Complete Solutions Guide

CALCULUS I WITH PRECALCULUS

A ONE-YEAR COURSE

SECOND EDITION

Larson / Hostetler / Edwards

Bruce H. Edwards
University of Florida

Dianna L. Zook
Indiana University, Purdue University at Fort Wayne

Houghton Mifflin Company Boston New York

Publisher: Jack Shira
Associate Sponsoring Editor: Cathy Cantin
Development Manager: Maureen Ross
Editorial Assistant: Elizabeth Kassab
Supervising Editor: Karen Carter
Senior Project Editor: Patty Bergin
Editorial Assistant: Julia Keller
Production Technology Supervisor: Gary Crespo
Executive Marketing Manager: Michael Busnach
Senior Marketing Manager: Danielle Potvin Curran
Marketing Coordinator: Nicole Mollica

Printed in the United States of America

ISBN: 0-618-56808-5

123456789-EB-09 08 07 06 05

PREFACE

The *Complete Solutions Guide* for *Calculus I with Precalculus: A One-Year Course,* Second Edition, is a supplement to the text by Ron Larson, Robert P. Hostetler, and Bruce H. Edwards. Solutions to every exercise in the text are given with all essential algebraic steps included.

We have made every effort to see that the solutions are correct. However, we would appreciate hearing about any errors or other suggestions for improvement.

We would like to thank the staff at Larson Texts, Inc. for their help in the production of this manual.

Bruce H. Edwards
Department of Mathematics
University of Florida

Dianna L. Zook
Indiana University
Purdue University
Fort Wayne, IN 46806

CONTENTS

Chapter P Prerequisites 1

Chapter 1 Functions and Their Graphs 90

Chapter 2 Polynomial and Rational Functions 158

Chapter 3 Limits and Their Properties 255

Chapter 4 Differentiation 295

Chapter 5 Applications of Differentiation 374

Chapter 6 Integration 476

Chapter 7 Exponential and Logarithmic Functions 546

Chapter 8 Exponential and Logarithmic Functions and Calculus 614

Chapter 9 Trigonometric Functions 658

Chapter 10 Analytic Trigonometry 743

Chapter 11 Trigonometric Functions and Calculus 824

Chapter 12 Topics in Analytic Geometry 886

Chapter 13 Additional Topics in Trigonometry 972

Chapter 14 Systems of Equations and Matrices 1038

CHAPTER P
Prerequisites

Section P.1 Solving Equations 2

Section P.2 Solving Inequalities 23

Section P.3 Graphical Representation of Data 44

Section P.4 Graphs of Equations 50

Section P.5 Linear Equations in Two Variables 58

Review Exercises 72

Problem Solving 86

CHAPTER P
Prerequisites

Section P.1 Solving Equations

1. $2(x - 1) = 2x - 2$ is an *identity* by the Distributive Property. It is true for all real values of x.

2. $3(x + 2) = 5x + 4$ is *conditional.* There are real values of x for which the equation is not true (for example, $x = 0$).

3. $-6(x - 3) + 5 = -2x + 10$ is *conditional.* There are real values of x for which the equation is not true.

4. $3(x + 2) - 5 = 3x + 1$ is an *identity* by simplification. It is true for all real values of x. $3(x + 2) - 5 = 3x + 6 - 5 = 3x + 1$

5. $4(x + 1) - 2x = 4x + 4 - 2x = 2x + 4 = 2(x + 2)$ is an *identity* by simplification. It is true for all real values of x.

6. $-7(x - 3) + 4x = 3(7 - x)$ is an *identity* by simplification. It is true for all real values of x.
$-7(x - 3) + 4x = -7x + 21 + 4x = 21 - 3x = 3(7 - x)$

7. $x^2 - 8x - 5 = (x - 4)^2 - 11$ is *conditional.* These are real values of x for which the equation is not true.

8. $x^2 + 2(3x - 2) = x^2 + 6x - 4$ is an *identity* by simplification. It is true for all real values of x.

9. $3 + \frac{1}{x + 1} = \frac{4x}{x + 1}$ is *conditional.* There are real values of x for which the equation is not true.

10. $\frac{5}{x} + \frac{3}{x} = 24$ is *conditional.* There are real values of x for which the equation is not true (for example, $x = 1$).

11.
$$\begin{aligned} x + 11 &= 15 \\ x + 11 - 11 &= 15 - 11 \\ x &= 4 \end{aligned}$$

12.
$$\begin{aligned} 7 - x &= 19 \\ 7 - x + x &= 19 + x \\ 7 &= 19 + x \\ 7 - 19 &= 19 + x - 19 \\ -12 &= x \end{aligned}$$

13.
$$\begin{aligned} 7 - 2x &= 25 \\ 7 - 7 - 2x &= 25 - 7 \\ -2x &= 18 \\ \frac{-2x}{-2} &= \frac{18}{-2} \\ x &= -9 \end{aligned}$$

14.
$$\begin{aligned} 7x + 2 &= 23 \\ 7x + 2 - 2 &= 23 - 2 \\ 7x &= 21 \\ \frac{7x}{7} &= \frac{21}{7} \\ x &= 3 \end{aligned}$$

15.
$$\begin{aligned} 8x - 5 &= 3x + 20 \\ 8x - 3x - 5 &= 3x - 3x + 20 \\ 5x - 5 &= 20 \\ 5x - 5 + 5 &= 20 + 5 \\ 5x &= 25 \\ \frac{5x}{5} &= \frac{25}{5} \\ x &= 5 \end{aligned}$$

16.
$$\begin{aligned} 7x + 3 &= 3x - 17 \\ 7x + 3 - 3 - 3x &= 3x - 17 - 3 - 3x \\ 4x &= -20 \\ x &= -5 \end{aligned}$$

17.
$$\begin{aligned} 2(x + 5) - 7 &= 3(x - 2) \\ 2x + 10 - 7 &= 3x - 6 \\ 2x + 3 &= 3x - 6 \\ -x &= -9 \\ x &= 9 \end{aligned}$$

18.
$$\begin{aligned} 3(x + 3) &= 5(1 - x) - 1 \\ 3x + 9 &= 5 - 5x - 1 \\ 3x + 9 &= 4 - 5x \\ 8x &= -5 \\ x &= -\tfrac{5}{8} \end{aligned}$$

19.
$$\begin{aligned} x - 3(2x + 3) &= 8 - 5x \\ x - 6x - 9 &= 8 - 5x \\ -5x - 9 &= 8 - 5x \\ -5x + 5x - 9 &= 8 - 5x + 5x \\ -9 &\neq 8 \end{aligned}$$

No solution

20.
$$\begin{aligned} 9x - 10 &= 5x + 2(2x - 5) \\ 9x - 10 &= 5x + 4x - 10 \\ 9x - 10 &= 9x - 10 \end{aligned}$$

The solution is the set of all real numbers.

21.
$$\begin{aligned} \frac{5x}{4} + \frac{1}{2} &= x - \frac{1}{2} \\ 4\left(\frac{5x}{4}\right) + 4\left(\frac{1}{2}\right) &= 4(x) - 4\left(\frac{1}{2}\right) \\ 5x + 2 &= 4x - 2 \\ x &= -4 \end{aligned}$$

22.
$$\begin{aligned} \frac{x}{5} - \frac{x}{2} &= 3 + \frac{3x}{10} \\ 10\left(\frac{x}{5} - \frac{x}{2}\right) &= 10\left(3 + \frac{3x}{10}\right) \\ 2x - 5x &= 30 + 3x \\ -6x &= 30 \\ x &= -5 \end{aligned}$$

23.
$$\begin{aligned} \frac{3}{2}(z + 5) - \frac{1}{4}(z + 24) &= 0 \\ 4\left(\frac{3}{2}\right)(z + 5) - 4\left(\frac{1}{4}\right)(z + 24) &= 4(0) \\ 6(z + 5) - (z + 24) &= 0 \\ 6z + 30 - z - 24 &= 0 \\ 5z &= -6 \\ z &= -\frac{6}{5} \end{aligned}$$

24.
$$\begin{aligned} \frac{3x}{2} + \frac{1}{4}(x - 2) &= 10 \\ (4)\left(\frac{3x}{2}\right) + (4)\frac{1}{4}(x - 2) &= (4)10 \\ 6x + (x - 2) &= 40 \\ 7x - 2 &= 40 \\ 7x &= 42 \\ x &= 6 \end{aligned}$$

25.
$$\begin{aligned} 0.25x + 0.75(10 - x) &= 3 \\ 4(0.25x) + 4(0.75)(10 - x) &= 4(3) \\ x + 3(10 - x) &= 12 \\ x + 30 - 3x &= 12 \\ -2x &= -18 \\ x &= 9 \end{aligned}$$

26.
$$\begin{aligned} 0.60x + 0.40(100 - x) &= 50 \\ 0.60x + 40 - 0.40x &= 50 \\ 0.20x &= 10 \\ x &= 50 \end{aligned}$$

27.
$$\begin{aligned} x + 8 &= 2(x - 2) - x \\ x + 8 &= 2x - 4 - x \\ x + 8 &= x - 4 \\ 8 &\neq -4 \end{aligned}$$

No solution

28.
$$\begin{aligned} 8(x + 2) - 3(2x + 1) &= 2(x + 5) \\ 8x + 16 - 6x - 3 &= 2x + 10 \\ 2x + 13 &= 2x + 10 \\ 13 &= 10 \end{aligned}$$

False equation $\Longrightarrow$ no solution

29. $$\frac{100-4x}{3} = \frac{5x+6}{4} + 6$$
$$12\left(\frac{100-4x}{3}\right) = 12\left(\frac{5x+6}{4}\right) + 12(6)$$
$$4(100-4x) = 3(5x+6) + 72$$
$$400 - 16x = 15x + 18 + 72$$
$$-31x = -310$$
$$x = 10$$

30. $$\frac{17+y}{y} + \frac{32+y}{y} = 100$$
$$(y)\frac{17+y}{y} + (y)\frac{32+y}{y} = 100(y)$$
$$17 + y + 32 + y = 100y$$
$$49 + 2y = 100y$$
$$49 = 98y$$
$$\frac{1}{2} = y$$

31. $$\frac{5x-4}{5x+4} = \frac{2}{3}$$
$$3(5x-4) = 2(5x+4)$$
$$15x - 12 = 10x + 8$$
$$5x = 20$$
$$x = 4$$

32. $$\frac{10x+3}{5x+6} = \frac{1}{2}$$
$$2(10x+3) = 1(5x+6)$$
$$20x + 6 = 5x + 6$$
$$15x = 0$$
$$x = 0$$

33. $$10 - \frac{13}{x} = 4 + \frac{5}{x}$$
$$\frac{10x-13}{x} = \frac{4x+5}{x}$$
$$10x - 13 = 4x + 5$$
$$6x = 18$$
$$x = 3$$

34. $$\frac{15}{x} - 4 = \frac{6}{x} + 3$$
$$\frac{15}{x} - \frac{6}{x} = 7$$
$$\frac{9}{x} = 7$$
$$9 = 7x$$
$$\frac{9}{7} = x$$

35. $$\frac{x}{x+4} + \frac{4}{x+4} + 2 = 0$$
$$\frac{x+4}{x+4} + 2 = 0$$
$$1 + 2 = 0$$
$$3 \neq 0$$

Contradiction : no solution
The variable is divided out.

36. $$3 = 2 + \frac{2}{z+2}$$
$$3(z+2) = \left(2 + \frac{2}{z+2}\right)(z+2)$$
$$3z + 6 = 2z + 4 + 2$$
$$z = 0$$

37. $$\frac{1}{x} + \frac{2}{x-5} = 0$$ Multiply both sides by $x(x-5)$.
$$1(x-5) + 2x = 0$$
$$3x - 5 = 0$$
$$3x = 5$$
$$x = \frac{5}{3}$$

38. $$\frac{7}{2x+1} - \frac{8x}{2x-1} = -4$$
$$(2x+1)(2x-1)\frac{7}{2x+1} - (2x+1)(2x-1)\frac{8x}{2x-1} = -4(2x+1)(2x-1)$$
$$7(2x-1) - 8x(2x+1) = -4(2x+1)(2x-1)$$
$$14x - 7 - 16x^2 - 8x = -16x^2 + 4$$
$$6x = 11$$
$$x = \frac{11}{6}$$

39. $\dfrac{2}{(x-4)(x-2)} = \dfrac{1}{x-4} + \dfrac{2}{x-2}$ Multiply both sides by $(x-4)(x-2)$.

$$2 = 1(x-2) + 2(x-4)$$
$$2 = x - 2 + 2x - 8$$
$$2 = 3x - 10$$
$$12 = 3x$$
$$4 = x$$

A check reveals that $x = 4$ is an extraneous solution–it makes the denominator zero. There is no real solution.

40. $$\frac{4}{x-1} + \frac{6}{3x+1} = \frac{15}{3x+1}$$

$$(x-1)(3x+1)\frac{4}{x-1} + (x-1)(3x+1)\frac{6}{3x+1} = (x-1)(3x+1)\frac{15}{3x+1}$$
$$4(3x+1) + 6(x-1) = 15(x-1)$$
$$12x + 4 + 6x - 6 = 15x - 15$$
$$18x - 2 = 15x - 15$$
$$3x = -13$$
$$x = -\frac{13}{3}$$

41. $$\frac{1}{x-3} + \frac{1}{x+3} = \frac{10}{x^2-9}$$

$$\frac{(x+3)+(x-3)}{x^2-9} = \frac{10}{x^2-9}$$
$$2x = 10$$
$$x = 5$$

42. $$\frac{1}{x-2} + \frac{3}{x+3} = \frac{4}{x^2+x-6}$$

$$(x^2+x-6)\frac{1}{x-2} + (x^2+x-6)\frac{3}{x+3} = (x^2+x-6)\frac{4}{x^2+x-6}$$
$$(x+3) + 3(x-2) = 4$$
$$x + 3 + 3x - 6 = 4$$
$$4x - 3 = 4$$
$$4x = 7$$
$$x = \frac{7}{4}$$

43. $\dfrac{3}{x^2-3x} + \dfrac{4}{x} = \dfrac{1}{x-3}$ Multiply both sides by $x(x-3)$.

$$3 + 4(x-3) = x$$
$$3 + 4x - 12 = x$$
$$3x = 9$$
$$x = 3$$

A check reveals that $x = 3$ is an extraneous solution, so there is no solution.

44. $$\frac{6}{x} - \frac{2}{x+3} = \frac{3(x+5)}{x^2+3x}$$

$$(x^2+3x)\frac{6}{x} - (x^2+3x)\frac{2}{x+3} = (x^2+3x)\frac{3(x+5)}{x^2+3x}$$

$$6(x+3) - 2x = 3(x+5)$$

$$6x + 18 - 2x = 3x + 15$$

$$4x + 18 = 3x + 15$$

$$x = -3$$

Check: $\frac{6}{-3} - \frac{2}{-3+3} = \frac{3(-3+5)}{[(-3)^2 + 3(-3)]}$

$$-2 - \frac{2}{0} = \frac{6}{0}$$

Division by zero is undefined. Thus, $x = -3$ is not a solution, and the original equation has no solution.

45. $$(x+2)^2 + 4 = (x+3)^2$$

$$x^2 + 4x + 4 + 4 = x^2 + 6x + 9$$

$$4x + 8 = 6x + 9$$

$$-2x = 1$$

$$x = -\tfrac{1}{2}$$

46. $$(x+1)^2 + 2(x-2) = (x+1)(x-2)$$

$$x^2 + 2x + 1 + 2x - 4 = x^2 - x - 2$$

$$5x = 1$$

$$x = \tfrac{1}{5}$$

47. $$(x+2)^2 - x^2 = 4(x+1)$$

$$x^2 + 4x + 4 - x^2 = 4x + 4$$

$$4 = 4$$

The equation is an identity; every real number is a solution.

48. $$(2x+1)^2 = 4(x^2 + x + 1)$$

$$4x^2 + 4x + 1 = 4x^2 + 4x + 4$$

$$1 = 4$$

This is a false equation. Thus, the equation has no solution.

49. $2x^2 = 3 - 8x$

General form: $2x^2 + 8x - 3 = 0$

50. $x^2 = 16x + 8$

General form: $x^2 - 16x - 8 = 0$

51. $(x-3)^2 = 3$

$x^2 - 6x + 9 = 3$

General form: $x^2 - 6x + 6 = 0$

52. $$13 - 3(x+7)^2 = 0$$

$$13 - 3(x^2 + 14x + 49) = 0$$

$$13 - 3x^2 - 42x - 147 = 0$$

General form: $-3x^2 - 42x - 134 = 0$

53. $\frac{1}{5}(3x^2 - 10) = 18x$

$3x^2 - 10 = 90x$

General form: $3x^2 - 90x - 10 = 0$

54. $$x(x+2) = 5x^2 + 1$$

$$x^2 + 2x = 5x^2 + 1$$

$$-4x^2 + 2x - 1 = 0$$

$$(-1)(-4x^2 + 2x - 1) = -1(0)$$

General form: $4x^2 - 2x + 1 = 0$

55. $6x^2 + 3x = 0$

$3x(2x+1) = 0$

$3x = 0$ or $2x + 1 = 0$

$x = 0$ or $x = -\frac{1}{2}$

56. $9x^2 - 18x = 0$

$9x(x-2) = 0$

$9x = 0 \Rightarrow x = 0$

$x - 2 = 0 \Rightarrow x = 2$

57. $x^2 - 2x - 8 = 0$

$(x-4)(x+2) = 0$

$x - 4 = 0$ or $x + 2 = 0$

$x = 4$ or $x = -2$

58. $x^2 - 10x + 9 = 0$

$(x-9)(x-1) = 0$

$x - 9 = 0 \Rightarrow x = 9$

$x - 1 = 0 \Rightarrow x = 1$

59. $x^2 + 10x + 25 = 0$

$(x+5)(x+5) = 0$

$x + 5 = 0$

$x = -5$

60. $4x^2 + 12x + 9 = 0$

$(2x+3)(2x+3) = 0$

$2x + 3 = 0 \Rightarrow x = -\frac{3}{2}$

61. $3 + 5x - 2x^2 = 0$

$(3-x)(1+2x) = 0$

$3 - x = 0$ or $1 + 2x = 0$

$x = 3$ or $x = -\frac{1}{2}$

62. $33 + 19x - 2x^2 = 0$
$2x^2 - 19x - 33 = 0$
$(2x + 3)(x - 11) = 0$
$2x + 3 = 0 \Rightarrow x = -\frac{3}{2}$
$x - 11 = 0 \Rightarrow x = 11$

63. $x^2 + 4x = 12$
$x^2 + 4x - 12 = 0$
$(x + 6)(x - 2) = 0$
$x + 6 = 0$ or $x - 2 = 0$
$x = -6$ or $x = 2$

64. $-x^2 + 8x = 12$
$-x^2 + 8x - 12 = 0$
$(-1)(-x^2 + 8x - 12) = (-1)(0)$
$x^2 - 8x + 12 = 0$
$(x - 6)(x - 2) = 0$
$x - 6 = 0 \Rightarrow x = 6$
$x - 2 = 0 \Rightarrow x = 2$

65. $\frac{3}{4}x^2 + 8x + 20 = 0$
$4(\frac{3}{4}x^2 + 8x + 20) = 4(0)$
$3x^2 + 32x + 80 = 0$
$(3x + 20)(x + 4) = 0$
$3x + 20 = 0$ or $x + 4 = 0$
$x = -\frac{20}{3}$ or $x = -4$

66. $\frac{1}{8}x^2 - x - 16 = 0$
$x^2 - 8x - 128 = 0$
$(x - 16)(x + 8) = 0$
$x - 16 = 0 \Rightarrow x = 16$
$x + 8 = 0 \Rightarrow x = -8$

67. $x^2 + 2ax + a^2 = 0$
$(x + a)^2 = 0$
$x + a = 0$
$x = -a$

68. $(x + a)^2 - b^2 = 0$
$x^2 + 2ax + a^2 - b^2 = 0$
$[x + (a + b)][x + (a - b)] = 0$
$x + (a + b) = 0 \Rightarrow x = -a - b$
$x + (a - b) = 0 \Rightarrow x = -a + b$

69. $x^2 = 49$
$x = \pm\sqrt{49}$
$= \pm 7$
$= \pm 7.00$

70. $x^2 = 169$
$x = \pm\sqrt{169} = \pm 13 = \pm 13.00$

71. $x^2 = 24$
$x = \pm\sqrt{24}$
$x = \pm 2\sqrt{6}$
$\approx \pm 4.90$

72. $x^2 = 32$
$x = \pm\sqrt{32} = \pm 4\sqrt{2} = \pm 5.66$

73. $3x^2 = 81$
$x^2 = 27$
$x = \pm\sqrt{27} = \pm 3\sqrt{3}$
$x \approx \pm 5.20$

74. $9x^2 = 36$
$x^2 = 4$
$x = \pm\sqrt{4} = \pm 2 = \pm 2.00$

75. $(x - 12)^2 = 16$
$x - 12 = \pm\sqrt{16}$
$x = 12 \pm 4$
$x = 16$ or $x = 8$
$x = 16.00$ or $x = 8.00$

76. $(x + 13)^2 = 25$
$x + 13 = \pm\sqrt{25}$
$x + 13 = \pm 5$
$x = -13 \pm 5 = -8, -18 = -8.00, -18.00$

77. $(x + 2)^2 = 14$
$x + 2 = \pm\sqrt{14}$
$x = -2 \pm \sqrt{14}$
$x \approx 1.74$ or $x \approx -5.74$

78. $(x - 5)^2 = 30$
$x - 5 = \pm\sqrt{30}$
$x = 5 \pm \sqrt{30} = -0.48, 10.48$

79. $(2x - 1)^2 = 18$
$2x - 1 = \pm\sqrt{18}$
$2x = 1 \pm 3\sqrt{2}$
$x = \dfrac{1 \pm 3\sqrt{2}}{2}$
$x \approx 2.62$ or $x \approx -1.62$

80. $(4x + 7)^2 = 44$

$$4x + 7 = \pm\sqrt{44}$$
$$4x = -7 \pm 2\sqrt{11}$$
$$x = \frac{-7 \pm 2\sqrt{11}}{4} = -\frac{7}{4} \pm \frac{\sqrt{11}}{2} = -3.41, -0.09$$

81. $(x - 7)^2 = (x + 3)^2$

$$x - 7 = \pm(x + 3)$$

$x - 7 = x + 3$ or $x - 7 = -x - 3$

$-7 \neq 3$ $\qquad$ $2x = 4$

No solution $\qquad$ $x = 2 = 2.00$

82. $(x + 5)^2 = (x + 4)^2$

$$x + 5 = \pm(x + 4)$$
$$x + 5 = +(x + 4)$$
$$5 = 4$$

No solution

$$x + 5 = -(x + 4)$$
$$x + 5 = -x - 4$$
$$2x = -9$$
$$x = -\tfrac{9}{2} = -4.50$$

83. $x^2 - 2x = 0$

$$x^2 - 2x + 1 = 0 + 1$$
$$(x - 1)^2 = 1$$
$$x - 1 = \pm\sqrt{1}$$
$$x = 1 \pm 1$$
$$x = 0 \quad \text{or} \quad x = 2$$

84. $x^2 + 4x = 0$

$$x^2 + 4x + 2^2 = 2^2$$
$$(x + 2)^2 = 4$$
$$x + 2 = \pm\sqrt{4}$$
$$x = -2 \pm 2 = 0, -4$$

85. $x^2 + 4x - 32 = 0$

$$x^2 + 4x = 32$$
$$x^2 + 4x + 2^2 = 32 + 2^2$$
$$(x + 2)^2 = 36$$
$$x + 2 = \pm\sqrt{36}$$
$$x = -2 \pm 6$$
$$x = 4 \quad \text{or} \quad x = -8$$

86. $x^2 - 2x - 3 = 0$

$$x^2 - 2x = 3$$
$$x^2 - 2x + (-1)^2 = 3 + 1$$
$$(x - 1)^2 = 4$$
$$x - 1 = \pm\sqrt{4}$$
$$x = 1 \pm 2 = 3, -1$$

87. $x^2 + 6x + 2 = 0$

$$x^2 + 6x = -2$$
$$x^2 + 6x + 3^2 = -2 + 3^2$$
$$(x + 3)^2 = 7$$
$$x + 3 = \pm\sqrt{7}$$
$$x = -3 \pm \sqrt{7}$$

88. $x^2 + 8x + 14 = 0$

$$x^2 + 8x = -14$$
$$x^2 + 8x + 4^2 = -14 + 16$$
$$(x + 4)^2 = 2$$
$$x + 4 = \pm\sqrt{2}$$
$$x = -4 \pm \sqrt{2}$$

89. $9x^2 - 18x = -3$

$$x^2 - 2x = -\frac{1}{3}$$
$$x^2 - 2x + 1 = -\frac{1}{3} + 1$$
$$(x - 1)^2 = \frac{2}{3}$$
$$x - 1 = \pm\sqrt{\frac{2}{3}}$$
$$x = 1 \pm \sqrt{\frac{6}{9}}$$
$$x = 1 \pm \frac{\sqrt{6}}{3}$$

90. $6x^2 - 12x = 15$

$$x^2 - 2x = \frac{5}{2}$$
$$x^2 - 2x + 1 = \frac{5}{2} + 1$$
$$(x - 1)^2 = \frac{7}{2}$$
$$x - 1 = \pm\sqrt{\frac{7}{2}}$$
$$x = 1 \pm \frac{\sqrt{14}}{2}$$

91. $8 + 4x - x^2 = 0$

$-x^2 + 4x + 8 = 0$

$x^2 - 4x - 8 = 0$

$x^2 - 4x = 8$

$x^2 - 4x + 2^2 = 8 + 2^2$

$(x - 2)^2 = 12$

$x - 2 = \pm\sqrt{12}$

$x = 2 \pm 2\sqrt{3}$

92. $4x^2 - 4x - 99 = 0$

$x^2 - x = \frac{99}{4}$

$x^2 - x + \left(\frac{1}{2}\right)^2 = \frac{99}{4} + \frac{1}{4}$

$\left(x - \frac{1}{2}\right)^2 = \frac{100}{4}$

$\left(x - \frac{1}{2}\right)^2 = 25$

$x - \frac{1}{2} = \pm\sqrt{25}$

$x = \frac{1}{2} \pm 5 = \frac{11}{2}, -\frac{9}{2}$

93. $2x^2 + x - 1 = 0$

$x = \dfrac{-b \pm \sqrt{b^2 - 4ac}}{2a}$

$= \dfrac{-1 \pm \sqrt{1^2 - 4(2)(-1)}}{2(2)}$

$= \dfrac{-1 \pm 3}{4} = \dfrac{1}{2}, -1$

94. $2x^2 - x - 1 = 0$

$x = \dfrac{-b \pm \sqrt{b^2 - 4ac}}{2a}$

$= \dfrac{-(-1) \pm \sqrt{(-1)^2 - 4(2)(-1)}}{2(2)}$

$= \dfrac{1 \pm \sqrt{1 + 8}}{4}$

$= \dfrac{1 \pm 3}{4} = 1, -\dfrac{1}{2}$

95. $16x^2 + 8x - 3 = 0$

$x = \dfrac{-b \pm \sqrt{b^2 - 4ac}}{2a}$

$= \dfrac{-8 \pm \sqrt{8^2 - 4(16)(-3)}}{2(16)}$

$= \dfrac{-8 \pm 16}{32} = \dfrac{1}{4}, -\dfrac{3}{4}$

96. $25x^2 - 20x + 3 = 0$

$x = \dfrac{-b \pm \sqrt{b^2 - 4ac}}{2a}$

$= \dfrac{-(-20) \pm \sqrt{(-20)^2 - 4(25)(3)}}{2(25)}$

$= \dfrac{20 \pm \sqrt{400 - 300}}{50}$

$= \dfrac{20 \pm 10}{50} = \dfrac{3}{5}, \dfrac{1}{5}$

97. $2 + 2x - x^2 = 0$

$x = \dfrac{-b \pm \sqrt{b^2 - 4ac}}{2a}$

$= \dfrac{-2 \pm \sqrt{2^2 - 4(-1)(2)}}{2(-1)}$

$= \dfrac{-2 \pm 2\sqrt{3}}{-2} = 1 \pm \sqrt{3}$

98. $x^2 - 10x + 22 = 0$

$x = \dfrac{-b \pm \sqrt{b^2 - 4ac}}{2a}$

$= \dfrac{-(-10) \pm \sqrt{(-10)^2 - 4(1)(22)}}{2(1)}$

$= \dfrac{10 \pm \sqrt{100 - 88}}{2}$

$= \dfrac{10 \pm 2\sqrt{3}}{2} = 5 \pm \sqrt{3}$

99. $14x = -44 - x^2$

$x = \dfrac{-b \pm \sqrt{b^2 - 4ac}}{2a}$

$= \dfrac{-14 \pm \sqrt{14^2 - 4(1)(44)}}{2(1)}$

$= \dfrac{-14 \pm 2\sqrt{5}}{2} = -7 \pm \sqrt{5}$

100. $6x = 4 - x^2$

$$x^2 + 6x - 4 = 0$$

$$x = \frac{-b \pm \sqrt{b^2 - 4ac}}{2a}$$

$$= \frac{-6 \pm \sqrt{6^2 - 4(1)(-4)}}{2(1)}$$

$$= \frac{-6 \pm \sqrt{36 + 16}}{2}$$

$$= \frac{-6 \pm 2\sqrt{13}}{2}$$

$$= -3 \pm \sqrt{13}$$

101. $x^2 + 8x - 4 = 0$

$$x = \frac{-b \pm \sqrt{b^2 - 4ac}}{2a}$$

$$= \frac{-8 \pm \sqrt{8^2 - 4(1)(-4)}}{2(1)}$$

$$= \frac{-8 \pm 4\sqrt{5}}{2}$$

$$= -4 \pm 2\sqrt{5}$$

102. $4x^2 - 4x - 4 = 0$

$$x^2 - x - 1 = 0$$

$$x = \frac{-b \pm \sqrt{b^2 - 4ac}}{2a}$$

$$= \frac{-(-1) \pm \sqrt{(-1)^2 - 4(1)(-1)}}{2(1)}$$

$$= \frac{1 \pm \sqrt{1 + 4}}{2}$$

$$= \frac{1}{2} \pm \frac{\sqrt{5}}{2}$$

103. $12x - 9x^2 = -3$

$$-9x^2 + 12x + 3 = 0$$

$$x = \frac{-b \pm \sqrt{b^2 - 4ac}}{2a}$$

$$= \frac{-12 \pm \sqrt{12^2 - 4(-9)(3)}}{2(-9)}$$

$$= \frac{-12 \pm 6\sqrt{7}}{-18} = \frac{2}{3} \pm \frac{\sqrt{7}}{3}$$

104. $16x^2 + 22 = 40x$

$$8x^2 - 20x + 11 = 0$$

$$x = \frac{-b \pm \sqrt{b^2 - 4ac}}{2a}$$

$$= \frac{-(-20) \pm \sqrt{(-20)^2 - 4(8)(11)}}{2(8)}$$

$$= \frac{20 \pm \sqrt{400 - 352}}{16} = \frac{20 \pm 4\sqrt{3}}{16}$$

$$= \frac{5}{4} \pm \frac{\sqrt{3}}{4}$$

105. $9x^2 + 24x + 16 = 0$

$$x = \frac{-b \pm \sqrt{b^2 - 4ac}}{2a}$$

$$= \frac{-24 \pm \sqrt{24^2 - 4(9)(16)}}{2(9)}$$

$$= \frac{-24 \pm 0}{18}$$

$$= -\frac{4}{3}$$

106. $36x^2 + 24x - 7 = 0$

$$x = \frac{-b \pm \sqrt{b^2 - 4ac}}{2a}$$

$$= \frac{-24 \pm \sqrt{24^2 - 4(36)(-7)}}{2(36)}$$

$$= \frac{-24 \pm \sqrt{576 + 1008}}{72}$$

$$= \frac{-24 \pm 12\sqrt{11}}{72}$$

$$= -\frac{1}{3} \pm \frac{\sqrt{11}}{6}$$

107. $4x^2 + 4x = 7$

$$4x^2 + 4x - 7 = 0$$

$$x = \frac{-b \pm \sqrt{b^2 - 4ac}}{2a}$$

$$= \frac{-4 \pm \sqrt{4^2 - 4(4)(-7)}}{2(4)}$$

$$= \frac{-4 \pm 8\sqrt{2}}{8} = -\frac{1}{2} \pm \sqrt{2}$$

108. $16x^2 - 40x + 5 = 0$

$$x = \frac{-b \pm \sqrt{b^2 - 4ac}}{2a}$$
$$= \frac{-(-40) \pm \sqrt{(-40)^2 - 4(16)(5)}}{2(16)}$$
$$= \frac{40 \pm \sqrt{1600 - 320}}{32}$$
$$= \frac{40 \pm 16\sqrt{5}}{32}$$
$$= \frac{5}{4} \pm \frac{\sqrt{5}}{2}$$

109. $28x - 49x^2 = 4$

$$-49x^2 + 28x - 4 = 0$$
$$x = \frac{-b \pm \sqrt{b^2 - 4ac}}{2a}$$
$$= \frac{-28 \pm \sqrt{28^2 - 4(-49)(-4)}}{2(-49)}$$
$$= \frac{-28 \pm 0}{-98} = \frac{2}{7}$$

110. $3x + x^2 - 1 = 0$

$$x^2 + 3x - 1 = 0$$
$$x = \frac{-b \pm \sqrt{b^2 - 4ac}}{2a}$$
$$= \frac{-3 \pm \sqrt{3^2 - 4(1)(-1)}}{2(1)}$$
$$= \frac{-3 \pm \sqrt{13}}{2}$$
$$= -\frac{3}{2} \pm \frac{\sqrt{13}}{2}$$

111. $8t = 5 + 2t^2$

$$-2t^2 + 8t - 5 = 0$$
$$t = \frac{-b \pm \sqrt{b^2 - 4ac}}{2a}$$
$$= \frac{-8 \pm \sqrt{8^2 - 4(-2)(-5)}}{2(-2)}$$
$$= \frac{-8 \pm 2\sqrt{6}}{-4} = 2 \pm \frac{\sqrt{6}}{2}$$

112. $25h^2 + 80h + 61 = 0$

$$h = \frac{-b \pm \sqrt{b^2 - 4ac}}{2a}$$
$$= \frac{-80 \pm \sqrt{80^2 - 4(25)(61)}}{2(25)}$$
$$= \frac{-80 \pm \sqrt{6400 - 6100}}{50}$$
$$= -\frac{8}{5} \pm \frac{10\sqrt{3}}{50}$$
$$= -\frac{8}{5} \pm \frac{\sqrt{3}}{5}$$

113. $(y - 5)^2 = 2y$

$$y^2 - 12y + 25 = 0$$
$$x = \frac{-b \pm \sqrt{b^2 - 4ac}}{2a}$$
$$= \frac{-(-12) \pm \sqrt{(-12)^2 - 4(1)(25)}}{2(1)}$$
$$= \frac{12 \pm 2\sqrt{11}}{2} = 6 \pm \sqrt{11}$$

114. $(z + 6)^2 = -2z$

$$z^2 + 12z + 36 = -2z$$
$$z^2 + 14z + 36 = 0$$
$$z = \frac{-b \pm \sqrt{b^2 - 4ac}}{2a}$$
$$= \frac{-14 \pm \sqrt{14^2 - 4(1)(36)}}{2(1)}$$
$$= \frac{-14 \pm \sqrt{52}}{2}$$
$$= -7 \pm \sqrt{13}$$

115. $\frac{1}{2}x^2 + \frac{3}{8}x = 2$

$$4x^2 + 3x = 16$$
$$4x^2 + 3x - 16 = 0$$
$$x = \frac{-b \pm \sqrt{b^2 - 4ac}}{2a}$$
$$= \frac{-3 \pm \sqrt{3^2 - 4(4)(-16)}}{2(4)}$$
$$= \frac{-3 \pm \sqrt{265}}{8} = -\frac{3}{8} \pm \frac{\sqrt{265}}{8}$$

116. $\left(\frac{5}{7}x - 14\right)^2 = 8x$

$\frac{25}{49}x^2 - 20x + 196 = 8x$

$\frac{25}{49}x^2 - 28x + 196 = 0$

$25x^2 - 1372x + 9604 = 0$

$$x = \frac{-b \pm \sqrt{b^2 - 4ac}}{2a} = \frac{-(-1372) \pm \sqrt{(-1372)^2 - 4(25)(9604)}}{2(25)}$$

$$= \frac{1372 \pm \sqrt{921{,}984}}{50} = \frac{686 \pm 196\sqrt{6}}{25} = \frac{686}{25} \pm \frac{196\sqrt{6}}{25}$$

117. $5.1x^2 - 1.7x - 3.2 = 0$

$$x = \frac{1.7 \pm \sqrt{(-1.7)^2 - 4(5.1)(-3.2)}}{2(5.1)}$$

$x \approx 0.976, -0.643$

118. $2x^2 - 2.50x - 0.42 = 0$

$$x = \frac{-b \pm \sqrt{b^2 - 4ac}}{2a}$$

$$= \frac{-(-2.50) \pm \sqrt{(-2.50)^2 - 4(2)(-0.42)}}{2(2)}$$

$= 1.400, -0.150$

119. $-0.067x^2 - 0.852x + 1.277 = 0$

$$x = \frac{-(-0.852) \pm \sqrt{(-0.852)^2 - 4(-0.067)(1.277)}}{2(-0.067)}$$

$x \approx -14.071, 1.355$

120. $-0.005x^2 + 0.101x - 0.193 = 0$

$$x = \frac{-b \pm \sqrt{b^2 - 4ac}}{2a}$$

$$= \frac{-0.101 \pm \sqrt{(0.101)^2 - 4(-0.005)(-0.193)}}{2(-0.005)}$$

$\approx 2.137, 18.063$

121. $422x^2 - 506x - 347 = 0$

$$x = \frac{506 \pm \sqrt{(-506)^2 - 4(422)(-347)}}{2(422)}$$

$x \approx 1.687, -0.488$

122. $1100x^2 + 326x - 715 = 0$

$$x = \frac{-b \pm \sqrt{b^2 - 4ac}}{2a}$$

$$= \frac{-326 \pm \sqrt{(326)^2 - 4(1100)(-715)}}{2(1100)}$$

$\approx 0.672, -0.968$

123. $12.67x^2 + 31.55x + 8.09 = 0$

$$x = \frac{-31.55 \pm \sqrt{(31.55)^2 - 4(12.67)(8.09)}}{2(12.67)}$$

$x \approx -2.200, -0.290$

124. $-3.22x^2 - 0.08x + 28.651 = 0$

$$x = \frac{-b \pm \sqrt{b^2 - 4ac}}{2a}$$

$$= \frac{-(-0.08) \pm \sqrt{(-0.08)^2 - 4(-3.22)(28.651)}}{2(-3.22)}$$

$\approx -2.995, 2.971$

125. $-16t^2 + 120t + 5 = 0$

$16t^2 - 120t - 5 = 0$

$$t = \frac{-(-120) \pm \sqrt{(-120)^2 - 4(16)(-5)}}{2(16)}$$

$$= \frac{120 \pm \sqrt{14,720}}{32}$$

$t \approx 7.541$ or $t \approx -0.041$

126. $-16t^2 + 48t + 3 = 0$

$16t^2 - 48t - 3 = 0$

$$t = \frac{-(-48) \pm \sqrt{(-48)^2 - 4(16)(-3)}}{2(16)}$$

$$= \frac{48 \pm \sqrt{2496}}{32}$$

$t \approx 3.061$ or $t \approx -0.061$

127. $-4.9t^2 + 46t + 1 = 0$

$4.9t^2 - 46t - 1 = 0$

$$t = \frac{-(-46) \pm \sqrt{(-46)^2 - 4(4.9)(-1)}}{2(4.9)}$$

$$= \frac{46 \pm \sqrt{2135.6}}{9.8}$$

$t \approx 9.409$ or $t \approx -0.022$

128. $-4.9t^2 + 30t + 2 = 0$

$4.9t^2 - 30t - 2 = 0$

$$t = \frac{-(-30) \pm \sqrt{(-30)^2 - 4(4.9)(-2)}}{2(4.9)}$$

$$= \frac{30 \pm \sqrt{939.2}}{9.8}$$

$t \approx 6.188$ or $t \approx -0.066$

129. $x^2 - 2x - 1 = 0$

$x^2 - 2x = 1$

$x^2 - 2x + 1^2 = 1 + 1^2$

$(x - 1)^2 = 2$

$x - 1 = \pm\sqrt{2}$

$x = 1 \pm \sqrt{2}$

130. $x^2 - 4x - 2 = 0$

$x^2 - 4x = 2$

$x^2 - 4x + 4 = 2 + 4$

$(x - 2)^2 = 6$

$x - 2 = \pm\sqrt{6}$

$x = 2 \pm \sqrt{6}$

131. $(x + 3)^2 = 81$

$x + 3 = \pm 9$

$x + 3 = 9$ or $x + 3 = -9$

$x = 6$ or $x = -12$

132. $x^2 - 14x + 49 = 0$

$(x - 7)^2 = 0$

$x - 7 = 0 \Longrightarrow x = 7$

133. $x^2 - x - \frac{11}{4} = 0$ Complete the Square

$x^2 - x = \frac{11}{4}$

$x^2 - x + \left(\frac{1}{2}\right)^2 = \frac{11}{4} + \left(\frac{1}{2}\right)^2$

$\left(x - \frac{1}{2}\right)^2 = \frac{12}{4}$

$x - \frac{1}{2} = \pm\sqrt{\frac{12}{4}}$

$x = \frac{1}{2} \pm \sqrt{3}$

134. $x^2 + 3x - \frac{3}{4} = 0$

$x^2 + 3x + \left(\frac{3}{2}\right)^2 = \frac{3}{4} + \frac{9}{4}$

$\left(x + \frac{3}{2}\right)^2 = 3$

$x + \frac{3}{2} = \pm\sqrt{3}$

$x = -\frac{3}{2} \pm \sqrt{3}$

135. $(x + 1)^2 = x^2$ Extract Square Roots

$x^2 = (x + 1)^2$

$x = \pm(x + 1)$

For $x = +(x + 1)$:

$0 \neq 1$ No solution

For $x = -(x + 1)$:

$2x = -1$

$x = -\frac{1}{2}$

136.
$$(x - 2)^2 = x^2 - 4$$
$$x^2 - 4x + 4 = x^2 - 4$$
$$-4x + 4 = -4$$
$$-4x = -8$$
$$x = 2$$

137. $3x + 4 = 2x^2 - 7$ Quadratic Formula
$$0 = 2x^2 - 3x - 11$$
$$x = \frac{-(-3) \pm \sqrt{(-3)^2 - 4(2)(-11)}}{2(2)}$$
$$= \frac{3 \pm \sqrt{97}}{4} = \frac{3}{4} \pm \frac{\sqrt{97}}{4}$$

138.
$$4x^2 + 2x + 4 = 2x + 8$$
$$4x^2 - 4 = 0$$
$$4(x^2 - 1) = 0$$
$$4(x + 1)(x - 1) = 0$$
$$x + 1 = 0 \Rightarrow x = -1$$
$$x - 1 = 0 \Rightarrow x = 1$$

139.
$$a^2x^2 - b^2 = 0$$
$$(ax + b)(ax - b) = 0$$
$$ax + b = 0 \Rightarrow x = -\frac{b}{a}$$
$$ax - b = 0 \Rightarrow x = \frac{b}{a}$$

140. $a^2x^2 - 2abx + b^2 = 0$
$$(ax - b)^2 = 0$$
$$ax - b = 0 \Rightarrow x = \frac{b}{a}$$

141.
$$4x^4 - 18x^2 = 0$$
$$2x^2(2x^2 - 9) = 0$$
$$2x^2 = 0 \Rightarrow x = 0$$
$$2x^2 - 9 = 0 \Rightarrow x = \pm\frac{3\sqrt{2}}{2}$$

142.
$$20x^3 - 125x = 0$$
$$5x(4x^2 - 25) = 0$$
$$5x(2x + 5)(2x - 5) = 0$$
$$5x = 0 \Rightarrow x = 0$$
$$2x + 5 = 0 \Rightarrow x = -\tfrac{5}{2}$$
$$2x - 5 = 0 \Rightarrow x = \tfrac{5}{2}$$

143.
$$x^4 - 81 = 0$$
$$(x^2 + 9)(x + 3)(x - 3) = 0$$
$x^2 + 9 = 0 \Rightarrow$ No real solutions
$$x + 3 = 0 \Rightarrow x = -3$$
$$x - 3 = 0 \Rightarrow x = 3$$

144.
$$x^6 - 64 = 0$$
$$(x^3 - 8)(x^3 + 8) = 0$$
$$(x - 2)(x^2 + 2x + 4)(x + 2)(x^2 - 2x + 4) = 0$$
$$x - 2 = 0 \Rightarrow x = 2$$
$x^2 + 2x + 4 = 0 \Rightarrow$ No real solutions
$$x + 2 = 0 \Rightarrow x = -2$$
$x^2 - 2x + 4 = 0 \Rightarrow$ No real solutions

Real solutions: $x = 2, -2$

145.
$$x^3 + 216 = 0$$
$$x^3 + 6^3 = 0$$
$$(x + 6)(x^2 - 6x + 36) = 0$$
$$x + 6 = 0 \Rightarrow x = -6$$
$x^2 - 6x + 36 = 0 \Rightarrow$ No real solutions

(By completing the square)

146.
$$27x^3 - 512 = 0$$
$$(3x - 8)(9x^2 + 24x + 64) = 0$$
$$3x - 8 = 0 \Rightarrow x = \frac{8}{3}$$
$9x^2 + 24x + 64 = 0 \Rightarrow$ No real solutions

Real solution: $x = \dfrac{8}{3}$

147. $5x^3 + 30x^2 + 45x = 0$

$5x(x^2 + 6x + 9) = 0$

$5x(x + 3)^2 = 0$

$5x = 0 \Rightarrow x = 0$

$x + 3 = 0 \Rightarrow x = -3$

148. $9x^4 - 24x^3 + 16x^2 = 0$

$x^2(9x^2 - 24x + 16) = 0$

$x^2(3x - 4)^2 = 0$

$x^2 = 0 \Rightarrow x = 0$

$3x - 4 = 0 \Rightarrow x = \frac{4}{3}$

149. $x^3 - 3x^2 - x + 3 = 0$

$x^2(x - 3) - (x - 3) = 0$

$(x - 3)(x^2 - 1) = 0$

$(x - 3)(x + 1)(x - 1) = 0$

$x - 3 = 0 \Rightarrow x = 3$

$x + 1 = 0 \Rightarrow x = -1$

$x - 1 = 0 \Rightarrow x = 1$

150. $x^3 + 2x^2 + 3x + 6 = 0$

$x^2(x + 2) + 3(x + 2) = 0$

$(x + 2)(x^2 + 3) = 0$

$x + 2 = 0 \Rightarrow x = -2$

$x^2 + 3 = 0 \Rightarrow$ No real solutions

Real solution: $x = -2$

151. $x^4 - x^3 + x - 1 = 0$

$x^3(x - 1) + (x - 1) = 0$

$(x - 1)(x^3 + 1) = 0$

$(x - 1)(x + 1)(x^2 - x + 1) = 0$

$x - 1 = 0 \Rightarrow x = 1$

$x + 1 = 0 \Rightarrow x = -1$

$x^2 - x + 1 = 0 \Rightarrow$ No real solutions (By the Quadratic Formula)

152. $x^4 + 2x^3 - 8x - 16 = 0$

$x^3(x + 2) - 8(x + 2) = 0$

$(x^3 - 8)(x + 2) = 0$

$(x - 2)(x^2 + 2x + 4)(x + 2) = 0$

$x - 2 = 0 \Rightarrow x = 2$

$x^2 + 2x + 4 = 0 \Rightarrow$ No real solutions

$x + 2 = 0 \Rightarrow x = -2$

Real solutions: $x = 2, -2$

153. $x^4 - 4x^2 + 3 = 0$

$(x^2 - 3)(x^2 - 1) = 0$

$(x + \sqrt{3})(x - \sqrt{3})(x + 1)(x - 1) = 0$

$x + \sqrt{3} = 0 \Rightarrow x = -\sqrt{3}$

$x - \sqrt{3} = 0 \Rightarrow x = \sqrt{3}$

$x + 1 = 0 \Rightarrow x = -1$

$x - 1 = 0 \Rightarrow x = 1$

154. $x^4 + 5x^2 - 36 = 0$

$(x^2 + 9)(x^2 - 4) = 0$

$(x^2 + 9)(x + 2)(x - 2) = 0$

$x^2 + 9 = 0 \Rightarrow$ No real solutions

$x + 2 = 0 \Rightarrow x = -2$

$x - 2 = 0 \Rightarrow x = 2$

Real solutions: $x = 2, -2$

155. $4x^4 - 65x^2 + 16 = 0$

$(4x^2 - 1)(x^2 - 16) = 0$

$(2x + 1)(2x - 1)(x + 4)(x - 4) = 0$

$2x + 1 = 0 \Rightarrow x = -\frac{1}{2}$

$2x - 1 = 0 \Rightarrow x = \frac{1}{2}$

$x + 4 = 0 \Rightarrow x = -4$

$x - 4 = 0 \Rightarrow x = 4$

156.

$$36t^4 + 29t^2 - 7 = 0$$
$$(36t^2 - 7)(t^2 + 1) = 0$$
$$(6t + \sqrt{7})(6t - \sqrt{7})(t^2 + 1) = 0$$
$$6t + \sqrt{7} = 0 \Rightarrow t = -\frac{\sqrt{7}}{6}$$
$$6t - \sqrt{7} = 0 \Rightarrow t = \frac{\sqrt{7}}{6}$$
$$t^2 + 1 = 0 \Rightarrow \text{No real solutions}$$

Real solutions: $t = \pm\frac{\sqrt{7}}{6}$

157.

$$x^6 + 7x^3 - 8 = 0$$
$$(x^3 + 8)(x^3 - 1) = 0$$
$$(x + 2)(x^2 - 2x + 4)(x - 1)(x^2 + x + 1) = 0$$
$$x + 2 = 0 \Rightarrow x = -2$$
$$x^2 - 2x + 4 = 0 \Rightarrow \text{No real solutions (By the Quadratic Formula)}$$
$$x - 1 = 0 \Rightarrow x = 1$$
$$x^2 + x + 1 = 0 \Rightarrow \text{No real solutions (By the Quadratic Formula)}$$

158.

$$x^6 + 3x^3 + 2 = 0$$
$$(x^3 + 2)(x^3 + 1) = 0$$
$$\left(x + \sqrt[3]{2}\right)\left[x^2 - \sqrt[3]{2}x + \left(\sqrt[3]{2}\right)^2\right](x + 1)(x^2 - x + 1) = 0$$
$$x + \sqrt[3]{2} = 0 \Rightarrow x = -\sqrt[3]{2}$$
$$x^2 - \sqrt[3]{2}x + \left(\sqrt[3]{2}\right)^2 = 0 \Rightarrow \text{No real solutions}$$
$$x + 1 = 0 \Rightarrow x = -1$$
$$x^2 - x + 1 = 0 \Rightarrow \text{No real solutions}$$

Real solutions: $x = -\sqrt[3]{2}, -1$

159. $\sqrt{2x} - 10 = 0$

$$\sqrt{2x} = 10$$
$$2x = 100$$
$$x = 50$$

160. $4\sqrt{x} - 3 = 0$

$$4\sqrt{x} = 3$$
$$16x = 9$$
$$x = \tfrac{9}{16}$$

161. $\sqrt{x - 10} - 4 = 0$

$$\sqrt{x - 10} = 4$$
$$x - 10 = 16$$
$$x = 26$$

162. $\sqrt{5 - x} - 3 = 0$

$$\sqrt{5 - x} = 3$$
$$5 - x = 9$$
$$x = -4$$

163. $\sqrt[3]{2x + 5} + 3 = 0$

$$\sqrt[3]{2x + 5} = -3$$
$$2x + 5 = -27$$
$$2x = -32$$
$$x = -16$$

164. $\sqrt[3]{3x + 1} - 5 = 0$

$$\sqrt[3]{3x + 1} = 5$$
$$3x + 1 = 125$$
$$3x = 124$$
$$x = \tfrac{124}{3}$$

165. $x + \sqrt{26 - 11x} = 4$

$$4 - x = \sqrt{26 - 11x}$$
$$16 - 8x + x^2 = 26 - 11x$$
$$x^2 + 3x - 10 = 0$$
$$(x + 5)(x - 2) = 0$$
$$x + 5 = 0 \Rightarrow x = -5$$
$$x - 2 = 0 \Rightarrow x = 2$$

166. $x + \sqrt{31 - 9x} = 5$

$$\sqrt{31 - 9x} = 5 - x$$
$$31 - 9x = 25 - 10x + x^2$$
$$0 = x^2 - x - 6$$
$$0 = (x - 3)(x + 2)$$
$$0 = x - 3 \Rightarrow x = 3$$
$$0 = x + 2 \Rightarrow x = -2$$

167. $\sqrt{x + 1} = \sqrt{3x + 1}$

$$x + 1 = 3x + 1$$
$$-2x = 0$$
$$x = 0$$

168. $\sqrt{x + 5} = \sqrt{x - 5}$

$$x + 5 = x - 5$$
$$5 = -5$$

No solution

169. $(x - 5)^{3/2} = 8$

$$(x - 5)^3 = 8^2$$
$$x - 5 = 8^{2/3}$$
$$x = 5 + 4$$
$$x = 9$$

170. $(x + 3)^{3/2} = 8$

$$(x + 3)^3 = 8^2$$
$$(x + 3)^3 = 64$$
$$x + 3 = \sqrt[3]{64}$$
$$x = -3 + 4 = 1$$

171. $(x + 3)^{2/3} = 8$

$$(x + 3)^2 = 8^3$$
$$x + 3 = \pm\sqrt{8^3}$$
$$x + 3 = \pm\sqrt{512}$$
$$x = -3 \pm 16\sqrt{2}$$

172. $(x + 2)^{2/3} = 9$

$$(x + 2)^2 = 9^3$$
$$x + 2 = \pm\sqrt{729}$$
$$x = -2 \pm 27 = -29, 25$$

173. $(x^2 - 5)^{3/2} = 27$

$$x^2 - 5 = 27^{2/3}$$
$$x^2 = 5 + 9$$
$$x^2 = 14$$
$$x = \pm\sqrt{14}$$

174. $(x^2 - x - 22)^{3/2} = 27$

$$x^2 - x - 22 = 27^{2/3}$$
$$x^2 - x - 22 = 9$$
$$x^2 - x - 31 = 0$$
$$x = \frac{-(-1) \pm \sqrt{(-1)^2 - 4(1)(-31)}}{2(1)}$$
$$= \frac{1 \pm 5\sqrt{5}}{2} = \frac{1}{2} \pm \frac{5\sqrt{5}}{2}$$

175. $3x(x - 1)^{1/2} + 2(x - 1)^{3/2} = 0$

$$(x - 1)^{1/2}[3x + 2(x - 1)] = 0$$
$$(x - 1)^{1/2}(5x - 2) = 0$$
$$(x - 1)^{1/2} = 0 \Rightarrow x - 1 = 0 \Rightarrow x = 1$$
$$5x - 2 = 0 \Rightarrow x = \tfrac{2}{5} \text{ which is extraneous.}$$

176. $4x^2(x - 1)^{1/3} + 6x(x - 1)^{4/3} = 0$

$$2x[2x(x - 1)^{1/3} + 3(x - 1)^{4/3}] = 0$$
$$2x(x - 1)^{1/3}[2x + 3(x - 1)] = 0$$
$$2x(x - 1)^{1/3}(5x - 3) = 0$$
$$2x = 0 \Rightarrow x = 0$$
$$x - 1 = 0 \Rightarrow x = 1$$
$$5x - 3 = 0 \Rightarrow x = \frac{3}{5}$$

177. $\dfrac{20 - x}{x} = x$

$$20 - x = x^2$$
$$0 = x^2 + x - 20$$
$$0 = (x + 5)(x - 4)$$
$$x + 5 = 0 \Rightarrow x = -5$$
$$x - 4 = 0 \Rightarrow x = 4$$

178. $\frac{4}{x} - \frac{5}{3} = \frac{x}{6}$

$(6x)\frac{4}{x} - (6x)\frac{5}{3} = (6x)\frac{x}{6}$

$24 - 10x = x^2$

$x^2 + 10x - 24 = 0$

$(x + 12)(x - 2) = 0$

$x + 12 = 0 \Longrightarrow x = -12$

$x - 2 = 0 \Longrightarrow x = 2$

179. $\frac{1}{x} - \frac{1}{x + 1} = 3$

$x(x + 1)\frac{1}{x} - x(x + 1)\frac{1}{x + 1} = x(x + 1)(3)$

$x + 1 - x = 3x(x + 1)$

$1 = 3x^2 + 3x$

$0 = 3x^2 + 3x - 1; \; a = 3, \; b = 3, \; c = -1$

$x = \frac{-3 \pm \sqrt{(3)^2 - 4(3)(-1)}}{2(3)} = \frac{-3 \pm \sqrt{21}}{6}$

180. $\frac{x}{x^2 - 4} + \frac{1}{x + 2} = 3$

$(x + 2)(x - 2)\frac{x}{x^2 - 4} + (x + 2)(x - 2)\frac{1}{x + 2} = 3(x + 2)(x - 2)$

$x + x - 2 = 3x^2 - 12$

$3x^2 - 2x - 10 = 0$

$a = 3, b = -2, c = -10$

$x = \frac{-(-2) \pm \sqrt{(-2)^2 - 4(3)(-10)}}{2(3)} = \frac{2 \pm \sqrt{124}}{6} = \frac{2 \pm 2\sqrt{31}}{6} = \frac{1 \pm \sqrt{31}}{3} = \frac{1}{3} \pm \frac{\sqrt{31}}{3}$

181. $x = \frac{3}{x} + \frac{1}{2}$

$(2x)(x) = (2x)\left(\frac{3}{x}\right) + (2x)\left(\frac{1}{2}\right)$

$2x^2 = 6 + x$

$2x^2 - x - 6 = 0$

$(2x + 3)(x - 2) = 0$

$2x + 3 = 0 \Longrightarrow x = -\frac{3}{2}$

$x - 2 = 0 \Longrightarrow x = 2$

182. $4x + 1 = \frac{3}{x}$

$(x)4x + (x)1 = (x)\frac{3}{x}$

$4x^2 + x = 3$

$4x^2 + x - 3 = 0$

$(4x - 3)(x + 1) = 0$

$4x - 3 = 0 \Longrightarrow x = \frac{3}{4}$

$x + 1 = 0 \Longrightarrow x = -1$

183. $\dfrac{4}{x+1} - \dfrac{3}{x+2} = 1$

$$4(x+2) - 3(x+1) = (x+1)(x+2),\ x \neq -2, -1$$
$$4x + 8 - 3x - 3 = x^2 + 3x + 2$$
$$x^2 + 2x - 3 = 0$$
$$(x-1)(x+3) = 0$$
$$x - 1 = 0 \Rightarrow x = 1$$
$$x + 3 = 0 \Rightarrow x = -3$$

184. $\dfrac{x+1}{3} - \dfrac{x+1}{x+2} = 0$

$$3(x+2)\frac{x+1}{3} - 3(x+2)\frac{x+1}{x+2} = 0$$
$$(x+2)(x+1) - 3(x+1) = 0$$
$$x^2 + 3x + 2 - 3x - 3 = 0$$
$$x^2 - 1 = 0$$
$$(x+1)(x-1) = 0$$
$$x + 1 = 0 \Rightarrow x = -1$$
$$x - 1 = 0 \Rightarrow x = 1$$

185. $|2x - 1| = 5$

$$2x - 1 = 5 \Rightarrow x = 3$$
$$-(2x - 1) = 5 \Rightarrow x = -2$$

186. $|3x + 2| = 7$

$$3x + 2 = 7 \Rightarrow x = \tfrac{5}{3}$$
$$-(3x + 2) = 7$$
$$-3x - 2 = 7 \Rightarrow x = -3$$

187. $|x| = x^2 + x - 3$

$x = x^2 + x - 3$ OR $-x = x^2 + x - 3$

$$x^2 - 3 = 0 \qquad\qquad x^2 + 2x - 3 = 0$$
$$x = \pm\sqrt{3} \qquad\qquad (x-1)(x+3) = 0$$
$$x - 1 = 0 \Rightarrow x = 1$$
$$x + 3 = 0 \Rightarrow x = -3$$

Only $x = \sqrt{3}$, and $x = -3$ are solutions to the original equation. $x = -\sqrt{3}$ and $x = 1$ are extraneous.

188. $|x^2 + 6x| = 3x + 18$

First equation:

$$x^2 + 6x = 3x + 18$$
$$x^2 + 3x - 18 = 0$$
$$(x-3)(x+6) = 0$$
$$x - 3 = 0 \Rightarrow x = 3$$
$$x + 6 = 0 \Rightarrow x = -6$$

Second equation:

$$-(x^2 + 6x) = 3x + 18$$
$$0 = x^2 + 9x + 18$$
$$0 = (x+3)(x+6)$$
$$0 = x + 3 \Rightarrow x = -3$$
$$0 = x + 6 \Rightarrow x = -6$$

189. $|x + 1| = x^2 - 5$

$x + 1 = x^2 - 5$ OR $-(x + 1) = x^2 - 5$

$$x^2 - x - 6 = 0 \qquad\qquad -x - 1 = x^2 - 5$$
$$(x-3)(x+2) = 0 \qquad\qquad x^2 + x - 4 = 0$$
$$x - 3 = 0 \Rightarrow x = 3 \qquad\qquad x = \frac{-1 \pm \sqrt{17}}{2}$$
$$x + 2 = 0 \Rightarrow x = -2$$

Only $x = 3$ and $x = \dfrac{-1 - \sqrt{17}}{2}$ are solutions to the original equation. $x = -2$ and $x = \dfrac{-1 + \sqrt{17}}{2}$ are extraneous.

190. $|x - 10| = x^2 - 10x$

First equation:

$x - 10 = x^2 - 10x$

$0 = x^2 - 11x + 10$

$0 = (x - 1)(x - 10)$

$0 = x - 1 \Rightarrow x = 1$, not a solution

$0 = x - 10 \Rightarrow x = 10$

Second equation:

$-(x - 10) = x^2 - 10x$

$0 = x^2 - 9x - 10$

$0 = (x - 10)(x + 1)$

$0 = x - 10 \Rightarrow x = 10$

$0 = x + 1 \Rightarrow x = -1$

191. The student should have subtracted $15x$ from both sides so that the equation is equal to zero. By factoring out an x, there are **two** solutions.

$x = 0$ or $x = 6$

192. The **zero-factor** property states that if the product of two factors is **zero,** then one (or both) of the factors much be zero. To solve $4x^2 + 4x = 15$, a student factors $4x$ from the left side of the equation and sets each factor equal to 15. The resulting **incorrect** solutions are $x = \frac{15}{4}$ and $x = 14$. To attempt to solve a quadratic equation by factoring, the equation should be in **general form** first. If it factors, then the zero-factor property can be employed to solve the equation.

$4x^2 + 4x = 15$

$4x^2 + 4x - 15 = 0$

$(2x - 3)(2x + 5) = 0$

$2x - 3 = 0 \Rightarrow x = \frac{3}{2}$

$2x + 5 = 0 \Rightarrow x = -\frac{5}{2}$

193. Answers will vary. For example:
Two equations are equivalent if they differ only by algebraic simplification steps and have the same solutions.

$x^2 + 3x + 4 = x + 1$ is equivalent to $x^2 + 2x + 3 = 0$.

194. Remove symbols of grouping, combine like terms, reduce fractions.

Add (or subtract) the same quantity to (from) both sides of the equation.

Multiply (or divide) both sides of the equation by the same nonzero quantity.

Interchange the two sides of the equation.

195. $C = 0.40m + 2750$

When $C = 10{,}000$ we have:

$10{,}000 = 0.40m + 2750$

$7250 = 0.40m$

$m = 18{,}125$ miles

196. Let n be the number of feet above flood level after t hours.

$n = -\frac{1}{4}t + 8$

$1 = -\frac{1}{4}t + 8 \Rightarrow t = 28$ hours

197. (a)

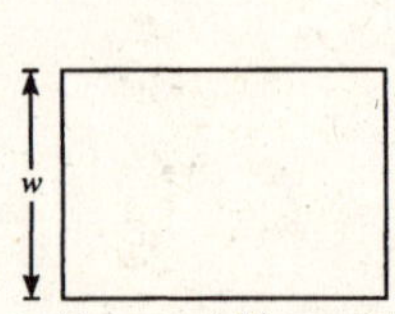

(b) $w(w + 14) = 1632$

$w^2 + 14w - 1632 = 0$

(c) $(w - 34)(w + 48) = 0$

$w = 34$ or $w = -48$ Extraneous

$l = 34 + 14 = 48$

width: 34 feet

length: 48 feet

198. $S = x^2 + 4xh$

$84 = x^2 + 4x(2)$

$0 = x^2 + 8x - 84$

$0 = (x + 14)(x - 6)$

$x = -14$ or $x = 6$

Since x must be positive, we have $x = 6$ inches. The dimensions of the base are 6 inches × 6 inches.

199. $16 = 0.432x - 10.44$

$26.44 = 0.432x$

$\frac{26.44}{0.432} = x$

$x \approx 61.2$ inches

200. $y = 0.449x - 12.15 = 19$

$x = \frac{19 + 12.15}{0.449} \approx 69.4$ inches, which is approximately 69 inches, so it is likely that both the foot bones and the thigh bone came from the same person.

201.

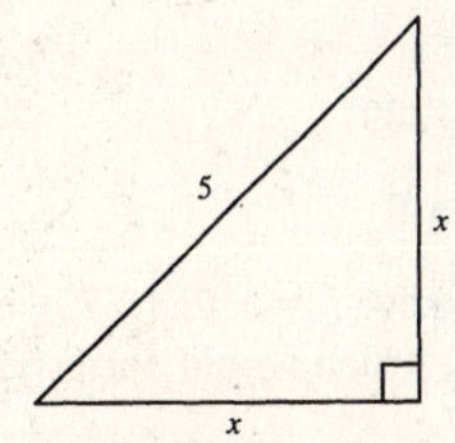

$x^2 + x^2 = 5^2$

$2x^2 = 25$

$x^2 = \frac{25}{2}$

$x = \sqrt{\frac{25}{2}} = \frac{5}{\sqrt{2}} = \frac{5\sqrt{2}}{2} \approx 3.54$ centimeters

202. $x^2 = 10^2 + \left(\frac{1}{2}x\right)^2$

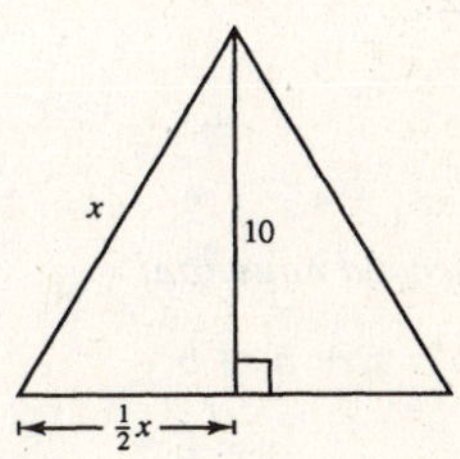

$x^2 = 100 + \frac{1}{4}x^2$

$\frac{3}{4}x^2 = 100$

$x^2 = \frac{400}{3}$

$x = \pm\sqrt{\frac{400}{3}} = \pm\frac{20}{\sqrt{3}} = \pm\frac{20\sqrt{3}}{3}$

Choose the positive value for x. So, the length of each side is $\frac{20\sqrt{3}}{3}$ inches.

203. Let r = speed of the eastbound plane and $r + 50$ = speed of the northbound plane. After 3 hours the eastbound plane has traveled $3r$ miles and the northbound plane has traveled $3(r + 50)$ miles.

$(3r)^2 + [3(r + 50)]^2 = 2440^2$

$9r^2 + 9(r^2 + 100r + 2500) = 5{,}953{,}600$

$18r^2 + 900r - 5{,}931{,}100 = 0$

By the Quadratic Formula, $r \approx 550$ (discard the negative value of r as extraneous).

Speed of the eastbound plane: 550 miles per hour

Speed of the northbound plane: 600 miles per hour

204. $C = \sqrt{0.2x + 1} = 2.5$

$0.2x + 1 = (2.5)^2$

$0.2x = 6.25 - 1$

$x = 26.25$

26,250 passengers

205. $x(20 - 0.0002x) = 500{,}000$

$20x - 0.0002x^2 = 500{,}000$

$0 = 0.0002x^2 - 20x + 500{,}000$

By the Quadratic Formula, $x = 50{,}000$ units.

206.
$$p = 40 - \sqrt{0.01x + 1}$$
$$37.55 = 40 - \sqrt{0.01x + 1}$$
$$40 - 37.55 = \sqrt{0.01x + 1}$$
$$(2.45)^2 = 0.01x + 1$$
$$x = 500.25$$

The demand is approximately 500 units per day.

207. $p = 0.0014t^2 + 0.004t + 0.23$ where $t = 0$ corresponds to 1940.

$0.0014t^2 + 0.004t + 0.23 > 8.00$

$0.0014t^2 + 0.004t - 7.77 > 0$

By the Quadratic Formula we have $t \approx -76$, which is extraneous, and $t \approx 73$, which would correspond to the year 2013.

208. $p = 0.0094t^2 + 1.078t + 78.31$ where $t = 0$ corresponds to 1900.

$0.0094t^2 + 1.078t + 78.31 > 300$

$0.0094t^2 + 1.078t - 221.69 > 0$

By the Quadratic Formula we have $t \approx -221.27$, which is extraneous, and $t \approx 106.59$, which would correspond to the early part of the year 2007.

209. False. $x(3 - x) = 10$ is a quadratic equation.

210. False. The product must equal **zero** for the Zero Factor Property to be used.

211. True. See Example 11. In solving the radical equation $\sqrt{2x + 7} - x = 2$, an extraneous solution of $x = -3$ was produced.

212. True, the expression inside the absolute value signs can be positive or negative. This results in two separate equations when solving an equation involving an absolute value.

213. $x + |x - a| = b$

Solving for x we have:

First equation:

$x + x - a = b$

$x = \dfrac{a + b}{2}$

Second equation:

$x - x + a = b$

$a = b$

Thus, $x = 9$ will be the only solution if $9 = \dfrac{a + b}{2}$ and $a \neq b$ and $9 > a$.

For example,

$a = 0, b = 18$

$a = 2, b = 16$, etc.

Answers will vary.

214. Isolate the absolute value by subtracting x from both sides of the equation. The expression inside the absolute value signs can be positive or negative, so two separate equations must be solved.

Section P.2 Solving Inequalities

1. Interval: $[-1, 5]$

Inequality: $-1 \le x \le 5$

The interval is bounded.

2. Interval: $(2, 10]$

Inequality: $2 < x \le 10$

The interval is bounded.

3. Interval: $(11, \infty)$

Inequality: $11 < x < \infty$

The interval is unbounded.

4. Interval: $[-5, \infty)$

Inequality:
$-5 \le x < \infty$ or $x \ge -5$

The interval is unbounded.

5. Interval: $(-\infty, -2)$

Inequality:
$-\infty < x < -2$ or $x < -2$

The interval is unbounded.

6. Interval: $(-\infty, 7]$

Inequality:
$-\infty < x \le 7$ or $x \le -7$

The interval is unbounded.

7. $x < 3$

Matches (b)

8. $x \ge 5$

Matches (f)

9. $-3 < x \le 4$

Matches (d)

10. $0 \le x \le \frac{9}{2}$

Matches (c)

11. $|x| < 3 \Rightarrow -3 < x < 3$

Matches (e)

12. $|x| > 4 \Rightarrow x > 4$ or $x < -4$

Matches (a)

13. (a) $x = 3$

$5(3) - 12 \overset{?}{>} 0$

$3 > 0$

Yes, $x = 3$ is a solution.

(b) $x = -3$

$5(-3) - 12 \overset{?}{>} 0$

$-27 \not> 0$

No, $x = -3$ is not a solution.

(c) $x = \frac{5}{2}$

$5(\frac{5}{2}) - 12 \overset{?}{>} 0$

$\frac{1}{2} > 0$

Yes, $x = \frac{5}{2}$ is a solution.

(d) $x = \frac{3}{2}$

$5(\frac{3}{2}) - 12 \overset{?}{>} 0$

$-\frac{9}{2} \not> 0$

No, $x = \frac{3}{2}$ is not a solution.

14. $2x + 1 < -3$

(a) $x = 0$

$2(0) + 1 \overset{?}{<} -3$

$1 \not< -3$

No, $x = 0$ is not a solution.

(b) $x = -\frac{1}{4}$

$2(-\frac{1}{4}) + 1 \overset{?}{<} -3$

$\frac{1}{2} \not< -3$

No, $x = -\frac{1}{4}$ is not a solution.

(c) $x = -4$

$2(-4) + 1 \overset{?}{<} -3$

$-7 < -3$

Yes, $x = -4$ is a solution.

(d) $x = -\frac{3}{2}$

$2(-\frac{3}{2}) + 1 \overset{?}{<} -3$

$-2 \not< -3$

No, $x = -\frac{3}{2}$ is not a solution.

15. (a) $x = 4$

$0 \overset{?}{<} \dfrac{4 - 2}{4} \overset{?}{<} 2$

$0 < \dfrac{1}{2} < 2$

Yes, $x = 4$ is a solution.

(b) $x = 10$

$0 \overset{?}{<} \dfrac{10 - 2}{4} \overset{?}{<} 2$

$0 < 2 \not< 2$

No, $x = 10$ is not a solution.

(c) $x = 0$

$0 \overset{?}{<} \dfrac{0 - 2}{4} \overset{?}{<} 2$

$0 \not< -\dfrac{1}{2} < 2$

No, $x = 0$ is not a solution.

(d) $x = \dfrac{7}{2}$

$0 \overset{?}{<} \dfrac{\frac{7}{2} - 2}{4} \overset{?}{<} 2$

$0 < \dfrac{3}{8} < 2$

Yes, $x = \frac{7}{2}$ is a solution.

16. $-1 < \dfrac{3-x}{2} \le 1$

(a) $x = 0$

$-1 \overset{?}{<} \dfrac{3-0}{2} \overset{?}{\le} 1$

$-1 < \dfrac{3}{2} \nleq 1$

No, $x = 0$ is not a solution.

(b) $x = -5$

$-1 \overset{?}{<} \dfrac{3+5}{2} \overset{?}{\le} 1$

$-1 < 4 \nleq 1$

No, $x = -5$ is not a solution.

(c) $x = 1$

$-1 \overset{?}{<} \dfrac{3-1}{2} \overset{?}{\le} 1$

$-1 < 1 \le 1$

Yes, $x = 1$ is a solution.

(d) $x = 5$

$-1 \overset{?}{<} \dfrac{3-5}{2} \overset{?}{\le} 1$

$-1 \nless -1 \le 1$

No, $x = 5$ is not a solution.

17. (a) $x = 13$

$|13 - 10| \overset{?}{\ge} 3$

$3 \ge 3$

Yes, $x = 13$ is a solution.

(b) $x = -1$

$|-1 - 10| \overset{?}{\ge} 3$

$11 \ge 3$

Yes, $x = -1$ is a solution.

(c) $x = 14$

$|14 - 10| \overset{?}{\ge} 3$

$4 \ge 3$

Yes, $x = 14$ is a solution.

(d) $x = 9$

$|9 - 10| \overset{?}{\ge} 3$

$1 \ngeq 3$

No, $x = 9$ is not a solution.

18. $|2x - 3| < 15$

(a) $x = -6$

$|2(-6) - 3| \overset{?}{<} 15$

$15 \nless 15$

No, $x = -6$ is not a solution.

(b) $x = 0$

$|2(0) - 3| \overset{?}{<} 15$

$3 < 15$

Yes, $x = 0$ is a solution.

(c) $x = 12$

$|2(12) - 3| \overset{?}{<} 15$

$21 \nless 15$

No, $x = 12$ is not a solution.

(d) $x = 7$

$|2(7) - 3| \overset{?}{<} 15$

$11 < 15$

Yes, $x = 7$ is a solution.

19. $4x < 12$

$\frac{1}{4}(4x) < \frac{1}{4}(12)$

$x < 3$

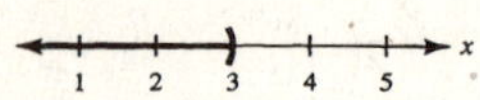

20. $-10x < 40$

$x > -4$

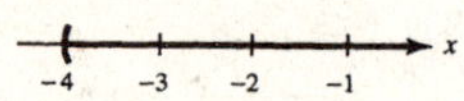

21. $2x > 3$

$x > \frac{3}{2}$

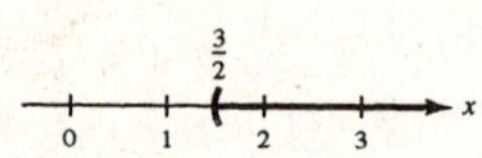

22. $-6x > 15$

$x < -\frac{15}{6}$

$x < -\frac{5}{2}$

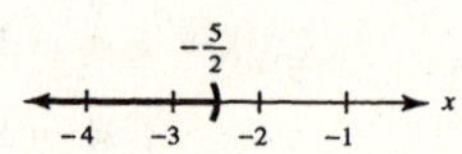

23. $x - 5 \ge 7$

$x \ge 12$

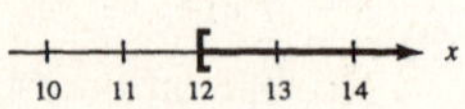

24. $x + 7 \le 12$

$x \le 5$

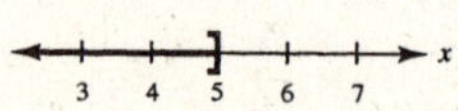

25. $2x + 7 < 3 + 4x$

$-2x < -4$

$x > 2$

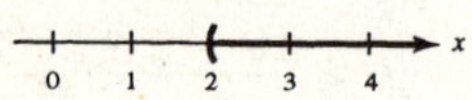

26. $3x + 1 \ge 2 + x$

$2x \ge 1$

$x \ge \frac{1}{2}$

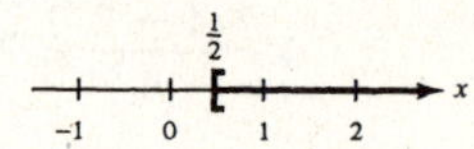

27. $2x - 1 \ge 1 - 5x$

$7x \ge 2$

$x \ge \frac{2}{7}$

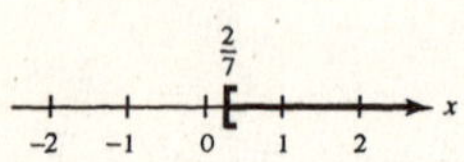

28. $6x - 4 \le 2 + 8x$

$-2x \le 6$

$x \ge -3$

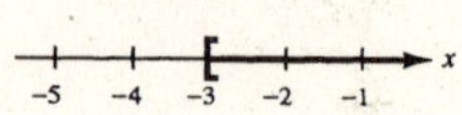

29. $4 - 2x < 3(3 - x)$

$4 - 2x < 9 - 3x$

$x < 5$

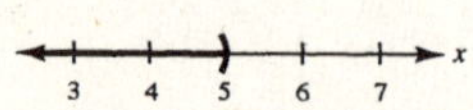

30. $4(x + 1) < 2x + 3$

$4x + 4 < 2x + 3$

$2x < -1$

$x < -\frac{1}{2}$

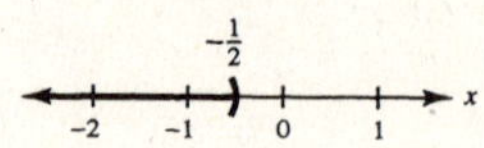

31. $\frac{3}{4}x - 6 \le x - 7$

$-\frac{1}{4}x \le -1$

$x \ge 4$

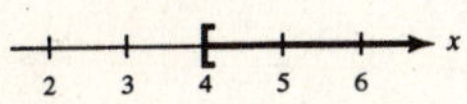

32. $3 + \frac{2}{7}x > x - 2$

$21 + 2x > 7x - 14$

$-5x > -35$

$x < 7$

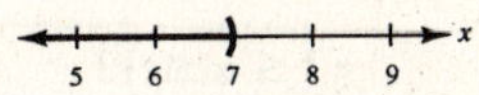

33. $\frac{1}{2}(8x + 1) \ge 3x + \frac{5}{2}$

$4x + \frac{1}{2} \ge 3x + \frac{5}{2}$

$x \ge 2$

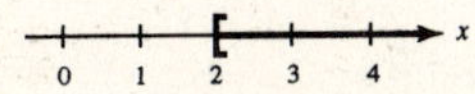

34. $9x - 1 < \frac{3}{4}(16x - 2)$

$36x - 4 < 48x - 6$

$-12x < -2$

$x > \frac{1}{6}$

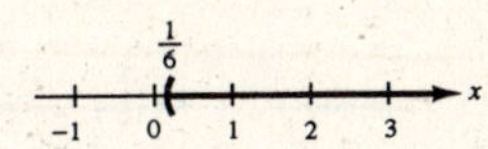

35. $3.6x + 11 \ge -3.4$

$3.6x \ge -14.4$

$x \ge -4$

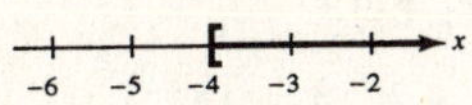

36. $15.6 - 1.3x < -5.2$

$-1.3x < -20.8$

$x > 16$

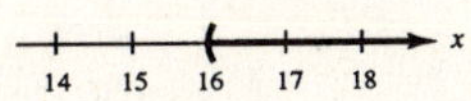

37. $1 < 2x + 3 < 9$

$-2 < 2x < 6$

$-1 < x < 3$

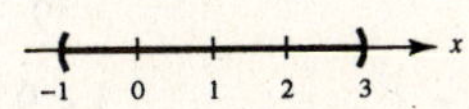

38. $-8 \le -(3x + 5) < 13$

$-8 \le -3x - 5 < 13$

$-3 \le -3x < 18$

$-6 < x \le 1$

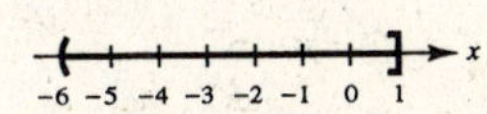

39. $-4 < \dfrac{2x - 3}{3} < 4$

$-12 < 2x - 3 < 12$

$-9 < 2x < 15$

$-\dfrac{9}{2} < x < \dfrac{15}{2}$

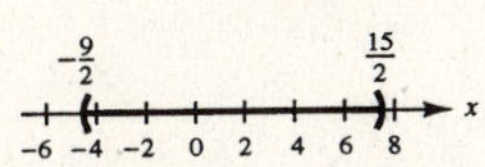

40. $0 \le \dfrac{x + 3}{2} < 5$

$0 \le x + 3 < 10$

$-3 \le x < 7$

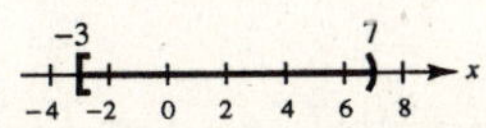

41. $\dfrac{3}{4} > x + 1 > \dfrac{1}{4}$

$-\dfrac{1}{4} > x > -\dfrac{3}{4}$

$-\dfrac{3}{4} < x < -\dfrac{1}{4}$

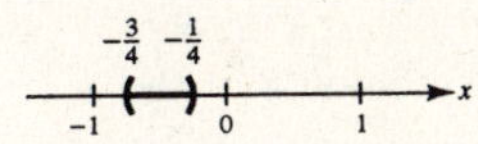

42. $-1 < 2 - \dfrac{x}{3} < 1$

$-3 < 6 - x < 3$

$-9 < -x < -3$

$3 < x < 9$

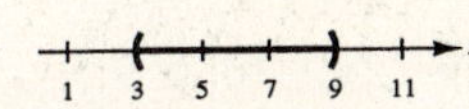

43. $3.2 \le 0.4x - 1 \le 4.4$

$4.2 \le 0.4x \le 5.4$

$10.5 \le x \le 13.5$

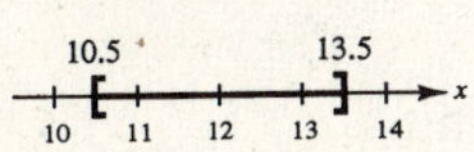

44. $4.5 > \dfrac{1.5x + 6}{2} > 10.5$

$9 > 1.5x + 6 > 21$

$3 > 1.5x > 15$

$2 > x > 10$

There is no solution.

45. $|x| < 6$

$-6 < x < 6$

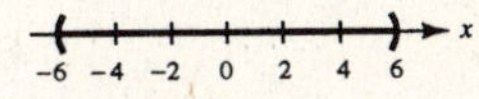

46. $|x| < 4$

$-4 < x < 4$

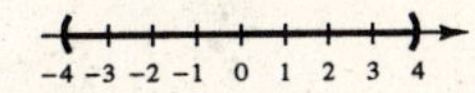

47. $\left|\dfrac{x}{2}\right| > 5$

$\dfrac{x}{2} < -5$ or $\dfrac{x}{2} > 5$

$x < -10$ $\quad$ $x > 10$

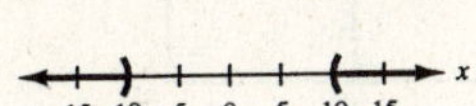

48. $\left|\dfrac{x}{5}\right| > 3$

$\dfrac{x}{5} < -3$ or $\dfrac{x}{5} > 3$

$x < -15$ or $x > 15$

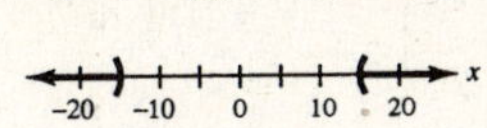

49. $|x - 5| < -1$

No solution. The absolute value of a number cannot be less than a negative number.

50. $|x - 5| \le -2$

No solution since $|x - 5|$ must be greater than or equal to zero.

51. $|x - 20| \le 6$

$-6 \le x - 20 \le 6$

$14 \le x \le 26$

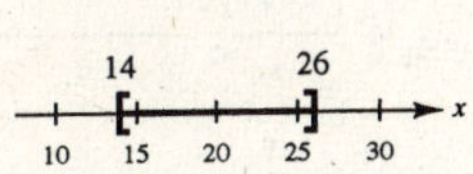

52. $|x - 7| < 5$

$-5 < x - 7 < 5$

$2 < x < 12$

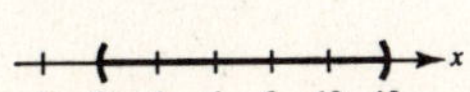

53. $|3 - 4x| \ge 9$

$3 - 4x \le -9$ or $3 - 4x \ge 9$

$-4x \le -12$ $\quad$ $-4x \ge 6$

$x \ge 3$ $\quad$ $x \le -\frac{3}{2}$

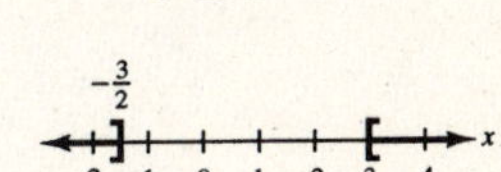

54. $|1 - 2x| > 5$

$1 - 2x < -5$ or $1 - 2x > 5$

$-2x < -6$ $\quad$ $-2x > 4$

$x > 3$ $\quad$ $x < -2$

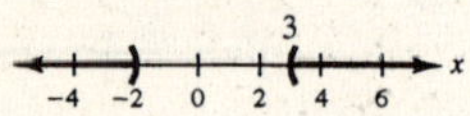

55. $\left|\frac{x - 3}{2}\right| \le 5$

$-5 \le \frac{x - 3}{2} \le 5$

$-10 \le x - 3 \le 10$

$-7 \le x \le 13$

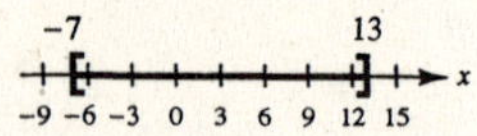

56. $\left|1 - \frac{2x}{3}\right| \le 1$

$-1 \le 1 - \frac{2x}{3} \le 1$

$-2 \le -\frac{2x}{3} \le 0$

$-6 \le -2x \le 0$

$3 \ge x \ge 0$ or $0 \le x \le 3$

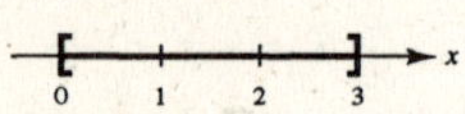

57. $|9 - 2x| - 2 < -1$

$|9 - 2x| < 1$

$-1 < 9 - 2x < 1$

$-10 < -2x < -8$

$5 > x > 4$

$4 < x < 5$

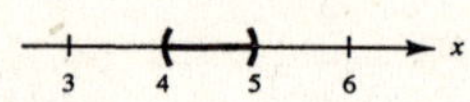

58. $|x + 14| + 3 < 17$

$|x + 14| < 14$

$-14 < x + 14 < 14$

$-28 < x < 0$

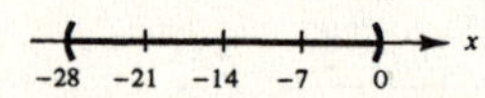

59. $2|x + 10| \ge 9$

$|x + 10| \ge \frac{9}{2}$

$x + 10 \le -\frac{9}{2}$ or $x + 10 \ge \frac{9}{2}$

$x \le -\frac{29}{2}$ $\quad$ $x \ge -\frac{11}{2}$

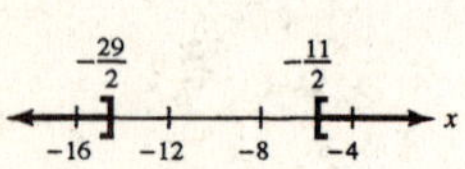

60. $3|4 - 5x| \ge 9$

$|4 - 5x| \ge 3$

$4 - 5x \le -3$ or $4 - 5x \ge 3$

$-5x \le -7$ $\quad$ $-5x \ge -1$

$x \ge \frac{7}{5}$ $\quad$ $x \le \frac{1}{5}$

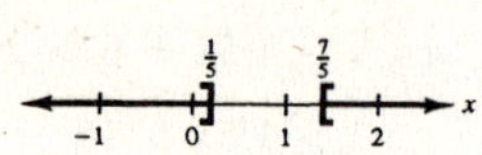

61. $6x > 12$

$x > 2$

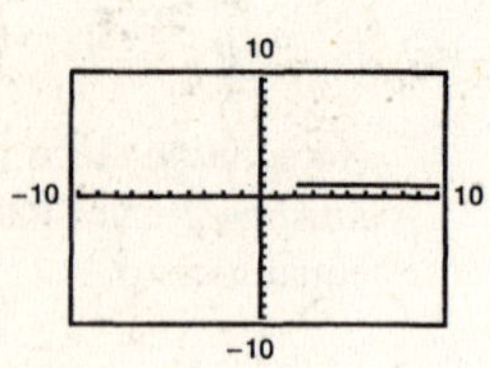

62. $3x - 1 \le 5$

$3x \le 6$

$x \le 2$

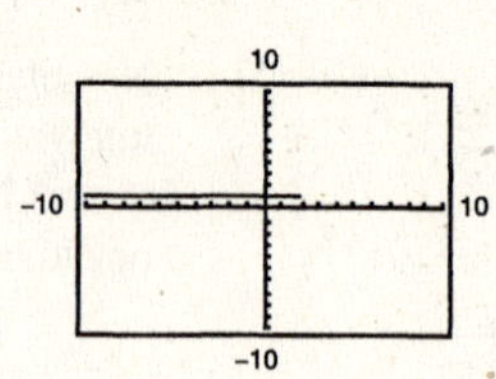

63. $5 - 2x \ge 1$

$-2x \ge -4$

$x \le 2$

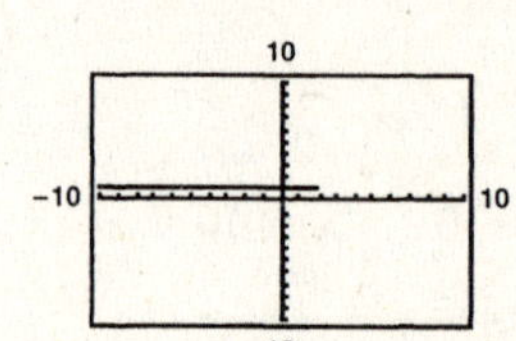

64. $3(x + 1) < x + 7$

$3x + 3 < x + 7$

$2x < 4$

$x < 2$

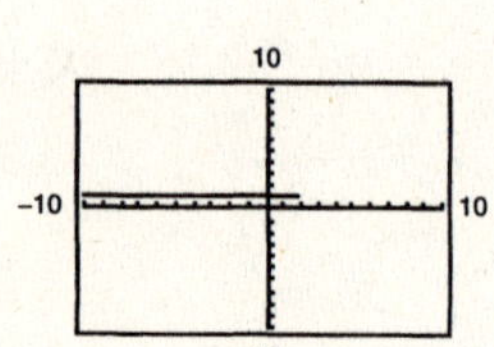

65. $|x - 8| \le 14$

$-14 \le x - 8 \le 14$

$-6 \le x \le 22$

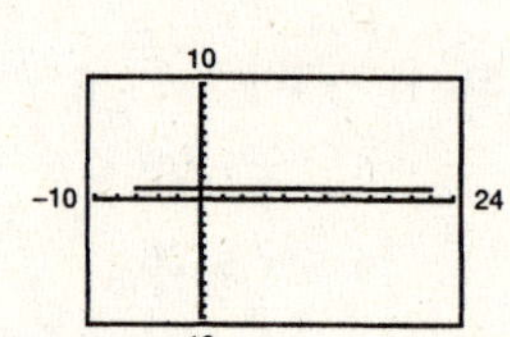

66. $|2x + 9| > 13$

$2x + 9 < -13$ or $2x + 9 > 13$

$2x < -22$ or $2x > 4$

$x < -11$ or $x > 2$

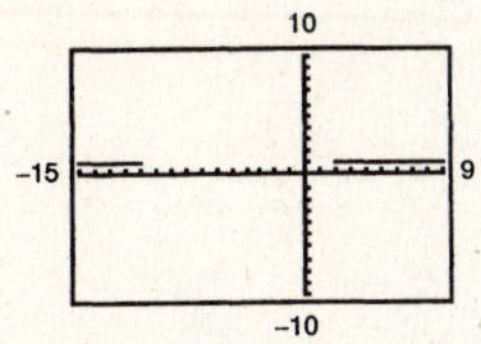

67. $2|x + 7| \ge 13$

$|x + 7| \ge \frac{13}{2}$

$x + 7 \le -\frac{13}{2}$ or $x + 7 \ge \frac{13}{2}$

$x \le -\frac{27}{2}$ $\quad$ $x \ge -\frac{1}{2}$

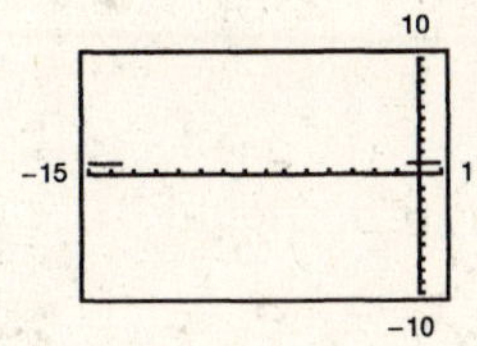

68. $\frac{1}{2}|x + 1| \le 3$

$|x + 1| \le 6$

$-6 \le x + 1 \le 6$

$-7 \le x \le 5$

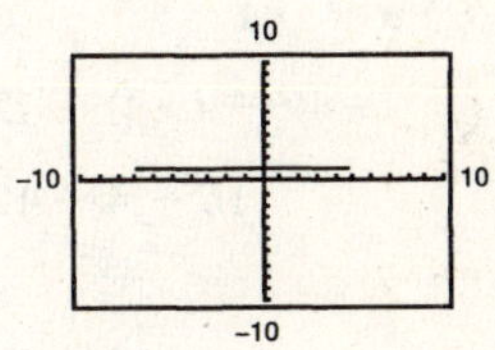

69. $x - 5 \ge 0$

$x \ge 5$

$[5, \infty)$

70. $\sqrt{x - 10}$

$x - 10 \ge 0$

$x \ge 10$

$[10, \infty)$

71. $x + 3 \ge 0$

$x \ge -3$

$[-3, \infty)$

72. $\sqrt{3 - x}$

$3 - x \ge 0$

$3 \ge x$

$(-\infty, 3]$

73. $7 - 2x \ge 0$

$-2x \ge -7$

$x \le \frac{7}{2}$

$\left(-\infty, \frac{7}{2}\right]$

74. $\sqrt[4]{6x + 15}$

$6x + 15 \ge 0$

$6x \ge -15$

$x \ge -\frac{5}{2}$

$\left[-\frac{5}{2}, \infty\right)$

75. $|x - 10| < 8$

All real numbers within 8 units of 10

76. $|x - 8| > 4$

All real numbers more than 4 units from 8

77. The midpoint of the interval $[-3, 3]$ is 0. The interval represents all real numbers x no more than 3 units from 0.

$|x - 0| \le 3$

$|x| \le 3$

78. The graph shows all real numbers more than 3 units from 0.

$|x - 0| > 3$

$|x| > 3$

79. The graph shows all real numbers at least 3 units from 7.

$|x - 7| \ge 3$

80. The graph shows all real numbers no more than 4 units from -1.

$|x + 1| \le 4$

81. All real numbers within 10 units of 12

$|x - 12| < 10$

82. All real numbers at least 5 units from 8

$|x - 8| \ge 5$

83. All real numbers more than 5 units from -3

$|x - (-3)| > 5$

$|x + 3| > 5$

84. All real numbers no more than 7 units from -6

$|x + 6| \le 7$

85. $x^2 - 3 < 0$

(a) $x = 3$

$(3)^2 - 3 \overset{?}{<} 0$

$6 \not< 0$

No, $x = 3$ is not a solution.

(b) $x = 0$

$(0)^2 - 3 \overset{?}{<} 0$

$-3 < 0$

Yes, $x = 0$ is a solution.

(c) $x = \frac{3}{2}$

$(\frac{3}{2})^2 - 3 \overset{?}{<} 0$

$-\frac{3}{4} < 0$

Yes, $x = \frac{3}{2}$ is a solution.

(d) $x = -5$

$(-5)^2 - 3 \overset{?}{<} 0$

$22 \not< 0$

No, $x = -5$ is not a solution.

86. $x^2 - x - 12 \geq 0$

(a) $x = 5$

$(5)^2 - (5) - 12 \overset{?}{\geq} 0$

$8 \geq 0$

Yes, $x = 5$ is a solution.

(b) $x = 0$

$(0)^2 - 0 - 12 \overset{?}{\geq} 0$

$-12 \not\geq 0$

No, $x = 0$ is not a solution.

(c) $x = -4$

$(-4)^2 - (-4) - 12 \overset{?}{\geq} 0$

$16 + 4 - 12 \overset{?}{\geq} 0$

$8 \geq 0$

Yes, $x = -4$ is a solution.

(d) $x = -3$

$(-3)^2 - (-3) - 12 \overset{?}{\geq} 0$

$9 + 3 - 12 \overset{?}{\geq} 0$

$0 \geq 0$

Yes, $x = -3$ is a solution.

87. $\dfrac{x + 2}{x - 4} \geq 3$

(a) $x = 5$

$\dfrac{5 + 2}{5 - 4} \overset{?}{\geq} 3$

$7 \geq 3$

Yes, $x = 5$ is a solution.

(b) $x = 4$

$\dfrac{4 + 2}{4 - 4} \overset{?}{\geq} 3$

$\dfrac{6}{0} \not\geq 3$; $\dfrac{6}{0}$ is undefined.

No, $x = 4$ is not a solution.

(c) $x = -\dfrac{9}{2}$

$\dfrac{-\frac{9}{2} + 2}{-\frac{9}{2} - 4} \overset{?}{\geq} 3$

$\dfrac{5}{17} \not\geq 3$

No, $x = -\dfrac{9}{2}$ is not a solution.

(d) $x = \dfrac{9}{2}$

$\dfrac{\frac{9}{2} + 2}{\frac{9}{2} - 4} \overset{?}{\geq} 3$

$13 \geq 3$

Yes, $x = \dfrac{9}{2}$ is a solution.

88. $\dfrac{3x^2}{x^2+4} < 1$

(a) $x = -2$

$$\frac{3(-2)^2}{(-2)^2+4} \overset{?}{<} 1$$

$$\frac{12}{8} \not< 1$$

No, $x = -2$ is not a solution.

(b) $x = -1$

$$\frac{3(-1)^2}{(-1)^2+4} \overset{?}{<} 1$$

$$\frac{3}{5} < 1$$

Yes, $x = -1$ is a solution.

(c) $x = 0$

$$\frac{3(0)^2}{(0)^2+4} \overset{?}{<} 1$$

$$0 < 1$$

Yes, $x = 0$ is a solution.

(d) $x = 3$

$$\frac{3(3)^2}{(3)^2+4} \overset{?}{<} 1$$

$$\frac{27}{13} \not< 1$$

No, $x = 3$ is not a solution.

89. $2x^2 - x - 6 = (2x+3)(x-2)$

$$2x + 3 = 0 \implies x = -\frac{3}{2}$$

$$x - 2 = 0 \implies x = 2$$

Critical numbers: $x = -\frac{3}{2}, x = 2$

90. $9x^3 - 25x^2 = 0$

$$x^2(9x - 25) = 0$$

$$x^2 = 0 \implies x = 0$$

$$9x - 25 = 0 \implies x = \frac{25}{9}$$

The critical numbers are 0 and $\frac{25}{9}$.

91. $2 + \dfrac{3}{x-5} = \dfrac{2(x-5)+3}{x-5} = \dfrac{2x-7}{x-5}$

$$2x - 7 = 0 \implies x = \frac{7}{2}$$

$$x - 5 = 0 \implies x = 5$$

Critical numbers: $x = \frac{7}{2}, x = 5$

92.

$$\frac{x}{x+2} - \frac{2}{x-1} = \frac{x(x-1) - 2(x+2)}{(x+2)(x-1)} = \frac{x^2 - x - 2x - 4}{(x+2)(x-1)} = \frac{(x-4)(x+1)}{(x+2)(x-1)}$$

$$(x-4)(x+1) = 0$$

$$x - 4 = 0 \implies x = 4$$

$$x + 1 = 0 \implies x = -1$$

$$(x+2)(x-1) = 0$$

$$x + 2 = 0 \implies x = -2$$

$$x - 1 = 0 \implies x = 1$$

The critical numbers are $-2, -1, 1, 4$.

93. $x^2 \le 9$

$x^2 - 9 \le 0$

$(x + 3)(x - 3) \le 0$

Critical numbers: $x = \pm 3$

Test intervals: $(-\infty, -3), (-3, 3), (3, \infty)$

Test: Is $(x + 3)(x - 3) \le 0$?

Interval	*x-value*	*Value of* $x^2 - 9$	*Conclusion*
$(-\infty, -3)$	$x = -4$	$16 - 9 = 7$	Positive
$(-3, 3)$	$x = 0$	$0 - 9 = -9$	Negative
$(3, \infty)$	$x = 4$	$16 - 9 = 7$	Positive

Solution set: $[-3, 3]$

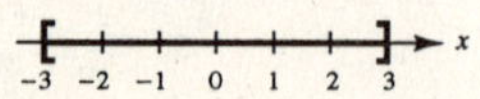

94. $x^2 < 5$

$x^2 - 5 < 0$

$(x + \sqrt{5})(x - \sqrt{5}) = 0$

Critical numbers: $x = \pm\sqrt{5}$

Test intervals: $(-\infty, -\sqrt{5}), (-\sqrt{5}, \sqrt{5}), (\sqrt{5}, \infty)$

Test: Is $x^2 - 5 < 0$?

Interval	*x-value*	*Value of* $x^2 - 5$	*Conclusion*
$(-\infty, -\sqrt{5})$	$x = -3$	$9 - 5 = 4$	Positive
$(-\sqrt{5}, \sqrt{5})$	$x = 0$	$0 - 5 = -5$	Negative
$(\sqrt{5}, \infty)$	$x = 3$	$9 - 5 = 4$	Positive

Solution interval: $(-\sqrt{5}, \sqrt{5})$

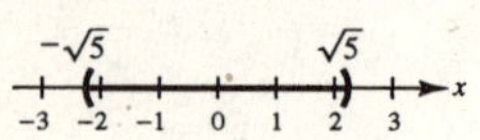

95. $(x + 2)^2 < 25$

$x^2 + 4x + 4 < 25$

$x^2 + 4x - 21 < 0$

$(x + 7)(x - 3) < 0$

Critical numbers: $x = -7, x = 3$

Test intervals: $(-\infty, -7), (-7, 3), (3, \infty)$

Test: Is $(x + 7)(x - 3) < 0$?

Interval	*x-value*	*Value of* $(x + 7)(x - 3)$	*Conclusion*
$(-\infty, -7)$	$x = -10$	$(-3)(-13) = 39$	Positive
$(-7, 3)$	$x = 0$	$(7)(-3) = -21$	Negative
$(3, \infty)$	$x = 5$	$(12)(2) = 24$	Positive

Solution set: $(-7, 3)$

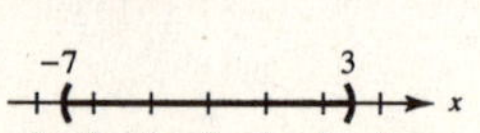

96. $(x - 3)^2 \le 1$

$x^2 - 6x + 9 \le 1$

$x^2 - 6x + 8 \le 0$

$(x - 2)(x - 4) \le 0$

Critical numbers: $x = 2, x = 4$

Test intervals: $(-\infty, 2), (2, 4), (4, \infty)$

Test: Is $(x - 2)(x - 4) \le 0$?

Interval	*x-value*	*Value of* $(x - 2)(x - 4)$	*Conclusion*
$(-\infty, 2)$	$x = 0$	$(-2)(-4) = 8$	Positive
$(2, 4)$	$x = 3$	$(1)(-1) = -1$	Negative
$(4, \infty)$	$x = 5$	$(3)(1) = 3$	Positive

Solution set: $[2, 4]$

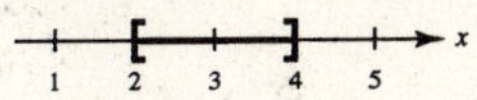

97. $x^2 + 4x + 4 \ge 9$

$x^2 + 4x - 5 \ge 0$

$(x + 5)(x - 1) \ge 0$

Critical numbers: $x = -5, x = 1$

Test intervals: $(-\infty, -5), (-5, 1), (1, \infty)$

Test: Is $(x + 5)(x - 1) \ge 0$?

Interval	*x-value*	*Value of* $(x + 5)(x - 1)$	*Conclusion*
$(-\infty, -5)$	$x = -6$	$(-1)(-7) = 7$	Positive
$(-5, 1)$	$x = 0$	$(5)(-1) = -5$	Negative
$(1, \infty)$	$x = 2$	$(7)(1) = 7$	Positive

Solution set: $(-\infty, -5] \cup [1, \infty)$

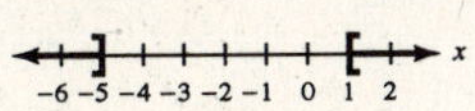

98. $x^2 - 6x + 9 \ge 16$

$x^2 - 6x - 7 \ge 0$

$(x - 7)(x + 1) \ge 0$

Critical numbers: $x = 7, x = -1$

Test intervals: $(-\infty, -1), (-1, 7), (7, \infty)$

Test: Is $(x - 7)(x + 1) \ge 0$?

Interval	*x-value*	*Value of* $(x - 7)(x + 1)$	*Conclusion*
$(-\infty, -1)$	$x = -2$	$(-9)(-1) = 9$	Positive
$(-1, 7)$	$x = 0$	$(-7)(1) = -7$	Negative
$(7, \infty)$	$x = 10$	$(3)(11) = 33$	Positive

Solution set: $(-\infty, -1] \cup [7, \infty)$

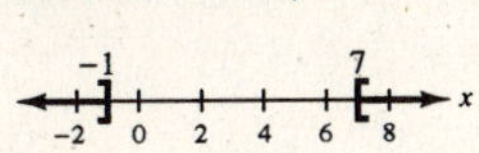

99. $x^2 + x < 6$

$x^2 + x - 6 < 0$

$(x + 3)(x - 2) < 0$

Critical numbers: $x = -3, x = 2$

Test intervals: $(-\infty, -3), (-3, 2), (2, \infty)$

Test: Is $(x + 3)(x - 2) < 0$?

Interval	*x-value*	*Value of* $(x + 3)(x - 2)$	*Conclusion*
$(-\infty, -3)$	$x = -4$	$(-1)(-6) = 6$	Positive
$(-3, 2)$	$x = 0$	$(3)(-2) = -6$	Negative
$(2, \infty)$	$x = 3$	$(6)(1) = 6$	Positive

Solution set: $(-3, 2)$

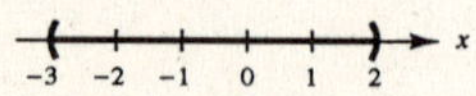

100. $x^2 + 2x > 3$

$x^2 + 2x - 3 > 0$

$(x + 3)(x - 1) > 0$

Critical numbers: $x = -3, x = 1$

Test intervals: $(-\infty, -3), (-3, 1), (1, \infty)$

Test: Is $(x + 3)(x - 1) > 0$?

Interval	*x-value*	*Value of* $(x + 3)(x - 1)$	*Conclusion*
$(-\infty, -3)$	$x = -4$	$(-1)(-5) = 5$	Positive
$(-3, 1)$	$x = 0$	$(3)(-1) = -3$	Negative
$(1, \infty)$	$x = 2$	$(5)(1) = 5$	Positive

Solution intervals: $(-\infty, -3) \cup (1, \infty)$

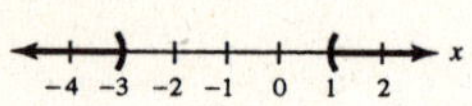

101. $x^2 + 2x - 3 < 0$

$(x + 3)(x - 1) < 0$

Critical numbers: $x = -3, x = 1$

Test intervals: $(-\infty, -3), (-3, 1), (1, \infty)$

Test: Is $(x + 3)(x - 1) < 0$?

Interval	*x-value*	*Value of* $(x + 3)(x - 1)$	*Conclusion*
$(-\infty, -3)$	$x = -4$	$(-1)(-5) = 5$	Positive
$(-3, 1)$	$x = 0$	$(3)(-1) = -3$	Negative
$(1, \infty)$	$x = 2$	$(5)(1) = 5$	Positive

Solution set: $(-3, 1)$

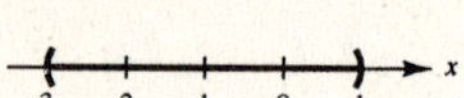

102. $x^2 - 4x - 1 > 0$

$$x = \frac{4 \pm \sqrt{16 + 4}}{2} = 2 \pm \sqrt{5}$$

Critical numbers: $x = 2 - \sqrt{5}, x = 2 + \sqrt{5}$

Test intervals: $(-\infty, 2 - \sqrt{5}), (2 - \sqrt{5}, 2 + \sqrt{5}), (2 + \sqrt{5}, \infty)$

Test: Is $x^2 - 4x - 1 > 0$?

Interval	*x-value*	*Value of* $x^2 - 4x - 1$	*Conclusion*
$(-\infty, 2 - \sqrt{5})$	$x = -1$	$(-1)^2 - 4(-1) - 1 = 4$	Positive
$(2 - \sqrt{5}, 2 + \sqrt{5})$	$x = 0$	$(0)^2 - 4(0) - 1 = -1$	Negative
$(2 + \sqrt{5}, \infty)$	$x = 5$	$(5)^2 - 4(5) - 1 = 4$	Positive

Solution intervals: $(-\infty, 2 - \sqrt{5}) \cup (2 + \sqrt{5}, \infty)$

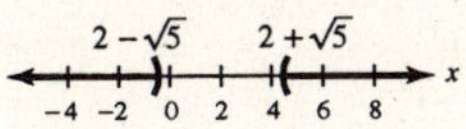

103. $x^2 + 8x - 5 \geq 0$

$x^2 + 8x - 5 = 0$ Complete the Square

$$x^2 + 8x + 16 = 5 + 16$$

$$(x + 4)^2 = 21$$

$$x + 4 = \pm\sqrt{21}$$

$$x = -4 \pm \sqrt{21}$$

Critical Numbers: $x = -4 \pm \sqrt{21}$

Test Intervals: $(-\infty, -4 - \sqrt{21}), (-4 - \sqrt{21}, -4 + \sqrt{21}), (-4 + \sqrt{21}, \infty)$

Test: Is $x^2 + 8x - 5 \geq 0$?

Interval	*x-value*	*Value of* $x^2 + 8x - 5$	*Conclusion*
$(-\infty, -4 - \sqrt{21})$	$x = -10$	$100 - 80 - 5 = 15$	Positive
$(-4 - \sqrt{21}, -4 + \sqrt{21})$	$x = 0$	$0 + 0 - 5 = -5$	Negative
$(-4 + \sqrt{21}, \infty)$	$x = 2$	$4 + 16 - 5 = 15$	Positive

Solution set: $(-\infty, -4 - \sqrt{21}] \cup [-4 + \sqrt{21}, \infty)$

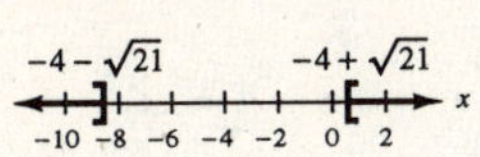

104. $-2x^2 + 6x + 15 \le 0$

$$2x^2 - 6x - 15 \ge 0$$

$$x = \frac{-(-6) \pm \sqrt{(-6)^2 - 4(2)(-15)}}{2(2)} = \frac{6 \pm \sqrt{156}}{4}$$

$$= \frac{6 \pm 2\sqrt{39}}{4} = \frac{3}{2} \pm \frac{\sqrt{39}}{2}$$

Critical numbers: $x = \frac{3}{2} - \frac{\sqrt{39}}{2},\ x = \frac{3}{2} + \frac{\sqrt{39}}{2}$

Test intervals: $\left(-\infty, \frac{3}{2} - \frac{\sqrt{39}}{2}\right), \left(\frac{3}{2} - \frac{\sqrt{39}}{2}, \frac{3}{2} + \frac{\sqrt{39}}{2}\right), \left(\frac{3}{2} + \frac{\sqrt{39}}{2}, \infty\right)$

Test: Is $2x^2 - 6x - 15 \ge 0$?

Interval	*x-value*	*Value of* $2x^2 - 6x - 15$	*Conclusion*
$\left(-\infty, \frac{3}{2} - \frac{\sqrt{39}}{2}\right)$	$x = -2$	$2(-2)^2 - 6(-2) - 15 = 5$	Positive
$\left(\frac{3}{2} - \frac{\sqrt{39}}{2}, \frac{3}{2} + \frac{\sqrt{39}}{2}\right)$	$x = 0$	$2(0)^2 - 6(0) - 15 = -15$	Negative
$\left(\frac{3}{2} + \frac{\sqrt{39}}{2}, \infty\right)$	$x = 5$	$2(5)^2 - 6(5) - 15 = 5$	Positive

Solution intervals: $\left(-\infty, \frac{3}{2} - \frac{\sqrt{39}}{2}\right] \cup \left[\frac{3}{2} + \frac{\sqrt{39}}{2}, \infty\right)$

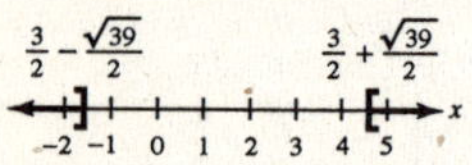

105. $x^3 - 3x^2 - x + 3 > 0$

$$x^2(x - 3) - 1(x - 3) > 0$$

$$(x^2 - 1)(x - 3) > 0$$

$$(x + 1)(x - 1)(x - 3) > 0$$

Critical Numbers: $x = \pm 1, x = 3$

Test Intervals: $(-\infty, -1), (-1, 1), (1, 3), (3, \infty)$

Test: Is $(x + 1)(x - 1)(x - 3) > 0$?

Interval	*x-value*	*Value of* $(x + 1)(x - 1)(x - 3)$	*Conclusion*
$(-\infty, -1)$	$x = -2$	$(-1)(-3)(-5) = -15$	Negative
$(-1, 1)$	$x = 0$	$(1)(-1)(-3) = 3$	Positive
$(1, 3)$	$x = 2$	$(3)(1)(-1) = -3$	Negative
$(3, \infty)$	$x = 4$	$(5)(3)(1) = 15$	Positive

Solution set: $(-1, 1) \cup (3, \infty)$

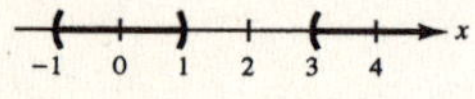

106. $x^3 + 2x^2 - 4x - 8 \le 0$

$x^2(x + 2) - 4(x + 2) \le 0$

$(x + 2)(x^2 - 4) \le 0$

$(x + 2)^2(x - 2) \le 0$

Critical numbers: $x = -2, x = 2$

Test intervals: $(-\infty, -2), (-2, 2), (2, \infty)$

Test: Is $(x + 2)^2(x - 2) \le 0$?

Interval	*x-value*	*Value of* $(x + 2)^2(x - 2)$	*Conclusion*
$(-\infty, -2)$	$x = -3$	$(-1)^2(-5) = -5$	Negative
$(-2, 2)$	$x = 0$	$(2)^2(-2) = -8$	Negative
$(2, \infty)$	$x = 3$	$(5)^2(1) = 25$	Positive

Solution interval: $(-\infty, 2]$

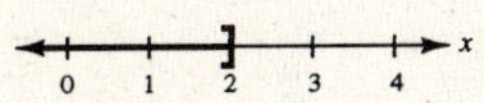

107. $x^3 - 2x^2 - 9x - 2 \ge -20$

$x^3 - 2x^2 - 9x + 18 \ge 0$

$x^2(x - 2) - 9(x - 2) \ge 0$

$(x - 2)(x^2 - 9) \ge 0$

$(x - 2)(x + 3)(x - 3) \ge 0$

Critical Numbers: $x = 2, x = \pm 3$

Test Intervals: $(-\infty, -3), (-3, 2), (2, 3), (3, \infty)$

Test: Is $(x - 2)(x + 3)(x - 3) \ge 0$?

Interval	*x-value*	*Value of* $(x - 2)(x + 3)(x - 3)$	*Conclusion*
$(-\infty, -3)$	$x = -4$	$(-6)(-1)(-7) = -42$	Negative
$(-3, 2)$	$x = 0$	$(-2)(3)(-3) = 18$	Positive
$(2, 3)$	$x = 2.5$	$(0.5)(5.5)(-0.5) = -1.375$	Negative
$(3, \infty)$	$x = 4$	$(2)(7)(1) = 14$	Positive

Solution set: $[-3, 2] \cup [3, \infty)$

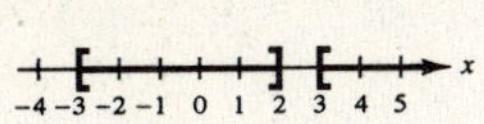

108. $2x^3 + 13x^2 - 8x - 46 \ge 6$

$2x^3 + 13x^2 - 8x - 52 \ge 0$

$2x(x^2 - 4) + 13(x^2 - 4) \ge 0$

$(2x + 13)(x + 2)(x - 2) \ge 0$

Critical numbers: $x = -\frac{13}{2}, x = -2, x = 2$

Test intervals: $\left(-\infty, -\frac{13}{2}\right), \left(-\frac{13}{2}, -2\right), (-2, 2), (2, \infty)$

Test: Is $(2x + 13)(x^2 - 4) \ge 0$?

Interval	*x-value*	*Value of* $(2x + 13)(x^2 - 4)$	*Conclusion*
$\left(-\infty, -\frac{13}{2}\right)$	$x = -7$	$(-1)(45) = -45$	Negative
$\left(-\frac{13}{2}, -2\right)$	$x = -3$	$(7)(5) = 35$	Positive
$(-2, 2)$	$x = 0$	$(13)(-4) = -52$	Negative
$(2, \infty)$	$x = 3$	$(19)(5) = 95$	Positive

Solution intervals: $\left[-\frac{13}{2}, -2\right] \cup [2, \infty)$

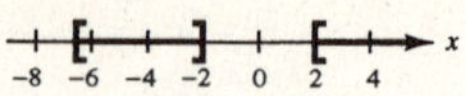

109. $4x^3 - 6x^2 < 0$

$2x^2(2x - 3) < 0$

Critical numbers: $x = 0, x = \frac{3}{2}$

Test intervals: $(-\infty, 0), \left(0, \frac{3}{2}\right), \left(\frac{3}{2}, \infty\right)$

Test: Is $2x^2(2x - 3) < 0$?

By testing an x-value in each test interval in the inequality, we see that the solution set is:

$(-\infty, 0) \cup \left(0, \frac{3}{2}\right)$

110. $4x^3 - 12x^2 > 0$

$4x^2(x - 3) > 0$

Critical numbers: $x = 0, x = 3$

Test intervals: $(-\infty, 0) \Longrightarrow 4x^2(x - 3) < 0$

$(0, 3) \Longrightarrow 4x^2(x - 3) < 0$

$(3, \infty) \Longrightarrow 4x^2(x - 3) > 0$

Solution interval: $(3, \infty)$

111. $x^3 - 4x \ge 0$

$x(x + 2)(x - 2) \ge 0$

Critical numbers: $x = 0, x = \pm 2$

Test intervals: $(-\infty, -2), (-2, 0), (0, 2), (2, \infty)$

Test: Is $x(x + 2)(x - 2) \ge 0$?

By testing an x-value in each test interval in the inequality, we see that the solution set is:

$[-2, 0] \cup [2, \infty)$

112. $2x^3 - x^4 \le 0$

$x^3(2 - x) \le 0$

Critical numbers: $x = 0, x = 2$

Test intervals: $(-\infty, 0) \Longrightarrow x^3(2 - x) < 0$

$(0, 2) \Longrightarrow x^3(2 - x) > 0$

$(2, \infty) \Longrightarrow x^3(2 - x) < 0$

Solution intervals: $(-\infty, 0] \cup [2, \infty)$

113. $(x - 1)^2(x + 2)^3 \ge 0$

Critical numbers: $x = 1, x = -2$

Test intervals: $(-\infty, -2), (-2, 1), (1, \infty)$

Test: Is $(x - 1)^2(x + 2)^3 \ge 0$?

By testing an x-value in each test interval in the inequality, we see that the solution set is: $[-2, \infty)$

114. $x^4(x - 3) \le 0$

Critical numbers: $x = 0, x = 3$

Test intervals: $(-\infty, 0) \Longrightarrow x^4(x - 3) < 0$

$(0, 3) \Longrightarrow x^4(x - 3) < 0$

$(3, \infty) \Longrightarrow x^4(x - 3) > 0$

Solution intervals: $(-\infty, 0] \cup [0, 3]$ or $(-\infty, 3]$

115. $\frac{1}{x} - x > 0$

$\frac{1 - x^2}{x} > 0$

Critical numbers: $x = 0, x = \pm 1$

Test intervals: $(-\infty, -1), (-1, 0), (0, 1), (1, \infty)$

Test: Is $\frac{1 - x^2}{x} > 0$?

By testing an x-value in each test interval in the inequality, we see that the solution set is:

$(-\infty, -1) \cup (0, 1)$

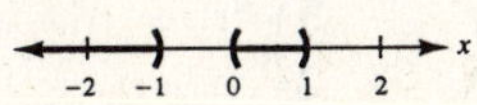

116. $\frac{1}{x} - 4 < 0$

$\frac{1 - 4x}{x} < 0$

Critical numbers: $x = 0, x = \frac{1}{4}$

Test intervals: $(-\infty, 0) \Rightarrow \frac{1 - 4x}{x} < 0$

$\left(0, \frac{1}{4}\right) \Rightarrow \frac{1 - 4x}{x} > 0$

$\left(\frac{1}{4}, \infty\right) \Rightarrow \frac{1 - 4x}{x} < 0$

Solution interval: $(-\infty, 0) \cup \left(\frac{1}{4}, \infty\right)$

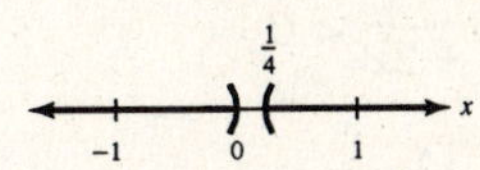

117. $\frac{x + 6}{x + 1} - 2 < 0$

$\frac{x + 6 - 2(x + 1)}{x + 1} < 0$

$\frac{4 - x}{x + 1} < 0$

Critical numbers: $x = -1, x = 4$

Test intervals: $(-\infty, -1), (-1, 4), (4, \infty)$

Test: Is $\frac{4 - x}{x + 1} < 0$?

By testing an x-value in each test interval in the inequality, we see that the solution set is:

$(-\infty, -1) \cup (4, \infty)$

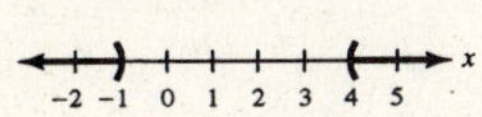

118. $\frac{x + 12}{x + 2} - 3 \ge 0$

$\frac{x + 12 - 3(x + 2)}{x + 2} \ge 0$

$\frac{6 - 2x}{x + 2} \ge 0$

Critical numbers: $x = -2, x = 3$

Test intervals: $(-\infty, -2) \Rightarrow \frac{6 - 2x}{x + 2} < 0$

$(-2, 3) \Rightarrow \frac{6 - 2x}{x + 2} > 0$

$(3, \infty) \Rightarrow \frac{6 - 2x}{x + 2} < 0$

Solution interval: $(-2, 3]$

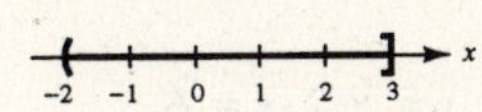

119. $\frac{3x - 5}{x - 5} > 4$

$\frac{3x - 5}{x - 5} - 4 > 0$

$\frac{3x - 5 - 4(x - 5)}{x - 5} > 0$

$\frac{15 - x}{x - 5} > 0$

Critical numbers: $x = 5, x = 15$

Test intervals: $(-\infty, 5), (5, 15), (15, \infty)$

Test: Is $\frac{15 - x}{x - 5} > 0$?

By testing an x-value in each test interval in the inequality, we see that the solution set is: $(5, 15)$

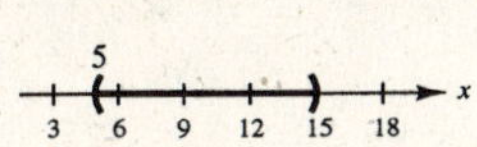

120. $$\frac{5+7x}{1+2x} < 4$$

$$\frac{5+7x-4(1+2x)}{1+2x} < 0$$

$$\frac{1-x}{1+2x} < 0$$

Critical numbers: $x = -\frac{1}{2}, x = 1$

Test intervals: $\left(-\infty, -\frac{1}{2}\right) \Rightarrow \frac{1-x}{1+2x} < 0$

$\left(-\frac{1}{2}, 1\right) \Rightarrow \frac{1-x}{1+2x} > 0$

$(1, \infty) \Rightarrow \frac{1-x}{1+2x} < 0$

Solution intervals: $\left(-\infty, -\frac{1}{2}\right) \cup (1, \infty)$

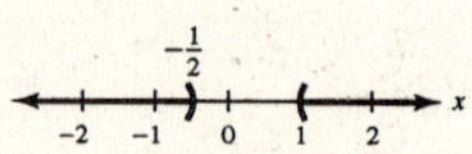

121. $$\frac{4}{x+5} > \frac{1}{2x+3}$$

$$\frac{4}{x+5} - \frac{1}{2x+3} > 0$$

$$\frac{4(2x+3)-(x+5)}{(x+5)(2x+3)} > 0$$

$$\frac{7x+7}{(x+5)(2x+3)} > 0$$

Critical numbers: $x = -1, x = -5, x = -\frac{3}{2}$

Test intervals: $(-\infty, -5), \left(-5, -\frac{3}{2}\right),$

$\left(-\frac{3}{2}, -1\right), (-1, \infty)$

Test: Is $\frac{7(x+1)}{(x+5)(2x+3)} > 0$?

By testing an x-value in each test interval in the inequality, we see that the solution set is:

$\left(-5, -\frac{3}{2}\right) \cup (-1, \infty)$

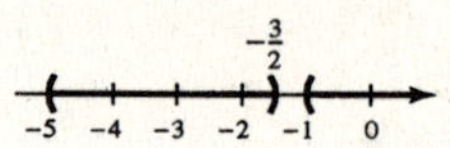

122. $$\frac{5}{x-6} > \frac{3}{x+2}$$

$$\frac{5(x+2)-3(x-6)}{(x-6)(x+2)} > 0$$

$$\frac{2x+28}{(x-6)(x+2)} > 0$$

Critical numbers: $x = -14, x = -2, x = 6$

Test intervals: $(-\infty, -14) \Rightarrow \frac{2x+28}{(x-6)(x+2)} < 0$

$(-14, -2) \Rightarrow \frac{2x+28}{(x-6)(x+2)} > 0$

$(-2, 6) \Rightarrow \frac{2x+28}{(x-6)(x+2)} < 0$

$(6, \infty) \Rightarrow \frac{2x+28}{(x-6)(x+2)} > 0$

Solution intervals: $(-14, -2) \cup (6, \infty)$

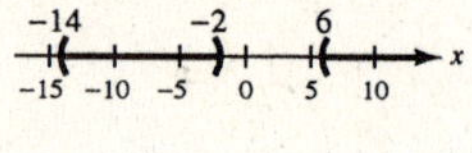

123. $$\frac{1}{x-3} \le \frac{9}{4x+3}$$

$$\frac{1}{x-3} - \frac{9}{4x+3} \le 0$$

$$\frac{4x+3-9(x-3)}{(x-3)(4x+3)} \le 0$$

$$\frac{30-5x}{(x-3)(4x+3)} \le 0$$

Critical numbers: $x = 3, x = -\frac{3}{4}, x = 6$

Test intervals: $\left(-\infty, -\frac{3}{4}\right), \left(-\frac{3}{4}, 3\right), (3, 6), (6, \infty)$

Test: Is $\frac{5(6-x)}{(x-3)(4x+3)} \le 0$?

By testing an x-value in each test interval in the inequality, we see that the solution set is:

$\left(-\frac{3}{4}, 3\right) \cup [6, \infty)$

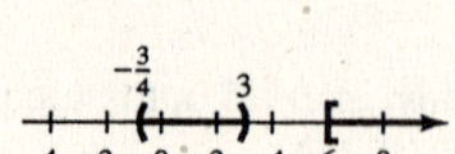

124. $$\frac{1}{x} \geq \frac{1}{x+3}$$

$$\frac{1(x+3) - 1(x)}{x(x+3)} \geq 0$$

$$\frac{3}{x(x+3)} \geq 0$$

Critical numbers: $x = -3, x = 0$

Test intervals: $(-\infty, -3) \Rightarrow \dfrac{3}{x(x+3)} > 0$

$(-3, 0) \Rightarrow \dfrac{3}{x(x+3)} < 0$

$(0, \infty) \Rightarrow \dfrac{3}{x(x+3)} > 0$

Solution intervals: $(-\infty, -3) \cup (0, \infty)$

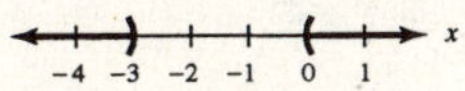

125. $$\frac{x^2 + 2x}{x^2 - 9} \leq 0$$

$$\frac{x(x+2)}{(x+3)(x-3)} \leq 0$$

Critical numbers: $x = 0, x = -2, x = \pm 3$

Test intervals: $(-\infty, -3), (-3, -2), (-2, 0),$
$(0, 3), (3, \infty)$

Test: Is $\dfrac{x(x+2)}{(x+3)(x-3)} \leq 0$?

By testing an x-value in each test interval in the inequality, we see that the solution set is: $(-3, -2] \cup [0, 3)$

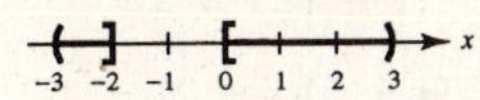

126. $$\frac{x^2 + x - 6}{x} \geq 0$$

$$\frac{(x+3)(x-2)}{x} \geq 0$$

Critical numbers: $x = -3, x = 0, x = 2$

Test intervals: $(-\infty, -3) \Rightarrow \dfrac{(x+3)(x-2)}{x} < 0$

$(-3, 0) \Rightarrow \dfrac{(x+3)(x-2)}{x} > 0$

$(0, 2) \Rightarrow \dfrac{(x+3)(x-2)}{x} < 0$

$(2, \infty) \Rightarrow \dfrac{(x+3)(x-2)}{x} > 0$

Solution intervals: $[-3, 0) \cup [2, \infty)$

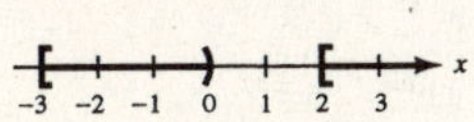

127. $$\frac{5}{x-1} - \frac{2x}{x+1} < 1$$

$$\frac{5}{x-1} - \frac{2x}{x+1} - 1 < 0$$

$$\frac{5(x+1) - 2x(x-1) - (x-1)(x+1)}{(x-1)(x+1)} < 0$$

$$\frac{5x + 5 - 2x^2 + 2x - x^2 + 1}{(x-1)(x+1)} < 0$$

$$\frac{-3x^2 + 7x + 6}{(x-1)(x+1)} < 0$$

$$\frac{-(3x+2)(x-3)}{(x-1)(x+1)} < 0$$

Critical Numbers: $x = -\frac{2}{3}, x = 3, x = \pm 1$

Test Intervals:

$$(-\infty, -1), \left(-1, -\frac{2}{3}\right), \left(-\frac{2}{3}, 1\right), (1, 3), (3, \infty)$$

Test: Is $\dfrac{-(3x+2)(x-3)}{(x-1)(x+1)} < 0$?

By testing an x-value in each test interval in the inequality, we see that the solution set is:

$$(-\infty, -1) \cup \left(-\frac{2}{3}, 1\right) \cup (3, \infty)$$

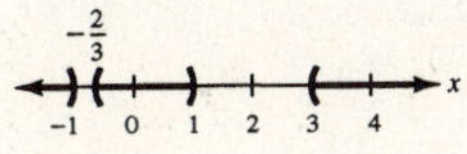

128. $$\frac{3x}{x-1} \le \frac{x}{x+4} + 3$$

$$\frac{3x(x+4) - x(x-1) - 3(x+4)(x-1)}{(x-1)(x+4)} \le 0$$

$$\frac{-x^2 + 4x + 12}{(x-1)(x+4)} \le 0$$

$$\frac{-(x-6)(x+2)}{(x-1)(x+4)} \le 0$$

Critical numbers: $x = -4, x = -2, x = 1, x = 6$

Test intervals: $(-\infty, -4) \Rightarrow \frac{-(x-6)(x+2)}{(x-1)(x+4)} < 0$

$(-4, -2) \Rightarrow \frac{-(x-6)(x+2)}{(x-1)(x+4)} > 0$

$(-2, 1) \Rightarrow \frac{-(x-6)(x+2)}{(x-1)(x+4)} < 0$

$(1, 6) \Rightarrow \frac{-(x-6)(x+2)}{(x-1)(x+4)} > 0$

$(6, \infty) \Rightarrow \frac{-(x-6)(x+2)}{(x-1)(x+4)} < 0$

Solution intervals: $(-\infty, -4), [-2, 1), [6, \infty)$

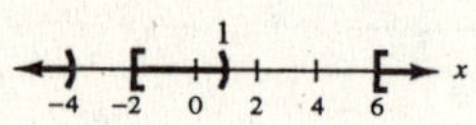

129. $4 - x^2 \ge 0$

$(2 + x)(2 - x) \ge 0$

Critical numbers: $x = \pm 2$

Test intervals: $(-\infty, -2), (-2, 2), (2, \infty)$

Test: Is $4 - x^2 \ge 0$?

By testing an x-value in each test interval in the inequality, we see that the domain is: $[-2, 2]$

130. $x^2 - 4 \ge 0$

$(x + 2)(x - 2) \ge 0$

Critical numbers: $x = -2, x = 2$

Test intervals: $(-\infty, -2) \Rightarrow (x+2)(x-2) > 0$

$(-2, 2) \Rightarrow (x+2)(x-2) < 0$

$(2, \infty) \Rightarrow (x+2)(x-2) > 0$

Domain: $(-\infty, -2] \cup [2, \infty)$

131. $x^2 - 7x + 12 \ge 0$

$(x - 3)(x - 4) \ge 0$

Critical numbers: $x = 3, x = 4$

Test intervals: $(-\infty, 3), (3, 4), (4, \infty)$

Test: Is $(x - 3)(x - 4) \ge 0$?

By testing an x-value in each test interval in the inequality, we see that the domain is: $(-\infty, 3] \cup [4, \infty)$

132. $x^2 - 5x - 6 \ge 0$

$(x + 1)(x - 6) \ge 0$

Critical numbers: $x = -1, x = 6$

Test intervals: $(-\infty, -1), (-1, 6)(6, \infty)$

Test: Is $(x + 1)(x - 6) \ge 0$?

By testing an x-value in each test interval in the inequality, we see that the domain set is:

$(-\infty, -1] \cup [6, \infty)$

133. $$\frac{x}{x^2 - 2x - 35} \ge 0$$

$$\frac{x}{(x+5)(x-7)} \ge 0$$

Critical Numbers: $x = 0, x = -5, x = 7$

Test Intervals: $(-\infty, -5), (-5, 0), (0, 7), (7, \infty)$

Test: Is $\frac{x}{(x+5)(x-7)} \ge 0$?

By testing an x-value in each test interval in the inequality, we see that the domain is: $(-5, 0] \cup (7, \infty)$

134. $\dfrac{x}{x^2 - 9} \geq 0$

$\dfrac{x}{(x + 3)(x - 3)} \geq 0$

Critical numbers: $x = -3, x = 0, x = 3$

Test intervals: $(-\infty, -3) \Rightarrow \dfrac{x}{(x + 3)(x - 3)} < 0$

$(-3, 0) \Rightarrow \dfrac{x}{(x + 3)(x - 3)} > 0$

$(0, 3) \Rightarrow \dfrac{x}{(x + 3)(x - 3)} < 0$

$(3, \infty) \Rightarrow \dfrac{x}{(x + 3)(x - 3)} > 0$

Domain: $(-3, 0] \cup (3, \infty)$

135. $0.4x^2 + 5.26 < 10.2$

$0.4x^2 - 4.94 < 0$

$0.4(x^2 - 12.35) < 0$

Critical numbers: $x \approx \pm 3.51$

Test intervals: $(-\infty, -3.51), (-3.51, 3.51), (3.51, \infty)$

By testing an x-value in each test interval in the inequality, we see that the solution set is: $(-3.51, 3.51)$

136. $-1.3x^2 + 3.78 > 2.12$

$-1.3x^2 + 1.66 > 0$

Critical numbers: $x \approx \pm 1.13$

Test intervals: $(-\infty, -1.13) \Rightarrow -1.3x^2 + 1.66 < 0$

$(-1.13, 1.13) \Rightarrow -1.3x^2 + 1.66 > 0$

$(1.13, \infty) \Rightarrow -1.3x^2 + 1.66 < 0$

Solution set: $(-1.13, 1.13)$

137. $-0.5x^2 + 12.5x + 1.6 > 0$

The zeros are $x = \dfrac{-12.5 \pm \sqrt{(12.5)^2 - 4(-0.5)(1.6)}}{2(-0.5)}$.

Critical numbers: $x \approx -0.13, x \approx 25.13$

Test intervals: $(-\infty, -0.13), (-0.13, 25.13), (25.13, \infty)$

By testing x-values in each test interval in the inequality, we see that the solution set is: $(-0.13, 25.13)$

138. $1.2x^2 + 4.8x + 3.1 < 5.3$

$1.2x^2 + 4.8x - 2.2 < 0$

Critical numbers: $x \approx -4.42, x \approx 0.42$

Test intervals:

$(-\infty, -4.42) \Rightarrow 1.2x^2 + 4.8x - 2.2 > 0$

$(-4.42, 0.42) \Rightarrow 1.2x^2 + 4.8x - 2.2 < 0$

$(0.42, \infty) \Rightarrow 1.2x^2 + 4.8x - 2.2 > 0$

Solution set: $(-4.42, 0.42)$

139. $\dfrac{1}{2.3x - 5.2} > 3.4$

$\dfrac{1}{2.3x - 5.2} - 3.4 > 0$

$\dfrac{-7.82x + 18.68}{2.3x - 5.2} > 0$

Critical numbers: $x \approx 2.39, x = 2.26$

Test intervals: $(-\infty, 2.26), (2.26, 2.39), (2.39, \infty)$

By testing x-values in each test interval in the inequality, we see that the solution set is: $(2.26, 2.39)$

140. $\dfrac{2}{3.1x - 3.7} > 5.8$

$\dfrac{2 - 5.8(3.1x - 3.7)}{3.1x - 3.7} > 0$

$\dfrac{23.46 - 17.98x}{3.1x - 3.7} > 0$

Critical numbers: $x \approx 1.19, x \approx 1.30$

Test intervals: $(-\infty, 1.19) \Rightarrow \dfrac{23.46 - 17.98x}{3.1x - 3.7} < 0$

$(1.19, 1.30) \Rightarrow \dfrac{23.46 - 17.98x}{3.1x - 3.7} > 0$

$(1.30, \infty) \Rightarrow \dfrac{23.46 - 17.98x}{3.1x - 3.7} < 0$

Solution interval: $(1.19, 1.30)$

141. $|x - a| \geq 2$

$x - a \leq -2$ or $x - a \geq 2$

$x \leq a - 2$ or $x \geq a + 2$

Matches graph (b)

142. $|x - b| < 4$

$-4 < x - b < 4$

$b - 4 < x < b + 4$

Matches graph (b)

143. $|ax - b| \leq c \Rightarrow c$ must be greater than or equal to zero.

$-c \leq ax - b \leq c$

$b - c \leq ax \leq b + c$

Let $a = 1$, then $b - c = 0$ and $b + c = 10$.
This is true when $b = c = 5$.
One set of values is: $a = 1, b = 5, c = 5$.

(Note: This solution is not unique. The following are also solutions.

$a = 2, b = c = 10$

$a = 3, b = c = 15$.)

In general, $a = k, b = c = 5k, k \geq 0$
or $a = k, b = 5k, c = -5k, k < 0$

144. $(x - a)(x - b)$

(a) The polynomial is zero when $x = a$ or $x = b$.

(b)

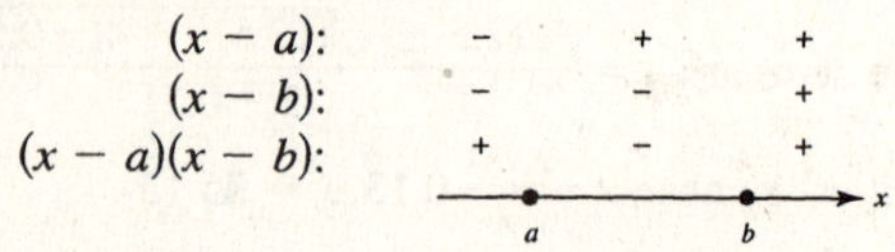

(c) A polynomial changes signs at its zeros.

145. Company A = Lemon Rental

Company B = Cars 'R Us

Company B fee > Company A fee

$150 + 0.25x > 250$

$0.25x > 100$

$x > 400$

If you drive more than 400 miles in a week, the rental fee for Company B is greater than the rental fee for Company A.

146. Let x be the number of copies.

$3000 + 0.03x < 0.10x$

$3000 < 0.07x$

$42{,}857.1 < x$

You must make at least 42,858 copies.

147. $1000(1 + r(2)) > 1062.50$

$1 + 2r > 1.0625$

$2r > 0.0625$

$r > 0.03125$

$r > 3.125\% \approx 3.1\%$

148. $1.266x - 35.766 > 200$

$1.266x > 235.766$

$x > 186.2$ pounds

149. $\left|\dfrac{h - 68.5}{2.7}\right| \leq 1$

$-1 \leq \dfrac{h - 68.5}{2.7} \leq 1$

$-2.7 \leq h - 68.5 \leq 2.7$

$65.8 \leq h \leq 71.2$ inches

150. $|h - 50| \leq 30$

$-30 \leq h - 50 \leq 30$

$20 \leq h \leq 80$

151. $2L + 2W = 100 \Rightarrow W = 50 - L$

$LW \geq 500$

$L(50 - L) \geq 500$

$-L^2 + 50L - 500 \geq 0$

By the Quadratic Formula we have:

Critical numbers: $L = 25 \pm 5\sqrt{5}$

Test: Is $-L^2 + 50L - 500 \geq 0$?

Solution set: $25 - 5\sqrt{5} \leq L \leq 25 + 5\sqrt{5}$

13.8 meters $\leq L \leq$ 36.2 meters

152. $2L + 2W = 440 \Longrightarrow W = 220 - L$

$$LW \geq 8000$$

$$L(220 - L) \geq 8000$$

$$-L^2 + 220L - 8000 \geq 0$$

By the Quadratic Formula we have:

Critical numbers: $L = 110 \pm 10\sqrt{41}$

Test intervals: $\left(-\infty, 110 - 10\sqrt{41}\right) \Longrightarrow -L^2 + 220L - 8000 < 0$

$\left(110 - 10\sqrt{41}, 110 + \sqrt{41}\right) \Longrightarrow -L^2 + 220L - 8000 > 0$

$\left(110 + 10\sqrt{41}, \infty\right) \Longrightarrow -L^2 + 220L - 8000 < 0$

Solution set: $110 - 10\sqrt{41} \leq L \leq 110 + 10\sqrt{41}$

$45.97 \text{ feet} \leq L \leq 174.03 \text{ feet}$

153.

$$1000(1 + r)^2 > 1100$$

$$(1 + r)^2 > 1.1$$

$$1 + 2r + r^2 - 1.1 > 0$$

$$r^2 + 2r - 0.1 > 0$$

By the Quadratic Formula we have:

Critical Numbers: $r = -1 \pm \sqrt{1.1}$

Since r cannot be negative, $r = -1 + \sqrt{1.1} \approx 0.0488$

$= 4.88\%.$

Thus, $r > 4.88\%$.

154.

$$P = R - C$$

$$1{,}650{,}000 \leq R - C$$

$$1{,}650{,}000 \leq x(50 - 0.0002x) - (12x + 150{,}000)$$

$$1{,}650{,}000 \leq -0.0002x^2 + 38x - 150{,}000$$

$$0.0002x^2 - 38x + 1{,}800{,}000 \leq 0$$

$$x^2 - 190{,}000x + 9{,}000{,}000{,}000 \leq 0$$

$$(x - 100{,}000)(x - 90{,}000) \leq 0$$

Critical numbers: $x = 90{,}000$ and $x = 100{,}000$

Test intervals: $(-\infty, 90{,}000) \Longrightarrow (x - 100{,}000)(x - 90{,}000) > 0$

$(90{,}000, 100{,}000) \Longrightarrow (x - 100{,}000)(x - 90{,}000) < 0$

$(100{,}000, \infty) \Longrightarrow (x - 100{,}000)(x - 90{,}000) > 0$

Solution interval: $[90{,}000, 100{,}000]$ or $90{,}000 \text{ units} \leq x \leq 100{,}000 \text{ units}$

155. $\frac{1}{R} = \frac{1}{R_1} + \frac{1}{2}$

$2R_1 = 2R + RR_1$

$2R_1 = R(2 + R_1)$

$\frac{2R_1}{2 + R_1} = R$

Since $R \geq 1$, we have

$\frac{2R_1}{2 + R_1} \geq 1$

$\frac{2R_1}{2 + R_1} - 1 \geq 0$

$\frac{R_1 - 2}{2 + R_1} \geq 0.$

Since $R_1 > 0$, the only critical number is $R_1 = 2$.

The inequality is satisfied when $R_1 \geq 2$ ohms.

156. (a)

d	4	6	8	10	12
Load	2223.9	5593.9	10,312	16,378	23,792

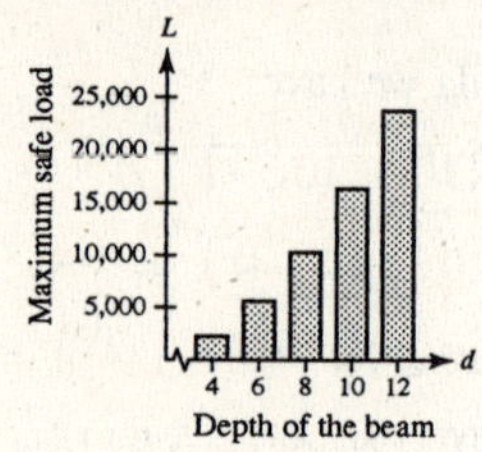

(b) $2000 \leq 168.5d^2 - 472.1$

$2472.1 \leq 168.5d^2$

$14.67 \leq d^2$

$3.83 \leq d$

157. False. c has to be greater than or equal to zero for $ac \leq bc$.

158. False. If $-10 \leq x \leq 8$, then $10 \geq -x$ and $-x \geq -8$.

159. True. $x^3 - 2x^2 - 11x + 12 = (x - 4)(x + 3)(x - 1)$.

The critical numbers are $x = 4$, $x = -3$, and $x = 1$.

The test intervals are $(-\infty, -3)$, $(-3, 1)$, $(1, 4)$ and $(4, \infty)$.

160. True.

$\frac{3}{2}x^2 + 3x + 6 = \frac{3}{2}(x^2 + 2x + 4)$

$= \frac{3}{2}(x^2 + 2x + 1 + 4 - 1)$

$= \frac{3}{2}(x + 1)^2 + \frac{9}{2}$

For any real number x, $\frac{3}{2}(x + 1)^2 + \frac{9}{2} \geq 0$.

Section P.3 Graphical Representation of Data

1.

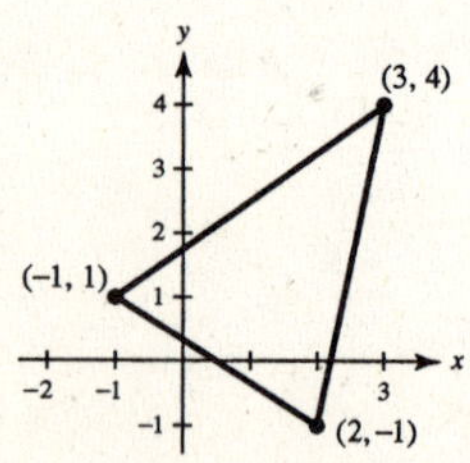

2.

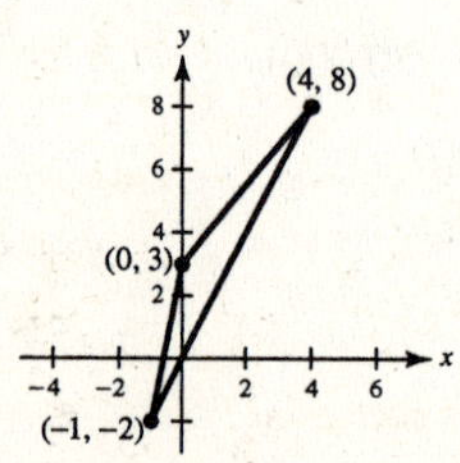

3.

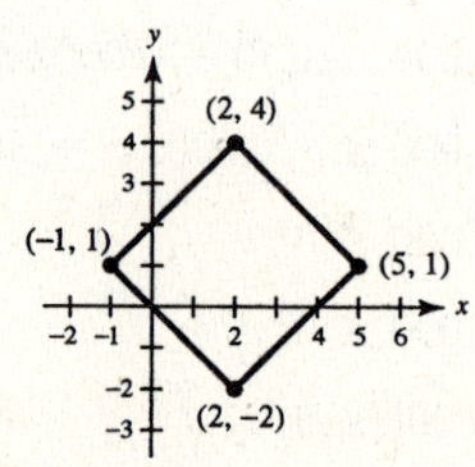

4.

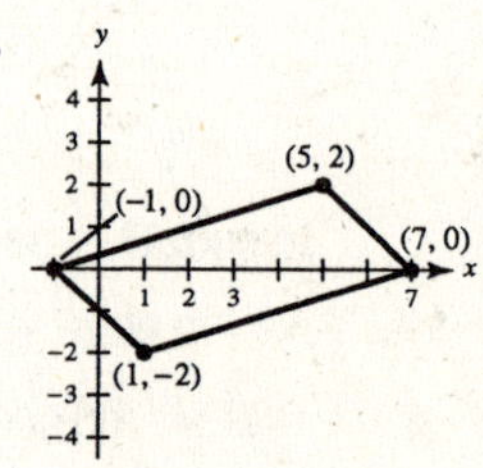

5. $A: (2, 6)$, $B: (-6, -2)$, $C: (4, -4)$, $D: (-3, 2)$

6. $A: (\frac{3}{2}, -4)$; $B: (0, -2)$; $C: (-3, \frac{5}{2})$, $D: (-6, 0)$

7. $(-3, 4)$

8. $(4, -8)$

9. $(-5, -5)$

10. $(-12, 0)$

11. $x > 0$ and $y < 0$ in Quadrant IV.

12. $x < 0$ and $y < 0$ in Quadrant III.

13. $x > 0$ and $y > 0$ means that (x, y) is in Quadrant I.

14. $x < 0$ and $y > 0$ means that (x, y) is in Quadrant II.

15. $x = -4$ and $y > 0$ in Quadrant II.

16. $x > 2$ and $y = 3$ in Quadrant I.

17. $y < -5$ in Quadrants III and IV.

18. $x > 4$ in Quadrants I and IV.

19. $(x, -y)$ is in the second Quadrant means that (x, y) is in Quadrant III.

20. If $(-x, y)$ is in Quadrant IV, then (x, y) must be in Quadrant III.

21. (x, y), $xy > 0$ means x and y have the same signs. This occurs in Quadrants I and III.

22. If $xy < 0$, then x and y have opposite signs. This happens in Quadrants II and IV.

23. The highest price of a stick of butter is 3.4 dollars. This occurred in 2001.

24. Percent increase: $\dfrac{3.4 - 1.5}{1.5} \approx 1.27$ or 127%.

25.

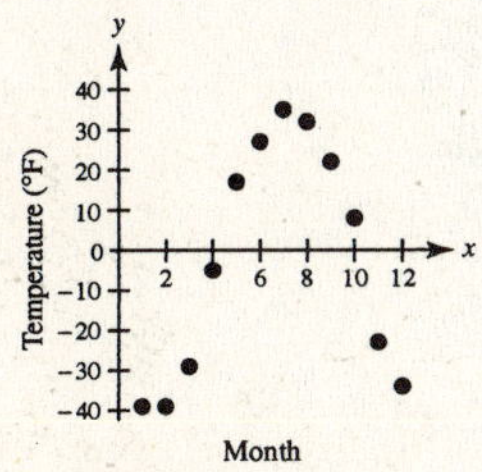

26.

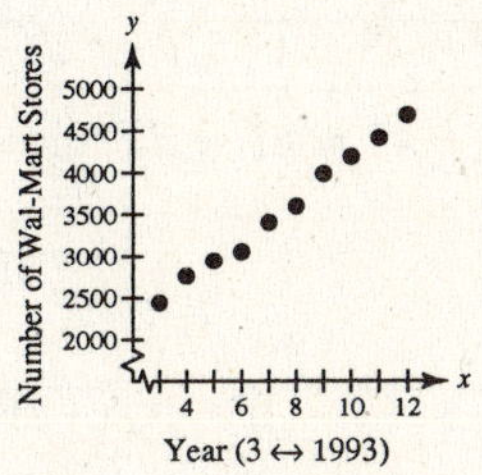

27. Percent increase:

$$\frac{2400 - 675}{675} \approx 2.56 = 256\%$$

28. (a) Percent increase $= \dfrac{850 - 675}{675} \approx 0.26 = 26\%$

(b) Percent increase $= \dfrac{2400 - 850}{850} \approx 1.82 = 182\%$

29. $(-2 + 2, -4 + 5) = (0, 1)$

$(2 + 2, -3 + 5) = (4, 2)$

$(-1 + 2, -1 + 5) = (1, 4)$

30. $(-3 + 6, 6 - 3) = (3, 3)$

$(-5 + 6, 3 - 3) = (1, 0)$

$(-3 + 6, 0 - 3) = (3, -3)$

$(-1 + 6, 3 - 3) = (5, 0)$

31. $(-7 + 4, -2 + 8) = (-3, 6)$

$(-2 + 4, 2 + 8) = (2, 10)$

$(-2 + 4, -4 + 8) = (2, 4)$

$(-7 + 4, -4 + 8) = (-3, 4)$

32. $(5 - 10, 8 - 6) = (-5, 2)$

$(3 - 10, 6 - 6) = (-7, 0)$

$(7 - 10, 6 - 6) = (-3, 0)$

$(5 - 10, 2 - 6) = (-5, -4)$

33. $(-5 + 12, 4 - 5) = (7, -1)$

$(-3 + 12, 0 - 5) = (9, -5)$

$(-1 + 12, 2 - 5) = (11, -3)$

34. $(12 - 8, -8 + 10) = (4, 2)$

$(7 - 8, -4 + 10) = (-1, 6)$

$(4 - 8, -6 + 10) = (-4, 4)$

35. To reflect the vertices in the y-axis, negate each x-coordinate.

Original Point	*Reflected Point*
(1, 5)	(−1, 5)
(5, 4)	(−5, 4)
(2, 2)	(−2, 2)

36. Negate each x-coordinate.

Original Point	*Reflected Point*
(−4, 5)	(4, 5)
(−2, 3)	(2, 3)
(−5, 1)	(5, 1)

37. Negate each x-coordinate.

Original Point	*Reflected Point*
(2, 1)	(−2, 1)
(5, 4)	(−5, 4)
(3, 6)	(−3, 6)

38. Negate each x-coordinate.

Original Point	*Reflected Point*
(−3, 2)	(3, 2)
(−5, 0)	(5, 0)
(−1, −3)	(1, −3)

39. Negate each x-coordinate.

Original Point	*Reflected Point*
(0, 3)	(0, 3)
(3, −2)	(−3, −2)
(6, 3)	(−6, 3)
(3, 8)	(−3, 8)

40. Negate each x-coordinate.

Original Point	*Reflected Point*
(−7, 1)	(7, 1)
(−5, 4)	(5, 4)
(−1, 4)	(1, 4)
(−3, 1)	(3, 1)

41. $(-1(1), -6) = (-1, -6)$

$(-1(6), -6) = (-6, -6)$

$(-1(6), -3) = (-6, -3)$

$(-1(3), -3) = (-3, -3)$

42. $(-1(-6), 0) = (6, 0)$

$(-1(-4), 3) = (4, 3)$

$(-1(-2), 0) = (2, 0)$

$(-1(-4), -5) = (4, -5)$

43.

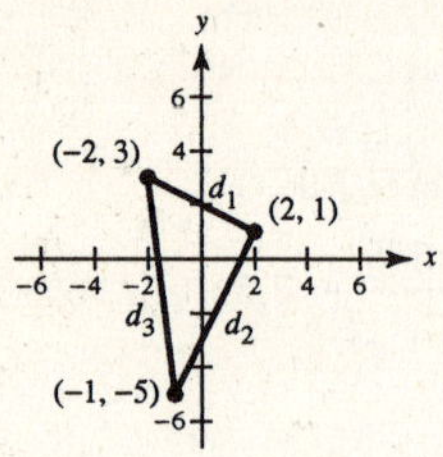

$d_1 = \sqrt{(2-(-2))^2 + (1-3)^2} = \sqrt{(4)^2 + (-2)^2} = \sqrt{20} = 2\sqrt{5}$

$d_2 = \sqrt{(-1-2)^2 + (-5-1)^2} = \sqrt{(-3)^2 + (-6)^2} = \sqrt{45} = 3\sqrt{5}$

$d_3 = \sqrt{(-2-(-1))^2 + (3-(-5))^2} = \sqrt{(-1)^2 + (8)^2} = \sqrt{65}$

Since $d_1^2 + d_2^2 = 20 + 45 = 65 = d_3^2$, the triangle is a right triangle.

44.

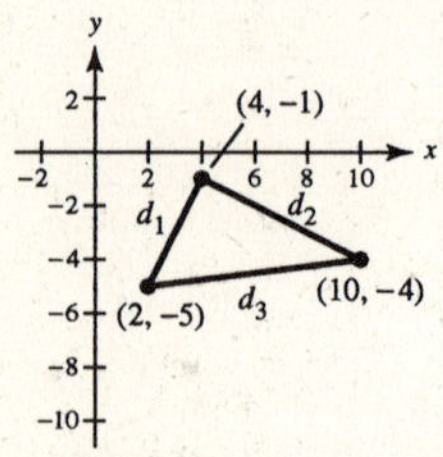

$d_1 = \sqrt{(2-4)^2 + (-5-(-1))^2} = \sqrt{(-2)^2 + (-4)^2} = \sqrt{20} = 2\sqrt{5}$

$d_2 = \sqrt{(10-4)^2 + (-4-(-1))^2} = \sqrt{(6)^2 + (-3)^2} = \sqrt{45} = 3\sqrt{5}$

$d_3 = \sqrt{(10-2)^2 + (-4-(-5))^2} = \sqrt{(8)^2 + (1)^2} = \sqrt{65}$

Since $d_1^2 + d_2^2 = 20 + 45 = 65 = d_3^2$, the triangle is a right triangle.

45.

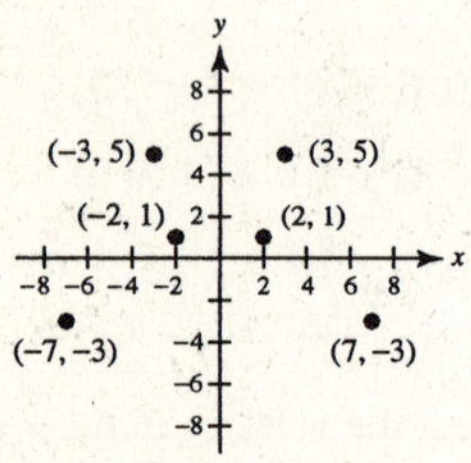

(a) The point is reflected through the y-axis.

(b) The point is reflected through the x-axis.

(c) The point is reflected through the origin.

46. On the x-axis, $y = 0$

On the y-axis, $x = 0$

47. Since (x_0, y_0) lies in Quadrant II, $(x_0, -y_0)$ must lie in Quadrant III. Matches (b)

48. Since (x_0, y_0) lies in Quadrant II, $(-2x_0, y_0)$ must lie in Quadrant I. Matches (c)

49. Since (x_0, y_0) lies in Quadrant II, $\left(x_0, \frac{1}{2}y_0\right)$ must lie in Quadrant II. Matches (d)

50. Since (x_0, y_0) lies in Quadrant II, $(-x_0, -y_0)$ must lie in Quadrant IV. Matches (a)

51. The decade that shows the greatest increase in minimum wage is 1990–2000.

52. (a) Minimum wage in 1990 = \$3.80

Minimum wage in 1995 = \$4.25

$$\text{Percent Increase} = \frac{\$4.25 - \$3.80}{\$3.80} = 0.12 \text{ or } 12\%$$

(b) Minimum wage in 1955 = \$0.75

$$\text{Percent increase} = \frac{\$4.25 - \$0.75}{\$0.75} = 4.67 \text{ or } 467\%$$

53. The point (65, 83) represents an entrance exam score of 65.

54. No. There are many variables that will affect the final exam score.

55. $d = |5 - (-3)| = 8$

56. $d = |1 - 8| = |-7| = 7$

57. $d = |2 - (-3)| = 5$

58. $d = |-4 - 6| = |-10| = 10$

59. (a) The distance between (0, 2) and (4, 2) is 4.

The distance between (4, 2) and (4, 5) is 3.

The distance between (0, 2) and (4, 5) is

$\sqrt{(4-0)^2 + (5-2)^2} = \sqrt{16+9} = \sqrt{25} = 5.$

(b) $4^2 + 3^2 = 16 + 9 = 25 = 5^2$

60. (a) $(1, 0), (13, 5)$

$$\text{Distance} = \sqrt{(13-1)^2 + (5-0)^2}$$
$$= \sqrt{12^2 + 5^2} = \sqrt{169} = 13$$

$(13, 5), (13, 0)$

$$\text{Distance} = |5 - 0| = |5| = 5$$

$(1, 0), (13, 0)$

$$\text{Distance} = |1 - 13| = |-12| = 12$$

(b) $5^2 + 12^2 = 25 + 144 = 169 = 13^2$

61. (a) The distance between $(-1, 1)$ and $(9, 1)$ is 10.

The distance between $(9, 1)$ and $(9, 4)$ is 3.

The distance between $(-1, 1)$ and $(9, 4)$ is

$$\sqrt{(9-(-1))^2 + (4-1)^2} = \sqrt{100+9} = \sqrt{109}.$$

(b) $10^2 + 3^2 = 109 = \left(\sqrt{109}\right)^2$

62. (a) $(1, 5), (5, -2)$

$$\text{Distance} = \sqrt{(1-5)^2 + (5-(-2))^2}$$
$$= \sqrt{(-4)^2 + (7)^2} = \sqrt{16+49} = \sqrt{65}$$

$(1, 5), (1, -2)$

$$\text{Distance} = |5 - (-2)| = |5 + 2| = |7| = 7$$

$(1, -2), (5, -2)$

$$\text{Distance} = |1 - 5| = |-4| = 4$$

(b) $4^2 + 7^2 = 16 + 49 = 65 = \left(\sqrt{65}\right)^2$

63. (a)

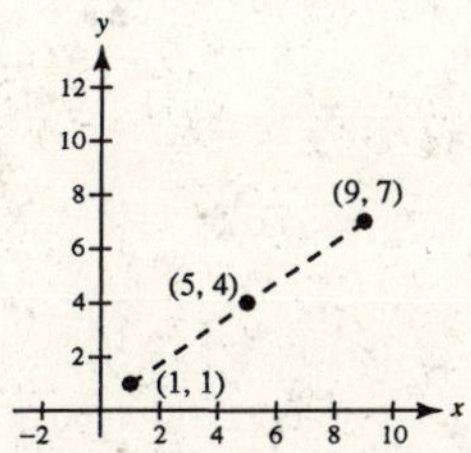

(b) $d = \sqrt{(9-1)^2 + (7-1)^2}$

$= \sqrt{64 + 36} = 10$

(c) $\left(\frac{9+1}{2}, \frac{7+1}{2}\right) = (5, 4)$

64. (a)

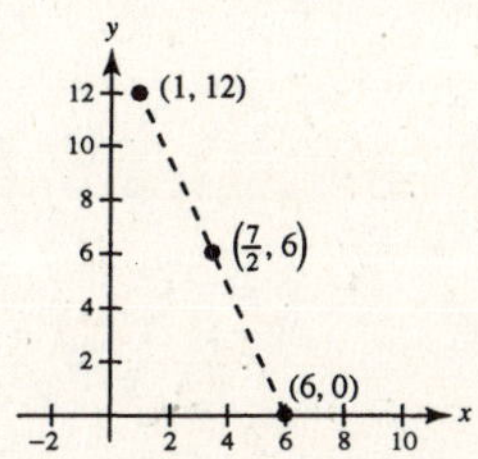

(b) $d = \sqrt{(1-6)^2 + (12-0)^2}$

$= \sqrt{25 + 144} = 13$

(c) $\left(\frac{1+6}{2}, \frac{12+0}{2}\right) = \left(\frac{7}{2}, 6\right)$

65. (a)

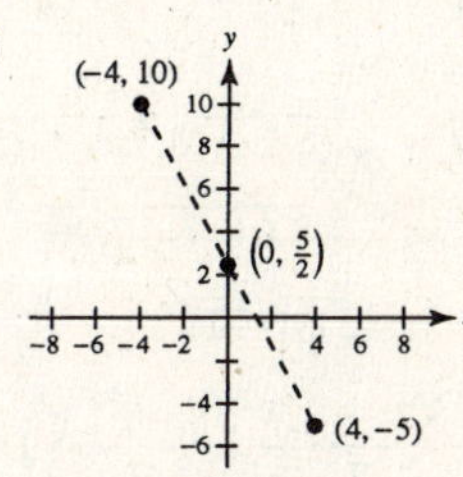

(b) $d = \sqrt{(4+4)^2 + (-5-10)^2}$

$= \sqrt{64 + 225} = 17$

(c) $\left(\frac{4-4}{2}, \frac{-5+10}{2}\right) = \left(0, \frac{5}{2}\right)$

66. (a)

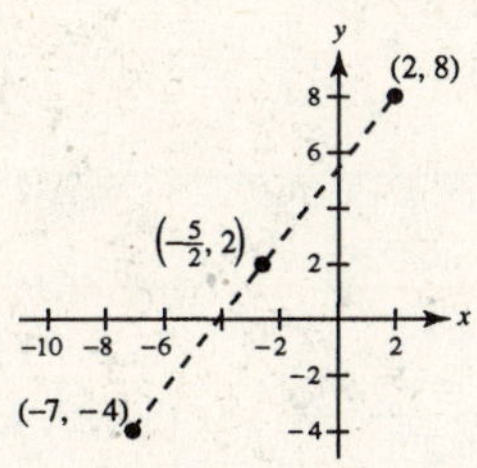

(b) $d = \sqrt{(-7-2)^2 + (-4-8)^2}$

$= \sqrt{81 + 144} = 15$

(c) $\left(\frac{-7+2}{2}, \frac{-4+8}{2}\right) = \left(-\frac{5}{2}, 2\right)$

67. (a)

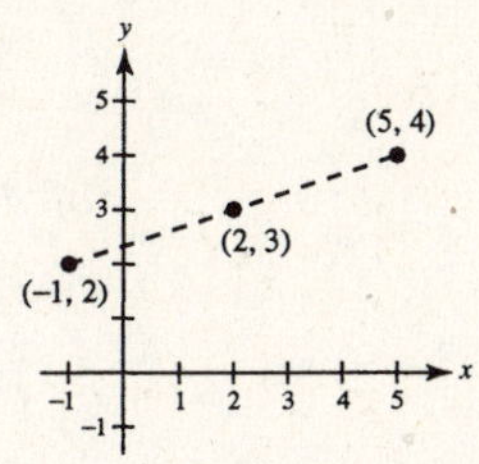

(b) $d = \sqrt{(5+1)^2 + (4-2)^2}$

$= \sqrt{36 + 4} = 2\sqrt{10}$

(c) $\left(\frac{-1+5}{2}, \frac{2+4}{2}\right) = (2, 3)$

68. (a)

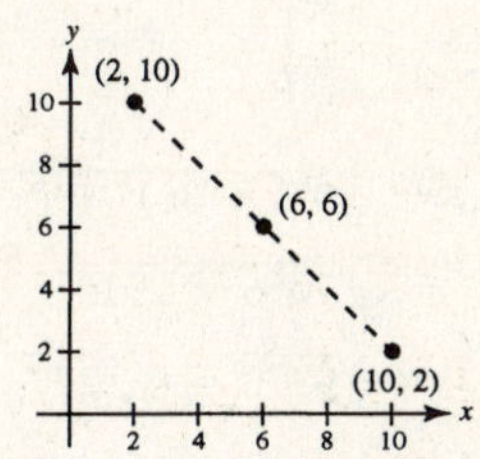

(b) $d = \sqrt{(2-10)^2 + (10-2)^2}$

$= \sqrt{64 + 64} = 8\sqrt{2}$

(c) $\left(\frac{2+10}{2}, \frac{10+2}{2}\right) = (6, 6)$

69. (a)

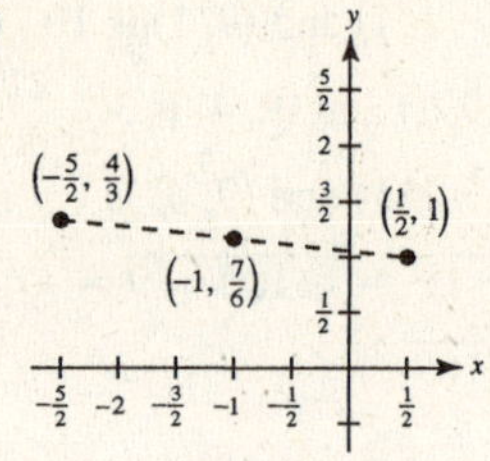

(b) $d = \sqrt{\left(\frac{1}{2} + \frac{5}{2}\right)^2 + \left(1 - \frac{4}{3}\right)^2}$

$d = \sqrt{9 + \frac{1}{9}} = \frac{\sqrt{82}}{3}$

(c) $\left(\frac{-\frac{5}{2} + \frac{1}{2}}{2}, \frac{\frac{4}{3} + 1}{2}\right) = \left(-1, \frac{7}{6}\right)$

70. (a)

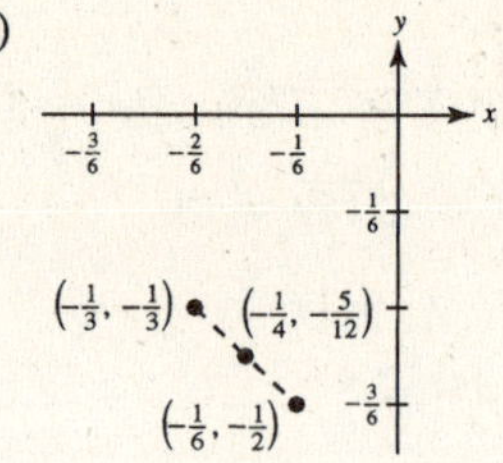

(b) $d = \sqrt{\left(-\frac{1}{3} + \frac{1}{6}\right)^2 + \left(-\frac{1}{3} + \frac{1}{2}\right)^2}$

$= \sqrt{\frac{1}{36} + \frac{1}{36}} = \frac{\sqrt{2}}{6}$

(c) $\left(\frac{-\frac{1}{3} + \left(-\frac{1}{6}\right)}{2}, \frac{-\frac{1}{3} + \left(-\frac{1}{2}\right)}{2}\right) = \left(-\frac{1}{4}, -\frac{5}{12}\right)$

71. (a)

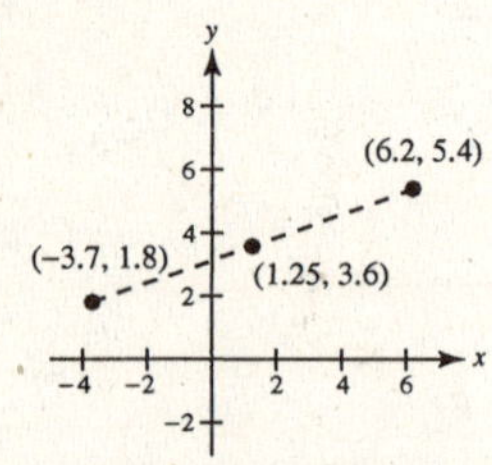

(b) $d = \sqrt{(6.2 + 3.7)^2 + (5.4 - 1.8)^2}$

$= \sqrt{98.01 + 12.96}$

$= \sqrt{110.97}$

(c) $\left(\frac{6.2 - 3.7}{2}, \frac{5.4 + 1.8}{2}\right) = (1.25, 3.6)$

72. (a)

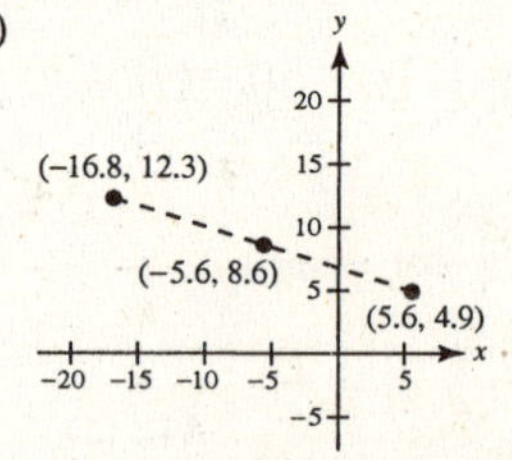

(b) $d = \sqrt{(-16.8 - 5.6)^2 + (12.3 - 4.9)^2}$

$= \sqrt{501.76 + 54.76} = \sqrt{556.52}$

(c) $\left(\frac{-16.8 + 5.6}{2}, \frac{12.3 + 4.9}{2}\right) = (-5.6, 8.6)$

73. (a)

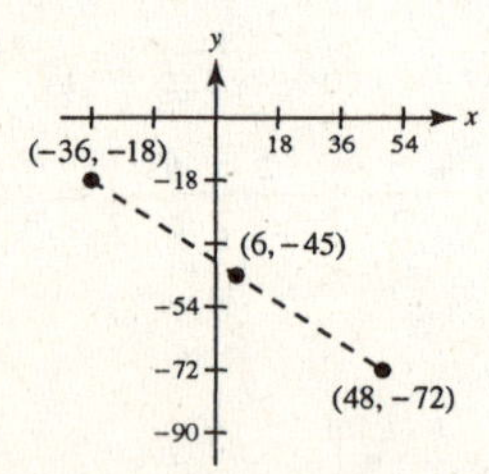

(b) $d = \sqrt{(48 + 36)^2 + (-72 + 18)^2}$

$= \sqrt{7056 + 2916}$

$= \sqrt{9972} = 6\sqrt{277}$

(c) $\left(\frac{-36 + 48}{2}, \frac{-18 - 72}{2}\right) = (6, -45)$

74. (a)

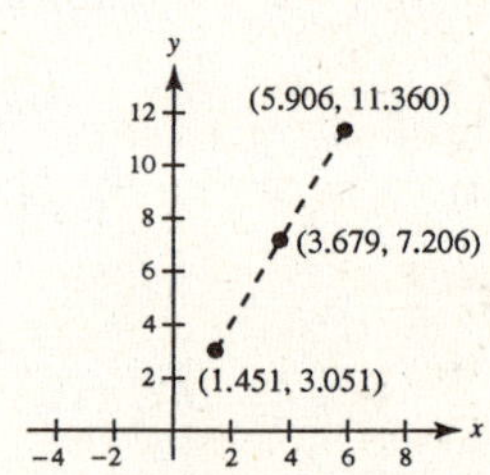

(b) $d = \sqrt{(1.451 - 5.906)^2 + (3.051 - 11.360)^2}$

$\approx \sqrt{88.887}$

(c) $\left(\frac{1.451 + 5.906}{2}, \frac{3.051 + 11.360}{2}\right) \approx (3.679, 7.206)$

75. Target: $\frac{25.371 + 43.917}{2} = 34.644$ billion dollars

76. Wal-Mart: $\frac{104.859 + 244.524}{2} = 174.6915$ billion dollars

77. $d_1 = \sqrt{(4 - 2)^2 + (0 - 1)^2} = \sqrt{5}$

$d_2 = \sqrt{(4 + 1)^2 + (0 + 5)^2} = \sqrt{50}$

$d_3 = \sqrt{(2 + 1)^2 + (1 + 5)^2} = \sqrt{45}$

$\left(\sqrt{5}\right)^2 + \left(\sqrt{45}\right)^2 = \left(\sqrt{50}\right)^2$

78. $d_1 = \sqrt{(1 - 3)^2 + (-3 - 2)^2} = \sqrt{4 + 25} = \sqrt{29}$

$d_2 = \sqrt{(3 + 2)^2 + (2 - 4)^2} = \sqrt{25 + 4} = \sqrt{29}$

$d_3 = \sqrt{(1 + 2)^2 + (-3 - 4)^2} = \sqrt{9 + 49} = \sqrt{58}$

$d_1 = d_2$

79. $d_1 = \sqrt{(0-2)^2 + (9-5)^2} = \sqrt{4+16} = \sqrt{20} = 2\sqrt{5}$

$d_2 = \sqrt{(-2-0)^2 + (0-9)^2} = \sqrt{4+81} = \sqrt{85}$

$d_3 = \sqrt{(0-(-2))^2 + (-4-0)^2} = \sqrt{4+16} = \sqrt{20} = 2\sqrt{5}$

$d_4 = \sqrt{(0-2)^2 + (-4-5)^2} = \sqrt{4+81} = \sqrt{85}$

Opposite sides have equal lengths of $2\sqrt{5}$ and $\sqrt{85}$.

80. $d_1 = \sqrt{(0-3)^2 + (1-7)^2} = \sqrt{9+36} = \sqrt{45} = 3\sqrt{5}$

$d_2 = \sqrt{(3-4)^2 + (7-4)^2} = \sqrt{1+9} = \sqrt{10}$

$d_3 = \sqrt{(4-1)^2 + (4+2)^2} = \sqrt{9+36} = \sqrt{45} = 3\sqrt{5}$

$d_4 = \sqrt{(0-1)^2 + (1+2)^2} = \sqrt{1+9} = \sqrt{10}$

Opposite sides have equal lengths of $3\sqrt{5}$ and $\sqrt{10}$.

81. $d = \sqrt{(45-10)^2 + (40-15)^2} = \sqrt{35^2 + 25^2} = \sqrt{1850} = 5\sqrt{74} \approx 43$ yards

82. Let (0, 0) represent the coordinates of the point of departure and let (100, 150) represent the coordinates of the destination.

(a)

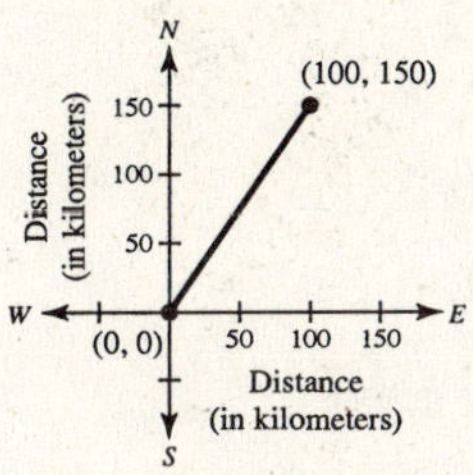

(b) Distance $= \sqrt{(0-100)^2 + (0-150)^2}$

$= \sqrt{10{,}000 + 22{,}500}$

$= \sqrt{32{,}500}$

$= 50\sqrt{13} \approx 180.28$ miles

83. Polo Ralph Lauren Corp: $\dfrac{81.3 + 183.7}{2} = 132.5$ million dollars

84. Russell Corp: $\dfrac{81.6 + 46.9}{2} = 64.25$ million dollars

85. False, you would have to use the Midpoint Formula 15 times.

86. True. Two sides of the triangle have lengths $\sqrt{149}$ and the third side has a length of $\sqrt{18}$.

87.

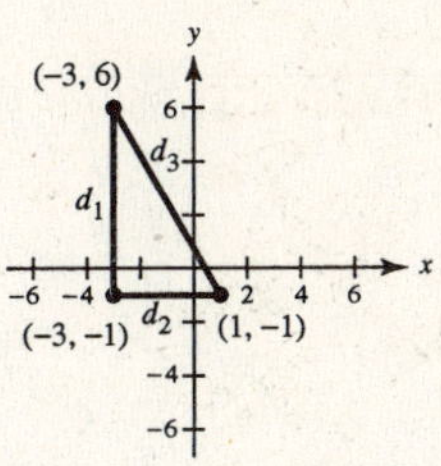

True. $d_1^2 + d_2^2 = 7^2 + 4^2 = 65 = d_3^2$

88.

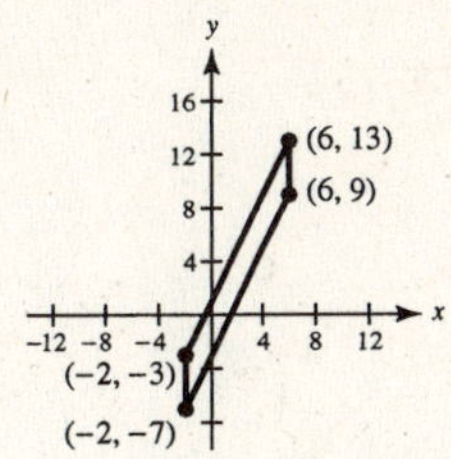

True. These points do **not** form the vertices of a square. From the graph we can see that the four sides do not all have the same length.

89. No. It depends on the magnitude of the quantities measured.

90. Use the Midpoint Formula to prove the diagonals of the parallelogram bisect each other.

$$\left(\frac{b+a}{2}, \frac{c+0}{2}\right) = \left(\frac{a+b}{2}, \frac{c}{2}\right)$$

$$\left(\frac{a+b+0}{2}, \frac{c+0}{2}\right) = \left(\frac{a+b}{2}, \frac{c}{2}\right)$$

Section P.4 Graphs of Equations

1. $y = \sqrt{x + 4}$

(a) $(0, 2)$: $2 \stackrel{?}{=} \sqrt{0 + 4}$

$2 = 2$

Yes, the point *is* on the graph.

(b) $(5, 3)$: $3 \stackrel{?}{=} \sqrt{5 + 4}$

$3 = \sqrt{9}$

Yes, the point *is* on the graph.

2. $y = x^2 - 3x + 2$

(a) $(2, 0)$: $(2)^2 - 3(2) + 2 \stackrel{?}{=} 0$

$4 - 6 + 2 \stackrel{?}{=} 0$

$0 = 0$

Yes, the point *is* on the graph.

(b) $(-2, 8)$: $(-2)^2 - 3(-2) + 2 \stackrel{?}{=} 8$

$4 + 6 + 2 \stackrel{?}{=} 8$

$12 \neq 8$

No, the point *is not* on the graph.

3. $y = 4 - |x - 2|$

(a) $(1, 5)$: $5 \stackrel{?}{=} 4 - |1 - 2|$

$5 \neq 4 - 1$

No, the point is *not* on the graph.

(b) $(6, 0)$: $0 \stackrel{?}{=} 4 - |6 - 2|$

$0 = 4 - 4$

Yes, the point *is* on the graph.

4. $y = \frac{1}{3}x^3 - 2x^2$

(a) $\left(2, -\frac{16}{3}\right)$: $\frac{1}{3}(2)^3 - 2(2)^2 \stackrel{?}{=} -\frac{16}{3}$

$\frac{1}{3} \cdot 8 - 2 \cdot 4 \stackrel{?}{=} -\frac{16}{3}$

$\frac{8}{3} - 8 \stackrel{?}{=} -\frac{16}{3}$

$\frac{8}{3} - \frac{24}{3} \stackrel{?}{=} -\frac{16}{3}$

$-\frac{16}{3} = -\frac{16}{3}$

Yes, the point *is* on the graph.

(b) $(-3, 9)$: $\frac{1}{3}(-3)^3 - 2(-3)^2 \stackrel{?}{=} 9$

$\frac{1}{3}(-27) - 2(9) \stackrel{?}{=} 9$

$-9 - 18 \stackrel{?}{=} 9$

$-27 \neq 9$

No, the point *is not* on the graph.

5. $y = -2x + 5$

x	-1	0	1	2	$\frac{5}{2}$
y	7	5	3	1	0

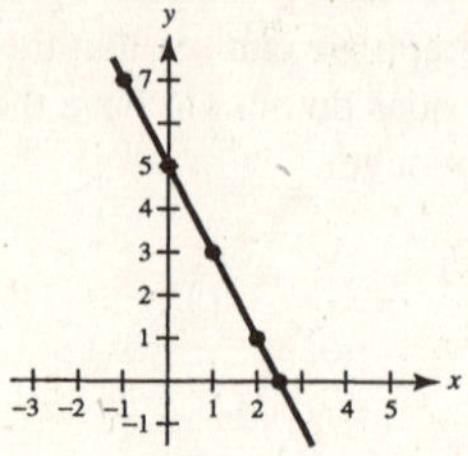

6. $y = \frac{3}{4}x - 1$

x	-2	0	1	$\frac{4}{3}$	2
y	$-\frac{5}{2}$	-1	$-\frac{1}{4}$	0	$\frac{1}{2}$

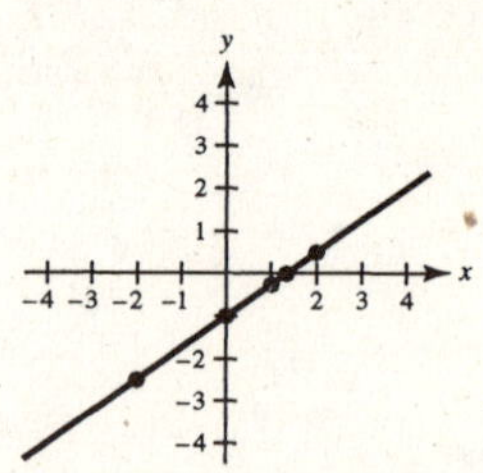

7. $y = x^2 - 3x$

x	-1	0	1	2	3
y	4	0	-2	-2	0

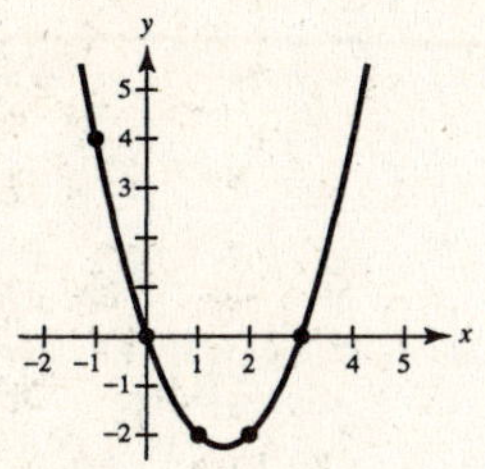

8. $5 - x^2$

x	-2	-1	0	1	2
y	1	4	5	4	1

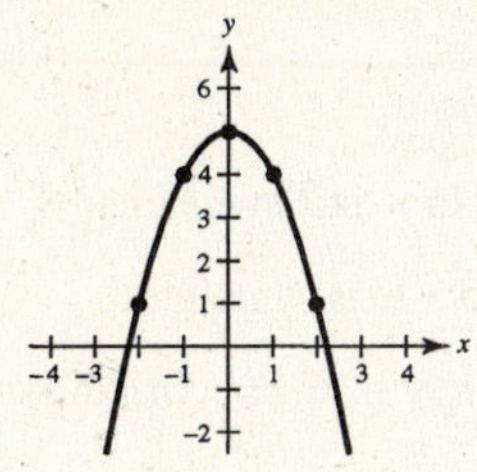

9. $y = 16 - 4x^2$

x-intercepts: $0 = 16 - 4x^2$

$4x^2 = 16$

$x^2 = 4$

$x = \pm 2$

$(-2, 0), (2, 0)$

y-intercept: $y = 16 - 4(0)^2 = 16$

$(0, 16)$

10. $y = (x + 2)^2$

x-intercept: $0 = (x + 2)^2$

$0 = x + 2$

$x = -2$

$(-2, 0)$

y-intercept: $y = (0 + 2)^2$

$y = 2^2$

$y = 4$

$(0, 4)$

11. $y = 2x^3 - 5x^2$

x-intercepts: $0 = 2x^3 - 5x^2$

$0 = x^2(2x - 5)$

$x = 0$ or $x = \frac{5}{2}$

$(0, 0), \left(\frac{5}{2}, 0\right)$

y-intercept: $y = 2(0)^3 - 5(0)^2$

$= 0$

$(0, 0)$

12. $y = 3x^4 - 2x^2$

x-intercepts: $0 = 3x^4 - 2x^2$

$0 = x^2(3x^2 - 2)$

$x = 0$ or $x = \pm\sqrt{\frac{2}{3}}$

$(0, 0), \left(-\sqrt{\frac{2}{3}}, 0\right), \left(\sqrt{\frac{2}{3}}, 0\right)$

y-intercept: $y = 3(0)^4 - 2(0)^2$

$= 0$

$(0, 0)$

13. $y^2 = x + 1$

x-intercept: $0 = x + 1$

$x = -1$

$(-1, 0)$

y-intercepts: $y^2 = 0 + 1$

$y = \pm 1$

$(0, 1), (0, -1)$

14. $y^2 - 4x + 9 = 0$

x-intercept: $(0)^2 - 4x + 9 = 0$

$-4x = -9$

$x = \frac{9}{4}$

$\left(\frac{9}{4}, 0\right)$

y-intercept: $y^2 - 4(0) + 9 = 0$

$y^2 = -9$

No real solution

No y-intercepts

15. $x^2 + y^2 = 25$

x-intercepts: $x^2 + (0)^2 = 25$

$x^2 = 25$

$x = \pm 5$

$(\pm 5, 0)$

y-intercepts: $(0)^2 + y^2 = 25$

$y^2 = 25$

$y = \pm 5$

$(0, \pm 5)$

16. $9x^2 + 9y^2 = 144$

x-intercepts: $9x^2 + 9(0)^2 = 144$

$9x^2 = 144$

$3x = \pm 12$

$x = \pm 4$

$(\pm 4, 0)$

y-intercepts: $9(0)^2 + 9y^2 = 144$

$9y^2 = 144$

$3y = \pm 12$

$y = \pm 4$

$(0, \pm 4)$

17. $x^2 - y = 0$

$(-x)^2 - y = 0 \Rightarrow x^2 - y = 0 \Rightarrow$ y-axis symmetry

$x^2 - (-y) = 0 \Rightarrow x^2 + y = 0 \Rightarrow$ No x-axis symmetry

$(-x)^2 - (-y) = 0 \Rightarrow x^2 + y = 0 \Rightarrow$ No origin symmetry

18. $x - (-y)^2 = 0 \Rightarrow x - y^2 = 0$

x-axis symmetry

19. $y = x^3$

$y = (-x)^3 \Rightarrow y = -x^3 \Rightarrow$ No y-axis symmetry

$-y = x^3 \Rightarrow y = -x^3 \Rightarrow$ No x-axis symmetry

$-y = (-x)^3 \Rightarrow -y = -x^3 \Rightarrow y = x^3 \Rightarrow$ Origin symmetry

20. $y = x^4 - x^2 + 3$

$y = (-x)^4 - (-x)^2 + 3 \Rightarrow$

$y = x^4 - x^2 + 3$

y-axis symmetry

21. $y = \dfrac{x}{x^2 + 1}$

$y = \dfrac{-x}{(-x)^2 + 1} \Rightarrow y = \dfrac{-x}{x^2 + 1} \Rightarrow$ No y-axis symmetry

$-y = \dfrac{x}{x^2 + 1} \Rightarrow y = \dfrac{-x}{x^2 + 1} \Rightarrow$ No x-axis symmetry

$-y = \dfrac{-x}{(-x)^2 + 1} \Rightarrow -y = \dfrac{-x}{x^2 + 1} \Rightarrow y = \dfrac{x}{x^2 + 1} \Rightarrow$ Origin symmetry

22. $y = \sqrt{9 - x^2}$

$y = \sqrt{9 - (-x)^2}$

$y = \sqrt{9 - x^2}$

y-axis symmetry

23. $xy^2 + 10 = 0$

$(-x)y^2 + 10 = 0 \Rightarrow -xy^2 + 10 = 0 \Rightarrow$ No y-axis symmetry

$x(-y)^2 + 10 = 0 \Rightarrow xy^2 + 10 = 0 \Rightarrow$ x-axis symmetry

$(-x)(-y)^2 + 10 = 0 \Rightarrow -xy^2 + 10 = 0 \Rightarrow$ No origin symmetry

24. $xy = 4$

$(-x)(-y) = 4$

$xy = 4$

Origin symmetry

25. $y = 1 - x$ has intercepts $(1, 0)$ and $(0, 1)$.

Matches graph (c)

26. $y = x^2 - 2x$ is a parabola.

Matches (a)

27. $y = x^3 - x + 1$ has a y-intercept of $(0, 1)$ and the points $(1, 1)$ and $(-2, -5)$ are on the graph.

Matches graph (b)

28. $y = |x| - 3$ involves an absolute value.

Matches (d)

29. $y = -3x + 1$

x-intercept: $\left(\frac{1}{3}, 0\right)$

y-intercept: $(0, 1)$

No symmetry

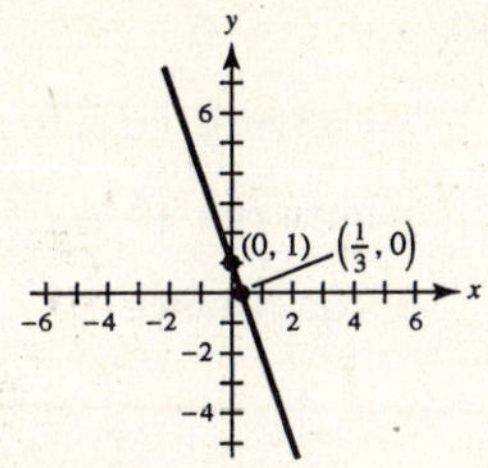

30. $y = 2x - 3$

x-intercept: $\left(\frac{3}{2}, 0\right)$

y-intercept: $(0, -3)$

No symmetry

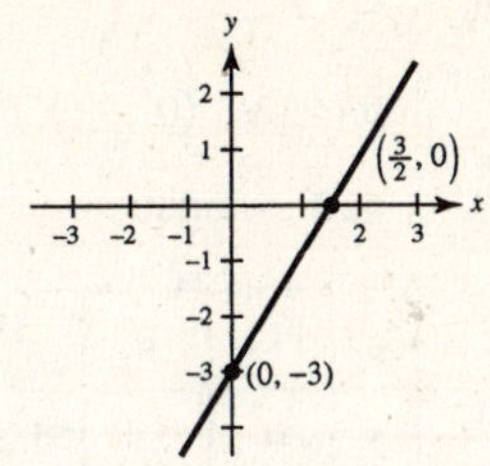

31. $y = x^2 - 2x$

Intercepts: $(0, 0), (2, 0)$

No symmetry

x	-1	0	1	2	3
y	3	0	-1	0	3

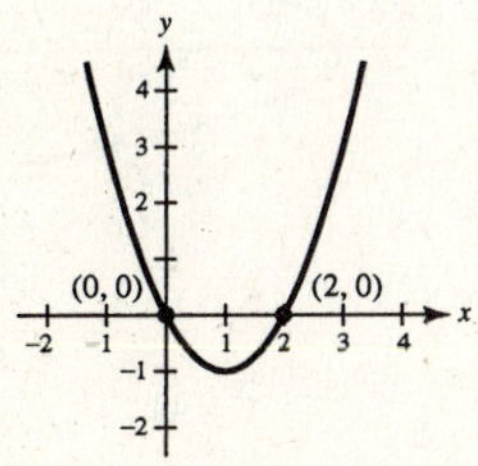

32. $y = -x^2 - 2x$

x-intercept: $(-2, 0), (0, 0)$

y-intercept: $(0, 0)$

No symmetry

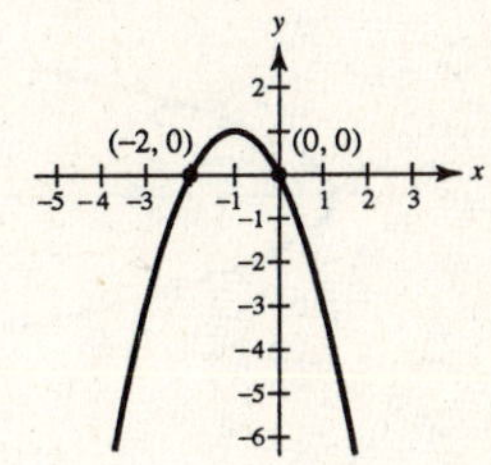

33. $y = x^3 + 3$

Intercepts: $(0, 3), \left(\sqrt[3]{-3}, 0\right)$

No symmetry

x	-2	-1	0	1	2
y	-5	2	3	4	11

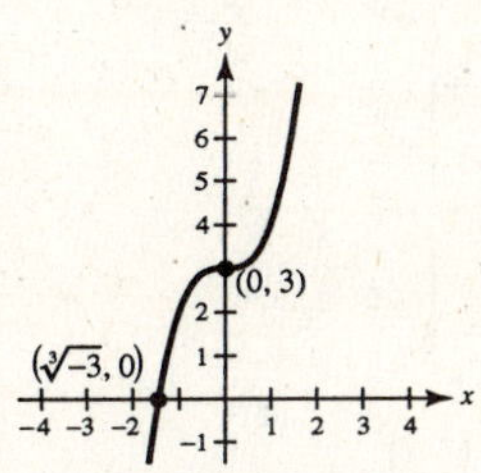

34. $y = x^3 - 1$

x-intercept: $(1, 0)$

y-intercept: $(0, -1)$

No symmetry

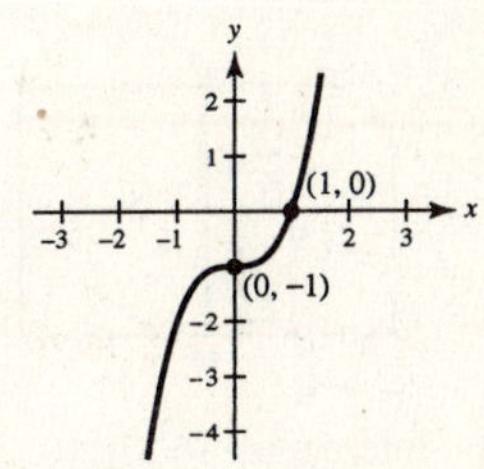

35. $y = \sqrt{x - 3}$

Domain: $[3, \infty)$

Intercept: $(3, 0)$

No symmetry

x	3	4	7	12
y	0	1	2	3

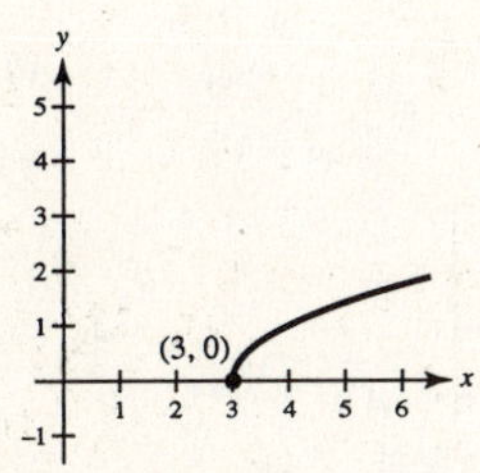

36. $y = \sqrt{1 - x}$

Domain: $(-\infty, 1]$

y-intercept: $(0, 1)$

No symmetry

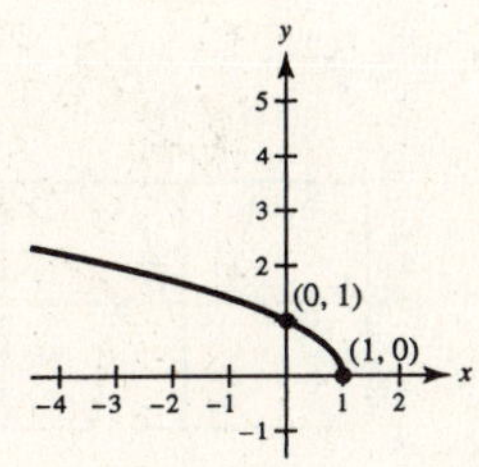

37. $y = |x - 6|$

Intercepts: $(0, 6), (6, 0)$

No symmetry

x	-2	0	2	4	6	8	10
y	8	6	4	2	0	2	4

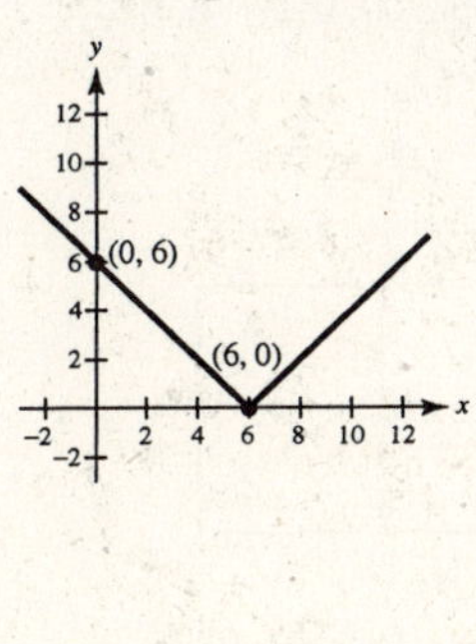

38. $y = 1 - |x|$

x-intercepts: $(\pm 1, 0)$

y-intercept: $(0, 1)$

y-axis symmetry

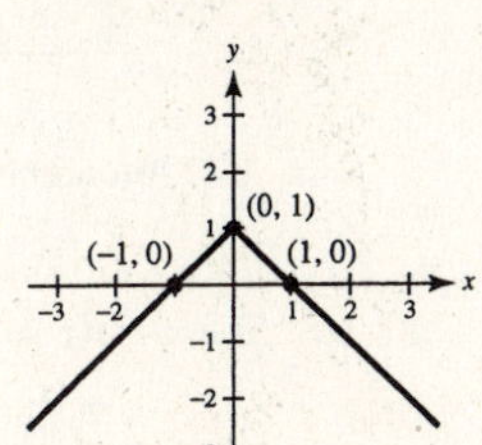

39. $x = y^2 - 1$

Intercepts: $(0, -1), (0, 1), (-1, 0)$

x-axis symmetry

x	-1	0	3
y	0	± 1	± 2

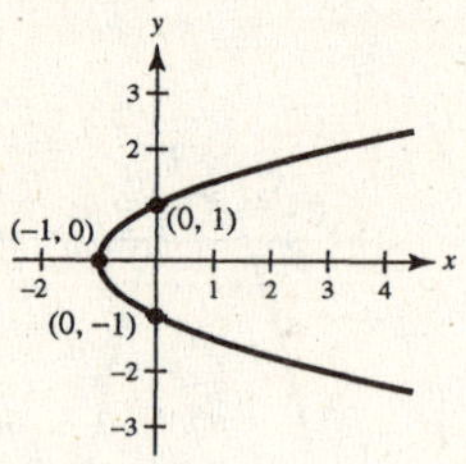

40. $x = y^2 - 5$

x-intercept: $(-5, 0)$

y-intercept: $(0, \pm\sqrt{5})$

x-axis symmetry

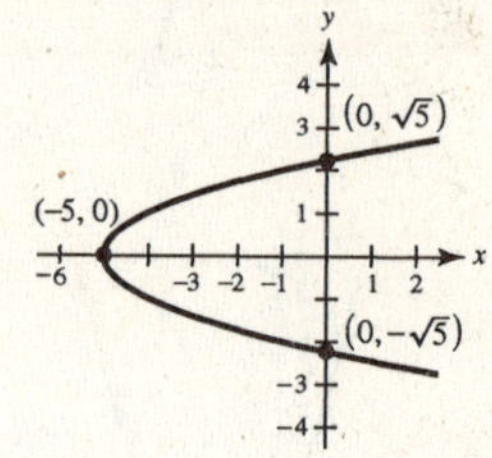

41. $y = 3 - \frac{1}{2}x$

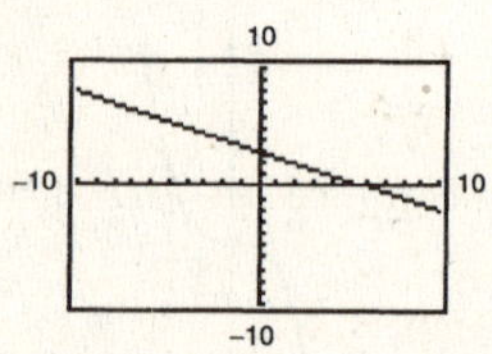

Intercepts: $(6, 0), (0, 3)$

42. $y = \frac{2}{3}x - 1$

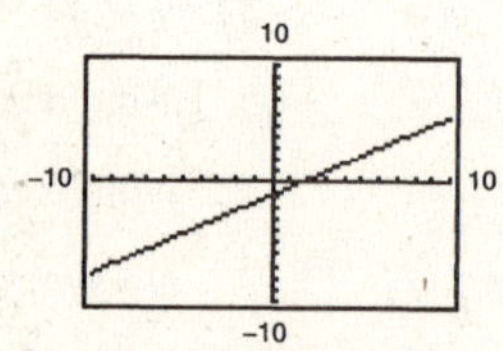

Intercepts: $(0, -1), \left(\frac{3}{2}, 0\right)$

43. $y = x^2 - 4x + 3$

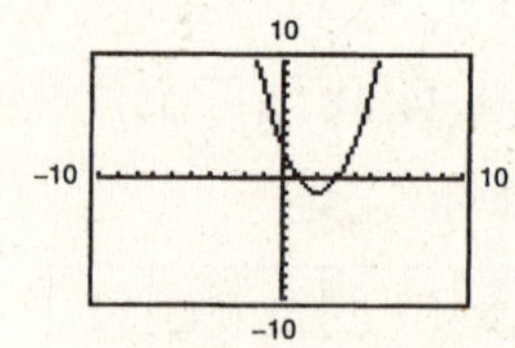

Intercepts: $(3, 0), (1, 0), (0, 3)$

44. $y = x^2 + x - 2$

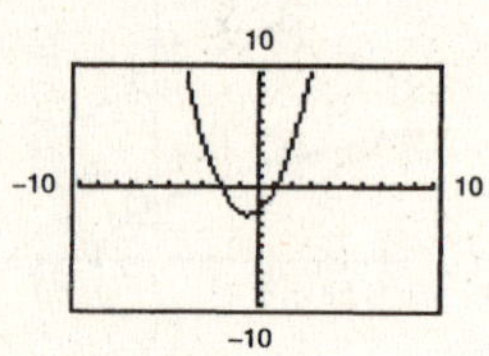

Intercepts: $(-2, 0), (1, 0), (0, -2)$

45. $y = \dfrac{2x}{x - 1}$

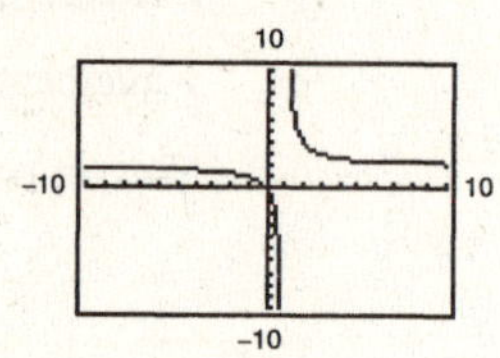

Intercept: $(0, 0)$

46. $y = \dfrac{4}{x^2 + 1}$

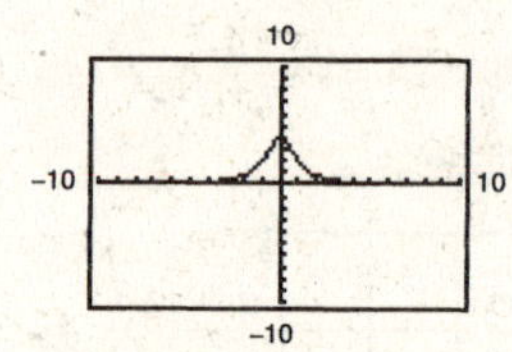

Intercept: $(0, 4)$

47. $y = \sqrt[3]{x}$

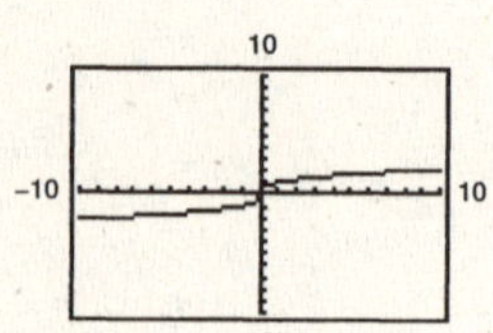

Intercept: $(0, 0)$

48. $y = \sqrt[3]{x + 1}$

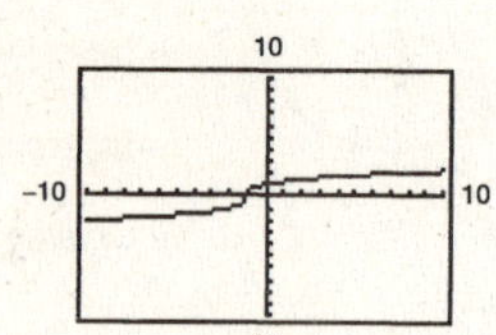

Intercepts: $(-1, 0), (0, 1)$

49. $y = x\sqrt{x + 6}$

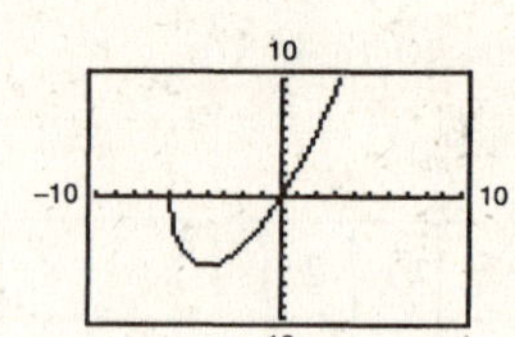

Intercepts: $(0, 0), (-6, 0)$

50. $y = (6 - x)\sqrt{x}$

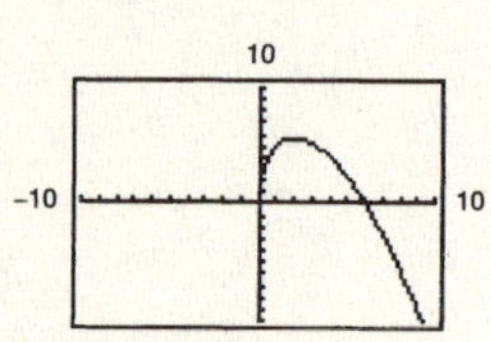

Intercepts: $(0, 0), (6, 0)$

51. $y = |x + 3|$

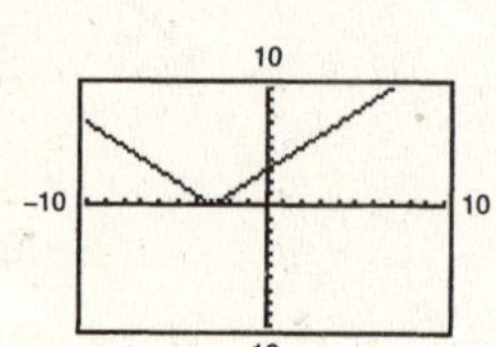

Intercepts: $(-3, 0), (0, 3)$

52. $y = 2 - |x|$

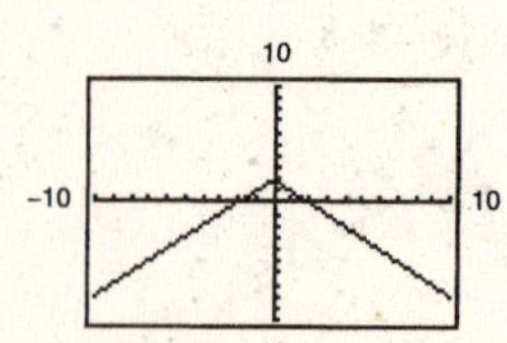

Intercepts: $(\pm 2, 0), (0, 2)$

53. Center: $(0, 0)$; radius: 4

Standard form: $(x - 0)^2 + (y - 0)^2 = 4^2$

$x^2 + y^2 = 16$

54. $(x - 0)^2 + (y - 0)^2 = 5^2$

$x^2 + y^2 = 25$

55. Center: $(2, -1)$; radius: 4

Standard form: $(x - 2)^2 + (y - (-1))^2 = 4^2$

$(x - 2)^2 + (y + 1)^2 = 16$

56. $(x - (-7))^2 + (y - (-4))^2 = 7^2$

$(x + 7)^2 + (y + 4)^2 = 49$

57. Center: $(-1, 2)$; solution point: $(0, 0)$

$(x - (-1))^2 + (y - 2)^2 = r^2$

$(0 + 1)^2 + (0 - 2)^2 = r^2 \Rightarrow 5 = r^2$

Standard form: $(x + 1)^2 + (y - 2)^2 = 5$

58. $r = \sqrt{(3 - (-1))^2 + (-2 - 1)^2}$

$= \sqrt{4^2 + (-3)^2} = \sqrt{25} = 5$

$(x - 3)^2 + (y - (-2))^2 = 5^2$

$(x - 3)^2 + (y + 2)^2 = 25$

59. Endpoints of a diameter: $(0, 0)$, $(6, 8)$

Center: $\left(\dfrac{0 + 6}{2}, \dfrac{0 + 8}{2}\right) = (3, 4)$

$(x - 3)^2 + (y - 4)^2 = r^2$

$(0 - 3)^2 + (0 - 4)^2 = r^2 \Rightarrow 25 = r^2$

Standard form: $(x - 3)^2 + (y - 4)^2 = 25$

60. $r = \dfrac{1}{2}\sqrt{(-4 - 4)^2 + (-1 - 1)^2}$

$= \dfrac{1}{2}\sqrt{(-8)^2 + (-2)^2}$

$= \dfrac{1}{2}\sqrt{64 + 4}$

$= \dfrac{1}{2}\sqrt{68} = \left(\dfrac{1}{2}\right)(2)\sqrt{17} = \sqrt{17}$

Midpoint of diameter (center of circle):

$\left(\dfrac{-4 + 4}{2}, \dfrac{-1 + 1}{2}\right) = (0, 0)$

$(x - 0)^2 + (y - 0)^2 = \left(\sqrt{17}\right)^2$

$x^2 + y^2 = 17$

61. $x^2 + y^2 = 25$

Center: $(0, 0)$

Radius: 5

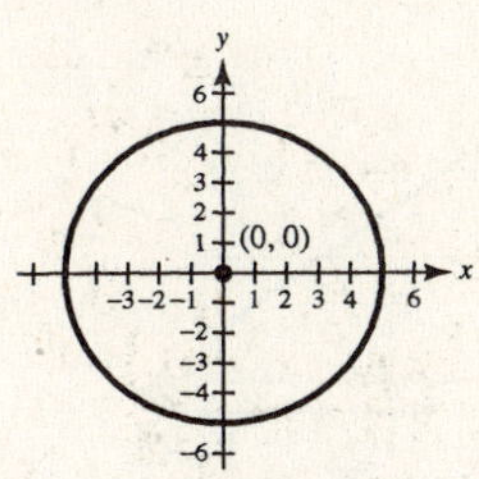

62. $x^2 + y^2 = 16$

Center: $(0, 0)$, radius: 4

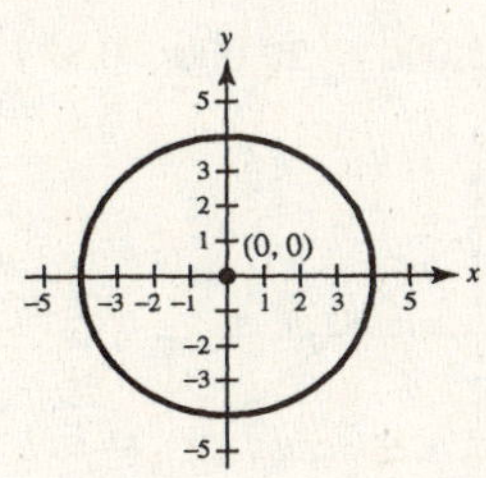

63. $(x - 1)^2 + (y + 3)^2 = 9$

Center: $(1, -3)$

Radius: 3

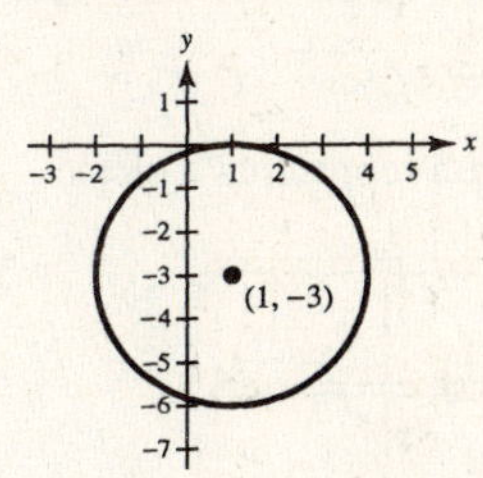

64. $x^2 + (y - 1)^2 = 1$

Center: $(0, 1)$, radius: 1

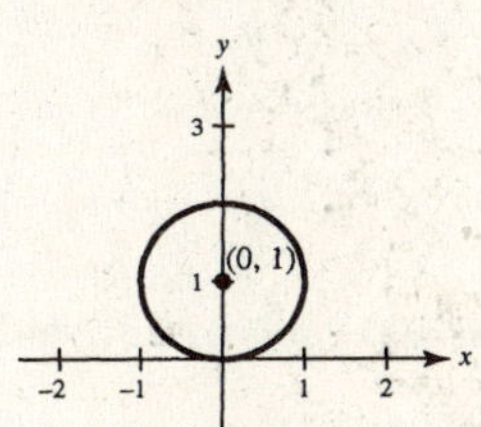

65. $\left(x - \frac{1}{2}\right)^2 + \left(y - \frac{1}{2}\right)^2 = \frac{9}{4}$

Center: $\left(\frac{1}{2}, \frac{1}{2}\right)$

Radius: $\frac{3}{2}$

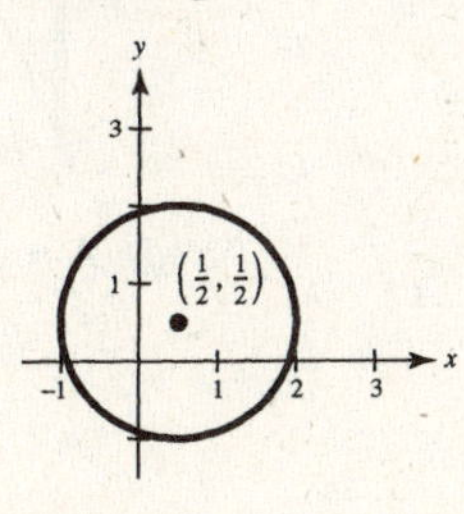

66. $(x - 2)^2 + (y + 1)^2 = 3$

Center: $(2, -1)$; radius: $\sqrt{3}$

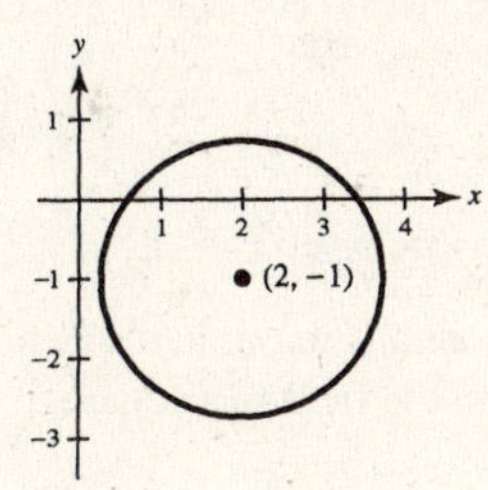

67. y-axis symmetry

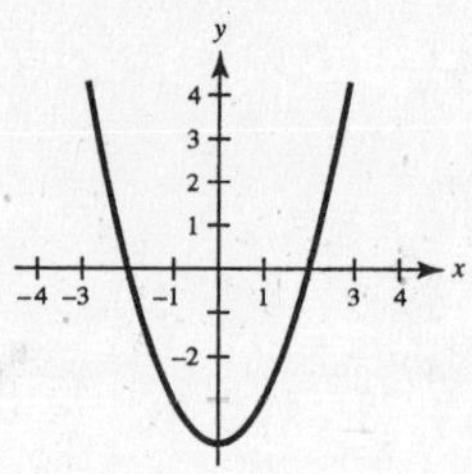

68.

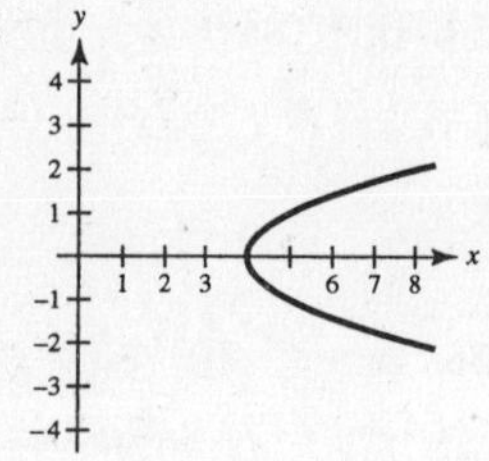

69. Origin symmetry

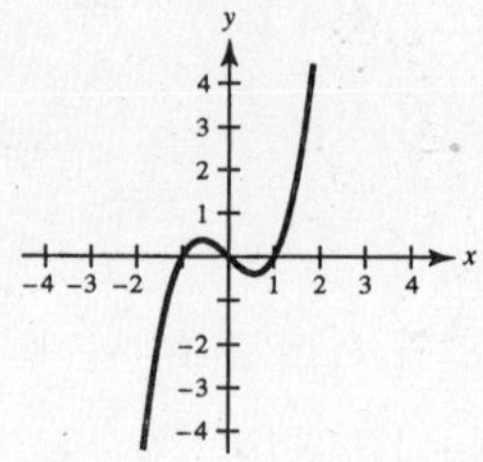

70.

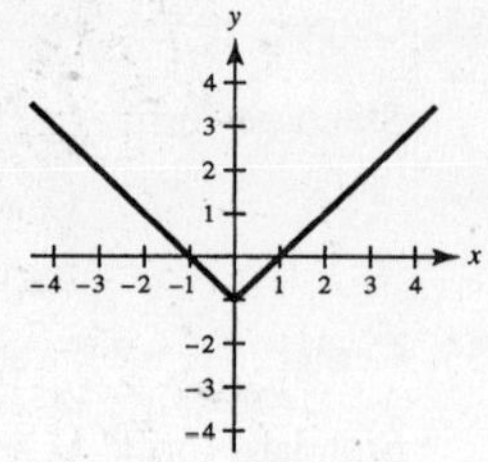

71. Answers will vary.

One possible equation with x-intercepts at $x = -2, x = 4$, and $x = 6$ is:

$$\begin{aligned} y &= (x + 2)(x - 4)(x - 6) \\ &= (x + 2)(x^2 - 10x + 24) \\ &= x^3 - 10x^2 + 24x + 2x^2 - 20x + 48 \\ &= x^3 - 8x^2 + 4x + 48 \end{aligned}$$

Any non-zero multiple of the right side of this equation, $y = k(x^3 - 8x^2 + 4x + 48)$, would also have these x-intercepts.

72. Answers will vary.

One possible equation with x-intercepts at $x = -\frac{5}{2}, x = 2$, and $x = \frac{3}{2}$ is:

$$\begin{aligned} y &= (2x + 5)(x - 2)(2x - 3) \\ &= (2x + 5)(2x^2 - 7x + 6) \\ &= 4x^3 - 14x^2 + 12x + 10x^2 - 35x + 30 \\ &= 4x^3 - 4x^2 - 23x + 30 \end{aligned}$$

Any non-zero multiple of the right side of this equation, $y = k(4x^3 - 4x^2 - 23x + 30)$, would also have these x-intercepts.

73. $y_1 = 4 + \sqrt{25 - x^2}$

$y_2 = 4 - \sqrt{25 - x^2}$

The graph represents a circle.

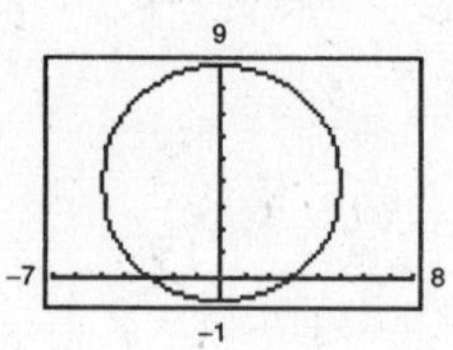

74. $y_1 = 1 - \sqrt{16 - x^2}$

$y_2 = 1 + \sqrt{16 - x^2}$

The graph represents a circle.

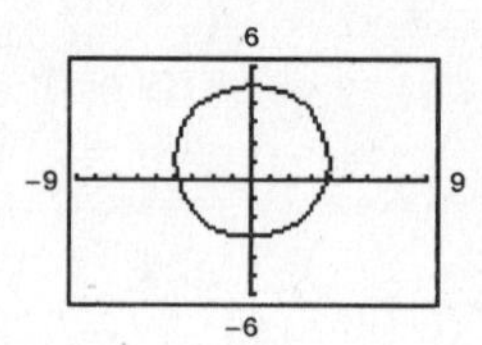

75. $y_1 = 2 + \sqrt{16 - (x - 1)^2}$

$y_2 = 2 - \sqrt{16 - (x - 1)^2}$

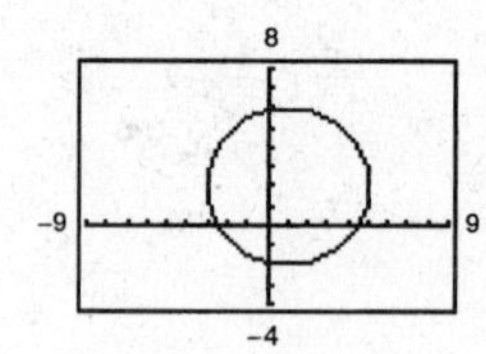

A circle is bounded by their graphs.

76. $y_1 = 3 + \sqrt{36 - (x + 1)^2}$

$y_2 = 3 - \sqrt{36 - (x + 1)^2}$

The graph represents a circle.

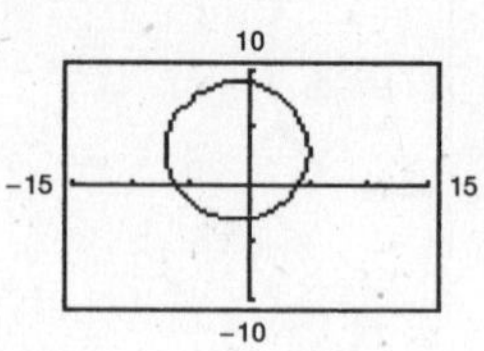

77. $y = 225{,}000 - 20{,}000t,\ 0 \le t \le 8$

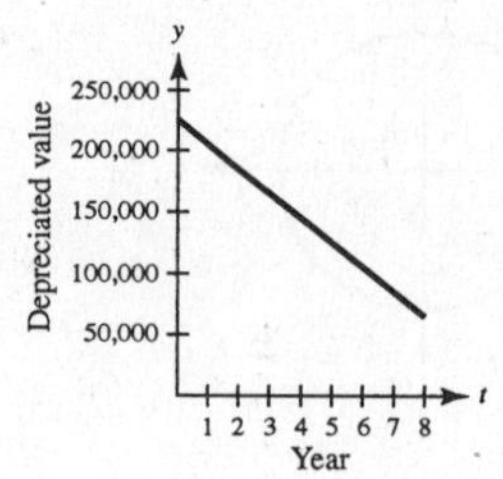

78. $y = 8100 - 929t, 0 \le t \le 6$

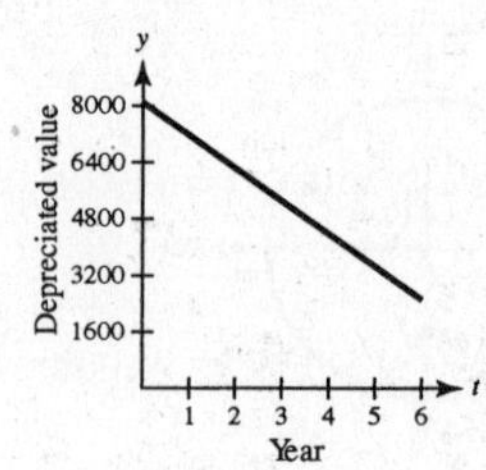

79. (a) $2x + 2w = \frac{920}{3}$

$x + w = \frac{460}{3}$

$w = \frac{460}{3} - x$

$A = xw = x\left(\frac{460}{3} - x\right)$

(b)

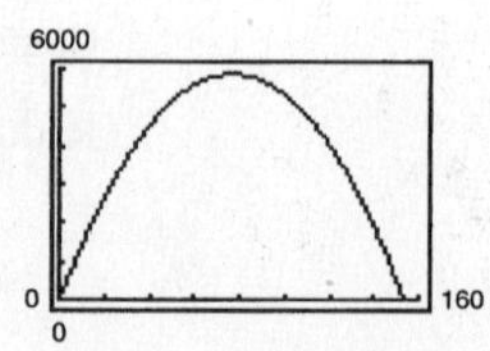

(c) The area is maximum when $x = w = 76\frac{2}{3}$ yards, or when the field is a square.

(d) A regulation football filed is 100 yards by $53\frac{1}{3}$ yards.

80. (a)

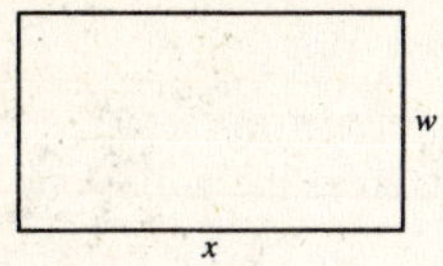

(b) $2x + 2w = 390$

$x + w = 195$

$w = 195 - x$

$A = xw = x(195 - x)$

(c)

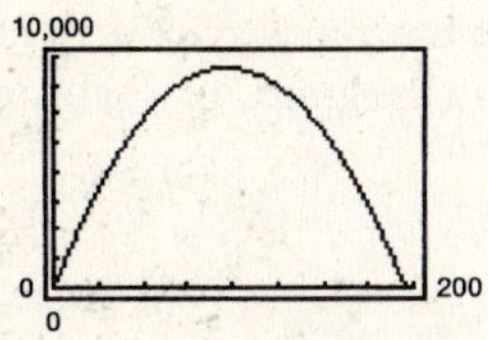

(d) The area is maximum when $x = w = 97.5$ yards, or when the field is a square.

(e) A regulation soccer field must be rectangular and the length of the touch line must be greater than the length of the goal line. The length can be between 100 yards and 130 yards. The width can be between 50 yards and 100 yards.

81. $y = -0.0026t^2 + 0.577t + 44.19$

(a) and (b)

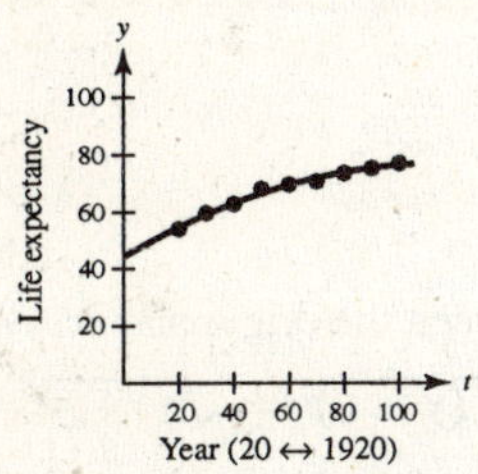

The curve seems to be a good fit for the data.

(c) For the year 2005, $t = 105$ and $y = 76.11$ years.

For the year 2010, $t = 110$ and $y = 76.2$ years.

(d) No. Because the model is quadratic, life expectancy decreases for $t > 111$ or 2011.

82. $y = -0.021t^2 + 0.49t + 9.3$

(a) and (b)

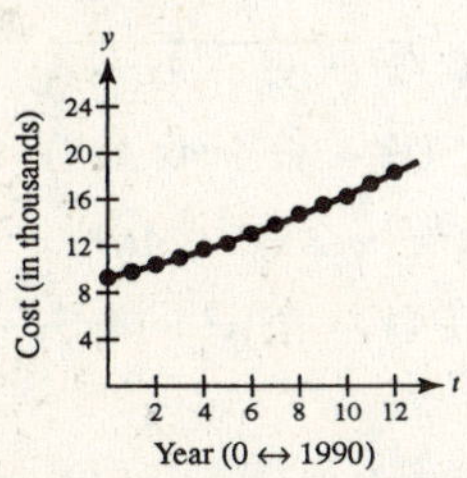

The curve seems to be a good fit for the data.

(c) For the year 2005, $t = 15$ and $y = \$21.375$ thousand.

For the year 2010, $t = 20$ and $y = \$27.5$ thousand.

(d) No. Since the model is quadratic, it increases without bound as t increases.

83. $y = \dfrac{10{,}770}{50^2} - 0.37 \approx 3.9$ ohms

84. (i) $y = kx + 5 \implies \dfrac{y - 5}{x} = k$

This matches table (b) when $k = 2$.

(ii) $y = x^2 + k \implies y - x^2 = k$

This matches table (d) when $k = -10$.

(iii) $y = kx^{3/2} \implies \dfrac{y}{x^{3/2}} = k$

This matches table (a) when $k = 3$.

(iv) $xy = k$

This matches table (c) when $k = 36$.

85. False.

To find y-intercepts, let x be zero and solve the equation for y.

86. True.

All linear equations of the form $y = mx + b$, which excludes vertical lines, cross the y-axis one time.

87. True.

A circle centered at the origin would have symmetries to the origin, the x-axis, and the y-axis.

88. False.

If the graph of an equation has x-axis symmetry, then for each point (x, y) on the graph, $(x, -y)$ is also on the graph.

89. False.

If $(1, -2)$ is a point on a graph that is symmetric with respect to the x-axis, then $(1, -(-2)) = (1, 2)$ is also a point on the graph.

90. True.

If $(1, -2)$ is a point on a graph that is symmetric with respect to the y-axis, then the reflected point is found by talking the negative of the x-coordinate. Thus, $(-1, -2)$ is also a point on the graph.

91. True.

If $b^2 - 4ac > 0$ and $a \neq 0$, then the graph of $y = ax^2 + bx + c$

has x-intercepts at $\left(\dfrac{-b - \sqrt{b^2 - 4ac}}{2a}, 0\right)$ and $\left(\dfrac{-b + \sqrt{b^2 - 4ac}}{2a}, 0\right)$.

92. True.

If $b^2 - 4ac = 0$ and $a \neq 0$, then the graph of

$y = ax^2 + bx + c$ has one x-intercept at $\left(\dfrac{-b}{2a}, 0\right)$.

This is also the vertex of the parabola.

93. The distance between the origin and (x, y) is:

$$\sqrt{(x - 0)^2 + (y - 0)^2} = \sqrt{x^2 + y^2}$$

The distance between $(2, 0)$ and (x, y) is:

$$\sqrt{(x - 2)^2 + (y - 0)^2} = \sqrt{x^2 + y^2 - 4x + 4}$$

Since the distance between (x, y) and the origin is K times the distance between (x, y) and $(2, 0)$, we have:

$$\sqrt{x^2 + y^2} = K\sqrt{x^2 + y^2 - 4x + 4}$$

$$x^2 + y^2 = K^2(x^2 + y^2 - 4x + 4)$$

$$x^2 + y^2 - K^2x^2 - K^2y^2 + 4K^2x - 4K^2 = 0$$

$$(1 - K^2)x^2 + (1 - K^2)y^2 + 4K^2x - 4K^2 = 0$$

94. The viewing rectangle is incorrect. Change the viewing rectangle. Answers will vary.

95. Assuming that the graph does not go beyond the vertical limits of the display, you will see the graph for the larger values of x.

Section P.5 Linear Equations in Two Variables

1.

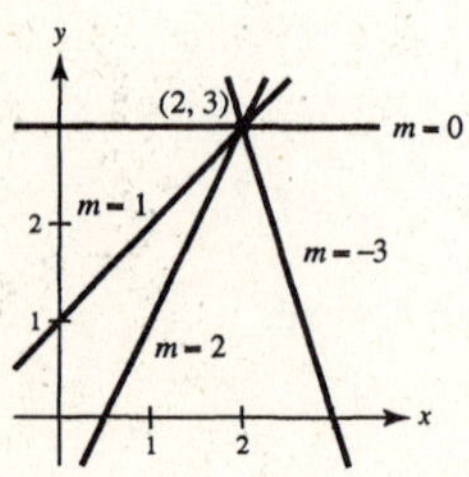

2.

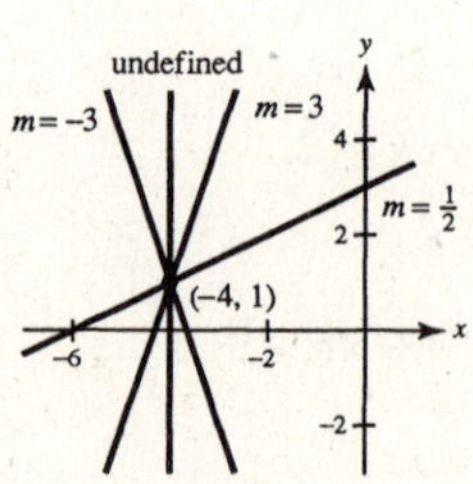

3. Two points on the line: (0, 0) and (5, 8)

$$\text{Slope} = \frac{\text{rise}}{\text{run}} = \frac{8}{5}$$

4. The line appears to go through (1, 0) and (4, 8).

$$\text{Slope} = \frac{y_2 - y_1}{x_2 - x_1} = \frac{8 - 0}{4 - 1} = \frac{8}{3}$$

5. Two points on the line: (0, 3) and (1, 3)

$$\text{Slope} = \frac{\text{rise}}{\text{run}} = \frac{0}{1} = 0$$

6. The line appears to go through (0, 7) and (7, 0).

$$\text{Slope} = \frac{y_2 - y_1}{x_2 - x_1} = \frac{0 - 7}{7 - 0} = -1$$

7. Two points on the line: (0, 8) and (2, 0)

$$\text{Slope} = \frac{\text{rise}}{\text{run}} = \frac{-8}{2} = -4$$

8. The line appears to go through (0, 1) and (6, 5).

$$\text{Slope} = \frac{y_2 - y_1}{x_2 - x_1} = \frac{5 - 1}{6 - 0} = \frac{4}{6} = \frac{2}{3}$$

9.

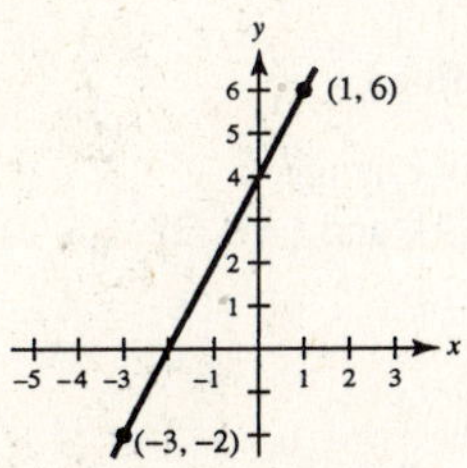

$$m = \frac{6 - (-2)}{1 - (-3)} = \frac{8}{4} = 2$$

10. $\text{Slope} = \dfrac{-4 - 4}{4 - 2} = -4$

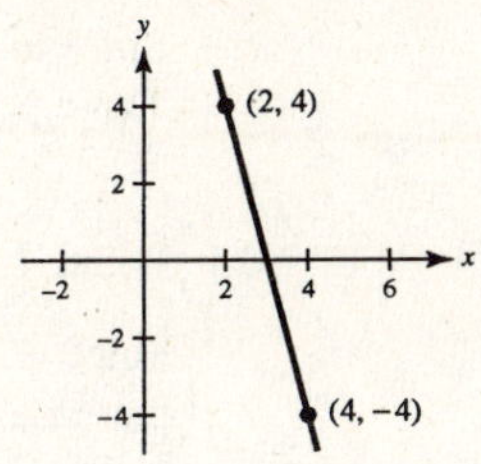

11.

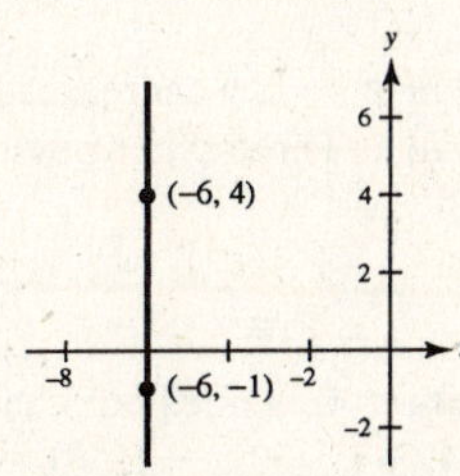

$$m = \frac{4 - (-1)}{-6 - (-6)} = \frac{5}{0}$$

m is undefined.

12. $\text{Slope} = \dfrac{0 - (-10)}{-4 - 0} = -\dfrac{5}{2}$

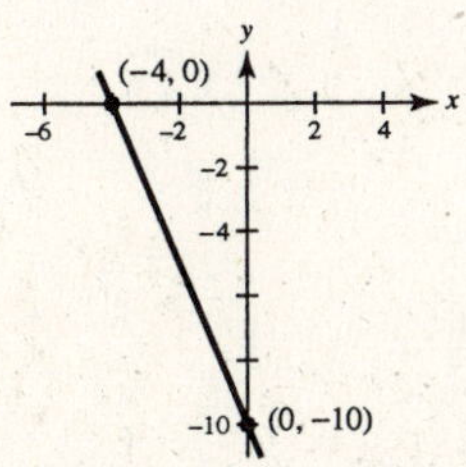

13.

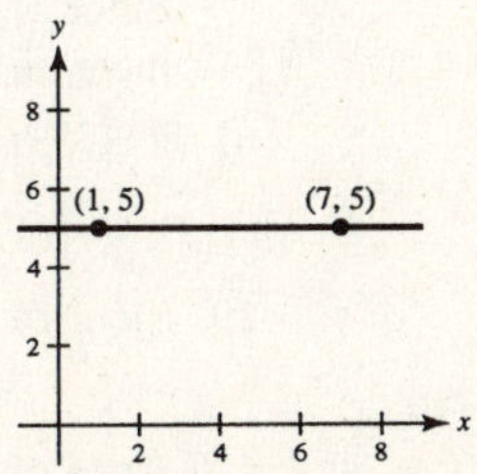

$$m = \frac{5 - 5}{7 - 1} = \frac{0}{6} = 0$$

14.

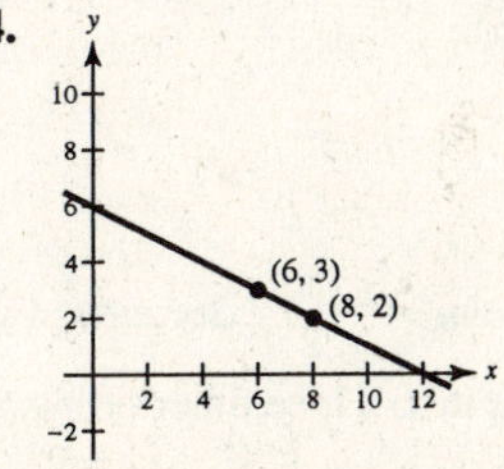

$$m = \frac{2 - 3}{8 - 6} = -\frac{1}{2}$$

15.

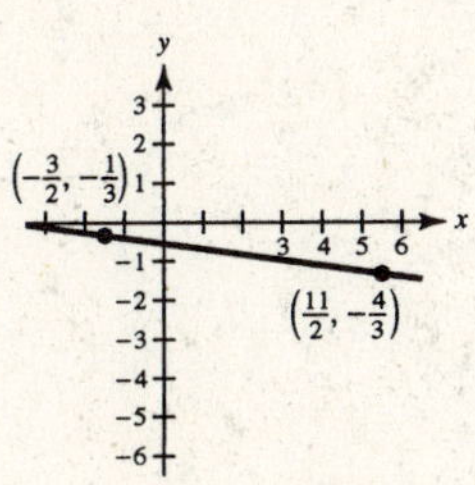

$$m = \frac{-\frac{1}{3} - \left(-\frac{4}{3}\right)}{-\frac{3}{2} - \frac{11}{2}} = -\frac{1}{7}$$

16. $\text{Slope} = \dfrac{-\frac{1}{4} - \frac{3}{4}}{\frac{5}{4} - \frac{7}{8}} = \dfrac{-1}{\frac{3}{8}} = -\dfrac{8}{3}$

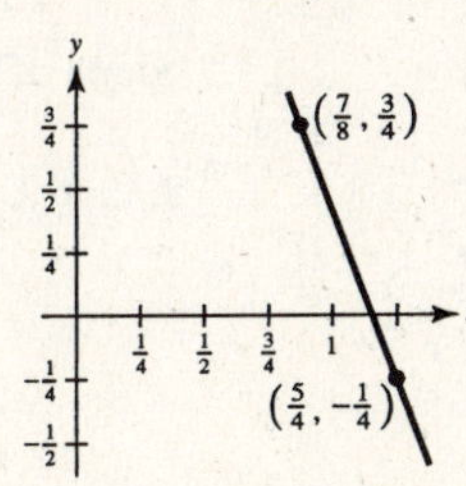

17.

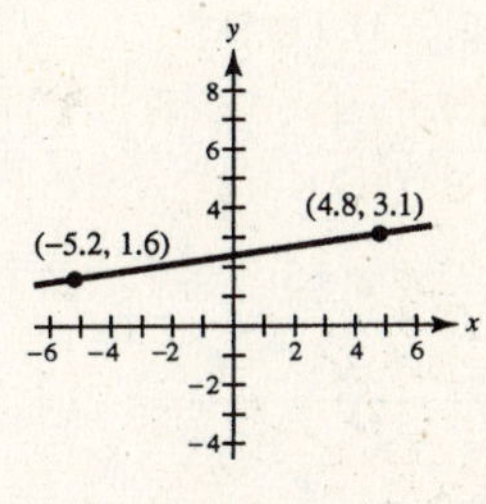

$$m = \frac{1.6 - 3.1}{-5.2 - 4.8} = \frac{-1.5}{-10} = 0.15$$

18. Slope $= \dfrac{-2.6 - (-8.3)}{2.25 - (-1.75)} = 1.425$

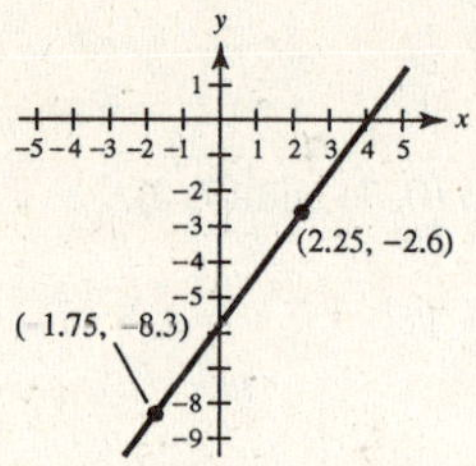

19. Point: (2, 1) Slope: $m = 0$

Since $m = 0$, y does not change.
Three points are (0, 1), (3, 1), and (−1, 1).

20. Because m is undefined, x does not change. Three other points are: (−4, 0), (−4, 3), (−4, 5).

21. Point: (5, −6) Slope: $m = 1$

Since $m = 1$, y increases by 1 for every one unit increase in x. Three points are (6, −5), (7, −4), and (8, −3).

22. Because $m = -1$, y decreases by 1 for every one unit increase in x. Three other points are: (0, 4), (9, −5), (11, −7).

23. Point: (−8, 1) Slope is undefined.

Since m is undefined, x does not change.
Three points are (−8, 0), (−8, 2), and (−8, 3).

24. Because $m = 0$, y does not change. Three other points are: (−4, −1), (−2, −1), (0, −1).

25. Point: (−5, 4) Slope: $m = 2$

Since $m = 2 = \frac{2}{1}$, y increases by 2 for every one unit increase in x. Three additional points are (−4, 6), (−3, 8), and (−2, 10).

26. Because $m = -2$, y decreases by 2 for every one unit increase in x. Three other points are: (−2, −5), (1, −11), (3, −15).

27. Point: (7, −2) Slope: $m = \frac{1}{2}$

Since $m = \frac{1}{2}$, y increases by 1 unit for every two unit increase in x. Three additional points are (9, −1), (11, 0), and (13, 1).

28. Because $m = -\dfrac{1}{2}$, y decreases by 1 for every 2 unit increase in x. Three other points are: (−3, −5), (1, −7), (5, −9).

29. Slope of L_1: $m = \dfrac{9+1}{5-0} = 2$

Slope of L_2: $m = \dfrac{1-3}{4-0} = -\dfrac{1}{2}$

L_1 and L_2 are perpendicular.

30. L_1: (−2, −1), (1, 5)

$$m_1 = \frac{5-(-1)}{1-(-2)} = \frac{6}{3} = 2$$

L_2: (1, 3), (5, −5)

$$m_2 = \frac{-5-3}{5-1} = \frac{-8}{4} = -2$$

The lines are neither parallel nor perpendicular.

31. Slope of L_1: $m = \dfrac{0-6}{-6-3} = \dfrac{2}{3}$

Slope of L_2: $m = \dfrac{\frac{7}{3}+1}{5-0} = \dfrac{2}{3}$

L_1 and L_2 are parallel.

32. L_1: (4, 8), (−4, 2)

$$m_1 = \frac{2-8}{-4-4} = \frac{-6}{-8} = \frac{3}{4}$$

L_2: $(3, -5), \left(-1, \frac{1}{3}\right)$

$$m_2 = \frac{\frac{1}{3}-(-5)}{-1-3} = \frac{\frac{16}{3}}{-4} = -\frac{4}{3}$$

The lines are perpendicular.

33. (a) $m = \frac{2}{3}$. Since the slope is positive, the line rises. Matches L_2.

(b) m is undefined. The line is vertical. Matches L_3.

(c) $m = -2$. The line falls. Matches L_1.

34. (a) $m = 0$. The line is horizontal. Matches L_2.

(b) $m = -\frac{3}{4}$. Because the slope is negative, the line falls. Matches L_1.

(c) $m = 1$. Because the slope is positive, the line rises. Matches L_3.

35. The slope is $m = -20$. This represents the decrease in the amount of the loan each week. Matches graph (b)

36. The y-intercept is 8.5 and the slope is 2, which represents the increase in hourly wage per unit produced. Matches graph (c)

37. The slope is $m = 0.32$. This represents the increase in travel cost for each mile driven. Matches graph (a)

38. The y-intercepts is 750 and the slope is −100, which represents the decrease in the value of the word processor each year. Matches graph (d)

39. (a) $m = 135$. The sales are increasing 135 units per year.

(b) $m = 0$. There is no change in sales.

(c) $m = -40$. The sales are decreasing 40 units per year.

40. (a) $m = 400$. The revenues are increasing \$400 per day.

(b) $m = 100$. The revenues are increasing \$100 per day.

(c) $m = 0$. There is no change in revenue. (Revenue remains constant.)

41. (a) and (b)

x	300	600	900	1200	1500	1800	2100
y	−25	−50	−75	−100	−125	−150	−175

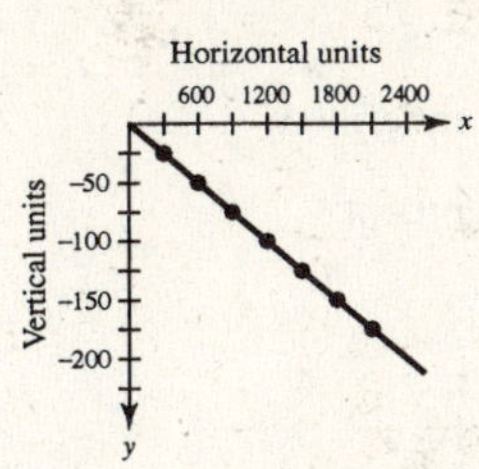

(c) $m = \frac{-50-(-25)}{600-300} = \frac{-25}{300} = -\frac{1}{12}$

$$y - (-50) = -\frac{1}{12}(x - 600)$$

$$y + 50 = -\frac{1}{12}x + 50$$

$$y = -\frac{1}{12}x$$

(d) Since $m = -\frac{1}{12}$, for every 12 horizontal measurements the vertical measurement decreases by 1.

(e) $\frac{1}{12} \approx 0.083 = 8.3\%$ grade

42. $y = \frac{6}{100}x$

$y = \frac{6}{100}(200) = 12$ feet

43. (a)

Years	Slope
1993–1994	0.91
1994–1995	1.47
1995–1996	0.90
1996–1997	0.72
1997–1998	1.01
1998–1999	0.84
1999–2000	0.98
2000–2001	0.63
2001–2002	0.54
2002–2003	1.29

The years 1994–1995 showed the greatest increase and the years 2001–2002 showed the smallest increase.

(b) $m = \dfrac{16.70 - 7.41}{13 - 3} = 0.929$

(c) On average, revenue increased 0.929 billion dollars each year.

44. (a)

Years	Slope
1993–1994	60.0
1994–1995	43.6
1995–1996	10.4
1996–1997	14.0
1997–1998	−1.2
1998–1999	−78.3
1999–2000	−20.0
2000–2001	49.7
2001–2002	76.7
2002–2003	72.2

The years 2001–2002 showed the greatest increase and the years 1998–1999 showed the greatest decrease.

(b) $m = \dfrac{350.0 - 122.9}{13 - 3} = 22.71$

(c) On average, revenue increased 22.71 million dollars each year.

45. $\dfrac{\text{rise}}{\text{run}} = \dfrac{3}{4} = \dfrac{x}{\frac{1}{2}(32)}$

$\dfrac{3}{4} = \dfrac{x}{16}$

$4x = 48$

$x = 12$

The maximum height in the attic is 12 feet.

46. Let x = the number of minutes.

Then $C = 0.01x + 10$.

The slope, $m = 0.01$, represents the per minute cost of the phone, while the C-intercept of 10 represents the basic monthly rate.

47. $y = x - 10$

Slope: $m = 1$

y-intercept: $(0, -10)$

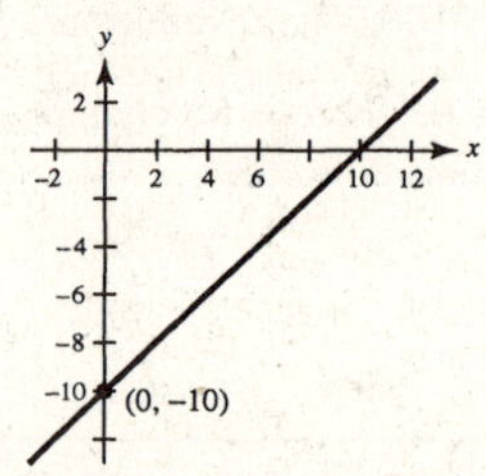

48. $y = 5x + 3$

Slope: $m = 5$

y-intercept: $(0, 3)$

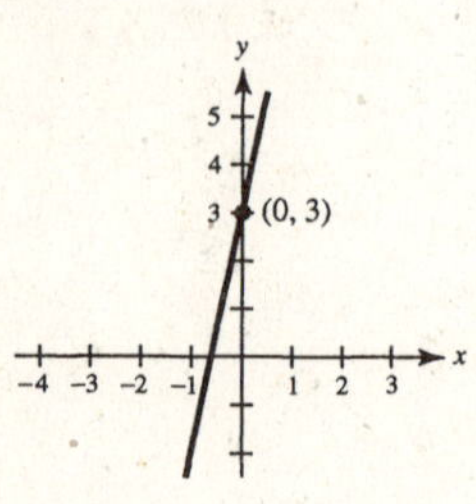

49. $3y + 5 = 0$

$3y = -5$

$y = -\frac{5}{3}$

Slope: $m = 0$

y-intercept: $\left(0, -\frac{5}{3}\right)$

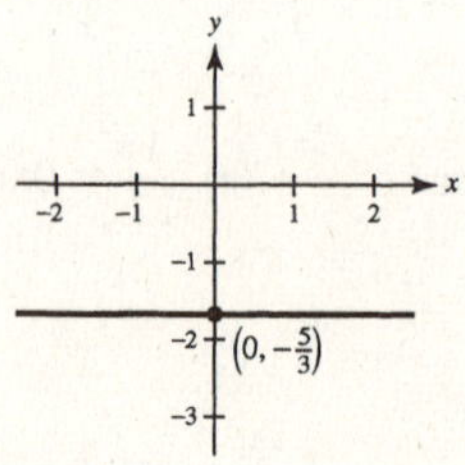

50. $5x - 2 = 0$

$x = \frac{2}{5}$

Slope: undefined

No y-intercept

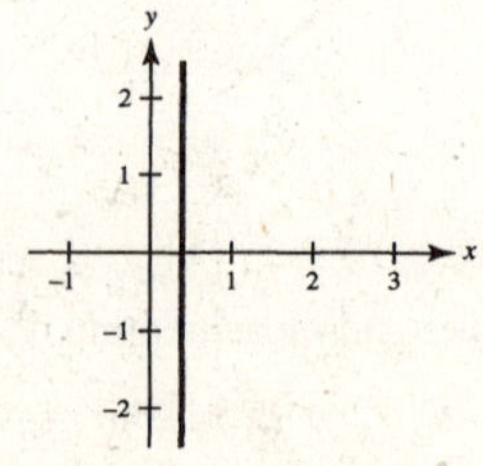

51. $2x + 3y = 9$

$$3y = -2x + 9$$

$$y = -\frac{2}{3}x + 3$$

Slope: $m = -\dfrac{2}{3}$

y-intercept: $(0, 3)$

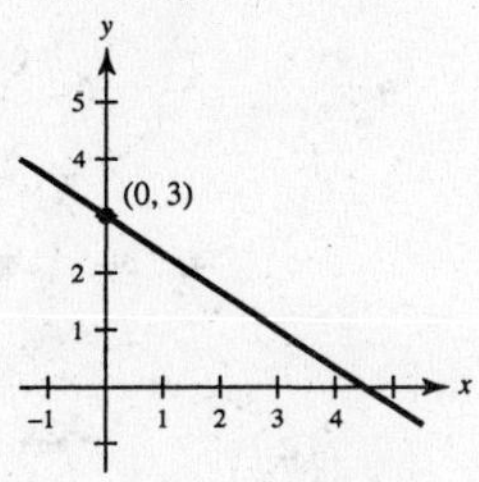

52. $7x + 6y = 30$

$$y = -\tfrac{7}{6}x + 5$$

Slope: $m = -\frac{7}{6}$

y-intercept: $(0, 5)$

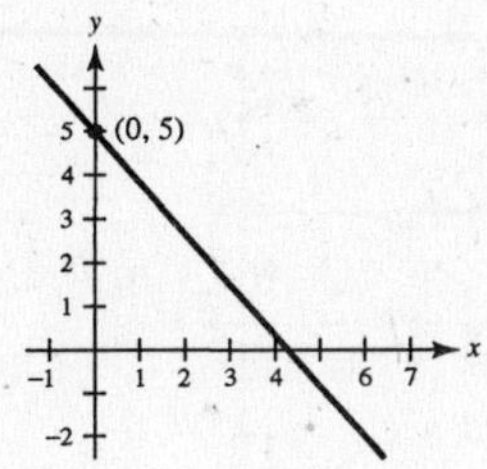

53. $-4x + 5y = -35$

$$5y = 4x - 35$$

$$y = \tfrac{4}{5}x - 7$$

Slope: $m = \frac{4}{5}$

y-intercept: $(0, -7)$

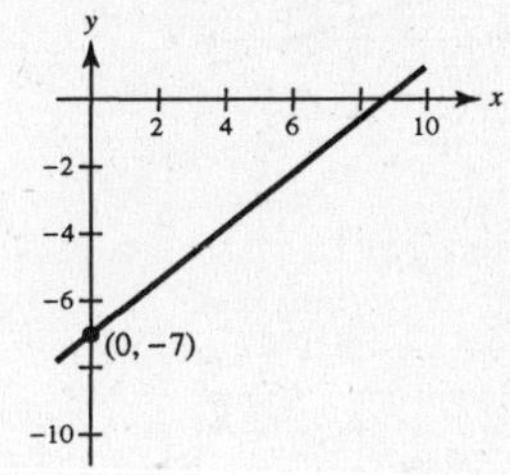

54. $-12x + 3y = 39$

$$3y = 12x + 39$$

$$y = 4x + 13$$

Slope: $m = 4$

y-intercept: $(0, 13)$

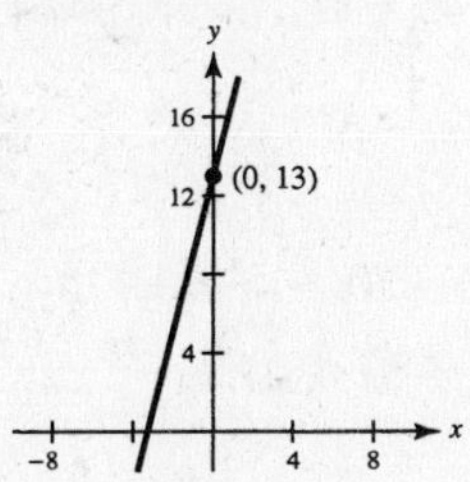

55. $m = -1, (0, 10)$

$$y - 10 = -1(x - 0)$$

$$y - 10 = -x$$

$$y = -x + 10$$

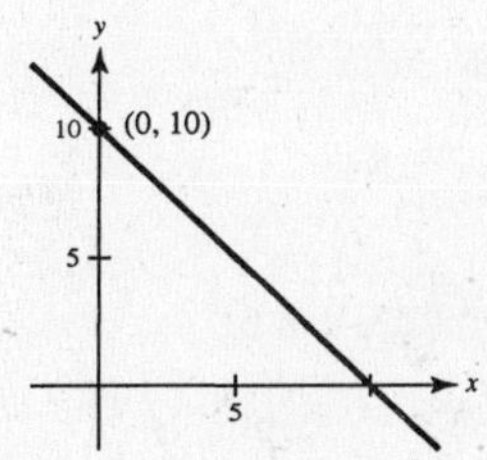

56. $y + 2 = 3(x - 0)$

$$y = 3x - 2$$

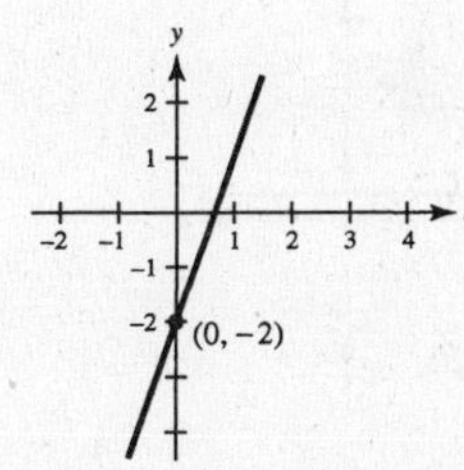

57. $m = 4, (0, 0)$

$$y - 0 = 4(x - 0)$$

$$y = 4x$$

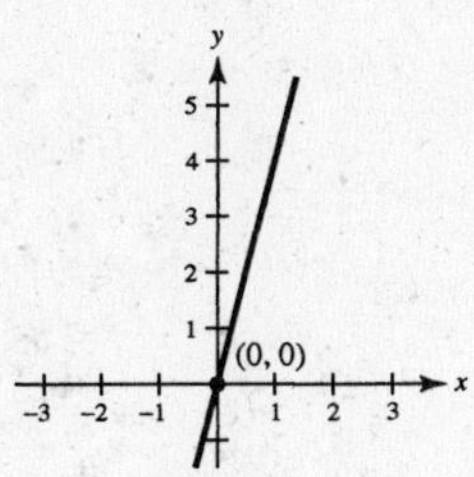

58. $y - 6 = -2(x + 3)$

$$y = -2x$$

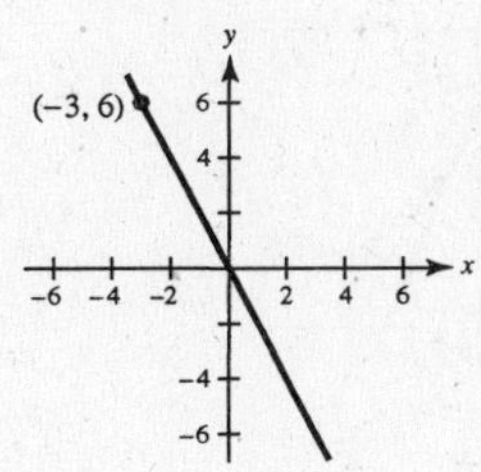

59. $m = \frac{3}{4}, (-2, -5)$

$$y + 5 = \tfrac{3}{4}(x + 2)$$

$$4y + 20 = 3x + 6$$

$$4y = 3x - 14$$

$$y = \tfrac{3}{4}x - \tfrac{7}{2}$$

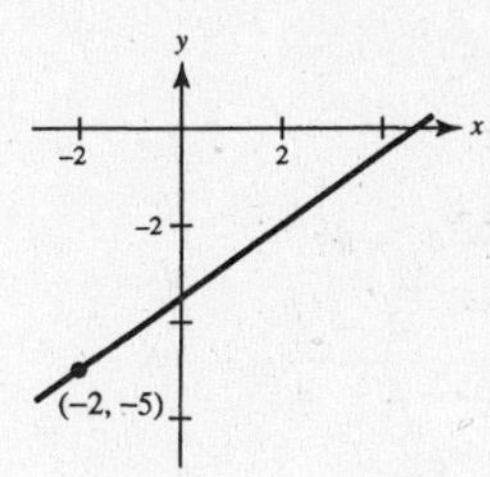

60. $y - 0 = -\frac{1}{3}(x - 4)$

$y = -\frac{1}{3}x + \frac{4}{3}$

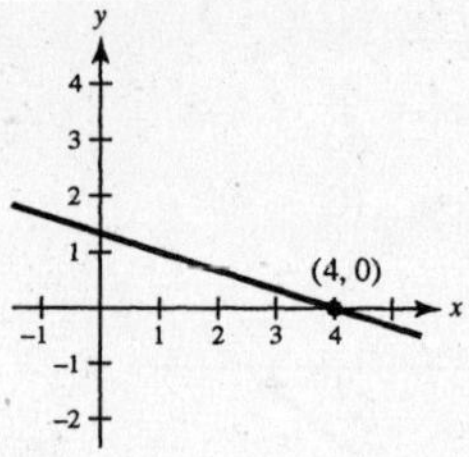

61. $m = 0, (-10, 4)$

$y - 4 = 0(x + 10)$

$y - 4 = 0$

$y = 4$

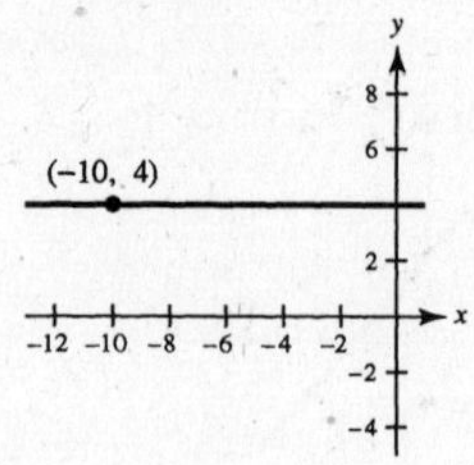

62. $x = 6$

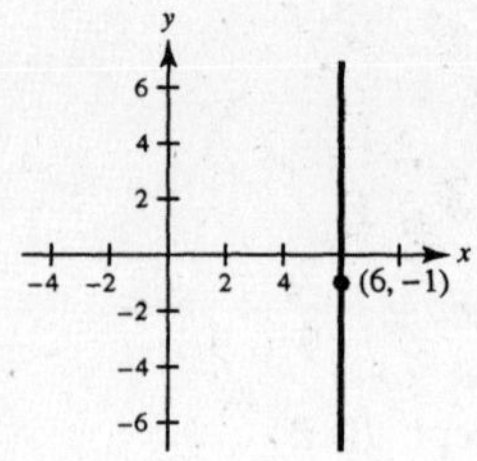

63. $m = 0, \left(-3, \frac{1}{3}\right)$

$y - \frac{1}{3} = 0(x + 3)$

$y - \frac{1}{3} = 0$

$y = \frac{1}{3}$

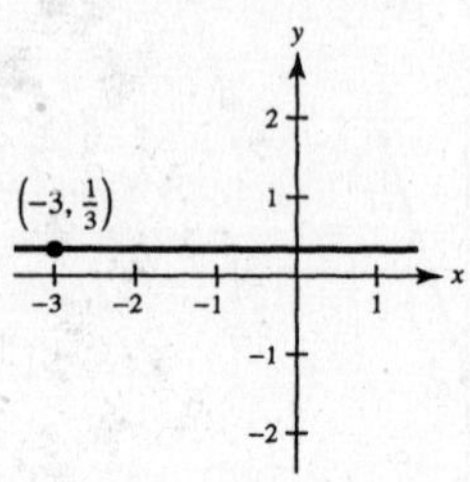

64. m is undefined, $\left(2, -\frac{1}{2}\right)$. The line is vertical.

$x = 2$

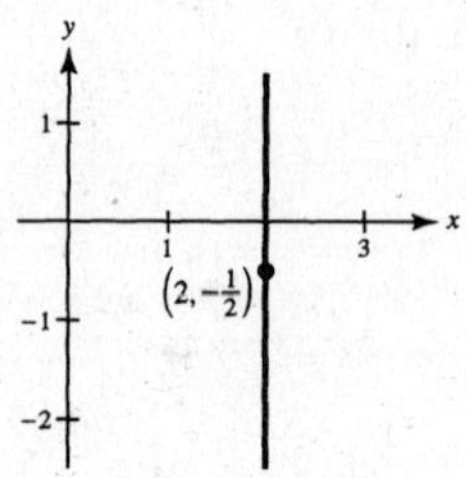

65. $m = -3, \left(-\frac{1}{2}, \frac{3}{2}\right)$

$y - \frac{3}{2} = -3\left(x + \frac{1}{2}\right)$

$y - \frac{3}{2} = -3x - \frac{3}{2}$

$y = -3x$

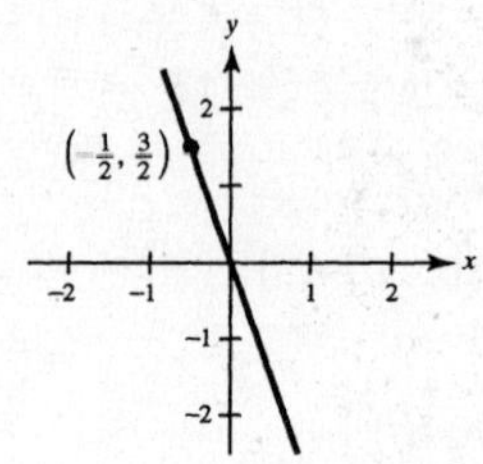

66. $y - \frac{5}{2} = \frac{4}{3}(x - 4)$

$y = \frac{4}{3}x - \frac{17}{6}$

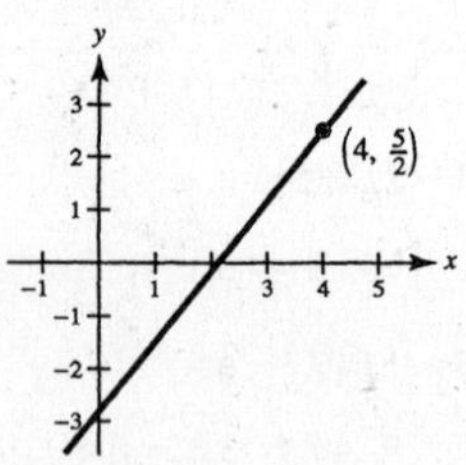

67. $m = -\frac{5}{2}, (2.3, -8.5)$

$y - (-8.5) = -\frac{5}{2}(x - 2.3)$

$y + 8.5 = -2.5x + 5.75$

$y = -2.5x - 2.75$

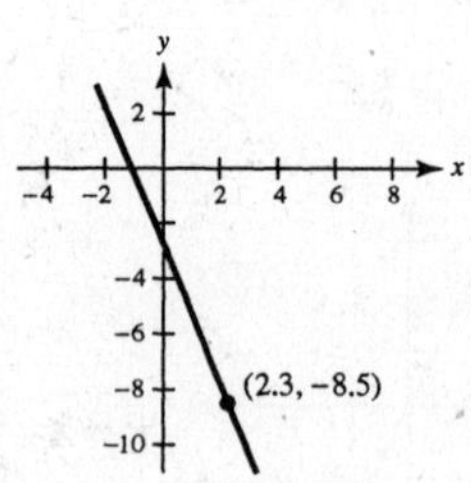

68. $y - 1.8 = 5(x - (-5.1))$

$y = 5x + 27.3$

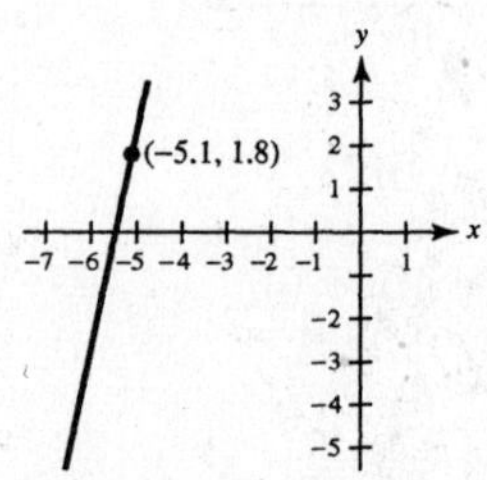

69. $(4, 3), (-4, -4)$

$y - 3 = \frac{-4 - 3}{-4 - 4}(x - 4)$

$y - 3 = \frac{7}{8}(x - 4)$

$y - 3 = \frac{7}{8}x - \frac{7}{2}$

$y = \frac{7}{8}x - \frac{1}{2}$

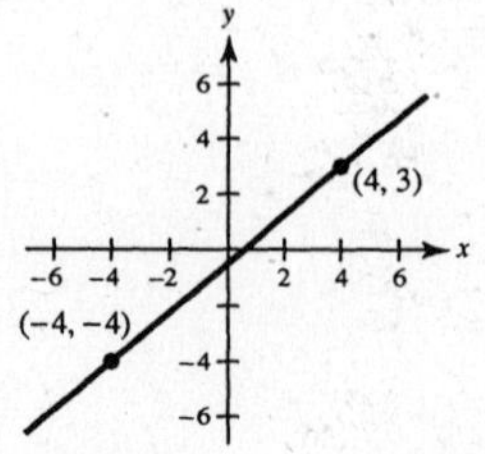

70. $y + 1 = \frac{5 + 1}{-5 - 5}(x - 5)$

$y = -\frac{3}{5}(x - 5) - 1$

$y = -\frac{3}{5}x + 2$

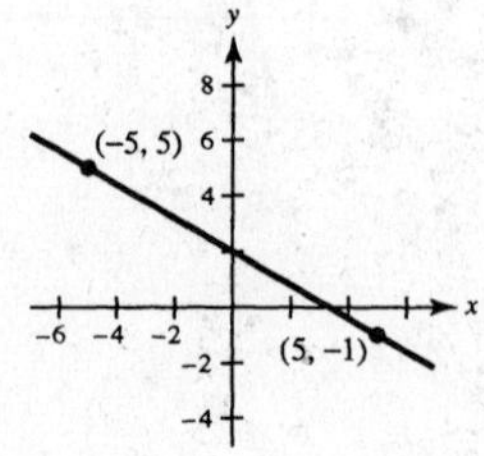

71. $(-1, 4), (6, 4)$

$$y - 4 = \frac{4 - 4}{6 - (-1)}(x + 1)$$

$$y - 4 = 0(x + 1)$$

$$y - 4 = 0$$

$$y = 4$$

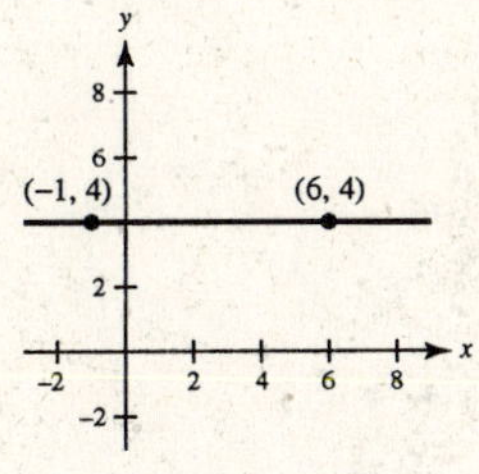

72. $(2, -3), (7, -3)$

$$y + 3 = \frac{-3 - (-3)}{7 - 2}(x - 2)$$

$$y + 3 = 0(x - 2)$$

$$y + 3 = 0$$

$$y = -3$$

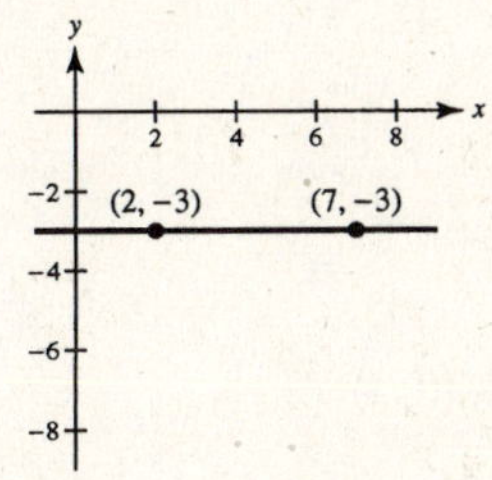

73. Since both points have $x = -8$, the slope is undefined.

$$x = -8$$

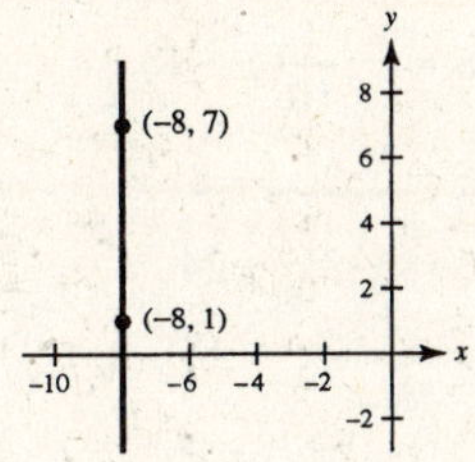

74. $(2, 5), (2, -6)$

$m = \dfrac{-6 - 5}{2 - 2} = -\dfrac{11}{0}$, which is undefined.

The line is vertical.

$$x = 2$$

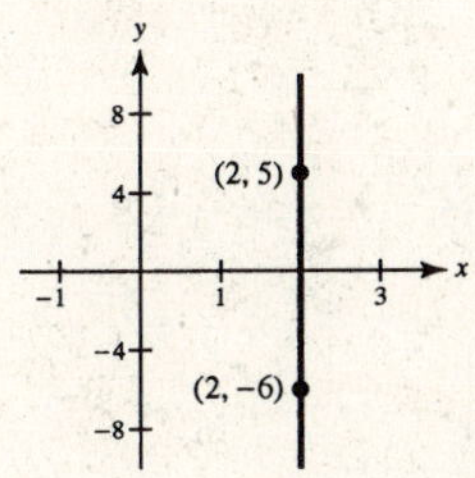

75. $(1, 1), \left(6, -\frac{2}{3}\right)$

$$y - 1 = \frac{-\frac{2}{3} - 1}{6 - 1}(x - 1)$$

$$y - 1 = -\frac{1}{3}(x - 1)$$

$$y - 1 = -\frac{1}{3}x + \frac{1}{3}$$

$$y = -\frac{1}{3}x + \frac{4}{3}$$

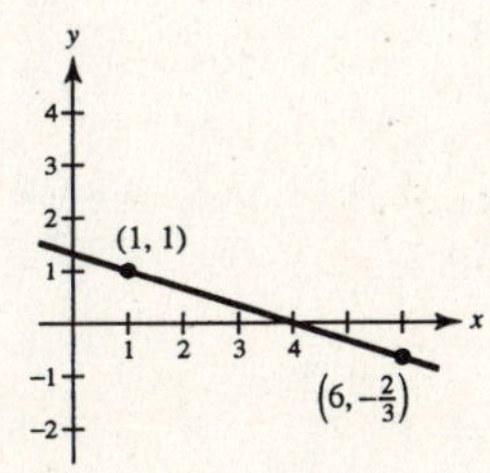

76. $y - \dfrac{1}{2} = \dfrac{\frac{5}{4} - \frac{1}{2}}{\frac{1}{2} - 2}(x - 2)$

$$y = -\frac{1}{2}(x - 2) + \frac{1}{2}$$

$$y = -\frac{1}{2}x + \frac{3}{2}$$

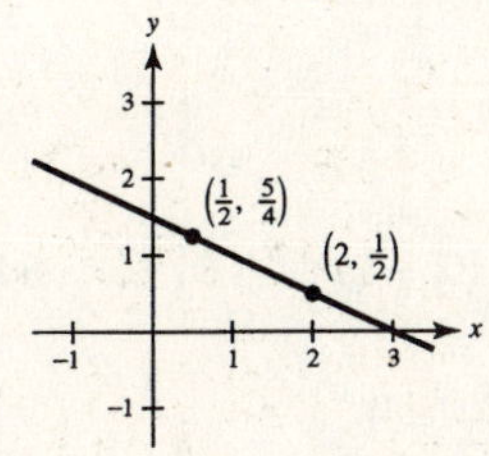

77. $\left(\frac{3}{4}, \frac{3}{2}\right), \left(-\frac{4}{3}, \frac{7}{4}\right)$

$$y - \frac{3}{2} = \frac{\frac{7}{4} - \frac{3}{2}}{-\frac{4}{3} - \frac{3}{4}}\left(x - \frac{3}{4}\right)$$

$$y - \frac{3}{2} = \frac{\frac{1}{4}}{-\frac{25}{12}}\left(x - \frac{3}{4}\right)$$

$$y - \frac{3}{2} = -\frac{3}{25}\left(x - \frac{3}{4}\right)$$

$$y - \frac{3}{2} = -\frac{3}{25}x + \frac{9}{100}$$

$$y = -\frac{3}{25}x + \frac{159}{100}$$

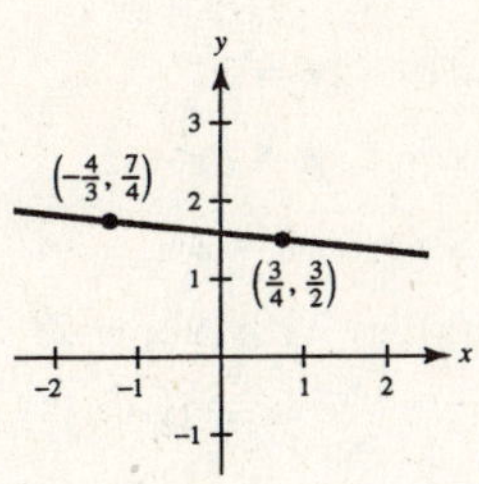

78. $y - \left(-\dfrac{3}{5}\right) = \dfrac{-\frac{9}{5} - \left(-\frac{3}{5}\right)}{\frac{9}{10} - \left(-\frac{1}{10}\right)}\left(x - \left(-\dfrac{1}{10}\right)\right)$

$$y = -\frac{6}{5}\left(x + \frac{1}{10}\right) - \frac{3}{5}$$

$$y = -\frac{6}{5}x - \frac{18}{25}$$

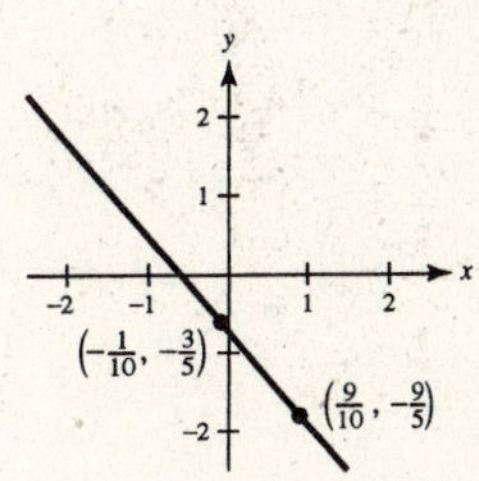

79. $(-8, 0.6), (2, -2.4)$

$$y - 0.6 = \frac{-2.4 - 0.6}{2 - (-8)}(x + 8)$$

$$y - 0.6 = -\frac{3}{10}(x + 8)$$

$$10y - 6 = -3(x + 8)$$

$$10y - 6 = -3x - 24$$

$$10y = -3x - 18$$

$$y = -\frac{3}{10}x - \frac{9}{5} \quad \text{or} \quad y = -0.3x - 1.8$$

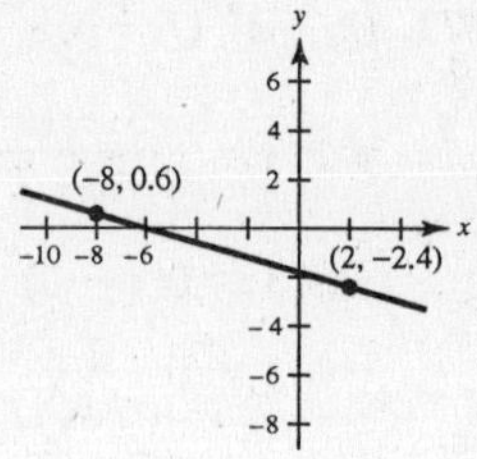

80. $y - 0.6 = \dfrac{-0.6 - 0.6}{-2 - 1}(x - 1)$

$$y = 0.4(x - 1) + 0.6$$

$$y = 0.4x + 0.2$$

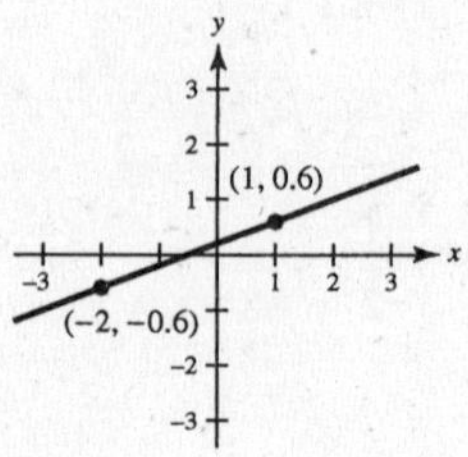

81. $(-3, 0), (0, 4)$

$$\frac{x}{-3} + \frac{y}{4} = 1$$

$$(-12)\frac{x}{-3} + (-12)\frac{y}{4} = (-12) \cdot 1$$

$$4x - 3y + 12 = 0$$

82. $\dfrac{x}{2} + \dfrac{y}{3} = 1$

$$3x + 2y - 6 = 0$$

83. $\left(\dfrac{2}{3}, 0\right), (0, -2)$

$$\frac{x}{\frac{2}{3}} + \frac{y}{-2} = 1$$

$$\frac{3x}{2} - \frac{y}{2} = 1$$

$$3x - y - 2 = 0$$

84. $\dfrac{x}{-\frac{1}{6}} + \dfrac{y}{-\frac{2}{3}} = 1$

$$6x + \frac{3}{2}y = -1$$

$$12x + 3y + 2 = 0$$

85. $(d, 0), (0, d), (-3, 4)$

$$\frac{x}{d} + \frac{y}{d} = 1$$

$$x + y = d$$

$$-3 + 4 = d$$

$$1 = d$$

$$x + y = 1$$

$$x + y - 1 = 0$$

86. $\dfrac{x}{c} + \dfrac{y}{c} = 1, \ c \neq 0$

$$x + y = c$$

$$1 + 2 = c$$

$$3 = c$$

$$x + y = 3$$

$$x + y - 3 = 0$$

87. $x + y = 7$

$$y = -x + 7$$

Slope: $m = -1$

(a) $m = -1, (-3, 2)$

$$y - 2 = -1(x + 3)$$

$$y - 2 = -x - 3$$

$$y = -x - 1$$

(b) $m = 1, (-3, 2)$

$$y - 2 = 1(x + 3)$$

$$y = x + 5$$

88. $4x - 2y = 3$

$$y = 2x - \tfrac{3}{2}$$

Slope: $m = 2$

(a) $y - 1 = 2(x - 2)$

$$y = 2x - 3$$

(b) $y - 1 = -\frac{1}{2}(x - 2)$

$$y = -\tfrac{1}{2}x + 2$$

89. $5x + 3y = 0$

$3y = -5x$

$y = -\frac{5}{3}x$

Slope: $m = -\frac{5}{3}$

(a) $m = -\frac{5}{3}, \left(\frac{7}{8}, \frac{3}{4}\right)$

$y - \frac{3}{4} = -\frac{5}{3}\left(x - \frac{7}{8}\right)$

$24y - 18 = -40\left(x - \frac{7}{8}\right)$

$24y - 18 = -40x + 35$

$24y = -40x + 53$

$y = -\frac{5}{3}x + \frac{53}{24}$

(b) $m = \frac{3}{5}, \left(\frac{7}{8}, \frac{3}{4}\right)$

$y - \frac{3}{4} = \frac{3}{5}\left(x - \frac{7}{8}\right)$

$40y - 30 = 24\left(x - \frac{7}{8}\right)$

$40y - 30 = 24x - 21$

$40y = 24x + 9$

$y = \frac{3}{5}x + \frac{9}{40}$

90. $3x + 4y = 7$

$y = -\frac{3}{4}x + \frac{7}{4}$

Slope: $m = -\frac{3}{4}$

(a) $y - \frac{7}{8} = -\frac{3}{4}\left(x - \left(-\frac{2}{3}\right)\right)$

$y = -\frac{3}{4}x + \frac{3}{8}$

(b) $y - \frac{7}{8} = \frac{4}{3}\left(x - \left(-\frac{2}{3}\right)\right)$

$y = \frac{4}{3}x + \frac{127}{72}$

91. $x = 4$

m is undefined.

(a) $(2, 5)$, m is undefined.

$x = 2$

(b) $(2, 5)$, $m = 0$

$y = 5$

92. $y = -3$

Slope: $m = 0$

(a) $y = 0$

(b) $x = -1 \Rightarrow x + 1 = 0$

93. $y = 2$, $m = 0$

(a) $(2, 1)$, $m = 0$

$y = 1$

(b) $(2, 1)$, m is undefined.

$x = 2$

94. $x = 2$, m is undefined.

(a) $(5, -3)$, m is undefined.

$x = 5$

(b) $(5, -3)$, $m = 0$

$y = -3$

95. $6x + 2y = 9$

$2y = -6x + 9$

$y = -3x + \frac{9}{2}$

Slope: $m = -3$

(a) $(-3.9, -1.4)$, $m = -3$

$y - (-1.4) = -3(x - (-3.9))$

$y + 1.4 = -3x - 11.7$

$y = -3x - 13.1$

(b) $(-3.9, -1.4)$, $m = \frac{1}{3}$

$y - (-1.4) = \frac{1}{3}(x - (-3.9))$

$y + 1.4 = \frac{1}{3}x + 1.3$

$y = \frac{1}{3}x - 0.1$

96. $x - y = 4$

$y = x - 4$

Slope: $m = 1$

(a) $y - 6.8 = 1(x - 2.5)$

$y = x + 4.3$

(b) $y - 6.8 = (-1)(x - 2.5)$

$y = -x + 9.3$

97. (a) $y = \frac{2}{3}x$ (b) $y = -\frac{3}{2}x$ (c) $y = \frac{2}{3}x + 2$

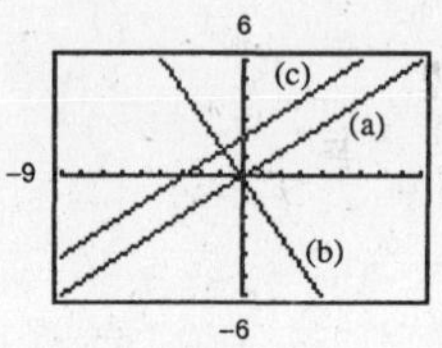

(a) is parallel to (c). (b) is perpendicular to (a) and (c).

98. (a) $y = 2x$ (b) $y = -2x$ (c) $y = \frac{1}{2}x$

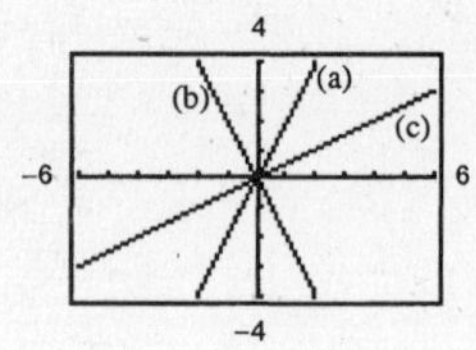

(b) and (c) are perpendicular.

99. (a) $y = x - 8$ (b) $y = x + 1$ (c) $y = -x + 3$

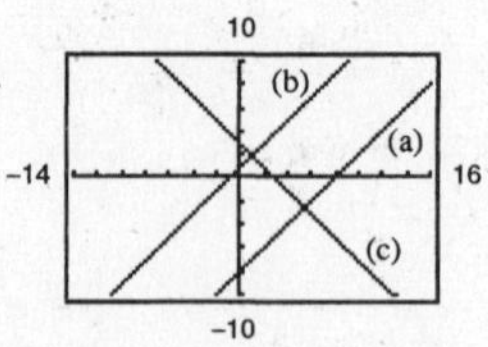

(a) is parallel to (b). (c) is perpendicular to (a) and (b).

100. (a) $y = -\frac{1}{2}x$ (b) $y = -\frac{1}{2}x + 3$ (c) $y = 2x - 4$

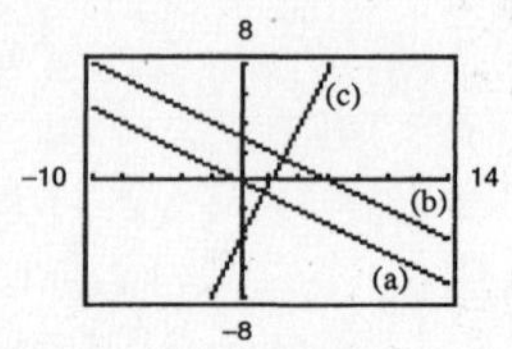

(a) and (b) are parallel.

(c) is perpendicular to (a) and (b).

101. $(5, 156), m = 4.50$

$$V - 156 = 4.50(t - 5)$$
$$V - 156 = 4.50t - 22.50$$
$$V = 4.5t + 133.50$$

102. $(5, 2540), m = 125$

$$V - 2540 = 125(t - 5)$$
$$V - 2540 = 125t - 625$$
$$V = 125t + 1915$$

103. Set the distance between $(6, 5)$ and (x, y) equal to the distance between $(1, -8)$ and (x, y).

$$\sqrt{(x-6)^2 + (y-5)^2} = \sqrt{(x-1)^2 + (y-(-8))^2}$$
$$(x-6)^2 + (y-5)^2 = (x-1)^2 + (y+8)^2$$
$$x^2 - 12x + 36 + y^2 - 10y + 25 = x^2 - 2x + 1 + y^2 + 16y + 64$$
$$x^2 + y^2 - 12x - 10y + 61 = x^2 + y^2 - 2x + 16y + 65$$
$$-12x - 10y + 61 = -2x + 16y + 65$$
$$-10x - 26y - 4 = 0$$
$$-2(5x + 13y + 2) = 0$$
$$5x + 13y + 2 = 0$$
$$13y = -5x - 2$$
$$y = -\frac{5}{13}x - \frac{2}{13}$$

104. Set the distance between $(4, -1)$ and (x, y) equal to the distance between $(-2, 3)$ and (x, y).

$$\sqrt{(x-4)^2 + [y-(-1)]^2} = \sqrt{[x-(-2)]^2 + (y-3)^2}$$
$$(x-4)^2 + (y+1)^2 = (x+2)^2 + (y-3)^2$$
$$x^2 - 8x + 16 + y^2 + 2y + 1 = x^2 + 4x + 4 + y^2 - 6y + 9$$
$$-8x + 2y + 17 = 4x - 6y + 13$$
$$0 = 12x - 8y - 4$$
$$0 = 4(3x - 2y - 1)$$
$$0 = 3x - 2y - 1$$

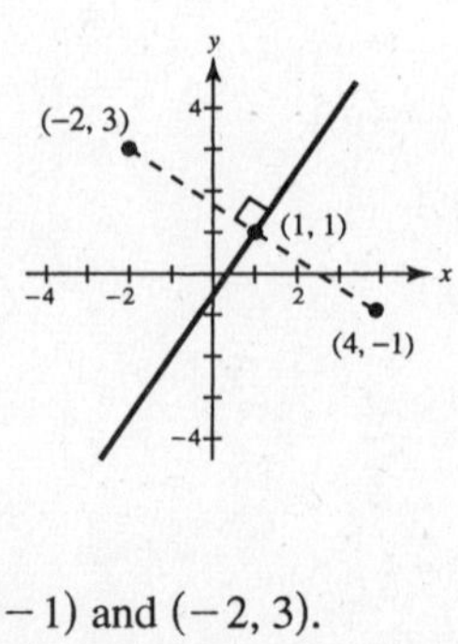

This line is the perpendicular bisector of the line segment connecting $(4, -1)$ and $(-2, 3)$.

105. Set the distance between $\left(-\frac{1}{2}, -4\right)$ and (x, y) equal to the distance between $\left(\frac{7}{2}, \frac{5}{4}\right)$ and (x, y).

$$\sqrt{\left(x - \left(-\tfrac{1}{2}\right)\right)^2 + (y - (-4))^2} = \sqrt{\left(x - \tfrac{7}{2}\right)^2 + \left(y - \tfrac{5}{4}\right)^2}$$

$$\left(x + \tfrac{1}{2}\right)^2 + (y + 4)^2 = \left(x - \tfrac{7}{2}\right)^2 + \left(y - \tfrac{5}{4}\right)^2$$

$$x^2 + x + \tfrac{1}{4} + y^2 + 8y + 16 = x^2 - 7x + \tfrac{49}{4} + y^2 - \tfrac{5}{2}y + \tfrac{25}{16}$$

$$x^2 + y^2 + x + 8y + \tfrac{65}{4} = x^2 + y^2 - 7x - \tfrac{5}{2}y + \tfrac{221}{16}$$

$$x + 8y + \tfrac{65}{4} = -7x - \tfrac{5}{2}y + \tfrac{221}{16}$$

$$8x + \tfrac{21}{2}y + \tfrac{39}{16} = 0$$

$$128x + 168y + 39 = 0$$

$$168y = -128x - 39$$

$$y = -\tfrac{16}{21}x - \tfrac{13}{56}$$

106. Set the distance between $\left(3, \frac{5}{2}\right)$ and (x, y) equal to the distance between $(-7, 1)$ and (x, y).

$$\sqrt{(x - 3)^2 + \left(y - \tfrac{5}{2}\right)^2} = \sqrt{[x - (-7)]^2 + (y - 1)^2}$$

$$(x - 3)^2 + \left(y - \tfrac{5}{2}\right)^2 = (x + 7)^2 + (y - 1)^2$$

$$x^2 - 6x + 9 + y^2 - 5y + \tfrac{25}{4} = x^2 + 14x + 49 + y^2 - 2y + 1$$

$$-6x - 5y + \tfrac{61}{4} = 14x - 2y + 50$$

$$-24x - 20y + 61 = 56x - 8y + 200$$

$$80x + 12y + 139 = 0$$

This line is the perpendicular bisector of the line segment connecting $\left(3, \frac{5}{2}\right)$ and $(-7, 1)$.

107. $t = 0$ represents 1999.

$(0, 0.28)$, $(5, 1.60)$

$$m = \frac{1.60 - 0.28}{5 - 0} = 0.264$$

$y = 0.264t + 0.28$

$t = 6$ represents 2005: $y = 0.264(6) + 0.28 \approx \$1.86$

$t = 11$ represents 2010: $y = 0.264(11) + 0.28 = \$3.18$

108. $t = 0$ represents 1996.

$(0, 397)$, $(8, 733)$

$$m = \frac{733 - 397}{8 - 0} = 42$$

$N = 42t + 397$

$t = 9$ represents 2005: $N = 42(9) + 397 = 775$ stores

$t = 16$ represents 2010: $N = 42(16) + 397 = 1069$ stores

109. Using the points $(0, 32)$ and $(100, 212)$, we have

$$m = \frac{212 - 32}{100 - 0} = \frac{180}{100} = \frac{9}{5}$$

$$F - 32 = \frac{9}{5}(C - 0)$$

$$F = \frac{9}{5}C + 32.$$

110. $F = \frac{9}{5}C + 32$

$F = 0°;\quad 0 = \frac{9}{5}C + 32$

$-32 = \frac{9}{5}C$

$-17.8 \approx C$

$C = -10°;\ F = \frac{9}{5}(-10) + 32$

$F = -18 + 32$

$F = 14$

$C = 10°;\ F = \frac{9}{5}(10) + 32$

$F = 18 + 32$

$F = 50$

$F = 68°;\ 68 = \frac{9}{5}C + 32$

$36 = \frac{9}{5}C$

$20 = C$

$F = 90°;\quad 90 = \frac{9}{5}C + 32$

$58 = \frac{9}{5}C$

$32.2 \approx C$

$C = 177°;\ F = \frac{9}{5}(177) + 32$

$F = 318.6 + 32$

$F = 350.6$

C	$-17.8°$	$-10°$	$10°$	$20°$	$32.2°$	$177°$
F	$0°$	$14°$	$50°$	$68°$	$90°$	$350.6°$

111. Let $t = 0$ represent 1998.

$(0, 28{,}500), (6, 36{,}400)$

$$m = \frac{36{,}400 - 28{,}500}{6 - 0} = \frac{3950}{3}$$

$$S = \frac{3950}{3}t + 28{,}500$$

$t = 10$ represents 2008: $S = \frac{3950}{3}(10) + 28{,}500$

$\approx \$41{,}666.67$

112. Let $t = 0$ represent 1998.

$(0, 2546), (6, 2902)$

$$m = \frac{2902 - 2546}{6 - 0} = \frac{178}{3}$$

$$N = \frac{178}{3}t + 2546$$

$t = 10$ represents 2008: $S = \frac{178}{3}(10) + 2546$

≈ 3139 students

113. Using the points $(0, 875)$ and $(5, 0)$, where the first coordinate represents the year t and the second coordinate represents the value V, we have

$$m = \frac{0 - 875}{5 - 0} = -175$$

$V = -175t + 875,\ 0 \le t \le 5.$

114. $(0, 25{,}000)$ and $(10, 2000)$

$$m = \frac{2000 - 25000}{10 - 0} = -2300$$

$V = -2300t + 25{,}000,\quad 0 \le t \le 10$

115. Sale price = List price − 15% of the list price

$S = L - 0.15L$

$S = 0.85L$

116. $W = 0.75x + 11.50$

117. (a) $C = 36{,}500 + 5.25t + 12.50t$

$= 17.75t + 36{,}500$

(b) $R = 27t$

(c) $P = R - C = 27t - (17.75t + 36{,}500)$

$= 9.25t - 36{,}500$

(d) $P = 0$

$9.25t - 36{,}500 = 0$

$$t = \frac{36{,}500}{9.25} \approx 3946 \text{ hours}$$

118. $(580, 50)$ and $(625, 47)$

(a) $m = \dfrac{47 - 50}{625 - 580} = \dfrac{-3}{45} = -\dfrac{1}{15}$

$$x - 50 = -\frac{1}{15}(p - 580)$$

$$x - 50 = -\frac{1}{15}p + \frac{116}{3}$$

$$x = -\frac{1}{15}p + \frac{266}{3}$$

(b) $x = -\frac{1}{15}(655) + \frac{266}{3} = 45$ units

(c) $x = -\frac{1}{15}(595) + \frac{266}{3} = 49$ units

119. (a)

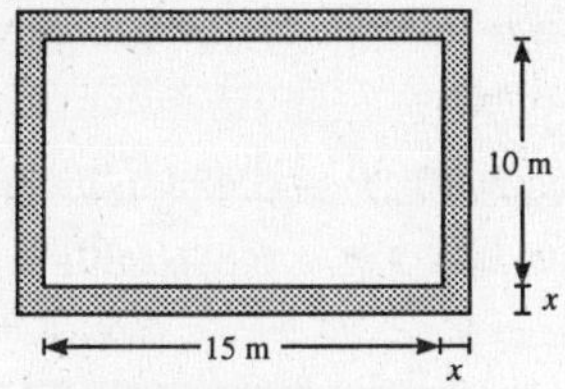

$y = 2(15 + 2x) + 2(10 + 2x)$

$= 8x + 50$

(b)

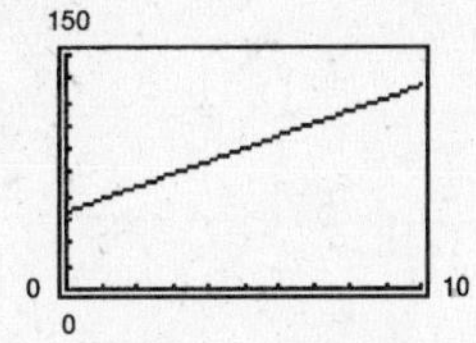

(c) Since $m = 8$, each 1 meter increase in x will increase y by 8 meters.

120. $W = 0.07S + 2500$

121. $C = 120 + 0.31x$

122. Let x = amount invested in the $2\frac{1}{2}\%$ fund and z = amount invested in the 4% fund.

(a) $x + z = 12{,}000 \Rightarrow z = 12{,}000 - x$ in the 4% fund.

(b) $y = 0.025x + 0.04(12000 - x)$

$= 0.025x + 480 - 0.04x$

$= -0.015x + 480$

(c)

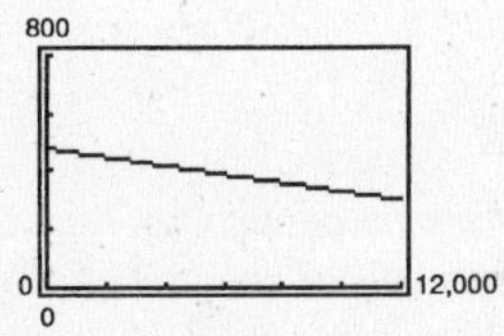

(d) As the amount invested at the lower interest rate increases, the annual interest decreases.

123. Answers will vary. One approximation is $y = 199.8t - 14.6$.

124.

x	18	10	19	16	13	15
y	87	55	96	79	76	82

(a) and (b)

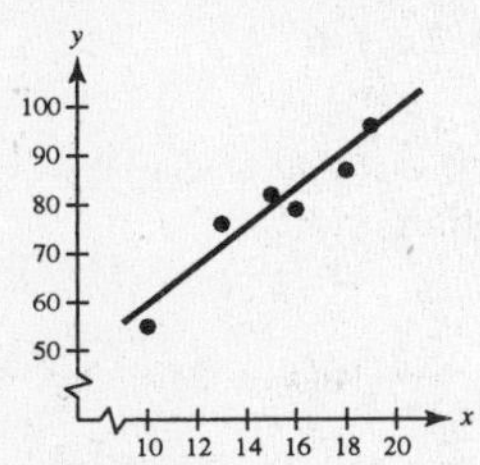

(c) Answers will vary.
One approximation is $y = 4x + 19$.

(d) Answers will vary, depending on the equation found in part (c).

$y = 4(17) + 19 = 87$

(e) If 4 points are added to each y-value, then each point would move up 4 units on the graph, and the y-intercept in the equation would increase by 4.

125. (a) and (b)

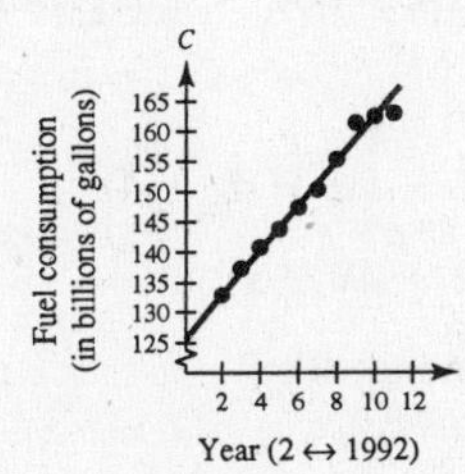

(c) Answers will vary.
One approximation is $C = 3.7t + 125.5$.
(Another is $C = 3.6t + 126.3$.)

(d) $m = 3.7$. This means that fuel consumption increases 3.7 billion gallons per year.

(e) $t = 20$ represents 2010: $C = 199.5$ billion gallons

$t = 30$ represents 2020: $C = 236.5$ billion gallons

126. (a) Let $t = 0$ represent 1990.

$(0, 3.022), (10, 4.240)$

$$m = \frac{4.240 - 3.022}{10 - 0} = 0.1218$$

$$P = 0.1218t + 3.022$$

(b)

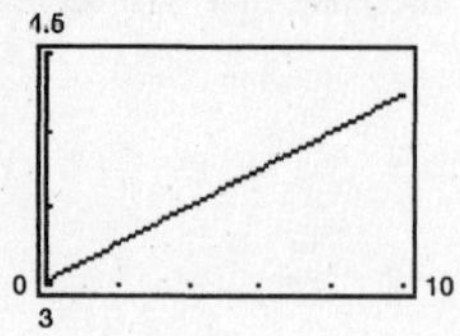

(c) $m = 0.1218$. The population is increasing by 0.1218 million per year.

(d) $t = 20$ represents 2010: $P = 5.458$ million

$t = 30$ represents 2020: $P = 6.676$ million

127. False. The slope with the greatest magnitude corresponds to the steepest line.

128. $(-8, 2)$ and $(-1, 4): m_1 = \dfrac{4 - 2}{-1 - (-8)} = \dfrac{2}{7}$

$(0, -4)$ and $(-7, 7): m_2 = \dfrac{7 - (-4)}{-7 - 0} = \dfrac{11}{-7}$

False, the lines are not parallel.

129. True. Given any two distinct points (x_1, y_1) and (x_2, y_2) on a nonvertical line, then the slope is $m = \dfrac{y_2 - y_1}{x_2 - x_1}$.

130. True

If two nonvertical lines are perpendicular, then their slopes are negative reciprocals of each other. If m_1 is positive then $m_2 = -\dfrac{1}{m_1}$ is negative, and vice versa.

Review Exercises for Chapter P

1. $6 - (x - 2)^2 = 6 - (x^2 - 4x + 4)$

$= 2 + 4x - x^2$

The equation is an indentity.

2. $3(x - 2) + 2x = 2(x + 3)$

$3x - 6 + 2x = 2x + 6$

$3x = 12$

Conditional equation

3. $-x^3 + x(7 - x) + 3 = -x^3 - x^2 + 7x + 3$

$x(-x^2 - x) + 7(x + 1) - 4 = -x^3 - x^2 + 7x + 3$

The equation is an identity.

4. $3(x^2 - 4x + 8) = -10(x + 2) - 3x^2 + 6$

$3x^2 - 12x + 24 = -10x - 20 - 3x^2 + 6$

$6x^2 - 2x + 38 = 0$

Conditional equation

5. $3x - 2(x + 5) = 10$

$3x - 2x - 10 = 10$

$x = 20$

6. $4x + 2(7 - x) = 5$

$4x + 14 - 2x = 5$

$2x = -9$

$x = -\dfrac{9}{2}$

7. $4(x + 3) - 3 = 2(4 - 3x) - 4$

$4x + 12 - 3 = 8 - 6x - 4$

$4x + 9 = -6x + 4$

$10x = -5$

$x = -\dfrac{1}{2}$

8. $\frac{1}{2}(x - 3) - 2(x + 1) = 5$

$$x - 3 - 4x - 4 = 10$$
$$-3x = 17$$
$$x = -\frac{17}{3}$$

9. $2x^2 - x - 28 = 0$

$$(2x + 7)(x - 4) = 0$$
$$2x + 7 = 0 \quad \text{or} \quad x - 4 = 0$$
$$x = -\tfrac{7}{2} \quad \text{or} \quad x = 4$$

10. $15 + x - 2x^2 = 0$

$$2x^2 - x - 15 = 0$$
$$(2x + 5)(x - 3) = 0$$
$$x = -\frac{5}{2}, 3$$

11. $16x^2 = 25$

$$x^2 = \frac{25}{16}$$
$$x = \pm\sqrt{\frac{25}{16}} = \pm\frac{5}{4}$$

12. $6 = 3x^2$

$$x^2 = 2$$
$$x = \pm\sqrt{2}$$

13. $(x - 8)^2 = 15$

$$x - 8 = \pm\sqrt{15}$$
$$x = 8 \pm \sqrt{15}$$

14. $(x + 4)^2 = 18$

$$x + 4 = \pm 3\sqrt{2}$$
$$x = -4 \pm 3\sqrt{2}$$

15. $x^2 + 6x - 3 = 0$

$$a = 1, \quad b = 6, \quad c = -3$$
$$x = \frac{-6 \pm \sqrt{6^2 - 4(1)(-3)}}{2(1)}$$
$$= \frac{-6 \pm \sqrt{48}}{2} = -3 \pm 2\sqrt{3}$$

16. $x^2 - 12x + 30 = 0$

$$x = \frac{-(-12) \pm \sqrt{(-12)^2 - 4(1)(30)}}{2(1)}$$
$$= 6 \pm \sqrt{6}$$

17. $-20 - 3x + 3x^2 = 0$

$$3x^2 - 3x - 20 = 0$$
$$a = 3, b = -3, c = -20$$
$$x = \frac{-(-3) \pm \sqrt{(-3)^2 - 4(3)(-20)}}{2(3)}$$
$$= \frac{3 \pm \sqrt{249}}{6} = \frac{1}{2} \pm \frac{\sqrt{249}}{6}$$

18. $-2x^2 - 5x + 27 = 0$

$$x = \frac{-(-5) \pm \sqrt{(-5)^2 - 4(-2)(27)}}{2(-2)}$$
$$= \frac{-5 \pm \sqrt{241}}{4}$$

19. $4x^3 - 6x^2 = 0$

$$x^2(4x - 6) = 0$$
$$x^2 = 0 \Longrightarrow x = 0$$
$$4x - 6 = 0 \Longrightarrow x = \tfrac{3}{2}$$

20. $5x^4 - 12x^3 = 0$

$$x^3(5x - 12) = 0$$
$$x = 0, \tfrac{12}{5}$$

21. $9x^4 + 27x^3 - 4x^2 - 12x = 0$

$$9x^3(x + 3) - 4x(x + 3) = 0$$
$$(9x^3 - 4x)(x + 3) = 0$$
$$x(9x^2 - 4)(x + 3) = 0$$
$$x(3x + 2)(3x - 2)(x + 3) = 0$$
$$x = 0$$
$$3x + 2 = 0 \Longrightarrow x = -\tfrac{2}{3}$$
$$3x - 2 = 0 \Longrightarrow x = \tfrac{2}{3}$$
$$x + 3 = 0 \Longrightarrow x = -3$$

22. $x^4 - 5x^2 + 6 = 0$

Let $u = x^2$.

$u^2 - 5u + 6 = 0$

$(u - 2)(u - 3) = 0$

$u = 2, 3$

$x^2 = 2, 3$

$x = \pm\sqrt{2}, \pm\sqrt{3}$

23. $\sqrt{x - 2} - 8 = 0$

$\sqrt{x - 2} = 8$

$x - 2 = 64$

$x = 66$

24. $\sqrt{x + 4} = 3$

$x + 4 = 9$

$x = 5$

25. $\sqrt{3x - 2} = 4 - x$

$3x - 2 = (4 - x)^2$

$3x - 2 = 16 - 8x + x^2$

$0 = 18 - 11x + x^2$

$0 = (x - 9)(x - 2)$

$0 = x - 9 \Rightarrow x = 9$, extraneous

$0 = x - 2 \Rightarrow x = 2$

26. $2\sqrt{x} - 5 = x$

$2\sqrt{x} = x + 5$

$4x = x^2 + 10x + 25$

$x^2 + 6x + 25 = 0$

$b^2 - 4ac = 6^2 - 4(1)(25) = -64 < 0 \Rightarrow$ no real solutions

Original equation has no real solutions.

27. $(x + 2)^{3/4} = 27$

$x + 2 = 27^{4/3}$

$x + 2 = 81$

$x = 79$

28. $(x - 1)^{2/3} - 25 = 0$

$x - 1 = 25^{3/2}$

$x - 1 = \pm 125$

$x = 126, -124$

29. $8x^2(x^2 - 4)^{1/3} + (x^2 - 4)^{4/3} = 0$

$(x^2 - 4)^{1/3}[8x^2 + x^2 - 4] = 0$

$(x^2 - 4)^{1/3}(9x^2 - 4) = 0$

$(x - 2)^{1/3}(x + 2)^{1/3}(3x - 2)(3x + 2) = 0$

$x - 2 = 0 \Rightarrow x = 2$

$x + 2 = 0 \Rightarrow x = -2$

$3x - 2 = 0 \Rightarrow x = \frac{2}{3}$

$3x + 2 = 0 \Rightarrow x = -\frac{2}{3}$

30. $(x + 4)^{1/2} + 5x(x + 4)^{3/2} = 0$

$(x + 4)^{1/2}[1 + 5x(x + 4)] = 0$

$(x + 4)^{1/2}[5x^2 + 20x + 1] = 0$

$(x + 4)^{1/2} = 0$

$x = -4$

$5x^2 + 20x + 1 = 0$

$$x = \frac{-20 \pm \sqrt{20^2 - 4(5)(1)}}{2(5)} = \frac{-10 \pm \sqrt{95}}{5}$$

$$x = -4, \frac{-10 \pm \sqrt{95}}{5}$$

31. $|2x + 3| = 7$

$2x + 3 = 7$ or $2x + 3 = -7$

$2x = 4$ or $2x = -10$

$x = 2$ $\qquad$ $x = -5$

32. $|x - 5| = 10$

First Equation	*Second Equation*
$x - 5 = 10$	$x - 5 = -10$
$x = 15$	$x = -5$

33. $|x^2 - 6| = x$

$x^2 - 6 = x$ $\qquad$ or $\qquad$ $-(x^2 - 6) = x$

$x^2 - x - 6 = 0$ $\qquad$ $x^2 + x - 6 = 0$

$(x - 3)(x + 2) = 0$ $\qquad$ $(x + 3)(x - 2) = 0$

$x - 3 = 0 \Rightarrow x = 3$ $\qquad$ $x - 2 = 0 \Rightarrow x = 2$

$x + 2 = 0 \Rightarrow x = -2$, extraneous $\qquad$ $x + 3 = 0 \Rightarrow x = -3$, extraneous

34. $|x^2 - 3| = 2x$

First Equation

$x^2 - 3 = 2x$

$x^2 - 2x - 3 = 0$

$(x - 3)(x + 1) = 0$

$x - 3 = 0 \Rightarrow x = 3$

$x + 1 = 0 \Rightarrow x = -1$, not a solution

Second equation

$-(x^2 - 3) = 2x$

$x^2 + 2x - 3 = 0$

$(x + 3)(x - 1) = 0$

$x + 3 = 0 \Rightarrow x = -3$, not a solution

$x - 1 = 0 \Rightarrow x = 1$

35. Let $x =$ the number of liters of pure antifreeze.

$$30\% \text{ of } (10 - x) + 100\% \text{ of } x = 50\% \text{ of } 10$$
$$0.30(10 - x) + 1.00x = 0.50(10)$$
$$3 - 0.30x + 1.00x = 5$$
$$0.70x = 2$$
$$x = \frac{2}{0.70} = \frac{20}{7} = 2\frac{6}{7} \text{ liters}$$

36. $42 - \sqrt{0.001x + 2} = 29.95$

$42 - 29.95 = \sqrt{0.001x + 2}$

$(12.05)^2 = 0.001x + 2$

$x = 143{,}202.5$

Number of units demanded per day: 143,203

37. $6x - 17 > 0$

(a) $x = 3$

$6(3) - 17 \overset{?}{>} 0$

$1 > 0$

Yes, $x = 3$ is a solution.

(b) $x = -4$

$6(-4) - 17 \overset{?}{>} 0$

$-41 \not> 0$

No, $x = -4$ is not a solution.

38. $-3 \le \frac{x - 3}{5} < 2$

(a) $x = 3$

$-3 \overset{?}{\le} \frac{3 - 3}{5} \overset{?}{<} 2$

$-3 \le 0 < 2$

Yes, $x = 3$ is a solution.

(b) $x = -12$

$-3 \overset{?}{\le} \frac{-12 - 3}{5} \overset{?}{<} 2$

$-3 \le -3 < 2$

Yes, $x = -12$ is a solution.

39. $9x - 8 \le 7x + 16$

$2x \le 24$

$x \le 12$

$(-\infty, 12]$

40. $4(5 - 2x) \le \frac{1}{2}(8 - x)$

$20 - 8x \le 4 - \frac{1}{2}x$

$-\frac{15}{2}x \le -16$

$x \ge \frac{32}{15}$

$\left[\frac{32}{15}, \infty\right)$

41. $\frac{15}{2}x + 4 > 3x - 5$

$15x + 8 > 6x - 10$

$9x > -18$

$x > -2$

$(-2, \infty)$

42. $\frac{1}{2}(3 - x) > \frac{1}{3}(2 - 3x)$

$3(3 - x) > 2(2 - 3x)$

$9 - 3x > 4 - 6x$

$3x > -5$

$x > -\frac{5}{3}$

$\left(-\frac{5}{3}, \infty\right)$

43. $-19 < \frac{3x - 17}{2} \le 34$

$-38 < 3x - 17 \le 68$

$-21 < 3x \le 85$

$-7 < x \le \frac{85}{3}$

$-7 < x \le 28\frac{1}{3}$ or $\left(-7, 28\frac{1}{3}\right]$

44. $-3 \le \frac{2x - 5}{3} < 5$

$-9 \le 2x - 5 < 15$

$-4 \le 2x < 20$

$-2 \le x < 10$

$[-2, 10)$

45. $|x + 1| \le 5$

$-5 \le x + 1 \le 5$

$-6 \le x \le 4$ or $[-6, 4]$

46. $|x - 2| < 1$

$-1 < x - 2 < 1$

$1 < x < 3$

$(1, 3)$

47. $|x - 3| > 4$

$x - 3 > 4$ or $x - 3 < -4$

$x > 7$ $\quad$ $x < -1$

$(-\infty, -1) \cup (7, \infty)$

48. $\left|x - \frac{3}{2}\right| \geq \frac{3}{2}$

$x - \frac{3}{2} \leq -\frac{3}{2}$ $\quad$ $x - \frac{3}{2} \geq \frac{3}{2}$

$x \leq 0$ $\quad$ $x \geq 3$

$(-\infty, 0], [3, \infty)$

49. $R > C$

$125.33x > 92x + 1200$

$33.33x > 1200$

$x > \dfrac{1200}{33.33} \approx 36$ units

50. $A = s^2$

$s = 19.3 \pm 0.5$ cm

$A = (19.3)^2 \approx 372 \text{ cm}^2$

Smallest area:
$A = (18.8)^2 \approx 353 \text{ cm}^2$

Largest area:
$A = (19.8)^2 \approx 392 \text{ cm}^2$

Interval containing area of square:
$353 \text{ cm}^2 < A < 392 \text{ cm}^2$

51. $x^2 - 6x - 27 < 0$

$(x + 3)(x - 9) < 0$

Critical numbers: $x = -3, x = 9$

Test intervals: $(-\infty, -3), (-3, 9), (9, \infty)$

Test: Is $(x + 3)(x - 9) < 0$?

Solution set: $(-3, 9)$

52. $x^2 - 2x \geq 3$

$x^2 - 2x - 3 \geq 0$

$(x - 3)(x + 1) \geq 0$

Critical numbers: $-1, 3$

Test intervals: $(-\infty, -1) \Rightarrow (x - 3)(x + 1) > 0$

$(-1, 3) \Rightarrow (x - 3)(x + 1) < 0$

$(3, \infty) \Rightarrow (x - 3)(x + 1) > 0$

Solution intervals: $(-\infty, -1] \cup [3, \infty)$

53. $6x^2 + 5x < 4$

$6x^2 + 5x - 4 < 0$

$(3x + 4)(2x - 1) < 0$

Critical numbers: $x = -\dfrac{4}{3}, x = \dfrac{1}{2}$

Test intervals: $\left(-\infty, -\dfrac{4}{3}\right), \left(-\dfrac{4}{3}, \dfrac{1}{2}\right), \left(\dfrac{1}{2}, \infty\right)$

Test: Is $(3x + 4)(2x - 1) < 0$?

Solution set: $\left(-\dfrac{4}{3}, \dfrac{1}{2}\right)$

54. $2x^2 + x \geq 15$

$2x^2 + x - 15 \geq 0$

$(2x - 5)(x + 3) \geq 0$

Critical numbers: $-3, \dfrac{5}{2}$

Test intervals: $(-\infty, -3) \Rightarrow (2x - 5)(x + 3) > 0$

$\left(-3, \dfrac{5}{2}\right) \Rightarrow (2x - 5)(x + 3) < 0$

$\left(\dfrac{5}{2}, \infty\right) \Rightarrow (2x - 5)(x + 3) > 0$

Solution intervals: $(-\infty, -3] \cup \left[\dfrac{5}{2}, \infty\right)$

55. $\dfrac{2}{x+1} \le \dfrac{3}{x-1}$

$\dfrac{2(x-1) - 3(x+1)}{(x+1)(x-1)} \le 0$

$\dfrac{2x - 2 - 3x - 3}{(x+1)(x-1)} \le 0$

$\dfrac{-(x+5)}{(x+1)(x-1)} \le 0$

Critical numbers: $x = -5, x = \pm 1$

Test intervals: $(-\infty, -5), (-5, -1), (-1, 1), (1, \infty)$

Test: Is $\dfrac{-(x+5)}{(x+1)(x-1)} \le 0$?

Solution set: $[-5, -1) \cup (1, \infty)$

56. $\dfrac{x-5}{3-x} < 0$

Critical numbers: 3, 5

Test intervals: $(-\infty, 3) \Rightarrow \dfrac{x-5}{3-x} < 0$

$(3, 5) \Rightarrow \dfrac{x-5}{3-x} > 0$

$(5, \infty) \Rightarrow \dfrac{x-5}{3-x} < 0$

Solution intervals: $(-\infty, 3) \cup (5, \infty)$

57. $\dfrac{x^2 + 7x + 12}{x} \ge 0$

$\dfrac{(x+4)(x+3)}{x} \ge 0$

Critical numbers: $x = -4, x = -3, x = 0$

Test intervals: $(-\infty, -4), (-4, -3), (-3, 0), (0, \infty)$

Test: Is $\dfrac{(x+4)(x+3)}{x} \ge 0$?

Solution set: $[-4, -3] \cup (0, \infty)$

58. $\dfrac{1}{x-2} > \dfrac{1}{x}$

$\dfrac{1}{x-2} - \dfrac{1}{x} > 0$

$\dfrac{2}{x(x-2)} > 0$

Critical numbers: 0, 2

Test intervals: $(-\infty, 0) \Rightarrow \dfrac{2}{x(x-2)} > 0$

$(0, 2) \Rightarrow \dfrac{2}{x(x-2)} < 0$

$(2, \infty) \Rightarrow \dfrac{2}{x(x-2)} > 0$

Solution intervals: $(-\infty, 0) \cup (2, \infty)$

59. $5000(1 + r)^2 > 5500$

$(1 + r)^2 > 1.1$

$1 + r > 1.0488$

$r > 0.0488$

$r > 4.9\%$

60. $P = \dfrac{1000(1 + 3t)}{5 + t} \ge 2000$

$\dfrac{1000(1 + 3t)}{5 + t} - 2000 \ge 0$

$\dfrac{1000t - 9000}{t + 5} \ge 0$

t must be greater than zero, so the critical value to check is $t = 9$.

$(0, 9) \Rightarrow \dfrac{1000t - 9000}{t + 5} < 0$

$(9, \infty) \Rightarrow \dfrac{1000t - 9000}{t + 5} > 0$

$t \ge 9$

So, the required time is at least 9 days.

61.

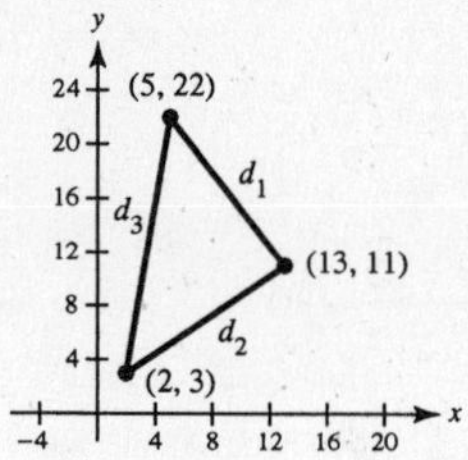

$d_1 = \sqrt{(13 - 5)^2 + (11 - 22)^2} = \sqrt{64 + 121} = \sqrt{185}$

$d_2 = \sqrt{(2 - 13)^2 + (3 - 11)^2} = \sqrt{121 + 64} = \sqrt{185}$

$d_3 = \sqrt{(2 - 5)^2 + (3 - 22)^2} = \sqrt{9 + 361} = \sqrt{370}$

$d_1^2 + d_2^2 = 185 + 185 = 370 = d_3^2$

Thus, the triangle is a right triangle.

62. Parallelogram: $(1, 2)$, $(8, 3)$, $(9, 6)$, $(2, 5)$

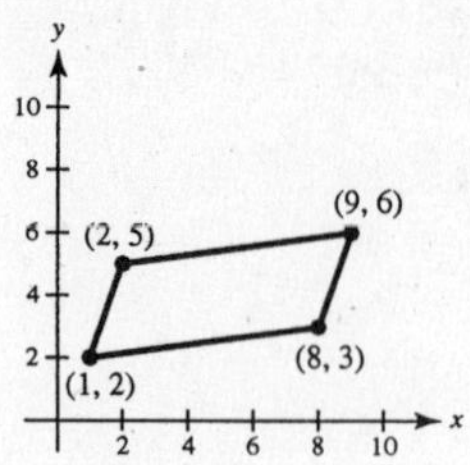

$(1, 2)$ and $(2, 5)$

$d = \sqrt{(2 - 1)^2 + (5 - 2)^2}$

$= \sqrt{1^2 + 3^2} = \sqrt{10}$

$(2, 5)$ and $(9, 6)$

$d = \sqrt{(9 - 2)^2 + (6 - 5)^2}$

$= \sqrt{7^2 + 1^2}$

$= \sqrt{50} = 5\sqrt{2}$

$(9, 6)$ and $(8, 3)$

$d = \sqrt{(8 - 9)^2 + (3 - 6)^2}$

$= \sqrt{(-1)^2 + (-3)^2} = \sqrt{10}$

$(8, 3)$ and $(1, 2)$

$d = \sqrt{(1 - 8)^2 + (2 - 3)^2}$

$= \sqrt{(-7)^2 + (-1)^2}$

$= \sqrt{50} = 5\sqrt{2}$

Opposite sides have equal lengths of $\sqrt{10}$ and $5\sqrt{2}$.

63. $x > 0$ and $y = -2$ in Quadrant IV.

64. $y > 0$

(x, y) lies in quadrants I or II.

65. $(-x, y)$ is in the third Quadrant means that (x, y) is in Quadrant IV.

66. $xy = 4$

x and y are both positive or both negative.
(x, y) lies in quadrants I or III.

67. (a)

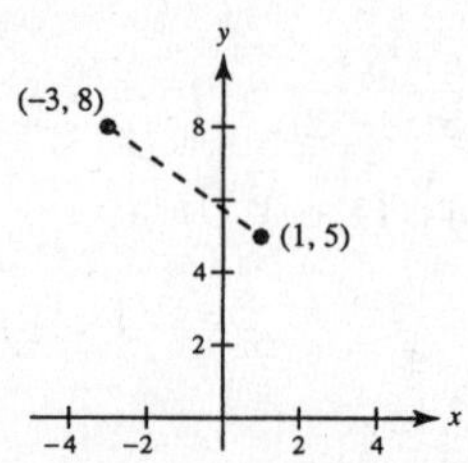

(b) $d = \sqrt{(-3 - 1)^2 + (8 - 5)^2} = \sqrt{16 + 9} = 5$

68. (a)

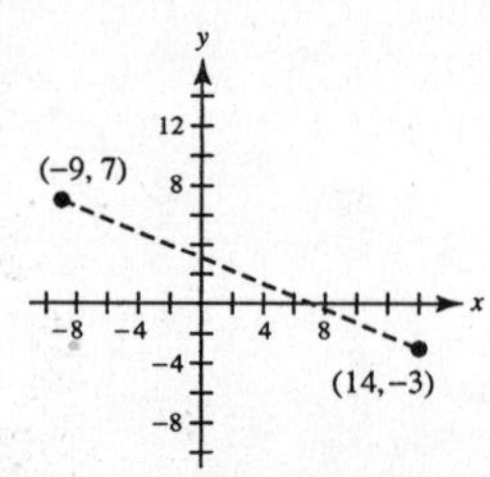

(b) $d = \sqrt{(-9 - 14)^2 + (7 - (-3))^2}$

$= \sqrt{629}$

69. (a)

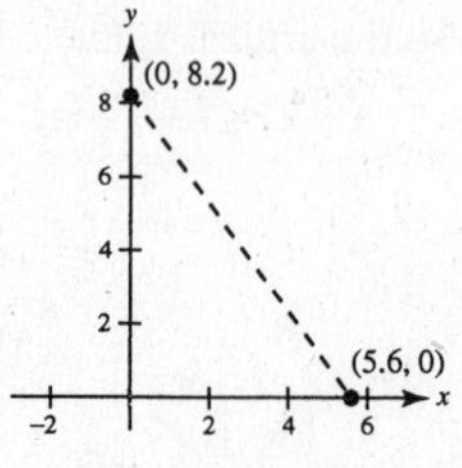

(b) $d = \sqrt{(5.6 - 0)^2 + (0 - 8.2)^2}$

$= \sqrt{31.36 + 67.24} = \sqrt{98.6} \approx 9.9$

70. (a)

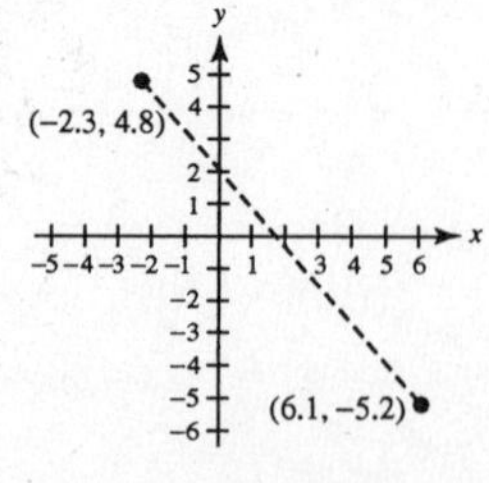

(b) $d = \sqrt{(-2.3 - 6.1)^2 + (4.8 - (-5.2))^2}$

$= \sqrt{170.56} \approx 13.06$

71. (a)

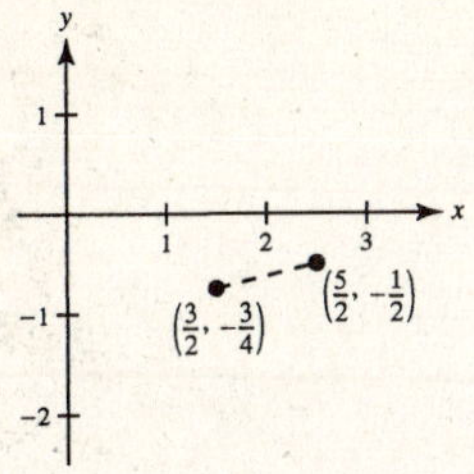

(b) $$d = \sqrt{\left(\frac{5}{2} - \frac{3}{2}\right)^2 + \left[-\frac{1}{2} - \left(-\frac{3}{4}\right)\right]^2}$$
$$= \sqrt{(1)^2 + \left(\frac{1}{4}\right)^2}$$
$$= \sqrt{1 + \frac{1}{16}}$$
$$= \sqrt{\frac{17}{16}}$$
$$= \frac{\sqrt{17}}{4}$$

72. (a)

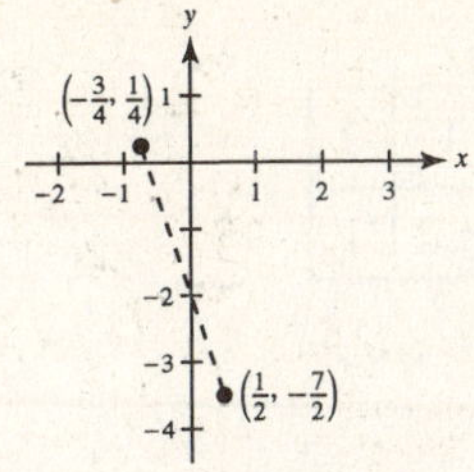

(b) $$d = \sqrt{\left(-\frac{3}{4} - \frac{1}{2}\right)^2 + \left[\frac{1}{4} - \left(-\frac{7}{2}\right)\right]^2}$$
$$= \sqrt{\left(-\frac{5}{4}\right)^2 + \left(\frac{15}{4}\right)^2}$$
$$= \sqrt{\frac{25}{16} + \frac{225}{16}}$$
$$= \sqrt{\frac{250}{16}}$$
$$= \frac{5\sqrt{10}}{4}$$

73. (a)

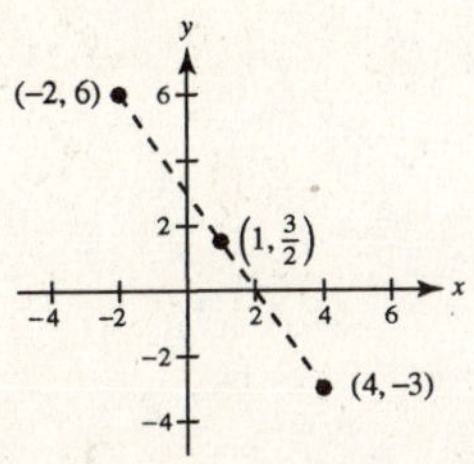

(b) Midpoint: $\left(\frac{-2 + 4}{2}, \frac{6 + (-3)}{2}\right) = \left(1, \frac{3}{2}\right)$

74. (a)

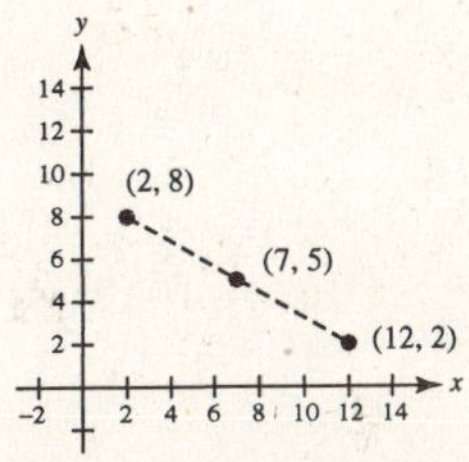

(b) Midpoint: $\left(\frac{2 + 12}{2}, \frac{8 + 2}{2}\right) = (7, 5)$

75. (a)

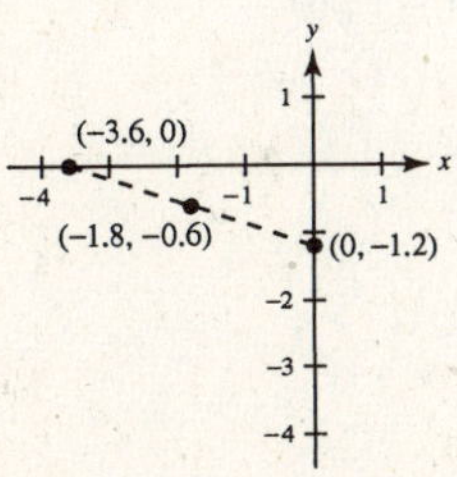

(b) Midpoint: $\left(\frac{0 + (-3.6)}{2}, \frac{-1.2 + 0}{2}\right) = (-1.8, -0.6)$

76. (a)

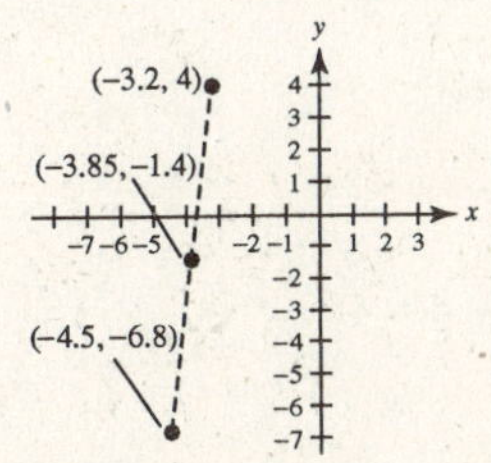

(b) Midpoint: $\left(\frac{-3.2 + (-4.5)}{2}, \frac{4 + (-6.8)}{2}\right)$

$= (-3.85, -1.4)$

77. Change in apparent temperature: 150° F − 70° F = 80° F

78.

Apparent Temperature	*Actual Temperature*
85°F	80°F
109°F	90°F

Change in actual temperature: 90°F − 80°F = 10°F

79. $y = 3x - 5$

x	-2	-1	0	1	2
y	-11	-8	-5	-2	1

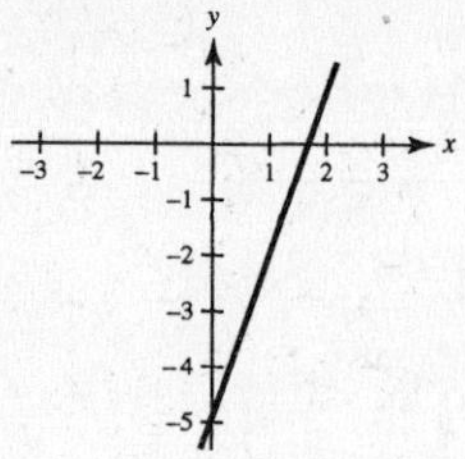

80. $y = -\frac{1}{2}x + 2$

x	-2	0	2
y	3	2	1

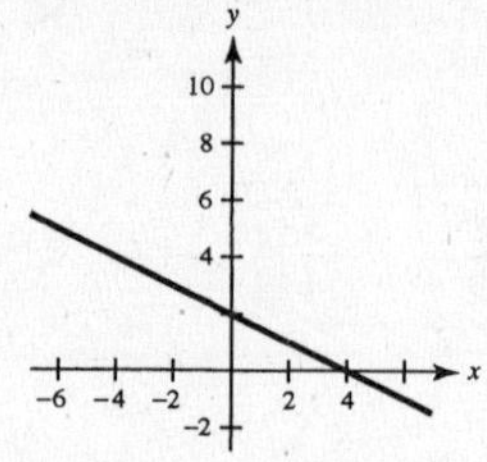

81. $y = x^2 - 3x$

x	-1	0	1	2	3	4
y	4	0	-2	-2	0	4

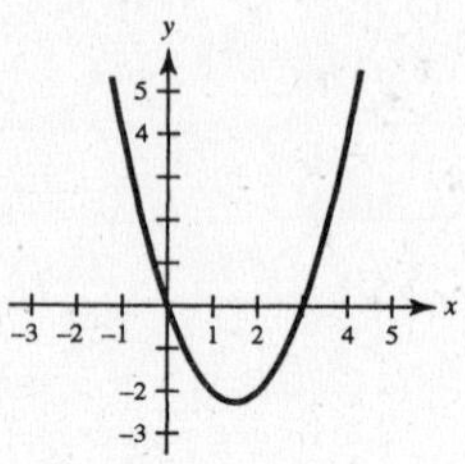

82. $y = 2x^2 - x - 9$

x	-2	-1	0	1	2
y	1	-6	-9	-8	-3

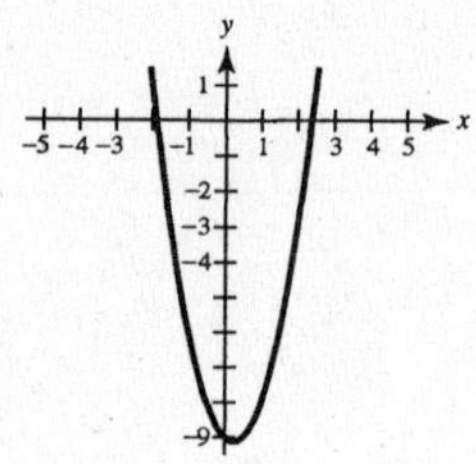

83. $y = 2x - 9$

x-intercept: $0 = 2x - 9 \Rightarrow x = \frac{9}{2}$

$\left(\frac{9}{2}, 0\right)$ is the *x*-intercept

y-intercept: $y = 2(0) - 9 = -9$

$(0, -9)$ is the *y*-intercept

84. $y = 3x - 10$

x-intercept: $0 = 3x - 10 \Rightarrow x = \frac{10}{3}$

$\left(\frac{10}{3}, 0\right)$

y-intercept: $y = 3(0) - 10 = -10$

$(0, -10)$

85. $y = |x - 4| - 4$

y-intercept: $y = |0 - 4| - 4 = 0$

$(0, 0)$

x-intercepts: $y = |x - 4| - 4 = 0$

$-x + 4 - 4 = 0 \qquad x - 4 - 4 = 0$

$x = 0 \qquad x = 8$

$(0, 0), (8, 0)$

86. $y = |x + 2| - 1$

x-intercepts: $0 = |x + 2| - 1$

$1 = |x + 2|$

$x + 2 = -1$ or $x + 2 = 1$

$x = -3 \qquad x = -1$

$(-3, 0), (-1, 0)$

y-intercept: $y = |0 + 2| - 1 = 1$

$(0, 1)$

87. $y = (x + 1)^2$

x-intercept: $0 = (x + 1)^2 \Rightarrow x = -1$

$(-1, 0)$ is the *x*-intercept.

y-intercept: $y = (0 + 1)^2 = 1$

$(0, 1)$ is the *y*-intercept.

88. $y = (x + 4)^2$

x-intercept: $0 = (x + 4)^2$

$0 = x + 4 \Rightarrow x = -4$

$(-4, 0)$

y-intercept: $y = (0 + 4)^2 = 16$

$(0, 16)$

89. $y = x\sqrt{9 - x^2}$

y-intercept: $y = 0\sqrt{9 - 0^2} = 0$

$(0, 0)$

x-intercepts: $y = x\sqrt{9 - x^2} = 0$

$x = 0 \qquad \sqrt{9 - x^2} = 0$

$x = \pm 3$

$(0, 0), (\pm 3, 0)$

90. $y = x\sqrt{4 - x^2}$

x-intercepts: $0 = x\sqrt{4 - x^2}$

$x = 0 \quad \text{or} \quad 4 - x^2 = 0 \Rightarrow x = \pm 2$

$(0, 0), (\pm 2, 0)$

y-intercept: $y = 0\sqrt{4 - 0^2} = 0$

$(0, 0)$

91. $y = 5 - x^2$

Intercepts: $(\pm\sqrt{5}, 0), (0, 5)$

y-axis symmetry

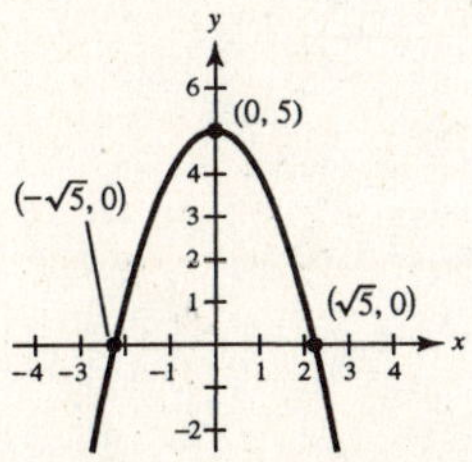

92. $y = 10 - x^2$

x-intercepts: $(\pm\sqrt{10}, 0)$

y-intercept: $(0, 10)$

y-axis symmetry

x	0	± 1	± 2	± 3	± 4
y	10	9	6	1	-6

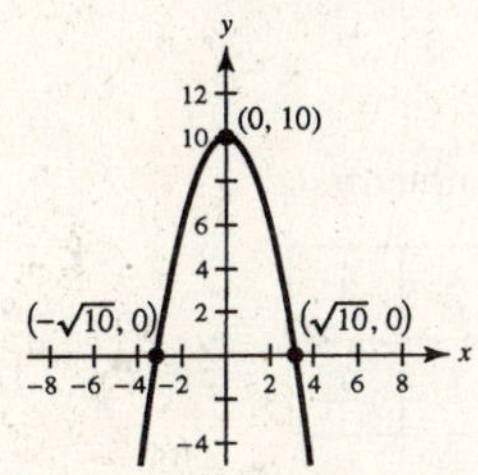

93. $y = x^3 - 3$

Intercepts: $(0, -3), (\sqrt[3]{3}, 0) \approx (1.44, 0)$

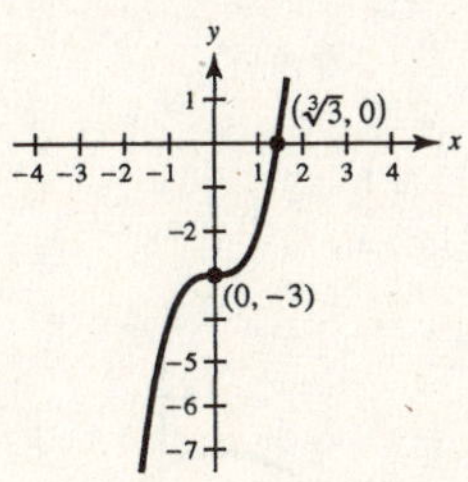

94. $y = x^3 - 2x$

x-intercepts: $(0, 0), (\pm\sqrt{2}, 0)$

y-intercept: $(0, 0)$

origin symmetry

x	-2	-1	0	1	2
y	-4	1	0	-1	4

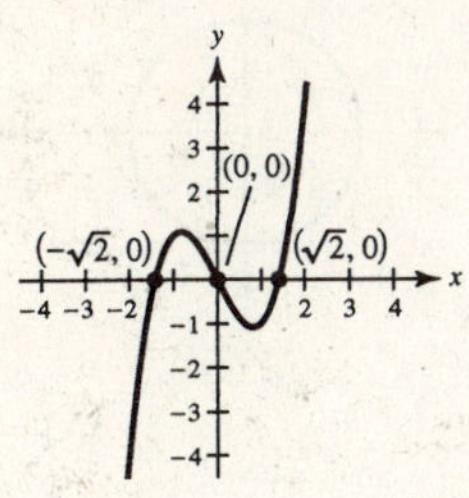

95. $y = \sqrt{x + 5}$

Domain: $[-5, \infty)$

Intercepts: $(-5, 0), (0, \sqrt{5})$

No axis or origin symmetry

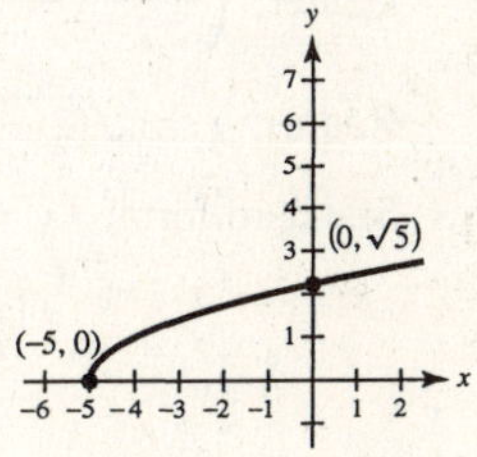

96. $y = \sqrt{x - 2}$

x-intercept: $(2, 0)$

y-intercept: None

No x-axis, y-axis, or origin symmetry.

x	2	3	6	11
y	0	1	2	3

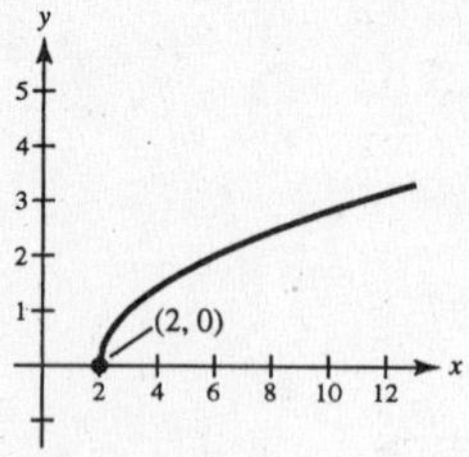

97. $y = 1 - |x|$

Intercepts: $(0, 1), (\pm 1, 0)$

Symmetry: y-axis

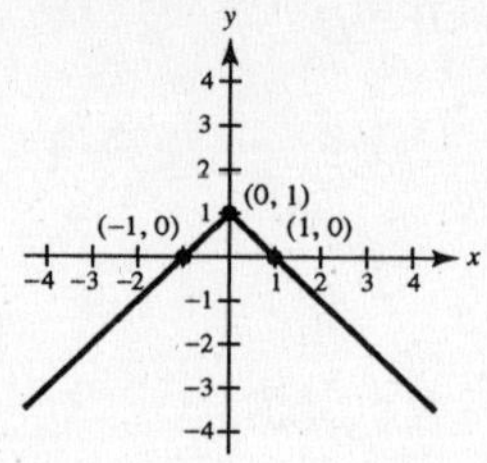

98. $y = |x - 1|$

x-intercept: $(1, 0)$

y-intercept: $(0, 1)$

No x-axis, y-axis, or origin symmetry.

x	-2	-1	0	1	2	3
y	3	2	1	0	1	2

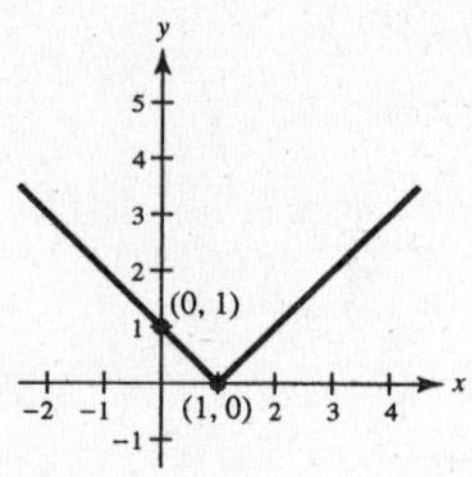

99. $x^2 + y^2 = 25$

Center: $(0, 0)$

Radius: 5

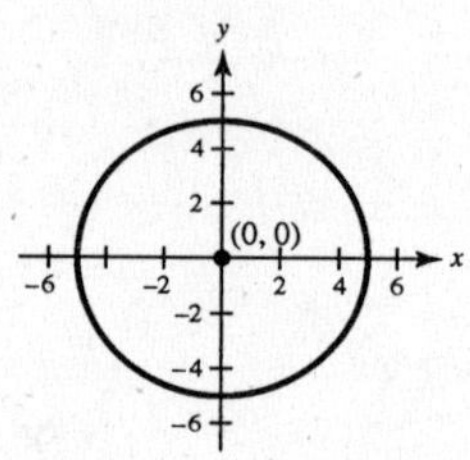

100. $x^2 + y^2 = 4$

Center: $(0, 0)$

Radius: $\sqrt{4} = 2$

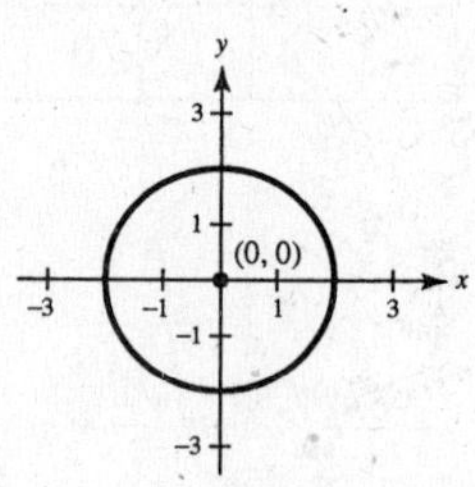

101. $(x + 2)^2 + y^2 = 16$

$(x - (-2))^2 + (y - 0)^2 = 4^2$

Center: $(-2, 0)$

Radius: 4

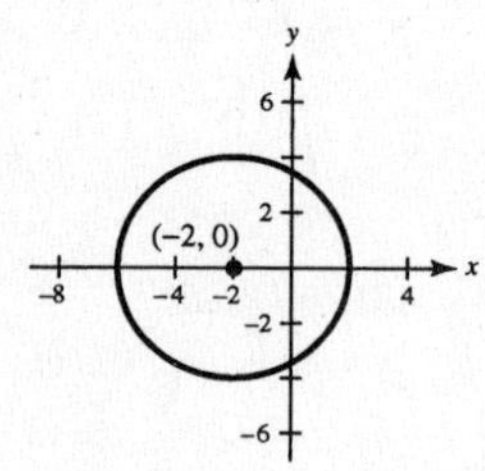

102. $x^2 + (y - 8)^2 = 81$

Center: $(0, 8)$

Radius: $\sqrt{81} = 9$

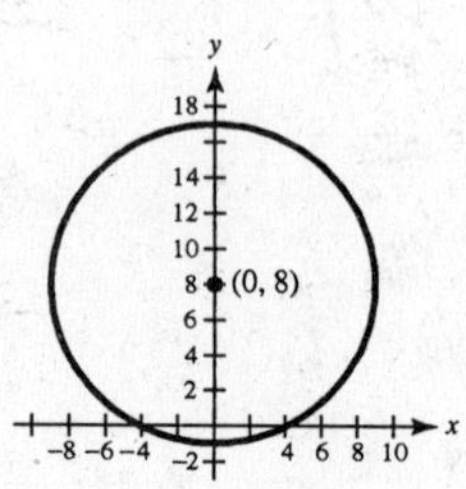

103. Endpoints of a diameter: $(0, 0)$ and $(4, -6)$

Center: $\left(\dfrac{0 + 4}{2}, \dfrac{0 + (-6)}{2}\right) = (2, -3)$

Radius: $r = \sqrt{(2 - 0)^2 + (-3 - 0)^2} = \sqrt{4 + 9} = \sqrt{13}$

Standard form: $(x - 2)^2 + (y - (-3))^2 = \left(\sqrt{13}\right)^2$

$$(x - 2)^2 + (y + 3)^2 = 13$$

104. Endpoints of diameter: $A(-2, -3), B(4, -10)$

Center: Midpoint of $\overline{AB} = \left(\frac{-2+4}{2}, \frac{-3+-10}{2}\right) = \left(1, -\frac{13}{2}\right)$

$$\text{Radius} = \sqrt{(1-(-2))^2 + \left(-\frac{13}{2} - (-3)\right)^2} = \sqrt{\frac{85}{4}}$$

$$(x-1)^2 + \left(y + \frac{13}{2}\right)^2 = \frac{85}{4}$$

105. $y = 0.070t + 0.37$,
$t = 0$ represents 1990.

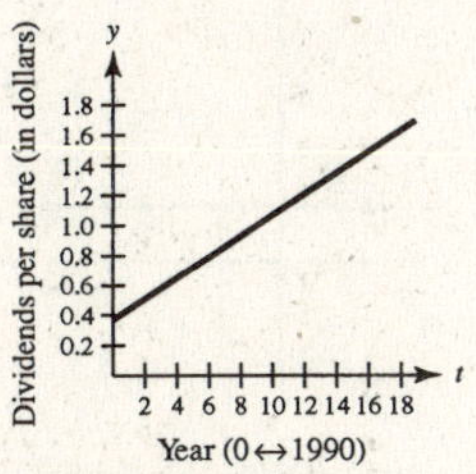

$0.070t + 0.37 = 1.50$

$0.070t = 1.13$

$t \approx 16$

This corresponds the year of 2006.

106. $A = -2x^2 + 100x$

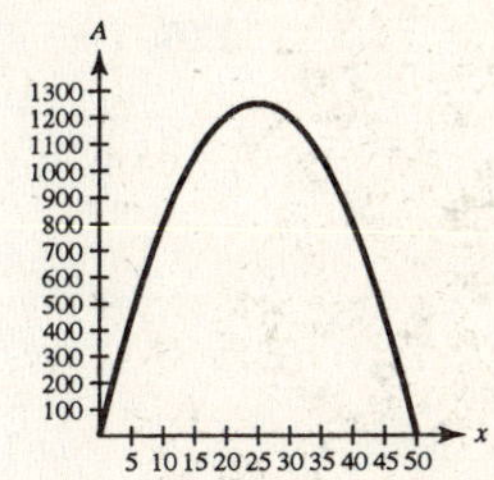

Maximum area: 1250 square feet
Length: 50 feet; Width: 25 feet

107. $y = 6$

Horizontal line

y-intercept: $(0, 6)$

Slope: $m = 0$

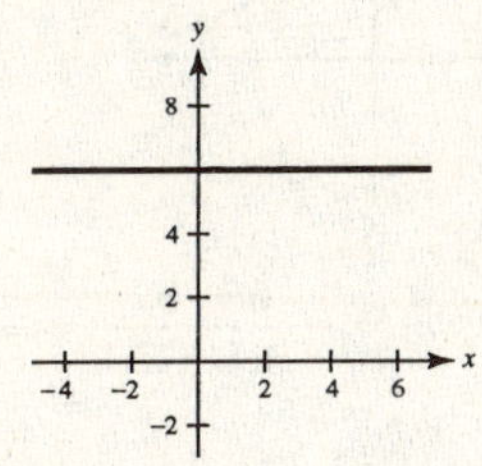

108. $x = -3$

Slope: undefined

y-intercept: none

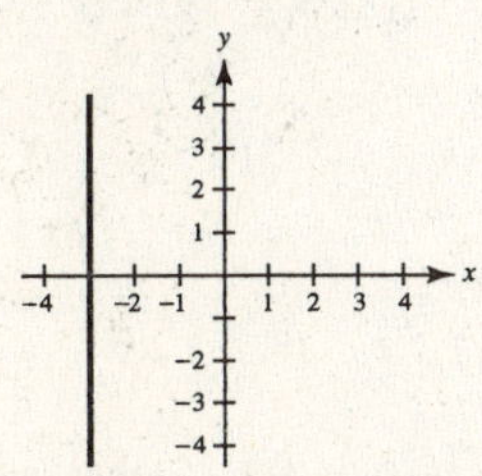

109. $y = 3x + 13$

y-intercept: $(0, 13)$

Slope: $m = 3 = \frac{3}{1}$

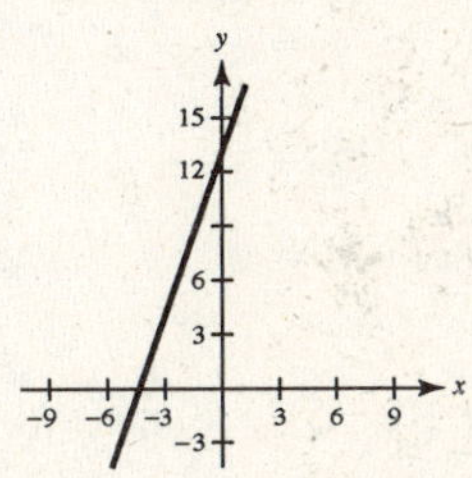

110. $y = -10x + 9$

Slope: -10

y-intercept: $(0, 9)$

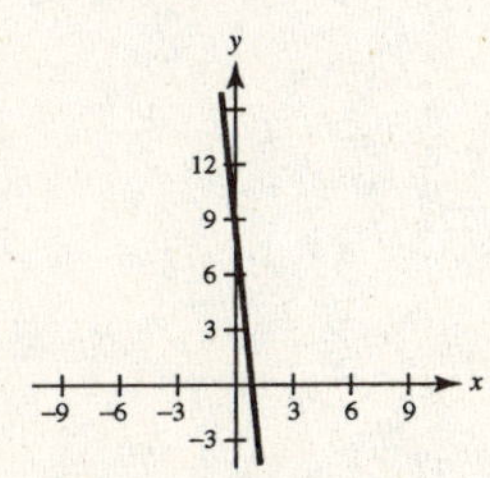

111. (a) $m = \frac{3}{2} > 0 \Rightarrow$ The line rises. Matches L_2.

(b) $m = 0 \Rightarrow$ The line is horizontal. Matches L_3.

(c) $m = -3 < 0 \Rightarrow$ The line falls. Matches L_1.

(d) $m = -\frac{1}{5} < 0 \Rightarrow$ The line gradually falls. Matches L_4.

112. (a) $m = -\frac{5}{2} \Rightarrow$ The line falls. Matches L_3.

(b) m is undefined. $\Rightarrow$ The line is vertical. Matches L_1.

(c) $m = 0 \Rightarrow$ The line is horizontal. Matches L_4.

(d) $m = \frac{1}{2} \Rightarrow$ The line rises. Matches L_2.

113.

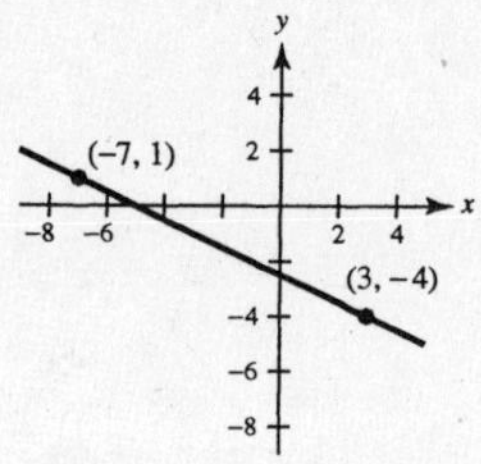

Slope: $m = \dfrac{1-(-4)}{-7-3} = \dfrac{5}{-10} = -\dfrac{1}{2}$

114. $(-1, 8), (6, 5)$

$m = \dfrac{5-8}{6-(-1)} = \dfrac{-3}{7}$

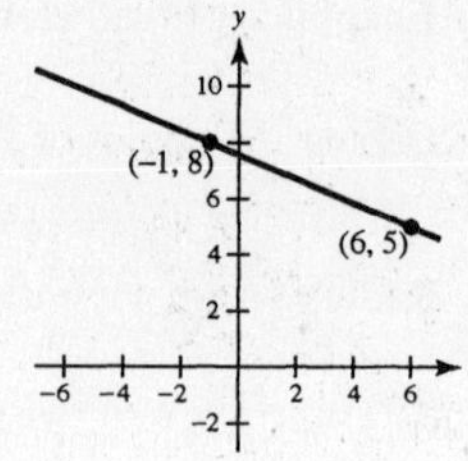

115.

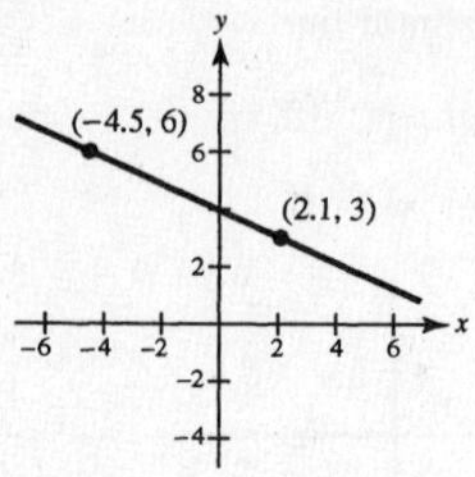

Slope: $m = \dfrac{6-3}{-4.5-2.1} = \dfrac{3}{-6.6} = -\dfrac{5}{11}$

116. $(-3, 2), (8, 2)$

$m = \dfrac{2-2}{-3-8} = \dfrac{0}{-11} = 0$

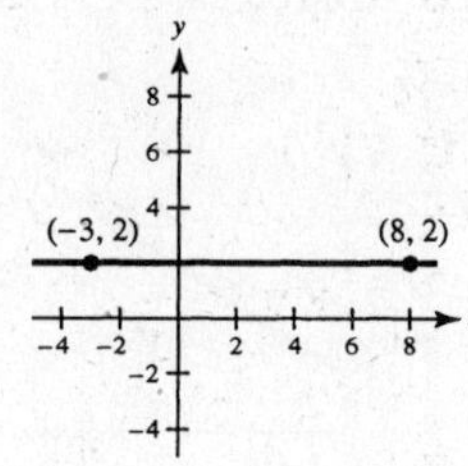

117. $(-5, 2), (0, 10)$

$m = \dfrac{10-2}{0+5} = \dfrac{8}{5}$

$y = \dfrac{8}{5}x + 10$

118. $(2, 5), (-2, -1)$

$m = \dfrac{-1-5}{-2-2} = \dfrac{-6}{-4} = \dfrac{3}{2}$

$y - 5 = \dfrac{3}{2}(x-2)$

$y - 5 = \dfrac{3}{2}x - 3$

$y = \dfrac{3}{2}x + 2$

$2y = 3x + 4$

$3x - 2y + 4 = 0$

119. $(-1, 4), (2, 0)$

Slope: $m = \dfrac{4-0}{-1-2} = -\dfrac{4}{3}$

$y - 0 = -\dfrac{4}{3}(x-2)$

$3y = -4x + 8$

$4x + 3y - 8 = 0$

120. $(11, -2), (6, -1)$

$m = \dfrac{-1-(-2)}{6-11} = -\dfrac{1}{5}$

$y - (-2) = -\dfrac{1}{5}(x - 11)$

$y + 2 = -\dfrac{1}{5}x + \dfrac{11}{5}$

$y = -\dfrac{1}{5}x + \dfrac{1}{5}$

$5y = -x + 1$

$x + 5y - 1 = 0$

121. Point: $(0, -5)$ Slope: $m = \frac{3}{2}$

$y - (-5) = \frac{3}{2}(x - 0)$

$2y + 10 = 3x$

$3x - 2y - 10 = 0$

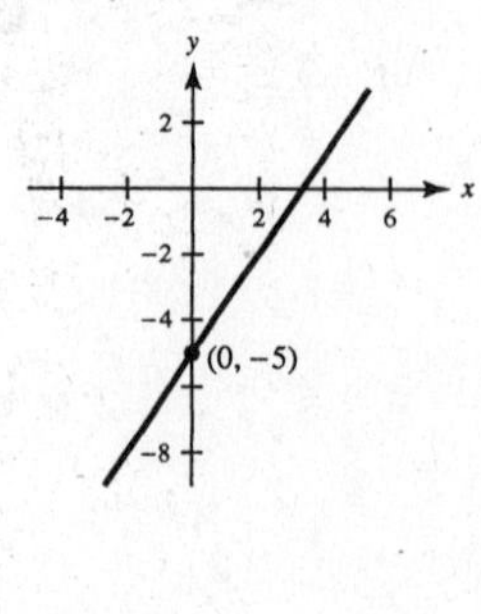

122. Point: $(10, -3)$ Slope: $m = -\frac{1}{2}$

$y - (-3) = -\frac{1}{2}(x - 10)$

$2y + 6 = -x + 10$

$x + 2y - 4 = 0$

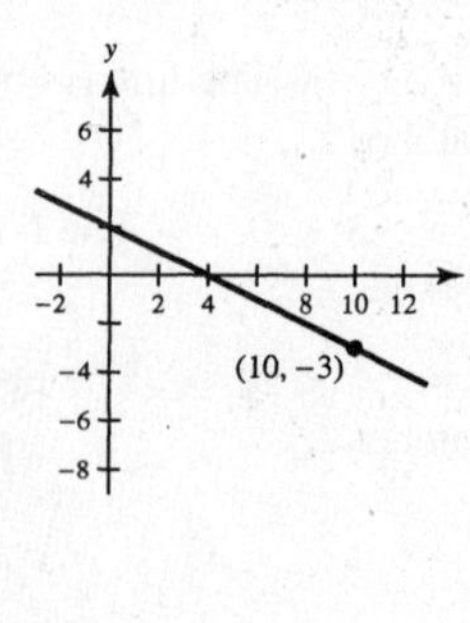

123. Point: $(-2, 6)$

Slope: $m = 0$

$y - 6 = 0(x - (-2))$

$y - 6 = 0$

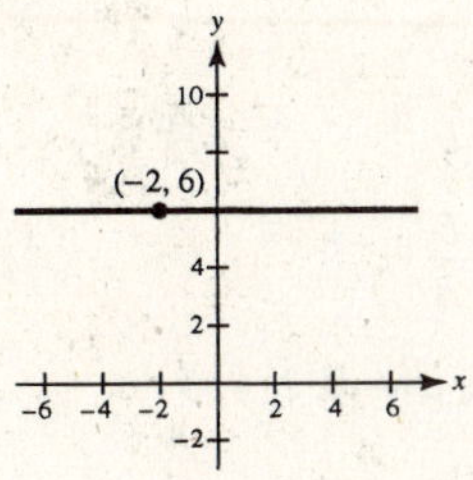

124. $(-3, 1), m = 0$

$y - 1 = 0(x + 3)$

$y - 1 = 0$ or $y = 1$

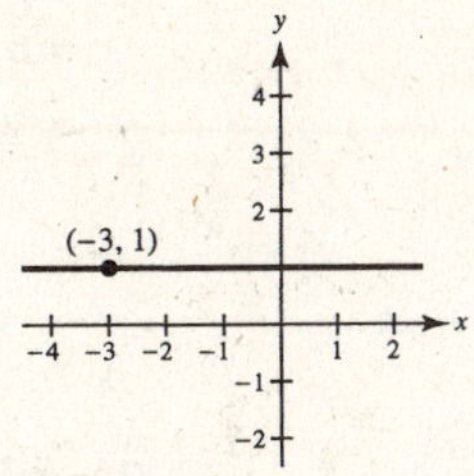

125. Point: $(-8, 5)$

Slope: Undefined

$x = -8$

$x + 8 = 0$

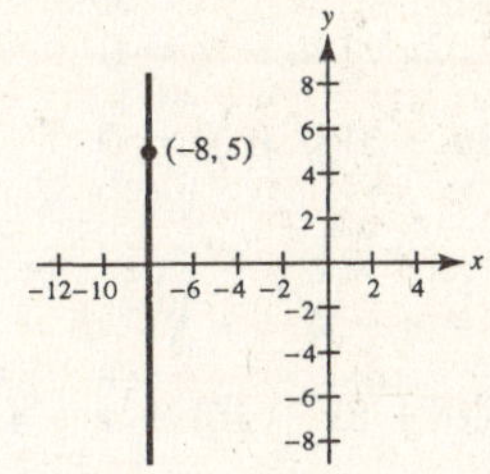

126. $(2, -3)$, m is undefined.

The line is vertical.

$x = 2$ or $x - 2 = 0$

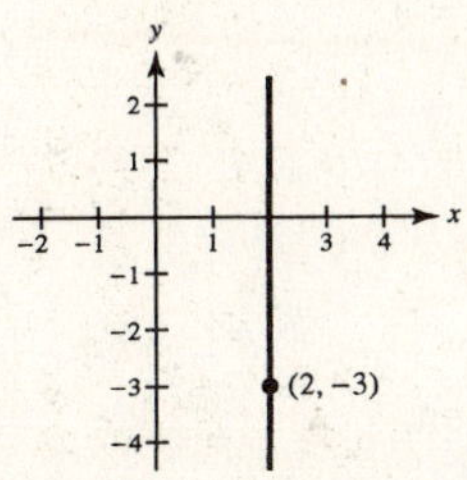

127. $x - 5 = 0 \Rightarrow x = 5$ and m is undefined.

(a) $(2, -1)$, m is undefined.

$x = 2$ or $x - 2 = 0$

(b) $(2, -1)$, $m = 0$

$y = -1$ or $y + 1 = 0$

128. $y - 1 = 0 \Rightarrow y = 1$ and $m = 0$

(a) $(3, 4)$, $m = 0$

$y = 4$ or $y - 4 = 0$

(b) $(3, 4)$, m is undefined.

$x = 3$ or $x - 3 = 0$

129. $x + 4 = 0 \Rightarrow x = -4$ and m is undefined.

(a) $(3, 2)$, m is undefined.

$x = 3$ or $x - 3 = 0$

(b) $(3, 2)$, $m = 0$

$y = 2$ or $y - 2 = 0$

130. $y + 6 = 0 \Rightarrow y = -6$ and $m = 0$

(a) $(-2, 1)$, $m = 0$

$y = 1$ or $y - 1 = 0$

(b) $(-2, 1)$, m is undefined.

$x = -2$ or $x + 2 = 0$

131. Point: $(3, -2)$, Line: $5x - 4y = 8$

(a) Parallel line: $m = \frac{5}{4}$

$$y - (-2) = \tfrac{5}{4}(x - 3)$$
$$4y + 8 = 5x - 15$$
$$5x - 4y - 23 = 0$$

(b) Perpendicular line: $m = -\frac{4}{5}$

$$y - (-2) = -\tfrac{4}{5}(x - 3)$$
$$5y + 10 = -4x + 12$$
$$4x + 5y - 2 = 0$$

132. $(-8, 3)$, $2x + 3y = 5$

$3y = 5 - 2x$

$y = \frac{5}{3} - \frac{2}{3}x$

(a) Parallel slope: $m = -\frac{2}{3}$

$$y - 3 = -\tfrac{2}{3}(x + 8)$$
$$3y - 9 = -2x - 16$$
$$2x + 3y + 7 = 0$$

(b) Perpendicular slope: $m = \frac{3}{2}$

$$y - 3 = \tfrac{3}{2}(x - (-8))$$
$$2y - 6 = 3x + 24$$
$$0 = 3x - 2y + 30$$

133. 2005 *Value*: \$12,500

Slope: \$850 increase per year

$V = 850(t - 5) + 12{,}500$

$V = 850t + 8250$

134. $(5, 72.95)$, $m = 5.15$

$$y - 72.95 = 5.15(x - 5)$$
$$y - 72.95 = 5.15x - 25.75$$
$$y = 5.15x + 42.20$$

Problem Solving for Chapter P

1. (a) $3(x + 4)^2 + (x + 4) - 2 = 0$

Let $u = x + 4$.

$3u^2 + u - 2 = 0$

$(3u - 2)(u + 1) = 0$

$u = \frac{2}{3}, -1$

$x = u - 4 = -\frac{10}{3}, -5$

(b) $3(x^2 + 8x + 16) + x + 4 - 2 = 0$

$3x^2 + 24x + 48 + x + 4 - 2 = 0$

$3x^2 + 25x + 50 = 0$

$(3x + 10)(x + 5) = 0$

$x = -\frac{10}{3}, -5$

(c) Answers will vary. Method (a) is slightly easier since there are less algebraic steps.

2. (a) $ax^2 + bx = 0, a \neq 0, b \neq 0$

$x(ax + b) = 0$

$x = 0$ or $ax + b = 0$

$ax = -b$

$x = -\frac{b}{a}$

(b) $ax^2 - (a - b)x - b = 0, a \neq 0, b \neq 0$

$ax^2 - ax + bx - b = 0$

$ax(x - 1) + b(x - 1) = 0$

$(ax + b)(x - 1) = 0$

$ax + b = 0$ or $x - 1 = 0$

$ax = -b$ $\quad x = 1$

$x = -\frac{b}{a}$

3. (a) $x^2 + bx + 4 = 0$

To have at least one real solution,

$b^2 - 4(1)(4) \geq 0$

$b^2 - 16 \geq 0.$

Critical numbers: $b = \pm 4$

Test intervals: $(-\infty, -4), (-4, 4), (4, \infty)$

Test: Is $b^2 - 16 \geq 0$?

By testing values in each test interval, we see that $b^2 - 16$ is greater than or equal to zero on the intervals $(-\infty, -4] \cup [4, \infty)$.

(b) $x^2 + bx - 4 = 0$

To have at least one real solution,

$b^2 - 4(1)(-4) \geq 0$

$b^2 + 16 \geq 0.$

This is true for all values of b:

$-\infty < b < \infty$

(c) $3x^2 + bx + 10 = 0$

To have at least one real solution,

$b^2 - 4(3)(10) \geq 0$

$b^2 - 120 \geq 0.$

Critical numbers: $b = \pm\sqrt{120} = \pm 2\sqrt{30}$

Test intervals:
$(-\infty, -2\sqrt{30}), (-2\sqrt{30}, 2\sqrt{30}), (2\sqrt{30}, \infty)$

Test: Is $b^2 - 120 \geq 0$?

By testing values in each test interval, we see that $b^2 - 120$ is greater than or equal to zero on the intervals $(-\infty, -2\sqrt{30}] \cup [2\sqrt{30}, \infty)$.

(d) $2x^2 + bx + 5 = 0$

To have at least one real solution,

$b^2 - 4(2)(5) \geq 0$

$b^2 - 40 \geq 0.$

This is true for $b \leq -2\sqrt{10}$ or $b \geq 2\sqrt{10}$,
$(-\infty, -2\sqrt{10}] \cup [2\sqrt{10}, \infty)$.

(e) If $a > 0$ and $c \leq 0$, then b can be any real number since $b^2 - 4ac$ would always be positive.

If $a > 0$ and $c > 0$, then $b \leq -2\sqrt{ac}$ or $b \geq 2\sqrt{ac}$, as in parts (a), (e), and (d).

(f) Since the intervals for b are symmetric about $b = 0$, the center of the interval is $b = 0$.

4. Since $x_m = \dfrac{x_1 + x_2}{2}$ and $y_m = \dfrac{y_1 + y_2}{2}$ we have:

(a) $2x_m = x_1 + x_2 \qquad 2y_m = y_1 + y_2$

$2x_m - x_1 = x_2 \qquad 2y_m - y_1 = y_2$

Thus, $(x_2, y_2) = (2x_m - x_1, 2y_m - y_1)$.

(b) (i) $(x_2, y_2) = (2x_m - x_1, 2y_m - y_1)$
$= (2 \cdot 4 - 1, 2(-1) - (-2)) = (7, 0)$

(ii) $(x_2, y_2) = (2x_m - x_1, 2y_m - y_1)$
$= (2 \cdot 2 - (-5), 2 \cdot 4 - 11) = (9, -3)$

(c) The midpoint of the given line segment is $\left(\dfrac{x_1 + x_2}{2}, \dfrac{y_1 + y_2}{2}\right)$.

The midpoint between (x_1, y_1) and $\left(\dfrac{x_1 + x_2}{2}, \dfrac{y_1 + y_2}{2}\right)$ is $\left(\dfrac{x_1 + \dfrac{x_1 + x_2}{2}}{2}, \dfrac{y_1 + \dfrac{y_1 + y_2}{2}}{2}\right) = \left(\dfrac{3x_1 + x_2}{4}, \dfrac{3y_1 + y_2}{4}\right)$.

The midpoint between $\left(\dfrac{x_1 + x_2}{2}, \dfrac{y_1 + y_2}{2}\right)$ and (x_2, y_2) is $\left(\dfrac{\dfrac{x_1 + x_2}{2} + x_2}{2}, \dfrac{\dfrac{y_1 + y_2}{2} + y_2}{2}\right) = \left(\dfrac{x_1 + 3x_2}{4}, \dfrac{y_1 + 3y_2}{4}\right)$.

Thus, the three points are $\left(\dfrac{3x_1 + x_2}{4}, \dfrac{3y_1 + y_2}{2}\right)$, $\left(\dfrac{x_1 + x_2}{2}, \dfrac{y_1 + y_2}{2}\right)$, and $\left(\dfrac{x_1 + 3x_2}{4}, \dfrac{y_1 + 3y_2}{4}\right)$.

(d) (i) $\left(\dfrac{3x_1 + x_2}{4}, \dfrac{3y_1 + y_2}{4}\right) = \left(\dfrac{3 \cdot 1 + 4}{4}, \dfrac{3(-2) - 1}{4}\right) = \left(\dfrac{7}{4}, -\dfrac{7}{4}\right)$

$\left(\dfrac{x_1 + x_2}{2}, \dfrac{y_1 + y_2}{2}\right) = \left(\dfrac{1 + 4}{2}, \dfrac{-2 - 1}{2}\right) = \left(\dfrac{5}{2}, -\dfrac{3}{2}\right)$

$\left(\dfrac{x_1 + 3x_2}{4}, \dfrac{y_1 + 3y_2}{4}\right) = \left(\dfrac{1 + 3 \cdot 4}{4}, \dfrac{-2 + 3(-1)}{4}\right) = \left(\dfrac{13}{4}, -\dfrac{5}{4}\right)$

(ii) $\left(\dfrac{3x_1 + x_2}{4}, \dfrac{3y_1 + y_2}{4}\right) = \left(\dfrac{3(-2) + 0}{4}, \dfrac{3(-3) + 0}{4}\right) = \left(-\dfrac{3}{2}, -\dfrac{9}{4}\right)$

$\left(\dfrac{x_1 + x_2}{2}, \dfrac{y_1 + y_2}{2}\right) = \left(\dfrac{-2 + 0}{2}, \dfrac{-3 + 0}{2}\right) = \left(-1, -\dfrac{3}{2}\right)$

$\left(\dfrac{x_1 + 3x_2}{4}, \dfrac{y_1 + 3y_2}{4}\right) = \left(\dfrac{-2 + 0}{4}, \dfrac{-3 + 0}{4}\right) = \left(-\dfrac{1}{2}, -\dfrac{3}{4}\right)$

5. $y = ax^2 + bx^3$

(a) If the graph is symmetric with respect to the y-axis, then you can replace x with $-x$ and the result is an equivalent equation. This happens when $b = 0$ and a is any real number.

(b) If the graph is symmetric with respect to the origin, then you can replace x with $-x$ and y with $-y$ and the result is an equivalent equation. This happens when $a = 0$ and b is any real number.

6. (a) Since there are three solutions, the equation is not linear nor is it quadratic. Neither

(b) Since there is only one solution, the equation could have been either linear or quadratic. Both

(c) Since there are two solutions, the equation is not linear but could be quadratic. Quadratic

(d) Since there are four solutions, the equation is not linear nor is it quadratic. Neither

7. No. The slope cannot be determined without knowing the scale on the y-axis. The slopes could be the same.

8. (a) $$x^2 - 6x + y^2 - 8y = 0$$
$$(x^2 - 6x + 9) + (y^2 - 8y + 16) = 9 + 16$$
$$(x - 3)^2 + (y - 4)^2 = 25$$

Center: $(3, 4)$ Radius: 5

(b) Slope of line from $(0, 0)$ to $(3, 4)$ is $\frac{4}{3}$. Slope of tangent line is $-\frac{3}{4}$. Hence,

$$y - 0 = -\frac{3}{4}(x - 0) \Rightarrow y = -\frac{3}{4}x. \quad \text{Tangent line}$$

(c) Slope of line from $(6, 0)$ to $(3, 4)$ is $\frac{4-0}{3-6} = -\frac{4}{3}$.

Slope of tangent line is $\frac{3}{4}$. Hence,

$$y - 0 = \frac{3}{4}(x - 6) \Rightarrow y = \frac{3}{4}x - \frac{9}{2}. \quad \text{Tangent line}$$

(d) $$-\frac{3}{4}x = \frac{3}{4}x - \frac{9}{2}$$
$$\frac{3}{2}x = \frac{9}{2}$$
$$x = 3$$

Intersection: $\left(3, -\frac{9}{4}\right)$

9. $$d_1d_2 = 1$$
$$[(x + 1)^2 + y^2][(x - 1)^2 + y^2] = 1$$
$$(x + 1)^2(x - 1)^2 + y^2[(x + 1)^2 + (x - 1)^2] + y^4 = 1$$
$$(x^2 - 1)^2 + y^2[2x^2 + 2] + y^4 = 1$$
$$x^4 - 2x^2 + 1 + 2x^2y^2 + 2y^2 + y^4 = 1$$
$$(x^4 + 2x^2y^2 + y^4) - 2x^2 + 2y^2 = 0$$
$$(x^2 + y^2)^2 = 2(x^2 - y^2)$$

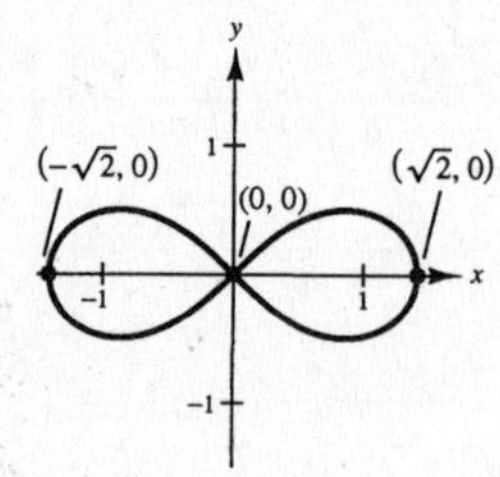

Let $y = 0$. Then $x^4 = 2x^2 \Rightarrow x = 0$ or $x^2 = 2$. Thus, $(0, 0)$, $(\sqrt{2}, 0)$ and $(-\sqrt{2}, 0)$ are on the curve.

10. The original point is (x, y).

(a) The transformed point $(-x, y)$ is a reflection through the y-axis.

(b) The transformed point $(x, -y)$ is a reflection through the x-axis.

(c) The transformed point $(-x, -y)$ is a reflection through the origin.

11. (a) $\frac{7000 - 5500}{10} = 150$ students per year

(b) 1993: $5500 + 3(150) = 5950$ students

1997: $5500 + 7(150) = 6550$ students

1999: $5500 + 9(150) = 6850$ students

(c) Equation: $y = 150x + 5500$ where $x = 0$ represents 1990.

Slope: $m = 150$

This means that enrollments increase by approximately 150 students per year.

12. Milk: $M = -0.289t + 25.58$

Coffee: $C = 1.13t + 15.0$

(a) $C = M$

$$1.13t + 15.0 = -0.289t + 25.58$$
$$1.419t = 10.58$$
$$t \approx 7.456$$

Point of intersection: $(7.456, 23.425)$

(b) and (c)

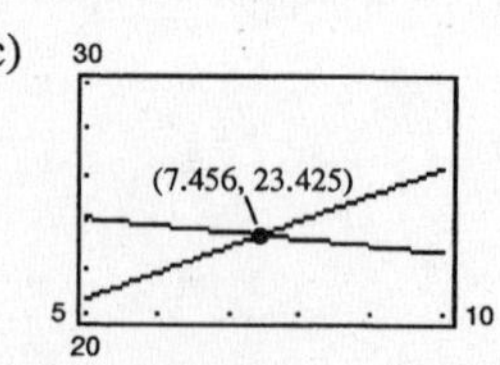

(d) Per capita consumption of milk was equal to per capita consumption of coffee in 1997.

13. (a) and (b)

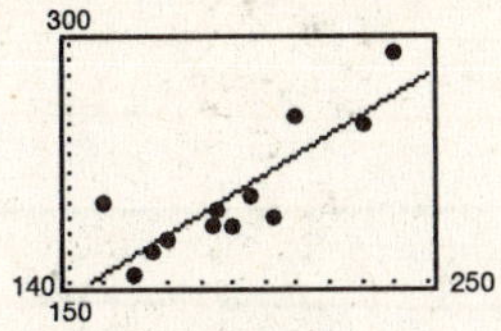

(c) $200 = 1.266x - 35.766$

$235.766 = 1.266x$

$x \approx 186.23$ pounds

(d) Answers will vary.

14. $S \approx 15.91t - 52.11$

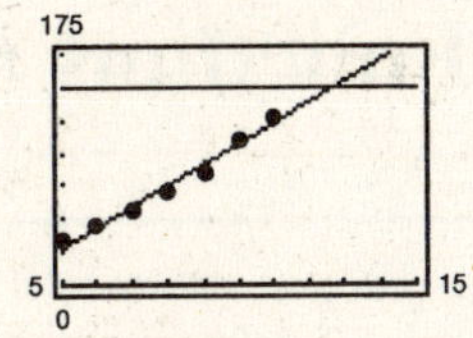

$S = 150$ when $t \approx 12.7 \approx 13$, which corresponds to 2003.

t	S
5	33.8
6	44.0
7	55.3
8	69.2
9	86.0
10	109.5
11	128.4

15. (a) Choice 1: $W = 3000 + 0.07s$

Choice 2: $W = 3400 + 0.05s$

(b) $3000 + 0.07s = 3400 + 0.05s$

$0.02s = 400$

$s = \$20{,}000$

The salaries are the same ($4400 per month) when sales equal $20,000.

(c) An ambitious salesperson who can sell more than $20,000 per month would be wise to select choice 1. A more conservative choice for a salesperson who is unsure of the market for this product would be choice 2.

CHAPTER 1
Functions and Their Graphs

Section 1.1 Functions **91**

Section 1.2 Analyzing Graphs of Functions **100**

Section 1.3 Shifting, Reflecting, and Stretching Graphs **112**

Section 1.4 Combinations of Functions **121**

Section 1.5 Inverse Functions **128**

Section 1.6 Mathematical Modeling **138**

Review Exercises **146**

Problem Solving **152**

CHAPTER 1
Functions and Their Graphs

Section 1.1 Functions

1. Yes, the relationship is a function. Each domain value is matched with only one range value.

2. No, it is not a function. The domain value of -1 is matched with two output values.

3. No, the relationship is not a function. The domain values are each matched with two range values.

4. Yes, it is a function. Each domain value is matched with only one range value.

5. No, the relationship is not a function. Each input value is matched with three output values.

6. Yes, the relationship is a function. Each input value is matched with only one output value.

7. Yes, it does represent a function. Each input value is matched with only one output value.

8. No, the table does not represent a function. The input values of 0 and 1 are each matched with two different output values.

9. No, it does not represent a function. The input values of 10 and 7 are each matched with two output values.

10. Yes, the table does represent a function. Each input value is matched with only one output value.

11. (a) Each element of A is matched with exactly one element of B, so it does represent a function.
    (b) The element 1 in A is matched with two elements, -2 and 1 of B, so it does not represent a function.
    (c) Each element of A is matched with exactly one element of B, so it does represent a function.
    (d) The element 2 in A is not matched with an element of B, so it does not represent a function.

12. (a) The element c in A is matched with two elements, 2 and 3 of B, so it is not a function.
    (b) Each element of A is matched with exactly one element of B, so it does represent a function.
    (c) This is not a function from A to B (it represents a function from B to A instead).
    (d) Each element of A is matched with exactly one element of B, so it does represent a function.

13. Each is a function. For each year there corresponds one and only one circulation.

14. From the graph we have $f(2002) \approx 9$ million.

15. $x^2 + y^2 = 4 \Longrightarrow y = \pm\sqrt{4 - x^2}$

    No, y *is not* a function of x.

16. $4x^2 + y^2 = 8 \Longrightarrow y = \pm\sqrt{8 - 4x^2}$

    No, y *is not* a function of x.

17. $x^2 + y = 4 \Longrightarrow y = 4 - x^2$

    Yes, y *is* a function of x.

18. $x + y^2 = 4 \Longrightarrow y = \pm\sqrt{4 - x}$

    Thus, y is *not* a function of x.

19. $2x + 3y = 4 \Longrightarrow y = \frac{1}{3}(4 - 2x)$

    Yes, y *is* a function of x.

20. $5x - 4y = 10 \Longrightarrow y = \frac{5x - 10}{4}$

    Yes, y *is* a function of x.

21. $y^2 = x^2 - 1 \Rightarrow y = \pm\sqrt{x^2 - 1}$

No, y *is not* a function of x.

22. $y^2 = x^2 + 4 \Rightarrow y = \pm\sqrt{x^2 + 4}$

No, y *is not* a function of x.

23. $y = |4 - x|$

Yes, y *is* a function of x.

24. $y = \sqrt{x - 5}$

Yes, y *is* a function of x.

25. $|x| - y = 1 \Rightarrow y = |x| - 1$

Yes, y *is* a function of x.

26. $|y| = 4 - x \Rightarrow y = \pm(4 - x)$

No, y *is not* a function of x.

27. $x = y^2 \Rightarrow y = \pm\sqrt{x}$

No, y *is not* a function of x.

28. $(x - 2)^2 + y^2 = 4 \Rightarrow y = \pm\sqrt{4 - (x - 2)^2}$

No, y *is not* a function of x.

29. $f(s) = \dfrac{1}{s + 1}$

(a) $f(4) = \dfrac{1}{(4) + 1}$ (b) $f(0) = \dfrac{1}{(0) + 1}$ (c) $f(4x) = \dfrac{1}{(4x) + 1}$ (d) $f(x + c) = \dfrac{1}{(x + c) + 1}$

30. $g(x) = x^2 - 2x$

(a) $g(2) = (2)^2 - 2(2)$

(b) $g(-3) = (-3)^2 - 2(-3)$

(c) $g(t + 1) = (t + 1)^2 - 2(t + 1)$

(d) $g(x + c) = (x + c)^2 - 2(x + c)$

31. $f(x) = 2x - 3$

(a) $f(1) = 2(1) - 3 = -1$

(b) $f(-3) = 2(-3) - 3 = -9$

(c) $f(x - 1) = 2(x - 1) - 3 = 2x - 5$

32. $g(y) = 7 - 3y$

(a) $g(0) = 7 - 3(0) = 7$

(b) $g(\frac{7}{3}) = 7 - 3(\frac{7}{3}) = 0$

(c) $g(s + 2) = 7 - 3(s + 2)$

$= 7 - 3s - 6 = 1 - 3s$

33. $V(r) = \frac{4}{3}\pi r^3$

(a) $V(3) = \frac{4}{3}\pi(3)^3 = \frac{4}{3}\pi(27) = 36\pi$

(b) $V(\frac{3}{2}) = \frac{4}{3}\pi(\frac{3}{2})^3 = \frac{4}{3}\pi(\frac{27}{8}) = \frac{9}{2}\pi$

(c) $V(2r) = \frac{4}{3}\pi(2r)^3 = \frac{4}{3}\pi(8r^3) = \frac{32}{3}\pi r^3$

34. $h(t) = t^2 - 2t$

(a) $h(2) = 2^2 - 2(2) = 0$

(b) $h(1.5) = (1.5)^2 - 2(1.5) = -0.75$

(c) $h(x + 2) = (x + 2)^2 - 2(x + 2) = x^2 + 2x$

35. $f(y) = 3 - \sqrt{y}$

(a) $f(4) = 3 - \sqrt{4} = 1$

(b) $f(0.25) = 3 - \sqrt{0.25} = 2.5$

(c) $f(4x^2) = 3 - \sqrt{4x^2} = 3 - 2|x|$

36. $f(x) = \sqrt{x + 8} + 2$

(a) $f(-8) = \sqrt{(-8) + 8} + 2 = 2$

(b) $f(1) = \sqrt{(1) + 8} + 2 = 5$

(c) $f(x - 8) = \sqrt{(x - 8) + 8} + 2 = \sqrt{x} + 2$

37. $q(x) = \dfrac{1}{x^2 - 9}$

(a) $q(0) = \dfrac{1}{0^2 - 9} = -\dfrac{1}{9}$

(b) $q(3) = \dfrac{1}{3^2 - 9}$ is undefined.

(c) $q(y + 3) = \dfrac{1}{(y + 3)^2 - 9} = \dfrac{1}{y^2 + 6y}$

38. $q(t) = \dfrac{2t^2 + 3}{t^2}$

(a) $q(2) = \dfrac{2(2)^2 + 3}{(2)^2} = \dfrac{8 + 3}{4} = \dfrac{11}{4}$

(b) $q(0) = \dfrac{2(0)^2 + 3}{(0)^2}$

Division by zero is undefined.

(c) $q(-x) = \dfrac{2(-x)^2 + 3}{(-x)^2} = \dfrac{2x^2 + 3}{x^2}$

39. $f(x) = \dfrac{|x|}{x}$

(a) $f(2) = \dfrac{|2|}{2} = 1$

(b) $f(-2) = \dfrac{|-2|}{-2} = -1$

(c) $f(x - 1) = \dfrac{|x - 1|}{x - 1}$

40. $f(x) = |x| + 4$

(a) $f(2) = |2| + 4 = 6$

(b) $f(-2) = |-2| + 4 = 6$

(c) $f(x^2) = |x^2| + 4 = x^2 + 4$

41. $f(x) = \begin{cases} 2x - 1, & x < 0 \\ 2x + 2, & x \geq 0 \end{cases}$

(a) $f(-1) = 2(-1) - 1 = -3$

(b) $f(0) = 2(0) + 2 = 2$

(c) $f(2) = 2(2) + 2 = 6$

42. $f(x) = \begin{cases} x^2 + 2, & x \leq 1 \\ 2x^2 + 1, & x > 1 \end{cases}$

(a) $f(-2) = (-2)^2 + 2 = 6$

(b) $f(1) = (1)^2 + 2 = 3$

(c) $f(2) = 2(2)^2 + 1 = 9$

43. $f(x) = x^2 - 3$

x	-2	-1	0	1	2
$f(x)$	1	-2	-3	-2	1

44. $g(x) = \sqrt{x - 3}$

$g(3) = \sqrt{3 - 3} = 0$

$g(4) = \sqrt{4 - 3} = 1$

$g(5) = \sqrt{5 - 3} = \sqrt{2}$

$g(6) = \sqrt{6 - 3} = \sqrt{3}$

$g(7) = \sqrt{7 - 3} = 2$

x	3	4	5	6	7
$g(x)$	0	1	$\sqrt{2}$	$\sqrt{3}$	2

45. $h(t) = \frac{1}{2}|t + 3|$

t	-5	-4	-3	-2	-1
$h(t)$	1	$\frac{1}{2}$	0	$\frac{1}{2}$	1

46. $f(s) = \dfrac{|s - 2|}{s - 2}$

$f(0) = \dfrac{|0 - 2|}{0 - 2} = \dfrac{2}{-2} = -1$

$f(1) = \dfrac{|1 - 2|}{1 - 2} = \dfrac{1}{-1} = -1$

$f\left(\dfrac{3}{2}\right) = \dfrac{\left|\frac{3}{2} - 2\right|}{\frac{3}{2} - 2} = \dfrac{\frac{1}{2}}{-\frac{1}{2}} = -1$

$f\left(\dfrac{5}{2}\right) = \dfrac{\left|\frac{5}{2} - 2\right|}{\frac{5}{2} - 2} = \dfrac{\frac{1}{2}}{\frac{1}{2}} = 1$

$f(4) = \dfrac{|4 - 2|}{4 - 2} = \dfrac{2}{2} = 1$

s	0	1	$\frac{3}{2}$	$\frac{5}{2}$	4
$f(s)$	-1	-1	-1	1	1

47. $f(x) = \begin{cases} -\frac{1}{2}x + 4, & x \leq 0 \\ (x - 2)^2, & x > 0 \end{cases}$

x	-2	-1	0	1	2
$f(x)$	5	$\frac{9}{2}$	4	1	0

48. $h(x) = \begin{cases} 9 - x^2, & x < 3 \\ x - 3, & x \geq 3 \end{cases}$

$h(1) = 9 - (1)^2 = 8$

$h(2) = 9 - (2)^2 = 5$

$h(3) = (3) - 3 = 0$

$h(4) = (4) - 3 = 1$

$h(5) = (5) - 3 = 2$

s	1	2	3	4	5
$h(x)$	8	5	0	1	2

49. $f(x) = 15 - 3x$

$$15 - 3x = 0$$
$$3x = 15$$
$$x = 5$$

50. $f(x) = 5x + 1$

$$5x + 1 = 0$$
$$x = -\frac{1}{5}$$

51. $f(x) = \dfrac{x^2 - 16}{x - 2}$

$$\frac{x^2 - 16}{x - 2} = 0$$
$$x^2 - 16 = 0$$
$$x^2 = 16$$
$$x = \pm 4$$

52. $f(x) = \dfrac{9 - x^2}{x + 1}$

$$\frac{9 - x^2}{x + 1} = 0$$
$$9 - x^2 = 0$$
$$x^2 = 9$$
$$x = \pm 3$$

53. $f(x) = x^2 - 5x + 4$

$$x^2 - 5x + 4 = 0$$
$$(x - 1)(x - 4) = 0$$
$$x - 1 = 0 \Rightarrow x = 1$$
$$x - 4 = 0 \Rightarrow x = 4$$

54. $f(x) = x^2 - 8x + 15$

$$x^2 - 8x + 15 = 0$$
$$(x - 5)(x - 3) = 0$$
$$x - 5 = 0 \Rightarrow x = 5$$
$$x - 3 = 0 \Rightarrow x = 3$$

55. $f(x) = x^3 - 4x^2 - x + 4$

$$x^3 - 4x^2 - x + 4 = 0$$
$$x^2(x - 4) - 1(x - 4) = 0$$
$$(x^2 - 1)(x - 4) = 0$$
$$(x + 1)(x - 1)(x - 4) = 0$$
$$x + 1 = 0 \Rightarrow x = -1$$
$$x - 1 = 0 \Rightarrow x = 1$$
$$x - 4 = 0 \Rightarrow x = 4$$

56. $f(x) = x^3 - x^2 - 4x + 4$

$$x^3 - x^2 - 4x + 4 = 0$$
$$x^2(x - 1) - 4(x - 1) = 0$$
$$(x - 1)(x^2 - 4) = 0$$
$$x - 1 = 0 \Rightarrow x = 1$$
$$x^2 - 4 = 0 \Rightarrow x = \pm 2$$

57. $f(x) = g(x)$

$$x^2 = x + 2$$
$$x^2 - x - 2 = 0$$
$$(x + 1)(x - 2) = 0$$
$$x + 1 = 0 \Rightarrow x = -1$$
$$x - 2 = 0 \Rightarrow x = 2$$

58. $f(x) = g(x)$

$$x^2 + 2x + 1 = 3x + 3$$
$$x^2 - x - 2 = 0$$
$$(x - 2)(x + 1) = 0$$
$$x - 2 = 0 \Rightarrow x = 2$$
$$x + 1 = 0 \Rightarrow x = -1$$

59. $f(x) = g(x)$

$$x^3 - 3x + 1 = x + 1$$
$$x^3 - 4x = 0$$
$$x(x^2 - 4) = 0$$
$$x\ (x + 2)(x - 2) = 0$$
$$x = 0$$
$$x + 2 = 0 \Rightarrow x = -2$$
$$x - 2 = 0 \Rightarrow x = 2$$

60. $f(x) = g(x)$

$$x^4 - 2x^2 = 2x^2$$
$$x^4 - 4x^2 = 0$$
$$x^2(x^2 - 4) = 0$$
$$x^2(x + 2)(x - 2) = 0$$
$$x^2 = 0 \Rightarrow x = 0$$
$$x + 2 = 0 \Rightarrow x = -2$$
$$x - 2 = 0 \Rightarrow x = 2$$

61. $f(x) = 5x^2 + 2x - 1$

Since $f(x)$ is a polynomial, the domain is all real numbers x.

62. $g(x) = 1 - 2x^2$

Because $g(x)$ is a polynomial, the domain is all real numbers x.

63. $h(t) = \dfrac{4t - 3}{t}$

Domain: All real numbers $t \neq 0$.

64. $s(y) = \dfrac{3y}{y + 5}$

$y + 5 \neq 0$

$y \neq -5$

The domain is all real numbers $y \neq -5$.

65. $g(y) = \sqrt{y - 10}$

Domain: $y - 10 \geq 0$

$y \geq 10$

66. $f(t) = \sqrt[3]{t + 4}$

Because $f(t)$ is a cube root, the domain is all real numbers t.

67. $f(x) = \sqrt[4]{1 - x^2}$

Domain: $1 - x^2 \geq 0$

$-x^2 \geq -1$

$x^2 \leq 1$

$x^2 - 1 \leq 0$

Critical Numbers: $x = \pm 1$

Test Intervals: $(-\infty, -1), (-1, 1), (1, \infty)$

Test: Is $x^2 - 1 \leq 0$?

Solution: $[-1, 1]$ or $-1 \leq x \leq 1$

68. $f(x) = \sqrt[4]{x^2 + 3x}$

$x^2 + 3x \geq 0$

$x(x + 3) \geq 0$

Test intervals: $(-\infty, -3), (-3, 0), (0, \infty)$

The domain is all real numbers $x \leq -3, x \geq 0$

69. $g(x) = \dfrac{1}{x} - \dfrac{3}{x + 2}$

Domain: All real numbers except $x = 0,\ x = -2$

70. $h(x) = \dfrac{2}{x} + \dfrac{5}{x - 2}$

Domain: All real numbers except $x = 0, x = 2$.

71. $f(s) = \dfrac{\sqrt{s - 1}}{s - 4}$

Domain: $s - 1 \geq 0 \Rightarrow s \geq 1$ and $s \neq 4$

The domain consists of all real numbers s, such that $s \geq 1$ and $s \neq 4$.

72. $f(x) = \dfrac{\sqrt{x + 6}}{6 + x}$

$x + 6 \geq 0 \Rightarrow x \geq -6$ and

$6 + x \neq 0 \Rightarrow x \neq -6$

The domain is all real numbers $x > -6$.

73. $f(x) = \dfrac{\sqrt[3]{x - 4}}{x}$

The domain is all real numbers except $x = 0$.

74. $f(x) = \dfrac{\sqrt[3]{x - 5}}{x^2 - 9}$

The domain is all real numbers except $x = \pm 3$.

75. $f(x) = x^2$

$\{(-2, 4), (-1, 1), (0, 0), (1, 1), (2, 4)\}$

76. $f(x) = \dfrac{2x}{x^2 + 1}$

$\left\{\left(-2, -\frac{4}{5}\right), (-1, -1), (0, 0), (1, 1), \left(2, \frac{4}{5}\right)\right\}$

77. $f(x) = \sqrt{x + 2}$

$\{(-2, 0), (-1, 1), (0, \sqrt{2}), (1, \sqrt{3}), (2, 2)\}$

78. $f(x) = |x + 1|$

$\{(-2, 1), (-1, 0), (0, 1), (1, 2), (2, 3)\}$

79. By plotting the points, we have a parabola, so $g(x) = cx^2$. Since $(-4, -32)$ is on the graph, we have $-32 = c(-4)^2 \Rightarrow c = -2$. Thus, $g(x) = -2x^2$.

80. By plotting the data, you can see that they represent a line, or $f(x) = cx$. Because $(0, 0)$ and $\left(1, \frac{1}{4}\right)$ are on the line, the slope is $\frac{1}{4}$. Thus, $f(x) = \frac{1}{4}x$.

81. Since the function is undefined at 0, we have $r(x) = c/x$. Since $(-4, -8)$ is on the graph, we have $-8 = c/-4 \Rightarrow c = 32$. Thus, $r(x) = 32/x$.

82. By plotting the data, you can see that they represent $h(x) = c\sqrt{|x|}$. Because $\sqrt{|-4|} = 2$ and $\sqrt{|-1|} = 1$ and the corresponding y values are 6 and 3, $c = 3$ and $h(x) = 3\sqrt{|x|}$.

83.

$$\begin{aligned}
f(x) &= x^2 - x + 1\\
f(2 + \Delta x) &= (2 + \Delta x)^2 - (2 + \Delta x) + 1\\
&= 4 + 4\Delta x + (\Delta x)^2 - 2 - \Delta x + 1\\
&= (\Delta x)^2 + 3\Delta x + 3\\
f(2) &= (2)^2 - 2 + 1 = 3\\
f(2 + \Delta x) - f(2) &= (\Delta x)^2 + 3\Delta x\\
\frac{f(2 + \Delta x) - f(2)}{\Delta x} &= \frac{(\Delta x)^2 + 3\Delta x}{\Delta x} = \Delta x + 3,\ \Delta x \neq 0
\end{aligned}$$

84.

$$\begin{aligned}
f(x) &= 5x - x^2\\
f(5 + \Delta x) &= 5(5 + \Delta x) - (5 + \Delta x)^2\\
&= 25 + 5\Delta x - (25 + 10\Delta x + (\Delta x)^2)\\
&= 25 + 5\Delta x - 25 - 10\Delta x - (\Delta x)^2\\
&= -(\Delta x)^2 - 5\Delta x\\
f(5) &= 5(5) - (5)^2\\
&= 25 - 25 = 0\\
\frac{f(5 + \Delta x) - f(5)}{\Delta x} &= \frac{-(\Delta x)^2 - 5\Delta x}{\Delta x}\\
&= \frac{-\Delta x(\Delta x + 5)}{\Delta x} = -(\Delta x + 5), \Delta x \neq 0
\end{aligned}$$

85.

$$\begin{aligned}
f(x) &= x^3 + 2x - 1\\
\frac{f(x + c) - f(x)}{c} &= \frac{[(x + c)^3 + 2(x + c) - 1] - (x^3 + 2x - 1)}{c}\\
&= \frac{x^3 + 3x^2c + 3xc^2 + c^3 + 2x + 2c - 1 - x^3 - 2x + 1}{c}\\
&= \frac{3x^2c + 3xc^2 + c^3 + 2c}{c} = \frac{c(3x^2 + 3xc + c^2 + 2)}{c}\\
&= 3x^2 + 3xc + c^2 + 2,\ c \neq 0
\end{aligned}$$

86.

$$\begin{aligned}
f(x) &= x^3 - x + 1\\
\frac{f(x + c) - f(x)}{c} &= \frac{[(x + c)^3 - (x + c) + 1] - (x^3 - x + 1)}{c}\\
&= \frac{x^3 + 3x^2c + 3xc^2 + c^3 - x - c + 1 - x^3 + x - 1}{c}\\
&= \frac{3x^2c + 3xc^2 + c^3 - c}{c} = \frac{c(3x^2 + 3xc + c^2 - 1)}{c}\\
&= 3x^2 + 3xc + c^2 - 1,\ c \neq 0
\end{aligned}$$

87. $g(x) = 3x - 1$

$$\frac{g(x) - g(3)}{x - 3} = \frac{(3x - 1) - 8}{x - 3} = \frac{3x - 9}{x - 3} = \frac{3(x - 3)}{x - 3} = 3,\ x \neq 3$$

88. $f(t) = \dfrac{1}{t}$

$f(1) = \dfrac{1}{1} = 1$

$$\frac{f(t) - f(1)}{t - 1} = \frac{\frac{1}{t} - 1}{t - 1} = \frac{\frac{1}{t} - \frac{t}{t}}{t - 1} = \frac{\frac{1 - t}{t}}{t - 1} = \frac{\left(-\frac{1}{t}\right)(t - 1)}{t - 1} = -\frac{1}{t}, \quad t \neq 1$$

89. $f(x) = \sqrt{5x}$

$$\frac{f(x) - f(5)}{x - 5} = \frac{\sqrt{5x} - 5}{x - 5}$$

90. $f(x) = x^{2/3} + 1$

$f(8) = 8^{2/3} + 1 = 5$

$$\frac{f(x) - f(8)}{x - 8} = \frac{x^{2/3} + 1 - 5}{x - 8} = \frac{x^{2/3} - 4}{x - 8}$$

91. No. The element 3 in the domain corresponds to two elements in the range.

92. The domain is the set of inputs of the function, and the range is the set of outputs.

93. An advantage to function notation is that it gives a name to the relationship so it can be easily referenced. When evaluating a function, you see both the input and output values.

94. $A = s^2$ and $P = 4s \Rightarrow \dfrac{P}{4} = s$

$$A = \left(\frac{P}{4}\right)^2 = \frac{P^2}{16}$$

95. $A = \pi r^2, \quad C = 2\pi r$

$r = \dfrac{C}{2\pi}$

$$A = \pi\left(\frac{C}{2\pi}\right)^2 = \frac{C^2}{4\pi}$$

96. $8^2 + \left(\dfrac{b}{2}\right)^2 = s^2$

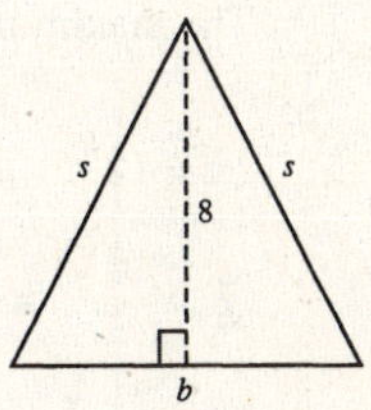

$\dfrac{b^2}{4} = s^2 - 64$

$b^2 = 4(s^2 - 64)$

$b = 2\sqrt{s^2 - 64}$

Thus, $A = \dfrac{1}{2}bh$

$= \dfrac{1}{2}\left(2\sqrt{s^2 - 64}\right)(8)$

$= 8\sqrt{s^2 - 64}$ square inches.

97. $A = \frac{1}{2}bh$, and in an equilateral triangle $b = s$

$s^2 = h^2 + \left(\dfrac{s}{2}\right)^2$

$$h = \sqrt{s^2 - \left(\frac{s}{2}\right)^2}$$

$$h = \sqrt{\frac{4s^2}{4} - \frac{s^2}{4}} = \frac{\sqrt{3}s}{2}$$

$$A = \frac{1}{2}s \cdot \frac{\sqrt{3}s}{2} = \frac{\sqrt{3}s^2}{4}$$

98. (a)

Height, x	Width	Volume, V
1	$24 - 2(1)$	$1[24 - 2(1)]^2 = 484$
2	$24 - 2(2)$	$2[24 - 2(2)]^2 = 800$
3	$24 - 2(3)$	$3[24 - 2(3)]^2 = 972$
4	$24 - 2(4)$	$4[24 - 2(4)]^2 = 1024$
5	$24 - 2(5)$	$5[24 - 2(5)]^2 = 980$
6	$24 - 2(6)$	$6[24 - 2(6)]^2 = 864$

The volume is maximum when $x = 4$.

(b)

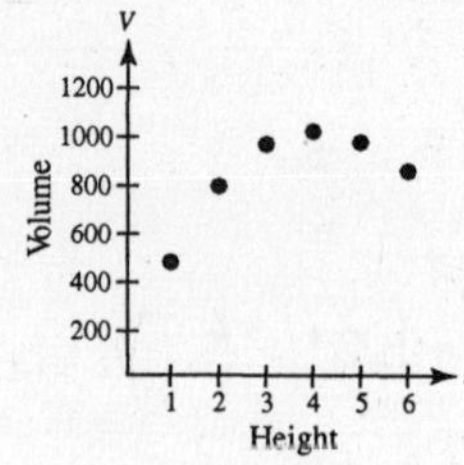

V is a function of x.

(c) $V = x(24 - 2x)^2$

Domain: $0 < x < 12$

99. (a)

Units x	*Price p*	*Profit P*
110	$90 - 10(0.15)$	$110[90 - 10(0.15)] - 110(60) = 3135$
120	$90 - 20(0.15)$	$120[90 - 20(0.15)] - 120(60) = 3240$
130	$90 - 30(0.15)$	$130[90 - 30(0.15)] - 130(60) = 3315$
140	$90 - 40(0.15)$	$140[90 - 40(0.15)] - 140(60) = 3360$
150	$90 - 50(0.15)$	$150[90 - 50(0.15)] - 150(60) = 3375$
160	$90 - 60(0.15)$	$160[90 - 60(0.15)] - 160(60) = 3360$

The maximum profit is \$3375.

(b)

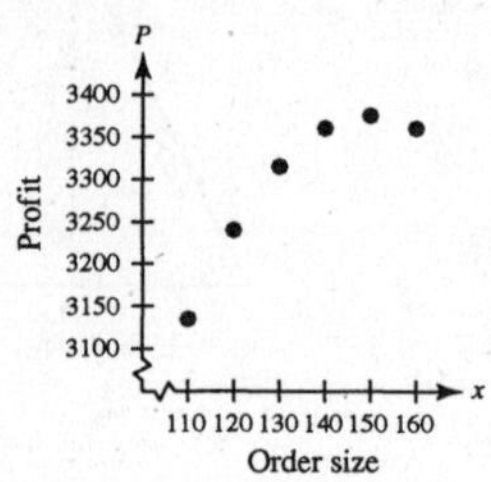

Yes

(c) Profit = Revenue − Cost

$= \text{(price per unit)(number of units)} - \text{(cost)(number of units)}$

$= [90 - (x - 100)(0.15)]x - 60x, \; x > 100$

$= (90 - 0.15x + 15)x - 60x$

$= (105 - 0.15x)x - 60x$

$= 105x - 0.15x^2 - 60x$

$= 45x - 0.15x^2, \; x > 100$

100. (a) Cost = variable costs + fixed costs

$C = 12.30x + 98{,}000$

(b) Revenue = price per unit × number of units

$R = 17.98x$

(c) Profit = Revenue − Cost

$P = 17.98x - (12.30x + 98{,}000)$

$P = 5.68x - 98{,}000$

101. (a) $V = l \cdot w \cdot h = x \cdot y \cdot x = x^2y$ where $4x + y = 108$.

Thus, $y = 108 - 4x$ and $V = x^2(108 - 4x) = 108x^2 - 4x^3$

(b) Domain: $0 < x < 27$

(c)

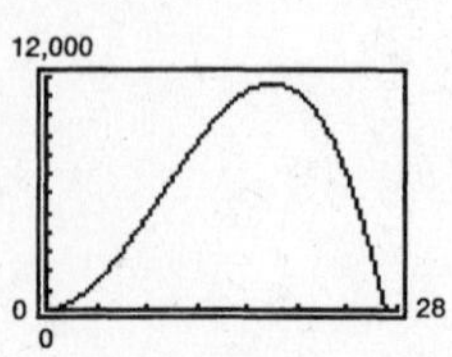

(d) The dimensions that will maximize the volume of the package are $18 \times 18 \times 36$. From the graph, the maximum volume occurs when $x = 18$. To find the dimension for y, use the equation $y = 108 - 4x$.

$y = 108 - 4x = 108 - 4(18) = 108 - 72 = 36.$

102. $p(t) = \begin{cases} -0.014t^2 + 2.35t + 23.7, & 2 \le t \le 6 \\ 0.193t^2 - 1.20t + 38.8, & 7 \le t \le 11 \end{cases}$

1993: $p(3) = 30.624$ thousand dollars

1997: $p(7) = 39.857$ thousand dollars

1999: $p(9) = 43.633$ thousand dollars

2001: $p(11) = 48.953$ thousand dollars

103. $C = 0.95x + 6000$

$$\overline{C} = \frac{C}{x} = \frac{0.95x + 6000}{x} = 0.95 + \frac{6000}{x}$$

104. (a) $R = n(\text{rate}) = n[8.00 - 0.05(n - 80)],\ n \ge 80$

$$R = 12.00n - 0.05n^2 = 12n - \frac{n^2}{20} = \frac{240n - n^2}{20},\ n \ge 80$$

(b)

n	90	100	110	120	130	140	150
$R(n)$	\$675	\$700	\$715	\$720	\$715	\$700	\$675

The revenue is maximum when 120 people take the trip.

105. $F(y) = 149.76\sqrt{10}y^{5/2}$

(a)

y	5	10	20	30	40
$F(y)$	26,474.08	149,760.00	847,170.49	2,334,527.36	4,792,320

The force, in tons, of the water against the dam increases with the depth of the water.

(b) It appears that approximately 21 feet of water would produce 1,000,000 tons of force. You can find a better estimate by creating a new table with y at smaller intervals between 20 and 25.

106. (a)

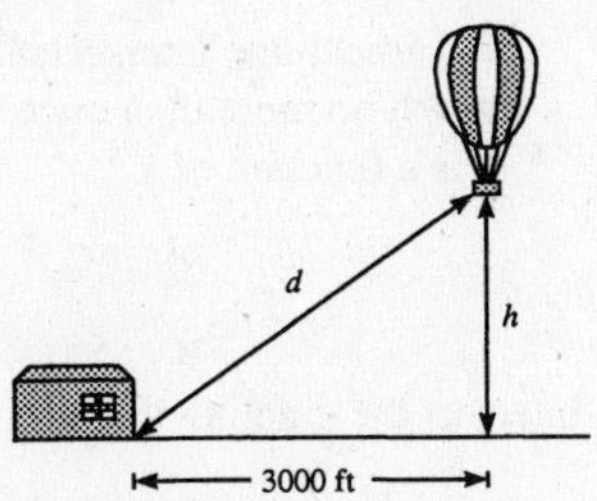

(b) $(3000)^2 + h^2 = d^2$

$$h = \sqrt{d^2 - (3000)^2}$$

Domain: $d \ge 3000$
(since both $d \ge 0$ and $d^2 - (3000)^2 \ge 0$)

107. Answers may vary slightly.

$$\frac{f(2001) - f(1993)}{2001 - 1993} \approx \frac{14{,}200 - 11{,}000}{8} = 400$$

The number of kidney transplants increased by approximately 400 transplants per year.

108. $y = -\frac{1}{10}x^2 + 3x + 6$

$$y(30) = -\frac{1}{10}(30)^2 + 3(30) + 6 = 6 \text{ feet}$$

If the child holds a glove at a height of 5 feet, then the ball *will* be over the child's head since it will be at a height of 6 feet.

109. False. The range is $[-1, \infty)$.

110. True, the set represents a function.
Each x-value corresponds to one y-value.

111. True.
As long as **all** elements in the domain are matched with elements in the range, even if it is the same element, then the relation is a function.

112. False.
Each element in the domain is matched with exactly **one** element in the range in a function.

Section 1.2 Analyzing Graphs of Functions

1. $f(x) = \frac{2}{3}x - 4$

Domain: All real numbers $(-\infty, \infty)$

Range: All real numbers $(-\infty, \infty)$

2. $f(x) = x^3 - 3x + 2$

Domain: All real numbers $(-\infty, \infty)$

Range: All real numbers $(-\infty, \infty)$

3. $f(x) = \sqrt{x^2 - 1}$

Domain: $(-\infty, -1] \cup [1, \infty)$

Range: $[0, \infty)$

4. $f(x) = \sqrt{x - 1}$

Domain: $x - 1 \geq 0$

$x \geq 1$

$[1, \infty)$

Range: $[0, \infty)$

5. $f(x) = 1 - x^2$

Domain: All real numbers

Range: $(-\infty, 1]$

6. $f(x) = \frac{1}{2}|x - 2|$

Domain: $(-\infty, \infty)$

Range: $[0, \infty)$

7. $h(x) = \sqrt{16 - x^2}$

Domain: $[-4, 4]$

Range: $[0, 4]$

8. $g(x) = \dfrac{|x - 1|}{x - 1}$

Domain: $x - 1 \neq 0$

$x \neq 1$

$(-\infty, 1), (1, \infty)$

Range: $\{-1, 1\}$

9. $f(x) = \dfrac{1}{x^2 + 1}$

Domain: All real numbers

Range: $(0, 1]$

10. $g(x) = \dfrac{x^2}{x^2 + 1}$

Domain: All real numbers

Range: $[0, 1)$

11. $y = \dfrac{1}{2}x^2$

A vertical line intersects the graph just once, so y is a function of x.

12. $y = \dfrac{1}{4}x^3$

A vertical line intersects the graph no more than once, so y is a function of x.

13. $x - y^2 = 1 \implies y = \pm\sqrt{x - 1}$

y is not a function of x.
Some vertical lines cross the graph twice.

14. $x^2 + y^2 = 25$

Some vertical lines intersect the graph twice, so y is not a function of x.

15. $x^2 = 2xy - 1$

A vertical line intersects the graph just once, so y is a function of x.

16. $x = |y + 2|$

Some vertical lines intersect the graph twice, so y is not a function of x.

17. $2x^2 - 7x - 30 = 0$

$(2x + 5)(x - 6) = 0$

$2x + 5 = 0$ or $x - 6 = 0$

$x = -\frac{5}{2}$ or $x = 6$

18. $f(x) = 3x^2 + 22x - 16$

$0 = (3x - 2)(x + 8)$

$3x - 2 = 0 \implies x = \frac{2}{3}$

$x + 8 = 0 \implies x = -8$

19. $f(x) = \dfrac{9x^2 - 4}{x}$

$$\frac{9x^2 - 4}{x} = 0$$
$$9x^2 - 4 = 0$$
$$x^2 = \frac{4}{9}$$
$$x = \pm\frac{2}{3}$$

20. $f(x) = \dfrac{x^2 - 9x + 14}{4x}$

$$0 = \frac{x^2 - 9x + 14}{4x}$$
$$0 = (x - 7)(x - 2)$$
$$x - 7 = 0 \Rightarrow x = 7$$
$$x - 2 = 0 \Rightarrow x = 2$$

21. $f(x) = \dfrac{1}{2}x^3 - 2x$

$$\frac{1}{2}x^3 - 2x = 0$$
$$x^3 - 4x = 0$$
$$x(x^2 - 4) = 0$$
$$x(x - 2)(x + 2) = 0$$
$$x = 0 \quad \text{or} \quad x = \pm 2$$

22. $f(x) = 9x^4 - 25x^2$

$$0 = 9x^4 - 25x^2$$
$$0 = x^2(9x^2 - 25)$$
$$x^2 = 0 \Rightarrow x = 0$$
$$9x^2 - 25 = 0 \Rightarrow x = \pm\tfrac{5}{3}$$

23. $f(x) = x^3 - 4x^2 - 9x + 36$

$$0 = x^3 - 4x^2 - 9x + 36$$
$$0 = x^2(x - 4) - 9(x - 4)$$
$$0 = (x - 4)(x^2 - 9)$$
$$x - 4 = 0 \Rightarrow x = 4$$
$$x^2 - 9 = 0 \Rightarrow x = \pm 3$$

24. $f(x) = 4x^3 - 24x^2 - x + 6$

$$4x^3 - 24x^2 - x + 6 = 0$$
$$4x^2(x - 6) - 1(x - 6) = 0$$
$$(x - 6)(4x^2 - 1) = 0$$
$$(x - 6)(2x + 1)(2x - 1) = 0$$
$$x - 6 = 0 \Rightarrow x = 6$$
$$2x + 1 = 0 \Rightarrow x = -\tfrac{1}{2}$$
$$2x - 1 = 0 \Rightarrow x = \tfrac{1}{2}$$

25. $3 + \dfrac{5}{x} = 0$

$$3x + 5 = 0$$
$$x = -\frac{5}{3}$$

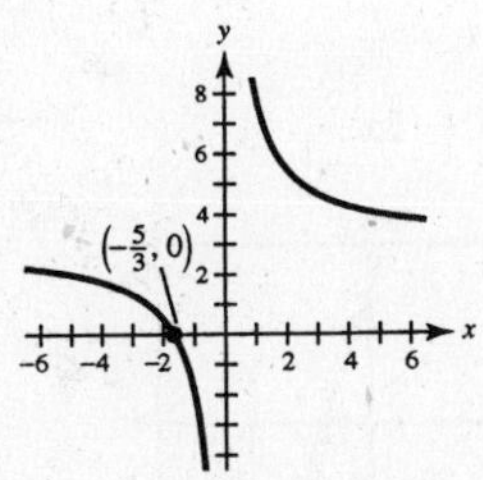

26. $2 - \dfrac{7}{x} = 0$

$$2x - 7 = 0$$
$$x = \frac{7}{2}$$

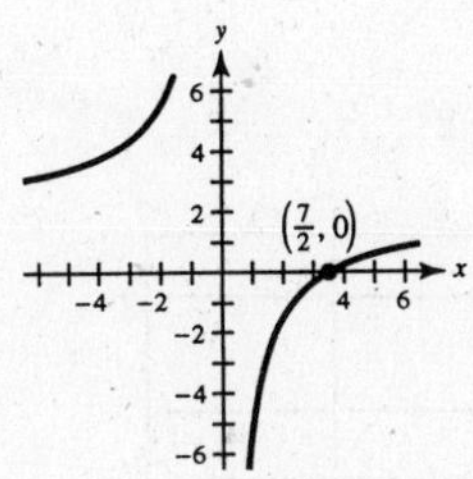

27. $\sqrt{2x + 11} = 0$

$$2x + 11 = 0$$
$$x = -\frac{11}{2}$$

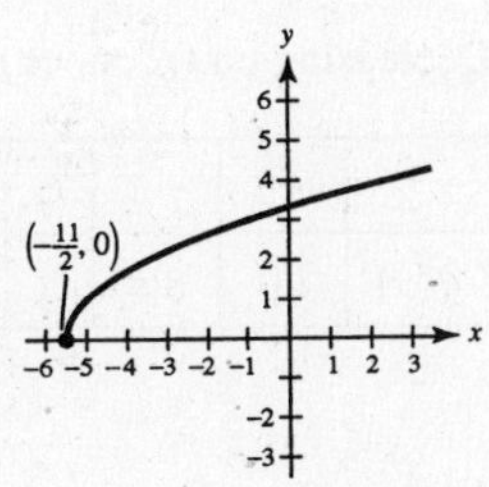

28. $0 = \sqrt{3x - 14} - 8$

$$8 = \sqrt{3x - 14}$$
$$64 = 3x - 14$$
$$x = 26$$

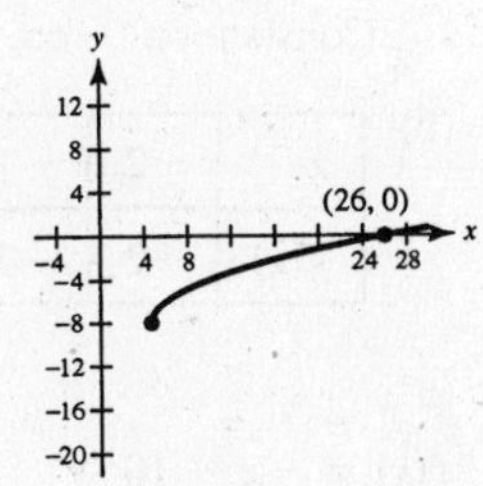

29. $\dfrac{3x^2 - 1}{x - 6} = 0$

$$3x^2 - 1 = 0$$
$$x^2 = \frac{1}{3}$$
$$x = \pm\frac{1}{\sqrt{3}}$$

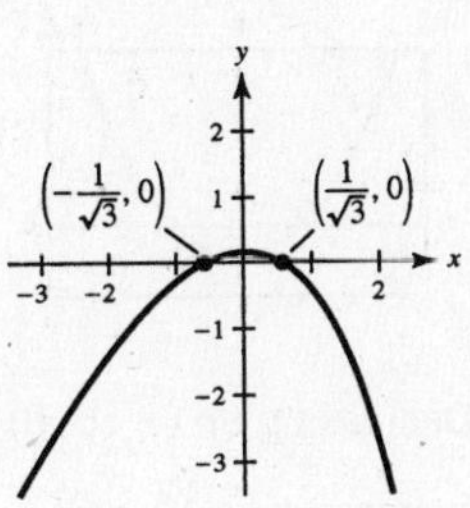

30. $0 = \dfrac{2x^2 - 9}{3 - x}$

$$2x^2 - 9 = 0 \Rightarrow x = \pm\frac{3\sqrt{2}}{2}$$

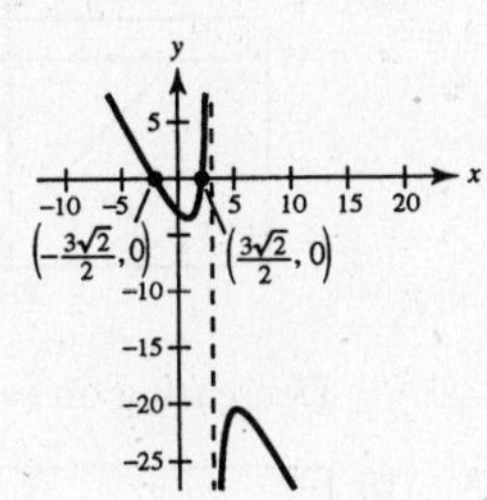

31. $f(x) = \frac{3}{2}x$

(a) f is increasing on $(-\infty, \infty)$.

(b) Since $f(-x) = -f(x)$, f is odd.

32. $f(x) = x^2 - 4x$

(a) The graph is decreasing on $(-\infty, 2)$ and increasing on $(2, \infty)$.

(b) $f(-x) = (-x)^2 - 4(-x) = x^2 + 4x$

$x^2 + 4x \neq f(x)$

$x^2 + 4x \neq -f(x)$

The function is neither odd nor even.

33. $f(x) = x^3 - 3x^2 + 2$

(a) f is increasing on $(-\infty, 0)$ and $(2, \infty)$.

f is decreasing on $(0, 2)$.

(b) $f(-x) = (-x)^3 - 3(-x)^2 + 2$

$= -x^3 - 3x^2 + 2$

$f(-x) \neq -f(x)$

$f(-x) \neq f(x)$

f is neither odd nor even.

34. $f(x) = \sqrt{x^2 - 1}$

(a) The graph is decreasing on $(-\infty, -1)$ and increasing on $(1, \infty)$.

(b) $f(-x) = \sqrt{(-x)^2 - 1} = \sqrt{x^2 - 1} = f(x)$

The function is even.

35. $f(x) = 3$

(a)

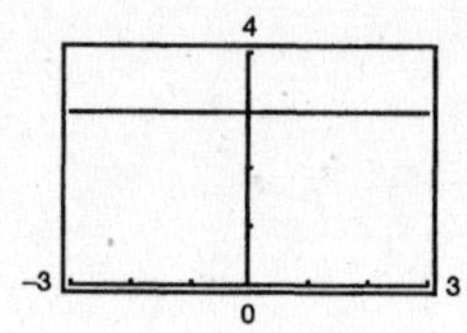

Constant on $(-\infty, \infty)$

(b)

x	-2	-1	0	1	2
$f(x)$	3	3	3	3	3

36. $g(x) = -2$

(a)

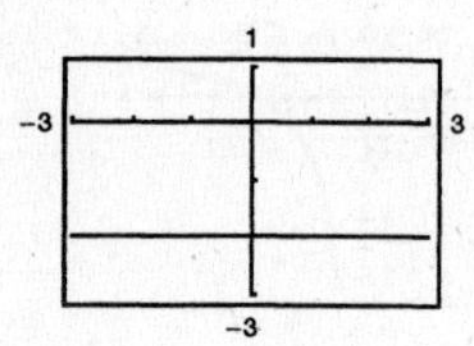

Constant on $(-\infty, \infty)$

(b)

x	-2	-1	0	1	2
$g(x)$	-2	-2	-2	-2	-2

37. $f(x) = 5 - 3x$

(a)

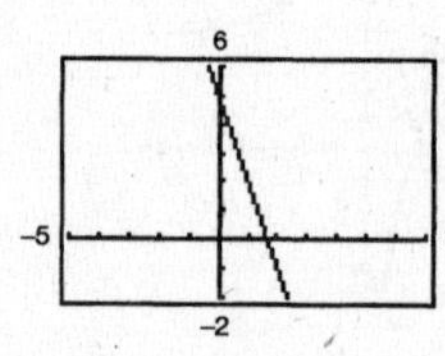

Decreasing on $(-\infty, \infty)$

(b)

x	-2	-1	0	1	2
$f(x)$	11	8	5	2	-1

38. $f(x) = -x - 10$

(a)

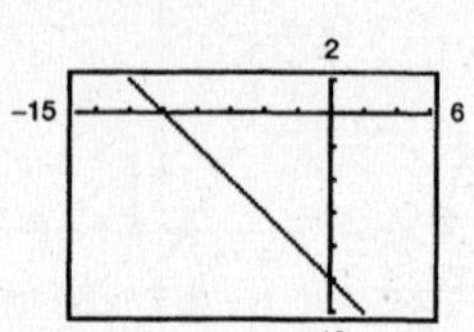

Decreasing on $(-\infty, \infty)$

(b)

x	-12	-8	-4	0	4
$f(x)$	2	-2	-6	-10	-14

39. $g(s) = \dfrac{s^2}{4} - 1$

(a)

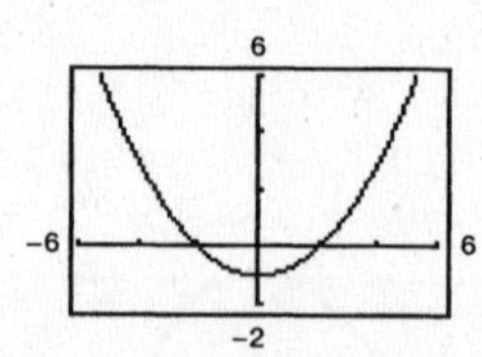

Decreasing on $(-\infty, 0)$

Increasing on $(0, \infty)$

(b)

s	-4	-2	0	2	4
$g(s)$	3	0	-1	0	3

40. $h(x) = x^2 - 4$

(a)

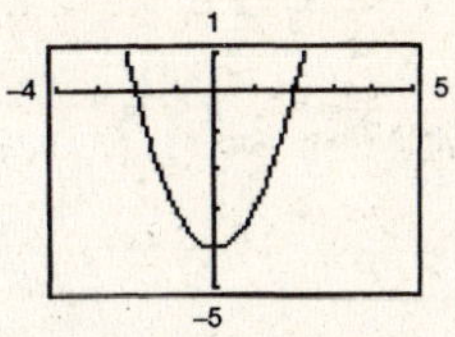

Decreasing on $(-\infty, 0)$

Increasing on $(0, \infty)$

(b)

x	-2	-1	0	1	2
$h(x)$	0	-3	-4	-3	0

41. $f(t) = -t^4 + 2t^2$

(a)

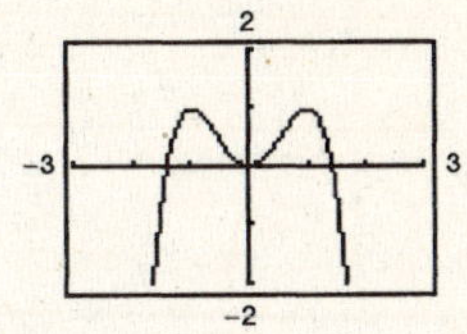

Increasing on $(-\infty, -1)$ and $(0, 1)$

Decreasing on $(-1, 0)$ and $(1, \infty)$

(b)

t	-2	-1	0	1	2
$f(t)$	-8	1	0	1	-8

42. $f(x) = 3x^4 - 6x^2$

(a)

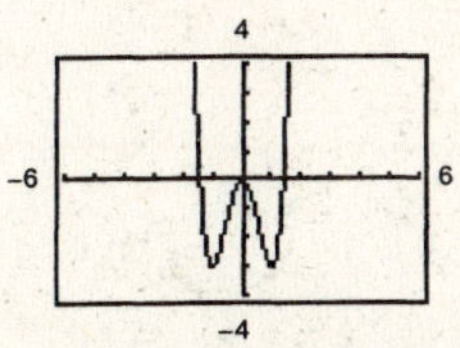

Increasing on $(-1, 0), (1, \infty)$

Decreasing on $(-\infty, -1), (0, 1)$

(b)

x	-2	-1	0	1	2
$f(x)$	24	-3	0	-3	24

43. $f(x) = x\sqrt{1 - x}$

(a)

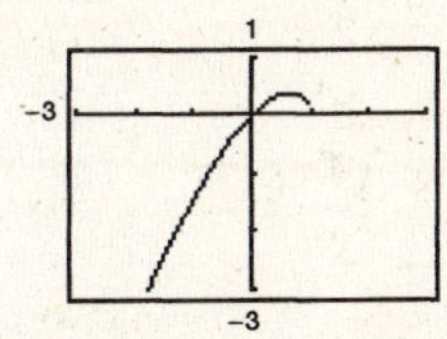

Increasing on $\left(-\infty, \frac{2}{3}\right)$

Decreasing on $\left(\frac{2}{3}, 1\right)$

(b)

x	-3	-2	-1	0	1
$f(x)$	-6	$-2\sqrt{3}$	$-\sqrt{2}$	0	0

44. $f(x) = x\sqrt{x + 3}$

(a)

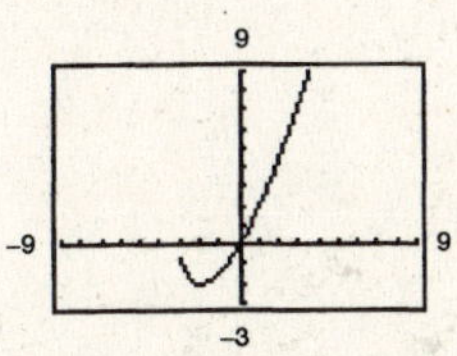

Increasing on $(-2, \infty)$

Decreasing on $(-3, -2)$

(b)

x	-3	-2	-1	0	1
$f(x)$	0	-2	$-\sqrt{2}$	0	2

45. $f(x) = x^{3/2} - 1$

(a)

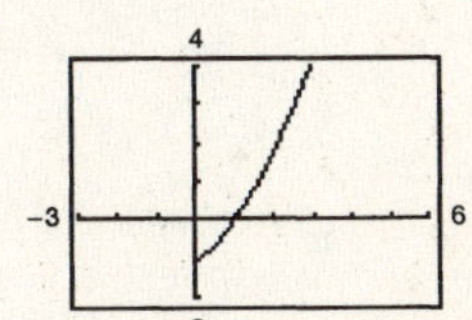

Increasing on $(0, \infty)$

(b)

x	0	1	2	3	4
$f(x)$	-1	0	1.83	4.2	7

46. $f(x) = x^{2/3} - 1$

(a)

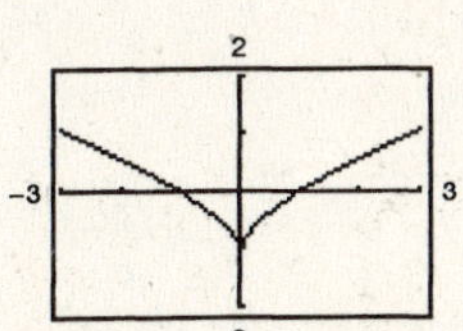

Increasing on $(0, \infty)$

Decreasing on $(-\infty, 0)$

(b)

x	-8	-1	0	1	8
$f(x)$	3	0	-1	0	3

47. $g(t) = \sqrt[3]{t-1}$

(a)

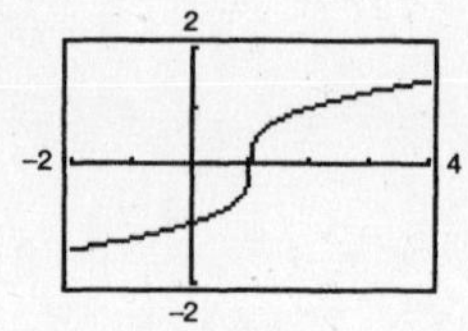

Increasing on $(-\infty, \infty)$

(b)

t	-2	-1	0	1	2
$g(t)$	-1.44	-1.26	-1	0	1

48. $f(x) = \sqrt[3]{x+5}$

(a)

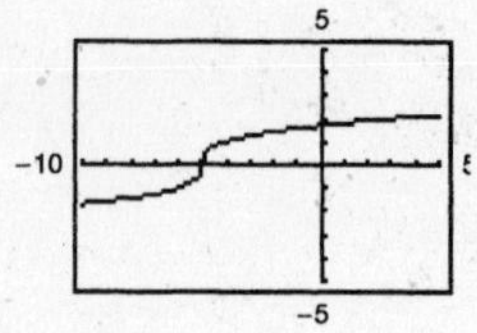

Increasing on $(-\infty, \infty)$

(b)

x	-13	-6	-5	-4	3
$f(x)$	-2	-1	0	1	2

49. $f(x) = |x+2| - |x-2|$

(a)

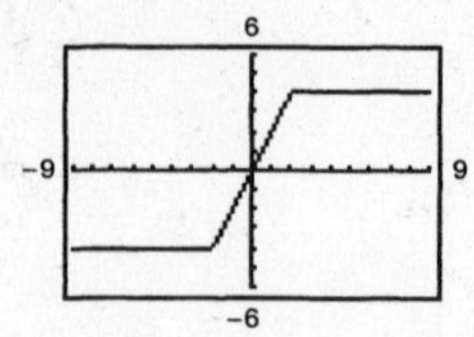

Increasing on $(-2, 2)$

Constant on $(-\infty, -2)$ and $(2, \infty)$

(b)

x	-2	-1	0	1	4
$f(x)$	-4	-2	0	2	4

50. $f(x) = |x+1| + |x-1|$

(a)

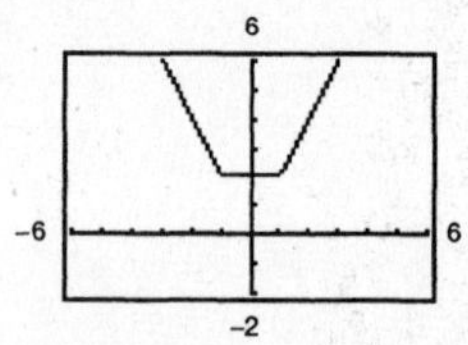

Decreasing on $(-\infty, -1)$

Constant on $(-1, 1)$

Increasing on $(1, \infty)$

(b)

x	-3	-2	-1	0	1	2	3
$f(x)$	6	4	2	2	2	4	6

51. $f(x) = \begin{cases} x+3, & x \le 0 \\ 3, & 0 < x \le 2 \\ 2x-1, & x > 2 \end{cases}$

(a)

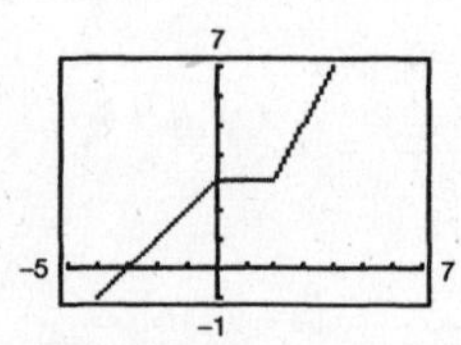

Increasing on $(-\infty, 0)$ and $(2, \infty)$

Constant on $(0, 2)$

(b)

x	-2	-1	0	1	2	3	4
$f(x)$	1	2	3	3	3	5	7

52. $f(x) = \begin{cases} 2x+1, & x \le -1 \\ x^2-2, & x > -1 \end{cases}$

(a)

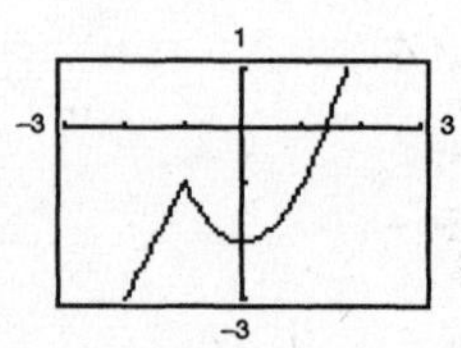

Increasing on $(-\infty, -1)$ and $(0, \infty)$

Decreasing on $(-1, 0)$

(b)

x	-2	-1	$-\frac{1}{2}$	0	1	2
$f(x)$	-3	-1	-1.75	-2	-1	2

53. $f(x) = (x-4)(x+2)$

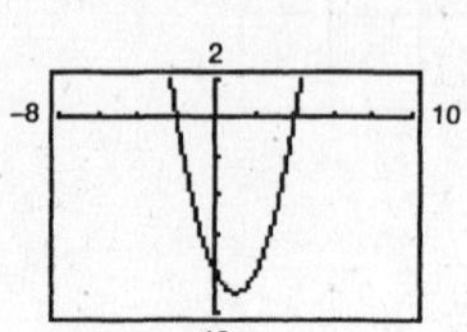

Relative Minimum at $(1, -9)$

54. $f(x) = 3x^2 - 2x - 5$

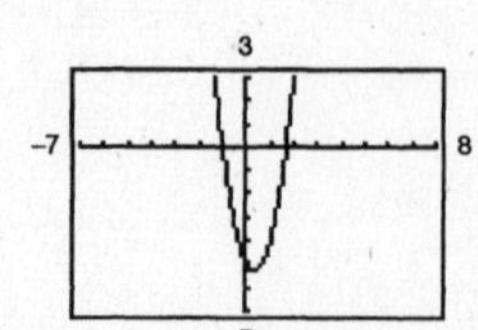

Relative minimum: $\left(\frac{1}{3}, -\frac{16}{3}\right)$

55. $f(x) = x(x - 2)(x + 3)$

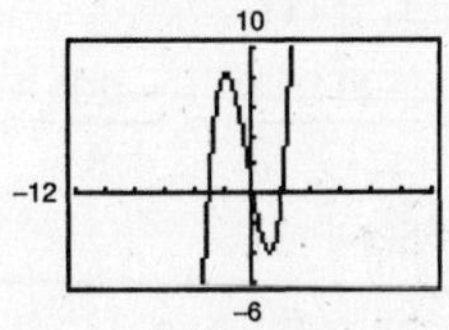

Relative minimum at $(1.12, -4.06)$

Relative maximum at $(-1.79, 8.21)$

56. $f(x) = x^3 - 3x^2 - x + 1$

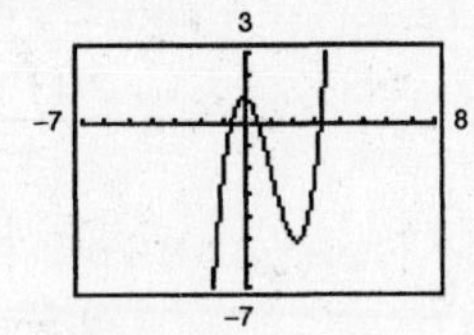

Relative maximum: $(-0.15, 1.08)$

Relative minimum: $(2.15, -5.08)$

57. $f(x) = 2x^3 - 5x^2 - 4x - 1$

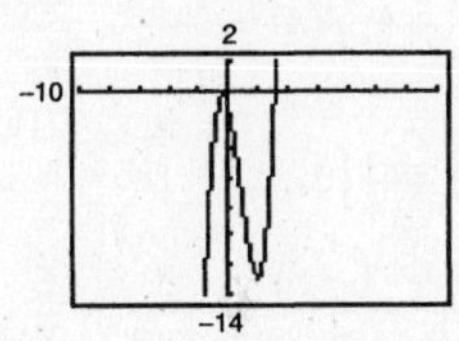

Relative minimum at $(2, -13)$

Relative maximum at $(-0.33, -0.30)$

58. $f(x) = 8x^4 - 3x - 1$

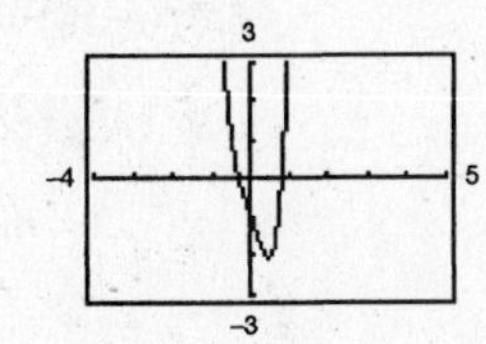

Relative minimum: $(0.45, -2.02)$

59. $f(x) = 1 - 2x$

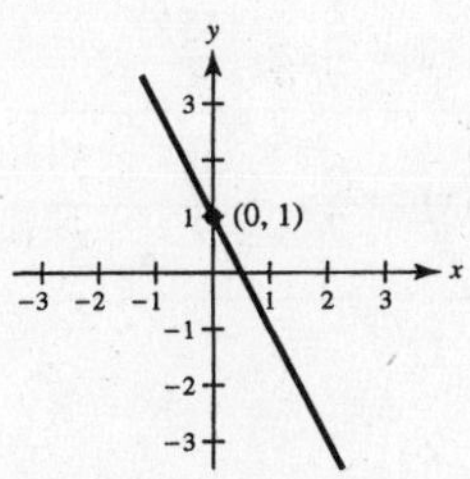

60. $f(x) = 3x - 11$

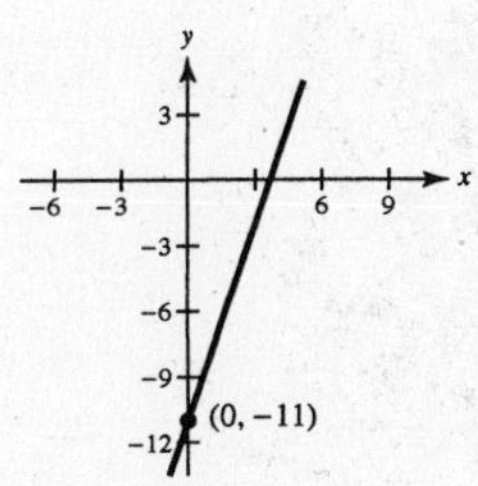

61. $f(x) = -x - \frac{3}{4}$

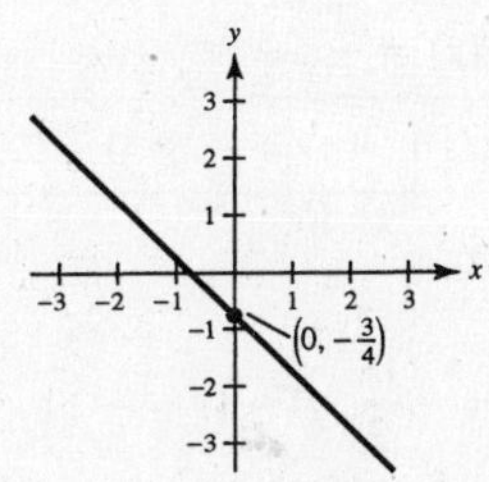

62. $f(x) = 3x - \frac{5}{2}$

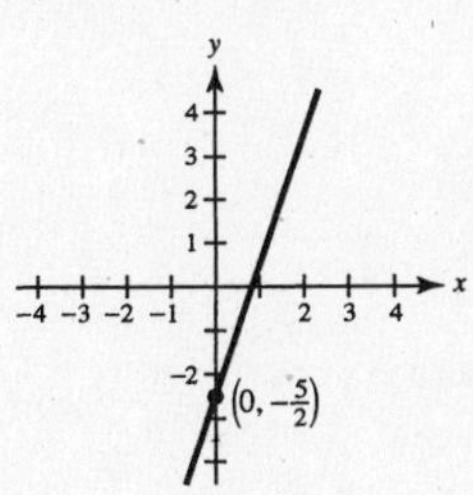

63. $f(x) = -\frac{1}{6}x - \frac{5}{2}$

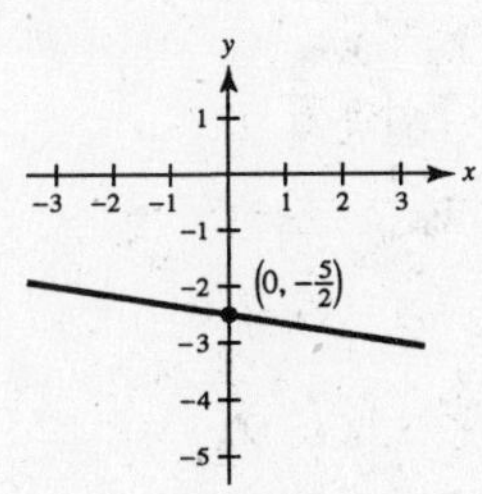

64. $f(x) = \frac{5}{6} - \frac{2}{3}x$

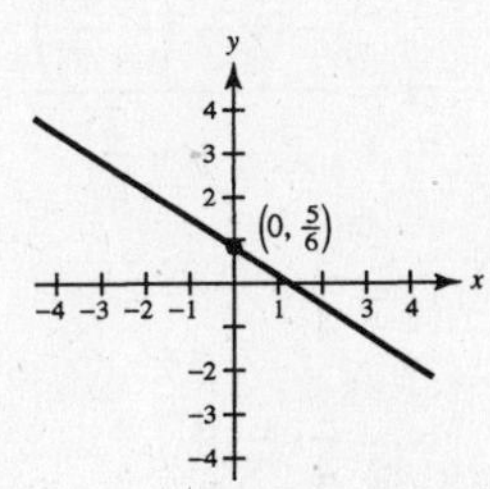

65. $f(x) = -1.8 + 2.5x$

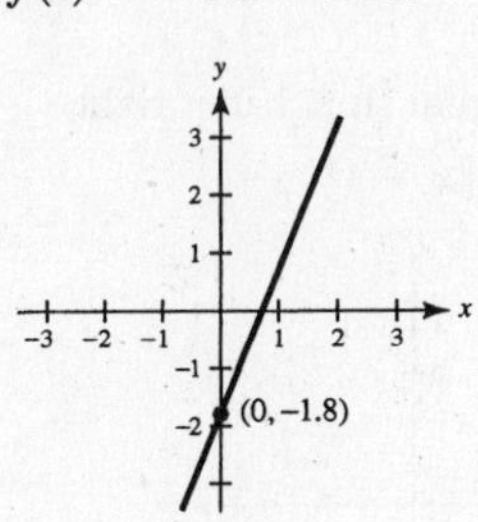

66. $f(x) = 10.2 + 3.1x$

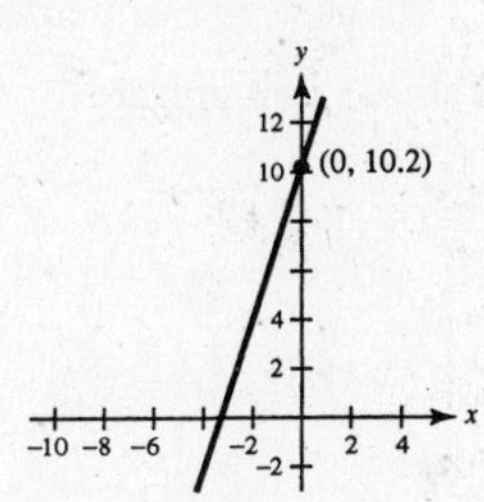

67. $f(1) = 4,\ f(0) = 6$

$(1, 4)$ and $(0, 6)$

$$m = \frac{6 - 4}{0 - 1} = -2$$

$$y - 6 = -2(x - 0)$$

$$y = -2x + 6$$

$$f(x) = -2x + 6$$

68. $f(-3) = -8, f(1) = 2$

$(-3, -8), (1, 2)$

$$m = \frac{2 - (-8)}{1 - (-3)} = \frac{10}{4} = \frac{5}{2}$$

$$y - 2 = \frac{5}{2}(x - 1)$$

$$y = \frac{5}{2}x - \frac{1}{2}$$

$$f(x) = \frac{5}{2}x - \frac{1}{2}$$

69. $f(5) = -4, f(-2) = 17$

$(5, -4)$ and $(-2, 17)$

$$m = \frac{17 - (-4)}{-2 - 5} = \frac{21}{-7} = -3$$

$$y - (-4) = -3(x - 5)$$

$$y + 4 = -3x + 15$$

$$y = -3x + 11$$

$$f(x) = -3x + 11$$

70. $f(3) = 9, f(-1) = -11$

$(3, 9), (-1, -11)$

$$m = \frac{-11 - 9}{-1 - 3} = \frac{-20}{-4} = 5$$

$$y - 9 = 5(x - 3)$$

$$y = 5x - 6$$

$$f(x) = 5x - 6$$

71. $f(-5) = -5, f(5) = -1$

$(-5, -5)$ and $(5, -1)$

$$m = \frac{-1 - (-5)}{5 - (-5)} = \frac{4}{10} = \frac{2}{5}$$

$$y - (-5) = \frac{2}{5}(x - (-5))$$

$$y + 5 = \frac{2}{5}x + 2$$

$$y = \frac{2}{5}x - 3$$

$$f(x) = \frac{2}{5}x - 3$$

72. $f(-10) = 12, f(16) = -1$

$(-10, 12), (16, -1)$

$$m = \frac{-1 - 12}{16 - (-10)} = \frac{-13}{26}$$

$$= -\frac{1}{2}$$

$$y - (-1) = -\frac{1}{2}(x - 16)$$

$$y = -\frac{1}{2}x + 7$$

$$f(x) = -\frac{1}{2}x + 7$$

73. $f\left(\frac{1}{2}\right) = -6, f(4) = -3$

$\left(\frac{1}{2}, -6\right)$ and $(4, -3)$

$$m = \frac{-3 - (-6)}{4 - \frac{1}{2}} = \frac{3}{7/2} = \frac{6}{7}$$

$$y - (-3) = \frac{6}{7}(x - 4)$$

$$y + 3 = \frac{6}{7}x - \frac{24}{7}$$

$$y = \frac{6}{7}x - \frac{45}{7}$$

$$f(x) = \frac{6}{7}x - \frac{45}{7}$$

74. $f\left(\frac{2}{3}\right) = -\frac{15}{2}, f(-4) = -11$

$\left(\frac{2}{3}, -\frac{15}{2}\right), (-4, -11)$

$$m = \frac{-11 - \left(-\frac{15}{2}\right)}{-4 - \frac{2}{3}} = \frac{-\frac{7}{2}}{-\frac{14}{3}} = \left(-\frac{7}{2}\right) \cdot \left(-\frac{3}{14}\right) = \frac{3}{4}$$

$$y - (-11) = \frac{3}{4}(x - (-4))$$

$$y = \frac{3}{4}x - 8$$

$$f(x) = \frac{3}{4}x - 8$$

75. Vertical shift 2 units downward.

$f(x) = [\![x]\!] - 2$

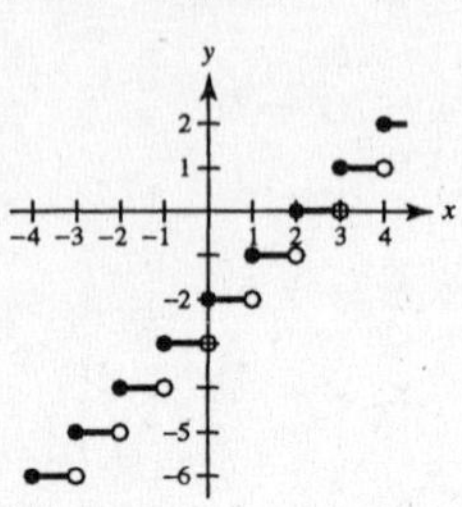

76. Vertical shift 1 unit upward.

$f(x) = [\![x]\!] + 1$

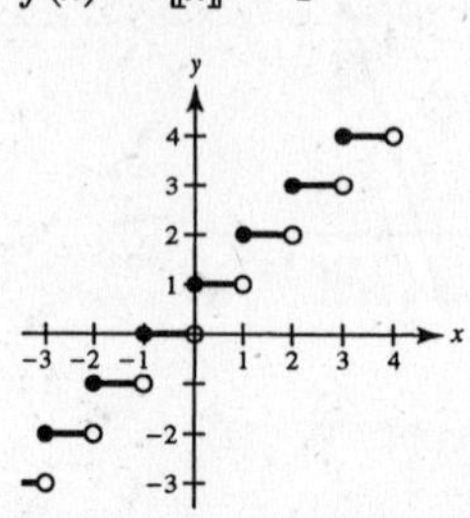

77. Horizontal shift 1 unit right.

$f(x) = [\![x - 1]\!]$

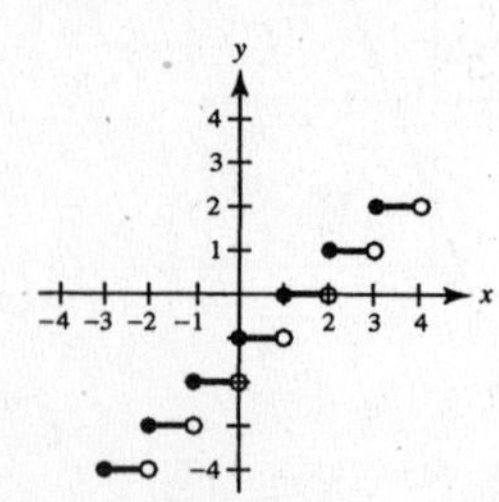

78. Horizontal shift 2 units left.

$f(x) = [\![x + 2]\!]$

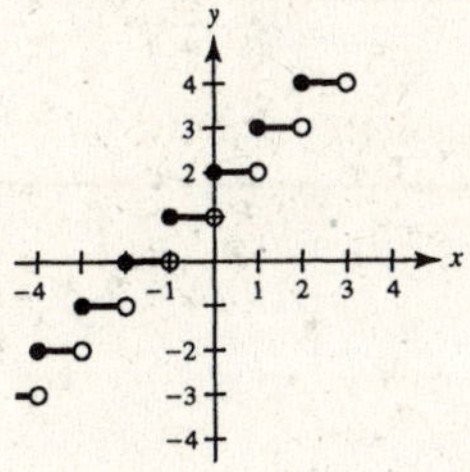

79. $f(x) = \begin{cases} 2x + 3, & x < 0 \\ 3 - x, & x \geq 0 \end{cases}$

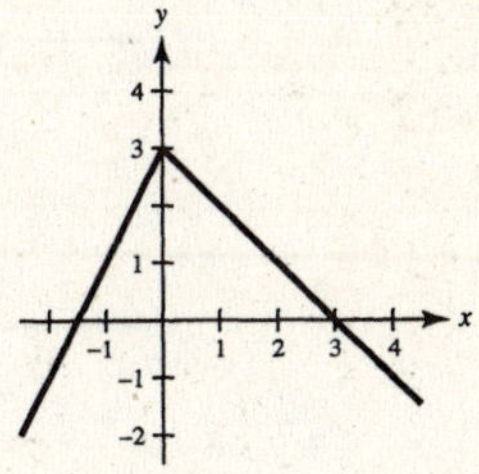

80. $f(x) = \begin{cases} \sqrt{4 + x}, & x < 0 \\ \sqrt{4 - x}, & x \geq 0 \end{cases}$

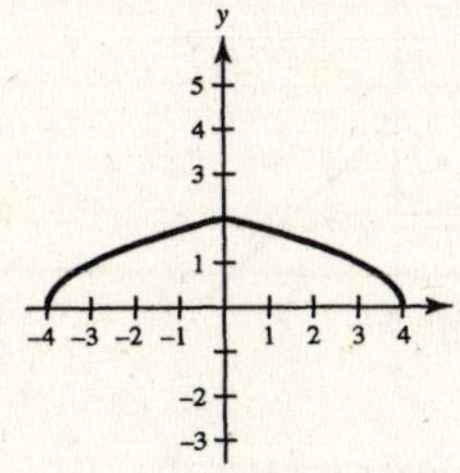

81. $f(x) = \begin{cases} x^2 + 5, & x \leq 1 \\ -x^2 + 4x + 3, & x > 1 \end{cases}$

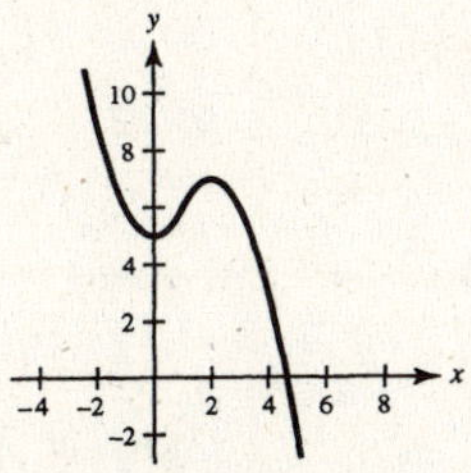

82. $f(x) = \begin{cases} 1 - (x - 1)^2, & x \leq 2 \\ \sqrt{x - 2}, & x > 2 \end{cases}$

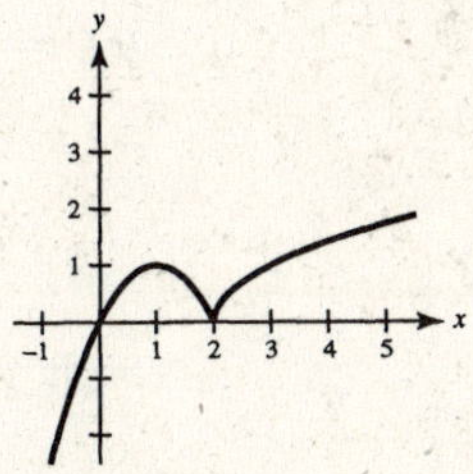

83. $f(x) = 4 - x$

$f(x) \geq 0$ on $(-\infty, 4]$.

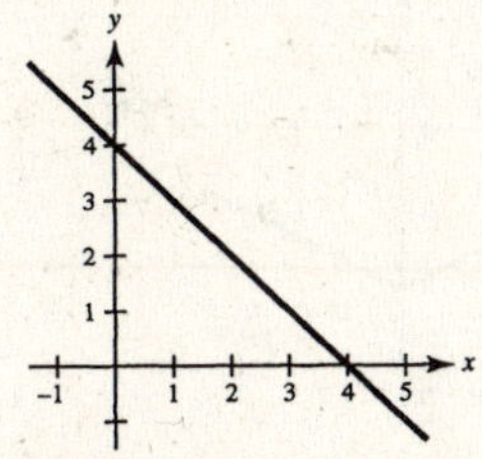

84. $f(x) = 4x + 2$

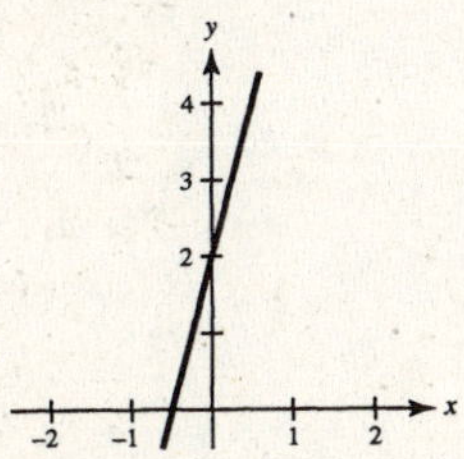

$$f(x) \geq 0$$
$$4x + 2 \geq 0$$
$$4x \geq -2$$
$$x \geq -\tfrac{1}{2}$$
$$\left[-\tfrac{1}{2}, \infty\right)$$

85. $f(x) = x^2 - 9$

$f(x) \geq 0$ on $(-\infty, -3]$ and $[3, \infty)$.

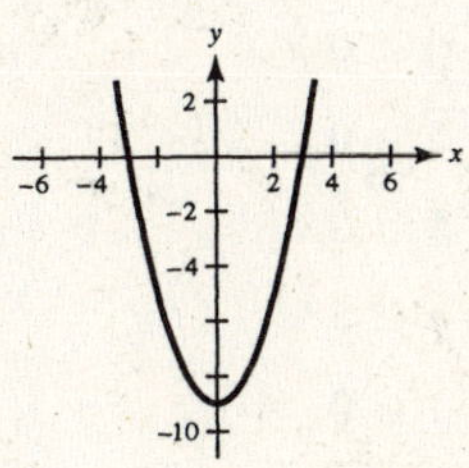

86. $f(x) = x^2 - 4x$

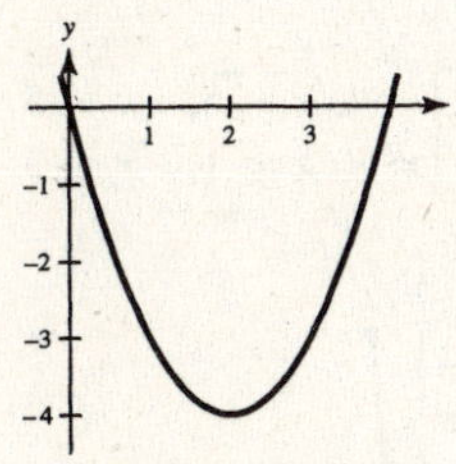

$$f(x) \geq 0$$
$$x^2 - 4x \geq 0$$
$$x(x - 4) \geq 0$$
$$(-\infty, 0], [4, \infty)$$

87. $f(x) = 1 - x^4$

$f(x) \geq 0$ on $[-1, 1]$.

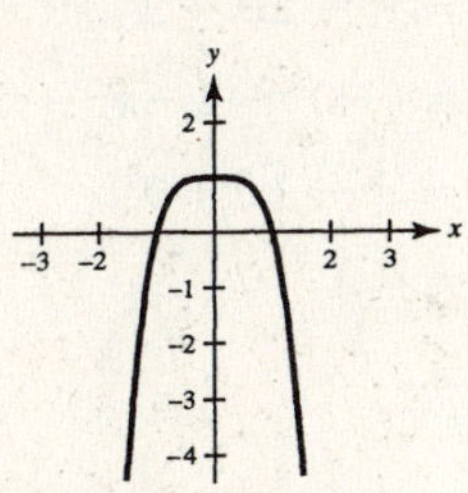

88. $f(x) = \sqrt{x + 2}$

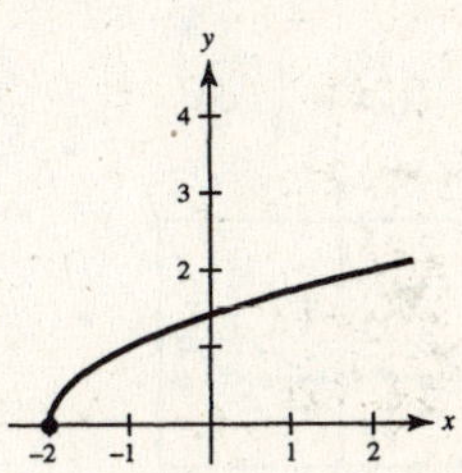

$$f(x) \geq 0$$
$$\sqrt{x + 2} \geq 0$$
$$x + 2 \geq 0$$
$$x \geq -2$$
$$[-2, \infty)$$

89. $f(x) = x^2 + 1$

$f(x) \geq 0$ on $(-\infty, \infty)$.

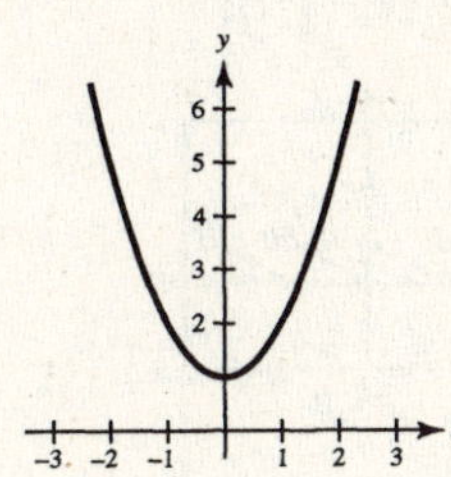

90. $f(x) = -(1 + |x|)$

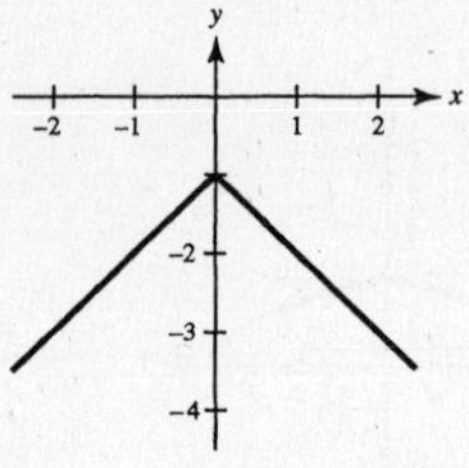

$f(x)$ is never greater than 0.
($f(x) < 0$ for all x.)

91. $f(x) = -5$, $f(x) < 0$ for all x.

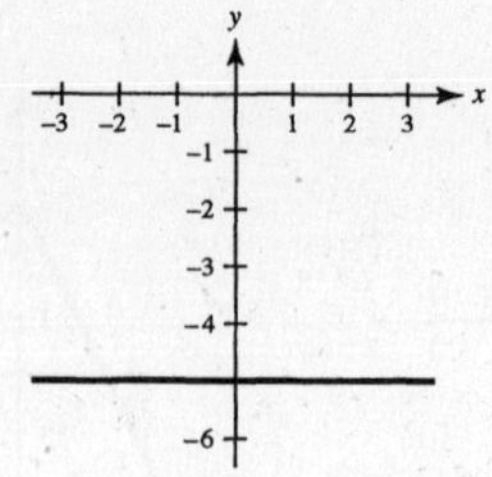

92. $f(x) = \frac{1}{2}(2 + |x|)$

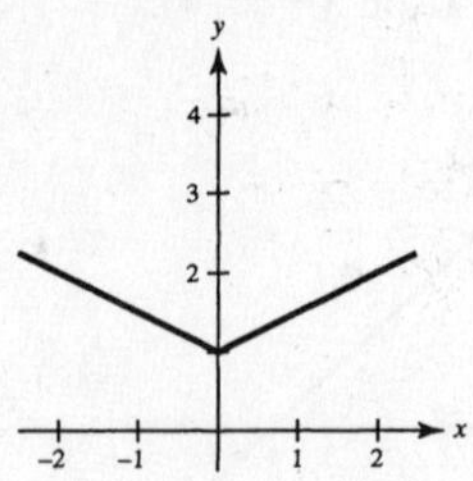

$f(x)$ is always greater than 0.
$(-\infty, \infty)$

93. $f(x) = \begin{cases} 1 - 2x^2, & x \le -2 \\ -x + 8, & x > -2 \end{cases}$

$f(x) \ge 0$ on $(-2, 8]$

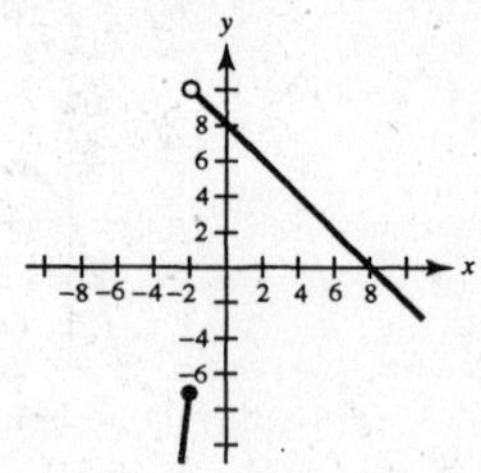

94. $f(x) = \begin{cases} \sqrt{x - 5}, & x > 5 \\ x^2 + x - 1, & x \le 5 \end{cases}$

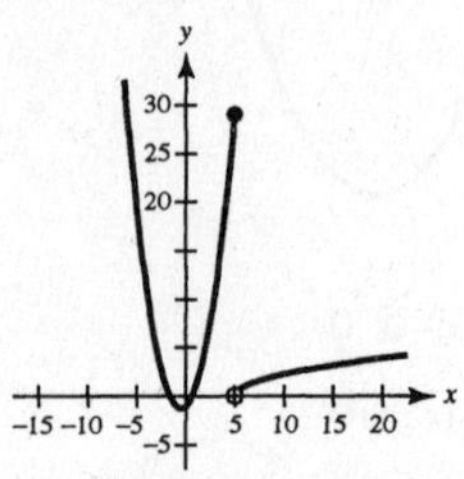

$$f(x) \ge 0$$
$$\sqrt{x - 5} \ge 0$$
$$x - 5 \ge 0$$
$$x \ge 5$$
$$x^2 + x - 1 \ge 0$$
$$x = \frac{-1 \pm \sqrt{1^2 - 4(1)(-1)}}{2(1)}$$
$$= \frac{-1 \pm \sqrt{5}}{2}$$
$$\left(-\infty, \frac{-1 - \sqrt{5}}{2}\right], \left[\frac{-1 + \sqrt{5}}{2}, \infty\right)$$

95. $s(x) = 2\left(\frac{1}{4}x - [\![\frac{1}{4}x]\!]\right)$

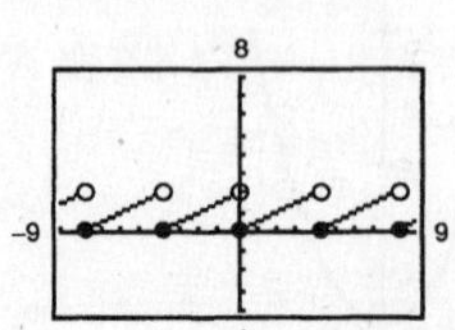

Domain: $(-\infty, \infty)$

Range: $[0, 2)$

Sawtooth pattern

96. $g(x) = 2\left(\frac{1}{4}x - [\![\frac{1}{4}x]\!]\right)^2$

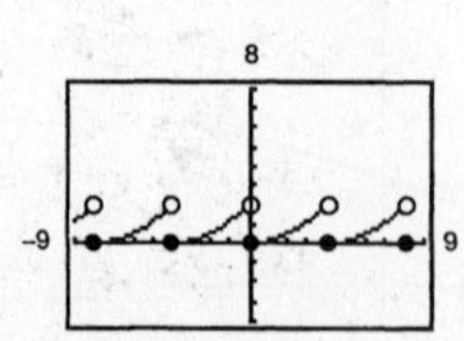

Domain: $(-\infty, \infty)$

Range: $[0, 2)$

Pattern: sawtooth

97. $f(x) = x^6 - 2x^2 + 3$

$$f(-x) = (-x)^6 - 2(-x)^2 + 3$$
$$= x^6 - 2x^2 + 3$$
$$= f(x)$$

f is even.

98. $h(x) = x^3 - 5x$

$h(-x) = (-x)^3 - 5(-x)$

$= -x^3 + 5x$

$= -(x^3 - 5x)$

$= -h(x)$

h is odd.

99. $g(x) = x^3 + 5x - 2$

$g(-x) = (-x)^3 + 5(-x) - 2$

$= -x^3 - 5x - 2$

$= -(x^3 + 5x + 2)$

$\neq g(x), \neq -g(x)$

g is neither even nor odd.

100. $f(x) = x\sqrt{1 - x^2}$

$f(-x) = -x\sqrt{1 - (-x)^2}$

$= -x\sqrt{1 - x^2}$

$= -f(x)$

The function is odd.

101. $f(t) = 2t^{3/2} = 2\sqrt{t^3}$ or $2(\sqrt{t})^3$

Domain: $t \geq 0$

$f(-t) = 2\sqrt{(-t)^3} = 2\sqrt{-t^3}, t \leq 0$

Note: $2\sqrt{-t^3} \neq -2\sqrt{t^3}$ thus $f(-t) \neq -f(t), \neq f(t)$

f is neither even nor odd

102. $g(s) = 4s^{2/3}$

$g(-s) = 4(-s)^{2/3}$

$= 4s^{2/3}$

$= g(s)$

The function is even.

103. $\left(-\frac{3}{2}, 4\right)$

(a) If f is even, another point is $\left(\frac{3}{2}, 4\right)$.

(b) If f is odd, another point is $\left(\frac{3}{2}, -4\right)$.

104. $\left(-\frac{5}{3}, -7\right)$

(a) If f is even, another point is $\left(\frac{5}{3}, -7\right)$.

(b) If f is odd, another point is $\left(\frac{5}{3}, 7\right)$.

105. $(4, 9)$

(a) If f is even, another point is $(-4, 9)$.

(b) If f is odd, another point is $(-4, -9)$.

106. $(5, -1)$

(a) If f is even, another point is $(-5, -1)$.

(b) If f is odd, another point is $(-5, 1)$.

107.

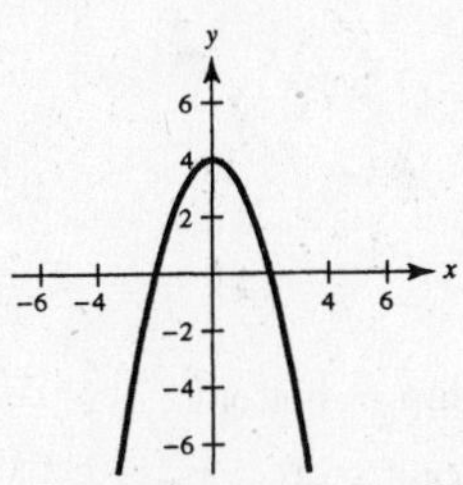

(a) Domain: All real numbers or $(-\infty, \infty)$

(b) Range: $(-\infty, 4]$

(c) Increasing on $(-\infty, 0)$
Decreasing on $(0, \infty)$

108.

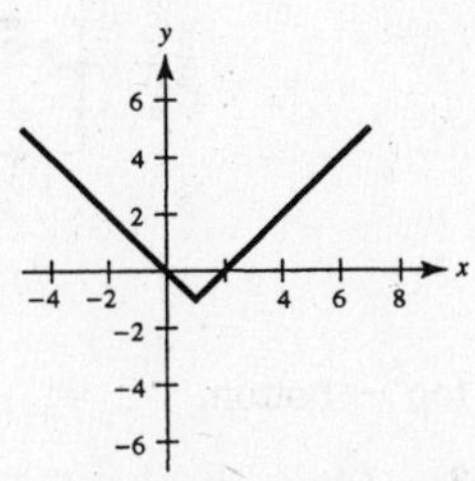

(a) Domain: All real numbers or $(-\infty, \infty)$

(b) Range: $[-1, \infty)$

(c) Increasing on $(1, \infty)$
Decreasing on $(-\infty, 1)$

109.

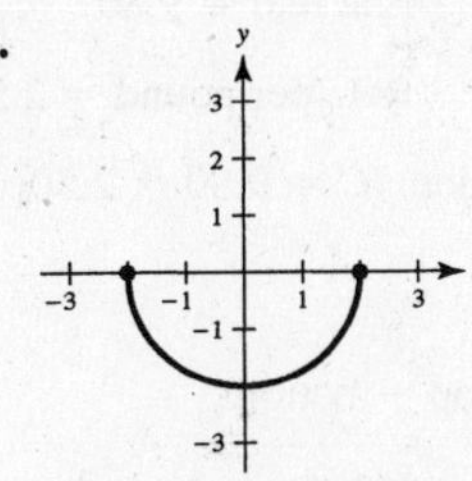

(a) Domain: $[-2, 2]$

(b) Range: $[-2, 0]$

(c) Increasing on $(0, 2)$
Decreasing on $(-2, 0)$

110.

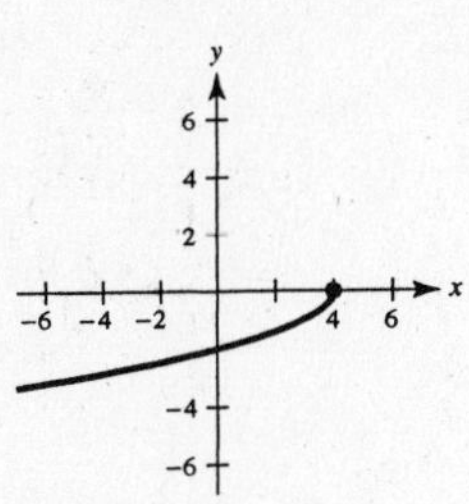

(a) Domain: $(-\infty, 4]$

(b) Range: $(-\infty, 0]$

(c) Increasing on $(-\infty, 4)$

111.

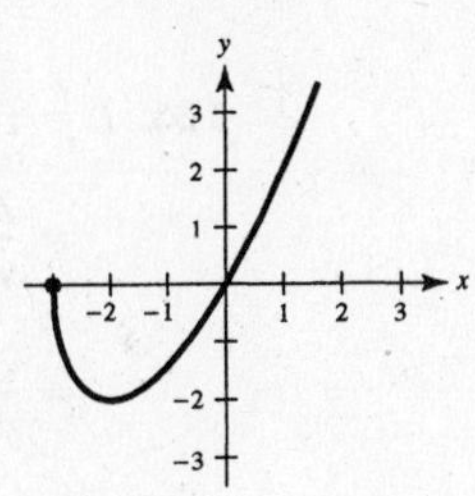

(a) Domain: $[-3, \infty)$

(b) Range: $[-2, \infty)$

(c) Increasing on $(-2, \infty)$
Decreasing on $(-3, -2)$

112.

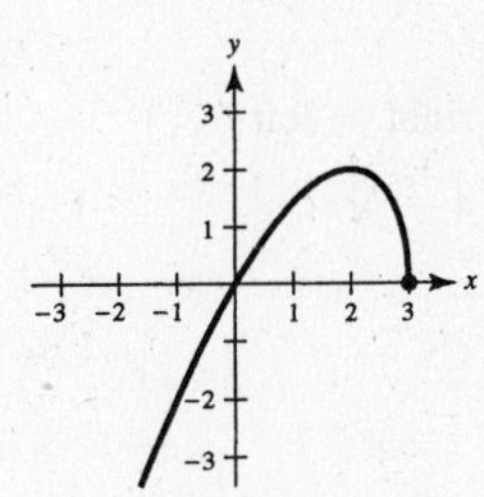

(a) Domain: $(-\infty, 3]$

(b) Range: $(-\infty, 2]$

(c) Increasing on $(-\infty, 2)$
Decreasing on $(2, 3)$

113.

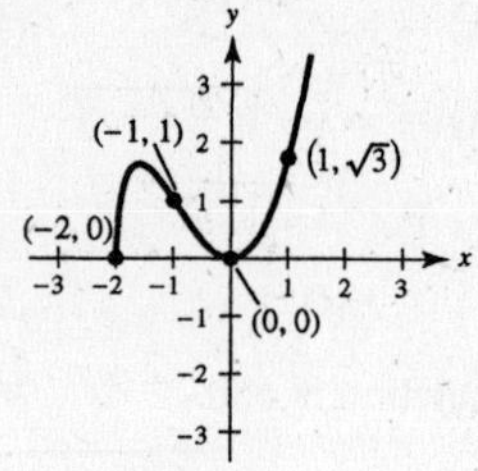

(a) $f(-1) = 1$

(b) $f(1) = \sqrt{3}$

(c) f is increasing on $(-2, -1.6)$ and $(0, \infty)$
f is decreasing on $(-1.6, 0)$.

114.

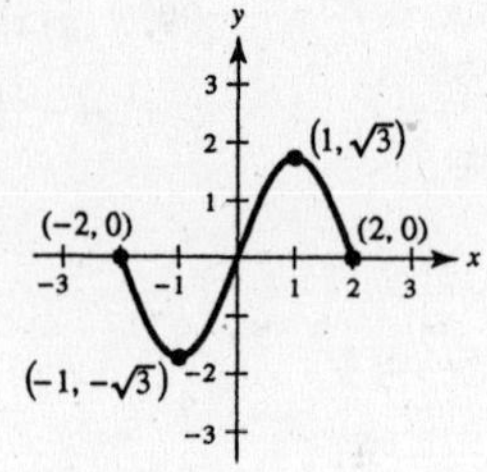

(a) $g(-1) = -\sqrt{3}$

(b) $g(1) = \sqrt{3}$

(c) g is increasing on $(-1, 1)$.
g is decreasing on $(-2, -1)$ and $(1, 2)$.

115. (a) $C_2(t) = 1.05 - 0.38[\![-(t-1)]\!]$ is the appropriate model since the cost does not increase until after the next minute of conversation has started.

(b)

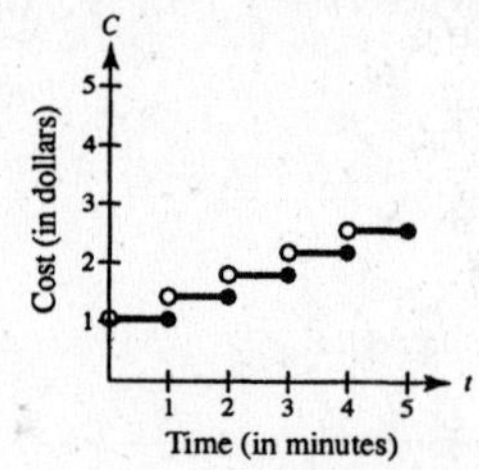

$C = 1.05 - 0.38[\![-17.75]\!] = \$7.89$

116. Model: (Total cost) = (Flat rate) + (Rate per pound)

Labels: Total cost $= C$

Flat rate $= 9.80$

Rate per pound $= 2.50[\![x]\!],\ x > 0$

Equation: $C = 9.80 + 2.50[\![x]\!],\ x > 0$

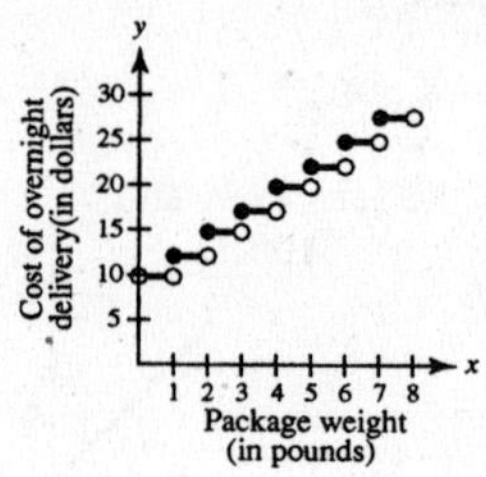

117. $h =$ top $-$ bottom

$= (-x^2 + 4x - 1) - 2$

$= -x^2 + 4x - 3$

118. $h =$ top $-$ bottom

$= 3 - (4x - x^2)$

$= 3 - 4x + x^2$

119. $h =$ top $-$ bottom

$= (4x - x^2) - 2x$

$= 2x - x^2$

120. $h =$ top $-$ bottom

$= 2 - \sqrt[3]{x}$

121. $L =$ right $-$ left

$= \frac{1}{2}y^2 - 0$

$= \frac{1}{2}y^2$

122. $L =$ right $-$ left

$= 2 - \sqrt[3]{2y}$

123. $L =$ right $-$ left

$= 4 - y^2$

124. $L =$ right $-$ left

$= \frac{2}{y} - 0$

$= \frac{2}{y}$

125.

Year	Doctors, y
1992	33.5
1993	33.4
1994	35.0
1995	35.7
1996	37.3
1997	38.9
1998	40.8
1999	43.5
2000	44.9
2001	47.0

(a) Let $t = 2$ corresponds to 1992.
$y \approx -0.01336t^3 + 0.3621t^2 - 1.234t + 34.55$

(b) Domain: All real numbers

(c) 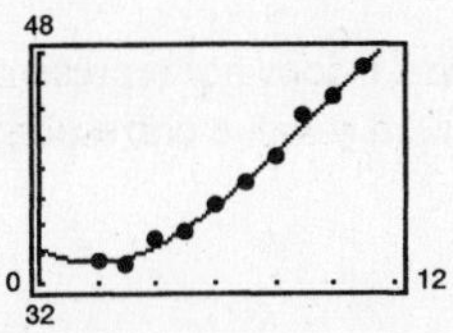

(d) The model is most accurate for the year 1996 and least accurate for the year 1999.

126. 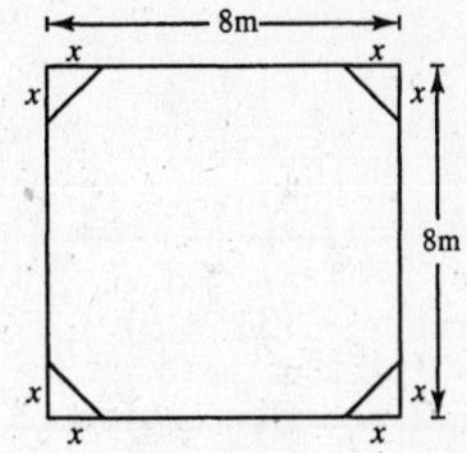

(a) $A = (8)(8) - 4(\frac{1}{2})(x)(x)$

$= 64 - 2x^2$

Domain: $0 \le x \le 4$

(b) 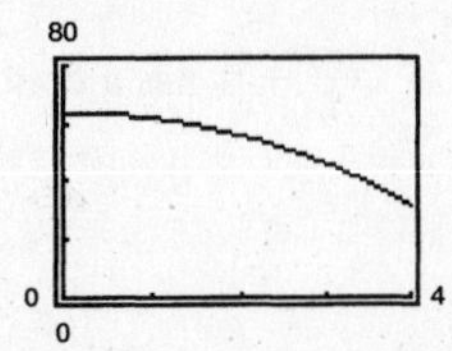

Range: $32 \le A \le 64$

(c) When $x = 4$, the resulting figure is a square.

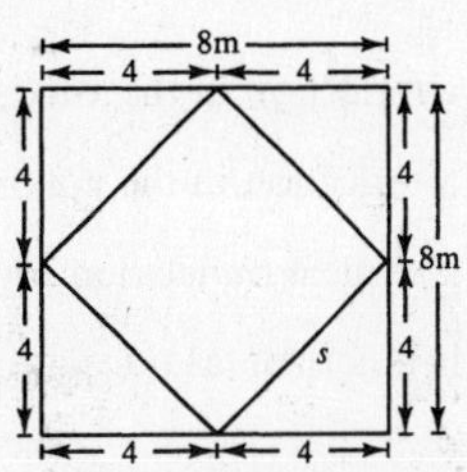

By the Pythagorean Theorem,
$4^2 + 4^2 = s^2 \implies s = \sqrt{32} = 4\sqrt{2}$ meters.

127. (a) For the average salaries of college professors, a scale of $10,000 would be appropriate.

(b) For the population of the United States, use a scale of 50,000,000.

(c) For the percent of the civilian workforce that is unemployed, use a scale of 1%.

128.

Interval	Intake Pipe	Drainpipe 1	Drainpipe 2
[0, 5]	Open	Closed	Closed
[5, 10]	Open	Open	Closed
[10, 20]	Closed	Closed	Closed
[20, 30]	Closed	Closed	Open
[30, 40]	Open	Open	Open
[40, 45]	Open	Closed	Open
[45, 50]	Open	Open	Open
[50, 60]	Open	Open	Closed

129. $N = -0.01842t^4 + 0.6098t^3 - 7.062t^2$
$+ 26.59t + 46.9$, $t = 0$ corresponds to 1990

(a) The numbers of new cases was increasing from 1990-1993 and decreasing from 1993-2001.

(b) There was a maximum number of approximately 78,000 new cases in 1993.

130. $f(x) = \begin{cases} x + 2, & x < -2 \\ 0, & -2 \le x \le 2 \\ ax + b, & x > 2 \end{cases}$

(a) If $f(x)$ is odd, then $f(-x) = -f(x)$

$f(-3) = -f(3) \Rightarrow -1 = -(3a + b)$

$f(-4) = -f(4) \Rightarrow -2 = -(4a + b)$

Solving the system: $4a + b = 2$

$3a + b = 1$

yields $a = 1$ and $b = -2$

(b) If $f(x)$ is even, then $f(-x) = f(x)$

$f(-3) = f(3) \Rightarrow -1 = 3a + b$

$f(-4) = f(4) \Rightarrow -2 = 4a + b$

Solving this system yields $a = -1$ and $b = 2$.

131. False. The function $f(x) = \sqrt{x^2 + 1}$ has a domain of all real numbers.

132. False. A piecewise-defined function is a function that is defined by two or more equations over a specified domain. That domain may or may not include x- and y-intercepts.

133. True. A horizontal line, $y = k$, k a constant, is even.

134. True.

The line $y = x$ is odd.

135. (a) Even. The graph is a reflection in the x-axis.

(b) Even. The graph is a reflection in the y-axis.

(c) Even. The graph is a vertical translation of f.

(d) Neither. The graph is a horizontal translation of f.

136. No, the graph in Exercise 11 does not represent x as a function of y. Each non zero y-value corresponds to two x-values.

137. Yes, each y-value corresponds to only one x-value, so the graph does represent y as a function of x. Some values of y near 0 "appear" to correspond to more than one value of x, but that is a result of the vertical scaling.

Section 1.3 Shifting, Reflecting, and Stretching Graphs

1. (a) $f(x) = x^3 + c$

$c = -2 : f(x) = x^3 - 2$ Vertical shift 2 units downward

$c = 0 : f(x) = x^3$ Basic cubic function

$c = 2 : f(x) = x^3 + 2$ Vertical shift 2 units upward

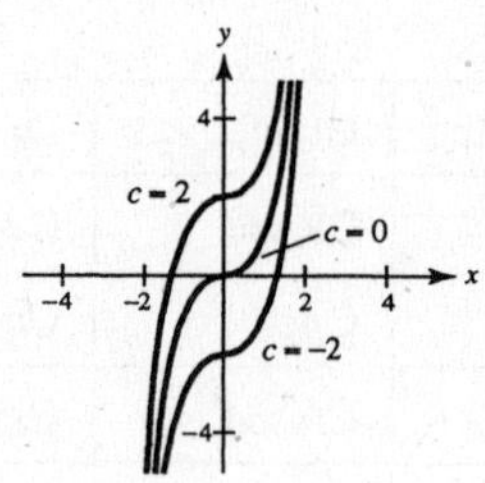

(b) $f(x) = (x - c)^3$

$c = -2 : f(x) = (x + 2)^3$ Horizontal shift 2 units to the left

$c = 0 : f(x) = x^3$ Basic cubic function

$c = 2 : f(x) = (x - 2)^3$ Horizontal shift 2 units to the right

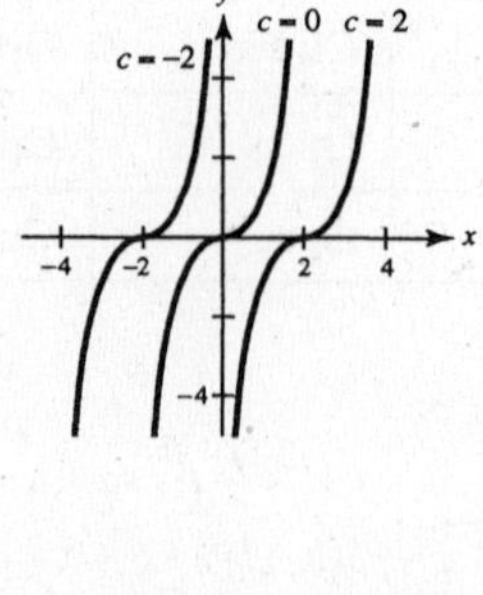

—CONTINUED—

1. —CONTINUED—

(c) $f(x) = (x + 1)^3 + c$ — Horizontal shift 1 unit to the left and a vertical shift

$c = -2$: $f(x) = (x + 1)^3 - 2$ — 2 units down

$c = 0$: $f(x) = (x + 1)^3$

$c = 2$: $f(x) = (x + 1)^3 + 2$ — 2 units up

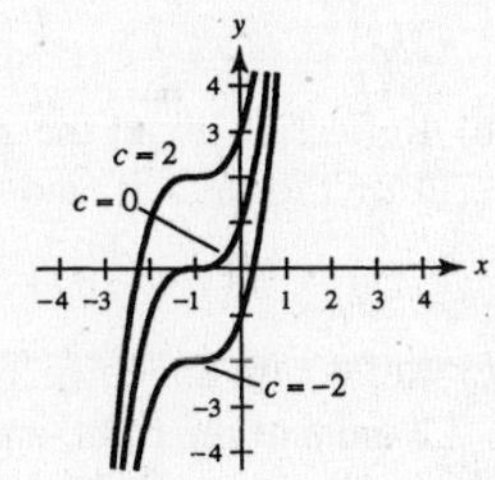

2. (a) $f(x) = |x| + c$

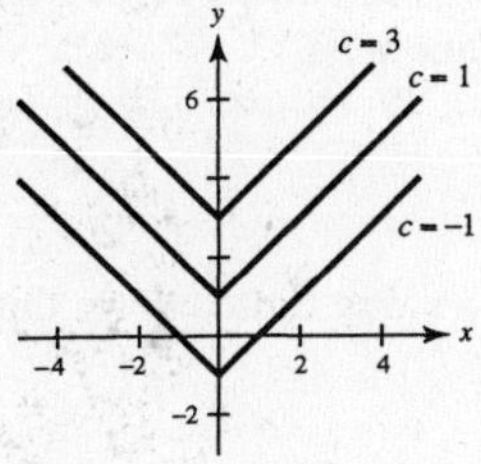

(b) $f(x) = |x - c|$

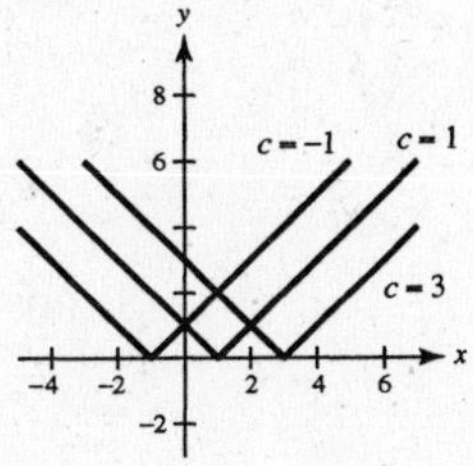

(c) $f(x) = |x + 4| + c$

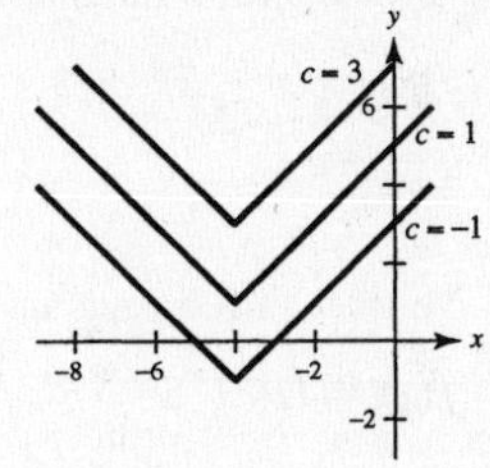

3. (a) $f(x) = \sqrt{x} + c$ — Vertical shifts

$c = -3 : f(x) = \sqrt{x} - 3$ — 3 units downward

$c = -1 : f(x) = \sqrt{x} - 1$ — 1 unit downward

$c = 1 : f(x) = \sqrt{x} + 1$ — 1 unit upward

$c = 3 : f(x) = \sqrt{x} + 3$ — 3 units upward

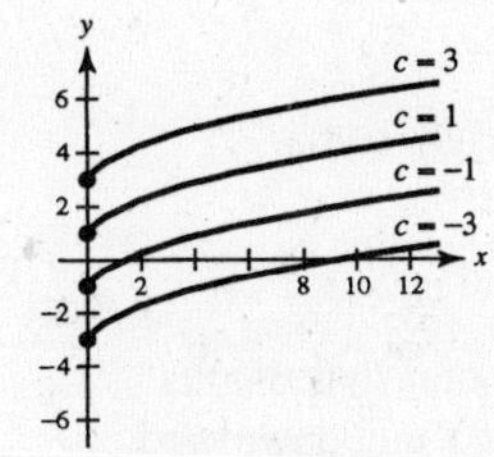

(b) $f(x) = \sqrt{x - c}$ — Horizontal shifts

$c = -3 : f(x) = \sqrt{x + 3}$ — 3 units to the left

$c = -1 : f(x) = \sqrt{x + 1}$ — 1 unit to the left

$c = 1 : f(x) = \sqrt{x - 1}$ — 1 unit to the right

$c = 3 : f(x) = \sqrt{x - 3}$ — 3 units to the right

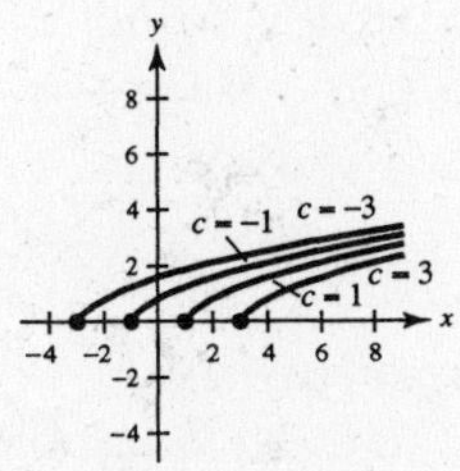

(c) $f(x) = \sqrt{x - 3} + c$ — Horizontal shift 3 units to the right and a vertical shift

$c = -3 : f(x) = \sqrt{x - 3} - 3$ — 3 units downward

$c = -1 : f(x) = \sqrt{x - 3} - 1$ — 1 unit downward

$c = 1 : f(x) = \sqrt{x - 3} + 1$ — 1 unit upward

$c = 3 : f(x) = \sqrt{x - 3} + 3$ — 3 units upward

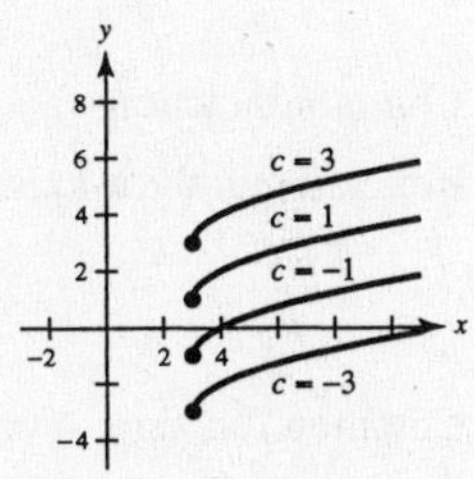

4. (a) $f(x) = \begin{cases} x^2 + c, & x < 0 \\ -x^2 + c, & x \geq 0 \end{cases}$

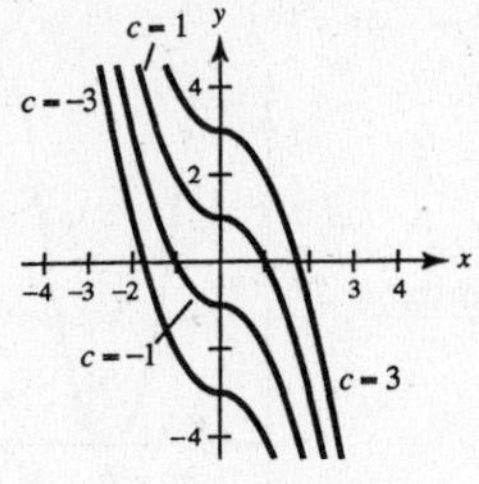

(b) $f(x) = \begin{cases} (x + c)^2, & x < 0 \\ -(x + c)^2, & x \geq 0 \end{cases}$

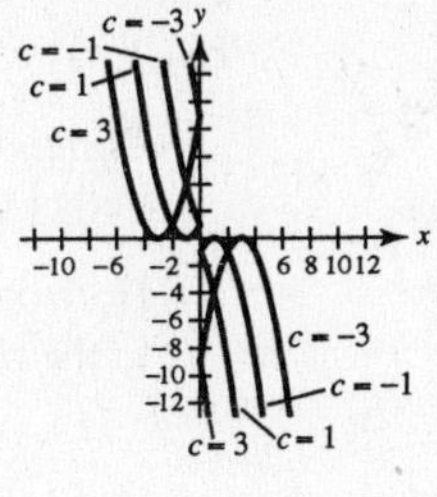

(c) $f(x) = \begin{cases} (x + 1)^2 + c, & x < 0 \\ -(x + 1)^2 + c, & x \geq 0 \end{cases}$

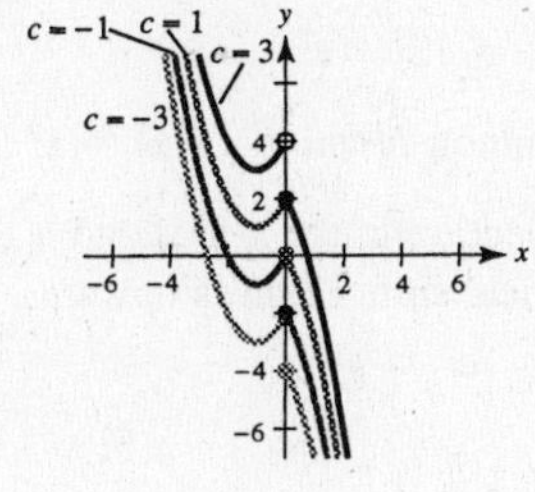

5. (a) Vertical shift 1 unit downward.

$f(x) = x^2 - 1$

(b) Reflection about the x-axis, horizontal shift 1 unit to the left, and a vertical shift 1 unit upward.

$f(x) = -(x + 1)^2 + 1$

(c) Reflection about the x-axis, horizontal shift 2 units to the right, and a vertical shift 6 units upward.

$f(x) = -(x - 2)^2 + 6$

(d) Horizontal shift 5 units to the right and a vertical shift 3 units downward.

$f(x) = (x - 5)^2 - 3$

6. (a) The graph of $f(x) = x^3$ was reflected in the x-axis and shifted upward one unit.

$y = -x^3 + 1 = 1 - x^3$

(b) The graph of $f(x) = x^3$ was shifted to the right one unit and upward one unit.

$y = (x - 1)^3 + 1$

(c) The graph of $f(x) = x^3$ was reflected in the x-axis and shifted to the left three units and downward one unit.

$y = -(x + 3)^3 - 1$

(d) The graph of $f(x) = x^3$ was shifted to the right ten units and downward four units.

$y = (x - 10)^3 - 4$

7. (a) Vertical shift 5 units upward.

$f(x) = |x| + 5$

(b) Reflection in the x-axis and a horizontal shift 3 units to the left.

$f(x) = -|x + 3|$

(c) Horizontal shift 2 units to the right and a vertical shift 4 units downward.

$f(x) = |x - 2| - 4$

(d) Reflection in the x-axis, horizontal shift 6 units to the right, and a vertical shift 1 unit downward.

$f(x) = -|x - 6| - 1$

8. (a) The graph of $f(x) = \sqrt{x}$ was shifted down three units.

$y = \sqrt{x} - 3$

(b) The graph of $f(x) = \sqrt{x}$ was shifted downward seven units and to the left one unit.

$y = \sqrt{x + 1} - 7$

(c) The graph of $f(x) = \sqrt{x}$ was reflected in the x-axis and shifted to the right five units and upward five units.

$y = -\sqrt{x - 5} + 5$

(d) The graph of $f(x) = \sqrt{x}$ was reflected about the x- and y-axis and shifted to the right three units and downward four units.

$y = -\sqrt{-x + 3} - 4 = -\sqrt{-(x - 3)} - 4$

9. Common function: $f(x) = x^3$

Horizontal shift 2 units to the right: $y = (x - 2)^3$

10. Common function: $y = x$

Transformation: vertical shrink

Formula: $y = \frac{1}{2}x$

11. Common function: $f(x) = x^2$

Reflection in the x-axis: $y = -x^2$

12. Common function: $y = 0$

Transformation: vertical shift 7 units upward

Formula: $y = 7$

13. Common function: $f(x) = \sqrt{x}$

Reflection in the x-axis and a vertical shift 1 unit upward: $y = -\sqrt{x} + 1$

14. Common function: $y = |x|$

Transformation: horizontal shift 2 units to the left

Formula: $y = |x + 2|$

15. $f(x) = 12 - x^2$

Common function: $g(x) = x^2$

Reflection in the x-axis and a vertical shift 12 units upward.

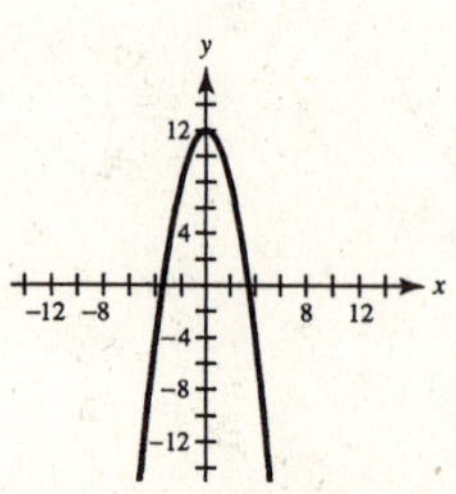

16. $f(x) = (x - 8)^2$

Common function: $g(x) = x^2$

Horizontal shift of eight units to the right.

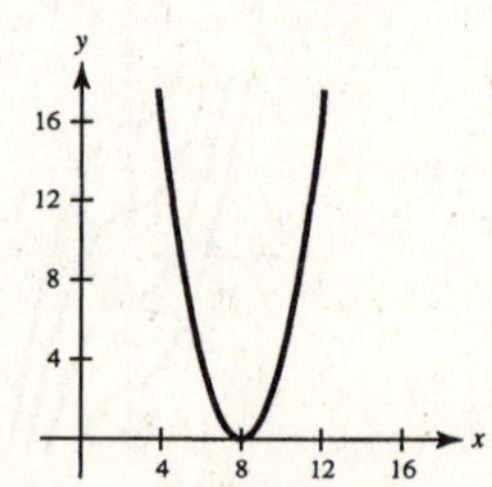

17. $f(x) = x^3 + 7$

Common function: $g(x) = x^3$

Vertical shift 7 units upward.

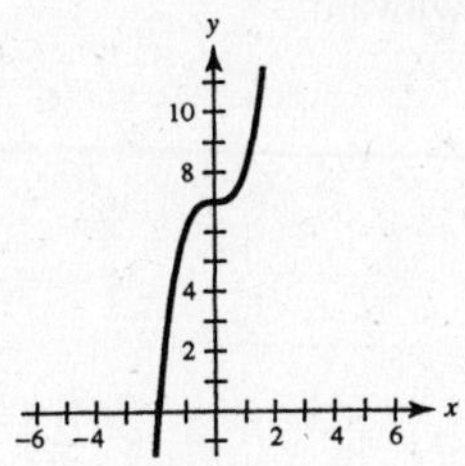

18. $f(x) = -x^3 - 1$

Common function: $g(x) = x^3$

Reflection in the x-axis; vertical shift of one unit downward.

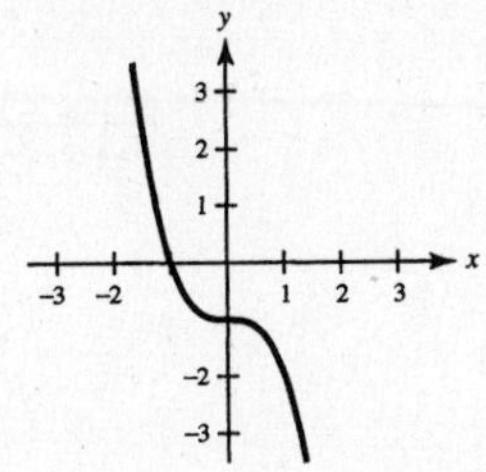

19. $f(x) = 2 - (x + 5)^2$

Common function: $g(x) = x^2$

Reflection in the x-axis, horizontal shift 5 units to the left, and a vertical shift 2 units upward.

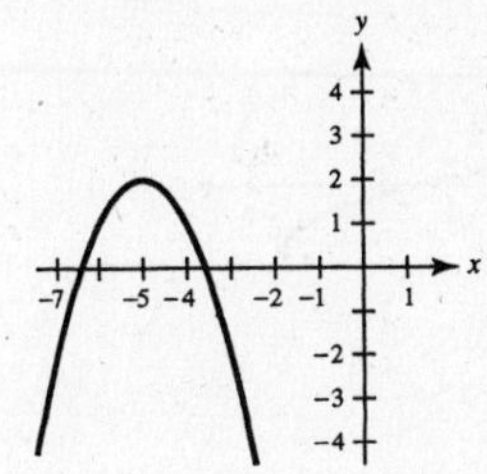

20. $f(x) = -(x + 10)^2 + 5$

Common function: $g(x) = x^2$

Reflection in the x-axis; horizontal shift of 10 units to the left; vertical shift of 5 units upward.

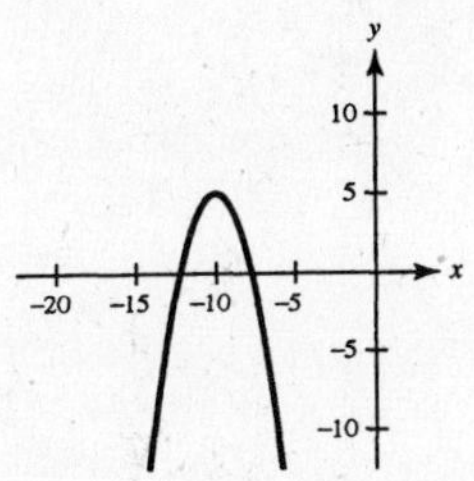

21. $f(x) = (x - 1)^3 + 2$

Common function: $g(x) = x^3$

Horizontal shift 1 unit to the right and a vertical shift 2 units upward.

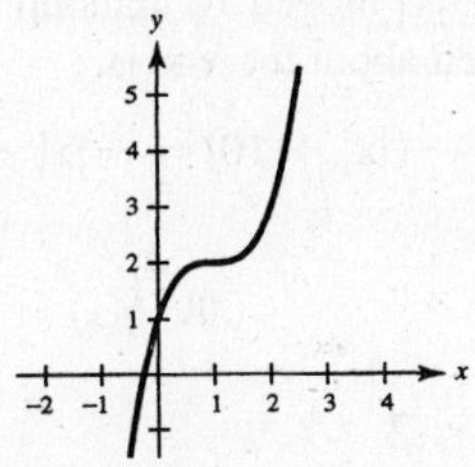

22. $f(x) = (x + 3)^3 - 10$

Common function: $g(x) = x^3$

Horizontal shift of 3 units to the left; vertical shift of 10 units downward.

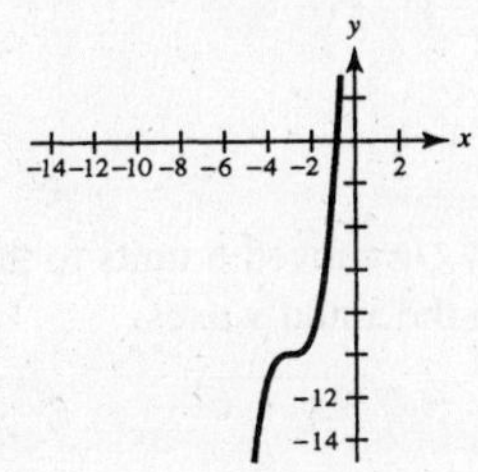

23. $f(x) = -|x| - 2$

Common function: $g(x) = |x|$

Reflection in the x-axis and a vertical shift 2 units downward.

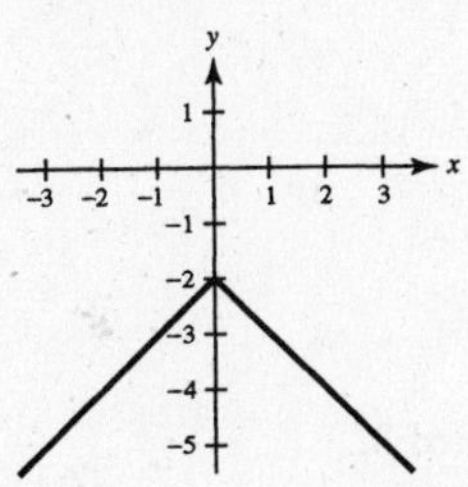

24. $f(x) = 6 - |x + 5|$

Common function: $g(x) = |x|$

Reflection in the x-axis; horizontal shift of 5 units to the left; vertical shift of 6 units upward.

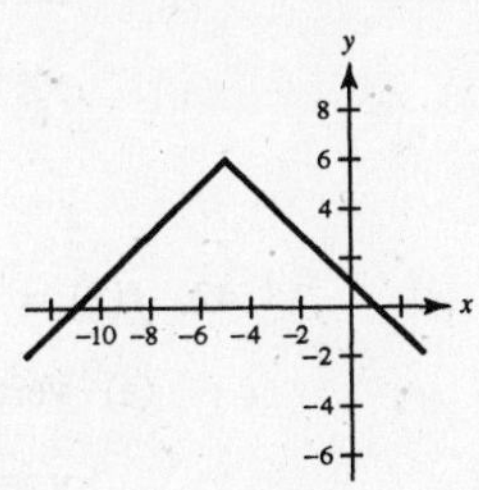

25. $f(x) = -|x + 4| + 8$

Common function: $g(x) = |x|$

Reflection in the x-axis, horizontal shift 4 units to the left, and a vertical shift 8 units upward.

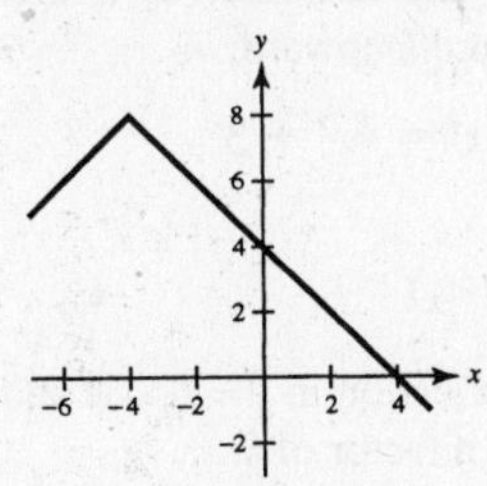

26. $f(x) = |-x + 3| + 9$

Common function: $g(x) = |x|$

Reflection in the y-axis; horizontal shift of 3 units to the right; vertical shift of 9 units upward.

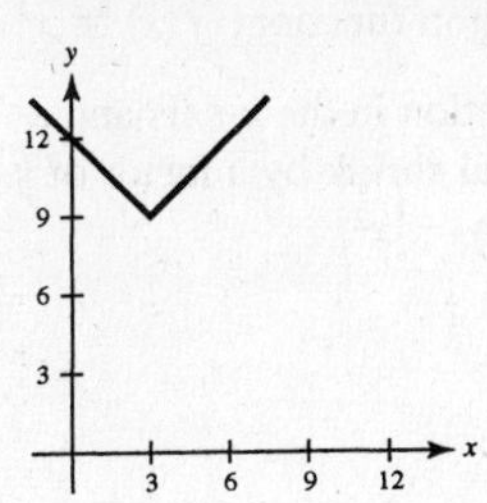

27. $f(x) = \sqrt{x - 9}$

Common function: $g(x) = \sqrt{x}$

Horizontal shift 9 units to the right.

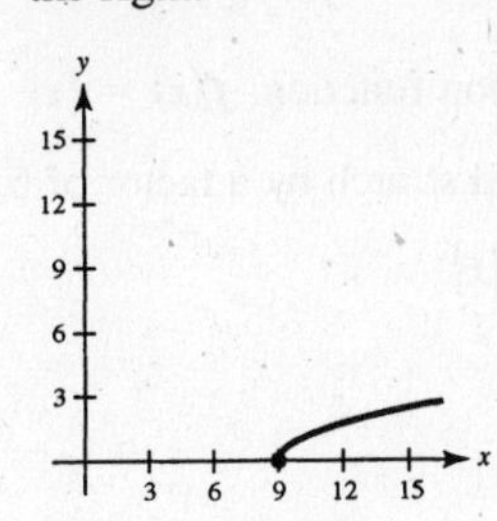

28. $f(x) = \sqrt{x + 4} + 8$

Common function: $g(x) = \sqrt{x}$

Horizontal shift of 4 units to the left; vertical shift of 8 units upward.

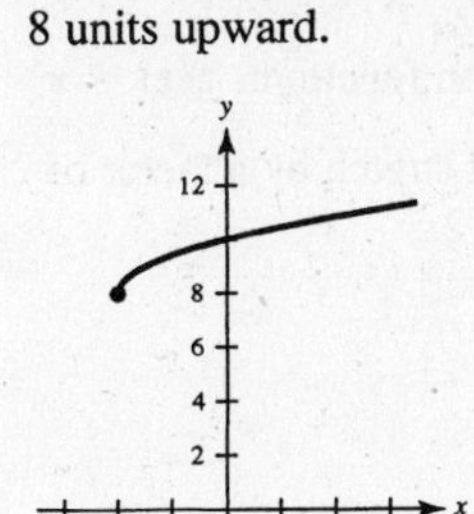

29. $f(x) = \sqrt{7 - x} - 2$ or $f(x) = \sqrt{-(x - 7)} - 2$

Common function: $g(x) = \sqrt{x}$

Reflection in the y-axis, horizontal shift 7 units to the right, and a vertical shift 2 units downward.

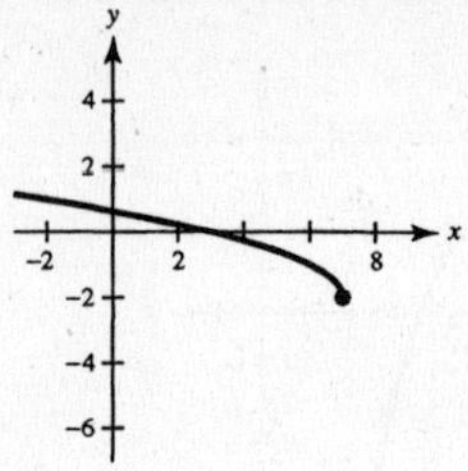

30. $f(x) = -\sqrt{x + 1} - 6$

Common function: $g(x) = \sqrt{x}$

Reflection in the x-axis; horizontal shift of 1 unit to the left; vertical shift of 6 units downward.

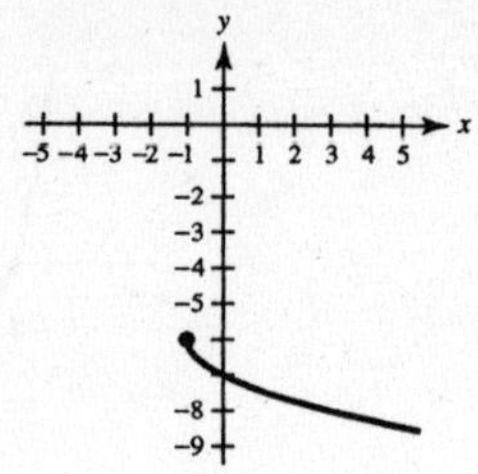

31. $f(x) = x^2$ moved 2 units to the right and 8 units down.

$g(x) = (x - 2)^2 - 8$

32. $g(x) = -[(x + 3)^2 + 7]$

$= -(x + 3)^2 - 7$

33. $f(x) = x^3$ moved 13 units to the right.

$g(x) = (x - 13)^3$

34. $g(x) = (-x + 6)^3 - 6$

35. $f(x) = |x|$ moved 10 units up and reflected about the x-axis.

$g(x) = -(|x| + 10) = -|x| - 10$

36. $g(x) = |x + 1| - 7$

37. $f(x) = \sqrt{x}$ moved 6 units to the left and reflected in both the x and y axes.

$g(x) = -\sqrt{-x + 6}$

38. $g(x) = -(\sqrt{-x} - 9)$

$= -\sqrt{-x} + 9$

39. $f(x) = x^2$

(a) Reflection in the x-axis and a vertical stretch by a factor of 3.

$g(x) = -3x^2$

(b) A vertical stretch by a factor of 4 and a vertical shift 3 units upward.

$g(x) = 4x^2 + 3$

40. $f(x) = x^3$

(a) Vertical shrink by a factor of $\frac{1}{4}$.

$y = \frac{1}{4}x^3$

(b) Reflection in the x-axis and a vertical stretch by a factor of 2.

$y = -2x^3$

41. $f(x) = |x|$

(a) Reflection in the x-axis and a vertical shrink by a factor of $\frac{1}{2}$.

$g(x) = -\frac{1}{2}|x|$

(b) Vertical stretch by a factor of 3 and a vertical shift 3 units downward.

$g(x) = 3|x| - 3$

42. $f(x) = \sqrt{x}$

(a) Vertical stretch by a factor of 8.

$y = 8\sqrt{x}$

(b) Reflection in the x-axis and a vertical shrink by a factor of $\frac{1}{4}$.

$y = -\frac{1}{4}\sqrt{x}$

43. Common function: $f(x) = x^3$

Vertical stretch by a factor of 2: $g(x) = 2x^3$

44. Common function: $f(x) = |x|$

Vertical stretch by a factor of 6.

$y = 6|x|$

45. Common function: $f(x) = x^2$

Reflection in the x-axis and a vertical shrink by a factor of $\frac{1}{2}$: $g(x) = -\frac{1}{2}x^2$

46. Common function: $f(x) = \sqrt{x}$

Vertical stretch by a factor of 3.

$y = 3\sqrt{x}$

47. Common function: $f(x) = \sqrt{x}$

Reflection in the y-axis and a vertical shrink by a factor of $\frac{1}{2}$: $g(x) = \frac{1}{2}\sqrt{-x}$

48. Common function: $f(x) = |x|$

Reflection in the x-axis; vertical stretch by a factor of 2; vertical shift of 2 units downward.

$y = -2|x| - 2$

49. Common function: $f(x) = x^3$

Reflection in the x-axis, horizontal shift 2 units to the right and a vertical shift 2 units upward: $g(x) = -(x - 2)^3 + 2$

50. The graph of $y = |x|$ has a horizontal shift of 4 units to the left and a vertical shift of 2 units downward.

$y = |x + 4| - 2$

51. Common function: $f(x) = \sqrt{x}$

Reflection in the x-axis and a vertical shift 3 units downward: $g(x) = -\sqrt{x} - 3$

52. The graph of $y = x^2$ has a horizontal shift of 2 units to the right and a vertical shift of 4 units upward.

$y = (x - 2)^2 + 4$

53. (a) $y = f(x) + 2$

Vertical shift 2 units upward.

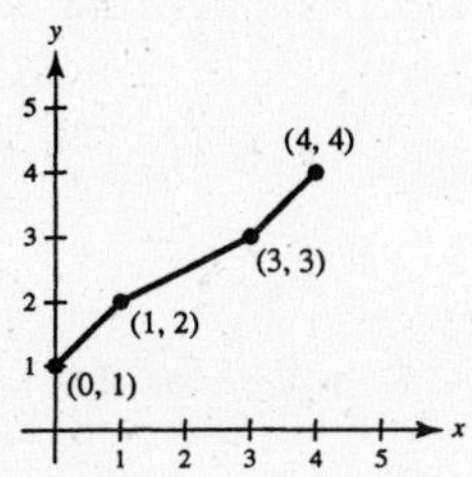

(b) $y = f(x - 2)$

Horizontal shift 2 units to the right.

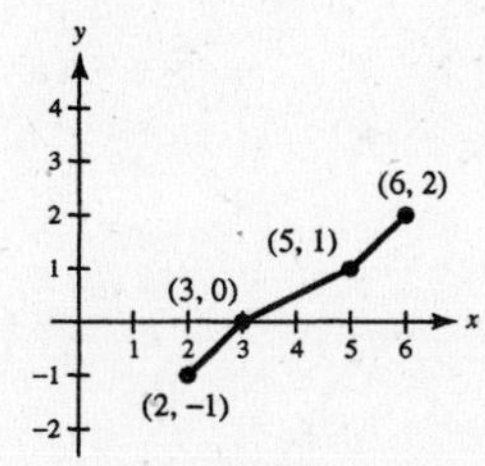

(c) $y = 2f(x)$

Vertical stretch by a factor of 2.

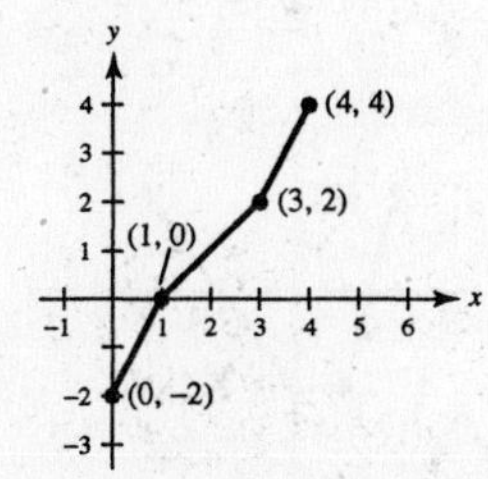

(d) $y = -f(x)$

Reflection in the x-axis.

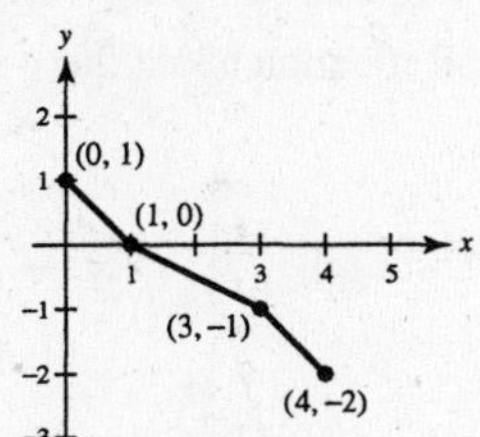

(e) $y = f(x + 3)$

Horizontal shift 3 units to the left.

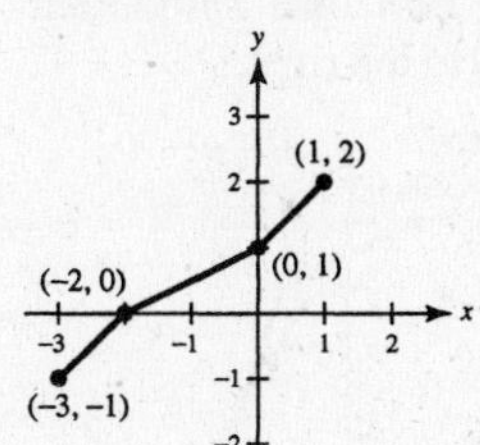

(f) $y = f(-x)$

Reflection in the y-axis.

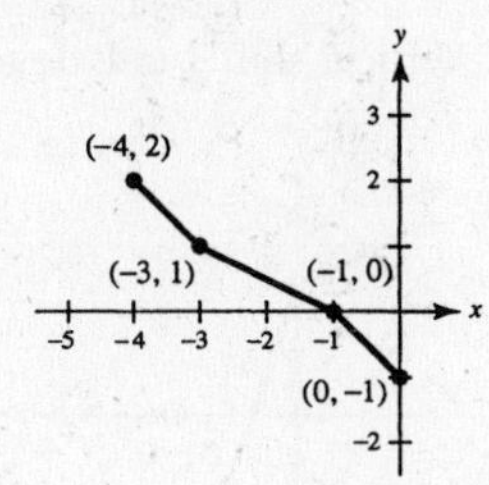

(g) $y = f(2x)$

Horizontal shrink. Each x-value is divided by 2.

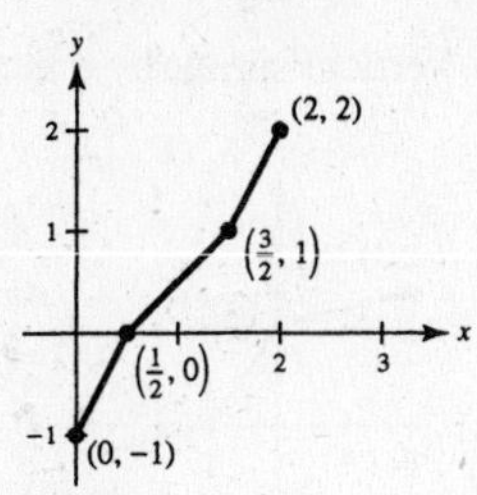

54. (a) $y = f(-x)$

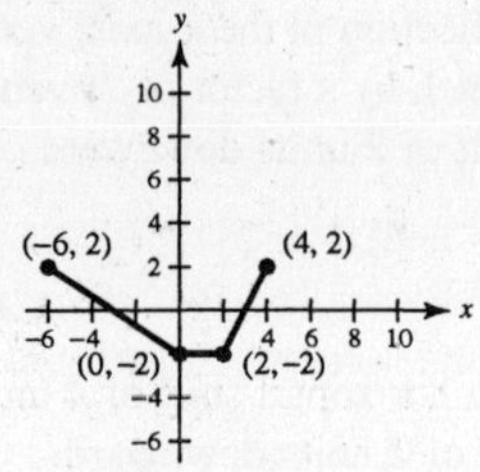

(b) $y = f(x) + 4$

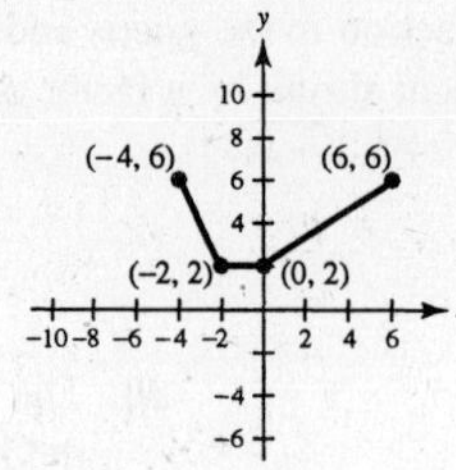

(c) $y = 2f(x)$

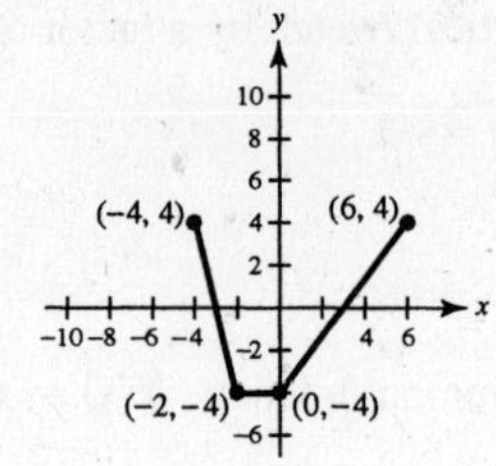

(d) $y = -f(x - 4)$

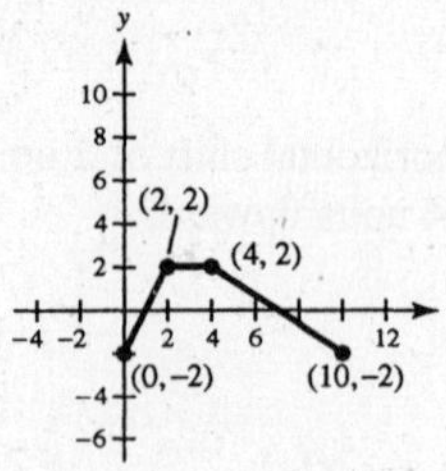

(e) $y = f(x) - 3$

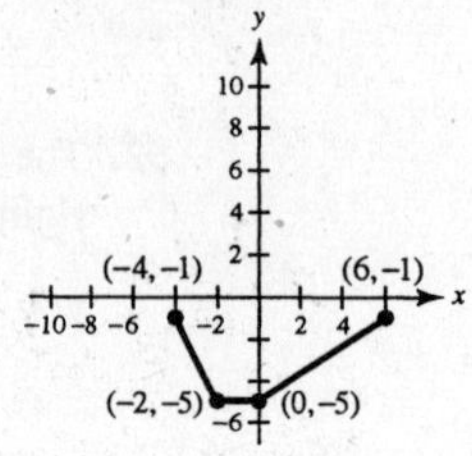

(f) $y = -f(x) - 1$

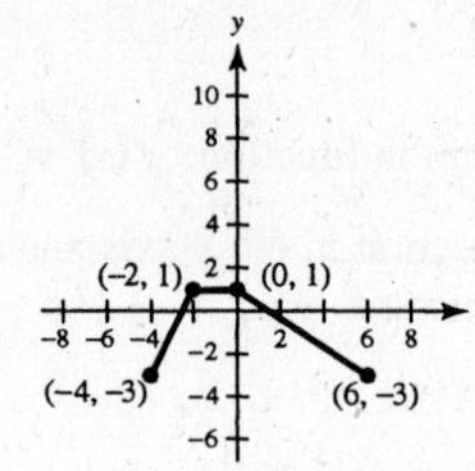

(g) $y = f(2x)$

Horizontal shrink. Each x-value is divided by 2.

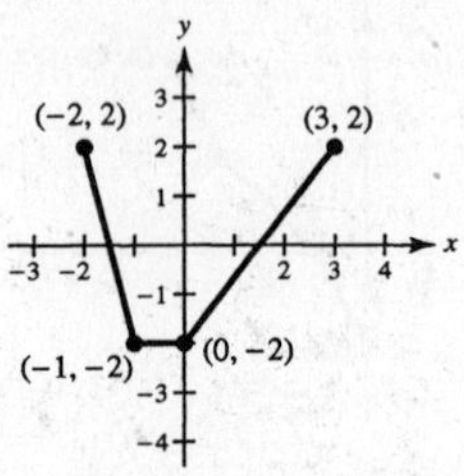

55. (a) $y = f(x) - 1$

Vertical shift 1 unit downward.

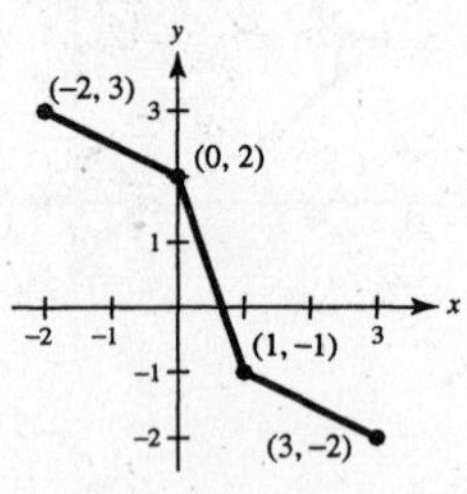

(b) $y = f(x - 1)$

Horizontal shift 1 unit to the right.

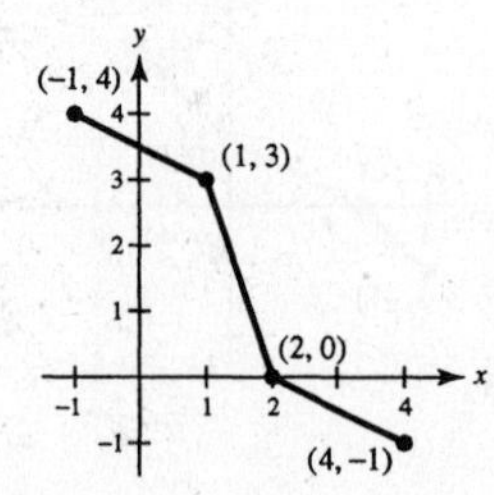

(c) $y = f(-x)$

Reflection about the y-axis.

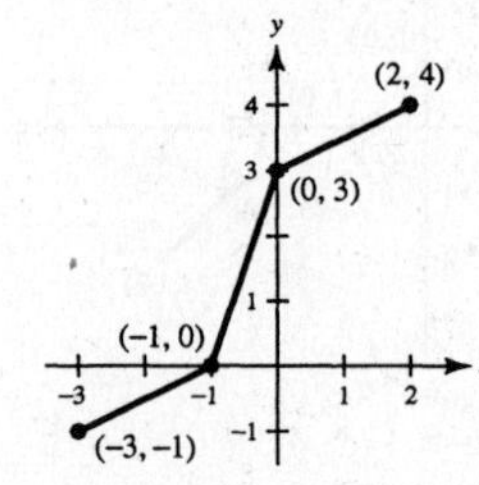

(d) $y = f(x + 1)$

Horizontal shift 1 unit to the left.

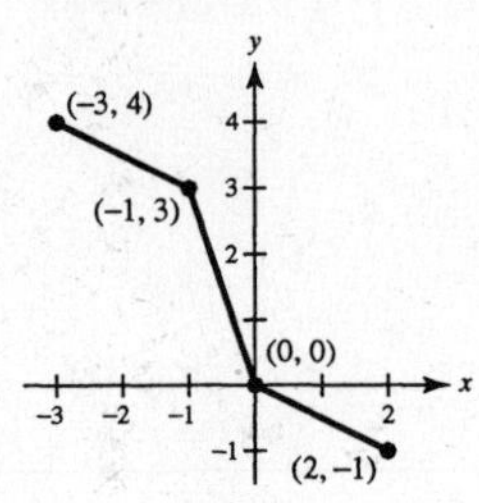

(e) $y = -f(x - 2)$

Reflection about the x-axis and a horizontal shift 2 units to the right.

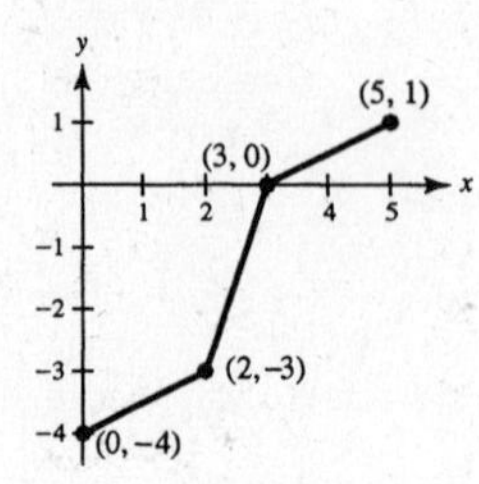

(f) $y = \frac{1}{2}f(x)$

Vertical shrink by a factor of $\frac{1}{2}$.

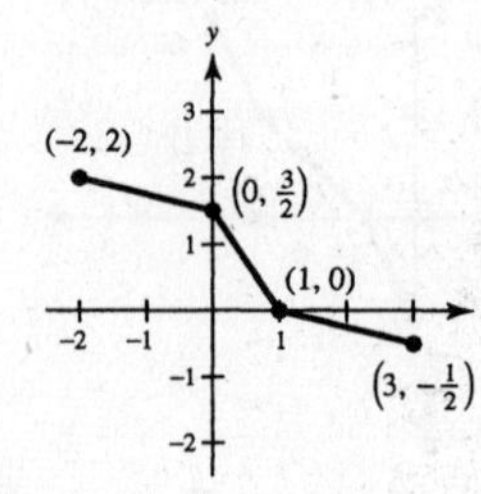

—CONTINUED—

55. —CONTINUED—

(g) $y = f\left(\frac{1}{2}x\right)$

Horizontal stretch. Each x-value is multiplied by 2.

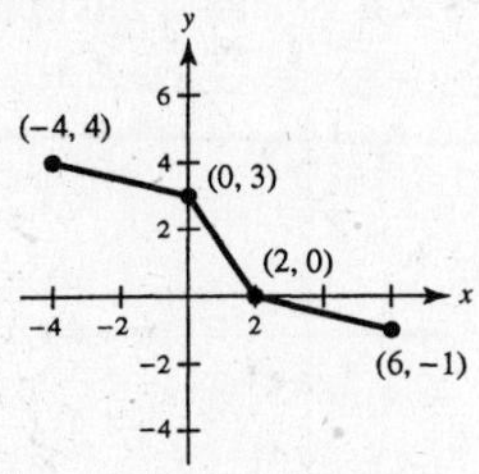

56. (a) $y = f(x - 5)$

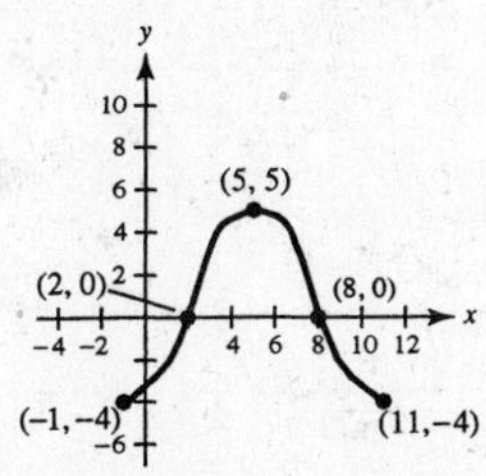

(b) $y = -f(x) + 3$

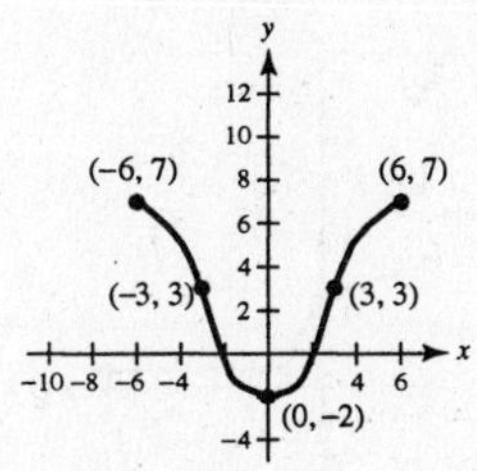

(c) $y = \frac{1}{3}f(x)$

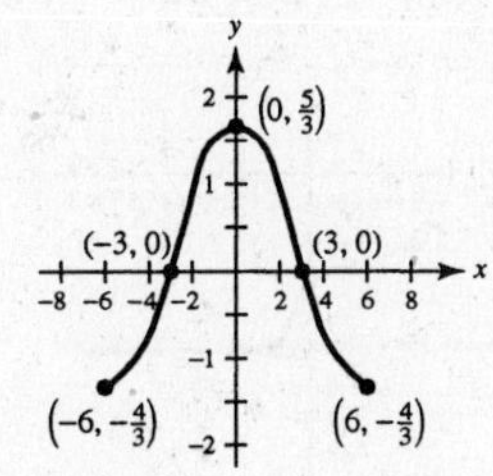

(d) $y = -f(x + 1)$

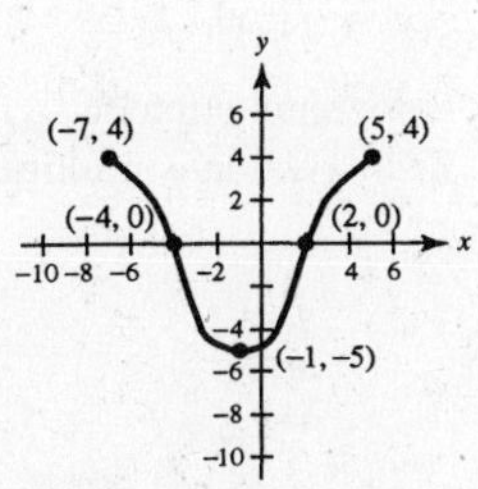

(e) $y = f(-x)$

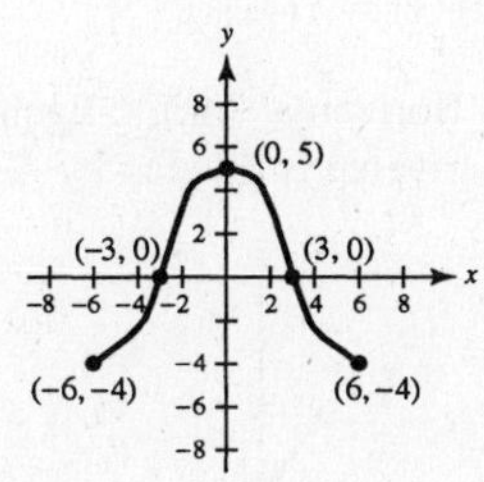

(f) $y = f(x) - 10$

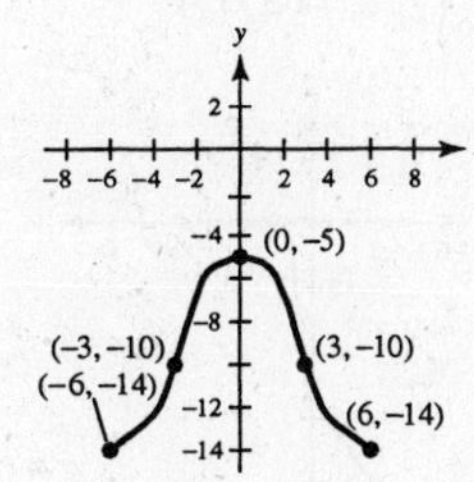

(g) $y = f\left(\frac{1}{2}x\right)$

Horizontal stretch. Each x-value is multiplied by 2.

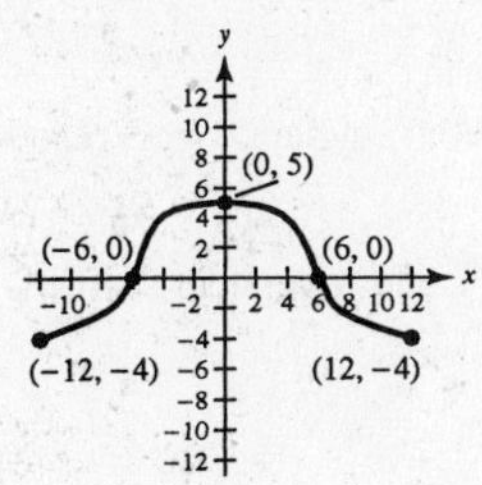

57. (a) $g(x) = f(x) + 2$

Vertical shift 2 units upward.

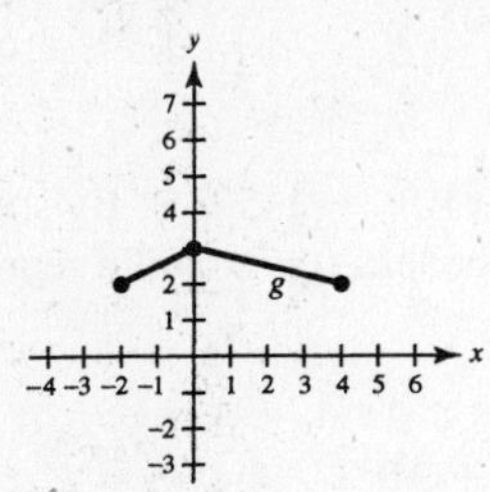

(b) $g(x) = f(x) - 1$

Vertical shift 1 unit downward.

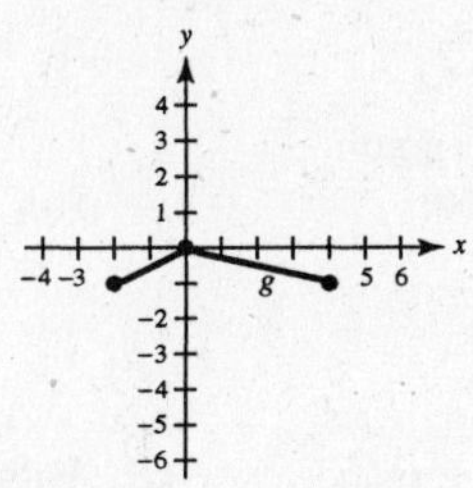

(c) $g(x) = f(-x)$

Reflection in the y-axis.

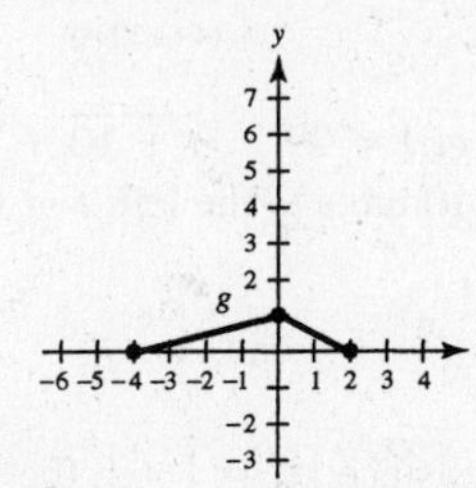

—CONTINUED—

57. —CONTINUED—

(d) $g(x) = -2f(x)$

Reflection in the x-axis and a vertical stretch by a factor of 2.

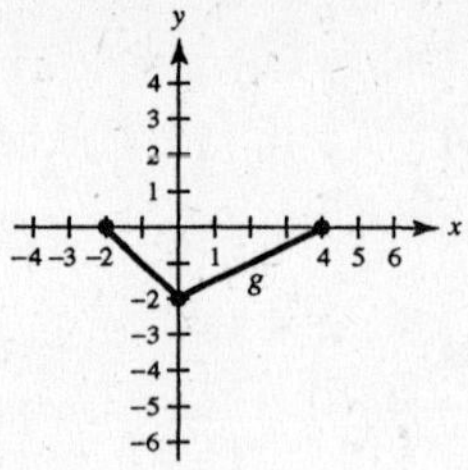

(e) $g(x) = f(2x)$

Horizontal shrink.
Each x-value is divided by 2.

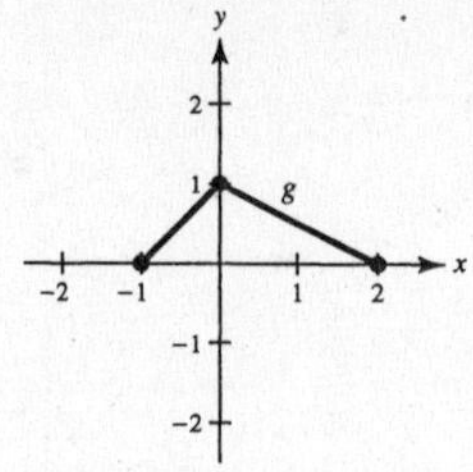

(f) $g(x) = f(\frac{1}{2}x)$

Horizontal stretch.
Each x-value is multiplied by 2.

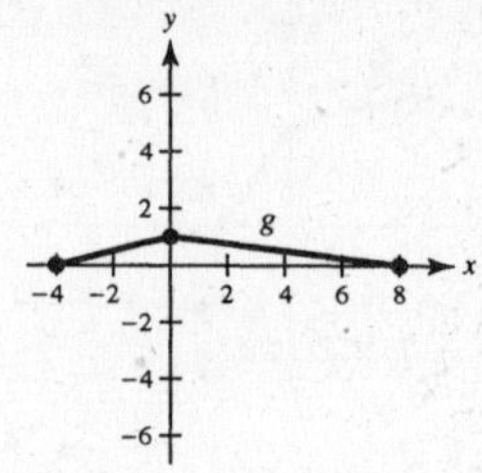

58. (a) $g(x) = f(x) - 5$

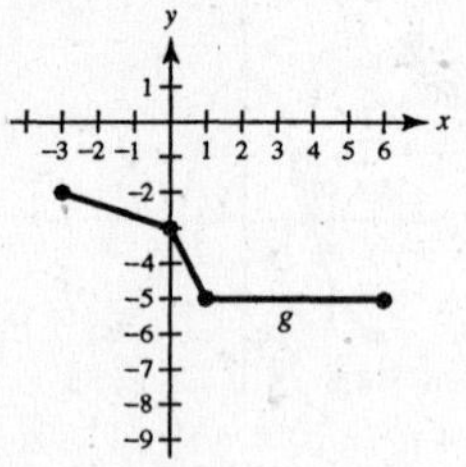

(b) $g(x) = f(x) + \frac{1}{2}$

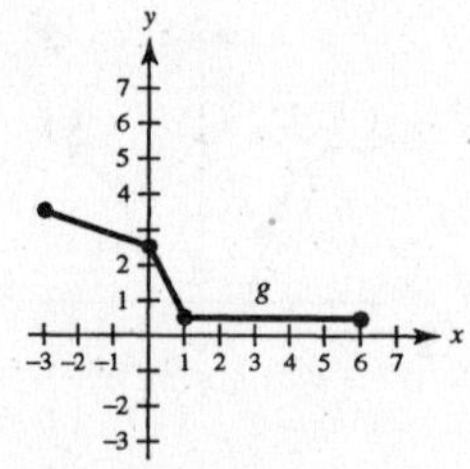

(c) $g(x) = f(-x)$

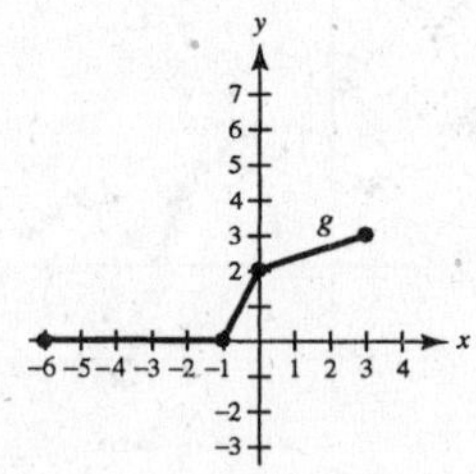

(d) $g(x) = -4f(x)$

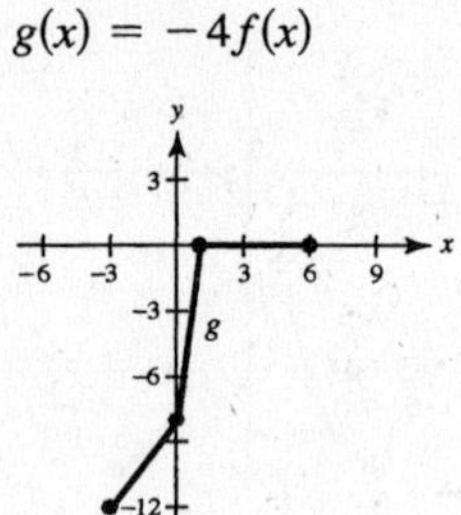

(e) $g(x) = f(4x)$

Horizontal shrink. Each x-value is divided by 4.

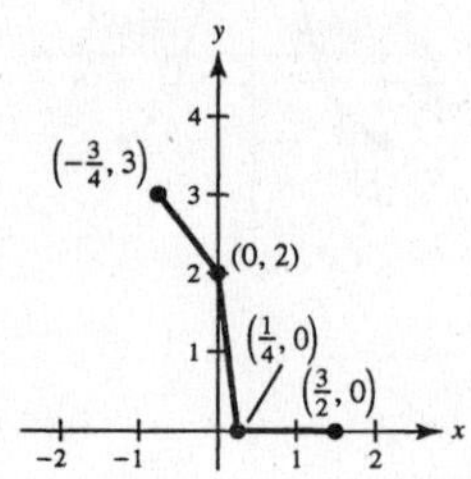

(f) $g(x) = f(\frac{1}{4}x)$

Horizontal stretch.
Each x-value is multiplied by 4.

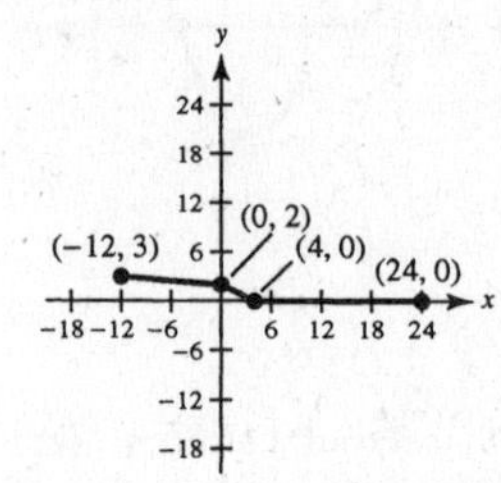

59. $T = f(t) = 29.1\sqrt{t} + 152.4$, $t = 4$ corresponds to 1994

(a) Vertical stretch of 29.1 and a vertical shift of 152.4 units upward

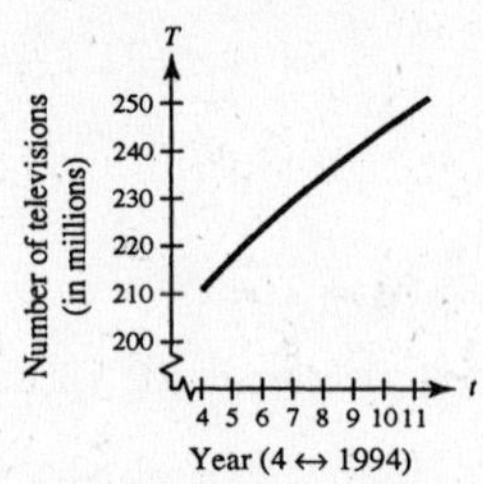

(b) $g(t) = 29.1\sqrt{t + 10} + 152.4$; By shifting the graph 10 units to the left, $t = 0$ would represent 2000.

60. $M = f(t) = 24.77t^2 + 3150.03$ where $t = 3$ represents 1993.

(a) Vertical stretch of 24.77 and a vertical shift of 3150.03 units upward

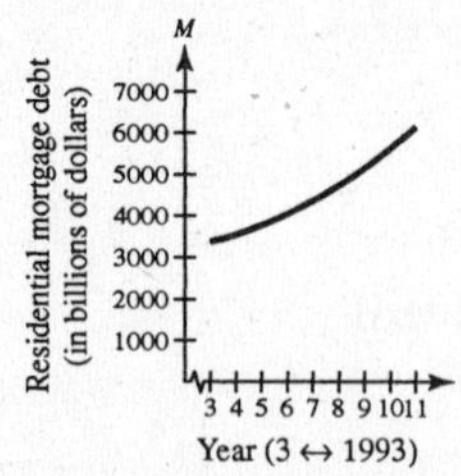

(b) $M = g(t) = 24.77(t + 10)^2 + 3150.03$; By shifting the graph 10 units to the left, $t = 0$ would represent 2000.

61. True, since $|x| = |-x|$, the graphs of $f(x) = |x| + 6$ and $f(x) = |-x| + 6$ are identical.

62. False. $f(x) = -[(x - 6)^2 + 3]$

The point $(-2, -67)$ lies on the transformation.

63. True

Let $g(x) = -f(x)$.

Then $g(-x) = -f(-x)$

$= -[-f(x)]$ since f is odd.

$= f(x)$

$= -g(x)$

Thus, $g(x) = -f(x)$ is also odd.

64. True.

Let $g(x) = f(x) + c$

Then $g(-x) = f(-x) + c$

$= f(x) + c$ Since f is even

$= g(x)$

Thus, $g(x) = f(x) + c$ is also even.

65. (a) and (b); Answers will vary.

Section 1.4 Combinations of Functions

1. $f(x) = x + 2, g(x) = x - 2$

(a) $(f + g)(x) = f(x) + g(x) = (x + 2) + (x - 2) = 2x$

(b) $(f - g)(x) = f(x) - g(x) = (x + 2) - (x - 2) = 4$

(c) $(fg)(x) = f(x) \cdot g(x) = (x + 2)(x - 2) = x^2 - 4$

(d) $\left(\frac{f}{g}\right)(x) = \frac{f(x)}{g(x)} = \frac{x + 2}{x - 2}$

Domain: all real numbers except $x = 2$

2. $f(x) = 2x - 5, g(x) = 2 - x$

(a) $(f + g)(x) = 2x - 5 + 2 - x = x - 3$

(b) $(f - g)(x) = 2x - 5 - (2 - x)$

$= 2x - 5 - 2 + x = 3x - 7$

(c) $(fg)(x) = (2x - 5)(2 - x)$

$= 4x - 2x^2 - 10 + 5x$

$= -2x^2 + 9x - 10$

(d) $\left(\frac{f}{g}\right)(x) = \frac{2x - 5}{2 - x}$

Domain: $2 - x \neq 0, x \neq 2$

3. $f(x) = x^2, g(x) = 2 - x$

$(f + g)(x) = f(x) + g(x) = x^2 + (2 - x) = x^2 - x + 2$

$(f - g)(x) = f(x) - g(x) = x^2 - (2 - x) = x^2 + x - 2$

$(fg)(x) = f(x) \cdot g(x) = x^2(2 - x) = 2x^2 - x^3$

$\left(\frac{f}{g}\right)(x) = \frac{f(x)}{g(x)} = \frac{x^2}{2 - x},$

Domain: all real numbers except $x = 2$

4. $f(x) = 2x - 5, g(x) = 4$

(a) $(f + g)(x) = 2x - 5 + 4 = 2x - 1$

(b) $(f - g)(x) = 2x - 5 - 4 = 2x - 9$

(c) $(fg)(x) = (2x - 5)(4) = 8x - 20$

(d) $\left(\frac{f}{g}\right)(x) = \frac{2x - 5}{4} = \frac{1}{2}x - \frac{5}{4}$

Domain: $-\infty < x < \infty$

5. $f(x) = x^2 + 6, g(x) = \sqrt{1 - x}$

$(f + g)(x) = f(x) + g(x) = (x^2 + 6) + \sqrt{1 - x}$

$(f - g)(x) = f(x) - g(x) = (x^2 + 6) - \sqrt{1 - x}$

$(fg)(x) = f(x) \cdot g(x) = (x^2 + 6)\sqrt{1 - x}$

$\left(\frac{f}{g}\right)(x) = \frac{f(x)}{g(x)} = \frac{x^2 + 6}{\sqrt{1 - x}},$

Domain: $x < 1$

6. $f(x) = \sqrt{x^2 - 4},\ g(x) = \frac{x^2}{x^2 + 1}$

(a) $(f + g)(x) = \sqrt{x^2 - 4} + \frac{x^2}{x^2 + 1}$

(b) $(f - g)(x) = \sqrt{x^2 - 4} - \frac{x^2}{x^2 + 1}$

(c) $(fg)(x) = \sqrt{x^2 - 4}\left(\frac{x^2}{x^2 + 1}\right) = \frac{x^2\sqrt{x^2 - 4}}{x^2 + 1}$

(d) $\left(\frac{f}{g}\right)(x) = \sqrt{x^2 - 4} \div \frac{x^2}{x^2 + 1} = \frac{(x^2 + 1)\sqrt{x^2 - 4}}{x^2}$

Domain: $x^2 - 4 \geq 0$

$x^2 \geq 4 \implies x \geq 2 \text{ or } x \leq -2$

Domain: $|x| \geq 2$

7. $f(x) = \frac{1}{x}, g(x) = \frac{1}{x^2}$

$(f + g)(x) = f(x) + g(x) = \frac{1}{x} + \frac{1}{x^2} = \frac{x + 1}{x^2}$

$(f - g)(x) = f(x) - g(x) = \frac{1}{x} - \frac{1}{x^2} = \frac{x - 1}{x^2}$

$(fg)(x) = f(x) \cdot g(x) = \frac{1}{x}\left(\frac{1}{x^2}\right) = \frac{1}{x^3}$

$\left(\frac{f}{g}\right)(x) = \frac{f(x)}{g(x)} = \frac{1/x}{1/x^2} = \frac{x^2}{x} = x, \ x \neq 0$

8. $f(x) = \frac{x}{x + 1}, \ g(x) = x^3$

(a) $(f + g)(x) = \frac{x}{x + 1} + x^3 = \frac{x + x^4 + x^3}{x + 1}$

(b) $(f - g)(x) = \frac{x}{x + 1} - x^3 = \frac{x - x^4 - x^3}{x + 1}$

(c) $(fg)(x) = \frac{x}{x + 1} \cdot x^3 = \frac{x^4}{x + 1}$

(d) $\left(\frac{f}{g}\right)(x) = \frac{x}{x + 1} \div x^3$

$= \frac{x}{x + 1} \cdot \frac{1}{x^3} = \frac{1}{x^2(x + 1)}$

Domain: $x \neq 0, -1$

For Exercises 9–18, $f(x) = x^2 + 1$ and $g(x) = x - 4$

9. $(f + g)(2) = f(2) + g(2) = (2^2 + 1) + (2 - 4) = 3$

10. $(f - g)(-1) = f(-1) - g(-1)$

$= (-1)^2 + 1 - (-1 - 4)$

$= 1 + 1 - (-5)$

$= 7$

11. $(f - g)(3t) = f(3t) - g(3t) = [(3t)^2 + 1] - (3t - 4)$

$= 9t^2 - 3t + 5$

12. $(f + g)(t - 2) = f(t - 2) + g(t - 2)$

$= (t - 2)^2 + 1 + (t - 2) - 4$

$= t^2 - 4t + 4 + 1 + t - 2 - 4$

$= t^2 - 3t - 1$

13. $(fg)(6) = f(6)g(6) = (6^2 + 1)(6 - 4) = 74$

14. $(fg)(-6) = f(-6) \cdot g(-6)$

$= [(-6)^2 + 1][(-6) - 4]$

$= (37)(-10)$

$= -370$

15. $\left(\frac{f}{g}\right)(5) = \frac{f(5)}{g(5)} = \frac{5^2 + 1}{5 - 4} = 26$

16. $\left(\frac{f}{g}\right)(0) = \frac{f(0)}{g(0)} = \frac{0^2 + 1}{0 - 4} = -\frac{1}{4}$

17. $\left(\frac{f}{g}\right)(-1) - 2g(3) = \frac{f(-1)}{g(-1)} - 2g(3)$

$= \frac{2}{-5} - 2(-1)$

$= \frac{8}{5}$

18. $(2f)(5) + g(3) = 2f(5) + g(3)$

$= 2(26) + (-1)$

$= 51$

19. $f(x) = \frac{1}{2}x$, $g(x) = x - 1$, $(f + g)(x) = \frac{3}{2}x - 1$

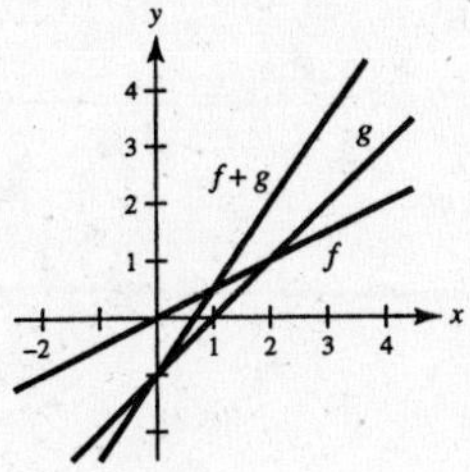

20. $f(x) = \frac{1}{3}x$, $g(x) = -x + 4$

$(f + g)(x) = \frac{1}{3}x - x + 4 = -\frac{2}{3}x + 4$

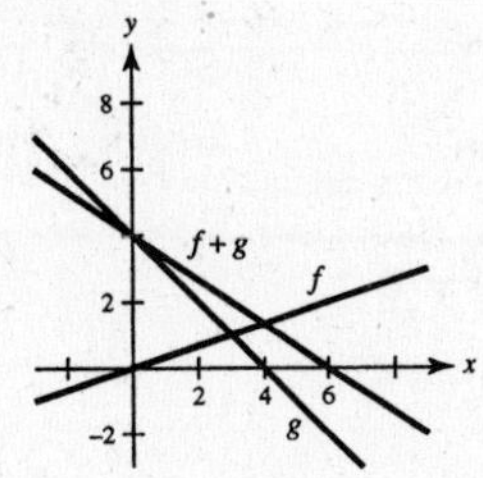

21. $f(x) = x^2$, $g(x) = -2x$, $(f + g)(x) = x^2 - 2x$

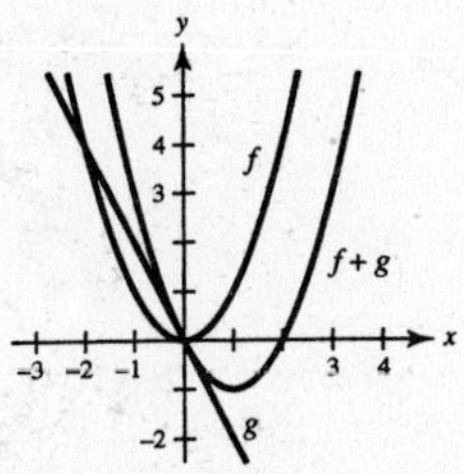

22. $f(x) = 4 - x^2$, $g(x) = x$

$(f + g)(x) = 4 - x^2 + x = 4 + x - x^2$

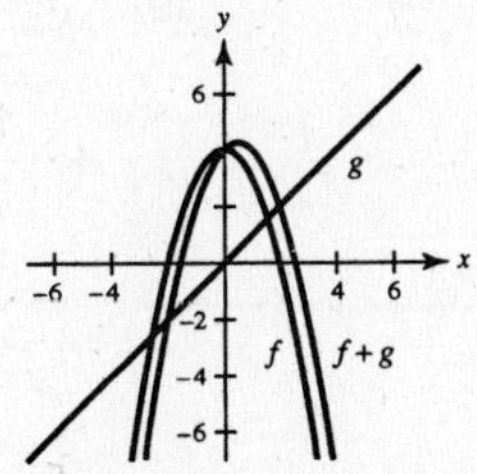

23.

x	0	1	2	3
f	2	3	1	2
g	-1	0	$\frac{1}{2}$	0
$f + g$	1	3	$\frac{3}{2}$	2

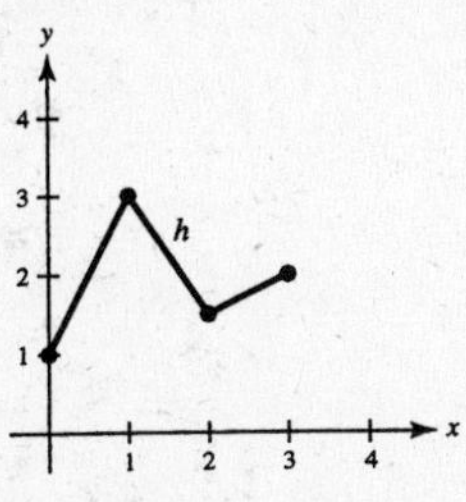

24.

x	-2	-1	0	1	2
$f(x)$	-2	0	-1	-1	1
$g(x)$	1	1	0	2	2
$h(x) = (f + g)(x)$	-1	1	-1	1	3

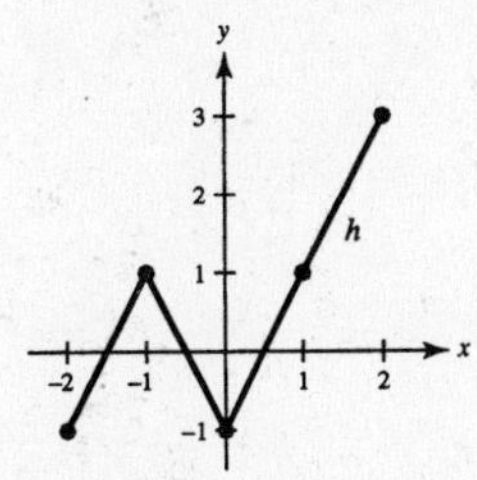

25.

x	-2	0	1	2	4
f	2	0	1	2	4
g	4	2	1	0	2
$f + g$	6	2	2	2	6

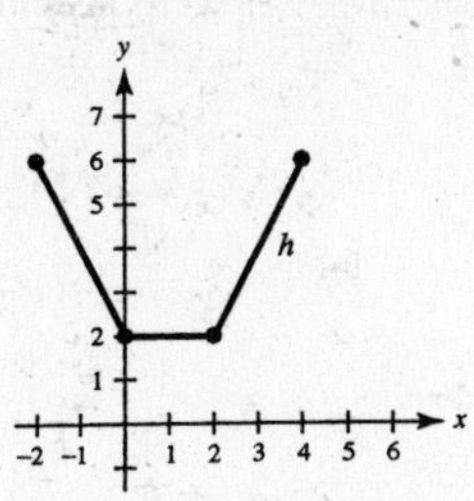

26.

x	0	1	2	3	4
$f(x)$	2	2	2	2	2
$g(x)$	0	1	$\sqrt{2}$	$\sqrt{3}$	2
$h(x) = (f + g)(x)$	2	3	$2 + \sqrt{2}$	$2 + \sqrt{3}$	4

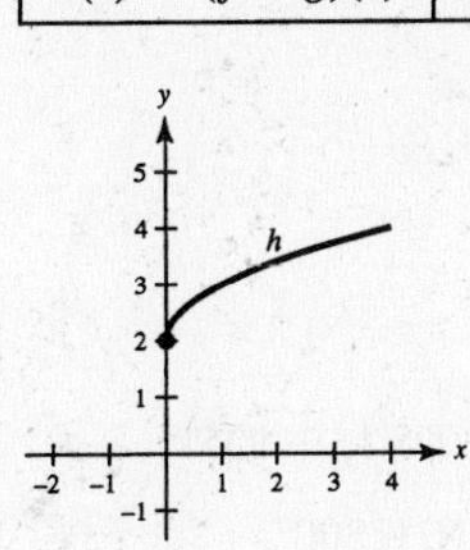

27. (a) $(f + g)(3) = f(3) + g(3) = 2 + 1 = 3$

(b) $\left(\dfrac{f}{g}\right)(2) = \dfrac{f(2)}{g(2)} = \dfrac{0}{2} = 0$

28. (a) $(f - g)(1) = f(1) - g(1)$

$= 2 - 3 = -1$

(b) $(fg)(4) = f(4) \cdot g(4) = 0$

$= 4 \cdot 0 = 0$

29. (a) $(f \circ g)(2) = f(g(2)) = f(2) = 0$

(b) $(g \circ f)(2) = g(f(2)) = g(0) = 4$

30. (a) $(f \circ g)(1) = f(g(1))$

$= f(3) = 2$

(b) $(g \circ f)(3) = g(f(3))$

$= g(2) = 2$

31. $f(x) = 3x,\ g(x) = -\dfrac{x^3}{10},\ (f + g)(x) = 3x - \dfrac{x^3}{10}$

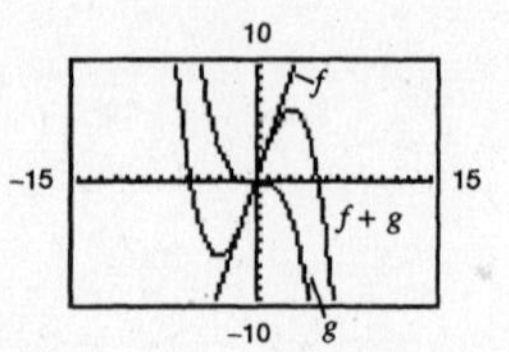

For $0 \le x \le 2$, $f(x)$ contributes most to the magnitude.
For $x > 6$, $g(x)$ contributes most to the magnitude.

32. $f(x) = \dfrac{x}{2},\ g(x) = \sqrt{x}$

$(f + g)(x) = \dfrac{x}{2} + \sqrt{x}$

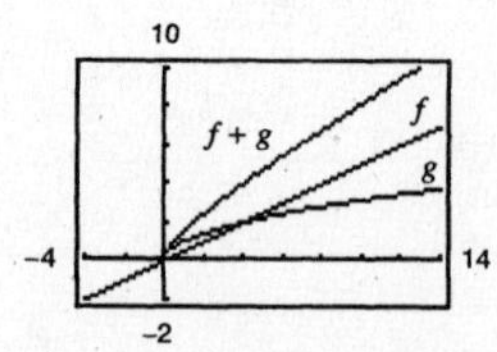

$g(x)$ contributes most to the magnitude of the sum for $0 \le x \le 2$. $f(x)$ contributes most to the magnitude of the sum for $x > 6$.

33. $T(x) = R(x) + B(x) = \frac{3}{4}x + \frac{1}{15}x^2$

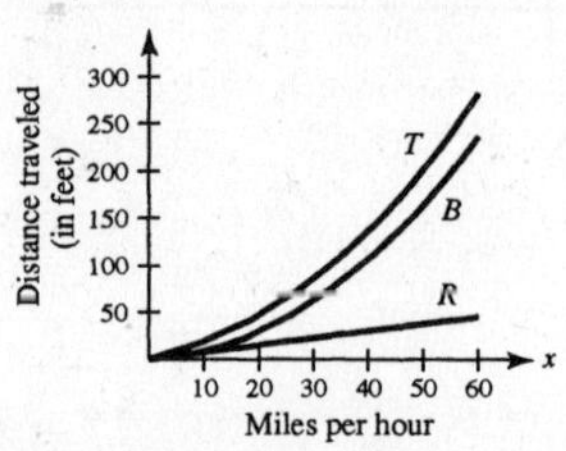

34. $R_1 = 480 - 8t - 0.8t^2,\ R_2 = 254 + 0.78t,\ t = 0, 1, 2, 3, 4$

$T = R_1 + R_2 = 734 - 7.22t - 0.8t^2,\ t = 0, 1, 2, 3, 4$

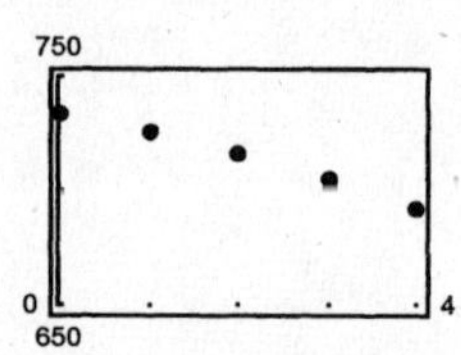

35.

Year	y_1	y_2
1995	1510	395
1996	1273	432
1997	1366	571
1998	1378	646
1999	1715	763
2000	2019	841
2001	2099	978

$y_1 \approx -14.056t^3 + 376.50t^2 - 3098.1t + 9351$

$y_2 \approx -0.500t^3 + 15.32t^2 - 47.1t + 300$

36.

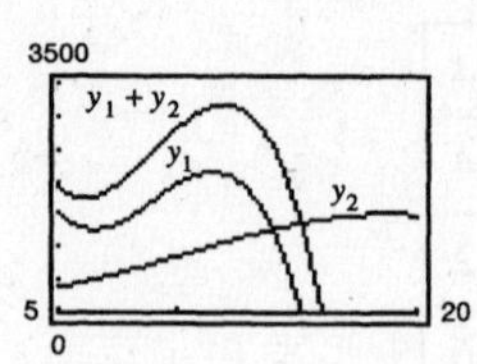

For 2005 use $t = 15$: $(y_1 + y_2)(15) \approx 1506$ thousand vehicles.

37. (a) T is a function of t since for each time t there corresponds one and only one temperature T.

(b) $T(4) = 60°$

$T(15) = 72°$

(c) $H(t) = T(t - 1)$; All the temperature changes would be one hour later.

(d) $H(t) = T(t) - 1$; The temperature would be decreased by one degree.

38. From 0 to 6, $T(t) = 60$.
From 6 to 6.5, $T(t)$ goes from 60 to 72.

$$T(t) - 60 = \frac{72 - 60}{6.5 - 6}(t - 6)$$

$$T(t) - 60 = 24t - 144$$

$$T(t) = 24t - 84$$

From 6.5 to 20.5, $T(t) = 72$.
From 20.5 to 21, $T(t)$ goes from 72 to 60.

$$T(t) - 60 = \frac{72 - 60}{20.5 - 21}(t - 21)$$

$$T(t) - 60 = -24(t - 21)$$

$$T(t) = -24t + 564$$

From 21 to 24, $T(t) = 60$.

$$T(t) = \begin{cases} 60, & 0 \le t \le 6 \\ 24t - 84, & 6 < t < 6.5 \\ 72, & 6.5 \le t \le 20.5 \\ -24t + 564, & 20.5 < t < 21 \\ 60, & 21 \le t \le 24 \end{cases}$$

T repeats every 24 hours.

39. $f(x) = x^2, g(x) = x - 1$

(a) $(f \circ g)(x) = f(g(x)) = f(x - 1) = (x - 1)^2$

(b) $(g \circ f)(x) = g(f(x)) = g(x^2) = x^2 - 1$

(c) $(f \circ f)(x) = f(f(x)) = f(x^2) = (x^2)^2 = x^4$

40. $f(x) = \sqrt[3]{x - 1},\ g(x) = x^3 + 1$

(a) $(f \circ g)(x) = f(g(x)) = f(x^3 + 1) = \sqrt[3]{(x^3 + 1) - 1} = \sqrt[3]{x^3} = x$

(b) $(g \circ f)(x) = g(f(x)) = g\left(\sqrt[3]{x - 1}\right) = \left(\sqrt[3]{x - 1}\right)^3 + 1 = (x - 1) + 1 = x$

(c) $(f \circ f)(x) = f(f(x)) = f\left(\sqrt[3]{x - 1}\right) = \sqrt[3]{\sqrt[3]{x - 1} - 1}$

41. $f(x) = 3x + 5, g(x) = 5 - x$

(a) $(f \circ g)(x) = f(g(x)) = f(5 - x) = 3(5 - x) + 5 = 20 - 3x$

(b) $(g \circ f)(x) = g(f(x)) = g(3x + 5) = 5 - (3x + 5) = -3x$

(c) $(f \circ f)(x) = f(f(x)) = f(3x + 5) = 3(3x + 5) + 5 = 9x + 20$

42. $f(x) = x^3,\ g(x) = \frac{1}{x}$

(a) $(f \circ g)(x) = f(g(x)) = f\left(\frac{1}{x}\right) = \left(\frac{1}{x}\right)^3 = \frac{1}{x^3}$

(b) $(g \circ f)(x) = g(f(x)) = g(x^3) = \frac{1}{x^3}$

(c) $(f \circ f)(x) = f(f(x)) = f(x^3) = (x^3)^3 = x^9$

43. $f(x) = \sqrt{x+4}$ Domain: $x \geq -4$

$g(x) = x^2$ Domain: all real numbers

(a) $(f \circ g)(x) = f(g(x)) = f(x^2) = \sqrt{x^2+4}$ Domain: all real numbers

(b) $(g \circ f)(x) = g(f(x)) = g(\sqrt{x+4}) = (\sqrt{x+4})^2 = x + 4$ Domain: $x \geq -4$

44. $f(x) = \sqrt[3]{x-5}$, $g(x) = x^3 + 1$

(a) $(f \circ g)(x) = f(g(x)) = f(x^3+1) = \sqrt[3]{x^3+1-5} = \sqrt[3]{x^3-4}$ Domain of $f, g, f \circ g, g \circ f$: all real numbers

(b) $(g \circ f)(x) = g(f(x)) = g(\sqrt[3]{x-5}) = (\sqrt[3]{x-5})^3 + 1 = x - 5 + 1 = x - 4$

45. $f(x) = \frac{1}{3}x - 3$ Domain: all real numbers

$g(x) = 3x + 1$ Domain: all real numbers

(a) $(f \circ g)(x) = f(g(x)) = f(3x+1) = \frac{1}{3}(3x+1) - 3 = x - \frac{8}{3}$ Domain: all real numbers

(b) $(g \circ f)(x) = g(f(x)) = g(\frac{1}{3}x - 3) = 3(\frac{1}{3}x - 3) + 1 = x - 8$ Domain: all real numbers

46. $f(x) = x^2 + 1, g(x) = \sqrt{x}$

(a) $(f \circ g)(x) = f(g(x)) = f(\sqrt{x}) = (\sqrt{x})^2 + 1 = x + 1$

(b) $(g \circ f)(x) = g(f(x)) = g(x^2+1) = \sqrt{x^2+1}$

Domain of $f, g \circ f$: all real numbers

Domain of $g, f \circ g$: $x \geq 0$

47. $f(x) = x^4$ Domain: all real numbers

$g(x) = x^4$ Domain: all real numbers

(a) and (b) $(f \circ g)(x) = (g \circ f)(x) = (x^4)^4 = x^{16}$

Domain: all real numbers

48. $f(x) = \sqrt{x}, g(x) = 2x - 3$

(a) $(f \circ g)(x) = f(g(x)) = f(2x-3) = \sqrt{2x-3}$

(b) $(g \circ f)(x) = g(f(x)) = g(\sqrt{x}) = 2\sqrt{x} - 3$

Domain of $f, g \circ f$: $x \geq 0$

Domain of g: all real numbers

Domain of $f \circ g$: $x \geq \frac{3}{2}$

49. $f(x) = |x|$ Domain: all real numbers

$g(x) = x + 6$ Domain: all real numbers

(a) $(f \circ g)(x) = f(g(x)) = f(x+6) = |x+6|$

Domain: all real numbers

(b) $(g \circ f)(x) = g(f(x)) = g(|x|) = |x| + 6$

Domain: all real numbers

50. $f(x) = x^{2/3}$, $g(x) = x^6$

(a) $(f \circ g)(x) = f(g(x)) = f(x^6) = (x^6)^{2/3} = x^4$

(b) $(g \circ f)(x) = g(f(x)) = g(x^{2/3}) = (x^{2/3})^6 = x^4$

Domain of $f, g, f \circ g, g \circ f$: all real numbers

51. $f(x) = \dfrac{1}{x}$ Domain: all real numbers except $x = 0$

$g(x) = x + 3$ Domain: all real numbers

(a) $(f \circ g)(x) = f(g(x)) = f(x+3) = \dfrac{1}{x+3}$

Domain: all real numbers except $x = -3$

(b) $(g \circ f)(x) = g(f(x)) = g\left(\dfrac{1}{x}\right) = \dfrac{1}{x} + 3$

Domain: all real numbers except $x = 0$

52. $f(x) = \dfrac{3}{x^2 - 1}$, $g(x) = x + 1$

(a) $(f \circ g)(x) = f(g(x)) = f(x + 1) = \dfrac{3}{(x + 1)^2 - 1} = \dfrac{3}{x^2 + 2x + 1 - 1} = \dfrac{3}{x^2 + 2x}$

(b) $(g \circ f)(x) = g(f(x)) = g\left(\dfrac{3}{x^2 - 1}\right) = \dfrac{3}{x^2 - 1} + 1 = \dfrac{3 + x^2 - 1}{x^2 - 1} = \dfrac{x^2 + 2}{x^2 - 1}$

Domain of f, $g \circ f$: all real numbers $x \neq \pm 1$

Domain of g: all real numbers

Domain of $f \circ g$: all real numbers $x \neq 0, -2$

53. Let $f(x) = x^2$ and $g(x) = 2x + 1$, then $(f \circ g)(x) = h(x)$.

This is not a unique solution.

For example, if $f(x) = (x + 1)^2$ and $g(x) = 2x$, then $(f \circ g)(x) = h(x)$ as well.

54. $h(x) = (1 - x)^3$

One possibility: Let $g(x) = 1 - x$ and $f(x) = x^3$.

$(f \circ g)(x) = f(1 - x) = (1 - x)^3 = h(x)$

55. Let $f(x) = \sqrt[3]{x}$ and $g(x) = x^2 - 4$, then $(f \circ g)(x) = h(x)$.

This answer is not unique. Other possibilities may be:

$f(x) = \sqrt[3]{x - 4}$ and $g(x) = x^2$

or $f(x) = \sqrt[3]{-x}$ and $g(x) = 4 - x^2$

or $f(x) = \sqrt[9]{x}$ and $g(x) = (x^2 - 4)^3$

56. $h(x) = \sqrt{9 - x}$

One possibility: Let $g(x) = 9 - x$ and $f(x) = \sqrt{x}$.

$(f \circ g)(x) = f(9 - x) = \sqrt{9 - x} = h(x)$

57. Let $f(x) = 1/x$ and $g(x) = x + 2$, then $(f \circ g)(x) = h(x)$.

This is not a unique solution.

Other possibilities may be:

$f(x) = \dfrac{1}{x + 2}$ and $g(x) = x$

or $f(x) = \dfrac{1}{x + 1}$ and $g(x) = x + 1$

58. $h(x) = \dfrac{4}{(5x + 2)^2}$

One possibility: Let $g(x) = 5x + 2$ and $f(x) = \dfrac{4}{x^2}$.

$(f \circ g)(x) = f(5x + 2) = \dfrac{4}{(5x + 2)^2} = h(x)$

59. Let $f(x) = \dfrac{x + 3}{4 + x}$ and $g(x) = -x^2$, then $(f \circ g)(x) = h(x)$.

This answer is not unique.

Another possibility may be:

$f(x) = \dfrac{x + 1}{x + 2}$ and $g(x) = -x^2 + 2$

60. $h(x) = \dfrac{27x^3 + 6x}{10 - 27x^3}$

One possibility: Let $g(x) = x^3$ and $f(x) = \dfrac{27x + 6\sqrt[3]{x}}{10 - 27x}$.

$(f \circ g)(x) = f(x^3) = \dfrac{27x^3 + 6\sqrt[3]{x^3}}{10 - 27x^3} = h(x)$

61. (a) $r(x) = \dfrac{x}{2}$

(b) $A(r) = \pi r^2$

(c) $(A \circ r)(x) = A(r(x)) = A\left(\dfrac{x}{2}\right) = \pi\left(\dfrac{x}{2}\right)^2$

$(A \circ r)(x)$ represents the area of the circular base of the tank on the square foundation with side length x.

62. $(A \circ r)(t) = A(r(t)) = A(0.6t) = \pi(0.6t)^2 = 0.36\pi t^2$

$A \circ r$ represents the area of the circle at time t.

63. $(C \circ x)(t) = C(x(t))$

$= 60(50t) + 750$

$= 3000t + 750$

$(C \circ x)(t)$ represents the cost after t production hours.

64. $P(n) = \sqrt{n^2 + 1}$

$n(t) = 8 + 0.1t$

$(P \circ n)(t) = P(n(t))$

$= \sqrt{(8 + 0.1t)^2 + 1}$

This function gives the average daily level of a certain pollutant in the water t years from now.

65. False. $(f \circ g)(x) = 6x + 1$ and $(g \circ f)(x) = 6x + 6$.

66. True. The range of g must be a subset of the domain of f for $(f \circ g)(x)$ to be defined. Since $(f \circ g)(x) = f(g(x))$ and since $g(x)$ represents the range of g, then f is being evaluated with values from g's range.

67. True

Let $h(x) = f(x) + g(x)$.

Then $h(-x) = f(-x) + g(-x)$

$= f(x) + g(x)$ since f and g are even

$= h(x)$.

Thus, $h(x) = f(x) + g(x)$ is also even.

68. False.

Let $h(x) = f(x) + g(x)$.

Then $h(-x) = f(-x) + g(-x)$

$= -f(x) - g(x)$ Since f and g are odd

$= -[f(x) + g(x)]$

$= -h(x)$.

Thus, $h(x) = f(x) + g(x)$ is also odd, not even.

69. (a) $f(g(x)) = f(0.03x) = 0.03x - 500{,}000$

(b) $g(f(x)) = g(x - 500{,}000) = 0.03(x - 500{,}000)$

$g(f(x))$ represents 3% of an amount over \$500,000.

Section 1.5 Inverse Functions

1. $f^{-1}(x) = \frac{x}{6} = \frac{1}{6}x$

$$f(f^{-1}(x)) = f\left(\frac{x}{6}\right) = 6\left(\frac{x}{6}\right) = x$$

$$f^{-1}(f(x)) = f^{-1}(6x) = \frac{6x}{6} = x$$

2. $f(x) = \frac{1}{3}x$

$f^{-1}(x) = 3x$

$$f(f^{-1}(x)) = f(3x) = \frac{1}{3}(3x) = x$$

$$f^{-1}(f(x)) = f^{-1}\left(\frac{1}{3}x\right) = 3\left(\frac{1}{3}x\right) = x$$

3. $f^{-1}(x) = x - 9$

$f(f^{-1}(x)) = f(x - 9) = (x - 9) + 9 = x$

$f^{-1}(f(x)) = f^{-1}(x + 9) = (x + 9) - 9 = x$

4. $f(x) = x - 4$

$f^{-1}(x) = x + 4$

$f(f^{-1}(x)) = f(x + 4) = x + 4 - 4 = x$

$f^{-1}(f(x)) = f^{-1}(x - 4) = x - 4 + 4 = x$

5. $f^{-1}(x) = \frac{x - 1}{3}$

$$f(f^{-1}(x)) = f\left(\frac{x - 1}{3}\right) = 3\left(\frac{x - 1}{3}\right) + 1 = x$$

$$f^{-1}(f(x)) = f^{-1}(3x + 1) = \frac{(3x + 1) - 1}{3} = x$$

6. $f(x) = -2x - 9$

$y = -2x - 9$

$x = -2y - 9$

$-\frac{1}{2}(x + 9) = y$

$f^{-1}(x) = -\frac{1}{2}(x + 9)$

$f(f^{-1}(x)) = f\left[-\frac{1}{2}(x + 9)\right] = -2\left[-\frac{1}{2}(x + 9)\right] - 9$

$= (x + 9) - 9 = x$

$f^{-1}(f(x)) = f^{-1}(-2x - 9) = -\frac{1}{2}[(-2x - 9) + 9]$

$= -\frac{1}{2}(-2x) = x$

7. $f(x) = \frac{x - 1}{5}$

$f^{-1}(x) = 5x + 1$

$f(f^{-1}(x)) = f(5x + 1) = \frac{5x + 1 - 1}{5} = \frac{5x}{5} = x$

$f^{-1}(f(x)) = f^{-1}\left(\frac{x - 1}{5}\right) = 5\left(\frac{x - 1}{5}\right) + 1 = x - 1 + 1 = x$

8. $f(x) = \frac{4x + 7}{2}$

$y = \frac{4x + 7}{2}$

$x = \frac{4y + 7}{2}$

$2x = 4y + 7$

$\frac{1}{4}(2x - 7) = y$

$f^{-1}(x) = \frac{1}{4}(2x - 7)$

$f(f^{-1}(x)) = f\left[\frac{1}{4}(2x - 7)\right] = \frac{4[1/4(2x - 7)] + 7}{2} = \frac{(2x - 7) + 7}{2} = x$

$f^{-1}(f(x)) = f^{-1}\left(\frac{4x + 7}{2}\right) = \frac{1}{4}\left[2\left(\frac{4x + 7}{2}\right) - 7\right] = \frac{1}{4}[(4x + 7) - 7] = x$

9. $f^{-1}(x) = x^3$

$f(f^{-1}(x)) = f(x^3) = \sqrt[3]{x^3} = x$

$f^{-1}(f(x)) = f^{-1}(\sqrt[3]{x}) = (\sqrt[3]{x})^3 = x$

10. $f(x) = x^5$

$f^{-1}(x) = \sqrt[5]{x}$

$f(f^{-1}(x)) = f(\sqrt[5]{x}) = (\sqrt[5]{x})^5 = x$

$f^{-1}(f(x)) = f^{-1}(x^5) = \sqrt[5]{x^5} = x$

11. (a) $f(g(x)) = f\left(\frac{x}{2}\right) = 2\left(\frac{x}{2}\right) = x$

$g(f(x)) = g(2x) = \frac{2x}{2} = x$

(b)

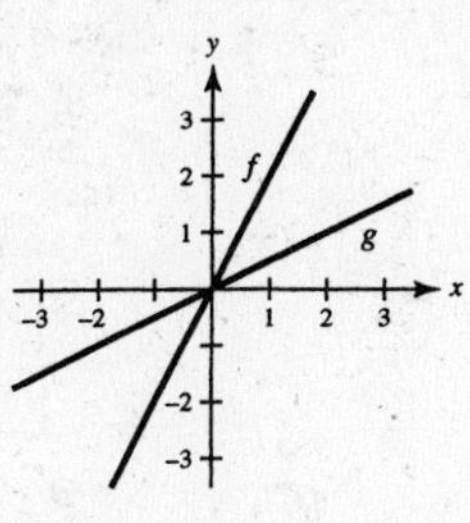

12. (a) $f(x) = x - 5,\ g(x) = x + 5$

$f(g(x)) = f(x + 5) = (x + 5) - 5 = x$

$g(f(x)) = g(x - 5) = (x - 5) + 5 = x$

(b)

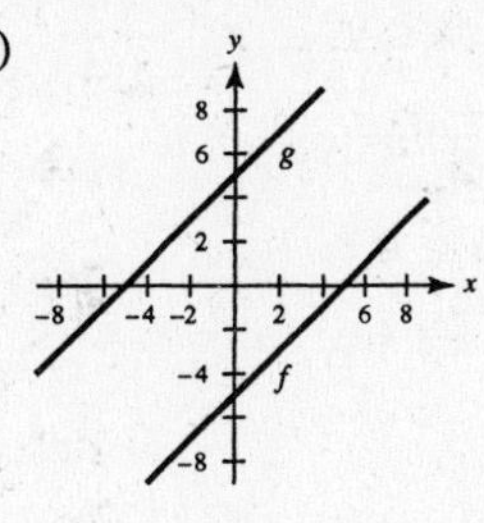

13. (a) $f(g(x)) = f\left(\dfrac{x-1}{5}\right) = 5\left(\dfrac{x-1}{5}\right) + 1 = x$

$g(f(x)) = g(5x + 1) = \dfrac{(5x+1) - 1}{5} = x$

(b)

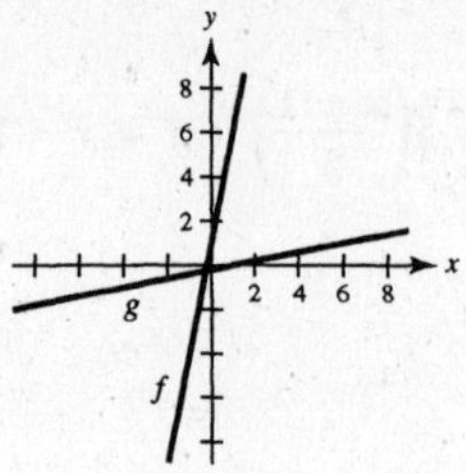

14. (a) $f(x) = 3 - 4x,\ g(x) = \dfrac{3-x}{4}$

$f(g(x)) = f\left(\dfrac{3-x}{4}\right) = 3 - 4\left(\dfrac{3-x}{4}\right) = 3 - (3 - x) = x$

$g(f(x)) = g(3 - 4x) = \dfrac{3 - (3 - 4x)}{4} = \dfrac{4x}{4} = x$

(b)

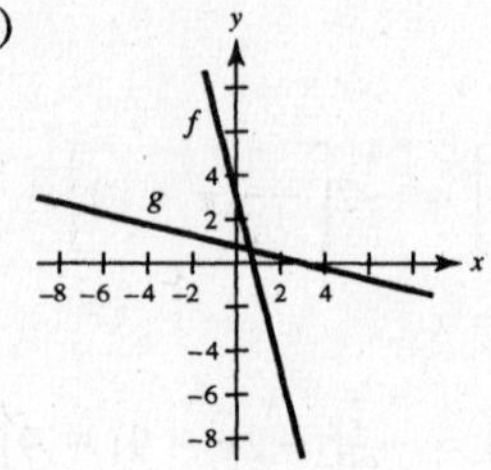

15. (a) $f(g(x)) = f(\sqrt[3]{x}) = (\sqrt[3]{x})^3 = x$

$g(f(x)) = g(x^3) = \sqrt[3]{x^3} = x$

(b)

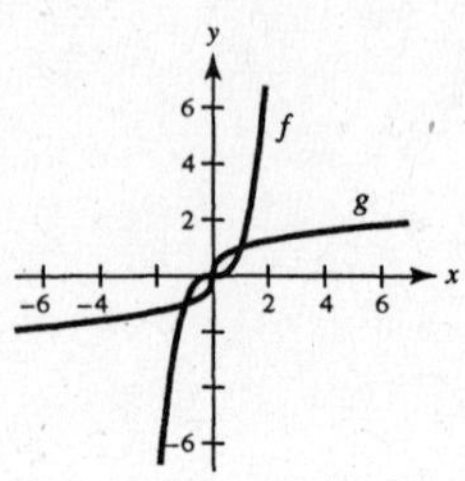

16. (a) $f(x) = \dfrac{1}{x},\ g(x) = \dfrac{1}{x}$

$f(g(x)) = f\left(\dfrac{1}{x}\right) = \dfrac{1}{1/x} = 1 \div \dfrac{1}{x} = 1 \cdot \dfrac{x}{1} = x$

$g(f(x)) = g\left(\dfrac{1}{x}\right) = \dfrac{1}{1/x} = 1 \div \dfrac{1}{x} = 1 \cdot \dfrac{x}{1} = x$

(b)

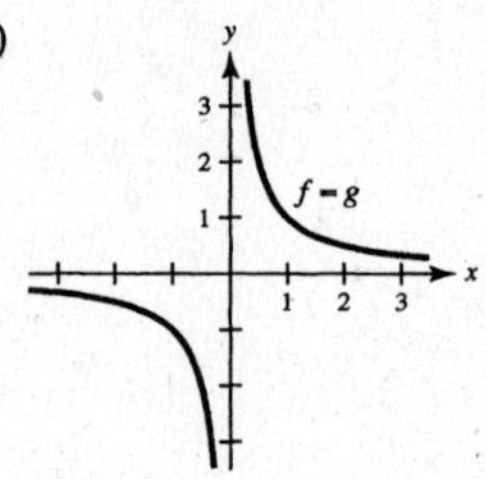

17. (a) $f(g(x)) = f(x^2 + 4),\ x \geq 0$

$= \sqrt{(x^2 + 4) - 4} = x$

$g(f(x)) = g(\sqrt{x - 4})$

$= (\sqrt{x-4})^2 + 4 = x$

(b)

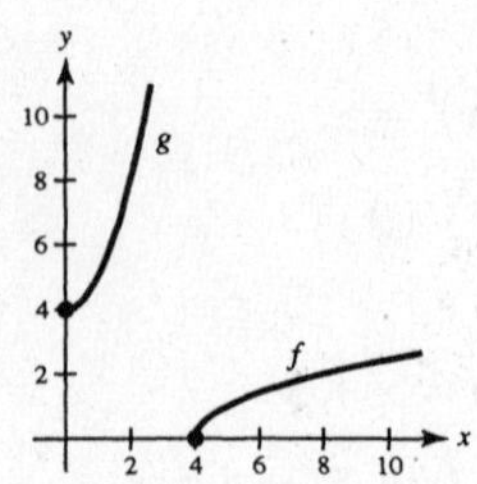

18. (a) $f(x) = 1 - x^3,\ g(x) = \sqrt[3]{1 - x}$

$f(g(x)) = f(\sqrt[3]{1-x}) = 1 - (\sqrt[3]{1-x})^3$

$= 1 - (1 - x) = x$

$g(f(x)) = g(1 - x^3) = \sqrt[3]{1 - (1 - x^3)}$

$= \sqrt[3]{x^3} = x$

(b)

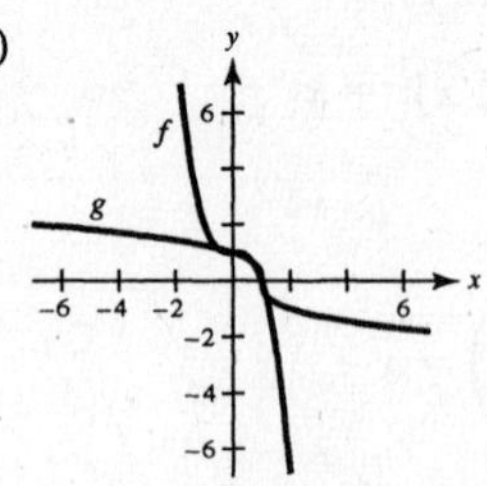

19. (a) $f(g(x)) = f(\sqrt{9 - x}),\ x \leq 9$

$= 9 - (\sqrt{9-x})^2 = x$

$g(f(x)) = g(9 - x^2),\ x \geq 0$

$= \sqrt{9 - (9 - x^2)} = x$

(b)

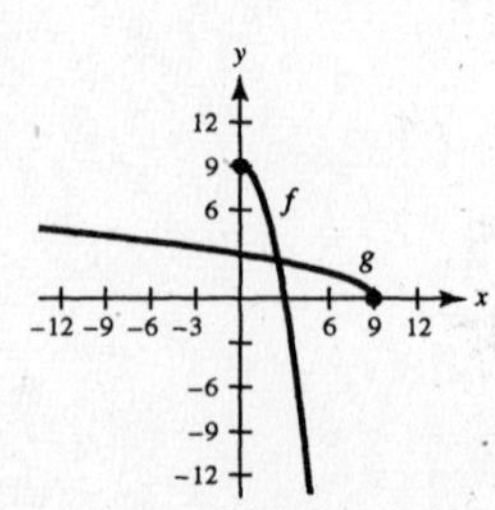

20. (a) $f(x) = \dfrac{1}{1+x},\ x \ge 0;\ g(x) = \dfrac{1-x}{x},\ 0 < x \le 1$

$$f(g(x)) = f\left(\frac{1-x}{x}\right) = \frac{1}{1 + \left(\dfrac{1-x}{x}\right)} = \frac{1}{\dfrac{x}{x} + \dfrac{1-x}{x}} = \frac{1}{\dfrac{1}{x}} = x$$

$$g(f(x)) = g\left(\frac{1}{1+x}\right) = \frac{1 - \left(\dfrac{1}{1+x}\right)}{\left(\dfrac{1}{1+x}\right)} = \frac{\dfrac{1+x}{1+x} - \dfrac{1}{1+x}}{\dfrac{1}{1+x}} = \frac{\dfrac{x}{1+x}}{\dfrac{1}{1+x}} = \frac{x}{1+x} \cdot \frac{x+1}{1} = x$$

(b)

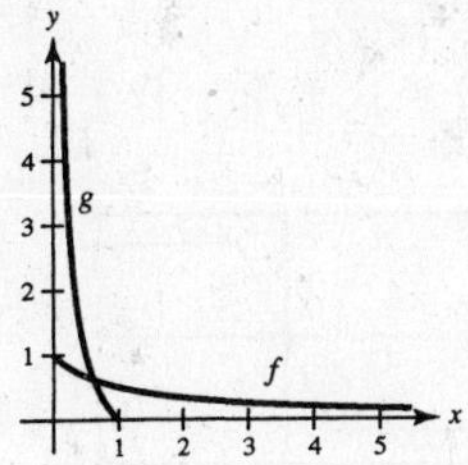

21. (a) $$f(g(x)) = f\left(-\frac{5x+1}{x-1}\right) = \frac{\left(-\dfrac{5x+1}{x-1} - 1\right)}{\left(-\dfrac{5x+1}{x-1} + 5\right)} \cdot \frac{x-1}{x-1} = \frac{-(5x+1) - (x-1)}{-(5x+1) + 5(x-1)} = \frac{-6x}{-6} = x$$

$$g(f(x)) = g\left(\frac{x-1}{x+5}\right) = -\frac{\left[5\left(\dfrac{x-1}{x+5}\right) + 1\right]}{\left[\dfrac{x-1}{x+5} - 1\right]} \cdot \frac{x+5}{x+5} = -\frac{5(x-1) + (x+5)}{(x-1) - (x+5)} = -\frac{6x}{-6} = x$$

(b)

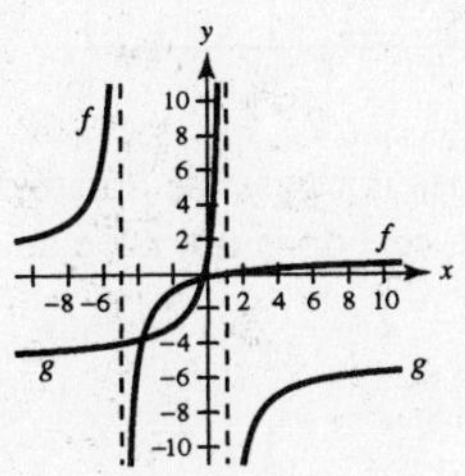

22. (a) $f(x) = \dfrac{x+3}{x-2},\ g(x) = \dfrac{2x+3}{x-1}$

$$f(g(x)) = f\left(\frac{2x+3}{x-1}\right) = \frac{\dfrac{2x+3}{x-1} + 3}{\dfrac{2x+3}{x-1} - 2} = \frac{\dfrac{2x+3+3x-3}{x-1}}{\dfrac{2x+3-2x+2}{x-1}} = \frac{5x}{5} = x$$

$$g(f(x)) = g\left(\frac{x+3}{x-2}\right) = \frac{2\left(\dfrac{x+3}{x-2}\right) + 3}{\dfrac{x+3}{x-2} - 1} = \frac{\dfrac{2x+6+3x-6}{x-2}}{\dfrac{x+3-x+2}{x-2}} = \frac{5x}{5} = x$$

(b)

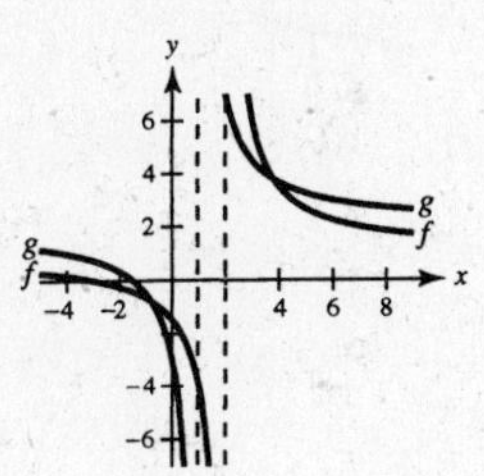

23. No, $\{(-2, -1), (1, 0), (2, 1), (1, 2), (-2, 3), (-6, 4)\}$ does not represent a function. -2 and 1 are paired with two different values.

24. Yes, $\{(10, -3), (6, -2), (4, -1), (1, 0), (-3, 2), (-10, 3)\}$ does represent a function.

$\{(10, -3), (6, -2), (4, -1), (1, 0), (-3, 2), (-10, 3)\}$

25. Since no horizontal line crosses the graph of f at more than one point, f **has** an inverse.

26. No, because some horizontal lines intersect the graph twice, f does **not** have an inverse.

27. Since some horizontal lines cross the graph of f twice, f does **not** have an inverse.

28. Yes, because no horizontal lines intersect the graph at more than one point, f **has** an inverse.

29. $g(x) = \dfrac{4 - x}{6}$

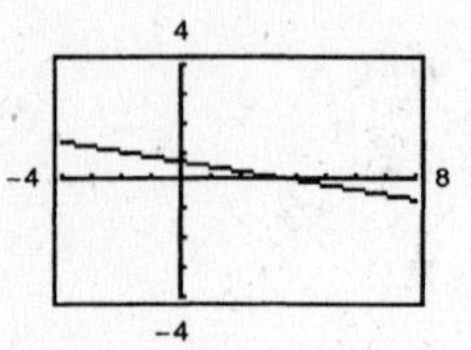

g passes the horizontal line test, so g **has** an inverse.

30. $f(x) = 10$

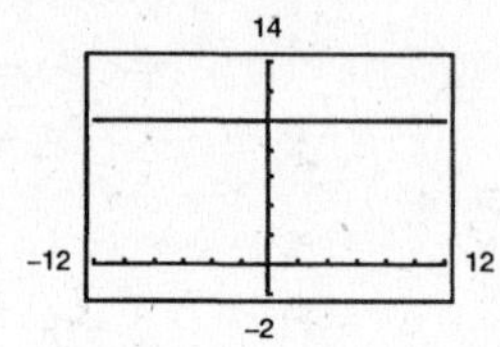

f does not pass the horizontal line test, so f does **not** have an inverse.

31. $h(x) = |x + 4| - |x - 4|$

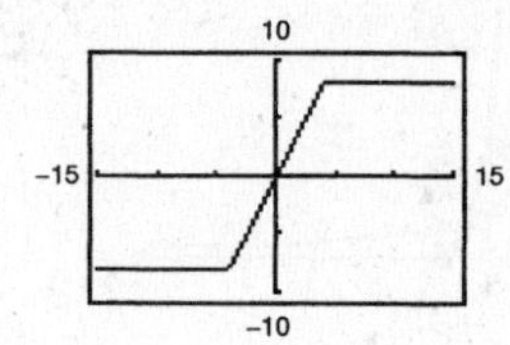

h does not pass the horizontal line test, so h does **not** have an inverse.

32. $g(x) = (x + 5)^3$

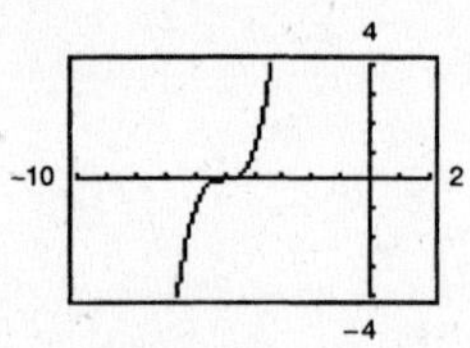

g passes the horizontal line test, so g **has** an inverse.

33. $f(x) = -2x\sqrt{16 - x^2}$

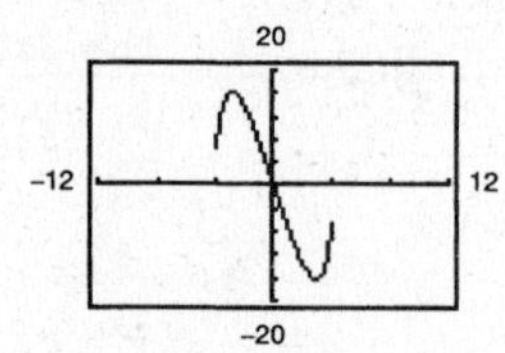

f does not pass the horizontal line test, so f does **not** have an inverse.

34. $f(x) = \frac{1}{8}(x + 2)^2 - 1$

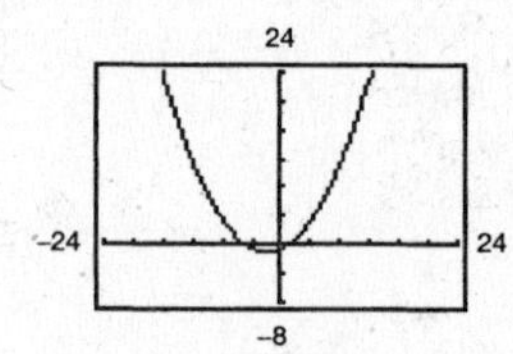

f does not pass the horizontal line test, so f does **not** have an inverse.

35.
$$f(x) = 2x - 3$$
$$y = 2x - 3$$
$$x = 2y - 3$$
$$y = \frac{x + 3}{2}$$
$$f^{-1}(x) = \frac{x + 3}{2}$$

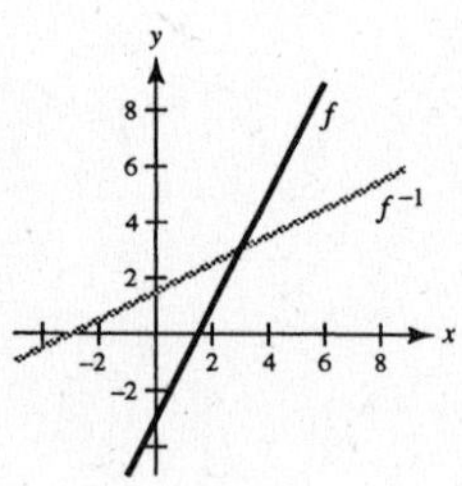

36.
$$f(x) = 3x + 1$$
$$y = 3x + 1$$
$$x = 3y + 1$$
$$\frac{x - 1}{3} = y$$
$$f^{-1}(x) = \frac{x - 1}{3}$$

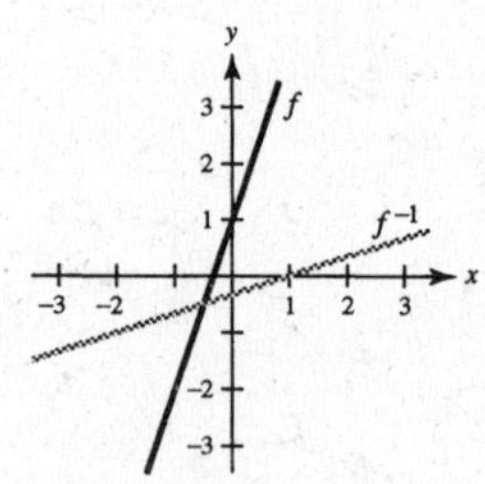

37.
$$f(x) = x^5 - 2$$
$$y = x^5 - 2$$
$$x = y^5 - 2$$
$$y = \sqrt[5]{x + 2}$$
$$f^{-1}(x) = \sqrt[5]{x + 2}$$

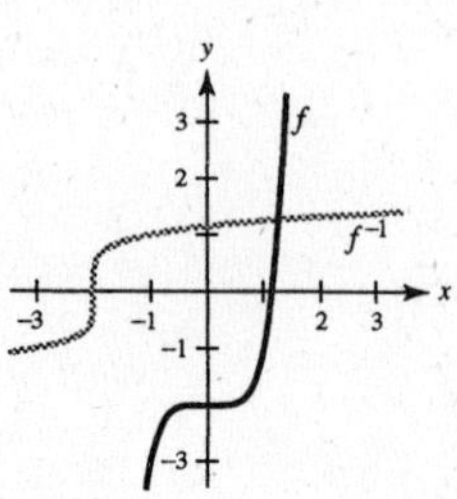

38.
$$f(x) = x^3 + 1$$
$$y = x^3 + 1$$
$$x = y^3 + 1$$
$$x - 1 = y^3$$
$$\sqrt[3]{x - 1} = y$$
$$f^{-1}(x) = \sqrt[3]{x - 1}$$

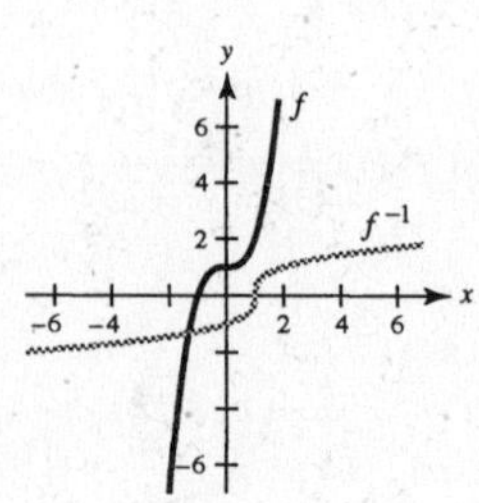

39. $f(x) = \sqrt{x}$

$y = \sqrt{x}$

$x = \sqrt{y}$

$y = x^2$

$f^{-1}(x) = x^2, \ x \geq 0$

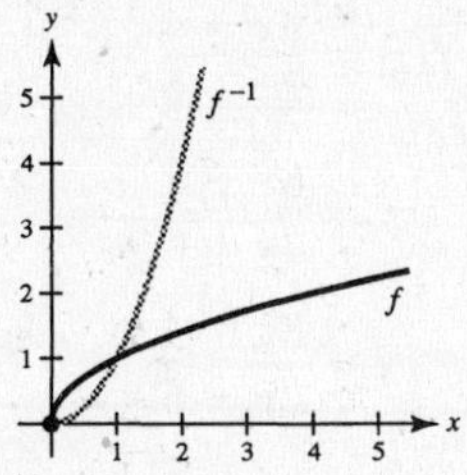

40. $f(x) = x^2, \ x \geq 0$

$y = x^2$

$x = y^2$

$\sqrt{x} = y$

$f^{-1}(x) = \sqrt{x}$

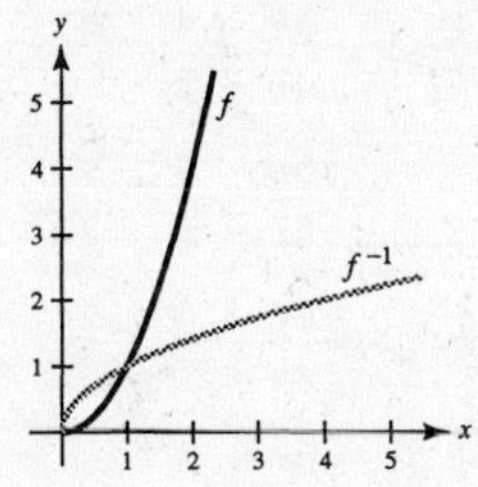

41. $f(x) = \sqrt{4 - x^2}, \ 0 \leq x \leq 2$

$y = \sqrt{4 - x^2}$

$x = \sqrt{4 - y^2}$

$f^{-1}(x) = \sqrt{4 - x^2}, \ 0 \leq x \leq 2$

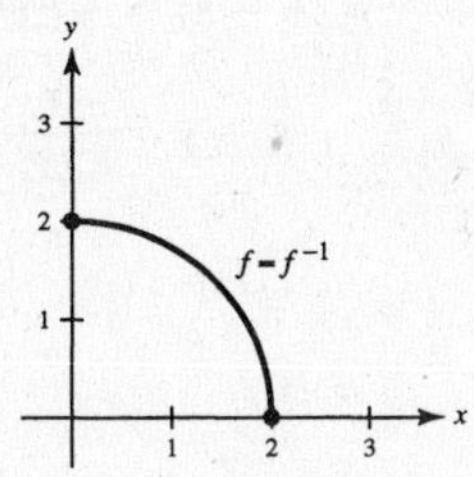

42. $f(x) = x^2 - 2, x \leq 0$

$y = x^2 - 2$

$x = y^2 - 2$

$\pm\sqrt{x + 2} = y$

$f^{-1}(x) = -\sqrt{x + 2}$

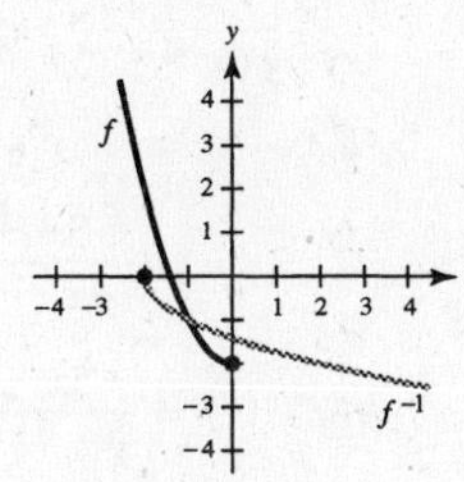

43. $f(x) = \dfrac{4}{x}$

$y = \dfrac{4}{x}$

$x = \dfrac{4}{y}$

$xy = 4$

$y = \dfrac{4}{x}$

$f^{-1}(x) = \dfrac{4}{x}$

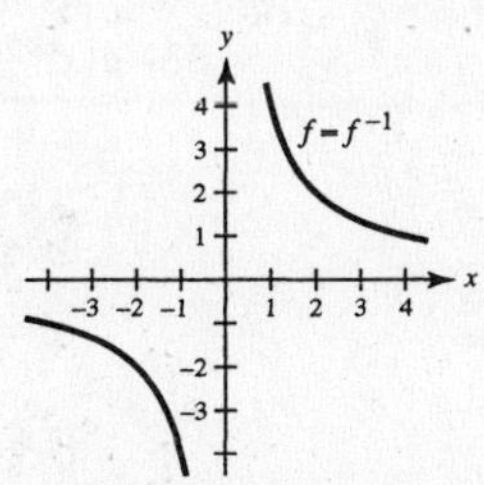

44. $f(x) = -\dfrac{2}{x}$

$y = -\dfrac{2}{x}$

$x = -\dfrac{2}{y}$

$y = -\dfrac{2}{x}$

$f^{-1}(x) = -\dfrac{2}{x}$

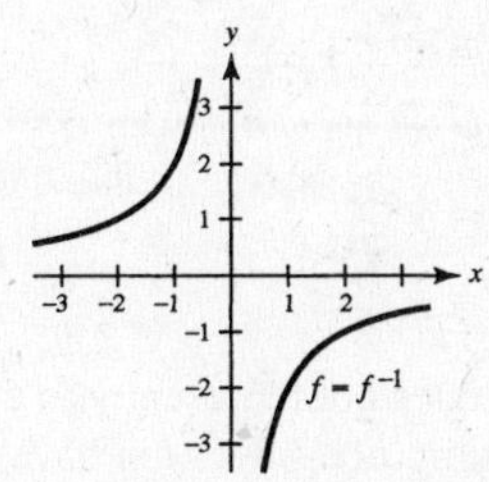

45. $f(x) = \dfrac{x + 1}{x - 2}$

$y = \dfrac{x + 1}{x - 2}$

$x = \dfrac{y + 1}{y - 2}$

$x(y - 2) = y + 1$

$xy - 2x = y + 1$

$xy - y = 2x + 1$

$y(x - 1) = 2x + 1$

$y = \dfrac{2x + 1}{x - 1}$

$f^{-1}(x) = \dfrac{2x + 1}{x - 1}$

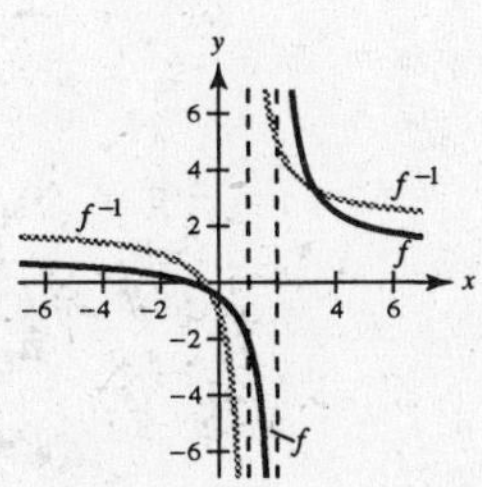

46. $f(x) = \dfrac{x - 3}{x + 2}$

$y = \dfrac{x - 3}{x + 2}$

$x = \dfrac{y - 3}{y + 2}$

$xy + 2x - y + 3 = 0$

$y(x - 1) = -2x - 3$

$y = \dfrac{-2x - 3}{x - 1}$

$f^{-1}(x) = \dfrac{-2x - 3}{x - 1}$

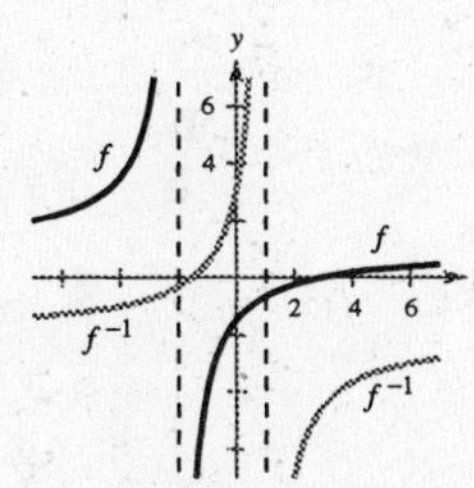

47. $f(x) = \sqrt[3]{x - 1}$

$y = \sqrt[3]{x - 1}$

$x = \sqrt[3]{y - 1}$

$x^3 = y - 1$

$y = x^3 + 1$

$f^{-1}(x) = x^3 + 1$

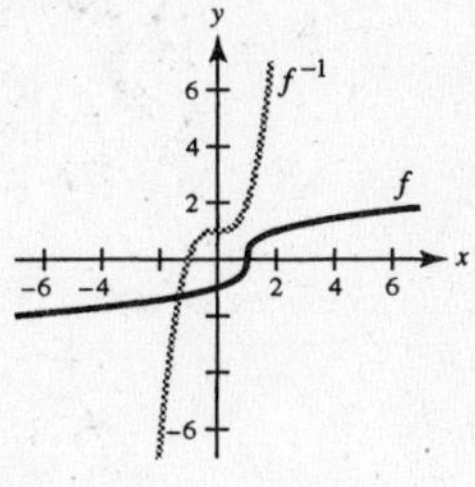

48. $f(x) = x^{3/5}$

$y = x^{3/5}$

$x = y^{3/5}$

$x^{5/3} = (y^{3/5})^{5/3}$

$x^{5/3} = y$

$f^{-1}(x) = x^{5/3}$

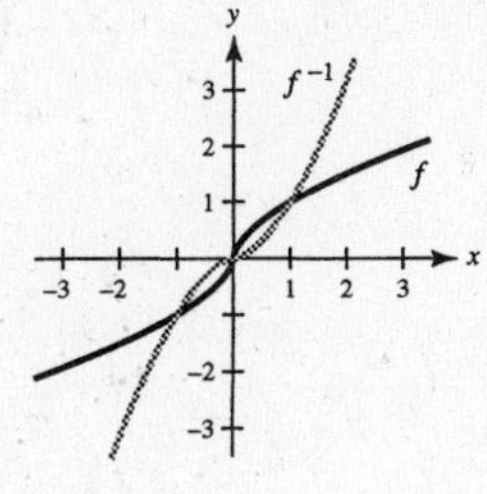

49. $f(x) = \dfrac{6x + 4}{4x + 5}$

$y = \dfrac{6x + 4}{4x + 5}$

$x = \dfrac{6y + 4}{4y + 5}$

$x(4y + 5) = 6y + 4$

$4xy + 5x = 6y + 4$

$4xy - 6y = -5x + 4$

$y(4x - 6) = -5x + 4$

$y = \dfrac{-5x + 4}{4x - 6}$

$f^{-1}(x) = \dfrac{-5x + 4}{4x - 6} = \dfrac{5x - 4}{6 - 4x}$

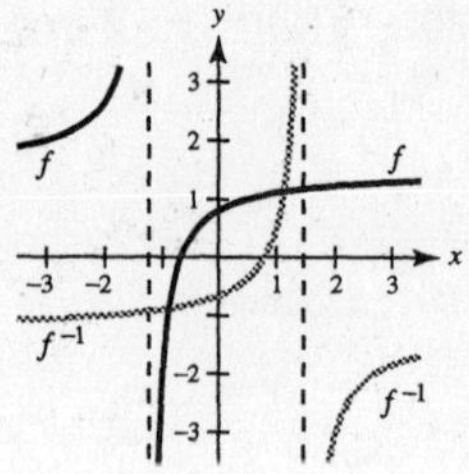

50. $f(x) = \dfrac{8x - 4}{2x + 6}$

$y = \dfrac{8x - 4}{2x + 6}$

$x = \dfrac{8y - 4}{2y + 6}$

$2xy + 6x = 8y - 4$

$y(2x - 8) = -6x - 4$

$y = \dfrac{-6x - 4}{2x - 8}$

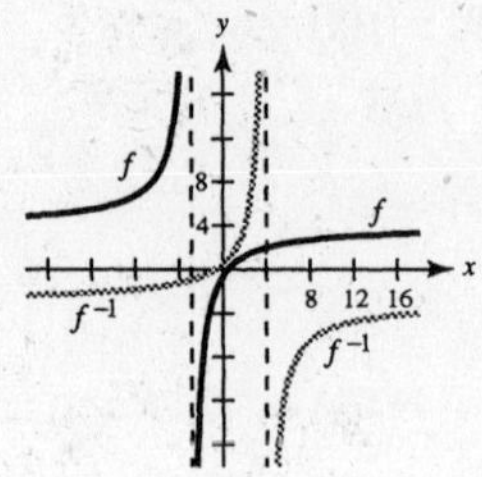

51. $f(x) = x^4$

$y = x^4$

$x = y^4$

$y = \pm\sqrt[4]{x}$

This does not represent y as a function of x. f does not have an inverse.

52. $f(x) = \dfrac{1}{x^2}$

$y = \dfrac{1}{x^2}$

$x = \dfrac{1}{y^2}$

$y^2 = \dfrac{1}{x}$

$y = \pm\sqrt{\dfrac{1}{x}}$

This does not represent y as a function of x. f does not have an inverse.

53. $g(x) = \dfrac{x}{8}$

$y = \dfrac{x}{8}$

$x = \dfrac{y}{8}$

$y = 8x$

This is a function of x, so g has an inverse.

$g^{-1}(x) = 8x$

54. $f(x) = 3x + 5$

$y = 3x + 5$

$x = 3y + 5$

$x - 5 = 3y$

$\dfrac{x - 5}{3} = y$

This is a function of x, so f has an inverse.

$f^{-1}(x) = \dfrac{x - 5}{3}$

55. $p(x) = -4$

$y = -4$

Since $y = -4$ for all x, the graph is a horizontal line and fails the horizontal line test. p does not have an inverse.

56. $f(x) = \dfrac{3x + 4}{5}$

$y = \dfrac{3x + 4}{5}$

$x = \dfrac{3y + 4}{5}$

$5x = 3y + 4$

$5x - 4 = 3y$

$\dfrac{5x - 4}{3} = y$

This is a function of x, so f has an inverse.

$f^{-1}(x) = \dfrac{5x - 4}{3}$

57. $f(x) = (x + 3)^2,\ x \geq -3 \implies y \geq 0$

$y = (x + 3)^2,\ x \geq -3,\ y \geq 0$

$x = (y + 3)^2,\ y \geq -3,\ x \geq 0$

$\sqrt{x} = y + 3,\ y \geq -3,\ x \geq 0$

$y = \sqrt{x} - 3,\ x \geq 0,\ y \geq -3$

This is a function of x, so f has an inverse.

$f^{-1}(x) = \sqrt{x} - 3,\ x \geq 0$

58. $q(x) = (x - 5)^2$

$y = (x - 5)^2$

$x = (y - 5)^2$

$\pm\sqrt{x} = y - 5$

$5 \pm \sqrt{x} = y$

This does not represent y as a function of x, so q does not have an inverse.

59. $f(x) = \begin{cases} x + 3, & x < 0 \\ 6 - x, & x \geq 0 \end{cases}$

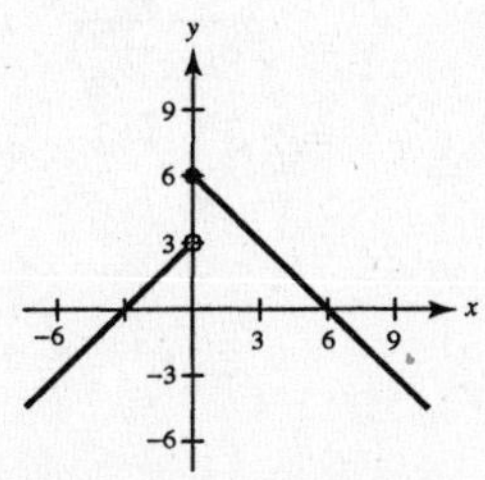

The graph fails the horizontal line test, so $f(x)$ does not have an inverse.

60. $f(x) = \begin{cases} -x, & x \leq 0 \\ x^2 - 3x, & x > 0 \end{cases}$

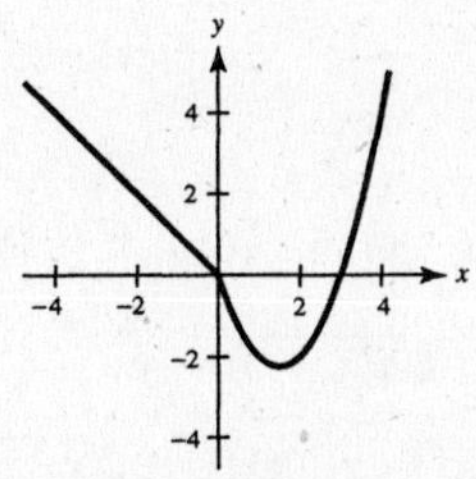

The graph fails the horizontal line test, so f does not have an inverse.

61. The graph represents a one-to-one function, so the function has an inverse.

$h(x) = \dfrac{1}{x}$

$y = \dfrac{1}{x}$

$xy = 1$

$y = \dfrac{1}{x}$

$h^{-1}(x) = \dfrac{1}{x}$

62. The graph represents a one-to-one function, so the function has an inverse.

$f(x) = |x - 2|,\ x \leq 2 \implies y \geq 0$

$y = |x - 2|,\ x \leq 2,\ y \geq 0$

$x = |y - 2|,\ y \leq 2,\ x \geq 0$

$x = y - 2$ or $-x = y - 2$

$2 + x = y$ or $2 - x = y$

The portion that satisfies the conditions $y \leq 2$ and $x \geq 0$ is $2 - x = y$.

$f^{-1}(x) = 2 - x,\ x \geq 0$

63. The graph represents a one-to-one function, so the function has an inverse.

$f(x) = \sqrt{2x + 3} \implies x \geq -\dfrac{3}{2},\ y \geq 0$

$y = \sqrt{2x + 3},\ x \geq -\dfrac{3}{2},\ y \geq 0$

$x = \sqrt{2y + 3},\ y \geq -\dfrac{3}{2},\ x \geq 0$

$x^2 = 2y + 3,\ x \geq 0,\ y \geq -\dfrac{3}{2}$

$y = \dfrac{x^2 - 3}{2},\ x \geq 0,\ y \geq -\dfrac{3}{2}$

$f^{-1}(x) = \dfrac{x^2 - 3}{2},\ x \geq 0$

64. The graph represents a one-to-one function, so the function has an inverse.

$f(x) = \sqrt{x - 2} \implies x \geq 2,\ y \geq 0$

$y = \sqrt{x - 2},\ x \geq 2,\ y \geq 0$

$x = \sqrt{y - 2},\ y \geq 2,\ x \geq 0$

$x^2 = y - 2,\ x \geq 0,\ y \geq 2$

$x^2 + 2 = y,\ x \geq 0,\ y \geq 2$

$f^{-1}(x) = x^2 + 2,\ x \geq 0$

In Exercises 65–70, $f(x) = \frac{1}{8}x - 3$, $f^{-1}(x) = 8(x + 3)$, $g(x) = x^3$, $g^{-1}(x) = \sqrt[3]{x}$.

65. $(f^{-1} \circ g^{-1})(1) = f^{-1}(g^{-1}(1)) = f^{-1}(\sqrt[3]{1}) = 8(\sqrt[3]{1} + 3) = 32$

66. $(g^{-1}f^{-1})(-3) = g^{-1}(f^{-1}(-3)) = g^{-1}(8(-3 + 3)) = g^{-1}(0) = \sqrt[3]{0} = 0$

67. $(f^{-1} \circ f^{-1})(6) = f^{-1}(f^{-1}(6)) = f^{-1}(8[6 + 3]) = 8[8(6 + 3) + 3] = 600$

68. $(g^{-1} \circ g^{-1})(-4) = g^{-1}(g^{-1}(-4)) = g^{-1}(\sqrt[3]{-4}) = \sqrt[3]{\sqrt[3]{-4}} = \sqrt[9]{-4}$

69.
$$(f \circ g)(x) = f(g(x)) = f(x^3) = \tfrac{1}{8}x^3 - 3$$
$$y = \tfrac{1}{8}x^3 - 3$$
$$x = \tfrac{1}{8}y^3 - 3$$
$$x + 3 = \tfrac{1}{8}y^3$$
$$8(x + 3) = y^3$$
$$\sqrt[3]{8(x + 3)} = y$$
$$(f \circ g)^{-1}(x) = 2\sqrt[3]{x + 3}$$

70.
$$g^{-1} \circ f^{-1} = g^{-1}(f^{-1}(x))$$
$$= g^{-1}(8(x + 3))$$
$$= \sqrt[3]{8(x + 3)}$$
$$= 2\sqrt[3]{x + 3}$$

In Exercises 71–74, $f(x) = x + 4$, $f^{-1}(x) = x - 4$, $g(x) = 2x - 5$, $g^{-1}(x) = \dfrac{x + 5}{2}$.

71. $(g^{-1} \circ f^{-1})(x) = g^{-1}(f^{-1}(x)) = g^{-1}(x - 4) = \dfrac{(x - 4) + 5}{2} = \dfrac{x + 1}{2}$

72. $f^{-1} \circ g^{-1}(x) = f^{-1}(g^{-1}(x)) = f^{-1}\left(\dfrac{x + 5}{2}\right) = \dfrac{x + 5}{2} - 4 = \dfrac{x + 5 - 8}{2} = \dfrac{x - 3}{2}$

73. $(f \circ g)(x) = f(g(x)) = f(2x - 5) = (2x - 5) + 4 = 2x - 1$

$$(f \circ g)^{-1}(x) = \frac{x + 1}{2}$$

Note: Comparing Exercises 71 and 73, we see that $(f \circ g)^{-1}(x) = (g^{-1} \circ f^{-1})(x)$.

74. $(g \circ f)(x) = g(f(x)) = g(x + 4) = 2(x + 4) - 5 = 2x + 8 - 5 = 2x + 3$
$$y = 2x + 3$$
$$x = 2y + 3$$
$$x - 3 = 2y$$
$$\frac{x - 3}{2} = y$$
$$(g \circ f)^{-1}(x) = \frac{x - 3}{2}$$

75. The inverse is a line through $(-1, 0)$.

Matches graph (c).

76. The inverse is a line through $(0, 6)$ and $(6, 0)$.

Matches graph (b).

77. The inverse is half a parabola starting at $(1, 0)$.

Matches graph (a).

78. The inverse is a cube root equation through $(0, 0)$.

Matches graph (d).

79.

x	-2	0	2	4	6	8
$f^{-1}(x)$	-2	-1	0	1	2	3

80.

x	-10	-7	-4	-1	2	5
$f^{-1}(x)$	-3	-2	-1	0	1	2

81. (a)

$$y = 8 + 0.75x$$
$$x = 8 + 0.75y$$
$$x - 8 = 0.75y$$
$$\frac{x-8}{0.75} = y$$
$$f^{-1}(x) = \frac{x-8}{0.75}$$

(b) $x =$ hourly wage

$y =$ number of units produced

(c) $y = \dfrac{22.25 - 8}{0.75} = 19$ units

82. (a) $(\text{Total cost}) = \begin{pmatrix}\text{Cost of}\\ \text{first commodity}\end{pmatrix} + \begin{pmatrix}\text{Cost of}\\ \text{second commodity}\end{pmatrix}$

Labels: Total cost $= y$

Amount of first commodity $= x$

Amount of second commodity $= 50 - x$

Cost of first commodity $= 1.25x$

Cost of second commodity $= 1.60(50 - x)$

Equation: $y = 1.25x + 1.60(50 - x)$

(b)

$$x = 1.25y + 1.60(50 - y)$$
$$x = 1.25y + 80 - 1.60y$$
$$x - 80 = -0.35y$$
$$\frac{x-80}{-0.35} = y$$
$$y = \frac{80-x}{0.35}$$

$x =$ total cost

$y =$ number of pounds of less expensive commodity

(c)

$$0 \le y \le 50$$
$$0 \le \frac{80-x}{0.35} \le 50$$
$$0 \le 80 - x \le 17.5$$
$$-80 \le -x \le -62.5$$
$$62.5 \le x \le 80$$

(d) $\dfrac{80 - 73}{0.35} = y = 20$ pounds

83. (a)

$$y = 0.03x^2 + 245.50,\ 0 < x < 100$$
$$x = 0.03y^2 + 245.50$$
$$x - 245.50 = 0.03y^2$$
$$\frac{x-245.50}{0.03} = y^2$$
$$\sqrt{\frac{x-245.50}{0.03}} = y,\ 245.50 < x < 545.50$$
$$f^{-1}(x) = \sqrt{\frac{x-245.50}{0.03}}$$

$x =$ temperature in degrees Fahrenheit

$y =$ percent load for a diesel engine

(b)

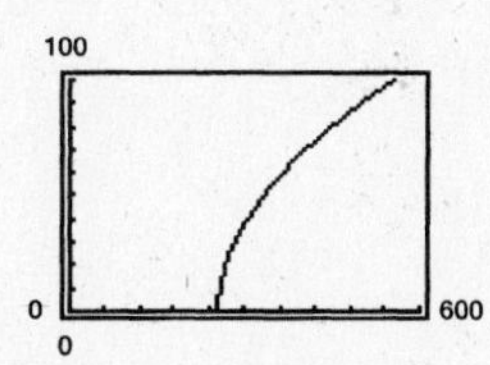

(c)

$$0.03x^2 + 245.50 < 500$$
$$0.03x^2 < 254.50$$
$$x^2 < 8483.333$$
$$x < 92.11$$

Thus, $0 < x < 92.11$.

84.

t	$f(t)$
5	1732
6	1770
7	1841
8	1903
9	1959
10	2066
11	2182

(a) f is one-to-one so f^{-1} *does* exist.

(b) f^{-1} represents the year for a given total out-of-pocket health care annual expenditure.

(c) $f^{-1}(1903) = 8$ which represents 1998.

(d) No. The function would no longer be one-to-one since $f(9) = f(12) = 1959$.

85. (a) $f^{-1}(59.196) = 10$

(b) f^{-1} yields the year for a given amount spent on wireless communications services in the United States.

(c) $f(t) = 12.016t - 57.366$

(d) $f^{-1}(t) = \dfrac{t + 57.366}{12.016}$

(e) $f^{-1}(170.938) = 19$ which represents 2009.

86. (a) $f^{-1}(103{,}874) = 9$ which represents 1999.

(b) f^{-1} yields the year for a given number of households.

(c) $f(t) \approx 1266.54t + 92{,}255.54$

(d) $f^{-1}(t) = \dfrac{t - 92{,}255.54}{1266.54}$

(e) $f^{-1}(111{,}254) \approx 15$ which represents 2005.

87. False. $f(x) = x^2$ is even and does not have an inverse.

88. True. If $f(x) = x - 6$ and $f^{-1}(x) = x + 6$, then the y-intercept of f is $(0, -6)$ and the x-intercept of f^{-1} is $(-6, 0)$.

89. True. If $f(x) = x^n$, then $f^{-1}(x) = \sqrt[n]{x}$.

90. False. Some examples:

$$f(x) = f^{-1}(x) = x$$

$$f(x) = f^{-1}(x) = \frac{1}{x}$$

$$f(x) = f^{-1}(x) = \sqrt{4 - x^2}, \quad 0 \le x \le 2$$

Section 1.6 Mathematical Modeling

1.

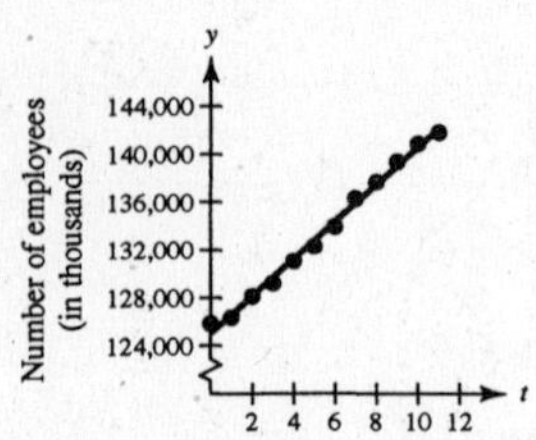

The model is a good fit for the actual data.

2.

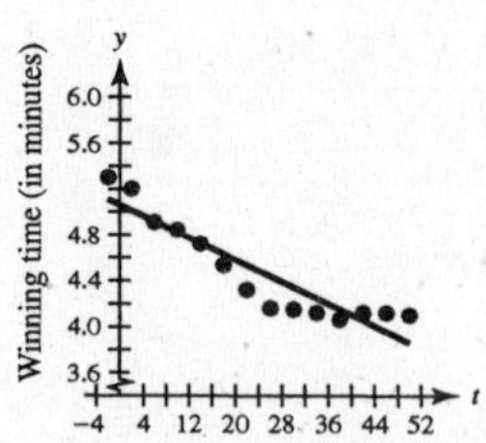

The model is a good fit for the actual data.

3. The graph appears to represent $y = 4/x$, so y varies inversely as x.

4. The graph appears to represent $y = \frac{3}{2}x$ which is a direct variation.

5. $k = 1$

x	2	4	6	8	10
$y = kx^2$	4	16	36	64	100

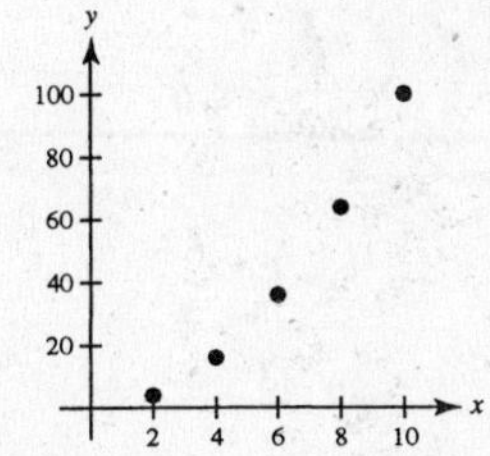

6. $k = 2$

x	2	4	6	8	10
$y = kx^2$	8	32	72	128	200

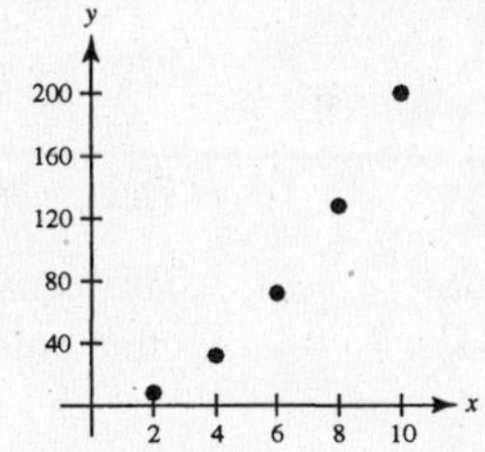

7. $k = \frac{1}{2}$

x	2	4	6	8	10
$y = kx^2$	2	8	18	32	50

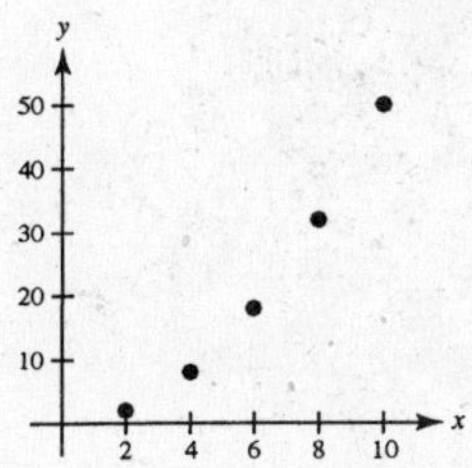

8. $k = \frac{1}{4}$

x	2	4	6	8	10
$y = kx^2$	1	4	9	16	25

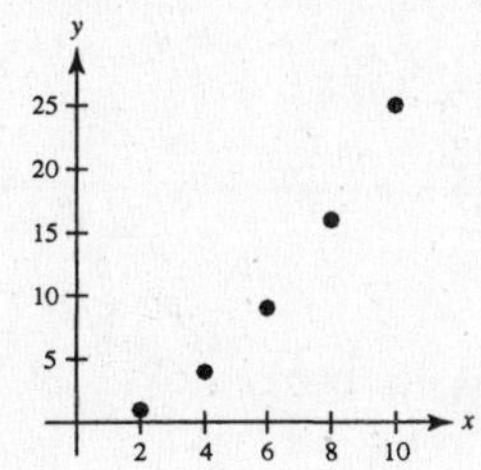

9. $k = 2$

x	2	4	6	8	10
$y = \frac{k}{x^2}$	$\frac{1}{2}$	$\frac{1}{8}$	$\frac{1}{18}$	$\frac{1}{32}$	$\frac{1}{50}$

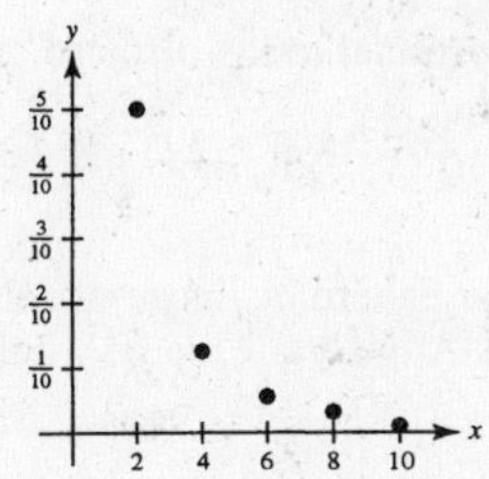

10. $k = 5$

x	2	4	6	8	10
$y = \frac{k}{x^2}$	$\frac{5}{4}$	$\frac{5}{16}$	$\frac{5}{36}$	$\frac{5}{64}$	$\frac{1}{20}$

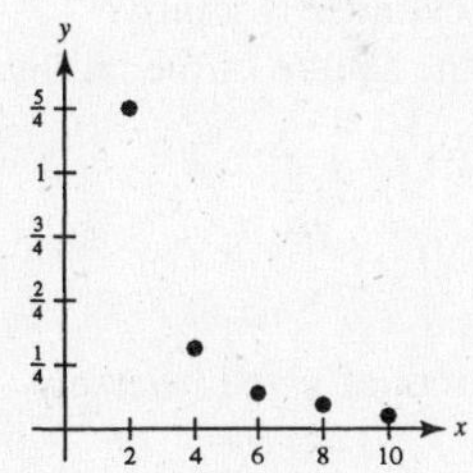

11. $k = 10$

x	2	4	6	8	10
$y = \frac{k}{x^2}$	$\frac{5}{2}$	$\frac{5}{8}$	$\frac{5}{18}$	$\frac{5}{32}$	$\frac{1}{10}$

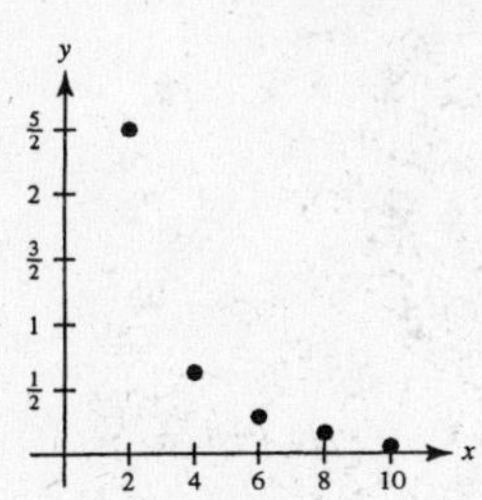

12. $k = 20$

x	2	4	6	8	10
$y = \frac{k}{x^2}$	5	$\frac{5}{4}$	$\frac{5}{9}$	$\frac{5}{16}$	$\frac{1}{5}$

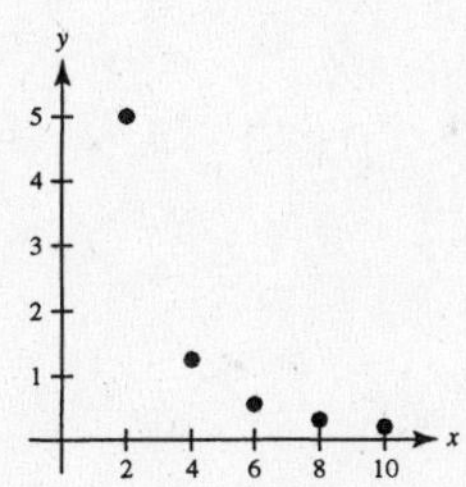

13. The table represents the equation $y = \frac{5}{x}$.

14. The table represents the equation $y = \frac{2}{5}x$.

15. $y = kx$

$-7 = k(10)$

$-\frac{7}{10} = k$

$y = -\frac{7}{10}x$

This equation checks with the other points given in the table.

16. $y = \frac{k}{x}$

$24 = \frac{k}{5}$

$120 = k$

Thus, $y = \frac{120}{x}$.

This equation checks with the other points given in the table.

17. $y = kx$

$12 = k(5)$

$\frac{12}{5} = k$

$y = \frac{12}{5}x$

18. $y = kx$

$14 = k(2)$

$7 = k$

$y = 7x$

19. $y = kx$

$2050 = k(10)$

$205 = k$

$y = 205x$

20. $y = kx$

$580 = k(6)$

$\frac{290}{3} = k$

$y = \frac{290}{3}x$

21. $A = kr^2$

22. $V = ke^3$

23. $y = \frac{k}{x^2}$

24. $h = \frac{k}{\sqrt{s}}$

25. $F = \frac{kg}{r^2}$

26. $z = kx^2y^3$

27. $A = \frac{1}{2}bh$

The area of a triangle is jointly proportional to its base and height.

28. $S = 4\pi r^2$

The surface area of a sphere varies directly as the square of the radius r.

29. $V = \frac{4}{3}\pi r^3$

The volume of a sphere varies directly as the cube of its radius.

30. $V = \pi r^2 h$

The volume of a right circular cylinder is jointly proportional to the height and the square of the radius.

31. $r = \frac{d}{t}$

Average speed is directly proportional to the distance and inversely proportional to the time.

32. $\omega = \sqrt{\frac{kg}{W}}$

ω varies directly as the square root of g and inversely as the square root of W.
(Note: The constant of proportionality is $\sqrt{k}$.)

33. The points do not follow a linear pattern. A linear model would be a poor approximation. A quadratic model would be better.

34. The data shown could be represented by a linear model which would be a good approximation.

35. The data shown could be represented by a linear model which would be a good approximation.

36. The points do not follow a linear pattern. A linear model would not be a good approximation.

37.

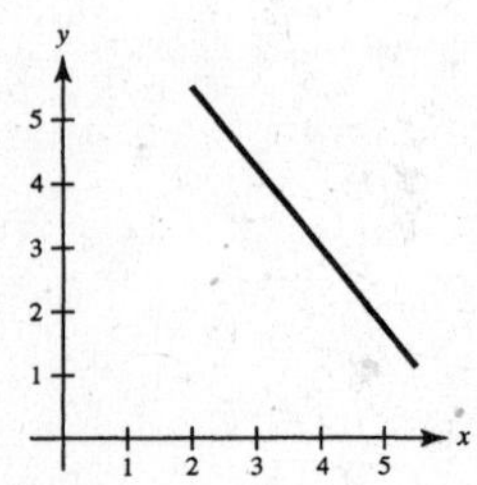

The line appears to pass through (2, 5.5) and (6, 0.5), so its equation is $y = -\frac{5}{4}x + 8$.

38.

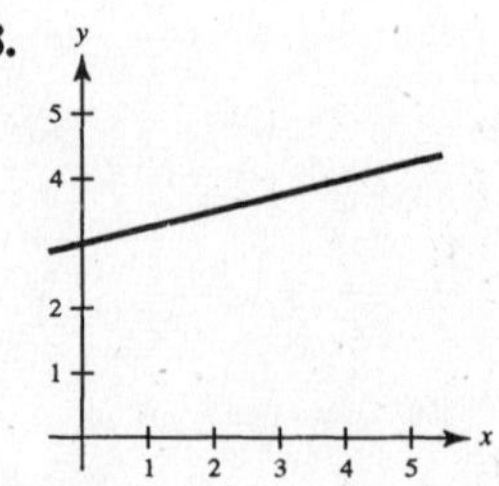

Using the points (0, 3) and (4, 4), we have $y = \frac{1}{4}x + 3$.

39.

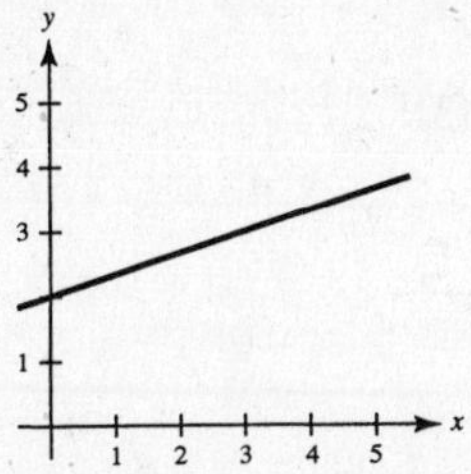

The line appears to pass through (0, 2) and (3, 3) so its equation is $y = \frac{1}{3}x + 2$.

40.

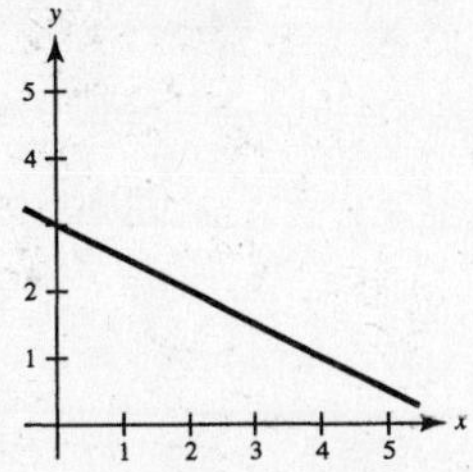

Using the points (2, 2) and (4, 1), we have $y = -\frac{1}{2}x + 3$.

41.
$$I = kP$$
$$187.50 = k(5000)$$
$$0.0375 = k$$
$$I = 0.0375P$$

42.
$$I = kP$$
$$87.50 = k(2500)$$
$$0.035 = k$$
$$I = 0.035P$$

43. $y = kx$
$$53 = k(14)$$
$$\frac{53}{14} = k$$
$$y = \frac{53}{14}x$$

Gallons	5	10	15	20	25	30
Liters	18.9	37.9	56.8	75.7	94.6	113.6

44. $y = kx$
$$33 = k(13)$$
$$\frac{33}{13} = k$$
$$y = \frac{33}{13}x$$

Inches	5	10	20	25	30
Centimeters	12.7	25.4	50.8	63.5	76.2

45. $d = kF$
$$0.15 = k(265)$$
$$\frac{3}{5300} = k$$
$$d = \frac{3}{5300}F$$

(a) $d = \frac{3}{5300}(90) \approx 0.05$ meter

(b) $0.1 = \frac{3}{5300}F$
$$\frac{530}{3} = F$$
$$F = 176\tfrac{2}{3} \text{ newtons}$$

46. $d = kF$
$$0.12 = k(220)$$
$$\frac{3}{5500} = k$$
$$d = \frac{3}{5500}F$$
$$0.16 = \frac{3}{5500}F$$
$$\frac{880}{3} = F$$

The required force is $293\frac{1}{3}$ newtons.

47. $d = kF$
$$1.9 = k(25) \implies k = 0.076$$
$$d = 0.076F$$

When the distance compressed is 3 inches, we have
$$3 = 0.076F$$
$$F \approx 39.47.$$

No child over 39.47 pounds should use the toy.

48. $d = kF$
$$1 = k(15)$$
$$k = \frac{1}{15}$$
$$d = \frac{1}{15}F$$
$$\frac{8}{2} = \frac{1}{15}F$$
$$F = 60 \text{ lb per spring}$$

Combined lifting force $= 2F = 120$ lbs.

49. $R = k(T - T_e)$

50. $P = \dfrac{k}{V}$

51. $F = \dfrac{km_1m_2}{r^2}$

52. $R = kS(S - L)$

53. $y = kx$

$10.22 = k(145.99)$

$0.07 \approx k$

$y = 0.07x$

$y = 0.07(540.50)$

$y \approx 37.84$

The sales tax is \$37.84.

54. $y = kx$

$5520 = k(150{,}000)$

$0.0368 = k$

$y = 0.0368x$

$y = 0.0368(200{,}000) = \$7360$

55. $A = kr^2$

$9\pi = k(3)^2$

$\pi = k$

$A = \pi r^2$

56. $y = \dfrac{k}{x}$

$3 = \dfrac{k}{25}$

$75 = k$

$y = \dfrac{75}{x}$

57. $y = \dfrac{k}{x}$

$7 = \dfrac{k}{4}$

$28 = k$

$y = \dfrac{28}{x}$

58. $z = kxy$

$64 = k(4)(8)$

$2 = k$

$z = 2xy$

59. $F = krs^3$

$4158 = k(11)(3)^3$

$k = 14$

$F = 14rs^3$

60. $P = \dfrac{kx}{y^2}$

$\dfrac{28}{3} = \dfrac{k(42)}{9^2}$

$\dfrac{28}{3} \cdot \dfrac{81}{42} = k$

$\dfrac{2 \cdot 27}{3} = k$

$18 = k$

$P = \dfrac{18x}{y^2}$

61. $z = \dfrac{kx^2}{y}$

$6 = \dfrac{k(6)^2}{4}$

$\dfrac{24}{36} = k$

$\dfrac{2}{3} = k$

$z = \dfrac{\frac{2}{3}x^2}{y} = \dfrac{2x^2}{3y}$

62. $v = \dfrac{kpq}{s^2}$

$1.5 = \dfrac{k(4.1)(6.3)}{(1.2)^2}$

$\dfrac{(1.5)(1.44)}{(4.1)(6.3)} = k$

$\dfrac{2.16}{25.83} = k$

$k = \dfrac{24}{287}$

$v = \dfrac{24pq}{287s^2}$

63. $d = kv^2$

$0.02 = k\left(\dfrac{1}{4}\right)^2$

$k = 0.32$

$d = 0.32v^2$

$0.12 = 0.32v^2$

$v^2 = \dfrac{0.12}{0.32} = \dfrac{3}{8}$

$v = \dfrac{\sqrt{3}}{2\sqrt{2}} = \dfrac{\sqrt{6}}{4} \approx 0.61$ mi/hr

64. $d = kv^2$

If the velocity is doubled:

$d = k(2v)^2$

$d = k \cdot 4v^2$

$\dfrac{4kv^2}{kv^2} = 4$

65. $r = \frac{kl}{A},\ A = \pi r^2 = \frac{\pi d^2}{4}$

$r = \frac{4kl}{\pi d^2}$

$64.9 = \frac{4(1000)k}{\pi\left(\frac{0.0126}{12}\right)^2}$

$k \approx 5.62 \times 10^{-8}$

$r = \frac{4(5.62 \times 10^{-8})l}{\pi\left(\frac{0.0126}{12}\right)^2}$

$33.5 = \frac{4(5.62 \times 10^{-8})l}{\pi\left(\frac{0.0126}{12}\right)^2}$

$\frac{33.5\pi\left(\frac{0.0126}{12}\right)^2}{4(5.62 \times 10^{-8})} = l$

$l \approx 516$ feet

66. From Exercise 65,

$k \approx 5.62 \times 10^{-8}$.

$r = \frac{4(5.62 \times 10^{-8})l}{\pi d^2}$

$d = \sqrt{\frac{4(5.62 \times 10^{-8})l}{\pi r}}$

$d = \sqrt{\frac{4(5.62 \times 10^{-8})(14)}{\pi(0.05)}}$

$d \approx 0.0045$ feet $= 0.054$ inch

67. $s = kt^2$

$144 = k(3)^2$

$16 = k$

$s = 16t^2$

$s = 16(5)^2 = 400$ feet

68. $d = ks^2$

$75 = k(30)^2$

$\frac{1}{12} = k$

$d = \frac{1}{12}s^2$

$d = \frac{1}{12}(50)^2$

$d = 208\frac{1}{3}$ feet

69. $P = kA = k(\pi r^2) = k\pi\left(\frac{d}{2}\right)^2$

$8.78 = k\pi\left(\frac{9}{2}\right)^2$

$\frac{4(8.78)}{81\pi} = k$

$k \approx 0.138$

However, we do not obtain \$11.78 when $d = 12$ inches.

$P = 0.138\pi\left(\frac{12}{2}\right)^2 \approx \$15.61$

Instead, $k = \frac{11.78}{36\pi} \approx 0.104$.

For the 15-inch pizza, we have $k = \frac{4(14.18)}{225\pi} \approx 0.080$.

The price is not directly proportional to the surface area. The best buy is the 15-inch pizza.

70. $d = \frac{k}{p}$

$500 = \frac{k}{3.75}$

$1875 = k$

$d = \frac{1875}{p}$

$d = \frac{1875}{4.25} \approx 441$ units

71. $v = \dfrac{k}{A}$

(a) $v = \dfrac{k}{0.75A} = \dfrac{4}{3}\left(\dfrac{k}{A}\right)$

The velocity is increased by one-third.

(b) $v = \dfrac{k}{\left(1 + \frac{1}{3}\right)A} = \dfrac{k}{\frac{4}{3}A} = \dfrac{3}{4}\left(\dfrac{k}{A}\right)$

The velocity is decreased by one-fourth.

72. $\text{Load} = \dfrac{kwd^2}{l}$

(a) $\text{load} = \dfrac{k(2w)d^2}{2l} = \dfrac{kwd^2}{l}$

The safe load is unchanged.

(b) $\text{load} = \dfrac{k(2w)(2d)^2}{l} = \dfrac{8kwd^2}{l}$

The safe load is 8 times as great.

(c) $\text{load} = \dfrac{k(2w)(2d)^2}{2l} = \dfrac{4kwd^2}{l}$

The safe load is 4 times as great.

(d) $\text{load} = \dfrac{kw\left(\frac{d}{2}\right)^2}{l} = \dfrac{\frac{1}{4}kwd^2}{l}$

The safe load is one-fourth as great.

73. (a)

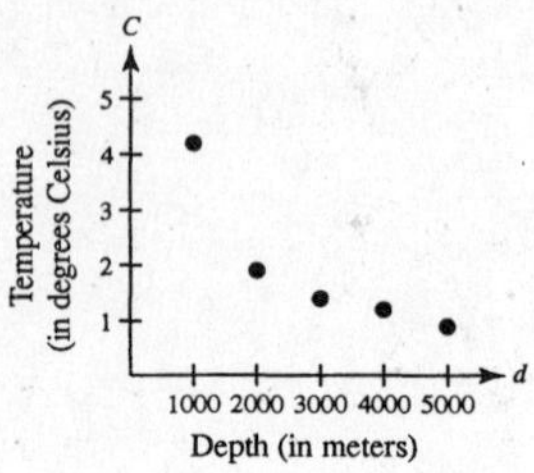

(b) $C = \dfrac{k}{d} \Rightarrow k = dC$

$(1000, 4.2) : k = 4200$

$(2000, 1.9) : k = 3800$

$(3000, 1.4) : k = 4200$

$(4000, 1.2) : k = 4800$

$(5000, 0.9) : k = 4500$

(c) $\dfrac{4200 + 3800 + 4200 + 4800 + 4500}{5} = 4300$

$C = \dfrac{4300}{d}$

(d)

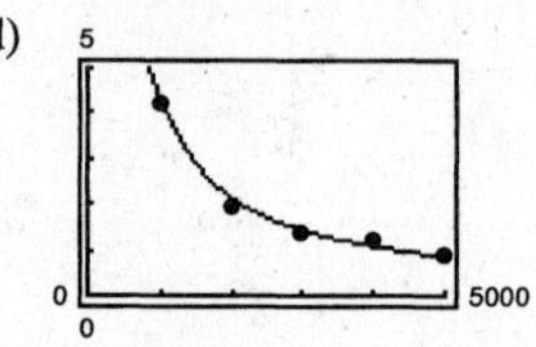

(e) $3 = \dfrac{4300}{d}$

$d = \dfrac{4300}{3} \approx 1433 \text{ meters}$

74. $y = \dfrac{262.76}{x^{2.12}}$

(a)

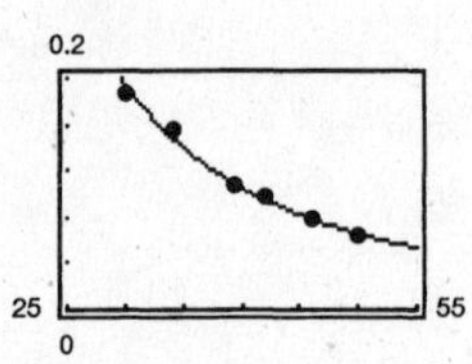

(b) $y = \dfrac{262.76}{(25)^{2.12}} \approx 0.2857$ microwatts per square centimeter

75. $y = \dfrac{k}{d^2}$

When d is doubled,

$$y = \frac{k}{(2d)^2} = \frac{k}{4d^2} = \frac{1}{4}\left(\frac{k}{d^2}\right)$$

Thus the illumination is $\frac{1}{4}$ of what it was before.

76. (a) $y = 1.08t + 127.7$

(b)

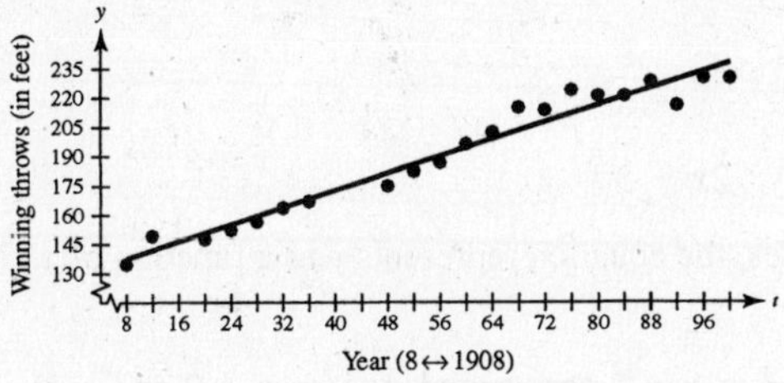

(c) $y(104) = 1.08(104) + 127.7 = 240.0$ feet

77. (a) $y = 2.1560t - 4.8904$

(b)

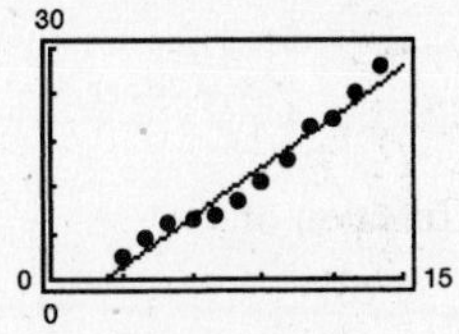

(c) $y(15) \approx 27.45$ billion dollars

(d) Answers will vary.

78. (a) and (c)

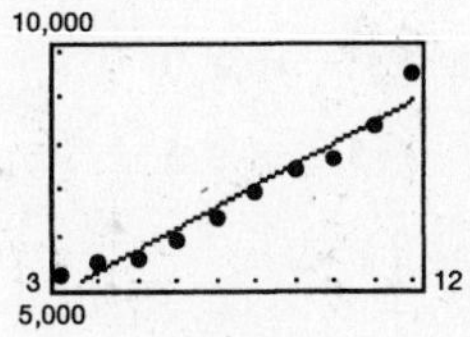

(b) $y \approx 463.27t + 3356.8$

(d) 2005 : $y(15) \approx 10{,}306$ million dollars

2010 : $y(20) \approx 12{,}622$ million dollars

(e) The receipts are increasing at a rate of 463.27 million dollars per year.

79. (a) $y \approx 1.63t + 54.2$

(b)

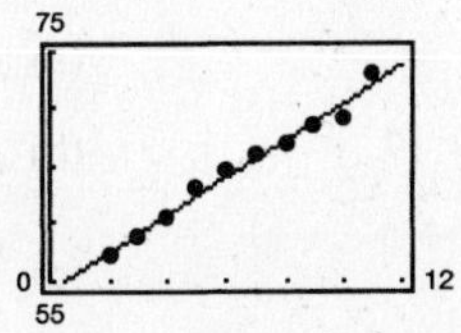

(c) 2005 : $y(15) = 78.65$ million households

2010 : $y(20) = 86.80$ million households

(d) The number of households with cable televisions is increasing at a rate of 1.63 million per year.

80. (a) $y \approx 1.1783x + 45.2935$

(b)

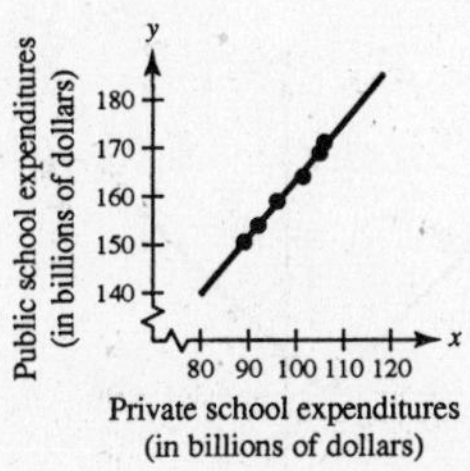

(c) $y \approx 1.1783(120) + 45.2935 \approx 186.7$ billion dollars

(d) For each billion dollars in private school expenditures, there was an expenditure of 1.1783 billion dollars for public schools (colleges and universities).

81. (a) $y \approx 0.45x - 4.9$

(b)

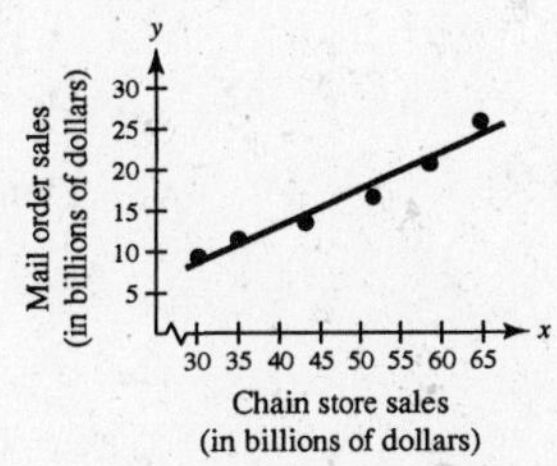

(c) $y \approx 0.45(75) - 4.9 \approx 28.85$ billion dollars

82. The accuracy of the model in predicting prize winnings is questionable because the model is based on limited data.

83. False. y will increase if k is positive and y will decrease if k is negative.

84. False. E is jointly proportional (not "directly proportional") to the mass of an object and the square of its velocity.

85. False.

"y varies directly as x" means $y = kx$ for some nonzero constant k.

"y is inversely proportional to x" means $y = \frac{k}{x}$ for some nonzero constant k.

86. False.

"a is jointly proportional to y and z with the constant of proportionality k" means $a = kyz$

Review Exercises for Chapter 1

1. $16x - y^4 = 0$

No. This determines x as a function of y.

2. $2x - y - 3 = 0$

$2x - 3 = y$

Yes, the equation represents y as a function of x.

3. $y = \sqrt{1 - x}$

Yes. This determines y as a function of x.

4. $|y| = x + 2$ corresponds to $y = x + 2$ or $-y = x + 2$. y is not a function of x. Each x-value, except $x = -2$, corresponds to two y-values.

5. $f(x) = x^2 + 1$

(a) $f(2) = 2^2 + 1 = 5$

(b) $f(-4) = (-4)^2 + 1 = 17$

(c) $f(t^2) = (t^2)^2 + 1 = t^4 + 1$

(d) $-f(x) = -(x^2 + 1)$

$= -x^2 - 1$

6. $g(x) = x^{4/3}$

(a) $g(8) = 8^{4/3} = 2^4 = 16$

(b) $g(t + 1) = (t + 1)^{4/3}$

(c) $\dfrac{g(8) - g(1)}{8 - 1} = \dfrac{16 - 1}{7} = \dfrac{15}{7}$

(d) $g(-x) = (-x)^{4/3} = x^{4/3}$

7. $f(x) = \sqrt{25 - x^2}$

Domain: $25 - x^2 \geq 0$

$-5 \leq x \leq 5$

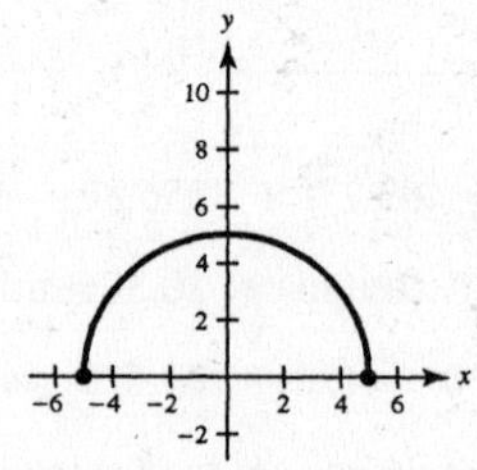

8. $f(x) = 3x + 4$

Domain: All real numbers

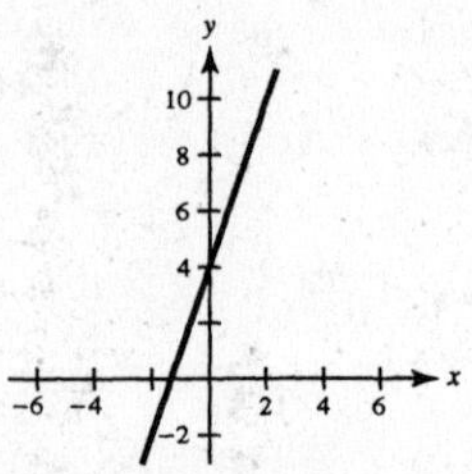

9. $h(x) = \dfrac{x}{x^2 - x - 6}$

$x^2 - x - 6 = 0$

$(x - 3)(x + 2) = 0$

Domain: All real numbers $x \neq -2, 3$

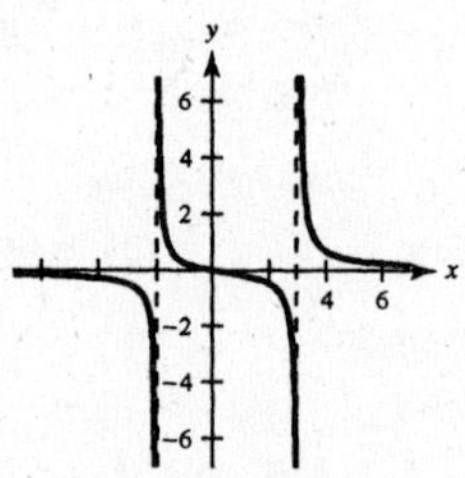

10. $h(t) = |t + 1|$

Domain: All real numbers

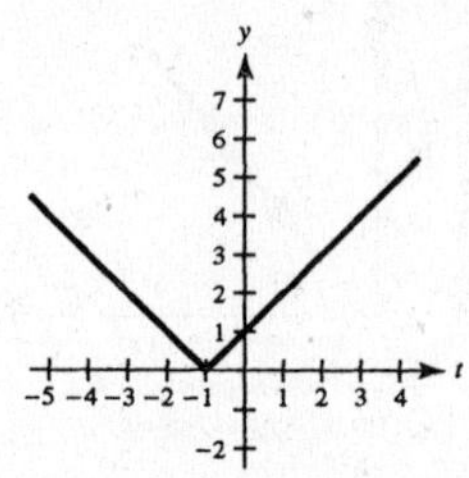

11. $v(t) = -32t + 48$

(a) $v(1) = -32(1) + 48 = 16$ feet per second

(b) $v(t) = 0$

$-32t + 48 = 0$

$t = \frac{48}{32} = 1.5$ seconds

(c) $v(2) = -32t + 48 = -32(2) + 48$

$= -16$ feet per second

12. (a) $f(x) = 0.40(50 - x) + 1.00x$

$= 0.4(50 - x) + x$

$= 20 + 0.6x$

(b) Domain: $0 \leq x \leq 50$

Range: When $x = 0$; $f(0) = 20$

When $x = 50$; $f(50) = 50$

$20 \leq y \leq 50$

(c) $20 + 0.6x = 0.5(50)$

$20 + 0.6x = 25$

$0.6x = 5$

$x = \dfrac{5}{0.6} = \dfrac{50}{6} = 8\frac{1}{3}$ liters

13. (a) Let x = the width of the rectangle and y = the length of the rectangle.

$$y = 4x + 2$$
$$P(x) = 2x + 2y$$
$$= 2x + 2(4x + 2)$$
$$= 10x + 4$$

(b) $A(x) = xy$
$$= x(4x + 2)$$
$$= 4x^2 + 2x$$

(c) Since x is the width of a rectangle, x must be positive.

Domain of $P(x)$: $x > 0$

Domain of $A(x)$: $x > 0$

(d)

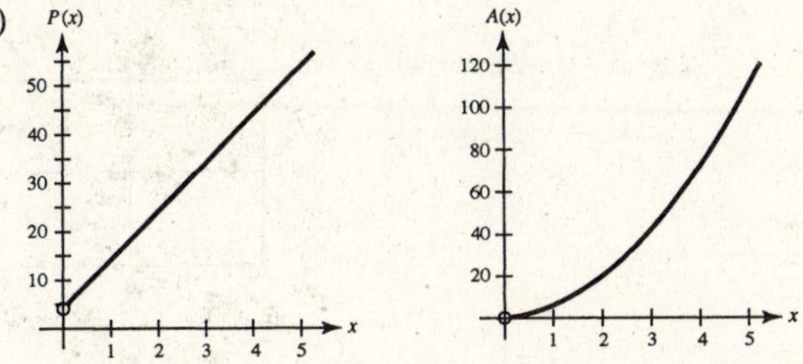

14. A function from a set A to a set B is a relation that assigns to each element x in set A exactly one element y in the set B.

15. $y = (x - 3)^2$

All vertical lines intersect graph exactly once $\Rightarrow$ y is a function of x.

16. $y = -\frac{3}{5}x^3 - 2x + 1$

The graph passes the Vertical Line Test. y **is** a function of x.

17. $x - 4 = y^2$

Some vertical lines intersect graph more than once $\Rightarrow$ y is **not** a function of x.

18. $x = -|4 - y|$

The graph fails the Vertical Line Test. Some vertical lines cross the graph twice. y is **not** a function of x.

19. $f(x) = 5x^2 + 4x - 1$

$$5x^2 + 4x - 1 = 0$$
$$(5x - 1)(x + 1) = 0$$

Zeros: $x = \frac{1}{5}, -1$

20. $f(x) = 3x^2 - 16x + 21$

$$3x^2 - 16x + 21 = 0$$
$$(3x - 7)(x - 3) = 0$$

Zeros: $x = \frac{7}{3}, 3$

21. $f(x) = \dfrac{8x + 3}{11 - x}$

$$\frac{8x + 3}{11 - x} = 0$$
$$8x + 3 = 0$$

Zero: $x = -\dfrac{3}{8}$

22. $f(x) = x^3 - x^2 - 25x + 25$

$$x^3 - x^2 - 25x + 25 = 0$$
$$x^2(x - 1) - 25(x - 1) = 0$$
$$(x^2 - 25)(x - 1) = 0$$
$$(x + 5)(x - 5)(x - 1) = 0$$

Zeros: $x = \pm 5, 1$

23. $g(x) = |x + 2| - |x - 2|$

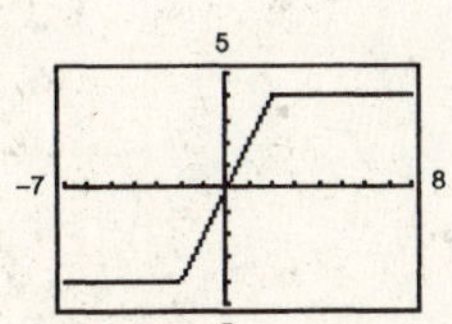

Increasing: $(-2, 2)$

Constant: $(-\infty, -2), (2, \infty)$

24. $f(x) = (x^2 - 4)^2$

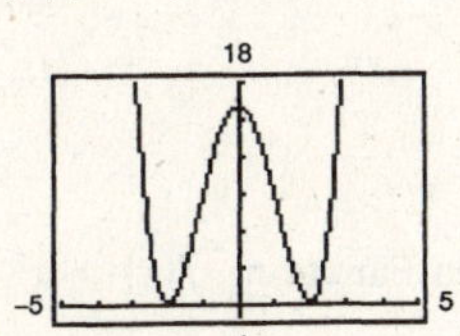

Increasing on $(-2, 0)$ and $(2, \infty)$

Decreasing on $(-\infty, -2)$ and $(0, 2)$

25. $h(x) = 4x^3 - x^4$

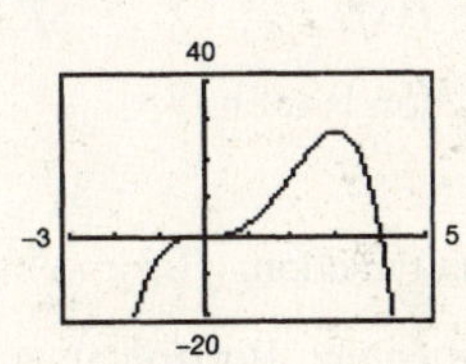

Increasing: $(-\infty, 3)$

Decreasing: $(3, \infty)$

26. $g(x) = \sqrt[3]{x(x + 3)^2}$

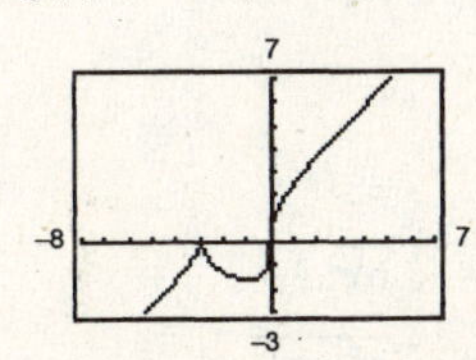

Increasing on $(-\infty, -3)$ and $(-1, \infty)$

Decreasing on $(-3, -1)$

27. $f(2) = -6, f(-1) = 3$

Two points: $(2, -6), (-1, 3)$

Slope: $m = \dfrac{-6 - 3}{2 - (-1)} = \dfrac{-9}{3} = -3$

$y - 3 = -3(x + 1)$

$y = -3x$

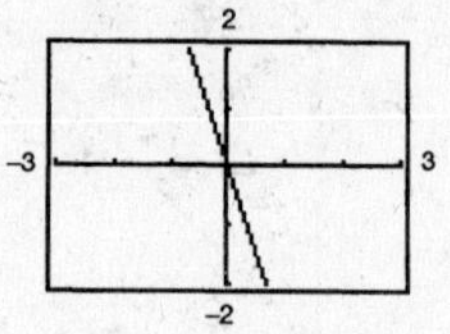

28. $f(0) = -5, f(4) = -8$

Two points: $(0, -5), (4, -8)$

$m = \dfrac{-8 - (-5)}{4 - 0} = -\dfrac{3}{4}$

$y - (-5) = -\dfrac{3}{4}(x - 0)$

$y = -\dfrac{3}{4}x - 5$

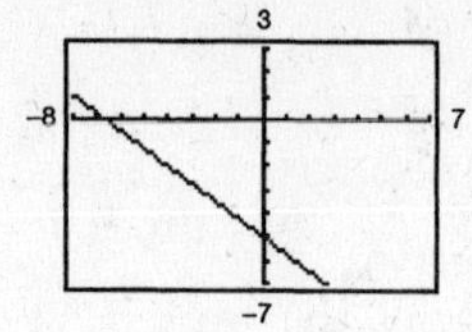

29. $f\left(-\dfrac{4}{5}\right) = 2, f\left(\dfrac{11}{5}\right) = 7$

Two points: $\left(-\dfrac{4}{5}, 2\right), \left(\dfrac{11}{5}, 7\right)$

Slope: $m = \dfrac{2 - 7}{-\frac{4}{5} - \frac{11}{5}} = \dfrac{-5}{-\frac{15}{5}} = \dfrac{5}{3}$

$y - 7 = \dfrac{5}{3}\left(x - \dfrac{11}{5}\right)$

$y = \dfrac{5}{3}x + \dfrac{10}{3}$

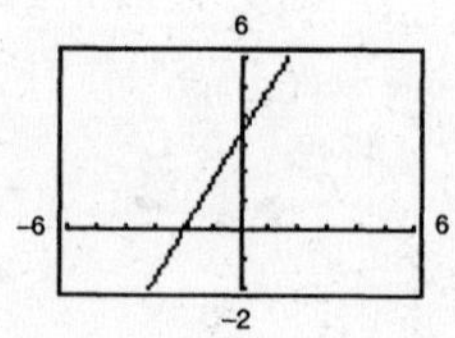

30. $f(3.3) = 5.6, f(-4.7) = -1.4$

Two points: $(3.3, 5.6), (-4.7, -1.4)$

$m = \dfrac{-1.4 - 5.6}{-4.7 - 3.3} = \dfrac{-7}{-8} = \dfrac{7}{8}$

$y - 5.6 = \dfrac{7}{8}(x - 3.3)$

$y - \dfrac{56}{10} = \dfrac{7}{8}\left(x - \dfrac{33}{10}\right)$

$y = \dfrac{7}{8}x - \dfrac{231}{80} + \dfrac{56}{10} \cdot \dfrac{8}{8}$

$y = \dfrac{7}{8}x + \dfrac{217}{80}$

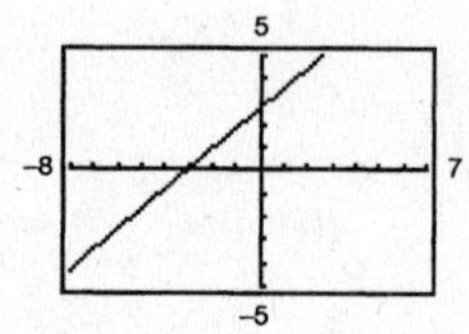

31. $f(x) = \begin{cases} 5x - 3, & x \geq -1 \\ -4x + 5, & x < -1 \end{cases}$

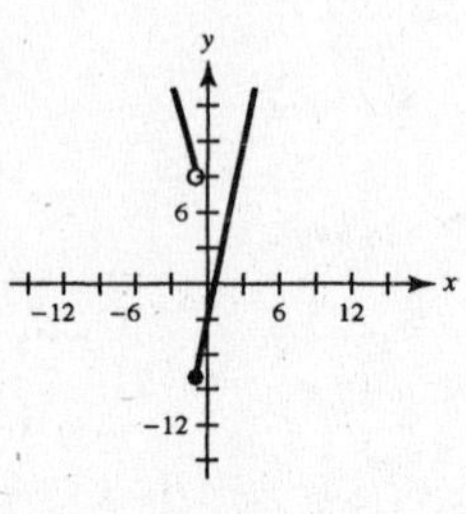

32. $f(x) = \begin{cases} x^2 - 2, & x < -2 \\ 5 \quad , & -2 \leq x \leq 0 \\ 8x - 5, & x > 0 \end{cases}$

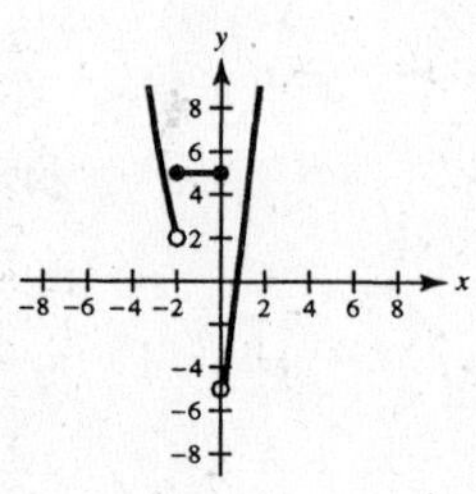

33. $f(x) = x^5 + 4x - 7$

$f(-x) = -x^5 - 4x - 7$

$f(-x) \neq f(x)$

$f(-x) \neq -f(x)$

Neither.

34. $f(x) = x^4 - 20x^2$

$f(-x) = (-x)^4 - 20(-x)^2$

$= x^4 - 20x^2$

$= f(x)$

The function is even.

35. $f(x) = 2x\sqrt{x^2 + 3}$

$f(-x) = -2x\sqrt{x^2 + 3}$

The function is odd.

36. $f(x) = \sqrt[5]{6x^2}$

$f(-x) = \sqrt[5]{6(-x)^2}$

$= \sqrt[5]{6x^2}$

The function is even.

37. Common Function: $f(x) = x^3$

Transformation: Horizontal shift 4 units to the left, and a vertical shift 4 units upward.

$g(x) = (x + 4)^3 + 4$

38. Common Function: $f(x) = x^2$

Transformation: Reflection in the x-axis and a horizontal shift 3 units to the left.

$h(x) = -(x + 3)^2$

39. Common Function: $f(x) = |x|$

Transformation: Reflection in the x-axis, a horizontal shift 2 units to the left, and a vertical shift 1 unit upward.

$$g(x) = -|x + 2| + 1$$

40. Common Function: $f(x) = \sqrt{x}$

Transformation: Reflection in the y-axis, a horizontal shift 2 units to the right, and a vertical shift 1 unit down.

$$g(x) = \sqrt{-(x - 2)} - 1$$

41. $f(x) = x^2$

$h(x) = x^2 - 9$

Transformation: Vertical shift 9 units downward.

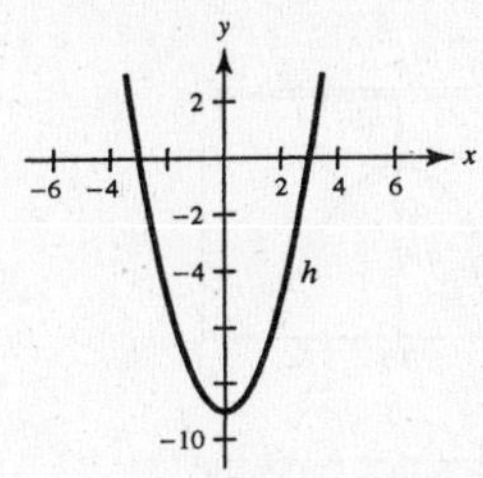

42. $f(x) = \sqrt{x}$

$h(x) = \sqrt{x - 7}$

Transformation: Horizontal shift 7 units to the right.

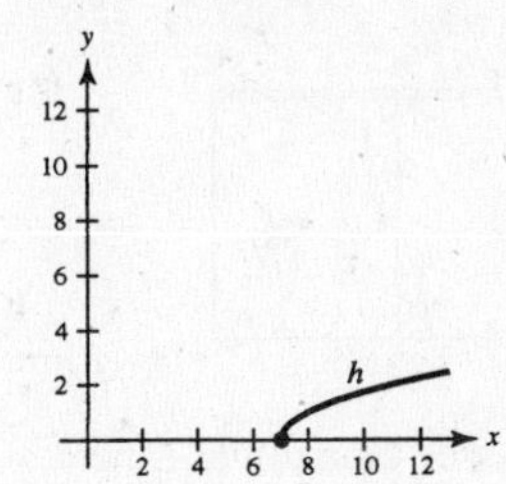

43. $f(x) = |x|$

$h(x) = |x + 3| - 5$

Transformation: Horizontal shift 3 units to the left and a vertical shift 5 units downward.

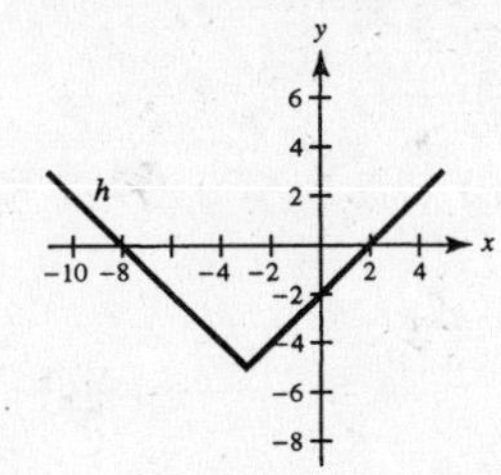

44. $f(x) = x^2$

$h(x) = -(x + 3)^2 + 1$

Transformation: Reflection in the x-axis, horizontal shift 3 units to the left, and a vertical shift 1 unit upward.

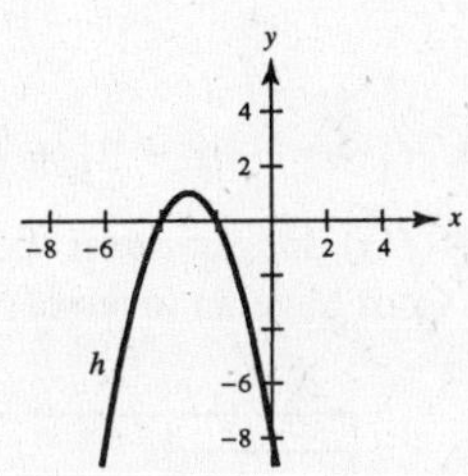

45. $f(x) = \sqrt{x}$

$h(x) = -\sqrt{x + 1} + 9$

Transformation: Reflection in the x-axis, horizontal shift 1 unit to the left, and a vertical shift 9 units upward.

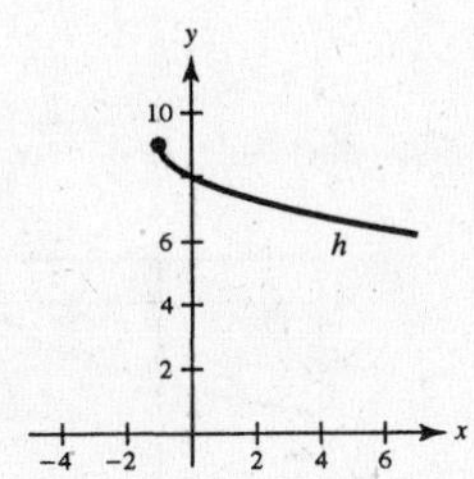

46. $f(x) = x^3$

$h(x) = -\frac{1}{3}x^3$

Transformation: Reflection in the x-axis and a vertical shrink by a factor of 1/3.

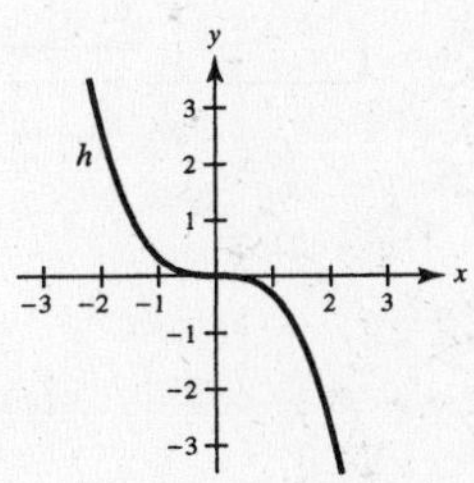

For Exercises 47–50, let $f(x) = 3 - 2x$, $g(x) = \sqrt{x}$, and $h(x) = 3x^2 + 2$.

47. $(f - g)(4) = f(4) - g(4)$

$= 3 - 2(4) - \sqrt{4}$

$= -7$

48. $\left(\dfrac{f}{h}\right)(0) = \dfrac{f(0)}{h(0)} = \dfrac{3 - 2(0)}{3(0)^2 + 2} = \dfrac{3}{2}$

49. $(h \circ g)(7) = 3(g(7))^2 + 2$

$= 3\left(\sqrt{7}\right)^2 + 2$

$= 23$

50. $(g \circ f)(-2) = g(f(-2)) = g(3 - 2(-2)) = g(7) = \sqrt{7}$

51. Let $t = 5$ represent 1995.

$y_1 \approx 1.5t^2 - 8.674t + 70.057$

$y_2 \approx 2.775t^2 - 25.468t + 205.371$

52.

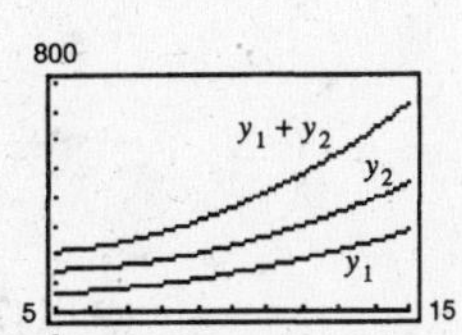

In 2005, let $t = 15$: $(y_1 + y_2)(15) = y_1(15) + y_2(15)$

≈ 725.2 billion dollars

53. $f(x) = 6x$

$f^{-1}(x) = \dfrac{x}{6}$

$f(f^{-1}(x)) = 6\left(\dfrac{x}{6}\right) = x$

$f^{-1}(f(x)) = \dfrac{6x}{6} = x$

54. $f(x) = \frac{1}{12}x$

$f^{-1}(x) = 12x$

$f(f^{-1}(x)) = \frac{1}{12}(12x) = x$

$f^{-1}(f(x)) = 12\left(\frac{1}{12}x\right) = x$

55. $f(x) = x - 7$

$f^{-1}(x) = x + 7$

$f(f^{-1}(x)) = (x + 7) - 7 = x$

$f^{-1}(f(x)) = (x - 7) + 7 = x$

56. $f(x) = x + 5$

$f^{-1}(x) = x - 5$

$f(f^{-1}(x)) = (x - 5) + 5 = x$

$f^{-1}(f(x)) = (x + 5) - 5 = x$

57. $f(x) = 3x^3 - 5$

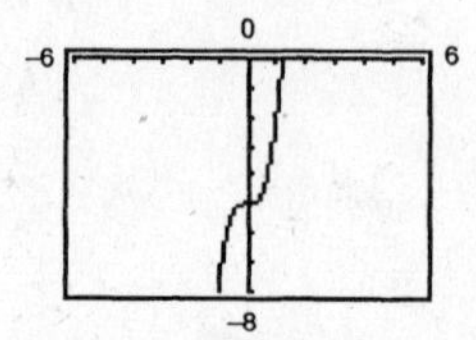

$f(x)$ passes Horizontal Line Test $\Rightarrow$ it has an inverse.

58. $f(x) = -\frac{1}{4}x^2 - 3$

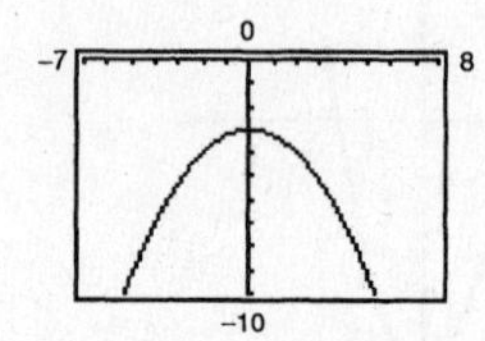

$f(x)$ does not pass the Horizontal Line Test, so $f(x)$ does **not** have an inverse.

59. $f(x) = -\sqrt{4 - x}$

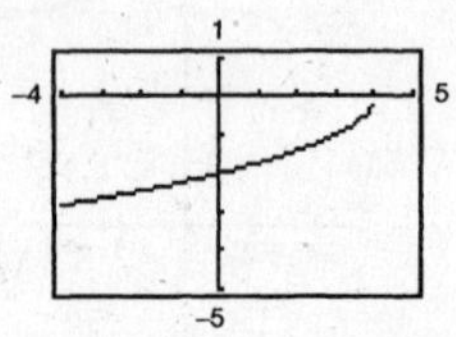

$f(x)$ passes Horizontal Line Test $\Rightarrow$ it has an inverse.

60. $f(x) = -|x + 2| + |7 - x|$

$f(x)$ does not pass the Horizontal Line Test, so $f(x)$ does **not** have an inverse.

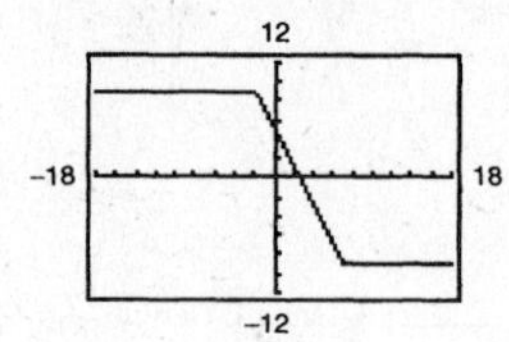

61. $f(x) = \frac{1}{2}x - 3$

(a) $y = \frac{1}{2}x - 3$

$x = \frac{1}{2}y - 3$

$2x + 6 = y$

$f^{-1}(x) = 2x + 6$

(b)

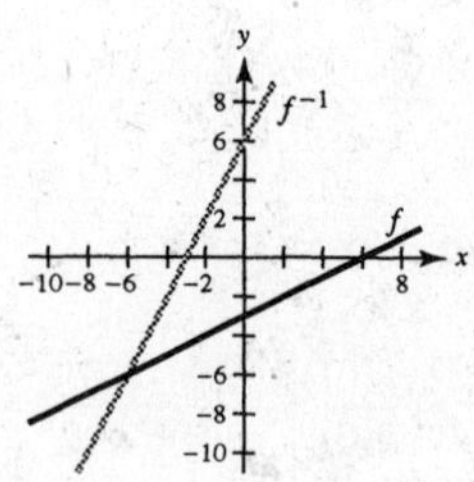

(c) $f^{-1}(f(x)) = 2\left[\frac{1}{2}x - 3\right] + 6 = x$

$f(f^{-1}(x)) = \frac{1}{2}[2x + 6] - 3 = x$

62. $f(x) = 5x - 7$

(a) $y = 5x - 7$

$x = 5y - 7$

$x + 7 = 5y$

$\dfrac{x + 7}{5} = y$

$f^{-1}(x) = \dfrac{x + 7}{5}$

(b)

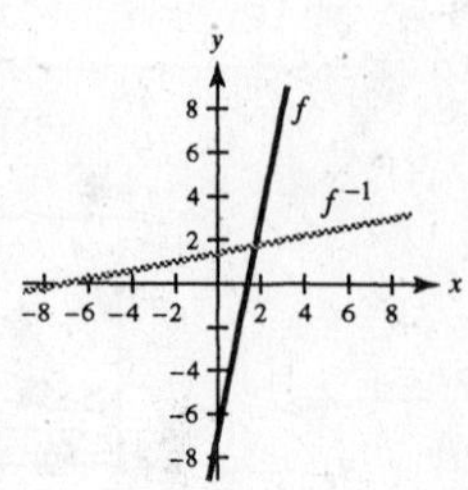

(c) $f^{-1}(f(x)) = \dfrac{(5x - 7) + 7}{5}$

$= \dfrac{5x}{5} = x$

$f(f^{-1}(x)) = 5\left(\dfrac{x + 7}{5}\right) - 7$

$= x + 7 - 7 = x$

63. $f(x) = \sqrt{x + 1}$

(a) $y = \sqrt{x + 1},\ x \ge -1,\ y \ge 0$

$x = \sqrt{y + 1},\ y \ge -1,\ x \ge 0$

$x^2 - 1 = y$

$f^{-1}(x) = x^2 - 1,\ x \ge 0$

(b)

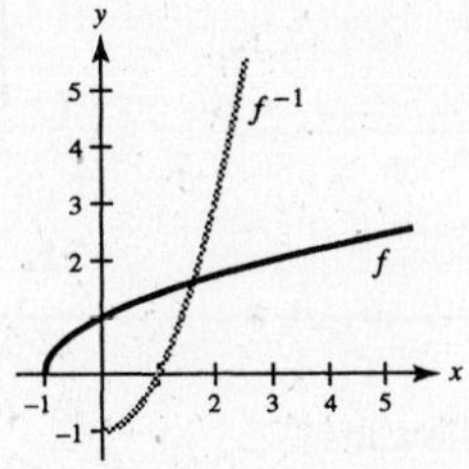

(c) $f^{-1}(f(x)) = \left[\sqrt{x + 1}\right]^2 - 1 = x$

$f(f^{-1}(x)) = \sqrt{[x^2 - 1] + 1} = x$

64. $f(x) = x^3 + 2$

(a) $y = x^3 + 2$

$x = y^3 + 2$

$x - 2 = y^3$

$\sqrt[3]{x - 2} = y$

$f^{-1}(x) = \sqrt[3]{x - 2}$

(b)

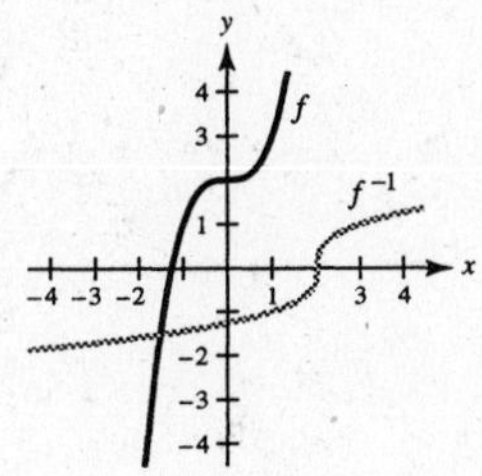

(c) $f^{-1}(f(x)) = \sqrt[3]{(x^3 + 2) - 2}$

$= \sqrt[3]{x^3} = x$

$f(f^{-1}(x)) = \left(\sqrt[3]{x - 2}\right)^3 + 2$

$= x - 2 + 2$

65. $f(x) = 2(x - 4)^2$

$f(x)$ is increasing on the interval $(4, \infty)$.

$y = 2(x - 4)^2$

$x = 2(y - 4)^2$

$\pm\sqrt{\dfrac{x}{2}} = y - 4$

Choose positive root, corresponding to $(4, \infty)$ as the domain of $f(x)$.

$4 + \sqrt{\dfrac{x}{2}} = y$

$f^{-1}(x) = \sqrt{\dfrac{x}{2}} + 4$

66. $f(x) = |x - 2| = \begin{cases} x - 2, & x \ge 2 \\ -(x - 2), & x < 2 \end{cases}$

$f(x)$ is increasing on $[2, \infty)$ or $x \ge 2$ so we will restrict the domain to this interval.

$f(x) = |x - 2| = x - 2,\ x \ge 2$ and $y \ge 0$

$f^{-1}(x) = x + 2,\ x \ge 0$ [Note: The domain of f^{-1} is the range of f.]

67. The Vertical Line Test is used to determine if a graph represents a function. The Horizontal Line Test is used to determine if a function has an inverse.

68. Let $t = 3$ represent 1993.

$S = -0.01911t^4 + 0.4830t^3 - 4.274t^2 + 16.24t - 16.9$

(a)

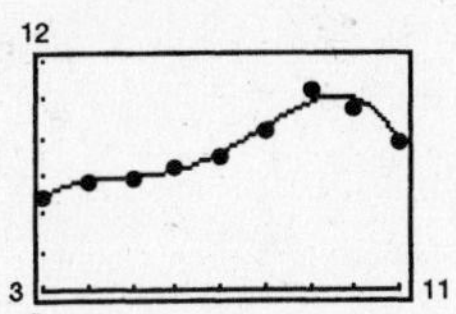

The model is a good fit to the data.

(b) The model also shows a downturn in sales. Answers will vary.

(c) $S(11) - S(10) \approx 7.67 - 10.00 = -2.33$ billion dollars

Actual decrease: $7.8 - 9.5 = -1.70$ billion dollars

The actual decrease was less than indicated by the model.

(d) $S(13) = -12.74$. Since the model decreases without bound when $t > 9.6$, it is not accurate in predicting future sales.

69. 2.5 miles = 4 kilometers $\Rightarrow$ 1 mile $= \frac{4}{2.5} = 1.6$ kilometers

$y = 1.6x$

Miles, x	2	5	10	12
Kilometers, $1.6x$	3.2	8	16	19.2

70. $P = kS^3$

$750 = k(27)^3 \Rightarrow k \approx 0.038104$

$P = k(40)^3 \approx (0.038104)(40)^3 \approx 2438.7$ kilowatts

71. $F = ks^2$

If s is doubled; $F = k(2s)^2 = 4ks^2$.

The force is changed by a factor of 4.

72. (a) and (c)

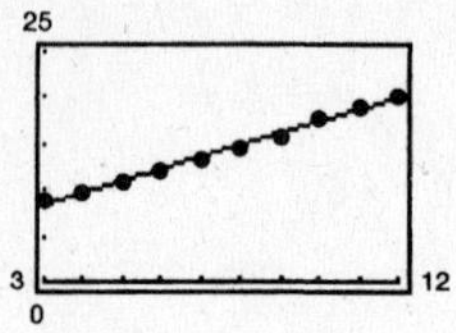

(b) $y \approx 1.2883t + 4.5547$

(d) 2000 : $y(10) \approx 17.44$ billion dollars

2003 : $y(13) \approx 21.30$ billion dollars

(e) The amount of child support collections is increasing at a rate of 1.2883 billion dollars per year.

73. y is inversely proportional to x: $y = \frac{k}{x}$

$y = 9$ when $x = 5.5$

$9 = \frac{k}{5.5} \Rightarrow k = (9)(5.5) = 49.5$

So, $y = \frac{49.5}{x}$

74. $F = cx\sqrt{y}$

$6 = c(9)\sqrt{4}$

$6 = 18c$

$\frac{1}{3} = c$

$F = \frac{1}{3}x\sqrt{y}$

75. If $y = kx$, then the y-intercept is $(0, 0)$.

Problem Solving for Chapter 1

1. $A = \frac{1}{2}bh = \frac{1}{2}xy$

Since $(0, y)$, $(2, 1)$, and $(x, 0)$ all lie on the same line, the slopes between any pair are equal.

$$\frac{1-y}{2-0} = \frac{0-1}{x-2}$$

$$\frac{1-y}{2} = \frac{-1}{x-2}$$

$$y = \frac{2}{x-2} + 1$$

$$y = \frac{x}{x-2}$$

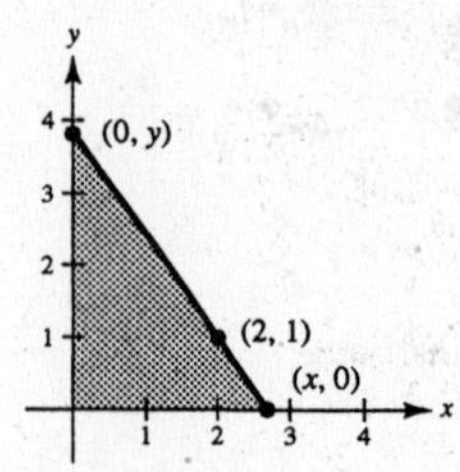

Therefore, $A = \frac{1}{2}x\left(\frac{x}{x-2}\right) = \frac{x^2}{2(x-2)}$.

The domain of A includes x-values such that $x^2/[2(x-2)] > 0$. The domain is $x > 2$.

2. $A = l \cdot w = (2x)y = 2xy$

But $y = \sqrt{36 - x^2}$, so $A = 2x\sqrt{36 - x^2}$, $0 < x < 6$.

3. $$f(x) = a_{2n+1}x^{2n+1} + a_{2n-1}x^{2n-1} + \cdots + a_3x^3 + a_1x$$
$$f(-x) = a_{2n+1}(-x)^{2n+1} + a_{2n-1}(-x)^{2n-1} + \cdots + a_3(-x)^3 + a_1(-x)$$
$$= -a_{2n+1}x^{2n+1} - a_{2n-1}x^{2n-1} - \cdots - a_3x^3 - a_1x = -f(x)$$

Therefore, $f(x)$ is odd.

4. $$f(x) = a_{2n}x^{2n} + a_{2n-2}x^{2n-2} + \cdots + a_2x^2 + a_0$$
$$f(-x) = a_{2n}(-x)^{2n} + a_{2n-2}(-x)^{2n-2} + \cdots + a_2(-x)^2 + a_0$$
$$= a_{2n}x^{2n} + a_{2n-2}x^{2n-2} + \cdots + a_2x^2 + a_0$$
$$= f(x)$$

Therefore, $f(x)$ is even.

5. (a) $y = x$

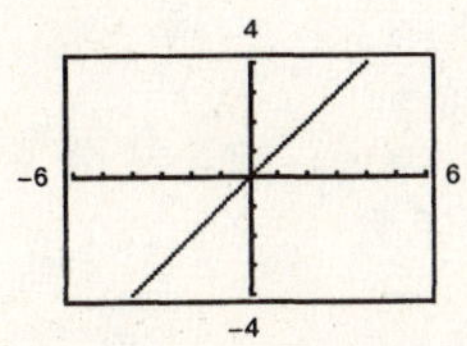

(b) $y = x^2$

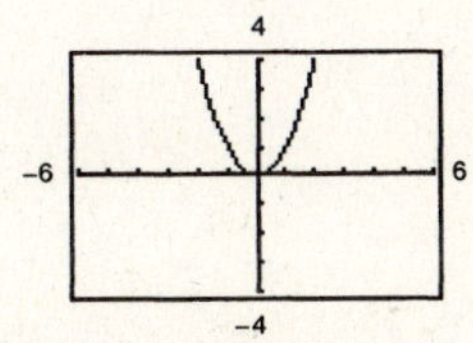

(c) $y = x^3$

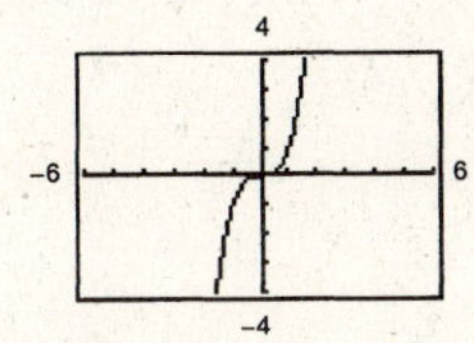

(d) $y = x^4$

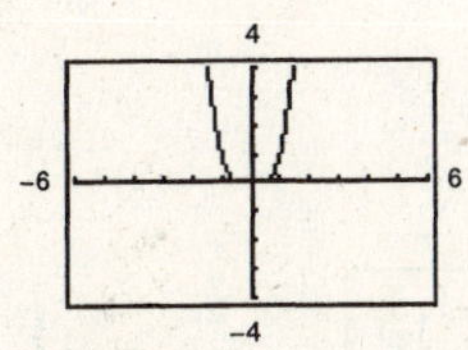

(e) $y = x^5$

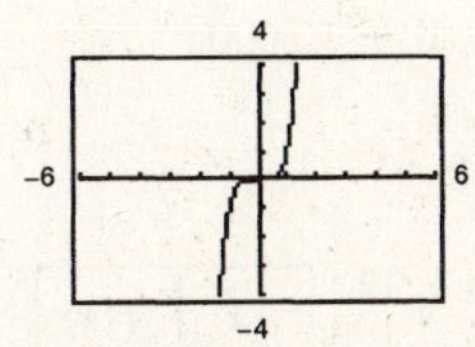

(f) $y = x^6$

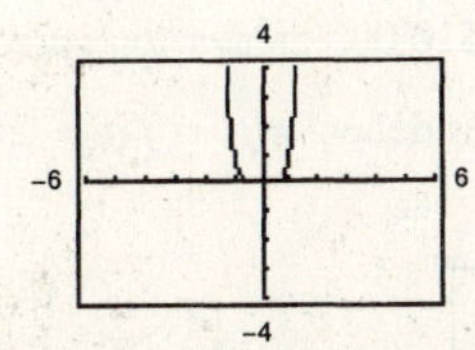

All the graphs pass through the origin. The graphs of the odd powers of x are symmetric with respect to the origin and the graphs of the even powers are symmetric with respect to the y-axis. As the powers increase, the graphs become flatter in the interval $-1 < x < 1$.

(g) The graph of $y = x^7$ will pass through the origin and will be symmetric with the origin.
The graph of $y = x^8$ will pass through the origin and will be symmetric with respect to the y-axis.

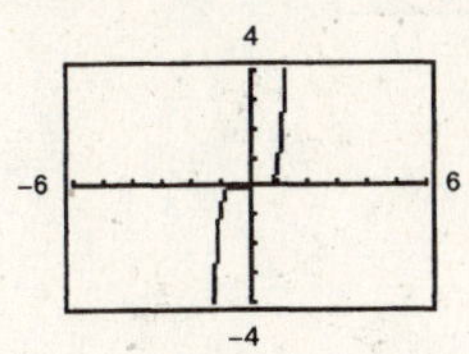

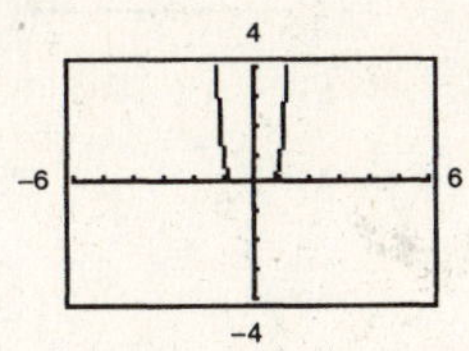

6. If you consider the x-axis to be a mirror, the graph of $y = -f(x)$ is the mirror image of the graph of $y = f(x)$.

7. $y = f(x + 2) - 1$

Horizontal shift 2 units to the left and a vertical shift 1 unit downward.

$(0, 1) \rightarrow (0 - 2, 1 - 1) = (-2, 0)$

$(1, 2) \rightarrow (1 - 2, 2 - 1) = (-1, 1)$

$(2, 3) \rightarrow (2 - 2, 3 - 1) = (0, 2)$

8. Let $f(x)$ and $g(x)$ be two odd functions and define $h(x) = f(x)g(x)$. Then

$$\begin{aligned} h(-x) &= f(-x)g(-x) \\ &= [-f(x)][-g(x)] \quad \text{Since } f \text{ and } g \text{ are odd} \\ &= f(x)g(x) \\ &= h(x) \end{aligned}$$

Thus, $h(x)$ is even.

Let $f(x)$ and $g(x)$ be two even functions and define $h(x) = f(x)g(x)$. Then

$$\begin{aligned} h(-x) &= f(-x)g(-x) \\ &= f(x)g(x) \quad \text{Since } f \text{ and } g \text{ are even} \\ &= h(x) \end{aligned}$$

Thus, $h(x)$ is even.

9. Let $f(x)$ be an odd function, $g(x)$ be an even function and define $h(x) = f(x)g(x)$. Then

$$\begin{aligned} h(-x) &= f(-x)g(-x) \\ &= [-f(x)]g(x) \quad \text{Since } f \text{ is odd and } g \text{ is even.} \\ &= -f(x)g(x) \\ &= -h(x) \end{aligned}$$

Thus, h is odd.

10. (a) The profits were only $\frac{3}{4}$ as large as expected: $g(t) = \frac{3}{4}f(t)$

(b) The profits were \$10,000 greater than predicted: $g(t) = f(t) + 10,000$

(c) There was a 2-year delay: $g(t) = f(t - 2)$

11.

x	1	3	4	6
f	1	2	6	7

x	1	2	6	7
$f^{-1}(x)$	1	3	4	6

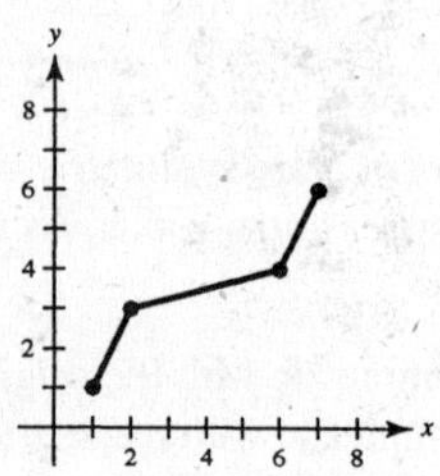

12.

x	$f(x)$
-2	-5
-1	-2
1	2
3	3

x	$f^{-1}(x)$
-5	-2
-2	-1
2	1
3	3

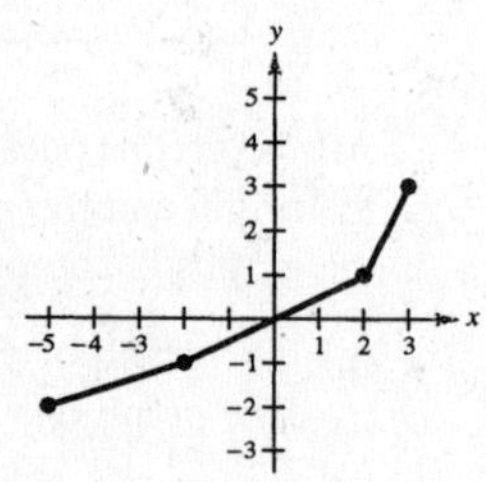

13.

x	-2	-1	3	4
f	6	0	-2	-3

x	-3	-2	0	6
$f^{-1}(x)$	4	3	-1	-2

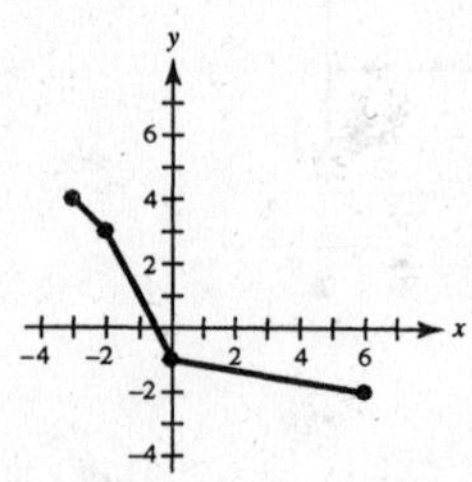

14.

x	$f(x)$
-4	3
-2	4
0	0
3	-1

The graph does not pass the Horizontal Line Test, so $f^{-1}(x)$ does not exist.

15. If $f(x) = k(2 - x - x^3)$ has an inverse and $f^{-1}(3) = -2$, then $f(-2) = 3$. Thus,

$$f(-2) = k(2 - (-2) - (-2)^3) = 3$$

$$k(2 + 2 + 8) = 3$$

$$12k = 3$$

$$k = \frac{3}{12} = \frac{1}{4}$$

Thus, $k = \frac{1}{4}$.

16. (a) The length of the trip in the water is $\sqrt{2^2 + x^2}$, and the length of the trip over land is $\sqrt{1 + (3 - x)^2}$. Hence, the total time is

$$T(x) = \frac{\sqrt{4 + x^2}}{2} + \frac{\sqrt{1 + (3 - x)^2}}{4} \text{ hours.}$$

(b) Domain of $T(x)$: $0 \le x \le 3$

(c)

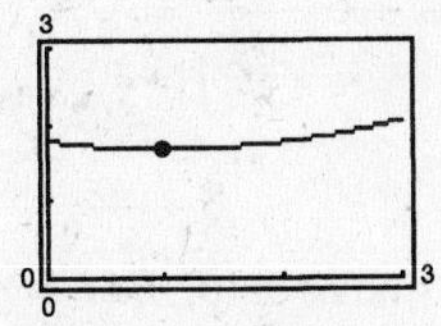

(d) $T(x)$ is a minimum when $x = 1$

(e) To reach point Q in the shortest amount of time, you should row to a point one mile down the coast, and then walk the rest of the way.

17. $H(x) = \begin{cases} 1, & x \ge 0 \\ 0, & x < 0 \end{cases}$

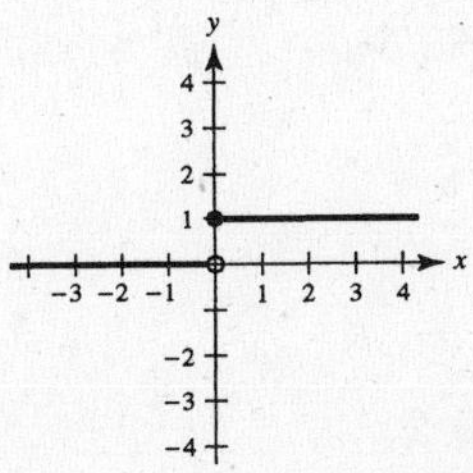

(a) $H(x) - 2$

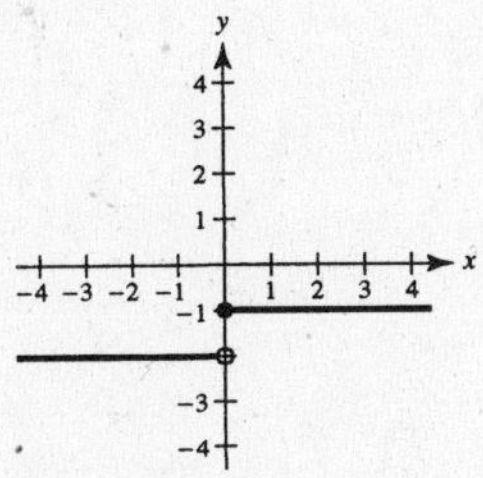

(b) $H(x - 2)$

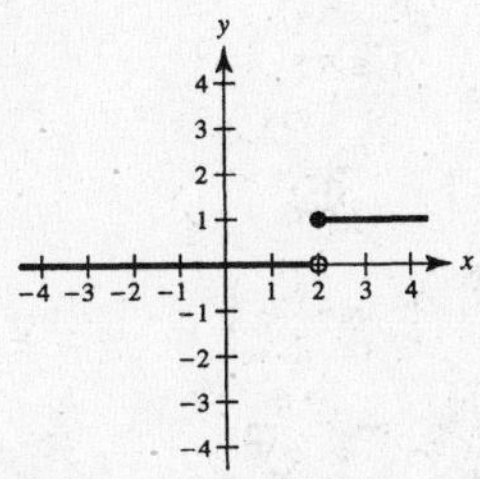

(c) $-H(x)$

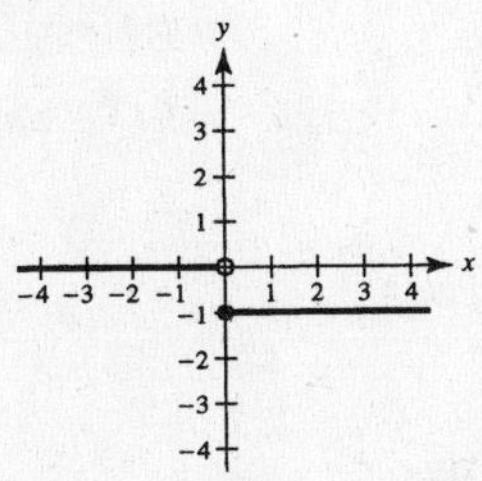

(d) $H(-x)$

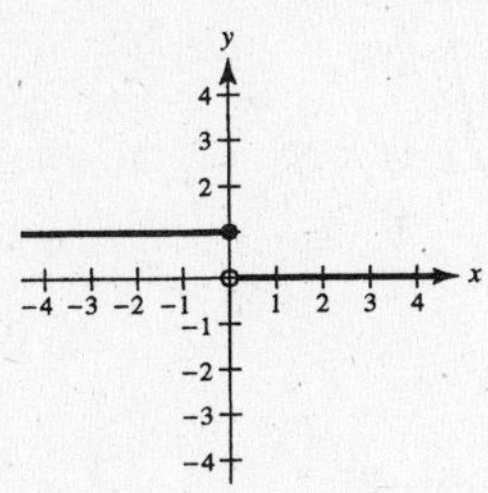

(e) $\frac{1}{2}H(x)$

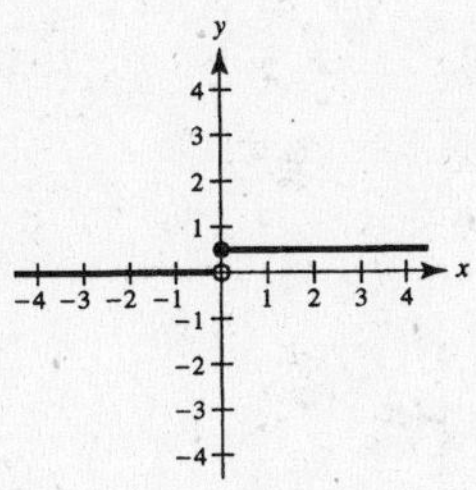

(f) $-H(x - 2) + 2$

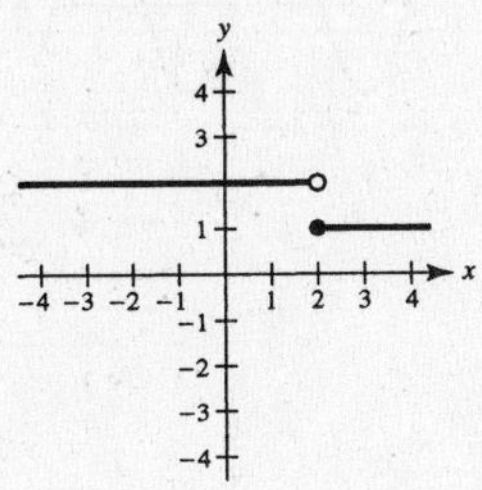

18. $f(x) = y = \dfrac{1}{1-x}$

(a) Domain: all $x \neq 1$

Range: all $y \neq 0$

(b) $f(f(x)) = f\left(\dfrac{1}{1-x}\right) = \dfrac{1}{1-\left(\dfrac{1}{1-x}\right)} = \dfrac{1}{\dfrac{1-x-1}{1-x}} = \dfrac{1-x}{-x} = \dfrac{x-1}{x}$

Domain: all $x \neq 0, 1$

(c) $f(f(f(x))) = f\left(\dfrac{x-1}{x}\right) = \dfrac{1}{1-\left(\dfrac{x-1}{x}\right)} = \dfrac{1}{\dfrac{1}{x}} = x$

Domain: all $x \neq 0, 1$

(d) The graph is not a line. It has holes at $(0, 0)$ and $(1, 1)$.

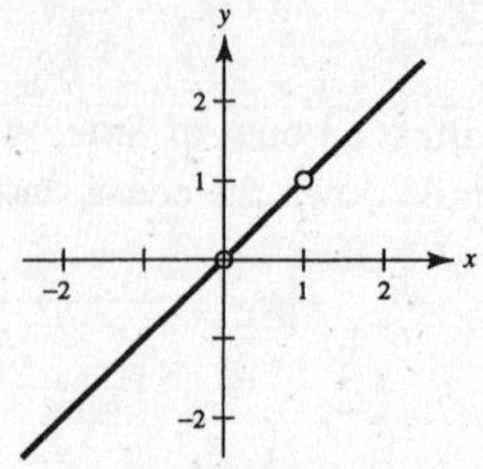

19. $(f \circ (g \circ h))(x) = f((g \circ h)(x))$

$= f(g(h(x)))$

$= (f \circ g \circ h)(x)$

$((f \circ g) \circ h)(x) = (f \circ g)(h(x))$

$= f(g(h(x)))$

$= (f \circ g \circ h)(x)$

20. (a) $f(x+1)$

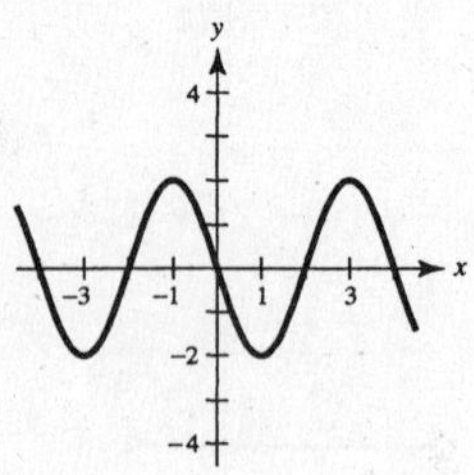

(b) $f(x) + 1$

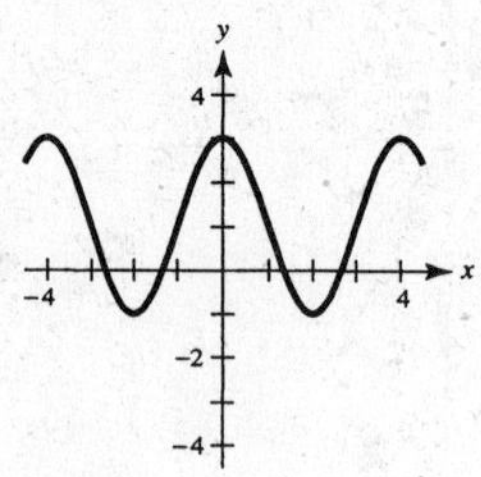

(c) $2f(x)$

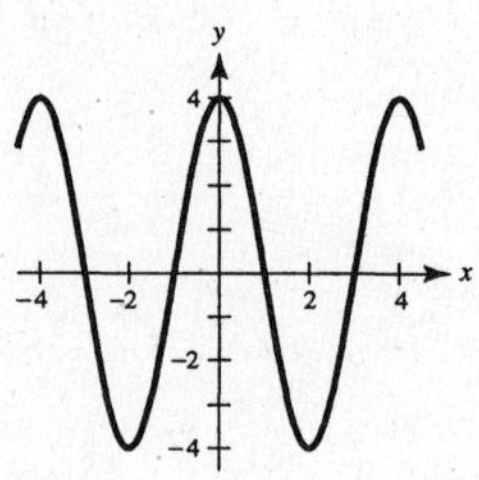

(d) $f(-x)$

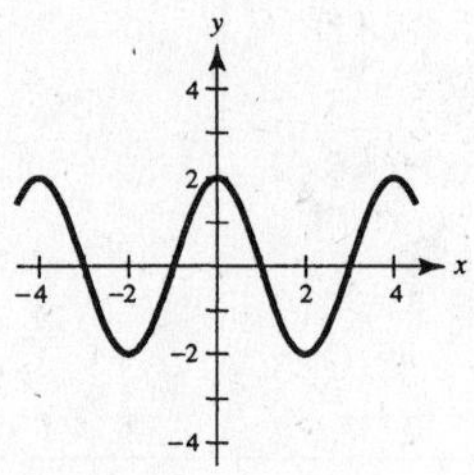

(e) $-f(x)$

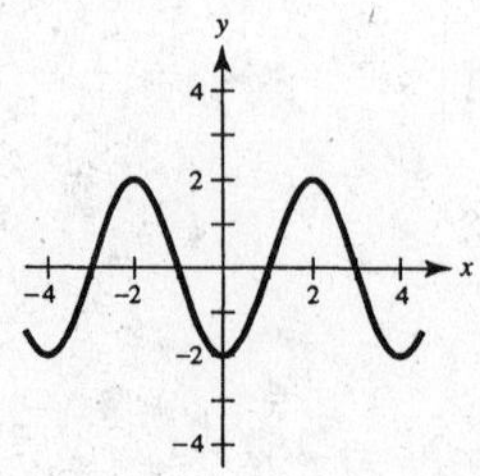

(f) $|f(x)|$

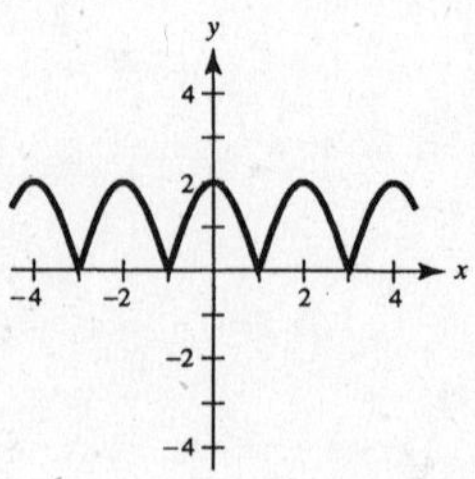

—CONTINUED—

20.—CONTINUED—

(g) $f(|x|)$

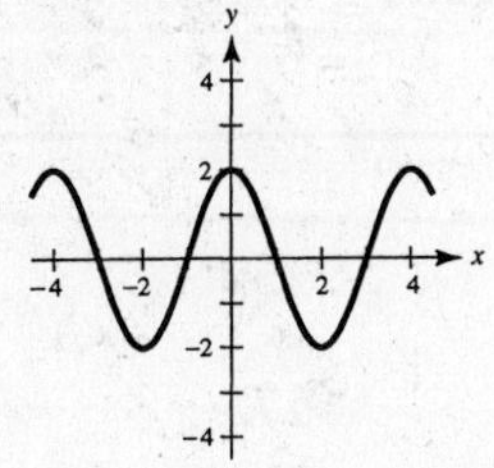

21.

x	$f(x)$	$(f+g)(x)$
-2	-2	0
-1	0	-2
0	1	0
1	0	1
2	1	1

This chart was derived from the graphs of f and $f + g$.

(a)

x	$g(x) = (f+g)(x) - f(x)$
-2	2
-1	-2
0	-1
1	1
2	0

(b)

x	$(g-f)(x) = g(x) - f(x)$
-2	4
-1	-2
0	-2
1	1
2	-1

(c)

x	$g(f(x))$
-2	$g(f(-2)) = g(-2) = 2$
-1	$g(f(-1)) = g(0) = -1$
0	$g(f(0)) = g(1) = 1$
1	$g(f(1)) = g(0) = -1$
2	$g(f(2)) = g(1) = 1$

(d)

x	$f(g(-x))$
-2	$f(g(2)) = f(0) = 1$
-1	$f(g(1)) = f(1) = 0$
0	$f(g(0)) = f(-1) = 0$
1	$f(g(-1)) = f(-2) = -2$
2	$f(g(-2)) = f(2) = 1$

CHAPTER 2
Polynomial and Rational Functions

Section 2.1 Quadratic Functions 159

Section 2.2 Polynomial Functions of Higher Degree 175

Section 2.3 Polynomial and Synthetic Division 186

Section 2.4 Complex Numbers 198

Section 2.5 The Fundamental Theorem of Algebra 204

Section 2.6 Rational Functions 220

Review Exercises 236

Problem Solving 250

CHAPTER 2
Polynomial and Rational Functions

Section 2.1 Quadratic Functions

1. $f(x) = (x - 2)^2$ opens upward and has vertex $(2, 0)$. Matches graph (g).

2. $f(x) = (x + 4)^2$ opens upward and has vertex $(-4, 0)$. Matches graph (c).

3. $f(x) = x^2 - 2$ opens upward and has vertex $(0, -2)$. Matches graph (b).

4. $f(x) = 3 - x^2$ opens downward and has vertex $(0, 3)$. Matches graph (h).

5. $f(x) = 4 - (x - 2)^2 = -(x - 2)^2 + 4$ opens downward and has vertex $(2, 4)$. Matches graph (f).

6. $f(x) = (x + 1)^2 - 2$ opens upward and has vertex $(-1, -2)$. Matches graph (a).

7. $f(x) = -(x - 3)^2 - 2$ opens downward and has vertex $(3, -2)$. Matches graph (e).

8. $f(x) = -(x - 4)^2$ opens downward and has vertex $(4, 0)$. Matches graph (d).

9. $y = x^2$

(a) $f(x) = \frac{1}{2}x^2$

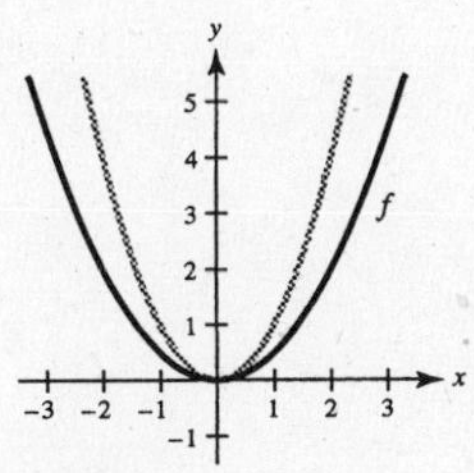

Vertical shrink by a factor of $\frac{1}{2}$.

(b) $g(x) = x^2 - 1$

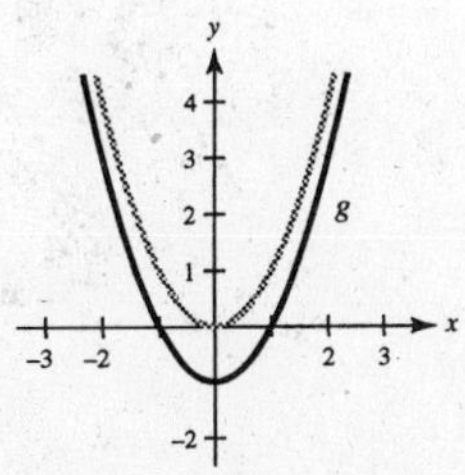

Vertical shift one unit down.

(c) $h(x) = (x - 1)^2$

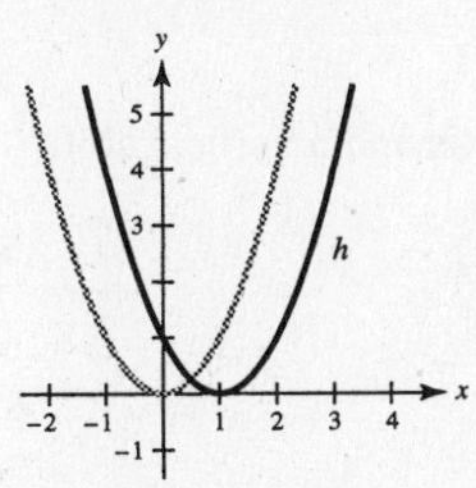

Horizontal shift one unit to the right.

(d) $k(x) = -\frac{1}{2}(x - 2)^2 - 1$

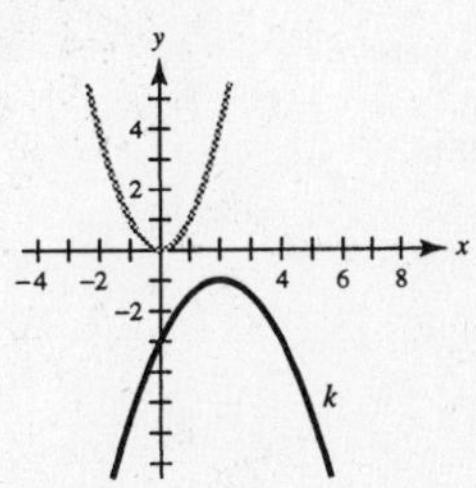

Vertical shrink by a factor of $\frac{1}{2}$, a reflection in the x-axis, a horizontal shift 2 units to the right, and a vertical shift one unit down.

10. $y = x^2$

(a) $f(x) = -3x^2$

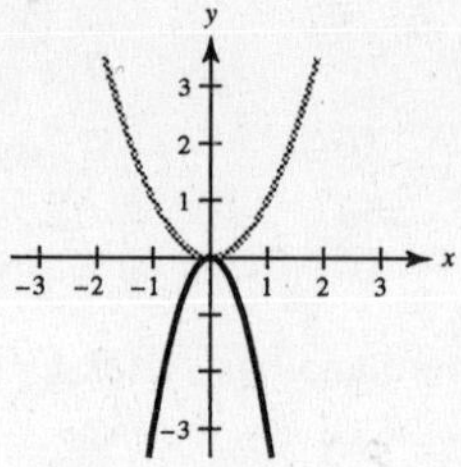

A reflection in the x-axis and a vertical stretch by a factor of 3.

(b) $g(x) = x^2 + 3$

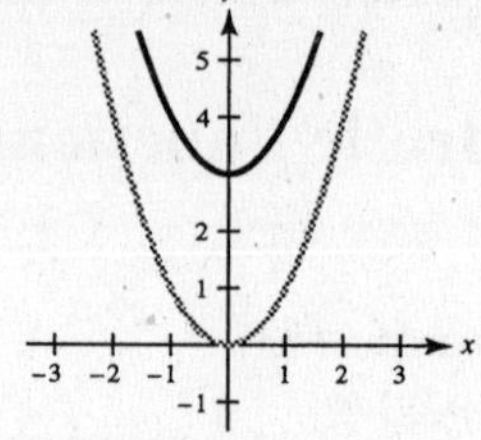

Vertical shift 3 units up.

(c) $h(x) = (x - 3)^2$

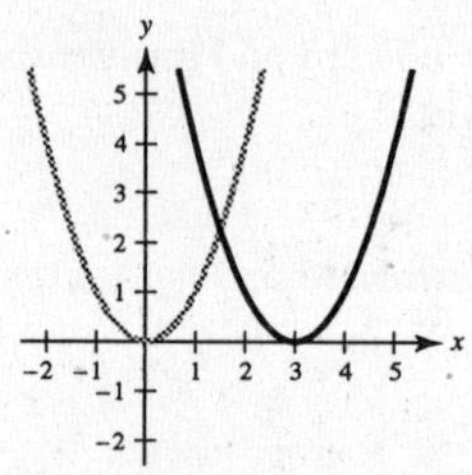

Horizontal shift 3 units to the right.

(d) $k(x) = \frac{1}{2}(x + 2)^2 + 1$

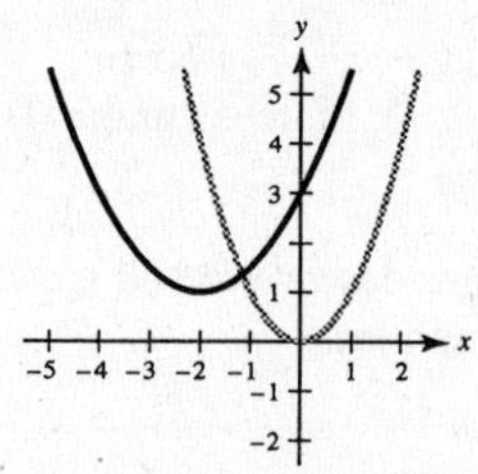

A vertical shrink by a factor of $\frac{1}{2}$, a horizontal shift 2 units to the left, and a vertical shift one unit up.

11. $y = x^2$

(a) $f(x) = -x^2$

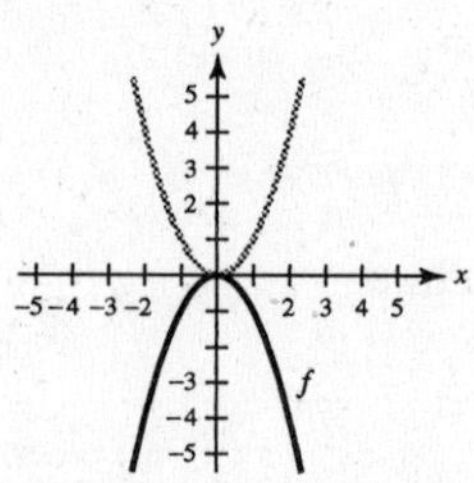

A reflection in the x-axis.

(b) $g(x) = -x^2 - 2$

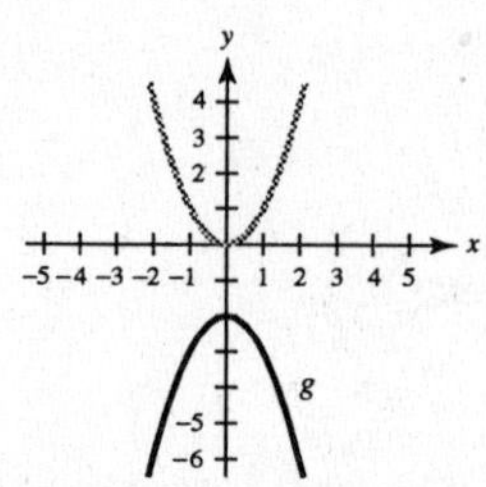

A reflection in the x-axis and a vertical shift 2 units down.

(c) $h(x) = -(x - 3)^2$

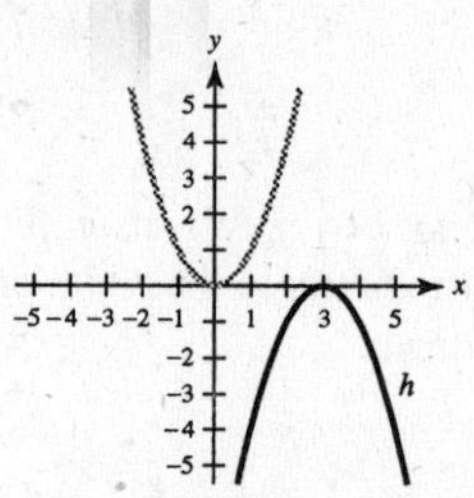

A reflection in the x-axis and a horizontal shift 3 units to the right.

(d) $k(x) = -\frac{1}{2}(x - 3)^2 - 2$

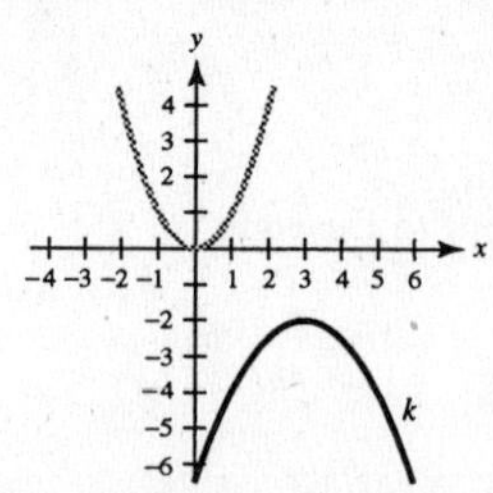

A reflection in the x-axis, a vertical shrink by a factor of $\frac{1}{2}$, a horizontal shift 3 units to the right and a vertical shift 2 units down.

12. $y = x^2$

(a) $f(x) = -\frac{1}{2}x^2$

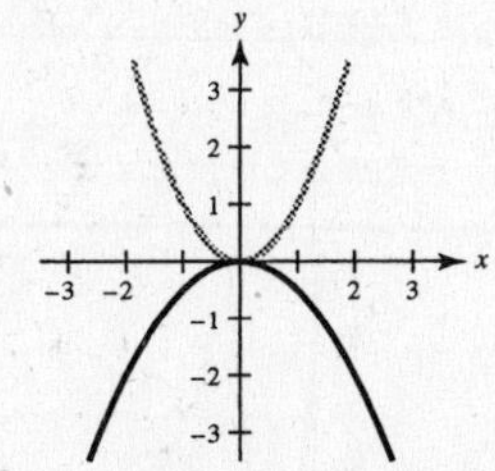

A reflection in the x-axis and a vertical shrink by a factor of $\frac{1}{2}$.

(b) $g(x) = x^2 - 4$

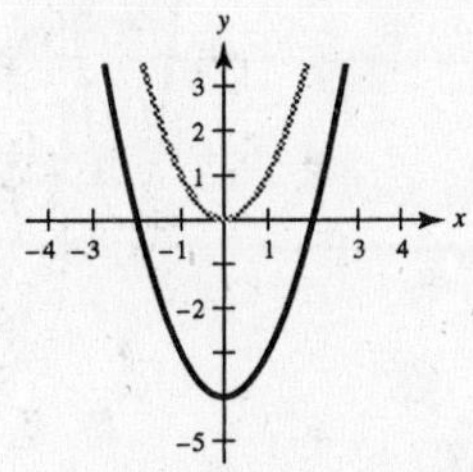

Vertical shift 4 units down.

(c) $h(x) = \left(x - \frac{1}{2}\right)^2$

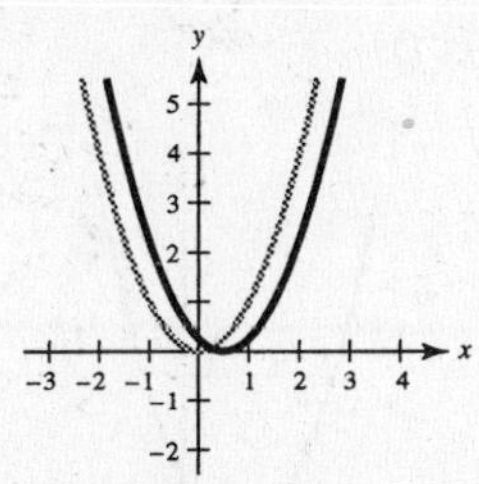

Horizontal shift $\frac{1}{2}$ unit to the right.

(d) $k(x) = \frac{1}{2}(x + 2)^2 - 1$

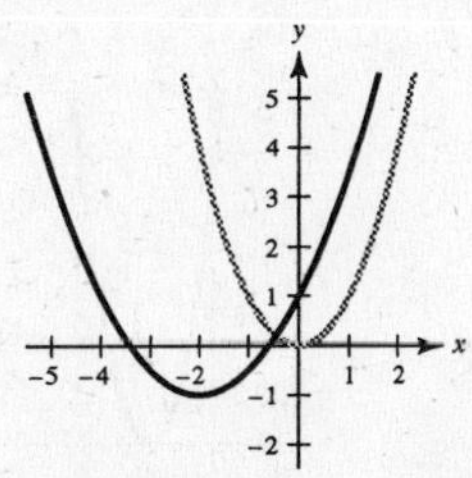

Vertical shrink by a factor of $\frac{1}{2}$, a horizontal shift 2 units to the left, and a vertical shift 1 unit down.

13. $f(x) = x^2 - 16$

Vertex: $(0, -16)$

Find x-intercepts: $x^2 - 16 = 0$

$$x^2 = 16$$
$$x = \pm 4$$

x-intercepts: $(\pm 4, 0)$

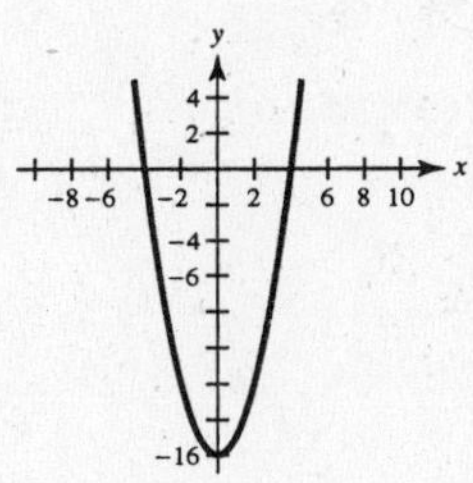

14. $h(x) = 25 - x^2$

Vertex: $(0, 25)$

Find x-intercepts: $25 - x^2 = 0$

$$x^2 = 25$$
$$x = \pm 5$$

x-intercepts: $(\pm 5, 0)$

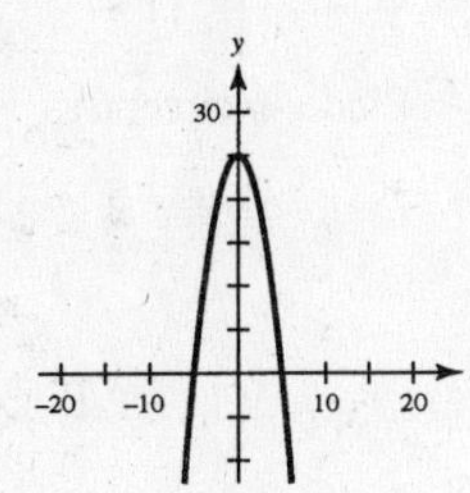

15. $f(x) = 16 - \frac{1}{4}x^2 = -\frac{1}{4}x^2 + 16$

Vertex: $(0, 16)$

Find x-intercepts: $16 - \frac{1}{4}x^2 = 0$

$$x^2 = 64$$
$$x = \pm 8$$

x-intercepts: $(\pm 8, 0)$

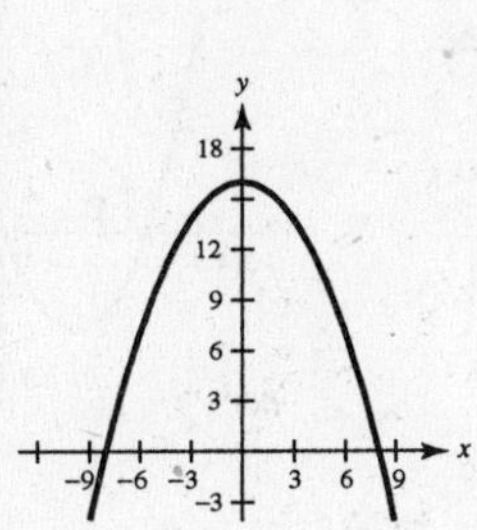

16. $f(x) = \frac{1}{2}x^2 - 4 = \frac{1}{2}(x - 0)^2 - 4$

Vertex: $(0, -4)$

Find x-intercepts:

$$\frac{1}{2}x^2 - 4 = 0$$
$$x^2 = 8$$
$$x = \pm\sqrt{8} = \pm 2\sqrt{2}$$

x-intercepts: $\left(-2\sqrt{2}, 0\right), \left(2\sqrt{2}, 0\right)$

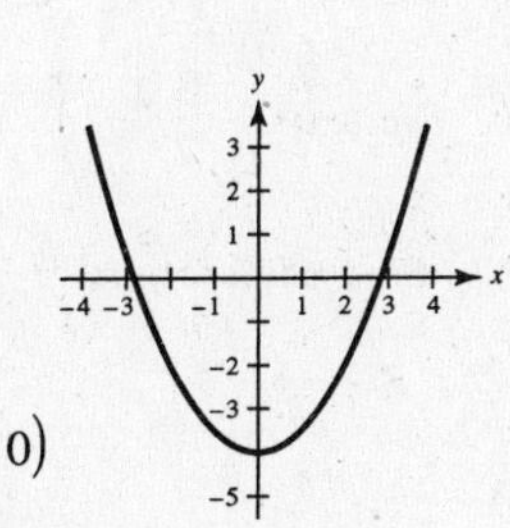

17. $f(x) = (x + 5)^2 - 6$

Vertex: $(-5, -6)$

Find x-intercepts:

$$(x + 5)^2 - 6 = 0$$
$$(x + 5)^2 = 6$$
$$x + 5 = \pm\sqrt{6}$$
$$x = -5 \pm \sqrt{6}$$

x-intercepts: $(-5 - \sqrt{6}, 0), (-5 + \sqrt{6}, 0)$

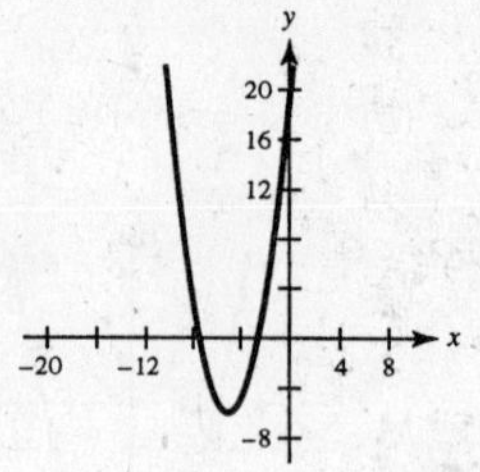

18. $f(x) = (x - 6)^2 + 3$

Vertex: $(6, 3)$

Find x-intercepts: $(x - 6)^2 + 3 = 0$

$$(x - 6)^2 = -3$$

Not possible for real x.

No x-intercepts

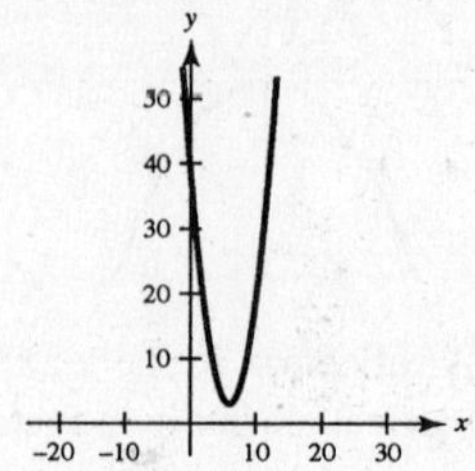

19. $h(x) = x^2 - 8x + 16 = (x - 4)^2$

Vertex: $(4, 0)$

x-intercept: $(4, 0)$

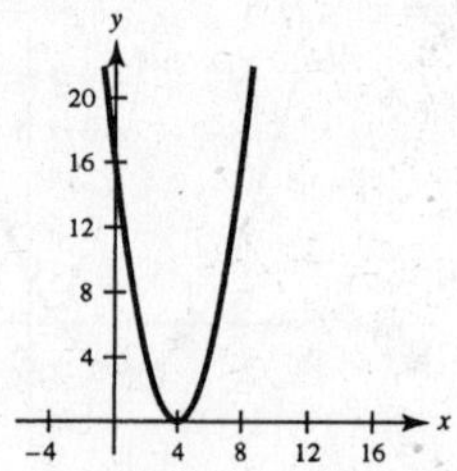

20. $g(x) = x^2 + 2x + 1 = (x + 1)^2$

Vertex: $(-1, 0)$

x-intercept: $(-1, 0)$

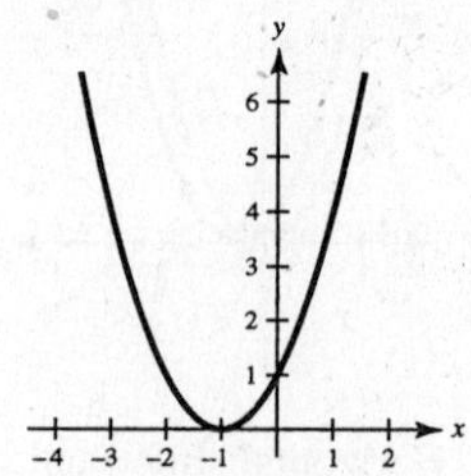

21. $f(x) = x^2 - x + \frac{5}{4}$

$$= \left(x^2 - x + \tfrac{1}{4}\right) - \tfrac{1}{4} + \tfrac{5}{4}$$
$$= \left(x - \tfrac{1}{2}\right)^2 + 1$$

Vertex: $\left(\frac{1}{2}, 1\right)$

Find x-intercepts:

$$x^2 - x + \frac{5}{4} = 0$$
$$x = \frac{1 \pm \sqrt{1 - 5}}{2}$$

Not a real number $\Longrightarrow$ No x-intercepts

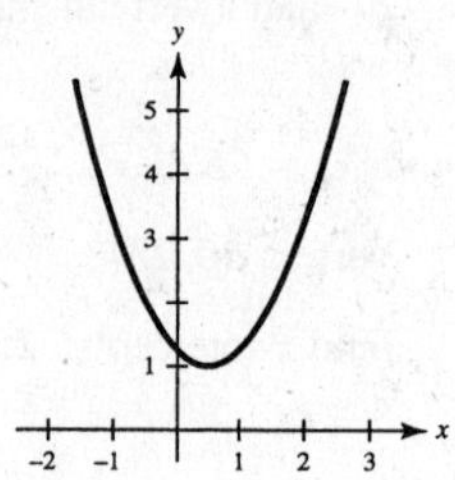

22. $f(x) = x^2 + 3x + \frac{1}{4} = \left(x^2 + 3x + \frac{9}{4}\right) - \frac{9}{4} + \frac{1}{4} = \left(x + \frac{3}{2}\right)^2 - 2$

Vertex: $\left(-\frac{3}{2}, -2\right)$

Find x-intercepts: $x^2 + 3x + \frac{1}{4} = 0$

$$x = \frac{-3 \pm \sqrt{9 - 1}}{2} = -\frac{3}{2} \pm \sqrt{2}$$

x-intercepts: $\left(-\frac{3}{2} \pm \sqrt{2}, 0\right)$

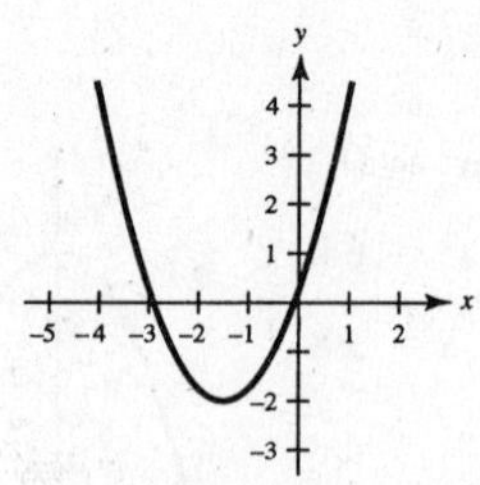

23. $f(x) = -x^2 + 2x + 5$

$= -(x^2 - 2x + 1) - (-1) + 5$

$= -(x - 1)^2 + 6$

Vertex: $(1, 6)$

Find x-intercepts:

$-x^2 + 2x + 5 = 0$

$x^2 - 2x - 5 = 0$

$$x = \frac{2 \pm \sqrt{4 + 20}}{2}$$

$= 1 \pm \sqrt{6}$

x-intercepts: $(1 - \sqrt{6}, 0), (1 + \sqrt{6}, 0)$

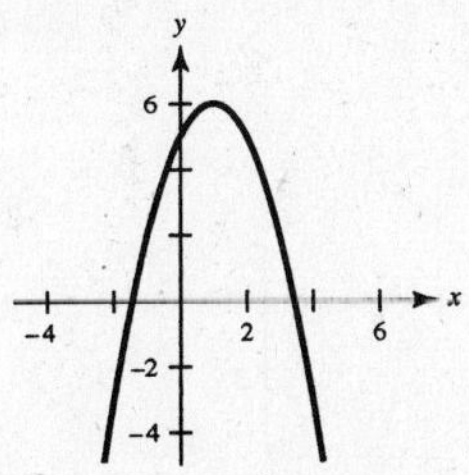

24. $f(x) = -x^2 - 4x + 1 = -(x^2 + 4x) + 1$

$= -(x^2 + 4x + 4) - (-4) + 1$

$= -(x + 2)^2 + 5$

Vertex: $(-2, 5)$

Find x-intercepts: $-x^2 - 4x + 1 = 0$

$x^2 + 4x - 1 = 0$

$$x = \frac{-4 \pm \sqrt{16 + 4}}{2}$$

$= -2 \pm \sqrt{5}$

x-intercepts: $(-2 \pm \sqrt{5}, 0)$

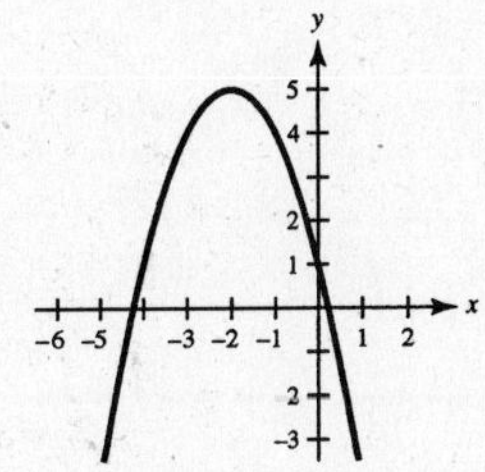

25. $h(x) = 4x^2 - 4x + 21$

$= 4\left(x^2 - x + \frac{1}{4}\right) - 4\left(\frac{1}{4}\right) + 21$

$= 4\left(x - \frac{1}{2}\right)^2 + 20$

Vertex: $\left(\frac{1}{2}, 20\right)$

Find x-intercepts:

$4x^2 - 4x + 21 = 0$

$$x = \frac{4 \pm \sqrt{16 - 336}}{2(4)}$$

Not a real number $\Longrightarrow$ No x-intercepts

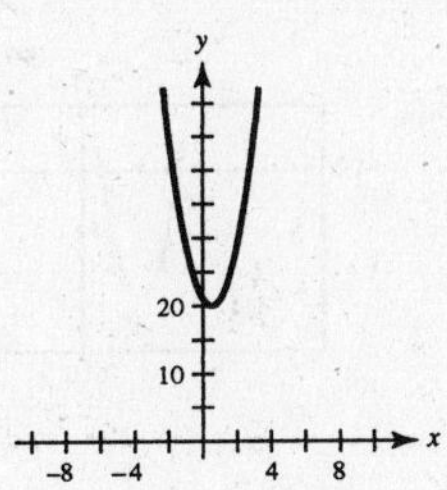

26. $f(x) = 2x^2 - x + 1$

$= 2\left(x^2 - \frac{1}{2}x\right) + 1$

$= 2\left(x - \frac{1}{4}\right)^2 - 2\left(\frac{1}{16}\right) + 1$

$= 2\left(x - \frac{1}{4}\right)^2 + \frac{7}{8}$

Vertex: $\left(\frac{1}{4}, \frac{7}{8}\right)$

Find x-intercepts: $2x^2 - x + 1 = 0$

$$x = \frac{1 \pm \sqrt{1 - 8}}{2(2)}$$

Not a real number

No x-intercepts

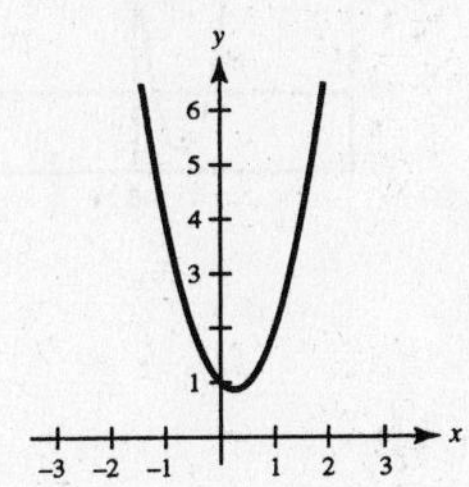

27. $f(x) = \frac{1}{4}x^2 - 2x - 12$

$= \frac{1}{4}(x^2 - 8x + 16) - \frac{1}{4}(16) - 12$

$= \frac{1}{4}(x - 4)^2 - 16$

Vertex: $(4, -16)$

Find x-intercepts:

$\frac{1}{4}x^2 - 2x - 12 = 0$

$x^2 - 8x - 48 = 0$

$(x + 4)(x - 12) = 0$

$x = -4$ or $x = 12$

x-intercepts: $(-4, 0), (12, 0)$

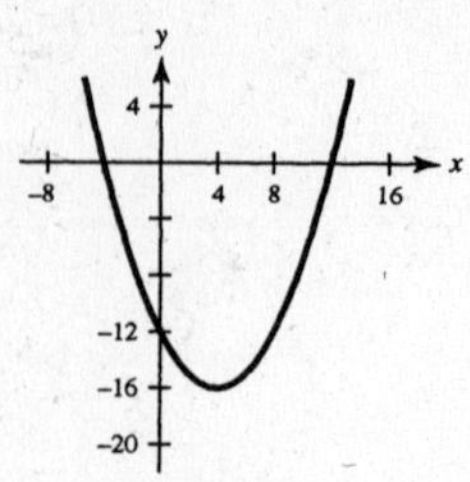

28. $f(x) = -\frac{1}{3}x^2 + 3x - 6$

$= -\frac{1}{3}(x^2 - 9x) - 6$

$= -\frac{1}{3}\left(x^2 - 9x + \frac{81}{4}\right) + \frac{1}{3}\left(\frac{81}{4}\right) - 6$

$= -\frac{1}{3}\left(x - \frac{9}{2}\right)^2 + \frac{3}{4}$

Vertex: $\left(\frac{9}{2}, \frac{3}{4}\right)$

Find x-intercepts: $-\frac{1}{3}x^2 + 3x - 6 = 0$

$x^2 - 9x + 18 = 0$

$(x - 3)(x - 6) = 0$

x-intercepts: $(3, 0), (6, 0)$

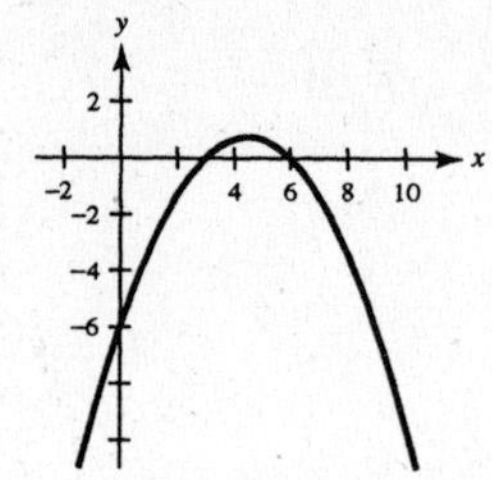

29. $f(x) = -(x^2 + 2x - 3) = -(x + 1)^2 + 4$

Vertex: $(-1, 4)$

x-intercepts: $(-3, 0), (1, 0)$

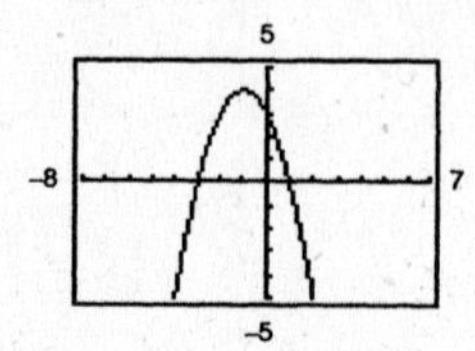

30. $f(x) = -(x^2 + x - 30)$

$= -(x^2 + x) + 30$

$= -\left(x^2 + x + \frac{1}{4}\right) + \frac{1}{4} + 30$

$= -\left(x + \frac{1}{2}\right)^2 + \frac{121}{4}$

Vertex: $\left(-\frac{1}{2}, \frac{121}{4}\right)$

x-intercepts: $(-6, 0), (5, 0)$

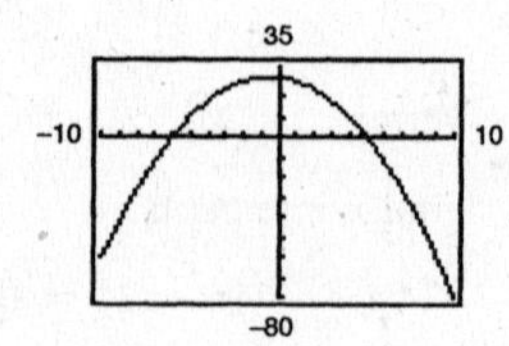

31. $g(x) = x^2 + 8x + 11 = (x + 4)^2 - 5$

Vertex: $(-4, -5)$

x-intercepts: $\left(-4 \pm \sqrt{5}, 0\right)$

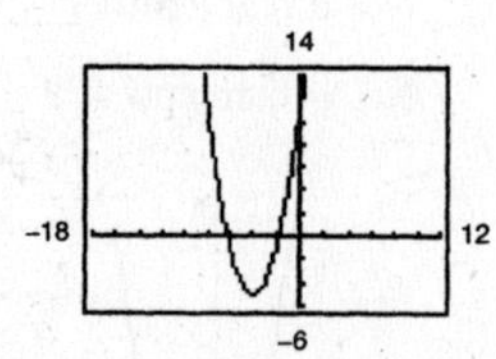

32. $f(x) = x^2 + 10x + 14$

$= (x^2 + 10x + 25) - 25 + 14$

$= (x + 5)^2 - 11$

Vertex: $(-5, -11)$

x-intercepts: $\left(-5 \pm \sqrt{11}, 0\right)$

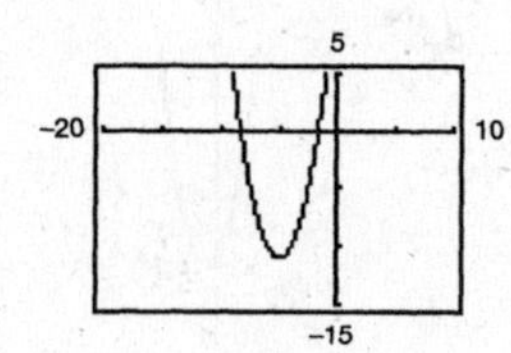

33. $f(x) = 2x^2 - 16x + 31$

$= 2(x - 4)^2 - 1$

Vertex: $(4, -1)$

x-intercepts: $\left(4 \pm \frac{1}{2}\sqrt{2}, 0\right)$

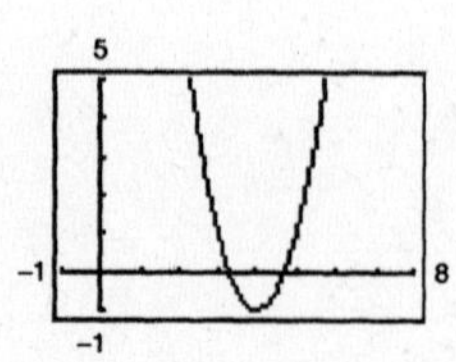

34. $f(x) = -4x^2 + 24x - 41$

$= -4(x^2 - 6x) - 41$

$= -4(x^2 - 6x + 9) + 36 - 41$

$= -4(x - 3)^2 - 5$

Vertex: $(3, -5)$

No x-intercepts

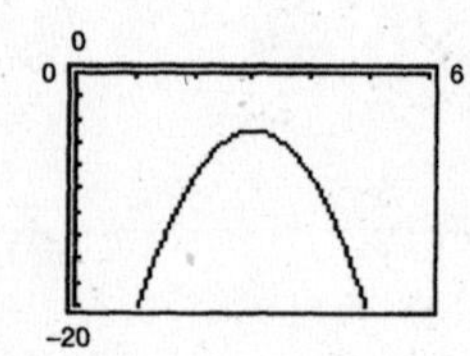

35. $g(x) = \frac{1}{2}(x^2 + 4x - 2) = \frac{1}{2}(x + 2)^2 - 3$

Vertex: $(-2, -3)$

x-intercepts: $(-2 \pm \sqrt{6}, 0)$

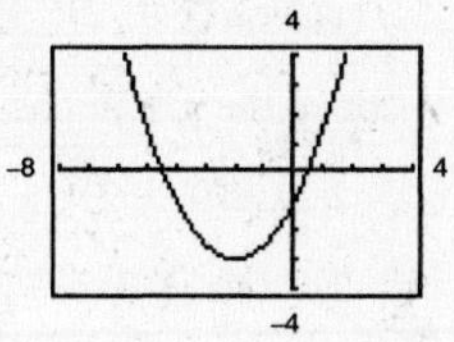

36. $f(x) = \frac{3}{5}(x^2 + 6x - 5)$

$= \frac{3}{5}(x^2 + 6x + 9) - \frac{27}{5} - 3$

$= \frac{3}{5}(x + 3)^2 - \frac{42}{5}$

Vertex: $\left(-3, -\frac{42}{5}\right)$

x-intercepts: $(-3 \pm \sqrt{14}, 0)$

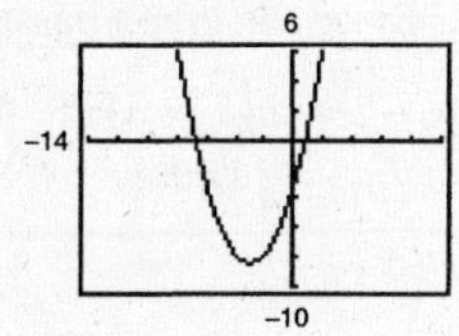

37. $(1, 0)$ is the vertex.

$y = a(x - 1)^2 + 0 = a(x - 1)^2$

Since the graph passes through the point $(0, 1)$, we have:

$1 = a(0 - 1)^2$

$1 = a$

$y = 1(x - 1)^2 = (x - 1)^2$

38. $(0, 1)$ is the vertex.

$f(x) = a(x - 0)^2 + 1 = ax^2 + 1$

Since the graph passes through $(1, 0)$,

$0 = a(1)^2 + 1$

$-1 = a.$

Thus, $y = -x^2 + 1.$

39. $(-1, 4)$ is the vertex.

$y = a(x + 1)^2 + 4$

Since the graph passes through the point $(1, 0)$, we have:

$0 = a(1 + 1)^2 + 4$

$-4 = 4a$

$-1 = a$

$y = -1(x + 1)^2 + 4 = -(x + 1)^2 + 4$

40. $(-2, -1)$ is the vertex.

$f(x) = a(x + 2)^2 - 1$

Since the graph passes through $(0, 3)$,

$3 = a(0 + 2)^2 - 1$

$3 = 4a - 1$

$4 = 4a$

$1 = a.$

Thus, $y = (x + 2)^2 - 1.$

41. $(-2, 2)$ is the vertex.

$y = a(x + 2)^2 + 2$

Since the graph passes through the point $(-1, 0)$, we have:

$0 = a(-1 + 2)^2 + 2$

$-2 = a$

$y = -2(x + 2)^2 + 2$

42. $(2, 0)$ is the vertex.

$f(x) = a(x - 2)^2 + 0 = a(x - 2)^2$

Since the graph passes through $(3, 2)$,

$2 = a(3 - 2)^2$

$2 = a.$

Thus, $y = 2(x - 2)^2.$

43. $(-2, 5)$ is the vertex.

$f(x) = a(x + 2)^2 + 5$

Since the graph passes through the point $(0, 9)$, we have:

$9 = a(0 + 2)^2 + 5$

$4 = 4a$

$1 = a$

$f(x) = 1(x + 2)^2 + 5 = (x + 2)^2 + 5$

44. $(4, -1)$ is the vertex.

$f(x) = a(x - 4)^2 - 1$

Since the graph passes through $(2, 3)$,

$3 = a(2 - 4)^2 - 1$

$3 = 4a - 1$

$4 = 4a$

$1 = a.$

Thus, $f(x) = (x - 4)^2 - 1.$

45. $(3, 4)$ is the vertex.

$f(x) = a(x - 3)^2 + 4$

Since the graph passes through the point $(1, 2)$, we have:

$2 = a(1 - 3)^2 + 4$

$-2 = 4a$

$-\frac{1}{2} = a$

$f(x) = -\frac{1}{2}(x - 3)^2 + 4$

46. $(2, 3)$ is the vertex.

$f(x) = a(x - 2)^2 + 3$

Since the graph passes through $(0, 2)$,

$2 = a(0 - 2)^2 + 3$

$2 = 4a + 3$

$-1 = 4a$

$-\frac{1}{4} = a.$

Thus, $f(x) = -\frac{1}{4}(x - 2)^2 + 3.$

47. $(5, 12)$ is the vertex.

$f(x) = a(x - 5)^2 + 12$

Since the graph passes through the point $(7, 15)$, we have:

$15 = a(7 - 5)^2 + 12$

$3 = 4a \Rightarrow a = \frac{3}{4}$

$f(x) = \frac{3}{4}(x - 5)^2 + 12$

48. $(-2, -2)$ is the vertex.

$f(x) = a(x + 2)^2 - 2$

Since the graph passes through $(-1, 0)$,

$0 = a(-1 + 2)^2 - 2$

$0 = a - 2$

$2 = a.$

Thus, $f(x) = 2(x + 2)^2 - 2.$

49. $\left(-\frac{1}{4}, \frac{3}{2}\right)$ is the vertex.

$f(x) = a\left(x + \frac{1}{4}\right)^2 + \frac{3}{2}$

Since the graph passes through the point $(-2, 0)$, we have:

$0 = a\left(-2 + \frac{1}{4}\right)^2 + \frac{3}{2}$

$-\frac{3}{2} = \frac{49}{16}a \Rightarrow a = -\frac{24}{49}$

$f(x) = -\frac{24}{49}\left(x + \frac{1}{4}\right)^2 + \frac{3}{2}$

50. $\left(\frac{5}{2}, -\frac{3}{4}\right)$ is the vertex.

$f(x) = a\left(x - \frac{5}{2}\right)^2 - \frac{3}{4}$

Since the graph passes through $(-2, 4)$,

$4 = a\left(-2 - \frac{5}{2}\right)^2 - \frac{3}{4}$

$4 = \frac{81}{4}a - \frac{3}{4}$

$\frac{19}{4} = \frac{81}{4}a$

$\frac{19}{81} = a$

Thus, $f(x) = \frac{19}{81}\left(x - \frac{5}{2}\right)^2 - \frac{3}{4}.$

51. $\left(-\frac{5}{2}, 0\right)$ is the vertex.

$f(x) = a\left(x + \frac{5}{2}\right)^2$

Since the graph passes through the point $\left(-\frac{7}{2}, -\frac{16}{3}\right)$, we have:

$-\frac{16}{3} = a\left(-\frac{7}{2} + \frac{5}{2}\right)^2$

$-\frac{16}{3} = a$

$f(x) = -\frac{16}{3}\left(x + \frac{5}{2}\right)^2$

52. $(6, 6)$ is the vertex.

$f(x) = a(x - 6)^2 + 6$

Since the graph passes through $\left(\frac{61}{10}, \frac{3}{2}\right)$,

$\frac{3}{2} = a\left(\frac{61}{10} - 6\right)^2 + 6$

$\frac{3}{2} = \frac{1}{100}a + 6$

$-\frac{9}{2} = \frac{1}{100}a$

$-450 = a$

Thus, $f(x) = -450(x - 6)^2 + 6.$

53. $y = x^2 - 16$

(a) x-intercepts: $(\pm 4, 0)$

(b) The x-intercepts and the solutions of the equation are the same.

(c) $0 = x^2 - 16$

$x^2 = 16$

$x = \pm 4 \Rightarrow$ The x-intercepts are $(\pm 4, 0)$.

54. $y = -x^2 + 4$

(a) From the graph it appears that the x-intercepts are $(\pm 2, 0)$.

(b) The x-intercepts and the solutions of $-x^2 + 4 = 0$ are the same.

(c) $-x^2 + 4 = 0$

$-x^2 = -4$

$x^2 = 4$

$x = \pm 2 \Rightarrow$ The x-intercepts are $(\pm 2, 0)$.

55. $y = -x^2 - 2x - 1$

(a) From the graph it appears that the x- intercept is $(-1, 0)$.

(b) The x-intercept and the solution to $-x^2 - 2x - 1 = 0$ are the same.

(c) $-x^2 - 2x - 1 = 0$

$x^2 + 2x + 1 = 0$

$(x + 1)^2 = 0$

$x + 1 = 0$

$x = -1 \Rightarrow$ The x-intercept is at $(-1, 0)$.

56. $y = x^2 - 6x + 9$

(a) From the graph it appears that the x-intercept is $(3, 0)$.

(b) The x-intercept and solution of $x^2 - 6x + 9 = 0$ are the same.

(c) $x^2 - 6x + 9 = 0$

$(x - 3)^2 = 0$

$x = 3 \Rightarrow$ The x-intercept is $(3, 0)$.

57. $y = x^2 - 4x - 5$

(a) x-intercepts: $(5, 0), (-1, 0)$

(b) The x-intercepts and the solutions of the equation are the same.

(c) $0 = x^2 - 4x - 5$

$0 = (x - 5)(x + 1)$

$x = 5$ or $x = -1$

The x-intercepts are $(5, 0)$ and $(-1, 0)$.

58. $y = -x^2 + 2x + 3$

(a) From the graph it appears that the x-intercepts are $(-1, 0)$ and $(3, 0)$.

(b) The x-intercepts and the solutions of the equation $-x^2 + 2x + 3 = 0$ are the same.

(c) $-x^2 + 2x + 3 = 0$

$x^2 - 2x - 3 = 0$

$(x + 1)(x - 3) = 0$

$x = -1$ or $x = 3 \Rightarrow$ The x-intercepts are $(-1, 0)$ and $(3, 0)$.

59. $y = -2x^2 + 3x + 2$

(a) From the graph it appears that the x-intercepts are $\left(-\frac{1}{2}, 0\right)$ and $(2, 0)$.

(b) The x-intercepts and the solutions of the equation are the same.

(c) $0 = -2x^2 + 3x + 2 \Rightarrow 2x^2 - 3x - 2 = 0$

$(2x + 1)(x - 2) = 0$

$x = -\frac{1}{2}$ or $x = 2 \Rightarrow$ The x-intercepts are $\left(-\frac{1}{2}, 0\right)$ and $(2, 0)$.

60. $y = 2x^2 + 5x - 3$

(a) From the graph it appears that the x-intercepts are $\left(\frac{1}{2}, 0\right)$ and $(-3, 0)$.

(b) The x-intercepts and solutions of $2x^2 + 5x - 3 = 0$ are the same.

(c) $2x^2 + 5x - 3 = 0$

$(2x - 1)(x + 3) = 0$

$x = \frac{1}{2}$ or $x = -3 \Rightarrow$ The x-intercepts are $\left(\frac{1}{2}, 0\right)$ and $(-3, 0)$.

61. $y = x^2 + 2x + 2$

(a) From the graph we see that there are no x-intercepts.

(b) $0 = x^2 + 2x + 2$ has no real solutions.

(c) Completing the square yields $(x + 1)^2 = -1$ which has no real solutions.

62. $y = -x^2 - 3x - 3$

(a) From the graph it appears that there are no x- intercepts.

(b) There are no x-intercepts and there are no real solutions to the equation $-x^2 - 3x - 3 = 0$.

(c) $-x^2 - 3x - 3 = 0$

$$x^2 + 3x + 3 = 0$$

$$x = \frac{-3 \pm \sqrt{3^2 - 4(1)(3)}}{2} = \frac{-3 \pm \sqrt{-3}}{2}$$

Not a real number $\Rightarrow$ No x-intercepts

63. $f(x) = ax^2 + bx + c$

$$= a\left(x^2 + \frac{b}{a}x\right) + c$$

$$= a\left(x^2 + \frac{b}{a}x + \frac{b^2}{4a^2} - \frac{b^2}{4a^2}\right) + c$$

$$= a\left(x^2 + \frac{b}{a}x + \frac{b^2}{4a^2}\right) - a\left(\frac{b^2}{4a^2}\right) + c$$

$$= a\left(x + \frac{b}{2a}\right)^2 - \frac{b^2}{4a} + \frac{4ac}{4a}$$

$$= a\left(x + \frac{b}{2a}\right)^2 + \frac{4ac - b^2}{4a}$$

The vertex is $\left(-\frac{b}{2a}, \frac{4ac - b^2}{4a}\right)$.

$$f\left(-\frac{b}{2a}\right) = a\left(-\frac{b}{2a}\right)^2 + b\left(-\frac{b}{2a}\right) + c$$

$$= a\left(\frac{b^2}{4a^2}\right) - \frac{b^2}{2a} + c$$

$$= \frac{b^2}{4a} - \frac{2b^2}{4a} + \frac{4ac}{4a}$$

$$= \frac{-b^2 + 4ac}{4a}$$

$$= \frac{4ac - b^2}{4a}$$

Thus, the vertex occurs at $\left(-\frac{b}{2a}, f\left(-\frac{b}{2a}\right)\right)$.

64. (a) Yes, it is possible for a quadratic equation to have only one x-intercept. That happens when the vertex is the x-intercept.

(b) Yes. If the vertex is above the x-axis and the parabola opens upward, or if the vertex is below the x-axis and the parabola opens downward, then the graph of the quadratic equation will have no x-intercepts.

Examples: $f(x) = x^2 + 4$; $g(x) = -x^2 - 1$

65. $f(x) = x^2 - 4x$

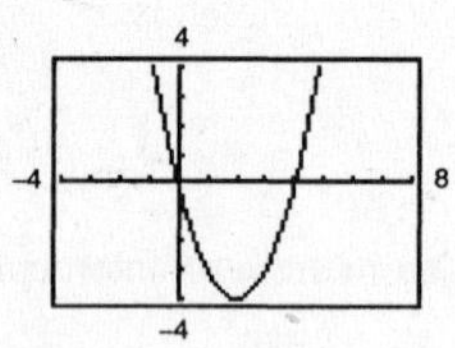

x-intercepts: $(0, 0)$, $(4, 0)$

$0 = x^2 - 4x$

$0 = x(x - 4)$

$x = 0$ or $x = 4$

66. $f(x) = -2x^2 + 10x$

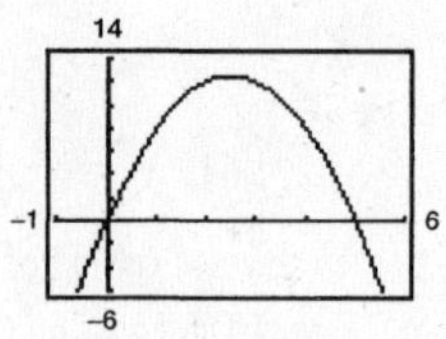

x-intercepts: $(0, 0)$, $(5, 0)$

$0 = -2x^2 + 10x$

$0 = -2x(x - 5)$

$-2x = 0 \Rightarrow x = 0$

$x - 5 = 0 \Rightarrow x = 5$

67. $f(x) = x^2 - 9x + 18$

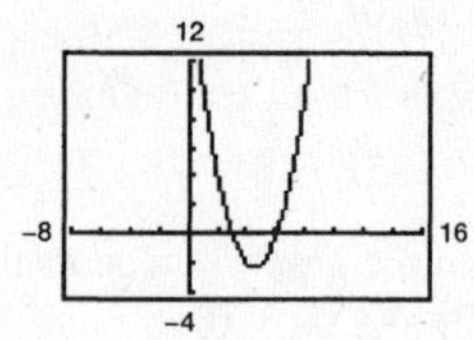

x-intercepts: $(3, 0)$, $(6, 0)$

$0 = x^2 - 9x + 18$

$0 = (x - 3)(x - 6)$

$x = 3$ or $x = 6$

68. $f(x) = x^2 - 8x - 20$

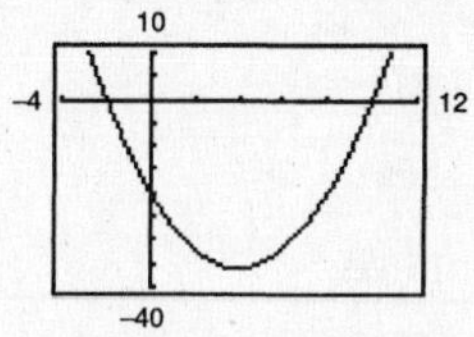

x-intercepts: $(-2, 0), (10, 0)$

$0 = x^2 - 8x - 20$

$0 = (x + 2)(x - 10)$

$x + 2 = 0 \Rightarrow x = -2$

$x - 10 = 0 \Rightarrow x = 10$

69. $f(x) = 2x^2 - 7x - 30$

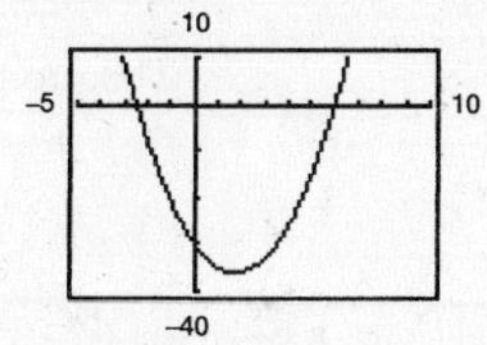

x-intercepts: $\left(-\frac{5}{2}, 0\right), (6, 0)$

$0 = 2x^2 - 7x - 30$

$0 = (2x + 5)(x - 6)$

$x = -\frac{5}{2}$ or $x = 6$

70. $f(x) = 4x^2 + 25x - 21$

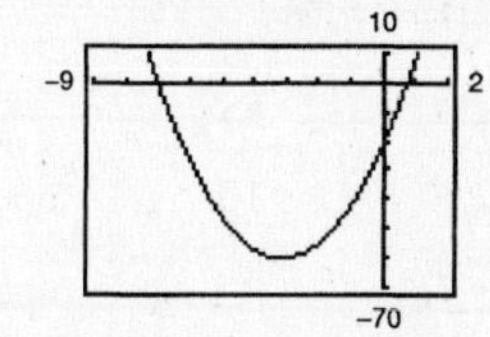

x-intercepts: $(-7, 0), \left(\frac{3}{4}, 0\right)$

$0 = 4x^2 + 25x - 21$

$0 = (x + 7)(4x - 3)$

$x + 7 = 0 \Rightarrow x = -7$

$4x - 3 = 0 \Rightarrow x = \frac{3}{4}$

71. $f(x) = -\frac{1}{2}(x^2 - 6x - 7)$

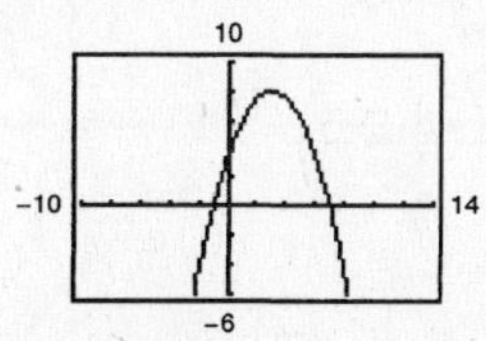

x-intercepts: $(-1, 0), (7, 0)$

$0 = -\frac{1}{2}(x^2 - 6x - 7)$

$0 = x^2 - 6x - 7$

$0 = (x + 1)(x - 7)$

$x = -1$ or $x = 7$

72. $f(x) = \frac{7}{10}(x^2 + 12x - 45)$

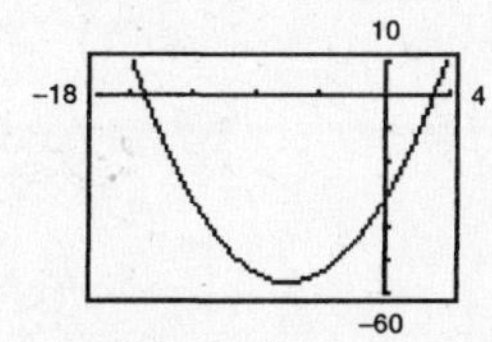

x-intercepts: $(-15, 0), (3, 0)$

$0 = \frac{7}{10}(x^2 + 12x - 45)$

$0 = (x + 15)(x - 3)$

$x + 15 = 0 \Rightarrow x = -15$

$x - 3 = 0 \Rightarrow x = 3$

73. $f(x) = [x - (-1)](x - 3)$
opens upward

$= (x + 1)(x - 3)$

$= x^2 - 2x - 3$

$g(x) = -[x - (-1)](x - 3)$
opens downward

$= -(x + 1)(x - 3)$

$= -(x^2 - 2x - 3)$

$= -x^2 + 2x + 3$

Note: $f(x) = a(x + 1)(x - 3)$ has x-intercepts $(-1, 0)$ and $(3, 0)$ for all real numbers $a \neq 0$.

74. $f(x) = [x - (-5)](x - 5)$

$= (x + 5)(x - 5)$

$= x^2 - 25$ opens upward

$g(x) = -f(x)$ opens downward

$g(x) = -x^2 + 25$

Note: $f(x) = a(x^2 - 25)$ has x-intercepts $(-5, 0)$ and $(5, 0)$ for all real numbers $a \neq 0$.

75. $f(x) = (x - 0)(x - 10)$ opens upward

$= x^2 - 10x$

$g(x) = -(x - 0)(x - 10)$ opens downward

$= -x^2 + 10x$

Note: $f(x) = a(x - 0)(x - 10) = ax(x - 10)$ has x-intercepts $(0, 0)$ and $(10, 0)$ for all real numbers $a \neq 0$.

76. $f(x) = (x - 4)(x - 8)$

$= x^2 - 12x + 32$ opens upward

$g(x) = -f(x)$ opens downward

$g(x) = -x^2 + 12x - 32$

Note: $f(x) = a(x - 4)(x - 8)$ has x-intercepts $(4, 0)$ and $(8, 0)$ for all real numbers $a \neq 0$.

77. $f(x) = [x - (-3)]\left[x - \left(-\frac{1}{2}\right)\right](2)$ opens upward

$= (x + 3)\left(x + \frac{1}{2}\right)(2)$

$= (x + 3)(2x + 1)$

$= 2x^2 + 7x + 3$

$g(x) = -(2x^2 + 7x + 3)$ opens downward

$= -2x^2 - 7x - 3$

Note: $f(x) = a(x + 3)(2x + 1)$ has x-intercepts $(-3, 0)$ and $\left(-\frac{1}{2}, 0\right)$ for all real numbers $a \neq 0$.

78. $f(x) = 2[x - (-\frac{5}{2})](x - 2)$

$= 2(x + \frac{5}{2})(x - 2)$

$= 2(x^2 + \frac{1}{2}x - 5)$

$= 2x^2 + x - 10$ opens upward

$g(x) = -f(x)$ opens downward

$g(x) = -2x^2 - x + 10$

Note: $f(x) = a(x + \frac{5}{2})(x - 2)$ has x-intercepts $(-\frac{5}{2}, 0)$ and $(2, 0)$ for all real numbers $a \neq 0$.

79. Let $x =$ the first number and $y =$ the second number. Then the sum is $x + y = 156 \Rightarrow y = 156 - x$.

The product is $P(x) = xy = x(156 - x)$

$= -x^2 + 156x$

$= -(x^2 - 156x)$

$= -(x^2 - 156x + 6084 - 6084)$

$= -(x - 78)^2 + 6084.$

The maximum value of the product occurs at the vertex and is 6084. This happens when $x = 78$ and $y = 156 - 78 = 78$. Thus, the numbers are 78 and 78.

80. Let $x =$ the first number and $y =$ the second number. Then the sum is

$x + y = 196 \Rightarrow y = 196 - x.$

The product is $P(x) = xy = x(196 - x) = 196x - x^2$.

$P(x) = -x^2 + 196x$

$= -(x^2 - 196x + 9604 - 9604)$

$= -[(x - 98)^2 - 9604]$

$= -(x - 98)^2 + 9604$

The maximum value of the product occurs at the vertex of $P(x)$ and is 9604. This happens when $x = y = 98$.

81. Let $x =$ the first number and $y =$ the second number. Then the sum is

$x + 2y = 24 \Rightarrow y = \dfrac{24 - x}{2}.$

The product is $P(x) = xy = x\left(\dfrac{24 - x}{2}\right)$.

$P(x) = \frac{1}{2}(-x^2 + 24x)$

$= -\frac{1}{2}(x^2 - 24x + 144 - 144)$

$= -\frac{1}{2}[(x - 12)^2 - 144] = -\frac{1}{2}(x - 12)^2 + 72$

The maximum value of the product occurs at the vertex of $P(x)$ and is 72. This happens when $x = 12$ and $y = (24 - 12)/2 = 6$. Thus, the numbers are 12 and 6.

82. Let $x =$ the first number and $y =$ the second number.

Then the sum is $x + 3y = 42 \Rightarrow y = \dfrac{42 - x}{3}.$

The product is $P(x) = xy = x\left(\dfrac{42 - x}{3}\right)$.

$P(x) = \frac{1}{3}(-x^2 + 42x)$

$= -\frac{1}{3}(x^2 - 42x + 441 - 441)$

$= -\frac{1}{3}[(x - 21)^2 - 441] = -\frac{1}{3}(x - 21)^2 + 147$

The maximum value of the product occurs at the vertex of $P(x)$ and is 147. This happens when $x = 21$ and $y = \dfrac{42 - 21}{3} = 7$. Thus, the numbers are 21 and 7.

83.

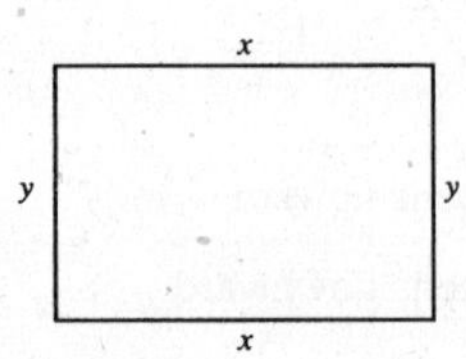

$2x + 2y = 100$

$y = 50 - x$

(a) $A(x) = xy = x(50 - x)$

Domain: $0 < x < 50$

(b)

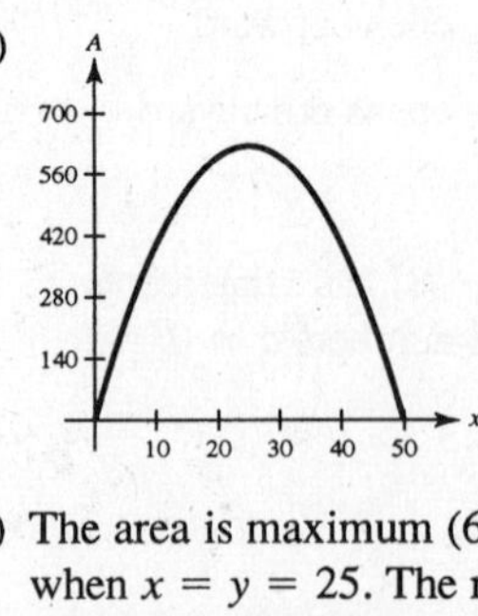

(c) The area is maximum (625 square feet) when $x = y = 25$. The rectangle has dimensions 25 ft × 25 ft.

84. Let x = length of rectangle and y = width of rectangle.

$2x + 2y = 36$

$y = 18 - x$

(a) $A(x) = xy = x(18 - x)$

Domain: $0 < x < 18$

(b)

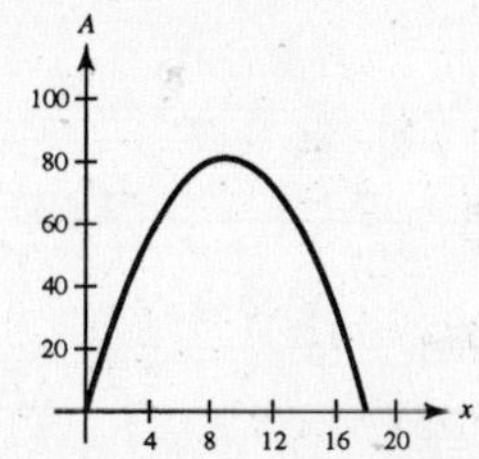

(c) The area is maximum (81 square meters) when $x = y = 9$ meters. The rectangle has dimensions 9 meters $\times$ 9 meters.

85. (a) $4x + 3y = 200 \implies y = \frac{1}{3}(200 - 4x)$

x	y	Area
2	$\frac{1}{3}[200 - 4(2)]$	$2xy = (2)(2)(\frac{1}{3})[200 - 4(2)] = 256$
4	$\frac{1}{3}[200 - 4(4)]$	$2xy = (2)(4)(\frac{1}{3})[200 - 4(4)] \approx 491$
6	$\frac{1}{3}[200 - 4(6)]$	$2xy = (2)(6)(\frac{1}{3})[200 - 4(6)] = 704$
8	$\frac{1}{3}[200 - 4(8)]$	$2xy = (2)(8)(\frac{1}{3})[200 - 4(8)] = 896$
10	$\frac{1}{3}[200 - 4(10)]$	$2xy = (2)(10)(\frac{1}{3})[200 - 4(10)] \approx 1067$
12	$\frac{1}{3}[200 - 4(12)]$	$2xy = (2)(12)(\frac{1}{3})[200 - 4(12)] = 1216$

(b)

x	y	Area
20	$\frac{1}{3}[200 - 4(20)]$	$2xy = (2)(20)(\frac{1}{3})[200 - 4(20)] = 1600$
22	$\frac{1}{3}[200 - 4(22)]$	$2xy = (2)(22)(\frac{1}{3})[200 - 4(22)] \approx 1643$
24	$\frac{1}{3}[200 - 4(24)]$	$2xy = (2)(24)(\frac{1}{3})[200 - 4(24)] = 1664$
26	$\frac{1}{3}[200 - 4(26)]$	$2xy = (2)(26)(\frac{1}{3})[200 - 4(26)] = 1664$
28	$\frac{1}{3}[200 - 4(28)]$	$2xy = (2)(28)(\frac{1}{3})[200 - 4(28)] \approx 1643$
30	$\frac{1}{3}[200 - 4(30)]$	$2xy = (2)(30)(\frac{1}{3})[200 - 4(30)] = 1600$

The table indicates that the area is maximum when x is between 24 and 26 feet and y is between $34\frac{2}{3}$ and 32 feet.

(c) $A = 2xy = 2x\left(\frac{200 - 4x}{3}\right) = \frac{2x(4)(50 - x)}{3}$

$= \frac{8x(50 - x)}{3}$

(d)

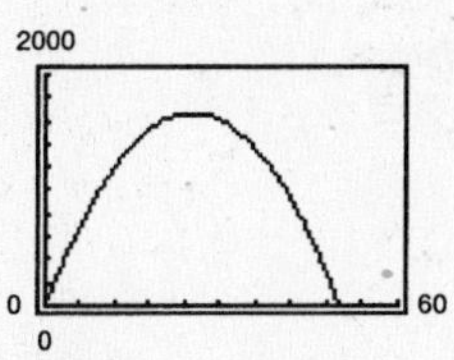

This area is maximum when $x = 25$ feet and $y = \frac{100}{3} = 33\frac{1}{3}$ feet.

—CONTINUED—

85. —CONTINUED—

(e) $A = \frac{8}{3}x(50 - x)$

$= -\frac{8}{3}(x^2 - 50x)$

$= -\frac{8}{3}(x^2 - 50x + 625 - 625)$

$= -\frac{8}{3}[(x - 25)^2 - 625]$

$= -\frac{8}{3}(x - 25)^2 + \frac{5000}{3}$

The maximum area occurs at the vertex and is 5000/3 square feet. This happens when $x = 25$ feet and $y = (200 - 4(25))/3 = 100/3$ feet. The dimensions are $2x = 50$ feet by $33\frac{1}{3}$ feet.

86. (a) Radius of semicircular ends of track: $r = \frac{1}{2}y$
distance around two semicircular parts of track:

$$d = 2\pi r = 2\pi\left(\frac{1}{2}y\right) = \pi y$$

(b) Distance traveled around track in one lap:

$$d = \pi y + 2x = 200$$

$$\pi y = 200 - 2x$$

$$y = \frac{200 - 2x}{\pi}$$

(c) Area of rectangular region:

$$A = xy = x\left(\frac{200 - 2x}{\pi}\right)$$

$$= \frac{1}{\pi}(200x - 2x^2)$$

$$= -\frac{2}{\pi}(x^2 - 100x)$$

$$= -\frac{2}{\pi}(x^2 - 100x + 2500 - 2500)$$

$$= -\frac{2}{\pi}(x - 50)^2 + \frac{5000}{\pi}$$

The area is maximum when $x = 50$ and

$$y = \frac{200 - 2(50)}{\pi} = \frac{100}{\pi}.$$

87. $R = 900x - 0.1x^2 = -0.1x^2 + 900x$

The vertex occurs at $x = -\frac{b}{2a} = -\frac{900}{2(-0.1)} = 4500.$

The revenue is maximum when $x = 4500$ units.

88. $R = 100x - 0.0002x^2 = -0.0002x^2 + 100x$

The vertex occurs at

$$x = -\frac{b}{2a} = -\frac{100}{2(-0.0002)} = 250{,}000.$$

The revenue is maximum when $x = 250{,}000$ units.

89. $C = 800 - 10x + 0.25x^2 = 0.25x^2 - 10x + 800$

The vertex occurs at $x = -\frac{b}{2a} = -\frac{-10}{2(0.25)} = 20.$

The cost is minimum when $x = 20$ fixtures.

90. $C = 100{,}000 - 110x + 0.045x^2$

The vertex occurs at

$$x = -\frac{-110}{2(0.045)} \approx 1222.$$

The cost is minimum when $x \approx 1222$ units.

91. $P = -0.0002x^2 + 140x - 250{,}000$

The vertex occurs at $x = -\frac{b}{2a} = -\frac{140}{2(-0.0002)} = 350{,}000.$

The profit is maximum when $x = 350{,}000$ units.

92. $P = 230 + 20x - 0.5x^2$

The vertex occurs at

$$x = -\frac{b}{2a} = -\frac{20}{2(-0.5)} = 20.$$

Because x is in hundreds of dollars, $20 \times 100 = 2000$ dollars is the amount spent on advertising that gives maximum profit.

93. $y = -\frac{1}{12}x^2 + 2x + 4$

(a) When $x = 0$, $y = 4$ feet.

(b) The vertex occurs at $x = -\frac{b}{2a} = -\frac{2}{2(-1/12)} = 12$. The maximum height is

$$y = -\frac{1}{12}(12)^2 + 2(12) + 4 = 16 \text{ feet.}$$

(c) When the ball strikes the ground, $y = 0$.

$$0 = -\frac{1}{12}x^2 + 2x + 4$$

$$0 = x^2 - 24x - 48 \qquad \text{Multiply both sides by } -12.$$

$$x = \frac{-(-24) \pm \sqrt{(-24)^2 - 4(1)(-48)}}{2(1)}$$

$$= \frac{24 \pm \sqrt{768}}{2} = \frac{24 \pm 16\sqrt{3}}{2} = 12 \pm 8\sqrt{3}$$

Using the positive value for x, we have $x = 12 + 8\sqrt{3} \approx 25.86$ feet.

94. $y = -\frac{4}{9}x^2 + \frac{24}{9}x + 12$

(a) When $x = 0$, $y = 12$ feet.

(b) The vertex occurs at $-\frac{b}{2a} = \frac{-24/9}{2(-4/9)} = 3$.

The maximum height is

$$y(3) = -\frac{4}{9}(3)^2 + \frac{24}{9}(3) + 12 = 16 \text{ feet.}$$

(c) $-\frac{4}{9}x^2 + \frac{24}{9}x + 12 = 0$

$$-4x^2 + 24x + 108 = 0$$

$$-4(x^2 - 6x - 27) = 0$$

$$-4(x + 3)(x - 9) = 0$$

$x = -3$ is extraneous, $x = 9$.

The diver hits the surface of the water 9 feet from the end of the diving board.

95. $V = 0.77x^2 - 1.32x - 9.31$, $5 \le x \le 40$

(a)

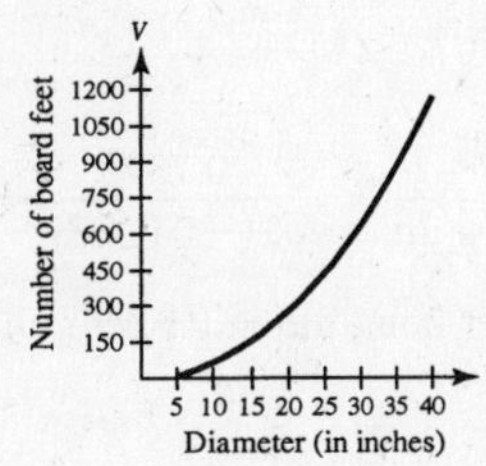

(b) $V(16) = 166.69$ board feet

(c) $500 = 0.77x^2 - 1.32x - 9.31$

$$0 = 0.77x^2 - 1.32x - 509.31$$

Using the Quadratic Formula and selecting the positive value for x, we have $x \approx 26.6$ inches in diameter.

96. (a)

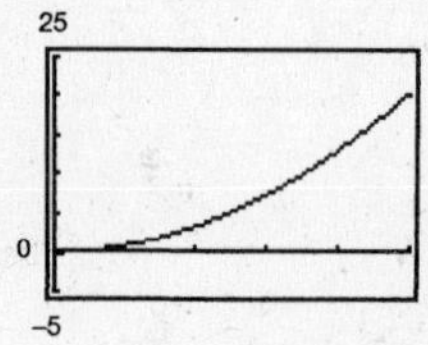

(b) $0.002s^2 + 0.005s - 0.029 = 10$

$$2s^2 + 5s - 29 = 10{,}000$$

$$2s^2 + 5s - 10{,}029 = 0$$

$a = 2, b = 5, c = -10{,}029$

$$s = \frac{-5 \pm \sqrt{5^2 - 4(2)(-10{,}029)}}{2(2)}$$

$$s = \frac{-5 \pm \sqrt{80{,}257}}{4}$$

$s \approx -72.1, 69.6$

The maximum speed if power is not to exceed 10 horsepower is 69.6 miles per hour.

97. Let $t = 0$ correspond to 1960.

$C = -1.43t^2 + 0.5t + 4287, \quad 0 \le t \le 42$

(a)

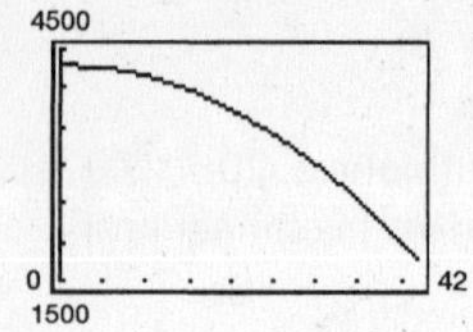

(b) The vertex occurs at $-\dfrac{b}{2a} = \dfrac{-0.5}{2(-1.43)} \approx 0.1748.$

$C(0.1748) \approx 4287$, which is the maximum average annual consumption. The warnings may not have had an immediate effect, and consumption was decreasing by 1966 without them, but over time they and other findings about the health risks of cigarettes have had an effect.

(c) $C(40) = 2019$

$\dfrac{209{,}128{,}094(2019)}{48{,}308{,}590} \approx 8740$ cigarettes per smoker per year.

$\dfrac{8740}{365} \approx 24$ cigarettes per smoker per day.

98. (a) $y \approx -0.0083x^2 + 0.7396x + 16.0303$

(b)

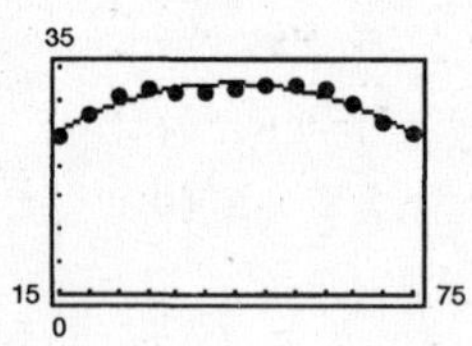

(c) The vertex occurs at $-\dfrac{b}{2a} = -\dfrac{0.7396}{2(-0.0083)} \approx 44.6.$

At $x \approx 44.6$ miles per hour, the miles per gallon are greatest.

99. (a) and (c)

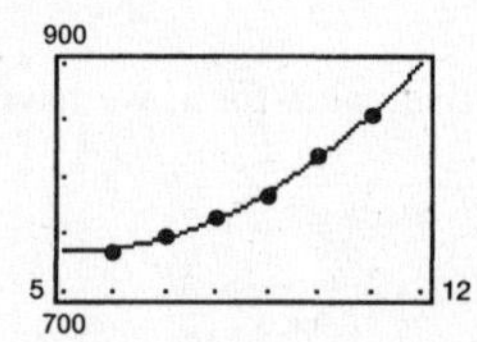

The model is a good fit for the data.

(b) $y \approx 3.55t^2 - 36.9t + 831$ where $t = 6$ corresponds to 1996.

(d) The number of hairdressers and cosmetologists is minimum when $t \approx 5$, which corresponds to 1995.

(e) $y(20) = 1513$ which corresponds to 1,513,000 hairdressers and cosmetologists in 2010.

100. True. The equation $f(x) = 0$ has no real solution, so the graph has no x-intercepts.

101. True. The vertex of $f(x)$ is $\left(-\frac{5}{4}, \frac{53}{4}\right)$ and the vertex of $g(x)$ is $\left(-\frac{5}{4}, -\frac{71}{4}\right)$.

102. True. The quadratic function, $f(x) = ax^2 + bx + c$, has a y-intercept at $(0, c)$ and this is the only y-intercept.

103. True. All quadratic functions of the form $y = a(x - 1)(x - 2)$, $a \ne 0$, have x-intercepts at $(1, 0)$ and $(2, 0)$.

Section 2.2 Polynomial Functions of Higher Degree

1. $f(x) = -2x + 3$ is a line with y-intercept $(0, 3)$.

Matches graph (c).

2. $f(x) = x^2 - 4x$ is a parabola with intercepts $(0, 0)$ and $(4, 0)$ and opens upward.

Matches graph (g).

3. $f(x) = -2x^2 - 5x$ is a parabola with x-intercepts $(0, 0)$ and $\left(-\frac{5}{2}, 0\right)$ and opens downward.

Matches graph (h).

4. $f(x) = 2x^3 - 3x + 1$ has intercepts $(0, 1)$, $(1, 0)$, $\left(-\frac{1}{2} - \frac{1}{2}\sqrt{3}, 0\right)$ and $\left(-\frac{1}{2} + \frac{1}{2}\sqrt{3}, 0\right)$.

Matches graph (f).

5. $f(x) = -\frac{1}{4}x^4 + 3x^2$ has intercepts $(0, 0)$ and $\left(\pm 2\sqrt{3}, 0\right)$.

Matches graph (a).

6. $f(x) = -\frac{1}{3}x^3 + x^2 - \frac{4}{3}$ has y-intercept $\left(0, -\frac{4}{3}\right)$.

Matches graph (e).

7. $f(x) = x^4 + 2x^3$ has intercepts $(0, 0)$ and $(-2, 0)$.

Matches graph (d).

8. $f(x) = \frac{1}{5}x^5 - 2x^3 + \frac{9}{5}x$ has intercepts $(0, 0)$, $(1, 0)$, $(-1, 0)$, $(3, 0)$, $(-3, 0)$.

Matches (b).

9. $y = x^3$

(a) $f(x) = (x - 2)^3$

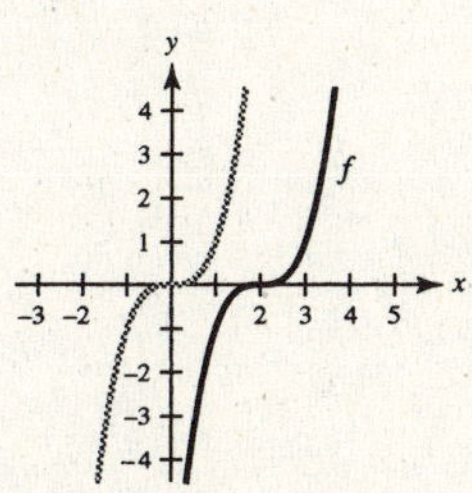

Horizontal shift two units to the right

(b) $f(x) = x^3 - 2$

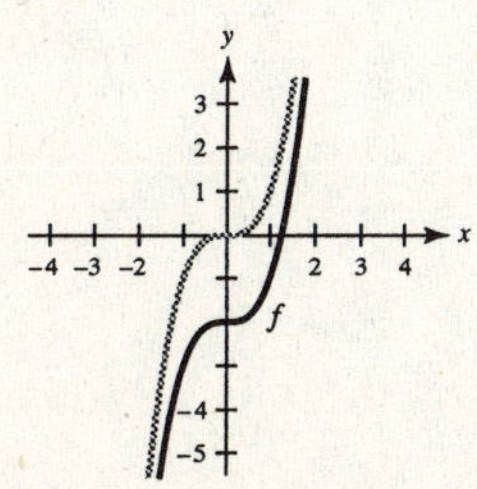

Vertical shift two units downward

(c) $f(x) = -\frac{1}{2}x^3$

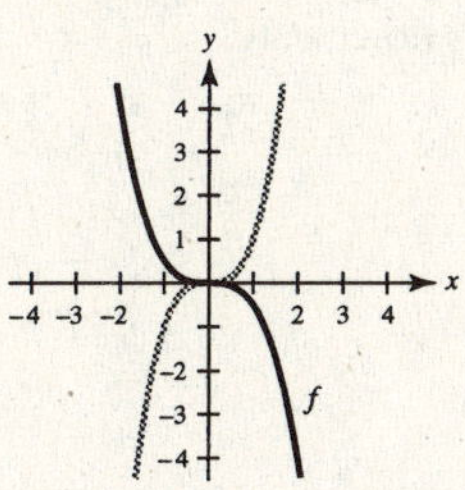

Reflection in the x-axis and a vertical shrink

(d) $f(x) = (x - 2)^3 - 2$

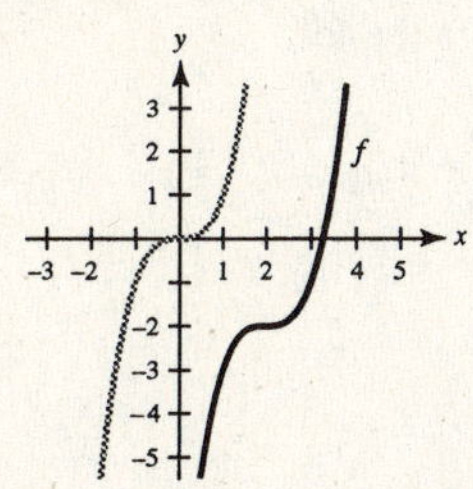

Horizontal shift two units to the right and a vertical shift two units downward

10. $y = x^5$

(a) $f(x) = (x + 1)^5$

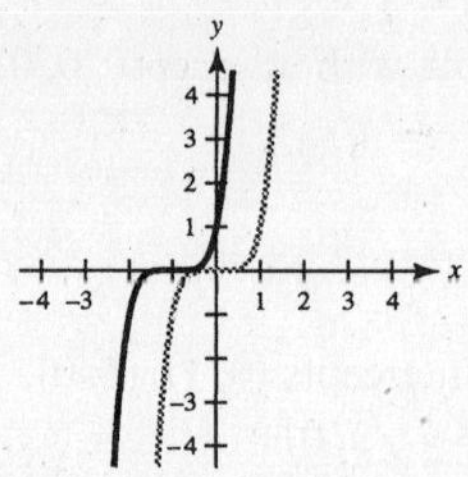

Horizontal shift 1 unit to the left

(b) $f(x) = x^5 + 1$

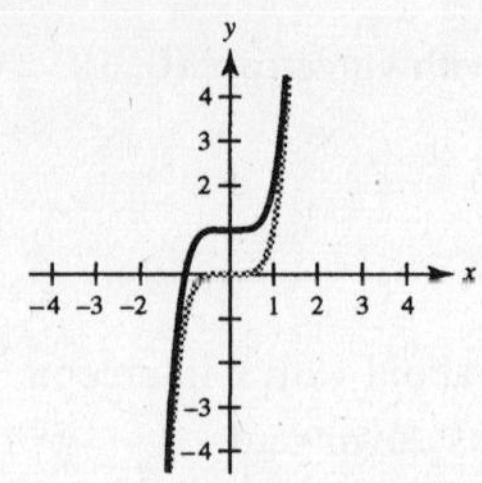

Vertical shift 1 unit upward

(c) $f(x) = 1 - \frac{1}{2}x^5$

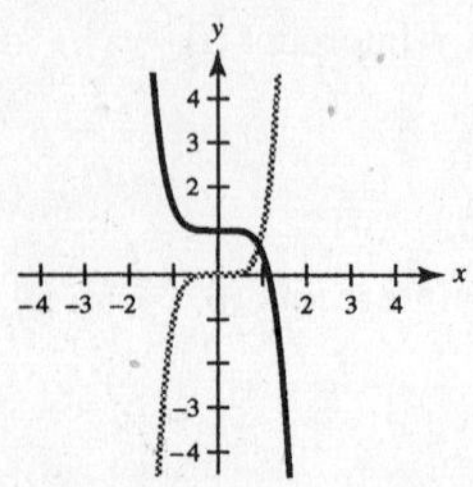

Reflection in the x-axis, vertical shrink by a factor of $\frac{1}{2}$, and vertical shift 1 unit upward

(d) $f(x) = -\frac{1}{2}(x + 1)^5$

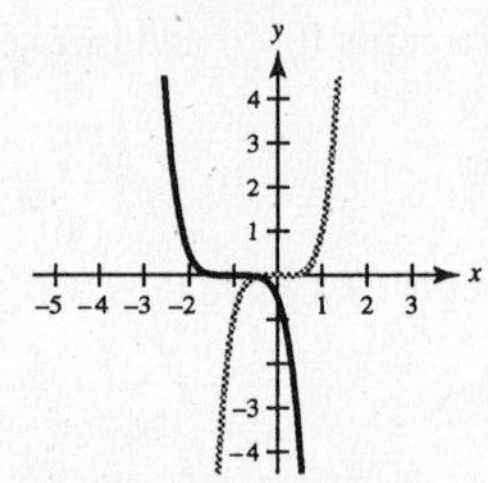

Reflection in the x-axis, vertical shrink by a factor of $\frac{1}{2}$, and horizontal shift 1 unit to the left

11. $y = x^4$

(a) $f(x) = (x + 3)^4$

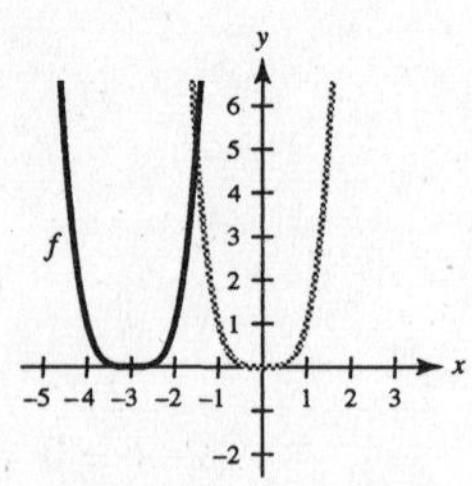

Horizontal shift three units to the left

(b) $f(x) = x^4 - 3$

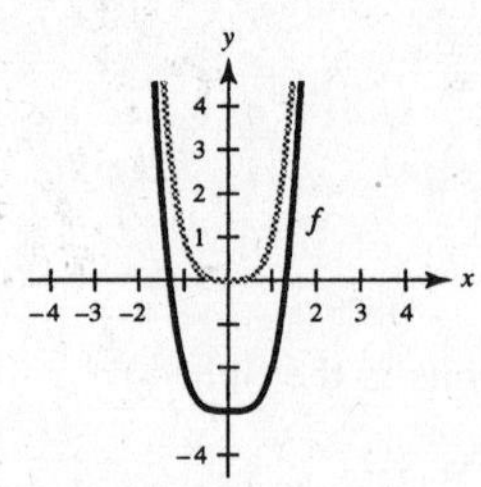

Vertical shift three units downward

(c) $f(x) = 4 - x^4$

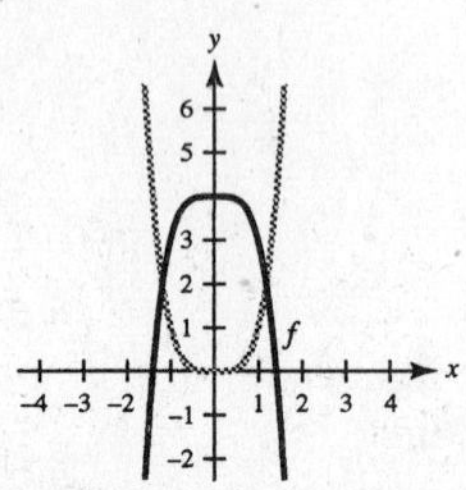

Reflection in the x-axis and then a vertical shift four units upward

(d) $f(x) = \frac{1}{2}(x - 1)^4$

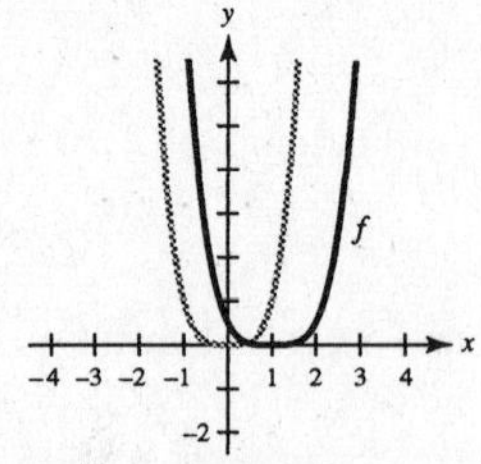

Horizontal shift one unit to the right and a vertical shrink

12. $y = x^6$

(a) $f(x) = -\frac{1}{8}x^6$

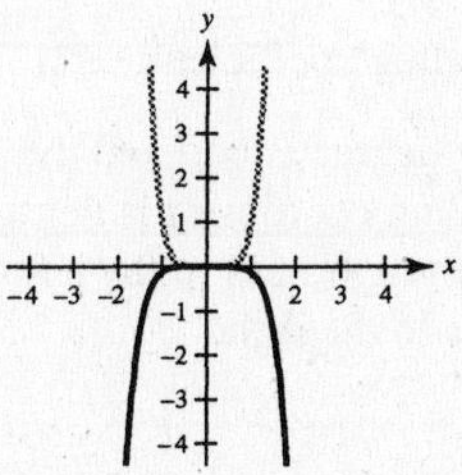

Vertical shrink by a factor of $\frac{1}{8}$ and reflection in the x-axis

(b) $f(x) = (x + 2)^6 - 4$

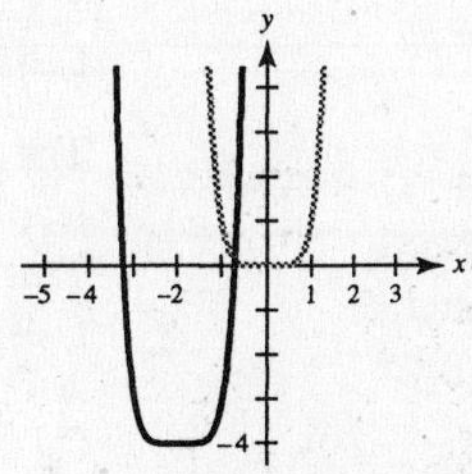

Horizontal shift 2 units to the left and vertical shift 4 units downward

(c) $f(x) = x^6 - 4$

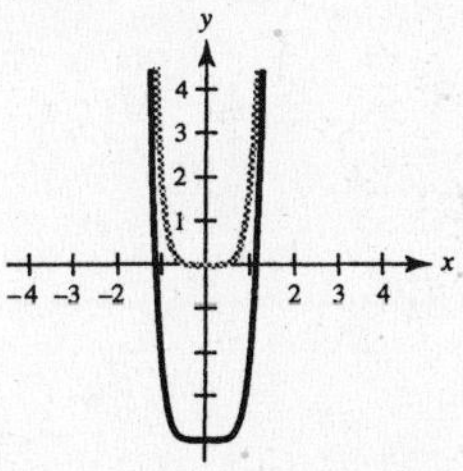

Vertical shift 4 units downward

(d) $f(x) = -\frac{1}{4}x^6 + 1$

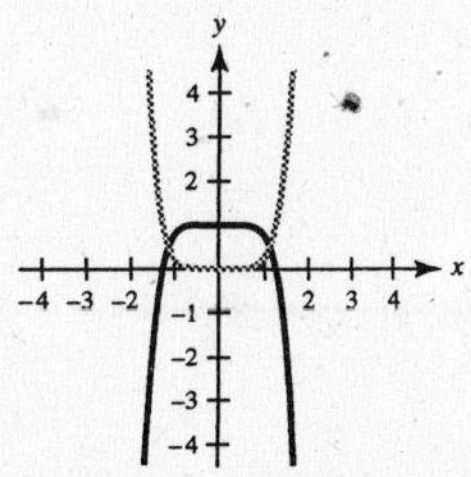

Reflection in the x-axis, vertical shrink by a factor of $\frac{1}{4}$, and vertical shift 1 unit upward

13. $f(x) = \frac{1}{3}x^3 + 5x$

Degree: 3

Leading coefficient: $\frac{1}{3}$

The degree is odd and the leading coefficient is positive. The graph falls to the left and rises to the right.

14. $f(x) = 2x^2 - 3x + 1$

Degree: 2

Leading coefficient: 2

The degree is even and the leading coefficient is positive. The graph rises to the left and rises to the right.

15. $g(x) = 5 - \frac{7}{2}x - 3x^2$

Degree: 2

Leading coefficient: -3

The degree is even and the leading coefficient is negative. The graph falls to the left and falls to the right.

16. $h(x) = 1 - x^6$

Degree: 6

Leading coefficient: -1

The degree is even and the leading coefficient is negative. The graph falls to the left and falls to the right.

17. $f(x) = -2.1x^5 + 4x^3 - 2$

Degree: 5

Leading coefficient: -2.1

The degree is odd and the leading coefficient is negative. The graph rises to the left and falls to the right.

18. $f(x) = 2x^5 - 5x + 7.5$

Degree: 5

Leading coefficient: 2

The degree is odd and the leading coefficient is positive. The graph falls to the left and rises to the right.

19. $f(x) = 6 - 2x + 4x^2 - 5x^3$

Degree: 3

Leading coefficient: -5

The degree is odd and the leading coefficient is negative. The graph rises to the left and falls to the right.

20. $f(x) = \dfrac{3x^4 - 2x + 5}{4}$

Degree: 4

Leading coefficient: $\frac{3}{4}$

The degree is even and the leading coefficient is positive. The graph rises to the left and rises to the right.

21. $h(t) = -\frac{2}{3}(t^2 - 5t + 3)$

Degree: 2

Leading coefficient: $-\frac{2}{3}$

The degree is even and the leading coefficient is negative. The graph falls to the left and falls to the right.

22. $f(s) = -\frac{7}{8}(s^3 + 5s^2 - 7s + 1)$

Degree: 3

Leading coefficient: $-\frac{7}{8}$

The degree is odd and the leading coefficient is negative. The graph rises to the left and falls to the right.

23. $f(x) = 3x^3 - 9x + 1;\ g(x) = 3x^3$

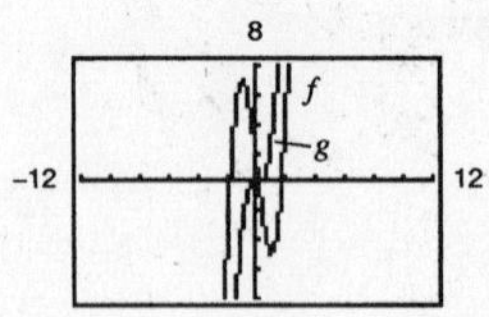

24. $f(x) = -\frac{1}{3}(x^3 - 3x + 2),\ g(x) = -\frac{1}{3}x^3$

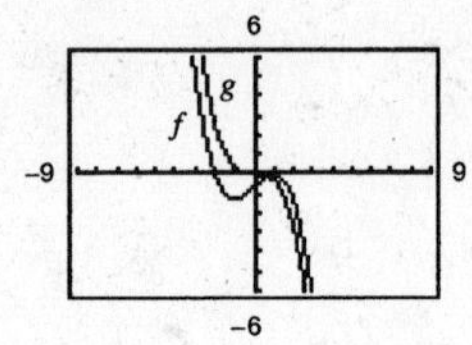

25. $f(x) = -(x^4 - 4x^3 + 16x);\ g(x) = -x^4$

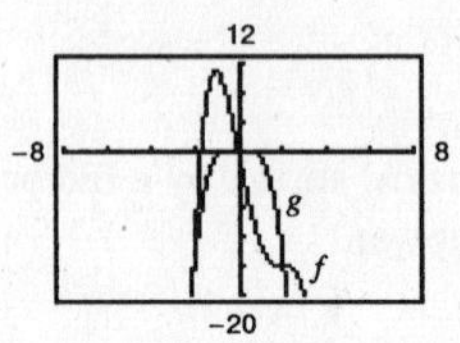

26. $f(x) = 3x^4 - 6x^2,\ g(x) = 3x^4$

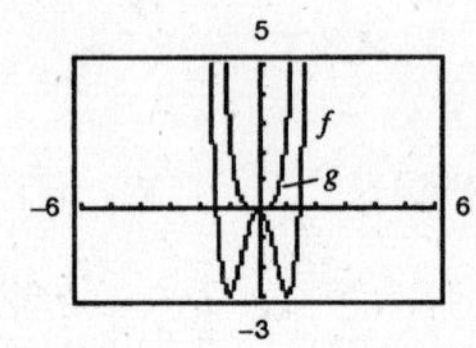

27. $f(x) = x^2 - 25$

$0 = (x + 5)(x - 5)$

$x = \pm 5$

28. $f(x) = 49 - x^2$

$= (7 - x)(7 + x)$

$x = \pm 7$

29. $h(t) = t^2 - 6t + 9$

$0 = (t - 3)^2$

$t = 3$

30. $f(x) = x^2 + 10x + 25$

$= (x + 5)^2$

$x = -5$

31. $f(x) = \frac{1}{3}x^2 + \frac{1}{3}x - \frac{2}{3}$

$0 = \frac{1}{3}(x + 2)(x - 1)$

$x = -2, 1$

32. $f(x) = \frac{1}{2}x^2 + \frac{5}{2}x - \frac{3}{2}$

$a = \frac{1}{2}, b = \frac{5}{2}, c = -\frac{3}{2}$

$$x = \frac{-\frac{5}{2} \pm \sqrt{\left(\frac{5}{2}\right)^2 - 4\left(\frac{1}{2}\right)\left(-\frac{3}{2}\right)}}{1}$$

$$= -\frac{5}{2} \pm \sqrt{\frac{37}{4}} = \frac{-5 \pm \sqrt{37}}{2}$$

33. $f(x) = 3x^2 - 12x + 3$

$0 = 3(x^2 - 4x + 1)$

$x = \dfrac{4 \pm \sqrt{16 - 4}}{2} = 2 \pm \sqrt{3}$

34. $g(x) = 5(x^2 - 2x - 1)$

$a = 1, b = -2, c = -1$

$x = \dfrac{-(-2) \pm \sqrt{(-2)^2 - 4(1)(-1)}}{2}$

$x = 1 \pm \sqrt{2}$

35. $f(t) = t^3 - 4t^2 + 4t$

$0 = t(t - 2)^2$

$t = 0, 2$

36. $f(x) = x^4 - x^3 - 20x^2$

$= x^2(x^2 - x - 20)$

$= x^2(x + 4)(x - 5)$

$x = 0, -4, 5$

37. $g(t) = \frac{1}{2}t^4 - \frac{1}{2}$

$0 = \frac{1}{2}(t + 1)(t - 1)(t^2 + 1)$

$t = \pm 1$

38. $f(x) = x^5 + x^3 - 6x$

$= x(x^4 + x^2 - 6)$

$= x(x^2 + 3)(x^2 - 2)$

$x = 0, \pm\sqrt{2}$

39. $g(t) = t^5 - 6t^3 + 9t$

$0 = t(t^2 - 3)^2$

$0 = t(t + \sqrt{3})^2(t - \sqrt{3})^2$

$t = 0, \pm\sqrt{3}$

40. $f(x) = 2x^4 - 2x^2 - 40$

$0 = 2(x^2 + 4)(x + \sqrt{5})(x - \sqrt{5})$

$x = \pm\sqrt{5}$

41. $f(x) = 5x^4 + 15x^2 + 10$

$0 = 5(x^4 + 3x^2 + 2)$

$0 = 5(x^2 + 2)(x^2 + 1)$

No real zeros

42. $f(x) = x^3 - 4x^2 - 25x + 100$

$= x^2(x - 4) - 25(x - 4)$

$= (x^2 - 25)(x - 4)$

$= (x + 5)(x - 5)(x - 4)$

$x = \pm 5, 4$

43. $y = 4x^3 - 20x^2 + 25x$

(a)

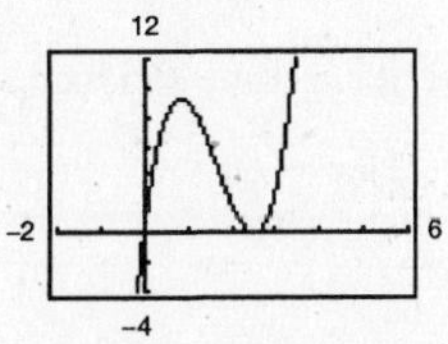

(b) x-intercepts: $(0, 0), \left(\frac{5}{2}, 0\right)$

(c) $0 = 4x^3 - 20x^2 + 25x$

$0 = x(2x - 5)^2$

$x = 0$ or $x = \frac{5}{2}$

The solutions are the same as the x-coordinates of the x-intercepts.

44. $y = 4x^3 + 4x^2 - 7x + 2$

(a)

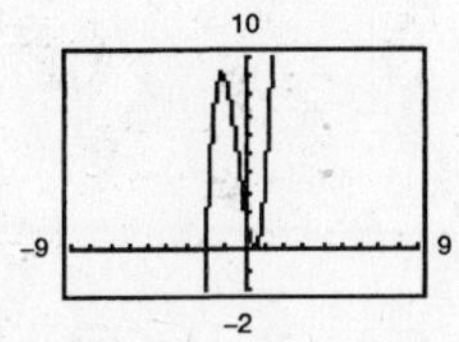

(b) x-intercepts: $(-2, 0), \left(\frac{1}{2}, 0\right)$

(c) $0 = 4x^3 + 4x^2 - 7x + 2$

$= (2x - 1)(2x^2 + 3x - 2)$

$= (2x - 1)(2x - 1)(x + 2)$

$x = -2, \frac{1}{2}$

The solutions are the same as the x-coordinates of the x-intercepts.

45. $y = x^5 - 5x^3 + 4x$

(a)

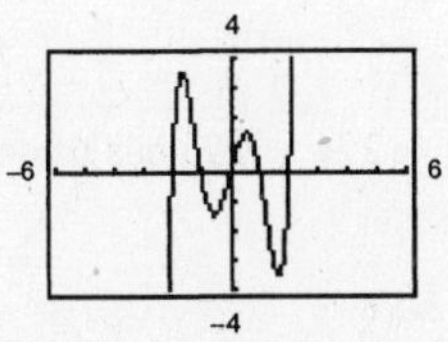

(b) x-intercepts: $(0, 0), (\pm 1, 0), (\pm 2, 0)$

(c) $0 = x^5 - 5x^3 + 4x$

$0 = x(x^2 - 1)(x^2 - 4)$

$0 = x(x + 1)(x - 1)(x + 2)(x - 2)$

$x = 0, \pm 1, \pm 2$

The solutions are the same as the x-coordinates of the x-intercepts.

46. $y = \frac{1}{4}x^3(x^2 - 9)$

(a)

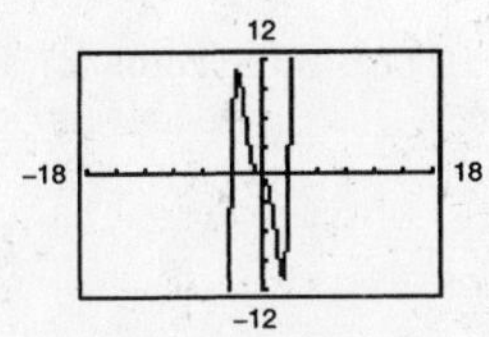

(b) x-intercepts: $(0, 0), (\pm 3, 0)$

(c) $0 = \frac{1}{4}x^3(x^2 - 9)$

$x = 0, \pm 3$

The solutions are the same as the x-coordinates of the x-intercepts.

47. $f(x) = (x - 0)(x - 10)$

$f(x) = x^2 - 10x$

Note: $f(x) = a(x - 0)(x - 10) = ax(x - 10)$ has zeros 0 and 10 for all real numbers $a \neq 0$.

48. $f(x) = (x - 0)(x - (-3))$

$= x(x + 3)$

$= x^2 + 3x$

Note: $f(x) = ax(x + 3)$ has zeros 0 and -3 for all real numbers a.

49. $f(x) = (x - 2)(x - (-6))$
$= (x - 2)(x + 6)$
$= x^2 + 4x - 12$

Note: $f(x) = a(x - 2)(x + 6)$ has zeros 2 and -6 for all real numbers $a \neq 0$.

50. $f(x) = (x - (-4))(x - 5)$
$= (x + 4)(x - 5)$
$= x^2 - x - 20$

Note: $f(x) = a(x + 4)(x - 5)$ has zeros -4 and 5 for all real numbers a.

51. $f(x) = (x - 0)(x - (-2))(x - (-3))$
$= x(x + 2)(x + 3)$
$= x^3 + 5x^2 + 6x$

Note: $f(x) = ax(x + 2)(x + 3)$ has zeros $0, -2, -3$ for all real numbers $a \neq 0$.

52. $f(x) = (x - 0)(x - 2)(x - 5)$
$= x(x - 2)(x - 5)$
$= x(x^2 - 7x + 10)$
$= x^3 - 7x^2 + 10x$

Note: $f(x) = ax(x - 2)(x - 5)$ has zeros 0, 2, 5 for all real numbers a.

53. $f(x) = (x - 4)(x + 3)(x - 3)(x - 0)$
$= (x - 4)(x^2 - 9)x$
$= x^4 - 4x^3 - 9x^2 + 36x$

Note: $f(x) = a(x^4 - 4x^3 - 9x^2 + 36x)$ has these zeros for all real numbers $a \neq 0$.

54. $f(x) = (x - (-2))(x - (-1))(x - 0)(x - 1)(x - 2)$
$= x(x + 2)(x + 1)(x - 1)(x - 2)$
$= x(x^2 - 4)(x^2 - 1)$
$= x(x^4 - 5x^2 + 4)$
$= x^5 - 5x^3 + 4x$

Note: $f(x) = ax(x + 2)(x + 1)(x - 1)(x - 2)$ has zeros $-2, -1, 0, 1, 2$ for all real numbers a.

55. $f(x) = \left[x - (1 + \sqrt{3})\right]\left[x - (1 - \sqrt{3})\right]$
$= \left[(x - 1) - \sqrt{3}\right]\left[(x - 1) + \sqrt{3}\right]$
$= (x - 1)^2 - (\sqrt{3})^2$
$= x^2 - 2x + 1 - 3$
$= x^2 - 2x - 2$

Note: $f(x) = a(x^2 - 2x - 2)$ has these zeros for all real numbers $a \neq 0$.

56. $f(x) = (x - 2)\left[x - (4 + \sqrt{5})\right]\left[x - (4 - \sqrt{5})\right]$
$= (x - 2)\left[(x - 4) + \sqrt{5}\right]\left[(x - 4) - \sqrt{5}\right]$
$= (x - 2)\left[(x - 4)^2 - 5\right]$
$= x(x - 4)^2 - 5x - 2(x - 4)^2 + 10$
$= x^3 - 8x^2 + 16x - 5x - 2x^2 + 16x - 32 + 10$
$= x^3 - 10x^2 + 27x - 22$

Note: $f(x) = a(x^3 - 10x^2 + 27x - 22)$ has these zeros for all real numbers a.

57. $f(x) = (x - (-2))(x - (-2)) = (x + 2)^2 = x^2 + 4x + 4$

Note: If $a \neq 0$, then $f(x) = a(x^2 + 4x + 4)$ has degree 2 and zero $x = -2$, as does $f(x) = a(x + 2)(x - b)$ for any b value.

58. $f(x) = (x - 9)^2 = x^2 - 18x + 81$

Note: $f(x) = a(x - 9)^2$ has a zero of $x = 9$ for all real numbers $a \neq 0$.

59. $f(x) = (x - (-3))(x - 0)(x - 1) = x(x + 3)(x - 1) = x^3 + 2x^2 - 3x$

Note: $f(x) = a(x^3 + 2x^2 - 3x)$, $a \neq 0$, has degree 3 and zeros $x = -3, 0, 1$.

60. $f(x) = (x + 2)(x - 4)(x - 7) = (x + 2)(x^2 - 11x + 28) = x^3 - 9x^2 + 6x + 56$

Note: $f(x) = a(x^3 - 9x^2 + 6x + 56)$, $a \neq 0$, has degree 3 and zeros $x = -2, 4$, and 7.

61. $f(x) = (x - 0)(x - \sqrt{3})(x - (-\sqrt{3}))$

$= x(x - \sqrt{3})(x + \sqrt{3})$

$= x^3 - 3x$

Note: $f(x) = a(x^3 - 3x)$, $a \neq 0$, has degree 3 and zeros $x = 0, \sqrt{3}, -\sqrt{3}$.

62. $f(x) = (x - 0)(x - 2\sqrt{2})[x - (-2\sqrt{2})]$

$= x(x - 2\sqrt{2})(x + 2\sqrt{2})$

$= x(x^2 - 8)$

$= x^3 - 8x$

Note: $f(x) = a(x^3 - 8x)$ has these zeros for all real numbers $a \neq 0$.

63. $f(x) = (x - 1)[x - (-2)][x - (1 + \sqrt{3})][x - (1 - \sqrt{3})]$

$= (x - 1)(x + 2)[(x - 1) - \sqrt{3}][(x - 1) + \sqrt{3}]$

$= (x^2 + x - 2)[(x - 1)^2 - 3]$

$= (x^2 + x - 2)(x^2 - 2x - 2)$

$= x^4 - x^3 - 6x^2 + 2x + 4$

Note: $f(x) = a(x^4 - x^3 - 6x^2 + 2x + 4)$ has these zeros for all real numbers $a \neq 0$.

64. $f(x) = (x - 3)[x - (-2)][x - (2 + \sqrt{5})][x - (2 - \sqrt{5})]$

$= (x - 3)(x + 2)[(x - 2) - \sqrt{5}][(x - 2) + \sqrt{5}]$

$= (x^2 - x - 6)[(x - 2)^2 - 5]$

$= (x^2 - x - 6)(x^2 - 4x - 1)$

$= x^4 - 5x^3 - 3x^2 + 25x + 6$

Note: $f(x) = a(x^4 - 5x^3 - 3x^2 + 25x + 6)$ has these zeros for all real numbers $a \neq 0$.

65. For $n = 5$ there are many possibilities.

$f(x) = (x - 0)^3(x + 4)(x - 3) = x^5 + x^4 - 12x^3$

$f(x) = x^2(x + 4)^2(x - 3) = x^5 + 5x^4 - 8x^3 - 48x^2$

$f(x) = x^2(x + 4)(x - 3)^2 = x^5 - 2x^4 - 15x^3 + 36x^2$

$f(x) = x(x + 4)^3(x - 3) = x^5 + 9x^4 + 12x^3 - 80x^2 - 192x$

$f(x) = x(x + 4)^2(x - 3)^2 = x^5 + 2x^4 - 23x^3 - 24x^2 + 144x$

$f(x) = x(x + 4)(x - 3)^3 = x^5 - 5x^4 - 9x^3 + 81x^2 - 108x$

Also, any non-zero multiple of the right side of these functions would have these zeros and a degree of 5.

66. $f(x) = (x + 3)^2(x - 1)(x - 5)(x - 6) = x^5 - 6x^4 - 22x^3 + 108x^2 + 189x - 270$

or $f(x) = (x + 3)(x - 1)^2(x - 5)(x - 6) = x^5 - 10x^4 + 14x^3 + 88x^2 - 183x + 90$

or $f(x) = (x + 3)(x - 1)(x - 5)^2(x - 6) = x^5 - 14x^4 + 50x^3 + 68x^2 - 555x + 450$

or $f(x) = (x + 3)(x - 1)(x - 5)(x - 6)^2 = x^5 - 15x^4 + 59x^3 + 63x^2 - 648x + 540$

Note: Any nonzero multiple of the right side of these functions would also have degree 5 and zeros $x = -3, 1, 5$, and 6.
Also, $f(x) = a(x + 3)(x - 1)(x - 5)(x - 6)(x - b)$, $a \neq 0$, has degree 5 and zeros $x = -3, 1, 5, 6$ (as well as $x = b$).

67. $f(x) = x^3 - 9x = x(x^2 - 9) = x(x + 3)(x - 3)$

(a) Falls to the left

Rises to the right

(b) Zeros: $0, -3, 3$

(c)

x	-3	-2	-1	0	1	2	3
$f(x)$	0	10	8	0	-8	-10	0

(d)

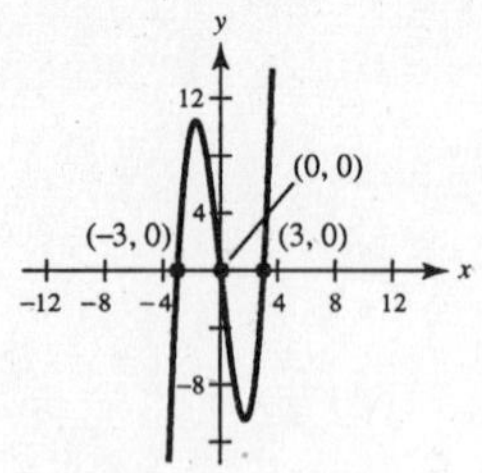

68. $g(x) = x^4 - 4x^2 = x^2(x + 2)(x - 2)$

(a) Rises to the left
Rises to the right

(b) Zeros: $-2, 0, 2$

(c)

x	±0.5	±1	±1.5	±2.5
$g(x)$	-0.94	-3	-3.94	14.1

(d)

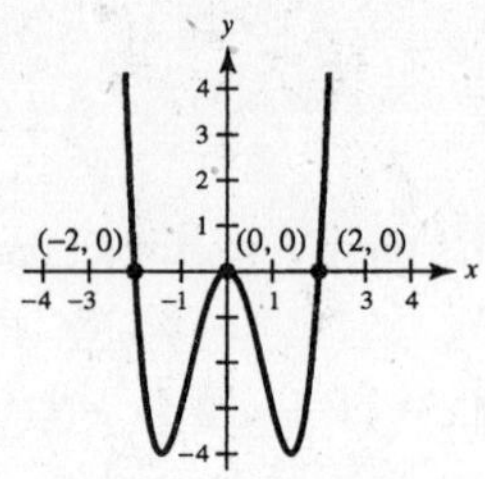

69. $f(t) = \frac{1}{4}(t^2 - 2t + 15) = \frac{1}{4}(t - 1)^2 + \frac{7}{2}$

(a) Rises to the left

Rises to the right

(b) No real zero (no x-intercepts)

(c)

t	-1	0	1	2	3
$f(t)$	4.5	3.75	3.5	3.75	4.5

(d) The graph is a parabola with vertex $\left(1, \frac{7}{2}\right)$.

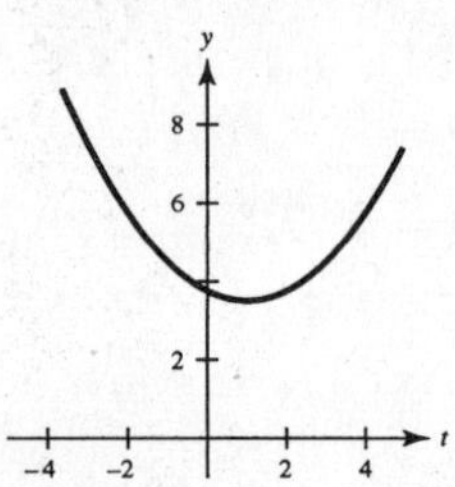

70. $g(x) = -x^2 + 10x - 16 = -(x - 2)(x - 8)$

(a) Falls to the left
Falls to the right

(b) Zeros: $2, 8$

(c)

x	1	3	5	7	9
$g(x)$	-7	5	9	5	-7

(d)

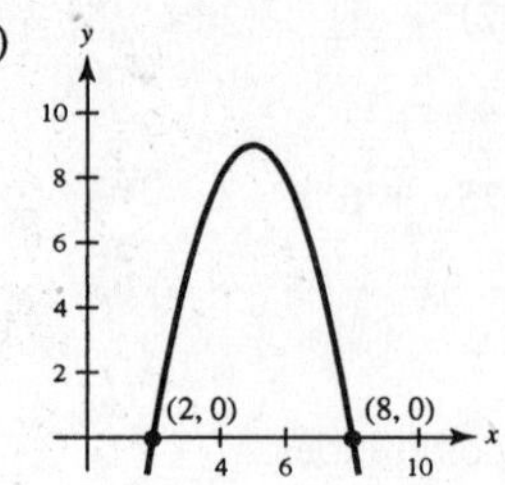

71. $f(x) = x^3 - 3x^2 = x^2(x - 3)$

(a) Falls to the left

Rises to the right

(b) Zeros: 0 and 3

(c)

x	-1	0	1	2	3
$f(x)$	-4	0	-2	-4	0

(d)

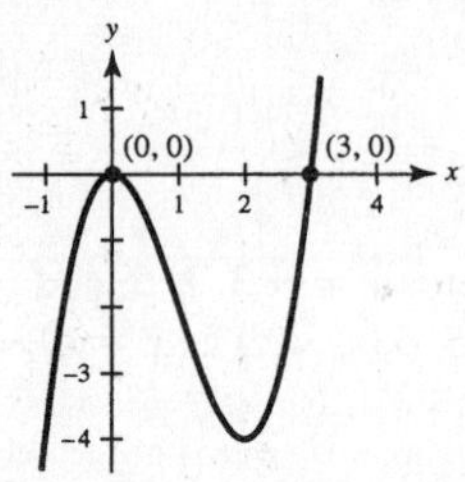

72. $f(x) = 1 - x^3$

(a) Rises to the left
Falls to the right

(b) Zero: 1

(c)

x	-2	-1	0	1	2
$f(x)$	9	2	1	0	-7

(d)

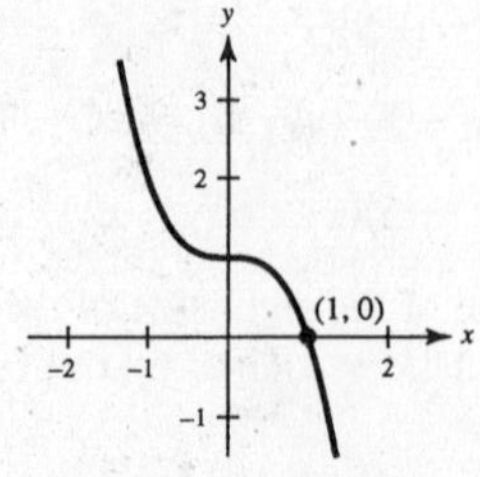

73. $f(x) = 3x^3 - 15x^2 + 18x = 3x(x - 2)(x - 3)$

(a) Falls to the left

Rises to the right

(b) Zeros: 0, 2, 3

(c)

x	0	1	2	2.5	3	3.5
$f(x)$	0	6	0	−1.875	0	7.875

(d)

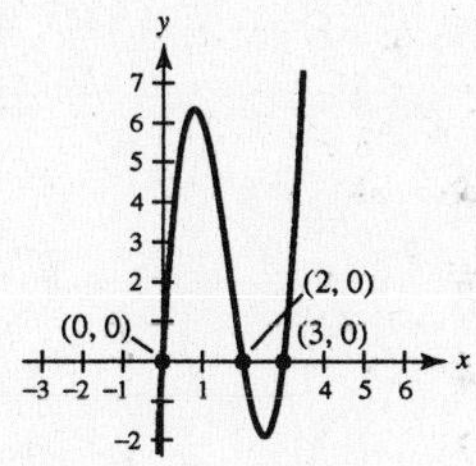

74. $f(x) = -4x^3 + 4x^2 + 15x$

$= -x(4x^2 - 4x - 15)$

$= -x(2x - 5)(2x + 3)$

(a) Rises to the left
Falls to the right

(b) Zeros: $-\frac{3}{2}, 0, \frac{5}{2}$

(c)

x	−3	−2	−1	0	1	2	3
$f(x)$	99	18	−7	0	15	14	−27

(d)

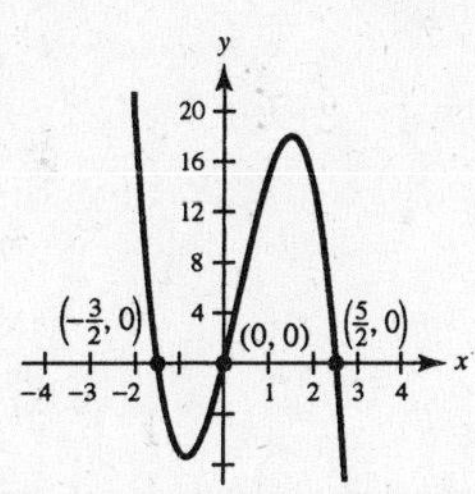

75. $f(x) = -5x^2 - x^3 = -x^2(5 + x)$

(a) Rises to the left

Falls to the right

(b) Zeros: 0, −5

(c)

x	−5	−4	−3	−2	−1	0	1
$f(x)$	0	−16	−18	−12	−4	0	−6

(d)

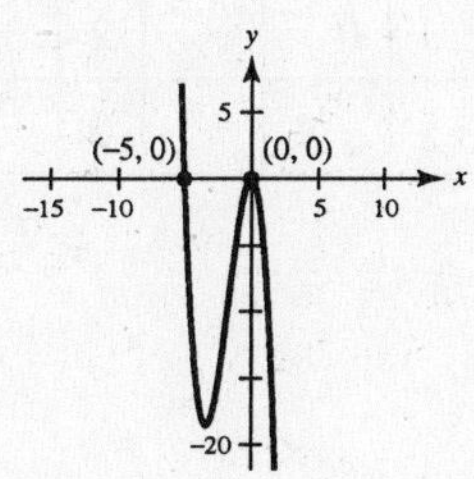

76. $f(x) = -48x^2 + 3x^4$

$= 3x^2(x^2 - 16)$

(a) Rises to the left
Rises to the right

(b) Zeros: $0, \pm 4$

(c)

x	−5	−4	−3	−2	−1	0	1	2	3	4	5
$f(x)$	675	0	−189	−144	−45	0	−45	−144	−189	0	675

(d)

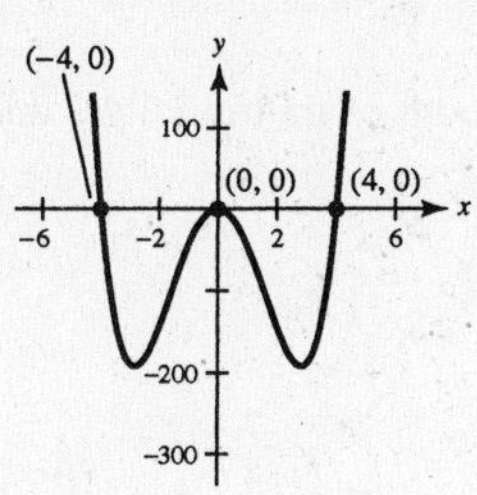

77. $f(x) = x^2(x - 4)$

(a) Falls to the left

Rises to the right

(b) Zeros: 0, 4

(c)

x	-1	0	1	2	3	4	5
$f(x)$	-5	0	-3	-8	-9	0	25

(d)

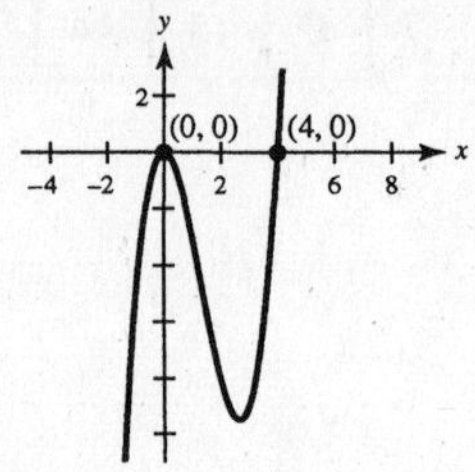

78. $h(x) = \frac{1}{3}x^3(x - 4)^2$

(a) Falls to the left
Rises to the right

(b) Zeros: 0 and 4

(c)

x	-1	0	1	2	3	4	5
$h(x)$	$-\frac{25}{3}$	0	3	$\frac{32}{3}$	9	0	$\frac{125}{3}$

(d)

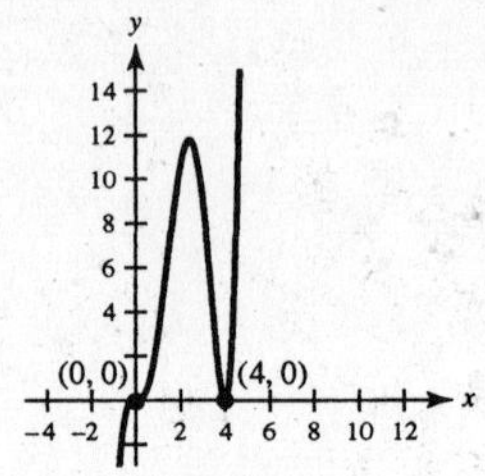

79. $g(t) = -\frac{1}{4}(t - 2)^2(t + 2)^2$

(a) Falls to the left

Falls to the right

(b) Zeros: 2 and -2

(c)

t	-3	-2	-1	0	1	2	3
$g(t)$	$-\frac{25}{4}$	0	$-\frac{9}{4}$	-4	$-\frac{9}{4}$	0	$-\frac{25}{4}$

(d)

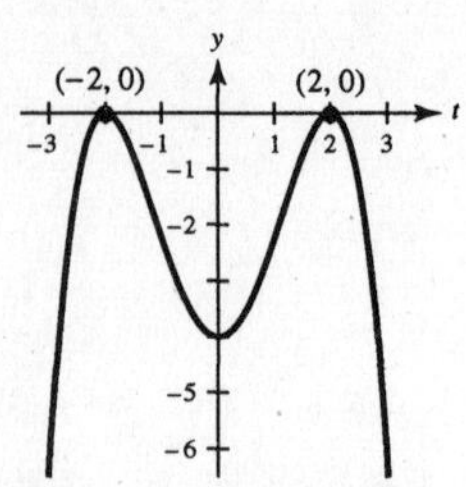

80. $g(x) = \frac{1}{10}(x + 1)^2(x - 3)^3$

(a) Falls to the left
Rises to the right

(b) Zeros: $-1, 3$

(c)

x	-2	-1	0	1	2
$f(x)$	-12.5	0	-2.7	-3.2	-0.9

(d)

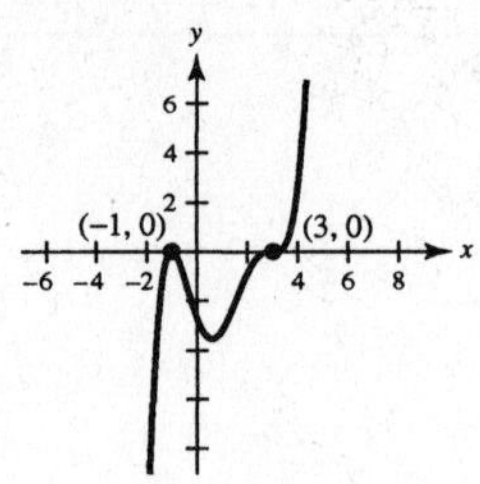

81. $f(x) = x^3 - 4x = x(x + 2)(x - 2)$

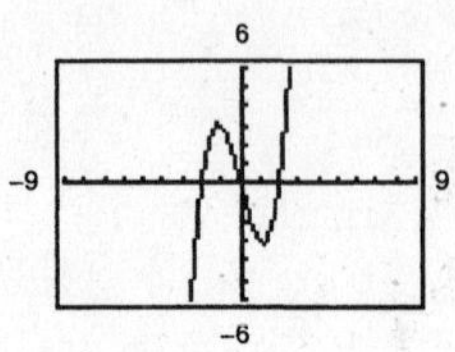

Zeros: $0, -2, 2$ all of multiplicity 1

82. $f(x) = \frac{1}{4}x^4 - 2x^2$

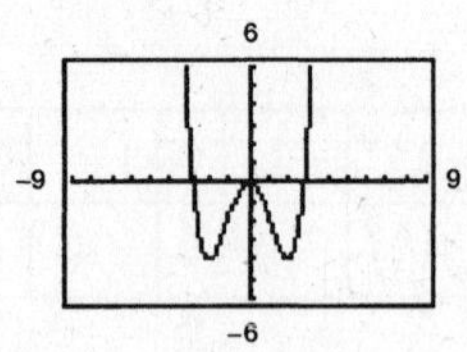

Zeros: -2.828 and 2.828 both of multiplicity 1 and 0 of multiplicity 2.

83. $g(x) = \frac{1}{5}(x + 1)^2(x - 3)(2x - 9)$

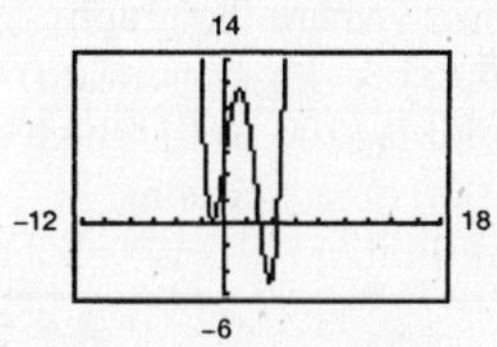

Zeros: -1 of multiplicity 2, 3 of multiplicity 1, and $\frac{9}{2}$ of multiplicity 1

84. $h(x) = \frac{1}{5}(x + 2)^2(3x - 5)^2$

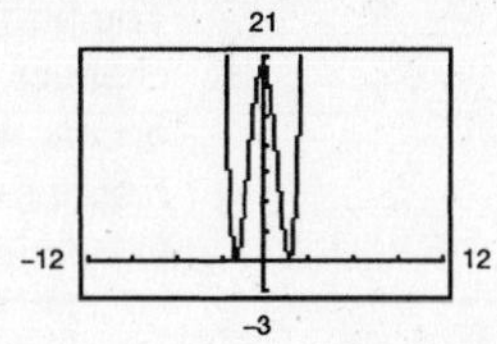

Zeros: $-2, \frac{5}{3}$, both multiplicity 2

85. (a) Degree: 3

Leading coefficient: Positive

(b) Degree: 2

Leading coefficient: Positive

(c) Degree: 4

Leading coefficient: Positive

(d) Degree: 5

Leading coefficient: Positive

86. $f(x) = x^4$; $f(x)$ is even.

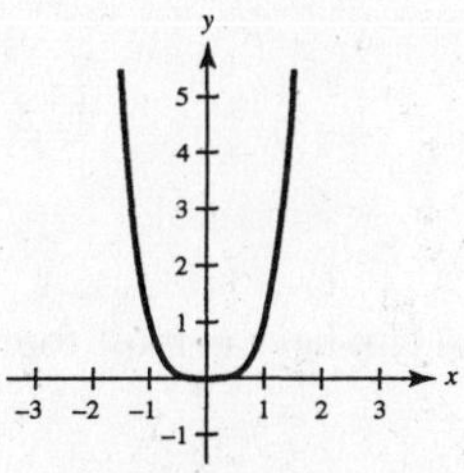

(a) $g(x) = f(x) + 2$

Vertical shift two units upward

$g(-x) = f(-x) + 2$

$= f(x) + 2$

$= g(x)$

Even

(b) $g(x) = f(x + 2)$

Horizontal shift two units to the left
Neither odd nor even

(c) $g(x) = f(-x) = (-x)^4 = x^4$

Reflection in the y-axis. The graph looks the same.
Even

(d) $g(x) = -f(x) = -x^4$

Reflection in the x-axis
Even

(e) $g(x) = f\left(\frac{1}{2}x\right) = \frac{1}{16}x^4$

Horizontal stretch
Even

(f) $g(x) = \frac{1}{2}f(x) = \frac{1}{2}x^4$

Vertical shrink
Even

(g) $g(x) = f(x^{3/4}) = (x^{3/4})^4 = \left(\sqrt[4]{x^3}\right)^4 = x^3,\ x \geq 0$

Neither

(h) $g(x) = (f \circ f)(x) = f(f(x)) = f(x^4) = (x^4)^4 = x^{16}$

Even

87. (a) and (b)

Box Height	Box Width	Box Volume, V
1	$36 - 2(1)$	$1[36 - 2(1)]^2 = 1156$
2	$36 - 2(2)$	$2[36 - 2(2)]^2 = 2048$
3	$36 - 2(3)$	$3[36 - 2(3)]^2 = 2700$
4	$36 - 2(4)$	$4[36 - 2(4)]^2 = 3136$
5	$36 - 2(5)$	$5[36 - 2(5)]^2 = 3380$
6	$36 - 2(6)$	$6[36 - 2(6)]^2 = 3456$
7	$36 - 2(7)$	$7[36 - 2(7)]^2 = 3388$

Volume is maximum at 3456 cubic inches when the height is 6 inches and the length and width are each 24 inches. So the dimensions are $6 \times 24 \times 24$ inches.

(c) Volume = length × width × height

height = x

length = width = $36 - 2x$

Thus, $V(x) = (36 - 2x)(36 - 2x)(x) = x(36 - 2x)^2$

Domain: $0 < x < 18$

(d)

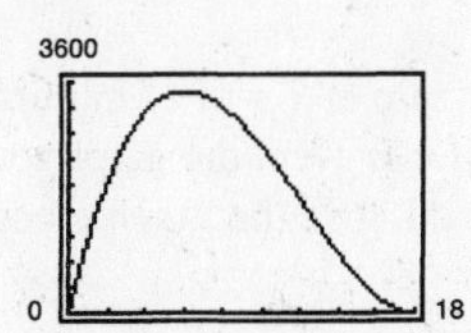

The maximum point on the graph occurs at $x = 6$. This agrees with the maximum found in part (b).

88. (a) Volume $= l \cdot w \cdot h = (24 - 2x)(24 - 4x)x$

$$= 2(12 - x) \cdot 4(6 - x)x$$

$$= 8x(12 - x)(6 - x)$$

(b) $x > 0, \quad 12 - x > 0, \quad 6 - x > 0$

$x < 12 \quad x < 6$

Domain: $0 < x < 6$

(c)

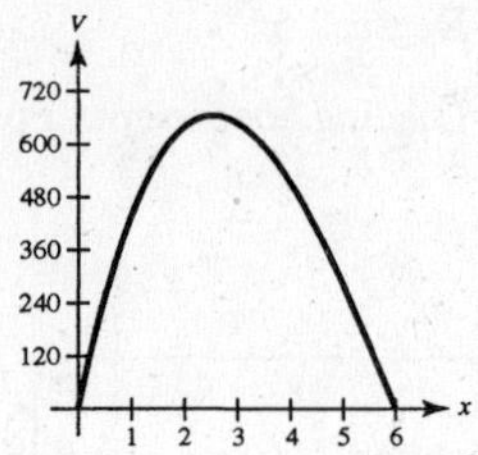

$x \approx 2.54$ when $V(x)$ is maximum.

89. $R = \frac{1}{100{,}000}(-x^3 + 600x^2)$

The point of diminishing returns (where the graph changes from curving upward to curving downward) occurs when $x = 200$. The point is $(200, 160)$ which corresponds to spending \$2,000,000 on advertising to obtain a revenue of \$160 million.

90. Let $t = 0$ correspond to 1990.

$N = -0.00082t^3 + 0.0161t^2 - 0.032t + 3.41$

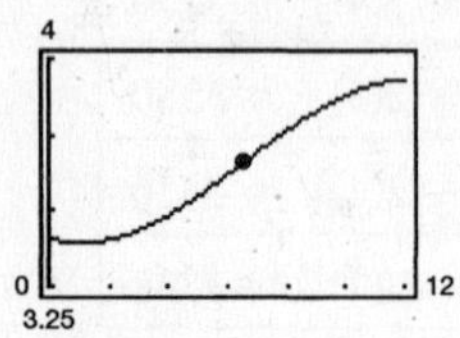

Participation is growing most rapidly when $t \approx 6.54$, or during the middle of 1996.

91. False. A fifth degree polynomial can have at most four turning points.

92. True. $f(x) = (x - 1)^6$ has one multiple zero.

93. False.

If the leading coefficient of a third-degree polynomial function is greater than zero, then the graph falls to the left and rises to the right. If the leading coefficient is less than zero, then the graph rises to the left and falls to the right.

94. True.

If a function has a repeated zero at $x = a$ of multiplicity k, $k > 1$, then $(x - a)^k$ is a factor of the function. If k is even, the graph touches, but does not cross, the x-axis at $x = a$. If k is odd, then the graph crosses the x-axis at $x = a$.

Section 2.3 Polynomial and Synthetic Division

1. $y_1 = \dfrac{4x}{x - 1}$ and $y_2 = 4 + \dfrac{4}{x - 1}$

$$\begin{array}{r} 4 \\ x - 1 \overline{) 4x + 0} \\ \underline{4x - 4} \\ 4 \end{array}$$

Thus, $\dfrac{4x}{x - 1} = 4 + \dfrac{4}{x - 1}$ and $y_1 = y_2$.

2. $y_1 = \dfrac{3x - 5}{x - 3}$ and $y_2 = 3 + \dfrac{4}{x - 3}$

$$\begin{array}{r} 3 \\ x - 3 \overline{) 3x - 5} \\ \underline{3x - 9} \\ 4 \end{array}$$

So, $y_1 = \dfrac{3x - 5}{x - 3} = 3 + \dfrac{4}{x - 3} = y_2$.

3. $y_1 = \dfrac{x^2}{x+2}$ and $y_2 = x - 2 + \dfrac{4}{x+2}$

$$\begin{array}{r} x - 2 \\ x + 2 \enclose{longdiv}{x^2 + 0x + 0} \\ \underline{x^2 + 2x} \\ -2x + 0 \\ \underline{-2x - 4} \\ 4 \end{array}$$

Thus, $\dfrac{x^2}{x+2} = x - 2 + \dfrac{4}{x+2}$ and $y_1 = y_2$.

4. $y_1 = \dfrac{x^4 - 3x^2 - 1}{x^2 + 5}$ and $y_2 = x^2 - 8 + \dfrac{39}{x^2 + 5}$

$$\begin{array}{r} x^2 - \quad 8 \\ x^2 + 0x + 5 \enclose{longdiv}{x^4 + 0x^3 - 3x^2 + 0x - 1} \\ \underline{x^4 + 0x^3 + 5x^2} \\ -8x^2 + 0x - 1 \\ \underline{-8x^2 + 0x - 40} \\ 39 \end{array}$$

Thus, $\dfrac{x^4 - 3x^2 - 1}{x^2 + 5} = x^2 - 8 + \dfrac{39}{x^2 + 5}$ and $y_1 = y_2$.

5. $y_1 = \dfrac{x^5 - 3x^3}{x^2 + 1}$ and $y_2 = x^3 - 4x + \dfrac{4x}{x^2 + 1}$

$$\begin{array}{r} x^3 - 4x \\ x^2 + 0x + 1 \enclose{longdiv}{x^5 + 0x^4 - 3x^3 + 0x^2 + 0x + 0} \\ \underline{x^5 + 0x^4 + x^3} \\ -4x^3 + 0x^2 + 0x \\ \underline{-4x^3 + 0x^2 - 4x} \\ 4x + 0 \end{array}$$

Thus, $\dfrac{x^5 - 3x^3}{x^2 + 1} = x^3 - 4x + \dfrac{4x}{x^2 + 1}$ and $y_1 = y_2$.

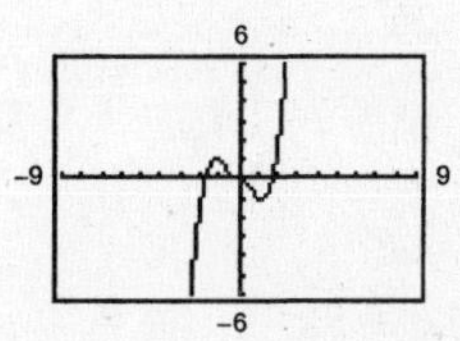

6.

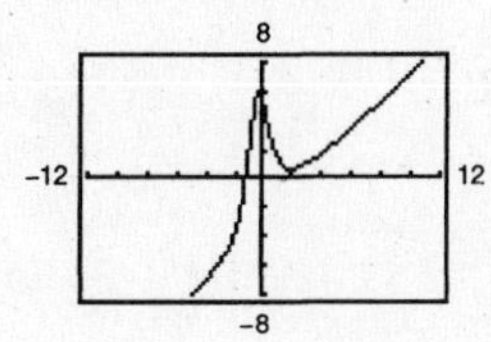

$y_1 = \dfrac{x^3 - 2x^2 + 5}{x^2 + x + 1}$ and $y_2 = x - 3 + \dfrac{2(x+4)}{x^2 + x + 1}$

$$\begin{array}{r} x - 3 \\ x^2 + x + 1 \enclose{longdiv}{x^3 - 2x^2 + 0x + 5} \\ \underline{x^3 + x^2 + x} \\ -3x^2 - x + 5 \\ \underline{-3x^2 - 3x - 3} \\ 2x + 8 \end{array}$$

So, $y_1 = \dfrac{x^3 - 2x^2 + 5}{x^2 + x + 1} = x - 3 + \dfrac{2(x+4)}{x^2 + x + 1} = y_2$.

7.

$$\begin{array}{r} 2x + 4 \\ x + 3 \enclose{longdiv}{2x^2 + 10x + 12} \\ \underline{2x^2 + 6x} \\ 4x + 12 \\ \underline{4x + 12} \\ 0 \end{array}$$

$\dfrac{2x^2 + 10x + 12}{x + 3} = 2x + 4$

8.

$$\begin{array}{r} 5x + 3 \\ x - 4 \enclose{longdiv}{5x^2 - 17x - 12} \\ \underline{5x^2 - 20x} \\ 3x - 12 \\ \underline{3x - 12} \\ 0 \end{array}$$

$\dfrac{5x^2 - 17x - 12}{x - 4} = 5x + 3$

9.

$$\begin{array}{r} x^2 - 3x + 1 \\ 4x + 5 \enclose{longdiv}{4x^3 - 7x^2 - 11x + 5} \\ \underline{4x^3 + 5x^2} \\ -12x^2 - 11x \\ \underline{-12x^2 - 15x} \\ 4x + 5 \\ \underline{4x + 5} \\ 0 \end{array}$$

$\dfrac{4x^3 - 7x^2 - 11x + 5}{4x + 5} = x^2 - 3x + 1$

10.

$$\begin{array}{r} 2x^2 - 4x + 3 \\ 3x - 2 \enclose{longdiv}{6x^3 - 16x^2 + 17x - 6} \\ \underline{6x^3 - 4x^2} \\ -12x^2 + 17x \\ \underline{-12x^2 + 8x} \\ 9x - 6 \\ \underline{9x - 6} \\ 0 \end{array}$$

$$\frac{6x^3 - 16x^2 + 17x - 6}{3x - 2} = 2x^2 - 4x + 3$$

11.

$$\begin{array}{r} x^3 + 3x^2 \quad - 1 \\ x + 2 \enclose{longdiv}{x^4 + 5x^3 + 6x^2 - x - 2} \\ \underline{x^4 + 2x^3} \\ 3x^3 + 6x^2 \\ \underline{3x^3 + 6x^2} \\ -x - 2 \\ \underline{-x - 2} \\ 0 \end{array}$$

$$\frac{x^4 + 5x^3 + 6x^2 - x - 2}{x + 2} = x^3 + 3x^2 - 1$$

12.

$$\begin{array}{r} x^2 + 7x + 18 \\ x - 3 \enclose{longdiv}{x^3 + 4x^2 - 3x - 12} \\ \underline{x^3 - 3x^2} \\ 7x^2 - 3x \\ \underline{7x^2 - 21x} \\ 18x - 12 \\ \underline{18x - 54} \\ 42 \end{array}$$

$$\frac{x^3 + 4x^2 - 3x - 12}{x - 3} = x^2 + 7x + 18 + \frac{42}{x - 3}$$

13.

$$\begin{array}{r} 7 \\ x + 2 \enclose{longdiv}{7x + 3} \\ \underline{7x + 14} \\ -11 \end{array}$$

$$\frac{7x + 3}{x + 2} = 7 - \frac{11}{x + 2}$$

14.

$$\begin{array}{r} 4 \\ 2x + 1 \enclose{longdiv}{8x - 5} \\ \underline{8x + 4} \\ -9 \end{array}$$

$$\frac{8x - 5}{2x + 1} = 4 - \frac{9}{2x + 1}$$

15.

$$\begin{array}{r} 3x + 5 \\ 2x^2 + 0x + 1 \enclose{longdiv}{6x^3 + 10x^2 + x + 8} \\ \underline{6x^3 + 0x^2 + 3x} \\ 10x^2 - 2x + 8 \\ \underline{10x^2 + 0x + 5} \\ -2x + 3 \end{array}$$

$$\frac{6x^3 + 10x^2 + x + 8}{2x^2 + 1} = 3x + 5 + \frac{-2x + 3}{2x^2 + 1}$$

$$= 3x + 5 - \frac{2x - 3}{2x^2 + 1}$$

16.

$$\begin{array}{r} x \\ x^2 + 0x + 1 \enclose{longdiv}{x^3 + 0x^2 + 0x - 9} \\ \underline{x^3 + 0x^2 + x} \\ -x - 9 \end{array}$$

$$\frac{x^3 - 9}{x^2 + 1} = x - \frac{x + 9}{x^2 + 1}$$

17.

$$\begin{array}{r} x^2 + 2x + 4 \\ x^2 - 2x + 3 \overline{)\, x^4 + 0x^3 + 3x^2 + 0x + 1} \\ \underline{x^4 - 2x^3 + 3x^2} \\ 2x^3 + 0x^2 + 0x \\ \underline{2x^3 - 4x^2 + 6x} \\ 4x^2 - 6x + 1 \\ \underline{4x^2 - 8x + 12} \\ 2x - 11 \end{array} \Rightarrow \quad \frac{x^4 + 3x^2 + 1}{x^2 - 2x + 3} = x^2 + 2x + 4 + \frac{2x - 11}{x^2 - 2x + 3}$$

18.

$$\begin{array}{r} x^2 \\ x^3 + 0x^2 + 0x - 1 \overline{)\, x^5 + 0x^4 + 0x^3 + 0x^2 + 0x + 7} \\ \underline{x^5 + 0x^4 + 0x^3 - x^2 \qquad\quad} \\ x^2 \qquad + 7 \end{array}$$

$$\frac{x^5 + 7}{x^3 - 1} = x^2 + \frac{x^2 + 7}{x^3 - 1}$$

19.

$$\begin{array}{r} x + 3 \\ x^3 - 3x^2 + 3x - 1 \overline{)\, x^4 + 0x^3 + 0x^2 + 0x + 0} \\ \underline{x^4 - 3x^3 + 3x^2 - x \quad} \\ 3x^3 - 3x^2 + x + 0 \\ \underline{3x^3 - 9x^2 + 9x - 3} \\ 6x^2 - 8x + 3 \end{array}$$

$$\frac{x^4}{(x - 1)^3} = x + 3 + \frac{6x^2 - 8x + 3}{(x - 1)^3}$$

20.

$$\begin{array}{r} 2x \\ x^2 - 2x + 1 \overline{)\, 2x^3 - 4x^2 - 15x + 5} \\ \underline{2x^3 - 4x^2 + 2x \quad} \\ -17x + 5 \end{array}$$

$$\frac{2x^3 - 4x^2 - 15x + 5}{(x - 1)^2} = 2x - \frac{17x - 5}{x^2 - 2x + 1}$$

21.

$$\begin{array}{r|rrrr} 5 & 3 & -17 & 15 & -25 \\ & & 15 & -10 & 25 \\ \hline & 3 & -2 & 5 & 0 \end{array}$$

$$\frac{3x^3 - 17x^2 + 15x - 25}{x - 5} = 3x^2 - 2x + 5$$

22.

$$\begin{array}{r|rrrr} -3 & 5 & 18 & 7 & -6 \\ & & -15 & -9 & 6 \\ \hline & 5 & 3 & -2 & 0 \end{array}$$

$$\frac{5x^3 + 18x^2 + 7x - 6}{x + 3} = 5x^2 + 3x - 2$$

23.

$$\begin{array}{r|rrrr} -2 & 4 & 8 & -9 & -18 \\ & & -8 & 0 & 18 \\ \hline & 4 & 0 & -9 & 0 \end{array}$$

$$\frac{4x^3 + 8x^2 - 9x - 18}{x + 2} = 4x^2 - 9$$

24.

$$\begin{array}{r|rrrr} 2 & 9 & -18 & -16 & 32 \\ & & 18 & 0 & -32 \\ \hline & 9 & 0 & -16 & 0 \end{array}$$

$$\frac{9x^3 - 18x^2 - 16x + 32}{x - 2} = 9x^2 - 16$$

25.

$$\begin{array}{r|rrrr} -10 & -1 & 0 & 75 & -250 \\ & & 10 & -100 & 250 \\ \hline & -1 & 10 & -25 & 0 \end{array}$$

$$\frac{-x^3 + 75x - 250}{x + 10} = -x^2 + 10x - 25$$

26.

$$\begin{array}{r|rrrr} 6 & 3 & -16 & 0 & -72 \\ & & 18 & 12 & 72 \\ \hline & 3 & 2 & 12 & 0 \end{array}$$

$$\frac{3x^3 - 16x^2 - 72}{x - 6} = 3x^2 + 2x + 12$$

27.

$$\begin{array}{r|rrrr} 4 & 5 & -6 & 0 & 8 \\ & & 20 & 56 & 224 \\ \hline & 5 & 14 & 56 & 232 \end{array}$$

$$\frac{5x^3 - 6x^2 + 8}{x - 4} = 5x^2 + 14x + 56 + \frac{232}{x - 4}$$

28. $$\begin{array}{r|rrrr} -2 & 5 & 0 & 6 & 8 \\ & & -10 & 20 & -52 \\ \hline & 5 & -10 & 26 & -44 \end{array}$$

$$\frac{5x^3 + 6x + 8}{x + 2} = 5x^2 - 10x + 26 - \frac{44}{x + 2}$$

29. $$\begin{array}{r|rrrrr} 6 & 10 & -50 & 0 & 0 & -800 \\ & & 60 & 60 & 360 & 2160 \\ \hline & 10 & 10 & 60 & 360 & 1360 \end{array}$$

$$\frac{10x^4 - 50x^3 - 800}{x - 6} = 10x^3 + 10x^2 + 60x + 360 + \frac{1360}{x - 6}$$

30. $$\begin{array}{r|rrrrrr} -3 & 1 & -13 & 0 & 0 & -120 & 80 \\ & & -3 & 48 & -144 & 432 & -936 \\ \hline & 1 & -16 & 48 & -144 & 312 & -856 \end{array}$$

$$\frac{x^5 - 13x^4 - 120x + 80}{x + 3} = x^4 - 16x^3 + 48x^2 - 144x + 312 - \frac{856}{x + 3}$$

31. $$\begin{array}{r|rrrr} -8 & 1 & 0 & 0 & 512 \\ & & -8 & 64 & -512 \\ \hline & 1 & -8 & 64 & 0 \end{array}$$

$$\frac{x^3 + 512}{x + 8} = x^2 - 8x + 64$$

32. $$\begin{array}{r|rrrr} 9 & 1 & 0 & 0 & -729 \\ & & 9 & 81 & 729 \\ \hline & 1 & 9 & 81 & 0 \end{array}$$

$$\frac{x^3 - 729}{x - 9} = x^2 + 9x + 81$$

33. $$\begin{array}{r|rrrrr} 2 & -3 & 0 & 0 & 0 & 0 \\ & & -6 & -12 & -24 & -48 \\ \hline & -3 & -6 & -12 & -24 & -48 \end{array}$$

$$\frac{-3x^4}{x - 2} = -3x^3 - 6x^2 - 12x - 24 - \frac{48}{x - 2}$$

34. $$\begin{array}{r|rrrrr} -2 & -3 & 0 & 0 & 0 & 0 \\ & & 6 & -12 & 24 & -48 \\ \hline & -3 & 6 & -12 & 24 & -48 \end{array}$$

$$\frac{-3x^4}{x + 2} = -3x^3 + 6x^2 - 12x + 24 - \frac{48}{x + 2}$$

35. $$\begin{array}{r|rrrrr} 6 & -1 & 0 & 0 & 180 & 0 \\ & & -6 & -36 & -216 & -216 \\ \hline & -1 & -6 & -36 & -36 & -216 \end{array}$$

$$\frac{180x - x^4}{x - 6} = -x^3 - 6x^2 - 36x - 36 - \frac{216}{x - 6}$$

36. $$\begin{array}{r|rrrr} -1 & -1 & 2 & -3 & 5 \\ & & 1 & -3 & 6 \\ \hline & -1 & 3 & -6 & 11 \end{array}$$

$$\frac{5 - 3x + 2x^2 - x^3}{x + 1} = -x^2 + 3x - 6 + \frac{11}{x + 1}$$

37. $$\begin{array}{r|rrrr} -\frac{1}{2} & 4 & 16 & -23 & -15 \\ & & -2 & -7 & 15 \\ \hline & 4 & 14 & -30 & 0 \end{array}$$

$$\frac{4x^3 + 16x^2 - 23x - 15}{x + \frac{1}{2}} = 4x^2 + 14x - 30$$

38. $$\begin{array}{r|rrrr} \frac{3}{2} & 3 & -4 & 0 & 5 \\ & & \frac{9}{2} & \frac{3}{4} & \frac{9}{8} \\ \hline & 3 & \frac{1}{2} & \frac{3}{4} & \frac{49}{8} \end{array}$$

$$\frac{3x^3 - 4x^2 + 5}{x - \frac{3}{2}} = 3x^2 + \frac{1}{2}x + \frac{3}{4} + \frac{49}{8x - 12}$$

39. $f(x) = x^3 - x^2 - 14x + 11,\ k = 4$

$$\begin{array}{r|rrrr} 4 & 1 & -1 & -14 & 11 \\ & & 4 & 12 & -8 \\ \hline & 1 & 3 & -2 & 3 \end{array}$$

$f(x) = (x - 4)(x^2 + 3x - 2) + 3$

$f(4) = 4^3 - 4^2 - 14(4) + 11 = 3$

40. $f(x) = x^3 - 5x^2 - 11x + 8,\ k = -2$

$$\begin{array}{r|rrrr} -2 & 1 & -5 & -11 & 8 \\ & & -2 & 14 & -6 \\ \hline & 1 & -7 & 3 & 2 \end{array}$$

$f(x) = (x + 2)(x^2 - 7x + 3) + 2$

$f(-2) = (-2)^3 - 5(-2)^2 - 11(-2) + 8$

$= -8 - 20 + 22 + 8 = 2$

41. $f(x) = 15x^4 + 10x^3 - 6x^2 + 14,\ k = -\frac{2}{3}$

$$\begin{array}{r|rrrrr} -\frac{2}{3} & 15 & 10 & -6 & 0 & 14 \\ & & -10 & 0 & 4 & -\frac{8}{3} \\ \hline & 15 & 0 & -6 & 4 & \frac{34}{3} \end{array}$$

$f(x) = \left(x + \frac{2}{3}\right)(15x^3 - 6x + 4) + \frac{34}{3}$

$f\left(-\frac{2}{3}\right) = 15\left(-\frac{2}{3}\right)^4 + 10\left(-\frac{2}{3}\right)^3 - 6\left(-\frac{2}{3}\right)^2 + 14 = \frac{34}{3}$

42. $f(x) = 10x^3 - 22x^2 - 3x + 4,\ k = \frac{1}{5}$

$$\begin{array}{r|rrrr} \frac{1}{5} & 10 & -22 & -3 & 4 \\ & & 2 & -4 & -\frac{7}{5} \\ \hline & 10 & -20 & -7 & \frac{13}{5} \end{array}$$

$f(x) = \left(x - \frac{1}{5}\right)(10x^2 - 20x - 7) + \frac{13}{5}$

$f\left(\frac{1}{5}\right) = 10\left(\frac{1}{5}\right)^3 - 22\left(\frac{1}{5}\right)^2 - 3\left(\frac{1}{5}\right) + 4$

$= \frac{2}{25} - \frac{22}{25} - \frac{3}{5} + 4 = \frac{65}{25} = \frac{13}{5}$

43. $f(x) = x^3 + 3x^2 - 2x - 14,\ k = \sqrt{2}$

$$\begin{array}{r|rrrr} \sqrt{2} & 1 & 3 & -2 & -14 \\ & & \sqrt{2} & 2 + 3\sqrt{2} & 6 \\ \hline & 1 & 3 + \sqrt{2} & 3\sqrt{2} & -8 \end{array}$$

$f(x) = \left(x - \sqrt{2}\right)\left[x^2 + \left(3 + \sqrt{2}\right)x + 3\sqrt{2}\right] - 8$

$f\left(\sqrt{2}\right) = \left(\sqrt{2}\right)^3 + 3\left(\sqrt{2}\right)^2 - 2\sqrt{2} - 14 = -8$

44. $f(x) = x^3 + 2x^2 - 5x - 4,\ k = -\sqrt{5}$

$$\begin{array}{r|rrrr} -\sqrt{5} & 1 & 2 & -5 & -4 \\ & & -\sqrt{5} & -2\sqrt{5} + 5 & 10 \\ \hline & 1 & 2 - \sqrt{5} & -2\sqrt{5} & 6 \end{array}$$

$f(x) = \left(x + \sqrt{5}\right)\left[x^2 + \left(2 - \sqrt{5}\right)x - 2\sqrt{5}\right] + 6$

$f\left(-\sqrt{5}\right) = \left(-\sqrt{5}\right)^3 + 2\left(-\sqrt{5}\right)^2 - 5\left(-\sqrt{5}\right) - 4$

$= -5\sqrt{5} + 10 + 5\sqrt{5} - 4 = 6$

45. $f(x) = -4x^3 + 6x^2 + 12x + 4,\ k = 1 - \sqrt{3}$

$$\begin{array}{r|rrrr} 1 - \sqrt{3} & -4 & 6 & 12 & 4 \\ & & -4 + 4\sqrt{3} & -10 + 2\sqrt{3} & -4 \\ \hline & -4 & 2 + 4\sqrt{3} & 2 + 2\sqrt{3} & 0 \end{array}$$

$f(x) = \left[x - \left(1 - \sqrt{3}\right)\right]\left[-4x^2 + \left(2 + 4\sqrt{3}\right)x + \left(2 + 2\sqrt{3}\right)\right]$

$f\left(1 - \sqrt{3}\right) = -4\left(1 - \sqrt{3}\right)^3 + 6\left(1 - \sqrt{3}\right)^2 + 12\left(1 - \sqrt{3}\right) + 4 = 0$

46. $f(x) = -3x^3 + 8x^2 + 10x - 8,\ k = 2 + \sqrt{2}$

$$\begin{array}{r|rrrr} 2 + \sqrt{2} & -3 & 8 & 10 & -8 \\ & & -6 - 3\sqrt{2} & -2 - 4\sqrt{2} & 8 \\ \hline & -3 & 2 - 3\sqrt{2} & 8 - 4\sqrt{2} & 0 \end{array}$$

$f(x) = \left(x - 2 - \sqrt{2}\right)\left[-3x^2 + \left(2 - 3\sqrt{2}\right)x + 8 - 4\sqrt{2}\right]$

$f\left(2 + \sqrt{2}\right) = -3\left(2 + \sqrt{2}\right)^3 + 8\left(2 + \sqrt{2}\right)^2 + 10\left(2 + \sqrt{2}\right) - 8$

$= -3\left(20 + 14\sqrt{2}\right) + 8\left(6 + 4\sqrt{2}\right) + 10\left(2 + \sqrt{2}\right) - 8$

$= -60 - 42\sqrt{2} + 48 + 32\sqrt{2} + 20 + 10\sqrt{2} - 8$

$= 0$

47. $f(x) = 4x^3 - 13x + 10$

(a) $$\begin{array}{r|rrrr} 1 & 4 & 0 & -13 & 10 \\ & & 4 & 4 & -9 \\ \hline & 4 & 4 & -9 & \underline{1} = f(1) \end{array}$$

(b) $$\begin{array}{r|rrrr} -2 & 4 & 0 & -13 & 10 \\ & & -8 & 16 & -6 \\ \hline & 4 & -8 & 3 & \underline{4} = f(-2) \end{array}$$

(c) $$\begin{array}{r|rrrr} \frac{1}{2} & 4 & 0 & -13 & 10 \\ & & 2 & 1 & -6 \\ \hline & 4 & 2 & -12 & \underline{4} = f\left(\frac{1}{2}\right) \end{array}$$

(d) $$\begin{array}{r|rrrr} 8 & 4 & 0 & -13 & 10 \\ & & 32 & 256 & 1944 \\ \hline & 4 & 32 & 243 & \underline{1954} = f(8) \end{array}$$

48. $g(x) = x^6 - 4x^4 + 3x^2 + 2$

(a) $$\begin{array}{r|rrrrrrr} 2 & 1 & 0 & -4 & 0 & 3 & 0 & 2 \\ & & 2 & 4 & 0 & 0 & 6 & 12 \\ \hline & 1 & 2 & 0 & 0 & 3 & 6 & 14 \end{array}$$

$g(2) = 14$

(b) $$\begin{array}{r|rrrrrrr} -4 & 1 & 0 & -4 & 0 & 3 & 0 & 2 \\ & & -4 & 16 & -48 & 192 & -780 & 3120 \\ \hline & 1 & -4 & 12 & -48 & 195 & -780 & 3122 \end{array}$$

$g(-4) = 3122$

(c) $$\begin{array}{r|rrrrrrr} 3 & 1 & 0 & -4 & 0 & 3 & 0 & 2 \\ & & 3 & 9 & 15 & 45 & 144 & 432 \\ \hline & 1 & 3 & 5 & 15 & 48 & 144 & 434 \end{array}$$

$g(3) = 434$

(d) $$\begin{array}{r|rrrrrrr} -1 & 1 & 0 & -4 & 0 & 3 & 0 & 2 \\ & & -1 & 1 & 3 & -3 & 0 & 0 \\ \hline & 1 & -1 & -3 & 3 & 0 & 0 & 2 \end{array}$$

$g(-1) = 2$

49. $h(x) = 3x^3 + 5x^2 - 10x + 1$

(a) $$\begin{array}{r|rrrr} 3 & 3 & 5 & -10 & 1 \\ & & 9 & 42 & 96 \\ \hline & 3 & 14 & 32 & \underline{97} = h(3) \end{array}$$

(b) $$\begin{array}{r|rrrr} \frac{1}{3} & 3 & 5 & -10 & 1 \\ & & 1 & 2 & -\frac{8}{3} \\ \hline & 3 & 6 & -8 & \underline{-\frac{5}{3}} = h\left(\frac{1}{3}\right) \end{array}$$

(c) $$\begin{array}{r|rrrr} -2 & 3 & 5 & -10 & 1 \\ & & -6 & 2 & 16 \\ \hline & 3 & -1 & -8 & \underline{17} = h(-2) \end{array}$$

(d) $$\begin{array}{r|rrrr} -5 & 3 & 5 & -10 & 1 \\ & & -15 & 50 & -200 \\ \hline & 3 & -10 & 40 & \underline{-199} = h(-5) \end{array}$$

50. $f(x) = 0.4x^4 - 1.6x^3 + 0.7x^2 - 2$

(a) $$\begin{array}{r|rrrrr} 1 & 0.4 & -1.6 & 0.7 & 0 & -2 \\ & & 0.4 & -1.2 & -0.5 & -0.5 \\ \hline & 0.4 & -1.2 & -0.5 & -0.5 & -2.5 \end{array}$$

$f(1) = -2.5$

(b) $$\begin{array}{r|rrrrr} -2 & 0.4 & -1.6 & 0.7 & 0 & -2 \\ & & -0.8 & 4.8 & -11 & 22 \\ \hline & 0.4 & -2.4 & 5.5 & -11 & 20 \end{array}$$

$f(-2) = 20$

(c) $$\begin{array}{r|rrrrr} 5 & 0.4 & -1.6 & 0.7 & 0 & -2 \\ & & 2.0 & 2.0 & 13.5 & 67.5 \\ \hline & 0.4 & 0.4 & 2.7 & 13.5 & 65.5 \end{array}$$

$f(5) = 65.5$

(d) $$\begin{array}{r|rrrrr} -10 & 0.4 & -1.6 & 0.7 & 0 & -2 \\ & & -4.0 & 56.0 & -567 & 5670 \\ \hline & 0.4 & -5.6 & 56.7 & -567 & 5668 \end{array}$$

$f(-10) = 5668$

51. $$\begin{array}{r|rrrr} 2 & 1 & 0 & -7 & 6 \\ & & 2 & 4 & -6 \\ \hline & 1 & 2 & -3 & 0 \end{array}$$

$$\begin{aligned} x^3 - 7x + 6 &= (x - 2)(x^2 + 2x - 3) \\ &= (x - 2)(x + 3)(x - 1) \end{aligned}$$

Zeros: $2, -3, 1$

52. $$\begin{array}{r|rrrr} -4 & 1 & 0 & -28 & -48 \\ & & -4 & 16 & 48 \\ \hline & 1 & -4 & -12 & 0 \end{array}$$

$$\begin{aligned} x^3 - 28x - 48 &= (x + 4)(x^2 - 4x - 12) \\ &= (x + 4)(x - 6)(x + 2) \end{aligned}$$

Zeros: $-4, -2, 6$

53. $$\begin{array}{r|rrrr} \frac{1}{2} & 2 & -15 & 27 & -10 \\ & & 1 & -7 & 10 \\ \hline & 2 & -14 & 20 & 0 \end{array}$$

$$\begin{aligned} &2x^3 - 15x^2 + 27x - 10 \\ &= \left(x - \tfrac{1}{2}\right)(2x^2 - 14x + 20) \\ &= (2x - 1)(x - 2)(x - 5) \end{aligned}$$

Zeros: $\frac{1}{2}, 2, 5$

54. $$\begin{array}{r|rrrr} \frac{2}{3} & 48 & -80 & 41 & -6 \\ & & 32 & -32 & 6 \\ \hline & 48 & -48 & 9 & 0 \end{array}$$

$$\begin{aligned} 48x^3 - 80x^2 + 41x - 6 &= \left(x - \tfrac{2}{3}\right)(48x^2 - 48x + 9) \\ &= \left(x - \tfrac{2}{3}\right)(4x - 3)(12x - 3) \\ &= (3x - 2)(4x - 3)(4x - 1) \end{aligned}$$

Zeros: $\frac{2}{3}, \frac{3}{4}, \frac{1}{4}$

55. $$\begin{array}{r|rrrr} \sqrt{3} & 1 & 2 & -3 & -6 \\ & & \sqrt{3} & 3 + 2\sqrt{3} & 6 \\ \hline & 1 & 2 + \sqrt{3} & 2\sqrt{3} & 0 \end{array}$$

$$\begin{array}{r|rrr} -\sqrt{3} & 1 & 2 + \sqrt{3} & 2\sqrt{3} \\ & & -\sqrt{3} & -2\sqrt{3} \\ \hline & 1 & 2 & 0 \end{array}$$

$x^3 + 2x^2 - 3x - 6 = \left(x - \sqrt{3}\right)\left(x + \sqrt{3}\right)(x + 2)$

Zeros: $\pm\sqrt{3}, -2$

56. $$\begin{array}{r|rrrr} \sqrt{2} & 1 & 2 & -2 & -4 \\ & & \sqrt{2} & 2\sqrt{2} + 2 & 4 \\ \hline & 1 & 2 + \sqrt{2} & 2\sqrt{2} & 0 \end{array}$$

$$\begin{aligned} x^3 + 2x^2 - 2x - 4 &= \left(x - \sqrt{2}\right)\left[x^2 + \left(2 + \sqrt{2}\right)x + 2\sqrt{2}\right] \\ &= \left(x - \sqrt{2}\right)(x + 2)\left(x + \sqrt{2}\right) \end{aligned}$$

Zeros: $-2, -\sqrt{2}, \sqrt{2}$

57. $$\begin{array}{r|rrrr} 1 + \sqrt{3} & 1 & -3 & 0 & 2 \\ & & 1 + \sqrt{3} & 1 - \sqrt{3} & -2 \\ \hline & 1 & -2 + \sqrt{3} & 1 - \sqrt{3} & 0 \end{array}$$

$$\begin{array}{r|rrr} 1 - \sqrt{3} & 1 & -2 + \sqrt{3} & 1 - \sqrt{3} \\ & & 1 - \sqrt{3} & -1 + \sqrt{3} \\ \hline & 1 & -1 & 0 \end{array}$$

$$\begin{aligned} x^3 - 3x^2 + 2 &= \left[x - \left(1 + \sqrt{3}\right)\right]\left[x - \left(1 - \sqrt{3}\right)\right](x - 1) \\ &= (x - 1)\left(x - 1 - \sqrt{3}\right)\left(x - 1 + \sqrt{3}\right) \end{aligned}$$

Zeros: $1, 1 \pm \sqrt{3}$

58. $$\begin{array}{r|rrrr} 2 - \sqrt{5} & 1 & -1 & -13 & -3 \\ & & 2 - \sqrt{5} & 7 - 3\sqrt{5} & 3 \\ \hline & 1 & 1 - \sqrt{5} & -6 - 3\sqrt{5} & 0 \end{array}$$

$$\begin{aligned} x^3 - x^2 - 13x - 3 &= \left[x - \left(2 - \sqrt{5}\right)\right]\left[x^2 + \left(1 - \sqrt{5}\right)x - \left(6 + 3\sqrt{5}\right)\right] \\ &= \left(x - 2 + \sqrt{5}\right)\left(x - 2 - \sqrt{5}\right)(x + 3) \end{aligned}$$

Zeros: $2 - \sqrt{5}, 2 + \sqrt{5}, -3$

59. $f(x) = 2x^3 + x^2 - 5x + 2;$

Factors: $(x + 2), (x - 1)$

(a) $$\begin{array}{r|rrrr} -2 & 2 & 1 & -5 & 2 \\ & & -4 & 6 & -2 \\ \hline & 2 & -3 & 1 & 0 \end{array}$$

$$\begin{array}{r|rrr} 1 & 2 & -3 & 1 \\ & & 2 & -1 \\ \hline & 2 & -1 & 0 \end{array}$$

Both are factors of $f(x)$ since the remainders are zero.

(b) The remaining factor of $f(x)$ is $(2x - 1)$.

(c) $f(x) = (2x - 1)(x + 2)(x - 1)$

(d) Zeros: $\frac{1}{2}, -2, 1$

(e)

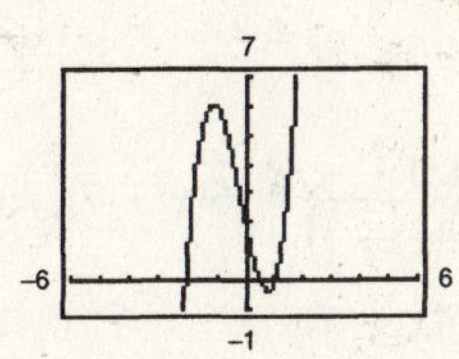

60. $f(x) = 3x^3 + 2x^2 - 19x + 6$; Factors: $(x + 3), (x - 2)$

(a) $\begin{array}{r|rrrr} -3 & 3 & 2 & -19 & 6 \\ & & -9 & 21 & -6 \\ \hline & 3 & -7 & 2 & 0 \end{array}$

$\begin{array}{r|rrr} 2 & 3 & -7 & 2 \\ & & 6 & -2 \\ \hline & 3 & -1 & 0 \end{array}$

(b) The remaining factor is $(3x - 1)$.

(c) $f(x) = 3x^3 + 2x^2 - 19x + 6$

$= (3x - 1)(x + 3)(x - 2)$

(d) Zeros: $\frac{1}{3}, -3, 2$

(e)

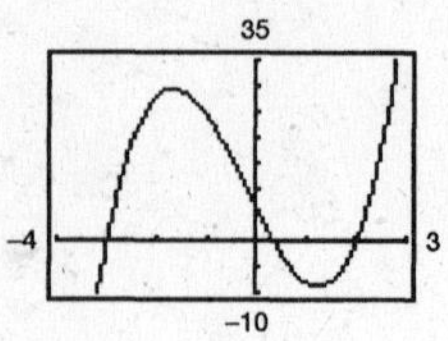

61. $f(x) = x^4 - 4x^3 - 15x^2 + 58x - 40$;

Factors: $(x - 5), (x + 4)$

(a) $\begin{array}{r|rrrrr} 5 & 1 & -4 & -15 & 58 & -40 \\ & & 5 & 5 & -50 & 40 \\ \hline & 1 & 1 & -10 & 8 & 0 \end{array}$

$\begin{array}{r|rrrr} -4 & 1 & 1 & -10 & 8 \\ & & -4 & 12 & -8 \\ \hline & 1 & -3 & 2 & 0 \end{array}$

Both are factors of $f(x)$ since the remainders are zero.

(b) $x^2 - 3x + 2 = (x - 1)(x - 2)$

The remaining factors are $(x - 1)$ and $(x - 2)$.

(c) $f(x) = (x - 1)(x - 2)(x - 5)(x + 4)$

(d) Zeros: $1, 2, 5, -4$

(e)

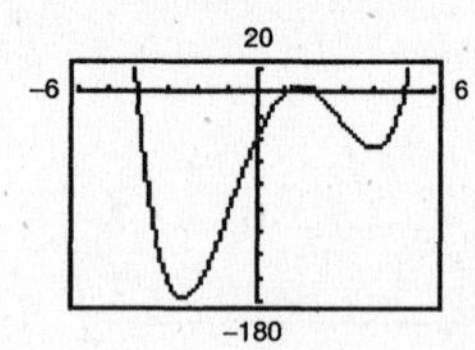

62. $f(x) = 8x^4 - 14x^3 - 71x^2 - 10x + 24$;
Factors: $(x + 2), (x - 4)$

(a) $\begin{array}{r|rrrrr} -2 & 8 & -14 & -71 & -10 & 24 \\ & & -16 & 60 & 22 & -24 \\ \hline & 8 & -30 & -11 & 12 & 0 \end{array}$

$\begin{array}{r|rrrr} 4 & 8 & -30 & -11 & 12 \\ & & 32 & 8 & -12 \\ \hline & 8 & 2 & -3 & 0 \end{array}$

(b) $8x^2 + 2x - 3 = (4x + 3)(2x - 1)$

The remaining factors are $(4x + 3)$ and $(2x - 1)$.

(c) $f(x) = (4x + 3)(2x - 1)(x + 2)(x - 4)$

(d) Zeros: $-\frac{3}{4}, \frac{1}{2}, -2, 4$

(e)

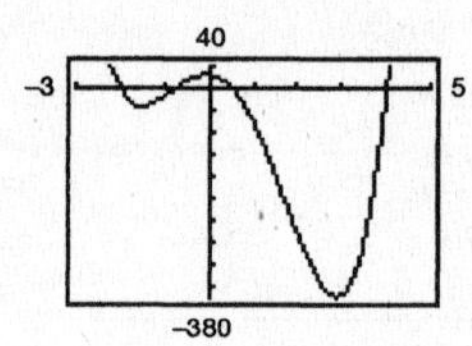

63. $f(x) = 6x^3 + 41x^2 - 9x - 14$; Factors: $(2x + 1), (3x - 2)$

(a) $\begin{array}{r|rrrr} -\frac{1}{2} & 6 & 41 & -9 & -14 \\ & & -3 & -19 & 14 \\ \hline & 6 & 38 & -28 & 0 \end{array}$

$\begin{array}{r|rrr} \frac{2}{3} & 6 & 38 & -28 \\ & & 4 & 28 \\ \hline & 6 & 42 & 0 \end{array}$

Both are factors since the remainders are zero.

(b) $6x + 42 = 6(x + 7)$

This shows that $\dfrac{f(x)}{\left(x + \frac{1}{2}\right)\left(x - \frac{2}{3}\right)} = 6(x + 7)$, so $\dfrac{f(x)}{(2x + 1)(3x - 2)} = x + 7$.

The remaining factor is $(x + 7)$.

(c) $f(x) = (x + 7)(2x + 1)(3x - 2)$

(d) Zeros: $-7, -\frac{1}{2}, \frac{2}{3}$

(e)

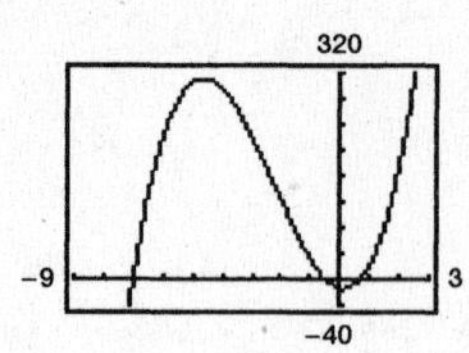

64. $f(x) = 10x^3 - 11x^2 - 72x + 45$;
Factors: $(2x + 5)$, $(5x - 3)$

(a)
$$\begin{array}{r|rrrr} -\frac{5}{2} & 10 & -11 & -72 & 45 \\ & & -25 & 90 & -45 \\ \hline & 10 & -36 & 18 & 0 \end{array}$$

$$\begin{array}{r|rrr} \frac{3}{5} & 10 & -36 & 18 \\ & & 6 & -18 \\ \hline & 10 & -30 & 0 \end{array}$$

(b) $10x - 30 = 10(x - 3)$

This shows that $\dfrac{f(x)}{\left(x + \frac{5}{2}\right)\left(x - \frac{3}{5}\right)} = 10(x - 3)$,

so $\dfrac{f(x)}{(2x + 5)(5x - 3)} = x - 3$.

The remaining factor is $(x - 3)$.

(c) $f(x) = (x - 3)(2x + 5)(5x - 3)$

(d) Zeros: $3, -\dfrac{5}{2}, \dfrac{3}{5}$

(e)

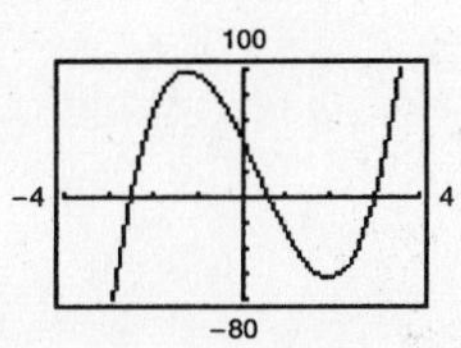

65. $f(x) = 2x^3 - x^2 - 10x + 5$;
Factors: $(2x - 1)$, $\left(x + \sqrt{5}\right)$

(a)
$$\begin{array}{r|rrrr} \frac{1}{2} & 2 & -1 & -10 & 5 \\ & & 1 & 0 & -5 \\ \hline & 2 & 0 & -10 & 0 \end{array}$$

$$\begin{array}{r|rrr} -\sqrt{5} & 2 & 0 & -10 \\ & & -2\sqrt{5} & 10 \\ \hline & 2 & -2\sqrt{5} & 0 \end{array}$$

Both are factors since the remainders are zero.

(b) $2x - 2\sqrt{5} = 2\left(x - \sqrt{5}\right)$

This shows that $\dfrac{f(x)}{\left(x - \frac{1}{2}\right)\left(x + \sqrt{5}\right)} = 2\left(x - \sqrt{5}\right)$,

so $\dfrac{f(x)}{(2x - 1)\left(x + \sqrt{5}\right)} = x - \sqrt{5}$.

The remaining factor is $\left(x - \sqrt{5}\right)$.

(c) $f(x) = \left(x + \sqrt{5}\right)\left(x - \sqrt{5}\right)(2x - 1)$

(d) Zeros: $-\sqrt{5}, \sqrt{5}, \frac{1}{2}$

(e)

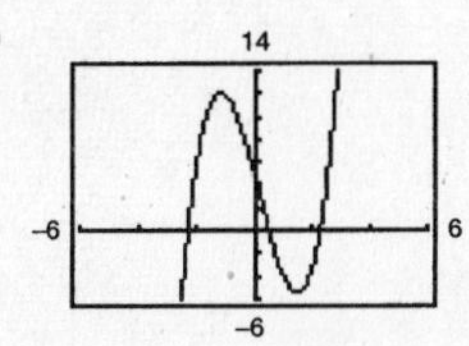

66. $f(x) = x^3 + 3x^2 - 48x - 144$;
Factors: $\left(x + 4\sqrt{3}\right)$, $(x + 3)$

(a)
$$\begin{array}{r|rrrr} -3 & 1 & 3 & -48 & -144 \\ & & -3 & 0 & 144 \\ \hline & 1 & 0 & -48 & 0 \end{array}$$

$$\begin{array}{r|rrr} -4\sqrt{3} & 1 & 0 & -48 \\ & & -4\sqrt{3} & 48 \\ \hline & 1 & -4\sqrt{3} & 0 \end{array}$$

(b) The remaining factor is $\left(x - 4\sqrt{3}\right)$.

(c) $f(x) = \left(x - 4\sqrt{3}\right)\left(x + 4\sqrt{3}\right)(x + 3)$

(d) Zeros: $\pm 4\sqrt{3}, -3$

(e)

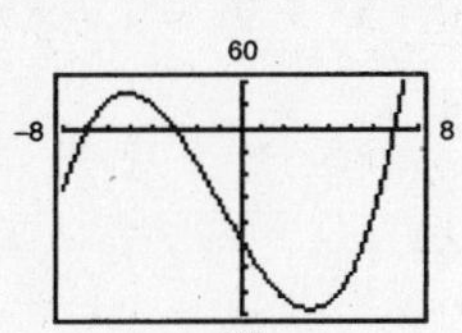

67. $f(x) = x^3 - 2x^2 - 5x + 10$

(a) The zeros of f are 2 and $\approx \pm 2.236$.

(b)
$$\begin{array}{r|rrrr} 2 & 1 & -2 & -5 & 10 \\ & & 2 & 0 & -10 \\ \hline & 1 & 0 & -5 & 0 \end{array}$$

(c) $f(x) = (x - 2)(x^2 - 5)$
$= (x - 2)\left(x - \sqrt{5}\right)\left(x + \sqrt{5}\right)$

68. $g(x) = x^3 - 4x^2 - 2x + 8$

(a) The zeros of g are $x = 4$, $x \approx -1.414$, $x \approx 1.414$.

(b)
$$\begin{array}{r|rrrr} 4 & 1 & -4 & -2 & 8 \\ & & 4 & 0 & -8 \\ \hline & 1 & 0 & -2 & 0 \end{array}$$

(c) $f(x) = (x - 4)(x^2 - 2)$
$= (x - 4)\left(x - \sqrt{2}\right)\left(x + \sqrt{2}\right)$

69. $h(t) = t^3 - 2t^2 - 7t + 2$

(a) The zeros of h are -2, ≈ 3.732, ≈ 0.268.

(b)
$$\begin{array}{r|rrrr} -2 & 1 & -2 & -7 & 2 \\ & & -2 & 8 & -2 \\ \hline & 1 & -4 & 1 & 0 \end{array}$$

$h(t) = (t + 2)(t^2 - 4t + 1)$

By the Quadratic Formula, the zeros of $t^2 - 4t + 1$ are $2 \pm \sqrt{3}$.

(c) Thus, $h(t) = (t + 2)[t - (2 + \sqrt{3})][t - (2 - \sqrt{3})]$.

70. $f(s) = s^3 - 12s^2 + 40s - 24$

(a) The zeros of f are $s = 6$, $s \approx 0.764$, $s \approx 5.236$.

(b)
$$\begin{array}{r|rrrr} 6 & 1 & -12 & 40 & -24 \\ & & 6 & -36 & 24 \\ \hline & 1 & -6 & 4 & 0 \end{array}$$

(c) $f(s) = (s - 6)(s^2 - 6s + 4)$
$= (s - 6)[s - (3 + \sqrt{5})][s - (3 - \sqrt{5})]$

71.
$$\begin{array}{r|rrrr} \frac{3}{2} & 4 & -8 & 1 & 3 \\ & & 6 & -3 & -3 \\ \hline & 4 & -2 & -2 & 0 \end{array}$$

$$\frac{4x^3 - 8x^2 + x + 3}{x - \frac{3}{2}} = 4x^2 - 2x - 2$$

Thus, $\dfrac{4x^3 - 8x^2 + x + 3}{2x - 3} = 2x^2 - x - 1, x \neq \dfrac{3}{2}$.

72. $\dfrac{x^3 + x^2 - 64x - 64}{x + 8}$

$$\begin{array}{r|rrrr} -8 & 1 & 1 & -64 & -64 \\ & & -8 & 56 & 64 \\ \hline & 1 & -7 & -8 & 0 \end{array}$$

$$\frac{x^3 + x^2 - 64x - 64}{x + 8} = x^2 - 7x - 8, \quad x \neq -8$$

73.
$$\begin{array}{r|rrrr} -1 & 1 & 3 & -1 & -3 \\ & & -1 & -2 & 3 \\ \hline & 1 & 2 & -3 & 0 \end{array}$$

$$\frac{x^3 + 3x^2 - x - 3}{x + 1} = x^2 + 2x - 3, x \neq -1$$

74. $\dfrac{2x^3 + 3x^2 - 3x - 2}{x - 1}$

$$\begin{array}{r|rrrr} 1 & 2 & 3 & -3 & -2 \\ & & 2 & 5 & 2 \\ \hline & 2 & 5 & 2 & 0 \end{array}$$

$$\frac{2x^3 + 3x^2 - 3x - 2}{x - 1} = 2x^2 + 5x + 2, \quad x \neq 1$$

75. Note that $x^2 + 3x + 2 = (x + 1)(x + 2)$.

$$\begin{array}{r|rrrrr} -1 & 1 & 6 & 11 & 6 & 0 \\ & & -1 & -5 & -6 & 0 \\ \hline & 1 & 5 & 6 & 0 & 0 \end{array}$$

$$\begin{array}{r|rrrr} -2 & 1 & 5 & 6 & 0 \\ & & -2 & -6 & 0 \\ \hline & 1 & 3 & 0 & 0 \end{array}$$

$$\frac{x^4 + 6x^3 + 11x^2 + 6x}{(x + 1)(x + 2)} = x^2 + 3x, x \neq -2, -1$$

76. $\dfrac{x^4 + 9x^3 - 5x^2 - 36x + 4}{(x + 2)(x - 2)}$

$$\begin{array}{r|rrrrr} 2 & 1 & 9 & -5 & -36 & 4 \\ & & 2 & 22 & 34 & -4 \\ \hline & 1 & 11 & 17 & -2 & 0 \end{array}$$

$$\begin{array}{r|rrrr} -2 & 1 & 11 & 17 & -2 \\ & & -2 & -18 & 2 \\ \hline & 1 & 9 & -1 & 0 \end{array}$$

$$\frac{x^4 + 9x^3 - 5x^2 - 36x + 4}{x^2 - 4} = x^2 + 9x - 1, \quad x \neq \pm 2$$

77.
$$\begin{array}{r l} & x^{2n} + 6x^n + 9 \\ x^n + 3 \,\big) & x^{3n} + 9x^{2n} + 27x^n + 27 \\ & \underline{x^{3n} + 3x^{2n}} \\ & \quad 6x^{2n} + 27x^n \\ & \quad \underline{6x^{2n} + 18x^n} \\ & \qquad 9x^n + 27 \\ & \qquad \underline{9x^n + 27} \\ & \qquad\qquad 0 \end{array}$$

$$\frac{x^{3n} + 9x^{2n} + 27x^n + 27}{x^n + 3} = x^{2n} + 6x^n + 9$$

78.

$$\begin{array}{r}
x^{2n} - x^n + 3 \\
x^n - 2 \overline{)\, x^{3n} - 3x^{2n} + 5x^n - 6} \\
\underline{x^{3n} - 2x^{2n} } \\
-x^{2n} + 5x^n \\
\underline{-x^{2n} + 2x^n } \\
3x^n - 6 \\
\underline{3x^n - 6} \\
0
\end{array}$$

$$\frac{x^{3n} - 3x^{2n} + 5x^n - 6}{x^n - 2} = x^{2n} - x^n + 3$$

79. A divisor divides evenly into a dividend if the remainder is zero.

80. You can check polynomial division by multiplying the quotient by the divisor. This should yield the original dividend if the multiplication was performed correctly.

81.
$$\begin{array}{r|rrrr}
5 & 1 & 4 & -3 & c \\
 & & 5 & 45 & 210 \\
\hline
 & 1 & 9 & 42 & c + 210
\end{array}$$

For $c + 210$ to equal zero, c must equal -210.

82.
$$\begin{array}{r|rrrrrr}
-2 & 1 & 0 & 0 & -2 & 1 & c \\
 & & -2 & 4 & -8 & 20 & -42 \\
\hline
 & 1 & -2 & 4 & -10 & 21 & c - 42
\end{array}$$

To divide evenly, $c - 42$ must equal zero.
Thus, c must equal 42.

83. $f(x) = (x + 3)^2(x - 3)(x + 1)^3$

The remainder when $k = -3$ is zero since $(x + 3)$ is a factor of $f(x)$.

84. In this case it is easier to evaluate $f(2)$ directly because $f(x)$ is in factored form. To evaluate using synthetic division you would have to expand each factor and then multiply it all out.

85. (a) and (b)

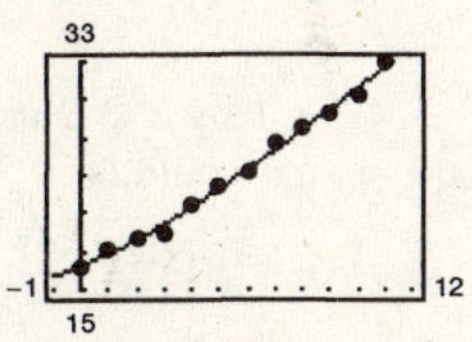

$y \approx -0.00320t^3 + 0.0856t^2 + 0.865t + 16.87$, where $t = 0$ corresponds to 1990.

(c)

t	0	1	2	3	4	5
R	16.87	17.82	18.92	20.15	21.49	22.94

t	6	7	8	9	10	11
R	24.45	26.02	27.63	29.26	30.88	32.48

The model is a good fit to the actual data.

(d)
$$\begin{array}{r|rrrr}
15 & -0.00320 & 0.0856 & 0.865 & 16.87 \\
 & & -0.048 & 0.564 & 21.435 \\
\hline
 & -0.00320 & 0.0376 & 1.429 & 38.305
\end{array}$$

Thus, $y(15) \approx \$38.31$.

No, because the model will eventually climb too fast.

86. (a) and (b)

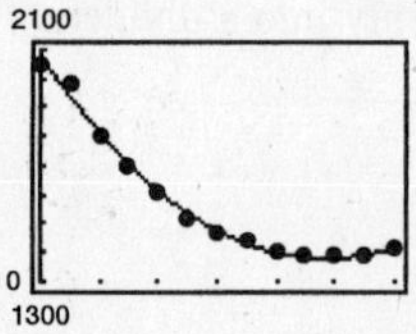

$y \approx -0.03001t^3 + 7.9653t^2 - 146.092t + 2073.64$ where $t = 0$ corresponds to 1990.

The model is a good fit to the actual data.

(c)

t	0	1	2	3	4	5	6
R	2074	1935	1813	1706	1615	1539	1477

t	7	8	9	10	11	12
R	1431	1399	1382	1379	1390	1416

(d)
$$\begin{array}{r|rrrr} 15 & -0.03001 & 7.9653 & -146.092 & 2073.64 \\ & & -0.45015 & 112.72725 & -500.47125 \\ \hline & -0.03001 & 7.51515 & -33.36475 & 1573.16875 \end{array}$$

$y(15) \approx 1573$ thousand

No, the model will eventually climb too fast.

87. False. If $(7x + 4)$ is a factor of f, then $-\frac{4}{7}$ is a zero of f.

88. True.

$$\begin{array}{r|rrrrrrr} \frac{1}{2} & 6 & 1 & -92 & 45 & 184 & 4 & -48 \\ & & 3 & 2 & -45 & 0 & 92 & 48 \\ \hline & 6 & 4 & -90 & 0 & 184 & 96 & 0 \end{array}$$

$f(x) = (2x - 1)(x + 1)(x - 2)(x - 3)(3x + 2)(x + 4)$

89. True. If $x = k$ is a zero of $f(x)$, then $(x - k)$ is a factor of $f(x)$, and $f(k) = 0$.

90. False.

To divide $x^4 - 3x^2 + 4x - 1$ by $x + 2$ using synthetic division, the set up would be:

$$\begin{array}{r|rrrrr} -2 & 1 & 0 & -3 & 4 & -1 \\ & & & & & \\ \hline \end{array}$$

A zero must be included for the missing x^3 term.

Section 2.4 Complex Numbers

1. $a + bi = -10 + 6i$

$a = -10$

$b = 6$

2. $a + bi = 13 + 4i$

$a = 13$

$b = 4$

3. $(a - 1) + (b + 3)i = 5 + 8i$

$a - 1 = 5 \Rightarrow a = 6$

$b + 3 = 8 \Rightarrow b = 5$

4. $(a + 6) + 2bi = 6 - 5i$

$2b = -5$

$b = -\frac{5}{2}$

$a + 6 = 6$

$a = 0$

5. $4 + \sqrt{-9} = 4 + 3i$

6. $3 + \sqrt{-16} = 3 + 4i$

7. $2 - \sqrt{-27} = 2 - \sqrt{27}i = 2 - 3\sqrt{3}i$

8. $1 + \sqrt{-8} = 1 + 2\sqrt{2}i$

9. $\sqrt{-75} = \sqrt{75}i = 5\sqrt{3}i$

10. $\sqrt{-4} = 2i$

11. $8 = 8 + 0i = 8$

12. 45

13. $-6i + i^2 = -6i - 1 = -1 - 6i$

14. $-4i^2 + 2i = -4(-1) + 2i$
$= 4 + 2i$

15. $\sqrt{-0.09} = \sqrt{0.09}i = 0.3i$

16. $\sqrt{-0.0004} = 0.02i$

17. $(5 + i) + (6 - 2i) = 11 - i$

18. $(13 - 2i) + (-5 + 6i) = 8 + 4i$

19. $(8 - i) - (4 - i) = 8 - i - 4 + i = 4$

20. $(3 + 2i) - (6 + 13i) = 3 + 2i - 6 - 13i = -3 - 11i$

21. $(-2 + \sqrt{-8}) + (5 - \sqrt{-50}) = -2 + 2\sqrt{2}i + 5 - 5\sqrt{2}i = 3 - 3\sqrt{2}i$

22. $(8 + \sqrt{-18}) - (4 + 3\sqrt{2}i) = 8 + 3\sqrt{2}i - 4 - 3\sqrt{2}i = 4$

23. $13i - (14 - 7i) = 13i - 14 + 7i = -14 + 20i$

24. $22 + (-5 + 8i) + 10i = 17 + 18i$

25. $-\left(\frac{3}{2} + \frac{5}{2}i\right) + \left(\frac{5}{3} + \frac{11}{3}i\right) = -\frac{3}{2} - \frac{5}{2}i + \frac{5}{3} + \frac{11}{3}i$
$= -\frac{9}{6} - \frac{15}{6}i + \frac{10}{6} + \frac{22}{6}i$
$= \frac{1}{6} + \frac{7}{6}i$

26. $(1.6 + 3.2i) + (-5.8 + 4.3i) = -4.2 + 7.5i$

27. $\sqrt{-6} \cdot \sqrt{-2} = (\sqrt{6}i)(\sqrt{2}i) = \sqrt{12}i^2 = (2\sqrt{3})(-1)$
$= -2\sqrt{3}$

28. $\sqrt{-5} \cdot \sqrt{-10} = (\sqrt{5}i)(\sqrt{10}i)$
$= \sqrt{50}i^2 = 5\sqrt{2}(-1) = -5\sqrt{2}$

29. $(\sqrt{-10})^2 = (\sqrt{10}i)^2 = 10i^2 = -10$

30. $(\sqrt{-75})^2 = (\sqrt{75}i)^2 = 75i^2 = -75$

31. $(1 + i)(3 - 2i) = 3 - 2i + 3i - 2i^2 = 3 + i + 2 = 5 + i$

32. $(6 - 2i)(2 - 3i) = 12 - 18i - 4i + 6i^2$
$= 12 - 22i - 6 = 6 - 22i$

33. $6i(5 - 2i) = 30i - 12i^2 = 30i + 12 = 12 + 30i$

34. $-8i(9 + 4i) = -72i - 32i^2$
$= 32 - 72i$

35. $(\sqrt{14} + \sqrt{10}i)(\sqrt{14} - \sqrt{10}i) = 14 - 10i^2$
$= 14 + 10 = 24$

36. $(3 + \sqrt{-5})(7 - \sqrt{-10}) = (3 + \sqrt{5}i)(7 - \sqrt{10}i)$
$= 21 - 3\sqrt{10}i + 7\sqrt{5}i - \sqrt{50}i^2$
$= 21 + \sqrt{50} + 7\sqrt{5}i - 3\sqrt{10}i$
$= (21 + 5\sqrt{2}) + (7\sqrt{5} - 3\sqrt{10})i$

37. $(4 + 5i)^2 = 16 + 40i + 25i^2$
$= 16 + 40i - 25$
$= -9 + 40i$

38. $(2 - 3i)^2 = 4 - 12i + 9i^2$
$= 4 - 9 - 12i$
$= -5 - 12i$

39. $(2 + 3i)^2 + (2 - 3i)^2 = 4 + 12i + 9i^2 + 4 - 12i + 9i^2$
$= 4 + 12i - 9 + 4 - 12i - 9$
$= -10$

40. $(1 - 2i)^2 - (1 + 2i)^2 = 1 - 4i + 4i^2 - (1 + 4i + 4i^2)$
$= 1 - 4i + 4i^2 - 1 - 4i - 4i^2$
$= -8i$

41. The complex conjugate of $6 + 3i$ is $6 - 3i$.
$(6 + 3i)(6 - 3i) = 36 - (3i)^2 = 36 + 9 = 45$

42. The complex conjugate of $7 - 12i$ is $7 + 12i$.
$(7 - 12i)(7 + 12i) = 49 - 144i^2$
$= 49 - (-144)$
$= 193$

43. The complex conjugate of $-1 - \sqrt{5}i$ is $-1 + \sqrt{5}i$.
$(-1 - \sqrt{5}i)(-1 + \sqrt{5}i) = (-1)^2 - (\sqrt{5}i)^2$
$= 1 + 5 = 6$

44. The complex conjugate of $-3 + \sqrt{2}i$ is $-3 - \sqrt{2}i$.
$(-3 + \sqrt{2}i)(-3 - \sqrt{2}i) = 9 - 2i^2$
$= 9 - (-2)$
$= 11$

45. The complex conjugate of $\sqrt{-20} = 2\sqrt{5}i$ is $-2\sqrt{5}i$.
$(2\sqrt{5}i)(-2\sqrt{5}i) = -20i^2 = 20$

46. The complex conjugate of $\sqrt{-15} = \sqrt{15}i$ is $-\sqrt{15}i$.
$(\sqrt{15}i)(-\sqrt{15}i) = -15i^2 = -(-15) = 15$

47. The complex conjugate of $\sqrt{8}$ is $\sqrt{8}$.
$(\sqrt{8})(\sqrt{8}) = 8$

48. The complex conjugate of $1 + \sqrt{8}$ is $1 + \sqrt{8}$.
$(1 + \sqrt{8})(1 + \sqrt{8}) = 1 + 2\sqrt{8} + 8$
$= 9 + 4\sqrt{2}$

49. $\dfrac{5}{i} = \dfrac{5}{i} \cdot \dfrac{-i}{-i} = \dfrac{-5i}{1} = -5i$

50. $-\dfrac{14}{2i} \cdot \dfrac{-2i}{-2i} = \dfrac{28i}{-4i^2} = \dfrac{28i}{4} = 7i$

51. $\dfrac{2}{4 - 5i} = \dfrac{2}{4 - 5i} \cdot \dfrac{4 + 5i}{4 + 5i} = \dfrac{2(4 + 5i)}{16 + 25} = \dfrac{8 + 10i}{41} = \dfrac{8}{41} + \dfrac{10}{41}i$

52. $\dfrac{5}{1 - i} \cdot \dfrac{1 + i}{1 + i} = \dfrac{5 + 5i}{1 - i^2} = \dfrac{5 + 5i}{2} = \dfrac{5}{2} + \dfrac{5}{2}i$

53. $\dfrac{3 + i}{3 - i} = \dfrac{3 + i}{3 - i} \cdot \dfrac{3 + i}{3 + i}$
$= \dfrac{9 + 6i + i^2}{9 + 1}$
$= \dfrac{8 + 6i}{10}$
$= \dfrac{4}{5} + \dfrac{3}{5}i$

54. $\dfrac{6 - 7i}{1 - 2i} \cdot \dfrac{1 + 2i}{1 + 2i} = \dfrac{6 + 12i - 7i - 14i^2}{1 - 4i^2}$
$= \dfrac{20 + 5i}{5} = \dfrac{20}{5} + \dfrac{5}{5}i = 4 + i$

55. $\dfrac{6-5i}{i} = \dfrac{6-5i}{i} \cdot \dfrac{-i}{-i}$

$= \dfrac{-6i+5i^2}{1}$

$= -5-6i$

56. $\dfrac{8+16i}{2i} \cdot \dfrac{-2i}{-2i} = \dfrac{-16i-32i^2}{-4i^2} = 8-4i$

57. $\dfrac{3i}{(4-5i)^2} = \dfrac{3i}{16-40i+25i^2} = \dfrac{3i}{-9-40i} \cdot \dfrac{-9+40i}{-9+40i}$

$= \dfrac{-27i+120i^2}{81+1600} = \dfrac{-120-27i}{1681}$

$= -\dfrac{120}{1681} - \dfrac{27}{1681}i$

58. $\dfrac{5i}{(2+3i)^2} = \dfrac{5i}{4+12i+9i^2} = \dfrac{5i}{-5+12i} \cdot \dfrac{-5-12i}{-5-12i}$

$= \dfrac{-25i-60i^2}{25-144i^2} = \dfrac{60-25i}{169} = \dfrac{60}{169} - \dfrac{25}{169}i$

59. $\dfrac{2}{1+i} - \dfrac{3}{1-i} = \dfrac{2(1-i)-3(1+i)}{(1+i)(1-i)}$

$= \dfrac{2-2i-3-3i}{1+1}$

$= \dfrac{-1-5i}{2}$

$= -\dfrac{1}{2} - \dfrac{5}{2}i$

60. $\dfrac{2i}{2+i} + \dfrac{5}{2-i} = \dfrac{2i(2-i)}{(2+i)(2-i)} + \dfrac{5(2+i)}{(2+i)(2-i)}$

$= \dfrac{4i-2i^2+10+5i}{4-i^2}$

$= \dfrac{12+9i}{5}$

$= \dfrac{12}{5} + \dfrac{9}{5}i$

61. $\dfrac{i}{3-2i} + \dfrac{2i}{3+8i} = \dfrac{i(3+8i)+2i(3-2i)}{(3-2i)(3+8i)}$

$= \dfrac{3i+8i^2+6i-4i^2}{9+24i-6i-16i^2}$

$= \dfrac{4i^2+9i}{9+18i+16}$

$= \dfrac{-4+9i}{25+18i} \cdot \dfrac{25-18i}{25-18i}$

$= \dfrac{-100+72i+225i-162i^2}{625+324}$

$= \dfrac{-100+297i+162}{949}$

$= \dfrac{62+297i}{949}$

$= \dfrac{62}{949} + \dfrac{297}{949}i$

62. $\dfrac{1+i}{i} - \dfrac{3}{4-i} = \dfrac{(1+i)(4-i)-3i}{i(4-i)}$

$= \dfrac{4-i+4i-i^2-3i}{4i-i^2}$

$= \dfrac{5}{1+4i} \cdot \dfrac{1-4i}{1-4i}$

$= \dfrac{5-20i}{1-16i^2}$

$= \dfrac{5}{17} - \dfrac{20}{17}i$

63. $x^2-2x+2=0;\ a=1,\ b=-2,\ c=2$

$x = \dfrac{-(-2) \pm \sqrt{(-2)^2-4(1)(2)}}{2(1)}$

$= \dfrac{2 \pm \sqrt{-4}}{2}$

$= \dfrac{2 \pm 2i}{2}$

$= 1 \pm i$

64. $x^2+6x+10=0;\quad a=1, b=6, c=10$

$x = \dfrac{-6 \pm \sqrt{6^2-4(1)(10)}}{2(1)}$

$= \dfrac{-6 \pm \sqrt{-4}}{2}$

$= -3 \pm i$

65. $4x^2 + 16x + 17 = 0;\ a = 4,\ b = 16,\ c = 17$

$$x = \frac{-16 \pm \sqrt{(16)^2 - 4(4)(17)}}{2(4)} = \frac{-16 \pm \sqrt{-16}}{8} = \frac{-16 \pm 4i}{8} = -2 \pm \frac{1}{2}i$$

66. $9x^2 - 6x + 37 = 0; a = 9, b = -6, c = 37$

$$x = \frac{-(-6) \pm \sqrt{(-6)^2 - 4(9)(37)}}{2(9)}$$

$$= \frac{6 \pm \sqrt{-1296}}{18}$$

$$= \frac{1}{3} \pm \frac{36i}{18}$$

$$= \frac{1}{3} \pm 2i$$

67. $4x^2 + 16x + 15 = 0;\ a = 4,\ b = 16,\ c = 15$

$$x = \frac{-16 \pm \sqrt{(16)^2 - 4(4)(15)}}{2(4)}$$

$$= \frac{-16 \pm \sqrt{16}}{8}$$

$$= \frac{-16 \pm 4}{8}$$

$$x = -\frac{12}{8} = -\frac{3}{2} \text{ or } x = -\frac{20}{8} = -\frac{5}{2}$$

68. $16t^2 - 4t + 3 = 0;\ a = 16,\ b = -4,\ c = 3$

$$t = \frac{-(-4) \pm \sqrt{(-4)^2 - 4(16)(3)}}{2(16)}$$

$$= \frac{4 \pm \sqrt{-176}}{32}$$

$$= \frac{4 \pm 4\sqrt{11}i}{32}$$

$$= \frac{1}{8} \pm \frac{\sqrt{11}}{8}i$$

69. $\frac{3}{2}x^2 - 6x + 9 = 0$ Multiply both sides by 2.

$$3x^2 - 12x + 18 = 0$$

$$x = \frac{-(-12) \pm \sqrt{(-12)^2 - 4(3)(18)}}{2(3)}$$

$$= \frac{12 \pm \sqrt{-72}}{6}$$

$$= \frac{12 \pm 6\sqrt{2}i}{6}$$

$$= 2 \pm \sqrt{2}i$$

70. $\frac{7}{8}x^2 - \frac{3}{4}x + \frac{5}{16} = 0$

$14x^2 - 12x + 5 = 0; a = 14, b = -12, c = 5$

$$x = \frac{-(-12) \pm \sqrt{(-12)^2 - 4(14)(5)}}{2(14)}$$

$$= \frac{12 \pm \sqrt{-136}}{28}$$

$$= \frac{12 \pm 2i\sqrt{34}}{28}$$

$$= \frac{3}{7} \pm \frac{\sqrt{34}}{14}i$$

71. $1.4x^2 - 2x - 10 = 0$ Multiply both sides by 5

$$7x^2 - 10x - 50 = 0$$

$$x = \frac{-(-10) \pm \sqrt{(-10)^2 - 4(7)(-50)}}{2(7)}$$

$$= \frac{10 \pm \sqrt{1500}}{14}$$

$$= \frac{10 \pm 10\sqrt{15}}{14}$$

$$= \frac{5 \pm 5\sqrt{15}}{7}$$

$$= \frac{5}{7} \pm \frac{5\sqrt{15}}{7}$$

72. $4.5x^2 - 3x + 12 = 0; a = 4.5, b = -3, c = 12$

$$x = \frac{-(-3) \pm \sqrt{(-3)^2 - 4(4.5)(12)}}{2(4.5)}$$

$$= \frac{3 \pm \sqrt{-207}}{9} = \frac{3 \pm 3i\sqrt{23}}{9} = \frac{1}{3} \pm \frac{\sqrt{23}}{3}i$$

73. $i^{40} = i^4 \cdot i^4 \cdot i^4 \cdot i^4 \cdot i^4 \cdot i^4 \cdot i^4 \cdot i^4 \cdot i^4 \cdot i^4$

$$= 1 \cdot 1 \cdot 1 \cdot 1 \cdot 1 \cdot 1 \cdot 1 \cdot 1 \cdot 1 \cdot 1$$

$$= 1$$

74. $i^{25} = i^4 \cdot i^4 \cdot i^4 \cdot i^4 \cdot i^4 \cdot i^4 \cdot i$

$= 1 \cdot 1 \cdot 1 \cdot 1 \cdot 1 \cdot 1 \cdot i$

$= i$

75. $i^{50} = i^{25} \cdot i^{25} = i \cdot i = i^2 = -1$

76. $i^{67} = i^{50} \cdot i^{17} = -1 \cdot i^4 \cdot i^4 \cdot i^4 \cdot i^4 \cdot i = -i$

77. $4i^2 - 2i^3 = -4 + 2i$

78. $-6i^3 + i^2 = -6i^2 i + i^2$

$= -6(-1)i + (-1)$

$= 6i - 1$

$= -1 + 6i$

79. $(-i)^3 = (-1)(i^3) = (-1)(-i) = i$

80. $-5i^5 = -5i^2 i^2 i$

$= -5(-1)(-1)i$

$= -5i$

81. $\left(\sqrt{-2}\right)^6 = \left(\sqrt{2}\,i\right)^6 = 8i^6 = 8i^4 i^2 = -8$

82. $\left(\sqrt{-75}\right)^3 = \left(5\sqrt{3}\,i\right)^3$

$= 5^3\left(\sqrt{3}\right)^3 i^3$

$= 125\left(3\sqrt{3}\right)(-1)i$

$= -375\sqrt{3}\,i$

83. $\dfrac{1}{(2i)^3} = \dfrac{1}{8i^3} = \dfrac{1}{-8i} \cdot \dfrac{8i}{8i} = \dfrac{8i}{-64i^2} = \dfrac{1}{8}i$

84. $\dfrac{1}{i^3} = \dfrac{1}{-i} = \dfrac{1}{-i} \cdot \dfrac{i}{i} = \dfrac{i}{-i^2} = \dfrac{i}{1} = i$

85. False, if $b = 0$ then $a + bi = a - bi = a$.
That is, if the complex number is real, the number equals its conjugate.

86. True.

$$x^4 - x^2 + 14 = 56$$

$$\left(-i\sqrt{6}\right)^4 - \left(-i\sqrt{6}\right)^2 + 14 \stackrel{?}{=} 56$$

$$36 + 6 + 14 \stackrel{?}{=} 56$$

$$56 = 56$$

87. False.

$i^{44} + i^{150} - i^{74} - i^{109} + i^{61} = (i^4)^{11} + (i^4)^{37}(i^2) - (i^4)^{18}(i^2) - (i^4)^{27}(i) + (i^4)^{15}(i)$

$= (1)^{11} + (1)^{37}(-1) - (1)^{18}(-1) - (1)^{27}(i) + (1)^{15}(i)$

$= 1 + (-1) + 1 - i + i = 1$

88. False.

$\sqrt{-2} \cdot \sqrt{-2} = \left(\sqrt{2}\,i\right)\left(\sqrt{2}\,i\right) = \sqrt{4}\,i^2 = 2(-1) = -2$

89. $(a + bi)(a - bi) = a^2 + abi - abi - b^2 i^2$

$= a^2 - b^2(-1)$

$= a^2 + b^2$

which is a real number since a and b are real numbers. Thus, the product of a complex number and its conjugate is a real number.

90. $\sqrt{-6}\sqrt{-6} = \sqrt{6}\,i\sqrt{6}\,i = 6i^2 = -6$

91. Answers will vary.

92. Answers will vary.

93. (a) $z_1 = 5 + 2i, z_2 = 3 - 4i$

$$\frac{1}{z} = \frac{1}{z_1} + \frac{1}{z_2}$$

$$= \frac{1}{5 + 2i} + \frac{1}{3 - 4i}$$

$$= \frac{3 - 4i + 5 + 2i}{(5 + 2i)(3 - 4i)}$$

$$= \frac{8 - 2i}{23 - 14i}$$

$$z = \left(\frac{23 - 14i}{8 - 2i}\right)\left(\frac{8 + 2i}{8 + 2i}\right)$$

$$= \frac{212 - 66i}{68} = \frac{53}{17} - \frac{33}{34}i$$

(b) $z_1 = 9 + 16i, z_2 = 20 - 10i$

$$\frac{1}{z} = \frac{1}{z_1} + \frac{1}{z_2}$$

$$= \frac{1}{9 + 16i} + \frac{1}{20 - 10i}$$

$$= \frac{20 - 10i + 9 + 16i}{(9 + 16i)(20 - 10i)}$$

$$= \frac{29 + 6i}{340 + 230i}$$

$$z = \left(\frac{340 + 230i}{29 + 6i}\right)\left(\frac{29 - 6i}{29 - 6i}\right)$$

$$= \frac{11{,}240 + 4630i}{877} = \frac{11{,}240}{877} + \frac{4630}{877}i$$

Section 2.5 The Fundamental Theorem of Algebra

1. $f(x) = x(x - 6)^2 = x(x - 6)(x - 6)$

The three zeros are: $x = 0, x = 6, x = 6$.

2. $f(x) = x^2(x + 3)(x^2 - 1) = x^2(x + 3)(x + 1)(x - 1)$

The five zeros are: $0, 0, -3, \pm 1$.

3. $g(x) = (x - 2)(x + 4)^3 = (x - 2)(x + 4)(x + 4)(x + 4)$

The four zeros are: $x = 2, x = -4, x = -4, x = -4$.

4. $f(x) = (x + 5)(x - 8)^2$

The three zeros are: $-5, 8, 8$.

5. $f(x) = (x + 6)(x + i)(x - i)$

The three zeros are: $x = -6, x = -i, x = i$.

6. $h(t) = (t - 3)(t - 2)(t - 3i)(t + 3i)$

The four zeros are: $3, 2, \pm 3i$.

7. $f(x) = x^3 + 3x^2 - x - 3$

Possible rational zeros: $\pm 1, \pm 3$

Zeros shown on graph: $-3, -1, 1$

8. $f(x) = x^3 - 4x^2 - 4x + 16$

$p =$ factor of 16

$q =$ factor of 1

Possible rational zeros: $\pm 1, \pm 2, \pm 4, \pm 8, \pm 16$

Zeros shown on graph: $-2, 2, 4$

9. $f(x) = 2x^4 - 17x^3 + 35x^2 + 9x - 45$

Possible rational zeros: $\pm 1, \pm 3, \pm 5, \pm 9, \pm 15, \pm 45, \pm\frac{1}{2}, \pm\frac{3}{2}, \pm\frac{5}{2}, \pm\frac{9}{2}, \pm\frac{15}{2}, \pm\frac{45}{2}$

Zeros shown on graph: $-1, \frac{3}{2}, 3, 5$

10. $f(x) = 4x^5 - 8x^4 - 5x^3 + 10x^2 + x - 2$

$p =$ factor of -2

$q =$ factor of 4

Possible rational zeros: $\pm 1, \pm 2, \pm\frac{1}{2}, \pm\frac{1}{4}$

Zeros shown on graph: $-1, -\frac{1}{2}, \frac{1}{2}, 1, 2$

11. $f(x) = x^3 - 6x^2 + 11x - 6$

Possible rational zeros: $\pm 1, \pm 2, \pm 3, \pm 6$

$$\begin{array}{r|rrrr} 1 & 1 & -6 & 11 & -6 \\ & & 1 & -5 & 6 \\ \hline & 1 & -5 & 6 & 0 \end{array}$$

$x^3 - 6x^2 + 11x - 6 = (x - 1)(x^2 - 5x + 6) = (x - 1)(x - 2)(x - 3)$

Thus, the real zeros are 1, 2, and 3.

12. $g(x) = x^3 - 4x^2 - x + 4 = x^2(x - 4) - 1(x - 4) = (x - 4)(x^2 - 1)$

$= (x - 4)(x - 1)(x + 1)$

Thus, the zeros of $g(x)$ are 4 and ± 1.

13. $f(x) = x^3 - 7x - 6$

Possible rational zeros: $\pm 1, \pm 2, \pm 3, \pm 6$

$$\begin{array}{r|rrrr} 3 & 1 & 0 & -7 & -6 \\ & & 3 & 9 & 6 \\ \hline & 1 & 3 & 2 & 0 \end{array}$$

$f(x) = (x - 3)(x^2 + 3x + 2) = (x - 3)(x + 2)(x + 1)$

Thus, the real zeros are $-2, -1, 3$.

14. $C(x) = 2x^3 + 3x^2 - 1$

Possible rational zeros: $\pm 1, \pm\frac{1}{2}$

$$\begin{array}{r|rrrr} -1 & 2 & 3 & 0 & -1 \\ & & -2 & -1 & 1 \\ \hline & 2 & 1 & -1 & 0 \end{array}$$

$$\begin{aligned} 2x^3 + 3x^2 - 1 &= (x + 1)(2x^2 + x - 1) \\ &= (x + 1)(x + 1)(2x - 1) \\ &= (x + 1)^2(2x - 1) \end{aligned}$$

Thus, the zeros are -1 and $\frac{1}{2}$.

15. $h(t) = t^3 + 12t^2 + 21t + 10$

Possible rational zeros: $\pm 1, \pm 2, \pm 5, \pm 10$

$$\begin{array}{r|rrrr} -1 & 1 & 12 & 21 & 10 \\ & & -1 & -11 & -10 \\ \hline & 1 & 11 & 10 & 0 \end{array}$$

$$\begin{aligned} t^3 + 12t^2 + 21t + 10 &= (t + 1)(t^2 + 11t + 10) \\ &= (t + 1)(t + 1)(t + 10) \\ &= (t + 1)^2(t + 10) \end{aligned}$$

Thus, the zeros are -1 and -10.

16. $p(x) = x^3 - 9x^2 + 27x - 27$

Possible rational zeros: $\pm 1, \pm 3, \pm 9, \pm 27$

$$\begin{array}{r|rrrr} 3 & 1 & -9 & 27 & -27 \\ & & 3 & -18 & 27 \\ \hline & 1 & -6 & 9 & 0 \end{array}$$

$$\begin{aligned} f(x) &= (x - 3)(x^2 - 6x + 9) \\ &= (x - 3)(x - 3)(x - 3) \end{aligned}$$

Thus, the real zero is 3.

17. $h(x) = x^3 - 9x^2 + 20x - 12$

Possible rational zeros: $\pm 1, \pm 2, \pm 3, \pm 4, \pm 6, \pm 12$

$$\begin{array}{r|rrrr} 1 & 1 & -9 & 20 & -12 \\ & & 1 & -8 & 12 \\ \hline & 1 & -8 & 12 & 0 \end{array}$$

$$\begin{aligned} h(x) &= (x - 1)(x^2 - 8x + 12) \\ &= (x - 1)(x - 2)(x - 6) \end{aligned}$$

Thus, the real zeros are 1, 2, 6.

18. $f(x) = 3x^3 - 19x^2 + 33x - 9$

Possible rational zeros: $\pm 1, \pm 3, \pm 9, \pm\frac{1}{3}$

$$\begin{array}{r|rrrr} 3 & 3 & -19 & 33 & -9 \\ & & 9 & -30 & 9 \\ \hline & 3 & -10 & 3 & 0 \end{array}$$

$$\begin{aligned} f(x) &= (x - 3)(3x^2 - 10x + 3) \\ &= (x - 3)(3x - 1)(x - 3) \end{aligned}$$

Thus, the real zeros are $3, \frac{1}{3}$.

19. $f(x) = 9x^4 - 9x^3 - 58x^2 + 4x + 24$

Possible rational zeros: $\pm 1, \pm 2, \pm 3, \pm 4, \pm 6, \pm 8, \pm 12, \pm 24, \pm\frac{1}{3}, \pm\frac{2}{3}, \pm\frac{4}{3}, \pm\frac{8}{3}, \pm\frac{1}{9}, \pm\frac{2}{9}, \pm\frac{4}{9}, \pm\frac{8}{9}$

$$\begin{array}{r|rrrrr} -2 & 9 & -9 & -58 & 4 & 24 \\ & & -18 & 54 & 8 & -24 \\ \hline & 9 & -27 & -4 & 12 & 0 \end{array}$$

$$\begin{array}{r|rrrr} 3 & 9 & -27 & -4 & 12 \\ & & 27 & 0 & -12 \\ \hline & 9 & 0 & -4 & 0 \end{array}$$

$$\begin{aligned} 9x^4 - 9x^3 - 58x^2 + 4x - 24 &= (x + 2)(x - 3)(9x^2 - 4) \\ &= (x + 2)(x - 3)(3x - 2)(3x + 2) \end{aligned}$$

Thus, the zeros are -2, 3, and $\pm\frac{2}{3}$.

20. $f(x) = 2x^4 - 15x^3 + 23x^2 + 15x - 25$

Possible rational zeros: $\pm 1, \pm 5, \pm 25, \pm\frac{1}{2}, \pm\frac{5}{2}, \pm\frac{25}{2}$

$$\begin{array}{r|rrrrr} 5 & 2 & -15 & 23 & 15 & -25 \\ & & 10 & -25 & -10 & 25 \\ \hline & 2 & -5 & -2 & 5 & 0 \end{array}$$

$$\begin{array}{r|rrrr} 1 & 2 & -5 & -2 & 5 \\ & & 2 & -3 & -5 \\ \hline & 2 & -3 & -5 & 0 \end{array}$$

$$\begin{array}{r|rrr} -1 & 2 & -3 & -5 \\ & & -2 & 5 \\ \hline & 2 & -5 & 0 \end{array}$$

$f(x) = (x - 5)(x - 1)(x + 1)(2x - 5)$

Thus, the real zeros are 5, 1, -1, $\frac{5}{2}$.

21. $z^4 - z^3 - 2z - 4 = 0$

Possible rational zeros: $\pm 1, \pm 2, \pm 4$

$$\begin{array}{r|rrrrr} -1 & 1 & -1 & 0 & -2 & -4 \\ & & -1 & 2 & -2 & 4 \\ \hline & 1 & -2 & 2 & -4 & 0 \end{array}$$

$$\begin{array}{r|rrrr} 2 & 1 & -2 & 2 & -4 \\ & & 2 & 0 & 4 \\ \hline & 1 & 0 & 2 & 0 \end{array}$$

$z^4 - z^3 - 2z - 4 = (z + 1)(z - 2)(z^2 + 2)$

The only real zeros are -1 and 2.

22. $x^4 - 13x^2 - 12x = 0$

$x(x^3 - 13x - 12) = 0$

$$\begin{array}{r|rrrr} -1 & 1 & 0 & -13 & -12 \\ & & -1 & 1 & 12 \\ \hline & 1 & -1 & -12 & 0 \end{array}$$

$x(x + 1)(x^2 - x - 12) = 0$

$x(x + 1)(x - 4)(x + 3) = 0$

The real zeros are 0, -1, 4, -3.

23. $2y^4 + 7y^3 - 26y^2 + 23y - 6 = 0$

Possible rational zeros: $\pm 1, \pm 2, \pm 3, \pm 6, \pm \frac{1}{2}, \pm \frac{3}{2}$

$$\begin{array}{r|rrrrr} 1 & 2 & 7 & -26 & 23 & -6 \\ & & 2 & 9 & -17 & 6 \\ \hline & 2 & 9 & -17 & 6 & 0 \end{array}$$

$$\begin{array}{r|rrrr} -6 & 2 & 9 & -17 & 6 \\ & & -12 & 18 & -6 \\ \hline & 2 & -3 & 1 & 0 \end{array}$$

$$\begin{aligned} 2y^4 + 7y^3 - 26y^2 + 23y - 6 &= (y - 1)(y + 6)(2y^2 - 3y + 1) \\ &= (y - 1)(y + 6)(2y - 1)(y - 1) \\ &= (y - 1)^2(y + 6)(2y - 1) \end{aligned}$$

The only real zeros are $1, -6$, and $\frac{1}{2}$.

24. $x^5 - x^4 - 3x^3 + 5x^2 - 2x = 0$

$x(x^4 - x^3 - 3x^2 + 5x - 2) = 0$

$$\begin{array}{r|rrrrr} 1 & 1 & -1 & -3 & 5 & -2 \\ & & 1 & 0 & -3 & 2 \\ \hline & 1 & 0 & -3 & 2 & 0 \end{array}$$

$$\begin{array}{r|rrrr} -2 & 1 & 0 & -3 & 2 \\ & & -2 & 4 & -2 \\ \hline & 1 & -2 & 1 & 0 \end{array}$$

$x(x - 1)(x + 2)(x^2 - 2x + 1) = 0$

$x(x - 1)(x + 2)(x - 1)(x - 1) = 0$

The real zeros are $-2, 0, 1$.

25. $f(x) = x^3 + x^2 - 4x - 4$

(a) Possible rational zeros: $\pm 1, \pm 2, \pm 4$

(b)

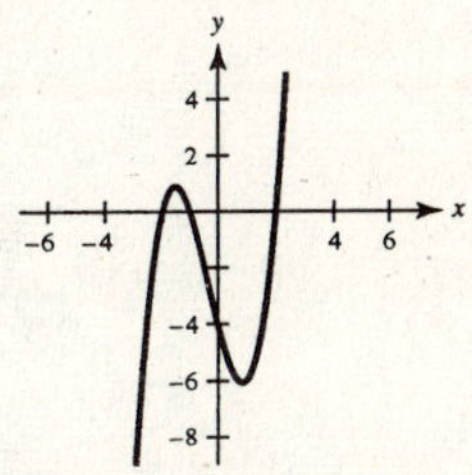

(c) The zeros are: $-2, -1, 2$.

26. $f(x) = -3x^3 + 20x^2 - 36x + 16$

(a) Possible rational zeros: $\pm 1, \pm 2, \pm 4, \pm 8, \pm 16, \pm \frac{1}{3}, \pm \frac{2}{3}, \pm \frac{4}{3}, \pm \frac{8}{3}, \pm \frac{16}{3}$

(b)

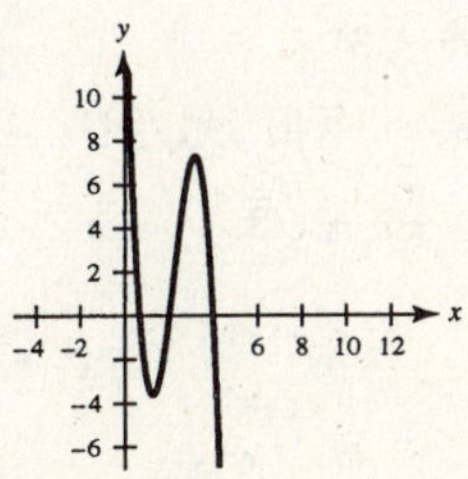

(c) Real zeros: $\frac{2}{3}, 2, 4$

27. $f(x) = -4x^3 + 15x^2 - 8x - 3$

(a) Possible rational zeros: $\pm 1, \pm 3, \pm \frac{1}{2}, \pm \frac{3}{2}, \pm \frac{1}{4}, \pm \frac{3}{4}$

(b)

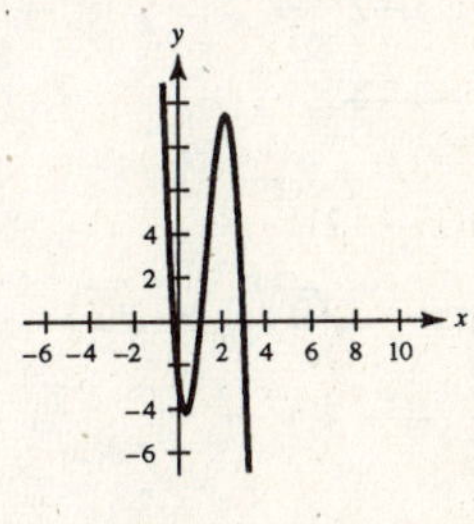

(c) The zeros are: $-\frac{1}{4}, 1, 3$.

28. $f(x) = 4x^3 - 12x^2 - x + 15$

(a) Possible rational zeros: $\pm 1, \pm 3, \pm 5, \pm 15, \pm \frac{1}{2}, \pm \frac{3}{2}, \pm \frac{5}{2}, \pm \frac{15}{2}, \pm \frac{1}{4}, \pm \frac{3}{4}, \pm \frac{5}{4}, \pm \frac{15}{4}$

(b)

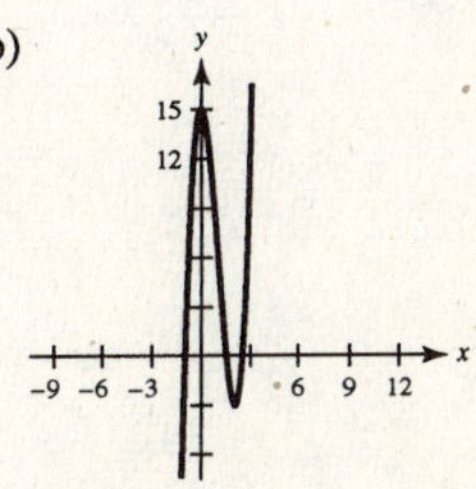

(c) Real zeros: $-1, \frac{3}{2}, \frac{5}{2}$

29. $f(x) = -2x^4 + 13x^3 - 21x^2 + 2x + 8$

(a) Possible rational zeros: $\pm 1, \pm 2, \pm 4, \pm 8, \pm \frac{1}{2}$

(b)

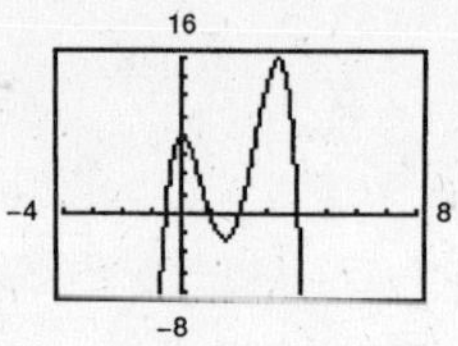

(c) The zeros are: $-\frac{1}{2}$, 1, 2, 4

30. $f(x) = 4x^4 - 17x^2 + 4$

(a) Possible rational zeros: $\pm 1, \pm 2, \pm 4, \pm \frac{1}{2}, \pm \frac{1}{4}$

(b)

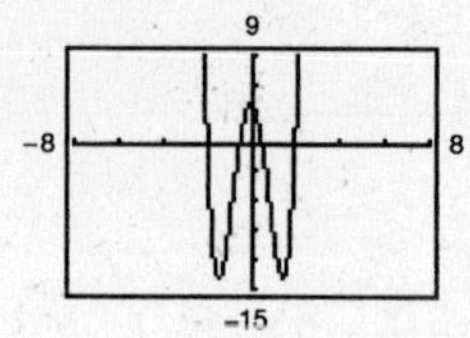

(c) Real zeros: $\pm 2, \pm \frac{1}{2}$

31. $f(x) = 32x^3 - 52x^2 + 17x + 3$

(a) Possible rational zeros: $\pm 1, \pm 3, \pm \frac{1}{2}, \pm \frac{3}{2}, \pm \frac{1}{4}, \pm \frac{3}{4}, \pm \frac{1}{8}, \pm \frac{3}{8}, \pm \frac{1}{16}, \pm \frac{3}{16}, \pm \frac{1}{32}, \pm \frac{3}{32}$

(b)

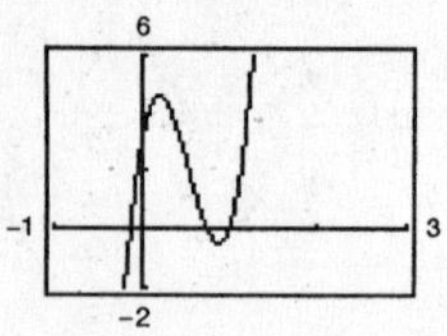

(c) The zeros are: $-\dfrac{1}{8}, \dfrac{3}{4}, 1$

32. $f(x) = 4x^3 + 7x^2 - 11x - 18$

(a) Possible rational zeros: $\pm 1, \pm 2, \pm 3, \pm 6, \pm 9, \pm 18, \pm \frac{1}{2}, \pm \frac{3}{2}, \pm \frac{9}{2}, \pm \frac{1}{4}, \pm \frac{3}{4}, \pm \frac{9}{4}$

(b)

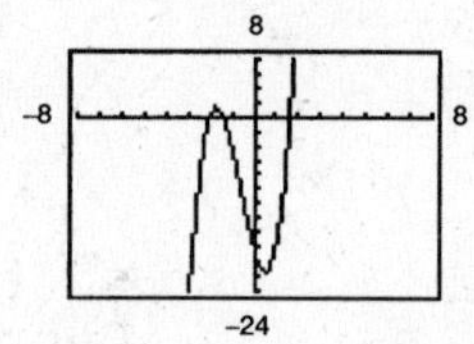

(c) Real zeros: $-2, \dfrac{1}{8} \pm \dfrac{\sqrt{145}}{8}$

33. $f(x) = x^4 - 3x^2 + 2$

(a) From the calculator we have

$x = \pm 1$ and $x \approx \pm 1.414$.

(b)
$$\begin{array}{r|rrrrr} 1 & 1 & 0 & -3 & 0 & 2 \\ & & 1 & 1 & -2 & -2 \\ \hline & 1 & 1 & -2 & -2 & 0 \end{array}$$

$$\begin{array}{r|rrrr} -1 & 1 & 1 & -2 & -2 \\ & & -1 & 0 & 2 \\ \hline & 1 & 0 & -2 & 0 \end{array}$$

$f(x) = (x - 1)(x + 1)(x^2 - 2)$

$= (x - 1)(x + 1)(x - \sqrt{2})(x + \sqrt{2})$

The exact roots are $x = \pm 1, \pm \sqrt{2}$.

34. $P(t) = t^4 - 7t^2 + 12$

(a) $t = \pm 2, \pm 1.732$

(b)
$$\begin{array}{r|rrrrr} 2 & 1 & 0 & -7 & 0 & 12 \\ & & 2 & 4 & -6 & -12 \\ \hline & 1 & 2 & -3 & -6 & 0 \end{array}$$

$$\begin{array}{r|rrrr} -2 & 1 & 2 & -3 & -6 \\ & & -2 & 0 & 6 \\ \hline & 1 & 0 & -3 & 0 \end{array}$$

$P(t) = (t - 2)(t + 2)(t^2 - 3)$

$= (t - 2)(t + 2)(t - \sqrt{3})(t + \sqrt{3})$

The exact roots are $x = \pm 2, \pm \sqrt{3}$.

35. $h(x) = x^5 - 7x^4 + 10x^3 + 14x^2 - 24x$

(a) $h(x) = x(x^4 - 7x^3 + 10x^2 + 14x - 24)$

From the calculator we have

$x = 0, 3, 4$ and $x \approx \pm 1.414$.

(b)
$$\begin{array}{r|rrrrr} 3 & 1 & -7 & 10 & 14 & -24 \\ & & 3 & -12 & -6 & 24 \\ \hline & 1 & -4 & -2 & 8 & 0 \end{array}$$

$$\begin{array}{r|rrrr} 4 & 1 & -4 & -2 & 8 \\ & & 4 & 0 & -8 \\ \hline & 1 & 0 & -2 & 0 \end{array}$$

$f(x) = x(x - 3)(x - 4)(x^2 - 2)$

$= x(x - 3)(x - 4)(x - \sqrt{2})(x + \sqrt{2})$

The exact roots are $x = 0, 3, 4, \pm \sqrt{2}$.

36. $g(x) = 6x^4 - 11x^3 - 51x^2 + 99x - 27$

(a) $x = \pm 3, 1.5, 0.333$

(b)
$$\begin{array}{r|rrrrr} 3 & 6 & -11 & -51 & 99 & -27 \\ & & 18 & 21 & -90 & 27 \\ \hline & 6 & 7 & -30 & 9 & 0 \end{array}$$

$$\begin{array}{r|rrrr} -3 & 6 & 7 & -30 & 9 \\ & & -18 & 33 & -9 \\ \hline & 6 & -11 & 3 & 0 \end{array}$$

$g(x) = (x - 3)(x + 3)(6x^2 - 11x + 3)$

$= (x - 3)(x + 3)(3x - 1)(2x - 3)$

The exact roots are $x = \pm 3, \frac{1}{3}, \frac{3}{2}$.

37. $f(x) = (x - 1)(x - 5i)(x + 5i)$

$= (x - 1)(x^2 + 25)$

$= x^3 - x^2 + 25x - 25$

Note: $f(x) = a(x^3 - x^2 + 25x - 25)$. where a is any nonzero real number, has the zeros 1 and $\pm 5i$.

38. $f(x) = (x - 4)(x - 3i)(x + 3i)$

$= (x - 4)(x^2 + 9)$

$= x^3 - 4x^2 + 9x - 36$

Note: $f(x) = a(x^3 - 4x^2 + 9x - 36)$, where a is any real number, has the zeros 4, $3i$ and $-3i$.

39. $f(x) = (x - 6)[x - (-5 + 2i)][x - (-5 - 2i)]$

$= (x - 6)[(x + 5) - 2i][(x + 5) + 2i]$

$= (x - 6)[(x + 5)^2 - (2i)^2]$

$= (x - 6)(x^2 + 10x + 25 + 4)$

$= (x - 6)(x^2 + 10x + 29)$

$= x^3 + 4x^2 - 31x - 174$

Note: $f(x) = a(x^3 + 4x^2 - 31x - 174)$, where a is any nonzero real number, has the zeros 6, and $-5 \pm 2i$.

40. $f(x) = (x - 2)(x - 4 - i)(x - 4 + i)$

$= (x - 2)(x^2 - 8x + 17)$

$= x^3 - 10x^2 + 33x - 34$

Note: $f(x) = a(x^3 - 10x^2 + 33x - 34)$ where a is any real number, has the zeros 2, $4 \pm i$.

41. If $3 + \sqrt{2}i$ is a zero, so is its conjugate, $3 - \sqrt{2}i$.

$f(x) = (3x - 2)(x + 1)\left[x - (3 + \sqrt{2}i)\right]\left[x - (3 - \sqrt{2}i)\right]$

$= (3x - 2)(x + 1)\left[(x - 3) - \sqrt{2}i\right]\left[(x - 3) + \sqrt{2}i\right]$

$= (3x^2 + x - 2)\left[(x - 3)^2 - (\sqrt{2}i)^2\right]$

$= (3x^2 + x - 2)(x^2 - 6x + 9 + 2)$

$= (3x^2 + x - 2)(x^2 - 6x + 11)$

$= 3x^4 - 17x^3 + 25x^2 + 23x - 22$

Note: $f(x) = a(3x^4 - 17x^3 + 25x^2 + 23x - 22)$, where a is any nonzero real number, has the zeros $\frac{2}{3}$, -1, and $3 \pm \sqrt{2}i$.

42. $f(x) = (x + 5)^2(x - 1 - \sqrt{3}i)(x - 1 + \sqrt{3}i)$

$= (x^2 + 10x + 25)(x^2 - 2x + 4)$

$= x^4 + 8x^3 + 9x^2 - 10x + 100$

Note: $f(x) = a(x^4 + 8x^3 + 9x^2 - 10x + 100)$, where a is any real number, has the zeros $-5, -5, 1 \pm \sqrt{3}i$.

43. $f(x) = x^4 + 6x^2 - 27$

(a) $f(x) = (x^2 + 9)(x^2 - 3)$

(b) $f(x) = (x^2 + 9)(x + \sqrt{3})(x - \sqrt{3})$

(c) $f(x) = (x + 3i)(x - 3i)(x + \sqrt{3})(x - \sqrt{3})$

44. $f(x) = x^4 - 2x^3 - 3x^2 + 12x - 18$

(a) $f(x) = (x^2 - 6)(x^2 - 2x + 3)$

(b) $f(x) = (x + \sqrt{6})(x - \sqrt{6})(x^2 - 2x + 3)$

(c) $f(x) = (x + \sqrt{6})(x - \sqrt{6})(x - 1 - \sqrt{2}i)(x - 1 + \sqrt{2}i)$

45.

$$
\begin{array}{r}
x^2 - 2x + 3 \\
x^2 - 2x - 2 \overline{)\, x^4 - 4x^3 + 5x^2 - 2x - 6} \\
\underline{x^4 - 2x^3 - 2x^2 } \\
-2x^3 + 7x^2 - 2x \\
\underline{-2x^3 + 4x^2 + 4x } \\
3x^2 - 6x - 6 \\
\underline{3x^2 - 6x - 6} \\
0
\end{array}
$$

$f(x) = (x^2 - 2x - 2)(x^2 - 2x + 3)$

(a) $f(x) = (x^2 - 2x - 2)(x^2 - 2x + 3)$

(b) $f(x) = \left(x - 1 + \sqrt{3}\right)\left(x - 1 - \sqrt{3}\right)(x^2 - 2x + 3)$

(c) $f(x) = \left(x - 1 + \sqrt{3}\right)\left(x - 1 - \sqrt{3}\right)\left(x - 1 + \sqrt{2}\,i\right)\left(x - 1 - \sqrt{2}\,i\right)$

Note: Use the Quadratic Formula for (b) and (c).

46. $f(x) = x^4 - 3x^3 - x^2 - 12x - 20$

(a) $f(x) = (x^2 + 4)(x^2 - 3x - 5)$

(b) $f(x) = (x^2 + 4)\left(x - \dfrac{3 + \sqrt{29}}{2}\right)\left(x - \dfrac{3 - \sqrt{29}}{2}\right)$

(c) $f(x) = (x + 2i)(x - 2i)\left(x - \dfrac{3 + \sqrt{29}}{2}\right)\left(x - \dfrac{3 - \sqrt{29}}{2}\right)$

47. $f(x) = 2x^3 + 3x^2 + 50x + 75$

Since $5i$ is a zero, so is $-5i$.

$$
\begin{array}{r|rrrr}
5i & 2 & 3 & 50 & 75 \\
 & & 10i & -50 + 15i & -75 \\
\hline
 & 2 & 3 + 10i & 15i & 0
\end{array}
$$

$$
\begin{array}{r|rrr}
-5i & 2 & 3 + 10i & 15i \\
 & & -10i & -15i \\
\hline
 & 2 & 3 & 0
\end{array}
$$

The zero of $2x + 3$ is $x = -\frac{3}{2}$.

The zeros of $f(x)$ are $x = -\frac{3}{2}$ and $x = \pm 5i$.

<u>Alternate Solution</u>

Since $x = \pm 5i$ are zeros of $f(x)$, $(x + 5i)(x - 5i) = x^2 + 25$ is a factor of $f(x)$.
By long division we have:

$$
\begin{array}{r}
2x + 3 \\
x^2 + 0x + 25 \overline{)\, 2x^3 + 3x^2 + 50x + 75} \\
\underline{2x^3 + 0x^2 + 50x } \\
3x^2 + 0x + 75 \\
\underline{3x^2 + 0x + 75} \\
0
\end{array}
$$

Thus, $f(x) = (x^2 + 25)(2x + 3)$ and the zeros of f are $x = \pm 5i$ and $x = -\frac{3}{2}$.

48. $f(x) = x^3 + x^2 + 9x + 9$

Since $3i$ is a zero, so is $-3i$.

$$\begin{array}{r|rrrr} 3i & 1 & 1 & 9 & 9 \\ & & 3i & -9+3i & -9 \\ \hline & 1 & 1+3i & 3i & 0 \end{array}$$

$$\begin{array}{r|rrr} -3i & 1 & 1+3i & 3i \\ & & -3i & -3i \\ \hline & 1 & 1 & 0 \end{array}$$

The zero of $x + 1$ is $x = -1$.

The zeros of f are $x = -1, \pm 3i$.

49. $f(x) = 2x^4 - x^3 + 7x^2 - 4x - 4$

Since $2i$ is a zero, so is $-2i$.

$$\begin{array}{r|rrrrr} 2i & 2 & -1 & 7 & -4 & -4 \\ & & 4i & -8-2i & 4-2i & 4 \\ \hline & 2 & -1+4i & -1-2i & -2i & 0 \end{array}$$

$$\begin{array}{r|rrrr} -2i & 2 & -1+4i & -1-2i & -2i \\ & & -4i & 2i & 2i \\ \hline & 2 & -1 & -1 & 0 \end{array}$$

The zeros of $2x^2 - x - 1 = (2x + 1)(x - 1)$ are $x = -\frac{1}{2}$ and $x = 1$.

The zeros of $f(x)$ are $x = \pm 2i$, $x = -\frac{1}{2}$, and $x = 1$.

Alternate Solution

Since $x = \pm 2i$ are zeros of $f(x)$, $(x + 2i)(x - 2i) = x^2 + 4$ is a factor of $f(x)$.

By long division we have:

$$\begin{array}{r} 2x^2 - x - 1 \\ x^2 + 0x + 4 \overline{)\, 2x^4 - x^3 + 7x^2 - 4x - 4} \\ \underline{2x^4 + 0x^3 + 8x^2 \qquad\qquad} \\ -x^3 - x^2 - 4x \qquad \\ \underline{-x^3 + 0x^2 - 4x \qquad} \\ -x^2 + 0x - 4 \\ \underline{-x^2 + 0x - 4} \\ 0 \end{array}$$

Thus, $f(x) = (x^2 + 4)(2x^2 - x - 1)$

$= (x + 2i)(x - 2i)(2x + 1)(x - 1)$

and the zeros of $f(x)$ are $x = \pm 2i$, $x = -\frac{1}{2}$, and $x = 1$.

50. $g(x) = x^3 - 7x^2 - x + 87$

Since $5 + 2i$ is a zero, so is $5 - 2i$.

$$\begin{array}{r|rrrr} 5+2i & 1 & -7 & -1 & 87 \\ & & 5+2i & -14+6i & -87 \\ \hline & 1 & -2+2i & -15+6i & 0 \end{array}$$

$$\begin{array}{r|rrr} 5-2i & 1 & -2+2i & -15+6i \\ & & 5-2i & 15-6i \\ \hline & 1 & 3 & 0 \end{array}$$

The zero of $x + 3$ is $x = -3$.

The zeros of f are $x = -3, 5 \pm 2i$.

51. $g(x) = 4x^3 + 23x^2 + 34x - 10$

Since $-3 + i$ is a zero, so is $-3 - i$.

$$\begin{array}{r|rrrr} -3+i & 4 & 23 & 34 & -10 \\ & & -12+4i & -37-i & 10 \\ \hline & 4 & 11+4i & -3-i & 0 \end{array}$$

$$\begin{array}{r|rrr} -3-i & 4 & 11+4i & -3-i \\ & & -12-4i & 3+i \\ \hline & 4 & -1 & 0 \end{array}$$

The zero of $4x - 1$ is $x = \frac{1}{4}$. The zeros of $g(x)$ are $x = -3 \pm i$ and $x = \frac{1}{4}$.

Alternate Solution

Since $-3 \pm i$ are zeros of $g(x)$,

$$\begin{aligned} [x - (-3 + i)][x - (-3 - i)] &= [(x + 3) - i][(x + 3) + i] \\ &= (x + 3)^2 - i^2 \\ &= x^2 + 6x + 10 \end{aligned}$$

is a factor of $g(x)$. By long division we have:

$$\begin{array}{r} 4x - 1 \\ x^2 + 6x + 10 \overline{)\, 4x^3 + 23x^2 + 34x - 10} \\ \underline{4x^3 + 24x^2 + 40x} \\ -x^2 - 6x - 10 \\ \underline{-x^2 - 6x - 10} \\ 0 \end{array}$$

Thus, $g(x) = (x^2 + 6x + 10)(4x - 1)$ and the zeros of $g(x)$ are $x = -3 \pm i$ and $x = \frac{1}{4}$.

52. $h(x) = 3x^3 - 4x^2 + 8x + 8$

Since $1 - \sqrt{3}i$ is a zero, so is $1 + \sqrt{3}i$.

$$\begin{array}{r|rrrr} 1-\sqrt{3}i & 3 & -4 & 8 & 8 \\ & & 3-3\sqrt{3}i & -10-2\sqrt{3}i & -8 \\ \hline & 3 & -1-3\sqrt{3}i & -2-2\sqrt{3}i & 0 \end{array}$$

$$\begin{array}{r|rrr} 1+\sqrt{3}i & 3 & -1-3\sqrt{3}i & -2-2\sqrt{3}i \\ & & 3+3\sqrt{3}i & 2+2\sqrt{3}i \\ \hline & 3 & 2 & 0 \end{array}$$

The zero of $3x + 2$ is $x = -\frac{2}{3}$.

The zeros of h are $x = -\frac{2}{3}, 1 \pm \sqrt{3}i$.

53. Since $-3 + \sqrt{2}\,i$ is a zero, so is $-3 - \sqrt{2}\,i$, and

$$\begin{aligned} &\left[x - \left(-3 + \sqrt{2}\,i\right)\right]\left[x - \left(-3 - \sqrt{2}\,i\right)\right] \\ &\quad = \left[(x + 3) - \sqrt{2}\,i\right]\left[(x + 3) + \sqrt{2}\,i\right] \\ &\quad = (x + 3)^2 - \left(\sqrt{2}\,i\right)^2 \\ &\quad = x^2 + 6x + 11 \end{aligned}$$

is a factor of $f(x)$. By long division, we have:

$$\begin{array}{r} x^2 - 3x + 2 \\ x^2 + 6x + 11 \overline{)\, x^4 + 3x^3 - 5x^2 - 21x + 22} \\ \underline{x^4 + 6x^3 + 11x^2} \\ -3x^3 - 16x^2 - 21x \\ \underline{-3x^3 - 18x^2 - 33x} \\ 2x^2 + 12x + 22 \\ \underline{2x^2 + 12x + 22} \\ 0 \end{array}$$

Thus,
$$\begin{aligned} f(x) &= (x^2 + 6x + 11)(x^2 - 3x + 2) \\ &= (x^2 + 6x + 11)(x - 1)(x - 2) \end{aligned}$$

and the zeros of f are $x = -3 \pm \sqrt{2}\,i$, $x = 1$, and $x = 2$.

54. $f(x) = x^3 + 4x^2 + 14x + 20$

Since $-1 - 3i$ is a zero, so is $-1 + 3i$.

$$\begin{array}{r|rrrr} -1-3i & 1 & 4 & 14 & 20 \\ & & -1-3i & -12-6i & -20 \\ \hline & 1 & 3-3i & 2-6i & 0 \end{array}$$

$$\begin{array}{r|rrr} -1+3i & 1 & 3-3i & 2-6i \\ & & -1+3i & -2+6i \\ \hline & 1 & 2 & 0 \end{array}$$

The zero of $x + 2$ is $x = -2$.

The zeros of f are $x = -2, -1 \pm 3i$.

55. $f(x) = x^2 + 25$

$= (x + 5i)(x - 5i)$

The zeros of $f(x)$ are $x = \pm 5i$.

56. $f(x) = x^2 - x + 56$

By the Quadratic Formula, the zeros of $f(x)$ are

$$x = \frac{1 \pm \sqrt{1 - 224}}{2} = \frac{1 \pm \sqrt{223}i}{2}.$$

$$f(x) = \left(x - \frac{1 - \sqrt{223}i}{2}\right)\left(x - \frac{1 + \sqrt{223}i}{2}\right)$$

57. $h(x) = x^2 - 4x + 1$

h has no rational zeros.

By the Quadratic Formula, the zeros are $x = \dfrac{4 \pm \sqrt{16 - 4}}{2} = 2 \pm \sqrt{3}$.

$h(x) = \left[x - \left(2 + \sqrt{3}\right)\right]\left[x - \left(2 - \sqrt{3}\right)\right] = \left(x - 2 - \sqrt{3}\right)\left(x - 2 + \sqrt{3}\right)$

58. $g(x) = x^2 + 10x + 23$

By the Quadratic Formula, the zeros of $f(x)$ are

$$x = \frac{-10 \pm \sqrt{100 - 92}}{2} = \frac{-10 \pm \sqrt{8}}{2} = -5 \pm \sqrt{2}.$$

$g(x) = \left(x + 5 + \sqrt{2}\right)\left(x + 5 - \sqrt{2}\right)$

59. $f(x) = x^4 - 81$

$= (x^2 - 9)(x^2 + 9)$

$= (x + 3)(x - 3)(x + 3i)(x - 3i)$

The zeros of $f(x)$ are $x = \pm 3$ and $x = \pm 3i$.

60. $f(y) = y^4 - 625$

$= (y^2 + 25)(y^2 - 25)$

Zeros: $y = \pm 5, \pm 5i$

$f(y) = (y + 5)(y - 5)(y + 5i)(y - 5i)$

61. $f(z) = z^2 - 2z + 2$

f has no rational zeros.

By the Quadratic Formula,

the zeros are $z = \dfrac{2 \pm \sqrt{4 - 8}}{2} = 1 \pm i$.

$f(z) = [z - (1 + i)][z - (1 - i)] = (z - 1 - i)(z - 1 + i)$

62. $h(x) = x^3 - 3x^2 + 4x - 2$

$$\begin{array}{r|rrrr} 1 & 1 & -3 & 4 & -2 \\ & & 1 & -2 & 2 \\ \hline & 1 & -2 & 2 & 0 \end{array}$$

By the Quadratic Formula, the zeros of $x^2 - 2x + 2$

are $x = \dfrac{2 \pm \sqrt{4 - 8}}{2} = 1 \pm i$.

Zeros: $x = 1, 1 \pm i$

$h(x) = (x - 1)(x - 1 - i)(x - 1 + i)$

63. $g(x) = x^3 - 6x^2 + 13x - 10$

Possible rational zeros: $\pm 1, \pm 2, \pm 5, \pm 10$

$$\begin{array}{r|rrrr} 2 & 1 & -6 & 13 & -10 \\ & & 2 & -8 & 10 \\ \hline & 1 & -4 & 5 & 0 \end{array}$$

By the Quadratic Formula, the zeros of

$x^2 - 4x + 5$ are $x = \dfrac{4 \pm \sqrt{16 - 20}}{2} = 2 \pm i$.

The zeros of $g(x)$ are $x = 2$ and $x = 2 \pm i$.

$g(x) = (x - 2)[x - (2 + i)][x - (2 - i)]$

$= (x - 2)(x - 2 - i)(x - 2 + i)$

64. $f(x) = x^3 - 2x^2 - 11x + 52$

$$\begin{array}{r|rrrr} -4 & 1 & -2 & -11 & 52 \\ & & -4 & 24 & -52 \\ \hline & 1 & -6 & 13 & 0 \end{array}$$

By the Quadratic Formula, the zeros of $x^2 - 6x + 13$ are $x = \dfrac{6 \pm \sqrt{36 - 52}}{2} = 3 \pm 2i.$

Zeros: $x = -4, 3 \pm 2i$

$f(x) = (x + 4)(x - 3 - 2i)(x - 3 + 2i)$

65. $h(x) = x^3 - x + 6$

Possible rational zeros: $\pm 1, \pm 2, \pm 3, \pm 6$

$$\begin{array}{r|rrrr} -2 & 1 & 0 & -1 & 6 \\ & & -2 & 4 & -6 \\ \hline & 1 & -2 & 3 & 0 \end{array}$$

By the Quadratic Formula, the zeros of $x^2 - 2x + 3$ are $x = \dfrac{2 \pm \sqrt{4 - 12}}{2} = 1 \pm \sqrt{2}\,i.$

The zeros of $h(x)$ are $x = -2$ and $x = 1 \pm \sqrt{2}\,i.$

$$\begin{aligned} h(x) &= [x - (-2)]\left[x - \left(1 + \sqrt{2}\,i\right)\right]\left[x - \left(1 - \sqrt{2}\,i\right)\right] \\ &= (x + 2)\left(x - 1 - \sqrt{2}\,i\right)\left(x - 1 + \sqrt{2}\,i\right) \end{aligned}$$

66. $h(x) = x^3 + 9x^2 + 27x + 35$

$$\begin{array}{r|rrrr} -5 & 1 & 9 & 27 & 35 \\ & & -5 & -20 & -35 \\ \hline & 1 & 4 & 7 & 0 \end{array}$$

By the Quadratic Formula, the zeros of $x^2 + 4x + 7$ are $x = \dfrac{-4 \pm \sqrt{16 - 28}}{2} = -2 \pm \sqrt{3}i.$

Zeros: $-5, -2 \pm \sqrt{3}i$

$h(x) = (x + 5)\left(x + 2 + \sqrt{3}i\right)\left(x + 2 - \sqrt{3}i\right)$

67. $f(x) = 5x^3 - 9x^2 + 28x + 6$

Possible rational zeros: $\pm 1, \pm 2, \pm 3, \pm 6, \pm\frac{1}{5}, \pm\frac{2}{5}, \pm\frac{3}{5}, \pm\frac{6}{5}$

$$\begin{array}{r|rrrr} -\frac{1}{5} & 5 & -9 & 28 & 6 \\ & & -1 & 2 & -6 \\ \hline & 5 & -10 & 30 & 0 \end{array}$$

By the Quadratic Formula, the zeros of $5x^2 - 10x + 30 = 5(x^2 - 2x + 6)$ are

$$x = \frac{2 \pm \sqrt{4 - 24}}{2} = 1 \pm \sqrt{5}\,i.$$

The zeros of $f(x)$ are $x = -\frac{1}{5}$ and $x = 1 \pm \sqrt{5}\,i.$

$$\begin{aligned} f(x) &= \left[x - \left(-\tfrac{1}{5}\right)\right](5)\left[x - \left(1 + \sqrt{5}\,i\right)\right]\left[x - \left(1 - \sqrt{5}\,i\right)\right] \\ &= (5x + 1)\left(x - 1 - \sqrt{5}\,i\right)\left(x - 1 + \sqrt{5}\,i\right) \end{aligned}$$

68. $g(x) = 3x^3 - 4x^2 + 8x + 8$

$$\begin{array}{r|rrrr} -\frac{2}{3} & 3 & -4 & 8 & 8 \\ & & -2 & 4 & -8 \\ \hline & 3 & -6 & 12 & 0 \end{array}$$

By the Quadratic Formula, the zeros of $3x^2 - 6x + 12 = 3(x^2 - 2x + 4)$ are

$$x = \frac{2 \pm \sqrt{4 - 16}}{2} = 1 \pm \sqrt{3}i$$

Zeros: $x = -\frac{2}{3}, 1 \pm \sqrt{3}i$

$g(x) = (3x + 2)\left(x - 1 + \sqrt{3}i\right)\left(x - 1 - \sqrt{3}i\right).$

69. $g(x) = x^4 - 4x^3 + 8x^2 - 16x + 16$

Possible rational zeros: $\pm 1, \pm 2, \pm 4, \pm 8, \pm 16$

$$\begin{array}{r|rrrrr} 2 & 1 & -4 & 8 & -16 & 16 \\ & & 2 & -4 & 8 & -16 \\ \hline & 1 & -2 & 4 & -8 & 0 \end{array}$$

$$\begin{array}{r|rrrr} 2 & 1 & -2 & 4 & -8 \\ & & 2 & 0 & 8 \\ \hline & 1 & 0 & 4 & 0 \end{array}$$

$g(x) = (x - 2)(x - 2)(x^2 + 4) = (x - 2)^2(x + 2i)(x - 2i)$

The zeros of $g(x)$ are 2 and $\pm 2i$.

70. $h(x) = x^4 + 6x^3 + 10x^2 + 6x + 9$

$$\begin{array}{r|rrrrr} -3 & 1 & 6 & 10 & 6 & 9 \\ & & -3 & -9 & -3 & -9 \\ \hline & 1 & 3 & 1 & 3 & 0 \end{array}$$

$$\begin{array}{r|rrrr} -3 & 1 & 3 & 1 & 3 \\ & & -3 & 0 & -3 \\ \hline & 1 & 0 & 1 & 0 \end{array}$$

The zeros of $x^2 + 1$ are $x = \pm i.$

Zeros: $x = -3, \pm i$

$h(x) = (x + 3)^2(x + i)(x - i)$

71. $f(x) = x^4 + 10x^2 + 9$

$$\begin{aligned} &= (x^2 + 1)(x^2 + 9) \\ &= (x + i)(x - i)(x + 3i)(x - 3i) \end{aligned}$$

The zeros of $f(x)$ are $x = \pm i$ and $x = \pm 3i.$

72. $f(x) = x^4 + 29x^2 + 100$

$= (x^2 + 25)(x^2 + 4)$

Zeros: $x = \pm 2i, \pm 5i$

$f(x) = (x + 2i)(x - 2i)(x + 5i)(x - 5i)$

73. $f(x) = x^3 + 24x^2 + 214x + 740$

Possible rational zeros: $\pm 1, \pm 2, \pm 4, \pm 5, \pm 10, \pm 20, \pm 37, \pm 74, \pm 148, \pm 185, \pm 370, \pm 740$

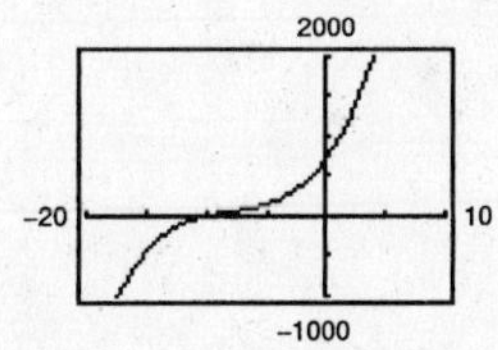

Based on the graph, try $x = -10$.

$$\begin{array}{r|rrrr} -10 & 1 & 24 & 214 & 740 \\ & & -10 & -140 & -740 \\ \hline & 1 & 14 & 74 & 0 \end{array}$$

By the Quadratic Formula, the zeros of $x^2 + 14x + 74$ are

$$x = \frac{-14 \pm \sqrt{196 - 296}}{2} = -7 \pm 5i.$$

The zeros of $f(x)$ are $x = -10$ and $x = -7 \pm 5i$.

74. $f(s) = 2s^3 - 5s^2 + 12s - 5$

Possible rational zeros: $\pm 1, \pm 5, \pm \frac{1}{2}, \pm \frac{5}{2}$

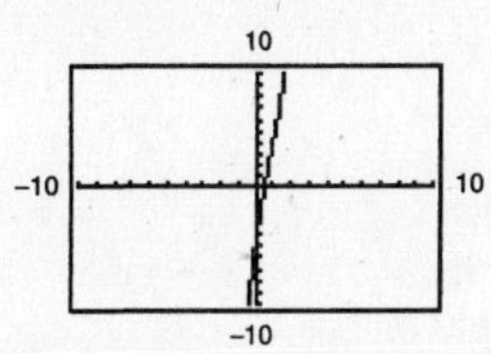

Based on the graph, try $s = \frac{1}{2}$.

$$\begin{array}{r|rrrr} \frac{1}{2} & 2 & -5 & 12 & -5 \\ & & 1 & -2 & 5 \\ \hline & 2 & -4 & 10 & 0 \end{array}$$

Zeros: $s = \dfrac{1}{2}, \dfrac{4 \pm \sqrt{64}i}{4} = 1 \pm 2i$

75. $f(x) = 16x^3 - 20x^2 - 4x + 15$

Possible rational zeros: $\pm 1, \pm 3, \pm 5, \pm 15, \pm \frac{1}{2}, \pm \frac{3}{2}, \pm \frac{5}{2}, \pm \frac{15}{2}, \pm \frac{1}{4}, \pm \frac{3}{4}, \pm \frac{5}{4}, \pm \frac{15}{4}, \pm \frac{1}{8}, \pm \frac{3}{8}, \pm \frac{5}{8}, \pm \frac{15}{8}, \pm \frac{1}{16}, \pm \frac{3}{16}, \pm \frac{5}{16}, \pm \frac{15}{16}$

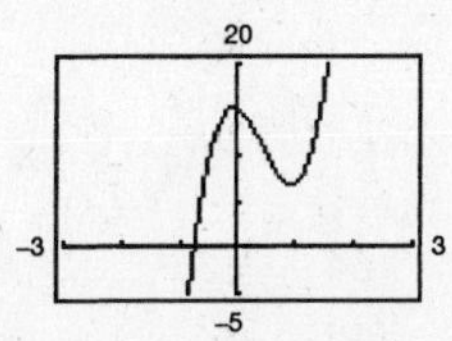

Based on the graph, try $x = -\frac{3}{4}$.

$$\begin{array}{r|rrrr} -\frac{3}{4} & 16 & -20 & -4 & 15 \\ & & -12 & 24 & -15 \\ \hline & 16 & -32 & 20 & 0 \end{array}$$

By the Quadratic Formula, the zeros of $16x^2 - 32x + 20 = 4(4x^2 - 8x + 5)$ are

$$x = \frac{8 \pm \sqrt{64 - 80}}{8} = 1 \pm \frac{1}{2}i.$$

The zeros of $f(x)$ are $x = -\frac{3}{4}$ and $x = 1 \pm \frac{1}{2}i$.

76. $f(x) = 9x^3 - 15x^2 + 11x - 5$

Possible rational zeros: $\pm 1, \pm 5, \pm \frac{1}{3}, \pm \frac{5}{3}, \pm \frac{1}{9}, \pm \frac{5}{9}$

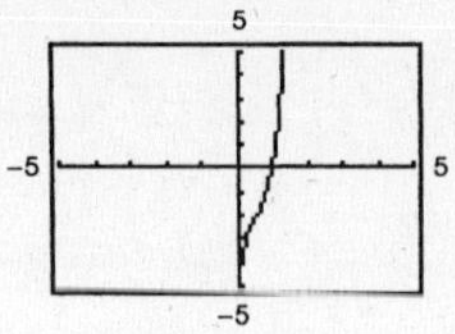

Based on the graph, try $x = 1$.

$$\begin{array}{r|rrrr} 1 & 9 & -15 & 11 & -5 \\ & & 9 & -6 & 5 \\ \hline & 9 & -6 & 5 & 0 \end{array}$$

Zeros: $x = 1, \dfrac{6 \pm \sqrt{144}i}{18} = \dfrac{1}{3} \pm \dfrac{2}{3}i$

77. $f(x) = 2x^4 + 5x^3 + 4x^2 + 5x + 2$

Possible rational zeros: $\pm 1, \pm 2, \pm \frac{1}{2}$

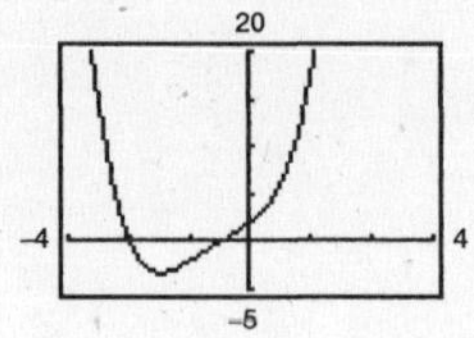

Based on the graph, try $x = -2$ and $x = -\frac{1}{2}$.

$$\begin{array}{r|rrrrr} -2 & 2 & 5 & 4 & 5 & 2 \\ & & -4 & -2 & -4 & -2 \\ \hline & 2 & 1 & 2 & 1 & 0 \end{array}$$

$$\begin{array}{r|rrrr} -\frac{1}{2} & 2 & 1 & 2 & 1 \\ & & -1 & 0 & -1 \\ \hline & 2 & 0 & 2 & 0 \end{array}$$

The zeros of $2x^2 + 2 = 2(x^2 + 1)$ are $x = \pm i$.

The zeros of $f(x)$ are $x = -2$, $x = -\frac{1}{2}$, and $x = \pm i$.

78. $g(x) = x^5 - 8x^4 + 28x^3 - 56x^2 + 64x - 32$

Possible rational zeros: $\pm 1, \pm 2, \pm 4, \pm 8, \pm 16, \pm 32$

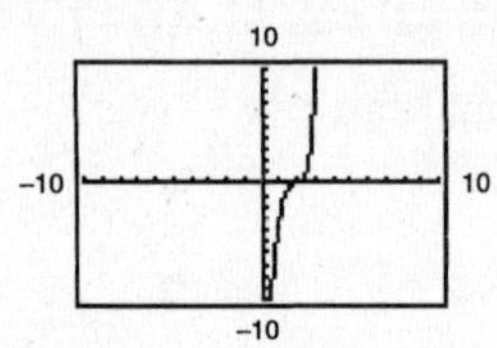

Based on the graph, try $x = 2$.

$$\begin{array}{r|rrrrrr} 2 & 1 & -8 & 28 & -56 & 64 & -32 \\ & & 2 & -12 & 32 & -48 & 32 \\ \hline & 1 & -6 & 16 & -24 & 16 & 0 \end{array}$$

$$\begin{array}{r|rrrrr} 2 & 1 & -6 & 16 & -24 & 16 \\ & & 2 & -8 & 16 & -16 \\ \hline & 1 & -4 & 8 & -8 & 0 \end{array}$$

$$\begin{array}{r|rrrr} 2 & 1 & -4 & 8 & -8 \\ & & 2 & -4 & 8 \\ \hline & 1 & -2 & 4 & 0 \end{array}$$

Zeros: $x = 2, \dfrac{2 \pm \sqrt{12}i}{2} = 1 \pm \sqrt{3}i$

79. $f(x) = 4x^3 - 3x - 1$

Possible rational zeros: $\pm 1, \pm \frac{1}{2}, \pm \frac{1}{4}$

$$\begin{array}{r|rrrr} 1 & 4 & 0 & -3 & -1 \\ & & 4 & 4 & 1 \\ \hline & 4 & 4 & 1 & 0 \end{array}$$

$4x^3 - 3x - 1 = (x - 1)(4x^2 + 4x + 1) = (x - 1)(2x + 1)^2$

Thus, the zeros are 1 and $-\frac{1}{2}$.

80. $f(z) = 12z^3 - 4z^2 - 27z + 9$

Possible rational zeros: $\pm 1, \pm 3, \pm 9, \pm \frac{1}{2}, \pm \frac{3}{2}, \pm \frac{9}{2}, \pm \frac{1}{3}, \pm \frac{1}{4}, \pm \frac{3}{4}, \pm \frac{9}{4}, \pm \frac{1}{6}, \pm \frac{1}{12}$

$$\begin{array}{r|rrrr} \frac{3}{2} & 12 & -4 & -27 & 9 \\ & & 18 & 21 & -9 \\ \hline & 12 & 14 & -6 & 0 \end{array}$$

$f(z) = 2\left(z - \frac{3}{2}\right)(6z^2 + 7z - 3)$

$= (2z - 3)(3z - 1)(2z + 3)$

Real zeros: $-\frac{3}{2}, \frac{1}{3}, \frac{3}{2}$

81. $f(y) = 4y^3 + 3y^2 + 8y + 6$

Possible rational zeros: $\pm 1, \pm 2, \pm 3, \pm 6, \pm \frac{1}{2}, \pm \frac{3}{2}, \pm \frac{1}{4}, \pm \frac{3}{4}$

$$\begin{array}{r|rrrr} -\frac{3}{4} & 4 & 3 & 8 & 6 \\ & & -3 & 0 & -6 \\ \hline & 4 & 0 & 8 & 0 \end{array}$$

$4y^3 + 3y^2 + 8y + 6 = \left(y + \frac{3}{4}\right)(4y^2 + 8) = \left(y + \frac{3}{4}\right)4(y^2 + 2) = (4y + 3)(y^2 + 2)$

Thus, the only real zero is $-\frac{3}{4}$.

82. $g(x) = 3x^3 - 2x^2 + 15x - 10$

Possible rational zeros: $\pm 1, \pm 2, \pm 5, \pm 10, \pm\frac{1}{3}, \pm\frac{2}{3}, \pm\frac{5}{3}, \pm\frac{10}{3}$

$$\begin{array}{r|rrrr} \frac{2}{3} & 3 & -2 & 15 & -10 \\ & & 2 & 0 & 10 \\ \hline & 3 & 0 & 15 & 0 \end{array}$$

$g(x) = \left(x - \frac{2}{3}\right)(3x^2 + 15) = (3x - 2)(x^2 + 5)$

Real zeros: $\frac{2}{3}$

83. $P(x) = x^4 - \frac{25}{4}x^2 + 9$

$= \frac{1}{4}(4x^4 - 25x^2 + 36)$

$= \frac{1}{4}(4x^2 - 9)(x^2 - 4)$

$= \frac{1}{4}(2x + 3)(2x - 3)(x + 2)(x - 2)$

The zeros are $\pm\frac{3}{2}$ and ± 2.

84. $f(x) = \frac{1}{2}(2x^3 - 3x^2 - 23x + 12)$

Possible rational zeros: $\pm 1, \pm 2, \pm 3, \pm 4, \pm 6, \pm 12, \pm\frac{1}{2}, \pm\frac{3}{2}$

$$\begin{array}{r|rrrr} 4 & 2 & -3 & -23 & 12 \\ & & 8 & 20 & -12 \\ \hline & 2 & 5 & -3 & 0 \end{array}$$

$f(x) = \frac{1}{2}(x - 4)(2x^2 + 5x - 3)$

$= \frac{1}{2}(x - 4)(2x - 1)(x + 3)$

Rational zeros: $-3, \frac{1}{2}, 4$

85. $f(x) = x^3 - \frac{1}{4}x^2 - x + \frac{1}{4}$

$= \frac{1}{4}(4x^3 - x^2 - 4x + 1)$

$= \frac{1}{4}[x^2(4x - 1) - 1(4x - 1)]$

$= \frac{1}{4}(4x - 1)(x^2 - 1)$

$= \frac{1}{4}(4x - 1)(x + 1)(x - 1)$

The zeros are $\frac{1}{4}$ and ± 1.

86. $f(z) = \frac{1}{6}(6z^3 + 11z^2 - 3z - 2)$

Possible rational zeros: $\pm 1, \pm 2, \pm\frac{1}{2}, \pm\frac{1}{3}, \pm\frac{2}{3}, \pm\frac{1}{6}$

$$\begin{array}{r|rrrr} -2 & 6 & 11 & -3 & -2 \\ & & -12 & 2 & 2 \\ \hline & 6 & -1 & -1 & 0 \end{array}$$

$f(x) = \frac{1}{6}(z + 2)(6z^2 - z - 1)$

$= \frac{1}{6}(z + 2)(3z + 1)(2z - 1)$

Rational zeros: $-2, -\frac{1}{3}, \frac{1}{2}$

87. $f(x) = x^3 - 1 = (x - 1)(x^2 + x + 1)$

Rational zeros: $1 \ (x = 1)$

Irrational zeros: 0

Matches (d).

88. $f(x) = x^3 - 2$

$= \left(x - \sqrt[3]{2}\right)\left(x^2 + \sqrt[3]{2}x + \sqrt[3]{4}\right)$

Rational zeros: 0

Irrational zeros: $1 \ \left(x = \sqrt[3]{2}\right)$

Matches (a).

89. $f(x) = x^3 - x = x(x + 1)(x - 1)$

Rational zeros: $3 \ (x = 0, \pm 1)$

Irrational zeros: 0

Matches (b).

90. $f(x) = x^3 - 2x$

$= x(x^2 - 2)$

$= x\left(x + \sqrt{2}\right)\left(x - \sqrt{2}\right)$

Rational zeros: $1 \ (x = 0)$

Irrational zeros: $2 \ \left(x = \pm\sqrt{2}\right)$

Matches (c).

91. Zeros: $-2, \frac{1}{2}, 3$

$f(x) = -(x + 2)(2x - 1)(x - 3)$

$= -2x^3 + 3x^2 + 11x - 6$

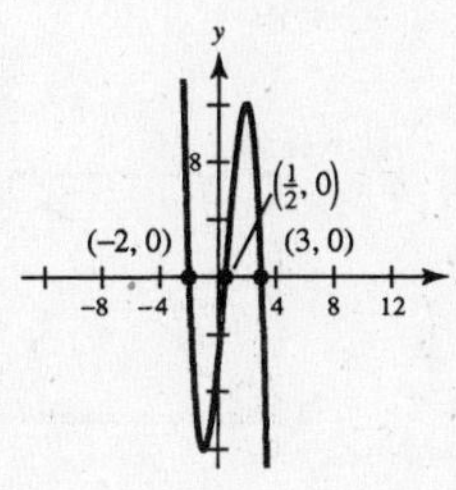

Any nonzero scalar multiple of f would have the same three zeros.

Let $g(x) = af(x),\ a > 0$.

There are infinitely many possible functions for f.

92.

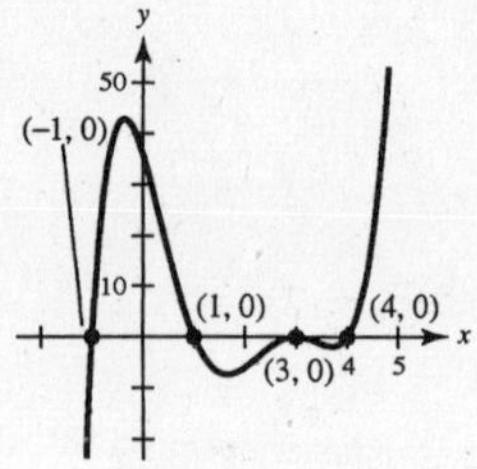

93. Interval: $(-\infty, -2)$, $(-2, 1)$, $(1, 4)$, $(4, \infty)$

Value of $f(x)$: Positive Negative Negative Positive

(a) Zeros of $f(x)$: $x = -2, x = 1, x = 4$.

(b) The graph touches the x-axis at $x = 1$.

(c) The least possible degree of the function is 4 because there are at least four real zeros (1 is repeated) and a function can have at most the number of real zeros equal to the degree of the function. The degree cannot be odd by the behavior at $\pm\infty$.

(d) The leading coefficient of f is positive. From the information in the table, you can conclude that the graph will eventually rise to the left and to the right.

(e) $f(x) = (x + 2)(x - 1)^2(x - 4)$

$$= x^4 - 4x^3 - 3x^2 + 14x - 8$$

(Any nonzero multiple of $f(x)$ is also a solution.)

(f)

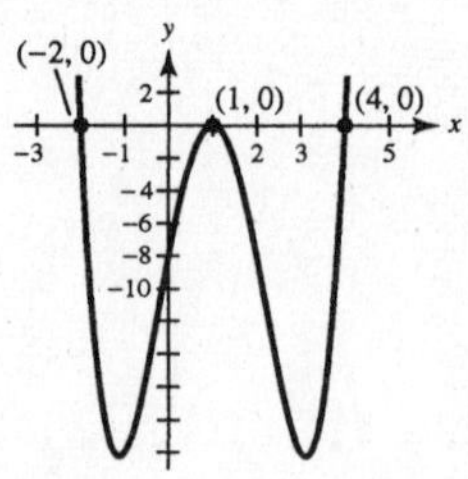

94. (a) $-2, 0, 2$

(b) The graph touches the x-axis at $x = 0$ and at $x = 2$.

(c) The least possible degree of f is 5 because there are at least 5 real zeros (0 and 2 repeated) and a function can have at most the number of real zeros equal to the degree of the function. The degree cannot be even by the definition of multiplicity.

(d) The leading coefficient of f is positive. From the information in the table, you can conclude that the graph will eventually fall to the left and rise to the right.

(e) $f(x) = (x - 0)^2[x - (-2)](x - 2)^2$

$$= x^2(x + 2)(x - 2)^2$$

$$= x^5 - 2x^4 - 4x^3 + 8x^2$$

(f)

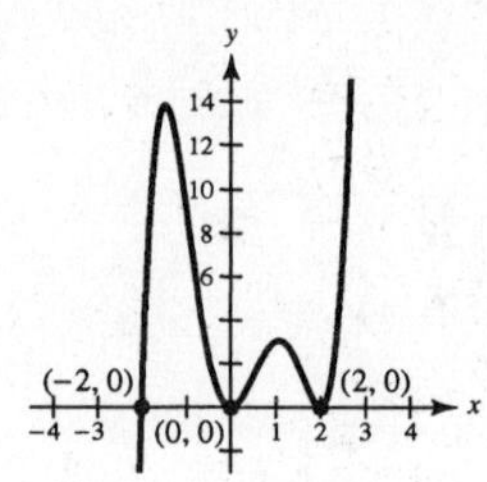

95. (a) Combined length and width:

$4x + y = 120 \Longrightarrow y = 120 - 4x$

$$\text{Volume} = l \cdot w \cdot h = x^2y$$
$$= x^2(120 - 4x)$$
$$= 4x^2(30 - x)$$

(b)

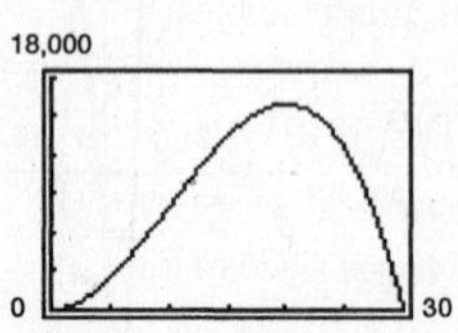

Dimensions with maximum volume:

20 in. × 20 in. × 40 in.

(c)

$$13{,}500 = 4x^2(30 - x)$$
$$4x^3 - 120x^2 + 13{,}500 = 0$$
$$x^3 - 30x^2 + 3375 = 0$$

15	1	−30	0	3375
		15	−225	−3375
	1	−15	−225	0

$(x - 15)(x^2 - 15x - 225) = 0$

Using the Quadratic Formula,

$$x = 15, \frac{15 \pm 15\sqrt{5}}{2}.$$

The value of $\dfrac{15 - 15\sqrt{5}}{2}$ is not possible because it is negative.

96. (a)

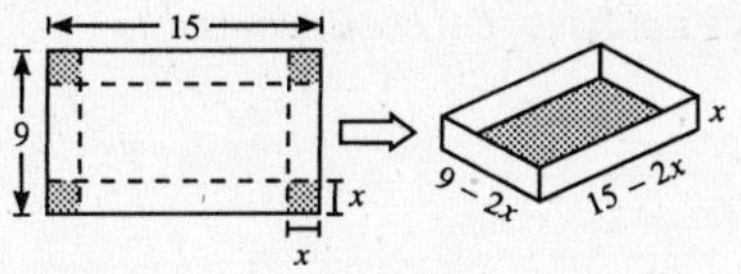

(b) $V = l \cdot w \cdot h = (15 - 2x)(9 - 2x)x$

$= x(9 - 2x)(15 - 2x)$

Since length, width, and height must be positive, we have $0 < x < \frac{9}{2}$ for the domain.

(c)

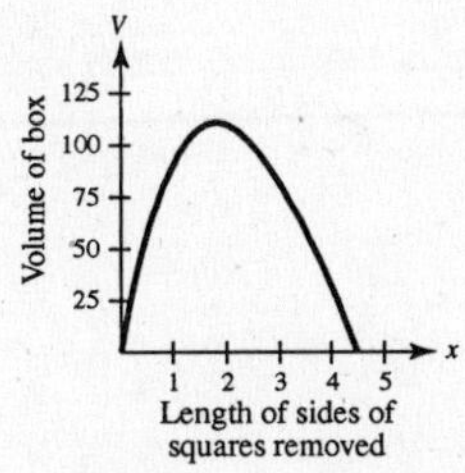

The volume is maximum when $x \approx 1.82$.

The dimensions are: length $\approx 15 - 2(1.82) = 11.36$

width $\approx 9 - 2(1.82) = 5.36$

height $= x \approx 1.82$

1.82 cm × 5.36 cm × 11.36 cm

(d) $56 = x(9 - 2x)(15 - 2x)$

$56 = 135x - 48x^2 + 4x^3$

$0 = 4x^3 - 48x^2 + 135x - 56$

The zeros of this polynomial are $\frac{1}{2}, \frac{7}{2}$, and 8. x cannot equal 8 since it is not in the domain of V. [The length cannot equal -1 and the width cannot equal -7.] The product of $(8)(-1)(-7) = 56$ so it showed up as an extraneous solution.]

97. $C = 100\left(\frac{200}{x^2} + \frac{x}{x + 30}\right),\ 1 \le x$

C is minimum when $3x^3 - 40x^2 - 2400x - 36000 = 0$. The only real zero is $x \approx 40$.

98.

$$P = -76x^3 + 4830x^2 - 320{,}000,\ 0 \le x \le 60$$

$$2{,}500{,}000 = -76x^3 + 4830x^2 - 320{,}000$$

$$76x^3 - 4830x^2 + 2{,}820{,}000 = 0$$

The zeros of this equation are $x \approx 46.1$, $x \approx 38.4$, and $x \approx -21.0$. Since $0 \le x \le 60$, we disregard $x \approx -21.0$. The smaller remaining solution is $x \approx 38.4$. The advertising amount is \$384,000.

99.

$$P = -45x^3 + 2500x^2 - 275{,}000$$

$$800{,}000 = -45x^3 + 2500x^2 - 275{,}000$$

$$0 = 45x^3 - 2500x^2 + 1{,}075{,}000$$

$$0 = 9x^3 - 500x^2 + 215{,}000$$

The zeros of this equation are $x \approx -18.0$, $x \approx 31.5$, and $x \approx 42.0$. Because $0 \le x \le 50$, disregard $x \approx -18.0$. The smaller remaining solution is $x \approx 31.5$, or \$315,000.

100. (a) $y \approx -0.00050t^3 + 0.03369t^2 - 0.26433t + 1.64329$

(b)

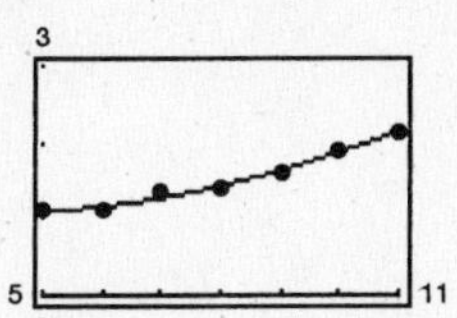

The model is a good fit to the actual data.

(c) $y(6.3958) \approx 1.2$. The average salary reached \$1.2 million in 1996.

(d) $y(10.4879) \approx 2$. The average salary reached \$2 million in 2000.

(e) No. Since the graph falls to the right (the degree is odd and the leading coefficient is negative), the model does not indicate that the average salaries of major league baseball players will continue to increase.

101. False. The most nonreal complex zeros it can have is two and the Linear Factorization Theorem guarantees that there are 3 linear factors, so one zero must be real.

102. False. f does not have real coefficients.

103. (a) $f(x) = (x - \sqrt{b}i)(x + \sqrt{b}i) = x^2 + b$

(b) $f(x) = [x - (a + bi)][x - (a - bi)]$

$= [(x - a) - bi][(x - a) + bi]$

$= (x - a)^2 - (bi)^2$

$= x^2 - 2ax + a^2 + b^2$

104. (a) No. The graph is not of $f(x) = x^2(x + 2)(x - 3.5)$ because this function has a $(0, 0)$ as an intercept and the given graph does not go through $(0, 0)$.

(b) No. The graph is not of $g(x) = (x + 2)(x - 3.5)$ because this function's graph is a parabola and the graph given is not. The function shown in the graph must be at least a fourth-degree polynomial.

(c) Yes. The function shown in the graph is $h(x) = (x + 2)(x - 3.5)(x^2 + 1)$.

(d) The graph is not of $k(x) = (x + 1)(x + 2)(x - 3.5)$ because this function has $(-1, 0)$ as an intercept and the given graph does not.

Section 2.6 Rational Functions

1. $f(x) = \dfrac{1}{x - 1}$

(a)

x	$f(x)$
0.5	-2
0.9	-10
0.99	-100
0.999	-1000

x	$f(x)$
1.5	2
1.1	10
1.01	100
1.001	1000

x	$f(x)$
5	0.25
10	$0.\overline{1}$
100	$0.\overline{01}$
1000	$0.\overline{001}$

x	$f(x)$
-5	$-0.1\overline{6}$
-10	$-0.\overline{09}$
-100	$-0.\overline{0099}$
-1000	$-0.\overline{000999}$

(b) The zero of the denominator is $x = 1$, so $x = 1$ is a vertical asymptote. The degree of the numerator is less than the degree of the denominator so the x-axis, or $y = 0$, is a horizontal asymptote.

(c) The domain is all real numbers except $x = 1$.

2. $f(x) = \dfrac{5x}{x - 1}$

(a)

x	$f(x)$
0.5	-5
0.9	-45
0.99	-495
0.999	-4995

x	$f(x)$
1.5	15
1.1	55
1.01	505
1.001	5005

x	$f(x)$
5	6.25
10	$5.\overline{55}$
100	$5.\overline{05}$
1000	$5.\overline{005}$

x	$f(x)$
-5	$4.1\overline{6}$
-10	$4.\overline{54}$
-100	$4.\overline{9504}$
-1000	$4.\overline{995004}$

(b) The zero of the denominator is $x = 1$, so $x = 1$ is a vertical asymptote. The degree of the numerator is equal to the degree of the denominator, so the line $y = \frac{5}{1} = 5$ is a horizontal asymptote.

(c) The domain is all real numbers except $x = 1$.

3. $f(x) = \dfrac{4x}{|x - 1|}$

(a)

x	$f(x)$
0.5	4
0.9	36
0.99	396
0.999	3996

x	$f(x)$
1.5	12
1.1	44
1.01	404
1.001	4004

x	$f(x)$
5	5
10	$4.\overline{4}$
100	$4.\overline{04}$
1000	$4.\overline{004}$

x	$f(x)$
-5	$-3.\overline{3}$
-10	$-3.\overline{63}$
-100	$-3.\overline{9603}$
-1000	$-3.\overline{996003}$

(b) The zero of the denominator is $x = 1$, so $x = 1$ is a vertical asymptote. Since $f(x) \longrightarrow 4$ as $x \longrightarrow \infty$ and $f(x) \longrightarrow -4$ as $x \longrightarrow -\infty$, both $y = 4$ and $y = -4$ are horizontal asymptotes.

(c) The domain is all real numbers except $x = 1$.

4. $f(x) = \dfrac{2}{|x - 1|}$

(a)

x	$f(x)$
0.5	4
0.9	20
0.99	200
0.999	2000

x	$f(x)$
1.5	4
1.1	20
1.01	200
1.001	2000

x	$f(x)$
5	0.50
10	$0.\overline{2}$
100	$0.\overline{02}$
1000	$0.\overline{002}$

x	$f(x)$
-5	$0.\overline{3}$
-10	$0.\overline{18}$
-100	$0.\overline{0198}$
-1000	$0.\overline{001998}$

(b) The zero of the denominator is $x = 1$, so $x = 1$ is a vertical asymptote. Because the degree of the numerator is less than the degree of the denominator, the x-axis or $y = 0$ is a horizontal asymptote.

(c) The domain is all real numbers except $x = 1$.

5. $f(x) = \dfrac{3x^2}{x^2 - 1}$

(a)

x	$f(x)$
0.5	-1
0.9	-12.79
0.99	-147.8
0.999	-1498

x	$f(x)$
1.5	5.4
1.1	17.29
1.01	152.3
1.001	1502

x	$f(x)$
5	3.125
10	$3.\overline{03}$
100	$3.\overline{0003}$
1000	3

x	$f(x)$
-5	3.125
-10	$3.\overline{03}$
-100	$3.\overline{0003}$
-1000	$3.\overline{000003}$

(b) The zeros of the denominator are $x = \pm 1$ so both $x = 1$ and $x = -1$ are vertical asymptotes. Since the degree of the numerator equals the degree of the denominator, $y = \frac{3}{1} = 3$ is a horizontal asymptote.

(c) The domain is all real numbers except $x = \pm 1$.

6. $f(x) = \dfrac{4x}{x^2 - 1}$

(a)

x	$f(x)$	x	$f(x)$	x	$f(x)$	x	$f(x)$
0.5	$-2.\overline{66}$	1.5	4.8	5	$0.8\overline{33}$	-5	$-0.8\overline{3}$
0.9	-18.95	1.1	20.95	10	$0.\overline{40}$	-10	$-0.\overline{40}$
0.99	-199	1.01	201	100	0.04	-100	$-0.04\overline{0004}$
0.999	-1999	1.001	2001	1000	0.004	-1000	$-0.004\overline{000004}$

(b) The zeros of the denominator are $x = \pm 1$ so both $x = 1$ and $x = -1$ are vertical asymptotes. Because the degree of the numerator is less than the degree of the denominator, the x-axis or $y = 0$ is a horizontal asymptote.

(c) The domain is all real numbers except $x = \pm 1$.

7. $f(x) = \dfrac{1}{x^2}$

Domain: all real numbers except $x = 0$

Vertical asymptote: $x = 0$

Horizontal asymptote: $y = 0$

[Degree of $N(x)$ < degree of $D(x)$]

8. $f(x) = \dfrac{4}{(x - 2)^3}$

Domain: all real numbers except $x = 2$

Vertical asymptote: $x = 2$

Horizontal asymptote: $y = 0$

[Degree of $N(x)$ < degree of $D(x)$]

9. $f(x) = \dfrac{2 + x}{2 - x} = \dfrac{x + 2}{-x + 2}$

Domain: all real numbers except $x = 2$

Vertical asymptote: $x = 2$

Horizontal asymptote: $y = -1$

[Degree of $N(x)$ = degree of $D(x)$]

10. $f(x) = \dfrac{1 - 5x}{1 + 2x} = \dfrac{-5x + 1}{2x + 1}$

Domain: all real numbers except $x = -\frac{1}{2}$

Vertical asymptote: $x = -\frac{1}{2}$

Horizontal asymptote: $y = -\frac{5}{2}$

[Degree of $N(x)$ = degree of $D(x)$]

11. $f(x) = \dfrac{x^3}{x^2 - 1}$

Domain: all real numbers except $x = \pm 1$

Vertical asymptotes: $x = \pm 1$

Horizontal asymptote: None

[Degree of $N(x)$ > degree of $D(x)$]

12. $f(x) = \dfrac{2x^2}{x + 1}$

Domain: all real numbers except $x = -1$

Vertical asymptote: $x = -1$

Horizontal asymptote: None

[Degree of $N(x)$ > degree of $D(x)$]

13. $f(x) = \dfrac{3x^2 + 1}{x^2 + x + 9}$

Domain: All real numbers. The denominator has no real zeros. [Try the Quadratic Formula on the denominator.]

Vertical asymptote: None

Horizontal asymptote: $y = 3$

[Degree of $N(x)$ = degree of $D(x)$]

14. $f(x) = \dfrac{3x^2 + x - 5}{x^2 + 1}$

Domain: all real numbers. (The denominator has no real zeros.)

Vertical asymptote: None

Horizontal asymptote: $y = 3$

[Degree of $N(x)$ = degree of $D(x)$]

15. $f(x) = \dfrac{2}{x+3}$

Vertical asymptote: $x = -3$

Matches graph (d).

16. $f(x) = \dfrac{1}{x-5}$

Vertical asymptote: $x = 5$

Horizontal asymptote: $y = 0$

Matches graph (a).

17. $f(x) = \dfrac{3x+1}{x}$

Horizontal asymptote: $y = 3$

Matches graph (f).

18. $f(x) = \dfrac{1-x}{x}$

Vertical asymptote: $x = 0$

Horizontal asymptote: $y = -1$

Matches graph (c).

19. $f(x) = \dfrac{x-1}{x-4}$

Vertical asymptote: $x = 4$

Horizontal asymptote: $y = 1$

Matches graph (e).

20. $f(x) = -\dfrac{x+2}{x+4}$

Vertical asymptote: $x = -4$

Horizontal asymptote: $y = -1$

Matches graph (b).

21. $g(x) = 1 - \dfrac{1}{x-1}$

$$1 - \frac{1}{x-1} = 0$$

$$1 = \frac{1}{x-1}$$

$$x - 1 = 1$$

$x = 2$ is a zero of $g(x)$.

22. $h(x) = 2 + \dfrac{5}{x^2+2}$

There are no real zeros.

23. $f(x) = 1 - \dfrac{3}{x-3}$

$$1 - \frac{3}{x-3} = 0$$

$$1 = \frac{3}{x-3}$$

$$x - 3 = 3$$

$x = 6$ is a zero of $f(x)$.

24. $g(x) = \dfrac{x^3-8}{x^2+1}$

The zero of g corresponds to the zero of the numerator and is $x = 2$.

25. $f(x) = \dfrac{1}{x+2}$

(a) y-intercept: $\left(0, \frac{1}{2}\right)$

(b) Vertical asymptote: $x = -2$
Horizontal asymptote: $y = 0$

(c) No origin or axis symmetry

(d)

x	-4	-3	-1	0	1
y	$-\frac{1}{2}$	-1	1	$\frac{1}{2}$	$\frac{1}{3}$

(e)

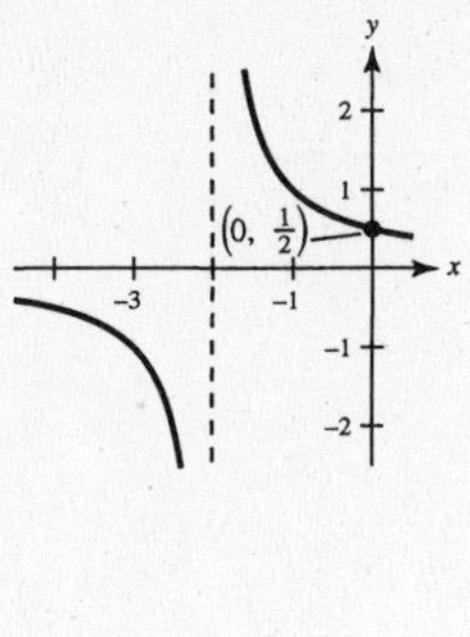

26. $f(x) = \dfrac{1}{x-3}$

(a) Intercept: $\left(0, -\frac{1}{3}\right)$

(b) Vertical asymptote: $x = 3$
Horizontal asymptote: $y = 0$

(c) No axis or origin symmetry

(d)

x	0	1	2	4	5	6
y	$-\frac{1}{3}$	$-\frac{1}{2}$	-1	1	$\frac{1}{2}$	$\frac{1}{3}$

(e)

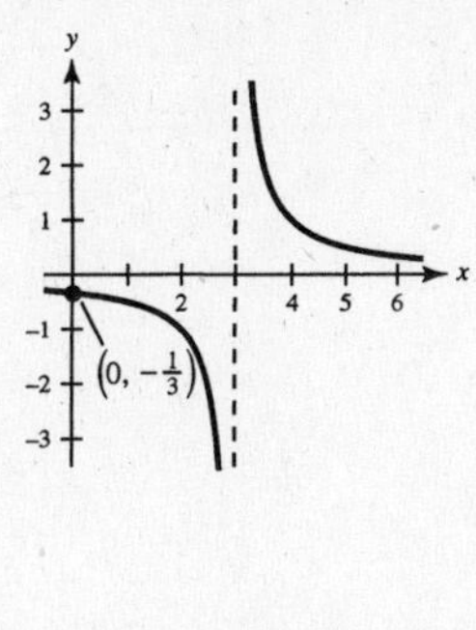

27. $h(x) = \dfrac{-1}{x+2}$

(a) y-intercept: $\left(0, -\frac{1}{2}\right)$

(b) Vertical asymptote: $x = -2$

Horizontal asymptote: $y = 0$

(c) No origin or axis symmetry

(d)

x	-4	-3	-1	0
y	$\frac{1}{2}$	1	-1	$-\frac{1}{2}$

(e)

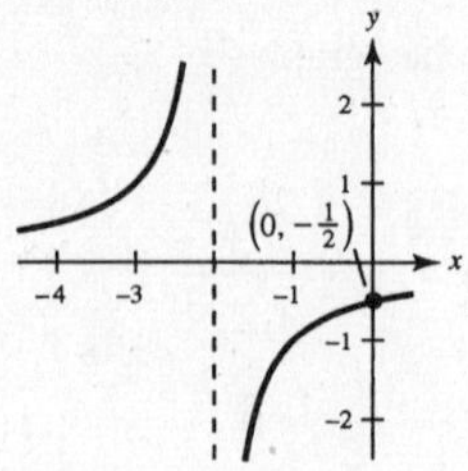

Note: This is the graph of $f(x) = \dfrac{1}{x+2}$ (Exercise 25) reflected about the x-axis.

28. $g(x) = \dfrac{1}{3-x} = -\dfrac{1}{x-3}$

(a) Intercept: $\left(0, \dfrac{1}{3}\right)$

(b) Vertical asymptote: $x = 3$
Horizontal asymptote: $y = 0$

(c) No axis or origin symmetry

(d)

x	0	1	2	4	5	6
y	$\frac{1}{3}$	$\frac{1}{2}$	1	-1	$-\frac{1}{2}$	$-\frac{1}{3}$

(e)

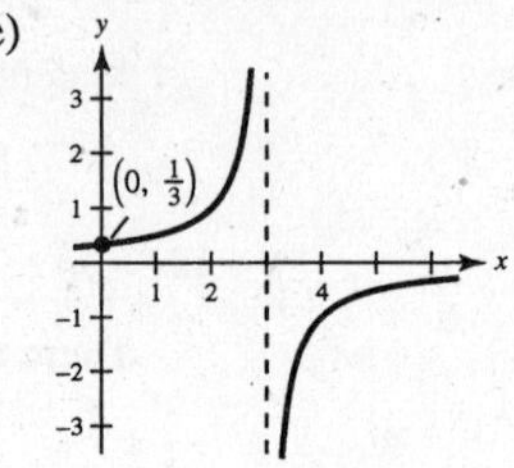

Note: This is the graph of $f(x) = \dfrac{1}{x-3}$ (Exercise 26) reflected about the x-axis.

29. $C(x) = \dfrac{5+2x}{1+x} = \dfrac{2x+5}{x+1}$

(a) x-intercept: $\left(-\frac{5}{2}, 0\right)$

y-intercept: $(0, 5)$

(b) Vertical asymptote: $x = -1$

Horizontal asymptote: $y = 2$

(c) No origin or axis symmetry

(d)

x	-4	-3	-2	0	1	2
$C(x)$	1	$\frac{1}{2}$	-1	5	$\frac{7}{2}$	3

(e)

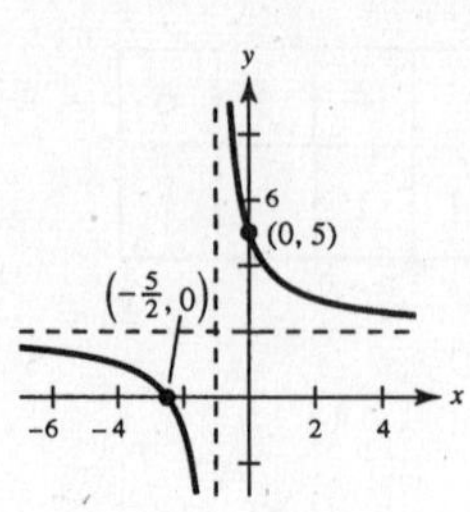

30. $P(x) = \dfrac{1-3x}{1-x} = \dfrac{3x-1}{x-1}$

(a) Intercepts: $\left(\dfrac{1}{3}, 0\right)$, $(0, 1)$

(b) Vertical asymptote: $x = 1$
Horizontal asymptote: $y = 3$

(c) No axis or origin symmetry

(d)

x	-1	0	2	3
y	2	1	5	4

(e)

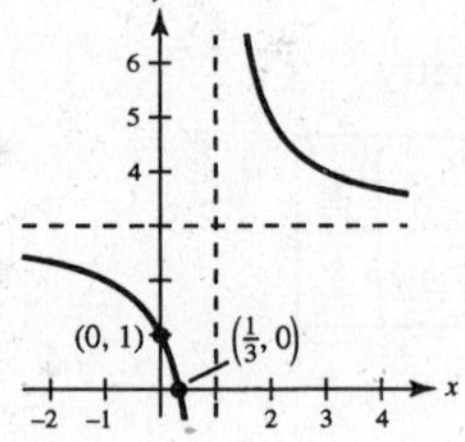

31. $g(x) = \dfrac{1}{x+2} + 2 = \dfrac{2x+5}{x+2}$

(a) x-intercept: $\left(-\frac{5}{2}, 0\right)$

y-intercept: $\left(0, \frac{5}{2}\right)$

(b) Vertical asymptote: $x = -2$

Horizontal asymptote: $y = 2$

(c) No origin or axis symmetry

(d)

x	-4	-3	-1	0	1
y	$\frac{3}{2}$	1	3	$\frac{5}{2}$	$\frac{7}{3}$

(e)

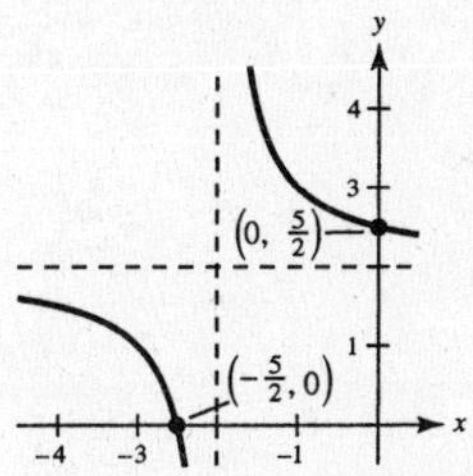

Note: This is the graph of $f(x) = \dfrac{1}{x+2}$ (Exercise 25) shifted upward two units.

32. $f(x) = 2 - \dfrac{3}{x^2}$

(a) Intercepts: $\left(-\dfrac{\sqrt{6}}{2}, 0\right), \left(\dfrac{\sqrt{6}}{2}, 0\right)$

(b) Vertical asymptote: $x = 0$
Horizontal asymptote: $y = 2$

(c) y-axis symmetry

(d)

x	-2	-1	$-\frac{1}{2}$	$\frac{1}{2}$	1	2
y	$\frac{5}{4}$	-1	-10	-10	-1	$\frac{5}{4}$

(e)

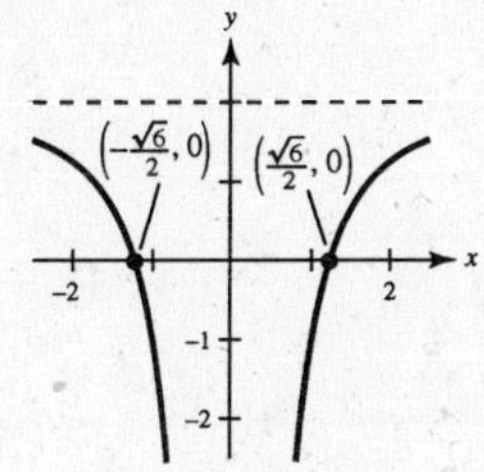

33. $f(x) = \dfrac{x^2}{x^2+9}$

(a) Intercept: $(0, 0)$

(b) Horizontal asymptote: $y = 1$

(c) y-axis symmetry

(d)

x	± 1	± 2	± 3
y	$\frac{1}{10}$	$\frac{4}{13}$	$\frac{1}{2}$

(e)

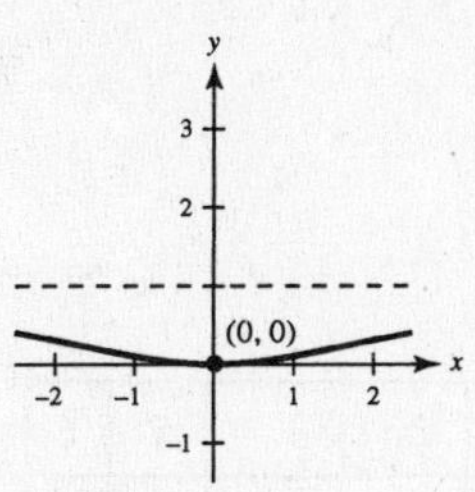

34. $f(t) = \dfrac{4t^2}{t^2+1}$

(a) Intercept: $(0, 0)$

(b) Vertical asymptote: None
Horizontal asymptote: $y = 4$

(c) y-axis symmetry

(d)

t	0	± 1	± 2	± 3
$f(t)$	0	2	$\frac{16}{5}$	$\frac{18}{5}$

(e)

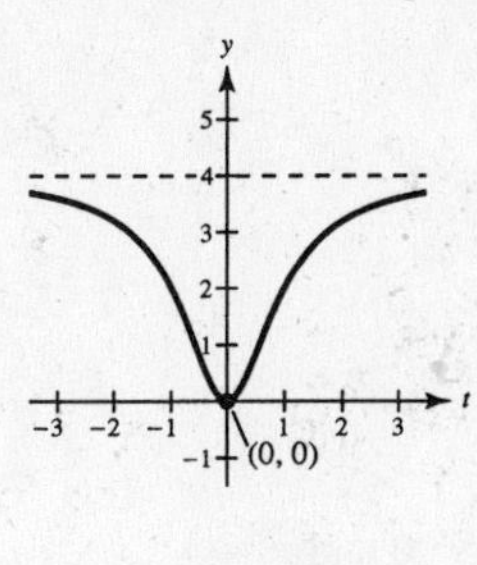

35. $h(x) = \dfrac{x^2}{x^2 - 9}$

(a) Intercept: $(0, 0)$

(b) Vertical asymptotes: $x = \pm 3$

Horizontal asymptote: $y = 1$

(c) y-axis symmetry

(d)

x	± 5	± 4	± 2	± 1	0
y	$\frac{25}{16}$	$\frac{16}{7}$	$-\frac{4}{5}$	$-\frac{1}{8}$	0

(e)

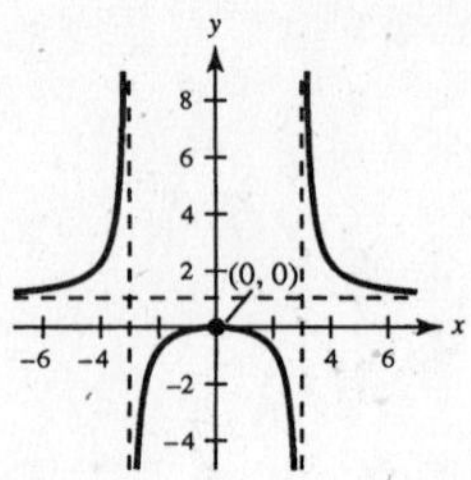

36. $g(x) = \dfrac{x}{x^2 - 9}$

(a) Intercept: $(0, 0)$

(b) Vertical asymptotes: $x = \pm 3$
Horizontal asymptote: $y = 0$

(c) Origin symmetry

(d)

x	-5	-4	-2	0	2	4	5
y	$-\frac{5}{16}$	$-\frac{4}{7}$	$\frac{2}{5}$	0	$-\frac{2}{5}$	$\frac{4}{7}$	$\frac{5}{16}$

(e)

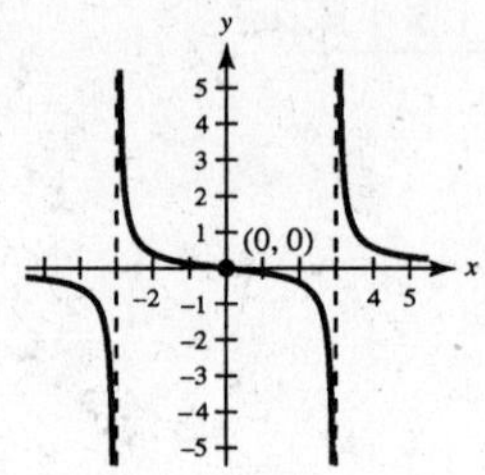

37. $g(s) = \dfrac{s}{s^2 + 1}$

(a) Intercept: $(0, 0)$

(b) Horizontal asymptote: $y = 0$

(c) Origin symmetry

(d)

s	-2	-1	0	1	2
$g(s)$	$-\frac{2}{5}$	$-\frac{1}{2}$	0	$\frac{1}{2}$	$\frac{2}{5}$

(e)

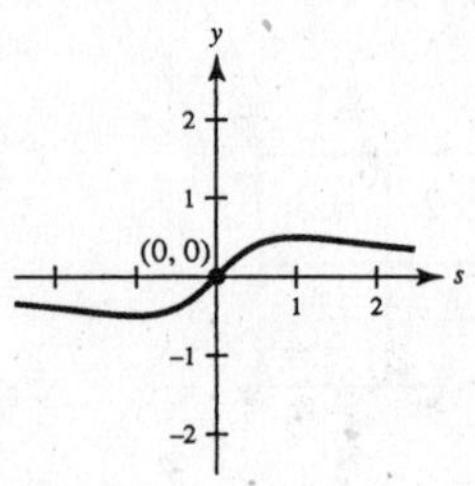

38. $f(x) = -\dfrac{1}{(x - 2)^2}$

(a) Intercept: $\left(0, -\dfrac{1}{4}\right)$

(b) Vertical asymptote: $x = 2$
Horizontal asymptote: $y = 0$

(c) No axis or origin symmetry

(d)

x	0	$\frac{1}{2}$	1	$\frac{3}{2}$	$\frac{5}{2}$	3	$\frac{7}{2}$	4
y	$-\frac{1}{4}$	$-\frac{4}{9}$	-1	-4	-4	-1	$-\frac{4}{9}$	$-\frac{1}{4}$

(e)

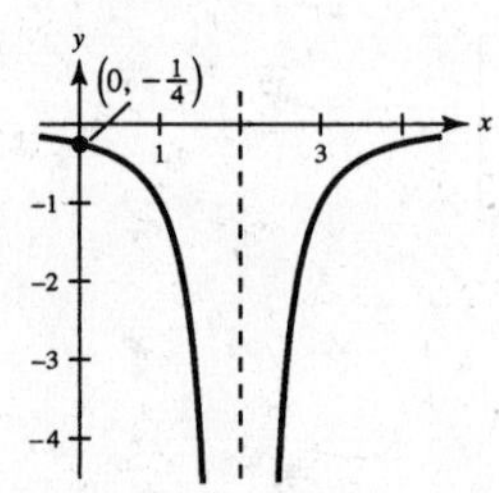

39. $g(x) = \dfrac{4(x + 1)}{x(x - 4)}$

(a) Intercept: $(-1, 0)$

(b) Vertical asymptotes: $x = 0$ and $x = 4$

Horizontal asymptote: $y = 0$

(c) No origin or axis symmetry

(d)

x	-2	-1	1	2	3	5	6
y	$-\frac{1}{3}$	0	$-\frac{8}{3}$	-3	$-\frac{16}{3}$	$\frac{24}{5}$	$\frac{7}{3}$

(e)

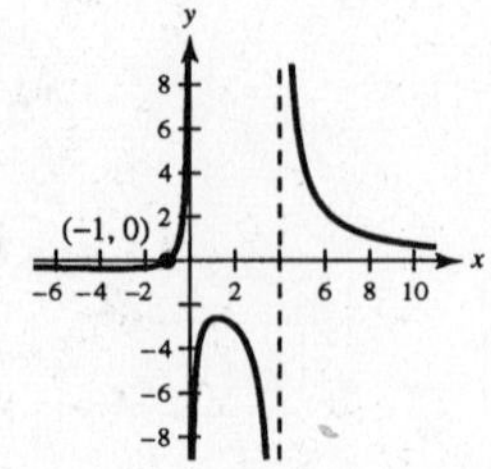

40. $h(x) = \dfrac{2}{x^2(x-2)}$

(a) No intercepts

(b) Vertical asymptotes: $x = 0, x = 2$

Horizontal asymptote: $y = 0$

(c) No axis or origin symmetry

(d)

x	-2	-1	$\frac{1}{2}$	1	$\frac{3}{2}$	$\frac{5}{2}$	3
y	$-\frac{1}{8}$	$-\frac{2}{3}$	$\frac{-16}{3}$	-2	$-\frac{16}{9}$	$\frac{16}{25}$	$\frac{2}{9}$

(e)

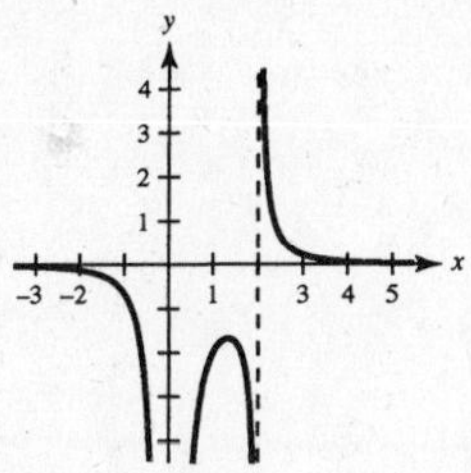

41. $f(x) = \dfrac{3x}{x^2 - x - 2} = \dfrac{3x}{(x+1)(x-2)}$

(a) Intercept: $(0, 0)$

(b) Vertical asymptotes: $x = -1$ and $x = 2$

Horizontal asymptote: $y = 0$

(c) No origin or axis symmetry

(d)

x	-3	0	1	3	4
y	$-\frac{9}{10}$	0	$-\frac{3}{2}$	$\frac{9}{4}$	$\frac{6}{5}$

(e)

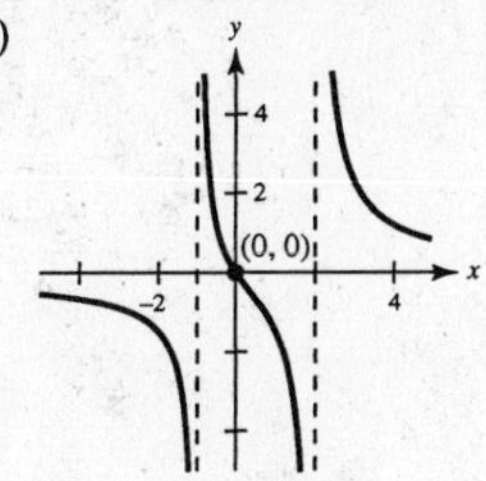

42. $f(x) = \dfrac{2x}{x^2 + x - 2} = \dfrac{2x}{(x+2)(x-1)}$

(a) Intercept: $(0, 0)$

(b) Vertical asymptotes: $x = -2, x = 1$

Horizontal asymptote: $y = 0$

(c) No axis or origin symmetry

(d)

x	-4	-3	-1	0	$\frac{1}{2}$	2	3
y	$-\frac{4}{5}$	$-\frac{3}{2}$	1	0	$-\frac{4}{5}$	1	$\frac{3}{5}$

(e)

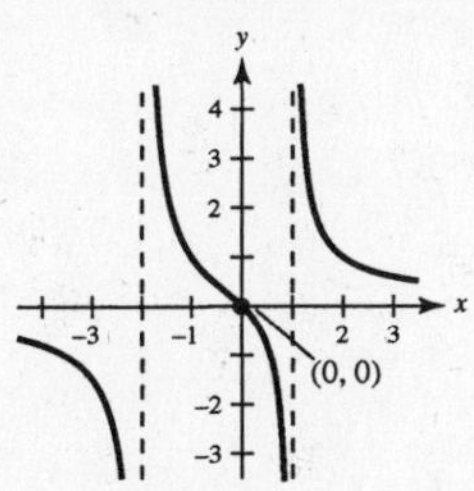

43. $f(x) = \dfrac{6x}{x^2 - 5x - 14} = \dfrac{6x}{(x+2)(x-7)}$

(a) Intercept: $(0, 0)$

(b) Vertical asymptotes: $x = -2$, and $x = 7$

Horizontal asymptotes: $y = 0$

(c) No origin or axis symmetry

(d)

x	-6	-4	0	2	4	6	8	10
$f(x)$	$-\frac{9}{13}$	$-\frac{12}{11}$	0	$-\frac{3}{5}$	$-\frac{4}{3}$	$-\frac{9}{2}$	$\frac{24}{5}$	$\frac{5}{3}$

(e)

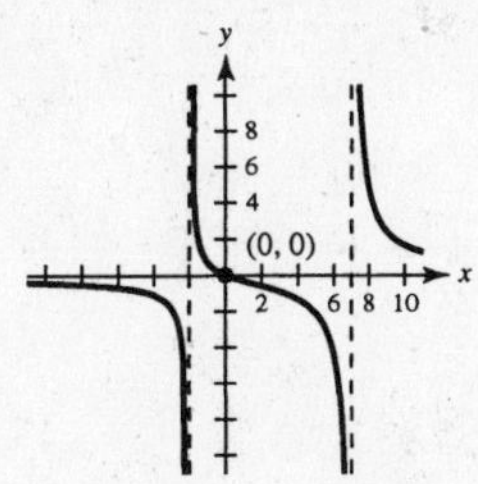

44. $f(x) = \dfrac{3(x^2+1)}{x^2 + 2x - 15} = \dfrac{3(x^2+1)}{(x-3)(x+5)}$

(a) Intercept: $\left(0, -\frac{1}{5}\right)$

(b) Vertical asymptotes: $x = -5, x = 3$

Horizontal asymptote: $y = 3$

(c) No axis or origin symmetry

(d)

x	-15	-12	-9	-6	-3
y	$\frac{113}{30}$	$\frac{29}{7}$	$\frac{41}{8}$	$\frac{37}{3}$	$-\frac{5}{2}$

x	0	6	9	12	15
y	$-\frac{1}{5}$	$\frac{37}{11}$	$\frac{41}{14}$	$\frac{145}{51}$	$\frac{113}{40}$

(e)

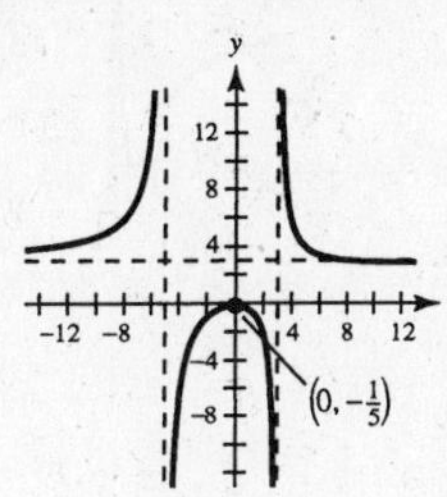

45. $h(t) = \dfrac{4}{t^2 + 1}$

Domain: all real numbers

Horizontal asymptote: $y = 0$

y-axis symmetry

t	± 2	± 1	0
$h(t)$	$\frac{4}{5}$	2	4

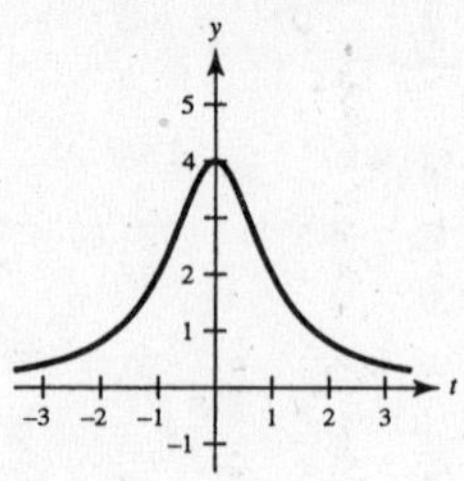

46. $g(x) = -\dfrac{x}{(x - 2)^2}$

Domain: all real numbers except $x = 2$

Vertical asymptote: $x = 2$

Horizontal asymptote: $y = 0$

x	-1	0	1	$\frac{3}{2}$	$\frac{5}{2}$	3	4
y	$\frac{1}{9}$	0	-1	-6	-10	-3	-1

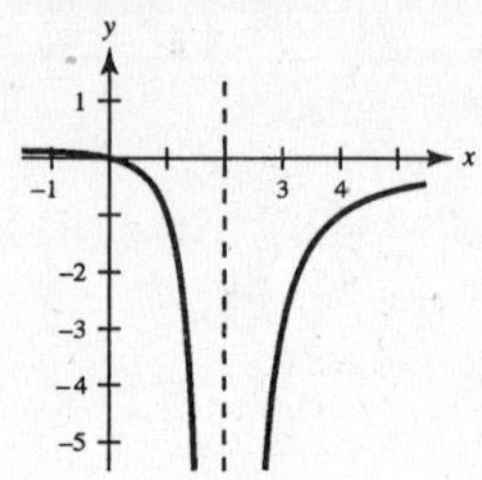

47. $f(t) = \dfrac{2t^2}{t^2 - 4}$

Domain: all real numbers except ± 2,

Vertical asymptotes: $t = \pm 2$

Horizontal asymptote: $y = 2$

y-axis symmetry

t	± 4	± 3	± 1	0
$f(t)$	$\frac{8}{3}$	$\frac{18}{5}$	$-\frac{2}{3}$	0

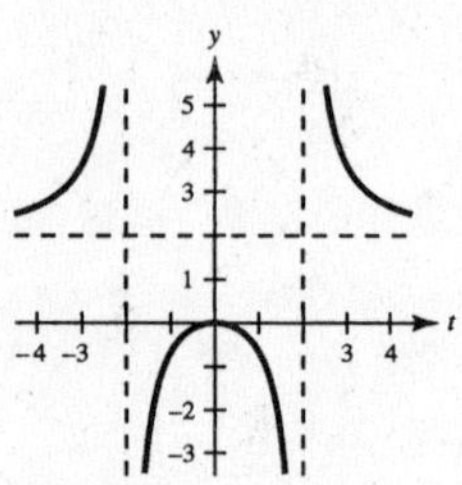

48. $f(x) = \dfrac{x + 4}{x^2 + x - 6}$

Domain: all real numbers except $x = -3$ and $x = 2$

Vertical asymptotes: $x = -3, x = 2$

Horizontal asymptote: $y = 0$

x	-6	-4	-2	-1	0	1	3	4
y	$-\frac{1}{12}$	0	$-\frac{1}{2}$	$-\frac{1}{2}$	$-\frac{2}{3}$	$-\frac{5}{4}$	$\frac{7}{6}$	$\frac{4}{7}$

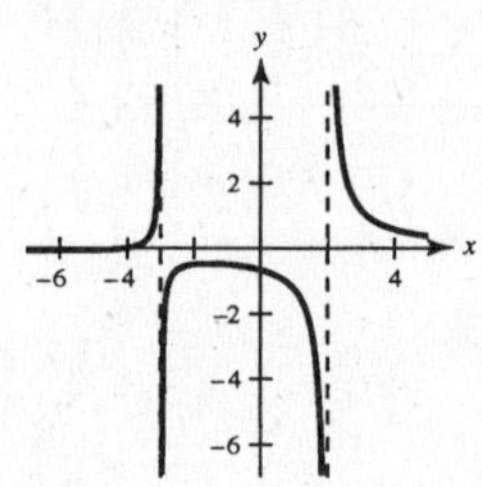

49. $f(x) = \dfrac{20x}{x^2 + 1} - \dfrac{1}{x} = \dfrac{19x^2 - 1}{x(x^2 + 1)}$

Domain: all real numbers except 0,

Vertical asymptote: $x = 0$

Horizontal asymptote: $y = 0$

Origin symmetry

x	-2	-1	1	2
y	$-\frac{15}{2}$	-9	9	$\frac{15}{2}$

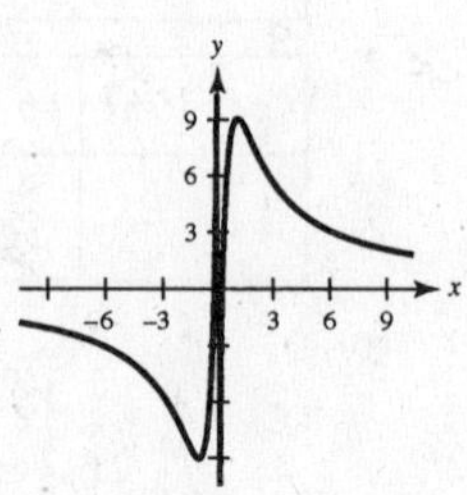

50. $f(x) = 5\left(\frac{1}{x-4} - \frac{1}{x+2}\right) = \frac{30}{(x-4)(x+2)}$

Domain: all real numbers except $x = -2$ and $x = 4$

Vertical asymptotes: $x = -2, x = 4$

Horizontal asymptote: $y = 0$

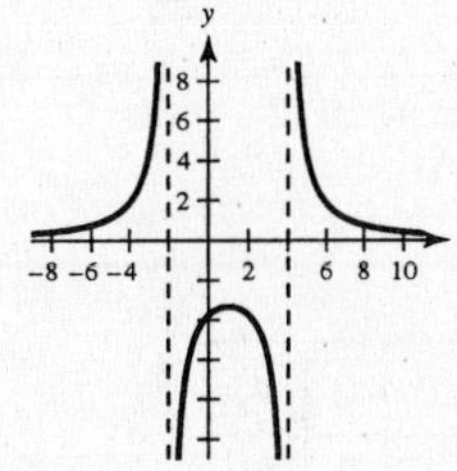

x	-4	-3	-1	0	1	2	3	5	6	7
y	$\frac{15}{8}$	$\frac{30}{7}$	-6	$-\frac{15}{4}$	$-\frac{10}{3}$	$-\frac{15}{4}$	-6	$\frac{30}{7}$	$\frac{15}{8}$	$\frac{10}{9}$

51. $f(x) = \frac{2x^2 + 1}{x} = 2x + \frac{1}{x}$

(a) No intercepts

(b) Vertical asymptote: $x = 0$
Slant asymptote: $y = 2x$

(c) Origin symmetry

(d)

x	-4	-2	2	4	6
$f(x)$	$-\frac{33}{4}$	$-\frac{9}{2}$	$\frac{9}{2}$	$\frac{33}{4}$	$\frac{73}{6}$

(e)

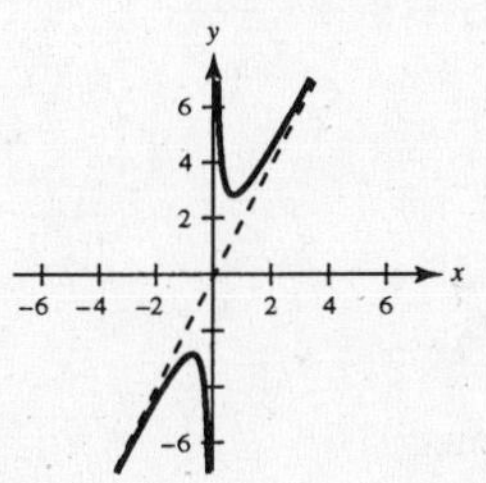

52. $f(x) = \frac{1 - x^2}{x} = -x + \frac{1}{x}$

(a) Intercepts: $(-1, 0), (1, 0)$

(b) Vertical asymptote: $x = 0$
Slant asymptote: $y = -x$

(c) Origin symmetry

(d)

x	-6	-4	-2	2	4	6
$f(x)$	$\frac{35}{6}$	$\frac{15}{4}$	$\frac{3}{2}$	$-\frac{3}{2}$	$-\frac{15}{4}$	$-\frac{35}{6}$

(e)

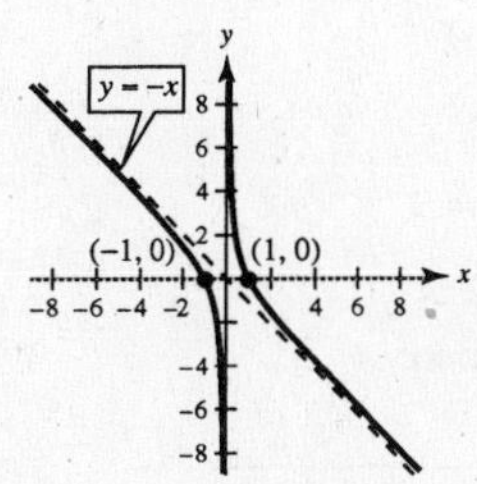

53. $g(x) = \frac{x^2 + 1}{x} = x + \frac{1}{x}$

(a) No intercepts

(b) Vertical asymptote: $x = 0$
Slant asymptote: $y = x$

(c) Origin symmetry

(d)

x	-4	-2	2	4	6
$g(x)$	$-\frac{17}{4}$	$-\frac{5}{2}$	$\frac{5}{2}$	$\frac{17}{4}$	$\frac{37}{6}$

(e)

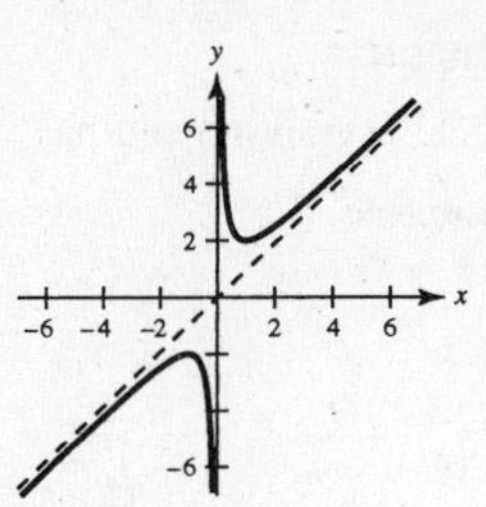

54. $h(x) = \frac{x^2}{x-1} = x + 1 + \frac{1}{x-1}$

(a) Intercept: $(0, 0)$

(b) Vertical asymptote: $x = 1$
Slant asymptote: $y = x + 1$

(c) No axis or origin symmetry

(d)

x	-4	-2	2	4	6
$h(x)$	$-\frac{16}{5}$	$-\frac{4}{3}$	4	$\frac{16}{3}$	$\frac{36}{5}$

(e)

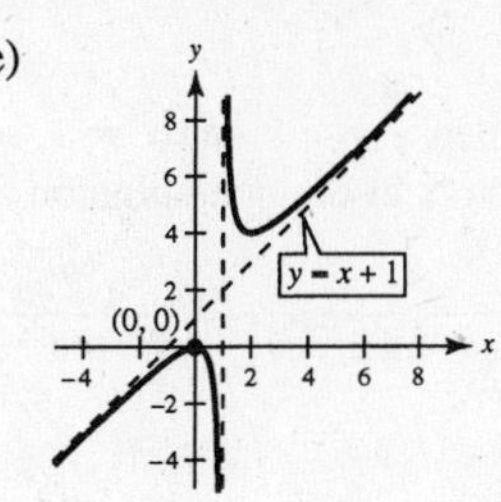

55. $f(x) = \dfrac{x^3}{x^2 - 1} = x + \dfrac{x}{x^2 - 1}$

(a) Intercept: $(0, 0)$

(b) Vertical asymptotes: $x = \pm 1$
Slant asymptote: $y = x$

(c) Origin symmetry

(d)

x	-4	-2	0	2	4
$f(x)$	$-\frac{64}{15}$	$-\frac{8}{3}$	0	$\frac{8}{3}$	$\frac{64}{15}$

(e)

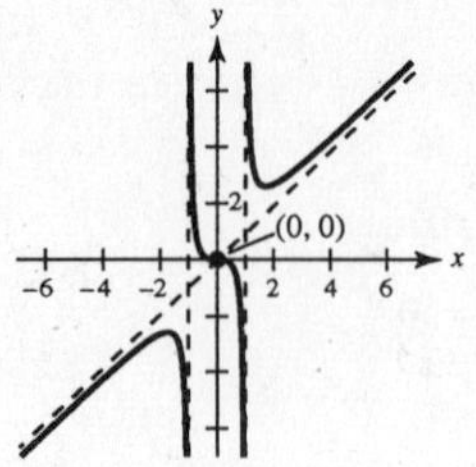

56. $g(x) = \dfrac{x^3}{2x^2 - 8} = \dfrac{1}{2}x + \dfrac{4x}{2x^2 - 8}$

(a) Intercept: $(0, 0)$

(b) Vertical asymptotes: $x = \pm 2$

Slant asymptote: $y = \dfrac{1}{2}x$

(c) Origin symmetry

(d)

x	-6	-4	-1	1	4	6
$g(x)$	$-\frac{27}{8}$	$-\frac{8}{3}$	$\frac{1}{6}$	$-\frac{1}{6}$	$\frac{8}{3}$	$\frac{27}{8}$

(e)

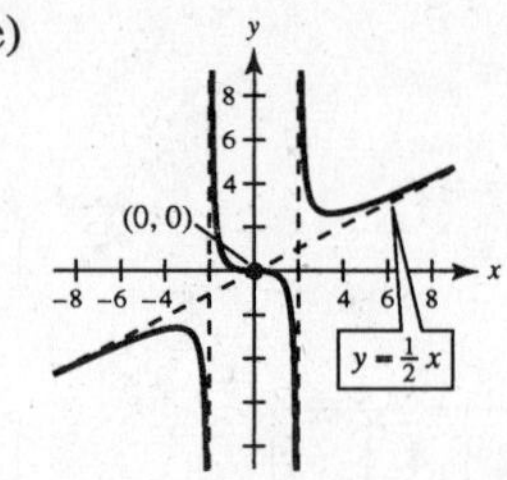

57. $f(x) = \dfrac{x^2 - x + 1}{x - 1} = x + \dfrac{1}{x - 1}$

(a) y-intercept: $(0, -1)$

(b) Vertical asymptote: $x = 1$
Slant asymptote: $y = x$

(c) No axis or origin symmetry

(d)

x	-4	-2	0	2	4
$f(x)$	$-\frac{21}{5}$	$-\frac{7}{3}$	-1	3	$\frac{13}{3}$

(e)

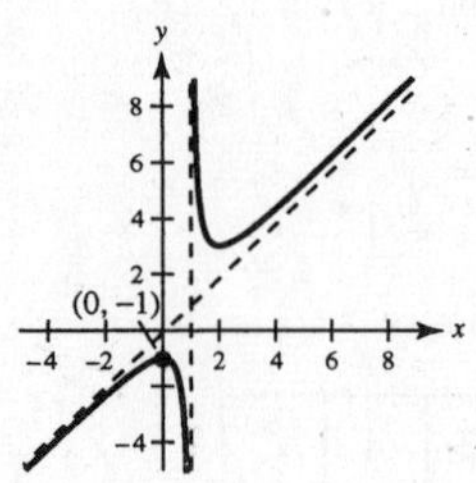

58. $f(x) = \dfrac{2x^2 - 5x + 5}{x - 2} = 2x - 1 + \dfrac{3}{x - 2}$

(a) Intercept: $\left(0, -\dfrac{5}{2}\right)$

(b) Vertical asymptote: $x = 2$
Slant asymptote: $y = 2x - 1$

(c) No axis or origin symmetry

(d)

x	-6	-4	-1	1	4	6
$f(x)$	$-\frac{107}{8}$	$-\frac{19}{2}$	-2	8	$\frac{47}{4}$	$\frac{68}{5}$

(e)

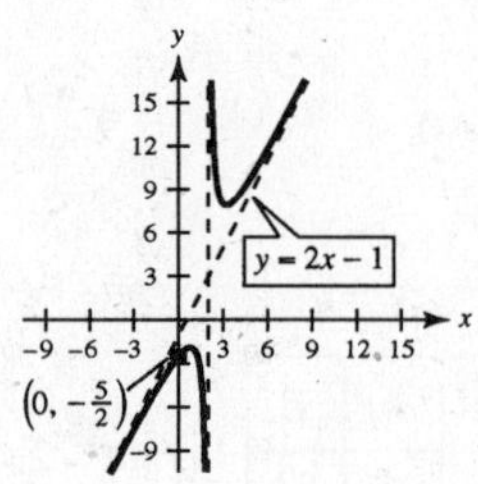

59. Vertical asymptotes: $x = -2, x = 1$

The denominator is zero when $x = -2$ and $x = 1$, so $(x + 2)$ and $(x - 1)$ are factors of the denominator.

One possibility: $f(x) = \dfrac{1}{(x + 2)(x - 1)} = \dfrac{1}{x^2 + x - 2}$

(Answers may vary).

60. Answers may vary. For example,
$f(x) = \dfrac{1}{x^2 + 1}$ has no vertical asymptote and has $y = 0$ as its horizontal asymptote.

61. Vertical asymptotes: None

Horizontal asymptote: $y = 2$

Since there are no vertical asymptotes, the denominator is never zero (if it is reduced), and since there is a horizontal asymptote, the degree of the numerator equals the degree of the denominator. Also, $\frac{a_n}{b_n} = 2.$

One possibility: $f(x) = \frac{2x^2}{x^2 + 1}$ (Answers may vary).

62. Answers will vary. For example,

$$f(x) = \frac{-6x^2 + 15x + 1}{x(2x - 5)}.$$

63. Domain: All real numbers

One possibility: $f(x) = \frac{1}{x^2 + 2}$

Domain: All real numbers except $x = 2$.

One possibility: $f(x) = \frac{1}{x - 2}$

(Answers are not unique).

64. An asymptote is a line to which a graph gets arbitrarily close to, but does not reach, as $|x|$ or $|y|$ increases without bound.

65. $f(x) = \frac{x^2 + 5x + 8}{x + 3} = x + 2 + \frac{2}{x + 3}$

Domain: all real numbers except -3

y-intercept: $\left(0, \frac{8}{3}\right)$

Vertical asymptote: $x = -3$

Slant asymptote: $y = x + 2$

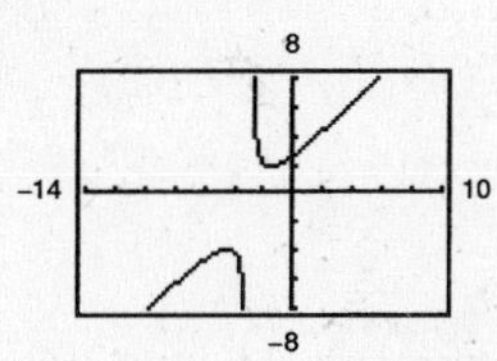

66. $f(x) = \frac{2x^2 + x}{x + 1} = 2x - 1 + \frac{1}{x + 1}$

Domain: all real numbers except $x = -1$

Intercepts: $\left(-\frac{1}{2}, 0\right)$, $(0, 0)$

Vertical asymptote: $x = -1$

Slant asymptote: $y = 2x - 1$

Line: $y = 2x - 1$

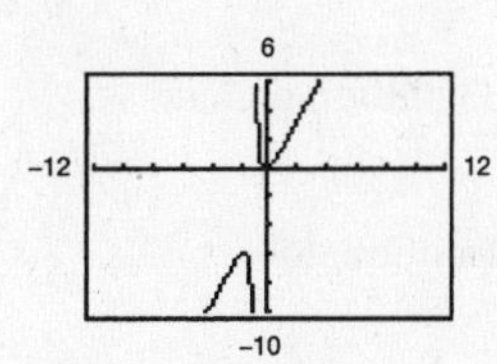

67. $g(x) = \frac{1 + 3x^2 - x^3}{x^2} = \frac{1}{x^2} + 3 - x = -x + 3 + \frac{1}{x^2}$

Domain: all real numbers except 0

Vertical asymptote: $x = 0$

Slant asymptote: $y = -x + 3$

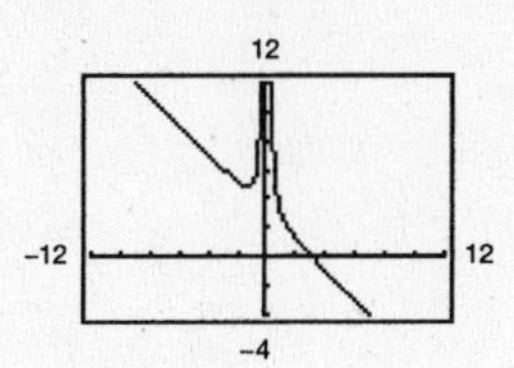

68. $h(x) = \dfrac{12 - 2x - x^2}{2(4 + x)} = -\dfrac{1}{2}x + 1 + \dfrac{2}{4 + x}$

Domain: all real numbers except $x = -4$

Intercepts: $(-4.61, 0), (2.61, 0), \left(0, \dfrac{3}{2}\right)$

Vertical asymptote: $x = -4$

Slant asymptote: $y = -\dfrac{1}{2}x + 1$

Line: $y = -\dfrac{1}{2}x + 1$

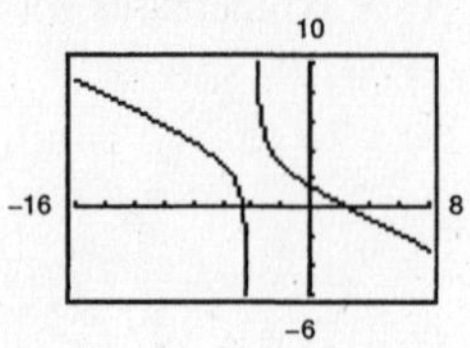

69. (a) x-intercept: $(-1, 0)$

(b) $0 = \dfrac{x + 1}{x - 3}$

$0 = x + 1$

$-1 = x$

70. (a) x-intercept: $(0, 0)$

(b) $0 = \dfrac{2x}{x - 3}$

$0 = 2x$

$0 = x$

71. (a) x-intercepts: $(\pm 1, 0)$

(b) $0 = \dfrac{1}{x} - x$

$x = \dfrac{1}{x}$

$x^2 = 1$

$x = \pm 1$

72. (a) x-intercepts: $(1, 0), (2, 0)$

(b) $0 = x - 3 + \dfrac{2}{x}$

$0 = x^2 - 3x + 2$

$0 = (x - 1)(x - 2)$

$x = 1, 2$

73. $C = \dfrac{255p}{100 - p}, \ 0 \le p < 100$

(a) $C(10) = \dfrac{255(10)}{100 - 10} \approx 28.33$ million dollars

(b) $C(40) = \dfrac{255(40)}{100 - 40} = 170$ million dollars

(c) $C(75) = \dfrac{255(75)}{100 - 75} = 765$ million dollars

(d) $C \longrightarrow \infty$ as $x \longrightarrow 100$. No, it would not be possible to remove 100% of the pollutants.

74. $C = \dfrac{25{,}000p}{100 - p}, \ 0 \le p < 100$

(a) $C = \dfrac{25{,}000(15)}{100 - 15} = \$4411.76$

(b) $C = \dfrac{25{,}000(50)}{100 - 50} = \$25{,}000$

(c) $C = \dfrac{25{,}000(90)}{100 - 90} = \$225{,}000$

(d) No. The function is undefined at $p = 100$ and the cost effectively grows without bound as 100% is approached.

75. $N = \dfrac{20(5 + 3t)}{1 + 0.04t}, \ 0 \le t$

(a) $N(5) \approx 333$ deer

$N(10) = 500$ deer

$N(25) = 800$ deer

(b) The herd is limited by the horizontal asymptote:

$N = \dfrac{60}{0.04} = 1500$ deer

76. (a) Let C be the fractional concentration of brine in final solution.

$$C = \frac{\text{(brine in 25\% solution)} + \text{(brine in 75\% solution)}}{\text{total amount of solution}}$$

$$= \frac{50 \cdot 0.25 + x \cdot 0.75}{50 + x} = \frac{12.5 + 0.75x}{x + 50} = \frac{3x + 50}{4(x + 50)}$$

(b) The amount x of added 75% solution can be any number of liters greater than or equal to zero that will fit in the 1000-liter tank:

$0 \le x \le 950$ or $[0, 950]$

(c) 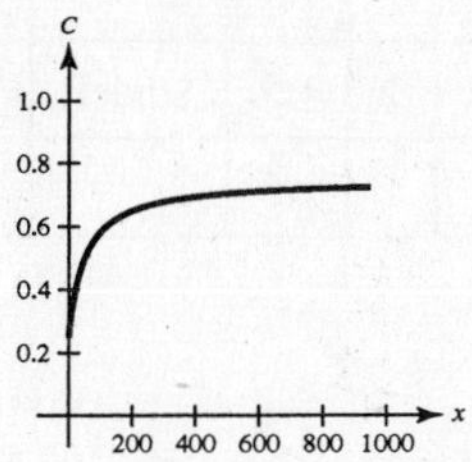

The concentration of brine increases at a slower rate.
The concentration of brine appears to be approaching 0.75 or 75%.

77. Let w be the width of the page and h be the height. The total of the horizontal margins is 3 inches and the total of the vertical margins is 2 inches. Thus,

64 square inches $= (w - 3)(h - 2)$.

Solving $64 = (w - 3)(h - 2)$ for h gives

$$h = 2 + \frac{64}{w - 3}.$$

The total area of the paper is

$$A = w \cdot h = w\left(2 + \frac{64}{w - 3}\right) = \frac{2w^2 + 58w}{w - 3}.$$

Plotting A versus w using a graphing utility, the graph has a minimum, for $w > 3$, at $w \approx 12.80$ inches.

$$h \approx 2 + \frac{64}{12.80 - 3} \approx 8.53 \text{ inches}$$

Width: 12.80 inches

Height: 8.53 inches

78. 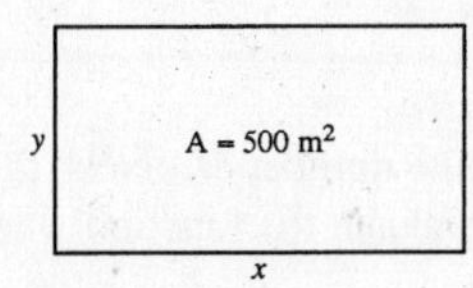

(a) $xy = 500$

$$y = \frac{500}{x}$$

(b) Domain: $x > 0$ or $(0, \infty)$

(c) When $x = 30$ m, $y = \dfrac{500}{30} = 16\dfrac{2}{3}$ m.

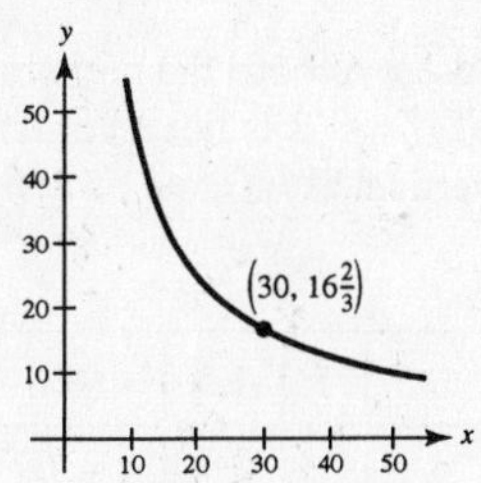

79. $C = \dfrac{3t^2 + t}{t^3 + 50}, t > 0$

(a) Since the degree of $N(t) <$ degree of $D(t)$, the horizontal asymptote is $C = 0$. The chemical will eventually dissipate.

(b) 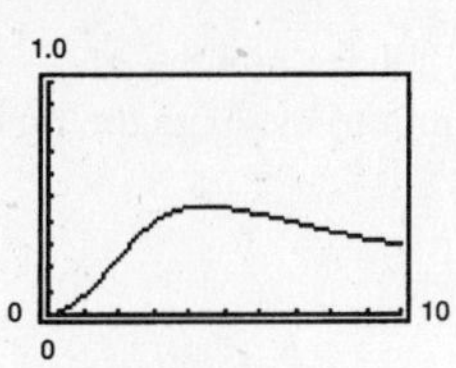

C is maximum when $t \approx 4.6$ hours

80. (a) Average speed $= \dfrac{\text{total distance}}{\text{total time}}$

$$50 = \frac{200}{\dfrac{100}{x} + \dfrac{100}{y}}$$

$$50 = \frac{2xy}{x + y}$$

$$50x + 50y = 2xy$$

$$y = \frac{25x}{x - 25}$$

(b) Vertical asymptote: $x = 25$

Horizontal asymptote: $y = 25$

81. $f(x) = \dfrac{x^2 - 1}{x + 1}$, $g(x) = x - 1$

(a) Domain of f: all real numbers except -1

Domain of g: all real numbers

(b) Because $(x + 1)$ is a factor of both the numerator and the denominator of f, $x = -1$ is not a vertical asymptote. f has no vertical asymptotes.

(c)

x	-3	-2	-1.5	-1	-0.5	0	1
$f(x)$	-4	-3	-2.5	Undef.	-1.5	-1	0
$g(x)$	-4	-3	-2.5	-2	-1.5	-1	0

(d)

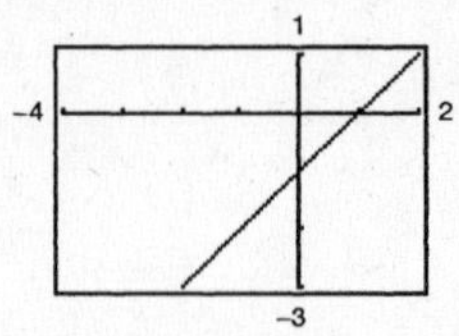

(e) Because there are only a finite number of pixels, the utility may not attempt to evaluate the function where it does not exist.

82. $f(x) = \dfrac{x^2(x - 2)}{x^2 - 2x}$, $g(x) = x$

(a) Domain of f: all real numbers except $x = 0$ and $x = 2$
Domain of g: all real numbers

(b) Since $x^2 - 2x$ is a factor of both the numerator and the denominator of f, neither $x = 0$ nor $x = 2$ is a vertical asymptote of f. Thus, f has no vertical asymptotes.

(c)

x	-1	0	1	1.5	2	2.5	3
$f(x)$	-1	Undef.	1	1.5	Undef.	2.5	3
$g(x)$	-1	0	1	1.5	2	2.5	3

(d)

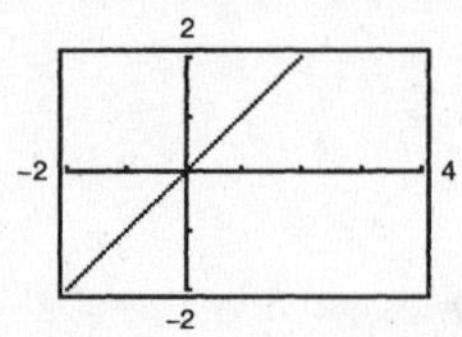

(e) Because there are only a finite number of pixels, the utility may not attempt to evaluate the function where it does not exist.

83. $f(x) = \dfrac{x - 2}{x^2 - 2x}$, $g(x) = \dfrac{1}{x}$

(a) Domain of f: all real numbers except 0 and 2

Domain of g: all real numbers except 0

(b) Because $(x - 2)$ is a factor of both the numerator and the denominator of f, $x = 2$ is not a vertical asymptote. The only vertical asymptote of f is $x = 0$.

(c)

x	-0.5	0	0.5	1	1.5	2	3
$f(x)$	-2	Undef.	2	1	$\frac{2}{3}$	Undef.	$\frac{1}{3}$
$g(x)$	-2	Undef.	2	1	$\frac{2}{3}$	$\frac{1}{2}$	$\frac{1}{3}$

(d)

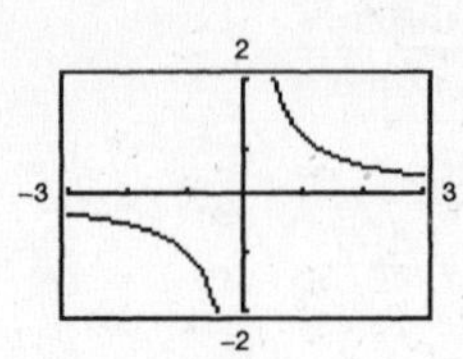

(e) Because there are only a finite number of pixels, the utility may not attempt to evaluate the function where it does not exist.

84. $f(x) = \dfrac{2x - 6}{x^2 - 7x + 12}$, $g(x) = \dfrac{2}{x - 4}$

(a) Domain of f: all real numbers except $x = 3$ and $x = 4$
Domain of g: all real numbers except $x = 4$

(b) Since $x - 3$ is a factor of both the numerator and the denominator of f, $x = 3$ is not a vertical asymptote of f. Thus, f has $x = 4$ as its only vertical asymptote.

(c)

x	0	1	2	3	4	5	6
$f(x)$	$-\frac{1}{2}$	$-\frac{2}{3}$	-1	Undef.	Undef.	2	1
$g(x)$	$-\frac{1}{2}$	$-\frac{2}{3}$	-1	-2	Undef.	2	1

(d)

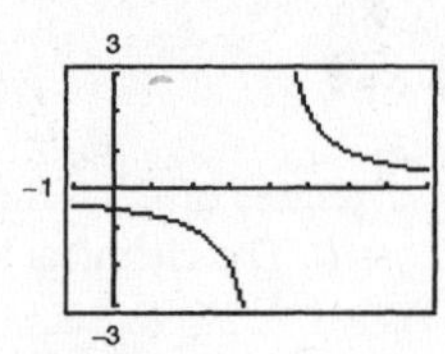

(e) Because there are only a finite number of pixels, the utility may not attempt to evaluate the function where it does not exist.

85. False. Polynomials do not have vertical asymptotes.

86. False. A rational function with the x-axis as a horizontal asymptote may cross the x-axis. For example, the graph of

$f(x) = \dfrac{x}{x^2 + 1}$ crosses $y = 0$, which is a horizontal asymptote.

87. True.

If $f(x) = \dfrac{N(x)}{D(x)}$, where $N(x)$ and $D(x)$ are polynomials, and $D(x) \neq 0$ for any real number, then $f(x)$ has no vertical asymptotes.

88. False.

$$f(x) = \frac{x^2 - 4}{x - 2} = \frac{(x + 2)(x - 2)}{x - 2} = x + 2, x \neq 2$$

Since $(x - 2)$ is a common factor of both the numerator and the denominator in this function, $x = 2$ is not a vertical asymptote. The graph has a "hole" at (2, 4).

89. $h(x) = \dfrac{6 - 2x}{3 - x}$

Although $x = 3$ is not in the domain of $h(x)$,

$h(x) = \dfrac{2(3 - x)}{3 - x} = 2$ for all real numbers except 3.

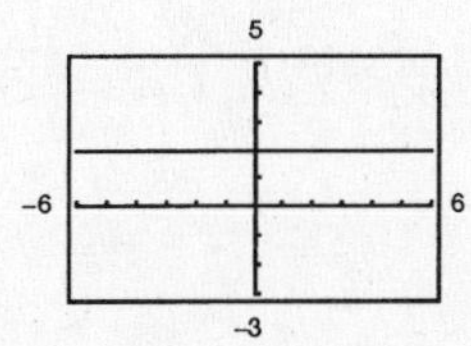

90. $g(x) = \dfrac{x^2 + x - 2}{x - 1} = \dfrac{(x + 2)(x - 1)}{x - 1}$

Since $N(x)/D(x)$ is not reduced, and $x - 1$ is a factor of both $N(x)$ and $D(x)$, $x = 1$ is **not** a vertical asymptote.

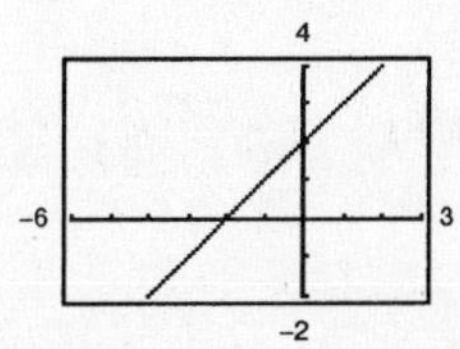

91. $P = \dfrac{1419.96 + 11.802t^2}{1.0 + 0.0027t^2}$, where $t = 7$ corresponds to 1997.

(a)

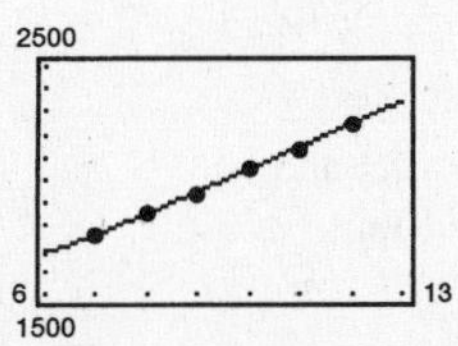

(b) $P(18) \approx 2797$ thousand

(c) Answers will vary. It will be somewhat useful since realistically, business partnerships will not continue to infinity. Likewise, the model levels off. However, the model is the best approximation only for the years in the data range, and as you choose a year farther from the data, the approximation is less accurate.

(d) $y \approx 95.2t + 1090$

(e) Answers will vary. The linear approximation fits the data better (it has an average error of 4.6 partnerships/year as opposed to the model given, which has an average error of 7 partnerships /year). However, the linear model does not interpolate past the data as well as the model given and the linear model continues to infinity as t increases.

92. $M = \dfrac{2252.5 + 34.743t^2}{1.0 + 0.035t^2}$, where $t = 5$ corresponds to 1995.

(a)

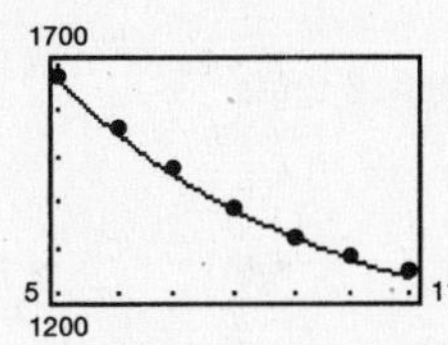

(b) $M(16) \approx 1119$ thousand

(c) $M \approx 0.55556t^3 - 4.7738t^2 - 105.710t + 2252.24$

(d) The cubic model is a better fit for the data, but would continue to infinity as t increases, whereas the rational model levels off.

Review Exercises for Chapter 2

1. (a) $y = 2x^2$

Vertical stretch

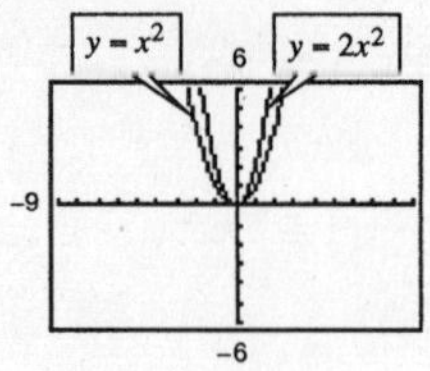

(b) $y = -2x^2$

Vertical stretch and a reflection in the x-axis

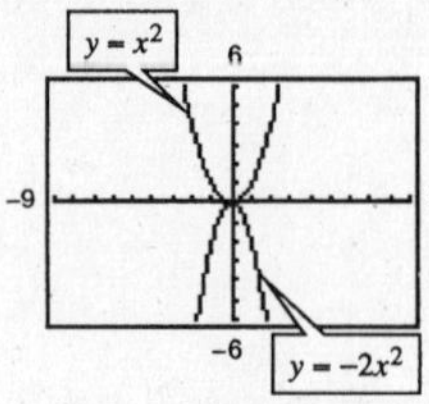

(c) $y = x^2 + 2$

Vertical shift two units upward

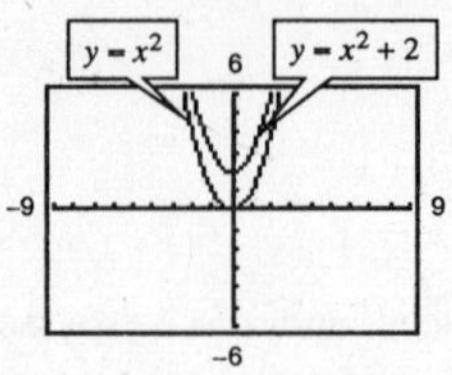

(d) $y = (x + 2)^2$

Horizontal shift two units to the left

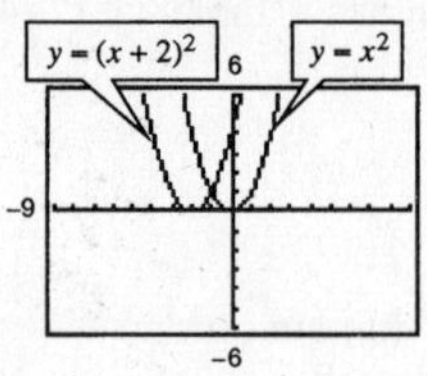

2. (a) $y = x^2 - 4$

Vertical shift downward 4 units

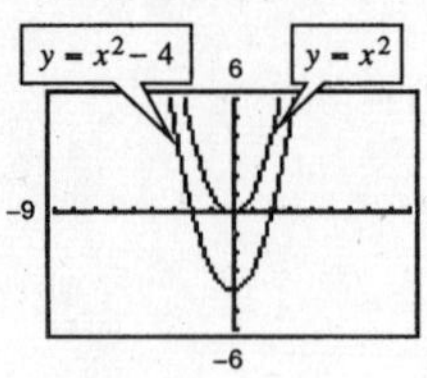

(b) $y = 4 - x^2$

Reflected in x-axis and vertical shift upward 4 units

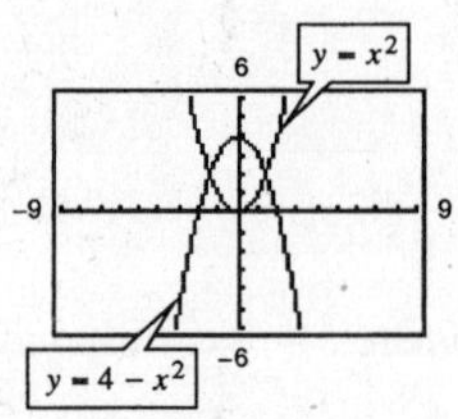

(c) $y = (x - 3)^2$

Horizontal shift to the right 3 units.

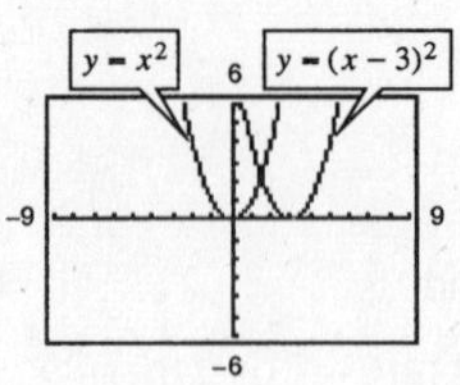

(d) $y = \frac{1}{2}x^2 - 1$

Vertical shrink by factor of $\frac{1}{2}$ and vertical shift downward 1 unit.

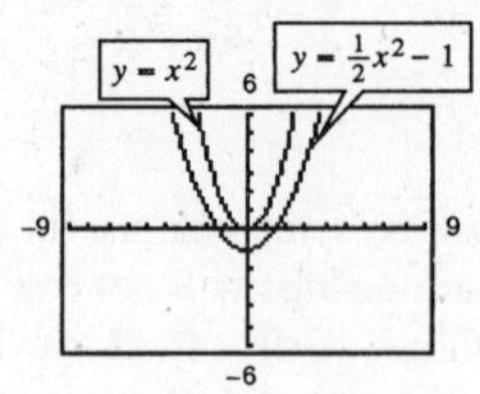

3. Vertical: $(4, 1) \Rightarrow f(x) = a(x - 4)^2 + 1$

Point: $(2, -1) \Rightarrow -1 = a(2 - 4)^2 + 1$

$-2 = 4a$

$-\frac{1}{2} = a$

Thus, $f(x) = -\frac{1}{2}(x - 4)^2 + 1$.

4. Vertex: $(2, 2) \Rightarrow f(x) = a(x - 2)^2 + 2$

Point: $(0, 3) \Rightarrow 3 = a(0 - 2)^2 + 2$

$3 = 4a + 2$

$1 = 4a$

$\frac{1}{4} = a$

$f(x) = \frac{1}{4}(x - 2)^2 + 2$

5. Vertex: $(1, -4) \implies f(x) = a(x-1)^2 - 4$

Point: $(2, -3) \implies -3 = a(2-1)^2 - 4$

$1 = a$

Thus, $f(x) = (x-1)^2 - 4$.

6. Vertex: $(2, 3) \implies f(x) = a(x-2)^2 + 3$

Point: $(-1, 6) \implies 6 = a(-1-2)^2 + 3$

$6 = 9a + 3$

$3 = 9a$

$\frac{1}{3} = a$

$f(x) = \frac{1}{3}(x-2)^2 + 3$

7. $g(x) = x^2 - 2x$

$= x^2 - 2x + 1 - 1$

$= (x-1)^2 - 1$

Vertex: $(1, -1)$

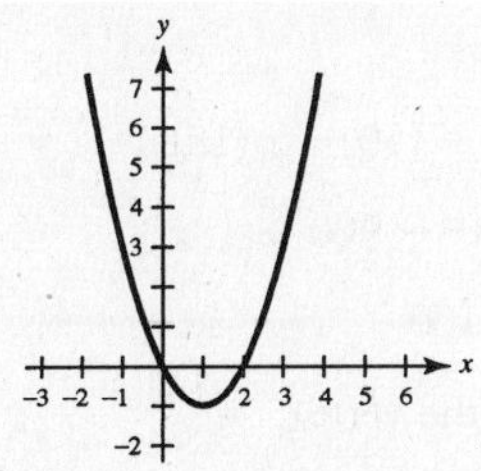

8. $f(x) = 6x - x^2$

$= -(x^2 - 6x + 9 - 9)$

$= -(x-3)^2 + 9$

Vertex $(3, 9)$

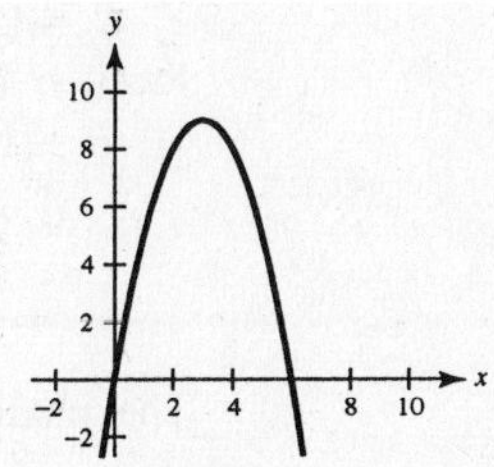

9. $f(x) = x^2 + 8x + 10$

$= x^2 + 8x + 16 - 16 + 10$

$= (x+4)^2 - 6$

Vertex: $(-4, -6)$

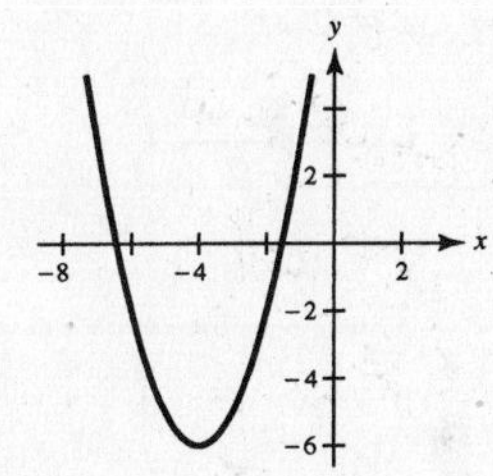

10. $h(x) = 3 + 4x - x^2$

$= -(x^2 - 4x - 3)$

$= -(x^2 - 4x + 4 - 4 - 3)$

$= -[(x-2)^2 - 7]$

$= -(x-2)^2 + 7$

Vertex $(2, 7)$

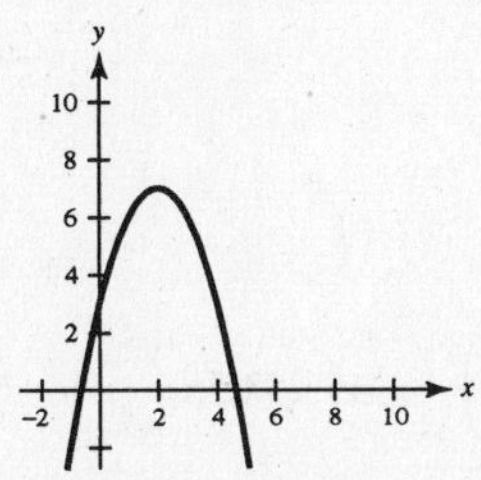

11. $f(t) = -2t^2 + 4t + 1$

$= -2(t^2 - 2t + 1 - 1) + 1$

$= -2[(t-1)^2 - 1] + 1$

$= -2(t-1)^2 + 3$

Vertex: $(1, 3)$

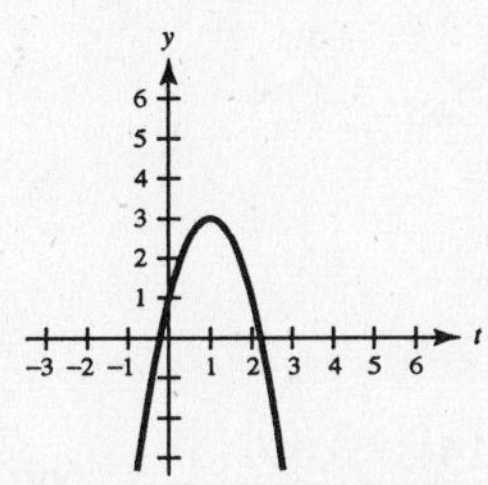

12. $f(x) = x^2 - 8x + 12$

$= x^2 - 8x + 16 - 16 + 12$

$= (x-4)^2 - 4$

Vertex $(4, -4)$

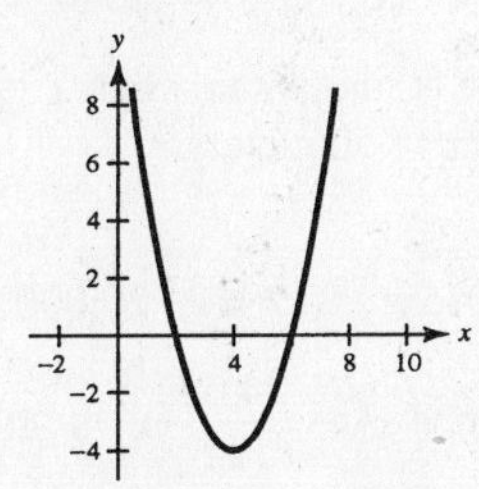

13. $h(x) = 4x^2 + 4x + 13$

$= 4(x^2 + x) + 13$

$= 4\left(x^2 + x + \frac{1}{4} - \frac{1}{4}\right) + 13$

$= 4\left(x^2 + x + \frac{1}{4}\right) - 1 + 13$

$= 4\left(x + \frac{1}{2}\right)^2 + 12$

Vertex: $\left(-\frac{1}{2}, 12\right)$

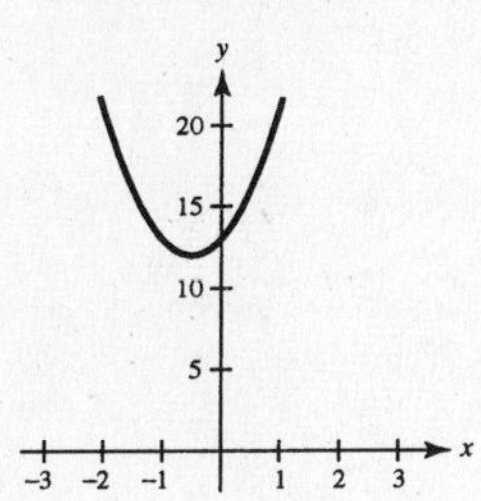

14. $h(x) = x^2 + 5x - 4$

$= x^2 + 5x + \frac{25}{4} - \frac{25}{4} - 4$

$= \left(x + \frac{5}{2}\right)^2 - \frac{25}{4} - \frac{16}{4}$

$= \left(x + \frac{5}{2}\right)^2 - \frac{41}{4}$

Vertex: $\left(-\frac{5}{2}, -\frac{41}{4}\right)$

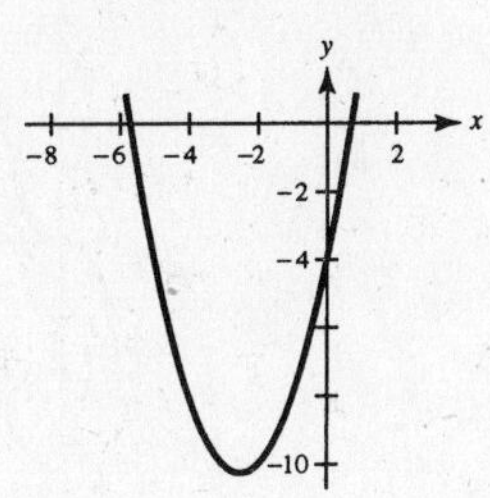

15. $f(x) = \frac{1}{3}(x^2 + 5x - 4)$

$= \frac{1}{3}\left(x^2 + 5x + \frac{25}{4} - \frac{25}{4} - 4\right)$

$= \frac{1}{3}\left[\left(x + \frac{5}{2}\right)^2 - \frac{41}{4}\right]$

$= \frac{1}{3}\left(x + \frac{5}{2}\right)^2 - \frac{41}{12}$

Vertex: $\left(-\frac{5}{2}, -\frac{41}{12}\right)$

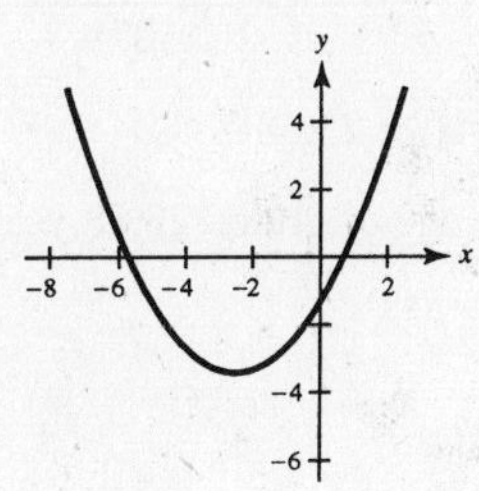

16. $f(x) = \frac{1}{2}(6x^2 - 24x + 22)$

$= 3x^2 - 12x + 11$

$= 3(x^2 - 4x + 4 - 4) + 11$

$= 3(x - 2)^2 + 3(-4) + 11$

$= 3(x - 2)^2 - 1$

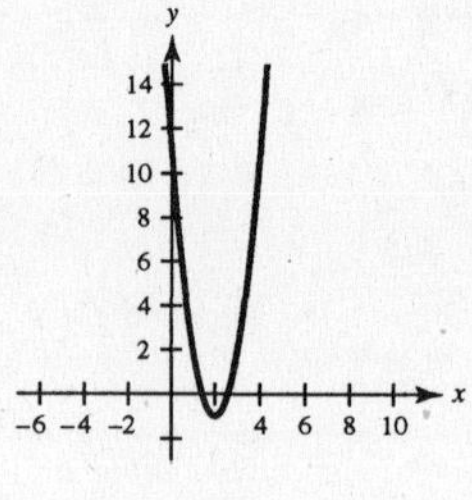

17. (a)

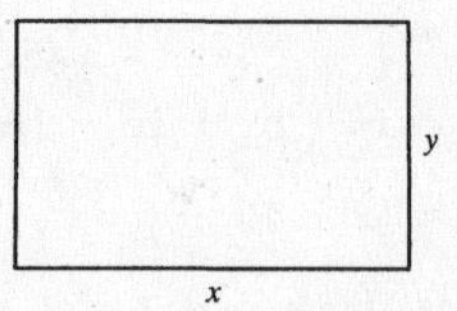

(b) $2x + 2y = 200$

$y = \dfrac{200 - 2x}{2}$

$y = 100 - x$

$A = xy = x(100 - x)$

$= 100x - x^2$

(c) $A = -x^2 + 100x$

$= -(x^2 - 100x)$

$= -(x^2 - 100x + 2500 - 2500)$

$= -[(x - 50)^2 - 2500]$

$= -(x - 50)^2 + 2500 \implies$ Vertex is $(50, 2500)$

Area is maximum when $x = 50$ meters and $y = 50$ meters.

18. Let $x =$ the number of \$30 increases in rent. Then,
Rent $= 540 + 30x$
Number of occupied units $= 50 - x$
Revenue $=$ (Number of occupied units)(rent)
$R = (50 - x)(540 + 30x)$
Cost $=$ (Number of occupied units)(\$18)
$C = (50 - x)(18)$
Profit $=$ Revenue $-$ Cost
$P = (50 - x)(540 + 30x) - (50 - x)(18)$

$= 27{,}000 + 960x - 30x^2 - 900 + 18x$

$= -30x^2 + 978x + 26{,}100$

The maximum profit occurs at the vertex.

$-\dfrac{b}{2a} = \dfrac{-978}{2(-30)} = 16.3 \approx 16$ increases

The corresponding rent is $540 + 30(16) = \$1020$.

19. $C = 0.055x^2 - 120x + 20{,}000$

$= 0.055\left(x^2 - \dfrac{24{,}000}{11}x\right) + 20{,}000$

$= 0.055\left(x^2 - \dfrac{24{,}000}{11}x + \dfrac{144{,}000{,}000}{121}\right) - 0.055\left(\dfrac{144{,}000{,}000}{121}\right) + 20{,}000 \approx 0.055\left(x - \dfrac{12{,}000}{11}\right)^2 - 45{,}454.55$

C is minimum when $x = \dfrac{12{,}000}{11} \approx 1091$ units

20. $y = -0.107x^2 + 5.68x - 48.5,\ 20 \le x \le 25$

$y = 26 \implies -0.107x^2 + 5.68x - 48.5 = 26$

$0 = 0.107x^2 - 5.68x + 74.5$

$x = \dfrac{5.68 \pm \sqrt{(-5.68)^2 - 4(0.107)(74.5)}}{2(0.107)}$

≈ 29.41 or 23.68

Average age of bride: 23.68 years

21. $y = x^3,\ f(x) = -(x - 4)^3$

$f(x)$ is a reflection in the x-axis and a horizontal shift 4 units to the right of the graph of $y = x^3$.

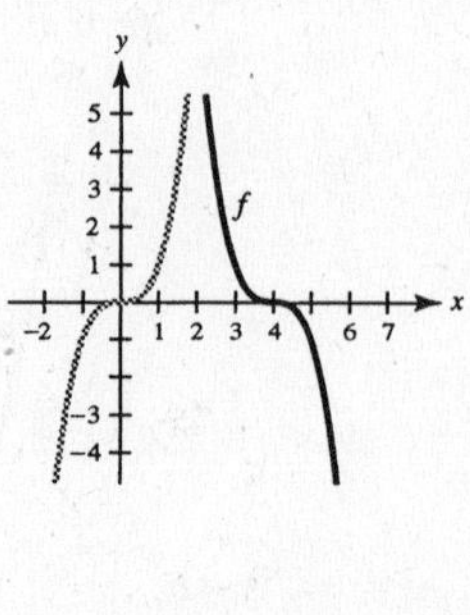

22. $y = x^3, f(x) = -4x^3$

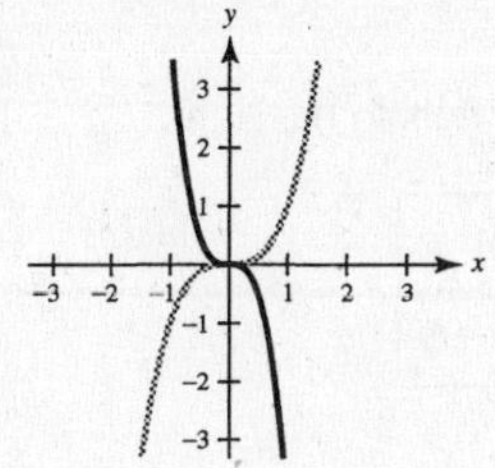

23. $y = x^4, f(x) = 2 - x^4$

$f(x)$ is a reflection in the x-axis and a vertical shift 2 units upward of the graph of $y = x^4$.

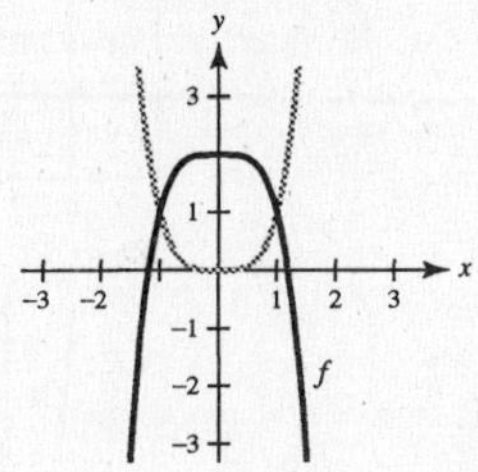

24. $y = x^4, f(x) = 2(x - 2)^4$

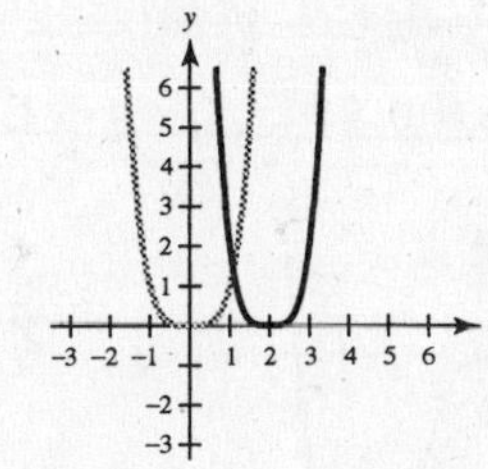

25. $y = x^5, f(x) = (x - 3)^5$

$f(x)$ is a horizontal shift 3 units to the right of the graph of $y = x^5$.

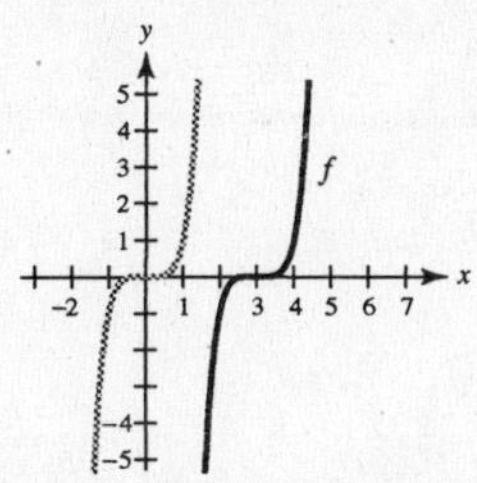

26. $y = x^5, f(x) = \frac{1}{2}x^5 + 3$

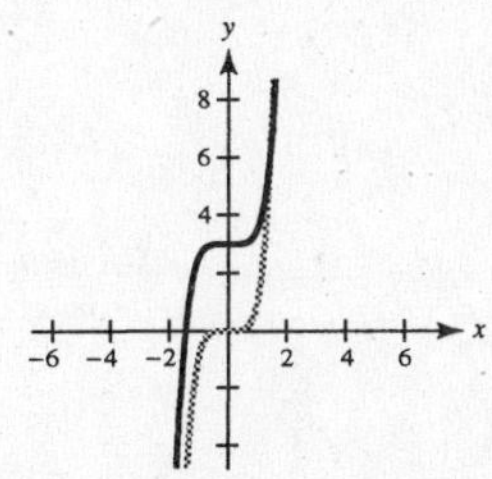

27. $f(x) = -x^2 + 6x + 9$

The degree is even and the leading coefficient is negative. The graph falls to the left and falls to the right.

28. $f(x) = \frac{1}{2}x^3 + 2x$

The degree is odd and the leading coefficient is positive. The graph falls to the left and rises to the right.

29. $g(x) = \frac{3}{4}(x^4 + 3x^2 + 2)$

The degree is even and the leading coefficient is positive. The graph rises to the left and rises to the right.

30. $h(x) = -x^5 - 7x^2 + 10x$

The degree is odd and the leading coefficient is negative. The graph rises to the left and falls to the right.

31. $f(x) = 2x^2 + 11x - 21$

$= 2\left(x + \frac{11}{4}\right)^2 - \frac{289}{8}$

Vertex: $(-2.75, -36.125)$

Zeros: $2x^2 + 11x - 21 = 0$

$(2x - 3)(x + 7) = 0$

$2x - 3 = 0 \Rightarrow x = \frac{3}{2}$

$x + 7 = 0 \Rightarrow x = -7$

The graph rises to the left and rises to the right.

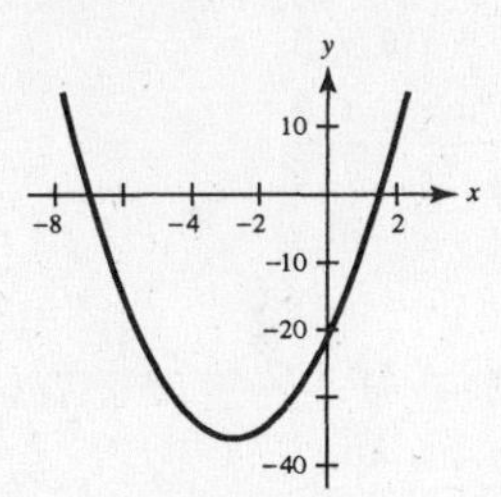

32. $f(x) = x(x + 3)^2$

$x(x + 3)^2 = 0$

Zeros: $x = -3, 0$

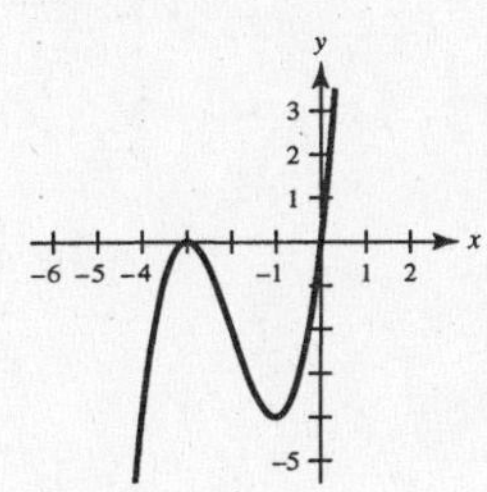

33. $f(t) = t^3 - 3t = t(t^2 - 3)$

Zeros: $t = 0, \pm\sqrt{3}$

The graph rises to the right and falls to the left.

x	-2	-1	0	1	2
y	-2	2	0	-2	2

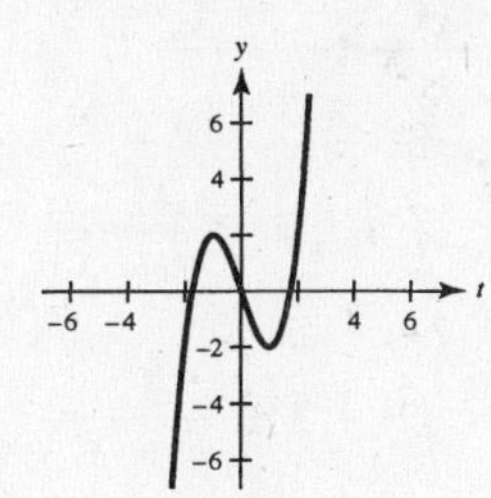

34. $f(x) = x^3 - 8x^2$

$x^3 - 8x^2 = x^2(x - 8) = 0$

Zeros: $x = 0, 8$

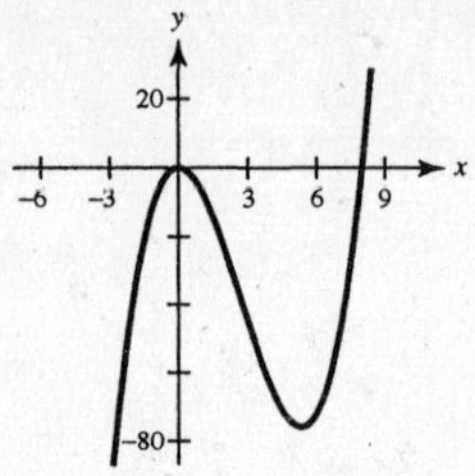

35. $f(x) = -12x^3 + 20x^2 = -4x^2(3x - 5)$

Zeros: $x = 0, \frac{5}{3}$

The graph rises to the left and falls to the right.

x	-2	-1	0	1	2
y	176	32	0	8	-16

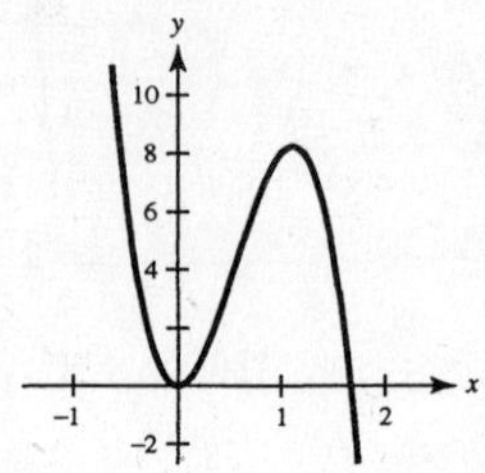

36. $g(x) = x^4 - x^3 - 2x^2$

$x^4 - x^3 - 2x^2 = x^2(x^2 - x - 2)$

$= x^2(x - 2)(x + 1)$

Zeros: $x = -1, 0, 2$

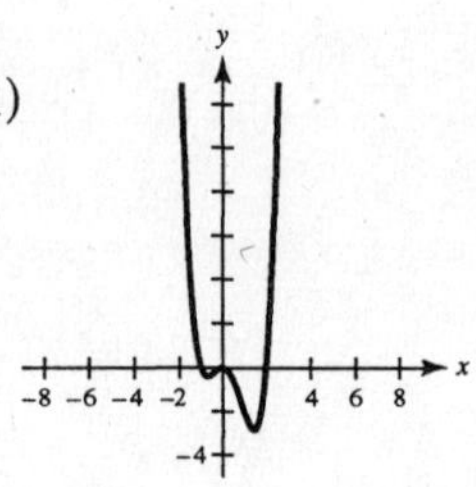

37.

$$\begin{array}{r} 8x + 5 \\ 3x - 2 \overline{) 24x^2 - x - 8} \\ \underline{24x^2 - 16x} \\ 15x - 8 \\ \underline{15x - 10} \\ 2 \end{array}$$

Thus, $\dfrac{24x^2 - x - 8}{3x - 2} = 8x + 5 + \dfrac{2}{3x - 2}$.

38.

$$\begin{array}{r} \frac{4}{3} \\ 3x - 2 \overline{) 4x + 7} \\ \underline{4x - \frac{8}{3}} \\ \frac{29}{3} \end{array}$$

$\dfrac{4x + 7}{3x - 2} = \dfrac{4}{3} + \dfrac{29}{3(3x - 2)}$

39.

$$\begin{array}{r} 5x + 2 \\ x^2 - 3x + 1 \overline{) 5x^3 - 13x^2 - x + 2} \\ \underline{5x^3 - 15x^2 + 5x} \\ 2x^2 - 6x + 2 \\ \underline{2x^2 - 6x + 2} \\ 0 \end{array}$$

Thus, $\dfrac{5x^3 - 13x^2 - x + 2}{x^2 - 3x + 1} = 5x + 2$.

40.

$$\begin{array}{r} 3x^2 + 3 \\ x^2 - 1 \overline{) 3x^4 + 0x^3 + 0x^2 + 0x + 0} \\ \underline{3x^4 - 3x^2} \\ 3x^2 \\ \underline{3x^2 - 3} \\ 3 \end{array}$$

$\dfrac{3x^4}{x^2 - 1} = 3x^2 + 3 + \dfrac{3}{x^2 - 1}$

41.

$$\begin{array}{r} x^2 - 3x + 2 \\ x^2 + 0x + 2 \overline{) x^4 - 3x^3 + 4x^2 - 6x + 3} \\ \underline{x^4 + 0x^3 + 2x^2} \\ -3x^3 + 2x^2 - 6x \\ \underline{-3x^3 + 0x^2 - 6x} \\ 2x^2 + 0x + 3 \\ \underline{2x^2 + 0x + 4} \\ -1 \end{array}$$

Thus,

$\dfrac{x^4 - 3x^3 + 4x^2 - 6x + 3}{x^2 + 2} = x^2 - 3x + 2 - \dfrac{1}{x^2 + 2}$.

42.

$$\begin{array}{r}
3x^2 + 5x + 8 \\
2x^2 - 1 \overline{\smash{)}\,6x^4 + 10x^3 + 13x^2 - 5x + 2} \\
\underline{6x^4 \qquad\quad - 3x^2 \qquad\qquad} \\
10x^3 + 16x^2 - 5x \qquad \\
\underline{10x^3 \qquad\quad - 5x \qquad} \\
16x^2 \qquad + 2 \\
\underline{16x^2 \qquad - 8} \\
10
\end{array}$$

$$\frac{6x^4 + 10x^3 + 13x^2 - 5x + 2}{2x^2 - 1} = 3x^2 + 5x + 8 + \frac{10}{2x^2 - 1}$$

43.

$$\begin{array}{r|rrrrr}
\frac{2}{3} & 6 & -4 & -27 & 18 & 0 \\
 & & 4 & 0 & -18 & 0 \\
\hline
 & 6 & 0 & -27 & 0 & 0
\end{array}$$

Thus, $\dfrac{6x^4 - 4x^3 - 27x^2 + 18x}{x - \frac{2}{3}} = 6x^3 - 27x.$

44.

$$\begin{array}{r|rrrr}
5 & 0.1 & 0.3 & 0 & -0.5 \\
 & & 0.5 & 4 & 20 \\
\hline
 & 0.1 & 0.8 & 4 & 19.5
\end{array}$$

$$\frac{0.1x^3 + 0.3x^2 - 0.5}{x - 5} = 0.1x^2 + 0.8x + 4 + \frac{19.5}{x - 5}$$

45.

$$\begin{array}{r|rrrr}
4 & 2 & -19 & 38 & 24 \\
 & & 8 & -44 & -24 \\
\hline
 & 2 & -11 & -6 & 0
\end{array}$$

Thus, $\dfrac{2x^3 - 19x^2 + 38x + 24}{x - 4} = 2x^2 - 11x - 6.$

46.

$$\begin{array}{r|rrrr}
-3 & 3 & 20 & 29 & -12 \\
 & & -9 & -33 & 12 \\
\hline
 & 3 & 11 & -4 & 0
\end{array}$$

$$\frac{3x^3 + 20x^2 + 29x - 12}{x + 3} = 3x^2 + 11x - 4$$

47. $f(x) = 20x^4 + 9x^3 - 14x^2 - 3x$

(a)
$$\begin{array}{r|rrrrr}
-1 & 20 & 9 & -14 & -3 & 0 \\
 & & -20 & 11 & 3 & 0 \\
\hline
 & 20 & -11 & -3 & 0 & 0
\end{array}$$

Yes, $x = -1$ is a zero of f.

(b)
$$\begin{array}{r|rrrrr}
\frac{3}{4} & 20 & 9 & -14 & -3 & 0 \\
 & & 15 & 18 & 3 & 0 \\
\hline
 & 20 & 24 & 4 & 0 & 0
\end{array}$$

Yes, $x = \frac{3}{4}$ is a zero of f.

(c)
$$\begin{array}{r|rrrrr}
0 & 20 & 9 & -14 & -3 & 0 \\
 & & 0 & 0 & 0 & 0 \\
\hline
 & 20 & 9 & -14 & -3 & 0
\end{array}$$

Yes, $x = 0$ is a zero of f.

(d)
$$\begin{array}{r|rrrrr}
1 & 20 & 9 & -14 & -3 & 0 \\
 & & 20 & 29 & 15 & 12 \\
\hline
 & 20 & 29 & 15 & 12 & 12
\end{array}$$

No, $x = 1$ is not a zero of f.

48. $f(x) = 3x^3 - 8x^2 - 20x + 16$

(a)
$$\begin{array}{r|rrrl}
4 & 3 & -8 & -20 & 16 \\
 & & 12 & 16 & -16 \\
\hline
 & 3 & 4 & -4 & 0 = f(4)
\end{array}$$

$x = 4$ is a zero of f.

(b)
$$\begin{array}{r|rrrl}
-4 & 3 & -8 & -20 & 16 \\
 & & -12 & 80 & -240 \\
\hline
 & 3 & -20 & 60 & -224 = f(-4)
\end{array}$$

$x = -4$ is not a zero of f.

(c)
$$\begin{array}{r|rrrl}
\frac{2}{3} & 3 & -8 & -20 & 16 \\
 & & 2 & -4 & -16 \\
\hline
 & 3 & -6 & -24 & 0 = f\left(\frac{2}{3}\right)
\end{array}$$

$x = \frac{2}{3}$ is a zero of f.

(d)
$$\begin{array}{r|rrrl}
-1 & 3 & -8 & -20 & 16 \\
 & & -3 & 11 & 9 \\
\hline
 & 3 & -11 & -9 & 25 = f(-1)
\end{array}$$

$x = -1$ is not a zero of f.

49. $f(x) = x^4 + 10x^3 - 24x^2 + 20x + 44$

(a)
$$\begin{array}{r|rrrrr} -3 & 1 & 10 & -24 & 20 & 44 \\ & & -3 & -21 & 135 & -465 \\ \hline & 1 & 7 & -45 & 155 & -421 \end{array}$$

Thus, $f(-3) = -421$.

(b)
$$\begin{array}{r|rrrrr} -1 & 1 & 10 & -24 & 20 & 44 \\ & & -1 & -9 & 33 & -53 \\ \hline & 1 & 9 & -33 & 53 & -9 \end{array}$$

Thus, $f(-1) = -9$.

50. $g(t) = 2t^5 - 5t^4 - 8t + 20$

(a)
$$\begin{array}{r|rrrrrr} -4 & 2 & -5 & 0 & 0 & -8 & 20 \\ & & -8 & 52 & -208 & 832 & -3296 \\ \hline & 2 & -13 & 52 & -208 & 824 & -3276 = g(-4) \end{array}$$

(b)
$$\begin{array}{r|rrrrrr} \sqrt{2} & 2 & -5 & 0 & 0 & -8 & 20 \\ & & 2\sqrt{2} & -5\sqrt{2}+4 & -10+4\sqrt{2} & -10\sqrt{2}+8 & -20 \\ \hline & 2 & -5+2\sqrt{2} & -5\sqrt{2}+4 & -10+4\sqrt{2} & -10\sqrt{2} & 0 = g(\sqrt{2}) \end{array}$$

51. $f(x) = x^3 + 4x^2 - 25x - 28$

(a)
$$\begin{array}{r} x^2 + 8x + 7 \\ x - 4 \overline{\smash{)}\, x^3 + 4x^2 - 25x - 28} \\ \underline{x^3 - 4x^2} \qquad\qquad \\ 8x^2 - 25x \qquad \\ \underline{8x^2 - 32x} \qquad \\ 7x - 28 \\ \underline{7x - 28} \\ 0 \end{array}$$

Since the remainder is 0, $(x - 4)$ is a factor of $f(x)$.

(b) $f(x) = (x - 4)(x^2 + 8x + 7)$

$= (x - 4)(x + 7)(x + 1)$

The remaining factors of $f(x)$ are $(x + 7)$ and $(x + 1)$.

(c) $f(x) = (x + 7)(x + 1)(x - 4)$

(d) The zeros of $f(x)$ are: $x = -7, -1, 4$.

(e)

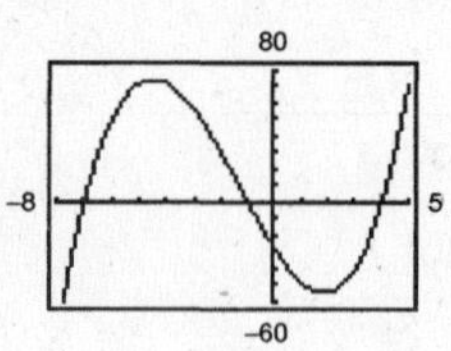

52. $f(x) = 2x^3 + 11x^2 - 21x - 90$; Factor: $(x + 6)$

(a)
$$\begin{array}{r|rrrr} -6 & 2 & 11 & -21 & -90 \\ & & -12 & 6 & 90 \\ \hline & 2 & -1 & -15 & 0 \end{array}$$

(b) $2x^2 - x - 15 = (2x + 5)(x - 3)$

(c) $f(x) = (x + 6)(2x + 5)(x - 3)$

(d) Zeros: $-6, -\frac{5}{2}, 3$

(e)

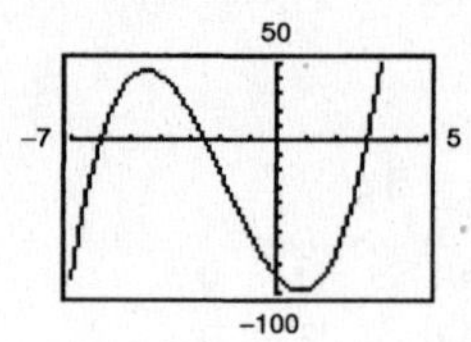

53. $f(x) = x^4 - 4x^3 - 7x^2 + 22x + 24$

(a)
$$\begin{array}{r|rrrrr} -2 & 1 & -4 & -7 & 22 & 24 \\ & & -2 & 12 & -10 & -24 \\ \hline & 1 & -6 & 5 & 12 & 0 \end{array}$$

$$\begin{array}{r|rrrr} 3 & 1 & -6 & 5 & 12 \\ & & 3 & -9 & -12 \\ \hline & 1 & -3 & -4 & 0 \end{array}$$

Since the remainders are zero, both $(x + 2)$ and $(x - 3)$ are factors of $f(x)$.

(b) $f(x) = (x + 2)(x - 3)(x^2 - 3x - 4)$

$= (x + 2)(x - 3)(x + 1)(x - 4)$

(c) $f(x) = (x + 1)(x - 4)(x + 2)(x - 3)$

(d) The zeros of $f(x)$ are: $x = -1, 4, -2, 3$

(e)

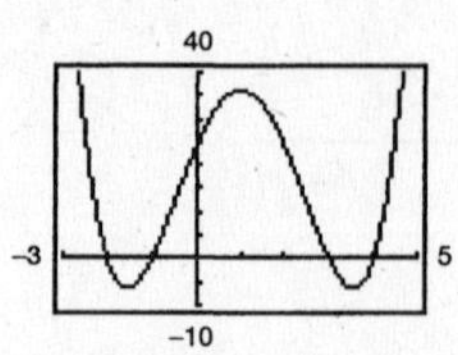

54. $f(x) = x^4 - 11x^3 + 41x^2 - 61x + 30$;
Factors: $(x - 2)(x - 5)$

(a)

2	1	−11	41	−61	30
		2	−18	46	−30
	1	−9	23	−15	0

5	1	−9	23	−15
		5	−20	15
	1	−4	3	0

(b) $x^2 - 4x + 3 = (x - 3)(x - 1)$

(c) $f(x) = (x - 1)(x - 2)(x - 3)(x - 5)$

(d) Zeros: 1, 2, 3, 5

(e)

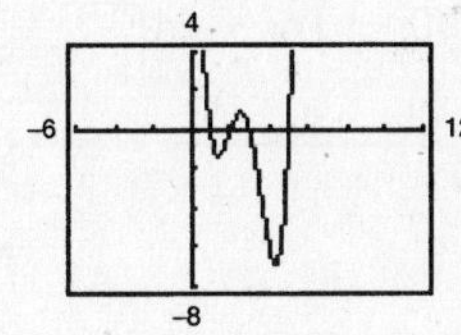

55. $A \approx -0.0022t^3 + 0.044t^2 + 0.17t + 2.3$

56.

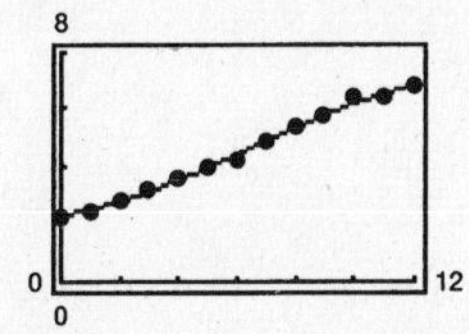

The model is a good fit to the actual data.

57.

t	A, actual	A, cubic model
0	2.3	2.3
1	2.4	2.5
2	2.9	2.8
3	3.2	3.1
4	3.6	3.5
5	4.0	4.0
6	4.2	4.4
7	4.9	4.9
8	5.4	5.3
9	5.8	5.8
10	6.4	6.2
11	6.5	6.6
12	6.9	6.9

58.

18	−0.0022	0.044	0.17	2.3
		−0.0396	0.0792	4.4856
	−0.0022	0.0044	0.2492	6.7856

$A(18) \approx 6.8$ million

No, the model falls to the right as t increases since the degree is odd and the leading coefficient is negative.

59. $3 - \sqrt{-25} = 3 - 5i$

60. $6 + \sqrt{-4} = 6 + 2i$

61. $-5i + i^2 = -5i - 1 = -1 - 5i$

62. $i^2 + 3i = -1 + 3i$

63. $\left(\frac{\sqrt{2}}{2} - \frac{\sqrt{2}}{2}i\right) - \left(\frac{\sqrt{2}}{2} + \frac{\sqrt{2}}{2}i\right) = \frac{\sqrt{2}}{2} - \frac{\sqrt{2}}{2}i - \frac{\sqrt{2}}{2} - \frac{\sqrt{2}}{2}i = -2\frac{\sqrt{2}}{2}i = -\sqrt{2}i$

64. $(7 + 5i) + (-4 + 2i) = 3 + 7i$

65. $(1 + 6i)(5 - 2i) = 5 - 2i + 30i - 12i^2 = 17 + 28i$

66. $5i(13 - 8i) = 40 + 65i$

67. $i(6 + i)(3 - 2i) = i(18 - 12i + 3i - 2i^2) = i(20 - 9i)$

$= 20i - 9i^2 = 9 + 20i$

68. $(10 - 8i)(2 - 3i) = 20 - 30i - 16i + 24i^2$

$= -4 - 46i$

69. $\dfrac{3 + 2i}{5 + i} = \dfrac{3 + 2i}{5 + i} \cdot \dfrac{5 - i}{5 - i} = \dfrac{15 - 3i + 10i - 2i^2}{25 - i^2}$

$= \dfrac{17 + 7i}{26} = \dfrac{17}{26} + \dfrac{7}{26}i$

70. $\dfrac{6 + i}{4 - i} = \dfrac{6 + i}{4 - i} \cdot \dfrac{4 + i}{4 + i} = \dfrac{23 + 10i}{17} = \dfrac{23}{17} + \dfrac{10}{17}i$

71. $\dfrac{1}{2 + i} - \dfrac{5}{1 + 4i} = \dfrac{1 + 4i - 5(2 + i)}{(2 + i)(1 + 4i)} = \dfrac{-9 - i}{-2 + 9i}$

$= \dfrac{9 + i}{2 - 9i} \cdot \dfrac{2 + 9i}{2 + 9i} = \dfrac{18 + 81i + 2i + 9i^2}{4 - 81i^2}$

$= \dfrac{9 + 83i}{85} = \dfrac{9}{85} + \dfrac{83}{85}i$

72. $\dfrac{4}{2 - 3i} + \dfrac{2}{1 + i} = \dfrac{4}{2 - 3i} \cdot \dfrac{2 + 3i}{2 + 3i} + \dfrac{2}{1 + i} \cdot \dfrac{1 - i}{1 - i}$

$= \dfrac{8 + 12i}{13} + \dfrac{2 - 2i}{2} = \dfrac{21 - i}{13}$

$= \dfrac{21}{13} - \dfrac{1}{13}i$

73. $2 + 8x^2 = 0$

$8x^2 = -2$

$x^2 = -\dfrac{1}{4}$

$x = \pm\sqrt{-\dfrac{1}{4}} = \pm\dfrac{1}{2}i$

74. $3x^2 + 1 = 0$

$x^2 = -\dfrac{1}{3}$

$x = \pm\dfrac{\sqrt{3}}{3}i$

75. $6x^2 + 3x + 27 = 0$

$3(2x^2 + x + 9) = 0$

By the Quadratic Formula: $x = \dfrac{-1 \pm \sqrt{-71}}{4}$

$= -\dfrac{1}{4} \pm \dfrac{\sqrt{71}}{4}i$

76. $x^2 - 2x + 10 = 0$

$x = \dfrac{-(-2) \pm \sqrt{(-2)^2 - 4(1)(10)}}{2(1)}$

$= \dfrac{2 \pm \sqrt{-36}}{2} = \dfrac{2 \pm 6i}{2}$

$= 1 \pm 3i$

77. $f(x) = (x - 4)(x + 9)^2$

There are three zeros (one repeating zero at $x = -9$).

The zeros are: $x = 4, -9$

78. $f(x) = 3x(x - 2)^2$

3 zeros

Zeros: 0, 2, 2

79. $f(x) = x^3 + 6x$

There are three zeros.

$f(x) = x(x^2 + 6) = x(x + \sqrt{6}i)(x - \sqrt{6}i)$

The zeros are: $x = 0, \pm\sqrt{6}i$

80. $f(x) = x^2 - 9x + 8$

2 zeros

$x^2 - 9x + 8 = (x - 1)(x - 8)$

Zeros: 1, 8

81. $f(x) = (x - 8)(x - 5)^2(x - 3 + i)(x - 3 - i)$

There are five zeros (one repeating zero).

The zeros are: $x = 8, 5, 3 \pm i$

82. $f(x) = (x + 4)(x - 6)(x - 2i)(x + 2i)$

4 zeros

Zeros: $-4, 6, 2i, -2i$

83. $f(x) = 3x^4 + 4x^3 - 5x^2 - 8$

Possible rational zeros: $\pm 1, \pm 2, \pm 4, \pm 8, \pm \frac{1}{3}, \pm \frac{2}{3}, \pm \frac{4}{3}, \pm \frac{8}{3}$

84. $f(x) = -4x^3 + 8x^2 - 3x + 15$

Possible rational zeros:
$\pm 1, \pm 3, \pm 5, \pm 15, \pm \frac{1}{2}, \pm \frac{3}{2}, \pm \frac{5}{2}, \pm \frac{15}{2}, \pm \frac{1}{4}, \pm \frac{3}{4}, \pm \frac{5}{4}, \pm \frac{15}{4}$

85. $f(x) = 3x^3 - 20x^2 + 7x + 30$

Possible rational zeros: $\pm 1, \pm 2, \pm 3, \pm 5, \pm 6, \pm 10, \pm 15, \pm 30, \pm \frac{1}{3}, \pm \frac{2}{3}, \pm \frac{5}{3}, \pm \frac{10}{3}$

$x = -1$ is a zero.

$$\begin{array}{r|rrrr} -1 & 3 & -20 & 7 & 30 \\ & & -3 & 23 & -30 \\ \hline & 3 & -23 & 30 & 0 \end{array}$$

$f(x) = (x + 1)(3x^2 - 23x + 30)$

$= (x + 1)(3x - 5)(x - 6)$

The zeros of $f(x)$ are: $x = -1, \frac{5}{3}, 6$.

86. $f(x) = x^3 + 9x^2 + 24x + 20$

Possible rational zeros: $\pm 1, \pm 2, \pm 4, \pm 5, \pm 10, \pm 20$

$x = -5$ is a zero.

$$\begin{array}{r|rrrr} -5 & 1 & 9 & 24 & 20 \\ & & -5 & -20 & -20 \\ \hline & 1 & 4 & 4 & 0 \end{array}$$

$f(x) = (x + 5)(x^2 + 4x + 4)$

$= (x + 5)(x + 2)^2$

The zeros of $f(x)$ are: $x = -5, -2$.

87. $f(x) = 25x^4 + 25x^3 - 154x^2 - 4x + 24$

Possible rational zeros:
$\pm 1, \pm 2, \pm 3, \pm 4, \pm 6, \pm 8, \pm 12, \pm 24, \pm \frac{1}{5}, \pm \frac{2}{5}, \pm \frac{3}{5}, \pm \frac{4}{5}, \pm \frac{6}{5}, \pm \frac{8}{5}, \pm \frac{12}{5}, \pm \frac{24}{5}, \pm \frac{1}{25}, \pm \frac{2}{25}, \pm \frac{3}{25}, \pm \frac{4}{25}, \pm \frac{6}{25}, \pm \frac{8}{25}, \pm \frac{12}{25}, \pm \frac{24}{25}$

$x = 2$ and $x = -3$ are zeros.

$$\begin{array}{r|rrrrr} 2 & 25 & 25 & -154 & -4 & 24 \\ & & 50 & 150 & -8 & -24 \\ \hline & 25 & 75 & -4 & -12 & 0 \end{array}$$

$$\begin{array}{r|rrrr} -3 & 25 & 75 & -4 & -12 \\ & & -75 & 0 & 12 \\ \hline & 25 & 0 & -4 & 0 \end{array}$$

$f(x) = (x - 2)(x + 3)(25x^2 - 4)$

$= (x - 2)(x + 3)(5x - 2)(5x + 2)$

The zeros of $f(x)$ are: $x = 2, -3, \pm \frac{2}{5}$.

88. $f(x) = x^4 + x^3 - 11x^2 + x - 12$

$$\begin{array}{r|rrrrr} 3 & 1 & 1 & -11 & 1 & -12 \\ & & 3 & 12 & 3 & 12 \\ \hline & 1 & 4 & 1 & 4 & 0 \end{array}$$

$$\begin{array}{r|rrrr} -4 & 1 & 4 & 1 & 4 \\ & & -4 & 0 & -4 \\ \hline & 1 & 0 & 1 & 0 \end{array}$$

$x^2 + 1$ has no real zeros.

Real zeros: $-4, 3$

89. Zeros: $2, -3, 1 - 2i$ (Since $1 - 2i$ is a zero, so is $1 + 2i$.)

$(x - 2)(x + 3)(x - (1 - 2i)[x - (1 + 2i)]$

$= (x^2 + x - 6)(x^2 - 2x + 5)$

$= x^4 - x^3 - 3x^2 + 17x - 30$

Any non-zero multiple of this polynomial would also have these zeros.

90. Zeros: $\frac{2}{3}, 4, \sqrt{3}i, -\sqrt{3}i$ (Since $\sqrt{3}i$ is a zero, so is $-\sqrt{3}i$)

$3\left(x - \frac{2}{3}\right)(x - 4)\left(x - \sqrt{3}i\right)\left(x + \sqrt{3}i\right) = (3x - 2)(x - 4)(x^2 + 3)$

$= 3x^4 - 14x^3 + 17x^2 - 42x + 24$

Any non-zero multiple of this polynomial would also have these zeros.

91. $f(x) = x^4 - 2x^3 - 2x^2 - 2x - 3$

$$\begin{array}{r} x^2 - 2x - 3 \\ x^2 + 0x + 1 \overline{)\, x^4 - 2x^3 - 2x^2 - 2x - 3} \\ \underline{x^4 + 0x^3 + x^2} \qquad\qquad \\ -2x^3 - 3x^2 - 2x \qquad \\ \underline{-2x^3 + 0x^2 - 2x} \qquad \\ -3x^2 + 0x - 3 \\ \underline{-3x^2 + 0x - 3} \\ 0 \end{array}$$

(a) Rationals: $f(x) = (x^2 + 1)(x^2 - 2x - 3)$

$= (x^2 + 1)(x + 1)(x - 3)$

(b) Linear and Quadratic Factors:

$f(x) = (x^2 + 1)(x + 1)(x - 3)$

(c) Completely factored:

$f(x) = (x + i)(x - i)(x + 1)(x - 3)$

92. $f(x) = x^4 - 2x^3 + 4x^2 + 2x - 5$

$$\begin{array}{r} x^2 - 2x + 5 \\ x^2 - 1 \overline{)\, x^4 - 2x^3 + 4x^2 + 2x - 5} \\ \underline{x^4 \qquad - x^2} \qquad\qquad \\ -2x^3 + 5x^2 + 2x \qquad \\ \underline{2x^3 \qquad + 2x} \qquad \\ 5x^2 \qquad -5 \\ \underline{5x^2 \qquad -5} \\ 0 \end{array}$$

(a) $f(x) = (x + 1)(x - 1)(x^2 - 2x + 5)$

(b) $f(x) = (x + 1)(x - 1)(x^2 - 2x + 5)$

(c) By the Quadratic Formula, the zeros of $x^2 - 2x + 5$ are

$$x = \frac{-(-2) \pm \sqrt{(-2)^2 - 4(1)(5)}}{2(1)}$$

$$= \frac{2 \pm \sqrt{-16}}{2} = 1 \pm 2i.$$

So, $f(x) = (x + 1)(x - 1)(x - 1 - 2i)(x - 1 + 2i)$.

93. Answers will vary.

$ax^2 + bx + c = 0$

(a) $b^2 - 4ac > 0$

$x^2 - 3x + 2 = 0$

$x = 1, 2$

(b) $b^2 - 4ac < 0$

$x^2 + 2x + 4 = 0$

$x = -1 \pm \sqrt{3}i$

(c) $b^2 - 4ac < 0$

$x^2 + 1 = 0$

$x = \pm i$

94. Exactly 2 real zeros and 2 complex zeros $\Longrightarrow$ 4 total zeros $\Longrightarrow$ function is of degree 4.

95. $f(x) = \dfrac{3x^2}{1 + 3x}$

$1 + 3x \neq 0$

$3x = -1$

$x = -\dfrac{1}{3}$

Domain: all real numbers except $x = -\frac{1}{3}$.

96. $f(x) = \dfrac{5x}{x + 12}$

$x + 12 = 0$

$x = -12$

Domain: all real numbers $x \neq -12$

97. $f(x) = \dfrac{x^2 + x - 2}{x^2 + 4}$

Domain: all real numbers

98. $f(x) = \dfrac{8}{x^2 - 10x + 24}$

$x^2 - 10x + 24 = (x - 6)(x - 4)$

Domain: all real numbers $x \neq 4, 6$

99. $f(x) = \dfrac{2x^2 + 5x - 3}{x^2 + 2}$

Vertical asymptote: none

Horizontal asymptote: $y = 2$

100. $f(x) = \dfrac{4}{x + 3}$

Horizontal asymptote: $y = 0$

Vertical asymptote: $x = -3$

101. $g(x) = \dfrac{1}{(x - 3)^2}$

Vertical asymptote: $x = 3$

Horizontal asymptote: $y = 0$

102. $g(x) = \dfrac{x^2}{x^2 - 4}$

Horizontal asymptote: $y = 1$

Vertical asymptotes: $x = \pm 2$

103. $f(x) = \dfrac{4}{x}$

No intercepts

Origin symmetry

Vertical asymptote: $x = 0$

Horizontal asymptote: $y = 0$

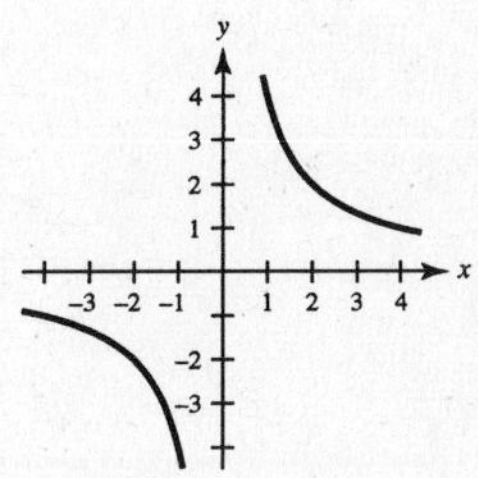

x	-3	-2	-1	1	2	3
y	$-\frac{4}{3}$	-2	-4	4	2	$\frac{4}{3}$

104. $f(x) = \dfrac{-5}{x^2}$

Intercepts: none

Symmetry: y-axis

Vertical asymptote: $x = 0$

Horizontal asymptote: $y = 0$

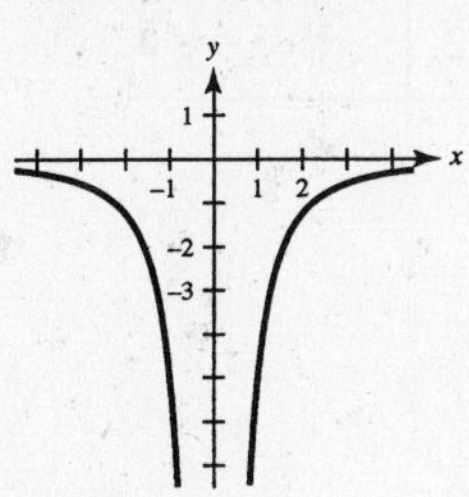

105. $h(x) = \dfrac{x - 3}{x - 2}$

x-intercept: $(3, 0)$

y-intercept: $\left(0, \dfrac{3}{2}\right)$

No axis or origin symmetry

Vertical asymptote: $x = 2$

Horizontal asymptote: $y = 1$

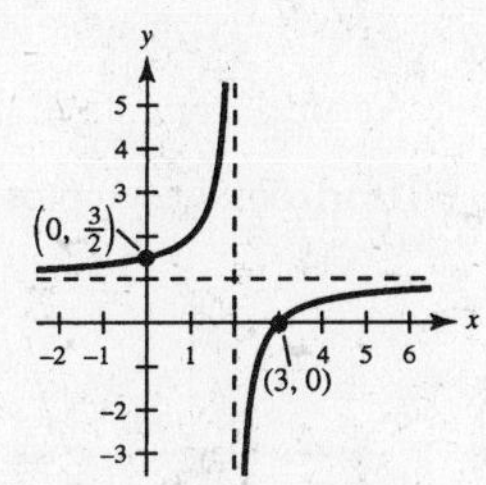

x	-1	0	1	3	4	5
y	$\frac{4}{3}$	$\frac{3}{2}$	2	0	$\frac{1}{2}$	$\frac{2}{3}$

106. $f(x) = \dfrac{2x}{x^2 + 4}$

Intercept: $(0, 0)$

Origin symmetry

Horizontal asymptote: $y = 0$

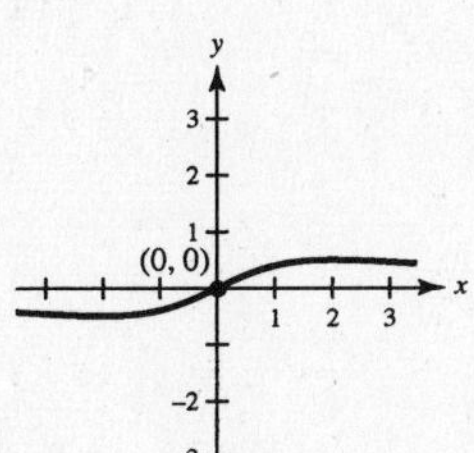

x	-2	-1	0	1	2
y	$-\frac{1}{2}$	$-\frac{2}{5}$	0	$\frac{2}{5}$	$\frac{1}{2}$

107. $p(x) = \dfrac{x^2}{x^2 + 1}$

Intercept: $(0, 0)$

Symmetry: y-axis

Vertical asymptote: none

Horizontal asymptote: $y = 1$

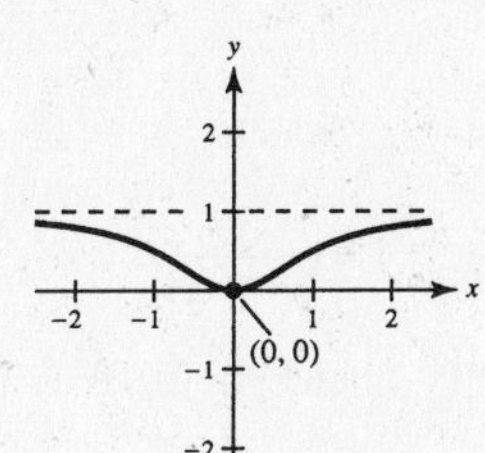

108. $h(x) = \dfrac{4}{(x-1)^2}$

y-intercept: $(0, 4)$

No axis or origin symmetry

Vertical asymptote: $x = 1$

Horizontal asymptote: $y = 0$

x	-2	-1	0	2	3	4
y	$\frac{4}{9}$	1	4	4	1	$\frac{4}{9}$

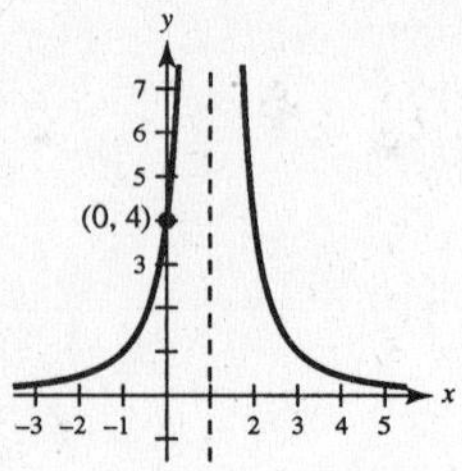

109. $y = \dfrac{2x^2}{x^2 - 4}$

Intercept: $(0, 0)$

y-axis symmetry

Vertical asymptotes: $x = 2, x = -2$

Horizontal asymptote: $y = 2$

x	± 5	± 4	± 3	± 1	0
y	$\frac{50}{21}$	$\frac{8}{3}$	$\frac{18}{5}$	$-\frac{2}{3}$	0

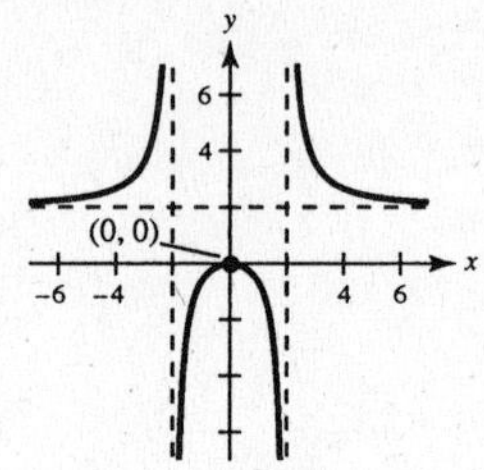

110. $f(x) = \dfrac{-6x^2}{x^2 + 1}$

Intercept: $(0, 0)$

Symmetry: y-axis

Vertical asymptote: none

Horizontal asymptote: $y = -6$

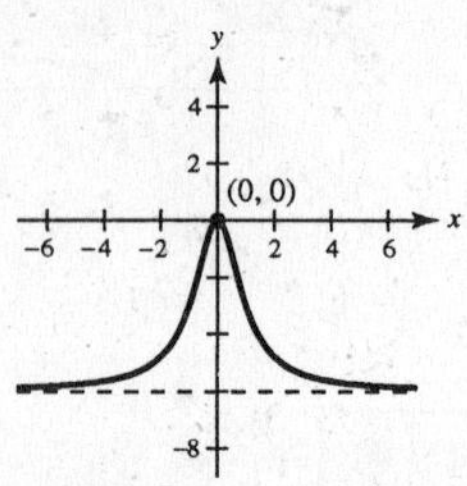

111. $g(x) = \dfrac{-2}{(x+3)^2}$

y-intercept: $\left(0, -\dfrac{2}{9}\right)$

No axis or origin symmetry

Vertical asymptote: $x = -3$

Horizontal asymptote: $y = 0$

x	-6	-5	-4	-2	-1	0
y	$-\frac{2}{9}$	$-\frac{1}{2}$	-2	-2	$-\frac{1}{2}$	$-\frac{2}{9}$

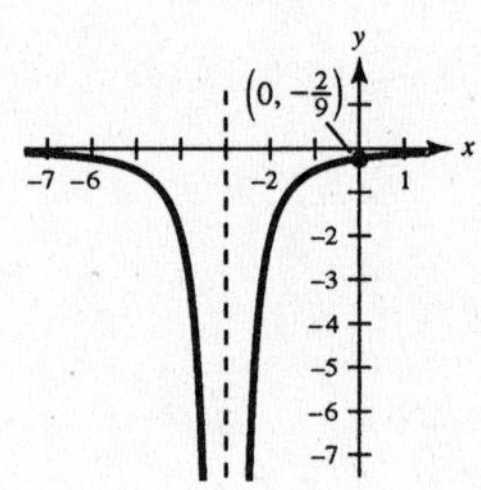

112. $y = \dfrac{x}{x^2 - 1}$

Intercept: $(0, 0)$

Symmetry: origin

Vertical asymptotes: $x = \pm 1$

Horizontal asymptote: $y = 0$

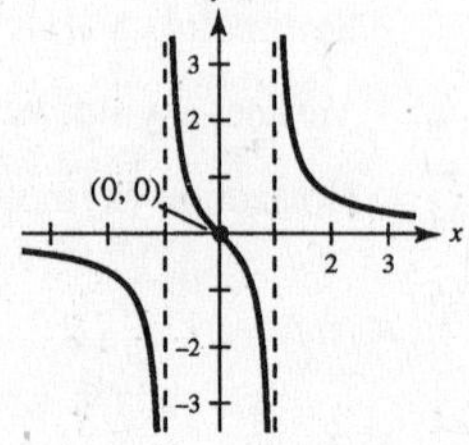

113. $f(x) = \dfrac{x^2 + 1}{x + 1}$

Using long division, $f(x) = \dfrac{x^2 + 1}{x + 1} = x - 1 + \dfrac{2}{x + 1}$

Slant asymptote: $y = x - 1$

Vertical asymptote: $x = -1$

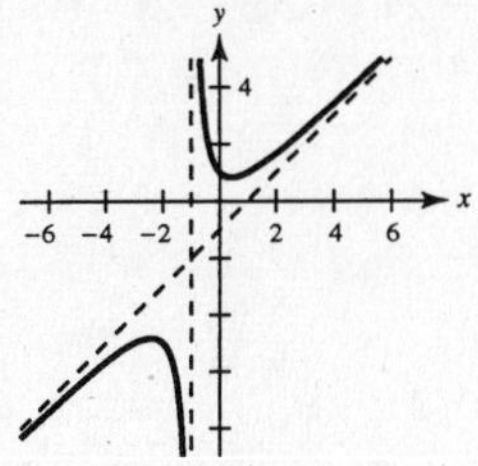

114. $f(x) = \dfrac{2x^3}{x^2 + 1} = 2x + \dfrac{-2x}{x^2 + 1}$

Slant asymptote: $y = 2x$

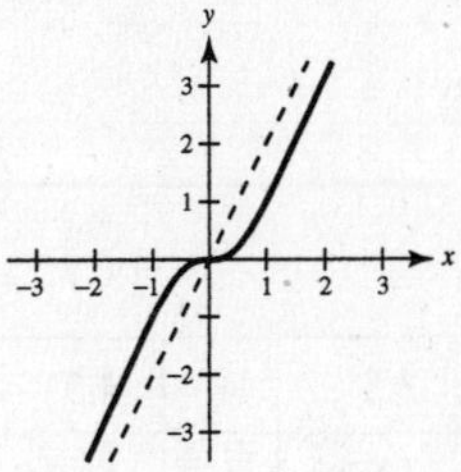

115. $f(x) = \dfrac{x^3}{x^2 - 4}$

Using long division, $f(x) = \dfrac{x^3}{x^2 - 4} = x + \dfrac{4x}{x^2 - 4}$

Slant asymptote: $y = x$

Vertical asymptotes: $x = \pm 2$

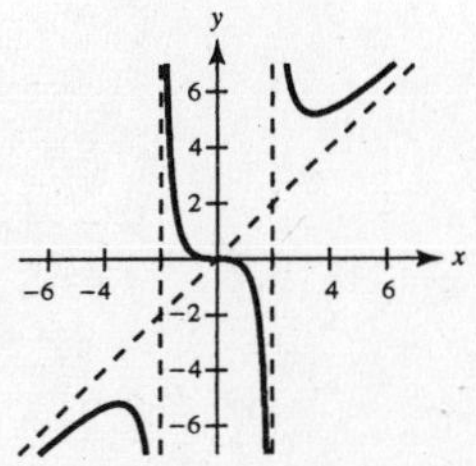

116. $f(x) = \dfrac{x^2 + 3x - 10}{x + 2} = x + 1 + \dfrac{-12}{x + 2}$

Slant asymptote: $y = x + 1$

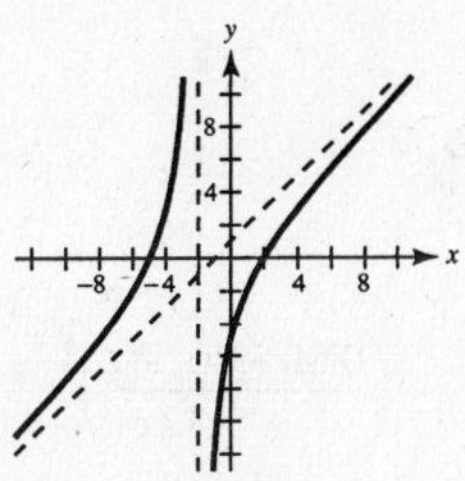

117. $\overline{C} = \dfrac{C}{x} = \dfrac{0.5x + 500}{x}, x > 0$

Horizontal asymptote: $y = 0.5$

Average cost per unit as the number of units increases without bound is 0.5.

118. $C = \dfrac{528p}{100 - p}, \quad 0 \le p < 100$

(a) When $p = 25$, $C = \dfrac{528(25)}{100 - 25} = \$176$ million.

(b) When $p = 50$, $C = \dfrac{528(50)}{100 - 50} = \$528$ million.

(c) When $p = 75$, $C = \dfrac{528(75)}{100 - 75} = \$1584$ million.

(d) As $p \longrightarrow 100$, $C \longrightarrow \infty$. No, it is not possible.

119. (a)

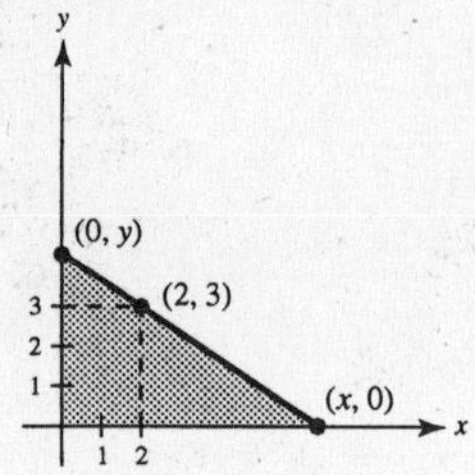

(b) $A = \frac{1}{2}bh$ where $b = x$ and $h = y$.

Slope: $\dfrac{y-0}{0-x} = \dfrac{0-3}{x-2}$

$$y = -x\left(\frac{-3}{x-2}\right) = \frac{3x}{x-2}$$

Area: $A = \dfrac{1}{2}xy = \dfrac{1}{2}x\left(\dfrac{3x}{x-2}\right) = \dfrac{3x^2}{2(x-2)},\ x > 2$

(c)

x	2.5	3	3.5	4	4.5
A	18.75	13.50	12.25	12	12.15

The area is minimum when

$x = 4$ and $y = \dfrac{3(4)}{4-2} = 6.$

(d)

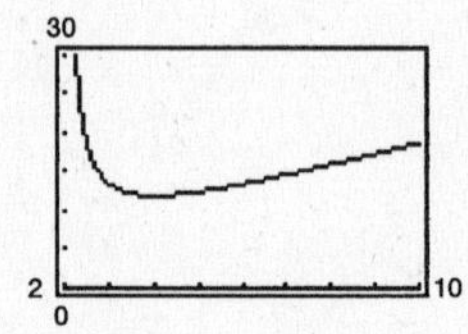

The area is minimum (12) when $x = 4$ and $y = 6$.

(e) $A = \dfrac{3x^2}{2x-4} = \dfrac{3}{2}x + 3 + \dfrac{12}{2x-4} = \dfrac{3}{2}(x+2) + \dfrac{6}{x-2}$

The slant asymptote is $y = \frac{3}{2}(x+2)$. The area increases without bound as x increases.

120. $y = \left(\dfrac{0.80 - 0.54x}{1 + 2.72x}\right)^2,\ x > 0$

$y = 0.1$ inch

$$0.1 = \left(\frac{0.80 - 0.54x}{1 + 2.72x}\right)^2$$

$$0.1(1 + 2.72x)^2 = (0.80 - 0.54x)^2$$

$$0.73984x^2 + 0.544x + 0.1 = 0.2916x^2 - 0.864x + 0.64$$

$$0.44824x^2 + 1.408x - 0.54 = 0$$

Using the Quadratic Formula, $x \approx -3.49, 0.346$.

Diameter of tube: 0.346 inches

Problem Solving for Chapter 2

1. Conditions (a) and (d) are preferable because profits would be increasing in the long run.

2. (a) If $f(x) = ax^2 + bx + c$ has two real zeros, then by the Quadratic Formula they are

$$x = \frac{-b \pm \sqrt{b^2 - 4ac}}{2a}.$$

The average of the zeros of f is

$$\frac{\dfrac{-b - \sqrt{b^2-4ac}}{2a} + \dfrac{-b + \sqrt{b^2-4ac}}{2a}}{2} = \frac{\dfrac{-2b}{2a}}{2} = -\frac{b}{2a}.$$

This is the x-coordinate of the vertex of the graph.

(b)

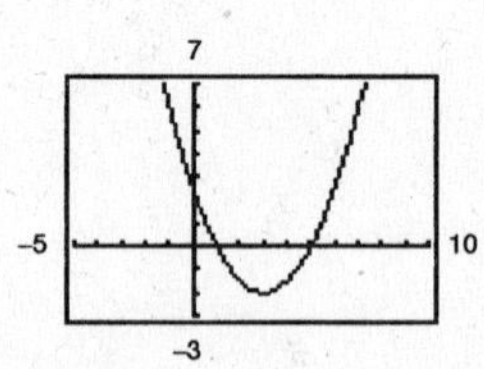

The x-coordinate of the vertex is 3, which is the average of the zeros 1 and 5.

3. If $h = 0$ and $k = 0$, then $a < -1$ produces a stretch that is reflected in the x-axis, and $-1 < a < 0$ produces a shrink that is reflected in the x-axis.

4. (a) $y_1 = -\frac{1}{3}(x - 2)^5 + 1$ is decreasing.

$y_2 = \frac{3}{5}(x + 2)^5 - 3$ is increasing.

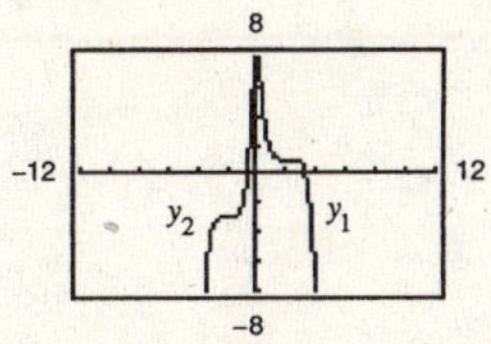

(b) The graph is either always increasing or always decreasing.

The behavior is determined by a.

If $a > 0$, $g(x)$ will always be increasing.

If $a < 0$, $g(x)$ will always be decreasing.

(c) $H(x) = x^5 - 3x^3 + 2x + 1$

Since $H(x)$ is not always increasing or always decreasing, $H(x) \neq a(x - h)^5 + k$.

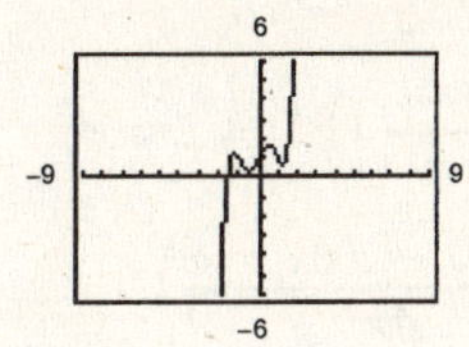

5. $f(x) = (x - k)q(x) + r$

(a) $k = 2,\ r = 5,\ q(x) =$ any quadratic $ax^2 + bx + c$ where $a > 0$.

One example: $f(x) = (x - 2)x^2 + 5 = x^3 - 2x^2 + 5$

(b) $k = -3,\ r = 1,\ q(x) =$ any quadratic $ax^2 + bx + c$ where $a < 0$.

One example:
$f(x) = (x + 3)(-x^2) + 1 = -x^3 - 3x^2 + 1$

6. $G = -0.003t^3 + 0.137t^2 + 0.458t - 0.839, \quad 2 \le t \le 34$

(a)

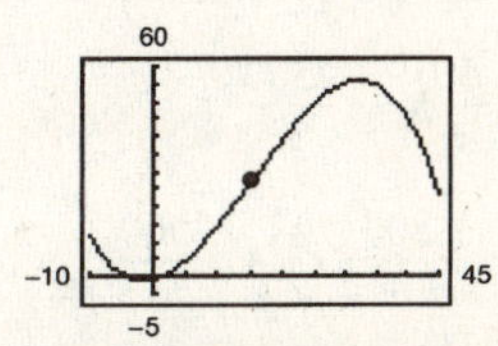

(b) The tree is growing most rapidly when it is approximately 15.2 years old.

(c) $y = -0.009t^2 + 0.274t + 0.458$

$$-\frac{b}{2a} = \frac{-0.274}{2(-0.009)} \approx 15.2222$$

$y(15.2222) \approx 2.5434$

Vertex: $(15.2222, 2.5434)$

(d) In both (b) and (c) the point of diminishing returns occured when $t \approx 15.2$.

7. $f(x) = \dfrac{2x^2 + x - 1}{x + 1}$

(a)

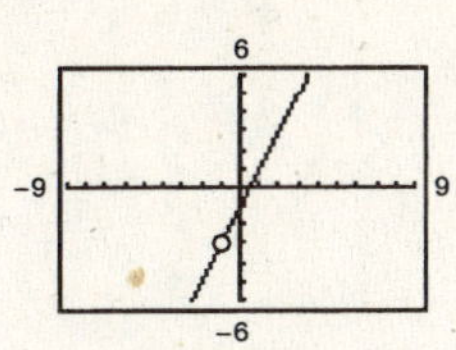

The graph has a "hole" when $x = -1$. There are no vertical asymptotes.

(b) $\dfrac{2x^2 + x - 1}{x + 1} = \dfrac{(2x - 1)(x + 1)}{x + 1} = 2x - 1, x \neq -1$

(c) As $x \to -1$, $\dfrac{2x^2 + x - 1}{x + 1} \to -3$

8. Let $x =$ length of the wire used to form the square.
Then $100 - x =$ length of wire used to form the circle.

(a) Let $s =$ the side of the square. Then $4s = x \Rightarrow s = \frac{x}{4}$ and the area of the square is $s^2 = \left(\frac{x}{4}\right)^2$.

Let $r =$ the radius of the circle. Then $2\pi r = 100 - x \Rightarrow r = \frac{100 - x}{2\pi}$ and the area of the circle is $\pi r^2 = \pi\left(\frac{100 - x}{2\pi}\right)^2$.

The combined area is:

$$\begin{aligned}
A(x) &= \left(\frac{x}{4}\right)^2 + \pi\left(\frac{100 - x}{2\pi}\right)^2 \\
&= \frac{x^2}{16} + \pi\left(\frac{10{,}000 - 200x + x^2}{4\pi^2}\right) \\
&= \frac{x^2}{16} + \frac{2500}{\pi} - \frac{50x}{\pi} + \frac{x^2}{4\pi} \\
&= \left(\frac{1}{16} + \frac{1}{4\pi}\right)x^2 - \frac{50x}{\pi} + \frac{2500}{\pi} \\
&= \left(\frac{\pi + 4}{16\pi}\right)x^2 - \frac{50}{\pi}x + \frac{2500}{\pi}
\end{aligned}$$

(b) Domain: Since the wire is 100 cm, $0 \le x \le 100$.

(c)
$$\begin{aligned}
A(x) &= \left(\frac{\pi + 4}{16\pi}\right)x^2 - \frac{50}{\pi}x + \frac{2500}{\pi} \\
&= \left(\frac{\pi + 4}{16\pi}\right)\left(x^2 - \frac{800}{\pi + 4}x\right) + \frac{2500}{\pi} \\
&= \left(\frac{\pi + 4}{16\pi}\right)\left[x^2 - \frac{800}{\pi + 4}x + \left(\frac{400}{\pi + 4}\right)^2 - \left(\frac{400}{\pi + 4}\right)^2\right] + \frac{2500}{\pi} \\
&= \left(\frac{\pi + 4}{16\pi}\right)\left[x - \left(\frac{400}{\pi + 4}\right)\right]^2 - \left(\frac{\pi + 4}{16\pi}\right)\left(\frac{400}{\pi + 4}\right)^2 + \frac{2500}{\pi} \\
&= \left(\frac{\pi + 4}{16\pi}\right)\left[x - \left(\frac{400}{\pi + 4}\right)\right]^2 - \frac{10{,}000}{\pi(\pi + 4)} + \frac{2500}{\pi} \\
&= \left(\frac{\pi + 4}{16\pi}\right)\left[x - \left(\frac{400}{\pi + 4}\right)\right]^2 + \frac{2500}{\pi + 4}
\end{aligned}$$

The minimum occurs at the vertex when $x = \frac{400}{\pi + 4} \approx 56$ cm and $A(x) \approx 350\text{ cm}^2$.

The maximum occurs at one of the endpoints of the domain.

When $x = 0$, $A(x) \approx 796\text{ cm}^2$.

When $x = 100$, $A(x) = 625\text{ cm}^2$.

Thus, the area is maximum when $x = 0$ cm.

(d) Answers will vary. Graph $A(x)$ to see where the minimum and maximum values occur.

9. $y = ax^2 + bx + c$

(a) $(0, -4)$: $-4 = c$

$(1, 0)$: $0 = a + b + c \Rightarrow a + b = 4$

$(2, 2)$: $2 = 4a + 2b + c \Rightarrow 4a + 2b = 6$

Solve the system of equations:

$$\begin{aligned} 4a + 2b = 6 &\Rightarrow \quad 2a + b = 3 \\ a + b = 4 &\Rightarrow \underline{-a - b = -4} \\ & \qquad a \qquad = -1 \\ & \quad -1 + b = 4 \\ & \qquad\quad b = 5 \end{aligned}$$

Thus, $y = -x^2 + 5x - 4$.

Check: $(4, 0)$: $0 = -(4)^2 + 5(4) - 4$ ✓

$(6, -10)$: $-10 = -(6)^2 + 5(6) - 4$ ✓

(b)

L_1	L_2
0	−4
1	0
2	2
4	0
6	−10

Use the "Quad Reg" feature of your graphing utility to obtain

$y = -x^2 + 5x - 4$

10.

$$\begin{array}{r} x^{n-1} + x^{n-2} + x^{n-3} + \cdots \quad \cdots + x + 1 \\ x - 1 \overline{\big) x^n + 0x^{n-1} + 0x^{n-2} + 0x^{n-3} + \cdots + 0x^2 + 0x - 1} \\ \underline{x^n - x^{n-1}} \qquad\qquad\qquad\qquad\qquad\qquad\qquad\qquad \\ x^{n-1} + 0x^{n-2} \qquad\qquad\qquad\qquad\qquad\qquad \\ \underline{x^{n-1} - x^{n-2}} \qquad\qquad\qquad\qquad\qquad\qquad \\ x^{n-2} + 0x^{n-3} \qquad\qquad\qquad\qquad \\ \underline{x^{n-2} - x^{n-3}} \qquad\qquad\qquad\qquad \\ x^{n-3} + 0x^{n-4} \qquad\qquad\qquad \\ \vdots \qquad\qquad\qquad\qquad \\ x^2 + 0x \qquad \\ \underline{x^2 - x} \qquad \\ x - 1 \\ \underline{x - 1} \\ 0 \end{array}$$

Thus, $\dfrac{x^n - 1}{x - 1} = x^{n-1} + x^{n-2} + x^{n-3} + \cdots + x + 1$.

11. (a) Slope $= \dfrac{9 - 4}{3 - 2} = 5$. Slope of tangent line is less than 5.

(b) Slope $= \dfrac{4 - 1}{2 - 1} = 3$. Slope of tangent line is greater than 3.

(c) Slope $= \dfrac{4.41 - 4}{2.1 - 2} = 4.1$. Slope of tangent line is less than 4.1.

(d)
$$\begin{aligned} \text{Slope} &= \frac{f(2 + h) - f(2)}{(2 + h) - 2} \\ &= \frac{(2 + h)^2 - 4}{h} \\ &= \frac{4h + h^2}{h} \\ &= 4 + h, h \neq 0 \end{aligned}$$

(e) Slope $= 4 + h$, $h \neq 0$

$4 + (-1) = 3$

$4 + 1 = 5$

$4 + 0.1 = 4.1$

The results are the same as in (a)–(c).

(f) Letting h get closer and closer to 0, the slope approaches 4. Hence, the slope at $(2, 4)$ is 4.

12. (a) $x + 2y = 100 \Rightarrow y = \dfrac{100 - x}{2}$

$A(x) = xy = x\left(\dfrac{100 - x}{2}\right) = -\dfrac{x^2}{2} + 50x$

Domain: $0 < x < 100$

(b)

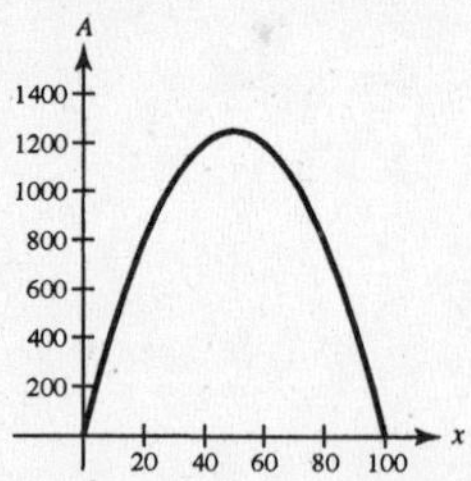

Maximum of 1250 m^2 at $x = 50$ m, $y = 25$ m

(c) $A(x) = -\dfrac{1}{2}(x^2 - 100x)$

$= -\dfrac{1}{2}(x^2 - 100x + 2500) + 1250$

$= -\dfrac{1}{2}(x - 50)^2 + 1250$

$A(50) = 1250 \text{ m}^2$ is the maximum.

$x = 50$ m, $y = 25$ m

13. $f(x) = \dfrac{ax + b}{cx + d}$

$f(x)$ has a vertical asymptote at $x = -\dfrac{d}{c}$ and a horizontal asymptote at $y = \dfrac{a}{c}$.

(i) $a > 0$

$b < 0$

$c > 0$

$d < 0$

$x = -\dfrac{d}{c}$ is positive.

$y = \dfrac{a}{c}$ is positive.

Both asymptotes are positive on graph (d).

(ii) $a > 0$

$b > 0$

$c < 0$

$d < 0$

$x = -\dfrac{d}{c}$ is negative.

$y = \dfrac{a}{c}$ is negative.

Both asymptotes are negative on graph (b).

(iii) $a < 0$

$b > 0$

$c > 0$

$d < 0$

$x = -\dfrac{d}{c}$ is positive.

$y = \dfrac{a}{c}$ is negative.

The vertical asymptote is positive and the horizontal asymptote is negative on graph (a).

(iv) $a > 0$

$b < 0$

$c > 0$

$d > 0$

$x = -\dfrac{d}{c}$ is negative.

$y = \dfrac{a}{c}$ is positive.

The vertical asymptote is negative and the horizontal asymptote is positive on graph (c).

CHAPTER 3
Limits and Their Properties

Section 3.1 A Preview of Calculus 256

Section 3.2 Finding Limits Graphically and Numerically 257

Section 3.3 Evaluating Limits Analytically 265

Section 3.4 Continuity and One-Sided Limits 272

Section 3.5 Infinite Limits 280

Review Exercises 286

Problem Solving 291

CHAPTER 3
Limits and Their Properties

Section 3.1 A Preview of Calculus

1. Precalculus: (20 ft/sec)(15 seconds) = 300 feet

2. Calculus: velocity is not constant

Distance $\approx$ (20 ft/sec)(15 seconds) = 300 feet

3. Calculus required: slope of tangent line at $x = 2$ is rate of change, and equals about 0.16.

4. Precalculus: rate of change = slope = 0.08

5. Precalculus: Area $= \frac{1}{2}bh = \frac{1}{2}(5)(3) = \frac{15}{2}$ sq. units

6. Precalculus: Area $= \pi(\sqrt{2})^2$

$= 2\pi$

7. $f(x) = 4x - x^2$

(a)

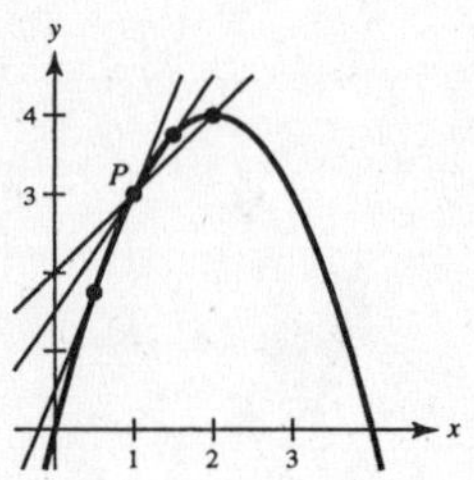

(b) $\text{slope} = m = \dfrac{(4x - x^2) - 3}{x - 1}$

$= \dfrac{(x - 1)(3 - x)}{x - 1} = 3 - x, \quad x \neq 1$

$x = 2$: $m = 3 - 2 = 1$

$x = 1.5$: $m = 3 - 1.5 = 1.5$

$x = 0.5$: $m = 3 - 0.5 = 2.5$

(c) At $P(1, 3)$ the slope is 2.

You can improve your approximation of the slope at $x = 1$ by considering x-values very close to 1.

8. $f(x) = \sqrt{x}$

(a)

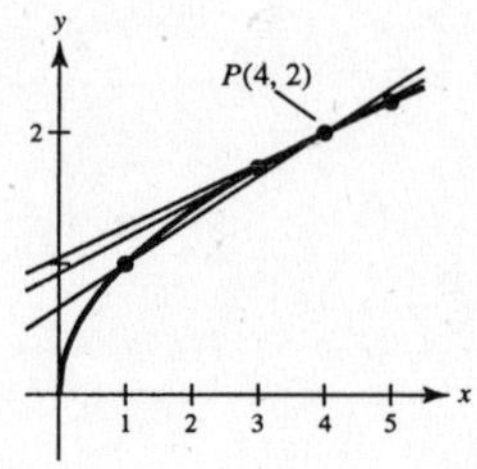

(b) $\text{slope} = m = \dfrac{\sqrt{x} - 2}{x - 4}$

$= \dfrac{\sqrt{x} - 2}{(\sqrt{x} + 2)(\sqrt{x} - 2)} = \dfrac{1}{\sqrt{x} + 2}, \quad x \neq 4$

$x = 1$: $m = \dfrac{1}{\sqrt{1} + 2} = \dfrac{1}{3}$

$x = 3$: $m = \dfrac{1}{\sqrt{3} + 2} \approx 0.2679$

$x = 5$: $m = \dfrac{1}{\sqrt{5} + 2} \approx 0.2361$

(c) At $P(4, 2)$ the slope is $\dfrac{1}{\sqrt{4} + 2} = \dfrac{1}{4} = 0.25$.

You can improve your approximation of the slope at $x = 4$ by considering x-values very close to 4.

9. $\text{Area} \approx 5 + \dfrac{5}{2} + \dfrac{5}{3} + \dfrac{5}{4} \approx 10.417$

$\text{Area} \approx \dfrac{1}{2}\left(5 + \dfrac{5}{1.5} + \dfrac{5}{2} + \dfrac{5}{2.5} + \dfrac{5}{3} + \dfrac{5}{3.5} + \dfrac{5}{4} + \dfrac{5}{4.5}\right) \approx 9.145$

10. For the graph on the left,

Area $\approx 3 + 4 + 3 + 0 = 10.$

For the graph on the right,

Area $\approx \frac{1}{2}[1.75 + 3 + 3.75 + 4 + 3.75 + 3 + 1.75 + 0]$

$= \frac{1}{2}[21] = 10.5$

[Note: The exact area is $10\frac{2}{3}$.

11. (a) $D_1 = \sqrt{(5-1)^2 + (1-5)^2} = \sqrt{16+16} \approx 5.66$

(b) $D_2 = \sqrt{1 + \left(\frac{5}{2}\right)^2} + \sqrt{1 + \left(\frac{5}{2} - \frac{5}{3}\right)^2} + \sqrt{1 + \left(\frac{5}{3} - \frac{5}{4}\right)^2} + \sqrt{1 + \left(\frac{5}{4} - 1\right)^2}$

$\approx 2.693 + 1.302 + 1.083 + 1.031 \approx 6.11$

(c) Increase the number of line segments.

Section 3.2 Finding Limits Graphically and Numerically

1.

x	1.9	1.99	1.999	2.001	2.01	2.1
$f(x)$	0.3448	0.3344	0.3334	0.3332	0.3322	0.3226

$\lim_{x\to 2} \frac{x-2}{x^2-x-2} \approx 0.3333$ (Actual limit is $\frac{1}{3}$.)

2.

x	1.9	1.99	1.999	2.001	2.01	2.1
$f(x)$	0.2564	0.2506	0.2501	0.2499	0.2494	0.2439

$\lim_{x\to 2} \frac{x-2}{x^2-4} \approx 0.25$ (Actual limit is $\frac{1}{4}$.)

3.

x	−0.1	−0.01	−0.001	0.001	0.01	0.1
$f(x)$	0.2911	0.2889	0.2887	0.2887	0.2884	0.2863

$\lim_{x\to 0} \frac{\sqrt{x+3} - \sqrt{3}}{x} \approx 0.2887$ (Actual limit is $1/(2\sqrt{3})$.)

4.

x	−3.1	−3.01	−3.001	−2.999	−2.99	−2.9
$f(x)$	−0.2485	−0.2498	−0.2500	−0.2500	−0.2502	−0.2516

$\lim_{x\to -3} \frac{\sqrt{1-x} - 2}{x+3} \approx -0.25$ (Actual limit is $-\frac{1}{4}$.)

5.

x	2.9	2.99	2.999	3.001	3.01	3.1
$f(x)$	−0.0641	−0.0627	−0.0625	−0.0625	−0.0623	−0.0610

$\lim_{x\to 3} \frac{[1/(x+1)] - (1/4)}{x-3} \approx -0.0625$ (Actual limit is $-\frac{1}{16}$.)

6.

x	3.9	3.99	3.999	4.001	4.01	4.1
$f(x)$	0.0408	0.0401	0.0400	0.0400	0.0399	0.0392

$\lim_{x\to 4} \frac{[x/(x+1)] - (4/5)}{x-4} \approx 0.04$ (Actual limit is $\frac{1}{25}$.)

7. $\lim_{x \to 3} (4 - x) = 1$

8. $\lim_{x \to 1} (x^2 + 2) = 3$

9. $\lim_{x \to 2} f(x) = \lim_{x \to 2} (4 - x) = 2$

10. $\lim_{x \to 1} f(x) = \lim_{x \to 1} (x^2 + 2) = 3$

11. $\lim_{x \to 5} \frac{|x - 5|}{x - 5}$ does not exist. For values of x to the left of 5, $|x - 5|/(x - 5)$ equals -1, whereas for values of x to the right of 5, $|x - 5|/(x - 5)$ equals 1.

12. $\lim_{x \to 3} \frac{1}{x - 3}$ does not exist since the function increases and decreases without bound as x approaches 3.

13. $C(t) = 0.75 - 0.50[\![-(t - 1)]\!]$

(a)

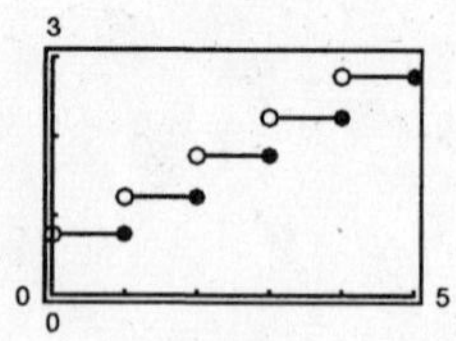

(b)

t	3	3.3	3.4	3.5	3.6	3.7	4
C	1.75	2.25	2.25	2.25	2.25	2.25	2.25

$\lim_{t \to 3.5} C(t) = 2.25$

(c)

t	2	2.5	2.9	3	3.1	3.5	4
C	1.25	1.75	1.75	1.75	2.25	2.25	2.25

$\lim_{t \to 3} C(t)$ does not exist. The values of C jump from 1.75 to 2.25 at $t = 3$.

14. $C(t) = 0.35 - 0.12[\![-(t - 1)]\!]$

(a)

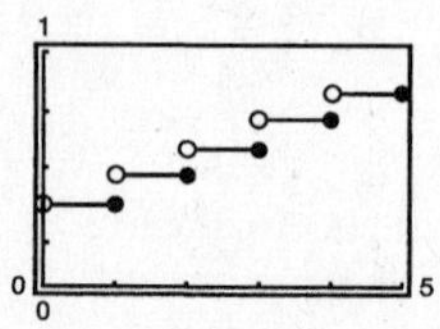

(b)

t	3	3.3	3.4	3.5	3.6	3.7	4
$C(t)$	0.59	0.71	0.71	0.71	0.71	0.71	0.71

$\lim_{t \to 3.5} C(t) = 0.71$

(c)

t	2	2.5	2.9	3	3.1	3.5	4
$C(t)$	0.47	0.59	0.59	0.59	0.71	0.71	0.71

$\lim_{t \to 3} C(t)$ does not exist. The values of C jump from 0.59 to 0.71 at $t = 3$.

15. You need to find δ such that $0 < |x - 1| < \delta$ implies $|f(x) - 1| = \left|\frac{1}{x} - 1\right| < 0.1$. That is,

$$-0.1 < \frac{1}{x} - 1 < 0.1$$

$$1 - 0.1 < \frac{1}{x} < 1 + 0.1$$

$$\frac{9}{10} < \frac{1}{x} < \frac{11}{10}$$

$$\frac{10}{9} > x > \frac{10}{11}$$

$$\frac{10}{9} - 1 > x - 1 > \frac{10}{11} - 1$$

$$\frac{1}{9} > x - 1 > -\frac{1}{11}.$$

So take $\delta = \frac{1}{11}$. Then $0 < |x - 1| < \delta$ implies

$$-\frac{1}{11} < x - 1 < \frac{1}{11}$$

$$-\frac{1}{11} < x - 1 < \frac{1}{9}.$$

Using the first series of equivalent inequalities, you obtain

$$|f(x) - 1| = \left|\frac{1}{x} - 1\right| < \varepsilon < 0.1.$$

16. You need to find δ such that $0 < |x - 2| < \delta$ implies $|f(x) - 3| = |x^2 - 1 - 3| = |x^2 - 4| < 0.2$. That is,

$$\begin{aligned} -0.2 < x^2 - 4 &< 0.2 \\ 4 - 0.2 < \quad x^2 \quad &< 4 + 0.2 \\ 3.8 < \quad x^2 \quad &< 4.2 \\ \sqrt{3.8} < \quad x \quad &< \sqrt{4.2} \\ \sqrt{3.8} - 2 < x - 2 &< \sqrt{4.2} - 2 \end{aligned}$$

So take $\delta = \sqrt{4.2} - 2 \approx 0.0494$.

Then $0 < |x - 2| < \delta$ implies

$$-\left(\sqrt{4.2} - 2\right) < x - 2 < \sqrt{4.2} - 2$$
$$\sqrt{3.8} - 2 < x - 2 < \sqrt{4.2} - 2.$$

Using the first series of equivalent inequalities, you obtain

$$|f(x) - 3| = |x^2 - 4| < \varepsilon = 0.2.$$

17. $\lim_{x\to 2} (3x + 2) = 8 = L$

$$|(3x + 2) - 8| < 0.01$$
$$|3x - 6| < 0.01$$
$$3|x - 2| < 0.01$$
$$0 < |x - 2| < \frac{0.01}{3} \approx 0.0033 = \delta$$

Hence, if $0 < |x - 2| < \delta = \frac{0.01}{3}$, you have

$$3|x - 2| < 0.01$$
$$|3x - 6| < 0.01$$
$$|(3x + 2) - 8| < 0.01$$
$$|f(x) - L| < 0.01.$$

18. $\lim_{x\to 4} \left(4 - \frac{x}{2}\right) = 2$

$$\left|\left(4 - \frac{x}{2}\right) - 2\right| < 0.01$$
$$\left|2 - \frac{x}{2}\right| < 0.01$$
$$\left|-\frac{1}{2}(x - 4)\right| < 0.01$$
$$0 < |x - 4| < 0.02 = \delta$$

Hence, if $0 < |x - 4| < \delta = 0.02$, you have

$$\left|-\frac{1}{2}(x - 4)\right| < 0.01$$
$$\left|2 - \frac{x}{2}\right| < 0.01$$
$$\left|\left(4 - \frac{x}{2}\right) - 2\right| < 0.01$$
$$|f(x) - L| < 0.01.$$

19. $\lim_{x\to 2} (x^2 - 3) = 1 = L$

$$|(x^2 - 3) - 1| < 0.01$$
$$|x^2 - 4| < 0.01$$
$$|(x + 2)(x - 2)| < 0.01$$
$$|x + 2|\,|x - 2| < 0.01$$
$$|x - 2| < \frac{0.01}{|x + 2|}$$

If we assume $1 < x < 3$, then $\delta = 0.01/5 = 0.002$.

Hence, if $0 < |x - 2| < \delta = 0.002$, you have

$$|x - 2| < 0.002 = \frac{1}{5}(0.01) < \frac{1}{|x + 2|}(0.01)$$
$$|x + 2||x - 2| < 0.01$$
$$|x^2 - 4| < 0.01$$
$$|(x^2 - 3) - 1| < 0.01$$
$$|f(x) - L| < 0.01.$$

20. $\lim_{x \to 5} (x^2 + 4) = 29$

$$|(x^2 + 4) - 29| < 0.01$$
$$|x^2 - 25| < 0.01$$
$$|(x + 5)(x - 5)| < 0.01$$
$$|x - 5| < \frac{0.01}{|x + 5|}$$

If we assume $4 < x < 6$, then $\delta = 0.01/11 \approx 0.0009$.

Hence, if $0 < |x - 5| < \delta = \frac{0.01}{11}$, you have

$$|x - 5| < \frac{0.01}{11} < \frac{1}{|x + 5|}(0.01)$$
$$|x - 5||x + 5| < 0.01$$
$$|x^2 - 25| < 0.01$$
$$|(x^2 + 4) - 29| < 0.01$$
$$|f(x) - L| < 0.01.$$

21. $\lim_{x \to 2} (x + 3) = 5$

Given $\varepsilon > 0$:

$$|(x + 3) - 5| < \varepsilon$$
$$|x - 2| < \varepsilon = \delta$$

Hence, let $\delta = \varepsilon$.

Hence, if $0 < |x - 2| < \delta = \varepsilon$, you have

$$|x - 2| < \varepsilon$$
$$|(x + 3) - 5| < \varepsilon$$
$$|f(x) - L| < \varepsilon.$$

22. $\lim_{x \to -3} (2x + 5) = -1$

Given $\varepsilon > 0$:

$$|(2x + 5) - (-1)| < \varepsilon$$
$$|2x + 6| < \varepsilon$$
$$2|x + 3| < \varepsilon$$
$$|x + 3| < \frac{\varepsilon}{2} = \delta$$

Hence, let $\delta = \varepsilon/2$.

Hence, if $0 < |x + 3| < \delta = \frac{\varepsilon}{2}$, you have

$$|x + 3| < \frac{\varepsilon}{2}$$
$$|2x + 6| < \varepsilon$$
$$|(2x + 5) - (-1)| < \varepsilon$$
$$|f(x) - L| < \varepsilon.$$

23. $\lim_{x \to -4} \left(\frac{1}{2}x - 1\right) = \frac{1}{2}(-4) - 1 = -3$

Given $\varepsilon > 0$:

$$\left|\left(\tfrac{1}{2}x - 1\right) - (-3)\right| < \varepsilon$$
$$\left|\tfrac{1}{2}x + 2\right| < \varepsilon$$
$$\tfrac{1}{2}|x - (-4)| < \varepsilon$$
$$|x - (-4)| < 2\varepsilon$$

Hence, let $\delta = 2\varepsilon$.

Hence, if $0 < |x - (-4)| < \delta = 2\varepsilon$, you have

$$|x - (-4)| < 2\varepsilon$$
$$\left|\tfrac{1}{2}x + 2\right| < \varepsilon$$
$$\left|\left(\tfrac{1}{2}x - 1\right) + 3\right| < \varepsilon$$
$$|f(x) - L| < \varepsilon.$$

24. $\lim_{x\to 1}\left(\frac{2}{3}x + 9\right) = \frac{2}{3}(1) + 9 = \frac{29}{3}$

Given $\varepsilon > 0$:

$$\left|\left(\tfrac{2}{3}x + 9\right) - \tfrac{29}{3}\right| < \varepsilon$$
$$\left|\tfrac{2}{3}x - \tfrac{2}{3}\right| < \varepsilon$$
$$\tfrac{2}{3}|x - 1| < \varepsilon$$
$$|x - 1| < \tfrac{3}{2}\varepsilon$$

Hence, let $\delta = (3/2)\varepsilon$.

Hence, if $0 < |x - 1| < \delta = \frac{3}{2}\varepsilon$, you have

$$|x - 1| < \tfrac{3}{2}\varepsilon$$
$$\left|\tfrac{2}{3}x - \tfrac{2}{3}\right| < \varepsilon$$
$$\left|\left(\tfrac{2}{3}x + 9\right) - \tfrac{29}{3}\right| < \varepsilon$$
$$|f(x) - L| < \varepsilon.$$

25. $\lim_{x\to 6} 3 = 3$

Given $\varepsilon > 0$:

$$|3 - 3| < \varepsilon$$
$$0 < \varepsilon$$

Hence, any $\delta > 0$ will work.

Hence, for any $\delta > 0$, you have

$$|3 - 3| < \varepsilon$$
$$|f(x) - L| < \varepsilon.$$

26. $\lim_{x\to 2}(-1) = -1$

Given $\varepsilon > 0$: $|-1 - (-1)| < \varepsilon$

$$0 < \varepsilon$$

Hence, any $\delta > 0$ will work.

Hence, for any $\delta > 0$, you have

$$|(-1) - (-1)| < \varepsilon$$
$$|f(x) - L| < \varepsilon.$$

27. $\lim_{x\to 0}\sqrt[3]{x} = 0$

Given $\varepsilon > 0$: $\left|\sqrt[3]{x} - 0\right| < \varepsilon$

$$\left|\sqrt[3]{x}\right| < \varepsilon$$
$$|x| < \varepsilon^3 = \delta$$

Hence, let $\delta = \varepsilon^3$.

Hence for $0 < |x - 0| < \delta = \varepsilon^3$, you have

$$|x| < \varepsilon^3$$
$$|\sqrt[3]{x}| < \varepsilon$$
$$|\sqrt[3]{x} - 0| < \varepsilon$$
$$|f(x) - L| < \varepsilon.$$

28. $\lim_{x\to 4}\sqrt{x} = \sqrt{4} = 2$

Given $\varepsilon > 0$:

$$\left|\sqrt{x} - 2\right| < \varepsilon$$
$$\left|\sqrt{x} - 2\right|\left|\sqrt{x} + 2\right| < \varepsilon\left|\sqrt{x} + 2\right|$$
$$\left|x - 4\right| < \varepsilon\left|\sqrt{x} + 2\right|$$

Assuming $1 < x < 9$, you can choose $\delta = 3\varepsilon$. Then,

$$0 < |x - 4| < \delta = 3\varepsilon \Rightarrow |x - 4| < \varepsilon\left|\sqrt{x} + 2\right|$$
$$\Rightarrow \left|\sqrt{x} - 2\right| < \varepsilon.$$

29. $\lim_{x\to -2}|x - 2| = |(-2) - 2| = 4$

Given $\varepsilon > 0$:

$$||x - 2| - 4| < \varepsilon$$
$$|-(x - 2) - 4| < \varepsilon$$

$(x - 2 < 0)$

$$|-x - 2| = |x + 2| = |x - (-2)| < \varepsilon$$

Hence, $\delta = \varepsilon$.

Hence for $0 < |x - (-2)| < \delta = \varepsilon$, you have

$$|x + 2| < \varepsilon$$
$$|-(x + 2)| < \varepsilon$$
$$|-(x - 2) - 4| < \varepsilon$$
$$||x - 2| - 4| < \varepsilon \quad \text{(because } x - 2 < 0\text{)}$$
$$|f(x) - L| < \varepsilon.$$

30. $\lim_{x \to 3} |x - 3| = 0$

Given $\varepsilon > 0$:

$$|(x - 3) - 0| < \varepsilon$$

$$|x - 3| < \varepsilon = \delta$$

Hence, let $\delta = \varepsilon$.

Hence for $0 < |x - 3| < \delta = \varepsilon$, you have

$$|x - 3| < \varepsilon$$

$$||x - 3| - 0| < \varepsilon$$

$$|f(x) - L| < \varepsilon.$$

31. $\lim_{x \to 1} (x^2 + 1) = 2$

Given $\varepsilon > 0$:

$$|(x^2 + 1) - 2| < \varepsilon$$

$$|x^2 - 1| < \varepsilon$$

$$|(x + 1)(x - 1)| < \varepsilon$$

$$|x - 1| < \frac{\varepsilon}{|x + 1|}$$

If we assume $0 < x < 2$, then $\delta = \varepsilon/3$.

Hence for $0 < |x - 1| < \delta = \frac{\varepsilon}{3}$, you have

$$|x - 1| < \frac{1}{3}\varepsilon < \frac{1}{|x + 1|}\varepsilon$$

$$|x^2 - 1| < \varepsilon$$

$$|(x^2 + 1) - 2| < \varepsilon$$

$$|f(x) - 2| < \varepsilon.$$

32. $\lim_{x \to -3} (x^2 + 3x) = 0$

Given $\varepsilon > 0$:

$$|(x^2 + 3x) - 0| < \varepsilon$$

$$|x(x + 3)| < \varepsilon$$

$$|x + 3| < \frac{\varepsilon}{|x|}$$

If we assume $-4 < x < -2$, then $\delta = \varepsilon/4$.

Hence for $0 < |x - (-3)| < \delta = \frac{\varepsilon}{4}$, you have

$$|x + 3| < \frac{1}{4}\varepsilon < \frac{1}{|x|}\varepsilon$$

$$|x(x + 3)| < \varepsilon$$

$$|x^2 + 3x - 0| < \varepsilon$$

$$|f(x) - L| < \varepsilon.$$

33. $f(x) = \dfrac{\sqrt{x + 5} - 3}{x - 4}$

$\lim_{x \to 4} f(x) = \frac{1}{6}$

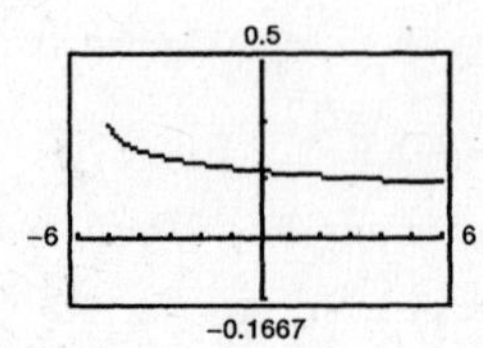

The domain is $[-5, 4) \cup (4, \infty)$.
The graphing utility does not show the hole at $\left(4, \frac{1}{6}\right)$.

34. $f(x) = \dfrac{x - 3}{x^2 - 4x + 3}$

$\lim_{x \to 3} f(x) = \frac{1}{2}$

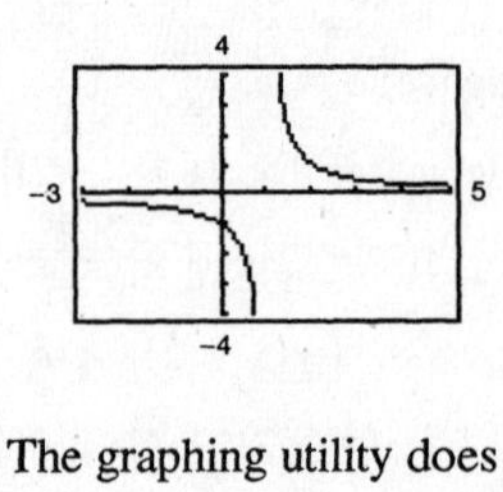

The domain is all $x \neq 1, 3$. The graphing utility does not show the hole at $\left(3, \frac{1}{2}\right)$.

35. $f(x) = \dfrac{x - 9}{\sqrt{x} - 3}$

$\lim_{x \to 9} f(x) = 6$

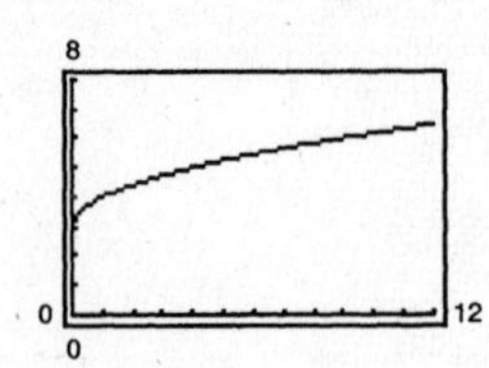

The domain is all $x \geq 0$ except $x = 9$. The graphing utility does not show the hole at $(9, 6)$.

36. $f(x) = \dfrac{x - 3}{x^2 - 9}$

$\lim_{x \to 3} f(x) = \dfrac{1}{6}$

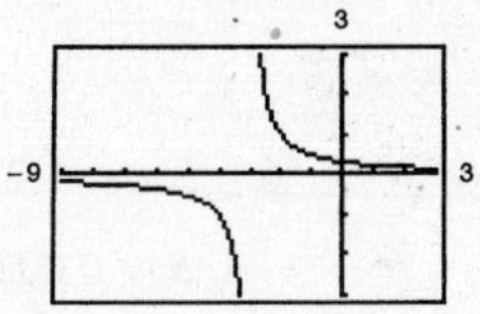

The domain is all $x \neq \pm 3$. The graphing utility does not show the hole at $\left(3, \frac{1}{6}\right)$.

37. $\lim_{x \to 8} f(x) = 25$ means that the values of f approach 25 as x gets closer and closer to 8.

38. (a) No. The fact that $f(2) = 4$ has no bearing on the existence of the limit of $f(x)$ as x approaches 2.

(b) No. The fact that $\lim_{x \to 2} f(x) = 4$ has no bearing on the value of f at 2.

39. (i) The values of f approach different numbers as x approaches c from different sides of c:

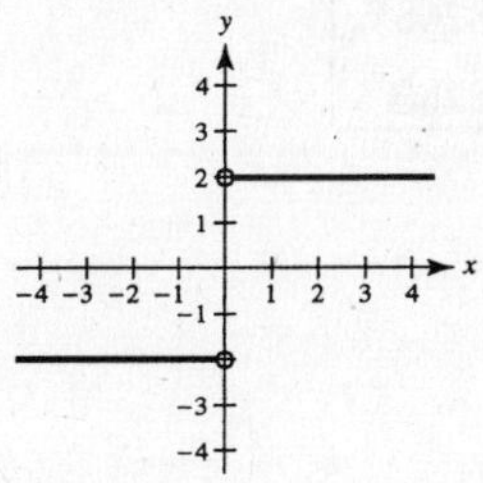

(ii) The values of f increase without bound as x approaches c:

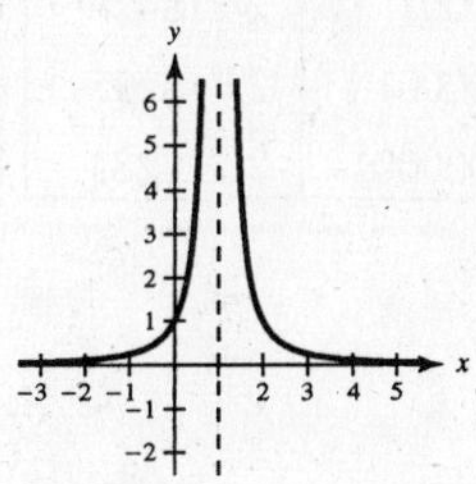

(iii) The values of f oscillate between two fixed numbers as x approaches c:

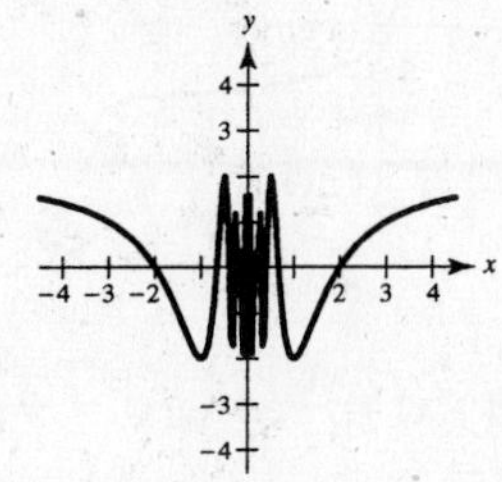

40. Let $p(x)$ be the atmospheric pressure in a plane at altitude x (in feet).

$\lim_{x \to 0^+} p(x) = 14.7 \text{ lb/in}^2$

41. False. For example, let $f(x) = \dfrac{x - 2}{x^2 - 4}$. Then $f(2)$ does not exist, but $\lim_{x \to 2} f(x) = \dfrac{1}{4}$ (See Exercise 2, Section 3.2.)

42. True

43. False. Let $f(x) = \begin{cases} 4 - x, & x \neq 2 \\ 0, & x = 2 \end{cases}$.

$\lim_{x \to 2} f(x) = 2$, but $f(2) = 0$ (See Exercise 9.)

44. False. Let

$$f(x) = \begin{cases} x^2 - 4x, & x \neq 4 \\ 10, & x = 4 \end{cases}.$$

$\lim_{x \to 4} f(x) = \lim_{x \to 4} (x^2 - 4x) = 0$ and $f(4) = 10 \neq 0$

45. (a) $C = 2\pi r$

$$r = \frac{C}{2\pi} = \frac{6}{2\pi} = \frac{3}{\pi} \approx 0.9549 \text{ cm}$$

(b) If $C = 5.5$, $r = \dfrac{5.5}{2\pi} \approx 0.87535$ cm.

If $C = 6.5$, $r = \dfrac{6.5}{2\pi} \approx 1.03451$ cm.

Thus, $0.87535 < r < 1.03451$.

(c) $\lim_{r \to 3/\pi} (2\pi r) = 6$; $\varepsilon = 0.5$; $\delta \approx 0.0796$

46. $V = \frac{4}{3}\pi r^3,\ V = 2.48$

(a) $2.48 = \frac{4}{3}\pi r^3$

$r^3 = \frac{1.86}{\pi}$

$r \approx 0.8397$ in.

(b) $2.45 \le V \le 2.51$

$2.45 \le \frac{4}{3}\pi r^3 \le 2.51$

$0.5849 \le r^3 \le 0.5992$

$0.8363 \le r \le 0.8431$

(c) For $\varepsilon = 2.51 - 2.48 = 0.03$,

$\delta \approx 0.003$.

47. $f(x) = (1 + x)^{1/x}$

$\lim_{x\to 0} (1 + x)^{1/x} = e \approx 2.71828$

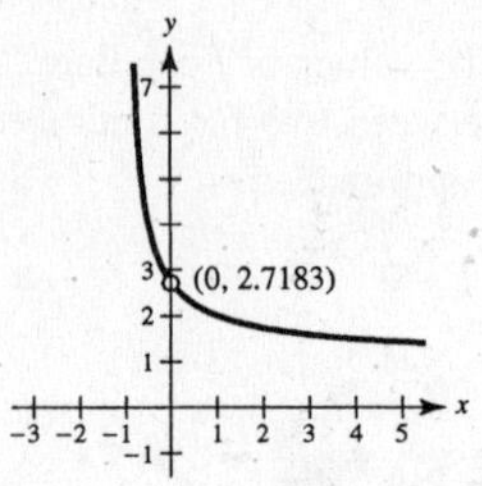

x	$f(x)$
-0.1	2.867972
-0.01	2.731999
-0.001	2.719642
-0.0001	2.718418
-0.00001	2.718295
-0.000001	2.718283

x	$f(x)$
0.1	2.593742
0.01	2.704814
0.001	2.716924
0.0001	2.718146
0.00001	2.718268
0.000001	2.718280

48. Let $f(x) = \frac{1}{x}$ and $g(x) = \frac{-1}{x}$.

Then $\lim_{x\to 0} f(x)$ and $\lim_{x\to 0} g(x)$ do not exist.

But, $\lim_{x\to 0} [f(x) + g(x)] = \lim_{x\to 0}\left[\frac{1}{x} - \frac{1}{x}\right] = 0$ does exist.

49. $f(x) = \sqrt{x}$

(a) $\lim_{x\to 0.25} \sqrt{x} = 0.5$ is true.

As x approaches $0.25 = \frac{1}{4}$, $f(x) = \sqrt{x}$ approaches $\frac{1}{2} = 0.5$

(b) $\lim_{x\to 0} \sqrt{x} = 0$ is false.

$f(x) = \sqrt{x}$ is not defined on an open interval containing 0 because the domain of f is $x \ge 0$.

50. The value of f at c has no bearing on the limit as x approaches c.

51. If $\lim_{x\to c} f(x) = L_1$ and $\lim_{x\to c} f(x) = L_2$, then for every $\varepsilon > 0$, there exists $\delta_1 > 0$ and $\delta_2 > 0$ such that $|x - c| < \delta_1 \Rightarrow |f(x) - L_1| < \varepsilon$ and $|x - c| < \delta_2 \Rightarrow |f(x) - L_2| < \varepsilon$. Let δ equal the smaller of δ_1 and δ_2. Then for $|x - c| < \delta$, we have

$$|L_1 - L_2| = |L_1 - f(x) + f(x) - L_2| \le |L_1 - f(x)| + |f(x) - L_2| < \varepsilon + \varepsilon.$$

Therefore, $|L_1 - L_2| < 2\varepsilon$. Since $\varepsilon > 0$ is arbitrary, it follows that $L_1 = L_2$.

52. $f(x) = mx + b,\ m \ne 0$. Let $\varepsilon > 0$ be given. Take $\delta = \frac{\varepsilon}{|m|}$.

If $0 < |x - c| < \delta = \frac{\varepsilon}{|m|}$, then

$$|m||x - c| < \varepsilon$$

$$|mx - mc| < \varepsilon$$

$$|(mx + b) - (mc + b)| < \varepsilon$$

which shows that $\lim_{x\to c} (mx + b) = mc + b$.

53. $\lim_{x\to c} [f(x) - L] = 0$ means that for every $\varepsilon > 0$ there exists $\delta > 0$ such that if

$$0 < |x - c| < \delta, \text{ then } |(f(x) - L) - 0| < \varepsilon.$$

This means the same as $|f(x) - L| < \varepsilon$ when $0 < |x - c| < \delta$.

Thus, $\lim_{x\to c} f(x) = L$.

54. (a) $(3x + 1)(3x - 1)x^2 + 0.01 = (9x^2 - 1)x^2 + \frac{1}{100}$

$$= 9x^4 - x^2 + \frac{1}{100}$$

$$= \frac{1}{100}(10x^2 - 1)(90x^2 - 1)$$

Thus, $(3x + 1)(3x - 1)x^2 + 0.01 > 0$ if $10x^2 - 1 < 0$ and $90x^2 - 1 < 0$.

Let $(a, b) = \left(-\frac{1}{\sqrt{90}}, \frac{1}{\sqrt{90}}\right)$.

For all $x \neq 0$ in (a, b), the graph is positive. You can verify this with a graphing utility.

(b) We are given $\lim_{x \to c} g(x) = L > 0$. Let $\varepsilon = \frac{1}{2}L$. There exists $\delta > 0$ such that $0 < |x - c| < \delta$ implies that $|g(x) - L| < \varepsilon = \frac{L}{2}$. That is,

$$-\frac{L}{2} < g(x) - L < \frac{L}{2}$$

$$\frac{L}{2} < g(x) \quad < \frac{3L}{2}.$$

For x in the interval $(c - \delta, c + \delta)$, $x \neq c$, we have $g(x) > \frac{L}{2} > 0$, as desired.

Section 3.3 Evaluating Limits Analytically

1. 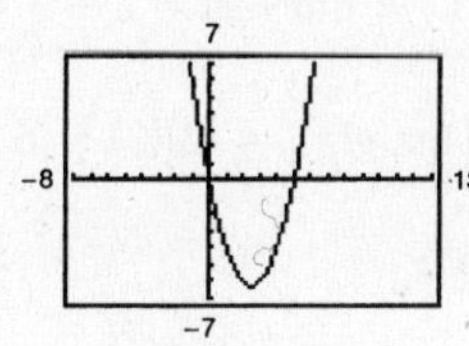

$h(x) = x^2 - 5x$

(a) $\lim_{x \to 5} h(x) = 0$

(b) $\lim_{x \to -1} h(x) = 6$

2. 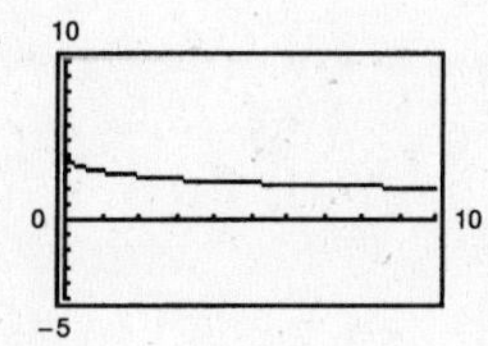

$g(x) = \frac{12(\sqrt{x} - 3)}{x - 9}$

(a) $\lim_{x \to 4} g(x) = 2.4$

(b) $\lim_{x \to 0} g(x) = 4$

3. 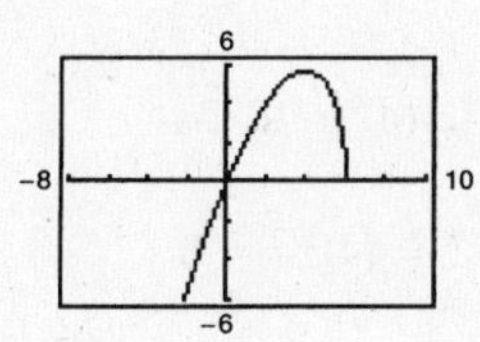

(a) $\lim_{x \to 0} f(x) = 0$

(b) $\lim_{x \to 2} f(x) = 4$

4. 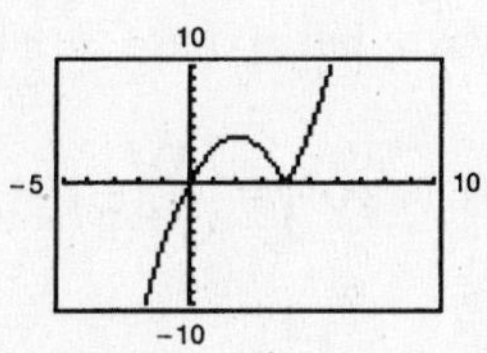

$f(t) = t|t - 4|$

(a) $\lim_{t \to 4} f(t) = 0$

(b) $\lim_{t \to -1} f(t) = -5$

5. $\lim_{x \to 2} x^4 = 2^4 = 16$

6. $\lim_{x \to -2} x^3 = (-2)^3 = -8$

7. $\lim_{x \to 0} (2x - 1) = 2(0) - 1 = -1$

8. $\lim_{x \to -3} (3x + 2) = 3(-3) + 2 = -7$

9. $\lim_{x \to -3} (x^2 + 3x) = (-3)^2 + 3(-3) = 9 - 9 = 0$

10. $\lim_{x \to 1} (-x^2 + 1) = -(1)^2 + 1 = 0$

11. $\lim_{x \to -3} (2x^2 + 4x + 1) = 2(-3)^2 + 4(-3) + 1 = 18 - 12 + 1 = 7$

12. $\lim_{x \to 1} (3x^3 - 2x^2 + 4) = 3(1)^3 - 2(1)^2 + 4 = 5$

13. $\lim_{x \to 2} \frac{1}{x} = \frac{1}{2}$

14. $\lim_{x \to -3} \frac{2}{x + 2} = \frac{2}{-3 + 2} = -2$

15. $\lim_{x \to 1} \frac{x - 3}{x^2 + 4} = \frac{1 - 3}{1^2 + 4} = \frac{-2}{5} = -\frac{2}{5}$

16. $\lim_{x \to 3} \frac{2x - 3}{x + 5} = \frac{2(3) - 3}{3 + 5} = \frac{3}{8}$

17. $\lim_{x \to 7} \frac{5x}{\sqrt{x + 2}} = \frac{5(7)}{\sqrt{7 + 2}} = \frac{35}{\sqrt{9}} = \frac{35}{3}$

18. $\lim_{x\to 3} \frac{\sqrt{x+1}}{x-4} = \frac{\sqrt{3+1}}{3-4} = -2$

19. $\lim_{x\to 3} \sqrt{x+1} = \sqrt{3+1} = 2$

20. $\lim_{x\to 4} \sqrt[3]{x+4} = \sqrt[3]{4+4} = 2$

21. $\lim_{x\to -4} (x+3)^2 = (-4+3)^2 = 1$

22. $\lim_{x\to 0} (2x-1)^3 = [2(0)-1]^3 = -1$

23. (a) $\lim_{x\to 1} f(x) = 5 - 1 = 4$

(b) $\lim_{x\to 4} g(x) = 4^3 = 64$

(c) $\lim_{x\to 1} g(f(x)) = g(f(1)) = g(4) = 64$

24. (a) $\lim_{x\to -3} f(x) = (-3) + 7 = 4$

(b) $\lim_{x\to 4} g(x) = 4^2 = 16$

(c) $\lim_{x\to -3} g(f(x)) = g(4) = 16$

25. (a) $\lim_{x\to 1} f(x) = 4 - 1 = 3$

(b) $\lim_{x\to 3} g(x) = \sqrt{3+1} = 2$

(c) $\lim_{x\to 1} g(f(x)) = g(3) = 2$

26. (a) $\lim_{x\to 4} f(x) = 2(4^2) - 3(4) + 1 = 21$

(b) $\lim_{x\to 21} g(x) = \sqrt[3]{21+6} = 3$

(c) $\lim_{x\to 4} g(f(x)) = g(21) = 3$

27. (a) $\lim_{x\to c} [5g(x)] = 5 \lim_{x\to c} g(x) = 5(3) = 15$

(b) $\lim_{x\to c} [f(x) + g(x)] = \lim_{x\to c} f(x) + \lim_{x\to c} g(x) = 2 + 3 = 5$

(c) $\lim_{x\to c} [f(x)g(x)] = \left[\lim_{x\to c} f(x)\right]\left[\lim_{x\to c} g(x)\right] = (2)(3) = 6$

(d) $\lim_{x\to c} \frac{f(x)}{g(x)} = \frac{\lim_{x\to c} f(x)}{\lim_{x\to c} g(x)} = \frac{2}{3}$

28. (a) $\lim_{x\to c} [4f(x)] = 4 \lim_{x\to c} f(x) = 4\left(\frac{3}{2}\right) = 6$

(b) $\lim_{x\to c} [f(x) + g(x)] = \lim_{x\to c} f(x) + \lim_{x\to c} g(x) = \frac{3}{2} + \frac{1}{2} = 2$

(c) $\lim_{x\to c} [f(x)g(x)] = \left[\lim_{x\to c} f(x)\right]\left[\lim_{x\to c} g(x)\right] = \left(\frac{3}{2}\right)\left(\frac{1}{2}\right) = \frac{3}{4}$

(d) $\lim_{x\to c} \frac{f(x)}{g(x)} = \frac{\lim_{x\to c} f(x)}{\lim_{x\to c} g(x)} = \frac{3/2}{1/2} = 3$

29. (a) $\lim_{x\to c} [f(x)]^3 = \left[\lim_{x\to c} f(x)\right]^3 = (4)^3 = 64$

(b) $\lim_{x\to c} \sqrt{f(x)} = \sqrt{\lim_{x\to c} f(x)} = \sqrt{4} = 2$

(c) $\lim_{x\to c} [3f(x)] = 3 \lim_{x\to c} f(x) = 3(4) = 12$

(d) $\lim_{x\to c} [f(x)]^{3/2} = \left[\lim_{x\to c} f(x)\right]^{3/2} = (4)^{3/2} = 8$

30. (a) $\lim_{x\to c} \sqrt[3]{f(x)} = \sqrt[3]{\lim_{x\to c} f(x)} = \sqrt[3]{27} = 3$

(b) $\lim_{x\to c} \frac{f(x)}{18} = \frac{\lim_{x\to c} f(x)}{\lim_{x\to c} 18} = \frac{27}{18} = \frac{3}{2}$

(c) $\lim_{x\to c} [f(x)]^2 = \left[\lim_{x\to c} f(x)\right]^2 = (27)^2 = 729$

(d) $\lim_{x\to c} [f(x)]^{2/3} = \left[\lim_{x\to c} f(x)\right]^{2/3} = (27)^{2/3} = 9$

31. $f(x) = -2x + 1$ and $g(x) = \frac{-2x^2 + x}{x}$ agree except at $x = 0$.

(a) $\lim_{x\to 0} g(x) = \lim_{x\to 0} f(x) = 1$

(b) $\lim_{x\to -1} g(x) = \lim_{x\to -1} f(x) = 3$

32. $f(x) = x - 3$ and $h(x) = \frac{x^2 - 3x}{x}$ agree except at $x = 0$.

(a) $\lim_{x\to -2} h(x) = \lim_{x\to -2} f(x) = -5$

(b) $\lim_{x\to 0} h(x) = \lim_{x\to 0} f(x) = -3$

33. $f(x) = x(x+1)$ and $g(x) = \frac{x^3 - x}{x - 1}$ agree except at $x = 1$.

(a) $\lim_{x\to 1} g(x) = \lim_{x\to 1} f(x) = 2$

(b) $\lim_{x\to -1} g(x) = \lim_{x\to -1} f(x) = 0$

34. $g(x) = \dfrac{1}{x-1}$ and $f(x) = \dfrac{x}{x^2 - x}$ agree except at $x = 0$.

(a) $\lim\limits_{x \to 1} f(x)$ does not exist.

(b) $\lim\limits_{x \to 0} f(x) = \lim\limits_{x \to 0} g(x) = -1$

35. $f(x) = \dfrac{x^2 - 1}{x + 1}$ and $g(x) = x - 1$ agree except at $x = -1$.

$\lim\limits_{x \to -1} f(x) = \lim\limits_{x \to -1} g(x) = -2$

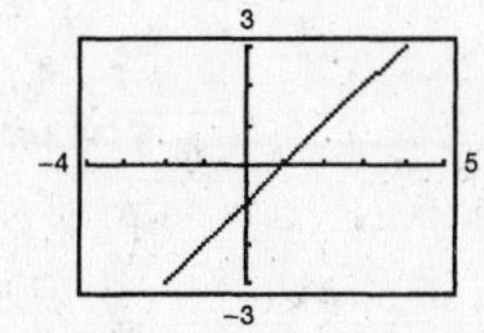

36. $f(x) = \dfrac{2x^2 - x - 3}{x + 1}$ and $g(x) = 2x - 3$ agree except at $x = -1$.

$\lim\limits_{x \to -1} f(x) = \lim\limits_{x \to -1} g(x) = -5$

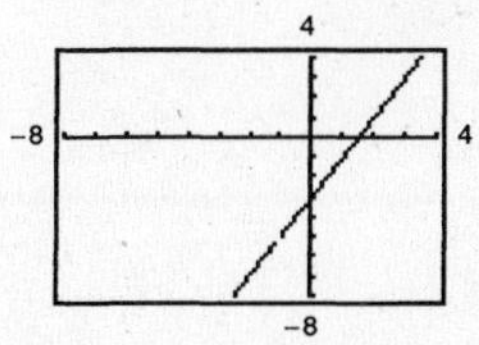

37. $f(x) = \dfrac{x^3 - 8}{x - 2}$ and $g(x) = x^2 + 2x + 4$ agree except at $x = 2$.

$\lim\limits_{x \to 2} f(x) = \lim\limits_{x \to 2} g(x) = 12$

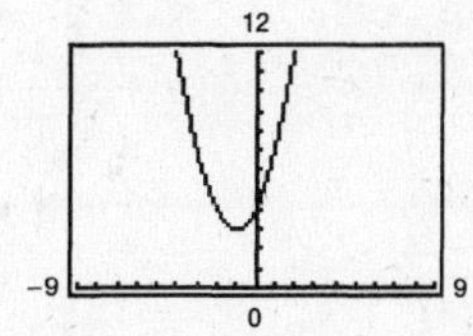

38. $f(x) = \dfrac{x^3 + 1}{x + 1}$ and $g(x) = x^2 - x + 1$ agree except at $x = -1$.

$\lim\limits_{x \to -1} f(x) = \lim\limits_{x \to -1} g(x) = 3$

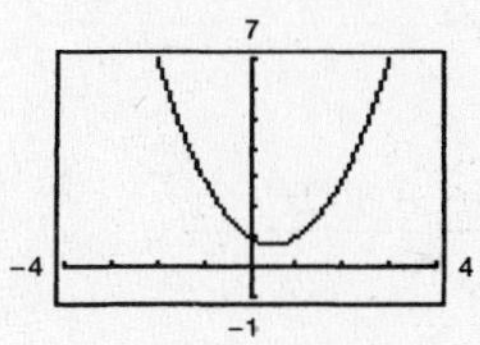

39. $\lim\limits_{x \to 5} \dfrac{x - 5}{x^2 - 25} = \lim\limits_{x \to 5} \dfrac{x - 5}{(x + 5)(x - 5)}$

$= \lim\limits_{x \to 5} \dfrac{1}{x + 5} = \dfrac{1}{10}$

40. $\lim\limits_{x \to 2} \dfrac{2 - x}{x^2 - 4} = \lim\limits_{x \to 2} \dfrac{-(x - 2)}{(x - 2)(x + 2)}$

$= \lim\limits_{x \to 2} \dfrac{-1}{x + 2} = -\dfrac{1}{4}$

41. $\lim\limits_{x \to -3} \dfrac{x^2 + x - 6}{x^2 - 9} = \lim\limits_{x \to -3} \dfrac{(x + 3)(x - 2)}{(x + 3)(x - 3)}$

$= \lim\limits_{x \to -3} \dfrac{x - 2}{x - 3} = \dfrac{-5}{-6} = \dfrac{5}{6}$

42. $\lim\limits_{x \to 4} \dfrac{x^2 - 5x + 4}{x^2 - 2x - 8} = \lim\limits_{x \to 4} \dfrac{(x - 4)(x - 1)}{(x - 4)(x + 2)}$

$= \lim\limits_{x \to 4} \dfrac{(x - 1)}{(x + 2)} = \dfrac{3}{6} = \dfrac{1}{2}$

43. $\lim\limits_{x \to 0} \dfrac{\sqrt{x + 5} - \sqrt{5}}{x} = \lim\limits_{x \to 0} \dfrac{\sqrt{x + 5} - \sqrt{5}}{x} \cdot \dfrac{\sqrt{x + 5} + \sqrt{5}}{\sqrt{x + 5} + \sqrt{5}}$

$= \lim\limits_{x \to 0} \dfrac{(x + 5) - 5}{x\left(\sqrt{x + 5} + \sqrt{5}\right)} = \lim\limits_{x \to 0} \dfrac{1}{\sqrt{x + 5} + \sqrt{5}} = \dfrac{1}{2\sqrt{5}} = \dfrac{\sqrt{5}}{10}$

44. $\lim_{x\to 0}\frac{\sqrt{2+x}-\sqrt{2}}{x} = \lim_{x\to 0}\frac{\sqrt{2+x}-\sqrt{2}}{x}\cdot\frac{\sqrt{2+x}+\sqrt{2}}{\sqrt{2+x}+\sqrt{2}}$

$$= \lim_{x\to 0}\frac{2+x-2}{\left(\sqrt{2+x}+\sqrt{2}\right)x} = \lim_{x\to 0}\frac{1}{\sqrt{2+x}+\sqrt{2}} = \frac{1}{2\sqrt{2}} = \frac{\sqrt{2}}{4}$$

45. $\lim_{x\to 4}\frac{\sqrt{x+5}-3}{x-4} = \lim_{x\to 4}\frac{\sqrt{x+5}-3}{x-4}\cdot\frac{\sqrt{x+5}+3}{\sqrt{x+5}+3}$

$$= \lim_{x\to 4}\frac{(x+5)-9}{(x-4)\left(\sqrt{x+5}+3\right)} = \lim_{x\to 4}\frac{1}{\sqrt{x+5}+3} = \frac{1}{\sqrt{9}+3} = \frac{1}{6}$$

46. $\lim_{x\to 3}\frac{\sqrt{x+1}-2}{x-3} = \lim_{x\to 3}\frac{\sqrt{x+1}-2}{x-3}\cdot\frac{\sqrt{x+1}+2}{\sqrt{x+1}+2} = \lim_{x\to 3}\frac{x-3}{(x-3)\left[\sqrt{x+1}+2\right]} = \lim_{x\to 3}\frac{1}{\sqrt{x+1}+2} = \frac{1}{4}$

47. $\lim_{x\to 0}\frac{\frac{1}{3+x}-\frac{1}{3}}{x} = \lim_{x\to 0}\frac{3-(3+x)}{3(3+x)x} = \lim_{x\to 0}\frac{-1}{3(3+x)} = \frac{-1}{9}$

48. $\lim_{x\to 0}\frac{\frac{1}{x+4}-\frac{1}{4}}{x} = \lim_{x\to 0}\frac{\frac{4-(x+4)}{4(x+4)}}{x} = \lim_{x\to 0}\frac{-1}{4(x+4)} = -\frac{1}{16}$

49. $\lim_{\Delta x\to 0}\frac{2(x+\Delta x)-2x}{\Delta x} = \lim_{\Delta x\to 0}\frac{2x+2\Delta x-2x}{\Delta x} = \lim_{\Delta x\to 0} 2 = 2$

50. $\lim_{\Delta x\to 0}\frac{(x+\Delta x)^2-x^2}{\Delta x} = \lim_{\Delta x\to 0}\frac{x^2+2x\,\Delta x+(\Delta x)^2-x^2}{\Delta x} = \lim_{\Delta x\to 0}\frac{\Delta x(2x+\Delta x)}{\Delta x} = \lim_{\Delta x\to 0}(2x+\Delta x) = 2x$

51. $\lim_{\Delta x\to 0}\frac{(x+\Delta x)^2-2(x+\Delta x)+1-(x^2-2x+1)}{\Delta x} = \lim_{\Delta x\to 0}\frac{x^2+2x\Delta x+(\Delta x)^2-2x-2\Delta x+1-x^2+2x-1}{\Delta x}$

$$= \lim_{\Delta x\to 0}(2x+\Delta x-2) = 2x-2$$

52. $\lim_{\Delta x\to 0}\frac{(x+\Delta x)^3-x^3}{\Delta x} = \lim_{\Delta x\to 0}\frac{x^3+3x^2\Delta x+3x(\Delta x)^2+(\Delta x)^3-x^3}{\Delta x}$

$$= \lim_{\Delta x\to 0}\frac{\Delta x(3x^2+3x\,\Delta x+(\Delta x)^2)}{\Delta x} = \lim_{\Delta x\to 0}(3x^2+3x\,\Delta x+(\Delta x)^2) = 3x^2$$

53. $\lim_{x\to 0}\frac{\sqrt{x+2}-\sqrt{2}}{x} \approx 0.354$

x	−0.1	−0.01	−0.001	0	0.001	0.01	0.1
$f(x)$	0.358	0.354	0.354	?	0.354	0.353	0.349

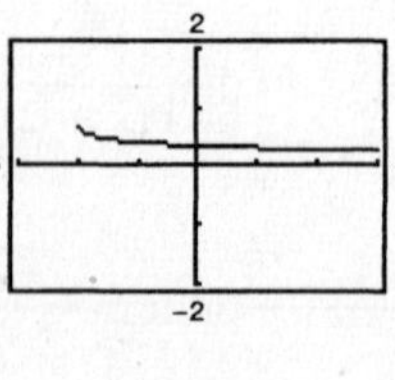

Analytically, $\lim_{x\to 0}\frac{\sqrt{x+2}-\sqrt{2}}{x} = \lim_{x\to 0}\frac{\sqrt{x+2}-\sqrt{2}}{x}\cdot\frac{\sqrt{x+2}+\sqrt{2}}{\sqrt{x+2}+\sqrt{2}}$

$$= \lim_{x\to 0}\frac{x+2-2}{x\left(\sqrt{x+2}+\sqrt{2}\right)} = \lim_{x\to 0}\frac{1}{\sqrt{x+2}+\sqrt{2}} = \frac{1}{2\sqrt{2}} = \frac{\sqrt{2}}{4} \approx 0.354.$$

54. $f(x) = \dfrac{4 - \sqrt{x}}{x - 16}$

x	15.9	15.99	15.999	16	16.001	16.01	16.1
$f(x)$	$-.1252$	$-.125$	$-.125$	?	$-.125$	$-.125$	$-.1248$

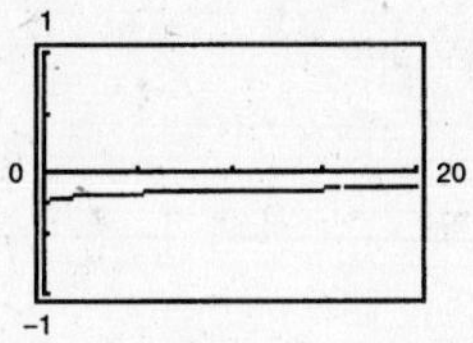

It appears that the limit is -0.125.

Analytically, $\displaystyle\lim_{x\to16} \frac{4 - \sqrt{x}}{x - 16} = \lim_{x\to16} \frac{(4 - \sqrt{x})}{(\sqrt{x} + 4)(\sqrt{x} - 4)}$

$$= \lim_{x\to16} \frac{-1}{\sqrt{x} + 4} = -\frac{1}{8}.$$

55. $\displaystyle\lim_{x\to0} \frac{\frac{1}{2 + x} - \frac{1}{2}}{x} = -\frac{1}{4}$

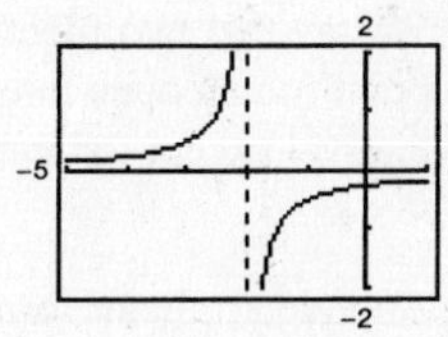

x	-0.1	-0.01	-0.001	0	0.001	0.01	0.1
$f(x)$	-0.263	-0.251	-0.250	?	-0.250	-0.249	-0.238

Analytically, $\displaystyle\lim_{x\to0} \frac{\frac{1}{2 + x} - \frac{1}{2}}{x} = \lim_{x\to0} \frac{2 - (2 + x)}{2(2 + x)} \cdot \frac{1}{x} = \lim_{x\to0} \frac{-x}{2(2 + x)} \cdot \frac{1}{x} = \lim_{x\to0} \frac{-1}{2(2 + x)} = -\frac{1}{4}.$

56. $\displaystyle\lim_{x\to2} \frac{x^5 - 32}{x - 2} = 80$

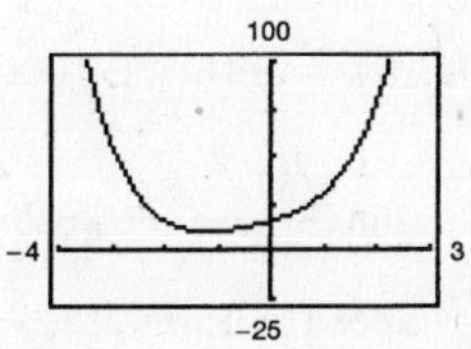

x	1.9	1.99	1.999	1.9999	2.0	2.0001	2.001	2.01	2.1
$f(x)$	72.39	79.20	79.92	79.99	?	80.01	80.08	80.80	88.41

Analytically, $\displaystyle\lim_{x\to2} \frac{x^5 - 32}{x - 2} = \lim_{x\to2} \frac{(x - 2)(x^4 + 2x^3 + 4x^2 + 8x + 16)}{x - 2}$

$$= \lim_{x\to2} (x^4 + 2x^3 + 4x^2 + 8x + 16) = 80.$$

(*Hint:* Use long division to factor $x^5 - 32$.)

57. $f(x) = 2x + 3$

$$\lim_{\Delta x\to0} \frac{f(x + \Delta x) - f(x)}{\Delta x} = \lim_{\Delta x\to0} \frac{2(x + \Delta x) + 3 - (2x + 3)}{\Delta x} = \lim_{\Delta x\to0} \frac{2x + 2\,\Delta x - 2x}{\Delta x} = \lim_{\Delta x\to0} \frac{2\,\Delta x}{\Delta x} = \lim_{\Delta x\to0} 2 = 2$$

58. $\displaystyle\lim_{\Delta x\to0} \frac{f(x + \Delta x) - f(x)}{\Delta x} = \lim_{\Delta x\to0} \frac{\sqrt{x + \Delta x} - \sqrt{x}}{\Delta x} = \lim_{\Delta x\to0} \frac{\sqrt{x + \Delta x} - \sqrt{x}}{\Delta x} \cdot \frac{\sqrt{x + \Delta x} + \sqrt{x}}{\sqrt{x + \Delta x} + \sqrt{x}}$

$$= \lim_{\Delta x\to0} \frac{x + \Delta x - x}{\Delta x(\sqrt{x + \Delta x} + \sqrt{x})} = \lim_{\Delta x\to0} \frac{1}{\sqrt{x + \Delta x} + \sqrt{x}} = \frac{1}{2\sqrt{x}}$$

59. $f(x) = \dfrac{4}{x}$

$$\lim_{\Delta x\to0} \frac{f(x + \Delta x) - f(x)}{\Delta x} = \lim_{\Delta x\to0} \frac{\frac{4}{x + \Delta x} - \frac{4}{x}}{\Delta x} = \lim_{\Delta x\to0} \frac{4x - 4(x + \Delta x)}{(x + \Delta x)x\,\Delta x}$$

$$= \lim_{\Delta x\to0} \frac{-4\,\Delta x}{(x + \Delta x)x\,\Delta x} = \lim_{\Delta x\to0} \frac{-4}{(x + \Delta x)x} = \frac{-4}{x^2}$$

60. $\lim_{\Delta x \to 0} \frac{f(x + \Delta x) - f(x)}{\Delta x} = \lim_{\Delta x \to 0} \frac{(x + \Delta x)^2 - 4(x + \Delta x) - (x^2 - 4x)}{\Delta x} = \lim_{\Delta x \to 0} \frac{x^2 + 2x\,\Delta x + \Delta x^2 - 4x - 4\,\Delta x - x^2 + 4x}{\Delta x}$

$= \lim_{\Delta x \to 0} \frac{\Delta x(2x + \Delta x - 4)}{\Delta x} = \lim_{\Delta x \to 0} (2x + \Delta x - 4) = 2x - 4$

61. $\lim_{x \to 0}(4 - x^2) = 4$

$\lim_{x \to 0}(4 + x^2) = 4$

Since $4 - x^2 \le f(x) \le 4 + x^2$ the Squeeze Theorem says $\lim_{x \to 0} f(x) = 4$.

62. $\lim_{x \to a} [b - |x - a|] \le \lim_{x \to a} f(x) \le \lim_{x \to a} [b + |x - a|]$

$b \le \lim_{x \to a} f(x) \le b$

Therefore, $\lim_{x \to a} f(x) = b$.

63. We say that two functions f and g agree at all but one point (on an open interval) if $f(x) = g(x)$ for all x in the interval except for $x = c$, where c is in the interval.

64. $f(x) = \frac{x^2 - 1}{x - 1}$ and $g(x) = x + 1$ agree at all points except $x = 1$.

65. An indeterminant form is obtained when evaluating a limit using direct substitution produces a meaningless fractional expression such as $0/0$. That is,

$$\lim_{x \to c} \frac{f(x)}{g(x)}$$

for which $\lim_{x \to c} f(x) = \lim_{x \to c} g(x) = 0$.

66. If a function f is squeezed between two functions h and g, $h(x) \le f(x) \le g(x)$, and h and g have the same limit L as $x \to c$, then $\lim_{x \to c} f(x)$ exists and equals L.

67. $s(t) = -16t^2 + 1000$

$\lim_{t \to 5} \frac{s(5) - s(t)}{5 - t} = \lim_{t \to 5} \frac{600 - (-16t^2 + 1000)}{5 - t} = \lim_{t \to 5} \frac{16(t + 5)(t - 5)}{-(t - 5)} = \lim_{t \to 5} -16(t + 5) = -160$ ft/sec.

Speed $= 160$ ft/sec

68. $s(t) = -16t^2 + 1000 = 0$ when $t = \sqrt{\frac{1000}{16}} = \frac{5\sqrt{10}}{2}$ seconds

$$\lim_{t \to 5\sqrt{10}/2} \frac{s\left(\frac{5\sqrt{10}}{2}\right) - s(t)}{\frac{5\sqrt{10}}{2} - t} = \lim_{t \to 5\sqrt{10}/2} \frac{0 - (-16t^2 + 1000)}{\frac{5\sqrt{10}}{2} - t}$$

$$= \lim_{t \to 5\sqrt{10}/2} \frac{16\left(t^2 - \frac{125}{2}\right)}{\frac{5\sqrt{10}}{2} - t} = \lim_{t \to 5\sqrt{10}/2} \frac{16\left(t + \frac{5\sqrt{10}}{2}\right)\left(t - \frac{5\sqrt{10}}{2}\right)}{-\left(t - \frac{5\sqrt{10}}{2}\right)}$$

$$= \lim_{t \to 5\sqrt{10}/2} -16\left(t + \frac{5\sqrt{10}}{2}\right) = -80\sqrt{10} \text{ ft/sec} \approx -253 \text{ ft/sec}$$

69. $s(t) = -4.9t^2 + 150$

$$\lim_{t \to 3} \frac{s(3) - s(t)}{3 - t} = \lim_{t \to 3} \frac{-4.9(3^2) + 150 - (-4.9t^2 + 150)}{3 - t}$$

$$= \lim_{t \to 3} \frac{-4.9(9 - t^2)}{3 - t}$$

$$= \lim_{t \to 3} \frac{-4.9(3 - t)(3 + t)}{3 - t}$$

$$= \lim_{t \to 3} -4.9(3 + t) = -29.4 \text{ m/sec}$$

70. $-4.9t^2 + 150 = 0$ when $t = \sqrt{\dfrac{150}{4.9}} = \sqrt{\dfrac{1500}{49}} \approx 5.53$ seconds.

The velocity at time $t = a$ is

$$\lim_{t\to a} \frac{s(a) - s(t)}{a - t} = \lim_{t\to a} \frac{(-4.9a^2 + 150) - (-4.9t^2 + 150)}{a - t} = \lim_{t\to a} \frac{-4.9(a - t)(a + t)}{a - t}$$

$$= \lim_{t\to a} -4.9(a + t) = -2a(4.9) = -9.8a \text{ m/sec.}$$

Hence, if $a = \sqrt{1500/49}$, the velocity is $-9.8\sqrt{1500/49} \approx -54.2$ m/sec.

71. False. As x approaches 0 from the left, $\dfrac{|x|}{x} = -1$.

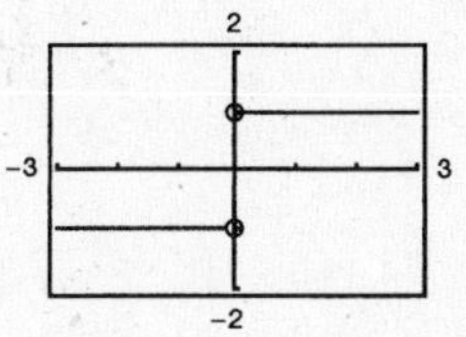

72. True. $\lim_{x\to 0} x^3 = 0^3 = 0$

73. True

74. False. Let $f(x) = \begin{cases} x & x \neq 1 \\ 3 & x = 1 \end{cases}$, $c = 1$.

Then $\lim_{x\to 1} f(x) = 1$ but $f(1) \neq 1$.

75. False. The limit does not exist.

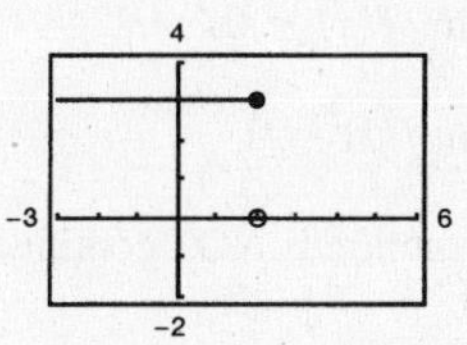

76. False. Let $f(x) = \frac{1}{2}x^2$ and $g(x) = x^2$. Then $f(x) < g(x)$ for all $x \neq 0$. But $\lim_{x\to 0} f(x) = \lim_{x\to 0} g(x) = 0$.

77. Let $f(x) = 1/x$ and $g(x) = -1/x$. $\lim_{x\to 0} f(x)$ and $\lim_{x\to 0} g(x)$ do not exist.

$$\lim_{x\to 0} [f(x) \div g(x)] = \lim_{x\to 0} \left[\frac{1}{x} \div \left(-\frac{1}{x}\right)\right]$$

$$= \lim_{x\to 0} (-1) = -1$$

78. Suppose, on the contrary, that $\lim_{x\to c} g(x)$ exists. Then, since $\lim_{x\to c} f(x)$ exists, so would $\lim_{x\to c} [f(x) + g(x)]$, which is a contradiction. Hence, $\lim_{x\to c} g(x)$ does not exist.

79. Given $f(x) = b$, show that for every $\varepsilon > 0$ there exists a $\delta > 0$ such that $|f(x) - b| < \varepsilon$ whenever $|x - c| < \delta$. Since $|f(x) - b| = |b - b| = 0 < \epsilon$ for any $\varepsilon > 0$, then any value of $\delta > 0$ will work.

80. $\lim_{x\to c} x^n = \lim_{x\to c} x \cdot x \cdot \cdots \cdot x$

$= \left(\lim_{x\to c} x\right)\left(\lim_{x\to c} x\right) \cdots \left(\lim_{x\to c}\right)$ by Property 3, Theorem 3.2

$= c \cdot c \cdot \cdots \cdot c$ by Property 2, Theorem 3.1

$= c^n$

81. If $b = 0$, then the property is true because both sides are equal to 0. If $b \neq 0$, let $\varepsilon > 0$ be given. Since $\lim_{x \to c} f(x) = L$, there exists $\delta > 0$ such that $|f(x) - L| < \varepsilon/|b|$ whenever $0 < |x - c| < \delta$. Hence, wherever $0 < |x - c| < \delta$, we have

$$|b||f(x) - L| < \varepsilon \quad \text{or} \quad |bf(x) - bL| < \varepsilon$$

which implies that $\lim_{x \to c} [bf(x)] = bL$.

82. $f(x) = \begin{cases} 0, & \text{if } x \text{ is rational} \\ 1, & \text{if } x \text{ is irrational} \end{cases}$

$g(x) = \begin{cases} 0, & \text{if } x \text{ is rational} \\ x, & \text{if } x \text{ is irrational} \end{cases}$

$\lim_{x \to 0} f(x)$ does not exist.

No matter how "close to" 0 x is, there are still an infinite number of rational and irrational numbers so that $\lim_{x \to 0} f(x)$ does not exist.

$\lim_{x \to 0} g(x) = 0.$

When x is "close to" 0, both parts of the function are "close to" 0.

Section 3.4 Continuity and One-Sided Limits

1. (a) $\lim_{x \to 3^+} f(x) = 1$

(b) $\lim_{x \to 3^-} f(x) = 1$

(c) $\lim_{x \to 3} f(x) = 1$

The function is continuous at $x = 3$.

2. (a) $\lim_{x \to -2^+} f(x) = -2$

(b) $\lim_{x \to -2^-} f(x) = -2$

(c) $\lim_{x \to -2} f(x) = -2$

The function is continuous at $x = -2$.

3. (a) $\lim_{x \to 3^+} f(x) = 0$

(b) $\lim_{x \to 3^-} f(x) = 0$

(c) $\lim_{x \to 3} f(x) = 0$

The function is NOT continuous at $x = 3$.

4. (a) $\lim_{x \to -2^+} f(x) = 2$

(b) $\lim_{x \to -2^-} f(x) = 2$

(c) $\lim_{x \to -2} f(x) = 2$

The function is NOT continuous at $x = -2$.

5. (a) $\lim_{x \to 4^+} f(x) = 2$

(b) $\lim_{x \to 4^-} f(x) = -2$

(c) $\lim_{x \to 4} f(x)$ does not exist

The function is NOT continuous at $x = 4$.

6. (a) $\lim_{x \to -1^+} f(x) = 0$

(b) $\lim_{x \to -1^-} f(x) = 2$

(c) $\lim_{x \to -1} f(x)$ does not exist.

The function is NOT continuous at $x = -1$.

7. $\displaystyle \lim_{x \to 5^+} \frac{x - 5}{x^2 - 25} = \lim_{x \to 5^+} \frac{1}{x + 5} = \frac{1}{10}$

8. $\displaystyle \lim_{x \to 2^+} \frac{2 - x}{x^2 - 4} = \lim_{x \to 2^+} -\frac{1}{x + 2} = -\frac{1}{4}$

9. $\displaystyle \lim_{x \to -3^-} \frac{x}{\sqrt{x^2 - 9}}$ does not exist because $\dfrac{x}{\sqrt{x^2 - 9}}$ grows without bound as $x \to -3^-$.

10.
$$\lim_{x \to 4^-} \frac{\sqrt{x} - 2}{x - 4} = \lim_{x \to 4^-} \frac{\sqrt{x} - 2}{x - 4} \cdot \frac{\sqrt{x} + 2}{\sqrt{x} + 2}$$
$$= \lim_{x \to 4^-} \frac{x - 4}{(x - 4)(\sqrt{x} + 2)}$$
$$= \lim_{x \to 4^-} \frac{1}{\sqrt{x} + 2} = \frac{1}{4}$$

11. $\lim_{x\to 0^-} \frac{|x|}{x} = \lim_{x\to 0^-} \frac{-x}{x} = -1$

12. $\lim_{x\to 2^+} \frac{|x-2|}{x-2} = \lim_{x\to 2^+} \frac{x-2}{x-2} = 1$

13. $$\lim_{\Delta x\to 0^-} \frac{\frac{1}{x+\Delta x} - \frac{1}{x}}{\Delta x} = \lim_{\Delta x\to 0^-} \frac{x-(x+\Delta x)}{x(x+\Delta x)} \cdot \frac{1}{\Delta x} = \lim_{\Delta x\to 0^-} \frac{-\Delta x}{x(x+\Delta x)} \cdot \frac{1}{\Delta x}$$
$$= \lim_{\Delta x\to 0^-} \frac{-1}{x(x+\Delta x)}$$
$$= \frac{-1}{x(x+0)} = -\frac{1}{x^2}$$

14. $$\lim_{\Delta x\to 0^+} \frac{(x+\Delta x)^2 + (x+\Delta x) - (x^2+x)}{\Delta x} = \lim_{\Delta x\to 0^+} \frac{x^2 + 2x(\Delta x) + (\Delta x)^2 + x + \Delta x - x^2 - x}{\Delta x}$$
$$= \lim_{\Delta x\to 0^+} \frac{2x(\Delta x) + (\Delta x)^2 + \Delta x}{\Delta x}$$
$$= \lim_{\Delta x\to 0^+} (2x + \Delta x + 1)$$
$$= 2x + 0 + 1 = 2x + 1$$

15. $\lim_{x\to 3^-} f(x) = \lim_{x\to 3^-} \frac{x+2}{2} = \frac{5}{2}$

16. $\lim_{x\to 2^+} f(x) = \lim_{x\to 2^+} (-x^2 + 4x - 2) = 2$

$\lim_{x\to 2^-} f(x) = \lim_{x\to 2^-} (x^2 - 4x + 6) = 2$

$\lim_{x\to 2} f(x) = 2$

17. $\lim_{x\to 1^+} f(x) = \lim_{x\to 1^+} (x+1) = 2$

$\lim_{x\to 1^-} f(x) = \lim_{x\to 1^-} (x^3+1) = 2$

$\lim_{x\to 1} f(x) = 2$

18. $\lim_{x\to 1^+} f(x) = \lim_{x\to 1^+} (1-x) = 0$

19. $\lim_{x\to 4^-} (3[\![x]\!] - 5) = 3(3) - 5 = 4$

($[\![x]\!] = 3$ for $3 < x < 4$)

20. $\lim_{x\to 2^+} (2x - [\![x]\!]) = 2(2) - 2 = 2$

21. $\lim_{x\to 3} (2 - [\![-x]\!])$ does not exist

because

$\lim_{x\to 3^-} (2 - [\![-x]\!]) = 2 - (-3) = 5$

and

$\lim_{x\to 3^+} (2 - [\![-x]\!]) = 2 - (-4) = 6.$

22. $\lim_{x\to 1} \left(1 - \left[\!\!\left[-\frac{x}{2}\right]\!\!\right]\right) = 1 - (-1) = 2$

23. $f(x) = \frac{1}{x^2 - 4}$

has discontinuities at $x = -2$ and $x = 2$ since $f(-2)$ and $f(2)$ are not defined.

24. $f(x) = \frac{x^2 - 1}{x + 1}$ has a discontinuity at $x = -1$ since $f(-1)$ is not defined.

25. $f(x) = \frac{[\![x]\!]}{2} + x$

has discontinuities at each integer k since $\lim_{x\to k^-} f(x) \neq \lim_{x\to k^+} f(x)$.

26. $f(x) = \begin{cases} x, & x < 1 \\ 2, & x = 1 \\ 2x - 1, & x > 1 \end{cases}$ has discontinuity at $x = 1$ since $f(1) = 2 \neq \lim_{x\to 1} f(x) = 1$.

27. $g(x) = \sqrt{25 - x^2}$ is continuous on $[-5, 5]$.

28. $f(t) = 3 - \sqrt{9 - t^2}$ is continuous on $[-3, 3]$.

29. $\lim_{x\to 0^-} f(x) = 3 = \lim_{x\to 0^+} f(x)$.
f is continuous on $[-1, 4]$.

30. $g(2)$ is not defined.
g is continuous on $[-1, 2)$.

31. $f(x) = x^2 - 2x + 1$ is continuous for all real x.

32. $f(x) = \dfrac{1}{x^2 + 1}$ is continuous for all real x.

33. $f(x) = \dfrac{x}{x^2 - x}$ is not continuous at $x = 0, 1$. Since

$\dfrac{x}{x^2 - x} = \dfrac{1}{x - 1}$ for $x \neq 0$, $x = 0$ is a removable discontinuity, whereas $x = 1$ is a nonremovable discontinuity.

34. $f(x) = \dfrac{x}{x^2 - 1}$ has nonremovable discontinuities at $x = 1$ and $x = -1$ since $\lim_{x\to 1} f(x)$ and $\lim_{x\to -1} f(x)$ do not exist.

35. $f(x) = \dfrac{x}{x^2 + 1}$ is continuous for all real x.

36. $f(x) = \dfrac{x - 3}{x^2 - 9}$ has a nonremovable discontinuity at $x = -3$ since $\lim_{x\to -3} f(x)$ does not exist, and has a removable discontinuity at $x = 3$ since

$$\lim_{x\to 3} f(x) = \lim_{x\to 3} \frac{1}{x + 3} = \frac{1}{6}.$$

37. $f(x) = \dfrac{x + 2}{(x + 2)(x - 5)}$

has a nonremovable discontinuity at $x = 5$ since $\lim_{x\to 5} f(x)$ does not exist, and has a removable discontinuity at $x = -2$ since

$$\lim_{x\to -2} f(x) = \lim_{x\to -2} \frac{1}{x - 5} = -\frac{1}{7}.$$

38. $f(x) = \dfrac{x - 1}{(x + 2)(x - 1)}$

has a nonremovable discontinuity at $x = -2$ since $\lim_{x\to -2} f(x)$ does not exist, and has a removable discontinuity at $x = 1$ since

$$\lim_{x\to 1} f(x) = \lim_{x\to 1} \frac{1}{x + 2} = \frac{1}{3}.$$

39. $f(x) = \dfrac{|x + 2|}{x + 2}$ has a nonremovable discontinuity at $x = -2$ since $\lim_{x\to -2} f(x)$ does not exist.

40. $f(x) = \dfrac{|x - 3|}{x - 3}$

has a nonremovable discontinuity at $x = 3$ since $\lim_{x\to 3} f(x)$ does not exist.

41. $f(x) = \begin{cases} x, & x \le 1 \\ x^2, & x > 1 \end{cases}$

has a **possible** discontinuity at $x = 1$.

1. $f(1) = 1$

2. $\left.\begin{aligned} \lim_{x\to 1^-} f(x) &= \lim_{x\to 1^-} x = 1 \\ \lim_{x\to 1^+} f(x) &= \lim_{x\to 1^+} x^2 = 1 \end{aligned}\right\} \lim_{x\to 1} f(x) = 1$

3. $f(1) = \lim_{x\to 1} f(x)$

f is continuous at $x = 1$, therefore, f is continuous for all real x.

42. $f(x) = \begin{cases} -2x + 3, & x < 1 \\ x^2, & x \ge 1 \end{cases}$

has a **possible** discontinuity at $x = 1$.

1. $f(1) = 1^2 = 1$

2. $\left.\begin{aligned} \lim_{x\to 1^-} f(x) &= \lim_{x\to 1^-} (-2x + 3) = 1 \\ \lim_{x\to 1^+} f(x) &= \lim_{x\to 1^+} x^2 = 1 \end{aligned}\right\} \lim_{x\to 1} f(x) = 1$

3. $f(1) = \lim_{x\to 1} f(x)$

f is continuous at $x = 1$, therefore, f is continuous for all real x.

43. $f(x) = \begin{cases} \frac{x}{2} + 1, & x \le 2 \\ 3 - x, & x > 2 \end{cases}$ has a **possible** discontinuity at $x = 2$.

1. $f(2) = \frac{2}{2} + 1 = 2$

2. $\left.\begin{array}{l} \lim_{x\to 2^-} f(x) = \lim_{x\to 2^-} \left(\frac{x}{2} + 1\right) = 2 \\ \lim_{x\to 2^+} f(x) = \lim_{x\to 2^+} (3 - x) = 1 \end{array}\right\} \lim_{x\to 2} f(x)$ does not exist.

Therefore, f has a nonremovable discontinuity at $x = 2$.

44. $f(x) = \begin{cases} -2x, & x \le 2 \\ x^2 - 4x + 1, & x > 2 \end{cases}$

has a possible discontinuity at $x = 2$.

1. $f(2) = -2(2) = -4$

2. $\left.\begin{array}{l} \lim_{x\to 2^-} f(x) = \lim_{x\to 2^-} (-2x) = -4 \\ \lim_{x\to 2^+} f(x) = \lim_{x\to 2^+} (x^2 - 4x + 1) = -3 \end{array}\right\} \lim_{x\to 2} f(x)$ does not exist.

Therefore, f has a nonremovable discontinuity at $x = 2$.

45. $f(x) = [\![x - 1]\!]$ has nonremovable discontinuities at each integer k.

46. $f(x) = 3 - [\![x]\!]$ has nonremovable discontinuities at each integer k.

47. $\lim_{x\to 0^+} f(x) = 0$

$\lim_{x\to 0^-} f(x) = 0$

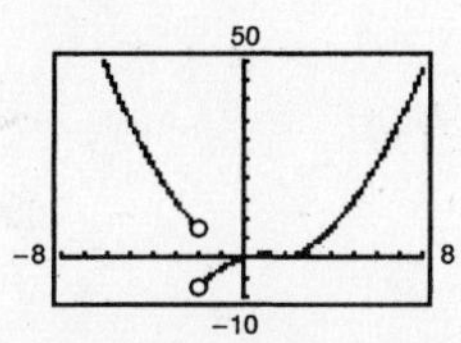

f is not continuous at $x = -2$.

48. $f(x) = \dfrac{|x^2 + 4x|(x + 2)}{x + 4}$

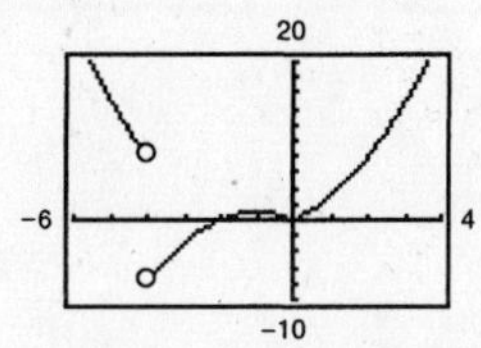

$\lim_{x\to 0^+} f(x) = 0$ and $\lim_{x\to 0^-} f(x) = 0$

The function is not continuous at $x = -4$.

49. Find a such that

$\lim_{x\to 2^+} ax^2 = 4a$

$= \lim_{x\to 2^-} x^3 = 8.$

$4a = 8$

$a = 2$

50. $\lim_{x\to a} g(x) = \lim_{x\to a} \dfrac{x^2 - a^2}{x - a}$

$= \lim_{x\to a} (x + a) = 2a$

Find a such that $2a = 8 \implies a = 4$.

51. $f(g(x)) = (x - 1)^2$

Continuous for all real x.

52. $h(x) = f(g(x)) = f(x^2) = \sqrt{x^2}$

Continuous for all real x.

53. $h(x) = f(g(x)) = f\left(\frac{1}{x}\right) = \dfrac{1}{\sqrt{1/x}} = \sqrt{x}, x \ne 0$

Continuous on $(0, \infty)$.

54. $h(x) = f(g(x)) = f(x - 1) = \dfrac{1}{\sqrt{x - 1}}$.

Continuous on $(1, \infty)$.

55. $f(g(x)) = \dfrac{1}{(x^2 + 5) - 6} = \dfrac{1}{x^2 - 1}$

Nonremovable discontinuities at $x = \pm 1$.

56. $h(x) = f(g(x)) = f\left(\dfrac{1}{x - 1}\right) = \dfrac{1}{\frac{1}{x - 1}} = x - 1, x \ne 1$

Continuous for all $x \ne 1$. Removable discontinuity at $x = 1$.

57. $f(x) = [\![x]\!] - x$

Nonremovable discontinuity at each integer.

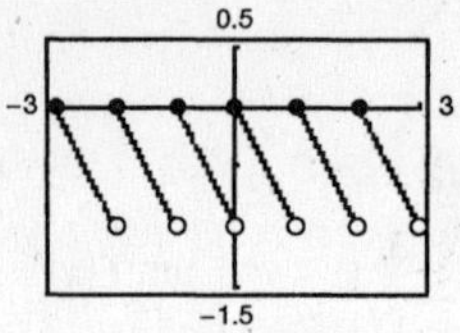

58. $h(x) = \dfrac{1}{(x+1)(x-2)}$

Nonremovable discontinuity at $x = -1$ and $x = 2$.

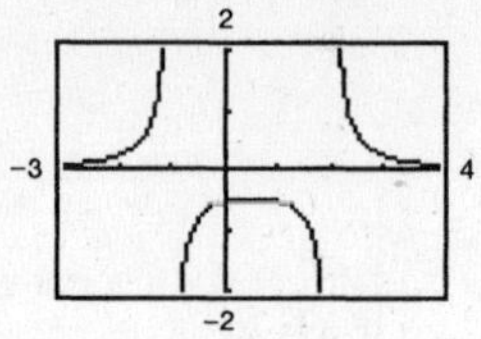

59. $g(x) = \begin{cases} 2x - 4, & x \le 3 \\ x^2 - 2x, & x > 3 \end{cases}$

Nonremovable discontinuity at $x = 3$.

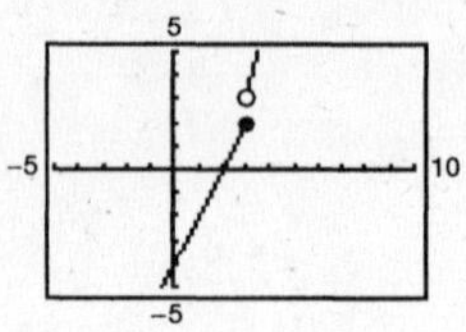

60. $f(x) = \begin{cases} x^2 - 2x + 2, & x < 2 \\ -x^2 + 6x - 6, & x \ge 2 \end{cases}$

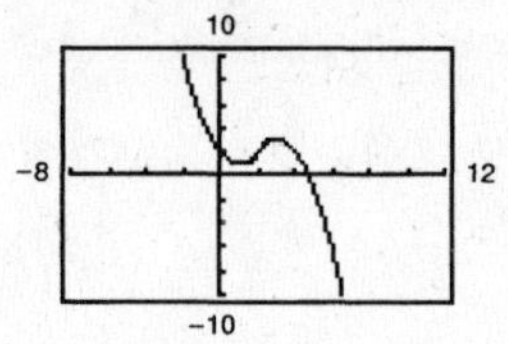

Continuous for all real x.

61. $f(x) = \dfrac{x}{x^2 + 1}$

Continuous on $(-\infty, \infty)$.

62. $f(x) = x\sqrt{x+3}$

Continuous on $[-3, \infty]$.

63. $f(x) = \dfrac{x^2}{x^2 - 36}$

Continuous on: $(-\infty, -6), (-6, 6), (6, \infty)$.

64. $f(x) = \dfrac{x+1}{\sqrt{x}}$

Continuous on $(0, \infty)$.

65. $f(x) = \dfrac{x^2 - x - 2}{x + 1}$

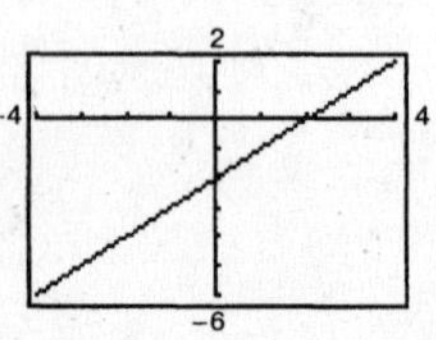

The graph **appears** to be continuous on the interval $[-4, 4]$. Since $f(-1)$ is not defined, we know that f has a discontinuity at $x = -1$. This discontinuity is removable so it does not show up on the graph.

66. $f(x) = \dfrac{x^3 - 8}{x - 2}$

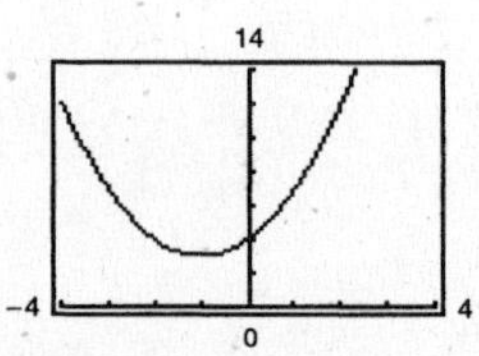

The graph **appears** to be continuous on the interval $[-4, 4]$. Since $f(2)$ is not defined, we know that f has a discontinuity at $x = 2$. This discontinuity is removable so it does not show up on the graph.

67. $f(x) = \frac{1}{16}x^4 - x^3 + 3$ is continuous on $[1, 2]$. $f(1) = \frac{33}{16}$ and $f(2) = -4$. By the Intermediate Value Theorem, $f(c) = 0$ for at least one value of c between 1 and 2.

68. $f(x) = x^3 + 3x - 2$ is continuous on $[0, 1]$.

$f(0) = -2$ and $f(1) = 2$.

By the Intermediate Value Theorem, $f(x) = 0$ for at least one value of c between 0 and 1.

69. $f(x) = x^3 + x - 1$

$f(x)$ is continuous on $[0, 1]$.

$f(0) = -1$ and $f(1) = 1$.

By the Intermediate Value Theorem, $f(x) = 0$ for at least one value of c between 0 and 1. Using a graphing utility, we find that $x \approx 0.6823$.

70. $f(x) = x^3 + 3x - 2$

$f(x)$ is continuous on $[0, 1]$.

$f(0) = -2$ and $f(1) = 2$.

By the Intermediate Value Theorem, $f(x) = 0$ for at least one value of c between 0 and 1. Using a graphing utility, we find that $x \approx 0.5961$.

71. $g(t) = 3\sqrt{t^2 + 1} - 4$. g is continuous on $[0, 1]$.

$g(0) = 3 - 4 = -1$ and $g(1) = 3\sqrt{2} - 4 \approx 0.2426$.

By the Intermediate Value Theorem, $g(t) = 0$ for at least one value of c between 0 and 1. Using the zoom feature, we find that $c \approx 0.88$. Using a graphing utility, we find that $c \approx 0.8819$.

72. $h(s) = 5 - \dfrac{2}{s^3}$

h is continuous on $\left[\frac{1}{2}, 1\right]$.

$h\left(\dfrac{1}{2}\right) = 5 - \dfrac{2}{\left(\dfrac{1}{8}\right)} = -11$ and $h(1) = 5 - 2 = 3$.

By the Intermediate Valve Theorem, $h(s) = 0$ for at least one value of c between $\frac{1}{2}$ and 1 (and hence in $[0, 1]$). Using the zoom feature, we find that $c \approx 0.74$. Using the graphing utility, we find that $c \approx 0.7368$.

73. $f(x) = x^2 + x - 1$

f is continuous on $[0, 5]$.

$f(0) = -1$ and $f(5) = 29$.

$-1 < 11 < 29$

The Intermediate Value Theorem applies.

$$x^2 + x - 1 = 11$$
$$x^2 + x - 12 = 0$$
$$(x + 4)(x - 3) = 0$$
$$x = -4 \text{ or } x = 3$$

$c = 3$ ($x = -4$ is not in the interval.)

Thus, $f(3) = 11$.

74. $f(x) = x^2 - 6x + 8$

f is continuous on $[0, 3]$.

$f(0) = 8$ and $f(3) = -1$.

$-1 < 0 < 8$

The Intermediate Value Theorem applies.

$$x^2 - 6x + 8 = 0$$
$$(x - 2)(x - 4) = 0$$
$$x = 2 \text{ or } x = 4$$

$c = 2$ ($x = 4$ is not in the interval.)

Thus, $f(2) = 0$.

75. $f(x) = x^3 - x^2 + x - 2$

f is continuous on $[0, 3]$.

$f(0) = -2$ and $f(3) = 19$.

$-2 < 4 < 19$

The Intermediate Value Theorem applies.

$$x^3 - x^2 + x - 2 = 4$$
$$x^3 - x^2 + x - 6 = 0$$
$$(x - 2)(x^2 + x + 3) = 0$$
$$x = 2$$

($x^2 + x + 3 = 0$ has no real solution.)

$$c = 2$$

Thus, $f(2) = 4$.

76. $f(x) = \dfrac{x^2 + x}{x - 1}$

f is continuous on $\left[\frac{5}{2}, 4\right]$. The nonremovable discontinuity, $x = 1$, lies outside the interval.

$f\left(\dfrac{5}{2}\right) = \dfrac{35}{6}$ and $f(4) = \dfrac{20}{3}$.

$\dfrac{35}{6} < 6 < \dfrac{20}{3}$

The Intermediate Value Theorem applies.

$$\frac{x^2 + x}{x - 1} = 6$$
$$x^2 + x = 6x - 6$$
$$x^2 - 5x + 6 = 0$$
$$(x - 2)(x - 3) = 0$$
$$x = 2 \text{ or } x = 3$$

$c = 3$ ($x = 2$ is not in the interval.)

Thus, $f(3) = 6$.

77. (a) The limit does not exist at $x = c$.

(b) The function is not defined at $x = c$.

(c) The limit exists at $x = c$, but it is not equal to the value of the function at $x = c$.

(d) The limit does not exist at $x = c$.

78. A discontinuity at $x = c$ is removable if you can define (or redefine) the function at $x = c$ in such a way that the new function is continuous at $x = c$. Answers will vary.

(a) $f(x) = \dfrac{|x - 2|}{x - 2}$

(b) $f(x) = \dfrac{x^2 + 2x}{x + 2}$

(c) $f(x) = \begin{cases} 1, & \text{if } x \geq 2 \\ 0, & \text{if } -2 < x < 2 \\ 1, & \text{if } x = -2 \\ 0, & \text{if } x < -2 \end{cases}$

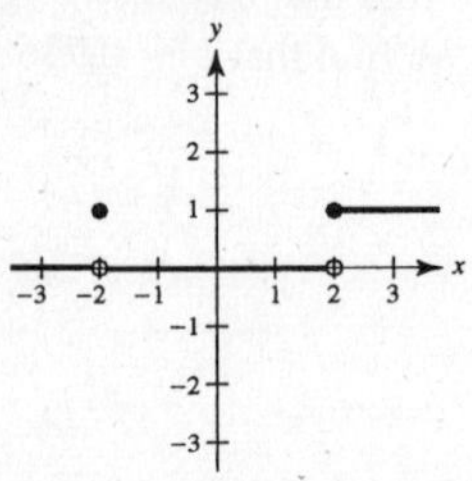

79.

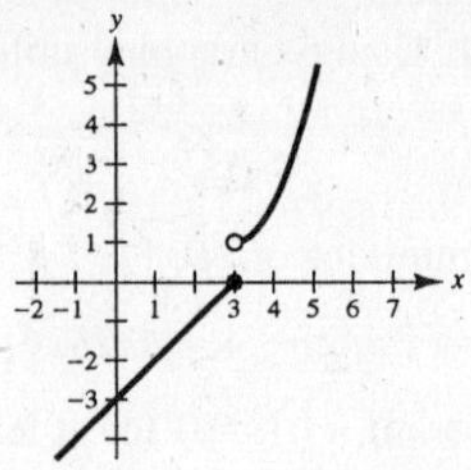

The function is not continuous at $x = 3$ because

$\lim_{x \to 3^+} f(x) = 1 \neq 0 = \lim_{x \to 3^-} f(x)$.

80. If f and g are continuous for all real x, then so is $f + g$ (Theorem 3.9, part 2). However, f/g might not be continuous if $g(x) = 0$. For example, let $f(x) = x$ and $g(x) = x^2 - 1$. Then f and g are continuous for all real x, but f/g is not continuous at $x = \pm 1$.

81. The functions agree for integer values of x:

$$\left.\begin{aligned} g(x) &= 3 - [\![-x]\!] = 3 - (-x) = 3 + x \\ f(x) &= 3 + [\![x]\!] = 3 + x \end{aligned}\right\} \text{ for } x \text{ an integer}$$

However, for non-integer values of x, the functions differ by 1.

$f(x) = 3 + [\![x]\!] = g(x) - 1 = 2 - [\![-x]\!]$.

For example, $f(\frac{1}{2}) = 3 + 0 = 3$, $g(\frac{1}{2}) = 3 - (-1) = 4$.

82. $C = \begin{cases} 1.04, & 0 < t \leq 2 \\ 1.04 + 0.36[\![t - 1]\!], & t > 2, t \text{ is not an integer} \\ 1.04 + 0.36(t - 2), & t > 2, t \text{ is an integer} \end{cases}$

Nonremovable discontinuity at each integer greater than 2.

You can also write C as

$$C = \begin{cases} 1.04, & 0 < t \leq 2 \\ 1.04 - 0.36[\![2 - t]\!], & t > 2 \end{cases}.$$

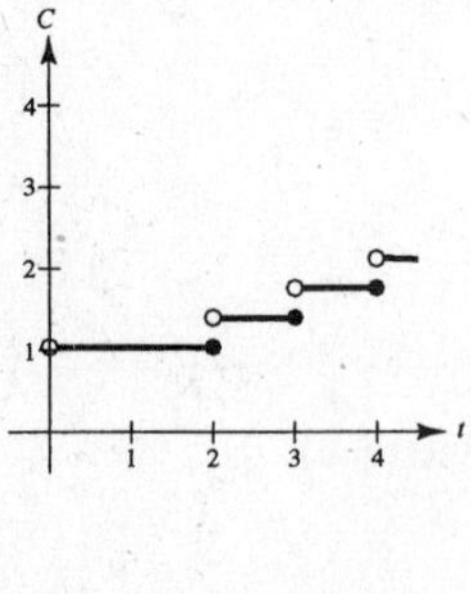

83. $N(t) = 25\left(2\left[\!\!\left[\dfrac{t + 2}{2}\right]\!\!\right] - t\right)$

t	0	1	1.8	2	3	3.8
$N(t)$	50	25	5	50	25	5

Discontinuous at every positive even integer. The company replenishes its inventory every two months.

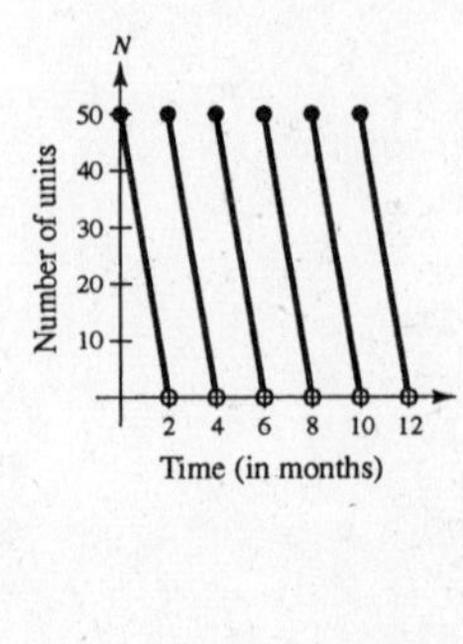

84. Let $s(t)$ be the position function for the run up to the campsite. $s(0) = 0$ ($t = 0$ corresponds to 8:00 A.M., $s(20) = k$ (distance to campsite)). Let $r(t)$ be the position function for the run back down the mountain: $r(0) = k$, $r(10) = 0$. Let $f(t) = s(t) - r(t)$.

When $t = 0$ (8:00 A.M.), $f(0) = s(0) - r(0) = 0 - k < 0$.

When $t = 10$ (8:10 A.M.), $f(10) = s(10) - r(10) > 0$.

Since $f(0) < 0$ and $f(10) > 0$, then there must be a value t in the interval $[0, 10]$ such that $f(t) = 0$. If $f(t) = 0$, then $s(t) - r(t) = 0$, which gives us $s(t) = r(t)$. Therefore, at some time t, where $0 \le t \le 10$, the position functions for the run up and the run down are equal.

85. Let $V = \frac{4}{3}\pi r^3$ be the volume of a sphere of radius r.

V is continuous on $[1, 5]$.

$V(1) = \frac{4}{3}\pi \approx 4.19$

$V(5) = \frac{4}{3}\pi(5^3) \approx 523.6$

Since $4.19 < 275 < 523.6$ the Intermediate Value Theorem implies that there is at least one value r between 1 and 5 such that $V(r) = 275$. (In fact, $r \approx 4.0341$.)

86. Suppose there exists x_1 in $[a, b]$ such that $f(x_1) > 0$ and there exists x_2 in $[a, b]$ such that $f(x_2) < 0$. Then by the Intermediate Value Theorem, $f(x)$ must equal zero for some value of x in $[x_1, x_2]$ (or $[x_2, x_1]$ if $x_2 < x_1$). Thus, f would have a zero in $[a, b]$, which is a contradiction. Therefore, $f(x) > 0$ for all x in $[a, b]$ or $f(x) < 0$ for all x in $[a, b]$.

87. Let c be any real number. Then $\lim_{x \to c} f(x)$ does not exist since there are both rational and irrational numbers arbitrarily close to c. Therefore, f is not continuous at c.

88. True

1. $f(c) = L$ is defined. **2.** $\lim_{x \to c} f(x) = L$ exists.

3. $f(c) = \lim_{x \to c} f(x)$

All of the conditions for continuity are met.

89. True; if $f(x) = g(x)$, $x \ne c$, then $\lim_{x \to c} f(x) = \lim_{x \to c} g(x)$ (if they exist) and at least one of these limits then does not equal the corresponding function value at $x = c$.

90. False; a rational function can be written as $P(x)/Q(x)$ where P and Q are polynomials of degree m and n, respectively. It can have, at most, n discontinuities.

91. False; $f(1)$ is not defined and $\lim_{x \to 1} f(x)$ does not exist.

92. (a)

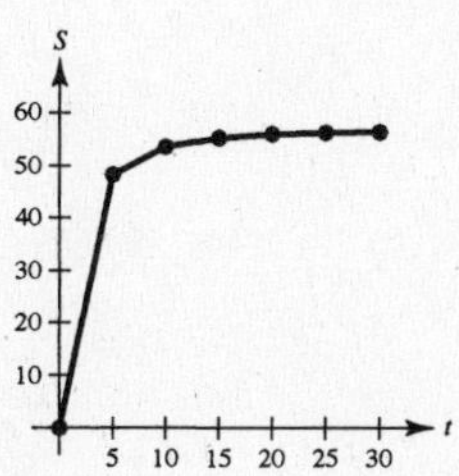

(b) There appears to be a limiting speed and a possible cause is air resistance.

93. (a) $f(x) = \begin{cases} 0 & 0 \le x < b \\ b & b < x \le 2b \end{cases}$

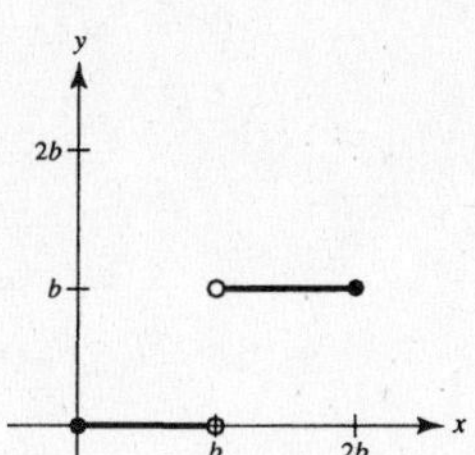

NOT continuous at $x = b$.

(b) $g(x) = \begin{cases} \dfrac{x}{2} & 0 \le x \le b \\ b - \dfrac{x}{2} & b < x \le 2b \end{cases}$

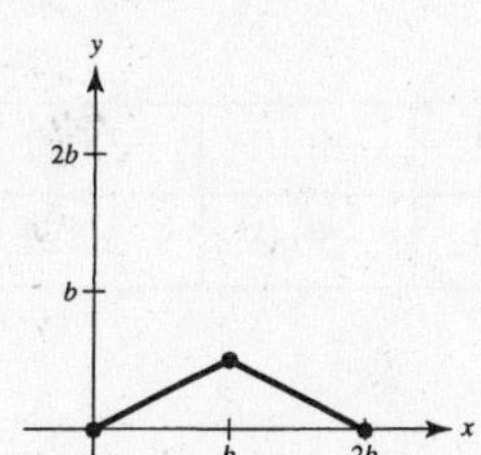

Continuous on $[0, 2b]$.

94. $h(x) = x[\![x]\!]$

h has nonremovable discontinuities at

$x = \pm 1, \pm 2, \pm 3, \ldots.$

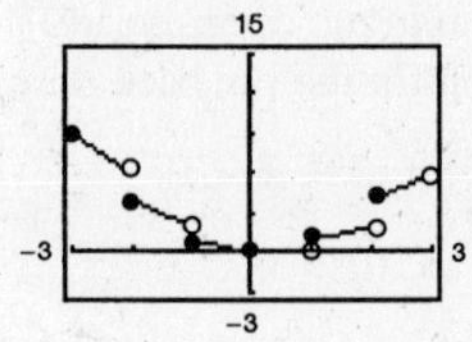

95. $f(x) = \dfrac{\sqrt{x + c^2} - c}{x}, \ c > 0$

Domain: $x + c^2 \geq 0 \Rightarrow x \geq -c^2$ and $x \neq 0$, $[-c^2, 0) \cup (0, \infty)$

$$\lim_{x\to 0} \frac{\sqrt{x + c^2} - c}{x} = \lim_{x\to 0} \frac{\sqrt{x + c^2} - c}{x} \cdot \frac{\sqrt{x + c^2} + c}{\sqrt{x + c^2} + c}$$

$$= \lim_{x\to 0} \frac{(x + c^2) - c^2}{x\left[\sqrt{x + c^2} + c\right]} = \lim_{x\to 0} \frac{1}{\sqrt{x + c^2} + c} = \frac{1}{2c}$$

Define $f(0) = 1/(2c)$ to make f continuous at $x = 0$.

96. Define $f(x) = f_2(x) - f_1(x)$. Since f_1 and f_2 are continuous on $[a, b]$, so is f.

$f(a) = f_2(a) - f_1(a) > 0$ and $f(b) = f_2(b) - f_1(b) < 0$.

By the Intermediate Value Theorem, there exists c in $[a, b]$ such that $f(c) = 0$.

$f(c) = f_2(c) - f_1(c) = 0 \Rightarrow f_1(c) = f_2(c)$

Section 3.5 Infinite Limits

1. $\displaystyle\lim_{x\to -2^+} 2\left|\frac{x}{x^2 - 4}\right| = \infty$

$\displaystyle\lim_{x\to -2^-} 2\left|\frac{x}{x^2 - 4}\right| = \infty$

2. $\displaystyle\lim_{x\to -2^+} \frac{1}{x + 2} = \infty$

$\displaystyle\lim_{x\to -2^-} \frac{1}{x + 2} = -\infty$

3. $f(x) = \dfrac{1}{x^2 - 9}$

x	-3.5	-3.1	-3.01	-3.001	-2.999	-2.99	-2.9	-2.5
$f(x)$	0.308	1.639	16.64	166.6	-166.7	-16.69	-1.695	-0.364

$\displaystyle\lim_{x\to -3^-} f(x) = \infty$

$\displaystyle\lim_{x\to -3^+} f(x) = -\infty$

4. $f(x) = \dfrac{x}{x^2 - 9}$

x	-3.5	-3.1	-3.01	-3.001	-2.999	-2.99	-2.9	-2.5
$f(x)$	-1.077	-5.082	-50.08	-500.1	499.9	49.92	4.915	0.9091

$\displaystyle\lim_{x\to -3^-} f(x) = -\infty$

$\displaystyle\lim_{x\to -3^+} f(x) = \infty$

5. $f(x) = \dfrac{x^2}{x^2 - 9}$

x	-3.5	-3.1	-3.01	-3.001	-2.999	-2.99	-2.9	-2.5
$f(x)$	3.769	15.75	150.8	1501	-1499	-149.3	-14.25	-2.273

$\lim_{x \to -3^-} f(x) = \infty$

$\lim_{x \to -3^+} f(x) = -\infty$

6. $f(x) = \dfrac{x^3}{x^2 - 9} = \dfrac{x^3}{(x - 3)(x + 3)}$

x	-3.5	-3.1	-3.01	-3.001	-2.999	-2.99	-2.9	-2.5
$f(x)$	-13.19	-48.84	-453.76	-4503.8	4496.3	446.26	41.34	5.68

$\lim_{x \to -3^-} f(x) = -\infty$

$\lim_{x \to -3^+} f(x) = \infty$

7. $\displaystyle\lim_{x \to 2^+} \frac{x^2 - 2}{(x - 2)(x + 1)} = \infty$

$\displaystyle\lim_{x \to 2^-} \frac{x^2 - 2}{(x - 2)(x + 1)} = -\infty$

Therefore, $x = 2$ is a vertical asymptote.

$\displaystyle\lim_{x \to -1^+} \frac{x^2 - 2}{(x - 2)(x + 1)} = \infty$

$\displaystyle\lim_{x \to -1^-} \frac{x^2 - 2}{(x - 2)(x + 1)} = -\infty$

Therefore, $x = -1$ is a vertical asymptote.

8. $f(x) = \dfrac{x^3}{x^2 - 1} = \dfrac{x^3}{(x - 1)(x + 1)}$

Vertical asymptotes: $x = 1, -1$

9. $\displaystyle\lim_{x \to 0^+} \frac{1}{x^2} = \infty = \lim_{x \to 0^-} \frac{1}{x^2}$

Therefore, $x = 0$ is a vertical asymptote.

10. $\displaystyle\lim_{x \to 2^+} \frac{4}{(x - 2)^3} = \infty$

$\displaystyle\lim_{x \to 2^-} \frac{4}{(x - 2)^3} = -\infty$

Therefore, $x = 2$ is a vertical asymptote.

11. $f(x) = \dfrac{x^2}{x^2 + x - 6} = \dfrac{x^2}{(x + 3)(x - 2)}$

Vertical asymptotes: $x = -3, 2$

12. $\displaystyle\lim_{x \to 0^-} \frac{2 + x}{x^2(1 - x)} = \lim_{x \to 0^+} \frac{2 + x}{x^2(1 - x)} = \infty$

Therefore, $x = 0$ is a vertical asymptote.

$\displaystyle\lim_{x \to 1^-} \frac{2 + x}{x^2(1 - x)} = \infty$

$\displaystyle\lim_{x \to 1^+} \frac{2 + x}{x^2(1 - x)} = -\infty$

Therefore, $x = 1$ is a vertical asymptote.

13. $\lim_{x\to-2^-} \dfrac{x^2}{x^2-4} = \infty$ and $\lim_{x\to-2^+} \dfrac{x^2}{x^2-4} = -\infty$.

Therefore, $x = -2$ is a vertical asymptote.

$\lim_{x\to2^-} \dfrac{x^2}{x^2-4} = -\infty$ and $\lim_{x\to2^+} \dfrac{x^2}{x^2-4} = \infty$.

Therefore, $x = 2$ is a vertical asymptote.

14. $f(x) = \dfrac{-4x}{x^2+4}$

$x^2 + 4 \neq 0$. No vertical asymptotes.

15. $g(t) = \dfrac{t-1}{t^2+1}$

$t^2 + 1 \neq 0$. No vertical asymptotes.

16. $h(s) = \dfrac{2s-3}{s^2-25} = \dfrac{2s-3}{(s-5)(s+5)}$

Vertical asymptotes: $s = 5, -5$

17. $\lim_{t\to0^+} \left(1 - \dfrac{4}{t^2}\right) = -\infty = \lim_{t\to0^-} \left(1 - \dfrac{4}{t^2}\right)$

Therefore, $t = 0$ is a vertical asymptote.

18. $g(x) = \dfrac{(1/2)x^3 - x^2 - 4x}{3x^2 - 6x - 24} = \dfrac{1}{6}\dfrac{x(x^2-2x-8)}{x^2-2x-8}$

$= \dfrac{1}{6}x,$

$x \neq -2, 4$

No vertical asymptotes. The graph has holes at $x = -2$ and $x = 4$.

19. $\lim_{x\to-2^+} \dfrac{x}{(x+2)(x-1)} = \infty$

$\lim_{x\to-2^-} \dfrac{x}{(x+2)(x-1)} = -\infty$

Therefore, $x = -2$ is a vertical asymptote.

$\lim_{x\to1^+} \dfrac{x}{(x+2)(x-1)} = \infty$

$\lim_{x\to1^-} \dfrac{x}{(x+2)(x-1)} = -\infty$

Therefore, $x = 1$ is a vertical asymptote.

20. $f(x) = \dfrac{4(x^2+x-6)}{x(x^3-2x^2-9x+18)} = \dfrac{4(x+3)(x-2)}{x(x-2)(x^2-9)} = \dfrac{4}{x(x-3)}, x \neq -3, 2$

Vertical asymptotes at $x = 0$ and $x = 3$. The graph has holes at $x = -3$ and $x = 2$.

21. $g(x) = \dfrac{x^3+1}{x+1} = \dfrac{(x+1)(x^2-x+1)}{x+1}$

has no vertical asymptote since

$\lim_{x\to-1} f(x) = \lim_{x\to-1} (x^2-x+1) = 3.$

22. $h(x) = \dfrac{x^2-4}{x^3+2x^2+x+2} = \dfrac{(x+2)(x-2)}{(x+2)(x^2+1)}$

has no vertical asymptote since

$\lim_{x\to-2} h(x) = \lim_{x\to-2} \dfrac{x-2}{x^2+1} = -\dfrac{4}{5}.$

23. $f(x) = \dfrac{(x-5)(x+3)}{(x-5)(x^2+1)} = \dfrac{x+3}{x^2+1}, x \neq 5$

No vertical asymptotes. The graph has a hole at $x = 5$.

24. $h(t) = \dfrac{t(t-2)}{(t-2)(t+2)(t^2+4)} = \dfrac{t}{(t+2)(t^2+4)}, t \neq 2$

Vertical asymptote at $t = -2$. The graph has a hole at $t = 2$.

25. $\lim_{x\to -1} \dfrac{x^2 - 1}{x + 1} = \lim_{x\to -1} (x - 1) = -2$

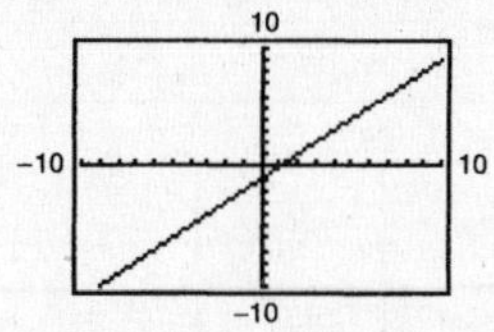

Removable discontinuity at $x = -1$.

26. $\lim_{x\to -1} \dfrac{x^2 - 6x - 7}{x + 1} = \lim_{x\to -1} (x - 7) = -8$

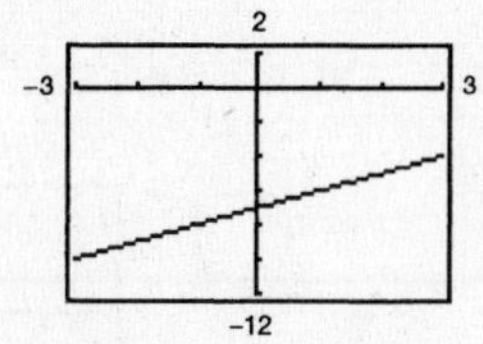

Removable discontinuity at $x = -1$.

27. $\lim_{x\to -1^+} \dfrac{x^2 + 1}{x + 1} = \infty$

$\lim_{x\to -1^-} \dfrac{x^2 + 1}{x + 1} = -\infty$

Vertical asymptote at $x = -1$.

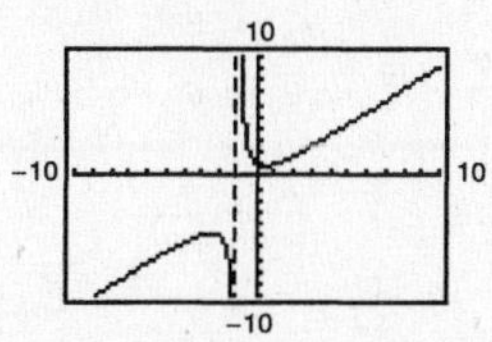

28. $\lim_{x\to -1^+} \dfrac{x - 1}{x + 1} = -\infty$

$\lim_{x\to -1^-} \dfrac{x - 1}{x + 1} = \infty$

Vertical asymptote at $x = -1$.

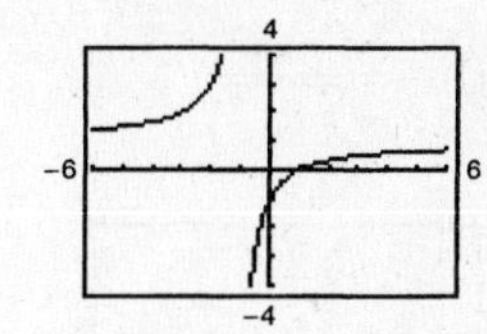

29. $\lim_{x\to 2^+} \dfrac{x - 3}{x - 2} = -\infty$

30. $\lim_{x\to 1^+} \dfrac{2 + x}{1 - x} = -\infty$

31. $\lim_{x\to 3^+} \dfrac{x^2}{(x - 3)(x + 3)} = \infty$

32. $\lim_{x\to 4^-} \dfrac{x^2}{x^2 + 16} = \dfrac{1}{2}$

33. $\lim_{x\to -3^-} \dfrac{x^2 + 2x - 3}{x^2 + x - 6} = \lim_{x\to -3^-} \dfrac{x - 1}{x - 2} = \dfrac{4}{5}$

34. $\lim_{x\to -(1/2)^+} \dfrac{6x^2 + x - 1}{4x^2 - 4x - 3} = \lim_{x\to -(1/2)^+} \dfrac{3x - 1}{2x - 3} = \dfrac{5}{8}$

35. $\lim_{x\to 1} \dfrac{x^2 - x}{(x^2 + 1)(x - 1)} = \lim_{x\to 1} \dfrac{x}{x^2 + 1} = \dfrac{1}{2}$

36. $\lim_{x\to 3} \dfrac{x - 2}{x^2} = \dfrac{1}{9}$

37. $\lim_{x\to 0^-} \left(1 + \dfrac{1}{x}\right) = -\infty$

38. $\lim_{x\to 0^-} \left(x^2 - \dfrac{1}{x}\right) = \infty$

39. $\lim_{x\to 4} f(x) = \lim_{x\to 4} \dfrac{1}{(x - 4)^2} = \infty$

40. $\lim_{x\to 4} g(x) = \lim_{x\to 4} (x^2 - 5x) = 16 - 20 = -4$

41. $\lim_{x\to 4} [f(x) + g(x)] = \lim_{x\to 4} \left[\dfrac{1}{(x - 4)^2} + x^2 - 5x\right] = \infty$

42. $\lim_{x\to 4} [f(x)g(x)] = \lim_{x\to 4} \dfrac{x^2 - 5x}{(x - 4)^2} = -\infty$

43. $\lim_{x\to 4} \left[\dfrac{f(x)}{g(x)}\right] = \lim_{x\to 4} \dfrac{1}{(x - 4)^2(x^2 - 5x)} = -\infty$

44. $\lim_{x\to 4} \left[\dfrac{g(x)}{f(x)}\right] = \lim_{x\to 4} (x^2 - 5x)(x - 4)^2 = 0$

45. $f(x) = \dfrac{x^2 + x + 1}{x^3 - 1}$

$\lim_{x \to 1^+} f(x) = \lim_{x \to 1^+} \dfrac{1}{x - 1} = \infty$

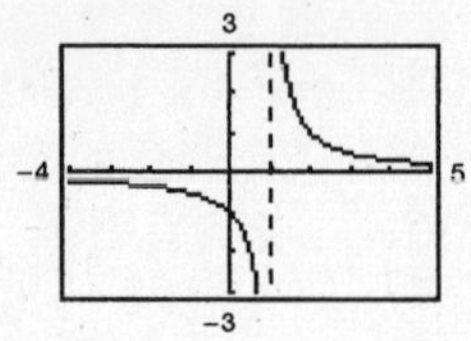

46. $f(x) = \dfrac{x^3 - 1}{x^2 + x + 1}$

$\lim_{x \to 1^-} f(x) = \lim_{x \to 1^-} (x - 1) = 0$

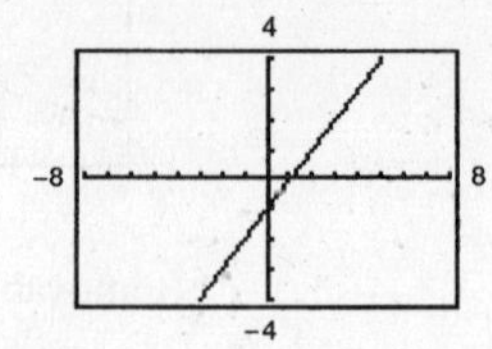

47. $f(x) = \dfrac{1}{x^2 - 25}$

$\lim_{x \to 5^-} f(x) = -\infty$

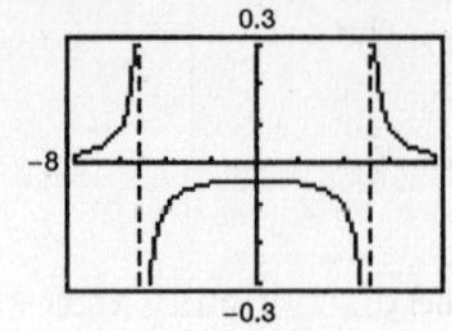

48. $f(x) = \dfrac{6 - x}{\sqrt{x - 3}}$

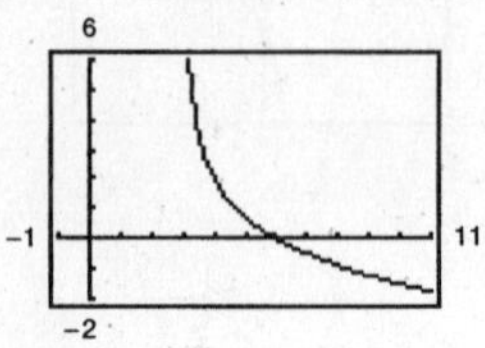

$\lim_{x \to 3^+} f(x) = \infty$

49. A limit in which $f(x)$ increases or decreases without bound as x approaches c is called an infinite limit. ∞ is not a number. Rather, the symbol

$$\lim_{x \to c} f(x) = \infty$$

says how the limit fails to exist.

50. The line $x = a$ is a vertical asymptote of the graph of f if $f(x)$ tends to ∞ or $-\infty$ as x tends to a from the left or right.

51. One answer is $f(x) = \dfrac{x - 3}{(x - 6)(x + 2)} = \dfrac{x - 3}{x^2 - 4x - 12}$.

52. No. For example, $f(x) = \dfrac{1}{x^2 + 1}$ has no vertical asymptote.

53.

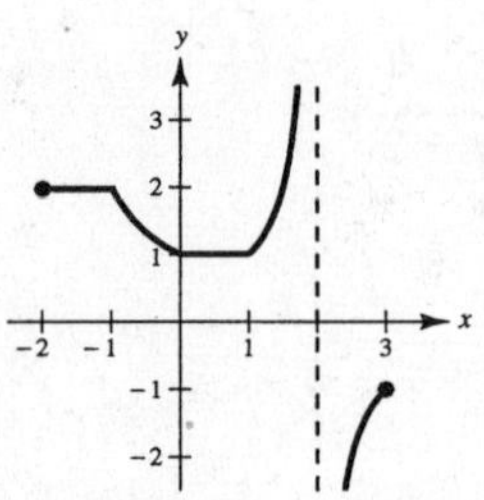

54. $P = \dfrac{k}{V}$

$\lim_{V \to 0^+} \dfrac{k}{V} = \infty$

(In this case we know that $k > 0$.)

55. $C = \dfrac{80{,}000p}{100 - p}, 0 \le p < 100$

(a) $C(15) \approx \$14{,}118$

(b) $C(50) = \$80{,}000$

(c) $C(90) = \$720{,}000$

(d) $\lim_{p \to 100^-} C = \infty$

The cost increases without bound.

56. $C = \dfrac{528x}{100 - x}, 0 \le x < 100$

(a) $C(25) = \$176$ million

(b) $C(50) = \$528$ million

(c) $C(75) = \$1584$ million

(d) $\lim_{x \to 100^-} \dfrac{528}{100 - x} = \infty$ Thus, it is not possible.

57. $m = \dfrac{m_0}{\sqrt{1 - (v^2/c^2)}}$

$\lim_{v \to c^-} m = \lim_{v \to c^-} \dfrac{m_0}{\sqrt{1 - (v^2/c^2)}} = \infty$

58. (a) $r = \dfrac{2(7)}{\sqrt{625 - 49}} = \dfrac{7}{12}$ ft/sec

(b) $r = \dfrac{2(15)}{\sqrt{625 - 225}} = \dfrac{3}{2}$ ft/sec

(c) $\lim_{x \to 25^-} \dfrac{2x}{\sqrt{625 - x^2}} = \infty$

59. (a) Average speed $= \dfrac{\text{Total distance}}{\text{Total time}}$

$$50 = \frac{2d}{(d/x) + (d/y)}$$

$$50 = \frac{2xy}{y + x}$$

$$50y + 50x = 2xy$$

$$50x = 2xy - 50y$$

$$50x = 2y(x - 25)$$

$$\frac{25x}{x - 25} = y$$

Domain: $x > 25$

(b)

x	30	40	50	60
y	150	66.667	50	42.857

(c) $\displaystyle\lim_{x\to 25^+} \frac{25x}{x - 25} = \infty$

As x gets close to 25 mph, y becomes larger and larger.

60. (a) $\dfrac{150}{x} + \dfrac{150}{y} = \dfrac{300}{60} = 5$

$$\frac{30}{x} + \frac{30}{y} = 1$$

$$30y + 30x = xy$$

$$30x = xy - 30y = (x - 30)y$$

$$y = \frac{30x}{x - 30}$$

(b) If $y \le 65$, then graph $y = \dfrac{30x}{x - 30}$ together with $y = 65$ to obtain $x \ge 55.7$ mph.

(c) $\displaystyle\lim_{x\to 30^+} y = \infty$

61. False; for instance, let

$$f(x) = \frac{x^2 - 1}{x - 1}.$$

The graph of f has a hole at $(1, 2)$, not a vertical asymptote.

62. False; for instance, let

$$f(x) = \frac{x^2 - 1}{x - 1}.$$

The graph of f has a hole at $(1, 2)$, not a vertical asymptote.

63. True

64. True

65. Let $f(x) = \dfrac{1}{x^2}$ and $g(x) = \dfrac{1}{x^4}$, and $c = 0$.

$$\lim_{x\to 0} \frac{1}{x^2} = \infty \text{ and } \lim_{x\to 0} \frac{1}{x^4} = \infty, \text{ but } \lim_{x\to 0}\left(\frac{1}{x^2} - \frac{1}{x^4}\right) = \lim_{x\to 0}\left(\frac{x^2 - 1}{x^4}\right) = -\infty \ne 0.$$

66. Given $\displaystyle\lim_{x\to c} f(x) = \infty$ and $\displaystyle\lim_{x\to c} g(x) = L$:

(2) Product:

If $L > 0$, then for $\varepsilon = L/2 > 0$ there exists $\delta_1 > 0$ such that $|g(x) - L| < L/2$ whenever $0 < |x - c| < \delta_1$. Thus, $L/2 < g(x) < 3L/2$. Since $\displaystyle\lim_{x\to c} f(x) = \infty$ then for $M > 0$, there exists $\delta_2 > 0$ such that $f(x) > M(2/L)$ whenever $|x - c| < \delta_2$. Let δ be the smaller of δ_1 and δ_2. Then for $0 < |x - c| < \delta$, we have $f(x)g(x) > M(2/L)(L/2) = M$. Therefore $\displaystyle\lim_{x\to c} f(x)g(x) = \infty$. The proof is similar for $L < 0$.

(3) Quotient: Let $\varepsilon > 0$ be given.

There exists $\delta_1 > 0$ such that $f(x) > 3L/2\epsilon$ whenever $0 < |x - c| < \delta_1$ and there exists $\delta_2 > 0$ such that $|g(x) - L| < L/2$ whenever $0 < |x - c| < \delta_2$. This inequality gives us $L/2 < g(x) < 3L/2$. Let δ be the smaller of δ_1 and δ_2. Then for $0 < |x - c| < \delta$, we have

$$\left|\frac{g(x)}{f(x)}\right| < \frac{3L/2}{3L/2\varepsilon} = \varepsilon.$$

Therefore, $\displaystyle\lim_{x\to c} \frac{g(x)}{f(x)} = 0$.

67. Given $\lim_{x\to c} f(x) = \infty$, let $g(x) = 1$. Then $\lim_{x\to c} \frac{g(x)}{f(x)} = 0$ by Theorem 3.13.

68. Given $\lim_{x\to c} \frac{1}{f(x)} = 0$.

Suppose $\lim_{x\to c} f(x)$ exists and equals L. Then,

$$\lim_{x\to c} \frac{1}{f(x)} = \frac{\lim_{x\to c} 1}{\lim_{x\to c} f(x)} = \frac{1}{L} = 0.$$

This is not possible. Thus, $\lim_{x\to c} f(x)$ does not exist.

69. $f(x) = \frac{1}{x-3}$ is defined for all $x > 3$. Let $M > 0$ be given. We need $\delta > 0$ such that $f(x) = \frac{1}{x-3} > M$ whenever $3 < x < 3 + \delta$.

Equivalently, $x - 3 < \frac{1}{M}$ whenever $|x-3| < \delta$, $x > 3$.

So take $\delta = \frac{1}{M}$. Then for $x > 3$ and $|x - 3| < \delta$,

$\frac{1}{x-3} > \frac{1}{\delta} = M$ and hence $f(x) > M$.

70. $f(x) = \frac{1}{x-4}$ is defined for all $x < 4$. Let $N < 0$ be given. We need $\delta > 0$ such that $f(x) = \frac{1}{x-4} < N$ whenever $4 - \delta < x < 4$.

Equivalently, $x - 4 > \frac{1}{N}$ whenever $|x-4| < \delta$, $x < 4$. Equivalently, $\frac{1}{|x-4|} < -\frac{1}{N}$ whenever $|x - 4| < \delta$, $x < 4$. So take $\delta = -\frac{1}{N}$. Note that $\delta > 0$ because $N < 0$. For $|x-4| < \delta$ and $x < 4$, $\frac{1}{|x-4|} > \frac{1}{\delta} = -N$, and $\frac{1}{x-4} = -\frac{1}{|x-4|} < N$.

Review Exercises for Chapter 3

1. Calculus required. Using a graphing utility, you can estimate the length to be 8.3.
Or, the length is slightly longer than the distance between the two points, approximately 8.25.

2. Precalculus. $L = \sqrt{(9-1)^2 + (3-1)^2} \approx 8.25$

3.

x	-0.1	-0.01	-0.001	0.001	0.01	0.1
$f(x)$	-1.053	-1.005	-1.001	-0.9995	-0.995	-0.9524

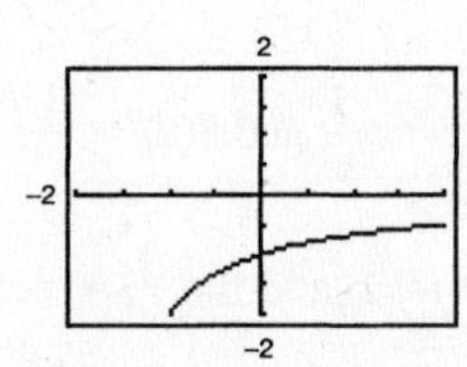

$\lim_{x\to 0} f(x) \approx -1$

4.

x	-0.1	-0.01	-0.001	0.001	0.01	0.1
$f(x)$	1.432	1.416	1.414	1.414	1.412	1.397

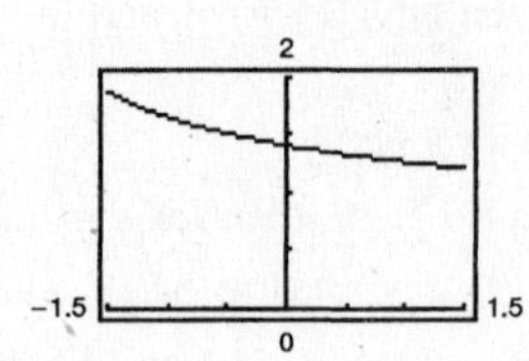

$\lim_{x\to 0} f(x) \approx 1.414$

5. $h(x) = \frac{x^2 - 2x}{x}$

(a) $\lim_{x\to 0} h(x) = -2$

(b) $\lim_{x\to -1} h(x) = -3$

6. $g(x) = \frac{3x}{x-2}$

(a) $\lim_{x\to 2} g(x)$ does not exist.

(b) $\lim_{x\to 0} g(x) = 0$

7. $\lim_{x\to1} (3 - x) = 3 - 1 = 2$

Let $\varepsilon > 0$ be given. Choose $\delta = \varepsilon$. Then for $0 < |x - 1| < \delta = \varepsilon$, you have

$$|x - 1| < \varepsilon$$
$$|1 - x| < \varepsilon$$
$$|(3 - x) - 2| < \varepsilon$$
$$|f(x) - L| < \varepsilon.$$

8. $\lim_{x\to9} \sqrt{x} = \sqrt{9} = 3.$

Let $\varepsilon > 0$ be given. We need

$$\left|\sqrt{x} - 3\right| < \varepsilon \Rightarrow \left|\sqrt{x} + 3\right|\left|\sqrt{x} - 3\right| < \varepsilon\left|\sqrt{x} + 3\right|$$
$$|x - 9| < \varepsilon\left|\sqrt{x} + 3\right|.$$

Assuming $4 < x < 16$, you can choose $\delta = 5\varepsilon$.

Hence, for $0 < |x - 9| < \delta = 5\varepsilon$, you have

$$|x - 9| < 5\varepsilon < \left|\sqrt{x} + 3\right|\varepsilon$$
$$\left|\sqrt{x} - 3\right| < \varepsilon$$
$$|f(x) - L| < \varepsilon.$$

9. $\lim_{x\to2} (x^2 - 3) = 1$

Let $\varepsilon > 0$ be given. We need $|x^2 - 3 - 1| < \varepsilon \Rightarrow |x^2 - 4| = |(x - 2)(x + 2)| < \varepsilon \Rightarrow |x - 2| < \frac{1}{|x + 2|}\varepsilon.$

Assuming, $1 < x < 3$, you can choose $\delta = \varepsilon/5$. Hence, for $0 < |x - 2| < \delta = \varepsilon/5$ you have

$$|x - 2| < \frac{\varepsilon}{5} < \frac{1}{|x + 2|}\varepsilon$$
$$|x - 2||x + 2| < \varepsilon$$
$$|x^2 - 4| < \varepsilon$$
$$|(x^2 - 3) - 1| < \varepsilon$$
$$|f(x) - L| < \varepsilon.$$

10. $\lim_{x\to5} 9 = 9$. Let $\varepsilon > 0$ be given. δ can be any positive number. Hence, for $0 < |x - 5| < \delta$, you have

$$|9 - 9| < \varepsilon$$
$$|f(x) - L| < \varepsilon.$$

11. $\lim_{t\to4} \sqrt{t + 2} = \sqrt{4 + 2} = \sqrt{6} \approx 2.45$

12. $\lim_{y\to4} 3|y - 1| = 3|4 - 1| = 9$

13. $\lim_{t\to-2} \frac{t + 2}{t^2 - 4} = \lim_{t\to-2} \frac{1}{t - 2} = -\frac{1}{4}$

14. $\lim_{t\to3} \frac{t^2 - 9}{t - 3} = \lim_{t\to3} (t + 3) = 6$

15. $\lim_{x\to4} \frac{\sqrt{x} - 2}{x - 4} = \lim_{x\to4} \frac{\sqrt{x} - 2}{(\sqrt{x} - 2)(\sqrt{x} + 2)}$

$$= \lim_{x\to4} \frac{1}{\sqrt{x} + 2} = \frac{1}{\sqrt{4} + 2} = \frac{1}{4}$$

16. $\lim_{x\to0} \frac{\sqrt{4 + x} - 2}{x} = \lim_{x\to0} \frac{\sqrt{4 + x} - 2}{x} \cdot \frac{\sqrt{4 + x} + 2}{\sqrt{4 + x} + 2}$

$$= \lim_{x\to0} \frac{1}{\sqrt{4 + x} + 2} = \frac{1}{4}$$

17. $\lim_{x\to0} \frac{[1/(x + 1)] - 1}{x} = \lim_{x\to0} \frac{1 - (x + 1)}{x(x + 1)}$

$$= \lim_{x\to0} \frac{-1}{x + 1} = -1$$

18. $\lim_{s\to 0}\frac{(1/\sqrt{1+s})-1}{s} = \lim_{s\to 0}\left[\frac{(1/\sqrt{1+s})-1}{s}\cdot\frac{(1/\sqrt{1+s})+1}{(1/\sqrt{1+s})+1}\right]$

$= \lim_{s\to 0}\frac{[1/(1+s)]-1}{s\left[(1/\sqrt{1+s})+1\right]} = \lim_{s\to 0}\frac{-1}{(1+s)\left[(1/\sqrt{1+s})+1\right]} = -\frac{1}{2}$

19. $\lim_{x\to -5}\frac{x^3+125}{x+5} = \lim_{x\to -5}\frac{(x+5)(x^2-5x+25)}{x+5}$

$= \lim_{x\to -5}(x^2-5x+25)$

$= 75$

20. $\lim_{x\to -2}\frac{x^2-4}{x^3+8} = \lim_{x\to -2}\frac{(x+2)(x-2)}{(x+2)(x^2-2x+4)}$

$= \lim_{x\to -2}\frac{x-2}{x^2-2x+4}$

$= -\frac{4}{12} = -\frac{1}{3}$

21. $\lim_{x\to c}[f(x)\cdot g(x)] = \left(-\frac{3}{4}\right)\left(\frac{2}{3}\right) = -\frac{1}{2}$

22. $\lim_{x\to c}[f(x)+2g(x)] = -\frac{3}{4}+2\left(\frac{2}{3}\right) = \frac{7}{12}$

23. $\lim_{t\to a}\frac{s(a)-s(t)}{a-t} = \lim_{t\to 4}\frac{(-4.9(4)^2+200)-(-4.9t^2+200)}{4-t}$

$= \lim_{t\to 4}\frac{4.9(t-4)(t+4)}{4-t}$

$= \lim_{t\to 4} -4.9(t+4) = -39.2 \text{ m/sec}$

24. $s(t) = 0 \implies -4.9t^2+200 = 0 \implies t^2 \approx 40.816 \implies t \approx 6.39$ sec

When $t = 6.39$, the velocity is approximately

$\lim_{t\to a}\frac{s(a)-s(t)}{a-t} = \lim_{t\to a} -4.9(a+t)$

$= \lim_{t\to 6.39} -4.9(6.39+6.39) = -62.6 \text{ m/sec.}$

25. $\lim_{x\to 3^-}\frac{|x-3|}{x-3} = \lim_{x\to 3^-}\frac{-(x-3)}{x-3} = -1$

26. $\lim_{x\to 4}[\![x-1]\!]$ does not exist.

The graph jumps from 2 to 3 at $x = 4$.

27. $\lim_{x\to 2}f(x) = 0$

28. $\lim_{x\to 1^+}g(x) = 1+1 = 2$

29. $\lim_{t\to 1}h(t)$ does not exist because $\lim_{t\to 1^-}h(t) = 1+1 = 2$ and $\lim_{t\to 1^+}h(t) = \frac{1}{2}(1+1) = 1$.

30. $\lim_{s\to -2}f(s) = 2$

31. $f(x) = [\![x+3]\!]$

$\lim_{x\to k^+}[\![x+3]\!] = k+3$ where k is an integer.

$\lim_{x\to k^-}[\![x+3]\!] = k+2$ where k is an integer.

Nonremovable discontinuity at each integer k.

Continuous on $(k, k+1)$ for all integers k.

32. $f(x) = \begin{cases}\frac{3x^2-x-2}{x-1}, & x \neq 1\\ 0, & x = 1\end{cases}$

$\lim_{x\to 1}f(x) = \lim_{x\to 1}\frac{3x^2-x-2}{x-1}$

$= \lim_{x\to 1}(3x+2) = 5 \neq 0$

Removable discontinuity at $x = 1$.

Continuous on $(-\infty, 1)\cup(1, \infty)$.

33. $f(x) = \dfrac{3x^2 - x - 2}{x - 1} = \dfrac{(3x + 2)(x - 1)}{x - 1}$

$\lim_{x \to 1} f(x) = \lim_{x \to 1} (3x + 2) = 5$

Removable discontinuity at $x = 1$.

Continuous on $(-\infty, 1) \cup (1, \infty)$.

34. $f(x) = \begin{cases} 5 - x, & x \le 2 \\ 2x - 3, & x > 2 \end{cases}$

$\lim_{x \to 2^-} (5 - x) = 3$

$\lim_{x \to 2^+} (2x - 3) = 1$

Nonremovable discontinuity at $x = 2$.

Continuous on $(-\infty, 2) \cup (2, \infty)$.

35. $f(x) = \dfrac{1}{(x - 2)^2}$

$\lim_{x \to 2} \dfrac{1}{(x - 2)^2} = \infty$

Nonremovable discontinuity at $x = 2$.

Continuous on $(-\infty, 2) \cup (2, \infty)$.

36. $f(x) = \sqrt{\dfrac{x + 1}{x}} = \sqrt{1 + \dfrac{1}{x}}$

$\lim_{x \to 0^+} \sqrt{1 + \dfrac{1}{x}} = \infty$

Domain: $(-\infty, -1], (0, \infty)$

Nonremovable discontinuity at $x = 0$.

Continuous on $(-\infty, -1] \cup (0, \infty)$.

37. $f(x) = \dfrac{3}{x + 1}$

$\lim_{x \to -1^-} f(x) = -\infty$

$\lim_{x \to -1^+} f(x) = \infty$

Nonremovable discontinuity at $x = -1$.

Continuous on $(-\infty, -1) \cup (-1, \infty)$.

38. $f(x) = \dfrac{x + 1}{2x + 2}$

$\lim_{x \to -1} \dfrac{x + 1}{2(x + 1)} = \dfrac{1}{2}$

Removable discontinuity at $x = -1$.

Continuous on $(-\infty, -1) \cup (-1, \infty)$.

39. $f(x) = \dfrac{\sqrt{2x + 1} - \sqrt{3}}{x - 1}$

(a)

x	1.1	1.01	1.001	1.0001
$f(x)$	0.5680	0.5764	0.5773	0.5773

$\lim_{x \to 1^+} \dfrac{\sqrt{2x + 1} - \sqrt{3}}{x - 1} \approx 0.577$ (Actual limit is $\sqrt{3}/3$.)

(b)

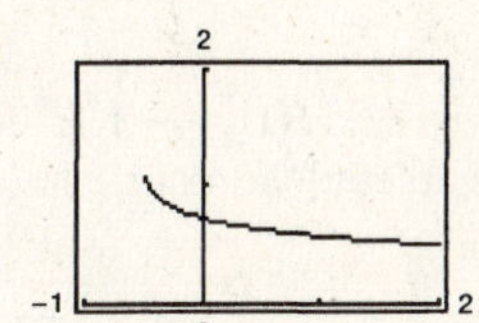

(c) $\lim_{x \to 1^+} \dfrac{\sqrt{2x + 1} - \sqrt{3}}{x - 1} = \lim_{x \to 1^+} \dfrac{\sqrt{2x + 1} - \sqrt{3}}{x - 1} \cdot \dfrac{\sqrt{2x + 1} + \sqrt{3}}{\sqrt{2x + 1} + \sqrt{3}}$

$$= \lim_{x \to 1^+} \frac{(2x + 1) - 3}{(x - 1)\left(\sqrt{2x + 1} + \sqrt{3}\right)}$$

$$= \lim_{x \to 1^+} \frac{2}{\sqrt{2x + 1} + \sqrt{3}}$$

$$= \frac{2}{2\sqrt{3}} = \frac{1}{\sqrt{3}} = \frac{\sqrt{3}}{3}$$

40. $f(x) = \dfrac{1 - \sqrt[3]{x}}{x - 1}$

(a)

x	1.1	1.01	1.001	1.0001
$f(x)$	−0.3228	−0.3322	−0.3332	−0.3333

$\lim_{x\to 1^+} \dfrac{1 - \sqrt[3]{x}}{x - 1} \approx -0.333$ (Actual limit is $-\frac{1}{3}$.)

(b)

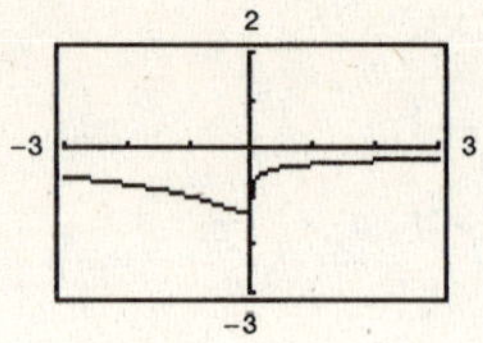

(c) $\lim_{x\to 1^+} \dfrac{1 - \sqrt[3]{x}}{x - 1} = \lim_{x\to 1^+} \dfrac{1 - \sqrt[3]{x}}{x - 1} \cdot \dfrac{1 + \sqrt[3]{x} + (\sqrt[3]{x})^2}{1 + \sqrt[3]{x} + (\sqrt[3]{x})^2}$

$$= \lim_{x\to 1^+} \frac{1 - x}{(x - 1)\left[1 + \sqrt[3]{x} + (\sqrt[3]{x})^2\right]}$$

$$= \lim_{x\to 1^+} \frac{-1}{1 + \sqrt[3]{x} + (\sqrt[3]{x})^2}$$

$$= -\frac{1}{3}$$

41. $f(2) = 5$

Find c so that $\lim_{x\to 2^+} (cx + 6) = 5$.

$c(2) + 6 = 5$

$2c = -1$

$c = -\frac{1}{2}$

42. $\lim_{x\to 1^+} (x + 1) = 2$

$\lim_{x\to 3^-} (x + 1) = 4$

Find b and c so that

$\lim_{x\to 1^-} (x^2 + bx + c) = 2$ and $\lim_{x\to 3^+} (x^2 + bx + c) = 4$.

Consequently we get

$1 + b + c = 2$ and $9 + 3b + c = 4$.

Solving simultaneously,

$b = -3$ and $c = 4$.

43. f is continuous on $[1, 2]$. $f(1) = -1 < 0$ and $f(2) = 13 > 0$. Therefore by the Intermediate Value Theorem, there is at least one value c in $(1, 2)$ such that $2c^3 - 3 = 0$.

44. $C = 9.80 + 2.50[-[\![-x]\!] - 1],\ x > 0$

$= 9.80 - 2.50[[\![-x]\!] + 1]$

C has a nonremovable discontinuity at each integer.

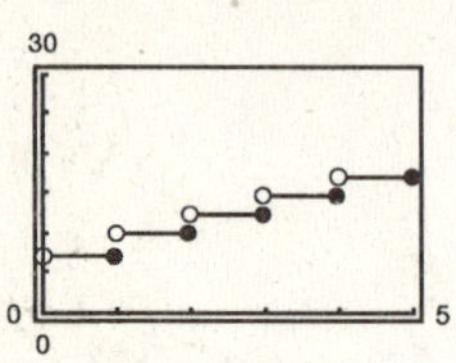

45. $f(x) = \dfrac{x^2 - 4}{|x - 2|} = (x + 2)\left[\dfrac{x - 2}{|x - 2|}\right]$

(a) $\lim_{x\to 2^-} f(x) = -4$

(b) $\lim_{x\to 2^+} f(x) = 4$

(c) $\lim_{x\to 2} f(x)$ does not exist.

46. $f(x) = \sqrt{(x - 1)x}$

(a) Domain: $(-\infty, 0] \cup [1, \infty)$

(b) $\lim_{x\to 0^-} f(x) = 0$

(c) $\lim_{x\to 1^+} f(x) = 0$

47. $g(x) = 1 + \dfrac{2}{x}$

Vertical asymptote at $x = 0$.

48. $h(x) = \dfrac{4x}{4 - x^2}$

Vertical asymptotes at $x = 2$ and $x = -2$.

49. $f(x) = \dfrac{8}{(x - 10)^2}$

Vertical asymptote at $x = 10$.

50. $f(x) = \dfrac{x + 3}{x(x^2 + 1)}$

$x^2 + 1 \neq 0$. Vertical asymptote: $x = 0$

51. $\displaystyle\lim_{x \to -2^-} \frac{2x^2 + x + 1}{x + 2} = -\infty$

52. $\displaystyle\lim_{x \to (1/2)^+} \frac{x}{2x - 1} = \infty$

53. $\displaystyle\lim_{x \to -1^+} \frac{x + 1}{x^3 + 1} = \lim_{x \to -1^+} \frac{1}{x^2 - x + 1} = \frac{1}{3}$

54. $\displaystyle\lim_{x \to -1^-} \frac{x + 1}{x^4 - 1} = \lim_{x \to -1^-} \frac{1}{(x^2 + 1)(x - 1)} = -\frac{1}{4}$

55. $\displaystyle\lim_{x \to 1^-} \frac{x^2 + 2x + 1}{x - 1} = -\infty$

56. $\displaystyle\lim_{x \to -1^+} \frac{x^2 - 2x + 1}{x + 1} = \infty$

57. $\displaystyle\lim_{x \to 0^+} \left(x - \frac{1}{x^3}\right) = -\infty$

58. $\displaystyle\lim_{x \to 2^-} \frac{1}{\sqrt[3]{x^2 - 4}} = -\infty$

59. $r = \dfrac{2L}{\sqrt{L^2 - 144}}$

(a) When $L = 25$, $r = \dfrac{2(25)}{\sqrt{25^2 - 144}} = \dfrac{50\sqrt{481}}{481} \approx 2.28$ ft/sec

(b) When $L = 13$, $r = \dfrac{2(13)}{\sqrt{13^2 - 144}} = \dfrac{26}{5} = 5.2$ ft/sec

(c) $\displaystyle\lim_{L \to 12^+} r = \infty$

Problem Solving for Chapter 3

1. (a) $$\text{Perimeter } \triangle PAO = \sqrt{x^2 + (y - 1)^2} + \sqrt{x^2 + y^2} + 1 = \sqrt{x^2 + (x^2 - 1)^2} + \sqrt{x^2 + x^4} + 1$$

$$\text{Perimeter } \triangle PBO = \sqrt{(x - 1)^2 + y^2} + \sqrt{x^2 + y^2} + 1 = \sqrt{(x - 1)^2 + x^4} + \sqrt{x^2 + x^4} + 1$$

(b) $$r(x) = \frac{\sqrt{x^2 + (x^2 - 1)^2} + \sqrt{x^2 + x^4} + 1}{\sqrt{(x - 1)^2 + x^4} + \sqrt{x^2 + x^4} + 1}$$

x	4	2	1	0.1	0.01
Perimeter $\triangle PAO$	33.02	9.08	3.41	2.10	2.01
Perimeter $\triangle PBO$	33.77	9.60	3.41	2.00	2.00
$r(x)$	0.98	0.95	1	1.05	1.005

(c) $\displaystyle\lim_{x \to 0^+} r(x) = \frac{1 + 0 + 1}{1 + 0 + 1} = \frac{2}{2} = 1$

2. (a) There are 6 triangles, each with a central angle of $60° = \pi/3$. Hence,

$$\text{Area hexagon} = 6\left[\frac{1}{2}bh\right] = 6\left[\frac{1}{2}(1)\sin\frac{\pi}{3}\right]$$

$$= \frac{3\sqrt{3}}{2} \approx 2.598.$$

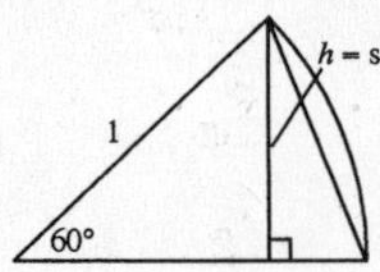

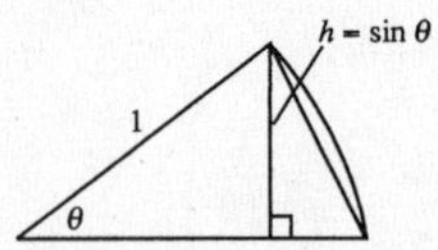

Error: $\pi - \dfrac{3\sqrt{3}}{2} \approx 0.5435.$

(b) There are n triangles, each with central angle of $\theta = 2\pi/n$. Hence,

$$A_n = n\left[\frac{1}{2}bh\right] = n\left[\frac{1}{2}(1)\sin\frac{2\pi}{n}\right] = \frac{n\sin(2\pi/n)}{2}.$$

(c)

n	6	12	24	48	96
A_n	2.598	3	3.106	3.133	3.139

(d) As n gets larger and larger, $2\pi/n$ approaches 0.

Letting $x = 2\pi/n$,

$$A_n = \frac{\sin(2\pi/n)}{2/n} = \frac{\sin(2\pi/n)}{(2\pi/n)}\pi = \frac{\sin x}{x}\pi$$

which approaches $(1)\pi = \pi$.

3. (a) $\text{Slope} = \dfrac{4-0}{3-0} = \dfrac{4}{3}$

(b) $\text{Slope} = -\dfrac{3}{4}$ Tangent line: $y - 4 = -\dfrac{3}{4}(x-3)$

$$y = -\frac{3}{4}x + \frac{25}{4}$$

(c) Let $Q = (x, y) = \left(x, \sqrt{25-x^2}\right)$

$$m_x = \frac{\sqrt{25-x^2}-4}{x-3}$$

(d) $$\lim_{x\to3} m_x = \lim_{x\to3}\frac{\sqrt{25-x^2}-4}{x-3}\cdot\frac{\sqrt{25-x^2}+4}{\sqrt{25-x^2}+4}$$

$$= \lim_{x\to3}\frac{25-x^2-16}{(x-3)\left(\sqrt{25-x^2}+4\right)}$$

$$= \lim_{x\to3}\frac{(3-x)(3+x)}{(x-3)\left(\sqrt{25-x^2}+4\right)}$$

$$= \lim_{x\to3}\frac{-(3+x)}{\sqrt{25-x^2}+4} = \frac{-6}{4+4} = -\frac{3}{4}$$

This is the slope of the tangent line at P.

4. (a) $\text{Slope} = -\dfrac{12}{5}$

(b) Slope of tangent line is $\dfrac{5}{12}$.

$$y + 12 = \frac{5}{12}(x-5)$$

$$y = \frac{5}{12}x - \frac{169}{12} \quad \text{Tangent line}$$

(c) $Q = (x, y) = \left(x, -\sqrt{169-x^2}\right)$

$$m_x = \frac{-\sqrt{169-x^2}+12}{x-5}$$

(d) $$\lim_{x\to5} m_x = \lim_{x\to5}\frac{12-\sqrt{169-x^2}}{x-5}\cdot\frac{12+\sqrt{169-x^2}}{12+\sqrt{169-x^2}}$$

$$= \lim_{x\to5}\frac{144-(169-x^2)}{(x-5)\left(12+\sqrt{169-x^2}\right)}$$

$$= \lim_{x\to5}\frac{x^2-25}{(x-5)\left(12+\sqrt{169-x^2}\right)}$$

$$= \lim_{x\to5}\frac{(x+5)}{12+\sqrt{169-x^2}}$$

$$= \frac{10}{12+12} = \frac{5}{12}$$

This is the same slope as part (b).

5. $$\frac{\sqrt{a+bx}-\sqrt{3}}{x} = \frac{\sqrt{a+bx}-\sqrt{3}}{x}\cdot\frac{\sqrt{a+bx}+\sqrt{3}}{\sqrt{a+bx}+\sqrt{3}}$$

$$= \frac{(a+bx)-3}{x\left(\sqrt{a+bx}+\sqrt{3}\right)}$$

Letting $a = 3$ simplifies the numerator.

Thus,

$$\lim_{x\to0}\frac{\sqrt{3+bx}-\sqrt{3}}{x} = \lim_{x\to0}\frac{bx}{x\left(\sqrt{3+bx}+\sqrt{3}\right)}$$

$$= \lim_{x\to0}\frac{b}{\sqrt{3+bx}+\sqrt{3}}.$$

Setting $\dfrac{b}{\sqrt{3}+\sqrt{3}} = \sqrt{3}$, you obtain $b = 6$.

Thus, $a = 3$ and $b = 6$.

6. (a) $3 + x^{1/3} \geq 0$

$x^{1/3} \geq -3$

$x \geq -27$

Domain: $x \geq -27, x \neq 1$

(b)

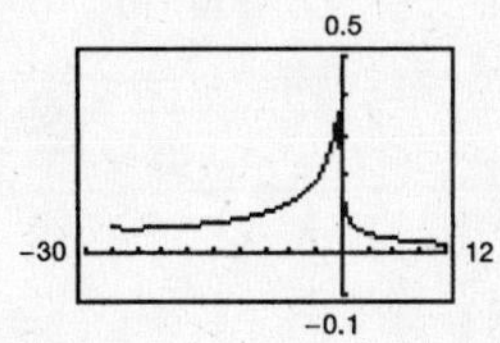

(c) $\lim_{x \to -27^+} f(x) = \dfrac{\sqrt{3 + (-27)^{1/3}} - 2}{-27 - 1}$

$= \dfrac{-2}{-28} = \dfrac{1}{14} \approx 0.0714$

(d) $\lim_{x \to 1} f(x) = \lim_{x \to 1} \dfrac{\sqrt{3 + x^{1/3}} - 2}{x - 1} \cdot \dfrac{\sqrt{3 + x^{1/3}} + 2}{\sqrt{3 + x^{1/3}} + 2}$

$= \lim_{x \to 1} \dfrac{3 + x^{1/3} - 4}{(x - 1)\left(\sqrt{3 + x^{1/3}} + 2\right)}$

$= \lim_{x \to 1} \dfrac{x^{1/3} - 1}{(x^{1/3} - 1)(x^{2/3} + x^{1/3} + 1)\left(\sqrt{3 + x^{1/3}} + 2\right)}$

$= \lim_{x \to 1} \dfrac{1}{(x^{2/3} + x^{1/3} + 1)\left(\sqrt{3 + x^{1/3}} + 2\right)}$

$= \dfrac{1}{(1 + 1 + 1)(2 + 2)} = \dfrac{1}{12}$

7. (a) $\lim_{x \to 2} f(x) = 3$: g_1, g_4

(b) f continuous at 2: g_1

(c) $\lim_{x \to 2^-} f(x) = 3$: g_1, g_3, g_4

8. (a)

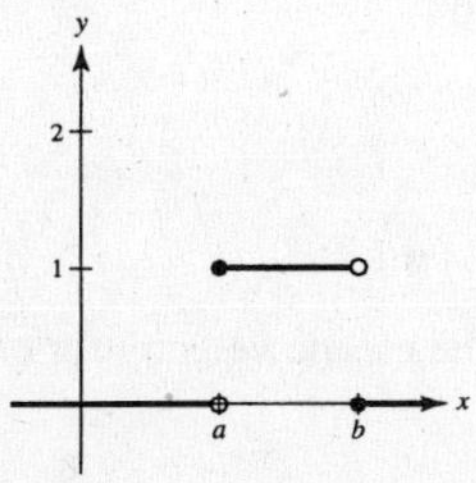

(b) (i) $\lim_{x \to a^+} P_{a,b}(x) = 1$

(ii) $\lim_{x \to a^-} P_{a,b}(x) = 0$

(iii) $\lim_{x \to b^+} P_{a,b}(x) = 0$

(iv) $\lim_{x \to b^-} P_{a,b}(x) = 1$

(c) $P_{a,b}$ is continuous for all positive real numbers except $x = a, b$.

(d) The area under the graph of U, and above the x-axis, is 1.

9.

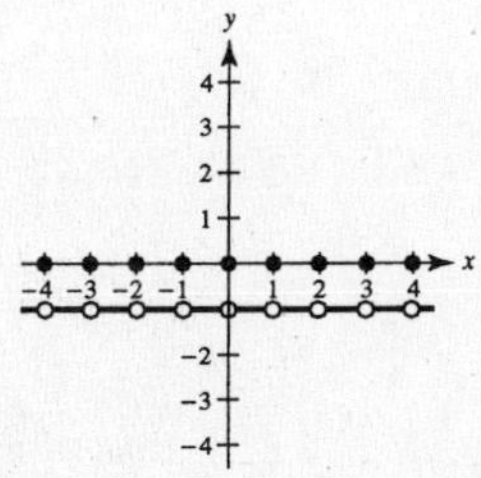

(a) $f(1) = [\![1]\!] + [\![-1]\!] = 1 + (-1) = 0$

$f(0) = 0$

$f\left(\frac{1}{2}\right) = 0 + (-1) = -1$

$f(-2.7) = -3 + 2 = -1$

(b) $\lim_{x \to 1^-} f(x) = -1$

$\lim_{x \to 1^+} f(x) = -1$

$\lim_{x \to 1/2} f(x) = -1$

(c) f is continuous for all real numbers except

$x = 0, \pm 1, \pm 2, \pm 3, \ldots$

10.

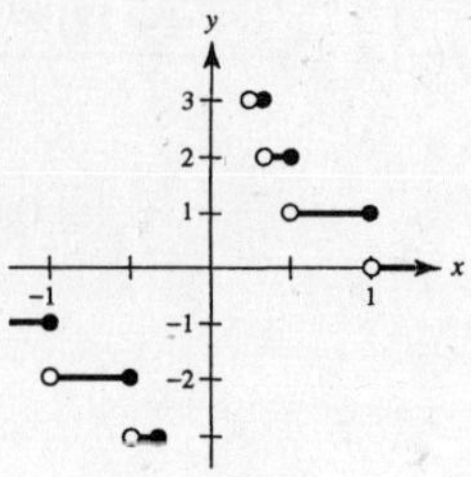

(a) $f\left(\frac{1}{4}\right) = [\![4]\!] = 4$

$f(3) = \left[\!\left[\frac{1}{3}\right]\!\right] = 0$

$f(1) = [\![1]\!] = 1$

(b) $\lim_{x\to 1^-} f(x) = 1$

$\lim_{x\to 1^+} f(x) = 0$

$\lim_{x\to 0^-} f(x) = -\infty$

$\lim_{x\to 0^+} f(x) = \infty$

(c) f is continuous for all real numbers except

$x = 0, \pm 1, \pm\frac{1}{2}, \pm\frac{1}{3}, \ldots$

11. (a) $$v^2 = \frac{192{,}000}{r} + v_0^2 - 48$$

$$\frac{192{,}000}{r} = v^2 - v_0^2 + 48$$

$$r = \frac{192{,}000}{v^2 - v_0^2 + 48}$$

$$\lim_{v\to 0} r = \frac{192{,}000}{48 - v_0^2}$$

Let $v_0 = \sqrt{48} = 4\sqrt{3}$ miles/sec.

(b) $$v^2 = \frac{1920}{r} + v_0^2 - 2.17$$

$$\frac{1920}{r} = v^2 - v_0^2 + 2.17$$

$$r = \frac{1920}{v^2 - v_0^2 + 2.17}$$

$$\lim_{v\to 0} r = \frac{1920}{2.17 - v_0^2}$$

Let $v_0 = \sqrt{2.17}$ mi/sec $(\approx 1.47$ mi/sec$)$.

(c) $$r = \frac{10{,}600}{v^2 - v_0^2 + 6.99}$$

$$\lim_{v\to 0} r = \frac{10{,}600}{6.99 - v_0^2}$$

Let $v_0 = \sqrt{6.99} \approx 2.64$ mi/sec.

Since this is smaller than the escape velocity for earth, the mass is less.

CHAPTER 4
Differentiation

Section 4.1 The Derivative and the Tangent Line Problem 296

Section 4.2 Basic Differentiation Rules and Rates of Change 308

Section 4.3 The Product and Quotient Rules and Higher-Order Derivatives 318

Section 4.4 The Chain Rule 329

Section 4.5 Implicit Differentiation 337

Section 4.6 Related Rates 349

Review Exercises 357

Problem Solving 368

CHAPTER 4
Differentiation

Section 4.1 The Derivative and the Tangent Line Problem

1. (a) At (x_1, y_1), slope $= 0$.

At (x_2, y_2), slope $\approx \frac{5}{2}$.

(b) At (x_1, y_1), slope $\approx -\frac{5}{2}$.

At (x_2, y_2), slope ≈ 2.

2. (a) At (x_1, y_1), slope $\approx \frac{2}{3}$.

At (x_2, y_2), slope $\approx -\frac{2}{5}$.

(b) At (x_1, y_1), slope $\approx \frac{4}{3}$.

At (x_2, y_2), slope $\approx \frac{5}{4}$.

3. (a), (b)

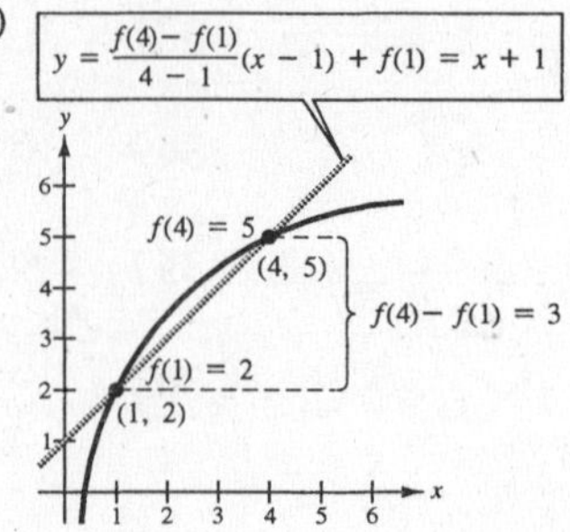

(c) $y = \dfrac{f(4) - f(1)}{4 - 1}(x - 1) + f(1)$

$= \dfrac{3}{3}(x - 1) + 2$

$= 1(x - 1) + 2$

$= x + 1$

4. (a) $\dfrac{f(4) - f(1)}{4 - 1} = \dfrac{5 - 2}{3} = 1$

$\dfrac{f(4) - f(3)}{4 - 3} \approx \dfrac{5 - 4.75}{1} = 0.25$

Thus, $\dfrac{f(4) - f(1)}{4 - 1} > \dfrac{f(4) - f(3)}{4 - 3}$.

(b) The slope of the tangent line at $(1, 2)$ equals $f'(1)$. This slope is steeper than the slope of the line through $(1, 2)$ and $(4, 5)$. Thus,

$\dfrac{f(4) - f(1)}{4 - 1} < f'(1)$.

5. $f(x) = 3 - 2x$ is a line. Slope $= -2$

6. $g(x) = \dfrac{3}{2}x + 1$ is a line. Slope $= \dfrac{3}{2}$

7. Slope at $(1, -3) = \lim\limits_{\Delta x \to 0} \dfrac{g(1 + \Delta x) - g(1)}{\Delta x}$

$= \lim\limits_{\Delta x \to 0} \dfrac{(1 + \Delta x)^2 - 4 - (-3)}{\Delta x}$

$= \lim\limits_{\Delta x \to 0} \dfrac{1 + 2(\Delta x) + (\Delta x)^2 - 1}{\Delta x}$

$= \lim\limits_{\Delta x \to 0} [2 + 2(\Delta x)] = 2$

8. Slope at $(2, 1) = \lim\limits_{\Delta x \to 0} \dfrac{g(2 + \Delta x) - g(2)}{\Delta x}$

$= \lim\limits_{\Delta x \to 0} \dfrac{5 - (2 + \Delta x)^2 - 1}{\Delta x}$

$= \lim\limits_{\Delta x \to 0} \dfrac{5 - 4 - 4(\Delta x) - (\Delta x)^2 - 1}{\Delta x}$

$= \lim\limits_{\Delta x \to 0} (-4 - \Delta x) = -4$

9. Slope at $(0, 0) = \lim\limits_{\Delta t \to 0} \dfrac{f(0 + \Delta t) - f(0)}{\Delta t}$

$= \lim\limits_{\Delta t \to 0} \dfrac{3(\Delta t) - (\Delta t)^2 - 0}{\Delta t}$

$= \lim\limits_{\Delta t \to 0} (3 - \Delta t) = 3$

10. Slope at $(-2, 7) = \lim\limits_{\Delta t \to 0} \dfrac{h(-2 + \Delta t) - h(-2)}{\Delta t}$

$= \lim\limits_{\Delta t \to 0} \dfrac{(-2 + \Delta t)^2 + 3 - 7}{\Delta t}$

$= \lim\limits_{\Delta t \to 0} \dfrac{4 - 4(\Delta t) + (\Delta t)^2 - 4}{\Delta t}$

$= \lim\limits_{\Delta t \to 0} (-4 + \Delta t) = -4$

11. $f(x) = 3$

$$f'(x) = \lim_{\Delta x \to 0} \frac{f(x + \Delta x) - f(x)}{\Delta x}$$
$$= \lim_{\Delta x \to 0} \frac{3 - 3}{\Delta x}$$
$$= \lim_{\Delta x \to 0} 0 = 0$$

12. $g(x) = -5$

$$g'(x) = \lim_{\Delta x \to 0} \frac{g(x + \Delta x) - g(x)}{\Delta x}$$
$$= \lim_{\Delta x \to 0} \frac{-5 - (-5)}{\Delta x}$$
$$= \lim_{\Delta x \to 0} \frac{0}{\Delta x} = 0$$

13. $f(x) = -5x$

$$f'(x) = \lim_{\Delta x \to 0} \frac{f(x + \Delta x) - f(x)}{\Delta x}$$
$$= \lim_{\Delta x \to 0} \frac{-5(x + \Delta x) - (-5x)}{\Delta x}$$
$$= \lim_{\Delta x \to 0} -5 = -5$$

14. $f(x) = 3x + 2$

$$f'(x) = \lim_{\Delta x \to 0} \frac{f(x + \Delta x) - f(x)}{\Delta x}$$
$$= \lim_{\Delta x \to 0} \frac{[3(x + \Delta x) + 2] - [3x + 2]}{\Delta x}$$
$$= \lim_{\Delta x \to 0} \frac{3\Delta x}{\Delta x}$$
$$= \lim_{\Delta x \to 0} 3 = 3$$

15. $h(s) = 3 + \frac{2}{3}s$

$$h'(s) = \lim_{\Delta s \to 0} \frac{h(s + \Delta s) - h(s)}{\Delta s}$$
$$= \lim_{\Delta s \to 0} \frac{3 + \frac{2}{3}(s + \Delta s) - \left(3 + \frac{2}{3}s\right)}{\Delta s}$$
$$= \lim_{\Delta s \to 0} \frac{\frac{2}{3}\Delta s}{\Delta s} = \frac{2}{3}$$

16. $f(x) = 9 - \frac{1}{2}x$

$$f'(x) = \lim_{\Delta x \to 0} \frac{f(x + \Delta x) - f(x)}{\Delta x}$$
$$= \lim_{\Delta x \to 0} \frac{[9 - (1/2)(x + \Delta x)] - [9 - (1/2)x]}{\Delta x}$$
$$= \lim_{\Delta x \to 0} \left(-\frac{1}{2}\right) = -\frac{1}{2}$$

17. $f(x) = 2x^2 + x - 1$

$$f'(x) = \lim_{\Delta x \to 0} \frac{f(x + \Delta x) - f(x)}{\Delta x}$$
$$= \lim_{\Delta x \to 0} \frac{[2(x + \Delta x)^2 + (x + \Delta x) - 1] - [2x^2 + x - 1]}{\Delta x}$$
$$= \lim_{\Delta x 0} \frac{(2x^2 + 4x\,\Delta x + 2(\Delta x)^2 + x + \Delta x - 1) - (2x^2 + x - 1)}{\Delta x}$$
$$= \lim_{\Delta x \to 0} \frac{4x\,\Delta x + 2(\Delta x)^2 + \Delta x}{\Delta x} = \lim_{\Delta x \to 0} (4x + 2\,\Delta x + 1) = 4x + 1$$

18. $f(x) = 1 - x^2$

$$f'(x) = \lim_{\Delta x \to 0} \frac{f(x + \Delta x) - f(x)}{\Delta x}$$
$$= \lim_{\Delta x \to 0} \frac{[1 - (x + \Delta x)^2] - [1 - x^2]}{\Delta x}$$
$$= \lim_{\Delta x \to 0} \frac{1 - x^2 - 2x\Delta x - (\Delta x)^2 - 1 + x^2}{\Delta x}$$
$$= \lim_{\Delta x \to 0} \frac{-2x\Delta x - (\Delta x)^2}{\Delta x} = \lim_{\Delta x \to 0} (-2x - \Delta x) = -2x$$

19. $f(x) = x^3 - 12x$

$$f'(x) = \lim_{\Delta x \to 0} \frac{f(x + \Delta x) - f(x)}{\Delta x}$$

$$= \lim_{\Delta x \to 0} \frac{[(x + \Delta x)^3 - 12(x + \Delta x)] - [x^3 - 12x]}{\Delta x}$$

$$= \lim_{\Delta x \to 0} \frac{x^3 + 3x^2\Delta x + 3x(\Delta x)^2 + (\Delta x)^3 - 12x - 12\,\Delta x - x^3 + 12x}{\Delta x}$$

$$= \lim_{\Delta x \to 0} \frac{3x^2\,\Delta x + 3x(\Delta x)^2 + (\Delta x)^3 - 12\,\Delta x}{\Delta x}$$

$$= \lim_{\Delta x \to 0} (3x^2 + 3x\,\Delta x + (\Delta x)^2 - 12) = 3x^2 - 12$$

20. $f(x) = x^3 + x^2$

$$f'(x) = \lim_{\Delta x \to 0} \frac{f(x + \Delta x) - f(x)}{\Delta x}$$

$$= \lim_{\Delta x \to 0} \frac{[(x + \Delta x)^3 + (x + \Delta x)^2] - [x^3 + x^2]}{\Delta x}$$

$$= \lim_{\Delta x \to 0} \frac{x^3 + 3x^2\,\Delta x + 3x(\Delta x)^2 + (\Delta x)^3 + x^2 + 2x\,\Delta x + (\Delta x)^2 - x^3 - x^2}{\Delta x}$$

$$= \lim_{\Delta x \to 0} \frac{3x^2\Delta x + 3x(\Delta x)^2 + (\Delta x)^3 + 2x\,\Delta x + (\Delta x)^2}{\Delta x}$$

$$= \lim_{\Delta x \to 0} (3x^2 + 3x\,\Delta x + (\Delta x)^2 + 2x + (\Delta x)) = 3x^2 + 2x$$

21. $f(x) = \dfrac{1}{x - 1}$

$$f'(x) = \lim_{\Delta x \to 0} \frac{f(x + \Delta x) - f(x)}{\Delta x}$$

$$= \lim_{\Delta x \to 0} \frac{\dfrac{1}{x + \Delta x - 1} - \dfrac{1}{x - 1}}{\Delta x}$$

$$= \lim_{\Delta x \to 0} \frac{(x - 1) - (x + \Delta x - 1)}{\Delta x(x + \Delta x - 1)(x - 1)}$$

$$= \lim_{\Delta x \to 0} \frac{-\Delta x}{\Delta x(x + \Delta x - 1)(x - 1)}$$

$$= \lim_{\Delta x \to 0} \frac{-1}{(x + \Delta x - 1)(x - 1)}$$

$$= -\frac{1}{(x - 1)^2}$$

22. $f(x) = \dfrac{1}{x^2}$

$$f'(x) = \lim_{\Delta x \to 0} \frac{f(x + \Delta x) - f(x)}{\Delta x}$$

$$= \lim_{\Delta x \to 0} \frac{\dfrac{1}{(x + \Delta x)^2} - \dfrac{1}{x^2}}{\Delta x}$$

$$= \lim_{\Delta x \to 0} \frac{x^2 - (x + \Delta x)^2}{\Delta x(x + \Delta x)^2 x^2}$$

$$= \lim_{\Delta x \to 0} \frac{-2x\,\Delta x - (\Delta x)^2}{\Delta x(x + \Delta x)^2 x^2}$$

$$= \lim_{\Delta x \to 0} \frac{-2x - \Delta x}{(x + \Delta x)^2 x^2}$$

$$= \frac{-2x}{x^4}$$

$$= -\frac{2}{x^3}$$

23. $f(x) = \sqrt{x+1}$

$$f'(x) = \lim_{\Delta x \to 0} \frac{f(x+\Delta x) - f(x)}{\Delta x}$$

$$= \lim_{\Delta x \to 0} \frac{\sqrt{x+\Delta x+1} - \sqrt{x+1}}{\Delta x} \cdot \left(\frac{\sqrt{x+\Delta x+1} + \sqrt{x+1}}{\sqrt{x+\Delta x+1} + \sqrt{x+1}}\right)$$

$$= \lim_{\Delta x \to 0} \frac{(x+\Delta x+1) - (x+1)}{\Delta x\left[\sqrt{x+\Delta x+1} + \sqrt{x+1}\right]}$$

$$= \lim_{\Delta x \to 0} \frac{1}{\sqrt{x+\Delta x+1} + \sqrt{x+1}}$$

$$= \frac{1}{\sqrt{x+1} + \sqrt{x+1}} = \frac{1}{2\sqrt{x+1}}$$

24. $f(x) = \dfrac{4}{\sqrt{x}}$

$$f'(x) = \lim_{\Delta x \to 0} \frac{f(x+\Delta x) - f(x)}{\Delta x}$$

$$= \lim_{\Delta x \to 0} \frac{\dfrac{4}{\sqrt{x+\Delta x}} - \dfrac{4}{\sqrt{x}}}{\Delta x}$$

$$= \lim_{\Delta x \to 0} \frac{4\sqrt{x} - 4\sqrt{x+\Delta x}}{\Delta x\sqrt{x}\sqrt{x+\Delta x}} \cdot \left(\frac{\sqrt{x} + \sqrt{x+\Delta x}}{\sqrt{x} + \sqrt{x+\Delta x}}\right)$$

$$= \lim_{\Delta x \to 0} \frac{4x - 4(x+\Delta x)}{\Delta x\sqrt{x}\sqrt{x+\Delta x}\left(\sqrt{x} + \sqrt{x+\Delta x}\right)}$$

$$= \lim_{\Delta x \to 0} \frac{-4}{\sqrt{x}\sqrt{x+\Delta x}\left(\sqrt{x} + \sqrt{x+\Delta x}\right)}$$

$$= \frac{-4}{\sqrt{x}\sqrt{x}\left(\sqrt{x} + \sqrt{x}\right)} = \frac{-2}{x\sqrt{x}}$$

25. (a) $f(x) = x^2 + 1$

$$f'(x) = \lim_{\Delta x \to 0} \frac{f(x+\Delta x) - f(x)}{\Delta x}$$

$$= \lim_{\Delta x \to 0} \frac{[(x+\Delta x)^2 + 1] - [x^2 + 1]}{\Delta x}$$

$$= \lim_{\Delta x \to 0} \frac{2x\,\Delta x + (\Delta x)^2}{\Delta x}$$

$$= \lim_{\Delta x \to 0} (2x + \Delta x) = 2x$$

At (2, 5), the slope of the tangent line is $m = 2(2) = 4$. The equation of the tangent line is

$$y - 5 = 4(x - 2)$$

$$y - 5 = 4x - 8$$

$$y = 4x - 3.$$

(b)

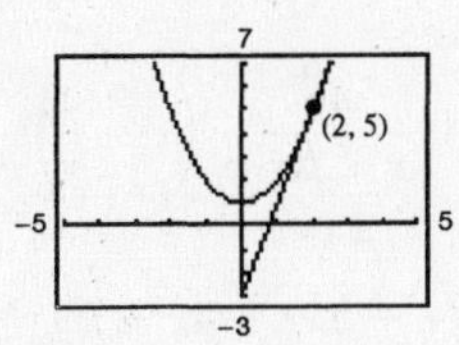

26. (a) $f(x) = x^2 + 2x + 1$

$$f'(x) = \lim_{\Delta x \to 0} \frac{f(x+\Delta x) - f(x)}{\Delta x}$$

$$= \lim_{\Delta x \to 0} \frac{[(x+\Delta x)^2 + 2(x+\Delta x) + 1] - [x^2 + 2x + 1]}{\Delta x}$$

$$= \lim_{\Delta x \to 0} \frac{2x\,\Delta x + (\Delta x)^2 + 2\,\Delta x}{\Delta x}$$

$$= \lim_{\Delta x \to 0} (2x + \Delta x + 2) = 2x + 2$$

At $(-3, 4)$, the slope of the tangent line is

$$m = 2(-3) + 2 = -4.$$

The equation of the tangent line is

$$y - 4 = -4(x + 3)$$

$$y = -4x - 8.$$

(b)

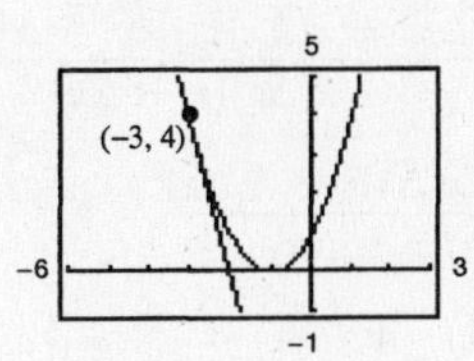

27. (a) $f(x) = x^3$

$$f'(x) = \lim_{\Delta x \to 0} \frac{f(x + \Delta x) - f(x)}{\Delta x}$$

$$= \lim_{\Delta x \to 0} \frac{(x + \Delta x)^3 - x^3}{\Delta x}$$

$$= \lim_{\Delta x \to 0} \frac{3x^2 \Delta x + 3x(\Delta x)^2 + (\Delta x)^3}{\Delta x}$$

$$= \lim_{\Delta x \to 0} (3x^2 + 3x\Delta x + (\Delta x)^2) = 3x^2$$

At $(2, 8)$, the slope of the tangent is $m = 3(2)^2 = 12$. The equation of the tangent line is

$$y - 8 = 12(x - 2)$$

$$y = 12x - 16.$$

(b)

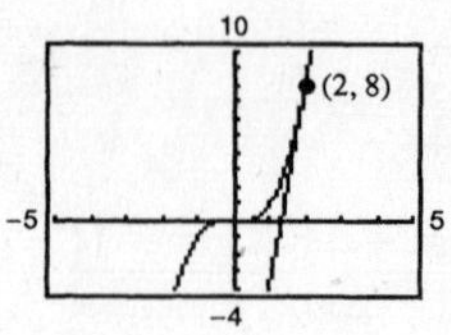

28. (a) $f(x) = \sqrt{x}$

$$f'(x) = \lim_{\Delta x \to 0} \frac{f(x + \Delta x) - f(x)}{\Delta x}$$

$$= \lim_{\Delta x \to 0} \frac{\sqrt{x + \Delta x} - \sqrt{x}}{\Delta x} \cdot \frac{\sqrt{x + \Delta x} + \sqrt{x}}{\sqrt{x + \Delta x} + \sqrt{x}}$$

$$= \lim_{\Delta x \to 0} \frac{(x + \Delta x) - x}{\Delta x\left(\sqrt{x + \Delta x} + \sqrt{x}\right)}$$

$$= \lim_{\Delta x \to 0} \frac{1}{\sqrt{x + \Delta x} + \sqrt{x}} = \frac{1}{2\sqrt{x}}$$

At $(1, 1)$, the slope of the tangent line is

$$m = \frac{1}{2\sqrt{1}} = \frac{1}{2}.$$

The equation of the tangent line is

$$y - 1 = \frac{1}{2}(x - 1)$$

$$y = \frac{1}{2}x + \frac{1}{2}.$$

(b)

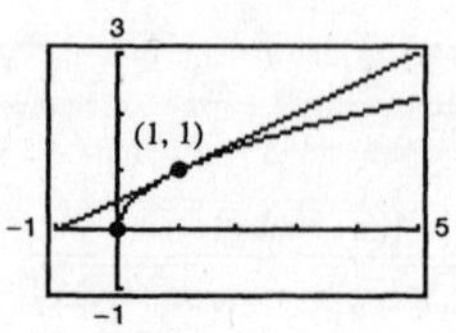

29. (a) $f(x) = x + \dfrac{4}{x}$

$$f'(x) = \lim_{\Delta x \to 0} \frac{f(x + \Delta x) - f(x)}{\Delta x}$$

$$= \lim_{\Delta x \to 0} \frac{(x + \Delta x) + \dfrac{4}{x + \Delta x} - \left(x + \dfrac{4}{x}\right)}{\Delta x}$$

$$= \lim_{\Delta x \to 0} \frac{x(x + \Delta x)(x + \Delta x) + 4x - x^2(x + \Delta x) - 4(x + \Delta x)}{x(\Delta x)(x + \Delta x)}$$

$$= \lim_{\Delta x \to 0} \frac{x^3 + 2x^2(\Delta x) + x(\Delta x)^2 - x^3 - x^2(\Delta x) - 4(\Delta x)}{x(\Delta x)(x + \Delta x)}$$

$$= \lim_{\Delta x \to 0} \frac{x^2(\Delta x) + x(\Delta x)^2 - 4(\Delta x)}{x(\Delta x)(x + \Delta x)}$$

$$= \lim_{\Delta x \to 0} \frac{x^2 + x(\Delta x) - 4}{x(x + \Delta x)}$$

$$= \frac{x^2 - 4}{x^2} = 1 - \frac{4}{x^2}$$

At $(4, 5)$, the slope of the tangent line is $m = 1 - \dfrac{4}{16} = \dfrac{3}{4}$.

The equation of the tangent line is $y - 5 = \dfrac{3}{4}(x - 4)$

$$y = \frac{3}{4}x + 2.$$

(b)

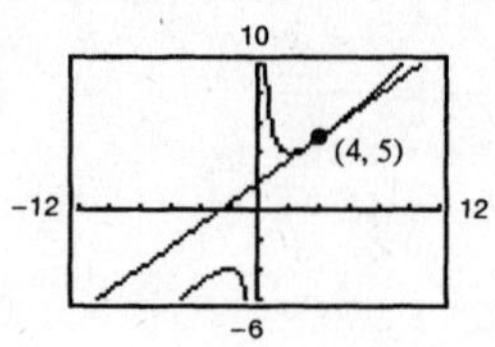

30. (a) $f(x) = \dfrac{1}{x+1}$

$$f'(x) = \lim_{\Delta x \to 0} \frac{f(x+\Delta x) - f(x)}{\Delta x}$$

$$= \lim_{\Delta x \to 0} \frac{\dfrac{1}{x+\Delta x+1} - \dfrac{1}{x+1}}{\Delta x}$$

$$= \lim_{\Delta x \to 0} \frac{(x+1) - (x+\Delta x+1)}{\Delta x(x+\Delta x+1)(x+1)}$$

$$= \lim_{\Delta x \to 0} -\frac{1}{(x+\Delta x+1)(x+1)}$$

$$= -\frac{1}{(x+1)^2}$$

At (0, 1), the slope of the tangent line is

$$m = \frac{-1}{(0+1)^2} = -1.$$

The equation of the tangent line is $y = -x + 1$.

(b)

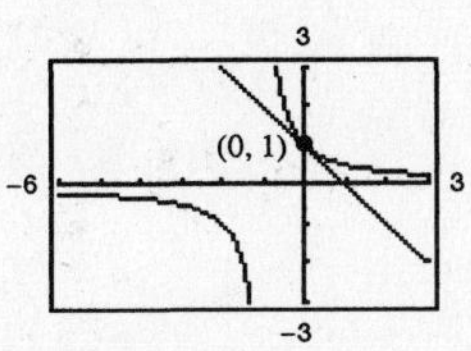

31. From Exercise 27 we know that $f'(x) = 3x^2$. Since the slope of the given line is 3, we have

$$3x^2 = 3$$

$$x = \pm 1.$$

Therefore, at the points $(1, 1)$ and $(-1, -1)$ the tangent lines are parallel to $3x - y + 1 = 0$. These lines have equations

$$y - 1 = 3(x - 1) \quad \text{and} \quad y + 1 = 3(x + 1)$$

$$y = 3x - 2 \qquad\qquad y = 3x + 2.$$

32. $f(x) = x^3 + 2$

$$f'(x) = \lim_{\Delta x \to 0} \frac{f(x+\Delta x) - f(x)}{\Delta x}$$

$$= \lim_{\Delta x \to 0} \frac{(x+\Delta x)^3 + 2 - (x^3+2)}{\Delta x}$$

$$= \lim_{\Delta x \to 0} \frac{3x^2\,\Delta x + 3x(\Delta x)^2 + (\Delta x)^3}{\Delta x}$$

$$= \lim_{\Delta x \to 0} [3x^2 + 3x\,\Delta x + (\Delta x)^2] = 3x^2$$

The slope of the line is 3. Thus,

$$3x^2 = 3$$

$$x^2 = 1$$

$$x = \pm 1.$$

At $(1, 3)$, $y - 3 = 3(x - 1)$

$$y = 3x.$$

At $(-1, 1)$, $y - 1 = 3(x + 1)$

$$y = 3x + 4.$$

33. $g(5) = 2$ because the tangent line passes through $(5, 2)$.

$$g'(5) = \frac{2-0}{5-9} = \frac{2}{-4} = -\frac{1}{2}$$

34. $h(-1) = 4$ because the tangent line passes through $(-1, 4)$.

$$h'(-1) = \frac{6-4}{3-(-1)} = \frac{2}{4} = \frac{1}{2}$$

35. $f(x) = x \Rightarrow f'(x) = 1$ (b)

36. $f(x) = x^2 \Rightarrow f'(x) = 2x$ (d)

37. $f(x) = \sqrt{x} \Rightarrow f'(x)$ matches (a)

(decreasing slope as $x \to \infty$)

38. f' does not exist at $x = 0$.

Matches (c)

39.

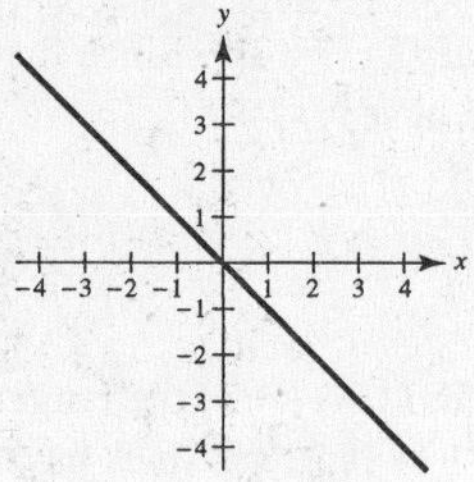

Answers will vary.

Sample answer: $y = -x$

40.

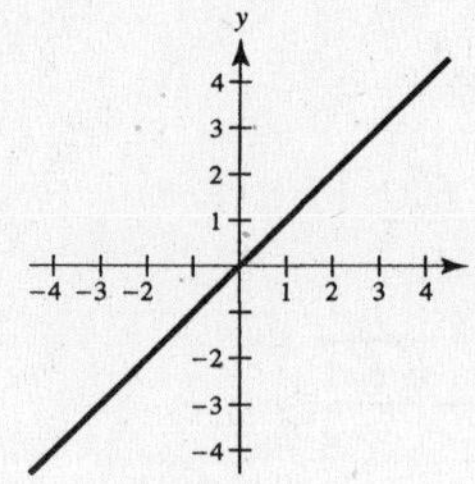

Answers will vary.

Sample answer: $y = x$

41. (a) If $f'(c) = 3$ and f is odd, then $f'(-c) = f'(c) = 3$.

(b) If $f'(c) = 3$ and f is even, then $f'(-c) = -f'(c) = -3$.

42. (a) Yes. $\lim_{\Delta x \to 0} \dfrac{f(x + 2\,\Delta x) - f(x)}{2\,\Delta x} = \lim_{\Delta x \to 0} \dfrac{f(x + \Delta x) - f(x)}{\Delta x} = f'(x)$

(b) No. The numerator does not approach zero.

(c) Yes.
$$\lim_{\Delta x \to 0} \frac{f(x + \Delta x) - f(x - \Delta x)}{2\,\Delta x} = \lim_{\Delta x \to 0} \frac{f(x + \Delta x) - f(x) - f(x - \Delta x) + f(x)}{2\,\Delta x}$$
$$= \lim_{\Delta x \to 0} \left[\frac{f(x + \Delta x) - f(x)}{2\,\Delta x} + \frac{f(x - \Delta x) - f(x)}{2(-\Delta x)}\right]$$
$$= \frac{1}{2}f'(x) + \frac{1}{2}f'(x) = f'(x)$$

(d) Yes. $\lim_{\Delta x \to 0} \dfrac{f(x + \Delta x) - f(x)}{\Delta x} = f'(x)$

43. The slope of the graph of f is $1 \implies f'(x) = 1$.

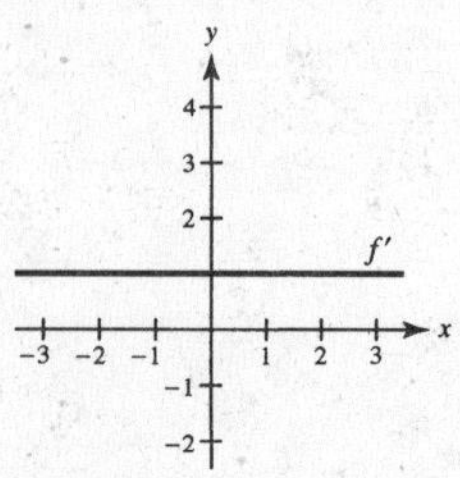

44. The slope of the graph of f is $0 \implies f'(x) = 0$.

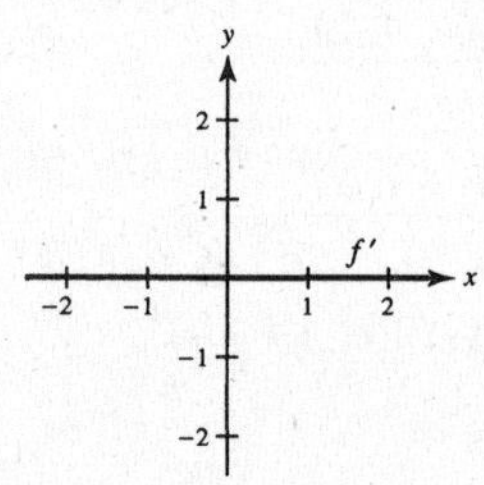

45. The slope of the graph of f is negative for $x < 4$, positive for $x > 4$, and 0 at $x = 4$.

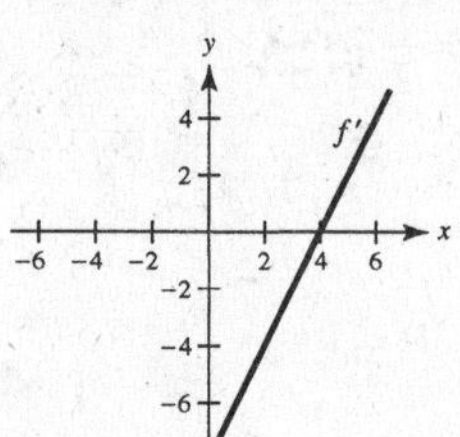

46. The slope of the graph of f is -1 for $x < 4$, 1 for $x > 4$, and undefined at $x = 4$.

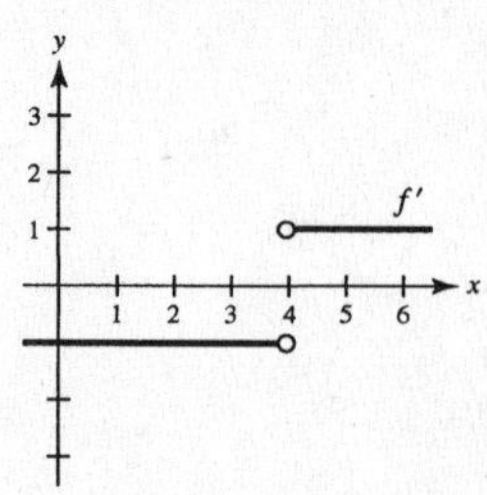

47. Let (x_0, y_0) be a point of tangency on the graph of f. By the limit definition for the derivative, $f'(x) = 4 - 2x$. The slope of the line through $(2, 5)$ and (x_0, y_0) equals the derivative of f at x_0:

$$\frac{5 - y_0}{2 - x_0} = 4 - 2x_0$$

$$5 - y_0 = (2 - x_0)(4 - 2x_0)$$

$$5 - (4x_0 - x_0^2) = 8 - 8x_0 + 2x_0^2$$

$$0 = x_0^2 - 4x_0 + 3$$

$$0 = (x_0 - 1)(x_0 - 3) \Rightarrow x_0 = 1, 3$$

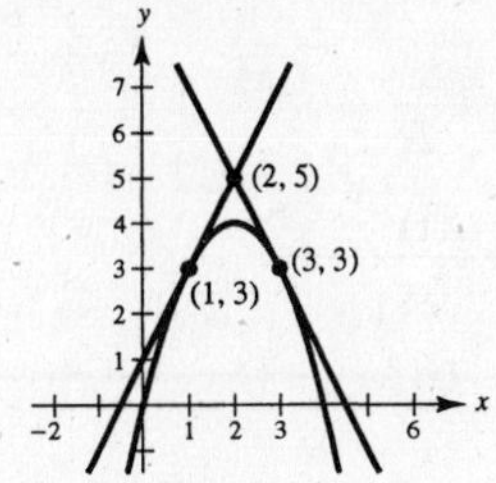

Therefore, the points of tangency are $(1, 3)$ and $(3, 3)$, and the corresponding slopes are 2 and -2. The equations of the tangent lines are

$$y - 5 = 2(x - 2) \qquad y - 5 = -2(x - 2)$$

$$y = 2x + 1 \qquad y = -2x + 9.$$

48. Let (x_0, y_0) be a point of tangency on the graph of f. By the limit definition for the derivative, $f'(x) = 2x$. The slope of the line through $(1, -3)$ and (x_0, y_0) equals the derivative of f at x_0:

$$\frac{-3 - y_0}{1 - x_0} = 2x_0$$

$$-3 - y_0 = (1 - x_0)2x_0$$

$$-3 - x_0^2 = 2x_0 - 2x_0^2$$

$$x_0^2 - 2x_0 - 3 = 0$$

$$(x_0 - 3)(x_0 + 1) = 0 \Rightarrow x_0 = 3, -1$$

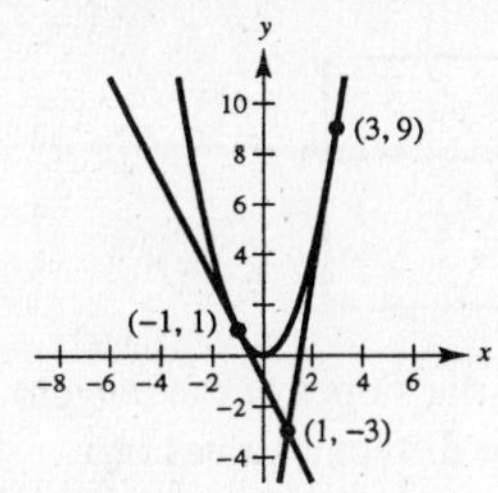

Therefore, the points of tangency are $(3, 9)$ and $(-1, 1)$, and the corresponding slopes are 6 and -2. The equations of the tangent lines are

$$y + 3 = 6(x - 1) \qquad y + 3 = -2(x - 1)$$

$$y = 6x - 9 \qquad y = -2x - 1.$$

49. (a) $g'(0) = -3$

(b) $g'(3) = 0$

(c) Because $g'(1) = -\frac{8}{3}$, g is decreasing (falling) at $x = 1$.

(d) Because $g'(-4) = \frac{7}{3}$, g is increasing (rising) at $x = -4$.

(e) Because $g'(x) > 0$ for all x between 4 and 6, $g(6)$ is greater than $g(4)$, and $g(6) - g(4) > 0$.

(f) No, it is not possible. All you can say is that g is decreasing (falling) at $x = 2$.

50. (a) $f(x) = x^2$

$$f'(x) = \lim_{\Delta x \to 0} \frac{f(x + \Delta x) - f(x)}{\Delta x}$$
$$= \lim_{\Delta x \to 0} \frac{(x + \Delta x)^2 - x^2}{\Delta x}$$
$$= \lim_{\Delta x \to 0} \frac{x^2 + 2x(\Delta x) + (\Delta x)^2 - x^2}{\Delta x}$$
$$= \lim_{\Delta x \to 0} \frac{\Delta x(2x + \Delta x)}{\Delta x}$$
$$= \lim_{\Delta x \to 0} (2x + \Delta x) = 2x$$

At $x = -1, f'(-1) = -2$ and the tangent line is

$y - 1 = -2(x + 1)$ or $y = -2x - 1.$

At $x = 0, f'(0) = 0$ and the tangent line is $y = 0$.

At $x = 1, f'(1) = 2$ and the tangent line is $y = 2x - 1$.

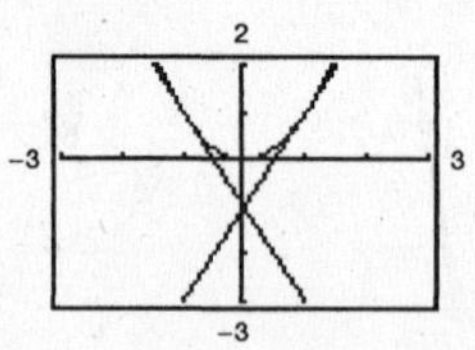

For this function, the slopes of the tangent lines are always distinct for different values of x.

(b) $$g'(x) = \lim_{\Delta x \to 0} \frac{g(x + \Delta x) - g(x)}{\Delta x}$$
$$= \lim_{\Delta x \to 0} \frac{(x + \Delta x)^3 - x^3}{\Delta x}$$
$$= \lim_{\Delta x \to 0} \frac{x^3 + 3x^2(\Delta x) + 3x(\Delta x)^2 + (\Delta x)^3 - x^3}{\Delta x}$$
$$= \lim_{\Delta x \to 0} \frac{\Delta x(3x^2 + 3x(\Delta x) + (\Delta x)^2)}{\Delta x}$$
$$= \lim_{\Delta x \to 0} (3x^2 + 3x(\Delta x) + (\Delta x)^2) = 3x^2$$

At $x = -1, g'(-1) = 3$ and the tangent line is

$y + 1 = 3(x + 1)$ or $y = 3x + 2.$

At $x = 0, g'(0) = 0$ and the tangent line is $y = 0$.

At $x = 1, g'(1) = 3$ and the tangent line is

$y - 1 = 3(x - 1)$ or $y = 3x - 2.$

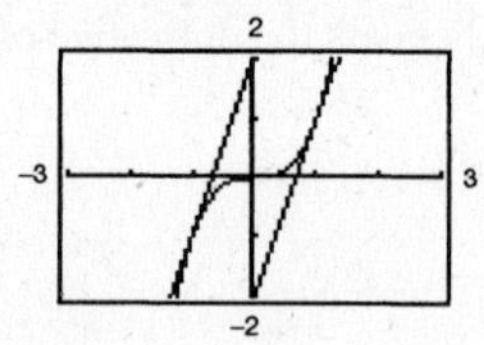

For this function, the slopes of the tangent lines are sometimes the same.

51. $f(x) = \frac{1}{4}x^3$

By the limit definition of the derivative we have $f'(x) = \frac{3}{4}x^2$.

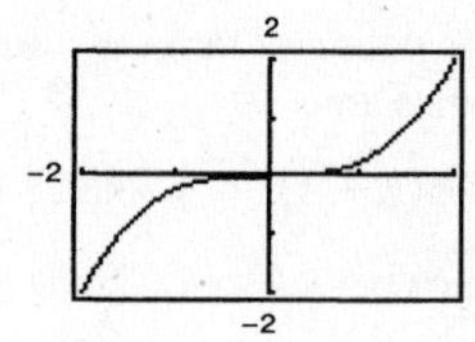

x	-2	-1.5	-1	-0.5	0	0.5	1	1.5	2
$f(x)$	-2	$-\frac{27}{32}$	$-\frac{1}{4}$	$-\frac{1}{32}$	0	$\frac{1}{32}$	$\frac{1}{4}$	$\frac{27}{32}$	2
$f'(x)$	3	$\frac{27}{16}$	$\frac{3}{4}$	$\frac{3}{16}$	0	$\frac{3}{16}$	$\frac{3}{4}$	$\frac{27}{16}$	3

52. $f(x) = \frac{1}{2}x^2$

By the limit definition of the derivative we have $f'(x) = x$.

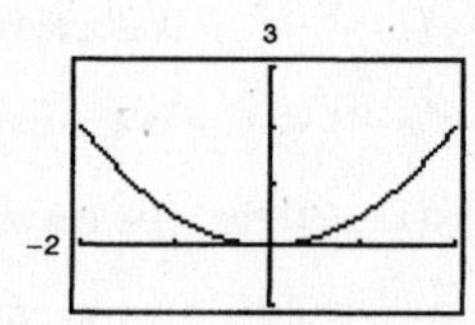

x	-2	-1.5	-1	-0.5	0	0.5	1	1.5	2
$f(x)$	2	1.125	0.5	0.125	0	0.125	0.5	1.125	2
$f'(x)$	-2	-1.5	-1	-0.5	0	0.5	1	1.5	2

53. $$g(x) = \frac{f(x + 0.01) - f(x)}{0.01}$$
$$= [2(x + 0.01) - (x + 0.01)^2 - 2x + x^2]100$$
$$= 1.99 - 2x$$

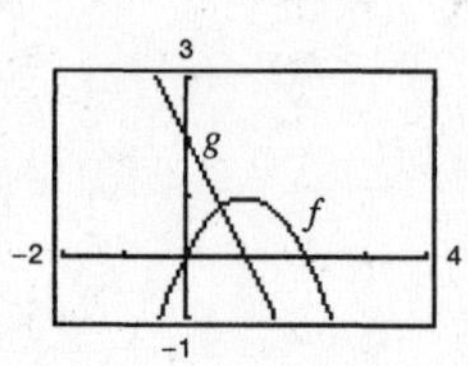

The graph of $g(x)$ is approximately the graph of $f'(x)$.

54. $$g(x) = \frac{f(x + 0.01) - f(x)}{0.01}$$
$$= (3\sqrt{x + 0.01} - 3\sqrt{x})100$$

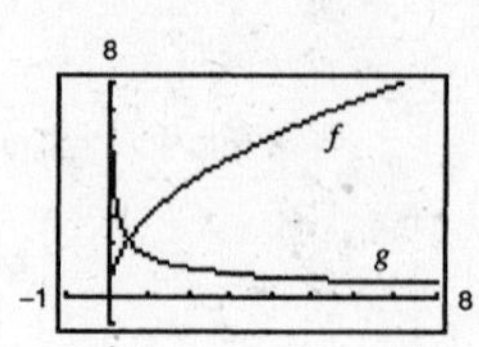

The graph of $g(x)$ is approximately the graph of $f'(x)$.

55. $f(2) = 2(4 - 2) = 4,\ f(2.1) = 2.1(4 - 2.1) = 3.99$

$$f'(2) \approx \frac{3.99 - 4}{2.1 - 2} = -0.1 \text{ [Exact: } f'(2) = 0\text{]}$$

56. $f(2) = \frac{1}{4}(2^3) = 2,\ f(2.1) = 2.31525$

$$f'(2) \approx \frac{2.31525 - 2}{2.1 - 2} = 3.1525 \text{ [Exact: } f'(2) = 3\text{]}$$

57. $f(x) = \dfrac{1}{\sqrt{x}}$ and $f'(x) = \dfrac{-1}{2x^{3/2}}$.

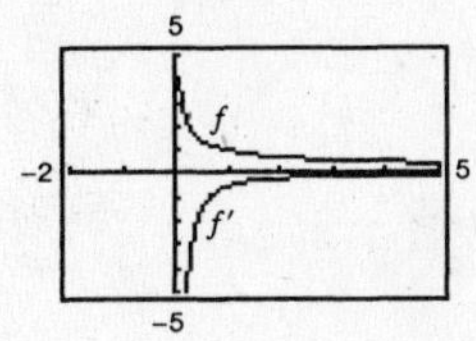

As $x \to \infty$, f is nearly horizontal and thus $f' \approx 0$.

58. $f(x) = \dfrac{x^3}{4} - 3x$ and $f'(x) = \dfrac{3}{4}x^2 - 3$

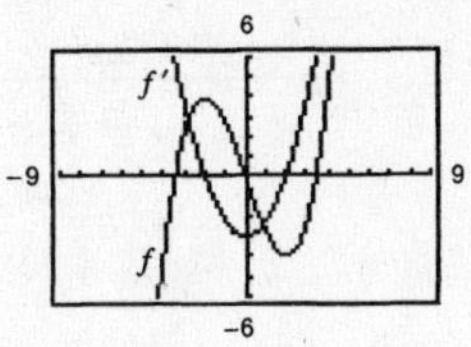

59. $f(x) = 4 - (x - 3)^2$

$$S_{\Delta x}(x) = \frac{f(2 + \Delta x) - f(2)}{\Delta x}(x - 2) + f(2)$$

$$= \frac{4 - (2 + \Delta x - 3)^2 - 3}{\Delta x}(x - 2) + 3 = \frac{1 - (\Delta x - 1)^2}{\Delta x}(x - 2) + 3 = (-\Delta x + 2)(x - 2) + 3$$

(a) $\Delta x = 1$: $S_{\Delta x} = (x - 2) + 3 = x + 1$

$\Delta x = 0.5$: $S_{\Delta x} = \left(\frac{3}{2}\right)(x - 2) + 3 = \frac{3}{2}x$

$\Delta x = 0.1$: $S_{\Delta x} = \left(\frac{19}{10}\right)(x - 2) + 3 = \frac{19}{10}x - \frac{4}{5}$

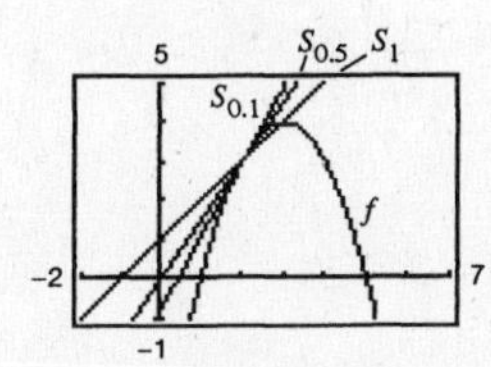

(b) As $\Delta x \to 0$, the line approaches the tangent line to f at $(2, 3)$.

60. $f(x) = x + \dfrac{1}{x}$

$$S_{\Delta x}(x) = \frac{f(2 + \Delta x) - f(2)}{\Delta x}(x - 2) + f(2) = \frac{(2 + \Delta x) + \frac{1}{2 + \Delta x} - \frac{5}{2}}{\Delta x}(x - 2) + \frac{5}{2}$$

$$= \frac{2(2 + \Delta x)^2 + 2 - 5(2 + \Delta x)}{2(2 + \Delta x)\,\Delta x}(x - 2) + \frac{5}{2} = \frac{(2\,\Delta x + 3)}{2(2 + \Delta x)}(x - 2) + \frac{5}{2}$$

(a) $\Delta x = 1$: $S_{\Delta x} = \frac{5}{6}(x - 2) + \frac{5}{2} = \frac{5}{6}x + \frac{5}{6}$

$\Delta x = 0.5$: $S_{\Delta x} = \frac{4}{5}(x - 2) + \frac{5}{2} = \frac{4}{5}x + \frac{9}{10}$

$\Delta x = 0.1$: $S_{\Delta x} = \frac{16}{21}(x - 2) + \frac{5}{2} = \frac{16}{21}x + \frac{41}{42}$

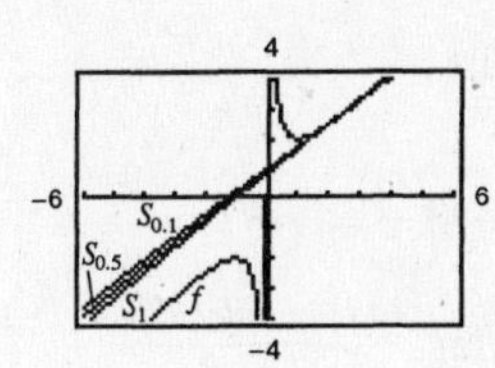

(b) As $\Delta x \to 0$, the line approaches the tangent line to f at $\left(2, \frac{5}{2}\right)$.

61. $f(x) = x^2 - 1,\ c = 2$

$$f'(2) = \lim_{x \to 2} \frac{f(x) - f(2)}{x - 2} = \lim_{x \to 2} \frac{(x^2 - 1) - 3}{x - 2} = \lim_{x \to 2} \frac{(x - 2)(x + 2)}{x - 2} = \lim_{x \to 2} (x + 2) = 4$$

62. $g(x) = x(x-1) = x^2 - x, c = 1$

$$g'(1) = \lim_{x\to1}\frac{g(x)-g(1)}{x-1} = \lim_{x\to1}\frac{x^2-x-0}{x-1} = \lim_{x\to1}\frac{x(x-1)}{x-1}$$

$$= \lim_{x\to1} x = 1$$

63. $f(x) = x^3 + 2x^2 + 1, c = -2$

$$f'(-2) = \lim_{x\to-2}\frac{f(x)-f(-2)}{x+2} = \lim_{x\to-2}\frac{(x^3+2x^2+1)-1}{x+2} = \lim_{x\to-2}\frac{x^2(x+2)}{x+2} = \lim_{x\to-2} x^2 = 4$$

64. $f(x) = x^3 + 2x, c = 1$

$$f'(1) = \lim_{x\to1}\frac{f(x)-f(1)}{x-1} = \lim_{x\to1}\frac{x^3+2x-3}{x-1} = \lim_{x\to1}\frac{(x-1)(x^2+x+3)}{x-1} = \lim_{x\to1}(x^2+x+3) = 5$$

65. $g(x) = \sqrt{|x|}, c = 0$

$$g'(0) = \lim_{x\to0}\frac{g(x)-g(0)}{x-0} = \lim_{x\to0}\frac{\sqrt{|x|}}{x}.$$ Does not exist.

As $x \to 0^-$, $\dfrac{\sqrt{|x|}}{x} = \dfrac{-1}{\sqrt{|x|}} \to -\infty$

As $x \to 0^+$, $\dfrac{\sqrt{|x|}}{x} = \dfrac{1}{\sqrt{x}} \to \infty$

66. $f(x) = \dfrac{1}{x}, c = 3$

$$f'(3) = \lim_{x\to3}\frac{f(x)-f(3)}{x-3} = \lim_{x\to3}\frac{(1/x)-(1/3)}{x-3} = \lim_{x\to3}\frac{3-x}{3x}\cdot\frac{1}{x-3} = \lim_{x\to3}\left(-\frac{1}{3x}\right) = -\frac{1}{9}$$

67. $f(x) = (x-6)^{2/3}, c = 6$

$$f'(6) = \lim_{x\to6}\frac{f(x)-f(6)}{x-6}$$

$$= \lim_{x\to6}\frac{(x-6)^{2/3}-0}{x-6}$$

$$= \lim_{x\to6}\frac{1}{(x-6)^{1/3}}$$

Does not exist.

68. $g(x) = (x+3)^{1/3}, c = -3$

$$g'(-3) = \lim_{x\to-3}\frac{g(x)-g(-3)}{x-(-3)}$$

$$= \lim_{x\to-3}\frac{(x+3)^{1/3}-0}{x+3}$$

$$= \lim_{x\to-3}\frac{1}{(x+3)^{2/3}}$$

Does not exist.

69. $h(x) = |x+5|, c = -5$

$$h'(-5) = \lim_{x\to-5}\frac{h(x)-h(-5)}{x-(-5)}$$

$$= \lim_{x\to-5}\frac{|x+5|-0}{x+5}$$

$$= \lim_{x\to-5}\frac{|x+5|}{x+5}$$

Does not exist.

70. $f(x) = |x-4|, c = 4$

$$f'(4) = \lim_{x\to4}\frac{f(x)-f(4)}{x-4} = \lim_{x\to4}\frac{|x-4|-0}{x-4} = \lim_{x\to4}\frac{|x-4|}{x-4}$$

Does not exist.

71. $f(x)$ is differentiable everywhere except at $x = -3$. (Sharp turn in the graph.)

72. $f(x)$ is differentiable everywhere except at $x = \pm 3$. (Sharp turns in the graph.)

73. $f(x)$ is differentiable everywhere except at $x = -1$. (Discontinuity)

74. $f(x)$ is differentiable everywhere except at $x = 1$. (Discontinuity)

75. $f(x)$ is differentiable everywhere except at $x = 3$. (Sharp turn in the graph.)

76. $f(x)$ is differentiable everywhere except at $x = 0$. (Sharp turn in the graph.)

77. $f(x)$ is differentiable on the interval $(1, \infty)$. (At $x = 1$ the tangent line is vertical.)

78. $f(x)$ is differentiable everywhere except at $x = \pm 2$. (Discontinuities)

79. $f(x)$ is differentiable everywhere except at $x = 0$. (Discontinuity)

80. $f(x)$ is differentiable everywhere except at $x = 1$. (Discontinuity)

81. $f(x) = |x - 1|$

The derivative from the left is $\lim_{x \to 1^-} \dfrac{f(x) - f(1)}{x - 1} = \lim_{x \to 1^-} \dfrac{|x - 1| - 0}{x - 1} = -1.$

The derivative from the right is $\lim_{x \to 1^+} \dfrac{f(x) - f(1)}{x - 1} = \lim_{x \to 1^+} \dfrac{|x - 1| - 0}{x - 1} = 1.$

The one-sided limits are not equal. Therefore, f is not differentiable at $x = 1$.

82. $f(x) = \sqrt{1 - x^2}$

The derivative from the left does not exist because

$$\lim_{x \to 1^-} \frac{f(x) - f(1)}{x - 1} = \lim_{x \to 1^-} \frac{\sqrt{1 - x^2} - 0}{x - 1} = \lim_{x \to 1^-} \frac{\sqrt{1 - x^2}}{x - 1} \cdot \frac{\sqrt{1 - x^2}}{\sqrt{1 - x^2}} = \lim_{x \to 1^-} -\frac{1 + x}{\sqrt{1 - x^2}} = -\infty. \text{ (Vertical tangent)}$$

The limit from the right does not exist since f is undefined for $x > 1$. Therefore, f is not differentiable at $x = 1$.

83. The function is continuous at $x = 1$.

$$f(x) = \begin{cases} (x - 1)^3, & x \le 1 \\ (x - 1)^2, & x > 1 \end{cases}$$

The derivative from the left is $\lim_{x \to 1^-} \dfrac{f(x) - f(1)}{x - 1} = \lim_{x \to 1^-} \dfrac{(x - 1)^3 - 0}{x - 1} = \lim_{x \to 1^-} (x - 1)^2 = 0.$

The derivative from the right is $\lim_{x \to 1^+} \dfrac{f(x) - f(1)}{x - 1} = \lim_{x \to 1^+} \dfrac{(x - 1)^2 - 0}{x - 1} = \lim_{x \to 1^+} (x - 1) = 0.$

These one-sided limits are equal. Therefore, f is differentiable at $x = 1$. $(f'(1) = 0)$

84. $f(x) = \begin{cases} x, & x \le 1 \\ x^2, & x > 1 \end{cases}$

The derivative from the left is $\lim_{x \to 1^-} \dfrac{f(x) - f(1)}{x - 1} = \lim_{x \to 1^-} \dfrac{x - 1}{x - 1} = \lim_{x \to 1^-} 1 = 1.$

The derivative from the right is $\lim_{x \to 1^+} = \dfrac{f(x) - f(1)}{x - 1} = \lim_{x \to 1^+} \dfrac{x^2 - 1}{x - 1} = \lim_{x \to 1^+} (x + 1) = 2.$

These one-sided limits are not equal. Therefore, f is not differentiable at $x = 1$.

85. Note that f is continuous at $x = 2$. $f(x) = \begin{cases} x^2 + 1, & x \le 2 \\ 4x - 3, & x > 2 \end{cases}$

The derivative from the left is $\lim_{x \to 2^-} \frac{f(x) - f(2)}{x - 2} = \lim_{x \to 2^-} \frac{(x^2 + 1) - 5}{x - 2} = \lim_{x \to 2^-} (x + 2) = 4$.

The derivative from the right is $\lim_{x \to 2^+} \frac{f(x) - f(2)}{x - 2} = \lim_{x \to 2^+} \frac{(4x - 3) - 5}{x - 2} = \lim_{x \to 2^+} 4 = 4$.

The one-sided limits are equal. Therefore, f is differentiable at $x = 2$. $(f'(2) = 4)$

86. Note that f is continuous at $x = 2$. $f(x) = \begin{cases} \frac{1}{2}x + 1, & x < 2 \\ \sqrt{2x}, & x \ge 2 \end{cases}$

The derivative from the left is $\lim_{x \to 2^-} \frac{f(x) - f(2)}{x - 2} = \lim_{x \to 2^-} \frac{\left(\frac{1}{2}x + 1\right) - 2}{x - 2} = \lim_{x \to 2^-} \frac{\frac{1}{2}(x - 2)}{x - 2} = \frac{1}{2}$.

The derivative from the right is $\lim_{x \to 2^+} \frac{f(x) - f(2)}{x - 2} = \lim_{x \to 2^+} \frac{\sqrt{2x} - 2}{x - 2} \cdot \frac{\sqrt{2x} + 2}{\sqrt{2x} + 2} = \lim_{x \to 2^+} \frac{2x - 4}{(x - 2)\left(\sqrt{2x} + 2\right)}$

$$= \lim_{x \to 2^+} \frac{2(x - 2)}{(x - 2)\left(\sqrt{2x} + 2\right)} = \lim_{x \to 2^+} \frac{2}{\sqrt{2x} + 2} = \frac{1}{2}.$$

The one-sided limits are equal. Therefore, f is differentiable at $x = 2$. $\left(f'(2) = \frac{1}{2}\right)$

87. False. the slope is $\lim_{\Delta x \to 0} \frac{f(2 + \Delta x) - f(2)}{\Delta x}$.

88. False. $y = |x - 2|$ is continuous at $x = 2$, but is not differentiable at $x = 2$. (Sharp turn in the graph.)

89. False. If the derivative from the left of a point does not equal the derivative from the right of a point, then the derivative does not exist at that point. For example, if $f(x) = |x|$, then the derivative from the left at $x = 0$ is -1 and the derivative from the right at $x = 0$ is 1. At $x = 0$, the derivative does not exist.

90. True—see Theorem 4.1.

91.

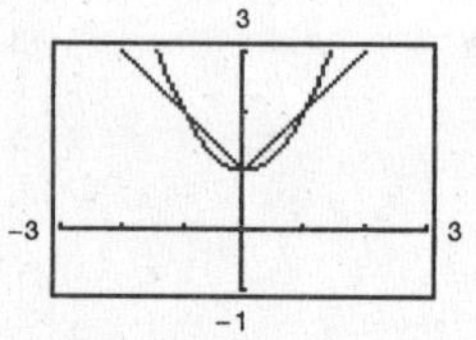

As you zoom in, the graph of $y_1 = x^2 + 1$ appears to be locally the graph of a horizontal line, whereas the graph of $y_2 = |x| + 1$ always has a sharp corner at $(0, 1)$. y_2 is not differentiable at $(0, 1)$.

Section 4.2 Basic Differentiation Rules and Rates of Change

1. (a) $y = x^{1/2}$

$y' = \frac{1}{2}x^{-1/2}$

$y'(1) = \frac{1}{2}$

(b) $y = x^3$

$y' = 3x^2$

$y'(1) = 3$

2. (a) $y = x^{-1/2}$

$y' = -\frac{1}{2}x^{-3/2}$

$y'(1) = -\frac{1}{2}$

(b) $y = x^{-1}$

$y' = -x^{-2}$

$y'(1) = -1$

3. $y = 8$

$y' = 0$

4. $f(x) = -2$

$f'(x) = 0$

5. $y = x^6$

$y' = 6x^5$

6. $y = x^8$

$y' = 8x^7$

7. $y = \frac{1}{x^7} = x^{-7}$

$y' = -7x^{-8} = \frac{-7}{x^8}$

8. $y = \frac{1}{x^8} = x^{-8}$

$y' = -8x^{-9} = \frac{-8}{x^9}$

9. $f(x) = \sqrt[5]{x} = x^{1/5}$

$f'(x) = \frac{1}{5}x^{-4/5} = \frac{1}{5x^{4/5}}$

10. $g(x) = \sqrt[4]{x} = x^{1/4}$

$g'(x) = \frac{1}{4}x^{-3/4} = \frac{1}{4x^{3/4}}$

11. $f(x) = x + 1$

$f'(x) = 1$

12. $g(x) = 3x - 1$

$g'(x) = 3$

13. $f(t) = -2t^2 + 3t - 6$

$f'(t) = -4t + 3$

14. $y = t^2 + 2t - 3$

$y' = 2t + 2$

15. $y = 16 - 3x - \frac{1}{2}x^2$

$y' = -3 - x$

16. $h(s) = 480 + 64s - 16s^2$

$h'(s) = 64 - 32s$

17. $g(x) = x^2 + 4x^3$

$g'(x) = 2x + 12x^2$

18. $y = 8 - x^3$

$y' = -3x^2$

19. $s(t) = t^3 - 2t + 4$

$s'(t) = 3t^2 - 2$

20. $f(x) = 2x^3 - x^2 + 3x$

$f'(x) = 6x^2 - 2x + 3$

	Function	*Rewrite*	*Derivative*	*Simplify*
21.	$y = \frac{5}{2x^2}$	$y = \frac{5}{2}x^{-2}$	$y' = -5x^{-3}$	$y' = \frac{-5}{x^3}$
22.	$y = \frac{2}{3x^2}$	$y = \frac{2}{3}x^{-2}$	$y' = -\frac{4}{3}x^{-3}$	$y' = -\frac{4}{3x^3}$
23.	$y = \frac{3}{(2x)^3}$	$y = \frac{3}{8}x^{-3}$	$y' = \frac{-9}{8}x^{-4}$	$y' = \frac{-9}{8x^4}$
24.	$y = \frac{\pi}{(3x)^2}$	$y = \frac{\pi}{9}x^{-2}$	$y' = -\frac{2\pi}{9}x^{-3}$	$y' = -\frac{2\pi}{9x^3}$
25.	$y = \frac{\sqrt{x}}{x}$	$y = x^{-1/2}$	$y' = -\frac{1}{2}x^{-3/2}$	$y' = -\frac{1}{2x^{3/2}}$
26.	$y = \frac{4}{x^{-3}}$	$y = 4x^3$	$y' = 12x^2$	$y' = 12x^2$

27. $f(x) = \frac{3}{x^2} = 3x^{-2}, (1, 3)$

$f'(x) = -6x^{-3} = \frac{-6}{x^3}$

$f'(1) = -6$

28. $f(t) = 3 - \frac{3}{5t}, \left(\frac{3}{5}, 2\right)$

$f'(t) = \frac{3}{5t^2}$

$f'\left(\frac{3}{5}\right) = \frac{5}{3}$

29. $f(x) = -\frac{1}{2} + \frac{7}{5}x^3, \left(0, -\frac{1}{2}\right)$

$f'(x) = \frac{21}{5}x^2$

$f'(0) = 0$

30. $y = 3x^3 - 6, (2, 18)$

$y' = 9x^2$

$y'(2) = 36$

31. $y = (2x + 1)^2, (0, 1)$

$= 4x^2 + 4x + 1$

$y' = 8x + 4$

$y'(0) = 4$

32. $f(x) = 3(5 - x)^2, (5, 0)$

$= 3x^2 - 30x + 75$

$f'(x) = 6x - 30$

$f'(5) = 0$

33. $f(x) = x^2 + 5 - 3x^{-2}$

$f'(x) = 2x + 6x^{-3} = 2x + \dfrac{6}{x^3}$

34. $f(x) = x^2 - 3x - 3x^{-2}$

$f'(x) = 2x - 3 + 6x^{-3}$

$= 2x - 3 + \dfrac{6}{x^3}$

35. $g(t) = t^2 - \dfrac{4}{t^3} = t^2 - 4t^{-3}$

$g'(t) = 2t + 12t^{-4} = 2t + \dfrac{12}{t^4}$

36. $f(x) = x + x^{-2}$

$f'(x) = 1 - 2x^{-3}$

$= 1 - \dfrac{2}{x^3}$

37. $f(x) = \dfrac{x^3 - 3x^2 + 4}{x^2} = x - 3 + 4x^{-2}$

$f'(x) = 1 - \dfrac{8}{x^3} = \dfrac{x^3 - 8}{x^3}$

38. $h(x) = \dfrac{2x^2 - 3x + 1}{x} = 2x - 3 + x^{-1}$

$h'(x) = 2 - \dfrac{1}{x^2} = \dfrac{2x^2 - 1}{x^2}$

39. $y = x(x^2 + 1) = x^3 + x$

$y' = 3x^2 + 1$

40. $y = 3x(6x - 5x^2) = 18x^2 - 15x^3$

$y' = 36x - 45x^2$

41. $f(x) = \sqrt{x} - 6\sqrt[3]{x} = x^{1/2} - 6x^{1/3}$

$f'(x) = \dfrac{1}{2}x^{-1/2} - 2x^{-2/3} = \dfrac{1}{2\sqrt{x}} - \dfrac{2}{x^{2/3}}$

42. $f(x) = \sqrt[3]{x} + \sqrt[5]{x} = x^{1/3} + x^{1/5}$

$f'(x) = \dfrac{1}{3}x^{-2/3} + \dfrac{1}{5}x^{-4/5} = \dfrac{1}{3x^{2/3}} + \dfrac{1}{5x^{4/5}}$

43. $h(s) = s^{4/5} - s^{2/3}$

$h'(s) = \dfrac{4}{5}s^{-1/5} - \dfrac{2}{3}s^{-1/3} = \dfrac{4}{5s^{1/5}} - \dfrac{2}{3s^{1/3}}$

44. $f(t) = t^{2/3} - t^{1/3} + 4$

$f'(t) = \dfrac{2}{3}t^{-1/3} - \dfrac{1}{3}t^{-2/3} = \dfrac{2}{3t^{1/3}} - \dfrac{1}{3t^{2/3}}$

45. (a) $y = x^4 - 3x^2 + 2$

$y' = 4x^3 - 6x$

At $(1, 0)$: $y' = 4(1)^3 - 6(1) = -2$

Tangent line: $y - 0 = -2(x - 1)$

$2x + y - 2 = 0$

(b)

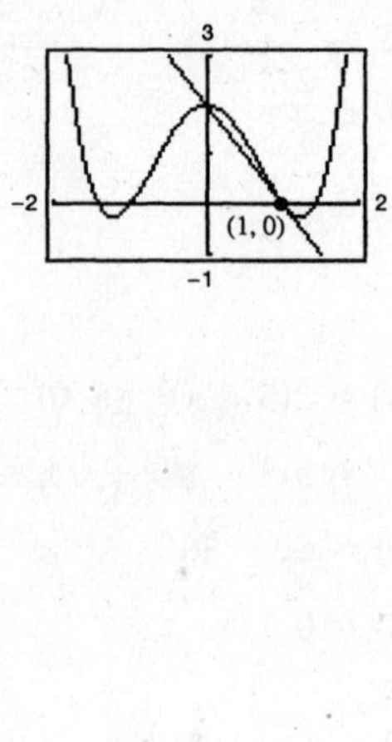

46. (a) $y = x^3 + x$

$y' = 3x^2 + 1$

At $(-1, -2)$: $y' = 3(-1)^2 + 1 = 4$.

Tangent line: $y + 2 = 4(x + 1)$

$4x - y + 2 = 0$

(b)

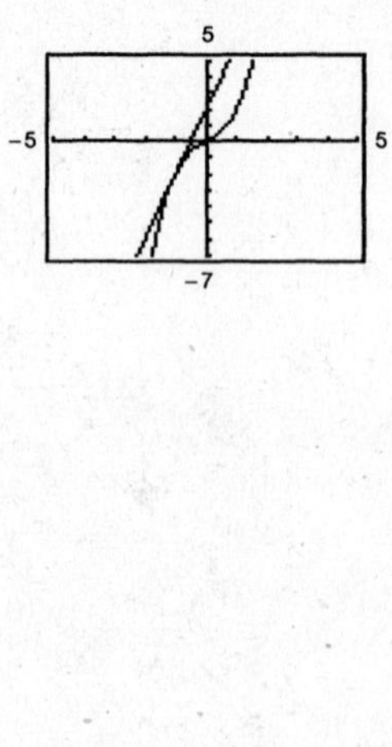

47. (a) $f(x) = \frac{2}{\sqrt[4]{x^3}} = 2x^{-3/4}$

$f'(x) = \frac{-3}{2}x^{-7/4} = \frac{-3}{2x^{7/4}}$

At $(1, 2)$: $f'(1) = \frac{-3}{2}$

Tangent line: $y - 2 = -\frac{3}{2}(x - 1)$

$y = -\frac{3}{2}x + \frac{7}{2}$

$3x + 2y - 7 = 0$

(b)

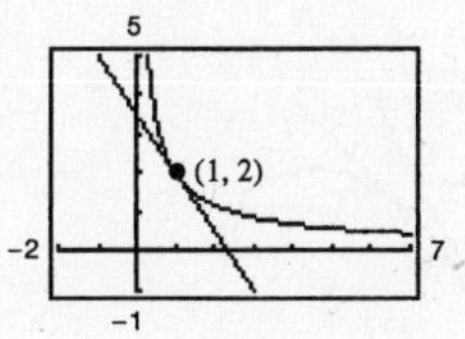

48. (a) $y = (x^2 + 2x)(x + 1)$

$= x^3 + 3x^2 + 2x$

$y' = 3x^2 + 6x + 2$

At $(1, 6)$: $y' = 3(1)^2 + 6(1) + 2 = 11.$

Tangent line: $y - 6 = 11(x - 1)$

$0 = 11x - y - 5$

(b)

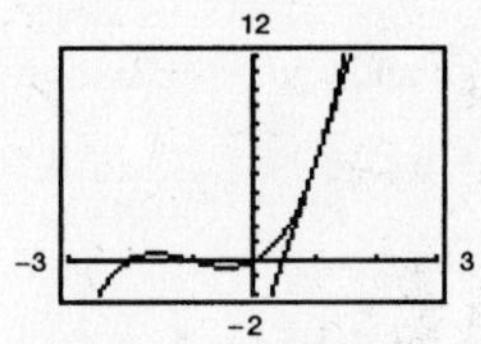

49. $y = x^4 - 8x^2 + 2$

$y' = 4x^3 - 16x$

$= 4x(x^2 - 4)$

$= 4x(x - 2)(x + 2)$

$y' = 0 \Rightarrow x = 0, \pm 2$

Horizontal tangents: $(0, 2)$, $(2, -14)$, $(-2, -14)$

50. $y = x^3 + x$

$y' = 3x^2 + 1 > 0$ for all x.

Therefore, there are no horizontal tangents.

51. $y = \frac{1}{x^2} = x^{-2}$

$y' = -2x^{-3} = \frac{-2}{x^3}$ cannot equal zero.

Therefore, there are no horizontal tangents.

52. $y = x^2 + 1$

$y' = 2x = 0 \Rightarrow x = 0$

At $x = 0$, $y = 1$.

Horizontal tangent: $(0, 1)$

53. $x^2 - kx = 4x - 9$ Equate functions

$2x - k = 4$ Equate derivatives

Hence, $k = 2x - 4$ and
$x^2 - (2x - 4)x = 4x - 9 \Rightarrow -x^2 = -9 \Rightarrow x = \pm 3.$

For $x = 3$, $k = 2$ and for $x = -3$, $k = -10$.

54. $k - x^2 = -4x + 7$ Equate functions

$-2x = -4$ Equate derivatives

Hence, $x = 2$ and $k - 4 = -8 + 7 \Rightarrow k = 3$

55. $\frac{k}{x} = -\frac{3}{4}x + 3$ Equate functions

$-\frac{k}{x^2} = -\frac{3}{4}$ Equate derivatives

Hence, $k = \frac{3}{4}x^2$ and $\frac{\frac{3}{4}x^2}{x} = \frac{-3}{4}x + 3 \Rightarrow \frac{3}{4}x = -\frac{3}{4}x + 3 \Rightarrow \frac{3}{2}x = 3 \Rightarrow x = 2 \Rightarrow k = 3.$

56. $k\sqrt{x} = x + 4$ Equate functions

$\dfrac{k}{2\sqrt{x}} = 1$ Equate derivatives

Hence, $k = 2\sqrt{x}$ and

$(2\sqrt{x})\sqrt{x} = x + 4 \Rightarrow 2x = x + 4 \Rightarrow x = 4 \Rightarrow k = 4.$

57. Let (x_1, y_1) and (x_2, y_2) be the points of tangency on $y = x^2$ and $y = -x^2 + 6x - 5$, respectively. The derivatives of these functions are

$$y' = 2x \Rightarrow m = 2x_1 \quad \text{and} \quad y' = -2x + 6 \Rightarrow m = -2x_2 + 6.$$

$$m = 2x_1 = -2x_2 + 6$$

$$x_1 = -x_2 + 3$$

Since $y_1 = x_1^2$ and $y_2 = -x_2^2 + 6x_2 - 5$,

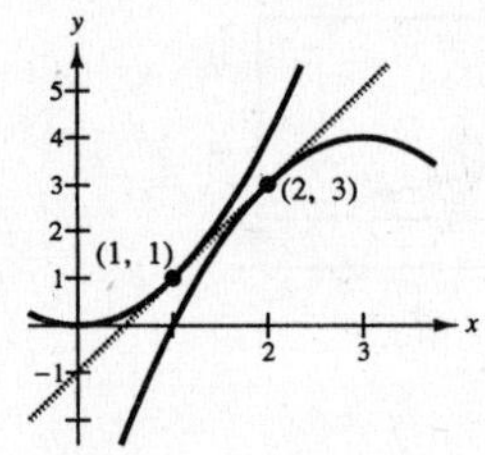

$$m = \frac{y_2 - y_1}{x_2 - x_1} = \frac{(-x_2^2 + 6x_2 - 5) - (x_1^2)}{x_2 - x_1} = -2x_2 + 6.$$

$$\frac{(-x_2^2 + 6x_2 - 5) - (-x_2 + 3)^2}{x_2 - (-x_2 + 3)} = -2x_2 + 6$$

$$(-x_2^2 + 6x_2 - 5) - (x_2^2 - 6x_2 + 9) = (-2x_2 + 6)(2x_2 - 3)$$

$$-2x_2^2 + 12x_2 - 14 = -4x_2^2 + 18x_2 - 18$$

$$2x_2^2 - 6x_2 + 4 = 0$$

$$2(x_2 - 2)(x_2 - 1) = 0$$

$$x_2 = 1 \text{ or } 2$$

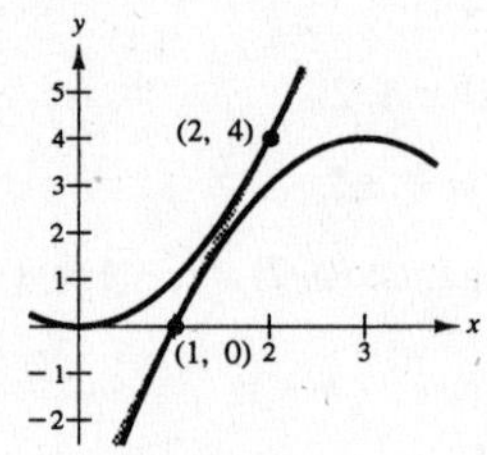

$x_2 = 1 \Rightarrow y_2 = 0, x_1 = 2$ and $y_1 = 4$

Thus, the tangent line through $(1, 0)$ and $(2, 4)$ is

$$y - 0 = \left(\frac{4 - 0}{2 - 1}\right)(x - 1) \Rightarrow y = 4x - 4.$$

$x_2 = 2 \Rightarrow y_2 = 3, x_1 = 1$ and $y_1 = 1$

Thus, the tangent line through $(2, 3)$ and $(1, 1)$ is

$$y - 1 = \left(\frac{3 - 1}{2 - 1}\right)(x - 1) \Rightarrow y = 2x - 1.$$

58. m_1 is the slope of the line tangent to $y = x$. m_2 is the slope of the line tangent to $y = 1/x$. Since

$$y = x \Rightarrow y' = 1 \Rightarrow m_1 = 1 \text{ and } y = \frac{1}{x} \Rightarrow y' = \frac{-1}{x^2} \Rightarrow m_2 = \frac{-1}{x^2}.$$

The points of intersection of $y = x$ and $y = 1/x$ are

$$x = \frac{1}{x} \Rightarrow x^2 = 1 \Rightarrow x = \pm 1.$$

At $x = \pm 1$, $m_2 = -1$. Since $m_2 = -1/m_1$, these tangent lines are perpendicular at the points intersection.

59. (a) The slope appears to be steepest between A and B.

(b) The average rate of change between A and B is **greater** than the instantaneous rate of change at B.

(c)

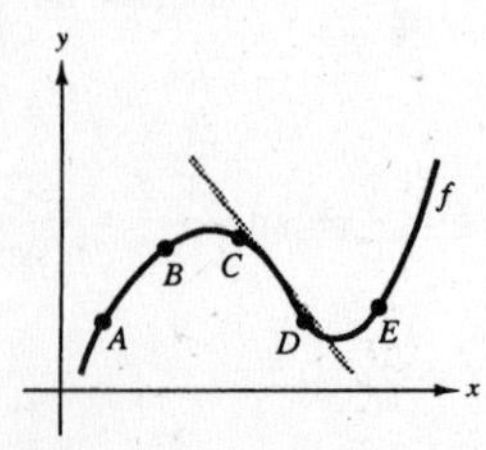

60. The graph of a function f such that $f' > 0$ for all x and the rate of change of the function is decreasing (i.e. $f'' < 0$) would, in general, look like the graph at the right.

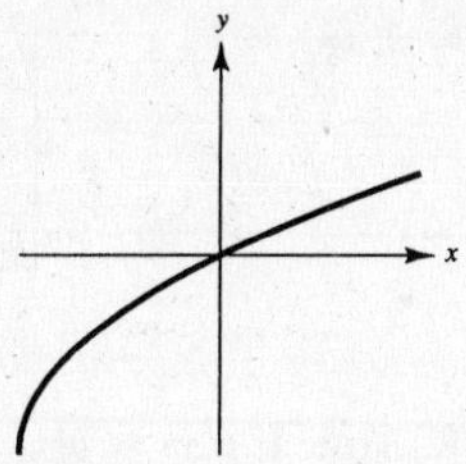

61. $g(x) = f(x) + 6 \Rightarrow g'(x) = f'(x)$

62. $g(x) = -5f(x) \Rightarrow g'(x) = -5f'(x)$

63.

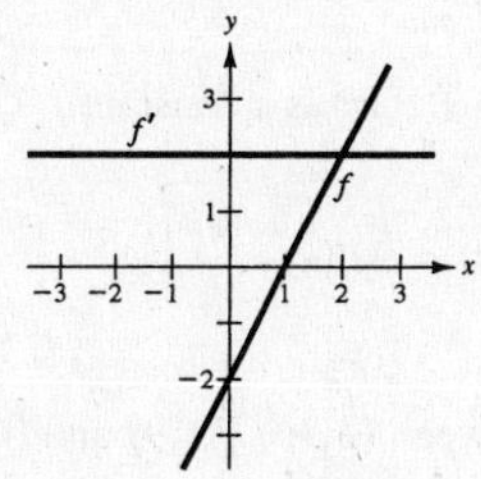

If f is linear then its derivative is a constant function.

$$f(x) = ax + b$$

$$f'(x) = a$$

64.

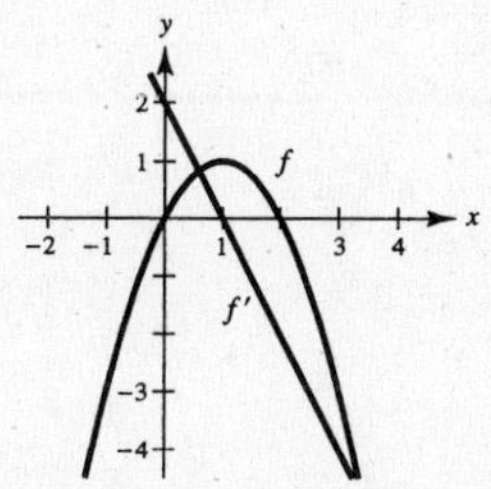

If f is quadratic, then its derivative is a linear function.

$$f(x) = ax^2 + bx + c$$

$$f'(x) = 2ax + b$$

65. $f(x) = \sqrt{x}$, $(-4, 0)$

$$f'(x) = \frac{1}{2}x^{-1/2} = \frac{1}{2\sqrt{x}}$$

$$\frac{1}{2\sqrt{x}} = \frac{0 - y}{-4 - x}$$

$$4 + x = 2\sqrt{x}y$$

$$4 + x = 2\sqrt{x}\sqrt{x}$$

$$4 + x = 2x$$

$$x = 4, y = 2$$

The point $(4, 2)$ is on the graph of f.

Tangent line: $y - 2 = \dfrac{0 - 2}{-4 - 4}(x - 4)$

$$4y - 8 = x - 4$$

$$0 = x - 4y + 4$$

66. $f(x) = \dfrac{2}{x}$, $(5, 0)$

$$f'(x) = -\frac{2}{x^2}$$

$$-\frac{2}{x^2} = \frac{0 - y}{5 - x}$$

$$-10 + 2x = -x^2y$$

$$-10 + 2x = -x^2\left(\frac{2}{x}\right)$$

$$-10 + 2x = -2x$$

$$4x = 10$$

$$x = \frac{5}{2}, y = \frac{4}{5}$$

The point $\left(\frac{5}{2}, \frac{4}{5}\right)$ is on the graph of f. The slope of the tangent line is $f'\left(\frac{5}{2}\right) = -\frac{8}{25}$.

Tangent line: $y - \dfrac{4}{5} = -\dfrac{8}{25}\left(x - \dfrac{5}{2}\right)$

$$25y - 20 = -8x + 20$$

$$8x + 25y - 40 = 0$$

67. $f'(1) = -1$

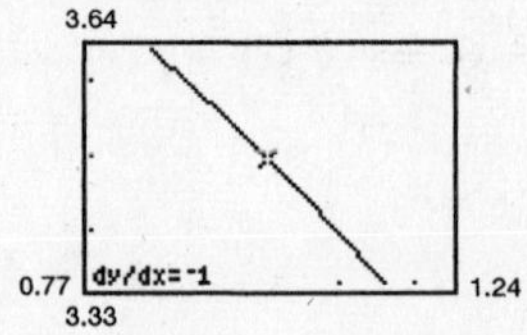

68. $f'(4) = 1$

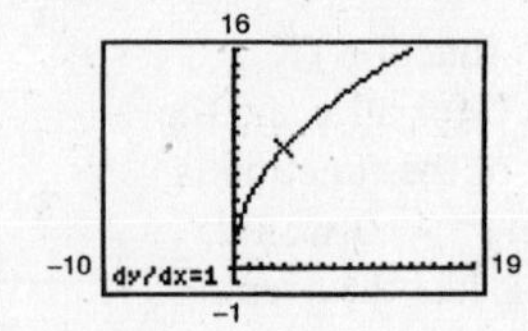

69. False. Let $f(x) = x^2$ and $g(x) = x^2 + 4$.
Then $f'(x) = g'(x) = 2x$, but $f(x) \neq g(x)$.

70. True. If $f(x) = g(x) + c$, then $f'(x) = g'(x) + 0 = g'(x)$.

71. False. If $y = \pi^2$, then $dy/dx = 0$. (π^2 is a constant.)

72. True. If $y = x/\pi = (1/\pi) \cdot x$, then $dy/dx = (1/\pi)(1) = 1/\pi$.

73. True. If $g(x) = 3f(x)$, then $g'(x) = 3f'(x)$.

74. False. If $f(x) = \dfrac{1}{x^n} = x^{-n}$, then $f'(x) = -nx^{-n-1} = \dfrac{-n}{x^{n+1}}$.

75. (a) One possible secant is between (3.9, 7.7019) and (4, 8):

$$y - 8 = \frac{8 - 7.7019}{4 - 3.9}(x - 4)$$

$$y - 8 = 2.981(x - 4)$$

$$y = S(x) = 2.981x - 3.924$$

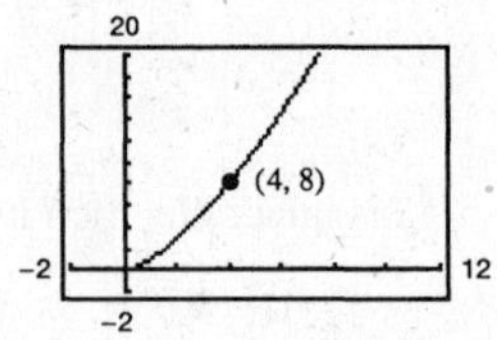

(b) $f'(x) = \frac{3}{2}x^{1/2} \Rightarrow f'(4) = \frac{3}{2}(2) = 3$

$T(x) = 3(x - 4) + 8 = 3x - 4$

$S(x)$ is an approximation of the tangent line $T(x)$.

(c) As you move further away from (4, 8), the accuracy of the approximation T gets worse.

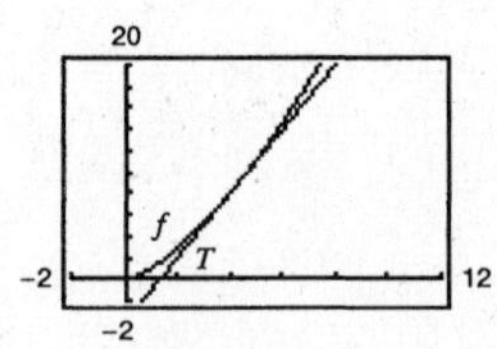

(d)

Δx	-3	-2	-1	-0.5	-0.1	0	0.1	0.5	1	2	3
$f(4 + \Delta x)$	1	2.828	5.196	6.548	7.702	8	8.302	9.546	11.180	14.697	18.520
$T(4 + \Delta x)$	-1	2	5	6.5	7.7	8	8.3	9.5	11	14	17

76. (a) Nearby point: (1.0073138, 1.0221024)

Secant line: $y - 1 = \dfrac{1.0221024 - 1}{1.0073138 - 1}(x - 1)$

$$y = 3.022(x - 1) + 1$$

(Answers will vary.)

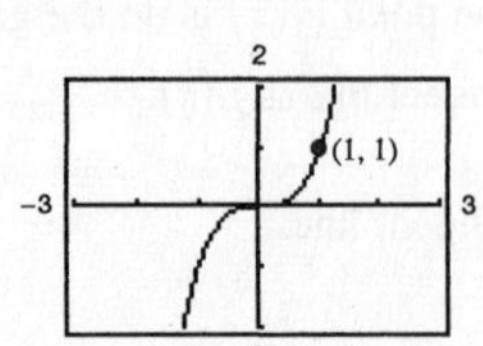

(b) $f'(x) = 3x^2$

$T(x) = 3(x - 1) + 1 = 3x - 2$

(c) The accuracy worsens as you move away from (1, 1).

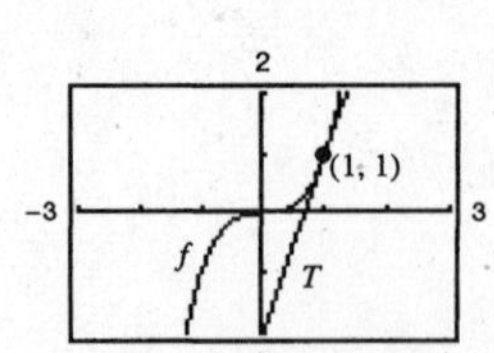

(d)

Δx	-3	-2	-1	-0.5	-0.1	0	0.1	0.5	1	2	3
$f(1 + \Delta x)$	-8	-1	0	0.125	0.729	1	1.331	3.375	8	27	64
$T(1 + \Delta x)$	-8	-5	-2	-0.5	0.7	1	1.3	2.5	4	7	10

The accuracy decreases more rapidly than in Exercise 75 because $y = x^3$ is less "linear" than $y = x^{3/2}$.

77. $f(t) = 2t + 7, [1, 2]$

$f'(t) = 2$

Instantaneous rate of change is the constant 2.
Average rate of change:

$$\frac{f(2) - f(1)}{2 - 1} = \frac{[2(2) + 7] - [2(1) + 7]}{1} = 2$$

(These are the same because f is a line of slope 2.)

78. $f(t) = t^2 - 3, [2, 2.1]$

$f'(t) = 2t$

Instantaneous rate of change:

$$(2, 1) \Rightarrow f'(2) = 2(2) = 4$$

$$(2.1, 1.41) \Rightarrow f'(2.1) = 4.2$$

Average rate of change:

$$\frac{f(2.1) - f(2)}{2.1 - 2} = \frac{1.41 - 1}{0.1} = 4.1$$

79. $f(x) = -\frac{1}{x}, [1, 2]$

$f'(x) = \frac{1}{x^2}$

Instantaneous rate of change:

$$(1, -1) \Rightarrow f'(1) = 1$$

$$\left(2, -\frac{1}{2}\right) \Rightarrow f'(2) = \frac{1}{4}$$

Average rate of change:

$$\frac{f(2) - f(1)}{2 - 1} = \frac{(-1/2) - (-1)}{2 - 1} = \frac{1}{2}$$

80. $f(x) = \frac{1}{x + 1}, [0, 3]$

$f'(x) = \frac{-1}{(x + 1)^2}$

Average rate of change: $\frac{f(3) - f(0)}{3 - 0} = \frac{\frac{1}{4} - 1}{3} = -\frac{1}{4}$

Instantaneous rates of change:

$(0, 1)$: $f'(0) = -1$

$\left(3, \frac{1}{4}\right)$: $f'(3) = \frac{-1}{16}$

81. (a) $s(t) = -16t^2 + 1362$

$v(t) = -32t$

(b) $\frac{s(2) - s(1)}{2 - 1} = 1298 - 1346 = -48$ ft/sec

(c) $v(t) = s'(t) = -32t$

When $t = 1$: $v(1) = -32$ ft/sec.

When $t = 2$: $v(2) = -64$ ft/sec.

(d) $-16t^2 + 1362 = 0$

$$t^2 = \frac{1362}{16} \Rightarrow t = \frac{\sqrt{1362}}{4} \approx 9.226 \text{ sec}$$

$$v\left(\frac{\sqrt{1362}}{4}\right) = -32\left(\frac{\sqrt{1362}}{4}\right)$$

$$= -8\sqrt{1362} \approx -295.242 \text{ ft/sec}$$

82.

$$s(t) = -16t^2 - 22t + 220$$

$$v(t) = -32t - 22$$

$$v(3) = -118 \text{ ft/sec}$$

$$s(t) = -16t^2 - 22t + 220$$

$$= 112 \text{ (height after falling 108 ft)}$$

$$-16t^2 - 22t + 108 = 0$$

$$-2(t - 2)(8t + 27) = 0$$

$$t = 2$$

$$v(2) = -32(2) - 22$$

$$= -86 \text{ ft/sec}$$

83. $s(t) = -4.9t^2 + v_0t + s_0$

$= -4.9t^2 + 120t$

$v(t) = -9.8t + 120$

$v(5) = -9.8(5) + 120 = 71$ m/sec

$v(10) = -9.8(10) + 120 = 22$ m/sec

84. $s(t) = -4.9t^2 + v_0t + s_0$

$= -4.9t^2 + s_0 = 0$ when $t = 6.8$.

$s_0 = 4.9t^2 = 4.9(6.8)^2 = 226.6$ m

85.

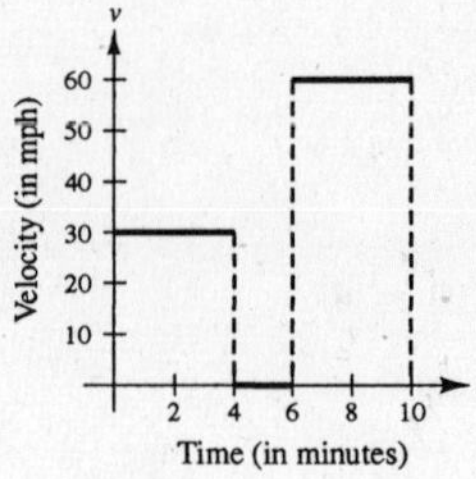

(The velocity has been converted to miles per hour.)

86.

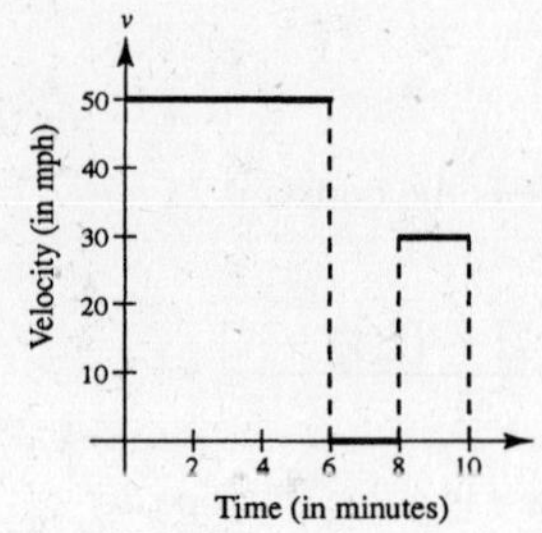

(The velocity has been converted to miles per hour.)

87. $v = 40 \text{ mph} = \frac{2}{3} \text{ mi/min}$

$\left(\frac{2}{3} \text{ mi/min}\right)(6 \text{ min}) = 4 \text{ mi}$

$v = 0 \text{ mph} = 0 \text{ mi/min}$

$(0 \text{ mi/min})(2 \text{ min}) = 0 \text{ mi}$

$v = 60 \text{ mph} = 1 \text{mi/min}$

$(1 \text{ mi/min})(2 \text{ min}) = 2 \text{ mi}$

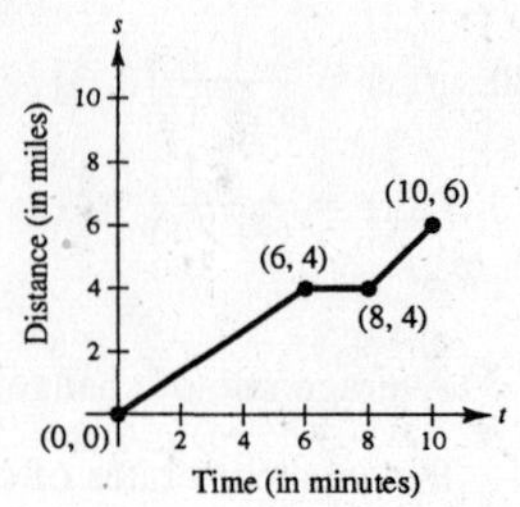

88.

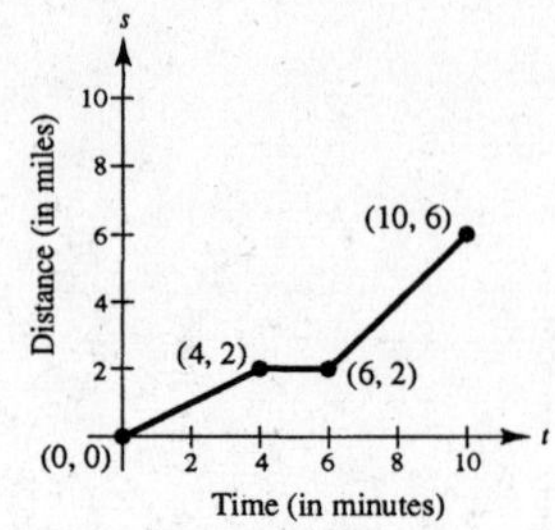

89. (a) $R(v) = 0.417v - 0.02$

(b) $B(v) = 0.00557v^2 + 0.0014v + 0.04$

(c) $T(v) = R(v) + B(v) = 0.00557v^2 + 0.418v + 0.02$

(d)

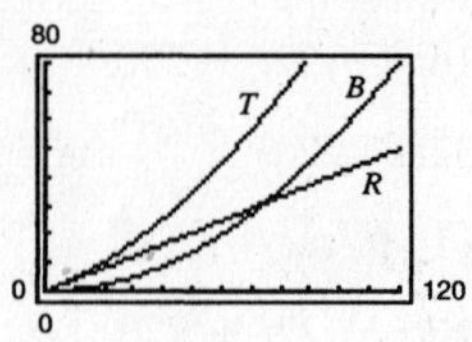

(e) $T'(v) = 0.01114v + 0.418$ (Answers will vary.)

$T'(40) = 0.01114(40) + 0.418 \approx 0.864$

$T'(80) = 0.01114(80) + 0.418 \approx 1.309$

$T'(100) = 0.01114(100) + 0.418 \approx 1.532$

(f) Stopping distances increase at an increasing rate.

90. $s(t) = -\frac{1}{2}at^2 + c$ and $s'(t) = -at.$

Average velocity:
$$\frac{s(t_0 + \Delta t) - s(t_0 - \Delta t)}{(t_0 + \Delta t) - (t_0 - \Delta t)} = \frac{[-(1/2)a(t_0 + \Delta t)^2 + c] - [-(1/2)a(t_0 - \Delta t)^2 + c)]}{2\,\Delta t}$$
$$= \frac{-(1/2)a(t_0^2 + 2t_0\,\Delta t + (\Delta t)^2) + (1/2)a(t_0^2 - 2t_0\,\Delta t + (\Delta t)^2)}{2\,\Delta t}$$
$$= \frac{-2at_0\,\Delta t}{2\,\Delta t}$$
$$= -at_0$$
$$= s'(t_0) \quad \text{Instantaneous velocity at } t = t_0$$

91. $A = s^2, \frac{dA}{ds} = 2s$

When $s = 4$ m,

$\frac{dA}{ds} = 8$ square meters per meter change in s.

92. $V = s^3, \frac{dV}{ds} = 3s^2$

When $s = 4$ cm, $\frac{dV}{ds} = 48 \text{ cm}^2$.

93. $$C = \frac{1{,}008{,}000}{Q} + 6.3Q$$

$$\frac{dC}{dQ} = -\frac{1{,}008{,}000}{Q^2} + 6.3$$

$C(351) - C(350) \approx 5083.095 - 5085 \approx -\$1.91$

When $Q = 350$, $\frac{dC}{dQ} \approx -\$1.93$.

94. $C =$ (gallons of fuel used)(cost per gallon)

$$= \left(\frac{15{,}000}{x}\right)(1.55) = \frac{23{,}250}{x}$$

$$\frac{dC}{dx} = -\frac{23{,}250}{x^2}$$

x	10	15	20	25	30	35	40
C	2325	1550	1162.5	930	775	664.3	581.3
$\frac{dC}{dx}$	-232.5	-103.3	-58.1	-37.2	-25.8	-19.0	-14.5

The driver who gets 15 miles per gallon would benefit more from a 1 mile per gallon increase in fuel efficiency. The rate of change is larger when $x = 15$.

95. (a) $f'(1.479)$ is the rate of change of the amount of gasoline sold when the price is \$1.479 per gallon.

(b) $f'(1.479)$ is usually negative. As prices go up, sales go down.

96. $$\frac{dT}{dt} = K(T - T_a)$$

97. $$f(x) = \begin{cases} ax^3, & x \le 2 \\ x^2 + b, & x > 2 \end{cases}$$

f must be continuous at $x = 2$ to be differentiable at $x = 2$.

$$\left.\begin{aligned} \lim_{x\to 2^-} f(x) &= \lim_{x\to 2^-} ax^3 = 8a \\ \lim_{x\to 2^+} f(x) &= \lim_{x\to 2^+} (x^2 + b) = 4 + b \end{aligned}\right\} \quad \begin{aligned} 8a &= 4 + b \\ 8a - 4 &= b \end{aligned}$$

$$f'(x) = \begin{cases} 3ax^2, & x < 2 \\ 2x, & x > 2 \end{cases}$$

For f to be differentiable at $x = 2$, the left derivative must equal the right derivative.

$$3a(2)^2 = 2(2)$$

$$12a = 4$$

$$a = \frac{1}{3}$$

$$b = 8a - 4 = -\frac{4}{3}$$

98. $y = \frac{1}{x}, \; x > 0$

$y' = -\frac{1}{x^2}$

At (a, b), the equation of the tangent line is

$y - \frac{1}{a} = -\frac{1}{a^2}(x - a)$ or $y = -\frac{x}{a^2} + \frac{2}{a}.$

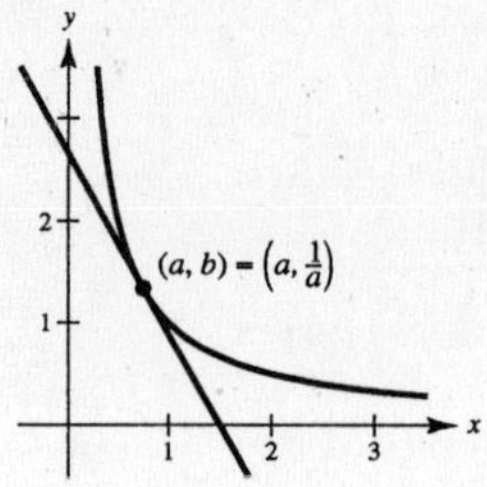

The x-intercept is $(2a, 0)$.

The y-intercept is $\left(0, \frac{2}{a}\right)$.

The area of the triangle is $A = \frac{1}{2}bh = \frac{1}{2}(2a)\left(\frac{2}{a}\right) = 2.$

99. $y = x^3 - 9x$

$y' = 3x^2 - 9$

Tangent lines through $(1, -9)$:

$$y + 9 = (3x^2 - 9)(x - 1)$$

$$(x^3 - 9x) + 9 = 3x^3 - 3x^2 - 9x + 9$$

$$0 = 2x^3 - 3x^2 = x^2(2x - 3)$$

$$x = 0 \text{ or } x = \tfrac{3}{2}$$

The points of tangency are $(0, 0)$ and $\left(\frac{3}{2}, -\frac{81}{8}\right)$. At $(0, 0)$ the slope is $y'(0) = -9$. At $\left(\frac{3}{2}, -\frac{81}{8}\right)$ the slope is $y'\left(\frac{3}{2}\right) = -\frac{9}{4}$.

Tangent lines:

$y - 0 = -9(x - 0)$ and $y + \frac{81}{8} = -\frac{9}{4}\left(x - \frac{3}{2}\right)$

$y = -9x \qquad y = -\frac{9}{4}x - \frac{27}{4}$

$9x + y = 0 \qquad 9x + 4y + 27 = 0$

100. $f(x) = x^5 + 3x^3 + 5x$

$f'(x) = 5x^4 + 9x^2 + 5$

Since $5x^4 + 9x^2 \geq 0$, $f'(x) \geq 5$. Thus, f does not have a tangent line with a slope of 3.

Section 4.3 The Product and Quotient Rules and Higher-Order Derivatives

1. $g(x) = (x^2 + 1)(x^2 - 2x)$

$g'(x) = (x^2 + 1)(2x - 2) + (x^2 - 2x)(2x)$

$= 2x^3 - 2x^2 + 2x - 2 + 2x^3 - 4x^2$

$= 4x^3 - 6x^2 + 2x - 2$

2. $f(x) = (6x + 5)(x^3 - 2)$

$f'(x) = (6x + 5)(3x^2) + (x^3 - 2)(6)$

$= 18x^3 + 15x^2 + 6x^3 - 12$

$= 24x^3 + 15x^2 - 12$

3. $h(t) = \sqrt[3]{t}(t^2 + 4) = t^{1/3}(t^2 + 4)$

$h'(t) = t^{1/3}(2t) + (t^2 + 4)\frac{1}{3}t^{-2/3}$

$= 2t^{4/3} + \frac{t^2 + 4}{3t^{2/3}}$

$= \frac{7t^2 + 4}{3t^{2/3}}$

4. $g(s) = \sqrt{s}(4 - s^2) = s^{1/2}(4 - s^2)$

$g'(s) = s^{1/2}(-2s) + (4 - s^2)\frac{1}{2}s^{-1/2} = -2s^{3/2} + \frac{4 - s^2}{2s^{1/2}}$

$= \frac{4 - 5s^2}{2s^{1/2}}$

5. $g(t) = (2t^2 - 3)(4 - t^2 - t^4)$

$$g'(t) = (2t^2 - 3)(-2t - 4t^3) + (4 - t^2 - t^4)(4t)$$
$$= -4t^3 + 6t - 8t^5 + 12t^3 + 16t - 4t^3 - 4t^5$$
$$= -2t(6t^4 - 2t^2 - 11)$$

6. $h(t) = (t^5 - 1)(4t^2 - 7t - 3)$

$$h'(t) = (t^5 - 1)(8t - 7) + (4t^2 - 7t - 3)(5t^4)$$
$$= 28t^6 - 42t^5 - 15t^4 - 8t + 7$$

7. $f(x) = \dfrac{x}{x^2 + 1}$

$$f'(x) = \frac{(x^2 + 1)(1) - x(2x)}{(x^2 + 1)^2} = \frac{1 - x^2}{(x^2 + 1)^2}$$

8. $g(t) = \dfrac{t^2 + 2}{2t - 7}$

$$g'(t) = \frac{(2t - 7)(2t) - (t^2 + 2)(2)}{(2t - 7)^2} = \frac{2t^2 - 14t - 4}{(2t - 7)^2}$$

9. $h(x) = \dfrac{\sqrt[3]{x}}{x^3 + 1} = \dfrac{x^{1/3}}{x^3 + 1}$

$$h'(x) = \frac{(x^3 + 1)\frac{1}{3}x^{-2/3} - x^{1/3}(3x^2)}{(x^3 + 1)^2}$$
$$= \frac{(x^3 + 1) - x(9x^2)}{3x^{2/3}(x^3 + 1)^2}$$
$$= \frac{1 - 8x^3}{3x^{2/3}(x^3 + 1)^2}$$

10. $h(s) = \dfrac{s}{\sqrt{s} - 1}$

$$h'(s) = \frac{(\sqrt{s} - 1)(1) - s\left(\frac{1}{2}s^{-1/2}\right)}{(\sqrt{s} - 1)^2}$$
$$= \frac{\sqrt{s} - 1 - \frac{1}{2}\sqrt{s}}{(\sqrt{s} - 1)^2} = \frac{\sqrt{s} - 2}{2(\sqrt{s} - 1)^2}$$

11. $f(x) = \dfrac{x^3 + 3x + 2}{x^2 - 1}$

$$f'(x) = \frac{(x^2 - 1)(3x^2 + 3) - (x^3 + 3x + 2)(2x)}{(x^2 - 1)^2}$$
$$= \frac{x^4 - 6x^2 - 4x - 3}{(x^2 - 1)^2}$$

12. $g(x) = \dfrac{3 - 2x - x^2}{x^2 - 1}$

$$g'(x) = \frac{(x^2 - 1)(-2 - 2x) - (3 - 2x - x^2)(2x)}{(x^2 - 1)^2}$$
$$= \frac{-2x^2 + 2 - 2x^3 + 2x - 6x + 4x^2 + 2x^3}{(x^2 - 1)^2}$$
$$= \frac{2x^2 - 4x + 2}{(x^2 - 1)^2}$$
$$= \frac{2(x - 1)^2}{(x - 1)^2(x + 1)^2}$$
$$= \frac{2}{(x + 1)^2}$$

Note: $g(x) = \dfrac{3 - 2x - x^2}{(x^2 - 1)} = \dfrac{-(x - 1)(x + 3)}{(x - 1)(x + 1)}$

$$= -\frac{(x + 3)}{x + 1}, x \neq 1$$

13. $f(x) = \dfrac{5}{x^2}(x + 3) = \dfrac{5}{x} + \dfrac{15}{x^2} = 5x^{-1} + 15x^{-2}$

$$f'(x) = -\frac{5}{x^2} - \frac{30}{x^3} = -\frac{5(x + 6)}{x^3}$$

$$f'(1) = -5 - 30 = -35$$

14. $f(x) = \dfrac{1}{7}(5 - 6x^2)$

$$f'(x) = \frac{1}{7}(-12x) = \frac{-12}{7}x$$

$$f'(1) = \frac{-12}{7}(1) = \frac{-12}{7}$$

15. $f(x) = \dfrac{x^2 - 4}{x - 3}$

$f'(x) = \dfrac{(x-3)(2x) - (x^2-4)(1)}{(x-3)^2} = \dfrac{2x^2 - 6x - x^2 + 4}{(x-3)^2}$

$= \dfrac{x^2 - 6x + 4}{(x-3)^2}$

$f'(1) = \dfrac{1 - 6 + 4}{(1-3)^2} = -\dfrac{1}{4}$

16. $f(x) = (x^2 - 2x + 1)(x^3 - 1)$

$f'(x) = (x^2 - 2x + 1)(3x^2) + (x^3 - 1)(2x - 2)$

$= 3x^2(x-1)^2 + 2(x-1)^2(x^2 + x + 1)$

$= (x-1)^2(5x^2 + 2x + 2)$

$f'(1) = 0$

17. $f(x) = (x^3 - 3x)(2x^2 + 3x + 5)$

$f'(x) = (x^3 - 3x)(4x + 3) + (2x^2 + 3x + 5)(3x^2 - 3)$

$= 10x^4 + 12x^3 - 3x^2 - 18x - 15$

$f'(0) = -15$

18. $f(x) = (x - 1)(x^2 - 3x + 2)$

$f'(x) = (x-1)(2x-3) + (x^2 - 3x + 2)(1)$

$= 2x^2 - 5x + 3 + x^2 - 3x + 2$

$= 3x^2 - 8x + 5$

$f'(0) = 5$

19. $f(x) = (x^5 - 3x)\left(\dfrac{1}{x^2}\right) = x^3 - \dfrac{3}{x}$

$f'(x) = 3x^2 + \dfrac{3}{x^2} = \dfrac{3(x^4 + 1)}{x^2}$

$f'(-1) = 3 + 3 = 6$

20. $f(x) = \dfrac{x + 1}{x - 1}$

$f'(x) = \dfrac{(x-1)(1) - (x+1)(1)}{(x-1)^2}$

$= \dfrac{x - 1 - x - 1}{(x-1)^2}$

$= -\dfrac{2}{(x-1)^2}$

$f'(2) = -\dfrac{2}{(2-1)^2} = -2$

	Function	Rewrite	Derivative	Simplify
21.	$y = \dfrac{x^2 + 2x}{3}$	$y = \dfrac{1}{3}x^2 + \dfrac{2}{3}x$	$y' = \dfrac{2}{3}x + \dfrac{2}{3}$	$y' = \dfrac{2x + 2}{3}$
22.	$y = \dfrac{7}{3x^3}$	$y = \dfrac{7}{3}x^{-3}$	$y' = -7x^{-4}$	$y' = -\dfrac{7}{x^4}$
23.	$y = \dfrac{4x^{3/2}}{x}$	$y = 4\sqrt{x}, x > 0$	$y' = 2x^{-1/2}$	$y' = \dfrac{2}{\sqrt{x}}$
24.	$y = \dfrac{3x^2 - 5}{7}$	$y = \dfrac{3}{7}x^2 - \dfrac{5}{7}$	$y' = \dfrac{6}{7}x$	$y' = \dfrac{6x}{7}$

25. $f(x) = \dfrac{3 - 2x - x^2}{x^2 - 1}$

$f'(x) = \dfrac{(x^2 - 1)(-2 - 2x) - (3 - 2x - x^2)(2x)}{(x^2 - 1)^2}$

$= \dfrac{2x^2 - 4x + 2}{(x^2 - 1)^2} = \dfrac{2(x-1)^2}{(x^2-1)^2}$

$= \dfrac{2}{(x+1)^2}, x \neq 1$

26. $f(x) = \dfrac{x^3 + 3x + 2}{x^2 - 1}$

$f'(x) = \dfrac{(x^2 - 1)(3x^2 + 3) - (x^3 + 3x + 2)(2x)}{(x^2 - 1)^2}$

$= \dfrac{x^4 - 6x^2 - 4x - 3}{(x^2 - 1)^2}$

27. $f(x) = x\left(1 - \frac{4}{x+3}\right) = x - \frac{4x}{x+3}$

$f'(x) = 1 - \frac{(x+3)4 - 4x(1)}{(x+3)^2} = \frac{(x^2+6x+9) - 12}{(x+3)^2}$

$= \frac{x^2+6x-3}{(x+3)^2}$

28. $f(x) = x^4\left[1 - \frac{2}{x+1}\right] = x^4\left[\frac{x-1}{x+1}\right]$

$f'(x) = x^4\left[\frac{(x+1)-(x-1)}{(x+1)^2}\right] + \left[\frac{x-1}{x+1}\right](4x^3)$

$= 2x^3\left[\frac{2x^2+x-2}{(x+1)^2}\right]$

29. $f(x) = \frac{2x+5}{\sqrt{x}} = 2x^{1/2} + 5x^{-1/2}$

$f'(x) = x^{-1/2} - \frac{5}{2}x^{-3/2} = x^{-3/2}\left[x - \frac{5}{2}\right]$

$= \frac{2x-5}{2x\sqrt{x}} = \frac{2x-5}{2x^{3/2}}$

30. $f(x) = \sqrt[3]{x}\left(\sqrt{x}+3\right) = x^{1/3}(x^{1/2}+3)$

$f'(x) = x^{1/3}\left(\frac{1}{2}x^{-1/2}\right) + (x^{1/2}+3)\left(\frac{1}{3}x^{-2/3}\right)$

$= \frac{5}{6}x^{-1/6} + x^{-2/3}$

$= \frac{5}{6x^{1/6}} + \frac{1}{x^{2/3}}$

Alternate solution:

$f(x) = \sqrt[3]{x}\left(\sqrt{x}+3\right)$

$= x^{5/6} + 3x^{1/3}$

$f'(x) = \frac{5}{6}x^{-1/6} + x^{-2/3}$

$= \frac{5}{6x^{1/6}} + \frac{1}{x^{2/3}}$

31. $h(s) = (s^3-2)^2 = s^6 - 4s^3 + 4$

$h'(s) = 6s^5 - 12s^2 = 6s^2(s^3-2)$

32. $h(x) = (x^2-1)^2 = x^4 - 2x^2 + 1$

$h'(x) = 4x^3 - 4x = 4x(x^2-1)$

33. $f(x) = \frac{2 - \frac{1}{x}}{x-3} = \frac{2x-1}{x(x-3)} = \frac{2x-1}{x^2-3x}$

$f'(x) = \frac{(x^2-3x)2 - (2x-1)(2x-3)}{(x^2-3x)^2} = \frac{2x^2-6x-4x^2+8x-3}{(x^2-3x)^2}$

$= \frac{-2x^2+2x-3}{(x^2-3x)^2} = -\frac{2x^2-2x+3}{x^2(x-3)^2}$

34. $g(x) = x^2\left(\frac{2}{x} - \frac{1}{x+1}\right) = 2x - \frac{x^2}{x+1}$

$g'(x) = 2 - \frac{(x+1)2x - x^2(1)}{(x+1)^2} = \frac{2(x^2+2x+1) - x^2 - 2x}{(x+1)^2} = \frac{x^2+2x+2}{(x+1)^2}$

35. $f(x) = (3x^3+4x)(x-5)(x+1)$

$f'(x) = (9x^2+4)(x-5)(x+1) + (3x^3+4x)(1)(x+1) + (3x^3+4x)(x-5)(1)$

$= (9x^2+4)(x^2-4x-5) + 3x^4 + 3x^3 + 4x^2 + 4x + 3x^4 - 15x^3 + 4x^2 - 20x$

$= 9x^4 - 36x^3 - 41x^2 - 16x - 20 + 6x^4 - 12x^3 + 8x^2 - 16x$

$= 15x^4 - 48x^3 - 33x^2 - 32x - 20$

36. $f(x) = (x^2 - x)(x^2 + 1)(x^2 + x + 1)$

$$f'(x) = (2x - 1)(x^2 + 1)(x^2 + x + 1) + (x^2 - x)(2x)(x^2 + x + 1) + (x^2 - x)(x^2 + 1)(2x + 1)$$
$$= (2x - 1)(x^4 + x^3 + 2x^2 + x + 1) + (x^2 - x)(2x^3 + 2x^2 + 2x) + (x^2 - x)(2x^3 + x^2 + 2x + 1)$$
$$= 2x^5 + x^4 + 3x^3 + x - 1 + 2x^5 - 2x^2 + 2x^5 - x^4 + x^3 - x^2 - x$$
$$= 6x^5 + 4x^3 - 3x^2 - 1$$

37. $f(x) = \dfrac{x^2 + c^2}{x^2 - c^2}$

$$f'(x) = \frac{(x^2 - c^2)(2x) - (x^2 + c^2)(2x)}{(x^2 - c^2)^2}$$
$$= \frac{-4xc^2}{(x^2 - c^2)^2}$$

38. $f(x) = \dfrac{c^2 - x^2}{c^2 + x^2}$

$$f'(x) = \frac{(c^2 + x^2)(-2x) - (c^2 - x^2)(2x)}{(c^2 + x^2)^2}$$
$$= \frac{-4xc^2}{(c^2 + x^2)^2}$$

39. $g(x) = \left(\dfrac{x + 1}{x + 2}\right)(2x - 5)$

$g'(x) = \dfrac{2x^2 + 8x - 1}{(x + 2)^2}$ (form of answer may vary)

40. $f(x) = \left(\dfrac{x^2 - x - 3}{x^2 + 1}\right)(x^2 + x + 1)$

$f'(x) = 2\dfrac{x^5 + 2x^3 + 2x^2 - 2}{(x^2 + 1)^2}$ (form of answer may vary)

41. (a) $f(x) = (x^3 - 3x + 1)(x + 2),\ (1, -3)$

$$f'(x) = (x^3 - 3x + 1)(1) + (x + 2)(3x^2 - 3)$$
$$= 4x^3 + 6x^2 - 6x - 5$$

$f'(1) = -1 =$ slope at $(1, -3)$.

Tangent line: $y + 3 = -1(x - 1) \Rightarrow y = -x - 2$

(b)

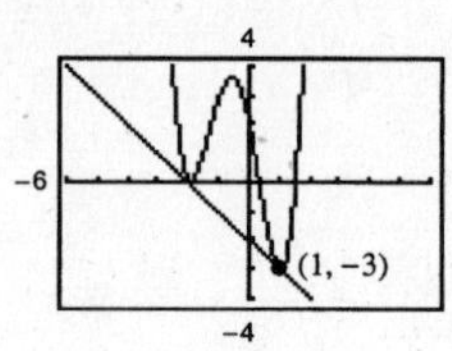

42. (a) $f(x) = (x - 1)(x^2 - 2),\ (0, 2)$

$f'(x) = (x - 1)(2x) + (x^2 - 2)(1) = 3x^2 - 2x - 2$

$f'(0) = -2 =$ slope at $(0, 2)$.

Tangent line: $y - 2 = -2x \Rightarrow y = -2x + 2$

(b)

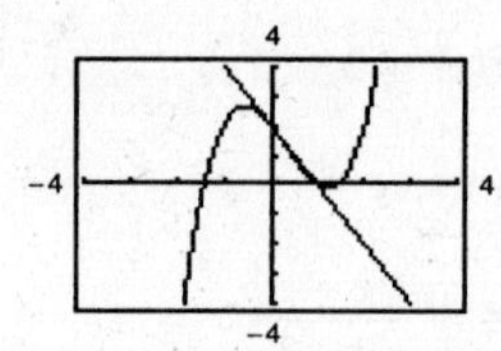

43. (a) $f(x) = \dfrac{x}{x - 1},\ (2, 2)$

$$f'(x) = \frac{(x - 1)(1) - x(1)}{(x - 1)^2} = \frac{-1}{(x - 1)^2}$$

$f'(2) = \dfrac{-1}{(2 - 1)^2} = -1 =$ slope at $(2, 2)$.

Tangent line: $y - 2 = -1(x - 2) \Rightarrow y = -x + 4$

(b)

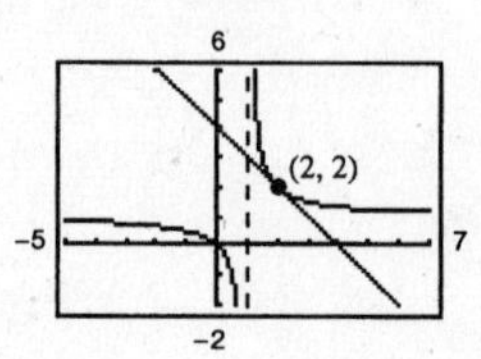

44. (a) $f(x) = \dfrac{x - 1}{x + 1},\ \left(2, \dfrac{1}{3}\right)$

$$f'(x) = \frac{(x + 1)(1) - (x - 1)(1)}{(x + 1)^2} = \frac{2}{(x + 1)^2}$$

$f'(2) = \dfrac{2}{9} =$ slope at $\left(2, \dfrac{1}{3}\right)$.

Tangent line: $y - \dfrac{1}{3} = \dfrac{2}{9}(x - 2) \Rightarrow y = \dfrac{2}{9}x - \dfrac{1}{9}$

(b)

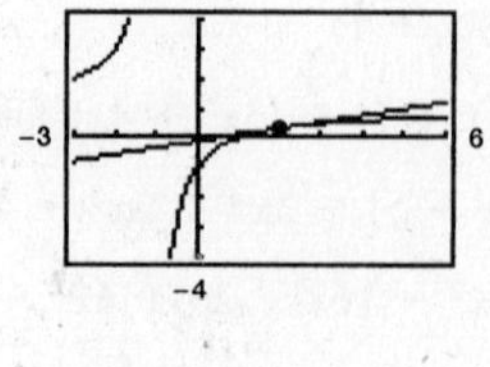

45. $f(x) = \dfrac{x^2}{x-1}$

$$f'(x) = \frac{(x-1)(2x) - x^2(1)}{(x-1)^2}$$

$$= \frac{x^2 - 2x}{(x-1)^2} = \frac{x(x-2)}{(x-1)^2}$$

$f'(x) = 0$ when $x = 0$ or $x = 2$.

Horizontal tangents are at $(0, 0)$ and $(2, 4)$.

46. $f(x) = \dfrac{x^2}{x^2+1}$

$$f'(x) = \frac{(x^2+1)(2x) - (x^2)(2x)}{(x^2+1)^2} = \frac{2x}{(x^2+1)^2}$$

$f'(x) = 0$ when $x = 0$.

Horizontal tangent is at $(0, 0)$.

47. $f(x) = \dfrac{8}{x^2+4}$; $(2, 1)$

$$f'(x) = \frac{(x^2+4)(0) - 8(2x)}{(x^2+4)^2} = \frac{-16x}{(x^2+4)^2}$$

$$f'(2) = \frac{-16(2)}{(4+4)^2} = -\frac{1}{2}$$

$$y - 1 = -\frac{1}{2}(x-2)$$

$$y = -\frac{1}{2}x + 2$$

$2y + x - 4 = 0$

48. $f(x) = \dfrac{27}{x^2+9}$; $\left(-3, \dfrac{3}{2}\right)$

$$f'(x) = \frac{(x^2+9)(0) - 27(2x)}{(x^2+9)^2} = \frac{-54x}{(x^2+9)^2}$$

$$f'(-3) = \frac{-54(-3)}{(9+9)^2} = \frac{1}{2}$$

$$y - \frac{3}{2} = \frac{1}{2}(x+3)$$

$$y = \frac{1}{2}x + 3$$

$2y - x - 6 = 0$

49. $f(x) = \dfrac{16x}{x^2+16}$; $\left(-2, -\dfrac{8}{5}\right)$

$$f'(x) = \frac{(x^2+16)(16) - 16x(2x)}{(x^2+16)^2} = \frac{256 - 16x^2}{(x^2+16)^2}$$

$$f'(-2) = \frac{256 - 16(4)}{20^2} = \frac{12}{25}$$

$$y + \frac{8}{5} = \frac{12}{25}(x+2)$$

$$y = \frac{12}{25}x - \frac{16}{25}$$

$25y - 12x + 16 = 0$

50. $f(x) = \dfrac{4x}{x^2+6}$; $\left(2, \dfrac{4}{5}\right)$

$$f'(x) = \frac{(x^2+6)(4) - 4x(2x)}{(x^2+6)^2} = \frac{24 - 4x^2}{(x^2+6)^2}$$

$$f'(2) = \frac{24 - 16}{10^2} = \frac{2}{25}$$

$$y - \frac{4}{5} = \frac{2}{25}(x-2)$$

$$y = \frac{2}{25}x + \frac{16}{25}$$

$25y - 2x - 16 = 0$

51. $f(x) = \dfrac{x+1}{x-1}$

$$f'(x) = \frac{(x-1) - (x+1)}{(x-1)^2} = \frac{-2}{(x-1)^2}$$

$2y + x = 6 \Rightarrow y = -\frac{1}{2}x + 3$; Slope: $-\frac{1}{2}$

$$\frac{-2}{(x-1)^2} = -\frac{1}{2}$$

$(x-1)^2 = 4$

$x - 1 = \pm 2$

$x = -1, 3;\quad f(-1) = 0,\ f(3) = 2$

$$y - 0 = -\frac{1}{2}(x+1) \Rightarrow y = -\frac{1}{2}x - \frac{1}{2}$$

$$y - 2 = -\frac{1}{2}(x-3) \Rightarrow y = -\frac{1}{2}x + \frac{7}{2}$$

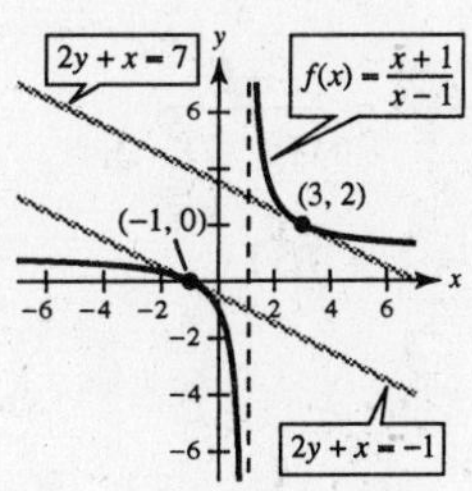

52. $f(x) = \dfrac{x}{x-1}$

$$f'(x) = \frac{(x-1) - x}{(x-1)^2} = \frac{-1}{(x-1)^2}$$

Let $(x, y) = (x, x/(x-1))$ be a point of tangency on the graph of f.

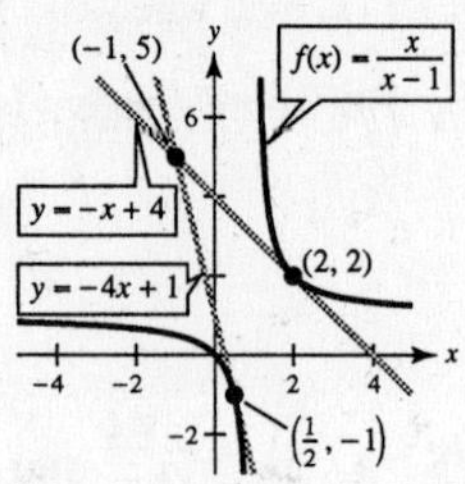

$$\frac{5 - (x/(x-1))}{-1 - x} = \frac{-1}{(x-1)^2}$$

$$\frac{4x - 5}{(x-1)(x+1)} = \frac{1}{(x-1)^2}$$

$$(4x - 5)(x - 1) = x + 1$$

$$4x^2 - 10x + 4 = 0$$

$$(x - 2)(2x - 1) = 0 \Rightarrow x = \frac{1}{2}, 2$$

$$f\left(\frac{1}{2}\right) = -1, f(2) = 2; f'\left(\frac{1}{2}\right) = -4, f'(2) = -1$$

Two tangent lines:

$$y + 1 = -4\left(x - \frac{1}{2}\right) \Rightarrow y = -4x + 1$$

$$y - 2 = -1(x - 2) \Rightarrow y = -x + 4$$

53. (a) $p'(x) = f'(x)g(x) + f(x)g'(x)$

$$p'(1) = f'(1)g(1) + f(1)g'(1) = 1(4) + 6\left(-\frac{1}{2}\right) = 1$$

(b) $q'(x) = \dfrac{g(x)f'(x) - f(x)g'(x)}{g(x)^2}$

$$q'(4) = \frac{3(-1) - 7(0)}{3^2} = -\frac{1}{3}$$

54. (a) $p'(x) = f'(x)g(x) + f(x)g'(x)$

$$p'(4) = \frac{1}{2}(8) + 1(0) = 4$$

(b) $q'(x) = \dfrac{g(x)f'(x) - f(x)g'(x)}{g(x)^2}$

$$q'(7) = \frac{4(2) - 4(-1)}{4^2} = \frac{12}{16} = \frac{3}{4}$$

55. $f'(x) = \dfrac{(x+2)3 - 3x(1)}{(x+2)^2} = \dfrac{6}{(x+2)^2}$

$$g'(x) = \frac{(x+2)5 - (5x+4)(1)}{(x+2)^2} = \frac{6}{(x+2)^2}$$

$$g(x) = \frac{5x+4}{(x+2)} = \frac{3x}{(x+2)} + \frac{2x+4}{(x+2)} = f(x) + 2$$

f and g differ by a constant.

56. $f(x) = \dfrac{5}{x-3}$, $f'(x) = \dfrac{(x-3)(0) - 5(1)}{(x-3)^2} = \dfrac{-5}{(x-3)^2}$

$$g(x) = \frac{x+2}{x-3}, g'(x) = \frac{(x-3)(1) - (x+2)(1)}{(x-3)^2} = \frac{-5}{(x-3)^2}$$

$$g(x) = \frac{x+2}{x-3} = \frac{x-3+5}{x-3} = 1 + \frac{5}{x-3} = 1 + f(x)$$

57. Area $= A(t) = (2t + 1)\sqrt{t} = 2t^{3/2} + t^{1/2}$

$$A'(t) = 2\left(\frac{3}{2}t^{1/2}\right) + \frac{1}{2}t^{-1/2}$$

$$= 3t^{1/2} + \frac{1}{2}t^{-1/2}$$

$$= \frac{6t + 1}{2\sqrt{t}} \text{ cm}^2\text{/sec}$$

58. $V = \pi r^2 h = \pi(t + 2)\left(\frac{1}{2}\sqrt{t}\right)$

$$= \frac{1}{2}(t^{3/2} + 2t^{1/2})\pi$$

$$V'(t) = \frac{1}{2}\left(\frac{3}{2}t^{1/2} + t^{-1/2}\right)\pi = \frac{3t + 2}{4t^{1/2}}\pi \text{ cubic inches/sec}$$

59. $C = 100\left(\frac{200}{x^2} + \frac{x}{x+30}\right),\ 1 \le x$

$\frac{dC}{dx} = 100\left(-\frac{400}{x^3} + \frac{30}{(x+30)^2}\right)$

(a) When $x = 10$: $\frac{dC}{dx} = -\$38.13$.

(b) When $x = 15$: $\frac{dC}{dx} = -\$10.37$.

(c) When $x = 20$: $\frac{dC}{dx} = -\$3.80$.

As the order size increases, the cost per item decreases.

60. $P = \frac{k}{V}$

$\frac{dP}{dV} = -\frac{k}{V^2}$

61. (a) $n(t) = -3.5806t^3 + 82.577t^2 - 603.60t + 1667.5$

$v(t) = -0.1361t^3 + 3.165t^2 - 23.02t + 59.8$

(b)

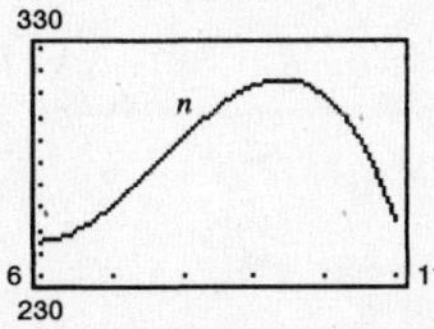

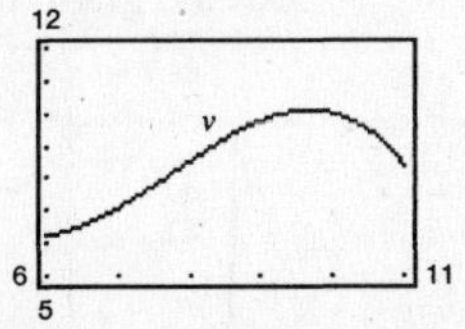

(c) $A(t) = \frac{v(t)}{n(t)} = \frac{-0.1361t^3 + 3.165t^2 - 23.02t + 59.8}{-3.5806t^3 + 82.577t^2 - 603.60t + 1667.5}$

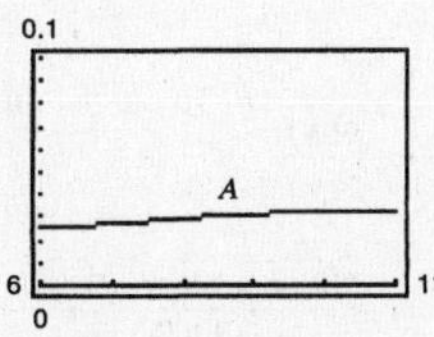

$A(t)$ represents the average retail value (in billions of dollars) per 1000 motor homes.

(d) $A'(t)$ is the rate of change of the average retail value (in billions of dollars) per 1000 motor homes, per year.

62. $P(t) = 500\left[1 + \frac{4t}{50+t^2}\right]$

$P'(t) = 500\left[\frac{(50+t^2)(4) - (4t)(2t)}{(50+t^2)^2}\right]$

$= 500\left[\frac{200 - 4t^2}{(50+t^2)^2}\right]$

$= 2000\left[\frac{50 - t^2}{(50+t^2)^2}\right]$

$P'(2) \approx 31.55$ bacteria per hour

63. $f(x) = 4x^{3/2}$

$f'(x) = 6x^{1/2}$

$f''(x) = 3x^{-1/2} = \frac{3}{\sqrt{x}}$

64. $f(x) = x + \frac{32}{x^2}$

$f'(x) = 1 - \frac{64}{x^3}$

$f''(x) = \frac{192}{x^4}$

65. $f(x) = \frac{x}{x-1}$

$f'(x) = \frac{(x-1)(1) - x(1)}{(x-1)^2} = \frac{-1}{(x-1)^2}$

$f''(x) = \frac{2}{(x-1)^3}$

66. $f(x) = \dfrac{x^2 + 2x - 1}{x} = x + 2 - \dfrac{1}{x}$

$f'(x) = 1 + \dfrac{1}{x^2}$

$f''(x) = -\dfrac{2}{x^3}$

67. $f'(x) = x^2$

$f''(x) = 2x$

68. $f''(x) = x^3$

$f'''(x) = 3x^2$

69. $f''(x) = 2 - 2x^{-1}$

$f'''(x) = 2x^{-2} = \dfrac{2}{x^2}$

70. $f'''(x) = 2\sqrt{x}$

$f^{(4)}(x) = \dfrac{1}{2}(2)x^{-1/2}$

$= \dfrac{1}{\sqrt{x}}$

71. $f(x) = 2x^2 - 2$

$f'(x) = 4x$

$f''(x) = 4$

72. $f^{(4)}(x) = 2x + 1$

$f^{(5)}(x) = 2$

$f^{(6)}(x) = 0$

73.

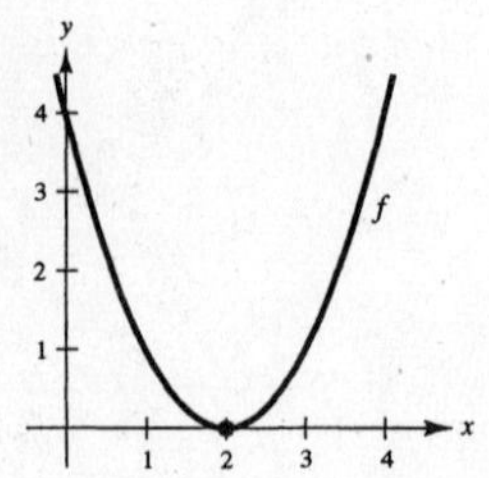

$f(2) = 0$

One such function is $f(x) = (x - 2)^2$.

74. The graph of a differentiable function f such that $f > 0$ and $f' < 0$ for all real numbers x would in general look like the graph below.

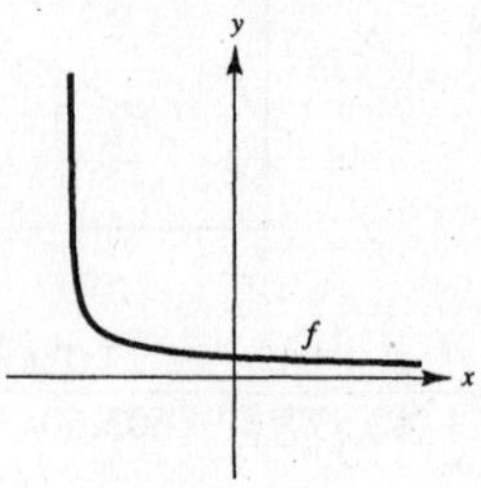

75. $f(x) = 2g(x) + h(x)$

$f'(x) = 2g'(x) + h'(x)$

$f'(2) = 2g'(2) + h'(2)$

$= 2(-2) + 4$

$= 0$

76. $f(x) = 4 - h(x)$

$f'(x) = -h'(x)$

$f'(2) = -h'(2) = -4$

77. $f(x) = \dfrac{g(x)}{h(x)}$

$f'(x) = \dfrac{h(x)g'(x) - g(x)h'(x)}{[h(x)]^2}$

$f'(2) = \dfrac{h(2)g'(2) - g(2)h'(2)}{[h(2)]^2}$

$= \dfrac{(-1)(-2) - (3)(4)}{(-1)^2}$

$= -10$

78. $f(x) = g(x)h(x)$

$f'(x) = g(x)h'(x) + h(x)g'(x)$

$f'(2) = g(2)h'(2) + h(2)g'(2)$

$= (3)(4) + (-1)(-2)$

$= 14$

79.

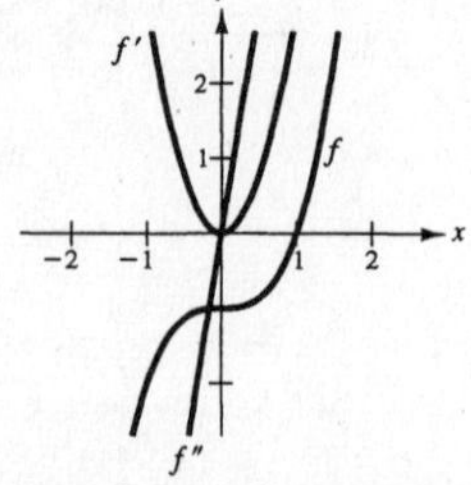

It appears that f is cubic; so f' would be quadratic and f'' would be linear.

80.

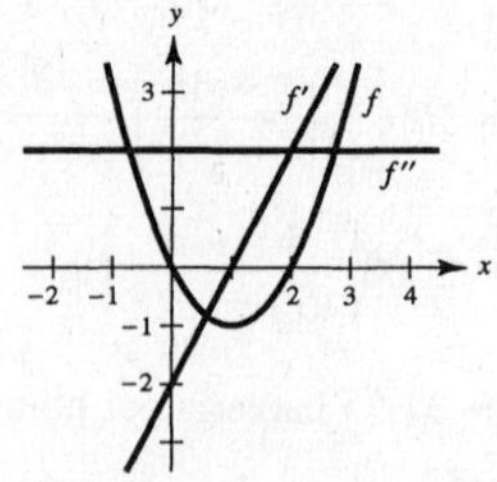

It appears that f is quadratic; so f' would be linear and f'' would be constant.

81.

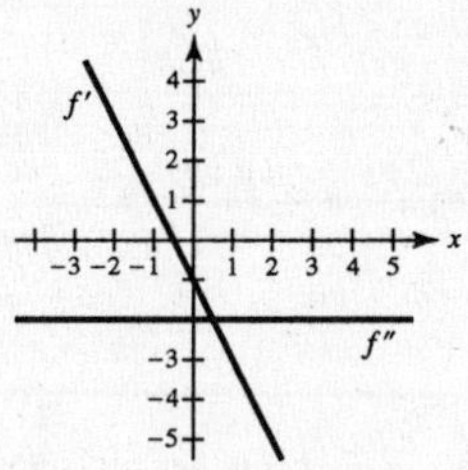

82.

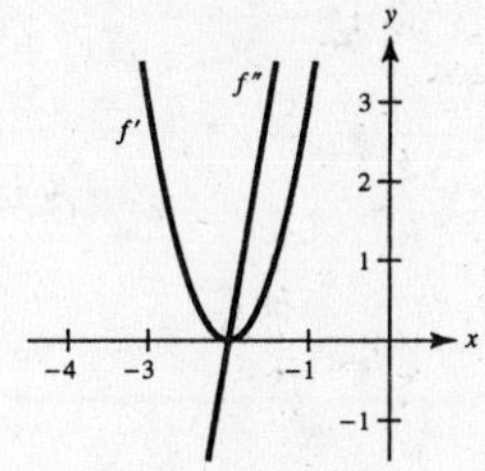

83.

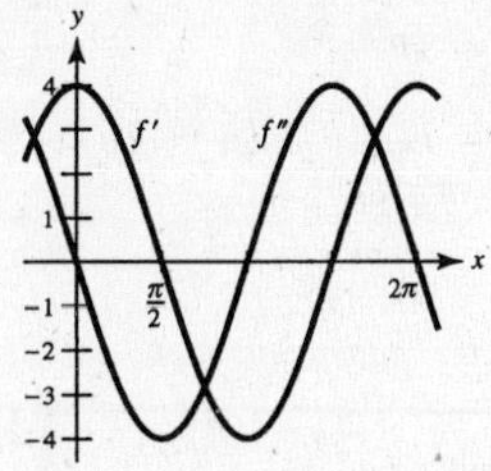

84.

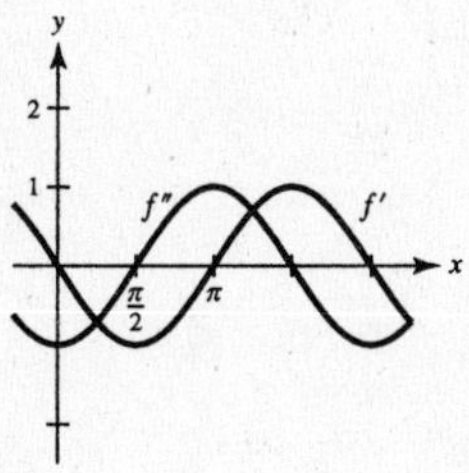

85. $y = \frac{1}{x}, y' = -\frac{1}{x^2}, y'' = \frac{2}{x^3}$

$$x^3y'' + 2x^2y' = x^3\left[\frac{2}{x^3}\right] + 2x^2\left[-\frac{1}{x^2}\right] = 2 - 2 = 0$$

86.

$$y = 2x^3 - 6x + 10$$
$$y' = 6x^2 - 6$$
$$y'' = 12x$$
$$y''' = 12$$
$$-y''' - xy'' - 2y' = -12 - x(12x) - 2(6x^2 - 6) = -24x^2$$

87. $v(t) = 36 - t^2, \ 0 \le t \le 6$

$a(t) = -2t$

$v(3) = 27 \text{ m/sec}$

$a(3) = -6 \text{ m/sec}^2$

The speed of the object is decreasing.

88. $s(t) = -8.25t^2 + 66t$

$v(t) = -16.50t + 66$

$a(t) = -16.50$

t(sec)	0	1	2	3	4
$s(t)$ (ft)	0	57.75	99	123.75	132
$v(t) = s'(t)$ (ft/sec)	66	49.5	33	16.5	0
$a(t) = v'(t)$ (ft/sec²)	−16.5	−16.5	−16.5	−16.5	−16.5

Average velocity on:

$[0, 1]$ is $\frac{57.75 - 0}{1 - 0} = 57.75$.

$[1, 2]$ is $\frac{99 - 57.75}{2 - 1} = 41.25$.

$[2, 3]$ is $\frac{123.75 - 99}{3 - 2} = 24.75$.

$[3, 4]$ is $\frac{132 - 123.75}{4 - 3} = 8.25$.

89. $v(t) = \frac{100t}{2t + 15}$

$$a(t) = \frac{(2t + 15)(100) - (100t)(2)}{(2t + 15)^2}$$
$$= \frac{1500}{(2t + 15)^2}$$

(a) $a(5) = \frac{1500}{[2(5) + 15]^2} = 2.4 \text{ ft/sec}^2$

(b) $a(10) = \frac{1500}{[2(10) + 15]^2} \approx 1.2 \text{ ft/sec}^2$

(c) $a(20) = \frac{1500}{[2(20) + 15]^2} \approx 0.5 \text{ ft/sec}^2$

90. (a)

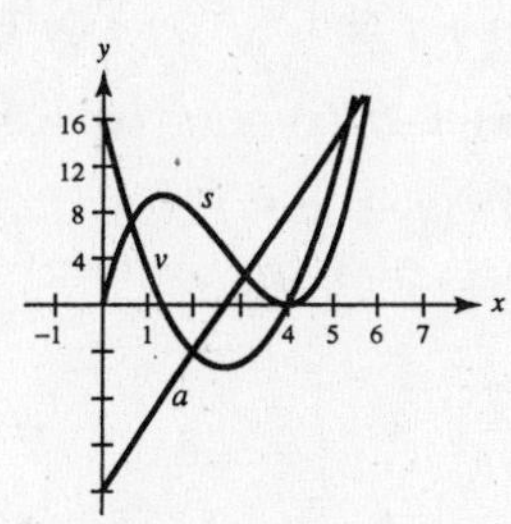

s position function

v velocity function

a acceleration function

(b) The particle speeds up (accelerates) when $a > 0$ and slows down when $a < 0$.

Answers will vary.

91. $f(x) = x^n$

$f^{(n)}(x) = n(n-1)(n-2)\cdots(2)(1) = n!$

92. $f(x) = \dfrac{1}{x}$

$$f^{(n)}(x) = \frac{(-1)^n(n)(n-1)(n-2)\cdots(2)(1)}{x^{n+1}}$$

$$= \frac{(-1)^n n!}{x^{n+1}}$$

Note: $n! = n(n-1)\cdots 3 \cdot 2 \cdot 1$ (read "*n* factorial.")

93. $f(x) = g(x)h(x)$

(a) $f'(x) = g(x)h'(x) + h(x)g'(x)$

$$f''(x) = g(x)h''(x) + g'(x)h'(x) + h(x)g''(x) + h'(x)g'(x)$$

$$= g(x)h''(x) + 2g'(x)h'(x) + h(x)g''(x)$$

$$f'''(x) = g(x)h'''(x) + g'(x)h''(x) + 2g'(x)h''(x) + 2g''(x)h'(x) + h(x)g'''(x) + h'(x)g''(x)$$

$$= g(x)h'''(x) + 3g'(x)h''(x) + 3g''(x)h'(x) + g'''(x)h(x)$$

$$f^{(4)}(x) = g(x)h^{(4)}(x) + g'(x)h'''(x) + 3g'(x)h'''(x) + 3g''(x)h''(x) + 3g''(x)h''(x) + 3g'''(x)h'(x)$$
$$+ g'''(x)h'(x) + g^{(4)}(x)h(x)$$

$$= g(x)h^{(4)}(x) + 4g'(x)h'''(x) + 6g''(x)h''(x) + 4g'''(x)h'(x) + g^{(4)}(x)h(x)$$

(b) $$f^{(n)}(x) = g(x)h^{(n)}(x) + \frac{n(n-1)(n-2)\cdots(2)(1)}{1[(n-1)(n-2)\cdots(2)(1)]}g'(x)h^{(n-1)}(x) + \frac{n(n-1)(n-2)\cdots(2)(1)}{(2)(1)[(n-2)(n-3)\cdots(2)(1)]}g''(x)h^{(n-2)}(x)$$

$$+ \frac{n(n-1)(n-2)\cdots(2)(1)}{(3)(2)(1)[(n-3)(n-4)\cdots(2)(1)]}g'''(x)h^{(n-3)}(x) + \cdots$$

$$+ \frac{n(n-1)(n-2)\cdots(2)(1)}{[(n-1)(n-2)\cdots(2)(1)](1)}g^{(n-1)}(x)h'(x) + g^{(n)}(x)h(x)$$

$$= g(x)h^{(n)}(x) + \frac{n!}{1!(n-1)!}g'(x)h^{(n-1)}(x) + \frac{n!}{2!(n-2)!}g''(x)h^{(n-2)}(x) + \cdots$$

$$+ \frac{n!}{(n-1)!1!}g^{(n-1)}(x)h'(x) + g^{(n)}(x)h(x)$$

Note: $n! = n(n-1) \ldots 3 \cdot 2 \cdot 1$ (read "*n* factorial.")

94. $[xf(x)]' = xf'(x) + f(x)$

$[xf(x)]'' = xf''(x) + f'(x) + f'(x) = xf''(x) + 2f'(x)$

$[xf(x)]''' = xf'''(x) + f''(x) + 2f''(x) = xf'''(x) + 3f''(x)$

In general, $[xf(x)]^{(n)} = xf^{(n)}(x) + nf^{(n-1)}(x)$.

95. (a) $f(x) = \sqrt{x} = x^{1/2}$ $\qquad f(1) = 1$

$$f'(x) = \frac{1}{2}x^{-1/2} = \frac{1}{2\sqrt{x}} \qquad f'(1) = \frac{1}{2}$$

$$f''(x) = -\frac{1}{4}x^{-3/2} = -\frac{1}{4x^{3/2}} \qquad f''(1) = -\frac{1}{4}$$

$$P_1(x) = f'(1)(x - 1) + f(1) = \frac{1}{2}(x - 1) + 1 = \frac{1}{2}x + \frac{1}{2}$$

$$P_2(x) = \frac{1}{2}f''(1)(x - 1)^2 + f'(1)(x - 1) + f(1)$$

$$= \frac{1}{2}\left(-\frac{1}{4}\right)(x - 1)^2 + \frac{1}{2}(x - 1) + 1$$

$$= -\frac{1}{8}(x - 1)^2 + \frac{1}{2}(x - 1) + 1$$

(b)

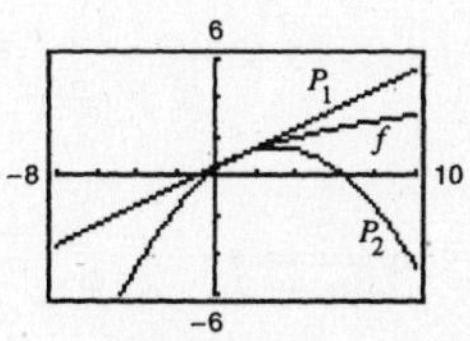

(c) P_2 is a better approximation.

(d) P_1 and P_2 become less accurate as you move farther away from $x = a = 1$.

96. (a) $f(x) = \frac{6}{x} = 6x^{-1}$ $\qquad f(2) = 3$

$$f'(x) = -6x^{-2} = \frac{-6}{x^2} \qquad f'(2) = -\frac{3}{2}$$

$$f''(x) = 12x^{-3} = \frac{12}{x^3} \qquad f''(2) = \frac{3}{2}$$

$$P_1(x) = f'(2)(x - 2) + f(2) = -\frac{3}{2}(x - 2) + 3$$

$$P_2(x) = \frac{1}{2}f''(2)(x - 2)^2 + f'(2)(x - 2) + f(2)$$

$$= \frac{1}{2}\left(\frac{3}{2}\right)(x - 2)^2 - \frac{3}{2}(x - 2) + 3$$

$$= \frac{3}{4}(x - 2)^2 - \frac{3}{2}(x - 2) + 3$$

(b)

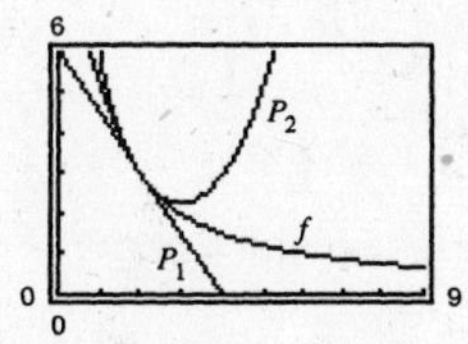

(c) P_2 is a better approximation than P_1.

(d) The accuracy worsens as you move farther away from $x = a = 2$.

97. False. If $y = f(x)g(x)$, then

$$\frac{dy}{dx} = f(x)g'(x) + g(x)f'(x).$$

98. True. y is a fourth-degree polynomial.

$$\frac{d^ny}{dx^n} = 0 \text{ when } n > 4.$$

99. True

$$h'(c) = f(c)g'(c) + g(c)f'(c)$$
$$= f(c)(0) + g(c)(0)$$
$$= 0$$

100. True

101. True

102. True. If $v(t) = c$ then $a(t) = v'(t) = 0$.

103. $f(x) = x|x| = \begin{cases} x^2, & \text{if } x \ge 0 \\ -x^2, & \text{if } x < 0 \end{cases}$

$$f'(x) = \left\{\begin{matrix} 2x, & \text{if } x \ge 0 \\ -2x, & \text{if } x < 0 \end{matrix}\right\} = 2|x|$$

$$f''(x) = \begin{cases} 2, & \text{if } x > 0 \\ -2, & \text{if } x < 0 \end{cases}$$

$f''(0)$ does not exist since the left and right derivatives are not equal.

104. (a) $(fg' - f'g)' = fg'' + f'g' - f'g' - f''g$

$$= fg'' - f''g \quad \text{True}$$

(b) $(fg)'' = (fg' + f'g)'$

$$= fg'' + f'g' + f'g' + f''g$$
$$= fg'' + 2f'g' + f''g$$
$$\ne fg'' + f''g \quad \text{False}$$

Section 4.4 The Chain Rule

$y = f(g(x))$	$u = g(x)$	$y = f(u)$
1. $y = (6x - 5)^4$	$u = 6x - 5$	$y = u^4$
2. $y = (5x - 2)^{3/2}$	$u = 5x - 2$	$y = u^{3/2}$

$y = f(g(x))$	$u = g(x)$	$y = f(u)$
3. $y = (x^2 - 3x + 4)^6$	$u = x^2 - 3x + 4$	$y = u^6$
4. $y = \dfrac{3}{x+2}$	$u = x + 2$	$y = \dfrac{3}{u}$ (answers may vary)
5. $y = \sqrt{x^2 - 1}$	$u = x^2 - 1$	$y = \sqrt{u}$
6. $y = \dfrac{1}{\sqrt{x+1}}$	$u = x + 1$	$y = u^{-1/2}$

7. $y = (2x - 7)^3$

$y' = 3(2x - 7)^2(2) = 6(2x - 7)^2$

8. $y = 3(4 - x^2)^5$

$y' = 15(4 - x^2)^4(-2x) = -30x(4 - x^2)^4$

9. $g(x) = 3(4 - 9x)^4$

$g'(x) = 12(4 - 9x)^3(-9) = -108(4 - 9x)^3$

10. $f(t) = (9t + 2)^{2/3}$

$f'(t) = \frac{2}{3}(9t + 2)^{-1/3}(9) = \dfrac{6}{\sqrt[3]{9t + 2}}$

11. $f(t) = (1 - t)^{1/2}$

$f'(t) = \frac{1}{2}(1 - t)^{-1/2}(-1) = -\dfrac{1}{2\sqrt{1 - t}}$

12. $g(x) = \sqrt{5 - 3x} = (5 - 3x)^{1/2}$

$g'(x) = \frac{1}{2}(5 - 3x)^{-1/2}(-3) = \dfrac{-3}{2\sqrt{5 - 3x}}$

13. $y = (9x^2 + 4)^{1/3}$

$y' = \frac{1}{3}(9x^2 + 4)^{-2/3}(18x) = \dfrac{6x}{(9x^2 + 4)^{2/3}}$

14. $g(x) = \sqrt{x^2 - 2x + 1} = \sqrt{(x - 1)^2} = |x - 1|$

$g'(x) = \begin{cases} 1, & x > 1 \\ -1, & x < 1 \end{cases}$

15. $y = 2(4 - x^2)^{1/4}$

$y' = 2\left(\frac{1}{4}\right)(4 - x^2)^{-3/4}(-2x)$

$= \dfrac{-x}{\sqrt[4]{(4 - x^2)^3}}$

16. $f(x) = -3(2 - 9x)^{1/4}$

$f'(x) = -\frac{3}{4}(2 - 9x)^{-3/4}(-9) = \dfrac{27}{4(2 - 9x)^{3/4}}$

17. $y = (x - 2)^{-1}$

$y' = -1(x - 2)^{-2}(1) = \dfrac{-1}{(x - 2)^2}$

18. $s(t) = (t^2 + 3t - 1)^{-1}$

$s'(t) = -1(t^2 + 3t - 1)^{-2}(2t + 3)$

$= \dfrac{-(2t + 3)}{(t^2 + 3t - 1)^2}$

19. $f(t) = (t - 3)^{-2}$

$f'(t) = -2(t - 3)^{-3} = \dfrac{-2}{(t - 3)^3}$

20. $y = -5(t + 3)^{-3}$

$y' = 15(t + 3)^{-4} = \dfrac{15}{(t + 3)^4}$

21. $y = (x + 2)^{-1/2}$

$\dfrac{dy}{dx} = -\frac{1}{2}(x + 2)^{-3/2} = -\dfrac{1}{2(x + 2)^{3/2}}$

22. $g(t) = (t^2 - 2)^{-1/2}$

$g'(t) = -\frac{1}{2}(t^2 - 2)^{-3/2}(2t) = -\dfrac{t}{(t^2 - 2)^{3/2}}$

23. $f(x) = x^2(x-2)^4$

$$f'(x) = x^2[4(x-2)^3(1)] + (x-2)^4(2x)$$
$$= 2x(x-2)^3[2x + (x-2)]$$
$$= 2x(x-2)^3(3x-2)$$

24. $f(x) = x(3x-9)^3$

$$f'(x) = x[3(3x-9)^2(3)] + (3x-9)^3(1)$$
$$= (3x-9)^2[9x + 3x - 9]$$
$$= 27(x-3)^2(4x-3)$$

25. $y = x\sqrt{1-x^2} = x(1-x^2)^{1/2}$

$$y' = x\left[\frac{1}{2}(1-x^2)^{-1/2}(-2x)\right] + (1-x^2)^{1/2}(1)$$
$$= -x^2(1-x^2)^{-1/2} + (1-x^2)^{1/2}$$
$$= (1-x^2)^{-1/2}[-x^2 + (1-x^2)]$$
$$= \frac{1-2x^2}{\sqrt{1-x^2}}$$

26. $y = \frac{1}{2}x^2\sqrt{16-x^2}$

$$y' = \frac{1}{2}x^2\left(\frac{1}{2}(16-x^2)^{-1/2}(-2x)\right) + x(16-x^2)^{1/2}$$
$$= \frac{-x^3}{2\sqrt{16-x^2}} + x\sqrt{16-x^2}$$
$$= \frac{-x(3x^2-32)}{2\sqrt{16-x^2}}$$

27. $y = \dfrac{x}{\sqrt{x^2+1}} = x(x^2+1)^{-1/2}$

$$y' = x\left[-\frac{1}{2}(x^2+1)^{-3/2}(2x)\right] + (x^2+1)^{-1/2}(1)$$
$$= -x^2(x^2+1)^{-3/2} + (x^2+1)^{-1/2}$$
$$= (x^2+1)^{-3/2}[-x^2 + (x^2+1)]$$
$$= \frac{1}{(x^2+1)^{3/2}}$$

28. $y = \dfrac{x}{\sqrt{x^4+4}}$

$$y' = \frac{(x^4+4)^{1/2}(1) - x\frac{1}{2}(x^4+4)^{-1/2}(4x^3)}{x^4+4}$$
$$= \frac{x^4+4-2x^4}{(x^4+4)^{3/2}} = \frac{4-x^4}{(x^4+4)^{3/2}}$$

29. $g(x) = \left(\dfrac{x+5}{x^2+2}\right)^2$

$$g'(x) = 2\left(\frac{x+5}{x^2+2}\right)\left(\frac{(x^2+2)-(x+5)(2x)}{(x^2+2)^2}\right)$$
$$= \frac{2(x+5)(2-10x-x^2)}{(x^2+2)^3}$$

30. $h(t) = \left(\dfrac{t^2}{t^3+2}\right)^2$

$$h'(t) = 2\left(\frac{t^2}{t^3+2}\right)\left(\frac{(t^3+2)(2t)-t^2(3t^2)}{(t^3+2)^2}\right)$$
$$= \frac{2t^2(4t-t^4)}{(t^3+2)^3} = \frac{2t^3(4-t^3)}{(t^3+2)^3}$$

31. $f(v) = \left(\dfrac{1-2v}{1+v}\right)^3$

$$f'(v) = 3\left(\frac{1-2v}{1+v}\right)^2\left(\frac{(1+v)(-2)-(1-2v)}{(1+v)^2}\right)$$
$$= \frac{-9(1-2v)^2}{(1+v)^4}$$

32. $g(x) = \left(\dfrac{3x^2-2}{2x+3}\right)^3$

$$g'(x) = 3\left(\frac{3x^2-2}{2x+3}\right)^2\left(\frac{(2x+3)(6x)-(3x^2-2)(2)}{(2x+3)^2}\right)$$
$$= \frac{3(3x^2-2)^2(6x^2+18x+4)}{(2x+3)^4}$$
$$= \frac{6(3x^2-2)^2(3x^2+9x+2)}{(2x+3)^4}$$

33. $y = \dfrac{\sqrt{x}+1}{x^2+1}$

$$y' = \frac{1-3x^2-4x^{3/2}}{2\sqrt{x}(x^2+1)^2}$$

The zero of y' corresponds to the point on the graph of y where the tangent line is horizontal.

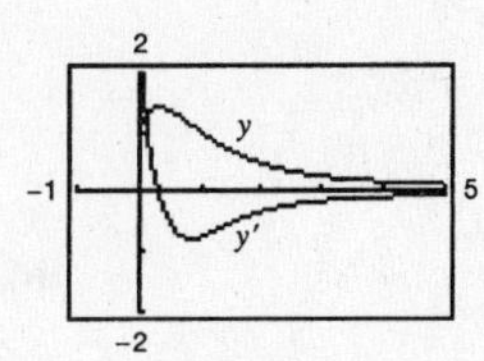

34. $y = \sqrt{\dfrac{2x}{x+1}}$

$y' = \dfrac{1}{\sqrt{2x}(x+1)^{3/2}}$

y' has no zeros.

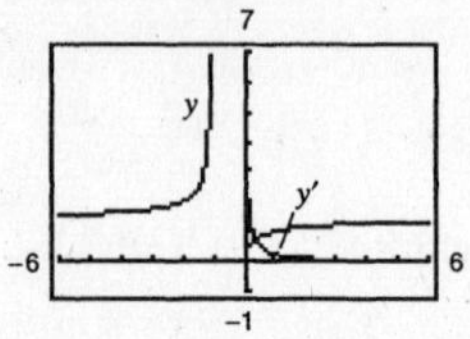

35. $g(t) = \dfrac{3t^2}{\sqrt{t^2 + 2t - 1}}$

$g'(t) = \dfrac{3t(t^2 + 3t - 2)}{(t^2 + 2t - 1)^{3/2}}$

The zeros of g' correspond to the points on the graph of g where the tangent lines are horizontal.

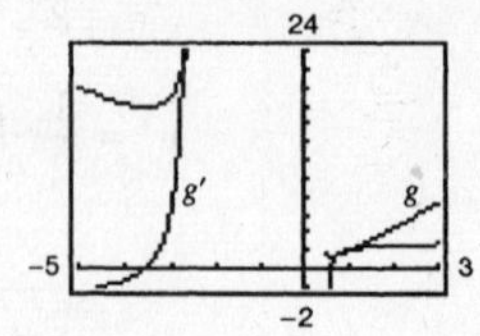

36. $f(x) = \sqrt{x}(2 - x)^2$

$f'(x) = \dfrac{(x - 2)(5x - 2)}{2\sqrt{x}}$

The zeros of f' correspond to the points on the graph of f where the tangent lines are horizontal.

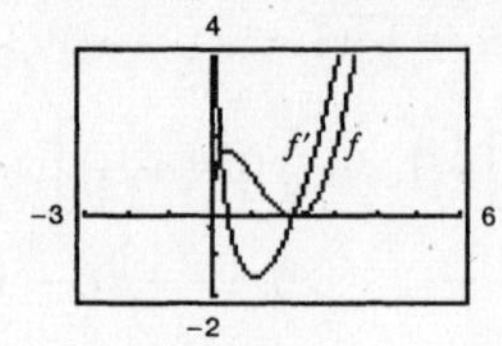

37. $y = \sqrt{\dfrac{x+1}{x}}$

$y' = -\dfrac{\sqrt{(x+1)/x}}{2x(x+1)}$

y' has no zeros.

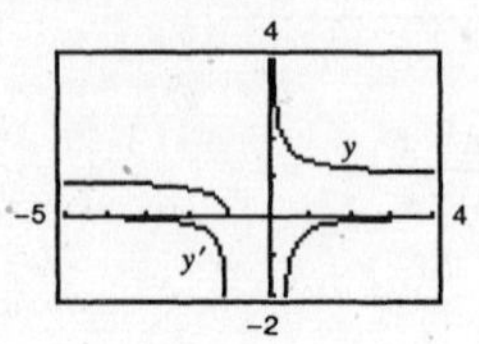

38. $y = (t^2 - 9)\sqrt{t+2}$

$y' = \dfrac{5t^2 + 8t - 9}{2\sqrt{t+2}}$

The zero of y' corresponds to the point on the graph of y where the tangent line is horizontal.

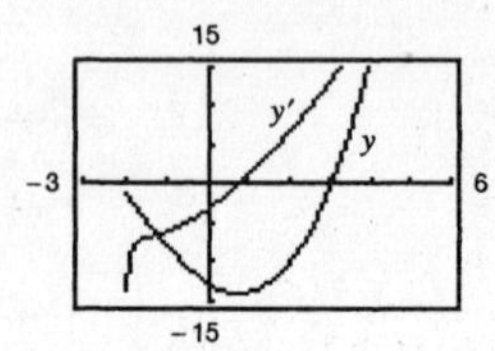

39. $s(t) = \dfrac{-2(2 - t)\sqrt{1 + t}}{3}$

$s'(t) = \dfrac{t}{\sqrt{1 + t}}$

The zero of $s'(t)$ corresponds to the point on the graph of $s(t)$ where the tangent line is horizontal.

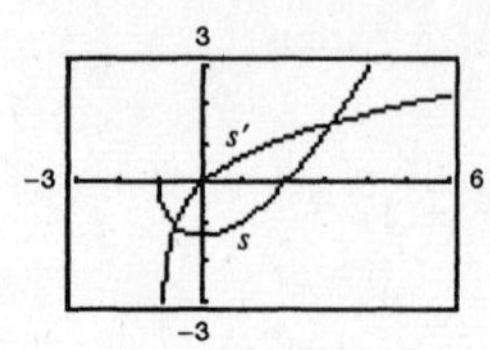

40. $g(x) = \sqrt{x-1} + \sqrt{x+1}$

$g'(x) = \dfrac{1}{2\sqrt{x-1}} + \dfrac{1}{2\sqrt{x+1}}$

g' has no zeros.

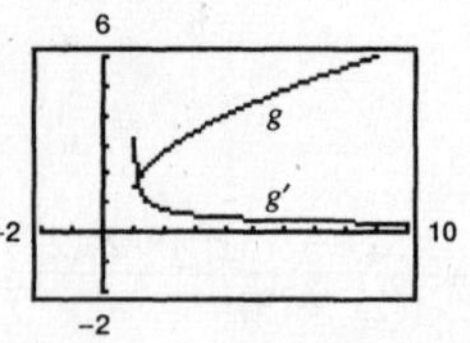

41. $s(t) = (t^2 + 2t + 8)^{1/2}, \ (2, 4)$

$s'(t) = \dfrac{1}{2}(t^2 + 2t + 8)^{-1/2}(2t + 2)$

$= \dfrac{t + 1}{\sqrt{t^2 + 2t + 8}}$

$s'(2) = \dfrac{3}{4}$

42. $y = (3x^3 + 4x)^{1/5}, \ (2, 2)$

$y' = \dfrac{1}{5}(3x^3 + 4x)^{-4/5}(9x^2 + 4)$

$= \dfrac{9x^2 + 4}{5(3x^3 + 4x)^{4/5}}$

$y'(2) = \dfrac{1}{2}$

43. $f(x) = \dfrac{3}{x^3 - 4} = 3(x^3 - 4)^{-1}, \ \left(-1, -\dfrac{3}{5}\right)$

$f'(x) = -3(x^3 - 4)^{-2}(3x^2) = -\dfrac{9x^2}{(x^3 - 4)^2}$

$f'(-1) = -\dfrac{9}{25}$

44. $f(x) = \dfrac{1}{(x^2 - 3x)^2} = (x^2 - 3x)^{-2}, \ \left(4, \dfrac{1}{16}\right)$

$f'(x) = -2(x^2 - 3x)^{-3}(2x - 3) = \dfrac{-2(2x - 3)}{(x^2 - 3x)^3}$

$f'(4) = -\dfrac{5}{32}$

45. $f(t) = \dfrac{3t + 2}{t - 1}, \ (0, -2)$

$f'(t) = \dfrac{(t - 1)(3) - (3t + 2)(1)}{(t - 1)^2} = \dfrac{-5}{(t - 1)^2}$

$f'(0) = -5$

46. $f(x) = \dfrac{x + 1}{2x - 3}, \ (2, 3)$

$f'(x) = \dfrac{(2x - 3)(1) - (x + 1)(2)}{(2x - 3)^2} = \dfrac{-5}{(2x - 3)^2}$

$f'(2) = -5$

47. (a) $f(x) = \sqrt{3x^2 - 2}, \ (3, 5)$

$$f'(x) = \frac{1}{2}(3x^2 - 2)^{-1/2}(6x)$$

$$= \frac{3x}{\sqrt{3x^2 - 2}}$$

$$f'(3) = \frac{9}{5}$$

Tangent line:

$$y - 5 = \frac{9}{5}(x - 3) \implies 9x - 5y - 2 = 0$$

(b)

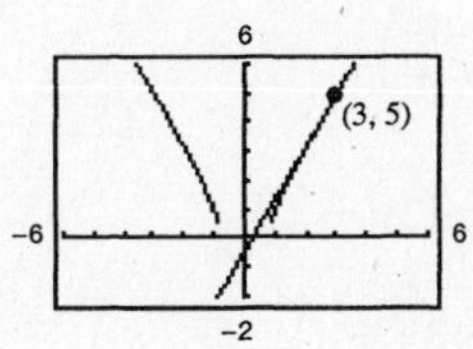

48. (a) $f(x) = \frac{1}{3}x\sqrt{x^2 + 5}, \ (2, 2)$

$$f'(x) = \frac{1}{3}x\left[\frac{1}{2}(x^2 + 5)^{-1/2}(2x)\right] + \frac{1}{3}(x^2 + 5)^{1/2}$$

$$= \frac{x^2}{3\sqrt{x^2 + 5}} + \frac{1}{3}\sqrt{x^2 + 5}$$

$$f'(2) = \frac{4}{3(3)} + \frac{1}{3}(3) = \frac{13}{9}$$

Tangent line:

$$y - 2 = \frac{13}{9}(x - 2) \implies 13x - 9y - 8 = 0$$

(b)

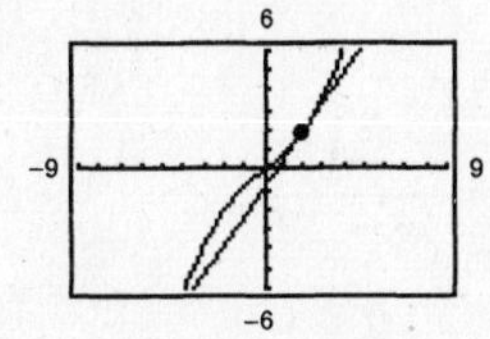

49. (a) $y = (2x^3 + 1)^2, \ (-1, 1)$

$$y' = 2(2x^3 + 1)(6x^2) = 12x^2(2x^3 + 1)$$

$$y'(-1) = 12(-1) = -12$$

Tangent line:

$$y - 1 = -12(x + 1)$$

$$y = -12x - 11$$

(b)

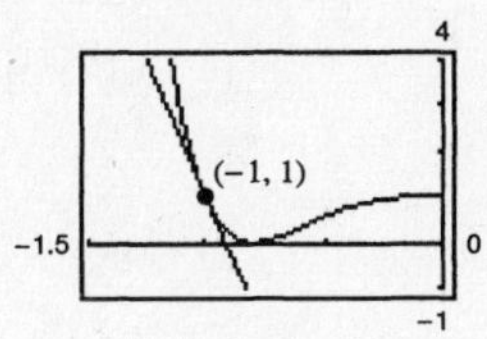

50. (a) $f(x) = (9 - x^2)^{2/3}, \ (1, 4)$

$$f'(x) = \frac{2}{3}(9 - x^2)^{-1/3}(-2x) = \frac{-4x}{3(9 - x^2)^{1/3}}$$

$$f'(1) = \frac{-4}{3(2)} = \frac{-2}{3}$$

Tangent Line:

$$y - 4 = \frac{-2}{3}(x - 1)$$

$$y = \frac{-2}{3}x + \frac{14}{3}$$

(b)

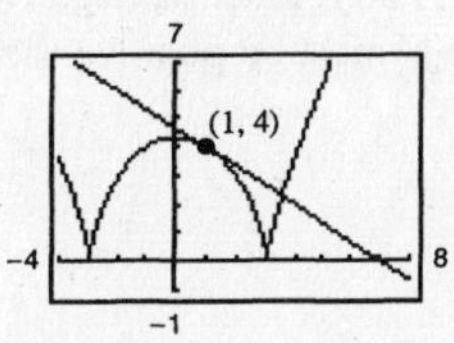

51. $f(x) = 2(x^2 - 1)^3$

$$f'(x) = 6(x^2 - 1)^2(2x)$$

$$= 12x(x^4 - 2x^2 + 1)$$

$$= 12x^5 - 24x^3 + 12x$$

$$f''(x) = 60x^4 - 72x^2 + 12$$

$$= 12(5x^2 - 1)(x^2 - 1)$$

52. $f(x) = \sqrt{x^2 + x + 1} = (x^2 + x + 1)^{1/2}$

$$f'(x) = \frac{1}{2}(x^2 + x + 1)^{-1/2}(2x + 1) = \frac{2x + 1}{2\sqrt{x^2 + x + 1}}$$

$$f''(x) = \frac{2\sqrt{x^2 + x + 1}(2) - (2x + 1)(x^2 + x + 1)^{-1/2}(2x + 1)}{4(x^2 + x + 1)}$$

$$= \frac{4(x^2 + x + 1)^{1/2} - (2x + 1)^2(x^2 + x + 1)^{-1/2}}{4(x^2 + x + 1)}$$

$$= \frac{4(x^2 + x + 1) - (2x + 1)^2}{4(x^2 + x + 1)^{3/2}}$$

$$= \frac{3}{4(x^2 + x + 1)^{3/2}}$$

53. $f(t) = \dfrac{\sqrt{t^2+1}}{t} = \dfrac{(t^2+1)^{1/2}}{t}$

$$f'(t) = \frac{t^2(t^2+1)^{-1/2} - (t^2+1)^{1/2}}{t^2} = (t^2+1)^{-1/2} - \frac{(t^2+1)^{1/2}}{t^2}$$

$$f''(t) = -(t^2+1)^{-3/2}(t) - \left[\frac{t^3(t^2+1)^{-1/2} - (t^2+1)^{1/2}(2t)}{t^4}\right]$$

$$= \frac{-t^4 - t^2(t^2+1) + 2(t^2+1)^2}{t^3(t^2+1)^{3/2}}$$

$$= \frac{3t^2+2}{t^3(t^2+1)^{3/2}}$$

54. $f(x) = (x-2)^{-1}$

$f'(x) = -(x-2)^{-2} = \dfrac{-1}{(x-2)^2}$

$f''(x) = 2(x-2)^{-3} = \dfrac{2}{(x-2)^3}$

55. $h(x) = \dfrac{1}{9}(3x+1)^3, \quad \left(1, \dfrac{64}{9}\right)$

$h'(x) = \dfrac{1}{9}3(3x+1)^2(3) = (3x+1)^2$

$h''(x) = 2(3x+1)(3) = 6(3x+1)$

$h''(1) = 24$

56. $f(x) = \dfrac{1}{\sqrt{x+4}} = (x+4)^{-1/2}, \quad \left(0, \dfrac{1}{2}\right)$

$f'(x) = -\dfrac{1}{2}(x+4)^{-3/2}$

$f''(x) = \dfrac{3}{4}(x+4)^{-5/2} = \dfrac{3}{4(x+4)^{5/2}}$

$f''(0) = \dfrac{3}{128}$

57. The zeros of f' correspond to the points where the graph of f has horizontal tangents.

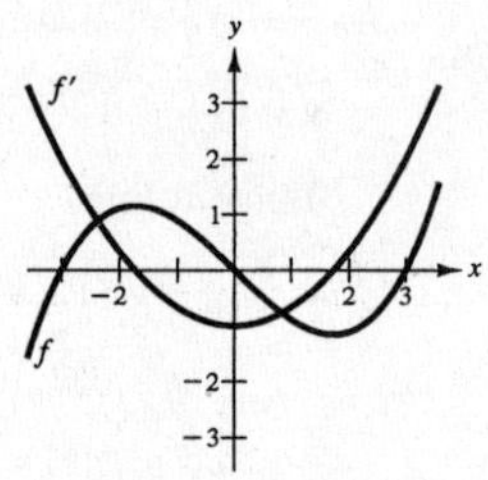

58. f is decreasing on $(-\infty, -1)$ so f' must be negative there. f is increasing on $(1, \infty)$ so f' must be positive there.

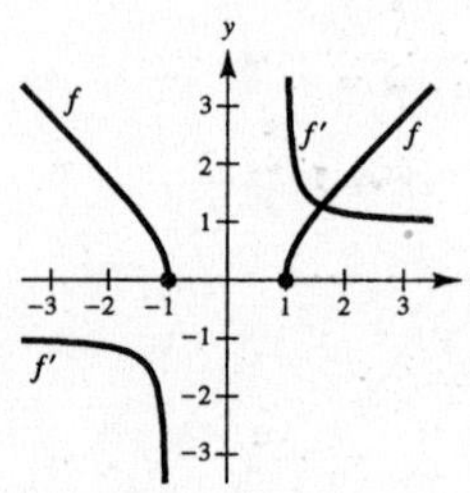

59. $g(x) = f(3x)$

$g'(x) = f'(3x)(3) \implies g'(x) = 3f'(3x)$

60. $g(x) = f(x^2)$

$g'(x) = f'(x^2)(2x) \implies g'(x) = 2xf'(x^2)$

61. (a) $f(x) = g(x)h(x)$

$f'(x) = g(x)h'(x) + g'(x)h(x)$

$f'(5) = (-3)(-2) + (6)(3) = 24$

(b) $f(x) = g(h(x))$

$f'(x) = g'(h(x))h'(x)$

$f'(5) = g'(3)(-2) = -2g'(3)$

Need $g'(3)$ to find $f'(5)$.

(c) $f(x) = \dfrac{g(x)}{h(x)}$

$f'(x) = \dfrac{h(x)g'(x) - g(x)h'(x)}{[h(x)]^2}$

$f'(5) = \dfrac{(3)(6) - (-3)(-2)}{(3)^2} = \dfrac{12}{9} = \dfrac{4}{3}$

(d) $f(x) = [g(x)]^3$

$f'(x) = 3[g(x)]^2g'(x)$

$f'(5) = 3(-3)^2(6) = 162$

62. To differentiate $y = x\sqrt{x^2 + 1}$, you would first apply the product rule.

63. (a) $h(x) = f(g(x)),\ g(1) = 4,\ g'(1) = -\frac{1}{2},\ f'(4) = -1,\ h'(x) = f'(g(x))g'(x)$

$$h'(1) = f'(g(1))g'(1)$$
$$= f'(4)g'(1)$$
$$= (-1)\left(-\tfrac{1}{2}\right) = \tfrac{1}{2}$$

(b) $s(x) = g(f(x)),\ f(5) = 6,\ f'(5) = -1,\ g'(6)$ does not exist.

$$s'(x) = g'(f(x))f'(x)$$
$$s'(5) = g'(f(5))f'(5) = g'(6)(-1)$$

Since $g'(6)$ does not exist, $s'(5)$ is not defined.

64. (a) $h(x) = f(g(x))$

$$h'(x) = f'(g(x))g'(x)$$
$$h'(3) = f'(g(3))g'(3)$$
$$= f'(5)(1)$$
$$= \tfrac{1}{2}$$

(b) $s(x) = g(f(x))$

$$s'(x) = g'(f(x))f'(x)$$
$$s'(9) = g'(f(9))f'(9)$$
$$= g'(8)(2)$$
$$= (-1)(2)$$
$$= -2$$

65. (a) $F = 132{,}400(331 - v)^{-1}$

$$F' = (-1)(132{,}400)(331 - v)^{-2}(-1)$$
$$= \frac{132{,}400}{(331 - v)^2}$$

When $v = 30$, $F' \approx 1.461$.

(b) $F = 132{,}400(331 + v)^{-1}$

$$F' = (-1)(132{,}400)(331 + v)^{-2}(1)$$
$$= \frac{-132{,}400}{(331 + v)^2}$$

When $v = 30$, $F' \approx -1.016$.

66. $S = C(R^2 - r^2) = CR^2 - Cr^2$

$$\frac{dS}{dt} = 2CR\frac{dR}{dt}$$
$$= 2(1.76 \cdot 10^5)(1.2 \cdot 10^{-2})(10^{-5})$$
$$= 4.224 \cdot 10^{-2} = 0.04224 \text{ cm/sec}^2$$

67. (a) $x = -1.6372t^3 + 19.3120t^2 - 0.5082t - 0.6162$

(b) $C = 60x + 1350$

$$= 60(-1.6372t^3 + 19.3120t^2 - 0.5082t - 0.6162) + 1350$$
$$\frac{dC}{dt} = 60(-4.9116t^2 + 38.624t - 0.5082)$$
$$= -294.696t^2 + 2317.44t - 30.492$$

(c) The function $\frac{dC}{dt}$ is quadratic, not linear. The cost function levels off at the end of the day, perhaps due to fatigue.

68. (a) $g(x) = f(x) - 2 \Rightarrow g'(x) = f'(x)$

(b) $h(x) = 2f(x) \Rightarrow h'(x) = 2f'(x)$

(c) $r(x) = f(-3x) \Rightarrow r'(x) = f'(-3x)(-3) = -3f'(-3x)$

Hence, you need to know $f'(-3x)$.

$r'(0) = -3f'(0) = (-3)\left(-\frac{1}{3}\right) = 1$

$r'(-1) = -3f'(3) = (-3)(-4) = 12$

(d) $s(x) = f(x + 2) \Rightarrow s'(x) = f'(x + 2)$

Hence, you need to know $f'(x + 2)$.

$s'(-2) = f'(0) = -\frac{1}{3}$, etc.

x	-2	-1	0	1	2	3
$f'(x)$	4	$\frac{2}{3}$	$-\frac{1}{3}$	-1	-2	-4
$g'(x)$	4	$\frac{2}{3}$	$-\frac{1}{3}$	-1	-2	-4
$h'(x)$	8	$\frac{4}{3}$	$-\frac{2}{3}$	-2	-4	-8
$r'(x)$		12	1			
$s'(x)$	$-\frac{1}{3}$	-1	-2	-4		

69. (a) $r'(x) = f'(g(x))g'(x)$

$r'(1) = f'(g(1))g'(1)$

Note that $g(1) = 4$ and $f'(4) = \dfrac{5 - 0}{6 - 2} = \dfrac{5}{4}$.

Also, $g'(1) = 0$. Thus, $r'(1) = 0$.

(b) $s'(x) = g'(f(x))f'(x)$

$s'(4) = g'(f(4))f'(4)$

Note that $f(4) = \dfrac{5}{2}$, $g'\left(\dfrac{5}{2}\right) = \dfrac{6 - 4}{6 - 2} = \dfrac{1}{2}$ and

$f'(4) = \dfrac{5}{4}$.

Thus, $s'(4) = \dfrac{1}{2}\left(\dfrac{5}{4}\right) = \dfrac{5}{8}$.

70. $|u| = \sqrt{u^2}$

$$\frac{d}{dx}[|u|] = \frac{d}{dx}\left[\sqrt{u^2}\right]$$

$$= \frac{1}{2}(u^2)^{-1/2}(2uu')$$

$$= \frac{uu'}{\sqrt{u^2}} = u'\frac{u}{|u|},\ u \neq 0$$

71. $g(x) = |2x - 3|$

$$g'(x) = 2\left(\frac{2x - 3}{|2x - 3|}\right),\quad x \neq \frac{3}{2}$$

72. $f(x) = |x^2 - 4|$

$$f'(x) = 2x\left(\frac{x^2 - 4}{|x^2 - 4|}\right),\ x \neq \pm 2$$

73. (a) $f(x) = \dfrac{1}{\sqrt{x^2 - 3}} = (x^2 - 3)^{-1/2}$, $\quad f(2) = 1$

$$f'(x) = -\frac{1}{2}(x^2 - 3)^{-3/2}(2x) = -\frac{x}{(x^2 - 3)^{3/2}},\ f'(2) = -2$$

$$f''(x) = \frac{(x^2 - 3)^{3/2}(-1) + x\left(\frac{3}{2}\right)(x^2 - 3)^{1/2}(2x)}{(x^2 - 3)^3}$$

$$= \frac{-(x^2 - 3) + 3x^2}{(x^2 - 3)^{5/2}} = \frac{2x^2 + 3}{(x^2 - 3)^{5/2}},\quad f''(2) = 11$$

$P_1(x) = f'(2)(x - 2) + f(2) = -2(x - 2) + 1 = -2x + 5$

$P_2(x) = \dfrac{1}{2}f''(2)(x - 2)^2 + f'(2)(x - 2) + f(2)$

$= \dfrac{1}{2}(11)(x - 2)^2 - 2(x - 2) + 1$

$= \dfrac{11}{2}(x - 2)^2 - 2(x - 2) + 1$

(b)

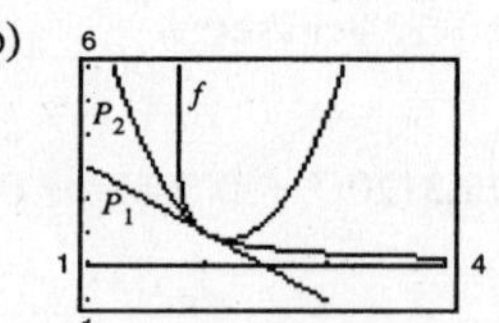

(c) P_2 is the better approximation.

(d) The accuracy worsens as you move away from $x = a = 2$.

74. (a) $f(x) = (x^2 - 3)^{1/2}$ $\quad f(2) = 1$

$f'(x) = x(x^2 - 3)^{-1/2}$ $\quad f'(2) = 2$

$$f''(x) = x^2\left(-(x^2 - 3)^{-3/2}\right) + (x^2 - 3)^{-1/2}$$

$$= \frac{-x^2 + (x^2 - 3)}{(x^2 - 3)^{3/2}}$$

$$= \frac{-3}{(x^2 - 3)^{3/2}} \qquad f''(2) = -3$$

$$P_1(x) = f'(2)(x - 2) + f(2) = 2(x - 2) + 1 = 2x - 3$$

$$P_2(x) = \frac{1}{2}f''(2)(x - 2)^2 + f'(2)(x - 2) + f(2)$$

$$= \frac{1}{2}(-3)(x - 2)^2 + 2(x - 2) + 1$$

$$= -\frac{3}{2}(x - 2)^2 + 2(x - 2) + 1$$

(b)

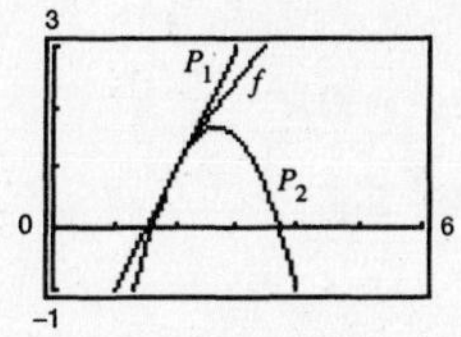

(c) P_2 is a better approximation.

(d) P_1 and P_2 become less accurate as you move further away from $x = a = 2$.

75. False. If $y = (1 - x)^{1/2}$, then $y' = \frac{1}{2}(1 - x)^{-1/2}(-1)$.

76. True. The chain rule implies that $\dfrac{dy}{dx} = \dfrac{dy}{du} \cdot \dfrac{du}{dv} \cdot \dfrac{dv}{dx}$

Section 4.5 Implicit Differentiation

1. $x^2 + y^2 = 36$

$$2x + 2yy' = 0$$

$$y' = \frac{-x}{y}$$

2. $x^2 - y^2 = 16$

$$2x - 2yy' = 0$$

$$y' = \frac{x}{y}$$

3. $x^{1/2} + y^{1/2} = 9$

$$\frac{1}{2}x^{-1/2} + \frac{1}{2}y^{-1/2}y' = 0$$

$$y' = -\frac{x^{-1/2}}{y^{-1/2}} = -\sqrt{\frac{y}{x}}$$

4. $x^3 + y^3 = 8$

$$3x^2 + 3y^2y' = 0$$

$$y' = -\frac{x^2}{y^2}$$

5. $x^3 - xy + y^2 = 4$

$$3x^2 - xy' - y + 2yy' = 0$$

$$(2y - x)y' = y - 3x^2$$

$$y' = \frac{y - 3x^2}{2y - x}$$

6. $x^2y + y^2x = -2$

$$x^2y' + 2xy + y^2 + 2yxy' = 0$$

$$(x^2 + 2xy)y' = -(y^2 + 2xy)$$

$$y' = \frac{-y(y + 2x)}{x(x + 2y)}$$

7. $x^3y^3 - y - x = 0$

$$3x^3y^2y' + 3x^2y^3 - y' - 1 = 0$$

$$(3x^3y^2 - 1)y' = 1 - 3x^2y^3$$

$$y' = \frac{1 - 3x^2y^3}{3x^3y^2 - 1}$$

8. $(xy)^{1/2} - x + 2y = 0$

$$\frac{1}{2}(xy)^{-1/2}(xy' + y) - 1 + 2y' = 0$$

$$\frac{x}{2\sqrt{xy}}y' + \frac{y}{2\sqrt{xy}} - 1 + 2y' = 0$$

$$xy' + y - 2\sqrt{xy} + 4\sqrt{xy}\,y' = 0$$

$$y' = \frac{2\sqrt{xy} - y}{4\sqrt{xy} + x}$$

9. $x^3 - 3x^2y + 2xy^2 = 12$

$$3x^2 - 3x^2y' - 6xy + 4xyy' + 2y^2 = 0$$

$$(4xy - 3x^2)y' = 6xy - 3x^2 - 2y^2$$

$$y' = \frac{6xy - 3x^2 - 2y^2}{4xy - 3x^2}$$

10. $x^3 - 2x^2y + 3xy^2 = 38$

$$3x^2 - 4xy - 2x^2\frac{dy}{dx} + 3y^2 + 6xy\frac{dy}{dx} = 0$$

$$(6xy - 2x^2)\frac{dy}{dx} = 4xy - 3x^2 - 3y^2$$

$$\frac{dy}{dx} = \frac{4xy - 3x^2 - 3y^2}{6xy - 2x^2}$$

11. (a) $x^2 + y^2 = 16$

$$y^2 = 16 - x^2$$

$$y = \pm\sqrt{16 - x^2}$$

(b)

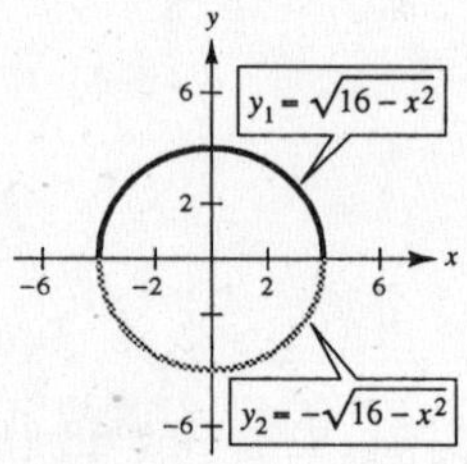

(c) Explicitly:

$$\frac{dy}{dx} = \pm\frac{1}{2}(16 - x^2)^{-1/2}(-2x)$$

$$= \frac{\mp x}{\sqrt{16 - x^2}} = \frac{-x}{\pm\sqrt{16 - x^2}} = \frac{-x}{y}$$

(d) Implicitly:

$$2x + 2yy' = 0$$

$$y' = -\frac{x}{y}$$

12. (a) $(x^2 - 4x + 4) + (y^2 + 6y + 9) = -9 + 4 + 9$

$$(x - 2)^2 + (y + 3)^2 = 4 \text{ (Circle)}$$

$$(y + 3)^2 = 4 - (x - 2)^2$$

$$y = -3 \pm \sqrt{4 - (x - 2)^2}$$

(b)

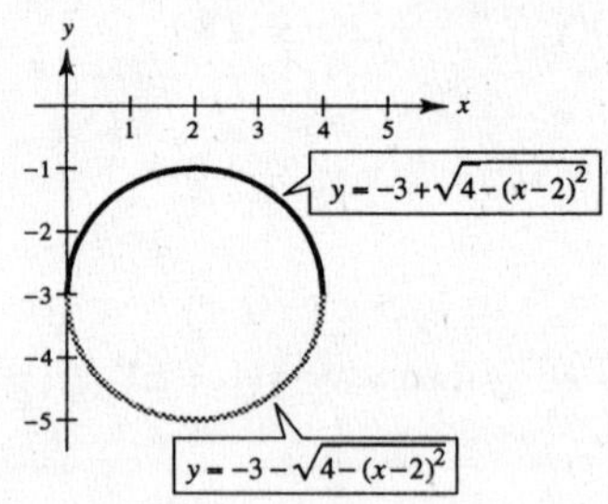

(c) Explicitly:

$$\frac{dy}{dx} = \pm\frac{1}{2}[4 - (x - 2)^2]^{-1/2}(-2)(x - 2)$$

$$= \frac{\mp(x - 2)}{\sqrt{4 - (x - 2)^2}}$$

$$= \frac{-(x - 2)}{\pm\sqrt{4 - (x - 2)^2}}$$

$$= \frac{-(x - 2)}{-3 \pm \sqrt{4 - (x - 2)^2} + 3}$$

$$= \frac{-(x - 2)}{y + 3}$$

(d) Implicitly:

$$2x + 2yy' - 4 + 6y' = 0$$

$$(2y + 6)y' = -2(x - 2)$$

$$y' = \frac{-(x - 2)}{y + 3}$$

13. (a) $16y^2 = 144 - 9x^2$

$$y^2 = \frac{1}{16}(144 - 9x^2) = \frac{9}{16}(16 - x^2)$$

$$y = \pm\frac{3}{4}\sqrt{16 - x^2}$$

(b)

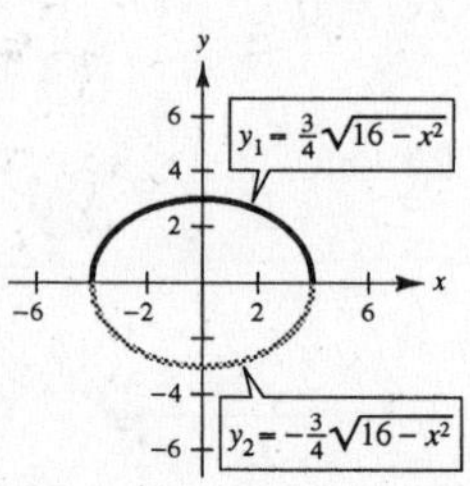

(c) Explicitly:

$$\frac{dy}{dx} = \pm\frac{3}{8}(16 - x^2)^{-1/2}(-2x)$$

$$= \mp\frac{3x}{4\sqrt{16 - x^2}} = \frac{-3x}{4(4/3)y} = \frac{-9x}{16y}$$

(d) Implicitly:

$$18x + 32yy' = 0$$

$$y' = \frac{-9x}{16y}$$

14. (a) $9y^2 = x^2 + 9$

$$y^2 = \frac{x^2}{9} + 1 = \frac{x^2 + 9}{9}$$

$$y = \frac{\pm\sqrt{x^2 + 9}}{3}$$

(b)

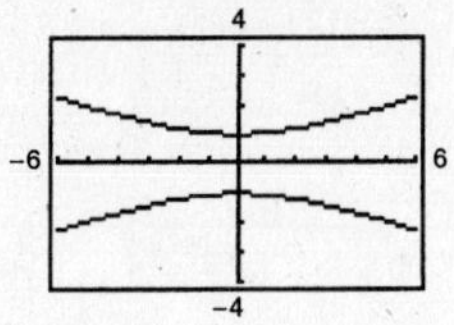

(c) Explicitly: $\frac{dy}{dx} = \frac{\pm\frac{1}{2}(x^2 + 9)^{-1/2}(2x)}{3} = \frac{\pm x}{3\sqrt{x^2 + 9}} = \frac{\pm x}{3(\pm 3y)} = \frac{x}{9y}$

(d) Implicitly: $9y^2 - x^2 = 9$

$$18yy' - 2x = 0$$

$$18yy' = 2x$$

$$y' = \frac{2x}{18y} = \frac{x}{9y}$$

15. $xy = 4$

$$xy' + y(1) = 0$$

$$xy' = -y$$

$$y' = \frac{-y}{x}$$

At $(-4, -1)$: $y' = -\frac{1}{4}$.

16. $x^2 - y^3 = 0$

$$2x - 3y^2y' = 0$$

$$y' = \frac{2x}{3y^2}$$

At $(1, 1)$: $y' = \frac{2}{3}$.

17. $y^2 = \frac{x^2 - 4}{x^2 + 4}$

$$2yy' = \frac{(x^2 + 4)(2x) - (x^2 - 4)(2x)}{(x^2 + 4)^2}$$

$$2yy' = \frac{16x}{(x^2 + 4)^2}$$

$$y' = \frac{8x}{y(x^2 + 4)^2}$$

At $(2, 0)$, y' is undefined.

18. $(x + y)^3 = x^3 + y^3$

$$x^3 + 3x^2y + 3xy^2 + y^3 = x^3 + y^3$$

$$3x^2y + 3xy^2 = 0$$

$$x^2y + xy^2 = 0$$

$$x^2y' + 2xy + 2xyy' + y^2 = 0$$

$$(x^2 + 2xy)y' = -(y^2 + 2xy)$$

$$y' = -\frac{y(y + 2x)}{x(x + 2y)}$$

At $(-1, 1)$: $y' = -1$.

19. $x^{2/3} + y^{2/3} = 5$

$$\frac{2}{3}x^{-1/3} + \frac{2}{3}y^{-1/3}y' = 0$$

$$y' = \frac{-x^{-1/3}}{y^{-1/3}} = -\sqrt[3]{\frac{y}{x}}$$

At $(8, 1)$: $y' = -\frac{1}{2}$.

20. $x^3 + y^3 = 4xy + 1$

$$3x^2 + 3y^2y' = 4xy' + 4y$$

$$(3y^2 - 4x)y' = 4y - 3x^2$$

$$y' = \frac{4y - 3x^2}{(3y^2 - 4x)}$$

At $(2, 1)$, $y' = \frac{4 - 12}{3 - 8} = \frac{8}{5}$.

21. $(x^2 + 4)y = 8$

$(x^2 + 4)y' + y(2x) = 0$

$$y' = \frac{-2xy}{x^2 + 4}$$

At $(2, 1)$: $y' = -\frac{1}{2}$

$\left(\text{Or, you could just solve for } y\text{: } y = \frac{8}{x^2 + 4}\right)$

22. $(4 - x)y^2 = x^3$

$(4 - x)(2yy') + y^2(-1) = 3x^2$

$$y' = \frac{3x^2 + y^2}{2y(4 - x)}$$

At $(2, 2)$: $y' = 2$.

23. $(x^2 + y^2)^2 = 4x^2y$

$2(x^2 + y^2)(2x + 2yy') = 4x^2y' + y(8x)$

$4x^3 + 4x^2yy' + 4xy^2 + 4y^3y' = 4x^2y' + 8xy$

$4x^2yy' + 4y^3y' - 4x^2y' = 8xy - 4x^3 - 4xy^2$

$4y'(x^2y + y^3 - x^2) = 4(2xy - x^3 - xy^2)$

$$y' = \frac{2xy - x^3 - xy^2}{x^2y + y^3 - x^2}$$

At $(1, 1)$: $y' = 0$.

24. $x^3 + y^3 - 6xy = 0$

$3x^2 + 3y^2y' - 6xy' - 6y = 0$

$y'(3y^2 - 6x) = 6y - 3x^2$

$$y' = \frac{6y - 3x^2}{3y^2 - 6x} = \frac{2y - x^2}{y^2 - 2x}$$

At $\left(\frac{4}{3}, \frac{8}{3}\right)$: $y' = \frac{(16/3) - (16/9)}{(64/9) - (8/3)} = \frac{32}{40} = \frac{4}{5}$.

25. $(y - 2)^2 = 4(x - 3)$, $(4, 0)$

$2(y - 2)y' = 4$

$$y' = \frac{2}{y - 2}$$

At $(4, 0)$: $y' = -1$.

Tangent line: $y - 0 = -1(x - 4)$

$y = -x + 4$

26. $(x + 1)^2 + (y - 2)^2 = 20$, $(3, 4)$

$2(x + 1) + 2(y - 2)y' = 0$

$2(y - 2)y' = -2(x + 1)$

$$y' = \frac{-(x + 1)}{y - 2}$$

At $(3, 4)$: $y' = -2$.

Tangent line: $y - 4 = -2(x - 3)$

$y = -2x + 10$

27. $xy = 1$, $(1, 1)$

$xy' + y = 0$

$$y' = \frac{-y}{x}$$

At $(1, 1)$: $y' = -1$.

Tangent line: $y - 1 = -1(x - 1)$

$y = -x + 2$

28. $7x^2 - 6\sqrt{3}xy + 13y^2 - 16 = 0$, $(\sqrt{3}, 1)$

$14x - 6\sqrt{3}xy' - 6\sqrt{3}y + 26yy' = 0$

$$y' = \frac{6\sqrt{3}y - 14x}{26y - 6\sqrt{3}x}$$

At $(\sqrt{3}, 1)$: $y' = \frac{6\sqrt{3} - 14\sqrt{3}}{26 - 6\sqrt{3}\sqrt{3}} = \frac{-8\sqrt{3}}{8} = -\sqrt{3}$.

Tangent line: $y - 1 = -\sqrt{3}(x - \sqrt{3})$

$y = -\sqrt{3}x + 4$

29. $x^2y^2 - 9x^2 - 4y^2 = 0, \quad (-4, 2\sqrt{3})$

$x^2\,2yy' + 2xy^2 - 18x - 8yy' = 0$

$$y' = \frac{18x - 2xy^2}{2x^2y - 8y}$$

At $(-4, 2\sqrt{3})$: $y' = \dfrac{18(-4) - 2(-4)(12)}{2(16)(2\sqrt{3}) - 16\sqrt{3}}$

$$= \frac{24}{48\sqrt{3}} = \frac{1}{2\sqrt{3}} = \frac{\sqrt{3}}{6}.$$

Tangent line: $y - 2\sqrt{3} = \dfrac{\sqrt{3}}{6}(x + 4)$

$$y = \frac{\sqrt{3}}{6}x + \frac{8}{3}\sqrt{3}$$

30. $x^{2/3} + y^{2/3} = 5, \quad (8, 1)$

$\dfrac{2}{3}x^{-1/3} + \dfrac{2}{3}y^{-1/3}y' = 0$

$$y' = \frac{-x^{-1/3}}{y^{-1/3}} = -\left(\frac{y}{x}\right)^{1/3}$$

At $(8, 1)$: $y' = -\dfrac{1}{2}$.

Tangent line: $y - 1 = -\dfrac{1}{2}(x - 8)$

$$y = -\frac{1}{2}x + 5$$

31. $3(x^2 + y^2)^2 = 100(x^2 - y^2), \quad (4, 2)$

$6(x^2 + y^2)(2x + 2yy') = 100(2x - 2yy')$

At $(4, 2)$:

$6(16 + 4)(8 + 4y') = 100(8 - 4y')$

$960 + 480y' = 800 - 400y'$

$880y' = -160$

$y' = -\frac{2}{11}$.

Tangent line: $y - 2 = -\frac{2}{11}(x - 4)$

$11y + 2x - 30 = 0$

$y = -\frac{2}{11}x + \frac{30}{11}$

32. $y^2(x^2 + y^2) = 2x^2, \quad (1, 1)$

$y^2x^2 + y^4 = 2x^2$

$2yy'x^2 + 2xy^2 + 4y^3y' = 4x$

At $(1, 1)$:

$2y' + 2 + 4y' = 4$

$6y' = 2$

$y' = \frac{1}{3}$.

Tangent line: $y - 1 = \frac{1}{3}(x - 1)$

$y = \frac{1}{3}x + \frac{2}{3}$

33. (a) $\dfrac{x^2}{2} + \dfrac{y^2}{8} = 1, \quad (1, 2)$

$x + \dfrac{yy'}{4} = 0$

$y' = -\dfrac{4x}{y}$

At $(1, 2)$: $y' = -2$.

Tangent line: $y - 2 = -2(x - 1)$

$y = -2x + 4$

Note: From part (a), $\dfrac{1(x)}{2} + \dfrac{2(y)}{8} = 1 \Rightarrow \dfrac{1}{4}y = -\dfrac{1}{2}x + 1 \Rightarrow y = -2x + 4$, Tangent line.

(b) $\dfrac{x^2}{a^2} + \dfrac{y^2}{b^2} = 1 \Rightarrow \dfrac{2x}{a^2} + \dfrac{2yy'}{b^2} = 0 \Rightarrow y' = \dfrac{-b^2x}{a^2y}$

$y - y_0 = \dfrac{-b^2x_0}{a^2y_0}(x - x_0)$, Tangent line at (x_0, y_0)

$\dfrac{y_0y}{b^2} - \dfrac{y_0^2}{b^2} = \dfrac{-x_0x}{a^2} + \dfrac{x_0^2}{a^2}$

Since $\dfrac{x_0^2}{a^2} + \dfrac{y_0^2}{b^2} = 1$, you have $\dfrac{y_0y}{b^2} + \dfrac{x_0x}{a^2} = 1$.

34. (a) $\frac{x^2}{6} - \frac{y^2}{8} = 1, \quad (3, -2)$

$$\frac{x}{3} - \frac{y}{4}y' = 0$$

$$\frac{y}{4}y' = \frac{x}{3}$$

$$y' = \frac{4x}{3y}$$

At $(3, -2)$: $y' = \frac{4(3)}{3(-2)} = -2.$

Tangent line: $y + 2 = -2(x - 3)$

$$y = -2x + 4$$

Note: From part (a), $\frac{3x}{6} - \frac{(-2)y}{8} = 1 \Rightarrow \frac{1}{2}x + \frac{y}{4} = 1 \Rightarrow y = -2x + 4,$ Tangent line.

(b) $\frac{x^2}{a^2} - \frac{y^2}{b^2} = 1 \Rightarrow \frac{2x}{a^2} - \frac{2yy'}{b^2} = 0 \Rightarrow y' = \frac{xb^2}{ya^2}$

$$y - y_0 = \frac{x_0 b^2}{y_0 a^2}(x - x_0), \quad \text{Tangent line at } (x_0, y_0)$$

$$\frac{yy_0}{b^2} - \frac{y_0^2}{b^2} = \frac{x_0 x}{a^2} - \frac{x_0^2}{a^2}$$

Since $\frac{x_0^2}{a^2} - \frac{y_0^2}{b^2} = 1$, you have $\frac{x_0 x}{a^2} - \frac{yy_0}{b^2} = 1.$

35. $x^2 + y^2 = 36$

$$2x + 2yy' = 0$$

$$y' = \frac{-x}{y}$$

$$y'' = \frac{y(-1) + xy'}{y^2} = \frac{-y + x\left(-\frac{x}{y}\right)}{y^2} = \frac{-y^2 - x^2}{y^3} = \frac{-36}{y^3}$$

36.

$$x^2y^2 - 2x = 3$$

$$2x^2yy' + 2xy^2 - 2 = 0$$

$$x^2yy' + xy^2 - 1 = 0$$

$$y' = \frac{1 - xy^2}{x^2y}$$

$$2xyy' + x^2(y')^2 + x^2yy'' + 2xyy' + y^2 = 0$$

$$4xyy' + x^2(y')^2 + x^2yy'' + y^2 = 0$$

$$\frac{4 - 4xy^2}{x} + \frac{(1 - xy^2)^2}{x^2y^2} + x^2yy'' + y^2 = 0$$

$$4xy^2 - 4x^2y^4 + 1 - 2xy^2 + x^2y^4 + x^4y^3y'' + x^2y^4 = 0$$

$$x^4y^3y'' = 2x^2y^4 - 2xy^2 - 1$$

$$y'' = \frac{2x^2y^4 - 2xy^2 - 1}{x^4y^3}$$

37.

$$x^2 - y^2 = 16$$
$$2x - 2yy' = 0$$
$$y' = \frac{x}{y}$$
$$x - yy' = 0$$
$$1 - yy'' - (y')^2 = 0$$
$$1 - yy'' - \left(\frac{x}{y}\right)^2 = 0$$
$$y^2 - y^3y'' = x^2$$
$$y'' = \frac{y^2 - x^2}{y^3} = \frac{-16}{y^3}$$

38.

$$1 - xy = x - y$$
$$y - xy = x - 1$$
$$y = \frac{x - 1}{1 - x} = -1$$
$$y' = 0$$
$$y'' = 0$$

39.

$$y^2 = x^3$$
$$2yy' = 3x^2$$
$$y' = \frac{3x^2}{2y} = \frac{3x^2}{2y} \cdot \frac{xy}{xy} = \frac{3y}{2x} \cdot \frac{x^3}{y^2} = \frac{3y}{2x}$$
$$y'' = \frac{2x(3y') - 3y(2)}{4x^2}$$
$$= \frac{2x[3 \cdot (3y/2x)] - 6y}{4x^2}$$
$$= \frac{3y}{4x^2} = \frac{3x}{4y}$$

40.

$$y^2 = 4x$$
$$2yy' = 4$$
$$y' = \frac{2}{y}$$
$$y'' = -2y^{-2}y' = \left[\frac{-2}{y^2}\right] \cdot \frac{2}{y} = \frac{-4}{y^3}$$

41.

$$\sqrt{x} + \sqrt{y} = 4$$
$$\frac{1}{2}x^{-1/2} + \frac{1}{2}y^{-1/2}y' = 0$$
$$y' = \frac{-\sqrt{y}}{\sqrt{x}}$$

At $(9, 1)$: $y' = -\frac{1}{3}$.

Tangent line: $y - 1 = -\frac{1}{3}(x - 9)$

$$y = -\frac{1}{3}x + 4$$
$$x + 3y - 12 = 0$$

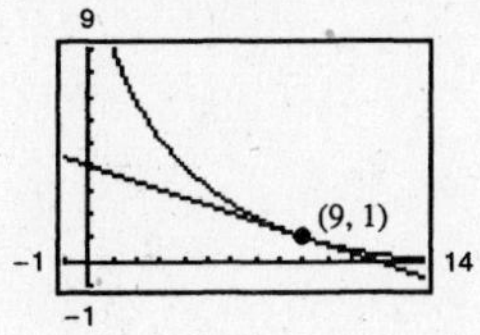

42.

$$y^2 = \frac{x - 1}{x^2 + 1}$$
$$2yy' = \frac{(x^2 + 1)(1) - (x - 1)(2x)}{(x^2 + 1)^2}$$
$$= \frac{x^2 + 1 - 2x^2 + 2x}{(x^2 + 1)^2}$$
$$y' = \frac{1 + 2x - x^2}{2y(x^2 + 1)^2}$$

At $\left(2, \frac{\sqrt{5}}{5}\right)$:

$$y' = \frac{1 + 4 - 4}{[(2\sqrt{5})/5](4 + 1)^2} = \frac{1}{10\sqrt{5}}.$$

Tangent line:

$$y - \frac{\sqrt{5}}{5} = \frac{1}{10\sqrt{5}}(x - 2)$$
$$10\sqrt{5}y - 10 = x - 2$$
$$x - 10\sqrt{5}y + 8 = 0$$

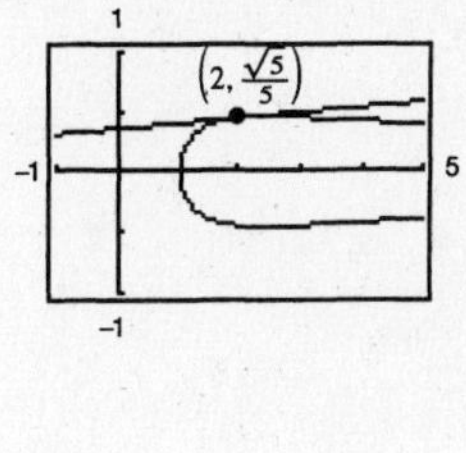

43. $x^2 + y^2 = 25$

$$2x + 2yy' = 0$$

$$y' = \frac{-x}{y}$$

At (4, 3):

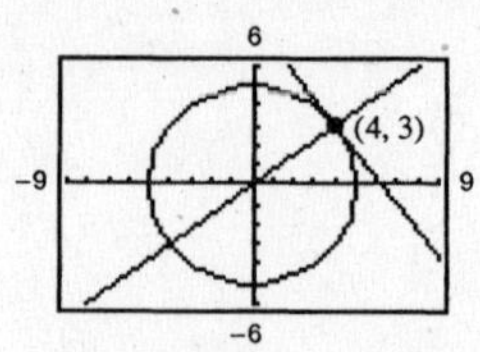

Tangent line: $y - 3 = \frac{-4}{3}(x - 4) \Rightarrow 4x + 3y - 25 = 0$

Normal line: $y - 3 = \frac{3}{4}(x - 4) \Rightarrow 3x - 4y = 0.$

At $(-3, 4)$:

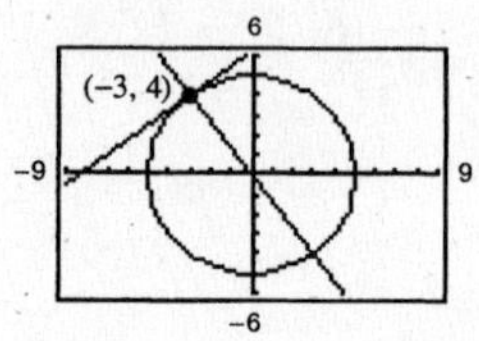

Tangent line: $y - 4 = \frac{3}{4}(x + 3) \Rightarrow 3x - 4y + 25 = 0$

Normal line: $y - 4 = \frac{-4}{3}(x + 3) \Rightarrow 4x + 3y = 0.$

44. $x^2 + y^2 = 9$

$$y' = \frac{-x}{y}$$

At (0, 3):

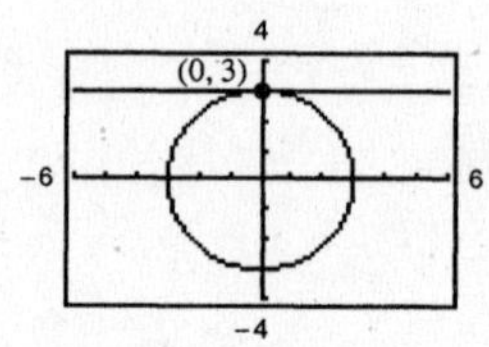

Tangent line: $y = 3$

Normal line: $x = 0.$

At $(2, \sqrt{5})$:

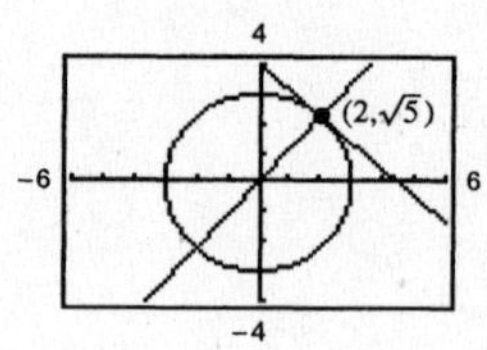

Tangent line: $y - \sqrt{5} = \frac{-2}{\sqrt{5}}(x - 2) \Rightarrow 2x + \sqrt{5}y - 9 = 0$

Normal line: $y - \sqrt{5} = \frac{\sqrt{5}}{2}(x - 2) \Rightarrow \sqrt{5}x - 2y = 0.$

45. $x^2 + y^2 = r^2$

$$2x + 2yy' = 0$$

$$y' = \frac{-x}{y} = \text{slope of tangent line}$$

$$\frac{y}{x} = \text{slope of normal line}$$

Let (x_0, y_0) be a point on the circle. If $x_0 = 0$, then the tangent line is horizontal, the normal line is vertical and, hence, passes through the origin. If $x_0 \neq 0$, then the equation of the normal line is

$$y - y_0 = \frac{y_0}{x_0}(x - x_0)$$

$$y = \frac{y_0}{x_0}x$$

which passes through the origin.

46. $y^2 = 4x$

$2yy' = 4$

$y' = \frac{2}{y} = 1$ at $(1, 2)$.

Equation of normal at $(1, 2)$ is $y - 2 = -1(x - 1)$, $y = 3 - x$. The centers of the circles must be on the normal and at a distance of 4 units from $(1, 2)$. Therefore,

$$(x - 1)^2 + [(3 - x) - 2]^2 = 16$$
$$2(x - 1)^2 = 16$$
$$x = 1 \pm 2\sqrt{2}.$$

Centers of the circles: $\left(1 + 2\sqrt{2}, 2 - 2\sqrt{2}\right)$ and $\left(1 - 2\sqrt{2}, 2 + 2\sqrt{2}\right)$

Equations: $\left(x - 1 - 2\sqrt{2}\right)^2 + \left(y - 2 + 2\sqrt{2}\right)^2 = 16$

$\left(x - 1 + 2\sqrt{2}\right)^2 + \left(y - 2 - 2\sqrt{2}\right)^2 = 16$

47. $25x^2 + 16y^2 + 200x - 160y + 400 = 0$

$$50x + 32yy' + 200 - 160y' = 0$$
$$y' = \frac{200 + 50x}{160 - 32y}$$

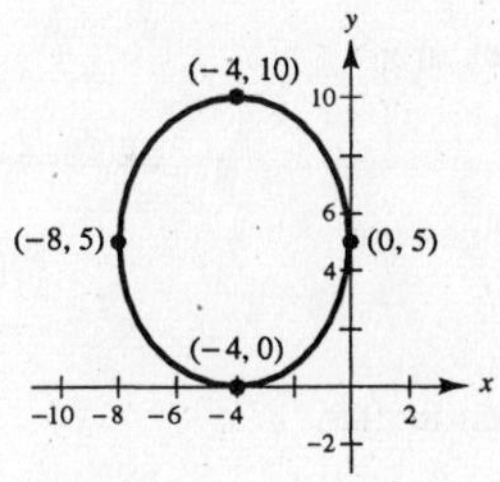

Horizontal tangents occur when $x = -4$:

$$25(16) + 16y^2 + 200(-4) - 160y + 400 = 0$$
$$y(y - 10) = 0 \Rightarrow y = 0, 10$$

Horizontal tangents: $(-4, 0), (-4, 10)$.

Vertical tangents occur when $y = 5$:

$$25x^2 + 400 + 200x - 800 + 400 = 0$$
$$25x(x + 8) = 0 \Rightarrow x = 0, -8$$

Vertical tangents: $(0, 5), (-8, 5)$

48. $4x^2 + y^2 - 8x + 4y + 4 = 0$

$$8x + 2yy' - 8 + 4y' = 0$$
$$y' = \frac{8 - 8x}{2y + 4} = \frac{4 - 4x}{y + 2}$$

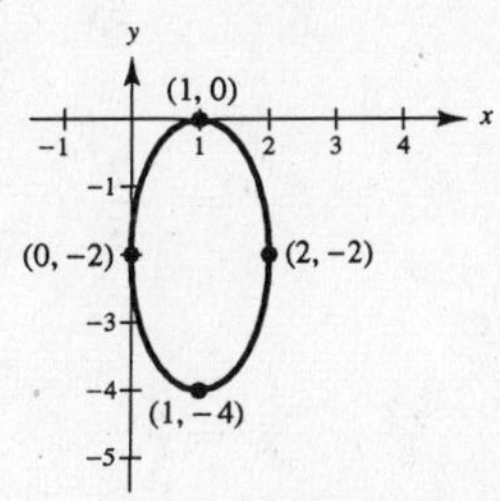

Horizontal tangents occur when $x = 1$:

$$4(1)^2 + y^2 - 8(1) + 4y + 4 = 0$$
$$y^2 + 4y = y(y + 4) = 0 \Rightarrow y = 0, -4$$

Horizontal tangents: $(1, 0), (1, -4)$.

Vertical tangents occur when $y = -2$:

$$4x^2 + (-2)^2 - 8x + 4(-2) + 4 = 0$$
$$4x^2 - 8x = 4x(x - 2) = 0 \Rightarrow x = 0, 2$$

Vertical tangents: $(0, -2), (2, -2)$

49. In the explicit form of a function, the variable is explicitly written as a function of x. In an implicit equation, the function is only implied by the equation. An example of an implicit function is $x^2 + xy = 5$. In explicit form, it would be

$$y = \frac{5 - x^2}{x}.$$

50. Given an implicit equation, first differentiate both sides with respect to x. Collect all terms involving y' on the left, and all other terms to the right. Factor out y' on the left side. Finally, divide both sides by the left-hand factor that does not contain y'.

51. Find the points of intersection by letting $y^2 = 4x$ in the equation $2x^2 + y^2 = 6$.

$2x^2 + 4x = 6$ and $(x + 3)(x - 1) = 0$

The curves intersect at $(1, \pm 2)$.

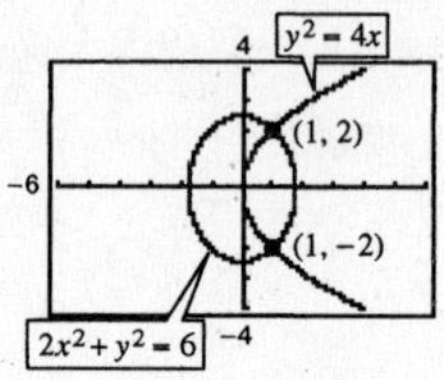

Ellipse:	Parabola:
$4x + 2yy' = 0$	$2yy' = 4$
$y' = -\frac{2x}{y}$	$y' = \frac{2}{y}$

At $(1, 2)$, the slopes are:

$y' = -1$ $\qquad$ $y' = 1.$

At $(1, -2)$, the slopes are:

$y' = 1$ $\qquad$ $y' = -1.$

Tangents are perpendicular.

52. Find the points of intersection by letting $y^2 = x^3$ in the equation $2x^2 + 3y^2 = 5$.

$2x^2 + 3x^3 = 5$ and $3x^3 + 2x^2 - 5 = 0$

Intersect when $x = 1$.

Points of intersection: $(1, \pm 1)$

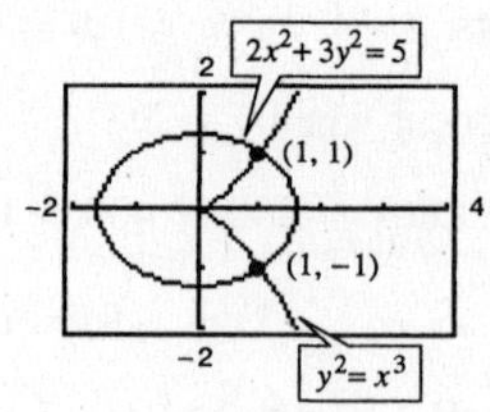

$y^2 = x^3$:	$2x^2 + 3y^2 = 5$:
$2yy' = 3x^2$	$4x + 6yy' = 0$
$y' = \frac{3x^2}{2y}$	$y' = -\frac{2x}{3y}$

At $(1, 1)$, the slopes are:

$y' = \frac{3}{2}$ $\qquad$ $y' = -\frac{2}{3}.$

At $(1, -1)$, the slopes are:

$y' = -\frac{3}{2}$ $\qquad$ $y' = \frac{2}{3}.$

Tangents are perpendicular.

53. $x + y = 0$

$x^2 + y^2 = 4$

The line and circle intersect at $(\sqrt{2}, -\sqrt{2})$ and $(-\sqrt{2}, \sqrt{2})$.

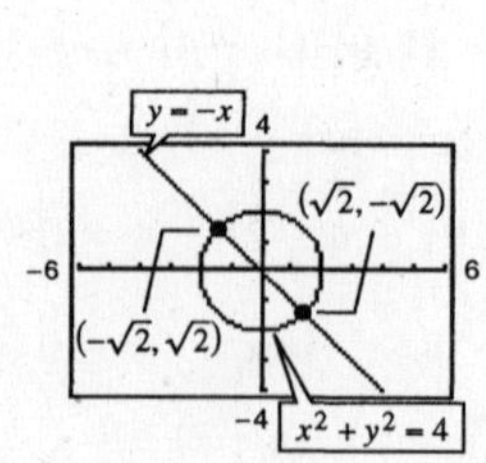

$x + y = 0 \Rightarrow y' = -1$

$$x^2 + y^2 = 4 \Rightarrow 2x + 2y\frac{dy}{dx} = 0 \Rightarrow \frac{dy}{dx} = -\frac{x}{y}$$

At $(-\sqrt{2}, \sqrt{2})$, slope of line is -1, slope of circle is 1.

At $(\sqrt{2}, -\sqrt{2})$, slope of line is -1, slope of circle is 1.

54. Rewriting each equation and differentiating,

$$x^3 = 3(y - 1) \qquad x(3y - 29) = 3$$

$$y = \frac{x^3}{3} + 1 \qquad y = \frac{1}{3}\left(\frac{3}{x} + 29\right)$$

$$y' = x^2 \qquad y' = -\frac{1}{x^2}.$$

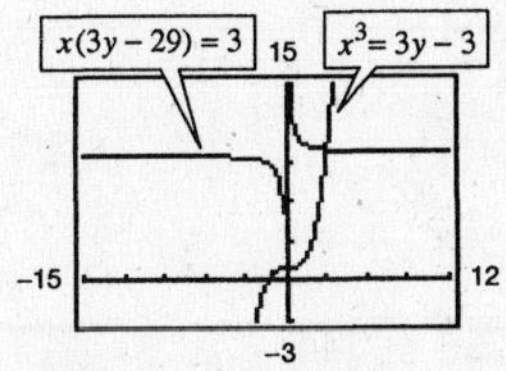

For each value of x, the derivatives are negative reciprocals of each other. Thus, the tangent lines are orthogonal at both points of intersection.

55. $xy = C \qquad x^2 - y^2 = K$

$$xy' + y = 0 \qquad 2x - 2yy' = 0$$

$$y' = -\frac{y}{x} \qquad y' = \frac{x}{y}$$

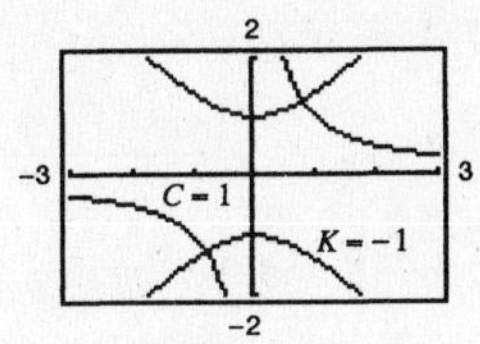

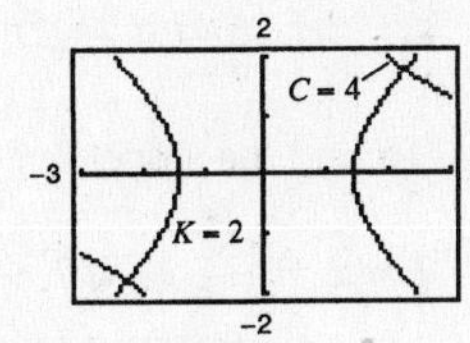

At any point of intersection (x, y) the product of the slopes is $(-y/x)(x/y) = -1$. The curves are orthogonal.

56. $x^2 + y^2 = C^2 \qquad y = Kx$

$$2x + 2yy' = 0 \qquad y' = K$$

$$y' = -\frac{x}{y}$$

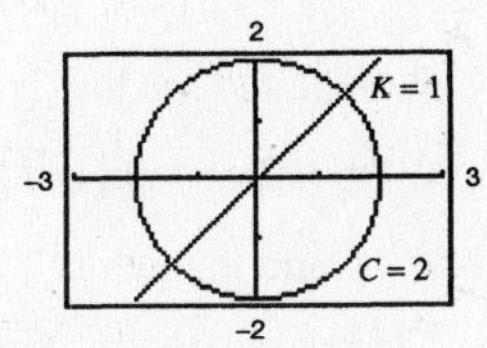

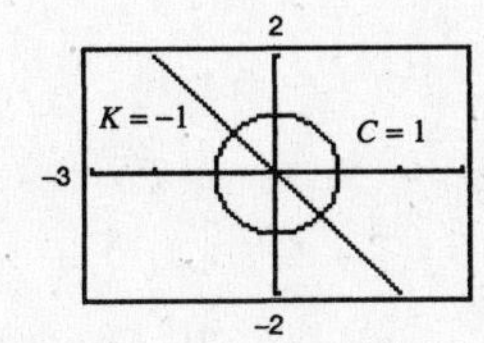

At the point of intersection (x, y), the product of the slopes is $(-x/y)(K) = (-x/Kx)(K) = -1$. The curves are orthogonal.

57. $2y^2 - 3x^4 = 0$

(a) $4yy' - 12x^3 = 0$

$$4yy' = 12x^3$$

$$y' = \frac{12x^3}{4y} = \frac{3x^3}{y}$$

(b) $4y\frac{dy}{dt} - 12x^3\frac{dx}{dt} = 0$

$$y\frac{dy}{dt} = 3x^3\frac{dx}{dt}$$

58. $x^2 - 3xy^2 + y^3 = 10$

(a) $2x - 3y^2 - 6xyy' + 3y^2y' = 0$

$$(-6xy + 3y^2)y' = 3y^2 - 2x$$

$$y' = \frac{3y^2 - 2x}{3y^2 - 6xy}$$

(b) $2x\frac{dx}{dt} - 3y^2\frac{dx}{dt} - 6xy\frac{dy}{dt} + 3y^2\frac{dy}{dt} = 0$

$$(2x - 3y^2)\frac{dx}{dt} = (6xy - 3y^2)\frac{dy}{dt}$$

59. (a) $x^4 = 4(4x^2 - y^2)$

$$4y^2 = 16x^2 - x^4$$

$$y^2 = 4x^2 - \frac{1}{4}x^4$$

$$y = \pm\sqrt{4x^2 - \frac{1}{4}x^4}$$

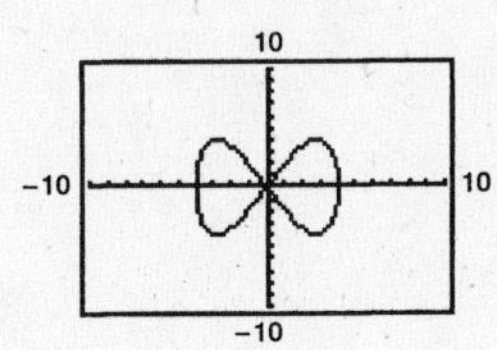

—CONTINUED—

59. —CONTINUED—

(b) $y = 3 \Rightarrow 9 = 4x^2 - \frac{1}{4}x^4$

$$36 = 16x^2 - x^4$$

$$x^4 - 16x^2 + 36 = 0$$

$$x^2 = \frac{16 \pm \sqrt{256 - 144}}{2} = 8 \pm \sqrt{28}$$

Note that $x^2 = 8 \pm \sqrt{28} = 8 \pm 2\sqrt{7} = (1 \pm \sqrt{7})^2$.

Hence, there are four values of x:

$$-1-\sqrt{7}, 1-\sqrt{7}, -1+\sqrt{7}, 1 + \sqrt{7}$$

To find the slope, $2yy' = 8x - x^3 \Rightarrow y' = \frac{x(8 - x^2)}{2(3)}$.

For $x = -1 - \sqrt{7}$, $y' = \frac{1}{3}(\sqrt{7} + 7)$, and the line is

$$y_1 = \tfrac{1}{3}(\sqrt{7} + 7)(x + 1 + \sqrt{7}) + 3 = \tfrac{1}{3}[(\sqrt{7} + 7)x + 8\sqrt{7} + 23)].$$

For $x = 1 - \sqrt{7}$, $y' = \frac{1}{3}(\sqrt{7} - 7)$, and the line is

$$y_2 = \tfrac{1}{3}(\sqrt{7} - 7)(x - 1 + \sqrt{7}) + 3 = \tfrac{1}{3}[(\sqrt{7} - 7)x + 23 - 8\sqrt{7}].$$

For $x = -1 + \sqrt{7}$, $y' = -\frac{1}{3}(\sqrt{7} - 7)$, and the line is

$$y_3 = -\tfrac{1}{3}(\sqrt{7} - 7)(x + 1 - \sqrt{7}) + 3 = -\tfrac{1}{3}[(\sqrt{7} - 7)x - (23 - 8\sqrt{7})].$$

For $x = 1 + \sqrt{7}$, $y' = -\frac{1}{3}(\sqrt{7} + 7)$, and the line is

$$y_4 = -\tfrac{1}{3}(\sqrt{7} + 7)(x - 1 - \sqrt{7}) + 3 = -\tfrac{1}{3}[(\sqrt{7} + 7)x - (8\sqrt{7} + 23)].$$

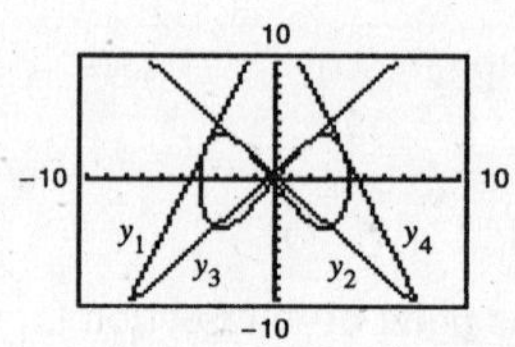

(c) Equating y_3 and y_4,

$$-\frac{1}{3}(\sqrt{7} - 7)(x + 1 - \sqrt{7}) + 3 = -\frac{1}{3}(\sqrt{7} + 7)(x - 1 - \sqrt{7}) + 3$$

$$(\sqrt{7} - 7)(x + 1 - \sqrt{7}) = (\sqrt{7} + 7)(x - 1 - \sqrt{7})$$

$$\sqrt{7}x + \sqrt{7} - 7 - 7x - 7 + 7\sqrt{7} = \sqrt{7}x - \sqrt{7} - 7 + 7x - 7 - 7\sqrt{7}$$

$$16\sqrt{7} = 14x$$

$$x = \frac{8\sqrt{7}}{7}$$

If $x = \frac{8\sqrt{7}}{7}$, then $y = 5$ and the lines intersect at $\left(\frac{8\sqrt{7}}{7}, 5\right)$.

60.

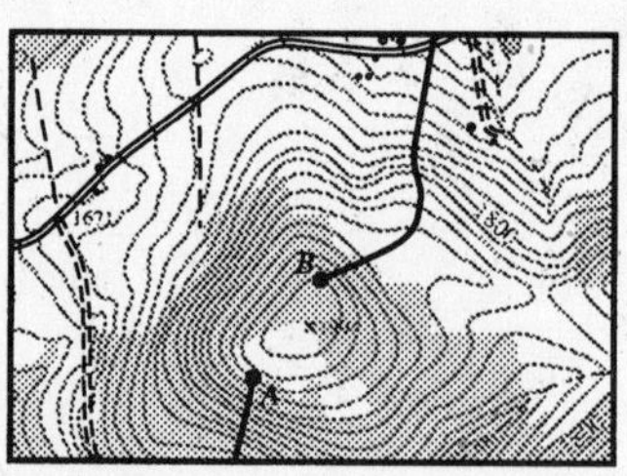

Use starting point B.

61. Suppose $n = \frac{p}{q}$ with p, q integers $q > 0$. If $q = 1$, then

$n = p$ and $\frac{d}{dx}[x^n] = nx^{n-1}$ by the integer case.

So suppose $q > 1$,

$y = x^{p/q}$ implies $y^q = x^p$ so

$$qy^{q-1}y' = px^{p-1}$$

$$y' = \frac{p}{q}\frac{x^{p-1}}{y^{q-1}}$$

$$= \frac{p}{q}\frac{x^{p-1}}{(x^{p/q})^{q-1}}$$

$$= \frac{p}{q}x^{p/q-1}$$

$$= nx^{n-1}.$$

62. $\sqrt{x} + \sqrt{y} = \sqrt{c}$

$$\frac{1}{2\sqrt{x}} + \frac{1}{2\sqrt{y}}\frac{dy}{dx} = 0$$

$$\frac{dy}{dx} = -\frac{\sqrt{y}}{\sqrt{x}}$$

Tangent line at (x_0, y_0):

$$y - y_0 = -\frac{\sqrt{y_0}}{\sqrt{x_0}}(x - x_0)$$

x-intercept: $\left(x_0 + \sqrt{x_0}\sqrt{y_0}, 0\right)$

y-intercept: $\left(0, y_0 + \sqrt{x_0}\sqrt{y_0}\right)$

Sum of intercepts:

$$\left(x_0 + \sqrt{x_0}\sqrt{y_0}\right) + \left(y_0 + \sqrt{x_0}\sqrt{y_0}\right)$$
$$= x_0 + 2\sqrt{x_0}\sqrt{y_0} + y_0 = \left(\sqrt{x_0} + \sqrt{y_0}\right)^2$$
$$= \left(\sqrt{c}\right)^2 = c.$$

Section 4.6 Related Rates

1. $y = \sqrt{x}$

$$\frac{dy}{dt} = \left(\frac{1}{2\sqrt{x}}\right)\frac{dx}{dt}$$

$$\frac{dx}{dt} = 2\sqrt{x}\frac{dy}{dt}$$

(a) When $x = 4$ and $dx/dt = 3$,

$$\frac{dy}{dt} = \frac{1}{2\sqrt{4}}(3) = \frac{3}{4}.$$

(b) When $x = 25$ and $dy/dt = 2$,

$$\frac{dx}{dt} = 2\sqrt{25}\,(2) = 20.$$

2. $y = 2(x^2 - 3x)$

$$\frac{dy}{dt} = (4x - 6)\frac{dx}{dt}$$

$$\frac{dx}{dt} = \frac{1}{4x - 6}\frac{dy}{dt}$$

(a) When $x = 3$ and $\frac{dx}{dt} = 2$, $\frac{dy}{dt} = [4(3) - 6](2) = 12$

(b) When $x = 1$ and $\frac{dy}{dt} = 5$, $\frac{dx}{dt} = \frac{1}{4(1) - 6}(5) = -\frac{5}{2}$

3. $xy = 4$

$$x\frac{dy}{dt} + y\frac{dx}{dt} = 0$$

$$\frac{dy}{dt} = \left(-\frac{y}{x}\right)\frac{dx}{dt}$$

$$\frac{dx}{dt} = \left(-\frac{x}{y}\right)\frac{dy}{dt}$$

(a) When $x = 8$, $y = 1/2$, and $dx/dt = 10$,

$$\frac{dy}{dt} = -\frac{1/2}{8}(10) = -\frac{5}{8}.$$

(b) When $x = 1$, $y = 4$, and $dy/dt = -6$,

$$\frac{dx}{dt} = -\frac{1}{4}(-6) = \frac{3}{2}.$$

4. $x^2 + y^2 = 25$

$$2x\frac{dx}{dt} + 2y\frac{dy}{dt} = 0$$

$$\frac{dy}{dt} = \left(-\frac{x}{y}\right)\frac{dx}{dt}$$

$$\frac{dx}{dt} = \left(-\frac{y}{x}\right)\frac{dy}{dt}$$

(a) When $x = 3$, $y = 4$, and $dx/dt = 8$,

$$\frac{dy}{dt} = -\frac{3}{4}(8) = -6$$

(b) When $x = 4$, $y = 3$, and $dy/dt = -2$,

$$\frac{dx}{dt} = -\frac{3}{4}(-2) = \frac{3}{2}.$$

5. $y = x^2 + 1$

$\frac{dx}{dt} = 2$

$\frac{dy}{dt} = 2x\frac{dx}{dt}$

(a) When $x = -1$,

$$\frac{dy}{dt} = 2(-1)(2) = -4 \text{ cm/sec.}$$

(b) When $x = 0$,

$$\frac{dy}{dt} = 2(0)(2) = 0 \text{ cm/sec.}$$

(c) When $x = 1$,

$$\frac{dy}{dt} = 2(1)(2) = 4 \text{ cm/sec.}$$

6. $y = \frac{1}{1 + x^2}$

$\frac{dx}{dt} = 2$

$\frac{dy}{dt} = \left[\frac{-2x}{(1 + x^2)^2}\right]\frac{dx}{dt}$

(a) When $x = -2$,

$$\frac{dy}{dt} = \frac{-2(-2)(2)}{25} = \frac{8}{25} \text{ cm/sec.}$$

(b) When $x = 0$,

$$\frac{dy}{dt} = 0 \text{ cm/sec.}$$

(c) When $x = 2$,

$$\frac{dy}{dt} = \frac{-2(2)(2)}{25} = \frac{-8}{25} \text{ cm/sec.}$$

7. $\sqrt{x} + \sqrt{y} = 4$

$\frac{1}{2\sqrt{x}}\frac{dx}{dt} + \frac{1}{2\sqrt{y}}\frac{dy}{dt} = 0$

$\frac{1}{2\sqrt{y}}\frac{dy}{dt} = \frac{-1}{2\sqrt{x}}\frac{dx}{dt}$

$\frac{dy}{dt} = -\frac{\sqrt{y}}{\sqrt{x}}\frac{dx}{dt} = -\frac{2\sqrt{y}}{\sqrt{x}}$

(a) $x = 1, y = 9, \frac{dy}{dt} = -\frac{2(3)}{1} = -6$ cm/sec

(b) $x = 4, y = 4, \frac{dy}{dt} = -2\left(\frac{2}{2}\right) = -2$ cm/sec

(c) $x = 9, y = 1, \frac{dy}{dt} = -\frac{2(1)}{3} = -\frac{2}{3}$ cm/sec

8. $x^2 + \sqrt{y} + y = 3$

$2x\frac{dx}{dt} + \frac{1}{2\sqrt{y}}\frac{dy}{dt} + \frac{dy}{dt} = 0$

$\left(1 + \frac{1}{2\sqrt{y}}\right)\frac{dy}{dt} = -2x\frac{dx}{dt} = -4x$

$\frac{dy}{dt} = \frac{-4x}{1 + \frac{1}{2\sqrt{y}}} = \frac{-8x\sqrt{y}}{1 + 2\sqrt{y}}$

(a) $x = -1, y = 1, \frac{dy}{dt} = \frac{8}{3}$

(b) $x = 0, \frac{dy}{dt} = 0$

(c) $x = 1, y = 1, \frac{dy}{dt} = \frac{-8}{3}$

9. (a) $\frac{dx}{dt}$ negative $\Rightarrow \frac{dy}{dt}$ positive

(b) $\frac{dy}{dt}$ positive $\Rightarrow \frac{dx}{dt}$ negative

10. (a) $\frac{dx}{dt}$ negative $\Rightarrow \frac{dy}{dt}$ negative

(b) $\frac{dy}{dt}$ positive $\Rightarrow \frac{dx}{dt}$ positive

11. Yes, y changes at a constant rate: $\frac{dy}{dt} = a \cdot \frac{dx}{dt}$.

No, the rate $\frac{dy}{dt}$ is a multiple of $\frac{dx}{dt}$.

12. Answers will vary.

13. $D = \sqrt{x^2 + y^2} = \sqrt{x^2 + (x^2 + 1)^2} = \sqrt{x^4 + 3x^2 + 1}$

$\frac{dx}{dt} = 2$

$$\frac{dD}{dt} = \frac{1}{2}(x^4 + 3x^2 + 1)^{-1/2}(4x^3 + 6x)\frac{dx}{dt} = \frac{2x^3 + 3x}{\sqrt{x^4 + 3x^2 + 1}}\frac{dx}{dt} = \frac{4x^3 + 6x}{\sqrt{x^4 + 3x^2 + 1}}$$

14. $D = \sqrt{x^2 + y^2} = \sqrt{x^2 + x}$

$$\frac{dx}{dt} = 2$$

$$\frac{dD}{dt} = \frac{1}{2}(x^2 + x)^{-1/2}(2x + 1)\frac{dx}{dt}$$

$$= \frac{2x + 1}{2\sqrt{x^2 + x}}\frac{dx}{dt}$$

$$= \frac{2x + 1}{\sqrt{x^2 + x}} \text{ cm/sec}$$

15. $A = \pi r^2$

$$\frac{dr}{dt} = 3$$

$$\frac{dA}{dt} = 2\pi r\frac{dr}{dt}$$

(a) When $r = 6$,

$$\frac{dA}{dt} = 2\pi(6)(3) = 36\pi \text{ cm}^2/\text{min}.$$

(b) When $r = 24$,

$$\frac{dA}{dt} = 2\pi(24)(3) = 144\pi \text{ cm}^2/\text{min}.$$

16. $A = \pi r^2$

$$\frac{dA}{dt} = 2\pi r\frac{dr}{dt}$$

If dr/dt is constant, dA/dt is not constant.

$\frac{dA}{dt}$ depends on r and $\frac{dr}{dt}$.

17. (a) $h^2 + \left(\frac{b}{2}\right)^2 = 30^2$

$$h = \sqrt{900 - b^2/4}$$

$$A = \frac{1}{2}bh = \frac{1}{2}b\sqrt{900 - b^2/4}$$

$$= \frac{b\sqrt{3600 - b^2}}{4}$$

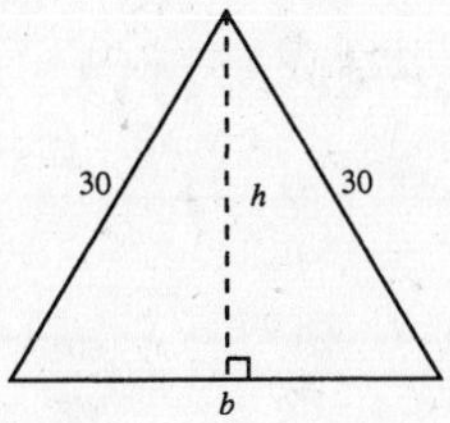

(b) $\frac{dA}{dt} = \frac{1}{4}\left[-b^2(3600 - b^2)^{-1/2} + (3600 - b^2)^{1/2}\right]\frac{db}{dt}$

$$= \frac{-b^2 + (3600 - b^2)}{4\sqrt{3600 - b^2}}(3)$$

$$= \frac{5400 - 3b^2}{2\sqrt{3600 - b^2}}$$

If $b = 20$, $\frac{dA}{dt} = \frac{105\sqrt{2}}{4} \approx 37.1$ cm/sec.

If $b = 56$, $\frac{dA}{dt} = -\frac{501\sqrt{29}}{29} \approx -93.0$ cm/sec.

(c) If $\frac{db}{dt}$ is constant, $\frac{dA}{dt}$ is a nonconstant function of b.

18. $V = \frac{4}{3}\pi r^3$

$$\frac{dr}{dt} = 2$$

$$\frac{dV}{dt} = 4\pi r^2\frac{dr}{dt}$$

(a) When $r = 6$, $\frac{dV}{dt} = 4\pi(6)^2(2) = 288\pi$ in^3/min.

When $r = 24$, $\frac{dV}{dt} = 4\pi(24)^2(2) = 4608\pi$ in^3/min.

(b) If dr/dt is constant, dV/dt is proportional to r^2.

19. $V = \frac{4}{3}\pi r^3$, $\frac{dV}{dt} = 800$

$$\frac{dV}{dt} = 4\pi r^2\frac{dr}{dt}$$

$$\frac{dr}{dt} = \frac{1}{4\pi r^2}\left(\frac{dV}{dt}\right) = \frac{1}{4\pi r^2}(800)$$

(a) When $r = 30$, $\frac{dr}{dt} = \frac{1}{4\pi(30)^2}(800) = \frac{2}{9\pi}$ cm/min.

(b) When $r = 60$, $\frac{dr}{dt} = \frac{1}{4\pi(60)^2}(800) = \frac{1}{18\pi}$ cm/min.

20. $V = x^3$

$\frac{dx}{dt} = 3$

$\frac{dV}{dt} = 3x^2 \frac{dx}{dt}$

(a) When $x = 1$,

$\frac{dV}{dt} = 3(1)^2(3) = 9 \text{ cm}^3/\text{sec.}$

(b) When $x = 10$,

$\frac{dV}{dt} = 3(10)^2(3) = 900 \text{ cm}^3/\text{sec.}$

21. $s = 6x^2$

$\frac{dx}{dt} = 3$

$\frac{ds}{dt} = 12x \frac{dx}{dt}$

(a) When $x = 1, \frac{ds}{dt} = 12(1)(3) = 36 \text{ cm}^2/\text{sec.}$

(b) When $x = 10, \frac{ds}{dt} = 12(10)(3) = 360 \text{ cm}^2/\text{sec.}$

22. $V = \frac{1}{3}\pi r^2 h = \frac{1}{3}\pi r^2(3r) = \pi r^3$

$\frac{dr}{dt} = 2$

$\frac{dV}{dt} = 3\pi r^2 \frac{dr}{dt}$

(a) When $r = 6$,

$\frac{dV}{dt} = 3\pi(6)^2(2) = 216\pi \text{ in}^3/\text{min.}$

(b) When $r = 24$,

$\frac{dV}{dt} = 3\pi(24)^2(2) = 3456\pi \text{ in}^3/\text{min.}$

23. $V = \frac{1}{3}\pi r^2 h = \frac{1}{3}\pi\left(\frac{9}{4}h^2\right)h$ [since $2r = 3h$]

$= \frac{3\pi}{4}h^3$

$\frac{dV}{dt} = 10$

$\frac{dV}{dt} = \frac{9\pi}{4}h^2 \frac{dh}{dt} \Rightarrow \frac{dh}{dt} = \frac{4(dV/dt)}{9\pi h^2}$

When $h = 15, \frac{dh}{dt} = \frac{4(10)}{9\pi(15)^2} = \frac{8}{405\pi}$ ft/min.

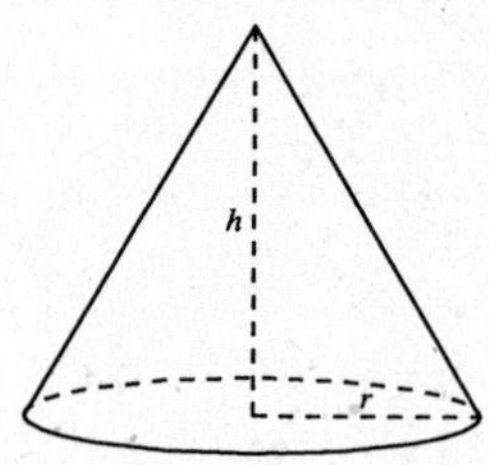

24. $V = \frac{1}{3}\pi r^2 h = \frac{1}{3}\pi \frac{25}{144}h^3 = \frac{25\pi}{3(144)}h^3$

$\left(\text{By similar triangles, } \frac{r}{5} = \frac{h}{12} \Rightarrow r = \frac{5}{12}h.\right)$

$\frac{dV}{dt} = 10$

$\frac{dV}{dt} = \frac{25\pi}{144}h^2 \frac{dh}{dt} \Rightarrow \frac{dh}{dt} = \left(\frac{144}{25\pi h^2}\right)\frac{dV}{dt}$

When $h = 8, \frac{dh}{dt} = \frac{144}{25\pi(64)}(10) = \frac{9}{10\pi}$ ft/min.

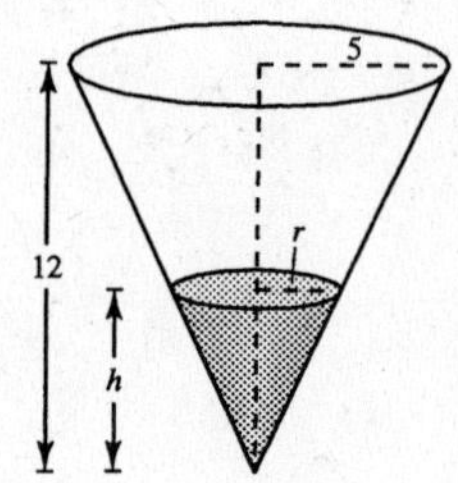

25.

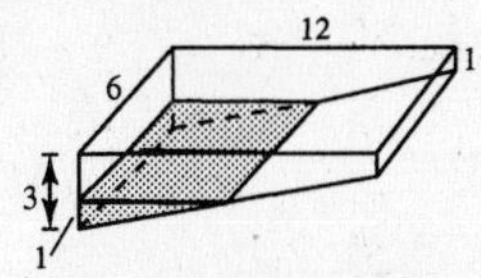

(a) Total volume of pool $= \frac{1}{2}(2)(12)(6) + (1)(6)(12) = 144 \text{ m}^3$

Volume of 1m. of water $= \frac{1}{2}(1)(6)(6) = 18 \text{ m}^3$

(see similar triangle diagram)

% pool filled $= \frac{18}{144}(100\%) = 12.5\%$

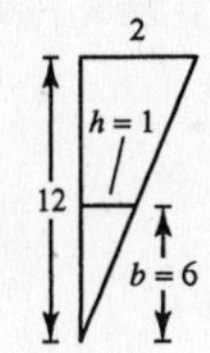

(b) Since for $0 \le h \le 2$, $b = 6h$, you have

$$V = \frac{1}{2}bh(6) = 3bh = 3(6h)h = 18h^2$$

$$\frac{dV}{dt} = 36h\frac{dh}{dt} = \frac{1}{4} \Rightarrow \frac{dh}{dt} = \frac{1}{144h} = \frac{1}{144(1)} = \frac{1}{144} \text{ m/min.}$$

26. $V = \frac{1}{2}bh(12) = 6bh = 6h^2$ (since $b = h$)

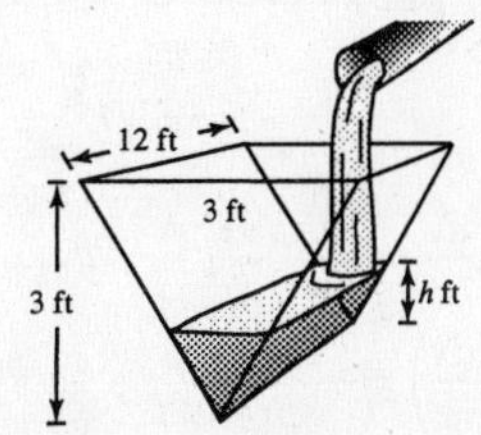

(a) $\frac{dV}{dt} = 12h\frac{dh}{dt} \Rightarrow \frac{dh}{dt} = \frac{1}{12h}\frac{dV}{dt}$

When $h = 1$ and $\frac{dV}{dt} = 2$, $\frac{dh}{dt} = \frac{1}{12(1)}(2) = \frac{1}{6}$ ft/min

(b) If $\frac{dh}{dt} = \frac{3}{8}$ inches/minute $= \frac{3}{8} \cdot \frac{1}{12} = \frac{1}{32}$ ft/minute and $h = 2$, then $\frac{dV}{dt} = 12(2)\left(\frac{1}{32}\right) = \frac{3}{4}$ ft³/min.

27. $x^2 + y^2 = 25^2$

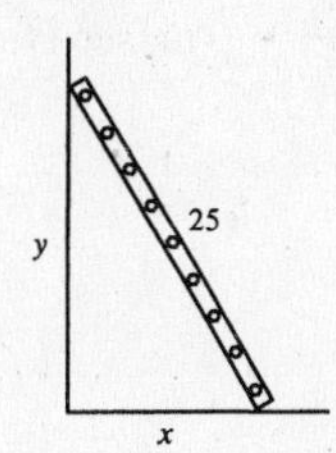

$$2x\frac{dx}{dt} + 2y\frac{dy}{dt} = 0$$

$$\frac{dy}{dt} = \frac{-x}{y} \cdot \frac{dx}{dt} = \frac{-2x}{y} \text{ since } \frac{dx}{dt} = 2.$$

(a) When $x = 7$, $y = \sqrt{576} = 24$, $\frac{dy}{dt} = \frac{-2(7)}{24} = \frac{-7}{12}$ ft/sec.

When $x = 15$, $y = \sqrt{400} = 20$, $\frac{dy}{dt} = \frac{-2(15)}{20} = \frac{-3}{2}$ ft/sec.

When $x = 24$, $y = 7$, $\frac{dy}{dt} = \frac{-2(24)}{7} = \frac{-48}{7}$ ft/sec.

(b) $A = \frac{1}{2}xy$

$$\frac{dA}{dt} = \frac{1}{2}\left(x\frac{dy}{dt} + y\frac{dx}{dt}\right)$$

From part (a) we have $x = 7$, $y = 24$, $\frac{dx}{dt} = 2$, and $\frac{dy}{dt} = -\frac{7}{12}$.

Thus, $\frac{dA}{dt} = \frac{1}{2}\left[7\left(-\frac{7}{12}\right) + 24(2)\right] = \frac{527}{24} \approx 21.96$ ft²/sec.

28. $x^2 + y^2 = 25$

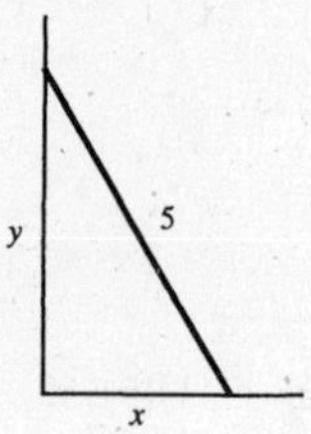

$$2x\frac{dx}{dt} + 2y\frac{dy}{dt} = 0$$

$$\frac{dx}{dt} = -\frac{y}{x}\cdot\frac{dy}{dt} = -\frac{0.15y}{x} \text{ since } \frac{dy}{dt} = 0.15.$$

When $x = 2.5$,

$$y = \sqrt{18.75}, \frac{dx}{dt} = -\frac{\sqrt{18.75}}{2.5}0.15 \approx -0.26 \text{ m/sec}$$

29. When $y = 6$, $x = \sqrt{12^2 - 6^2} = 6\sqrt{3}$, and

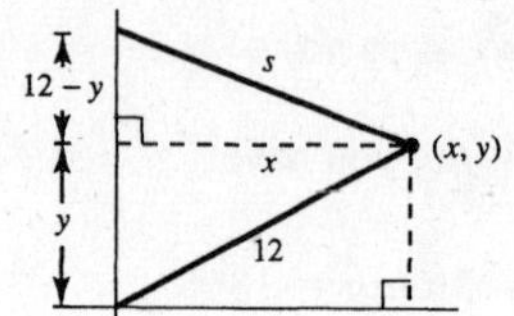

$$s = \sqrt{x^2 + (12 - y)^2}$$
$$= \sqrt{108 + 36} = 12.$$

$$x^2 + (12 - y)^2 = s^2$$

$$2x\frac{dx}{dt} + 2(12 - y)(-1)\frac{dy}{dt} = 2s\frac{ds}{dt}$$

$$x\frac{dx}{dt} + (y - 12)\frac{dy}{dt} = s\frac{ds}{dt}$$

Also, $x^2 + y^2 = 12^2$

$$2x\frac{dx}{dt} + 2y\frac{dy}{dt} = 0 \Rightarrow \frac{dy}{dt} = \frac{-x}{y}\frac{dx}{dt}.$$

Thus, $x\dfrac{dx}{dt} + (y - 12)\left(\dfrac{-x}{y}\dfrac{dx}{dt}\right) = s\dfrac{ds}{dt}$

$$\frac{dx}{dt}\left[x - x + \frac{12x}{y}\right] = s\frac{ds}{dt} \Rightarrow \frac{dx}{dt} = \frac{sy}{12x}\cdot\frac{ds}{dt} = \frac{(12)(6)}{(12)(6\sqrt{3})}(-0.2) = \frac{-1}{5\sqrt{3}} = \frac{-\sqrt{3}}{15} \text{ m/sec (horizontal)}$$

$$\frac{dy}{dt} = \frac{-x}{y}\frac{dx}{dt} = \frac{-6\sqrt{3}}{6}\cdot\frac{(-\sqrt{3})}{15} = \frac{1}{5} \text{ m/sec (vertical).}$$

30. Let L be the length of the rope.

(a) $L^2 = 144 + x^2$

$$2L\frac{dL}{dt} = 2x\frac{dx}{dt}$$

$$\frac{dx}{dt} = \frac{L}{x}\cdot\frac{dL}{dt} = -\frac{4L}{x} \text{ since } \frac{dL}{dt} = -4 \text{ ft/sec.}$$

When $L = 13$,

$$x = \sqrt{L^2 - 144} = \sqrt{169 - 144} = 5$$

$$\frac{dx}{dt} = -\frac{4(13)}{5} = -\frac{52}{5} = -10.4 \text{ ft/sec.}$$

Speed of the boat increases as it approaches the dock.

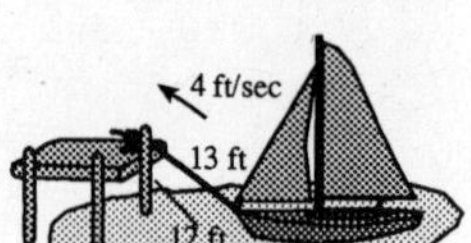

(b) If $\dfrac{dx}{dt} = -4$, and $L = 13$,

$$\frac{dL}{dt} = \frac{x}{L}\frac{dx}{dt}$$
$$= \frac{5}{13}(-4)$$
$$= \frac{-20}{13} \text{ ft/sec}$$

As $x \to 0$, $\dfrac{dL}{dt}$ decreases.

31. (a) $s^2 = x^2 + y^2$

$$\frac{dx}{dt} = -450$$

$$\frac{dy}{dt} = -600$$

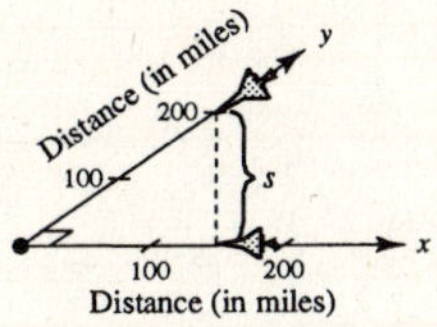

$$2s\frac{ds}{dt} = 2x\frac{dx}{dt} + 2y\frac{dy}{dt}$$

$$\frac{ds}{dt} = \frac{x(dx/dt) + y(dy/dt)}{s}$$

When $x = 150$ and $y = 200$, $s = 250$ and

$$\frac{ds}{dt} = \frac{150(-450) + 200(-600)}{250} = -750 \text{ mph.}$$

(b) $t = \dfrac{250}{750} = \dfrac{1}{3}\text{ hr} = 20\text{ min}$

32. $x^2 + y^2 = s^2$

$$2x\frac{dx}{dt} + 0 = 2s\frac{ds}{dt} \qquad \left(\text{since } \frac{dy}{dt} = 0\right)$$

$$\frac{dx}{dt} = \frac{s}{x}\frac{ds}{dt}$$

When $s = 10$, $x = \sqrt{100 - 25} = \sqrt{75} = 5\sqrt{3}$

$$\frac{dx}{dt} = \frac{10}{5\sqrt{3}}(-240) = \frac{-480}{\sqrt{3}} = -160\sqrt{3} \approx -277.13 \text{ mph.}$$

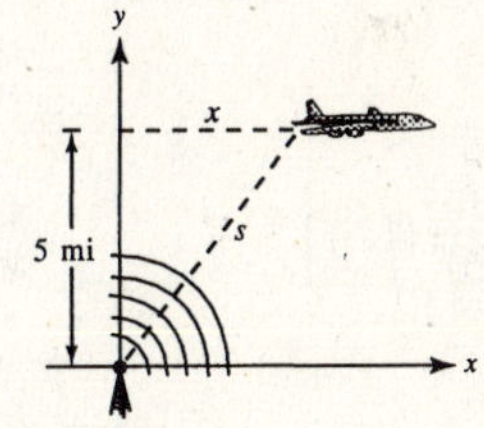

33. $s^2 = 90^2 + x^2$

$$x = 30$$

$$\frac{dx}{dt} = -28$$

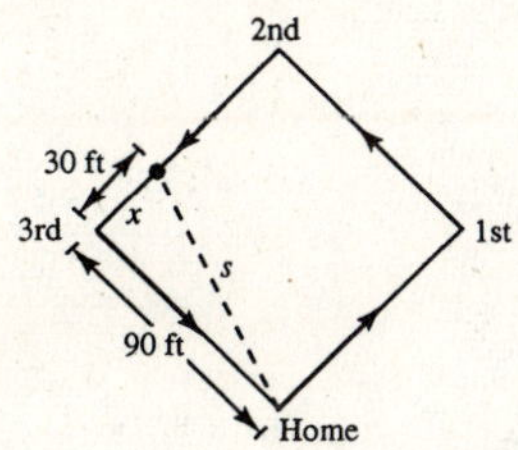

$$2s\frac{ds}{dt} = 2x\frac{dx}{dt} \Rightarrow \frac{ds}{dt} = \frac{x}{s}\cdot\frac{dx}{dt}$$

When $x = 30$,

$$s = \sqrt{90^2 + 30^2} = 30\sqrt{10}$$

$$\frac{ds}{dt} = \frac{30}{30\sqrt{10}}(-28) = \frac{-28}{\sqrt{10}} \approx -8.85 \text{ ft/sec.}$$

34. $s^2 = 90^2 + x^2$

$$x = 60$$

$$\frac{dx}{dt} = 28$$

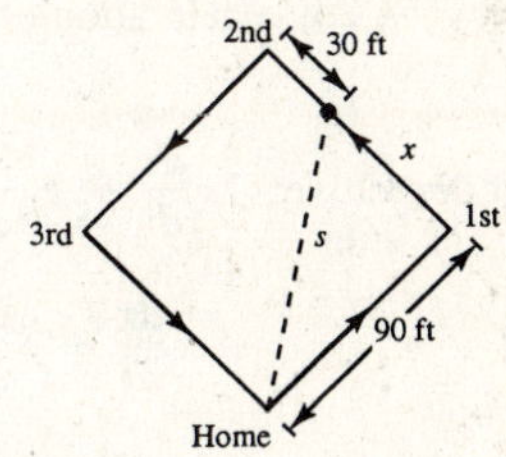

$$\frac{ds}{dt} = \frac{x}{s}\cdot\frac{dx}{dt}$$

When $x = 60$,

$$s = \sqrt{90^2 + 60^2} = 30\sqrt{13}$$

$$\frac{ds}{dt} = \frac{60}{30\sqrt{13}}(28) = \frac{56}{\sqrt{13}} \approx 15.53 \text{ ft/sec.}$$

35. (a) $\dfrac{15}{6} = \dfrac{y}{y - x} \Rightarrow 15y - 15x = 6y$

$$y = \frac{5}{3}x$$

$$\frac{dx}{dt} = 5$$

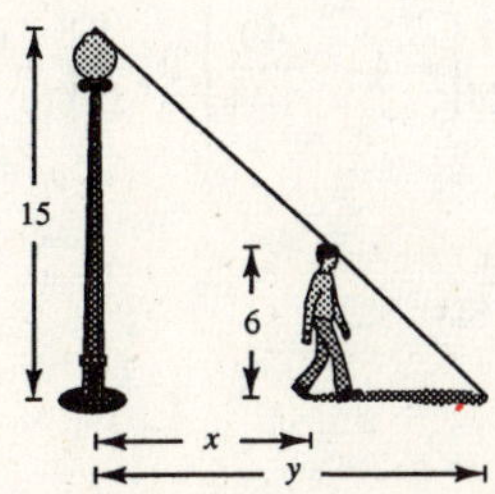

$$\frac{dy}{dt} = \frac{5}{3}\cdot\frac{dx}{dt} = \frac{5}{3}(5) = \frac{25}{3}\text{ ft/sec}$$

(b) $\dfrac{d(y - x)}{dt} = \dfrac{dy}{dt} - \dfrac{dx}{dt} = \dfrac{25}{3} - 5 = \dfrac{10}{3}\text{ ft/sec}$

36. $pV^{1.3} = k$

$$1.3\,pV^{0.3}\frac{dV}{dt} + V^{1.3}\frac{dp}{dt} = 0$$

$$V^{0.3}\left(1.3p\frac{dV}{dt} + V\frac{dp}{dt}\right) = 0$$

$$1.3p\frac{dV}{dt} = -V\frac{dp}{dt}$$

37. Since the evaporation rate is proportional to the surface area, $dV/dt = k(4\pi r^2)$. However, since $V = (4/3)\pi r^3$, we have

$$\frac{dV}{dt} = 4\pi r^2\frac{dr}{dt}.$$

Therefore,

$$k(4\pi r^2) = 4\pi r^2\frac{dr}{dt} \Rightarrow k = \frac{dr}{dt}.$$

38. $$\frac{1}{R} = \frac{1}{R_1} + \frac{1}{R_2}$$

$$\frac{dR_1}{dt} = 1$$

$$\frac{dR_2}{dt} = 1.5$$

$$\frac{1}{R^2} \cdot \frac{dR}{dt} = \frac{1}{R_1^2} \cdot \frac{dR_1}{dt} + \frac{1}{R_2^2} \cdot \frac{dR_2}{dt}$$

When $R_1 = 50$ and $R_2 = 75$,

$$R = 30$$

$$\frac{dR}{dt} = (30)^2\left[\frac{1}{(50)^2}(1) + \frac{1}{(75)^2}(1.5)\right]$$

$$= 0.6 \text{ ohms/sec.}$$

39. $x^2 + y^2 = 25$; acceleration of the top of the ladder $= \dfrac{d^2y}{dt^2}$

First derivative: $2x\dfrac{dx}{dt} + 2y\dfrac{dy}{dt} = 0$

$$x\frac{dx}{dt} + y\frac{dy}{dt} = 0$$

Second derivative: $x\dfrac{d^2x}{dt^2} + \dfrac{dx}{dt} \cdot \dfrac{dx}{dt} + y\dfrac{d^2y}{dt^2} + \dfrac{dy}{dt} \cdot \dfrac{dy}{dt} = 0$

$$\frac{d^2y}{dt^2} = \left(\frac{1}{y}\right)\left[-x\frac{d^2x}{dt^2} - \left(\frac{dx}{dt}\right)^2 - \left(\frac{dy}{dt}\right)^2\right]$$

When $x = 7$, $y = 24$, $\dfrac{dy}{dt} = -\dfrac{7}{12}$, and $\dfrac{dx}{dt} = 2$ (see Exercise 27). Since $\dfrac{dx}{dt}$ is constant, $\dfrac{d^2x}{dt^2} = 0$.

$$\frac{d^2y}{dt^2} = \frac{1}{24}\left[-7(0) - (2)^2 - \left(-\frac{7}{12}\right)^2\right] = \frac{1}{24}\left[-4 - \frac{49}{144}\right] = \frac{1}{24}\left[-\frac{625}{144}\right] \approx -0.1808 \text{ ft/sec}^2$$

40. $L^2 = 144 + x^2$; acceleration of the boat $= \dfrac{d^2x}{dt^2}$.

First derivative: $2L\dfrac{dL}{dt} = 2x\dfrac{dx}{dt}$

$$L\frac{dL}{dt} = x\frac{dx}{dt}$$

Second derivative: $L\dfrac{d^2L}{dt^2} + \dfrac{dL}{dt} \cdot \dfrac{dL}{dt} = x\dfrac{d^2x}{dt^2} + \dfrac{dx}{dt} \cdot \dfrac{dx}{dt}$

$$\frac{d^2x}{dt^2} = \left(\frac{1}{x}\right)\left[L\frac{d^2L}{dt^2} + \left(\frac{dL}{dt}\right)^2 - \left(\frac{dx}{dt}\right)^2\right]$$

When $L = 13$, $x = 5$, $\dfrac{dx}{dt} = -10.4$, and $\dfrac{dL}{dt} = -4$ (see Exercise 30). Since $\dfrac{dL}{dt}$ is constant, $\dfrac{d^2L}{dt^2} = 0$.

$$\frac{d^2x}{dt^2} = \frac{1}{5}[13(0) + (-4)^2 - (-10.4)^2]$$

$$= \frac{1}{5}[16 - 108.16] = \frac{1}{5}[-92.16] = -18.432 \text{ ft/sec}^2$$

41. $y(t) = -4.9t^2 + 20$

$$\frac{dy}{dt} = -9.8t$$

$y(1) = -4.9 + 20 = 15.1$

$y'(1) = -9.8$

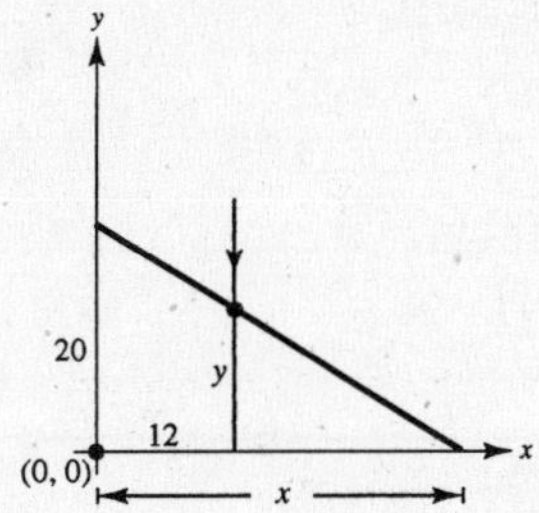

By similar triangles, $\dfrac{20}{x} = \dfrac{y}{x - 12}$

$$20x - 240 = xy.$$

When $y = 15.1$, $20x - 240 = x(15.1)$

$$(20 - 15.1)x = 240$$

$$x = \frac{240}{4.9}.$$

$$20x - 240 = xy$$

$$20\frac{dx}{dt} = x\frac{dy}{dt} + y\frac{dx}{dt}$$

$$\frac{dx}{dt} = \frac{x}{20 - y}\frac{dy}{dt}$$

At $t = 1$, $\dfrac{dx}{dt} = \dfrac{240/4.9}{20 - 15.1}(-9.8) \approx -97.96$ m/sec.

Review Exercises for Chapter 4

1. $f(x) = x^2 - 2x + 3$

$$f'(x) = \lim_{\Delta x \to 0} \frac{f(x + \Delta x) - f(x)}{\Delta x}$$

$$= \lim_{\Delta x \to 0} \frac{[(x + \Delta x)^2 - 2(x + \Delta x) + 3] - [x^2 - 2x + 3]}{\Delta x}$$

$$= \lim_{\Delta x \to 0} \frac{(x^2 + 2x(\Delta x) + (\Delta x)^2 - 2x - 2(\Delta x) + 3) - (x^2 - 2x + 3)}{\Delta x}$$

$$= \lim_{\Delta x \to 0} \frac{2x(\Delta x) + (\Delta x)^2 - 2(\Delta x)}{\Delta x} = \lim_{\Delta x \to 0} (2x + \Delta x - 2) = 2x - 2$$

2. $f(x) = \dfrac{x + 1}{x - 1}$

$$f'(x) = \lim_{\Delta x \to 0} \frac{f(x + \Delta x) - f(x)}{\Delta x} = \lim_{\Delta x \to 0} \frac{\dfrac{x + \Delta x + 1}{x + \Delta x - 1} - \dfrac{x + 1}{x - 1}}{\Delta x}$$

$$= \lim_{\Delta x \to 0} \frac{(x + \Delta x + 1)(x - 1) - (x + \Delta x - 1)(x + 1)}{\Delta x(x + \Delta x - 1)(x - 1)}$$

$$= \lim_{\Delta x \to 0} \frac{(x^2 + x\,\Delta x + x - x - \Delta x - 1) - (x^2 + x\,\Delta x - x + x + \Delta x - 1)}{\Delta x(x + \Delta x - 1)(x - 1)}$$

$$= \lim_{\Delta x \to 0} \frac{-2\Delta x}{\Delta x(x + \Delta x - 1)(x - 1)} = \lim_{\Delta x \to 0} \frac{-2}{(x + \Delta x - 1)(x - 1)} = \frac{-2}{(x - 1)^2}$$

3. $f(x) = \sqrt{x} + 1$

$$f'(x) = \lim_{\Delta x \to 0} \frac{f(x + \Delta x) - f(x)}{\Delta x}$$
$$= \lim_{\Delta x \to 0} \frac{\left(\sqrt{x + \Delta x} + 1\right) - \left(\sqrt{x} + 1\right)}{\Delta x}$$
$$= \lim_{\Delta x \to 0} \frac{\sqrt{x + \Delta x} - \sqrt{x}}{\Delta x} \cdot \frac{\sqrt{x + \Delta x} + \sqrt{x}}{\sqrt{x + \Delta x} + \sqrt{x}}$$
$$= \lim_{\Delta x \to 0} \frac{(x + \Delta x) - x}{\Delta x\left(\sqrt{x + \Delta x} + \sqrt{x}\right)}$$
$$= \lim_{\Delta x \to 0} \frac{1}{\sqrt{x + \Delta x} + \sqrt{x}} = \frac{1}{2\sqrt{x}}$$

4. $f(x) = \dfrac{2}{x}$

$$f'(x) = \lim_{\Delta x \to 0} \frac{f(x + \Delta x) - f(x)}{\Delta x}$$
$$= \lim_{\Delta x \to 0} \frac{\dfrac{2}{x + \Delta x} - \dfrac{2}{x}}{\Delta x}$$
$$= \lim_{\Delta x \to 0} \frac{2x - (2x + 2\Delta x)}{\Delta x(x + \Delta x)x}$$
$$= \lim_{\Delta x \to 0} \frac{-2\Delta x}{\Delta x(x + \Delta x)x}$$
$$= \lim_{\Delta x \to 0} \frac{-2}{(x + \Delta x)x} = \frac{-2}{x^2}$$

5. f is differentiable for all $x \neq -1$.

6. f is differentiable for all $x \neq -3$.

7. $f(x) = 4 - |x - 2|$

(a) Continuous at $x = 2$.

(b) Not differentiable at $x = 2$ because of the sharp turn in the graph.

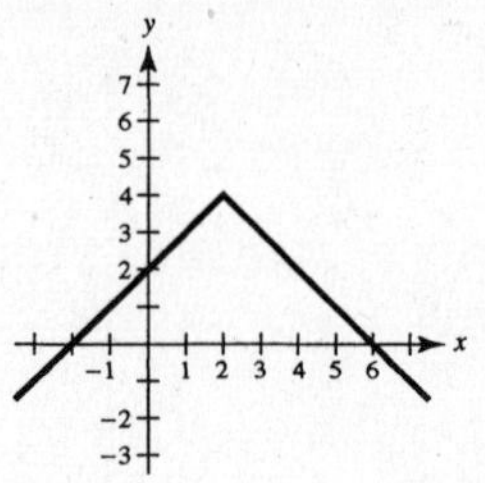

8. $f(x) = \begin{cases} x^2 + 4x + 2, & \text{if } x < -2 \\ 1 - 4x - x^2, & \text{if } x \geq -2 \end{cases}$

(a) Nonremovable discontinuity at $x = -2$.

(b) Not differentiable at $x = -2$ because the function is discontinuous there.

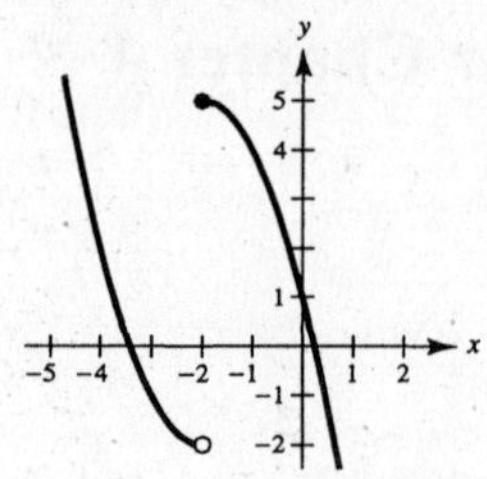

9. Using the limit definition, you obtain $g'(x) = \frac{4}{3}x - \frac{1}{6}$.

At $x = -1$, $g'(-1) = -\frac{4}{3} - \frac{1}{6} = \frac{-3}{2}$.

10. Using the limit definition, you obtain

$$h'(x) = \frac{3}{8} - 4x.$$

At $x = -2$, $h'(-2) = \frac{3}{8} - 4(-2) = \frac{67}{8}$.

11. (a) Using the limit definition, $f'(x) = 3x^2$.

At $x = -1$, $f'(-1) = 3$. The tangent line is

$$y - (-2) = 3(x - (-1))$$
$$y = 3x + 1.$$

(b)

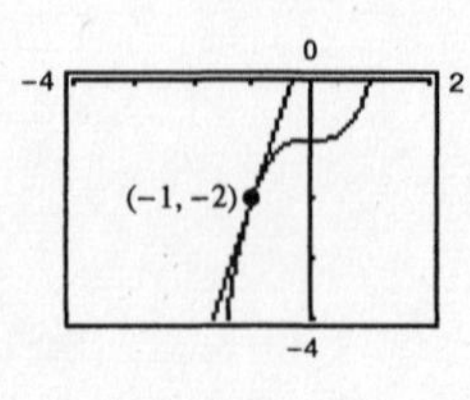

12. (a) Using the limit definition,

$$f'(x) = \frac{-2}{(x + 1)^2}.$$

At $x = 0$, $f'(0) = -2$.

The tangent line is

$$y - 2 = -2(x - 0)$$
$$y = -2x + 2.$$

(b)

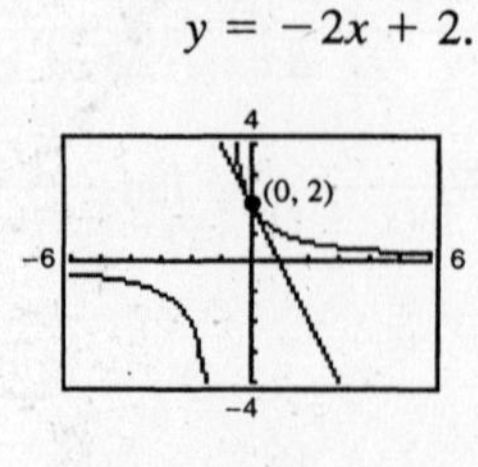

13. $g'(2) = \lim_{x\to 2} \frac{g(x) - g(2)}{x - 2}$

$= \lim_{x\to 2} \frac{x^2(x-1) - 4}{x - 2}$

$= \lim_{x\to 2} \frac{x^3 - x^2 - 4}{x - 2}$

$= \lim_{x\to 2} \frac{(x-2)(x^2 + x + 2)}{x - 2}$

$= \lim_{x\to 2} (x^2 + x + 2) = 8$

14. $f'(2) = \lim_{x\to 2} \frac{f(x) - f(2)}{x - 2}$

$= \lim_{x\to 2} \frac{\frac{1}{x+1} - \frac{1}{3}}{x - 2}$

$= \lim_{x\to 2} \frac{3 - x - 1}{(x-2)(x+1)3}$

$= \lim_{x\to 2} \frac{-1}{(x+1)3} = \frac{-1}{9}$

15. $y = 25$

$y' = 0$

16. $y = -12$

$y' = 0$

17. $f(x) = x^8$

$f'(x) = 8x^7$

18. $g(x) = x^{12}$

$g'(x) = 12x^{11}$

19. $h(t) = 3t^4$

$h'(t) = 12t^3$

20. $f(t) = -8t^5$

$f'(t) = -40t^4$

21. $f(x) = x^3 - 3x^2$

$f'(x) = 3x^2 - 6x = 3x(x - 2)$

22. $g(s) = 4s^4 - 5s^2$

$g'(s) = 16s^3 - 10s$

23. $h(x) = 6\sqrt{x} + 3\sqrt[3]{x} = 6x^{1/2} + 3x^{1/3}$

$h'(x) = 3x^{-1/2} + x^{-2/3} = \frac{3}{\sqrt{x}} + \frac{1}{\sqrt[3]{x^2}}$

24. $f(x) = x^{1/2} - x^{-1/2}$

$f'(x) = \frac{1}{2}x^{-1/2} + \frac{1}{2}x^{-3/2} = \frac{x+1}{2x^{3/2}}$

25. $g(t) = \frac{2}{3}t^{-2}$

$g'(t) = \frac{-4}{3}t^{-3} = \frac{-4}{3t^3}$

26. $h(x) = \frac{2}{9}x^{-2}$

$h'(x) = \frac{-4}{9}x^{-3} = \frac{-4}{9x^3}$

27. f is always increasing so $f' > 0$.

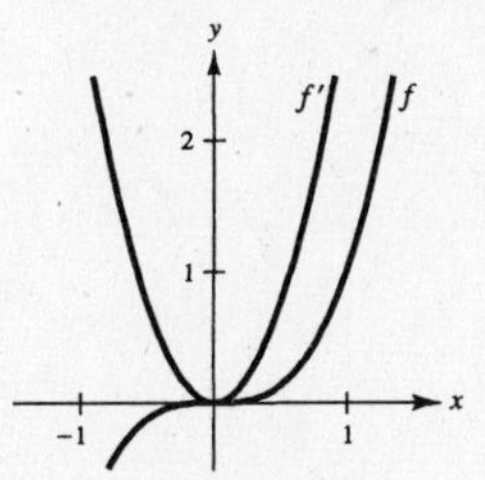

28.

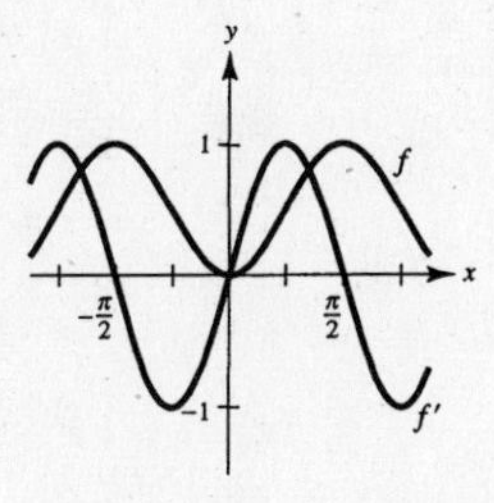

29. $F = 200\sqrt{T}$

$F'(T) = \frac{100}{\sqrt{T}}$

(a) When $T = 4$, $F'(4) = 50$ vibrations/sec/lb.

(b) When $T = 9$, $F'(9) = 33\frac{1}{3}$ vibrations/sec/lb.

30. $s = -16t^2 + s_0$

First ball:

$-16t^2 + 100 = 0$

$t = \sqrt{\dfrac{100}{16}} = \dfrac{10}{4} = 2.5$ seconds to hit ground

Second ball:

$-16t^2 + 75 = 0$

$t^2 = \sqrt{\dfrac{75}{16}} = \dfrac{5\sqrt{3}}{4} \approx 2.165$ seconds to hit ground

Since the second ball was released one second after the first ball, the first ball will hit the ground first. The second ball will hit the ground $3.165 - 2.5 = 0.665$ second later.

31. $s(t) = -16t^2 + s_0$

$s(9.2) = -16(9.2)^2 + s_0 = 0$

$s_0 = 1354.24$

The building is approximately 1354 feet high (or 415 m).

32. $s(t) = -16t^2 + 14{,}400 = 0$

$16t^2 = 14{,}400$

$t = 30$ sec

Since 600 mph $= \frac{1}{6}$ mi/sec, in 30 seconds the bomb will move horizontally $(\frac{1}{6})(30) = 5$ miles.

33. (a)

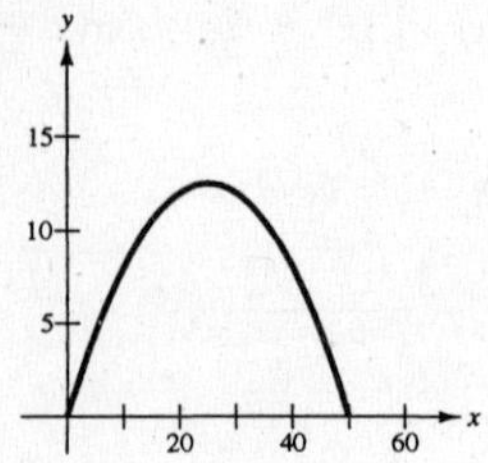

(b) $0 = x - 0.02x^2$

$0 = x\left(1 - \dfrac{x}{50}\right)$ implies $x = 50$.

Total horizontal distance: 50

(c) Ball reaches maximum height when $x = 25$.

(d) $y = x - 0.02x^2$

$y' = 1 - 0.04x$

$y'(0) = 1$

$y'(10) = 0.6$

$y'(25) = 0$

$y'(30) = -0.2$

$y'(50) = -1$

(e) $y'(25) = 0$

34.

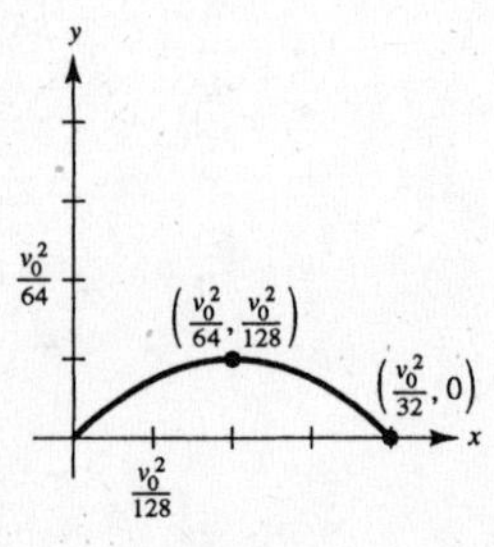

(a) $y = x - \dfrac{32}{v_0^2}x^2 = x\left(1 - \dfrac{32}{v_0^2}x\right)$

$= 0$ if $x = 0$ or $x = \dfrac{v_0^2}{32}$.

Projectile strikes the ground when $x = v_0^2/32$.

Projectile reaches its maximum height at $x = v_0^2/64$. (one-half the distance)

(b) $y' = 1 - \dfrac{64}{v_0^2}x$

When $x = \dfrac{v_0^2}{64}$, $y' = 1 - \dfrac{64}{v_0^2}\left(\dfrac{v_0^2}{64}\right) = 0$.

—CONTINUED—

34. —CONTINUED—

(c) $y = x - \dfrac{32}{v_0^2}x^2 = x\left(1 - \dfrac{32}{v_0^2}x\right) = 0$

when $x = 0$ and $x = v_0^2/32$. Therefore, the range is $x = v_0^2/32$. When the initial velocity is doubled the range is

$$x = \frac{(2v_0)^2}{32} = \frac{4v_0^2}{32}$$

or four times the initial range. From part (a), the maximum height occurs when $x = v_0^2/64$. The maximum height is

$$y\left(\frac{v_0^2}{64}\right) = \frac{v_0^2}{64} - \frac{32}{v_0^2}\left(\frac{v_0^2}{64}\right)^2 = \frac{v_0^2}{64} - \frac{v_0^2}{128} = \frac{v_0^2}{128}.$$

If the initial velocity is doubled, the maximum height is

$$y\left[\frac{(2v_0)^2}{64}\right] = \frac{(2v_0)^2}{128} = 4\left(\frac{v_0^2}{128}\right)$$

or four times the original maximum height.

(d) $v_0 = 70$ ft/sec

Range: $x = \dfrac{v_0^2}{32} = \dfrac{(70)^2}{32} = 153.125$ ft

Maximum height: $y = \dfrac{v_0^2}{128} = \dfrac{(70)^2}{128} \approx 38.28$ ft

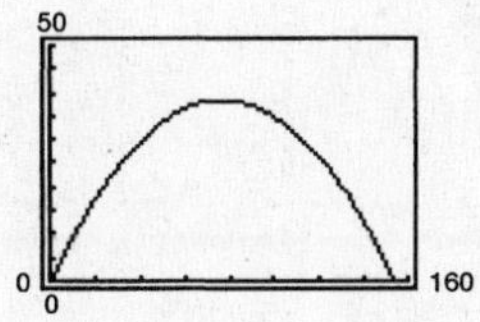

35. $x(t) = t^2 - 3t + 2 = (t - 2)(t - 1)$

(a) $v(t) = x'(t) = 2t - 3$

(b) $v(t) < 0$ for $t < \frac{3}{2}$.

(c) $v(t) = 0$ for $t = \frac{3}{2}$.

$x = \left(\frac{3}{2} - 2\right)\left(\frac{3}{2} - 1\right) = \left(-\frac{1}{2}\right)\left(\frac{1}{2}\right) = -\frac{1}{4}$

(d) $x(t) = 0$ for $t = 1, 2$.

$|v(1)| = |2(1) - 3| = 1$

$|v(2)| = |2(2) - 3| = 1$

The speed is 1 when the position is 0.

36. (a) $y = 0.14x^2 - 4.43x + 58.4$

(b)

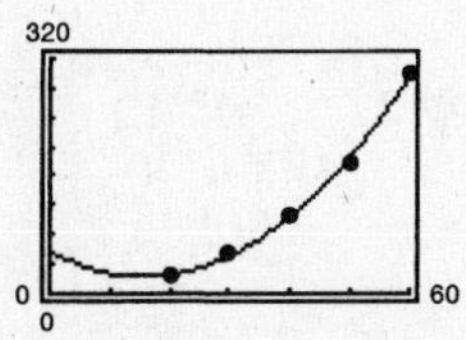

(c)

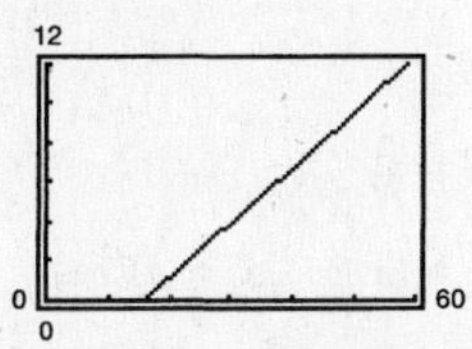

(d) If $x = 65$, $y \approx 362$ feet.

(e) As the speed increases, the stopping distance increases at an increasing rate.

37. $f(x) = (3x^2 + 7)(x^2 - 2x + 3)$

$f'(x) = (3x^2 + 7)(2x - 2) + (x^2 - 2x + 3)(6x)$

$= 2(6x^3 - 9x^2 + 16x - 7)$

38. $g(x) = (x^3 - 3x)(x + 2)$

$g'(x) = (x^3 - 3x)(1) + (x + 2)(3x^2 - 3)$

$= x^3 - 3x + 3x^3 + 6x^2 - 3x - 6$

$= 4x^3 + 6x^2 - 6x - 6$

39. $h(t) = \sqrt{t}(t^3 + 4t - 1) = t^{7/2} + 4t^{3/2} - t^{1/2}$

$h'(t) = \dfrac{7}{2}t^{5/2} + 6t^{1/2} - \dfrac{1}{2t^{1/2}}$

$= \dfrac{7t^3 + 12t - 1}{2\sqrt{t}}$

40. $f(z) = z^{1/3}(z^2 + 5z) = z^{7/3} + 5z^{4/3}$

$f'(z) = \dfrac{7}{3}z^{4/3} + \dfrac{20}{3}z^{1/3}$

$= \dfrac{\sqrt[3]{z}}{3}(7z + 20)$

41. $f(x) = \dfrac{x^2 + x - 1}{x^2 - 1}$

$f'(x) = \dfrac{(x^2 - 1)(2x + 1) - (x^2 + x - 1)(2x)}{(x^2 - 1)^2}$

$= \dfrac{-(x^2 + 1)}{(x^2 - 1)^2}$

42. $f(x) = \dfrac{6x - 5}{x^2 + 1}$

$f'(x) = \dfrac{(x^2 + 1)(6) - (6x - 5)(2x)}{(x^2 + 1)^2}$

$= \dfrac{2(3 + 5x - 3x^2)}{(x^2 + 1)^2}$

43. $f(x) = (4 - 3x^2)^{-1}$

$f'(x) = -(4 - 3x^2)^{-2}(-6x) = \dfrac{6x}{(4 - 3x^2)^2}$

44. $f(x) = 9(3x^2 - 2x)^{-1}$

$f'(x) = -9(3x^2 - 2x)^{-2}(6x - 2) = \dfrac{18(1 - 3x)}{(3x^2 - 2x)^2}$

45. $f(x) = \dfrac{2x^3 - 1}{x^2} = 2x - x^{-2}, (1, 1)$

$f'(x) = 2 + 2x^{-3} = 2 + \dfrac{2}{x^3}$

$f'(1) = 4$

Tangent line: $y - 1 = 4(x - 1)$

$y = 4x - 3$

46. $f(x) = \dfrac{x + 1}{x - 1}, \left(\dfrac{1}{2}, -3\right)$

$f'(x) = \dfrac{(x - 1) - (x + 1)}{(x - 1)^2} = \dfrac{-2}{(x - 1)^2}$

$f'\left(\dfrac{1}{2}\right) = \dfrac{-2}{\left(\dfrac{1}{2} - 1\right)^2} = -8$

Tangent line: $y + 3 = -8\left(x - \dfrac{1}{2}\right)$

$y = -8x + 1$

47. $g(t) = t^3 - 3t + 2$

$g'(t) = 3t^2 - 3$

$g''(t) = 6t$

48. $f(x) = 12x^{1/4}$

$f'(x) = 3x^{-3/4}$

$f''(x) = \dfrac{-9}{4}x^{-7/4} = \dfrac{-9}{4x^{7/4}}$

49. $f(x) = (3x^2 + 7)(x^2 - 2x + 3)$

$f'(x) = (3x^2 + 7)(2x - 2) + (x^2 - 2x + 3)(6x)$

$= 6x^3 - 6x^2 + 14x - 14 + 6x^3 - 12x^2 + 18x$

$= 12x^3 - 18x^2 + 32x - 14$

$f''(x) = 36x^2 - 36x + 32$

$= 4(9x^2 - 9x + 8)$

50. $g(t) = \dfrac{t + 3}{t - 4}$

$g'(t) = \dfrac{(t - 4) - (t + 3)}{(t - 4)^2} = -7(t - 4)^{-2}$

$g''(t) = 14(t - 4)^{-3} = \dfrac{14}{(t - 4)^3}$

51. $h(x) = (x^2 - 4x)^3$

$h'(x) = 3(x^2 - 4x)^2(2x - 4)$

$= 6x^2(x - 4)^2(x - 2)$

52. $g(x) = (5 - 3x)^4$

$g'(x) = 4(5 - 3x)^3(-3)$

$= -12(5 - 3x)^3$

$= 12(3x - 5)^3$

53. $f(x) = -2(1 - 4x^2)^2$

$f'(x) = -4(1 - 4x^2)(-8x)$

$= 32x(1 - 4x^2)$

$= 32x - 128x^3$

54. $f(x) = \left[\sqrt{x}(x - 3)\right]^2 = x(x^2 - 6x + 9) = x^3 - 6x^2 + 9x$

$f'(x) = 3x^2 - 12x + 9 = 3(x^2 - 4x + 3) = 3(x - 3)(x - 1)$

55. $h(x) = \left(\dfrac{x-3}{x^2+1}\right)^2$

$$h'(x) = 2\left(\frac{x-3}{x^2+1}\right)\left(\frac{(x^2+1)(1)-(x-3)(2x)}{(x^2+1)^2}\right)$$

$$= \frac{2(x-3)(-x^2+6x+1)}{(x^2+1)^3}$$

56. $f(x) = \left(x^2 + \dfrac{1}{x}\right)^5$

$$f'(x) = 5\left(x^2+\frac{1}{x}\right)^4\left(2x - \frac{1}{x^2}\right)$$

57. $f(s) = (s^2-1)^{5/2}(s^3+5)$

$$f'(s) = (s^2-1)^{5/2}(3s^2) + (s^3+5)\left(\tfrac{5}{2}\right)(s^2-1)^{3/2}(2s)$$

$$= s(s^2-1)^{3/2}[3s(s^2-1)+5(s^3+5)]$$

$$= s(s^2-1)^{3/2}(8s^3-3s+25)$$

58. $h(t) = \dfrac{t}{(1-t)^3}$

$$h'(t) = \frac{(1-t)^3-(t)(3)(1-t)^2}{(1-t)^6}$$

$$= \frac{(1-t)-3t}{(1-t)^4}$$

$$= \frac{-4t+1}{(1-t)^4}$$

59. $f(x) = \sqrt{1-x^3} = (1-x^3)^{1/2}$

$$f'(x) = \frac{1}{2}(1-x^3)^{-1/2}(-3x^2) = \frac{-3x^2}{2\sqrt{1-x^3}}$$

$$f'(-2) = -\frac{-3(-2)^2}{2\sqrt{9}} = -2$$

60. $f(x) = \sqrt[3]{x^2-1}$

$$f'(x) = \frac{1}{3}(x^2-1)^{-2/3}(2x) = \frac{2x}{3(x^2-1)^{2/3}}$$

$$f'(3) = \frac{6}{3(4)} = \frac{1}{2}$$

61. $f(t) = t^2(t-1)^5$

$$f'(t) = t(t-1)^4(7t-2)$$

The zeros of f' correspond to the points on the graph of f where the tangent line is horizontal.

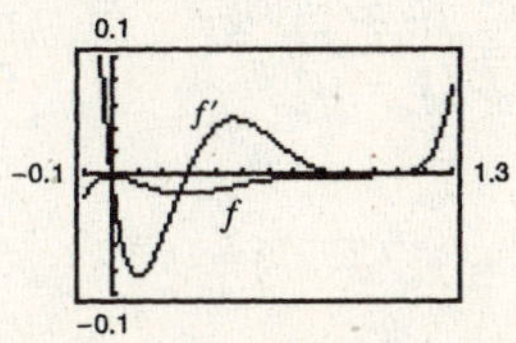

62. $f(x) = [(x-2)(x+4)]^2 = (x^2+2x-8)^2$

$$f'(x) = 4(x^3+3x^2-6x-8)$$

$$= 4(x-2)(x+1)(x+4)$$

The zeros of f' correspond to the points on the graph of f where the tangent line is horizontal.

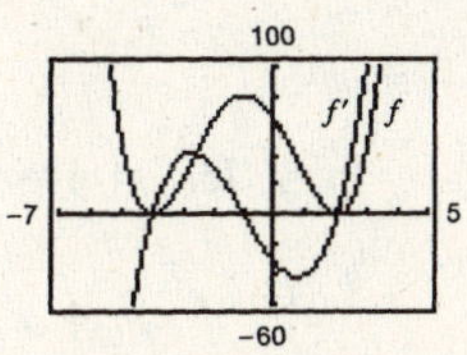

63. $g(x) = 2x(x+1)^{-1/2}$

$$g'(x) = \frac{x+2}{(x+1)^{3/2}}$$

g' does not equal zero for any value of x in the domain. The graph of g has no horizontal tangent lines.

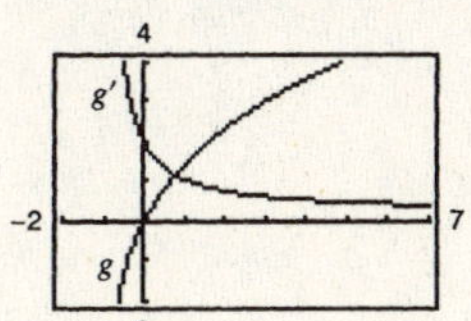

64. $g(x) = x(x^2+1)^{1/2}$

$$g'(x) = \frac{2x^2+1}{\sqrt{x^2+1}}$$

g' does not equal zero for any value of x. The graph of g has no horizontal tangent lines.

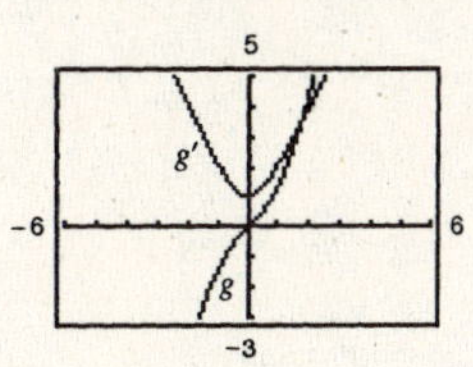

65. $f(t) = \sqrt{t+1}\,\sqrt[3]{t+1}$

$f(t) = (t+1)^{1/2}(t+1)^{1/3} = (t+1)^{5/6}$

$f'(t) = \dfrac{5}{6(t+1)^{1/6}}$

f' does not equal zero for any t in the domain. The graph of f has no horizontal tangent lines.

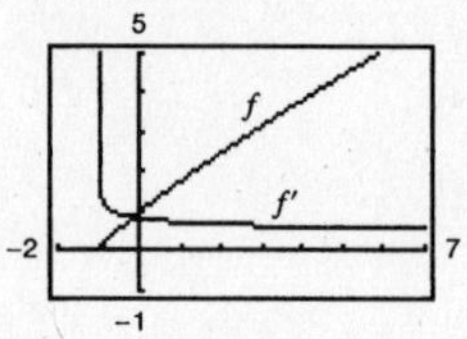

66. $y = \sqrt{3x}(x+2)^3$

$y' = \dfrac{3(x+2)^2(7x+2)}{2\sqrt{3x}}$

y' does not equal zero for any x in the domain. The graph has no horizontal tangent lines.

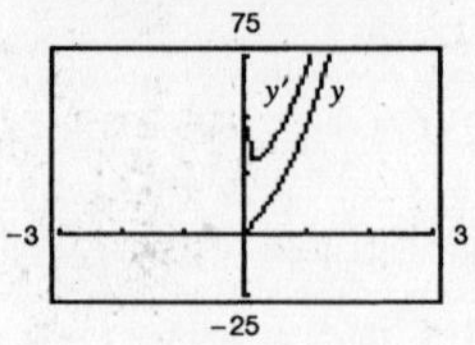

67. $f(x) = \sqrt{x^2+9} = (x^2+9)^{1/2}$

$f'(x) = \dfrac{1}{2}(x^2+9)^{-1/2}(2x) = \dfrac{x}{(x^2+9)^{1/2}}$

$f''(x) = \dfrac{(x^2+9)^{1/2} - x\frac{1}{2}(x^2+9)^{-1/2}(2x)}{(x^2+9)}$

$= \dfrac{(x^2+9) - x^2}{(x^2+9)^{3/2}} = \dfrac{9}{(x^2+9)^{3/2}}$

68. $g(x) = (5-2x)^{1/2}$

$g'(x) = \dfrac{1}{2}(5-2x)^{-1/2}(-2) = -(5-2x)^{-1/2}$

$g''(x) = \dfrac{1}{2}(5-2x)^{-3/2}(-2) = \dfrac{-1}{(5-2x)^{3/2}}$

69. $y = \dfrac{4}{(x-2)^2} = 4(x-2)^{-2}$

$y' = -8(x-2)^{-3}$

$y'' = 24(x-2)^{-4} = \dfrac{24}{(x-2)^4}$

70. $y = 3(x-1)^{-1/2}$

$y' = -\dfrac{3}{2}(x-1)^{-3/2}$

$y'' = \dfrac{9}{4}(x-1)^{-5/2} = \dfrac{9}{4(x-1)^{5/2}}$

71. $f(t) = \dfrac{t}{(1-t)^2}$

$f'(t) = \dfrac{t+1}{(1-t)^3}$

$f''(t) = \dfrac{2(t+2)}{(1-t)^4}$

72. $g(x) = \dfrac{6x-5}{x^2+1}$

$g'(x) = \dfrac{2(-3x^2+5x+3)}{(x^2+1)^2}$

$g''(x) = \dfrac{2(6x^3-15x^2-18x+5)}{(x^2+1)^3}$

73. $g(x) = \dfrac{x}{\sqrt{x+3}}$

$g'(x) = \dfrac{x+6}{2(x+3)^{3/2}}$

$g''(x) = -\dfrac{x+12}{4(x+3)^{5/2}}$

74. $h(x) = x\sqrt{x^2-1}$

$h'(x) = \dfrac{2x^2-1}{\sqrt{x^2-1}}$

$h''(x) = \dfrac{x(2x^2-3)}{(x^2-1)^{3/2}}$

75. (a) $y = (x+3)^3, \quad (-2, 1)$

$y' = 3(x+3)^2, \quad y'(-2) = 3$

$y - 1 = 3(x+2)$

$3x - y + 7 = 0$ or $y = 3x + 7$

(b)

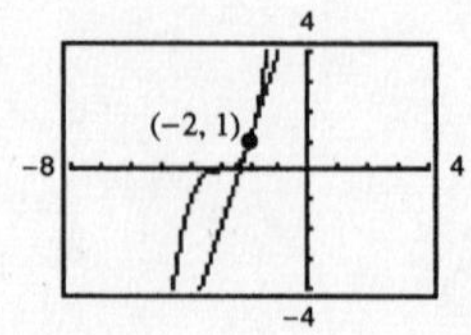

76. (a) $y = (x-2)^2, \quad (2, 0)$

$y' = 2(x-2), \quad y'(2) = 0$

$y - 0 = 0(x-2)$

$y = 0$

(tangent line is the x-axis)

(b)

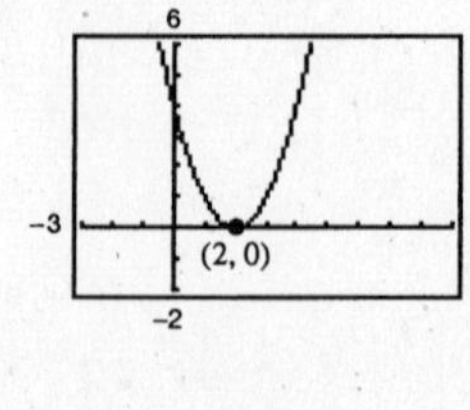

77. (a) $y = \sqrt[3]{(x-2)^2} = (x-2)^{2/3}, \quad (3, 1)$

$y' = \frac{2}{3}(x-2)^{-1/3}, \quad y'(3) = \frac{2}{3}$

$y - 1 = \frac{2}{3}(x-3)$

$3y - 3 = 2x - 6$

$2x - 3y - 3 = 0$ or $y = \frac{2}{3}x - 1$

(b)

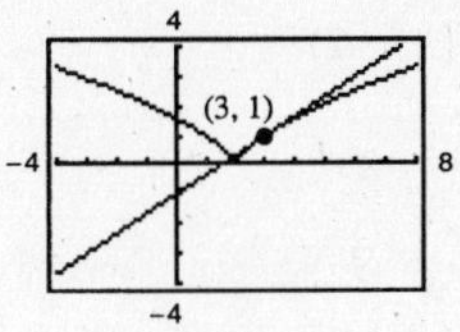

78. (a) $y = \frac{2x}{1-x^2}, \quad (0, 0)$

$y' = \frac{(1-x^2)(2) - 2x(-2x)}{(1-x^2)^2} = \frac{2+2x^2}{(1-x^2)^2}$

$y'(0) = 2$

$y - 0 = 2(x - 0)$

$y = 2x$

(b)

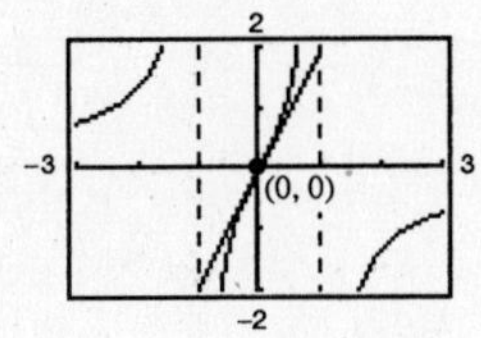

79. $T = \frac{700}{t^2 + 4t + 10} = 700(t^2 + 4t + 10)^{-1}$

$T' = \frac{-1400(t+2)}{(t^2 + 4t + 10)^2}$

(a) When $t = 1$,

$T' = \frac{-1400(1+2)}{(1+4+10)^2} \approx -18.667$ deg/hr.

(b) When $t = 3$,

$T' = \frac{-1400(3+2)}{(9+12+10)^2} \approx -7.284$ deg/hr.

(c) When $t = 5$,

$T' = \frac{-1400(5+2)}{(25+20+10)^2} \approx -3.240$ deg/hr.

(d) When $t = 10$,

$T' = \frac{-1400(10+2)}{(100+40+10)^2} \approx -0.747$ deg/hr.

80. $v = \sqrt{2gh} = \sqrt{2(32)h} = 8\sqrt{h}$

$\frac{dv}{dh} = \frac{4}{\sqrt{h}}$

(a) When $h = 9$, $\frac{dv}{dh} = \frac{4}{3}$ ft/sec.

(b) When $h = 4$, $\frac{dv}{dh} = 2$ ft/sec.

81. $x^2 + 3xy + y^3 = 10$

$2x + 3xy' + 3y + 3y^2y' = 0$

$3(x + y^2)y' = -(2x + 3y)$

$y' = \frac{-(2x + 3y)}{3(x + y^2)}$

82. $x^2 + 9y^2 - 4x + 3y = 0$

$2x + 18yy' - 4 + 3y' = 0$

$3(6y + 1)y' = 4 - 2x$

$y' = \frac{4 - 2x}{3(6y + 1)}$

83. $y\sqrt{x} - x\sqrt{y} = 16$

$y\left(\frac{1}{2}x^{-1/2}\right) + x^{1/2}y' - x\left(\frac{1}{2}y^{-1/2}y'\right) - y^{1/2} = 0$

$\left(\sqrt{x} - \frac{x}{2\sqrt{y}}\right)y' = \sqrt{y} - \frac{y}{2\sqrt{x}}$

$\frac{2\sqrt{xy} - x}{2\sqrt{y}}y' = \frac{2\sqrt{xy} - y}{2\sqrt{x}}$

$y' = \frac{2\sqrt{xy} - y}{2\sqrt{x}} \cdot \frac{2\sqrt{y}}{2\sqrt{xy} - x} = \frac{2y\sqrt{x} - y\sqrt{y}}{2x\sqrt{y} - x\sqrt{x}}$

84.
$$\begin{aligned} y^2 &= x^3 - x^2y + xy - y^2 \\ 0 &= x^3 - x^2y + xy - 2y^2 \\ 0 &= 3x^2 - x^2y' - 2xy + xy' + y - 4yy' \\ (x^2 - x + 4y)y' &= 3x^2 - 2xy + y \\ y' &= \frac{3x^2 - 2xy + y}{x^2 - x + 4y} \end{aligned}$$

85. $x^2 + y^2 = 20$

$2x + 2yy' = 0$

$y' = -\dfrac{x}{y}$

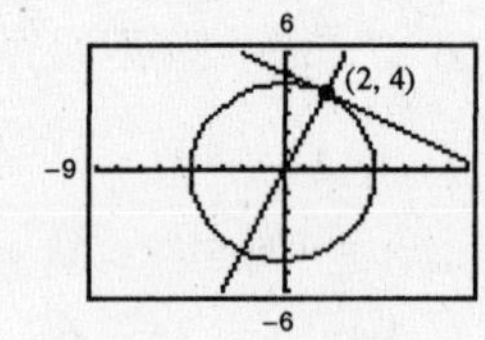

At (2, 4): $y' = -\dfrac{1}{2}$.

Tangent line: $y - 4 = -\dfrac{1}{2}(x - 2)$

$x + 2y - 10 = 0$

Normal line: $y - 4 = 2(x - 2)$

$2x - y = 0$

86. $x^2 - y^2 = 16$

$2x - 2yy' = 0$

$y' = \dfrac{x}{y}$

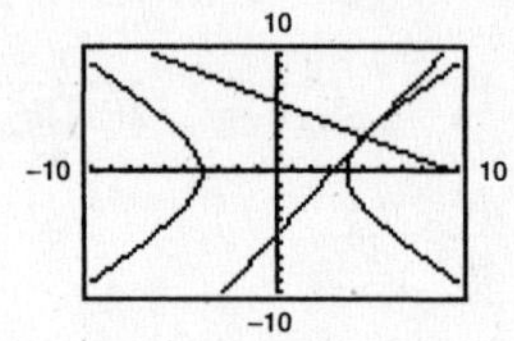

At (5, 3): $y' = \dfrac{5}{3}$.

Tangent line: $y - 3 = \dfrac{5}{3}(x - 5)$

$5x - 3y - 16 = 0$

Normal line: $y - 3 = -\dfrac{3}{5}(x - 5)$

$3x + 5y - 30 = 0$

87. $y = \sqrt{x}$

$\dfrac{dy}{dt} = 2$ units/sec

$\dfrac{dy}{dt} = \dfrac{1}{2\sqrt{x}}\dfrac{dx}{dt} \Rightarrow \dfrac{dx}{dt} = 2\sqrt{x}\dfrac{dy}{dt} = 4\sqrt{x}$

(a) When $x = \dfrac{1}{2}$, $\dfrac{dx}{dt} = 2\sqrt{2}$ units/sec.

(b) When $x = 1$, $\dfrac{dx}{dt} = 4$ units/sec.

(c) When $x = 4$, $\dfrac{dx}{dt} = 8$ units/sec.

88. $y = \sqrt{x}$, $\dfrac{dy}{dt} = 2$

$D = \sqrt{x^2 + y^2} = \sqrt{y^4 + y^2}$

$\dfrac{dD}{dt} = \dfrac{1}{2}(y^4 + y^2)^{-1/2}(4y^3 + 2y)\dfrac{dy}{dt}$

$= \dfrac{4y^3 + 2y}{\sqrt{y^4 + y^2}}$

(a) $x = \dfrac{1}{2}$, $y = \dfrac{\sqrt{2}}{2}$, $\dfrac{dD}{dt} = \dfrac{\sqrt{2} + \sqrt{2}}{\sqrt{3/2}}$

$= \dfrac{4\sqrt{2}}{\sqrt{3}} = \dfrac{4\sqrt{6}}{3} \approx 3.3$ units/sec

(b) $x = 1$, $y = 1$, $\dfrac{dD}{dt} = \dfrac{6}{\sqrt{2}} = 3\sqrt{2} \approx 4.2$ units/sec

(c) $x = 4$, $y = 2$, $\dfrac{dD}{dt} = \dfrac{36}{\sqrt{20}} = \dfrac{18}{\sqrt{5}}$

$= \dfrac{18\sqrt{5}}{5} \approx 8.0$ units/sec

89. Surface area $= A = 6x^2$, x length of edge.

$\dfrac{dx}{dt} = 5$

$\dfrac{da}{dt} = 12x\dfrac{dx}{dt} = 12(4.5)(5) = 270\ \text{cm}^2/\text{sec}$

90. $\dfrac{s}{h} = \dfrac{1/2}{2}$

$s = \dfrac{1}{4}h$

$\dfrac{dV}{dt} = 1$

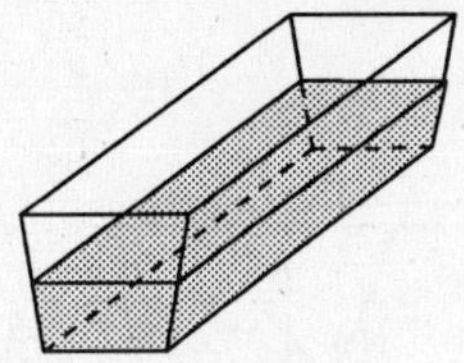

Width of water at depth h:

$w = 2 + 2s = 2 + 2\left(\dfrac{1}{4}h\right) = \dfrac{4+h}{2}$

$V = \dfrac{5}{2}\left(2 + \dfrac{4+h}{2}\right)h = \dfrac{5}{4}(8+h)h$

$\dfrac{dV}{dt} = \dfrac{5}{2}(4+h)\dfrac{dh}{dt}$

$\dfrac{dh}{dt} = \dfrac{2(dV/dt)}{5(4+h)}$

When $h = 1$, $\dfrac{dh}{dt} = \dfrac{2}{25}$ m/min.

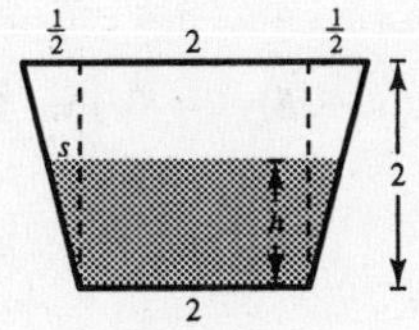

91. $D = \sqrt{x^2 + y^2}$

$\dfrac{dD}{dt} = \dfrac{1}{2}(x^2+y^2)^{-1/2}\left(2x\dfrac{dx}{dt} + 2y\dfrac{dy}{dt}\right)$

$= \dfrac{x\dfrac{dx}{dt} + y\dfrac{dy}{dt}}{\sqrt{x^2+y^2}}$

When $t = \dfrac{1}{2}$, $x = 32.5$ and $y = 25$.

$\dfrac{dD}{dt} = \dfrac{(32.5)(65) + (25)(50)}{\sqrt{32.5^2 + 25^2}}$

$= \dfrac{3362.5}{\sqrt{1681.25}} \approx 82$ mph

92. $V = \pi r^2(0.08)$

$\dfrac{dV}{dt} = 2\pi r(0.08)\dfrac{dr}{dt}$

When $r = 750$ and $\dfrac{dr}{dt} = 0.5$,

$\dfrac{dV}{dt} = 2\pi(750)(0.08)(0.5)$

$= 60\pi$ cubic feet/minute.

93. For $0 \le h \le 5$,

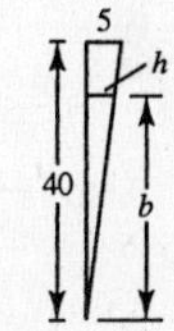

$\dfrac{b}{40} = \dfrac{h}{5}$

$b = 8h.$

$V = \dfrac{1}{2}bh(20) = 10bh = 10(8h)h = 80h^2$

$\dfrac{dV}{dt} = 160h\dfrac{dh}{dt} = 10 \Rightarrow \dfrac{dh}{dt} = \dfrac{1}{16h}$

When $h = 4$, $\dfrac{dh}{dt} = \dfrac{1}{16(4)} = \dfrac{1}{64}$ ft/min

94. (a) $\dfrac{20}{6} = \dfrac{y}{y-x}$

$20y - 20x = 6y$

$14y = 20x$

$y = \dfrac{10}{7}x$

$\dfrac{dx}{dt} = -5$

$\dfrac{dy}{dt} = \dfrac{10}{7}\dfrac{dx}{dt} = \dfrac{10}{7}(-5) = \dfrac{-50}{7}$ ft/sec

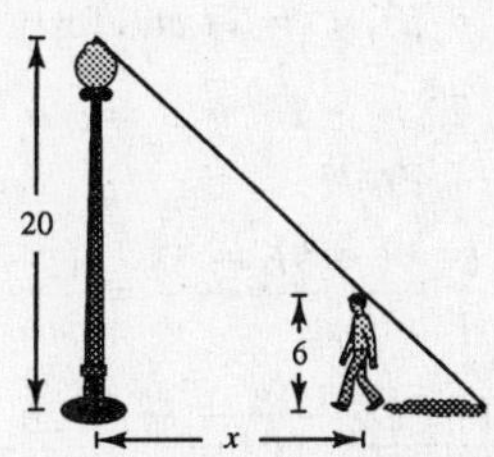

(b) $\dfrac{d(y-x)}{dt} = \dfrac{dy}{dt} - \dfrac{dx}{dt} = \dfrac{-50}{7} - (-5) = \dfrac{-50}{7} + \dfrac{35}{7} = \dfrac{-15}{7}$ ft/sec

Problem Solving for Chapter 4

1. (a) $x^2 + (y - r)^2 = r^2$ Circle

$x^2 = y$ Parabola

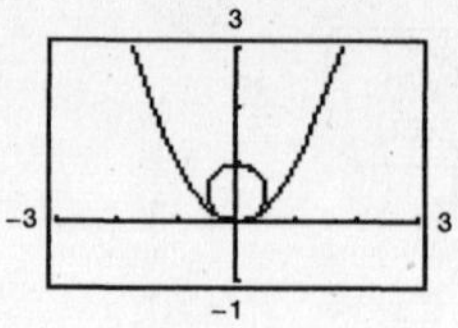

Substituting,

$$(y - r)^2 = r^2 - y$$

$$y^2 - 2ry + r^2 = r^2 - y$$

$$y^2 - 2ry + y = 0$$

$$y(y - 2r + 1) = 0$$

Since you want only one solution, let $1 - 2r = 0 \Rightarrow r = \frac{1}{2}$.

Graph $y = x^2$ and $x^2 + \left(y - \frac{1}{2}\right)^2 = \frac{1}{4}$.

(b) Let (x, y) be a point of tangency: $x^2 + (y - b)^2 = 1 \Rightarrow 2x + 2(y - b)y' = 0 \Rightarrow y' = \dfrac{x}{b - y}$ (circle).

$y = x^2 \Rightarrow y' = 2x$ (parabola). Equating,

$$2x = \frac{x}{b - y}$$

$$2(b - y) = 1$$

$$b - y = \frac{1}{2} \Rightarrow b = y + \frac{1}{2}.$$

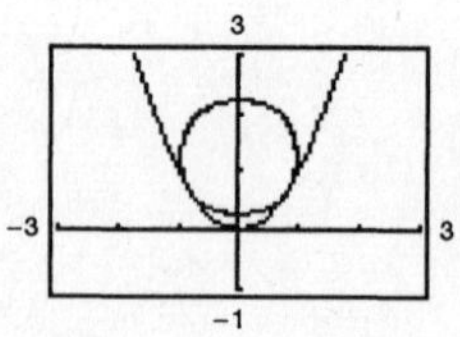

Also, $x^2 + (y - b)^2 = 1$ and $y = x^2$ imply

$$y + (y - b)^2 = 1 \Rightarrow y + \left[y - \left(y + \frac{1}{2}\right)\right]^2 = 1 \Rightarrow y + \frac{1}{4} = 1 \Rightarrow y = \frac{3}{4} \text{ and } b = \frac{5}{4}.$$

Center: $\left(0, \dfrac{5}{4}\right)$

Graph $y = x^2$ and $x^2 + \left(y - \dfrac{5}{4}\right)^2 = 1$.

2.

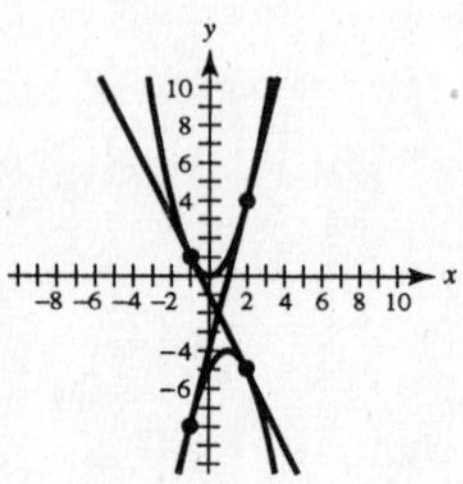

Let (a, a^2) and $(b, -b^2 + 2b - 5)$ be the points of tangency.

For $y = x^2$, $y' = 2x$ and for $y = -x^2 + 2x - 5$, $y' = -2x + 2$.

Thus, $2a = -2b + 2 \Rightarrow a + b = 1$, or $a = 1 - b$. Furthermore, the slope of the common tangent line is

$$\frac{a^2 - (-b^2 + 2b - 5)}{a - b} = \frac{(1 - b)^2 + b^2 - 2b + 5}{(1 - b) - b} = -2b + 2$$

$$\Rightarrow \frac{1 - 2b + b^2 + b^2 - 2b + 5}{1 - 2b} = -2b + 2$$

$$\Rightarrow 2b^2 - 4b + 6 = 4b^2 - 6b + 2$$

$$\Rightarrow 2b^2 - 2b - 4 = 0$$

$$\Rightarrow b^2 - b - 2 = 0$$

$$\Rightarrow (b - 2)(b + 1) = 0$$

$$b = 2, -1.$$

For $b = 2$, $a = 1 - b = -1$ and the points of tangency are $(-1, 1)$, $(2, -5)$. The tangent line has slope -2:

$$y - 1 = -2(x + 1) \Rightarrow y = -2x - 1.$$

For $b = -1$, $a = 1 - b = 2$ and the points of tangency are $(2, 4)$ and $(-1, -8)$. The tangent line has slope 4:

$$y - 4 = 4(x - 2) \Rightarrow y = 4x - 4.$$

3. (a) $f(x) = \sqrt{x+1}, f'(x) = \dfrac{1}{2\sqrt{x+1}}, f(0) = 1, f'(0) = \dfrac{1}{2}$

$y - 1 = \dfrac{1}{2}(x - 0) \Rightarrow y = \dfrac{1}{2}x + 1$

$P_1(x) = \dfrac{1}{2}x + 1$

(b) $f''(x) = -\dfrac{1}{4\sqrt{(x+1)^3}}, \quad f''(0) = -\dfrac{1}{4}$

$P_2(x) = a_0 + a_1x + a_2x^2, \; P_2'(x) = a_1 + 2a_2x, \; P_2''(x) = 2a_2 = -\dfrac{1}{4}$

Hence, $a_2 = -\dfrac{1}{8}$ and $P_2(x) = 1 + \dfrac{1}{2}x - \dfrac{1}{8}x^2$.

(c)

x	-1.0	-0.1	-0.001	0	0.001	0.1	1.0
$\sqrt{x+1}$	0	0.9487	0.9995	1	1.0005	1.0488	1.4142
$P_2(x)$	0.3750	0.9488	0.9995	1	1.0005	1.0488	1.3750

The accuracy is better near $x = 0$.

(d)

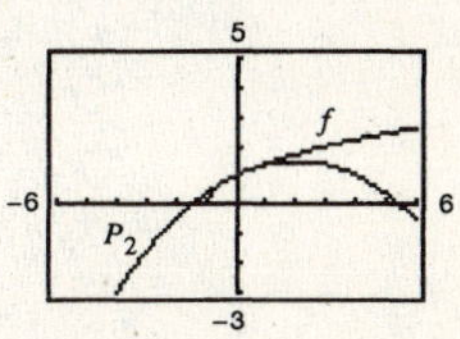

The graphs are close together near $x = 0$.

4. At the point $(1, 4)$

$a + b = 4$

$\dfrac{b}{2\sqrt{4}} = \dfrac{3}{2} \Rightarrow b = 6 \Rightarrow a = 4 - 6 = -2$

$f(x) = 6\sqrt{x} - 2$

5. Let $p(x) = Ax^3 + Bx^2 + Cx + D$.

$p'(x) = 3Ax^2 + 2Bx + C$

At $(1, 1)$: $A + B + C + D = 1$ Equation 1

$3A + 2B + C = 14$ Equation 2

At $(-1, -3)$: $-A + B - C + D = -3$ Equation 3

$3A - 2B + C = -2$ Equation 4

Adding Equations 1 and 3: $2B + 2D = -2$

Subtracting Equations 1 and 3: $2A + 2C = 4$

Adding Equations 2 and 4: $6A + 2C = 12$

Subtracting Equations 2 and 4: $4B = 16$

Hence, $B = 4$ and $D = \frac{1}{2}(-2 - 2B) = -5$.

Subtracting $2A + 2C = 4$ and $6A + 2C = 12$, you obtain $4A = 8 \Rightarrow A = 2$. Finally, $C = \frac{1}{2}(4 - 2A) = 0$.

Thus, $p(x) = 2x^3 + 4x^2 - 5$.

6. (a) $y = x^2, y' = 2x$. Slope $= 4$ at $(2, 4)$.

Tangent line: $y - 4 = 4(x - 2)$

$$y = 4x - 4$$

(b) Slope of normal line: $-\frac{1}{4}$.

Normal line: $y - 4 = -\frac{1}{4}(x - 2)$

$$y = -\frac{1}{4}x + \frac{9}{2}$$

$$y = -\frac{1}{4}x + \frac{9}{2} = x^2 \Rightarrow 4x^2 + x - 18 = 0 \Rightarrow (4x + 9)(x - 2) = 0$$

$x = 2, -\frac{9}{4}$. Second intersection point: $\left(-\frac{9}{4}, \frac{81}{16}\right)$

(c) Tangent line: $y = 0$

Normal line: $x = 0$

(d) Let (a, a^2), $a \neq 0$, be a point on the parabola $y = x^2$. Tangent line at (a, a^2) is $y = 2a(x - a) + a^2$.

Normal line at (a, a^2) is $y = -\frac{1}{2a}(x - a) + a^2$. To find points of intersection, solve

$$x^2 = -\frac{1}{2a}(x - a) + a^2$$

$$x^2 + \frac{1}{2a}x = a^2 + \frac{1}{2}$$

$$x^2 + \frac{1}{2a}x + \frac{1}{16a^2} = a^2 + \frac{1}{2} + \frac{1}{16a^2}$$

$$\left(x + \frac{1}{4a}\right)^2 = \left(a + \frac{1}{4a}\right)^2$$

$$x + \frac{1}{4a} = \pm\left(a + \frac{1}{4a}\right)$$

$$x + \frac{1}{4a} = a + \frac{1}{4a} \Rightarrow x = a \text{ (Point of tangency)}$$

$$x + \frac{1}{4a} = -\left(a + \frac{1}{4a}\right) \Rightarrow x = -a - \frac{1}{2a} = -\frac{2a^2 + 1}{2a}$$

The normal line intersects a second time at $x = -\frac{2a^2 + 1}{2a}$.

7. (a) $x^4 = a^2x^2 - a^2y^2$

$a^2y^2 = a^2x^2 - x^4$

$$y = \frac{\pm\sqrt{a^2x^2 - x^4}}{a}$$

Graph: $y_1 = \frac{\sqrt{a^2x^2 - x^4}}{a}$ and $y_2 = -\frac{\sqrt{a^2x^2 - x^4}}{a}$.

(b)

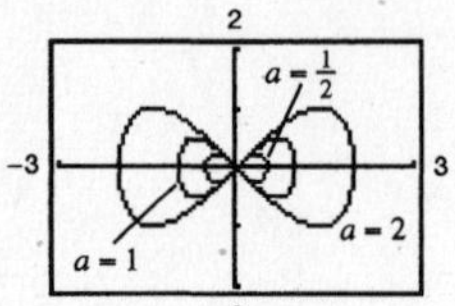

$(\pm a, 0)$ are the x-intercepts, along with $(0, 0)$.

—CONTINUED—

7. —CONTINUED—

(c) Differentiating implicitly,

$$4x^3 = 2a^2x - 2a^2yy'$$

$$y' = \frac{2a^2x - 4x^3}{2a^2y} = \frac{x(a^2 - 2x^2)}{a^2y} = 0 \Rightarrow 2x^2 = a^2 \Rightarrow x = \frac{\pm a}{\sqrt{2}}.$$

$$\left(\frac{a^2}{2}\right)^2 = a^2\left(\frac{a^2}{2}\right) - a^2y^2$$

$$\frac{a^4}{4} = \frac{a^4}{2} - a^2y^2$$

$$a^2y^2 = \frac{a^4}{4}$$

$$y^2 = \frac{a^2}{4}$$

$$y = \pm\frac{a}{2}$$

Four points: $\left(\frac{a}{\sqrt{2}}, \frac{a}{2}\right), \left(\frac{a}{\sqrt{2}}, -\frac{a}{2}\right), \left(-\frac{a}{\sqrt{2}}, \frac{a}{2}\right), \left(\frac{-a}{\sqrt{2}}, -\frac{a}{2}\right)$

8. (a) $b^2y^2 = x^3(a - x)$; $a, b > 0$

$$y^2 = \frac{x^3(a - x)}{b^2}$$

Graph $y_1 = \frac{\sqrt{x^3(a - x)}}{b}$ and $y_2 = -\frac{\sqrt{x^3(a - x)}}{b}$.

(b) a determines the x-intercept on the right: $(a, 0)$. b affects the height.

(c) Differentiating implicitly,

$$2b^2yy' = 3x^2(a - x) - x^3 = 3ax^2 - 4x^3$$

$$y' = \frac{(3ax^2 - 4x^3)}{2b^2y} = 0 \Rightarrow 3ax^2 = 4x^3$$

$$3a = 4x$$

$$x = \frac{3a}{4}.$$

$$b^2y^2 = \left(\frac{3a}{4}\right)^3\left(a - \frac{3a}{4}\right) = \frac{27a^3}{64}\left(\frac{1}{4}a\right)$$

$$y^2 = \frac{27a^4}{256b^2} \Rightarrow y = \pm\frac{3\sqrt{3}a^2}{16b}$$

Two points: $\left(\frac{3a}{4}, \frac{3\sqrt{3}a^2}{16b}\right), \left(\frac{3a}{4}, \frac{-3\sqrt{3}a^2}{16b}\right)$

9. (a)

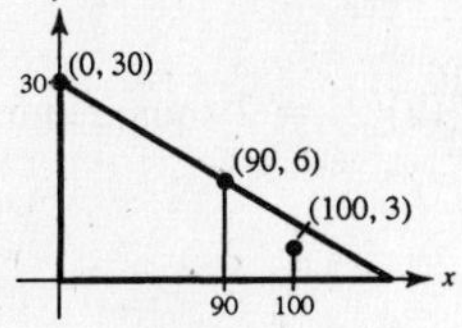

Not drawn to scale

Line determined by $(0, 30)$ and $(90, 6)$:

$$y - 30 = \frac{30 - 6}{0 - 90}(x - 0) = -\frac{24}{90}x = -\frac{4}{15}x \Rightarrow y = -\frac{4}{15}x + 30$$

When $x = 100$, $y = \frac{-4}{15}(100) + 30 = \frac{10}{3} > 3 \Rightarrow$ Shadow determined by man.

—CONTINUED—

9. —CONTINUED—

(b)

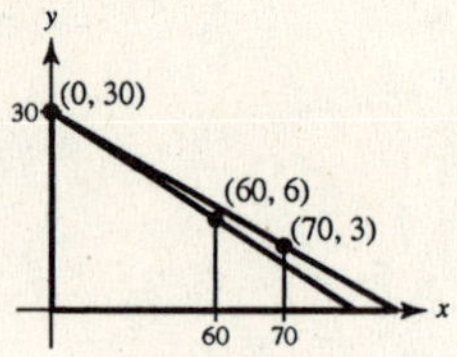

Not drawn to scale

Line determined by (0, 30) and (60, 6):

$$y - 30 = \frac{30 - 6}{0 - 60}(x - 0) = -\frac{2}{5}x \Rightarrow y = -\frac{2}{5}x + 30$$

When $x = 70$, $y = \frac{-2}{5}(70) + 30 = 2 < 3 \Rightarrow$ Shadow determined by child.

(c) Need (0, 30), $(d, 6)$, $(d + 10, 3)$ collinear.

$$\frac{30 - 6}{0 - d} = \frac{6 - 3}{d - (d + 10)} \Rightarrow \frac{24}{d} = \frac{3}{10} \Rightarrow d = 80 \text{ feet}$$

(d) Let y be the distance from the base of the street light to the tip of the shadow. We know that $\frac{dx}{dt} = -5$.

For $x > 80$, the shadow is determined by the man.

$$\frac{y}{30} = \frac{y - x}{6} \Rightarrow y = \frac{5}{4}x \text{ and } \frac{dy}{dt} = \frac{5}{4}\frac{dx}{dt} = \frac{-25}{4}.$$

For $x < 80$, the shadow is determined by the child.

$$\frac{y}{30} = \frac{y - x - 10}{3} \Rightarrow y = \frac{10}{9}x + \frac{100}{9} \text{ and } \frac{dy}{dt} = \frac{10}{9}\frac{dx}{dt} = \frac{-50}{9}.$$

Therefore,

$$\frac{dy}{dt} = \begin{cases} \dfrac{-25}{4} & x > 80 \\ \dfrac{-50}{9} & 0 < x < 80 \end{cases}.$$

$\frac{dy}{dt}$ is not continuous at $x = 80$.

10. (a) $y = x^{1/3} \Rightarrow \frac{dy}{dt} = \frac{1}{3}x^{-2/3}\frac{dx}{dt}$

$$1 = \frac{1}{3}(8)^{-2/3}\frac{dx}{dt}$$

$$\frac{dx}{dt} = 12 \text{ cm/sec}$$

(b) $D = \sqrt{x^2 + y^2} \Rightarrow \frac{dD}{dt} = \frac{1}{2}(x^2 + y^2)\left(2x\frac{dx}{dt} + 2y\frac{dy}{dt}\right)$

$$= \frac{x\frac{dx}{dt} + y\frac{dy}{dt}}{\sqrt{x^2 + y^2}}$$

$$= \frac{8(12) + 2(1)}{\sqrt{64 + 4}} = \frac{98}{\sqrt{68}} = \frac{49}{\sqrt{17}} \text{ cm/sec.}$$

11. (a) $y = 1/x$, $y' = -1/x^2$. At (1, 1), slope $= -1$.

Tangent line: $y - 1 = -1(x - 1)$ or $y = -x + 2$

Point A: $y = 0, x = 2$: $A = (2, 0)$

Point B $= (0, 2)$

Distance AP = Distance BP $= \sqrt{2}$

—CONTINUED—

(b) Area $= \frac{1}{2}(\text{base})(\text{height}) = \frac{1}{2}(2)(2) = 2$ square units

11. —CONTINUED—

(c) At $\left(2, \frac{1}{2}\right)$, slope $= -\frac{1}{4}$.

Tangent line $y - \frac{1}{2} = -\frac{1}{4}(x - 2)$ or $y = -\frac{1}{4}x + 1$

Point A: $y = 0, x = 4$: $A = (4, 0)$

Point B $= (0, 1)$

Distance AP $= \sqrt{2^2 + (1/2)^2} = \sqrt{17}/2$

Distance BP $= \sqrt{2^2 + (1/2)^2} = \sqrt{17}/2$

Area $= \frac{1}{2}(4)(1) = 2$ square units

(d) $y = a/x, y' = -a/x^2$. $P = (c, a/c)$ on graph.

Tangent line $y - \frac{a}{c} = \frac{-a}{c^2}(x - c)$ or $y = \frac{-a}{c^2}x + \frac{2a}{c}$

Point A: $y = 0, x = 2c$. $A = (2c, 0)$

Point B $= (0, 2a/c)$

Distance AP $= \sqrt{c^2 + (a/c)^2}$ = Distance BP

Area $= \frac{1}{2}$(base)(height) $= \frac{1}{2}(2c)\left(\frac{2a}{c}\right) = 2a$, independent of point.

12. $L'(x) = \lim_{\Delta x \to 0} \frac{L(x + \Delta x) - L(x)}{\Delta x}$

$= \lim_{\Delta x \to 0} \frac{L(x) + L(\Delta x) - L(x)}{\Delta x}$

$= \lim_{\Delta x \to 0} \frac{L(\Delta x)}{\Delta x}$

Also, $L'(0) = \lim_{\Delta x \to 0} \frac{L(\Delta x) - L(0)}{\Delta x}$.

But, $L(0) = 0$ because
$L(0) = L(0 + 0) = L(0) + L(0) \Rightarrow L(0) = 0$.

Thus, $L'(x) = L'(0)$, for all x.

The graph of L is a line through the origin of slope $L'(0)$.

13. (a) $v(t) = -\frac{27}{5}t + 27$ ft/sec

$a(t) = -\frac{27}{5}$ ft/sec^2

(b) $v(t) = -\frac{27}{5}t + 27 = 0 \Rightarrow \frac{27}{5}t$
$= 27 \Rightarrow t = 5$ seconds

$s(5) = -\frac{27}{10}(5)^2 + 27(5) + 6 = 73.5$ feet

(c) The acceleration due to gravity on Earth is greater in magnitude than that on the moon.

14. $j(t) = a'(t)$

(a) $j(t)$ is the rate of change of the acceleration.

(b) From Exercise 88 in Section 4.3,

$s(t) = -8.25t^2 + 66t$

$v(t) = -16.5t + 66$

$a(t) = -16.5$

$a'(t) = j(t) = 0$

(c) Position function: a

Velocity function: d

Acceleration function: b

Jerk function: c

CHAPTER 5
Applications of Differentiation

Section 5.1 Extrema on an Interval 375

Section 5.2 Rolle's Theorem and the Mean Value Theorem . 380

Section 5.3 Increasing and Decreasing Functions and the First Derivative Test 388

Section 5.4 Concavity and the Second Derivative Test 401

Section 5.5 Limits at Infinity 413

Section 5.6 A Summary of Curve Sketching 423

Section 5.7 Optimization Problems 438

Section 5.8 Differentials 452

Review Exercises 457

Problem Solving 470

CHAPTER 5
Applications of Differentiation

Section 5.1 Extrema on an Interval

1. A: neither
B: absolute maximum (and relative maximum)
C: neither
D: neither
E: relative maximum
F: relative minimum
G: neither

2. A: absolute minimum
B: relative maximum
C: neither
D: relative minimum
E: relative maximum
F: relative minimum
G: neither

3. $f(x) = \dfrac{x^2}{x^2 + 4}$

$$f'(x) = \frac{(x^2 + 4)(2x) - (x^2)(2x)}{(x^2 + 4)^2} = \frac{8x}{(x^2 + 4)^2}$$

$f'(0) = 0$

4. $f(x) = \dfrac{1}{2}x^3 - \dfrac{3}{2}x^2 + 1$

$f'(x) = \dfrac{3}{2}x^2 - 3x$

$f'(0) = 0; f'(2) = \dfrac{3}{2}(2)^2 - 3(2) = 0$

5. $f(x) = x + \dfrac{27}{2x^2} = x + \dfrac{27}{2}x^{-2}$

$f'(x) = 1 - 27x^{-3} = 1 - \dfrac{27}{x^3}$

$f'(3) = 1 - \dfrac{27}{3^3} = 1 - 1 = 0$

6. $f(x) = -3x\sqrt{x + 1}$

$$\begin{aligned} f'(x) &= -3x\left[\frac{1}{2}(x + 1)^{-1/2}\right] + \sqrt{x + 1}(-3) \\ &= -\frac{3}{2}(x + 1)^{-1/2}[x + 2(x + 1)] \\ &= -\frac{3}{2}(x + 1)^{-1/2}(3x + 2) \end{aligned}$$

$f'\left(-\dfrac{2}{3}\right) = 0$

7. $f(x) = (x + 2)^{2/3}$

$f'(x) = \dfrac{2}{3}(x + 2)^{-1/3}$

$f'(-2)$ is undefined.

8. Using the limit definition of the derivative,

$$\lim_{x \to 0^-} \frac{f(x) - f(0)}{x - 0} = \lim_{x \to 0^-} \frac{(4 - |x|) - 4}{x} = 1$$

$$\lim_{x \to 0^+} \frac{f(x) - f(0)}{x - 0} = \lim_{x \to 0^+} \frac{(4 - |x|) - 4}{x - 0} = -1$$

$f'(0)$ does not exist, since the one-sided derivatives are not equal.

9. Critical numbers: $x = 2$

$x = 2$: absolute maximum

10. Critical number: 0

The function has neither a maximum nor a minimum at $x = 0$.

11. Critical numbers: $x = 1, 2, 3$

$x = 1, 3$: absolute maximum

$x = 2$: absolute minimum

12. Critical numbers: 2, 5

At $x = 5$, the function has an absolute maximum at $x = 2$, none of those.

13. $f(x) = x^2(x - 3) = x^3 - 3x^2$

$f'(x) = 3x^2 - 6x = 3x(x - 2)$

Critical numbers: $x = 0, x = 2$

14. $g(x) = x^2(x^2 - 4) = x^4 - 4x^2$

$g'(x) = 4x^3 - 8x = 4x(x^2 - 2)$

Critical numbers: $x = 0, x = \pm\sqrt{2}$

15. $g(t) = t\sqrt{4 - t},\ t < 3$

$$g'(t) = t\left[\frac{1}{2}(4 - t)^{-1/2}(-1)\right] + (4 - t)^{1/2}$$

$$= \frac{1}{2}(4 - t)^{-1/2}[-t + 2(4 - t)]$$

$$= \frac{8 - 3t}{2\sqrt{4 - t}}$$

Critical number is $t = \frac{8}{3}$.

16. $f(x) = \dfrac{4x}{x^2 + 1}$

$$f'(x) = \frac{(x^2 + 1)(4) - (4x)(2x)}{(x^2 + 1)^2} = \frac{4(1 - x^2)}{(x^2 + 1)^2}$$

Critical numbers: $x = \pm 1$

17. $f(x) = 2(3 - x), [-1, 2]$

$f'(x) = -2 \Rightarrow$ No critical numbers

Left endpoint: $(-1, 8)$ Maximum

Right endpoint: $(2, 2)$ Minimum

18. $f(x) = \dfrac{2x + 5}{3}, [0, 5]$

$f'(x) = \dfrac{2}{3} \Rightarrow$ No critical numbers

Left endpoint: $\left(0, \dfrac{5}{3}\right)$ Minimum

Right endpoint: $(5, 5)$ Maximum

19. $f(x) = -x^2 + 3x, [0, 3]$

$f'(x) = -2x + 3$

Left endpoint: $(0, 0)$ Minimum

Critical number: $\left(\frac{3}{2}, \frac{9}{4}\right)$ Maximum

Right endpoint: $(3, 0)$ Minimum

20. $f(x) = x^2 + 2x - 4, [-1, 1]$

$f'(x) = 2x + 2 = 2(x + 1)$

Left endpoint: $(-1, -5)$ Minimum

Right endpoint: $(1, -1)$ Maximum

21. $f(x) = x^3 - \dfrac{3}{2}x^2, [-1, 2]$

$f'(x) = 3x^2 - 3x = 3x(x - 1)$

Left endpoint: $\left(-1, -\dfrac{5}{2}\right)$ Minimum

Right endpoint: $(2, 2)$ Maximum

Critical number: $(0, 0)$

Critical number: $\left(1, -\dfrac{1}{2}\right)$

22. $f(x) = x^3 - 12x, [0, 4]$

$f'(x) = 3x^2 - 12 = 3(x^2 - 4)$

Left endpoint: $(0, 0)$

Critical number: $(2, -16)$ Minimum

Right endpoint: $(4, 16)$ Maximum

Note: $x = -2$ is not in the interval.

23. $f(x) = 3x^{2/3} - 2x, [-1, 1]$

$f'(x) = 2x^{-1/3} - 2 = \dfrac{2(1 - \sqrt[3]{x})}{\sqrt[3]{x}}$

Left endpoint: $(-1, 5)$ Maximum

Critical number: $(0, 0)$ Minimum

Right endpoint: $(1, 1)$

24. $g(x) = \sqrt[3]{x}, [-1, 1]$

$g'(x) = \dfrac{1}{3x^{2/3}}$

Left endpoint: $(-1, -1)$ Minimum

Critical number: $(0, 0)$

Right endpoint: $(1, 1)$ Maximum

25. $y = 3 - |t - 3|, [-1, 5]$

From the graph, you see that $t = 3$ is a critical number.

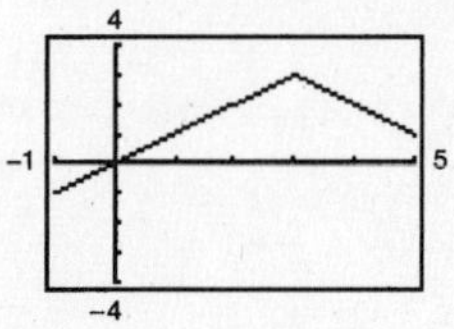

Left endpoint: $(-1, -1)$ Minimum

Right endpoint: $(5, 1)$

Critical number: $(3, 3)$ Maximum

26. $g(t) = \dfrac{t^2}{t^2 + 3}, [-1, 1]$

$$g'(t) = \frac{6t}{(t^2 + 3)^2}$$

Left endpoint: $\left(-1, \dfrac{1}{4}\right)$ Maximum

Critical number: $(0, 0)$ Minimum

Right endpoint: $\left(1, \dfrac{1}{4}\right)$ Maximum

27. $h(s) = \dfrac{1}{s - 2}, [0, 1]$

$$h'(s) = \frac{-1}{(s - 2)^2}$$

Left endpoint: $\left(0, -\dfrac{1}{2}\right)$ Maximum

Right endpoint: $(1, -1)$ Minimum

28. $h(t) = \dfrac{t}{t - 2}, [3, 5]$

$$h'(t) = \frac{-2}{(t - 2)^2}$$

Left endpoint: $(3, 3)$ Maximum

Right endpoint: $\left(5, \dfrac{5}{3}\right)$ Minimum

29. (a) Minimum: $(0, -3)$

Maximum: $(2, 1)$

(b) Minimum: $(0, -3)$

(c) Maximum: $(2, 1)$

(d) No extrema

30. (a) Minimum: $(4, 1)$

Maximum: $(1, 4)$

(b) Maximum: $(1, 4)$

(c) Minimum: $(4, 1)$

(d) No extrema

31. $f(x) = x^2 - 2x$

(a) Minimum: $(1, -1)$

Maximum: $(-1, 3)$

(b) Maximum: $(3, 3)$

(c) Minimum: $(1, -1)$

(d) Minimum: $(1, -1)$

32. (a) Minima: $(-2, 0)$ and $(2, 0)$

Maximum: $(0, 2)$

(b) Minimum: $(-2, 0)$

(c) Maximum: $(0, 2)$

(d) Maximum: $\left(1, \sqrt{3}\right)$

33. $f(x) = \begin{cases} 2x + 2, & 0 \le x \le 1 \\ 4x^2, & 1 < x \le 3 \end{cases}$

Left endpoint: $(0, 2)$ Minimum

Right endpoint: $(3, 36)$ Maximum

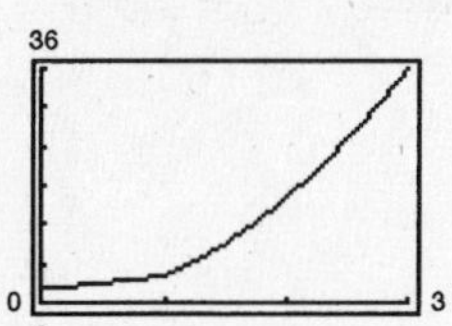

34. $f(x) = \begin{cases} 2 - x^2, & 1 \le x < 3 \\ 2 - 3x, & 3 \le x \le 5 \end{cases}$

Left endpoint: $(1, 1)$ Maximum

Right endpoint: $(5, -13)$ Minimum

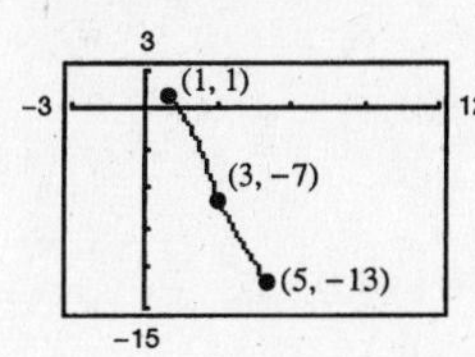

35. $f(x) = \dfrac{3}{x - 1}, (1, 4]$

Right endpoint: $(4, 1)$ Minimum

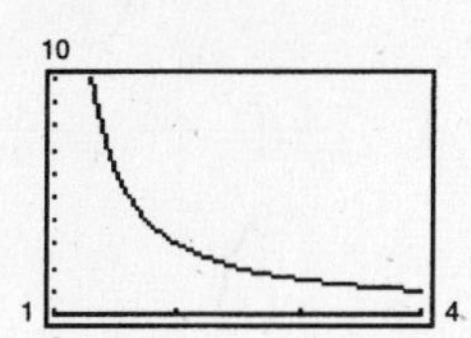

36. $f(x) = \dfrac{2}{2 - x}, [0, 2)$

Left endpoint: $(0, 1)$ Minimum

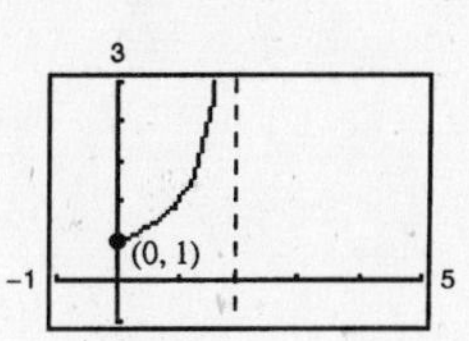

37. (a)

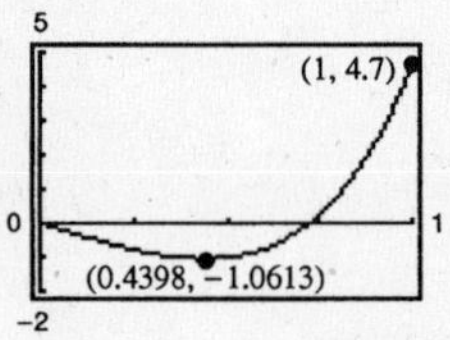

Maximum: $(1, 4.7)$ (endpoint)

Minimum: $(0.4398, -1.0613)$

(b)
$$f(x) = 3.2x^5 + 5x^3 - 3.5x, [0, 1]$$
$$f'(x) = 16x^4 + 15x^2 - 3.5$$
$$16x^4 + 15x^2 - 3.5 = 0$$
$$x^2 = \frac{-15 \pm \sqrt{(15)^2 - 4(16)(-3.5)}}{2(16)}$$
$$= \frac{-15 \pm \sqrt{449}}{32}$$
$$x = \sqrt{\frac{-15 + \sqrt{449}}{32}} \approx 0.4398$$
$$f(0) = 0$$
$$f(1) = 4.7 \text{ Maximum (endpoint)}$$
$$f\left(\sqrt{\frac{-15 + \sqrt{449}}{32}}\right) \approx -1.0613$$

Minimum: $(0.4398, -1.0613)$

38. (a)

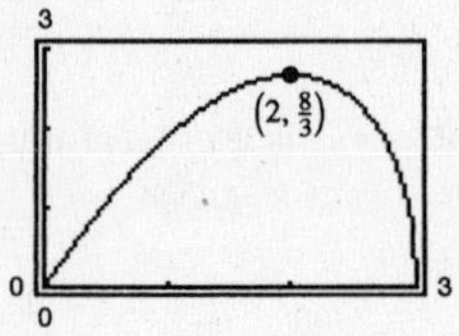

Maximum: $\left(2, \frac{8}{3}\right)$

Minimum: $(0, 0), (3, 0)$

(b) $f(x) = \frac{4}{3}x\sqrt{3 - x}, [0, 3]$

$$f'(x) = \frac{4}{3}\left[x\left(\frac{1}{2}\right)(3 - x)^{-1/2}(-1) + (3 - x)^{1/2}(1)\right]$$
$$= \frac{4}{3}(3 - x)^{-1/2}\left(\frac{1}{2}\right)[-x + 2(3 - x)]$$
$$= \frac{2(6 - 3x)}{3\sqrt{3 - x}} = \frac{6(2 - x)}{3\sqrt{3 - x}} = \frac{2(2 - x)}{\sqrt{3 - x}}$$

Critical number: $x = 2$

$f(0) = 0$ Minimum

$f(3) = 0$ Minimum

$f(2) = \frac{8}{3}$

Maximum: $\left(2, \frac{8}{3}\right)$

39. $f(x) = (1 + x^3)^{1/2}, [0, 2]$

$$f'(x) = \frac{3}{2}x^2(1 + x^3)^{-1/2}$$
$$f''(x) = \frac{3}{4}(x^4 + 4x)(1 + x^3)^{-3/2}$$
$$f'''(x) = -\frac{3}{8}(x^6 + 20x^3 - 8)(1 + x^3)^{-5/2}$$

Setting $f''' = 0$, we have $x^6 + 20x^3 - 8 = 0$.

$$x^3 = \frac{-20 \pm \sqrt{400 - 4(1)(-8)}}{2}$$
$$x = \sqrt[3]{-10 + \sqrt{108}} = \sqrt{3} - 1$$

In the interval $[0, 2]$, choose

$$x = \sqrt[3]{-10 + \sqrt{108}} = \sqrt{3} - 1 \approx 0.732.$$

$\left|f''\left(\sqrt[3]{-10 + \sqrt{108}}\right)\right| \approx 1.47$ is the maximum value.

40. $f(x) = \frac{1}{x^2 + 1}, \left[\frac{1}{2}, 3\right]$

$$f'(x) = \frac{-2x}{(x^2 + 1)^2}$$
$$f''(x) = \frac{-2(1 - 3x^2)}{(x^2 + 1)^3}$$
$$f'''(x) = \frac{24x - 24x^3}{(x^2 + 1)^4}$$

Setting $f''' = 0$, we have $x = 0, \pm 1$.

$|f''(1)| = \frac{1}{2}$ is the maximum value.

41. The function f is continuous on $[3, 5]$, so it has a maximum by Theorem 5.1. The function is not continuous on $[1, 3]$.

42. Let $f(x) = 1/x$. f is continuous on $(0, 1)$ but does not have a maximum. f is also continuous on $(-1, 0)$ but does not have a minimum. This can occur if one of the endpoints is an infinite discontinuity.

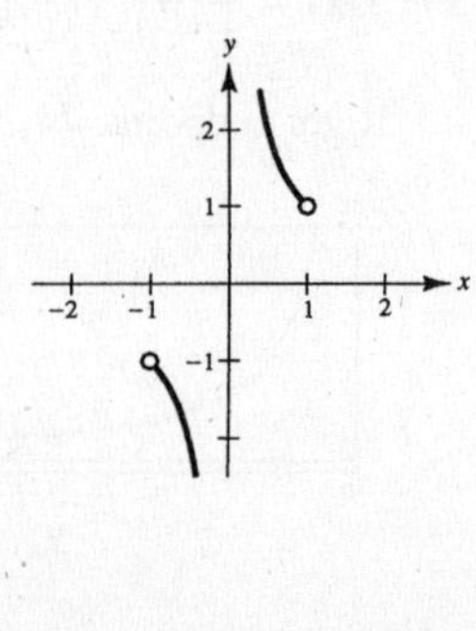

43. $f(x) = (x + 1)^{2/3}, [0, 2]$

$f'(x) = \frac{2}{3}(x + 1)^{-1/3}$

$f''(x) = -\frac{2}{9}(x + 1)^{-4/3}$

$f'''(x) = \frac{8}{27}(x + 1)^{-7/3}$

$f^{(4)}(x) = -\frac{56}{81}(x + 1)^{-10/3}$

$f^{(5)}(x) = \frac{560}{243}(x + 1)^{-13/3}$

$|f^{(4)}(0)| = \frac{56}{81}$ is the maximum value.

44. $f(x) = \frac{1}{x^2 + 1}, [-1, 1]$

$f'''(x) = \frac{24x - 24x^3}{(x^2 + 1)^4}$ (See Exercise 40.)

$f^{(4)}(x) = \frac{24(5x^4 - 10x^2 + 1)}{(x^2 + 1)^5}$

$f^{(5)}(x) = \frac{-240x(3x^4 - 10x^2 + 3)}{(x^2 + 1)^6}$

$|f^{(4)}(0)| = 24$ is the maximum value.

45.

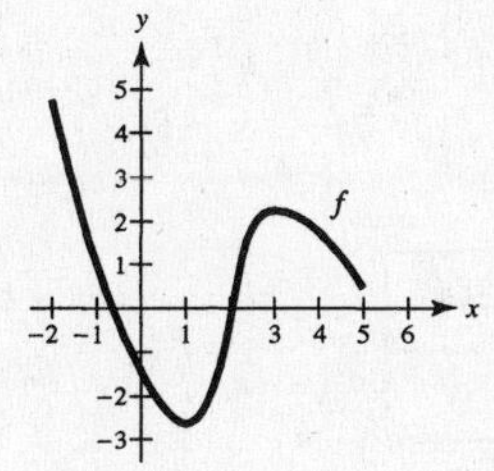

46.

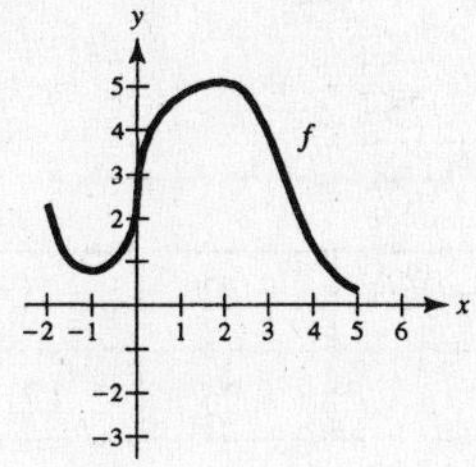

47. (a) Yes

(b) No

48. (a) No

(b) Yes

49. (a) No

(b) Yes

50. (a) No

(b) Yes

51. $P = VI - RI^2 = 12I - 0.5I^2, 0 \le I \le 15$

$P = 0$ when $I = 0$.

$P = 67.5$ when $I = 15$.

$P' = 12 - I = 0$

Critical number: $I = 12$ amps

When $I = 12$ amps, $P = 72$, the maximum output.

No, a 20-amp fuse would not increase the power output. P is decreasing for $I > 12$.

52. $C = 2x + \frac{300,000}{x}, 1 \le x \le 300$

$C(1) = 300,002$

$C(300) = 1600$

$C' = 2 - \frac{300,000}{x^2} = 0$

$2x^2 = 300,000$

$x^2 = 150,000$

$x = 100\sqrt{15} \approx 387 > 300$ (outside of interval)

C is minimized when $x = 300$ units.

Yes, if $1 \le x \le 400$, then $x = 387$ would minimize C.

53. (a) $y = ax^2 + bx + c$

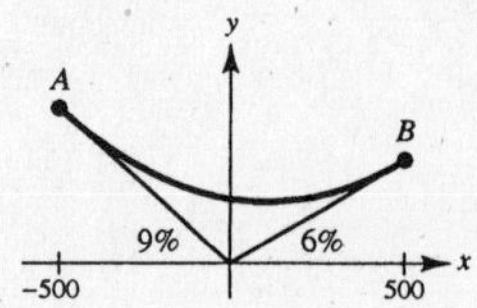

$y' = 2ax + b$

The coordinates of B are (500, 30), and those of A are (−500, 45). From the slopes at A and B,

$$-1000a + b = -0.09$$
$$1000a + b = 0.06.$$

Solving these two equations, you obtain $a = 3/40000$ and $b = -3/200$. From the points (500, 30) and (−500, 45), you obtain

$$30 = \frac{3}{40000}500^2 + 500\left(\frac{-3}{200}\right) + c$$

$$45 = \frac{3}{40000}500^2 - 500\left(\frac{-3}{200}\right) + c.$$

In both cases, $c = 18.75 = \frac{75}{4}$. Thus,

$$y = \frac{3}{40000}x^2 - \frac{3}{200}x + \frac{75}{4}.$$

(b)

x	−500	−400	−300	−200	−100	0	100	200	300	400	500
d	0	.75	3	6.75	12	18.75	12	6.75	3	.75	0

For $-500 \le x \le 0$, $d = (ax^2 + bx + c) - (-0.09x)$.

For $0 \le x \le 500$, $d = (ax^2 + bx + c) - (0.06x)$.

(c) The lowest point on the highway is (100, 18), which is not directly over the point where the two hillsides come together.

54. $f(x) = [\![x]\!]$

The derivative of f is undefined at every integer and is zero at any noninteger real number. All real numbers are critical numbers.

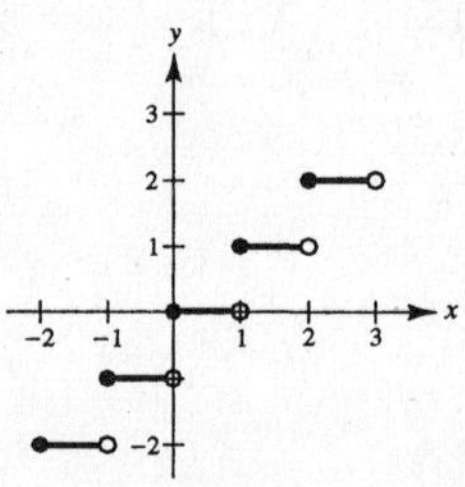

55. True.

56. True. This is stated in the Extreme Value Theorem.

57. True.

58. False. Let $f(x) = x^2$. $x = 0$ is a critical number of f.

$$g(x) = f(x - k)$$
$$= (x - k)^2$$

$x = k$ is a critical number of g.

Section 5.2 Rolle's Theorem and the Mean Value Theorem

1. Rolle's Theorem does not apply to $f(x) = 1 - |x - 1|$ over $[0, 2]$ since f is not differentiable at $x = 1$.

2. $f(-2) = f(2) = 0$. But, f is not continuous on $(-2, 2)$ because f is not defined at $x = 0$.

3. $f(x) = \left|\frac{1}{x}\right|$

$f(-1) = f(1) = 1$. But, f is not continuous on $[-1, 1]$.

4. $f(x) = \sqrt{(2 - x^{2/3})^3}$

$f(-1) = f(1) = 1$

$f'(x) = \dfrac{-\sqrt{(2 - x^{2/3})}}{x^{1/3}}$

f is not differentiable at $x = 0$.

5. $f(x) = x^2 - x - 2 = (x - 2)(x + 1)$

x-intercepts: $(-1, 0), (2, 0)$

$f'(x) = 2x - 1 = 0$ at $x = \frac{1}{2}$.

6. $f(x) = x(x - 3)$

x-intercepts: $(0, 0), (3, 0)$

$f'(x) = 2x - 3 = 0$ at $x = \frac{3}{2}$.

7. $f(x) = x\sqrt{x + 4}$

x-intercepts: $(-4, 0), (0, 0)$

$$f'(x) = x\frac{1}{2}(x + 4)^{-1/2} + (x + 4)^{1/2}$$

$$= (x + 4)^{-1/2}\left(\frac{x}{2} + (x + 4)\right)$$

$$f'(x) = \left(\frac{3}{2}x + 4\right)(x + 4)^{-1/2} = 0 \text{ at } x = -\frac{8}{3}$$

8. $f(x) = -3x\sqrt{x + 1}$

x-intercepts: $(-1, 0), (0, 0)$

$$f'(x) = -3x\frac{1}{2}(x + 1)^{-1/2} - 3(x + 1)^{1/2}$$

$$= -3(x + 1)^{-1/2}\left(\frac{x}{2} + (x + 1)\right)$$

$$f'(x) = -3(x + 1)^{-1/2}\left(\frac{3}{2}x + 1\right) = 0 \text{ at } x = -\frac{2}{3}.$$

9. $f(x) = x^2 - 2x, [0, 2]$

$f(0) = f(2) = 0$

f is continuous on $[0, 2]$. f is differentiable on $(0, 2)$. Rolle's Theorem applies.

$$f'(x) = 2x - 2$$

$$2x - 2 = 0 \implies x = 1$$

c value: 1

10. $f(x) = x^2 - 5x + 4, [1, 4]$

$f(1) = f(4) = 0$

f is continuous on $[1, 4]$. f is differentiable on $(1, 4)$.

Rolle's Theorem applies.

$$f'(x) = 2x - 5$$

$$2x - 5 = 0 \implies x = \frac{5}{2}$$

c value: $\frac{5}{2}$

11. $f(x) = (x - 1)(x - 2)(x - 3), [1, 3]$

$f(1) = f(3) = 0$

f is continuous on $[1, 3]$. f is differentiable on $(1, 3)$. Rolle's Theorem applies.

$$f(x) = x^3 - 6x^2 + 11x - 6$$

$$f'(x) = 3x^2 - 12x + 11$$

$$3x^2 - 12x + 11 = 0 \implies x = \frac{6 \pm \sqrt{3}}{3}$$

$c = \dfrac{6 - \sqrt{3}}{3}, c = \dfrac{6 + \sqrt{3}}{3}$

12. $f(x) = (x - 3)(x + 1)^2, [-1, 3]$

$f(-1) = f(3) = 0$

f is continuous on $[-1, 3]$. f is differentiable on $(-1, 3)$. Rolle's Theorem applies.

$$f'(x) = (x - 3)(2)(x + 1) + (x + 1)^2$$

$$= (x + 1)[2x - 6 + x + 1]$$

$$= (x + 1)(3x - 5)$$

c value: $\frac{5}{3}$

13. $f(x) = x^{2/3} - 1, [-8, 8]$

$f(-8) = f(8) = 3$

f is continuous on $[-8, 8]$. f is not differentiable on $(-8, 8)$ since $f'(0)$ does not exist. Rolle's Theorem does not apply.

14. $f(x) = 3 - |x - 3|, [0, 6]$

$f(0) = f(6) = 0$

f is continuous on $[0, 6]$. f is not differentiable on $(0, 6)$ since $f'(3)$ does not exist. Rolle's Theorem does not apply.

15. $f(x) = \dfrac{x^2 - 2x - 3}{x + 2}$, $[-1, 3]$

$f(-1) = f(3) = 0$

f is continuous on $[-1, 3]$. (**Note:** The discontinuity, $x = -2$, is not in the interval.) f is differentiable on $(-1, 3)$. Rolle's Theorem applies.

$$f'(x) = \frac{(x + 2)(2x - 2) - (x^2 - 2x - 3)(1)}{(x + 2)^2} = 0$$

$$\frac{x^2 + 4x - 1}{(x + 2)^2} = 0$$

$$x = \frac{-4 \pm 2\sqrt{5}}{2} = -2 \pm \sqrt{5}$$

c value: $-2 + \sqrt{5}$

16. $f(x) = \dfrac{x^2 - 1}{x}$, $[-1, 1]$

$f(-1) = f(1) = 0$

f is not continuous on $[-1, 1]$ since $f(0)$ does not exist. Rolle's Theorem does not apply.

17. $f(x) = |x| - 1$, $[-1, 1]$

$f(-1) = f(1) = 0$

f is continuous on $[-1, 1]$. f is not differentiable on $(-1, 1)$ since $f'(0)$ does not exist. Rolle's Theorem does not apply.

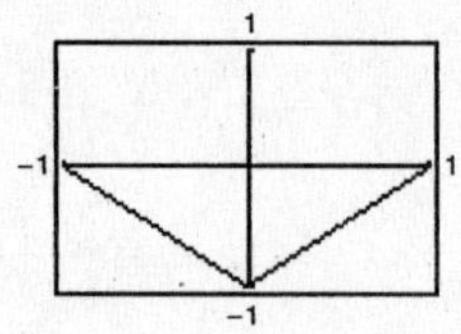

18. $f(x) = x - x^{1/3}$, $[0, 1]$

$f(0) = f(1) = 0$

f is continuous on $[0, 1]$. f is differentiable on $(0, 1)$. (**Note:** f is not differentiable at $x = 0$.) Rolle's Theorem applies.

$$f'(x) = 1 - \frac{1}{3\sqrt[3]{x^2}} = 0$$

$$1 = \frac{1}{3\sqrt[3]{x^2}}$$

$$\sqrt[3]{x^2} = \frac{1}{3}$$

$$x^2 = \frac{1}{27}$$

$$x = \sqrt{\frac{1}{27}} = \frac{\sqrt{3}}{9}$$

c value: $\dfrac{\sqrt{3}}{9} \approx 0.1925$

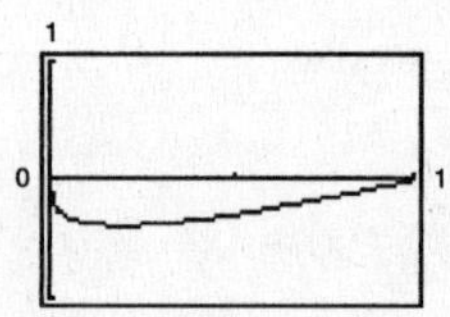

19. $C(x) = 10\left(\dfrac{1}{x} + \dfrac{x}{x + 3}\right)$

(a) $C(3) = C(6) = \dfrac{25}{3}$

(b) $C'(x) = 10\left(-\dfrac{1}{x^2} + \dfrac{3}{(x + 3)^2}\right) = 0$

$$\frac{3}{x^2 + 6x + 9} = \frac{1}{x^2}$$

$$2x^2 - 6x - 9 = 0$$

$$x = \frac{6 \pm \sqrt{108}}{4}$$

$$= \frac{6 \pm 6\sqrt{3}}{4} = \frac{3 \pm 3\sqrt{3}}{2}$$

In the interval $(3, 6)$: $c = \dfrac{3 + 3\sqrt{3}}{2} \approx 4.098$.

20. $f(t) = -16t^2 + 48t + 32$

(a) $f(1) = f(2) = 64$

(b) $v = f'(t)$ must be 0 at some time in $(1, 2)$.

$$f'(t) = -32t + 48 = 0$$

$$t = \frac{3}{2} \text{ seconds}$$

21. $f(x) = x^2 + 1$

(a) slope $= \dfrac{5-2}{2+1} = 1$

secant line: $y - 2 = 1(x + 1)$

$$y = x + 3$$

(b) $f'(x) = 2x = 1 \Rightarrow c = \dfrac{1}{2}$

$f'\left(\dfrac{1}{2}\right) = 1, \quad f\left(\dfrac{1}{2}\right) = \dfrac{5}{4}$

(c) Tangent line: $y - \dfrac{5}{4} = 1\left(x - \dfrac{1}{2}\right)$

$$y = x + \frac{3}{4}$$

(d)

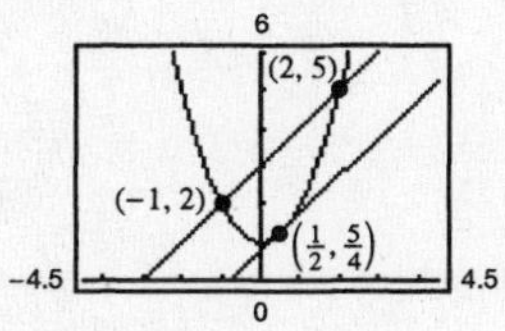

22. $f(x) = -x^2 - x + 6$

(a) slope $= \dfrac{4-0}{-2-2} = -1$

secant line: $y - 0 = -1(x - 2)$

$$y = -x + 2$$

(b) $f'(x) = -2x - 1 = -1 \Rightarrow c = 0$ and $f(c) = 6$

(c) Tangent line: $y - 6 = -1(x - 0)$

$$y = -x + 6$$

(d)

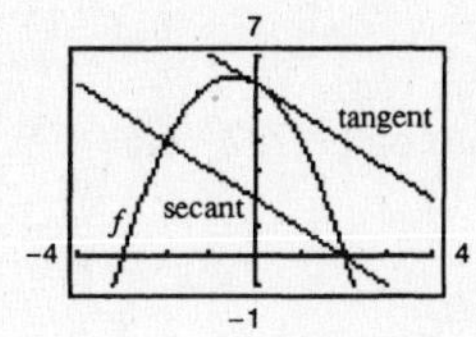

23.

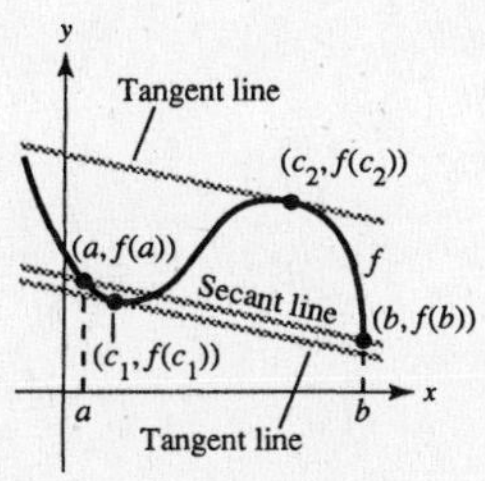

24.

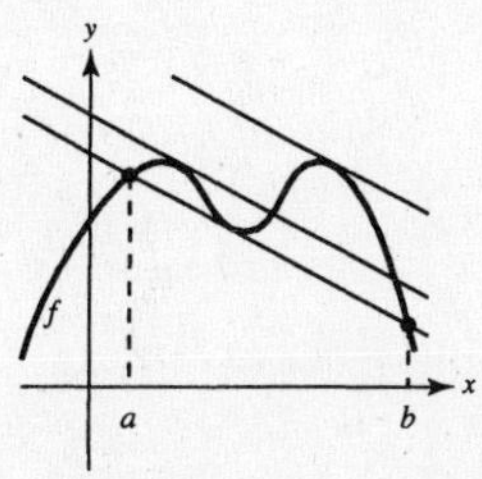

25. $f(x) = \dfrac{1}{x-3}, \; [0, 6]$

f has a discontinuity at $x = 3$.

26. $f(x) = |x - 3|, \; [0, 6]$

f is not differentiable at $x = 3$.

27. $f(x) = x^2$ is continuous on $[-2, 1]$ and differentiable on $(-2, 1)$.

$$\frac{f(1) - f(-2)}{1 - (-2)} = \frac{1-4}{3} = -1$$

$f'(x) = 2x = -1$ when $x = -\dfrac{1}{2}$. Therefore,

$$c = -\frac{1}{2}.$$

28. $f(x) = x(x^2 - x - 2)$ is continuous on $[-1, 1]$ and differentiable on $(-1, 1)$.

$$\frac{f(1) - f(-1)}{1 - (-1)} = -1$$

$$f'(x) = 3x^2 - 2x - 2 = -1$$

$$(3x + 1)(x - 1) = 0$$

$$c = -\frac{1}{3}$$

29. $f(x) = x^{2/3}$ is continuous on $[0, 1]$ and differentiable on $(0, 1)$.

$$\frac{f(1) - f(0)}{1 - 0} = 1$$

$$f'(x) = \frac{2}{3}x^{-1/3} = 1$$

$$x = \left(\frac{2}{3}\right)^3 = \frac{8}{27}$$

$$c = \frac{8}{27}$$

30. $f(x) = (x + 1)/x$ is continuous on $[1/2, 2]$ and differentiable on $(1/2, 2)$.

$$\frac{f(2) - f(1/2)}{2 - (1/2)} = \frac{(3/2) - 3}{3/2} = -1$$

$$f'(x) = \frac{-1}{x^2} = -1$$

$$x^2 = 1$$

$$c = 1$$

31. $f(x) = \sqrt{2 - x}$ is continuous on $[-7, 2]$ and differentiable on $(-7, 2)$.

$$\frac{f(2) - f(-7)}{2 - (-7)} = \frac{0 - 3}{9} = -\frac{1}{3}$$

$$f'(x) = \frac{-1}{2\sqrt{2 - x}} = -\frac{1}{3}$$

$$2\sqrt{2 - x} = 3$$

$$\sqrt{2 - x} = \frac{3}{2}$$

$$2 - x = \frac{9}{4}$$

$$x = -\frac{1}{4}$$

$$c = -\frac{1}{4}$$

32. $f(x) = x^3$ is continuous on $[0, 1]$ and differentiable on $(0, 1)$.

$$\frac{f(1) - f(0)}{1 - 0} = \frac{1 - 0}{1} = 1$$

$$f'(x) = 3x^2 = 1$$

$$x = \pm\frac{\sqrt{3}}{3}$$

In the interval $(0, 1)$: $c = \frac{\sqrt{3}}{3}$.

33. $f(x) = \frac{x}{x + 1}$ on $\left[-\frac{1}{2}, 2\right]$.

(a)

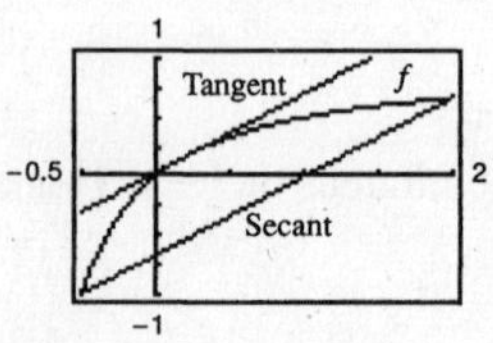

(b) Secant line:

$$\text{slope} = \frac{f(2) - f(-1/2)}{2 - (-1/2)} = \frac{2/3 - (-1)}{5/2} = \frac{2}{3}$$

$$y - \frac{2}{3} = \frac{2}{3}(x - 2)$$

$$3y - 2 = 2x - 4$$

$$3y - 2x + 2 = 0$$

(c) $f'(x) = \frac{1}{(x + 1)^2} = \frac{2}{3}$

$$(x + 1)^2 = \frac{3}{2}$$

$$x = -1 \pm \sqrt{\frac{3}{2}} = -1 \pm \frac{\sqrt{6}}{2}$$

In the interval $[-1/2, 2]$, $c = -1 + (\sqrt{6}/2)$.

$$f(c) = \frac{-1 + (\sqrt{6}/2)}{[-1 + (\sqrt{6}/2)] + 1} = \frac{-2 + \sqrt{6}}{\sqrt{6}} = \frac{-2}{\sqrt{6}} + 1$$

Tangent line: $y - 1 + \frac{2}{\sqrt{6}} = \frac{2}{3}\left(x - \frac{\sqrt{6}}{2} + 1\right)$

$$y - 1 + \frac{\sqrt{6}}{3} = \frac{2}{3}x - \frac{\sqrt{6}}{3} + \frac{2}{3}$$

$$3y - 2x - 5 + 2\sqrt{6} = 0$$

34. (a) $f(x) = -x^2 + x^3, \quad [0, 1]$

(b) $f(0) = 0, \quad f(1) = 0 \quad$ Slope $= 0$

Secant line: $y - 0 = 0(x - 0)$

$$y = 0$$

(c) $f'(x) = -2x + 3x^2$

$$0 = -2x + 3x^2$$

$$0 = x(-2 + 3x)$$

$$x = 0, \frac{2}{3} \quad \text{Use } x = \frac{2}{3} \text{ in } (0, 1).$$

$$f\left(\frac{2}{3}\right) = -\frac{4}{27}$$

Tangent line: $y + \dfrac{4}{27} = 0\left(x - \dfrac{2}{3}\right)$

$$y = -\frac{4}{27}$$

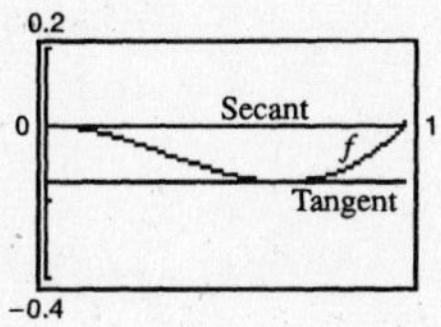

35. $f(x) = \sqrt{x}, \; [1, 9]$

$(1, 1), (9, 3)$

$$m = \frac{3 - 1}{9 - 1} = \frac{1}{4}$$

(a)

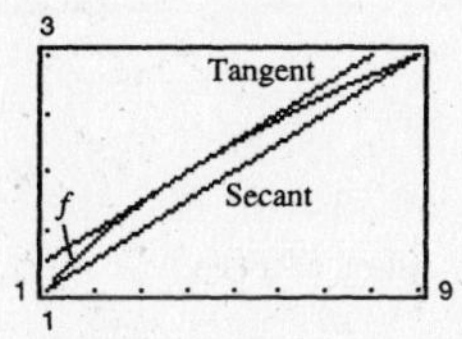

(b) Secant line: $y - 1 = \dfrac{1}{4}(x - 1)$

$$y = \frac{1}{4}x + \frac{3}{4}$$

$$0 = x - 4y + 3$$

(c) $$f'(x) = \frac{1}{2\sqrt{x}}$$

$$\frac{f(9) - f(1)}{9 - 1} = \frac{1}{4}$$

$$\frac{1}{2\sqrt{c}} = \frac{1}{4}$$

$$\sqrt{c} = 2$$

$$c = 4$$

$$(c, f(c)) = (4, 2)$$

$$m = f'(4) = \frac{1}{4}$$

Tangent line: $y - 2 = \dfrac{1}{4}(x - 4)$

$$y = \frac{1}{4}x + 1$$

$$0 = x - 4y + 4$$

36. (a) $f(x) = \sqrt{x + 1}, \quad [0, 8]$

(b) $f(0) = 1, \quad f(8) = \sqrt{9} = 3 \quad$ Slope $= \dfrac{3 - 1}{8 - 0} = \dfrac{1}{4}$

Secant line: $y - 1 = \dfrac{1}{4}(x - 0)$

$$y = \frac{1}{4}x + 1$$

(c) $$f'(x) = \frac{1}{2\sqrt{x + 1}}$$

$$\frac{1}{4} = \frac{1}{2\sqrt{x + 1}}$$

$$\sqrt{x + 1} = 2$$

$$x + 1 = 4$$

$$x = 3$$

$$f(3) = 2$$

Tangent line: $y - 2 = \dfrac{1}{4}(x - 3)$

$$y = \frac{1}{4}x + \frac{5}{4}$$

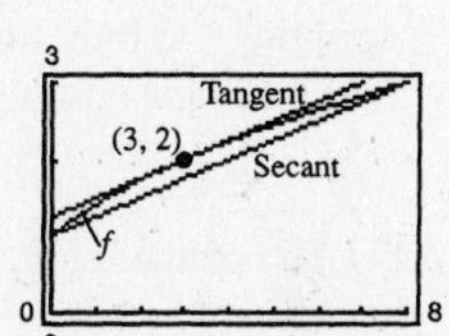

37. $s(t) = -4.9t^2 + 500$

(a) $v_{avg} = \dfrac{s(3) - s(0)}{3 - 0} = \dfrac{455.9 - 500}{3} = -14.7 \text{ m/sec}$

(b) $s(t)$ is continuous on $[0, 3]$ and differentiable on $(0, 3)$. Therefore, the Mean Value Theorem applies.

$$v(t) = s'(t) = -9.8t = -14.7 \text{ m/sec}$$

$$t = \frac{-14.7}{-9.8} = 1.5 \text{ seconds}$$

38. $S(t) = 200\left(5 - \dfrac{9}{2 + t}\right)$

(a) $\dfrac{S(12) - S(0)}{12 - 0} = \dfrac{200[5 - (9/14)] - 200[5 - (9/2)]}{12}$

$$= \frac{450}{7}$$

(b) $S'(t) = 200\left(\dfrac{9}{(2 + t)^2}\right) = \dfrac{450}{7}$

$$\frac{1}{(2 + t)^2} = \frac{1}{28}$$

$$2 + t = 2\sqrt{7}$$

$$t = 2\sqrt{7} - 2 \approx 3.2915 \text{ months}$$

$S'(t)$ is equal to the average value in April.

39. No. Let $f(x) = x^2$ on $[-1, 2]$.

$$f'(x) = 2x$$

$f'(0) = 0$ and zero is in the interval $(-1, 2)$ but $f(-1) \neq f(2)$.

40. $f(a) = f(b)$ and $f'(c) = 0$ where c is in the interval (a, b).

(a) $g(x) = f(x) + k$

$g(a) = g(b) = f(a) + k$

$g'(x) = f'(x) \Rightarrow g'(c) = 0$

Interval: $[a, b]$

Critical number of g: c

(b) $g(x) = f(x - k)$

$g(a + k) = g(b + k) = f(a)$

$g'(x) = f'(x - k)$

$g'(c + k) = f'(c) = 0$

Interval: $[a + k, b + k]$

Critical number of g: $c + k$

(c) $g(x) = f(kx)$

$g\left(\dfrac{a}{k}\right) = g\left(\dfrac{b}{k}\right) = f(a)$

$g'(x) = kf'(kx)$

$g'\left(\dfrac{c}{k}\right) = kf'(c) = 0$

Interval: $\left[\dfrac{a}{k}, \dfrac{b}{k}\right]$

(Assume $k > 0$)

Critical number of g: $\dfrac{c}{k}$

41. Let $S(t)$ be the position function of the plane. If $t = 0$ corresponds to 2 P.M., $S(0) = 0$, $S(5.5) = 2500$ and the Mean Value Theorem says that there exists a time t_0, $0 < t_0 < 5.5$, such that

$$S'(t_0) = v(t_0) = \frac{2500 - 0}{5.5 - 0} \approx 454.55.$$

Applying the Intermediate Value Theorem to the velocity function on the intervals $[0, t_0]$ and $[t_0, 5.5]$, you see that there are at least two times during the flight when the speed was 400 miles per hour. $(0 < 400 < 454.54)$

42. Let $T(t)$ be the temperature of the core of the object. Then $T(0) = 1500°$ and $T(5) = 390°$. The average temperature over the interval $[0, 5]$ is

$$\frac{390 - 1500}{5 - 0} = -222° \text{ F/hr.}$$

By the Mean Value Theorem, there exists a time to, $0 < t_0 < 5$, such that $T'(t_0) = -222$.

43. (a) $f(x) = |9 - x^2|$

$$f(x) = \begin{cases} 9 - x^2, & -3 \le x \le 3 \\ x^2 - 9, & x > 3 \text{ or } x < -3 \end{cases}$$

$$f'(x) = \begin{cases} -2x, & -3 < x < 3 \\ 2x, & x > 3 \text{ or } x < -3 \end{cases}$$

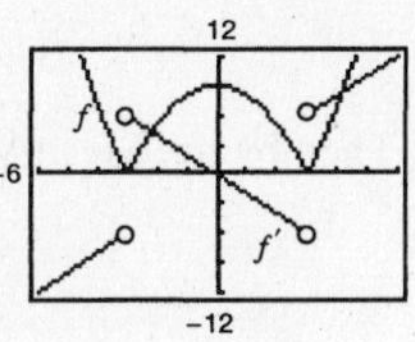

(b) f is continuous.

f' is not continuous.

(c) Since $f(-1) = f(1) = 8$ and f is differentiable on $(-1, 1)$, Rolle's Theorem applies on $[-1, 1]$. Since $f(2) = 5$ and $f(4) = 7$, and f is not differentiable at $x = 3$, Rolle's Theorem does not apply on $[2, 4]$.

(d) $\lim_{x \to 3^-} f'(x) = -6$ and $\lim_{x \to 3^+} f'(x) = 6$.

44. (a) f is continuous on $[-10, 4]$ and changes sign, $(f(-8) > 0,\ f(3) < 0)$. By the Intermediate Value Theorem, there exists at least one value of x in $[-10, 4]$ satisfying $f(x) = 0$.

(b) There exist real numbers a and b such that $-10 < a < b < 4$ and $f(a) = f(b) = 2$. Therefore, by Rolle's Theorem there exists at least one number c in $(-10, 4)$ such that $f'(c) = 0$. This is called a critical number.

(c)

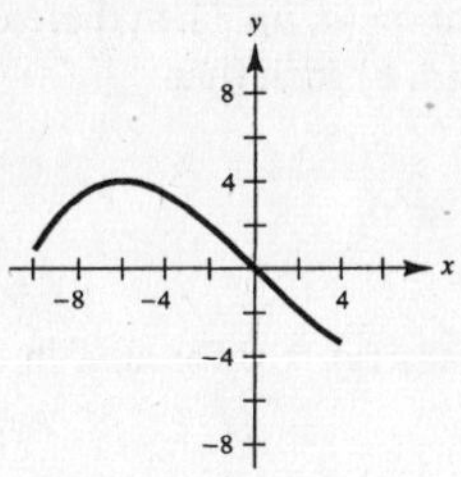

(d)

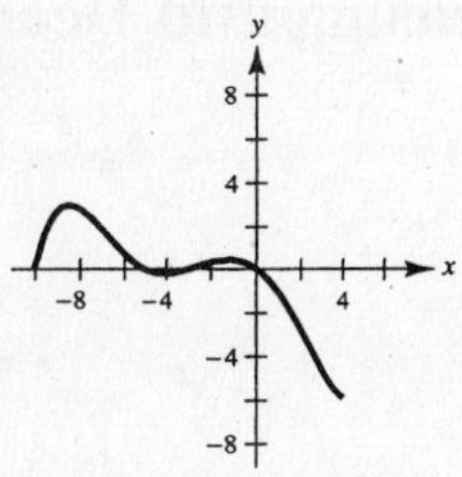

(e) No, f' did not have to be continuous on $[-10, 4]$.

45. $f'(x) = 0$

$f(x) = c$

$f(2) = 5$

Hence, $f(x) = 5$.

46. $f'(x) = 4$

$f(x) = 4x + c$

$f(0) = 1 \Rightarrow c = 1$

Hence, $f(x) = 4x + 1$.

47. $f'(x) = 2x$

$f(x) = x^2 + c$

$f(1) = 0 \Rightarrow 0 = 1 + c \Rightarrow c = -1$

Hence, $f(x) = x^2 - 1$.

48. $f'(x) = 2x + 3$

$f(x) = x^2 + 3x + c$

$f(1) = 0 \Rightarrow 0 = 1 + 3 + c \Rightarrow c = -4$

Hence, $f(x) = x^2 + 3x - 4$.

49. f is continuous on $[-5, 5]$ and does not satisfy the conditions of the Mean Value Theorem.
$\Rightarrow f$ is not differentiable on $(-5, 5)$.
Example: $f(x) = |x|$

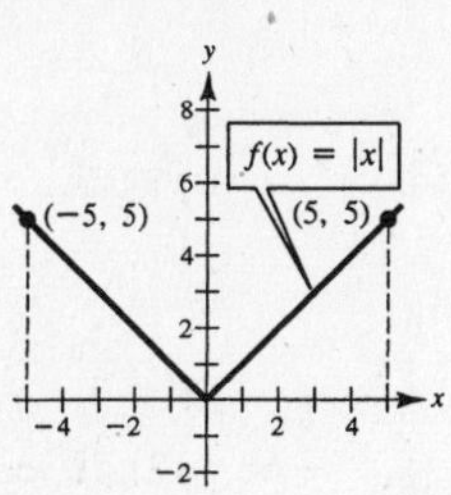

50. f is not continuous on $[-5, 5]$.

Example: $f(x) = \begin{cases} 1/x, & x \ne 0 \\ 0, & x = 0 \end{cases}$

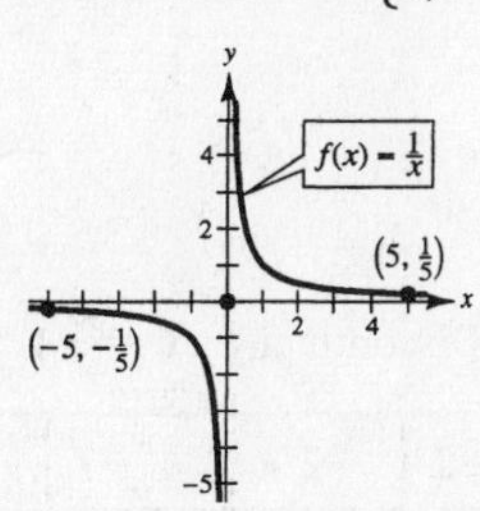

51. False. $f(x) = 1/x$ has a discontinuity at $x = 0$.

52. False. f must also be continuous *and* differentiable on each interval. Let

$$f(x) = \frac{x^3 - 4x}{x^2 - 1}.$$

53. True. A polynomial is continuous and differentiable everywhere.

54. True

55. Suppose $f(x)$ is not constant on (a, b). Then there exists x_1 and x_2 in (a, b) such that $f(x_1) \neq f(x_2)$. Then by the Mean Value Theorem, there exists c in (a, b) such that

$$f'(c) = \frac{f(x_2) - f(x_1)}{x_2 - x_1} \neq 0.$$

This contradicts the fact that $f'(x) = 0$ for all x in (a, b).

Section 5.3 Increasing and Decreasing Functions and the First Derivative Test

1. $f(x) = x^2 - 6x + 8$

Increasing on: $(3, \infty)$

Decreasing on: $(-\infty, 3)$

2. $y = -(x + 1)^2$

Increasing on: $(-\infty, -1)$

Decreasing on: $(-1, \infty)$

3. $y = \dfrac{x^3}{4} - 3x$

Increasing on: $(-\infty, -2), (2, \infty)$

Decreasing on: $(-2, 2)$

4. $f(x) = x^4 - 2x^2$

Increasing on: $(-1, 0), (1, \infty)$

Decreasing on: $(-\infty, -1), (0, 1)$

5. $f(x) = \dfrac{1}{x^2} = x^{-2}$

$$f'(x) = \frac{-2}{x^3}$$

Discontinuity: $x = 0$

Test intervals:	$-\infty < x < 0$	$0 < x < \infty$
Sign of $f'(x)$:	$f' > 0$	$f' < 0$
Conclusion:	Increasing	Decreasing

Increasing on: $(-\infty, 0)$

Decreasing on: $(0, \infty)$

6. $y = \dfrac{x^2}{x + 1}$

$$y' = \frac{x(x + 2)}{(x + 1)^2}$$

Critical numbers: $x = 0, -2$ Discontinuity: $x = -1$

Test intervals:	$-\infty < x < -2$	$-2 < x < -1$	$-1 < x < 0$	$0 < x < \infty$
Sign of $f'(x)$:	$y' > 0$	$y' < 0$	$y' < 0$	$y' > 0$
Conclusion:	Increasing	Decreasing	Decreasing	Increasing

Increasing on: $(-\infty, -2), (0, \infty)$

Decreasing on: $(-2, -1), (-1, 0)$

7. $g(x) = x^2 - 2x - 8$

$g'(x) = 2x - 2$

Critical number: $x = 1$

Test intervals:	$-\infty < x < 1$	$1 < x < \infty$
Sign of $g'(x)$:	$g' < 0$	$g' > 0$
Conclusion:	Decreasing	Increasing

Increasing on: $(1, \infty)$

Decreasing on: $(-\infty, 1)$

8. $h(x) = 27x - x^3$

$h'(x) = 27 - 3x^2 = 3(3 - x)(3 + x)$

$h'(x) = 0$

Critical numbers: $x = \pm 3$

Test intervals:	$-\infty < x < -3$	$-3 < x < 3$	$3 < x < \infty$
Sign of $h'(x)$:	$h' < 0$	$h' > 0$	$h' < 0$
Conclusion:	Decreasing	Increasing	Decreasing

Increasing on: $(-3, 3)$

Decreasing on: $(-\infty, -3), (3, \infty)$

9. $y = x\sqrt{16 - x^2}$ Domain: $[-4, 4]$

$$y' = \frac{-2(x^2 - 8)}{\sqrt{16 - x^2}} = \frac{-2}{\sqrt{16 - x^2}}\left(x - 2\sqrt{2}\right)\left(x + 2\sqrt{2}\right)$$

Critical numbers: $x = \pm 2\sqrt{2}$

Test intervals:	$-4 < x < -2\sqrt{2}$	$-2\sqrt{2} < x < 2\sqrt{2}$	$2\sqrt{2} < x < 4$
Sign of y':	$y' < 0$	$y' > 0$	$y' < 0$
Conclusion:	Decreasing	Increasing	Decreasing

Increasing on: $\left(-2\sqrt{2}, 2\sqrt{2}\right)$

Decreasing on: $\left(-4, -2\sqrt{2}\right), \left(2\sqrt{2}, 4\right)$

10. $y = x + \dfrac{4}{x}$

$$y' = \frac{(x - 2)(x + 2)}{x^2}$$

Critical numbers: $x = \pm 2$ Discontinuity: 0

Test intervals:	$-\infty < x < -2$	$-2 < x < 0$	$0 < x < 2$	$2 < x < \infty$
Sign of y':	$y' > 0$	$y' < 0$	$y' < 0$	$y' > 0$
Conclusion:	Increasing	Decreasing	Decreasing	Increasing

Increasing: $(-\infty, -2), (2, \infty)$

Decreasing: $(-2, 0), (0, 2)$

11. $f(x) = x^2 - 6x$

$f'(x) = 2x - 6 = 0$

Critical number: $x = 3$

Test intervals:	$-\infty < x < 3$	$3 < x < \infty$
Sign of $f'(x)$:	$f' < 0$	$f' > 0$
Conclusion:	Decreasing	Increasing

Increasing on: $(3, \infty)$

Decreasing on: $(-\infty, 3)$

Relative minimum: $(3, -9)$

12. $f(x) = x^2 + 8x + 10$

$f'(x) = 2x + 8 = 0$

Critical number: $x = -4$

Test intervals:	$-\infty < x < -4$	$-4 < x < \infty$
Sign of $f'(x)$:	$f' < 0$	$f' > 0$
Conclusion:	Decreasing	Increasing

Increasing on: $(-4, \infty)$

Decreasing on: $(-\infty, -4)$

Relative minimum: $(-4, -6)$

13. $f(x) = -2x^2 + 4x + 3$

$f'(x) = -4x + 4 = 0$

Critical number: $x = 1$

Test intervals:	$-\infty < x < 1$	$1 < x < \infty$
Sign of $f'(x)$:	$f' > 0$	$f' < 0$
Conclusion:	Increasing	Decreasing

Increasing on: $(-\infty, 1)$

Decreasing on: $(1, \infty)$

Relative maximum: $(1, 5)$

14. $f(x) = -(x^2 + 8x + 12)$

$f'(x) = -2x - 8 = 0$

Critical number: $x = -4$

Test intervals:	$-\infty < x < -4$	$-4 < x < \infty$
Sign of $f'(x)$:	$f' > 0$	$f' < 0$
Conclusion:	Increasing	Decreasing

Increasing on: $(-\infty, -4)$

Decreasing on: $(-4, \infty)$

Relative maximum: $(-4, 4)$

15. $f(x) = x^3 - 6x^2 + 15$

$f'(x) = 3x^2 - 12x = 3x(x - 4)$

Critical numbers: $x = 0, 4$

Test intervals:	$-\infty < x < 0$	$0 < x < 4$	$4 < x < \infty$
Sign of $f'(x)$:	$f' > 0$	$f' < 0$	$f' > 0$
Conclusion:	Increasing	Decreasing	Increasing

Increasing on: $(-\infty, 0), (4, \infty)$

Decreasing on: $(0, 4)$

Relative maximum: $(0, 15)$

Relative minimum: $(4, -17)$

16. $f(x) = x^4 - 2x^3$

$f'(x) = 4x^3 - 6x^2 = 2x^2(2x - 3)$

Critical numbers: $x = 0, \frac{3}{2}$

Interval:	$-\infty < x < 0$	$0 < x < 3/2$	$3/2 < x < \infty$
Test value:	-1	1	2
Sign of $f'(x)$:	$f'(-1) < 0$	$f'(1) < 0$	$f'(2) > 0$
Conclusion:	Decreasing	Decreasing	Increasing

Increasing on: $\left(\frac{3}{2}, \infty\right)$

Decreasing on: $\left(-\infty, \frac{3}{2}\right)$

By the First Derivative Test, f has one relative minimum at the point $\left(\frac{3}{2}, \frac{-27}{16}\right)$.

17. $f(x) = x^2(3 - x) = 3x^2 - x^3$

$f'(x) = 6x - 3x^2 = 3x(2 - x)$

Critical numbers: $x = 0, 2$

Test intervals:	$-\infty < x < 0$	$0 < x < 2$	$2 < x < \infty$
Sign of $f'(x)$:	$f' < 0$	$f' > 0$	$f' < 0$
Conclusion:	Decreasing	Increasing	Decreasing

Increasing on: $(0, 2)$

Decreasing on: $(-\infty, 0), (2, \infty)$

Relative maximum: $(2, 4)$

Relative minimum: $(0, 0)$

18. $f(x) = (x + 2)^2(x - 1)$

$f'(x) = 3x(x + 2)$

Critical numbers: $x = -2, 0$

Test intervals:	$-\infty < x < -2$	$-2 < x < 0$	$0 < x < \infty$
Sign of $f'(x)$:	$f' > 0$	$f' < 0$	$f' > 0$
Conclusion:	Increasing	Decreasing	Increasing

Increasing on: $(-\infty, -2), (0, \infty)$

Decreasing on: $(-2, 0)$

Relative maximum: $(-2, 0)$

Relative minimum: $(0, -4)$

19. $f(x) = 2x^3 + 3x^2 - 12x$

$f'(x) = 6x^2 + 6x - 12 = 6(x + 2)(x - 1) = 0$

Critical numbers: $x = -2, 1$

Test intervals:	$-\infty < x < -2$	$-2 < x < 1$	$1 < x < \infty$
Sign of $f'(x)$:	$f' > 0$	$f' < 0$	$f' > 0$
Conclusion:	Increasing	Decreasing	Increasing

Increasing on: $(-\infty, -2), (1, \infty)$

Decreasing on: $(-2, 1)$

Relative maximum: $(-2, 20)$

Relative minimum: $(1, -7)$

20. $f(x) = (x - 3)^3$

$f'(x) = 3(x - 3)^2$

Critical number: $f'(x) = 0 \Rightarrow x = 3$

Interval:	$-\infty < x < 3$	$3 < x < \infty$
Test value:	$x = 0$	$x = 4$
Sign of $f'(x)$:	$f'(0) > 0$	$f'(4) > 0$
Conclusion:	Increasing	Increasing

Increasing on $(-\infty, \infty)$

No relative extrema.

21. $f(x) = \dfrac{x^5 - 5x}{5}$

$f'(x) = x^4 - 1$

Critical numbers: $x = -1, 1$

Test intervals:	$-\infty < x < -1$	$-1 < x < 1$	$1 < x < \infty$
Sign of $f'(x)$:	$f' > 0$	$f' < 0$	$f' > 0$
Conclusion:	Increasing	Decreasing	Increasing

Increasing on: $(-\infty, -1), (1, \infty)$

Decreasing on: $(-1, 1)$

Relative maximum: $\left(-1, \frac{4}{5}\right)$

Relative minimum: $\left(1, -\frac{4}{5}\right)$

22. $f(x) = x^4 - 32x + 4$

$f'(x) = 4x^3 - 32 = 4(x^3 - 8)$

Critical number: $x = 2$

Test intervals:	$-\infty < x < 2$	$2 < x < \infty$
Sign of $f'(x)$:	$f' < 0$	$f' > 0$
Conclusion:	Decreasing	Increasing

Increasing on: $(2, \infty)$

Decreasing on: $(-\infty, 2)$

Relative minimum: $(2, -44)$

23. $f(x) = x^{1/3} + 1$

$f'(x) = \frac{1}{3}x^{-2/3} = \frac{1}{3x^{2/3}}$

Critical number: $x = 0$

Test intervals:	$-\infty < x < 0$	$0 < x < \infty$
Sign of $f'(x)$:	$f' > 0$	$f' > 0$
Conclusion:	Increasing	Increasing

Increasing on: $(-\infty, \infty)$

No relative extrema.

24. $f(x) = x^{2/3} - 4$

$f'(x) = \frac{2}{3}x^{-1/3} = \frac{2}{3x^{1/3}}$

Critical number: $x = 0$

Test intervals:	$-\infty < x < 0$	$0 < x < \infty$
Sign of $f'(x)$:	$f' < 0$	$f' > 0$
Conclusion:	Decreasing	Increasing

Increasing on: $(0, \infty)$

Decreasing on: $(-\infty, 0)$

Relative minimum: $(0, -4)$

25. $f(x) = (x - 1)^{2/3}$

$f'(x) = \frac{2}{3(x - 1)^{1/3}}$

Critical number: $x = 1$

Test intervals:	$-\infty < x < 1$	$1 < x < \infty$
Sign of $f'(x)$:	$f' < 0$	$f' > 0$
Conclusion:	Decreasing	Increasing

Increasing on: $(1, \infty)$

Decreasing on: $(-\infty, 1)$

Relative minimum: $(1, 0)$

26. $f(x) = (x - 1)^{1/3}$

$f'(x) = \frac{1}{3(x - 1)^{2/3}}$

Critical number: $x = 1$

Test intervals:	$-\infty < x < 1$	$1 < x < \infty$
Sign of $f'(x)$:	$f' > 0$	$f' > 0$
Conclusion:	Increasing	Increasing

Increasing on: $(-\infty, \infty)$

No relative extrema.

27. $f(x) = 5 - |x - 5|$

$$f'(x) = -\frac{x - 5}{|x - 5|} = \begin{cases} 1, & x < 5 \\ -1, & x > 5 \end{cases}$$

Critical number: $x = 5$

Test intervals:	$-\infty < x < 5$	$5 < x < \infty$
Sign of $f'(x)$:	$f' > 0$	$f' < 0$
Conclusion:	Increasing	Decreasing

Increasing on: $(-\infty, 5)$

Decreasing on: $(5, \infty)$

Relative maximum: $(5, 5)$

28. $f(x) = |x + 3| - 1$

$$f'(x) = \frac{x + 3}{|x + 3|} = \begin{cases} 1, & x > -3 \\ -1, & x < -3 \end{cases}$$

Critical number: $x = -3$

Test intervals:	$-\infty < x < -3$	$-3 < x < \infty$
Sign of $f'(x)$:	$f' < 0$	$f' > 0$
Conclusion:	Decreasing	Increasing

Increasing on: $(-3, \infty)$

Decreasing on: $(-\infty, -3)$

Relative minimum: $(-3, -1)$

29. $f(x) = x + \dfrac{1}{x}$

$$f'(x) = 1 - \frac{1}{x^2} = \frac{x^2 - 1}{x^2}$$

Critical numbers: $x = -1, 1$

Discontinuity: $x = 0$

Test intervals:	$-\infty < x < -1$	$-1 < x < 0$	$0 < x < 1$	$1 < x < \infty$
Sign of $f'(x)$:	$f' > 0$	$f' < 0$	$f' < 0$	$f' > 0$
Conclusion:	Increasing	Decreasing	Decreasing	Increasing

Increasing on: $(-\infty, -1), (1, \infty)$

Decreasing on: $(-1, 0), (0, 1)$

Relative maximum: $(-1, -2)$

Relative minimum: $(1, 2)$

30. $f(x) = \dfrac{x}{x + 1}$

$$f'(x) = \frac{(x + 1)(1) - (x)(1)}{(x + 1)^2} = \frac{1}{(x + 1)^2}$$

Discontinuity: $x = -1$

Test intervals:	$-\infty < x < -1$	$-1 < x < \infty$
Sign of $f'(x)$:	$f' > 0$	$f' > 0$
Conclusion:	Increasing	Increasing

Increasing on: $(-\infty, -1), (-1, \infty)$

No relative extrema.

31. $f(x) = \dfrac{x^2}{x^2 - 9}$

$f'(x) = \dfrac{(x^2 - 9)(2x) - (x^2)(2x)}{(x^2 - 9)^2} = \dfrac{-18x}{(x^2 - 9)^2}$

Critical number: $x = 0$

Discontinuities: $x = -3, 3$

Test intervals:	$-\infty < x < -3$	$-3 < x < 0$	$0 < x < 3$	$3 < x < \infty$
Sign of $f'(x)$:	$f' > 0$	$f' > 0$	$f' < 0$	$f' < 0$
Conclusion:	Increasing	Increasing	Decreasing	Decreasing

Increasing on: $(-\infty, -3), (-3, 0)$

Decreasing on: $(0, 3), (3, \infty)$

Relative maximum: $(0, 0)$

32. $f(x) = \dfrac{x + 3}{x^2} = \dfrac{1}{x} + \dfrac{3}{x^2}$

$f'(x) = -\dfrac{1}{x^2} - \dfrac{6}{x^3} = \dfrac{-(x + 6)}{x^3}$

Critical number: $x = -6$

Discontinuity: $x = 0$

Test intervals:	$-\infty < x < -6$	$-6 < x < 0$	$0 < x < \infty$
Sign of $f'(x)$:	$f' < 0$	$f' > 0$	$f' < 0$
Conclusion:	Decreasing	Increasing	Decreasing

Increasing on: $(-6, 0)$

Decreasing on: $(-\infty, -6), (0, \infty)$

Relative minimum: $\left(-6, -\dfrac{1}{12}\right)$

33. $f(x) = \dfrac{x^2 - 2x + 1}{x + 1}$

$f'(x) = \dfrac{(x + 1)(2x - 2) - (x^2 - 2x + 1)(1)}{(x + 1)^2} = \dfrac{x^2 + 2x - 3}{(x + 1)^2} = \dfrac{(x + 3)(x - 1)}{(x + 1)^2}$

Critical numbers: $x = -3, 1$

Discontinuity: $x = -1$

Test intervals:	$-\infty < x < -3$	$-3 < x < -1$	$-1 < x < 1$	$1 < x < \infty$
Sign of $f'(x)$:	$f' > 0$	$f' < 0$	$f' < 0$	$f' > 0$
Conclusion:	Increasing	Decreasing	Decreasing	Increasing

Increasing on: $(-\infty, -3), (1, \infty)$

Decreasing on: $(-3, -1), (-1, 1)$

Relative maximum: $(-3, -8)$

Relative minimum: $(1, 0)$

34. $f(x) = \dfrac{x^2 - 3x - 4}{x - 2}$

$$f'(x) = \frac{(x-2)(2x-3) - (x^2 - 3x - 4)(1)}{(x-2)^2} = \frac{x^2 - 4x + 10}{(x-2)^2}$$

Discontinuity: $x = 2$

Test intervals:	$-\infty < x < 2$	$2 < x < \infty$
Sign of $f'(x)$:	$f' > 0$	$f' > 0$
Conclusion:	Increasing	Increasing

Increasing on: $(-\infty, 2), (2, \infty)$

No relative extrema.

35. $f(x) = 2x\sqrt{9 - x^2}, [-3, 3]$

(a) $f'(x) = \dfrac{2(9 - 2x^2)}{\sqrt{9 - x^2}}$

(b)

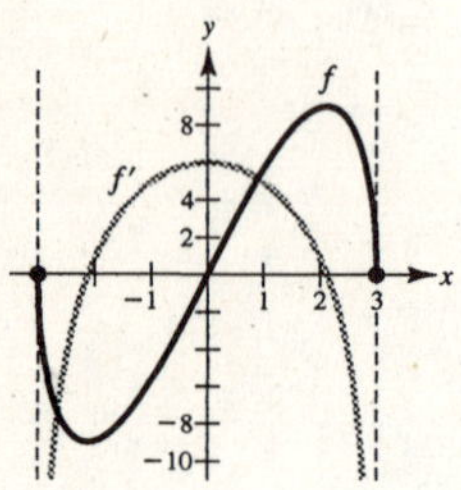

(c) $\dfrac{2(9 - 2x^2)}{\sqrt{9 - x^2}} = 0$

Critical numbers: $x = \pm\dfrac{3}{\sqrt{2}} = \pm\dfrac{3\sqrt{2}}{2}$

(d) Intervals:

$\left(-3, -\frac{3\sqrt{2}}{2}\right)$	$\left(-\frac{3\sqrt{2}}{2}, \frac{3\sqrt{2}}{2}\right)$	$\left(\frac{3\sqrt{2}}{2}, 3\right)$
$f'(x) < 0$	$f'(x) > 0$	$f'(x) < 0$
Decreasing	Increasing	Decreasing

f is increasing when f' is positive and decreasing when f' is negative.

36. $f(x) = 10\left(5 - \sqrt{x^2 - 3x + 16}\right), [0, 5]$

(a) $f'(x) = -\dfrac{5(2x - 3)}{\sqrt{x^2 - 3x + 16}}$

(b)

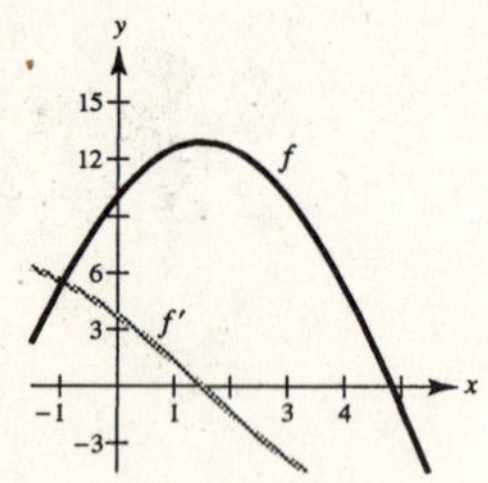

(c) $-\dfrac{5(2x - 3)}{\sqrt{x^2 - 3x + 16}} = 0$

Critical number: $x = \dfrac{3}{2}$

(d) Intervals:

$\left(0, \frac{3}{2}\right)$	$\left(\frac{3}{2}, 5\right)$
$f'(x) > 0$	$f'(x) < 0$
Increasing	Decreasing

f is increasing when f' is positive and decreasing when f' is negative.

37. $f(x) = \dfrac{x^5 - 4x^3 + 3x}{x^2 - 1} = \dfrac{(x^2 - 1)(x^3 - 3x)}{x^2 - 1} = x^3 - 3x, \ x \neq \pm 1$

$f(x) = g(x) = x^3 - 3x$ for all $x \neq \pm 1$.

$f'(x) = 3x^2 - 3 = 3(x^2 - 1), x \neq \pm 1 \quad f'(x) \neq 0$

f symmetric about origin

zeros of f: $(0, 0), (\pm\sqrt{3}, 0)$

No relative extrema.

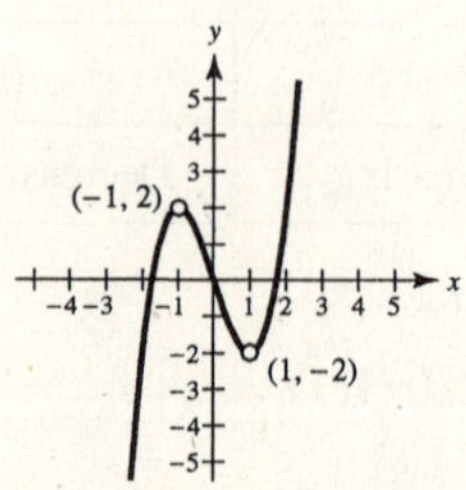

Holes at $(-1, 2)$ and $(1, -2)$.

38. $f(x)$ is symmetric to the y-axis.

Zeros: $(0, 0)$, $(\pm\sqrt{3}, 0)$

Relative maximum at $(0, 0)$

Relative minima at $\left(\dfrac{\pm\sqrt{6}}{2}, -\dfrac{9}{4}\right)$

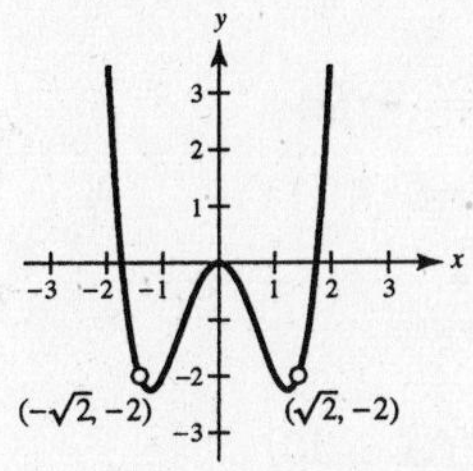

g is continuous on $(-\infty, \infty)$ and $f(x)$ has holes at $x = \pm\sqrt{2}$. Otherwise, the graphs of f and g are the same.

39. $f(x) = c$ is constant $\Rightarrow f'(x) = 0$.

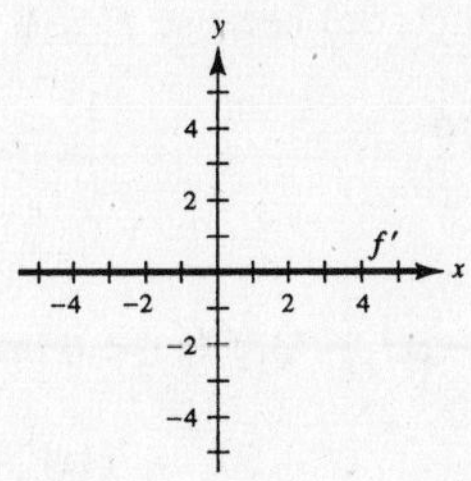

40. $f(x)$ is a line of slope $\approx 2 \Rightarrow f'(x) = 2$.

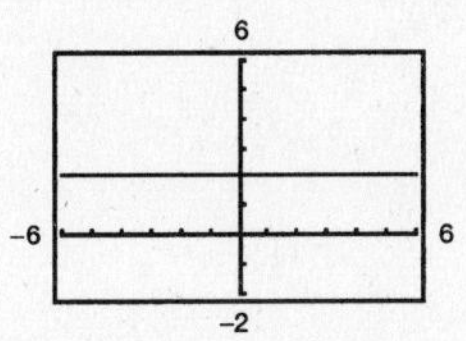

41. f is quadratic $\Rightarrow f'$ is a line.

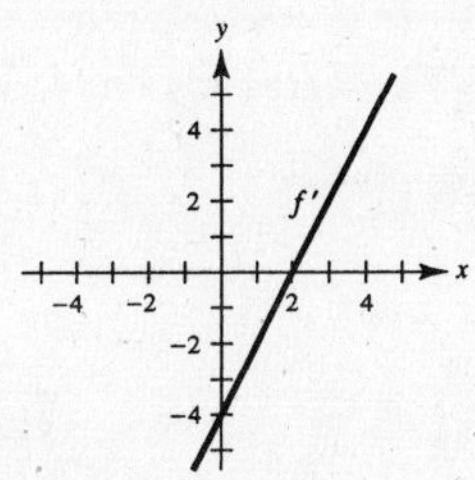

42. f is a 4th degree polynomial $\Rightarrow f'$ is a cubic polynomial.

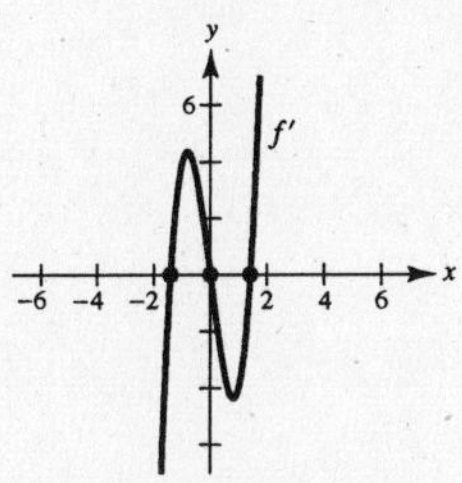

43. f has positive, but decreasing slope.

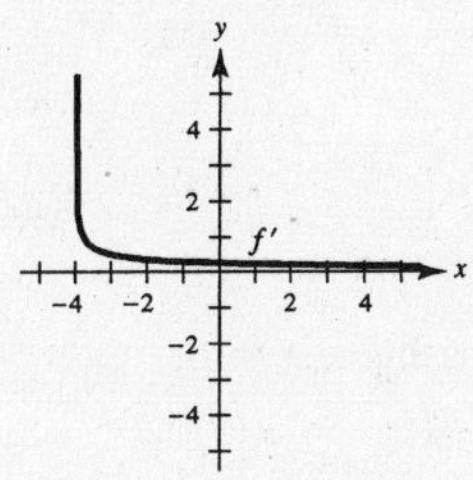

44. f has positive slope.

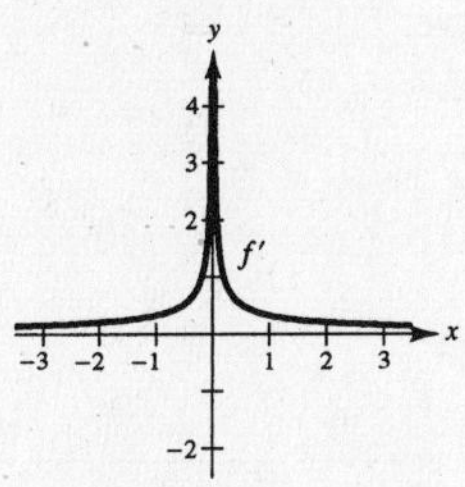

45. (a) f increasing on $(2, \infty)$ because $f' > 0$ on $(2, \infty)$.

f decreasing on $(-\infty, 2)$ because $f' < 0$ on $(-\infty, 2)$.

(b) f has a relative minimum at $x = 2$.

46. (a) f increasing on $(-\infty, -1)$ because $f' > 0$ on $(-\infty, -1)$.

f decreasing on $(-1, \infty)$ because $f' < 0$ on $(-1, \infty)$.

(b) f has a relative maximum at $x = -1$.

47. (a) f increasing on $(-\infty, 0)$ and $(1, \infty)$ because $f' > 0$ there

f decreasing on $(0, 1)$ because $f' < 0$ there

(b) f has a relative maximum at $x = 0$, and a relative minimum at $x = 1$.

48. (a) f increasing on $(-\infty, -1)$ and $(0, 1)$ because $f' > 0$ there

f decreasing on $(-1, 0)$ and $(1, \infty)$ because $f' < 0$ there

(b) f has a relative maximum at $x = -1$ and $x = 1$.

f has a relative minimum at $x = 0$.

In Exercises 49–54, $f'(x) > 0$ on $(-\infty, -4)$, $f'(x) < 0$ on $(-4, 6)$ and $f'(x) > 0$ on $(6, \infty)$.

49. $g(x) = f(x) + 5$

$g'(x) = f'(x)$

$g'(0) = f'(0) < 0$

50. $g(x) = 3f(x) - 3$

$g'(x) = 3f'(x)$

$g'(-5) = 3f'(-5) > 0$

51. $g(x) = -f(x)$

$g'(x) = -f'(x)$

$g'(-6) = -f'(-6) < 0$

52. $g(x) = -f(x)$

$g'(x) = -f'(x)$

$g'(0) = -f'(0) > 0$

53. $g(x) = f(x - 10)$

$g'(x) = f'(x - 10)$

$g'(0) = f'(-10) > 0$

54. $g(x) = f(x - 10)$

$g'(x) = f'(x - 10)$

$g'(8) = f'(-2) < 0$

55. $f'(x) = \begin{cases} > 0, & x < 4 \Rightarrow f \text{ is increasing on } (-\infty, 4). \\ \text{undefined}, & x = 4 \\ < 0, & x > 4 \Rightarrow f \text{ is decreasing on } (4, \infty). \end{cases}$

Two possibilities for $f(x)$ are given below.

(a) (b)

56. Critical number: $x = 5$

$f'(4) = -2.5 \Rightarrow f$ is decreasing at $x = 4$.

$f'(6) = 3 \Rightarrow f$ is increasing at $x = 6$.

$(5, f(5))$ is a relative minimum.

57. The critical numbers are in intervals $(-0.50, -0.25)$ and $(0.25, 0.50)$ since the sign of f' changes in these intervals. f is decreasing on approximately $(-1, -0.40)$, $(0.48, 1)$, and increasing on $(-0.40, 0.48)$.

Relative minimum when $x \approx -0.40$.

Relative maximum when $x \approx 0.48$.

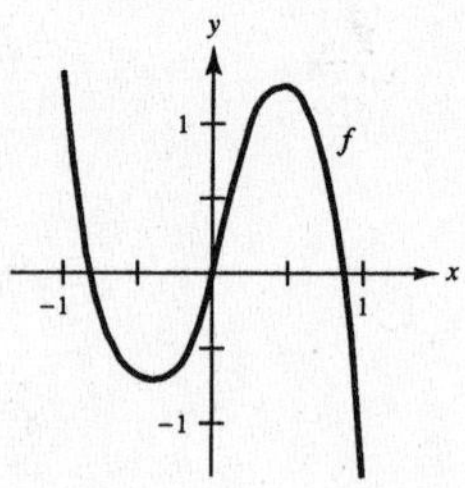

One possible answer. Others could be vertical shifts of this curve.

58. $P = 2.44x - \dfrac{x^2}{20{,}000} - 5000, \; 0 \le x \le 35{,}000$

$P' = 2.44 - \dfrac{x}{10{,}000} = 0$

$x = 24{,}400$

Test intervals:	$0 < x < 24{,}400$	$24{,}400 < x < 35{,}000$
Sign of P':	$P' > 0$	$P' < 0$

Increasing when $0 < x < 24{,}400$ hamburgers.

Decreasing when $24{,}400 < x < 35{,}000$ hamburgers.

59. $v = k(R - r)r^2 = k(Rr^2 - r^3)$

$v' = k(2Rr - 3r^2)$

$= kr(2R - 3r) = 0$

$r = 0 \text{ or } \frac{2}{3}R$

Maximum when $r = \frac{2}{3}R$.

60. $C = \dfrac{3t}{27 + t^3}, t \geq 0$

(a)

t	0	0.5	1	1.5	2	2.5	3
$C(t)$	0	0.055	0.107	0.148	0.171	0.176	0.167

The concentration seems greater near $t = 2.5$ hours.

(b)

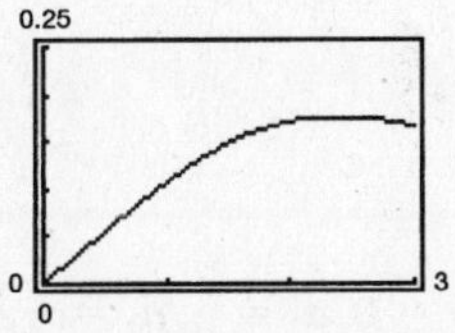

The concentration is greatest when $t \approx 2.38$ hours.

(c) $C' = \dfrac{(27 + t^3)(3) - (3t)(3t^2)}{(27 + t^3)^2}$

$= \dfrac{3(27 - 2t^3)}{(27 + t^3)^2}$

$C' = 0$ when $t = 3/\sqrt[3]{2} \approx 2.38$ hours.

By the First Derivative Test, this is a maximum.

61. $P = \dfrac{vR_1R_2}{(R_1 + R_2)^2}$, v and R_1 are constant.

$\dfrac{dP}{dR_2} = \dfrac{(R_1 + R_2)^2(vR_1) - vR_1R_2[2(R_1 + R_2)(1)]}{(R_1 + R_2)^4}$

$= \dfrac{vR_1(R_1 - R_2)}{(R_1 + R_2)^3} = 0 \Rightarrow R_2 = R_1$

Maximum when $R_1 = R_2$.

62. $R = \sqrt{0.001T^4 - 4T + 100}$

(a) $R' = \dfrac{0.004T^3 - 4}{2\sqrt{0.001T^4 - 4T + 100}} = 0$

$T = 10°, R \approx 8.3666\Omega$

(b)

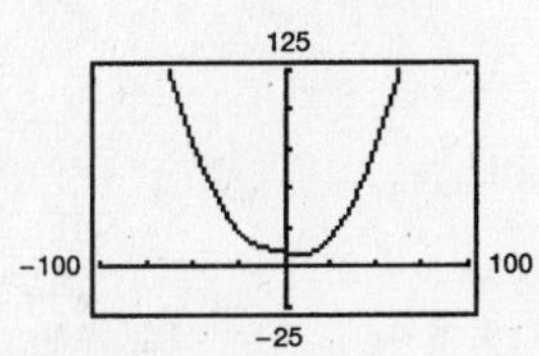

The minimum resistance is approximately $R \approx 8.37\Omega$ at $T = 10°$.

63. (a) $B = -0.14649t^4 + 8.5979t^3 - 182.185t^2 + 1710.31t - 5250.7$

(b)

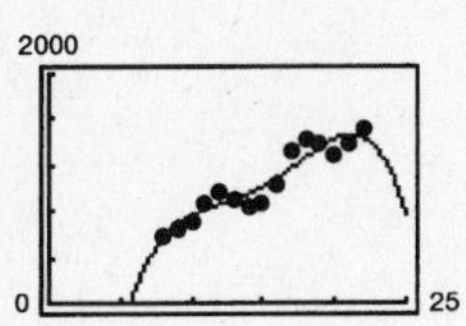

(c) $B' = 0$ when $t \approx 21.1$

Maximum is (21.1, 1458).

64. Assume that $f'(x) < 0$ for all x in the interval (a, b) and let $x_1 < x_2$ be any two points in the interval. By the Mean Value Theorem, we know there exists a number c such that $x_1 < c < x_2$, and

$$f'(c) = \frac{f(x_2) - f(x_1)}{x_2 - x_1}.$$

Since $f'(c) < 0$ and $x_2 - x_1 > 0$, then $f(x_2) - f(x_1) < 0$, which implies that $f(x_2) < f(x_1)$. Thus, f is decreasing on the interval.

65. Suppose $f'(x)$ changes from positive to negative at c. Then there exists a and b in I such that $f'(x) > 0$ for all x in (a, c) and $f'(x) < 0$ for all x in (c, b). By Theorem 5.5, f is increasing on (a, c) and decreasing on (c, b). Therefore, $f(c)$ is a maximum of f on (a, b) and thus, a relative maximum of f.

66. Let $f(x) = (1 + x)^n - nx - 1$. Then

$$f'(x) = n(1 + x)^{n-1} - n$$
$$= n[(1 + x)^{n-1} - 1] > 0 \text{ since } x > 0 \text{ and } n > 1.$$

Thus, $f(x)$ is increasing on $(0, \infty)$. Since $f(0) = 0 \Rightarrow f(x) > 0$ on $(0, \infty)$

$$(1 + x)^n - nx - 1 > 0 \Rightarrow (1 + x)^n > 1 + nx.$$

67. (a) Use a cubic polynomial $f(x) = a_3x^3 + a_2x^2 + a_1x + a_0$.

(b) $f'(x) = 3a_3x^2 + 2a_2x + a_1$

$(0, 0)$:	$0 = a_0$	$(f(0) = 0)$
	$0 = a_1$	$(f'(0) = 0)$
$(2, 2)$:	$2 = 8a_3 + 4a_2$	$(f(2) = 2)$
	$0 = 12a_3 + 4a_2$	$(f'(2) = 0)$

(c) The solution is $a_0 = a_1 = 0, a_2 = \frac{3}{2}, a_3 = -\frac{1}{2}$:

$$f(x) = -\frac{1}{2}x^3 + \frac{3}{2}x^2.$$

(d)

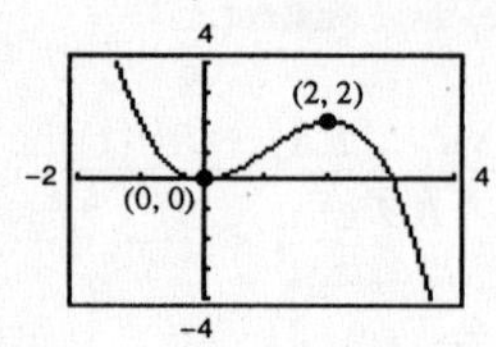

68. (a) Use a cubic polynomial $f(x) = a_3x^3 + a_2x^2 + a_1x + a_0$.

(b) $f'(x) = 3a_3x^2 + 2a_2x + a_1$

$(0, 0)$:	$0 = a_0$	$(f(0) = 0)$
	$0 = a_1$	$(f'(0) = 0)$
$(4, 1000)$:	$1000 = 64a_3 + 16a_2$	$(f(4) = 1000)$
	$0 = 48a_3 + 8a_2$	$(f'(4) = 0)$

(c) The solution is $a_0 = a_1 = 0, a_2 = \frac{375}{2}, a_3 = \frac{-125}{4}$

$$f(x) = \frac{-125}{4}x^3 + \frac{375}{2}x^2.$$

(d)

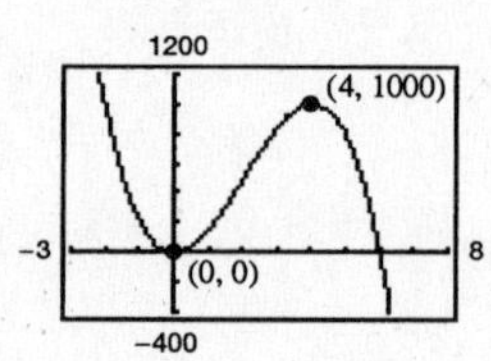

69. (a) Use a fourth degree polynomial $f(x) = a_4x^4 + a_3x^3 + a_2x^2 + a_1x + a_0$.

(b) $f'(x) = 4a_4x^3 + 3a_3x^2 + 2a_2x + a_1$

$(0, 0)$:	$0 = a_0$	$(f(0) = 0)$
	$0 = a_1$	$(f'(0) = 0)$
$(4, 0)$:	$0 = 256a_4 + 64a_3 + 16a_2$	$(f(4) = 0)$
	$0 = 256a_4 + 48a_3 + 8a_2$	$(f'(4) = 0)$
$(2, 4)$:	$4 = 16a_4 + 8a_3 + 4a_2$	$(f(2) = 4)$
	$0 = 32a_4 + 12a_3 + 4a_2$	$(f'(2) = 0)$

—CONTINUED—

69. —CONTINUED—

(c) The solution is $a_0 = a_1 = 0, a_2 = 4, a_3 = -2, a_4 = \frac{1}{4}$.

$$f(x) = \frac{1}{4}x^4 - 2x^3 + 4x^2$$

(d)

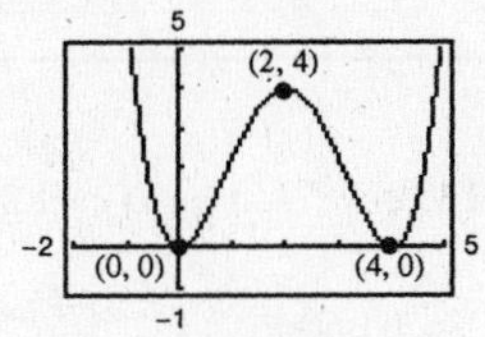

70. (a) Use a fourth degree polynomial $f(x) = a_4x^4 + a_3x^3 + a_2x^2 + a_1x + a_0$.

(b) $f'(x) = 4a_4x^3 + 3a_3x^2 + 2a_2x + a_1$

$(1, 2)$:	$2 = a_4 + a_3 + a_2 + a_1 + a_0$	$(f(1) = 2)$
	$0 = 4a_4 + 3a_3 + 2a_2 + a_1$	$(f'(1) = 0)$
$(-1, 4)$:	$4 = a_4 - a_3 + a_2 - a_1 + a_0$	$(f(-1) = 4)$
	$0 = -4a_4 + 3a_3 - 2a_2 + a_1$	$(f'(-1) = 0)$
$(3, 4)$:	$4 = 81a_4 + 27a_3 + 9a_2 + 3a_1 + a_0$	$(f(3) = 4)$
	$0 = 108a_4 + 27a_3 + 6a_2 + a_1$	$(f'(3) = 0)$

(c) The solution is $a_0 = \frac{23}{8}, a_1 = -\frac{3}{2}, a_2 = \frac{1}{4}, a_3 = \frac{1}{2}, a_4 = -\frac{1}{8}$

$$f(x) = -\frac{1}{8}x^4 + \frac{1}{2}x^3 + \frac{1}{4}x^2 - \frac{3}{2}x + \frac{23}{8}.$$

(d)

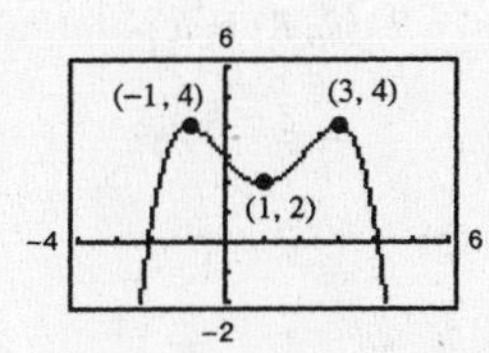

71. True.

Let $h(x) = f(x) + g(x)$ where f and g are increasing. Then $h'(x) = f'(x) + g'(x) > 0$ since $f'(x) > 0$ and $g'(x) > 0$.

72. False.

Let $h(x) = f(x)g(x)$ where $f(x) = g(x) = x$. Then $h(x) = x^2$ is decreasing on $(-\infty, 0)$.

73. False.

Let $f(x) = x^3$, then $f'(x) = 3x^2$ and f only has one critical number. Or, let $f(x) = x^3 + 3x + 1$, then $f'(x) = 3(x^2 + 1)$ has no critical numbers.

74. True.

If $f(x)$ is an nth-degree polynomial, then the degree of $f'(x)$ is $n - 1$.

75. False. For example, $f(x) = x^3$ does not have a relative extrema at the critical number $x = 0$.

76. False.

The function might not be continuous.

Section 5.4 Concavity and the Second Derivative Test

1. $y = x^2 - x - 2, y'' = 2$

Concave upward: $(-\infty, \infty)$

2. $y = -x^3 + 3x^2 - 2, y'' = -6x + 6$

Concave upward: $(-\infty, 1)$

Concave downward: $(1, \infty)$

3. $f(x) = \dfrac{24}{x^2 + 12}, y'' = \dfrac{-144(4 - x^2)}{(x^2 + 12)^3}$

Concave upward: $(-\infty, -2), (2, \infty)$

Concave downward: $(-2, 2)$

4. $f(x) = \dfrac{x^2 - 1}{2x + 1}, y'' = \dfrac{-6}{(2x + 1)^3}$

Concave upward: $\left(-\infty, -\dfrac{1}{2}\right)$

Concave downward: $\left(-\dfrac{1}{2}, \infty\right)$

5. $f(x) = \dfrac{x^2 + 1}{x^2 - 1}, y'' = \dfrac{4(3x^2 + 1)}{(x^2 - 1)^3}$

Concave upward: $(-\infty, -1), (1, \infty)$

Concave downward: $(-1, 1)$

6. $y = \dfrac{1}{270}(-3x^5 + 40x^3 + 135x), y'' = \dfrac{-2}{9}x(x - 2)(x + 2)$

Concave upward: $(-\infty, -2), (0, 2)$

Concave downward: $(-2, 0), (2, \infty)$

7. $g(x) = 3x^2 - x^3$

$g'(x) = 6x - 3x^2$

$g''(x) = 6 - 6x$

Concave upward: $(-\infty, 1)$

Concave downward: $(1, \infty)$

8. $h(x) = x^5 - 5x + 2$

$h'(x) = 5x^4 - 5$

$h''(x) = 20x^3$

Concave upward: $(0, \infty)$

Concave downward: $(-\infty, 0)$

9. $y = 2x$

$y' = 2$

$y'' = 0$

No concavity

10. $y = x + \dfrac{2}{x} = x + 2x^{-1}$

$y' = 1 - \dfrac{2}{x^2} = 1 - 2x^{-2}$

$y'' = 4x^{-3} = \dfrac{4}{x^3}$

Concave upward: $(0, \infty)$

Concave downward: $(-\infty, 0)$

11. $f(x) = x^3 - 6x^2 + 12x$

$f'(x) = 3x^2 - 12x + 12$

$f''(x) = 6(x - 2) = 0$ when $x = 2$.

The concavity changes at $x = 2$. $(2, 8)$ is a point of inflection.

Concave upward: $(2, \infty)$

Concave downward: $(-\infty, 2)$

12. $f(x) = 2x^3 - 3x^2 - 12x + 5$

$f'(x) = 6x^2 - 6x - 12$

$f''(x) = 12x - 6$

$f''(x) = 12x - 6 = 0$ when $x = \dfrac{1}{2}$.

Test interval	$-\infty < x < \frac{1}{2}$	$\frac{1}{2} < x < \infty$
Sign of $f''(x)$	$f''(x) < 0$	$f''(x) > 0$
Conclusion	Concave downward	Concave upward

Point of inflection: $\left(\dfrac{1}{2}, \dfrac{-3}{2}\right)$

13. $f(x) = \frac{1}{4}x^4 - 2x^2$

$f'(x) = x^3 - 4x$

$f''(x) = 3x^2 - 4$

$f''(x) = 3x^2 - 4 = 0$ when $x = \pm\frac{2}{\sqrt{3}}$.

Test interval:	$-\infty < x < -\frac{2}{\sqrt{3}}$	$-\frac{2}{\sqrt{3}} < x < \frac{2}{\sqrt{3}}$	$\frac{2}{\sqrt{3}} < x < \infty$
Sign of $f''(x)$:	$f''(x) > 0$	$f''(x) < 0$	$f''(x) > 0$
Conclusion:	Concave upward	Concave downward	Concave upward

Points of inflection: $\left(\pm\frac{2}{\sqrt{3}}, -\frac{20}{9}\right)$

14. $f(x) = 2x^4 - 8x + 3$

$f'(x) = 8x^3 - 8$

$f''(x) = 24x^2 = 0$ when $x = 0$.

However, $(0, 3)$ is not a point of inflection since $f''(x) \geq 0$ for all x.

Concave upward on $(-\infty, \infty)$.

15. $f(x) = x(x - 4)^3$

$f'(x) = x[3(x - 4)^2] + (x - 4)^3$

$= (x - 4)^2(4x - 4)$

$f''(x) = 4(x - 1)[2(x - 4)] + 4(x - 4)^2$

$= 4(x - 4)[2(x - 1) + (x - 4)]$

$= 4(x - 4)(3x - 6) = 12(x - 4)(x - 2)$

$f''(x) = 12(x - 4)(x - 2) = 0$ when $x = 2, 4$.

Test interval:	$-\infty < x < 2$	$2 < x < 4$	$4 < x < \infty$
Sign of $f''(x)$:	$f''(x) > 0$	$f''(x) < 0$	$f''(x) > 0$
Conclusion:	Concave upward	Concave downward	Concave upward

Points of inflection: $(2, -16), (4, 0)$

16. $f(x) = x^3(x - 4)$

$f'(x) = x^3 + 3x^2(x - 4)$

$= x^2[x + 3(x - 4)] = 4x^2(x - 3)$

$f''(x) = 4x^2 + 8x(x - 3) = 4x[x + 2(x - 3)] = 12x(x - 2) = 0$

$f''(x) = 12x(x - 2) = 0$ when $x = 0, 2$.

Test interval	$-\infty < x < 0$	$0 < x < 2$	$2 < x < \infty$
Sign of $f''(x)$	$f''(x) > 0$	$f''(x) < 0$	$f''(x) > 0$
Conclusion	Concave upward	Concave downward	Concave upward

Points of inflection: $(0, 0), (2, -16)$

17. $f(x) = x\sqrt{x+3}$, Domain: $[-3, \infty)$

$$f'(x) = x\left(\frac{1}{2}\right)(x+3)^{-1/2} + \sqrt{x+3} = \frac{3(x+2)}{2\sqrt{x+3}}$$

$$f''(x) = \frac{6\sqrt{x+3} - 3(x+2)(x+3)^{-1/2}}{4(x+3)} = \frac{3(x+4)}{4(x+3)^{3/2}}$$

$f''(x) > 0$ on the entire domain of f (except for $x = -3$, for which $f''(x)$ is undefined). There are no points of inflection. Concave upward on $(-3, \infty)$.

18. $f(x) = x\sqrt{x+1}$, Domain: $[-1, \infty)$

$$f'(x) = (x)\frac{1}{2}(x+1)^{-1/2} + \sqrt{x+1} = \frac{3x+2}{2\sqrt{x+1}}$$

$$f''(x) = \frac{6\sqrt{x+1} - (3x+2)(x+1)^{-1/2}}{4(x+1)} = \frac{3x+4}{4(x+1)^{3/2}}$$

$f''(x) > 0$ on the entire domain of f (except for $x = -1$, for which $f''(x)$ is undefined). There are no points of inflection.

Concave upward on $(-1, \infty)$.

19. $f(x) = \dfrac{x}{x^2+1}$

$$f'(x) = \frac{1-x^2}{(x^2+1)^2}$$

$$f''(x) = \frac{2x(x^2-3)}{(x^2+1)^3} = 0 \text{ when } x = 0, \pm\sqrt{3}.$$

Test intervals:	$-\infty < x < -\sqrt{3}$	$-\sqrt{3} < x < 0$	$0 < x < \sqrt{3}$	$\sqrt{3} < x < \infty$
Sign of $f'(x)$:	$f'' < 0$	$f'' > 0$	$f'' < 0$	$f'' > 0$
Conclusion:	Concave downward	Concave upward	Concave downward	Concave upward

Points of inflection: $\left(-\sqrt{3}, -\frac{\sqrt{3}}{4}\right), (0, 0), \left(\sqrt{3}, \frac{\sqrt{3}}{4}\right)$

20. $f(x) = \dfrac{x+1}{\sqrt{x}}$ Domain: $x > 0$

$$f'(x) = \frac{x-1}{2x^{3/2}}$$

$$f''(x) = \frac{3-x}{4x^{5/2}}$$

Test intervals	$0 < x < 3$	$3 < x < \infty$
Sign of $f''(x)$	$f'' > 0$	$f'' < 0$
Conclusion	Concave upward	Concave downward

Point of inflection: $\left(3, \frac{4}{\sqrt{3}}\right) = \left(3, \frac{4\sqrt{3}}{3}\right)$

21. $f(x) = 6x - x^2$

$f'(x) = 6 - 2x$ Critical number: $x = 3$

$f''(x) = -2$

$f''(3) = -2 < 0$

f has a relative maximum at $(3, 9)$.

22. $f(x) = x^2 + 3x - 8$

$f'(x) = 2x + 3$

$f''(x) = 2$

Critical number: $x = -\frac{3}{2}$

$f''\left(-\frac{3}{2}\right) > 0$

Therefore, $\left(-\frac{3}{2}, -\frac{41}{4}\right)$ is a relative minimum.

23. $f(x) = (x - 5)^2$

$f'(x) = 2(x - 5)$

$f''(x) = 2$

Critical number: $x = 5$

$f''(5) > 0$

Therefore, $(5, 0)$ is a relative minimum.

24. $f(x) = -(x - 5)^2$

$f'(x) = -2(x - 5)$

$f''(x) = -2$

Critical number: $x = 5$

$f''(5) < 0$

Therefore, $(5, 0)$ is a relative maximum.

25. $g(x) = x^2(6 - x) = 6x^2 - x^3$

$g'(x) = 12x - 3x^2 = 3x(4 - x)$ Critical numbers: $x = 0, 4$

$g''(x) = 12 - 6x$

$g''(0) = 12 > 0$ Relative minimum at $(0, 0)$.

$g''(4) = -12 < 0$ Relative maximum at $(4, 32)$.

26. $f(x) = 5 + 3x^2 - x^3$

$f'(x) = 6x - 3x^2 = 3x(2 - x)$ Critical numbers: $x = 0, 2$

$f''(x) = 6 - 6x$

$f''(0) = 6 > 0$ $(0, 5)$ is relative minimum.

$f''(2) = -6 < 0$ $(2, 9)$ is relative maximum.

27. $f(x) = x^3 - 3x^2 + 3$

$f'(x) = 3x^2 - 6x = 3x(x - 2)$

$f''(x) = 6x - 6 = 6(x - 1)$

Critical numbers: $x = 0, x = 2$

$f''(0) = -6 < 0$

Therefore, $(0, 3)$ is a relative maximum.

$f''(2) = 6 > 0$

Therefore, $(2, -1)$ is a relative minimum.

28. $f(x) = x^3 - 9x^2 + 27x$

$f'(x) = 3x^2 - 18x + 27 = 3(x - 3)^2$

$f''(x) = 6(x - 3)$

Critical number: $x = 3$

However, $f''(3) = 0$, so we must use the First Derivative Test. $f'(x) \geq 0$ for all x and, therefore, there are no relative extrema.

29. $f(x) = x^4 - 4x^3 + 2$

$f'(x) = 4x^3 - 12x^2 = 4x^2(x - 3)$

$f''(x) = 12x^2 - 24x = 12x(x - 2)$

Critical numbers: $x = 0, x = 3$

However, $f''(0) = 0$, so we must use the First Derivative Test. $f'(x) < 0$ on the intervals $(-\infty, 0)$ and $(0, 3)$; hence, $(0, 2)$ is not an extremum. $f''(3) > 0$ so $(3, -25)$ is a relative minimum.

30. $g(x) = -\frac{1}{8}(x + 2)^2(x - 4)^2$

$$g'(x) = \frac{-(x - 4)(x - 1)(x + 2)}{2}$$

$g''(x) = 3 + 3x - \frac{3}{2}x^2$

Critical numbers: $x = -2, 1, 4$

$g''(-2) = -9 < 0$

$(-2, 0)$ is a relative maximum.

$g''(1) = 9/2 > 0$

$(1, -10.125)$ is a relative minimum.

$g''(4) = -9 < 0$

$(4, 0)$ is a relative maximum.

31. $f(x) = x^{2/3} - 3$

$$f'(x) = \frac{2}{3x^{1/3}}$$

$$f''(x) = \frac{-2}{9x^{4/3}}$$

Critical number: $x = 0$

However, $f''(0)$ is undefined, so we must use the First Derivative Test. Since $f'(x) < 0$ on $(-\infty, 0)$ and $f'(x) > 0$ on $(0, \infty)$, $(0, -3)$ is a relative minimum.

32. $f(x) = \sqrt{x^2 + 1}$

$$f'(x) = \frac{x}{\sqrt{x^2 + 1}}$$

Critical number: $x = 0$

$$f''(x) = \frac{1}{(x^2 + 1)^{3/2}}$$

$f''(0) = 1 > 0$

Therefore, $(0, 1)$ is a relative minimum.

33. $f(x) = x + \dfrac{4}{x}$

$$f'(x) = 1 - \frac{4}{x^2} = \frac{x^2 - 4}{x^2}$$

$$f''(x) = \frac{8}{x^3}$$

Critical numbers: $x = \pm 2$

$f''(-2) < 0$

Therefore, $(-2, -4)$ is a relative maximum.

$f''(2) > 0$

Therefore, $(2, 4)$ is a relative minimum.

34. $f(x) = \dfrac{x}{x - 1}$

$$f'(x) = \frac{-1}{(x - 1)^2}$$

There are no critical numbers and $x = 1$ is not in the domain. There are no relative extrema.

35. $f(x) = 0.2x^2(x - 3)^3, [-1, 4]$

(a) $f'(x) = 0.2x(5x - 6)(x - 3)^2$

$f''(x) = (x - 3)(4x^2 - 9.6x + 3.6)$

$= 0.4(x - 3)(10x^2 - 24x + 9)$

(b) $f''(0) < 0 \Rightarrow (0, 0)$ is a relative maximum.

$f''(\frac{6}{5}) > 0 \Rightarrow (1.2, -1.6796)$ is a relative minimum.

Points of inflection:

$(3, 0), (0.4652, -0.7048), (1.9348, -0.9048)$

(c)

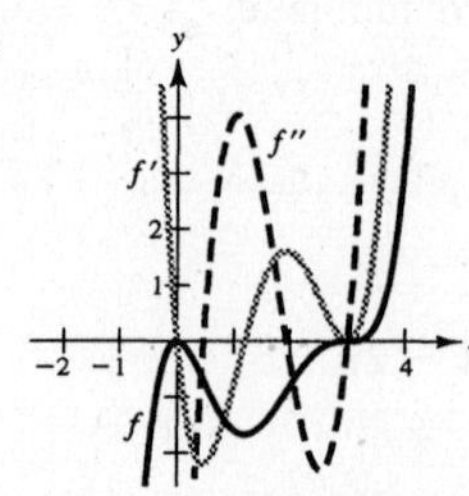

f is increasing when $f' > 0$ and decreasing when $f' < 0$. f is concave upward when $f'' > 0$ and concave downward when $f'' < 0$.

36. $f(x) = x^2\sqrt{6 - x^2}, \left[-\sqrt{6}, \sqrt{6}\right]$

(a) $f'(x) = \dfrac{3x(4 - x^2)}{\sqrt{6 - x^2}}$

$f'(x) = 0$ when $x = 0, x = \pm 2$.

$$f''(x) = \frac{6(x^4 - 9x^2 + 12)}{(6 - x^2)^{3/2}}$$

$f''(x) = 0$ when $x = \pm\sqrt{\dfrac{9 - \sqrt{33}}{2}}$.

(b) $f''(0) > 0 \Rightarrow (0, 0)$ is a relative minimum.

$f''(\pm 2) < 0 \Rightarrow \left(\pm 2, 4\sqrt{2}\right)$ are relative maxima.

Points of inflection: $(\pm 1.2758, 3.4036)$

(c)

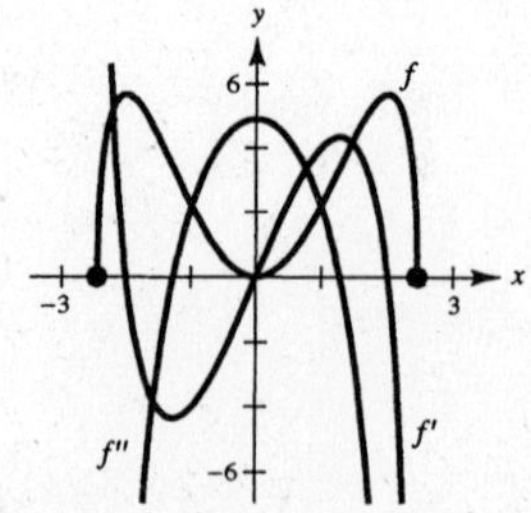

The graph of f is increasing when $f' > 0$ and decreasing when $f' < 0$. f is concave upward when $f'' > 0$ and concave downward when $f'' < 0$.

37. (a)

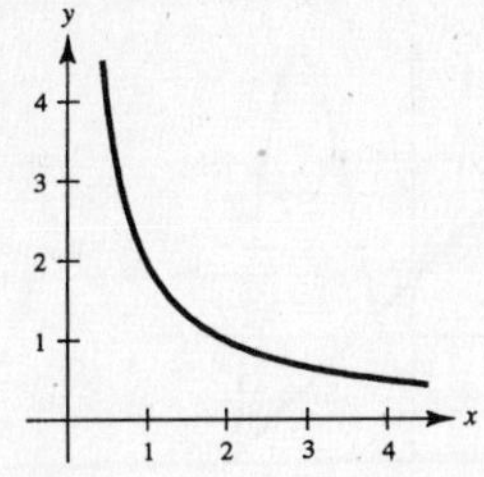

$f' < 0$ means f decreasing

f' increasing means concave upward

(b)

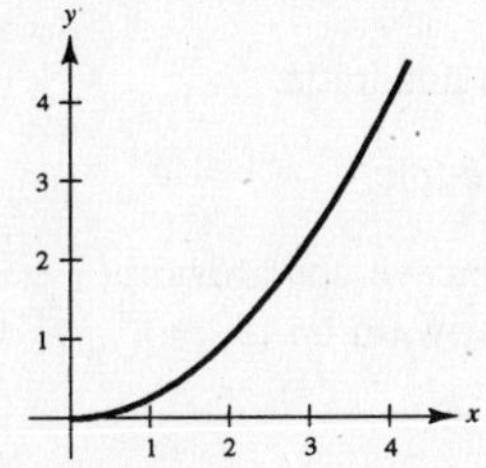

$f' > 0$ means f increasing

f' increasing means concave upward

38. (a)

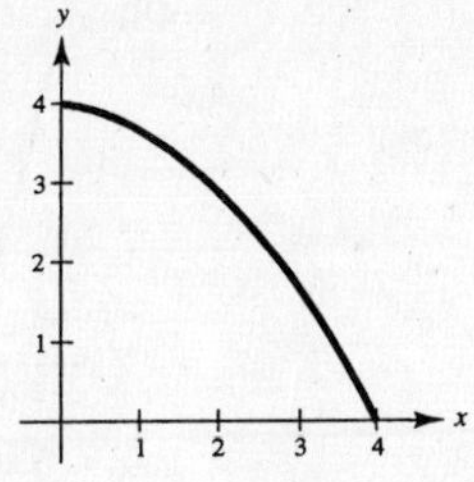

$f' < 0$ means f decreasing

f' decreasing means concave downward

(b)

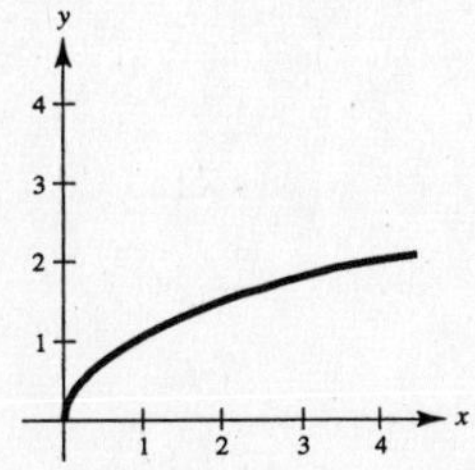

$f' > 0$ means f increasing

f' decreasing means concave downward

39. Let $f(x) = x^4$.

$f''(x) = 12x^2$

$f''(0) = 0$, but $(0, 0)$ is not a point of inflection.

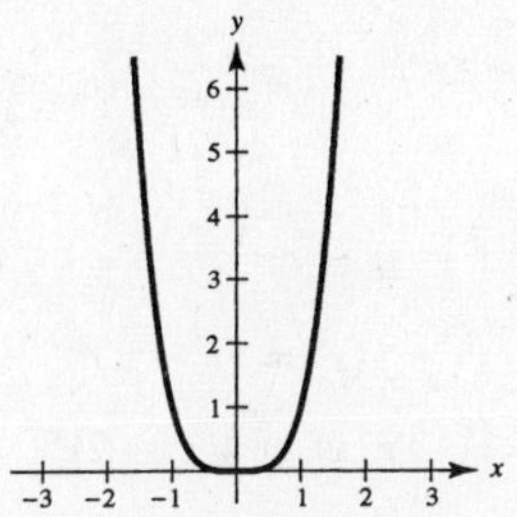

40. (a) The rate of change of sales is increasing.
$S'' > 0$

(b) The rate of change of sales is decreasing.
$S' > 0, S'' < 0$

(c) The rate of change of sales is constant.
$S' = C, S'' = 0$

(d) Sales are steady.
$S = C, S' = 0, S'' = 0$

(e) Sales are declining, but at a lower rate.
$S' < 0, S'' > 0$

(f) Sales have bottomed out and have started to rise.
$S' > 0$

41.

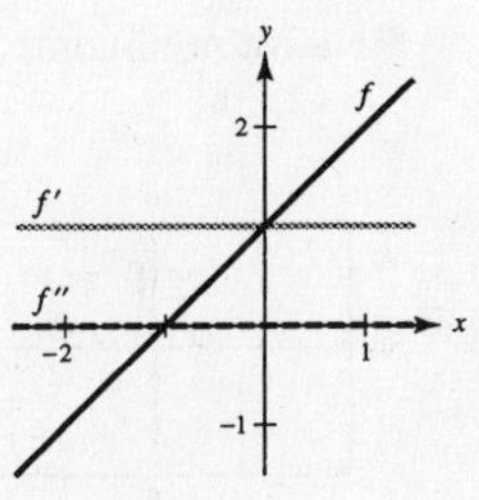

42.

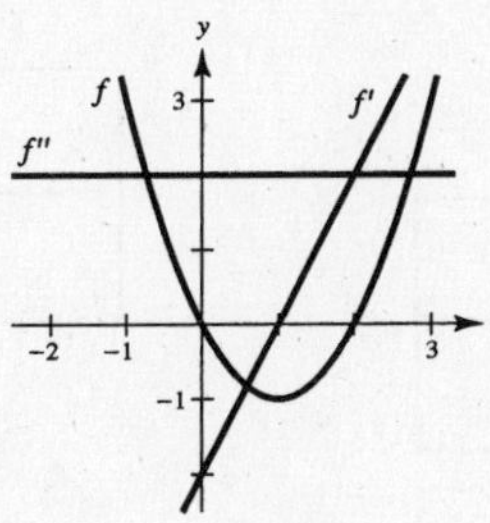

43.

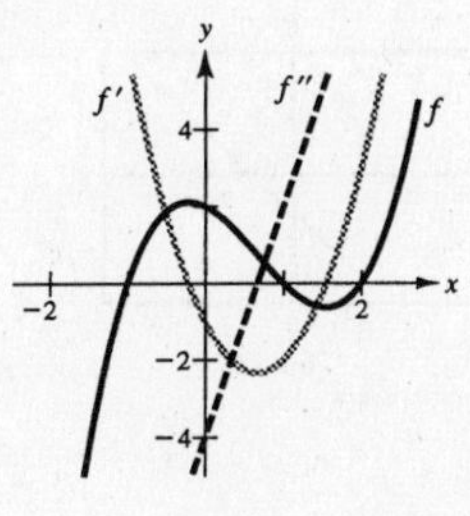

44.

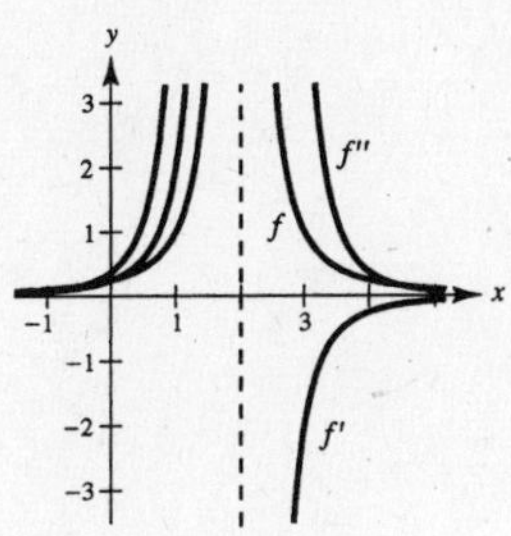

45.

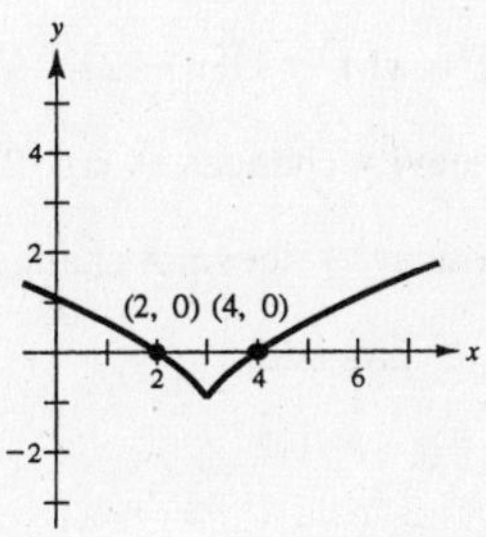

46.

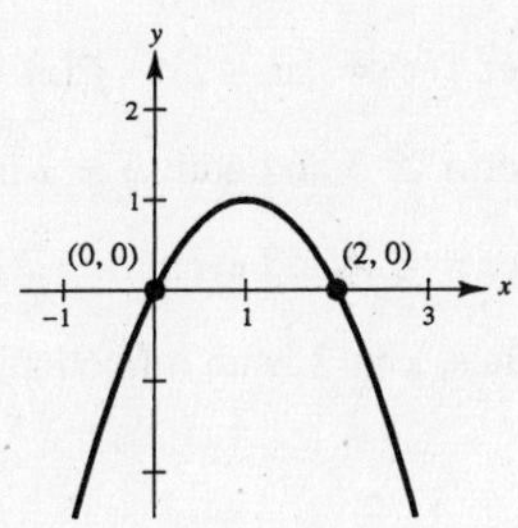

47.

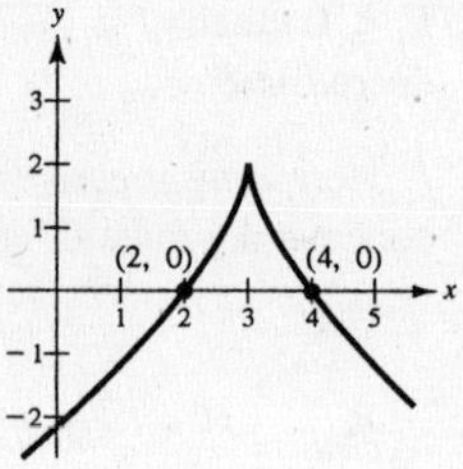

48.

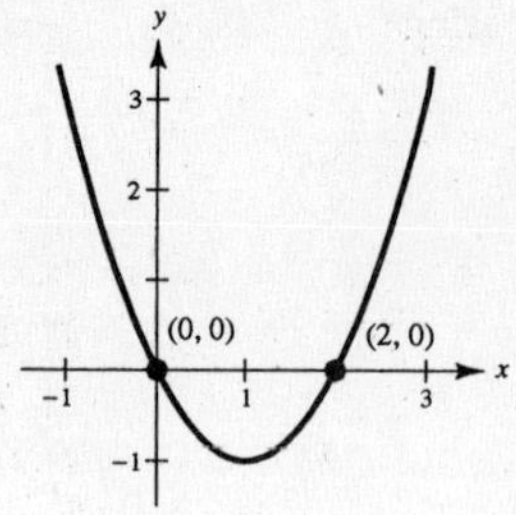

49.

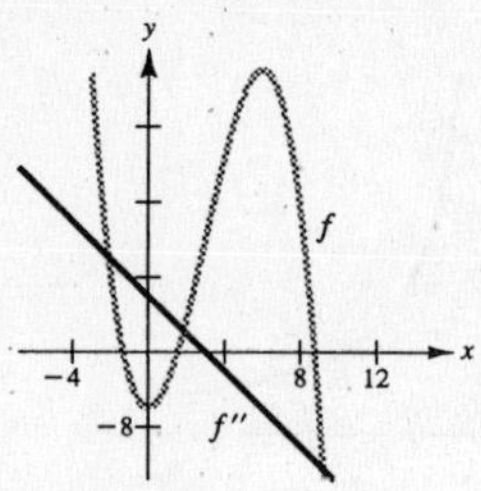

f'' is linear.

f' is quadratic.

f is cubic.

f concave upwards on $(-\infty, 3)$, downward on $(3, \infty)$.

50. (a)

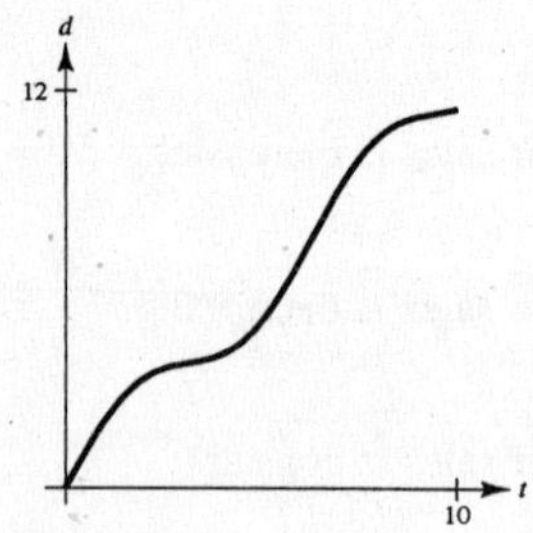

(b) Since the depth d is always increasing, there are no relative extrema. $f'(x) > 0$

(c) The rate of change of d is decreasing until you reach the widest point of the jug, then the rate increases until you reach the narrowest part of the jug's neck, then the rate decreases until you reach the top of the jug.

51. (a) $n = 1$:

$f(x) = x - 2$

$f'(x) = 1$

$f''(x) = 0$

No inflection points

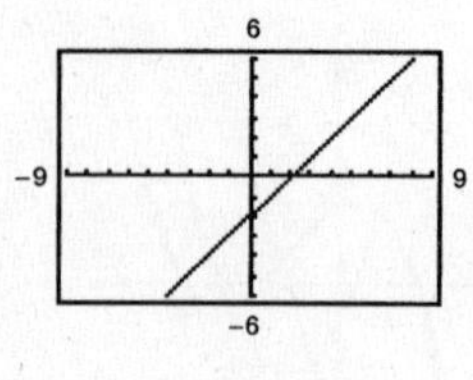

$n = 2$:

$f(x) = (x - 2)^2$

$f'(x) = 2(x - 2)$

$f''(x) = 2$

No inflection points

Relative minimum: $(2, 0)$

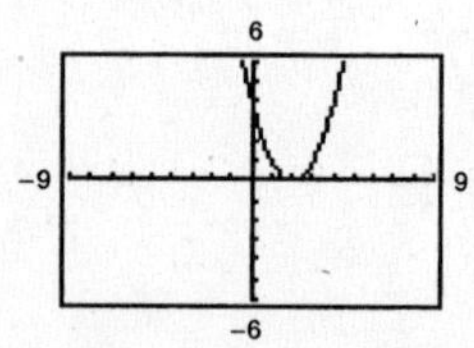

$n = 3$:

$f(x) = (x - 2)^3$

$f'(x) = 3(x - 2)^2$

$f''(x) = 6(x - 2)$

Inflection point: $(2, 0)$

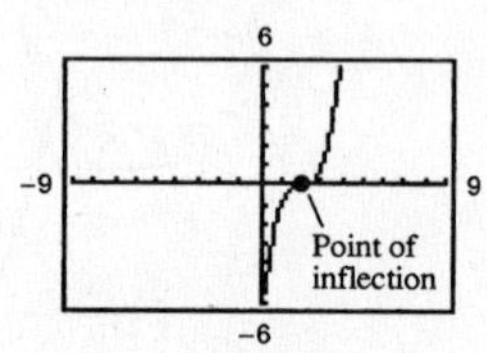

$n = 4$:

$f(x) = (x - 2)^4$

$f'(x) = 4(x - 2)^3$

$f''(x) = 12(x - 2)^2$

No inflection points:

Relative minimum: $(2, 0)$

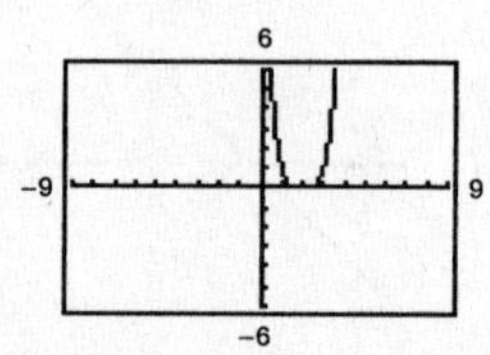

Conclusion: If $n \geq 3$ and n is odd, then $(2, 0)$ is an inflection point. If $n \geq 2$ and n is even, then $(2, 0)$ is a relative minimum.

(b) Let $f(x) = (x - 2)^n$, $f'(x) = n(x - 2)^{n-1}$, $f''(x) = n(n - 1)(x - 2)^{n-2}$.

For $n \geq 3$ and odd, $n - 2$ is also odd and the concavity changes at $x = 2$.

For $n \geq 4$ and even, $n - 2$ is also even and the concavity does not change at $x = 2$.

Thus, $x = 2$ is an inflection point if and only if $n \geq 3$ is odd.

52. (a) $f(x) = \sqrt[3]{x}$

$f'(x) = \frac{1}{3}x^{-2/3}$

$f''(x) = -\frac{2}{9}x^{-5/3}$

Inflection point: $(0, 0)$

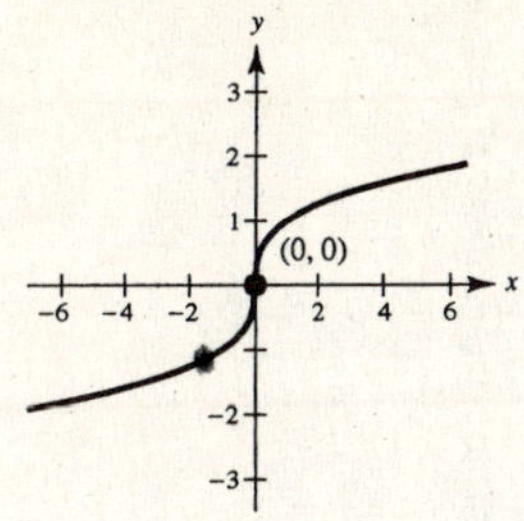

(b) $f''(x)$ does not exist at $x = 0$.

53. $f(x) = ax^3 + bx^2 + cx + d$

Relative maximum: $(3, 3)$

Relative minimum: $(5, 1)$

Point of inflection: $(4, 2)$

$f'(x) = 3ax^2 + 2bx + c, f''(x) = 6ax + 2b$

$$\left.\begin{array}{l} f(3) = 27a + 9b + 3c + d = 3 \\ f(5) = 125a + 25b + 5c + d = 1 \end{array}\right\} 98a + 16b + 2c = -2 \Rightarrow 49a + 8b + c = -1$$

$f'(3) = 27a + 6b + c = 0,\ f''(4) = 24a + 2b = 0$

$$\begin{array}{llll} 49a + 8b + c = -1 & & 24a + 2b = 0 \\ 27a + 6b + c = 0 & & 22a + 2b = -1 \\ 22a + 2b = -1 & & 2a = 1 \end{array}$$

$a = \frac{1}{2}, b = -6, c = \frac{45}{2}, d = -24$

$f(x) = \frac{1}{2}x^3 - 6x^2 + \frac{45}{2}x - 24$

54. $f(x) = ax^3 + bx^2 + cx + d$

Relative maximum: $(2, 4)$

Relative minimum: $(4, 2)$

Point of inflection: $(3, 3)$

$f'(x) = 3ax^2 + 2bx + c, f''(x) = 6ax + 2b$

$$\left.\begin{array}{l} f(2) = 8a + 4b + 2c + d = 4 \\ f(4) = 64a + 16b + 4c + d = 2 \end{array}\right\} 56a + 12b + 2c = -2 \Rightarrow 28a + 6b + c = -1$$

$f'(2) = 12a + 4b + c = 0,\ f'(4) = 48a + 8b + c = 0,\ f''(3) = 18a + 2b = 0$

$$\begin{array}{llll} 28a + 6b + c = -1 & & 18a + 2b = 0 \\ 12a + 4b + c = 0 & & 16a + 2b = -1 \\ 16a + 2b = -1 & & 2a = 1 \end{array}$$

$a = \frac{1}{2}, b = -\frac{9}{2}, c = 12, d = -6$

$f(x) = \frac{1}{2}x^3 - \frac{9}{2}x^2 + 12x - 6$

55. $f(x) = ax^3 + bx^2 + cx + d$

Maximum: $(-4, 1)$

Minimum: $(0, 0)$

(a) $f'(x) = 3ax^2 + 2bx + c, \quad f''(x) = 6ax + 2b$

$$f(0) = 0 \Rightarrow d = 0$$

$$f(-4) = 1 \Rightarrow \quad -64a + 16b - 4c = 1$$

$$f'(-4) = 0 \Rightarrow \quad 48a - 8b + c = 0$$

$$f'(0) = 0 \Rightarrow \quad c = 0$$

Solving this system yields $a = \frac{1}{32}$ and $b = 6a = \frac{3}{16}$.

$$f(x) = \tfrac{1}{32}x^3 + \tfrac{3}{16}x^2$$

(b) The plane would be descending at the greatest rate at the point of inflection.

$$f''(x) = 6ax + 2b = \frac{3}{16}x + \frac{3}{8} = 0 \Rightarrow x = -2$$

Two miles from touchdown.

56. $D = 2x^4 - 5Lx^3 + 3L^2x^2$

$$D' = 8x^3 - 15Lx^2 + 6L^2x = x(8x^2 - 15Lx + 6L^2) = 0$$

$$x = 0 \text{ or } x = \frac{15L \pm \sqrt{33}L}{16} = \left(\frac{15 \pm \sqrt{33}}{16}\right)L$$

By the Second Derivative Test, the deflection is maximum when

$$x = \left(\frac{15 - \sqrt{33}}{16}\right)L \approx 0.578L.$$

57. (a) line OA: $y = -0.06x$ slope: -0.06

line CB: $y = 0.04x + 50$ slope: 0.04

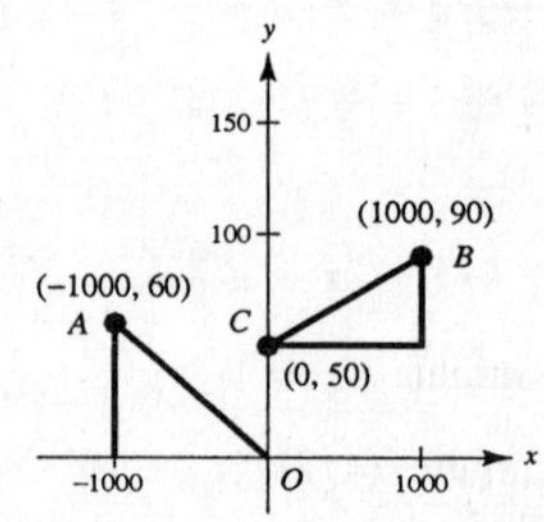

$$f(x) = ax^3 + bx^2 + cx + d$$

$$f'(x) = 3ax^2 + 2bx + c$$

$(-1000, 60)$: $\quad 60 = (-1000)^3a + (1000)^2b - 1000c + d$

$$-0.06 = (1000)^2 3a - 2000b + c$$

$(1000, 90)$: $\quad 90 = (1000)^3a + (1000)^2b + 1000c + d$

$$0.04 = (1000)^2 3a + 2000b + c$$

The solution to this system of 4 equations is $a = -1.25 \times 10^{-8}$, $b = 0.000025$, $c = 0.0275$, and $d = 50$.

(b) $y = -1.25 \times 10^{-8}x^3 + 0.000025x^2 + 0.0275x + 50$

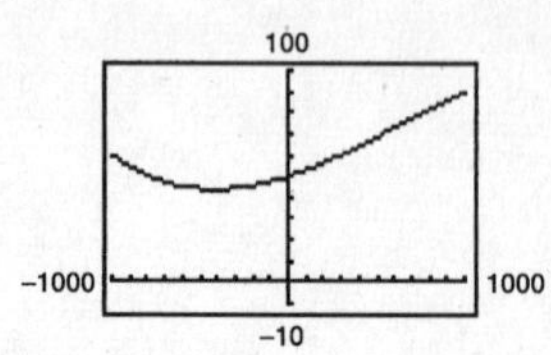

(c)

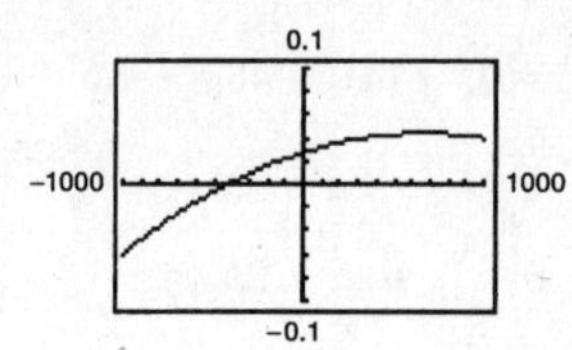

(d) The steepest part of the road is 6% at the point A.

58. $S = \dfrac{5.755T^3}{10^8} - \dfrac{8.521T^2}{10^6} + \dfrac{6.540T}{10^5} + 0.99987, \quad 0 < T < 25$

(a) The maximum occurs when $T \approx 4°$ and $S \approx 0.999999$.

(b)

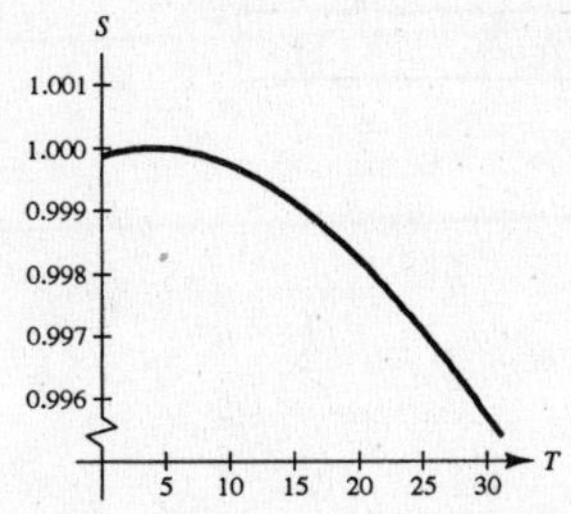

(c) $S(20°) \approx 0.9982$

59. $C = 0.5x^2 + 15x + 5000$

$\overline{C} = \dfrac{C}{x} = 0.5x + 15 + \dfrac{5000}{x}$

$\overline{C}$ = average cost per unit

$\dfrac{d\overline{C}}{dx} = 0.5 - \dfrac{5000}{x^2} = 0$ when $x = 100$.

By the First Derivative Test, $\overline{C}$ is minimized when $x = 100$ units.

60. $C = 2x + \dfrac{300{,}000}{x}$

$C' = 2 - \dfrac{300{,}000}{x^2} = 0$ when $x = 100\sqrt{15} \approx 387$.

By the First Derivative Test, C is minimized when $x \approx 387$ units.

61. $f(x) = \sqrt{1 - x}, \qquad f(0) = 1$

$f'(x) = -\dfrac{1}{2\sqrt{1 - x}}, \qquad f'(0) = -\dfrac{1}{2}$

$f''(x) = -\dfrac{1}{4(1 - x)^{3/2}}, \qquad f''(0) = -\dfrac{1}{4}$

$P_1(x) = 1 + \left(-\dfrac{1}{2}\right)(x - 0) = 1 - \dfrac{x}{2}$

$P_1{}'(x) = -\dfrac{1}{2}$

$P_2(x) = 1 + \left(-\dfrac{1}{2}\right)(x - 0) + \dfrac{1}{2}\left(-\dfrac{1}{4}\right)(x - 0)^2 = 1 - \dfrac{x}{2} - \dfrac{x^2}{8}$

$P_2{}'(x) = -\dfrac{1}{2} - \dfrac{x}{4}$

$P_2{}''(x) = -\dfrac{1}{4}$

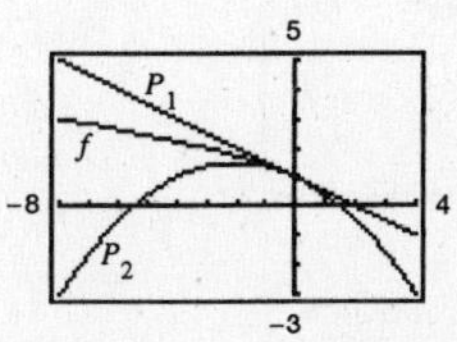

The values of f, P_1, P_2, and their first derivatives are equal at $x = 0$. The values of the second derivatives of f and P_2 are equal at $x = 0$. The approximations worsen as you move away from $x = 0$.

62. $f(x) = \dfrac{\sqrt{x}}{x-1}$, $\quad f(2) = \sqrt{2}$

$f'(x) = \dfrac{-(x+1)}{2\sqrt{x}(x-1)^2}$, $\quad f'(2) = -\dfrac{3}{2\sqrt{2}} = -\dfrac{3\sqrt{2}}{4}$

$f''(x) = \dfrac{3x^2+6x-1}{4x^{3/2}(x-1)^3}$, $\quad f''(2) = \dfrac{23}{8\sqrt{2}} = \dfrac{23\sqrt{2}}{16}$

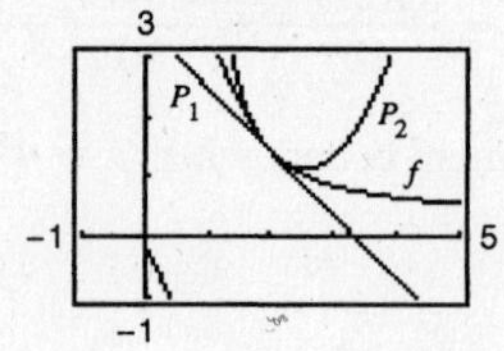

$$P_1(x) = \sqrt{2} + \left(-\frac{3\sqrt{2}}{4}\right)(x-2) = -\frac{3\sqrt{2}}{4}x + \frac{5\sqrt{2}}{2}$$

$$P_1'(x) = -\frac{3\sqrt{2}}{4}$$

$$P_2(x) = \sqrt{2} + \left(-\frac{3\sqrt{2}}{4}\right)(x-2) + \frac{1}{2}\left(\frac{23\sqrt{2}}{16}\right)(x-2)^2 = \sqrt{2} - \frac{3\sqrt{2}}{4}(x-2) + \frac{23\sqrt{2}}{32}(x-2)^2$$

$$P_2'(x) = -\frac{3\sqrt{2}}{4} + \frac{23\sqrt{2}}{16}(x-2)$$

$$P_2''(x) = \frac{23\sqrt{2}}{16}$$

The values of f, P_1, P_2 and their first derivatives are equal at $x = 2$. The values of the second derivatives of f and P_2 are equal at $x = 2$. The approximations worsen as you move away from $x = 2$.

63. $S = \dfrac{100t^2}{65+t^2}$, $t > 0$

(a)

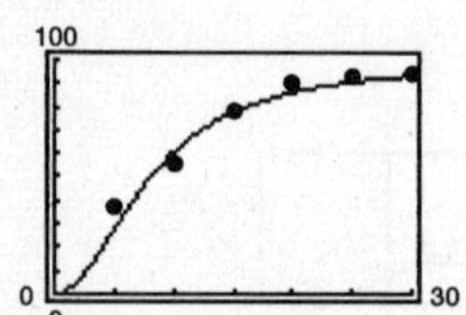

(b) $S'(t) = \dfrac{13{,}000t}{(65+t^2)^2}$

$S''(t) = \dfrac{13{,}000(65-3t^2)}{(65+t^2)^3} = 0 \Rightarrow t = 4.65$

S is concave upward on $(0, 4.65)$, concave downward on $(4.65, 30)$.

(c) $S'(t) > 0$ for $t > 0$.

As t increases, the speed increases, but at a slower rate.

64. $f(x) = x(x-6)^2 = x^3 - 12x^2 + 36x$

$f'(x) = 3x^2 - 24x + 36 = 3(x-2)(x-6) = 0$

$f''(x) = 6x - 24 = 6(x-4) = 0$

Relative extrema: $(2, 32)$ and $(6, 0)$

Point of inflection $(4, 16)$ is midway between the relative extrema of f.

65. True. Let $y = ax^3 + bx^2 + cx + d$, $a \neq 0$. Then $y'' = 6ax + 2b = 0$ when $x = -(b/3a)$, and the concavity changes at this point.

66. False. $f(x) = 1/x$ has a discontinuity at $x = 0$.

67. False. Concavity is determined by f''.

68. False. For example, let $f(x) = (x-2)^4$.

Section 5.5 Limits at Infinity

1. $f(x) = \dfrac{3x^2}{x^2 + 2}$

No vertical asymptotes

Horizontal asymptote: $y = 3$

Matches (f)

2. $f(x) = \dfrac{2x}{\sqrt{x^2 + 2}}$

No vertical asymptotes

Horizontal asymptotes: $y = \pm 2$

Matches (c)

3. $f(x) = \dfrac{x}{x^2 + 2}$

No vertical asymptotes

Horizontal asymptote: $y = 0$

Matches (d)

4. $f(x) = 2 + \dfrac{x^2}{x^4 + 1}$

No vertical asymptotes

Horizontal asymptote: $y = 2$

Matches (a)

5. Horizontal asymptote: $y = 0$

Intercept: $(0, 4)$

Matches (b)

6. $f(x) = \dfrac{2x^2 - 3x + 5}{x^2 + 1}$

No vertical asymptotes

Horizontal asymptote: $y = 2$

Matches (e)

7. $f(x) = \dfrac{4x + 3}{2x - 1}$

x	10^0	10^1	10^2	10^3	10^4	10^5	10^6
$f(x)$	7	2.26	2.025	2.0025	2.0003	2	2

$\lim_{x \to \infty} f(x) = 2$

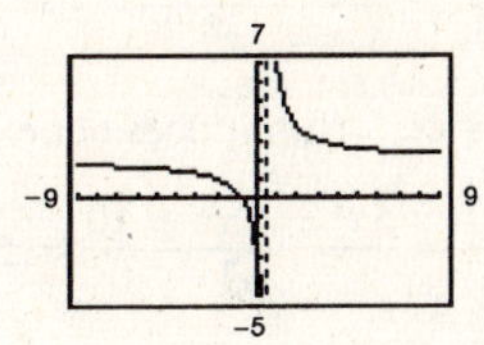

8. $f(x) = \dfrac{2x^2}{x + 1}$

x	10^0	10^1	10^2	10^3	10^4	10^5	10^6
$f(x)$	1	18.18	198.02	1998.002	19,998	199,998	1,999,998

$\lim_{x \to \infty} f(x) = \infty$ (Limit does not exist.)

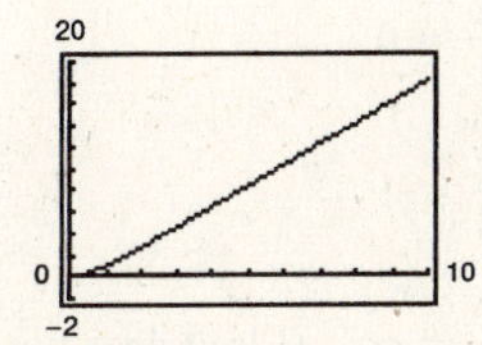

9. $f(x) = \dfrac{-6x}{\sqrt{4x^2 + 5}}$

x	10^0	10^1	10^2	10^3	10^4	10^5	10^6
$f(x)$	-2	-2.98	-2.9998	-3	-3	-3	-3

$\lim_{x \to \infty} f(x) = -3$

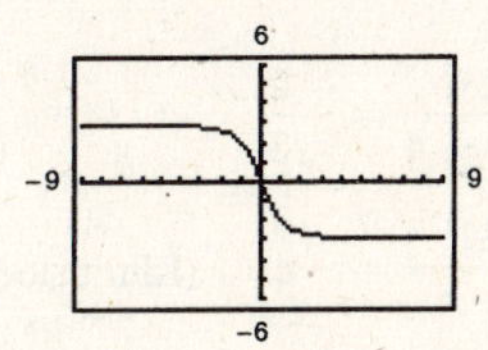

10. $f(x) = \dfrac{8x}{\sqrt{x^2 - 3}}$

x	10^1	10^2	10^3	10^4	10^5	10^6	10^7
$f(x)$	8.12	8.001	8	8	8	8	8

$\lim_{x \to \infty} f(x) = 8$

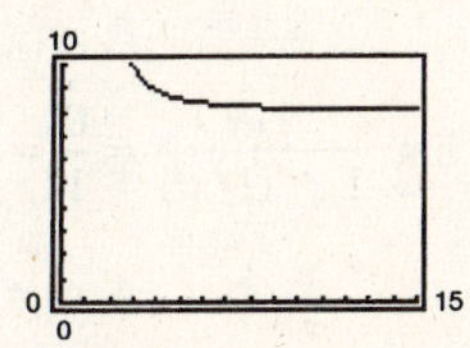

11. $f(x) = 5 - \dfrac{1}{x^2 + 1}$

x	10^0	10^1	10^2	10^3	10^4	10^5	10^6
$f(x)$	4.5	4.99	4.9999	4.999999	5	5	5

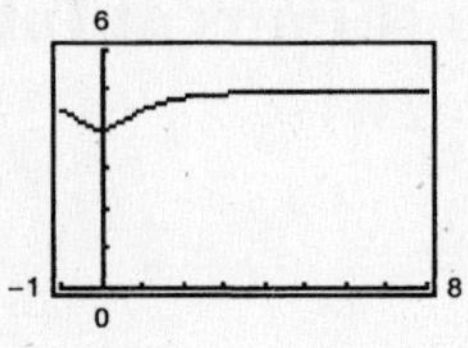

$\lim\limits_{x\to\infty} f(x) = 5$

12. $f(x) = 4 + \dfrac{3}{x^2 + 2}$

x	10^0	10^1	10^2	10^3	10^4	10^5	10^6
$f(x)$	5	4.03	4.0003	4.0	4.0	4	4

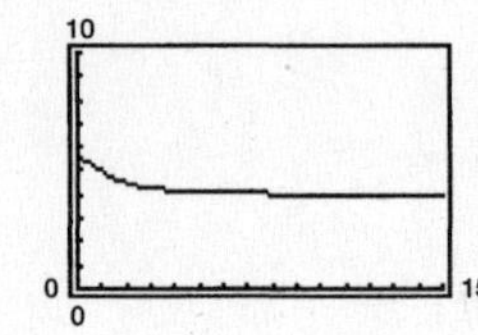

$\lim\limits_{x\to\infty} f(x) = 4$

13. (a) $h(x) = \dfrac{f(x)}{x^2} = \dfrac{5x^3 - 3x^2 + 10}{x^2} = 5x - 3 + \dfrac{10}{x^2}$

$\lim\limits_{x\to\infty} h(x) = \infty$ (Limit does not exist)

(b) $h(x) = \dfrac{f(x)}{x^3} = \dfrac{5x^3 - 3x^2 + 10}{x^3} = 5 - \dfrac{3}{x} + \dfrac{10}{x^3}$

$\lim\limits_{x\to\infty} h(x) = 5$

14. (a) $h(x) = \dfrac{f(x)}{x} = \dfrac{5x^2 - 3x + 7}{x} = 5x - 3 + \dfrac{7}{x}$

$\lim\limits_{x\to\infty} h(x) = \infty$ (Limit does not exist.)

(b) $h(x) = \dfrac{f(x)}{x^2} = \dfrac{5x^2 - 3x + 7}{x^2} = 5 - \dfrac{3}{x} + \dfrac{7}{x^2}$

$\lim\limits_{x\to\infty} h(x) = 5$

15. (a) $\lim\limits_{x\to\infty} \dfrac{x^2 + 2}{x^3 - 1} = 0$

(b) $\lim\limits_{x\to\infty} \dfrac{x^2 + 2}{x^2 - 1} = 1$

(c) $\lim\limits_{x\to\infty} \dfrac{x^2 + 2}{x - 1} = \infty$ (Limit does not exist.)

16. (a) $\lim\limits_{x\to\infty} \dfrac{3 - 2x}{3x^3 - 1} = 0$

(b) $\lim\limits_{x\to\infty} \dfrac{3 - 2x}{3x - 1} = -\dfrac{2}{3}$

(c) $\lim\limits_{x\to\infty} \dfrac{3 - 2x^2}{3x - 1} = -\infty$ (Limit does not exist.)

17. (a) $\lim\limits_{x\to\infty} \dfrac{5 - 2x^{3/2}}{3x^2 - 4} = 0$

(b) $\lim\limits_{x\to\infty} \dfrac{5 - 2x^{3/2}}{3x^{3/2} - 4} = -\dfrac{2}{3}$

(c) $\lim\limits_{x\to\infty} \dfrac{5 - 2x^{3/2}}{3x - 4} = -\infty$ (Limit does not exist.)

18. (a) $\lim\limits_{x\to\infty} \dfrac{5x^{3/2}}{4x^2 + 1} = 0$

(b) $\lim\limits_{x\to\infty} \dfrac{5x^{3/2}}{4x^{3/2} + 1} = \dfrac{5}{4}$

(c) $\lim\limits_{x\to\infty} \dfrac{5x^{3/2}}{4\sqrt{x} + 1} = \infty$ (Limit does not exist.)

19. $\lim\limits_{x\to\infty} \dfrac{2x - 1}{3x + 2} = \lim\limits_{x\to\infty} \dfrac{2 - (1/x)}{3 + (2/x)} = \dfrac{2 - 0}{3 + 0} = \dfrac{2}{3}$

20. $\lim\limits_{x\to\infty} \dfrac{3x^3 + 2}{9x^3 - 2x^2 + 7} = \dfrac{3}{9} = \dfrac{1}{3}$

21. $\lim\limits_{x\to\infty} \dfrac{x}{x^2 - 1} = \lim\limits_{x\to\infty} \dfrac{1/x}{1 - (1/x^2)} = \dfrac{0}{1} = 0$

22. $\lim\limits_{x\to-\infty} \dfrac{4x^2 + 3}{2x^2 - 1} = \dfrac{4}{2} = 2$

23. $\lim\limits_{x\to\infty} \left(10 - \dfrac{2}{x^2}\right) = 10 - 0 = 10$

24. $\lim\limits_{x\to\infty} \left(4 + \dfrac{3}{x}\right) = 4 + 0 = 4$

25. $\lim_{x \to -\infty} \frac{5x^2}{x + 3} = \lim_{x \to -\infty} \frac{5x}{1 + (3/x)} = -\infty$

Limit does not exist.

26. $\lim_{x \to -\infty} \left(\frac{1}{2}x - \frac{4}{x^2}\right) = -\infty$ (Limit does not exist.)

27. $\lim_{x \to -\infty} \frac{x}{\sqrt{x^2 - x}} = \lim_{x \to -\infty} \frac{1}{\frac{\sqrt{x^2 - x}}{-\sqrt{x^2}}}$, $\left(\text{for } x < 0 \text{ we have } x = -\sqrt{x^2}\right)$

$= \lim_{x \to -\infty} \frac{-1}{\sqrt{1 - (1/x)}} = -1$

28. $\lim_{x \to -\infty} \frac{x}{\sqrt{x^2 + 1}} = \lim_{x \to -\infty} \frac{1}{\left(\frac{\sqrt{x^2 + 1}}{-\sqrt{x^2}}\right)}$, $\left(\text{for } x < 0, x = -\sqrt{x^2}\right)$

$= \lim_{x \to -\infty} \frac{-1}{\sqrt{x + (1/x^2)}} = -1$

29. $\lim_{x \to \infty} \frac{2x + 1}{\sqrt{x^2 - x}} = \lim_{x \to \infty} \frac{2 + (1/x)}{\sqrt{1 - (1/x)}} = \frac{2}{1} = 2$

30. $\lim_{x \to -\infty} \frac{-3x + 1}{\sqrt{x^2 + x}} = \lim_{x \to -\infty} \frac{-3 + (1/x)}{\frac{\sqrt{x^2 + x}}{-\sqrt{x^2}}}$, $\left(\text{for } x < 0 \text{ we have } -\sqrt{x^2} = x\right)$

$= \lim_{x \to -\infty} \frac{3 - (1/x)}{\sqrt{1 + (1/x)}} = 3$

31. (a) $f(x) = \frac{|x|}{x + 1}$

$\lim_{x \to \infty} \frac{|x|}{x + 1} = 1$

$\lim_{x \to -\infty} \frac{|x|}{x + 1} = -1$

Therefore, $y = 1$ and $y = -1$ are both horizontal asymptotes.

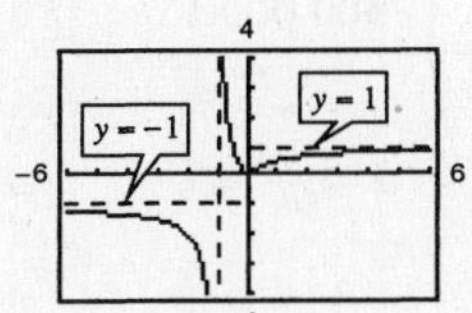

32. $f(x) = \frac{3x}{\sqrt{x^2 + 2}}$

$\lim_{x \to \infty} f(x) = 3$

$\lim_{x \to -\infty} f(x) = -3$

Therefore, $y = 3$ and $y = -3$ are both horizontal asymptotes.

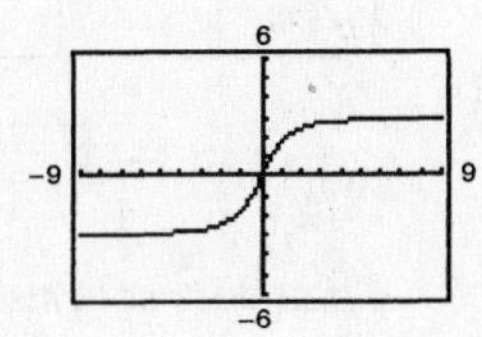

33. $\lim_{x \to -\infty} \left(x + \sqrt{x^2 + 3}\right) = \lim_{x \to -\infty} \left[\left(x + \sqrt{x^2 + 3}\right) \cdot \frac{x - \sqrt{x^2 + 3}}{x - \sqrt{x^2 + 3}}\right] = \lim_{x \to -\infty} \frac{-3}{x - \sqrt{x^2 + 3}} = 0$

34. $\lim_{x \to \infty} \left(2x - \sqrt{4x^2 + 1}\right) = \lim_{x \to \infty} \left[\left(2x - \sqrt{4x^2 + 1}\right) \cdot \frac{2x + \sqrt{4x^2 + 1}}{2x + \sqrt{4x^2 + 1}}\right] = \lim_{x \to \infty} \frac{-1}{2x + \sqrt{4x^2 + 1}} = 0$

35. $\lim_{x \to \infty} \left(x - \sqrt{x^2 + x}\right) = \lim_{x \to \infty} \left[\left(x - \sqrt{x^2 + x}\right) \cdot \frac{x + \sqrt{x^2 + x}}{x + \sqrt{x^2 + x}}\right]$

$= \lim_{x \to \infty} \frac{-x}{x + \sqrt{x^2 + x}} = \lim_{x \to \infty} \frac{-1}{1 + \sqrt{1 + (1/x)}} = -\frac{1}{2}$

36. $\lim_{x \to -\infty} \left(3x + \sqrt{9x^2 - x}\right) = \lim_{x \to -\infty} \left[\left(3x + \sqrt{9x^2 - x}\right) \cdot \frac{3x - \sqrt{9x^2 - x}}{3x - \sqrt{9x^2 - x}}\right]$

$$= \lim_{x \to -\infty} \frac{x}{3x - \sqrt{9x^2 - x}}$$

$$= \lim_{x \to -\infty} \frac{1}{3 - \dfrac{\sqrt{9x^2 - x}}{-\sqrt{x^2}}} \quad \left(\text{for } x < 0 \text{ we have } x = -\sqrt{x^2}\right)$$

$$= \lim_{x \to -\infty} \frac{1}{3 + \sqrt{9 - (1/x)}} = \frac{1}{6}$$

37.

x	10^0	10^1	10^2	10^3	10^4	10^5	10^6
$f(x)$	1	0.513	0.501	0.500	0.500	0.500	0.500

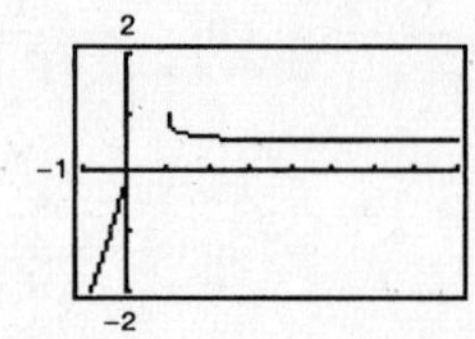

$$\lim_{x \to \infty} \left(x - \sqrt{x(x-1)}\right) = \lim_{x \to \infty} \frac{x - \sqrt{x^2 - x}}{1} \cdot \frac{x + \sqrt{x^2 - x}}{x + \sqrt{x^2 - x}}$$

$$= \lim_{x \to \infty} \frac{x}{x + \sqrt{x^2 - x}}$$

$$= \lim_{x \to \infty} \frac{1}{1 + \sqrt{1 - (1/x)}}$$

$$= \frac{1}{2}$$

38.

x	10^0	10^1	10^2	10^3	10^4	10^5	10^6
$f(x)$	1.0	5.1	50.1	500.1	5000.1	50,000.1	500,000.1

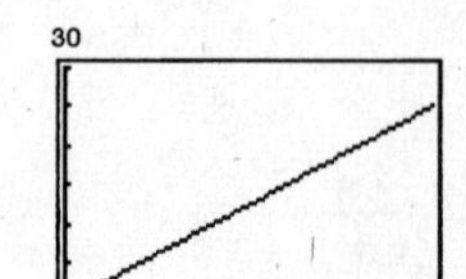

$$\lim_{x \to \infty} \frac{x^2 - x\sqrt{x^2 - x}}{1} \cdot \frac{x^2 + x\sqrt{x^2 - x}}{x^2 + x\sqrt{x^2 - x}} = \lim_{x \to \infty} \frac{x^3}{x^2 + x\sqrt{x^2 - x}} = \infty$$

Limit does not exist.

39.

x	10^0	10^1	10^2	10^3	10^4	10^5	10^6
$f(x)$	-0.236	-0.025	-0.002	-2.5×10^{-4}	-2.5×10^{-5}	-2.5×10^{-6}	0

$$2x - \sqrt{4x^2 + 1} = 2x - \sqrt{4x^2 + 1} \left(\frac{2x + \sqrt{4x^2 + 1}}{2x + \sqrt{4x^2 + 1}}\right)$$

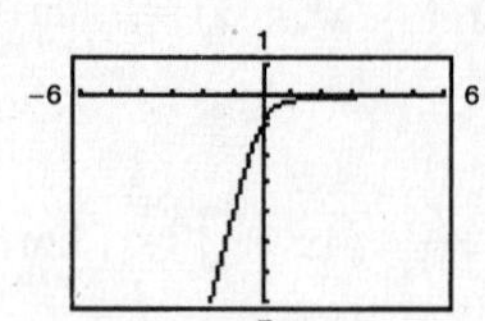

$$= \frac{4x^2 - (4x^2 + 1)}{2x + \sqrt{4x^2 + 1}}$$

$$= \frac{-1}{2x + \sqrt{4x^2 + 1}}$$

Hence, $\lim_{x \to \infty} f(x) = \lim_{x \to \infty} \frac{-1}{2x + \sqrt{4x^2 + 1}} = 0$

40.

x	10^0	10^1	10^2	10^3	10^4	10^5	10^6
$f(x)$	2.000	0.348	0.101	0.032	0.010	0.003	0.001

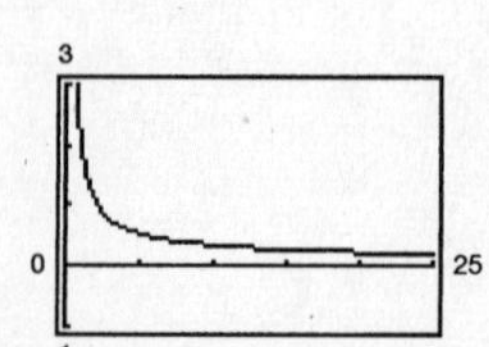

$$\lim_{x \to \infty} \frac{x + 1}{x\sqrt{x}} = 0$$

41. (a)

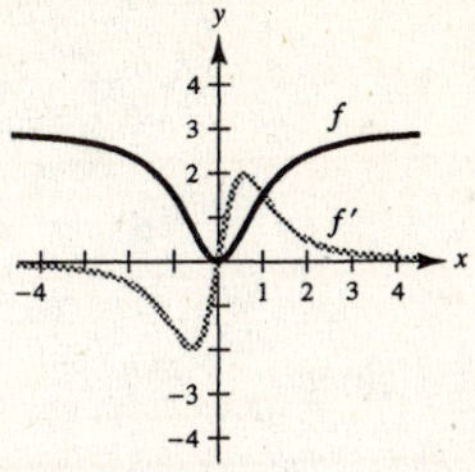

(b) $\lim_{x\to\infty} f(x) = 3 \qquad \lim_{x\to\infty} f'(x) = 0$

(c) Since $\lim_{x\to\infty} f(x) = 3$, the graph approaches that of a horizontal line, $\lim_{x\to\infty} f'(x) = 0$.

42. $x = 2$ is a critical number.

$f'(x) < 0$ for $x < 2$.

$f'(x) > 0$ for $x > 2$.

$\lim_{x\to-\infty} f(x) = \lim_{x\to\infty} f(x) = 6$

For example, let $f(x) = \dfrac{-6}{0.1(x-2)^2 + 1} + 6$.

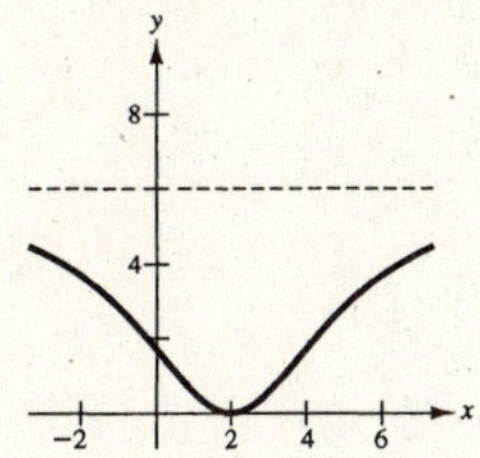

43. Yes. For example, let $f(x) = \dfrac{6|x-2|}{\sqrt{(x-2)^2+1}}$.

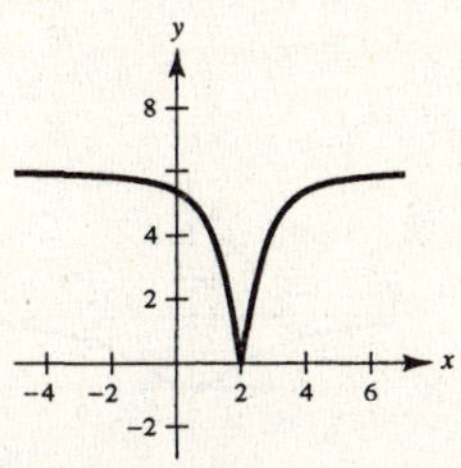

44. (a) The function is even: $\lim_{x\to-\infty} f(x) = 5$

(b) The function is odd: $\lim_{x\to-\infty} f(x) = -5$

45. $y = \dfrac{2+x}{1-x}$

Intercepts: $(-2, 0), (0, 2)$

Symmetry: none

Horizontal asymptote: $y = -1$ since

$$\lim_{x\to-\infty} \frac{2+x}{1-x} = -1 = \lim_{x\to\infty} \frac{2+x}{1-x}.$$

Discontinuity: $x = 1$ (Vertical asymptote)

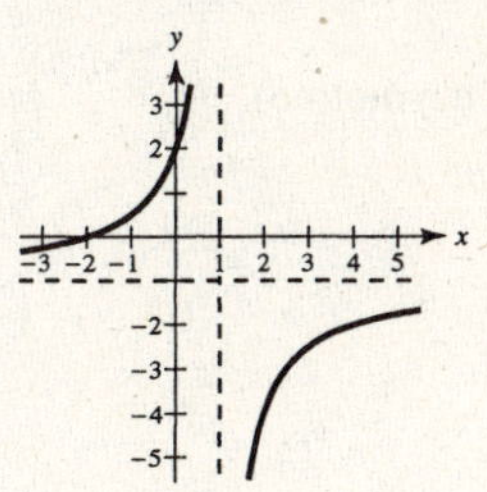

46. $y = \dfrac{x-3}{x-2}$

Intercepts: $(3, 0), \left(0, \dfrac{3}{2}\right)$

Symmetry: none

Horizontal asymptote: $y = 1$ since

$$\lim_{x\to-\infty} \frac{x-3}{x-2} = 1 = \lim_{x\to\infty} \frac{x-3}{x-2}.$$

Discontinuity: $x = 2$ (Vertical asymptote)

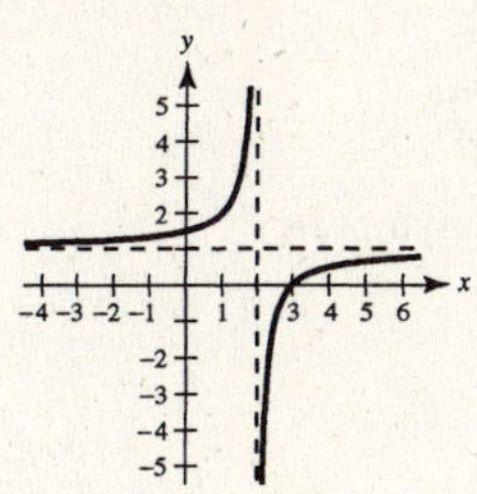

47. $y = \dfrac{x}{x^2-4}$

Intercept: $(0, 0)$

Symmetry: origin

Horizontal asymptote: $y = 0$

Vertical asymptote: $x = \pm 2$

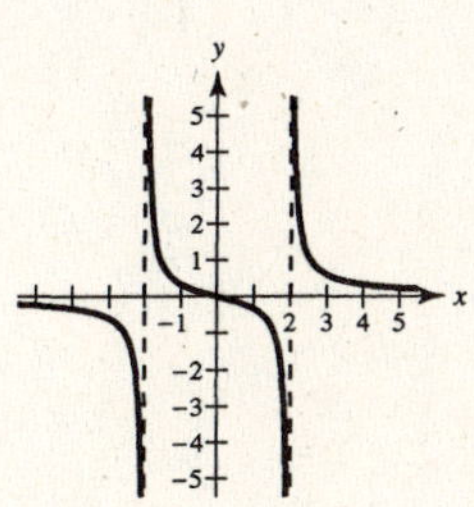

48. $y = \dfrac{2x}{9-x^2}$

Intercept: $(0, 0)$

Symmetry: origin

Horizontal asymptote: $y = 0$

Vertical asymptote: $x = \pm 3$

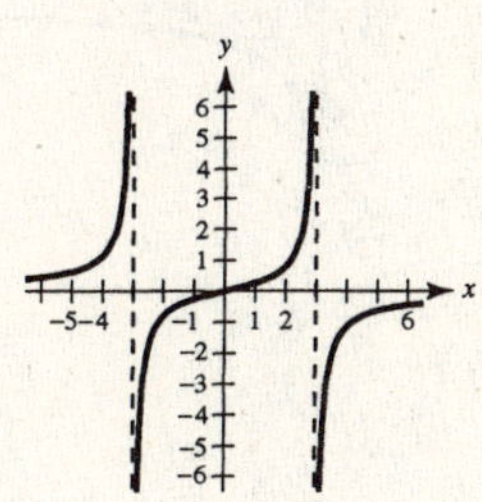

49. $y = \dfrac{x^2}{x^2 + 9}$

Intercept: (0, 0)

Symmetry: y-axis

Horizontal asymptote: $y = 1$ since

$$\lim_{x \to -\infty} \frac{x^2}{x^2 + 9} = 1 = \lim_{x \to \infty} \frac{x^2}{x^2 + 9}.$$

Relative minimum: (0, 0)

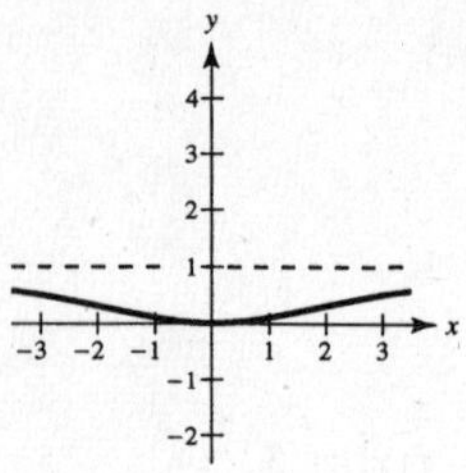

50. $y = \dfrac{x^2}{x^2 - 9}$

Intercept: (0, 0)

Symmetry: y-axis

Horizontal asymptote: $y = 1$ since

$$\lim_{x \to -\infty} \frac{x^2}{x^2 - 9} = 1 = \lim_{x \to \infty} \frac{x^2}{x^2 - 9}.$$

Discontinuities: $x = \pm 3$ (Vertical asymptotes)

Relative maximum: (0, 0)

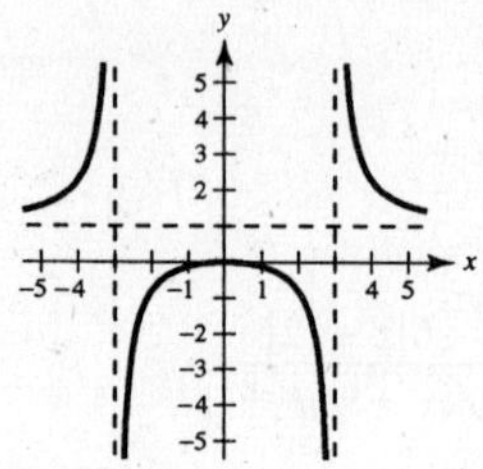

51. $y = \dfrac{2x^2}{x^2 - 4}$

Intercept: (0, 0)

Symmetry: y-axis

Horizontal asymptote: $y = 2$

Vertical asymptotes: $x = \pm 2$

Relative maximum: (0, 0)

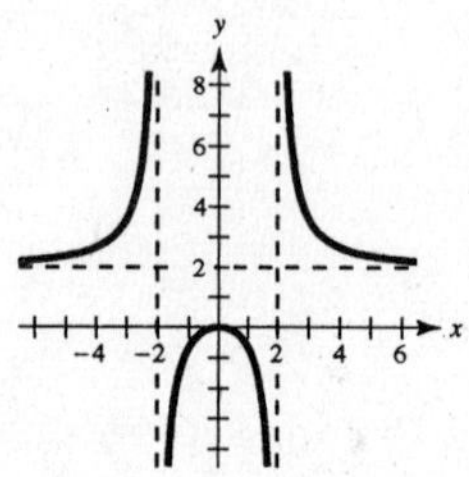

52. $y = \dfrac{2x^2}{x^2 + 4}$

Intercept: (0, 0)

Symmetry: y-axis

Horizontal asymptote: $y = 2$

Relative minimum: (0, 0)

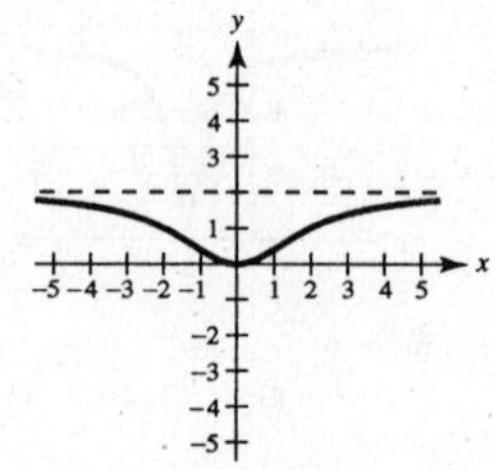

53. $xy^2 = 4$

Domain: $x > 0$

Intercepts: none

Symmetry: x-axis

Horizontal asymptote: $y = 0$ since

$$\lim_{x \to \infty} \frac{2}{\sqrt{x}} = 0 = \lim_{x \to \infty} -\frac{2}{\sqrt{x}}.$$

Discontinuity: $x = 0$ (Vertical asymptote)

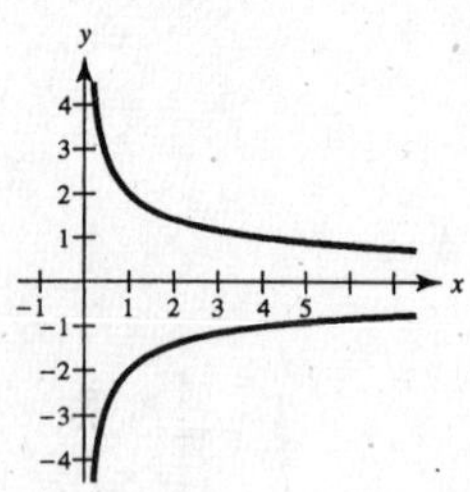

54. $x^2y = 4$

Intercepts: none

Symmetry: y-axis

Horizontal asymptote: $y = 0$ since

$$\lim_{x \to -\infty} \frac{4}{x^2} = 0 = \lim_{x \to \infty} \frac{4}{x^2}.$$

Discontinuity: $x = 0$ (Vertical asymptote)

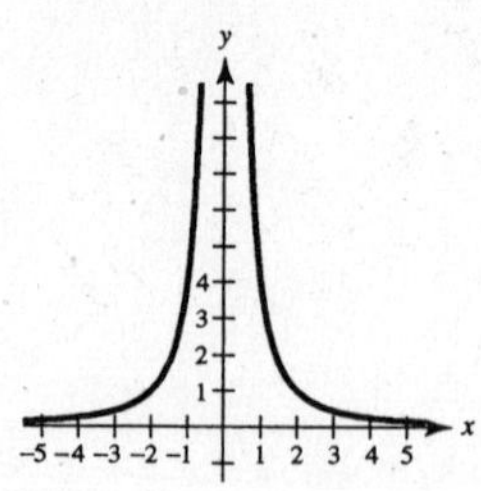

55. $y = \dfrac{2x}{1 - x}$

Intercept: $(0, 0)$

Symmetry: none

Horizontal asymptote: $y = -2$ since

$$\lim_{x \to -\infty} \frac{2x}{1 - x} = -2 = \lim_{x \to \infty} \frac{2x}{1 - x}.$$

Discontinuity: $x = 1$ (Vertical asymptote)

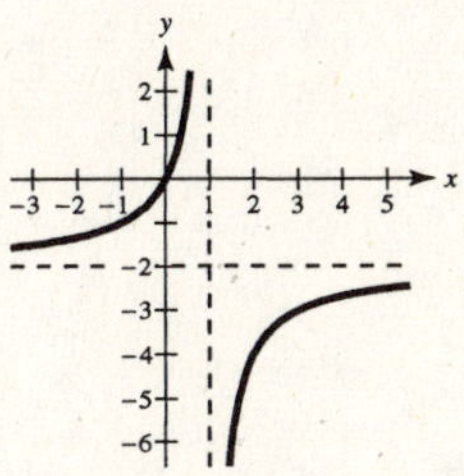

56. $y = \dfrac{2x}{1 - x^2}$

Intercept: $(0, 0)$

Symmetry: origin

Horizontal asymptote: $y = 0$ since

$$\lim_{x \to -\infty} \frac{2x}{1 - x^2} = 0 = \lim_{x \to \infty} \frac{2x}{1 - x^2}.$$

Discontinuities: $x = \pm 1$ (Vertical asymptotes)

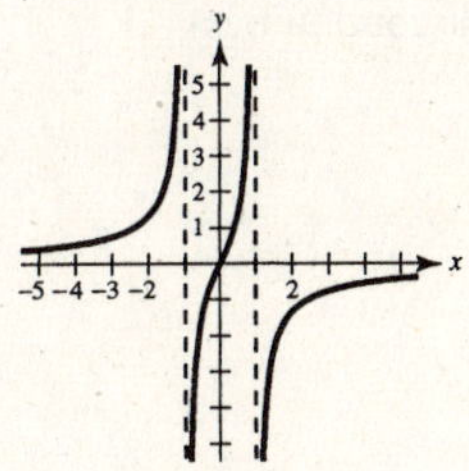

57. $y = 2 - \dfrac{3}{x^2}$

Intercepts: $\left(\pm\sqrt{3/2}, 0\right)$

Symmetry: y-axis

Horizontal asymptote: $y = 2$ since

$$\lim_{x \to -\infty} \left(2 - \frac{3}{x^2}\right) = 2 = \lim_{x \to \infty} \left(2 - \frac{3}{x^2}\right).$$

Discontinuity: $x = 0$ (Vertical asymptote)

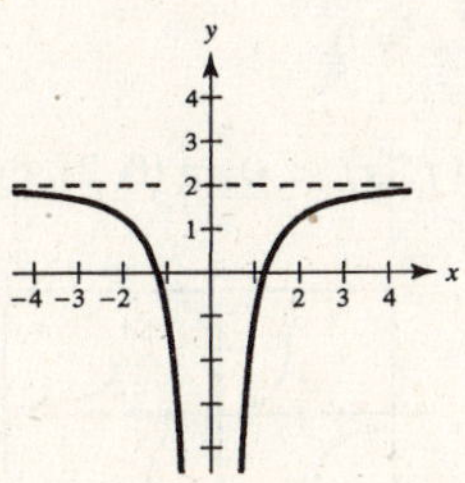

58. $y = 1 + \dfrac{1}{x}$

Intercept: $(-1, 0)$

Symmetry: none

Horizontal asymptote: $y = 1$ since

$$\lim_{x \to -\infty} \left(1 + \frac{1}{x}\right) = 1 = \lim_{x \to \infty} \left(1 + \frac{1}{x}\right).$$

Discontinuity: $x = 0$ (Vertical asymptote)

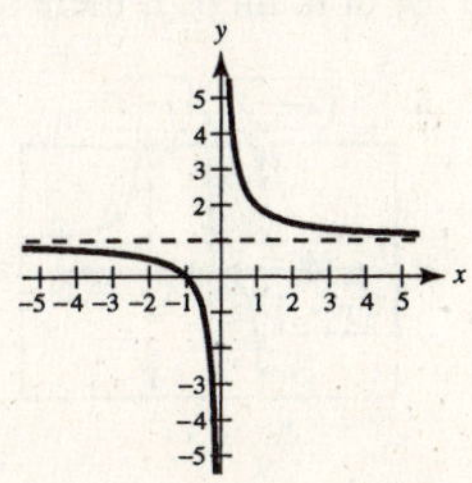

59. $y = \dfrac{x^3}{\sqrt{x^2 - 4}}$

Domain: $(-\infty, -2), (2, \infty)$

Intercepts: none

Symmetry: origin

Horizontal asymptote: none

Vertical asymptotes: $x = \pm 2$ (discontinuities)

Relative minimum: $\left(\sqrt{6}, 6\sqrt{3}\right)$

Relative maximum: $\left(-\sqrt{6}, -6\sqrt{3}\right)$

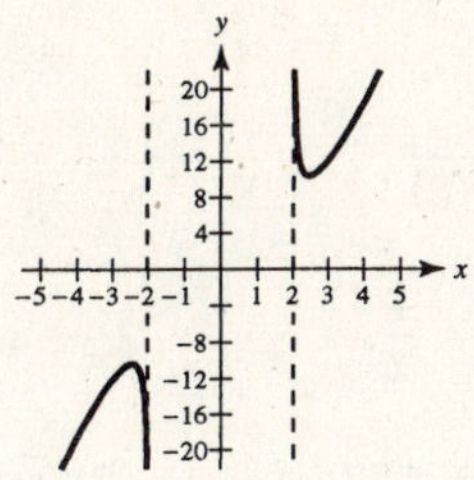

60. $y = \dfrac{x}{\sqrt{x^2 - 4}}$

Domain: $(-\infty, -2), (2, \infty)$

Intercepts: none

Symmetry: origin

Horizontal asymptotes: $y = \pm 1$ since

$$\lim_{x \to \infty} \frac{x}{\sqrt{x^2 - 4}} = 1, \lim_{x \to -\infty} \frac{x}{\sqrt{x^2 - 4}} = -1.$$

Vertical asymptotes: $x = \pm 2$ (discontinuities)

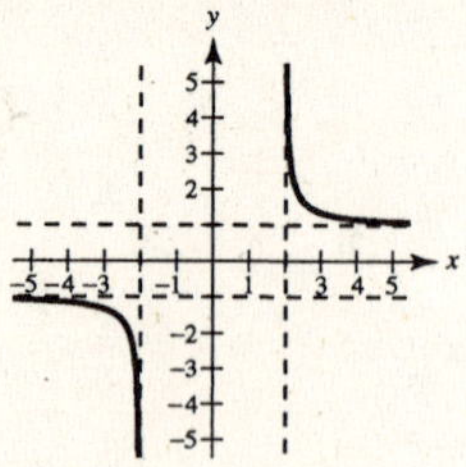

61. $f(x) = 5 - \dfrac{1}{x^2} = \dfrac{5x^2 - 1}{x^2}$

Domain: $(-\infty, 0), (0, \infty)$

$f'(x) = \dfrac{2}{x^3} \Rightarrow$ No relative extrema

$f''(x) = -\dfrac{6}{x^4} \Rightarrow$ No points of inflection

Vertical asymptote: $x = 0$

Horizontal asymptote: $y = 5$

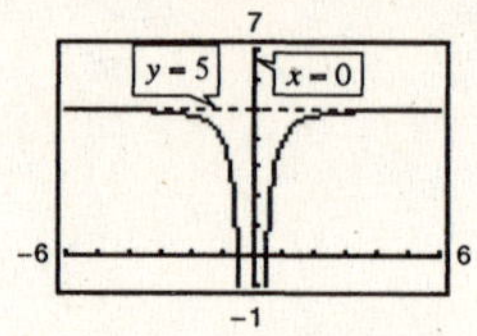

62. $f(x) = \dfrac{x^2}{x^2 - 1}$

$$f'(x) = \frac{(x^2 - 1)(2x) - x^2(2x)}{(x^2 - 1)^2} = \frac{-2x}{(x^2 - 1)^2} = 0 \text{ when } x = 0.$$

$$f''(x) = \frac{(x^2 - 1)^2(-2) + 2x(2)(x^2 - 1)(2x)}{(x^2 - 1)^4} = \frac{2(3x^2 + 1)}{(x^2 - 1)^3}$$

Since $f''(0) < 0$, then $(0, 0)$ is a relative maximum. Since $f''(x) \neq 0$, nor is it undefined in the domain of f, there are no points of inflection.

Vertical asymptotes: $x = \pm 1$

Horizontal asymptote: $y = 1$

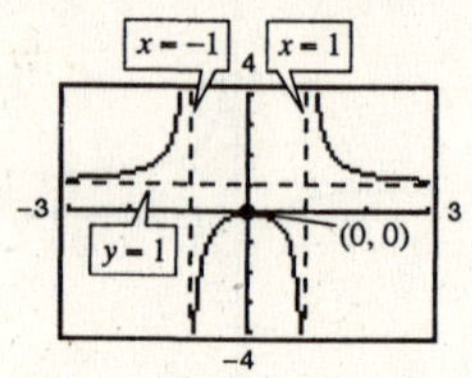

63. $f(x) = \dfrac{x}{x^2 - 4}$

$$f'(x) = \frac{(x^2 - 4) - x(2x)}{(x^2 - 4)^2}$$

$$= \frac{-(x^2 + 4)}{(x^2 - 4)^2} \neq 0 \text{ for any } x \text{ in the domain of } f.$$

$$f''(x) = \frac{(x^2 - 4)^2(-2x) + (x^2 + 4)(2)(x^2 - 4)(2x)}{(x^2 - 4)^4}$$

$$= \frac{2x(x^2 + 12)}{(x^2 - 4)^3} = 0 \text{ when } x = 0.$$

Since $f''(x) > 0$ on $(-2, 0)$ and $f''(x) < 0$ on $(0, 2)$, then $(0, 0)$ is a point of inflection.

Vertical asymptotes: $x = \pm 2$

Horizontal asymptote: $y = 0$

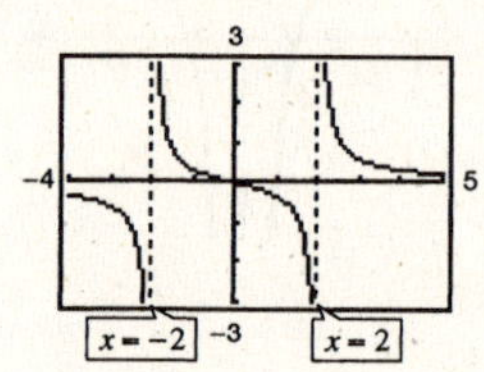

64. $f(x) = \dfrac{1}{x^2 - x - 2} = \dfrac{1}{(x + 1)(x - 2)}$

$$f'(x) = \frac{-(2x - 1)}{(x^2 - x - 2)^2} = 0 \text{ when } x = \frac{1}{2}.$$

$$f''(x) = \frac{(x^2 - x - 2)^2(-2) + (2x - 1)(2)(x^2 - x - 2)(2x - 1)}{(x^2 - x - 2)^4}$$

$$= \frac{6(x^2 - x + 1)}{(x^2 - x - 2)^3}$$

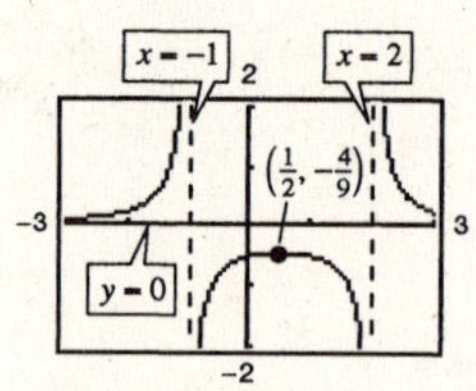

Since $f''\left(\frac{1}{2}\right) < 0$, then $\left(\frac{1}{2}, -\frac{4}{9}\right)$ is a relative maximum. Since $f''(x) \neq 0$, nor is it undefined in the domain of f, there are no points of inflection.

Vertical asymptotes: $x = -1, x = 2$ Horizontal asymptote: $y = 0$

65. $f(x) = \dfrac{x-2}{x^2-4x+3} = \dfrac{x-2}{(x-1)(x-3)}$

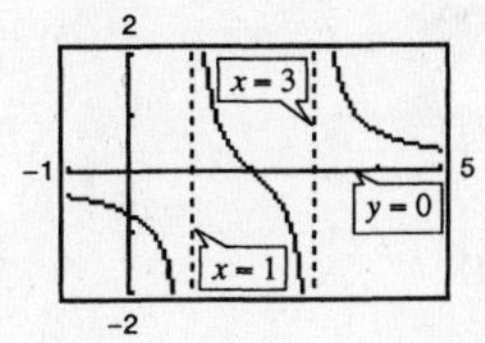

$$f'(x) = \frac{(x^2-4x+3)-(x-2)(2x-4)}{(x^2-4x+3)^2} = \frac{-x^2+4x-5}{(x^2-4x+3)^2} \neq 0$$

$$f''(x) = \frac{(x^2-4x+3)^2(-2x+4)-(-x^2+4x-5)(2)(x^2-4x+3)(2x-4)}{(x^2-4x+3)^4}$$

$$= \frac{2(x^3-6x^2+15x-14)}{(x^2-4x+3)^3} = 0 \text{ when } x = 2.$$

Since $f''(x) > 0$ on $(1, 2)$ and $f''(x) < 0$ on $(2, 3)$, then $(2, 0)$ is a point of inflection.

Vertical asymptote: $x = 1, x = 3$

Horizontal asymptote: $y = 0$

66. $f(x) = \dfrac{x+1}{x^2+x+1}$

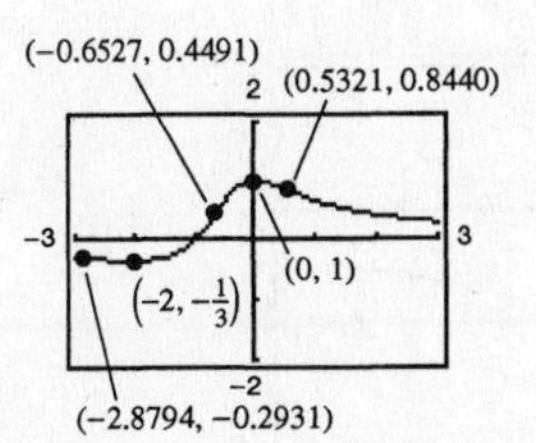

$$f'(x) = \frac{-x(x+2)}{(x^2+x+1)^2} = 0 \text{ when } x = 0, -2.$$

$$f''(x) = \frac{2(x^3+3x^2-1)}{(x^2+x+1)^3} = 0 \text{ when } x \approx 0.5321, -0.6527, -2.8794.$$

$f''(0) < 0$

Therefore, $(0, 1)$ is a relative maximum.

$f''(-2) > 0$

Therefore,

$$\left(-2, -\frac{1}{3}\right)$$

is a relative minimum.

Points of inflection: $(0.5321, 0.8440)$, $(-0.6527, 0.4491)$ and $(-2.8794, -0.2931)$ Horizontal asymptote: $y = 0$

67. $f(x) = \dfrac{x^3-3x^2+2}{x(x-3)}$, $g(x) = x + \dfrac{2}{x(x-3)}$

(a)

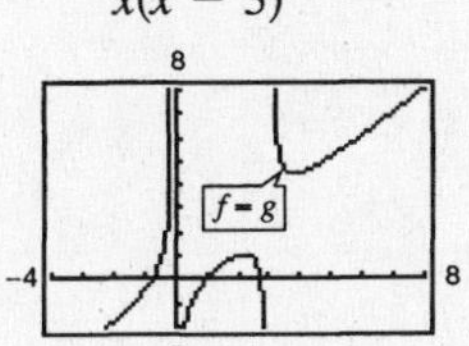

(b) $f(x) = \dfrac{x^3-3x^2+2}{x(x-3)}$

$$= \frac{x^2(x-3)}{x(x-3)} + \frac{2}{x(x-3)}$$

$$= x + \frac{2}{x(x-3)} = g(x)$$

(c)

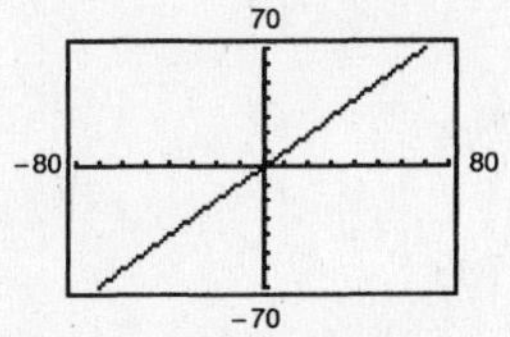

The graph appears as the slant asymptote $y = x$.

68. $f(x) = -\dfrac{x^3-2x^2+2}{2x^2}$, $g(x) = -\dfrac{1}{2}x + 1 - \dfrac{1}{x^2}$

(a)

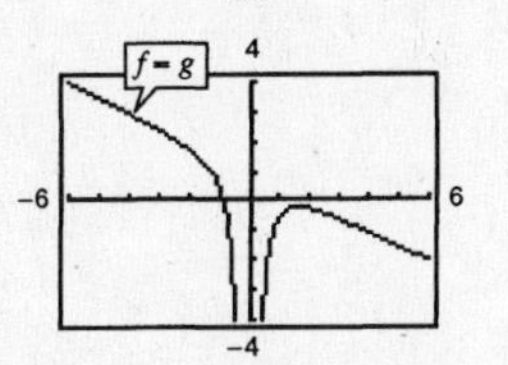

(b) $f(x) = -\dfrac{x^3-2x^2+2}{2x^2}$

$$= -\left[\frac{x^3}{2x^2} - \frac{2x^2}{2x^2} + \frac{2}{2x^2}\right]$$

$$= -\frac{1}{2}x + 1 - \frac{1}{x^2} = g(x)$$

(c)

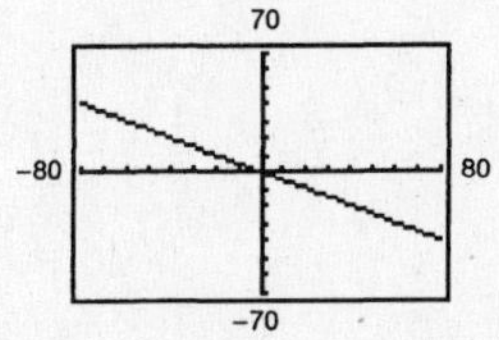

The graph appears as the slant asymptote $y = -\frac{1}{2}x + 1$.

69. $C = 0.5x + 500$

$\overline{C} = \dfrac{C}{x}$

$\overline{C} = 0.5 + \dfrac{500}{x}$

$\lim_{x \to \infty} \left(0.5 + \dfrac{500}{x}\right) = 0.5$

70. $\lim_{v_1/v_2 \to \infty} 100\left[1 - \dfrac{1}{(v_1/v_2)^c}\right] = 100[1 - 0] = 100\%$

71. (a) $T_1(t) = -0.003t^2 + 0.677t + 26.564$

(b)

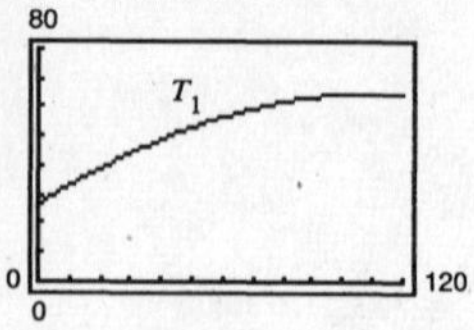

(c)

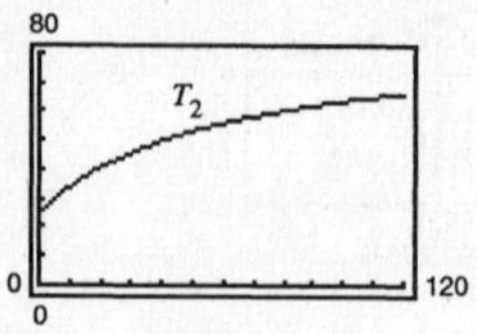

$T_2 = \dfrac{1451 + 86t}{58 + t}$

(d) $T_1(0) \approx 26.6$

$T_2(0) \approx 25.0$

(e) $\lim_{t \to \infty} T_2 = \dfrac{86}{1} = 86$

(f) The limiting temperature is 86. T_1 has no horizontal asymptote.

72. line: $mx - y + 4 = 0$

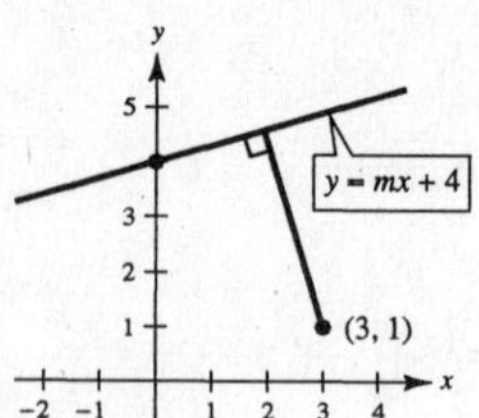

(a) $d = \dfrac{|Ax_1 + By_1 + C|}{\sqrt{A^2 + B^2}} = \dfrac{|m(3) - 1(1) + 4|}{\sqrt{m^2 + 1}}$

$= \dfrac{|3m + 3|}{\sqrt{m^2 + 1}}$

(b)

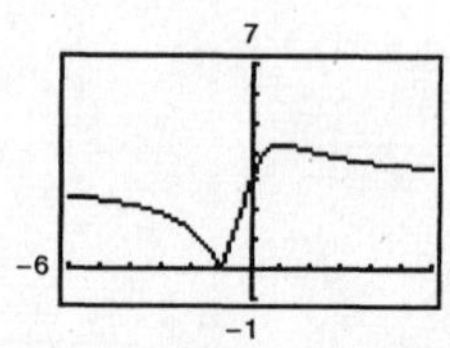

(c) $\lim_{m \to \infty} d(m) = 3 = \lim_{m \to -\infty} d(m)$

The line approaches the vertical line $x = 0$. Hence, the distance approaches 3.

73. $y = \dfrac{3.351t^2 + 42.461t - 543.730}{t^2}$

(a)

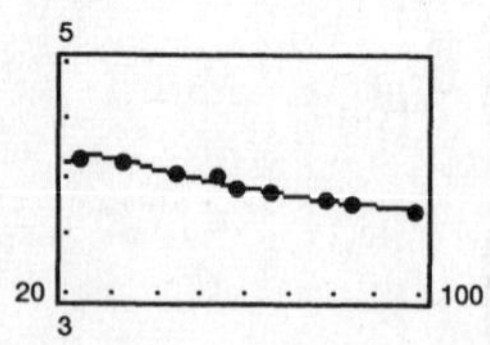

(b) Yes. $\lim_{t \to \infty} y = 3.351$

74. $S = \dfrac{100t^2}{65 + t^2},\ t > 0$

(a)

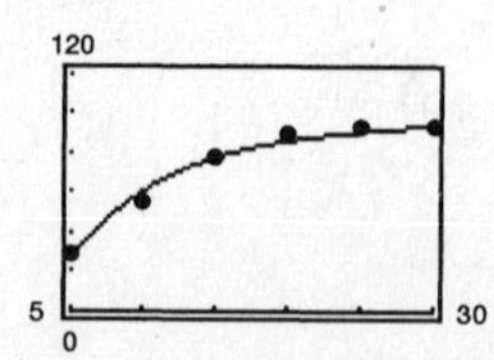

(b) Yes. $\lim_{t \to \infty} S = \dfrac{100}{1} = 100$

75. False. Let $f(x) = \dfrac{2x}{\sqrt{x^2 + 2}}$. (See Exercise 2.)

76. False. Let $y_1 = \sqrt{x + 1}$, then $y_1(0) = 1$. Thus, $y_1' = 1/(2\sqrt{x+1})$ and $y_1'(0) = 1/2$. Finally,

$$y_1'' = -\frac{1}{4(x+1)^{3/2}} \text{ and } y_1''(0) = -\frac{1}{4}.$$

Let $p = ax^2 + bx + 1$, then $p(0) = 1$. Thus, $p' = 2ax + b$ and $p'(0) = \frac{1}{2} \Rightarrow b = \frac{1}{2}$. Finally, $p'' = 2a$ and $p''(0) = -\frac{1}{4} \Rightarrow a = -\frac{1}{8}$. Therefore,

$$f(x) = \begin{cases} (-1/8)x^2 + (1/2)x + 1, & x < 0 \\ \sqrt{x+1}, & x \ge 0 \end{cases} \text{ and } f(0) = 1,$$

$$f'(x) = \begin{cases} (1/2) - (1/4)x, & x < 0 \\ 1/(2\sqrt{x+1}), & x > 0 \end{cases} \text{ and } f'(0) = \frac{1}{2}, \text{ and}$$

$$f''(x) = \begin{cases} (-1/4), & x < 0 \\ -1/(4(x+1)^{3/2}), & x > 0 \end{cases} \text{ and } f''(0) = -\frac{1}{4}.$$

$f''(x) < 0$ for all real x, but $f(x)$ *increases* without bound.

77. $\displaystyle\lim_{x\to\infty} \frac{p(x)}{q(x)} = \lim_{x\to\infty} \frac{a_n x^n + \cdots + a_1 x + a_0}{b_m x^m + \cdots + b_1 x + b_0}$

Divide $p(x)$ and $q(x)$ by x^m.

Case 1: If $n < m$: $\displaystyle\lim_{x\to\infty} \frac{p(x)}{q(x)} = \lim_{x\to\infty} \frac{\dfrac{a_n}{x^{m-n}} + \cdots + \dfrac{a_1}{x^{m-1}} + \dfrac{a_0}{x^m}}{b_m + \cdots + \dfrac{b_1}{x^{m-1}} + \dfrac{b_0}{x^m}} = \frac{0 + \cdots + 0 + 0}{b_m + \cdots + 0 + 0} = \frac{0}{b_m} = 0.$

Case 2: If $m = n$: $\displaystyle\lim_{x\to\infty} \frac{p(x)}{q(x)} = \lim_{x\to\infty} \frac{a_n + \cdots + \dfrac{a_1}{x^{m-1}} + \dfrac{a_0}{x^m}}{b_m + \cdots + \dfrac{b_1}{x^{m-1}} + \dfrac{b_0}{x^m}} = \frac{a_n + \cdots + 0 + 0}{b_m + \cdots + 0 + 0} = \frac{a_n}{b_m}.$

Case 3: If $n > m$: $\displaystyle\lim_{x\to\infty} \frac{p(x)}{q(x)} = \lim_{x\to\infty} \frac{a_n x^{n-m} + \cdots + \dfrac{a_1}{x^{m-1}} + \dfrac{a_0}{x^m}}{b_m + \cdots + \dfrac{b_1}{x^{m-1}} + \dfrac{b_0}{x^m}} = \frac{\pm\infty + \cdots + 0}{b_m + \cdots + 0} = \pm\infty.$

Section 5.6 A Summary of Curve Sketching

1. f has constant negative slope. Matches (d)

2. The slope of f approaches ∞ as $x \to 0^-$, and approaches $-\infty$ as $x \to 0^+$. Matches (c)

3. $f'(x) = 0$ at $x \approx \pm\dfrac{1}{2}$

Matches graph (a)

4. The slope is positive up to approximately $x = 1.5$. Matches (b)

5. (a) $f'(x) = 0$ for $x = -2$ and $x = 2$.

f' is negative for $-2 < x < 2$ (decreasing function).

f' is positive for $x > 2$ and $x < -2$ (increasing function).

(b) $f''(x) = 0$ at $x = 0$ (Inflection point).

f'' is positive for $x > 0$ (Concave upward).

f'' is negative for $x < 0$ (Concave downward).

(c) f' is increasing on $(0, \infty)$. $(f'' > 0)$

(d) $f'(x)$ is minimum at $x = 0$. The rate of change of f at $x = 0$ is less than the rate of change of f for all other values of x.

6. (a) x_0, x_2, x_4

(b) x_2, x_3

(c) x_1

(d) x_1

(e) x_2, x_3

7. $y = \dfrac{x^2}{x^2 + 3}$

$$y' = \frac{6x}{(x^2 + 3)^2} = 0 \text{ when } x = 0.$$

$$y'' = \frac{18(1 - x^2)}{(x^2 + 3)^3} = 0 \text{ when } x = \pm 1.$$

Horizontal asymptote: $y = 1$

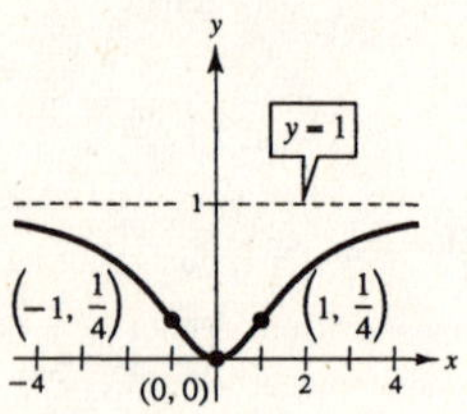

	y	y'	y''	Conclusion
$-\infty < x < -1$		$-$	$-$	Decreasing, concave down
$x = -1$	$\frac{1}{4}$	$-$	0	Point of inflection
$-1 < x < 0$		$-$	$+$	Decreasing, concave up
$x = 0$	0	0	$+$	Relative minimum
$0 < x < 1$		$+$	$+$	Increasing, concave up
$x = 1$	$\frac{1}{4}$	$+$	0	Point of inflection
$1 < x < \infty$		$+$	$-$	Increasing, concave down

8. $y = \dfrac{x}{x^2 + 1}$

$y' = \dfrac{1 - x^2}{(x^2 + 1)^2} = \dfrac{(1 - x)(x + 1)}{(x^2 + 1)^2} = 0$ when $x = \pm 1$.

$y'' = -\dfrac{2x(3 - x^2)}{(x^2 + 1)^3} = 0$ when $x = 0, \pm\sqrt{3}$.

Horizontal asymptote: $y = 0$

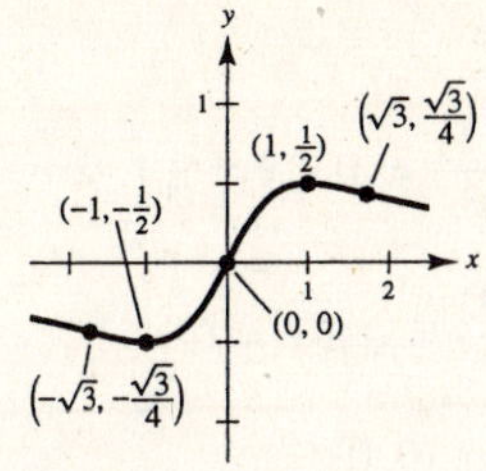

	y	y'	y''	Conclusion
$-\infty < x < -\sqrt{3}$		$-$	$-$	Decreasing, concave down
$x = -\sqrt{3}$	$-\dfrac{\sqrt{3}}{4}$	$-$	0	Point of inflection
$-\sqrt{3} < x < -1$		$-$	$+$	Decreasing, concave up
$x = -1$	$-\dfrac{1}{2}$	0	$+$	Relative minimum
$-1 < x < 0$		$+$	$+$	Increasing, concave up
$x = 0$	0	$+$	0	Point of inflection
$0 < x < 1$		$+$	$-$	Increasing, concave down
$x = 1$	$\dfrac{1}{2}$	0	$-$	Relative maximum
$1 < x < \sqrt{3}$		$-$	$-$	Decreasing, concave down
$x = \sqrt{3}$	$\dfrac{\sqrt{3}}{4}$	$-$	0	Point of inflection
$\sqrt{3} < x < \infty$		$-$	$+$	Decreasing, concave up

9. $y = \dfrac{1}{x - 2} - 3$

$y' = -\dfrac{1}{(x - 2)^2} < 0$ when $x \neq 2$.

$y'' = \dfrac{2}{(x - 2)^3}$

No relative extrema, no points of inflection

Intercepts: $\left(\dfrac{7}{3}, 0\right), \left(0, -\dfrac{7}{2}\right)$

Vertical asymptote: $x = 2$

Horizontal asymptote: $y = -3$

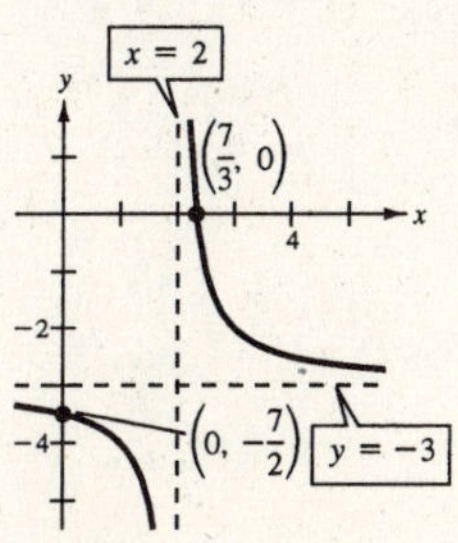

10. $y = \dfrac{x^2 + 1}{x^2 - 9}$

$y' = \dfrac{-20x}{(x^2 - 9)^2} = 0$ when $x = 0$.

$y'' = \dfrac{60(x^2 + 3)}{(x^2 - 9)^3} < 0$ when $x = 0$.

Therefore, $\left(0, -\dfrac{1}{9}\right)$ is a relative maximum.

Intercept: $\left(0, -\dfrac{1}{9}\right)$

Vertical asymptotes: $x = \pm 3$

Horizontal asymptote: $y = 1$

Symmetric about y-axis

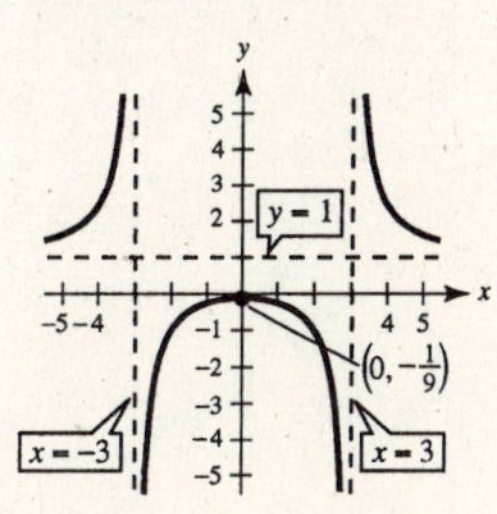

11. $y = \dfrac{2x}{x^2 - 1}$

$y' = \dfrac{-2(x^2 + 1)}{(x^2 - 1)^2} < 0$ if $x \neq \pm 1$.

$y'' = \dfrac{4x(x^2 + 3)}{(x^2 - 1)^3} = 0$ if $x = 0$.

Inflection point: $(0, 0)$

Intercept: $(0, 0)$

Vertical asymptote: $x = \pm 1$

Horizontal asymptote: $y = 0$

Symmetry with respect to the origin

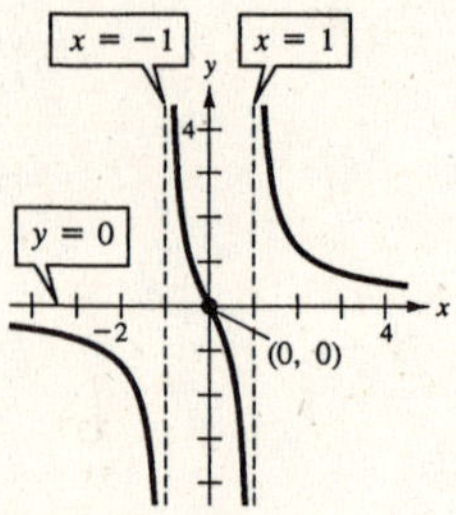

12. $f(x) = \dfrac{x + 2}{x} = 1 + \dfrac{2}{x}$

$f'(x) = \dfrac{-2}{x^2} < 0$ when $x \neq 0$.

$f''(x) = \dfrac{4}{x^3} \neq 0$

Intercept: $(-2, 0)$

Vertical asymptote: $x = 0$

Horizontal asymptote: $y = 1$

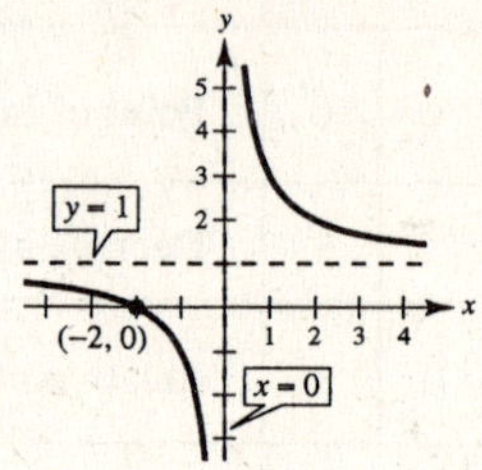

13. $g(x) = x + \dfrac{4}{x^2 + 1}$

$g'(x) = 1 - \dfrac{8x}{(x^2 + 1)^2} = \dfrac{x^4 + 2x^2 - 8x + 1}{(x^2 + 1)^2} = 0$ when $x \approx 0.1292, 1.6085$.

$g''(x) = \dfrac{8(3x^2 - 1)}{(x^2 + 1)^3} = 0$ when $x = \pm\dfrac{\sqrt{3}}{3}$.

$g''(0.1292) < 0$, therefore, $(0.1292, 4.064)$ is relative maximum.

$g''(1.6085) > 0$, therefore, $(1.6085, 2.724)$ is a relative minimum.

Points of inflection: $\left(-\dfrac{\sqrt{3}}{3}, 2.423\right), \left(\dfrac{\sqrt{3}}{3}, 3.577\right)$

Intercepts: $(0, 4)$, $(-1.3788, 0)$

Slant asymptote: $y = x$

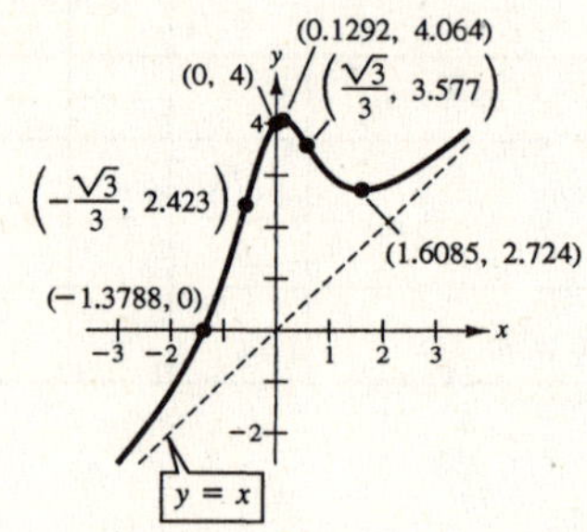

14. $f(x) = x + \dfrac{32}{x^2}$

$f'(x) = 1 - \dfrac{64}{x^3} = \dfrac{(x - 4)(x^2 + 4x + 16)}{x^3} = 0$ when $x = 4$.

$f''(x) = \dfrac{192}{x^4} > 0$ if $x \neq 0$.

Therefore, $(4, 6)$ is a relative minimum.

Intercept; $\left(-2\sqrt[3]{4}, 0\right)$

Vertical asymptote: $x = 0$

Slant asymptote: $y = x$

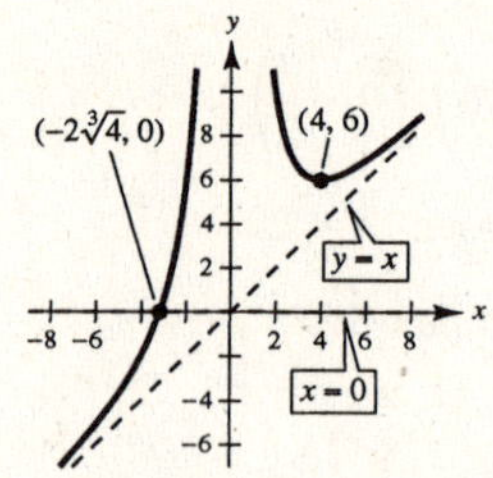

15. $f(x) = \dfrac{x^2 + 1}{x} = x + \dfrac{1}{x}$

$f'(x) = 1 - \dfrac{1}{x^2} = 0$ when $x = \pm 1$.

$f''(x) = \dfrac{2}{x^3} \neq 0$

Relative maximum: $(-1, -2)$

Relative minimum: $(1, 2)$

Vertical asymptote: $x = 0$

Slant asymptote: $y = x$

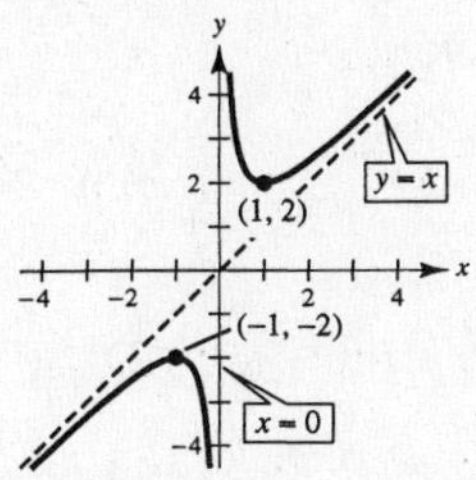

16. $f(x) = \dfrac{x^3}{x^2 - 4} = x + \dfrac{4x}{x^2 - 4}$

$f'(x) = \dfrac{x^2(x^2 - 12)}{(x^2 - 4)^2} = 0$ when $x = 0, \pm 2\sqrt{3}$.

$f''(x) = \dfrac{8x(x^2 + 12)}{(x^2 - 4)^3} = 0$ when $x = 0$.

Intercept: $(0, 0)$

Relative maximum: $\left(-2\sqrt{3}, -3\sqrt{3}\right)$

Relative minimum: $\left(2\sqrt{3}, 3\sqrt{3}\right)$

Inflection point: $(0, 0)$

Vertical asymptotes: $x = \pm 2$

Slant asymptote: $y = x$

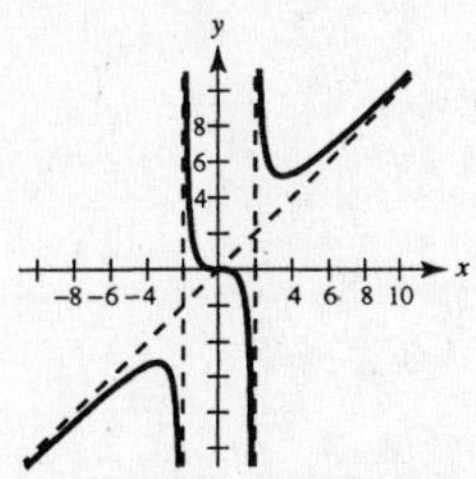

17. $y = \dfrac{x^2 - 6x + 12}{x - 4} = x - 2 + \dfrac{4}{x - 4}$

$y' = 1 - \dfrac{4}{(x - 4)^2}$

$= \dfrac{(x - 2)(x - 6)}{(x - 4)^2} = 0$ when $x = 2, 6$.

$y'' = \dfrac{8}{(x - 4)^3}$

$y'' < 0$ when $x = 2$.

Therefore, $(2, -2)$ is a relative maximum.

$y'' > 0$ when $x = 6$.

Therefore, $(6, 6)$ is a relative minimum.

Vertical asymptote: $x = 4$

Slant asymptote: $y = x - 2$

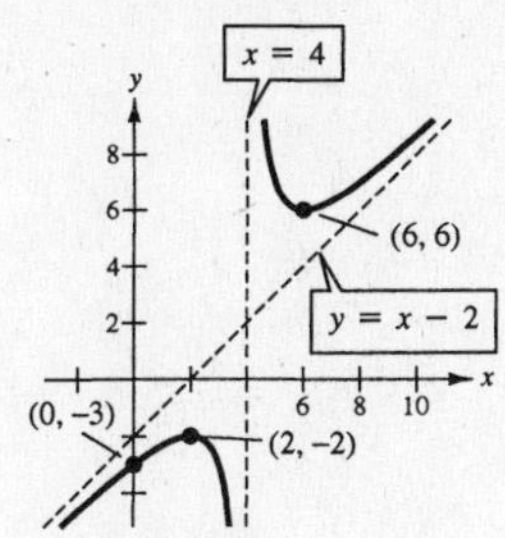

18. $y = \dfrac{2x^2 - 5x + 5}{x - 2} = 2x - 1 + \dfrac{3}{x - 2}$

$y' = 2 - \dfrac{3}{(x - 2)^2} = \dfrac{2x^2 - 8x + 5}{(x - 2)^2} = 0$ when $x = \dfrac{4 \pm \sqrt{6}}{2}$.

$y'' = \dfrac{6}{(x - 2)^3} \neq 0$

Relative maximum: $\left(\dfrac{4 - \sqrt{6}}{2}, -1.8990\right)$

Relative minimum: $\left(\dfrac{4 + \sqrt{6}}{2}, 7.8990\right)$

Intercept: $(0, -5/2)$

Vertical asymptote: $x = 2$

Slant asymptote: $y = 2x - 1$

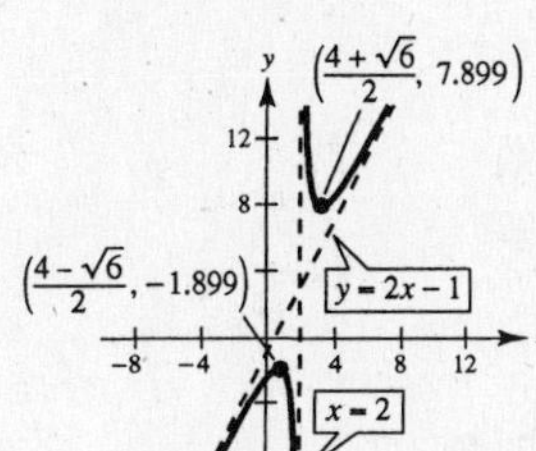

19. $y = x\sqrt{4-x}$

$y' = \dfrac{8-3x}{2\sqrt{4-x}}$

$y'' = \dfrac{3x-16}{4(4-x)^{3/2}}$

Domain: $x \le 4$

Intercepts: $(0, 0)$, $(4, 0)$

Relative maximum: $\left(\dfrac{8}{3}, \dfrac{16}{3\sqrt{3}}\right) \approx \left(\dfrac{8}{3}, 3.079\right)$

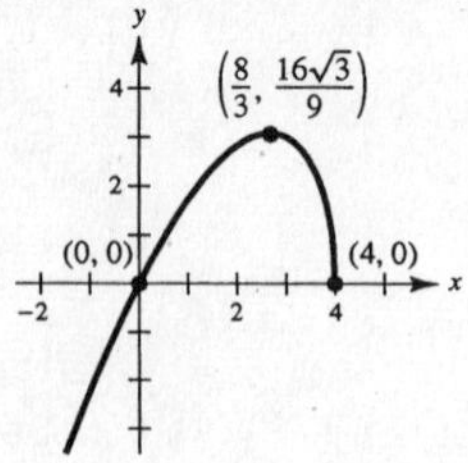

20. $g(x) = x\sqrt{9-x}$ Domain: $x \le 9$

$g'(x) = \dfrac{3(6-x)}{2\sqrt{9-x}} = 0$ when $x = 6$

$g''(x) = \dfrac{3(x-12)}{4(9-x)^{3/2}} < 0$ when $x = 6$

Relative maximum: $(6, 6\sqrt{3})$

Intercepts: $(0, 0)$, $(9, 0)$

Concave downward on $(-\infty, 9)$

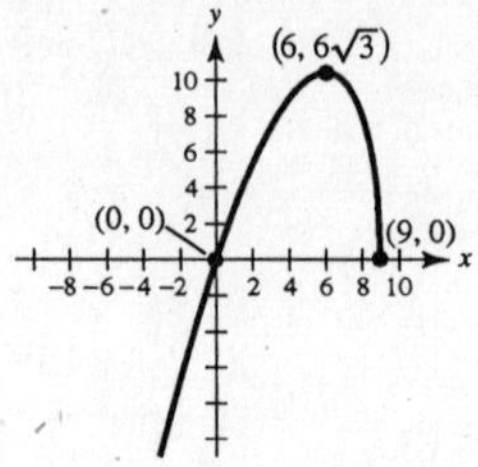

21. $h(x) = x\sqrt{9-x^2}$ Domain: $-3 \le x \le 3$

$h'(x) = \dfrac{9-2x^2}{\sqrt{9-x^2}} = 0$ when $x = \pm\dfrac{3}{\sqrt{2}} = \pm\dfrac{3\sqrt{2}}{2}$.

$h''(x) = \dfrac{x(2x^2-27)}{(9-x^2)^{3/2}} = 0$ when $x = 0$.

Relative maximum: $\left(\dfrac{3\sqrt{2}}{2}, \dfrac{9}{2}\right)$

Relative minimum: $\left(-\dfrac{3\sqrt{2}}{2}, -\dfrac{9}{2}\right)$

Intercepts: $(0, 0)$, $(\pm 3, 0)$

Symmetric with respect to the origin

Point of inflection: $(0, 0)$

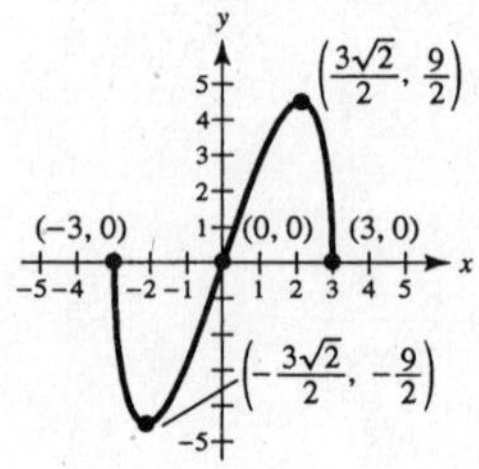

22. $y = x\sqrt{16-x^2}$ Domain: $-4 \le x \le 4$

$y' = \dfrac{2(8-x^2)}{\sqrt{16-x^2}} = 0$ when $x = \pm 2\sqrt{2}$.

$y'' = \dfrac{2x(x^2-24)}{(16-x^2)^{3/2}} = 0$ when $x = 0$.

Relative maximum: $(2\sqrt{2}, 8)$

Relative minimum: $(-2\sqrt{2}, -8)$

Intercepts: $(0, 0)$, $(\pm 4, 0)$

Symmetric with respect to the origin

Point of inflection: $(0, 0)$

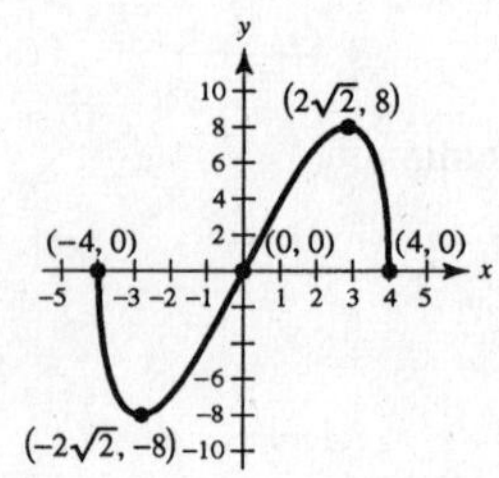

23. $y = 3x^{2/3} - 2x$

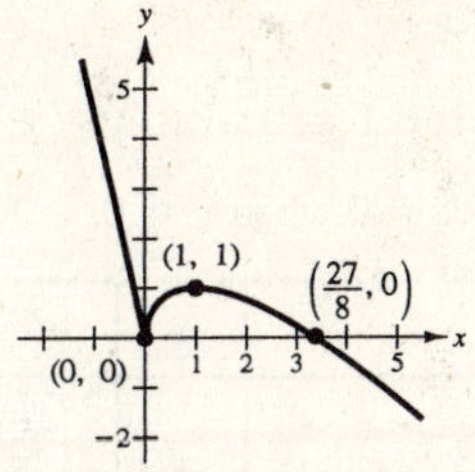

$y' = 2x^{-1/3} - 2 = \dfrac{2(1 - x^{1/3})}{x^{1/3}}$

$= 0$ when $x = 1$ and undefined when $x = 0$.

$y'' = \dfrac{-2}{3x^{4/3}} < 0$ when $x \neq 0$.

	y	y'	y''	Conclusion
$-\infty < x < 0$		$-$	$-$	Decreasing, concave down
$x = 0$	0	Undefined	Undefined	Relative minimum
$0 < x < 1$		$+$	$-$	Increasing, concave down
$x = 1$	1	0	$-$	Relative maximum
$1 < x < \infty$		$-$	$-$	Decreasing, concave down

24. $y = 3(x - 1)^{2/3} - (x - 1)^2$

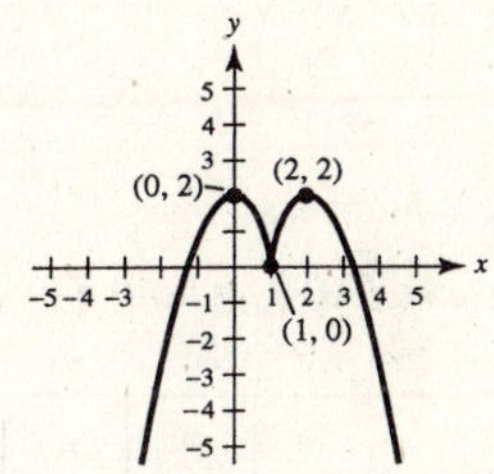

$y' = \dfrac{2}{(x-1)^{1/3}} - 2(x - 1) = \dfrac{2 - 2(x-1)^{4/3}}{(x-1)^{1/3}} = 0$ when $x = 0, 2$.

(y' undefined for $x = 1$)

$y'' = \dfrac{-2}{3(x-1)^{4/3}} - 2 < 0$ for all $x \neq 1$.

Concave downward on $(-\infty, 1)$ and $(1, \infty)$

Relative maximum: $(0, 2), (2, 2)$

Relative minimum: $(1, 0)$

Intercepts: $(0, 2), (1, 0), (-1.280, 0), (3.280, 0)$

25. $y = x^3 - 3x^2 + 3$

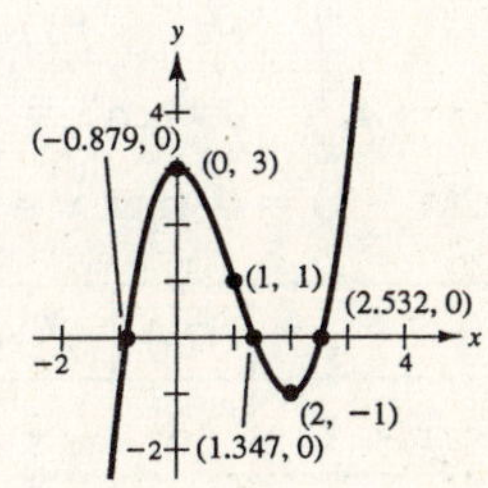

$y' = 3x^2 - 6x = 3x(x - 2) = 0$ when $x = 0, x = 2$.

$y'' = 6x - 6 = 6(x - 1) = 0$ when $x = 1$.

	y	y'	y''	Conclusion
$-\infty < x < 0$		$+$	$-$	Increasing, concave down
$x = 0$	3	0	$-$	Relative maximum
$0 < x < 1$		$-$	$-$	Decreasing, concave down
$x = 1$	1	$-$	0	Point of inflection
$1 < x < 2$		$-$	$+$	Decreasing, concave up
$x = 2$	-1	0	$+$	Relative minimum
$2 < x < \infty$		$+$	$+$	Increasing, concave up

26. $y = -\frac{1}{3}(x^3 - 3x + 2)$

$y' = -x^2 + 1 = 0$ when $x = \pm 1$.

$y'' = -2x = 0$ when $x = 0$.

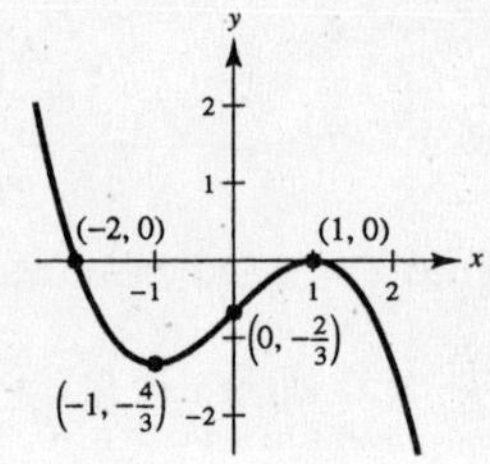

	y	y'	y''	Conclusion
$-\infty < x < -1$		$-$	$+$	Decreasing, concave up
$x = -1$	$-\frac{4}{3}$	0	$+$	Relative minimum
$-1 < x < 0$		$+$	$+$	Increasing, concave up
$x = 0$	$-\frac{2}{3}$	$+$	0	Point of inflection
$0 < x < 1$		$+$	$-$	Increasing, concave down
$x = 1$	0	0	$-$	Relative maximum
$1 < x < \infty$		$-$	$-$	Decreasing, concave down

27. $y = 2 - x - x^3$

$y' = -1 - 3x^2$

No critical numbers

$y'' = -6x = 0$ when $x = 0$.

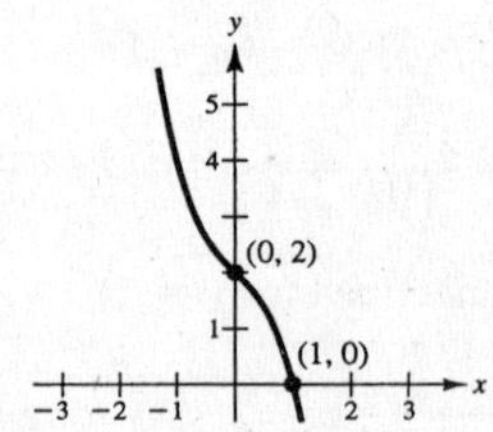

	y	y'	y''	Conclusion
$-\infty < x < 0$		$-$	$+$	Decreasing, concave up
$x = 0$	2	$-$	0	Point of inflection
$0 < x < \infty$		$-$	$-$	Decreasing, concave down

28. $f(x) = \frac{1}{3}(x - 1)^3 + 2$

$f'(x) = (x - 1)^2 = 0$ when $x = 1$.

$f''(x) = 2(x - 1) = 0$ when $x = 1$.

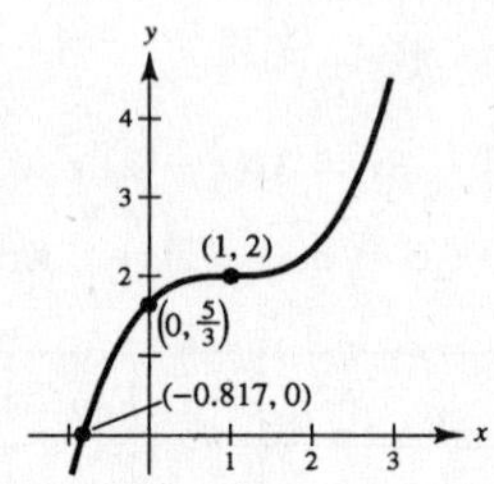

	$f(x)$	$f'(x)$	$f''(x)$	Conclusion
$-\infty < x < 1$		$+$	$-$	Increasing, concave down
$x = 1$	2	0	0	Point of inflection
$1 < x < \infty$		$+$	$+$	Increasing, concave up

29. $f(x) = 3x^3 - 9x + 1$

$f'(x) = 9x^2 - 9 = 9(x^2 - 1) = 0$ when $x = \pm 1$.

$f''(x) = 18x = 0$ when $x = 0$.

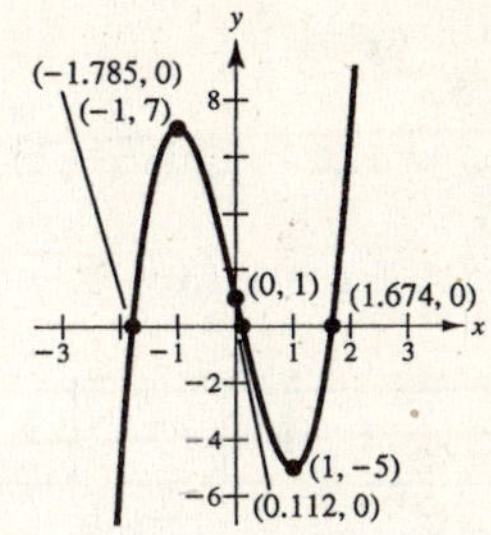

	$f(x)$	$f'(x)$	$f''(x)$	Conclusion
$-\infty < x < -1$		+	−	Increasing, concave down
$x = -1$	7	0	−	Relative maximum
$-1 < x < 0$		−	−	Decreasing, concave down
$x = 0$	1	−	0	Point of inflection
$0 < x < 1$		−	+	Decreasing, concave up
$x = 1$	−5	0	+	Relative minimum
$1 < x < \infty$		+	+	Increasing, concave up

30. $f(x) = (x + 1)(x - 2)(x - 5)$

$f'(x) = (x + 1)(x - 2) + (x + 1)(x - 5) + (x - 2)(x - 5)$

$= 3(x^2 - 4x + 1) = 0$ when $x = 2 \pm \sqrt{3}$.

$f''(x) = 6(x - 2) = 0$ when $x = 2$.

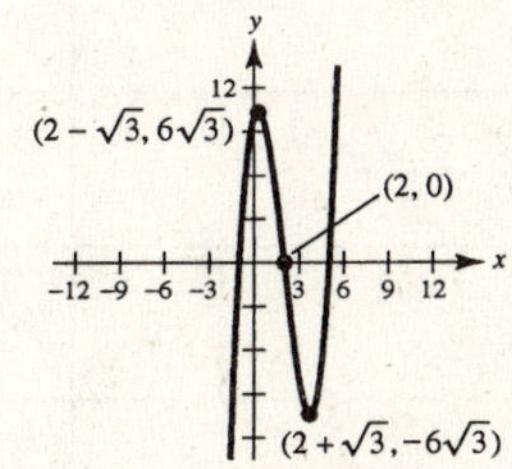

	$f(x)$	$f'(x)$	$f''(x)$	Conclusion
$-\infty < x < 2 - \sqrt{3}$		+	−	Increasing, concave down
$x = 2 - \sqrt{3}$	$6\sqrt{3}$	0	−	Relative maximum
$2 - \sqrt{3} < x < 2$		−	−	Decreasing, concave down
$x = 2$	0	−	0	Point of inflection
$2 < x < 2 + \sqrt{3}$		−	+	Decreasing, concave up
$x = 2 + \sqrt{3}$	$-6\sqrt{3}$	0	+	Relative minimum
$2 + \sqrt{3} < x < \infty$		+	+	Increasing, concave up

Intercepts: (0, 10), (−1, 0), (2, 0), (5, 0)

31. $y = 3x^4 + 4x^3$

$y' = 12x^3 + 12x^2 = 12x^2(x + 1) = 0$ when $x = 0, x = -1$.

$y'' = 36x^2 + 24x = 12x(3x + 2) = 0$ when $x = 0, x = -\frac{2}{3}$.

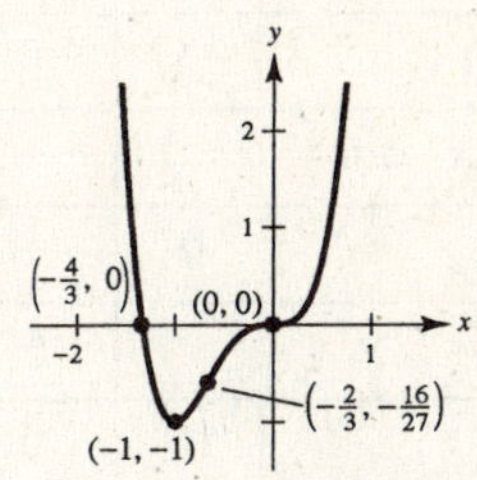

	y	y'	y''	Conclusion
$-\infty < x < -1$		−	+	Decreasing, concave up
$x = -1$	−1	0	+	Relative minimum
$-1 < x < -\frac{2}{3}$		+	+	Increasing, concave up
$x = -\frac{2}{3}$	$-\frac{16}{27}$	+	0	Point of inflection
$-\frac{2}{3} < x < 0$		+	−	Increasing, concave down
$x = 0$	0	0	0	Point of inflection
$0 < x < \infty$		+	+	Increasing, concave up

32. $y = 3x^4 - 6x^2 + \dfrac{5}{3}$

$y' = 12x^3 - 12x = 12x(x^2 - 1) = 0$ when $x = 0, x = \pm 1$.

$y'' = 36x^2 - 12 = 12(3x^2 - 1) = 0$ when $x = \pm\dfrac{\sqrt{3}}{3}$.

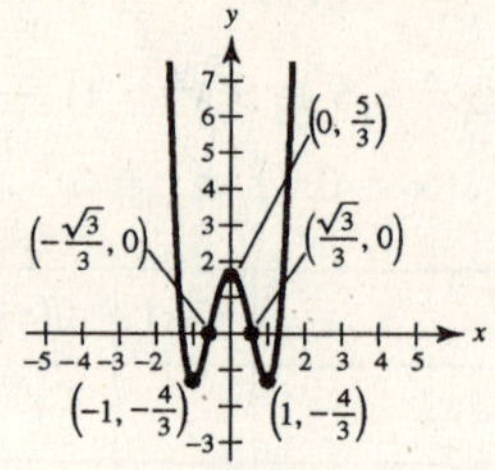

	y	y'	y''	Conclusion
$-\infty < x < -1$		$-$	$+$	Decreasing, concave up
$x = -1$	$-4/3$	0	$+$	Relative minimum
$-1 < x < -\dfrac{\sqrt{3}}{3}$		$+$	$+$	Increasing, concave up
$x = -\dfrac{\sqrt{3}}{3}$	0	$+$	0	Point of inflection
$-\dfrac{\sqrt{3}}{3} < x < 0$		$+$	$-$	Increasing, concave down
$x = 0$	$5/3$	0	$-$	Relative maximum
$0 < x < \dfrac{\sqrt{3}}{3}$		$-$	$-$	Decreasing, concave down
$x = \dfrac{\sqrt{3}}{3}$	0	$-$	0	Point of inflection
$\dfrac{\sqrt{3}}{3} < x < 1$		$-$	$+$	Decreasing, concave up
$x = 1$	$-4/3$	0	$+$	Relative minimum
$1 < x < \infty$		$+$	$+$	Increasing, concave up

33. $f(x) = x^4 - 4x^3 + 16x$

$f'(x) = 4x^3 - 12x^2 + 16 = 4(x + 1)(x - 2)^2 = 0$ when $x = -1, x = 2$.

$f''(x) = 12x^2 - 24x = 12x(x - 2) = 0$ when $x = 0, x = 2$.

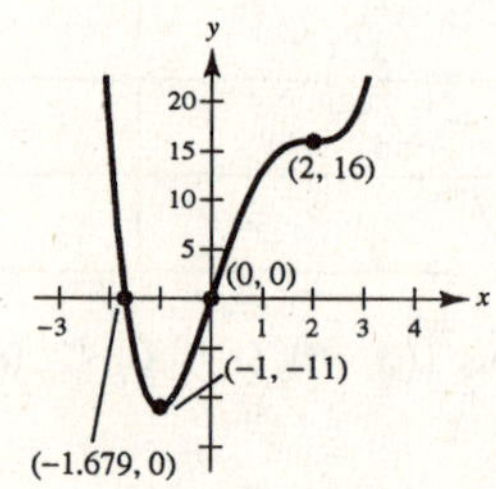

	$f(x)$	$f'(x)$	$f''(x)$	Conclusion
$-\infty < x < -1$		$-$	$+$	Decreasing, concave up
$x = -1$	-11	0	$+$	Relative minimum
$-1 < x < 0$		$+$	$+$	Increasing, concave up
$x = 0$	0	$+$	0	Point of inflection
$0 < x < 2$		$+$	$-$	Increasing, concave down
$x = 2$	16	0	0	Point of inflection
$2 < x < \infty$		$+$	$+$	Increasing, concave up

34. $f(x) = x^4 - 8x^3 + 18x^2 - 16x + 5$

$f'(x) = 4x^3 - 24x^2 + 36x - 16 = 4(x - 4)(x - 1)^2 = 0$ when $x = 1, x = 4$.

$f''(x) = 12x^2 - 48x + 36 = 12(x - 3)(x - 1) = 0$ when $x = 3, x = 1$.

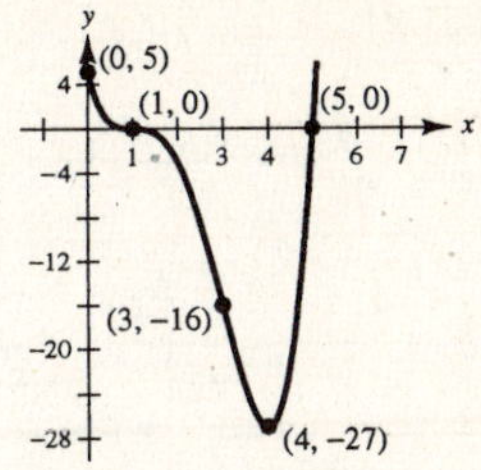

	$f(x)$	$f'(x)$	$f''(x)$	Conclusion
$-\infty < x < 1$		$-$	$+$	Decreasing, concave up
$x = 1$	0	0	0	Point of inflection
$1 < x < 3$		$-$	$-$	Decreasing, concave down
$x = 3$	-16	$-$	0	Point of inflection
$3 < x < 4$		$-$	$+$	Decreasing, concave up
$x = 4$	-27	0	$+$	Relative minimum
$4 < x < \infty$		$+$	$+$	Increasing, concave up

35. $y = x^5 - 5x$

$y' = 5x^4 - 5 = 5(x^4 - 1) = 0$ when $x = \pm 1$.

$y'' = 20x^3 = 0$ when $x = 0$.

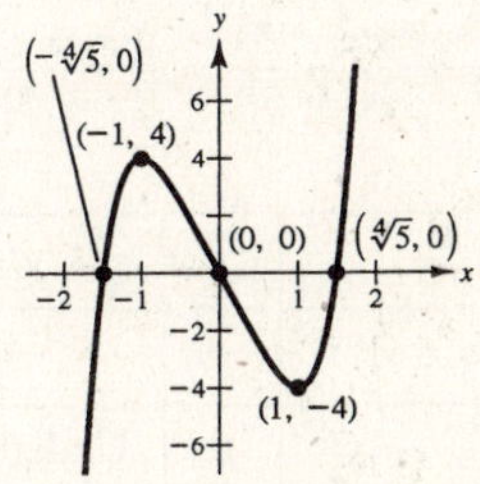

	y	y'	y''	Conclusion
$-\infty < x < -1$		$+$	$-$	Increasing, concave down
$x = -1$	4	0	$-$	Relative maximum
$-1 < x < 0$		$-$	$-$	Decreasing, concave down
$x = 0$	0	$-$	0	Point of inflection
$0 < x < 1$		$-$	$+$	Decreasing, concave up
$x = 1$	-4	0	$+$	Relative minimum
$1 < x < \infty$		$+$	$+$	Increasing, concave up

36. $y = (x - 1)^5$

$y' = 5(x - 1)^4 = 0$ when $x = 1$.

$y'' = 20(x - 1)^3 = 0$ when $x = 1$.

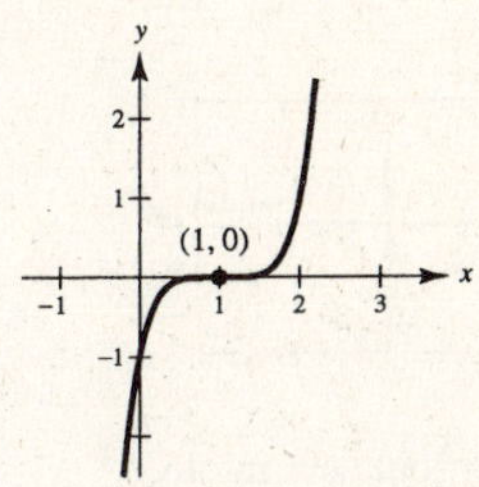

	y	y'	y''	Conclusion
$-\infty < x < 1$		$+$	$-$	Increasing, concave down
$x = 1$	0	0	0	Point of inflection
$1 < x < \infty$		$+$	$+$	Increasing, concave up

37. $y = |2x - 3|$

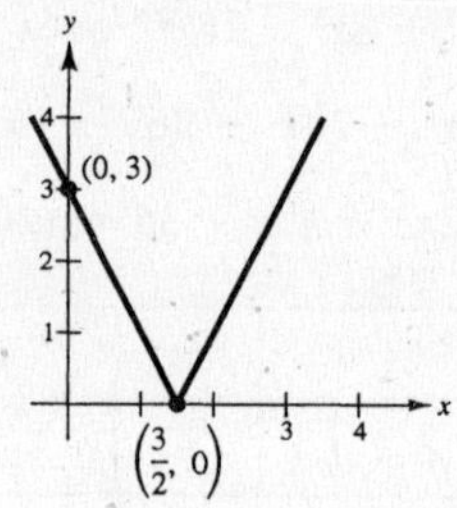

$y' = \dfrac{2(2x - 3)}{|2x - 3|}$ undefined at $x = \dfrac{3}{2}$.

$y'' = 0$

	y	y'	Conclusion
$-\infty < x < \frac{3}{2}$		$-$	Decreasing
$x = \frac{3}{2}$	0	Undefined	Relative minimum
$\frac{3}{2} < x < \infty$		$+$	Increasing

38. $y = |x^2 - 6x + 5|$

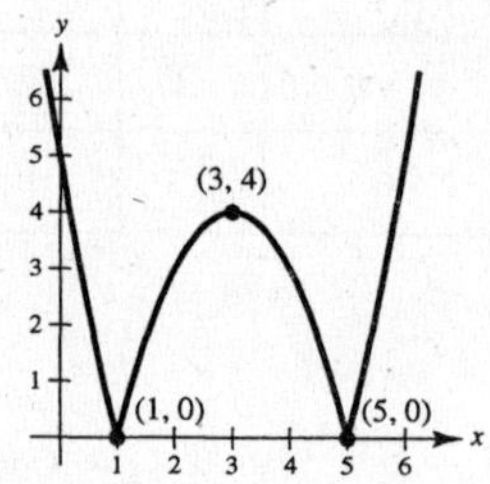

$y' = \dfrac{2(x - 3)(x^2 - 6x + 5)}{|x^2 - 6x + 5|} = \dfrac{2(x - 3)(x - 5)(x - 1)}{|(x - 5)(x - 1)|}$

$= 0$ when $x = 3$ and undefined when $x = 1, x = 5$.

$y'' = \dfrac{2(x^2 - 6x + 5)}{|x^2 - 6x + 5|} = \dfrac{2(x - 5)(x - 1)}{|(x - 5)(x - 1)|}$ undefined when $x = 1, x = 5$.

	y	y'	y''	Conclusion
$-\infty < x < 1$		$-$	$+$	Decreasing, concave up
$x = 1$	0	Undefined	Undefined	Relative minimum, point of inflection
$1 < x < 3$		$+$	$-$	Increasing, concave down
$x = 3$	4	0	$-$	Relative maximum
$3 < x < 5$		$-$	$-$	Decreasing, concave down
$x = 5$	0	Undefined	Undefined	Relative minimum, point of inflection
$5 < x < \infty$		$+$	$+$	Increasing, concave up

39. $f(x) = \dfrac{20x}{x^2 + 1} - \dfrac{1}{x} = \dfrac{19x^2 - 1}{x(x^2 + 1)}$

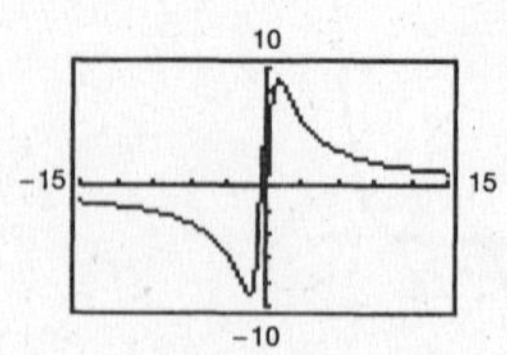

$x = 0$ vertical asymptote

$y = 0$ horizontal asymptote

Minimum: $(-1.10, -9.05)$

Maximum: $(1.10, 9.05)$

Points of inflection: $(-1.84, -7.86)$, $(1.84, 7.86)$

40. $f(x) = 5\left(\dfrac{1}{x - 4} - \dfrac{1}{x + 2}\right)$

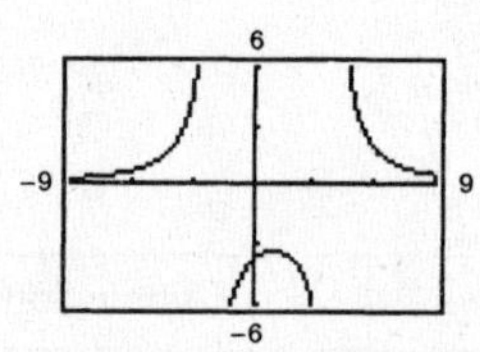

$x = -2, 4$ vertical asymptote

$y = 0$ horizontal asymptote

41. $y = \dfrac{x}{\sqrt{x^2 + 7}}$

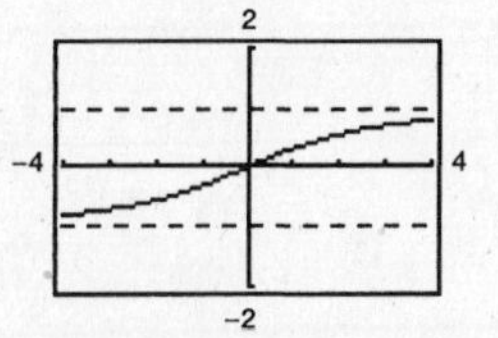

(0, 0) point of inflection

$y = \pm 1$ horizontal asymptotes

42. $f(x) = \dfrac{4x}{\sqrt{x^2 + 15}}$

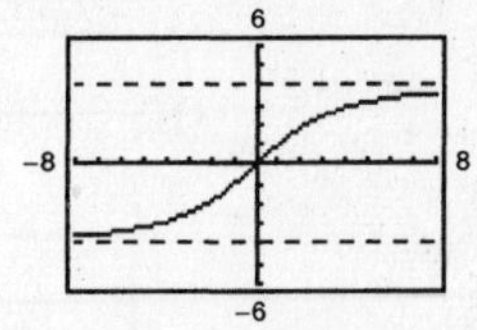

$y = \pm 4$ horizontal asymptotes

(0, 0) point of inflection

43. f is cubic.

f' is quadratic.

f'' is linear.

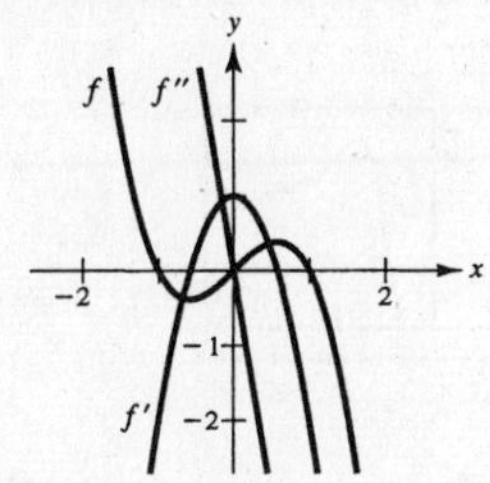

44. f'' is constant.

f' is linear.

f is quadratic.

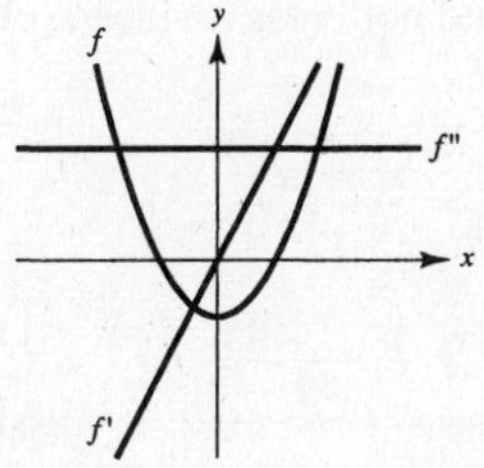

45.

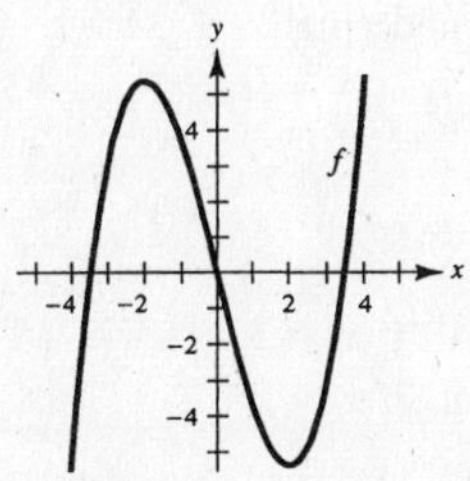

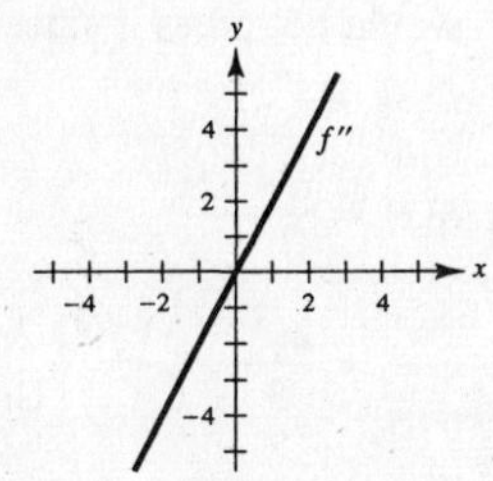

(any vertical translate of f will do)

46.

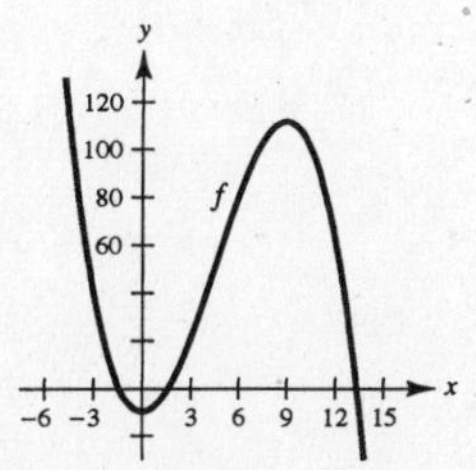

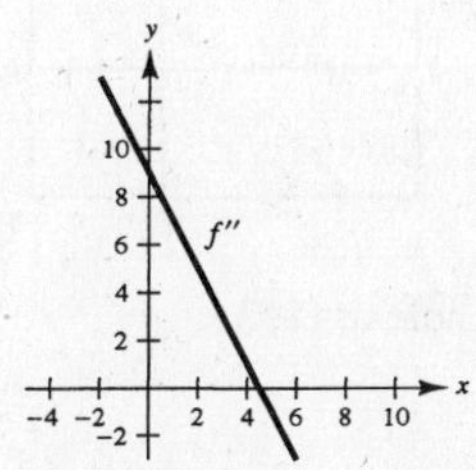

(any vertical translate of f will do)

47.

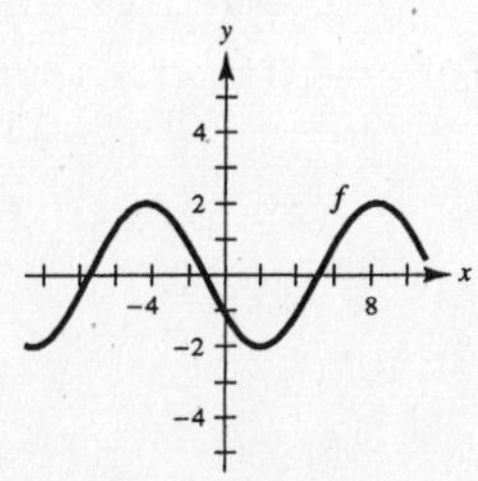

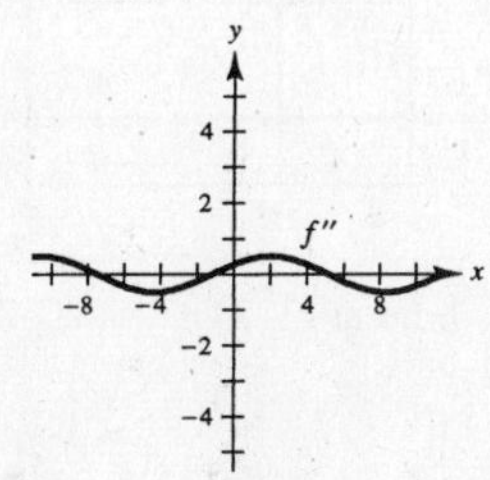

(any vertical translate of f will do)

48.

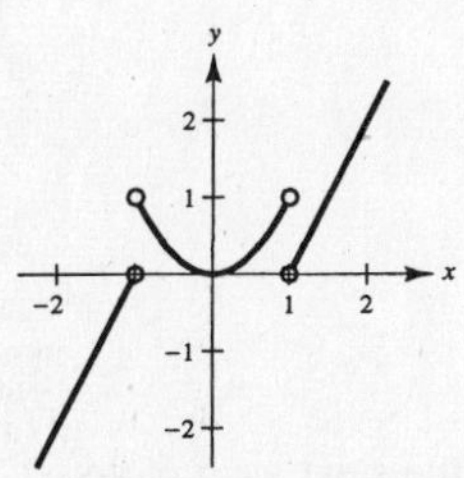

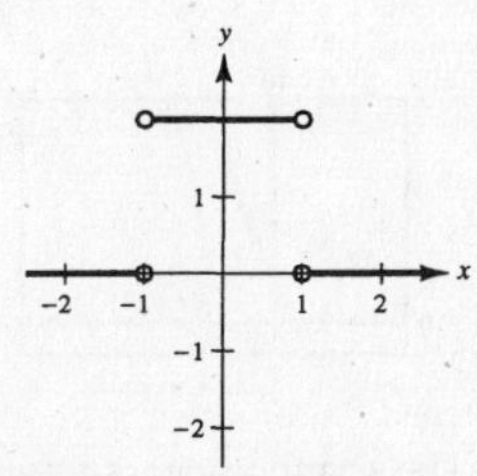

(any vertical translate of the 3 segments of f will do)

49. Since the slope is negative, the function is decreasing on $(2, 8)$, and hence $f(3) > f(5)$.

50. If $f'(x) = 2$ in $[-5, 5]$, then $f(x) = 2x + 3$ and $f(2) = 7$ is the least possible value of $f(2)$. If $f'(x) = 4$ in $[-5, 5]$, then $f(x) = 4x + 3$ and $f(2) = 11$ is the greatest possible value of $f(2)$.

51. $f(x) = \dfrac{4(x-1)^2}{x^2 - 4x + 5}$

Vertical asymptote: none

Horizontal asymptote: $y = 4$

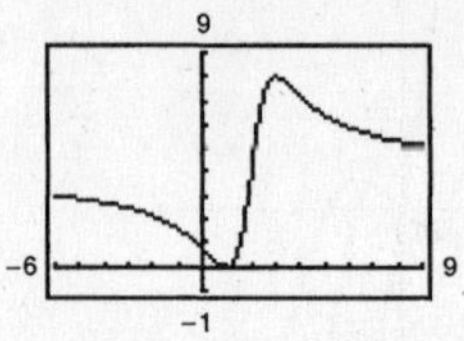

The graph crosses the horizontal asymptote $y = 4$. If a function has a vertical asymptote at $x = c$, the graph would not cross it since $f(c)$ is undefined.

52. $g(x) = \dfrac{3x^4 - 5x + 3}{x^4 + 1}$

Vertical asymptote: none

Horizontal asymptote: $y = 3$

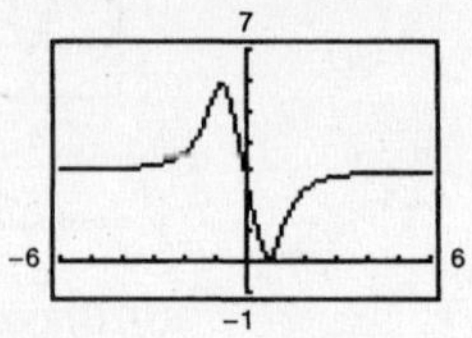

The graph crosses the horizontal asymptote $y = 3$. If a function has a vertical asymptote at $x = c$, the graph would not cross it since $f(c)$ is undefined.

53. $h(x) = \dfrac{6 - 2x}{3 - x}$

$$= \frac{2(3-x)}{3-x} = \begin{cases} 2, & \text{if } x \neq 3 \\ \text{Undefined}, & \text{if } x = 3 \end{cases}$$

The rational function is not reduced to lowest terms.

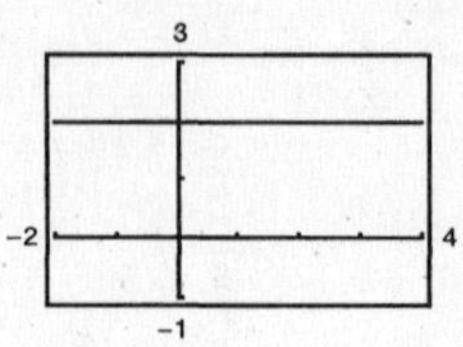

hole at $(3, 2)$

54. $g(x) = \dfrac{x^2 + x - 2}{x - 1}$

$$= \frac{(x+2)(x-1)}{x-1} = \begin{cases} x + 2, & \text{if } x \neq 1 \\ \text{Undefined}, & \text{if } x = 1 \end{cases}$$

The rational function is not reduced to lowest terms.

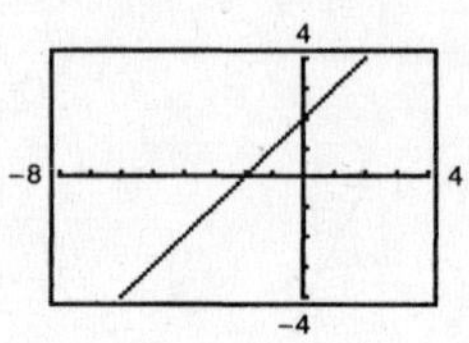

hole at $(1, 3)$

55. $f(x) = -\dfrac{x^2 - 3x - 1}{x - 2} = -x + 1 + \dfrac{3}{x - 2}$

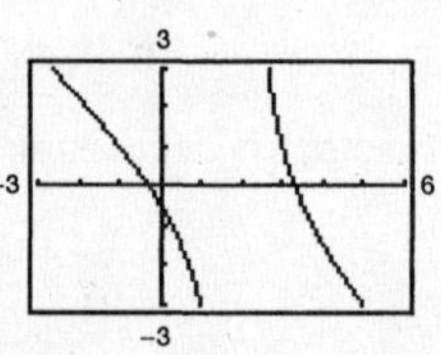

The graph appears to approach the slant asymptote $y = -x + 1$.

56. $g(x) = \dfrac{2x^2 - 8x - 15}{x - 5}$

$$= 2x + 2 - \frac{5}{x - 5}$$

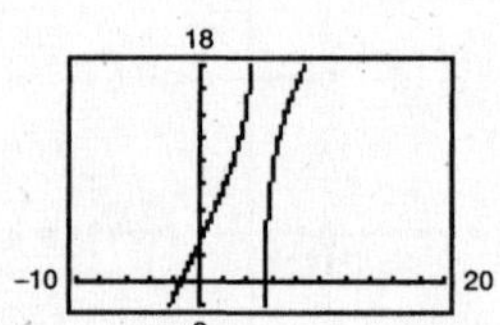

The graph appears to approach the slant asymptote $y = 2x + 2$.

57. $f(x) = \dfrac{x^3}{x^2 + 1} = x - \dfrac{x}{x^2 + 1}$

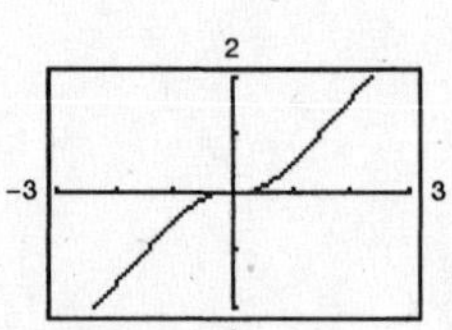

The graph appears to approach the slant asymptote $y = x$.

58. $h(x) = \dfrac{-x^3 + x^2 + 4}{x^2} = -x + 1 + \dfrac{4}{x^2}$

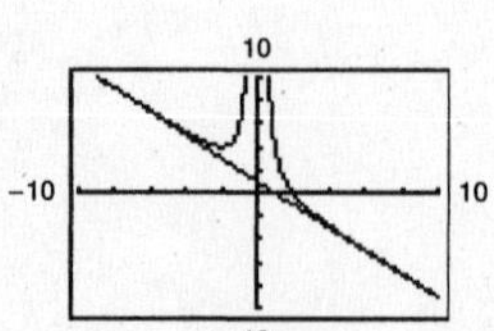

The graph appears to approach the slant asymptote $y = -x + 1$.

59. Vertical asymptote: $x = 5$

Horizontal asymptote: $y = 0$

$$y = \frac{1}{x - 5}$$

60. Vertical asymptote: $x = -3$

Horizontal asymptote: none

$$y = \frac{x^2}{x + 3}$$

61. Vertical asymptote: $x = 5$

Slant asymptote: $y = 3x + 2$

$$y = 3x + 2 + \frac{1}{x - 5} = \frac{3x^2 - 13x - 9}{x - 5}$$

62. Vertical asymptote: $x = 0$

Slant asymptote: $y = -x$

$$y = -x + \frac{1}{x} = \frac{1 - x^2}{x}$$

63. $f(x) = \dfrac{ax}{(x - b)^2}$

(a) The graph has a vertical asymptote at $x = b$. If $a > 0$, the graph approaches ∞ as $x \to b$. If $a < 0$, the graph approaches $-\infty$ as $x \to b$. The graph approaches its vertical asymptote faster as $|a| \to 0$.

(b) As b varies, the position of the vertical asymptote changes: $x = b$. Also, the coordinates of the minimum $(a > 0)$ or maximum $(a < 0)$ are changed.

64. $f(x) = \frac{1}{2}(ax)^2 - (ax) = \frac{1}{2}(ax)(ax - 2), a \neq 0$

$f'(x) = a^2x - a = a(ax - 1) = 0$ when $x = \dfrac{1}{a}$.

$f''(x) = a^2 > 0$ for all x.

(a) Intercepts: $(0, 0), \left(\frac{2}{a}, 0\right)$

Relative minimum: $\left(\frac{1}{a}, -\frac{1}{2}\right)$

Points of inflection: none

(b)

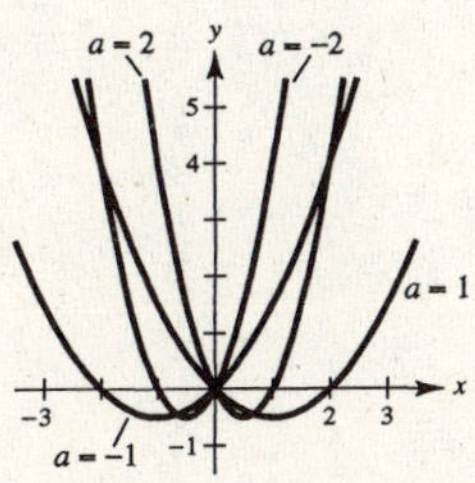

65. $f(x) = \dfrac{3x^n}{x^4 + 1}$

(a) For n even, f is symmetric about the y-axis. For n odd, f is symmetric about the origin.

(b) The x-axis will be the horizontal asymptote if the degree of the numerator is less than 4. That is, $n = 0, 1, 2, 3$.

(c) $n = 4$ gives $y = 3$ as the horizontal asymptote.

(d) There is a slant asymptote $y = 3x$ if $n = 5$:

$$\frac{3x^5}{x^4 + 1} = 3x - \frac{3x}{x^4 + 1}.$$

(e)

n	0	1	2	3	4	5
M	1	2	3	2	1	0
N	2	3	4	5	2	3

66. Tangent line at P: $y - y_0 = f'(x_0)(x - x_0)$

(a) Let $y = 0$: $-y_0 = f'(x_0)(x - x_0)$

$$f'(x_0)x = x_0 f'(x_0) - y_0$$

$$x = x_0 - \frac{y_0}{f'(x_0)} = x_0 - \frac{f(x_0)}{f'(x_0)}$$

x-intercept: $\left(x_0 - \dfrac{f(x_0)}{f'(x_0)}, 0\right)$

(b) Let $x = 0$: $y - y_0 = f'(x_0)(-x_0)$

$$y = y_0 - x_0 f'(x_0)$$

$$y = f(x_0) - x_0 f'(x_0)$$

y-intercept: $(0, f(x_0) - x_0 f'(x_0))$

—CONTINUED—

66. —CONTINUED—

(c) Normal line: $y - y_0 = -\frac{1}{f'(x_0)}(x - x_0)$

Let $y = 0$: $-y_0 = -\frac{1}{f'(x_0)}(x - x_0)$

$-y_0 f'(x_0) = -x + x_0$

$x = x_0 + y_0 f'(x_0) = x_0 + f(x_0)f'(x_0)$

x-intercept: $(x_0 + f(x_0)f'(x_0), 0)$

(d) Let $x = 0$: $y - y_0 = \frac{-1}{f'(x_0)}(-x_0)$

$y = y_0 + \frac{x_0}{f'(x_0)}$

y-intercept: $\left(0, y_0 + \frac{x_0}{f'(x_0)}\right)$

(e) $|BC| = \left|x_0 - \frac{f(x_0)}{f'(x_0)} - x_0\right| = \left|\frac{f(x_0)}{f'(x_0)}\right|$

(f) $|PC|^2 = y_0^2 + \left(\frac{f(x_0)}{f'(x_0)}\right)^2 = \frac{f(x_0)^2 f'(x_0)^2 + f(x_0)^2}{f'(x_0)^2}$

$|PC| = \left|\frac{f(x_0)\sqrt{1 + [f'(x_0)]^2}}{f'(x_0)}\right|$

(g) $|AB| = |x_0 - (x_0 + f(x_0)f'(x_0))| = |f(x_0)f'(x_0)|$

(h) $|AP|^2 = f(x_0)^2 f'(x_0)^2 + y_0^2$

$|AP| = |f(x_0)|\sqrt{1 + [f'(x_0)]^2}$

67. (a)

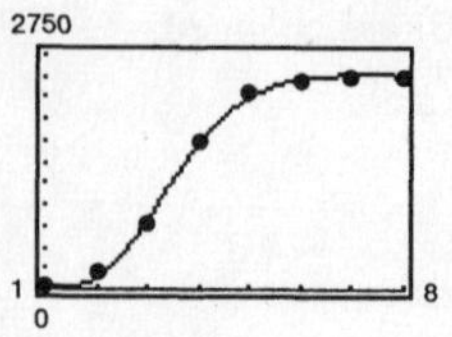

(b) When $t = 10$, $N(10) \approx 2434$ bacteria.

(c) N is a maximum when $t \approx 7.2$ (seventh day).

(d) $N''(t) = 0$ for $t \approx 3.2$

(e) $\lim_{t\to\infty} N(t) = \frac{13{,}250}{7} \approx 1892.86$

68. $y = \sqrt{4 + 16x^2}$

As $x \to \infty$, $y \to 4x$. As $x \to -\infty$, $y \to -4x$.

Slant asymptotes: $y = \pm 4x$

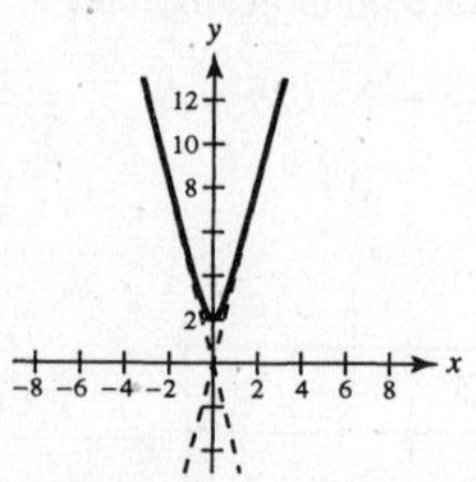

69. $y = \sqrt{x^2 + 6x} = \sqrt{(x + 3)^2 - 9}$

$y \to x + 3$ as $x \to \infty$, and $y \to -x - 3$ as $x \to -\infty$.

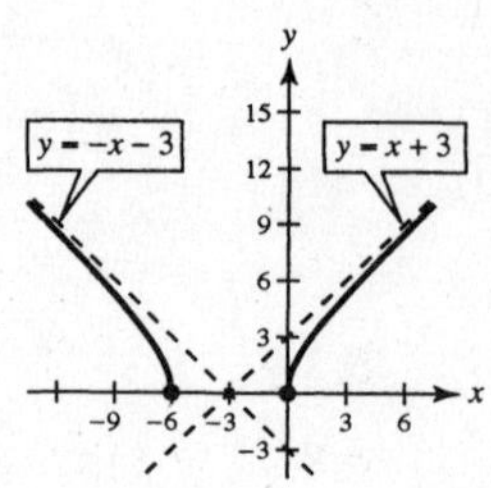

Section 5.7 Optimization Problems

1. (a)

First Number, x	Second Number	Product, P
10	110 − 10	10(110 − 10) = 1000
20	110 − 20	20(110 − 20) = 1800
30	110 − 30	30(110 − 30) = 2400
40	110 − 40	40(110 − 40) = 2800
50	110 − 50	50(110 − 50) = 3000
60	110 − 60	60(110 − 60) = 3000

—CONTINUED—

1. —CONTINUED—

(b)

First Number, x	Second Number	Product, P
10	$110 - 10$	$10(110 - 10) = 1000$
20	$110 - 20$	$20(110 - 20) = 1800$
30	$110 - 30$	$30(110 - 30) = 2400$
40	$110 - 40$	$40(110 - 40) = 2800$
50	$110 - 50$	$50(110 - 50) = 3000$
60	$110 - 60$	$60(110 - 60) = 3000$
70	$110 - 70$	$70(110 - 70) = 2800$
80	$110 - 80$	$80(110 - 80) = 2400$
90	$110 - 90$	$90(110 - 90) = 1800$
100	$110 - 100$	$100(110 - 100) = 1000$

The maximum is attained near $x = 50$ and 60.

(c) $P = x(110 - x) = 110x - x^2$

(d)

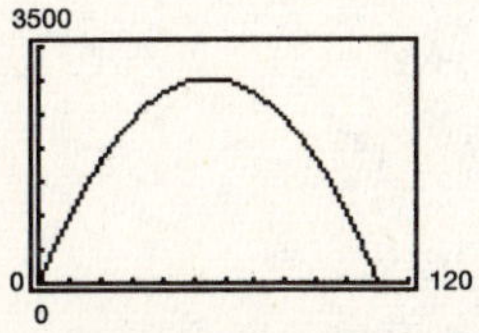

The solution appears to be $x = 55$.

(e) $\frac{dP}{dx} = 110 - 2x = 0$ when $x = 55$.

$\frac{d^2P}{dx^2} = -2 < 0$

P is a maximum when $x = 110 - x = 55$.

The two numbers are 55 and 55.

2. (a)

Height, x	Length & Width	Volume
1	$24 - 2(1)$	$1[24 - 2(1)]^2 = 484$
2	$24 - 2(2)$	$2[24 - 2(2)]^2 = 800$
3	$24 - 2(3)$	$3[24 - 2(3)]^2 = 972$
4	$24 - 2(4)$	$4[24 - 2(4)]^2 = 1024$
5	$24 - 2(5)$	$5[24 - 2(5)]^2 = 980$
6	$24 - 2(6)$	$6[24 - 2(6)]^2 = 864$

(b) $V = x(24 - 2x)^2, 0 < x < 12$

(c) $\frac{dV}{dx} = 2x(24 - 2x)(-2) + (24 - 2x)^2 = (24 - 2x)(24 - 6x)$

$= 12(12 - x)(4 - x) = 0$ when $x = 12, 4$ (12 is not in the domain).

$\frac{d^2V}{dx^2} = 12(2x - 16)$

$\frac{d^2V}{dx^2} < 0$ when $x = 4$.

When $x = 4$, $V = 1024$ is maximum.

(d)

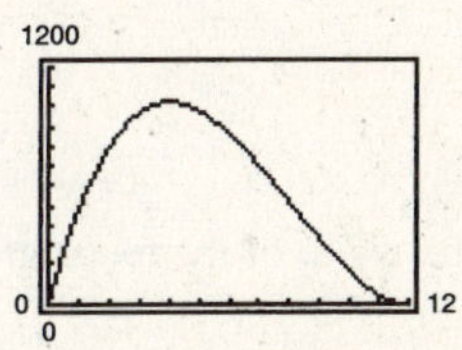

The maximum volume seems to be 1024.

3. Let x and y be two positive numbers such that $x + y = S$.

$$P = xy = x(S - x) = Sx - x^2$$

$$\frac{dP}{dx} = S - 2x = 0 \text{ when } x = \frac{S}{2}.$$

$$\frac{d^2P}{dx^2} = -2 < 0 \text{ when } x = \frac{S}{2}.$$

P is a maximum when $x = y = S/2$.

4. Let x and y be two positive numbers such that $xy = 192$.

$$S = x + y = x + \frac{192}{x}$$

$$\frac{dS}{dx} = 1 - \frac{192}{x^2} = 0 \text{ when } x = \sqrt{192}.$$

$$\frac{d^2S}{dx^2} = \frac{384}{x^3} > 0 \text{ when } x = \sqrt{192}.$$

S is a minimum when $x = y = \sqrt{192}$.

5. Let x and y be two positive numbers such that $xy = 192$.

$$S = x + 3y = \frac{192}{y} + 3y$$

$$\frac{dS}{dy} = 3 - \frac{192}{y^2} = 0 \text{ when } y = 8.$$

$$\frac{d^2S}{dy^2} = \frac{384}{y^3} > 0 \text{ when } y = 8.$$

S is minimum when $y = 8$ and $x = 24$.

6. Let x be a positive number.

$$S = x + \frac{1}{x}$$

$$\frac{dS}{dx} = 1 - \frac{1}{x^2} = 0 \text{ when } x = 1.$$

$$\frac{d^2S}{dx^2} = \frac{2}{x^3} > 0 \text{ when } x = 1.$$

The sum is a minimum when $x = 1$ and $1/x = 1$.

7. Let x and y be two positive numbers such that $x + 2y = 100$.

$$P = xy = y(100 - 2y) = 100y - 2y^2$$

$$\frac{dP}{dy} = 100 - 4y = 0 \text{ when } y = 25.$$

$$\frac{d^2P}{dy^2} = -4 < 0 \text{ when } y = 25.$$

P is a maximum when $x = 50$ and $y = 25$.

8. $x^2 + y = 27 \Rightarrow y = 27 - x^2$

Maximize: $f = xy = x(27 - x^2)$

$$f(x) = 27x - x^3$$

$$f'(x) = 27 - 3x^2 = 0$$

$$x = 3$$

$$f''(x) = -6x$$

$$f''(3) = -18 \quad \text{Maximum}$$

$x = 3, y = 18$

9. Let x be the length and y the width of the rectangle.

$$2x + 2y = 100$$

$$y = 50 - x$$

$$A = xy = x(50 - x)$$

$$\frac{dA}{dx} = 50 - 2x = 0 \text{ when } x = 25.$$

$$\frac{d^2A}{dx^2} = -2 < 0 \text{ when } x = 25.$$

A is maximum when $x = y = 25$ meters.

10. Let x be the length and y the width of the rectangle.

$$2x + 2y = P$$

$$y = \frac{P - 2x}{2} = \frac{P}{2} - x$$

$$A = xy = x\left(\frac{P}{2} - x\right) = \frac{P}{2}x - x^2$$

$$\frac{dA}{dx} = \frac{P}{2} - 2x = 0 \text{ when } x = \frac{P}{4}.$$

$$\frac{d^2A}{dx^2} = -2 < 0 \text{ when } x = \frac{P}{4}.$$

A is maximum when $x = y = P/4$ units. (A square!)

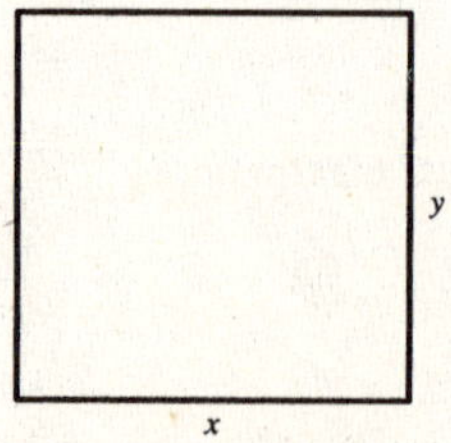

11. Let x be the length and y the width of the rectangle.

$$xy = 64$$

$$y = \frac{64}{x}$$

$$P = 2x + 2y = 2x + 2\left(\frac{64}{x}\right) = 2x + \frac{128}{x}$$

$$\frac{dP}{dx} = 2 - \frac{128}{x^2} = 0 \text{ when } x = 8.$$

$$\frac{d^2P}{dx^2} = \frac{256}{x^3} > 0 \text{ when } x = 8.$$

P is minimum when $x = y = 8$ feet.

12. Let x be the length and y the width of the rectangle.

$$xy = A$$

$$y = \frac{A}{x}$$

$$P = 2x + 2y = 2x + 2\left(\frac{A}{x}\right) = 2x + \frac{2A}{x}$$

$$\frac{dP}{dx} = 2 - \frac{2A}{x^2} = 0 \text{ when } x = \sqrt{A}.$$

$$\frac{d^2P}{dx^2} = \frac{4A}{x^3} > 0 \text{ when } x = \sqrt{A}.$$

P is minimum when $x = y = \sqrt{A}$ centimeters. (A square!)

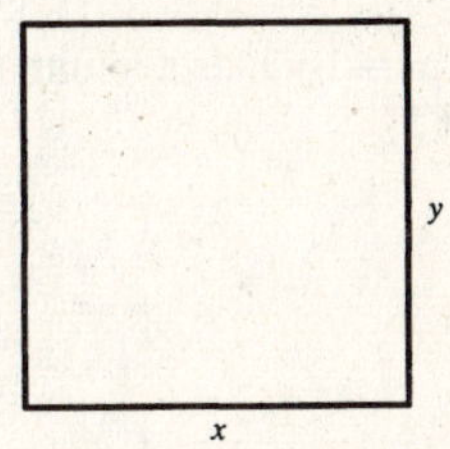

13. $d = \sqrt{(x - 4)^2 + \left(\sqrt{x} - 0\right)^2}$

$= \sqrt{x^2 - 7x + 16}$

Since d is smallest when the expression inside the radical is smallest, you need only find the critical numbers of

$$f(x) = x^2 - 7x + 16.$$

$$f'(x) = 2x - 7 = 0$$

$$x = \tfrac{7}{2}$$

By the First Derivative Test, the point nearest to $(4, 0)$ is $\left(7/2, \sqrt{7/2}\right)$.

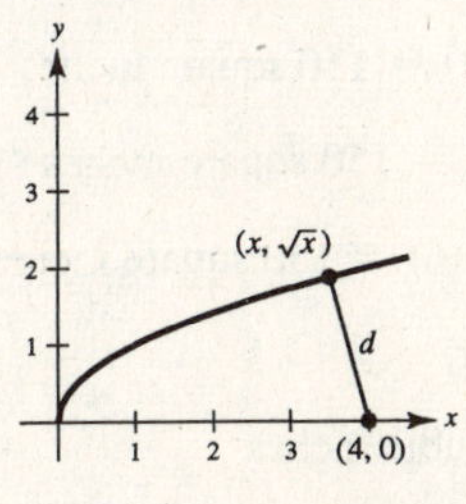

14. $f(x) = \sqrt{x - 8}$, $(2, 0)$

From the graph, it is clear that $(8, 0)$ is the closest point on the graph of f to $(2, 0)$.

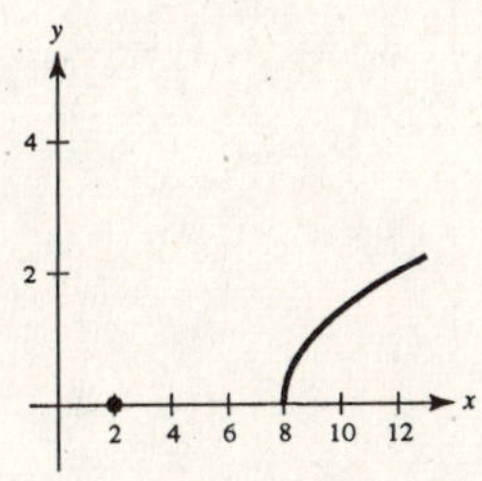

15. $d = \sqrt{(x - 2)^2 + [x^2 - (1/2)]^2}$

$= \sqrt{x^4 - 4x + (17/4)}$

Since d is smallest when the expression inside the radical is smallest, you need only find the critical numbers of

$$f(x) = x^4 - 4x + \tfrac{17}{4}.$$

$$f'(x) = 4x^3 - 4 = 0$$

$$x = 1$$

By the First Derivative Test, the point nearest to $\left(2, \frac{1}{2}\right)$ is $(1, 1)$.

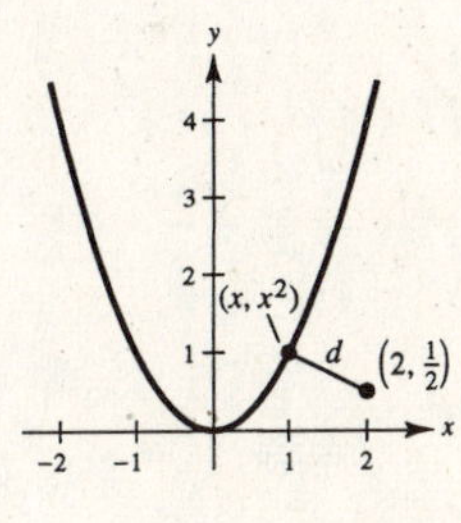

16. $f(x) = (x + 1)^2, (5, 3)$

$$d = \sqrt{(x - 5)^2 + [(x + 1)^2 - 3]^2}$$
$$= \sqrt{x^2 - 10x + 25 + (x^2 + 2x - 2)^2}$$
$$= \sqrt{x^2 - 10x + 25 + x^4 + 4x^3 - 8x + 4}$$
$$= \sqrt{x^4 + 4x^3 + x^2 - 18x + 29}$$

Since d is smallest when the expression inside the radical is smallest, you need to find the critical numbers of

$$g(x) = x^4 + 4x^3 + x^2 - 18x + 29$$
$$g'(x) = 4x^3 + 12x^2 + 2x - 18$$
$$= 2(x - 1)(2x^2 + 8x + 9).$$

By the First Derivative Test, $x = 1$ yields a minimum. Hence, $(1, 4)$ is closest to $(5, 3)$.

17. $$\frac{dQ}{dx} = kx(Q_0 - x) = kQ_0x - kx^2$$

$$\frac{d^2Q}{dx^2} = kQ_0 - 2kx$$
$$= k(Q_0 - 2x) = 0 \text{ when } x = \frac{Q_0}{2}.$$

$$\frac{d^3Q}{dx^3} = -2k < 0 \text{ when } x = \frac{Q_0}{2}.$$

dQ/dx is maximum when $x = Q_0/2$.

18. $$F = \frac{v}{22 + 0.02v^2}$$

$$\frac{dF}{dv} = \frac{22 - 0.02v^2}{(22 + 0.02v^2)^2}$$
$$= 0 \text{ when } v = \sqrt{1100} \approx 33.166.$$

By the First Derivative Test, the flow rate on the road is maximized when $v \approx 33$ mph.

19. $xy = 180{,}000$ (see figure)

$S = x + 2y = \left(x + \dfrac{360{,}000}{x}\right)$ where S is the length of fence needed.

$$\frac{dS}{dx} = 1 - \frac{360{,}000}{x^2} = 0 \text{ when } x = 600.$$

$$\frac{d^2S}{dx^2} = \frac{720{,}000}{x^3} > 0 \text{ when } x = 600.$$

S is a minimum when $x = 600$ meters and $y = 300$ meters.

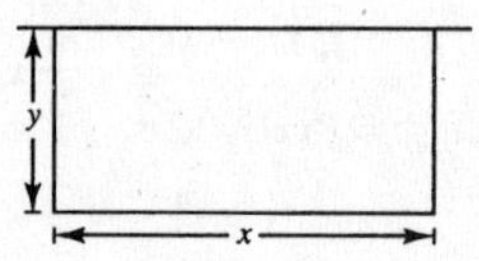

20. $4x + 3y = 200$ is the perimeter. (see figure)

$$A = 2xy = 2x\left(\frac{200 - 4x}{3}\right) = \frac{8}{3}(50x - x^2)$$

$$\frac{dA}{dx} = \frac{8}{3}(50 - 2x) = 0 \text{ when } x = 25.$$

$$\frac{d^2A}{dx^2} = -\frac{16}{3} < 0 \text{ when } x = 25.$$

A is a maximum when $x = 25$ feet and $y = \frac{100}{3}$ feet.

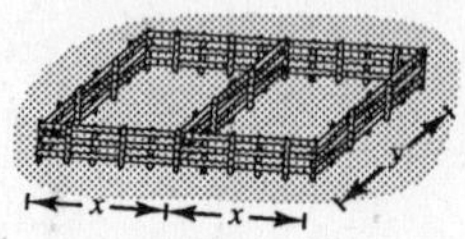

21. (a) $A = 4$(area of side) $+ 2$(area of Top)

(a) $A = 4(3)(11) + 2(3)(3) = 150$ square inches

(b) $A = 4(5)(5) + 2(5)(5) = 150$ square inches

(c) $A = 4(3.25)(6) + 2(6)(6) = 150$ square inches

(b) $V =$ (length)(width)(height)

(a) $V = (3)(3)(11) = 99$ cubic inches

(b) $V = (5)(5)(5) = 125$ cubic inches

(c) $V = (6)(6)(3.25) = 117$ cubic inches

(c) $S = 4xy + 2x^2 = 150 \Rightarrow y = \dfrac{150 - 2x^2}{4x}$

$$V = x^2y = x^2\left(\frac{150 - 2x^2}{4x}\right) = \frac{75}{2}x - \frac{1}{2}x^3$$

$$V' = \frac{75}{2} - \frac{3}{2}x^2 = 0 \Rightarrow x = \pm 5$$

By the First Derivative Test, $x = 5$ yields the maximum volume. Dimensions: $5 \times 5 \times 5$. (A cube!)

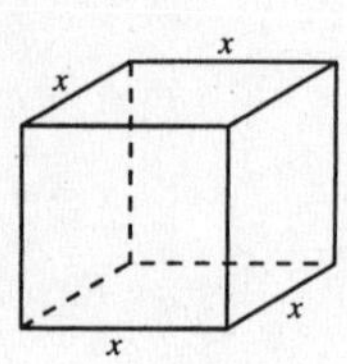

22. Surface Area $= 337.5$

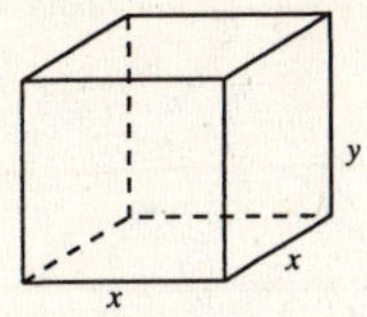

$$2x^2 + 4xy = 337.5$$

$$y = \frac{337.5 - 2x^2}{4x}$$

$$V = x^2y = x^2\left[\frac{337.5 - 2x^2}{4x}\right] = \frac{1}{4}[337.5x - 2x^3]$$

$$V'(x) = \frac{1}{4}[337.5 - 6x^2] = 0$$

$$6x^2 = 337.5$$

$$x = 7.5$$

$$y = 7.5$$

Dimensions: $7.5 \times 7.5 \times 7.5$ [cube!]

23. $16 = 2y + x + \pi\left(\frac{x}{2}\right)$

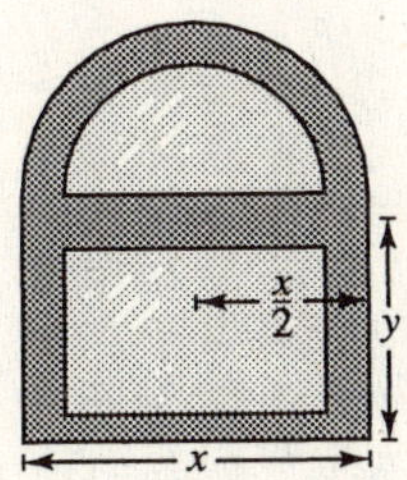

$$32 = 4y + 2x + \pi x$$

$$y = \frac{32 - 2x - \pi x}{4}$$

$$A = xy + \frac{\pi}{2}\left(\frac{x}{2}\right)^2 = \left(\frac{32 - 2x - \pi x}{4}\right)x + \frac{\pi x^2}{8}$$

$$= 8x - \frac{1}{2}x^2 - \frac{\pi}{4}x^2 + \frac{\pi}{8}x^2$$

$$\frac{dA}{dx} = 8 - x - \frac{\pi}{2}x + \frac{\pi}{4}x = 8 - x\left(1 + \frac{\pi}{4}\right)$$

$$= 0 \text{ when } x = \frac{8}{1 + (\pi/4)} = \frac{32}{4 + \pi}.$$

$$\frac{d^2A}{dx^2} = -\left(1 + \frac{\pi}{4}\right) < 0 \text{ when } x = \frac{32}{4 + \pi}$$

$$y = \frac{32 - 2[32/(4 + \pi)] - \pi[32/(4 + \pi)]}{4} = \frac{16}{4 + \pi}$$

The area is maximum when $y = \dfrac{16}{4 + \pi}$ feet and $x = \dfrac{32}{4 + \pi}$ feet.

24. You can see from the figure that $A = xy$ and $y = \dfrac{6 - x}{2}$.

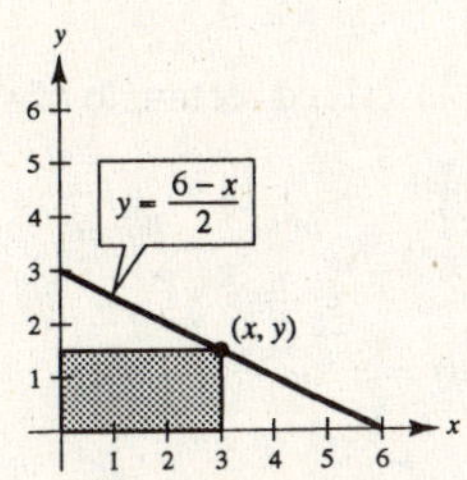

$$A = x\left(\frac{6 - x}{2}\right) = \frac{1}{2}(6x - x^2).$$

$$\frac{dA}{dx} = \frac{1}{2}(6 - 2x) = 0 \text{ when } x = 3.$$

$$\frac{d^2A}{dx^2} = -1 < 0 \text{ when } x = 3.$$

A is a maximum when $x = 3$ and $y = 3/2$.

25. (a) $\dfrac{y-2}{0-1} = \dfrac{0-2}{x-1}$

$$y = 2 + \frac{2}{x-1}$$

$$L = \sqrt{x^2+y^2} = \sqrt{x^2 + \left(2 + \frac{2}{x-1}\right)^2}$$

$$= \sqrt{x^2 + 4 + \frac{8}{x-1} + \frac{4}{(x-1)^2}}, \quad x > 1$$

(b)

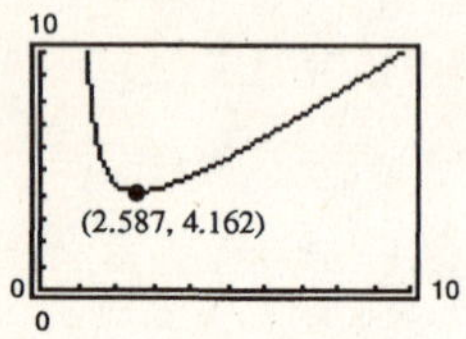

L is minimum when $x \approx 2.587$ and $L \approx 4.162$.

(c) Area $= A(x) = \dfrac{1}{2}xy = \dfrac{1}{2}x\left(2 + \dfrac{2}{x-1}\right) = x + \dfrac{x}{x-1}$

$$A'(x) = 1 + \frac{(x-1) - x}{(x-1)^2} = 1 - \frac{1}{(x-1)^2} = 0$$

$$(x-1)^2 = 1$$

$$x - 1 = \pm 1$$

$$x = 0, 2 \text{ (select } x = 2)$$

Then $y = 4$ and $A = 4$.

Vertices: $(0, 0)$, $(2, 0)$, $(0, 4)$

26. (a) $A = \dfrac{1}{2}$ base $\times$ height

$$= \frac{1}{2}\left(2\sqrt{16-h^2}\right)(4+h)$$

$$= \sqrt{16-h^2}(4+h)$$

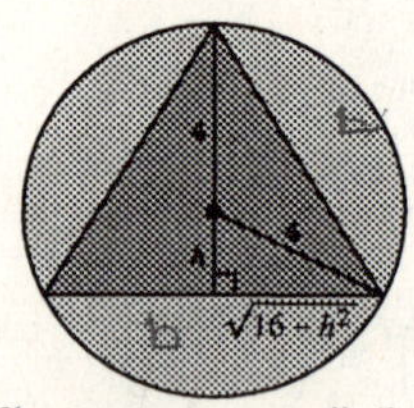

$$\frac{dA}{dh} = \frac{1}{2}(16-h^2)^{-1/2}(-2h)(4+h) + (16-h^2)^{1/2}$$

$$= (16-h^2)^{-1/2}[-h(4+h) + (16-h^2)]$$

$$= \frac{-2(h^2+2h-8)}{\sqrt{16-h^2}} = \frac{-2(h+4)(h-2)}{\sqrt{16-h^2}}$$

$\dfrac{dA}{dh} = 0$ when $h = 2$, which is a maximum by the First Derivative Test.

Hence, the sides are $2\sqrt{16-h^2} = 4\sqrt{3}$, an equilateral triangle. Area $= 12\sqrt{3}$ sq. units.

(b) Equilateral triangle

27. $A = 2xy = 2x\sqrt{25-x^2}$ (see figure)

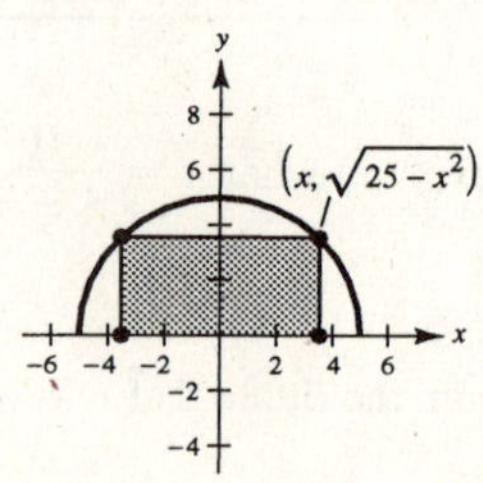

$$\frac{dA}{dx} = 2x\left(\frac{1}{2}\right)\left(\frac{-2x}{\sqrt{25-x^2}}\right) + 2\sqrt{25-x^2}$$

$$= 2\left(\frac{25-2x^2}{\sqrt{25-x^2}}\right) = 0 \text{ when } x = y = \frac{5\sqrt{2}}{2} \approx 3.54.$$

By the First Derivative Test, the inscribed rectangle of maximum area has vertices

$$\left(\pm\frac{5\sqrt{2}}{2}, 0\right), \left(\pm\frac{5\sqrt{2}}{2}, \frac{5\sqrt{2}}{2}\right).$$

Width: $\dfrac{5\sqrt{2}}{2}$; Length: $5\sqrt{2}$

28. $A = 2xy = 2x\sqrt{r^2-x^2}$ (see figure)

$$\frac{dA}{dx} = \frac{2(r^2-2x^2)}{\sqrt{r^2-x^2}} = 0 \text{ when } x = \frac{\sqrt{2}r}{2}.$$

By the First Derivative Test, A is maximum when the rectangle has dimensions $\sqrt{2}r$ by $(\sqrt{2}r)/2$.

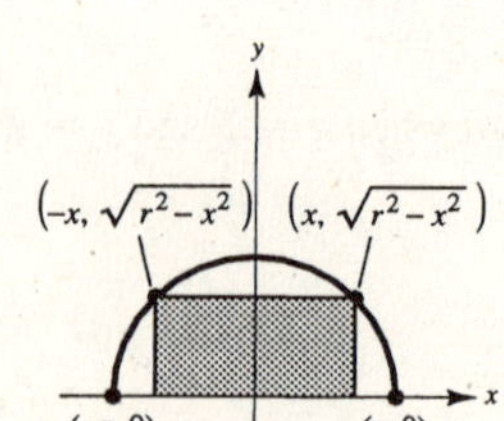

29. $xy = 30 \Rightarrow y = \dfrac{30}{x}$

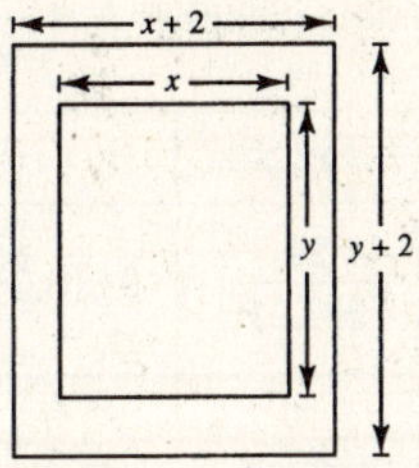

$$A = (x+2)\left(\frac{30}{x} + 2\right) \quad \text{(see figure)}$$

$$\frac{dA}{dx} = (x+2)\left(\frac{-30}{x^2}\right) + \left(\frac{30}{x} + 2\right) = \frac{2(x^2 - 30)}{x^2} = 0 \text{ when } x = \sqrt{30}.$$

$$y = \frac{30}{\sqrt{30}} = \sqrt{30}$$

By the First Derivative Test, the dimensions $(x + 2)$ by $(y + 2)$ are $\left(2 + \sqrt{30}\right)$ by $\left(2 + \sqrt{30}\right)$ (approximately 7.477 by 7.477). These dimensions yield a minimum area.

30. $xy = 36 \Rightarrow y = \dfrac{36}{x}$

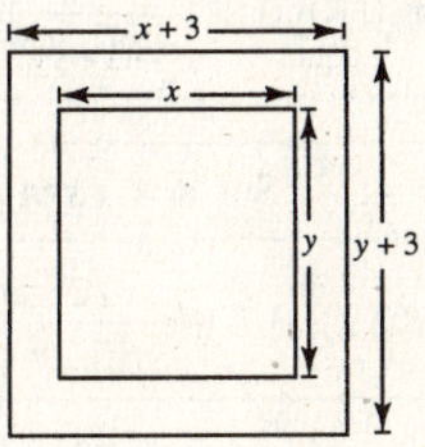

$$A = (x+3)(y+3) = (x+3)\left(\frac{36}{x} + 3\right)$$

$$= 36 + \frac{108}{x} + 3x + 9$$

$$\frac{dA}{dx} = \frac{-108}{x^2} + 3 = 0 \Rightarrow 3x^2 = 108 \Rightarrow x = 6, y = 6$$

Dimensions: 9×9

31. (a) $P = 2x + 2\pi r$

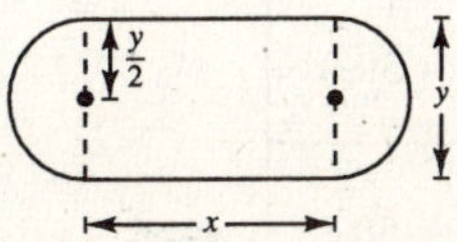

$$= 2x + 2\pi\left(\frac{y}{2}\right)$$

$$= 2x + \pi y = 200$$

$$\Rightarrow y = \frac{200 - 2x}{\pi} = \frac{2}{\pi}(100 - x)$$

(b)

Length, x	Width, y	Area, xy
10	$\frac{2}{\pi}(100 - 10)$	$(10)\frac{2}{\pi}(100 - 10) \approx 573$
20	$\frac{2}{\pi}(100 - 20)$	$(20)\frac{2}{\pi}(100 - 20) \approx 1019$
30	$\frac{2}{\pi}(100 - 30)$	$(30)\frac{2}{\pi}(100 - 30) \approx 1337$
40	$\frac{2}{\pi}(100 - 40)$	$(40)\frac{2}{\pi}(100 - 40) \approx 1528$
50	$\frac{2}{\pi}(100 - 50)$	$(50)\frac{2}{\pi}(100 - 50) \approx 1592$
60	$\frac{2}{\pi}(100 - 60)$	$(60)\frac{2}{\pi}(100 - 60) = 1528$

The maximum area of the rectangle is approximately 1592 m^2.

(c) $A = xy = x\dfrac{2}{\pi}(100 - x) = \dfrac{2}{\pi}(100x - x^2)$

(d) $A' = \dfrac{2}{\pi}(100 - 2x)$

$A' = 0$ when $x = 50$.

Maximum area $= A(50) = \dfrac{5000}{\pi}$

(e)

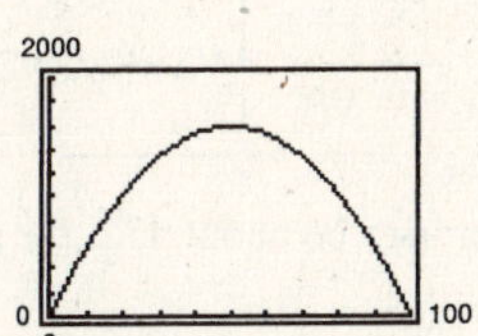

Maximum area is approximately

1591.55 m^2 ($x = 50$ m).

32. $V = \pi r^2 h = 22$ cubic inches or $h = \dfrac{22}{\pi r^2}$

(a)

Radius, r	Height	Surface Area
0.2	$\dfrac{22}{\pi(0.2)^2}$	$2\pi(0.2)\left[0.2 + \dfrac{22}{\pi(0.2)^2}\right] \approx 220.3$
0.4	$\dfrac{22}{\pi(0.4)^2}$	$2\pi(0.4)\left[0.4 + \dfrac{22}{\pi(0.4)^2}\right] \approx 111.0$
0.6	$\dfrac{22}{\pi(0.6)^2}$	$2\pi(0.6)\left[0.6 + \dfrac{22}{\pi(0.6)^2}\right] \approx 75.6$
0.8	$\dfrac{22}{\pi(0.8)^2}$	$2\pi(0.8)\left[0.8 + \dfrac{22}{\pi(0.8)^2}\right] \approx 59.0$

(b)

Radius, r	Height	Surface Area
0.2	$\dfrac{22}{\pi(0.2)^2}$	$2\pi(0.2)\left[0.2 + \dfrac{22}{\pi(0.2)^2}\right] \approx 220.3$
0.4	$\dfrac{22}{\pi(0.4)^2}$	$2\pi(0.4)\left[0.4 + \dfrac{22}{\pi(0.4)^2}\right] \approx 111.0$
0.6	$\dfrac{22}{\pi(0.6)^2}$	$2\pi(0.6)\left[0.6 + \dfrac{22}{\pi(0.6)^2}\right] \approx 75.6$
0.8	$\dfrac{22}{\pi(0.8)^2}$	$2\pi(0.8)\left[0.8 + \dfrac{22}{\pi(0.8)^2}\right] \approx 59.0$
1.0	$\dfrac{22}{\pi(1.0)^2}$	$2\pi(1.0)\left[1.0 + \dfrac{22}{\pi(1.0)^2}\right] \approx 50.3$
1.2	$\dfrac{22}{\pi(1.2)^2}$	$2\pi(1.2)\left[1.2 + \dfrac{22}{\pi(1.2)^2}\right] \approx 45.7$
1.4	$\dfrac{22}{\pi(1.4)^2}$	$2\pi(1.4)\left[1.4 + \dfrac{22}{\pi(1.4)^2}\right] \approx 43.7$
1.6	$\dfrac{22}{\pi(1.6)^2}$	$2\pi(1.6)\left[1.6 + \dfrac{22}{\pi(1.6)^2}\right] \approx 43.6$
1.8	$\dfrac{22}{\pi(1.8)^2}$	$2\pi(1.8)\left[1.8 + \dfrac{22}{\pi(1.8)^2}\right] \approx 44.8$
2.0	$\dfrac{22}{\pi(2.0)^2}$	$2\pi(2.0)\left[2.0 + \dfrac{22}{\pi(2.0)^2}\right] \approx 47.1$

The minimum seems to be about 43.6 for $r = 1.6$.

(c) $S = 2\pi r^2 + 2\pi rh$

$$= 2\pi r(r + h) = 2\pi r\left[r + \frac{22}{\pi r^2}\right] = 2\pi r^2 + \frac{44}{r}$$

(d)

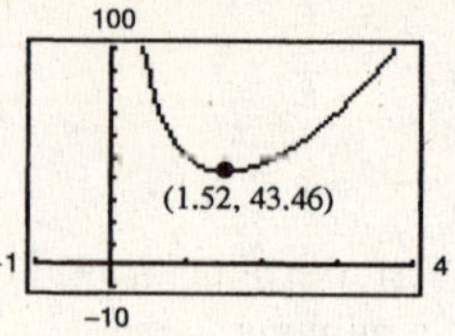

The minimum seems to be 43.46 for $r \approx 1.52$.

(e) $\dfrac{dS}{dr} = 4\pi r - \dfrac{44}{r^2} = 0$ when $r = \sqrt[3]{11/\pi} \approx 1.52$ in.

$h = \dfrac{22}{\pi r^2} \approx 3.04$ in.

Note: Notice that

$$h = \frac{22}{\pi r^2} = \frac{22}{\pi(11/\pi)^{2/3}} = 2\left(\frac{11^{1/3}}{\pi^{1/3}}\right) = 2r.$$

33. $V = \pi r^2 h = V_0$ cubic units or $h = \dfrac{V_0}{\pi r^2}$

$$S = 2\pi r^2 + 2\pi rh = 2\left(\pi r^2 + \frac{V_0}{r}\right)$$

$\dfrac{dS}{dr} = 2\left(2\pi r - \dfrac{V_0}{r^2}\right) = 0$ when $r = \sqrt[3]{\dfrac{V_0}{2\pi}}$ units.

$$h = \frac{V_0}{\pi\left(\sqrt[3]{V_0/2\pi}\right)^2} = \frac{V_0(2\pi)^{2/3}}{\pi V_0^{2/3}} = \frac{2V_0^{1/3}}{(2\pi)^{1/3}} = 2r$$

By the First Derivative Test, this will yield the minimum surface area.

34. Let x be the sides of the square ends and y the length of the package.

$$P = 4x + y = 108 \Rightarrow y = 108 - 4x$$
$$V = x^2y = x^2(108 - 4x) = 108x^2 - 4x^3$$
$$\frac{dV}{dx} = 216x - 12x^2$$
$$= 12x(18 - x) = 0 \text{ when } x = 18.$$
$$\frac{d^2V}{dx^2} = 216 - 24x = -216 < 0 \text{ when } x = 18.$$

The volume is maximum when $x = 18$ inches and $y = 108 - 4(18) = 36$ inches.

35. $V = \pi r^2 x$

$$x + 2\pi r = 108 \Rightarrow x = 108 - 2\pi r \quad \text{(see figure)}$$
$$V = \pi r^2(108 - 2\pi r) = \pi(108r^2 - 2\pi r^3)$$
$$\frac{dV}{dr} = \pi(216r - 6\pi r^2) = 6\pi r(36 - \pi r)$$
$$= 0 \text{ when } r = \frac{36}{\pi} \text{ and } x = 36.$$
$$\frac{d^2V}{dr^2} = \pi(216 - 12\pi r) < 0 \text{ when } r = \frac{36}{\pi}.$$

Volume is maximum when $x = 36$ inches and $r = 36/\pi \approx 11.459$ inches.

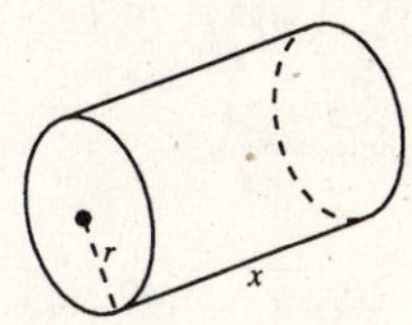

36. $V = \frac{1}{3}\pi x^2 h = \frac{1}{3}\pi x^2\left(r + \sqrt{r^2 - x^2}\right)$ (see figure)

$$\frac{dV}{dx} = \frac{1}{3}\pi\left[\frac{-x^3}{\sqrt{r^2 - x^2}} + 2x\left(r + \sqrt{r^2 - x^2}\right)\right] = \frac{\pi x}{3\sqrt{r^2 - x^2}}\left(2r^2 + 2r\sqrt{r^2 - x^2} - 3x^2\right) = 0$$
$$2r^2 + 2r\sqrt{r^2 - x^2} - 3x^2 = 0$$
$$2r\sqrt{r^2 - x^2} = 3x^2 - 2r^2$$
$$4r^2(r^2 - x^2) = 9x^4 - 12x^2r^2 + 4r^4$$
$$0 = 9x^4 - 8x^2r^2 = x^2(9x^2 - 8r^2)$$
$$x = 0, \frac{2\sqrt{2}r}{3}$$

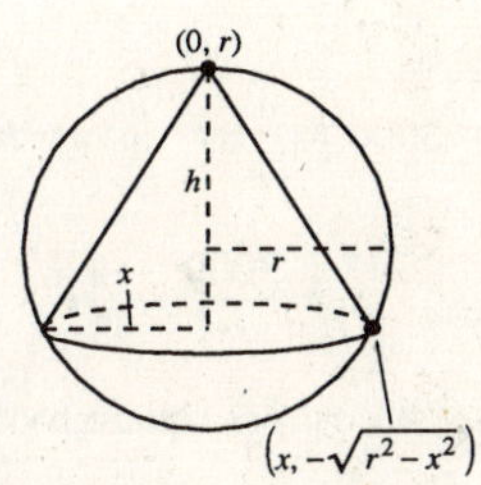

By the First Derivative Test, the volume is a maximum when

$$x = \frac{2\sqrt{2}r}{3} \text{ and } h = r + \sqrt{r^2 - x^2} = \frac{4r}{3}.$$

Thus, the maximum volume is

$$V = \frac{1}{3}\pi\left(\frac{8r^2}{9}\right)\left(\frac{4r}{3}\right) = \frac{32\pi r^3}{81} \text{ cubic units.}$$

37. $V = \pi x^2 h = \pi x^2\left(2\sqrt{r^2 - x^2}\right) = 2\pi x^2\sqrt{r^2 - x^2}$ (see figure)

$$\frac{dV}{dx} = 2\pi\left[x^2\left(\frac{1}{2}\right)(r^2 - x^2)^{-1/2}(-2x) + 2x\sqrt{r^2 - x^2}\right] = \frac{2\pi x}{\sqrt{r^2 - x^2}}(2r^2 - 3x^2)$$
$$= 0 \text{ when } x = 0 \text{ and } x^2 = \frac{2r^2}{3} \Rightarrow x = \frac{\sqrt{6}r}{3}.$$

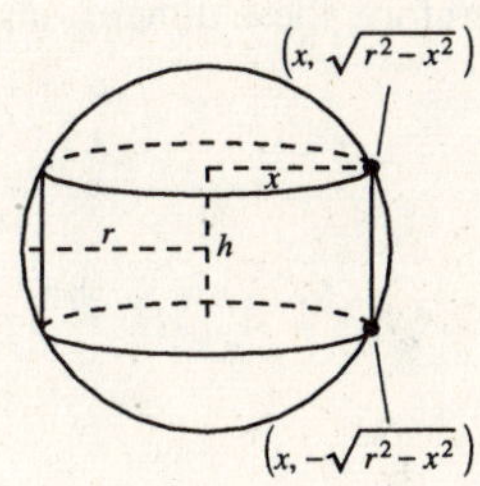

By the First Derivative Test, the volume is a maximum when

$$x = \frac{\sqrt{6}r}{3} \text{ and } h = \frac{2r}{\sqrt{3}}.$$

Thus, the maximum volume is

$$V = \pi\left(\frac{2}{3}r^2\right)\left(\frac{2r}{\sqrt{3}}\right) = \frac{4\pi r^3}{3\sqrt{3}}.$$

38. Answers will vary. See the text Guidelines on page 365.

39. No, there is no minimum area. If the sides are x and y, then $2x + 2y = 20 \Rightarrow y = 10 - x$.
The area is $A(x) = x(10 - x) = 10x - x^2$. This can be made arbitrarily small by selecting $x \approx 0$.

40. No. The volume will change because the shape of the container changes when squeezed.

41. $V = 12 = \frac{4}{3}\pi r^3 + \pi r^2 h$

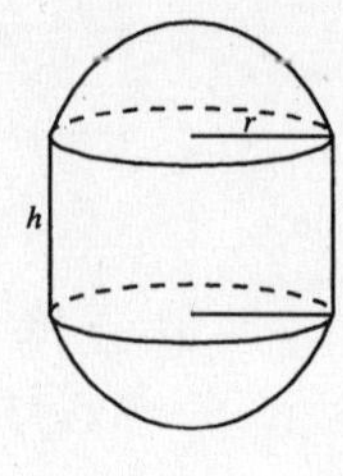

$$h = \frac{12 - (4/3)\pi r^3}{\pi r^2} = \frac{12}{\pi r^2} - \frac{4}{3}r$$

$$S = 4\pi r^2 + 2\pi r h = 4\pi r^2 + 2\pi r\left(\frac{12}{\pi r^2} - \frac{4}{3}r\right)$$

$$= 4\pi r^2 + \frac{24}{r} - \frac{8}{3}\pi r^2 = \frac{4}{3}\pi r^2 + \frac{24}{r}$$

$$\frac{dS}{dr} = \frac{8}{3}\pi r - \frac{24}{r^2} = 0 \text{ when } r = \sqrt[3]{9/\pi} \approx 1.42 \text{ cm.}$$

$$\frac{d^2S}{dr^2} = \frac{8}{3}\pi + \frac{48}{r^3} > 0 \text{ when } r = \sqrt[3]{9/\pi} \text{ cm.}$$

The surface area is minimum when $r = \sqrt[3]{9/\pi}$ cm and $h = 0$. The resulting solid is a sphere of radius $r \approx 1.42$ cm.

42. $V = 3000 = \frac{4}{3}\pi r^3 + \pi r^2 h$

$$h = \frac{3000}{\pi r^2} - \frac{4}{3}r$$

Let k = cost per square foot of the surface area of the sides, then $2k$ = cost per square foot of the hemispherical ends.

$$C = 2k(4\pi r^2) + k(2\pi r h) = k\left[8\pi r^2 + 2\pi r\left(\frac{3000}{\pi r^2} - \frac{4}{3}r\right)\right] = k\left[\frac{16}{3}\pi r^2 + \frac{6000}{r}\right]$$

$$\frac{dC}{dr} = k\left[\frac{32}{3}\pi r - \frac{6000}{r^2}\right] = 0 \text{ when } r = \sqrt[3]{\frac{1125}{2\pi}} \approx 5.636 \text{ feet and } h \approx 22.545 \text{ feet.}$$

By the Second Derivative Test, we have

$$\frac{d^2C}{dr^2} = k\left[\frac{32}{3}\pi + \frac{12{,}000}{r^3}\right] > 0 \text{ when } r = \sqrt[3]{\frac{1125}{2\pi}}.$$

Therefore, these dimensions will produce a minimum cost.

43. Let x be the length of a side of the square and y the length of a side of the triangle.

$$4x + 3y = 10$$

$$A = x^2 + \frac{1}{2}y\left(\frac{\sqrt{3}}{2}y\right)$$

$$= \frac{(10 - 3y)^2}{16} + \frac{\sqrt{3}}{4}y^2$$

$$\frac{dA}{dy} = \frac{1}{8}(10 - 3y)(-3) + \frac{\sqrt{3}}{2}y = 0$$

$$-30 + 9y + 4\sqrt{3}y = 0$$

$$y = \frac{30}{9 + 4\sqrt{3}}$$

$$\frac{d^2A}{dy^2} = \frac{9 + 4\sqrt{3}}{8} > 0$$

A is minimum when

$$y = \frac{30}{9 + 4\sqrt{3}} \text{ and } x = \frac{10\sqrt{3}}{9 + 4\sqrt{3}}.$$

44. Let x be the side of the triangle and y the side of the square.

$$A = \frac{3}{4}\left(\cot\frac{\pi}{3}\right)x^2 + \frac{4}{4}\left(\cot\frac{\pi}{4}\right)y^2 \text{ where } 3x + 4y = 20$$

$$= \frac{\sqrt{3}}{4}x^2 + \left(5 - \frac{3}{4}x\right)^2, \quad 0 \le x \le \frac{20}{3}.$$

$$A' = \frac{\sqrt{3}}{2}x + 2\left(5 - \frac{3}{4}x\right)\left(-\frac{3}{4}\right) = 0$$

$$x = \frac{60}{4\sqrt{3} + 9}$$

When $x = 0$, $A = 25$, when $x = 60/(4\sqrt{3} + 9)$, $A \approx 10.874$, and when $x = 20/3$, $A \approx 19.245$. Area is maximum when all 20 feet are used on the square.

45. Let S be the strength and k the constant of proportionality. Given $h^2 + w^2 = 24^2$, $h^2 = 24^2 - w^2$,

$$S = kwh^2$$

$$S = kw(576 - w^2) = k(576w - w^3)$$

$$\frac{dS}{dw} = k(576 - 3w^2) = 0 \text{ when } w = 8\sqrt{3}, h = 8\sqrt{6}.$$

$$\frac{d^2S}{dw^2} = -6kw < 0 \text{ when } w = 8\sqrt{3}.$$

These values yield a maximum.

46.

$$S = \sqrt{x^2 + 4}, L = \sqrt{1 + (3 - x)^2}$$

$$\text{Time} = T = \frac{\sqrt{x^2 + 4}}{2} + \frac{\sqrt{x^2 - 6x + 10}}{4}$$

$$\frac{dT}{dx} = \frac{x}{2\sqrt{x^2 + 4}} + \frac{x - 3}{4\sqrt{x^2 - 6x + 10}} = 0$$

$$\frac{x^2}{x^2 + 4} = \frac{9 - 6x + x^2}{4(x^2 - 6x + 10)}$$

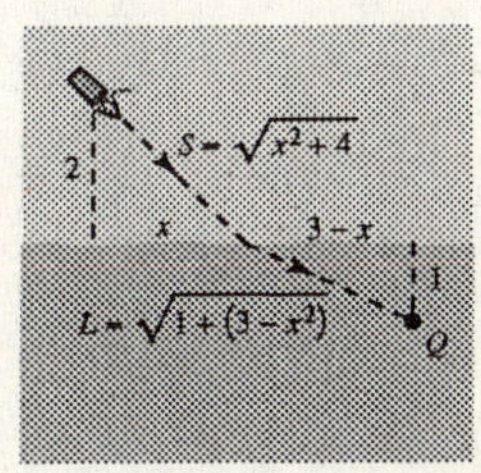

$$x^4 - 6x^3 + 9x^2 + 8x - 12 = 0$$

You need to find the roots of this equation in the interval $[0, 3]$. By using a computer or graphics calculator, you can determine that this equation has only one root in this interval ($x = 1$). Testing at this value and at the endpoints, you see that $x = 1$ yields the minimum time. Thus, the man should row to a point 1 mile from the nearest point on the coast.

47. $f(x) = \frac{1}{2}x^2 \qquad g(x) = \frac{1}{16}x^4 - \frac{1}{2}x^2$ on $[0, 4]$

(a) 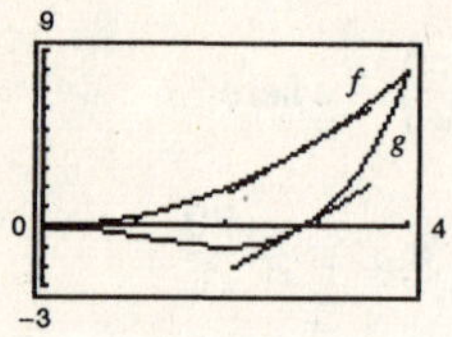

(b) $d(x) = f(x) - g(x) = \frac{1}{2}x^2 - \left(\frac{1}{16}x^4 - \frac{1}{2}x^2\right) = x^2 - \frac{1}{16}x^4$

$d'(x) = 2x - \frac{1}{4}x^3 = 0 \implies 8x = x^3$

$\implies x = 0, 2\sqrt{2}$ (in $[0, 4]$)

The maximum distance is $d = 4$ when $x = 2\sqrt{2}$.

(c) $f'(x) = x$, Tangent line at $(2\sqrt{2}, 4)$ is

$$y - 4 = 2\sqrt{2}(x - 2\sqrt{2})$$
$$y = 2\sqrt{2}x - 4.$$

$g'(x) = \frac{1}{4}x^3 - x$, Tangent line at $(2\sqrt{2}, 0)$ is

$$y - 0 = \left(\tfrac{1}{4}(2\sqrt{2})^3 - 2\sqrt{2}\right)(x - 2\sqrt{2})$$
$$y = 2\sqrt{2}x - 8.$$

The tangent lines are parallel and 4 vertical units apart.

(d) The tangent lines will be parallel. If $d(x) = f(x) - g(x)$, then $d'(x) = 0 = f'(x) - g'(x)$ implies that $f'(x) = g'(x)$ at the point x where the distance is maximum.

48. Let A be the amount of the power line.

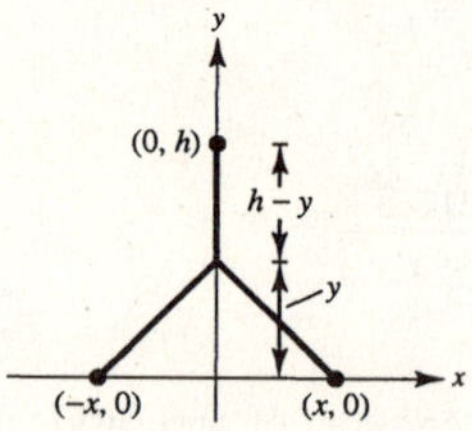

$$A = h - y + 2\sqrt{x^2 + y^2}$$

$$\frac{dA}{dy} = -1 + \frac{2y}{\sqrt{x^2 + y^2}} = 0 \text{ when } y = \frac{x}{\sqrt{3}}.$$

$$\frac{d^2A}{dy^2} = \frac{2x^2}{(x^2 + y^2)^{3/2}} > 0 \text{ for } y = \frac{x}{\sqrt{3}}.$$

The amount of power line is minimum when $y = x/\sqrt{3}$.

49. Let F be the illumination at point P which is x units from source 1.

$$F = \frac{kI_1}{x^2} + \frac{kI_2}{(d - x)^2}$$

$$\frac{dF}{dx} = \frac{-2kI_1}{x^3} + \frac{2kI_2}{(d - x)^3} = 0 \text{ when } \frac{2kI_1}{x^3} = \frac{2kI_2}{(d - x)^3}.$$

$$\frac{\sqrt[3]{I_1}}{\sqrt[3]{I_2}} = \frac{x}{d - x}$$

$$(d - x)\sqrt[3]{I_1} = x\sqrt[3]{I_2}$$

$$d\sqrt[3]{I_1} = x\left(\sqrt[3]{I_1} + \sqrt[3]{I_2}\right)$$

$$x = \frac{d\sqrt[3]{I_1}}{\sqrt[3]{I_1} + \sqrt[3]{I_2}}$$

$$\frac{d^2F}{dx^2} = \frac{6kI_1}{x^4} + \frac{6kI_2}{(d - x)^4} > 0 \text{ when } x = \frac{d\sqrt[3]{I_1}}{\sqrt[3]{I_1} + \sqrt[3]{I_2}}.$$

This is the minimum point.

50. $C(x) = 2k\sqrt{x^2 + 4} + k(4 - x)$

$$C'(x) = \frac{2xk}{\sqrt{x^2 + 4}} - k = 0$$

$$2x = \sqrt{x^2 + 4}$$

$$4x^2 = x^2 + 4$$

$$3x^2 = 4$$

$$x = \frac{2}{\sqrt{3}}$$

51. Let d be the amount deposited in the bank, i be the interest rate paid by the bank, and P be the profit.

$$P = (0.12)d - id$$

$d = ki^2$ (since d is proportional to i^2)

$$P = (0.12)(ki^2) - i(ki^2) = k(0.12i^2 - i^3)$$

$$\frac{dP}{di} = k(0.24i - 3i^2) = 0 \text{ when } i = \frac{0.24}{3} = 0.08.$$

$\frac{d^2P}{di^2} = k(0.24 - 6i) < 0$ when $i = 0.08$ (**Note:** $k > 0$).

The profit is a maximum when $i = 8\%$.

52. $C = 100\left(\dfrac{200}{x^2} + \dfrac{x}{x+30}\right),\ 1 \le x$

$C' = 100\left(-\dfrac{400}{x^3} + \dfrac{30}{(x+30)^2}\right)$

Approximation: $x \approx 40.45$ units, or 4045 units

53. $P = -\dfrac{1}{10}s^3 + 6s^2 + 400$

(a) $\dfrac{dP}{ds} = -\dfrac{3}{10}s^2 + 12s = -\dfrac{3}{10}s(s - 40) = 0$

when $x = 0$, $s = 40$.

$\dfrac{d^2P}{ds^2} = -\dfrac{3}{5}s + 12$

$\dfrac{d^2P}{ds^2}(0) > 0 \Rightarrow s = 0$ yields a minimum.

$\dfrac{d^2P}{ds^2}(40) < 0 \Rightarrow s = 40$ yields a maximum.

The maximum profit occurs when $s = 40$, which corresponds to \$40,000 ($P =$ \$3,600,000).

(b) $\dfrac{d^2P}{ds^2} = -\dfrac{3}{5}s + 12 = 0$ when $s = 20$.

The point of diminishing returns occurs when $s = 20$, which corresponds to \$20,000 being spent on advertising.

54. Label the figure so that $r^2 = x^2 + h^2$.
Then, the area A is 8 times the area of the region given by $OPQR$:

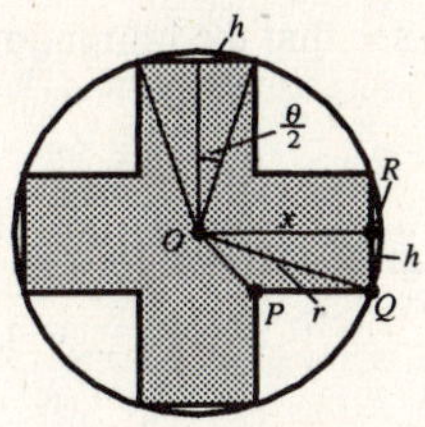

$A = 8\left[\dfrac{1}{2}h^2 + (x - h)h\right]$

$= 8\left[\dfrac{1}{2}(r^2 - x^2) + \left(x - \sqrt{r^2 - x^2}\right)\sqrt{r^2 - x^2}\right]$

$= 8x\sqrt{r^2 - x^2} + 4x^2 - 4r^2$

$A'(x) = 8\sqrt{r^2 - x^2} - \dfrac{8x^2}{\sqrt{r^2 - x^2}} + 8x = 0$

$\dfrac{8x^2}{\sqrt{r^2 - x^2}} = 8x + 8\sqrt{r^2 - x^2}$

$x^2 = x\sqrt{r^2 - x^2} + (r^2 - x^2)$

$2x^2 - r^2 = x\sqrt{r^2 - x^2}$

$4x^4 - 4x^2r^2 + r^4 = x^2(r^2 - x^2)$

$5x^4 - 5x^2r^2 + r^4 = 0$ Quadratic in x^2.

$x^2 = \dfrac{5r^2 \pm \sqrt{25r^4 - 20r^4}}{10} = \dfrac{r^2}{10}[5 \pm \sqrt{5}]$.

Take larger value.

$x = r\sqrt{\dfrac{5 + \sqrt{5}}{10}} \approx 0.85065r$ Critical number

55. $S_1 = (4m - 1)^2 + (5m - 6)^2 + (10m - 3)^2$

$\dfrac{dS_1}{dm} = 2(4m - 1)(4) + 2(5m - 6)(5) + 2(10m - 3)(10) = 282m - 128 = 0$ when $m = \dfrac{64}{141}$.

Line: $y = \dfrac{64}{141}x$

$S = \left|4\left(\dfrac{64}{141}\right) - 1\right| + \left|5\left(\dfrac{64}{141}\right) - 6\right| + \left|10\left(\dfrac{64}{141}\right) - 3\right|$

$= \left|\dfrac{256}{141} - 1\right| + \left|\dfrac{320}{141} - 6\right| + \left|\dfrac{640}{141} - 3\right| = \dfrac{858}{141} \approx 6.1$ mi

56. $S_2 = |4m - 1| + |5m - 6| + |10m - 3|$

Using a graphing utility, you can see that the minimum occurs when $m = 0.3$.

Line $y = 0.3x$

$$S_2 = |4(0.3) - 1| + |5(0.3) - 6| + |10(0.3) - 3|$$

$$= 4.7 \text{ mi.}$$

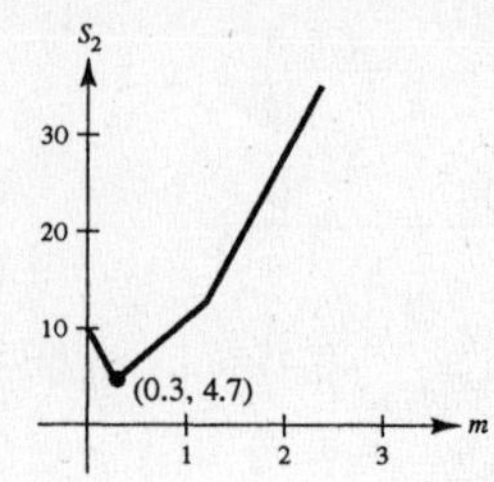

57. $S_3 = \dfrac{|4m - 1|}{\sqrt{m^2 + 1}} + \dfrac{|5m - 6|}{\sqrt{m^2 + 1}} + \dfrac{|10m - 3|}{\sqrt{m^2 + 1}}$

Using a graphing utility, you can see that the minimum occurs when $m \approx 0.3$.

Line: $y \approx 0.3x$

$$S_3 = \frac{|4(0.3) - 1| + |5(0.3) - 6| + |10(0.3) - 3|}{\sqrt{(0.3)^2 + 1}} \approx 4.5 \text{ mi.}$$

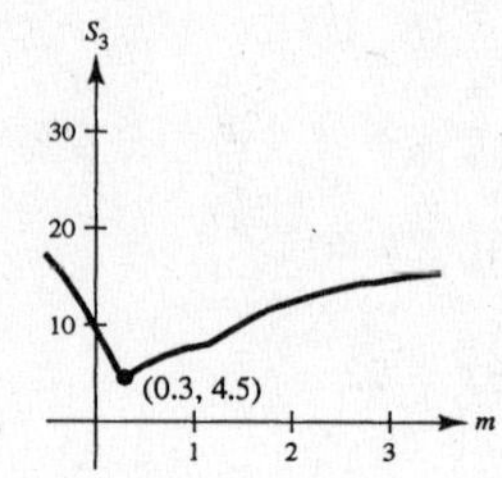

Section 5.8 Differentials

1. $f(x) = x^2$

$f'(x) = 2x$

Tangent line at (2, 4): $y - f(2) = f'(2)(x - 2)$

$$y - 4 = 4(x - 2)$$

$$y = 4x - 4$$

x	1.9	1.99	2	2.01	2.1
$f(x) = x^2$	3.6100	3.9601	4	4.0401	4.4100
$T(x) = 4x - 4$	3.6000	3.9600	4	4.0400	4.4000

2. $f(x) = \dfrac{6}{x^2} = 6x^{-2}$

$f'(x) = -12x^{-3} = \dfrac{-12}{x^3}$

Tangent line at $\left(2, \dfrac{3}{2}\right)$:

$$y - \frac{3}{2} = \frac{-12}{8}(x - 2) = \frac{-3}{2}(x - 2)$$

$$y = -\frac{3}{2}x + \frac{9}{2}$$

x	1.9	1.99	2	2.01	2.1
$f(x) = \dfrac{6}{x^2}$	1.6620	1.5151	1.5	1.4851	1.3605
$T(x) = -\dfrac{3}{2}x + \dfrac{9}{2}$	1.65	1.515	1.5	1.485	1.35

3. $f(x) = x^5$

$f'(x) = 5x^4$

Tangent line at (2, 32): $y - f(2) = f'(2)(x - 2)$

$$y - 32 = 80(x - 2)$$

$$y = 80x - 128$$

x	1.9	1.99	2	2.01	2.1
$f(x) = x^5$	24.7610	31.2080	32	32.8080	40.8410
$T(x) = 80x - 128$	24.0000	31.2000	32	32.8000	40.0000

4. $f(x) = \sqrt{x}$

$f'(x) = \dfrac{1}{2\sqrt{x}}$

Tangent line at $(2, \sqrt{2})$:

$y - f(2) = f'(2)(x - 2)$

$y - \sqrt{2} = \dfrac{1}{2\sqrt{2}}(x - 2)$

$y = \dfrac{x}{2\sqrt{2}} + \dfrac{1}{\sqrt{2}}$

x	1.9	1.99	2	2.01	2.1
$f(x) = \sqrt{x}$	1.3784	1.4107	1.4142	1.4177	1.4491
$T(x) = \dfrac{x}{\sqrt{2}} + \dfrac{1}{\sqrt{2}}$	1.3789	1.4107	1.4142	1.4177	1.4496

5. $y = f(x) = \frac{1}{2}x^3, f'(x) = \frac{3}{2}x^2, x = 2, \Delta x = dx = 0.1$

$\Delta y = f(x + \Delta x) - f(x)$
$= f(2.1) - f(2)$
$= 0.6305$

$dy = f'(x)dx$
$= f'(2)(0.1)$
$= 6(0.1) = 0.6$

6. $y = f(x) = 1 - 2x^2, f'(x) = -4x, x = 0, \Delta x = dx = -0.1$

$\Delta y = f(x + \Delta x) - f(x)$
$= f(-0.1) - f(0)$
$= [1 - 2(-0.1)^2] - [1 - 2(0)^2] = -0.02$

$dy = f'(x)\, dx$
$= f'(0)(-0.1)$
$= (0)(-0.1) = 0$

7. $y = f(x) = x^4 + 1, f'(x) = 4x^3, x = -1, \Delta x = dx = 0.01$

$\Delta y = f(x + \Delta x) - f(x)$
$= f(-0.99) - f(-1)$
$= [(-0.99)^4 + 1] - [(-1)^4 + 1] \approx -0.0394$

$dy = f'(x)\, dx$
$= f'(-1)(0.01)$
$= (-4)(0.01) = -0.04$

8. $y = f(x) = 2x + 1, f'(x) = 2, x = 2, \Delta x = dx = 0.01$

$\Delta y = f(x + \Delta x) - f(x)$
$= f(2.01) - f(2)$
$= [2(2.01) + 1] - [2(2) + 1] = 0.02$

$dy = f'(x)\, dx$
$= f'(2)(0.01)$
$= 2(0.01) = 0.02$

9. $y = 3x^2 - 4$

$dy = 6x\, dx$

10. $y = 3x^{2/3}$

$dy = 2x^{-1/3}dx = \dfrac{2}{x^{1/3}}dx$

11. $y = 4x^3$

$\dfrac{dy}{dx} = 12x^2$

$dy = 12x^2\, dx$

12. $y = 3$

$\dfrac{dy}{dx} = 0$

$dy = 0\, dx = 0$

13. $y = \dfrac{x + 1}{2x - 1}$

$dy = \dfrac{-3}{(2x - 1)^2}dx$

14. $y = \dfrac{x}{x + 5}$

$\dfrac{dy}{dx} = \dfrac{(x + 5) - x}{(x + 5)^2} = \dfrac{5}{(x + 5)^2}$

$dy = \dfrac{5}{(x + 5)^2}dx$

15. $y = \sqrt{x} = x^{1/2}$

$\dfrac{dy}{dx} = \dfrac{1}{2\sqrt{x}}$

$dy = \dfrac{1}{2\sqrt{x}}dx$

16. $y = \sqrt{9 - x^2}$

$$dy = \frac{1}{2}(9 - x^2)^{-1/2}(-2x)dx = \frac{-x}{\sqrt{9 - x^2}}\,dx$$

17. $y = x\sqrt{1 - x^2}$

$$dy = \left(x\frac{-x}{\sqrt{1 - x^2}} + \sqrt{1 - x^2}\right)dx = \frac{1 - 2x^2}{\sqrt{1 - x^2}}\,dx$$

18. $y = \sqrt{x} + \dfrac{1}{\sqrt{x}}$

$$dy = \left(\frac{1}{2\sqrt{x}} - \frac{1}{2x\sqrt{x}}\right)dx = \frac{x - 1}{2x\sqrt{x}}\,dx$$

19. (a) $f(1.9) = f(2 - 0.1) \approx f(2) + f'(2)(-0.1)$

$\approx 1 + (1)(-0.1) = 0.9$

(b) $f(2.04) = f(2 + 0.04) \approx f(2) + f'(2)(0.04)$

$\approx 1 + (1)(0.04) = 1.04$

20. $f(2) = 1$ and $f'(2) = -1$

(a) $f(1.9) \approx f(2) + f'(2)\,dx$

$= 1 + (-1)(-0.1) = 1.1$

(b) $f(2.04) \approx f(2) + f'(2)\,dx$

$= 1 + (-1)(0.04) = 0.96$

21. (a) $f(1.9) = f(2 - 0.1) \approx f(2) + f'(2)(-0.1)$

$\approx 1 + \left(-\frac{1}{2}\right)(-0.1) = 1.05$

(b) $f(2.04) = f(2 + 0.04) \approx f(2) + f'(2)(0.04)$

$\approx 1 + \left(-\frac{1}{2}\right)(0.04) = 0.98$

22. (a) $f(1.9) = f(2 - 0.1) \approx f(2) + f'(2)(-0.1)$

$\approx 1 + 0(-0.1) = 1$

(b) $f(2.04) = f(2 + 0.04) \approx f(2) + f'(2)(0.04)$

$\approx 1 + 0(0.04) = 1$

23. (a) $g(2.93) = g(3 - 0.07) \approx g(3) + g'(3)(-0.07)$

$\approx 8 + \left(-\frac{1}{2}\right)(-0.07) = 8.035$

(b) $g(3.1) = g(3 + 0.1) \approx g(3) + g'(3)(0.1)$

$\approx 8 + \left(-\frac{1}{2}\right)(0.1) = 7.95$

24. (a) $g(2.93) = g(3 - 0.07) \approx g(3) + g'(3)(-0.07)$

$\approx 8 + (3)(-0.07) = 7.79$

(b) $g(3.1) = g(3 + 0.1) \approx g(3) + g'(3)(0.1)$

$\approx 8 + (3)(0.1) = 8.3$

25. $A = x^2$

$x = 12$

$\Delta x = dx = \pm\frac{1}{64}$

$dA = 2x\,dx$

$\Delta A \approx dA = 2(12)\left(\pm\frac{1}{64}\right)$

$= \pm\frac{3}{8}$ square inches

26. $A = \frac{1}{2}bh,\ b = 36,\ h = 50$

$db = dh = \pm 0.25$

$dA = \frac{1}{2}b\,dh + \frac{1}{2}h\,db$

$\Delta A \approx dA = \frac{1}{2}(36)(\pm 0.25) + \frac{1}{2}(50)(\pm 0.25)$

$= \pm 10.75$ square centimeters

27. $A = \pi r^2$

$r = 14$

$\Delta r = dr = \pm\frac{1}{4}$

$\Delta A \approx dA = 2\pi r\,dr = \pi(28)\left(\pm\frac{1}{4}\right)$

$= \pm 7\pi$ square inches

28. $x = 12$ inches

$\Delta x = dx = \pm 0.03$ inch

(a) $V = x^3$

$dV = 3x^2\,dx = 3(12)^2(\pm 0.03)$

$= \pm 12.96$ cubic inches

(b) $S = 6x^2$

$dS = 12x\,dx = 12(12)(\pm 0.03)$

$= \pm 4.32$ square inches

29. $r = 6$ inches

$\Delta r = dr = \pm 0.02$ inches

(a) $V = \frac{4}{3}\pi r^3$

$dV = 4\pi r^2\,dr = 4\pi(6)^2(\pm 0.02) = \pm 2.88\pi$ cubic inches

(b) $S = 4\pi r^2$

$dS = 8\pi r\,dr = 8\pi(6)(\pm 0.02) = \pm 0.96\pi$ square inches

(c) Relative error: $\dfrac{dV}{V} = \dfrac{4\pi r^2\,dr}{(4/3)\pi r^3} = \dfrac{3dr}{r}$

$= \frac{3}{6}(0.02) = 0.01 = 1\%$

Relative error: $\dfrac{dS}{S} = \dfrac{8\pi r\,dr}{4\pi r^2} = \dfrac{2dr}{r}$

$= \dfrac{2(0.02)}{6} = 0.00666\ldots = \frac{2}{3}\%$

30. (a) $x = 15$ centimeters

$\Delta x = dx = \pm 0.05$ centimeters

$A = x^2$

$dA = 2x\,dx = 2(15)(\pm 0.05)$

$= \pm 1.5$ square centimeters

Percentage error:

$$\frac{dA}{A} = \frac{1.5}{(15)^2} = 0.00666\ldots = \frac{2}{3}\%$$

(b) $\dfrac{dA}{A} = \dfrac{2x\,dx}{x^2} = \dfrac{2\,dx}{x} \le 0.025$

$\dfrac{dx}{x} \le \dfrac{0.025}{2} = 0.0125 = 1.25\%$

31. (a) $C = 56$ centimeters

$\Delta C = dC = \pm 1.2$ centimeters

$C = 2\pi r \Rightarrow r = \dfrac{C}{2\pi}$

$A = \pi r^2 = \pi\left(\dfrac{C}{2\pi}\right)^2 = \dfrac{1}{4\pi}C^2$

$dA = \dfrac{1}{2\pi}C\,dC = \dfrac{1}{2\pi}(56)(\pm 1.2) = \dfrac{33.6}{\pi}$

$\dfrac{dA}{A} = \dfrac{33.6/\pi}{[1/(4\pi)](56)^2} \approx 0.042857 = 4.2857\%$

(b) $\dfrac{dA}{A} = \dfrac{(1/2\pi)C\,dC}{(1/4\pi)C^2} = \dfrac{2dC}{C} \le 0.03$

$\dfrac{dC}{C} \le \dfrac{0.03}{2} = 0.015 = 1.5\%$

32. $P = (500x - x^2) - \left(\frac{1}{2}x^2 - 77x + 3000\right)$, x changes from 115 to 120

$dP = (500 - 2x - x + 77)dx = (577 - 3x)\,dx = [577 - 3(115)](120 - 115) = 1160$

Approximate percentage change: $\dfrac{dP}{P}(100) = \dfrac{1160}{43517.50}(100) \approx 2.7\%$

33. $V = \pi r^2 h = 40\pi r^2$, $r = 5$ cm, $h = 40$ cm, $dr = 0.2$ cm

$\Delta V \approx dV = 80\pi r\,dr = 80\pi(5)(0.2) = 80\pi$ cm^3

34. $V = \frac{4}{3}\pi r^3$, $r = 100$ cm, $dr = 0.2$ cm

$\Delta V \approx dV = 4\pi r^2\,dr = 4\pi(100)^2(0.2) = 8000\pi$ cm^3

35. $E = IR$

$R = \dfrac{E}{I}$

$dR = -\dfrac{E}{I^2}dI$

$\dfrac{dR}{R} = \dfrac{-(E/I^2)dI}{E/I} = -\dfrac{dI}{I}$

$\left|\dfrac{dR}{R}\right| = \left|-\dfrac{dI}{I}\right| = \left|\dfrac{dI}{I}\right|$

36. (a) $T = 2\pi\sqrt{L/g}$

$$dT = \frac{\pi}{g\sqrt{L/g}}\,dL$$

Relative error:

$$\frac{dT}{T} = \frac{(\pi\,dL)/(g\sqrt{L/g})}{2\pi\sqrt{L/g}}$$

$$= \frac{dL}{2L}$$

$$= \frac{1}{2}\,(\text{relative error in } L)$$

$$= \frac{1}{2}(0.005) = 0.0025$$

Percentage error: $\frac{dT}{T}(100) = 0.25\% = \frac{1}{4}\%$

(b) $(0.0025)(3600)(24) = 216$ seconds

$= 3.6$ minutes

37. Let $f(x) = \sqrt{x}, x = 100, dx = -0.6.$

$$f(x + \Delta x) \approx f(x) + f'(x)\,dx$$

$$= \sqrt{x} + \frac{1}{2\sqrt{x}}\,dx$$

$$f(x + \Delta x) = \sqrt{99.4}$$

$$\approx \sqrt{100} + \frac{1}{2\sqrt{100}}(-0.6) = 9.97$$

Using a calculator: $\sqrt{99.4} \approx 9.96995$

38. Let $f(x) = \sqrt[3]{x}, x = 27, dx = -1$

$$f(x + \Delta x) \approx f(x) + f'(x)dx = \sqrt[3]{x} + \frac{1}{3\sqrt[3]{x^2}}dx$$

$$\sqrt[3]{26} \approx \sqrt[3]{27} + \frac{1}{3\sqrt[3]{27^2}}(-1) = 3 - \frac{1}{27} \approx 2.9630$$

Using a calculator, $\sqrt[3]{26} \approx 2.9625$

39. Let $f(x) = \sqrt[4]{x}, x = 625, dx = -1.$

$$f(x + \Delta x) \approx f(x) + f'(x)\,dx = \sqrt[4]{x} + \frac{1}{4\sqrt[4]{x^3}}\,dx$$

$$f(x + \Delta x) = \sqrt[4]{624} \approx \sqrt[4]{625} + \frac{1}{4(\sqrt[4]{625})^3}(-1)$$

$$= 5 - \frac{1}{500} = 4.998$$

Using a calculator, $\sqrt[4]{624} \approx 4.9980.$

40. Let $f(x) = x^3, x = 3, dx = -0.01.$

$$f(x + \Delta x) \approx f(x) + f'(x)\,dx$$

$$= x^3 + 3x^2\,dx$$

$$f(x + \Delta x) = (2.99)^3 \approx 3^3 + 3(3)^2(-0.01)$$

$$= 27 - 0.27 = 26.73$$

Using a calculator: $(2.99)^3 \approx 26.7309$

41. Let $f(x) = \sqrt{x}, x = 4, dx = 0.02, f'(x) = 1/(2\sqrt{x}).$

Then

$$f(4.02) \approx f(4) + f'(4)\,dx$$

$$\sqrt{4.02} \approx \sqrt{4} + \frac{1}{2\sqrt{4}}(0.02) = 2 + \frac{1}{4}(0.02).$$

42. Let $f(x) = -\frac{1}{x}, x = 2, dx = 0.04, f'(x) = \frac{-1}{x^2}.$

Then

$$f(2.04) = \frac{1}{2.04} \approx f(2) + f'(2)\,dx$$

$$= \frac{1}{2} - \frac{1}{4}(0.04).$$

43. $f(x) = \sqrt{x + 4}$

$$f'(x) = \frac{1}{2\sqrt{x + 4}}$$

At $(0, 2)$, $f(0) = 2, f'(0) = \frac{1}{4}.$

Tangent line: $y - 2 = \frac{1}{4}(x - 0)$

$$y = \frac{1}{4}x + 2$$

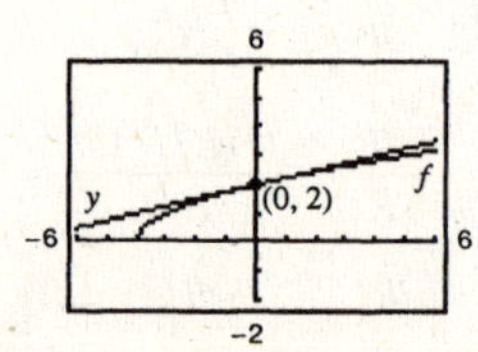

44. $f(x) = \dfrac{1}{1 - x} = (1 - x)^{-1}$

$f'(x) = \dfrac{1}{(1 - x)^2}$

$f(0) = 1$

$f'(0) = 1$

Tangent line at $(0, 1)$: $y - 1 = 1(x - 0)$

$y = x + 1$

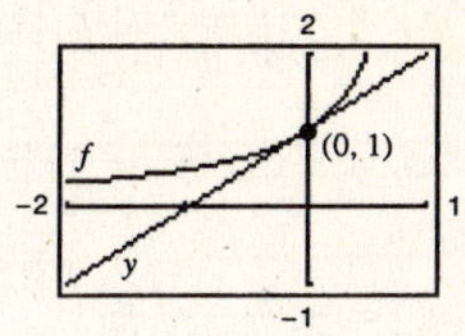

45. In general, when $\Delta x \to 0$, dy approaches Δy.

46. Propagated error $= f(x + \Delta x) - f(x)$,

relative error $= \left|\dfrac{dy}{y}\right|$, and the percent error $= \left|\dfrac{dy}{y}\right| \times 100$.

47. True

48. True, $\dfrac{\Delta y}{\Delta x} = \dfrac{dy}{dx} = a$

49. True

50. False

Let $f(x) = \sqrt{x}$, $x = 1$, and $\Delta x = dx = 3$. Then

$$\Delta y = f(x + \Delta x) - f(x) = f(4) - f(1) = 1$$

and

$$dy = f'(x)\,dx = \frac{1}{2\sqrt{1}}(3) = \frac{3}{2}.$$

Thus, $dy > \Delta y$ in this example.

Review Exercises for Chapter 5

1. A number c in the domain of f is a critical number if $f'(c) = 0$ or f' is undefined at c.

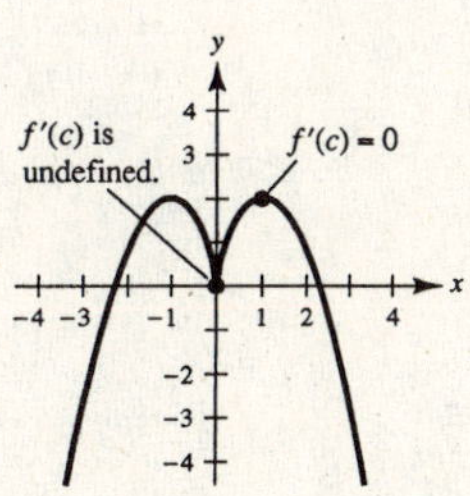

2. (a) $f(4) = -f(-4) = -3$

(b) $f(-3) = -f(3) = -(-4) = 4$

(c)

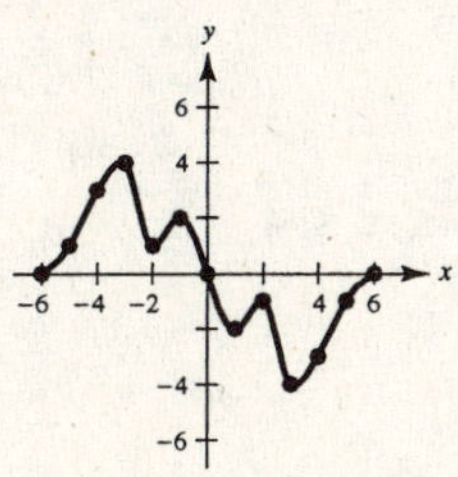

At least six critical numbers on $(-6, 6)$.

(d) Yes. Since $f(-2) = -f(2) = -(-1) = 1$ and $f(1) = -f(-1) = -2$, the Mean Value says that there exists at least one value c in $(-2, 1)$ such that

$$f'(c) = \frac{f(1) - f(-2)}{1 - (-2)} = \frac{-2 - 1}{1 + 2} = -1.$$

(e) No, $\lim_{x \to 0} f(x)$ exists because f is continuous at $(0, 0)$.

(f) Yes, f is differentiable at $x = 2$.

3. $f(x) = x\sqrt{5 - x}, \quad [0, 4]$

$f'(x) = \dfrac{10 - 3x}{2\sqrt{5 - x}}$ Critical number: $x = \dfrac{10}{3}$

$f(0) = 0$ Minimum

$f\left(\dfrac{10}{3}\right) = \dfrac{10\sqrt{15}}{9} \approx 4.303$ Maximum

$f(4) = 4$

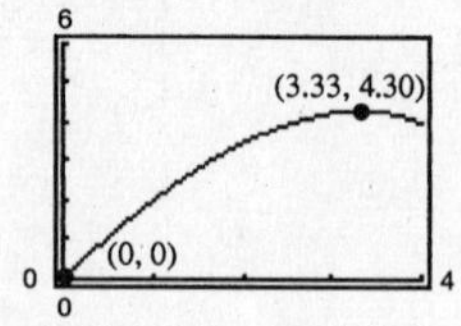

4. $f(x) = \dfrac{x}{\sqrt{x^2 + 1}}, [0, 2]$

$f'(x) = x\left[-\dfrac{1}{2}(x^2 + 1)^{-3/2}(2x)\right] + (x^2 + 1)^{-1/2}$

$= \dfrac{1}{(x^2 + 1)^{3/2}}$

No critical numbers

Left endpoint: $(0, 0)$ Minimum

Right endpoint: $(2, 2/\sqrt{5})$ Maximum

5. Yes. $f(-3) = f(2) = 0$. f is continuous on $[-3, 2]$, differentiable on $(-3, 2)$.

$f'(x) = (x + 3)(3x - 1) = 0$ for $x = \frac{1}{3}$.

$c = \frac{1}{3}$ satisfies $f'(c) = 0$.

6. No. f is not differentiable at $x = 2$.

7. $f(x) = 3 - |x - 4|$

(a)

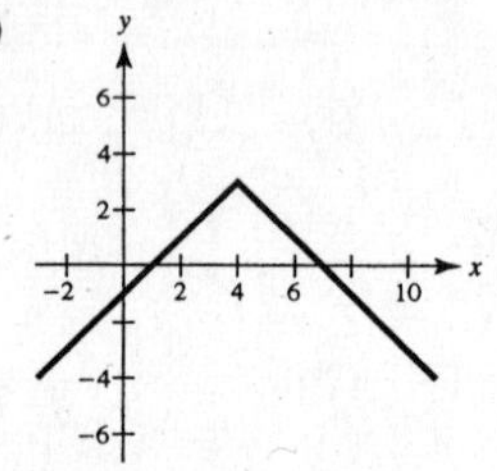

$f(1) = f(7) = 0$

(b) f is not differentiable at $x = 4$.

8. No; the function is discontinuous at $x = 0$ which is in the interval $[-2, 1]$.

9.

$f(x) = x^{2/3}, 1 \le x \le 8$

$f'(x) = \dfrac{2}{3}x^{-1/3}$

$\dfrac{f(b) - f(a)}{b - a} = \dfrac{4 - 1}{8 - 1} = \dfrac{3}{7}$

$f'(c) = \dfrac{2}{3}c^{-1/3} = \dfrac{3}{7}$

$c = \left(\dfrac{14}{9}\right)^3 = \dfrac{2744}{729} \approx 3.764$

10.

$f(x) = \dfrac{1}{x}, \quad 1 \le x \le 4$

$f'(x) = -\dfrac{1}{x^2}$

$\dfrac{f(b) - f(a)}{b - a} = \dfrac{(1/4) - 1}{4 - 1} = \dfrac{-3/4}{3} = -\dfrac{1}{4}$

$f'(c) = \dfrac{-1}{c^2} = -\dfrac{1}{4}$

$c = 2$

11. On $[0, 2]$, $f(x) = 9 - x^2$, $f'(x) = -2x$.

$f'(x) = \dfrac{f(2) - f(0)}{2 - 0} = \dfrac{5 - 9}{2} = -2$

$-2x = -2$

$x = 1$

$f'(1) = -2$

12.

$f(x) = \sqrt{x} - 2x, 0 \le x \le 4$

$f'(x) = \dfrac{1}{2\sqrt{x}} - 2$

$\dfrac{f(b) - f(a)}{b - a} = \dfrac{-6 - 0}{4 - 0} = -\dfrac{3}{2}$

$f'(c) = \dfrac{1}{2\sqrt{c}} - 2 = -\dfrac{3}{2}$

$c = 1$

13. $f(x) = Ax^2 + Bx + C$

$$f'(x) = 2Ax + B$$

$$\frac{f(x_2) - f(x_1)}{x_2 - x_1} = \frac{A(x_2^2 - x_1^2) + B(x_2 - x_1)}{x_2 - x_1}$$

$$= A(x_1 + x_2) + B$$

$$f'(c) = 2Ac + B = A(x_1 + x_2) + B$$

$$2Ac = A(x_1 + x_2)$$

$$c = \frac{x_1 + x_2}{2} = \text{Midpoint of } [x_1, x_2]$$

14. $f(x) = 2x^2 - 3x + 1$

$$f'(x) = 4x - 3$$

$$\frac{f(b) - f(a)}{b - a} = \frac{21 - 1}{4 - 0} = 5$$

$$f'(c) = 4c - 3 = 5$$

$$c = 2 = \text{Midpoint of } [0, 4]$$

15. $f(x) = (x - 1)^2(x - 3)$

$$f'(x) = (x - 1)^2(1) + (x - 3)(2)(x - 1)$$

$$= (x - 1)(3x - 7)$$

Critical numbers: $x = 1$ and $x = \frac{7}{3}$

Interval:	$-\infty < x < 1$	$1 < x < \frac{7}{3}$	$\frac{7}{3} < x < \infty$
Sign of $f'(x)$:	$f'(x) > 0$	$f'(x) < 0$	$f'(x) > 0$
Conclusion:	Increasing	Decreasing	Increasing

16. $g(x) = (x + 1)^3$

$$g'(x) = 3(x + 1)^2$$

Critical number: $x = -1$

Interval	$-\infty < x < -1$	$-1 < x < \infty$
Sign of $g'(x)$	$g'(x) > 0$	$g'(x) > 0$
Conclusion	Increasing	Increasing

17. $h(x) = \sqrt{x}(x - 3) = x^{3/2} - 3x^{1/2}$

Domain: $(0, \infty)$

$$h'(x) = \frac{3}{2}x^{1/2} - \frac{3}{2}x^{-1/2}$$

$$= \frac{3}{2}x^{-1/2}(x - 1) = \frac{3(x - 1)}{2\sqrt{x}}$$

Critical number: $x = 1$

Interval:	$0 < x < 1$	$1 < x < \infty$
Sign of $h'(x)$:	$h'(x) < 0$	$h'(x) > 0$
Conclusion:	Decreasing	Increasing

18. $y = 3(x - 2)^{2/3} - 2x + 4$

$$y' = 2(x - 2)^{-1/3} - 2$$

$$2(x - 2)^{-1/3} - 2 = 0$$

$$\frac{1}{(x - 2)^{1/3}} = 1$$

$$x = 3$$

Critical numbers: $x = 2, 3$

Increasing on $(2, 3)$

Decreasing on $(-\infty, 2), (3, \infty)$

Interval:	$-\infty < x < 2$	$2 < x < 3$	$3 < x < \infty$
Test value:	$x = 0$	$x = 2.5$	$x = 4$
Sign of $f'(x)$:	$f'(0) < 0$	$f'(2.5) > 0$	$f'(4) < 0$
Conclusion:	Decreasing	Increasing	Decreasing

19. $h(t) = \frac{1}{4}t^4 - 8t$

$h'(t) = t^3 - 8 = 0$ when $t = 2$.

Relative minimum: $(2, -12)$

Test Interval:	$-\infty < t < 2$	$2 < t < \infty$
Sign of $h'(t)$:	$h'(t) < 0$	$h'(t) > 0$
Conclusion:	Decreasing	Increasing

20. $g(x) = \frac{1}{27}(x^4 + 4x^3)$

$g'(x) = \frac{1}{27}(4x^3 + 12x^2) = \frac{4x^2}{27}(x + 3)$

Critical numbers: $x = 0, -3$

Interval:	$-\infty < x < -3$	$-3 < x < 0$	$0 < x < \infty$
Test value:	$x = -4$	$x = -1$	$x = 1$
Sign of $f'(x)$:	$g'(-4) < 0$	$g'(-1) > 0$	$g'(1)0$
Conclusion:	Decreasing	Increasing	Increasing

Minimum: $(-3, -1)$

21. $f(x) = -x^3 + 6x^2 - 9x + 1$

$f'(x) = -3x^2 + 12x - 9$

$f''(x) = -6x + 12 = 0 \Rightarrow x = 2.$

Because f'' changes sign at $x = 2$, $(2, -1)$ is a point of inflection.

22. $f(x) = (x + 2)^2(x - 4) = x^3 - 12x - 16$

$f'(x) = 3x^2 - 12$

$f''(x) = 6x = 0$ when $x = 0.$

Point of inflection: $(0, -16)$

Test Interval	$-\infty < x < 0$	$0 < x < \infty$
Sign of $f''(x)$	$f''(x) < 0$	$f''(x) > 0$
Conclusion	Concave downward	Concave upward

23. $g(x) = 2x^2(1 - x^2)$

$g'(x) = -4x(2x^2 - 1)$ Critical numbers: $x = 0, \pm\frac{1}{\sqrt{2}}$

$g''(x) = 4 - 24x^2$

$g''(0) = 4 > 0$ Relative minimum at $(0, 0)$

$g''\left(\pm\frac{1}{\sqrt{2}}\right) = -8 < 0$ Relative maximums at $\left(\pm\frac{1}{\sqrt{2}}, \frac{1}{2}\right)$

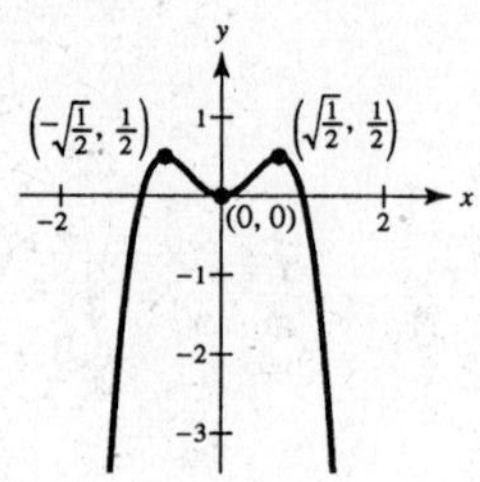

24. $h(t) = t - 4\sqrt{t + 1}$ Domain: $[-1, \infty)$

$h'(t) = 1 - \frac{2}{\sqrt{t + 1}} = 0 \Rightarrow t = 3$

$h''(t) = \frac{1}{(t + 1)^{3/2}}$

$h''(3) = \frac{1}{8} > 0$ $(3, -5)$ is a relative minimum.

25.

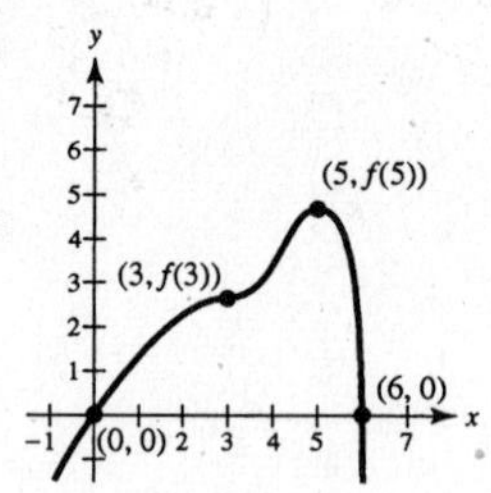

26.

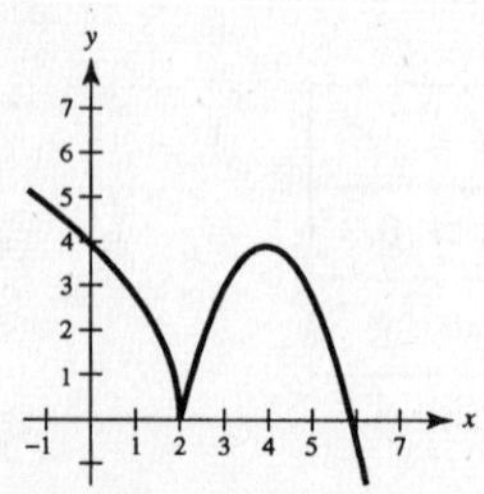

27. The first derivative is positive and the second derivative is negative. The graph is increasing and is concave down.

28. $C = \left(\frac{Q}{x}\right)s + \left(\frac{x}{2}\right)r$

$$\frac{dC}{dx} = -\frac{Qs}{x^2} + \frac{r}{2} = 0$$

$$\frac{Qs}{x^2} = \frac{r}{2}$$

$$x^2 = \frac{2Qs}{r}$$

$$x = \sqrt{\frac{2Qs}{r}}$$

29. (a) $D = 0.00375t^4 - 0.2517t^3 + 5.179t^2 - 21.89t + 94.6$

(b)

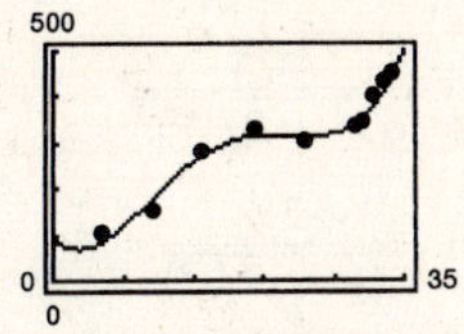

(c) Maximum in 2004

Minimum in 1972 ($t \approx 2.6$)

(d) The increase is greatest in 2004.

30. (a) $S = -0.1222t^3 + 3.565t^2 - 30.49t + 86.8$

(b)

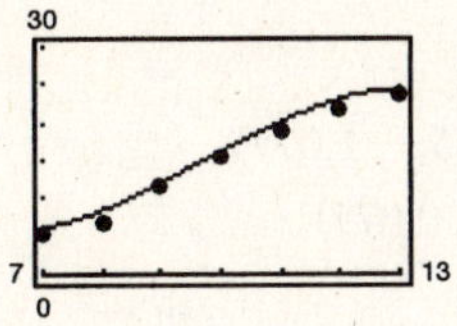

(c) $S'(t) = -0.3666t^2 + 7.13t - 30.49$

$S''(t) = -0.7332t + 7.13 = 0 \Rightarrow t \approx 9.7.$

This is a maximum by the First Derivative Test.

(d) No, because the t^3 coefficient is negative.

31. $\lim_{x\to\infty} \frac{2x^2}{3x^2 + 5} = \lim_{x\to\infty} \frac{2}{3 + 5/x^2} = \frac{2}{3}$

32. $\lim_{x\to\infty} \frac{2x}{3x^2 + 5} = \lim_{x\to\infty} \frac{2/x}{3 + 5/x^2} = 0$

33. $\lim_{x\to-\infty} \left(7 + \frac{1}{x}\right) = 7 + 0 = 7$

34. $\lim_{x\to\infty} \frac{3x}{\sqrt{x^2 + 4}} = \lim_{x\to\infty} \frac{3}{\sqrt{1 + 4/x^2}} = 3$

35. $h(x) = \frac{2x + 3}{x - 4}$

Discontinuity: $x = 4$

$$\lim_{x\to\infty} \frac{2x + 3}{x - 4} = \lim_{x\to\infty} \frac{2 + (3/x)}{1 - (4/x)} = 2$$

Vertical asymptote: $x = 4$

Horizontal asymptote: $y = 2$

36. $g(x) = \frac{5x^2}{x^2 + 2}$

$$\lim_{x\to\infty} \frac{5x^2}{x^2 + 2} = \lim_{x\to\infty} \frac{5}{1 + (2/x^2)} = 5$$

Horizontal asymptote: $y = 5$

37. $f(x) = \frac{3}{x} - 2$

Discontinuity: $x = 0$

$$\lim_{x\to\infty} \left(\frac{3}{x} - 2\right) = -2$$

Vertical asymptote: $x = 0$

Horizontal asymptote: $y = -2$

38. $f(x) = \dfrac{3x}{\sqrt{x^2 + 2}}$

$$\lim_{x \to \infty} \frac{3x}{\sqrt{x^2 + 2}} = \lim_{x \to \infty} \frac{3x/x}{\sqrt{x^2 + 2}/\sqrt{x^2}}$$

$$= \lim_{x \to \infty} \frac{3}{\sqrt{1 + (2/x^2)}} = 3$$

$$\lim_{x \to -\infty} \frac{3x}{\sqrt{x^2 + 2}} = \lim_{x \to -\infty} \frac{3x/x}{\sqrt{x^2 + 2}/(-\sqrt{x^2})}$$

$$= \lim_{x \to -\infty} \frac{3}{-\sqrt{1 + (2/x^2)}} = -3$$

Horizontal asymptotes: $y = \pm 3$

39. $f(x) = x^3 + \dfrac{243}{x}$

Relative minimum: $(3, 108)$

Relative maximum: $(-3, -108)$

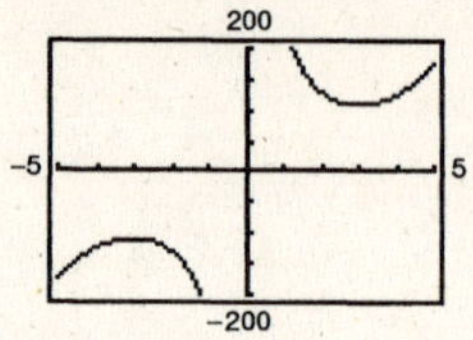

Vertical asymptote: $x = 0$

40. $f(x) = |x^3 - 3x^2 + 2x| = |x(x - 1)(x - 2)|$

Relative minima: $(0, 0), (1, 0), (2, 0)$

Relative maxima: $(1.577, 0.38), (0.423, 0.38)$

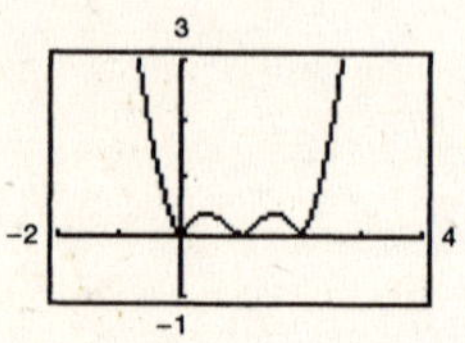

41. $f(x) = \dfrac{x - 1}{1 + 3x^2}$

Relative minimum: $(-0.155, -1.077)$

Relative maximum: $(2.155, 0.077)$

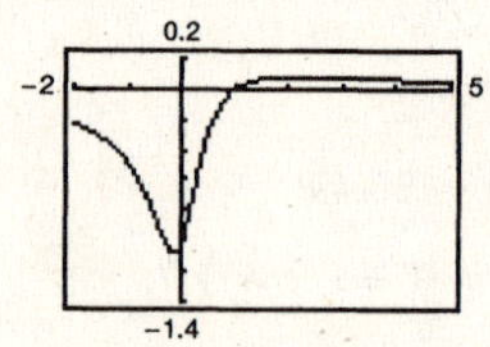

Horizontal asymptote: $y = 0$

42.

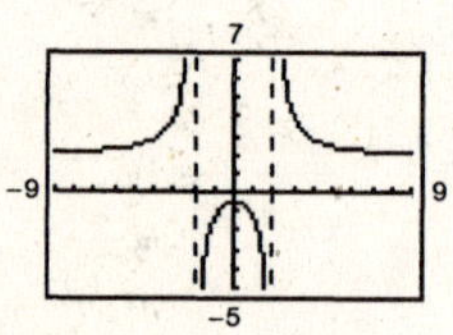

Horizontal asymptote: $y = 2$

Vertical asymptotes: $x = \pm 2$

Relative maximum: $\left(0, -\dfrac{1}{2}\right)$

43. $f(x) = 4x - x^2 = x(4 - x)$

Domain: $(-\infty, \infty)$; Range: $(-\infty, 4]$

$f'(x) = 4 - 2x = 0$ when $x = 2$.

$f''(x) = -2$

Therefore, $(2, 4)$ is a relative maximum.

Intercepts: $(0, 0), (4, 0)$

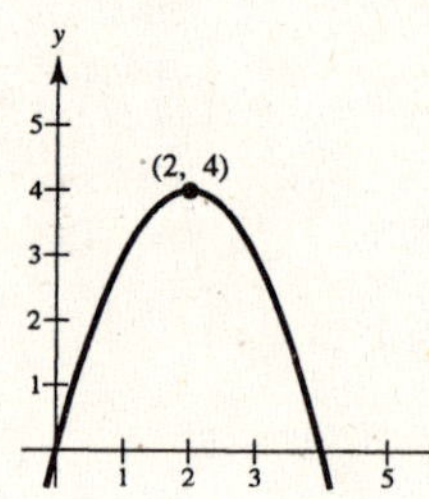

44. $f(x) = 4x^3 - x^4 = x^3(4 - x)$

Domain: $(-\infty, \infty)$; Range: $(-\infty, 27)$

$f'(x) = 12x^2 - 4x^3 = 4x^2(3 - x) = 0$ when $x = 0, 3$.

$f''(x) = 24x - 12x^2 = 12x(2 - x) = 0$ when $x = 0, 2$.

$f''(3) < 0$

Therefore, $(3, 27)$ is a relative maximum.

Points of inflection: $(0, 0), (2, 16)$

Intercepts: $(0, 0), (4, 0)$

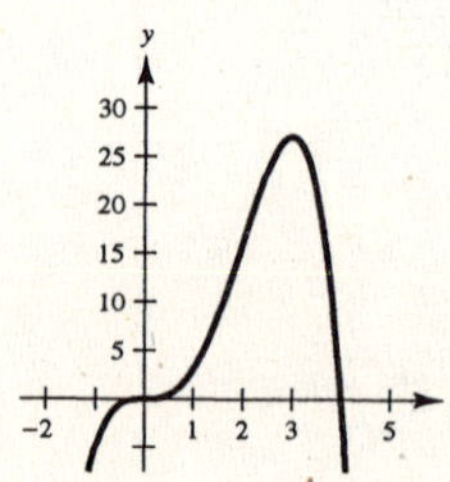

45. $f(x) = x\sqrt{16 - x^2}$, Domain: $[-4, 4]$, Range: $[-8, 8]$

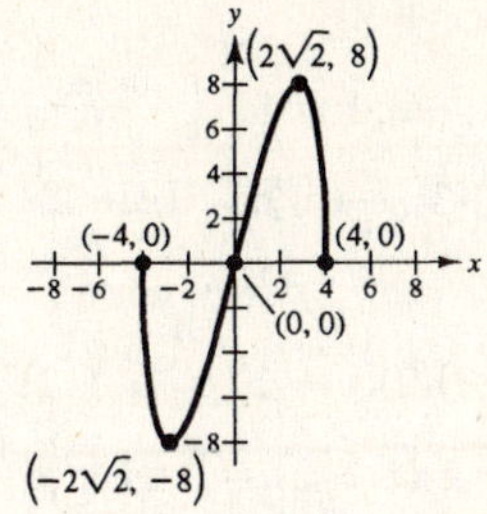

Domain: $[-4, 4]$; Range: $[-8, 8]$

$$f'(x) = \frac{16 - 2x^2}{\sqrt{16 - x^2}} = 0 \text{ when } x = \pm 2\sqrt{2} \text{ and undefined when } x = \pm 4.$$

$$f''(x) = \frac{2x(x^2 - 24)}{(16 - x^2)^{3/2}}$$

$f''(-2\sqrt{2}) > 0$

Therefore, $(-2\sqrt{2}, -8)$ is a relative minimum.

$f''(2\sqrt{2}) < 0$

Therefore, $(2\sqrt{2}, 8)$ is a relative maximum.

Point of inflection: $(0, 0)$

Intercepts: $(-4, 0), (0, 0), (4, 0)$

Symmetry with respect to origin

46. $f(x) = (x^2 - 4)^2$

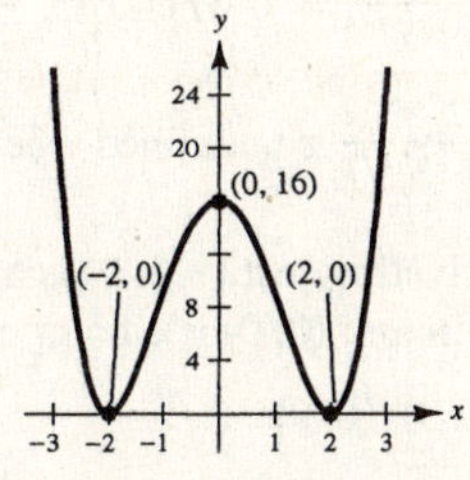

Domain: $(-\infty, \infty)$; Range: $[0, \infty)$

$f'(x) = 4x(x^2 - 4) = 0$ when $x = 0, \pm 2$.

$f''(x) = 4(3x^2 - 4) = 0$ when $x = \pm\dfrac{2\sqrt{3}}{3}$.

$f''(0) < 0$

Therefore, $(0, 16)$ is a relative maximum.

$f''(\pm 2) > 0$

Therefore, $(\pm 2, 0)$ are relative minima.

Points of inflection: $(\pm 2\sqrt{3}/3, 64/9)$

Intercepts: $(-2, 0), (0, 16), (2, 0)$

Symmetry with respect to y-axis

47. $f(x) = (x - 1)^3(x - 3)^2$

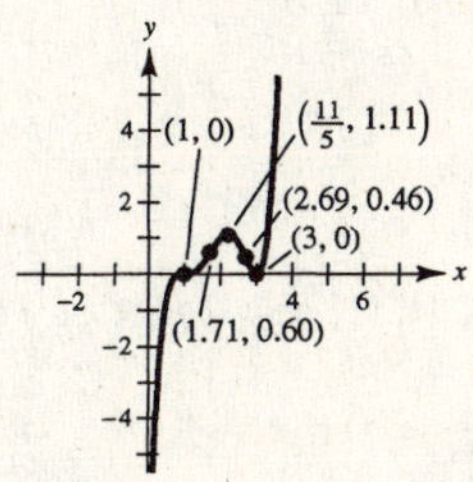

Domain: $(-\infty, \infty)$; Range: $(-\infty, \infty)$

$f'(x) = (x - 1)^2(x - 3)(5x - 11) = 0$ when $x = 1, \dfrac{11}{5}, 3$.

$f''(x) = 4(x - 1)(5x^2 - 22x + 23) = 0$ when $x = 1, \dfrac{11 \pm \sqrt{6}}{5}$.

$f''(3) > 0$

Therefore, $(3, 0)$ is a relative minimum.

$f''\left(\dfrac{11}{5}\right) < 0$

Therefore, $\left(\dfrac{11}{5}, \dfrac{3456}{3125}\right)$ is a relative maximum.

Points of inflection: $(1, 0), \left(\dfrac{11 - \sqrt{6}}{5}, 0.60\right), \left(\dfrac{11 + \sqrt{6}}{5}, 0.46\right)$

Intercepts: $(0, -9), (1, 0), (3, 0)$

48. $f(x) = (x - 3)(x + 2)^3$

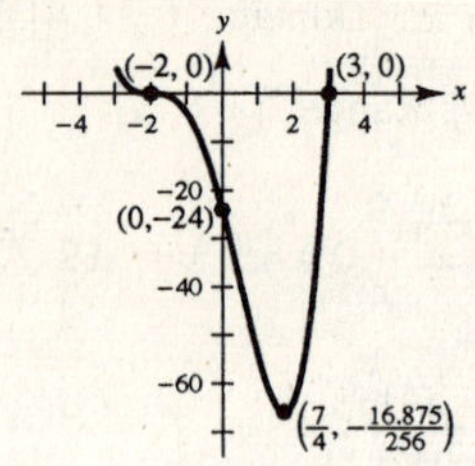

Domain: $(-\infty, \infty)$; Range: $\left[-\frac{16,875}{256}, \infty\right)$

$$f'(x) = (x - 3)(3)(x + 2)^2 + (x + 2)^3$$
$$= (4x - 7)(x + 2)^2 = 0 \text{ when } x = -2, \tfrac{7}{4}.$$

$$f''(x) = (4x - 7)(2)(x + 2) + (x + 2)^2(4)$$
$$= 6(2x - 1)(x + 2) = 0 \text{ when } x = -2, \tfrac{1}{2}.$$

$f''\left(\frac{7}{4}\right) > 0$

Therefore, $\left(\frac{7}{4}, -\frac{16,875}{256}\right)$ is a relative minimum.

Points of inflection: $(-2, 0)$, $\left(\frac{1}{2}, -\frac{625}{16}\right)$

Intercepts: $(-2, 0)$, $(0, -24)$, $(3, 0)$

49. $f(x) = x^{1/3}(x + 3)^{2/3}$

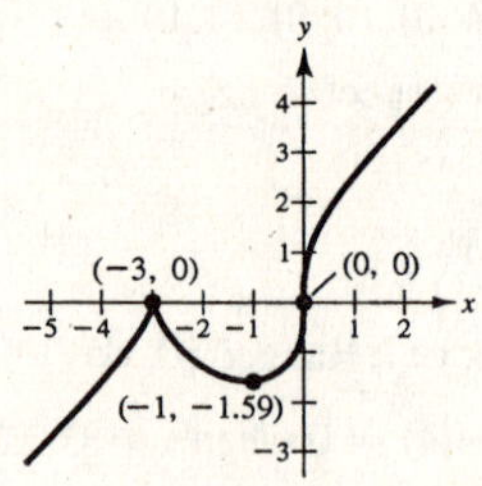

Domain: $(-\infty, \infty)$; Range: $(-\infty, \infty)$

$$f'(x) = \frac{x + 1}{(x + 3)^{1/3}x^{2/3}} = 0 \text{ when } x = -1 \text{ and undefined when } x = -3, 0.$$

$$f''(x) = \frac{-2}{x^{5/3}(x + 3)^{4/3}} \text{ is undefined when } x = 0, -3.$$

By the First Derivative Test $(-3, 0)$ is a relative maximum and $\left(-1, -\sqrt[3]{4}\right)$ is a relative minimum. $(0, 0)$ is a point of inflection.

Intercepts: $(-3, 0)$, $(0, 0)$

50. $f(x) = (x - 2)^{1/3}(x + 1)^{2/3}$

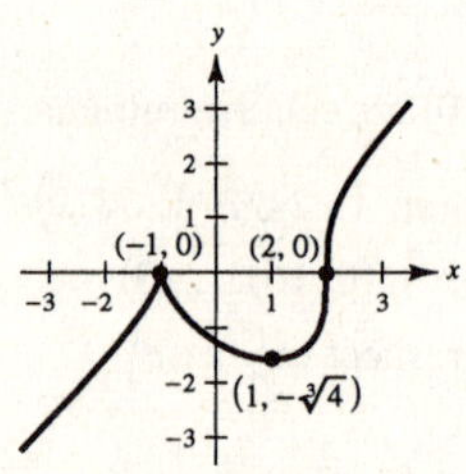

Graph of Exercise 49 translated 2 units to the right (x replaces by $x - 2$).

$(-1, 0)$ is a relative maximum.

$\left(1, -\sqrt[3]{4}\right)$ is a relative minimum.

$(2, 0)$ is a point of inflection.

Intercepts: $(-1, 0)$, $(2, 0)$

51. $f(x) = \dfrac{x + 1}{x - 1}$

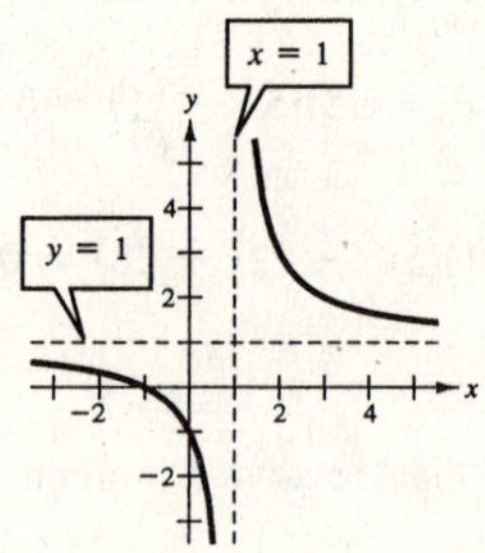

Domain: $(-\infty, 1)$, $(1, \infty)$; Range: $(-\infty, 1)$, $(1, \infty)$

$$f'(x) = \frac{-2}{(x - 1)^2} < 0 \text{ if } x \neq 1.$$

$$f''(x) = \frac{4}{(x - 1)^3}$$

Horizontal asymptote: $y = 1$

Vertical asymptote: $x = 1$

Intercepts: $(-1, 0)$, $(0, -1)$

52. $f(x) = \dfrac{2x}{1 + x^2}$

Domain: $(-\infty, \infty)$; Range: $[-1, 1]$

$f'(x) = \dfrac{2(1 - x)(1 + x)}{(1 + x^2)^2} = 0$ when $x = \pm 1$.

$f''(x) = \dfrac{-4x(3 - x^2)}{(1 + x^2)^3} = 0$ when $x = 0, \pm\sqrt{3}$.

$f''(1) < 0$

Therefore, $(1, 1)$ is a relative maximum.

$f''(-1) > 0$

Therefore, $(-1, -1)$ is a relative minimum.

Points of inflection: $\left(-\sqrt{3}, -\sqrt{3}/2\right)$, $(0, 0)$, $\left(\sqrt{3}, \sqrt{3}/2\right)$

Intercept: $(0, 0)$

Symmetric with respect to the origin

Horizontal asymptote: $y = 0$

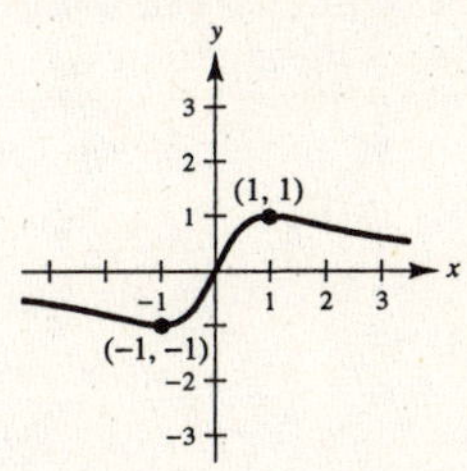

53. $f(x) = \dfrac{4}{1 + x^2}$

Domain: $(-\infty, \infty)$; Range: $(0, 4]$

$f'(x) = \dfrac{-8x}{(1 + x^2)^2} = 0$ when $x = 0$.

$f''(x) = \dfrac{-8(1 - 3x^2)}{(1 + x^2)^3} = 0$ when $x = \pm\dfrac{\sqrt{3}}{3}$.

$f''(0) < 0$

Therefore, $(0, 4)$ is a relative maximum.

Points of inflection: $\left(\pm\sqrt{3}/3, 3\right)$

Intercept: $(0, 4)$

Symmetric to the y-axis

Horizontal asymptote: $y = 0$

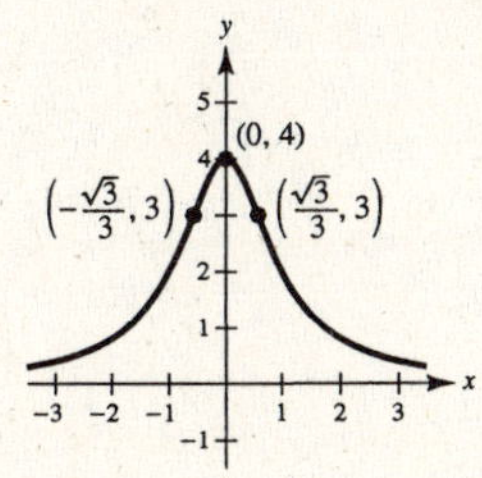

54. $f(x) = \dfrac{x^2}{1 + x^4}$

Domain: $(-\infty, \infty)$; Range: $\left[0, \dfrac{1}{2}\right]$

$f'(x) = \dfrac{(1 + x^4)(2x) - x^2(4x^3)}{(1 + x^4)^2} = \dfrac{2x(1 - x)(1 + x)(1 + x^2)}{(1 + x^4)^2} = 0$ when $x = 0, \pm 1$.

$f''(x) = \dfrac{(1 + x^4)^2(2 - 10x^4) - (2x - 2x^5)(2)(1 + x^4)(4x^3)}{(1 + x^4)^4} = \dfrac{2(1 - 12x^4 + 3x^8)}{(1 + x^4)^3} = 0$ when $x = \pm\sqrt[4]{\dfrac{6 \pm \sqrt{33}}{3}}$.

$f''(\pm 1) < 0$

Therefore, $\left(\pm 1, \dfrac{1}{2}\right)$ are relative maxima.

$f''(0) > 0$

Therefore, $(0, 0)$ is a relative minimum.

Points of inflection: $\left(\pm\sqrt[4]{\dfrac{6 - \sqrt{33}}{3}}, 0.27\right)$, $\left(\pm\sqrt[4]{\dfrac{6 + \sqrt{33}}{3}}, 0.40\right)$

Intercept: $(0, 0)$

Symmetric to the y-axis

Horizontal asymptote: $y = 0$

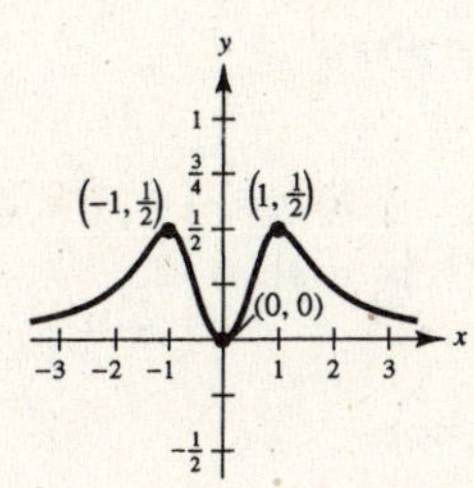

55. $f(x) = x^3 + x + \dfrac{4}{x}$

Domain: $(-\infty, 0), (0, \infty)$; Range: $(-\infty, -6], [6, \infty)$

$f'(x) = 3x^2 + 1 - \dfrac{4}{x^2} = \dfrac{3x^4 + x^2 - 4}{x^2} = 0$ when $x = \pm 1$.

$f''(x) = 6x + \dfrac{8}{x^3} = \dfrac{6x^4 + 8}{x^3} \neq 0$

$f''(-1) < 0$

Therefore, $(-1, -6)$ is a relative maximum.

$f''(1) > 0$

Therefore, $(1, 6)$ is a relative minimum.

Vertical asymptote: $x = 0$

Symmetric with respect to origin

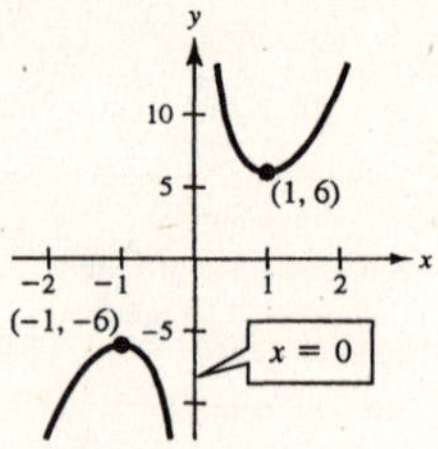

56. $f(x) = x^2 + \dfrac{1}{x} = \dfrac{x^3 + 1}{x}$

Domain: $(-\infty, 0), (0, \infty)$; Range: $(-\infty, \infty)$

$f'(x) = 2x - \dfrac{1}{x^2} = \dfrac{2x^3 - 1}{x^2} = 0$ when $x = \dfrac{1}{\sqrt[3]{2}}$.

$f''(x) = 2 + \dfrac{2}{x^3} = \dfrac{2(x^3 + 1)}{x^3} = 0$ when $x = -1$.

$f''\left(\dfrac{1}{\sqrt[3]{2}}\right) > 0$

Therefore, $\left(\dfrac{1}{\sqrt[3]{2}}, \dfrac{3}{\sqrt[3]{4}}\right)$ is a relative minimum.

Point of inflection: $(-1, 0)$

Intercept: $(-1, 0)$

Vertical asymptote: $x = 0$

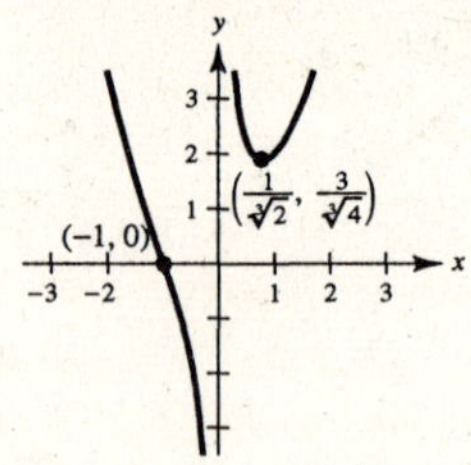

57. $f(x) = |x^2 - 9|$

Domain: $(-\infty, \infty)$; Range: $[0, \infty)$

$f'(x) = \dfrac{2x(x^2 - 9)}{|x^2 - 9|} = 0$ when $x = 0$ and is undefined when $x = \pm 3$.

$f''(x) = \dfrac{2(x^2 - 9)}{|x^2 - 9|}$ is undefined at $x = \pm 3$.

$f''(0) < 0$

Therefore, $(0, 9)$ is a relative maximum.

Relative minima: $(\pm 3, 0)$

Intercepts: $(\pm 3, 0), (0, 9)$

Symmetric to the y-axis

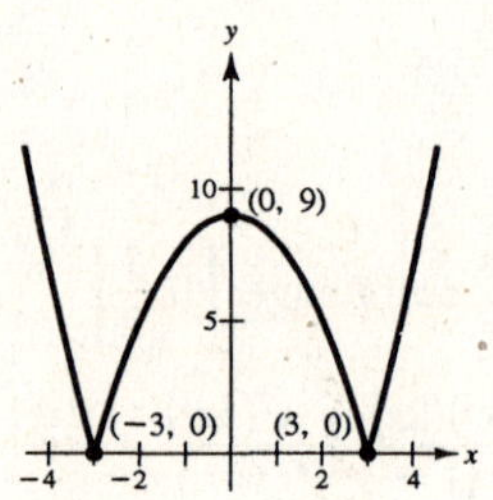

58. $f(x) = |x - 1| + |x - 3| = \begin{cases} -2x + 4, & x \le 1 \\ 2, & 1 < x \le 3 \\ 2x - 4, & x > 3 \end{cases}$

Domain: $(-\infty, \infty)$

Range: $[2, \infty)$

Intercept: $(0, 4)$

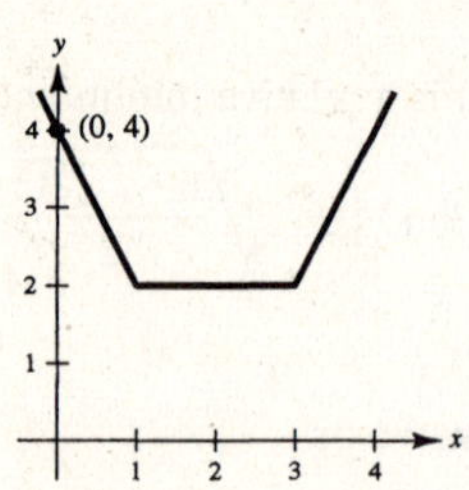

59. $x^2 + 4y^2 - 2x - 16y + 13 = 0$

(a) $(x^2 - 2x + 1) + 4(y^2 - 4y + 4) = -13 + 1 + 16$

$$(x-1)^2 + 4(y-2)^2 = 4$$

$$\frac{(x-1)^2}{4} + \frac{(y-2)^2}{1} = 1$$

The graph is an ellipse:

Maximum: $(1, 3)$

Minimum: $(1, 1)$

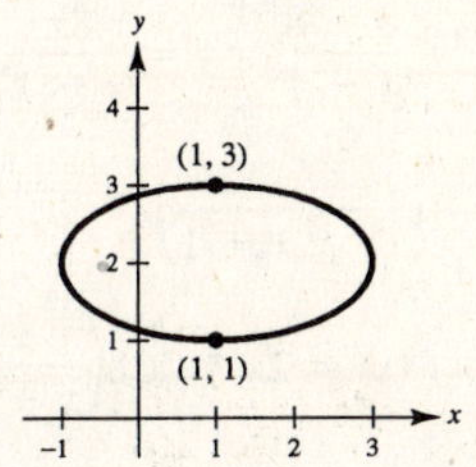

(b) $x^2 + 4y^2 - 2x - 16y + 13 = 0$

$$2x + 8y\frac{dy}{dx} - 2 - 16\frac{dy}{dx} = 0$$

$$\frac{dy}{dx}(8y - 16) = 2 - 2x$$

$$\frac{dy}{dx} = \frac{2 - 2x}{8y - 16} = \frac{1 - x}{4y - 8}$$

The critical numbers are $x = 1$ and $y = 2$. These correspond to the points $(1, 1)$, $(1, 3)$, $(2, -1)$, and $(2, 3)$. Hence, the maximum is $(1, 3)$ and the minimum is $(1, 1)$.

60. $f(x) = x^n$, n is a positive integer.

(a) $f'(x) = nx^{n-1}$

The function has a relative minimum at $(0, 0)$ when n is even.

(b) $f''(x) = n(n-1)x^{n-2}$

The function has a point of inflection at $(0, 0)$ when n is odd and $n \geq 3$.

61. Let $t = 0$ at noon.

$$L = d^2 = (100 - 12t)^2 + (-10t)^2 = 10{,}000 - 2400t + 244t^2$$

$$\frac{dL}{dt} = -2400 + 488t = 0 \text{ when } t = \frac{300}{61} \approx 4.92 \text{ hr.}$$

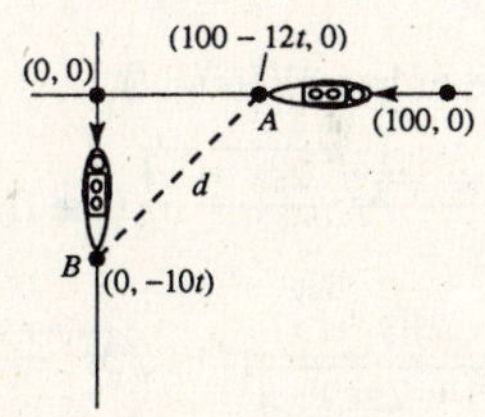

Ship A at $(40.98, 0)$; Ship B at $(0, -49.18)$

$$d^2 = 10{,}000 - 2400t + 244t^2$$

$$\approx 4098.36 \text{ when } t \approx 4.92 \approx \text{4:55 P.M.}.$$

$d \approx 64$ km

62. Ellipse: $\dfrac{x^2}{144} + \dfrac{y^2}{16} = 1$, $y = \dfrac{1}{3}\sqrt{144 - x^2}$

$$A = (2x)\left(\frac{2}{3}\sqrt{144 - x^2}\right) = \frac{4}{3}x\sqrt{144 - x^2}$$

$$\frac{dA}{dx} = \frac{4}{3}\left[\frac{-x^2}{\sqrt{144 - x^2}} + \sqrt{144 - x^2}\right]$$

$$= \frac{4}{3}\left[\frac{144 - 2x^2}{\sqrt{144 - x^2}}\right] = 0 \text{ when } x = \sqrt{72} = 6\sqrt{2}.$$

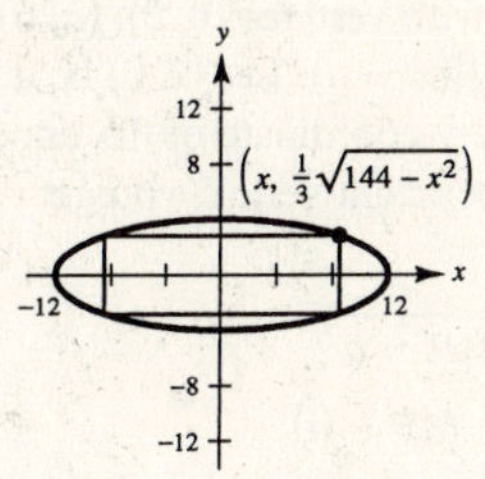

The dimensions of the rectangle are $2x = 12\sqrt{2}$ by $y = \dfrac{2}{3}\sqrt{144 - 72} = 4\sqrt{2}$.

63. We have points $(0, y)$, $(x, 0)$, and $(1, 8)$. Thus,

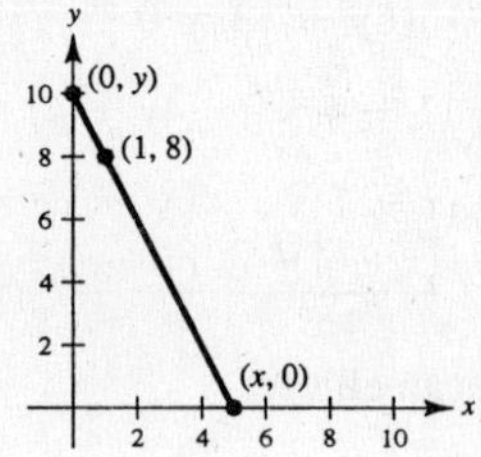

$$m = \frac{y-8}{0-1} = \frac{0-8}{x-1} \text{ or } y = \frac{8x}{x-1}.$$

Let $f(x) = L^2 = x^2 + \left(\frac{8x}{x-1}\right)^2.$

$$f'(x) = 2x + 128\left(\frac{x}{x-1}\right)\left[\frac{(x-1)-x}{(x-1)^2}\right] = 0$$

$$x - \frac{64x}{(x-1)^3} = 0$$

$$x[(x-1)^3 - 64] = 0 \text{ when } x = 0, 5 \text{ (minimum)}.$$

Vertices of triangle: $(0, 0)$, $(5, 0)$, $(0, 10)$

64. We have points $(0, y)$, $(x, 0)$, and $(4, 5)$. Thus,

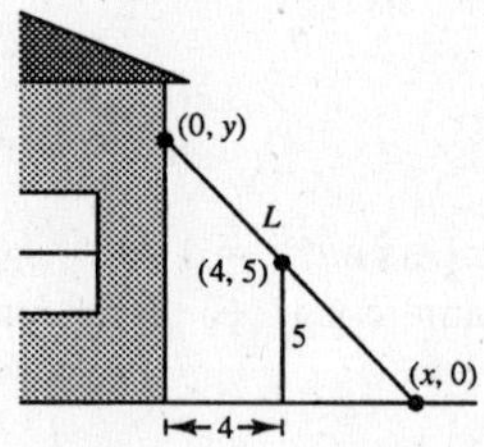

$$m = \frac{y-5}{0-4} = \frac{5-0}{4-x} \text{ or } y = \frac{5x}{x-4}.$$

Let $f(x) = L^2 = x^2 + \left(\frac{5x}{x-4}\right)^2.$

$$f'(x) = 2x + 50\left(\frac{x}{x-4}\right)\left[\frac{x-4-x}{(x-4)^2}\right] = 0$$

$$x - \frac{100x}{(x-4)^3} = 0$$

$$x[(x-4)^3 - 100] = 0 \text{ when } x = 0 \text{ or } x = 4 + \sqrt[3]{100}.$$

$$L = \sqrt{x^2 + \frac{25x^2}{(x-4)^2}} = \frac{x}{x-4}\sqrt{(x-4)^2 + 25} = \frac{\sqrt[3]{100} + 4}{\sqrt[3]{100}}\sqrt{100^{2/3} + 25} \approx 12.7 \text{ feet}$$

65. $A = \text{(Average of bases)(Height)}$

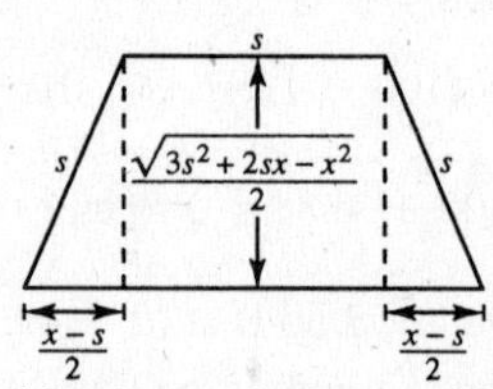

$$= \left(\frac{x+s}{2}\right)\frac{\sqrt{3s^2 + 2sx - x^2}}{2} \text{ (see figure)}$$

$$\frac{dA}{dx} = \frac{1}{4}\left[\frac{(s-x)(s+x)}{\sqrt{3s^2 + 2sx - x^2}} + \sqrt{3s^2 + 2sx - x^2}\right]$$

$$= \frac{2(2s-x)(s+x)}{4\sqrt{3s^2 + 2sx - x^2}} = 0 \text{ when } x = 2s.$$

A is a maximum when $x = 2s$.

66. Label triangle with vertices $(0, 0)$, $(a, 0)$, and (b, c). The equations of the sides of the triangle are $y = (c/b)x$ and $y = [c/(b-a)](x-a)$. Let $(x, 0)$ be a vertex of the inscribed rectangle. The coordinates of the upper left vertex are $(x, (c/b)x)$. The y-coordinate of the upper right vertex of the rectangle is $(c/b)x$. Solving for the x-coordinate $\bar{x}$ of the rectangle's upper right vertex, you get

$$\frac{c}{b}x = \frac{c}{b-a}(\bar{x} - a)$$

$$(b-a)x = b(\bar{x} - a)$$

$$\bar{x} = \frac{b-a}{b}x + a = a - \frac{a-b}{b}x.$$

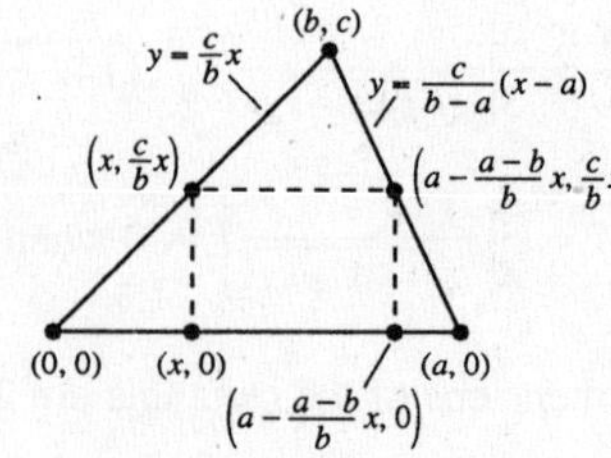

—CONTINUED—

66. —CONTINUED—

Finally, the lower right vertex is

$$\left(a - \frac{a-b}{b}x, 0\right).$$

Width of rectangle: $a - \dfrac{a-b}{b}x - x$

Height of rectangle: $\dfrac{c}{b}x$ (see figure)

$$A = (\text{Width})(\text{Height}) = \left(a - \frac{a-b}{b}x - x\right)\left(\frac{c}{b}x\right) = \left(a - \frac{a}{b}x\right)\frac{c}{b}x$$

$$\frac{dA}{dx} = \left(a - \frac{a}{b}x\right)\frac{c}{b} + \left(\frac{c}{b}x\right)\left(-\frac{a}{b}\right) = \frac{ac}{b} - \frac{2ac}{b^2}x = 0 \text{ when } x = \frac{b}{2}.$$

$$A\left(\frac{b}{2}\right) = \left(a - \frac{a}{b}\frac{b}{2}\right)\left(\frac{c}{b}\frac{b}{2}\right) = \left(\frac{a}{2}\right)\left(\frac{c}{2}\right) = \frac{1}{4}ac = \frac{1}{2}\left(\frac{1}{2}ac\right) = \frac{1}{2}(\text{Area of triangle})$$

67. Total cost = (Cost per hour)(Number of hours)

$$T = \left(\frac{v^2}{600} + 5\right)\left(\frac{110}{v}\right) = \frac{11v}{60} + \frac{550}{v}$$

$$\frac{dT}{dv} = \frac{11}{60} - \frac{550}{v^2} = \frac{11v^2 - 33{,}000}{60v^2}$$

$$= 0 \text{ when } v = \sqrt{3000} = 10\sqrt{30} \approx 54.8 \text{ mph.}$$

$$\frac{d^2T}{dv^2} = \frac{1100}{v^3} > 0 \text{ when } v = 10\sqrt{30} \text{ so this value yields a minimum.}$$

68. Total cost = (Cost per hour)(Number of hours)

$$T = \left(\frac{v^2}{500} + 7.50\right)\left(\frac{110}{v}\right) = \frac{11v}{50} + \frac{825}{v}$$

$$\frac{dT}{dv} = \frac{11}{50} - \frac{825}{v^2} = \frac{11v^2 - 41{,}250}{50v^2}$$

$$= 0 \text{ when } v = \sqrt{3750} = 25\sqrt{6} \approx 61.2 \text{ mph.}$$

$$\frac{d^2T}{dv^2} = \frac{1650}{v^3} > 0 \text{ when } v = 25\sqrt{6} \text{ so this value yields a minimum.}$$

69. $y = (3x^2 - 2)^3$

$$\frac{dy}{dx} = 3(3x^2 - 2)^2(6x)$$

$$dy = 18x(3x^2 - 2)^2\,dx$$

70. $y = \sqrt{36 - x^2}$

$$\frac{dy}{dx} = \frac{1}{2}(36 - x^2)^{-1/2}(-2x) = \frac{-x}{\sqrt{36 - x^2}}$$

$$dy = \frac{-x}{\sqrt{36 - x^2}}dx$$

71. $S = 4\pi r^2$. $dr = \Delta r = \pm 0.025$

$$dS = 8\pi r\, dr = 8\pi(9)(\pm 0.025)$$

$$= \pm 1.8\pi \text{ square cm}$$

$$\frac{dS}{S}(100) = \frac{8\pi r\, dr}{4\pi r^2}(100) = \frac{2\, dr}{r}(100)$$

$$= \frac{2(\pm 0.025)}{9}(100) \approx \pm 0.56\%$$

$$V = \frac{4}{3}\pi r^3$$

$$dV = 4\pi r^2\, dr = 4\pi(9)^2(\pm 0.025)$$

$$= \pm 8.1\pi \text{ cubic cm}$$

$$\frac{dV}{V}(100) = \frac{4\pi r^2\, dr}{(4/3)\pi r^3}(100) = \frac{3\, dr}{r}(100)$$

$$= \frac{3(\pm 0.025)}{9}(100) \approx \pm 0.83\%$$

72. $p = 75 - \frac{1}{4}x$

$$\Delta p = p(8) - p(7)$$

$$= \left(75 - \frac{8}{4}\right) - \left(75 - \frac{7}{4}\right) = -\frac{1}{4}$$

$$dp = -\frac{1}{4}dx = -\frac{1}{4}(1) = -\frac{1}{4}$$

[$\Delta p = dp$ because p is linear]

Problem Solving for Chapter 5

1. Assume $y_1 < d < y_2$. Let $g(x) = f(x) - d(x - a)$. g is continuous on $[a, b]$ and therefore has a minimum $(c, g(c))$ on $[a, b]$. The point c cannot be an endpoint of $[a, b]$ because

$$g'(a) = f'(a) - d = y_1 - d < 0$$

$$g'(b) = f'(b) - d = y_2 - d > 0$$

Hence, $a < c < b$ and $g'(c) = 0 \Rightarrow f'(c) = d$.

2. (a) $$dV = 3x^2\, dx = 3x^2\,\Delta x$$

$$\Delta V = (x + \Delta x)^3 - x^3 = 3x^2\,\Delta x + 3x(\Delta x)^2 + (\Delta x)^3$$

$$\Delta V - dV = 3x(\Delta x)^2 + (\Delta x)^3 = \underbrace{[3x\,\Delta x + (\Delta x)^2]}_{\varepsilon}\,\Delta x$$

$$= \varepsilon\,\Delta x, \text{ where } \varepsilon \to 0 \text{ as } \Delta x \to 0.$$

(b) Let $\varepsilon = \dfrac{\Delta y}{\Delta x} - f'(x)$. Then $\varepsilon \to 0$ as $\Delta x \to 0$.

Furthermore, $\Delta y - dy = \Delta y - f'(x)\, dx = \varepsilon\,\Delta x$.

3. (a) For $a = -3, -2, -1, 0$, p has a relative maximum at $(0, 0)$.

For $a = 1, 2, 3$, p has a relative maximum at $(0, 0)$ and 2 relative minima.

(b) $p'(x) = 4ax^3 - 12x = 4x(ax^2 - 3) = 0 \Rightarrow x = 0, \pm\sqrt{\dfrac{3}{a}}$

$p''(x) = 12ax^2 - 12 = 12(ax^2 - 1)$

For $x = 0$, $p''(0) = -12 < 0 \Rightarrow p$ has a relative maximum at $(0, 0)$.

(c) If $a > 0$, $x = \pm\sqrt{\dfrac{3}{a}}$ are the remaining critical numbers.

$$p''\left(\pm\sqrt{\frac{3}{a}}\right) = 12a\left(\frac{3}{a}\right) - 12 = 24 > 0 \Rightarrow p \text{ has relative minima for } a > 0.$$

(d) $(0, 0)$ lies on $y = -3x^2$.

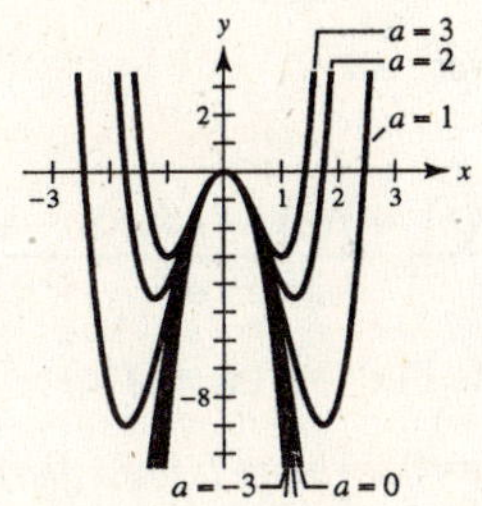

Let $x = \pm\sqrt{\dfrac{3}{a}}$. Then

$$p(x) = a\left(\frac{3}{a}\right)^2 - 6\left(\frac{3}{a}\right) = \frac{9}{a} - \frac{18}{a} = -\frac{9}{a}.$$

Thus, $y = -\dfrac{9}{a} = -3\left(\pm\sqrt{\dfrac{3}{a}}\right)^2 = -3x^2$ is satisfied by all the relative extrema of p.

4. Let $h(x) = g(x) - f(x)$, which is continuous on $[a, b]$ and differentiable on (a, b). $h(a) = 0$ and $h(b) = g(b) - f(b)$.

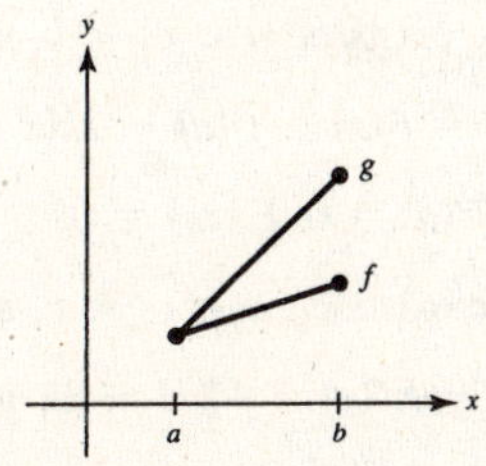

By the Mean Value Theorem, there exists c in (a, b) such that

$$h'(c) = \frac{h(b) - h(a)}{b - a} = \frac{g(b) - f(b)}{b - a}.$$

Since $h'(c) = g'(c) - f'(c) > 0$ and $b - a > 0$,

$g(b) - f(b) > 0 \Rightarrow g(b) > f(b)$.

5. $p(x) = x^4 + ax^2 + 1$

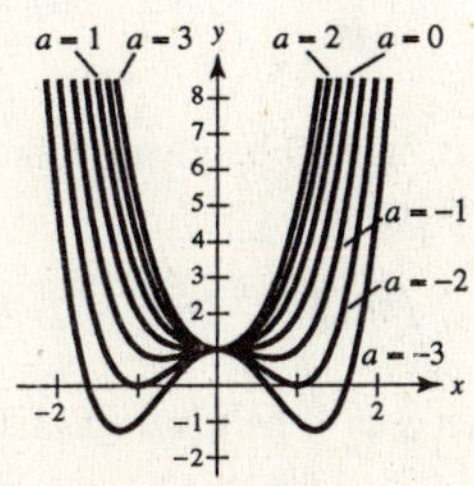

(a) $p'(x) = 4x^3 + 2ax = 2x(2x^2 + a)$

$p''(x) = 12x^2 + 2a$

For $a \geq 0$, there is one relative minimum at $(0, 1)$.

(b) For $a < 0$, there is a relative maximum at $(0, 1)$.

(c) For $a < 0$, there are two relative minima at $x = \pm\sqrt{-\dfrac{a}{2}}$.

(d) There are either 1 or 3 critical points. The above analysis shows that there cannot be exactly two relative extrema.

6. (a) $f' = 2ax + b$, $f'' = 2a \neq 0$. No points of inflection.

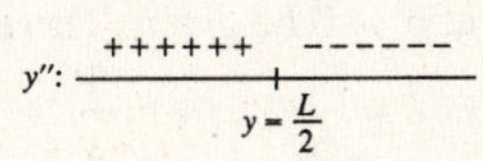

(b) $f' = 3ax^2 + 2bx + c$, $f'' = 6ax + 2b = 0 \Rightarrow x = \dfrac{-b}{3a}$. One point of inflection.

(c) $y' = ky(L - y) = kLy - ky^2$

$y'' = kLy' - 2kyy' = ky'(L - 2y)$

If $y = \dfrac{L}{2}$, then $y'' = 0$ and this is a point of inflection because of the analysis above.

7. $f(x) = \frac{c}{x} + x^2$

$f'(x) = -\frac{c}{x^2} + 2x = 0 \Rightarrow \frac{c}{x^2} = 2x \Rightarrow x^3 = \frac{c}{2} \Rightarrow x = \sqrt[3]{\frac{c}{2}}$

$f''(x) = \frac{2c}{x^3} + 2$

If $c = 0, f(x) = x^2$ has a relative minimum, but no relative maximum.

If $c > 0, x = \sqrt[3]{\frac{c}{2}}$ is a relative minimum, because $f''\left(\sqrt[3]{\frac{c}{2}}\right) > 0$.

If $c < 0, x = \sqrt[3]{\frac{c}{2}}$ is a relative minimum too.

Answer: all c.

8. Set $\frac{f(b) - f(a) - f'(a)(b - a)}{(b - a)^2} = k$.

Define $F(x) = f(x) - f(a) - f'(a)(x - a) - k(x - a)^2$.

$F(a) = 0, F(b) = f(b) - f(a) - f'(a)(b - a) - k(b - a)^2 = 0$

F is continuous on $[a, b]$ and differentiable on (a, b).

There exists $c_1, a < c_1 < b$, satisfying $F'(c_1) = 0$.

$F'(x) = f'(x) - f'(a) - 2k(x - a)$ satisfies the hypothesis of Rolle's Theorem on $[a, c_1]$:

$F'(a) = 0, F'(c_1) = 0.$

There exists $c_2, a < c_2 < c_1$ satisfying $F''(c_2) = 0$.

Finally, $F''(x) = f''(x) - 2k$ and $F''(c_2) = 0$ implies that

$k = \frac{f''(c_2)}{2}$.

Thus, $k = \frac{f(b) - f(a) - f'(a)(b - a)}{(b - a)^2} = \frac{f''(c_2)}{2} \Rightarrow f(b) = f(a) + f'(a)(b - a) + \frac{1}{2}f''(c_2)(b - a)^2$.

9. (a) Let $M > 0$ be given. Take $N = \sqrt{M}$. Then whenever $x > N = \sqrt{M}$, you have

$f(x) = x^2 > M.$

(b) Let $\varepsilon > 0$ be given. Let $M = \sqrt{\frac{1}{\varepsilon}}$. Then whenever $x > M = \sqrt{\frac{1}{\varepsilon}}$, you have

$x^2 > \frac{1}{\varepsilon} \Rightarrow \frac{1}{x^2} < \varepsilon \Rightarrow \left|\frac{1}{x^2} - 0\right| < \varepsilon.$

(c) Let $\varepsilon > 0$ be given. There exists $N > 0$ such that $|f(x) - L| < \varepsilon$ whenever $x > N$.

Let $\delta = \frac{1}{N}$. Let $x = \frac{1}{y}$.

If $0 < y < \delta = \frac{1}{N}$, then $\frac{1}{x} < \frac{1}{N} \Rightarrow x > N$ and

$|f(x) - L| = \left|f\left(\frac{1}{y}\right) - L\right| < \varepsilon.$

10. Let T be the intersection of PQ and RS. Let MN be the perpendicular to SQ and PR passing through T.

Let $TM = x$ and $TN = b - x$.

$$\frac{SN}{b-x} = \frac{MR}{x} \Rightarrow SN = \frac{b-x}{x}MR$$

$$\frac{NQ}{b-x} = \frac{PM}{x} \Rightarrow NQ = \frac{b-x}{x}PM$$

$$SQ = \frac{b-x}{x}(MR + PM) = \frac{b-x}{x}d$$

$$A(x) = \text{Area} = \frac{1}{2}dx + \frac{1}{2}\left(\frac{b-x}{x}d\right)(b-x) = \frac{1}{2}d\left[x + \frac{(b-x)^2}{x}\right] = \frac{1}{2}d\left[\frac{2x^2 - 2bx + b^2}{x}\right]$$

$$A'(x) = \frac{1}{2}d\left[\frac{x(4x-2b) - (2x^2 - 2bx + b^2)}{x^2}\right]$$

$$A'(x) = 0 \Rightarrow 4x^2 - 2xb = 2x^2 - 2bx + b^2$$

$$2x^2 = b^2$$

$$x = \frac{b}{\sqrt{2}}$$

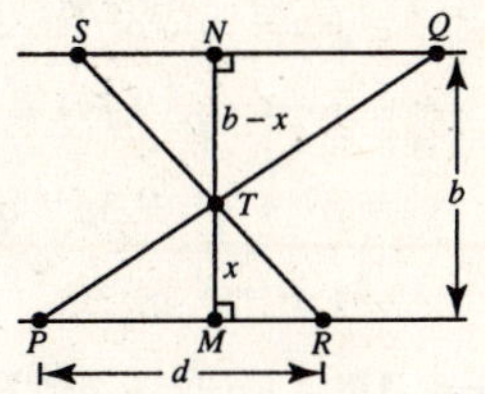

Hence, we have $SQ = \dfrac{b-x}{x}d = \dfrac{b - (b/\sqrt{2})}{b/\sqrt{2}}d = (\sqrt{2} - 1)d.$

Using the Second Derivative Test, this is a minimum. There is no maximum.

11. Let m be the slope of the line, $m < 0$.

$y - q = m(x - p)$

$y = mx + q - mp$

Intercepts: $y = 0 \Rightarrow x = \dfrac{mp - q}{m} = p - \dfrac{q}{m}$ $\quad A = \left(p - \dfrac{q}{m}, 0\right)$

$x = 0 \Rightarrow y = q - mp \quad B = (0, q - mp)$

(a) $f(m) = OA + OB = p - \dfrac{q}{m} + q - mp$

$f'(m) = \dfrac{q}{m^2} - p = 0 \Rightarrow m^2 = \dfrac{q}{p} \Rightarrow m = -\sqrt{(q/p)}.$

Thus $f\left(-\sqrt{(q/p)}\right) = p + q\sqrt{(p/q)} + q + p\sqrt{(q/p)}$

$= p + q + 2\sqrt{pq}.$

(b) $g(m) = OA \cdot OB = \left(p - \dfrac{q}{m}\right)(q - mp)$

$= pq - mp^2 - \dfrac{q^2}{m} + pq$

$g'(m) = -p^2 + \dfrac{q^2}{m^2} = 0 \Rightarrow m^2 = \dfrac{q^2}{p^2} \Rightarrow m = -\dfrac{q}{p}.$

$g\left(-\dfrac{q}{p}\right) = pq + \left(\dfrac{q}{p}\right)p^2 + q^2\left(\dfrac{p}{q}\right) + pq$

$= 4pq$

—CONTINUED—

11. —CONTINUED—

(c) Let $h(m) = (AB)^2 = \left(p - \frac{q}{m}\right)^2 + (q - mp)^2$

$$= p^2 - \frac{2pq}{m} + \frac{q^2}{m^2} + q^2 - 2pqm + m^2p^2$$

$$h'(m) = \frac{2pq}{m^2} - \frac{2q^2}{m^3} - 2pq + 2mp^2 = 0$$

$$2pqm - 2q^2 - 2pqm^3 + 2m^4p^2 = 0$$

$$(pm - q)(pm^3 + q) = 0$$

Thus, $m = \left(\frac{-q}{p}\right)^{1/3}$ and

$$AB^2 = \left(p - \frac{q}{m}\right)^2 + (q - mp)^2$$

$$= (p + p^{1/3}q^{2/3})^2 + (q + p^{2/3}q^{1/3})^2$$

$$= p^2 + 2p^{4/3}q^{2/3} + p^{2/3}q^{4/3} + q^2 + 2p^{2/3}q^{4/3} + p^{4/3}q^{2/3}$$

$$= p^2 + q^2 + 3p^{4/3}q^{2/3} + 3p^{2/3}q^{4/3}$$

$$= (p^{2/3} + q^{2/3})^3$$

$$\Rightarrow AB = (p^{2/3} + q^{2/3})^{3/2}.$$

12. $f(x) = |8 - x^3|$

(a) $f(x) = |(2 - x)(4 + 2x + x^2)|$

$$= \begin{cases} x^3 - 8, & x \ge 2 \\ 8 - x^3, & x < 2 \end{cases}$$

(b) $f'(x) = \begin{cases} 3x^2, & x > 2 \\ -3x^2, & x < 2 \end{cases}$

(c) f increasing on $(2, \infty)$

f decreasing on $(-\infty, 2)$

(d) $f''(x) = \begin{cases} 6x, & x > 2 \\ -6x, & x < 2 \end{cases}$

f concave upward on $(-\infty, 0)$ and $(2, \infty)$

f concave downward on $(0, 2)$

(e) $(0, 8)$ is the only point of inflection. $(2, 0)$ is not a point of inflection because f does not have a tangent line at $(2, 0)$.

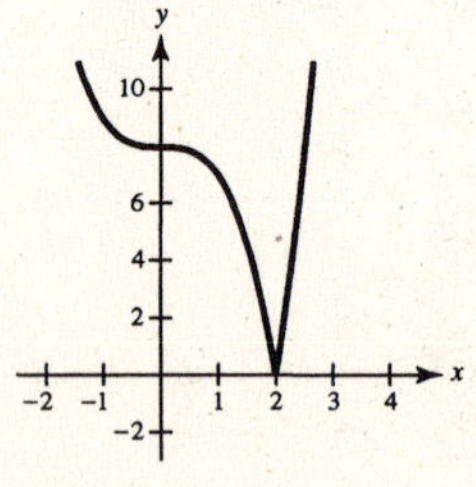

13. Place the triangle as indicated in the figure. Let (x_0, y_0) be between A and B.

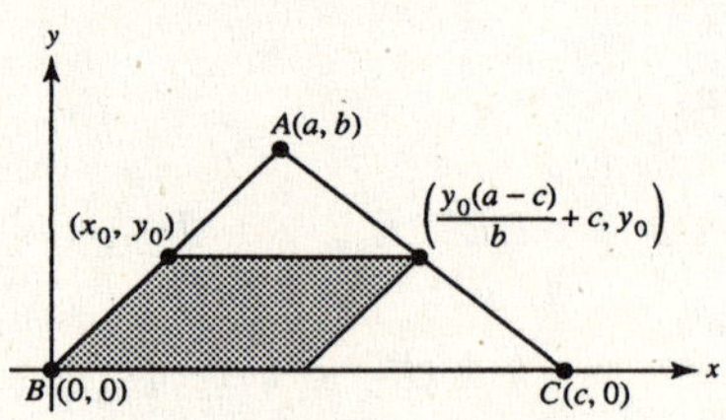

$$\text{Area} = f(y_0) = \text{(base)(height)}$$

$$= \left(\frac{y_0(a - c)}{b} + c - x_0\right)(y_0)$$

$$= \left(\frac{y_0(a - c)}{b} + c - \frac{ay_0}{b}\right)y_0$$

$$= \left(c - \frac{cy_0}{b}\right)y_0$$

$$= cy_0 - \frac{cy_0^2}{b}$$

$$f'(y_0) = c - \frac{2cy_0}{b} = 0 \Rightarrow y_0 = \frac{b}{2}, \quad x_0 = \frac{a}{2}$$

and area $= \frac{1}{4}bc = \frac{1}{2}$ area $\triangle ABC$

Without calculus, observe from the figure that the maximum area is obtained by using the midpoints of each side.

14. Distance $= \sqrt{4^2 + x^2} + \sqrt{(4 - x)^2 + 4^2} = f(x)$

$$f'(x) = \frac{x}{\sqrt{4^2 + x^2}} - \frac{4 - x}{\sqrt{(4 - x)^2 + 4^2}} = 0$$

$$x\sqrt{(4 - x)^2 + 4^2} = (4 - x)\sqrt{4^2 + x^2}$$

$$x^2[16 - 8x + x^2 + 16] = (x^2 - 8x + 16)(16 + x^2)$$

$$32x^2 - 8x^3 + x^4 = x^4 - 8x^3 + 32x^2 - 128x + 256$$

$$128x = 256$$

$$x = 2$$

The bug should head towards the midpoint of the opposite side.

Without Calculus: Imagine opening up the cube:

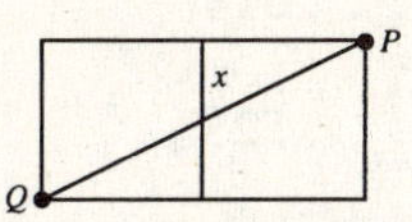

The shortest distance is the line PQ, passing through the midpoint.

15. The line has equation $\frac{x}{3} + \frac{y}{4} = 1$ or $y = -\frac{4}{3}x + 4$.

Rectangle:

$$\text{Area} = A = xy = x\left(-\frac{4}{3}x + 4\right) = -\frac{4}{3}x^2 + 4x.$$

$$A'(x) = -\frac{8}{3}x + 4 = 0 \Rightarrow \frac{8}{3}x = 4 \Rightarrow x = \frac{3}{2}$$

Dimensions: $\frac{3}{2} \times 2$ Calculus was helpful.

Circle: The distance from the center (r, r) to the line $\frac{x}{3} + \frac{y}{4} - 1 = 0$ must be r:

$$r = \frac{\left|\frac{r}{3} + \frac{r}{4} - 1\right|}{\sqrt{\frac{1}{9} + \frac{1}{16}}} = \frac{12}{5}\left|\frac{7r - 12}{12}\right| = \frac{|7r - 12|}{5}$$

$$5r = |7r - 12| \Rightarrow r = 1 \text{ or } r = 6.$$

Clearly, $r = 1$.

Semicircle: The center lies on the line $\frac{x}{3} + \frac{y}{4} = 1$ and satisfies $x = y = r$.

Thus $\frac{r}{3} + \frac{r}{4} = 1 \Rightarrow \frac{7}{12}r = 1 \Rightarrow r = \frac{12}{7}$.

No calculus necessary.

16. (a) $s = \dfrac{v\,\frac{\text{km}}{\text{hr}}\left(1000\,\frac{\text{m}}{\text{km}}\right)}{\left(3600\,\frac{\text{sec}}{\text{hr}}\right)} = \frac{5}{18}v$

v	20	40	60	80	100
s	5.56	11.11	16.67	22.22	27.78
d	5.1	13.7	27.2	44.2	66.4

$$d(s) = 0.071s^2 + 0.389s + 0.727$$

(b) The distance between the back of the first vehicle and the front of the second vehicle is $d(s)$, the safe stopping distance. The first vehicle passes the given point in $5.5/s$ seconds, and the second vehicle takes $d(s)/s$ more seconds. Hence,

$$T = \frac{d(s)}{s} + \frac{5.5}{s}.$$

(c)

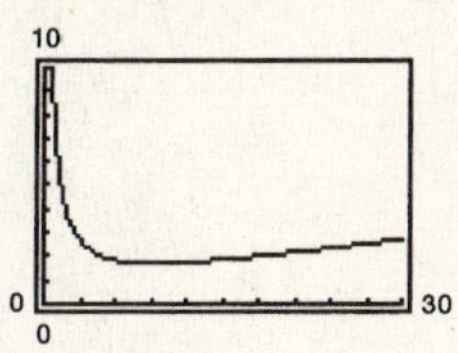

$$T = \frac{1}{s}(0.071s^2 + 0.389s + 0.727) + \frac{5.5}{s}$$

The minimum is attained when $s \approx 9.365$ m/sec.

(d) $$T(s) = 0.071s + 0.389 + \frac{6.227}{s}$$

$$T'(s) = 0.071 - \frac{6.227}{s^2} \Rightarrow s^2 = \frac{6.227}{0.071}$$

$$\Rightarrow s \approx 9.365 \text{ m/sec}$$

$T(9.365) \approx 1.719$ seconds

$$9.365 \text{ m/sec} \cdot \frac{3600}{1000} = 33.7 \text{ km/hr}$$

(e) $d(9.365) = 10.597$ m

17. (a)

x	0	0.5	1	2
$\sqrt{1 + x}$	1	1.2247	1.4142	1.7321
$\frac{1}{2}x + 1$	1	1.25	1.5	2

(b) Let $f(x) = \sqrt{1 + x}$. Using the Mean Value Theorem on the interval $[0, x]$,there exists c, $0 < c < x$, satisfying

$$f'(c) = \frac{1}{2\sqrt{1 + c}} = \frac{f(x) - f(0)}{x - 0} = \frac{\sqrt{1 + x} - 1}{x}.$$

Thus $\sqrt{1 + x} = \dfrac{x}{2\sqrt{1 + c}} + 1 < \dfrac{x}{2} + 1$

$\left(\text{because } \sqrt{1 + c} > 1\right)$.

CHAPTER 6
Integration

Section 6.1 Antiderivatives and Indefinite Integration **477**

Section 6.2 Area **486**

Section 6.3 Riemann Sums and Definite Integrals **498**

Section 6.4 The Fundamental Theorem of Calculus **505**

Section 6.5 Integration by Substitution **514**

Section 6.6 Numerical Integration **525**

Review Exercises **532**

Problem Solving **541**

CHAPTER 6
Integration

Section 6.1 Antiderivatives and Indefinite Integration

1. $\frac{d}{dx}\left(\frac{3}{x^3} + C\right) = \frac{d}{dx}(3x^{-3} + C) = -9x^{-4} = \frac{-9}{x^4}$

2. $\frac{d}{dx}\left(x^4 + \frac{1}{x} + C\right) = 4x^3 - \frac{1}{x^2}$

3. $\frac{d}{dx}\left(\frac{1}{3}x^3 - 4x + C\right) = x^2 - 4 = (x - 2)(x + 2)$

4. $\frac{d}{dx}\left(\frac{2(x^2 + 3)}{3\sqrt{x}} + C\right) = \frac{d}{dx}\left(\frac{2}{3}x^{3/2} + 2x^{-1/2} + C\right)$

$= x^{1/2} - x^{-3/2} = \frac{x^2 - 1}{x^{3/2}}$

5. $\frac{dy}{dt} = 3t^2$

$y = t^3 + C$

Check: $\frac{d}{dt}[t^3 + C] = 3t^2$

6. $\frac{dr}{d\theta} = \pi$

$r = \pi\theta + C$

Check: $\frac{d}{d\theta}[\pi\theta + C] = \pi$

7. $\frac{dy}{dx} = x^{3/2}$

$y = \frac{2}{5}x^{5/2} + C$

Check: $\frac{d}{dx}\left[\frac{2}{5}x^{5/2} + C\right] = x^{3/2}$

8. $\frac{dy}{dx} = 2x^{-3}$

$y = \frac{2x^{-2}}{-2} + C = \frac{-1}{x^2} + C$

Check: $\frac{d}{dx}\left[\frac{-1}{x^2} + C\right] = 2x^{-3}$

	Given	Rewrite	Integrate	Simplify
9.	$\int \sqrt[3]{x}\,dx$	$\int x^{1/3}\,dx$	$\frac{x^{4/3}}{4/3} + C$	$\frac{3}{4}x^{4/3} + C$
10.	$\int \frac{1}{x^2}\,dx$	$\int x^{-2}\,dx$	$\frac{x^{-1}}{-1} + C$	$-\frac{1}{x} + C$
11.	$\int \frac{1}{x\sqrt{x}}\,dx$	$\int x^{-3/2}\,dx$	$\frac{x^{-1/2}}{-1/2} + C$	$-\frac{2}{\sqrt{x}} + C$
12.	$\int x(x^2 + 3)\,dx$	$\int (x^3 + 3x)\,dx$	$\frac{x^4}{4} + 3\left(\frac{x^2}{2}\right) + C$	$\frac{1}{4}x^4 + \frac{3}{2}x^2 + C$
13.	$\int \frac{1}{2x^3}\,dx$	$\frac{1}{2}\int x^{-3}\,dx$	$\frac{1}{2}\left(\frac{x^{-2}}{-2}\right) + C$	$-\frac{1}{4x^2} + C$
14.	$\int \frac{1}{(3x)^2}\,dx$	$\frac{1}{9}\int x^{-2}\,dx$	$\frac{1}{9}\left(\frac{x^{-1}}{-1}\right) + C$	$\frac{-1}{9x} + C$

15. $\int (x+3)\,dx = \frac{x^2}{2} + 3x + C$

Check: $\frac{d}{dx}\left[\frac{x^2}{2} + 3x + C\right] = x + 3$

16. $\int (5 - x)\,dx = 5x - \frac{x^2}{2} + C$

Check: $\frac{d}{dx}\left[5x - \frac{x^2}{2} + C\right] = 5 - x$

17. $\int (2x - 3x^2)\,dx = x^2 - x^3 + C$

Check: $\frac{d}{dx}[x^2 - x^3 + C] = 2x - 3x^2$

18. $\int (4x^3 + 6x^2 - 1)\,dx = x^4 + 2x^3 - x + C$

Check: $\frac{d}{dx}[x^4 + 2x^3 - x + C] = 4x^3 + 6x^2 - 1$

19. $\int (x^3 + 2)\,dx = \frac{1}{4}x^4 + 2x + C$

Check: $\frac{d}{dx}\left(\frac{1}{4}x^4 + 2x + C\right) = x^3 + 2$

20. $\int (x^3 - 4x + 2)\,dx = \frac{x^4}{4} - 2x^2 + 2x + C$

Check: $\frac{d}{dx}\left[\frac{x^4}{4} - 2x^2 + 2x + C\right] = x^3 - 4x + 2$

21. $\int (x^{3/2} + 2x + 1)\,dx = \frac{2}{5}x^{5/2} + x^2 + x + C$

Check: $\frac{d}{dx}\left(\frac{2}{5}x^{5/2} + x^2 + x + C\right) = x^{3/2} + 2x + 1$

22. $\int \left(\sqrt{x} + \frac{1}{2\sqrt{x}}\right) dx = \int \left(x^{1/2} + \frac{1}{2}x^{-1/2}\right) dx = \frac{x^{3/2}}{3/2} + \frac{1}{2}\left(\frac{x^{1/2}}{1/2}\right) + C = \frac{2}{3}x^{3/2} + x^{1/2} + C$

Check: $\frac{d}{dx}\left(\frac{2}{3}x^{3/2} + x^{1/2} + C\right) = x^{1/2} + \frac{1}{2}x^{-1/2} = \sqrt{x} + \frac{1}{2\sqrt{x}}$

23. $\int \sqrt[3]{x^2}\,dx = \int x^{2/3}\,dx = \frac{x^{5/3}}{5/3} + C = \frac{3}{5}x^{5/3} + C$

Check: $\frac{d}{dx}\left(\frac{3}{5}x^{5/3} + C\right) = x^{2/3} = \sqrt[3]{x^2}$

24. $\int \left(\sqrt[4]{x^3} + 1\right) dx = \int (x^{3/4} + 1)\,dx = \frac{4}{7}x^{7/4} + x + C$

Check: $\frac{d}{dx}\left(\frac{4}{7}x^{7/4} + x + C\right) = x^{3/4} + 1 = \sqrt[4]{x^3} + 1$

25. $\int \frac{1}{x^3}\,dx = \int x^{-3}\,dx = \frac{x^{-2}}{-2} + C = -\frac{1}{2x^2} + C$

Check: $\frac{d}{dx}\left(-\frac{1}{2x^2} + C\right) = \frac{1}{x^3}$

26. $\int \frac{1}{x^4}\,dx = \int x^{-4}\,dx = \frac{x^{-3}}{-3} + C = -\frac{1}{3x^3} + C$

Check: $\frac{d}{dx}\left(-\frac{1}{3x^3} + C\right) = \frac{1}{x^4}$

27. $\int \frac{x^2 + x + 1}{\sqrt{x}}\,dx = \int (x^{3/2} + x^{1/2} + x^{-1/2})\,dx = \frac{2}{5}x^{5/2} + \frac{2}{3}x^{3/2} + 2x^{1/2} + C = \frac{2}{15}x^{1/2}(3x^2 + 5x + 15) + C$

Check: $\frac{d}{dx}\left(\frac{2}{5}x^{5/2} + \frac{2}{3}x^{3/2} + 2x^{1/2} + C\right) = x^{3/2} + x^{1/2} + x^{-1/2} = \frac{x^2 + x + 1}{\sqrt{x}}$

28. $\int \frac{x^2 + 2x - 3}{x^4}\,dx = \int (x^{-2} + 2x^{-3} - 3x^{-4})\,dx = \frac{x^{-1}}{-1} + \frac{2x^{-2}}{-2} - \frac{3x^{-3}}{-3} + C = \frac{-1}{x} - \frac{1}{x^2} + \frac{1}{x^3} + C$

Check: $\frac{d}{dx}\left[-\frac{1}{x} - \frac{1}{x^2} + \frac{1}{x^3} + C\right] = x^{-2} + 2x^{-3} - 3x^{-4} = \frac{x^2 + 2x - 3}{x^4}$

29. $\int (x+1)(3x-2)\,dx = \int (3x^2 + x - 2)\,dx$

$$= x^3 + \frac{1}{2}x^2 - 2x + C$$

Check: $\frac{d}{dx}\left(x^3 + \frac{1}{2}x^2 - 2x + C\right) = 3x^2 + x - 2$

$$= (x+1)(3x-2)$$

30. $\int (2t^2 - 1)^2\,dt = \int (4t^4 - 4t^2 + 1)\,dt$

$$= \frac{4}{5}t^5 - \frac{4}{3}t^3 + t + C$$

Check: $\frac{d}{dt}\left(\frac{4}{5}t^5 - \frac{4}{3}t^3 + t + C\right) = 4t^4 - 4t^2 + 1$

$$= (2t^2 - 1)^2$$

31. $\int y^2\sqrt{y}\,dy = \int y^{5/2}\,dy = \frac{2}{7}y^{7/2} + C$

Check: $\frac{d}{dy}\left(\frac{2}{7}y^{7/2} + C\right) = y^{5/2} = y^2\sqrt{y}$

32. $\int (1 + 3t)t^2\,dt = \int (t^2 + 3t^3)\,dt = \frac{1}{3}t^3 + \frac{3}{4}t^4 + C$

Check: $\frac{d}{dt}\left(\frac{1}{3}t^3 + \frac{3}{4}t^4 + C\right) = t^2 + 3t^3 = (1 + 3t)t^2$

33. $\int dx = \int 1\,dx = x + C$

Check: $\frac{d}{dx}(x + C) = 1$

34. $\int 3\,dt = 3t + C$

Check: $\frac{d}{dt}(3t + C) = 3$

35.

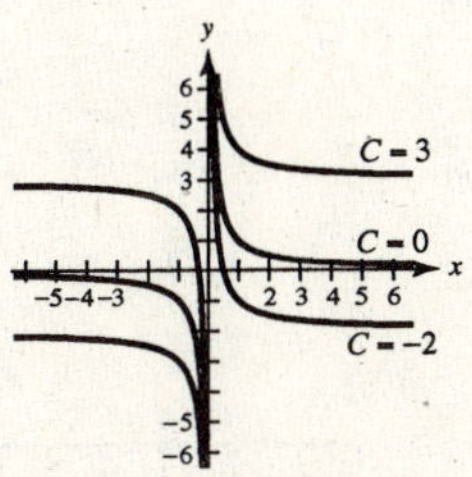

36. $f(x) = \sqrt{x}$

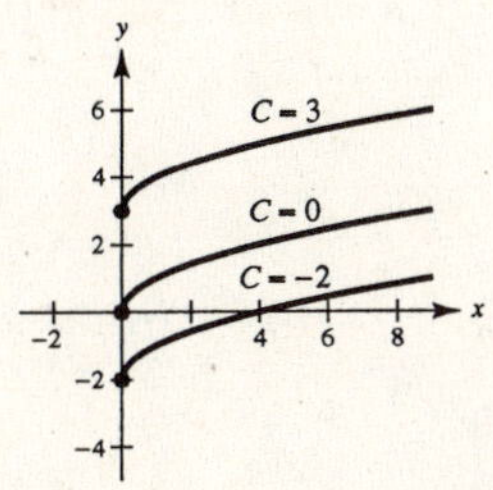

37. $f'(x) = 2$

$f(x) = 2x + C$

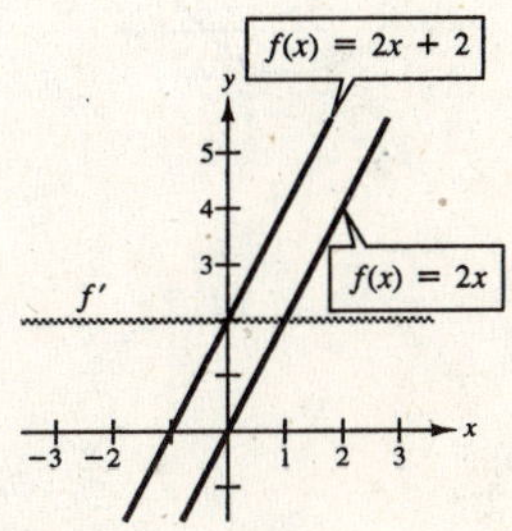

Answers will vary.

38. $f'(x) = x$

$f(x) = \frac{x^2}{2} + C$

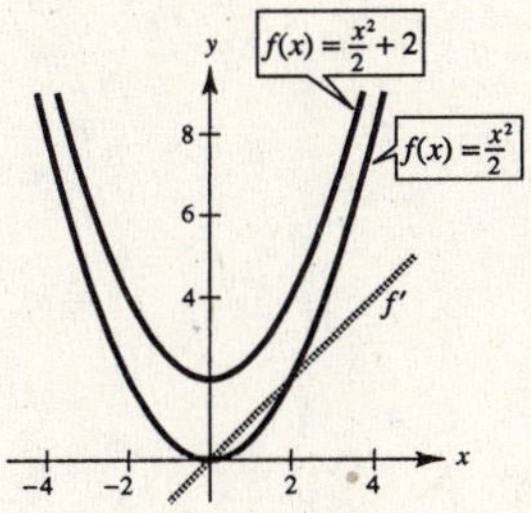

39. $f'(x) = 1 - x^2$

$f(x) = x - \frac{x^3}{3} + C$

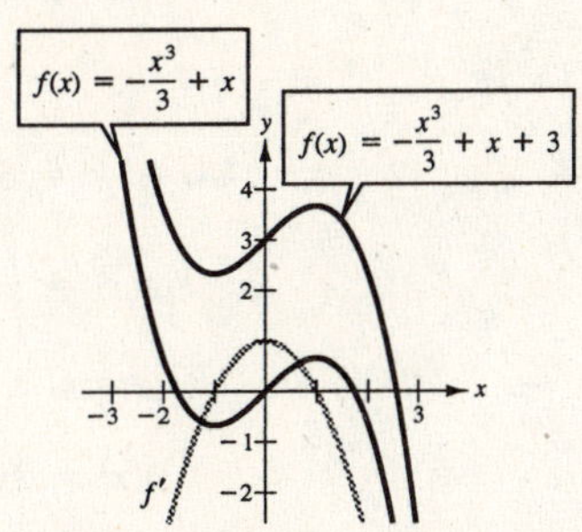

40. $f'(x) = \frac{1}{x^2}$

$f(x) = -\frac{1}{x} + C$

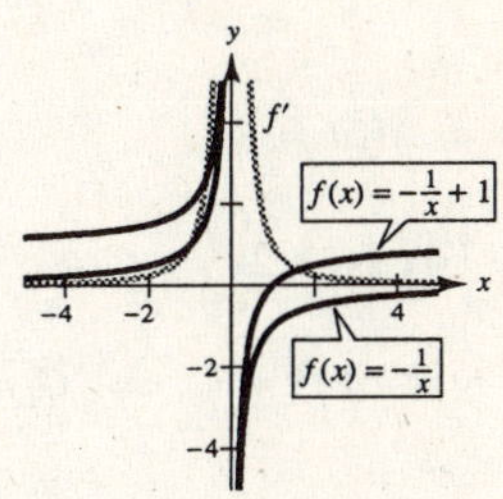

Answers will vary.

41. $\dfrac{dy}{dx} = 2x - 1, \ (1, 1)$

$$y = \int (2x - 1)\, dx = x^2 - x + C$$

$$1 = (1)^2 - (1) + C \Rightarrow C = 1$$

$$y = x^2 - x + 1$$

42. $\dfrac{dy}{dx} = 2(x - 1) = 2x - 2, \ (3, 2)$

$$y = \int 2(x - 1)\, dx = x^2 - 2x + C$$

$$2 = (3)^2 - 2(3) + C \Rightarrow C = -1$$

$$y = x^2 - 2x - 1$$

43. $\dfrac{dy}{dx} = 3x^2 - 1$

$$y = \int (3x^2 - 1)\, dx = x^3 - x + C$$

$$2 = 0^3 - 0 + C \Rightarrow C = 2$$

$$y = x^3 - x + 2$$

44. $\dfrac{dy}{dx} = -\dfrac{1}{x^2} = -x^{-2}, \ (1,3)$

$$y = \int -x^{-2}\, dx = \frac{1}{x} + C$$

$$3 = \frac{1}{1} + C \Rightarrow C = 2$$

$$y = \frac{1}{x} + 2, \ x > 0$$

45. (a) Answers will vary.

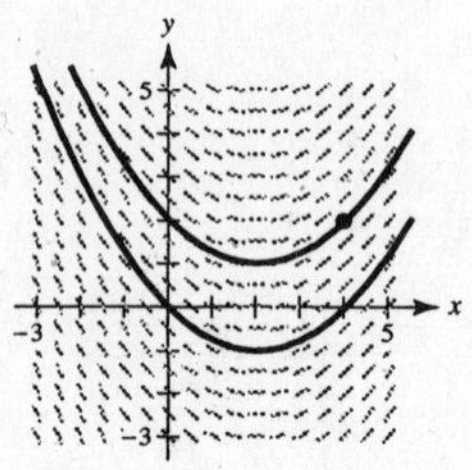

(b) $\dfrac{dy}{dx} = \dfrac{1}{2}x - 1, \ (4, 2)$

$$y = \frac{x^2}{4} - x + C$$

$$2 = \frac{4^2}{4} - 4 + C$$

$$2 = C$$

$$y = \frac{x^2}{4} - x + 2$$

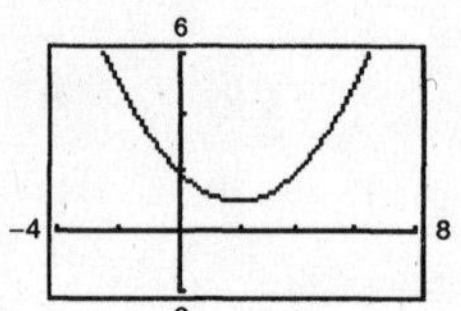

46. (a)

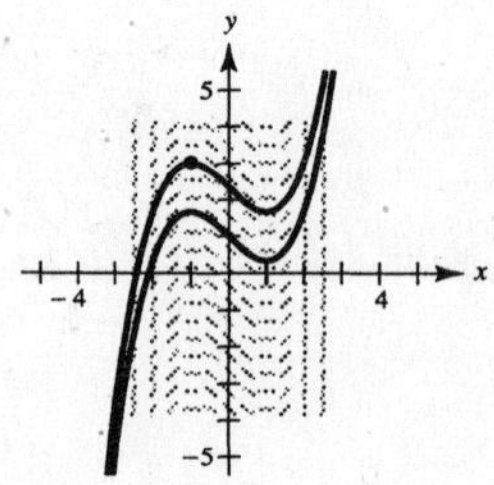

(b) $\dfrac{dy}{dx} = x^2 - 1, \ (-1, 3)$

$$y = \frac{x^3}{3} - x + C$$

$$3 = \frac{(-1)^3}{3} - (-1) + C$$

$$3 = -\frac{1}{3} + 1 + C$$

$$C = \frac{7}{3}$$

$$y = \frac{x^3}{3} - x + \frac{7}{3}$$

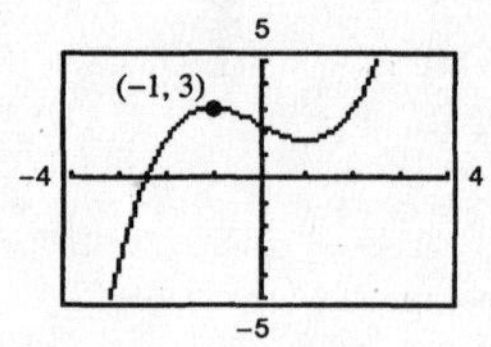

47. $f'(x) = 4x, f(0) = 6$

$$f(x) = \int 4x\, dx = 2x^2 + C$$

$$f(0) = 6 = 2(0)^2 + C \Rightarrow C = 6$$

$$f(x) = 2x^2 + 6$$

48. $g'(x) = 6x^2, g(0) = -1$

$$g(x) = \int 6x^2\, dx = 2x^3 + C$$

$$g(0) = -1 = 2(0)^3 + C \Rightarrow C = -1$$

$$g(x) = 2x^3 - 1$$

49. $h'(t) = 8t^3 + 5, h(1) = -4$

$$h(t) = \int (8t^3 + 5) dt = 2t^4 + 5t + C$$

$$h(1) = -4 = 2 + 5 + C \Rightarrow C = -11$$

$$h(t) = 2t^4 + 5t - 11$$

50. $f'(s) = 6s - 8s^3, f(2) = 3$

$$f(s) = \int (6s - 8s^3)\, ds = 3s^2 - 2s^4 + C$$

$$f(2) = 3 = 3(2)^2 - 2(2)^4 + C = 12 - 32 + C \Rightarrow C = 23$$

$$f(s) = 3s^2 - 2s^4 + 23$$

51. $f''(x) = 2$

$f'(2) = 5$

$f(2) = 10$

$f'(x) = \int 2\,dx = 2x + C_1$

$f'(2) = 4 + C_1 = 5 \Rightarrow C_1 = 1$

$f'(x) = 2x + 1$

$f(x) = \int (2x + 1)\,dx = x^2 + x + C_2$

$f(2) = 6 + C_2 = 10 \Rightarrow C_2 = 4$

$f(x) = x^2 + x + 4$

52. $f''(x) = x^2$

$f'(0) = 6$

$f(0) = 3$

$f'(x) = \int x^2\,dx = \frac{1}{3}x^3 + C_1$

$f'(0) = 0 + C_1 = 6 \Rightarrow C_1 = 6$

$f'(x) = \frac{1}{3}x^3 + 6$

$f(x) = \int \left(\frac{1}{3}x^3 + 6\right) dx = \frac{1}{12}x^4 + 6x + C_2$

$f(0) = 0 + 0 + C_2 = 3 \Rightarrow C_2 = 3$

$f(x) = \frac{1}{12}x^4 + 6x + 3$

53. (a) $h(t) = \int (1.5t + 5)\,dt = 0.75t^2 + 5t + C$

$h(0) = 0 + 0 + C = 12 \Rightarrow C = 12$

$h(t) = 0.75t^2 + 5t + 12$

(b) $h(6) = 0.75(6)^2 + 5(6) + 12 = 69$ cm

54. $\frac{dP}{dt} = k\sqrt{t},\ 0 \le t \le 10$

$P(t) = \int kt^{1/2}\,dt = \frac{2}{3}kt^{3/2} + C$

$P(0) = 0 + C = 500 \Rightarrow C = 500$

$P(1) = \frac{2}{3}k + 500 = 600 \Rightarrow k = 150$

$P(t) = \frac{2}{3}(150)t^{3/2} + 500 = 100t^{3/2} + 500$

$P(7) = 100(7)^{3/2} + 500 \approx 2352$ bacteria

55. $f(0) = -4$. Graph of f' is given.

(a) $f'(4) \approx -1.0$

(b) No. The slopes of the tangent lines are greater than 2 on [0, 2]. Therefore, f must increase more than 4 units on [0, 2].

(c) No, $f(5) < f(4)$ because f is decreasing on [4, 5].

(d) f has a relative minimum at $x \approx -0.75$ and $x \approx 6.25$. $f' = 0$ at both these values of x, and f' is negative to the left and positive to the right of each.

(e) f is concave upward when f' is increasing on $(-\infty, 1)$ and $(5, \infty)$. f is concave downward on (1, 5). Points of inflection at $x = 1, 5$.

(f) f'' is a minimum at $x = 3$.

(g)

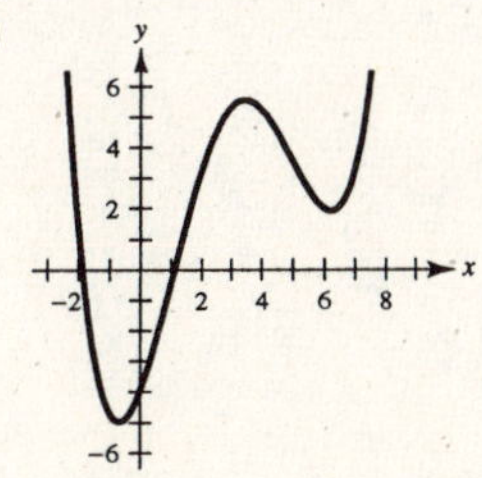

56. Since f'' is negative on $(-\infty, 0)$, f' is decreasing on $(-\infty, 0)$. Since f'' is positive on $(0, \infty)$, f' is increasing on $(0, \infty)$. f' has a relative minimum at (0, 0). Since f' is positive on $(-\infty, \infty)$, f is increasing on $(-\infty, \infty)$.

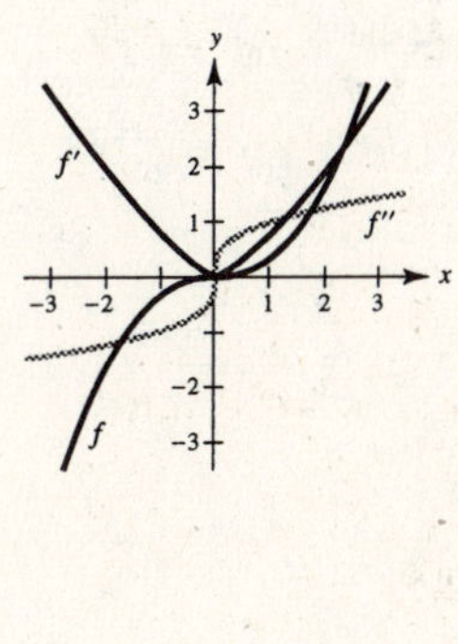

57. $a(t) = -32$ ft/sec^2

$$v(t) = \int -32\,dt = -32t + C_1$$

$$v(0) = 60 = C_1$$

$$s(t) = \int (-32t + 60)dt = -16t^2 + 60t + C_2$$

$$s(0) = 6 = C_2$$

$s(t) = -16t^2 + 60t + 6$ Position function

The ball reaches its maximim height when

$$v(t) = -32t + 60 = 0$$

$$32t = 60$$

$$t = \frac{15}{8} \text{ seconds.}$$

$$s\left(\frac{15}{8}\right) = -16\left(\frac{15}{8}\right)^2 + 60\left(\frac{15}{8}\right) + 6 = 62.25 \text{ feet}$$

58. $f''(t) = a(t) = -32$ ft/sec^2

$$f'(0) = v_0$$

$$f(0) = s_0$$

$$f'(t) = v(t) = \int -32\,dt = -32t + C_1$$

$$f'(0) = 0 + C_1 = v_0 \Rightarrow C_1 = v_0$$

$$f'(t) = -32t + v_0$$

$$f(t) = s(t) = \int (-32t + v_0)\,dt = -16t^2 + v_0t + C_2$$

$$f(0) = 0 + 0 + C_2 = s_0 \Rightarrow C_2 = s_0$$

$$f(t) = -16t^2 + v_0t + s_0$$

59. From Exercise 58, we have:

$$s(t) = -16t^2 + v_0t$$

$s'(t) = -32t + v_0 = 0$ when $t = \frac{v_0}{32}$ = time to reach maximum height.

$$s\left(\frac{v_0}{32}\right) = -16\left(\frac{v_0}{32}\right)^2 + v_0\left(\frac{v_0}{32}\right) = 550$$

$$-\frac{v_0^2}{64} + \frac{v_0^2}{32} = 550$$

$$v_0^2 = 35{,}200$$

$$v_0 \approx 187.617 \text{ ft/sec}$$

60. $v_0 = 16$ ft/sec

$s_0 = 64$ ft

(a) $$s(t) = -16t^2 + 16t + 64 = 0$$

$$-16(t^2 - t - 4) = 0$$

$$t = \frac{1 \pm \sqrt{17}}{2}$$

Choosing the positive value,

$$t = \frac{1 + \sqrt{17}}{2} \approx 2.562 \text{ seconds.}$$

(b) $$v(t) = s'(t) = -32t + 16$$

$$v\left(\frac{1 + \sqrt{17}}{2}\right) = -32\left(\frac{1 + \sqrt{17}}{2}\right) + 16$$

$$= -16\sqrt{17} \approx -65.970 \text{ ft/sec}$$

61. $a(t) = -9.8$

$$v(t) = \int -9.8\,dt = -9.8t + C_1$$

$$v(0) = v_0 = C_1 \Rightarrow v(t) = -9.8t + v_0$$

$$f(t) = \int (-9.8t + v_0)\,dt = -4.9t^2 + v_0t + C_2$$

$$f(0) = s_0 = C_2 \Rightarrow f(t) = -4.9t^2 + v_0t + s_0$$

62. From Exercise 61, $f(t) = -4.9t^2 + 1800$. (Using the canyon floor as position 0.)

$$f(t) = 0 = -4.9t^2 + 1800$$

$$4.9t^2 = 1800$$

$$t^2 = \frac{1800}{4.9} \Rightarrow t \approx \sqrt{367.35} \approx 19.2 \text{ sec}$$

63. From Exercise 61, $f(t) = -4.9t^2 + 10t + 2$.

$v(t) = -9.8t + 10 = 0$ (Maximum height when $v = 0$.)

$9.8t = 10$

$$t = \frac{10}{9.8}$$

$$f\left(\frac{10}{9.8}\right) \approx 7.1 \text{ m}$$

64. From Exercise 61, $f(t) = -4.9t^2 + v_0 t + 2$. If

$f(t) = 200 = -4.9t^2 + v_0 t + 2$,

then

$v(t) = -9.8t + v_0 = 0$

for this t value. Hence, $t = v_0/9.8$ and we solve

$$-4.9\left(\frac{v_0}{9.8}\right)^2 + v_0\left(\frac{v_0}{9.8}\right) + 2 = 200$$

$$\frac{-4.9\, v_0^2}{(9.8)^2} + \frac{v_0^2}{9.8} = 198$$

$$-4.9\, v_0^2 + 9.8\, v_0^2 = (9.8)^2\, 198$$

$$4.9\, v_0^2 = (9.8)^2\, 198$$

$$v_0^2 = 3880.8 \implies v_0 \approx 62.3 \text{ m/sec.}$$

65. $a = -1.6$

$$v(t) = \int -1.6\, dt = -1.6t + v_0 = -1.6t,$$

since the stone was dropped, $v_0 = 0$.

$$s(t) = \int (-1.6t)\, dt = -0.8t^2 + s_0$$

$$s(20) = 0 \implies -0.8(20)^2 + s_0 = 0$$

$$s_0 = 320$$

Thus, the height of the cliff is 320 meters.

$$v(t) = -1.6t$$

$$v(20) = -32 \text{ m/sec}$$

66. $$\int v\, dv = -GM\int \frac{1}{y^2}\, dy$$

$$\frac{1}{2}v^2 = \frac{GM}{y} + C$$

When $y = R$, $v = v_0$.

$$\frac{1}{2}v_0^2 = \frac{GM}{R} + C$$

$$C = \frac{1}{2}v_0^2 - \frac{GM}{R}$$

$$\frac{1}{2}v^2 = \frac{GM}{y} + \frac{1}{2}v_0^2 - \frac{GM}{R}$$

$$v^2 = \frac{2GM}{y} + v_0^2 - \frac{2GM}{R}$$

$$v^2 = v_0^2 + 2GM\left(\frac{1}{y} - \frac{1}{R}\right)$$

67. $x(t) = t^3 - 6t^2 + 9t - 2 \quad 0 \le t \le 5$

(a) $v(t) = x'(t) = 3t^2 - 12t + 9$

$= 3(t^2 - 4t + 3) = 3(t - 1)(t - 3)$

$a(t) = v'(t) = 6t - 12 = 6(t - 2)$

(b) $v(t) > 0$ when $0 < t < 1$ or $3 < t < 5$.

(c) $a(t) = 6(t - 2) = 0$ when $t = 2$.

$v(2) = 3(1)(-1) = -3$

68. $x(t) = (t - 1)(t - 3)^2 \quad 0 \le t \le 5$

$= t^3 - 7t^2 + 15t - 9$

(a) $v(t) = x'(t) = 3t^2 - 14t + 15 = (3t - 5)(t - 3)$

$a(t) = v'(t) = 6t - 14$

(b) $v(t) > 0$ when $0 < t < \frac{5}{3}$ and $3 < t < 5$.

(c) $a(t) = 6t - 14 = 0$ when $t = \frac{7}{3}$.

$$v\left(\frac{7}{3}\right) = \left(3\left(\frac{7}{3}\right) - 5\right)\left(\frac{7}{3} - 3\right) = 2\left(-\frac{2}{3}\right) = -\frac{4}{3}$$

69. $v(t) = \dfrac{1}{\sqrt{t}} = t^{-1/2} \quad t > 0$

$x(t) = \int v(t)\,dt = 2t^{1/2} + C$

$x(1) = 4 = 2(1) + C \Rightarrow C = 2$

$x(t) = 2t^{1/2} + 2$ position function

$a(t) = v'(t) = -\dfrac{1}{2}t^{-3/2} = \dfrac{-1}{2t^{3/2}}$ acceleration

70. (a) $v(0) = 25 \text{ km/hr} = 25 \cdot \dfrac{1000}{3600} = \dfrac{250}{36} \text{ m/sec}$

$v(13) = 80 \text{ km/hr} = 80 \cdot \dfrac{1000}{3600} = \dfrac{800}{36} \text{ m/sec}$

$a(t) = a$ (constant acceleration)

$v(t) = at + C$

$v(0) = \dfrac{250}{36} \Rightarrow v(t) = at + \dfrac{250}{36}$

$v(13) = \dfrac{800}{36} = 13a + \dfrac{250}{36}$

$\dfrac{550}{36} = 13a$

$a = \dfrac{550}{468} = \dfrac{275}{234} \approx 1.175 \text{ m/sec}^2$

(b) $s(t) = a\dfrac{t^2}{2} + \dfrac{250}{36}t \qquad (s(0) = 0)$

$s(13) = \dfrac{275}{234}\dfrac{(13)^2}{2} + \dfrac{250}{36}(13) \approx 189.58 \text{ m}$

71. $v(0) = 45 \text{ mph} = 66 \text{ ft/sec}$

$30 \text{ mph} = 44 \text{ ft/sec}$

$15 \text{ mph} = 22 \text{ ft/sec}$

$a(t) = -a$

$v(t) = -at + 66$

$s(t) = -\dfrac{a}{2}t^2 + 66t$ (Let $s(0) = 0$.)

$v(t) = 0$ after car moves 132 ft.

$-at + 66 = 0$ when $t = \dfrac{66}{a}$.

$s\left(\dfrac{66}{a}\right) = -\dfrac{a}{2}\left(\dfrac{66}{a}\right)^2 + 66\left(\dfrac{66}{a}\right)$

$= 132$ when $a = \dfrac{33}{2} = 16.5$.

$a(t) = -16.5$

$v(t) = -16.5t + 66$

$s(t) = -8.25t^2 + 66t$

(a) $-16.5t + 66 = 44$

$t = \dfrac{22}{16.5} \approx 1.333$

$s\left(\dfrac{22}{16.5}\right) \approx 73.33 \text{ ft}$

(b) $-16.5t + 66 = 22$

$t = \dfrac{44}{16.5} \approx 2.667$

$s\left(\dfrac{44}{16.5}\right) \approx 117.33 \text{ ft}$

(c)

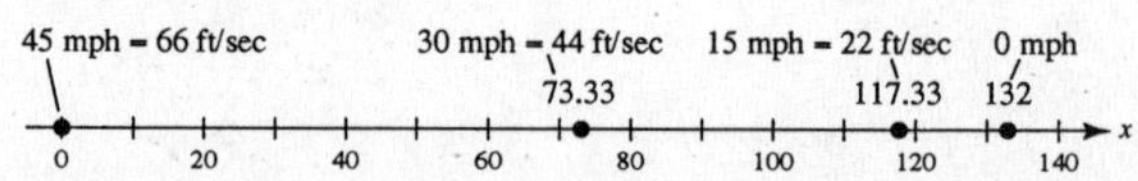

It takes 1.333 seconds to reduce the speed from 45 mph to 30 mph, 1.333 seconds to reduce the speed from 30 mph to 15 mph, and 1.333 seconds to reduce the speed from 15 mph to 0 mph. Each time, less distance is needed to reach the next speed reduction.

72. Truck: $v(t) = 30$

$s(t) = 30t$ (Let $s(0) = 0$.)

Automobile: $a(t) = 6$

$v(t) = 6t$ (Let $v(0) = 0$.)

$s(t) = 3t^2$ (Let $s(0) = 0$.)

At the point where the automobile overtakes the truck:

$30t = 3t^2$

$0 = 3t^2 - 30t$

$0 = 3t(t - 10)$ when $t = 10$ sec.

(a) $s(10) = 3(10)^2 = 300$ ft

(b) $v(10) = 6(10) = 60 \text{ ft/sec} \approx 41 \text{ mph}$

73. (a) $v = 0.6139t^3 - 5.525t^2 + 0.0492t + 65.9881$

(b) $s(t) = \int v(t)dt = \frac{0.6139t^4}{4} - \frac{5.525t^3}{3} + \frac{0.0492t^2}{2} + 65.9881t$

(**Note:** Assume $s(0) = 0$ is initial position)

$s(6) \approx 197.9$ feet

74. $\frac{(1 \text{ mi/hr})(5280 \text{ ft/mi})}{(3600 \text{ sec/hr})} = \frac{22}{15}$ ft/sec

(a)

t	0	5	10	15	20	25	30
v_1(ft/sec)	0	3.67	10.27	23.47	42.53	66	95.33
v_2(ft/sec)	0	30.8	55.73	74.8	88	93.87	95.33

(b) $v_1(t) = 0.1068t^2 - 0.0416t + 0.3679$

$v_2(t) = -0.1208t^2 + 6.7991t - 0.0707$

(c) $s_1(t) = \int v_1(t)\,dt = \frac{0.1068}{3}t^3 - \frac{0.0416}{2}t^2 + 0.3679t$

$s_2(t) = \int v_2(t)\,dt = -\frac{0.1208t^3}{3} + \frac{6.7991t^2}{2} - 0.0707t$

[In both cases, the constant of integration is 0 because $s_1(0) = s_2(0) = 0$.]

$s_1(30) \approx 953.5$ feet

$s_2(30) \approx 1970.3$ feet

The second car was going faster than the first until the end.

75. True

76. True

77. True

78. True

79. False. For example, $\int x \cdot x\,dx \neq \int x\,dx \cdot \int x\,dx$ because $\frac{x^3}{3} + C \neq \left(\frac{x^2}{2} + C_1\right)\left(\frac{x^2}{2} + C_2\right)$.

80. False. f has an infinite number of antiderivatives, each differing by a constant.

81. $f''(x) = 2x$

$f'(x) = x^2 + C$

$f'(2) = 0 \Rightarrow 4 + C = 0 \Rightarrow C = -4$

$f(x) = \frac{x^3}{3} - 4x + C_1$

$f(2) = 0 \Rightarrow \frac{8}{3} - 8 + C_1 = 0 \Rightarrow C_1 = \frac{16}{3}$

Answer: $f(x) = \frac{x^3}{3} - 4x + \frac{16}{3}$

82. $f'(x) = \begin{cases} -1, & 0 \le x < 2 \\ 2, & 2 < x < 3 \\ 0, & 3 < x \le 4 \end{cases}$

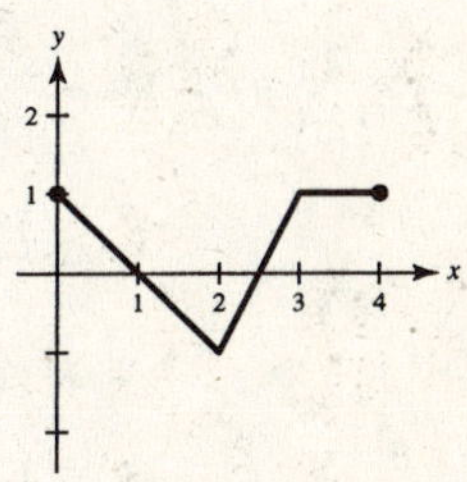

$f(x) = \begin{cases} -x + C_1, & 0 \le x < 2 \\ 2x + C_2, & 2 < x < 3 \\ C_3, & 3 < x \le 4 \end{cases}$

$f(0) = 1 \Rightarrow C_1 = 1$

f continuous at $x = 2 \Rightarrow -2 + 1 = 4 + C_2 \Rightarrow C_2 = -5$

f continuous at $x = 3 \Rightarrow 6 - 5 = C_3 = 1$

$f(x) = \begin{cases} -x + 1, & 0 \le x < 2 \\ 2x - 5, & 2 \le x < 3 \\ 1, & 3 \le x \le 4 \end{cases}$

Section 6.2 Area

1. $\sum_{i=1}^{5}(2i+1) = 2\sum_{i=1}^{5} i + \sum_{i=1}^{5} 1 = 2(1+2+3+4+5)+5 = 35$

2. $\sum_{k=3}^{6} k(k-2) = 3(1)+4(2)+5(3)+6(4) = 50$

3. $\sum_{k=0}^{4} \frac{1}{k^2+1} = 1 + \frac{1}{2} + \frac{1}{5} + \frac{1}{10} + \frac{1}{17} = \frac{158}{85}$

4. $\sum_{j=3}^{5} \frac{1}{j} = \frac{1}{3} + \frac{1}{4} + \frac{1}{5} = \frac{47}{60}$

5. $\sum_{k=1}^{4} c = c + c + c + c = 4c$

6. $\sum_{i=1}^{4}[(i-1)^2 + (i+1)^3] = (0+8)+(1+27)+(4+64)+(9+125) = 238$

7. $\sum_{i=1}^{9} \frac{1}{3i}$

8. $\sum_{i=1}^{15} \frac{5}{1+i}$

9. $\sum_{j=1}^{8}\left[5\left(\frac{j}{8}\right)+3\right]$

10. $\sum_{j=1}^{4}\left[1-\left(\frac{j}{4}\right)^2\right]$

11. $\frac{2}{n}\sum_{i=1}^{n}\left[\left(\frac{2i}{n}\right)^3 - \left(\frac{2i}{n}\right)\right]$

12. $\frac{2}{n}\sum_{i=1}^{n}\left[1-\left(\frac{2i}{n}-1\right)^2\right]$

13. $\frac{3}{n}\sum_{i=1}^{n}\left[2\left(1+\frac{3i}{n}\right)^2\right]$

14. $\frac{1}{n}\sum_{i=0}^{n-1}\sqrt{1-\left(\frac{i}{n}\right)^2}$

15. $\sum_{i=1}^{20} 2i = 2\sum_{i=1}^{20} i = 2\left[\frac{20(21)}{2}\right] = 420$

16. $\sum_{i=1}^{15}(2i-3) = 2\sum_{i=1}^{15} i - 3(15)$

$= 2\left[\frac{15(16)}{2}\right] - 45 = 195$

17. $\sum_{i=1}^{20}(i-1)^2 = \sum_{i=1}^{19} i^2$

$= \left[\frac{19(20)(39)}{6}\right] = 2470$

18. $\sum_{i=1}^{10}(i^2-1) = \sum_{i=1}^{10} i^2 - \sum_{i=1}^{10} 1$

$= \left[\frac{10(11)(21)}{6}\right] - 10 = 375$

19. $\sum_{i=1}^{15} i(i-1)^2 = \sum_{i=1}^{15} i^3 - 2\sum_{i=1}^{15} i^2 + \sum_{i=1}^{15} i$

$= \frac{15^2(16)^2}{4} - 2\frac{15(16)(31)}{6} + \frac{15(16)}{2}$

$= 14{,}400 - 2{,}480 + 120$

$= 12{,}040$

20. $\sum_{i=1}^{10} i(i^2+1) = \sum_{i=1}^{10} i^3 + \sum_{i=1}^{10} i$

$= \frac{10^2(11)^2}{4} + \left[\frac{10(11)}{2}\right] = 3080$

21. sum seq(x ^ 2 + 3, x, 1, 20, 1) = 2930 (*TI-82*)

$\sum_{i=1}^{20}(i^2+3) = \frac{20(20+1)(2(20)+1)}{6} + 3(20)$

$= \frac{(20)(21)(41)}{6} + 60 = 2930$

22. sum seq(x ^ 3 − 2x, x, 1, 15, 1) = 14,160 (*TI-82*)

$\sum_{i=1}^{15}(i^3-2i) = \frac{(15)^2(15+1)^2}{4} - 2\frac{15(15+1)}{2}$

$= \frac{(15)^2(16)^2}{4} - 15(16) = 14{,}160$

23. $S = \left[3 + 4 + \frac{9}{2} + 5\right](1) = \frac{33}{2} = 16.5$

$s = \left[1 + 3 + 4 + \frac{9}{2}\right](1) = \frac{25}{2} = 12.5$

24. $S = [5 + 5 + 4 + 2](1) = 16$

$s = [4 + 4 + 2 + 0](1) = 10$

25. $S = [3 + 3 + 5](1) = 11$

$s = [2 + 2 + 3](1) = 7$

26. $S = \left[5 + 2 + 1 + \frac{2}{3} + \frac{1}{2}\right] = \frac{55}{6}$

$s = \left[2 + 1 + \frac{2}{3} + \frac{1}{2} + \frac{1}{3}\right] = \frac{9}{2} = 4.5$

27. $S(4) = \sqrt{\frac{1}{4}}\left(\frac{1}{4}\right) + \sqrt{\frac{1}{2}}\left(\frac{1}{4}\right) + \sqrt{\frac{3}{4}}\left(\frac{1}{4}\right) + \sqrt{1}\left(\frac{1}{4}\right) = \frac{1 + \sqrt{2} + \sqrt{3} + 2}{8} \approx 0.768$

$s(4) = 0\left(\frac{1}{4}\right) + \sqrt{\frac{1}{4}}\left(\frac{1}{4}\right) + \sqrt{\frac{1}{2}}\left(\frac{1}{4}\right) + \sqrt{\frac{3}{4}}\left(\frac{1}{4}\right) = \frac{1 + \sqrt{2} + \sqrt{3}}{8} \approx 0.518$

28.
$$S(8) = \left(\sqrt{\frac{1}{4}} + 2\right)\frac{1}{4} + \left(\sqrt{\frac{1}{2}} + 2\right)\frac{1}{4} + \left(\sqrt{\frac{3}{4}} + 2\right)\frac{1}{4} + \left(\sqrt{1} + 2\right)\frac{1}{4}$$
$$+ \left(\sqrt{\frac{5}{4}} + 2\right)\frac{1}{4} + \left(\sqrt{\frac{3}{2}} + 2\right)\frac{1}{4} + \left(\sqrt{\frac{7}{4}} + 2\right)\frac{1}{4} + \left(\sqrt{2} + 2\right)\frac{1}{4}$$
$$= \frac{1}{4}\left(16 + \frac{1}{2} + \frac{\sqrt{2}}{2} + \frac{\sqrt{3}}{2} + 1 + \frac{\sqrt{5}}{2} + \frac{\sqrt{6}}{2} + \frac{\sqrt{7}}{2} + \sqrt{2}\right) \approx 6.038$$
$$s(8) = (0 + 2)\frac{1}{4} + \left(\sqrt{\frac{1}{4}} + 2\right)\frac{1}{4} + \left(\sqrt{\frac{1}{2}} + 2\right)\frac{1}{4} + \cdots + \left(\sqrt{\frac{7}{4}} + 2\right)\frac{1}{4} \approx 5.685$$

29. $S(5) = 1\left(\frac{1}{5}\right) + \frac{1}{6/5}\left(\frac{1}{5}\right) + \frac{1}{7/5}\left(\frac{1}{5}\right) + \frac{1}{8/5}\left(\frac{1}{5}\right) + \frac{1}{9/5}\left(\frac{1}{5}\right) = \frac{1}{5} + \frac{1}{6} + \frac{1}{7} + \frac{1}{8} + \frac{1}{9} \approx 0.746$

$s(5) = \frac{1}{6/5}\left(\frac{1}{5}\right) + \frac{1}{7/5}\left(\frac{1}{5}\right) + \frac{1}{8/5}\left(\frac{1}{5}\right) + \frac{1}{9/5}\left(\frac{1}{5}\right) + \frac{1}{2}\left(\frac{1}{5}\right) = \frac{1}{6} + \frac{1}{7} + \frac{1}{8} + \frac{1}{9} + \frac{1}{10} \approx 0.646$

30.
$$S(5) = 1\left(\frac{1}{5}\right) + \sqrt{1 - \left(\frac{1}{5}\right)^2}\left(\frac{1}{5}\right) + \sqrt{1 - \left(\frac{2}{5}\right)^2}\left(\frac{1}{5}\right) + \sqrt{1 - \left(\frac{3}{5}\right)^2}\left(\frac{1}{5}\right) + \sqrt{1 - \left(\frac{4}{5}\right)^2}\left(\frac{1}{5}\right)$$
$$= \frac{1}{5}\left[1 + \frac{\sqrt{24}}{5} + \frac{\sqrt{21}}{5} + \frac{\sqrt{16}}{5} + \frac{\sqrt{9}}{5}\right] \approx 0.859$$
$$s(5) = \sqrt{1 - \left(\frac{1}{5}\right)^2}\left(\frac{1}{5}\right) + \sqrt{1 - \left(\frac{2}{5}\right)^2}\left(\frac{1}{5}\right) + \sqrt{1 - \left(\frac{3}{5}\right)^2}\left(\frac{1}{5}\right) + \sqrt{1 - \left(\frac{4}{5}\right)^2}\left(\frac{1}{5}\right) + 0 \approx 0.659$$

31. $\lim_{n\to\infty}\left[\left(\frac{4}{3n^3}\right)(2n^3 + 3n^2 + n)\right] = \frac{4}{3}\lim_{n\to\infty}\left[\frac{2n^3 + 3n^2 + n}{n^3}\right]$

$= \frac{4}{3}(2) = \frac{8}{3}$

32. $\lim_{n\to\infty}\left(\frac{8}{3} + \frac{4}{n} + \frac{4}{3n^2}\right) = \frac{8}{3} + 0 + 0 = \frac{8}{3}$

33. $\lim_{n\to\infty}\left[\left(\frac{81}{n^4}\right)\frac{n^2(n + 1)^2}{4}\right] = \frac{81}{4}\lim_{n\to\infty}\left[\frac{n^4 + 2n^3 + n^2}{n^4}\right]$

$= \frac{81}{4}(1) = \frac{81}{4}$

34. $\lim_{n\to\infty}\left[\left(\frac{64}{n^3}\right)\frac{n(n + 1)(2n + 1)}{6}\right] = \frac{64}{6}\lim_{n\to\infty}\left[\frac{2n^3 + 3n^2 + n}{n^3}\right]$

$= \frac{64}{6}(2) = \frac{64}{3}$

35. $\lim_{n\to\infty}\left[\left(\frac{18}{n^2}\right)\frac{n(n+1)}{2}\right] = \frac{18}{2}\lim_{n\to\infty}\left[\frac{n^2+n}{n^2}\right] = \frac{18}{2}(1) = 9$

36. $\lim_{n\to\infty}\left[\left(\frac{1}{n^2}\right)\frac{n(n+1)}{2}\right] = \frac{1}{2}\lim_{n\to\infty}\left[\frac{n^2+n}{n^2}\right] = \frac{1}{2}(1) = \frac{1}{2}$

37. $\sum_{i=1}^{n}\frac{2i+1}{n^2} = \frac{1}{n^2}\sum_{i=1}^{n}(2i+1) = \frac{1}{n^2}\left[2\frac{n(n+1)}{2}+n\right] = \frac{n+2}{n} = S(n)$

$$\begin{aligned} S(10) &= \frac{12}{10} = 1.2 \\ S(100) &= 1.02 \\ S(1000) &= 1.002 \\ S(10{,}000) &= 1.0002 \end{aligned}$$

38. $\sum_{j=1}^{n}\frac{4j+3}{n^2} = \frac{1}{n^2}\sum_{j=1}^{n}(4j+3) = \frac{1}{n^2}\left[\frac{4n(n+1)}{2}+3n\right] = \frac{2n+5}{n} = S(n)$

$$\begin{aligned} S(10) &= \frac{25}{10} = 2.5 \\ S(100) &= 2.05 \\ S(1000) &= 2.005 \\ S(10{,}000) &= 2.0005 \end{aligned}$$

39. $\sum_{k=1}^{n}\frac{6k(k-1)}{n^3} = \frac{6}{n^3}\sum_{k=1}^{n}(k^2-k) = \frac{6}{n^3}\left[\frac{n(n+1)(2n+1)}{6}-\frac{n(n+1)}{2}\right]$

$= \frac{6}{n^2}\left[\frac{2n^2+3n+1-3n-3}{6}\right] = \frac{1}{n^2}[2n^2-2] = S(n)$

$$\begin{aligned} S(10) &= 1.98 \\ S(100) &= 1.9998 \\ S(1000) &= 1.999998 \\ S(10{,}000) &= 1.99999998 \end{aligned}$$

40. $\sum_{i=1}^{n}\frac{4i^2(i-1)}{n^4} = \frac{4}{n^4}\sum_{i=1}^{n}(i^3-i^2) = \frac{4}{n^4}\left[\frac{n^2(n+1)^2}{4}-\frac{n(n+1)(2n+1)}{6}\right]$

$= \frac{4}{n^3}\left[\frac{n^3+2n^2+n}{4}-\frac{2n^2+3n+1}{6}\right]$

$= \frac{1}{3n^3}[3n^3+6n^2+3n-4n^2-6n-2]$

$= \frac{1}{3n^3}[3n^3+2n^2-3n-2] = S(n)$

$$\begin{aligned} S(10) &= 1.056 \\ S(100) &= 1.006566 \\ S(1000) &= 1.00066567 \\ S(10{,}000) &= 1.000066657 \end{aligned}$$

41. $\lim_{n\to\infty}\sum_{i=1}^{n}\left(\frac{16i}{n^2}\right) = \lim_{n\to\infty}\frac{16}{n^2}\sum_{i=1}^{n}i = \lim_{n\to\infty}\frac{16}{n^2}\left(\frac{n(n+1)}{2}\right) = \lim_{n\to\infty}\left[8\left(\frac{n^2+n}{n^2}\right)\right] = 8\lim_{n\to\infty}\left(1+\frac{1}{n}\right) = 8$

42. $\lim_{n\to\infty} \sum_{i=1}^{n} \left(\frac{2i}{n}\right)\left(\frac{2}{n}\right) = \lim_{n\to\infty} \frac{4}{n^2}\sum_{i=1}^{n} i = \lim_{n\to\infty} \frac{4}{n^2}\left(\frac{n(n+1)}{2}\right) = \lim_{n\to\infty} \frac{4}{2}\left(1 + \frac{1}{n}\right) = 2$

43. $\lim_{n\to\infty} \sum_{i=1}^{n} \frac{1}{n^3}(i-1)^2 = \lim_{n\to\infty} \frac{1}{n^3}\sum_{i=1}^{n-1} i^2 = \lim_{n\to\infty} \frac{1}{n^3}\left[\frac{(n-1)(n)(2n-1)}{6}\right]$

$$= \lim_{n\to\infty} \frac{1}{6}\left[\frac{2n^3 - 3n^2 + n}{n^3}\right] = \lim_{n\to\infty}\left[\frac{1}{6}\left(\frac{2 - (3/n) + (1/n^2)}{1}\right)\right] = \frac{1}{3}$$

44. $\lim_{n\to\infty} \sum_{i=1}^{n} \left(1 + \frac{2i}{n}\right)^2\left(\frac{2}{n}\right) = \lim_{n\to\infty} \frac{2}{n^3}\sum_{i=1}^{n} (n+2i)^2$

$$= \lim_{n\to\infty} \frac{2}{n^3}\left[\sum_{i=1}^{n} n^2 + 4n\sum_{i=1}^{n} i + 4\sum_{i=1}^{n} i^2\right]$$

$$= \lim_{n\to\infty} \frac{2}{n^3}\left[n^3 + (4n)\left(\frac{n(n+1)}{2}\right) + \frac{4(n)(n+1)(2n+1)}{6}\right]$$

$$= 2\lim_{n\to\infty}\left[1 + 2 + \frac{2}{n} + \frac{4}{3} + \frac{2}{n} + \frac{2}{3n^2}\right]$$

$$= 2\left(1 + 2 + \frac{4}{3}\right) = \frac{26}{3}$$

45. $\lim_{n\to\infty} \sum_{i=1}^{n} \left(1 + \frac{i}{n}\right)\left(\frac{2}{n}\right) = 2\lim_{n\to\infty} \frac{1}{n}\left[\sum_{i=1}^{n} 1 + \frac{1}{n}\sum_{i=1}^{n} i\right] = 2\lim_{n\to\infty} \frac{1}{n}\left[n + \frac{1}{n}\left(\frac{n(n+1)}{2}\right)\right] = 2\lim_{n\to\infty}\left[1 + \frac{n^2+n}{2n^2}\right] = 2\left(1 + \frac{1}{2}\right) = 3$

46. $\lim_{n\to\infty} \sum_{i=1}^{n} \left(1 + \frac{2i}{n}\right)^3\left(\frac{2}{n}\right) = 2\lim_{n\to\infty} \frac{1}{n^4}\sum_{i=1}^{n} (n+2i)^3$

$$= 2\lim_{n\to\infty} \frac{1}{n^4}\sum_{i=1}^{n}(n^3 + 6n^2 i + 12ni^2 + 8i^3)$$

$$= 2\lim_{n\to\infty} \frac{1}{n^4}\left[n^4 + 6n^2\left(\frac{n(n+1)}{2}\right) + 12n\left(\frac{n(n+1)(2n+1)}{6}\right) + 8\left(\frac{n^2(n+1)^2}{4}\right)\right]$$

$$= 2\lim_{n\to\infty}\left(1 + 3 + \frac{3}{n} + 4 + \frac{6}{n} + \frac{2}{n^2} + 2 + \frac{4}{n} + \frac{2}{n^2}\right)$$

$$= 2\lim_{n\to\infty}\left(10 + \frac{13}{n} + \frac{4}{n^2}\right) = 20$$

47. (a)

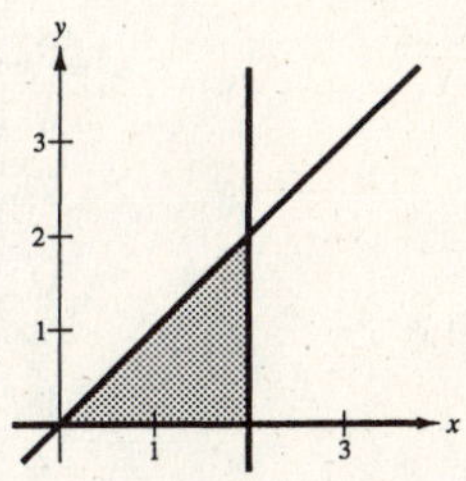

(b) $\Delta x = \frac{2-0}{n} = \frac{2}{n}$

Endpoints:

$$0 < 1\left(\frac{2}{n}\right) < 2\left(\frac{2}{n}\right) < \cdots < (n-1)\left(\frac{2}{n}\right) < n\left(\frac{2}{n}\right) = 2$$

(c) Since $y = x$ is increasing, $f(m_i) = f(x_{i-1})$ on $[x_{i-1}, x_i]$.

$$s(n) = \sum_{i=1}^{n} f(x_{i-1})\,\Delta x$$

$$= \sum_{i=1}^{n} f\left(\frac{2i-2}{n}\right)\left(\frac{2}{n}\right) = \sum_{i=1}^{n}\left[(i-1)\left(\frac{2}{n}\right)\right]\left(\frac{2}{n}\right)$$

(d) $f(M_i) = f(x_i)$ on $[x_{i-1}, x_i]$

$$S(n) = \sum_{i=1}^{n} f(x_i)\,\Delta x = \sum_{i=1}^{n} f\left(\frac{2i}{n}\right)\frac{2}{n} = \sum_{i=1}^{n}\left[i\left(\frac{2}{n}\right)\right]\left(\frac{2}{n}\right)$$

—CONTINUED—

47. —CONTINUED—

(e)

x	5	10	50	100
$s(n)$	1.6	1.8	1.96	1.98
$S(n)$	2.4	2.2	2.04	2.02

(f) $$\lim_{n\to\infty}\sum_{i=1}^{n}\left[(i-1)\left(\frac{2}{n}\right)\right]\left(\frac{2}{n}\right) = \lim_{n\to\infty}\frac{4}{n^2}\sum_{i=1}^{n}(i-1)$$
$$= \lim_{n\to\infty}\frac{4}{n^2}\left[\frac{n(n+1)}{2} - n\right]$$
$$= \lim_{n\to\infty}\left[\frac{2(n+1)}{n} - \frac{4}{n}\right] = 2$$

$$\lim_{n\to\infty}\sum_{i=1}^{n}\left[i\left(\frac{2}{n}\right)\right]\left(\frac{2}{n}\right) = \lim_{n\to\infty}\frac{4}{n^2}\sum_{i=1}^{n} i$$
$$= \lim_{n\to\infty}\left(\frac{4}{n^2}\right)\frac{n(n+1)}{2}$$
$$= \lim_{n\to\infty}\frac{2(n+1)}{n} = 2$$

48. (a)

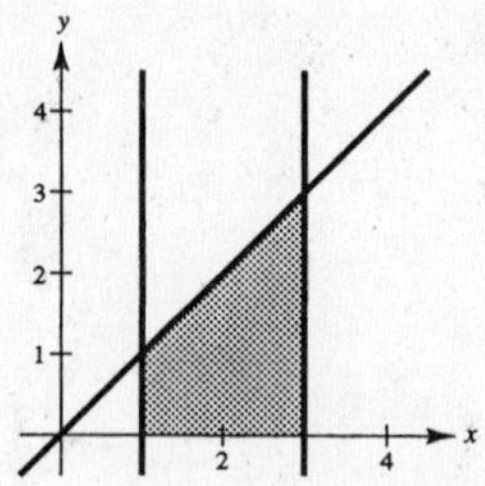

(b) $\Delta x = \dfrac{3-1}{n} = \dfrac{2}{n}$

Endpoints:

$$1 < 1 + \frac{2}{n} < 1 + \frac{4}{n} < \cdots < 1 + \frac{2n}{n} = 3$$
$$1 < 1 + 1\left(\frac{2}{n}\right) < 1 + 2\left(\frac{2}{n}\right) < \cdots < 1 + (n-1)\left(\frac{2}{n}\right) < 1 + n\left(\frac{2}{n}\right)$$

(c) Since $y = x$ is increasing, $f(m_i) = f(x_{i-1})$ on $[x_{i-1}, x_i]$.

$$s(n) = \sum_{i=1}^{n} f(x_{i-1})\,\Delta x$$
$$= \sum_{i=1}^{n} f\left[1 + (i-1)\left(\frac{2}{n}\right)\right]\left(\frac{2}{n}\right) = \sum_{i=1}^{n}\left[1 + (i-1)\left(\frac{2}{n}\right)\right]\left(\frac{2}{n}\right)$$

(d) $f(M_i) = f(x_i)$ on $[x_{i-1}, x_i]$

$$S(n) = \sum_{i=1}^{n} f(x_i)\,\Delta x = \sum_{i=1}^{n} f\left[1 + i\left(\frac{2}{n}\right)\right]\left(\frac{2}{n}\right) = \sum_{i=1}^{n}\left[1 + i\left(\frac{2}{n}\right)\right]\left(\frac{2}{n}\right)$$

(e)

x	5	10	50	100
$s(n)$	3.6	3.8	3.96	3.98
$S(n)$	4.4	4.2	4.04	4.02

(f) $$\lim_{n\to\infty}\sum_{i=1}^{n}\left[1 + (i-1)\left(\frac{2}{n}\right)\right]\left(\frac{2}{n}\right) = \lim_{n\to\infty}\left(\frac{2}{n}\right)\left[n + \frac{2}{n}\left(\frac{n(n+1)}{2} - n\right)\right]$$
$$= \lim_{n\to\infty}\left[2 + \frac{2n+2}{n} - \frac{4}{n}\right] = \lim_{n\to\infty}\left[4 - \frac{2}{n}\right] = 4$$

$$\lim_{n\to\infty}\sum_{i=1}^{n}\left[1 + i\left(\frac{2}{n}\right)\right]\left(\frac{2}{n}\right) = \lim_{n\to\infty}\frac{2}{n}\left[n + \left(\frac{2}{n}\right)\frac{n(n+1)}{2}\right]$$
$$= \lim_{n\to\infty}\left[2 + \frac{2(n+1)}{n}\right] = \lim_{n\to\infty}\left[4 + \frac{2}{n}\right] = 4$$

49. $y = -2x + 3$ on $[0, 1]$. $\left(\textbf{Note: } \Delta x = \dfrac{1-0}{n} = \dfrac{1}{n}\right)$

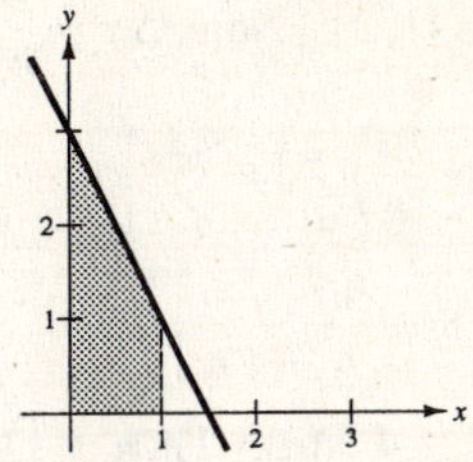

$$s(n) = \sum_{i=1}^{n} f\left(\frac{i}{n}\right)\left(\frac{1}{n}\right) = \sum_{i=1}^{n}\left[-2\left(\frac{i}{n}\right) + 3\right]\left(\frac{1}{n}\right)$$

$$= 3 - \frac{2}{n^2}\sum_{i=1}^{n} i = 3 - \frac{2(n+1)n}{2n^2} = 2 - \frac{1}{n}$$

$$\text{Area} = \lim_{n\to\infty} s(n) = 2$$

50. $y = 3x - 4$ on $[2, 5]$. $\left(\textbf{Note: } \Delta x = \dfrac{5-2}{n} = \dfrac{3}{n}\right)$

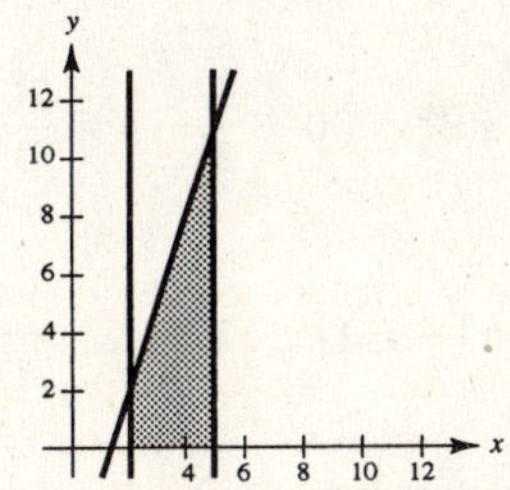

$$S(n) = \sum_{i=1}^{n} f\left(2 + \frac{3i}{n}\right)\left(\frac{3}{n}\right)$$

$$= \sum_{i=1}^{n}\left[3\left(2 + \frac{3i}{n}\right) - 4\right]\left(\frac{3}{n}\right) = 18 + 3\left(\frac{3}{n}\right)^2 \sum_{i=1}^{n} i - 12$$

$$= 6 + \frac{27}{n^2}\left(\frac{(n+1)n}{2}\right) = 6 + \frac{27}{2}\left(1 + \frac{1}{n}\right)$$

$$\text{Area} = \lim_{n\to\infty} S(n) = 6 + \frac{27}{2} = \frac{39}{2}$$

51. $y = x^2 + 2$ on $[0, 1]$. $\left(\textbf{Note: } \Delta x = \dfrac{1}{n}\right)$

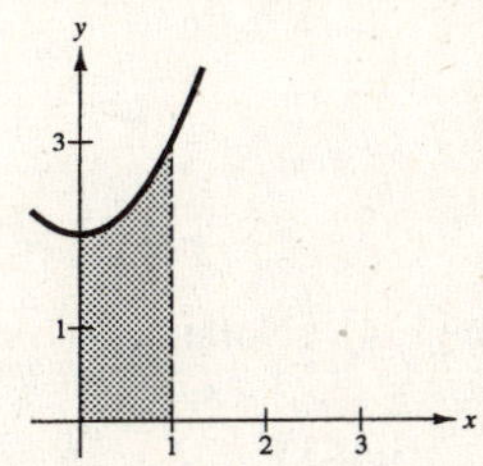

$$S(n) = \sum_{i=1}^{n} f\left(\frac{i}{n}\right)\left(\frac{1}{n}\right) = \sum_{i=1}^{n}\left[\left(\frac{i}{n}\right)^2 + 2\right]\left(\frac{1}{n}\right)$$

$$= \left[\frac{1}{n^3}\sum_{i=1}^{n} i^2\right] + 2 = \frac{n(n+1)(2n+1)}{6n^3} + 2 = \frac{1}{6}\left(2 + \frac{3}{n} + \frac{1}{n^2}\right) + 2$$

$$\text{Area} = \lim_{n\to\infty} S(n) = \frac{7}{3}$$

52. $y = x^2 + 1$ on $[0, 3]$. $\left(\textbf{Note: } \Delta x = \dfrac{3-0}{n} = \dfrac{3}{n}\right)$

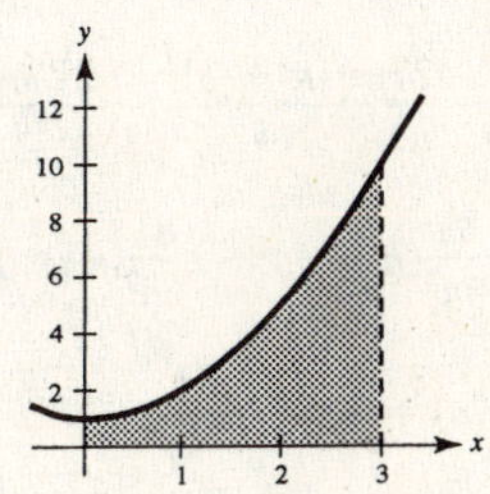

$$S(n) = \sum_{i=1}^{n} f\left(\frac{3i}{n}\right)\left(\frac{3}{n}\right) = \sum_{i=1}^{n}\left[\left(\frac{3i}{n}\right)^2 + 1\right]\left(\frac{3}{n}\right)$$

$$= \frac{27}{n^3}\sum_{l=1}^{n} l^2 + \frac{3}{n}\sum_{l=1}^{n} 1$$

$$= \frac{27}{n^3}\,\frac{n(n+1)(2n+1)}{6} + \frac{3}{n}(n) = \frac{9}{2}\,\frac{2n^2 + 3n + 1}{n^2} + 3$$

$$\text{Area} = \lim_{n\to\infty} S(n) = \frac{9}{2}(2) + 3 = 12$$

53. $y = 16 - x^2$ on $[1, 3]$. $\left(\textbf{Note: } \Delta x = \frac{2}{n}\right)$

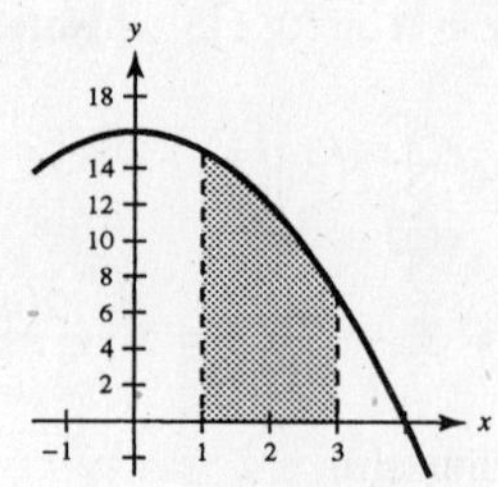

$$s(n) = \sum_{i=1}^{n} f\left(1 + \frac{2i}{n}\right)\left(\frac{2}{n}\right) = \sum_{i=1}^{n}\left[16 - \left(1 + \frac{2i}{n}\right)^2\right]\left(\frac{2}{n}\right)$$

$$= \frac{2}{n}\sum_{i=1}^{n}\left[15 - \frac{4i^2}{n^2} - \frac{4i}{n}\right]$$

$$= \frac{2}{n}\left[15n - \frac{4}{n^2}\frac{n(n+1)(2n+1)}{6} - \frac{4}{n}\frac{n(n+1)}{2}\right]$$

$$= 30 - \frac{8}{6n^2}(n+1)(2n+1) - \frac{4}{n}(n+1)$$

$$\text{Area} = \lim_{n\to\infty} s(n) = 30 - \frac{8}{3} - 4 = \frac{70}{3} = 23\frac{1}{3}$$

54. $y = 1 - x^2$ on $[-1, 1]$. Find area of region over the interval $[0, 1]$. $\left(\textbf{Note: } \Delta x = \frac{1}{n}\right)$

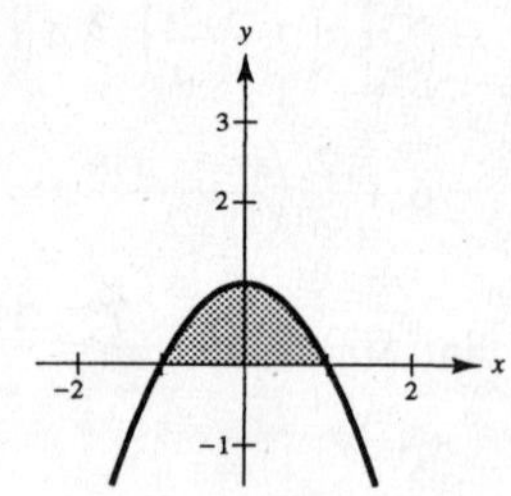

$$s(n) = \sum_{i=1}^{n} f\left(\frac{i}{n}\right)\left(\frac{1}{n}\right) = \sum_{i=1}^{n}\left[1 - \left(\frac{i}{n}\right)^2\right]\left(\frac{1}{n}\right)$$

$$= 1 - \frac{1}{n^3}\sum_{i=1}^{n} i^2 = 1 - \frac{n(n+1)(2n+1)}{6n^3} = 1 - \frac{1}{6}\left(2 + \frac{3}{n} + \frac{1}{n^2}\right)$$

$$\frac{1}{2}\text{Area} = \lim_{n\to\infty} s(n) = 1 - \frac{1}{3} = \frac{2}{3}$$

$$\text{Area} = \frac{4}{3}$$

55. $y = 64 - x^3$ on $[1, 4]$. $\left(\textbf{Note: } \Delta x = \frac{4-1}{n} = \frac{3}{n}\right)$

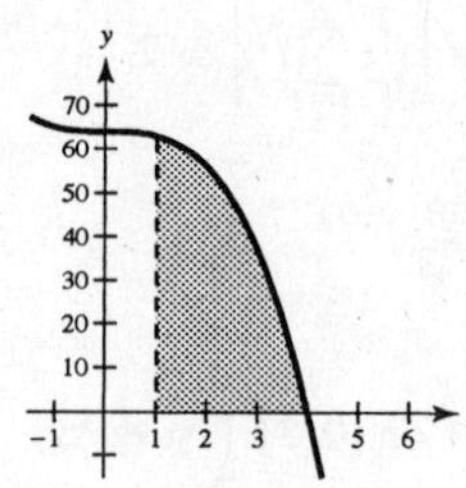

$$s(n) = \sum_{i=1}^{n} f\left(1 + \frac{3i}{n}\right)\left(\frac{3}{n}\right) = \sum_{i=1}^{n}\left[64 - \left(1 + \frac{3i}{n}\right)^3\right]\left(\frac{3}{n}\right)$$

$$= \frac{3}{n}\sum_{i=1}^{n}\left[63 - \frac{27i^3}{n^3} - \frac{27i^2}{n^2} - \frac{9i}{n}\right]$$

$$= \frac{3}{n}\left[63n - \frac{27}{n^3}\frac{n^2(n+1)^2}{4} - \frac{27}{n^2}\frac{n(n+1)(2n+1)}{6} - \frac{9}{n}\frac{n(n+1)}{2}\right]$$

$$= 189 - \frac{81}{4n^2}(n+1)^2 - \frac{81}{6n^2}(n+1)(2n+1) - \frac{27}{2}\frac{n+1}{n}$$

$$\text{Area} = \lim_{n\to\infty} s(n) = 189 - \frac{81}{4} - 27 - \frac{27}{2} = \frac{513}{4} = 128.25$$

56. $y = 2x - x^3$ on $[0, 1]$. $\left(\textbf{Note: } \Delta x = \frac{1-0}{n} = \frac{1}{n}\right)$

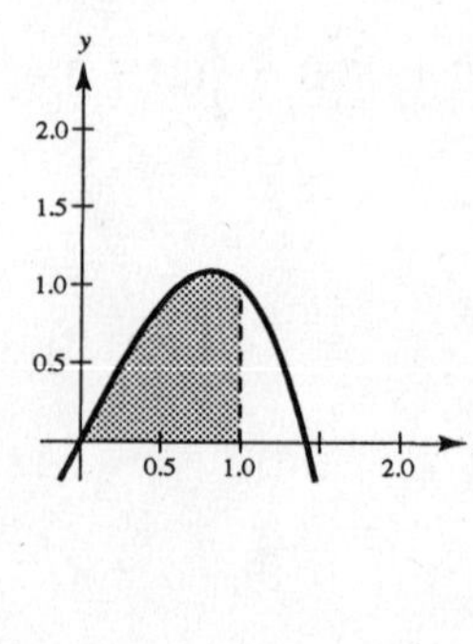

Since y both increases and decreases on $[0, 1]$, $T(n)$ is neither an upper nor lower sum.

$$T(n) = \sum_{i=1}^{n} f\left(\frac{i}{n}\right)\left(\frac{1}{n}\right) = \sum_{i=1}^{n}\left[2\left(\frac{i}{n}\right) - \left(\frac{i}{n}\right)^3\right]\left(\frac{1}{n}\right)$$

$$= \frac{2}{n^2}\sum_{i=1}^{n} i - \frac{1}{n^4}\sum_{i=1}^{n} i^3 = \frac{n(n+1)}{n^2} - \frac{1}{n^4}\left[\frac{n^2(n+1)^2}{4}\right]$$

$$= 1 + \frac{1}{n} - \frac{1}{4} - \frac{2}{4n} - \frac{1}{4n^2}$$

$$\text{Area} = \lim_{n\to\infty} T(n) = 1 - \frac{1}{4} = \frac{3}{4}$$

57. $y = x^2 - x^3$ on $[-1, 1]$. $\left(\textbf{Note: } \Delta x = \dfrac{1-(-1)}{n} = \dfrac{2}{n}\right)$

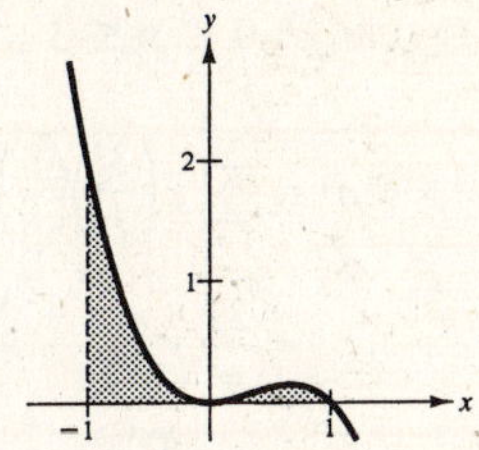

Again, $T(n)$ is neither an upper nor a lower sum.

$$T(n) = \sum_{i=1}^{n} f\left(-1 + \frac{2i}{n}\right)\left(\frac{2}{n}\right) = \sum_{i=1}^{n}\left[\left(-1 + \frac{2i}{n}\right)^2 - \left(-1 + \frac{2i}{n}\right)^3\right]\left(\frac{2}{n}\right)$$

$$= \sum_{i=1}^{n}\left[\left(1 - \frac{4i}{n} + \frac{4i^2}{n^2}\right) - \left(-1 + \frac{6i}{n} - \frac{12i^2}{n^2} + \frac{8i^3}{n^3}\right)\right]\left(\frac{2}{n}\right)$$

$$= \sum_{i=1}^{n}\left[2 - \frac{10i}{n} + \frac{16i^2}{n^2} - \frac{8i^3}{n^3}\right]\left(\frac{2}{n}\right) = \frac{4}{n}\sum_{i=1}^{n} 1 - \frac{20}{n^2}\sum_{i=1}^{n} i + \frac{32}{n^3}\sum_{i=1}^{n} i^2 - \frac{16}{n^4}\sum_{i=1}^{n} i^3$$

$$= \frac{4}{n}(n) - \frac{20}{n^2}\cdot\frac{n(n+1)}{2} + \frac{32}{n^3}\cdot\frac{n(n+1)(2n+1)}{6} - \frac{16}{n^4}\cdot\frac{n^2(n+1)^2}{4}$$

$$= 4 - 10\left(1 + \frac{1}{n}\right) + \frac{16}{3}\left(2 + \frac{3}{n} + \frac{1}{n^2}\right) - 4\left(1 + \frac{2}{n} + \frac{1}{n^2}\right)$$

$$\text{Area} = \lim_{n\to\infty} T(n) = 4 - 10 + \frac{32}{3} - 4 = \frac{2}{3}$$

58. $y = x^2 - x^3$ on $[-1, 0]$. $\left(\textbf{Note: } \Delta x = \dfrac{0-(-1)}{n} = \dfrac{1}{n}\right)$

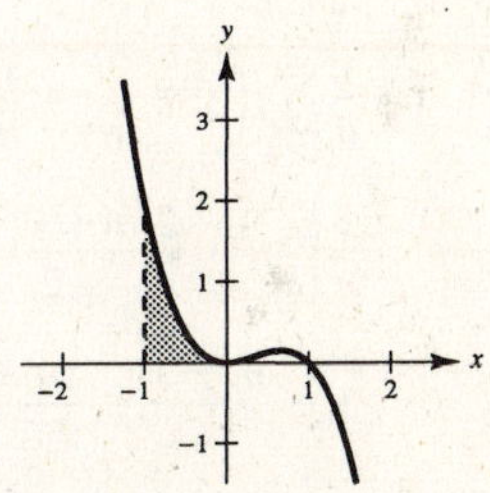

$$s(n) = \sum_{i=1}^{n} f\left(-1 + \frac{i}{n}\right)\left(\frac{1}{n}\right) = \sum_{i=1}^{n}\left[\left(-1 + \frac{i}{n}\right)^2 - \left(-1 + \frac{i}{n}\right)^3\right]\left(\frac{1}{n}\right)$$

$$= \sum_{i=1}^{n}\left[\left(2 - \frac{5i}{n} + \frac{4i^2}{n^2} - \frac{i^3}{n^3}\right)\left(\frac{1}{n}\right)\right] = 2 - \frac{5}{n^2}\sum_{i=1}^{n} i + \frac{4}{n^3}\sum_{i=1}^{n} i^2 - \frac{1}{n^4}\sum_{i=1}^{n} i^3$$

$$= 2 - \frac{5}{2} - \frac{5}{2n} + \frac{4}{3} + \frac{2}{n} + \frac{2}{3n^3} - \frac{1}{4} - \frac{1}{2n} - \frac{1}{4n^2}$$

$$\text{Area} = \lim_{n\to\infty} s(n) = 2 - \frac{5}{2} + \frac{4}{3} - \frac{1}{4} = \frac{7}{12}$$

59. $f(y) = 3y,\ 0 \le y \le 2$ $\left(\textbf{Note: } \Delta y = \dfrac{2-0}{n} = \dfrac{2}{n}\right)$

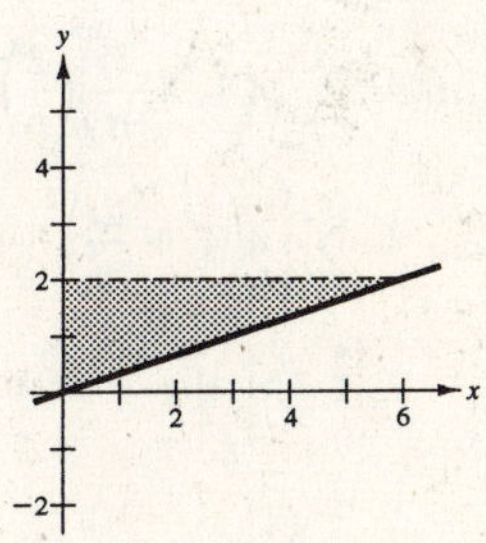

$$S(n) = \sum_{i=1}^{n} f(m_i)\,\Delta y = \sum_{i=1}^{n} f\left(\frac{2i}{n}\right)\left(\frac{2}{n}\right) = \sum_{i=1}^{n} 3\left(\frac{2i}{n}\right)\left(\frac{2}{n}\right)$$

$$= \frac{12}{n^2}\sum_{i=1}^{n} i = \left(\frac{12}{n^2}\right)\cdot\frac{n(n+1)}{2} = \frac{6(n+1)}{n} = 6 + \frac{6}{n}$$

$$\text{Area} = \lim_{n\to\infty} S(n) = \lim_{n\to\infty}\left(6 + \frac{6}{n}\right) = 6$$

60. $g(y) = \frac{1}{2}y,\ 2 \le y \le 4.$ $\left(\textbf{Note: } \Delta y = \dfrac{4-2}{n} = \dfrac{2}{n}\right)$

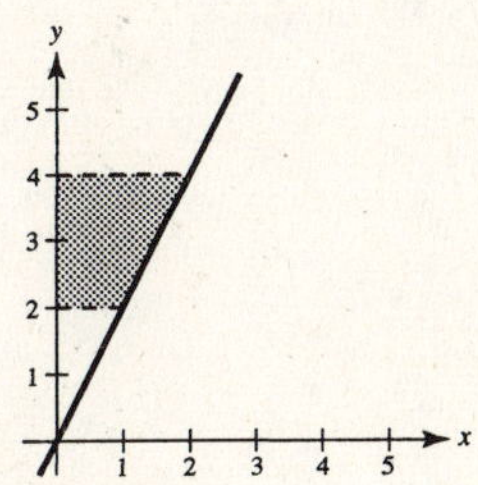

$$S(n) = \sum_{i=1}^{n} g\left(2 + \frac{2i}{n}\right)\left(\frac{2}{n}\right)$$

$$= \sum_{i=1}^{n} \frac{1}{2}\left(2 + \frac{2i}{n}\right)\left(\frac{2}{n}\right) = \frac{2}{n}\sum_{i=1}^{n}\left(1 + \frac{i}{n}\right)$$

$$= \frac{2}{n}\left[n + \frac{1}{n}\frac{n(n+1)}{2}\right] = 2 + \frac{n+1}{n}$$

$$\text{Area} = \lim_{n\to\infty} S(n) = 2 + 1 = 3$$

61. $f(y) = y^2, 0 \le y \le 3$ $\left(\textbf{Note: } \Delta y = \frac{3-0}{n} = \frac{3}{n}\right)$

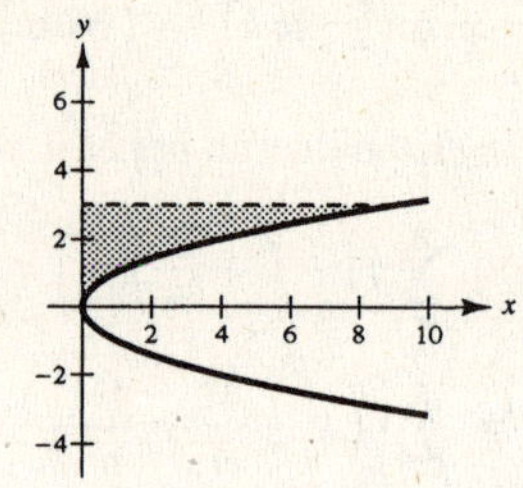

$$S(n) = \sum_{i=1}^{n} f\left(\frac{3i}{n}\right)\left(\frac{3}{n}\right) = \sum_{i=1}^{n} \left(\frac{3i}{n}\right)^2\left(\frac{3}{n}\right) = \frac{27}{n^3}\sum_{i=1}^{n} i^2$$

$$= \frac{27}{n^3} \cdot \frac{n(n+1)(2n+1)}{6} = \frac{9}{n^2}\left(\frac{2n^2+3n+1}{2}\right) = 9 + \frac{27}{2n} + \frac{9}{2n^2}$$

$$\text{Area} = \lim_{n\to\infty} S(n) = \lim_{n\to\infty}\left(9 + \frac{27}{2n} + \frac{9}{2n^2}\right) = 9$$

62. $f(y) = 4y - y^2, 1 \le y \le 2.$ $\left(\textbf{Note: } \Delta y = \frac{2-1}{n} = \frac{1}{n}\right)$

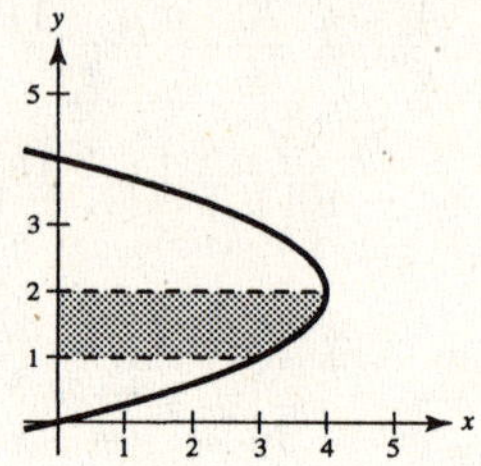

$$S(n) = \sum_{i=1}^{n} f\left(1 + \frac{i}{n}\right)\left(\frac{1}{n}\right)$$

$$= \frac{1}{n}\sum_{i=1}^{n}\left[4\left(1 + \frac{i}{n}\right) - \left(1 + \frac{i}{n}\right)^2\right]$$

$$= \frac{1}{n}\sum_{i=1}^{n}\left(4 + \frac{4i}{n} - 1 - \frac{2i}{n} - \frac{i^2}{n^2}\right)$$

$$= \frac{1}{n}\sum_{i=1}^{n}\left(3 + \frac{2i}{n} - \frac{i^2}{n^2}\right)$$

$$= \frac{1}{n}\left[3n + \frac{2}{n}\frac{n(n+1)}{2} - \frac{1}{n^2}\frac{n(n+1)(2n+1)}{6}\right]$$

$$= 3 + \frac{n+1}{n} - \frac{(n+1)(2n+1)}{6n^2}$$

$$\text{Area} = \lim_{n\to\infty} S(n) = 3 + 1 - \frac{1}{3} = \frac{11}{3}$$

63. $g(y) = 4y^2 - y^3, 1 \le y \le 3.$ $\left(\textbf{Note: } \Delta y = \frac{3-1}{n} = \frac{2}{n}\right)$

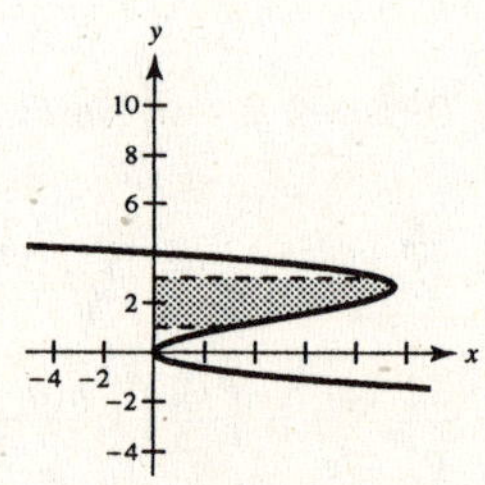

$$S(n) = \sum_{i=1}^{n} g\left(1 + \frac{2i}{n}\right)\left(\frac{2}{n}\right)$$

$$= \sum_{i=1}^{n}\left[4\left(1 + \frac{2i}{n}\right)^2 - \left(1 + \frac{2i}{n}\right)^3\right]\frac{2}{n}$$

$$= \frac{2}{n}\sum_{i=1}^{n}\left[4\left[1 + \frac{4i}{n} + \frac{4i^2}{n^2}\right] - \left[1 + \frac{6i}{n} + \frac{12i^2}{n^2} + \frac{8i^3}{n^3}\right]\right]$$

$$= \frac{2}{n}\sum_{i=1}^{n}\left[3 + \frac{10i}{n} + \frac{4i^2}{n^2} - \frac{8i^3}{n^3}\right] = \frac{2}{n}\left[3n + \frac{10}{n}\frac{n(n+1)}{2} + \frac{4}{n^2}\frac{n(n+1)(2n+1)}{6} - \frac{8}{n^3}\frac{n^2(n+1)^2}{4}\right]$$

$$\text{Area} = \lim_{n\to\infty} S(n) = 6 + 10 + \frac{8}{3} - 4 = \frac{44}{3}$$

64. $h(y) = y^3 + 1, 1 \le y \le 2 \left(\textbf{Note: } \Delta y = \frac{1}{n}\right)$

$$S(n) = \sum_{i=1}^{n} h\left(1 + \frac{i}{n}\right)\left(\frac{1}{n}\right)$$
$$= \sum_{i=1}^{n}\left[\left(1 + \frac{i}{n}\right)^3 + 1\right]\frac{1}{n}$$
$$= \frac{1}{n}\sum_{i=1}^{n}\left(2 + \frac{i^3}{n^3} + \frac{3i^2}{n^2} + \frac{3i}{n}\right)$$
$$= \frac{1}{n}\left[2n + \frac{1}{n^3}\frac{n^2(n+1)^2}{4} + \frac{3}{n^2}\frac{n(n+1)(2n+1)}{6} + \frac{3}{n}\frac{n(n+1)}{2}\right]$$
$$= 2 + \frac{(n+1)^2}{4n^2} + \frac{1}{2}\frac{(n+1)(2n+1)}{n^2} + \frac{3(n+1)}{2n}$$

$$\text{Area} = \lim_{n \to \infty} S(n) = 2 + \frac{1}{4} + 1 + \frac{3}{2} = \frac{19}{4}$$

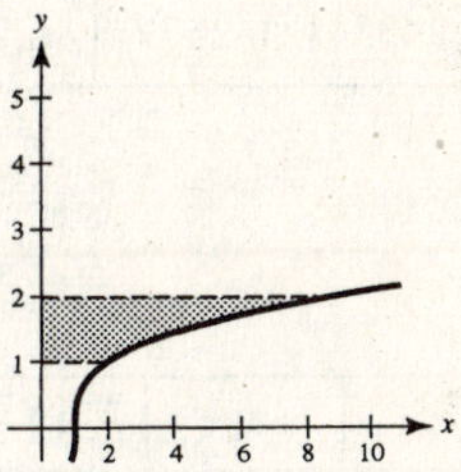

65. $f(x) = x^2 + 3, 0 \le x \le 2, n = 4$

Let $c_i = \dfrac{x_i + x_{i-1}}{2}$.

$$\Delta x = \frac{1}{2}, c_1 = \frac{1}{4}, c_2 = \frac{3}{4}, c_3 = \frac{5}{4}, c_4 = \frac{7}{4}$$

$$\text{Area} \approx \sum_{i=1}^{n} f(c_i)\,\Delta x = \sum_{i=1}^{4}[c_i^2 + 3]\left(\frac{1}{2}\right)$$
$$= \frac{1}{2}\left[\left(\frac{1}{16} + 3\right) + \left(\frac{9}{16} + 3\right) + \left(\frac{25}{16} + 3\right) + \left(\frac{49}{16} + 3\right)\right]$$
$$= \frac{69}{8}$$

66. $f(x) = x^2 + 4x, 0 \le x \le 4, n = 4$

Let $c_i = \dfrac{x_i + x_{i-1}}{2}$.

$$\Delta x = 1, c_1 = \frac{1}{2}, c_2 = \frac{3}{2}, c_3 = \frac{5}{2}, c_4 = \frac{7}{2}$$

$$\text{Area} \approx \sum_{i=1}^{n} f(c_i)\,\Delta x = \sum_{i=1}^{4}[c_i^2 + 4c_i](1)$$
$$= \left[\left(\frac{1}{4} + 2\right) + \left(\frac{9}{4} + 6\right) + \left(\frac{25}{4} + 10\right) + \left(\frac{49}{4} + 14\right)\right]$$
$$= 53$$

67. $f(x) = \sqrt{x - 1},\ [1, 2], n = 4$

Let $c_i = \dfrac{x_i + x_{i-1}}{2}$.

$$\Delta x = \frac{1}{4}, c_1 = \frac{9}{8}, c_2 = \frac{11}{8}, c_3 = \frac{13}{8}, c_4 = \frac{15}{8}$$

$$\text{Area} \approx \sum_{i=1}^{n} f(c_i)\,\Delta x = \sum_{i=1}^{4}\sqrt{c_i - 1}\left(\frac{1}{4}\right)$$
$$= \frac{1}{4}\left[\sqrt{1/8} + \sqrt{3/8} + \sqrt{5/8} + \sqrt{7/8}\right]$$
$$\approx 0.6730$$

68. $f(x) = \dfrac{1}{x^2 + 1},\ [0, 2], n = 4$

Let $c_i = \dfrac{x_i + x_{i-1}}{2}$

$$\Delta x = \frac{1}{2}, c_1 = \frac{1}{4}, c_2 = \frac{3}{4}, c_3 = \frac{5}{4}, c_4 = \frac{7}{4}$$

$$\text{Area} \approx \sum_{i=1}^{n} f(c_i)\Delta x = \sum_{i=1}^{4}\frac{1}{c_i^2 + 1}\left(\frac{1}{2}\right)$$
$$= \frac{1}{2}\left[\frac{1}{\frac{1}{16} + 1} + \frac{1}{\frac{9}{16} + 1} + \frac{1}{\frac{25}{16} + 1} + \frac{1}{\frac{49}{16} + 1}\right] \approx 1.1088$$

69. $f(x) = \sqrt{x}$ on $[0, 4]$.

n	4	8	12	16	20
Approximate area	5.3838	5.3523	5.3439	5.3403	5.3384

(Exact value is 16/3.)

70. $f(x) = x^{3/2} + 2$, $[0, 2]$.

n	4	8	12	16	20
Approximate area	6.2435	6.2577	6.2605	6.2615	6.2619

71. $f(x) = \dfrac{5x}{x^2 + 1}$, $[1, 3]$

n	4	8	12	16	20
Approximate Area	4.0272	4.0246	4.0241	4.0239	4.0238

72. $f(x) = \dfrac{8}{x^2 + 1}$ on $[2, 6]$.

n	4	8	12	16	20
Approximate area	2.3397	2.3755	2.3824	2.3848	2.3860

73. We can use the line $y = x$ bounded by $x = a$ and $x = b$. The sum of the areas of these inscribed rectangles is the lower sum.

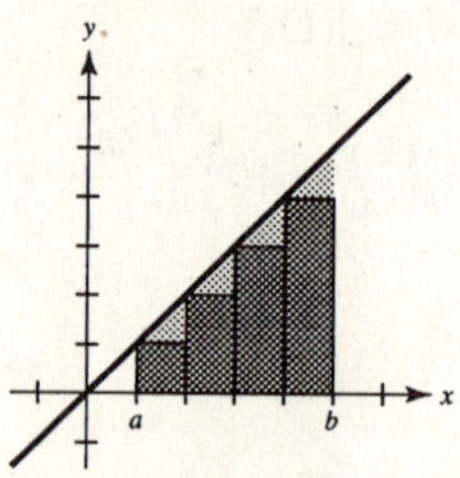

The sum of the areas of these circumscribed rectangles is the upper sum.

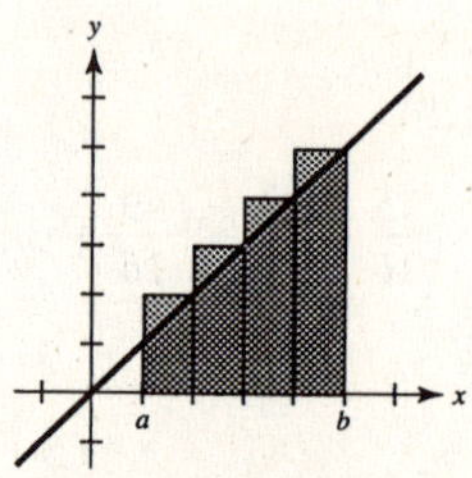

We can see that the rectangles do not contain all of the area in the first graph and the rectangles in the second graph cover more than the area of the region.

The exact value of the area lies between these two sums.

74. See the Definition of Area. Page 404.

75. (a) In each case, $\Delta x = 4/n$. The lower sum uses left endpoints, $(i - 1)(4/n)$. The upper sum uses right endpoints, $(i)(4/n)$. The Midpoint Rule uses midpoints, $\left(i - \frac{1}{2}\right)(4/n)$.

(b)

n	4	8	20	100	200
$s(n)$	15.333	17.368	18.459	18.995	19.06
$S(n)$	21.733	20.568	19.739	19.251	19.188
$M(n)$	19.403	19.201	19.137	19.125	19.125

(c) $s(n)$ increases because the lower sum approaches the exact value as n increases. $S(n)$ decreases because the upper sum approaches the exact value as n increases. Because of the shape of the graph, the lower sum is always smaller than the exact value, whereas the upper sum is always larger.

76. $f(x) = \sqrt[3]{x}, 0 \le x \le 8$

n	10	20	50	100	200
$s(n)$	10.998	11.519	11.816	11.910	11.956
$S(n)$	12.598	12.319	12.136	12.070	12.036
$M(n)$	12.040	12.016	12.005	12.002	12.001

(**Note:** exact answer is 12.)

77.

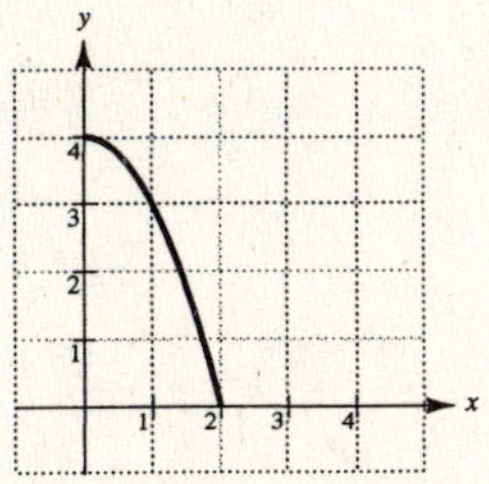

b. $A \approx 6$ square units

78.

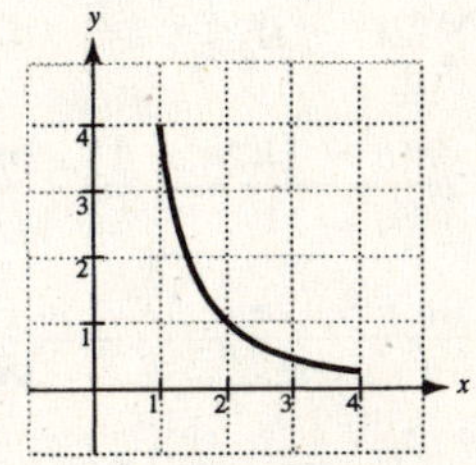

a. $A \approx 3$ square units

79. True. (Theorem 6.2 (2))

80. True. (Theorem 6.3)

81. (a) $y = (-4.09 \times 10^{-5})x^3 + 0.016x^2 - 2.67x + 452.9$

(b)

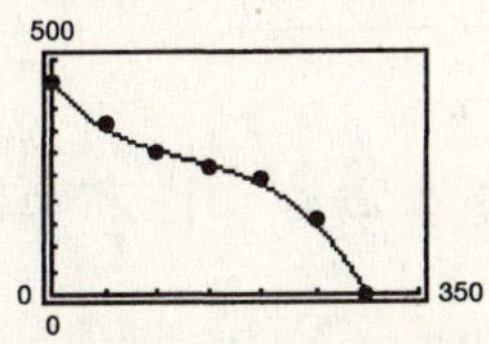

(c) Using the integration capability of a graphing utility, you obtain

$$A \approx 76{,}897.5 \text{ ft}^2.$$

82. Answers will vary.

83. Suppose there are n rows and $n + 1$ columns in the figure. The stars on the left total $1 + 2 + \cdots + n$, as do the stars on the right. There are $n(n + 1)$ stars in total, hence

$$2[1 + 2 + \cdots + n] = n(n + 1)$$

$$1 + 2 + \cdots + n = \tfrac{1}{2}(n)(n + 1).$$

84. (a) $\sum_{i=1}^{n} 2i = n(n + 1)$

The formula is true for $n = 1$: $2 = 1(1 + 1) = 2$

Assume that the formula is true for $n = k$:

$$\sum_{i=1}^{k} 2i = k(k + 1).$$

Then we have
$$\sum_{i=1}^{k+1} 2i = \sum_{i=1}^{k} 2i + 2(k + 1)$$
$$= k(k + 1) + 2(k + 1)$$
$$= (k + 1)(k + 2).$$

Which shows that the formula is true for $n = k + 1$.

(b) $\sum_{i=1}^{n} i^3 = \dfrac{n^2(n + 1)^2}{4}$

The formula is true for $n = 1$ because

$$1^3 = \frac{1^2(1 + 1)^2}{4} = \frac{4}{4} = 1.$$

Assume that the formula is true for $n = k$:

$$\sum_{i=1}^{k} i^3 = \frac{k^2(k + 1)^2}{4}.$$

Then we have
$$\sum_{i=1}^{k+1} i^3 = \sum_{i=1}^{k} i^3 + (k + 1)^3$$
$$= \frac{k^2(k + 1)^2}{4} + (k + 1)^3$$
$$= \frac{(k + 1)^2}{4}[k^2 + 4(k + 1)]$$
$$= \frac{(k + 1)^2}{4}(k + 2)^2$$

which shows that the formula is true for $n = k + 1$.

Section 6.3 Riemann Sums and Definite Integrals

1. $f(x) = \sqrt{x}, y = 0, x = 0, x = 3, c_i = \dfrac{3i^2}{n^2}$

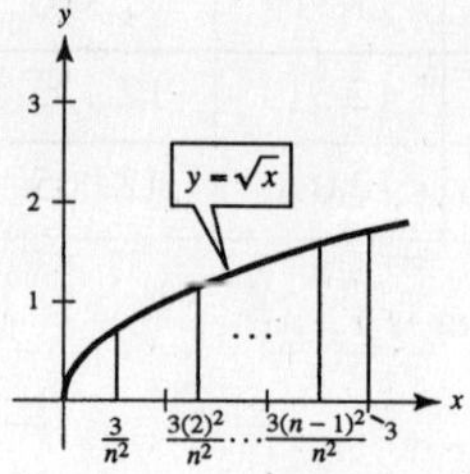

$$\Delta x_i = \frac{3i^2}{n^2} - \frac{3(i-1)^2}{n^2} = \frac{3}{n^2}(2i-1)$$

$$\lim_{n\to\infty}\sum_{i=1}^{n} f(c_i)\Delta x_i = \lim_{n\to\infty}\sum_{i=1}^{n}\sqrt{\frac{3i^2}{n^2}}\,\frac{3}{n^2}(2i-1)$$

$$= \lim_{n\to\infty}\frac{3\sqrt{3}}{n^3}\sum_{i=1}^{n}(2i^2 - i)$$

$$= \lim_{n\to\infty}\frac{3\sqrt{3}}{n^3}\left[2\frac{n(n+1)(2n+1)}{6} - \frac{n(n+1)}{2}\right]$$

$$= \lim_{n\to\infty} 3\sqrt{3}\left[\frac{(n+1)(2n+1)}{3n^2} - \frac{n+1}{2n^2}\right]$$

$$= 3\sqrt{3}\left[\frac{2}{3} - 0\right] = 2\sqrt{3} \approx 3.464$$

2. $f(x) = \sqrt[3]{x}, y = 0, x = 0, x = 1, c_i = \dfrac{i^3}{n^3}$

$$\Delta x_i = \frac{i^3}{n^3} - \frac{(i-1)^3}{n^3} = \frac{3i^2 - 3i + 1}{n^3}$$

$$\lim_{n\to\infty}\sum_{i=1}^{n} f(c_i)\,\Delta x_i = \lim_{n\to\infty}\sum_{i=1}^{n}\sqrt[3]{\frac{i^3}{n^3}}\left[\frac{3i^2 - 3i + 1}{n^3}\right]$$

$$= \lim_{n\to\infty}\frac{1}{n^4}\sum_{i=1}^{n}(3i^3 - 3i^2 + i)$$

$$= \lim_{n\to\infty}\frac{1}{n^4}\left[3\left(\frac{n^2(n+1)^2}{4}\right) - 3\left(\frac{n(n+1)(2n+1)}{6}\right) + \frac{n(n+1)}{2}\right]$$

$$= \lim_{n\to\infty}\frac{1}{n^4}\left[\frac{3n^4 + 6n^3 + 3n^2}{4} - \frac{2n^3 + 3n^2 + n}{2} + \frac{n^2 + n}{2}\right]$$

$$= \lim_{n\to\infty}\frac{1}{n^4}\left[\frac{3n^4}{4} + \frac{n^3}{2} - \frac{n^2}{4}\right] = \lim_{n\to\infty}\left[\frac{3}{4} + \frac{1}{2n} - \frac{1}{4n^2}\right] = \frac{3}{4}$$

3. $y = 6$ on $[4, 10]$. $\left(\textbf{Note: } \Delta x = \dfrac{10-4}{n} = \dfrac{6}{n}, \|\Delta\| \to 0 \text{ as } n \to \infty\right)$

$$\sum_{i=1}^{n} f(c_i)\,\Delta x_i = \sum_{i=1}^{n} f\left(4 + \frac{6i}{n}\right)\left(\frac{6}{n}\right) = \sum_{i=1}^{n} 6\left(\frac{6}{n}\right) = \sum_{i=1}^{n}\frac{36}{n} = 36$$

$$\int_4^{10} 6\,dx = \lim_{n\to\infty} 36 = 36$$

4. $y = x$ on $[-2, 3]$. $\left(\textbf{Note: } \Delta x = \dfrac{3-(-2)}{n} = \dfrac{5}{n}, \|\Delta\| \to 0 \text{ as } n \to \infty\right)$

$$\sum_{i=1}^{n} f(c_i)\,\Delta x_i = \sum_{i=1}^{n} f\left(-2 + \frac{5i}{n}\right)\left(\frac{5}{n}\right) = \sum_{i=1}^{n}\left(-2 + \frac{5i}{n}\right)\left(\frac{5}{n}\right) = -10 + \frac{25}{n^2}\sum_{i=1}^{n} i$$

$$= -10 + \left(\frac{25}{n^2}\right)\frac{n(n+1)}{2} = -10 + \frac{25}{2}\left(1 + \frac{1}{n}\right) = \frac{5}{2} + \frac{25}{2n}$$

$$\int_{-2}^{3} x\,dx = \lim_{n\to\infty}\left(\frac{5}{2} + \frac{25}{2n}\right) = \frac{5}{2}$$

5. $y = x^3$ on $[-1, 1]$. $\left(\textbf{Note: } \Delta x = \frac{1-(-1)}{n} = \frac{2}{n}, \|\Delta\| \to 0 \text{ as } n \to \infty\right)$

$$\sum_{i=1}^{n} f(c_i)\,\Delta x_i = \sum_{i=1}^{n} f\left(-1 + \frac{2i}{n}\right)\left(\frac{2}{n}\right) = \sum_{i=1}^{n}\left(-1 + \frac{2i}{n}\right)^3\left(\frac{2}{n}\right) = \sum_{i=1}^{n}\left[-1 + \frac{6i}{n} - \frac{12i^2}{n^2} + \frac{8i^3}{n^3}\right]\left(\frac{2}{n}\right)$$

$$= -2 + \frac{12}{n^2}\sum_{i=1}^{n} i - \frac{24}{n^3}\sum_{i=1}^{n} i^2 + \frac{16}{n^4}\sum_{i=1}^{n} i^3$$

$$= -2 + 6\left(1 + \frac{1}{n}\right) - 4\left(2 + \frac{3}{n} + \frac{1}{n^2}\right) + 4\left(1 + \frac{2}{n} + \frac{1}{n^2}\right) = \frac{2}{n}$$

$$\int_{-1}^{1} x^3\,dx = \lim_{n\to\infty} \frac{2}{n} = 0$$

6. $y = 3x^2$ on $[1, 3]$. $\left(\textbf{Note: } \Delta x = \frac{3-1}{n} = \frac{2}{n}, \|\Delta\| \to 0 \text{ as } n \to \infty\right)$

$$\sum_{i=1}^{n} f(c_i)\Delta x_i = \sum_{i=1}^{n} f\left(1 + \frac{2i}{n}\right)\left(\frac{2}{n}\right)$$

$$= \sum_{i=1}^{n} 3\left(1 + \frac{2i}{n}\right)^2\left(\frac{2}{n}\right)$$

$$= \frac{6}{n}\sum_{i=1}^{n}\left(1 + \frac{4i}{n} + \frac{4i^2}{n^2}\right)$$

$$= \frac{6}{n}\left[n + \frac{4}{n}\frac{n(n+1)}{2} + \frac{4}{n^2}\frac{n(n+1)(2n+1)}{6}\right]$$

$$= 6 + 12\frac{n+1}{n} + 4\frac{(n+1)(2n+1)}{n^2}$$

$$\int_{1}^{3} 3x^2\,dx = \lim_{n\to\infty}\left[6 + \frac{12(n+1)}{n} + \frac{4(n+1)(2n+1)}{n^2}\right]$$

$$= 6 + 12 + 8 = 26$$

7. $y = x^2 + 1$ on $[1, 2]$. $\left(\textbf{Note: } \Delta x = \frac{2-1}{n} = \frac{1}{n}, \|\Delta\| \to 0 \text{ as } n \to \infty\right)$

$$\sum_{i=1}^{n} f(c_i)\,\Delta x_i = \sum_{i=1}^{n} f\left(1 + \frac{i}{n}\right)\left(\frac{1}{n}\right) = \sum_{i=1}^{n}\left[\left(1 + \frac{i}{n}\right)^2 + 1\right]\left(\frac{1}{n}\right) = \sum_{i=1}^{n}\left[1 + \frac{2i}{n} + \frac{i^2}{n^2} + 1\right]\left(\frac{1}{n}\right)$$

$$= 2 + \frac{2}{n^2}\sum_{i=1}^{n} i + \frac{1}{n^3}\sum_{i=1}^{n} i^2 = 2 + \left(1 + \frac{1}{n}\right) + \frac{1}{6}\left(2 + \frac{3}{n} + \frac{1}{n^2}\right) = \frac{10}{3} + \frac{3}{2n} + \frac{1}{6n^2}$$

$$\int_{1}^{2} (x^2 + 1)\,dx = \lim_{n\to\infty}\left(\frac{10}{3} + \frac{3}{2n} + \frac{1}{6n^2}\right) = \frac{10}{3}$$

8. $y = 3x^2 + 2$ on $[-1, 2]$. $\left(\textbf{Note: } \Delta x = \frac{2-(-1)}{n} = \frac{3}{n}; \|\Delta\| \to 0 \text{ as } n \to \infty\right)$

$$\begin{aligned}\sum_{i=1}^{n} f(c_i)\Delta x_i &= \sum_{i=1}^{n} f\left(-1 + \frac{3i}{n}\right)\left(\frac{3}{n}\right) \\ &= \frac{3}{n}\sum_{i=1}^{n}\left[3\left(-1 + \frac{3i}{n}\right)^2 + 2\right] \\ &= \frac{3}{n}\sum_{i=1}^{n}\left[3\left(1 - \frac{6i}{n} + \frac{9i^2}{n^2}\right) + 2\right] \\ &= \frac{3}{n}\left[3n - \frac{18}{n}\frac{n(n+1)}{2} + \frac{27}{n^2}\frac{n(n+1)(2n+1)}{6} + 2n\right] \\ &= 15 - \frac{27(n+1)}{n} + \frac{27}{2}\frac{(n+1)(2n+1)}{n^2}\end{aligned}$$

$$\begin{aligned}\int_{-1}^{2}(3x^2+2)\,dx &= \lim_{n\to\infty}\left[15 - 27\frac{(n+1)}{n} + \frac{27}{2}\frac{(n+1)(2n+1)}{n^2}\right] \\ &= 15 - 27 + 27 = 15\end{aligned}$$

9. $\displaystyle\lim_{\|\Delta\|\to 0}\sum_{i=1}^{n}(3c_i + 10)\,\Delta x_i = \int_{-1}^{5}(3x+10)\,dx$

on the interval $[-1, 5]$.

10. $\displaystyle\lim_{\|\Delta\|\to 0}\sum_{i=1}^{n}6c_i(4 - c_i)^2\,\Delta x_i = \int_{0}^{4}6x(4-x)^2\,dx$

on the interval $[0, 4]$.

11. $\displaystyle\lim_{\|\Delta\|\to 0}\sum_{i=1}^{n}\sqrt{c_i^2+4}\,\Delta x_i = \int_{0}^{3}\sqrt{x^2+4}\,dx$

on the interval $[0, 3]$.

12. $\displaystyle\lim_{\|\Delta\|\to 0}\sum_{i=1}^{n}\left(\frac{3}{c_i^2}\right)\Delta x_i = \int_{1}^{3}\frac{3}{x^2}\,dx$

on the interval $[1, 3]$.

13. $\displaystyle\int_0^5 3\,dx$

14. $\displaystyle\int_0^2 (4-2x)\,dx$

15. $\displaystyle\int_{-4}^{4}(4 - |x|)\,dx$

16. $\displaystyle\int_0^2 x^2\,dx$

17. $\displaystyle\int_{-2}^{2}(4-x^2)\,dx$

18. $\displaystyle\int_{-1}^{1}\frac{1}{x^2+1}\,dx$

19. $\displaystyle\int_0^2\sqrt{x+1}\,dx$

20. $\displaystyle\int_{-1}^{1}(x^2+1)^3\,dx$

21. $\displaystyle\int_0^2 y^3\,dy$

22. $\displaystyle\int_0^2 (y-2)^2\,dy$

23. Rectangle

$A = bh = 3(4)$

$A = \displaystyle\int_0^3 4\,dx = 12$

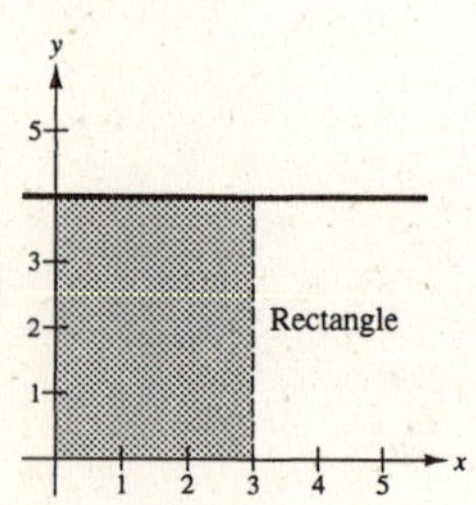

24. Rectangle

$A = bh = 2(4)(a)$

$A = \displaystyle\int_{-a}^{a}4\,dx = 8a$

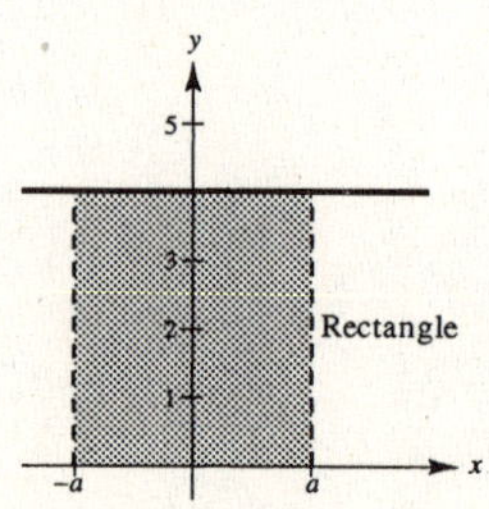

25. Triangle

$A = \frac{1}{2}bh = \frac{1}{2}(4)(4)$

$A = \displaystyle\int_0^4 x\,dx = 8$

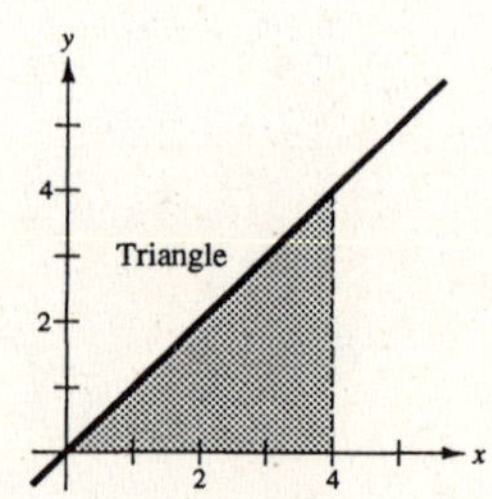

26. Triangle

$$A = \frac{1}{2}bh = \frac{1}{2}(4)(2)$$

$$A = \int_0^4 \frac{x}{2}\,dx = 4$$

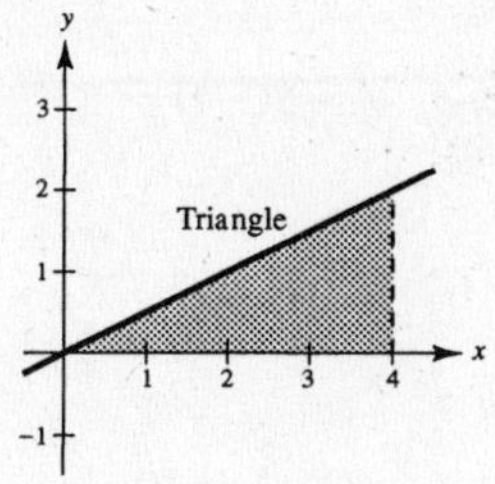

27. Trapezoid

$$A = \frac{b_1 + b_2}{2}h = \left(\frac{5 + 9}{2}\right)2$$

$$A = \int_0^2 (2x + 5)\,dx = 14$$

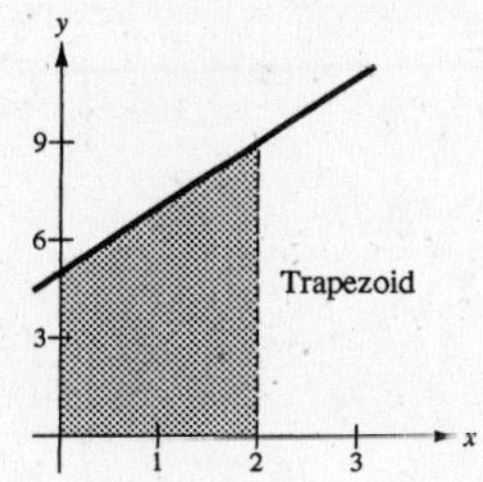

28. Triangle

$$A = \frac{1}{2}bh = \frac{1}{2}(8)(8) = 32$$

$$A = \int_0^8 (8 - x)\,dx = 32$$

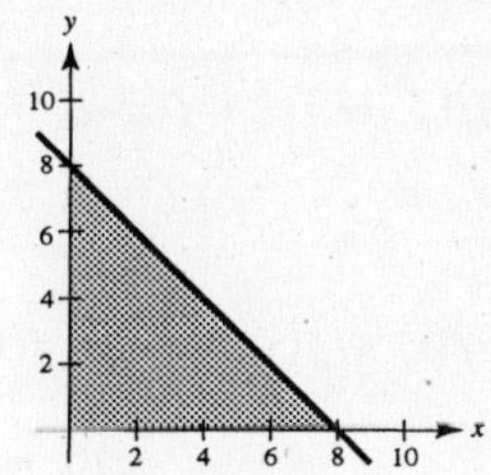

29. Triangle

$$A = \frac{1}{2}bh = \frac{1}{2}(2)(1)$$

$$A = \int_{-1}^1 (1 - |x|)\,dx = 1$$

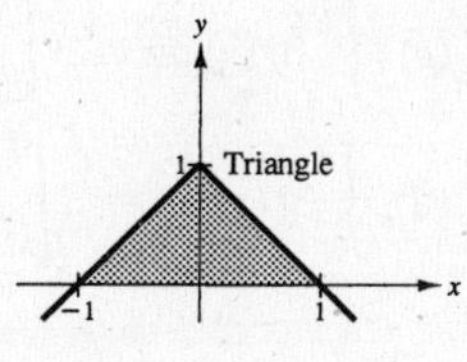

30. Triangle

$$A = \frac{1}{2}bh = \frac{1}{2}(2a)a$$

$$A = \int_{-a}^a (a - |x|)\,dx = a^2$$

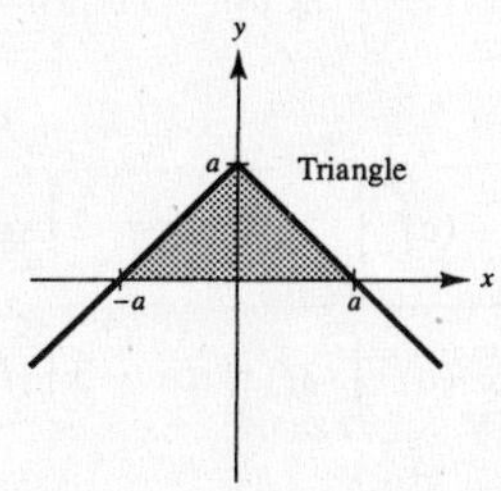

31. Semicircle

$$A = \frac{1}{2}\pi r^2 = \frac{1}{2}\pi(3)^2$$

$$A = \int_{-3}^3 \sqrt{9 - x^2}\,dx = \frac{9\pi}{2}$$

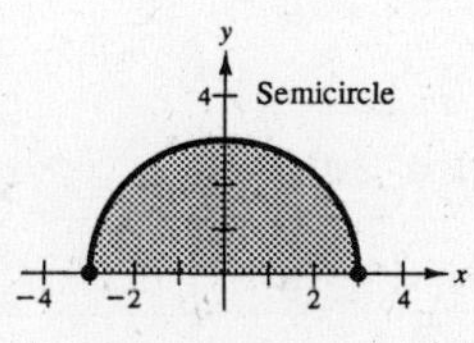

32. Semicircle

$$A = \frac{1}{2}\pi r^2$$

$$A = \int_{-r}^r \sqrt{r^2 - x^2}\,dx = \frac{1}{2}\pi r^2$$

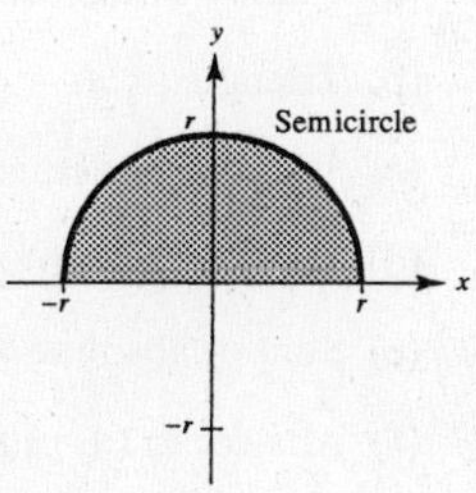

In Exercises 33–40, $\int_2^4 x^3\,dx = 60$, $\int_2^4 x\,dx = 6$, $\int_2^4 dx = 2$

33. $\int_4^2 x\,dx = -\int_2^4 x\,dx = -6$

34. $\int_2^2 x^3\,dx = 0$

35. $\int_2^4 4x\,dx = 4\int_2^4 x\,dx = 4(6) = 24$

36. $\int_2^4 15\,dx = 15\int_2^4 dx = 15(2) = 30$

37. $\int_2^4 (x - 8)\,dx = \int_2^4 x\,dx - 8\int_2^4 dx = 6 - 8(2) = -10$

38. $\int_2^4 (x^3 + 4)\,dx = \int_2^4 x^3\,dx + 4\int_2^4 dx = 60 + 4(2) = 68$

39. $\int_2^4 \left(\frac{1}{2}x^3 - 3x + 2\right)dx = \frac{1}{2}\int_2^4 x^3\,dx - 3\int_2^4 x\,dx + 2\int_2^4 dx$

$= \frac{1}{2}(60) - 3(6) + 2(2) = 16$

40. $\int_2^4 (6 + 2x - x^3)\,dx = 6\int_2^4 dx + 2\int_2^4 x\,dx - \int_2^4 x^3\,dx$

$= 6(2) + 2(6) - 60 = -36$

41. (a) $\int_0^7 f(x)\,dx = \int_0^5 f(x)\,dx + \int_5^7 f(x)\,dx = 10 + 3 = 13$

(b) $\int_5^0 f(x)\,dx = -\int_0^5 f(x)\,dx = -10$

(c) $\int_5^5 f(x)\,dx = 0$

(d) $\int_0^5 3f(x)\,dx = 3\int_0^5 f(x)\,dx = 3(10) = 30$

42. (a) $\int_0^6 f(x)\,dx = \int_0^3 f(x)\,dx + \int_3^6 f(x)\,dx = 4 + (-1) = 3$

(b) $\int_6^3 f(x)\,dx = -\int_3^6 f(x)\,dx = -(-1) = 1$

(c) $\int_3^3 f(x)\,dx = 0$

(d) $\int_3^6 -5f(x)\,dx = -5\int_3^6 f(x)\,dx = -5(-1) = 5$

43. (a) $\int_2^6 [f(x) + g(x)]\,dx = \int_2^6 f(x)\,dx + \int_2^6 g(x)\,dx$

$= 10 + (-2) = 8$

(b) $\int_2^6 [g(x) - f(x)]\,dx = \int_2^6 g(x)\,dx - \int_2^6 f(x)\,dx$

$= -2 - 10 = -12$

(c) $\int_2^6 2g(x)\,dx = 2\int_2^6 g(x)\,dx = 2(-2) = -4$

(d) $\int_2^6 3f(x)\,dx = 3\int_2^6 f(x)\,dx = 3(10) = 30$

44. (a) $\int_{-1}^0 f(x)\,dx = \int_{-1}^1 f(x)\,dx - \int_0^1 f(x)\,dx = 0 - 5 = -5$

(b) $\int_0^1 f(x)\,dx - \int_{-1}^0 f(x)\,dx = 5 - (-5) = 10$

(c) $\int_{-1}^1 3f(x)\,dx = 3\int_{-1}^1 f(x)\,dx = 3(0) = 0$

(d) $\int_0^1 3f(x)\,dx = 3\int_0^1 f(x)\,dx = 3(5) = 15$

45. (a) Quarter circle below x-axis: $-\frac{1}{4}\pi r^2 = -\frac{1}{4}\pi(2)^2 = -\pi$

(b) Triangle: $\frac{1}{2}bh = \frac{1}{2}(4)(2) = 4$

(c) Triangle + Semicircle below x-axis: $-\frac{1}{2}(2)(1) - \frac{1}{2}\pi(2)^2 = -(1 + 2\pi)$

(d) Sum of parts (b) and (c): $4 - (1 + 2\pi) = 3 - 2\pi$

(e) Sum of absolute values of (b) and (c): $4 + (1 + 2\pi) = 5 + 2\pi$

(f) Answer to (d) plus $2(10) = 20$: $(3 - 2\pi) + 20 = 23 - 2\pi$

46. (a) $\int_0^1 -f(x)dx = -\int_0^1 f(x)dx = \frac{1}{2}$

(b) $\int_3^4 3f(x)dx = 3(2) = 6$

(c) $\int_0^7 f(x)dx = -\frac{1}{2} + \frac{1}{2}(2)(2) + 2 + \frac{1}{2}(2)(2) - \frac{1}{2} = 5$

(d) $\int_5^{11} f(x)dx = \frac{1}{2} - \frac{1}{2}(4)(2) + \frac{1}{2} = -3$

(e) $\int_0^{11} f(x)dx = -\frac{1}{2} + 2 + 2 + 2 - 4 + \frac{1}{2} = 2$

(f) $\int_4^{10} f(x)dx = 2 - 4 = -2$

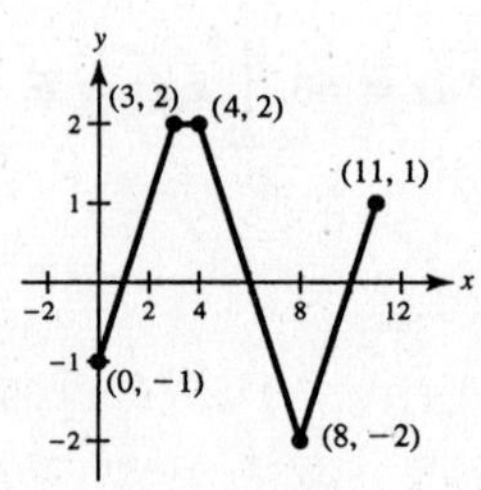

47. (a) $\int_0^5 [f(x) + 2]\,dx = \int_0^5 f(x)\,dx + \int_0^5 2\,dx = 4 + 10 = 14$

(b) $\int_{-2}^3 f(x + 2)\,dx = \int_0^5 f(x)\,dx = 4$ (Let $u = x + 2$.)

(c) $\int_{-5}^5 f(x)\,dx = 2\int_0^5 f(x)\,dx = 2(4) = 8$ (f even)

(d) $\int_{-5}^5 f(x)\,dx = 0$ (f odd)

48. $f(x) = \begin{cases} 4, & x < 4 \\ x, & x \geq 4 \end{cases}$

$$\int_0^8 f(x)\,dx = 4(4) + 4(4) + \frac{1}{2}(4)(4) = 40$$

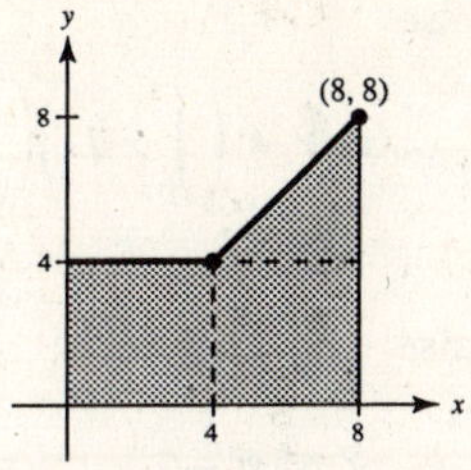

49. The left endpoint approximation will be greater than the actual area: >

50. The right endpoint approximation will be less than the actual area: <

51. $f(x) = \dfrac{1}{x - 4}$

is not integrable on the interval $[3, 5]$ and f has a discontinuity at $x = 4$.

52. $f(x) = |x|/x$ is integrable on $[-1, 1]$, but is not continuous on $[-1, 1]$. There is discontinuity at $x = 0$. To see that

$$\int_{-1}^{1} \frac{|x|}{x}\,dx$$

is integrable, sketch a graph of the region bounded by $f(x) = |x|/x$ and the x-axis for $-1 \leq x \leq 1$. You see that the integral equals 0.

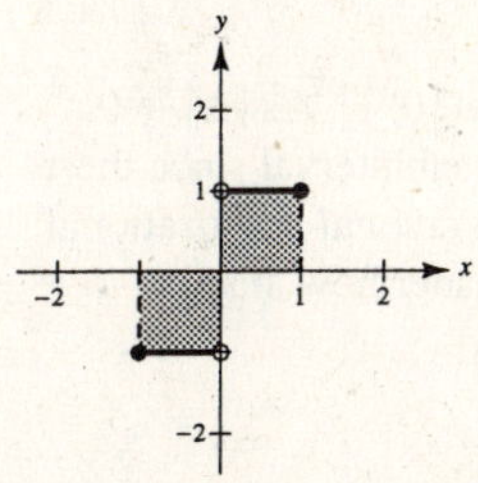

53.

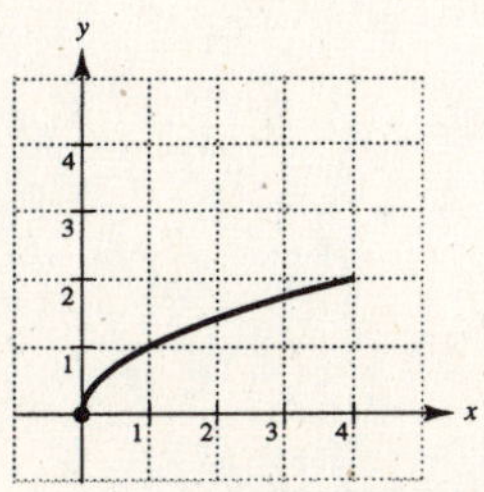

a. $A \approx 5$ square units

54.

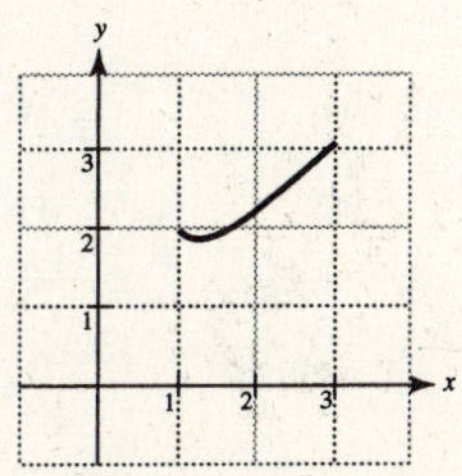

d. $\displaystyle\int_1^3 \frac{x^3 + 1}{x^2}\,dx \approx 4\frac{1}{2}$

55. $\displaystyle\int_0^3 x\sqrt{3 - x}\,dx$

n	4	8	12	16	20
$L(n)$	3.6830	3.9956	4.0707	4.1016	4.1177
$M(n)$	4.3082	4.2076	4.1838	4.1740	4.1690
$R(n)$	3.6830	3.9956	4.0707	4.1016	4.1177

56. $\displaystyle\int_0^3 \frac{5}{x^2 + 1}\,dx$

n	4	8	12	16	20
$L(n)$	7.9224	7.0855	6.8062	6.6662	6.5822
$M(n)$	6.2485	6.2470	7.2460	6.2457	6.2455
$R(n)$	4.5474	5.3980	5.6812	5.8225	5.9072

57. True

58. False

$$\int_0^1 x\sqrt{x}\,dx \neq \left(\int_0^1 x\,dx\right)\left(\int_0^1 \sqrt{x}\,dx\right)$$

59. True

60. True

61. False

$$\int_0^2 (-x)\,dx = -2$$

62. False

$$\int_{-2}^4 x\,dx = 6$$

63. $f(x) = x^2 + 3x$, $[0, 8]$

$x_0 = 0, x_1 = 1, x_2 = 3, x_3 = 7, x_4 = 8$

$\Delta x_1 = 1, \Delta x_2 = 2, \Delta x_3 = 4, \Delta x_4 = 1$

$c_1 = 1, c_2 = 2, c_3 = 5, c_4 = 8$

$$\sum_{i=1}^4 f(c_i)\,\Delta x = f(1)\,\Delta x_1 + f(2)\,\Delta x_2 + f(5)\,\Delta x_3 + f(8)\,\Delta x_4$$

$$= (4)(1) + (10)(2) + (40)(4) + (88)(1) = 272$$

64. $f(x) = \begin{cases} 1, & x \text{ is rational} \\ 0, & x \text{ is irrational} \end{cases}$

is not integrable on the interval $[0, 1]$. As $\|\Delta\| \to 0$, $f(c_i) = 1$ or $f(c_i) = 0$ in each subinterval since there are an infinite number of both rational and irrational numbers in any interval, no matter how small.

65. $f(x) = \begin{cases} 0, & x = 0 \\ \dfrac{1}{x}, & 0 < x \leq 1 \end{cases}$

The limit

$$\lim_{\|\Delta\| \to 0} \sum_{i=1}^n f(c_i)\Delta x_i$$

does not exist. This does not contradict Theorem 6.4 because f is not continuous on $[0, 1]$.

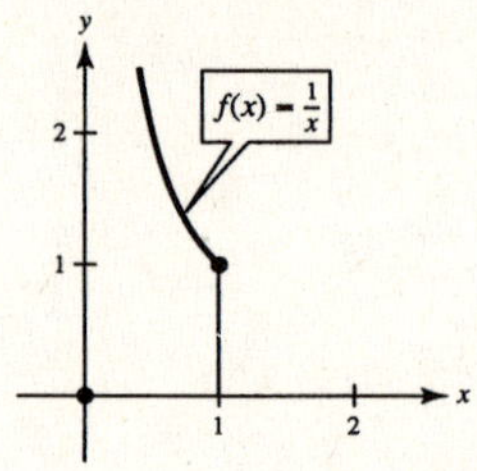

66. The function f is nonnegative between $x = -1$ and $x = 1$.

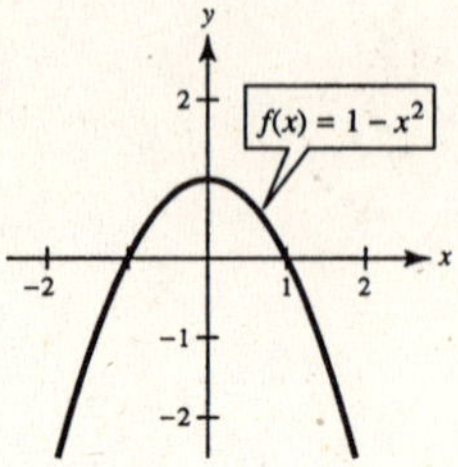

Hence,

$$\int_a^b (1 - x^2)dx$$

is a maximum for $a = -1$ and $b = 1$.

67. To find $\int_0^2 [\![x]\!]\,dx$, use a geometric approach.

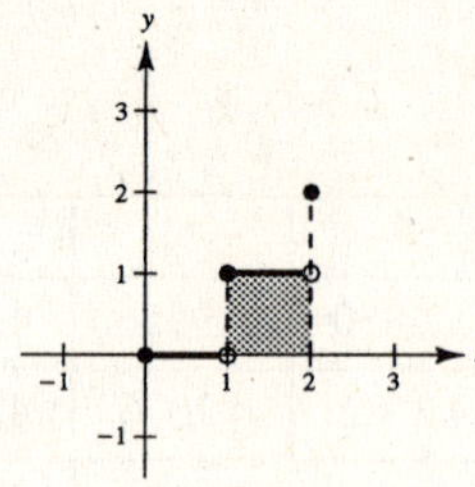

Thus,

$$\int_0^2 [\![x]\!]\,dx = 1(2 - 1) = 1.$$

Section 6.4 The Fundamental Theorem of Calculus

1. $f(x) = \dfrac{4}{x^2 + 1}$

$\displaystyle\int_0^{\pi} \frac{4}{x^2 + 1}\, dx$ is positive.

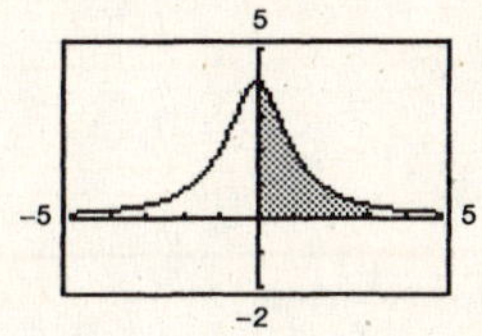

2. $f(x) = \sqrt[3]{x - 3}$

$\displaystyle\int_1^5 \sqrt[3]{x - 3}\, dx = 0$

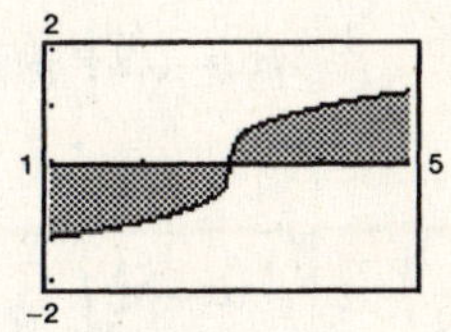

3. $f(x) = x\sqrt{x^2 + 1}$

$\displaystyle\int_{-2}^{2} x\sqrt{x^2 + 1}\, dx = 0$

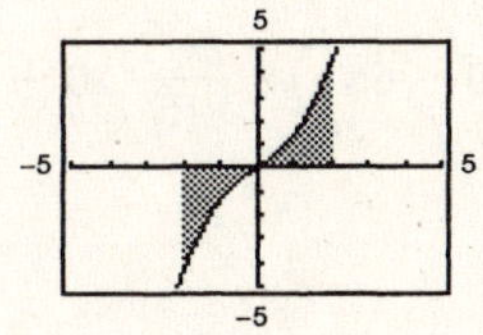

4. $f(x) = x\sqrt{2 - x}$

$\displaystyle\int_{-2}^{2} x\sqrt{2 - x}\, dx$ is negative.

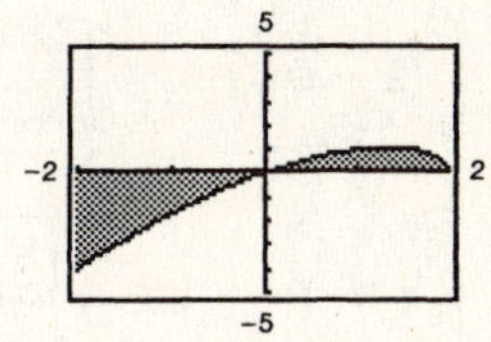

5. $\displaystyle\int_0^1 2x\, dx = \Big[x^2\Big]_0^1 = 1 - 0 = 1$

6. $\displaystyle\int_2^7 3\, dv = \Big[3v\Big]_2^7 = 3(7) - 3(2) = 15$

7. $\displaystyle\int_{-1}^{0} (x - 2)\, dx = \left[\frac{x^2}{2} - 2x\right]_{-1}^{0} = 0 - \left(\frac{1}{2} + 2\right) = -\frac{5}{2}$

8. $\displaystyle\int_2^5 (-3v + 4)\, dv = \left[-\frac{3}{2}v^2 + 4v\right]_2^5 = \left(-\frac{75}{2} + 20\right) - (-6 + 8) = -\frac{39}{2}$

9. $\displaystyle\int_{-1}^{1} (t^2 - 2)\, dt = \left[\frac{t^3}{3} - 2t\right]_{-1}^{1} = \left(\frac{1}{3} - 2\right) - \left(-\frac{1}{3} + 2\right) = -\frac{10}{3}$

10. $\displaystyle\int_1^3 (3x^2 + 5x - 4)\, dx = \left[x^3 + \frac{5x^2}{2} - 4x\right]_1^3 = \left(27 + \frac{45}{2} - 12\right) - \left(1 + \frac{5}{2} - 4\right) = 38$

11. $\displaystyle\int_0^1 (2t - 1)^2\, dt = \int_0^1 (4t^2 - 4t + 1)\, dt = \left[\frac{4}{3}t^3 - 2t^2 + t\right]_0^1 = \frac{4}{3} - 2 + 1 = \frac{1}{3}$

12. $\displaystyle\int_{-1}^{1} (t^3 - 9t)\, dt = \left[\frac{1}{4}t^4 - \frac{9}{2}t^2\right]_{-1}^{1} = \left(\frac{1}{4} - \frac{9}{2}\right) - \left(\frac{1}{4} - \frac{9}{2}\right) = 0$

13. $\displaystyle\int_1^2 \left(\frac{3}{x^2} - 1\right) dx = \left[-\frac{3}{x} - x\right]_1^2 = \left(-\frac{3}{2} - 2\right) - (-3 - 1) = \frac{1}{2}$

14. $\displaystyle\int_{-2}^{-1} \left(u - \frac{1}{u^2}\right) du = \left[\frac{u^2}{2} + \frac{1}{u}\right]_{-2}^{-1} = \left(\frac{1}{2} - 1\right) - \left(2 - \frac{1}{2}\right) = -2$

15. $\displaystyle\int_1^4 \frac{u - 2}{\sqrt{u}}\, du = \int_1^4 (u^{1/2} - 2u^{-1/2})\, du = \left[\frac{2}{3}u^{3/2} - 4u^{1/2}\right]_1^4 = \left[\frac{2}{3}(\sqrt{4})^3 - 4\sqrt{4}\right] - \left[\frac{2}{3} - 4\right] = \frac{2}{3}$

16. $\displaystyle\int_{-3}^{3} v^{1/3}\, dv = \left[\frac{3}{4}v^{4/3}\right]_{-3}^{3} = \frac{3}{4}\left[(\sqrt[3]{3})^4 - (\sqrt[3]{-3})^4\right] = 0$

17. $\displaystyle\int_{-1}^{1} (\sqrt[3]{t} - 2)\, dt = \left[\frac{3}{4}t^{4/3} - 2t\right]_{-1}^{1} = \left(\frac{3}{4} - 2\right) - \left(\frac{3}{4} + 2\right) = -4$

18. $\displaystyle\int_{1}^{8} \sqrt{\frac{2}{x}}\, dx = \sqrt{2}\int_{1}^{8} x^{-1/2}\, dx = \left[\sqrt{2}(2)x^{1/2}\right]_{1}^{8} = \left[2\sqrt{2x}\right]_{1}^{8} = 8 - 2\sqrt{2}$

19. $\displaystyle\int_{0}^{1} \frac{x - \sqrt{x}}{3}\, dx = \frac{1}{3}\int_{0}^{1} (x - x^{1/2})\, dx = \frac{1}{3}\left[\frac{x^2}{2} - \frac{2}{3}x^{3/2}\right]_{0}^{1} = \frac{1}{3}\left(\frac{1}{2} - \frac{2}{3}\right) = -\frac{1}{18}$

20. $\displaystyle\int_{0}^{2} (2 - t)\sqrt{t}\, dt = \int_{0}^{2} (2t^{1/2} - t^{3/2})\, dt = \left[\frac{4}{3}t^{3/2} - \frac{2}{5}t^{5/2}\right]_{0}^{2} = \left[\frac{t\sqrt{t}}{15}(20 - 6t)\right]_{0}^{2} = \frac{2\sqrt{2}}{15}(20 - 12) = \frac{16\sqrt{2}}{15}$

21. $\displaystyle\int_{-1}^{0} (t^{1/3} - t^{2/3})\, dt = \left[\frac{3}{4}t^{4/3} - \frac{3}{5}t^{5/3}\right]_{-1}^{0} = 0 - \left(\frac{3}{4} + \frac{3}{5}\right) = -\frac{27}{20}$

22. $\displaystyle\int_{-8}^{-1} \frac{x - x^2}{2\sqrt[3]{x}}\, dx = \frac{1}{2}\int_{-8}^{-1} (x^{2/3} - x^{5/3})\, dx$

$\displaystyle= \frac{1}{2}\left[\frac{3}{5}x^{5/3} - \frac{3}{8}x^{8/3}\right]_{-8}^{-1} = \left[\frac{x^{5/3}}{80}(24 - 15x)\right]_{-8}^{-1} = -\frac{1}{80}(39) + \frac{32}{80}(144) = \frac{4569}{80}$

23. $\displaystyle\int_{0}^{3} |2x - 3|\, dx = \int_{0}^{3/2} (3 - 2x)\, dx + \int_{3/2}^{3} (2x - 3)\, dx$ $\left(\text{split up the integral at the zero } x = \frac{3}{2}\right)$

$\displaystyle= \left[3x - x^2\right]_{0}^{3/2} + \left[x^2 - 3x\right]_{3/2}^{3} = \left(\frac{9}{2} - \frac{9}{4}\right) - 0 + (9 - 9) - \left(\frac{9}{4} - \frac{9}{2}\right) = 2\left(\frac{9}{2} - \frac{9}{4}\right) = \frac{9}{2}$

24. $\displaystyle\int_{1}^{4} (3 - |x - 3|)\, dx = \int_{1}^{3} [3 + (x - 3)]\, dx + \int_{3}^{4} [3 - (x - 3)]\, dx$

$\displaystyle= \int_{1}^{3} x\, dx + \int_{3}^{4} (6 - x)\, dx$

$\displaystyle= \left[\frac{x^2}{2}\right]_{1}^{3} + \left[6x - \frac{x^2}{2}\right]_{3}^{4}$

$\displaystyle= \left(\frac{9}{2} - \frac{1}{2}\right) + \left[(24 - 8) - \left(18 - \frac{9}{2}\right)\right]$

$\displaystyle= 4 + 16 - 18 + \frac{9}{2} = \frac{13}{2}$

25. $\displaystyle\int_{0}^{3} |x^2 - 4|\, dx = \int_{0}^{2} (4 - x^2)\, dx + \int_{2}^{3} (x^2 - 4)\, dx = \left[4x - \frac{x^3}{3}\right]_{0}^{2} + \left[\frac{x^3}{3} - 4x\right]_{2}^{3} = \left(8 - \frac{8}{3}\right) + (9 - 12) - \left(\frac{8}{3} - 8\right) = \frac{23}{3}$

26. $\displaystyle\int_{0}^{4} |x^2 - 4x + 3|\, dx = \int_{0}^{1} (x^2 - 4x + 3)\, dx - \int_{1}^{3} (x^2 - 4x + 3)\, dx + \int_{3}^{4} (x^2 - 4x + 3)\, dx$

$\displaystyle= \left[\frac{x^3}{3} - 2x^2 + 3x\right]_{0}^{1} - \left[\frac{x^3}{3} - 2x^2 + 3x\right]_{1}^{3} + \left[\frac{x^3}{3} - 2x^2 + 3x\right]_{3}^{4}$

$\displaystyle= \left(\frac{1}{3} - 2 + 3\right) - (9 - 18 + 9) + \left(\frac{1}{3} - 2 + 3\right) + \left(\frac{64}{3} - 32 + 12\right) - (9 - 18 + 9)$

$\displaystyle= \frac{4}{3} - 0 + \frac{4}{3} + \frac{4}{3} - 0 = 4$

27. $A = \int_0^1 (x - x^2)\,dx = \left[\frac{x^2}{2} - \frac{x^3}{3}\right]_0^1 = \frac{1}{6}$

28. $A = \int_{-1}^{3} (-x^2 + 2x + 3)\,dx = \left[-\frac{x^3}{3} + x^2 + 3x\right]_{-1}^{3} = (-9 + 9 + 9) - \left(\frac{1}{3} + 1 - 3\right) = \frac{32}{3}$

29. $A = \int_{-1}^{1} (1 - x^4)\,dx = \left[x - \frac{1}{5}x^5\right]_{-1}^{1} = \frac{8}{5}$

30. $A = \int_1^2 \frac{1}{x^2}\,dx = -\frac{1}{x}\Big]_1^2 = -\frac{1}{2} - (-1) = \frac{1}{2}$

31. $A = \int_0^4 (2x)^{1/3}\,dx = 2^{1/3}\int_0^4 x^{1/3}\,dx$

$= 2^{1/3}\left[\frac{3}{4}x^{4/3}\right]_0^4$

$= 2^{1/3}\frac{3}{4}4 \cdot 4^{1/3}$

$= 6$

32. $A = \int_0^3 (3 - x)\sqrt{x}\,dx = \int_0^3 (3x^{1/2} - x^{3/2})\,dx$

$= \left[2x^{3/2} - \frac{2}{5}x^{5/2}\right]_0^3$

$= \left[\frac{x\sqrt{x}}{5}(10 - 2x)\right]_0^3$

$= \frac{12\sqrt{3}}{5}$

33. Since $y \ge 0$ on $[0, 2]$,

$A = \int_0^2 (3x^2 + 1)\,dx = \left[x^3 + x\right]_0^2 = 8 + 2 = 10.$

34. Since $y \ge 0$ on $[0. 8]$,

$\text{Area} = \int_0^8 (1 + x^{1/3})\,dx = \left[x + \frac{3}{4}x^{4/3}\right]_0^8$

$= 8 + \frac{3}{4}(16) = 20$

35. Since $y \ge 0$ on $[0, 2]$,

$A = \int_0^2 (x^3 + x)\,dx = \left[\frac{x^4}{4} + \frac{x^2}{2}\right]_0^2 = 4 + 2 = 6.$

36. Since $y \ge 0$ on $[0, 3]$,

$A = \int_0^3 (3x - x^2)\,dx = \left[\frac{3}{2}x^2 - \frac{x^3}{3}\right]_0^3 = \frac{9}{2}.$

37. $\int_0^2 x^3\,dx = \frac{x^4}{4}\Big]_0^2 = 4$

$f(c)(2 - 0) = 4$

$c^3 = 2$

$c = 2^{1/3} \approx 1.2599$

38. $\int_1^3 \frac{9}{x^3}\,dx = \left[-\frac{9}{2x^2}\right]_1^3 = -\frac{1}{2} + \frac{9}{2} = 4$

$f(c)(3 - 1) = 4$

$\frac{9}{c^3} = 2$

$c^3 = \frac{9}{2}$

$c = \sqrt[3]{\frac{9}{2}} \approx 1.6510$

39. $\int_0^3 (-x^2 + 4x)\,dx = \left[-\frac{x^3}{3} + 2x^2\right]_0^3 = -9 + 18 = 9$

$f(c)(3 - 0) = 9$

$-c^2 + 4c = 3$

$c^2 - 4c + 3 = 0$

$(c - 1)(c - 3) = 0$

$c = 1, 3$

40. $\displaystyle\int_0^2 (x - 2\sqrt{x})\,dx = \left[\frac{x^2}{2} - \frac{4x^{3/2}}{3}\right]_0^2 = 2 - \frac{8\sqrt{2}}{3}$

$$f(c)(2 - 0) = \frac{6 - 8\sqrt{2}}{3}$$

$$c - 2\sqrt{c} = \frac{3 - 4\sqrt{2}}{3}$$

$$c - 2\sqrt{c} + 1 = \frac{3 - 4\sqrt{2}}{3} + 1$$

$$\left(\sqrt{c} - 1\right)^2 = \frac{6 - 4\sqrt{2}}{3}$$

$$\sqrt{c} - 1 = \pm\sqrt{\frac{6 - 4\sqrt{2}}{3}}$$

$$c = \left[1 \pm \sqrt{\frac{6 - 4\sqrt{2}}{3}}\right]^2$$

$c \approx 0.4380$ or $c \approx 1.7908$

41. $\displaystyle\frac{1}{2 - (-2)}\int_{-2}^2 (4 - x^2)\,dx = \frac{1}{4}\left[4x - \frac{1}{3}x^3\right]_{-2}^2$

$$= \frac{1}{4}\left[\left(8 - \frac{8}{3}\right) - \left(-8 + \frac{8}{3}\right)\right]$$

$$= \frac{8}{3}$$

Average value $= \dfrac{8}{3}$

$4 - x^2 = \dfrac{8}{3}$ when $x^2 = 4 - \dfrac{8}{3}$ or $x = \pm\dfrac{2\sqrt{3}}{3} \approx \pm 1.155.$

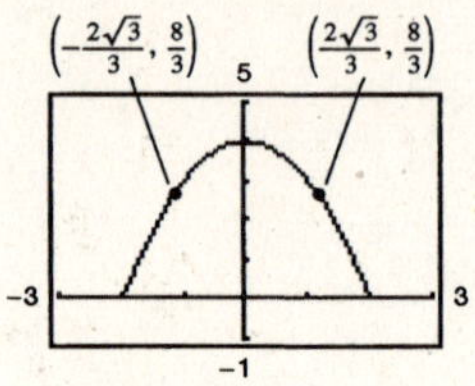

42. $\displaystyle\frac{1}{3 - 1}\int_1^3 \frac{4(x^2 + 1)}{x^2}\,dx = 2\int_1^3 (1 + x^{-2})\,dx = 2\left[x - \frac{1}{x}\right]_1^3$

$$= 2\left(3 - \frac{1}{3}\right) = \frac{16}{3}$$

Average value: $= \dfrac{16}{3}$

$\dfrac{4(x^2 + 1)}{x^2} = \dfrac{16}{3}$ when $12(x^2 + 1) = 16x^2$

or $x = \sqrt{3} \approx 1.732$

43. $\displaystyle\frac{1}{4 - 0}\int_0^4 (x - 2x^{1/2})\,dx = \frac{1}{4}\left[\frac{x^2}{2} - \frac{4}{3}x^{3/2}\right]_0^4$

$$= \frac{1}{4}\left[8 - \frac{4}{3}(8)\right] = -\frac{2}{3}$$

$$x - 2x^{1/2} = -\frac{2}{3}$$

$$3x - 6x^{1/2} + 2 = 0$$

$$\sqrt{x} = \frac{6 \pm \sqrt{36 - 24}}{6} = 1 \pm \frac{\sqrt{3}}{3}$$

$$x = \frac{4}{3} \pm \frac{2\sqrt{3}}{3} \qquad (x \approx 0.179, 2.488)$$

44. $\displaystyle\frac{1}{2 - 0}\int_0^2 (x - 3)^{-2}\,dx = \frac{1}{2}\left[\frac{-1}{x - 3}\right]_0^2 = \frac{1}{2}\left(1 - \frac{1}{3}\right) = \frac{1}{3}$

$$\frac{1}{(x - 3)^2} = \frac{1}{3}$$

$$(x - 3)^2 = 3$$

$$x - 3 = \pm\sqrt{3}$$

$$x = 3 \pm \sqrt{3}$$

Take $x = 3 - \sqrt{3} \approx 1.268$.

45. The distance traveled is $\int_0^8 v(t)\,dt$. The area under the curve from $0 \le t \le 8$ is approximately (18 squares) (30) ≈ 540 ft.

46. The distance traveled is $\int_0^5 v(t)\,dt$. The area under the curve from $0 \le t \le 5$ is approximately (29 squares) (5) $= 145$ ft.

47. If f is continuous on $[a, b]$ and $F'(x) = f(x)$ on $[a, b]$,

then $\displaystyle\int_a^b f(x)\,dx = F(b) - F(a)$.

48. (a) $\displaystyle\int_1^7 f(x)\,dx$ = Sum of the areas

$= A_1 + A_2 + A_3 + A_4$

$= \frac{1}{2}(3 + 1) + \frac{1}{2}(1 + 2) + \frac{1}{2}(2 + 1) + (3)(1)$

$= 8$

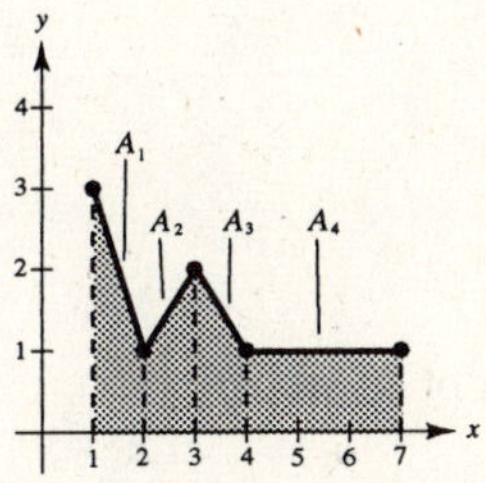

(b) Average value $= \dfrac{\int_1^7 f(x)\,dx}{7 - 1} = \dfrac{8}{6} = \dfrac{4}{3}$

(c) $A = 8 + (6)(2) = 20$

Average value $= \dfrac{20}{6} = \dfrac{10}{3}$

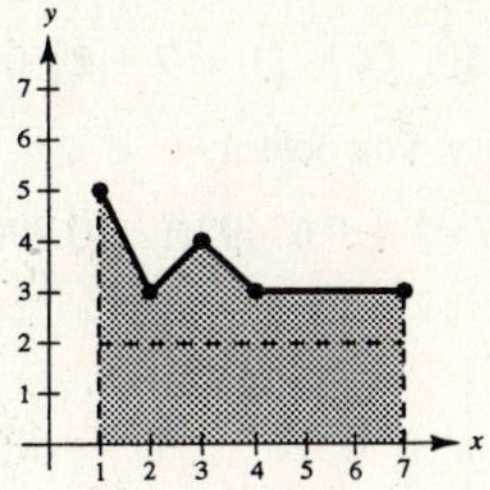

49. $\displaystyle\int_0^2 f(x)\,dx = -(\text{area of region } A) = -1.5$

50. $\displaystyle\int_2^6 f(x)\,dx = (\text{area of region } B) = \int_0^6 f(x)\,dx - \int_0^2 f(x)\,dx$

$= 3.5 - (-1.5) = 5.0$

51. $\displaystyle\int_0^6 |f(x)|\,dx = -\int_0^2 f(x)\,dx + \int_2^6 f(x)\,dx = 1.5 + 5.0 = 6.5$

52. $\displaystyle\int_0^2 -2f(x)\,dx = -2\int_0^2 f(x)\,dx = -2(-1.5) = 3.0$

53. $\displaystyle\int_0^6 [2 + f(x)]\,dx = \int_0^6 2\,dx + \int_0^6 f(x)\,dx$

$= 12 + 3.5 = 15.5$

54. Average value $= \dfrac{1}{6}\displaystyle\int_0^6 f(x)\,dx = \frac{1}{6}(3.5) = 0.5833$

55. $\dfrac{1}{5 - 0}\displaystyle\int_0^5 (0.1729t + 0.1522t^2 - 0.0374t^3)\,dt \approx \frac{1}{5}\Big[0.08645t^2 + 0.05073t^3 - 0.00935t^4\Big]_0^5 \approx 0.5318$ liter

56. $\dfrac{1}{R - 0}\displaystyle\int_0^R k(R^2 - r^2)\,dr = \frac{k}{R}\left[R^2 r - \frac{r^3}{3}\right]_0^R = \frac{2kR^2}{3}$

57. (a) $v = -8.61 \times 10^{-4}t^3 + 0.0782t^2 - 0.208t + 0.0952$

(b)

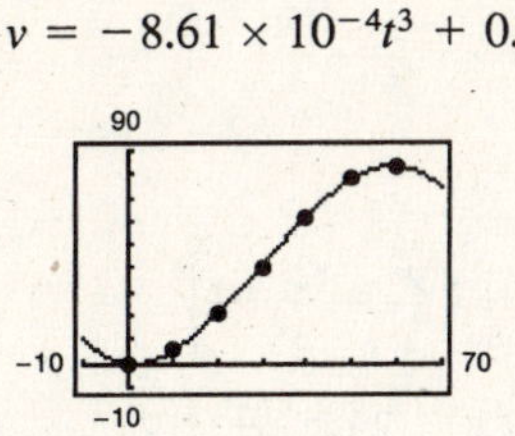

(c) $\displaystyle\int_0^{60} v(t)\,dt = \left[\frac{-8.61 \times 10^{-4}t^4}{4} + \frac{0.0782t^3}{3} - \frac{0.208t^2}{2} + 0.0952t\right]_0^{60} \approx 2472$ meters

58. (a) histogram

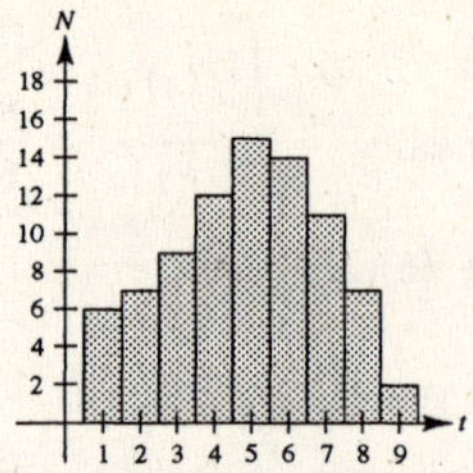

(b) $[6 + 7 + 9 + 12 + 15 + 14 + 11 + 7 + 2]60 = (83)60 = 4980$ customers

(c) Using a graphing utility, you obtain

$$N(t) = -0.084175t^3 + 0.63492t^2 + 0.79052 + 4.10317.$$

(d)

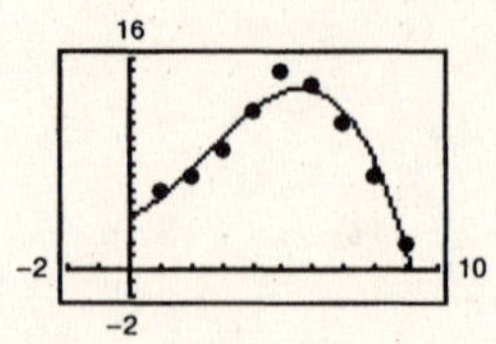

(e) $\int_0^9 N(t)\,dt \approx 85.162$

The estimated number of customers is $(85.162)(60) \approx 5110$.

(f) Between 3 P.M. and 7 P.M., the number of customers is approximately $\left(\int_3^7 N(t)\,dt\right)(60) \approx (50.28)(60) \approx 3017$.

Hence, $3017/240 \approx 12.6$ per minute.

59. $F(x) = \int_0^x (t - 5)\,dt = \left[\frac{t^2}{2} - 5t\right]_0^x = \frac{x^2}{2} - 5x$

$F(2) = \frac{4}{2} - 5(2) = -8$

$F(5) = \frac{25}{2} - 5(5) = -\frac{25}{2}$

$F(8) = \frac{64}{2} - 5(8) = -8$

60. $F(x) = \int_2^x (t^3 + 2t - 2)\,dt = \left[\frac{t^4}{4} + t^2 - 2t\right]_2^x = \left(\frac{x^4}{4} + x^2 - 2x\right) - (4 + 4 - 4) = \frac{x^4}{4} + x^2 - 2x - 4$

$F(2) = 4 + 4 - 4 - 4 = 0$ $\left[\textbf{Note: } F(2) = \int_2^2 (t^3 + 2t - 2)\,dt = 0\right]$

$F(5) = \frac{625}{4} + 25 - 10 - 4 = 167.25$

$F(8) = \frac{8^4}{4} + 64 - 16 - 4 = 1068$

61. $F(x) = \int_1^x \frac{10}{v^2}\,dv = \int_1^x 10v^{-2}\,dv = \left.\frac{-10}{v}\right]_1^x$

$= -\frac{10}{x} + 10 = 10\left(1 - \frac{1}{x}\right)$

$F(2) = 10\left(\frac{1}{2}\right) = 5$

$F(5) = 10\left(\frac{4}{5}\right) = 8$

$F(8) = 10\left(\frac{7}{8}\right) = \frac{35}{4}$

62. $F(x) = \int_1^x \left(y - \sqrt{y}\right)dy = \left[\frac{y^2}{2} - \frac{2}{3}y^{3/2}\right]_1^x$

$= \left(\frac{x^2}{2} - \frac{2}{3}x^{3/2}\right) - \left(\frac{1}{2} - \frac{2}{3}\right) = \frac{x^2}{2} - \frac{2}{3}x^{3/2} + \frac{1}{6}$

$F(2) = \frac{2^2}{2} - \frac{2}{3}2^{3/2} + \frac{1}{6} = \frac{13}{6} - \frac{4}{3}\sqrt{2}$

$F(5) = \frac{38}{3} - \frac{10\sqrt{5}}{3}$

$F(8) = \frac{193}{6} - \frac{32\sqrt{2}}{3}$

63. $g(x) = \int_0^x f(t)dt$

(a) $g(0) = \int_0^0 f(t)dt = 0$

$g(2) = \int_0^2 f(t)dt \approx 4 + 2 + 1 = 7$

$g(4) = \int_0^4 f(t)dt \approx 7 + 2 = 9$

$g(6) = \int_0^6 f(t)dt \approx 9 + (-1) = 8$

$g(8) = \int_0^8 f(t)dt \approx 8 - 3 = 5$

(b) g increasing on $(0, 4)$ and decreasing on $(4, 8)$

(c) g is a maximum of 9 at $x = 4$.

(d)

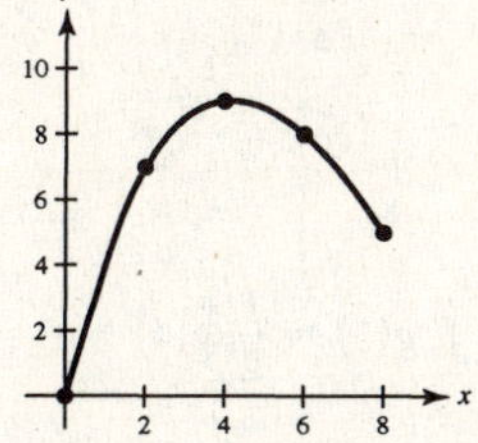

64. $g(x) = \int_0^x f(t)dt$

(a) $g(0) = \int_0^0 f(t)dt = 0$

$g(2) = \int_0^2 f(t)dt = -\frac{1}{2}(2)(4) = -4$

$g(4) = \int_0^4 f(t)dt = -\frac{1}{2}(4)(4) = -8$

$g(6) = \int_0^6 f(t)dt = -8 + 2 + 4 = -2$

$g(8) = \int_0^8 f(t)dt = -2 + 6 = 4$

(b) g decreasing on $(0, 4)$ and increasing on $(4, 8)$

(c) g is a minimum of -8 at $x = 4$.

(d)

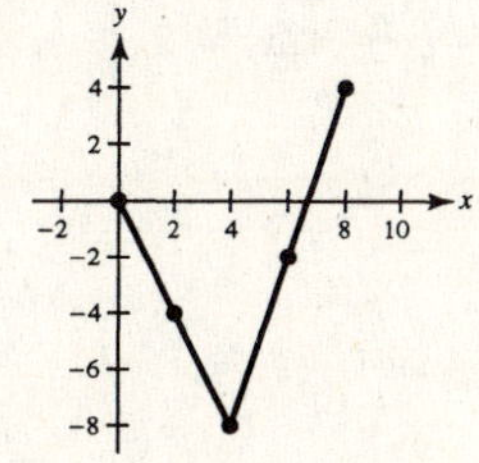

65. (a) $\int_0^x (t + 2)\, dt = \left[\frac{t^2}{2} + 2t\right]_0^x = \frac{1}{2}x^2 + 2x$

(b) $\frac{d}{dx}\left[\frac{1}{2}x^2 + 2x\right] = x + 2$

66. (a) $\int_0^x t(t^2 + 1)\, dt = \int_0^x (t^3 + t)\, dt = \left[\frac{1}{4}t^4 + \frac{1}{2}t^2\right]_0^x$

$= \frac{1}{4}x^4 + \frac{1}{2}x^2 = \frac{x^2}{4}(x^2 + 2)$

(b) $\frac{d}{dx}\left[\frac{1}{4}x^4 + \frac{1}{2}x^2\right] = x^3 + x = x(x^2 + 1)$

67. (a) $\int_8^x \sqrt[3]{t}\, dt = \left[\frac{3}{4}t^{4/3}\right]_8^x = \frac{3}{4}(x^{4/3} - 16) = \frac{3}{4}x^{4/3} - 12$

(b) $\frac{d}{dx}\left[\frac{3}{4}x^{4/3} - 12\right] = x^{1/3} = \sqrt[3]{x}$

68. (a) $\int_4^x \sqrt{t}\, dt = \left[\frac{2}{3}t^{3/2}\right]_4^x = \frac{2}{3}x^{3/2} - \frac{16}{3} = \frac{2}{3}(x^{3/2} - 8)$

(b) $\frac{d}{dx}\left[\frac{2}{3}x^{3/2} - \frac{16}{3}\right] = x^{1/2} = \sqrt{x}$

69. (a) $F(x) = \int_1^x \frac{1}{t^2}\, dt = -\frac{1}{t}\Big]_1^x = 1 - \frac{1}{x}$

(b) $F'(x) = \frac{d}{dx}\left(1 - \frac{1}{x}\right) = \frac{1}{x^2}$

70. (a) $F(x) = \int_0^x t^{3/2}\, dt = \frac{2}{5}t^{5/2}\Big]_0^x = \frac{2}{5}x^{5/2}$

(b) $F'(x) = \frac{2}{5}\left(\frac{5}{2}\right)x^{3/2} = x^{3/2}$

71. $F(x) = \int_{-2}^x (t^2 - 2t)\, dt$

$F'(x) = x^2 - 2x$

72. $F(x) = \int_1^x \sqrt[4]{t}\, dt$

$F'(x) = \sqrt[4]{x}$

73. $F(x) = \int_{-1}^x \sqrt{t^4 + 1}\, dt$

$F'(x) = \sqrt{x^4 + 1}$

74. $F(x) = \int_1^x \frac{t^2}{t^2 + 1}\, dt$

$F'(x) = \frac{x^2}{x^2 + 1}$

75. $F(x) = \int_x^{x+2} (4t + 1)\,dt = \left[2t^2 + t\right]_x^{x+2}$

$F'(x) = 8$

Alternate solution:

$F(x) = \int_x^{x+2} (4t + 1)\,dt$

$= \int_x^0 (4t + 1)\,dt + \int_0^{x+2} (4t + 1)\,dt$

$= -\int_0^x (4t + 1)\,dt + \int_0^{x+2} (4t + 1)\,dt$

$F'(x) = -(4x + 1) + 4(x + 2) + 1 = 8$

76. $F(x) = \int_{-x}^{x} t^3\,dt = \left[\frac{t^4}{4}\right]_{-x}^{x} = 0$

$F'(x) = 0$

Alternate solution:

$F(x) = \int_{-x}^{x} t^3\,dt$

$= \int_{-x}^{0} t^3\,dt + \int_0^x t^3\,dt$

$= -\int_0^{-x} t^3\,dt + \int_0^x t^3\,dt$

$F'(x) = -(-x)^3(-1) + (x^3) = 0$

77. $F(x) = \int_2^{x^2} t^{-3}\,dt = \left[\frac{t^{-2}}{-2}\right]_2^{x^2} = \left[-\frac{1}{2t^2}\right]_2^{x^2} = \frac{-1}{2x^4} + \frac{1}{8} \Rightarrow F'(x) = 2x^{-5}$

Alternate solution: $F'(x) = (x^2)^{-3}(2x) = 2x^{-5}$

78. $F(x) = \int_0^{3x} \sqrt{1 + t^3}\,dt$

$F'(x) = \sqrt{1 + (3x)^3}\,(3)$

$= 3\sqrt{1 + 27x^3}$

79. $g(x) = \int_0^x f(t)\,dt$

$g(0) = 0, g(1) \approx \frac{1}{2}, g(2) \approx 1, g(3) \approx \frac{1}{2}, g(4) = 0$

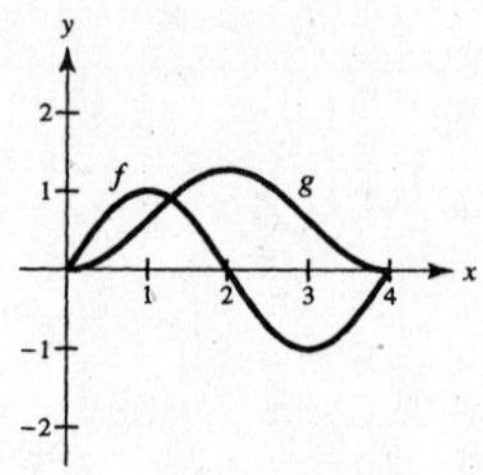

g has a relative maximum at $x = 2$.

80. (a)

x	1	2	3	4	5	6	7	8	9	10
$g(x)$	1	2	0	−2	−4	−6	−3	0	3	6

(b)

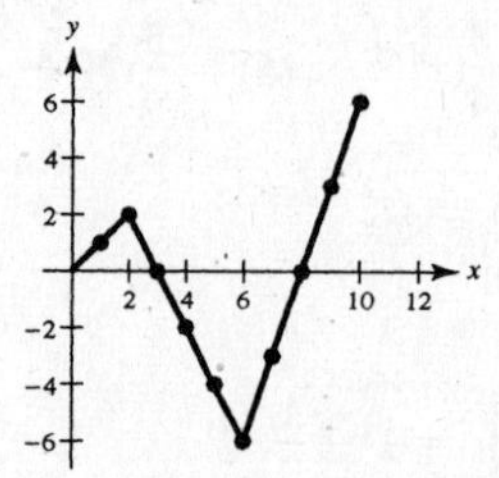

(c) Minimum of g at $(6, -6)$.

(d) Maximum at $(10, 6)$. Relative maximum at $(2, 2)$.

(e) On $[6, 10]$ g increases at a rate of $\frac{12}{4} = 3$.

(f) Zeros of g: $x = 3, x = 8$.

81. (a) $C(x) = 5000\left(25 + 3\int_0^x t^{1/4}\,dt\right)$

$= 5000\left(25 + 3\left[\frac{4}{5}t^{5/4}\right]_0^x\right)$

$= 5000\left(25 + \frac{12}{5}x^{5/4}\right) = 1000(125 + 12x^{5/4})$

(b) $C(1) = 1000(125 + 12(1)) = \$137{,}000$

$C(5) = 1000(125 + 12(5)^{5/4}) \approx \$214{,}721$

$C(10) = 1000(125 + 12(10)^{5/4}) \approx \$338{,}394$

82. (a) $g(t) = 4 - \dfrac{4}{t^2}$

$\lim_{t\to\infty} g(t) = 4$

Horizontal asymptote: $y = 4$

(b) $A(x) = \displaystyle\int_1^x \left(4 - \frac{4}{t^2}\right) dt$

$= \left[4t + \dfrac{4}{t}\right]_1^x = 4x + \dfrac{4}{x} - 8$

$= \dfrac{4x^2 - 8x + 4}{x} = \dfrac{4(x-1)^2}{x}$

$\lim_{x\to\infty} A(x) = \lim_{x\to\infty}\left(4x + \dfrac{4}{x} - 8\right) = \infty$

The graph of $A(x)$ does not have a horizontal asymptote.

83. True

84. True

85. The function $f(x) = \dfrac{1}{x^2}$ is not continuous at $x = 0$. You cannot simply integrate from -1 to 1.

86. Let $F(t)$ be an antiderivative of $f(t)$. Then,

$$\int_{u(x)}^{v(x)} f(t)\,dt = \Big[F(t)\Big]_{u(x)}^{v(x)} = F(v(x)) - F(u(x))$$

$$\frac{d}{dx}\left[\int_{u(x)}^{v(x)} f(t)\,dt\right] = \frac{d}{dx}[F(v(x)) - F(u(x))]$$
$$= F'(v(x))v'(x) - F'(u(x))u'(x)$$
$$= f(v(x))v'(x) - f(u(x))u'(x).$$

87. $f(x) = \displaystyle\int_0^{1/x} \frac{1}{t^2+1}\,dt + \int_0^x \frac{1}{t^2+1}\,dt$

By the Second Fundamental Theorem of Calculus, we have

$$f'(x) = \frac{1}{(1/x)^2 + 1}\left(-\frac{1}{x^2}\right) + \frac{1}{x^2+1}$$
$$= -\frac{1}{1+x^2} + \frac{1}{x^2+1} = 0.$$

Since $f'(x) = 0$, $f(x)$ must be constant.

88. $G(x) = \displaystyle\int_0^x \left[s\int_0^s f(t)\,dt\right] ds$

(a) $G(0) = \displaystyle\int_0^0 \left[s\int_0^s f(t)\,dt\right] ds = 0$

(b) Let $F(s) = s\displaystyle\int_0^s f(t)\,dt$.

$G(x) = \displaystyle\int_0^x F(s)\,ds$

$G'(x) = F(x) = x\displaystyle\int_0^x f(t)\,dt$

$G'(0) = 0\displaystyle\int_0^0 f(t)\,dt = 0$

(c) $G''(x) = x \cdot f(x) + \displaystyle\int_0^x f(t)\,dt$

(d) $G''(0) = 0 \cdot f(0) + \displaystyle\int_0^0 f(t)\,dt = 0$

89. $x(t) = t^3 - 6t^2 + 9t - 2$

$x'(t) = 3t^2 - 12t + 9$

$= 3(t^2 - 4t + 3)$

$= 3(t-3)(t-1)$

Total distance $= \displaystyle\int_0^5 |x'(t)|\,dt$

$= \displaystyle\int_0^5 3|(t-3)(t-1)|\,dt$

$= 3\displaystyle\int_0^1 (t^2 - 4t + 3)\,dt - 3\int_1^3 (t^2 - 4t + 3)\,dt + 3\int_3^5 (t^2 - 4t + 3)\,dt$

$= 4 + 4 + 20$

$= 28$ units

90. $x(t) = (t-1)(t-3)^2 = t^3 - 7t^2 + 15t - 9$

$x'(t) = 3t^2 - 14t + 15$

Using a graphing utility,

$$\text{Total distance} = \int_0^5 |x'(t)|\,dt \approx 27.37 \text{ units.}$$

91. $\text{Total distance} = \int_1^4 |x'(t)|\,dt$

$= \int_1^4 |v(t)|\,dt$

$= \int_1^4 \frac{1}{\sqrt{t}}\,dt$

$= 2t^{1/2}\Big]_1^4$

$= 2(2-1) = 2$ units

Section 6.5 Integration by Substitution

$\int f(g(x))g'(x)\,dx$	$u = g(x)$	$du = g'(x)\,dx$
1. $\int (5x^2+1)^2(10x)\,dx$	$5x^2+1$	$10x\,dx$
2. $\int (x^3+3)\,3x^2\,dx$	x^3+3	$3x^2\,dx$
3. $\int \frac{x}{\sqrt{x^2+1}}\,dx$	x^2+1	$2x\,dx$
4. $\int x^2\sqrt{x^3+1}\,dx$	x^3+1	$3x^2\,dx$

5. $\int (1+2x)^4\,2\,dx = \frac{(1+2x)^5}{5} + C$

Check: $\frac{d}{dx}\left[\frac{(1+2x)^5}{5} + C\right] = 2(1+2x)^4$

6. $\int (x^2-9)^3(2x)\,dx = \frac{(x^2-9)^4}{4} + C$

Check: $\frac{d}{dx}\left[\frac{(x^2-9)^4}{4} + C\right] = \frac{4(x^2-9)^3}{4}(2x) = (x^2-9)^3(2x)$

7. $\int (9-x^2)^{1/2}(-2x)\,dx = \frac{(9-x^2)^{3/2}}{3/2} + C = \frac{2}{3}(9-x^2)^{3/2} + C$

Check: $\frac{d}{dx}\left[\frac{2}{3}(9-x^2)^{3/2} + C\right] = \frac{2}{3}\cdot\frac{3}{2}(9-x^2)^{1/2}(-2x) = \sqrt{9-x^2}(-2x)$

8. $\int (1-2x^2)^{1/3}(-4x)\,dx = \frac{3}{4}(1-2x^2)^{4/3} + C$

Check: $\frac{d}{dx}\left[\frac{3}{4}(1-2x^2)^{4/3} + C\right] = \frac{3}{4}\cdot\frac{4}{3}(1-2x^2)^{1/3}(-4x) = (1-2x^2)^{1/3}(-4x)$

9. $\displaystyle\int x^3(x^4+3)^2\,dx = \frac{1}{4}\int (x^4+3)^2(4x^3)\,dx = \frac{1}{4}\frac{(x^4+3)^3}{3} + C = \frac{(x^4+3)^3}{12} + C$

Check: $\displaystyle\frac{d}{dx}\left[\frac{(x^4+3)^3}{12} + C\right] = \frac{3(x^4+3)^2}{12}(4x^3) = (x^4+3)^2(x^3)$

10. $\displaystyle\int x^2(x^3+5)^4\,dx = \frac{1}{3}\int (x^3+5)^4(3x^2)\,dx = \frac{1}{3}\frac{(x^3+5)^5}{5} + C = \frac{(x^3+5)^5}{15} + C$

Check: $\displaystyle\frac{d}{dx}\left[\frac{(x^3+5)^5}{15} + C\right] = \frac{5(x^3+5)^4(3x^2)}{15} = (x^3+5)^4x^2$

11. $\displaystyle\int x^2(x^3-1)^4\,dx = \frac{1}{3}\int (x^3-1)^4(3x^2)\,dx = \frac{1}{3}\left[\frac{(x^3-1)^5}{5}\right] + C = \frac{(x^3-1)^5}{15} + C$

Check: $\displaystyle\frac{d}{dx}\left[\frac{(x^3-1)^5}{15} + C\right] = \frac{5(x^3-1)^4(3x^2)}{15} = x^2(x^3-1)^4$

12. $\displaystyle\int x(4x^2+3)^3\,dx = \frac{1}{8}\int (4x^2+3)^3(8x)\,dx = \frac{1}{8}\left[\frac{(4x^2+3)^4}{4}\right] + C = \frac{(4x^2+3)^4}{32} + C$

Check: $\displaystyle\frac{d}{dx}\left[\frac{(4x^2+3)^4}{32} + C\right] = \frac{4(4x^2+3)^3(8x)}{32} = x(4x^2+3)^3$

13. $\displaystyle\int t\sqrt{t^2+2}\,dt = \frac{1}{2}\int (t^2+2)^{1/2}(2t)\,dt = \frac{1}{2}\frac{(t^2+2)^{3/2}}{3/2} + C = \frac{(t^2+2)^{3/2}}{3} + C$

Check: $\displaystyle\frac{d}{dt}\left[\frac{(t^2+2)^{3/2}}{3} + C\right] = \frac{3/2(t^2+2)^{1/2}(2t)}{3} = (t^2+2)^{1/2}t$

14. $\displaystyle\int t^3\sqrt{t^4+5}\,dt = \frac{1}{4}\int (t^4+5)^{1/2}(4t^3)\,dt = \frac{1}{4}\frac{(t^4+5)^{3/2}}{3/2} + C = \frac{1}{6}(t^4+5)^{3/2} + C$

Check: $\displaystyle\frac{d}{dt}\left[\frac{1}{6}(t^4+5)^{3/2} + C\right] = \frac{1}{6}\cdot\frac{3}{2}(t^4+5)^{1/2}(4t^3) = (t^4+5)^{1/2}(t^3)$

15. $\displaystyle\int 5x(1-x^2)^{1/3}\,dx = -\frac{5}{2}\int (1-x^2)^{1/3}(-2x)\,dx = -\frac{5}{2}\cdot\frac{(1-x^2)^{4/3}}{4/3} + C = -\frac{15}{8}(1-x^2)^{4/3} + C$

Check: $\displaystyle\frac{d}{dx}\left[-\frac{15}{8}(1-x^2)^{4/3} + C\right] = -\frac{15}{8}\cdot\frac{4}{3}(1-x^2)^{1/3}(-2x) = 5x(1-x^2)^{1/3} = 5x\sqrt[3]{1-x^2}$

16. $\displaystyle\int u^2\sqrt{u^3+2}\,du = \frac{1}{3}\int (u^3+2)^{1/2}(3u^2)\,du = \frac{1}{3}\frac{(u^3+2)^{3/2}}{3/2} + C = \frac{2(u^3+2)^{3/2}}{9} + C$

Check: $\displaystyle\frac{d}{du}\left[\frac{2(u^3+2)^{3/2}}{9} + C\right] = \frac{2}{9}\cdot\frac{3}{2}(u^3+2)^{1/2}(3u^2) = (u^3+2)^{1/2}(u^2)$

17. $\displaystyle\int \frac{x}{(1-x^2)^3}\,dx = -\frac{1}{2}\int (1-x^2)^{-3}(-2x)\,dx = -\frac{1}{2}\frac{(1-x^2)^{-2}}{-2} + C = \frac{1}{4(1-x^2)^2} + C$

Check: $\displaystyle\frac{d}{dx}\left[\frac{1}{4(1-x^2)^2} + C\right] = \frac{1}{4}(-2)(1-x^2)^{-3}(-2x) = \frac{x}{(1-x^2)^3}$

18. $\displaystyle\int \frac{x^3}{(1+x^4)^2}\,dx = \frac{1}{4}\int (1+x^4)^{-2}(4x^3)\,dx = -\frac{1}{4}(1+x^4)^{-1} + C = \frac{-1}{4(1+x^4)} + C$

Check: $\displaystyle\frac{d}{dx}\left[\frac{-1}{4(1+x^4)} + C\right] = \frac{1}{4}(1+x^4)^{-2}(4x^3) = \frac{x^3}{(1+x^4)^2}$

19. $\displaystyle\int \frac{x^2}{(1+x^3)^2}\,dx = \frac{1}{3}\int (1+x^3)^{-2}(3x^2)\,dx = \frac{1}{3}\left[\frac{(1+x^3)^{-1}}{-1}\right] + C = -\frac{1}{3(1+x^3)} + C$

Check: $\displaystyle\frac{d}{dx}\left[-\frac{1}{3(1+x^3)} + C\right] = -\frac{1}{3}(-1)(1+x^3)^{-2}(3x^2) = \frac{x^2}{(1+x^3)^2}$

20. $\displaystyle\int \frac{x^2}{(16-x^3)^2}\,dx = -\frac{1}{3}\int (16-x^3)^{-2}(-3x^2)\,dx = -\frac{1}{3}\left[\frac{(16-x^3)^{-1}}{-1}\right] + C = \frac{1}{3(16-x^3)} + C$

Check: $\displaystyle\frac{d}{dx}\left[\frac{1}{3(16-x^3)} + C\right] = \frac{1}{3}(-1)(16-x^3)^{-2}(-3x^2) = \frac{x^2}{(16-x^3)^2}$

21. $\displaystyle\int \frac{x}{\sqrt{1-x^2}}\,dx = -\frac{1}{2}\int (1-x^2)^{-1/2}(-2x)\,dx = -\frac{1}{2}\frac{(1-x^2)^{1/2}}{1/2} + C = -\sqrt{1-x^2} + C$

Check: $\displaystyle\frac{d}{dx}[-(1-x^2)^{1/2} + C] = -\frac{1}{2}(1-x^2)^{-1/2}(-2x) = \frac{x}{\sqrt{1-x^2}}$

22. $\displaystyle\int \frac{x^3}{\sqrt{1+x^4}}\,dx = \frac{1}{4}\int (1+x^4)^{-1/2}(4x^3)\,dx = \frac{1}{4}\frac{(1+x^4)^{1/2}}{1/2} + C = \frac{\sqrt{1+x^4}}{2} + C$

Check: $\displaystyle\frac{d}{dx}\left[\frac{\sqrt{1+x^4}}{2} + C\right] = \frac{1}{2}\cdot\frac{1}{2}(1+x^4)^{-1/2}(4x^3) = \frac{x^3}{\sqrt{1+x^4}}$

23. $\displaystyle\int \left(1+\frac{1}{t}\right)^3\left(\frac{1}{t^2}\right)dt = -\int \left(1+\frac{1}{t}\right)^3\left(-\frac{1}{t^2}\right)dt = -\frac{[1+(1/t)]^4}{4} + C$

Check: $\displaystyle\frac{d}{dt}\left[-\frac{[1+(1/t)]^4}{4} + C\right] = -\frac{1}{4}(4)\left(1+\frac{1}{t}\right)^3\left(-\frac{1}{t^2}\right) = \frac{1}{t^2}\left(1+\frac{1}{t}\right)^3$

24. $\displaystyle\int \left[x^2 + \frac{1}{(3x)^2}\right]dx = \int \left(x^2 + \frac{1}{9}x^{-2}\right)dx = \frac{x^3}{3} + \frac{1}{9}\left(\frac{x^{-1}}{-1}\right) + C = \frac{x^3}{3} - \frac{1}{9x} + C = \frac{3x^4-1}{9x} + C$

Check: $\displaystyle\frac{d}{dx}\left[\frac{1}{3}x^3 - \frac{1}{9}x^{-1} + C\right] = x^2 + \frac{1}{9}x^{-2} = x^2 + \frac{1}{(3x)^2}$

25. $\displaystyle\int \frac{1}{\sqrt{2x}}\,dx = \frac{1}{2}\int (2x)^{-1/2}\,2\,dx = \frac{1}{2}\left[\frac{(2x)^{1/2}}{1/2}\right] + C = \sqrt{2x} + C$

Check: $\displaystyle\frac{d}{dx}[\sqrt{2x} + C] = \frac{1}{2}(2x)^{-1/2}(2) = \frac{1}{\sqrt{2x}}$

26. $\displaystyle\int \frac{1}{2\sqrt{x}}\,dx = \frac{1}{2}\int x^{-1/2}\,dx = \frac{1}{2}\left(\frac{x^{1/2}}{1/2}\right) + C = \sqrt{x} + C$

Check: $\displaystyle\frac{d}{dx}[\sqrt{x} + C] = \frac{1}{2\sqrt{x}}$

27. $\displaystyle\int \frac{x^2 + 3x + 7}{\sqrt{x}}\,dx = \int (x^{3/2} + 3x^{1/2} + 7x^{-1/2})\,dx = \frac{2}{5}x^{5/2} + 2x^{3/2} + 14x^{1/2} + C = \frac{2}{5}\sqrt{x}(x^2 + 5x + 35) + C$

Check: $\displaystyle\frac{d}{dx}\left[\frac{2}{5}x^{5/2} + 2x^{3/2} + 14x^{1/2} + C\right] = \frac{x^2 + 3x + 7}{\sqrt{x}}$

28. $\displaystyle\int \frac{t + 2t^2}{\sqrt{t}}\,dt = \int (t^{1/2} + 2t^{3/2})\,dt = \frac{2}{3}t^{3/2} + \frac{4}{5}t^{5/2} + C = \frac{2}{15}t^{3/2}(5 + 6t) + C$

Check: $\displaystyle\frac{d}{dt}\left[\frac{2}{3}t^{3/2} + \frac{4}{5}t^{5/2} + C\right] = t^{1/2} + 2t^{3/2} = \frac{t + 2t^2}{\sqrt{t}}$

29. $\displaystyle\int t^2\left(t - \frac{2}{t}\right)dt = \int (t^3 - 2t)\,dt = \frac{1}{4}t^4 - t^2 + C$

Check: $\displaystyle\frac{d}{dt}\left[\frac{1}{4}t^4 - t^2 + C\right] = t^3 - 2t = t^2\left(t - \frac{2}{t}\right)$

30. $\displaystyle\int \left(\frac{t^3}{3} + \frac{1}{4t^2}\right)dt = \int \left(\frac{1}{3}t^3 + \frac{1}{4}t^{-2}\right)dt = \frac{1}{3}\left(\frac{t^4}{4}\right) + \frac{1}{4}\left(\frac{t^{-1}}{-1}\right) + C = \frac{1}{12}t^4 - \frac{1}{4t} + C$

Check: $\displaystyle\frac{d}{dt}\left[\frac{1}{12}t^4 - \frac{1}{4t} + C\right] = \frac{1}{3}t^3 + \frac{1}{4t^2}$

31. $\displaystyle\int (9 - y)\sqrt{y}\,dy = \int (9y^{1/2} - y^{3/2})\,dy = 9\left(\frac{2}{3}y^{3/2}\right) - \frac{2}{5}y^{5/2} + C = \frac{2}{5}y^{3/2}(15 - y) + C$

Check: $\displaystyle\frac{d}{dy}\left[\frac{2}{5}y^{3/2}(15 - y) + C\right] = \frac{d}{dy}\left[6y^{3/2} - \frac{2}{5}y^{5/2} + C\right] = 9y^{1/2} - y^{3/2} = (9 - y)\sqrt{y}$

32. $\displaystyle\int 2\pi y(8 - y^{3/2})\,dy = 2\pi\int (8y - y^{5/2})\,dy = 2\pi\left(4y^2 - \frac{2}{7}y^{7/2}\right) + C = \frac{4\pi y^2}{7}(14 - y^{3/2}) + C$

Check: $\displaystyle\frac{d}{dy}\left[\frac{4\pi y^2}{7}(14 - y^{3/2}) + C\right] = \frac{d}{dy}\left[2\pi\left(4y^2 - \frac{2}{7}y^{7/2}\right) + C\right] = 16\pi y - 2\pi y^{5/2} = (2\pi y)(8 - y^{3/2})$

33. $$\begin{aligned} y &= \int \left[4x + \frac{4x}{\sqrt{16 - x^2}}\right]dx \\ &= 4\int x\,dx - 2\int (16 - x^2)^{-1/2}(-2x)\,dx \\ &= 4\left(\frac{x^2}{2}\right) - 2\left[\frac{(16 - x^2)^{1/2}}{1/2}\right] + C \\ &= 2x^2 - 4\sqrt{16 - x^2} + C \end{aligned}$$

34. $$\begin{aligned} y &= \int \frac{10x^2}{\sqrt{1 + x^3}}\,dx \\ &= \frac{10}{3}\int (1 + x^3)^{-1/2}(3x^2)\,dx \\ &= \frac{10}{3}\left[\frac{(1 + x^3)^{1/2}}{1/2}\right] + C \\ &= \frac{20}{3}\sqrt{1 + x^3} + C \end{aligned}$$

35. $$\begin{aligned} y &= \int \frac{x + 1}{(x^2 + 2x - 3)^2}\,dx \\ &= \frac{1}{2}\int (x^2 + 2x - 3)^{-2}(2x + 2)\,dx \\ &= \frac{1}{2}\left[\frac{(x^2 + 2x - 3)^{-1}}{-1}\right] + C \\ &= -\frac{1}{2(x^2 + 2x - 3)} + C \end{aligned}$$

36. $$\begin{aligned} y &= \int \frac{x - 4}{\sqrt{x^2 - 8x + 1}}\,dx \\ &= \frac{1}{2}\int (x^2 - 8x + 1)^{-1/2}(2x - 8)\,dx \\ &= \frac{1}{2}\left[\frac{(x^2 - 8x + 1)^{1/2}}{1/2}\right] + C \\ &= \sqrt{x^2 - 8x + 1} + C \end{aligned}$$

37. (a)

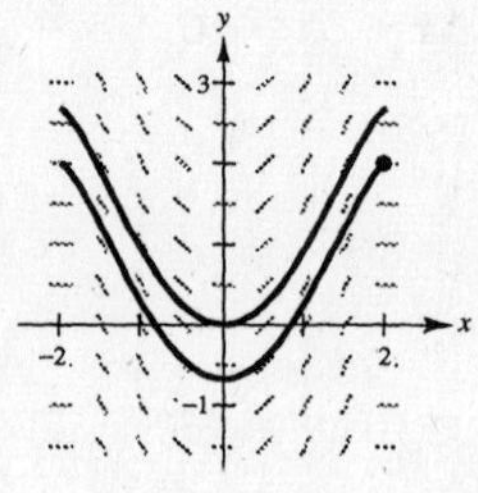

(b) $\dfrac{dy}{dx} = x\sqrt{4 - x^2}$, $(2, 2)$

$$y = \int x\sqrt{4 - x^2}\,dx = -\frac{1}{2}\int (4 - x^2)^{1/2}(-2x\,dx)$$

$$= -\frac{1}{2}\cdot\frac{2}{3}(4 - x^2)^{3/2} + C = -\frac{1}{3}(4 - x^2)^{3/2} + C$$

$$(2, 2):\; 2 = -\frac{1}{3}(4 - 2^2)^{3/2} + C \Rightarrow C = 2$$

$$y = -\frac{1}{3}(4 - x^2)^{3/2} + 2$$

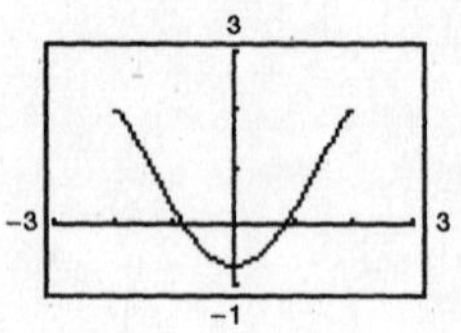

38. $\dfrac{dy}{dx} = \dfrac{x}{\sqrt{x^2 + 1}}$

(a)

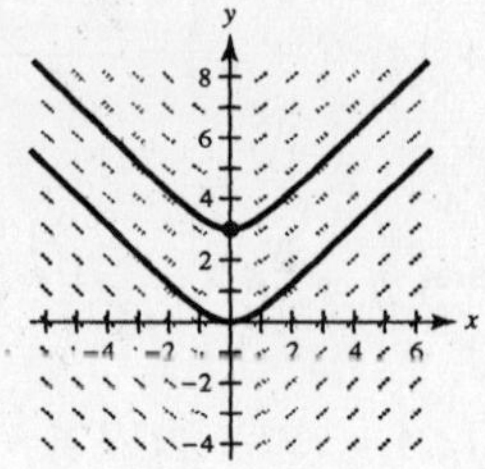

Slope field of $y' = \dfrac{x}{\sqrt{x^2 + 1}}$ with two solutions shown,one through $(0, 3)$ and one though $(0, 0)$.

(b) $$\int dy = \int \frac{x}{(x^2 + 1)^{1/2}}\,dx = \frac{1}{2}\int (x^2 + 1)^{-1/2}(2x)\,dx$$

$$= \frac{1}{2}\frac{(x^2 + 1)^{1/2}}{1/2} + C = \sqrt{x^2 + 1} + C$$

$$3 = \sqrt{0^2 + 1} + C \Rightarrow C = 2$$

$$y = \sqrt{x^2 + 1} + 2$$

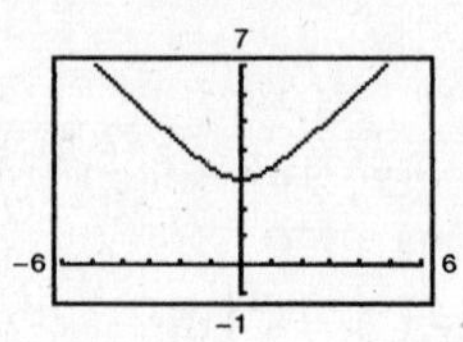

39. $u = x + 2$, $x = u - 2$, $dx = du$

$$\int x\sqrt{x + 2}\,dx = \int (u - 2)\sqrt{u}\,du$$

$$= \int (u^{3/2} - 2u^{1/2})\,du$$

$$= \frac{2}{5}u^{5/2} - \frac{4}{3}u^{3/2} + C$$

$$= \frac{2u^{3/2}}{15}(3u - 10) + C$$

$$= \frac{2}{15}(x + 2)^{3/2}[3(x + 2) - 10] + C$$

$$= \frac{2}{15}(x + 2)^{3/2}(3x - 4) + C$$

40. $u = 2x + 1$, $x = \frac{1}{2}(u - 1)$, $dx = \frac{1}{2}\,du$

$$\int x\sqrt{2x + 1}\,dx = \int \frac{1}{2}(u - 1)\sqrt{u}\,\frac{1}{2}\,du$$

$$= \frac{1}{4}\int (u^{3/2} - u^{1/2})\,du$$

$$= \frac{1}{4}\left(\frac{2}{5}u^{5/2} - \frac{2}{3}u^{3/2}\right) + C$$

$$= \frac{u^{3/2}}{30}(3u - 5) + C$$

$$= \frac{1}{30}(2x + 1)^{3/2}[3(2x + 1) - 5] + C$$

$$= \frac{1}{30}(2x + 1)^{3/2}(6x - 2) + C$$

$$= \frac{1}{15}(2x + 1)^{3/2}(3x - 1) + C$$

41. $u = 1 - x, x = 1 - u, dx = -du$

$$\int x^2\sqrt{1-x}\,dx = -\int (1-u)^2\sqrt{u}\,du$$

$$= -\int (u^{1/2} - 2u^{3/2} + u^{5/2})\,du$$

$$= -\left(\frac{2}{3}u^{3/2} - \frac{4}{5}u^{5/2} + \frac{2}{7}u^{7/2}\right) + C$$

$$= -\frac{2u^{3/2}}{105}(35 - 42u + 15u^2) + C$$

$$= -\frac{2}{105}(1-x)^{3/2}[35 - 42(1-x) + 15(1-x)^2] + C$$

$$= -\frac{2}{105}(1-x)^{3/2}(15x^2 + 12x + 8) + C$$

42. $u = 2 - x, x = 2 - u, dx = -du$

$$\int (x+1)\sqrt{2-x}\,dx = -\int (3-u)\sqrt{u}\,du$$

$$= -\int (3u^{1/2} - u^{3/2})\,du$$

$$= -\left(2u^{3/2} - \frac{2}{5}u^{5/2}\right) + C$$

$$= -\frac{2u^{3/2}}{5}(5 - u) + C$$

$$= -\frac{2}{5}(2-x)^{3/2}[5 - (2-x)] + C$$

$$= -\frac{2}{5}(2-x)^{3/2}(x+3) + C$$

43. $u = 2x - 1, x = \frac{1}{2}(u+1), dx = \frac{1}{2}\,du$

$$\int \frac{x^2 - 1}{\sqrt{2x-1}}\,dx = \int \frac{[(1/2)(u+1)]^2 - 1}{\sqrt{u}}\frac{1}{2}\,du$$

$$= \frac{1}{8}\int u^{-1/2}[(u^2 + 2u + 1) - 4]\,du$$

$$= \frac{1}{8}\int (u^{3/2} + 2u^{1/2} - 3u^{-1/2})\,du$$

$$= \frac{1}{8}\left(\frac{2}{5}u^{5/2} + \frac{4}{3}u^{3/2} - 6u^{1/2}\right) + C$$

$$= \frac{u^{1/2}}{60}(3u^2 + 10u - 45) + C$$

$$= \frac{\sqrt{2x-1}}{60}[3(2x-1)^2 + 10(2x-1) - 45] + C$$

$$= \frac{1}{60}\sqrt{2x-1}(12x^2 + 8x - 52) + C$$

$$= \frac{1}{15}\sqrt{2x-1}(3x^2 + 2x - 13) + C$$

44. Let $u = x + 4$, $x = u - 4$, $du = dx$.

$$\int \frac{2x+1}{\sqrt{x+4}}\,dx = \int \frac{2(u-4)+1}{\sqrt{u}}\,du$$
$$= \int (2u^{1/2} - 7u^{-1/2})\,du$$
$$= \frac{4}{3}u^{3/2} - 14u^{1/2} + C$$
$$= \frac{2}{3}u^{1/2}(2u - 21) + C$$
$$= \frac{2}{3}\sqrt{x+4}[2(x+4) - 21] + C$$
$$= \frac{2}{3}\sqrt{x+4}(2x - 13) + C$$

45. $u = x + 1$, $x = u - 1$, $dx = du$

$$\int \frac{-x}{(x+1) - \sqrt{x+1}}\,dx = \int \frac{-(u-1)}{u - \sqrt{u}}\,du$$
$$= -\int \frac{(\sqrt{u}+1)(\sqrt{u}-1)}{\sqrt{u}(\sqrt{u}-1)}\,du$$
$$= -\int (1 + u^{-1/2})\,du$$
$$= -(u + 2u^{1/2}) + C$$
$$= -u - 2\sqrt{u} + C$$
$$= -(x+1) - 2\sqrt{x+1} + C$$
$$= -x - 2\sqrt{x+1} - 1 + C$$
$$= -\left(x + 2\sqrt{x+1}\right) + C_1$$

where $C_1 = -1 + C$.

46. $u = t - 4$, $t = u + 4$, $dt = du$

$$\int t\sqrt[3]{t-4}\,dt = \int (u+4)u^{1/3}\,du$$
$$= \int (u^{4/3} + 4u^{1/3})\,du$$
$$= \frac{3}{7}u^{7/3} + 3u^{4/3} + C$$
$$= \frac{3u^{4/3}}{7}(u + 7) + C$$
$$= \frac{3}{7}(t-4)^{4/3}[(t-4) + 7] + C$$
$$= \frac{3}{7}(t-4)^{4/3}(t+3) + C$$

47. Let $u = x^2 + 1$, $du = 2x\,dx$.

$$\int_{-1}^{1} x(x^2+1)^3\,dx = \frac{1}{2}\int_{-1}^{1}(x^2+1)^3(2x)\,dx$$
$$= \left[\frac{1}{8}(x^2+1)^4\right]_{-1}^{1} = 0$$

48. Let $u = x^3 + 8$, $du = 3x^2\,dx$.

$$\int_{-2}^{4} x^2(x^3+8)^2\,dx = \frac{1}{3}\int_{-2}^{4}(x^3+8)^2(3x^2)\,dx$$
$$= \left[\frac{1}{3}\frac{(x^3+8)^3}{3}\right]_{-2}^{4}$$
$$= \frac{1}{9}[(64+8)^3 - (-8+8)^3] = 41{,}472$$

49. Let $u = x^3 + 1$, $du = 3x^2\,dx$

$$\int_{1}^{2} 2x^2\sqrt{x^3+1}\,dx = 2\cdot\frac{1}{3}\int_{1}^{2}(x^3+1)^{1/2}(3x^2)\,dx$$
$$= \left[\frac{2}{3}\frac{(x^3+1)^{3/2}}{3/2}\right]_{1}^{2}$$
$$= \frac{4}{9}\left[(x^3+1)^{3/2}\right]_{1}^{2}$$
$$= \frac{4}{9}[27 - 2\sqrt{2}] = 12 - \frac{8}{9}\sqrt{2}$$

50. Let $u = 1 - x^2$, $du = -2x\,dx$.

$$\int_0^1 x\sqrt{1-x^2}\,dx = -\frac{1}{2}\int_0^1 (1-x^2)^{1/2}(-2x)\,dx = \left[-\frac{1}{3}(1-x^2)^{3/2}\right]_0^1 = 0 + \frac{1}{3} = \frac{1}{3}$$

51. Let $u = 2x + 1$, $du = 2\,dx$.

$$\int_0^4 \frac{1}{\sqrt{2x+1}}\,dx = \frac{1}{2}\int_0^4 (2x+1)^{-1/2}(2)\,dx = \left[\sqrt{2x+1}\right]_0^4 = \sqrt{9} - \sqrt{1} = 2$$

52. Let $u = 1 + 2x^2$, $du = 4x\,dx$.

$$\int_0^2 \frac{x}{\sqrt{1+2x^2}}\,dx = \frac{1}{4}\int_0^2 (1+2x^2)^{-1/2}(4x)\,dx = \left[\frac{1}{2}\sqrt{1+2x^2}\right]_0^2 = \frac{3}{2} - \frac{1}{2} = 1$$

53. Let $u = 1 + \sqrt{x}$, $du = \frac{1}{2\sqrt{x}}\,dx$.

$$\int_1^9 \frac{1}{\sqrt{x}(1+\sqrt{x})^2}\,dx = 2\int_1^9 (1+\sqrt{x})^{-2}\left(\frac{1}{2\sqrt{x}}\right)dx = \left[-\frac{2}{1+\sqrt{x}}\right]_1^9 = -\frac{1}{2} + 1 = \frac{1}{2}$$

54. Let $u = 4 + x^2$, $du = 2x\,dx$.

$$\int_0^2 x\sqrt[3]{4+x^2}\,dx = \frac{1}{2}\int_0^2 (4+x^2)^{1/3}(2x)\,dx = \left[\frac{3}{8}(4+x^2)^{4/3}\right]_0^2 = \frac{3}{8}(8^{4/3} - 4^{4/3}) = 6 - \frac{3}{2}\sqrt[3]{4} \approx 3.619$$

55. $u = 2 - x$, $x = 2 - u$, $dx = -du$

When $x = 1$, $u = 1$. When $x = 2$, $u = 0$.

$$\int_1^2 (x-1)\sqrt{2-x}\,dx = \int_1^0 -[(2-u)-1]\sqrt{u}\,du = \int_1^0 (u^{3/2} - u^{1/2})\,du = \left[\frac{2}{5}u^{5/2} - \frac{2}{3}u^{3/2}\right]_1^0 = -\left[\frac{2}{5} - \frac{2}{3}\right] = \frac{4}{15}$$

56. Let $u = 2x - 1$, $du = 2\,dx$, $x = \frac{1}{2}(u+1)$.

When $x = 1$, $u = 1$. When $x = 5$, $u = 9$.

$$\begin{aligned}\int_1^5 \frac{x}{\sqrt{2x-1}}\,dx &= \int_1^9 \frac{1/2(u+1)}{\sqrt{u}}\frac{1}{2}\,du = \frac{1}{4}\int_1^9 (u^{1/2} + u^{-1/2})\,du \\ &= \frac{1}{4}\left[\frac{2}{3}u^{3/2} + 2u^{1/2}\right]_1^9 \\ &= \frac{1}{4}\left[\left(\frac{2}{3}(27) + 2(3)\right) - \left(\frac{2}{3} + 2\right)\right] \\ &= \frac{16}{3}\end{aligned}$$

57. Let $u = \sqrt{x-5}$, $x = u^2 + 5$, $dx = 2u\,du$.

$$\begin{aligned}\int_5^{14} x\sqrt{x-5}\,dx &= \int_0^3 (u^2+5)\,u\,(2u\,du) \\ &= \int_0^3 (2u^4 + 10u^2)\,du \\ &= \left[\frac{2u^5}{5} + \frac{10u^3}{3}\right]_0^3 \\ &= \frac{2(243)}{5} + 10(9) = \frac{936}{5}\end{aligned}$$

58. Let $u = x + 1$, $du = dx$.

$$\int_0^1 \frac{1}{\sqrt{x+1}}\,dx = \int_1^2 u^{-1/2}\,du = 2\sqrt{u}\Big]_1^2 = 2\sqrt{2} - 2$$

59. $u = x + 1$, $x = u - 1$, $dx = du$

When $x = 0$, $u = 1$. When $x = 7$, $u = 8$.

$$\begin{aligned}\text{Area} &= \int_0^7 x\sqrt[3]{x+1}\,dx = \int_1^8 (u-1)\sqrt[3]{u}\,du \\ &= \int_1^8 (u^{4/3} - u^{1/3})\,du = \left[\frac{3}{7}u^{7/3} - \frac{3}{4}u^{4/3}\right]_1^8 = \left(\frac{384}{7} - 12\right) - \left(\frac{3}{7} - \frac{3}{4}\right) = \frac{1209}{28}\end{aligned}$$

60. $u = x + 2, x = u - 2, dx = du$

When $x = -2, u = 0$. When $x = 6, u = 8$.

$$\text{Area} = \int_{-2}^{6} x^2 \sqrt[3]{x+2}\,dx = \int_0^8 (u-2)^2 \sqrt[3]{u}\,du = \int_0^8 (u^{7/3} - 4u^{4/3} + 4u^{1/3})\,du = \left[\frac{3}{10}u^{10/3} - \frac{12}{7}u^{7/3} + 3u^{4/3}\right]_0^8 = \frac{4752}{35}$$

61. $\displaystyle\int_0^4 \frac{x}{\sqrt{2x+1}}\,dx \approx 3.333 = \frac{10}{3}$

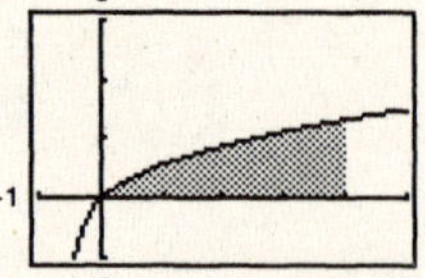

62. $\displaystyle\int_0^2 x^3\sqrt{x+2}\,dx \approx 7.581$

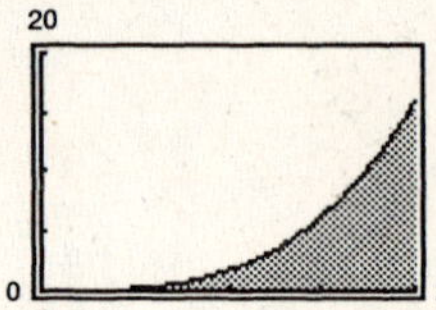

63. $\displaystyle\int_3^7 x\sqrt{x-3}\,dx \approx 28.8 = \frac{144}{5}$

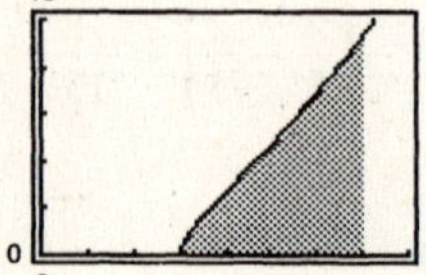

64. $\displaystyle\int_1^5 x^2\sqrt{x-1}\,dx \approx 67.505$

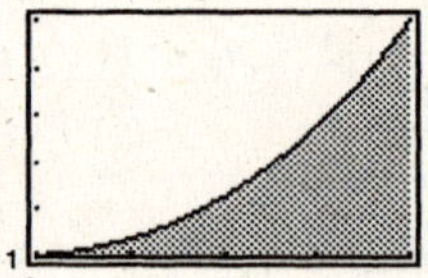

65. $\displaystyle\int (2x-1)^2\,dx = \frac{1}{2}\int (2x-1)^2\,2\,dx = \frac{1}{6}(2x-1)^3 + C_1 = \frac{4}{3}x^3 - 2x^2 + x - \frac{1}{6} + C_1$

$\displaystyle\int (2x-1)^2\,dx = \int (4x^2 - 4x + 1)\,dx = \frac{4}{3}x^3 - 2x^2 + x + C_2$

They differ by a constant: $C_2 = C_1 - \dfrac{1}{6}$.

66. $\displaystyle\int x(x^2-1)^2\,dx = \frac{1}{2}\int (x^2-1)^2(2x)\,dx$

$$= \frac{1}{2}\frac{(x^2-1)^3}{3} + C = \frac{1}{6}(x^2-1)^3 + C$$

$\displaystyle\int x(x^2-1)^2\,dx = \int (x^5 - 2x^3 + x)\,dx$

$$= \frac{x^6}{6} - \frac{x^4}{2} + \frac{x^2}{2} + C_1$$

$$= \frac{1}{6}(x^6 - 3x^4 + 3x^2) + C_1$$

$$= \frac{1}{6}(x^6 - 3x^4 + 3x^2 - 1) + C_1 + \frac{1}{6}$$

$$= \frac{1}{6}(x^2-1)^3 + C_1$$

67. $f(x) = x^2(x^2+1)$ is even.

$$\int_{-2}^{2} x^2(x^2+1)\,dx = 2\int_0^2 (x^4 + x^2)\,dx = 2\left[\frac{x^5}{5} + \frac{x^3}{3}\right]_0^2$$

$$= 2\left[\frac{32}{5} + \frac{8}{3}\right] = \frac{272}{15}$$

68. $\int_{-3}^{3}(9-x^2)\,dx = 2\int_0^3 (9-x^2)\,dx$

$= 2\left[9x - \frac{x^3}{3}\right]_0^3$

$= 2[27-9] = 36$

Even function

69. $f(x) = x(x^2+1)^3$ is odd.

$\int_{-2}^{2} x(x^2+1)^3\,dx = 0$

70. $\int_{-3}^{3} x\sqrt{9-x^2}\,dx = 0$

Odd function

71. $\int_0^2 x^2\,dx = \left[\frac{x^3}{3}\right]_0^2 = \frac{8}{3}$; the function x^2 is an even function.

(a) $\int_{-2}^{0} x^2\,dx = \int_0^2 x^2\,dx = \frac{8}{3}$

(b) $\int_{-2}^{2} x^2\,dx = 2\int_0^2 x^2\,dx = \frac{16}{3}$

(c) $\int_0^2 (-x^2)\,dx = -\int_0^2 x^2\,dx = -\frac{8}{3}$

(d) $\int_{-2}^{0} 3x^2\,dx = 3\int_0^2 x^2\,dx = 8$

72. (a) $\int_{-1}^{1} x^2(x^2+1)\,dx = 2\int_0^1 (x^4+x^2)\,dx = 2\left[\frac{x^5}{5} + \frac{x^3}{3}\right]_0^1$

$= 2\left[\frac{1}{5} + \frac{1}{3}\right] = \frac{16}{15}$

(b) $\int_{-2}^{2} x^3(x^2+1)\,dx = 0$ Odd function

(c) $\int_{-5}^{5} \frac{x}{\sqrt{x^2+1}}\,dx = 0$ Odd function

(d) $\int_{-3}^{3} |x|(x^2+1)\,dx = 2\int_0^3 x(x^2+1)\,dx = 2\int_0^3 (x^3+x)\,dx$

$= 2\left[\frac{x^4}{4} + \frac{x^2}{2}\right]_0^3 = 2\left[\frac{81}{4} + \frac{9}{2}\right] = \frac{99}{2}$

73. $\int_{-4}^{4}(x^3+6x^2-2x-3)\,dx = \int_{-4}^{4}(x^3-2x)\,dx + \int_{-4}^{4}(6x^2-3)\,dx = 0 + 2\int_0^4 (6x^2-3)\,dx = 2\left[2x^3-3x\right]_0^4 = 232$

74. $\int_{-2}^{2}(x^4-3x+5)\,dx = \int_{-2}^{2}(x^4+5)\,dx + \int_{-2}^{2}(-3x)\,dx$

$= 2\int_0^2 (x^4+5)\,dx + 0$

$= 2\left[\frac{x^5}{5} + 5x\right]_0^2$

$= 2\left[\frac{32}{5} + 10\right] = \frac{164}{5}$

75. If $u = 5 - x^2$, then $du = -2x\,dx$ and $\int x(5-x^2)^3\,dx = -\frac{1}{2}\int (5-x^2)^3(-2x)\,dx = -\frac{1}{2}\int u^3\,du.$

76. $f(x) = x(x^2+1)^2$ is odd. Hence, $\int_{-2}^{2} x(x^2+1)^2\,dx = 0.$

77. $\dfrac{dQ}{dt} = k(100 - t)^2$

$$Q(t) = \int k(100 - t)^2\,dt = -\frac{k}{3}(100 - t)^3 + C$$

$$Q(100) = C = 0$$

$$Q(t) = -\frac{k}{3}(100 - t)^3$$

$$Q(0) = -\frac{k}{3}(100)^3 = 2{,}000{,}000 \Rightarrow k = -6$$

Thus, $Q(t) = 2(100 - t)^3$. When $t = 50$, $Q(50) =$ \$250,000.

78. $\dfrac{dV}{dt} = \dfrac{k}{(t + 1)^2}$

$$V(t) = \int \frac{k}{(t + 1)^2}\,dt = -\frac{k}{t + 1} + C$$

$$V(0) = -k + C = 500{,}000$$

$$V(1) = -\frac{1}{2}k + C = 400{,}000$$

Solving this system yields $k = -200{,}000$ and $C = 300{,}000$. Thus,

$$V(t) = \frac{200{,}000}{t + 1} + 300{,}000.$$

When $t = 4$, $V(4) =$ \$340,000.

79. Let $u = x + h$, then $du = dx$. When $x = a$, $u = a + h$. When $x = b$, $u = b + h$. Thus,

$$\int_a^b f(x + h)\,dx = \int_{a+h}^{b+h} f(u)\,du = \int_{a+h}^{b+h} f(x)\,dx.$$

80. (a) Let $u = 1 - x$, $du = -dx$, $x = 1 - u$.

$x = 0 \Rightarrow u = 1$, $x = 1 \Rightarrow u = 0$

$$\int_0^1 x^2(1 - x)^5 dx = \int_1^0 (1 - u)^2 u^5(-du) = \int_0^1 u^5(1 - u)^2 du = \int_0^1 x^5(1 - x)^2 dx$$

(b) Let $u = 1 - x$, $du = -dx$, $x = 1 - u$.

$x = 0 \Rightarrow u = 1$, $x = 1 \Rightarrow u = 0$

$$\int_0^1 x^a(1 - x)^b dx = \int_1^0 (1 - u)^a u^b(-du) = \int_0^1 u^b(1 - u)^a du = \int_0^1 x^b(1 - x)^a dx$$

81. False

$$\int (2x + 1)^2\,dx = \frac{1}{2}\int (2x + 1)^2\,2\,dx = \frac{1}{6}(2x + 1)^3 + C$$

82. False

$$\int x(x^2 + 1)\,dx = \frac{1}{2}\int (x^2 + 1)(2x)\,dx = \frac{1}{4}(x^2 + 1)^2 + C$$

83. True

$$\int_{-10}^{10} (ax^3 + bx^2 + cx + d)\,dx = \underbrace{\int_{-10}^{10} (ax^3 + cx)\,dx}_{\text{Odd}} + \underbrace{\int_{-10}^{10} (bx^2 + d)\,dx}_{\text{Even}} = 0 + 2\int_0^{10} (bx^2 + d)\,dx$$

84. Let $u = cx$, $du = c\,dx$:

$$c\int_a^b f(cx)dx = c\int_{ca}^{cb} f(u)\frac{du}{c} = \int_{ca}^{cb} f(u)du = \int_{ca}^{cb} f(x)dx$$

85. Let $f(x)$ be an odd function and let $u = -x$. Then,

$$\int_{-a}^a f(x)\,dx = \int_a^{-a} f(-u)(-du) = \int_a^{-a} f(u)\,du = -\int_{-a}^a f(u)\,du$$

Thus,

$$2\int_{-a}^a f(x)\,dx = 0 \Rightarrow \int_{-a}^a f(x)\,dx = 0$$

Section 6.6 Numerical Integration

1. Exact: $\displaystyle\int_0^2 x^2\,dx = \left[\frac{1}{3}x^3\right]_0^2 = \frac{8}{3} \approx 2.6667$

Trapezoidal: $\displaystyle\int_0^2 x^2\,dx \approx \frac{1}{4}\left[0 + 2\left(\frac{1}{2}\right)^2 + 2(1)^2 + 2\left(\frac{3}{2}\right)^2 + (2)^2\right] = \frac{11}{4} = 2.7500$

Simpson's: $\displaystyle\int_0^2 x^2\,dx \approx \frac{1}{6}\left[0 + 4\left(\frac{1}{2}\right)^2 + 2(1)^2 + 4\left(\frac{3}{2}\right)^2 + (2)^2\right] = \frac{8}{3} \approx 2.6667$

2. Exact: $\displaystyle\int_0^1 \left(\frac{x^2}{2} + 1\right) dx = \left[\frac{x^3}{6} + x\right]_0^1 = \frac{7}{6} \approx 1.1667$

Trapezoidal: $\displaystyle\int_0^1 \left(\frac{x^2}{2} + 1\right) dx \approx \frac{1}{8}\left[1 + 2\left(\frac{(1/4)^2}{2} + 1\right) + 2\left(\frac{(1/2)^2}{2} + 1\right) + 2\left(\frac{(3/4)^2}{2} + 1\right) + \left(\frac{1^2}{2} + 1\right) = \frac{75}{64} \approx 1.1719$

Simpson's: $\displaystyle\int_0^1 \left(\frac{x^2}{2} + 1\right) dx \approx \frac{1}{12}\left[1 + 4\left(\frac{(1/4)^2}{2} + 1\right) + 2\left(\frac{(1/2)^2}{2} + 1\right) + 4\left(\frac{(3/4)^2}{2} + 1\right) + \left(\frac{1^2}{2} + 1\right)\right] = \frac{7}{6} \approx 1.1667$

3. Exact: $\displaystyle\int_0^2 x^3\,dx = \left[\frac{x^4}{4}\right]_0^2 = 4.000$

Trapezoidal: $\displaystyle\int_0^2 x^3\,dx \approx \frac{1}{4}\left[0 + 2\left(\frac{1}{2}\right)^3 + 2(1)^3 + 2\left(\frac{3}{2}\right)^3 + (2)^3\right] = \frac{17}{4} = 4.2500$

Simpson's: $\displaystyle\int_0^2 x^3\,dx \approx \frac{1}{6}\left[0 + 4\left(\frac{1}{2}\right)^3 + 2(1)^3 + 4\left(\frac{3}{2}\right)^3 + (2)^3\right] = \frac{24}{6} = 4.0000$

4. Exact: $\displaystyle\int_1^2 \frac{1}{x^2}\,dx = \left[\frac{-1}{x}\right]_1^2 = 0.5000$

Trapezoidal: $\displaystyle\int_1^2 \frac{1}{x^2}\,dx \approx \frac{1}{8}\left[1 + 2\left(\frac{4}{5}\right)^2 + 2\left(\frac{4}{6}\right)^2 + 2\left(\frac{4}{7}\right)^2 + \frac{1}{4}\right] \approx 0.5090$

Simpson's: $\displaystyle\int_1^2 \frac{1}{x^2}\,dx \approx \frac{1}{12}\left[1 + 4\left(\frac{4}{5}\right)^2 + 2\left(\frac{4}{6}\right)^2 + 4\left(\frac{4}{7}\right)^2 + \frac{1}{4}\right] \approx 0.5004$

5. Exact: $\displaystyle\int_0^2 x^3\,dx = \left[\frac{1}{4}x^4\right]_0^2 = 4.0000$

Trapezoidal: $\displaystyle\int_0^2 x^3\,dx \approx \frac{1}{8}\left[0 + 2\left(\frac{1}{4}\right)^3 + 2\left(\frac{2}{4}\right)^3 + 2\left(\frac{3}{4}\right)^3 + 2(1)^3 + 2\left(\frac{5}{4}\right)^3 + 2\left(\frac{6}{4}\right)^3 + 2\left(\frac{7}{4}\right)^3 + 8\right] = 4.0625$

Simpson's: $\displaystyle\int_0^2 x^3\,dx \approx \frac{1}{12}\left[0 + 4\left(\frac{1}{4}\right)^3 + 2\left(\frac{2}{4}\right)^3 + 4\left(\frac{3}{4}\right)^3 + 2(1)^3 + 4\left(\frac{5}{4}\right)^3 + 2\left(\frac{6}{4}\right)^3 + 4\left(\frac{7}{4}\right)^3 + 8\right] = 4.0000$

6. Exact: $\displaystyle\int_0^8 \sqrt[3]{x}\,dx = \left[\frac{3}{4}x^{4/3}\right]_0^8 = 12.0000$

Trapezoidal: $\displaystyle\int_0^8 \sqrt[3]{x}\,dx \approx \frac{1}{2}[0 + 2 + 2\sqrt[3]{2} + 2\sqrt[3]{3} + 2\sqrt[3]{4} + 2\sqrt[3]{5} + 2\sqrt[3]{6} + 2\sqrt[3]{7} + 2] \approx 11.7296$

Simpson's: $\displaystyle\int_0^8 \sqrt[3]{x}\,dx \approx \frac{1}{3}[0 + 4 + 2\sqrt[3]{2} + 4\sqrt[3]{3} + 2\sqrt[3]{4} + 4\sqrt[3]{5} + 2\sqrt[3]{6} + 4\sqrt[3]{7} + 2] \approx 11.8632$

7. Exact: $\int_4^9 \sqrt{x}\,dx = \left[\frac{2}{3}x^{3/2}\right]_4^9 = 18 - \frac{16}{3} = \frac{38}{3} \approx 12.6667$

Trapezoidal: $\int_4^9 \sqrt{x}\,dx \approx \frac{5}{16}\left[2 + 2\sqrt{\frac{37}{8}} + 2\sqrt{\frac{21}{4}} + 2\sqrt{\frac{47}{8}} + 2\sqrt{\frac{26}{4}} + 2\sqrt{\frac{57}{8}} + 2\sqrt{\frac{31}{4}} + 2\sqrt{\frac{67}{8}} + 3\right]$

≈ 12.6640

Simpson's: $\int_4^9 \sqrt{x}\,dx \approx \frac{5}{24}\left[2 + 4\sqrt{\frac{37}{8}} + \sqrt{21} + 4\sqrt{\frac{47}{8}} + \sqrt{26} + 4\sqrt{\frac{57}{8}} + \sqrt{31} + 4\sqrt{\frac{67}{8}} + 3\right] \approx 12.6667$

8. Exact: $\int_1^3 (4 - x^2)\,dx = \left[4x - \frac{x^3}{3}\right]_1^3 = 3 - \frac{11}{3} = -\frac{2}{3} \approx -0.6667$

Trapezoidal: $\int_1^3 (4 - x^2)\,dx \approx \frac{1}{4}\left\{3 + 2\left[4 - \left(\frac{3}{2}\right)^2\right] + 2(0) + 2\left[4 - \left(\frac{5}{2}\right)^2\right] - 5\right\} = -0.7500$

Simpson's: $\int_1^3 (4 - x^2)\,dx \approx \frac{1}{6}\left[3 + 4\left(4 - \frac{9}{4}\right) + 0 + 4\left(4 - \frac{25}{4}\right) - 5\right] \approx -0.6667$

9. Exact: $\int_1^2 \frac{1}{(x+1)^2}\,dx = \left[-\frac{1}{x+1}\right]_1^2 = -\frac{1}{3} + \frac{1}{2} = \frac{1}{6} \approx 0.1667$

Trapezoidal: $\int_1^2 \frac{1}{(x+1)^2}\,dx \approx \frac{1}{8}\left[\frac{1}{4} + 2\left(\frac{1}{((5/4)+1)^2}\right) + 2\left(\frac{1}{((3/2)+1)^2}\right) + 2\left(\frac{1}{((7/4)+1)^2}\right) + \frac{1}{9}\right]$

$= \frac{1}{8}\left(\frac{1}{4} + \frac{32}{81} + \frac{8}{25} + \frac{32}{121} + \frac{1}{9}\right) \approx 0.1676$

Simpson's: $\int_1^2 \frac{1}{(x+1)^2}\,dx \approx \frac{1}{12}\left[\frac{1}{4} + 4\left(\frac{1}{((5/4)+1)^2}\right) + 2\left(\frac{1}{((3/2)+1)^2}\right) + 4\left(\frac{1}{((7/4)+1)^2}\right) + \frac{1}{9}\right]$

$= \frac{1}{12}\left(\frac{1}{4} + \frac{64}{81} + \frac{8}{25} + \frac{64}{121} + \frac{1}{9}\right) \approx 0.1667$

10. Exact: $\int_0^2 x\sqrt{x^2+1}\,dx = \frac{1}{3}\left[(x^2+1)^{3/2}\right]_0^2 = \frac{1}{3}(5^{3/2} - 1) \approx 3.393$

Trapezoidal: $\int_0^2 x\sqrt{x^2+1}\,dx \approx \frac{1}{4}\left[0 + 2\left(\frac{1}{2}\right)\sqrt{(1/2)^2+1} + 2(1)\sqrt{1^2+1} + 2\left(\frac{3}{2}\right)\sqrt{(3/2)^2+1} + 2\sqrt{2^2+1}\right] \approx 3.457$

Simpson's: $\int_0^2 x\sqrt{x^2+1}\,dx \approx \frac{1}{6}\left[0 + 4\left(\frac{1}{2}\right)\sqrt{(1/2)^2+1} + 2(1)\sqrt{1^2+1} + 4\left(\frac{3}{2}\right)\sqrt{(3/2)^2+1} + 2\sqrt{2^2+1}\right] \approx 3.392$

11. $\int_0^4 \frac{1}{x+1}\,dx,\ n = 4$

Trapezoidal Rule: 1.6833

Simpson's Rule: 1.6222

Graphing utility: 1.6094

12. Trapezoidal: $\int_0^2 \frac{1}{\sqrt{1+x^3}}\,dx \approx \frac{1}{4}\left[1 + 2\left(\frac{1}{\sqrt{1+(1/2)^3}}\right) + 2\left(\frac{1}{\sqrt{1+1^3}}\right) + 2\left(\frac{1}{\sqrt{1+(3/2)^3}}\right) + \frac{1}{3}\right] \approx 1.397$

Simpson's: $\int_0^2 \frac{1}{\sqrt{1+x^3}}\,dx \approx \frac{1}{6}\left[1 + 4\left(\frac{1}{\sqrt{1+(1/2)^3}}\right) + 2\left(\frac{1}{\sqrt{1+1^3}}\right) + 4\left(\frac{1}{\sqrt{1+(3/2)^3}}\right) + \frac{1}{3}\right] \approx 1.405$

Graphing utility: 1.402

13. Trapezoidal: $\int_0^2 \sqrt{1+x^3}\,dx \approx \frac{1}{4}[1 + 2\sqrt{1+(1/8)} + 2\sqrt{2} + 2\sqrt{1+(27/8)} + 3] \approx 3.2833$

Simpson's: $\int_0^2 \sqrt{1+x^3}\,dx \approx \frac{1}{6}[1 + 4\sqrt{1+(1/8)} + 2\sqrt{2} + 4\sqrt{1+(27/8)} + 3] \approx 3.2396$

Graphing utility: 3.2413

14. $\int_0^4 \sqrt{x^2+1}\,dx,\ n = 4$

Trapezoidal Rule: 9.3741

Simpson's Rule: 9.3004

Graphing Utility: 9.2936

15. $\int_0^1 \sqrt{x}\sqrt{1-x}\,dx = \int_0^1 \sqrt{x(1-x)}\,dx$

Trapezoidal: $\int_0^1 \sqrt{x(1-x)}\,dx \approx \frac{1}{8}\left[0 + 2\sqrt{\frac{1}{4}\left(1-\frac{1}{4}\right)} + 2\sqrt{\frac{1}{2}\left(1-\frac{1}{2}\right)} + 2\sqrt{\frac{3}{4}\left(1-\frac{3}{4}\right)}\right] \approx 0.3415$

Simpson's: $\int_0^1 \sqrt{x(1-x)}\,dx \approx \frac{1}{12}\left[0 + 4\sqrt{\frac{1}{4}\left(1-\frac{1}{4}\right)} + 2\sqrt{\frac{1}{2}\left(1-\frac{1}{2}\right)} + 4\sqrt{\frac{3}{4}\left(1-\frac{3}{4}\right)}\right] \approx 0.3720$

Graphing utility: 0.3927

16. $\int_0^1 \frac{1}{x^2+1}\,dx,\ n = 2$

Trapezoidal Rule: 0.775

Simpson's Rule: 0.7833

Graphing Utility: 0.7854

17. $\int_{-2}^2 \frac{1}{x^2+1}\,dx \quad n = 8$

Trapezoidal Rule: 2.2077

Simpson's Rule: 2.2103

Graphing utility: 2.2143

18. $\int_{-1}^1 x\sqrt{x+1}\,dx,\ n = 4$

Trapezoidal Rule: 0.4830

Simpson's Rule: 0.4082

Graphing Utility: 0.3771

19. $\int_1^7 \frac{\sqrt{x-1}}{x}\,dx,\ n = 6$

Trapezoidal Rule: 2.3521

Simpson's Rule: 2.4385

Graphing utility: 2.5326

20. $\int_3^6 \frac{1}{1-\sqrt{x-1}}\,dx,\ n = 6$

Trapezoidal Rule: -3.8693

Simpson's Rule: -3.8331

Graphing Utility: -3.8303

21.

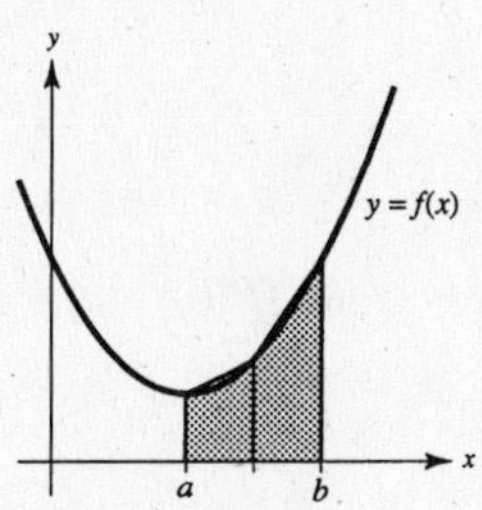

The Trapezoidal Rule overestimates the area if the graph of the integrand is concave up.

22. Trapezoidal: Linear polynomials

Simpson's: Quadratic polynomials

23. $f(x) = 2x + 3$

$f'(x) = 2$

$f''(x) = 0$

The error is 0 for both rules.

24. $f(x) = (x - 1)^{-2}$

$f'(x) = -2(x - 1)^{-3}$

$f''(x) = 6(x - 1)^{-4}$

$f'''(x) = -24(x - 1)^{-5}$

$f^{(4)}(x) = 120(x - 1)^{-6}$

(a) Trapezoidal: Error $\leq \frac{(4 - 2)^3}{12(4^2)}(6) = \frac{1}{4}$,

since $|f''(x)|$ is a maximum of 6 at $x = 2$.

(b) Simpson's: Error $\leq \frac{(4 - 2)^5}{180(4^4)}(120) = \frac{1}{12}$

since $|f^{(4)}(x)|$ is a maximum of 120 at $x = 2$.

25. $f(x) = x^3$

$f'(x) = 3x^2$

$f''(x) = 6x$

$f'''(x) = 6$

$f^{(4)}(x) = 0$

(a) Trapezoidal: Error $\leq \frac{(2 - 0)^3}{12(4^2)}(12) = 0.5$ since

$f''(x)$ is maximum in $[0, 2]$ when $x = 2$.

(b) Simpson's: Error $\leq \frac{(2 - 0)^5}{180(4^4)}(0) = 0$ since

$f^{(4)}(x) = 0$.

26. $f(x) = \frac{1}{x + 1}$

$f'(x) = \frac{-1}{(x + 1)^2}$

$f''(x) = \frac{2}{(x + 1)^3}$

$f'''(x) = \frac{-6}{(x + 1)^4}$

$f^{(4)}(x) = \frac{24}{(x + 1)^5}$

(a) Trapezoidal: Error $\leq \frac{(1 - 0)^2}{12(4^2)}(2) = \frac{1}{96} \approx 0.01$

since $f''(x)$ is maximum in $[0, 1]$ when $x = 0$.

(b) Simpson's: Error $\leq \frac{(1 - 0)^5}{180(4^4)}(24) = \frac{1}{1920} \approx 0.0005$

since $f^{(4)}(x)$ is maximum in $[0, 1]$ when $x = 0$.

27. $f''(x) = \frac{2}{x^3}$ in $[1, 3]$.

(a) $|f''(x)|$ is maximum when $x = 1$ and $|f''(1)| = 2$.

Trapezoidal: Error $\leq \frac{2^3}{12n^2}(2) < 0.00001$, $n^2 > 133{,}333.33$, $n > 365.15$; let $n = 366$.

$f^{(4)}(x) = \frac{24}{x^5}$ in $[1, 3]$

(b) $|f^{(4)}(x)|$ is maximum when $x = 1$ and when $|f^{(4)}(1)| = 24$.

Simpson's: Error $\leq \frac{2^5}{180n^4}(24) < 0.00001$, $n^4 > 426{,}666.67$, $n > 25.56$; let $n = 26$.

28. $f''(x) = \frac{2}{(1 + x)^3}$ in $[0, 1]$.

(a) $|f''(x)|$ is maximum when $x = 0$ and $|f''(0)| = 2$.

Trapezoidal: Error $\leq \frac{1}{12n^2}(2) < 0.00001$, $n^2 > 16{,}666.67$, $n > 129.10$; let $n = 130$.

$f^{(4)}(x) = \frac{24}{(1 + x)^5}$ in $[0, 1]$

(b) $|f^{(4)}(x)|$ is maximum when $x = 0$ and $|f^{(4)}(0)| = 24$.

Simpson's: Error $\leq \frac{1}{180n^4}(24) < 0.00001$, $n^4 > 13{,}333.33$, $n > 10.75$; let $n = 12$. (In Simpson's Rule n must be even.)

29. $f(x) = x^{-2}, [1, 3]$

(a) $f'(x) = -2x^{-3}, f''(x) = 6x^{-4}$

$|f''(x)|$ is maximum when $x = 1$ and $|f''(1)| = 6$.

Trapezoidal: Error $\leq \dfrac{8}{12n^2}(6) < 0.00001,\ n^2 > 400{,}000,\ \ n > 632.5$; Let $n = 633$

(b) $f'''(x) = -24x^{-5}, f^{(4)}(x) = 120x^{-6}$

$|f^{(4)}(x)|$ is maximum when $x = 1$ and $|f^{(4)}(1)| = 120$.

Simpson's Rule: Error $\leq \dfrac{32}{180n^4}(120) < 0.00001,\ n^4 > 2{,}133{,}333.3,\ n > 38.2$; Let $n = 40$ (must be even)

30. $f(x) = (x + 2)^{-1}, [0, 1]$

(a) $f'(x) = -(x + 2)^{-2}, f''(x) = 2(x + 2)^{-3} = \dfrac{2}{(x + 2)^3}$

$|f''(x)|$ is maximum when $x = 0$ and $|f''(0)| = \dfrac{1}{4}$

Trapezoidal: Error $\leq \dfrac{1}{12n^2}\left(\dfrac{1}{4}\right) < 0.00001$

$n^2 > 2083.3$

$n > 45.6$

Let $n = 46$.

(b) $f'''(x) = -6(x + 2)^{-4}, f^{(4)}(x) = 24(x + 2)^{-5}$

$|f^{(4)}(x)|$ is maximum when $x = 0$ and $|f^{(4)}(0)| = \dfrac{24}{32} = \dfrac{3}{4}$.

Simpson's Rule: Error $\leq \dfrac{1}{180n^4}\left(\dfrac{3}{4}\right) < 0.00001$

$n^4 > 416.67$

$n > 4.5$

Let $n = 6$. (must be even)

31. $f(x) = \sqrt{1 + x}$

(a) $f''(x) = -\dfrac{1}{4(1 + x)^{3/2}}$ in $[0, 2]$.

$|f''(x)|$ is maximum when $x = 0$ and $|f''(0)| = \dfrac{1}{4}$.

Trapezoidal: Error $\leq \dfrac{8}{12n^2}\left(\dfrac{1}{4}\right) < 0.00001,\ n^2 > 16{,}666.67,\ n > 129.10$; let $n = 130$.

(b) $f^{(4)}(x) = \dfrac{-15}{16(1 + x)^{7/2}}$ in $[0, 2]$

$|f^{(4)}(x)|$ is maximum when $x = 0$ and $|f^{(4)}(0)| = \dfrac{15}{16}$.

Simpson's: Error $\leq \dfrac{32}{180n^4}\left(\dfrac{15}{16}\right) < 0.00001,\ n^4 > 16{,}666.67,\ n > 11.36$; let $n = 12$.

32. $f(x) = (x + 1)^{2/3}$

(a) $f''(x) = -\dfrac{2}{9(x + 1)^{4/3}}$ in $[0, 2]$.

$|f''(x)|$ is maximum when $x = 0$ and $|f''(0)| = \dfrac{2}{9}$.

Trapezoidal: Error $\leq \dfrac{8}{12n^4}\left(\dfrac{2}{9}\right) < 0.00001,\ n^2 > 14{,}814.81,\ n > 121.72$; let $n = 122$.

(b) $f^{(4)}(x) = -\dfrac{56}{81(x + 1)^{10/3}}$ in $[0, 2]$

$|f^{(4)}(x)|$ is maximum when $x = 0$ and $|f^{(4)}(0)| = \dfrac{56}{81}$.

Simpson's: Error $\leq \dfrac{32}{180n^4}\left(\dfrac{56}{81}\right) < 0.00001,\ n^4 > 12{,}290.81,\ n > 10.530$; let $n = 12$. (In Simpson's Rule n must be even.)

33. Let $f(x) = Ax^3 + Bx^2 + Cx + D$. Then $f^{(4)}(x) = 0$.

Simpson's: Error $\leq \dfrac{(b-a)^5}{180n^4}(0) = 0$

Therefore, Simpson's Rule is exact when approximating the integral of a cubic polynomial.

Example: $\displaystyle\int_0^1 x^3\,dx = \frac{1}{6}\left[0 + 4\left(\frac{1}{2}\right)^3 + 1\right] = \frac{1}{4}$

This is the exact value of the integral.

34. The program will vary depending upon the computer or programmable calculator that you use.

35. $f(x) = \sqrt{2 + 3x^2}$ on $[0, 4]$.

n	$L(n)$	$M(n)$	$R(n)$	$T(n)$	$S(n)$
4	12.7771	15.3965	18.4340	15.6055	15.4845
8	14.0868	15.4480	16.9152	15.5010	15.4662
10	14.3569	15.4544	16.6197	15.4883	15.4658
12	14.5386	15.4578	16.4242	15.4814	15.4657
16	14.7674	15.4613	16.1816	15.4745	15.4657
20	14.9056	15.4628	16.0370	15.4713	15.4657

36. $f(x) = \sqrt{1 - x^2}$ on $[0, 1]$.

n	$L(n)$	$M(n)$	$R(n)$	$T(n)$	$S(n)$
4	0.8739	0.7960	0.6239	0.7489	0.7709
8	0.8350	0.7892	0.7100	0.7725	0.7803
10	0.8261	0.7881	0.7261	0.7761	0.7818
12	0.8200	0.7875	0.7367	0.7783	0.7826
16	0.8121	0.7867	0.7496	0.7808	0.7836
20	0.8071	0.7864	0.7571	0.7821	0.7841

37. $W = \displaystyle\int_0^5 100x\sqrt{125 - x^3}\,dx$

Simpson's Rule: $n = 12$

$$\int_0^5 100x\sqrt{125 - x^3}\,dx \approx \frac{5}{3(12)}\left[0 + 400\left(\frac{5}{12}\right)\sqrt{125 - \left(\frac{5}{12}\right)^3} + 200\left(\frac{10}{12}\right)\sqrt{125 - \left(\frac{10}{12}\right)^3}\right.$$
$$\left. + 400\left(\frac{15}{12}\right)\sqrt{125 - \left(\frac{15}{12}\right)^3} + \cdots + 0\right] \approx 10{,}233.58 \text{ ft} \cdot \text{lb}$$

38. (a) Trapezoidal:

$$\int_0^2 f(x)\,dx \approx \frac{2}{2(8)}[4.32 + 2(4.36) + 2(4.58) + 2(5.79) + 2(6.14) + 2(7.25) + 2(7.64) + 2(8.08) + 8.14] \approx 12.518$$

(b) Simpson's:

$$\int_0^2 f(x)\,dx \approx \frac{2}{3(8)}[4.32 + 4(4.36) + 2(4.58) + 4(5.79) + 2(6.14) + 4(7.25) + 2(7.64) + 4(8.08) + 8.14] \approx 12.592$$

(c) Using a graphing utility,

$$y = -1.3727x^3 + 4.0092x^2 - 0.6202x + 4.2844$$

Integrating, $\int_0^2 y\,dx \approx 12.53$

39. $\int_0^{1/2} \frac{6}{\sqrt{1-x^2}}\,dx$ Simpson's Rule, $n = 6$

$$\pi \approx \frac{\left(\frac{1}{2} - 0\right)}{3(6)}[6 + 4(6.0209) + 2(6.0851) + 4(6.1968) + 2(6.3640) + 4(6.6002) + 6.9282]$$

$$\approx \frac{1}{36}[113.098] \approx 3.1416$$

40. Simpson's Rule: $n = 6$

$$\pi = 4\int_0^1 \frac{1}{1+x^2}\,dx \approx \frac{4}{3(6)}\left[1 + \frac{4}{1+(1/6)^2} + \frac{2}{1+(2/6)^2} + \frac{4}{1+(3/6)^2} + \frac{2}{1+(4/6)^2} + \frac{4}{1+(5/6)^2} + \frac{1}{2}\right]$$

$$\approx 3.14159$$

41. Area $\approx \frac{1000}{2(10)}[125 + 2(125) + 2(120) + 2(112) + 2(90) + 2(90) + 2(95) + 2(88) + 2(75) + 2(35)] = 89{,}250$ sq m

42. Area $\approx \frac{120}{2(12)}[75 + 2(81) + 2(84) + 2(76) + 2(67) + 2(68) + 2(69) + 2(72) + 2(68) + 2(56) + 2(42) + 2(23) + 0]$

$= 7435$ sq m

43. $f(x) = \frac{1}{x+1}$, $n = 10$

Use a computer algebra system to determine t such that

$$\frac{t}{30}\left[\frac{1}{0+1} + 4\left(\frac{1}{\frac{t}{10}+1}\right) + 2\left(\frac{1}{\frac{2t}{10}+1}\right) + 4\left(\frac{1}{\frac{3t}{10}+1}\right) + \cdots + 4\left(\frac{1}{\frac{9t}{10}+1}\right) + \frac{1}{t+1}\right] = 2$$

to find $t = 6.366$.

44. The quadratic polynomial

$$p(x) = \frac{(x-x_2)(x-x_3)}{(x_1-x_2)(x_1-x_3)}y_1 + \frac{(x-x_1)(x-x_3)}{(x_2-x_1)(x_2-x_3)}y_2 + \frac{(x-x_1)(x-x_2)}{(x_3-x_1)(x_3-x_2)}y_3$$

passes through the three points.

Review Exercises for Chapter 6

1.

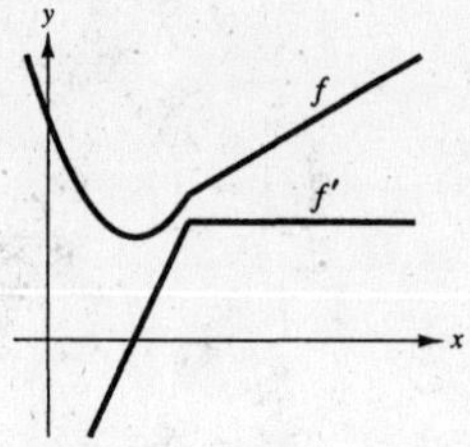

2.

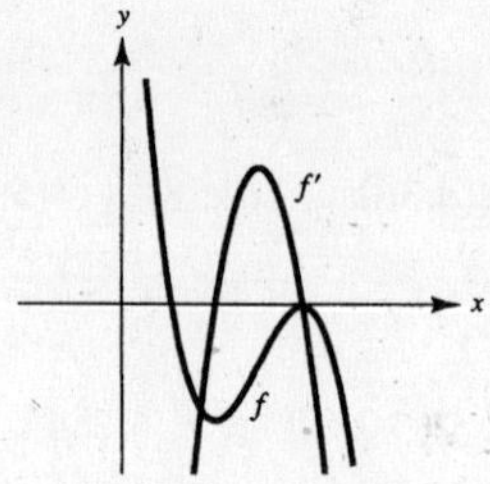

3. $\int (2x^2 + x - 1)\,dx = \frac{2}{3}x^3 + \frac{1}{2}x^2 - x + C$

4. $u = 3x$

$du = 3\,dx$

$\int \frac{2}{\sqrt[3]{3x}}\,dx = \frac{2}{3}\int (3x)^{-1/3}(3)\,dx = (3x)^{2/3} + C$

5. $\int \frac{x^3 + 1}{x^2}\,dx = \int \left(x + \frac{1}{x^2}\right)dx = \frac{1}{2}x^2 - \frac{1}{x} + C$

6. $\int \frac{x^3 - 2x^2 + 1}{x^2}\,dx = \int (x - 2 + x^{-2})\,dx$

$= \frac{1}{2}x^2 - 2x - \frac{1}{x} + C$

7. $\int \sqrt[3]{x}\,(x + 3)\,dx = \int (x^{4/3} + 3x^{1/3})\,dx$

$= \frac{3}{7}x^{7/3} + \frac{9}{4}x^{4/3} + C$

8. $\int x^2(x + 5)^2\,dx = \int x^2(x^2 + 10x + 25)\,dx$

$= \int x^4 + 10x^3 + 25x^2\,dx$

$= \frac{x^5}{5} + \frac{5x^4}{2} + \frac{25x^3}{3} + C$

9. $f'(x) = -2x,\ (-1, 1)$

$f(x) = \int -2x\,dx = -x^2 + C$

When $x = -1$:

$y = -1 + C = 1$

$C = 2$

$y = 2 - x^2$

10. $f''(x) = 6(x - 1)$

$f'(x) = \int 6(x - 1)\,dx = 3(x - 1)^2 + C_1$

Since the slope of the tangent line at $(2, 1)$ is 3, it follows that $f'(2) = 3 + C_1 = 3$ when $C_1 = 0$.

$f'(x) = 3(x - 1)^2$

$f(x) = \int 3(x - 1)^2\,dx = (x - 1)^3 + C_2$

$f(2) = 1 + C_2 = 1$ when $C_2 = 0$.

$f(x) = (x - 1)^3$

11. (a)

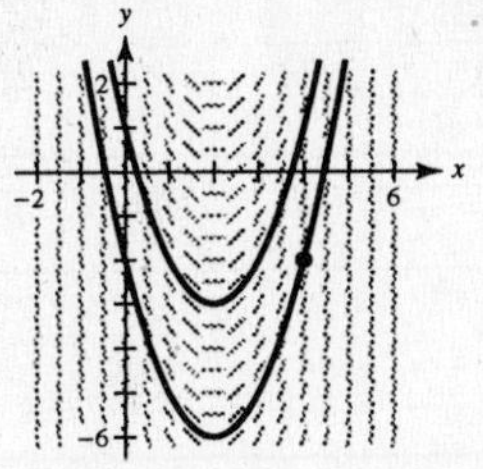

(b) $\dfrac{dy}{dx} = 2x - 4, \quad (4, -2)$

$$y = \int (2x - 4)dx = x^2 - 4x + C$$

$$-2 = 16 - 16 + C \Rightarrow C = -2$$

$$y = x^2 - 4x - 2$$

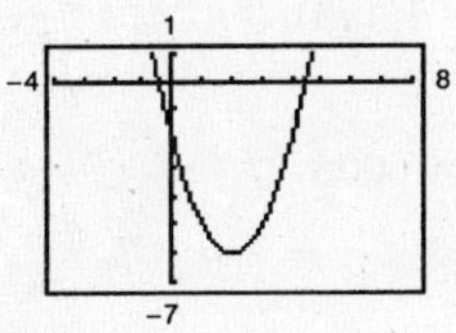

12. (a)

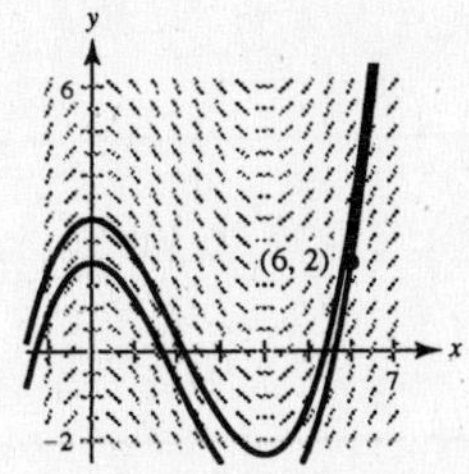

(b) $\dfrac{dy}{dx} = \dfrac{1}{2}x^2 - 2x, \quad (6, 2)$

$$y = \int \left(\frac{1}{2}x^2 - 2x\right)dx = \frac{1}{6}x^3 - x^2 + C$$

$$2 = \frac{1}{6}(6^3) - 6^2 + C \Rightarrow C = 2$$

$$y = \frac{1}{6}x^3 - x^2 + 2$$

13. $a(t) = a$

$v(t) = \int a\,dt = at + C_1$

$v(0) = 0 + C_1 = 0$ when $C_1 = 0$.

$v(t) = at$

$s(t) = \int at\,dt = \dfrac{a}{2}t^2 + C_2$

$s(0) = 0 + C_2 = 0$ when $C_2 = 0$.

$s(t) = \dfrac{a}{2}t^2$

$s(30) = \dfrac{a}{2}(30)^2 = 3600$ or

$a = \dfrac{2(3600)}{(30)^2} = 8 \text{ ft/sec}^2.$

$v(30) = 8(30) = 240$ ft/sec

14. 45 mph = 66 ft/sec

30 mph = 44 ft/sec

$a(t) = -a$

$v(t) = -at + 66$ since $v(0) = 66$ ft/sec.

$s(t) = -\dfrac{a}{2}t^2 + 66t$ since $s(0) = 0$.

Solving the system

$v(t) = -at + 66 = 44$

$s(t) = -\dfrac{a}{2}t^2 + 66t = 264$

we obtain $t = 24/5$ and $a = 55/12$. We now solve $-(55/12)t + 66 = 0$ and get $t = 72/5$. Thus,

$$s\left(\frac{72}{5}\right) = -\frac{55/12}{2}\left(\frac{72}{5}\right)^2 + 66\left(\frac{72}{5}\right) \approx 475.2 \text{ ft.}$$

Stopping distance from 30 mph to rest is

$475.2 - 264 = 211.2$ ft.

15. $a(t) = -32$

$v(t) = -32t + 96$

$s(t) = -16t^2 + 96t$

(a) $v(t) = -32t + 96 = 0$ when $t = 3$ sec.

(b) $s(3) = -144 + 288 = 144$ ft

(c) $v(t) = -32t + 96 = \dfrac{96}{2}$ when $t = \dfrac{3}{2}$ sec.

(d) $s\left(\dfrac{3}{2}\right) = -16\left(\dfrac{9}{4}\right) + 96\left(\dfrac{3}{2}\right) = 108$ ft

16. $a(t) = -9.8 \text{ m/sec}^2$

$v(t) = -9.8t + v_0 = -9.8t + 40$

$s(t) = -4.9t^2 + 40t \quad (s(0) = 0)$

(a) $v(t) = -9.8t + 40 = 0$ when $t = \dfrac{40}{9.8} \approx 4.08$ sec.

(b) $s(4.08) \approx 81.63$ m

(c) $v(t) = -9.8t + 40 = 20$ when $t = \dfrac{20}{9.8} \approx 2.04$ sec.

(d) $s(2.04) \approx 61.2$ m

17. $\displaystyle\sum_{i=1}^{8}\frac{1}{4i} = \frac{1}{4(1)} + \frac{1}{4(2)} + \cdots + \frac{1}{4(8)}$

18. $\displaystyle\sum_{i=1}^{12}\frac{i+2}{2i} = \frac{1+2}{2(1)} + \frac{2+2}{2(2)} + \frac{3+2}{2(3)} + \cdots + \frac{12+2}{2(12)}$

19. $\displaystyle\sum_{i=1}^{n}\left(\frac{3}{n}\right)\left(\frac{i+1}{n}\right)^2 = \frac{3}{n}\left(\frac{1+1}{n}\right)^2 + \frac{3}{n}\left(\frac{2+1}{n}\right)^2 + \cdots + \frac{3}{n}\left(\frac{n+1}{n}\right)^2$

20. $\displaystyle\sum_{i=1}^{n}3i\left[2 + \frac{(i+1)^2}{n}\right] = 3\left[2 + \frac{4}{n}\right] + 6\left[2 + \frac{9}{n}\right] + \cdots + 3n\left[2 + \frac{(n+1)^2}{n}\right]$

21. $\displaystyle\sum_{i=1}^{10}3i = 3\left(\frac{10(11)}{2}\right) = 165$

22. $\displaystyle\sum_{i=1}^{20}(4i-1) = 4\frac{20(21)}{2} - 20 = 820$

23. $\displaystyle\sum_{i=1}^{20}(i+1)^2 = \sum_{i=1}^{20}(i^2+2i+1)$

$\displaystyle= \frac{20(21)(41)}{6} + 2\frac{20(21)}{2} + 20$

$= 2870 + 420 + 20 = 3310$

24. $\displaystyle\sum_{i=1}^{12}i(i^2-1) = \sum_{i=1}^{12}(i^3-i)$

$\displaystyle= \frac{(12^2)(13^2)}{4} - \frac{12(13)}{2}$

$= 6084 - 78 = 6006$

25. (a) $\displaystyle\sum_{i=1}^{10}(2i-1)$ (b) $\displaystyle\sum_{i=1}^{n}i^3$ (c) $\displaystyle\sum_{i=1}^{10}(4i+2)$

26. $x_1 = 2, x_2 = -1, x_3 = 5, x_4 = 3, x_5 = 7$

(a) $\displaystyle\frac{1}{5}\sum_{i=1}^{5}x_i = \frac{1}{5}(2 - 1 + 5 + 3 + 7) = \frac{16}{5}$

(b) $\displaystyle\sum_{i=1}^{5}\frac{1}{x_i} = \frac{1}{2} - 1 + \frac{1}{5} + \frac{1}{3} + \frac{1}{7} = \frac{37}{210}$

(c) $\displaystyle\sum_{i=1}^{5}(2x_i - x_i^2) = [2(2) - (2)^2] + [2(-1) - (-1)^2] + [2(5) - (5)^2] + [2(3) - (3)^2] + [2(7) - (7)^2] = -56$

(d) $\displaystyle\sum_{i=2}^{5}(x_i - x_{i-1}) = (-1 - 2) + [5 - (-1)] + (3 - 5) + (7 - 3) = 5$

27. $\displaystyle y = \frac{10}{x^2+1}, \Delta x = \frac{1}{2}, n = 4$

$\displaystyle S(n) = S(4) = \frac{1}{2}\left[\frac{10}{1} + \frac{10}{(1/2)^2+1} + \frac{10}{(1)^2+1} + \frac{10}{(3/2)^2+1}\right]$

≈ 13.0385

$\displaystyle s(n) = s(4) = \frac{1}{2}\left[\frac{10}{(1/2)^2+1} + \frac{10}{1+1} + \frac{10}{(3/2)^2+1} + \frac{10}{2^2+1}\right]$

≈ 9.0385

$9.0385 <$ Area of Region < 13.0385

28. $y = 9 - \frac{1}{4}x^2, \Delta x = 1, n = 4$

$$S(4) = 1\left[\left(9 - \frac{1}{4}(4)\right) + \left(9 - \frac{1}{4}(9)\right) + \left(9 - \frac{1}{4}(16)\right) + \left(9 - \frac{1}{4}(25)\right)\right]$$

$$\approx 22.5$$

$$s(4) = 1\left[\left(9 - \frac{1}{4}(9)\right) + \left(9 - \frac{1}{4}(16)\right) + \left(9 - \frac{1}{4}(25)\right) + (9 - 9)\right]$$

$$\approx 14.5$$

29. $y = 6 - x, \Delta x = \frac{4}{n}$, right endpoints

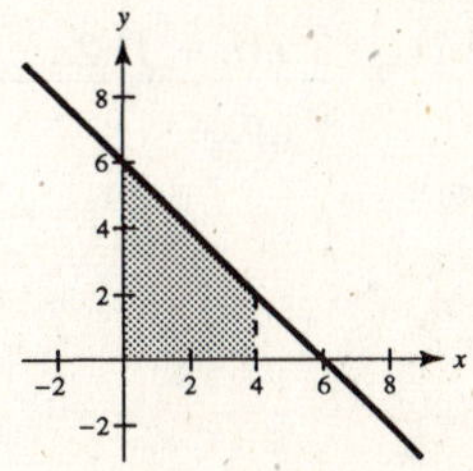

$$\text{Area} = \lim_{n\to\infty} \sum_{i=1}^{n} f(c_i)\,\Delta x$$

$$= \lim_{n\to\infty} \sum_{i=1}^{n} \left(6 - \frac{4i}{n}\right)\frac{4}{n}$$

$$= \lim_{n\to\infty} \frac{4}{n}\left[6n - \frac{4}{n}\frac{n(n+1)}{2}\right]$$

$$= \lim_{n\to\infty} \left[24 - 8\frac{n+1}{n}\right] = 24 - 8 = 16$$

30. $y = x^2 + 3, \Delta x = \frac{2}{n}$ right endpoints

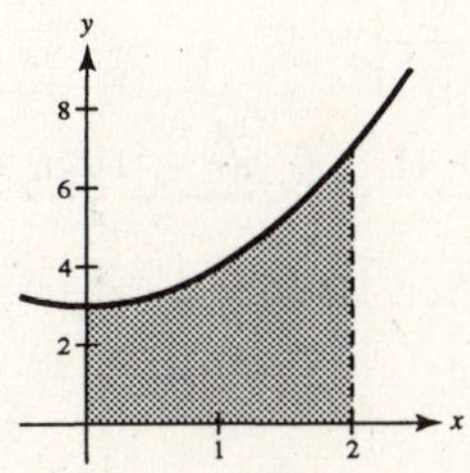

$$\text{Area} = \lim_{n\to\infty} \sum_{i=1}^{n} f(c_i)\,\Delta x$$

$$= \lim_{n\to\infty} \sum_{i=1}^{n} \left[\left(\frac{2i}{n}\right)^2 + 3\right]\left(\frac{2}{n}\right)$$

$$= \lim_{n\to\infty} \frac{2}{n} \sum_{i=1}^{n} \left[\frac{4i^2}{n^2} + 3\right]$$

$$= \lim_{n\to\infty} \frac{2}{n}\left[\frac{4}{n^2}\frac{n(n+1)(2n+1)}{6} + 3n\right]$$

$$= \lim_{n\to\infty} \left[\frac{4}{3}\frac{(n+1)(2n+1)}{n^2} + 6\right] = \frac{8}{3} + 6 = \frac{26}{3}$$

31. $y = 5 - x^2, \Delta x = \frac{3}{n}$

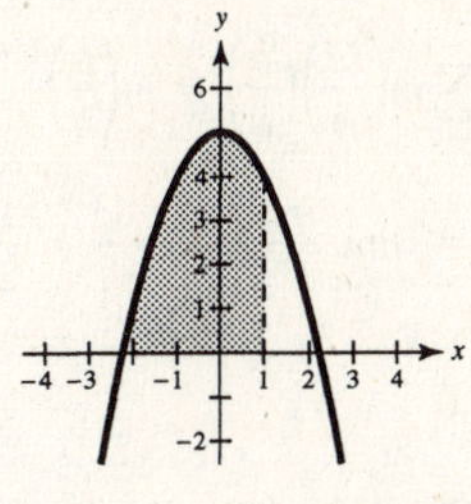

$$\text{Area} = \lim_{n\to\infty} \sum_{i=1}^{n} f(c_i)\,\Delta x$$

$$= \lim_{n\to\infty} \sum_{i=1}^{n} \left[5 - \left(-2 + \frac{3i}{n}\right)^2\right]\left(\frac{3}{n}\right)$$

$$= \lim_{n\infty} \frac{3}{n} \sum_{i=1}^{n} \left[1 + \frac{12i}{n} - \frac{9i^2}{n^2}\right]$$

$$= \lim_{n\to\infty} \frac{3}{n}\left[n + \frac{12}{n}\frac{n(n+1)}{2} - \frac{9}{n^2}\frac{n(n+1)(2n+1)}{6}\right]$$

$$= \lim_{n\to\infty} \left[3 + 18\frac{n+1}{n} - \frac{9}{2}\frac{(n+1)(2n+1)}{n^2}\right]$$

$$= 3 + 18 - 9 = 12$$

32. $y = \frac{1}{4}x^3, \Delta x = \frac{2}{n}$

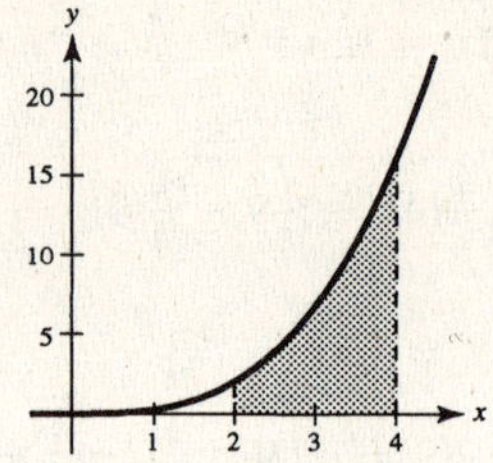

$$\begin{aligned}
\text{Area} &= \lim_{n\to\infty} \sum_{i=1}^{n} f(c_i)\,\Delta x \\
&= \lim_{n\to\infty} \sum_{i=1}^{n} \frac{1}{4}\left(2 + \frac{2i}{n}\right)^3\left(\frac{2}{n}\right) \\
&= \lim_{n\to\infty} \frac{1}{2n} \sum_{i=1}^{n}\left[8 + \frac{24i}{n} + \frac{24i^2}{n^2} + \frac{8i^3}{n^3}\right] \\
&= \lim_{n\to\infty} \frac{4}{n} \sum_{i=1}^{n}\left[1 + \frac{3i}{n} + \frac{3i^2}{n^2} + \frac{i^3}{n^3}\right] \\
&= \lim_{n\to\infty} \frac{4}{n}\left[n + \frac{3}{n}\frac{n(n+1)}{2} + \frac{3}{n^2}\frac{n(n+1)(2n+1)}{6} + \frac{1}{n^3}\frac{n^2(n+1)^2}{4}\right] \\
&= 4 + 6 + 4 + 1 = 15
\end{aligned}$$

33. $x = 5y - y^2, 2 \le y \le 5, \Delta y = \frac{3}{n}$

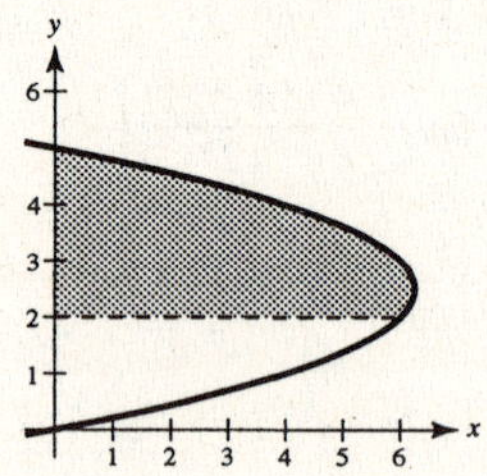

$$\begin{aligned}
\text{Area} &= \lim_{n\to\infty} \sum_{i=1}^{n}\left[5\left(2 + \frac{3i}{n}\right) - \left(2 + \frac{3i}{n}\right)^2\right]\left(\frac{3}{n}\right) \\
&= \lim_{n\to\infty} \frac{3}{n} \sum_{i=1}^{n}\left[10 + \frac{15i}{n} - 4 - 12\frac{i}{n} - \frac{9i^2}{n^2}\right] \\
&= \lim_{n\to\infty} \frac{3}{n} \sum_{i=1}^{n}\left[6 + \frac{3i}{n} - \frac{9i^2}{n^2}\right] \\
&= \lim_{n\to\infty} \frac{3}{n}\left[6n + \frac{3}{n}\frac{n(n+1)}{2} - \frac{9}{n^2}\frac{n(n+1)(2n+1)}{6}\right] \\
&= \left[18 + \frac{9}{2} - 9\right] = \frac{27}{2}
\end{aligned}$$

34. (a) $S = m\left(\frac{b}{4}\right)\left(\frac{b}{4}\right) + m\left(\frac{2b}{4}\right)\left(\frac{b}{4}\right) + m\left(\frac{3b}{4}\right)\left(\frac{b}{4}\right) + m\left(\frac{4b}{4}\right)\left(\frac{b}{4}\right) = \frac{mb^2}{16}(1 + 2 + 3 + 4) = \frac{5mb^2}{8}$

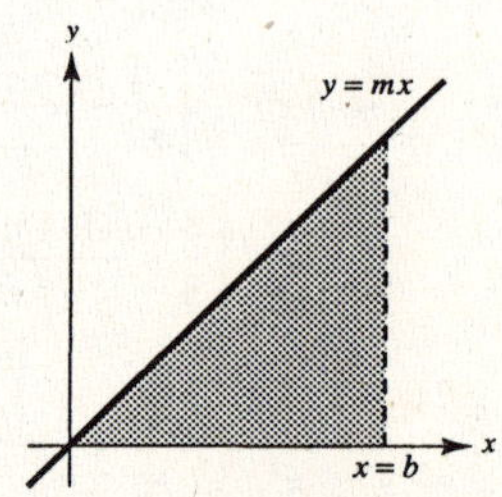

$s = m(0)\left(\frac{b}{4}\right) + m\left(\frac{b}{4}\right)\left(\frac{b}{4}\right) + m\left(\frac{2b}{4}\right)\left(\frac{b}{4}\right) + m\left(\frac{3b}{4}\right)\left(\frac{b}{4}\right) = \frac{mb^2}{16}(1 + 2 + 3) = \frac{3mb^2}{8}$

(b) $S(n) = \sum_{i=1}^{n} f\left(\frac{bi}{n}\right)\left(\frac{b}{n}\right) = \sum_{i=1}^{n}\left(\frac{mbi}{n}\right)\left(\frac{b}{n}\right) = m\left(\frac{b}{n}\right)^2 \sum_{i=1}^{n} i = \frac{mb^2}{n^2}\left(\frac{n(n+1)}{2}\right) = \frac{mb^2(n+1)}{2n}$

$s(n) = \sum_{i=0}^{n-1} f\left(\frac{bi}{n}\right)\left(\frac{b}{n}\right) = \sum_{i=0}^{n-1} m\left(\frac{bi}{n}\right)\left(\frac{b}{n}\right) = m\left(\frac{b}{n}\right)^2 \sum_{i=0}^{n-1} i = \frac{mb^2}{n^2}\left(\frac{(n-1)n}{2}\right) = \frac{mb^2(n-1)}{2n}$

(c) $\text{Area} = \lim_{n\to\infty} \frac{mb^2(n+1)}{2n} = \lim_{n\to\infty} \frac{mb^2(n-1)}{2n} = \frac{1}{2}mb^2 = \frac{1}{2}(b)(mb) = \frac{1}{2}(\text{base})(\text{height})$

35. $\lim_{\|\Delta\|\to\infty} \sum_{i=1}^{n} (2c_i - 3)\,\Delta xi = \int_4^6 (2x - 3)\,dx$

36. $\lim_{\|\Delta\|\to\infty} \sum_{i=1}^{n} 3c_i(9 - c_i^2)\,\Delta xi = \int_1^3 3x(9 - x^2)\,dx$

37.

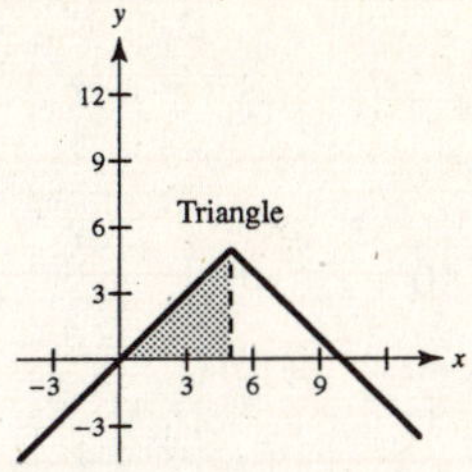

$$\text{Area} = \frac{1}{2}(\text{base})(\text{height})$$
$$= \frac{1}{2}(5)(5) = \frac{25}{2}$$

38.

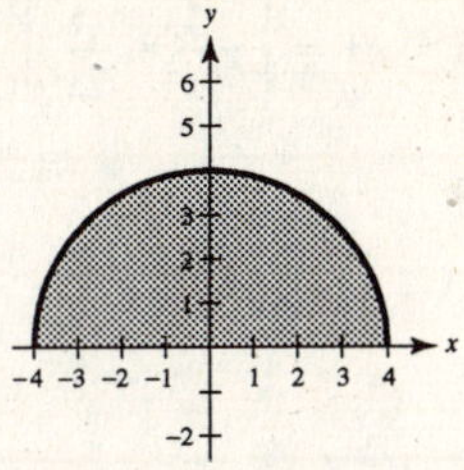

$$\int_{-4}^{4} \sqrt{16 - x^2}\,dx = \frac{1}{2}\pi(4)^2 = 8\pi \quad \text{(semicircle)}$$

39. (a) $\displaystyle\int_2^6 [f(x) + g(x)]\,dx = \int_2^6 f(x)\,dx + \int_2^6 g(x)\,dx = 10 + 3 = 13$

(b) $\displaystyle\int_2^6 [f(x) - g(x)]\,dx = \int_2^6 f(x)\,dx - \int_2^6 g(x)\,dx = 10 - 3 = 7$

(c) $\displaystyle\int_2^6 [2f(x) - 3g(x)]\,dx = 2\int_2^6 f(x)\,dx - 3\int_2^6 g(x)\,dx = 2(10) - 3(3) = 11$

(d) $\displaystyle\int_2^6 5f(x)\,dx = 5\int_2^6 f(x)\,dx = 5(10) = 50$

40. (a) $\displaystyle\int_0^6 f(x)\,dx = \int_0^3 f(x)\,dx + \int_3^6 f(x)\,dx = 4 + (-1) = 3$

(b) $\displaystyle\int_6^3 f(x)\,dx = -\int_3^6 f(x)\,dx = -(-1) = 1$

(c) $\displaystyle\int_4^4 f(x)\,dx = 0$

(d) $\displaystyle\int_3^6 -10\,f(x)\,dx = -10\int_3^6 f(x)\,dx = -10(-1) = 10$

41. $\displaystyle\int_1^8 (\sqrt[3]{x} + 1)\,dx = \left[\frac{3}{4}x^{4/3} + x\right]_1^8 = \left[\frac{3}{4}(16) + 8\right] - \left[\frac{3}{4} + 1\right] = \frac{73}{4}$ (c)

42. $\displaystyle\int_1^3 \frac{12}{x^3}\,dx = \left[\frac{12x^{-2}}{-2}\right]_1^3 = \left[\frac{-6}{x^2}\right]_1^3 = \frac{-6}{9} + 6 = \frac{16}{3}$, (d)

43. $\displaystyle\int_0^4 (2 + x)\,dx = \left[2x + \frac{x^2}{2}\right]_0^4 = 8 + \frac{16}{2} = 16$

44. $\displaystyle\int_{-1}^1 (t^2 + 2)\,dt = \left[\frac{t^3}{3} + 2t\right]_{-1}^1 = \frac{14}{3}$

45. $\displaystyle\int_{-1}^1 (4t^3 - 2t)\,dt = \left[t^4 - t^2\right]_{-1}^1 = 0$

46. $$\int_{-2}^2 (x^4 + 2x^2 - 5)\,dx = \left[\frac{x^5}{5} + \frac{2x^3}{3} - 5x\right]_{-2}^2$$
$$= \left(\frac{32}{5} + \frac{16}{3} - 10\right) - \left(-\frac{32}{5} - \frac{16}{3} + 10\right)$$
$$= \frac{52}{15}$$

47. $\displaystyle\int_4^9 x\sqrt{x}\,dx = \int_4^9 x^{3/2}\,dx = \left[\frac{2}{5}x^{5/2}\right]_4^9 = \frac{2}{5}[(\sqrt{9})^5 - (\sqrt{4})^5] = \frac{2}{5}(243 - 32) = \frac{422}{5}$

48. $\displaystyle\int_1^2\left(\frac{1}{x^2}-\frac{1}{x^3}\right)dx = \int_1^2 (x^{-2}-x^{-3})\,dx = \left[-\frac{1}{x}+\frac{1}{2x^2}\right]_1^2 = \left(-\frac{1}{2}+\frac{1}{8}\right)-\left(-1+\frac{1}{2}\right)=\frac{1}{8}$

49. $\displaystyle\int_1^3 (2x-1)\,dx = \left[x^2-x\right]_1^3 = 6$

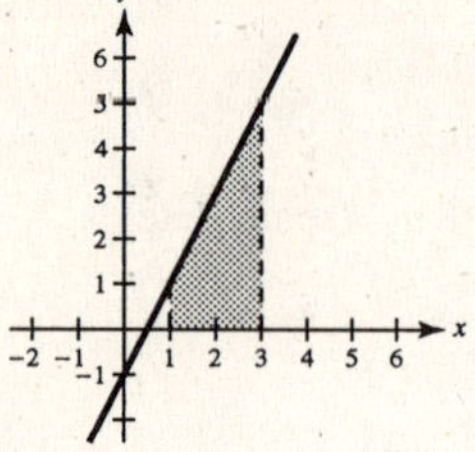

50. $\displaystyle\int_0^2 (x+4)\,dx = \left[\frac{x^2}{2}+4x\right]_0^2 = 10$

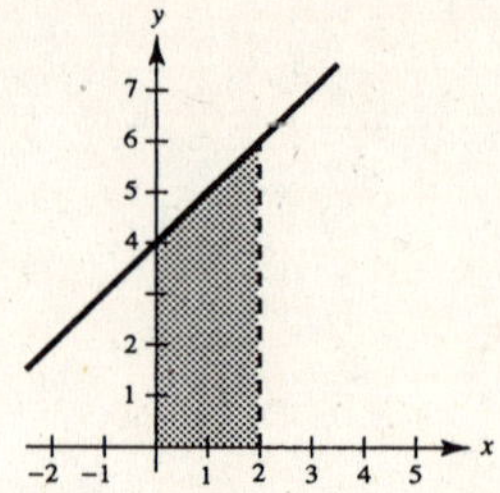

51. $\displaystyle\int_3^4 (x^2-9)\,dx = \left[\frac{x^3}{3}-9x\right]_3^4$

$\displaystyle = \left(\frac{64}{3}-36\right)-(9-27)$

$\displaystyle = \frac{64}{3}-\frac{54}{3}=\frac{10}{3}$

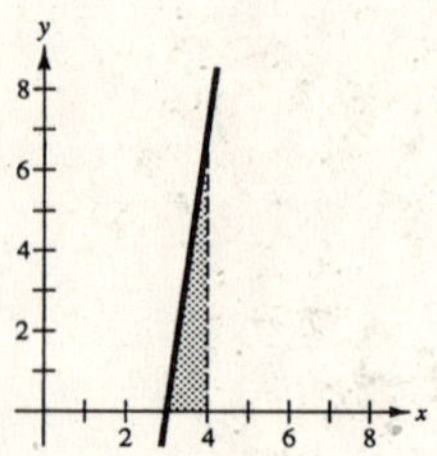

52. $\displaystyle\int_{-1}^2 (-x^2+x+2)\,dx = \left[-\frac{x^3}{3}+\frac{x^2}{2}+2x\right]_{-1}^2$

$\displaystyle = \left(-\frac{8}{3}+2+4\right)-\left(\frac{1}{3}+\frac{1}{2}-2\right)$

$\displaystyle = \frac{10}{3}+\frac{7}{6}=\frac{9}{2}$

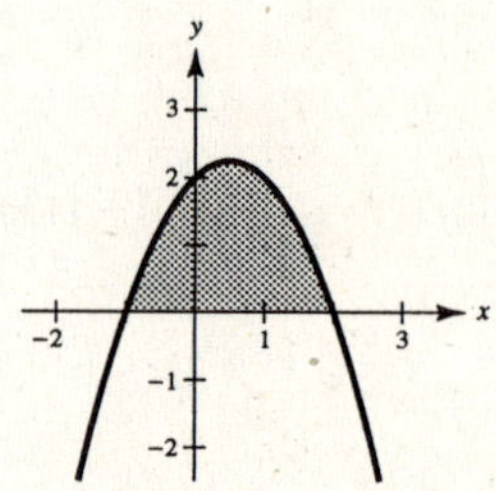

53. $\displaystyle\int_0^1 (x-x^3)\,dx = \left[\frac{x^2}{2}-\frac{x^4}{4}\right]_0^1 = \frac{1}{2}-\frac{1}{4}=\frac{1}{4}$

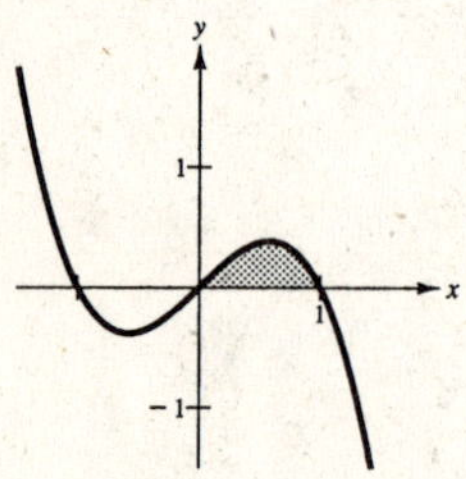

54. $\displaystyle\int_0^1 \sqrt{x}(1-x)\,dx = \int_0^1 (x^{1/2}-x^{3/2})\,dx$

$\displaystyle = \left[\frac{2}{3}x^{3/2}-\frac{2}{5}x^{5/2}\right]_0^1$

$\displaystyle = \frac{2}{3}-\frac{2}{5}=\frac{4}{15}$

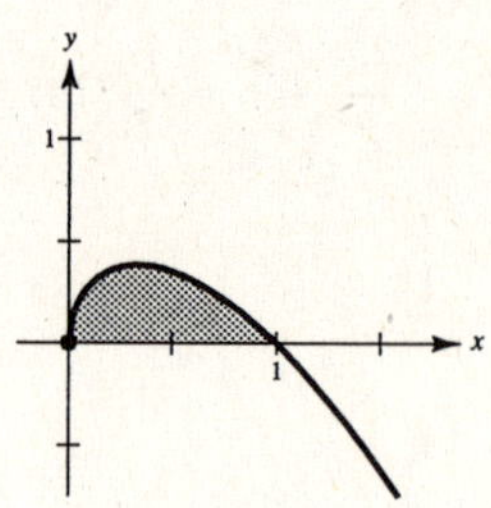

55. Area $\displaystyle = \int_1^9 \frac{4}{\sqrt{x}}\,dx = \left[\frac{4x^{1/2}}{(1/2)}\right]_1^9 = 8(3-1)=16$

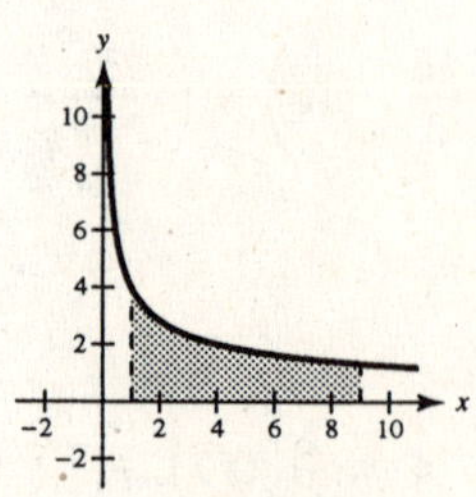

56. Area $= \int_0^1 (x - x^5)\,dx = \left[\frac{x^2}{2} - \frac{x^6}{6}\right]_0^1 = \frac{1}{3}$

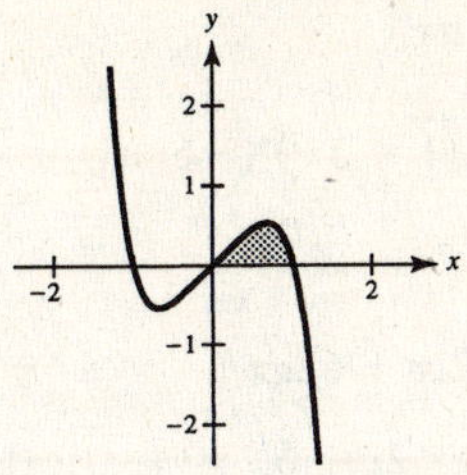

57. $\frac{1}{9-4}\int_4^9 \frac{1}{\sqrt{x}}\,dx = \left[\frac{1}{5}2\sqrt{x}\right]_4^9 = \frac{2}{5}(3-2) = \frac{2}{5}$ Average value

$\frac{2}{5} = \frac{1}{\sqrt{x}}$

$\sqrt{x} = \frac{5}{2}$

$x = \frac{25}{4}$

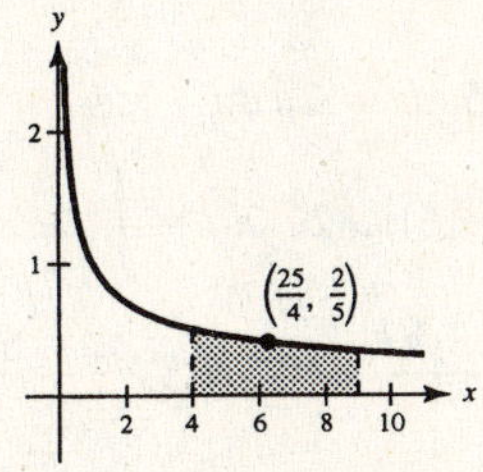

58. $\frac{1}{2-0}\int_0^2 x^3\,dx = \left[\frac{x^4}{8}\right]_0^2 = 2$

$x^3 = 2$

$x = \sqrt[3]{2}$

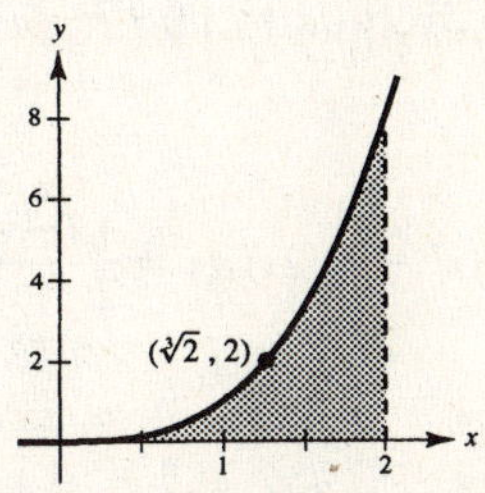

59. $F'(x) = x^2\sqrt{1 + x^3}$

60. $F'(x) = x^2 + 3x + 2$

61. $\int (x^2 + 1)^3\,dx = \int (x^6 + 3x^4 + 3x^2 + 1)\,dx = \frac{x^7}{7} + \frac{3}{5}x^5 + x^3 + x + C$

62. $\int \left(x + \frac{1}{x}\right)^2 dx = \int (x^2 + 2 + x^{-2})\,dx$

$= \frac{x^3}{3} + 2x - \frac{1}{x} + C$

63. $\int x(x^2 + 1)^3\,dx = \frac{1}{2}\int (x^2 + 1)^3 (2x)\,dx = \frac{(x^2 + 1)^4}{8} + C$

64. $u = x^3 + 3,\ du = 3x^2\,dx$

$\int x^2\sqrt{x^3 + 3}\,dx = \frac{1}{3}\int (x^3 + 3)^{1/2}\,3x^2\,dx = \frac{2}{9}(x^3 + 3)^{3/2} + C$

65. $u = x^3 + 3,\ du = 3x^2\,dx$

$\int \frac{x^2}{\sqrt{x^3 + 3}}\,dx = \int (x^3 + 3)^{-1/2}\,x^2\,dx = \frac{1}{3}\int (x^3 + 3)^{-1/2}\,3x^2\,dx = \frac{2}{3}(x^3 + 3)^{1/2} + C$

66. $u = 25 - 9x^2,\ du = -18x\,dx$

$\int \frac{x}{\sqrt{25 - 9x^2}}\,dx = -\frac{1}{18}\int \frac{-18x}{\sqrt{25 - 9x^2}}\,dx = -\frac{1}{18}2\sqrt{25 - 9x^2} + C = -\frac{1}{9}\sqrt{25 - 9x^2} + C$

67. $u = 1 - 3x^2, du = -6x\,dx$

$$\int x(1 - 3x^2)^4\,dx = -\frac{1}{6}\int (1 - 3x^2)^4(-6x\,dx) = -\frac{1}{30}(1 - 3x^2)^5 + C = \frac{1}{30}(3x^2 - 1)^5 + C$$

68. $u = x^2 + 6x - 5, du = (2x + 6)\,dx$

$$\int \frac{x + 3}{(x^2 + 6x - 5)^2} = \frac{1}{2}\int \frac{2x + 6}{(x^2 + 6x - 5)^2}\,dx = \frac{-1}{2}(x^2 + 6x - 5)^{-1} + C = \frac{-1}{2(x^2 + 6x - 5)} + C$$

69. $u = \sqrt{x + 5}, x = u^2 - 5, dx = 2u\,du$

$$\int x^2\sqrt{x + 5}\,dx = \int (u^2 - 5)^2\,u(2u)\,du = \int (2u^6 - 20u^4 + 50u^2)\,du$$

$$= \frac{2(x + 5)^{7/2}}{7} - 4(x + 5)^{5/2} + \frac{50}{3}(x + 5)^{3/2} + C$$

70. $u = x + 5, x = u - 5, dx = du$

$$\int x\sqrt{x + 5}\,dx = \int (u - 5)\sqrt{u}\,du = \int (u^{3/2} - 5u^{1/2})\,du = \frac{2}{5}(x + 5)^{5/2} - \frac{10}{3}(x + 5)^{3/2} + C$$

71. $\displaystyle\int_{-1}^{2} x(x^2 - 4)\,dx = \frac{1}{2}\int_{-1}^{2} (x^2 - 4)(2x)\,dx = \frac{1}{2}\frac{(x^2 - 4)^2}{2}\Bigg]_{-1}^{2} = \frac{1}{4}[0 - 9] = -\frac{9}{4}$

72. $\displaystyle\int_0^1 x^2(x^3 + 1)^3\,dx = \frac{1}{3}\int_0^1 (x^3 + 1)^3(3x^2)\,dx = \frac{1}{12}\Big[(x^3 + 1)^4\Big]_0^1 = \frac{1}{12}(16 - 1) = \frac{5}{4}$

73. $\displaystyle\int_0^3 \frac{1}{\sqrt{1 + x}}\,dx = \int_0^3 (1 + x)^{-1/2}\,dx = \Big[2(1 + x)^{1/2}\Big]_0^3 = 4 - 2 = 2$

74. $\displaystyle\int_3^6 \frac{x}{3\sqrt{x^2 - 8}}\,dx = \frac{1}{6}\int_3^6 (x^2 - 8)^{-1/2}(2x)\,dx = \left[\frac{1}{3}(x^2 - 8)^{1/2}\right]_3^6 = \frac{1}{3}(2\sqrt{7} - 1)$

75. $u = 1 - y, y = 1 - u, dy = -du$

When $y = 0, u = 1$. When $y = 1, u = 0$.

$$2\pi\int_0^1 (y + 1)\sqrt{1 - y}\,dy = 2\pi\int_1^0 -[(1 - u) + 1]\sqrt{u}\,du$$

$$= 2\pi\int_1^0 (u^{3/2} - 2u^{1/2})\,du = 2\pi\left[\frac{2}{5}u^{5/2} - \frac{4}{3}u^{3/2}\right]_1^0 = \frac{28\pi}{15}$$

76. $u = x + 1, x = u - 1, dx = du$

When $x = -1, u = 0$. When $x = 0, u = 1$.

$$2\pi\int_{-1}^0 x^2\sqrt{x + 1}\,dx = 2\pi\int_0^1 (u - 1)^2\sqrt{u}\,du$$

$$= 2\pi\int_0^1 (u^{5/2} - 2u^{3/2} + u^{1/2})\,du = 2\pi\left[\frac{2}{7}u^{7/2} - \frac{4}{5}u^{5/2} + \frac{2}{3}u^{3/2}\right]_0^1 = \frac{32\pi}{105}$$

77. $C = \frac{15,000}{M}\int_t^{t+1} p\,ds = \frac{15,000}{M}\int_t^{t+1}(1.20 + 0.04s)\,ds$

$= \frac{15,000}{M}\left[1.20s + 0.02s^2\right]_t^{t+1}$

(a) 2005 corresponds to $t = 15$.

$$C = \frac{15,000}{M}\left[1.20s + 0.02s^2\right]_{15}^{16} = \frac{27,300}{M} \text{ dollars}$$

(b) 2007 corresponds to $t = 17$.

$$C = \frac{15,000}{M}\left[1.20s + 0.02s^2\right]_{17}^{18} = \frac{28,500}{M} \text{ dollars}$$

78. Trapezoidal Rule ($n = 4$): $\int_1^2 \frac{1}{1+x^3}\,dx \approx \frac{1}{8}\left[\frac{1}{1+1^3} + \frac{2}{1+(1.25)^3} + \frac{2}{1+(1.5)^3} + \frac{2}{1+(1.75)^3} + \frac{1}{1+2^3}\right] \approx 0.257$

Simpson's Rule ($n = 4$): $\int_1^2 \frac{1}{1+x^3}\,dx \approx \frac{1}{12}\left[\frac{1}{1+1^3} + \frac{4}{1+(1.25)^3} + \frac{2}{1+(1.5)^3} + \frac{4}{1+(1.75)^3} + \frac{1}{1+2^3}\right] \approx 0.254$

Graphing utility: 0.254

79. Trapezoidal Rule ($n = 4$): $\int_0^1 \frac{x^{3/2}}{3-x^2}\,dx \approx \frac{1}{8}\left[0 + \frac{2(1/4)^{3/2}}{3-(1/4)^2} + \frac{2(1/2)^{3/2}}{3-(1/2)^2} + \frac{2(3/4)^{3/2}}{3-(3/4)^2} + \frac{1}{2}\right] \approx 0.172$

Simpson's Rule ($n = 4$): $\int_0^1 \frac{x^{3/2}}{3-x^2}\,dx \approx \frac{1}{12}\left[0 + \frac{4(1/4)^{3/2}}{3-(1/4)^2} + \frac{2(1/2)^{3/2}}{3-(1/2)^2} + \frac{4(3/4)^{3/2}}{3-(3/4)^2} + \frac{1}{2}\right] \approx 0.166$

Graphing utility: 0.166

80. (a) $R < I < T < L$

(b) $S(4) = \frac{4-0}{3(4)}[f(0) + 4f(1) + 2f(2) + 4f(3) + f(4)]$

$\approx \frac{1}{3}\left[4 + 4(2) + 2(1) + 4\left(\frac{1}{2}\right) + \frac{1}{4}\right] \approx 5.417$

Problem Solving for Chapter 6

1. (a) $L(1) = \int_1^1 \frac{1}{t}\,dt = 0$

(b) $L'(x) = \frac{1}{x}$ by the Second Fundamental Theorem of Calculus.

$L'(1) = 1$

(c) $L(x) = 1 = \int_1^x \frac{1}{t}\,dt$ for $x \approx 2.718$

$\int_1^{2.718} \frac{1}{t}\,dt = 0.999896$

(Note: The exact value of x is e, the base of the natural logarithm function.)

(d) We first show that $\int_1^{x_1} \frac{1}{t}\,dt = \int_{1/x_1}^1 \frac{1}{t}\,dt.$

To see this, let $u = \frac{t}{x_1}$ and $du = \frac{1}{x_1}\,dt.$

Then $\int_1^{x_1} \frac{1}{t}\,dt = \int_{1/x_1}^1 \frac{1}{ux_1}(x_1\,du) = \int_{1/x_1}^1 \frac{1}{u}\,du = \int_{1/x_1}^1 \frac{1}{t}\,dt.$

Now, $L(x_1x_2) = \int_1^{x_1x_2} \frac{1}{t}\,dt = \int_{1/x_1}^{x_2} \frac{1}{u}\,du \left(\text{using } u = \frac{t}{x_1}\right)$

$= \int_{1/x_1}^1 \frac{1}{u}\,du + \int_1^{x_2} \frac{1}{u}\,du$

$= \int_1^{x_1} \frac{1}{u}\,du + \int_1^{x_2} \frac{1}{u}\,du$

$= L(x_1) + L(x_2).$

2. (a) $F(x) = \int_2^x \sqrt{1 + t^3}\, dt$

x	0	1.0	1.5	1.9	2.0	2.1	2.5	3.0	4.0	5.0
$F(x)$	-3.241	-2.130	-1.265	-0.290	0	0.310	1.763	4.100	10.742	20.355

(b) $G(x) = \dfrac{1}{x - 2}\int_2^x \sqrt{1 + t^3}\, dt$

x	1.9	1.95	1.99	2.01	2.1
$G(x)$	2.9011	2.9503	2.9900	3.0100	3.1011

$\lim_{x\to 2} G(x) \approx 3$

(c) $F'(2) = \lim_{x\to 2} \dfrac{F(x) - F(2)}{x - 2} = \lim_{x\to 2} \dfrac{1}{x - 2}\int_2^x \sqrt{1 + t^3}\, dt = \lim_{x\to 2} G(x)$

Since $F'(x) = \sqrt{1 + x^3}$, $F'(2) = \sqrt{1 + 8} = 3 = \lim_{x\to 2} G(x)$.

3. (a) Slope $= \dfrac{9 - 0}{3 - 0} = 3$

(b) $f'(x) = 2x$

$$\text{Average value} = \frac{1}{3 - 0}\int_0^3 f'(x)\, dx = \frac{1}{3}\int_0^3 2x\, dx$$

$$= \frac{1}{3}\Big[x^2\Big]_0^3 = 3$$

(c) $\text{Average value} = \dfrac{1}{b - a}\int_a^b f'(x)\, dx = \dfrac{f(b) - f(a)}{b - a}$

4. Let d be the distance traversed and a be the uniform acceleration. We can assume that $v(0) = 0$ and $s(0) = 0$. Then

$$a(t) = a$$

$$v(t) = at$$

$$s(t) = \frac{1}{2}at^2.$$

$s(t) = d$ when $t = \sqrt{\dfrac{2d}{a}}$.

The highest speed is $v = a\sqrt{\dfrac{2d}{a}} = \sqrt{2ad}$.

The lowest speed is $v = 0$.

The mean speed is $\dfrac{1}{2}\left(\sqrt{2ad} + 0\right) = \sqrt{\dfrac{ad}{2}}$.

The time necessary to traverse the distance d at the mean speed is

$$t = \frac{d}{\sqrt{ad/2}} = \sqrt{\frac{2d}{a}}$$

which is the same as the time calculated above.

5. (a)

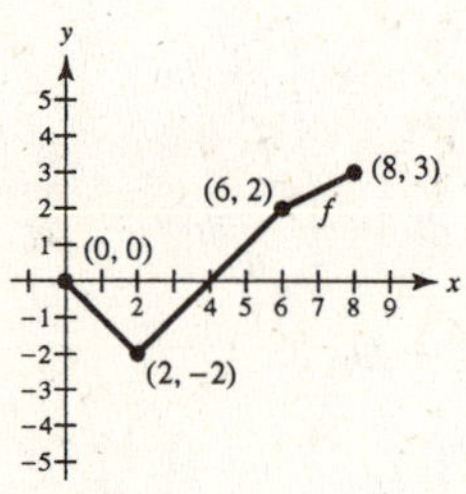

(b)

x	0	1	2	3	4	5	6	7	8
$F(x)$	0	$-\frac{1}{2}$	-2	$-\frac{7}{2}$	-4	$-\frac{7}{2}$	-2	$\frac{1}{4}$	3

—CONTINUED—

5. —CONTINUED—

(c) $f(x) = \begin{cases} -x, & 0 \le x < 2 \\ x - 4, & 2 \le x < 6 \\ \frac{1}{2}x - 1, & 6 \le x \le 8 \end{cases}$

$$F(x) = \int_0^x f(t)\,dt = \begin{cases} (-x^2/2), & 0 \le x < 2 \\ (x^2/2) - 4x + 4, & 2 \le x < 6 \\ (1/4)x^2 - x - 5, & 6 \le x \le 8 \end{cases}$$

$F'(x) = f(x)$. F is decreasing on $(0, 4)$ and increasing on $(4, 8)$. Therefore, the minimum is -4 at $x = 4$, and the maximum is 3 at $x = 8$.

(d) $F''(x) = f'(x) = \begin{cases} -1, & 0 < x < 2 \\ 1, & 2 < x < 6 \\ \frac{1}{2}, & 6 < x < 8 \end{cases}$

$x = 2$ is a point of inflection, whereas $x = 6$ is not.

6. (a)

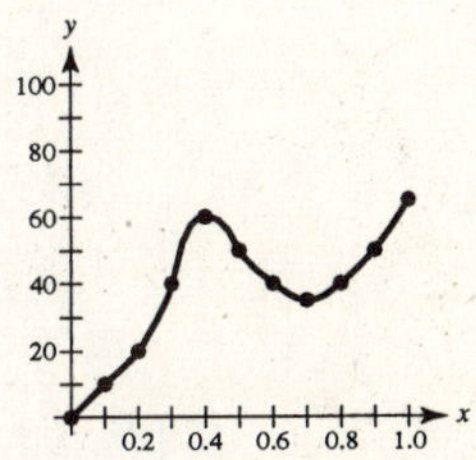

(b) v is increasing (positive acceleration) on $(0, 0.4)$ and $(0.7, 1.0)$.

(c) Average acceleration $= \dfrac{v(0.4) - v(0)}{0.4 - 0} = \dfrac{60 - 0}{0.4} = 150 \text{ mi/hr}^2$

(d) This integral is the total distance traveled in miles.

$$\int_0^1 v(t)\,dt \approx \frac{1}{10}[0 + 2(20) + 2(60) + 2(40) + 2(40) + 65] = \frac{385}{10} = 38.5 \text{ miles}$$

(e) One approximation is

$$a(0.8) \approx \frac{v(0.9) - v(0.8)}{0.9 - 0.8} = \frac{50 - 40}{0.1} = 100 \text{ mi/hr}^2.$$

(other answers possible)

7. (a) $\displaystyle\int_{-1}^{1} \sqrt{x + 2}\,dx \approx f\left(-\frac{1}{\sqrt{3}}\right) + f\left(\frac{1}{\sqrt{3}}\right)$

$\approx 1.1927 + 1.6054 = 2.7981$

Exact: 2.7974

Error: ≈ 0.0007

(c) Let $p(x)\,dx = ax^3 + bx^2 + cx + d$.

$$\int_{-1}^{1} p(x)\,dx = \left[\frac{ax^4}{4} + \frac{bx^3}{3} + \frac{cx^2}{2} + dx\right]_{-1}^{1} = \frac{2b}{3} + 2d$$

$$p\left(-\frac{1}{\sqrt{3}}\right) + p\left(\frac{1}{\sqrt{3}}\right) = \left(\frac{b}{3} + d\right) + \left(\frac{b}{3} + d\right) = \frac{2b}{3} + 2d$$

(b) $\displaystyle\int_{-1}^{1} \frac{1}{1 + x^2}\,dx \approx \frac{1}{1 + \left(\frac{1}{3}\right)} + \frac{1}{1 + \frac{1}{3}} = \frac{3}{2}$

$\left(\textbf{Note: } \text{Exact answer is } \dfrac{\pi}{2} \approx 1.5708.\right)$

8. $\displaystyle\int_0^x f(t)(x - t)\,dt = \int_0^x xf(t)\,dt - \int_0^x tf(t)\,dt = x\int_0^x f(t)\,dt - \int_0^x tf(t)\,dt$

Thus, $\displaystyle\frac{d}{dx}\int_0^x f(t)(x - t)\,dt = xf(x) + \int_0^x f(t)\,dt - xf(x) = \int_0^x f(t)\,dt.$

Differentiating the other integral,

$$\frac{d}{dx}\int_0^x \left(\int_0^t f(v)\,dv\right)dt = \int_0^x f(v)\,dv.$$

Thus, the two original integrals have equal derivatives,

$$\int_0^x f(t)(x - t)\,dt = \int_0^t \left(\int_0^t f(v)\,dv\right)dt + C.$$

Letting $x = 0$, we see that $C = 0$.

9. Consider $F(x) = [f(x)]^2 \Rightarrow F'(x) = 2f(x)f'(x)$. Thus,

$$\int_a^b f(x)f'(x)\,dx = \int_a^b \frac{1}{2}F'(x)\,dx$$

$$= \left[\frac{1}{2}F(x)\right]_a^b$$

$$= \frac{1}{2}[F(b) - F(a)]$$

$$= \frac{1}{2}[f(b)^2 - f(a)^2].$$

10. Consider $\int_0^1 \sqrt{x}\,dx = \frac{2}{3}x^{3/2}\Big]_0^1 = \frac{2}{3}$. The corresponding Riemann Sum using right-hand endpoints is

$$S(n) = \frac{1}{n}\left[\sqrt{\frac{1}{n}} + \sqrt{\frac{2}{n}} + \cdots + \sqrt{\frac{n}{n}}\right]$$

$$= \frac{1}{n^{3/2}}[\sqrt{1} + \sqrt{2} + \cdots + \sqrt{n}].$$

Thus, $\lim_{n\to\infty} \frac{\sqrt{1} + \sqrt{2} + \cdots + \sqrt{n}}{n^{3/2}} = \frac{2}{3}$.

11. Consider $\int_0^1 x^5\,dx = \frac{x^6}{6}\Big]_0^1 = \frac{1}{6}$.

The corresponding Riemann Sum using right endpoints is

$$S(n) = \frac{1}{n}\left[\left(\frac{1}{n}\right)^5 + \left(\frac{2}{n}\right)^5 + \cdots + \left(\frac{n}{n}\right)^5\right]$$

$$= \frac{1}{n^6}[1^5 + 2^5 + \cdots + n^5].$$

Thus, $\lim_{n\to\infty} S(n) = \lim_{n\to\infty} \frac{1^5 + 2^5 + \cdots + n^5}{n^6} = \frac{1}{6}$.

12. (a) Area $= \int_{-3}^{3} (9 - x^2)\,dx = 2\int_0^3 (9 - x^2)\,dx$

$$= 2\left[9x - \frac{x^3}{3}\right]_0^3$$

$$= 2[27 - 9] = 36$$

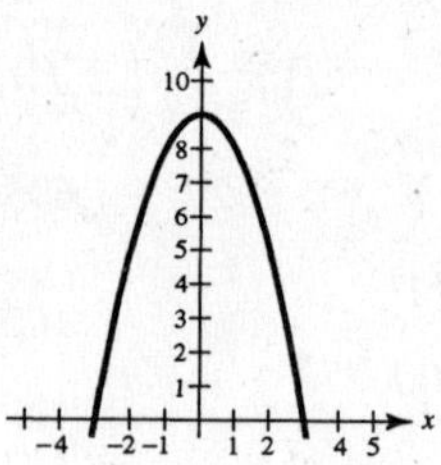

(b) Base = 6, height = 9. Area $= \frac{2}{3}bh = \frac{2}{3}(6)(9) = 36$.

(c) Let the parabola be given by $y = b^2 - a^2x^2,\ a, b > 0$.

Area $= 2\int_0^{b/a} (b^2 - a^2x^2)\,dx$

$$= 2\left[b^2x - a^2\frac{x^3}{3}\right]_0^{b/a}$$

$$= 2\left[b^2\left(\frac{b}{a}\right) - \frac{a^2}{3}\left(\frac{b}{a}\right)^3\right]$$

$$= 2\left[\frac{b^3}{a} - \frac{1}{3}\frac{b^3}{a}\right] = \frac{4}{3}\frac{b^3}{a}$$

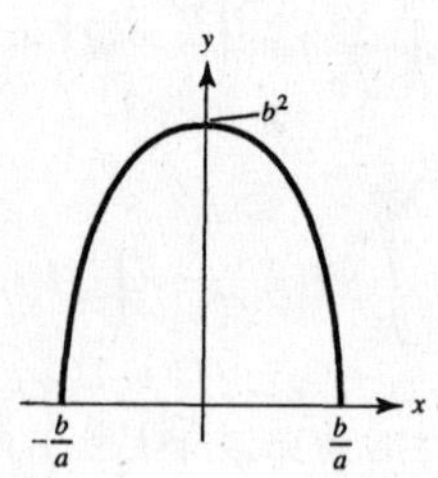

Base $= \frac{2b}{a}$, height $= b^2$

Archimedes' Formula: Area $= \frac{2}{3}\left(\frac{2b}{a}\right)(b^2) = \frac{4}{3}\frac{b^3}{a}$

13. By Theorem 6.8, $0 < f(x) \le M \Rightarrow \int_a^b f(x)\,dx \le \int_a^b M\,dx = M(b-a)$.

Similarly, $m \le f(x) \Rightarrow m(b-a) = \int_a^b m\,dx \le \int_a^b f(x)\,dx$. Thus, $m(b-a) \le \int_a^b f(x)\,dx \le M(b-a)$.

On the interval $[0, 1]$, $1 \le \sqrt{1+x^4} \le \sqrt{2}$ and $b - a = 1$. Thus, $1 \le \int_0^1 \sqrt{1+x^4}\,dx \le \sqrt{2}$.

$\left(\textbf{Note: } \int_0^1 \sqrt{1+x^4}\,dx \approx 1.0894\right)$

14. (a) $A = S^2 = (1 + 2 + \cdots + n)^2$

$$= \left[\frac{n(n+1)}{2}\right]^2$$

by the formula $1 + 2 + \cdots + n = \dfrac{n(n+1)}{2}$

(b) $A_k = k[1 + 2 + \cdots + k] + k[1 + 2 + \cdots + (k-1)]$

$$= k\frac{k(k+1)}{2} + k\frac{k(k-1)}{2}$$

$$= \frac{k^2}{2}[2k] = k^3$$

(c) $A = A_1 + A_2 + \cdots + A_n = 1^3 + 2^3 + \cdots + n^3$ by part (b)

$A = \left[\dfrac{n(n+1)}{2}\right]^2$ by part (a).

Hence, $1^3 + 2^3 + \cdots + n^3 = \left[\dfrac{n(n+1)}{2}\right]^2$.

CHAPTER 7
Exponential and Logarithmic Functions

Section 7.1 Exponential Functions and Their Graphs 547

Section 7.2 Logarithmic Functions and Their Graphs 557

Section 7.3 Using Properties of Logarithms 566

Section 7.4 Exponential and Logarithmic Equations 573

Section 7.5 Exponential and Logarithmic Models 588

Review Exercises 598

Problem Solving 610

CHAPTER 7
Exponential and Logarithmic Functions

Section 7.1 Exponential Functions and Their Graphs

1. $(3.4)^{5.6} \approx 946.852$

2. $5000(2^{-1.5}) \approx 1767.767$

3. $(1.005)^{400} \approx 7.352$

4. $8^{2\pi} \approx 472{,}369.379$

5. $5^{-\pi} \approx 0.006$

6. $\sqrt[3]{4395} \approx 16.380$

7. $100^{\sqrt{2}} \approx 673.639$

8. $500\left(\frac{3}{4}\right)^{10} \approx 28.157$

9. $f(x) = 3^x$

$g(x) = 3^{x-4}$

Because $g(x) = f(x - 4)$, the graph of g can be obtained by shifting the graph of f four units to the right.

10. $f(x) = 4^x, g(x) = 4^x + 1$

Because $g(x) = f(x) + 1$, the graph of g can be obtained by shifting the graph of f one unit upward.

11. $f(x) = -2^x$

$g(x) = 5 - 2^x$

Because $g(x) = 5 + f(x)$, the graph of g can be obtained by shifting the graph of f five units upward.

12. $f(x) = 10^x, g(x) = 10^{-x+3}$

Because $g(x) = f(-x + 3)$, the graph of g can be obtained by reflecting the graph of f in the y-axis and shifting the graph three units to the right. (Note: This is equivalent to shifting the graph three units to the left and then reflecting the graph in the y-axis.)

13. $f(x) = \left(\frac{3}{5}\right)^x$

$g(x) = -\left(\frac{3}{5}\right)^{x+4}$

Because $g(x) = -f(x + 4)$, the graph of g can be obtained by reflecting the graph of f in the x-axis and shifting f four units to the left.

14. $f(x) = \left(\frac{7}{2}\right)^x, g(x) = -\left(\frac{7}{2}\right)^{-x+6}$

Because $g(x) = -f(-x + 6)$, the graph of g can be obtained by reflecting the graph of f in the x-axis and y-axis and shifting the graph six units to the right. (Note: This is equivalent to shifting the graph six units to the left and then reflecting the graph in the x-axis and y-axis.)

15. $f(x) = 0.3^x$

$g(x) = -0.3^x + 5$

Because $g(x) = -f(x) + 5$, the graph of g can be obtained by reflecting the graph of f in the x-axis and shifting the graph five units upward.

16. $f(x) = 3.6^x, g(x) = -3.6^{-x} + 8$

Because $g(x) = -f(-x) + 8$, the graph of g can be obtained by reflecting the graph of f in the x-axis and y-axis and shifting the graph eight units upward.

17. $f(x) = 2^x$

Increasing

Asymptote: $y = 0$

Intercept: $(0, 1)$

Matches graph (d)

18. $f(x) = 2^x + 1$ rises to the right.

Asymptote: $y = 1$

Intercept: $(0, 2)$

Matches graph (c)

19. $f(x) = 2^{-x}$

Decreasing

Asymptote: $y = 0$

Intercept: $(0, 1)$

Matches graph (a)

20. $f(x) = 2^{x-2}$ rises to the right.

Asymptote: $y = 0$

Intercept: $\left(0, \frac{1}{4}\right)$

Matches graph (b)

21. $f(x) = \left(\frac{1}{2}\right)^x$

x	-2	-1	0	1	2
$f(x)$	4	2	1	0.5	0.25

Asymptote: $y = 0$

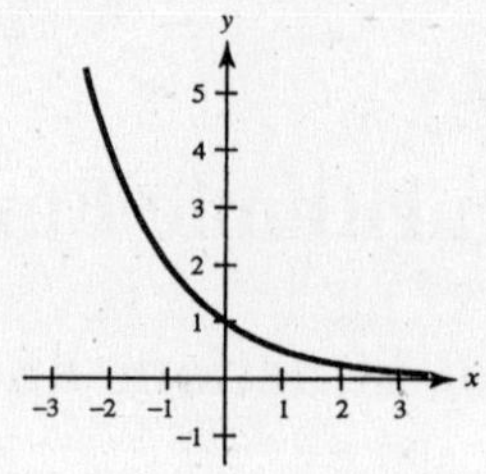

22. $f(x) = 6^x$

x	-2	-1	0	1	2
$f(x)$	0.028	0.167	1	6	36

Asymptote: $y = 0$

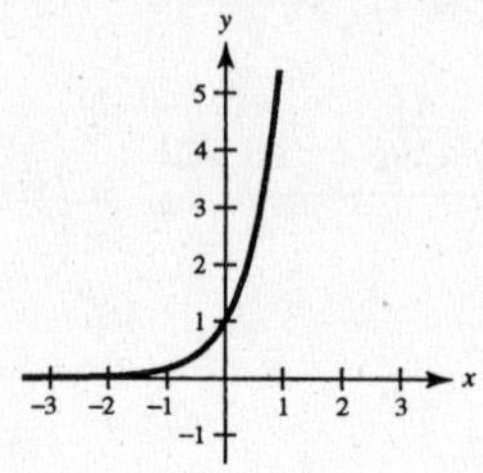

23. $f(x) = \left(\frac{1}{2}\right)^{-x} = 2^x$

x	-2	-1	0	1	2
$f(x)$	0.25	0.5	1	2	4

Asymptote: $y = 0$

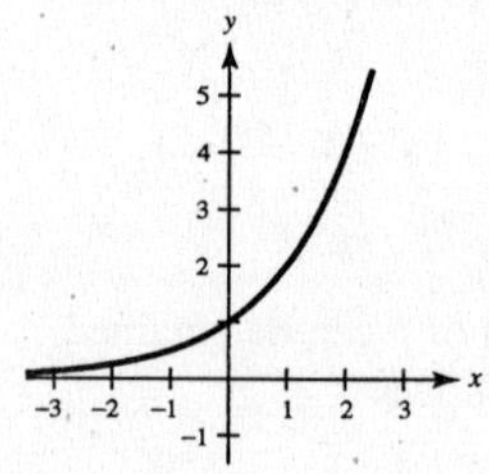

24. $f(x) = 6^{-x}$

x	-2	-1	0	1	2
$f(x)$	36	6	1	0.167	0.028

Asymptote: $y = 0$

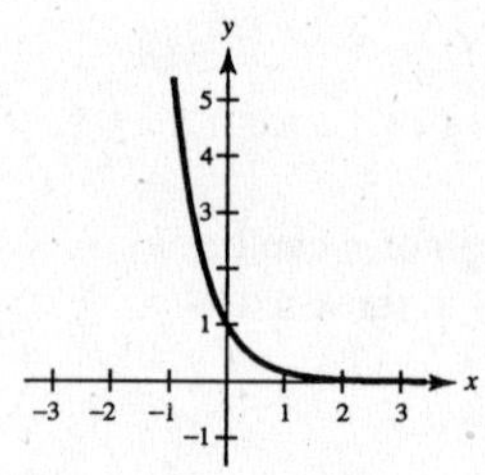

25. $f(x) = 2^{x-1}$

x	-2	-1	0	1	2
$f(x)$	0.125	0.25	0.5	1	2

Asymptote: $y = 0$

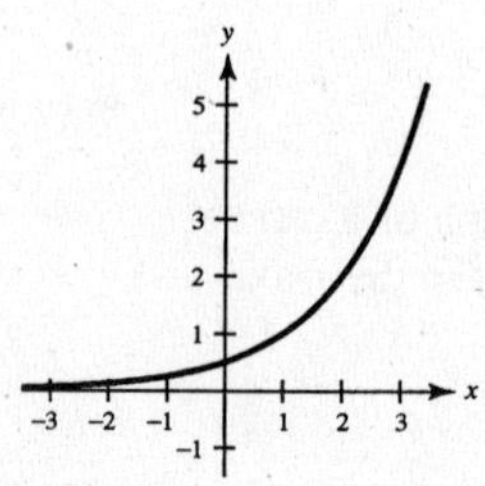

26. $f(x) = 3^{x+2}$

x	-4	-3	-2	-1	0
$f(x)$	0.111	0.333	1	3	9

Asymptote: $y = 0$

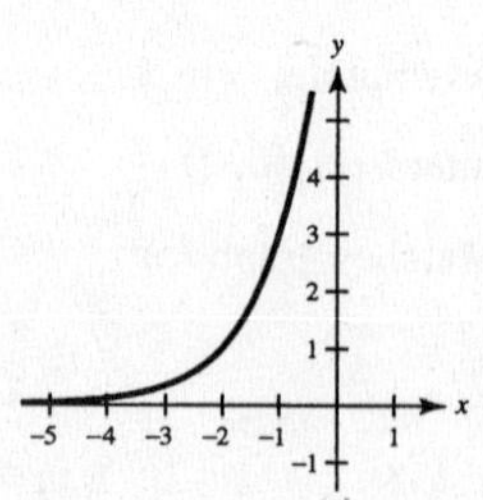

27. $f(x) = 4^{x-3} + 3$

x	-1	0	1	2	3
$f(x)$	3.004	3.016	3.063	3.25	4

Asymptote: $y = 3$

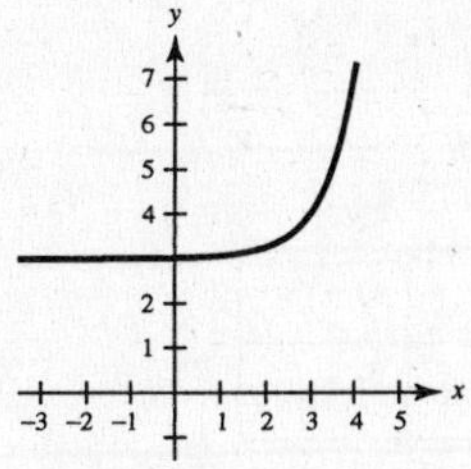

28. $f(x) = -4^{x-3} - 3$

x	-1	0	1	3	4
$f(x)$	-3.004	-3.016	-3.063	-4	-7

Asymptote: $y = -3$

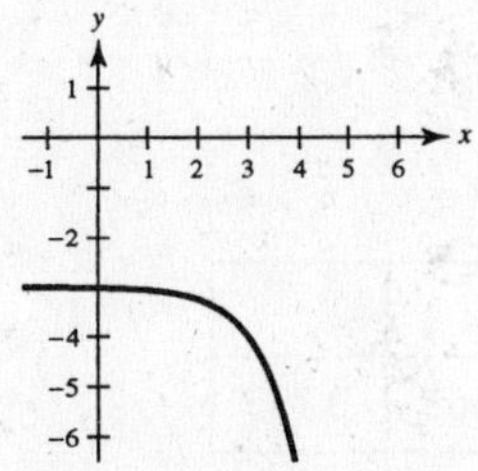

29. $g(x) = 5^x$

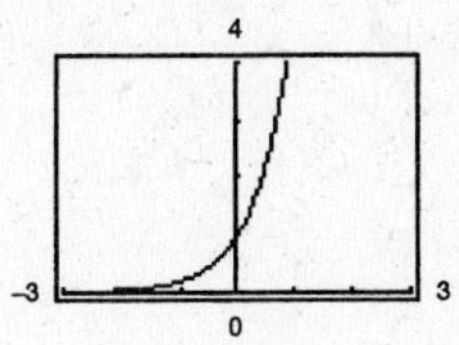

30. $f(x) = \left(\frac{3}{2}\right)^x$

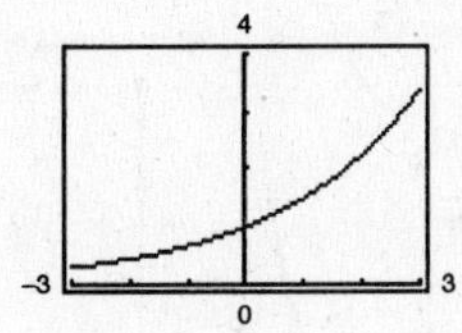

31. $f(x) = \left(\frac{1}{5}\right)^x = 5^{-x}$

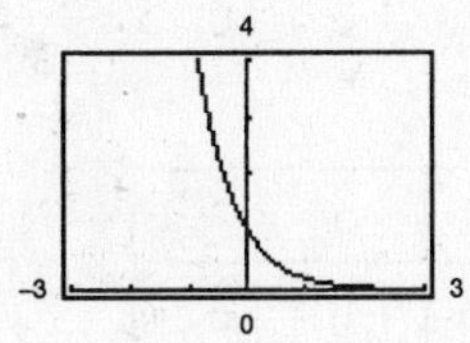

32. $h(x) = \left(\frac{3}{2}\right)^{-x}$

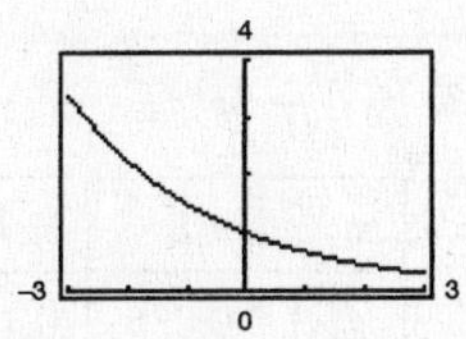

33. $h(x) = 5^{x-2}$

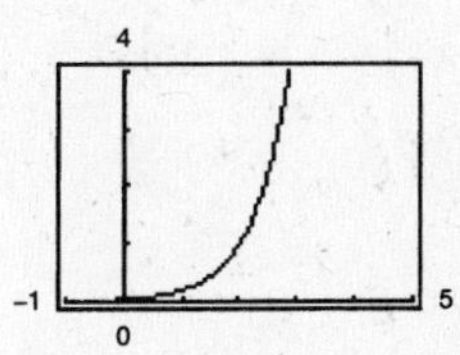

34. $g(x) = \left(\frac{3}{2}\right)^{x+2}$

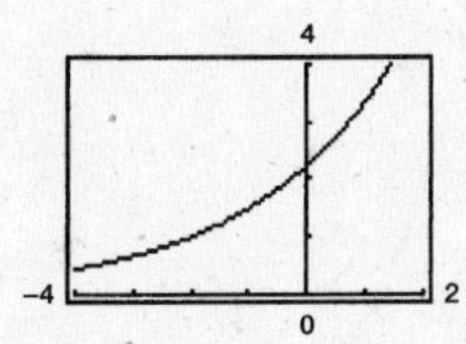

35. $g(x) = 5^{-x} - 3$

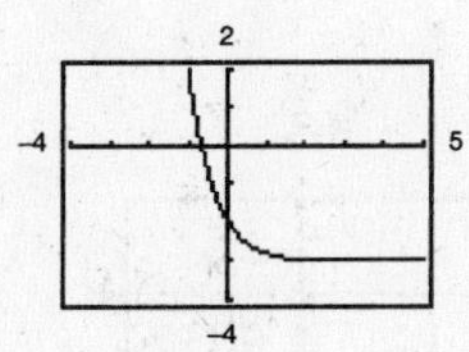

36. $f(x) = \left(\frac{3}{2}\right)^{-x} + 2$

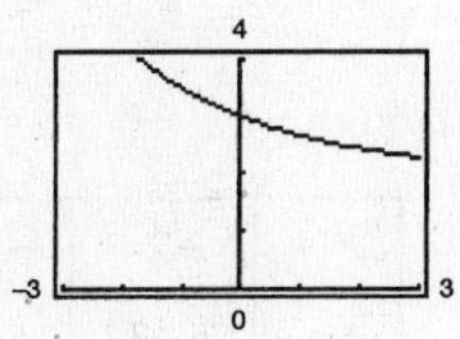

37. $y = 2^{-x^2}$

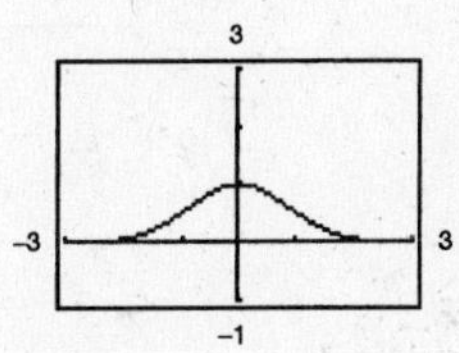

38. $y = 3^{-|x|}$

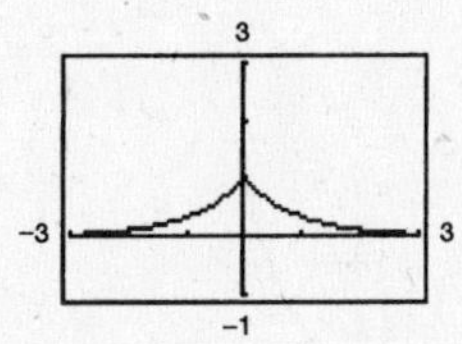

39. $y = 3^{x-2} + 1$

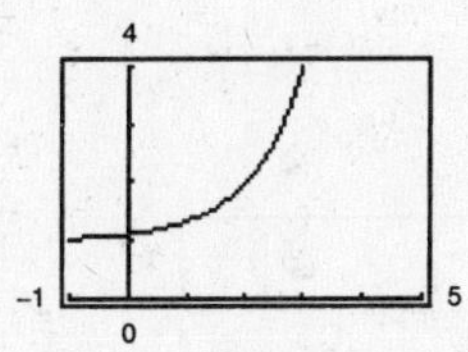

40. $y = 4^{x+1} - 2$

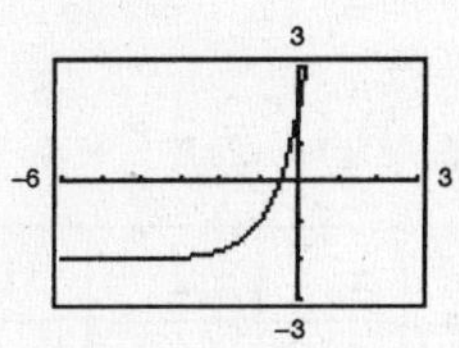

41. $y = 1.08^{-5x}$

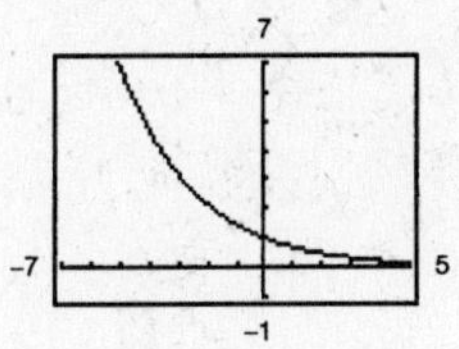

42. $y = 1.08^{5x}$

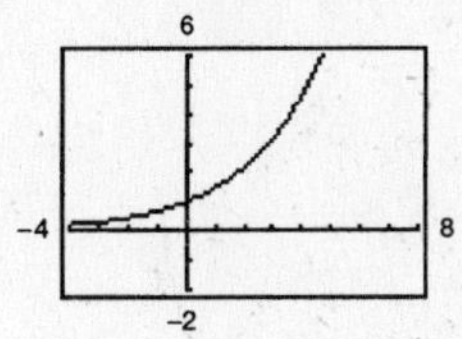

43. $e^{1/2} \approx 1.649$

44. $e^{-2} \approx 0.135$

45. $e^{-3/4} \approx 0.472$

46. $e^{3.2} \approx 24.533$

47. $40e^{0.012} \approx 40.483$

48. $2.5e^{0.001} \approx 2.503$

49. $500e^{-1.250} \approx 143.252$

50. $1000e^{-2.5} \approx 82.085$

51. $f(x) = e^x$

x	-2	-1	0	1	2
$f(x)$	0.135	0.368	1	2.718	7.389

Asymptote: $y = 0$

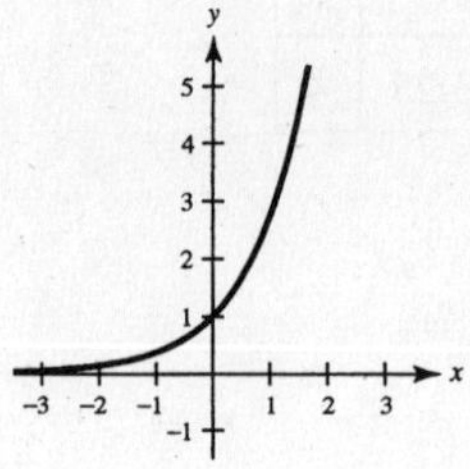

52. $f(x) = e^{-x}$

x	-2	-1	0	1	2
$f(x)$	7.389	2.718	1	0.368	0.135

Asymptote: $y = 0$

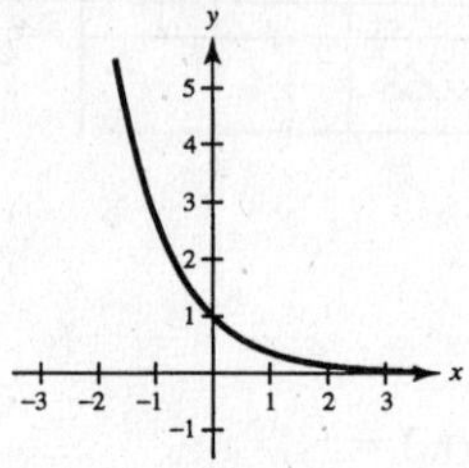

53. $f(x) = 3e^{x+4}$

x	-8	-7	-6	-5	-4
$f(x)$	0.055	0.149	0.406	1.104	3

Asymptote: $y = 0$

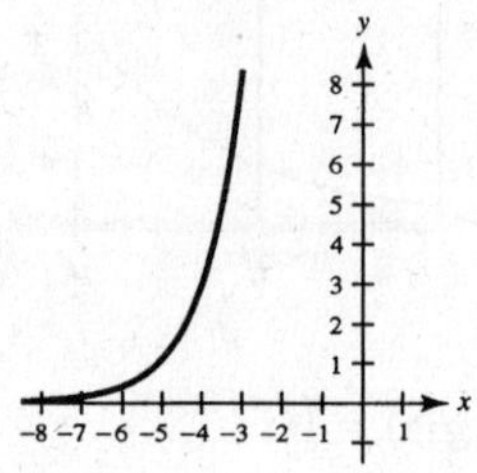

54. $f(x) = 2e^{-0.5x}$

x	-2	-1	0	1	2
$f(x)$	5.437	3.297	2	1.213	0.736

Asymptote: $y = 0$

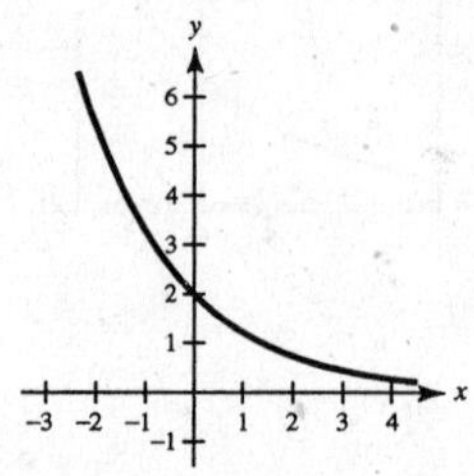

55. $f(x) = 2e^{x-2} + 4$

x	-2	-1	0	1	2
$f(x)$	4.037	4.100	4.271	4.736	6

Asymptote: $y = 4$

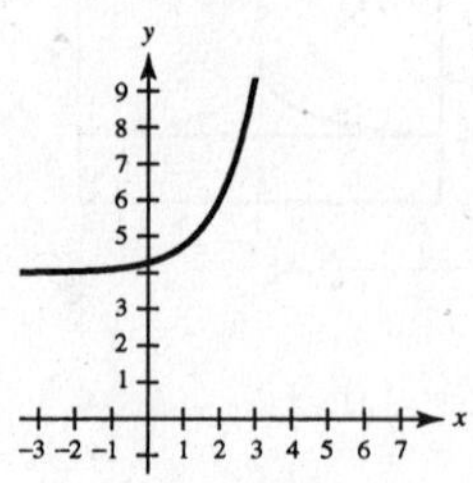

56. $f(x) = 2 + e^{x-5}$

x	0	2	4	5	6
$f(x)$	2.007	2.050	2.368	3	4.718

Asymptote: $y = 2$

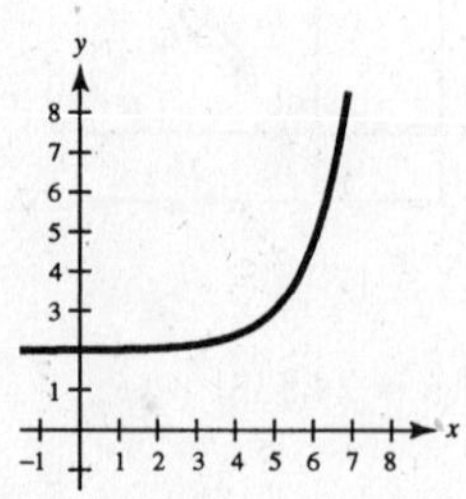

57. $s(t) = 2e^{0.12t}$

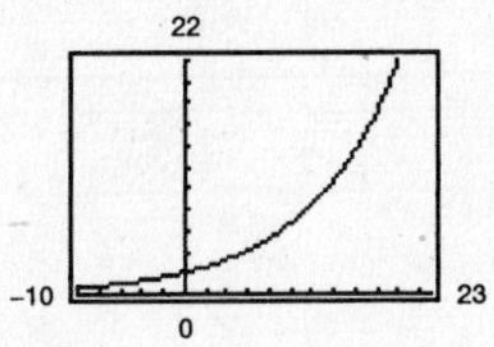

58. $s(t) = 3e^{-0.2t}$

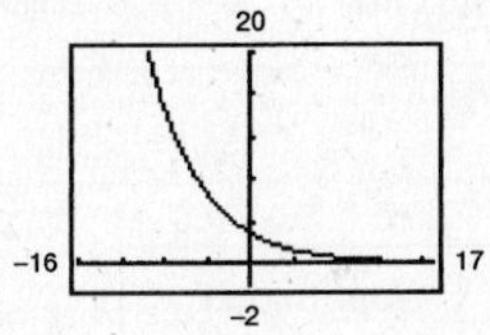

59. $g(x) = 1 + e^{-x}$

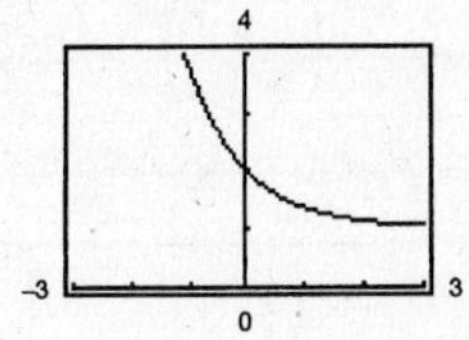

60. $h(x) = e^{x-2}$

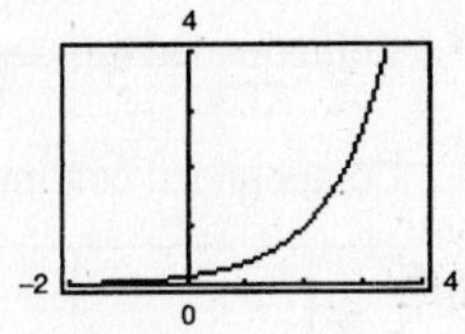

61. $f(x) = 1000e^{0.01x}$

x	-300	-200	-100	-50	0
$f(x)$	49.79	135.34	367.88	606.53	1000

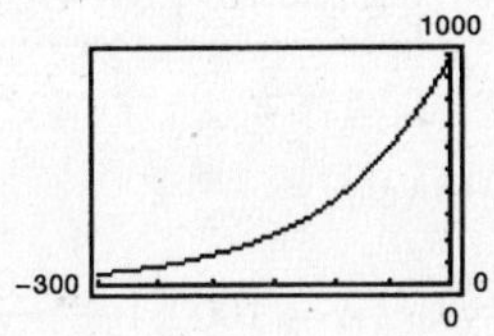

62. $f(x) = 2500e^{0.02x}$

x	-400	-300	-200	-100	0	50
$f(x)$	0.84	6.20	45.79	338.34	2500	6795.70

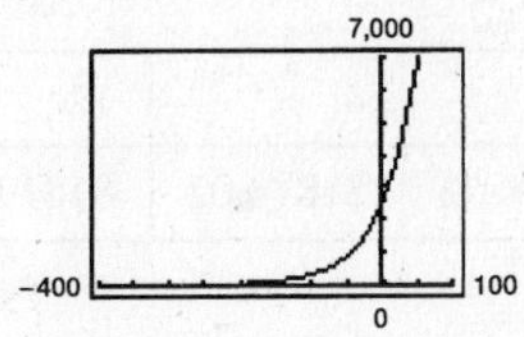

63. $f(x) = 50e^{-0.0025x}$

x	-1000	-750	-500	-250	0
$f(x)$	609.12	326.04	174.52	93.41	50

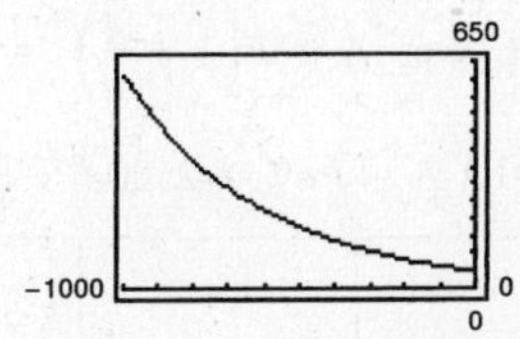

64. $f(x) = 150e^{-0.001x}$

x	-1000	-750	-500	-250	0	250
$f(x)$	407.74	317.55	247.31	192.60	150	116.82

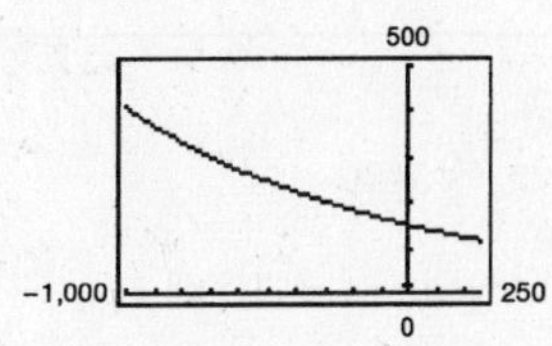

65. $f(x) = \dfrac{250}{1 + 2e^{-0.01x}}$

x	-400	0	200	400	600
$f(x)$	2.27	83.33	196.75	241.17	248.77

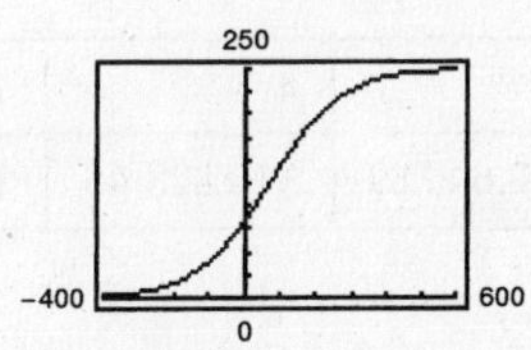

66. $f(x) = \dfrac{100}{1 + 5e^{-0.02x}}$

x	-200	0	200	400	600
$f(x)$	0.36	16.67	91.61	99.83	99.997

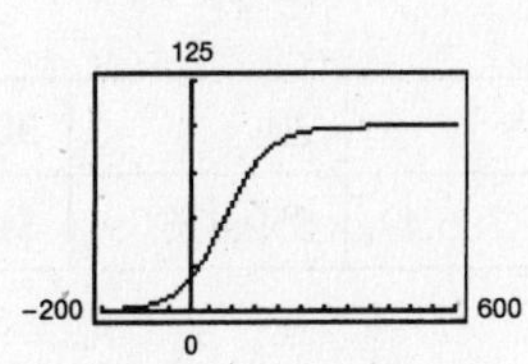

67. $P = \$2500,\ r = 8\%,\ t = 10$ years

Compounded n times per year: $A = P\left(1 + \frac{r}{n}\right)^{nt} = 2500\left(1 + \frac{0.08}{n}\right)^{10n}$

Compounded continuously: $A = Pe^{rt} = 2500e^{0.08(10)}$

n	1	2	4	12	365	Continuous Compounding
A	\$5397.31	\$5477.81	\$5520.10	\$5549.10	\$5563.36	\$5563.85

68. $P = \$1000,\ r = 6\%,\ t = 10$ years

Compounded n times per year: $A = 1000\left(1 + \frac{0.06}{n}\right)^{10n}$

Compounded continuously: $A = 1000e^{0.06(10)}$

n	1	2	4	12	365	Continuous
A	\$1790.85	\$1806.11	\$1814.02	\$1819.40	\$1822.03	\$1822.12

69. $P = \$2500,\ r = 8\%,\ t = 20$ years

Compounded n times per year: $A = P\left(1 + \frac{r}{n}\right)^{nt} = 2500\left(1 + \frac{0.08}{n}\right)^{20n}$

Compounded continuously: $A = Pe^{rt} = 2500e^{0.08(20)}$

n	1	2	4	12	365	Continuous Compounding
A	\$11,652.39	\$12,002.55	\$12,188.60	\$12,317.01	\$12,380.41	\$12,382.58

70. $P = \$1000,\ r = 6\%,\ t = 40$ years

Compounded n times per year: $A = 1000\left(1 + \frac{0.06}{n}\right)^{40n}$

Compounded continuously: $A = 1000e^{0.06(40)}$

n	1	2	4	12	365	Continuous
A	\$10,285.72	\$10,640.89	\$10,828.46	\$10,957.45	\$11,021.00	\$11,023.18

71. $A = Pe^{rt}$

$A = 12{,}000e^{0.08t}$

t	1	10	20	30	40	50
A	\$12,999.44	\$26,706.49	\$59,436.39	\$132,278.12	\$294,390.36	\$655,177.80

72. $A = Pe^{rt}$

$A = 12{,}000e^{0.06t}$

t	1	10	20	30	40	50
A	\$12,742.04	\$21,865.43	\$39,841.40	\$72,595.77	\$132,278.12	\$241,026.44

73. $A = Pe^{rt}$

$A = 12{,}000e^{0.065t}$

t	1	10	20	30	40	50
A	\$12,805.91	\$22,986.49	\$44,031.56	\$84,344.25	\$161,564.86	\$309,484.08

74. $A = Pe^{rt}$

$A = 12{,}000e^{0.075t}$

t	1	10	20	30	40	50
A	\$12,934.61	\$25,404.00	\$53,780.27	\$113,852.83	\$241,026.44	\$510,252.98

75. $f(x) = 3^{x-2}$

$= 3^x 3^{-2}$

$= 3^x\left(\frac{1}{3^2}\right)$

$= \frac{1}{9}(3^x)$

$= h(x)$

Thus, $f(x) \neq g(x)$, but $f(x) = h(x)$.

76. $g(x) = 2^{2x+6}$

$= 2^{2x} \cdot 2^6$

$= 64(2^{2x})$

$= 64(2^2)^x$

$= 64(4^x)$

$= h(x)$

Thus, $g(x) = h(x)$ but $g(x) \neq f(x)$.

77. $f(x) = 16(4^{-x})$

$= 4^2(4^{-x})$

$= 4^{2-x}$

$= \left(\frac{1}{4}\right)^{-(2-x)}$

$= \left(\frac{1}{4}\right)^{x-2}$

$= g(x)$

and $f(x) = 16(4^{-x})$

$= 16(2^2)^{-x}$

$= 16(2^{-2x})$

$= h(x)$

Thus, $f(x) = g(x) = h(x)$.

78. $f(x) = 5^{-x} + 3$

$g(x) = 5^{3-x} = 5^3 \cdot 5^{-x}$

$h(x) = -5^{x-3} = -(5^x \cdot 5^{-3})$

Thus, none are equal.

79. $y = 3^x$ and $y = 4^x$

x	-2	-1	0	1	2
3^x	$\frac{1}{9}$	$\frac{1}{3}$	1	3	9
4^x	$\frac{1}{16}$	$\frac{1}{4}$	1	4	16

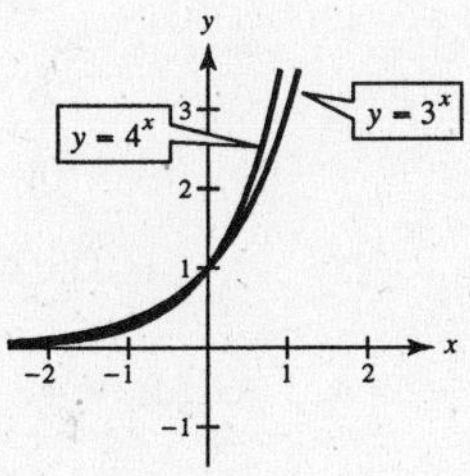

(a) $4^x < 3^x$ when $x < 0$.

(b) $4^x > 3^x$ when $x > 0$.

80. $y = \left(\frac{1}{2}\right)^x$ and $y = \left(\frac{1}{4}\right)^x$

x	-2	-1	0	1	2
$\left(\frac{1}{2}\right)^x$	4	2	1	$\frac{1}{2}$	$\frac{1}{4}$
$\left(\frac{1}{4}\right)^x$	16	4	1	$\frac{1}{4}$	$\frac{1}{16}$

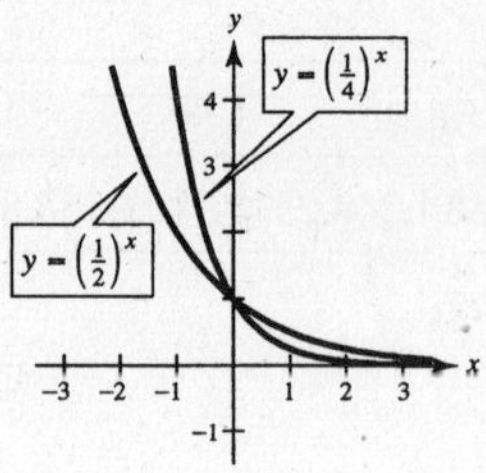

(a) $\left(\frac{1}{4}\right)^x < \left(\frac{1}{2}\right)^x$ when $x > 0$.

(b) $\left(\frac{1}{4}\right)^x > \left(\frac{1}{2}\right)^x$ when $x < 0$.

81.

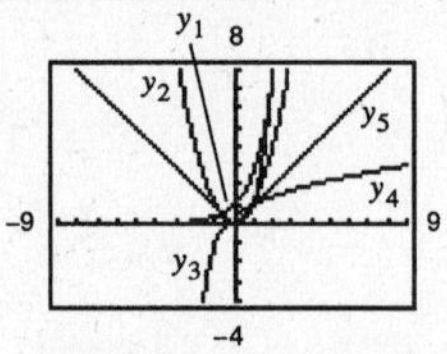

$y_1 = e^x$

$y_2 = x^2$

$y_3 = x^3$

$y_4 = \sqrt{x}$

$y_5 = |x|$

The function that increases at the fastest rate for "large" values of x is $y_1 = e^x$. (Note: One of the intersection points of $y = e^x$ and $y = x^3$ is approximately (4.536, 93) and past this point $e^x > x^3$. This is not shown on the graph above.)

82. The exponential function, $y = e^x$, increases at a faster rate than the polynomial function, $y = x^n$.

83. It usually implies rapid growth.

84. The functions (c) 3^x and (d) 2^{-x} are exponential.

85. $A = 25{,}000e^{(0.0875)(25)} \approx \$222{,}822.57$

86. $A = 5000e^{(0.075)(50)} \approx \$212{,}605.41$

87. (a) The steeper curve represents the investment earning compound interest, because compound interest earns more than simple interest. With simple interest there is no compounding so the growth is linear.

(b) Compound interest formula: $A = 500\left(1 + \dfrac{0.07}{1}\right)^{(1)t}$

$= 500(1.07)^t$

Simple interest formula: $A = Prt + P$

$= 500(0.07)t + 500$

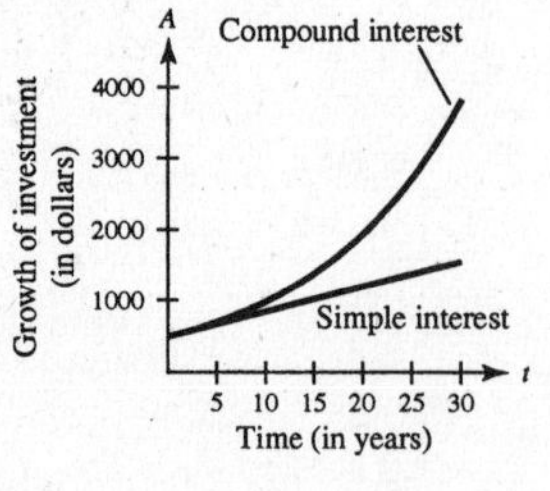

88. $V(t) = 20{,}000\left(\dfrac{3}{4}\right)^t$

(a)

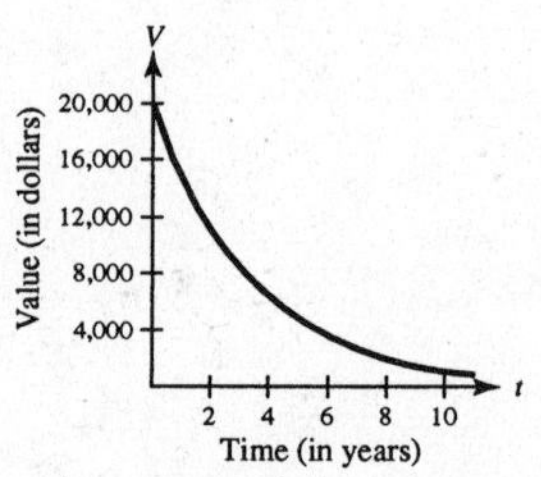

(b) $V(0) = 20{,}000$

The original value was \$20,000

(c) $V(4) = 20{,}000\left(\dfrac{3}{4}\right)^4$

$\approx \$6328.13$

89. $C(10) = 23.95(1.04)^{10} \approx \$35.45$

90. $p = 5000\left(1 - \dfrac{4}{4 + e^{-0.002x}}\right)$

(a)

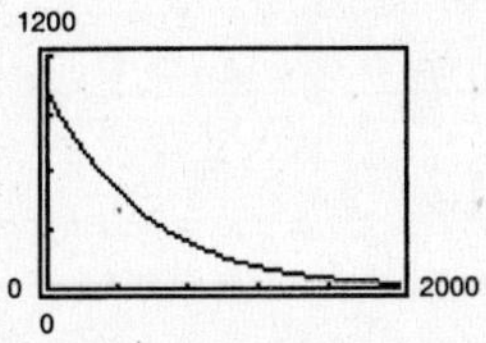

(b) When $x = 500$:

$$p = 5000\left(1 - \frac{4}{4 + e^{-0.002(500)}}\right) \approx \$421.12$$

(c) Since (600, 303.07) is on the graph in part (a), it appears that the greatest price that will still yield a demand of at least 600 units is about $303.

91. $P(t) = 100e^{0.2197t}$

(a) $P(0) = 100$

(b) $P(5) \approx 300$

(c) $P(10) \approx 900$

92. $P(t) = 2500e^{0.0293t}$

(a) $P(10) = 2500e^{0.0293(10)} \approx 3351$

(b) $P(20) = 2500e^{0.0293(20)} \approx 4492$

93. $Q = 25\left(\frac{1}{2}\right)^{t/1599}$

(a) $Q(0) = 25$ grams

(b) $Q(1000) = 25\left(\frac{1}{2}\right)^{1000/1599} \approx 16.21$ grams

(c)

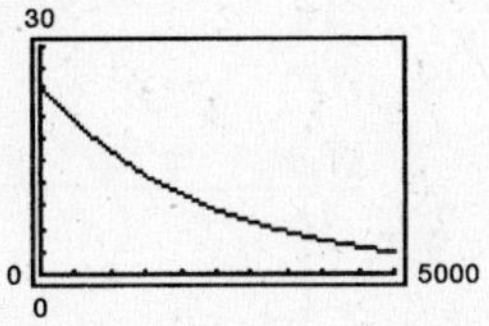

94. $Q = 10\left(\frac{1}{2}\right)^{t/5715}$

(a) $Q(0) = 10$ grams

(b) $Q(2000) = 10\left(\frac{1}{2}\right)^{2000/5715} \approx 7.85$ grams

(c)

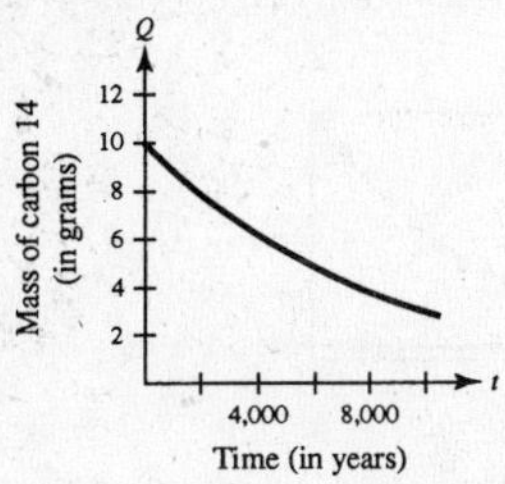

95. $P = 107{,}428e^{-0.15h}$

(a)

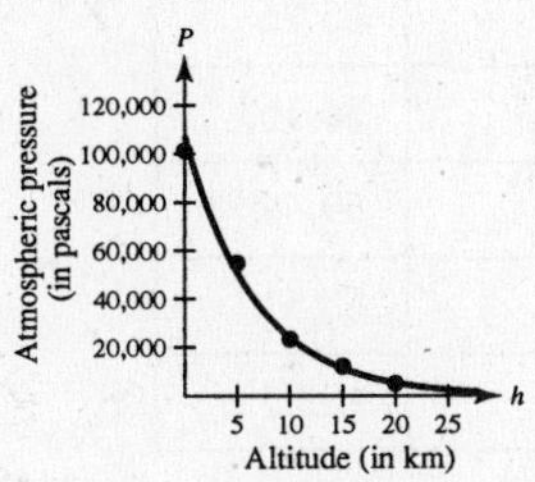

(b)

h	0	5	10	15	20
P (data)	101,293	54,735	23,294	12,157	5,069
P (model)	107,428	50,745	23,970	11,323	5,349

(c) $P(8) \approx 32{,}357$ pascals

(d) $P = 21{,}000$ when $h \approx 10.88$ kilometers.

96. $y = \dfrac{300}{3 + 17e^{-0.065x}}$

(a)

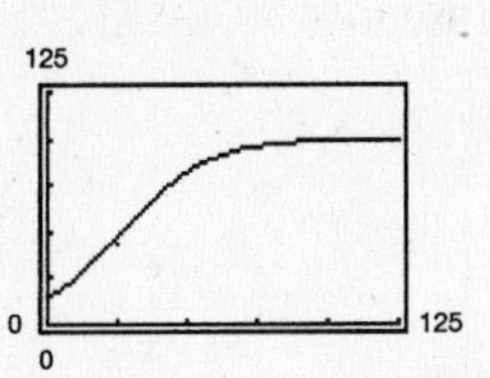

(b)

x	0	25	50	75	100
y	15	47	82	96	99

The model is a "good fit."

(c) When $x = 36$: $y = \dfrac{300}{3 + 17e^{-0.065(36)}} \approx 64.7\%$

(d) $\dfrac{200}{3} = \dfrac{300}{3 + 17e^{-0.065x}}$ when $x \approx 37$.

97. $P = 115.5e^{0.0445t}$ where $t = 0$ corresponds to 1970 and P is in thousands.

(a)

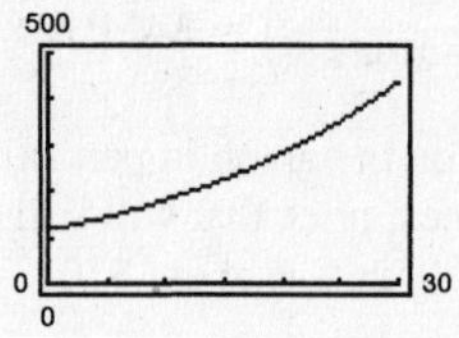

(b) For 1970 use $t = 0$: $P = 115{,}500$ people

For 1980 use $t = 10$: $P \approx 180{,}240$ people

For 1990 use $t = 20$: $P \approx 281{,}260$ people

For 2000 use $t = 30$: $P \approx 438{,}900$ people

The rate at which the population is increasing is increasing, therefore the population is growing exponentially and does not fit a linear model.

(c) $P = 750$ when $t \approx 42$ which corresponds to 2012.

98. $M = 45.7e^{-0.03139t}$ where $t = 0$ corresponds to 1940.

(a)

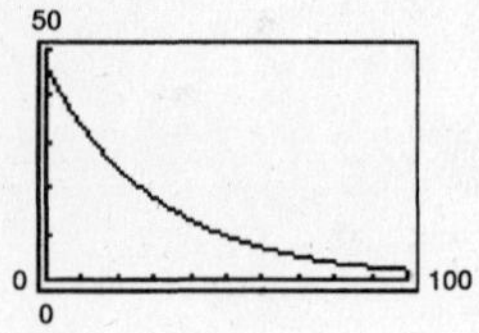

(b) $M = 10$ when $t \approx 48.4$ which corresponds to 1988.

(c) $M = 5$ when $t \approx 70.5$ which corresponds to 2010.

(d) No. For the graph to have a t-intercept, there would have to be a time when there was no infant mortality.

99. $P = 1.879e^{0.01451t}$ where $t = 50$ corresponds to 1850.

(a) and (b)

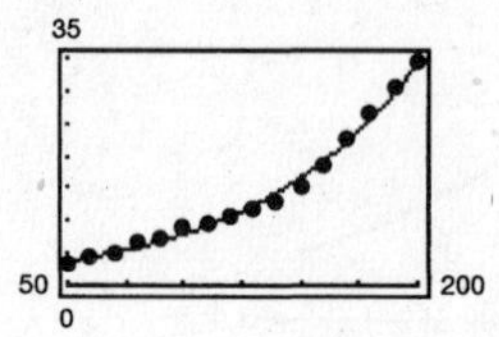

(c) $P = 40$ when $t \approx 211$ which corresponds to 2011.

100. $P = 4.623e^{0.02084t}$ where $t = 50$ corresponds to 1950.

(a) and (b)

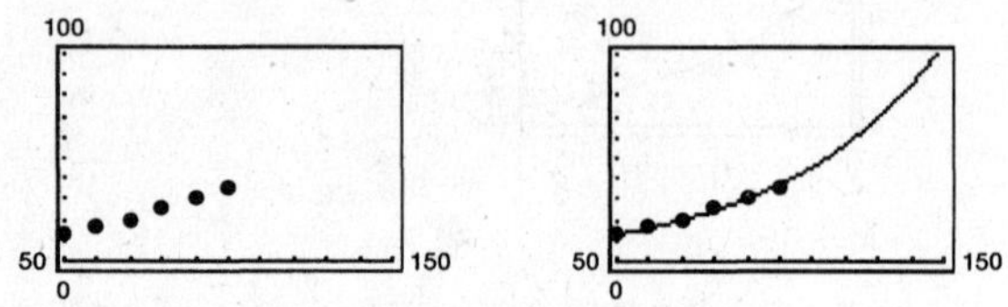

(c)

t	P(projections)	P(model)
110	40.24	45.76
120	54.63	56.36
130	71.45	69.42
140	80.05	85.51
150	86.71	105.32

Since the model increases very rapidly, it eventually becomes too large to use for projections.

101. $D = -182.08 + 677.14e^{-0.0142t}$, $t = 0 \leftrightarrow 1970$

(a) $D(0) \approx 495.06$ deaths per 100,000 people in 1970.

$D(10) \approx 405.42$ deaths per 100,000 people in 1980.

$D(20) \approx 327.65$ deaths per 100,000 people in 1990.

$D(30) \approx 260.17$ deaths per 100,000 people in 2000.

(b) No. The model decreases rapidly and eventually becomes negative as t increases.

102. $C = -31{,}933.90 + 33{,}188.79e^{0.1404t}$, $0 \leftrightarrow 1990$

(a)

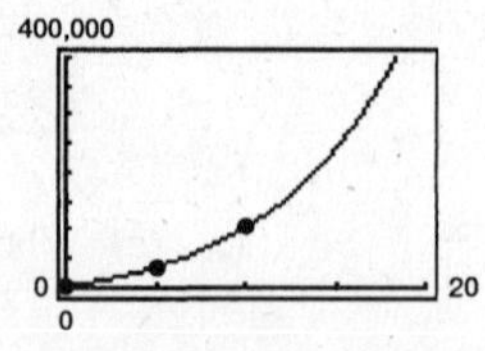

(b) $C = 350{,}000$ when $t \approx 17.4$ which corresponds to the year 2007.

103. True

104. True, if $a > 0, a \neq 1$.

105. False, $e \neq \dfrac{271{,}801}{99{,}990}$.

e is an irrational number.

106. True. By properties of exponents $\dfrac{e^x}{e^2} = e^{x-2}$.

107. (a) $f(x) = \dfrac{8}{1 + e^{-0.5x}}$

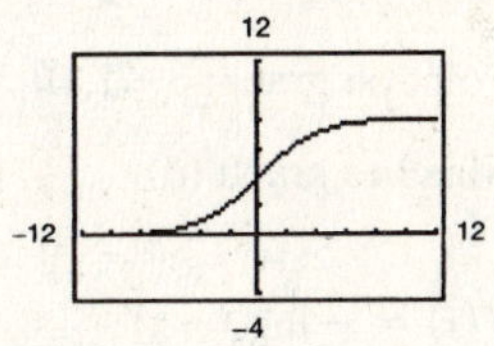

Horizontal asymptotes: $y = 0$ and $y = 8$

(b) $g(x) = \dfrac{8}{1 + e^{-0.5/x}}$

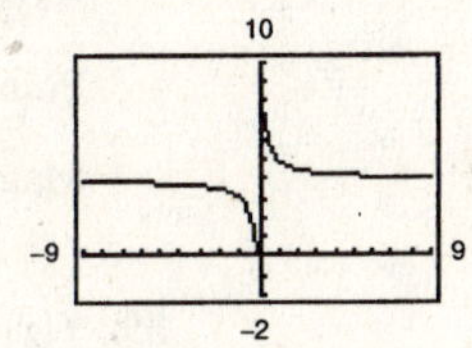

Horizontal asymptote: $y = 4$

Vertical asymptote: $x = 0$

108. (a) $f(x) = x^2e^{-x}$

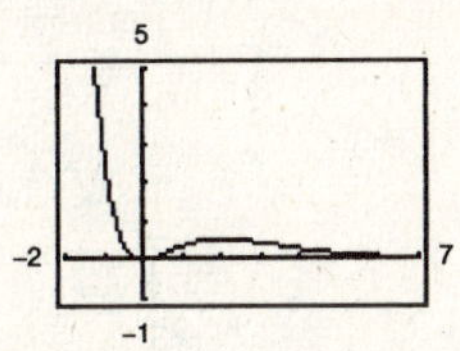

Decreasing: $(-\infty, 0)$, $(2, \infty)$

Increasing: $(0, 2)$

Relative maximum: $(2, 4e^{-2})$

Relative minimum: $(0, 0)$

(b) $g(x) = x2^{3-x}$

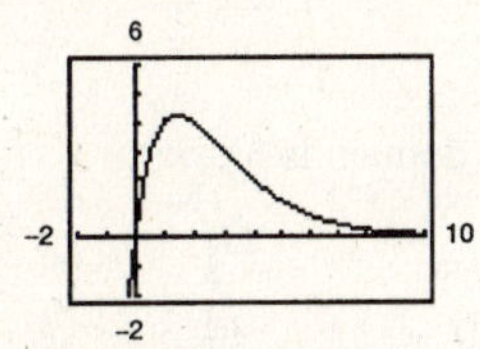

Decreasing: $(1.44, \infty)$

Increasing: $(-\infty, 1.44)$

Relative maximum: $(1.44, 4.25)$

Section 7.2 Logarithmic Functions and Their Graphs

1. $\log_4 64 = 3 \Rightarrow 4^3 = 64$

2. $\log_3 81 = 4 \Rightarrow 3^4 = 81$

3. $\log_7 \frac{1}{49} = -2 \Rightarrow 7^{-2} = \frac{1}{49}$

4. $\log_{10} \frac{1}{1000} = -3 \Rightarrow 10^{-3} = \frac{1}{1000}$

5. $\log_{32} 4 = \frac{2}{5} \Rightarrow 32^{2/5} = 4$

6. $\log_{16} 8 = \frac{3}{4} \Rightarrow 16^{3/4} = 8$

7. $\log_{1/4} 64 = -3 \Rightarrow \left(\frac{1}{4}\right)^{-3} = 64$

8. $\log_{1/2} 128 = -7 \Rightarrow \left(\frac{1}{2}\right)^{-7} = 128$

9. $5^3 = 125 \Rightarrow \log_5 125 = 3$

10. $8^2 = 64 \Rightarrow \log_8 64 = 2$

11. $81^{1/4} = 3 \Rightarrow \log_{81} 3 = \frac{1}{4}$

12. $9^{3/2} = 27 \Rightarrow \log_9 27 = \frac{3}{2}$

13. $6^{-2} = \frac{1}{36} \Rightarrow \log_6 \frac{1}{36} = -2$

14. $10^{-3} = 0.001 \Rightarrow \log_{10} 0.001 = -3$

15. $\left(\frac{2}{3}\right)^{-2} = \frac{9}{4} \Rightarrow \log_{2/3} \frac{9}{4} = -2$

16. $\left(\frac{3}{5}\right)^{-3} = \frac{125}{27} \Rightarrow \log_{3/5} \frac{125}{27} = -3$

17. $a^k = b \Rightarrow \log_a b = k$

18. $u^v = w \Rightarrow \log_u w = v$

19. $\log_2 16 = \log_2 2^4 = 4$

20. $\log_2 \frac{1}{8} = \log_2 2^{-3} = -3$

21. $\log_{16} 4 = \log_{16} 16^{1/2} = \frac{1}{2}$

22. $\log_{27} 9 = \log_{27} 27^{2/3} = \frac{2}{3}$

23. $\log_7 1 = \log_7 7^0 = 0$

24. $\log_{10} 1000 = \log_{10} 10^3 = 3$

25. $\log_{10} 0.01 = \log_{10} 10^{-2} = -2$

26. $\log_{10} 10 = \log_{10} 10^1 = 1$

27. $\log_8 32 = \log_8 8^{5/3} = \frac{5}{3}$

28. $\log_9 243 = \log_9 9^{5/2} = \frac{5}{2}$

29. $\log_a a^2 = 2$

30. $\log_b b^{-3} = -3$

31. $\log_{10} 345 \approx 2.538$

32. $\log_{10} 145 \approx 2.161$

33. $\log_{10} \frac{4}{5} \approx -0.097$

34. $\log_{10} 12.5 \approx 1.097$

35. $f(x) = \log_3 x + 2$

Asymptote: $x = 0$

Point on graph: $(1, 2)$

Matches graph (c)

36. $f(x) = -\log_3 x$

Asymptote: $x = 0$

Point on graph: $(1, 0)$

Matches graph (f)

37. $f(x) = -\log_3(x + 2)$

Asymptote: $x = -2$

Point on graph: $(-1, 0)$

Matches graph (d)

38. $f(x) = \log_3(x - 1)$

Asymptote: $x = 1$

Point on graph: $(2, 0)$

Matches graph (e)

39. $f(x) = \log_3(1 - x)$

Asymptote: $x = 1$

Point on graph: $(0, 0)$

Matches graph (b)

40. $f(x) = -\log_3(-x)$

Asymptote: $x = 0$

Point on graph: $(-1, 0)$

Matches graph (a)

41. $f(x) = \log_4 x$

Domain: $x > 0 \implies$ The domain is $(0, \infty)$.

x-intercept: $(1, 0)$

Vertical asymptote: $x = 0$

$y = \log_4 x \implies 4^y = x$

x	$\frac{1}{4}$	1	2	4
$f(x)$	-1	0	$\frac{1}{2}$	1

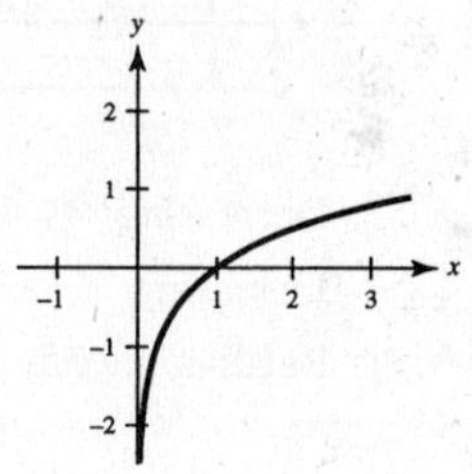

42. $g(x) = \log_6 x$

Domain: $(0, \infty)$

x-intercept: $(1, 0)$

Vertical asymptote: $x = 0$

$y = \log_6 x \implies 6^y = x$

x	$\frac{1}{6}$	1	$\sqrt{6}$	6
y	-1	0	$\frac{1}{2}$	1

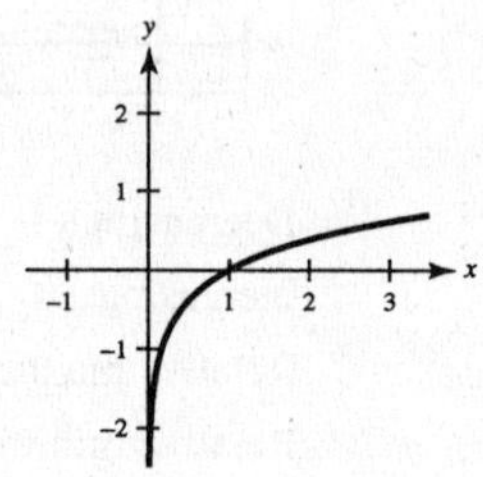

43. $y = -\log_3 x + 2$

Domain: $(0, \infty)$

x-intercept:

$$-\log_3 x + 2 = 0$$
$$2 = \log_3 x$$
$$3^2 = x$$
$$9 = x$$

The x-intercept is $(9, 0)$.

Vertical asymptote: $x = 0$

$y = -\log_3 x + 2$

$\log_3 x = 2 - y \implies 3^{2-y} = x$

x	27	9	3	1	$\frac{1}{3}$
y	-1	0	1	2	3

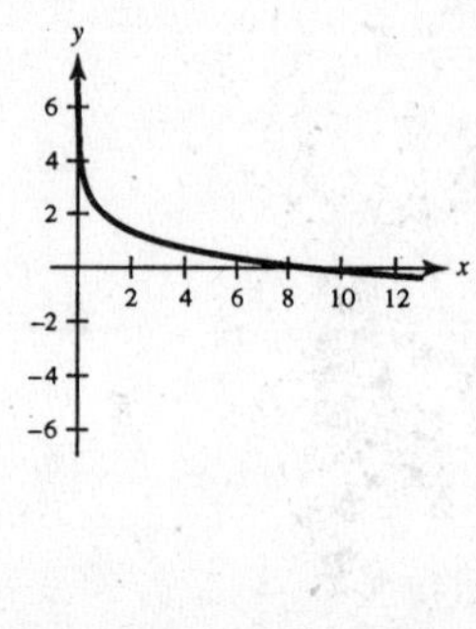

44. $h(x) = \log_4(x - 3)$

Domain: $x - 3 > 0 \implies x > 3$

The domain is $(3, \infty)$.

x-intercept: $\log_4(x - 3) = 0$

$$4^0 = x - 3$$
$$1 = x - 3$$
$$4 = x$$

The x-intercept is $(4, 0)$.

Vertical asymptote: $x - 3 = 0 \implies x = 3$

$y = \log_4(x - 3) \implies 4^y + 3 = x$

x	$3\frac{1}{4}$	4	7	19
y	-1	0	1	2

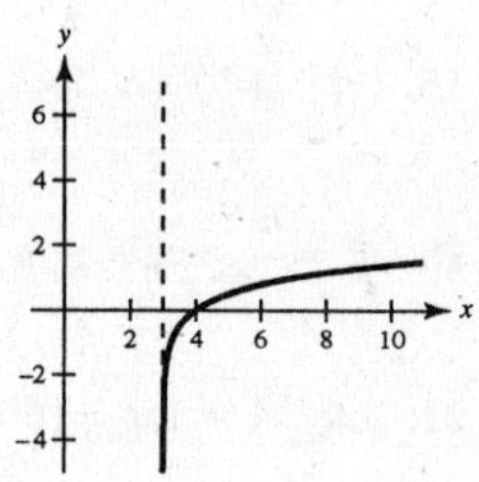

45. $f(x) = -\log_6(x + 2)$

Domain: $x + 2 > 0 \Rightarrow x > -2$

The domain is $(-2, \infty)$.

x-intercept:

$$0 = -\log_6(x + 2)$$
$$0 = \log_6(x + 2)$$
$$6^0 = x + 2$$
$$1 = x + 2$$
$$-1 = x$$

The x-intercept is $(-1, 0)$.

Vertical asymptote: $x + 2 = 0 \Rightarrow x = -2$

$$y = -\log_6(x + 2)$$
$$-y = \log_6(x + 2)$$
$$6^{-y} - 2 = x$$

x	4	-1	$-1\frac{5}{6}$	$-1\frac{35}{36}$
$f(x)$	-1	0	1	2

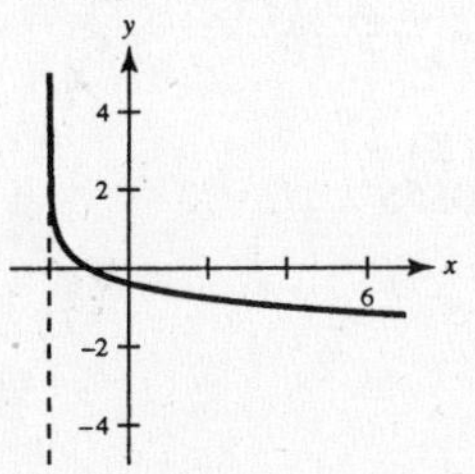

46. $y = \log_5(x - 1) + 4$

Domain: $x - 1 > 0 \Rightarrow x > 1$

The domain is $(1, \infty)$.

x-intercept: $\log_5(x - 1) + 4 = 0$

$$\log_5(x - 1) = -4$$
$$5^{-4} = x - 1$$
$$\frac{1}{625} = x - 1$$
$$\frac{626}{625} = x$$

The x-intercept is $\left(\frac{626}{625}, 0\right)$.

Vertical asymptote: $x - 1 = 0 \Rightarrow x = 1$

$y = \log_5(x - 1) + 4 \Rightarrow 5^{y-4} + 1 = x$

x	1.00032	1.0016	1.008	1.04	1.2
y	-1	0	1	2	3

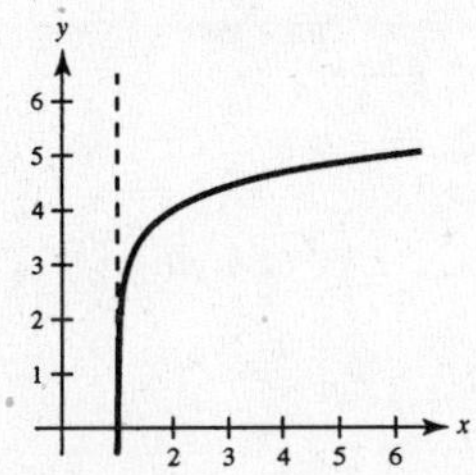

47. $y = \log_{10}\left(\frac{x}{5}\right)$

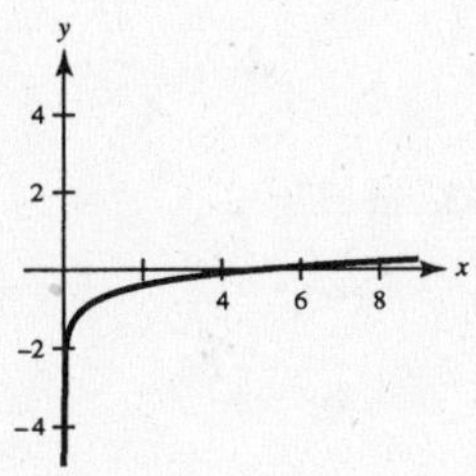

Domain: $\frac{x}{5} > 0 \Rightarrow x > 0$

The domain is $(0, \infty)$.

x-intercept:

$$\log_{10}\left(\frac{x}{5}\right) = 0$$
$$\frac{x}{5} = 10^0$$
$$\frac{x}{5} = 1 \Rightarrow x = 5$$

The x-intercept is $(5, 0)$.

Vertical asymptote: $\frac{x}{5} = 0 \Rightarrow x = 0$

The vertical asymptote is the y-axis.

x	1	2	3	4	5	6	7
y	-0.70	-0.40	-0.22	-0.10	0	0.08	0.15

48. $y = \log_{10}(-x)$

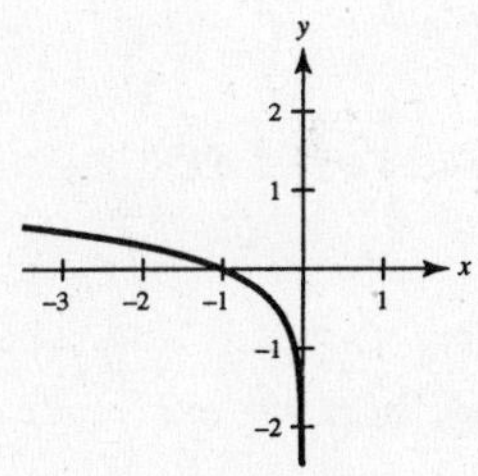

Domain: $-x > 0 \Rightarrow x < 0$

The domain is $(-\infty, 0)$.

x-intercept: $\log_{10}(-x) = 0$

$$10^0 = -x$$
$$-1 = x$$

The x-intercept is $(-1, 0)$.

Vertical asymptote: $x = 0$

$y = \log_{10}(-x) \Rightarrow -10^y = x$

x	$-\frac{1}{100}$	$-\frac{1}{10}$	-1	-10
y	-2	-1	0	1

49. $f(x) = \log_{10}(x + 1)$

Domain: $x + 1 > 0 \Rightarrow x > -1$

The domain is $(-1, \infty)$.

x-intercept:

$$\log_{10}(x + 1) = 0$$
$$x + 1 = 10^0$$
$$x + 1 = 1$$
$$x = 0$$

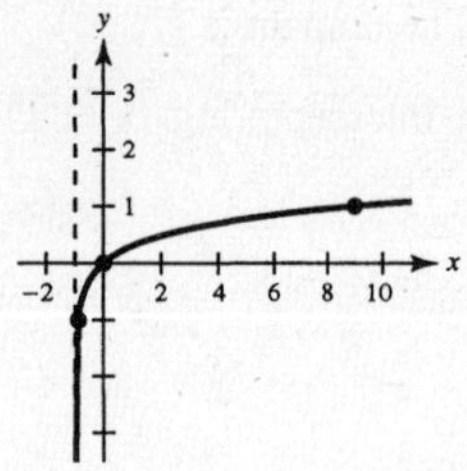

The x-intercept is $(0, 0)$.

Vertical asymptote: $x + 1 = 0 \Rightarrow x = -1$

$y = \log_{10}(x + 1) \Rightarrow 10^y = x + 1 \Rightarrow 10^y - 1 = x$

x	-0.9	0	9
y	-1	0	1

50. $f(x) = \log_{10}(x - 1)$

Domain: $x - 1 > 0 \Rightarrow x > 1$

The domain is $(1, \infty)$.

x-intercept:

$$\log_{10}(x - 1) = 0$$
$$x - 1 = 10^0$$
$$x - 1 = 1$$
$$x = 2$$

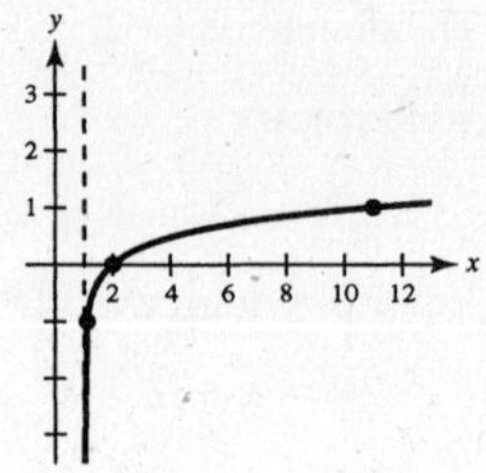

The x-intercept is $(2, 0)$.

Vertical asymptote: $x - 1 = 0 \Rightarrow x = 1$

$y = \log_{10}(x - 1) \Rightarrow 10^y = x - 1 \Rightarrow 10^y + 1 = x$

x	1.1	2	11
y	-1	0	1

51. $\ln 4 = 1.386\ldots \Rightarrow e^{1.386\ldots} = 4$

52. $\ln 2 = 0.6931\ldots \Rightarrow e^{0.6931\ldots} = 2$

53. $\ln 1 = 0 \Rightarrow e^0 = 1$

54. $\ln e^2 = 2 \Rightarrow e^2 = e^2$

55. $\ln e^{-3} = -3 \Rightarrow e^{-3} = e^{-3}$

56. $\ln \sqrt{e} = \frac{1}{2} \Rightarrow e^{1/2} = \sqrt{e}$

57. $e^3 = 20.0855\ldots \Rightarrow \ln 20.0855\ldots = 3$

58. $e^2 = 7.3890\ldots \Rightarrow \ln 7.3890\ldots = 2$

59. $e^0 = 1 \Rightarrow \ln 1 = 0$

60. $e^{-1} = 0.3678\ldots \Rightarrow \ln 0.3678\ldots = -1$

61. $\sqrt{e} = 1.6487\ldots \Rightarrow e^{1/2} = 1.6487\ldots$
$\Rightarrow \ln 1.6487\ldots = \frac{1}{2}$

62. $\sqrt[3]{e} = 1.3956\ldots \Rightarrow e^{1/3} = 1.3956\ldots$
$\Rightarrow \ln 1.3956\ldots = \frac{1}{3}$

63. $\ln e^3 = 3$

64. $\ln e^{-2} = -2$

65. $\ln e^{-2.1} = -2.1$

66. $\ln e^{4.5} = 4.5$

67. $\ln\left(\frac{1}{e^3}\right) = \ln e^{-3} = -3$

68. $\ln\left(\frac{1}{e}\right) = \ln e^{-1} = -1$

69. $\ln 18.42 \approx 2.913$

70. $\ln \sqrt{42} \approx 1.869$

71. $3 \ln 0.32 \approx -3.418$

72. $2 \ln 0.75 = -0.575$

73. $\ln(1 + \sqrt{3}) \approx 1.005$

74. $\ln(\sqrt{5} - 2) \approx -1.444$

75. $\ln \frac{2}{3} \approx -0.405$

76. $\ln\left(\frac{1}{2}\right) = -0.693$

77. $f(x) = -\ln x$

The graph of f can be obtained by reflecting the graph of $y = \ln x$ about the x-axis.

Matches graph (d)

78. $f(x) = \ln(x - 1)$

The graph of f can be obtained by shifting the graph of $y = \ln x$ one unit to the right.

Matches graph (e)

79. $f(x) = \ln x - 1$

The graph of f can be obtained by shifting the graph of $y = \ln x$ one unit downward.

Matches graph (a)

80. $f(x) = -\ln x + 1$

The graph of f can be obtained by reflecting the graph of $y = \ln x$ about the x-axis and shifting it one unit upward.

Matches graph (b)

81. $f(x) = \ln(1 - x) = \ln(-x + 1)$

The graph of f can be obtained by shifting the graph of $y = \ln x$ one unit to the left and then reflecting it about the y-axis.

Matches graph (c)

82. $f(x) = -\ln(1 - x)$

The graph of f can be obtained by shifting the graph of $y = \ln x$ one unit to the left and then reflecting the graph about both the x-axis and the y-axis. (or by reflecting the graph obtained in exercise 81 about the x-axis.)

Matches graph (f)

83. $f(x) = \ln(x - 2)$

Domain: $x - 2 > 0 \Rightarrow x > 2$

The domain is $(2, \infty)$.

x-intercept: $0 = \ln(x - 2)$

$$e^0 = x - 2$$

$$3 = x$$

The x-intercept is $(3, 0)$.

Vertical asymptote: $x - 2 = 0 \Rightarrow x = 2$

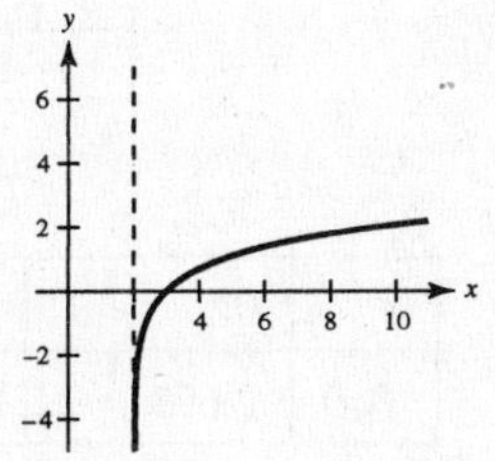

x	2.5	3	4	5
$f(x)$	-0.69	0	0.69	1.10

84. $h(x) = \ln(x + 1)$

Domain: $x + 1 > 0 \Rightarrow x > -1$

The domain is $(-1, \infty)$.

x-intercept: $\ln(x + 1) = 0$

$$e^0 = x + 1$$

$$1 = x + 1$$

$$0 = x$$

The x-intercept is $(0, 0)$.

Vertical asymptote: $x + 1 = 0 \Rightarrow x = -1$

$y = \ln(x + 1) \Rightarrow e^y - 1 = x$

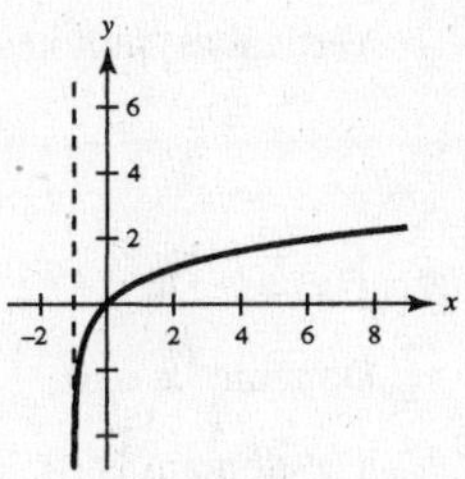

x	-0.39	0	1.72	6.39	19.09
y	$-\frac{1}{2}$	0	1	2	3

85. $g(x) = \ln(-x)$

Domain: $-x > 0 \Rightarrow x < 0$

The domain is $(-\infty, 0)$.

x-intercept: $0 = \ln(-x)$

$$e^0 = -x$$

$$-1 = x$$

The x-intercept is $(-1, 0)$.

Vertical asymptote: $-x = 0 \Rightarrow x = 0$

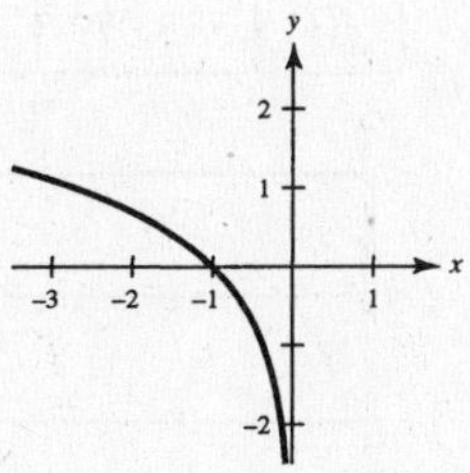

x	-0.5	-1	-2	-3
$g(x)$	-0.69	0	0.69	1.10

86. $f(x) = \ln(3 - x)$

Domain: $3 - x > 0 \Rightarrow x < 3$

The domain is $(-\infty, 3)$.

x-intercept: $\ln(3 - x) = 0$

$$e^0 = 3 - x$$

$$1 = 3 - x$$

$$2 = x$$

The x-intercept is $(2, 0)$.

Vertical asymptote: $3 - x = 0 \Rightarrow x = 3$

$y = \ln(3 - x) \Rightarrow 3 - e^y = x$

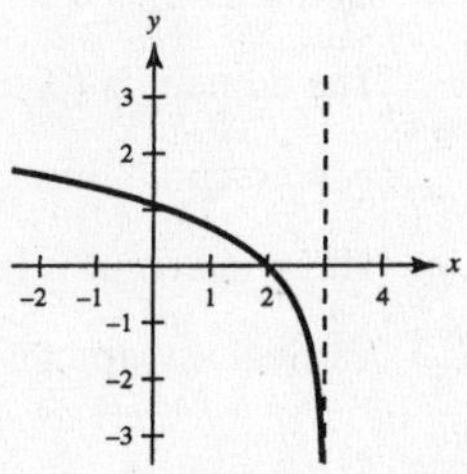

x	2.95	2.86	2.63	2	0.28
y	-3	-2	-1	0	1

87. $f(x) = \ln(x^2 - 1)$

Domain: $x^2 - 1 > 0 \Rightarrow x^2 > 1 \Rightarrow x < -1$ or $x > 1$

The domain is $(-\infty, -1) \cup (1, \infty)$.

x-intercepts: $\ln(x^2 - 1) = 0 \Rightarrow x^2 - 1 = 1 \Rightarrow x^2 = 2 \Rightarrow x = \pm\sqrt{2}$

The x-intercepts are $(\pm\sqrt{2}, 0)$.

Vertical asymptotes: $x^2 - 1 = 0 \Rightarrow x^2 = 1 \Rightarrow x = \pm 1$

x	± 2	± 3	± 4
$f(x)$	1.099	2.079	2.708

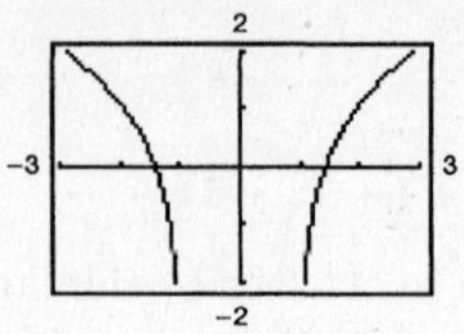

88. $f(x) = \ln(x^2 - 4)$

Domain: $x^2 - 4 > 0 \Rightarrow x^2 > 4 \Rightarrow x < -2$ or $x > 2$

The domain is $(-\infty, -2) \cup (2, \infty)$.

x-intercepts: $\ln(x^2 - 4) = 0 \Rightarrow x^2 - 4 = 1 \Rightarrow x^2 = 5 \Rightarrow x = \pm\sqrt{5}$

The x-intercepts are $(\pm\sqrt{5}, 0)$.

Vertical asymptotes: $x^2 - 4 = 0 \Rightarrow x^2 = 4 \Rightarrow x = \pm 2$

x	± 3	± 4	± 5
$f(x)$	1.609	2.485	3.045

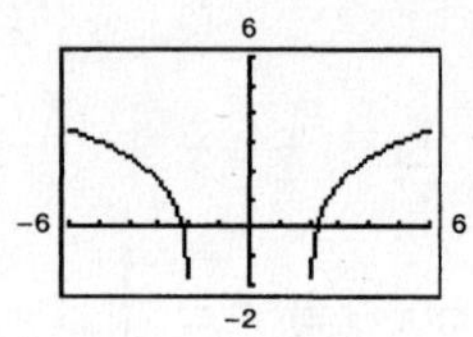

89. $f(x) = -\ln x + 2$

Domain: $x > 0$

The domain is $(0, \infty)$.

x-intercept: $-\ln x + 2 = 0 \Rightarrow \ln x = 2 \Rightarrow x = e^2$

The x-intercept is $(e^2, 0)$.

Vertical asymptote: $x = 0$

x	0.5	1	3	5	7	9
$f(x)$	2.693	2	0.901	0.391	0.054	−0.197

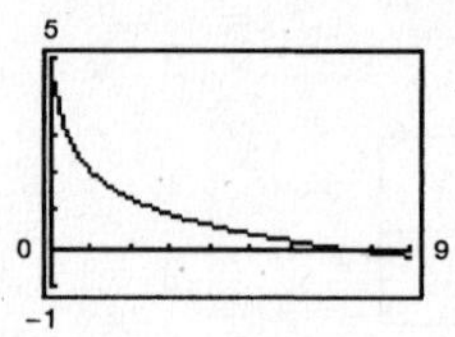

90. $f(x) = 3 \ln x - 1$

Domain: $x > 0$

The domain is $(0, \infty)$.

x-intercept: $3 \ln x - 1 = 0 \Rightarrow \ln x = \frac{1}{3} \Rightarrow x = e^{1/3}$

The x-intercept is $(e^{1/3}, 0)$.

Vertical asymptote: $x = 0$

x	0.5	1	2	3	4
$f(x)$	−3.079	−1	1.079	2.296	3.159

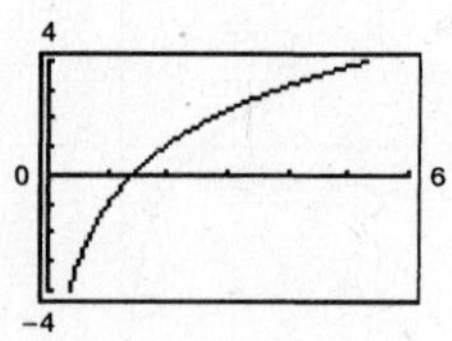

91. $f(x) = |\ln x|$

(a)

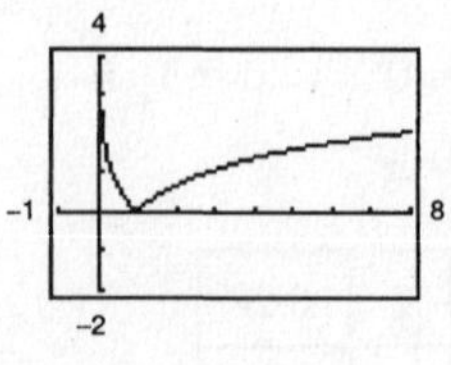

(b) Increasing on $(1, \infty)$

Decreasing on $(0, 1)$

(c) Relative minimum: $(1, 0)$

92. (a) $h(x) = \ln(x^2 + 1)$

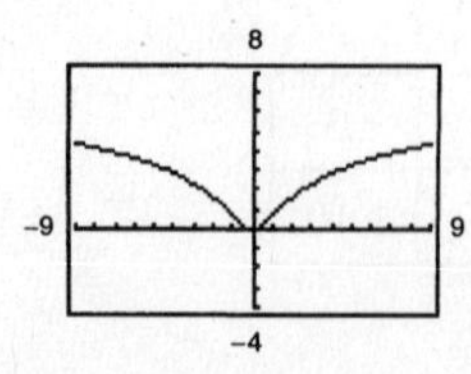

(b) Increasing on $(0, \infty)$.

Decreasing on $(-\infty, 0)$.

(c) Relative minimum: $(0, 0)$

93. $f(x) = \frac{x}{2} - \ln \frac{x}{4}$

(a)

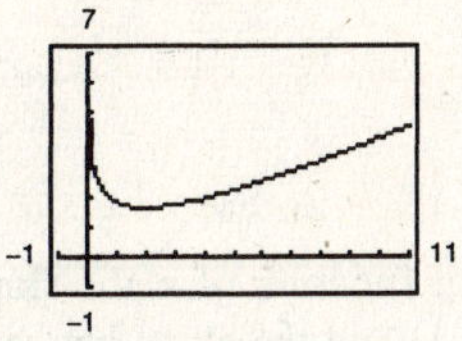

(b) Increasing on $(2, \infty)$.

Decreasing on $(0, 2)$.

(c) Relative minimum: $\left(2, 1 - \ln \frac{1}{2}\right)$

94. $g(x) = \dfrac{12 \ln x}{x}$

(a)

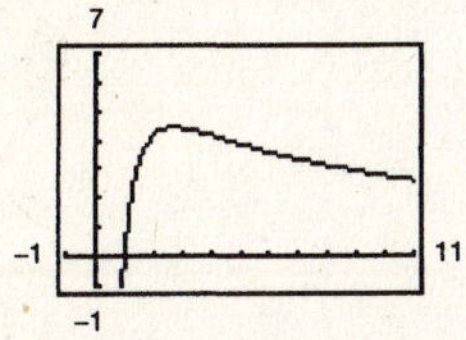

(b) Increasing on $(0, 2.72)$.

Decreasing on $(2.72, \infty)$.

(c) Relative maximum: $(2.72, 4.41)$

95. $f(x) = 3^x$, $g(x) = \log_3 x$

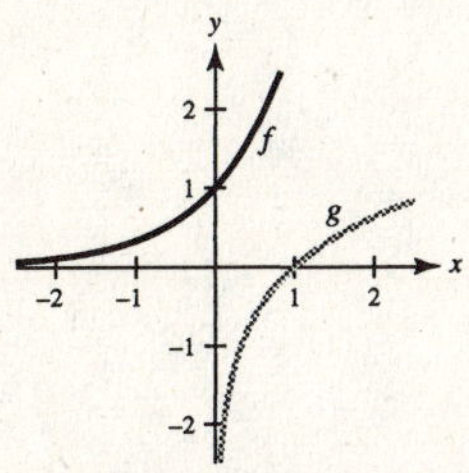

f and g are inverses. Their graphs are reflected about the line $y = x$.

96. $f(x) = 5^x$, $g(x) = \log_5 x$

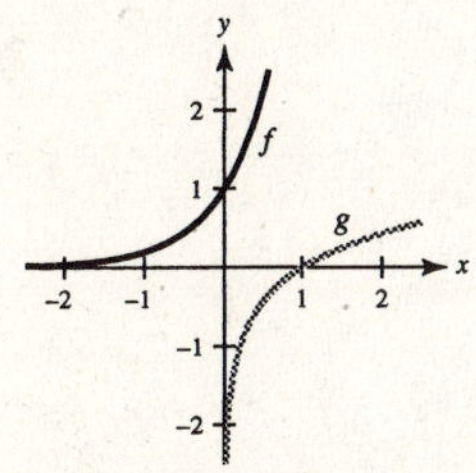

f and g are inverses. Their graphs are reflected about the line $y = x$.

97. $f(x) = e^x$, $g(x) = \ln x$

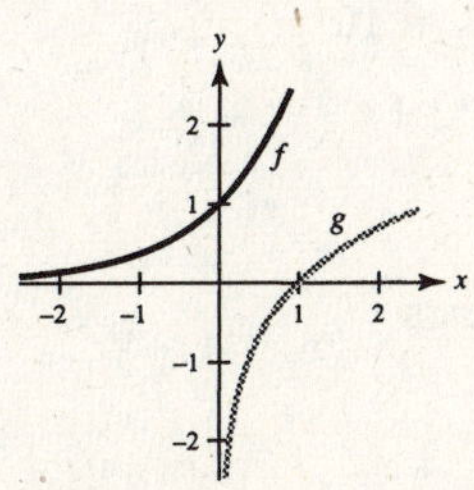

f and g are inverses. Their graphs are reflected about the line $y = x$.

98. $f(x) = 10^x$, $g(x) = \log_{10} x$

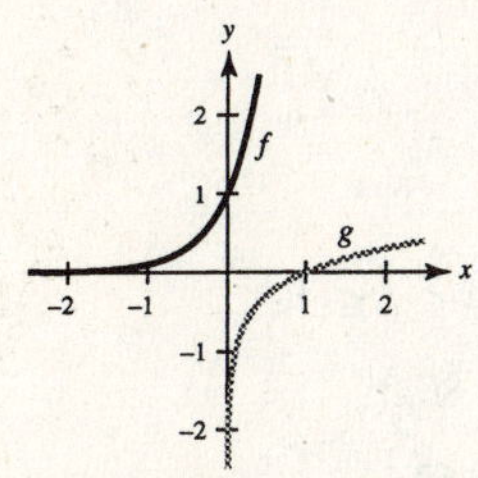

f and g are inverses. Their graphs are reflected about the line $y = x$.

99. (a) $f(x) = \ln x$

$g(x) = \sqrt{x}$

The natural log function grows at a slower rate than the square root function.

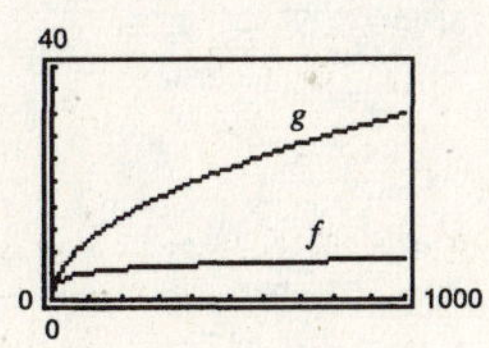

(b) $f(x) = \ln x$

$g(x) = \sqrt[4]{x}$

The natural log function grows at a slower rate than the fourth root function.

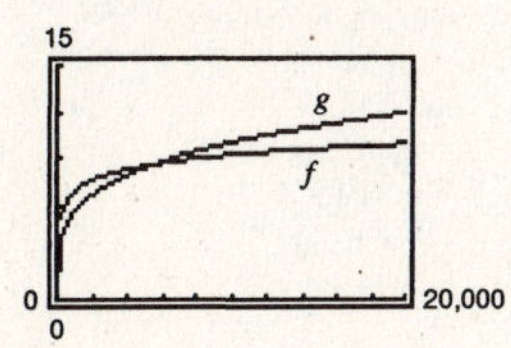

100. (a) False. If y were an exponential function of x, then $y = a^x$, but $a^1 = a$, not 0. Because one point is $(1, 0)$, y is not an exponential function of x.

(b) True. $y = \log_a x$

For $a = 2$, $y = \log_2 x$.

$x = 1, \log_2 1 = 0$

$x = 2, \log_2 2 = 1$

$x = 8, \log_2 8 = 3$

(c) True. $x = a^y$

For $a = 2$, $x = 2^y$.

$y = 0, 2^0 = 1$

$y = 1, 2^1 = 2$

$y = 3, 2^3 = 8$

(d) False. If y were a linear function of x, the slope between $(1, 0)$ and $(2, 1)$ and the slope between $(2, 1)$ and $(8, 3)$ would be the same.

However,

$$m_1 = \frac{1-0}{2-1} = 1 \text{ and } m_2 = \frac{3-1}{8-2} = \frac{2}{6} = \frac{1}{3}.$$

Therefore, y is not a linear function of x.

101. $f(x) = \log_{10} x$

(a) Domain: $(0, \infty)$

(b)
$$y = \log_{10} x$$
$$x = \log_{10} y$$
$$10^x = y$$
$$f^{-1}(x) = 10^x$$

(c) Since $\log_{10} 1000 = 3$ and $\log_{10} 10{,}000 = 4$, the interval in which $f(x)$ will be found is $(3, 4)$.

(d) When $f(x)$ is negative, x is in the interval $(0, 1)$.

(e) $0 = \log_{10} 1$

$1 = \log_{10} 10$

$2 = \log_{10} 100$

$3 = \log_{10} 1000$

When $f(x)$ is increased by one unit, x is increased by a factor of 10.

(f) $f(x_1) = 3n$ $\qquad$ $f(x_2) = n$

$\log_{10} x_1 = 3n$ $\qquad$ $\log_{10} x_2 = n$

$x_1 = 10^{3n}$ $\qquad$ $x_2 = 10^n$

$x_1 : x_2 = 10^{3n} : 10^n = 10^{2n} : 1$

102. $f(t) = 80 - 17\log_{10}(t + 1), 0 \le t \le 12$

(a) $f(0) = 80 - 17\log_{10} 1 = 80.0$

(b) $f(4) = 80 - 17\log_{10} 5 \approx 68.1$

(c) $f(10) = 80 - 17\log_{10} 11 \approx 62.3$

103. $t = \dfrac{10 \ln 2}{\ln 67 - \ln 50} \approx 23.68$ years

104. $t = \dfrac{\ln 2}{r}$

(a)

r	0.005	0.01	0.015	0.02	0.025	0.03
t	138.6	69.3	46.2	34.7	27.7	23.1

(b) Answers will vary.

105. $t = \dfrac{\ln K}{0.095}$

(a)

K	1	2	4	6	8	10	12
t	0	7.3	14.6	18.9	21.9	24.2	26.2

The number of years required to multiply the original investment by K increases with K. However, the larger the value of K, the fewer the years required to increase the value of the investment by an additional multiple of the original investment.

(b)

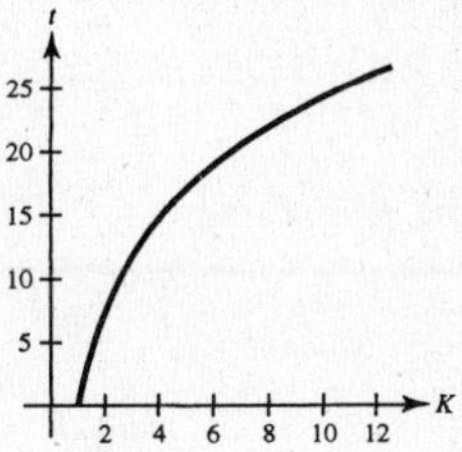

106. $W = 19{,}440(\ln 9 - \ln 3) \approx 21{,}357$ ft-lb

107. $y = 80.4 - 11 \ln x$

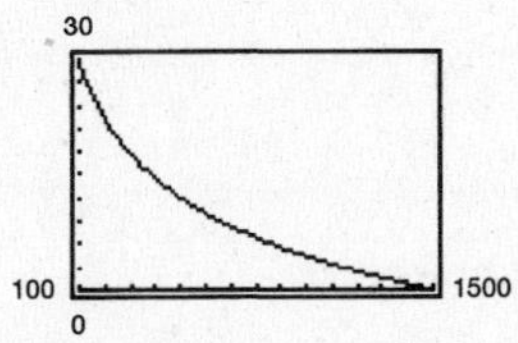

$y(300) = 80.4 - 11 \ln 300 \approx 17.66 \text{ ft}^3/\text{min}$

108. (a) $\dfrac{450}{30} = 15$ cubic feet per minute

(b) 380 cubic feet of air space per child

(c) Total air space required $= 380(30)$

$= 11{,}400$cubic feet

Let $x =$ square feet of floor space and $h = 30$ feet.

$$V = xh$$
$$11{,}400 = x(30)$$
$$380 = x$$

If the ceiling height is 30 feet, the minimum number of square feet of floor space required is 380 square feet.

109. $f(x) = \dfrac{\ln x}{x}$

(a)

x	1	5	10	10^2	10^4	10^6
$f(x)$	0	0.322	0.230	0.046	0.00092	0.0000138

(b) As $x \to \infty$, $f(x) \to 0$.

(c)

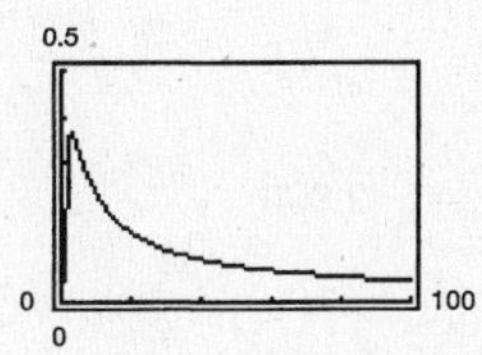

110 $\beta = 10 \log_{10}\left(\dfrac{I}{10^{-12}}\right)$

(a) $\beta = 10 \log_{10}\left(\dfrac{1}{10^{-12}}\right) = 10 \log_{10}(10^{12}) = 120$ decibels

(b) $\beta = 10 \log_{10}\left(\dfrac{10^{-2}}{10^{-12}}\right) = 10 \log_{10}(10^{10}) = 100$ decibels

(c) No, the difference is due to the logarithmic relationship between intensity and number of decibels.

111. $t = 12.542 \ln\left(\dfrac{1100.65}{1100.65 - 1000}\right) \approx 30$ years

112. $t = 12.542 \ln\left(\dfrac{1254.68}{1254.68 - 1000}\right) \approx 20$ years

113. Total amount $= (1100.65)(12)(30) = \$396{,}234$

Interest $= 396{,}234 - 150{,}000 = \$246{,}234$

114. Total amount $= (1254.68)(12)(20) \approx \$301{,}123.20$

Interest $= \$301{,}123.20 - 150{,}000 = \$151{,}123.20$

115. $y = 33.06 + 0.312\sqrt{t}\ln t,\ 1 \leftrightarrow 1991$

(a)

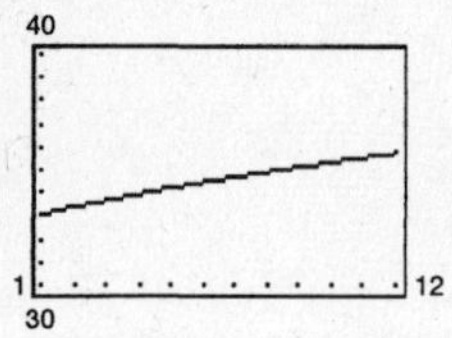

(b) $y(1) \approx 33.06$ years

$y(12) \approx 35.75$ years

(c) The model would probably work for a few years after 2002 since the population of the United States is gradually aging.

(d) The model is undefined for $t \leq 0$ so cannot be used for years prior to 1991.

116. $f(t) = 95 - 12\log_{10}(t + 1),\ 0 \leq t \leq 12$

(a) $f(0) = 95$

(b) $f(6) \approx 84.96$

(c)

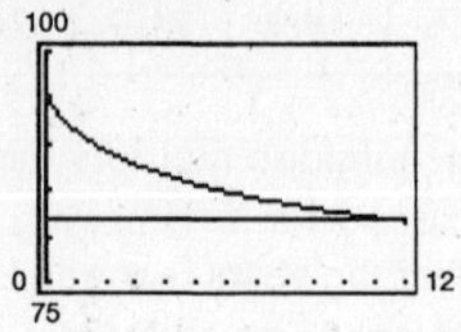

$f(t) = 82$ when $t \approx 11.11$ which corresponds to eleven months after the study.

117. False. You can determine the graph of $f(x) = \log_6 x$ by graphing $g(x) = 6^x$ and reflecting it about the line $y = x$.

118. True since $3^3 = 27$.

119. False. To find the domain of $f(x) = \log_2(bx + c)$ where b and c are nonzero constants, set $bx + c > 0$ and solve.

$$bx + c > 0 \Rightarrow x > -\frac{c}{b}$$

The domain of $f(x) = \log_2(bx + c)$ is $\left(-\frac{c}{b}, \infty\right)$.

120. True. These are the inverse properties.

Section 7.3 Using Properties of Logarithms

1. $\log_3 7 = \dfrac{\log_{10} 7}{\log_{10} 3} = \dfrac{\ln 7}{\ln 3} \approx 1.771$

2. $\log_7 4 = \dfrac{\log_{10} 4}{\log_{10} 7} = \dfrac{\ln 4}{\ln 7} \approx 0.712$

3. $\log_{1/2} 4 = \dfrac{\log_{10} 4}{\log_{10} \frac{1}{2}} = \dfrac{\ln 4}{\ln \frac{1}{2}} = -2.000$

4. $\log_{1/4} 5 = \dfrac{\log_{10} 5}{\log_{10} \frac{1}{4}} = \dfrac{\ln 5}{\ln \frac{1}{4}} \approx -1.161$

5. $\log_9(0.4) = \dfrac{\log_{10} 0.4}{\log_{10} 9} = \dfrac{\ln 0.4}{\ln 9} \approx -0.417$

6. $\log_{20} 0.125 = \dfrac{\log_{10} 0.125}{\log_{10} 20} = \dfrac{\ln 0.125}{\ln 20} \approx -0.694$

7. $\log_{15} 1250 = \dfrac{\log_{10} 1250}{\log_{10} 15} = \dfrac{\ln 1250}{\ln 15} \approx 2.633$

8. $\log_3 0.015 = \dfrac{\log_{10} 0.015}{\log_{10} 3} = \dfrac{\ln 0.015}{\ln 3} \approx -3.823$

9. (a) $\log_5 x = \dfrac{\log_{10} x}{\log_{10} 5}$

(b) $\log_5 x = \dfrac{\ln x}{\ln 5}$

10. (a) $\log_3 x = \dfrac{\log_{10} x}{\log_{10} 3}$

(b) $\log_3 x = \dfrac{\ln x}{\ln 3}$

11. (a) $\log_{1/5} x = \dfrac{\log_{10} x}{\log_{10}\left(\frac{1}{5}\right)}$

(b) $\log_{1/5} x = \dfrac{\ln x}{\ln\left(\frac{1}{5}\right)}$

12. (a) $\log_{1/3} x = \dfrac{\log_{10} x}{\log_{10} \frac{1}{3}}$

(b) $\log_{1/3} x = \dfrac{\ln x}{\ln \frac{1}{3}}$

13. (a) $\log_x \frac{3}{10} = \frac{\log_{10}\left(\frac{3}{10}\right)}{\log_{10} x}$

(b) $\log_x \frac{3}{10} = \frac{\ln\left(\frac{3}{10}\right)}{\ln x}$

14. (a) $\log_x \frac{3}{4} = \frac{\log_{10} \frac{3}{4}}{\log_{10} x}$

(b) $\log_x \frac{3}{4} = \frac{\ln \frac{3}{4}}{\ln x}$

15. (a) $\log_{2.6} x = \frac{\log_{10} x}{\log_{10} 2.6}$

(b) $\log_{2.6} x = \frac{\ln x}{\ln 2.6}$

16. (a) $\log_{7.1} x = \frac{\log_{10} x}{\log_{10} 7.1}$

(b) $\log_{7.1} x = \frac{\ln x}{\ln 7.1}$

17. $f(x) = \log_2 x = \frac{\log_{10} x}{\log_{10} 2} = \frac{\ln x}{\ln 2}$

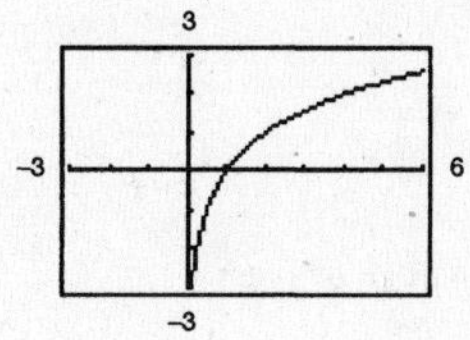

18. $f(x) = \log_4 x = \frac{\log_{10} x}{\log_{10} 4} = \frac{\ln x}{\ln 4}$

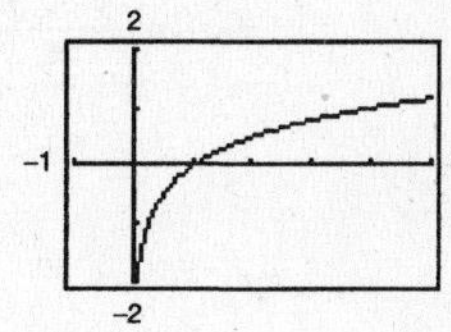

19. $f(x) = \log_{1/2} x = \frac{\log_{10} x}{\log_{10} \frac{1}{2}} = \frac{\ln x}{\ln\left(\frac{1}{2}\right)}$

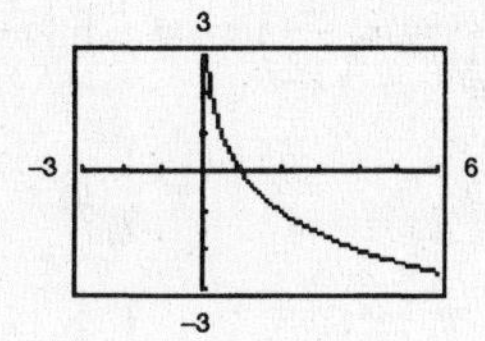

20. $f(x) = \log_{1/4} x = \frac{\log_{10} x}{\log_{10} \frac{1}{4}} = \frac{\ln x}{\ln \frac{1}{4}}$

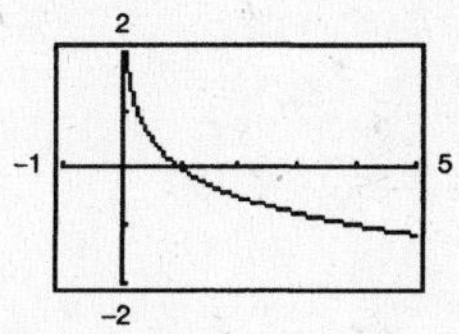

21. $f(x) = \log_{11.8} x = \frac{\log_{10} x}{\log_{10} 11.8} = \frac{\ln x}{\ln 11.8}$

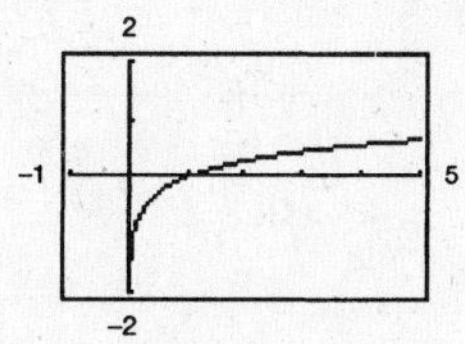

22. $f(x) = \log_{12.4} x = \frac{\log_{10} x}{\log_{10} 12.4} = \frac{\ln x}{\ln 12.4}$

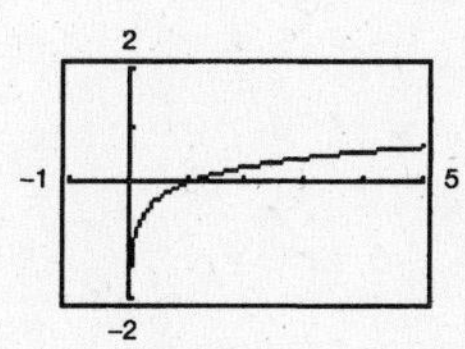

23. $\log_{10} 5x = \log_{10} 5 + \log_{10} x$

24. $\log_{10} 10z = \log_{10} 10 + \log_{10} z$

(Note: This is equivalent to $1 + \log_{10} z$.)

25. $\log_{10} \frac{5}{x} = \log_{10} 5 - \log_{10} x$

26. $\log_{10} \frac{y}{2} = \log_{10} y - \log_{10} 2$

27. $\log_8 x^4 = 4 \log_8 x$

28. $\log_6 z^{-3} = -3 \log_6 z$

29. $\ln \sqrt{z} = \ln z^{1/2} = \frac{1}{2} \ln z$

30. $\ln \sqrt[3]{t} = \ln t^{1/3} = \frac{1}{3} \ln t$

31. $\ln xyz = \ln x + \ln y + \ln z$

32. $\ln \frac{xy}{z} = \ln x + \ln y - \ln z$

33. $\ln \sqrt{a - 1} = \frac{1}{2} \ln(a - 1),\ a > 1$

34. $\ln\left(\frac{x^2 - 1}{x^3}\right) = \ln(x^2 - 1) - \ln x^3$

$= \ln[(x + 1)(x - 1)] - \ln x^3$

$= \ln(x + 1) + \ln(x - 1) - 3 \ln x$

35. $\ln z(z - 1)^2 = \ln z + \ln(z - 1)^2$

$= \ln z + 2 \ln(z - 1),\ z > 1$

36. $\ln \dfrac{x}{\sqrt{x^2+1}} = \ln x - \ln\sqrt{x^2+1}$

$= \ln x - \ln(x^2+1)^{1/2}$

$= \ln x - \dfrac{1}{2}\ln(x^2+1)$

37. $\ln \sqrt[3]{\dfrac{x}{y}} = \dfrac{1}{3}\ln\dfrac{x}{y}$

$= \dfrac{1}{3}[\ln x - \ln y]$

$= \dfrac{1}{3}\ln x - \dfrac{1}{3}\ln y$

38. $\ln \sqrt{\dfrac{x^2}{y^3}} = \ln\left(\dfrac{x^2}{y^3}\right)^{1/2} = \dfrac{1}{2}\ln\left(\dfrac{x^2}{y^3}\right)$

$= \dfrac{1}{2}(\ln x^2 - \ln y^3)$

$= \dfrac{1}{2}(2\ln x - 3\ln y)$

39. $\ln\left(\dfrac{x^4\sqrt{y}}{z^5}\right) = \ln x^4\sqrt{y} - \ln z^5$

$= \ln x^4 + \ln\sqrt{y} - \ln z^5$

$= 4\ln x + \dfrac{1}{2}\ln y - 5\ln z$

40. $\ln\sqrt{x^2(x+2)} = \ln[x^2(x+2)]^{1/2}$

$= \ln[x(x+2)^{1/2}]$

$= \ln x + \ln(x+2)^{1/2}$

$= \ln x + \dfrac{1}{2}\ln(x+2)$

41. $\log_b\left(\dfrac{x^2}{y^2z^3}\right) = \log_b x^2 - \log_b y^2z^3$

$= \log_b x^2 - [\log_b y^2 + \log_b z^3]$

$= 2\log_b x - 2\log_b y - 3\log_b z$

42. $\log_b \dfrac{\sqrt{x}\,y^4}{z^4} = \log_b \sqrt{x}\,y^4 - \log_b z^4$

$= \log_b x^{1/2} + \log_b y^4 - \log_b z^4$

$= \dfrac{1}{2}\log_b x + 4\log_b y - 4\log_b z$

43. $\ln x + \ln 3 = \ln 3x$

44. $\ln y + \ln t = \ln yt = \ln ty$

45. $\log_4 z - \log_4 y = \log_4 \dfrac{z}{y}$

46. $\log_5 8 - \log_5 t = \log_5 \dfrac{8}{t}$

47. $2\log_2(x+4) = \log_2(x+4)^2$

48. $-4\log_6 2x = \log_6(2x)^{-4} = \log_6 \dfrac{1}{16x^4}$

49. $\frac{1}{4}\log_3 5x = \log_3(5x)^{1/4} = \log_3 \sqrt[4]{5x}$

50. $\frac{2}{3}\log_7(z-2) = \log_7(z-2)^{2/3}$

51. $\ln x - 3\ln(x+1) = \ln x - \ln(x+1)^3$

$= \ln \dfrac{x}{(x+1)^3}$

52. $2\ln 8 + 5\ln z = \ln 8^2 + \ln z^5$

$= \ln 64z^5$

53. $\ln(x-2) - \ln(x+2) = \ln\left(\dfrac{x-2}{x+2}\right)$

54. $3\ln x + 4\ln y - 4\ln z = \ln x^3 + \ln y^4 - \ln z^4$

$= \ln x^3y^4 - \ln z^4$

$= \ln \dfrac{x^3y^4}{z^4}$

55. $\ln x - 4[\ln(x+2) + \ln(x-2)] = \ln x - 4\ln(x+2)(x-2)$

$= \ln x - 4\ln(x^2-4)$

$= \ln x - \ln(x^2-4)^4$

$= \ln \dfrac{x}{(x^2-4)^4}$

56. $4[\ln z + \ln(z+5)] - 2\ln(z-5) = 4[\ln z(z+5)] - \ln(z-5)^2$

$= \ln[z(z+5)]^4 - \ln(z-5)^2$

$= \ln \dfrac{z^4(z+5)^4}{(z-5)^2}$

57. $\frac{1}{3}[2\ln(x+3) + \ln x - \ln(x^2-1)] = \frac{1}{3}[\ln(x+3)^2 + \ln x - \ln(x^2-1)]$

$= \frac{1}{3}[\ln x(x+3)^2 - \ln(x^2-1)]$

$= \dfrac{1}{3}\ln \dfrac{x(x+3)^2}{x^2-1}$

$= \ln \sqrt[3]{\dfrac{x(x+3)^2}{x^2-1}}$

58. $2[\ln x - \ln(x+1) - \ln(x-1)] = 2\left[\ln \dfrac{x}{x+1} - \ln(x-1)\right]$

$= 2\left[\ln \dfrac{x}{(x+1)(x-1)}\right]$

$= 2\left[\ln \dfrac{x}{x^2-1}\right]$

$= \ln\left(\dfrac{x}{x^2-1}\right)^2$

59. $\frac{1}{3}[\ln y + 2\ln(y+4)] - \ln(y-1) = \frac{1}{3}[\ln y + \ln(y+4)^2] - \ln(y-1)$

$= \dfrac{1}{3}\ln y(y+4)^2 - \ln(y-1)$

$= \ln \sqrt[3]{y(y+4)^2} - \ln(y-1)$

$= \ln \dfrac{\sqrt[3]{y(y+4)^2}}{y-1}$

60. $\frac{1}{2}[\ln(x+1) + 2\ln(x-1)] + 6\ln x = \frac{1}{2}[\ln(x+1) + \ln(x-1)^2] + \ln x^6$

$= \frac{1}{2}[\ln(x+1)(x-1)^2] + \ln x^6$

$= \ln[(x+1)(x-1)^2]^{1/2} + \ln x^6$

$= \ln[(x+1)^{1/2}(x-1)] + \ln x^6$

$= \ln\left[x^6(x-1)\sqrt{x+1}\right]$

61. $\frac{1}{2}[\ln(x+3) + \ln(x-3)] - \ln(x^2-9) = \frac{1}{2}\ln[(x+3)(x-3)] - \ln[(x+3)(x-3)]$

$$= -\frac{1}{2}\ln[(x+3)(x-3)]$$
$$= \ln[(x^2-9)^{-1/2}]$$
$$= \ln\left[\frac{1}{\sqrt{x^2-9}}\right]$$
$$= \ln\left[1/\sqrt{(x+3)(x-3)}\right]$$

62. $\frac{1}{2}\ln(x^2-16) - \ln(x+4) - \ln(x-4) = \frac{1}{2}\ln[(x+4)(x-4)] - [\ln(x+4) + \ln(x-4)]$

$$= \frac{1}{2}\ln[(x+4)(x-4)] - \ln[(x+4)(x-4)]$$
$$= -\frac{1}{2}\ln[(x+4)(x-4)]$$
$$= \ln[(x^2-16)^{-1/2}]$$
$$= \ln\left[\frac{1}{\sqrt{x^2-16}}\right]$$
$$= \ln\left[1/\sqrt{(x+4)(x-4)}\right]$$

63. $\ln(x+2) + \ln(x-2) - \ln\sqrt{x^2-4} = \ln[(x+2)(x-2)] - \frac{1}{2}\ln(x^2-4)$

$$= \ln(x^2-4) - \frac{1}{2}\ln(x^2-4)$$
$$= \frac{1}{2}\ln(x^2-4)$$
$$= \ln\sqrt{x^2-4}$$
$$= \ln\sqrt{(x+2)(x-2)}$$

64. $\ln(x+1) + \ln(x-1) - 2\ln(x^2-1) = \ln[(x+1)(x-1)] - 2\ln(x^2-1)$

$$= \ln(x^2-1) - 2\ln(x^2-1)$$
$$= -1\ln(x^2-1)$$
$$= \ln[(x^2-1)^{-1}]$$
$$= \ln\left(\frac{1}{x^2-1}\right)$$
$$= \ln\left[\frac{1}{(x+1)(x-1)}\right]$$

65. $\log_2\frac{32}{4} = \log_2 32 - \log_2 4 \neq \frac{\log_2 32}{\log_2 4}$

The first two expressions are equal by Property 2.

66. $\log_7\sqrt{70} = \frac{1}{2}\log_7 70 = \frac{1}{2}[\log_7 7 + \log_7 10] = \frac{1}{2}[1 + \log_7 10] = \frac{1}{2} + \frac{1}{2}\log_7 10$

$= \frac{1}{2} + \log_7\sqrt{10}$ by Property 1 and Property 3

67. $\log_3 9 = 2\log_3 3 = 2$

68. $\log_6\sqrt[3]{6} = \log_6 6^{1/3} = \frac{1}{3}\log_6 6 = \frac{1}{3}(1) = \frac{1}{3}$

69. $\log_4 16^{1.2} = 1.2(\log_4 16) = 1.2 \log_4 4^2 = 1.2(2) = 2.4$

70. $\log_5 \frac{1}{125} = \log_5 5^{-3} = -3 \log_5 5 = -3(1) = -3$

71. $\log_3(-9)$ is undefined. -9 is not in the domain of $\log_3 x$.

72. $\log_2(-16)$ is undefined because -16 is not in the domain of $\log_2 x$.

73.
$$\begin{aligned}\log_5 75 - \log_5 3 &= \log_5 \tfrac{75}{3}\\ &= \log_5 25\\ &= \log_5 5^2\\ &= 2 \log_5 5\\ &= 2\end{aligned}$$

74.
$$\begin{aligned}\log_4 2 + \log_4 32 &= \log_4 4^{1/2} + \log_4 4^{5/2}\\ &= \tfrac{1}{2}\log_4 4 + \tfrac{5}{2}\log_4 4\\ &= \tfrac{1}{2}(1) + \tfrac{5}{2}(1)\\ &= 3\end{aligned}$$

75. $\ln e^2 - \ln e^5 = 2 - 5 = -3$

76. $3 \ln e^4 = (3)(4) \ln e = 12(1) = 12$

77. $\log_{10} 0$ is undefined. 0 is not in the domain of $\log_{10} x$.

78. $\ln 1 = 0$

79. $\ln e^{4.5} = 4.5$

80. $\ln \sqrt[4]{e^3} = \ln e^{3/4} = \frac{3}{4} \ln e = \frac{3}{4}(1) = \frac{3}{4}$

81. $\log_4 8 = \dfrac{\log_2 8}{\log_2 4} = \dfrac{\log_2 2^3}{\log_2 2^2} = \dfrac{3}{2}$

82.
$$\begin{aligned}\log_2(4^2 \cdot 3^4) &= \log_2 4^2 + \log_2 3^4\\ &= 2\log_2 4 + 4\log_2 3\\ &= 2\log_2 2^2 + 4\log_2 3\\ &= 4\log_2 2 + 4\log_2 3\\ &= 4 + 4\log_2 3\end{aligned}$$

83.
$$\begin{aligned}\log_5 \tfrac{1}{250} &= \log_5\left(\tfrac{1}{125}\cdot\tfrac{1}{2}\right) = \log_5 \tfrac{1}{125} + \log_5 \tfrac{1}{2}\\ &= \log_5 5^{-3} + \log_5 2^{-1}\\ &= -3 - \log_5 2\end{aligned}$$

84.
$$\begin{aligned}\log_{10} \tfrac{9}{300} &= \log_{10} \tfrac{3}{100}\\ &= \log_{10} 3 - \log_{10} 100\\ &= \log_{10} 3 - \log_{10} 10^2\\ &= \log_{10} 3 - 2\log_{10} 10\\ &= \log_{10} 3 - 2\end{aligned}$$

85. $\ln(5e^6) = \ln 5 + \ln e^6 = \ln 5 + 6 = 6 + \ln 5$

86. $\ln \dfrac{6}{e^2} = \ln 6 - \ln e^2 = \ln 6 - 2\ln e = \ln 6 - 2$

87. $\log_b 10 = \log_b(2 \cdot 5) = \log_b 2 + \log_b 5 = 0.2891 + 0.6712 = 0.9603$

88. $\log_b 6 = \log_b(2 \cdot 3) = \log_b 2 + \log_b 3 = 0.2891 + 0.4582 = 0.7473$

89.
$$\begin{aligned}\log_b 2.5 = \log_b \tfrac{5}{2} = \log_b 5 - \log_b 2 &= 0.6712 - 0.2891\\ &= 0.3821\end{aligned}$$

90.
$$\begin{aligned}\log_b 0.4 = \log_b \tfrac{2}{5} = \log_b 2 - \log_b 5 &= 0.2891 - 0.6712\\ &= -0.3821\end{aligned}$$

91.
$$\begin{aligned}\log_b \sqrt{15b} &= \tfrac{1}{2}\log_b(3 \cdot 5 \cdot b)\\ &= \tfrac{1}{2}[\log_b 3 + \log_b 5 + \log_b b]\\ &= \tfrac{1}{2}[0.4582 + 0.6712 + 1]\\ &= 1.0647\end{aligned}$$

92.
$$\begin{aligned}\log_b \sqrt{3b^3} &= \tfrac{1}{2}\log_b(3b^3)\\ &= \tfrac{1}{2}[\log_b 3 + \log_b b^3]\\ &= \tfrac{1}{2}[0.4582 + 3]\\ &= 1.7291\end{aligned}$$

93. $\log_b\left(\frac{25}{3b}\right) = \log_b 25 - \log_b 3b$

$= \log_b 5^2 - [\log_b 3 + \log_b b]$

$= 2\log_b 5 - \log_b 3 - \log_b b$

$= 2(0.6712) - 0.4582 - 1$

$= -0.1158$

94. $\log_b\left(\frac{32b^2}{5}\right) = \log_b(32b^2) - \log_b 5$

$= \log_b 32 + \log_b b^2 - \log_b 5$

$= \log_b 2^5 + \log_b b^2 - \log_b 5$

$= 5\log_b 2 + 2\log_b b - \log_b 5$

$= 5(0.2891) + 2(1) - 0.6712$

$= 2.7743$

95. $f(x) = \log_{10} x$

$g(x) = \frac{\ln x}{\ln 10}$

$f(x) = g(x)$

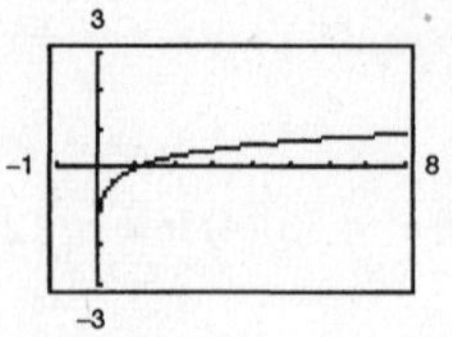

96. $f(x) = \ln x$

$g(x) = \frac{\log_{10} x}{\log_{10} e}$

$f(x) = g(x)$

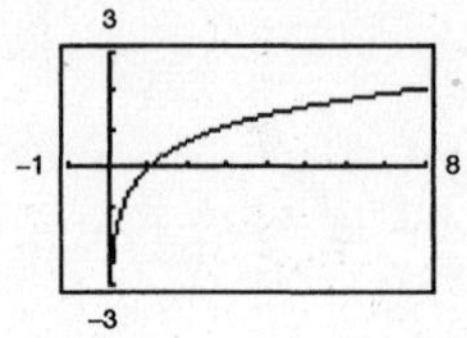

97. $f(x) = \ln\frac{x}{2},\ g(x) = \frac{\ln x}{\ln 2},\ h(x) = \ln x - \ln 2$

$f(x) = h(x)$ by Property 2.

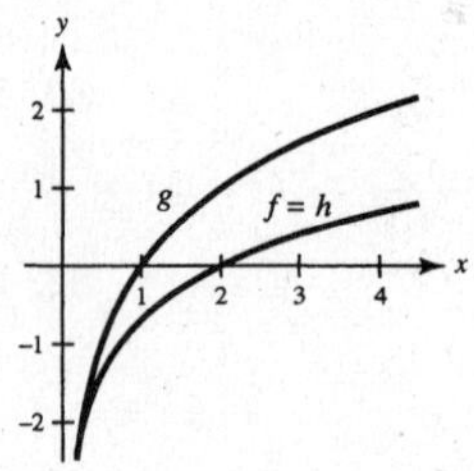

98. $f(t) = 90 - 15\log_{10}(t+1),\ 0 \le t \le 12$

(a) $f(0) = 90$

(b) $f(6) \approx 77$

(c) $f(12) \approx 73$

(d) $75 = 90 - 15\log_{10}(t+1)$

$-15 = -15\log_{10}(t+1)$

$1 = \log_{10}(t+1)$

$10^1 = t + 1$

$t = 9$ months

(e) $f(t) = 90 - \log_{10}(t+1)^{15}$

(f)

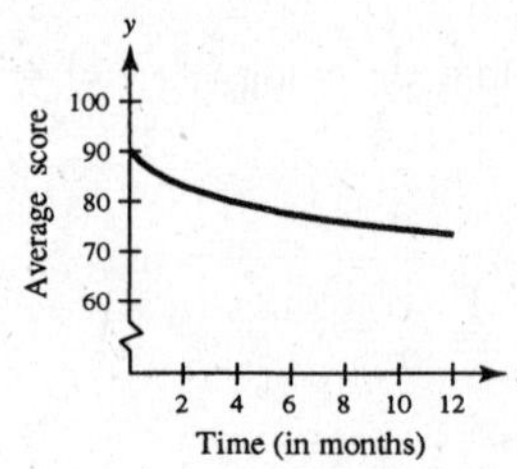

99. $\beta = 10\log_{10}\left(\frac{I}{10^{-12}}\right)$

$= 10[\log_{10} I - \log_{10} 10^{-12}]$

$= 10[\log_{10} I - (-12)\log_{10} 10]$

$= 10[\log_{10} I + 12]$

For $I = 10^{-6}$, $\beta = 10[\log_{10} 10^{-6} + 12] = 60$ decibels

100. (a) $f(x) = \ln x^2$

$g(x) = 2\ln x$

The graphs are identical for $x > 0$, $f(x) = \ln x^2$ is also defined for $x < 0$, whereas $g(x) = 2\ln x$ is not defined for $x < 0$, so the graphs are not identical here.

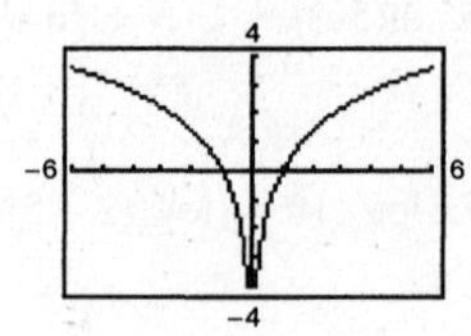

(b) $f(x) = \ln\sqrt{x^2+1}$

$g(x) = \frac{1}{2}\ln(x^2+1)$

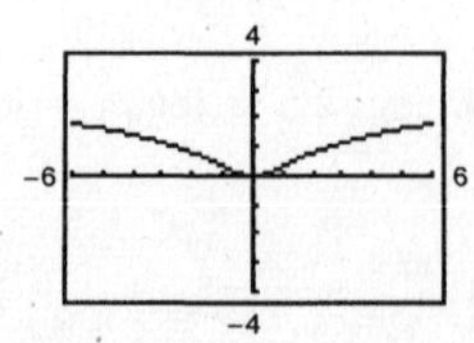

The graphs are identical. The domains of $f(x)$ and $g(x)$ are both all real numbers and $f(x) = g(x)$ by properties of logarithms.

101. $f(x) = \ln x$

False, $f(0) \neq 0$ since 0 is not in the domain of $f(x)$.
$f(1) = \ln 1 = 0$

102. $f(ax) = f(a) + f(x), a > 0, x > 0$

True, because

$$f(ax) = \ln ax = \ln a + \ln x = f(a) + f(x).$$

103. False. $f(x) - f(2) = \ln x - \ln 2 = \ln \frac{x}{2} \neq \ln(x - 2)$

104. $\sqrt{f(x)} = \frac{1}{2} f(x)$; False.

$\sqrt{f(x)} = \sqrt{\ln x}$ can't be simplified further.

$$f(\sqrt{x}) = \ln \sqrt{x} = \ln x^{1/2} = \frac{1}{2} \ln x = \frac{1}{2} f(x)$$

105. False.

$$f(u) = 2f(v) \Rightarrow \ln u = 2 \ln v \Rightarrow \ln u = \ln v^2 \Rightarrow u = v^2$$

106. If $f(x) < 0$, then $0 < x < 1$.

True.

107. Let $x = \log_b u$ and $y = \log_b v$, then $b^x = u$ and $b^y = v$.

$$\frac{u}{v} = \frac{b^x}{b^y} = b^{x-y}$$

Then $\log_b\left(\frac{u}{v}\right) = \log_b(b^{x-y}) = x - y = \log_b u - \log_b v$.

108. Let $x = \log_b u$, then $u = b^x$ and $u^n = b^{nx}$.

$\log_b u^n = \log_b b^{nx} = nx = n \log_b u$

Section 7.4 Exponential and Logarithmic Equations

1. $4^{2x-7} = 64$

(a) $x = 5$

$4^{2(5)-7} = 4^3 = 64$

Yes, $x = 5$ is a solution.

(b) $x = 2$

$4^{2(2)-7} = 4^{-3} = \frac{1}{64} \neq 64$

No, $x = 2$ is not a solution.

2. $2^{3x+1} = 32$

(a) $x = -1$

$2^{3(-1)+1} = 2^{-2} = \frac{1}{4}$

No, $x = -1$ is not a solution.

(b) $x = 2$

$2^{3(2)+1} = 2^7 = 128$

No, $x = 2$ is not a solution.

3. $3e^{x+2} = 75$

(a) $x = -2 + e^{25}$

$3e^{(-2+e^{25})+2} = 3e^{e^{25}} \neq 75$

No, $x = -2 + e^{25}$ is not a solution.

(b) $x = -2 + \ln 25$

$3e^{(-2+\ln 25)+2} = 3e^{\ln 25} = 3(25) = 75$

Yes, $x = -2 + \ln 25$ is a solution.

(c) $x \approx 1.219$

$3e^{1.219+2} = 3e^{3.219} \approx 75$

Yes, $x \approx 1.219$ is a solution.

4. $5^{2x+3} = 812$

(a) $x = -1.5 + \log_5 \sqrt{812}$

$$5^{2(-1.5+\log_5\sqrt{812})+3} = 5^{-3+2\log_5\sqrt{812}+3}$$
$$= 5^{2\log_5 812^{1/2}}$$
$$= 5^{\log_5(812^{1/2})^2}$$
$$= 5^{\log_5 812} = 812$$

Yes, $x = -1.5 + \log_5\sqrt{812}$ is a solution.

(b) $x \approx 0.581$

$5^{2(0.581)+3} = 5^{4.162} \approx 811$

No, $x \approx 0.581$ is not a solution.

(c) $x = \dfrac{1}{2}\left(-3 + \dfrac{\ln 812}{\ln 5}\right)$

$$5^{2[1/2(-3+(\ln 812/\ln 5))]+3} = 5^{-3+(\ln 812/\ln 5)+3}$$
$$= 5^{\ln 812/\ln 5}$$
$$= 5^{\log_5 812}$$
$$= 812$$

Yes, $x = \dfrac{1}{2}\left(-3 + \dfrac{\ln 812}{\ln 5}\right)$ is a solution.

5. $\log_4(3x) = 3 \Rightarrow 3x = 4^3 \Rightarrow 3x = 64$

(a) $x \approx 21.3560$

$3(21.3560) = 64.0680 \neq 64$

No, $x \approx 21.3560$ is not a solution.

(b) $x = -4$

$3(-4) = -12 \neq 64$

No, $x = -4$ is not a solution.

(c) $x = \dfrac{64}{3}$

$3\left(\dfrac{64}{3}\right) = 64$

Yes, $x = \dfrac{64}{3}$ is a solution.

6. $\ln(x - 1) = 3.8$

(a) $x = 1 + e^{3.8}$

$\ln(1 + e^{3.8} - 1) = \ln e^{3.8} = 3.8$

Yes, $x = 1 + e^{3.8}$ is a solution.

(b) $x \approx 45.701$

$\ln(45.701 - 1) = \ln(44.701) \approx 3.8$

Yes, $x \approx 45.701$ is a solution.

(c) $x = 1 + \ln 3.8$

$\ln(1 + \ln 3.8 - 1) = \ln(\ln 3.8) \approx 0.289$

No, $x = 1 + \ln 3.8$ is not a solution.

7. $4^x = 16$

$4^x = 4^2$

$x = 2$

8. $3^x = 243$

$3^x = 3^5$

$x = 5$

9. $5^x = \dfrac{1}{625}$

$5^x = \dfrac{1}{5^4}$

$5^x = 5^{-4}$

$x = -4$

10. $3^x = \dfrac{1}{81}$

$3^x = \dfrac{1}{3^4}$

$3^x = 3^{-4}$

$x = -4$

11. $27^x = 9$

$(3^3)^x = 3^2$

$3^{3x} = 3^2$

$3x = 2$

$x = \frac{2}{3}$

12. $8^x = 4$

$8^x = 8^{2/3}$

$x = \frac{2}{3}$

13. $\left(\frac{1}{2}\right)^x = 32$

$2^{-x} = 2^5$

$-x = 5$

$x = -5$

14. $\left(\frac{1}{4}\right)^x = 64$

$4^{-x} = 4^3$

$-x = 3$

$x = -3$

15. $\left(\frac{3}{4}\right)^x = \frac{27}{64}$

$\left(\frac{3}{4}\right)^x = \left(\frac{3}{4}\right)^3$

$x = 3$

16. $\left(\frac{2}{3}\right)^x = \frac{4}{9}$

$\left(\frac{2}{3}\right)^x = \left(\frac{2}{3}\right)^2$

$x = 2$

17. $3^{x-1} = 27$

$3^{x-1} = 3^3$

$x - 1 = 3$

$x = 4$

18. $2^{x-3} = 32$

$2^{x-3} = 2^5$

$x - 3 = 5$

$x = 8$

19. $\ln x - \ln 2 = 0$

$\ln x = \ln 2$

$x = 2$

20. $\ln x - \ln 5 = 0$

$\ln x = \ln 5$

$x = 5$

21. $e^x = 2$

$\ln e^x = \ln 2$

$x = \ln 2$

$x \approx 0.693$

22. $e^x = 4$

$\ln e^x = \ln 4$

$x = \ln 4$

$x \approx 1.386$

23. $\ln x = -1$

$e^{\ln x} = e^{-1}$

$x = e^{-1}$

$x \approx 0.368$

24. $\ln x = -7$

$e^{\ln x} = e^{-7}$

$x = e^{-7}$

$x \approx 0.000912$

25. $\log_x 16 = 4$

$x^4 = 16$

$x = \sqrt[4]{16}$

$x = 2$

26. $\log_x 625 = 4$

$x^{\log_x 625} = x^4$

$625 = x^4$

$\sqrt[4]{625} = x$

$5 = x$

27. $\log_{10} x - 2 = 0$

$\log_{10} x = 2$

$10^{\log_{10} x} = 10^2$

$x = 10^2$

$x = 100$

28. $\log_{10} x + 3 = 0$

$\log_{10} x = -3$

$10^{\log_{10} x} = 10^{-3}$

$x = 10^{-3}$

$x = 0.001$

29. $\log_{1/2} x = -3$

$\left(\frac{1}{2}\right)^{-3} = x$

$2^3 = x$

$8 = x$

30. $\log_{1/3} x = -2$

$\left(\frac{1}{3}\right)^{-2} = x$

$3^2 = x$

$9 = x$

31. $f(x) = g(x)$

$2^x = 8$

$2^x = 2^3$

$x = 3$

Point of intersection: $(3, 8)$

32. $f(x) = g(x)$

$27^x = 9$

$27^x = 27^{2/3}$

$x = \frac{2}{3}$

Point of intersection: $\left(\frac{2}{3}, 9\right)$

33. $f(x) = g(x)$

$\log_3 x = 2$

$x = 3^2$

$x = 9$

Point of intersection: $(9, 2)$

34. $f(x) = g(x)$

$\ln(x - 4) = 0$

$e^{\ln(x-4)} = e^0$

$x - 4 = 1$

$x = 5$

Point of intersection: $(5, 0)$

35. $\log_{10} 10^{x^2} = x^2$

36. $\log_6 6^{2x-1} = 2x - 1$

37. $8^{\log_8(x-2)} = x - 2$

38. $4^{\log_4 x^3} = x^3$

39. $\ln e^{7x+2} = 7x + 2$

40. $\ln e^{x^4} = x^4$

41. $e^{\ln(5x+2)} = 5x + 2$

42. $e^{\ln x^2} = x^2$

43. $-1 + \ln e^{2x} = -1 + 2x = 2x - 1$

44. $-8 + e^{\ln x^3} = -8 + x^3 = x^3 - 8$

45. $e^x = 10$

$x = \ln 10 \approx 2.303$

46. $4e^x = 91$

$e^x = \dfrac{91}{4}$

$\ln e^x = \ln \dfrac{91}{4}$

$x = \ln \dfrac{91}{4} \approx 3.125$

47. $7 - 2e^x = 5$

$-2e^x = -2$

$e^x = 1$

$x = \ln 1 = 0$

48. $-14 + 3e^x = 11$

$3e^x = 25$

$e^x = \dfrac{25}{3}$

$\ln e^x = \ln \dfrac{25}{3}$

$x = \ln \dfrac{25}{3} \approx 2.120$

49. $e^{3x} = 12$

$3x = \ln 12$

$x = \dfrac{\ln 12}{3} \approx 0.828$

50. $e^{2x} = 50$

$\ln e^{2x} = \ln 50$

$2x = \ln 50$

$x = \dfrac{\ln 50}{2} \approx 1.956$

51. $500e^{-x} = 300$

$e^{-x} = \dfrac{3}{5}$

$-x = \ln \dfrac{3}{5}$

$x = -\ln \dfrac{3}{5} = \ln \dfrac{5}{3} \approx 0.511$

52. $1000e^{-4x} = 75$

$e^{-4x} = \frac{3}{40}$

$\ln e^{-4x} = \ln \frac{3}{40}$

$-4x = \ln \frac{3}{40}$

$x = -\frac{1}{4} \ln \frac{3}{40} \approx 0.648$

53. $e^{2x} - 4e^x - 5 = 0$

$(e^x + 1)(e^x - 5) = 0$

$e^x = -1$ or $e^x = 5$

(No solution) $x = \ln 5 \approx 1.609$

54. $e^{2x} - 5e^x + 6 = 0$

$(e^x - 2)(e^x - 3) = 0$

$e^x = 2$ or $e^x = 3$

$x = \ln 2 \approx 0.693$ or $x = \ln 3 \approx 1.099$

55. $20(100 - e^{x/2}) = 500$

$100 - e^{x/2} = 25$

$-e^{x/2} = -75$

$e^{x/2} = 75$

$\dfrac{x}{2} = \ln 75$

$x = 2 \ln 75 \approx 8.635$

56. $\dfrac{400}{1 + e^{-x}} = 350$

$400 = 350(1 + e^{-x})$

$\dfrac{8}{7} = 1 + e^{-x}$

$\dfrac{8}{7} - 1 = e^{-x}$

$\dfrac{1}{7} = e^{-x}$

$\ln \dfrac{1}{7} = \ln e^{-x}$

$-x = \ln \dfrac{1}{7}$

$-x = \ln 7^{-1}$

$-x = -\ln 7$

$x = \ln 7 \approx 1.946$

57. $10^x = 42$

$x = \log_{10} 42 \approx 1.623$

58. $10^x = 570$

$\log_{10} 10^x = \log_{10} 570$

$x = \log_{10} 570$

≈ 2.756

59. $3^{2x} = 80$

$\ln 3^{2x} = \ln 80$

$2x \ln 3 = \ln 80$

$x = \dfrac{\ln 80}{2 \ln 3} \approx 1.994$

60.
$$\begin{aligned} 6^{5x} &= 3000 \\ \ln 6^{5x} &= \ln 3000 \\ (5x)\ln 6 &= \ln 3000 \\ 5x &= \frac{\ln 3000}{\ln 6} \\ x &= \frac{\ln 3000}{5 \ln 6} \\ &\approx 0.894 \end{aligned}$$

61.
$$\begin{aligned} 5^{-t/2} &= 0.20 \\ 5^{-t/2} &= \frac{1}{5} \\ 5^{-t/2} &= 5^{-1} \\ -\frac{t}{2} &= -1 \\ t &= 2 \end{aligned}$$

62.
$$\begin{aligned} 4^{-3t} &= 0.10 \\ \ln 4^{-3t} &= \ln 0.10 \\ (-3t)\ln 4 &= \ln 0.10 \\ -3t &= \frac{\ln 0.10}{\ln 4} \\ t &= -\frac{\ln 0.10}{3 \ln 4} \approx 0.554 \end{aligned}$$

63.
$$\begin{aligned} 2^{3-x} &= 565 \\ \ln 2^{3-x} &= \ln 565 \\ (3-x)\ln 2 &= \ln 565 \\ 3\ln 2 - x\ln 2 &= \ln 565 \\ -x\ln 2 &= \ln 565 - \ln 2^3 \\ x\ln 2 &= \ln 8 - \ln 565 \\ x &= \frac{\ln 8 - \ln 565}{\ln 2} \approx -6.142 \end{aligned}$$

64.
$$\begin{aligned} 8^{-2-x} &= 431 \\ \ln 8^{-2-x} &= \ln 431 \\ (-2-x)\ln 8 &= \ln 431 \\ -2\ln 8 - x\ln 8 &= \ln 431 \\ -x\ln 8 &= \ln 431 + \ln 8^2 \\ x\ln 8 &= -\ln 431 - \ln 64 \\ x &= \frac{-\ln 431 - \ln 64}{\ln 8} \approx -4.917 \end{aligned}$$

65. $g(x) = 6e^{1-x} - 25$

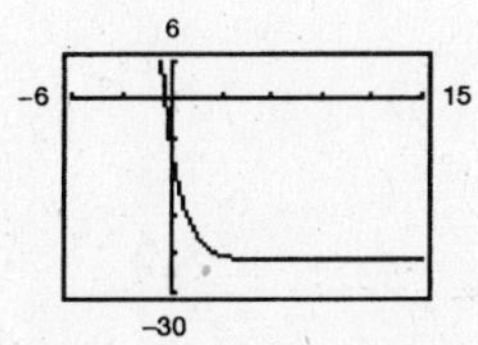

The zero is $x \approx -0.427$.

66. $f(x) = -4e^{-x-1} + 15$

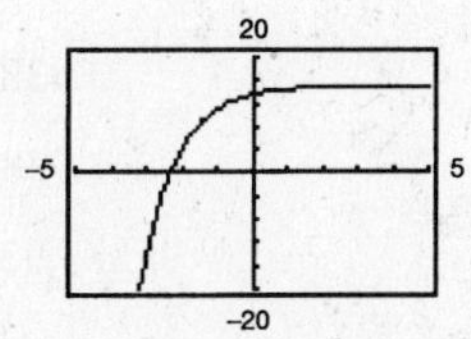

The zero is -2.322.

67. $f(x) = 3e^{3x/2} - 962$

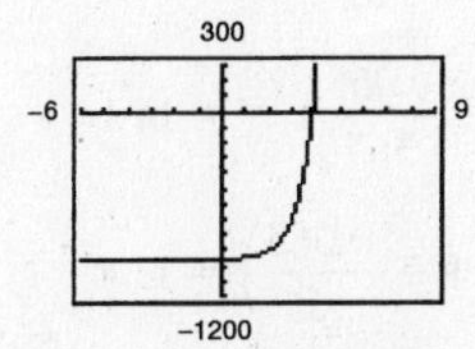

The zero is $x \approx 3.847$.

68. $g(x) = 8e^{-2x/3} - 11$

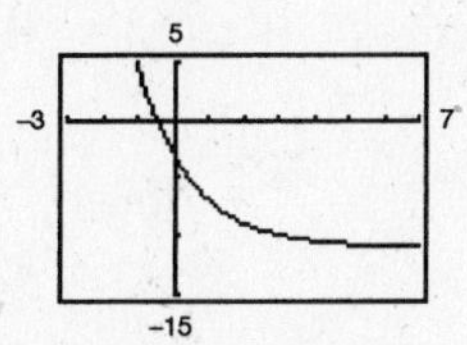

The zero is -0.478.

69. $g(t) = e^{0.09t} - 3$

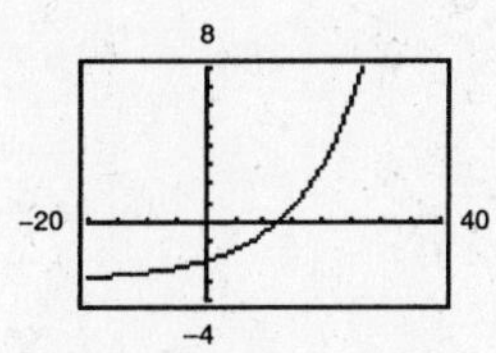

The zero is $t \approx 12.207$.

70. $f(x) = -e^{1.8x} + 7$

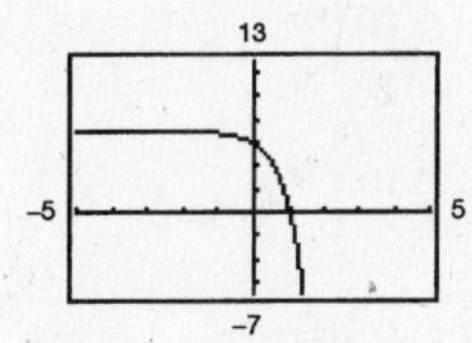

The zero is 1.081.

71. $h(t) = e^{0.125t} - 8$

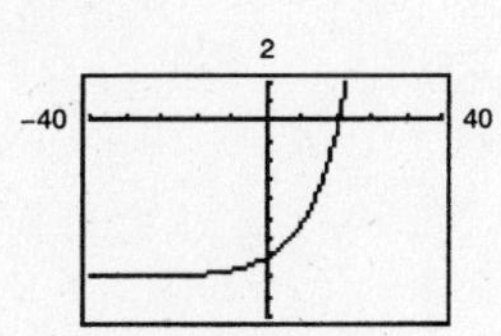

The zero is $t \approx 16.636$.

72. $f(x) = e^{2.724x} - 29$

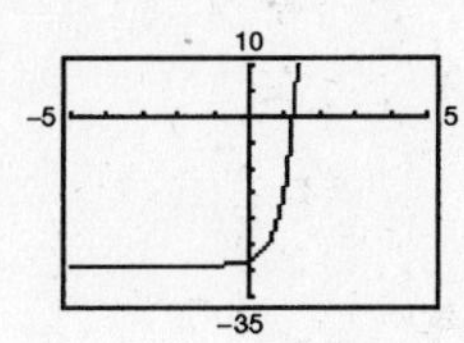

The zero is 1.236.

73.
$$8(10^{3x}) = 12$$
$$10^{3x} = \tfrac{12}{8}$$
$$\log_{10} 10^{3x} = \log_{10}\left(\tfrac{3}{2}\right)$$
$$3x = \log_{10}\left(\tfrac{3}{2}\right)$$
$$x = \tfrac{1}{3}\log_{10}\left(\tfrac{3}{2}\right) \approx 0.059$$

74.
$$5(10^{x-6}) = 7$$
$$10^{x-6} = \tfrac{7}{5}$$
$$\log_{10} 10^{x-6} = \log_{10} \tfrac{7}{5}$$
$$x - 6 = \log_{10} \tfrac{7}{5}$$
$$x = 6 + \log_{10} \tfrac{7}{5} \approx 6.146$$

75.
$$3(5^{x-1}) = 21$$
$$5^{x-1} = 7$$
$$\ln 5^{x-1} = \ln 7$$
$$(x-1)\ln 5 = \ln 7$$
$$x - 1 = \frac{\ln 7}{\ln 5}$$
$$x = 1 + \frac{\ln 7}{\ln 5} \approx 2.209$$

76.
$$8(3^{6-x}) = 40$$
$$3^{6-x} = 5$$
$$\ln 3^{6-x} = \ln 5$$
$$(6 - x)\ln 3 = \ln 5$$
$$6 - x = \frac{\ln 5}{\ln 3}$$
$$-x = \frac{\ln 5}{\ln 3} - 6$$
$$x = 6 - \frac{\ln 5}{\ln 3}$$
$$\approx 4.535$$

77.
$$\left(1 + \frac{0.065}{365}\right)^{365t} = 4$$
$$\ln\left(1 + \frac{0.065}{365}\right)^{365t} = \ln 4$$
$$365t \ln\left(1 + \frac{0.065}{365}\right) = \ln 4$$
$$t = \frac{\ln 4}{365 \ln\left(1 + \frac{0.065}{365}\right)} \approx 21.330$$

78.
$$\left(4 - \frac{2.471}{40}\right)^{9t} = 21$$
$$3.938225^{9t} = 21$$
$$\ln 3.938225^{9t} = \ln 21$$
$$9t \ln 3.938225 = \ln 21$$
$$t = \frac{\ln 21}{9 \ln 3.938225}$$
$$\approx 0.247$$

79.
$$\left(1 + \frac{0.10}{12}\right)^{12t} = 2$$
$$\ln\left(1 + \frac{0.10}{12}\right)^{12t} = \ln 2$$
$$12t \ln\left(1 + \frac{0.10}{12}\right) = \ln 2$$
$$t = \frac{\ln 2}{12 \ln\left(1 + \frac{0.10}{12}\right)} \approx 6.960$$

80.
$$\left(16 - \frac{0.878}{26}\right)^{3t} = 30$$
$$\ln\left(16 - \frac{0.878}{26}\right)^{3t} = \ln 30$$
$$3t \ln\left(16 - \frac{0.878}{26}\right) = \ln 30$$
$$t = \frac{\ln 30}{3 \ln\left(16 - \frac{0.878}{26}\right)} \approx 0.409$$

81. $\dfrac{3000}{2 + e^{2x}} = 2$

$$3000 = 2(2 + e^{2x})$$
$$1500 = 2 + e^{2x}$$
$$1498 = e^{2x}$$
$$\ln 1498 = 2x$$
$$x = \frac{\ln 1498}{2} \approx 3.656$$

82. $\dfrac{119}{e^{6x} - 14} = 7$

$$119 = 7(e^{6x} - 14)$$
$$17 = e^{6x} - 14$$
$$31 = e^{6x}$$
$$\ln 31 = \ln e^{6x}$$
$$\ln 31 = 6x$$
$$x = \frac{\ln 31}{6} \approx 0.572$$

83. $\ln x = -3$

$$x = e^{-3} \approx 0.050$$

84. $\ln x = 2$

$$e^{\ln x} = e^2$$
$$x = e^2 \approx 7.389$$

85. $\ln 2x = 2.4$

$$2x = e^{2.4}$$
$$x = \frac{e^{2.4}}{2} \approx 5.512$$

86. $\ln 4x = 1$

$$e^{\ln 4x} = e^1$$
$$4x = e$$
$$x = \frac{e}{4} \approx 0.680$$

87. $3 \ln 5x = 10$

$$\ln 5x = \frac{10}{3}$$
$$5x = e^{10/3}$$
$$x = \frac{e^{10/3}}{5} \approx 5.606$$

88. $2 \ln x = 7$

$$\ln x = \frac{7}{2}$$
$$e^{\ln x} = e^{7/2}$$
$$x = e^{7/2} \approx 33.115$$

89. $\ln \sqrt{x + 2} = 1$

$$\sqrt{x + 2} = e^1$$
$$x + 2 = e^2$$
$$x = e^2 - 2 \approx 5.389$$

90. $\ln \sqrt{x - 8} = 5$

$$e^{\ln \sqrt{x-8}} = e^5$$
$$\sqrt{x - 8} = e^5$$
$$x - 8 = e^{10}$$
$$x = e^{10} + 8 \approx 22{,}034.466$$

91. $\ln(x - 2) - \ln 3 = 2$

$$\ln\left(\frac{x - 2}{3}\right) = 2$$
$$\frac{x - 2}{3} = e^2$$
$$x - 2 = 3e^2$$
$$x = 3e^2 + 2 \approx 24.167$$

92. $\ln(x + 1) - \ln 2 = 3$

$$\ln\left(\frac{x + 1}{2}\right) = 3$$
$$\frac{x + 1}{2} = e^3$$
$$x + 1 = 2e^3$$
$$x = 2e^3 - 1 \approx 39.171$$

93. $\ln(2x - 1) + \ln 2 = -2$

$$\ln[2(2x - 1)] = -2$$
$$2(2x - 1) = e^{-2}$$
$$4x - 2 = e^{-2}$$
$$4x = e^{-2} + 2$$
$$x = \frac{e^{-2} + 2}{4} \approx 0.534$$

94. $\ln(3x + 1) + \ln 3 = -7$

$$\ln[3(3x + 1)] = -7$$
$$3(3x + 1) = e^{-7}$$
$$9x + 3 = e^{-7}$$
$$9x = e^{-7} - 3$$
$$x = \frac{e^{-7} - 3}{9} \approx -0.333$$

95. $\ln x + \ln(x - 1) = 1$

$\ln[x(x - 1)] = 1$

$x(x - 1) = e^1$

$x^2 - x - e = 0$

By the Quadratic Formula we have:

$x = \dfrac{1 \pm \sqrt{1 + 4e}}{2}$

$x = \dfrac{1 - \sqrt{1 + 4e}}{2} \approx -1.223$ is extraneous.

$x = \dfrac{1 + \sqrt{1 + 4e}}{2} \approx 2.223$ is the only solution.

96. $\ln x + \ln(x + 1) = 1$

$\ln[x(x + 1)] = 1$

$e^{\ln[x(x+1)]} = e^1$

$x(x + 1) = e^1$

$x^2 + x - e = 0$

$x = \dfrac{-1 + \sqrt{1 + 4e}}{2}$

The only solution is $x = \dfrac{-1 + \sqrt{1 + 4e}}{2} \approx 1.223.$

97. $\ln x + \ln(x - 2) = 1$

$\ln[x(x - 2)] = 1$

$x(x - 2) = e^1$

$x^2 - 2x - e = 0$

$x = \dfrac{2 \pm \sqrt{4 + 4e}}{2}$

$= \dfrac{2 \pm 2\sqrt{1 + e}}{2}$

$= 1 \pm \sqrt{1 + e}$

The negative value is extraneous. The only solution is $x = 1 + \sqrt{1 + e} \approx 2.928.$

98. $\ln x + \ln(x + 3) = 1$

$\ln[x(x + 3)] = 1$

$e^{\ln[x(x+3)]} = e^1$

$x(x + 3) = e^1$

$x^2 + 3x - e = 0$

$x = \dfrac{-3 \pm \sqrt{9 + 4e}}{2}$

The only solution is $x = \dfrac{-3 + \sqrt{9 + 4e}}{2} \approx 0.729.$

99. $\ln(x + 5) = \ln(x - 1) - \ln(x + 1)$

$\ln(x + 5) = \ln\left(\dfrac{x - 1}{x + 1}\right)$

$x + 5 = \dfrac{x - 1}{x + 1}$

$(x + 5)(x + 1) = x - 1$

$x^2 + 6x + 5 = x - 1$

$x^2 + 5x + 6 = 0$

$(x + 2)(x + 3) = 0$

$x = -2$ or $x = -3$

Both of these solutions are extraneous, so the equation has no solution.

100. $\ln(x + 1) - \ln(x - 2) = \ln x^2$

$\ln\left(\dfrac{x + 1}{x - 2}\right) = \ln x^2$

$\dfrac{x + 1}{x - 2} = x^2$

$x + 1 = x^3 - 2x^2$

$0 = x^3 - 2x^2 - x - 1$

From the graph, we have $x \approx 2.547.$

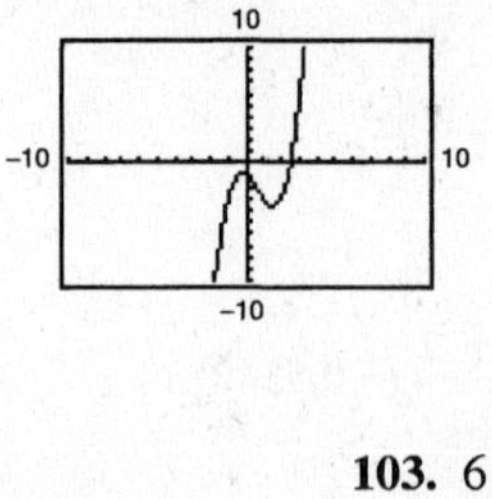

101. $\log_{10}(z - 3) = 2$

$10^{\log_{10}(z-3)} = 10^2$

$z - 3 = 10^2$

$z = 10^2 + 3 = 103$

102. $\log_{10}(x + 2) = 3$

$10^{\log_{10}(x+2)} = 10^3$

$x + 2 = 1000$

$x = 998$

103. $6 \log_3(0.5x) = 11$

$\log_3(0.5x) = \frac{11}{6}$

$3^{\log_3(0.5x)} = 3^{11/6}$

$0.5x = 3^{11/6}$

$x = 2(3^{11/6}) \approx 14.988$

104. $5 \log_{10}(x - 2) = 11$

$$\log_{10}(x - 2) = \tfrac{11}{5}$$

$$10^{\log_{10}(x-2)} = 10^{11/5}$$

$$x - 2 = 10^{11/5}$$

$$x = 10^{11/5} + 2 \approx 160.489$$

105. $\log_{10}(x + 4) - \log_{10} x = \log_{10}(x + 2)$

$$\log_{10}\left(\frac{x + 4}{x}\right) = \log_{10}(x + 2)$$

$$\frac{x + 4}{x} = x + 2$$

$$x + 4 = x^2 + 2x$$

$$0 = x^2 + x - 4$$

$$x = \frac{-1 \pm \sqrt{17}}{2} \quad \text{Quadratic Formula}$$

Choosing the positive value of x (the negative value is extraneous), we have

$$x = \frac{-1 + \sqrt{17}}{2} \approx 1.562.$$

106. $\log_2 x + \log_2(x + 2) = \log_2(x + 6)$

$$\log_2[x(x + 2)] = \log_2(x + 6)$$

$$x(x + 2) = x + 6$$

$$x^2 + x - 6 = 0$$

$$(x + 3)(x - 2) = 0$$

$x = -3$ or $x = 2$

The value $x = -3$ is extraneous.

The only solution is $x = 2$.

107. $\log_4 x - \log_4(x - 1) = \frac{1}{2}$

$$\log_4\left(\frac{x}{x - 1}\right) = \frac{1}{2}$$

$$4^{\log_4\left(\frac{x}{x-1}\right)} = 4^{1/2}$$

$$\frac{x}{x - 1} = 4^{1/2}$$

$$x = 2(x - 1)$$

$$x = 2x - 2$$

$$-x = -2$$

$$x = 2$$

108. $\log_3 x + \log_3(x - 8) = 2$

$$\log_3[x(x - 8)] = 2$$

$$3^{\log_3(x^2 - 8x)} = 3^2$$

$$x^2 - 8x = 9$$

$$x^2 - 8x - 9 = 0$$

$$(x - 9)(x + 1) = 0$$

$x = 9$ or $x = -1$

The value $x = -1$ is extraneous.

The only solution is $x = 9$.

109. $\log_{10} 8x - \log_{10}(1 + \sqrt{x}) = 2$

$$\log_{10} \frac{8x}{1 + \sqrt{x}} = 2$$

$$\frac{8x}{1 + \sqrt{x}} = 10^2$$

$$8x = 100(1 + \sqrt{x})$$

$$2x = 25(1 + \sqrt{x})$$

$$2x = 25 + 25\sqrt{x}$$

$$2x - 25 = 25\sqrt{x}$$

$$(2x - 25)^2 = (25\sqrt{x})^2$$

$$4x^2 - 100x + 625 = 625x$$

$$4x^2 - 725x + 625 = 0$$

$$x = \frac{725 \pm \sqrt{725^2 - 4(4)(625)}}{2(4)}$$

$$x = \frac{725 \pm \sqrt{515625}}{8}$$

$$x = \frac{25(29 \pm 5\sqrt{33})}{8}$$

$x \approx 0.866$ (extraneous) or $x \approx 180.384$

The only solution is $x = \dfrac{25(29 + 5\sqrt{33})}{8} \approx 180.384$.

110. $\log_{10} 4x - \log_{10}(12 + \sqrt{x}) = 2$

$$\log_{10}\left(\frac{4x}{12 + \sqrt{x}}\right) = 2$$

$$10^{\log_{10}(4x/(12 + \sqrt{x}))} = 10^2$$

$$\frac{4x}{12 + \sqrt{x}} = 100$$

$$4x = 100(12 + \sqrt{x})$$

$$4x = 1200 + 100\sqrt{x}$$

$$4x - 1200 = 100\sqrt{x}$$

$$x - 300 = 25\sqrt{x}$$

$$(x - 300)^2 = (25\sqrt{x})^2$$

$$x^2 - 600x + 90{,}000 = 625x$$

$$x^2 - 1225x + 90{,}000 = 0$$

$$x = \frac{1225 \pm \sqrt{(-1225)^2 - 4(1)(90{,}000)}}{2}$$

$$x = \frac{1225 \pm \sqrt{1{,}140{,}625}}{2}$$

$$x = \frac{1225 \pm 125\sqrt{73}}{2}$$

$x \approx 78.500$ (extraneous) or $x \approx 1146.500$

The only solution is $x = \dfrac{1225 + 125\sqrt{73}}{2} \approx 1146.500$.

111. $y_1 = 7$

$y_2 = 2^x$

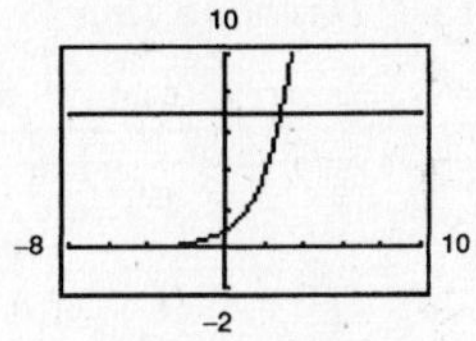

From the graph we have $x \approx 2.807$ when $y = 7$.

The point of intersection is approximately (2.807, 7).

112. $y_1 = 500$

$y_2 = 1500e^{-x/2}$

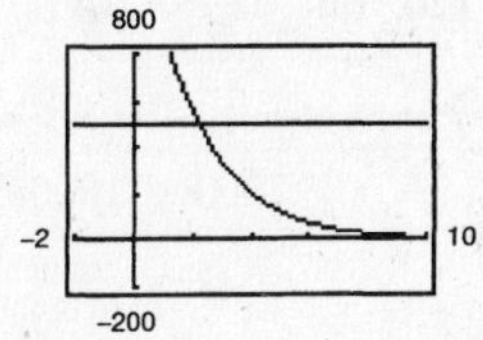

From the graph, we have $x \approx 2.197$.

The point of intersection is approximately (2.197, 500).

113. $y_1 = 3$

$y_2 = \ln x$

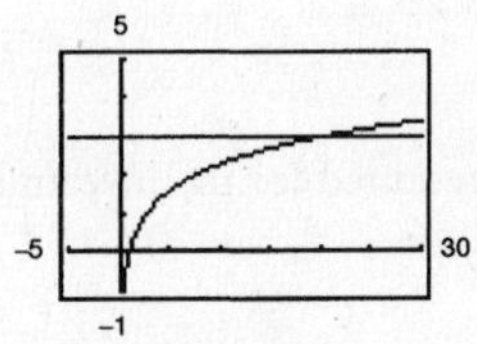

From the graph we have $x \approx 20.086$ when $y = 3$.

The point of intersection is approximately (20.086, 3).

114. $y_1 = 10$

$y_2 = 4\ln(x - 2)$

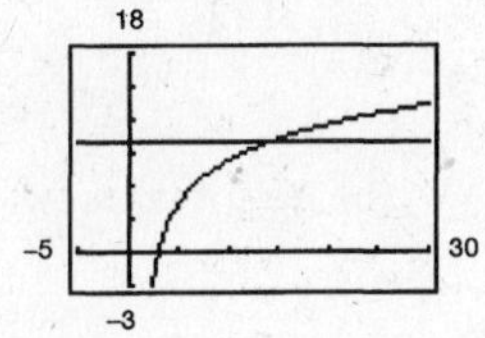

From the graph, we have $x \approx 14.182$.

The point of intersection is approximately (14.182, 10).

115. Rewrite the given equation in a form that allows use of the one-to-one properties of exponential functions, or rewrite the exponential equation in logarithmic form and use the properties of logarithms. Be sure to check for extraneous solutions.

116. Rewrite the equation in a form that allows the use of the one-to-one properties of logarithmic functions, or rewrite the logarithmic equation in exponential form and use the properties of exponents. Be sure to check for extraneous solutions.

117. $A = Pe^{rt}$

(a) $A = (2P)e^{rt} = 2(Pe^{rt})$ This doubles your money.

(b) $A = Pe^{(2r)t} = Pe^{rt}e^{rt} = e^{rt}(Pe^{rt})$

(c) $A = Pe^{r(2t)} = Pe^{rt}e^{rt} = e^{rt}(Pe^{rt})$

Doubling the interest rate yields the same result as doubling the number of years.

If $2 > e^{rt}$ (i.e., $rt < \ln 2$), then doubling your investment would yield the most money. If $rt > \ln 2$, then doubling either the interest rate or the number of years would yield more money.

118. No. Doubling time does not depend on the amount of the investment, but depends on the interest rate, r.

$$2P = Pe^{rt}$$
$$2 = e^{rt}$$
$$\ln 2 = rt$$
$$\frac{\ln 2}{r} = t$$

119. (a)

$$A = Pe^{rt}$$
$$2000 = 1000e^{0.085t}$$
$$2 = e^{0.085t}$$
$$\ln 2 = 0.085t$$
$$\frac{\ln 2}{0.085} = t$$
$$t \approx 8.2 \text{ years}$$

(b) Yes, it takes twice as long to quadruple.

(c) and (d)

Doubling Time

$$2P = Pe^{rt}$$
$$2 = e^{rt}$$
$$\ln 2 = rt$$
$$\frac{\ln 2}{r} = t$$

Quadrupling Time

$$4P = Pe^{rt}$$
$$4 = e^{rt}$$
$$\ln 4 = rt$$
$$\frac{\ln 4}{r} = t$$
$$\frac{\ln 2^2}{r} = t$$
$$\frac{2\ln 2}{r} = t$$
$$2\left(\frac{\ln 2}{r}\right) = t$$

The time required for the investment to quadruple is twice the doubling time.

120. (a)

$$r = 0.12$$
$$A = Pe^{rt}$$
$$2000 = 1000e^{0.12t}$$
$$2 = e^{0.12t}$$
$$\ln 2 = \ln e^{0.12t}$$
$$\ln 2 = 0.12t$$
$$\frac{\ln 2}{0.12} = t$$
$$t \approx 5.8 \text{ years}$$

(b) Yes, it takes twice as long to quadruple.

(c) and (d)

Doubling Time

$$2P = Pe^{rt}$$
$$2 = e^{rt}$$
$$\ln 2 = rt$$
$$\frac{\ln 2}{r} = t$$

Quadrupling Time

$$4P = Pe^{rt}$$
$$4 = e^{rt}$$
$$\ln 4 = rt$$
$$\ln 2^2 = rt$$
$$\frac{2 \ln 2}{r} = t$$
$$2\left(\frac{\ln 2}{r}\right) = t$$

The time required for the investment to quadruple is twice the doubling time.

121.

$$A = Pe^{rt}$$
$$3000 = 1000e^{0.085t}$$
$$3 = e^{0.085t}$$
$$\ln 3 = 0.085t$$
$$\frac{\ln 3}{0.085} = t$$
$$t \approx 12.9 \text{ years}$$

122.

$$r = 0.12$$
$$A = Pe^{rt}$$
$$3000 = 1000e^{0.12t}$$
$$3 = e^{0.12t}$$
$$\ln 3 = \ln e^{0.12t}$$
$$\ln 3 = 0.12t$$
$$\frac{\ln 3}{0.12} = t$$
$$t = 9.2 \text{ years}$$

123. $p = 7000\left(1 - \dfrac{6}{6 + e^{-0.0012x}}\right)$

(a) When $p = \$350$:

$$350 = 7000\left(1 - \frac{6}{6 + e^{-0.0012x}}\right)$$
$$0.05 = 1 - \frac{6}{6 + e^{-0.0012x}}$$
$$\frac{6}{6 + e^{-0.0012x}} = 0.95$$
$$6 = 0.95(6 + e^{-0.0012x})$$
$$6 = 5.7 + 0.95e^{-0.0012x}$$
$$\frac{0.3}{0.95} = e^{-0.0012x}$$
$$\ln\left(\frac{6}{19}\right) = -0.0012x$$
$$x = \frac{\ln(6/19)}{-0.0012} \approx 961 \text{ units}$$

(b) When $p = \$300$

$$300 = 7000\left(1 - \frac{6}{6 + e^{-0.0012x}}\right)$$
$$\frac{3}{70} = 1 - \frac{6}{6 + e^{-0.0012x}}$$
$$\frac{6}{6 + e^{-0.0012x}} = \frac{67}{70}$$
$$6 = \frac{67}{70}(6 + e^{-0.0012x})$$
$$\frac{420}{67} = 6 + e^{-0.0012x}$$
$$\frac{18}{67} = e^{-0.0012x}$$
$$\ln\left(\frac{18}{67}\right) = -0.0012x$$
$$x = \frac{\ln(18/67)}{-0.0012} \approx 1095 \text{ units}$$

124. $p = 5000\left(1 - \dfrac{4}{4 + e^{-0.002x}}\right)$

(a) When $p = \$600$:

$$600 = 5000\left(1 - \frac{4}{4 + e^{-0.002x}}\right)$$

$$0.12 = 1 - \frac{4}{4 + e^{-0.002x}}$$

$$\frac{4}{4 + e^{-0.002x}} = 0.88$$

$$4 = 3.52 + 0.88e^{-0.002x}$$

$$0.48 = 0.88e^{-0.002x}$$

$$\frac{6}{11} = e^{-0.002x}$$

$$\ln\frac{6}{11} = \ln e^{-0.002x}$$

$$\ln\frac{6}{11} = -0.002x$$

$$x = -\frac{\ln(6/11)}{0.002} \approx 303 \text{ units}$$

(b) When $p = \$400$:

$$400 = 5000\left(1 - \frac{4}{4 + e^{-0.002x}}\right)$$

$$0.08 = 1 - \frac{4}{4 + e^{-0.002x}}$$

$$\frac{4}{4 + e^{-0.002x}} = 0.92$$

$$4 = 3.68 + 0.92e^{-0.002x}$$

$$0.32 = 0.92e^{-0.002x}$$

$$\frac{8}{23} = e^{-0.002x}$$

$$\ln\frac{8}{23} = \ln e^{-0.002x}$$

$$\ln\frac{8}{23} = -0.002x$$

$$x = -\frac{\ln(8/23)}{0.002} \approx 528 \text{ units}$$

125. $V = 6.7e^{-48.1/t}$, $t \geq 0$

(a)

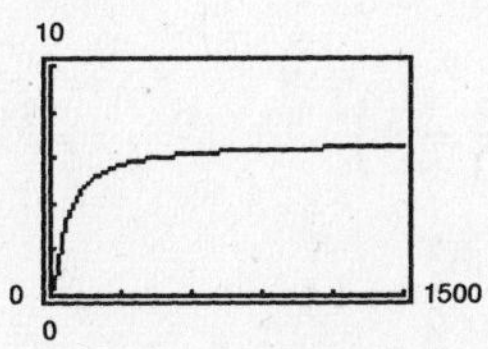

(b) As $t \to \infty$, $V \to 6.7$.

Horizontal asymptote: $V = 6.7$

The yield will approach 6.7 million cubic feet per acre.

(c)

$$1.3 = 6.7e^{-48.1/t}$$

$$\frac{1.3}{6.7} = e^{-48.1/t}$$

$$\ln\left(\frac{13}{67}\right) = \frac{-48.1}{t}$$

$$t = \frac{-48.1}{\ln(13/67)} \approx 29.3 \text{ years}$$

126. $N = 68(10^{-0.04x})$

When $N = 21$:

$$21 = 68(10^{-0.04x})$$

$$\frac{21}{68} = 10^{-0.04x}$$

$$\log_{10}\frac{21}{68} = -0.04x$$

$$x = -\frac{\log_{10}(21/68)}{0.04} \approx 12.76 \text{ inches}$$

127. (a) From the graph shown in the textbook, we see horizontal asymptotes at $y = 0$ and $y = 100$. These represent the lower and upper percent bounds; the range falls between 0% and 100%.

(b) Males

$$50 = \frac{100}{1 + e^{-0.6114(x-69.71)}}$$
$$1 + e^{-0.6114(x-69.71)} = 2$$
$$e^{-0.6114(x-69.71)} = 1$$
$$-0.6114(x - 69.71) = \ln 1$$
$$-0.6114(x - 69.71) = 0$$
$$x = 69.71 \text{ inches}$$

Females

$$50 = \frac{100}{1 + e^{-0.66607(x-64.51)}}$$
$$1 + e^{-0.66607(x-64.51)} = 2$$
$$e^{-0.66607(x-64.51)} = 1$$
$$-0.66607(x - 64.51) = \ln 1$$
$$-0.66607(x - 64.51) = 0$$
$$x = 64.51 \text{ inches}$$

128. $P = \dfrac{0.83}{1 + e^{-0.2n}}$

(a)

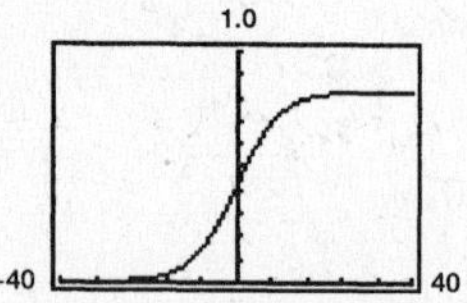

(b) Horizontal asymptotes: $P = 0, P = 0.83$
The upper asymptote, $P = 0.83$, indicates that the proportion of correct responses will approach 0.83 as the number of trials increases.

(c) When $P = 60\%$ or $P = 0.60$:

$$0.60 = \frac{0.83}{1 + e^{-0.2n}}$$
$$1 + e^{-0.2n} = \frac{0.83}{0.60}$$
$$e^{-0.2n} = \frac{0.83}{0.60} - 1$$
$$\ln e^{-0.2n} = \ln\left(\frac{0.83}{0.60} - 1\right)$$
$$-0.2n = \ln\left(\frac{0.83}{0.60} - 1\right)$$
$$n = -\frac{\ln\left(\frac{0.83}{0.60} - 1\right)}{0.2} \approx 5 \text{ trials}$$

129. $T = 20[1 + 7(2^{-h})]$

(a) From the graph in the textbook we see a horizontal asymptote at $T = 20$.
This represents the room temperature.

(b)
$$100 = 20[1 + 7(2^{-h})]$$
$$5 = 1 + 7(2^{-h})$$
$$4 = 7(2^{-h})$$
$$\frac{4}{7} = 2^{-h}$$
$$\ln\left(\frac{4}{7}\right) = \ln 2^{-h}$$
$$\ln\left(\frac{4}{7}\right) = -h \ln 2$$
$$\frac{\ln\left(\frac{4}{7}\right)}{-\ln 2} = h$$
$$h \approx 0.81 \text{ hour}$$

130. $E = 20.39e^{0.1066t}$ where $t = 0$ corresponds to 1960.

(a)

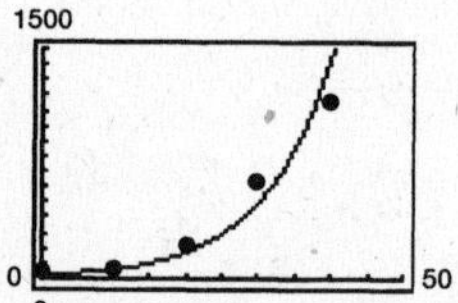

The model is a fairly good fit to the data.

(b) For 2005, let $t = 45$: $E \approx 2470.2$ billion dollars

For 2010, let $t = 50$: $E \approx 4209.3$ billion dollars

For 2015, let $t = 55$: $E \approx 7172.8$ billion dollars

(c)
$$20.39e^{0.1066t} = 1000$$
$$e^{0.1066t} = \frac{1000}{20.39}$$
$$0.1066t = \ln\left(\frac{1000}{20.39}\right)$$
$$t = \frac{1}{0.1066}\ln\left(\frac{1000}{20.39}\right) \approx 36.5$$

which corresponds to the middle of the year 1996.

(d) No. The model grows very rapidly as t increases.

131. $N = 17.02 - 3.096 \ln t$,

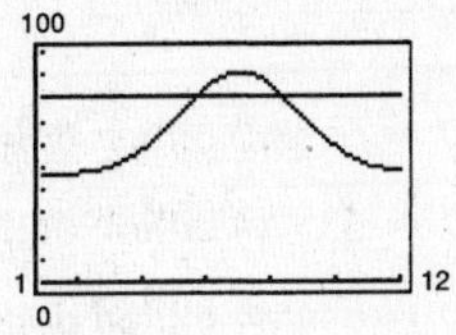

The high temperature is likely to be above 80° in the months of June, July, and August.

132. $N = 17.02 - 3.096 \ln t$ where $t = 5$ corresponds to 1995.

(a) $17.02 - 3.096 \ln t = 10$

$$7.02 = 3.096 \ln t$$

$$\frac{7.02}{3.096} = \ln t$$

$$t = e^{(7.02/3.096)} \approx 9.65 \approx 10 \text{ which corresponds with the year 2000.}$$

(b)

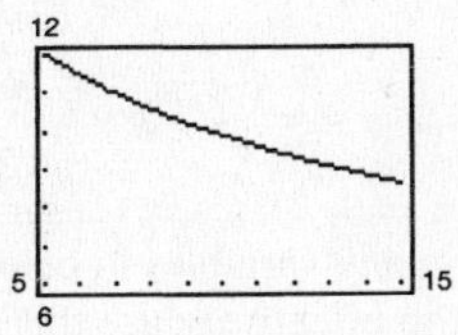

(c)

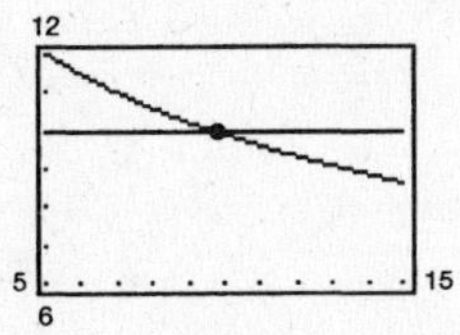

The graphs of $y_1 = 17.02 - 3.096 \ln t$ and $y_2 = 10$ intersect when $t \approx 9.65 \approx 10$.

133. $P = \dfrac{8.617}{1 + 4.65e^{-0.0380t}} + 1.0005$ where $t = 0$ corresponds to 1950.

(a)

Year	Population	Model
1950	2.526	2.526
1960	3.058	3.062
1970	3.722	3.715
1980	4.454	4.465
1990	5.276	5.273
2000	6.079	6.083

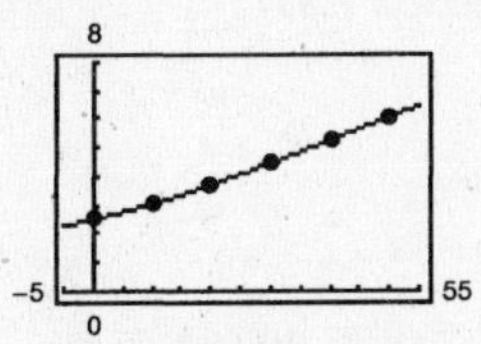

The model is a good fit.

(b) For 2010: $P(60) \approx 6.840$

For 2020: $P(70) \approx 7.503$

For 2030: $P(80) \approx 8.050$

(c) $\dfrac{8.617}{1 + 4.65e^{-0.0380t}} + 1.0005 = 8$

$$8.617 = 6.9995(1 + 4.65e^{-0.0380t})$$

$$\frac{86{,}170}{69{,}995} = 1 + 4.65e^{-0.0380t}$$

$$\frac{1}{4.65}\left[\frac{86{,}170}{69{,}995} - 1\right] = e^{-0.0380t}$$

$$t = \frac{\ln\left\{\dfrac{1}{4.65}\left(\dfrac{86{,}170}{69{,}995} - 1\right)\right\}}{-0.0380} \approx 79$$

According to the model, the world population would reach 8 billion in the year 2029.

(d)

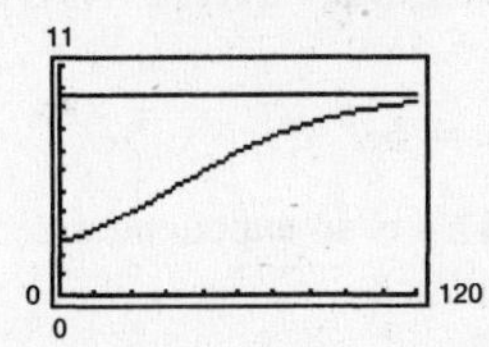

The horizontal asymptote is at $y = 8.617 + 1.0005 = 9.6175$. This would imply that the world population maximum would be 9.6175 billion.

(e) World population models can be used for short-range predictions, but there are too many unknown factors to use them for long-range predicitons.

134. $D = \dfrac{6296.7}{1 + 16{,}945.6e^{-0.1928t}} + 204.25$

where $t = 0$ corresponds to 1940.

(a) 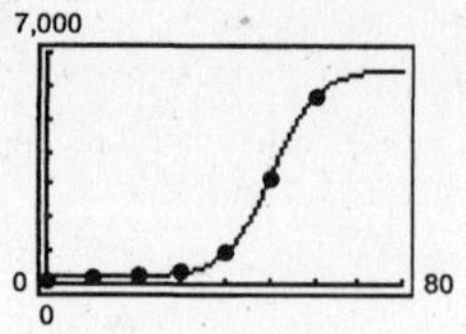

The model is a good fit to the actual data.

(b) For 2005, use $t = 65$: $D \approx 6138.0$ billion dollars

For 2010, use $t = 70$: $D \approx 6357.4$ billion dollars

For 2015, use $t = 75$: $D \approx 6445.4$ billion dollars

(c) $\dfrac{6296.7}{1 + 16{,}945.6e^{-0.1928t}} + 204.25 = 1000$

$6296.7 = 795.75(1 + 16{,}945.6e^{-0.1928t})$

$\dfrac{6296.7}{795.75} - 1 = 16{,}945.6e^{-0.1928t}$

$\ln\left[\dfrac{(6296.7/795.75) - 1}{16{,}945.6}\right] = -0.1928t$

$t \approx 40.5$ which corresponds to the middle of the year 1980.

(d) As $t \to \infty, D \to 6296.7 + 204.25 = 6500.95$
According to the model, the gross federal debt has an upper limit of 6500.95 billion dollars.

(e) $\dfrac{6296.7}{1 + 16{,}945.6e^{-0.1928t}} + 204.25 = 10{,}000$

$6296.7 = 9795.75(1 + 16{,}945.6e^{-0.1928t})$

$\dfrac{6296.7}{9795.75} - 1 = 16{,}945.6e^{-0.1928t}$

$\ln\left[\dfrac{(6296.7/9795.75) - 1}{16{,}945.6}\right] = -0.1928t$

On the left we have $\ln(-0.0000211)$ which is not a real number. The equation has no real solution. This agrees with part (d) since the model has an upper limit of 6500.95 billion dollars.

(f) No. There are far too many variables in determining the gross federal debt to effectively use this model, especially since it contains an upper limit of 6500.95 billion dollars.

135. $\log_a(uv) = \log_a u + \log_a v$

True by Property 1 in Section 7.3.

136. $\log_a(u + v) = (\log_a u)(\log_a v)$

False. $2.04 \approx \log_{10}(10 + 100) \neq (\log_{10} 10)(\log_{10} 100) = 2$

137. $\log_a(u - v) = \log_a u - \log_a v$

False. $1.95 \approx \log_{10}(100 - 10) \neq \log_{10} 100 - \log_{10} 10 = 1$

138. $\log_a\left(\dfrac{u}{v}\right) = \log_a u - \log_a v$

True by Property 2 in Section 7.3.

Section 7.5 Exponential and Logarithmic Models

1. $y = 2e^{x/4}$

This is an exponential growth model.

Matches graph (c)

2. $y = 6e^{-x/4}$

This is an exponential decay model.

Matches graph (e).

3. $y = 6 + \log_{10}(x + 2)$

This is a logarithmic function shifted up 6 units and left 2 units.

Matches graph (b)

4. $y = 3e^{-(x-2)^2/5}$

This is a Gaussian model.

Matches graph (a)

5. $y = \ln(x + 1)$

This is a logarithmic model.

Matches graph (d)

6. $y = \dfrac{4}{1 + e^{-2x}}$

This is a logistic growth model.

Matches graph (f)

7.

$$y = ae^{bx}$$
$$1 = ae^{b(0)} \Rightarrow 1 = a$$
$$10 = e^{b(3)}$$
$$\ln 10 = 3b$$
$$\frac{\ln 10}{3} = b \Rightarrow b \approx 0.7675$$

Thus, $y = e^{0.7675x}$.

8.

$$y = ae^{bx}$$
$$\frac{1}{2} = ae^{b(0)} \Rightarrow a = \frac{1}{2}$$
$$5 = \frac{1}{2}e^{b(4)}$$
$$10 = e^{4b}$$
$$\ln 10 = \ln e^{4b}$$
$$\ln 10 = 4b$$
$$\frac{\ln 10}{4} = b \Rightarrow b \approx 0.5756$$

Thus, $y = \frac{1}{2}e^{0.5756x}$.

9.

$$y = ae^{bx}$$
$$5 = ae^{b(0)} \Rightarrow 5 = a$$
$$1 = 5e^{b(4)}$$
$$\frac{1}{5} = e^{4b}$$
$$\ln\left(\frac{1}{5}\right) = 4b$$
$$\frac{\ln\left(\frac{1}{5}\right)}{4} = b \Rightarrow b \approx -0.4024$$

Thus, $y = 5e^{-0.4024x}$.

10.

$$y = ae^{bx}$$
$$1 = ae^{b(0)} \Rightarrow 1 = a$$
$$\frac{1}{4} = e^{b(3)}$$
$$\ln\left(\frac{1}{4}\right) = \ln e^{3b}$$
$$\ln\left(\frac{1}{4}\right) = 3b$$
$$\frac{\ln\left(\frac{1}{4}\right)}{3} = b \Rightarrow b \approx -0.4621$$

Thus, $y = e^{-0.4621x}$.

11. (a) Logarithmic

(b) Logistic

(c) Exponential (decay)

(d) Linear

(e) None of the above (appears to be a combination of a linear and a quadratic)

(f) Exponential (growth)

12. $y = \dfrac{a}{1 + be^{-xt}}$

This curve would have a vertical asymptote if $1 + be^{-xt} = 0$. Solving for b yields

$$be^{-xt} = -1$$
$$b = -e^{xt}.$$

Since $e^{xt} > 0$ for all x and t, $b = -e^{xt} \Rightarrow b < 0$. So, if $b < 0$, there will be a vertical asymptote when $b = -e^{xt}$.

13. $y = \dfrac{a}{1 + be^{-xt}}$

This curve does not have a vertical asymptote as long as $1 + be^{-xt} \neq 0$. This means $b \neq -e^{xt}$. Since $-e^{xt} < 0$ for all x and t, b cannot be negative. Thus, if $b > 0$, the curve has no vertical asymptotes. Also, if $b = 0$, $y = a$ is a horizontal line and the graph would not have a vertical asymptote.

14. $y = \dfrac{1}{3\sqrt{2\pi}}e^{-(x-70)^2/18}$

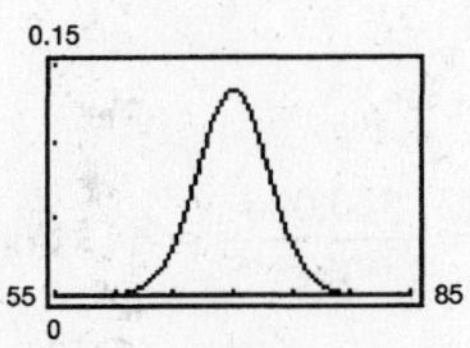

This is a bell shaped curve with a maximum when $x = 70$. The average height of American men between 18 and 24 years old is 70 inches.

15. Since $A = 1000e^{0.12t}$, the time to double is given by $2000 = 1000e^{0.12t}$ and we have

$$2000 = 1000e^{0.12t}$$
$$2 = e^{0.12t}$$
$$\ln 2 = \ln e^{0.12t}$$
$$\ln 2 = 0.12t$$
$$t = \frac{\ln 2}{0.12} \approx 5.78 \text{ years.}$$

Amount after 10 years: $A = 1000e^{1.2} \approx \$3320.12$

16. Since $A = 20{,}000e^{0.105t}$, the time to double is given by $40{,}000 = 20{,}000e^{0.105t}$, and we have

$$40{,}000 = 20{,}000e^{0.105t}$$
$$2 = e^{0.105t}$$
$$\ln 2 = \ln e^{0.105t}$$
$$\ln 2 = 0.105t$$
$$t = \frac{\ln 2}{0.105} \approx 6.60 \text{ years.}$$

Amount after 10 years: $A = 20{,}000e^{0.105(10)} \approx \$57{,}153.02$

17. Since $A = 750e^{rt}$ and $A = 1500$ when $t = 7.75$, we have the following.

$$1500 = 750e^{7.75r}$$
$$2 = e^{7.75r}$$
$$\ln 2 = \ln e^{7.75r}$$
$$\ln 2 = 7.75r$$
$$r = \frac{\ln 2}{7.75} \approx 0.089438 = 8.9438\%$$

Amount after 10 years: $A = 750e^{0.089438(10)} \approx \$1834.37$

18. Since $A = 10{,}000e^{rt}$ and $A = 20{,}000$ when $t = 12$, we have

$$20{,}000 = 10{,}000e^{12r}$$
$$2 = e^{12r}$$
$$\ln 2 = \ln e^{12r}$$
$$\ln 2 = 12r$$
$$r = \frac{\ln 2}{12} \approx 0.057762 = 5.7762\%$$

Amount after 10 years:

$$A = 10{,}000e^{0.057762(10)} \approx \$17{,}817.97$$

19. Since $A = 500e^{rt}$ and $A = \$1505.00$ when $t = 10$, we have the following.

$$1505.00 = 500e^{10r}$$
$$r = \frac{\ln(1505.00/500)}{10} \approx 0.110 = 11.0\%$$

The time to double is given by

$$1000 = 500e^{0.110t}$$
$$t = \frac{\ln 2}{0.110} \approx 6.3 \text{ years.}$$

20. Since $A = 600e^{rt}$ and $A = 19{,}205$ when $t = 10$, we have

$$19{,}205 = 600e^{10r}$$
$$\frac{19{,}205}{600} = e^{10r}$$
$$\ln\left(\frac{19{,}205}{600}\right) = \ln e^{10r}$$
$$\ln\left(\frac{19{,}205}{600}\right) = 10r$$
$$r = \frac{\ln(19{,}205/600)}{10} \approx 0.3466 \text{ or } 34.66\%.$$

The time to double is given by

$$1200 = 600e^{0.3466t}$$
$$t = \frac{\ln 2}{0.3466} \approx 2 \text{ years.}$$

21. Since $A = Pe^{0.045t}$ and $A = 10{,}000.00$ when $t = 10$, we have the following.

$$10{,}000.00 = Pe^{0.045(10)}$$
$$\frac{10{,}000.00}{e^{0.045(10)}} = P \approx \$6376.28$$

The time to double is given by

$$t = \frac{\ln 2}{0.045} \approx 15.40 \text{ years.}$$

22. Since $A = Pe^{0.08t}$ and $A = 20{,}000$ when $t = 10$, we have

$$20{,}000 = Pe^{0.08(10)}$$
$$P = \frac{20{,}000}{e^{0.08(10)}} \approx \$8986.58.$$

The time to double is given by

$$t = \frac{\ln 2}{0.08} \approx 8.66 \text{ years.}$$

23. $500{,}000 = P\left(1 + \frac{0.075}{12}\right)^{12(20)}$

$$P = \frac{500{,}000}{\left(1 + \frac{0.075}{12}\right)^{12(20)}} = \frac{500{,}000}{1.00625^{240}} \approx \$112{,}087.09$$

24. $$A = P\left(1 + \frac{r}{n}\right)^{nt}$$

$$500{,}000 = P\left(1 + \frac{0.12}{12}\right)^{12(40)}$$

$$P = \$4214.16$$

25. $P = 1000, r = 11\%$

(a) $n = 1$

$$(1 + 0.11)^t = 2$$

$$t \ln 1.11 = \ln 2$$

$$t = \frac{\ln 2}{\ln 1.11} \approx 6.642 \text{ years}$$

(b) $n = 12$

$$\left(1 + \frac{0.11}{12}\right)^{12t} = 2$$

$$12t \ln\left(1 + \frac{0.11}{12}\right) = \ln 2$$

$$t = \frac{\ln 2}{12 \ln\left(1 + \frac{0.11}{12}\right)} \approx 6.330 \text{ years}$$

(c) $n = 365$

$$\left(1 + \frac{0.11}{365}\right)^{365t} = 2$$

$$365t \ln\left(1 + \frac{0.11}{365}\right) = \ln 2$$

$$t = \frac{\ln 2}{365 \ln\left(1 + \frac{0.11}{365}\right)} \approx 6.302 \text{ years}$$

(d) Continuously

$$e^{0.11t} = 2$$

$$0.11t = \ln 2$$

$$t = \frac{\ln 2}{0.11} \approx 6.301 \text{ years}$$

26. $P = 1000, r = 10.5\% = 0.105$

(a) $n = 1$

$$t = \frac{\ln 2}{\ln(1 + 0.105)} \approx 6.94 \text{ years}$$

(b) $n = 12$

$$t = \frac{\ln 2}{12 \ln\left(1 + \frac{0.105}{12}\right)} \approx 6.63 \text{ years}$$

(c) $n = 365$

$$t = \frac{\ln 2}{365 \ln\left(1 + \frac{0.105}{365}\right)} \approx 6.602 \text{ years}$$

(d) Compounded continuously

$$t = \frac{\ln 2}{0.105} \approx 6.601 \text{ years}$$

27. $$3P = Pe^{rt}$$

$$3 = e^{rt}$$

$$\ln 3 = rt$$

$$\frac{\ln 3}{r} = t$$

r	2%	4%	6%	8%	10%	12%
$t = \frac{\ln 3}{r}$ (years)	54.93	27.47	18.31	13.73	10.99	9.16

28.

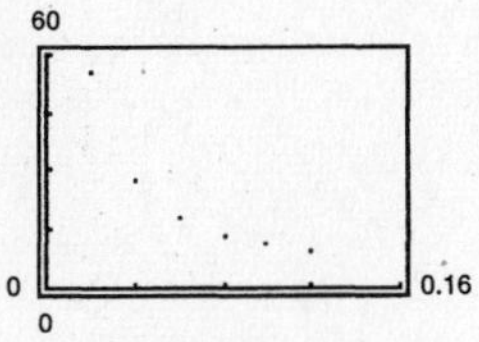

Using the power regression feature of a graphing utility,

$t = 1.099r^{-1}$.

29.

$$3P = P(1+r)^t$$
$$3 = (1+r)^t$$
$$\ln 3 = \ln(1+r)^t$$
$$\ln 3 = t\ln(1+r)$$
$$\frac{\ln 3}{\ln(1+r)} = t$$

r	2%	4%	6%	8%	10%	12%
$t = \frac{\ln 3}{\ln(1+r)}$ (years)	55.48	28.01	18.85	14.27	11.53	9.69

30.

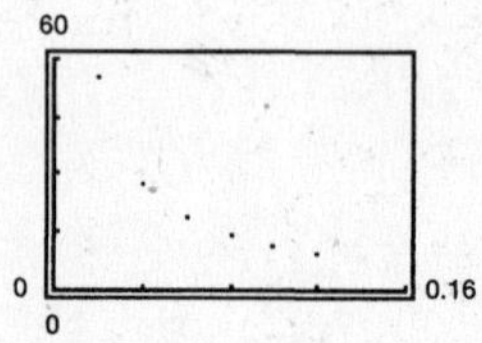

Using the power regression feature of a graphing utility,

$t = 1.222r^{-0.97}$.

31. Continuous compounding results in faster growth.

$A = 1 + 0.075[\![t]\!]$ and $A = e^{0.07t}$

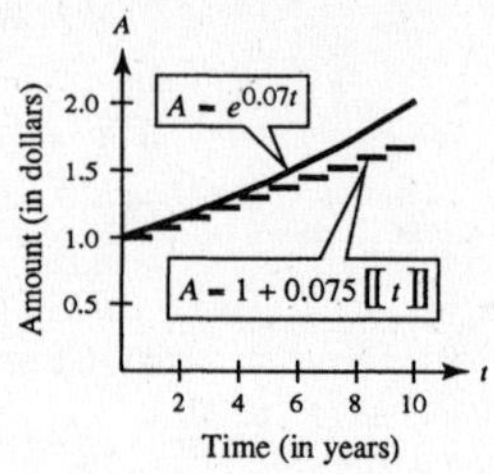

32.

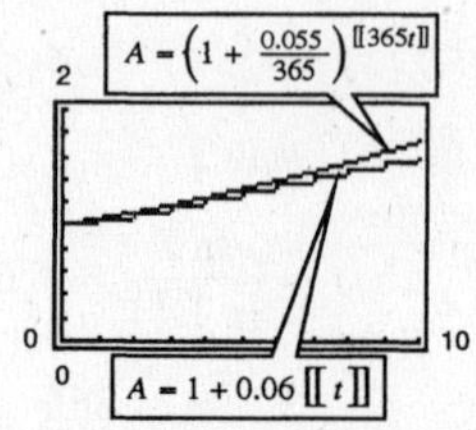

From the graph, $5\frac{1}{2}\%$ compounded daily grows faster than 6% simple interest.

33. $P = 73{,}796e^{0.030t}$ where $t = 0$ corresponds to 1970.

$$73{,}796e^{0.030t} = 350{,}000$$
$$0.030t = \ln\left(\frac{350{,}000}{73{,}796}\right)$$

$t \approx 52$ which corresponds to the year 2022.

34. $P = 101{,}659e^{0.021t}$ where $t = 0$ corresponds to 1970.

$$101{,}659e^{0.021t} = 225{,}000$$
$$0.021t = \ln\left(\frac{225{,}000}{101{,}659}\right)$$

$t \approx 38$ which corresponds to the year 2008.

35. $P = 1{,}458{,}000e^{-0.016t}$ where $t = 0$ corresponds to 1970.

$$1{,}458{,}000e^{-0.016t} = 800{,}000$$
$$-0.016t = \ln\left(\frac{800{,}000}{1{,}458{,}000}\right)$$

$t \approx 38$ which corresponds to the year 2008.

36.

$$N = 250e^{kt}$$
$$280 = 250e^{k(10)}$$
$$1.12 = e^{10k}$$
$$k = \frac{\ln 1.12}{10}$$
$$N = 250e^{[(\ln 1.12)/10]t}$$
$$500 = 250e^{[(\ln 1.12)/10]t}$$
$$2 = e^{[(\ln 1.12)/10]t}$$
$$\ln 2 = [(\ln 1.12)/10]t$$
$$t = \frac{\ln 2}{(\ln 1.12)/10} \approx 61.16 \text{ hours}$$

37.

$$N = 100e^{kt}$$
$$300 = 100e^{5k}$$
$$3 = e^{5k}$$
$$\ln 3 = \ln e^{5k}$$
$$\ln 3 = 5k$$
$$k = \frac{\ln 3}{5} \approx 0.2197$$
$$N = 100e^{0.2197t}$$
$$200 = 100e^{0.2197t}$$
$$t = \frac{\ln 2}{0.2197} \approx 3.15 \text{ hours}$$

38. (a) Russia: $(0, 145.0), (8, 142.3)$

$a = 145.0$

$142.3 = 145.0e^{b(8)}$

$\ln\left(\frac{142.3}{145.0}\right) = 8b$

$b \approx -0.0023$

For 2020, use $t = 18$.

$y = 145.0e^{-0.0023(18)} \approx 139.1$ million

Mali: $(0, 11.3), (8, 14.0)$

$a = 11.3$

$14.0 = 11.3e^{b(8)}$

$\ln\left(\frac{14.0}{11.3}\right) = 8b$

$b \approx 0.0268$

For 2020, use $t = 18$.

$y = 11.3e^{0.0268(18)} \approx 18.3$ million

Singapore: $(0, 4.5), (8, 5.8)$

$a = 4.5$

$5.8 = 4.5e^{b(8)}$

$\ln\left(\frac{5.8}{4.5}\right) = 8b$

$b \approx 0.0317$

For 2020, use $t = 18$.

$y = 4.5e^{0.0317(18)} \approx 8.0$ million

Sweden: $(0, 9.0), (8, 9.1)$

$a = 9.0$

$9.1 = 9.0e^{b(8)}$

$\ln\left(\frac{9.1}{9.0}\right) = 8b$

$b \approx 0.0014$

For 2020, use $t = 18$.

$y = 9.0e^{0.0014(18)} \approx 9.2$ million

(b) The constant b determines the growth rates. The greater the rate of growth, the greater the value of b.

(c) The constant b determines whether the population is increasing $(b > 0)$ or decreasing $(b < 0)$.

39. $y = Ce^{kt}$

$\frac{1}{2}C = Ce^{(1599)k}$

$\frac{1}{2} = e^{(1599)k}$

$\ln\frac{1}{2} = \ln e^{(1599)k}$

$\ln\frac{1}{2} = 1599k$

$k = \frac{\ln(1/2)}{1599}$

When $t = 100$, we have

$y = Ce^{[\ln(1/2)/1599](100)} \approx 0.957C = 95.7\%C$.

After 100 years, approximately 95.7% of the radioactive radium will remain.

40. $y = Ce^{kt}$

$\frac{1}{2}C = Ce^{5715k}$

$\ln\frac{1}{2} = 5715k$

$k = \frac{\ln(1/2)}{5715}$

The ancient charcoal has only 15% as much radioactive carbon.

$0.15C = Ce^{[(\ln 0.5)/5715]t}$

$\ln 0.15 = \frac{\ln 0.5}{5715}t$

$t = \frac{5715 \ln 0.15}{\ln 0.5} \approx 15{,}642$ years

41. $(0, 22{,}000), (2, 13{,}000)$

(a) $m = \frac{13{,}000 - 22{,}000}{2 - 0} = -4500$

$b = 22{,}000$

Thus, $V = -4500t + 22{,}000$.

(b) $a = 22{,}000$

$13{,}000 = 22{,}000e^{k(2)}$

$\frac{13}{22} = e^{2k}$

$\ln\left(\frac{13}{22}\right) = \ln e^{2k}$

$\ln\left(\frac{13}{22}\right) = 2k \Rightarrow k \approx -0.263$

Thus, $V = 22{,}000e^{-0.263t}$.

(c) The exponential model depreciates faster in the first two years.

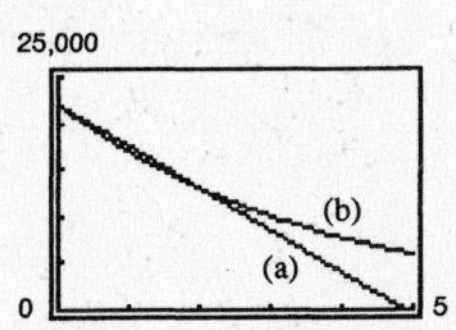

—CONTINUED—

41. —CONTINUED—

(d)

t	1	3
$V = -4500t + 22{,}000$	\$17,500	\$8500
$V = 22{,}000e^{-0.263t}$	\$16,912	\$9995

(e) The slope of the linear model means that the car depreciates \$4500 per year.

42. (0, 2000), (2, 500)

(a) $m = \dfrac{500 - 2000}{2 - 0} = -750$

$V = -750t + 2000$

(b) $500 = 2000e^{k(2)}$

$\ln \dfrac{1}{4} = 2k \Rightarrow k \approx -0.6931$

$V = 2000e^{-0.6931t}$

(c) The exponential model depreciates faster in the first two years.

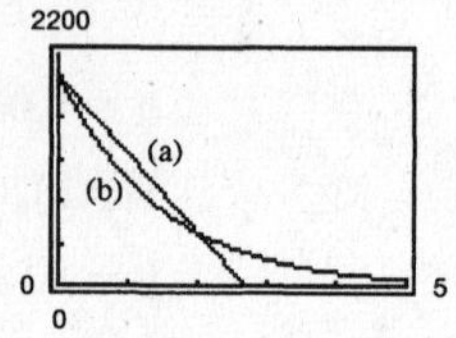

(d)

t	1	3
$V = -750t + 2000$	\$1250	$-$\$250
$V = 2000e^{-0.6931t}$	\$1000	\$250

(e) The slope of the linear model means that the computer depreciates \$750 per year.

43. $S(t) = 100(1 - e^{kt})$

(a) $15 = 100(1 - e^{k(1)})$

$-85 = -100e^k$

$\frac{85}{100} = e^k$

$0.85 = e^k$

$\ln 0.85 = \ln e^k$

$k = \ln 0.85$

$k \approx -0.1625$

$S(t) = 100(1 - e^{-0.1625t})$

(b)

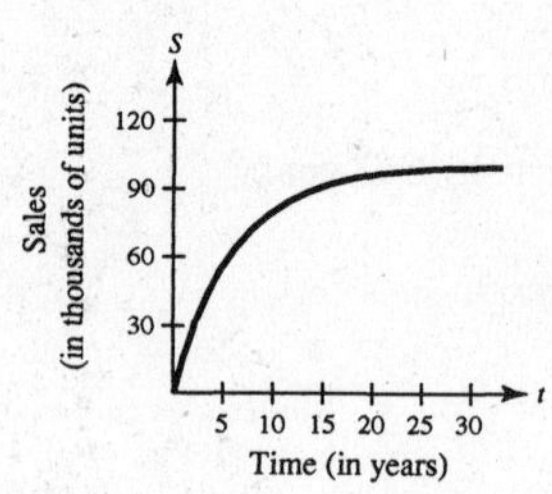

(c) $S(5) = 100(1 - e^{-0.1625(5)})$

$\approx 55.625 = 55{,}625$ units

44. $S = \dfrac{500{,}000}{1 + 0.6e^{kt}}$

(a) $300{,}000 = \dfrac{500{,}000}{1 + 0.6e^{2k}}$

$1 + 0.6e^{2k} = \dfrac{5}{3}$

$0.6e^{2k} = \dfrac{2}{3}$

$e^{2k} = \dfrac{10}{9}$

$2k = \ln\left(\dfrac{10}{9}\right)$

$k = \dfrac{1}{2}\ln\left(\dfrac{10}{9}\right) \approx 0.053$

$S = \dfrac{500{,}000}{1 + 0.6e^{0.053t}}$

(b) When $t = 5$:

$S = \dfrac{500{,}000}{1 + 0.6e^{[0.5\ln(10/9)](5)}} \approx 280{,}771$ units

45. $S = 10(1 - e^{kx})$

$x = 5$ (in hundreds), $S = 2.5$ (in thousands)

(a) $2.5 = 10(1 - e^{k(5)})$

$0.25 = 1 - e^{5k}$

$e^{5k} = 0.75$

$\ln e^{5k} = \ln 0.75$

$5k = \ln 0.75$

$k \approx -0.0575$

$S = 10(1 - e^{-0.0575x})$

(b) When $x = 7$, $S = 10(1 - e^{-0.0575(7)}) \approx 3.314$ which corresponds to 3314 units.

46. $y = ae^{bt}$

$632{,}000 = 742{,}000e^{b(2)}$

$\frac{632}{742} = e^{2b}$

$b = \frac{1}{2}\ln\left(\frac{632}{742}\right)$

$y = 742{,}000e^{0.5[\ln(632/742)](5)} \approx \$496{,}806$

47. $N = 30(1 - e^{kt})$

(a) $N = 19,\ t = 20$

$19 = 30(1 - e^{20k})$

$30e^{20k} = 11$

$e^{20k} = \dfrac{11}{30}$

$\ln e^{20k} = \ln\left(\dfrac{11}{30}\right)$

$20k = \ln \dfrac{11}{30}$

$k \approx -0.050$

$N = 30(1 - e^{-0.050t})$

(b) $N = 25$

$25 = 30(1 - e^{-0.050t})$

$\dfrac{5}{30} = e^{-0.050t}$

$\ln\left(\dfrac{5}{30}\right) = \ln e^{-0.050t}$

$\ln\left(\dfrac{5}{30}\right) = -0.050t$

$t = -\dfrac{\ln(5/30)}{-0.050} \approx 36$ days

(c) No. It is not a linear function.

48. $p(t) = \dfrac{1000}{1 + 9e^{-0.1656t}}$

(a)

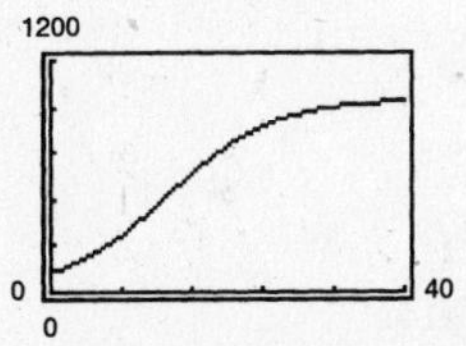

The horizontal asymptotes are $p = 0$ and $p = 1000$. The asymptote with the larger p-value, $p = 1000$, indicates that the population size will approach 1000 as time increases.

(b) $p(5) = \dfrac{1000}{1 + 9e^{-0.1656(5)}} \approx 203$ animals

(c) $500 = \dfrac{1000}{1 + 9e^{-0.1656t}}$

$1 + 9e^{-0.1656t} = 2$

$9e^{-0.1656t} = 1$

$e^{-0.1656t} = \dfrac{1}{9}$

$t = -\dfrac{\ln(1/9)}{0.1656} \approx 13$ months

49. $\frac{1}{2}C = Ce^{k(1599)}$

$0.5 = e^{k(1599)}$

$\ln 0.5 = \ln e^{k(1599)}$

$\ln 0.5 = k(1599)$

$k = \frac{\ln 0.5}{1599}$

Given $C = 10$ grams after 1000, years we have

$y = 10e^{[(\ln 0.5)/1599](1000)}$

≈ 6.48 grams.

50. $\frac{1}{2}C = Ce^{k(1599)}$

$\frac{1}{2} = e^{k(1599)}$

$\ln \frac{1}{2} = \ln e^{k(1599)}$

$\ln \frac{1}{2} = k(1599)$

$k = \frac{\ln \frac{1}{2}}{1599}$

Given $y = 1.5$ grams after 1000 years, we have

$1.5 = Ce^{[\ln(1/2)/1599](1000)}$

$C \approx 2.31$ grams.

51. $\frac{1}{2}C = Ce^{k(5715)}$

$0.5 = e^{k(5715)}$

$\ln 0.5 = \ln e^{k(5715)}$

$\ln 0.5 = k(5715)$

$k = \frac{\ln 0.5}{5715}$

Given $y = 2$ grams after 1000 years, we have

$2 = Ce^{[(\ln 0.5)/5715](1000)}$

$C \approx 2.26$ grams.

52. $\frac{1}{2}C = Ce^{k(5715)}$

$\frac{1}{2} = e^{k(5715)}$

$\ln \frac{1}{2} = \ln e^{k(5715)}$

$\ln \frac{1}{2} = k(5715)$

$k = \frac{\ln \frac{1}{2}}{5715}$

Given $C = 3$ grams, after 1000 years we have

$y = 3e^{[(\ln 1/2)/5715](1000)}$

$y \approx 2.66$ grams.

53. $\frac{1}{2}C = Ce^{k(24,100)}$

$0.5 = e^{k(24,100)}$

$\ln 0.5 = \ln e^{k(24,100)}$

$\ln 0.5 = k(24,100)$

$k = \frac{\ln 0.5}{24,100}$

Given $y = 2.1$ grams after 1000 years, we have

$2.1 = Ce^{[(\ln 0.5)/24,100](1000)}$

$C \approx 2.16$ grams.

54. $\frac{1}{2}C = Ce^{k(24,100)}$

$\frac{1}{2} = e^{k(24,100)}$

$\ln \frac{1}{2} = \ln e^{k(24,100)}$

$\ln \frac{1}{2} = k(24,100)$

$k = \frac{\ln \frac{1}{2}}{24,100}$

Given $y = 0.4$ grams after 1000 years, we have

$0.4 = Ce^{[(\ln 1/2)/24,100](1000)}$

$C \approx 0.41$ grams.

55. $R = \log_{10} \frac{I}{I_0} = \log_{10} I$ since $I_0 = 1$.

(a) $R = \log_{10} 80{,}500{,}000 \approx 7.91$

(b) $R = \log_{10} 48{,}275{,}000 \approx 7.68$

(c) $R = \log_{10} 251{,}200 \approx 5.40$

56. $R = \log_{10} \frac{I}{I_0} = \log_{10} I$ since $I_0 = 1$.

(a) $8.6 = \log_{10} I$

$10^{8.6} = I \approx 398{,}107{,}171$

(b) $6.7 = \log_{10} I$

$10^{6.7} = I \approx 5{,}011{,}872$

(c) $7.6 = \log_{10} I$

$10^{7.6} = I \approx 39{,}810{,}717$

57. $\beta = 10 \log_{10} \frac{I}{I_0}$ where $I_0 = 10^{-12}$ watt/m^2

(a) $\beta = 10 \log_{10} \frac{10^{-10}}{10^{-12}} = 10 \log_{10} 10^2 = 20$ decibels

(b) $\beta = 10 \log_{10} \frac{10^{-5}}{10^{-12}} = 10 \log_{10} 10^7 = 70$ decibels

(c) $\beta = 10 \log_{10} \frac{10^{-2.5}}{10^{-12}} = 10 \log_{10} 10^{9.5} = 95$ decibels

(d) $\beta = 10 \log_{10} \frac{1}{10^{-12}} = 10 \log_{10} 10^{12} = 120$ decibels

58. $\beta(I) = 10\log_{10}\frac{I}{I_0}$ where $I_0 = 10^{-12}$ watt/m^2

(a) $\beta(10^{-9}) = 10\log_{10}\frac{10^{-9}}{10^{-12}} = 10\log_{10}10^{3} = 30$ decibels

(b) $\beta(10^{-3.5}) = 10\log_{10}\frac{10^{-3.5}}{10^{-12}} = 10\log_{10}10^{8.5} = 85$ decibels

(c) $\beta(10^{-3}) = 10\log_{10}\frac{10^{-3}}{10^{-12}} = 10\log_{10}10^{9} = 90$ decibels

(d) $\beta(10^{-0.5}) = 10\log_{10}\frac{10^{-0.5}}{10^{-12}} = 10\log_{10}10^{11.5} = 115$ decibels

59. $\beta = 10\log_{10}\frac{I}{I_0}$

$\frac{\beta}{10} = \log_{10}\frac{I}{I_0}$

$10^{\beta/10} = 10^{\log_{10}I/I_0}$

$10^{\beta/10} = \frac{I}{I_0}$

$I = I_0 10^{\beta/10}$

$\%\text{ decrease} = \frac{I_0 10^{9.3} - I_0 10^{8.0}}{I_0 10^{9.3}} \times 100 \approx 95\%$

60. $\beta = 10\log_{10}\frac{I}{I_0}$

$10^{\beta/10} = \frac{I}{I_0}$

$I = I_0\, 10^{\beta/10}$

$\%\text{ decrease} = \frac{I_0\, 10^{8.8} - I_0\, 10^{7.2}}{I_0\, 10^{8.8}} \times 100 \approx 97\%$

61. $\text{pH} = -\log_{10}[\text{H}^+] = -\log_{10}(2.3 \times 10^{-5}) \approx 4.64$

62. $\text{pH} = -\log_{10}[\text{H}^+] = -\log_{10}[11.3 \times 10^{-6}] \approx 4.95$

63. $5.8 = -\log_{10}[\text{H}^+]$

$-5.8 = \log_{10}[\text{H}^+]$

$10^{-5.8} = 10^{\log_{10}[\text{H}^+]}$

$10^{-5.8} = [\text{H}^+]$

$[\text{H}^+] \approx 1.58 \times 10^{-6}$ moles per liter

64. $3.2 = -\log_{10}[\text{H}^+]$

$10^{-3.2} = [\text{H}^+]$

$[\text{H}^+] \approx 6.3 \times 10^{-4}$ moles per liter

65. $2.5 = -\log_{10}[\text{H}^+]$

$-2.5 = \log_{10}[\text{H}^+]$

$10^{-2.5} = 10^{\log_{10}[\text{H}^+]}$

$10^{-2.5} = [\text{H}^+]$ for the fruit.

$9.5 = -\log_{10}[\text{H}^+]$

$10^{-9.5} = [\text{H}^+]$ for the antacid tablet.

$\frac{10^{-2.5}}{10^{-9.5}} = 10^{7}$

66. $\text{pH} - 1 = -\log_{10}[\text{H}^+]$

$-(\text{pH} - 1) = \log_{10}[\text{H}^+]$

$10^{-(\text{pH}-1)} = [\text{H}^+]$

$10^{-\text{pH}+1} = [\text{H}^+]$

$10^{-\text{pH}} \cdot 10 = [\text{H}^+]$

The hydrogen ion concentration is increased by a factor of 10.

67. False. The domain can be the set of real numbers for a logistics growth function.

68. False. A logistics growth function never has an x-intercept.

69. True. $e^{-x^2} > 0$ for all x, so $y \neq 0$ and there are no x-intercepts.

70. True. $y = e^{-x^2}$ is a Gaussian model with $a = c = 1$ and $b = 0$. The graph of a Gaussian model is a bell-shaped curve.

71. Answers will vary.

72. At 9:00 A.M. we have:

$$t = -10 \ln\left(\frac{85.7 - 70}{98.6 - 70}\right) \approx 6 \text{ hours.}$$

From this we can conclude that the person died at 3:00 A.M.

Review Exercises for Chapter 7

1. $(6.1)^{2.4} \approx 76.699$

2. $-14(5^{-0.8}) \approx -3.863$

3. $2^{-0.5\pi} \approx 0.337$

4. $\sqrt[5]{1278} \approx 4.181$

5. $60^{\sqrt{3}} \approx 1201.845$

6. $7^{-\sqrt{11}} \approx 0.002$

7. $f(x) = 4^x$

Intercept: $(0, 1)$

Horizontal asymptote: x-axis

Increasing on: $(-\infty, \infty)$

Matches graph (c)

8. $f(x) = 4^{-x}$

Intercept: $(0, 1)$

Horizontal asymptote: $y = 0$

Decreasing on: $(-\infty, \infty)$

Matches graph (d)

9. $f(x) = -4^x$

Intercept: $(0, -1)$

Horizontal asymptote: x-axis

Decreasing on: $(-\infty, \infty)$

Matches graph (a)

10. $f(x) = 4^x + 1$

Intercept: $(0, 2)$

Horizontal asymptote: $y = 1$

Increasing on: $(-\infty, \infty)$

Matches graph (b)

11. $f(x) = 4^{-x} + 4$

Horizontal asymptote: $y = 4$

x	-1	0	1	2	3
$f(x)$	8	5	4.25	4.0625	4.016

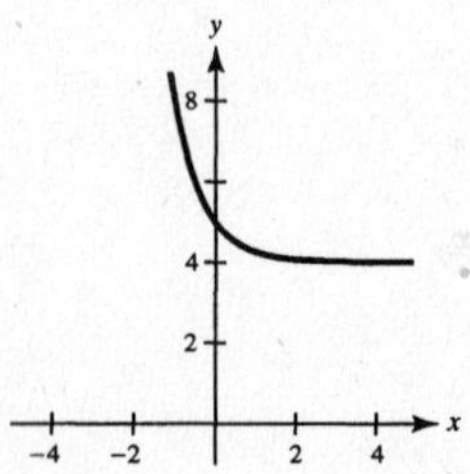

12. $f(x) = -4^x - 3$

Horizontal asymptote: $y = -3$

x	-2	-1	0	1	2
$f(x)$	-3.063	-3.25	-4	-7	-19

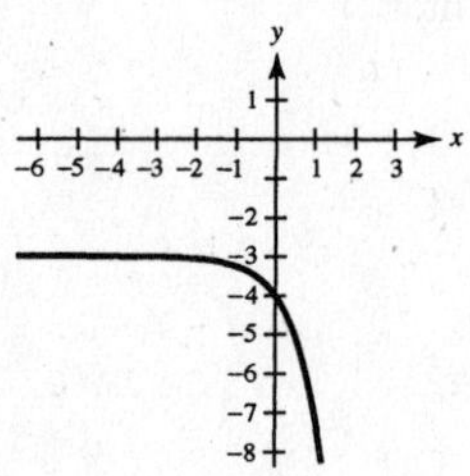

13. $f(x) = -2.65^{x+1}$

Horizontal asymptote: $y = 0$

x	-2	-1	0	1	2
$f(x)$	-0.377	-1	-2.65	-7.023	-18.61

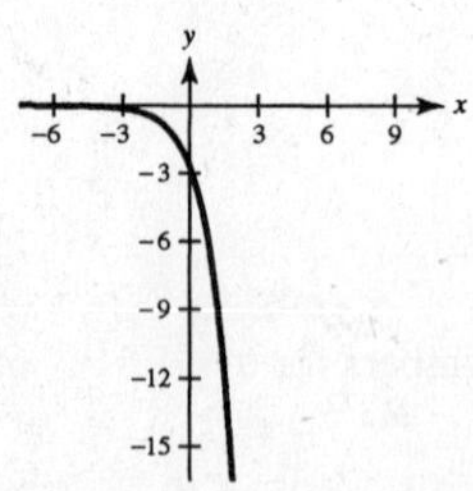

14. $f(x) = 2.65^{x-1}$

Horizontal asymptote: $y = 0$

x	-3	-1	0	1	3
$f(x)$	0.020	0.142	0.377	1	7.023

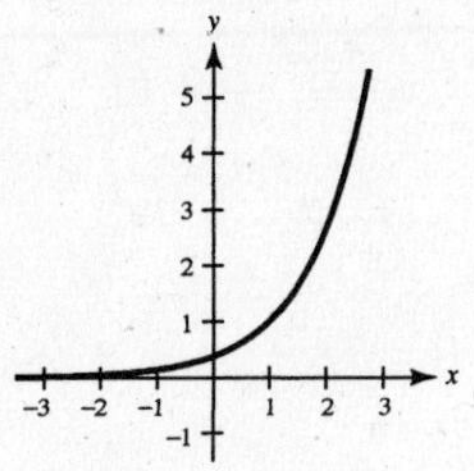

15. $f(x) = 5^{x-2} + 4$

Horizontal asymptote: $y = 4$

x	-1	0	1	2	3
$f(x)$	4.008	4.04	4.2	5	9

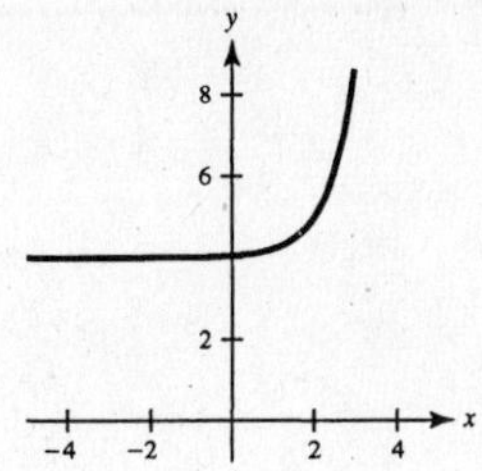

16. $f(x) = 2^{x-6} - 5$

Horizontal asymptote: $y = -5$

x	0	5	6	7	8	9
$f(x)$	-4.984	-4.5	-4	-3	-1	3

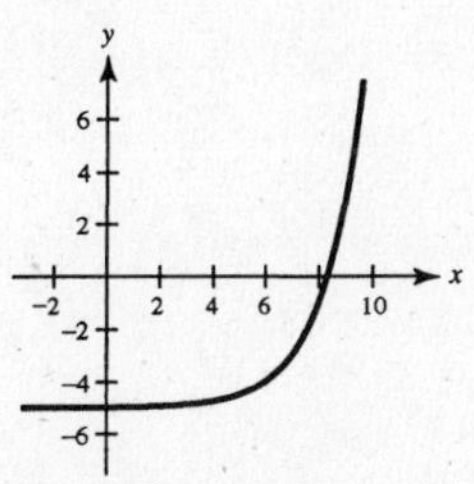

17. $f(x) = \left(\frac{1}{2}\right)^{-x} + 3 = 2^x + 3$

Horizontal asymptote: $y = 3$

x	-2	-1	0	1	2
$f(x)$	3.25	3.5	4	5	7

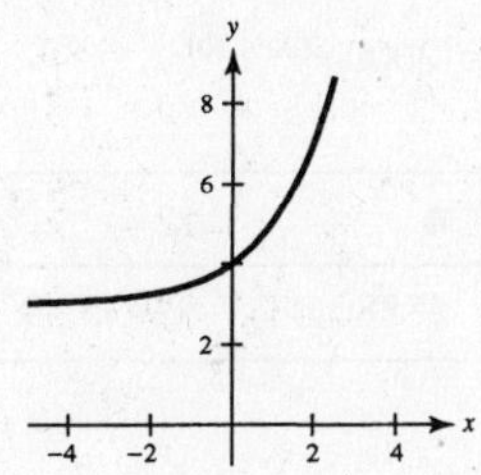

18. $f(x) = \left(\frac{1}{8}\right)^{x+2} - 5$

Horizontal asymptote: $y = -5$

x	-3	-2	-1	0	2
$f(x)$	3	-4	-4.875	-4.984	-5

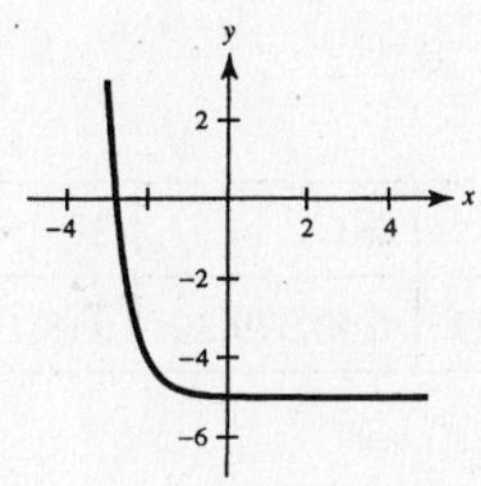

19. $e^8 \approx 2980.958$

20. $e^{5/8} \approx 1.868$

21. $e^{-1.7} \approx 0.183$

22. $e^{0.278} = 1.320$

23. $h(x) = e^{-x/2}$

x	-2	-1	0	1	2
$h(x)$	2.72	1.65	1	0.61	0.37

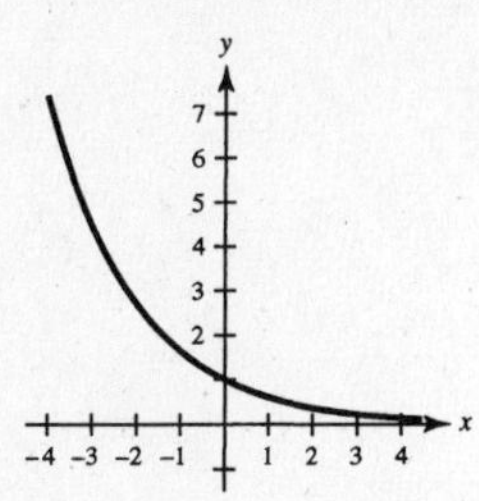

24. $h(x) = 2 - e^{-x/2}$

x	-2	-1	0	1	2
y	-0.72	0.35	1	1.39	1.63

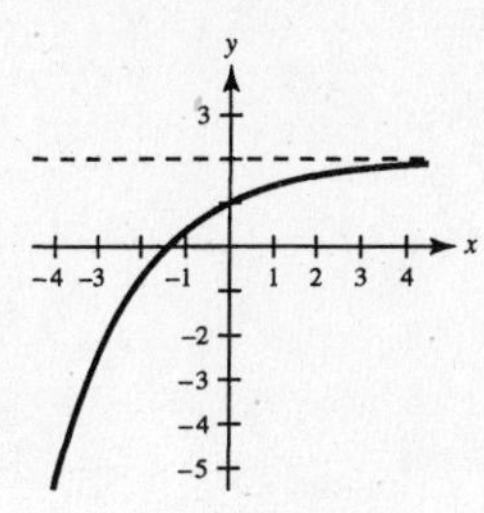

25. $f(x) = e^{x+2}$

x	-3	-2	-1	0	1
$f(x)$	0.37	1	2.72	7.39	20.09

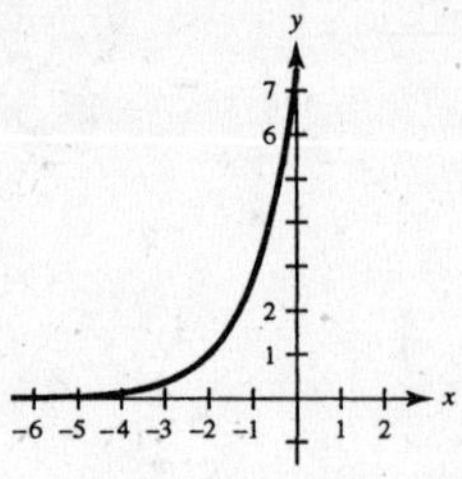

26. $s(t) = 4e^{-2/t}, t > 0$

t	$\frac{1}{2}$	1	2	3	4
y	0.07	0.54	1.47	2.05	2.43

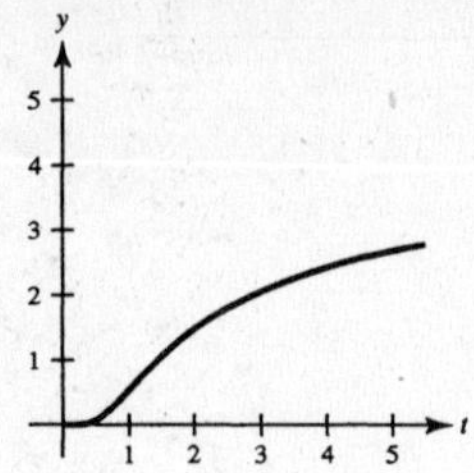

27. $A = 3500\left(1 + \frac{0.065}{n}\right)^{10n}$ or $A = 3500e^{(0.065)(10)}$

n	1	2	4	12	365	Continuous Compounding
A	\$6569.98	\$6635.43	\$6669.46	\$6692.64	\$6704.00	\$6704.39

28. $A = 2000\left(1 + \frac{0.05}{n}\right)^{30n}$ or $A = 2000e^{(0.05)(30)}$

n	1	2	4	12	365	Continuous
A	\$8643.88	\$8799.58	\$8880.43	\$8935.49	\$8962.46	\$8963.38

29. $200{,}000 = Pe^{0.08t}$

$$P = \frac{200{,}000}{e^{0.08t}}$$

t	1	10	20	30	40	50
P	\$184,623.27	\$89,865.79	\$40,379.30	\$18,143.59	\$8,152.44	\$3,663.13

30. $200{,}000 = P\left(1 + \frac{0.06}{12}\right)^{12t} = P(1.005)^{12t}$

$$P = \frac{200{,}000}{(1.005)^{12t}}$$

t	1	10	20	30	40	50
P	\$188,381.07	\$109,926.55	\$60,419.23	\$33,208.39	\$18,252.42	\$10,032.13

31. $F(t) = 1 - e^{-t/3}$

(a) $F\left(\frac{1}{2}\right) \approx 0.154$

(b) $F(2) \approx 0.487$

(c) $F(5) \approx 0.811$

32. $V(t) = 14{,}000\left(\frac{3}{4}\right)^t$

(a)

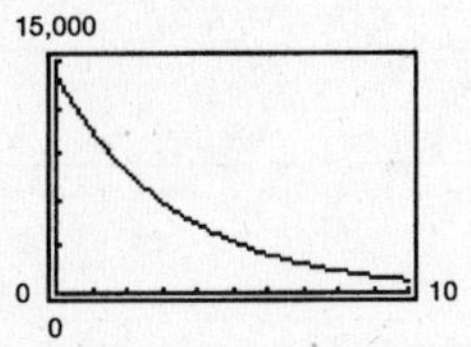

(b) $V(2) = 14{,}000\left(\frac{3}{4}\right)^2 = \$7875$

(c) According to the model, the car depreciates most rapidly at the beginning. Yes, this is realistic.

33. (a) $A = 50{,}000e^{(0.0875)(35)} \approx \$1{,}069{,}047.14$

(b) The doubling time is

$$\frac{\ln 2}{0.0875} \approx 7.9 \text{ years.}$$

34. $y = 28e^{0.6-0.012s}$, $s \geq 50$

s	50	55	60	65	70
y	28	26.4	24.8	23.4	22.0

35. Order: $b < d < a < c$

36. Since graphs (b) and (d) represent exponential decay, b and d are negative.

Since graph (a) and (c) represent exponential growth, a and c are positive.

37. $4^3 = 64$

$\log_4 64 = 3$

38. $25^{3/2} = 125$

$\log_{25} 125 = \frac{3}{2}$

39. $\log_{10} 1000 = \log_{10} 10^3 = 3$

40. $\log_9 3 = \log_9 9^{1/2} = \frac{1}{2}$

41. $\log_2 \dfrac{1}{8} = \log_2 2^{-3} = -3$

42. $\log_a \dfrac{1}{a} = \log_a a^{-1} = -1$

43. $g(x) = \log_7 x \Rightarrow x = 7^y$

Vertical asymptote: $x = 0$

x	$\frac{1}{7}$	1	7	49
$g(x)$	-1	0	1	2

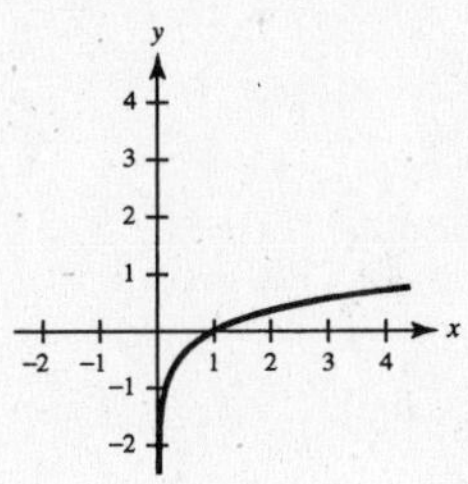

44. $g(x) = \log_5 x \Rightarrow 5^y = x$

Domain: $(0, \infty)$

Vertical asymptote: $x = 0$

x	$\frac{1}{25}$	$\frac{1}{5}$	1	5	25
$g(x)$	-2	-1	0	1	2

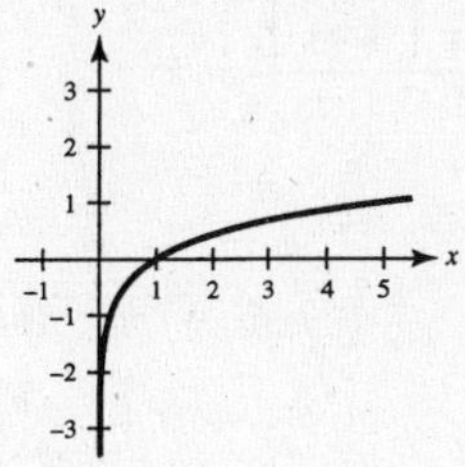

45. $f(x) = \log_{10}\left(\dfrac{x}{3}\right) \Rightarrow \dfrac{x}{3} = 10^y \Rightarrow x = 3(10^y)$

Vertical asymptote: $x = 0$

x	0.03	0.3	3	30
$f(x)$	-2	-1	0	1

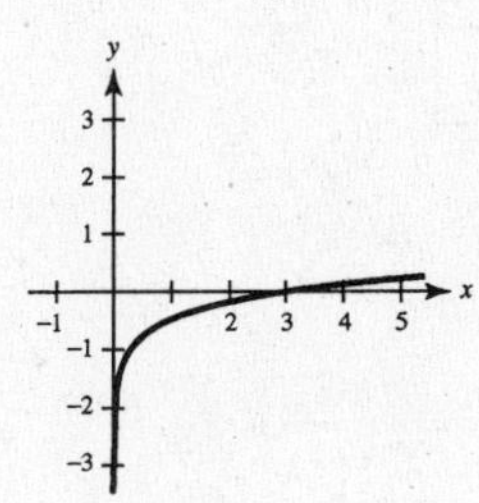

46. $f(x) = 6 + \log_{10} x$

Domain: $(0, \infty)$

Vertical asymptote: $x = 0$

x	1	2	4	6	8	10
$f(x)$	6	6.3	6.6	6.8	6.9	7

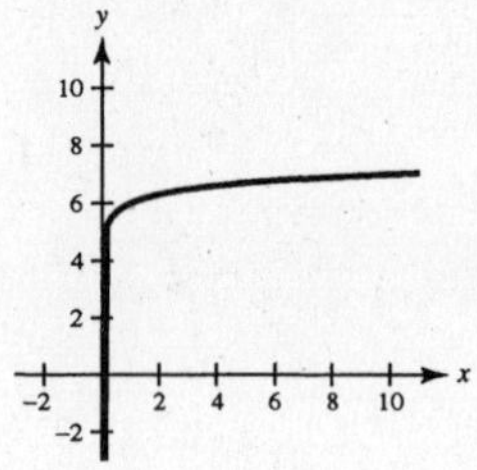

47. $f(x) = 4 - \log_{10}(x + 5)$

Vertical asymptote: $x = -5$

x	-4	-3	-2	-1	0	1
$f(x)$	4	3.70	3.52	3.40	3.30	3.22

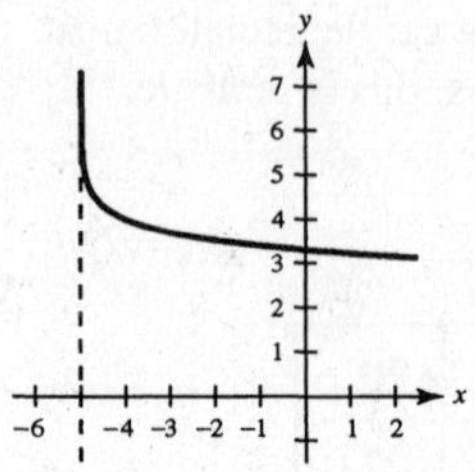

48. $f(x) = \log_{10}(x - 3) + 1$

Domain: $(3, \infty)$

Vertical asymptote: $x = 3$

x	4	5	6	7	8
$f(x)$	1	1.3	1.5	1.6	1.7

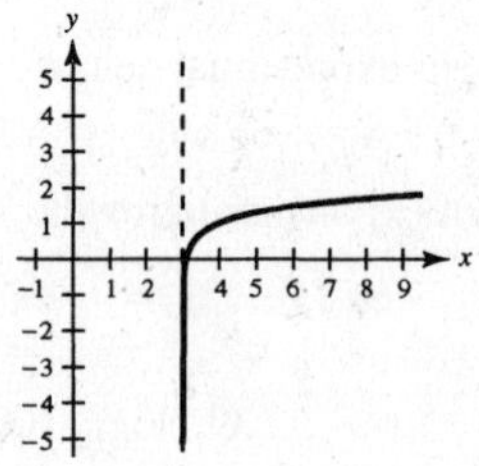

49. $\ln 22.6 \approx 3.118$

50. $\ln 0.98 \approx -0.020$

51. $\ln e^{-12} = -12$

52. $\ln e^7 = 7$

53. $\ln(\sqrt{7} + 5) \approx 2.034$

54. $\ln\left(\frac{\sqrt{3}}{8}\right) \approx -1.530$

55. $f(x) = \ln x + 3$

Domain: $(0, \infty)$

Vertical asymptote: $x = 0$

x	1	2	3	$\frac{1}{2}$	$\frac{1}{4}$
$f(x)$	3	3.69	4.10	2.31	1.61

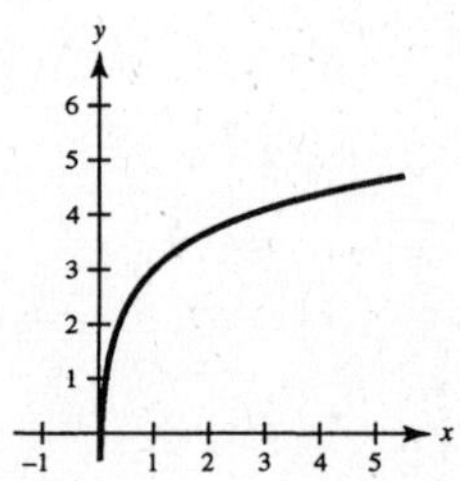

56. $f(x) = \ln(x - 3)$

Domain: $(3, \infty)$

Vertical asymptote: $x = 3$

x	3.5	4	4.5	5	5.5
y	-0.69	0	0.41	0.69	0.92

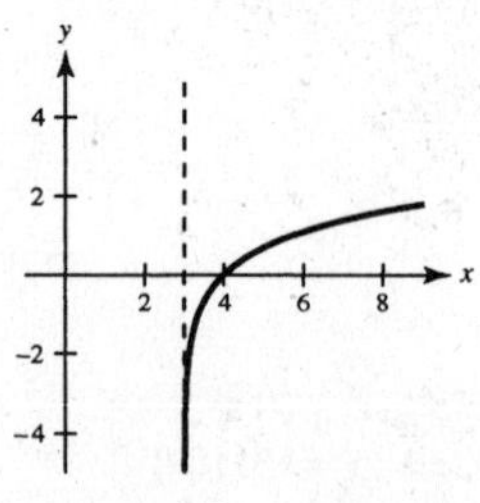

57. $h(x) = \ln(x^2) = 2\ln|x|$

Vertical asymptote: $x = 0$

x	± 0.5	± 1	± 2	± 3	± 4
y	-1.39	0	1.39	2.20	2.77

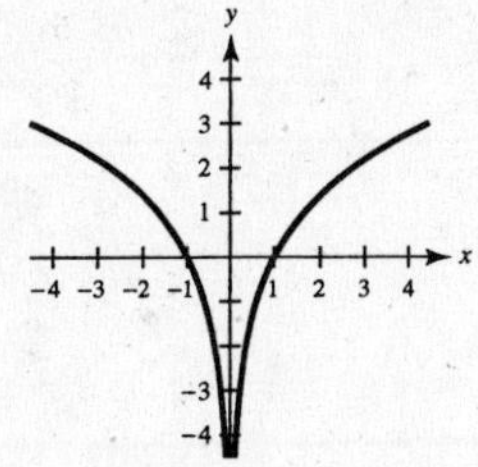

58. $f(x) = \frac{1}{4}\ln x$

Domain: $(0, \infty)$

Vertical asymptote: $x = 0$

x	$\frac{1}{2}$	1	$\frac{3}{2}$	2	$\frac{5}{2}$	3
y	-0.17	0	0.10	0.17	0.23	0.27

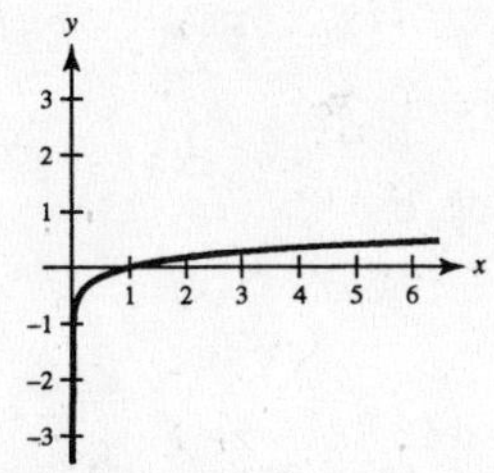

59. $s = 25 - \dfrac{13\ln(10/12)}{\ln 3} \approx 27.16$ miles

60. $T = 87.97 + 34.96\ln p + 7.91\sqrt{p}$

(a)

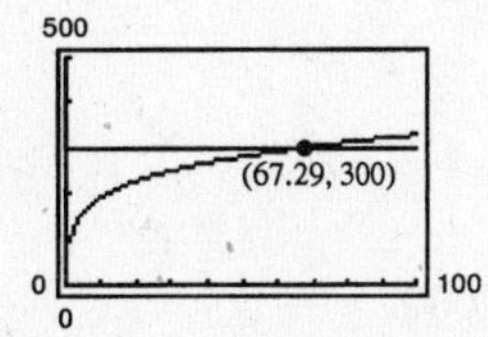

(b) $T > 300°\text{F}$ when $p > 67.29$ pounds per square inch.

(c) $T(74) \approx 306.48°\text{F}$

61. $\log_4 9 = \dfrac{\log_{10} 9}{\log_{10} 4} \approx 1.585$

$\log_4 9 = \dfrac{\ln 9}{\ln 4} \approx 1.585$

62. $\log_{12} 200 = \dfrac{\log_{10} 200}{\log_{10} 12} \approx 2.132$

$\log_{12} 200 = \dfrac{\ln 200}{\ln 12} \approx 2.132$

63. $\log_{1/2} 5 = \dfrac{\log_{10} 5}{\log_{10}(1/2)} \approx -2.322$

$\log_{1/2} 5 = \dfrac{\ln 5}{\ln(1/2)} \approx -2.322$

64. $\log_3 0.28 = \dfrac{\log_{10} 0.28}{\log_{10} 3} \approx -1.159$

$\log_3 0.28 = \dfrac{\ln 0.28}{\ln 3} \approx -1.159$

65. $\ln 8 + \ln 5 = \ln[(8)(5)] = \ln 40$

66. $\ln\sqrt[4]{\dfrac{x}{y}} = \ln\left(\dfrac{x}{y}\right)^{1/4} = \ln\left(\dfrac{x^{1/4}}{y^{1/4}}\right)$

$= \ln x^{1/4} - \ln y^{1/4} = \dfrac{1}{4}\ln x - \dfrac{1}{4}\ln y$

67. $\log_8\left(\dfrac{\sqrt{x}}{y^3}\right) = \log_8\sqrt{x} - \log_8 y^3 = \dfrac{1}{2}\log_8 x - 3\log_8 y$

68. $\log_{10}\left(\dfrac{p^2q^3}{r}\right) = \log_{10}(p^2q^3) - \log_{10} r$

$= \log_{10} p^2 + \log_{10} q^3 - \log_{10} r$

$= 2\log_{10} p + 3\log_{10} q - \log_{10} r$

69. $\log_5 5x^2 = \log_5 5 + \log_5 x^2$

$= 1 + 2\log_5|x|$

70. $\log_7 \frac{\sqrt{x}}{4} = \log_7 \sqrt{x} - \log_7 4$

$= \log_7 x^{1/2} - \log_7 4$

$= \frac{1}{2}\log_7 x - \log_7 4$

71. $\log_{10} \frac{5\sqrt{y}}{x^2} = \log_{10} 5\sqrt{y} - \log_{10} x^2$

$= \log_{10} 5 + \log_{10} \sqrt{y} - \log_{10} x^2$

$= \log_{10} 5 + \frac{1}{2}\log_{10} y - 2\log_{10}|x|$

72. $\ln\left|\frac{x-1}{x+1}\right| = \ln|x-1| - \ln|x+1|$

73. $\log_2 5 + \log_2 x = \log_2 5x$

74. $\log_6 y - 2\log_6 z = \log_6 y - \log_6 z^2$

$= \log_6 \frac{y}{z^2}$

75. $\frac{1}{2}\ln|2x-1| - 2\ln|x+1| = \ln\sqrt{|2x-1|} - \ln|x+1|^2$

$= \ln \frac{\sqrt{|2x-1|}}{(x+1)^2}$

76. $5\ln|x-2| - \ln|x+2| - 3\ln|x| = \ln|x-2|^5 - \ln|x+2| - \ln|x|^3$

$= \ln\left|\frac{(x-2)^5}{(x+2)x^3}\right|$

77. $t = 50\log_{10}\frac{18{,}000}{18{,}000 - h}$

(a) Domain: $0 \le h < 18{,}000$

(b)

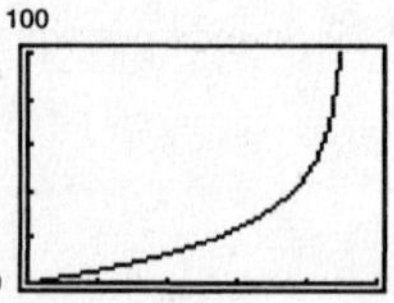

Vertical asymptote: $h = 18{,}000$

(c) As the plane approaches its absolute ceiling, it climbs at a slower rate, so the time required increases.

(d) $50\log_{10}\frac{18{,}000}{18{,}000 - 4000} \approx 5.46$ minutes

78. $R = \log_{10}\frac{I}{I_0} = \log_{10} I - \log_{10} I_0$

If $I_0 = 1$, then $R = \log_{10} I - \log_{10} 1 = \log_{10} I$

79. $8^x = 512$

$8^x = 8^3$

$x = 3$

80. $3^x = 729$

$3^x = 3^6$

$x = 6$

81. $6^x = \frac{1}{216}$

$6^x = 6^{-3}$

$x = -3$

82. $6^{x-2} = 1296$

$6^{x-2} = 6^4$

$x - 2 = 4$

$x = 6$

83. $\log_7 x = 4$

$7^{\log_7 x} = 7^4$

$x = 7^4$

$x = 2401$

84. $\log_x 243 = 5$

$x^5 = 243$

$x^5 = 3^5$

$x = 3$

85. $e^x = 12$

$\ln e^x = \ln 12$

$x = \ln 12 \approx 2.485$

86. $e^{3x} = 25$

$\ln e^{3x} = \ln 25$

$3x = \ln 25$

$x = \frac{\ln 25}{3} \approx 1.073$

87. $3e^{-5x} = 132$

$e^{-5x} = 44$

$\ln e^{-5x} = \ln 44$

$-5x = \ln 44$

$x = \dfrac{\ln 44}{-5} \approx -0.757$

88. $14e^{3x+2} = 560$

$e^{3x+2} = 40$

$\ln e^{3x+2} = \ln 40$

$3x + 2 = \ln 40$

$x = \dfrac{(\ln 40) - 2}{3} \approx 0.563$

89. $e^x + 13 = 35$

$e^x = 22$

$\ln e^x = \ln 22$

$x = \ln 22 \approx 3.091$

90. $e^x - 28 = -8$

$e^x = 20$

$x = \ln 20 \approx 2.996$

91. $-4(5^x) = -68$

$5^x = 17$

$\ln 5^x = \ln 17$

$x \ln 5 = \ln 17$

$x = \dfrac{\ln 17}{\ln 5} \approx 1.760$

92. $2(12^x) = 190$

$12^x = 95$

$\ln 12^x = \ln 95$

$x \ln 12 = \ln 95$

$x = \dfrac{\ln 95}{\ln 12} \approx 1.833$

93. $e^{2x} - 7e^x + 10 = 0$

$(e^x - 2)(e^x - 5) = 0$

$e^x = 2$ $\qquad$ $e^x = 5$

$\ln e^x = \ln 2$ $\qquad$ $\ln e^x = \ln 5$

$x = \ln 2 \approx 0.693$ $\qquad$ $x = \ln 5 \approx 1.609$

94. $e^{2x} - 6e^x + 8 = 0$

$(e^x - 2)(e^x - 4) = 0$

$e^x = 2$ or $e^x = 4$

$x = \ln 2$ $\qquad$ $x = \ln 4$

$x \approx 0.693$ $\qquad$ $x \approx 1.386$

95. $2^{0.6x} - 3x = 0$

Graph $y_1 = 2^{0.6x} - 3x$.

The x-intercepts are at $x \approx 0.39$ and at $x \approx 7.48$.

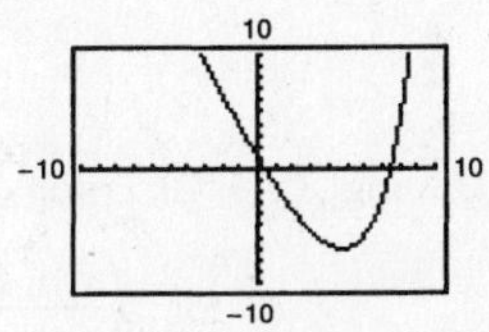

96. $4^{-0.2x} + x = 0$

Graph $y_1 = 4^{-0.2x} + x$.

The x-intercepts are at $x \approx -7.04$ and $x \approx -1.53$.

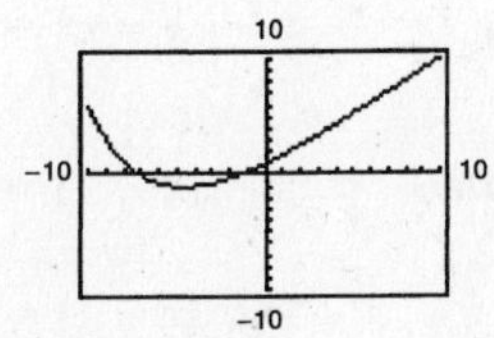

97. $25e^{-0.3x} = 12$

Graph $y_1 = 25e^{-0.3x}$ and $y_2 = 12$.

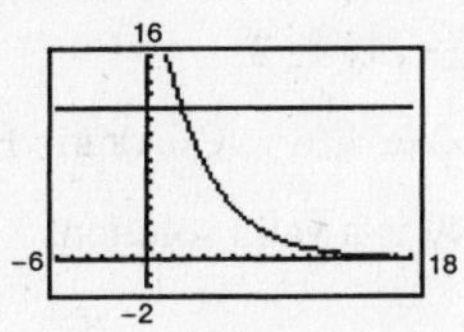

The graphs intersect at $x \approx 2.45$.

98. $4e^{1.2x} = 9$

Graph $y_1 = 4e^{1.2x}$ and $y_2 = 9$.

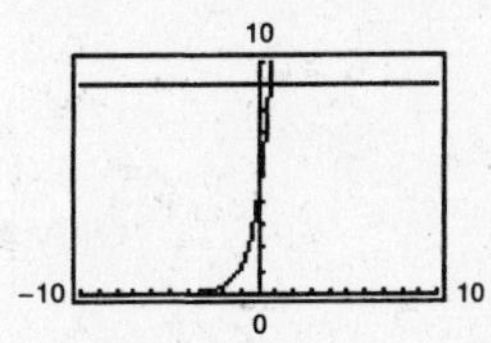

The intersection is at $x \approx 0.68$.

99. $2^{x-1} - x^2 = 0$

$2^{x-1} = x^2$

Graph $y_1 = 2^{x-1}$ and $y_2 = x^2$ and find the points of intersection.

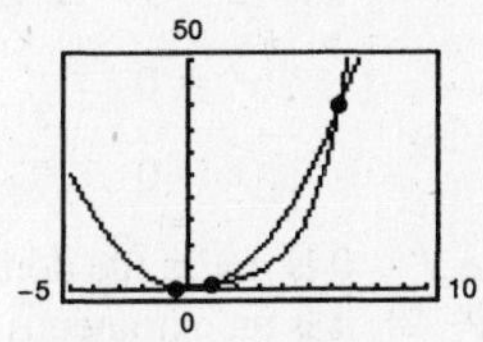

There are three points of intersection. $x \approx -0.58, x = 1, x \approx 6.32$

100. $e^{2x} - 10x = 4$

$e^{2x} = 10x + 4$

Graph $y_1 = e^{2x}$ and $y_2 = 10x + 4$ and find the points of intersection.

There are two points of intersection.

$x \approx -0.35$ and $x \approx 1.46$

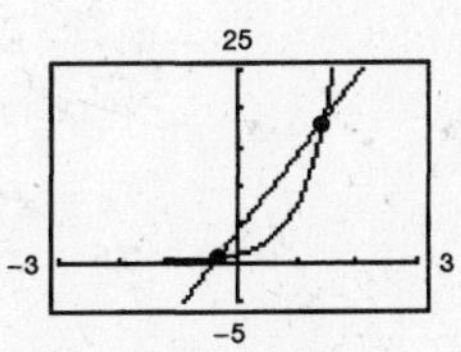

101. $\ln 3x = 8.2$

$e^{\ln 3x} = e^{8.2}$

$3x = e^{8.2}$

$x = \dfrac{e^{8.2}}{3} \approx 1213.650$

102. $\ln 5x = 7.2$

$5x = e^{7.2}$

$x = \dfrac{e^{7.2}}{5} \approx 267.886$

103. $2 \ln 4x = 15$

$\ln 4x = \frac{15}{2}$

$e^{\ln 4x} = e^{7.5}$

$4x = e^{7.5}$

$x = \frac{1}{4}e^{7.5} \approx 452.011$

104. $4 \ln 3x = 15$

$\ln 3x = \dfrac{15}{4}$

$3x = e^{15/4}$

$x = \dfrac{e^{15/4}}{3} \approx 14.174$

105. $\ln x - \ln 3 = 2$

$\ln \dfrac{x}{3} = 2$

$e^{(\ln x/3)} = e^2$

$\dfrac{x}{3} = e^2$

$x = 3e^2 \approx 22.167$

106. $\ln\sqrt{x + 8} = 3$

$\dfrac{1}{2}\ln(x + 8) = 3$

$\ln(x + 8) = 6$

$x + 8 = e^6$

$x = e^6 - 8 \approx 395.429$

107. $\ln\sqrt{x + 1} = 2$

$\dfrac{1}{2}\ln(x + 1) = 2$

$\ln(x + 1) = 4$

$e^{\ln(x+1)} = e^4$

$x + 1 = e^4$

$x = e^4 - 1 \approx 53.598$

108. $\ln x - \ln 5 = 4$

$\ln \dfrac{x}{5} = 4$

$\dfrac{x}{5} = e^4$

$x = 5e^4 \approx 272.991$

109. $\log_{10}(x - 1) = \log_{10}(x - 2) - \log_{10}(x + 2)$

$\log_{10}(x - 1) = \log_{10}\left(\dfrac{x - 2}{x + 2}\right)$

$10^{\log_{10}(x-1)} = 10^{\log_{10}(x-2/x+2)}$

$x - 1 = \dfrac{x - 2}{x + 2}$

$(x - 1)(x + 2) = x - 2$

$x^2 + x - 2 = x - 2$

$x^2 = 0$

$x = 0$

Since $x = 0$ is not in the domain of $\ln(x - 1)$ or of $\ln(x - 2)$, it is an extraneous solution. The equation has no solution.

110. $\log_{10}(x + 2) - \log_{10} x = \log_{10}(x + 5)$

$\log_{10}\left(\dfrac{x + 2}{x}\right) = \log_{10}(x + 5)$

$\dfrac{x + 2}{x} = x + 5$

$x + 2 = x^2 + 5x$

$0 = x^2 + 4x - 2$

$x = -2 \pm \sqrt{6}$ Quadratic Formula

Only $x = -2 + \sqrt{6} \approx 0.449$ is a valid solution.

111. $\log_3(1 - x) = -1$

$1 - x = 3^{-1}$

$1 - \frac{1}{3} = x$

$x = \frac{2}{3} \approx 0.667$

112. $\log_7(-x - 4) = 2$

$-x - 4 = 7^2$

$-x = 49 + 4$

$x = -53 = -53.000$

113. $2 \ln(x + 3) + 3x = 8$

Graph $y_1 = 2 \ln(x + 3) + 3x$ and $y_2 = 8$.

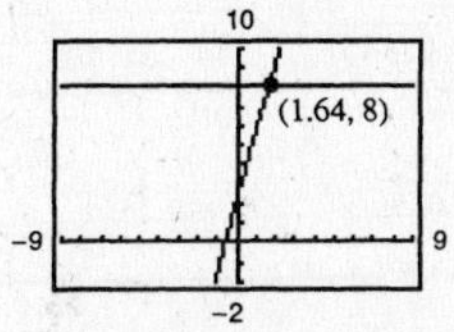

The graphs intersect at approximately $(1.64, 8)$.
The solution of the equation is $x \approx 1.64$.

114. $6 \log_{10}(x^2 + 1) - x = 0$

Graph

$$y_1 = 6 \log_{10}(x^2 + 1) - x.$$

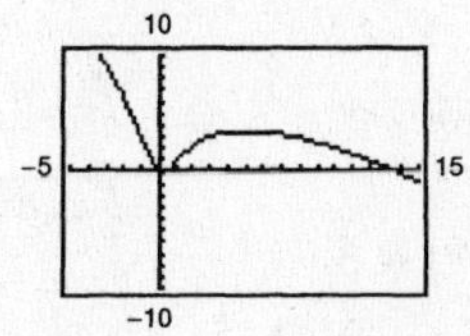

Zoom in to see the behavior near the origin.

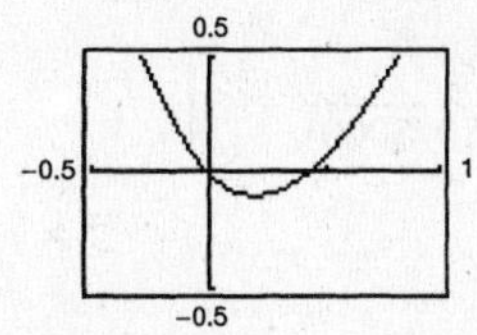

The solutions of the equation occur at the x-intercepts, which are at $x = 0$, $x \approx 0.42$, and $x \approx 13.63$.

115. $4 \ln(x + 5) - x = 10$

Graph $y_1 = 4 \ln(x + 5) - x$ and $y_2 = 10$.

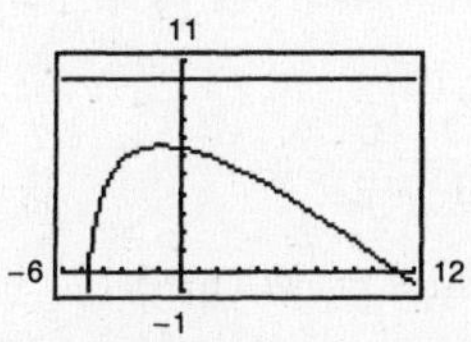

The graphs do not intersect.
The equation has no solution.

116. $x - 2 \log_{10}(x + 4) = 0$

Graph

$$y_1 = x - 2 \log_{10}(x + 4).$$

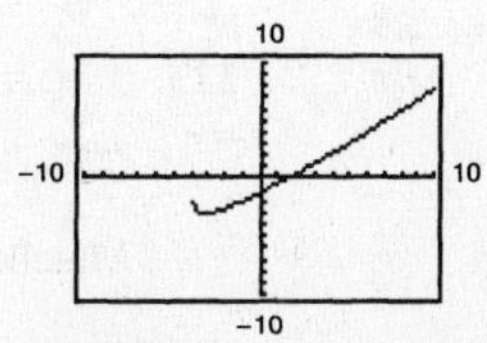

Note that $x = -4$ is a vertical asymptote, but the graph's behavior near $x = -4$ is not visible in most typical viewing windows. Zoom in to see the behavior near $x = -4$.

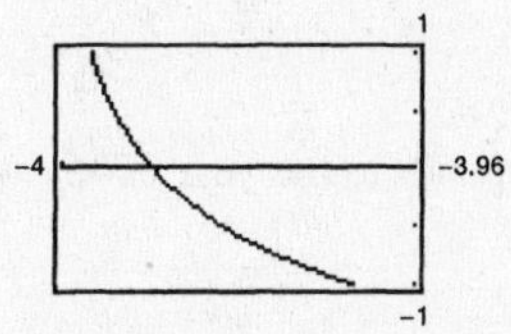

The solutions of the equation occur at the x-intercepts, which are at $x \approx -3.99$ and $x \approx 1.48$.

117.
$$\begin{aligned} 3(7550) &= 7550e^{0.0725t} \\ 3 &= e^{0.0725t} \\ \ln 3 &= \ln e^{0.0725t} \\ \ln 3 &= 0.0725t \\ t &= \frac{\ln 3}{0.0725} \approx 15.2 \text{ years} \end{aligned}$$

118. $p = 5000\left(1 - \dfrac{3}{3 + e^{-0.0015x}}\right)$

(a) $450 = 5000\left(1 - \dfrac{3}{3 + e^{-0.0015x}}\right)$

$\dfrac{450}{5000} - 1 = -\dfrac{3}{3 + e^{-0.0015x}}$

$\left(1 - \dfrac{450}{5000}\right)(3 + e^{-0.0015x}) = 3$

$e^{-0.0015x} = \dfrac{3}{1 - (450/5000)} - 3$

$-0.0015x = \ln\left[\dfrac{3}{1 - (450/5000)} - 3\right]$

$x \approx 810$ units

(b) $400 = 5000\left(1 - \dfrac{3}{3 + e^{-0.0015x}}\right)$

$\dfrac{400}{5000} - 1 = -\dfrac{3}{3 + e^{-0.0015x}}$

$\left(1 - \dfrac{2}{25}\right)(3 + e^{-0.0015x}) = 3$

$e^{-0.0015x} = \dfrac{3}{23/25} - 3$

$-0.0015x = \ln\left[3\left(\dfrac{25}{23}\right) - 3\right]$

$x \approx 896$ units

119. $y = e^{-2x/3}$

Exponential decay model

Matches graph (e)

120. $y = 4e^{2x/3}$

Exponential growth model

Matches graph (b)

121. $y = \ln(x + 3)$

Logarithmic model

Vertical asymptote: $x = -3$

Graph includes $(-2, 0)$

Matches graph (f)

122. $y = 7 - \log_{10}(x + 3)$

Logarithmic model

Vertical asymptote: $x = -3$

Matches graph (d)

123. $y = 2e^{-(x+4)^2/3}$

Gaussian model

Matches graph (a)

124. $y = \dfrac{6}{1 + 2e^{-2x}}$

Logistics growth model

Matches graph (c)

125. $P = 895{,}397e^{-0.011t}$ where $t = 0$ corresponds to 1970.

$895{,}397e^{-0.011t} = 600{,}000$

$-0.011t = \ln\left(\dfrac{600{,}000}{895{,}397}\right)$

$t \approx 36.4$ years which corresponds to the year 2006.

126. $y = Ce^{kt}$

$\dfrac{1}{2}C = Ce^{k(245{,}500)}$

$\ln\left(\dfrac{1}{2}\right) = 245{,}500k$

$k = \dfrac{\ln(1/2)}{245{,}500}$

When $t = 5000$, we have:
$y = Ce^{[\ln(1/2)/245{,}500](5000)} \approx 0.986C = 98.6\%C.$

After 5000 years, approximately 98.6% of the radioactive uranium II will remain.

127. $20{,}000 = 10{,}000e^{r(5)}$

$2 = e^{5r}$

$\ln 2 = 5r$

$\dfrac{\ln 2}{5} = r$

$r \approx 0.138629 = 13.8629\%$

128. $A = 10{,}000e^{0.138629}$

$\approx \$11{,}486.98$

129. $y = ae^{bx}$

$2 = ae^{b(0)} \implies a = 2$

$3 = 2e^{b(4)}$

$1.5 = e^{4b}$

$\ln 1.5 = 4b \implies b \approx 0.1014$

Thus, $y \approx 2e^{0.1014x}$

130. $y = ae^{bx}$

$\frac{1}{2} = ae^{b(0)} \implies a = \frac{1}{2}$

$5 = \frac{1}{2}e^{b(5)}$

$10 = e^{5b}$

$\ln 10 = 5b$

$\frac{\ln 10}{5} = b$

$b \approx 0.4605$

$y = \frac{1}{2}e^{0.4605x}$

131. $y = 0.0499e^{-(x-71)^2/128}$

(a) Graph $y_1 = 0.0499e^{-(x-71)^2/128}$.

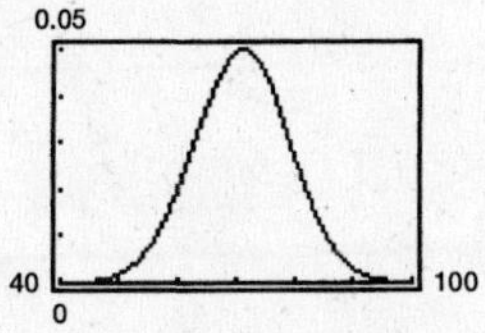

(b) The average test score is 71.

132. $N = \frac{157}{1 + 5.4e^{-0.12t}}$

(a) When $N = 50$:

$$50 = \frac{157}{1 + 5.4e^{-0.12t}}$$

$$1 + 5.4e^{-0.12t} = \frac{157}{50}$$

$$5.4e^{-0.12t} = \frac{107}{50}$$

$$e^{-0.12t} = \frac{107}{270}$$

$$-0.12t = \ln\frac{107}{270}$$

$$t = \frac{\ln(107/270)}{-0.12} \approx 7.7 \text{ weeks}$$

(b) When $N = 75$:

$$75 = \frac{157}{1 + 5.4e^{-0.12t}}$$

$$1 + 5.4e^{-0.12t} = \frac{157}{75}$$

$$5.4e^{-0.12t} = \frac{82}{75}$$

$$e^{-0.12t} = \frac{82}{405}$$

$$-0.12t = \ln\frac{82}{405}$$

$$t = \frac{\ln(82/405)}{-0.12} \approx 13.3 \text{ weeks}$$

133. $\beta = 10 \log_{10}\left(\frac{I}{10^{-16}}\right)$

$125 = 10 \log_{10}\left(\frac{I}{10^{-16}}\right)$

$12.5 = \log_{10}\left(\frac{I}{10^{-16}}\right)$

$10^{12.5} = \frac{I}{10^{-16}}$

$I = 10^{-3.5} \text{ watt/cm}^2$

134. $R = \log_{10} I$ since $I_0 = 1$.

(a) $\log_{10} I = 8.4$

$I = 10^{8.4} \approx 251{,}188{,}643$

(b) $\log_{10} I = 6.85$

$I = 10^{6.85} \approx 7{,}079{,}458$

(c) $\log_{10} I = 9.1$

$I = 10^{9.1} \approx 1{,}258{,}925{,}412$

Problem Solving for Chapter 7

1.

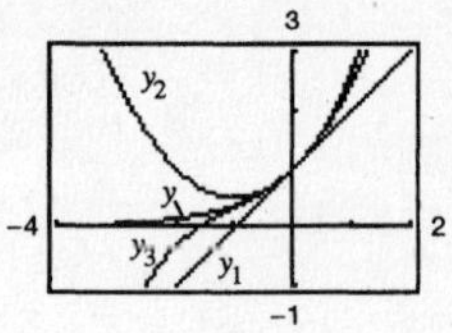

2. $y_4 = 1 + \dfrac{x}{1!} + \dfrac{x^2}{2!} + \dfrac{x^3}{3!} + \dfrac{x^4}{4!}$

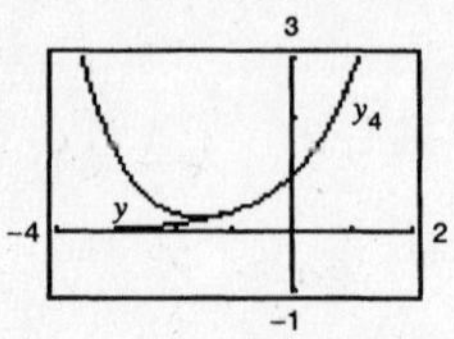

As more terms are added, the polynomial approaches e^x.

$$e^x = 1 + \frac{x}{1!} + \frac{x^2}{2!} + \frac{x^3}{3!} + \frac{x^4}{4!} + \frac{x^5}{5!} + \cdots$$

3. (a) $f(u + v) = a^{u+v}$

$= a^u \cdot a^v$

$= f(u) \cdot f(v)$

(b) $f(2x) = a^{2x}$

$= (a^x)^2$

$= [f(x)]^2$

4. $y = \ln x$

$y_1 = x - 1$

$y_2 = (x - 1) - \frac{1}{2}(x - 1)^2$

$y_3 = (x - 1) - \frac{1}{2}(x - 1)^2 + \frac{1}{3}(x - 1)^3$

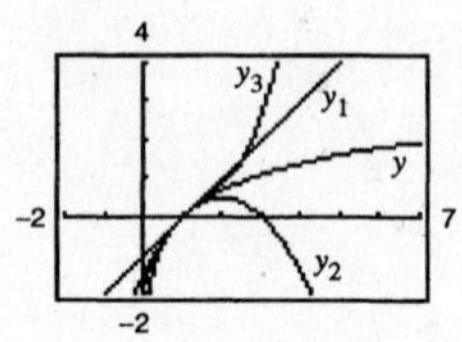

5. $y_4 = (x - 1) - \frac{1}{2}(x - 1)^2 + \frac{1}{3}(x - 1)^3 - \frac{1}{4}(x - 1)^4$

The pattern implies that

$$\ln x = (x - 1) - \tfrac{1}{2}(x - 1)^2 + \tfrac{1}{3}(x - 1)^3 - \tfrac{1}{4}(x - 1)^4 + \cdots.$$

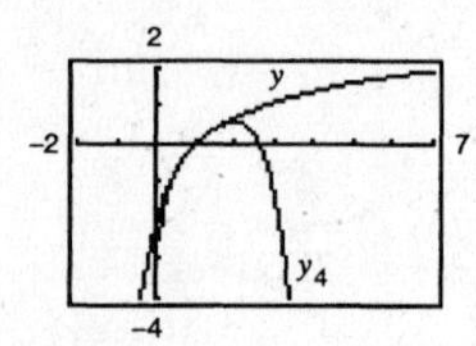

6. $\ln 2 \approx 0.6931$, $\ln 3 \approx 1.0986$, $\ln 5 \approx 1.6094$

$\ln 2 \approx 0.6931$

$\ln 3 \approx 1.0986$

$\ln 4 = \ln(2 \cdot 2) = \ln 2 + \ln 2 \approx 0.6931 + 0.6931 = 1.3862$

$\ln 5 \approx 1.6094$

$\ln 6 = \ln(2 \cdot 3) = \ln 2 + \ln 3 \approx 0.6931 + 1.0986 = 1.7917$

$\ln 8 = \ln 2^3 = 3 \ln 2 \approx 3(0.6931) = 2.0793$

$\ln 9 = \ln 3^2 = 2 \ln 3 \approx 2(1.0986) = 2.1972$

$\ln 10 = \ln(5 \cdot 2) = \ln 5 + \ln 2 \approx 1.6094 + 0.6931 = 2.3025$

$\ln 12 = \ln(2^2 \cdot 3) = \ln 2^2 + \ln 3 = 2 \ln 2 + \ln 3 \approx 2(0.6931) + 1.0986 = 2.4848$

$\ln 15 = \ln(5 \cdot 3) = \ln 5 + \ln 3 \approx 1.6094 + 1.0986 = 2.7080$

$\ln 16 = \ln 2^4 = 4 \ln 2 \approx 4(0.6931) = 2.7724$

$\ln 18 = \ln(3^2 \cdot 2) = \ln 3^2 + \ln 2 = 2 \ln 3 + \ln 2 \approx 2(1.0986) + 0.6931 = 2.8903$

$\ln 20 = \ln(5 \cdot 2^2) = \ln 5 + \ln 2^2 = \ln 5 + 2 \ln 2 \approx 1.6094 + 2(0.6931) = 2.9956$

7. $y = (1 + x)^{1/x}$

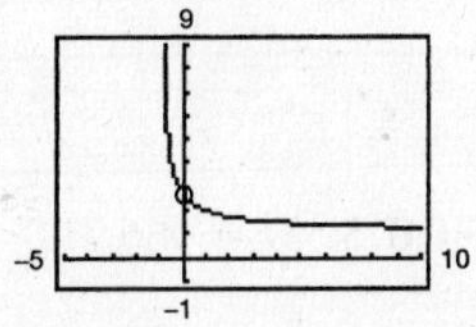

Near $x = 0$ the graph approaches e.

There is no y-intercept.

x	$(1 + x)^{1/x}$
-1	undefined
-0.1	2.867972
-0.01	2.731999
-0.001	2.719642
-0.0001	2.718418
-0.00001	2.718295
↓	↓
0^-	e

x	$(1 + x)^{1/x}$
1	2
0.1	2.593743
0.01	2.704814
0.001	2.716924
0.0001	2.718146
0.00001	2.718268
↓	↓
0^+	e

8. $t_1 = 40.757 + 0.556s - 15.817 \ln s$

$t_2 = 1.2259 + 0.0023s^2$

(a) Linear model: $t_3 = 0.2729s - 6.0143$

Exponential model: $t_4 = 1.5385e^{0.02913s}$ or $t_4 = 1.5385(1.0296)^s$

(b)

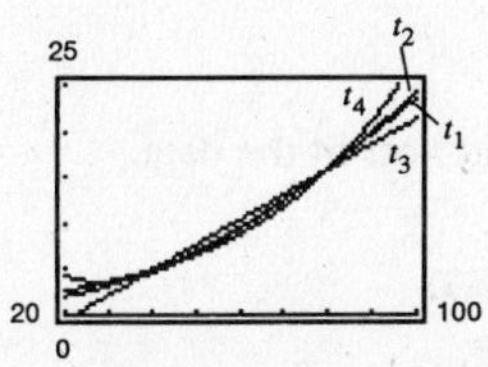

(c)

s	30	40	50	60	70	80	90
t_1	3.6	4.6	6.7	9.4	12.5	15.9	19.6
t_2	3.3	4.9	7.0	9.5	12.5	15.9	19.9
t_3	2.2	4.9	7.6	10.4	13.1	15.8	18.5
t_4	3.7	4.9	6.6	8.8	11.8	15.8	21.2

(d) Model t_1: $S_1 = |3.4 - 3.6| + |5 - 4.6| + |7 - 6.7| + |9.3 - 9.4| + |12 - 12.5| + |15.8 - 15.9| + |20 - 19.6| = 2.0$

Model t_2: $S_2 = |3.4 - 3.3| + |5 - 4.9| + |7 - 7| + |9.3 - 9.5| + |12 - 12.5| + |15.8 - 15.9| + |20 - 19.9| = 1.1$

Model t_3: $S_3 = |3.4 - 2.2| + |5 - 4.9| + |7 - 7.6| + |9.3 - 10.4| + |12 - 13.1| + |15.8 - 15.8| + |20 - 18.5| = 5.6$

Model t_4: $S_4 = |3.4 - 3.7| + |5 - 4.9| + |7 - 6.6| + |9.3 - 8.8| + |12 - 11.8| + |15.8 - 15.8| + |20 - 21.2| = 2.7$

The quadratic model, t_2, best fits the data.

9. Interest: $u = M - \left(M - \frac{Pr}{12}\right)\left(1 + \frac{r}{12}\right)^{12t}$

Principal: $v = \left(M - \frac{Pr}{12}\right)\left(1 + \frac{r}{12}\right)^{12t}$

(a) $P = 120{,}000,\ t = 35,\ r = 0.075,\ M = 809.09$

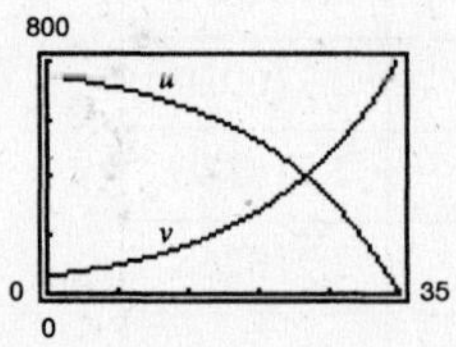

(b) In the early years of the mortgage, the majority of the monthly payment goes toward interest. The principal and interest are nearly equal when $t \approx 26$ years.

(c) $P = 120{,}000,\ t = 20,\ r = 0.075,\ M = 966.71$

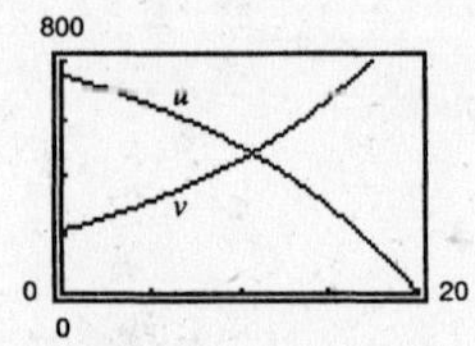

The interest is still the majority of the monthly payment in the early years. Now the principal and interest are nearly equal when $t \approx 10.729 \approx 11$ years.

10.

t	y
4	131.5
5	120.7
6	93.9
7	82.0
8	65.6
9	36.6
10	27.6

t	y
11	22.0
12	19.8
13	15.1
14	11.7
15	10.2
16	10.1
17	7.5

t	y
18	5.4
19	5.3
20	4.8
21	5.5
22	4.4

(a)

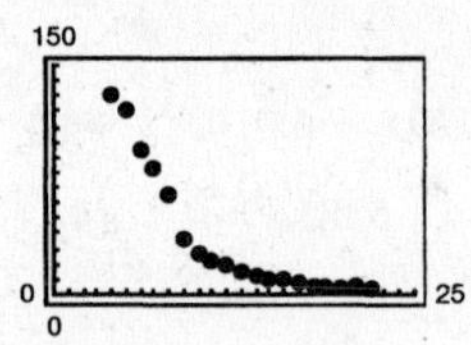

(b) A logistic model would best fit the data.

(c) $y = \frac{233.11080}{1 + 0.17995e^{0.34649x}}$

(d) Answers will vary.

11. $F(t) = 1 - e^{-t/3}$

(a) $F(1) \approx 0.283$

(b) $F(2.5) \approx 0.565$

(c) $F(4) \approx 0.736$

12. $p(t) = \frac{1200}{1 + 3e^{-t/5}}$

(a) $p(0) = 300$

(b) $p(5) = \frac{1200}{1 + 3e^{-1}} \approx 570$

(c)
$$800 = \frac{1200}{1 + 3e^{-t/5}}$$
$$1 + 3e^{-t/5} = \frac{1200}{800}$$
$$3e^{-t/5} = 0.5$$
$$-\frac{t}{5} = \ln\left(\frac{0.5}{3}\right)$$
$$t = -5\ln\left(\frac{0.5}{3}\right)$$
$$t \approx 8.96 \text{ years}$$

13. (c)

The graph passes through $(0, 0)$. It is symmetric to the y-axis. $\lim_{x \to \pm\infty} y = 6 \Rightarrow y = 6$ is a horizontal asymptote.

14. $(\ln x)^2 = \ln x^2$

$(\ln x)^2 - 2\ln x = 0$

$\ln x(\ln x - 2) = 0$

$\ln x = 0$ or $\ln x = 2$

$x = 1$ or $x = e^2$

15. $y_1 = c_1\left(\frac{1}{2}\right)^{t/k_1}$ and $y_2 = c_2\left(\frac{1}{2}\right)^{t/k_2}$

$$c_1\left(\frac{1}{2}\right)^{t/k_1} = c_2\left(\frac{1}{2}\right)^{t/k_2}$$

$$\frac{c_1}{c_2} = \left(\frac{1}{2}\right)^{(t/k_2 - t/k_1)}$$

$$\ln\left(\frac{c_1}{c_2}\right) = \left(\frac{t}{k_2} - \frac{t}{k_1}\right)\ln\left(\frac{1}{2}\right)$$

$$\ln\left(\frac{c_1}{c_2}\right) = t\left(\frac{k_1 - k_2}{k_1 k_2}\right)(-\ln 2)$$

$$k_1 k_2 \ln\left(\frac{c_1}{c_2}\right) = t(k_2 - k_1)\ln 2$$

$$t = \frac{k_1 k_2 \ln(c_1/c_2)}{(k_2 - k_1)\ln 2}$$

16. Let $\log_a x = m$ and $\log_{a/b} x = n$. Then $x = a^m$ and $x = \left(\frac{a}{b}\right)^n$.

$$a^m = \left(\frac{a}{b}\right)^n$$

$$a^{m/n} = \frac{a}{b}$$

$$a^{m/n - 1} = \frac{1}{b}$$

$$\log_a \frac{1}{b} = \frac{m}{n} - 1$$

$$1 + \log_a \frac{1}{b} = \frac{m}{n}$$

$$1 + \log_a \frac{1}{b} = \frac{\log_a x}{\log_{a/b} x}$$

17. $f(x) = e^x - e^{-x}$

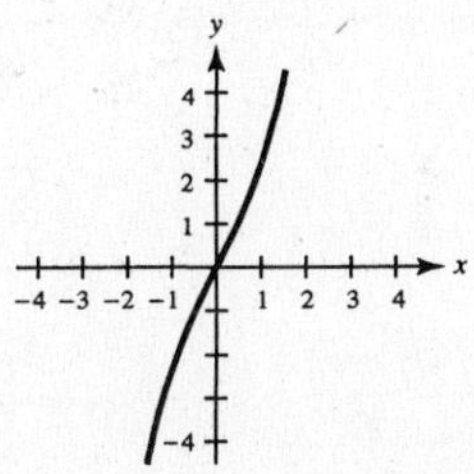

$$y = e^x - e^{-x}$$

$$x = e^y - e^{-y}$$

$$x = \frac{e^{2y} - 1}{e^y}$$

$$xe^y = e^{2y} - 1$$

$$e^{2y} - xe^y - 1 = 0$$

$$e^y = \frac{x \pm \sqrt{x^2 + 4}}{2} \quad \text{Quadratic Formula}$$

Choosing the positive quantity for e^y we have

$$y = \ln\left(\frac{x + \sqrt{x^2 + 4}}{2}\right). \text{ Thus,}$$

$$f^{-1}(x) = \ln\left(\frac{x + \sqrt{x^2 + 4}}{2}\right).$$

18. $[f(x)]^2 - [g(x)]^2 = \left(\frac{e^x + e^{-x}}{2}\right)^2 - \left(\frac{e^x - e^{-x}}{2}\right)^2$

$$= \left(\frac{e^{2x} + 2 + e^{-2x}}{4}\right) - \left(\frac{e^{2x} - 2 + e^{-2x}}{4}\right)$$

$$= \frac{4}{4}$$

$$= 1$$

19. $f(x) = \frac{a^x + 1}{a^x - 1}$, $a > 0$, $a \neq 1$

$$x = \frac{a^y + 1}{a^y - 1}$$

$$x(a^y - 1) = a^y + 1$$

$$xa^y - a^y = x + 1$$

$$a^y(x - 1) = x + 1$$

$$a^y = \frac{x + 1}{x - 1}$$

$$y = \log_a\left(\frac{x + 1}{x - 1}\right) = \frac{\ln\left(\frac{x + 1}{x - 1}\right)}{\ln a} = f^{-1}(x)$$

20. $B = B_0 a^{kt}$ through $(0, 500)$ and $(2, 200)$

$$B_0 = 500$$

$$200 = 500a^{k(2)}$$

$$\frac{2}{5} = a^{2k}$$

$$\log_a\left(\frac{2}{5}\right) = 2k$$

$$\frac{1}{2}\log_a\left(\frac{2}{5}\right) = k$$

$$B = 500a^{[1/2 \log_a(2/5)]t}$$

$$= 500\left[a^{\log_a(2/5)}\right]^{t/2}$$

$$= 500\left(\frac{2}{5}\right)^{t/2}$$

CHAPTER 8
Exponential and Logarithmic Functions and Calculus

Section 8.1 Exponential Functions: Differentiation and Integration . . 615

Section 8.2 Logarithmic Functions and Differentiation 624

Section 8.3 Logarithmic Functions and Integration 633

Section 8.4 Differential Equations: Growth and Decay 641

Review Exercises 647

Problem Solving 652

CHAPTER 8
Exponential and Logarithmic Functions and Calculus

Section 8.1 Exponential Functions: Differentiation and Integration

1. (a) $y = e^{3x}$

$y' = 3e^{3x}$

At $(0, 1)$, $y' = 3$.

(b) $y = e^{-3x}$

$y' = -3e^{-3x}$

At $(0, 1)$, $y' = -3$.

2. (a) $y = e^{2x}$

$y' = 2e^{2x}$

At $(0, 1)$, $y' = 2$.

(b) $y = e^{-2x}$

$y' = -2e^{-2x}$

At $(0, 1)$, $y' = -2$.

3. $f(x) = e^{2x}$

$f'(x) = 2e^{2x}$

4. $f(x) = e^{1-x}$

$f'(x) = -e^{1-x}$

5. $y = e^{-2x+x^2}$

$\dfrac{dy}{dx} = 2(x-1)e^{-2x+x^2}$

6. $y = e^{-x^2}$

$\dfrac{dy}{dx} = -2xe^{-x^2}$

7. $f(x) = e^{1/x} = e^{x^{-1}}$

$f'(x) = (-x^{-2})e^{x^{-1}} = -\dfrac{e^{1/x}}{x^2}$

8. $f(x) = e^{-1/x^2}$

$f'(x) = \dfrac{2e^{-1/x^2}}{x^3}$

9. $y = e^{\sqrt{x}}$

$\dfrac{dy}{dx} = \dfrac{e^{\sqrt{x}}}{2\sqrt{x}}$

10. $g(x) = e^{3\sqrt{x}}$

$g'(x) = e^{3\sqrt{x}}\left(\dfrac{3}{2}x^{-1/2}\right)$

$= \dfrac{3e^{3\sqrt{x}}}{2\sqrt{x}}$

11. $f(x) = (x+1)e^{3x}$

$f'(x) = (x+1)(3e^{3x}) + e^{3x}(1)$

$= e^{3x}(3x+4)$

12. $y = x^2e^{-x}$

$\dfrac{dy}{dx} = x^2(-e^{-x}) + 2xe^{-x}$

$= (2x - x^2)e^{-x}$

13. $f(x) = \dfrac{e^{x^2}}{x}$

$f'(x) = \dfrac{x(2xe^{x^2}) - e^{x^2}}{x^2}$

$= \dfrac{e^{x^2}(2x^2-1)}{x^2}$

14. $f(x) = \dfrac{e^{x/2}}{\sqrt{x}}$

$f'(x) = \dfrac{\sqrt{x}[(1/2)e^{x/2}] - e^{x/2}[1/(2\sqrt{x})]}{x} \cdot \dfrac{2\sqrt{x}}{2\sqrt{x}}$

$= \dfrac{xe^{x/2} - e^{x/2}}{2x\sqrt{x}} = \dfrac{e^{x/2}(x-1)}{2x\sqrt{x}}$

15. $g(t) = (e^{-t} + e^{t})^3$

$g'(t) = 3(e^{-t} + e^{t})^2(e^{t} - e^{-t})$

16. $y = (1 - e^{-x})^2$

$y' = 2(1 - e^{-x})(e^{-x}) = 2e^{-x}(1 - e^{-x})$

17. $y = \dfrac{2}{e^x + e^{-x}} = 2(e^x + e^{-x})^{-1}$

$\dfrac{dy}{dx} = -2(e^x + e^{-x})^{-2}(e^x - e^{-x})$

$= \dfrac{-2(e^x - e^{-x})}{(e^x + e^{-x})^2}$

18. $f(x) = \dfrac{e^x - e^{-x}}{2} = \dfrac{1}{2}(e^x - e^{-x})$

$f'(x) = \dfrac{1}{2}(e^x + e^{-x}) = \dfrac{e^x + e^{-x}}{2}$

19. $y = x^2e^x - 2xe^x + 2e^x = e^x(x^2 - 2x + 2)$

$\dfrac{dy}{dx} = e^x(2x - 2) + e^x(x^2 - 2x + 2) = x^2e^x$

20. $y = xe^x - e^x = e^x(x - 1)$

$\dfrac{dy}{dx} = e^x + e^x(x - 1) = xe^x$

21. $xe^y - 10x + 3y = 0$

$xe^y\dfrac{dy}{dx} + e^y - 10 + 3\dfrac{dy}{dx} = 0$

$\dfrac{dy}{dx}(xe^y + 3) = 10 - e^y$

$\dfrac{dy}{dx} = \dfrac{10 - e^y}{xe^y + 3}$

22. $e^{xy} + x^2 - y^2 = 10$

$\left(x\dfrac{dy}{dx} + y\right)e^{xy} + 2x - 2y\dfrac{dy}{dx} = 0$

$\dfrac{dy}{dx}(xe^{xy} - 2y) = -ye^{xy} - 2x$

$\dfrac{dy}{dx} = -\dfrac{ye^{xy} + 2x}{xe^{xy} - 2y}$

23. $f(x) = 2e^{3x} + 3e^{-2x}$

$f'(x) = 6e^{3x} - 6e^{-2x}$

$f''(x) = 18e^{3x} + 12e^{-2x}$

24. $f(x) = 5e^{-x} - 2e^{-5x}$

$f'(x) = -5e^{-x} + 10e^{-5x}$

$f''(x) = 5e^{-x} - 50e^{-5x}$

25. $f(x) = (3 + 2x)e^{-3x}$

$f'(x) = (3 + 2x)(-3e^{-3x}) + e^{-3x}(2)$

$= e^{-3x}(-6x - 7)$

$f''(x) = e^{-3x}(-6) + (-6x - 7)(-3e^{-3x})$

$= 3e^{-3x}(6x + 5)$

26. $g(x) = (1 + 2x)e^{4x}$

$g'(x) = (1 + 2x)(4e^{4x}) + e^{4x}(2) = 2e^{4x}(4x + 3)$

$g''(x) = 2e^{4x}(4) + (4x + 3)(8e^{4x}) = 8e^{4x}(4x + 4) = 32e^{4x}(x + 1)$

27. $f(x) = \dfrac{e^x + e^{-x}}{2}$

$f'(x) = \dfrac{e^x - e^{-x}}{2} = 0$ when $x = 0$.

$f''(x) = \dfrac{e^x + e^{-x}}{2} > 0$

Relative minimum: (0, 1)

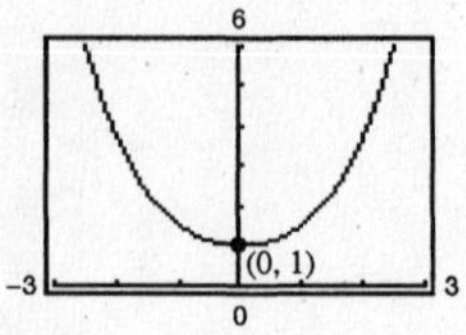

28. $f(x) = \dfrac{e^x - e^{-x}}{2}$

$f'(x) = \dfrac{e^x + e^{-x}}{2} > 0$

$f''(x) = \dfrac{e^x - e^{-x}}{2} = 0$ when $x = 0$.

Point of inflection: (0, 0)

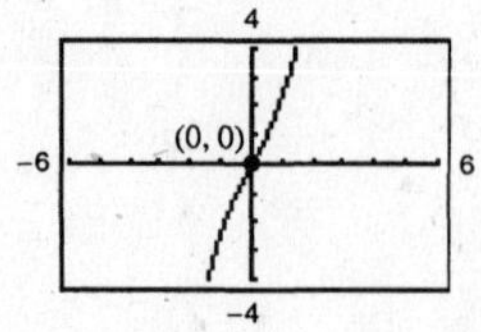

29. $g(x) = \dfrac{1}{\sqrt{2\pi}} e^{-(x-2)^2/2}$

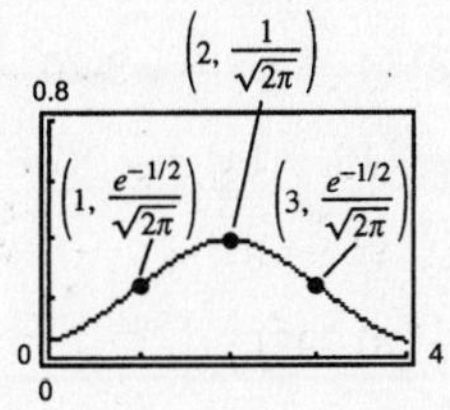

$g'(x) = \dfrac{-1}{\sqrt{2\pi}}(x-2)e^{-(x-2)^2/2}$

$g''(x) = \dfrac{1}{\sqrt{2\pi}}(x-1)(x-3)e^{-(x-2)^2/2}$

Relative maximum: $\left(2, \dfrac{1}{\sqrt{2\pi}}\right) \approx (2, 0.399)$

Points of inflection: $\left(1, \dfrac{1}{\sqrt{2\pi}}e^{-1/2}\right), \left(3, \dfrac{1}{\sqrt{2\pi}}e^{-1/2}\right) \approx (1, 0.242), (3, 0.242)$

30. $g(x) = \dfrac{1}{\sqrt{2\pi}} e^{-(x-3)^2/2}$

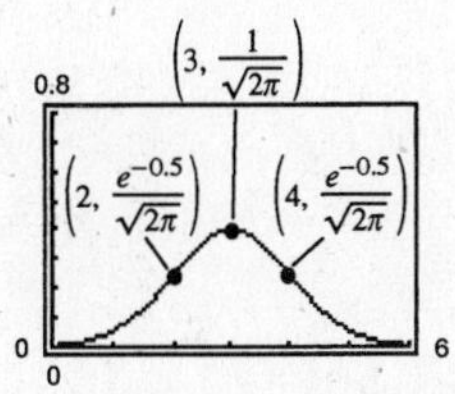

$g'(x) = \dfrac{-1}{\sqrt{2\pi}}(x-3)e^{-(x-3)^2/2}$

$g''(x) = \dfrac{1}{\sqrt{2\pi}}(x-2)(x-4)e^{-(x-3)^2/2}$

Relative maximum: $\left(3, \dfrac{1}{\sqrt{2\pi}}\right) \approx (3, 0.399)$

Points of inflection: $\left(2, \dfrac{1}{\sqrt{2\pi}}e^{-1/2}\right), \left(4, \dfrac{1}{\sqrt{2\pi}}e^{-1/2}\right) \approx (2, 0.242), (4, 0.242)$

31. $f(x) = x^2e^{-x}$

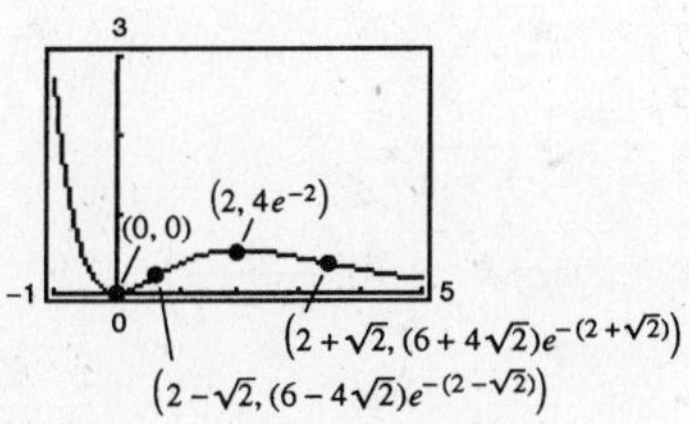

$f'(x) = -x^2e^{-x} + 2xe^{-x} = xe^{-x}(2-x) = 0$ when $x = 0, 2$.

$f''(x) = -e^{-x}(2x - x^2) + e^{-x}(2 - 2x)$

$= e^{-x}(x^2 - 4x + 2) = 0$ when $x = 2 \pm \sqrt{2}$.

Relative minimum: $(0, 0)$

Relative maximum: $(2, 4e^{-2})$

$x = 2 \pm \sqrt{2}$

$y = (2 \pm \sqrt{2})^2 e^{-(2 \pm \sqrt{2})}$

Point of inflection: $(3.414, 0.384), (0.586, 0.191)$

32. $f(x) = xe^{-x}$

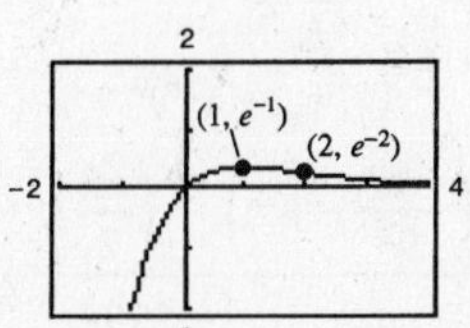

$f'(x) = -xe^{-x} + e^{-x} = e^{-x}(1-x) = 0$ when $x = 1$.

$f''(x) = -e^{-x} + (-e^{-x})(1-x) = e^{-x}(x-2) = 0$ when $x = 2$.

Relative maximum: $(1, e^{-1})$

Point of inflection: $(2, 2e^{-2})$

33. $g(t) = 1 + (2+t)e^{-t}$

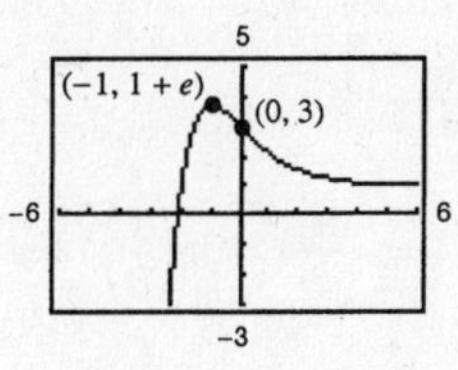

$g'(t) = -(1+t)e^{-t}$

$g''(t) = te^{-t}$

Relative maximum: $(-1, 1+e) \approx (-1, 3.718)$

Point of inflection: $(0, 3)$

34. $f(x) = -2 + e^{3x}(4 - 2x)$

$f'(x) = e^{3x}(-2) + 3e^{3x}(4 - 2x) = e^{3x}(10 - 6x) = 0$ when $x = \frac{5}{3}$.

$f''(x) = e^{3x}(-6) + 3e^{3x}(10 - 6x) = e^{3x}(24 - 18x) = 0$ when $x = \frac{4}{3}$.

Relative maximum: $\left(\frac{5}{3}, 96.942\right)$

Point of inflection: $\left(\frac{4}{3}, 70.798\right)$

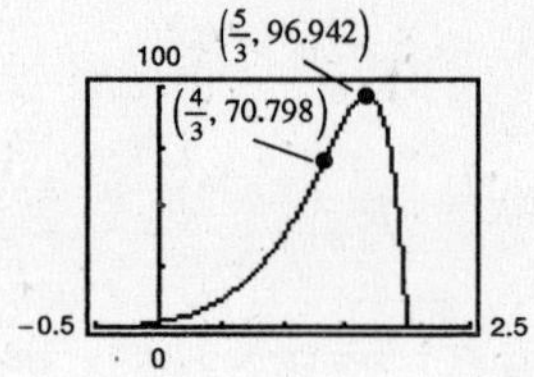

35. $A = 2xe^{-x^2}$

$$\frac{dA}{dx} = -4x^2e^{-x^2} + 2e^{-x^2}$$

$$= 2e^{-x^2}(1 - 2x^2) = 0 \text{ when } x = \frac{\sqrt{2}}{2}.$$

$A = \sqrt{2}e^{-1/2}$

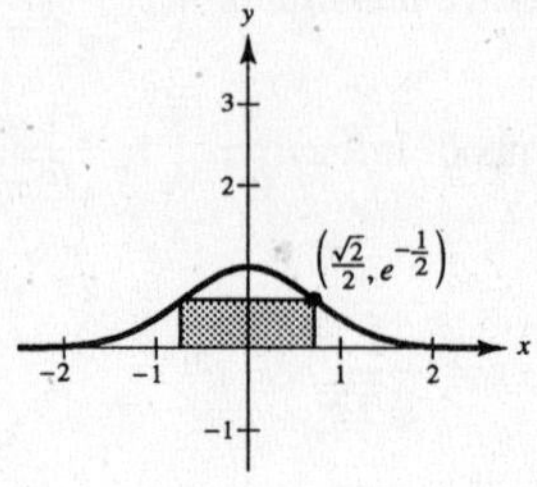

36. (a)

$$f(c) = f(c + x)$$

$$10ce^{-c} = 10(c + x)e^{-(c+x)}$$

$$\frac{c}{e^c} = \frac{c + x}{e^{c+x}}$$

$$ce^{c+x} = (c + x)e^c$$

$$ce^x = c + x$$

$$ce^x - c = x$$

$$c = \frac{x}{e^x - 1}$$

(b) $A(x) = xf(c) = x\left[10\left(\frac{x}{e^x - 1}\right)e^{-(x/(e^x-1))}\right]$

$$= \frac{10x^2}{e^x - 1}e^{x/(1-e^x)}$$

(c) $A(x) = \dfrac{10x^2}{e^x - 1}e^{x/(1-e^x)}$

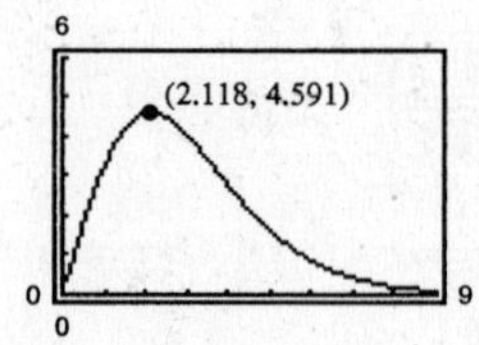

The maximum area is 4.591 for $x = 2.118$ and $f(x) = 2.547$.

(d) $c = \dfrac{x}{e^x - 1}$

$\lim_{x \to 0^+} c = 1$

$\lim_{x \to \infty} c = 0$

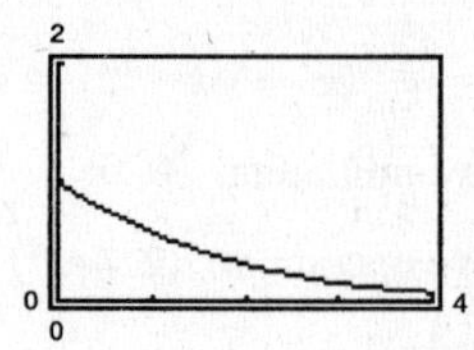

37. $y = \dfrac{L}{1 + ae^{-x/b}}, a > 0, b > 0, L > 0$

$$y' = \frac{-L\left(-\frac{a}{b}e^{-x/b}\right)}{(1 + ae^{-x/b})^2} = \frac{\frac{aL}{b}e^{-x/b}}{(1 + ae^{-x/b})^2}$$

$$y'' = \frac{(1 + ae^{-x/b})^2\left(\frac{-aL}{b^2}e^{-x/b}\right) - \left(\frac{aL}{b}e^{-x/b}\right)2(1 + ae^{-x/b})\left(\frac{-a}{b}e^{-x/b}\right)}{(1 + ae^{-x/b})^4}$$

$$= \frac{(1 + ae^{-x/b})\left(\frac{-aL}{b^2}e^{-x/b}\right) + 2\left(\frac{aL}{b}e^{-x/b}\right)\left(\frac{a}{b}e^{-x/b}\right)}{(1 + ae^{-x/b})^3}$$

$$= \frac{Lae^{-x/b}[ae^{-x/b} - 1]}{(1 + ae^{-x/b})^3 b^2}$$

$y'' = 0$ if $ae^{-x/b} = 1 \Rightarrow \dfrac{-x}{b} = \ln\left(\dfrac{1}{a}\right) \Rightarrow x = b \ln a$

$$y(b \ln a) = \frac{L}{1 + ae^{-(b \ln a)/b}} = \frac{L}{1 + a(1/a)} = \frac{L}{2}$$

Therefore, the y-coordinate of the inflection point is $L/2$.

38. $y = e^{-x}$

$y' = -e^{-x}$ (Slope of tangent line)

$-\dfrac{1}{y'} = e^{x}$ (Slope of normal line)

$y - e^{-x_0} = e^{x_0}(x - x_0)$

We want (0, 0) to satisfy the equation:

$-e^{-x_0} = -x_0e^{x_0}$

$1 = x_0e^{2x_0}$

$x_0e^{2x_0} - 1 = 0$ when $x_0 \approx 0.4263$.

$(0.4263, e^{-0.4263})$

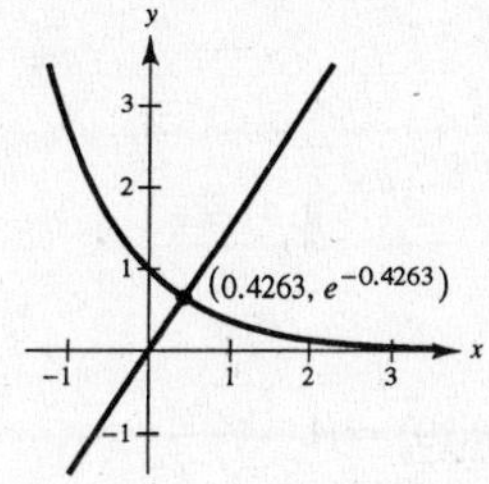

39. $y = e^{-x}$ Point: (0, 1)

$y' = -e^{-x}$ (Slope of tangent line)

$-\dfrac{1}{y'} = e^{x}$ (Slope of normal line)

At (0, 1), $e^{x} = 1$, $y = x + 1$.

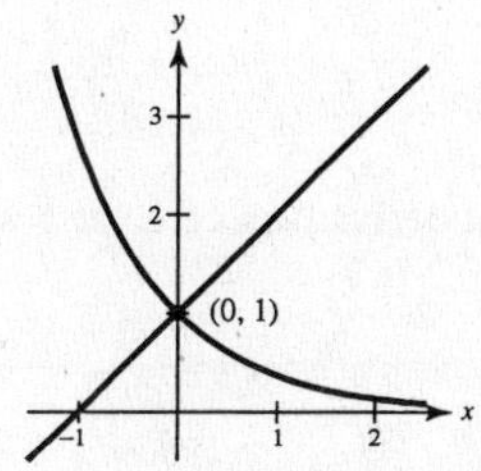

40. $V = 15{,}000e^{-0.6286t}, 0 \le t \le 10$

(a) 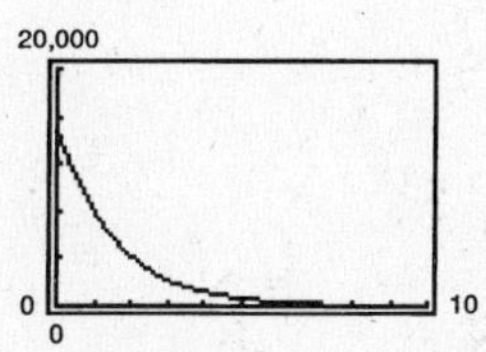

(b) $\dfrac{dV}{dt} = -9429e^{-0.6286t}$

When $t = 1$, $\dfrac{dV}{dt} \approx -5028.84$.

When $t = 5$, $\dfrac{dV}{dt} \approx -406.89$.

(c) 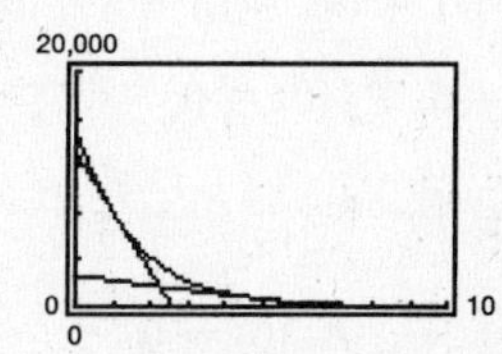

41. (a) 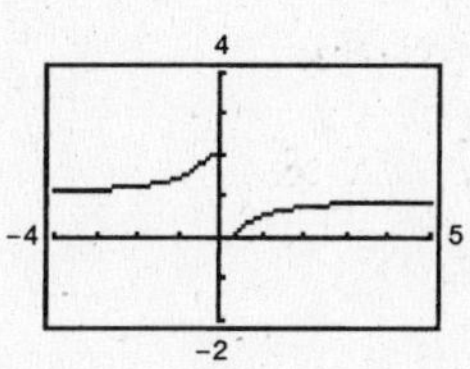

(b) When x increases without bound, $1/x$ approaches zero, and $e^{1/x}$ approaches 1. Therefore, $f(x)$ approaches $2/(1 + 1) = 1$. Thus, $f(x)$ has a horizontal asymptote at $y = 1$. As x approaches zero from the right, $1/x$ approaches ∞, $e^{1/x}$ approaches ∞ and $f(x)$ approaches zero. As x approaches zero from the left, $1/x$ approaches $-\infty$, $e^{1/x}$ approaches zero, and $f(x)$ approaches 2. The limit does not exist since the left limit does not equal the right limit. Therefore, $x = 0$ is a nonremovable discontinuity.

42. (a) 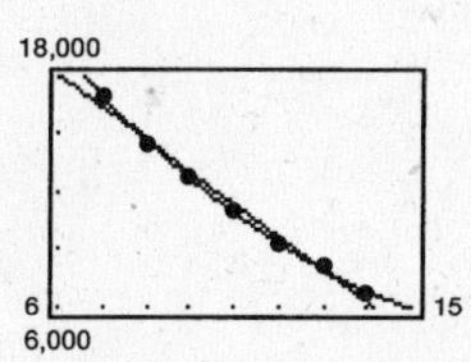

Linear:
$V \approx -1686.7857t + 28{,}242.1429$

Quadratic:
$V \approx 109.5238t^2 - 3877.2619t + 38{,}756.4286$

(b) The slope in the linear model represents the average amount of depreciation per year. The car is worth \$1686.79 less each year.

(c) Exponential: $V \approx 49{,}591.0596(0.8592^t)$

$= 49{,}591.0596e^{(\ln 0.8592)t}$

(d) $\lim_{t\to\infty} 49{,}591.0596(0.8592^t) = 0$

Eventually, the car will have no value.

(e) $V = 49{,}591.0596e^{(\ln 0.8592)t}$

$\dfrac{dV}{dt} = 49{,}591.0596(\ln 0.8592)e^{(\ln 0.8592)t}$

When $t = 8$: $\dfrac{dV}{dt} \approx -2235.10$ dollars per year.

When $t = 12$: $\dfrac{dV}{dt} \approx -1218.07$ dollars per year.

43. $f(x) = e^{x/2}, f(0) = 1$

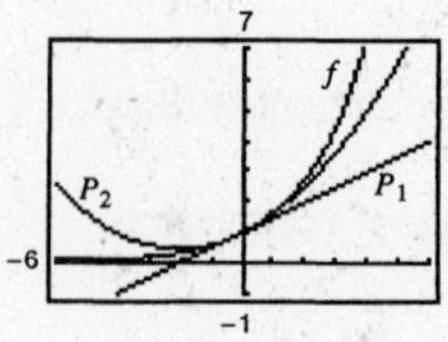

$f'(x) = \frac{1}{2}e^{x/2}, f'(0) = \frac{1}{2}$

$f''(x) = \frac{1}{4}e^{x/2}, f''(0) = \frac{1}{4}$

$P_1(x) = 1 + \frac{1}{2}(x - 0) = \frac{x}{2} + 1, P_1(0) = 1$

$P_1'(x) = \frac{1}{2}, P_1'(0) = \frac{1}{2}$

$P_2(x) = 1 + \frac{1}{2}(x - 0) + \frac{1}{8}(x - 0)^2 = \frac{x^2}{8} + \frac{x}{2} + 1, P_2(0) = 1$

$P_2'(x) = \frac{1}{4}x + \frac{1}{2}, P_2'(0) = \frac{1}{2}$

$P_2''(x) = \frac{1}{4}, P_2''(0) = \frac{1}{4}$

The values of f, P_1, P_2 and their first derivatives agree at $x = 0$. The values of the second derivatives of f and P_2 agree at $x = 0$.

44. $f(x) = e^{-x^2/2}, f(0) = 1$

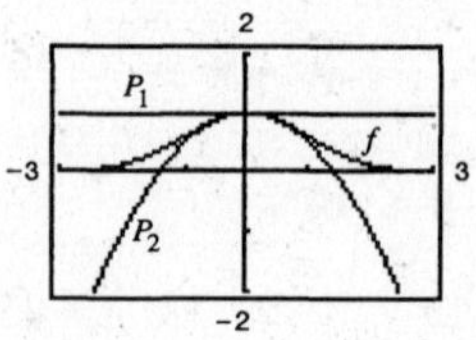

$f'(x) = -xe^{-x^2/2}, f'(0) = 0$

$f''(x) = x^2e^{-x^2/2} - e^{-x^2/2} = e^{-x^2/2}(x^2 - 1), f''(0) = -1$

$P_1(x) = 1 + 0(x - 0) = 1, P_1(0) = 1$

$P_1'(x) = 0,\ P_1'(0) = 0$

$P_2(x) = 1 + 0(x - 0) - \frac{1}{2}(x - 0)^2 = 1 - \frac{x^2}{2}, P_2(0) = 1$

$P_2'(x) = -x,\ P_2'(0) = 0$

$P_2''(x) = -1,\ P_2''(0) = -1$

The values of f, P_1, P_2 and their first derivatives agree at $x = 0$. The values of the second derivatives of f and P_2 agree at $x = 0$.

45. Let $u = 5x$, $du = 5\,dx$.

$$\int e^{5x}\,5dx = e^{5x} + C$$

46. Let $u = -x^4$, $du = -4x^3\,dx$.

$$\int e^{-x^4}(-4x^3)\,dx = e^{-x^4} + C$$

47. Let $u = -2x$, $du = -2\,dx$.

$$\int_0^1 e^{-2x}dx = -\frac{1}{2}\int_0^1 e^{-2x}(-2)\,dx = -\frac{1}{2}e^{-2x}\Big]_0^1$$

$$= \frac{1}{2}(1 - e^{-2}) = \frac{e^2 - 1}{2e^2}$$

48. $\displaystyle\int_3^4 e^{3-x}dx = \Big[-e^{3-x}\Big]_3^4 = -e^{-1} + 1 = 1 - \frac{1}{e}$

49. $\displaystyle\int xe^{-x^2}dx = -\frac{1}{2}\int e^{-x^2}(-2x)dx = -\frac{1}{2}e^{-x^2} + C$

50. $\displaystyle\int x^2e^{x^3/2}\,dx = \frac{2}{3}\int e^{x^3/2}\left(\frac{3x^2}{2}\right)dx = \frac{2}{3}e^{x^3/2} + C$

51. $\displaystyle\int \frac{e^{\sqrt{x}}}{\sqrt{x}}dx = 2\int e^{\sqrt{x}}\left(\frac{1}{2\sqrt{x}}\right)dx = 2e^{\sqrt{x}} + C$

52. $\displaystyle\int \frac{e^{1/x^2}}{x^3}\,dx = -\frac{1}{2}\int e^{1/x^2}\left(\frac{-2}{x^3}\right)dx = -\frac{1}{2}e^{1/x^2} + C$

53. Let $u = \frac{3}{x}$, $du = -\frac{3}{x^2}\,dx$.

$$\int_1^3 \frac{e^{3/x}}{x^2}\,dx = -\frac{1}{3}\int_1^3 e^{3/x}\left(-\frac{3}{x^2}\right)dx = -\frac{1}{3}e^{3/x}\Big]_1^3 = \frac{e}{3}(e^2 - 1)$$

54. Let $u = \frac{-x^2}{2}$, $du = -x\,dx$.

$$\int_0^{\sqrt{2}} xe^{-x^2/2}\,dx = -\int_0^{\sqrt{2}} e^{-x^2/2}(-x)\,dx = \Big[-e^{-x^2/2}\Big]_0^{\sqrt{2}} = 1 - e^{-1} = \frac{e-1}{e}$$

55. $\displaystyle\int (1 + e^x)^2\,dx = \int (1 + 2e^x + e^{2x})\,dx = x + 2e^x + \frac{1}{2}e^{2x} + C$

56. $\displaystyle\int (e^x - e^{-x})^2\,dx = \int (e^{2x} - 2 + e^{-2x})\,dx = \frac{1}{2}e^{2x} - 2x + \left(-\frac{1}{2}\right)e^{-2x} + C$

57. $\displaystyle\int e^{-x}(1 + e^{-x})^2\,dx = -\int (1 + e^{-x})^2(-e^{-x})\,dx = -\frac{1}{3}(1 + e^{-x})^3 + C$

58. $\displaystyle\int e^{2x}(1 - 3e^{2x})^2\,dx = -\frac{1}{6}\int (1 - 3e^{2x})^2(-6e^{2x})\,dx = -\frac{1}{6}\frac{(1 - 3e^{2x})^3}{3} + C = -\frac{1}{18}(1 - 3e^{2x})^3 + C$

59. Let $u = 1 - e^x$, $du = -e^x\,dx$.

$$\int e^x\sqrt{1 - e^x}\,dx = -\int (1 - e^x)^{1/2}(-e^x)\,dx = -\frac{2}{3}(1 - e^x)^{3/2} + C$$

60. $\displaystyle\int e^x(e^x - e^{-x})\,dx = \int (e^{2x} - 1)\,dx = \frac{1}{2}e^{2x} - x + C$

61. $\displaystyle\int \frac{e^{-x}}{(1 + e^{-x})^2}\,dx = -\int (1 + e^{-x})^{-2}(-e^{-x})\,dx = (1 + e^{-x})^{-1} + C = \frac{1}{1 + e^{-x}} + C$

62. $\displaystyle\int \frac{e^{2x}}{(1 + e^{2x})^2}\,dx = \frac{1}{2}\int (1 + e^{2x})^{-2}(2e^{2x})\,dx = \frac{1}{2}\left[\frac{(1 + e^{2x})^{-1}}{-1}\right] + C = -\frac{1}{2(1 + e^{2x})} + C$

63. $\displaystyle\int \frac{e^x + e^{-x}}{\sqrt{e^x - e^{-x}}}\,dx = \int (e^x - e^{-x})^{-1/2}(e^x + e^{-x})\,dx = 2(e^x - e^{-x})^{1/2} + C = 2\sqrt{e^x - e^{-x}} + C$

64. Let $u = e^x + e^{-x}$, $du = (e^x - e^{-x})\,dx$.

$$\int \frac{2e^x - 2e^{-x}}{(e^x + e^{-x})^2}\,dx = 2\int (e^x + e^{-x})^{-2}(e^x - e^{-x})\,dx = \frac{-2}{e^x + e^{-x}} + C$$

65. $\displaystyle\int \frac{5 - e^x}{e^{2x}}\,dx = \int 5e^{-2x}\,dx - \int e^{-x}\,dx = -\frac{5}{2}e^{-2x} + e^{-x} + C$

66. $\displaystyle\int \frac{e^{2x} + 2e^x + 1}{e^x}\,dx = \int (e^x + 2 + e^{-x})\,dx = e^x + 2x - e^{-x} + C$

67. Let $u = ax^2$, $du = 2ax\,dx$. (Assume $a \neq 0$)

$$y = \int xe^{ax^2}\,dx = \frac{1}{2a}\int e^{ax^2}(2ax)\,dx = \frac{1}{2a}e^{ax^2} + C$$

68. $y = \int (e^x - e^{-x})^2\,dx$

$$= \int (e^{2x} - 2 + e^{-2x})\,dx$$

$$= \frac{1}{2}e^{2x} - 2x - \frac{1}{2}e^{-2x} + C$$

69. $f'(x) = \int \frac{1}{2}(e^x + e^{-x})\,dx = \frac{1}{2}(e^x - e^{-x}) + C_1$

$f'(0) = C_1 = 0$

$f(x) = \int \frac{1}{2}(e^x - e^{-x})\,dx = \frac{1}{2}(e^x + e^{-x}) + C_2$

$f(0) = 1 + C_2 = 1 \Rightarrow C_2 = 0$

$f(x) = \frac{1}{2}(e^x + e^{-x})$

70. $f'(x) = \int (x + e^{2x})\,dx = \frac{x^2}{2} = \frac{e^{2x}}{2} + C_1$

$f'(0) = 0 + \frac{1}{2} + C_1 = \frac{1}{2} \Rightarrow C_1 = 0$

$f'(x) = \frac{1}{2}(x^2 - e^{2x})$

$f(x) = \int \frac{1}{2}(x^2 + e^{2x})\,dx = \frac{1}{2}\left(\frac{x^3}{3} + \frac{e^{2x}}{2}\right) + C_2$

$f(0) = \frac{1}{2}\left(0 + \frac{1}{2}\right) + C_2 = \frac{1}{4} \Rightarrow C_2 = 0$

$f(x) = \frac{1}{12}(2x^3 + 3e^{2x})$

71. (a)

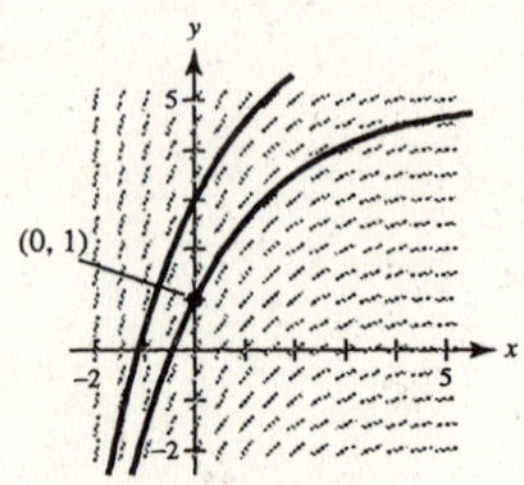

(b) $\frac{dy}{dx} = 2e^{-x/2}$, $(0, 1)$

$$y = \int 2e^{-x/2}\,dx = -4\int e^{-x/2}\left(-\frac{1}{2}dx\right)$$

$$= -4e^{-x/2} + C$$

$(0, 1)$: $1 = -4e^0 + C = -4 + C \Rightarrow C = 5$

$y = -4e^{-x/2} + 5$

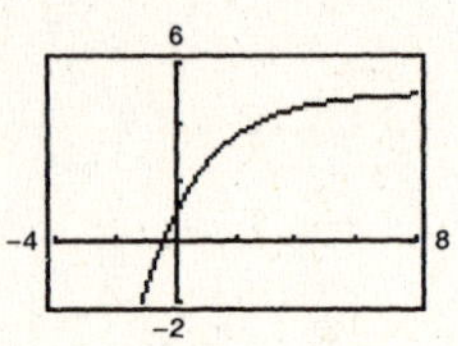

72. (a)

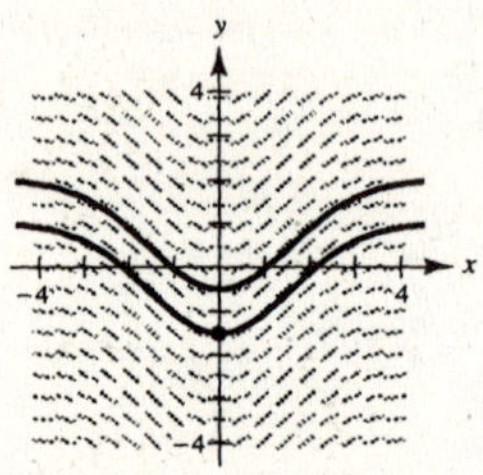

(b) $\frac{dy}{dx} = xe^{-0.2x^2}$, $\left(0, -\frac{3}{2}\right)$

$$y = \int xe^{-0.2x^2}\,dx = \frac{1}{-0.4}\int e^{-0.2x^2}(-0.4x)\,dx$$

$$= -\frac{1}{0.4}e^{-0.2x^2} + C = -2.5e^{-0.2x^2} + C$$

$\left(0, -\frac{3}{2}\right)$: $-\frac{3}{2} = -2.5e^0 + C = -2.5 + C \Rightarrow C = 1$

$y = -2.5e^{-0.2x^2} + 1$

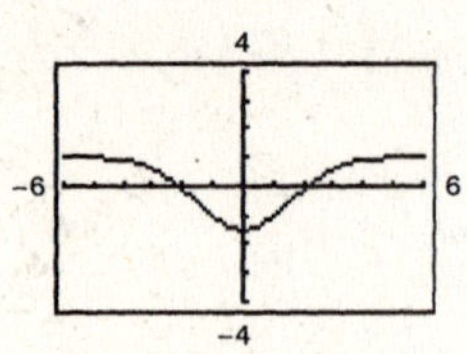

73. $\int_0^5 e^x\,dx = e^x\Big]_0^5 = e^5 - 1 \approx 147.413$

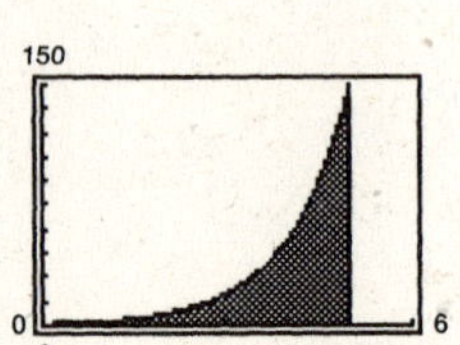

74. $\int_a^b e^{-x}\,dx = -e^{-x}\Big]_a^b = e^{-a} - e^{-b}$

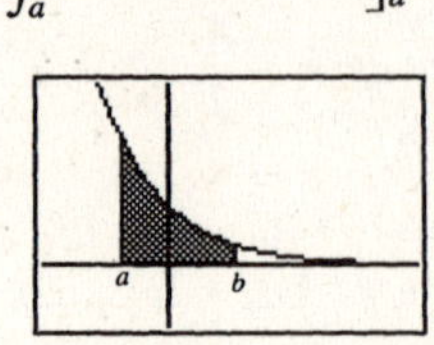

75. $\displaystyle\int_0^{\sqrt{6}} xe^{-(x^2/4)}\,dx = \left[-2e^{-(x^2/4)}\right]_0^{\sqrt{6}}$

$= 2 - 2e^{-3/2} = 2(1 - e^{-3/2}) \approx 1.554$

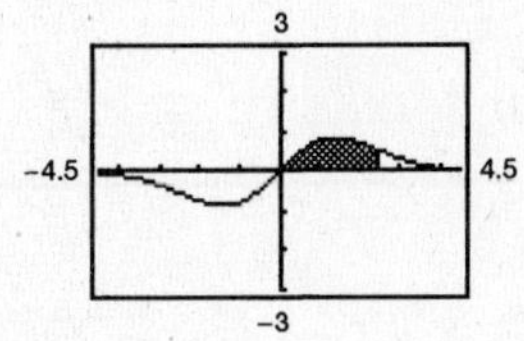

76. $\displaystyle\int_0^{2} (e^{-2x} + 2)\,dx = \left[-\frac{1}{2}e^{-2x} + 2x\right]_0^{2}$

$\displaystyle= -\frac{1}{2}e^{-4} + 4 + \frac{1}{2} \approx 4.4908$

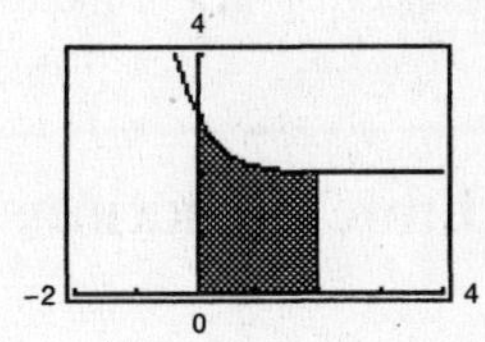

77. $\displaystyle\int_0^4 \sqrt{x}\,e^x\,dx,\ n = 12$

Midpoint Rule: 92.1898

Trapezoidal Rule: 93.8371

Simpson's Rule: 92.7385

78. $\displaystyle\int_0^2 2xe^{-x}\,dx,\ n = 12$

Midpoint Rule: 1.1906

Trapezoidal Rule: 1.1827

Simpson's Rule: 1.1880

79. $\displaystyle 0.0665\int_{48}^{60} e^{-0.0139(t-48)^2}\,dt$

$n = 24$

Simpson's Rule: $0.4772 = 47.72\%$

80. $\displaystyle\int_0^x 0.3^{-0.3t}\,dt = \frac{1}{2}$

$\displaystyle\left[-e^{-0.3t}\right]_0^x = \frac{1}{2}$

$\displaystyle -e^{-0.3x} + 1 = \frac{1}{2}$

$\displaystyle e^{-0.3x} = \frac{1}{2}$

$\displaystyle -0.3x = \ln\frac{1}{2} = -\ln 2$

$\displaystyle x = \frac{\ln 2}{0.3} \approx 2.31 \text{ minutes}$

81. $\displaystyle\int_0^x e^t\,dt \ge \int_0^x 1\,dt$

$\displaystyle\left[e^t\right]_0^x \ge \left[t\right]_0^x$

$e^x - 1 \ge x \Longrightarrow e^x \ge 1 + x$ for $x \ge 0$

82.

t	0	1	2	3	4
R	425	240	118	71	36
$\ln R$	6.052	5.481	4.771	4.263	3.584

(a) $\ln R = -0.6155t + 6.0609$

$R = e^{-0.6155t + 6.0609} = 428.78e^{-0.6155t}$

(b)

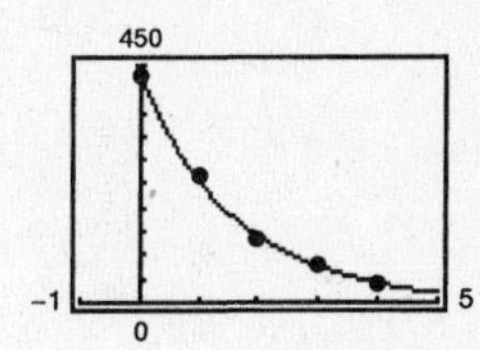

(c) $\displaystyle\int_0^4 R(t)\,dt = \int_0^4 428.78e^{-0.6155t}\,dt \approx 637.2 \text{ liters}$

83. Yes. $f(x) = Ce^x$, C a constant.

84. (a) Power Rule: $u = e^x + 1,\ du = e^x\,dx$

(b) Exponential Rule: $u = x^2,\ du = 2x\,dx$

85. $\displaystyle e^{-x} > 0 \Longrightarrow \int_0^2 e^{-x}dx > 0.$

86. $\ln \frac{e^a}{e^b} = \ln e^a - \ln e^b = a - b$

$\ln e^{a-b} = a - b$

Therefore, $\ln \frac{e^a}{e^b} = \ln e^{a-b}$ and since $y = \ln x$ is one-to-one, we have $\frac{e^a}{e^b} = e^{a-b}$.

Section 8.2 Logarithmic Functions and Differentiation

1. $\lim_{x \to 3^+} \ln(x - 3) = -\infty$

2. $\lim_{x \to 6^-} \ln(6 - x) = -\infty$

3. $\lim_{x \to 2^-} \ln[x^2(3 - x)] = \ln 4 \approx 1.3863$

4. $\lim_{x \to 5^+} \ln \frac{x}{\sqrt{x - 4}} = \ln 5 \approx 1.6094$

5. $y = \ln x^3 = 3 \ln x$

$y' = \frac{3}{x}$

At $(1, 0)$, $y' = 3$.

6. $y = \ln x^{3/2} = \frac{3}{2} \ln x$

$y' = \frac{3}{2x}$

At $(1, 0)$, $y' = \frac{3}{2}$.

7. $y = \ln x^2 = 2 \ln x$

$y' = \frac{2}{x}$

At $(1, 0)$, $y' = 2$.

8. $y = \ln x^{1/2} = \frac{1}{2} \ln x$

$y' = \frac{1}{2x}$

At $(1, 0)$, $y' = \frac{1}{2}$.

9. $g(x) = \ln x^2 = 2 \ln x$

$g'(x) = \frac{2}{x}$

10. $h(x) = \ln(2x^2 + 1)$

$h'(x) = \frac{1}{2x^2 + 1}(4x) = \frac{4x}{2x^2 + 1}$

11. $y = \ln \sqrt{x^4 - 4x} = \frac{1}{2} \ln(x^4 - 4x)$

$\frac{dy}{dx} = \frac{(4x^3 - 4)}{2(x^4 - 4x)} = \frac{2(x^3 - 1)}{x(x^3 - 4)}$

12. $y = \ln(1 - x)^{3/2} = \frac{3}{2} \ln(1 - x)$

$\frac{dy}{dx} = \frac{3}{2}\left(\frac{-1}{1 - x}\right) = \frac{3}{2(x - 1)}$

13. $y = (\ln x)^4$

$\frac{dy}{dx} = 4(\ln x)^3\left(\frac{1}{x}\right) = \frac{4(\ln x)^3}{x}$

14. $y = x \ln x$

$\frac{dy}{dx} = x\left(\frac{1}{x}\right) + \ln x = 1 + \ln x$

15. $y = \ln x\sqrt{x^2 - 1} = \ln x + \frac{1}{2} \ln(x^2 - 1)$

$\frac{dy}{dx} = \frac{1}{x} + \frac{1}{2}\left(\frac{2x}{x^2 - 1}\right) = \frac{2x^2 - 1}{x(x^2 - 1)}$

16. $y = \ln \sqrt{x^2 - 4} = \frac{1}{2} \ln(x^2 - 4)$

$\frac{dy}{dx} = \frac{1}{2}\left(\frac{2x}{x^2 - 4}\right)(2x) = \frac{x}{x^2 - 4}$

17. $f(x) = \ln \frac{x}{x^2 + 1} = \ln x - \ln(x^2 + 1)$

$f'(x) = \frac{1}{x} - \frac{2x}{x^2 + 1} = \frac{1 - x^2}{x(x^2 + 1)}$

18. $f(x) = \ln\left(\frac{2x}{x + 3}\right) = \ln 2x - \ln(x + 3)$

$f'(x) = \frac{1}{x} - \frac{1}{x + 3} = \frac{3}{x(x + 3)}$

19. $g(t) = \dfrac{\ln t}{t^2}$

$g'(t) = \dfrac{t^2(1/t) - 2t\ln t}{t^4} = \dfrac{1 - 2\ln t}{t^3}$

20. $h(t) = \dfrac{\ln t}{t}$

$h'(t) = \dfrac{t(1/t) - \ln t}{t^2} = \dfrac{1 - \ln t}{t^2}$

21. $y = \ln(\ln x^2)$

$\dfrac{dy}{dx} = \dfrac{(2x/x^2)}{\ln x^2} = \dfrac{2}{x\ln x^2} = \dfrac{1}{x\ln x}$

22. $y = \ln(\ln x)$

$\dfrac{dy}{dx} = \dfrac{1/x}{\ln x} = \dfrac{1}{x\ln x}$

23. $y = \ln\sqrt{\dfrac{x+1}{x-1}} = \dfrac{1}{2}[\ln(x+1) - \ln(x-1)]$

$\dfrac{dy}{dx} = \dfrac{1}{2}\left[\dfrac{1}{x+1} - \dfrac{1}{x-1}\right] = \dfrac{1}{1-x^2}$

24. $y = \ln\sqrt[3]{\dfrac{x-1}{x+1}} = \dfrac{1}{3}[\ln(x-1) - \ln(x+1)]$

$y' = \dfrac{1}{3}\left[\dfrac{1}{x-1} - \dfrac{1}{x+1}\right] = \dfrac{1}{3}\dfrac{2}{x^2-1} = \dfrac{2}{3(x^2-1)}$

25. $y = \ln e^{x^2} = x^2$

$\dfrac{dy}{dx} = 2x$

26. $y = \ln e^{-x/2} = -\dfrac{x}{2}$

$y' = -\dfrac{1}{2}$

27. $y = \ln\left(\dfrac{1+e^x}{1-e^x}\right)$

$= \ln(1+e^x) - \ln(1-e^x)$

$\dfrac{dy}{dx} = \dfrac{e^x}{1+e^x} + \dfrac{e^x}{1-e^x} = \dfrac{2e^x}{1-e^{2x}}$

28. $y = \ln\left(\dfrac{e^x + e^{-x}}{2}\right)$

$= \ln(e^x + e^{-x}) - \ln 2$

$\dfrac{dy}{dx} = \dfrac{e^x - e^{-x}}{e^x + e^{-x}}$

$= \dfrac{e^{2x} - 1}{e^{2x} + 1}$

29. $f(x) = e^{-x}\ln x$

$f'(x) = e^{-x}\left(\dfrac{1}{x}\right) - e^{-x}\ln x$

$= e^{-x}\left(\dfrac{1}{x} - \ln x\right)$

30. $g(x) = e^3\ln x$

$g'(x) = \dfrac{e^3}{x}$

31. $f(x) = \ln\dfrac{\sqrt{4+x^2}}{x} = \dfrac{1}{2}\ln(4+x^2) - \ln x$

$f'(x) = \dfrac{x}{4+x^2} - \dfrac{1}{x} = \dfrac{-4}{x(x^2+4)}$

32. $f(x) = \ln\left(x + \sqrt{4+x^2}\right)$

$f'(x) = \dfrac{1}{x + \sqrt{4+x^2}}\left(1 + \dfrac{x}{\sqrt{4+x^2}}\right)$

$= \dfrac{1}{\sqrt{4+x^2}}$

33. $y = \dfrac{-\sqrt{x^2+1}}{x} + \ln\left(x + \sqrt{x^2+1}\right)$

$\dfrac{dy}{dx} = \dfrac{-x\left(x/\sqrt{x^2+1}\right) + \sqrt{x^2+1}}{x^2} + \left(\dfrac{1}{x + \sqrt{x^2+1}}\right)\left(1 + \dfrac{x}{\sqrt{x^2+1}}\right)$

$= \dfrac{1}{x^2\sqrt{x^2+1}} + \left(\dfrac{1\left(x - \sqrt{x^2+1}\right)}{-1}\right)\left(\dfrac{\sqrt{x^2+1} + x}{\sqrt{x^2+1}}\right) = \dfrac{\sqrt{x^2+1}}{x^2}$

34. $y = \dfrac{-\sqrt{x^2+4}}{2x^2} - \dfrac{1}{4}\ln\left(\dfrac{2 + \sqrt{x^2+4}}{x}\right) = \dfrac{-\sqrt{x^2+4}}{2x^2} - \dfrac{1}{4}\ln\left(2 + \sqrt{x^2+4}\right) + \dfrac{1}{4}\ln x$

$\dfrac{dy}{dx} = \dfrac{-2x^2\left(x/\sqrt{x^2+4}\right) + 4x\sqrt{x^2+4}}{4x^4} - \dfrac{1}{4}\left(\dfrac{1}{2 + \sqrt{x^2+4}}\right)\left(\dfrac{x}{\sqrt{x^2+4}}\right) + \dfrac{1}{4x} = \dfrac{\sqrt{x^2+4}}{x^3}$

35. $f(x) = \ln|x^2 - 1|$

$f'(x) = \dfrac{2x}{x^2 - 1}$

36. $f(x) = \ln\left|\dfrac{x+5}{x}\right| = \ln|x + 5| - \ln|x|$

$f'(x) = \dfrac{1}{x+5} - \dfrac{1}{x} = \dfrac{-5}{x(x+5)}$

37. $f(x) = 4^x$

$f'(x) = (\ln 4)\, 4^x$

38. $g(x) = 2^{-x}$

$g'(x) = -(\ln 2)2^{-x}$

39. $y = 5^{x-2}$

$\dfrac{dy}{dx} = (\ln 5)\, 5^{x-2}$

40. $y = x(6^{-2x})$

$\dfrac{dy}{dx} = x[-2(\ln 6)6^{-2x}] + 6^{-2x}$

$= 6^{-2x}[-2x(\ln 6) + 1]$

$= 6^{-2x}(1 - 2x \ln 6)$

41. $g(t) = t^2\, 2^t$

$g'(t) = t^2(\ln 2)2^t + (2t)\, 2^t$

$= t\,2^t(t \ln 2 + 2)$

$= 2^t\, t(2 + t \ln 2)$

42. $f(t) = \dfrac{3^{2t}}{t}$

$f'(t) = \dfrac{t(2 \ln 3)\, 3^{2t} - 3^{2t}}{t^2}$

$= \dfrac{3^{2t}\,(2t \ln 3 - 1)}{t^2}$

43. $y = \log_3 x$

$\dfrac{dy}{dx} = \dfrac{1}{x \ln 3}$

44. $y = \log_{10}(2x) = \log_{10}2 + \log_{10} x$

$\dfrac{dy}{dx} = 0 + \dfrac{1}{x \ln 10} = \dfrac{1}{x \ln 10}$

45. $f(x) = \log_2 \dfrac{x^2}{x-1}$

$= 2 \log_2 x - \log_2 (x - 1)$

$f'(x) = \dfrac{2}{x \ln 2} - \dfrac{1}{(x-1) \ln 2}$

$= \dfrac{x-2}{(\ln 2)x(x-1)}$

46. $h(x) = \log_3 \dfrac{x\sqrt{x-1}}{2}$

$= \log_3 x + \dfrac{1}{2} \log_3 (x - 1) - \log_3 2$

$h'(x) = \dfrac{1}{x \ln 3} + \dfrac{1}{2} \cdot \dfrac{1}{(x-1) \ln 3} - 0$

$= \dfrac{1}{\ln 3}\left[\dfrac{1}{x} + \dfrac{1}{2(x-1)}\right]$

$= \dfrac{1}{\ln 3}\left[\dfrac{3x-2}{2x(x-1)}\right]$

47. $y = \log_5 \sqrt{x^2 - 1} = \dfrac{1}{2} \log_5 (x^2 - 1)$

$\dfrac{dy}{dx} = \dfrac{1}{2} \cdot \dfrac{2x}{(x^2-1)\ln 5} = \dfrac{x}{(x^2-1)\ln 5}$

48. $y = \log_{10} \dfrac{x^2 - 1}{x}$

$= \log_{10}(x^2 - 1) - \log_{10} x$

$\dfrac{dy}{dx} = \dfrac{2x}{(x^2-1) \ln 10} - \dfrac{1}{x \ln 10}$

$= \dfrac{1}{\ln 10}\left[\dfrac{2x}{x^2-1} - \dfrac{1}{x}\right]$

$= \dfrac{1}{\ln 10}\left[\dfrac{x^2+1}{x(x^2-1)}\right]$

49. $g(t) = \dfrac{10 \log_4 t}{t} = \dfrac{10(\ln t)/(\ln 4)}{t} = \dfrac{10}{\ln 4}\left(\dfrac{\ln t}{t}\right)$

$g'(t) = \dfrac{10}{2 \ln 2}\left(\dfrac{t(1/t) - \ln t}{t^2}\right)$

$= \dfrac{5}{\ln 2}\left(\dfrac{1 - \ln t}{t^2}\right) = \dfrac{5(1 - \ln t)}{t^2 \ln 2}$

50. $f(t) = t^{3/2}\log_2\sqrt{t+1} = t^{3/2}\frac{1}{2}\frac{\ln(t+1)}{\ln 2}$

$$f'(t) = \frac{1}{2\ln 2}\left[t^{3/2}\frac{1}{t+1} + \frac{3}{2}t^{1/2}\ln(t+1)\right]$$

51. $y = x\sqrt{x^2-1}$

$$\ln y = \ln x + \frac{1}{2}\ln(x^2-1)$$

$$\frac{1}{y}\left(\frac{dy}{dx}\right) = \frac{1}{x} + \frac{x}{x^2-1}$$

$$\frac{dy}{dx} = y\left[\frac{2x^2-1}{x(x^2-1)}\right] = \frac{2x^2-1}{\sqrt{x^2-1}}$$

52. $y = \sqrt{(x-1)(x-2)(x-3)}$

$$\ln y = \frac{1}{2}[\ln(x-1) + \ln(x-2) + \ln(x-3)]$$

$$\frac{1}{y}\left(\frac{dy}{dx}\right) = \frac{1}{2}\left[\frac{1}{x-1} + \frac{1}{x-2} + \frac{1}{x-3}\right]$$

$$= \frac{1}{2}\left[\frac{3x^2-12x+11}{(x-1)(x-2)(x-3)}\right]$$

$$\frac{dy}{dx} = \frac{3x^2-12x+11}{2y}$$

53. $y = \frac{x^2\sqrt{3x-2}}{(x-1)^2}$

$$\ln y = 2\ln x + \frac{1}{2}\ln(3x-2) - 2\ln(x-1)$$

$$\frac{1}{y}\left(\frac{dy}{dx}\right) = \frac{2}{x} + \frac{3}{2(3x-2)} - \frac{2}{x-1}$$

$$\frac{dy}{dx} = y\left[\frac{3x^2-15x+8}{2x(3x-2)(x-1)}\right]$$

$$= \frac{3x^3-15x^2+8x}{2(x-1)^3\sqrt{3x-2}}$$

54. $y = \sqrt{\frac{x^2-1}{x^2+1}}$

$$\ln y = \frac{1}{2}[\ln(x^2-1) - \ln(x^2+1)]$$

$$\frac{1}{y}\frac{dy}{dx} = \frac{1}{2}\left[\frac{2x}{x^2-1} - \frac{2x}{x^2+1}\right]$$

$$\frac{dy}{dx} = \sqrt{\frac{x^2-1}{x^2+1}}\left[\frac{2x}{x^4-1}\right]$$

$$= \frac{(x^2-1)^{1/2}2x}{(x^2+1)^{1/2}(x^2-1)(x^2+1)}$$

$$= \frac{2x}{(x^2+1)^{3/2}(x^2-1)^{1/2}}$$

55. $y = \frac{x(x-1)^{3/2}}{\sqrt{x+1}}$

$$\ln y = \ln x + \frac{3}{2}\ln(x-1) - \frac{1}{2}\ln(x+1)$$

$$\frac{1}{y}\left(\frac{dy}{dx}\right) = \frac{1}{x} + \frac{3}{2}\left(\frac{1}{x-1}\right) - \frac{1}{2}\left(\frac{1}{x+1}\right)$$

$$\frac{dy}{dx} = \frac{y}{2}\left[\frac{2}{x} + \frac{3}{x-1} - \frac{1}{x+1}\right]$$

$$= \frac{y}{2}\left[\frac{4x^2+4x-2}{x(x^2-1)}\right]$$

$$= \frac{(2x^2+2x-1)\sqrt{x-1}}{(x+1)^{3/2}}$$

56. $y = \frac{(x+1)(x+2)}{(x-1)(x-2)}$

$$\ln y = \ln(x+1) + \ln(x+2) - \ln(x-1) - \ln(x-2)$$

$$\frac{1}{y}\left(\frac{dy}{dx}\right) = \frac{1}{x+1} + \frac{1}{x+2} - \frac{1}{x-1} - \frac{1}{x-2}$$

$$\frac{dy}{dx} = y\left[\frac{-2}{x^2-1} + \frac{-4}{x^2-4}\right] = y\left[\frac{-6x^2+12}{(x^2-1)(x^2-4)}\right]$$

$$= \frac{6y(2-x^2)}{(x^2-1)(x^2-4)}$$

57. $y = x^{2/x}$

$$\ln y = \frac{2}{x}\ln x$$

$$\frac{1}{y}\left(\frac{dy}{dx}\right) = \frac{2}{x}\left(\frac{1}{x}\right) + \ln x\left(-\frac{2}{x^2}\right)$$

$$\frac{dy}{dx} = \frac{2y}{x^2}(1-\ln x) = 2x^{(2/x)-2}(1-\ln x)$$

58. $$y = x^{x-1}$$
$$\ln y = (x-1)(\ln x)$$
$$\frac{1}{y}\left(\frac{dy}{dx}\right) = (x-1)\left(\frac{1}{x}\right) + \ln x$$
$$\frac{dy}{dx} = y\left[\frac{x-1}{x} + \ln x\right]$$
$$= x^{x-2}(x - 1 + x\ln x)$$

59. $$y = (x-2)^{x+1}$$
$$\ln y = (x+1)\ln(x-2)$$
$$\frac{1}{y}\left(\frac{dy}{dx}\right) = (x+1)\left(\frac{1}{x-2}\right) + \ln(x-2)$$
$$\frac{dy}{dx} = y\left[\frac{x+1}{x-2} + \ln(x-2)\right]$$
$$= (x-2)^{x+1}\left[\frac{x+1}{x-2} + \ln(x-2)\right]$$

60. $$y = (1+x)^{1/x}$$
$$\ln y = \frac{1}{x}\ln(1+x)$$
$$\frac{1}{y}\left(\frac{dy}{dx}\right) = \frac{1}{x}\left(\frac{1}{1+x}\right) + \ln(1+x)\left(-\frac{1}{x^2}\right)$$
$$\frac{dy}{dx} = \frac{y}{x}\left[\frac{1}{x+1} - \frac{\ln(x+1)}{x}\right]$$

61. (a) $y = 3x^2 - \ln x,\ (1, 3)$
$$\frac{dy}{dx} = 6x - \frac{1}{x}$$
When $x = 1, \frac{dy}{dx} = 5.$

Tangent line: $y - 3 = 5(x - 1)$
$$y = 5x - 2$$

(b)

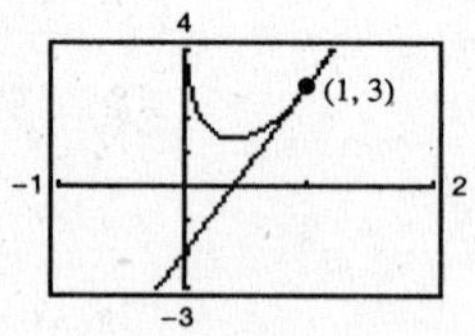

62. (a) $y = 4 - x^2 - \ln\left(\frac{1}{2}x + 1\right),\ (0, 4)$
$$\frac{dy}{dx} = -2x - \frac{1}{(1/2)x + 1}\left(\frac{1}{2}\right)$$
$$= -2x - \frac{1}{x+2}$$
When $x = 0, \frac{dy}{dx} = -\frac{1}{2}.$

Tangent line: $y - 4 = -\frac{1}{2}(x - 0)$
$$y = -\frac{1}{2}x + 4$$

(b)

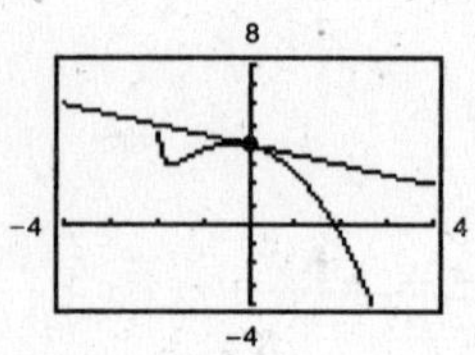

63. $$x^2 - 3\ln y + y^2 = 10$$
$$2x - \frac{3}{y}\frac{dy}{dx} + 2y\frac{dy}{dx} = 0$$
$$2x = \frac{dy}{dx}\left(\frac{3}{y} - 2y\right)$$
$$\frac{dy}{dx} = \frac{2x}{(3/y) - 2y} = \frac{2xy}{3 - 2y^2}$$

64. $$\ln(xy) + 5x = 30$$
$$\ln x + \ln y + 5x = 30$$
$$\frac{1}{x} + \frac{1}{y}\frac{dy}{dx} + 5 = 0$$
$$\frac{1}{y}\frac{dy}{dx} = -\frac{1}{x} - 5$$
$$\frac{dy}{dx} = -\frac{y}{x} - 5y = -\left(\frac{y + 5xy}{x}\right)$$

65. $$y = 2(\ln x) + 3$$
$$y' = \frac{2}{x}$$
$$y'' = -\frac{2}{x^2}$$
$$xy'' + y' = x\left(-\frac{2}{x^2}\right) + \frac{2}{x} = 0$$

66. $y = x(\ln x) - 4x$

$$y' = x\left(\frac{1}{x}\right) + \ln x - 4 = -3 + \ln x$$

$$(x + y) - xy' = x + x\ln x - 4x - x(-3 + \ln x) = 0$$

67. $y = \dfrac{x^2}{2} - \ln x$, Domain: $(0, \infty)$

$$y' = x - \frac{1}{x} = \frac{(x+1)(x-1)}{x} = 0 \text{ when } x = \pm 1.$$

$$y'' = 1 + \frac{1}{x^2} > 0$$

Relative minimum: $\left(1, \frac{1}{2}\right)$

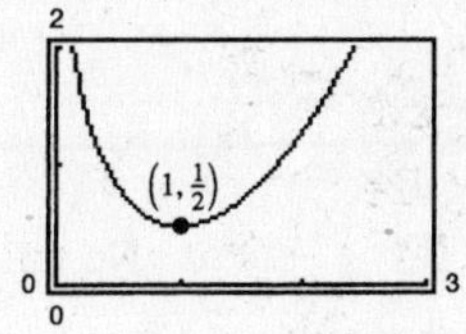

68. $y = x - \ln x$

Domain: $(0, \infty)$

$$y' = 1 - \frac{1}{x} = 0 \text{ when } x = 1.$$

$$y'' = \frac{1}{x^2} > 0$$

Relative minimum: $(1, 1)$

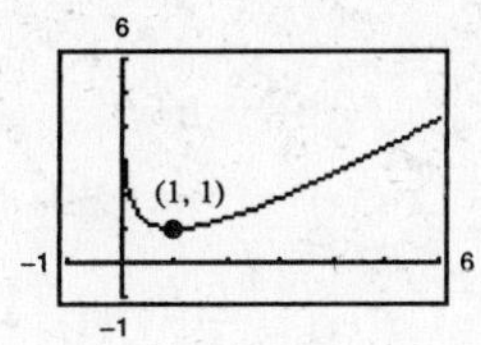

69. $y = x\ln x$, Domain: $(0, \infty)$

$$y' = x\left(\frac{1}{x}\right) + \ln x = 1 + \ln x = 0 \text{ when } x = e^{-1}.$$

$$y'' = \frac{1}{x} > 0$$

Relative minimum: $(e^{-1}, -e^{-1})$

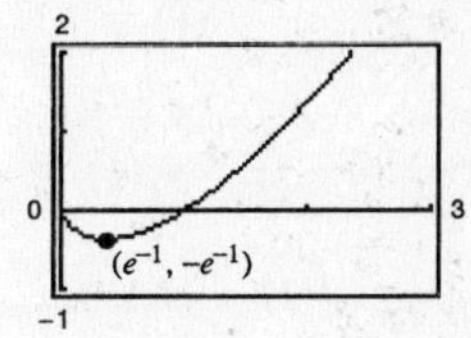

70. $y = \dfrac{\ln x}{x}$, Domain: $(0, \infty)$

$$y' = \frac{x(1/x) - \ln x}{x^2} = \frac{1 - \ln x}{x^2} = 0 \text{ when } x = e.$$

$$y'' = \frac{x^2(-1/x) - (1 - \ln x)(2x)}{x^4} = \frac{2(\ln x) - 3}{x^3} = 0 \text{ when } x = e^{3/2}.$$

Relative maximum: (e, e^{-1})

Point of inflection: $\left(e^{3/2}, \frac{3}{2}e^{-3/2}\right)$

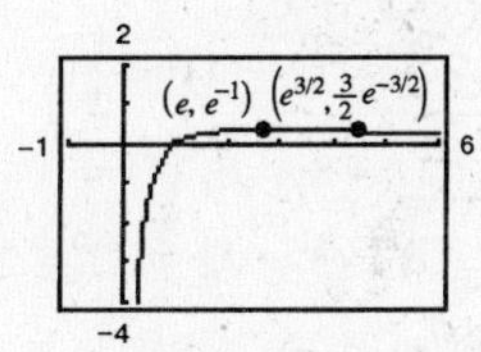

71. $y = \dfrac{x}{\ln x}$, Domain: $(0, 1) \cup (1, \infty)$

$$y' = \frac{(\ln x)(1) - (x)(1/x)}{(\ln x)^2} = \frac{\ln x - 1}{(\ln x)^2} = 0 \text{ when } x = e.$$

$$y'' = \frac{2 - \ln x}{x(\ln x)^3} = 0 \text{ when } x = e^2.$$

Relative minimum: (e, e)

Point of inflection: $(e^2, e^2/2)$

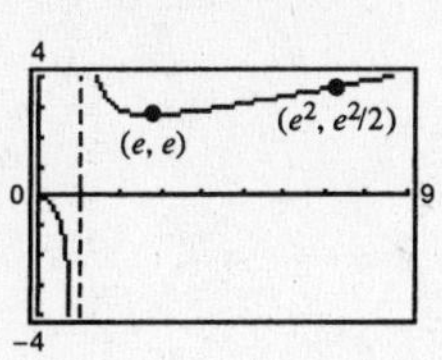

72. $y = x^2 \ln \dfrac{x}{4}$. Domain $x > 0$

$$y' = x^2\left(\frac{1}{x}\right) + 2x\ln\frac{x}{4} = x\left(1 + 2\ln\frac{x}{4}\right) = 0 \text{ when}$$

$$-1 = 2\ln\frac{x}{4} \Rightarrow \ln\frac{x}{4} = -\frac{1}{2} \Rightarrow x = 4e^{-1/2}.$$

$$y'' = 1 + 2\ln\frac{x}{4} + 2x\left(\frac{1}{x}\right) = 3 + 2\ln\frac{x}{4}$$

$y'' = 0$ when $x = 4e^{-3/2}$.

Relative minimum: $(4e^{-1/2}, -8e^{-1})$

Point of inflection: $(4e^{-3/2}, -24e^{-3})$

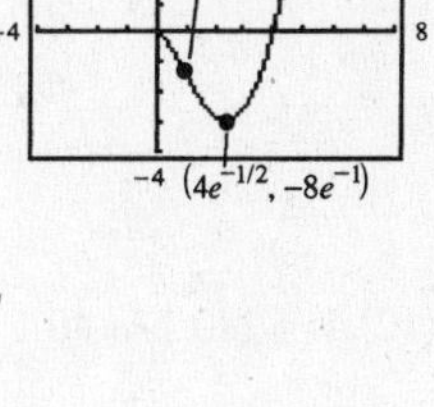

73. $y = x^2 - \ln x$, Domain: $x > 0$

$y' = 2x - \dfrac{1}{x} = 0$

$2x^2 = 1$

$x = \dfrac{1}{\sqrt{2}} = \dfrac{\sqrt{2}}{2}$

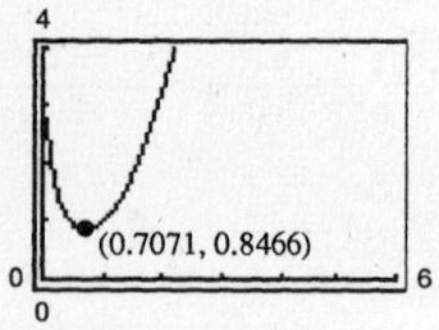

$y'' = 2 + \dfrac{1}{x^2} \neq 0$ No inflection points

Relative minimum: $\left(\dfrac{\sqrt{2}}{2}, \dfrac{1}{2} + \dfrac{1}{2}\ln 2\right) \approx (0.7071, 0.8466)$

74. $y = (\ln x)^2$, Domain: $x > 0$

$y' = 2\ln x\left(\dfrac{1}{x}\right) = \dfrac{2\ln x}{x}$

$y' = 0$ for $x = 1$.

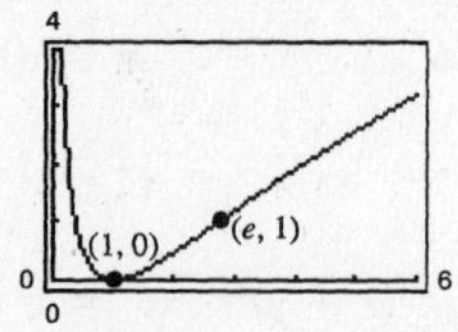

$y'' = \dfrac{x(2/x) - 2\ln x}{x^2} = \dfrac{2 - 2\ln x}{x^2} = 0$ for $x = e$.

Relative minimum: $(1, 0)$

Inflection point: $(e, 1)$

75. $f(x) = \ln x$, $f(1) = 0$

$f'(x) = \dfrac{1}{x}$, $f'(1) = 1$

$f''(x) = -\dfrac{1}{x^2}$, $f''(1) = -1$

$P_1(x) = f(1) + f'(1)(x - 1) = x - 1$, $P_1(1) = 0$

$P_2(x) = f(1) + f'(1)(x - 1) + \dfrac{1}{2}f''(1)(x - 1)^2$

$= (x - 1) - \dfrac{1}{2}(x - 1)^2$, $P_2(1) = 0$

$P_1'(x) = 1$, $P_1'(1) = 1$

$P_2'(x) = 1 - (x - 1) = 2 - x$, $P_2'(1) = 1$

$P_2''(x) = -1$, $P_2''(1) = -1$

The values of f, P_1, P_2, and their first derivatives agree at $x = 1$. The values of the second derivatives of f and P_2 agree at $x = 1$.

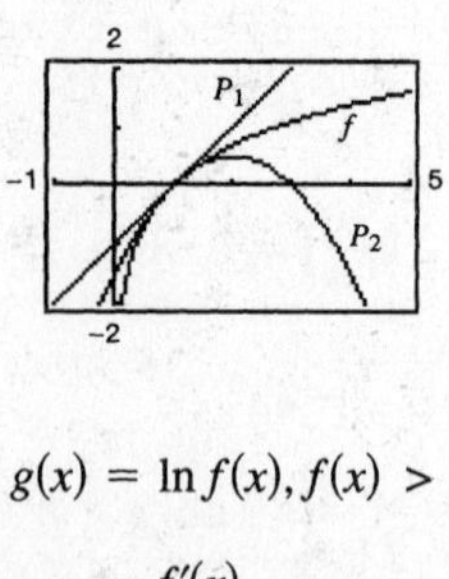

76. $f(x) = x\ln x$, $f(1) = 0$

$f'(x) = 1 + \ln x$, $f(1) = 1$

$f''(x) = \dfrac{1}{x}$, $f''(1) = 1$

$P_1(x) = f(1) + f'(1)(x - 1) = x - 1$, $P_1(1) = 0$

$P_2(x) = f(1) + f'(1)(x - 1) + \dfrac{1}{2}f''(1)(x - 1)^2$

$= (x - 1) + \dfrac{1}{2}(x - 1)^2$, $P_2(1) = 0$

$P_1'(x) = 1$, $P_1'(1) = 1$

$P_2'(x) = 1 + (x - 1) = x$, $P_2'(1) = 1$

$P_2''(x) = x$, $P_2''(1) = 1$

The values of f, P_1, P_2, and their first derivatives agree at $x = 1$. The values of the second derivatives of f and P_2 agree at $x = 1$.

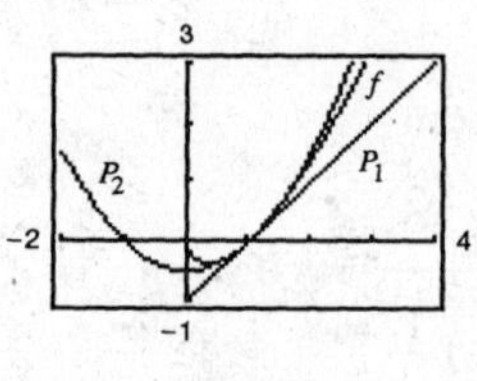

77. $g(x) = \ln f(x)$, $f(x) > 0$

$g'(x) = \dfrac{f'(x)}{f(x)}$

(a) Yes. If the graph of g is increasing, then $g'(x) > 0$. Since $f(x) > 0$, you know that $f'(x) = g'(x)f(x)$ and thus, $f'(x) > 0$. Therefore, the graph of f is increasing.

(b) No. Let $f(x) = x^2 + 1$ (positive and concave up). $g(x) = \ln(x^2 + 1)$ is not concave up.

78. (a) $f(1) \neq f(3)$

(b) $f'(x) = 1 - \dfrac{2}{x} = 0$ for $x = 2$.

79. $t = \dfrac{5.315}{-6.7968 + \ln x}$, $1000 < x$

(a) $t(1167.41) \approx 20$ years

$T = (1167.41)(20)(12) = \$280{,}178.40$

(b) $t(1068.45) \approx 30$ years

$T = (1068.45)(30)(12) = \$384{,}642.00$

—CONTINUED—

79. —CONTINUED—

(c) $\dfrac{dt}{dx} = -5.315(-6.7968 + \ln x)^{-2}\left(\dfrac{1}{x}\right)$

$= -\dfrac{5.315}{x(-6.7968 + \ln x)^2}$

When $x = 1167.41$, $dt/dx \approx -0.0645$.

When $x = 1068.45$, $dt/dx \approx -0.1585$.

(d) There are two obvious benefits to paying a higher monthly payment:

1. The term is lower.

2. The total amount paid is lower.

80. (a) You get an error message because $\ln h$ does not exist for $h = 0$.

(b) Reversing the data, you obtain

$h = 0.8627 - 6.4474 \ln p$.

(c)

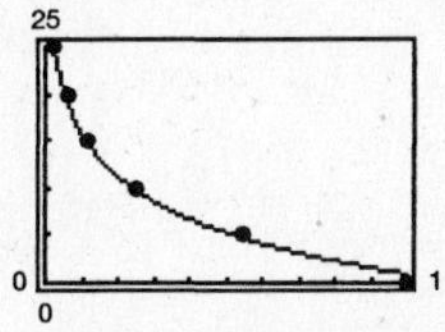

(d) If $p = 0.75$, $h \approx 2.72$ km.

(e) If $h = 13$ km, $p \approx 0.15$ atmosphere.

(f) $h = 0.8627 - 6.4474 \ln p$

$1 = -6.4474\dfrac{1}{p}\dfrac{dp}{dh}$ (implicit differentiation)

$\dfrac{dp}{dh} = \dfrac{p}{-6.4474}$

For $h = 5$, $p = 0.55$ and $dp/dh = -0.0853$ atmos/km.

For $h = 20$, $p = 0.06$ and $dp/dh = -0.00931$ atmos/km.

As the altitude increases, the rate of change of pressure decreases.

81. (a)

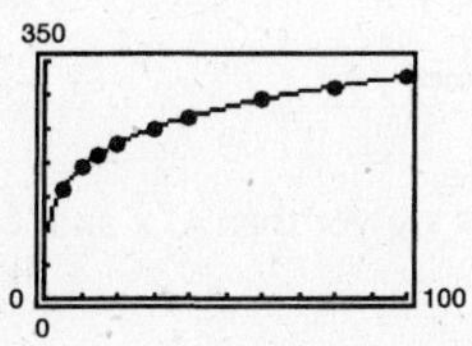

(b) $T'(p) = \dfrac{34.96}{p} + \dfrac{3.955}{\sqrt{p}}$

$T'(10) \approx 4.75$ deg/lb/in^2

$T'(70) \approx 0.97$ deg/lb/in^2

(c)

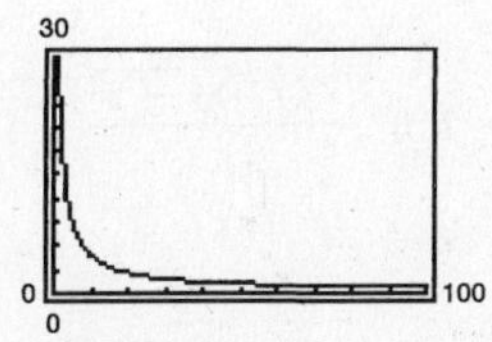

$\lim_{p\to\infty} T'(p) = 0$

As the pressure increases, the rate of change of temperature approaches zero.

82. $p = 91.6 - 15.6 \ln(t + 1),\ 0 \le t \le 48$

$p' = \dfrac{-15.6}{t + 1}$

$p'(12) = \dfrac{-15.6}{13} = -1.2\%$ per month.

83. $C = 6000 + 300x + 300x \ln x$

Average cost: $\overline{C} = \dfrac{C}{x} = \dfrac{6000}{x} + 300 + 300 \ln x$

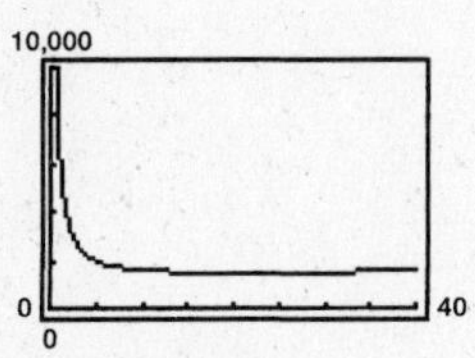

Minimum average cost: $1498.72

Analytically,

$\overline{C} = \dfrac{6000}{x} + 300 + 300 \ln x$

$\dfrac{d\overline{C}}{dx} = -\dfrac{6000}{x^2} + \dfrac{300}{x} = \dfrac{300x - 6000}{x^2}$

Critical Number: $x = 20$

On $(0, 20)$, $\dfrac{d\overline{C}}{dx} < 0 \Rightarrow \overline{C}$ is decreasing.

On $(20, \infty)$, $\dfrac{d\overline{C}}{dx} > 0 \Rightarrow \overline{C}$ is increasing.

Thus, $\overline{C}(20) \approx \$1498.72$ is the minimum value.

84. $P = \dfrac{0.86}{1 + 0.779^n}$

(a) $\lim_{n\to\infty} \dfrac{0.86}{1 + 0.779^n} = \dfrac{0.86}{1 + 0} = 0.86$

(b) $\dfrac{dP}{dn} = \dfrac{(1 + 0.779^n)(0) - 0.86(\ln 0.779)(0.779^n)}{(1 + 0.779^n)^2}$

$= -\dfrac{0.86(\ln 0.779)(0.779^n)}{(1 + 0.779^n)^2}$

When $n = 3: \dfrac{dP}{dn} \approx 0.047.$

When $n = 10: \dfrac{dP}{dn} \approx 0.015.$

85. (a) $\lim_{t\to\infty} 6.7e^{(-48.1)/t} = 6.7e^0 = 6.7$ million ft^3

(b) $V' = \dfrac{322.27}{t^2} e^{-(48.1)/t}$

$V'(20) \approx 0.073$ million ft^3/yr

$V'(60) \approx 0.040$ million ft^3/yr

86. (a)

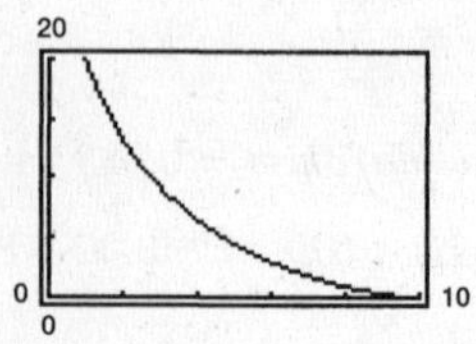

(b) $y'(5) = -\sqrt{3} \approx -1.7321$

$y'(9) = -\dfrac{\sqrt{19}}{9} \approx -0.4843$

(c) $\lim_{x\to 10^-} -\dfrac{\sqrt{100 - x^2}}{x} = 0$

87. (a)

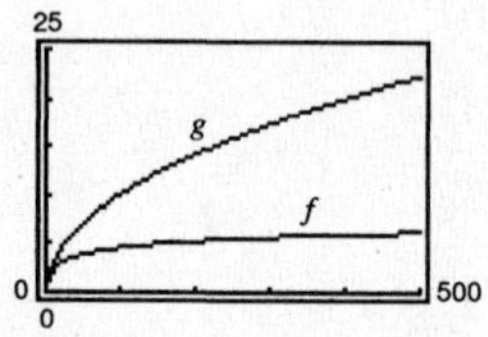

(b)

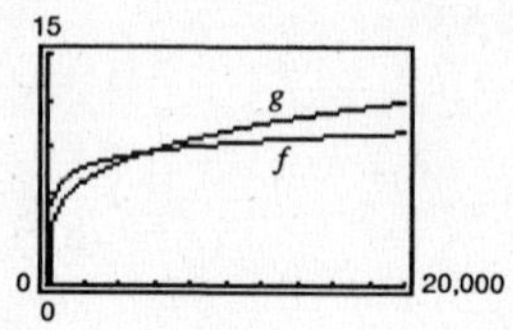

The rate of growth of $\ln x$ is slower than $\sqrt{x}$ and $\sqrt[4]{x}$ for large x.

88. $f(x) = \log_2 x \Rightarrow f'(x) = \dfrac{1}{x \ln 2}$

$g(x) = x^x \Rightarrow g'(x) = x^x(1 + \ln x)$

[**Note:** Let $y = g(x)$. Then: $\ln y = \ln x^x = x \ln x$

$\dfrac{1}{y}y' = x \cdot \dfrac{1}{x} + \ln x$

$y' = y(1 + \ln x)$

$y' = x^x(1 + \ln x) = g'(x)$.]

$h(x) = x^2 \Rightarrow h'(x) = 2x$

$k(x) = 2^x \Rightarrow k'(x) = (\ln 2)2^x$

From greatest to smallest rate of growth:

$g(x), k(x), h(x), f(x)$

89. $y = \ln x$

$y' = \dfrac{1}{x} > 0$ for $x > 0$.

Since $\ln x$ is increasing on its entire domain $(0, \infty)$, it is a strictly monotonic function and therefore, is one-to-one.

90. False

$\dfrac{d}{dx}[\ln(x^2 + 5x)] = \dfrac{1}{x^2 + 5x}(2x + 5) \neq \dfrac{2}{x} + \dfrac{1}{x}$

91. False.

π is a constant.

$\dfrac{d}{dx}[\ln \pi] = 0$

92. $f(x) = \dfrac{\ln x}{x}$

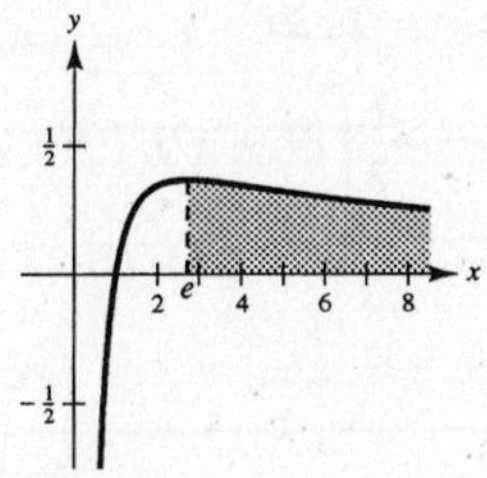

(a) $f'(x) = \dfrac{1 - \ln x}{x^2} = 0$ when $x = e$.

On $(0, e), f'(x) > 0 \Rightarrow f$ is increasing.

On $(e, \infty), f'(x) < 0 \Rightarrow f$ is decreasing.

(b) For $e \le A < B$, we have:

$$\frac{\ln A}{A} > \frac{\ln B}{B}$$

$$B \ln A > A \ln B$$

$$\ln A^B > \ln B^A$$

$$A^B > B^A.$$

(c) Since $e < \pi$, from part (b) we have $e^\pi > \pi^e$.

93. Let $y = \ln x$

$$e^y = x$$

$$e^y \frac{dy}{dx} = 1$$

$$\frac{dy}{dx} = \frac{1}{e^y} = \frac{1}{x}.$$

Section 8.3 Logarithmic Functions and Integration

1. $\displaystyle\int \frac{3}{x}\,dx = 3\int \frac{1}{x}\,dx = 3 \ln|x| + C$

2. $\displaystyle\int \frac{10}{x}\,dx = 10\int \frac{1}{x}\,dx = 10 \ln|x| + C$

3. $u = x + 1, du = dx$

$$\int \frac{1}{x+1}\,dx = \ln|x + 1| + C$$

4. $u = x - 5, du = dx$

$$\int \frac{1}{x-5}\,dx = \ln|x - 5| + C$$

5. $u = 3 - 2x, du = -2\,dx$

$$\int \frac{1}{3-2x}\,dx = -\frac{1}{2}\int \frac{1}{3-2x}(-2)\,dx$$

$$= -\frac{1}{2}\ln|3 - 2x| + C$$

6. $\displaystyle\int \frac{1}{3x+2}\,dx = \frac{1}{3}\int \frac{1}{3x+2}(3)\,dx$

$$= \frac{1}{3}\ln|3x + 2| + C$$

7. $u = x^2 + 1, du = 2x\,dx$

$$\int \frac{x}{x^2+1}\,dx = \frac{1}{2}\int \frac{1}{x^2+1}(2x)\,dx$$

$$= \frac{1}{2}\ln(x^2 + 1) + C$$

$$= \ln\sqrt{x^2 + 1} + C$$

8. $u = 3 - x^3, du = -3x^2\,dx$

$$\int \frac{x^2}{3-x^3}\,dx = -\frac{1}{3}\int \frac{1}{3-x^3}(-3x^2)\,dx$$

$$= -\frac{1}{3}\ln|3 - x^3| + C$$

9. $\displaystyle\int \frac{x^2-4}{x}\,dx = \int\left(x - \frac{4}{x}\right)dx$

$\displaystyle = \frac{x^2}{2} - 4\ln|x| + C$

10. $u = 9 - x^2$, $du = -2x\,dx$

$\displaystyle\int \frac{x}{\sqrt{9-x^2}}\,dx = -\frac{1}{2}\int (9-x^2)^{-1/2}(-2x)\,dx$

$\displaystyle = -\sqrt{9-x^2} + C$

11. $u = x^3 + 3x^2 + 9x$, $du = 3(x^2 + 2x + 3)\,dx$

$\displaystyle\int \frac{x^2+2x+3}{x^3+3x^2+9x}\,dx = \frac{1}{3}\int \frac{3(x^2+2x+3)}{x^3+3x^2+9x}\,dx = \frac{1}{3}\ln|x^3+3x^2+9x| + C$

12. $\displaystyle\int \frac{x(x+2)}{x^3+3x^2-4}\,dx = \frac{1}{3}\int \frac{3x^2+6x}{x^3+3x^2-4}\,dx \quad (u = x^3+3x^2-4)$

$\displaystyle = \frac{1}{3}\ln|x^3+3x^2-4| + C$

13. $\displaystyle\int \frac{x^2-3x+2}{x+1}\,dx = \int\left(x - 4 + \frac{6}{x+1}\right)dx$

$\displaystyle = \frac{x^2}{2} - 4x + 6\ln|x+1| + C$

14. $\displaystyle\int \frac{2x^2+7x-3}{x-2}\,dx = \int\left(2x + 11 + \frac{19}{x-2}\right)dx$

$\displaystyle = x^2 + 11x + 19\ln|x-2| + C$

15. $\displaystyle\int \frac{x^3-3x^2+5}{x-3}\,dx = \int\left(x^2 + \frac{5}{x-3}\right)dx$

$\displaystyle = \frac{x^3}{3} + 5\ln|x-3| + C$

16. $\displaystyle\int \frac{x^3-6x-20}{x+5}\,dx = \int\left(x^2 - 5x + 19 - \frac{115}{x+5}\right)dx$

$\displaystyle = \frac{x^3}{3} - \frac{5x^2}{2} + 19x - 115\ln|x+5| + C$

17. $\displaystyle\int \frac{x^4+x-4}{x^2+2}\,dx = \int\left(x^2 - 2 + \frac{x}{x^2+2}\right)dx$

$\displaystyle = \frac{x^3}{3} - 2x + \frac{1}{2}\ln(x^2+2) + C$

18. $\displaystyle\int \frac{x^3-3x^2+4x-9}{x^2+3}\,dx = \int\left(-3 + x + \frac{x}{x^2+3}\right)dx$

$\displaystyle = -3x + \frac{x^2}{2} + \frac{1}{2}\ln(x^2+3) + C$

19. $u = \ln x$, $du = \dfrac{1}{x}\,dx$

$\displaystyle\int \frac{(\ln x)^2}{x}\,dx = \frac{1}{3}(\ln x)^3 + C$

20. $\displaystyle\int \frac{1}{x\ln(x^3)}\,dx = \frac{1}{3}\int \frac{1}{\ln x}\cdot\frac{1}{x}\,dx$

$\displaystyle = \frac{1}{3}\ln|\ln|x|| + C$

21. Let $u = 1 + e^{-x}$, $du = -e^{-x}dx$.

$\displaystyle\int \frac{e^{-x}}{1+e^{-x}}\,dx = -\int \frac{-e^{-x}}{1+e^{-x}}\,dx = -\ln(1+e^{-x}) + C$

$\displaystyle = \ln\left(\frac{e^x}{e^x+1}\right) + C = x - \ln(e^x+1) + C$

22. Let $u = 1 + e^{2x}$, $du = 2e^{2x}\,dx$.

$\displaystyle\int \frac{e^{2x}}{1+e^{2x}}\,dx = \frac{1}{2}\int \frac{2e^{2x}}{1+e^{2x}}\,dx = \frac{1}{2}\ln(1+e^{2x}) + C$

23. Let $u = e^x - e^{-x}$, $du = (e^x + e^{-x})\,dx$.

$\displaystyle\int \frac{e^x+e^{-x}}{e^x-e^{-x}}\,dx = \ln|e^x - e^{-x}| + C$

24. $\displaystyle\int \ln(e^{2x-1})\,dx = \int (2x-1)\,dx$

$\displaystyle = x^2 - x + C$

25. $u = x + 1,\ du = dx$

$$\int \frac{1}{\sqrt{x+1}}\,dx = \int (x+1)^{-1/2}\,dx$$
$$= 2(x+1)^{1/2} + C$$
$$= 2\sqrt{x+1} + C$$

26. $u = 1 + x^{1/3},\ du = \dfrac{1}{3x^{2/3}}\,dx$

$$\int \frac{1}{x^{2/3}(1+x^{1/3})}\,dx = 3\int \frac{1}{1+x^{1/3}}\left(\frac{1}{3x^{2/3}}\right)dx$$
$$= 3\ln\left|1 + x^{1/3}\right| + C$$

27.
$$\int \frac{2x}{(x-1)^2}\,dx = \int \frac{2x-2+2}{(x-1)^2}\,dx$$
$$= \int \frac{2(x-1)}{(x-1)^2}\,dx + 2\int \frac{1}{(x-1)^2}\,dx$$
$$= 2\int \frac{1}{x-1}\,dx + 2\int \frac{1}{(x-1)^2}\,dx$$
$$= 2\ln|x-1| - \frac{2}{(x-1)} + C$$

28.
$$\int \frac{x(x-2)}{(x-1)^3}\,dx = \int \frac{x^2-2x+1-1}{(x-1)^3}\,dx$$
$$= \int \frac{(x-1)^2}{(x-1)^3}\,dx - \int \frac{1}{(x-1)^3}\,dx$$
$$= \int \frac{1}{x-1}\,dx - \int \frac{1}{(x-1)^3}\,dx$$
$$= \ln|x-1| + \frac{1}{2(x-1)^2} + C$$

29. $u = 1 + \sqrt{2x},\ du = \dfrac{1}{\sqrt{2x}}\,dx \Rightarrow (u-1)\,du = dx$

$$\int \frac{1}{1+\sqrt{2x}}\,dx = \int \frac{(u-1)}{u}\,du = \int \left(1 - \frac{1}{u}\right)du$$
$$= u - \ln|u| + C_1$$
$$= \left(1+\sqrt{2x}\right) - \ln\left|1+\sqrt{2x}\right| + C_1$$
$$= \sqrt{2x} - \ln\left(1+\sqrt{2x}\right) + C$$

where $C = C_1 + 1$.

30. $u = 1 + \sqrt{3x},\ du = \dfrac{3}{2\sqrt{3x}}dx \Rightarrow dx = \dfrac{2}{3}(u-1)\,du$

$$\int \frac{1}{1+\sqrt{3x}}\,dx = \int \frac{1}{u}\frac{2}{3}(u-1)\,du$$
$$= \frac{2}{3}\int \left(1 - \frac{1}{u}\right)du$$
$$= \frac{2}{3}\left[u - \ln|u|\right] + C$$
$$= \frac{2}{3}\left[1 + \sqrt{3x} - \ln\left(1+\sqrt{3x}\right)\right] + C$$
$$= \frac{2}{3}\sqrt{3x} - \frac{2}{3}\ln\left(1+\sqrt{3x}\right) + C_1$$

where $C_1 = C + \dfrac{2}{3}$

31. $u = \sqrt{x} - 3,\ du = \dfrac{1}{2\sqrt{x}}\,dx \Rightarrow 2(u+3)\,du = dx$

$$\int \frac{\sqrt{x}}{\sqrt{x}-3}\,dx = 2\int \frac{(u+3)^2}{u}\,du = 2\int \frac{u^2+6u+9}{u}\,du = 2\int \left(u + 6 + \frac{9}{u}\right)du$$
$$= 2\left[\frac{u^2}{2} + 6u + 9\ln|u|\right] + C_1 = u^2 + 12u + 18\ln|u| + C_1$$
$$= \left(\sqrt{x}-3\right)^2 + 12\left(\sqrt{x}-3\right) + 18\ln\left|\sqrt{x}-3\right| + C_1$$
$$= x + 6\sqrt{x} + 18\ln\left|\sqrt{x}-3\right| + C \text{ where } C = C_1 - 27.$$

32. $u = x^{1/3} - 1,\ du = \dfrac{1}{3x^{2/3}}\,dx \Rightarrow dx = 3(u+1)^2\,du$

$$\int \frac{\sqrt[3]{x}}{\sqrt[3]{x}-1}\,dx = \int \frac{u+1}{u}3(u+1)^2\,du$$
$$= 3\int \frac{u+1}{u}(u^2+2u+1)\,du$$
$$= 3\int \left(u^2+3u+3+\frac{1}{u}\right)du$$
$$= 3\left[\frac{u^3}{3}+\frac{3u^2}{2}+3u+\ln|u|\right]+C$$
$$= 3\left[\frac{(x^{1/3}-1)^3}{3}+\frac{3(x^{1/3}-1)^2}{2}+3(x^{1/3}-1)+\ln|x^{1/3}-1|\right]+C$$
$$= 3\ln|x^{1/3}-1| + \frac{3x^{2/3}}{2} + 3x^{1/3} + x + C_1$$

where $C_1 = C - \dfrac{11}{2}$

33. $\displaystyle\int 3^x\,dx = \frac{3^x}{\ln 3} + C$

34. $\displaystyle\int 5^{-x}\,dx = \frac{-5^{-x}}{\ln 5} + C$

35. $\displaystyle\int x5^{-x^2}\,dx = -\frac{1}{2}\int 5^{-x^2}(-2x)\,dx$
$$= -\left(\frac{1}{2}\right)\frac{5^{-x^2}}{\ln 5} + C$$
$$= \frac{-1}{2\ln 5}(5^{-x^2}) + C$$

36. $\displaystyle\int (3-x)7^{(3-x)^2}\,dx = -\frac{1}{2}\int -2(3-x)7^{(3-x)^2}\,dx$
$$= -\frac{1}{2\ln 7}\left[7^{(3-x)^2}\right] + C$$

37. $\displaystyle\int \frac{3^{2x}}{1+3^{2x}}\,dx,\ u = 1+3^{2x},\ du = 2(\ln 3)3^{2x}\,dx$
$$\frac{1}{2\ln 3}\int \frac{(2\ln 3)3^{2x}}{1+3^{2x}}\,dx = \frac{1}{2\ln 3}\ln(1+3^{2x}) + C$$

38. $\displaystyle\int \frac{2^{-x}}{1+2^{-x}}\,dx = \frac{-1}{\ln 2}\int \frac{1}{1+2^{-x}}(-\ln 2)2^{-x}\,dx$
$$= \frac{-1}{\ln 2}\ln(1+2^{-x}) + C$$
$$= \frac{-\ln(1+2^{-x})}{\ln 2} + C$$

39. $\displaystyle\int_{-1}^{2} 2^x\,dx = \left[\left(\frac{1}{\ln 2}\right)2^x\right]_{-1}^{2}$
$$= \left(\frac{1}{\ln 2}\right)\left[4-\frac{1}{2}\right]$$
$$= \frac{7}{2\ln 2} = \frac{7}{\ln 4}$$

40. $\displaystyle\int_{-2}^{0} (3^3-5^2)\,dx = \int_{-2}^{0}(27-25)\,dx$
$$= \int_{-2}^{0} 2\,dx = \Big[2x\Big]_{-2}^{0} = 0-(-4) = 4$$

41. $\displaystyle\int_0^4 \frac{5}{3x+1}\,dx = \left[\frac{5}{3}\ln|3x+1|\right]_0^4$
$$= \frac{5}{3}\ln 13 \approx 4.275$$

42. $\displaystyle\int_{-1}^{1} \frac{1}{x+2}\,dx = \Big[\ln|x+2|\Big]_{-1}^{1}$
$$= \ln 3 - \ln 1 = \ln 3$$

43. $u = 1+\ln x,\ du = \dfrac{1}{x}\,dx$
$$\int_1^e \frac{(1+\ln x)^2}{x}\,dx = \left[\frac{1}{3}(1+\ln x)^3\right]_1^e = \frac{7}{3}$$

44. $u = \ln x,\ du = \dfrac{1}{x}\,dx$
$$\int_e^{e^2} \frac{1}{x\ln x}\,dx = \int_e^{e^2}\left(\frac{1}{\ln x}\right)\frac{1}{x}\,dx = \Big[\ln|\ln|x||\Big]_e^{e^2} = \ln 2$$

45. $\displaystyle\int_0^1 \frac{x^2-2}{x+1}\,dx = \int_0^1 \left(x - 1 - \frac{1}{x+1}\right) dx$

$\displaystyle = \left[\frac{1}{2}x^2 - x - \ln|x+1|\right]_0^1 = -\frac{1}{2} - \ln 2$

≈ -1.193

46. $\displaystyle\int_0^1 \frac{x-1}{x+1}\,dx = \int_0^1 1\,dx + \int_0^1 \frac{-2}{x+1}\,dx = \Big[x - 2\ln|x+1|\Big]_0^1 = 1 - 2\ln 2$

47. $\displaystyle\int \frac{1}{1+\sqrt{x}}\,dx = 2(1+\sqrt{x}) - 2\ln(1+\sqrt{x}) + C_1$

$= 2[\sqrt{x} - \ln(1+\sqrt{x})] + C$ where $C = C_1 + 2$.

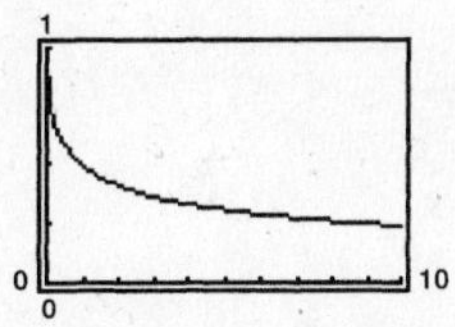

48. $\displaystyle\int \frac{1-\sqrt{x}}{1+\sqrt{x}}\,dx = -(1+\sqrt{x})^2 + 6(1+\sqrt{x}) - 4\ln(1+\sqrt{x}) + C_1$

$= 4\sqrt{x} - x - 4\ln(1+\sqrt{x}) + C$ where $C = C_1 + 5$.

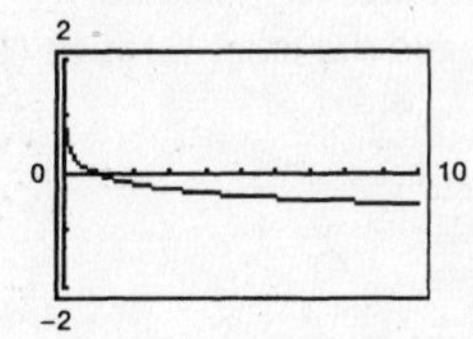

49. $\displaystyle y = \int \frac{3}{2-x}\,dx$

$\displaystyle = -3\int \frac{1}{x-2}\,dx$

$= -3\ln|x-2| + C$

$(1, 0)$: $0 = -3\ln|1-2| + C \Rightarrow C = 0$

$y = -3\ln|x-2|$

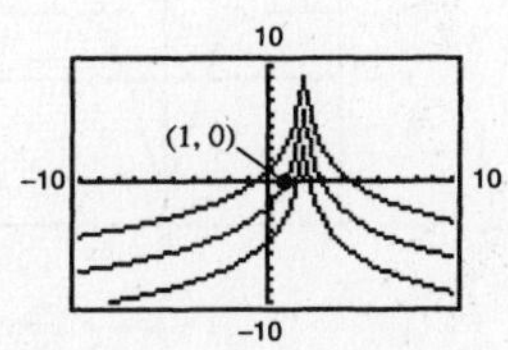

50. $\displaystyle y = \int \frac{2x}{x^2-9}\,dx$

$= \ln|x^2-9| + C$

$(0, 4)$: $4 = \ln|0-9| + C \Rightarrow C = 4 - \ln 9$

$y = \ln|x^2-9| + 4 - \ln 9$

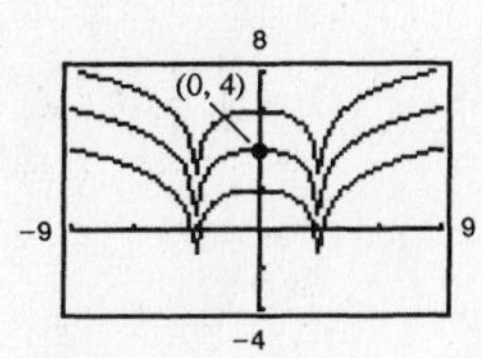

51. $\displaystyle\frac{dy}{dx} = \frac{1}{x+2}$, $(0, 1)$

(a)

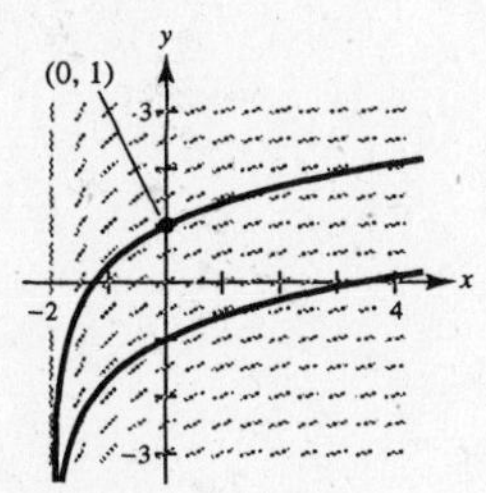

(b) $\displaystyle y = \int \frac{1}{x+2}\,dx = \ln|x+2| + C$

$y(0) = 1 \Rightarrow 1 = \ln 2 + C \Rightarrow C = 1 - \ln 2$

Hence, $\displaystyle y = \ln|x+2| + 1 - \ln 2 = \ln\left|\frac{x+2}{2}\right| + 1.$

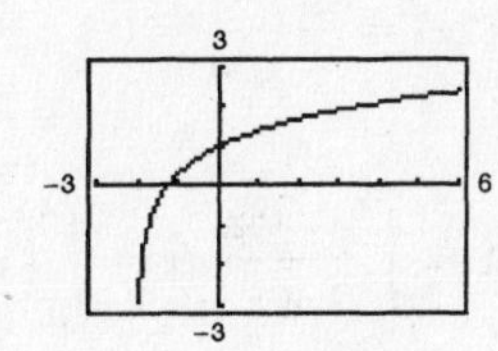

52. $\displaystyle\frac{dy}{dx} = \frac{\ln x}{x}$, $(1, -2)$

(a)

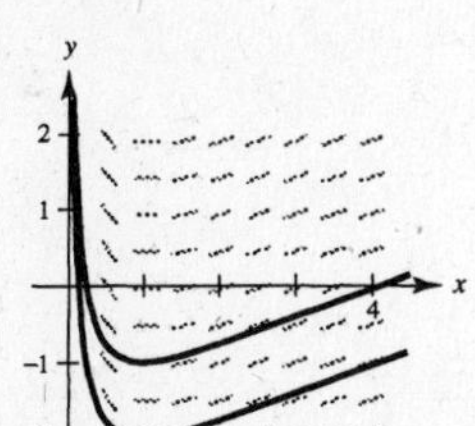

(b) $\displaystyle y = \int \frac{\ln x}{x}\,dx = \frac{1}{2}(\ln x)^2 + C$

$(1, -2)$: $-2 = \frac{1}{2}(\ln 1)^2 + C = C$

$y = \frac{1}{2}(\ln x)^2 - 2$

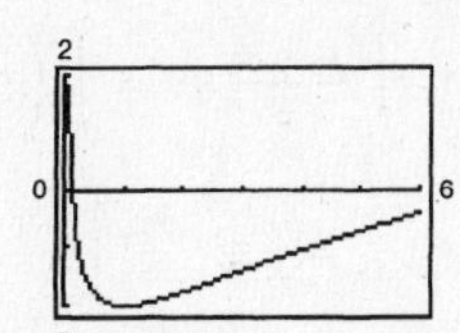

53. $\frac{dy}{dx} = 0.4^{x/3}, \left(0, \frac{1}{2}\right)$

$$y = \int 0.4^{x/3}\,dx = 3\int 0.4^{x/3}\left(\frac{1}{3}\,dx\right)$$

$$= \frac{3}{\ln 0.4}0.4^{x/3} + C = \frac{-3}{\ln 2.5}(0.4)^{x/3} + C$$

$$\frac{1}{2} = -\frac{3}{\ln 2.5} + C \Rightarrow C = \frac{1}{2} + \frac{3}{\ln 2.5}$$

$$y = \frac{-3}{\ln 2.5}(0.4)^{x/3} + \frac{1}{2} + \frac{3}{\ln 2.5}$$

$$= \frac{3(1 - 0.4^{x/3})}{\ln 2.5} + \frac{1}{2}$$

(a)

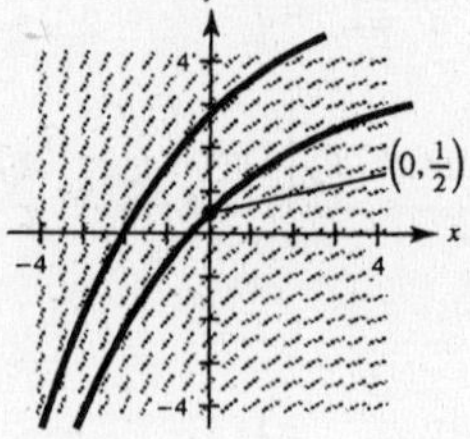

(b)

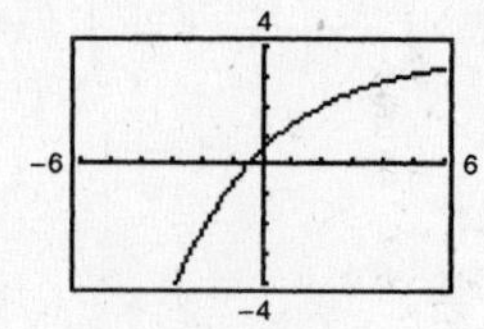

54. $\frac{dy}{dx} = 5x(2^{-x^2/2}), (2, 1)$

(a) Slope field, with 2 solutions indicated.
One through (2, 1) and the other through (2, − 1).

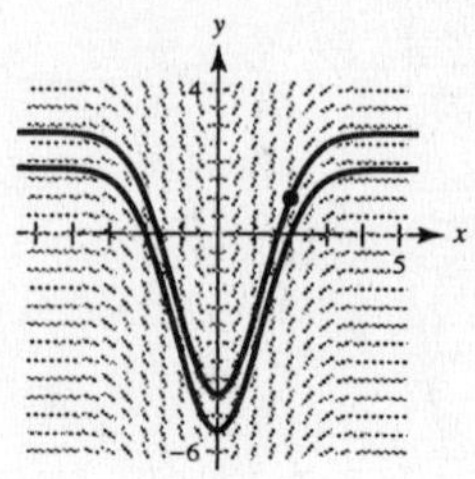

(b) $y = \int 5x(2^{-x^2/2})\,dx = \frac{(-5)2^{-x^2/2}}{\ln 2} + C$

$$1 = \frac{(-5)2^{-2}}{\ln 2} + C = \frac{-5}{4\ln 2} + C \Rightarrow C = 1 + \frac{5}{4\ln 2}$$

$$y = \frac{-5}{\ln 2}2^{-x^2/2} + 1 + \frac{5}{4\ln 2}$$

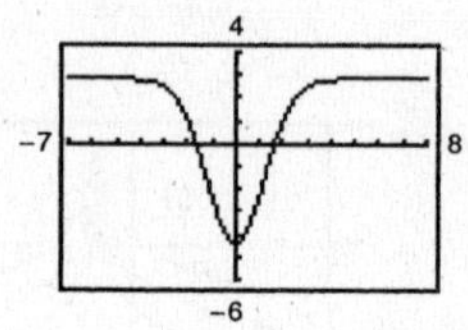

55. $F(x) = \int_1^x \frac{1}{t}\,dt$

$F'(x) = \frac{1}{x}$

56. $F(x) = \int_0^x \frac{1}{t+1}\,dt = \ln|t+1|\Big]_0^x = \ln|x+1|$

$F'(x) = \frac{1}{x+1}$

57. $F(x) = \int_x^{3x} \frac{1}{t}\,dt = \int_1^{3x} \frac{1}{t}\,dt - \int_1^x \frac{1}{t}\,dt$

$F'(x) = \frac{3}{3x} - \frac{1}{x} = 0$

58. $F(x) = \int_1^{x^2} \frac{1}{t}\,dt$

$F'(x) = \frac{2x}{x^2} = \frac{2}{x}$

59. $A = \int_1^4 \frac{x^2+4}{x}\,dx = \int_1^4\left(x + \frac{4}{x}\right)dx$

$$= \left[\frac{x^2}{2} + 4\ln x\right]_1^4 = (8 + 4\ln 4) - \frac{1}{2}$$

$$= \frac{15}{2} + 8\ln 2 \approx 13.045 \text{ square units}$$

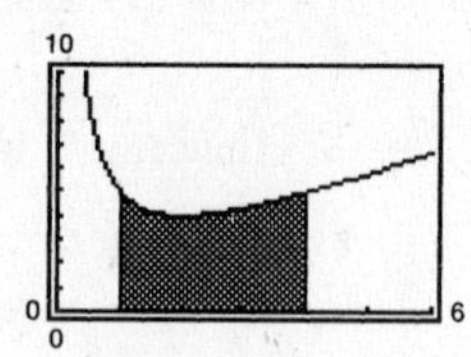

60. $A = \int_1^4 \frac{x+4}{x}\,dx = \int_1^4\left(1 + \frac{4}{x}\right)dx$

$$= \Big[x + 4\ln x\Big]_1^4$$

$$= 4 + 4\ln 4 - 1$$

$$= 3 + 4\ln 4 \approx 8.5452$$

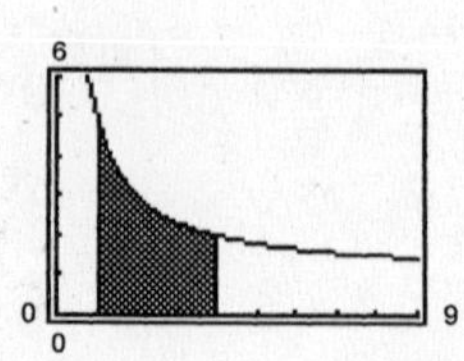

61. $A = \int_0^3 3^x\,dx = \left[\frac{3^x}{\ln 3}\right]_0^3 = \frac{26}{\ln 3} \approx 23.666$

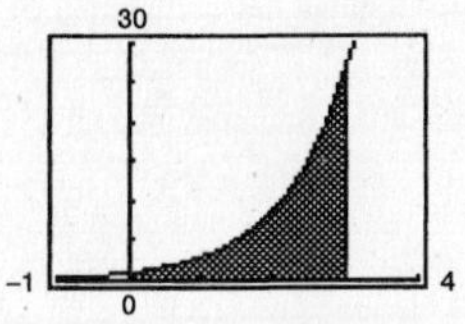

62. $A = \int_0^4 2^{-x}\,dx = \left[\frac{-2^{-x}}{\ln 2}\right]_0^4 = \frac{1}{\ln 2}[1 - 2^{-4}] = \frac{15}{16 \ln 2}$

≈ 1.3525

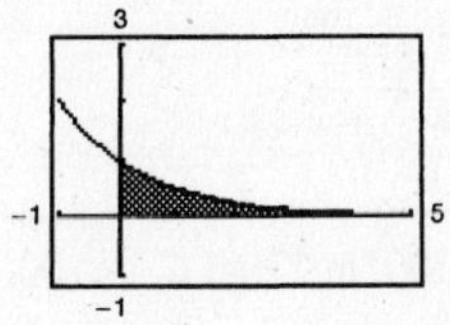

63. Power Rule

64. Substitution: $(u = x^2 + 4)$ and Power Rule

65. Substitution: $(u = x^2 + 4)$ and Log Rule

66. Exponential Rule:

$u = -\frac{x^2}{2},\ du = -x\,dx$

67. Divide the polynomials:

$\frac{x^2}{x+1} = x - 1 + \frac{1}{x+1}$

68. Answers will vary.

69. Average value $= \frac{1}{4-2}\int_2^4 \frac{8}{x^2}\,dx = 4\int_2^4 x^{-2}\,dx$

$= \left[-4\frac{1}{x}\right]_2^4$

$= -4\left(\frac{1}{4} - \frac{1}{2}\right) = 1$

70. Average value $= \frac{1}{4-2}\int_2^4 \frac{4(x+1)}{x^2}\,dx$

$= 2\int_2^4 \left(\frac{1}{x} + \frac{1}{x^2}\right) dx$

$= 2\left[\ln x - \frac{1}{x}\right]_2^4$

$= 2\left[\ln 4 - \frac{1}{4} - \ln 2 + \frac{1}{2}\right]$

$= 2\left[\ln 2 + \frac{1}{4}\right] = \ln 4 + \frac{1}{2} \approx 1.8863$

71. Average value $= \frac{1}{e-1}\int_1^e \frac{\ln x}{x}\,dx = \frac{1}{e-1}\left[\frac{(\ln x)^2}{2}\right]_1^e = \frac{1}{e-1}\left(\frac{1}{2}\right) = \frac{1}{2e-2} \approx 0.291$

72. $\frac{1}{6-0}\int_0^6 \frac{8}{x+2}\,dx = \frac{1}{6}8 \ln|x+2|\Big]_0^6 = \frac{4}{3}[\ln 8 - \ln 2] = \frac{4}{3}\ln 4$

73. $P = \int \frac{3000}{1 + 0.25t}\,dt = (3000)(4)\int \frac{0.25}{1 + 0.25t}\,dt = 12{,}000 \ln|1 + 0.25t| + C$

$P(0) = 12{,}000 \ln|1 + 0.25(0)| + C = 1000$

$C = 1000$

$P = 12{,}000 \ln|1 + 0.25t| + 1000 = 1000[12 \ln|1 + 0.25t| + 1]$

$P(3) = 1000[12(\ln 1.75) + 1] \approx 7715$

74. $t = \frac{10}{\ln 2}\int_{250}^{300} \frac{1}{T - 100}\,dT$

$= \frac{10}{\ln 2}\left[\ln(T - 100)\right]_{250}^{300} = \frac{10}{\ln 2}[\ln 200 - \ln 150] = \frac{10}{\ln 2}\left[\ln\left(\frac{4}{3}\right)\right] \approx 4.1504$ minutes

75. $\frac{1}{50-40}\int_{40}^{50} \frac{90{,}000}{400 + 3x}\,dx = \left[3000 \ln|400 + 3x|\right]_{40}^{50} \approx \$168.27$

76. $\dfrac{dS}{dt} = \dfrac{k}{t}$

$S(t) = \displaystyle\int \frac{k}{t}\,dt = k\ln|t| + C = k\ln t + C$ since $t > 1$.

$S(2) = k\ln 2 + C = 200$

$S(4) = k\ln 4 + C = 300$

Solving this system yields $k = 100/\ln 2$ and $C = 100$. Thus,

$$S(t) = \frac{100\ln t}{\ln 2} + 100 = 100\left[\frac{\ln t}{\ln 2} + 1\right].$$

77. (a) $\displaystyle\int_0^4 f(t)\,dt \approx 5.67$

$\displaystyle\int_0^4 g(t)\,dt \approx 5.67$

$\displaystyle\int_0^4 h(t)\,dt \approx 5.67$

(b)

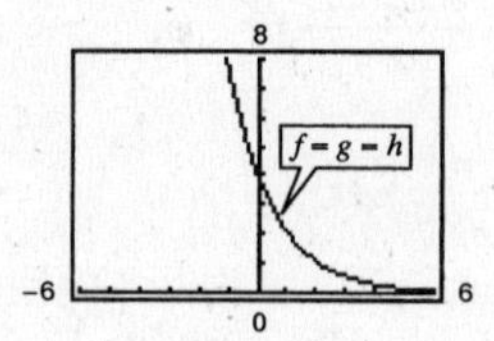

(c) The functions appear to be equal: $f(t) = g(t) = h(t)$

Analytically,

$$f(t) = 4\left(\frac{3}{8}\right)^{2t/3} = 4\left[\left(\frac{3}{8}\right)^{2/3}\right]^t = 4\left(\frac{9^{1/3}}{4}\right)^t = g(t)$$

$$h(t) = 4e^{-0.653886t} = 4\left[e^{-0.653886}\right]^t = 4(0.52002)^t$$

$$g(t) = 4\left(\frac{9^{1/3}}{4}\right)^t = 4(0.52002)^t.$$

No. The definite integrals over a given interval may be equal when the functions are not equal.

78. False

$\frac{1}{2}(\ln x) = \ln(x^{1/2}) \neq (\ln x)^{1/2}$

79. False

$\dfrac{d}{dx}[\ln x] = \dfrac{1}{x}$

80. True

$\displaystyle\int \frac{1}{x}\,dx = \ln|x| + c_1$

$= \ln|x| + \ln|c| = \ln|cx|,\ c \neq 0$

81. False; the integrand has a nonremovable discontinuity at $x = 0$.

82. $f(x) = \dfrac{x}{1 + x^2}$

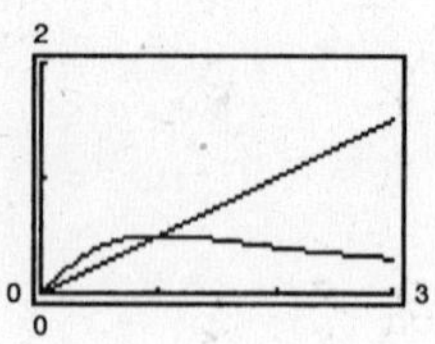

(a) $f(x) = \dfrac{x}{1 + x^2}$ and $y = \dfrac{1}{2}x$ intersect at $(0, 0)$ and $\left(1, \dfrac{1}{2}\right)$

$$A = \int_0^1 \left(\frac{x}{1 + x^2} - \frac{1}{2}x\right) dx = \left[\frac{1}{2}\ln(1 + x^2) - \frac{1}{4}x^2\right]_0^1$$

$$= \left(\frac{1}{2}\ln 2 - \frac{1}{4}\right) - 0 = \frac{1}{4}(2\ln 2 - 1)$$

$$\approx 0.097$$

—CONTINUED—

82. —CONTINUED—

(b) First m must be positive for $y = mx$ to intersect $f(x)$ in Quadrant I.

$$\frac{x}{1+x^2} = mx$$

$$x = mx(1 + x^2)$$

$$x = mx + mx^3$$

$$0 = x(m + mx^2 - 1) \Rightarrow x = 0 \quad \text{or}$$

$$mx^2 + m - 1 = 0 \Rightarrow x^2 = \frac{1-m}{m} \Rightarrow \frac{1-m}{m} \ge 0$$

Thus, we have $0 < m \le 1$.

(c) $$A = \int_0^{\sqrt{(1-m)/m}} \left(\frac{x}{1+x^2} - mx\right) dx$$

$$= \left[\frac{1}{2}\ln(1 + x^2) - \frac{1}{2}mx^2\right]_0^{\sqrt{(1-m)/m}}$$

$$= \frac{1}{2}\left[\ln\left(1 + \frac{1-m}{m}\right) - m\left(\frac{1-m}{m}\right)\right]$$

$$= \frac{1}{2}\left[\ln\left(\frac{1}{m}\right) - 1 + m\right]$$

83. $$F(x) = \int_x^{2x} \frac{1}{t}\,dt$$

$$= \Big[\ln|t|\Big]_x^{2x}$$

$= \ln(2x) - \ln(x)$ since $x > 0$ on $(0, \infty)$.

$$= \ln\left(\frac{2x}{x}\right)$$

$= \ln 2$ which is constant on the interval $(0, \infty)$.

Section 8.4 Differential Equations: Growth and Decay

1. $$\frac{dy}{dx} = x + 2$$

$$y = \int (x+2)\,dx = \frac{x^2}{2} + 2x + C$$

2. $$\frac{dy}{dx} = 4 - x$$

$$y = \int (4 - x)\,dx = 4x - \frac{x^2}{2} + C$$

3. $$\frac{dy}{dx} = y + 2$$

$$\frac{dy}{y+2} = dx$$

$$\int \frac{1}{y+2}\,dy = \int dx$$

$$\ln|y + 2| = x + C_1$$

$$y + 2 = e^{x+C_1} = Ce^x$$

$$y = Ce^x - 2$$

4. $$\frac{dy}{dx} = 4 - y$$

$$\frac{dy}{4-y} = dx$$

$$\int \frac{-1}{4-y}\,dy = \int -dx$$

$$\ln|4 - y| = -x + C_1$$

$$4 - y = e^{-x+C_1} = Ce^{-x}$$

$$y = 4 - Ce^{-x}$$

5. $$y' = \frac{5x}{y}$$

$$yy' = 5x$$

$$\int yy'\,dx = \int 5x\,dx$$

$$\int y\,dy = \int 5x\,dx$$

$$\frac{1}{2}y^2 = \frac{5}{2}x^2 + C_1$$

$$y^2 - 5x^2 = C$$

6. $$y' = \frac{\sqrt{x}}{3y}$$

$$3yy' = \sqrt{x}$$

$$\int 3yy'\,dx = \int \sqrt{x}\,dx$$

$$\frac{3y^2}{2} = \frac{2}{3}x^{3/2} + C_1$$

$$9y^2 - 4x^{3/2} = C$$

7.
$$y' = \sqrt{x}\,y$$
$$\frac{y'}{y} = \sqrt{x}$$
$$\int \frac{y'}{y}\,dx = \int \sqrt{x}\,dx$$
$$\int \frac{dy}{y} = \int \sqrt{x}\,dx$$
$$\ln y = \frac{2}{3}x^{3/2} + C_1$$
$$y = e^{(2/3)x^{3/2} + C_1}$$
$$= e^{C_1}\,e^{(2/3)x^{3/2}}$$
$$= Ce^{(2/3)x^{3/2}}$$

8.
$$y' = x(1 + y)$$
$$\frac{y'}{1 + y} = x$$
$$\int \frac{y'}{1 + y}\,dx = \int x\,dx$$
$$\int \frac{dy}{1 + y} = \int x\,dx$$
$$\ln(1 + y) = \frac{x^2}{2} + C_1$$
$$1 + y = e^{(x^2/2) + C_1}$$
$$y = e^{C_1}\,e^{x^2/2} - 1$$
$$= Ce^{x^2/2} - 1$$

9. $(1 + x^2)y' - 2xy = 0$
$$y' = \frac{2xy}{1 + x^2}$$
$$\frac{y'}{y} = \frac{2x}{1 + x^2}$$
$$\int \frac{y'}{y}\,dx = \int \frac{2x}{1 + x^2}\,dx$$
$$\int \frac{dy}{y} = \int \frac{2x}{1 + x^2}\,dx$$
$$\ln y = \ln(1 + x^2) + C_1$$
$$\ln y = \ln(1 + x^2) + \ln C$$
$$\ln y = \ln C(1 + x^2)$$
$$y = C(1 + x^2)$$

10.
$$xy + y' = 100x$$
$$y' = 100x - xy = x(100 - y)$$
$$\frac{y'}{100 - y} = x$$
$$\int \frac{y'}{100 - y}\,dx = \int x\,dx$$
$$\int \frac{1}{100 - y}\,dy = \int x\,dx$$
$$-\ln(100 - y) = \frac{x^2}{2} + C_1$$
$$\ln(100 - y) = -\frac{x^2}{2} - C_1$$
$$100 - y = e^{-(x^2/2) - C_1}$$
$$-y = e^{-C_1}\,e^{-x^2/2} - 100$$
$$y = 100 - Ce^{-x^2/2}$$

11.
$$\frac{dQ}{dt} = \frac{k}{t^2}$$
$$\int \frac{dQ}{dt}\,dt = \int \frac{k}{t^2}\,dt$$
$$\int dQ = -\frac{k}{t} + C$$
$$Q = -\frac{k}{t} + C$$

12.
$$\frac{dP}{dt} = k(10 - t)$$
$$\int \frac{dP}{dt}\,dt = \int k(10 - t)\,dt$$
$$\int dP = -\frac{k}{2}(10 - t)^2 + C$$
$$P = -\frac{k}{2}(10 - t)^2 + C$$

13.
$$\frac{dN}{ds} = k(250 - s)$$
$$\int \frac{dN}{ds}\,ds = \int k(250 - s)\,ds$$
$$\int dN = -\frac{k}{2}(250 - s)^2 + C$$
$$N = -\frac{k}{2}(250 - s)^2 + C$$

14. $$\frac{dy}{dx} = kx(L - y)$$

$$\frac{1}{L - y}\frac{dy}{dx} = kx$$

$$\int \frac{1}{L - y}\frac{dy}{dx}\,dx = \int kx\,dx$$

$$\int \frac{1}{L - y}\,dy = \frac{kx^2}{2} + C_1$$

$$-\ln(L - y) = \frac{kx^2}{2} + C_1$$

$$L - y = e^{-(kx^2/2) - C_1}$$

$$-y = -L + e^{-C_1}e^{-kx^2/2}$$

$$y = L - Ce^{-kx^2/2}$$

15. (a)

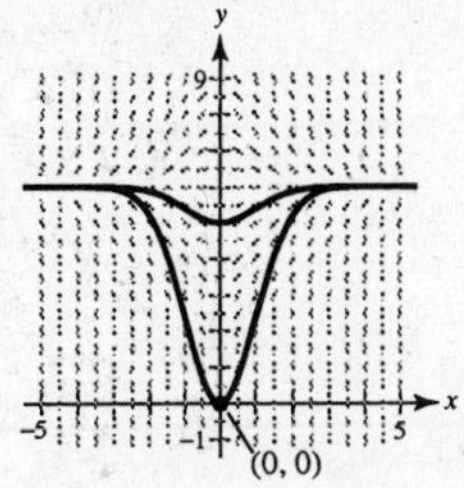

(b) $$\frac{dy}{dx} = x(6 - y), \quad (0, 0)$$

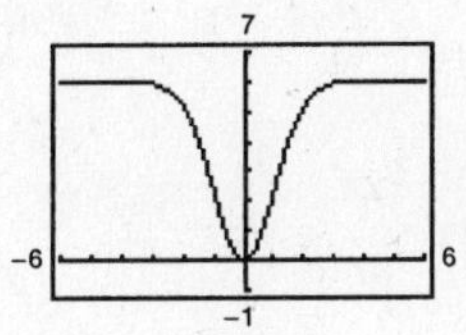

$$\frac{dy}{y - 6} = -x$$

$$\ln|y - 6| = \frac{-x^2}{2} + C$$

$$y - 6 = e^{-x^2/2 + C} = C_1e^{-x^2/2}$$

$$y = 6 + C_1e^{-x^2/2}$$

$$(0, 0):\ 0 = 6 + C_1 \Rightarrow C_1 = -6 \Rightarrow y = 6 - 6e^{-x^2/2}$$

16. (a)

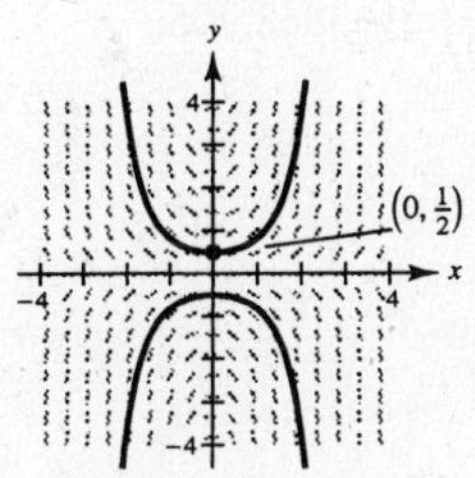

(b) $$\frac{dy}{dx} = xy, \quad \left(0, \frac{1}{2}\right)$$

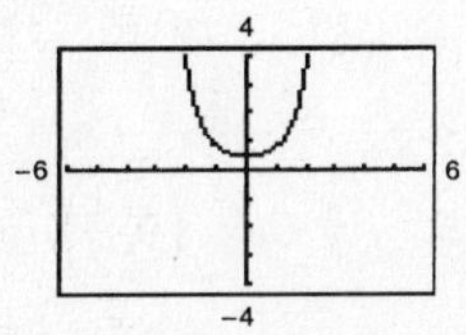

$$\frac{dy}{y} = x\,dx$$

$$\ln|y| = \frac{x^2}{2} + C$$

$$y = e^{x^2/2 + C} = C_1e^{x^2/2}$$

$$\left(0, \frac{1}{2}\right):\ \frac{1}{2} = C_1e^0 \Rightarrow C_1 = \frac{1}{2} \Rightarrow y = \frac{1}{2}e^{x^2/2}$$

17. $$\frac{dy}{dt} = \frac{1}{2}t, \quad (0, 10)$$

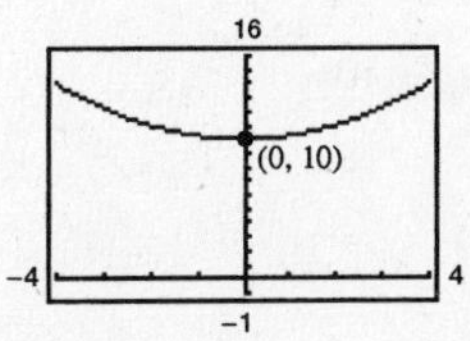

$$\int dy = \int \frac{1}{2}t\,dt$$

$$y = \frac{1}{4}t^2 + C$$

$$10 = \frac{1}{4}(0)^2 + C \Rightarrow C = 10$$

$$y = \frac{1}{4}t^2 + 10$$

18. $$\frac{dy}{dt} = -\frac{3}{4}\sqrt{t}, \quad (0, 10)$$

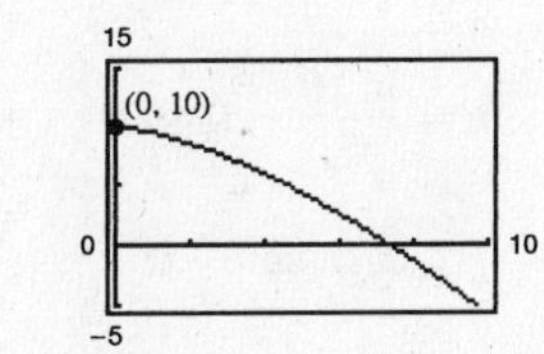

$$\int dy = \int -\frac{3}{4}\sqrt{t}\,dt$$

$$y = -\frac{1}{2}t^{3/2} + C$$

$$10 = -\frac{1}{2}(0)^{3/2} + C \Rightarrow C = 10$$

$$y = -\frac{1}{2}t^{3/2} + 10$$

19. $$\frac{dy}{dt} = -\frac{1}{2}y, \quad (0, 10)$$

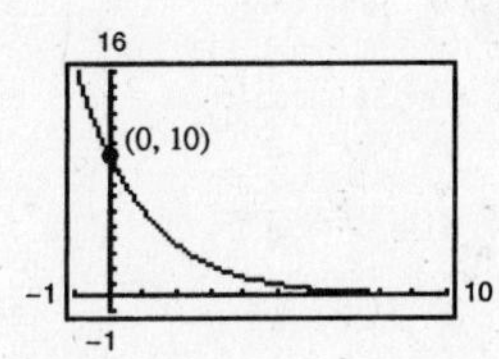

$$\int \frac{dy}{y} = \int -\frac{1}{2}\,dt$$

$$\ln y = -\frac{1}{2}t + C_1$$

$$y = e^{-(t/2) + C_1} = e^{C_1}e^{-t/2} = Ce^{-t/2}$$

$$10 = Ce^0 \Rightarrow C = 10$$

$$y = 10e^{-t/2}$$

20. $$\frac{dy}{dt} = \frac{3}{4}y, \quad (0, 10)$$

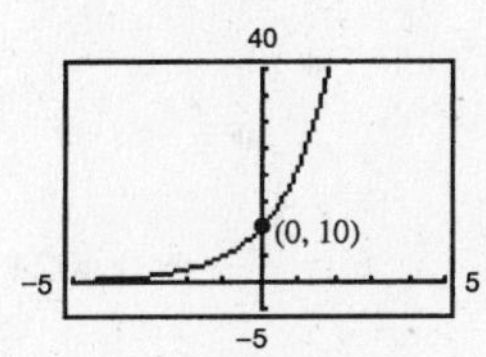

$$\int \frac{dy}{y} = \int \frac{3}{4}\,dt$$

$$\ln y = \frac{3}{4}t + C_1$$

$$y = e^{(3/4)t + C_1}$$

$$= e^{C_1}e^{(3/4)t} = Ce^{3t/4}$$

$$10 = Ce^0 \Rightarrow C = 10$$

$$y = 10e^{3t/4}$$

21. $\frac{dy}{dx} = ky$

$y = Ce^{kx}$ (Theorem 8.8)

$(0, 4)$: $4 = Ce^0 = C$

$(3, 10)$: $10 = 4e^{3k} \Rightarrow k = \frac{1}{3}\ln\left(\frac{5}{2}\right)$

When $x = 6$, $y = 4e^{1/3\ln(5/2)(6)} = 4e^{\ln(5/2)^2}$

$= 4\left(\frac{5}{2}\right)^2 = 25.$

22. $\frac{dN}{dt} = kN$

$N = Ce^{kt}$ (Theorem 8.8)

$(0, 250)$: $C = 250$

$(1, 400)$: $400 = 250e^k \Rightarrow k = \ln\frac{400}{250} = \ln\frac{8}{5}$

When $t = 4$, $N = 250e^{4\ln(8/5)} = 250e^{\ln(8/5)^4}$

$= 250\left(\frac{8}{5}\right)^4 = \frac{8192}{5}.$

23. $\frac{dV}{dt} = kV$

$V = Ce^{kt}$ (Theorem 8.8)

$(0, 20{,}000)$: $C = 20{,}000$

$(4, 12{,}500)$: $12{,}500 = 20{,}000e^{4k} \Rightarrow k = \frac{1}{4}\ln\left(\frac{5}{8}\right)$

When $t = 6$, $V = 20{,}000e^{1/4\ln(5/8)(6)} = 20{,}000e^{\ln(5/8)^{3/2}}$

$= 20{,}000\left(\frac{5}{8}\right)^{3/2} \approx 9882.118.$

24. $\frac{dP}{dt} = kP$

$P = Ce^{kt}$ (Theorem 8.8)

$(0, 5000)$: $C = 5000$

$(1, 4750)$: $4750 = 5000e^k \Rightarrow k = \ln\left(\frac{19}{20}\right)$

When $t = 5$, $P = 5000e^{\ln(19/20)(5)}$

$= 5000\left(\frac{19}{20}\right)^5 \approx 3868.905.$

25. $y = Ce^{kt}$, $\left(0, \frac{1}{2}\right)$, $(5, 5)$

$C = \frac{1}{2}$

$y = \frac{1}{2}e^{kt}$

$5 = \frac{1}{2}e^{5k}$

$k = \frac{\ln 10}{5} \approx 0.4605$

$y = \frac{1}{2}e^{0.4605t}$

26. $y = Ce^{kt}$, $(0, 4)$, $\left(5, \frac{1}{2}\right)$

$C = 4$

$y = 4e^{kt}$

$\frac{1}{2} = 4e^{5k}$

$k = \frac{\ln(1/8)}{5} \approx -0.4159$

$y = 4e^{-0.4159t}$

27. $y = Ce^{kt}$, $(1, 1)$, $(5, 5)$

$1 = Ce^k$

$5 = Ce^{5k}$

$5Ce^k = Ce^{5k}$

$5e^k = e^{5k}$

$5 = e^{4k}$

$k = \frac{\ln 5}{4} \approx 0.4024$

$y = Ce^{0.4024t}$

$1 = Ce^{0.4024}$

$C \approx 0.6687$

$y = 0.6687e^{0.4024t}$

28. $y = Ce^{kt}$, $\left(3, \frac{1}{2}\right)$, $(4, 5)$

$\frac{1}{2} = Ce^{3k}$

$5 = Ce^{4k}$

$2Ce^{3k} = \frac{1}{5}Ce^{4k}$

$10e^{3k} = e^{4k}$

$10 = e^k$

$k = \ln 10 \approx 2.3026$

$y = Ce^{2.3026t}$

$5 = Ce^{2.3026(4)}$

$C \approx 0.0005$

$y = 0.0005e^{2.3026t}$

29. $\dfrac{dy}{dx} = \dfrac{1}{2}xy$

$\dfrac{dy}{dx} > 0$ when $xy > 0$. Quadrants I and III.

30. $\dfrac{dy}{dx} = \dfrac{1}{2}x^2y$

$\dfrac{dy}{dx} > 0$ when $y > 0$. Quadrants I and II.

31. $\dfrac{1}{2}C = Ce^{k(1599)}$

$0.5 = e^{k(1599)}$

$\ln 0.5 = k(1599)$

$k = \dfrac{\ln 0.5}{1599}$

Given $C = 20$, after 1000 years we have:

$y = 20e^{[(\ln 0.5)/1599](1000)} \approx 12.96$ grams,

After 10,000 years we have:

$y \approx 20e^{[(\ln 0.5)/1599](10{,}000)} \approx 0.26$ grams.

32. $\dfrac{1}{2}C = Ce^{k(1599)}$

$0.5 = e^{k(1599)}$

$\ln 0.5 = k(1599)$

$k = \dfrac{\ln 0.5}{1599}$

Given $y = 3$ grams after 1000 years, we have:

$3 = Ce^{[(\ln 0.5)/1599](1000)} \implies C \approx 4.63$ grams.

After 10,000 years we have:

$y \approx 4.63e^{[(\ln 0.5)/1599](10{,}000)} \approx 0.60$ gram.

33. $\dfrac{1}{2}C = Ce^{k(5715)}$

$0.5 = e^{k(5715)}$

$\ln 0.5 = k(5715)$

$k = \dfrac{\ln 0.5}{5715}$

Given $y = 2$ grams after 10,000 years, we have:

$2 = Ce^{[(\ln 0.5)/5715](10{,}000)} \implies C \approx 6.73$ grams.

After 1000 years we have:

$y \approx 6.73e^{[(\ln 0.5)/5715](1000)} \approx 5.96$ grams.

34. $\dfrac{1}{2}C = Ce^{k(5715)}$

$0.5 = e^{k(5715)}$

$\ln 0.5 = k(5715)$

$k = \dfrac{\ln 0.5}{5715}$

Since $C = 5$ grams, after 1000 years we have:

$y = 5e^{[(\ln 0.5)/5715](1000)} \approx 4.43$ grams,

After 10,000 years we have:

$y \approx 5e^{[(\ln 0.5)/5715](10{,}000)} \approx 1.49$ grams.

35. $\dfrac{1}{2}C = Ce^{k(24{,}100)}$

$0.5 = e^{k(24{,}100)}$

$\ln 0.5 = k(24{,}100)$

$k = \dfrac{\ln 0.5}{24{,}100}$

Given $y = 2.5$ grams after 1000 years, we have:

$2.5 = Ce^{[(\ln 0.5)/24{,}100](1000)} \implies C \approx 2.57$ grams.

After 10,000 years we have:

$y \approx 2.57e^{[(\ln 0.5)/24{,}100](10{,}000)} \approx 1.93$ grams.

36. $\dfrac{1}{2}C = Ce^{k(24{,}100)}$

$0.5 = e^{k(24{,}100)}$

$\ln 0.5 = k(24{,}100)$

$k = \dfrac{\ln 0.5}{24{,}100}$

Since $y = 0.4$ grams after 10,000 years, we have:

$0.4 = Ce^{[(\ln 0.5)/24{,}100](10{,}000)} \implies C \approx 0.53$ gram.

After 1000 years we have:

$y \approx 0.53e^{[(\ln 0.5)/24{,}100](1000)} \approx 0.51$ gram.

37. (a) $7.7 = Ce^{-0.009(1)}$

$C = \dfrac{7.7}{e^{-0.009}} \approx 7.77$

$P = 7.77e^{-0.009t}$

(b) $P = 7.77e^{-0.009(15)} \approx 6.79$ million people.

38. (a) $12.7 = Ce^{0.018(1)}$

$C = \dfrac{12.7}{e^{0.018}} \approx 12.47$

$P = 12.47e^{0.018t}$

(b) $P = 12.47e^{0.018(15)} \approx 16.34$ million people.

39. (a) $5.2 = Ce^{0.026(1)}$

$C = \dfrac{5.2}{e^{0.026}} \approx 5.07$

$P = 5.07e^{0.026t}$

(b) $P = 5.07e^{0.026(15)} \approx 7.49$ million people.

40. (a) $3.6 = Ce^{-0.002(1)}$

$C = \dfrac{3.6}{e^{-0.002}} \approx 3.61$

$P = 3.61e^{-0.002t}$

(b) $P = 3.61e^{-0.002(15)} \approx 3.50$ million people.

41. $S = Ce^{k/t}$

(a) $S = 5$ when $t = 1$.

$5 = Ce^{k}$

$\lim_{t \to \infty} Ce^{k/t} = C = 30$

$5 = 30e^{k}$

$k = \ln \frac{1}{6} \approx -1.7918$

$S \approx 30e^{-1.7918/t}$

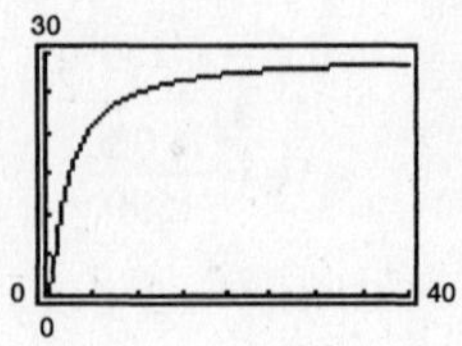

(b) When $t = 5$, $S \approx 20.9646$ which is 20,965 units.

42. $S = 25(1 - e^{kt})$

(a) $4 = 25(1 - e^{k(1)}) \Rightarrow 1 - e^{k} = \dfrac{4}{25} \Rightarrow e^{k} = \dfrac{21}{25} \Rightarrow k = \ln\left(\dfrac{21}{25}\right) \approx -0.1744$

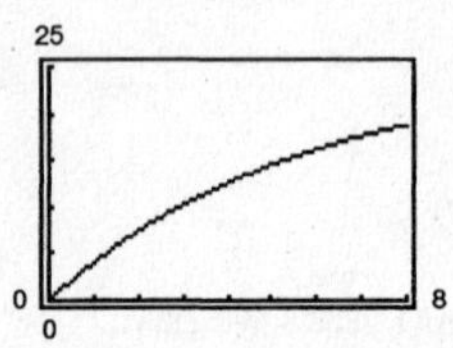

(b) 25,000 units $\left(\lim_{t \to \infty} S = 25\right)$

(c) When $t = 5$, $S \approx 14.547$ which is 14,547 units.

43. $R = \dfrac{\ln I - 0}{\ln 10}$, $I = e^{R \ln 10} = 10^{R}$

(a) $8.3 = \dfrac{\ln I - 0}{\ln 10}$

$I = 10^{8.3} \approx 199{,}526{,}231.5$

(b) $2R = \dfrac{\ln I - 0}{\ln 10}$

$I = e^{2R \ln 10} = e^{2R \ln 10} = (e^{R \ln 10})^2 = (10^{R})^2$

Increases by a factor of $e^{R \ln 10}$ or 10^{R}.

(c) $\dfrac{dR}{dI} = \dfrac{1}{I \ln 10}$

44. Since $\dfrac{dy}{dt} = k(y - 80)$

$\displaystyle\int \frac{1}{y - 80}\, dy = \int k\, dt$

$\ln(y - 80) = kt + C.$

When $t = 0$, $y = 1500$. Thus, $C = \ln 1420$.

When $t = 1$, $y = 1120$. Thus,

$k(1) + \ln 1420 = \ln(1120 - 80)$

$k = \ln 1040 - \ln 1420 = \ln \dfrac{104}{142}.$

Thus, $y = 1420e^{[\ln(104/142)]t} + 80$.

When $t = 5$, $y \approx 379.2°$.

45. $A(t) = V(t)e^{-0.10t} = 100{,}000e^{0.8\sqrt{t}}\, e^{-0.10t} = 100{,}000e^{0.8\sqrt{t} - 0.10t}$

$\dfrac{dA}{dt} = 100{,}000\left(\dfrac{0.4}{\sqrt{t}} - 0.10\right)e^{0.8\sqrt{t} - 0.10t} = 0$ when 16.

The timber should be harvested in the year 2014, (1998 + 16). **Note:** You could also use a graphing utility to graph $A(t)$ and find the maximum of $A(t)$. Use the viewing rectangle $0 \le x \le 30$ and $0 \le y \le 600{,}000$.

46. (a) Receipts: $R = 941.6088(1.0756)^t \approx 941.6088e^{0.0729t}$

Interest: $I = 0.1416t^4 - 3.9288t^3 + 36.5993t^2 - 120.8152t + 417.0364$

(b)

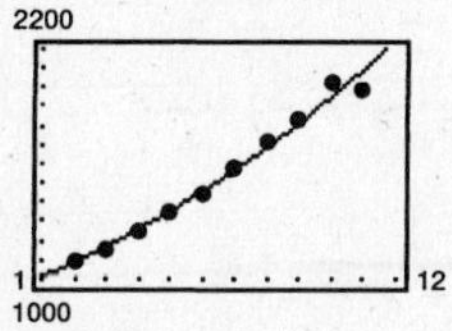

$1.0756 = e^k \Rightarrow k = \ln(1.0756) \approx 0.0729$

(c)

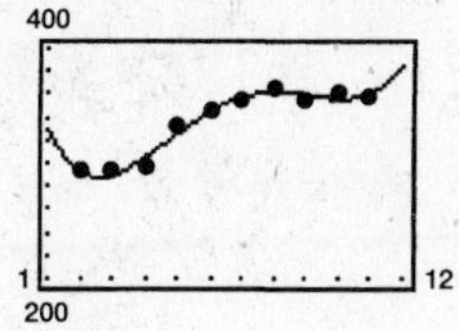

(d) Let $P = \dfrac{I}{R} \times 100.$

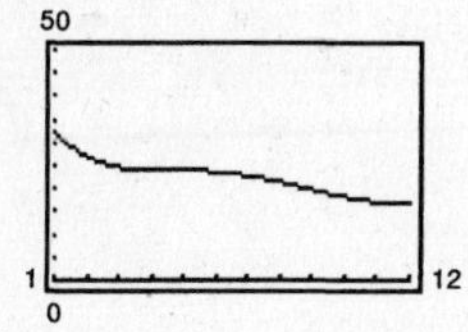

47. False. If $y = Ce^{kt}$, $y' = Cke^{kt} \neq$ constant.

48. True

49. True

50. True, if $k > 0$.

Review Exercises for Chapter 8

1. $y = e^{-x/2}$

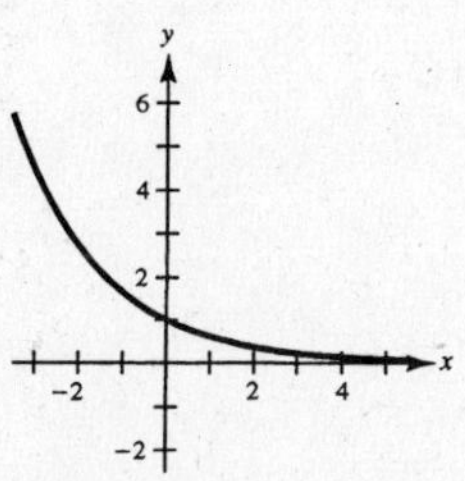

2. $y = 4e^{-x^2}$

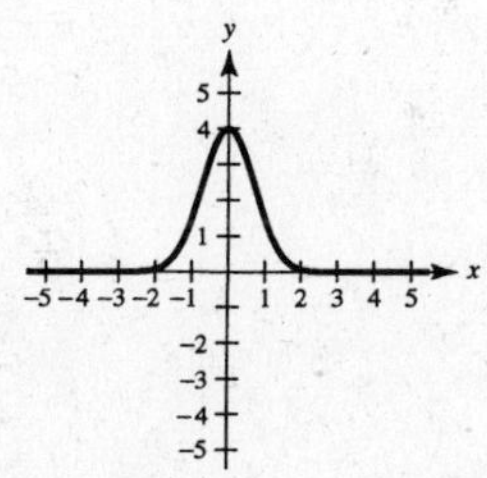

3. $f(x) = e^{-x^3}$

$f'(x) = e^{-x^3}(-3x^2) = -3x^2e^{-x^3}$

4. $g(x) = xe^x$

$g'(x) = xe^x + e^x = (x + 1)e^x$

5. $g(t) = t^2e^t$

$g'(x) = t^2e^t + 2te^t = te^t(t + 2)$

6. $h(z) = e^{-z^2/2}$

$h'(z) = -ze^{-z^2/2}$

7. $y = \sqrt{e^{2x} + e^{-2x}}$

$y' = \dfrac{1}{2}(e^{2x} + e^{-2x})^{-1/2}(2e^{2x} - 2e^{-2x})$

$= \dfrac{e^{2x} - e^{-2x}}{\sqrt{e^{2x} + e^{-2x}}}$

8. $y = 3e^{-3/t}$

$y' = 3e^{-3/t}(3t^{-2}) = \dfrac{9e^{-3/t}}{t^2}$

9. $g(x) = \dfrac{x^2}{e^x}$

$g'(x) = \dfrac{e^x(2x) - x^2e^x}{e^{2x}} = \dfrac{x(2 - x)}{e^x}$

10. $f(t) = \dfrac{e^t}{1 + e^t}$

$f'(t) = \dfrac{(1 + e^t)e^t - e^t(e^t)}{(1 + e^t)^2}$

$= \dfrac{e^t}{(1 + e^t)^2}$

11. $e^x + y^2 = 0$

$e^x + 2yy' = 0$

$y' = \dfrac{-e^x}{2y}$

12. $x + y = xe^y$

$1 + y' = xe^y y' + e^y$

$y' - xe^y y' = e^y - 1$

$y' = \dfrac{e^y - 1}{1 - xe^y}$

13. Let $u = -3x^2$, $du = -6x\,dx$.

$\displaystyle\int xe^{-3x^2}\,dx = -\frac{1}{6}\int e^{-3x^2}(-6x)\,dx = -\frac{1}{6}e^{-3x^2} + C$

14. Let $u = \dfrac{1}{x}$, $du = \dfrac{-1}{x^2}\,dx$.

$\displaystyle\int \frac{e^{1/x}}{x^2}\,dx = -\int e^{1/x}\left(-\frac{1}{x^2}\right)dx = -e^{1/x} + C$

15. $\int \frac{e^{4x} - e^{2x} + 1}{e^x}\,dx = \int (e^{3x} - e^x + e^{-x})\,dx$

$= \frac{1}{3}e^{3x} - e^x - e^{-x} + C$

$= \frac{e^{4x} - 3e^{2x} - 3}{3e^x} + C$

16. Let $u = e^{2x} + e^{-2x}$, $du = 2(e^{2x} - e^{-2x})\,dx$.

$\int \frac{e^{2x} - e^{-2x}}{(e^{2x} + e^{-2x})^2}\,dx = \frac{1}{2}\int (e^{2x} + e^{-2x})^{-2}\,2(e^{2x} - e^{-2x})\,dx$

$= \frac{-1}{2}(e^{2x} + e^{-2x})^{-1} + C$

$= \frac{-1}{2(e^{2x} + e^{-2x})} + C$

17. $\int xe^{1-x^2}\,dx = -\frac{1}{2}\int e^{1-x^2}(-2x)\,dx$

$= -\frac{1}{2}e^{1-x^2} + C$

18. Let $u = x^3 + 1$, $du = 3x^2\,dx$.

$\int x^2e^{x^3+1}\,dx = \frac{1}{3}\int e^{x^3+1}(3x^2)\,dx = \frac{1}{3}e^{x^3+1} + C$

19. Let $u = e^x - 1$, $du = e^x\,dx$.

$\int \frac{e^x}{(e^x - 1)^{3/2}}\,dx = \int (e^x - 1)^{-3/2}\,e^x\,dx$

$= -2(e^x - 1)^{-1/2} + C$

$= \frac{-2}{\sqrt{e^x - 1}} + C$

20. Let $u = e^{2x} + 1$, $du = 2e^{2x}\,dx$.

$\int \frac{e^{2x}}{\sqrt{e^{2x} + 1}}\,dx = \frac{1}{2}\int (e^{2x} + 1)^{-1/2}\,2e^{2x}\,dx$

$= \frac{1}{2}\frac{(e^{2x} + 1)^{1/2}}{1/2} + C$

$= \sqrt{e^{2x} + 1} + C$

21.

$$y = 5e^{2x} - 12e^{3x}$$
$$y' = 10e^{2x} - 36e^{3x}$$
$$y'' = 20e^{2x} - 108e^{3x}$$
$$y'' - 5y' + 6y = (20e^{2x} - 108e^{3x}) - 5(10e^{2x} - 36e^{3x}) + 6(5e^{2x} - 12e^{3x})$$
$$= 0$$

22. (a), (c)

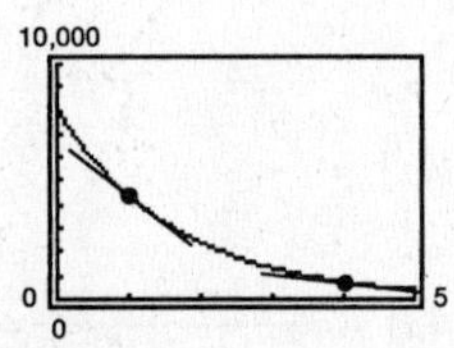

(b) $V = 8000e^{-0.6t},\ 0 \le t \le 5$

$V'(t) = -4800e^{-0.6t}$

$V'(1) = -2634.3$ dollars/year

$V'(4) = -435.4$ dollars/year

23. Area $= \int_0^4 xe^{-x^2}\,dx = \left[-\frac{1}{2}e^{-x^2}\right]_0^4$

$= -\frac{1}{2}(e^{-16} - 1) \approx 0.500$

24. Area $= \int_0^2 2e^{-x}\,dx = \left[-2e^{-x}\right]_0^2$

$= -2e^{-2} + 2 = 2 - \frac{2}{e^2} \approx 1.729$

25. $g(x) = \ln\sqrt{x} = \frac{1}{2}\ln x$

$g'(x) = \frac{1}{2x}$

26. $h(x) = \ln\frac{x(x-1)}{x-2} = \ln x + \ln(x-1) - \ln(x-2)$

$h'(x) = \frac{1}{x} + \frac{1}{x-1} - \frac{1}{x-2} = \frac{x^2 - 4x + 2}{x^3 - 3x^2 + 2x}$

27. $f(x) = x\sqrt{\ln x}$

$f'(x) = \left(\frac{x}{2}\right)(\ln x)^{-1/2}\left(\frac{1}{x}\right) + \sqrt{\ln x}$

$= \frac{1}{2\sqrt{\ln x}} + \sqrt{\ln x} = \frac{1 + 2\ln x}{2\sqrt{\ln x}}$

28. $f(x) = \ln[x(x^2 - 2)^{2/3}] = \ln x + \frac{2}{3}\ln(x^2 - 2)$

$f'(x) = \frac{1}{x} + \frac{2}{3}\left(\frac{2x}{x^2 - 2}\right) = \frac{7x^2 - 6}{3x^3 - 6x}$

29. $y = \frac{1}{b^2}\left[\ln(a + bx) + \frac{a}{a + bx}\right]$

$\frac{dy}{dx} = \frac{1}{b^2}\left[\frac{b}{a + bx} - \frac{ab}{(a + bx)^2}\right] = \frac{x}{(a + bx)^2}$

30. $y = \frac{1}{b^2}[a + bx - a\ln(a + bx)]$

$\frac{dy}{dx} = \frac{1}{b^2}\left(b - \frac{ab}{a + bx}\right) = \frac{x}{a + bx}$

31. $y = -\frac{1}{a}\ln\left(\frac{a + bx}{x}\right) = -\frac{1}{a}[\ln(a + bx) - \ln x]$

$\frac{dy}{dx} = -\frac{1}{a}\left(\frac{b}{a + bx} - \frac{1}{x}\right) = \frac{1}{x(a + bx)}$

32. $y = -\frac{1}{ax} + \frac{b}{a^2}\ln\frac{a + bx}{x}$

$= -\frac{1}{ax} + \frac{b}{a^2}[\ln(a + bx) - \ln x]$

$\frac{dy}{dx} = -\frac{1}{a}\left(-\frac{1}{x^2}\right) + \frac{b}{a^2}\left[\frac{b}{a + bx} - \frac{1}{x}\right]$

$= \frac{1}{ax^2} + \frac{b}{a^2}\left[\frac{-a}{x(a + bx)}\right] = \frac{1}{ax^2} - \frac{b}{ax(a + bx)}$

$= \frac{(a + bx) - bx}{ax^2(a + bx)} = \frac{1}{x^2(a + bx)}$

33. $f(x) = 3^{x-1}$

$f'(x) = 3^{x-1}\ln 3$

34. $f(x) = 4^x e^x$

$f'(x) = 4^x e^x + (\ln 4)4^x e^x = 4^x e^x(1 + \ln 4)$

35. $y = x^{2x+1}$

$\ln y = (2x + 1)\ln x$

$\frac{y'}{y} = \frac{2x + 1}{x} + 2\ln x$

$y' = y\left(\frac{2x + 1}{x} + 2\ln x\right) = x^{2x+1}\left(\frac{2x + 1}{x} + 2\ln x\right)$

36. $y = x(4^{-x})$

$y' = 4^{-x} - x \cdot 4^{-x}\ln 4$

37. $g(x) = \log_3\sqrt{1 - x} = \frac{1}{2}\log_3(1 - x)$

$g'(x) = \frac{1}{2}\frac{-1}{(1 - x)\ln 3} = \frac{1}{2(x - 1)\ln 3}$

38. $h(x) = \log_5\frac{x}{x - 1} = \log_5 x - \log_5(x - 1)$

$h'(x) = \frac{1}{\ln 5}\left[\frac{1}{x} - \frac{1}{x - 1}\right] = \frac{1}{\ln 5}\left[\frac{-1}{x(x - 1)}\right]$

39. $\ln x + y^2 = 0$

$\frac{1}{x} + 2yy' = 0$

$2yy' = \frac{-1}{x}$

$y' = \frac{-1}{2xy}$

40. $x\ln y - 3xy = 4$

$x\frac{y'}{y} + \ln y - 3xy' - 3y = 0$

$\left(\frac{x}{y} - 3x\right)y' = 3y - \ln y$

$y' = \frac{3y - \ln y}{x/y - 3x}$

$= \frac{3y^2 - y\ln y}{x - 3xy}$

41. $y = \left(\dfrac{6x}{x^2+1}\right)^{1/2}$

$$\ln y = \frac{1}{2}[\ln(6x) - \ln(x^2+1)]$$

$$\frac{y'}{y} = \frac{1}{2}\left[\frac{1}{x} - \frac{2x}{x^2+1}\right]$$

$$y' = \frac{1}{2}\left(\frac{6x}{x^2+1}\right)^{1/2}\left(\frac{x^2+1-2x^2}{x(x^2+1)}\right)$$

$$= \frac{\sqrt{6}(1-x^2)}{2\sqrt{x}(x^2+1)^{3/2}}$$

42. $y = [x(x+4)(x-3)]^{1/2}$

$$\ln y = \frac{1}{2}[\ln x + \ln(x+4) + \ln(x-3)]$$

$$\frac{y'}{y} = \frac{1}{2}\left[\frac{1}{x} + \frac{1}{x+4} + \frac{1}{x-3}\right]$$

$$y' = \frac{\sqrt{x(x+4)(x-3)}}{2}\left(\frac{1}{x} + \frac{1}{x+4} + \frac{1}{x-3}\right)$$

43. $y \ln|1-x| = 1$

$$y\left(\frac{-1}{1-x}\right) + y'\ln|1-x| = 0$$

$$\frac{-y}{1-x} + y'\left(\frac{1}{y}\right) = 0$$

$$y' = \frac{y^2}{1-x}$$

44. (a) $y = x^a$

$y' = ax^{a-1}$

(b) $y = a^x$

$y' = (\ln a)a^x$

(c) $y = x^x$

$\ln y = x \ln x$

$\dfrac{1}{y}y' = x \cdot \dfrac{1}{x} + (1)\ln x$

$y' = y(1 + \ln x)$

$y' = x^x(1 + \ln x)$

(d) $y = a^a$

$y' = 0$

45.

h	0	5	10	15	20
P	10,332	5583	2376	1240	517
$\ln P$	9.243	8.627	7.773	7.123	6.248

(a)

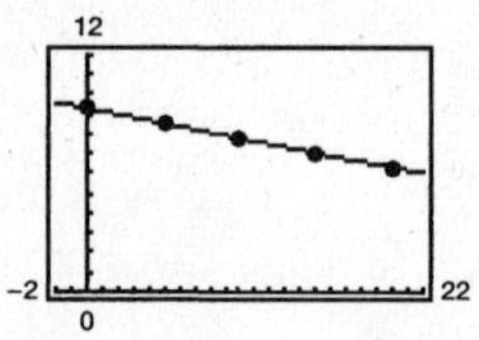

$y = -0.1499h + 9.3018$ is the regression line for data $(h, \ln P)$.

(b) $\ln P = ah + b$

$P = e^{ah+b} = e^b e^{ah}$

$P = Ce^{ah}, C = e^b$

For our data, $a = -0.1499$ and $C = e^{9.3018} = 10{,}957.7$.

$P = 10{,}957.7e^{-0.1499h}$

(c)

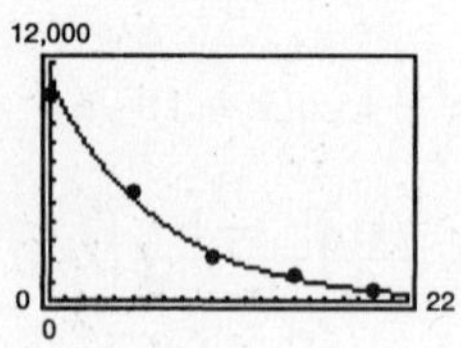

(d) $\dfrac{dP}{dh} = (10{,}957.7)(-0.1499)e^{-0.1499h}$

$= -1642.56e^{-0.1499h}$

For $h = 5$, $\dfrac{dP}{dh} = -776.3$. For $h = 18$, $\dfrac{dP}{dh} \approx -110.6$.

46. (a)

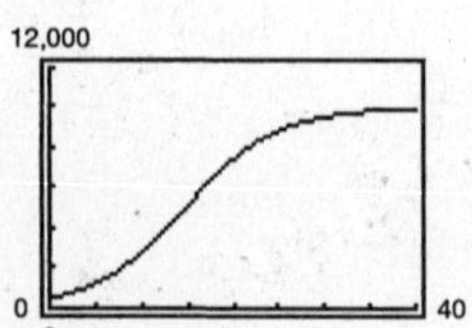

(b) Limiting size: 10,000 fish

(c) $p(t) = \dfrac{10{,}000}{1 + 19e^{-t/5}}$

$$p'(t) = \frac{e^{-t/5}}{(1+19e^{-t/5})^2}\left(\frac{19}{5}\right)(10{,}000)$$

$$= \frac{38{,}000e^{-t/5}}{(1+19e^{-t/5})^2}$$

$p'(1) \approx 113.5$ fish/month

$p'(10) \approx 403.2$ fish/month

(d) $p''(t) = -\dfrac{38{,}000}{5}(e^{-t/5})\left[\dfrac{1 - 19e^{-t/5}}{(1+19e^{-t/5})^3}\right] = 0$

$19e^{-t/5} = 1$

$\dfrac{t}{5} = \ln 19$

$t = 5 \ln 19 \approx 14.72$

47. $u = 7x - 2, du = 7dx$

$$\int \frac{1}{7x-2}\,dx = \frac{1}{7}\int \frac{1}{7x-2}(7)\,dx = \frac{1}{7}\ln|7x-2| + C$$

48. $u = x^2 - 1, du = 2x\,dx$

$$\int \frac{x}{x^2-1}\,dx = \frac{1}{2}\int \frac{2x}{x^2-1}\,dx = \frac{1}{2}\ln|x^2-1| + C$$

49. $$\int_1^4 \frac{x+1}{x}\,dx = \int_1^4 \left(1 + \frac{1}{x}\right)dx = \Big[x + \ln|x|\Big]_1^4 = 3 + \ln 4$$

50. $$\int_1^e \frac{\ln x}{x}\,dx = \int_1^e (\ln x)^1\left(\frac{1}{x}\right)dx = \left[\frac{1}{2}(\ln x)^2\right]_1^e = \frac{1}{2}$$

51. Let $u = e^{2x} + e^{-2x}, du = (2e^{2x} - e^{-2x})\,dx.$

$$\int \frac{e^{2x} - e^{-2x}}{e^{2x} + e^{-2x}}\,dx = \frac{1}{2}\int \frac{2e^{2x} - 2e^{-2x}}{e^{2x} + e^{-2x}}\,dx$$

$$= \frac{1}{2}\ln(e^{2x} + e^{-2x}) + C$$

52. $u = \ln x, du = \frac{1}{x}\,dx$

$$\int \frac{\ln\sqrt{x}}{x}\,dx = \frac{1}{2}\int (\ln x)\left(\frac{1}{x}\right)dx = \frac{1}{4}(\ln x)^2 + C$$

53. Let $u = e^x - 1, du = e^x\,dx.$

$$\int \frac{e^x}{e^x - 1}\,dx = \ln|e^x - 1| + C$$

54. $$\int \frac{e^{2x}}{e^{2x}+1}\,dx = \frac{1}{2}\int \frac{1}{e^{2x}+1}2e^{2x}\,dx$$

$$= \frac{1}{2}\ln(e^{2x} + 1) + C$$

55. $$\int (x+1)5^{(x+1)^2}\,dx = \frac{1}{2}\frac{1}{\ln 5}5^{(x+1)^2} + C$$

56. $$\int \frac{2^{-1/t}}{t^2}\,dt = \frac{1}{\ln 2}2^{-1/t} + C$$

57. $$\frac{dy}{dx} = \frac{x^2+3}{x}$$

$$\int dy = \int \left(x + \frac{3}{x}\right)dx$$

$$y = \frac{x^2}{2} + 3\ln|x| + C$$

58. $$\frac{dy}{dx} = \frac{e^{-2x}}{1 + e^{-2x}}$$

$$\int dy = \int \frac{e^{-2x}}{1+e^{-2x}}\,dx = -\frac{1}{2}\int \frac{-2e^{-2x}}{1+e^{-2x}}\,dx$$

$$y = -\frac{1}{2}\ln(1 + e^{-2x}) + C$$

59. (a) $$P = \frac{1}{100}\left(25 + \int_{2.5}^{10}\frac{25}{x}\,dx\right) = \frac{1}{100}\left(25 + \Big[25\ln x\Big]_{2.5}^{10}\right) = \frac{1}{4}(1 + \ln 4) \approx 0.60$$

(b) $$P = \frac{1}{100}\left(50 + \int_5^{10}\frac{50}{x}\,dx\right) = \frac{1}{100}\left(50 + \Big[50\ln x\Big]_5^{10}\right) = \frac{1}{2}(1 + \ln 2) \approx 0.85$$

60. $$\text{Average value} = \frac{1}{b-a}\int_a^b f(x)\,dx$$

$$= \frac{1}{10-5}\int_5^{10}\frac{1}{x-1}\,dx = \frac{1}{5}\Big[\ln|x-1|\Big]_5^{10}$$

$$= \frac{1}{5}(\ln 9 - \ln 4) = \frac{1}{5}\ln\frac{9}{4}$$

61. $$P(h) = 30e^{kh}$$

$$P(18{,}000) = 30e^{18{,}000k} = 15$$

$$k = \frac{\ln(1/2)}{18{,}000} = \frac{-\ln 2}{18{,}000}$$

$$P(h) = 30e^{-(h\ln 2)/18{,}000}$$

$$P(35{,}000) = 30e^{-(35{,}000\ln 2)/18{,}000} \approx 7.79 \text{ inches}$$

62.

$$y = Ce^{kt}$$

$$\frac{1}{2}C = Ce^{k(1599)}$$

$$\frac{1}{2} = e^{k(1599)}$$

$$\ln\left(\frac{1}{2}\right) = k(1599)$$

$$k = \frac{\ln(1/2)}{1599}$$

$$y = Ce^{[\ln(1/2)/1599]t}$$

When $C = 5$ and $t = 600$:

$$y = 5e^{[\ln(1/2)/1599](600)} \approx 3.855 \text{ grams}$$

63.

$$P = Ce^{0.015t}$$

$$2C = Ce^{0.015t}$$

$$2 = e^{0.015t}$$

$$\ln 2 = 0.015t$$

$$t = \frac{\ln 2}{0.015} \approx 46.21 \text{ years}$$

64. (a)

$$\frac{dy}{ds} = -0.012y,\ s > 50$$

$$\frac{-1}{0.012}\int\frac{dy}{y} = \int ds$$

$$\frac{-1}{0.012}\ln y = s + C_1$$

$$y = Ce^{-0.012s}$$

When $s = 50$, $y = 28 = Ce^{-0.012(50)} \Rightarrow C = 28e^{0.6}$.

$$y = 28e^{0.6-0.012s},\ s > 50$$

(b)

Speed(s)	50	55	60	65	70
Miles per Gallon (y)	28	26.4	24.8	23.4	22.0

Problem Solving for Chapter 8

1. $y = e^{-x}$, $y' = -e^{-x} \Rightarrow y - b = -e^{-a}(x - a)$ is equation of tangent line.

Simplifying

$$y = -e^{-a}x + b + ae^{-a}$$

$$y = -e^{-a}x + e^{-a} + ae^{-a}$$

Let $x = 0$: $y = e^{-a}(1 + a)$. y-intercept is $R = (0, (1 + a)e^{-a})$.

Let $y = 0$: $-x + 1 + a = 0$. x-intercept is $Q = (a + 1, 0)$.

Area of $\Delta ORQ = A(a) = \frac{1}{2}(a + 1)(1 + a)e^{-a} = \frac{1}{2}e^{-a}(1 + a)^2$

$$A'(a) = \frac{1}{2}e^{-a}2(1 + a) - \frac{1}{2}e^{-a}(1 + a)^2 = 0$$

$$\frac{1}{2}e^{-a}2(1 + a) = \frac{1}{2}e^{-a}(1 + a)^2$$

$$2 = 1 + a$$

$$a = 1,\ b = e^{-1}$$

$$P = (1, e^{-1}) \text{ or } P = \left(1, \frac{1}{e}\right)$$

Maximum area $= \frac{1}{2}e^{-1}(2)^2 = 2e^{-1} = \frac{2}{e} \approx 0.7358$ square units

2. (a) $f(x) = \ln(x + 1), f(0) = 0, P_1(x) = a_0 + a_1x, P_1(0) = a_0 \Rightarrow a_0 = 0$

$f'(x) = \frac{1}{x+1}, f'(0) = 1, P_1{}'(x) = a_1, P_1{}'(0) = a_1 \Rightarrow a_1 = 1$

Thus, $P_1(x) = x$.

(b) $f(x) = \ln(x + 1),\ f(0) = 0, P_2(x) = a_0 + a_1x + a_2x^2, P_2(0) = a_0 \Rightarrow a_0 = 0$

$f'(x) = \frac{1}{x+1}, f'(0) = 1, P_2{}'(x) = a_1 + 2a_2x, P_2(0) = a_1 \Rightarrow a_1 = 1$

$f''(x) = \frac{-1}{(x+1)^2}, f''(0) = -1, P_2{}''(x) = 2a_2, P_2{}''(0) = 2a_2 \Rightarrow a_2 = -\frac{1}{2}$

Thus, $P_2(x) = x - \frac{x^2}{2}$.

(c)

x	-1.0	-0.01	-0.0001	0	0.0001	0.01	1.0
$\ln(x+1)$	undefined	-0.01005	≈ 0	0	≈ 0	0.00995	0.69315
$P_2(x)$	-1.5	-0.01005	≈ 0	0	≈ 0	0.00995	0.05

$P_2(x)$ and $\ln(x + 1)$ are nearly equal for x-values near 0.

(d)

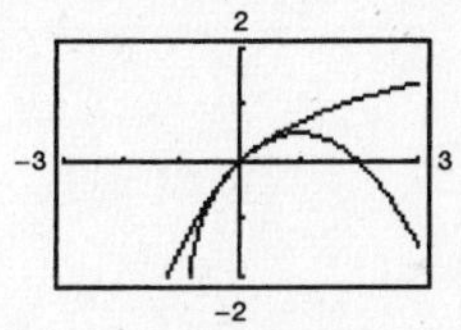

P_2 closely approximates $f(x) = \ln(x + 1)$ near $x = 0$. The approximation is poor if x is not close to 0.

3. (a) $\int_0^2 f(x)\,dx = \int_0^1 f(x)\,dx + \int_1^2 f(x)\,dx.$

Use substitution for the integral on the right.

$x = u + 1, dx = du$. Then

$$\int_0^2 f(x)\,dx = \int_0^1 f(x)\,dx + \int_0^1 f(u+1)\,du$$

$$= \int_0^1 [f(x) + f(x+1)]\,dx.$$

(b) $\int_0^1 \left(\sqrt{x} + \sqrt{x+1}\right) dx = \int_0^2 \sqrt{x}\,dx = \frac{2}{3}x^{3/2}\Big]_0^2 = \frac{4}{3}\sqrt{2}$

(c) $\int_0^1 (e^x + e^{x+1})\,dx = \int_0^2 e^x\,dx = e^x\Big]_0^2 = e^2 - 1$

4. (a) $y = \ln x \Rightarrow x = e^y$

$$\text{Area } A = \int_0^1 e^y\,dy$$

$$= e^y\Big]_0^1 = e - 1$$

(b) Area $B = \int_1^e \ln x\,dx =$ (Area rectangle) $-$ (Area A)

$= (1)(e) - (e - 1) = 1$

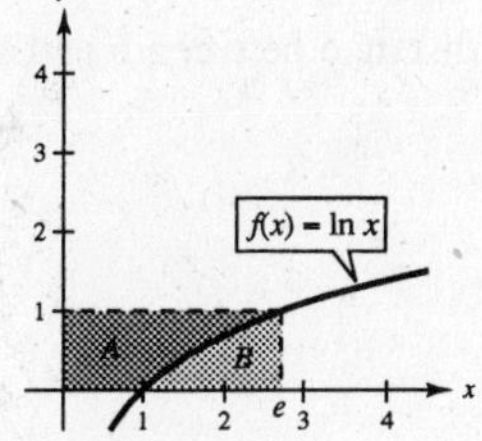

5. (a) $e^x > 1$

$$\int_0^x e^t\,dt > \int_0^x 1\,dt$$

$$e^x - 1 > x$$

$$e^x > 1 + x$$

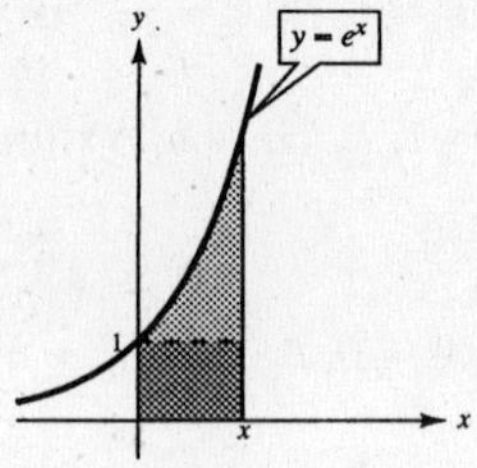

(b) $\displaystyle\int_0^x e^t\,dt > \int_0^x (1 + t)\,dt$

$$e^x - 1 > x + \frac{x^2}{2}$$

$$e^x > 1 + x + \frac{x^2}{2}$$

(c) Continuing this pattern,

$$e^x > 1 + x + \frac{x^2}{2!} + \frac{x^3}{3!}$$

$$\vdots$$

$$e^x > 1 + x + \frac{x^2}{2!} + \cdots + \frac{x^n}{n!}.$$

6. $y = 0.5^x$ and $y = 1.2^x$ intersect $y = x$.

$y = 2^x$ does not intersect $y = x$.

Suppose $y = x$ is tangent to $y = a^x$ at (x, y).

$$a^x = x \Rightarrow a = x^{1/x}.$$

$$y' = a^x \ln a = 1 \Rightarrow x \ln x^{1/x} = 1 \Rightarrow \ln x = 1 \Rightarrow x = e, a = e^{1/e}$$

For $0 < a \le e^{1/e} \approx 1.445$, the curve $y = a^x$ intersects $y = x$.

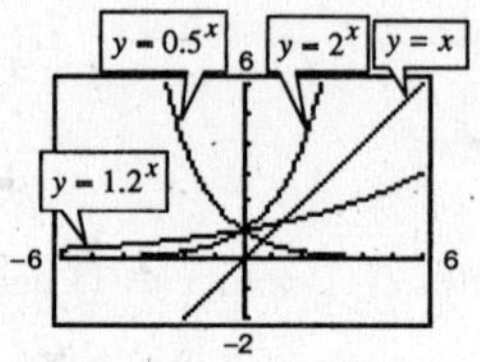

7.

$$y = \ln x$$

$$y' = \frac{1}{x}$$

$$y - b = \frac{1}{a}(x - a)$$

$$y = \frac{1}{a}x + b - 1 \quad \text{Tangent line}$$

If $x = 0$, $c = b - 1$. Thus, the distance between b and c is

$$b - c = b - (b - 1) = 1.$$

8.

$$y = e^x$$

$$y' = e^x$$

$$y - b = e^a(x - a)$$

$$y = e^a x - ae^a + b \quad \text{Tangent line}$$

If $y = 0$,

$$e^a x = ae^a - b$$

$$bx = ab - b \quad (b = e^a)$$

$$x = a - 1$$

$$c = a - 1.$$

Thus, $a - c = a - (a - 1) = 1.$

9. (a) $$\frac{dy}{dt} = y^{1.01}$$

$$\int y^{-1.01}\,dy = \int dt$$

$$\frac{y^{-0.01}}{-0.01} = t + C_1$$

$$\frac{1}{y^{0.01}} = -0.01t + C$$

$$y^{0.01} = \frac{1}{C - 0.01t}$$

$$y = \frac{1}{(C - 0.01t)^{100}}$$

$$y(0) = 1:\ 1 = \frac{1}{C^{100}} \Rightarrow C = 1$$

Hence, $y = \dfrac{1}{(1 - 0.01t)^{100}}$.

$\lim_{t\to T^-} y = \infty$ when $T = 100$.

(b) $$\int y^{-(1+\varepsilon)}\,dy = \int k\,dt$$

$$\frac{y^{-\varepsilon}}{-\varepsilon} = kt + C_1$$

$$y^{-\varepsilon} = -\varepsilon kt + C$$

$$y = \frac{1}{(C - \varepsilon kt)^{1/\varepsilon}}$$

$$y(0) = y_0 = \frac{1}{C^{1/\varepsilon}} \Rightarrow C^{1/\varepsilon} = \frac{1}{y_0} \Rightarrow C = \left(\frac{1}{y_0}\right)^{\varepsilon}$$

Hence, $y = \dfrac{1}{\left(\left(\dfrac{1}{y_0}\right)^{\varepsilon} - \varepsilon kt\right)^{1/\varepsilon}}$.

For $t \to \dfrac{1}{y_0^{\varepsilon}\varepsilon k}$, $y \to \infty$.

10. $f(x) = \begin{cases} |x|^x, & x \neq 0 \\ 1, & x = 0 \end{cases}$

(a)

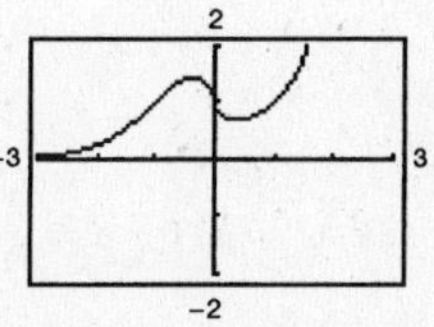

Domain: all x

(b) Using the zoom and trace feature, the limit appears to be 1.

(c) Clearly, f is continuous for all $x \neq 0$. For $x = 0$, $\lim_{x\to 0} f(x) = f(0) = 1$. Hence, f is continuous for all x.

(d) Visually, the slope appears to be -1.

(e) $f'(x) \approx \dfrac{f(x + \Delta x) - f(x)}{\Delta x}$ and $f'(x) \approx \dfrac{f(x) - f(x - \Delta x)}{\Delta x}$

Adding, $2f'(x) \approx \dfrac{f(x + \Delta x) - f(x - \Delta x)}{\Delta x}$

which yields $f'(x) \approx \dfrac{f(x + \Delta x) - f(x - \Delta x)}{2\,\Delta x}$

$$f'(0) \approx \frac{f(0 + \Delta x) - f(0 - \Delta x)}{2\,\Delta x}$$

$$= \frac{f(\Delta x) - f(-\Delta x)}{2\,\Delta x}$$

$$= \frac{|\Delta x|^{\Delta x} - |-\Delta x|^{-\Delta x}}{2\,\Delta x}$$

For $\Delta x \approx 0$, $f'(0)$ approaches $-\infty$.

$f'(0) = -\infty$ (undefined)

(f) For $x > 0$, $\quad y = x^x$.

$$\ln y = x \ln x$$

$$\frac{y'}{y} = 1 + \ln x$$

$$y' = x^x(1 + \ln x)$$

$y' \to -\infty$ as $x \to 0^+$

Similarly, for $x < 0$, $y' \to -\infty$ as $x \to 0^-$.

(g) $f'(x) = 0$ for $x = \pm\dfrac{1}{e}$.

By the first Derivative Test

$\left(-\dfrac{1}{e}, e^{1/e}\right)$ relative maximum

$\left(\dfrac{1}{e}, \left(\dfrac{1}{e}\right)^{1/e}\right)$ relative minimum

11. Let $u = 1 + \sqrt{x}$, $\sqrt{x} = u - 1$, $x = u^2 - 2u + 1$,

$dx = (2u - 2)\,du$.

$$\text{Area} = \int_1^4 \frac{1}{\sqrt{x} + x}\,dx = \int_2^3 \frac{2u - 2}{(u - 1) + (u^2 - 2u + 1)}\,du$$

$$= \int_2^3 \frac{2(u - 1)}{u^2 - u}\,du$$

$$= \int_2^3 \frac{2}{u}\,du$$

$$= \Big[2 \ln |u|\Big]_2^3$$

$$= 2 \ln 3 - 2 \ln 2 = 2 \ln\left(\frac{3}{2}\right)$$

$$\approx 0.8109$$

12. (a) $\dfrac{dy}{dt} = y(1 - y)$, $y(0) = \dfrac{1}{4}$

$$\int \left(\frac{1}{y} + \frac{1}{1 - y}\right) dy = \int dt$$

$$\ln|y| - \ln|1 - y| = t + C$$

$$\ln\left|\frac{y}{1 - y}\right| = t + C$$

$$\frac{y}{1 - y} = e^{t+C} = C_1 e^t$$

$$y = C_1 e^t - y C_1 e^t$$

$$y = \frac{C_1 e^t}{1 + C_1 e^t} = \frac{1}{1 + C_2 e^{-t}}$$

$$y(0) = \frac{1}{4} = \frac{1}{1 + C_2} \Rightarrow C_2 = 3$$

Hence, $y = \dfrac{1}{1 + 3e^{-t}}$.

(b)

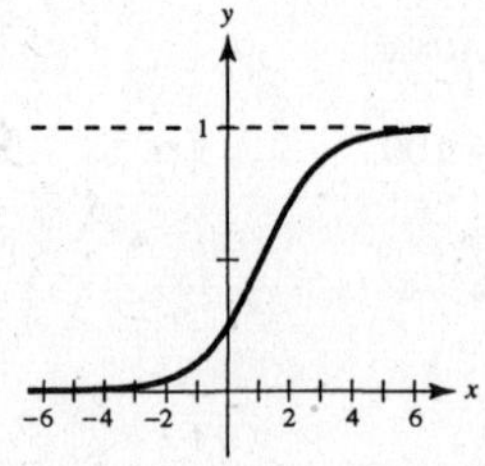

$$\frac{dy}{dt} = y(1 - y) = y - y^2$$

$$\frac{d^2y}{dt^2} = y'' = y' - 2yy' \Rightarrow y'' = 0 \text{ for } y = \frac{1}{2}$$

$$\frac{d^2y}{dt^2} > 0 \text{ if } 0 < y < \frac{1}{2} \text{ and } \frac{d^2y}{dt^2} < 0 \text{ if } \frac{1}{2} < y < 1.$$

Thus, the rate of growth is maximum at $y = \dfrac{1}{2}$, the point of inflection.

(c) $y' = y(1 - y)$, $y(0) = 2$

As before, $y = \dfrac{1}{1 + C_2 e^{-t}}$.

$$y(0) = 2 = \frac{1}{1 + C_2} \Rightarrow C_2 = -\frac{1}{2}$$

Thus, $y = \dfrac{1}{1 - \frac{1}{2}e^{-t}} = \dfrac{2}{2 - e^{-t}}$.

The graph is different:

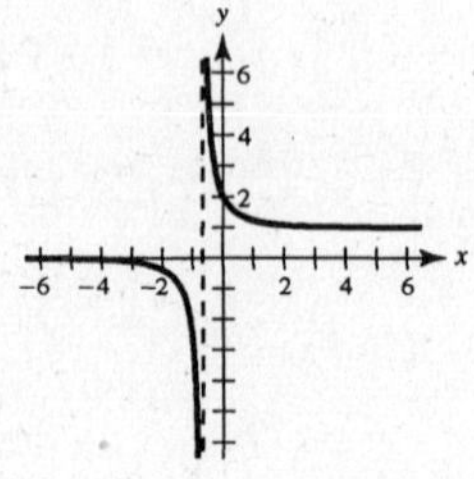

13. (a) $\dfrac{dS}{dt} = k_1 S(L - S)$

$S = \dfrac{L}{1 + Ce^{-kt}}$ is a solution because

$$\frac{dS}{dt} = -L(1 + Ce^{-kt})^{-2}(-C\,ke^{-kt})$$

$$= \frac{LC\,ke^{-kt}}{(1 + Ce^{-kt})^2}$$

$$= \left(\frac{k}{L}\right)\frac{L}{1 + Ce^{-kt}} \cdot \frac{C\,Le^{-kt}}{1 + Ce^{-kt}}$$

$$= \left(\frac{k}{L}\right)\frac{L}{1 + Ce^{-kt}} \cdot \left(L - \frac{L}{1 + Ce^{-kt}}\right)$$

$= k_1 S(L - S)$, where $k_1 = \dfrac{k}{L}$.

$L = 100$. Also, $S = 10$ when $t = 0 \Rightarrow C = 9$. And, $S = 20$ when $t = 1 \Rightarrow k = -\ln(4/9)$.

Particular Solution. $S = \dfrac{100}{1 + 9e^{\ln(4/9)t}}$

$$= \frac{100}{1 + 9e^{-0.8109t}}$$

(c)

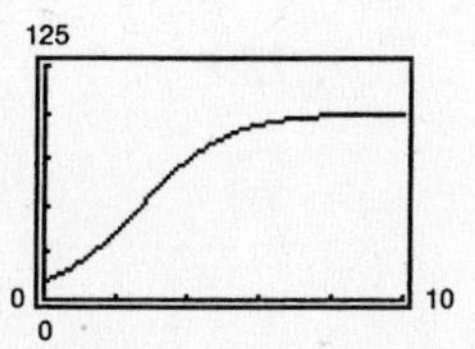

(b) $\dfrac{dS}{dt} = k_1 S(100 - S)$

$$\frac{d^2S}{dt^2} = k_1\left[S\left(-\frac{dS}{dt}\right) + (100 - S)\frac{dS}{dt}\right]$$

$$= k_1(100 - 2S)\frac{dS}{dt}$$

$= 0$ when $S = 50$ or $\dfrac{dS}{dt} = 0$.

Choosing $S = 50$, we have:

$$50 = \frac{100}{1 + 9e^{\ln(4/9)t}}$$

$$2 = 1 + 9e^{\ln(4/9)t}$$

$$\frac{\ln(1/9)}{\ln(4/9)} = t$$

$t \approx 2.7$ months (This is the inflection point)

(d)

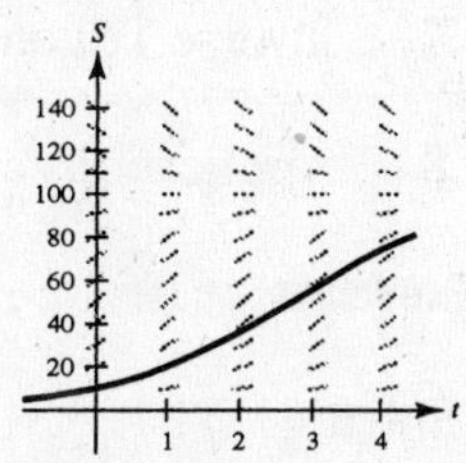

(e) Sales will decrease toward the line $S = L$.

CHAPTER 9
Trigonometric Functions

Section 9.1 Radian and Degree Measure **659**

Section 9.2 Trigonometric Functions: The Unit Circle **666**

Section 9.3 Right Triangle Trigonometry **671**

Section 9.4 Trigonometric Functions of Any Angle **681**

Section 9.5 Graphs of Sine and Cosine Functions **692**

Section 9.6 Graphs of Other Trigonometric Functions **702**

Section 9.7 Inverse Trigonometric Functions **712**

Section 9.8 Applications and Models **720**

Review Exercises **729**

Problem Solving **739**

CHAPTER 9
Trigonometric Functions

Section 9.1 Radian and Degree Measure

1.

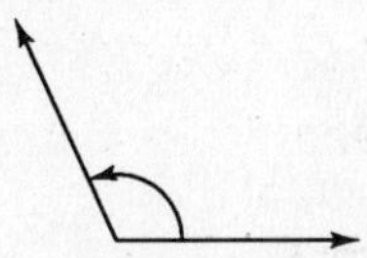

The angle shown is approximately 2 radians.

2. 5.5 radians

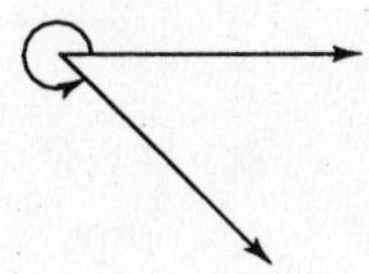

3.

The angle shown is approximately -3 radians.

4. -4 radians

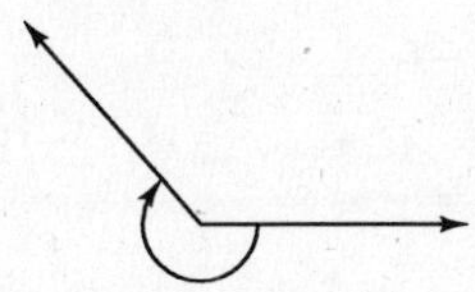

5.

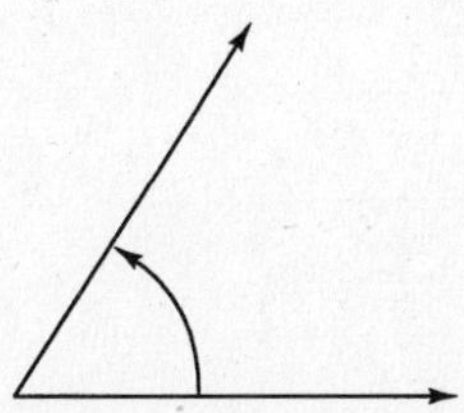

The angle shown is approximately 1 radian.

6. 6.5 radians

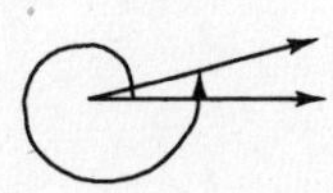

7. (a) Since $0 < \frac{\pi}{5} < \frac{\pi}{2}$; $\frac{\pi}{5}$ lies in Quadrant I.

(b) Since $\pi < \frac{7\pi}{5} < \frac{3\pi}{2}$; $\frac{7\pi}{5}$ lies in Quadrant III.

8. (a) Since $\pi < \frac{11\pi}{8} < \frac{3\pi}{2}$; $\frac{11\pi}{8}$ lies in Quadrant III.

(b) Since $\pi < \frac{9\pi}{8} < \frac{3\pi}{2}$; $\frac{9\pi}{8}$ lies in Quadrant III.

9. (a) Since $-\frac{\pi}{2} < -\frac{\pi}{12} < 0$; $-\frac{\pi}{12}$ lies in Quadrant IV.

(b) Since $-\frac{3\pi}{2} < -\frac{11\pi}{9} < -\pi$; $-\frac{11\pi}{9}$ lies in Quadrant II.

10. (a) Since $-\frac{\pi}{2} < -1 < 0$; -1 lies in Quadrant IV.

(b) Since $-\pi < -2 < -\frac{\pi}{2}$; -2 lies in Quadrant III.

11. (a) Since $\pi < 3.5 < \frac{3\pi}{2}$; 3.5 lies in Quadrant III.

(b) Since $\frac{\pi}{2} < 2.25 < \pi$; 2.25 lies in Quadrant II.

12. (a) Since $\frac{3\pi}{2} < 6.02 < 2\pi$; 6.02 lies in Quadrant IV.

(b) Since $-\frac{3\pi}{2} < -4.25 < -\pi$; 4.25 lies in Quadrant II.

13. (a)

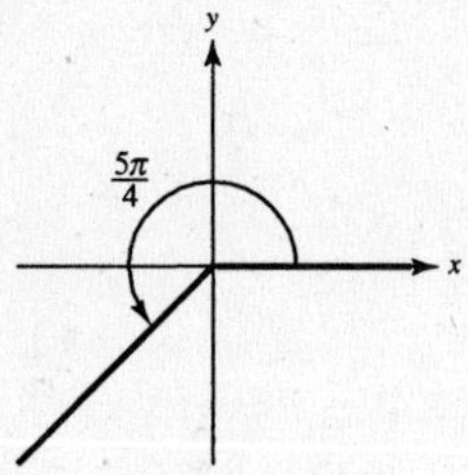

(b)

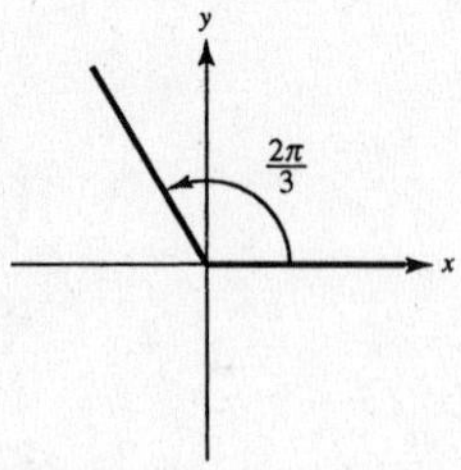

14. (a) $-\frac{7\pi}{4}$

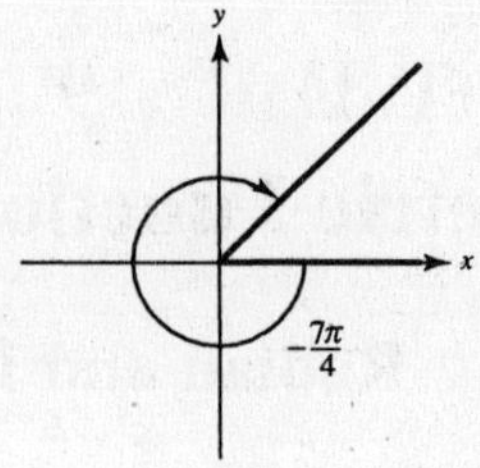

(b) $-\frac{5\pi}{2}$

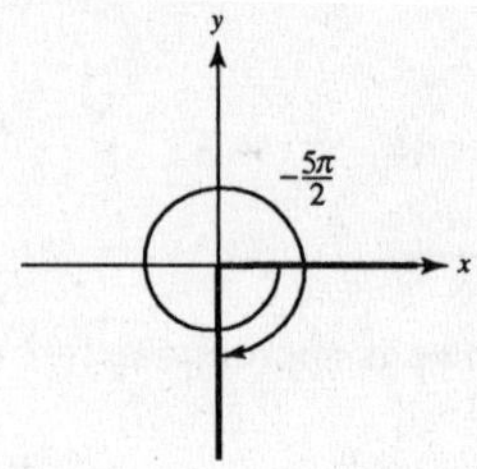

15. (a)

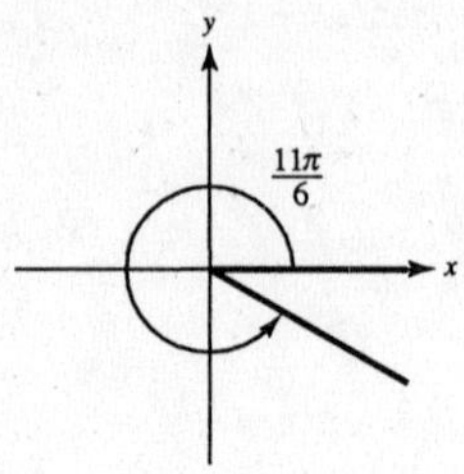

(b)

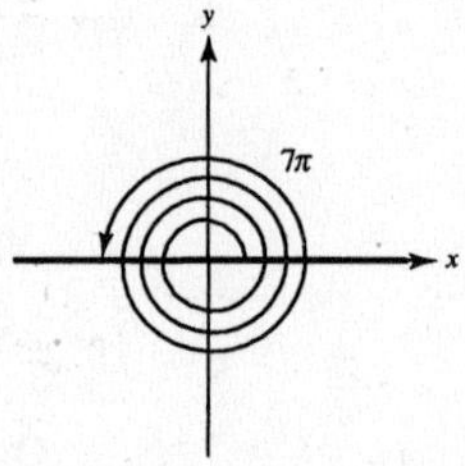

16. (a) 4

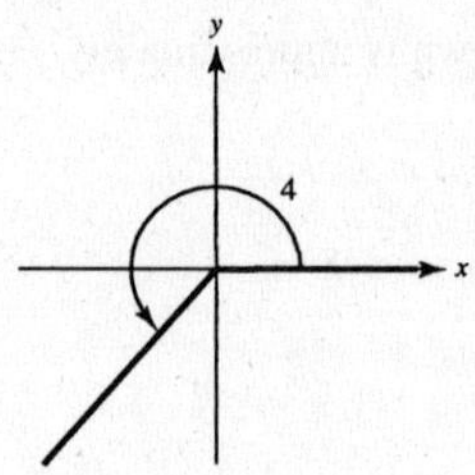

(b) -3

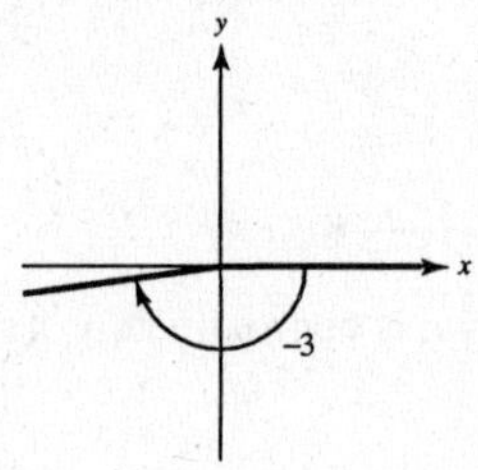

17. (a) Coterminal angles for $\frac{\pi}{6}$

$\frac{\pi}{6} + 2\pi = \frac{13\pi}{6}$

$\frac{\pi}{6} - 2\pi = -\frac{11\pi}{6}$

(b) Coterminal angles for $\frac{5\pi}{6}$

$\frac{5\pi}{6} + 2\pi = \frac{17\pi}{6}$

$\frac{5\pi}{6} - 2\pi = -\frac{7\pi}{6}$

18. (a) $\frac{7\pi}{6} + 2\pi = \frac{19\pi}{6}$

$\frac{7\pi}{6} - 2\pi = -\frac{5\pi}{6}$

(b) $-\frac{11\pi}{6} + 2\pi = \frac{\pi}{6}$

$-\frac{11\pi}{6} - 2\pi = -\frac{23\pi}{6}$

19. (a) Coterminal angles for $\frac{2\pi}{3}$

$\frac{2\pi}{3} + 2\pi = \frac{8\pi}{3}$

$\frac{2\pi}{3} - 2\pi = -\frac{4\pi}{3}$

(b) Coterminal angles for $\frac{\pi}{12}$

$\frac{\pi}{12} + 2\pi = \frac{25\pi}{12}$

$\frac{\pi}{12} - 2\pi = -\frac{23\pi}{12}$

20. (a) $-\frac{9\pi}{4} + 2\pi = -\frac{\pi}{4}$

$-\frac{9\pi}{4} + 4\pi = \frac{7\pi}{4}$

(b) $-\frac{2\pi}{15} + 2\pi = \frac{28\pi}{15}$

$-\frac{2\pi}{15} - 2\pi = -\frac{32\pi}{15}$

21. (a) Complement: $\frac{\pi}{2} - \frac{\pi}{3} = \frac{\pi}{6}$

Supplement: $\pi - \frac{\pi}{3} = \frac{2\pi}{3}$

(b) Complement: Not possible; $\frac{3\pi}{4}$ is greater than $\frac{\pi}{2}$.

Supplement: $\pi - \frac{3\pi}{4} = \frac{\pi}{4}$

22. (a) Complement: $\frac{\pi}{2} - \frac{\pi}{12} = \frac{5\pi}{12}$

Supplement: $\pi - \frac{\pi}{12} = \frac{11\pi}{12}$

(b) Complement: none $\left(\frac{11\pi}{12} > \frac{\pi}{2}\right)$

Supplement: $\pi - \frac{11\pi}{12} = \frac{\pi}{12}$

23. (a) Complement: $\frac{\pi}{2} - 1 \approx 0.57$

Supplement: $\pi - 1 \approx 2.14$

(b) Complement: Not possible. 2 is greater than $\frac{\pi}{2}$.

Supplement: $\pi - 2 \approx 1.14$

24. (a) Complement: none $\left(3 > \frac{\pi}{2}\right)$

Supplement: $\pi - 3 \approx 0.14$

(b) Complement: $\frac{\pi}{2} - 1.5 \approx 0.07$

Supplement: $\pi - 1.5 \approx 1.64$

25.

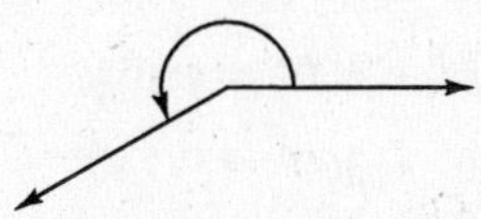

The angle shown is approximately 210°.

26. 120°

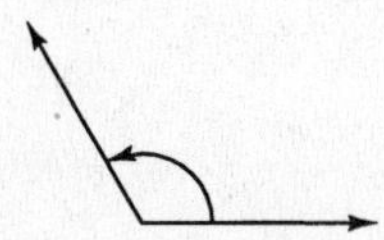

27.

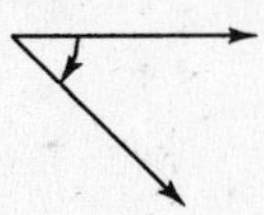

The angle shown is approximately $-60°$.

28. $-330°$

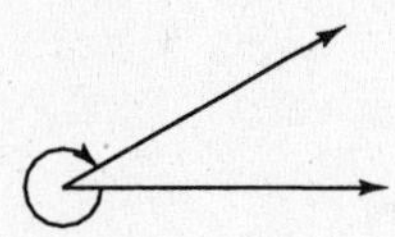

29.

The angle shown is approximately 165°.

30. 10°

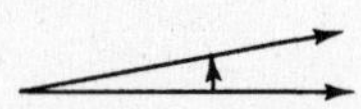

31. (a) Since $90° < 130° < 180°$; 130° lies in Quadrant II.

(b) Since $270° < 285° < 360°$; 285° lies in Quadrant IV.

32. (a) Since $0° < 8.3° < 90°$; 8.3° lies in Quadrant I.

(b) Since $180° < 257° 30' < 270°$; $257° 30'$ lies in Quadrant III.

33. (a) Since $-180° < -132°50' < -90°$; $-132° 50'$ lies in Quadrant III.

(b) Since $-360° < -336° < -270°$; $-336°$ lies in Quadrant I.

34. (a) Since $-270° < -260° < -180°$; $-260°$ lies in Quadrant II.

(b) Since $-90° < -3.4° < 0°$; $-3.4°$ lies in Quadrant IV.

35. (a)

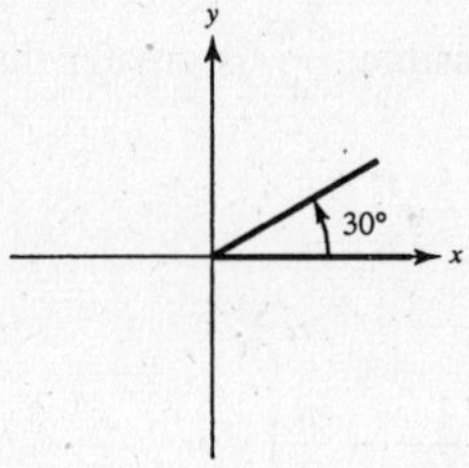

(b)

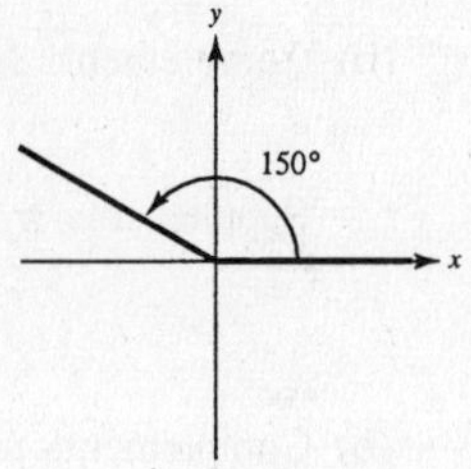

36. (a) $-270°$

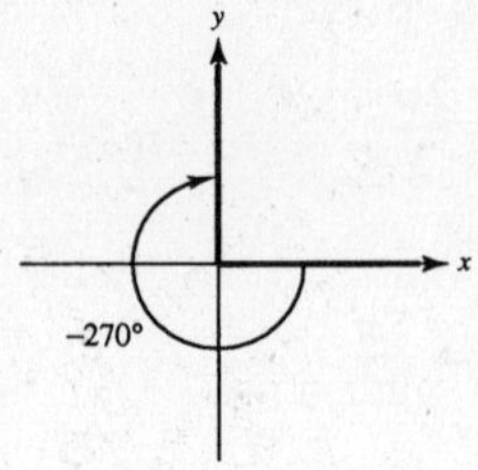

(b) $-120°$

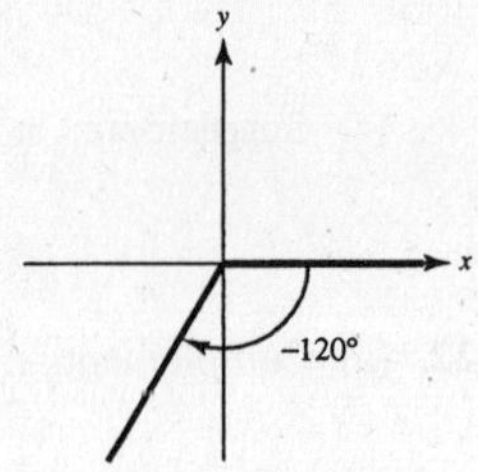

37. (a)

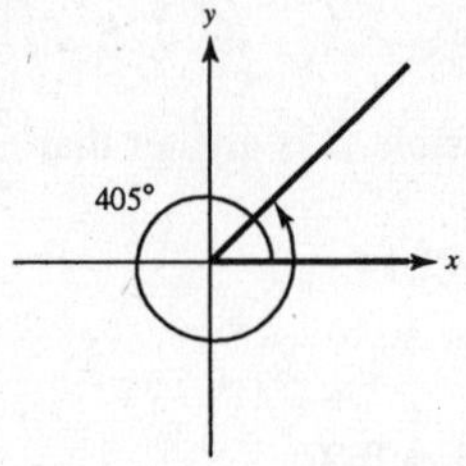

(b)

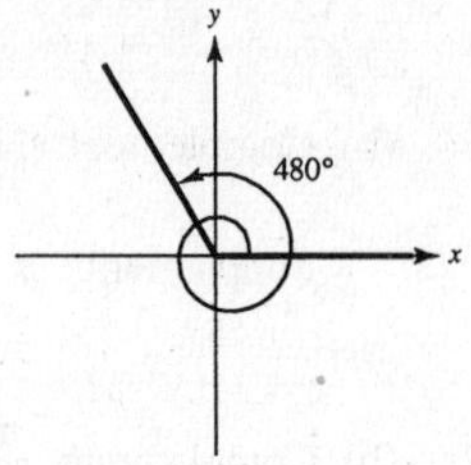

38. (a) $-750°$

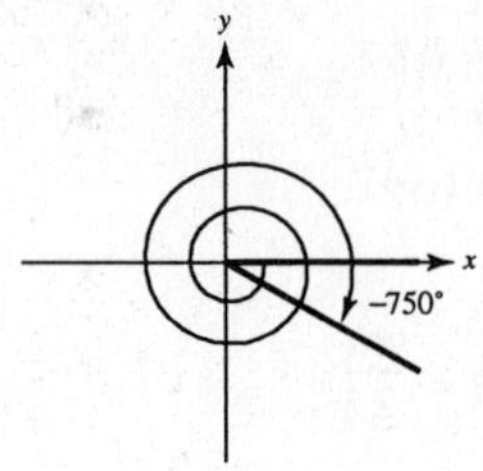

(b) $-600°$

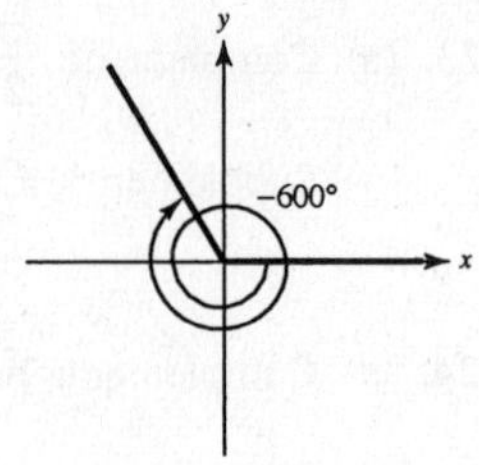

39. (a) Coterminal angles for 45°

$45° + 360° = 405°$

$45° - 360° = -315°$

(b) Coterminal angles for $-36°$

$-36° + 360° = 324°$

$-36 - 360° = -396°$

40. (a) Coterminal angles for 240°

$240° + 360° = 600°$

$240° - 360° = -120°$

(b) $-420° + 720° = 300°$

$-420° + 360° = -60°$

41. (a) Coterminal angles for 120°

$120° + 360° = 480°$

$120° - 360° = -240°$

(b) Coterminal angles for $-180°$

$-180° + 360° = 180°$

$-180° - 360° = -540°$

42. (a) $-510° + 720° = 210°$

$-510° + 360° = -150°$

(b) $230° + 360° = 590°$

$230° - 360° = -130°$

43. (a) Complement: $90° - 18° = 72°$

Supplement: $180° - 18° = 162°$

(b) Complement: Not possible; 115° is greater than 90°.

Supplement: $180° - 115° = 65°$

44. (a) Complement: $90° - 3° = 87°$

Supplement: $180° - 3° = 177°$

(b) Complement: $90° - 64° = 26°$

Supplement: $180° - 64° = 116°$

45. (a) Complement: $90° - 79° = 11°$

Supplement: $180° - 79° = 101°$

(b) Complement: Not possible. 150° is greater than 90°.

Supplement: $180° - 150° = 30°$

46. (a) Complement: none $(130° > 90°)$

Supplement: $180° - 130° = 50°$

(b) Complement: none $(170° > 90°)$

Supplement: $180° - 170° = 10°$

47. (a) $30° = 30°\left(\frac{\pi}{180°}\right) = \frac{\pi}{6}$

(b) $150° = 150°\left(\frac{\pi}{180°}\right) = \frac{5\pi}{6}$

48. (a) $315° = 315°\left(\frac{\pi}{180°}\right) = \frac{7\pi}{4}$

(b) $120° = 120°\left(\frac{\pi}{180°}\right) = \frac{2\pi}{3}$

49. (a) $-120° = -120°\left(\frac{\pi}{180°}\right) = -\frac{2\pi}{3}$

(b) $-240° = -240°\left(\frac{\pi}{180°}\right) = -\frac{4\pi}{3}$

50. (a) $-270° = -270°\left(\dfrac{\pi}{180°}\right) = -\dfrac{3\pi}{2}$

(b) $-90° = -90°\left(\dfrac{\pi}{180°}\right) = -\dfrac{\pi}{2}$

51. $115° = 115°\left(\dfrac{\pi}{180°}\right) \approx 2.007$ radians

52. $87.4° = 87.4°\left(\dfrac{\pi}{180°}\right) \approx 1.525$ radians

53. $-216.35° = -216.35°\left(\dfrac{\pi}{180°}\right) \approx -3.776$ radians

54. $-48.27° = -48.27°\left(\dfrac{\pi}{180°}\right) \approx -0.842$ radians

55. $532° = 532°\left(\dfrac{\pi}{180°}\right) \approx 9.285$ radians

56. $345° = 345°\left(\dfrac{\pi}{180°}\right) \approx 6.021$ radians

57. $-0.83° = -0.83°\left(\dfrac{\pi}{180°}\right) \approx -0.014$ radian

58. $0.54° = 0.54°\left(\dfrac{\pi}{180°}\right) \approx 0.009$ radians

59. (a) $\dfrac{3\pi}{2} = \dfrac{3\pi}{2}\left(\dfrac{180}{\pi}\right)^{\circ} = 270°$

(b) $\dfrac{7\pi}{6} = \dfrac{7\pi}{6}\left(\dfrac{180}{\pi}\right)^{\circ} = 210°$

60. (a) $\dfrac{2\pi}{3} = \dfrac{2\pi}{3}\left(\dfrac{180°}{\pi}\right) = 120°$

(b) $\dfrac{5\pi}{6} = \dfrac{5\pi}{6}\left(\dfrac{180°}{\pi}\right) = 150°$

61. (a) $-\dfrac{3\pi}{4} = -\dfrac{3\pi}{4}\left(\dfrac{180°}{\pi}\right) = -135°$

(b) $-\dfrac{7\pi}{6} = -\dfrac{7\pi}{6}\left(\dfrac{180°}{\pi}\right) = -210°$

62. (a) $\dfrac{11\pi}{6} = \dfrac{11\pi}{6}\left(\dfrac{180°}{\pi}\right) = 330°$

(b) $\dfrac{7\pi}{4} = \dfrac{7\pi}{4}\left(\dfrac{180°}{\pi}\right) = 315°$

63. $\dfrac{\pi}{7} = \dfrac{\pi}{7}\left(\dfrac{180°}{\pi}\right) \approx 25.714°$

64. $\dfrac{5\pi}{11} = \dfrac{5\pi}{11}\left(\dfrac{180°}{\pi}\right) \approx 81.818°$

65. $\dfrac{15\pi}{8} = \dfrac{15\pi}{8}\left(\dfrac{180°}{\pi}\right) = 337.5°$

66. $\dfrac{13\pi}{2} = \dfrac{13\pi}{2}\left(\dfrac{180°}{\pi}\right) = 1170°$

67. $-4.2\pi = -4.2\pi\left(\dfrac{180°}{\pi}\right) = -756°$

68. $4.8\pi = 4.8\pi\left(\dfrac{180°}{\pi}\right) = 864°$

69. $-2 = -2\left(\dfrac{180°}{\pi}\right) \approx -114.592°$

70. $-0.57 = -0.57\left(\dfrac{180°}{\pi}\right) \approx -32.659°$

71. (a) $54°\,45' = 54° + \left(\frac{45}{60}\right)^{\circ} = 54.75°$

(b) $-128°\,30' = -128° - \left(\frac{30}{60}\right)^{\circ} = -128.5°$

72. (a) $245°10' = 245° + \left(\frac{10}{60}\right)^{\circ}$

$\approx 245° + 0.167°$

$= 245.167°$

(b) $2°12' = 2° + \left(\frac{12}{60}\right)^{\circ}$

$= 2° + 0.2° = 2.2°$

73. (a) $85°\,18'\,30'' = \left(85 + \frac{18}{60} + \frac{30}{3600}\right)^{\circ} \approx 85.308°$

(b) $330°\,25'' = \left(330 + \frac{25}{3600}\right)^{\circ} \approx 330.007°$

74. (a) $-135°36'' = -135° - \left(\frac{36}{3600}\right)^{\circ}$

$= -135° - 0.01° = -135.01°$

(b) $-408°16'20'' = -\left(408° + \left(\frac{16}{60}\right)^{\circ} + \left(\frac{20}{3600}\right)^{\circ}\right)$

$\approx -(408° + 0.2667° + 0.0056°)$

$= -408.272°$

75. (a) $240.6° = 240° + 0.6(60)' = 240° 36'$

(b) $-145.8° = -[145° + 0.8(60')] = -145° 48'$

76. (a) $-345.12° = -(345° + (0.12)(60'))$

$= -(345° + 7' + 0.2(60''))$

$= -345° 7' 12''$

(b) $0.45° = 0° + (0.45)(60')$

$= 0° + 27'$

$= 0°27'$

77. (a) $2.5° = 2° 30'$

(b) $-3.58° = -3° 34' 48''$

78. (a) $-0.355° = -(0° + (0.355)(60'))$

$= -(0° + 21' + (0.3)(60''))$

$= -(0° + 21' + 18'')$

$= -0°21'18''$

(b) $0.7865 = 0° + (0.7865)(60')$

$= 0° + 47' + (0.19)(60'')$

$= 0° + 47' + 11.4''$

$= 0°47'11.4''$

79. (a) An angle is in standard position if its vertex is at the origin and its initial side is on the positive x-axis.

(b) A negative angle is generated by a clockwise rotation of the terminal side.

(c) Two angles in standard position with the same terminal sides are coterminal.

(d) An obtuse angle measures between 90° and 180°.

80. Increases, since the linear speed is proportional to the radius.

81. 1 Radian $= \left(\frac{180}{\pi}\right)^{\circ} \approx 57.3°$, so one radian is much larger than one degree.

82. The arc length is increasing. In order for the angle θ to remain constant as the radius r increases, the arc length s must increase in proportion to r, as can be seen from the formula $s = r\theta$.

83. $s = r\theta$

$6 = 5\theta$

$\theta = \frac{6}{5}$ radians

84. $s = r\theta$

$29 = 10\,\theta$

$\theta = \frac{29}{10}$ radians

85. $s = r\theta$

$32 = 7\theta$

$\theta = \frac{32}{7} = 4\frac{4}{7}$ radians

86. $s = r\theta$

$60 = 75\theta$

$\theta = \frac{60}{75} = \frac{4}{5}$ radians

Because the angle represented is clockwise, this angle is $-\frac{4}{5}$ radians.

87. $s = r\theta$

$6 = 27\theta$

$\theta = \frac{6}{27} = \frac{2}{9}$ radian

88. $r = 14$ feet, $s = 8$ feet

$\theta = \frac{s}{r} = \frac{8}{14} = \frac{4}{7}$ radians

89. $s = r\theta$

$25 = 14.5\theta$

$\theta = \frac{25}{14.5} = \frac{50}{29}$ radians

90. $r = 80$ kilometers,

$s = 160$ kilometers

$\theta = \frac{s}{r} = \frac{160}{80} = 2$ radians

91. $s = r\theta$, θ in radians

$s = 15(180)\left(\frac{\pi}{180}\right) = 15\pi$ inches

≈ 47.12 inches

92. $r = 9$ feet, $\theta = 60° = \dfrac{\pi}{3}$

$s = r\theta = 9\left(\dfrac{\pi}{3}\right) = 3\pi$ feet

93. $s = r\theta$, θ in radians

$s = 3(1) = 3$ meters

94. $r = 20$ centimeters, $\theta = \dfrac{\pi}{4}$

$s = r\theta = 20\left(\dfrac{\pi}{4}\right) = 5\pi$ centimeters

95. $\theta = 41°\,15'31'' - 32°\,47'0'' = 8°\,28'31'' \approx 8.47528° \approx 0.14792$ radian

$s = r\theta = 4000(0.14797) \approx 591.08$ miles

96. $r = 4000$ miles

$\theta = 47°\,36'23'' - 37°\,46'30'' = 9°\,49'53'' \approx 0.1716$ radian

$s = r\theta \approx (4000)(0.1716) \approx 686.4$ miles

97. $\theta = 42°\,7'45'' - 25°\,46'26'' = 16°\,21'19'' \approx 0.285453$ radian

$s = r\theta = 4000(0.285453) \approx 1141.81$ miles

98. $r = 4000$ miles

$\theta = 31°\,47' + 26°\,11' = 57°\,58' \approx 1.0117$ radians

$s = r\theta \approx (4000)(1.0117) \approx 4046.8$ miles

99. (a) 65 miles per hour $= \dfrac{65(5280)}{60} = 5720$ feet per minute

The circumference of the tire is $C = 2.5\pi$ feet.

The number of revolutions per minute is $r = \dfrac{5720}{2.5\pi} \approx 728.3$ rev/min.

(b) The angular speed is $\dfrac{\theta}{t}$.

$\theta = \dfrac{5720}{2.5\pi}(2\pi) = 4576$ radians

Angular speed $= \dfrac{4576 \text{ radians}}{1 \text{ minute}} = 4576$ rad/min

100. Linear velocity for either pulley: $1700(2\pi) = 3400\pi$ in/min

(a) Angular speed of motor pulley: $\omega = \dfrac{v}{r} = \dfrac{3400\pi}{1} = 3400\pi$ rad/min

Angular speed of the saw arbor: $\omega = \dfrac{v}{r} = \dfrac{3400\pi}{2} = 1700\pi$ rad/min

(b) Revolutions per minute of the saw arbor: $\dfrac{1700\pi}{2\pi} = 850$ rev/min

101. Circumference: $C = 2\pi(1.68) = 3.36\pi$ inches

360 rev/min = 6 rev/sec

Linear speed: $(3.36\pi)(6) = 20.16\pi$ inches/sec $= 1209.6\pi$ inches/minute

102. (a) Arc length of larger sprocket in feet: $s = r\theta$

$$s = \frac{1}{3}(2\pi) = \frac{2\pi}{3} \text{ feet}$$

Therefore, the chain moves $\frac{2\pi}{3}$ feet as does the smaller rear sprocket.

Thus, the angle θ of the smaller sprocket is $\left(r = 2 \text{ inches} = \frac{2}{12} \text{ feet}\right)$.

$$\theta = \frac{s}{r} = \frac{\frac{2\pi}{3} \text{ ft}}{\frac{2}{12} \text{ ft}} = 4\pi$$

and the arc length of the tire in feet is: $s = \theta r$

$$s = (4\pi)\left(\frac{14}{12}\right) = \frac{14\pi}{3} \text{ feet.}$$

$$\text{Speed} = \frac{s}{t} = \frac{\frac{14\pi}{3}}{1 \text{ sec}} = \frac{14\pi}{3} \text{ feet per second}$$

(b) $\dfrac{14\pi \text{ feet}}{3 \text{ seconds}} \times \dfrac{3600 \text{ seconds}}{1 \text{ hour}} \times \dfrac{1 \text{ mile}}{5280 \text{ feet}} \approx 10 \text{ miles per hour}$

103. False. A measurement of 4π radians corresponds to two complete revolutions from the initial to the terminal side of an angle.

104. True. If α and β are coterminal angles, then

$$\alpha = \beta + n(360°)$$

where n is an integer.

The difference between α and β is $\alpha - \beta = n(360°)$

$$= 2\pi n$$

105. False. $1° = \frac{\pi}{180}$ radian

106. False.

If $0 < \theta < \frac{\pi}{2}$, then θ is acute. θ is obtuse if $\frac{\pi}{2} < \theta < \pi$.

Section 9.2 Trigonometric Functions: The Unit Circle

1. $x = -\frac{8}{17}, \quad y = \frac{15}{17}$

$\sin\theta = y = \frac{15}{17}$ $\qquad \csc\theta = \frac{1}{y} = \frac{17}{15}$

$\cos\theta = x = -\frac{8}{17}$ $\qquad \sec\theta = \frac{1}{x} = -\frac{17}{8}$

$\tan\theta = \frac{y}{x} = -\frac{15}{8}$ $\qquad \cot\theta = \frac{x}{y} = -\frac{8}{15}$

2. $x = \frac{12}{13}, \quad y = \frac{5}{13}$

$\sin\theta = y = \frac{5}{13}$ $\qquad \csc\theta = \frac{1}{y} = \frac{13}{5}$

$\cos\theta = x = \frac{12}{13}$ $\qquad \sec\theta = \frac{1}{x} = \frac{13}{12}$

$\tan\theta = \frac{y}{x} = \frac{5}{12}$ $\qquad \cot\theta = \frac{x}{y} = \frac{12}{5}$

3. $x = \dfrac{12}{13}, \quad y = -\dfrac{5}{13}$

$\sin\theta = y = -\dfrac{5}{13} \qquad \csc\theta = \dfrac{1}{y} = -\dfrac{13}{5}$

$\cos\theta = x = \dfrac{12}{13} \qquad \sec\theta = \dfrac{1}{x} = \dfrac{13}{12}$

$\tan\theta = \dfrac{y}{x} = -\dfrac{5}{12} \qquad \cot\theta = \dfrac{x}{y} = -\dfrac{12}{5}$

4. $x = -\dfrac{4}{5}, y = -\dfrac{3}{5}$

$\sin\theta = y = -\dfrac{3}{5} \qquad \csc\theta = \dfrac{1}{y} = -\dfrac{5}{3}$

$\cos\theta = x = -\dfrac{4}{5} \qquad \sec\theta = \dfrac{1}{x} = -\dfrac{5}{4}$

$\tan\theta = \dfrac{y}{x} = \dfrac{3}{4} \qquad \cot\theta = \dfrac{x}{y} = \dfrac{4}{3}$

5. $t = \dfrac{\pi}{4}$ corresponds to $\left(\dfrac{\sqrt{2}}{2}, \dfrac{\sqrt{2}}{2}\right)$.

6. $t = \dfrac{\pi}{3}, (x\ y) = \left(\dfrac{1}{2}, \dfrac{\sqrt{3}}{2}\right)$

7. $t = \dfrac{7\pi}{6}$ corresponds to $\left(-\dfrac{\sqrt{3}}{2}, -\dfrac{1}{2}\right)$.

8. $t = \dfrac{5\pi}{4}, (x, y) = \left(-\dfrac{\sqrt{2}}{2}, -\dfrac{\sqrt{2}}{2}\right)$

9. $t = \dfrac{4\pi}{3}$ corresponds to $\left(-\dfrac{1}{2}, -\dfrac{\sqrt{3}}{2}\right)$.

10. $t = \dfrac{5\pi}{3}, (x, y) = \left(\dfrac{1}{2}, -\dfrac{\sqrt{3}}{2}\right)$

11. $t = \dfrac{3\pi}{2}$ corresponds to $(0, -1)$.

12. $t = \pi, (x, y) = (-1, 0)$

13. $t = \dfrac{\pi}{4}$ corresponds to $\left(\dfrac{\sqrt{2}}{2}, \dfrac{\sqrt{2}}{2}\right)$.

$\sin t = y = \dfrac{\sqrt{2}}{2}$

$\cos t = x = \dfrac{\sqrt{2}}{2}$

$\tan t = \dfrac{y}{x} = 1$

14. $t = \dfrac{\pi}{3}, (x, y) = \left(\dfrac{1}{2}, \dfrac{\sqrt{3}}{2}\right)$

$\sin\dfrac{\pi}{3} = \dfrac{\sqrt{3}}{2}$

$\cos\dfrac{\pi}{3} = \dfrac{1}{2}$

$\tan\dfrac{\pi}{3} = \dfrac{\dfrac{\sqrt{3}}{2}}{\dfrac{1}{2}} = \sqrt{3}$

15. $t = -\dfrac{\pi}{6}$ corresponds to $\left(\dfrac{\sqrt{3}}{2}, -\dfrac{1}{2}\right)$.

$\sin t = y = -\dfrac{1}{2}$

$\cos t = x = \dfrac{\sqrt{3}}{2}$

$\tan t = \dfrac{y}{x} = -\dfrac{1}{\sqrt{3}} = -\dfrac{\sqrt{3}}{3}$

16. $t = -\dfrac{\pi}{4}, (x, y) = \left(\dfrac{\sqrt{2}}{2}, -\dfrac{\sqrt{2}}{2}\right)$

$\sin\left(-\dfrac{\pi}{4}\right) = -\dfrac{\sqrt{2}}{2}$

$\cos\left(-\dfrac{\pi}{4}\right) = \dfrac{\sqrt{2}}{2}$

$\tan\left(-\dfrac{\pi}{4}\right) = \dfrac{-\dfrac{\sqrt{2}}{2}}{\dfrac{\sqrt{2}}{2}} = -1$

17. $t = -\frac{7\pi}{4}$ corresponds to $\left(\frac{\sqrt{2}}{2}, \frac{\sqrt{2}}{2}\right)$.

$\sin t = y = \frac{\sqrt{2}}{2}$

$\cos t = x = \frac{\sqrt{2}}{2}$

$\tan t = \frac{y}{x} = 1$

18. $t = -\frac{4\pi}{3}$, $(x, y) = \left(-\frac{1}{2}, \frac{\sqrt{3}}{2}\right)$

$\sin\left(-\frac{4\pi}{3}\right) = \frac{\sqrt{3}}{2}$

$\cos\left(-\frac{4\pi}{3}\right) = -\frac{1}{2}$

$\tan\left(-\frac{4\pi}{3}\right) = \frac{\frac{\sqrt{3}}{2}}{-\frac{1}{2}} = -\sqrt{3}$

19. $t = \frac{11\pi}{6}$ corresponds to $\left(\frac{\sqrt{3}}{2}, -\frac{1}{2}\right)$.

$\sin t = y = -\frac{1}{2}$

$\cos t = x = \frac{\sqrt{3}}{2}$

$\tan t = \frac{y}{x} = -\frac{1}{\sqrt{3}} = -\frac{\sqrt{3}}{3}$

20. $t = \frac{5\pi}{3}$, $(x, y) = \left(\frac{1}{2}, -\frac{\sqrt{3}}{2}\right)$

$\sin \frac{5\pi}{3} = -\frac{\sqrt{3}}{2}$

$\cos \frac{5\pi}{3} = \frac{1}{2}$

$\tan \frac{5\pi}{3} = \frac{-\frac{\sqrt{3}}{2}}{\frac{1}{2}} = -\sqrt{3}$

21. $t = -\frac{3\pi}{2}$ corresponds to $(0, 1)$.

$\sin t = y = 1$

$\cos t = x = 0$

$\tan t = \frac{y}{x}$ is undefined.

22. $t = -2\pi$, $(x, y) = (1, 0)$

$\sin(-2\pi) = 0$

$\cos(-2\pi) = 1$

$\tan(-2\pi) = \frac{0}{1} = 0$

23. $t = \pi$ corresponds to the point $(x, y) = (-1, 0)$.

$\sin \pi = y = 0$

$\cos \pi = x = -1$

$\tan \pi = \frac{y}{x} = \frac{0}{-1} = 0$

24. $t = -\frac{\pi}{2}$ corresponds to the point $(x, y) = (0, -1)$.

$\sin\left(-\frac{\pi}{2}\right) = y = -1$

$\cos\left(-\frac{\pi}{2}\right) = x = 0$

$\tan\left(-\frac{\pi}{2}\right) = \frac{y}{x}$ is undefined.

25. $t = \frac{3\pi}{4}$ corresponds to $\left(-\frac{\sqrt{2}}{2}, \frac{\sqrt{2}}{2}\right)$.

$\sin t = y = \frac{\sqrt{2}}{2}$ $\qquad \csc t = \frac{1}{y} = \sqrt{2}$

$\cos t = x = -\frac{\sqrt{2}}{2}$ $\qquad \sec t = \frac{1}{x} = -\sqrt{2}$

$\tan t = \frac{y}{x} = -1$ $\qquad \cot t = \frac{x}{y} = -1$

26. $t = \frac{5\pi}{6}$, $(x, y) = \left(-\frac{\sqrt{3}}{2}, \frac{1}{2}\right)$

$\sin \frac{5\pi}{6} = \frac{1}{2}$ $\qquad \csc \frac{5\pi}{6} = \frac{1}{\sin t} = 2$

$\cos \frac{5\pi}{6} = -\frac{\sqrt{3}}{2}$ $\qquad \sec \frac{5\pi}{6} = \frac{1}{\cos t} = -\frac{2\sqrt{3}}{3}$

$\tan \frac{5\pi}{6} = \frac{\frac{1}{2}}{-\frac{\sqrt{3}}{2}} = -\frac{\sqrt{3}}{3}$ $\qquad \cot \frac{5\pi}{6} = \frac{1}{\tan t} = -\sqrt{3}$

27. $t = \frac{\pi}{2}$ corresponds to $(0, 1)$.

$\sin t = y = 1$ $\quad$ $\csc t = \frac{1}{y} = 1$

$\cos t = x = 0$ $\quad$ $\sec t = \frac{1}{x}$ is undefined.

$\tan t = \frac{y}{x}$ is undefined. $\quad$ $\cot t = \frac{x}{y} = 0$

28. $t = \frac{3\pi}{2}, (x, y) = (0, -1)$

$\sin \frac{3\pi}{2} = -1$ $\quad$ $\csc \frac{3\pi}{2} = \frac{1}{\sin t} = -1$

$\cos \frac{3\pi}{2} = 0$ $\quad$ $\sec \frac{3\pi}{2}$ is undefined.

$\tan \frac{3\pi}{2}$ is undefined. $\quad$ $\cot \frac{3\pi}{2} = \frac{0}{-1} = 0$

29. $t = -\frac{\pi}{3}$ corresponds to $\left(\frac{1}{2}, -\frac{\sqrt{3}}{2}\right)$.

$\sin t = y = -\frac{\sqrt{3}}{2}$ $\quad$ $\csc t = \frac{1}{y} = -\frac{2\sqrt{3}}{3}$

$\cos t = x = \frac{1}{2}$ $\quad$ $\sec t = \frac{1}{x} = 2$

$\tan t = \frac{y}{x} = -\sqrt{3}$ $\quad$ $\cot t = \frac{x}{y} = -\frac{\sqrt{3}}{3}$

30. $t = -\frac{3\pi}{2}, (x, y) = (0, 1)$

$\sin\left(-\frac{3\pi}{2}\right) = 1$ $\quad$ $\csc\left(-\frac{3\pi}{2}\right) = \frac{1}{\sin t} = 1$

$\cos\left(-\frac{3\pi}{2}\right) = 0$ $\quad$ $\sec\left(-\frac{3\pi}{2}\right)$ is undefined.

$\tan\left(-\frac{3\pi}{2}\right)$ is undefined. $\quad$ $\cot\left(-\frac{3\pi}{2}\right) = \frac{0}{1} = 0$

31. $\sin 5\pi = \sin \pi = 0$

32. $\cos 5\pi = \cos \pi = -1$

33. $\cos \frac{8\pi}{3} = \cos \frac{2\pi}{3} = -\frac{1}{2}$

34. $\sin \frac{9\pi}{4} = \sin \frac{\pi}{4} = \frac{\sqrt{2}}{2}$

35. $\cos(-3\pi) = \cos \pi = -1$

36. $\sin (-3\pi) = \sin(-\pi) = 0$

37. $\sin\left(-\frac{9\pi}{4}\right) = \sin\left(\frac{7\pi}{4}\right) = -\frac{\sqrt{2}}{2}$

38. $\cos\left(-\frac{8\pi}{3}\right) = \cos \frac{4\pi}{3} = -\frac{1}{2}$

39. $\sin t = \frac{1}{3}$

(a) $\sin(-t) = -\sin t = -\frac{1}{3}$

(b) $\csc(-t) = -\csc t = -3$

40. $\sin(-t) = \frac{3}{8}$

(a) $\sin t = -\sin(-t) = -\frac{3}{8}$

(b) $\csc t = \frac{1}{\sin t} = -\frac{8}{3}$

41. $\cos(-t) = -\frac{1}{5}$

(a) $\cos t = \cos(-t) = -\frac{1}{5}$

(b) $\sec(-t) = \frac{1}{\cos(-t)} = -5$

42. $\cos t = -\frac{3}{4}$

(a) $\cos(-t) = \cos t = -\frac{3}{4}$

(b) $\sec(-t) = \sec t = \frac{1}{\cos t} = -\frac{4}{3}$

43. $\sin t = \frac{4}{5}$

(a) $\sin(\pi - t) = \sin t = \frac{4}{5}$

(b) $\sin(t + \pi) = -\sin t = -\frac{4}{5}$

44. $\cos t = \frac{4}{5}$

(a) $\cos(\pi - t) = -\cos t = -\frac{4}{5}$

(b) $\cos(t + \pi) = -\cos t = -\frac{4}{5}$

45. $\sin \frac{\pi}{4} \approx 0.7071$

46. $\tan \frac{\pi}{3} \approx 1.7321$

47. $\csc 1.3 = \frac{1}{\sin 1.3} \approx 1.0378$

48. $\cot 1 \approx 0.6421$

49. $\cos(-1.7) \approx -0.1288$

50. $\cos(-2.5) \approx -0.8011$

51. $\csc 0.8 = \dfrac{1}{\sin 0.8} \approx 1.3940$

52. $\sec 1.8 \approx -4.4014$

53. $\sec 22.8 = \dfrac{1}{\cos 22.8} \approx -1.4486$

54. $\sin(-0.9) \approx -0.7833$

55. (a) $\sin 5 \approx -1$

(b) $\cos 2 \approx -0.4$

56. (a) $\sin 0.75 = y \approx 0.7$

(b) $\cos 2.5 = x \approx -0.8$

57. (a) $\sin t = 0.25$

$t \approx 0.25$ or 2.89

(b) $\cos t = -0.25$

$t \approx 1.82$ or 4.46

58. (a) $\sin t = -0.75$

$t \approx 4.0$ or $t \approx 5.4$

(b) $\cos t = 0.75$

$t \approx 0.7$ or $t \approx 5.6$

59. $\cos 1.5 \approx 0.0707$

$2 \cos 0.75 \approx 1.4634$

$\cos 2t \neq 2 \cos t$

60. $\sin 1 = \sin(0.25 + 0.75) \approx 0.8415$

$\sin 0.25 + \sin 0.75 \approx 0.2474 + 0.6816$

≈ 0.9290

61. (a) The points have y-axis symmetry.

(b) $\sin t_1 = \sin(\pi - t_1)$ since they have the same y-value.

(c) $\cos(\pi - t_1) = -\cos t_1$ since the x-values have the opposite signs.

62. (a) (x_1, y_1) and (x_2, y_2) are symmetric about the origin.

(b) Since $(x_2, y_2) = (-x_1, -y_1)$, $\sin(t_1 + \pi) = -\sin t_1$.

(c) Since $(x_2, y_2) = (-x_1, -y_1)$, $\cos(t_1 + \pi) = -\cos t_1$.

63. $\cos \theta = x = \cos(-\theta)$

$\sec \theta = \dfrac{1}{x} = \sec(-\theta)$

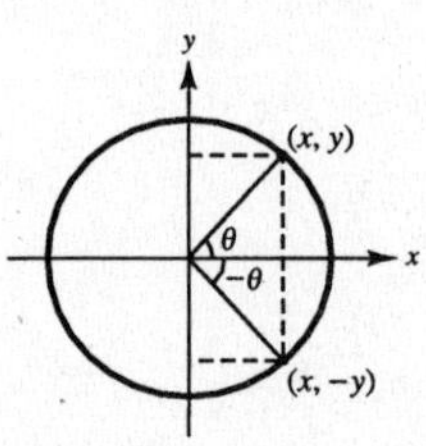

64. $\sin \theta = y$

$\sin(-\theta) = -y = -\sin \theta$

$\csc \theta = \dfrac{1}{y}$

$\csc(-\theta) = -\dfrac{1}{y} = -\csc \theta$

$\tan \theta = \dfrac{y}{x}$

$\tan(-\theta) = \dfrac{-y}{x} = -\tan \theta$

$\cot \theta = \dfrac{x}{y}$

$\cot(-\theta) = \dfrac{x}{-y} = -\cot \theta$

65. $h(t) = f(t)g(t)$ is an odd function, since

$h(-t) = f(-t)g(-t)$

$= -f(t)g(t)$

$= -h(t)$.

66. $f(t) = \sin t$ and $g(t) = \tan t$

Both f and g are odd functions.

$h(t) = f(t)g(t) = \sin t \tan t$

$h(-t) = \sin(-t)\tan(-t)$

$= (-\sin t)(-\tan t)$

$= \sin t \tan t = h(t)$

The function $h(t) = f(t)g(t)$ is even.

67. $y(t) = \frac{1}{4}\cos 6t$

(a) $y(0) = \frac{1}{4}\cos 0 = 0.2500$ feet

(b) $y\left(\frac{1}{4}\right) = \frac{1}{4}\cos\frac{3}{2} \approx 0.0177$ feet

(c) $y\left(\frac{1}{2}\right) = \frac{1}{4}\cos 3 \approx -0.2475$ feet

68. $y(t) = \dfrac{1}{4}e^{-t}\cos 6t$

(a) $t = 0$ $\quad y(0) = \dfrac{1}{4}e^{-0}\cos 0 = \dfrac{1}{4} = 0.2500$ foot

(b) $t = \dfrac{1}{4}$ $\quad y\left(\dfrac{1}{4}\right) = \dfrac{1}{4}e^{-1/4}\cos\left(6 \cdot \dfrac{1}{4}\right) \approx 0.0138$ foot

(c) $t = \dfrac{1}{2}$ $\quad y\left(\dfrac{1}{2}\right) = \dfrac{1}{4}e^{-1/2}\cos\left(6 \cdot \dfrac{1}{2}\right) \approx -0.1501$ foot

69. False. $\sin(-t) = -\sin t$ means the function is odd, not that the sine of a negative angle is a negative number.

For example: $\sin\left(-\dfrac{3\pi}{2}\right) = -\sin\left(\dfrac{3\pi}{2}\right) = -(-1) = 1.$

Even though the angle is negative, the sine value is positive.

70. True. $\tan a = \tan(a - 6\pi)$ since the period of tan is π.

71. False. Because the sine function is odd, its graph is symmetric with respect to the origin, not the y-axis.

72. True. Because the cosine function is even, $\cos(-x) = \cos x$ and the graph is symmetric with respect to the y-axis.

Section 9.3 Right Triangle Trigonometry

1. $\text{hyp} = \sqrt{6^2 + 8^2} = \sqrt{36 + 64} = \sqrt{100} = 10$

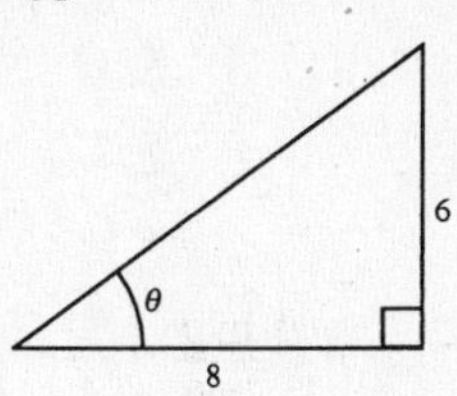

$\sin\theta = \dfrac{\text{opp}}{\text{hyp}} = \dfrac{6}{10} = \dfrac{3}{5}$ $\qquad \csc\theta = \dfrac{\text{hyp}}{\text{opp}} = \dfrac{10}{6} = \dfrac{5}{3}$

$\cos\theta = \dfrac{\text{adj}}{\text{hyp}} = \dfrac{8}{10} = \dfrac{4}{5}$ $\qquad \sec\theta = \dfrac{\text{hyp}}{\text{adj}} = \dfrac{10}{8} = \dfrac{5}{4}$

$\tan\theta = \dfrac{\text{opp}}{\text{adj}} = \dfrac{6}{8} = \dfrac{3}{4}$ $\qquad \cot\theta = \dfrac{\text{adj}}{\text{opp}} = \dfrac{8}{6} = \dfrac{4}{3}$

2.

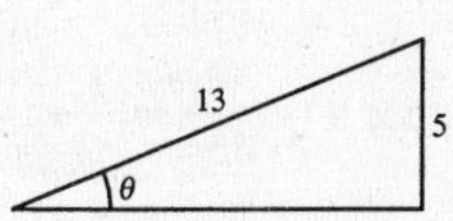

$\text{adj} = \sqrt{13^2 - 5^2} = \sqrt{169 - 25} = 12$

$\sin\theta = \dfrac{\text{opp}}{\text{hyp}} = \dfrac{5}{13}$ $\qquad \csc\theta = \dfrac{\text{hyp}}{\text{opp}} = \dfrac{13}{5}$

$\cos\theta = \dfrac{\text{adj}}{\text{hyp}} = \dfrac{12}{13}$ $\qquad \sec\theta = \dfrac{\text{hyp}}{\text{adj}} = \dfrac{13}{12}$

$\tan\theta = \dfrac{\text{opp}}{\text{adj}} = \dfrac{5}{12}$ $\qquad \cot\theta = \dfrac{\text{adj}}{\text{opp}} = \dfrac{12}{5}$

3. $\text{adj} = \sqrt{41^2 - 9^2} = \sqrt{1681 - 81} = \sqrt{1600} = 40$

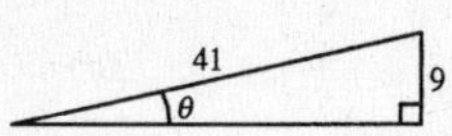

$\sin\theta = \dfrac{\text{opp}}{\text{hyp}} = \dfrac{9}{41}$ $\qquad \csc\theta = \dfrac{\text{hyp}}{\text{opp}} = \dfrac{41}{9}$

$\cos\theta = \dfrac{\text{adj}}{\text{hyp}} = \dfrac{40}{41}$ $\qquad \sec\theta = \dfrac{\text{hyp}}{\text{adj}} = \dfrac{41}{40}$

$\tan\theta = \dfrac{\text{opp}}{\text{adj}} = \dfrac{9}{40}$ $\qquad \cot\theta = \dfrac{\text{adj}}{\text{opp}} = \dfrac{40}{9}$

4.

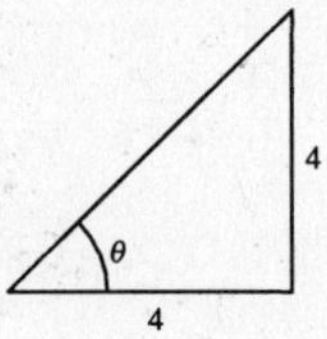

$$\text{hyp} = \sqrt{4^2 + 4^2} = \sqrt{32} = 4\sqrt{2}$$

$$\sin\theta = \frac{\text{opp}}{\text{hyp}} = \frac{4}{4\sqrt{2}} = \frac{1}{\sqrt{2}} = \frac{\sqrt{2}}{2} \qquad \csc\theta = \frac{\text{hyp}}{\text{opp}} = \frac{4\sqrt{2}}{4} = \sqrt{2}$$

$$\cos\theta = \frac{\text{adj}}{\text{hyp}} = \frac{4}{4\sqrt{2}} = \frac{1}{\sqrt{2}} = \frac{\sqrt{2}}{2} \qquad \sec\theta = \frac{\text{hyp}}{\text{adj}} = \frac{4\sqrt{2}}{4} = \sqrt{2}$$

$$\tan\theta = \frac{\text{opp}}{\text{adj}} = \frac{4}{4} = 1 \qquad \cot\theta = \frac{\text{adj}}{\text{opp}} = \frac{4}{4} = 1$$

5. $\text{adj} = \sqrt{3^2 - 1^2} = \sqrt{8} = 2\sqrt{2}$

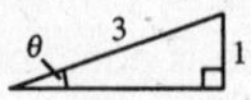

$$\sin\theta = \frac{\text{opp}}{\text{hyp}} = \frac{1}{3} \qquad \csc\theta = \frac{\text{hyp}}{\text{opp}} = 3$$

$$\cos\theta = \frac{\text{adj}}{\text{hyp}} = \frac{2\sqrt{2}}{3} \qquad \sec\theta = \frac{\text{hyp}}{\text{adj}} = \frac{3}{2\sqrt{2}} = \frac{3\sqrt{2}}{4}$$

$$\tan\theta = \frac{\text{opp}}{\text{adj}} = \frac{1}{2\sqrt{2}} = \frac{\sqrt{2}}{4} \qquad \cot\theta = \frac{\text{adj}}{\text{opp}} = 2\sqrt{2}$$

$\text{adj} = \sqrt{6^2 - 2^2} = \sqrt{32} = 4\sqrt{2}$

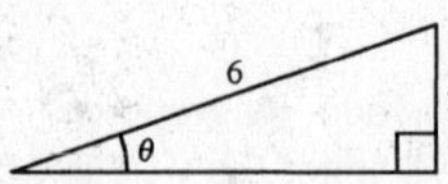

$$\sin\theta = \frac{\text{opp}}{\text{hyp}} = \frac{2}{6} = \frac{1}{3} \qquad \csc\theta = \frac{\text{hyp}}{\text{opp}} = \frac{6}{2} = 3$$

$$\cos\theta = \frac{\text{adj}}{\text{hyp}} = \frac{4\sqrt{2}}{6} = \frac{2\sqrt{2}}{3} \qquad \sec\theta = \frac{\text{hyp}}{\text{adj}} = \frac{6}{4\sqrt{2}} = \frac{3}{2\sqrt{2}} = \frac{3\sqrt{2}}{4}$$

$$\tan\theta = \frac{\text{opp}}{\text{adj}} = \frac{2}{4\sqrt{2}} = \frac{1}{2\sqrt{2}} = \frac{\sqrt{2}}{4} \qquad \cot\theta = \frac{\text{adj}}{\text{opp}} = \frac{4\sqrt{2}}{2} = 2\sqrt{2}$$

The function values are the same since the triangles are similar and the corresponding sides are proportional.

6.

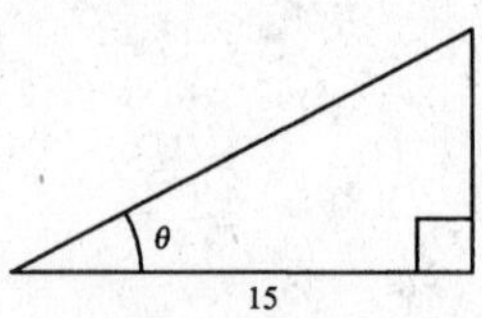

$$\text{hyp} = \sqrt{15^2 + 8^2} = \sqrt{289} = 17$$

$$\sin\theta = \frac{\text{opp}}{\text{hyp}} = \frac{8}{17} \qquad \csc\theta = \frac{\text{hyp}}{\text{opp}} = \frac{17}{8}$$

$$\cos\theta = \frac{\text{adj}}{\text{hyp}} = \frac{15}{17} \qquad \sec\theta = \frac{\text{hyp}}{\text{adj}} = \frac{17}{15}$$

$$\tan\theta = \frac{\text{opp}}{\text{adj}} = \frac{8}{15} \qquad \cot\theta = \frac{\text{adj}}{\text{opp}} = \frac{15}{8}$$

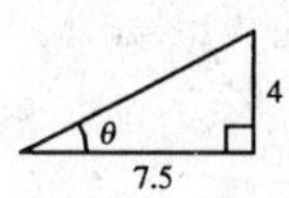

$$\text{hyp} = \sqrt{7.5^2 + 4^2} = \frac{17}{2}$$

$$\sin\theta = \frac{\text{opp}}{\text{hyp}} = \frac{4}{\frac{17}{2}} = \frac{8}{17} \qquad \csc\theta = \frac{\text{hyp}}{\text{opp}} = \frac{\frac{17}{2}}{4} = \frac{17}{8}$$

$$\cos\theta = \frac{\text{adj}}{\text{hyp}} = \frac{7.5}{\frac{17}{2}} = \frac{15}{17} \qquad \sec\theta = \frac{\text{hyp}}{\text{adj}} = \frac{\frac{17}{2}}{7.5} = \frac{17}{15}$$

$$\tan\theta = \frac{\text{opp}}{\text{adj}} = \frac{4}{7.5} = \frac{8}{15} \qquad \cot\theta = \frac{\text{adj}}{\text{opp}} = \frac{7.5}{4} = \frac{15}{8}$$

The function values are the same because the triangles are similar, and corresponding sides are proportional.

7. $\text{opp} = \sqrt{5^2 - 4^2} = 3$

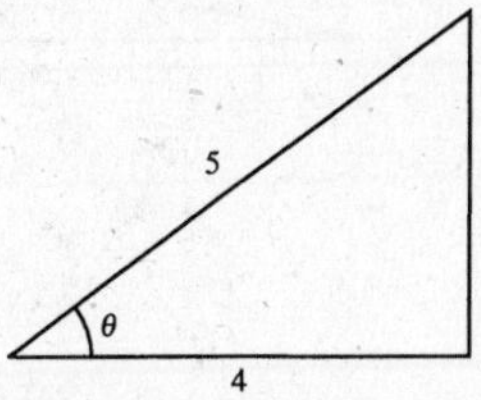

$\sin\theta = \dfrac{\text{opp}}{\text{hyp}} = \dfrac{3}{5}$ $\qquad \csc\theta = \dfrac{\text{hyp}}{\text{opp}} = \dfrac{5}{3}$

$\cos\theta = \dfrac{\text{adj}}{\text{hyp}} = \dfrac{4}{5}$ $\qquad \sec\theta = \dfrac{\text{hyp}}{\text{adj}} = \dfrac{5}{4}$

$\tan\theta = \dfrac{\text{opp}}{\text{adj}} = \dfrac{3}{4}$ $\qquad \cot\theta = \dfrac{\text{adj}}{\text{opp}} = \dfrac{4}{3}$

$\text{opp} = \sqrt{1.25^2 - 1^2} = 0.75$

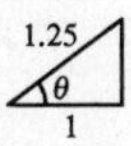

$\sin\theta = \dfrac{\text{opp}}{\text{hyp}} = \dfrac{0.75}{1.25} = \dfrac{3}{5}$ $\qquad \csc\theta = \dfrac{\text{hyp}}{\text{opp}} = \dfrac{1.25}{0.75} = \dfrac{5}{3}$

$\cos\theta = \dfrac{\text{adj}}{\text{hyp}} = \dfrac{1}{1.25} = \dfrac{4}{5}$ $\qquad \sec\theta = \dfrac{\text{hyp}}{\text{adj}} = \dfrac{1.25}{1} = \dfrac{5}{4}$

$\tan\theta = \dfrac{\text{opp}}{\text{adj}} = \dfrac{0.75}{1} = \dfrac{3}{4}$ $\qquad \cot\theta = \dfrac{\text{adj}}{\text{opp}} = \dfrac{1}{0.75} = \dfrac{4}{3}$

The function values are the same since the triangles are similar and the corresponding sides are proportional.

8.

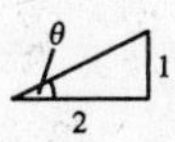

$\text{hyp} = \sqrt{1^2 + 2^2} = \sqrt{5}$

$\sin\theta = \dfrac{\text{opp}}{\text{hyp}} = \dfrac{1}{\sqrt{5}} = \dfrac{\sqrt{5}}{5}$ $\qquad \csc\theta = \dfrac{\text{hyp}}{\text{opp}} = \dfrac{\sqrt{5}}{1} = \sqrt{5}$

$\cos\theta = \dfrac{\text{adj}}{\text{hyp}} = \dfrac{2}{\sqrt{5}} = \dfrac{2\sqrt{5}}{5}$ $\qquad \sec\theta = \dfrac{\text{hyp}}{\text{adj}} = \dfrac{\sqrt{5}}{2}$

$\tan\theta = \dfrac{\text{opp}}{\text{adj}} = \dfrac{1}{2}$ $\qquad \cot\theta = \dfrac{\text{adj}}{\text{opp}} = \dfrac{2}{1} = 2$

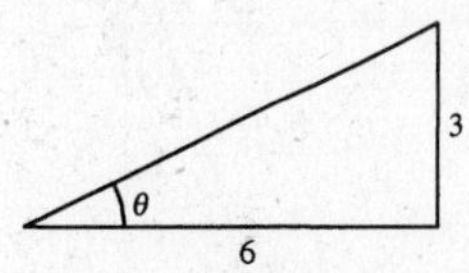

$\text{hyp} = \sqrt{3^2 + 6^2} = 3\sqrt{5}$

$\sin\theta = \dfrac{3}{3\sqrt{5}} = \dfrac{1}{\sqrt{5}} = \dfrac{\sqrt{5}}{5}$ $\qquad \csc\theta = \dfrac{3\sqrt{5}}{3} = \sqrt{5}$

$\cos\theta = \dfrac{6}{3\sqrt{5}} = \dfrac{2}{\sqrt{5}} = \dfrac{2\sqrt{5}}{5}$ $\qquad \sec\theta = \dfrac{3\sqrt{5}}{6} = \dfrac{\sqrt{5}}{2}$

$\tan\theta = \dfrac{3}{6} = \dfrac{1}{2}$ $\qquad \cot\theta = \dfrac{6}{3} = 2$

The function values are the same because the triangles are similar, and corresponding sides are proportional.

9. Given: $\sin\theta = \dfrac{3}{4} = \dfrac{\text{opp}}{\text{hyp}}$

$3^2 + (\text{adj})^2 = 4^2$

$\text{adj} = \sqrt{7}$

$\cos\theta = \dfrac{\sqrt{7}}{4}$

$\tan\theta = \dfrac{3\sqrt{7}}{7}$

$\cot\theta = \dfrac{\sqrt{7}}{3}$

$\sec\theta = \dfrac{4\sqrt{7}}{7}$

$\csc\theta = \dfrac{4}{3}$

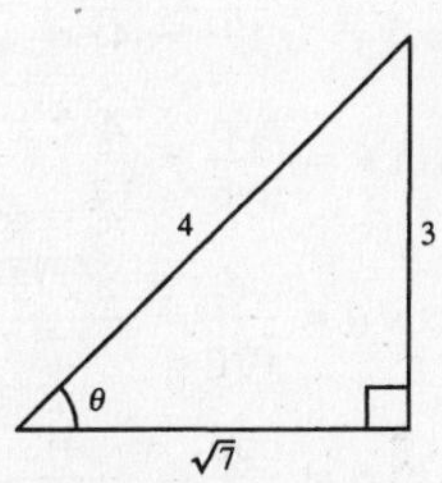

10. $\text{hyp} = \sqrt{7^2 - 5^2} = \sqrt{24} = 2\sqrt{6}$

$\sin\theta = \dfrac{\text{opp}}{\text{hyp}} = \dfrac{2\sqrt{6}}{7}$

$\tan\theta = \dfrac{\text{opp}}{\text{adj}} = \dfrac{2\sqrt{6}}{5}$

$\csc\theta = \dfrac{1}{\sin\theta} = \dfrac{7}{2\sqrt{6}} = \dfrac{7\sqrt{6}}{12}$

$\sec\theta = \dfrac{1}{\cos\theta} = \dfrac{7}{5}$

$\cot\theta = \dfrac{1}{\tan\theta} = \dfrac{5}{2\sqrt{6}} = \dfrac{5\sqrt{6}}{12}$

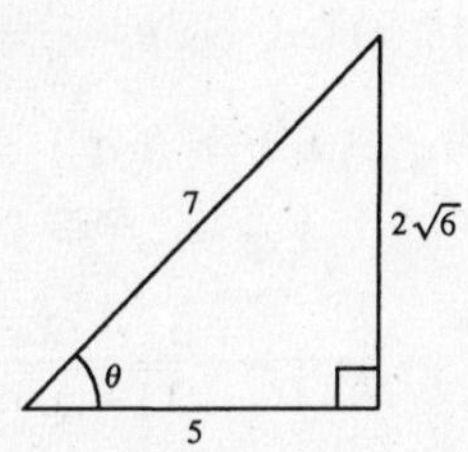

11. Given: $\sec\theta = 2 = \frac{2}{1} = \frac{\text{hyp}}{\text{adj}}$

$(\text{opp})^2 + 1^2 = 2^2$

$\text{opp} = \sqrt{3}$

$\sin\theta = \frac{\sqrt{3}}{2}$

$\cos\theta = \frac{1}{2}$

$\tan\theta = \sqrt{3}$

$\cot\theta = \frac{\sqrt{3}}{3}$

$\csc\theta = \frac{2\sqrt{3}}{3}$

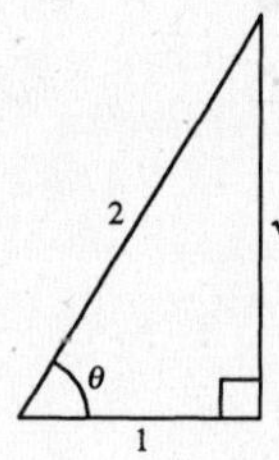

12. $\text{hyp} = \sqrt{5^2 + 1^2} = \sqrt{26}$

$\sin\theta = \frac{\text{opp}}{\text{hyp}} = \frac{1}{\sqrt{26}} = \frac{\sqrt{26}}{26}$

$\cos\theta = \frac{\text{adj}}{\text{hyp}} = \frac{5}{\sqrt{26}} = \frac{5\sqrt{26}}{26}$

$\tan\theta = \frac{\text{opp}}{\text{adj}} = \frac{1}{5}$

$\csc\theta = \frac{\text{hyp}}{\text{opp}} = \frac{\sqrt{26}}{1} = \sqrt{26}$

$\sec\theta = \frac{\text{hyp}}{\text{adj}} = \frac{\sqrt{26}}{5}$

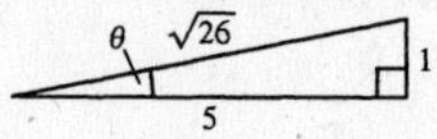

13. Given: $\tan\theta = 3 = \frac{3}{1} = \frac{\text{opp}}{\text{adj}}$

$3^2 + 1^2 = (\text{hyp})^2$

$\text{hyp} = \sqrt{10}$

$\sin\theta = \frac{3\sqrt{10}}{10}$

$\cos\theta = \frac{\sqrt{10}}{10}$

$\cot\theta = \frac{1}{3}$

$\sec\theta = \sqrt{10}$

$\csc\theta = \frac{\sqrt{10}}{3}$

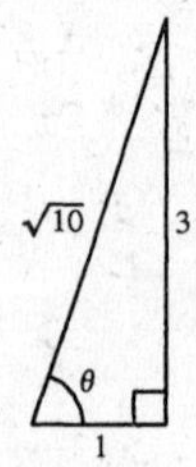

14. $\text{opp} = \sqrt{6^2 - 1^2} = \sqrt{35}$

$\sin\theta = \frac{\text{opp}}{\text{hyp}} = \frac{\sqrt{35}}{6}$

$\cos\theta = \frac{\text{adj}}{\text{hyp}} = \frac{1}{6}$

$\tan\theta = \frac{\text{opp}}{\text{adj}} = \frac{\sqrt{35}}{1} = \sqrt{35}$

$\csc\theta = \frac{\text{hyp}}{\text{opp}} = \frac{6}{\sqrt{35}} = \frac{6\sqrt{35}}{35}$

$\cot\theta = \frac{\text{adj}}{\text{opp}} = \frac{1}{\sqrt{35}} = \frac{\sqrt{35}}{35}$

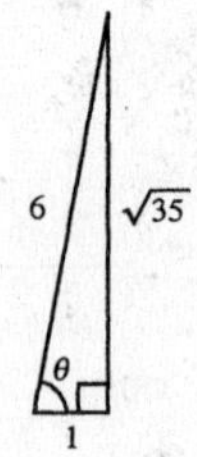

15. Given: $\cot\theta = \frac{3}{2} = \frac{\text{adj}}{\text{opp}}$

$2^2 + 3^2 = (\text{hyp})^2$

$\text{hyp} = \sqrt{13}$

$\sin\theta = \frac{2}{\sqrt{13}} = \frac{2\sqrt{13}}{13}$

$\cos\theta = \frac{3}{\sqrt{13}} = \frac{3\sqrt{13}}{13}$

$\tan\theta = \frac{2}{3}$

$\csc\theta = \frac{\sqrt{13}}{2}$

$\sec\theta = \frac{\sqrt{13}}{3}$

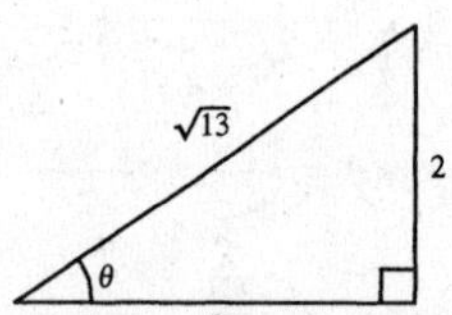

16. $\text{adj} = \sqrt{17^2 - 4^2} = \sqrt{273}$

$\sin\theta = \frac{\text{opp}}{\text{hyp}} = \frac{4}{17}$

$\cos\theta = \frac{\text{adj}}{\text{hyp}} = \frac{\sqrt{273}}{17}$

$\tan\theta = \frac{\text{opp}}{\text{adj}} = \frac{4}{\sqrt{273}} = \frac{4\sqrt{273}}{273}$

$\sec\theta = \frac{1}{\cos\theta} = \frac{17}{\sqrt{273}} = \frac{17\sqrt{273}}{273}$

$\cot\theta = \frac{1}{\tan\theta} = \frac{\sqrt{273}}{4}$

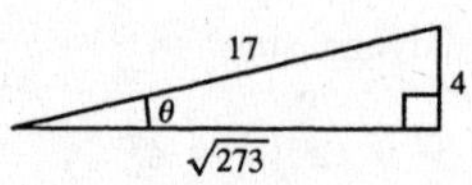

17.

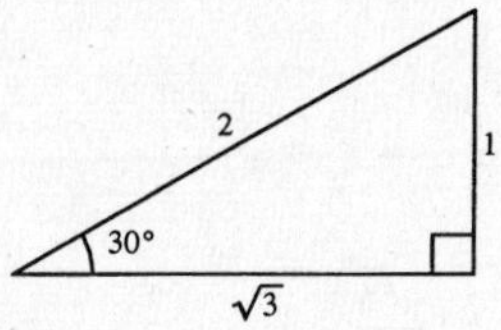

Function	θ(deg)	θ(rad)	Function Value
sin	30°	$\frac{\pi}{6}$	$\frac{1}{2}$

18.

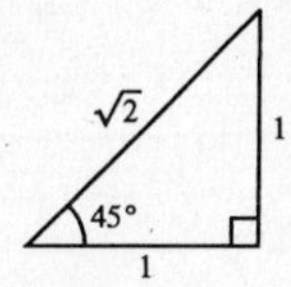

Function	θ(deg)	θ(rad)	Function Value
cos	45°	$\frac{\pi}{4}$	$\frac{\sqrt{2}}{2}$

19.

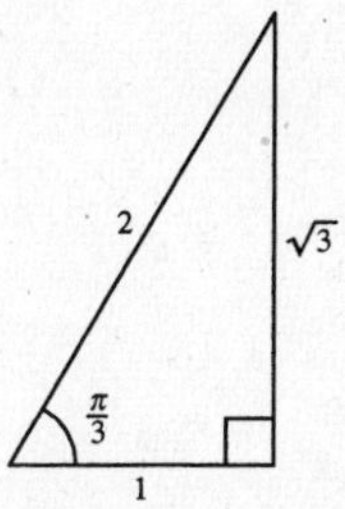

Function	θ(deg)	θ(rad)	Function Value
tan	60°	$\frac{\pi}{3}$	$\sqrt{3}$

20.

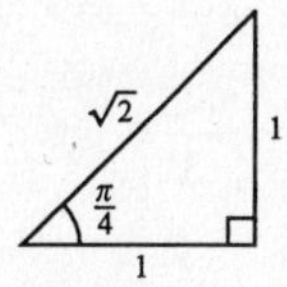

Function	θ(deg)	θ(rad)	Function Value
cot	45°	$\frac{\pi}{4}$	1

21.

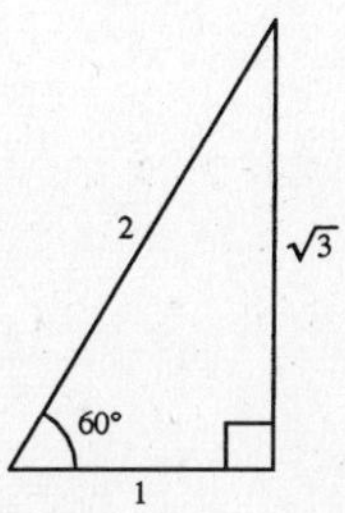

Function	θ(deg)	θ(rad)	Function Value
sec	60°	$\frac{\pi}{3}$	2

22.

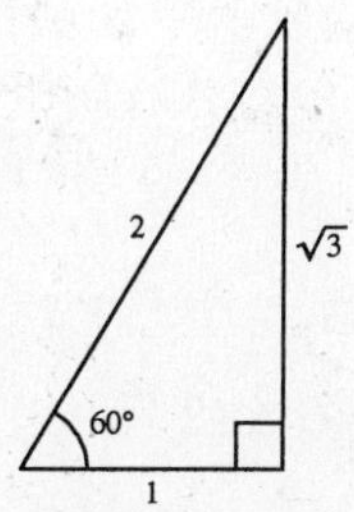

Function	θ(deg)	θ(rad)	Function Value
csc	60°	$\frac{\pi}{3}$	$\frac{2\sqrt{3}}{3}$

23.

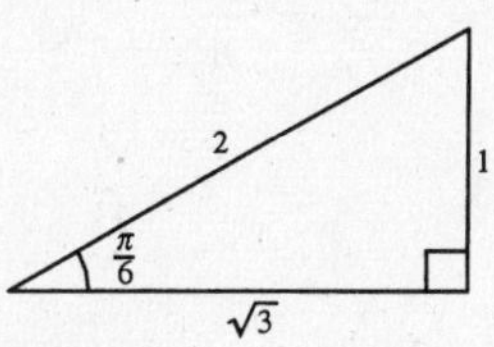

Function	θ(deg)	θ(rad)	Function Value
cos	30°	$\frac{\pi}{6}$	$\frac{\sqrt{3}}{2}$

24.

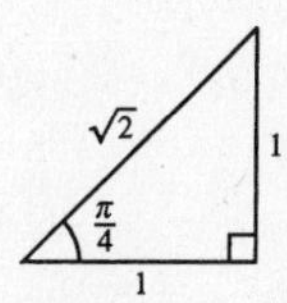

Function	θ(deg)	θ(rad)	Function Value
sin	45°	$\frac{\pi}{4}$	$\frac{\sqrt{2}}{2}$

25.

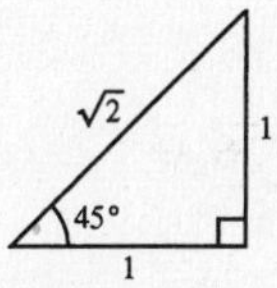

Function	θ(deg)	θ(rad)	Function Value
tan	45°	$\frac{\pi}{4}$	1

26.

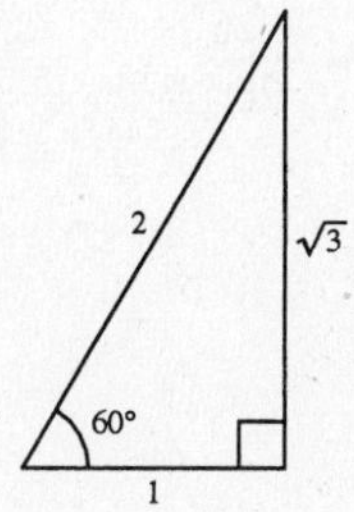

Function	θ(deg)
cot	60°

θ(rad)	Function Value
$\frac{\pi}{3}$	$\frac{\sqrt{3}}{3}$

27. $\sin 60° = \dfrac{\sqrt{3}}{2}$, $\cos 60° = \dfrac{1}{2}$

(a) $\tan 60° = \dfrac{\sin 60°}{\cos 60°} = \sqrt{3}$

(b) $\sin 30° = \cos 60° = \dfrac{1}{2}$

(c) $\cos 30° = \sin 60° = \dfrac{\sqrt{3}}{2}$

(d) $\cot 60° = \dfrac{\cos 60°}{\sin 60°} = \dfrac{1}{\sqrt{3}} = \dfrac{\sqrt{3}}{3}$

28. $\sin 30° = \dfrac{1}{2}$, $\tan 30° = \dfrac{\sqrt{3}}{3}$

(a) $\csc 30° = \dfrac{1}{\sin 30°} = 2$

(b) $\cot 60° = \tan(90° - 60°) = \tan 30° = \dfrac{\sqrt{3}}{3}$

(c) $\cos 30° = \dfrac{\sin 30°}{\tan 30°} = \dfrac{\frac{1}{2}}{\frac{\sqrt{3}}{3}} = \dfrac{3}{2\sqrt{3}} = \dfrac{\sqrt{3}}{2}$

(d) $\cot 30° = \dfrac{1}{\tan 30°} = \dfrac{3}{\sqrt{3}} = \dfrac{3\sqrt{3}}{3} = \sqrt{3}$

29. $\csc\theta = \dfrac{\sqrt{13}}{2}$, $\sec\theta = \dfrac{\sqrt{13}}{3}$

(a) $\sin\theta = \dfrac{1}{\csc\theta} = \dfrac{2}{\sqrt{13}} = \dfrac{2\sqrt{13}}{13}$

(b) $\cos\theta = \dfrac{1}{\sec\theta} = \dfrac{3}{\sqrt{13}} = \dfrac{3\sqrt{13}}{13}$

(c) $\tan\theta = \dfrac{\sin\theta}{\cos\theta} = \dfrac{\frac{2\sqrt{13}}{13}}{\frac{3\sqrt{13}}{13}} = \dfrac{2}{3}$

(d) $\sec(90° - \theta) = \csc\theta = \dfrac{\sqrt{13}}{2}$

30. $\sec\theta = 5$, $\tan\theta = 2\sqrt{6}$

(a) $\cos\theta = \dfrac{1}{\sec\theta} = \dfrac{1}{5}$

(b) $\cot\theta = \dfrac{1}{\tan\theta} = \dfrac{1}{2\sqrt{6}} = \dfrac{\sqrt{6}}{12}$

(c) $\cot(90° - \theta) = \tan\theta = 2\sqrt{6}$

(d) $\sin\theta = \tan\theta\cos\theta = \left(2\sqrt{6}\right)\left(\dfrac{1}{5}\right) = \dfrac{2\sqrt{6}}{5}$

31. $\cos\alpha = \dfrac{1}{3}$

(a) $\sec\alpha = \dfrac{1}{\cos\alpha} = 3$

(b) $\sin^2\alpha + \cos^2\alpha = 1$

$$\sin^2\alpha + \left(\frac{1}{3}\right)^2 = 1$$

$$\sin^2\alpha = \frac{8}{9}$$

$$\sin\alpha = \frac{2\sqrt{2}}{3}$$

(c) $\cot\alpha = \dfrac{\cos\alpha}{\sin\alpha} = \dfrac{\frac{1}{3}}{2\sqrt{2}/3} = \dfrac{1}{2\sqrt{2}} = \dfrac{\sqrt{2}}{4}$

(d) $\sin(90° - \alpha) = \cos\alpha = \dfrac{1}{3}$

32. $\tan\beta = 5$

(a) $\cot\beta = \dfrac{1}{\tan\beta} = \dfrac{1}{5}$

(b) $\cos\beta = \dfrac{1}{\sqrt{1 + \tan^2\beta}} = \dfrac{1}{\sqrt{1 + 5^2}}$

$$= \frac{1}{\sqrt{26}} = \frac{\sqrt{26}}{26}$$

(c) $\tan(90° - \beta) = \cot\beta = \dfrac{1}{\tan\beta} = \dfrac{1}{5}$

(d) $\csc\beta = \sqrt{1 + \cot^2\beta}$

$$= \sqrt{1 + \left(\frac{1}{5}\right)^2}$$

$$= \sqrt{1 + \frac{1}{25}} = \sqrt{\frac{26}{25}} = \frac{\sqrt{26}}{5}$$

33. (a) $\sin 10° \approx 0.1736$

(b) $\cos 80° \approx 0.1736$

Note: $\cos 80° = \sin(90° - 80°) = \sin 10°$

34. (a) $\tan 23.5° \approx 0.4348$

(b) $\cot 66.5° = \dfrac{1}{\tan 66.5°} \approx 0.4348$

35. (a) $\sin 16.35° \approx 0.2815$

(b) $\csc 16.35° = \dfrac{1}{\sin 16.35°} \approx 3.5523$

36. (a) $\cos 16° 18' = \cos\left(16 + \dfrac{18}{60}\right)^\circ \approx 0.9598$

(b) $\sin 73° 56' = \sin\left(73 + \dfrac{56}{60}\right)^\circ \approx 0.9609$

37. (a) $\sec 42°12' = \sec 42.2° = \dfrac{1}{\cos 42.2°} \approx 1.3499$

(b) $\csc 48°7' = \dfrac{1}{\sin\left(48 + \frac{7}{60}\right)^\circ} \approx 1.3432$

38. (a) $\cos 4° 50' 15'' = \cos\left(4 + \dfrac{50}{60} + \dfrac{15}{3600}\right)^\circ$

≈ 0.9964

(b) $\sec 4° 50' 15'' = \dfrac{1}{\cos 4° 50' 15''}$

≈ 1.0036

39. (a) $\cot 11°15' = \dfrac{1}{\tan 11.25°} \approx 5.0273$

(b) $\tan 11° 15' = \tan 11.25° \approx 0.1989$

40. (a) $\sec 56° 8' 10'' = \sec\left(56 + \dfrac{8}{60} + \dfrac{10}{3600}\right)^\circ \approx 1.7946$

(b) $\cos 56° 8' 10'' = \cos\left(56 + \dfrac{8}{60} + \dfrac{10}{3600}\right)^\circ \approx 0.5572$

41. (a) $\csc 32° 40' 3'' = \dfrac{1}{\sin 32.6675°} \approx 1.8527$

(b) $\tan 44° 28' 16'' \approx \tan 44.4711° \approx 0.9817$

42. (a) $\sec\left(\dfrac{9}{5} \cdot 20 + 32\right)^\circ \approx 2.6695$

(b) $\cot\left(\dfrac{9}{5} \cdot 30 + 32\right)^\circ \approx 0.0699$

43. (a) $\sin\theta = \dfrac{1}{2} \Rightarrow \theta = 30° = \dfrac{\pi}{6}$

(b) $\csc\theta = 2 \Rightarrow \theta = 30° = \dfrac{\pi}{6}$

44. (a) $\cos\theta = \dfrac{\sqrt{2}}{2} \Rightarrow \theta = 45° = \dfrac{\pi}{4}$

(b) $\tan\theta = 1 \Rightarrow \theta = 45° = \dfrac{\pi}{4}$

45. (a) $\sec\theta = 2 \Rightarrow \theta = 60° = \dfrac{\pi}{3}$

(b) $\cot\theta = 1 \Rightarrow \theta = 45° = \dfrac{\pi}{4}$

46. (a) $\tan\theta = \sqrt{3} \Rightarrow \theta = 60° = \dfrac{\pi}{3}$

(b) $\cos\theta = \dfrac{1}{2} \Rightarrow \theta = 60° = \dfrac{\pi}{3}$

47. (a) $\csc\theta = \dfrac{2\sqrt{3}}{3} \Rightarrow \theta = 60° = \dfrac{\pi}{3}$

(b) $\sin\theta = \dfrac{\sqrt{2}}{2} \Rightarrow \theta = 45° = \dfrac{\pi}{4}$

48. (a) $\cot\theta = \dfrac{\sqrt{3}}{3}$

$\tan\theta = \dfrac{3}{\sqrt{3}} = \sqrt{3} \Rightarrow \theta = 60° = \dfrac{\pi}{3}$

(b) $\sec\theta = \sqrt{2}$

$\cos\theta = \dfrac{1}{\sqrt{2}} = \dfrac{\sqrt{2}}{2} \Rightarrow \theta = 45° = \dfrac{\pi}{4}$

49. (a) $\sin\theta = 0.0145 \Rightarrow \theta \approx 0.83° \approx 0.015$ radian

(b) $\sin\theta = 0.4565 \Rightarrow \theta \approx 27° \approx 0.474$ radian

50. (a) $\cos\theta = 0.9848 \Rightarrow \theta \approx 10° \approx 0.175$

(b) $\cos\theta = 0.8746 \Rightarrow \theta \approx 29° \approx 0.506$

51. (a) $\tan\theta = 0.0125 \Rightarrow \theta \approx 0.72° \approx 0.012$ radian

(b) $\tan\theta = 2.3545 \Rightarrow \theta \approx 67° \approx 1.169$ radians

52. (a) $\sin\theta = 0.3746 \Rightarrow \theta \approx 22° \approx 0.384$

(b) $\cos\theta = 0.3746 \Rightarrow \theta \approx 68° \approx 1.187$

53.

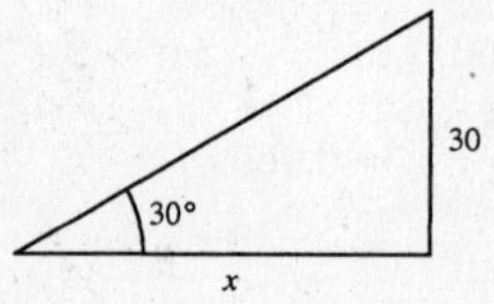

$$\tan 30° = \frac{30}{x}$$
$$\frac{1}{\sqrt{3}} = \frac{30}{x}$$
$$x = 30\sqrt{3}$$

54. $\sin 60° = \frac{y}{18}$

$$y = 18 \sin 60° = 18\frac{\sqrt{3}}{2}$$
$$= 9\sqrt{3}$$

55.

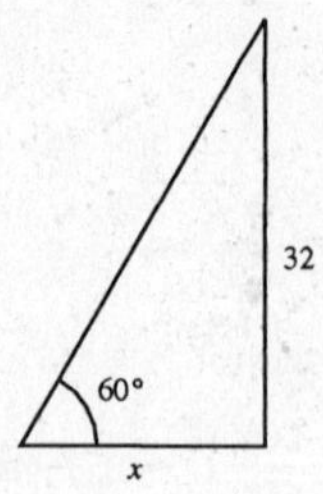

$$\tan 60° = \frac{32}{x}$$
$$\sqrt{3} = \frac{32}{x}$$
$$\sqrt{3}x = 32$$
$$x = \frac{32}{\sqrt{3}} = \frac{32\sqrt{3}}{3}$$

56. $\sin 45° = \frac{20}{r}$

$$r = \frac{20}{\sin 45°} = \frac{20}{\frac{\sqrt{2}}{2}} = 20\sqrt{2}$$

57. $\tan\theta\cot\theta = \tan\theta\left(\frac{1}{\tan\theta}\right) = 1$

58. $\cos\theta\sec\theta = \cos\theta\frac{1}{\cos\theta} = 1$

59. $\tan\alpha\cos\alpha = \left(\frac{\sin\alpha}{\cos\alpha}\right)\cos\alpha = \sin\alpha$

60. $\cot\alpha\sin\alpha = \frac{\cos\alpha}{\sin\alpha}\sin\alpha = \cos\alpha$

61. $(1 + \cos\theta)(1 - \cos\theta) = 1 - \cos^2\theta$

$$= (\sin^2\theta + \cos^2\theta) - \cos^2\theta$$
$$= \sin^2\theta$$

62. $(1 + \sin\theta)(1 - \sin\theta) = 1 - \sin^2\theta = \cos^2\theta$

63. $(\sec\theta + \tan\theta)(\sec\theta - \tan\theta) = \sec^2\theta - \tan^2\theta$

$$= (1 + \tan^2\theta) - \tan^2\theta$$
$$= 1$$

64. $\sin^2\theta - \cos^2\theta = \sin^2\theta - (1 - \sin^2\theta)$

$$= \sin^2\theta - 1 + \sin^2\theta$$
$$= 2\sin^2\theta - 1$$

65. $\frac{\sin\theta}{\cos\theta} + \frac{\cos\theta}{\sin\theta} = \frac{\sin^2\theta + \cos^2\theta}{\sin\theta\cos\theta}$

$$= \frac{1}{\sin\theta\cos\theta}$$
$$= \frac{1}{\sin\theta}\cdot\frac{1}{\cos\theta}$$
$$= \csc\theta\sec\theta$$

66. $\frac{\tan\beta + \cot\beta}{\tan\beta} = \frac{\tan\beta}{\tan\beta} + \frac{\cot\beta}{\tan\beta}$

$$= 1 + \frac{\cot\beta}{\left(\frac{1}{\cot\beta}\right)}$$
$$= 1 + \cot^2\beta = \csc^2\beta$$

67. This is true because the corresponding sides of similar triangles are proportional.

68. Yes. Given $\tan\theta$, $\sec\theta$ can be found from the identity $1 + \tan^2\theta = \sec^2\theta$.

69. (a)

θ	0.1	0.2	0.3	0.4	0.5
$\sin\theta$	0.0998	0.1987	0.2955	0.3894	0.4794

(b) As $\theta \to 0$, $\sin\theta \to 0$.

70. (a)

θ	$0°$	$18°$	$36°$	$54°$	$72°$	$90°$
$\sin \theta$	0	0.3090	0.5878	0.8090	0.9511	1
$\cos \theta$	1	0.9511	0.8090	0.5878	0.3090	0

(b) $\sin \theta$ increases from 0 to 1.

(c) $\cos \theta$ decreases from 1 to 0.

(d) As the angle increases the length of the side opposite the angle increases relative to the length of the hypotenuse and the length of the side adjacent to the angle decreases relative to the length of the hypotenuse. Thus the sine increases and the cosine decreases.

71. (a)

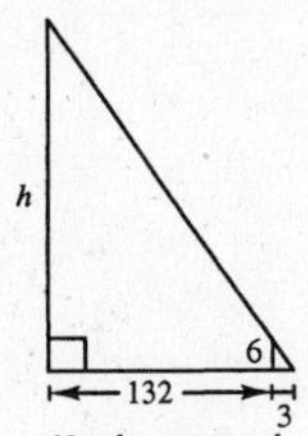

Not drawn to scale

(b) $\dfrac{6}{3} = \dfrac{h}{135}$

(c) $2(135) = h$

$h = 270$ feet

72. $\dfrac{h}{30} = \dfrac{6}{10}$

$h = 30\dfrac{6}{10} = 18$ feet

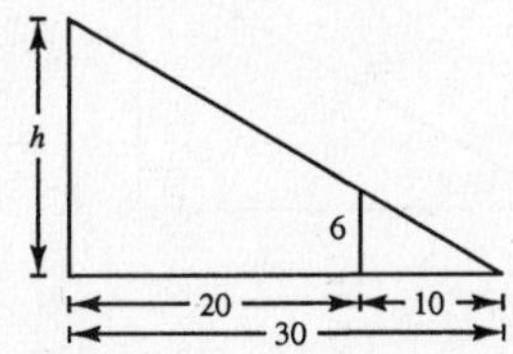

73. (a)

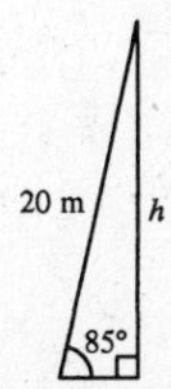

Not drawn to scale

(b) $\sin 85° = \dfrac{h}{20}$

(c) $h = 20 \sin 85° \approx 19.9$ meters

(d) As θ decreases, the height decreases and the base increases. The hypotenuse is fixed at 20 meters.

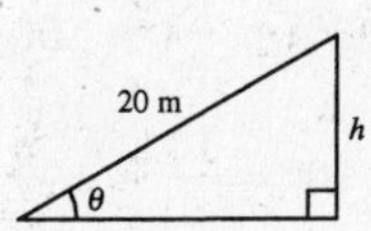

(e) $h = 20 \sin \theta$

θ	h(meters)
80°	19.7
70°	18.8
60°	17.3
50°	15.3
40°	12.9
30°	10.0
20°	6.8
10°	3.5

(f) As θ approaches zero, h, the height, approaches zero.

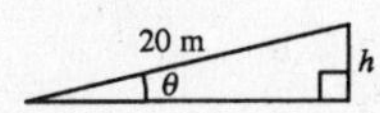

74. $\tan \theta = \dfrac{\text{opp}}{\text{adj}}$

$\tan 54° = \dfrac{w}{100}$

$w = 100 \tan 54° \approx 137.6$ feet

75. $\tan 3.5° = \dfrac{h}{x + 13}$ and $\tan 9° = \dfrac{h}{x}$

$h = (x + 13) \tan 3.5°$ and $h = x \tan 9°$

$x \tan 3.5° + 13 \tan 3.5° = x \tan 9°$

$13 \tan 3.5° = x(\tan 9° - \tan 3.5°)$

$\dfrac{13 \tan 3.5°}{\tan 9° - \tan 3.5°} = x$

$x \approx 8.1783$ miles

$h = x \tan 9° \approx 1.3$ miles

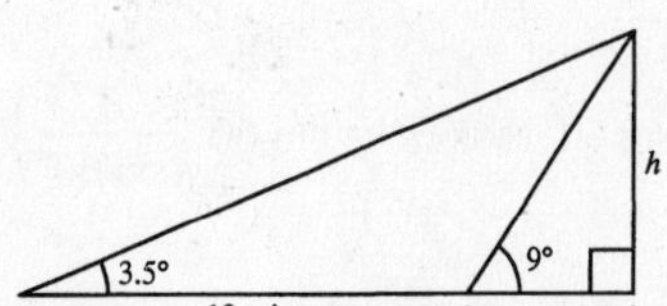

Not drawn to scale

76. Let $x =$ distance from the boat to the shoreline

$$\tan 4° = \frac{40}{x}$$

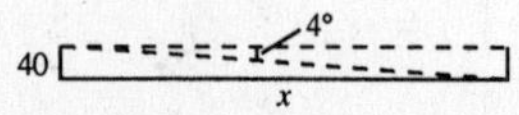

$$x = \frac{40}{\tan 4°} \approx 572 \text{ feet}$$

77. (a)

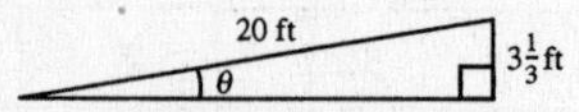

(b) $\sin\theta = \frac{\text{opp}}{\text{hyp}}$

$\sin\theta = \frac{3\frac{1}{3}}{20}$

(c) $\sin\theta = \frac{1}{6} \Longrightarrow \theta = 9.59°$

78.

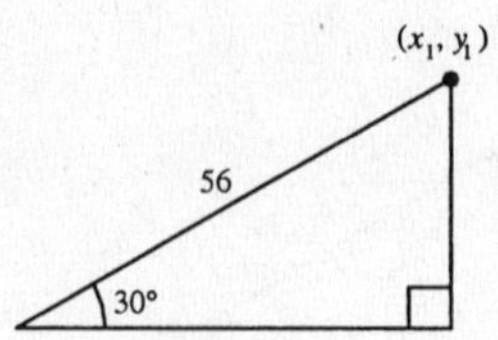

$$\sin 30° = \frac{y_1}{56}$$

$$y_1 = (\sin 30°)(56) = \left(\frac{1}{2}\right)(56) = 28$$

$$\cos 30° = \frac{x_1}{56}$$

$$x_1 = \cos 30°(56) = \frac{\sqrt{3}}{2}(56) = 28\sqrt{3}$$

$$(x_1, y_1) = \left(28\sqrt{3}, 28\right)$$

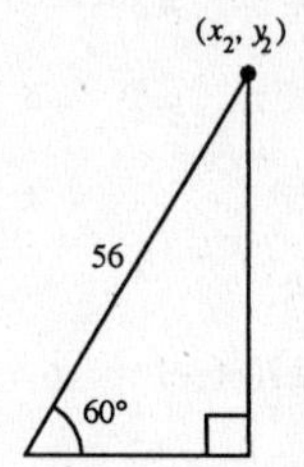

$$\sin 60° = \frac{y_2}{56}$$

$$y_2 = \sin 60°(56) = \left(\frac{\sqrt{3}}{2}\right)(56) = 28\sqrt{3}$$

$$\cos 60° = \frac{x_2}{56}$$

$$x_2 = (\cos 60°)(56) = \left(\frac{1}{2}\right)(56) = 28$$

$$(x_2, y_2) = \left(28, 28\sqrt{3}\right)$$

79. $\tan 3° = \frac{x}{15}$

$x = 15\tan 3°$

$d = 5 + 2x = 5 + 2(15\tan 3°) \approx 6.57$ centimeters

80. $x \approx 9.397,\quad y \approx 3.420$

$\sin 20° = \frac{y}{10} \approx 0.34$ $\qquad$ $\cot 20° = \frac{x}{y} \approx 2.75$

$\cos 20° = \frac{x}{10} \approx 0.94$ $\qquad$ $\sec 20° = \frac{10}{x} \approx 1.06$

$\tan 20° = \frac{y}{x} \approx 0.36$ $\qquad$ $\csc 20° = \frac{10}{y} \approx 2.92$

81. $x \approx 2.588,\quad y \approx 9.659$

$\sin 75° = \frac{y}{10} \approx 0.97$ $\qquad$ $\cot 75° = \frac{x}{y} \approx 0.27$

$\cos 75° = \frac{x}{10} \approx 0.26$ $\qquad$ $\sec 75° = \frac{10}{x} \approx 3.86$

$\tan 75° = \frac{y}{x} \approx 3.73$ $\qquad$ $\csc 75° = \frac{10}{y} \approx 1.04$

82. $x \approx 6.428,\quad y \approx 7.660$

$\sin 50° = \frac{y}{10} \approx 0.77$ $\qquad$ $\cot 50° = \frac{x}{y} \approx 0.84$

$\cos 50° = \frac{x}{10} \approx 0.64$ $\qquad$ $\sec 50° = \frac{10}{x} \approx 1.56$

$\tan 50° = \frac{y}{x} \approx 1.19$ $\qquad$ $\csc 50° = \frac{10}{y} \approx 1.31$

83. True, $\csc x = \frac{1}{\sin x} \Longrightarrow \sin 60° \csc 60° = \sin 60°\left(\frac{1}{\sin 60°}\right) = 1.$

84. $\sec 30° = \csc 60°$

True, because $\sec(90° - \theta) = \csc\theta$.

85. False, $\frac{\sqrt{2}}{2} + \frac{\sqrt{2}}{2} = \sqrt{2} \neq 1.$

86. $\cot^2 10° - \csc^2 10° = -1$

True, because

$$1 + \cot^2 \theta = \csc^2 \theta$$
$$\cot^2\theta = \csc^2 \theta - 1$$
$$\cot^2 \theta - \csc^2 \theta = -1.$$

87. False,

$$\frac{\sin 60°}{\sin 30°} = \frac{\cos 30°}{\sin 30°} = \cot 30° \approx 1.7321;\ \sin 2° \approx 0.0349.$$

88. $\tan[(5°)^2] = \tan^2(5°)$

False,

$\tan[(5°)^2] = \tan 25° \approx 0.4663$

$\tan^2(5°) = (\tan 5°)(\tan 5°)$

$\approx 0.0077.$

89. False,

$$2 \sin 30° = 1;\ \sin 60° = \frac{\sqrt{3}}{2}.$$

90. False,

$$\frac{1}{2}\cos 60° = \frac{1}{4};\ \cos 30° = \frac{\sqrt{3}}{2}.$$

91. False,

$$\sin 30° = \frac{1}{2};\ 1 - \cos 30° = 1 - \frac{\sqrt{3}}{2} = \frac{2 - \sqrt{3}}{2}.$$

92. False,

$$\cos 60° = \frac{1}{2};\ 1 - \cos 30° = 1 - \frac{\sqrt{3}}{2} = \frac{2 - \sqrt{3}}{2}.$$

93. False,

$$\frac{1}{2}\tan\frac{\pi}{3} = \frac{\sqrt{3}}{2};\ \tan\frac{\pi}{6} = \frac{\sqrt{3}}{3} = \frac{1}{\sqrt{3}}.$$

94. False,

$$2\cot\frac{\pi}{6} = 2\sqrt{3};\ \cot\frac{\pi}{3} = \frac{\sqrt{3}}{3} = \frac{1}{\sqrt{3}}.$$

95. False,

$$\sin\frac{6}{\pi} \approx 0.94;\ \csc\frac{\pi}{6} = 2.$$

96. False,

$$\sec\frac{\pi}{3} = 2;\ \cos\frac{3\pi}{4} = -\frac{\sqrt{2}}{2}.$$

Section 9.4 Trigonometric Functions of Any Angle

1. (a) $(x, y) = (4, 3)$

$r = \sqrt{16 + 9} = 5$

$\sin\theta = \frac{y}{r} = \frac{3}{5}$ $\qquad \csc\theta = \frac{r}{y} = \frac{5}{3}$

$\cos\theta = \frac{x}{r} = \frac{4}{5}$ $\qquad \sec\theta = \frac{r}{x} = \frac{5}{4}$

$\tan\theta = \frac{y}{x} = \frac{3}{4}$ $\qquad \cot\theta = \frac{x}{y} = \frac{4}{3}$

(b) $(x, y) = (8, -15)$

$r = \sqrt{64 + 225} = 17$

$\sin\theta = \frac{y}{r} = -\frac{15}{17}$ $\qquad \csc\theta = \frac{r}{y} = -\frac{17}{15}$

$\cos\theta = \frac{x}{r} = \frac{8}{17}$ $\qquad \sec\theta = \frac{r}{x} = \frac{17}{8}$

$\tan\theta = \frac{y}{x} = -\frac{15}{8}$ $\qquad \cot\theta = \frac{x}{y} = -\frac{8}{15}$

2. (a) $x = -12, y = -5$

$r = \sqrt{(-12)^2 + (-5)^2} = 13$

$\sin\theta = \frac{y}{r} = -\frac{5}{13}$

$\cos\theta = \frac{x}{r} = -\frac{12}{13}$

$\tan\theta = \frac{y}{x} = \frac{-5}{-12} = \frac{5}{12}$

$\csc\theta = \frac{r}{y} = \frac{13}{-5} = -\frac{13}{5}$

$\sec\theta = \frac{r}{x} = \frac{13}{-12} = -\frac{13}{12}$

$\cot\theta = \frac{x}{y} = \frac{-12}{-5} = \frac{12}{5}$

(b) $x = -1, y = 1$

$r = \sqrt{(-1)^2 + 1^2} = \sqrt{2}$

$\sin\theta = \frac{y}{r} = \frac{1}{\sqrt{2}} = \frac{\sqrt{2}}{2}$

$\cos\theta = \frac{x}{r} = \frac{-1}{\sqrt{2}} = -\frac{\sqrt{2}}{2}$

$\tan\theta = \frac{y}{x} = \frac{1}{-1} = -1$

$\csc\theta = \frac{r}{y} = \frac{\sqrt{2}}{1} = \sqrt{2}$

$\sec\theta = \frac{r}{x} = \frac{\sqrt{2}}{-1} = -\sqrt{2}$

$\cot\theta = \frac{x}{y} = \frac{-1}{1} = -1$

3. (a) $(x, y) = \left(-\sqrt{3}, -1\right)$

$r = \sqrt{3 + 1} = 2$

$\sin\theta = \frac{y}{r} = -\frac{1}{2}$

$\cos\theta = \frac{x}{r} = -\frac{\sqrt{3}}{2}$

$\tan\theta = \frac{y}{x} = \frac{\sqrt{3}}{3}$

$\csc\theta = \frac{r}{y} = -2$

$\sec\theta = \frac{r}{x} = -\frac{2\sqrt{3}}{3}$

$\cot\theta = \frac{x}{y} = \sqrt{3}$

(b) $(x, y) = (-4, 1)$

$r = \sqrt{16 + 1} = \sqrt{17}$

$\sin\theta = \frac{y}{r} = \frac{\sqrt{17}}{17}$

$\cos\theta = \frac{x}{r} = -\frac{4\sqrt{17}}{17}$

$\tan\theta = \frac{y}{x} = -\frac{1}{4}$

$\csc\theta = \frac{r}{y} = \sqrt{17}$

$\sec\theta = \frac{r}{x} = -\frac{\sqrt{17}}{4}$

$\cot\theta = \frac{x}{y} = -4$

4. (a) $x = 3, y = 1$

$r = \sqrt{3^2 + 1^2} = \sqrt{10}$

$\sin\theta = \frac{y}{r} = \frac{1}{\sqrt{10}} = \frac{\sqrt{10}}{10}$

$\cos\theta = \frac{x}{r} = \frac{3}{\sqrt{10}} = \frac{3\sqrt{10}}{10}$

$\tan\theta = \frac{y}{x} = \frac{1}{3}$

$\csc\theta = \frac{r}{y} = \frac{\sqrt{10}}{1} = \sqrt{10}$

$\sec\theta = \frac{r}{x} = \frac{\sqrt{10}}{3}$

$\cot\theta = \frac{x}{y} = \frac{3}{1} = 3$

(b) $x = 4, y = -4$

$r = \sqrt{4^2 + (-4)^2} = 4\sqrt{2}$

$\sin\theta = \frac{y}{r} = \frac{-4}{4\sqrt{2}} = -\frac{\sqrt{2}}{2}$

$\cos\theta = \frac{x}{r} = \frac{4}{4\sqrt{2}} = \frac{\sqrt{2}}{2}$

$\tan\theta = \frac{y}{x} = \frac{-4}{4} = -1$

$\csc\theta = \frac{r}{y} = \frac{4\sqrt{2}}{-4} = -\sqrt{2}$

$\sec\theta = \frac{r}{x} = \frac{4\sqrt{2}}{4} = \sqrt{2}$

$\cot\theta = \frac{x}{y} = \frac{4}{-4} = -1$

5. $(x, y) = (7, 24)$

$r = \sqrt{49 + 576} = 25$

$\sin\theta = \frac{y}{r} = \frac{24}{25}$ $\qquad \csc\theta = \frac{r}{y} = \frac{25}{24}$

$\cos\theta = \frac{x}{r} = \frac{7}{25}$ $\qquad \sec\theta = \frac{r}{x} = \frac{25}{7}$

$\tan\theta = \frac{y}{x} = \frac{24}{7}$ $\qquad \cot\theta = \frac{x}{y} = \frac{7}{24}$

6. $x = 8, y = 15$

$r = \sqrt{8^2 + 15^2} = 17$

$\sin\theta = \frac{y}{r} = \frac{15}{17}$ $\qquad \csc\theta = \frac{r}{y} = \frac{17}{15}$

$\cos\theta = \frac{x}{r} = \frac{8}{17}$ $\qquad \sec\theta = \frac{r}{x} = \frac{17}{8}$

$\tan\theta = \frac{y}{x} = \frac{15}{8}$ $\qquad \cot\theta = \frac{x}{y} = \frac{8}{15}$

7. $(x, y) = (-4, 10)$

$r = \sqrt{16 + 100} = 2\sqrt{29}$

$\sin\theta = \frac{y}{r} = \frac{5\sqrt{29}}{29}$

$\cos\theta = \frac{x}{r} = -\frac{2\sqrt{29}}{29}$

$\tan\theta = \frac{y}{x} = -\frac{5}{2}$

$\csc\theta = \frac{r}{y} = \frac{\sqrt{29}}{5}$

$\sec\theta = \frac{r}{x} = -\frac{\sqrt{29}}{2}$

$\cot\theta = \frac{x}{y} = -\frac{2}{5}$

8. $x = -5, y = -2$

$r = \sqrt{(-5)^2 + (-2)^2} = \sqrt{29}$

$\sin\theta = \frac{y}{r} = \frac{-2}{\sqrt{29}} = -\frac{2\sqrt{29}}{29}$

$\cos\theta = \frac{x}{r} = \frac{-5}{\sqrt{29}} = -\frac{5\sqrt{29}}{29}$

$\tan\theta = \frac{y}{x} = \frac{-2}{-5} = \frac{2}{5}$

$\csc\theta = \frac{r}{y} = \frac{\sqrt{29}}{-2} = -\frac{\sqrt{29}}{2}$

$\sec\theta = \frac{r}{x} = \frac{\sqrt{29}}{-5} = -\frac{\sqrt{29}}{5}$

$\cot\theta = \frac{x}{y} = \frac{-5}{-2} = \frac{5}{2}$

9. $(x, y) = \left(-\frac{5}{12}, 1\right)$

$r = \sqrt{\frac{25}{144} + 1} = \sqrt{\frac{169}{144}} = \frac{13}{12}$

$\sin\theta = \frac{y}{r} = \frac{1}{13/12} = \frac{12}{13}$

$\cos\theta = \frac{x}{r} = \frac{-5/12}{13/12} = -\frac{5}{13}$

$\tan\theta = \frac{y}{x} = \frac{1}{-5/12} = -\frac{12}{5}$

$\csc\theta = \frac{r}{y} = \frac{13/12}{1} = \frac{13}{12}$

$\sec\theta = \frac{r}{x} = \frac{13/12}{-5/12} = -\frac{13}{5}$

$\cot\theta = \frac{x}{y} = \frac{-5/12}{1} = -\frac{5}{12}$

10. $(x, y) = \left(\frac{1}{2}, -\frac{12}{7}\right)$

$r = \sqrt{\frac{1}{4} + \frac{144}{49}} = \sqrt{\frac{625}{196}} = \frac{25}{14}$

$\sin\theta = \frac{y}{r} = \frac{-12/7}{25/14} = -\frac{24}{25}$

$\cos\theta = \frac{x}{r} = \frac{1/2}{25/14} = \frac{7}{25}$

$\tan\theta = \frac{y}{x} = \frac{-12/7}{1/2} = -\frac{24}{7}$

$\csc\theta = \frac{r}{y} = \frac{25/14}{-12/7} = -\frac{25}{24}$

$\sec\theta = \frac{r}{x} = \frac{25/14}{1/2} = \frac{25}{7}$

$\cot\theta = \frac{x}{y} = \frac{1/2}{-12/7} = -\frac{7}{24}$

11. $\sin\theta < 0 \implies \theta$ lies in Quadrant III or in Quadrant IV.

$\cos\theta < 0 \implies \theta$ lies in Quadrant II or in Quadrant III.

$\sin\theta < 0$ *and* $\cos\theta < 0 \implies \theta$ lies in Quadrant III.

12. $\sin\theta > 0$ and $\cos\theta > 0$

$\frac{y}{r} > 0$ and $\frac{x}{r} > 0$

Quadrant I

13. $\sin\theta > 0 \Rightarrow \theta$ lies in Quadrant I or in Quadrant II.

$\tan\theta < 0 \Rightarrow \theta$ lies in Quadrant II or in Quadrant IV.

$\sin\theta > 0$ *and* $\tan\theta < 0 \Rightarrow \theta$ lies in Quadrant II.

14. $\sec\theta > 0$ and $\cot\theta < 0$

$\frac{r}{x} > 0$ and $\frac{x}{y} < 0$

Quadrant IV

15. $\sin\theta = \frac{y}{r} = \frac{3}{5} \Rightarrow x^2 = 25 - 9 = 16$

θ in Quadrant II $\Rightarrow x = -4$

$\sin\theta = \frac{y}{r} = \frac{3}{5}$ $\qquad \csc\theta = \frac{r}{y} = \frac{5}{3}$

$\cos\theta = \frac{x}{r} = -\frac{4}{5}$ $\qquad \sec\theta = \frac{r}{x} = -\frac{5}{4}$

$\tan\theta = \frac{y}{x} = -\frac{3}{4}$ $\qquad \cot\theta = \frac{x}{y} = -\frac{4}{3}$

16. $\cos\theta = \frac{x}{r} = \frac{-4}{5} \Rightarrow y = |3|$

θ in Quadrant III $\Rightarrow y = -3$

$\sin\theta = \frac{y}{r} = -\frac{3}{5}$ $\qquad \csc\theta = -\frac{5}{3}$

$\cos\theta = \frac{x}{r} = -\frac{4}{5}$ $\qquad \sec\theta = -\frac{5}{4}$

$\tan\theta = \frac{y}{x} = \frac{3}{4}$ $\qquad \cot\theta = \frac{4}{3}$

17. $\tan\theta = \frac{y}{x} = \frac{-15}{8}$

$\sin\theta < 0$ and $\tan\theta < 0 \Rightarrow \theta$ is in Quadrant IV $\Rightarrow$
$y < 0$ and $x > 0$

$x = 8, y = -15, r = 17$

$\sin\theta = \frac{y}{r} = -\frac{15}{17}$ $\qquad \csc\theta = \frac{r}{y} = -\frac{17}{15}$

$\cos\theta = \frac{x}{r} = \frac{8}{17}$ $\qquad \sec\theta = \frac{r}{x} = \frac{17}{8}$

$\tan\theta = \frac{y}{x} = -\frac{15}{8}$ $\qquad \cot\theta = \frac{x}{y} = -\frac{8}{15}$

18. $\cos\theta = \frac{x}{r} = \frac{8}{17} \Rightarrow y = |15|$

$\tan\theta < 0 \Rightarrow y = -15$

$\sin\theta = \frac{y}{r} = \frac{-15}{17} = -\frac{15}{17}$ $\qquad \csc\theta = -\frac{17}{15}$

$\cos\theta = \frac{x}{r} = \frac{8}{17}$ $\qquad \sec\theta = \frac{17}{8}$

$\tan\theta = \frac{y}{x} = \frac{-15}{8} = -\frac{15}{8}$ $\qquad \cot\theta = -\frac{8}{15}$

19. $\cot\theta = \frac{x}{y} = -\frac{3}{1} = \frac{3}{-1}$

$\cos\theta > 0 \Rightarrow \theta$ is in Quadrant IV $\Rightarrow x$ is positive;
$x = 3, y = -1, r = \sqrt{10}$

$\sin\theta = \frac{y}{r} = -\frac{\sqrt{10}}{10}$ $\qquad \csc\theta = \frac{r}{y} = -\sqrt{10}$

$\cos\theta = \frac{x}{r} = \frac{3\sqrt{10}}{10}$ $\qquad \sec\theta = \frac{r}{x} = \frac{\sqrt{10}}{3}$

$\tan\theta = \frac{y}{x} = -\frac{1}{3}$ $\qquad \cot\theta = \frac{x}{y} = -3$

20. $\csc\theta = \frac{r}{y} = \frac{4}{1} \Rightarrow x = \left|\sqrt{15}\right|$

$\cot\theta < 0 \Rightarrow x = -\sqrt{15}$

$\sin\theta = \frac{y}{r} = \frac{1}{4}$ $\qquad \csc\theta = 4$

$\cos\theta = \frac{x}{r} = -\frac{\sqrt{15}}{4}$ $\qquad \sec\theta = -\frac{4\sqrt{15}}{15}$

$\tan\theta = \frac{y}{x} = -\frac{\sqrt{15}}{15}$ $\qquad \cot\theta = -\sqrt{15}$

21. $\sec\theta = \frac{r}{x} = \frac{2}{-1} \Rightarrow y^2 = 4 - 1 = 3$

$\sin\theta > 0 \Rightarrow \theta$ is in Quadrant II $\Rightarrow y = \sqrt{3}$

$\sin\theta = \frac{y}{r} = \frac{\sqrt{3}}{2}$ $\qquad \csc\theta = \frac{r}{y} = \frac{2\sqrt{3}}{3}$

$\cos\theta = \frac{x}{r} = -\frac{1}{2}$ $\qquad \sec\theta = \frac{r}{x} = -2$

$\tan\theta = \frac{y}{x} = -\sqrt{3}$ $\qquad \cot\theta = \frac{x}{y} = -\frac{\sqrt{3}}{3}$

22. $\sin\theta = 0 \Rightarrow \theta = 0 + \pi n$

$\sec\theta = -1 \Rightarrow \theta = \pi + 2\pi n$

$y = 0, x = -r$

$\sin\theta = 0$ $\qquad \csc\theta = \frac{r}{y}$ is undefined

$\cos\theta = \frac{x}{r} = \frac{-r}{r} = -1$ $\qquad \sec\theta = \frac{r}{x} = -1$

$\tan\theta = \frac{y}{x} = 0$ $\qquad \cot\theta = \frac{x}{y}$ is undefined

23. $\cot\theta$ is undefined, $\frac{\pi}{2} \le \theta \le \frac{3\pi}{2} \implies y = 0 \implies \theta = \pi$

$\sin\pi = 0$ $\quad$ $\csc\pi$ is undefined.

$\cos\pi = -1$ $\quad$ $\sec\pi = -1$

$\tan\pi = 0$ $\quad$ $\cot\pi$ is undefined.

24. $\tan\theta$ is undefined $\implies \theta = n\pi + \frac{\pi}{2}$

$\pi \le \theta \le 2\pi \implies \theta = \frac{3\pi}{2}, x = 0, y = -r$

$\sin\theta = \frac{y}{r} = \frac{-r}{r} = -1$ $\quad$ $\csc\theta = \frac{r}{y} = -1$

$\cos\theta = \frac{x}{r} = \frac{0}{r} = 0$ $\quad$ $\sec\theta = \frac{r}{x}$ is undefined.

$\tan\theta = \frac{y}{x}$ is undefined. $\quad$ $\cot\theta = \frac{x}{y} = \frac{0}{y} = 0$

25. To find a point on the terminal side of θ, use any point on the line $y = -x$ that lies in Quadrant II. $(-1, 1)$ is one such point.

$x = -1, y = 1, r = \sqrt{2}$

$\sin\theta = \frac{1}{\sqrt{2}} = \frac{\sqrt{2}}{2}$

$\cos\theta = -\frac{1}{\sqrt{2}} = -\frac{\sqrt{2}}{2}$

$\tan\theta = -1$

$\csc\theta = \sqrt{2}$

$\sec\theta = -\sqrt{2}$

$\cot\theta = -1$

26. Let $x > 0$.

$\left(-x, -\frac{1}{3}x\right)$ Quadrant III

$r = \sqrt{x^2 + \frac{1}{9}x^2} = \frac{\sqrt{10}x}{3}$

$\sin\theta = \frac{y}{r} = \frac{(-1/3)x}{\sqrt{10}x/3} = -\frac{\sqrt{10}}{10}$

$\cos\theta = \frac{x}{r} = \frac{-x}{\sqrt{10}x/3} = -\frac{3\sqrt{10}}{10}$

$\tan\theta = \frac{y}{x} = \frac{(-1/3)x}{-x} = \frac{1}{3}$

$\csc\theta = \frac{r}{y} = \frac{\sqrt{10}x/3}{(-1/3)x} = -\sqrt{10}$

$\sec\theta = \frac{r}{x} = \frac{\sqrt{10}x/3}{-x} = -\frac{\sqrt{10}}{3}$

$\cot\theta = \frac{x}{y} = \frac{-x}{(-1/3)x} = 3$

27. To find a point on the terminal side of θ, use any point on the line $y = 2x$ that lies in Quadrant III. $(-1, -2)$ is one such point.

$x = -1, y = -2, r = \sqrt{5}$

$\sin\theta = -\frac{2}{\sqrt{5}} = -\frac{2\sqrt{5}}{5}$ $\quad$ $\csc\theta = \frac{\sqrt{5}}{-2} = -\frac{\sqrt{5}}{2}$

$\cos\theta = -\frac{1}{\sqrt{5}} = -\frac{\sqrt{5}}{5}$ $\quad$ $\sec\theta = \frac{\sqrt{5}}{-1} = -\sqrt{5}$

$\tan\theta = \frac{-2}{-1} = 2$ $\quad$ $\cot\theta = \frac{-1}{-2} = \frac{1}{2}$

28. Let $x > 0$.

$4x + 3y = 0 \implies y = -\frac{4}{3}x$

$\left(x, -\frac{4}{3}x\right)$ Quadrant IV

$r = \sqrt{x^2 + \frac{16}{9}x^2} = \frac{5}{3}x$

$\sin\theta = \frac{y}{r} = \frac{(-4/3)x}{5/3x} = -\frac{4}{5}$ $\quad$ $\csc\theta = -\frac{5}{4}$

$\cos\theta = \frac{x}{r} = \frac{x}{5/3x} = \frac{3}{5}$ $\quad$ $\sec\theta = \frac{5}{3}$

$\tan\theta = \frac{y}{x} = \frac{(-4/3)x}{x} = -\frac{4}{3}$ $\quad$ $\tan\theta = -\frac{3}{4}$

29. $(x, y) = (-1, 0), r = 1$

$\cos\pi = \frac{x}{r} = \frac{-1}{1} = -1$

30. $\cos\frac{3\pi}{2} = \frac{x}{r} = \frac{0}{1} = 0$

since $\frac{3\pi}{2}$ corresponds to $(0, -1)$.

31. $(x, y) = (-1, 0), r = 1$

$\sec \pi = \frac{r}{x} = \frac{1}{-1} = -1$

32. $\sec \frac{3\pi}{2} = \frac{r}{x} = \frac{1}{0} \Rightarrow$ undefined

since $\frac{3\pi}{2}$ corresponds to $(0, -1)$.

33. $(x, y) = (0, 1), r = 1$

$\tan \frac{\pi}{2} = \frac{y}{x} = \frac{1}{0}$ undefined.

34. $\tan \pi = \frac{y}{x} = \frac{0}{-1} = 0$

since π corresponds to $(-1, 0)$.

35. $(x, y) = (0, 1)$

$\cot \frac{\pi}{2} = \frac{x}{y} = \frac{0}{1} = 0$

36. $\csc \pi = \frac{r}{y} = \frac{1}{0} \Rightarrow$ undefined

since π corresponds to $(-1, 0)$.

37. $\theta = \frac{2\pi}{3}$

$\theta' = \pi - \frac{2\pi}{3} = \frac{\pi}{3}$

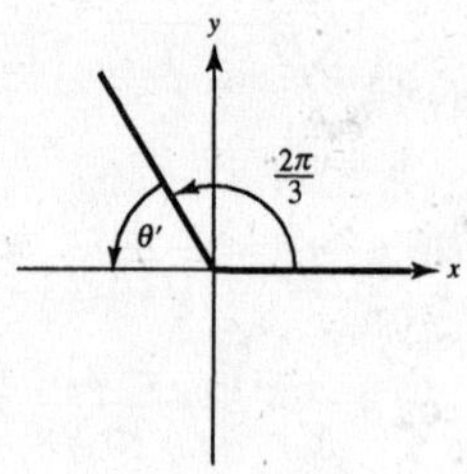

38. $\theta = \frac{7\pi}{4}$

$\theta' = 2\pi - \frac{7\pi}{4} = \frac{\pi}{4}$

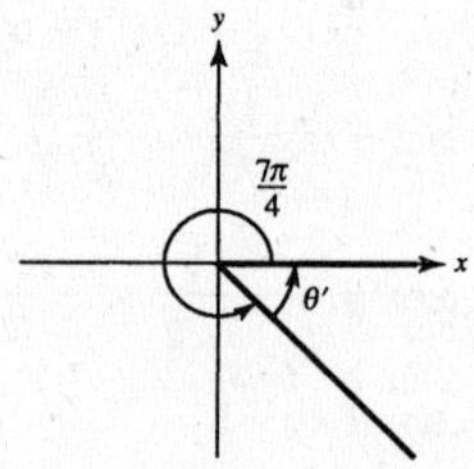

39. $\theta = \frac{5\pi}{6}$

$\theta' = \pi - \frac{5\pi}{6} = \frac{\pi}{6}$

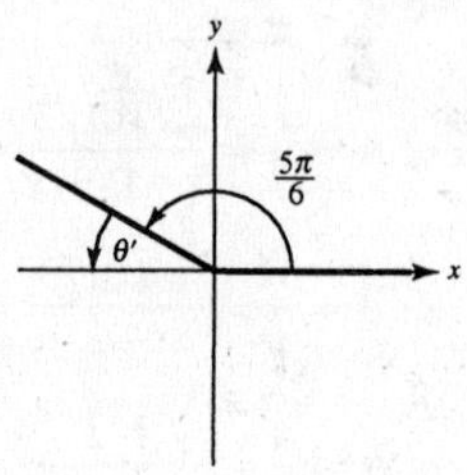

40. $\theta = \frac{9\pi}{4}$

$\theta' = \frac{9\pi}{4} - 2\pi = \frac{\pi}{4}$

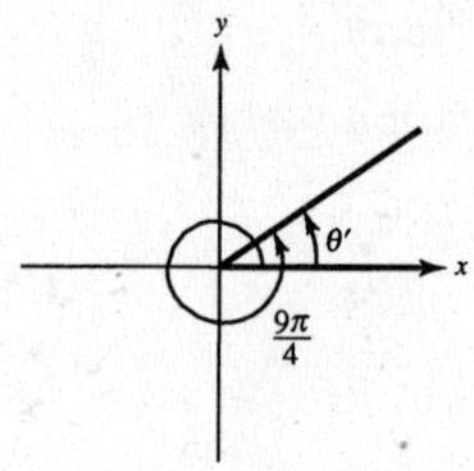

41. $\theta = -\frac{2\pi}{3}$

$\theta' = -\frac{2\pi}{3} + \pi = \frac{\pi}{3}$

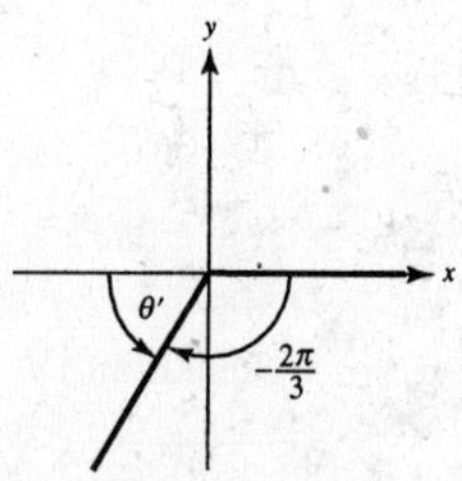

42. $\theta = -\frac{5\pi}{6}$

$\theta' = -\frac{5\pi}{6} + \pi = \frac{\pi}{6}$

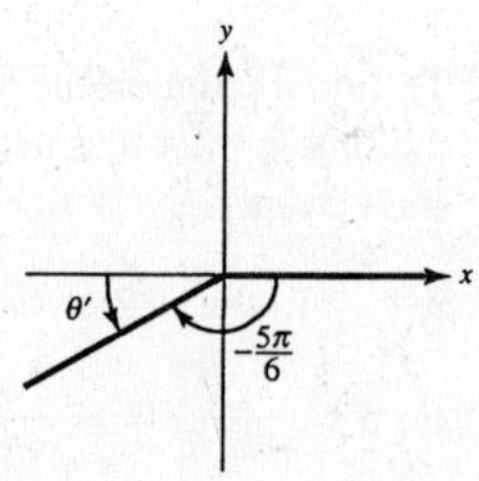

43. $\theta = 120°$

$\theta' = 180° - 120° = 60°$

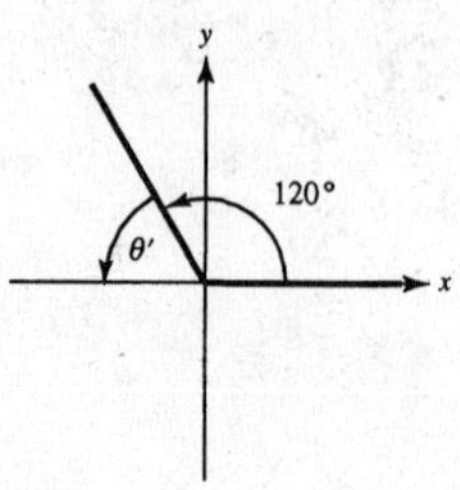

44. $\theta = 135°$

$\theta' = 180° - 135° = 45°$

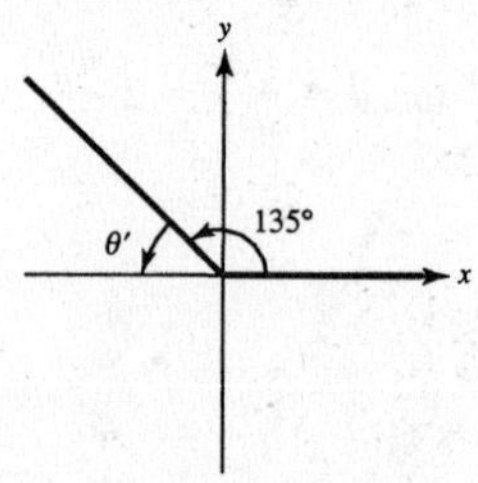

45. $\theta = 225°$

$\theta' = 225° - 180° = 45°$

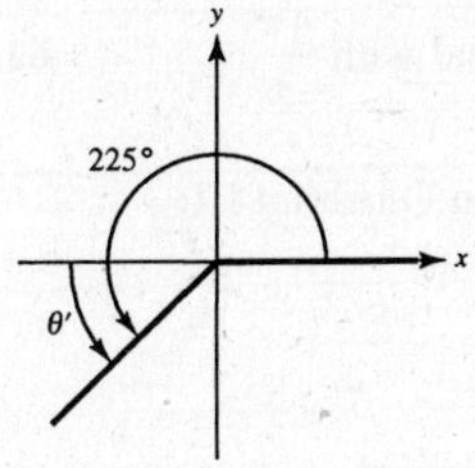

46. $\theta = 330°$

$\theta' = 360° - 330° = 30°$

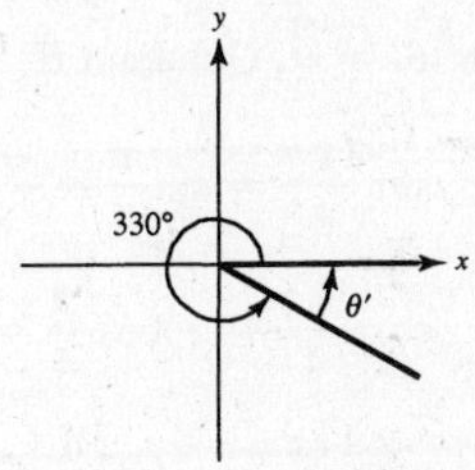

47. $\theta = -135°$

$\theta' = -135° + 180° = 45°$

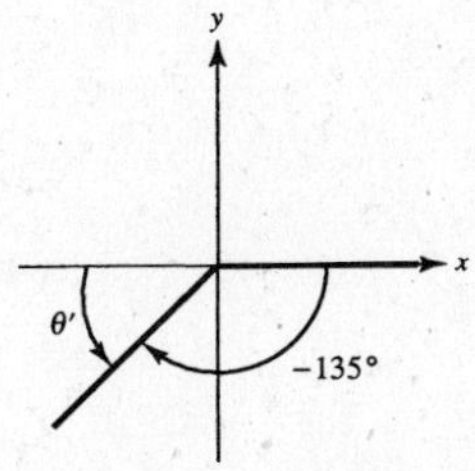

48. $\theta = -240°$

$-240°$ is coterminal with $120°$

$\theta' = 180° - 120° = 60°$

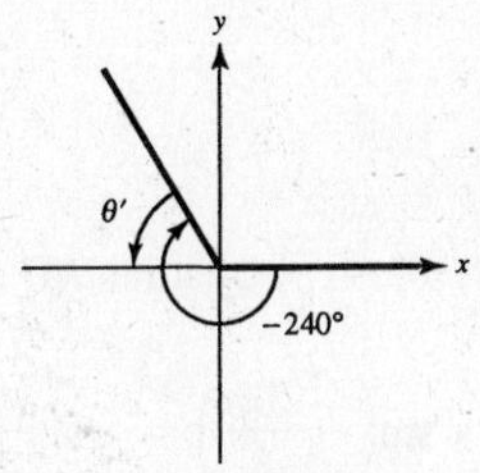

49. $\theta' = 45°$, Quadrant III

$\sin 225° = -\sin 45° = -\frac{\sqrt{2}}{2}$

$\cos 225° = -\cos 45° = -\frac{\sqrt{2}}{2}$

$\tan 225° = \tan 45° = 1$

50. $\theta = 300°$, $\theta' = 360° - 300° = 60°$, Quadrant IV

$\sin 300° = -\sin 60° = -\frac{\sqrt{3}}{2}$

$\cos 300° = \cos 60° = \frac{1}{2}$

$\tan 300° = -\tan 60° = -\sqrt{3}$

51. $\theta' = 30°$, Quadrant I

$\sin 750° = \sin 30° = \frac{1}{2}$

$\cos 750° = \cos 30° = \frac{\sqrt{3}}{2}$

$\tan 750° = \tan 30° = \frac{\sqrt{3}}{3}$

52. $\theta = -405°$ is coterminal with $315°$.

$\theta' = 360° - 315° = 45°$ in Quadrant IV.

$\sin(-405°) = -\sin 45° = -\frac{\sqrt{2}}{2}$

$\cos(-405°) = \cos 45° = \frac{\sqrt{2}}{2}$

$\tan(-405°) = -\tan 45° = -1$

53. $\theta' = 30°$, Quadrant III

$\sin(-150°) = -\sin 30° = -\frac{1}{2}$

$\cos(-150°) = -\cos 30° = -\frac{\sqrt{3}}{2}$

$\tan(-150°) = \tan 30° = \frac{\sqrt{3}}{3}$

54. $\theta = -840°$ is coterminal with $240°$.

$\theta' = 240° - 180° = 60°$in Quadrant III.

$\sin(-840°) = -\sin 60° = -\frac{\sqrt{3}}{2}$

$\cos(-840°) = -\cos 60° = -\frac{1}{2}$

$\tan(-840°) = \tan 60° = \sqrt{3}$

55. $\theta' = \frac{\pi}{3}$, Quadrant III

$\sin\frac{4\pi}{3} = -\sin\frac{\pi}{3} = -\frac{\sqrt{3}}{2}$

$\cos\frac{4\pi}{3} = -\cos\frac{\pi}{3} = -\frac{1}{2}$

$\tan\frac{4\pi}{3} = \tan\frac{\pi}{3} = \sqrt{3}$

56. $\theta' = \frac{\pi}{4}$, Quadrant II

$\sin\frac{3\pi}{4} = \sin\frac{\pi}{4} = \frac{\sqrt{2}}{2}$

$\cos\frac{3\pi}{4} = -\cos\frac{\pi}{4} = -\frac{\sqrt{2}}{2}$

$\tan\frac{3\pi}{4} = -\tan\frac{\pi}{4} = -1$

57. $\theta' = \frac{\pi}{6}$, Quadrant II

$\sin\left(-\frac{7\pi}{6}\right) = \sin\frac{\pi}{6} = \frac{1}{2}$

$\cos\left(-\frac{7\pi}{6}\right) = -\cos\frac{\pi}{6} = -\frac{\sqrt{3}}{2}$

$\tan\left(-\frac{7\pi}{6}\right) = -\tan\frac{\pi}{6} = -\frac{\sqrt{3}}{3}$

58. $\theta' = \frac{\pi}{3}$, Quadrant II

$\sin\left(-\frac{4\pi}{3}\right) = \sin\frac{\pi}{3} = \frac{\sqrt{3}}{2}$

$\cos\left(-\frac{4\pi}{3}\right) = -\cos\frac{\pi}{3} = -\frac{1}{2}$

$\tan\left(-\frac{4\pi}{3}\right) = -\tan\frac{\pi}{3} = -\sqrt{3}$

59. $\theta' = \frac{\pi}{4}$, Quadrant II

$\sin \frac{11\pi}{4} = \sin \frac{\pi}{4} = \frac{\sqrt{2}}{2}$

$\cos \frac{11\pi}{4} = -\cos \frac{\pi}{4} = -\frac{\sqrt{2}}{2}$

$\tan \frac{11\pi}{4} = -\tan \frac{\pi}{4} = -1$

60. $\theta = \frac{10\pi}{3}$ is coterminal with $\frac{4\pi}{3}$.

$\theta' = \frac{4\pi}{3} - \pi = \frac{\pi}{3}$ in Quadrant III.

$\sin \frac{10\pi}{3} = -\sin \frac{\pi}{3} = -\frac{\sqrt{3}}{2}$

$\cos \frac{10\pi}{3} = -\cos \frac{\pi}{3} = -\frac{1}{2}$

$\tan \frac{10\pi}{3} = \tan \frac{\pi}{3} = \sqrt{3}$

61. $\theta' = \frac{\pi}{6}$, Quadrant III

$\sin\left(-\frac{17\pi}{6}\right) = -\sin \frac{\pi}{6} = -\frac{1}{2}$

$\cos\left(-\frac{17\pi}{6}\right) = -\cos \frac{\pi}{6} = -\frac{\sqrt{3}}{2}$

$\tan\left(-\frac{17\pi}{6}\right) = \tan \frac{\pi}{6} = \frac{\sqrt{3}}{3}$

62. $\theta' = \frac{\pi}{3}$, Quadrant III

$\sin\left(-\frac{20\pi}{3}\right) = -\sin \frac{\pi}{3} = -\frac{\sqrt{3}}{2}$

$\cos\left(-\frac{20\pi}{3}\right) = -\cos \frac{\pi}{3} = -\frac{1}{2}$

$\tan\left(-\frac{20\pi}{3}\right) = \tan \frac{\pi}{3} = \sqrt{3}$

63. $\sin \theta = -\frac{3}{5}$

$\sin^2 \theta + \cos^2 \theta = 1$

$\cos^2 \theta = 1 - \sin^2 \theta$

$\cos^2 \theta = 1 - \left(-\frac{3}{5}\right)^2$

$\cos^2 \theta = 1 - \frac{9}{25}$

$\cos^2 \theta = \frac{16}{25}$

$\cos \theta > 0$ in Quadrant IV.

$\cos \theta = \frac{4}{5}$

64. $\cot \theta = -3$

$1 + \cot^2 \theta = \csc^2 \theta$

$1 + (-3)^2 = \csc^2 \theta$

$10 = \csc^2 \theta$

$\csc \theta > 0$ in Quadrant II.

$\sqrt{10} = \csc \theta$

$\csc \theta = \frac{1}{\sin \theta}$

$\sin \theta = \frac{1}{\csc \theta} = \frac{1}{\sqrt{10}} = \frac{\sqrt{10}}{10}$

65. $\tan \theta = \frac{3}{2}$

$\sec^2 \theta = 1 + \tan^2 \theta$

$\sec^2 \theta = 1 + \left(\frac{3}{2}\right)^2$

$\sec^2 \theta = 1 + \frac{9}{4}$

$\sec^2 \theta = \frac{13}{4}$

$\sec \theta < 0$ in Quadrant III.

$\sec \theta = -\frac{\sqrt{13}}{2}$

66. $\csc \theta = -2$

$1 + \cot^2 \theta = \csc^2 \theta$

$\cot^2 \theta = \csc^2 \theta - 1$

$\cot^2 \theta = (-2)^2 - 1$

$\cot^2 \theta = 3$

$\cot \theta < 0$ in Quadrant IV.

$\cot \theta = -\sqrt{3}$

67. $\cos \theta = \frac{5}{8}$

$\cos \theta = \frac{1}{\sec \theta} \Rightarrow \sec \theta = \frac{1}{\cos \theta}$

$\sec \theta = \frac{1}{5/8} = \frac{8}{5}$

68. $\sec \theta = -\frac{9}{4}$

$1 + \tan^2 \theta = \sec^2 \theta$

$\tan^2 \theta = \sec^2 \theta - 1$

$\tan^2 \theta = \left(-\frac{9}{4}\right)^2 - 1$

$\tan^2 \theta = \frac{65}{16}$

$\tan \theta > 0$ in Quadrant III.

$\tan \theta = \frac{\sqrt{65}}{4}$

69. $\sin 10° \approx 0.1736$

70. $\sec 235° = \frac{1}{\cos 235°} \approx -1.7434$

$\sec 235° = \frac{1}{\cos 235°}$

≈ -1.7434

71. $\cos(-110°) \approx -0.3420$

72. $\csc(-320°) = \dfrac{1}{\sin(-320°)}$

$= 1.5557$

73. $\tan 4.5 \approx 4.6373$

74. $\cot 1.35 = \dfrac{1}{\tan 1.35} \approx 0.2245$

75. $\tan\left(\dfrac{5\pi}{9}\right) \approx -5.6713$

76. $\tan\left(-\dfrac{11\pi}{9}\right) \approx -0.8391$

77. $\sin(-0.65) \approx -0.6052$

78. $\sin 0.65 \approx 0.6052$

79. (a) $\sin\theta = \dfrac{1}{2} \Rightarrow$ reference angle is 30° or $\dfrac{\pi}{6}$ and θ is in Quadrant I or Quadrant II.

Values in degrees: 30°, 150°

Values in radians: $\dfrac{\pi}{6}, \dfrac{5\pi}{6}$

(b) $\sin\theta = -\dfrac{1}{2} \Rightarrow$ reference angle is 30° or $\dfrac{\pi}{6}$ and θ is in Quadrant III or Quadrant IV.

Values in degrees: 210°, 330°

Values in radians: $\dfrac{7\pi}{6}, \dfrac{11\pi}{6}$

80. (a) $\cos\theta = \dfrac{\sqrt{2}}{2} \Rightarrow$ reference angle is 45° or $\dfrac{\pi}{4}$ and θ is in Quadrant I or IV.

Values in degrees: 45°, 315°

Values in radians: $\dfrac{\pi}{4}, \dfrac{7\pi}{4}$

(b) $\cos\theta = -\dfrac{\sqrt{2}}{2} \Rightarrow$ reference angle is 45° or $\dfrac{\pi}{4}$ and θ is in Quadrant II or III.

Values in degrees: 135°, 225°

Values in radians: $\dfrac{3\pi}{4}, \dfrac{5\pi}{4}$

81. (a) $\csc\theta = \dfrac{2\sqrt{3}}{3} \Rightarrow$ reference angle is 60° or $\dfrac{\pi}{3}$ and θ is in Quadrant I or Quadrant II.

Values in degrees: 60°, 120°

Values in radians: $\dfrac{\pi}{3}, \dfrac{2\pi}{3}$

(b) $\cot\theta = -1 \Rightarrow$ reference angle is 45° or $\dfrac{\pi}{4}$ and θ is in Quadrant II or Quadrant IV.

Values in degrees: 135°, 315°

Values in radians: $\dfrac{3\pi}{4}, \dfrac{7\pi}{4}$

82. (a) $\sec\theta = 2 \Rightarrow$ reference angle is 60° or $\dfrac{\pi}{3}$ and θ is in Quadrant I or IV.

Values in degrees: 60°, 300°

Values in radians: $\dfrac{\pi}{3}, \dfrac{5\pi}{3}$

(b) $\sec\theta = -2 \Rightarrow$ reference angle is 60° or $\dfrac{\pi}{3}$ and θ is in Quadrant II or III.

Values in degrees: 120°, 240°

Values in radians: $\dfrac{2\pi}{3}, \dfrac{4\pi}{3}$

83. (a) $\tan\theta = 1 \Rightarrow$ reference angle is 45° or $\dfrac{\pi}{4}$ and θ is in Quadrant I or Quadrant III.

Values in degrees: 45°, 225°

Values in radians: $\dfrac{\pi}{4}, \dfrac{5\pi}{4}$

(b) $\cot\theta = -\sqrt{3} \Rightarrow$ reference angle is 30° or $\dfrac{\pi}{6}$ and θ is in Quadrant II or Quadrant IV.

Values in degrees: 150°, 330°

Values in radians: $\dfrac{5\pi}{6}, \dfrac{11\pi}{6}$

84. (a) $\sin\theta = \frac{\sqrt{3}}{2} \Rightarrow$ reference angle is $60°$ or $\frac{\pi}{3}$ and θ is in Quadrant I or II.

Values in degrees: $60°, 120°$

Values in radians: $\frac{\pi}{3}, \frac{2\pi}{3}$

(b) $\sin\theta = -\frac{\sqrt{3}}{2} \Rightarrow$ reference angle is $60°$ or $\frac{\pi}{3}$ and θ is in Quadrant III or IV.

Values in degrees: $240°, 300°$

Values in radians: $\frac{4\pi}{3}, \frac{5\pi}{3}$

85. $\sin\theta = 0.8191$

Quadrant I: $\theta = \sin^{-1} 0.8191 \approx 54.99°$

Quadrant II: $\theta = 180° - \sin^{-1} 0.8191 \approx 125.01°$

86. $\cos\theta = 0.8746$

Quadrant I: $\theta = \cos^{-1} 0.8746 \approx 29.00°$

Quadrant IV: $\theta = 360° - 29.00° = 331.00°$

87. $\cos\theta = -0.4367 \Rightarrow \theta' \approx 64.11°$

Quadrant II: $\theta \approx 180° - 64.11° = 115.89°$

Quadrant III: $\theta \approx 180° + 64.11° = 244.11°$

88. $\sin\theta = -0.6514$

$\sin^{-1}(-0.6514) \approx -40.65°$

Quadrant III: $\theta = 180° + 40.65° = 220.65°$

Quadrant IV: $\theta = 360° - 40.65° = 319.35°$

89. $\tan\theta = 2.521$

Quadrant I: $\theta = \tan^{-1} 2.521 \approx 68.36°$

Quadrant III: $\theta = 180° + \tan^{-1} 2.521 \approx 248.36°$

90. $\tan\theta = 3.262$

Quadrant I: $\theta = \tan^{-1} 3.262 \approx 72.96°$

Quadrant III: $\theta = 180° + \tan^{-1} 3.262 \approx 252.96°$

91. $\sec\theta = -4.156 \Rightarrow \theta' = \cos^{-1}\left(\frac{1}{4.156}\right) \approx 76.08°$

Quadrant II: $\theta = 180° - \theta' \approx 103.92°$

Quadrant III: $\theta = 180° + \theta' \approx 256.08°$

92. $\csc\theta = -6.125 \Rightarrow \theta' = \sin^{-1}\left(\frac{1}{6.125}\right) \approx 9.40°$

Quadrant III: $\theta = 180° + \theta' \approx 189.40°$

Quadrant IV: $\theta = 360° - \theta' \approx 350.60°$

93. $\cot\theta = 15.254 \Rightarrow \theta' = \tan^{-1}\left(\frac{1}{15.254}\right) \approx 3.75°$

Quadrant I: $\theta = \theta' \approx 3.75°$

Quadrant III: $\theta = 180° + \theta' \approx 183.75°$

94. $\sec\theta = 10.542 \Rightarrow \theta' = \cos^{-1}\left(\frac{1}{10.542}\right) \approx 84.56°$

Quadrant I: $\theta = \theta' \approx 84.56°$

Quadrant IV: $\theta = 360° - \theta' \approx 275.44°$

95. $\cos\theta = 0.9848 \Rightarrow \theta' \approx 0.175$

Quadrant I: $\theta = \cos^{-1}(0.9848) \approx 0.175$

Quadrant IV: $\theta = 2\pi - \theta' \approx 6.109$

96. $\sin\theta = 0.0175$

Quadrant I: $\theta = \sin^{-1} 0.0175 \approx 0.018$

Quadrant II: $\theta = \pi - 0.018 = 3.124$

97. $\tan\theta = 1.192 \Rightarrow \theta' \approx 0.873$

Quadrant I: $\theta = \tan^{-1} 1.192 \approx 0.873$

Quadrant III: $\theta = \pi + \theta' \approx 4.014$

98. $\cot\theta = 5.671 \Rightarrow \frac{1}{\tan\theta} = 5.671$

Quadrant I: $\theta = \tan^{-1}\left(\frac{1}{5.671}\right) \approx 0.1745$

Quadrant III: $\theta = \pi + 0.1745 \approx 3.316$

99. $\sec\theta = -2.6667 \Rightarrow \theta' = \cos^{-1}\left(\frac{1}{2.6667}\right) \approx 1.1864$

Quadrant II: $\theta = \pi - 1.1864 \approx 1.955$

Quadrant III: $\theta = \pi + 1.1864 \approx 4.328$

100. $\cos\theta = -0.3214$

Quadrant II: $\theta = \cos^{-1}(-0.3214) \approx 1.898$

Quadrant III: $\theta = 2\pi - 1.898 \approx 4.385$

101. As θ increases from 0° to 90°, x decreases from 12 cm to 0 cm and y increases from 0 cm to 12 cm.

Therefore, $\sin\theta = \frac{y}{12}$ increases from 0 to 1 and $\cos\theta = \frac{x}{12}$ decreases from 1 to 0. Thus,

$\tan\theta = \frac{y}{x}$ increases without bound, and when $\theta = 90°$ the tangent is undefined.

102. Determine the trigonometric function of the reference angle and prefix the appropriate sign.

103. First, determine a positive coterminal angle. Then determine the trigonometric function of the reference angle and prefix the appropriate sign.

104. $T = 49.5 + 20.5\cos\left(\frac{\pi t}{6} - \frac{7\pi}{6}\right)$

(a) January: $T(1) = 29°$

(b) July: $T(7) = 70°$

(c) December: $T(12) \approx 31.7°$

105. $S = 23.1 + 0.442t + 4.3\sin\frac{\pi t}{6}$

(a) February 2005 $\Rightarrow t = 2$

$S = 23.1 + 0.442(2) + 4.3\sin\frac{2\pi}{6}$

≈ 27.7 thousand or 27,700 units

(b) February 2006 $\Rightarrow t = 14$

$S = 23.1 + 0.442(14) + 4.3\sin\frac{14\pi}{6}$

≈ 33.0 thousand or 33,000 units

(c) September 2005 $\Rightarrow t = 9$

$S = 23.1 + 0.442(9) + 4.3\sin\frac{9\pi}{6}$

≈ 22.8 thousand or 22,800 units

(d) September 2006 $\Rightarrow t = 21$

$S = 23.1 + 0.442(21) + 4.3\sin\frac{21\pi}{6}$

≈ 28.1 thousand or 28,100 units

106. $y(t) = 2\cos 6t$

(a) $y(0) = 2\cos 0 = 2$ centimeters

(b) $y\left(\frac{1}{4}\right) = 2\cos\left(\frac{3}{2}\right) \approx 0.14$ centimeter

(c) $y\left(\frac{1}{2}\right) = 2\cos 3 \approx -1.98$ centimeters

107. $y(t) = 2e^{-t}\cos 6t$

(a) $t = 0$

$y(0) = 2e^{-0}\cos 0 = 2$ centimeters

(b) $t = \frac{1}{4}$

$y\left(\frac{1}{4}\right) = 2e^{-1/4}\cos\left(6 \cdot \frac{1}{4}\right) \approx 0.11$ centimeter

(c) $t = \frac{1}{2}$

$y\left(\frac{1}{2}\right) = 2e^{-1/2}\cos\left(6 \cdot \frac{1}{2}\right) \approx -1.2$ centimeters

108. $I = 5e^{-2t}\sin t$

$I(0.7) = 5e^{-1.4}\sin 0.7 \approx 0.79$

109. $\sin\theta = \frac{6}{d} \Rightarrow d = \frac{6}{\sin\theta}$

(a) $\theta = 30°$

$d = \frac{6}{\sin 30°} = \frac{6}{1/2} = 12$ miles

(b) $\theta = 90°$

$d = \frac{6}{\sin 90°} = \frac{6}{1} = 6$ miles

(c) $\theta = 120°$

$d = \frac{6}{\sin 120°} \approx 6.9$ miles

110. $h(t) = f(t)g(t) = \sin t\cos t$

$$\begin{aligned} h(-t) = f(-t)g(-t) &= \sin(-t)\cos(-t) \\ &= (-\sin t)(\cos t) \quad \text{since } f \text{ is odd and } g \text{ is even.} \\ &= -\sin t\cos t \\ &= -h(t) \end{aligned}$$

Thus, $h(t)$ is an odd function.

111. False. In each of the four quadrants, the sign of the secant function and the cosine function will be the same since they are reciprocals of each other.

112. False. Let $n = 1$ and $\theta = 225°$. $0 \le 135 \le 360$, but $360n - \theta = 135$ is not the reference angle. The reference angle would be 45°.

113. True. If $\sin \theta = \frac{1}{4}$ and $\cos \theta < 0$, then θ lies in Quadrant II. Let $y = 1$ and $r = 4$, then $x = -\sqrt{4^2 - 1^2} = -\sqrt{15}$ and $\tan \theta = \frac{1}{-\sqrt{15}} = -\frac{\sqrt{15}}{15}$.

114. False. If $\tan \theta = \frac{4}{3}$ and $\sin \theta < 0$, then θ lies in Quadrant III. Let $y = -4$ and $x = -3$, then $r = \sqrt{(-3)^2 + (-4)^2} = 5$ and $\cos \theta = -\frac{3}{5}$.

Section 9.5 Graphs of Sine and Cosine Functions

1. $y = 3 \sin 2x$

Period: $\frac{2\pi}{2} = \pi$

Amplitude: $|3| = 3$

2. $y = 2 \cos 3x$

Period: $\frac{2\pi}{b} = \frac{2\pi}{3}$

Amplitude: $|a| = 2$

3. $y = \frac{5}{2} \cos \frac{x}{2}$

Period: $\frac{2\pi}{\frac{1}{2}} = 4\pi$

Amplitude: $\left|\frac{5}{2}\right| = \frac{5}{2}$

4. $y = -3 \sin \frac{x}{3}$

Period: $\frac{2\pi}{b} = \frac{2\pi}{\frac{1}{3}} = 6\pi$

Amplitude: $|a| = |-3| = 3$

5. $y = \frac{1}{2} \sin \frac{\pi x}{3}$

Period: $\frac{2\pi}{\frac{\pi}{3}} = 6$

Amplitude: $\left|\frac{1}{2}\right| = \frac{1}{2}$

6. $y = \frac{3}{2} \cos \frac{\pi x}{2}$

Period: $\frac{2\pi}{b} = \frac{2\pi}{\frac{\pi}{2}} = 4$

Amplitude: $|a| = \frac{3}{2}$

7. $y = -2 \sin x$

Period: $\frac{2\pi}{1} = 2\pi$

Amplitude: $|-2| = 2$

8. $y = -\cos \frac{2x}{3}$

Period: $\frac{2\pi}{b} = \frac{2\pi}{\frac{2}{3}} = 3\pi$

Amplitude: $|a| = |-1| = 1$

9. $y = 3 \sin 10x$

Period: $\frac{2\pi}{10} = \frac{\pi}{5}$

Amplitude: $|3| = 3$

10. $y = \frac{1}{3} \sin 8x$

Period: $\frac{2\pi}{b} = \frac{2\pi}{8} = \frac{\pi}{4}$

Amplitude: $|a| = \frac{1}{3}$

11. $y = \frac{1}{2} \cos \frac{2x}{3}$

Period: $\frac{2\pi}{\frac{2}{3}} = 3\pi$

Amplitude: $\left|\frac{1}{2}\right| = \frac{1}{2}$

12. $y = \frac{5}{2} \cos \frac{x}{4}$

Period: $\frac{2\pi}{b} = \frac{2\pi}{\frac{1}{4}} = 8\pi$

Amplitude: $|a| = \frac{5}{2}$

13. $y = \frac{1}{4}\sin 2\pi x$

Period: $\frac{2\pi}{2\pi} = 1$

Amplitude: $\left|\frac{1}{4}\right| = \frac{1}{4}$

14. $y = \frac{2}{3}\cos\frac{\pi x}{10}$

Period: $\frac{2\pi}{b} = \frac{2\pi}{\frac{\pi}{10}} = 20$

Amplitude: $|a| = \frac{2}{3}$

15. $f(x) = \sin x$

$g(x) = \sin(x - \pi)$

The graph of g is a horizontal shift to the right π units of the graph of f (a phase shift).

16. $f(x) = \cos x$, $g(x) = \cos(x + \pi)$
g is a horizontal shift of f π units to the left.

17. $f(x) = \cos 2x$

$g(x) = -\cos 2x$

The graph of g is a reflection in the x-axis of the graph of f.

18. $f(x) = \sin 3x$, $g(x) = \sin(-3x)$

g is a reflection of f about the y-axis.

19. $f(x) = \cos x$

$g(x) = \cos 2x$

The period of f is twice that of g.

20. $f(x) = \sin x$, $g(x) = \sin 3x$
The period of g is one-third the period of f.

21. $f(x) = \sin 2x$

$g(x) = 3 + \sin 2x$

The graph of g is a vertical shift 3 units upward of the graph of f.

22. $f(x) = \cos 4x$, $g(x) = -2 + \cos 4x$
g is a vertical shift of f two units downward.

23. The graph of g has twice the amplitude as the graph of f. The period is the same.

24. The period of g is one-third the period of f.

25. The graph of g is a horizontal shift π units to the right of the graph of f.

26. Shift the graph of f two units upward to obtain the graph of g.

27. $f(x) = -2\sin x$

Period: 2π

Amplitude: 2

$g(x) = 4\sin x$

Period: 2π

Amplitude: 4

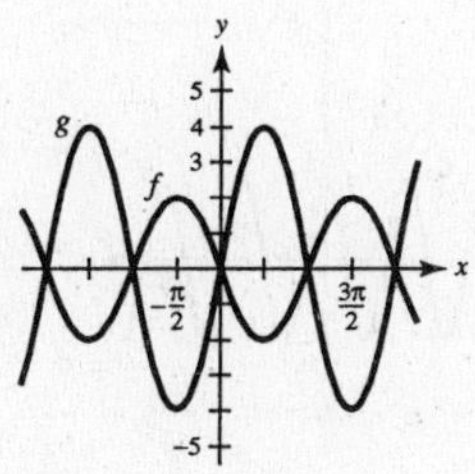

28. $f(x) = \sin x$

Period: 2π

Amplitude: 1

$g(x) = 4\sin\left(\frac{x}{3}\right)$

Period: 6π

Amplitude: 1

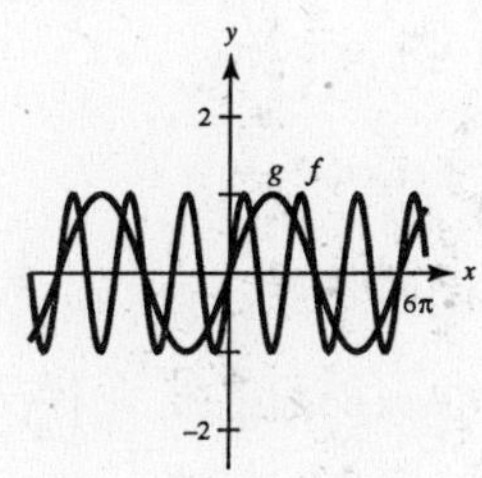

29. $f(x) = \cos x$

Period: 2π

Amplitude: 1

$g(x) = 1 + \cos x$
is a vertical shift of the graph of $f(x)$ one unit upward.

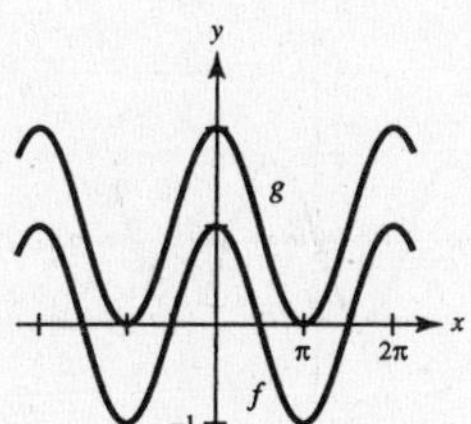

30. $f(x) = 2\cos 2x$

Period: π

Amplitude: 1

$g(x) = -\cos 4x$

Period: $\frac{\pi}{2}$

Amplitude: 1

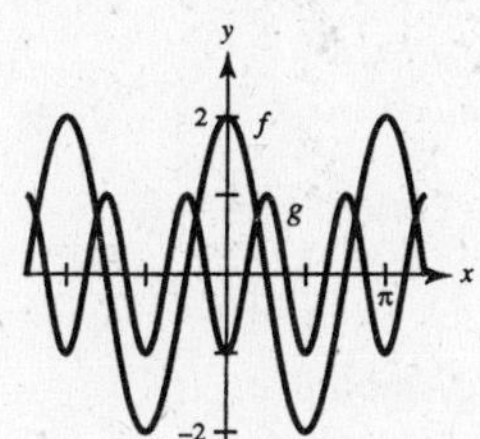

31. $f(x) = -\frac{1}{2}\sin\frac{x}{2}$

Period: 4π

Amplitude: $\frac{1}{2}$

$g(x) = 3 - \frac{1}{2}\sin\frac{x}{2}$ is the graph of $f(x)$ shifted vertically three units upward.

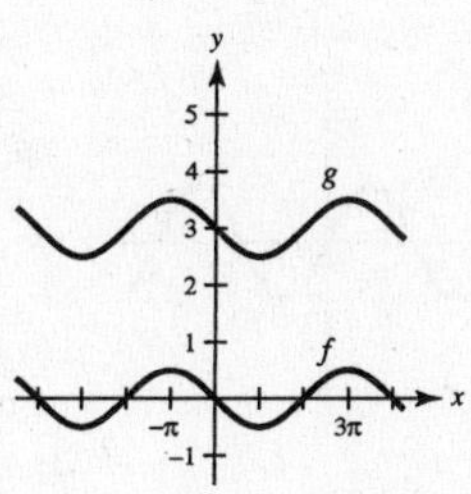

32. $f(x) = 4\sin\pi x$

Period: 2

Amplitude: 1

$g(x) = 4\sin\pi x - 3$ is the graph of $f(x)$ shifted vertically three units downward.

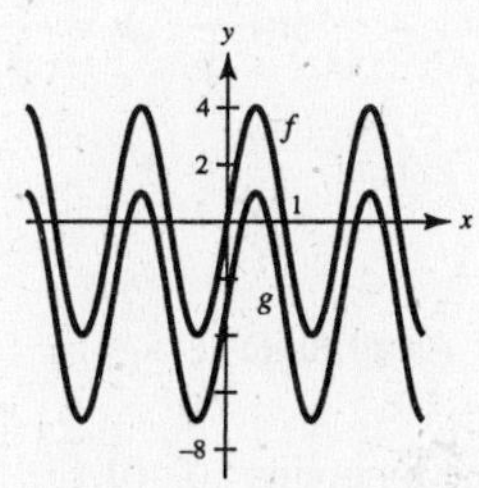

33. $f(x) = 2\cos x$

Period: 2π

Amplitude: 2

$g(x) = 2\cos(x + \pi)$ is the graph of $f(x)$ shifted π units to the left.

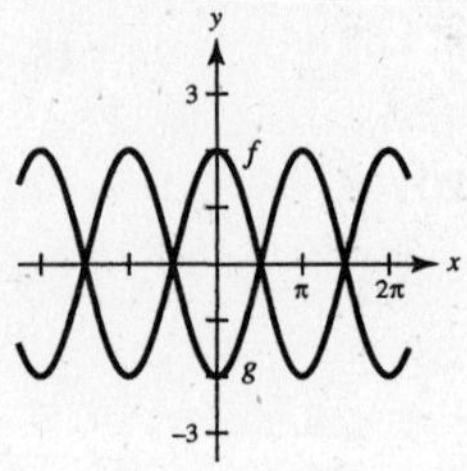

34. $f(x) = -\cos x$

Period: 2π

Amplitude: 1

$g(x) = -\cos(x - \pi)$ is the graph of $f(x)$ shifted π units to the right.

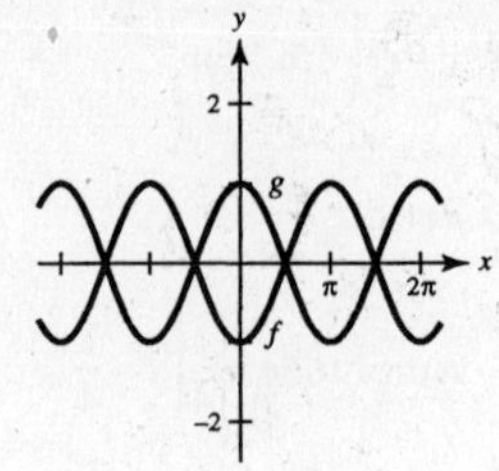

35. $y = -2\sin 6x$; $a = -2$, $b = 6$, $c = 0$

Period: $\dfrac{2\pi}{6} = \dfrac{\pi}{3}$

Amplitude: $|-2| = 2$

Key points: $(0, 0), \left(\dfrac{\pi}{12}, -2\right), \left(\dfrac{\pi}{6}, 0\right), \left(\dfrac{\pi}{4}, 2\right), \left(\dfrac{\pi}{3}, 0\right)$

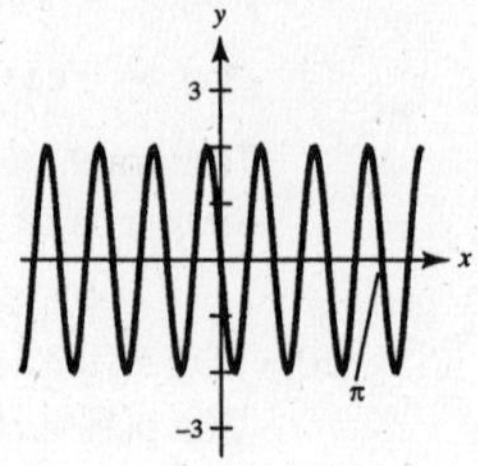

36. $y = -3\cos 4x$

Period: $\dfrac{2\pi}{4} = \dfrac{\pi}{2}$

Amplitude: 3

Key points: $(0, -3), \left(\dfrac{\pi}{8}, 0\right), \left(\dfrac{\pi}{4}, 3\right), \left(\dfrac{3\pi}{8}, 0\right), \left(\dfrac{\pi}{2}, -3\right)$

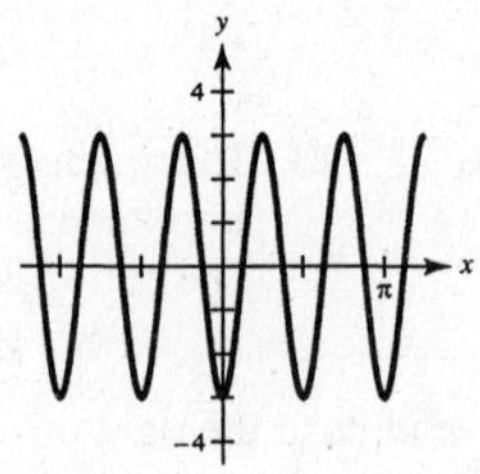

37. $y = \cos 2\pi x$

Period: $\dfrac{2\pi}{2\pi} = 1$

Amplitude: 1

Key points: $(0, 1), \left(\dfrac{1}{4}, 0\right), \left(\dfrac{1}{2}, -1\right), \left(\dfrac{3}{4}, 0\right)$

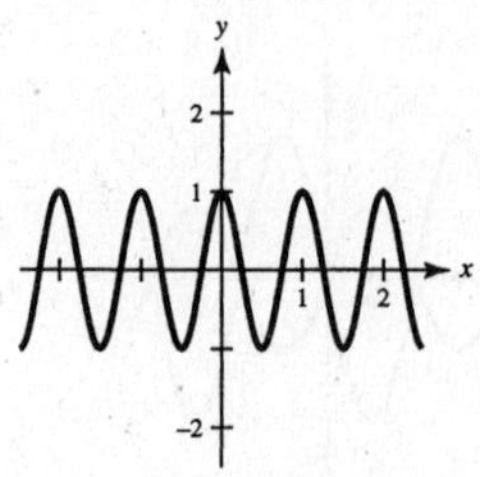

38. $y = \sin\dfrac{\pi x}{4}$

Period: $\dfrac{2\pi}{\frac{\pi}{4}} = 8$

Amplitude: 1

Key points: $(0, 0), (2, 1), (4, 0), (6, -1), (8, 0)$

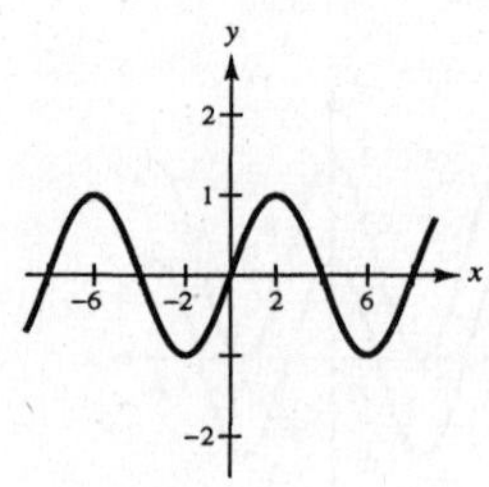

39. $y = -\sin\dfrac{2\pi x}{3}$; $a = -1, b = \dfrac{2\pi}{3}, c = 0$

Period: $\dfrac{2\pi}{\frac{2\pi}{3}} = 3$

Amplitude: 1

Key points: $(0, 0), \left(\dfrac{3}{4}, -1\right), \left(\dfrac{3}{2}, 0\right), \left(\dfrac{9}{4}, 1\right), (3, 0)$

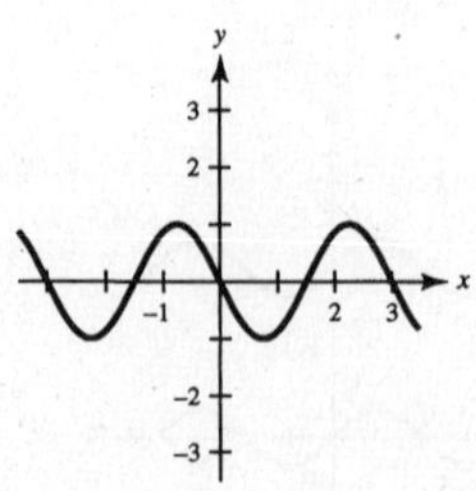

40. $y = -10 \cos \dfrac{\pi x}{6}$

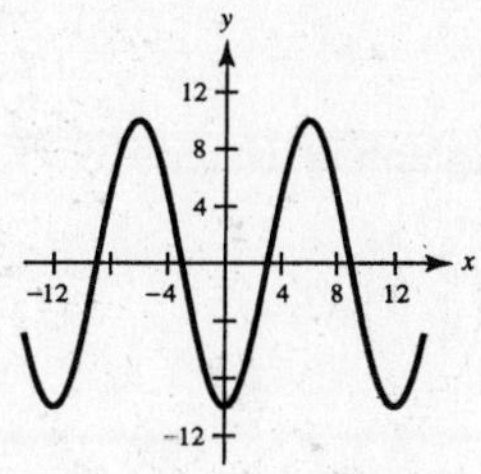

Period: $\dfrac{2\pi}{\frac{\pi}{6}} = 12$

Amplitude: 10

Key points: $(0, -10), (3, 0), (6, 10), (9, 0), (12, -10)$

41. $y = \sin\left(x - \dfrac{\pi}{4}\right)$; $a = 1,\ b = 1,\ c = \dfrac{\pi}{4}$

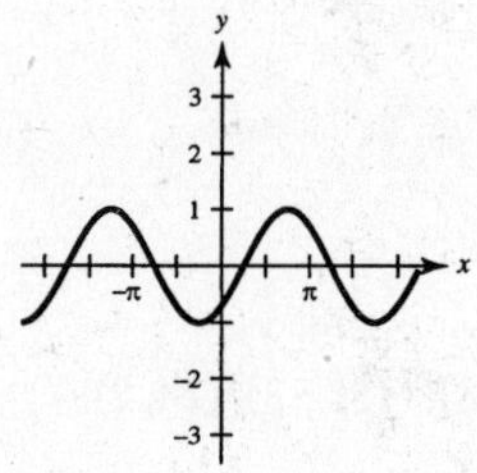

Period: 2π

Amplitude: 1

Shift: Set $x - \dfrac{\pi}{4} = 0$ and $x - \dfrac{\pi}{4} = 2\pi$

$x = \dfrac{\pi}{4}$ $\qquad x = \dfrac{9\pi}{4}$

Key points: $\left(\dfrac{\pi}{4}, 0\right), \left(\dfrac{3\pi}{4}, 1\right), \left(\dfrac{5\pi}{4}, 0\right), \left(\dfrac{7\pi}{4}, -1\right), \left(\dfrac{9\pi}{4}, 0\right)$

42. $y = \sin(x - \pi)$

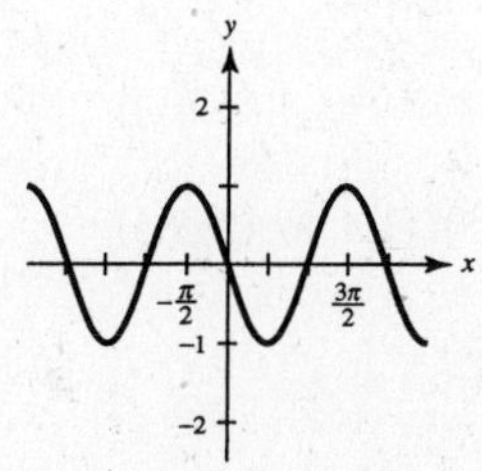

Period: 2π

Amplitude: 1

Key points: $(0, 0), \left(\dfrac{\pi}{2}, -1\right), (\pi, 0), \left(\dfrac{3\pi}{2}, 1\right), (2\pi, 0)$

43. $y = 3 \cos (x + \pi)$

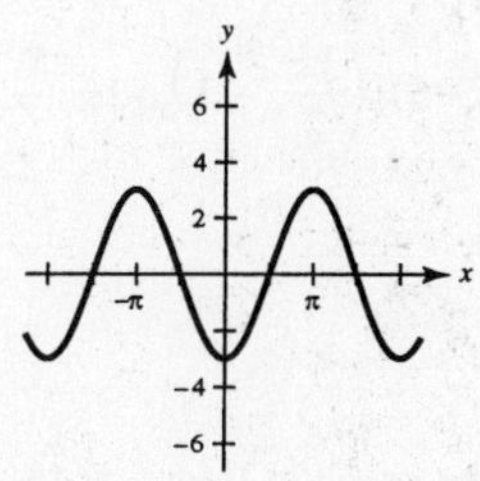

Period: 2π

Amplitude: 3

Shift: Set $x + \pi = 0$ and $x + \pi = 2\pi$

$x = -\pi$ $\qquad x = \pi$

Key points: $(-\pi, 3), \left(-\dfrac{\pi}{2}, 0\right), (0, -3), \left(\dfrac{\pi}{2}, 0\right), (\pi, 3)$

44. $y = 4 \cos\left(x + \dfrac{\pi}{4}\right)$

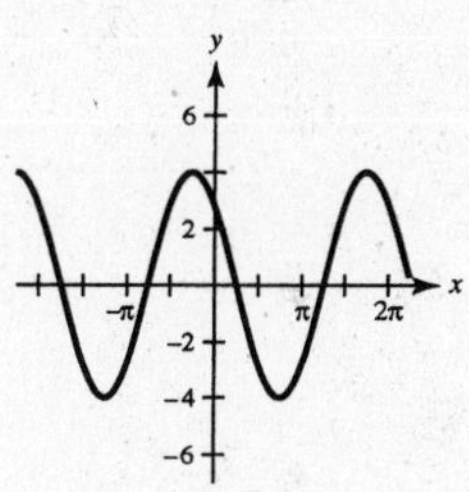

Period: 2π

Amplitude: 4

Shift: Set $x + \dfrac{\pi}{4} = 0$ and $x + \dfrac{\pi}{4} = 2\pi$

$x = -\dfrac{\pi}{4}$ $\qquad x = \dfrac{7\pi}{4}$

Key points: $\left(-\dfrac{\pi}{4}, 4\right), \left(\dfrac{\pi}{4}, 0\right), \left(\dfrac{3\pi}{4}, -4\right), \left(\dfrac{5\pi}{4}, 0\right), \left(\dfrac{7\pi}{4}, 4\right)$

45. $y = 2 - \sin \dfrac{2\pi x}{3}$

Vertical shift 2 units upward of the graph in Exercise 39.

Key points: $(0, 2), \left(\dfrac{3}{4}, 1\right), \left(\dfrac{3}{2}, 2\right), \left(\dfrac{9}{4}, 3\right), (3, 2)$

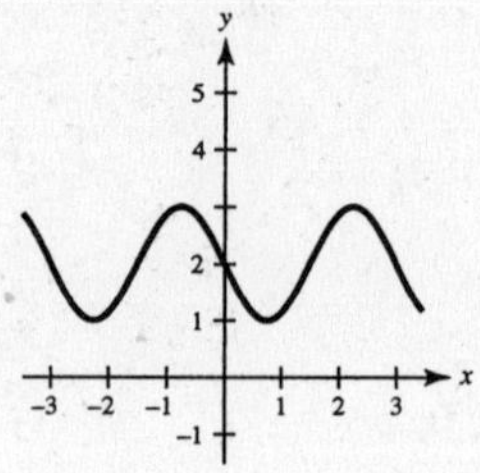

46. $y = -3 + 5 \cos \dfrac{\pi t}{12}$

Period: $\dfrac{2\pi}{\dfrac{\pi}{12}} = 24$

Amplitude: 5

Key points: $(0, 2), (6, -3), (12, -8), (18, -3), (24, 2)$

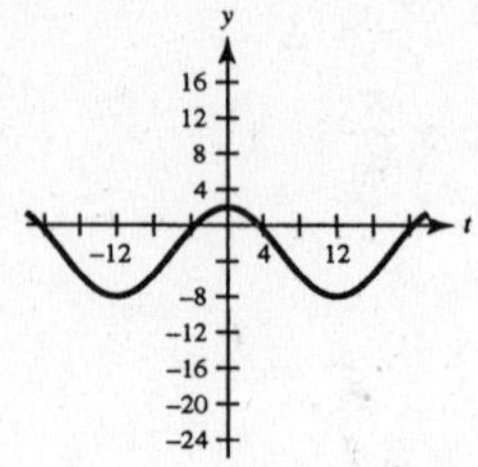

47. $y = 2 + \dfrac{1}{2}\cos(\pi x)$

Period: $\dfrac{2\pi}{\pi} = 2$

Amplitude: $\dfrac{1}{2}$

Vertical shift 2 units upward.

Key points: $(0, 2.5), (0.5, 2), (1, 1.5), (1.5, 2), (2, 2.5)$

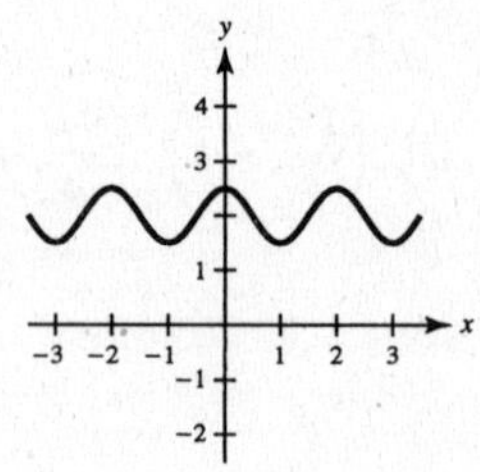

48. $y = 2 \cos x - 3$

Period: 2π

Amplitude: 2

Key points: $(0, -1), \left(\dfrac{\pi}{2}, -3\right), (\pi, -5), \left(\dfrac{3\pi}{2}, -3\right), (2\pi, -1)$

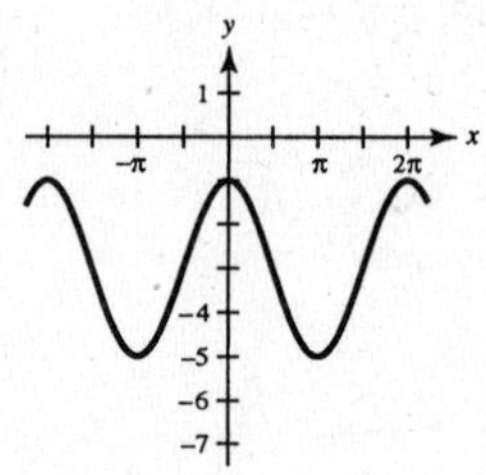

49. $y = 3 \cos(x + \pi) - 3$

Vertical shift 3 units downward of the graph in Exercise 43.

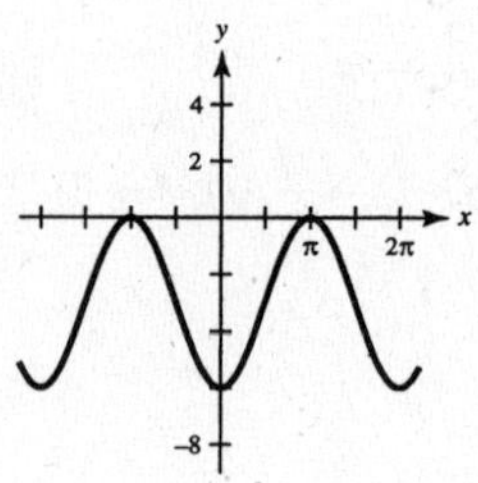

50. $y = 4 \cos\left(x + \dfrac{\pi}{4}\right) + 4$

Period: 2π

Amplitude: 4

Shift: $x + \dfrac{\pi}{4} = 0$ and $x + \dfrac{\pi}{4} = 2\pi$

$x = -\dfrac{\pi}{4}$ $\qquad x = \dfrac{7\pi}{4}$

Key points: $\left(-\dfrac{\pi}{4}, 8\right), \left(\dfrac{\pi}{4}, 4\right), \left(\dfrac{3\pi}{4}, 0\right), \left(\dfrac{5\pi}{4}, 4\right), \left(\dfrac{7\pi}{4}, 8\right)$

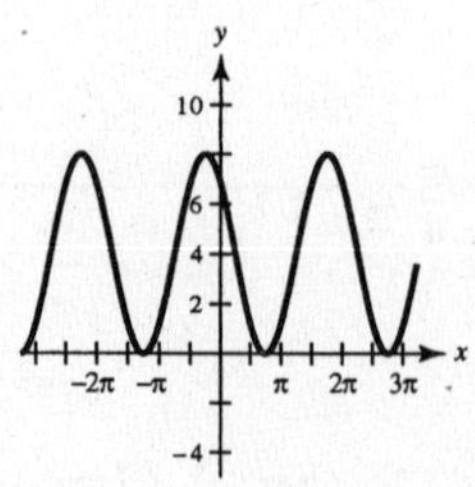

51. $y = \frac{2}{3}\cos\left(\frac{x}{2} - \frac{\pi}{4}\right)$; $a = \frac{2}{3}$, $b = \frac{1}{2}$, $c = \frac{\pi}{4}$

Period: 4π

Amplitude: $\frac{2}{3}$

Shift: $\frac{x}{2} - \frac{\pi}{4} = 0$ and $\frac{x}{2} - \frac{\pi}{4} = 2\pi$

$x = \frac{\pi}{2}$ $\qquad x = \frac{9\pi}{2}$

Key points: $\left(\frac{\pi}{2}, \frac{2}{3}\right), \left(\frac{3\pi}{2}, 0\right), \left(\frac{5\pi}{2}, \frac{-2}{3}\right), \left(\frac{7\pi}{2}, 0\right), \left(\frac{9\pi}{2}, \frac{2}{3}\right)$

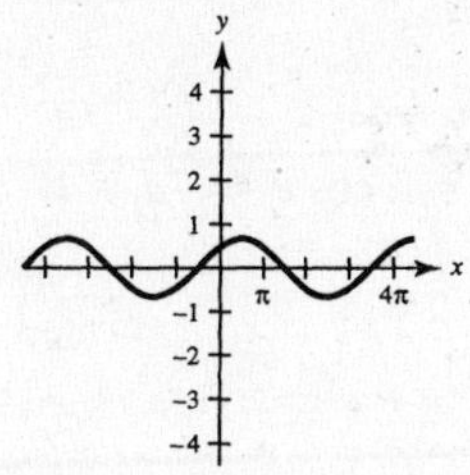

52. $y = -3\cos(6x + \pi)$

Period: $\frac{2\pi}{6} = \frac{\pi}{3}$

Amplitude: 3

Shift: $6x + \pi = 0$ and $6x + \pi = 2\pi$

$x = -\frac{\pi}{6}$ $\qquad x = \frac{\pi}{6}$

Key points: $\left(-\frac{\pi}{6}, -3\right), \left(-\frac{\pi}{12}, 0\right), (0, 3), \left(\frac{\pi}{12}, 0\right), \left(\frac{\pi}{6}, -3\right)$

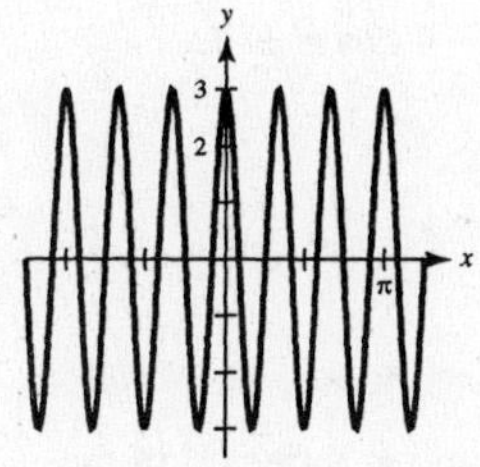

53. $y = -2\sin(4x + \pi)$

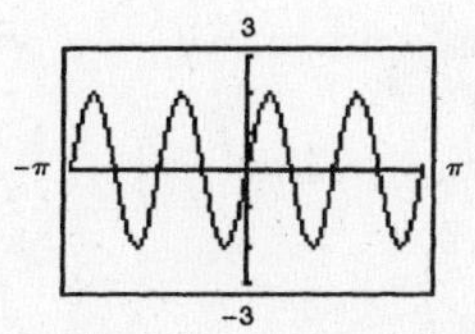

54. $y = -4\sin\left(\frac{2}{3}x - \frac{\pi}{3}\right)$

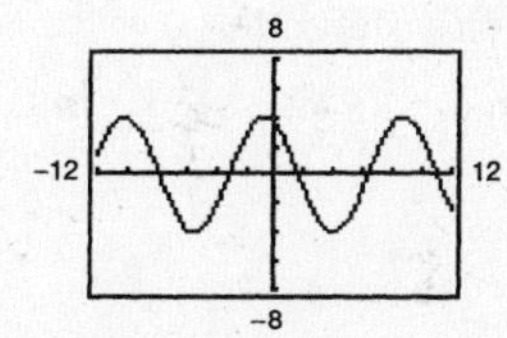

55. $y = \cos\left(2\pi x - \frac{\pi}{2}\right) + 1$

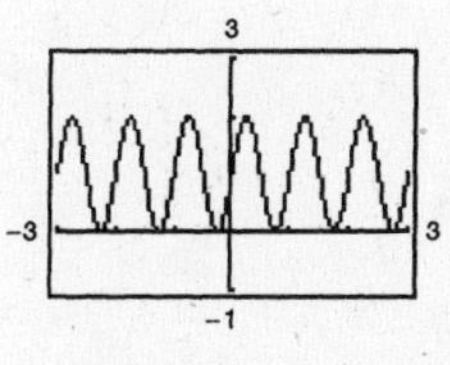

56. $y = 3\cos\left(\frac{\pi x}{2} + \frac{\pi}{2}\right) - 2$

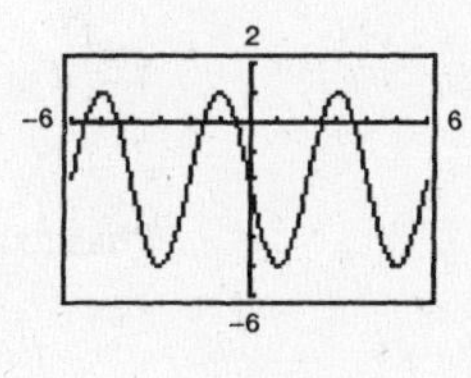

57. $y = 5\sin(\pi - 2x) + 10$

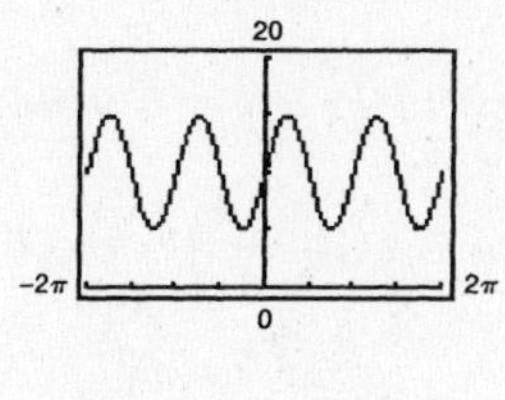

58. $y = 5\cos(\pi - 2x) + 2$

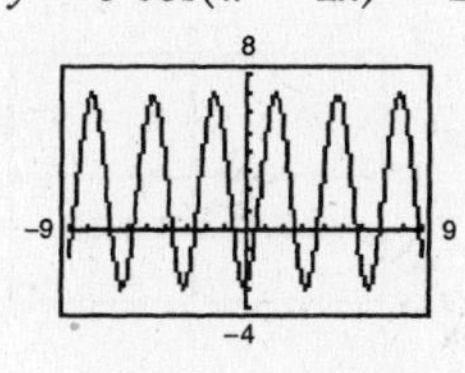

59. $y = -0.1\sin\left(\frac{\pi x}{10} + \pi\right)$

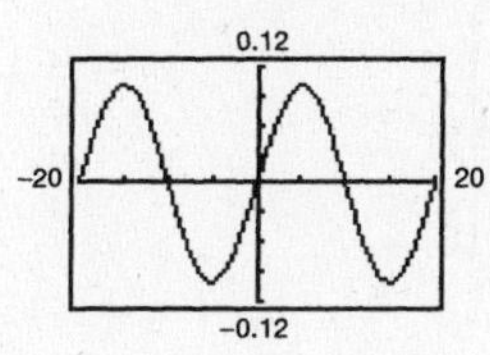

60. $y = \frac{1}{100}\sin 120\pi t$

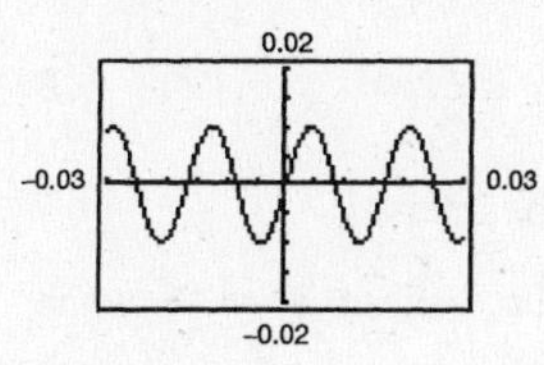

61. $f(x) = a \cos x + d$

Amplitude: $\frac{1}{2}[3 - (-1)] = 2 \Rightarrow a = 2$

Vertical shift 1 unit upward of $g(x) = 2 \cos x \Rightarrow d = 1$.

Thus, $f(x) = 2 \cos x + 1$.

62. $f(x) = a \cos x + d$

Amplitude: $\dfrac{1 - (-3)}{2} = 2$

$1 = 2 \cos 0 + d$

$d = 1 - 2 = -1$

$a = 2, d = -1$

63. $f(x) = a \cos x + d$

Amplitude: $\frac{1}{2}[8 - 0] = 4$

Since $f(x)$ is the graph of $g(x) = 4 \cos x$ reflected in the x-axis and shifted vertically 4 units upward, we have $a = -4$ and $d = 4$. Thus, $f(x) = -4 \cos x + 4$.

64. $f(x) = a \cos x + d$

Amplitude: $\dfrac{-2 - (-4)}{2} = 1$

Reflected in the x-axis: $a = -1$

$-4 = -1 \cos 0 + d$

$d = -3$

Thus, $f(x) = -\cos x - 3$

65. $y = a \sin(bx - c)$

Amplitude: $|a| = |3|$ Since the graph is reflected in the x-axis, we have $a = -3$.

Period: $\dfrac{2\pi}{b} = \pi \Rightarrow b = 2$

Phase shift: $c = 0$

Thus, $y = -3 \sin 2x$.

66. $y = a \sin(bx - c)$

Amplitude: $2 \Rightarrow a = 2$

Period: 4π

$\dfrac{2\pi}{b} = 4\pi \Rightarrow b = \dfrac{1}{2}$

Phase shift: $c = 0$

Thus, $y = 2 \sin\left(\dfrac{1}{2}x\right)$

67. $y = a \sin(bx - c)$

Amplitude: $a = 2$

Period: $2\pi \Rightarrow b = 1$

Phase shift: $bx - c = 0$ when $x = -\dfrac{\pi}{4}$

$$(1)\left(\frac{-\pi}{4}\right) - c = 0 \Rightarrow c = -\frac{\pi}{4}$$

Thus, $y = 2 \sin\left(x + \dfrac{\pi}{4}\right)$.

68. $y = a \sin(bx - c)$

Amplitude: $2 \Rightarrow a = 2$

Period: 4

$\dfrac{2\pi}{b} = 4 \Rightarrow b = \dfrac{\pi}{2}$

Phase shift: $\dfrac{c}{b} = -1 \Rightarrow c = -\dfrac{\pi}{2}$

Thus, $y = 2 \sin\left(\dfrac{\pi}{2}x - \dfrac{\pi}{2}\right)$

69. $y_1 = \sin x$

$y_2 = -\dfrac{1}{2}$

In the interval $[-2\pi, 2\pi]$, $\sin x = -\dfrac{1}{2}$ when

$x = -\dfrac{5\pi}{6}, -\dfrac{\pi}{6}, \dfrac{7\pi}{6}, \dfrac{11\pi}{6}$.

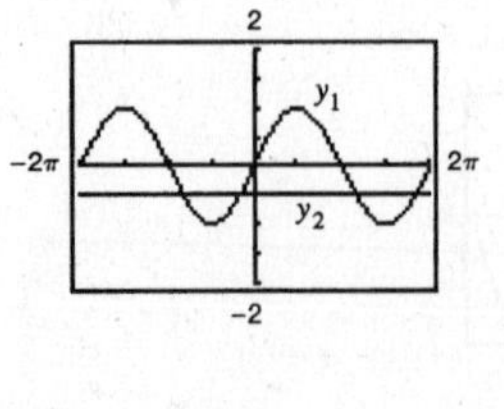

70. $y_1 = \cos x$

$y_2 = -1$

$y_1 = y_2$ when $x = \pi, -\pi$

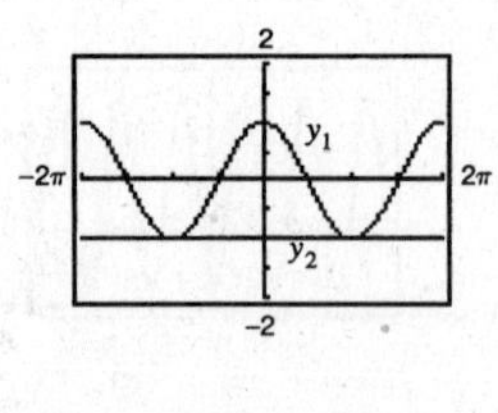

71. $y_1 = \cos x$

$y_2 = \dfrac{\sqrt{2}}{2}$

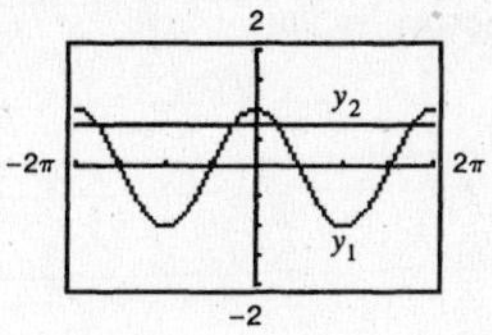

In the interval

$[-2\pi, 2\pi]$, $\cos x = \dfrac{\sqrt{2}}{2}$

when $x = \pm\dfrac{\pi}{4}, \pm\dfrac{7\pi}{4}$.

72. $y_1 = \sin x$

$y_2 = \dfrac{\sqrt{3}}{2}$

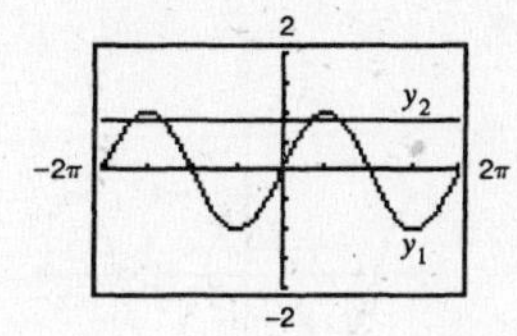

$y_1 = y_2$

when $x = \dfrac{\pi}{3}, \dfrac{2\pi}{3}, -\dfrac{4\pi}{3}, -\dfrac{5\pi}{3}$.

73.

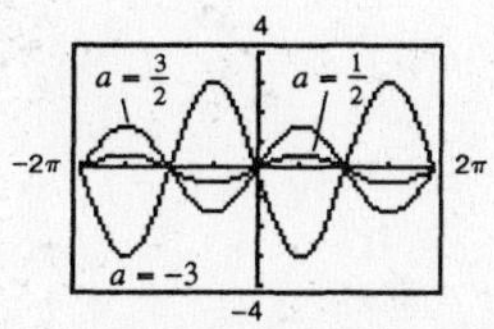

As a changes from $\frac{1}{2}$ to $\frac{3}{2}$, the amplitude increases. When $a = -3$, the amplitude again increases but the graph is reflected in the x-axis also.

74. $y = 2 + \sin x$

$y = 3.5 + \sin x$

$y = -2 + \sin x$

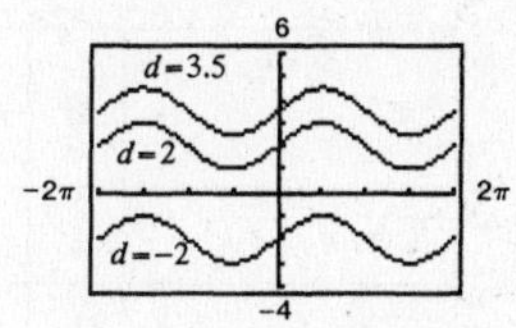

Each value of d produces a vertical shift of $y = \sin x$ upward (or downward) by d units.

75.

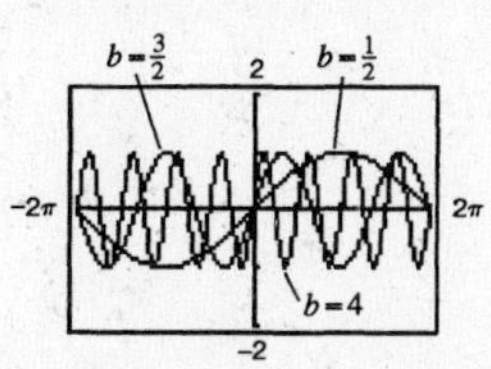

As b increases the period decreases.

76. $y = \sin(x - 1)$

$y = \sin(x - 3)$

$y = \sin(x - (-2)) = \sin(x + 2)$

Each value of c produces a horizontal shift of $y = \sin x$ to the left (or right) by c units.

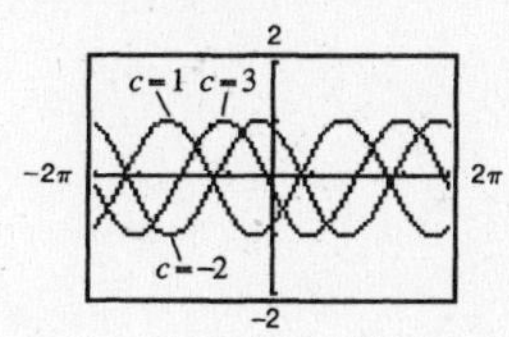

77.

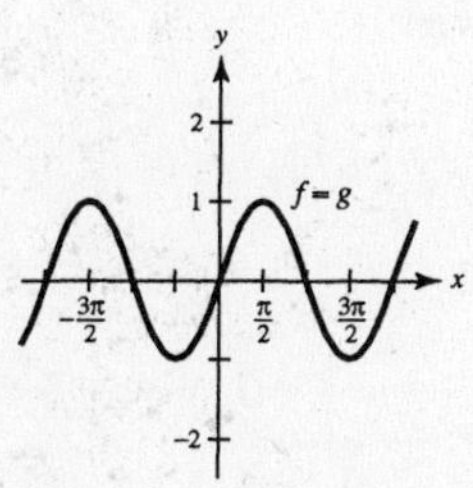

Since the graphs are the same, the conjecture is that

$\sin(x) = \cos\left(x - \dfrac{\pi}{2}\right)$.

78. $f(x) = \sin x$, $g(x) = -\cos\left(x + \dfrac{\pi}{2}\right)$

x	0	$\frac{\pi}{2}$	π	$\frac{3\pi}{2}$	2π
$\sin x$	0	1	0	-1	0
$-\cos\left(x + \frac{\pi}{2}\right)$	0	1	0	-1	0

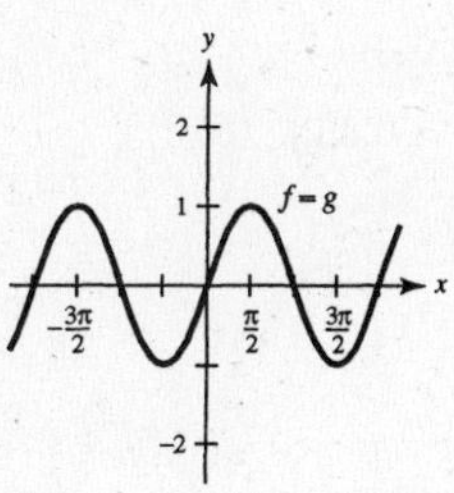

Conjecture: $\sin x = -\cos\left(x + \dfrac{\pi}{2}\right)$

79.

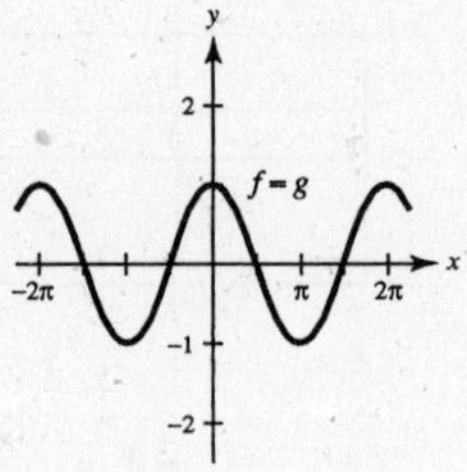

Since the graphs are the same, the conjecture is that

$$\cos x = -\sin\left(x - \frac{\pi}{2}\right).$$

80. $f(x) = \cos x$, $g(x) = -\cos(x - \pi)$

x	0	$\frac{\pi}{2}$	π	$\frac{3\pi}{2}$	2π
$\cos x$	1	0	-1	0	1
$-\cos(x - \pi)$	1	0	-1	0	1

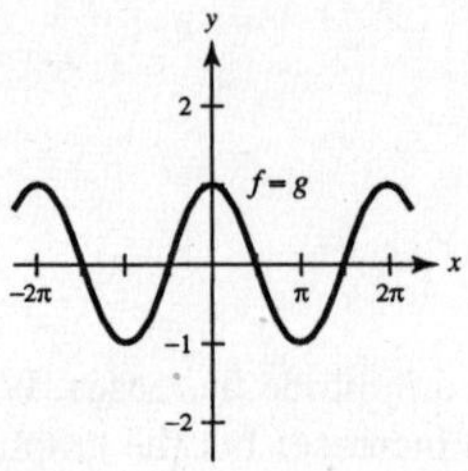

Conjecture: $\cos x = -\cos(x - \pi)$

81. (a)

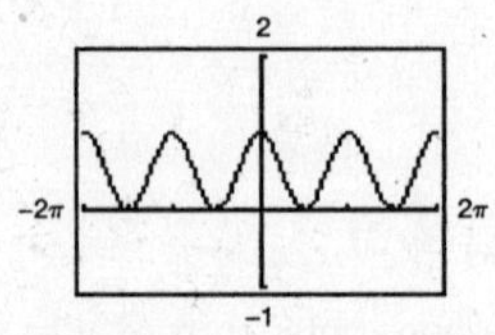

h is even

(b)

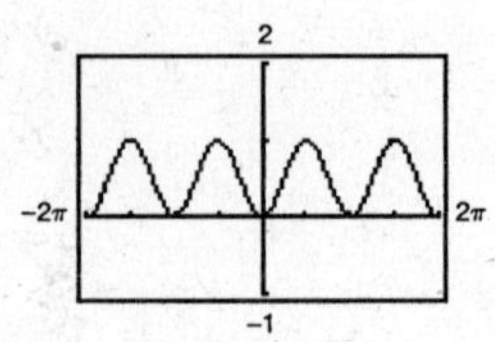

h is even

82. $f(x)$ is even $\Rightarrow f(-x) = f(x)$

$g(x)$ is odd $\Rightarrow g(-x) = -g(x)$

(a) $h(x) = [f(x)]^2$

$h(-x) = [f(-x)]^2$

$= [f(x)]^2$

$= h(x) \Rightarrow h(x)$ is even

(b) $h(x) = [g(x)]^2$

$h(-x) = [g(-x)]^2$

$= [-g(x)]^2$

$= [g(x)]^2$

$= h(x) \Rightarrow h(x)$ is even

83. $y = 0.85 \sin \frac{\pi t}{3}$

(a) Time for one cycle = one period $= \frac{2\pi}{\pi/3} = 6$ sec

(b) Cycles per min $= \frac{60}{6} = 10$ cycles per min

(c) Amplitude: 0.85

Period: 6

Key points: $(0, 0)$, $\left(\frac{3}{2}, 0.85\right)$, $(3, 0)$, $\left(\frac{9}{2}, -0.85\right)$, $(6, 0)$

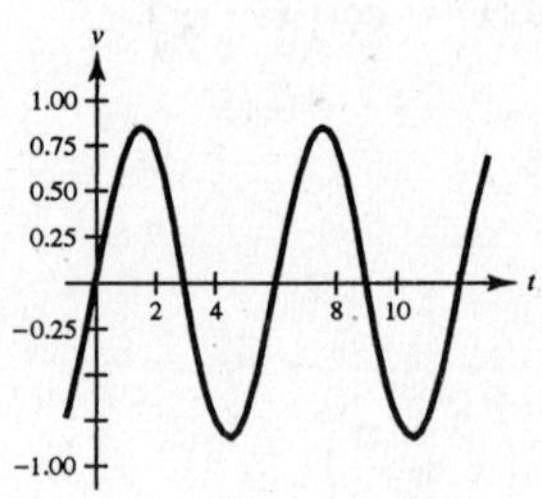

84. $v = 1.75 \sin \dfrac{\pi t}{2}$

(a) Period: $\dfrac{2\pi}{\pi/2} = 4$ seconds

(b) $\dfrac{1 \text{ cycle}}{4 \text{ seconds}} \cdot \dfrac{60 \text{ seconds}}{1 \text{ minute}} = 15$ cycles per minute

(c)

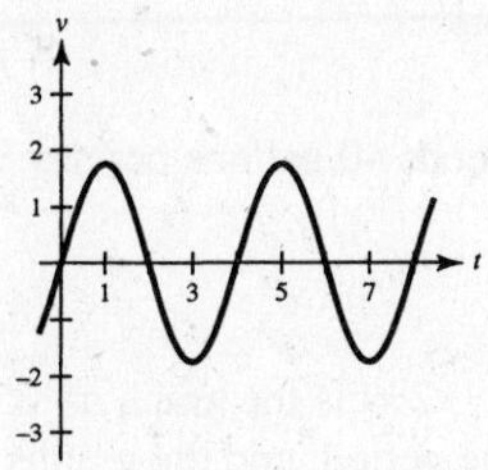

85. $y = 0.001 \sin 880\pi t$

(a) Period: $\dfrac{2\pi}{880\pi} = \dfrac{1}{440}$ seconds

(b) $f = \dfrac{1}{p} = 440$ cycles per second

86. $P = 100 - 20 \cos \dfrac{5\pi t}{3}$

(a) Period: $\dfrac{2\pi}{5\pi/3} = \dfrac{6}{5}$ seconds

(b) $\dfrac{1 \text{ heartbeat}}{6/5 \text{ seconds}} \cdot \dfrac{60 \text{ seconds}}{1 \text{ minute}} = 50$ heartbeats per minute

87. (a) $A(t) = 19.7 \sin(0.47t - 1.74) + 70.24$

(b)

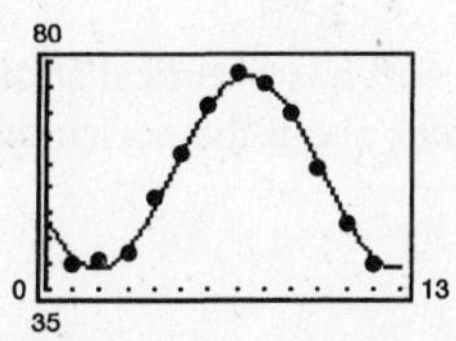

The model is a good fit for most months.

(c)

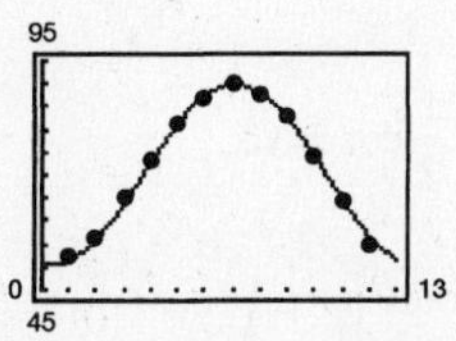

The model is a good fit.

(d) Use the constant term of each model to estimate the average annual temperature.

Nantucket: 58°

Athens: 70.24°

(e) $N(t)$ has a period of 11 months, $A(t)$ has a period of about 13 months. The periods should both be 12 months to correspond with one year.

(f) Athens has slightly more variability in temperature since its amplitude is greater than the amplitude in the Nantucket model.

88.

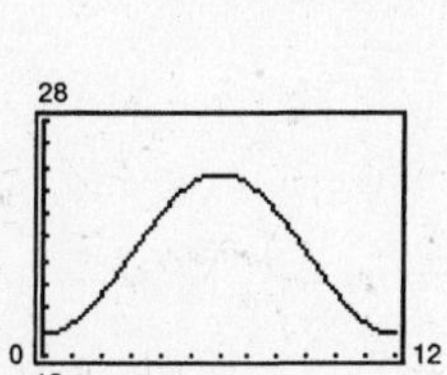

89. $S = 74.50 + 43.75 \sin \dfrac{\pi t}{6}$

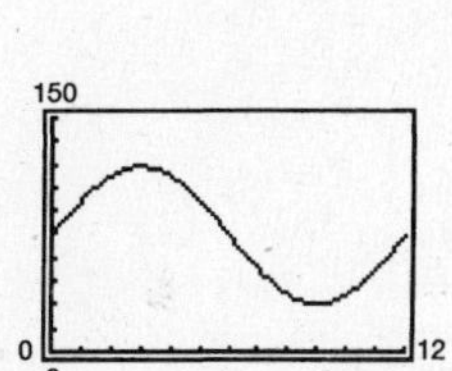

90. $C = 30.3 + 21.6 \sin\left(\frac{2\pi t}{365} + 10.9\right)$

(a) Period $= \dfrac{2\pi}{\frac{2\pi}{365}} = 365$

Yes, this is what is expected because there are 365 days in a year.

(b) The average daily fuel consumption is given by the amount of the vertical shift (from 0) which is given by the constant 30.3.

(c)

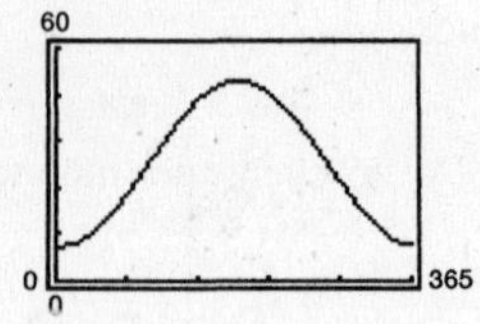

The consumption exceeds 40 gallons per day when $124 < x < 252$.

91. (a) and (c)

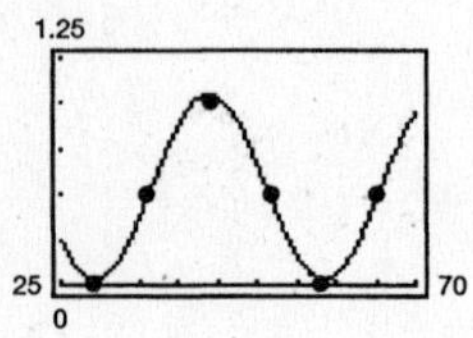

(b) $y = 0.51 \sin(0.216x - 1.62) + 0.54$

(d) For June 27, 2007, use $x = 543$.

$y(543) \approx 0.816$

92. False. The graph of $\sin(x + 2\pi)$ is the graph of $\sin(x)$ translated to the *left* by one period, and the graphs are indeed identical.

93. False. $y = \frac{1}{2} \cos 2x$ has an amplitude that is **half** that of $y = \cos x$. For $y = a \cos bx$, the amplitude is $|a|$.

94. True. $f(x) = 1 + 2 \sin(4x - \pi)$ has a vertical shift one unit upward so it oscillates about the horizontal line $y = 1$.

Section 9.6 Graphs of Other Trigonometric Functions

1. $y = \sec 3x$

Period: $\dfrac{2\pi}{3}$

Matches graph (e).

2. $y = \tan \dfrac{x}{3}$

Period: $\dfrac{\pi}{1/3} = 3\pi$

Matches graph (c).

3. $y = \dfrac{1}{2} \cot \pi x$

Period: $\dfrac{\pi}{\pi} = 1$

Matches graph (a).

4. $y = \dfrac{1}{2} \sec \dfrac{\pi x}{2}$

Period: $\dfrac{2\pi}{b} = \dfrac{2\pi}{\pi/2} = 4$

Asymptotes: $x = -1, x = 1$

Matches graph (f).

5. $y = -\csc x$

Period: 2π

Matches graph (d).

6. $y = -2 \sec \dfrac{\pi x}{2}$

Period: $\dfrac{2\pi}{b} = \dfrac{2\pi}{\pi/2} = 4$

Asymptotes: $x = -1, x = 1$

Reflected in x-axis

Matches graph (b).

7. $y = \frac{1}{3}\tan x$

Period: π

Two consecutive asymptotes:

$x = -\frac{\pi}{2}$ and $x = \frac{\pi}{2}$

x	$-\frac{\pi}{4}$	0	$\frac{\pi}{4}$
y	$-\frac{1}{3}$	0	$\frac{1}{3}$

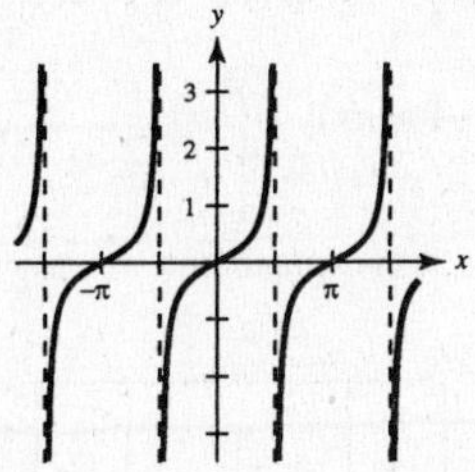

8. $y = \frac{1}{4}\tan x$

Period: π

Asymptotes: $x = -\frac{\pi}{2}, x = \frac{\pi}{2}$

x	$-\frac{\pi}{4}$	0	$\frac{\pi}{4}$
y	$-\frac{1}{4}$	0	$\frac{1}{4}$

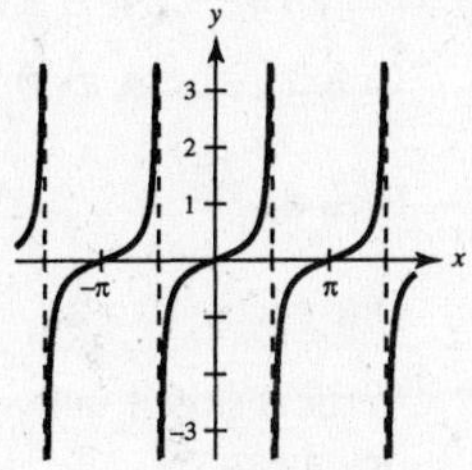

9. $y = \tan 3x$

Period: $\frac{\pi}{3}$

Two consecutive asymptotes:

$3x = -\frac{\pi}{2} \Rightarrow x = -\frac{\pi}{6}$

$3x = \frac{\pi}{2} \Rightarrow x = \frac{\pi}{6}$

x	$-\frac{\pi}{12}$	0	$\frac{\pi}{12}$
y	-1	0	1

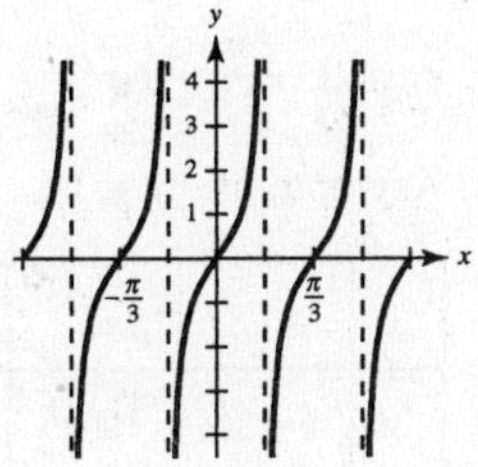

10. $y = -3\tan \pi x$

Period: $\frac{\pi}{\pi} = 1$

Asymptotes: $x = -\frac{1}{2}, x = \frac{1}{2}$

x	$-\frac{1}{4}$	0	$\frac{1}{4}$
y	3	0	-3

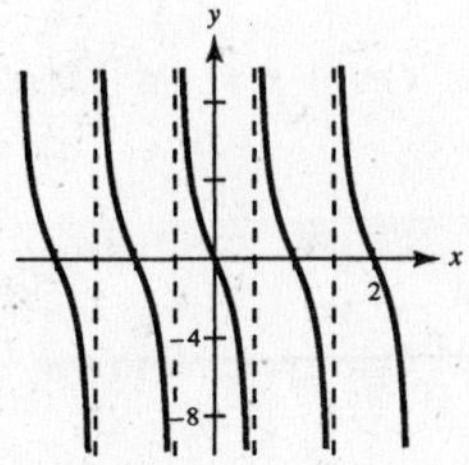

11. $y = -\frac{1}{2}\sec x$

Graph $y = -\frac{1}{2}\cos x$ first.

Period: 2π

One cycle: 0 to 2π

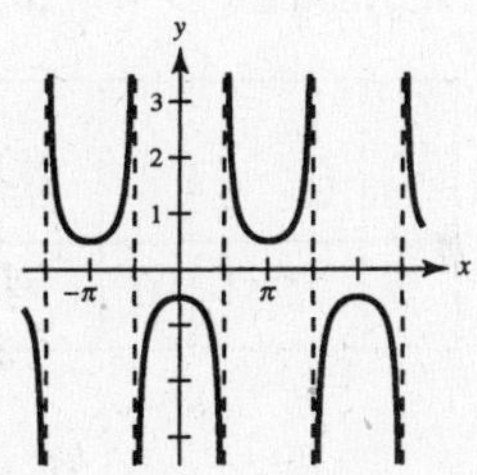

12. $y = \frac{1}{4}\sec x$

Period: 2π

Asymptotes: $x = -\frac{\pi}{2}, x = \frac{\pi}{2}$

x	$-\frac{\pi}{4}$	0	$\frac{\pi}{4}$
y	0.354	$\frac{1}{4}$	0.354

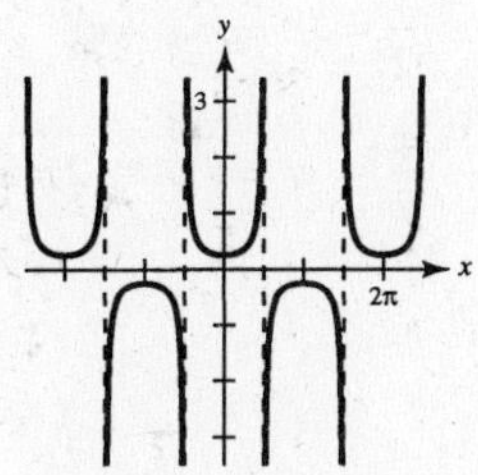

13. $y = \csc \pi x$

Graph $y = \sin \pi x$ first.

Period: $\frac{2\pi}{\pi} = 2$

One cycle: 0 to 2

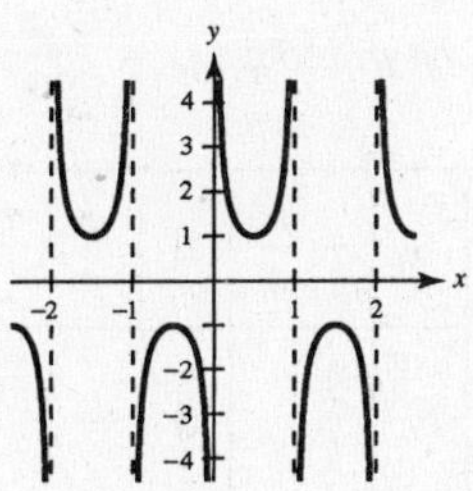

14. $y = 3\csc 4x$

Period: $\frac{2\pi}{4} = \frac{\pi}{2}$

Asymptotes: $x = 0, x = \frac{\pi}{4}$

x	$\frac{\pi}{16}$	$\frac{\pi}{8}$	$\frac{3\pi}{16}$
y	4.243	3	4.243

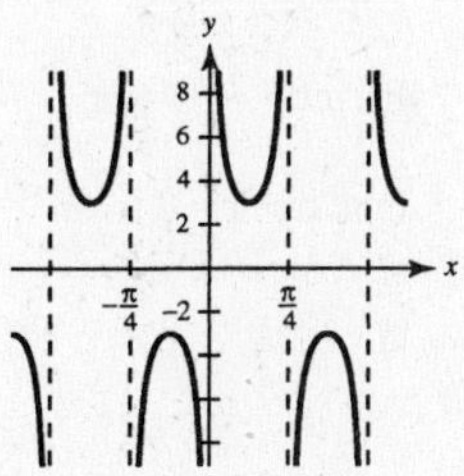

15. $y = \sec \pi x - 1$

Graph $y = \cos \pi x$ first

Period: $\dfrac{2\pi}{\pi} = 2$

One cycle: 0 to 2

Vertical shift 1 unit downward.

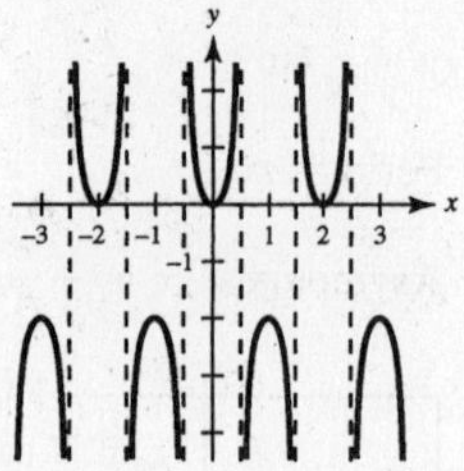

16. $y = -2 \sec 4x + 2$

Period: $\dfrac{2\pi}{4} = \dfrac{\pi}{2}$

Asymptotes: $x = -\dfrac{\pi}{8}, x = \dfrac{\pi}{8}$

x	$-\dfrac{\pi}{16}$	0	$\dfrac{\pi}{16}$
y	-0.828	0	-0.828

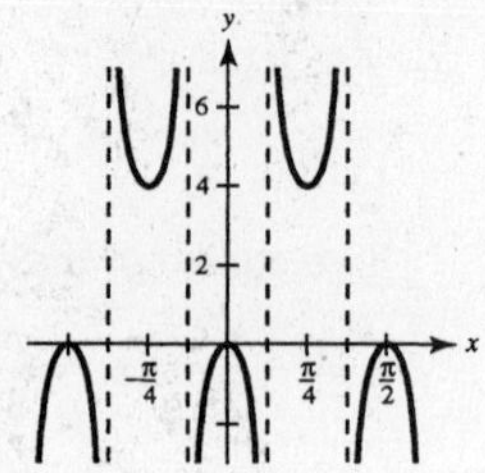

17. $y = \csc \dfrac{x}{2}$

Graph $y = \sin \dfrac{x}{2}$ first.

Period: $\dfrac{2\pi}{\frac{1}{2}} = 4\pi$

One cycle: 0 to 4π

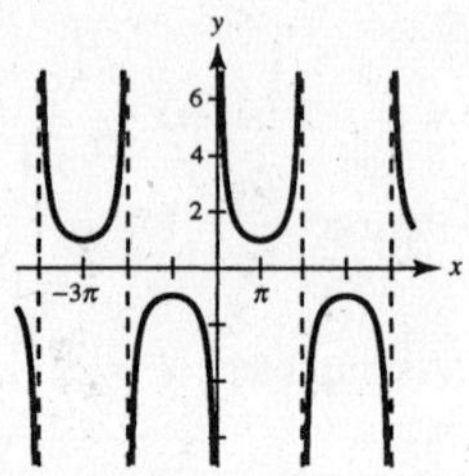

18. $y = \csc \dfrac{x}{3}$

Period: $\dfrac{2\pi}{\frac{1}{3}} = 6\pi$

Asymptotes: $x = 0, x = 3\pi$

x	π	2π	4π
y	1.155	1.155	-1.155

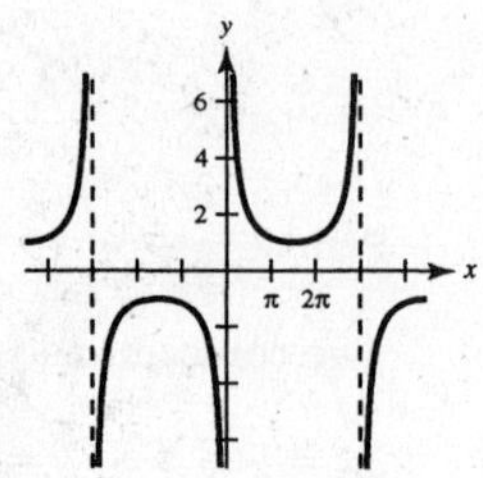

19. $y = \cot \dfrac{x}{2}$

Period: $\dfrac{\pi}{\frac{1}{2}} = 2\pi$

Two consecutive asymptotes:

$\dfrac{x}{2} = 0 \Longrightarrow x = 0$

$\dfrac{x}{2} = \pi \Longrightarrow x = 2\pi$

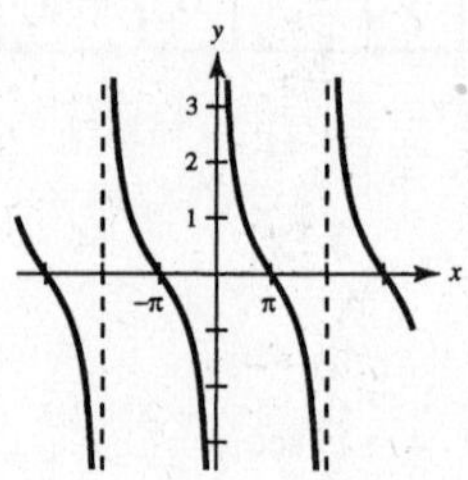

x	$\dfrac{\pi}{2}$	π	$\dfrac{3\pi}{2}$
y	1	0	-1

20. $y = 3 \cot \dfrac{\pi x}{2}$

Period: $\dfrac{\pi}{\frac{\pi}{2}} = 2$

Asymptotes: $x = 0, x = 2$

x	$\dfrac{1}{4}$	1	$\dfrac{3}{2}$
y	7.243	0	-3

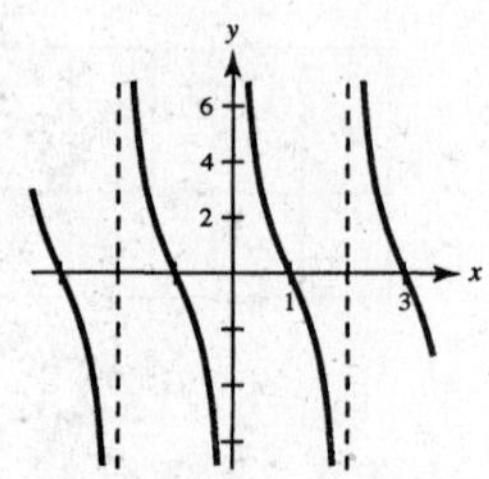

21. $y = \dfrac{1}{2} \sec 2x$

Graph $y = \dfrac{1}{2} \cos 2x$ first.

Period: $\dfrac{2\pi}{2} = \pi$

One cycle: 0 to π

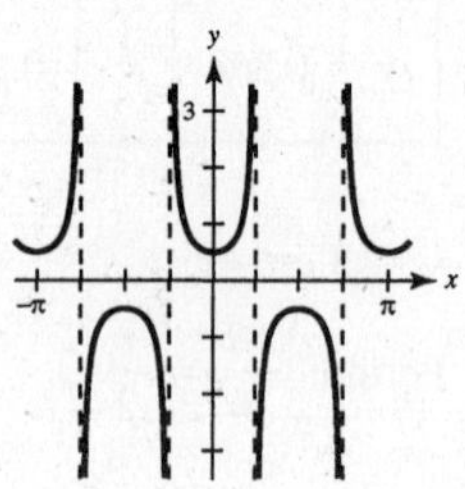

22. $y = -\dfrac{1}{2} \tan x$

Period: π

Asymptotes: $x = -\dfrac{\pi}{2}, x = \dfrac{\pi}{2}$

x	$-\dfrac{\pi}{4}$	0	$\dfrac{\pi}{4}$
y	$\dfrac{1}{2}$	0	$-\dfrac{1}{2}$

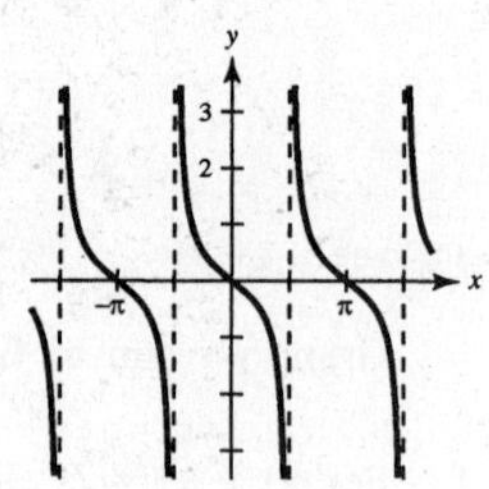

23. $y = \tan \dfrac{\pi x}{4}$

Period: $\dfrac{\pi}{\frac{\pi}{4}} = 4$

Two consecutive asymptotes:

$\dfrac{\pi x}{4} = -\dfrac{\pi}{2} \Rightarrow x = -2$

$\dfrac{\pi x}{4} = \dfrac{\pi}{2} \Rightarrow x = 2$

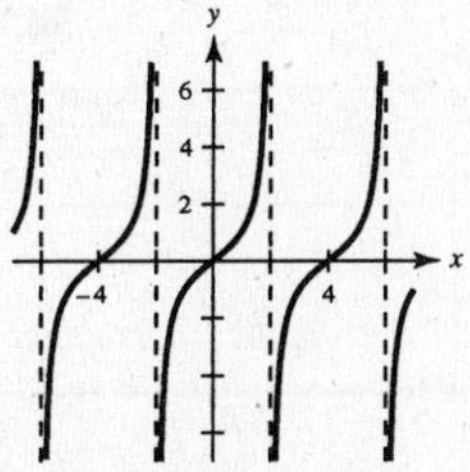

x	-1	0	1
y	-1	0	1

24. $y = \tan(x + \pi)$

Period: π

Asymptotes: $x = -\dfrac{\pi}{2}, x = \dfrac{\pi}{2}$

x	$-\dfrac{\pi}{4}$	0	$\dfrac{\pi}{4}$
y	-1	0	1

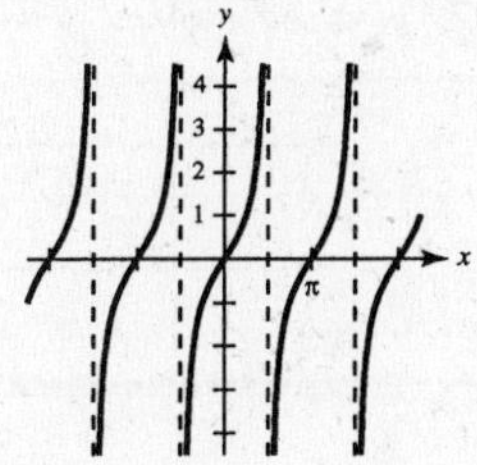

25. $y = \csc(\pi - x)$

Graph $y = \sin(\pi - x)$ first.

Period: 2π

Shift: Set $\pi - x = 0$ and $\pi - x = 2\pi$

$x = \pi$ $\qquad$ $x = -\pi$

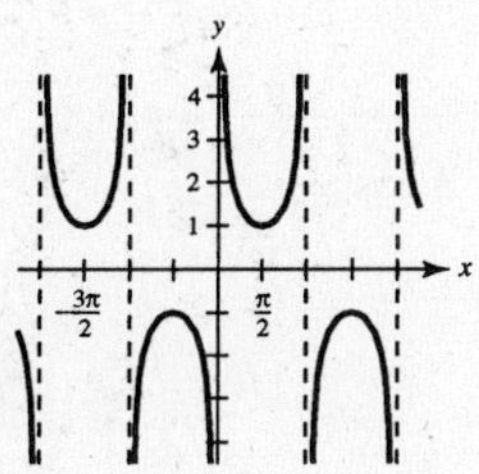

26. $y = \sec(\pi - x)$

Period: 2π

Asymptotes: $x = -\dfrac{\pi}{2}, x = \dfrac{\pi}{2}$

x	$-\dfrac{\pi}{4}$	0	$\dfrac{\pi}{4}$
y	-1.414	-1	-1.414

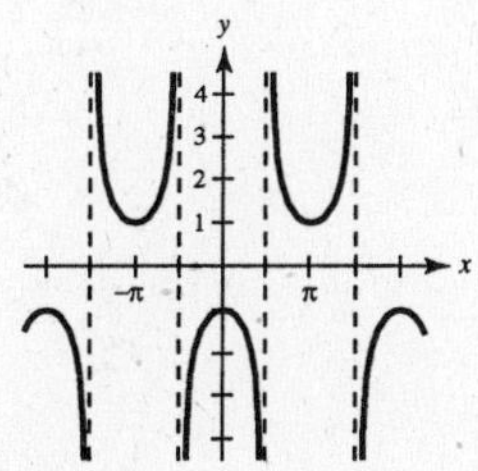

27. $y = \dfrac{1}{4}\csc\left(x + \dfrac{\pi}{4}\right)$

Graph $y = \dfrac{1}{4}\sin\left(x + \dfrac{\pi}{4}\right)$ first.

Period: 2π

Shift:

Set $x + \dfrac{\pi}{4} = 0$ and $x + \dfrac{\pi}{4} = 2\pi$

$x = -\dfrac{\pi}{4}$ to $x = \dfrac{7\pi}{4}$

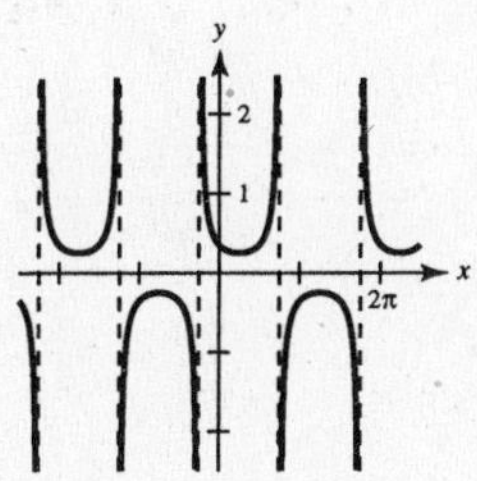

28. $y = 2\cot\left(x + \dfrac{\pi}{2}\right)$

Period: π

Asymptotes: $x = -\dfrac{\pi}{2}, x = \dfrac{\pi}{2}$

x	$-\dfrac{\pi}{4}$	0	$\dfrac{\pi}{4}$
y	2	0	-2

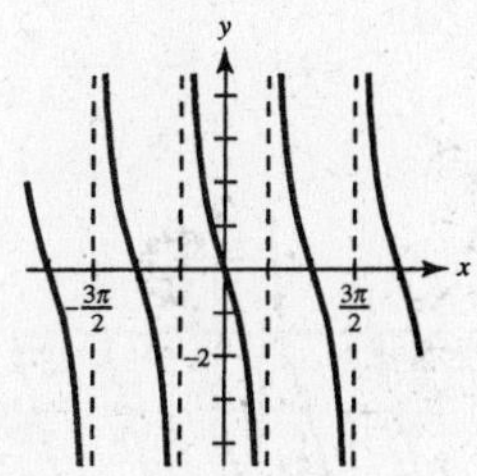

29. $y = \tan \dfrac{x}{3}$

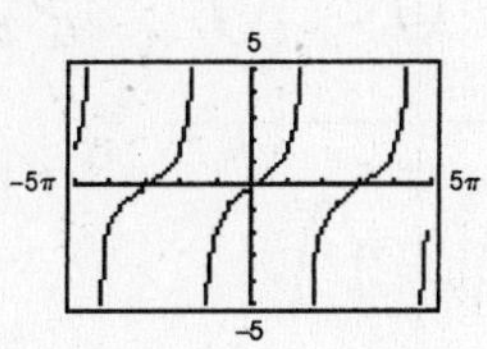

30. $y = -\tan 2x$

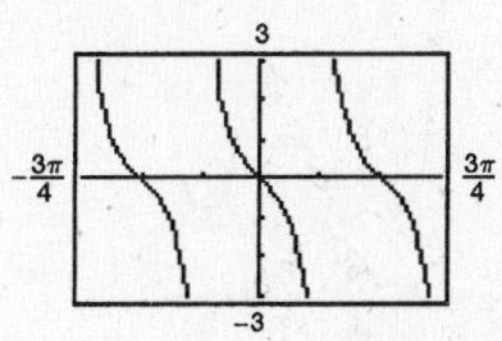

31. $y = -2\sec 4x$

$= \dfrac{-2}{\cos 4x}$

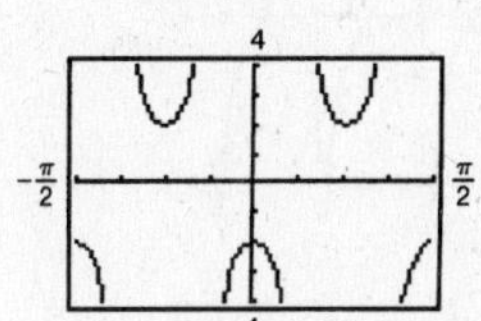

32. $y = \sec \pi x \Rightarrow y = \dfrac{1}{\cos(\pi x)}$

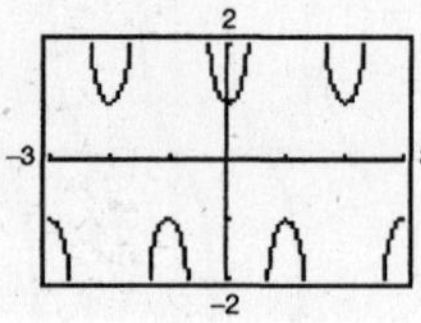

33. $y = \tan\left(x - \dfrac{\pi}{4}\right)$

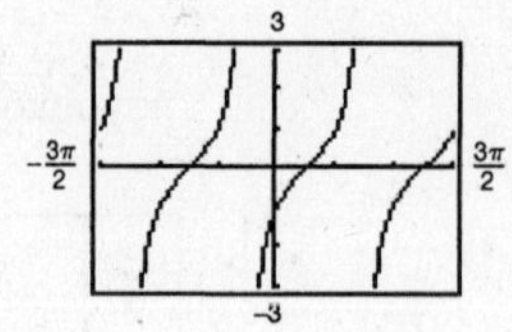

34. $y = -\csc(4x - \pi)$

$y = \dfrac{-1}{\sin(4x - \pi)}$

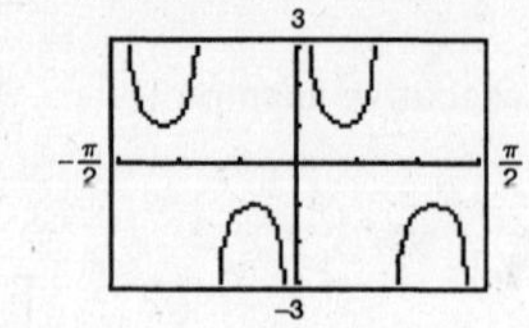

35. $y = \dfrac{1}{4}\cot\left(x - \dfrac{\pi}{2}\right) = \dfrac{1}{4\tan\left(x - \dfrac{\pi}{2}\right)}$

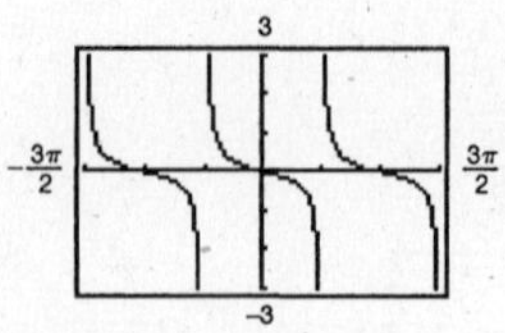

36. $y = 2\sec(2x - \pi) \Rightarrow y = \dfrac{2}{\cos(2x - \pi)}$

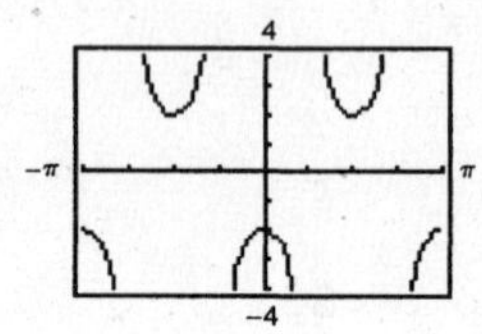

37. $y = 0.1\tan\left(\dfrac{\pi x}{4} + \dfrac{\pi}{4}\right)$

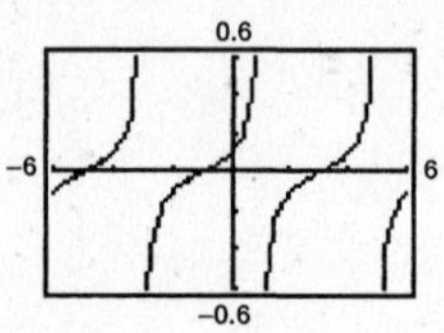

38. $y = \dfrac{1}{3}\sec\left(\dfrac{\pi x}{2} + \dfrac{\pi}{2}\right) \Rightarrow y = \dfrac{1}{3\cos\left(\dfrac{\pi x}{2} + \dfrac{\pi}{2}\right)}$

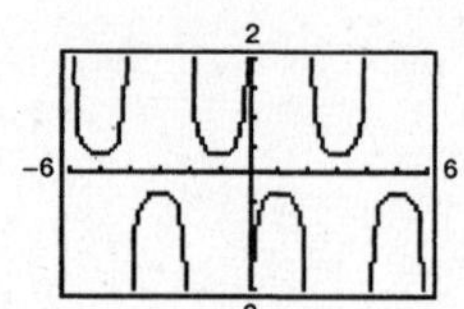

39. $\tan x = 1$

$x = -\dfrac{7\pi}{4}, -\dfrac{3\pi}{4}, \dfrac{\pi}{4}, \dfrac{5\pi}{4}$

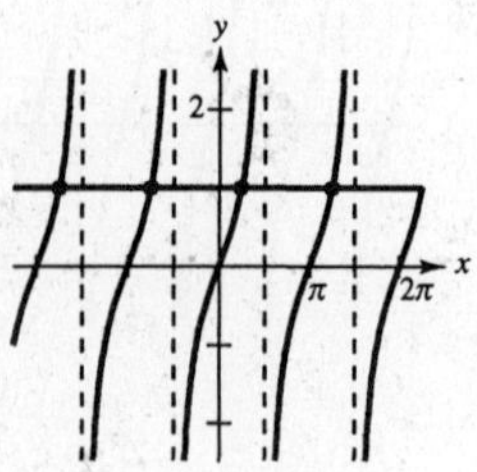

40. $\tan x = \sqrt{3}$

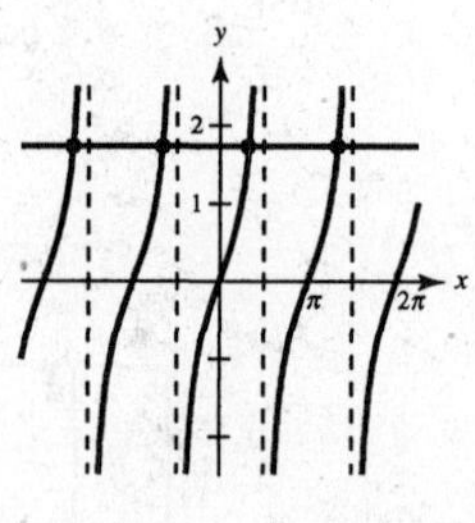

$x = -\dfrac{5\pi}{3}, -\dfrac{2\pi}{3}, \dfrac{\pi}{3}, \dfrac{4\pi}{3}$

41. $\cot x = -\dfrac{\sqrt{3}}{3}$

$x = -\dfrac{4\pi}{3}, -\dfrac{\pi}{3}, \dfrac{2\pi}{3}, \dfrac{5\pi}{3}$

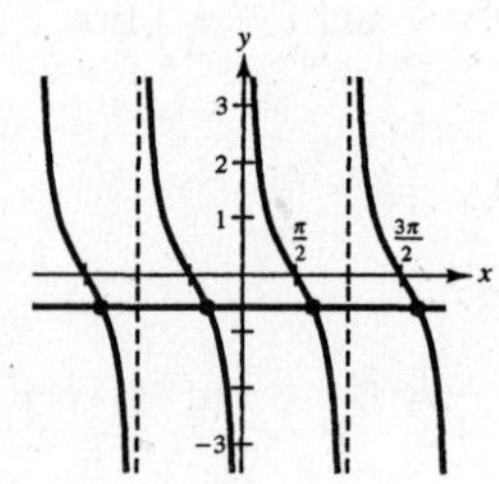

42. $\cot x = 1$

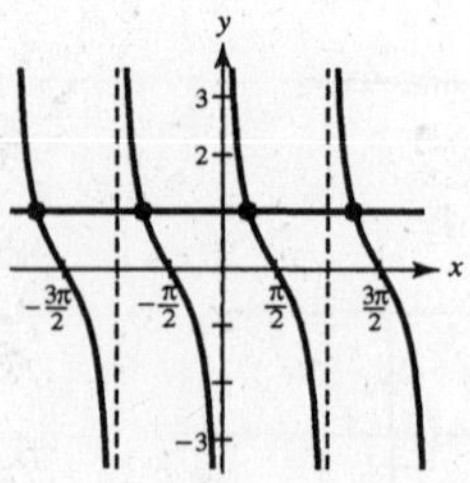

$x = -\dfrac{7\pi}{4}, -\dfrac{3\pi}{4}, \dfrac{\pi}{4}, \dfrac{5\pi}{4}$

43. $\sec x = -2$

$x = \pm\dfrac{2\pi}{3}, \pm\dfrac{4\pi}{3}$

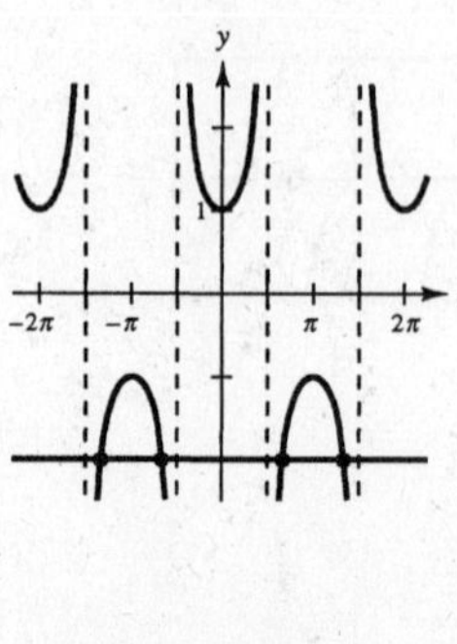

44. $\sec x = 2$

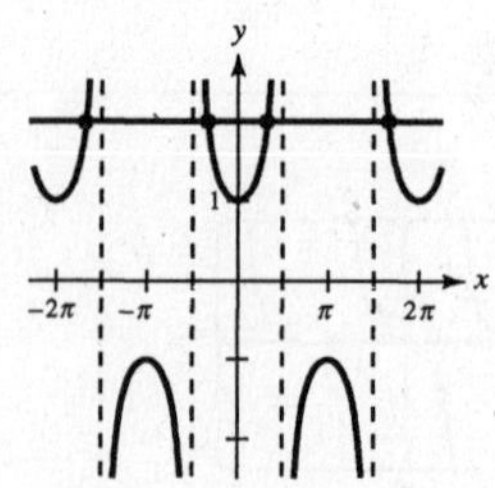

$x = -\dfrac{5\pi}{3}, -\dfrac{\pi}{3}, \dfrac{\pi}{3}, \dfrac{5\pi}{3}$

45. $\csc x = \sqrt{2}$

$x = -\frac{7\pi}{4}, -\frac{5\pi}{4}, \frac{\pi}{4}, \frac{3\pi}{4}$

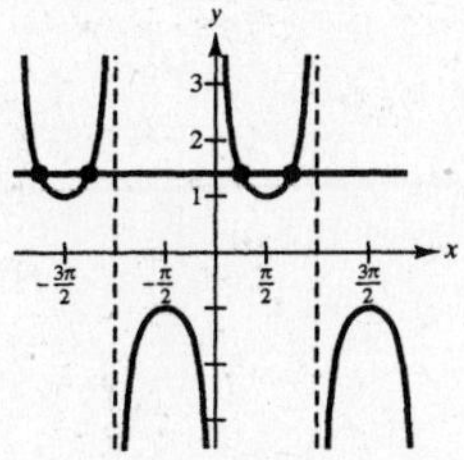

46. $\csc x = -\frac{2\sqrt{3}}{3}$

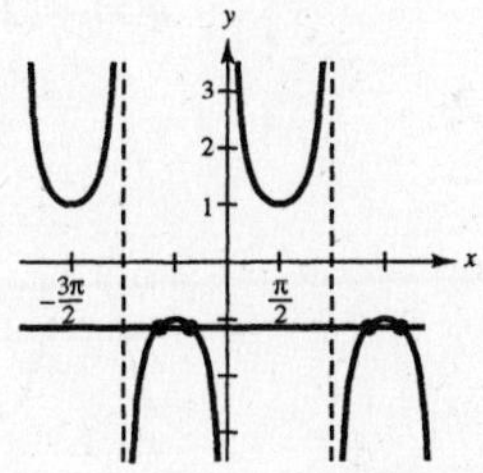

$x = -\frac{2\pi}{3}, -\frac{\pi}{3}, \frac{4\pi}{3}, \frac{5\pi}{3}$

47. $f(x) = \sec x$

$\sec(-x) = \sec x$

Thus, the function is even and the graph of $y = \sec x$ has y-axis symmetry.

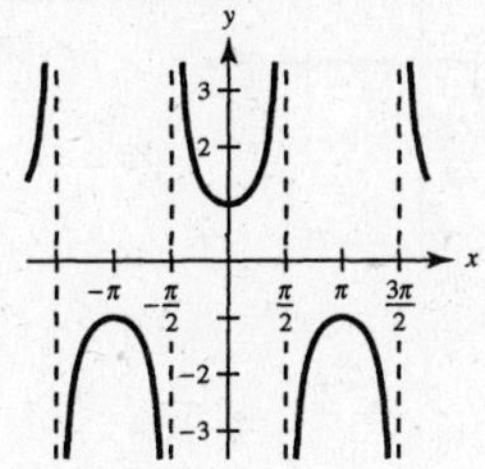

48. $f(x) = \tan x$

$\tan(-x) = -\tan x$

Thus, the function is odd and the graph of $y = \tan x$ is symmetric with the origin.

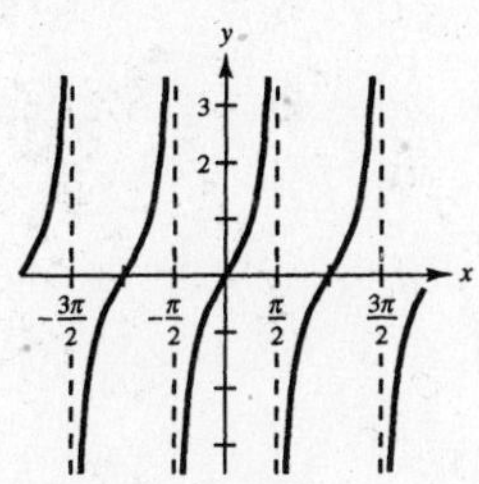

49. $g(x) = \sin(x + \pi)$

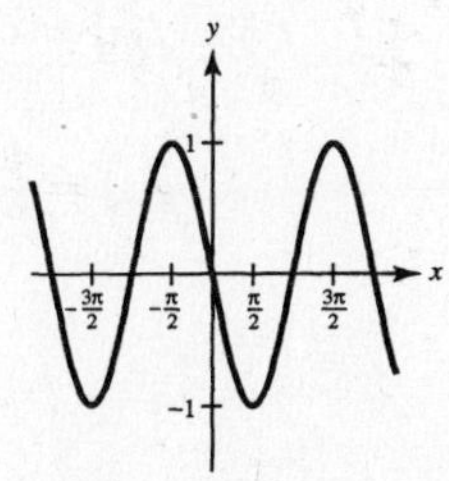

The graph has origin symmetry, so the function is odd.

50. $g(x) = \cos(x - \pi)$

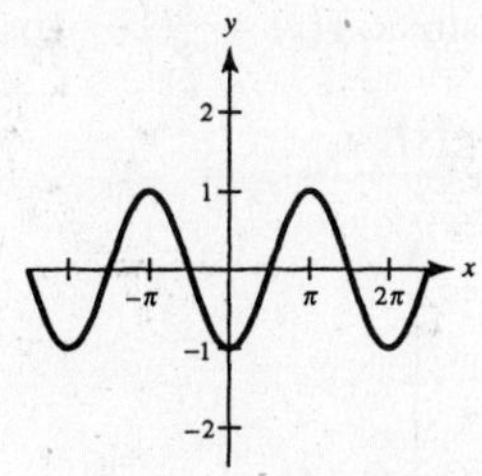

The graph has y-axis symmetry, so the function is even.

51. $y_1 = \sin x \csc x$ and $y_2 = 1$

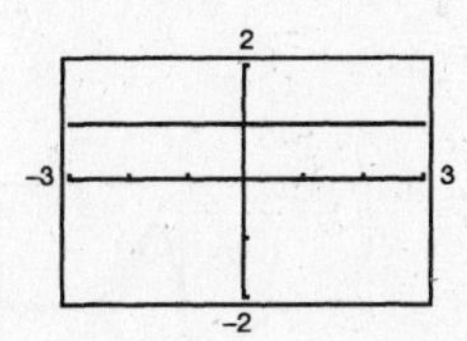

$\sin x \csc x = \sin x\left(\frac{1}{\sin x}\right) = 1, \sin x \neq 0$

The expressions are equivalent except when $\sin x = 0$ and y_1 is undefined.

52. $y_1 = \sin x \sec x, y_2 = \tan x$

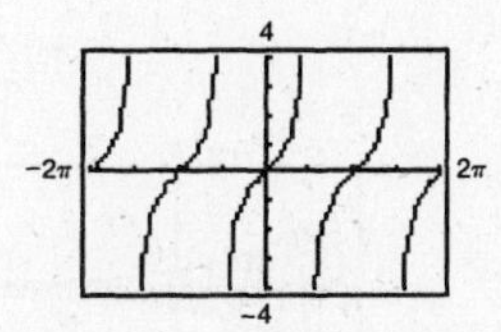

It appears that $y_1 = y_2$.

$\sin x \sec x = \sin x \frac{1}{\cos x} = \frac{\sin x}{\cos x} = \tan x$

53. $y_1 = \frac{\cos x}{\sin x}$ and $y_2 = \cot x = \frac{1}{\tan x}$

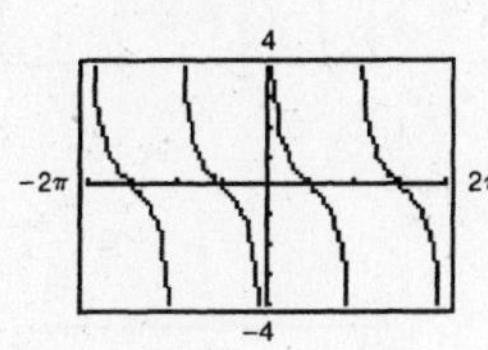

$\cot x = \frac{\cos x}{\sin x}$

The expressions are equivalent.

54. $y_1 = \sec^2 x - 1, y_2 = \tan^2 x$

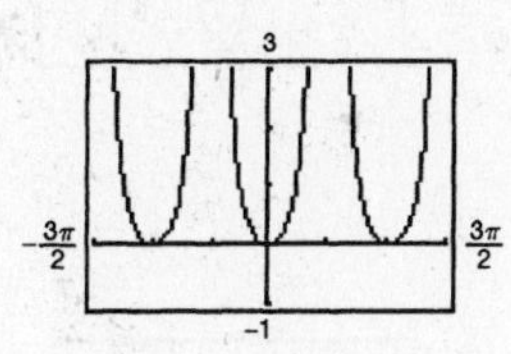

It appears that $y_1 = y_2$.

$1 + \tan^2 x = \sec^2 x$

$\tan^2 x = \sec^2 x - 1$

55. $f(x) = \sin x + \cos\left(x + \dfrac{\pi}{2}\right), g(x) = 0$

$f(x) = g(x)$ The graph is the line $y = 0$.

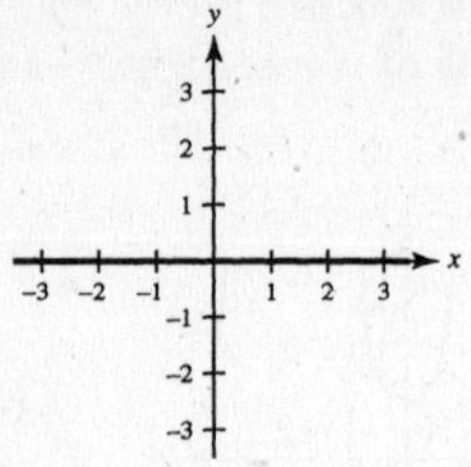

56. $f(x) = \sin x - \cos\left(x + \dfrac{\pi}{2}\right)$

$g(x) = 2 \sin x$

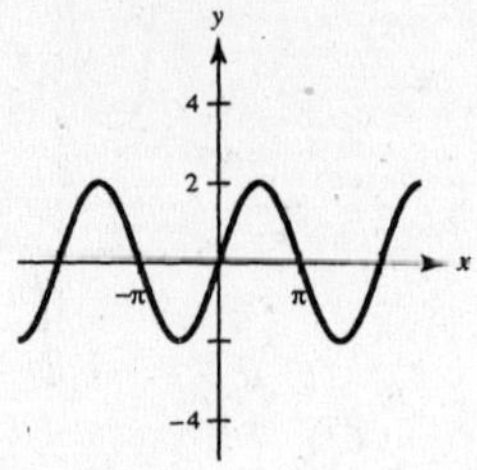

It appears that $f(x) = g(x)$. That is,

that $\sin x - \cos\left(x + \dfrac{\pi}{2}\right) = 2 \sin x$.

57. $f(x) = \sin^2 x, g(x) = \dfrac{1}{2}(1 - \cos 2x)$

$f(x) = g(x)$

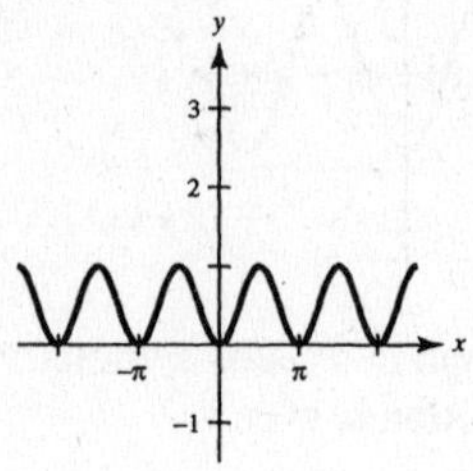

58. $f(x) = \cos^2 \dfrac{\pi x}{2}$

$g(x) = \dfrac{1}{2}(1 + \cos \pi x)$

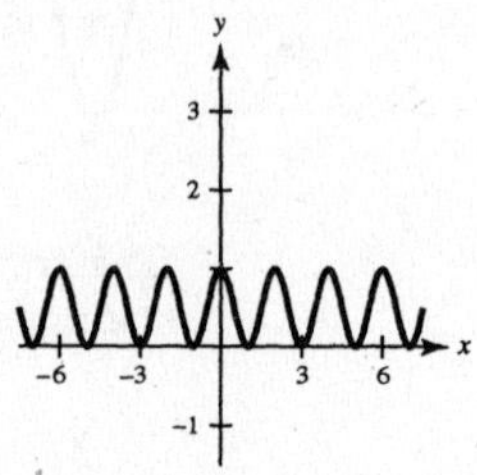

It appears that $f(x) = g(x)$. That is,

that $\cos^2 \dfrac{\pi x}{2} = \dfrac{1}{2}(1 + \cos \pi x)$.

59. $f(x) = 2^{-x/4} \cos \pi x$

$-2^{-x/4} \leq f(x) \leq 2^{-x/4}$

The damping factor is $y = 2^{-x/4}$.

As $x \to \infty, f(x) \to 0$

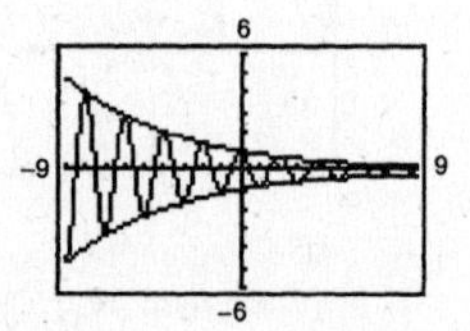

60. $h(x) = 2^{-x^2/4} \sin x$

Damping factor: $2^{-x^2/4}$

As $x \longrightarrow \infty, h(x) \longrightarrow 0$.

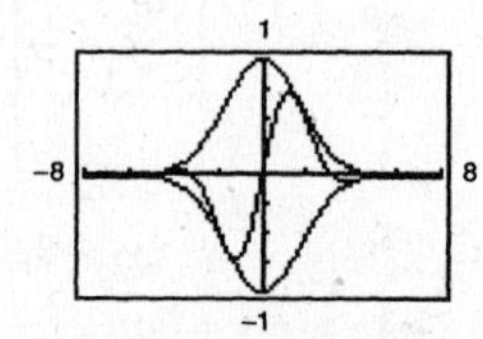

61. $g(x) = e^{-x^2/2} \sin x$

$-e^{-x^2/2} \leq g(x) \leq e^{-x^2/2}$

The damping factor is $y = e^{-x^2/2}$.

As $x \to \infty, g(x) \to 0$

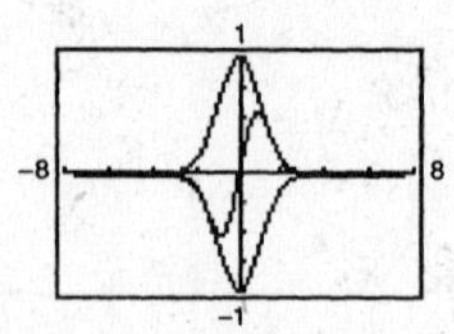

62. $f(x) = e^{-x} \cos x$

Damping factor: e^{-x}

As $x \longrightarrow \infty, f(x) \longrightarrow 0$.

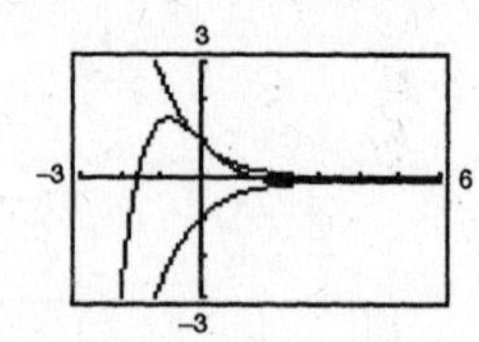

63. $y = \dfrac{6}{x} + \cos x, x > 0$

As $x \longrightarrow 0, y \longrightarrow \infty$.

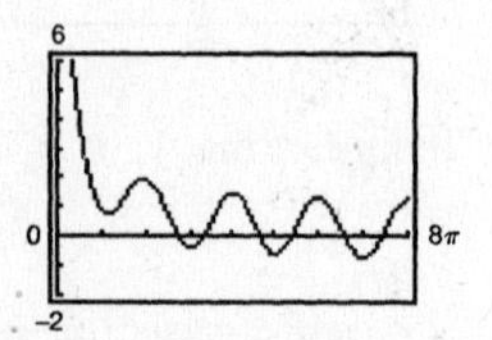

64. $y = \dfrac{4}{x} + \sin 2x, x > 0$

As $x \longrightarrow 0, y \longrightarrow \infty$.

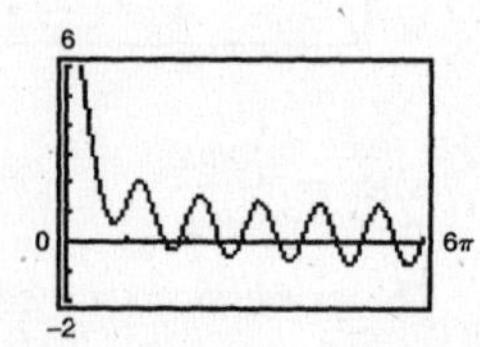

65. $g(x) = \dfrac{\sin x}{x}$

As $x \longrightarrow 0,\ g(x) \longrightarrow 1$.

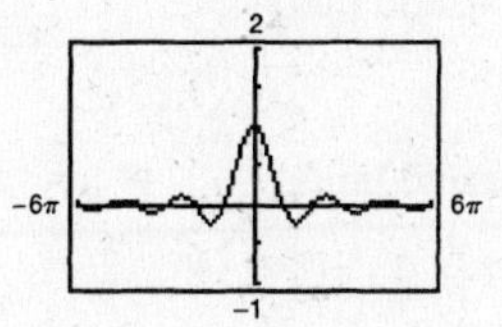

66. $f(x) = \dfrac{1 - \cos x}{x}$

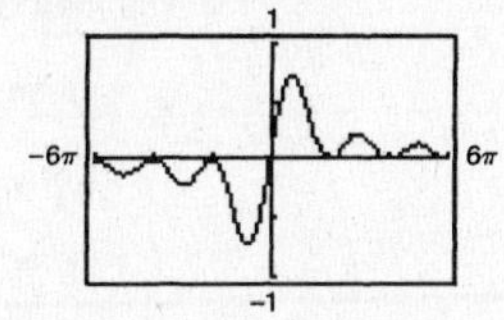

As $x \longrightarrow 0, f(x) \longrightarrow 0$.

67. $f(x) = \sin \dfrac{1}{x}$

As $x \longrightarrow 0, f(x)$ oscillates between -1 and 1.

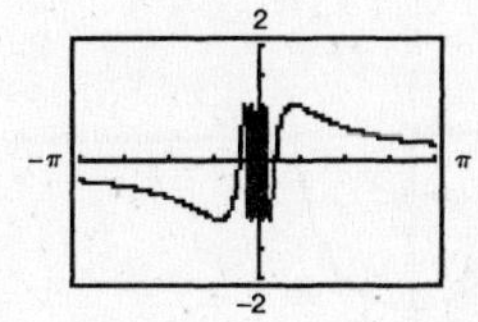

68. $h(x) = x \sin \dfrac{1}{x}$

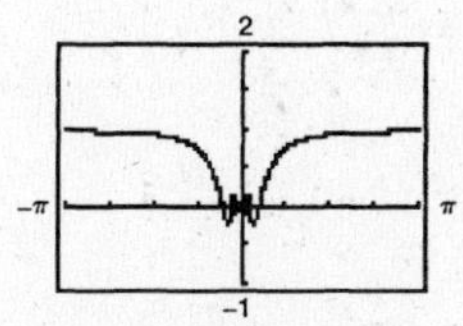

As $x \longrightarrow 0$, $h(x)$ oscillates.

69. $f(x) = |x \cos x|$

As $x \longrightarrow 0, f(x) \longrightarrow 0$.

Matches graph (d).

70. $f(x) = x \sin x$

Matches graph (a) as $x \longrightarrow 0, f(x) \longrightarrow 0$.

71. $g(x) = |x| \sin x$

As $x \longrightarrow 0, g(x) \longrightarrow 0$.

Matches graph (b).

72. $g(x) = |x| \cos x$

Matches graph (c) as $x \longrightarrow 0,\ g(x) \longrightarrow 0$.

73. $f(x) = 2 \sin x$

$g(x) = \dfrac{1}{2} \csc x$

(a)

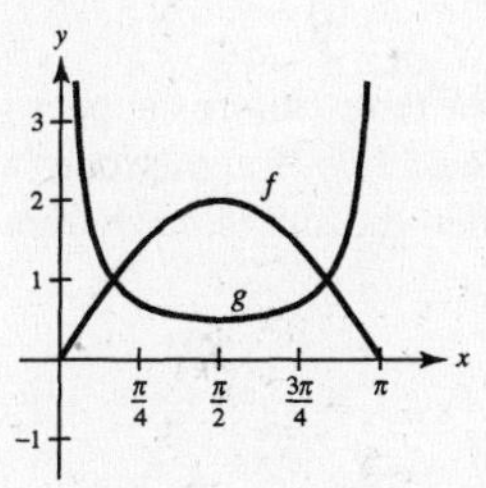

(b) $f > g$ on the interval, $\dfrac{\pi}{6} < x < \dfrac{5\pi}{6}$

(c) As $x \longrightarrow \pi$, $f(x) = 2 \sin x \longrightarrow 0$ and $g(x) = \dfrac{1}{2} \csc x \longrightarrow \infty$ if $x \longrightarrow \pi^+$ and $-\infty$ if $x \longrightarrow \pi^-$ since $g(x)$ is the reciprocal of $f(x)$.

74. $f(x) = \tan \dfrac{\pi x}{2},\ g(x) = \dfrac{1}{2} \sec \dfrac{\pi x}{2}$

(a)

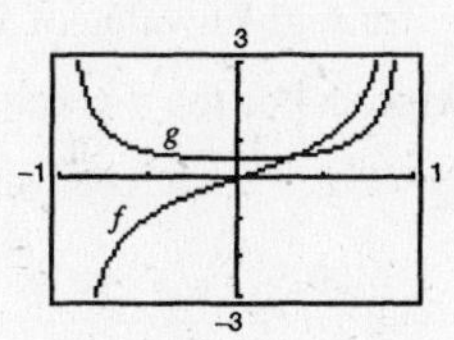

(b) The interval in which $f < g$ is $\left(-1, \frac{1}{3}\right)$.

(c) The interval in which $2f < 2g$ is $\left(-1, \frac{1}{3}\right)$, which is the same interval as part (b).

75. $\tan x = \dfrac{7}{d}$

$d = \dfrac{7}{\tan x} = 7 \cot x$

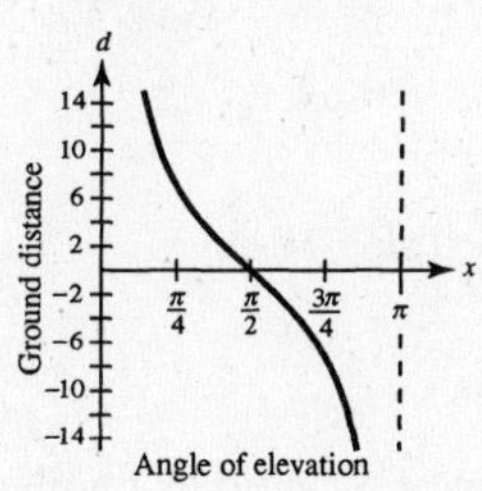

76. $\cos x = \dfrac{27}{d}$

$d = \dfrac{27}{\cos x} = 27 \sec x, \; -\dfrac{\pi}{2} < x < \dfrac{\pi}{2}$

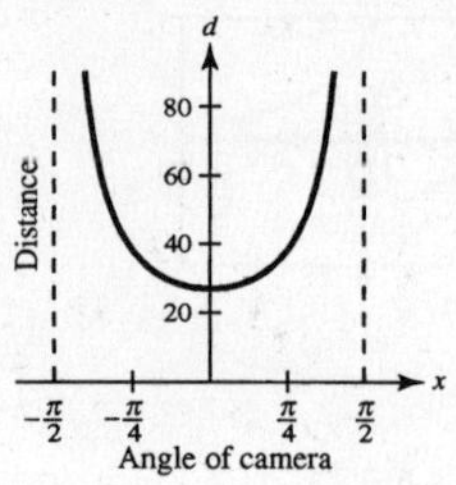

77. (a) $S = 74 + 3t + 40 \sin \dfrac{\pi t}{6}$

(b) Maximum sales: March

($t \approx 3.27$, $S \approx 123.41$ thousand units)

Minimum sales: August

($t \approx 8.73$, $S \approx 60.59$ thousand units)

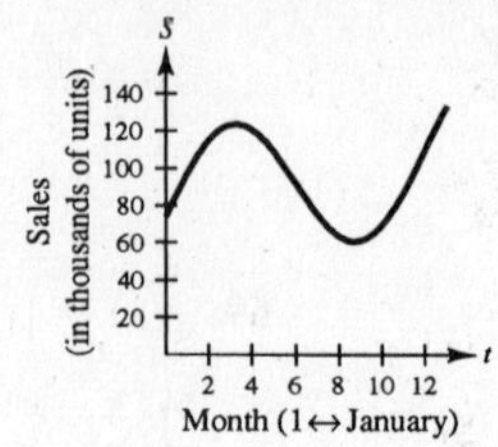

78. $C = 10{,}000 + 3000 \sin \dfrac{2\pi t}{24}$

$R = 15{,}000 + 5000 \cos \dfrac{2\pi t}{24}$

(a)

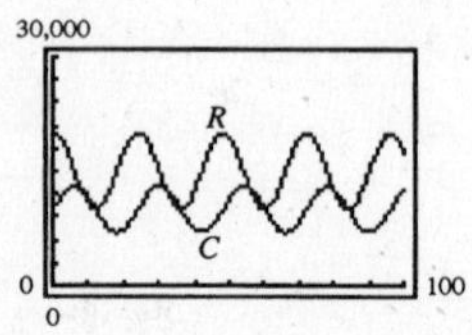

(b) When the prey population is highest, the predator population is increasing.
When the prey population is lowest, the predator population is decreasing.
When the predator population is lowest, the prey population is increasing.
When the predator population is highest, the prey population is decreasing.

(c) Period (for both): $\dfrac{2\pi}{2\pi/24} = 24$ months

Answers will vary. Some factors may include food sources for the prey, weather, disease, birth rates, and hunters.

79. $H(t) = 54.33 - 20.38 \cos \dfrac{\pi t}{6} - 15.69 \sin \dfrac{\pi t}{6}$

$L(t) = 39.36 - 15.70 \cos \dfrac{\pi t}{6} - 14.16 \sin \dfrac{\pi t}{6}$

(a) Period of $\cos \dfrac{\pi t}{6}$: $\dfrac{2\pi}{\pi/6} = 12$

Period of $\sin \dfrac{\pi t}{6}$: $\dfrac{2\pi}{\pi/6} = 12$

Period of $H(t)$: 12

Period of $L(t)$: 12

(b) From the graph, it appears that the greatest difference between high and low temperatures occurs in summer. The smallest difference occurs in winter.

(c) The highest high and low temperatures appear to occur around the middle of July, roughly one month after the time when the sun is northernmost in the sky.

80. (a)

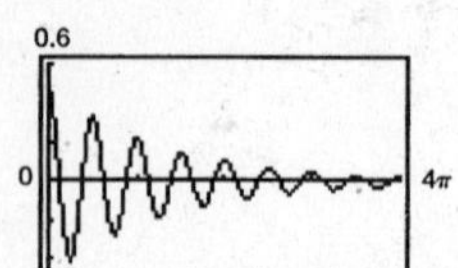

$y = \frac{1}{2}e^{-t/4} \cos 4t$

(b) The displacement is a damped sine wave.
$y \longrightarrow 0$ as t increases.

81. True. Since $y = \csc x = \dfrac{1}{\sin x}$, for a given value of x, the y-coordinate of $\csc x$ is the reciprocal of the y-coordinate of $\sin x$.

82. True. The graph of $y = \sec x$ can be obtained by translating the graph of $y = \csc x$ to the left by $\dfrac{\pi}{2}$ units.
Thus, $y = \sec x = \csc\left(x + \dfrac{\pi}{2}\right) = \dfrac{1}{\sin\left(x + \dfrac{\pi}{2}\right)}$.

83. True. $f(x) = \tan 2x$ has a period of $\frac{\pi}{2}$ and has consecutive asymptotes at:

$$2x = -\frac{\pi}{2} \quad \text{and} \quad 2x = \frac{\pi}{2}$$

$$x = -\frac{\pi}{4} \qquad\qquad x = \frac{\pi}{4}$$

All the vertical asymptotes can be written as $x = \frac{\pi}{4} + n\left(\frac{\pi}{2}\right)$ where n is an integer.

84. True. $g(x) = \cot 4x$ has a period of $\frac{\pi}{4}$ and has consecutive asymptotes at:

$$4x = 0 \quad \text{and} \quad 4x = \pi$$

$$x = 0 \qquad\qquad x = \frac{\pi}{4}$$

All the vertical asymptotes can be written as $x = 0 + n\left(\frac{\pi}{4}\right) = \frac{n\pi}{4}$ where n is an integer.

85. As $x \longrightarrow \frac{\pi}{2}$ from the left, $f(x) = \tan x \longrightarrow \infty$.

As $x \longrightarrow \frac{\pi}{2}$ from the right, $f(x) = \tan x \longrightarrow -\infty$.

86. As $x \longrightarrow \pi$ from the left, $f(x) = \csc x \longrightarrow \infty$.

As $x \longrightarrow \pi$ from the right, $f(x) = \csc x \longrightarrow -\infty$.

87.

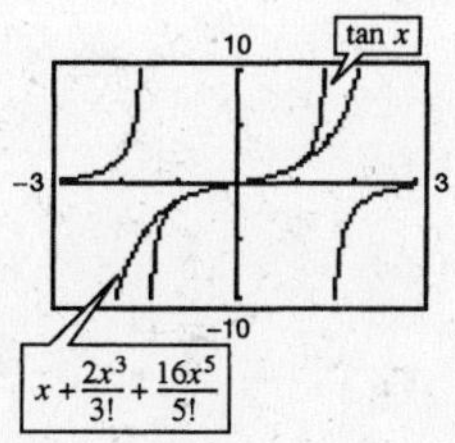

The polynomial approximation is best near the origin and becomes worse as x moves away from 0.

88.

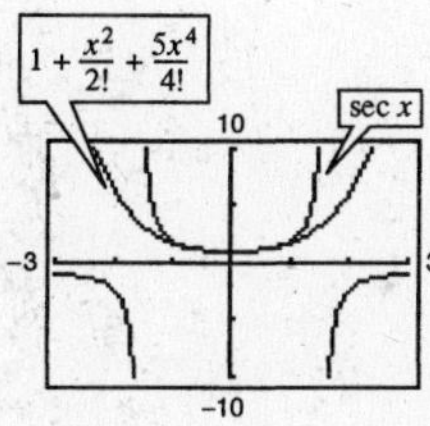

The polynomial approximation is best near the origin and becomes worse as x moves away from 0.

89. (a) $y_1 = \frac{4}{\pi}\left(\sin \pi x + \frac{1}{3}\sin 3\pi x\right)$

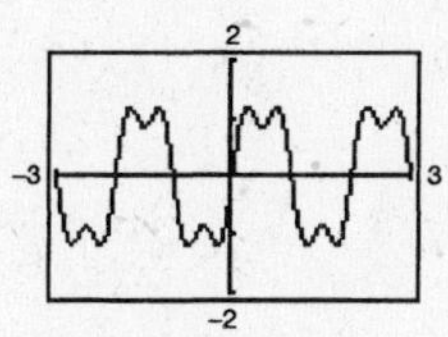

$y_2 = \frac{4}{\pi}\left(\sin \pi x + \frac{1}{3}\sin 3\pi x + \frac{1}{5}\sin 5\pi x\right)$

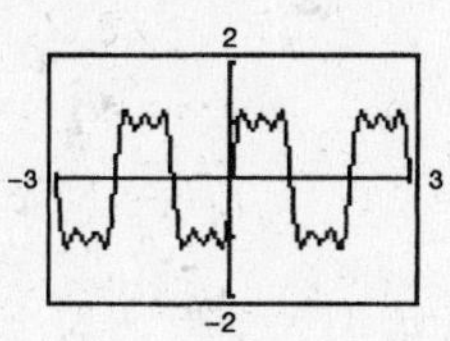

(b) $y_3 = \frac{4}{\pi}\left(\sin \pi x + \frac{1}{3}\sin 3\pi x + \frac{1}{5}\sin 5\pi x + \frac{1}{7}\sin 7\pi x\right)$

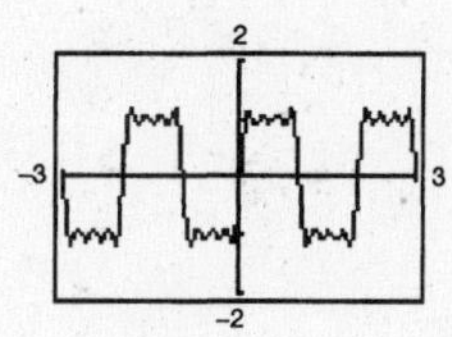

(c) $y_4 = \frac{4}{\pi}\left(\sin \pi x + \frac{1}{3}\sin 3\pi x + \frac{1}{5}\sin 5\pi x + \frac{1}{7}\sin 7\pi x + \frac{1}{9}\sin 9\pi x\right)$

Section 9.7 Inverse Trigonometric Functions

1. $y = \arcsin \frac{1}{2} \Rightarrow \sin y = \frac{1}{2}$ for

$-\frac{\pi}{2} \le y \le \frac{\pi}{2} \Rightarrow y = \frac{\pi}{6}$

2. $y = \arcsin 0 \Rightarrow \sin y = 0$ for

$-\frac{\pi}{2} \le y \le \frac{\pi}{2} \Rightarrow y = 0$

3. $y = \arccos \frac{1}{2} \Rightarrow \cos y = \frac{1}{2}$ for

$0 \le y \le \pi \Rightarrow y = \frac{\pi}{3}$

4. $y = \arccos 0 \Rightarrow \cos y = 0$ for

$0 \le y \le \pi \Rightarrow y = \frac{\pi}{2}$

5. $y = \arctan \frac{\sqrt{3}}{3} \Rightarrow \tan y = \frac{\sqrt{3}}{3}$ for

$-\frac{\pi}{2} < y < \frac{\pi}{2} \Rightarrow y = \frac{\pi}{6}$

6. $y = \arctan(\sqrt{3}) \Rightarrow \tan y = \sqrt{3}$ for

$-\frac{\pi}{2} < y < \frac{\pi}{2} \Rightarrow y = \frac{\pi}{3}$

7. $y = \arccos\left(-\frac{\sqrt{3}}{2}\right) \Rightarrow \cos y = -\frac{\sqrt{3}}{2}$ for

$0 \le y \le \pi \Rightarrow y = \frac{5\pi}{6}$

8. $y = \arcsin\left(-\frac{\sqrt{2}}{2}\right) \Rightarrow \sin y = -\frac{\sqrt{2}}{2}$ for

$-\frac{\pi}{2} \le y \le \frac{\pi}{2} \Rightarrow y = -\frac{\pi}{4}$

9. $y = \arctan(-\sqrt{3}) \Rightarrow \tan y = -\sqrt{3}$ for

$-\frac{\pi}{2} < y < \frac{\pi}{2} \Rightarrow y = -\frac{\pi}{3}$

10. $y = \arctan(-1) \Rightarrow \tan y = -1$ for

$-\frac{\pi}{2} < y < \frac{\pi}{2} \Rightarrow y = -\frac{\pi}{4}$

11. $y = \arccos\left(-\frac{1}{2}\right) \Rightarrow \cos y = -\frac{1}{2}$ for

$0 \le y \le \pi \Rightarrow y = \frac{2\pi}{3}$

12. $y = \arcsin \frac{\sqrt{2}}{2} \Rightarrow \sin y = \frac{\sqrt{2}}{2}$ for

$-\frac{\pi}{2} \le y \le \frac{\pi}{2} \Rightarrow y = \frac{\pi}{4}$

13. $y = \arcsin \frac{\sqrt{3}}{2} \Rightarrow \sin y = \frac{\sqrt{3}}{2}$ for

$-\frac{\pi}{2} \le y \le \frac{\pi}{2} \Rightarrow y = \frac{\pi}{3}$

14. $y = \arctan\left(-\frac{\sqrt{3}}{3}\right) \Rightarrow \tan y = -\frac{\sqrt{3}}{3}$ for

$-\frac{\pi}{2} < y < \frac{\pi}{2} \Rightarrow y = -\frac{\pi}{6}$

15. $y = \arctan 0 \Rightarrow \tan y = 0$ for

$-\frac{\pi}{2} < y < \frac{\pi}{2} \Rightarrow y = 0$

16. $y = \arccos 1 \Rightarrow \cos y = 1$ for

$0 \le y \le \pi \Rightarrow y = 0$

17. $\arccos 0.28 = \cos^{-1} 0.28 \approx 1.29$

18. $\arcsin 0.45 \approx 0.47$

19. $\arcsin(-0.75) = \sin^{-1}(-0.75) \approx -0.85$

20. $\arccos(-0.7) \approx 2.35$

21. $\arctan(-3) = \tan^{-1}(-3) \approx -1.25$

22. $\arctan 15 \approx 1.50$

23. $\arcsin 0.31 = \sin^{-1} 0.31 \approx 0.32$

24. $\arccos 0.26 \approx 1.31$

25. $\arccos(-0.41) = \cos^{-1}(-0.41)$

≈ 1.99

26. $\arcsin(-0.125) \approx -0.13$

27. $\arctan 0.92 = \tan^{-1} 0.92 \approx 0.74$

28. $\arctan 2.8 \approx 1.23$

29. $\arcsin\left(\frac{3}{4}\right) = \sin^{-1}(0.75) \approx 0.85$

30. $\arccos\left(-\frac{1}{3}\right) \approx 1.91$

31. $\arctan\left(\frac{7}{2}\right) = \tan^{-1}(3.5) \approx 1.29$

32. $\arctan\left(-\frac{95}{7}\right) \approx -1.50$

33. This is the graph of $y = \arctan x$. The coordinates are

$$\left(-\sqrt{3}, -\frac{\pi}{3}\right), \left(-\frac{\sqrt{3}}{3}, -\frac{\pi}{6}\right), \text{ and } \left(1, \frac{\pi}{4}\right).$$

34. $\arccos(-1) = \pi$

$$\arccos\left(-\frac{1}{2}\right) = \frac{2\pi}{3}$$

$$\cos\left(\frac{\pi}{6}\right) = \frac{\sqrt{3}}{2}$$

35. $f(x) = \tan x$ and $g(x) = \arctan x$

Graph $y_1 = \tan x$

Graph $y_2 = \tan^{-1} x$

Graph $y_3 = x$

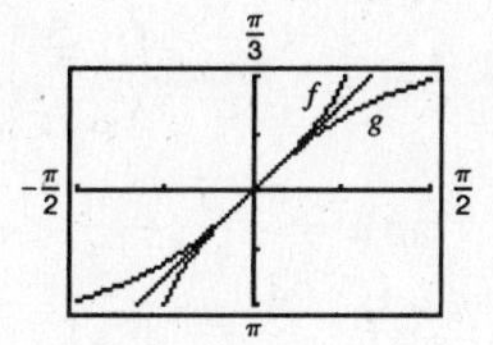

36. $f(x) = \sin x$

$g(x) = \arcsin x$

$y = x$

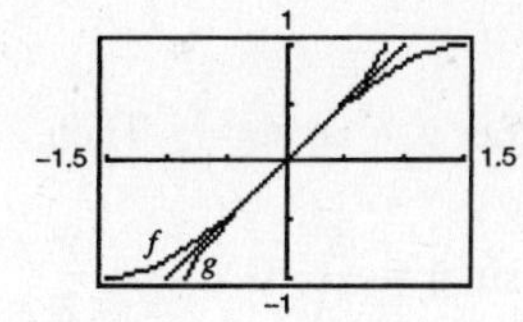

37. $\tan \theta = \dfrac{x}{4}$

$\theta = \arctan \dfrac{x}{4}$

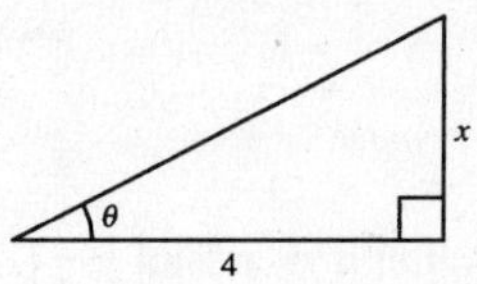

38. $\cos \theta = \dfrac{4}{x}$

$\theta = \arccos \dfrac{4}{x}$

39. $\sin \theta = \dfrac{x+2}{5}$

$\theta = \arcsin\left(\dfrac{x+2}{5}\right)$

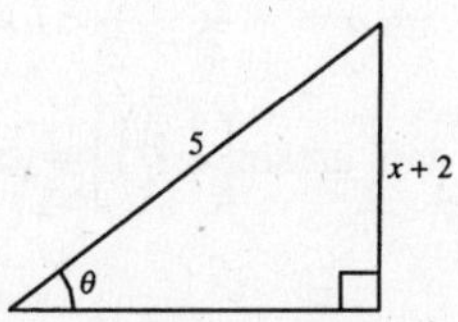

40. $\tan \theta = \dfrac{x+1}{10}$

$\theta = \arctan\left(\dfrac{x+1}{10}\right)$

41. $\cos \theta = \dfrac{x+3}{2x}$

$\theta = \arccos\left(\dfrac{x+3}{2x}\right)$

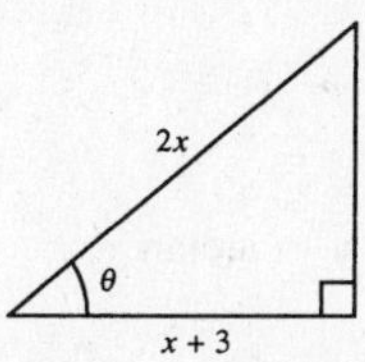

42. $\tan \theta = \dfrac{x-1}{x^2-1} = \dfrac{1}{x+1}$

$\theta = \arctan \dfrac{1}{x+1}$

$x \neq 1$

43. $\sin(\arcsin 0.3) = 0.3$

44. $\tan(\arctan 25) = 25$

45. $\cos[\arccos(-0.1)] = -0.1$

46. $\sin[\arcsin(-0.2)] = -0.2$

47. $\arcsin(\sin 3\pi) = \arcsin(0) = 0$

Note: 3π is not in the range of the arcsine function.

48. $\arccos\left(\cos \dfrac{7\pi}{2}\right) = \arccos 0 = \dfrac{\pi}{2}$

Note: $\dfrac{7\pi}{2}$ is not in the range of the arccosine function.

49. Let $y = \arctan \dfrac{3}{4}$. Then,

$\tan y = \dfrac{3}{4}, \ 0 < y < \dfrac{\pi}{2}$

and $\sin y = \dfrac{3}{5}$.

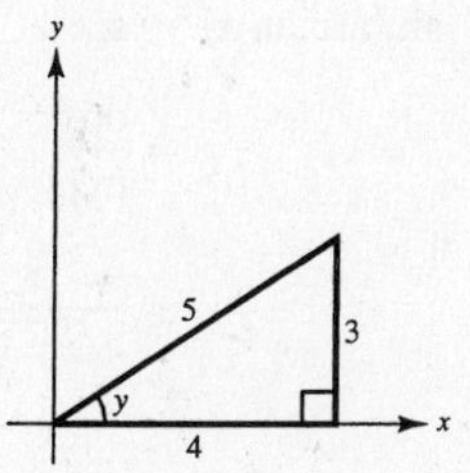

50. Let $u = \arcsin \dfrac{4}{5}$,

$\sin u = \dfrac{4}{5}, 0 < u < \dfrac{\pi}{2}$.

$\sec\left(\arcsin \dfrac{4}{5}\right) = \sec u = \dfrac{\text{hyp}}{\text{adj}} = \dfrac{5}{3}$

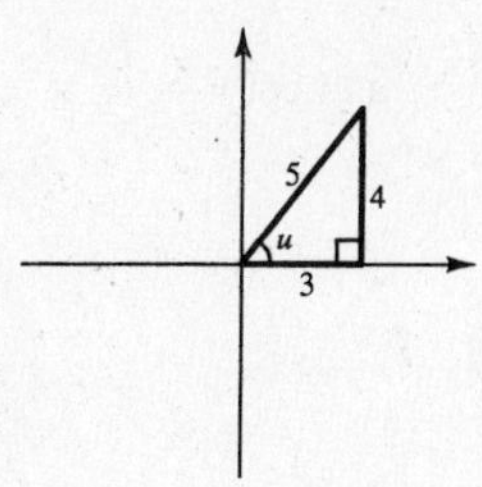

51. Let $y = \arctan 2$. Then,

$$\tan y = 2 = \frac{2}{1},\ 0 < y < \frac{\pi}{2}$$

and $\cos y = \dfrac{1}{\sqrt{5}} = \dfrac{\sqrt{5}}{5}$.

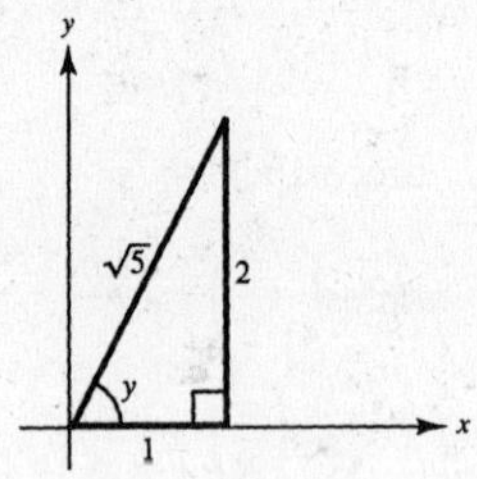

52. Let $u = \arccos \dfrac{\sqrt{5}}{5}$,

$$\cos u = \frac{\sqrt{5}}{5},\ 0 < u < \frac{\pi}{2}.$$

$$\sin\left(\arccos \frac{\sqrt{5}}{5}\right) = \sin u$$

$$= \frac{2}{\sqrt{5}} = \frac{2\sqrt{5}}{5}$$

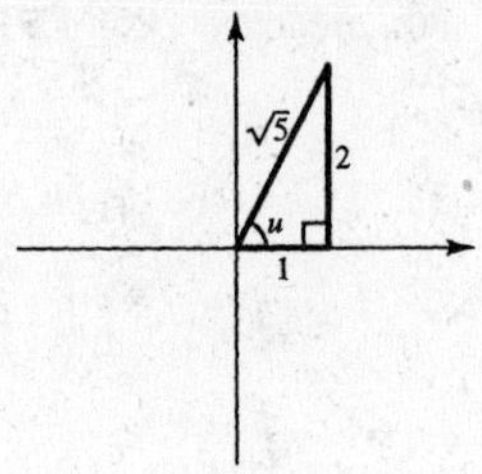

53. Let $y = \arcsin \dfrac{5}{13}$. Then,

$$\sin y = \frac{5}{13},\ 0 < y < \frac{\pi}{2}$$

and $\cos y = \dfrac{12}{13}$.

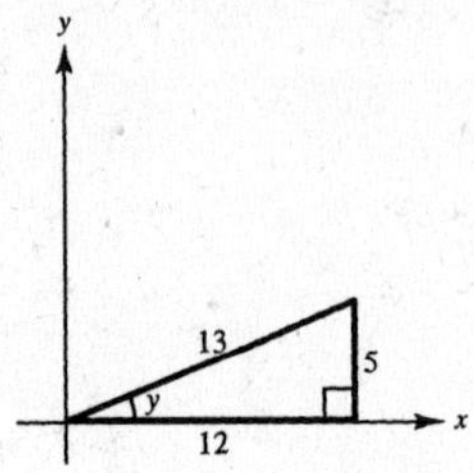

54. Let $u = \arctan\left(-\dfrac{5}{12}\right)$,

$$\tan u = -\frac{5}{12},\ -\frac{\pi}{2} < u < 0.$$

$$\csc\left[\arctan\left(-\frac{5}{12}\right)\right] = \csc u$$

$$= \frac{\text{hyp}}{\text{opp}} = -\frac{13}{5}$$

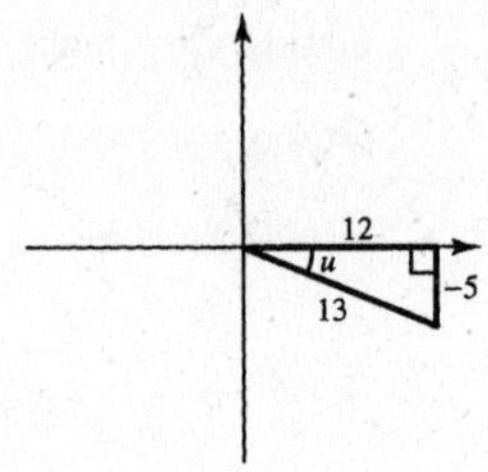

55. Let $y = \arctan\left(-\dfrac{3}{5}\right)$. Then,

$$\tan y = -\frac{3}{5},\ -\frac{\pi}{2} < y < 0$$

and $\sec y = \dfrac{\sqrt{34}}{5}$.

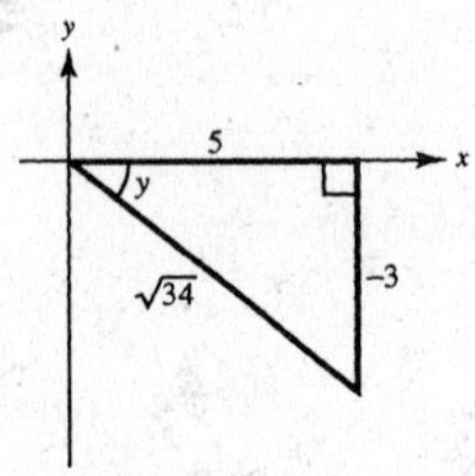

56. Let $u = \arcsin\left(-\dfrac{3}{4}\right)$,

$$\sin u = -\frac{3}{4},\ -\frac{\pi}{2} < u < 0.$$

$$\tan\left[\arcsin\left(-\frac{3}{4}\right)\right] = \tan u$$

$$= -\frac{3}{\sqrt{7}}$$

$$= -\frac{3\sqrt{7}}{7}$$

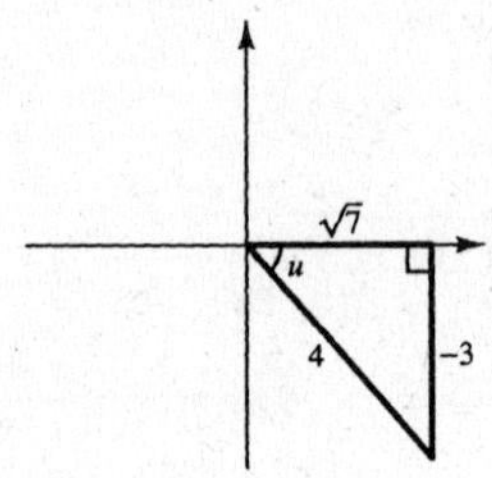

57. Let $y = \arccos\left(-\dfrac{2}{3}\right)$. Then,

$$\cos y = -\frac{2}{3},\ \frac{\pi}{2} < y < \pi$$

and $\sin y = \dfrac{\sqrt{5}}{3}$.

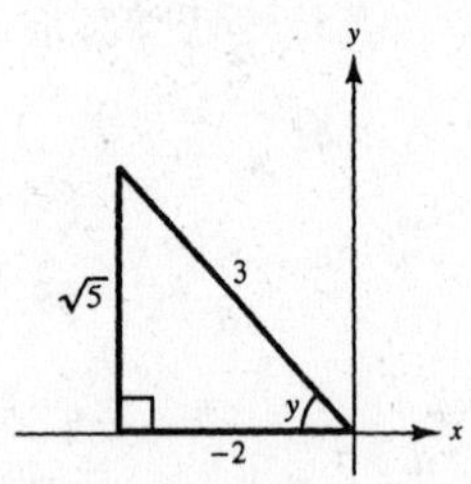

58. Let $u = \arctan \dfrac{5}{8}$,

$$\tan u = \frac{5}{8},\ 0 < u < \frac{\pi}{2}.$$

$$\cot\left(\arctan \frac{5}{8}\right) = \cot u$$

$$= \frac{\text{adj}}{\text{opp}} = \frac{8}{5}$$

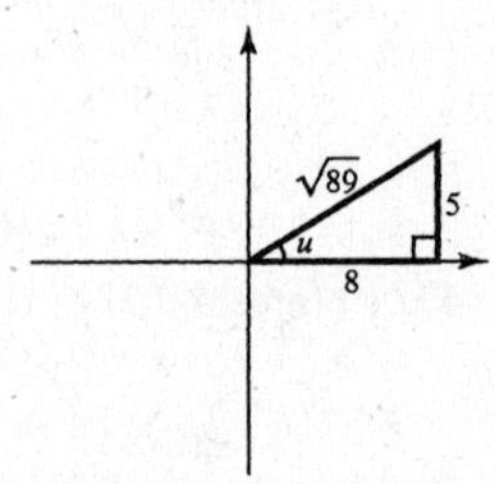

59. Let $y = \arctan x$. Then,

$$\tan y = x = \frac{x}{1}$$

and $\cot y = \dfrac{1}{x}$.

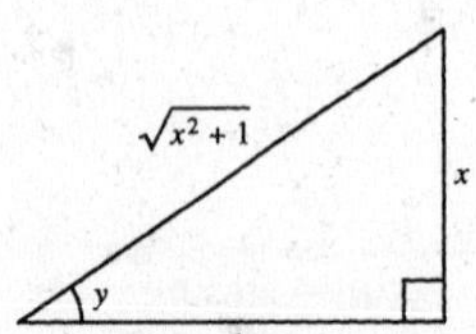

60. Let $u = \arctan x$,

$$\tan u = x = \frac{x}{1}.$$

$$\sin(\arctan x) = \sin u$$

$$= \frac{\text{opp}}{\text{hyp}}$$

$$= \frac{x}{\sqrt{x^2 + 1}}$$

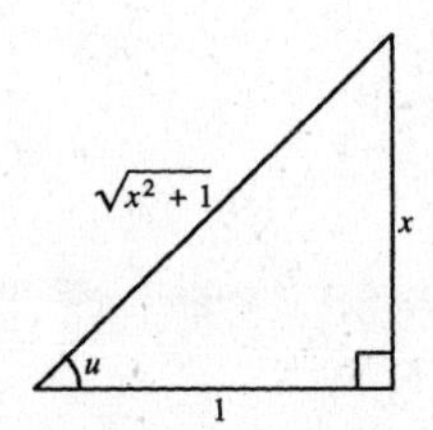

61. Let $y = \arcsin(2x)$. Then,

$\sin y = 2x = \dfrac{2x}{1}$

and $\cos y = \sqrt{1 - 4x^2}$.

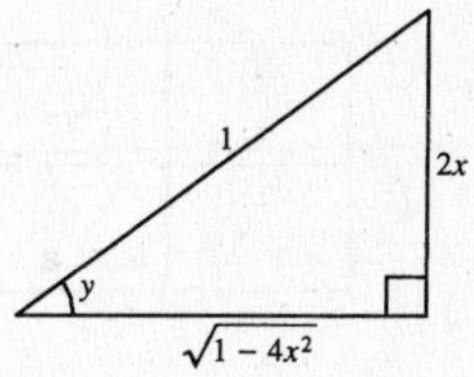

62. Let $u = \arctan 3x$,

$\tan u = 3x = \dfrac{3x}{1}$.

$\sec(\arctan 3x) = \sec u$

$= \dfrac{\text{hyp}}{\text{adj}}$

$= \sqrt{9x^2 + 1}$

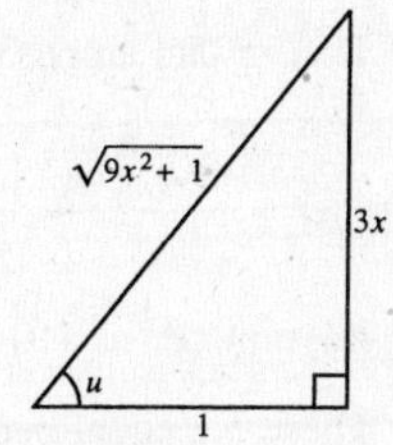

63. Let $y = \arccos x$. Then,

$\cos y = x = \dfrac{x}{1}$

and $\sin y = \sqrt{1 - x^2}$.

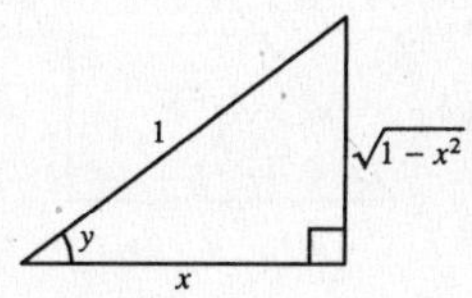

64. Let $u = \arcsin(x - 1)$,

$\sin u = x - 1 = \dfrac{x - 1}{1}$.

$\sec[\arcsin(x - 1)] = \sec u$

$= \dfrac{\text{hyp}}{\text{adj}}$

$= \dfrac{1}{\sqrt{2x - x^2}}$

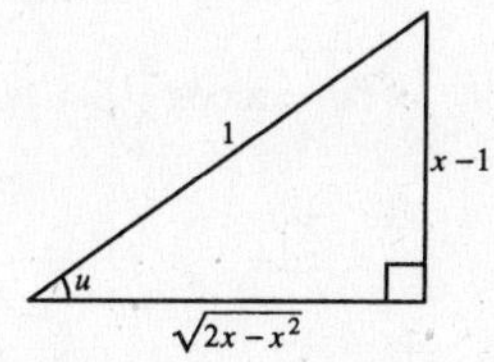

65. Let $y = \arccos\left(\dfrac{x}{3}\right)$. Then,

$\cos y = \dfrac{x}{3}$

and $\tan y = \dfrac{\sqrt{9 - x^2}}{x}$.

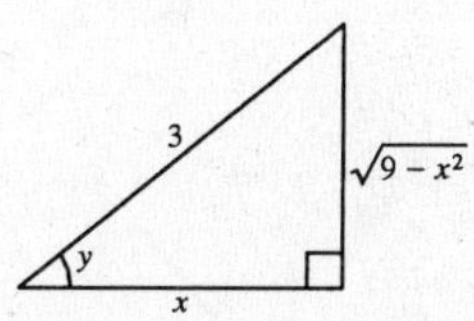

66. Let $u = \arctan \dfrac{1}{x}$,

$\tan u = \dfrac{1}{x}$.

$\cot\left(\arctan \dfrac{1}{x}\right) = \cot u$

$= \dfrac{\text{adj}}{\text{opp}} = x$

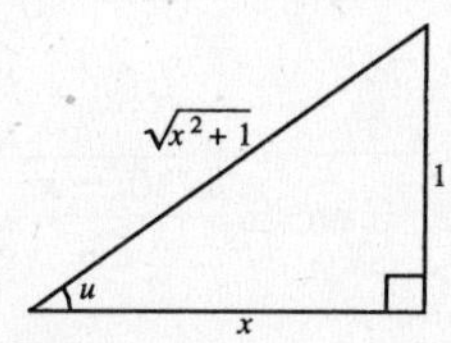

67. Let $y = \arctan \dfrac{x}{\sqrt{2}}$. Then,

$\tan y = \dfrac{x}{\sqrt{2}}$

and $\csc y = \dfrac{\sqrt{x^2 + 2}}{x}$.

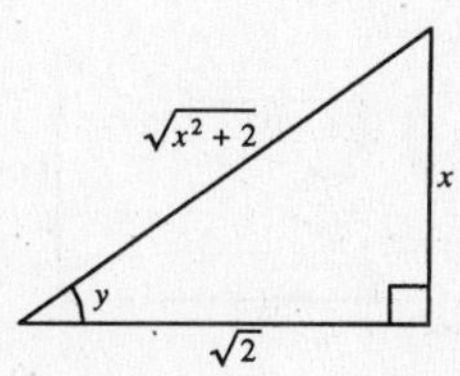

68. Let $u = \arcsin \dfrac{x - h}{r}$,

$\sin u = \dfrac{x - h}{r}$.

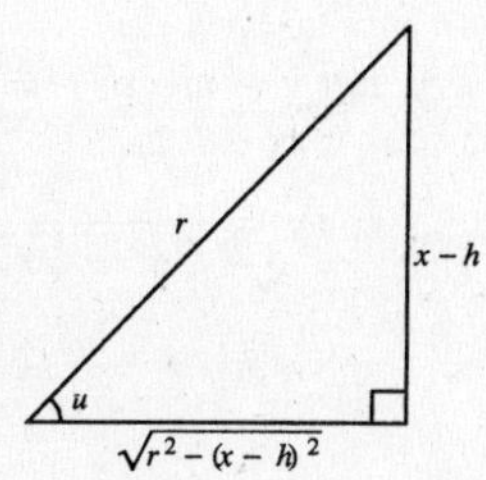

$\cos\left(\arcsin \dfrac{x - h}{r}\right) = \cos u = \dfrac{\sqrt{r^2 - (x - h)^2}}{r}$

69. $f(x) = \sin(\arctan 2x)$, $g(x) = \dfrac{2x}{\sqrt{1 + 4x^2}}$

Let $y = \arctan 2x$. Then,

$\tan y = 2x = \dfrac{2x}{1}$

and $\sin y = \dfrac{2x}{\sqrt{1 + 4x^2}}$.

$g(x) = \dfrac{2x}{\sqrt{1 + 4x^2}} = f(x)$

The graph has horizontal asymptotes at $y = \pm 1$.

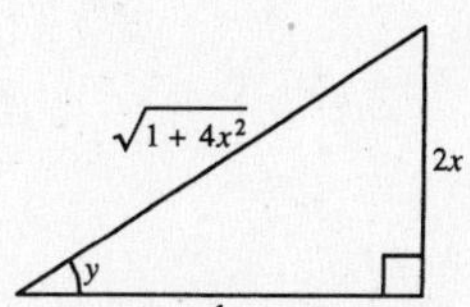

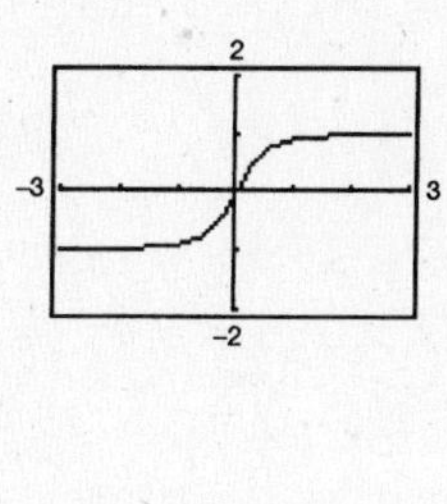

70. $f(x) = \tan\left(\arccos \frac{x}{2}\right)$

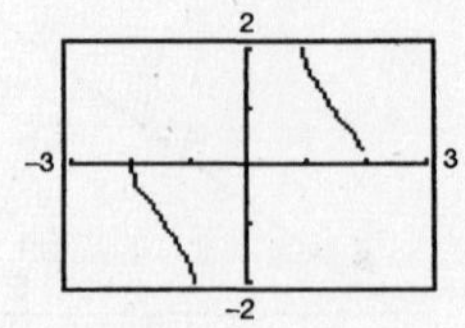

$g(x) = \frac{\sqrt{4 - x^2}}{x}$

Asymptote: $x = 0$

These are equal because:

Let $u = \arccos \frac{x}{2}$.

$\tan\left(\arccos \frac{x}{2}\right) = \tan u = \frac{\sqrt{4 - x^2}}{x}$

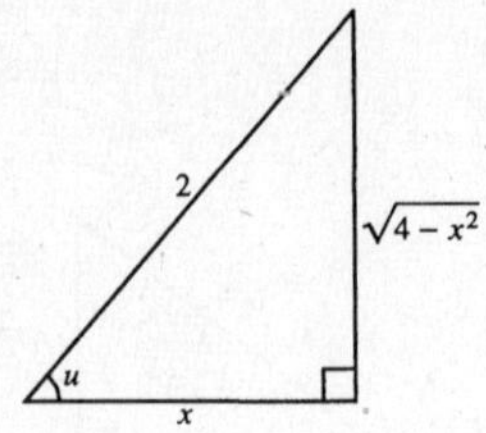

71. Let $y = \arctan \frac{9}{x}$. Then,

$\tan y = \frac{9}{x}$ and $\sin y = \frac{9}{\sqrt{x^2 + 81}}, x > 0; \frac{-9}{\sqrt{x^2 + 81}}, x < 0.$

Thus, $y = \arcsin \frac{9}{\sqrt{x^2 + 81}}, x > 0; y = \arcsin \frac{-9}{\sqrt{x^2 + 81}}, x < 0.$

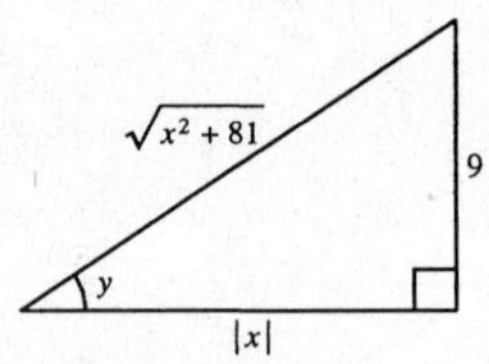

72. If $\arcsin \frac{\sqrt{36 - x^2}}{6} = u$,

then $\sin u = \frac{\sqrt{36 - x^2}}{6}$.

$\arcsin \frac{\sqrt{36 - x^2}}{6} = \arccos \frac{x}{6}$

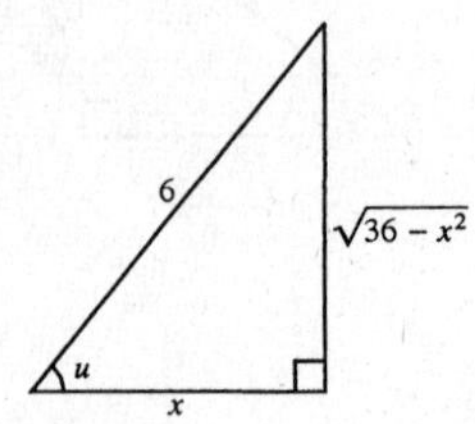

73. Let $y = \arccos \frac{3}{\sqrt{x^2 - 2x + 10}}$. Then,

$\cos y = \frac{3}{\sqrt{x^2 - 2x + 10}} = \frac{3}{\sqrt{(x - 1)^2 + 9}}$

and $\sin y = \frac{|x - 1|}{\sqrt{(x - 1)^2 + 9}}$.

Thus, $y = \arcsin \frac{|x - 1|}{\sqrt{(x - 1)^2 + 9}} = \arcsin \frac{|x - 1|}{\sqrt{x^2 - 2x + 10}}$.

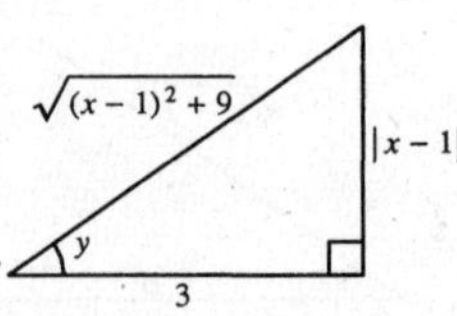

74. If $\arccos \frac{x - 2}{2} = u$,

then $\cos u = \frac{x - 2}{2}$.

$\arccos \frac{x - 2}{2} = \arctan \frac{\sqrt{4x - x^2}}{x - 2}$

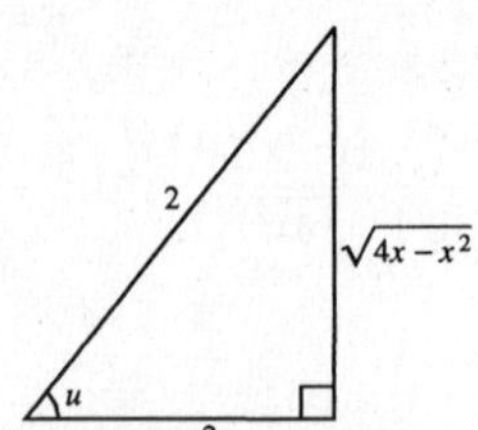

75. $y = 2 \arccos x$

Domain: $-1 \le x \le 1$

Range: $0 \le y \le 2\pi$

Vertical stretch of $f(x) = \arccos x$

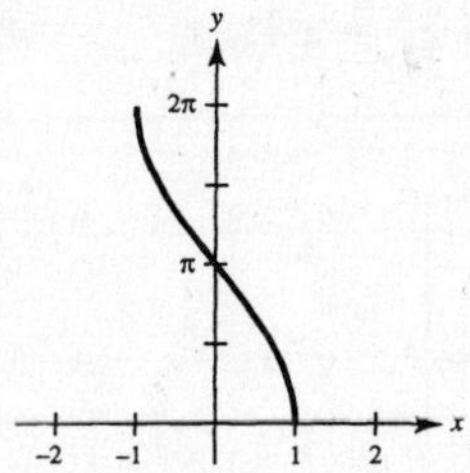

76. $y = \arcsin \frac{x}{2}$

Domain: $[-2, 2]$

Range: $\left[-\frac{\pi}{2}, \frac{\pi}{2}\right]$

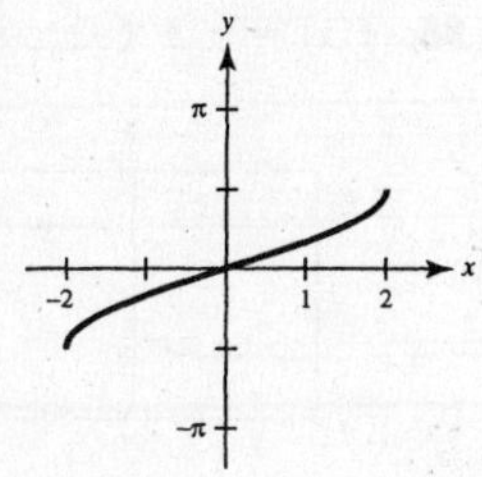

77. The graph of $f(x) = \arcsin(x - 1)$ is a horizontal translation of the graph of $y = \arcsin x$ by one unit.

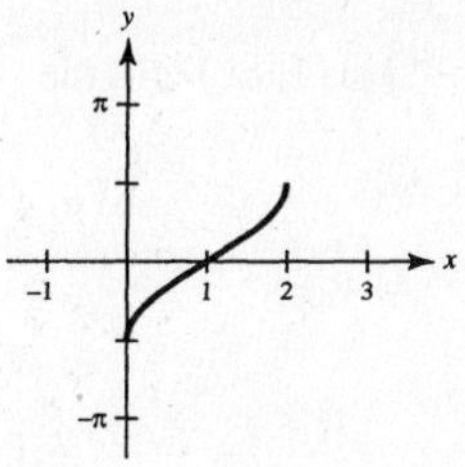

78. $g(t) = \arccos(t + 2)$

Domain: $[-3, -1]$,

Range: $[0, \pi]$

This is the graph of $y = \arccos t$ shifted two units to the left.

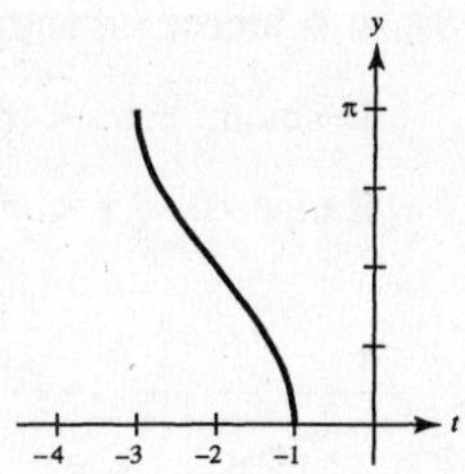

79. $f(x) = \arctan 2x$

Domain: all real numbers

Range: $-\frac{\pi}{2} < y < \frac{\pi}{2}$

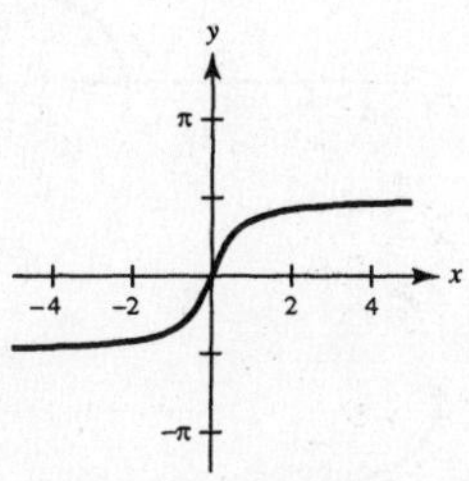

80. $f(x) = \frac{\pi}{2} + \arctan x$

Domain: $(-\infty, \infty)$

Range: $(0, \pi)$

This is the graph of $y = \arctan x$

shifted upward $\frac{\pi}{2}$ units.

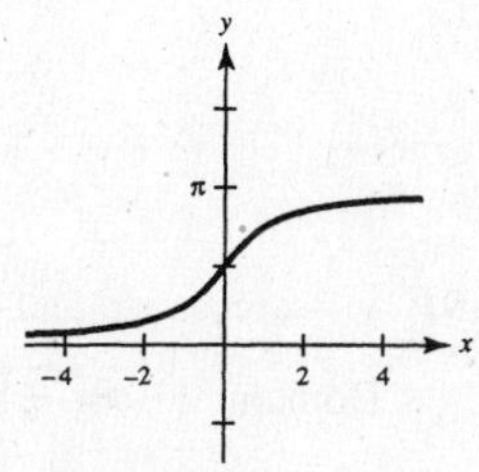

81. $h(v) = \tan(\arccos v) = \frac{\sqrt{1 - v^2}}{v}$

Domain: $-1 \le v \le 1, v \ne 0$

Range: all real numbers

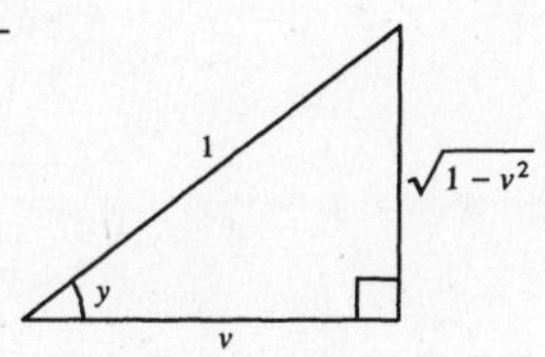

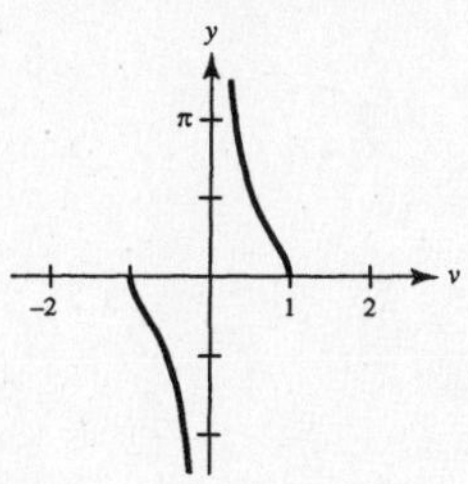

82. $f(x) = \arccos \frac{x}{4}$

Domain: $[-4, 4]$

Range: $[0, \pi]$

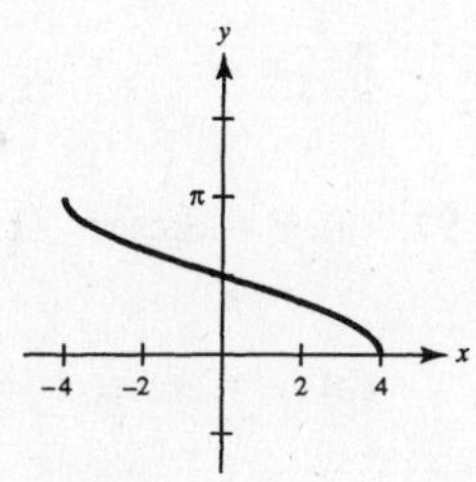

83. $f(x) = 2 \arccos (2x)$

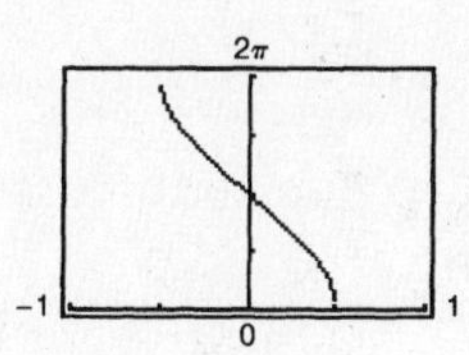

84. $f(x) = \pi \arcsin (4x)$

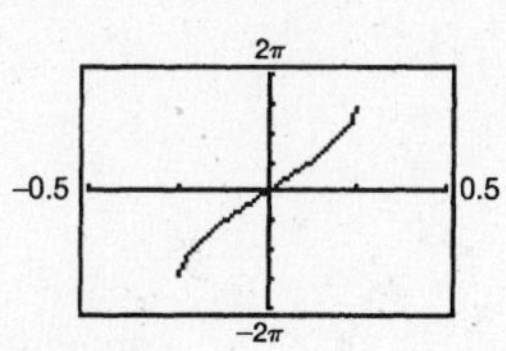

85. $f(x) = \arctan (2x - 3)$

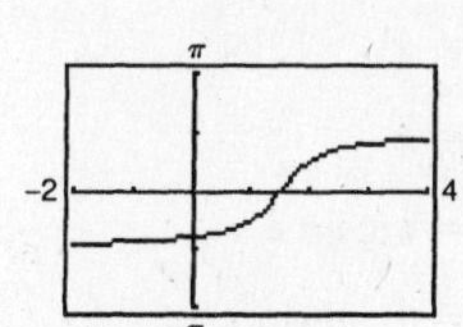

86. $f(x) = -3 + \arctan(\pi x)$

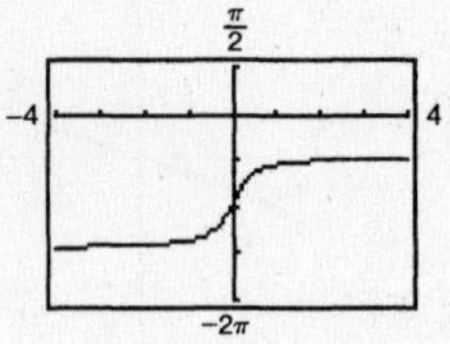

87. $f(x) = \pi - \arcsin\left(\frac{2}{3}\right) \approx 2.412$

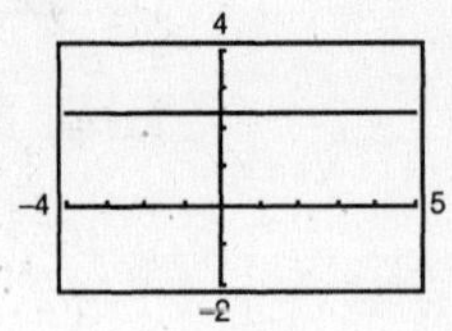

88. $f(x) = \frac{\pi}{2} + \arccos\left(\frac{1}{\pi}\right) \approx 2.82$

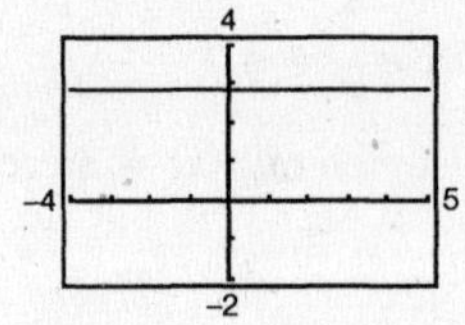

89. $y = \operatorname{arccot} x$ if and only if $\cot y = x$.

Domain: $-\infty < x < \infty$

Range: $0 < x < \pi$

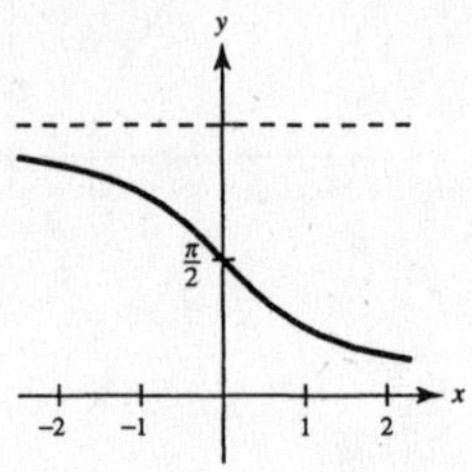

90. $y = \operatorname{arcsec} x$ if and only if $\sec y = x$ where $x \le -1 \cup x \ge 1$ and $0 \le y < \frac{\pi}{2}$ and $\frac{\pi}{2} < y \le \pi$. The domain of y arcsec x is $(-\infty, -1] \cup [1, \infty)$ and the range is $\left[0, \frac{\pi}{2}\right) \cup \left(\frac{\pi}{2}, \pi\right]$.

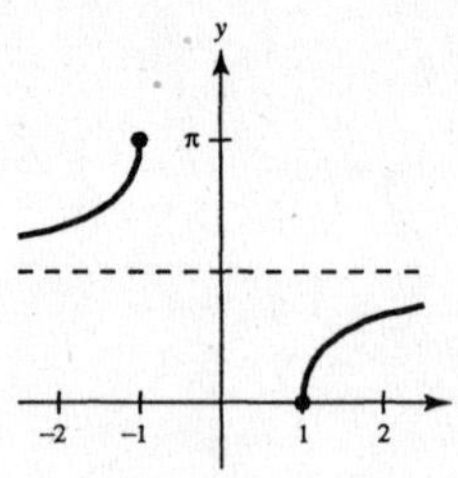

91. $y = \operatorname{arccsc} x$ if and only if $\csc y = x$.

Domain: $(-\infty, -1] \cup [1, \infty)$

Range: $\left[-\frac{\pi}{2}, 0\right) \cup \left(0, \frac{\pi}{2}\right]$

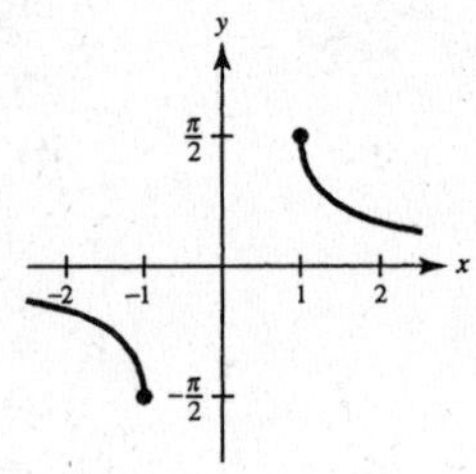

92. (a) $y = \operatorname{arcsec} \sqrt{2} \Rightarrow \sec y = \sqrt{2}$ and $0 \le y < \frac{\pi}{2} \cup \frac{\pi}{2} < y \le \pi \Rightarrow y = \frac{\pi}{4}$

(b) $y = \operatorname{arcsec} 1 \Rightarrow \sec y = 1$ and $0 \le y < \frac{\pi}{2} \cup \frac{\pi}{2} < y \le \pi \Rightarrow y = 0$

(c) $y = \operatorname{arccot}(-\sqrt{3}) \Rightarrow \cot y = -\sqrt{3}$ and $0 < y < \pi \Rightarrow y = \frac{5\pi}{6}$

(d) $y = \operatorname{arccsc} 2 \Rightarrow \csc y = 2$ and $-\frac{\pi}{2} \le y < 0 \cup 0 < y \le \frac{\pi}{2} \Rightarrow y = \frac{\pi}{6}$

93. Let $y = \arcsin(-x)$. Then,

$$\sin y = -x$$
$$-\sin y = x$$
$$\sin(-y) = x$$
$$-y = \arcsin x$$
$$y = -\arcsin x.$$

Therefore, $\arcsin(-x) = -\arcsin x$.

94.

$$y = \arctan(-x)$$
$$\tan y = -x, \; -\frac{\pi}{2} < y < \frac{\pi}{2}$$
$$-\tan y = x$$
$$\tan(-y) = x, \; -\frac{\pi}{2} < -y < \frac{\pi}{2}$$
$$\arctan(\tan(-y)) = \arctan x$$
$$-y = \arctan x$$
$$y = -\arctan x$$

95. $y = \pi - \arccos x$

$\cos y = \cos(\pi - \arccos x)$

$\cos y = \cos \pi \cos(\arccos x) + \sin \pi \sin(\arccos x)$

$\cos y = -x$

$y = \arccos(-x)$

Therefore, $\arccos(-x) = \pi - \arccos x$.

96. (a) $f \circ f^{-1} = \sin(\arcsin x)$

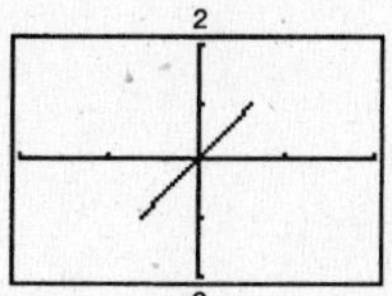

$f^{-1} \circ f = \arcsin(\sin x)$

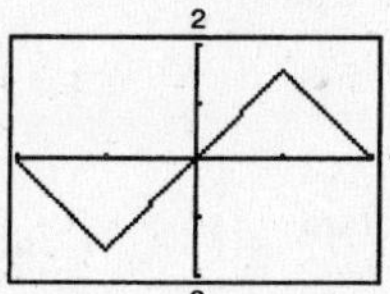

(b) The graphs coincide with the graph of $y = x$ only for certain values of x.

$f \circ f^{-1} = x$ over its entire domain, $-1 \le x \le 1$.

$f^{-1} \circ f = x$ over the region $-\dfrac{\pi}{2} \le x \le \dfrac{\pi}{2}$,

corresponding to the region where $\sin x$ is one-to-one and thus has an inverse.

97. $f(t) = 3\cos 2t + 3\sin 2t = \sqrt{3^2 + 3^2}\sin\left(2t + \arctan\dfrac{3}{3}\right)$

$= 3\sqrt{2}\sin(2t + \arctan 1)$

$= 3\sqrt{2}\sin\left(2t + \dfrac{\pi}{4}\right)$

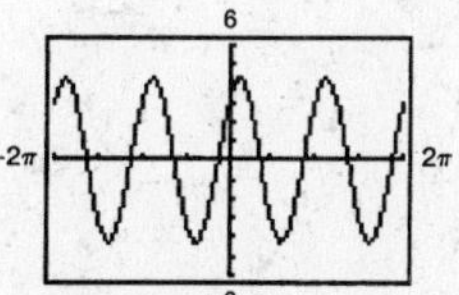

The graphs are the same.

98. $f(t) = 4\cos \pi t + 3\sin \pi t$

$= \sqrt{4^2 + 3^2}\sin\left(\pi t + \arctan\dfrac{4}{3}\right)$

$= 5\sin\left(\pi t + \arctan\dfrac{4}{3}\right)$

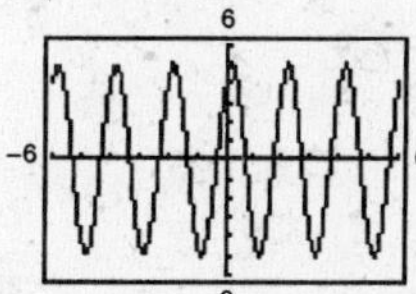

The graph implies that

$A\cos \omega t + B\sin \omega t = \sqrt{A^2 + B^2}\sin\left(\omega t + \arctan\dfrac{A}{B}\right)$ is true.

99. (a) $\sin\theta = \dfrac{5}{s}$

$\theta = \arcsin\dfrac{5}{s}$

(b) $s = 40$: $\theta = \arcsin\dfrac{5}{40} \approx 0.13$

$s = 20$: $\theta = \arcsin\dfrac{5}{20} \approx 0.25$

100. (a) $\tan\theta = \dfrac{s}{750}$

$\theta = \arctan\dfrac{s}{750}$

(b) When $s = 300$,

$\theta = \arctan\dfrac{300}{750} \approx 0.38 \approx 21.8°$.

When $s = 1200$,

$\theta = \arctan\dfrac{1200}{750} \approx 1.01 \approx 58.0°$.

101. $\beta = \arctan \frac{3x}{x^2 + 4}$

(a) 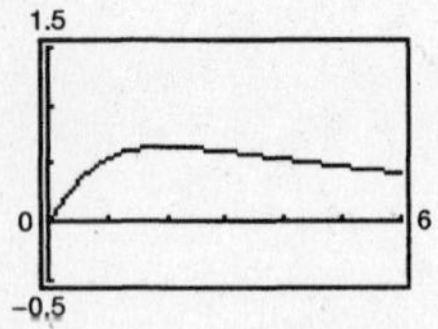

(b) β is maximum when $x = 2$.

(c) The graph has a horizontal asymptote at $\beta = 0$. As x increases, β decreases. This means the size of the picture in the photo gets smaller.

102. (a) $\tan \theta = \frac{6}{x}$

$\theta = \arctan \frac{6}{x}$

(b) $x = 7$ miles

$\theta = \arctan \frac{6}{7} \approx 0.71 \approx 40.6°$

$x = 1$ mile

$\theta = \arctan \frac{6}{1} \approx 1.41 \approx 80.5°$

103. (a) $\tan\theta = \frac{x}{20}$

$\theta = \arctan \frac{x}{20}$

(b) $x = 5$: $\theta = \arctan \frac{5}{20} \approx 0.24$

$x = 12$: $\theta = \arctan \frac{12}{20} \approx 0.54$

104. Area $= \arctan b - \arctan a$

(a) $a = 0, b = 1$

Area $= \arctan 1 - \arctan 0 = \frac{\pi}{4} - 0 = \frac{\pi}{4}$

(b) $a = -1, b = 1$

Area $= \arctan 1 - \arctan(-1)$

$= \frac{\pi}{4} - \left(-\frac{\pi}{4}\right) = \frac{\pi}{2}$

(c) $a = 0, b = 3$

Area $= \arctan 3 - \arctan 0$

$\approx 1.25 - 0 = 1.25$

$= 1.25$

(d) $a = -1, b = 3$

Area $= \arctan 3 - \arctan(-1)$

$\approx 1.25 - \left(-\frac{\pi}{4}\right) \approx 2.03$

105. False.

$\frac{5\pi}{6}$ is not in the range of arcsin (x).

$\arcsin \frac{1}{2} = \frac{\pi}{6}$

106. False; $\arctan 1 = \frac{\pi}{4}$. $\frac{5\pi}{4}$ is not in the range of the arctangent function.

107. False.

$\sin \theta = 0.2$ has two solutions for $0 \le \theta < 2\pi$:

$\theta = \arcsin 0.2$ and $\theta = \pi - \arcsin 0.2$

108. False. $\cos \theta = -0.2$ has two solutions for $0 \le \theta < 2\pi$: $\theta = \arccos(-0.2)$ and $\theta = \pi + \arccos(0.2)$.

Section 9.8 Applications and Models

1. Given: $A = 20°$, $b = 10$

$\tan A = \frac{a}{b} \Rightarrow a = b \tan A = 10 \tan 20° \approx 3.64$

$\cos A = \frac{b}{c} \Rightarrow c = \frac{b}{\cos A} = \frac{10}{\cos 20°} \approx 10.64$

$B = 90° - 20° = 70°$

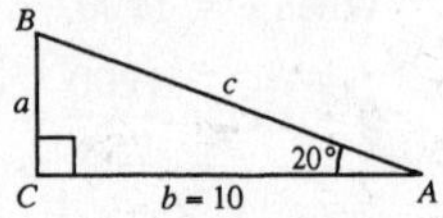

2. $B = 54°, c = 15$

$A = 90° - B = 90° - 54° = 36°$

$\sin B = \frac{b}{c} \Rightarrow b = c \sin B$

$= 15 \sin 54° \approx 12.14$

$\cos B = \frac{a}{c} \Rightarrow a = c \cos B$

$= 15 \cos 54° \approx 8.82$

3. Given: $B = 71°,\ b = 24$

$$\tan B = \frac{b}{a} \Rightarrow a = \frac{b}{\tan B} = \frac{24}{\tan 71°} \approx 8.26$$

$$\sin B = \frac{b}{c} \Rightarrow c = \frac{b}{\sin B} = \frac{24}{\sin 71°} \approx 25.38$$

$$A = 90° - 71° = 19°$$

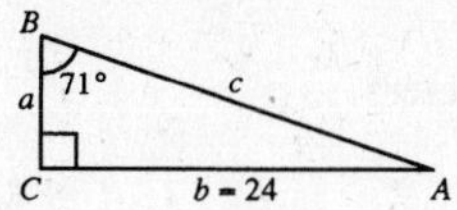

4. $A = 8.4°, a = 40.5$

$$B = 90° - A = 90° - 8.4° = 81.6°$$

$$\tan A = \frac{a}{b} \Rightarrow b = \frac{a}{\tan A}$$

$$= \frac{40.5}{\tan 8.4°} \approx 274.27$$

$$\sin A = \frac{a}{c} \Rightarrow c = \frac{a}{\sin A}$$

$$= \frac{40.5}{\sin 8.4°} \approx 277.24$$

5. Given: $a = 6,\ b = 10$

$$c^2 = a^2 + b^2 \Rightarrow c = \sqrt{36 + 100}$$

$$= 2\sqrt{34} \approx 11.66$$

$$\tan A = \frac{a}{b} = \frac{6}{10} \Rightarrow A = \arctan \frac{3}{5} \approx 30.96°$$

$$B = 90° - 30.96° = 59.04°$$

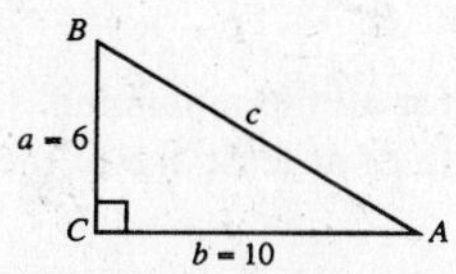

6. $a = 25, c = 35$

$$b = \sqrt{c^2 - a^2} = \sqrt{35^2 - 25^2} = \sqrt{600} \approx 24.49$$

$$\sin A = \frac{a}{c} \Rightarrow A = \arcsin \frac{a}{c}$$

$$= \arcsin \frac{25}{35} \approx 45.58°$$

$$\cos B = \frac{a}{c} \Rightarrow B = \arccos \frac{a}{c}$$

$$= \arccos \frac{25}{35} \approx 44.42°$$

7. $b = 16,\ c = 52$

$$a = \sqrt{52^2 - 16^2}$$

$$= \sqrt{2448} = 12\sqrt{17} \approx 49.48$$

$$\cos A = \frac{16}{52}$$

$$A = \arccos \frac{16}{52} \approx 72.08°$$

$$B = 90° - 72.08° \approx 17.92°$$

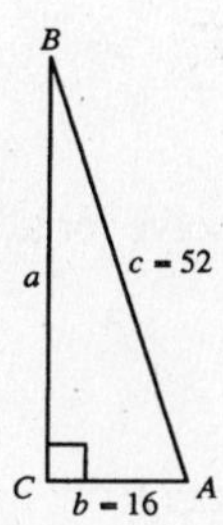

8. $b = 1.32, c = 9.45$

$$a = \sqrt{c^2 - b^2} = \sqrt{87.5601} \approx 9.36$$

$$\cos A = \frac{b}{c} \Rightarrow A = \arccos \frac{b}{c}$$

$$= \arccos \frac{1.32}{9.45} \approx 81.97°$$

$$\sin B = \frac{b}{c} \Rightarrow B = \arcsin \frac{b}{c}$$

$$= \arcsin \frac{1.32}{9.45} \approx 8.03°$$

9. $A = 12°15',\ c = 430.5$

$$B = 90° - 12°15' = 77°45'$$

$$\sin 12°15' = \frac{a}{430.5}$$

$$a = 430.5 \sin 12°15' \approx 91.34$$

$$\cos 12°15' = \frac{b}{430.5}$$

$$b = 430.5 \cos 12°15' \approx 420.70$$

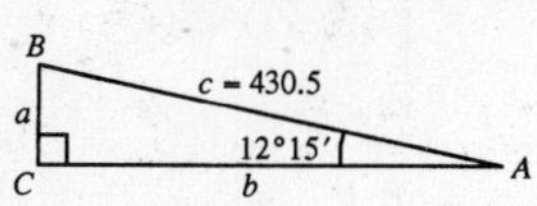

10. $B = 65°\ 12', a = 14.2$

$$A = 90° - B = 90° - 65°\ 12' = 24°\ 48'$$

$$\cos B = \frac{a}{c} \Rightarrow c = \frac{a}{\cos B}$$

$$= \frac{14.2}{\cos 65°\ 12'} \approx 33.85$$

$$\tan B = \frac{b}{a} \Rightarrow b = a \tan B$$

$$= 14.2 \tan 65°\ 12'$$

$$\approx 30.73$$

11. $\tan\theta = \dfrac{h}{\frac{1}{2}b} \Rightarrow h = \dfrac{1}{2}b\tan\theta$

$h = \dfrac{1}{2}(4)\tan 52° \approx 2.56$ inches

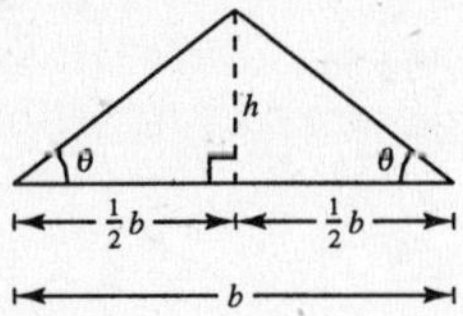

12. $\theta = 18°, b = 10$ meters

$\tan\theta = \dfrac{\text{altitude}}{\frac{b}{2}}$

$\text{altitude} = \dfrac{b}{2}\tan\theta$

$= \dfrac{10}{2}\tan 18° \approx 1.62$ meters

13. $\tan\theta = \dfrac{h}{\frac{1}{2}b} \Rightarrow h = \dfrac{1}{2}b\tan\theta$

$h = \dfrac{1}{2}(46)\tan 41° \approx 19.99$ inches

14. $\theta = 27°, b = 11$ feet

$\tan\theta = \dfrac{\text{altitude}}{\frac{b}{2}}$

$\text{altitude} = \dfrac{b}{2}\tan\theta$

$= \dfrac{11}{2}\tan 27° \approx 2.80$ feet

15. Given A, C, a, we can solve for all the unknown parts.

$B = 180° - A - C$ or $B = 90° - A$

$\tan B = \dfrac{b}{a} \Rightarrow b = a\tan B$

$c = \sqrt{a^2 + b^2}$

16. Given A, B, C, we cannot solve for all the remaining parts. We need to know the length of at least one of the sides.

17. Given C, a, c, we can solve for all the remaining parts.

$b = \sqrt{c^2 - a^2}$

$\sin A = \dfrac{a}{c} \Rightarrow A = \sin^{-1}\left(\dfrac{a}{c}\right)$

$B = 90° - A$

18. Given B, C, a, we can solve for all the remaining parts.

$A = 90° - B$

$\tan B = \dfrac{b}{a} \Rightarrow b = a\tan B$

$c = \sqrt{a^2 + b^2}$

19. N 45° W

45° west of north

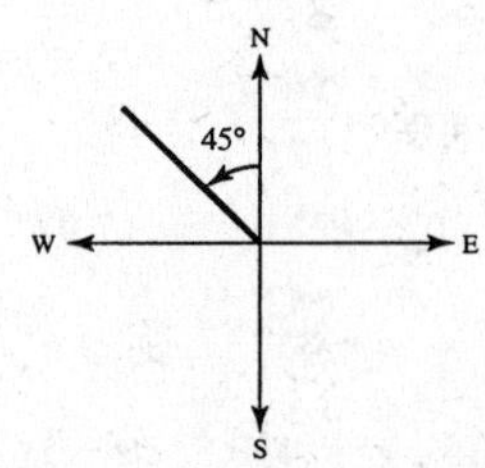

20. S 60° E

60° east of south

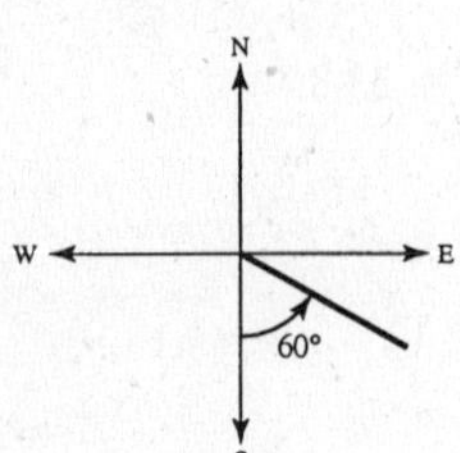

21. S 75° W

75° west of south

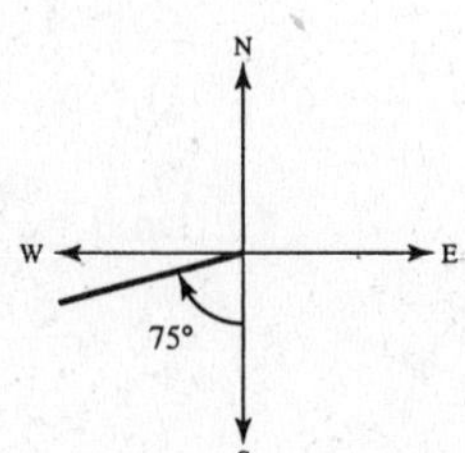

22. N 30° E

30° east of north

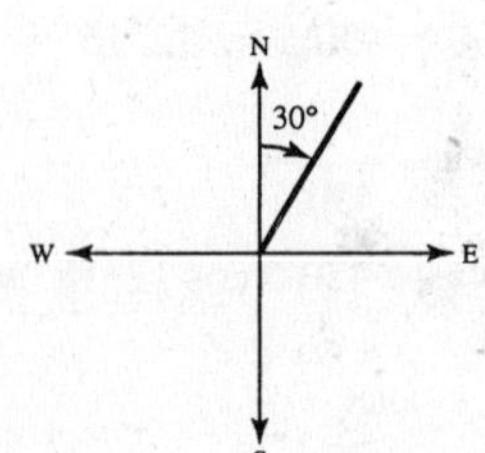

23. $d = a \sin \omega t$

Amplitude: $|a|$

24. $d = a \cos \omega t$

Period: $\dfrac{2\pi}{\omega}$

25. $\tan 25° = \dfrac{50}{x}$

$x = \dfrac{50}{\tan 25°} \approx 107.2$ feet

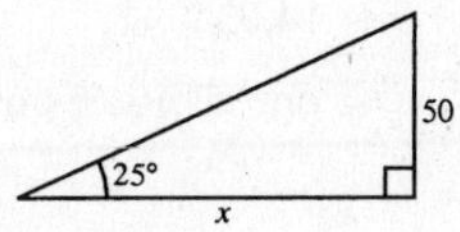

26. $\tan 20° = \dfrac{600}{x}$

$x = \dfrac{600}{\tan 20°} \approx 1648.5$ feet

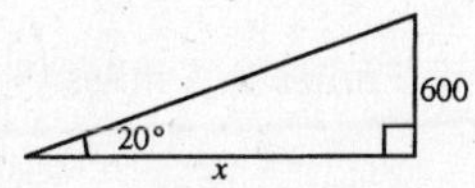

27. (a)

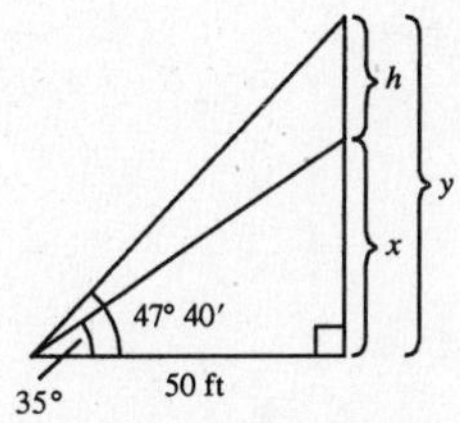

(b) Let the height of the church $= x$ and the height of the church and steeple $= y$. Then,

$\tan 35° = \dfrac{x}{50}$ and $\tan 47°40' = \dfrac{y}{50}$

$x = 50 \tan 35°$ and $y = 50 \tan 47°40'$

$h = y - x = 50\,(\tan 47°40' - \tan 35°)$.

(c) $h \approx 19.9$ feet

28. $\tan 33° = \dfrac{h}{125}$

$h = 125 \tan 33° \approx 81.2$ feet

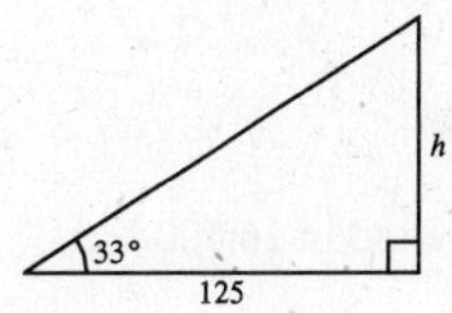

29. $\sin 80° = \dfrac{h}{20}$

$20 \sin 80° = h$

$h \approx 19.7$ feet

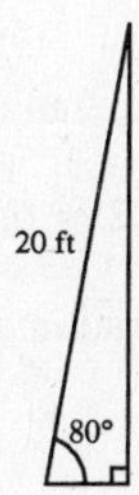

30. $\tan 51° = \dfrac{h}{100}$

$h = 100 \tan 51°$

≈ 123.5 feet

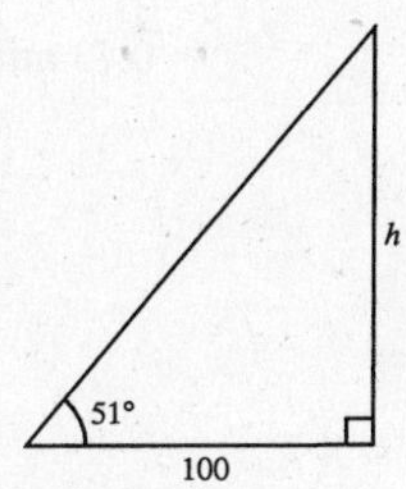

31. $\sin 34° = \dfrac{x}{4000}$

$x = 4000 \sin 34°$

≈ 2236.8 feet

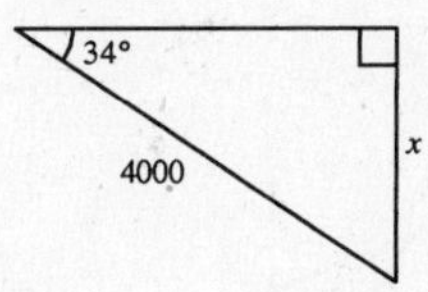

32. $\sin 60° = \dfrac{h}{75}$

$h = 75 \sin 60° \approx 65$ feet

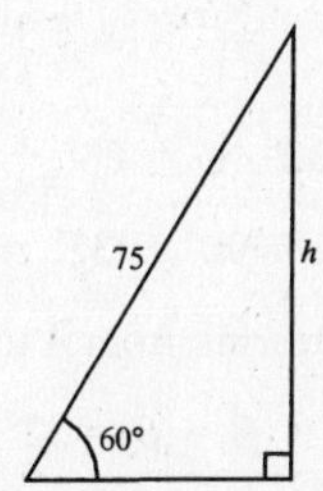

33. $\tan \theta = \dfrac{75}{50}$

$\theta = \arctan \dfrac{3}{2} \approx 56.3°$

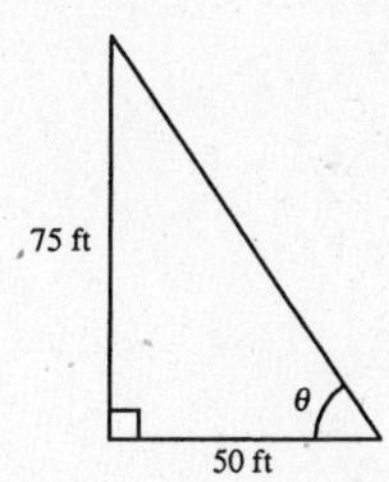

34. (a)

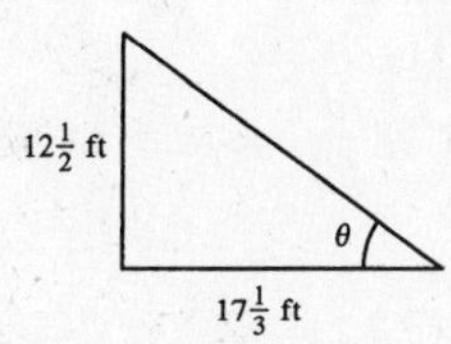

(b) $\tan \theta = \dfrac{12\frac{1}{2}}{17\frac{1}{3}}$

(c) $\theta = \arctan \dfrac{12\frac{1}{2}}{17\frac{1}{3}} \approx 35.8$

35.

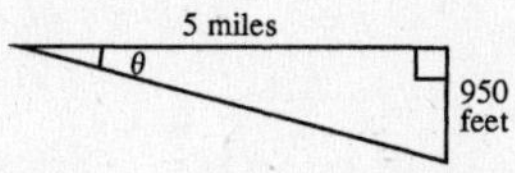

Not drawn to scale

$$5 \text{ miles} = 5 \text{ miles}\left(\frac{5280 \text{ ft}}{1 \text{ mile}}\right) = 26{,}400 \text{ feet}$$

$$\tan\theta = \frac{950}{26{,}400}$$

$$\theta = \arctan\left(\frac{950}{26{,}400}\right) \approx 2.06°$$

36. $\sin\theta = \dfrac{4000}{16{,}500}$

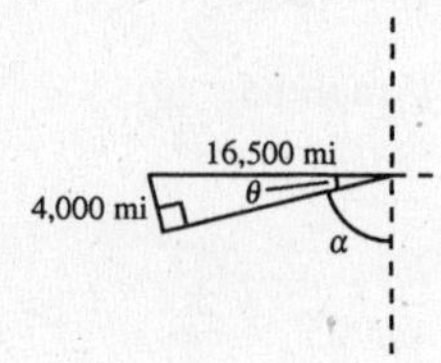

$$\theta = \arcsin\left(\frac{4000}{16{,}500}\right)$$

$$\theta \approx 14.03°$$

$$\alpha = 90° - \theta \approx 75.97°$$

37. Since the airplane speed is

$$\left(275\frac{\text{ft}}{\text{sec}}\right)\left(60\frac{\text{sec}}{\text{min}}\right) = 16{,}500\frac{\text{ft}}{\text{min}},$$

after one minute its distance travelled in 16,500 feet.

$$\sin 18° = \frac{a}{16{,}500}$$

$$a = 16{,}500 \sin 18°$$

$$\approx 5099 \text{ ft}$$

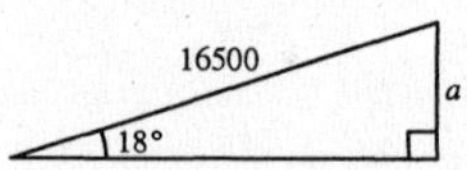

38.

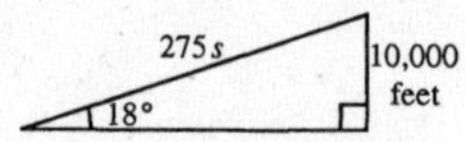

$$\sin 18° = \frac{10{,}000}{275\,s}$$

$$s = \frac{10{,}000}{275(\sin 18°)} \approx 117.7 \text{ seconds}$$

39. $\sin 10.5° = \dfrac{x}{4}$

$$x = 4 \sin 10.5°$$

$$\approx 0.73 \text{ mile}$$

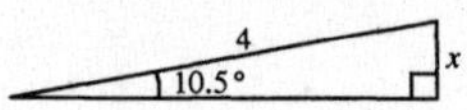

40.

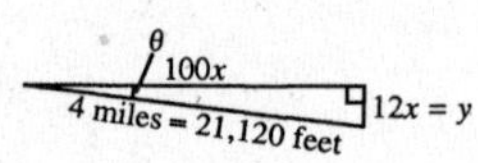

Angle of grade: $\tan\theta = \dfrac{12x}{100x}$

$$\theta = \arctan 0.12 \approx 6.8°$$

Change in elevation: $\sin\theta = \dfrac{y}{21{,}120}$

$$y = 21{,}120 \sin\theta$$

$$= 21{,}120 \sin(\arctan 0.12)$$

$$\approx 2516.3 \text{ feet}$$

41. $\theta = 32°$, $\phi = 68°$

(a) $\alpha = 90° - 32° = 58°$

Bearing from A to C: N 58° E

(b) $\beta = \theta = 32°$

$\gamma = 90° - \phi = 22°$

$C = \beta + \gamma = 54°$

$$\tan C = \frac{d}{50} \Rightarrow \tan 54° = \frac{d}{50} \Rightarrow d \approx 68.82 \text{ meters}$$

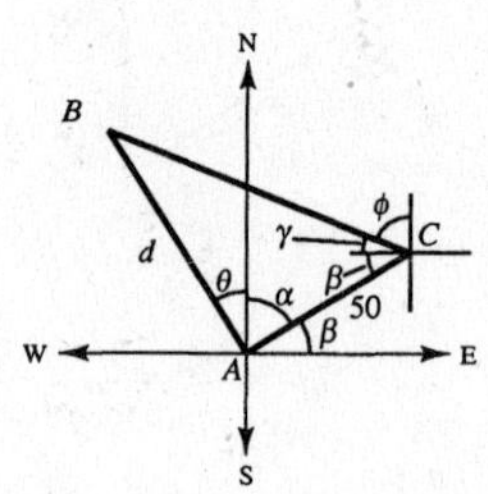

42.

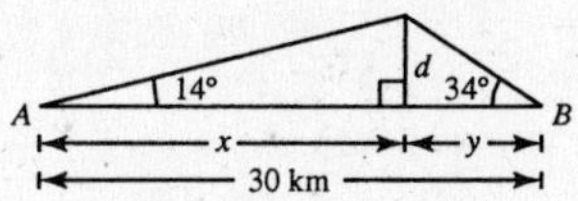

$$\tan 14° = \frac{d}{x} \implies x = d \cot 14°$$

$$\tan 34° = \frac{d}{y} \implies \frac{d}{30 - x} = \frac{d}{30 - d \cot 14°}$$

$$\cot 34° = \frac{30 - d \cot 14°}{d}$$

$$d \cot 34° = 30 - d \cot 14°$$

$$d = \frac{30}{\cot 34° + \cot 14°}$$

$$\approx 5.46 \text{ kilometers}$$

43. $\tan \theta = \dfrac{45}{30} \implies \theta \approx 56.3°$

Bearing: N 56.3° W

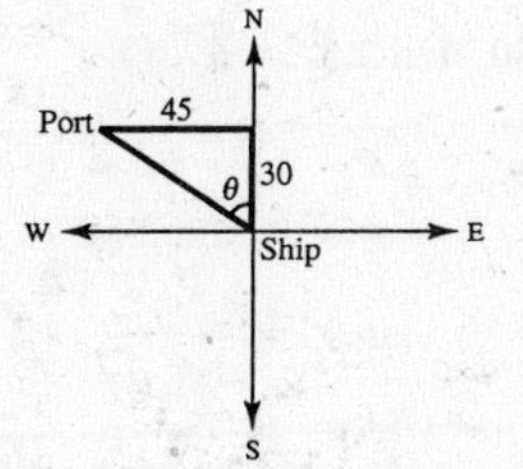

44. $\tan \theta = \dfrac{85}{120} \implies \theta = 35.3°$

Bearing: S 35.3° W

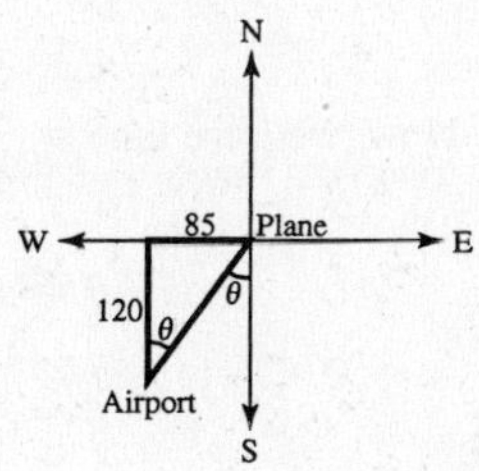

45. $\tan 6.5° = \dfrac{350}{d} \implies d \approx 3071.91$ ft

$\tan 4° = \dfrac{350}{D} \implies D \approx 5005.23$ ft

Distance between ships: $D - d \approx 1933.32$ ft

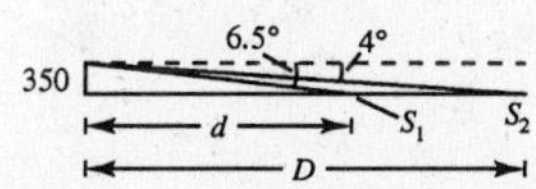

46.

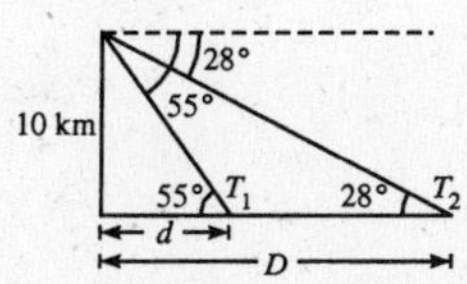

$$\cot 55° = \frac{d}{10} \implies d \approx 7 \text{ kilometers}$$

$$\cot 28° = \frac{D}{10} \implies D \approx 18.8 \text{ kilometers}$$

Distance between towns:

$D - d = 18.8 - 7 = 11.8$ kilometers

47. The plane has traveled $1.5\,(600) = 900$ miles.

$\sin 38° = \dfrac{a}{900} \implies a \approx 554$ miles north

$\cos 38° = \dfrac{b}{900} \implies b \approx 709$ miles east

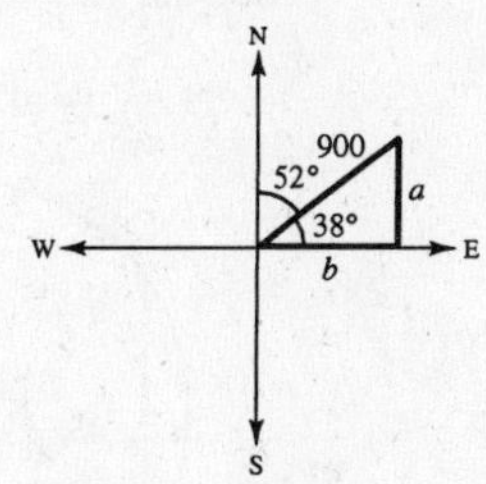

48. $\sin 63° = \dfrac{a}{120} \implies a \approx 107$ nautical miles south

$\cos 63° = \dfrac{b}{120} \implies b \approx 54.5$ nautical miles west

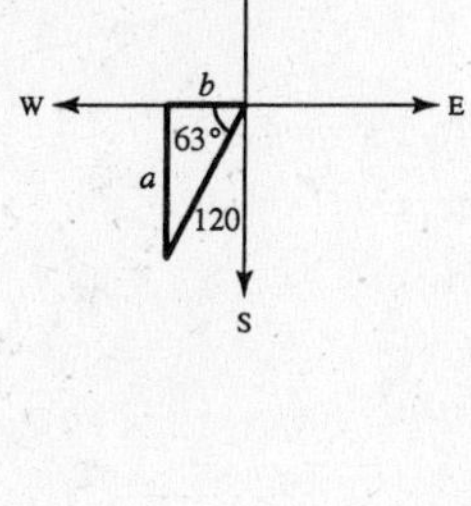

49. $\tan 57° = \dfrac{a}{x} \implies x = a \cot 57°$

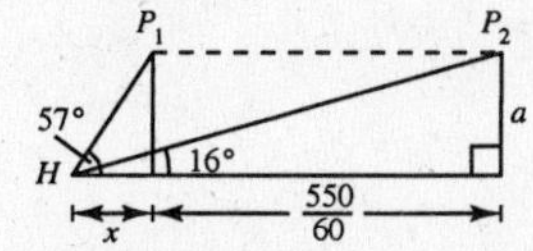

$$\tan 16° = \frac{a}{x + \frac{55}{6}}$$

$$\tan 16° = \frac{a}{a \cot 57° + \frac{55}{6}}$$

$$\cot 16° = \frac{a \cot 57° + \frac{55}{6}}{a}$$

$$a \cot 16° - a \cot 57° = \frac{55}{6} \implies a \approx 3.23 \text{ miles}$$

$$\approx 17{,}054 \text{ ft}$$

50. $\tan 2.5° = \dfrac{h}{x}$

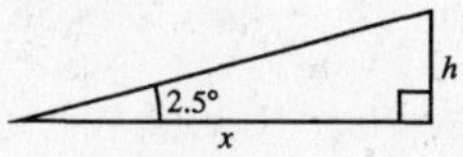

Not drawn to scale

$x = \dfrac{h}{\tan 2.5°}$

$\tan 9° = \dfrac{h}{x - 17}$

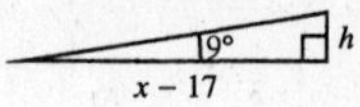

$x = \dfrac{h}{\tan 9°} + 17$

$\dfrac{h}{\tan 2.5°} = \dfrac{h}{\tan 9°} + 17$

$h = \dfrac{17}{\left(\dfrac{1}{\tan 2.5°} - \dfrac{1}{\tan 9°}\right)} \approx 1.025 \text{ miles}$

$\approx 5410 \text{ feet}$

51. L_1: $3x - 2y = 5 \Rightarrow y = \dfrac{3}{2}x - \dfrac{5}{2} \Rightarrow m_1 = \dfrac{3}{2}$

L_2: $x + y = 1 \Rightarrow y = -x + 1 \Rightarrow m_2 = -1$

$\tan \alpha = \left|\dfrac{-1 - \frac{3}{2}}{1 + (-1)\left(\frac{3}{2}\right)}\right| = \left|\dfrac{-\frac{5}{2}}{-\frac{1}{2}}\right| = 5$

$\alpha = \arctan 5 \approx 78.7°$

52. $L_1 = 2x - y = 8 \Rightarrow m_1 = 2$

$L_2 = x - 5y = -4 \Rightarrow m_2 = \dfrac{1}{5}$

$\tan \alpha = \left|\dfrac{m_2 - m_1}{1 + m_2 m_1}\right|$

$\alpha = \arctan\left|\dfrac{m_2 - m_1}{1 + m_2 m_1}\right|$

$= \arctan\left|\dfrac{\frac{1}{5} - 2}{1 + \frac{1}{5}(2)}\right|$

$= \arctan\left(\dfrac{9}{7}\right) \approx 52.1°$

53. The diagonal of the base has a length of $\sqrt{a^2 + a^2} = \sqrt{2}a$.

Now, we have $\tan\theta = \dfrac{a}{\sqrt{2}a} = \dfrac{1}{\sqrt{2}}$

$\theta = \arctan \dfrac{1}{\sqrt{2}}$

$\theta \approx 35.3°$.

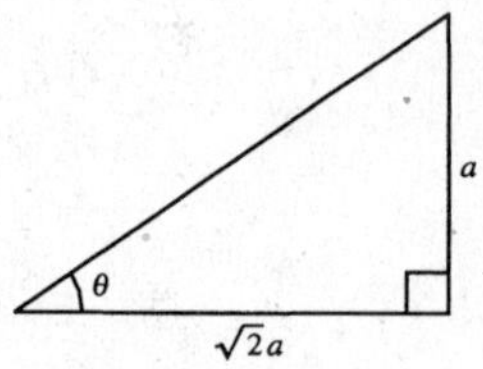

54.

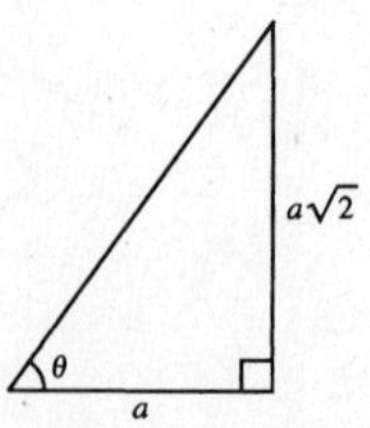

$\tan \theta = \dfrac{a\sqrt{2}}{a} = \sqrt{2}$

$\theta = \arctan \sqrt{2} \approx 54.7°$

55. $\sin 36° = \dfrac{d}{25} \Rightarrow d \approx 14.69$

Length of side: $2d \approx 29.4$ inches

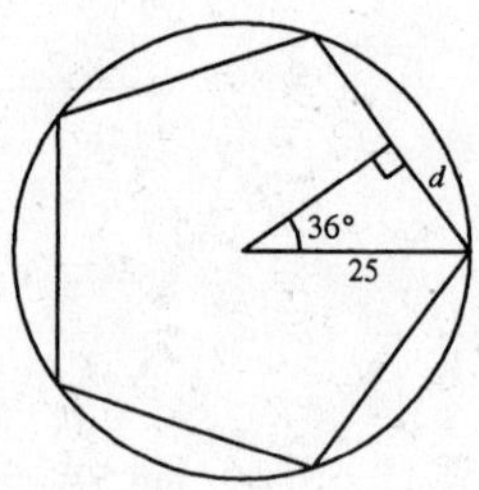

56. $\sin 30° = \dfrac{a}{25}$

$a = 25 \sin 30° = 12.5$

Length of side $= 2a = 2(12.5)$

$= 25$ inches

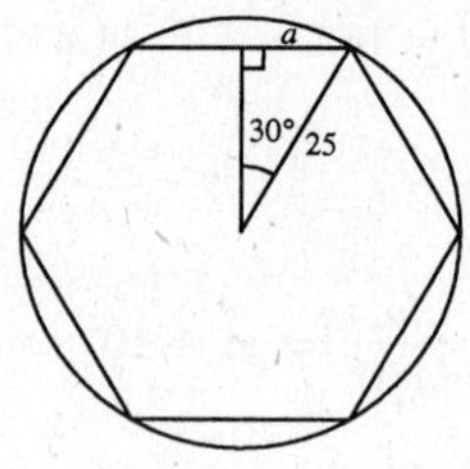

57. $\cos 30° = \dfrac{b}{r}$

$b = r \cos 30°$

$b = \dfrac{\sqrt{3}r}{2}$

$y = 2b = 2\left(\dfrac{\sqrt{3}r}{2}\right) = \sqrt{3}r$

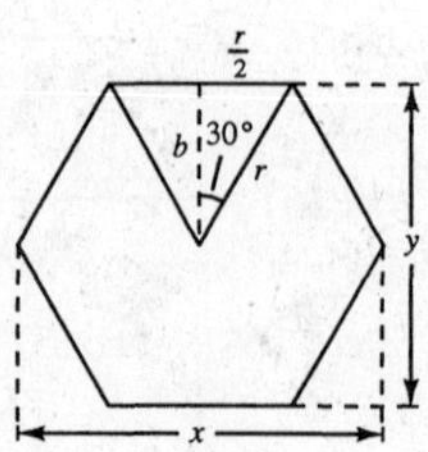

58. $c = \dfrac{35}{2} = 17.5$

$\sin 15° = \dfrac{a}{c}$

$a = c \sin 15° = 17.5 \sin 15° \approx 4.53$

Distance $= 2a \approx 9.06$ centimeters

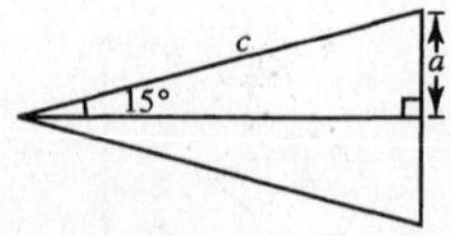

59.

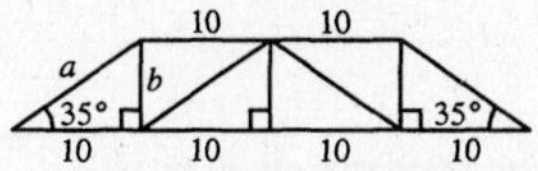

$$\tan 35° = \frac{b}{10}$$

$$b = 10 \tan 35° \approx 7$$

$$\cos 35° = \frac{10}{a}$$

$$a = \frac{10}{\cos 35°} \approx 12.2$$

60.

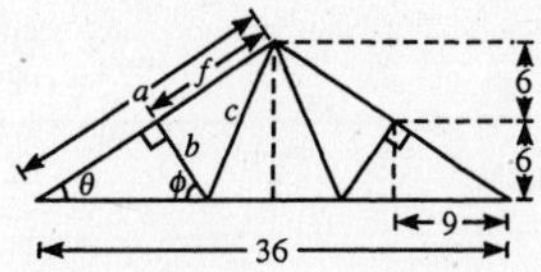

$$\tan \theta = \frac{12}{18}$$

$$\theta = \arctan \frac{2}{3} = 0.588 \text{ rad} \approx 33.7°$$

$$\cos \theta = \frac{18}{a}$$

$$a = \frac{18}{\cos \theta} \approx 21.6 \text{ feet}$$

$$f \approx \frac{21.6}{2} = 10.8$$

$$\phi \approx 90° - 33.7° = 56.3°$$

$$\sin \phi = \frac{6}{b}$$

$$b = \frac{6}{\sin \phi} \approx 7.2 \text{ feet}$$

$$c = \sqrt{10.8^2 + 7.2^2} \approx 13 \text{ feet}$$

61. $d = 4 \cos 8\pi t$

(a) Maximum displacement = amplitude = 4

(b) Frequency $= \frac{\omega}{2\pi} = \frac{8\pi}{2\pi}$

$= 4$ cycles per unit of time

(c) $8\pi t = \frac{\pi}{2} \Rightarrow t = \frac{1}{16}$

62. $d = \frac{1}{2} \cos 20\pi t$

(a) Maximum displacement: $|a| = \left|\frac{1}{2}\right| = \frac{1}{2}$

(b) Frequency

$$\frac{\omega}{2\pi} = \frac{20\pi}{2\pi} = 10$$

(c) Least positive value for t for which $d = 0$

$$\frac{1}{2} \cos 20\pi t = 0$$

$$\cos 20\pi t = 0$$

$$20\pi t = \arccos 0$$

$$20\pi t = \frac{\pi}{2}$$

$$t = \frac{\pi}{2} \cdot \frac{1}{20\pi} = \frac{1}{40}$$

63. $d = \frac{1}{16} \sin 120\pi t$

(a) Maximum displacement = amplitude $= \frac{1}{16}$

(b) Frequency $= \frac{\omega}{2\pi} = \frac{120\pi}{2\pi}$

$= 60$ cycles per unit of time

(c) $120\pi t = \pi \Rightarrow t = \frac{1}{120}$

64. $d = \frac{1}{64} \sin 792\,\pi t$

(a) Maximum displacement:

$$|a| = \left|\frac{1}{64}\right| = \frac{1}{64}$$

(b) Frequency

$$\frac{\omega}{2\pi} = \frac{792\pi}{2\pi} = 396$$

(c) Least positive value for t for which $d = 0$

$$\frac{1}{64} \sin 792\,\pi t = 0$$

$$\sin 792\,\pi t = 0$$

$$792\,\pi t = \arcsin 0 + \pi$$

$$792\,\pi t = \pi$$

$$t = \frac{\pi}{792\,\pi} = \frac{1}{792}$$

65. $d = 0$ when $t = 0$, $a = 4$, Period $= 2$

Use $d = a \sin \omega t$ since $d = 0$ when $t = 0$.

$$\frac{2\pi}{\omega} = 2 \Rightarrow \omega = \pi$$

Thus, $d = 4 \sin \pi t$.

66. Displacement at $t = 0$ is $0 \Rightarrow d = a \sin \omega t$

Amplitude: $|a| = 3$

Period: $\frac{2\pi}{\omega} = 6 \Rightarrow \omega = \frac{\pi}{3}$

$$d = 3 \sin\left(\frac{\pi t}{3}\right)$$

67. $d = 3$ when $t = 0$, $a = 3$, Period $= 1.5$

Use $d = a \cos \omega t$ since $d = 3$ when $t = 0$.

$$\frac{2\pi}{\omega} = 1.5 \Rightarrow \omega = \frac{4\pi}{3}$$

Thus, $d = 3 \cos\left(\frac{4\pi}{3}t\right) = 3 \cos\left(\frac{4\pi t}{3}\right)$.

68. Displacement at $t = 0$ is $2 \Rightarrow d = a \cos \omega t$

Amplitude: $|a| = 2$

Period: $\frac{2\pi}{\omega} = 10 \Rightarrow \omega = \frac{\pi}{5}$

$$d = 2 \cos\left(\frac{\pi t}{5}\right)$$

69. $d = a \sin \omega t$

$$\text{Period} = \frac{2\pi}{\omega} = \frac{1}{\text{frequency}}$$

$$\frac{2\pi}{\omega} = \frac{1}{264}$$

$$\omega = 2\pi(264) = 528\pi$$

70. At $t = 0$, buoy is at its high point $\Rightarrow d = a \cos \omega t$.

Distance from high to low $= 2|a| = 3.5$

$$|a| = \frac{7}{4}$$

Returns to high point every 10 seconds:

$$\text{Period} = \frac{2\pi}{\omega} = 10 \Rightarrow \omega = \frac{\pi}{5}$$

$$d = \frac{7}{4} \cos \frac{\pi t}{5}$$

71. $y = \frac{1}{4} \cos 16t,\ t > 0$

(a)

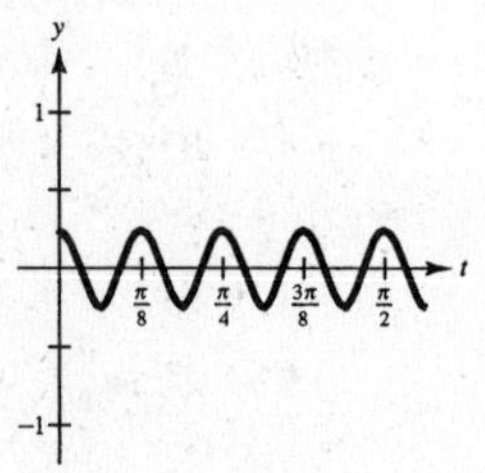

(b) Period: $\frac{2\pi}{16} = \frac{\pi}{8}$

(c) $\frac{1}{4} \cos 16t = 0$ when $16t = \frac{\pi}{2} \Rightarrow t = \frac{\pi}{32}$

72. (a)

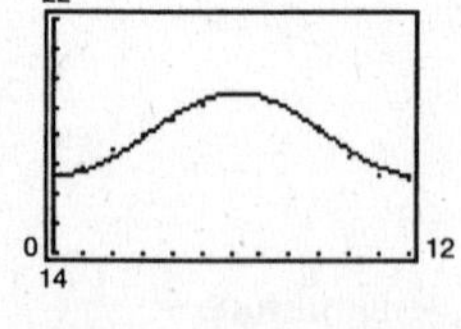

(b) $\text{Period} = \frac{2\pi}{b} = \frac{2\pi}{\pi/6} = 12$

Yes, this is what is expected there are 12 months in one year.

(c) Amplitude: $|a| = |1.41| = 1.41$, which represents the maximum change in time of sunset from the average time ($d = 18.09$) of sunset.

73. False. Since the tower is not exactly vertical, a right triangle with sides 191 ft and d is not formed.

74. False. One period is the time for one complete cycle of the motion.

75. False. $\tan \beta = \frac{h}{d} \Rightarrow h = d \tan \beta$

76. False. $\cos \alpha = \frac{d}{g} \Rightarrow g = \frac{d}{\cos \alpha}$

Review Exercises for Chapter 9

1. $\theta \approx 1$ radian

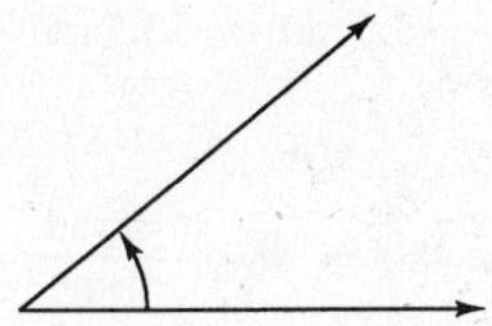

2. 2 radians

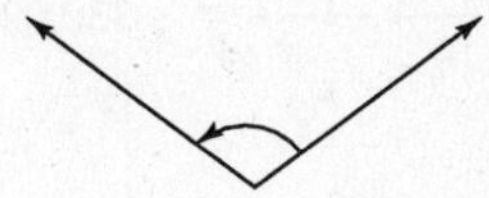

3. $\theta \approx 5$ radians

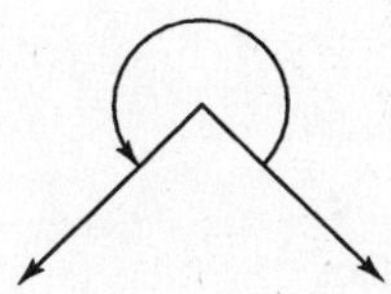

4. -3.5 radians

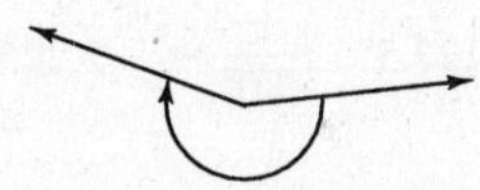

5. $\theta = \frac{11\pi}{4}$

Coterminal angles:

$\frac{11\pi}{4} - 2\pi = \frac{3\pi}{4}$

$\frac{3\pi}{4} - 2\pi = -\frac{5\pi}{4}$

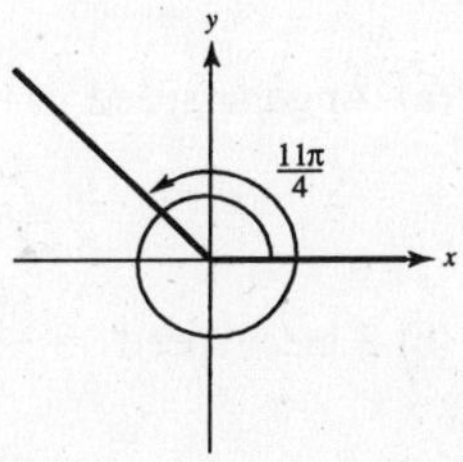

6. $\frac{2\pi}{9} + 2\pi = \frac{20\pi}{9}$

$\frac{2\pi}{9} - 2\pi = -\frac{16\pi}{9}$

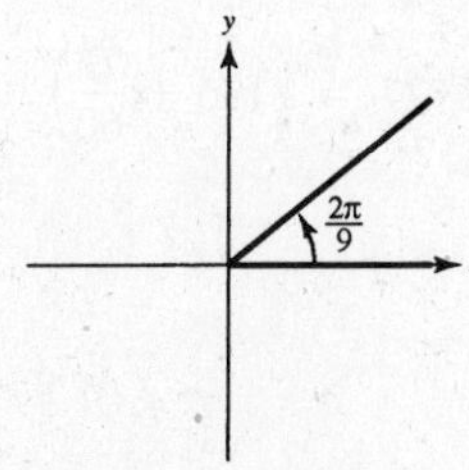

7. $\theta = -\frac{4\pi}{3}$

Coterminal angles:

$-\frac{4\pi}{3} + 2\pi = \frac{2\pi}{3}$

$-\frac{4\pi}{3} - 2\pi = -\frac{10\pi}{3}$

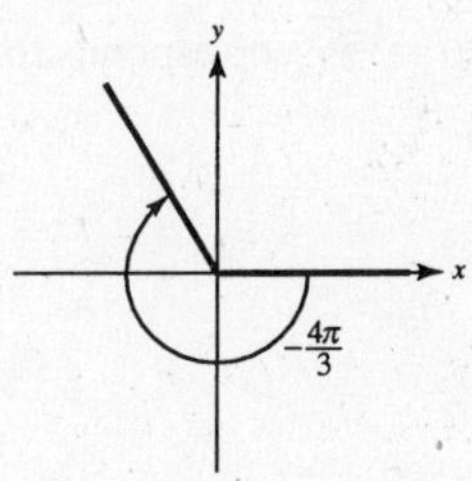

8. $-\frac{23\pi}{3} + 8\pi = \frac{\pi}{3}$

$-\frac{23\pi}{3} + 2\pi = -\frac{17\pi}{3}$

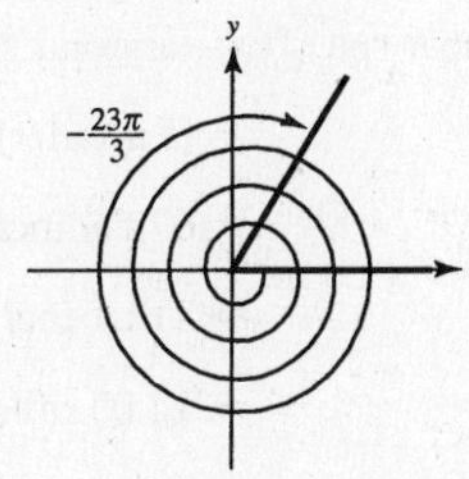

9. $\theta = 70°$

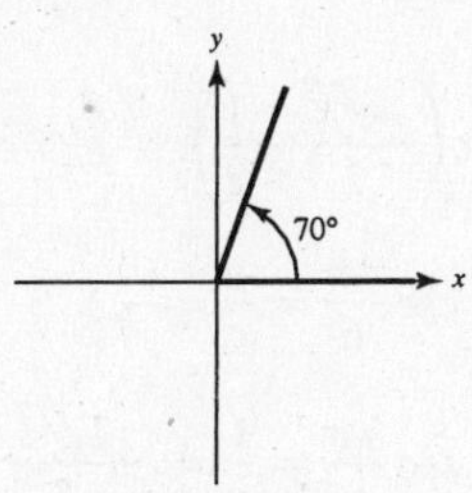

Coterminal angles:

$70° + 360° = 430°$

$70° - 360° = -290°$

10.

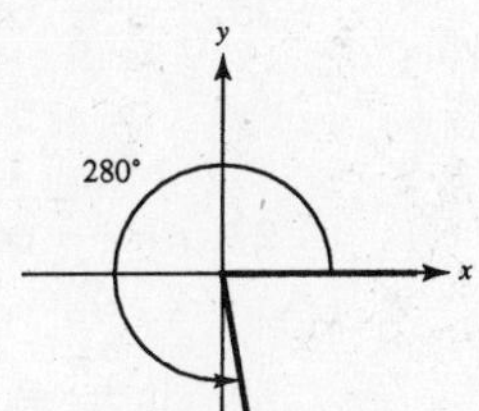

$280° + 360° = 640°$

$280° - 360° = -80°$

11. $\theta = -110°$

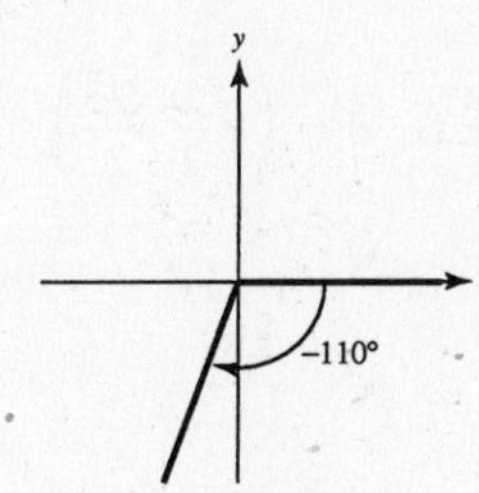

Coterminal angles:

$-110° + 360° = 250°$

$-110° - 360° = -470°$

12. $-405° + 720° = 315°$

$-405° + 360° = -45°$

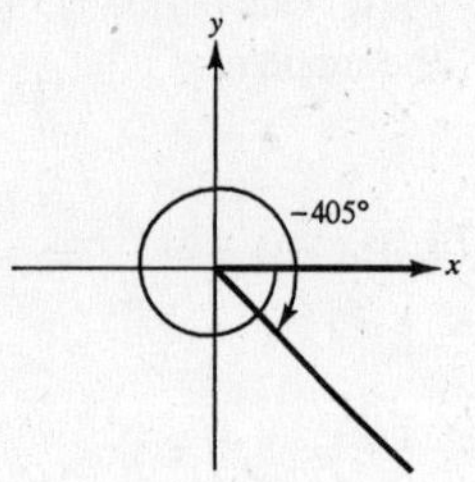

13. $\dfrac{5\pi \text{ rad}}{7} = \dfrac{5\pi \text{ rad}}{7} \cdot \dfrac{180°}{\pi \text{ rad}} \approx 128.57°$

14. $-\dfrac{11\pi}{6} \cdot \dfrac{180°}{\pi} = -330.00°$

15. $-3.5 \text{ rad} = -3.5 \text{ rad} \cdot \dfrac{180°}{\pi \text{ rad}} \approx -200.54°$

16. $5.7 \cdot \dfrac{180°}{\pi} \approx 326.59°$

17. $480° = 480° \cdot \dfrac{\pi \text{ rad}}{180°} = \dfrac{8\pi}{3} \text{ radians} \approx 8.3776 \text{ radians}$

18. $-127.5° \cdot \dfrac{\pi}{180°} \approx -2.2253$

19. $-33°45' = -33.75° = -33.75° \cdot \dfrac{\pi \text{ rad}}{180°} = -\dfrac{3\pi}{16} \text{ radian} \approx -0.5890 \text{ radian}$

20. $196°\,77' = \left(196 + \dfrac{77}{60}\right)^{\circ} \cdot \dfrac{\pi}{180°} \approx 3.4432$

21. (a) Angular speed $= \dfrac{(33\frac{1}{3})(2\pi) \text{ radians}}{1 \text{ minute}}$

$= 66\frac{2}{3}\pi \text{ radians per minute}$

(b) Linear speed $= \dfrac{6(66\frac{2}{3}\pi) \text{ inches}}{1 \text{ minute}}$

$= 400\pi \text{ inches per minute}$

22. (linear speed) = (angular speed) · (radius)

$= (5\pi \text{ rad}/s) \cdot (13.5 \text{ inches})$

$= 67.5\pi$ inches per second

≈ 212.1 inches per second

≈ 12.05 miles per hour

23. $t = \dfrac{2\pi}{3}$ corresponds to the point $\left(-\dfrac{1}{2}, \dfrac{\sqrt{3}}{2}\right)$.

24. $t = \dfrac{3\pi}{4}, (x, y) = \left(-\dfrac{\sqrt{2}}{2}, \dfrac{\sqrt{2}}{2}\right)$

25. $t = \dfrac{5\pi}{6}$ corresponds to the point $\left(-\dfrac{\sqrt{3}}{2}, \dfrac{1}{2}\right)$.

26. $t = -\dfrac{4\pi}{3}, (x, y) = \left(-\dfrac{1}{2}, \dfrac{\sqrt{3}}{2}\right)$

27. $t = \dfrac{7\pi}{6}$ corresponds to the point $\left(-\dfrac{\sqrt{3}}{2}, -\dfrac{1}{2}\right)$.

$\sin\dfrac{7\pi}{6} = y = -\dfrac{1}{2}$ $\qquad \csc\dfrac{7\pi}{6} = \dfrac{1}{y} = -2$

$\cos\dfrac{7\pi}{6} = x = -\dfrac{\sqrt{3}}{2}$ $\qquad \sec\dfrac{7\pi}{6} = \dfrac{1}{x} = -\dfrac{2\sqrt{3}}{3}$

$\tan\dfrac{7\pi}{6} = \dfrac{y}{x} = \dfrac{1}{\sqrt{3}} = \dfrac{\sqrt{3}}{3}$ $\qquad \cot\dfrac{7\pi}{6} = \dfrac{x}{y} = \sqrt{3}$

28. $t = \frac{5\pi}{4}$

$\sin\frac{5\pi}{4} = \frac{-\sqrt{2}}{2}$ $\quad \csc\frac{5\pi}{4} = -\sqrt{2}$

$\cos\frac{5\pi}{4} = \frac{-\sqrt{2}}{2}$ $\quad \sec\frac{5\pi}{4} = -\sqrt{2}$

$\tan\frac{5\pi}{4} = 1$ $\quad \cot\frac{5\pi}{4} = 1$

29. $t = -\frac{2\pi}{3}$ corresponds to the point $\left(-\frac{1}{2}, -\frac{\sqrt{3}}{2}\right)$.

$\sin\left(-\frac{2\pi}{3}\right) = y = -\frac{\sqrt{3}}{2}$ $\quad \csc\left(-\frac{2\pi}{3}\right) = \frac{1}{y} = -\frac{2\sqrt{3}}{3}$

$\cos\left(-\frac{2\pi}{3}\right) = x = -\frac{1}{2}$ $\quad \sec\left(-\frac{2\pi}{3}\right) = \frac{1}{x} = -2$

$\tan\left(-\frac{2\pi}{3}\right) = \frac{y}{x} = \sqrt{3}$ $\quad \cot\left(-\frac{2\pi}{3}\right) = \frac{x}{y} = \frac{\sqrt{3}}{3}$

30. $t = \frac{-11\pi}{6}$

$\sin\frac{-11\pi}{6} = \frac{1}{2}$ $\quad \csc\frac{-11\pi}{6} = 2$

$\cos\frac{-11\pi}{6} = \frac{\sqrt{3}}{2}$ $\quad \sec\frac{-11\pi}{6} = \frac{2\sqrt{3}}{3}$

$\tan\frac{-11\pi}{6} = \frac{\sqrt{3}}{3}$ $\quad \cot\frac{-11\pi}{6} = \sqrt{3}$

31. $\sin\frac{11\pi}{4} = \sin\frac{3\pi}{4} = \frac{\sqrt{2}}{2}$

32. $\cos 4\pi = \cos 0 = 1$

33. $\sin\left(-\frac{17\pi}{6}\right) = \sin\left(-\frac{5\pi}{6}\right) = -\frac{1}{2}$

34. $\cos\left(-\frac{13\pi}{3}\right) = \cos\left(\frac{5\pi}{3}\right) = \frac{1}{2}$

35. $\cos 84° \approx 0.10$

36. $\sin 6° \approx 0.10$

37. $\sec\frac{12\pi}{5} = \frac{1}{\cos\left(\frac{12\pi}{5}\right)} \approx 3.24$

38. $\sin\left(-\frac{\pi}{9}\right) \approx -0.34$

39. opp = 4, adj = 5, hyp = $\sqrt{4^2 + 5^2} = \sqrt{41}$

$\sin\theta = \frac{\text{opp}}{\text{hyp}} = \frac{4}{\sqrt{41}} = \frac{4\sqrt{41}}{41}$ $\quad \csc\theta = \frac{\text{hyp}}{\text{opp}} = \frac{\sqrt{41}}{4}$

$\cos\theta = \frac{\text{adj}}{\text{hyp}} = \frac{5}{\sqrt{41}} = \frac{5\sqrt{41}}{41}$ $\quad \sec\theta = \frac{\text{hyp}}{\text{adj}} = \frac{\sqrt{41}}{5}$

$\tan\theta = \frac{\text{opp}}{\text{adj}} = \frac{4}{5}$ $\quad \cot\theta = \frac{\text{adj}}{\text{opp}} = \frac{5}{4}$

40. adj = 6, opp = 6

hyp = $\sqrt{6^2 + 6^2} = 6\sqrt{2}$

$\sin\theta = \frac{6}{6\sqrt{2}} = \frac{\sqrt{2}}{2}$ $\quad \csc\theta = \frac{6\sqrt{2}}{6} = \sqrt{2}$

$\cos\theta = \frac{6}{6\sqrt{2}} = \frac{\sqrt{2}}{2}$ $\quad \sec\theta = \frac{6\sqrt{2}}{6} = \sqrt{2}$

$\tan\theta = \frac{6}{6} = 1$ $\quad \cot\theta = \frac{6}{6} = 1$

41. opp = 4, hyp = 8, adj = $\sqrt{8^2 - 4^2} = \sqrt{48} = 4\sqrt{3}$

$\sin\theta = \frac{\text{opp}}{\text{hyp}} = \frac{4}{8} = \frac{1}{2}$ $\quad \csc\theta = \frac{\text{hyp}}{\text{opp}} = \frac{8}{4} = 2$

$\cos\theta = \frac{\text{adj}}{\text{hyp}} = \frac{4\sqrt{3}}{8} = \frac{\sqrt{3}}{2}$ $\quad \sec\theta = \frac{\text{hyp}}{\text{adj}} = \frac{8}{4\sqrt{3}} = \frac{2\sqrt{3}}{3}$

$\tan\theta = \frac{\text{opp}}{\text{adj}} = \frac{4}{4\sqrt{3}} = \frac{\sqrt{3}}{3}$ $\quad \cot\theta = \frac{\text{adj}}{\text{opp}} = \frac{4\sqrt{3}}{4} = \sqrt{3}$

42. opp = 5, hyp = 9

adj = $\sqrt{9^2 - 5^2} = 2\sqrt{14}$

$\sin\theta = \frac{5}{9}$ $\quad \csc\theta = \frac{9}{5}$

$\cos\theta = \frac{2\sqrt{14}}{9}$ $\quad \sec\theta = \frac{9}{2\sqrt{14}} = \frac{9\sqrt{14}}{28}$

$\tan\theta = \frac{5}{2\sqrt{14}} = \frac{5\sqrt{14}}{28}$ $\quad \cot = \frac{2\sqrt{14}}{5}$

43. $\sin\theta = \dfrac{1}{3}$

(a) $\csc\theta = \dfrac{1}{\sin\theta} = 3$

(b) $\sin^2\theta + \cos^2\theta = 1$

$$\left(\frac{1}{3}\right)^2 + \cos^2\theta = 1$$

$$\cos^2\theta = 1 - \frac{1}{9}$$

$$\cos^2\theta = \frac{8}{9}$$

$$\cos\theta = \sqrt{\frac{8}{9}}$$

$$\cos\theta = \frac{2\sqrt{2}}{3}$$

(c) $\sec\theta = \dfrac{1}{\cos\theta} = \dfrac{3}{2\sqrt{2}} = \dfrac{3\sqrt{2}}{4}$

(d) $\tan\theta = \dfrac{\sin\theta}{\cos\theta} = \dfrac{\frac{1}{3}}{\frac{2\sqrt{2}}{3}} = \dfrac{1}{2\sqrt{2}} = \dfrac{\sqrt{2}}{4}$

44. $\tan\theta = 4$

(a) $\cot\theta = \dfrac{1}{\tan\theta} = \dfrac{1}{4}$

(b) $\sec\theta = \sqrt{1 + \tan^2\theta} = \sqrt{1 + 16} = \sqrt{17}$

(c) $\cos\theta = \dfrac{1}{\sec\theta} = \dfrac{1}{\sqrt{17}} = \dfrac{\sqrt{17}}{17}$

(d) $\csc\theta = \sqrt{1 + \cot^2\theta} = \sqrt{1 + \dfrac{1}{16}} = \dfrac{\sqrt{17}}{4}$

45. $\csc\theta = \dfrac{6}{5}$

(a) $\sin\theta = \dfrac{1}{\csc\theta} = \dfrac{5}{6}$

(b) $\sin^2\theta + \cos^2\theta = 1$

$$\left(\frac{5}{6}\right)^2 + \cos^2\theta = 1$$

$$\cos^2\theta = 1 - \frac{25}{36}$$

$$\cos^2\theta = \frac{11}{36}$$

$$\cos\theta = \sqrt{\frac{11}{36}}$$

$$\cos\theta = \frac{\sqrt{11}}{6}$$

(c) $\sec\theta = \dfrac{1}{\cos\theta} = \dfrac{6}{\sqrt{11}} = \dfrac{6\sqrt{11}}{11}$

(d) $\tan\theta = \dfrac{\sin\theta}{\cos\theta} = \dfrac{\frac{5}{6}}{\frac{\sqrt{11}}{6}} = \dfrac{5}{\sqrt{11}} = \dfrac{5\sqrt{11}}{11}$

46. $\csc\theta = \dfrac{5}{3}$

(a) $\sin\theta = \dfrac{1}{\csc\theta} = \dfrac{3}{5}$

(b) $\cot^2\theta + 1 = \csc^2\theta$

$$\cot^2\theta = \left(\frac{5}{3}\right)^2 - 1 = \frac{25}{9} - 1 = \frac{16}{9}$$

$$\cot\theta = \sqrt{\frac{16}{9}} = \frac{4}{3}$$

(c) $\tan\theta = \dfrac{1}{\cot\theta} = \dfrac{3}{4}$

(d) $\sec(90° - \theta) = \csc\theta = \dfrac{5}{3}$

47. $\tan 33° \approx 0.65$

48. $\csc 11° \approx 5.24$

49. $\sin 34.2° \approx 0.56$

50. $\sec 79.3° \approx 5.39$

51. $\cot 15°14' \approx \cot 15.2333° = \dfrac{1}{\tan 15.2333°} \approx 3.67$

52. $\cos 78°\,11'58'' = \cos\left(78 + \dfrac{11}{60} + \dfrac{58}{3600}\right)^\circ \approx 0.20$

53. $\sin 1°10' = \dfrac{x}{3.5}$

$x = 3.5 \sin 1°10' \approx 0.07$ Kilometer

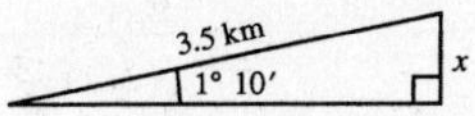

Not drawn to scale

54.

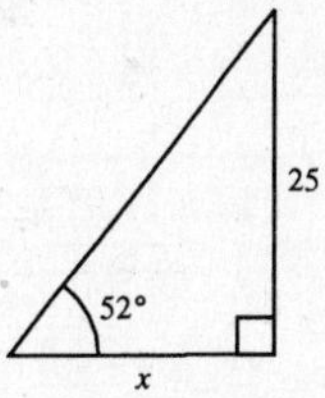

$\tan 52° = \dfrac{25}{x}$

$x = \dfrac{25}{\tan 52°} \approx 19.5$ feet

55. $x = 12, y = 16, r = \sqrt{144 + 256} = \sqrt{400} = 20$

$\sin\theta = \dfrac{y}{r} = \dfrac{4}{5}$ $\quad \csc\theta = \dfrac{r}{y} = \dfrac{5}{4}$

$\cos\theta = \dfrac{x}{r} = \dfrac{3}{5}$ $\quad \sec\theta = \dfrac{r}{x} = \dfrac{5}{3}$

$\tan\theta = \dfrac{y}{x} = \dfrac{4}{3}$ $\quad \cot\theta = \dfrac{x}{y} = \dfrac{3}{4}$

56. $(x, y) = (3, -4)$

$r = \sqrt{3^2 + (-4)^2} = 5$

$\sin\theta = \dfrac{y}{r} = -\dfrac{4}{5}$ $\quad \csc\theta = \dfrac{r}{y} = -\dfrac{5}{4}$

$\cos\theta = \dfrac{x}{r} = \dfrac{3}{5}$ $\quad \sec\theta = \dfrac{r}{x} = \dfrac{5}{3}$

$\tan\theta = \dfrac{y}{x} = -\dfrac{4}{3}$ $\quad \cot\theta = \dfrac{x}{y} = -\dfrac{3}{4}$

57. $x = \dfrac{2}{3}, y = \dfrac{5}{2}$

$r = \sqrt{\left(\dfrac{2}{3}\right)^2 + \left(\dfrac{5}{2}\right)^2} = \dfrac{\sqrt{241}}{6}$

$\sin\theta = \dfrac{y}{r} = \dfrac{\frac{5}{2}}{\frac{\sqrt{241}}{6}} = \dfrac{15}{\sqrt{241}} = \dfrac{15\sqrt{241}}{241}$ $\quad \csc\theta = \dfrac{r}{y} = \dfrac{\frac{\sqrt{241}}{6}}{\frac{5}{2}} = \dfrac{2\sqrt{241}}{30} = \dfrac{\sqrt{241}}{15}$

$\cos\theta = \dfrac{x}{r} = \dfrac{\frac{2}{3}}{\frac{\sqrt{241}}{6}} = \dfrac{4}{\sqrt{241}} = \dfrac{4\sqrt{241}}{241}$ $\quad \sec\theta = \dfrac{r}{x} = \dfrac{\frac{\sqrt{241}}{6}}{\frac{2}{3}} = \dfrac{\sqrt{241}}{4}$

$\tan\theta = \dfrac{y}{x} = \dfrac{\frac{5}{2}}{\frac{2}{3}} = \dfrac{15}{4}$ $\quad \cot\theta = \dfrac{x}{y} = \dfrac{\frac{2}{3}}{\frac{5}{2}} = \dfrac{4}{15}$

58. $(x, y) = \left(-\dfrac{10}{3}, -\dfrac{2}{3}\right)$

$r = \sqrt{\left(-\dfrac{10}{3}\right)^2 + \left(-\dfrac{2}{3}\right)^2} = \dfrac{2\sqrt{26}}{3}$

$\sin\theta = \dfrac{y}{r} = \dfrac{-\frac{2}{3}}{\frac{2\sqrt{26}}{3}} = -\dfrac{\sqrt{26}}{26}$ $\quad \csc\theta = \dfrac{r}{y} = \dfrac{\frac{2\sqrt{26}}{3}}{-\frac{2}{3}} = -\sqrt{26}$

$\cos\theta = \dfrac{x}{r} = \dfrac{-\frac{10}{3}}{\frac{2\sqrt{26}}{3}} = -\dfrac{5\sqrt{26}}{26}$ $\quad \sec\theta = \dfrac{r}{x} = \dfrac{\frac{2\sqrt{26}}{3}}{-\frac{10}{3}} = -\dfrac{\sqrt{26}}{5}$

$\tan\theta = \dfrac{y}{x} = \dfrac{-\frac{2}{3}}{-\frac{10}{3}} = \dfrac{1}{5}$ $\quad \cot\theta = \dfrac{x}{y} = \dfrac{-\frac{10}{3}}{-\frac{2}{3}} = 5$

59. $x = -0.5, y = 4.5$

$r = \sqrt{(-0.5)^2 + 4.5^2} = \sqrt{20.5}$

$\sin\theta = \frac{y}{r} = \frac{4.5}{\sqrt{20.5}} \approx 1$ $\qquad \csc\theta = \frac{r}{y} = \frac{\sqrt{20.5}}{4.5} \approx 1$

$\cos\theta = \frac{x}{r} = \frac{-0.5}{\sqrt{20.5}} \approx -0.1$ $\qquad \sec\theta = \frac{r}{x} = \frac{\sqrt{20.5}}{-0.5} \approx -9$

$\tan\theta = \frac{y}{x} = -9$ $\qquad \cot\theta = \frac{x}{y} = -\frac{1}{9} \approx -0.1$

60. $x = 0.3, y = -0.4$

$r = \sqrt{0.09 + 0.16} = 0.5$

$\sin\theta = \frac{y}{r} = \frac{-0.4}{0.5} = -\frac{4}{5}$ $\qquad \csc\theta = \frac{r}{y} = \frac{0.5}{-0.4} = -\frac{5}{4}$

$\cos\theta = \frac{x}{r} = \frac{0.3}{0.5} = \frac{3}{5}$ $\qquad \sec\theta = \frac{r}{x} = \frac{0.5}{0.3} = \frac{5}{3}$

$\tan\theta = \frac{y}{x} = \frac{-0.4}{0.3} = -\frac{4}{3}$ $\qquad \cot\theta = \frac{x}{y} = \frac{0.3}{-0.4} = -\frac{3}{4}$

61. $(x, 4x), x > 0$

$x = x, y = 4x$

$r = \sqrt{x^2 + (4x)^2} = \sqrt{17}x$

$\sin\theta = \frac{y}{r} = \frac{4x}{\sqrt{17}x} = \frac{4\sqrt{17}}{17}$

$\cos\theta = \frac{x}{r} = \frac{x}{\sqrt{17}x} = \frac{\sqrt{17}}{17}$

$\tan\theta = \frac{y}{x} = \frac{4x}{x} = 4$

$\csc\theta = \frac{r}{y} = \frac{\sqrt{17}x}{4x} = \frac{\sqrt{17}}{4}$

$\sec\theta = \frac{r}{x} = \frac{\sqrt{17}x}{x} = \sqrt{17}$

$\cot\theta = \frac{x}{y} = \frac{x}{4x} = \frac{1}{4}$

62. $(x', y') = (-2x, -3x), x > 0$

$r = \sqrt{(-2x)^2 + (-3x)^2} = \sqrt{13}x$

$\sin\theta = \frac{y'}{r} = \frac{-3x}{\sqrt{13}x} = -\frac{3\sqrt{13}}{13}$

$\cos\theta = \frac{x'}{r} = \frac{-2x}{\sqrt{13}x} = -\frac{2\sqrt{13}}{13}$

$\tan\theta = \frac{y'}{x'} = \frac{-3x}{-2x} = \frac{3}{2}$

$\csc\theta = \frac{r}{y'} = \frac{\sqrt{13}x}{-3x} = -\frac{\sqrt{13}}{3}$

$\sec\theta = \frac{r}{x'} = \frac{\sqrt{13}x}{-2x} = -\frac{\sqrt{13}}{2}$

$\cot\theta = \frac{x'}{y'} = \frac{-2x}{-3x} = \frac{2}{3}$

63. $\sec\theta = \frac{6}{5}$, $\tan\theta < 0 \implies \theta$ is in Quadrant IV.

$r = 6, x = 5, y = -\sqrt{36 - 25} = -\sqrt{11}$

$\sin\theta = \frac{y}{r} = -\frac{\sqrt{11}}{6}$ $\qquad \csc\theta = \frac{r}{y} = -\frac{6\sqrt{11}}{11}$

$\cos\theta = \frac{x}{r} = \frac{5}{6}$ $\qquad \sec\theta = \frac{r}{x} = \frac{6}{5}$

$\tan\theta = \frac{y}{x} = -\frac{\sqrt{11}}{5}$ $\qquad \cot\theta = \frac{x}{y} = -\frac{5\sqrt{11}}{11}$

64. $\csc\theta = \frac{3}{2}$, $\cos\theta < 0$

θ is in Quadrant II.

$\sin\theta = \frac{1}{\csc\theta} = \frac{2}{3}$

$\cos\theta = -\sqrt{1 - \sin^2\theta} = -\frac{\sqrt{5}}{3}$

$\tan\theta = \frac{\sin\theta}{\cos\theta} = -\frac{2\sqrt{5}}{5}$

$\sec\theta = \frac{1}{\cos\theta} = -\frac{3\sqrt{5}}{5}$

$\cot\theta = \frac{1}{\tan\theta} = -\frac{\sqrt{5}}{2}$

65. $\sin\theta = \frac{3}{8}$, $\cos\theta < 0 \implies \theta$ is in Quadrant II.

$y = 3, r = 8, x = -\sqrt{55}$

$\sin\theta = \frac{y}{r} = \frac{3}{8}$ $\qquad \csc\theta = \frac{8}{3}$

$\cos\theta = \frac{x}{r} = -\frac{\sqrt{55}}{8}$ $\qquad \sec\theta = -\frac{8}{\sqrt{55}} = -\frac{8\sqrt{55}}{55}$

$\tan\theta = \frac{y}{x} = -\frac{3}{\sqrt{55}} = -\frac{3\sqrt{55}}{55}$ $\qquad \cot\theta = -\frac{\sqrt{55}}{3}$

66. $\tan\theta = \frac{5}{4}, \cos\theta < 0$

θ is in Quadrant III.

$$\sec\theta = -\sqrt{1+\tan^2\theta} = -\sqrt{1+\left(\frac{25}{16}\right)} = -\frac{\sqrt{41}}{4}$$

$$\cos\theta = \frac{1}{\sec\theta} = -\frac{4\sqrt{41}}{41}$$

$$\sin\theta = -\sqrt{1-\cos^2\theta} = -\sqrt{1-\frac{16}{41}} = -\frac{5\sqrt{41}}{41}$$

$$\csc\theta = \frac{1}{\sin\theta} = -\frac{\sqrt{41}}{5}$$

$$\cot\theta = \frac{1}{\tan\theta} = \frac{4}{5}$$

67. $\cos\theta = \frac{x}{r} = \frac{-2}{5} \Rightarrow y^2 = 21$

$\sin\theta > 0 \Rightarrow \theta$ is in Quadrant II $\Rightarrow y = \sqrt{21}$.

$$\sin\theta = \frac{y}{r} = \frac{\sqrt{21}}{5}$$

$$\tan\theta = \frac{y}{x} = -\frac{\sqrt{21}}{2}$$

$$\csc\theta = \frac{r}{y} = \frac{5}{\sqrt{21}} = \frac{5\sqrt{21}}{21}$$

$$\sec\theta = \frac{r}{x} = \frac{5}{-2} = -\frac{5}{2}$$

$$\cot\theta = \frac{x}{y} = \frac{-2}{\sqrt{21}} = -\frac{2\sqrt{21}}{21}$$

68. $\sin\theta = -\frac{2}{4} = -\frac{1}{2}, \cos\theta > 0$

θ is in Quadrant IV.

$$\csc\theta = \frac{1}{\sin\theta} = -2$$

$$\cos\theta = \sqrt{1-\sin^2\theta} = \sqrt{1-\left(\frac{1}{4}\right)} = \frac{\sqrt{3}}{2} \qquad \sec\theta = \frac{1}{\cos\theta} = \frac{2\sqrt{3}}{3}$$

$$\tan\theta = \frac{\sin\theta}{\cos\theta} = -\frac{\sqrt{3}}{3} \qquad \cot\theta = \frac{1}{\tan\theta} = -\sqrt{3}$$

69. $\tan\frac{\pi}{3} = \sqrt{3}$

70. $\sec\frac{\pi}{4} = \frac{1}{\left(\cos\frac{\pi}{4}\right)} = \frac{1}{\left(\frac{1}{\sqrt{2}}\right)} = \sqrt{2}$

71. $\cos\left(-\frac{7\pi}{3}\right) = \cos\frac{\pi}{3} = \frac{1}{2}$

72. $\cot\left(-\frac{5\pi}{4}\right) = \cot\left(\frac{3\pi}{4}\right) = \frac{\cos\left(\frac{3\pi}{4}\right)}{\sin\left(\frac{3\pi}{4}\right)} = -1$

73. $\cos 495° = -\cos 45° = -\frac{\sqrt{2}}{2}$

74. $\sin(-150°) = \sin(210°) = -\frac{1}{2}$

75. $\sin 4 \approx -0.76$

76. $\tan 3 \approx -0.14$

77. $\sin(-3.2) \approx 0.06$

78. $\cot(-4.8) \approx 0.09$

79. $\sin 3\pi = 0$

80. $\cot(1.5\pi) = 0$

81. $\sec\frac{12\pi}{5} = \frac{1}{\cos\left(\frac{12\pi}{5}\right)} \approx 3.24$

82. $\tan\left(\frac{-25\pi}{7}\right) \approx 4.38$

83. $y = \sin x$

Amplitude: 1

Period: 2π

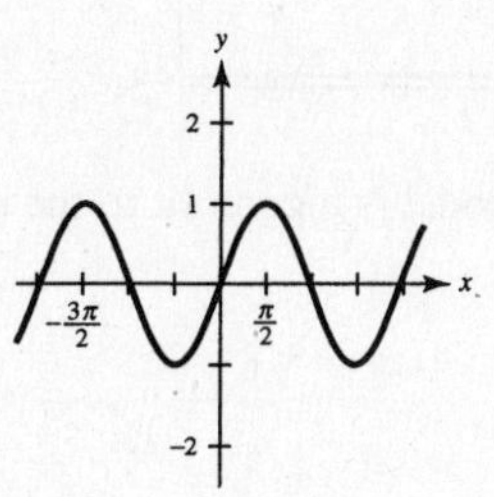

84. $y = \cos x$

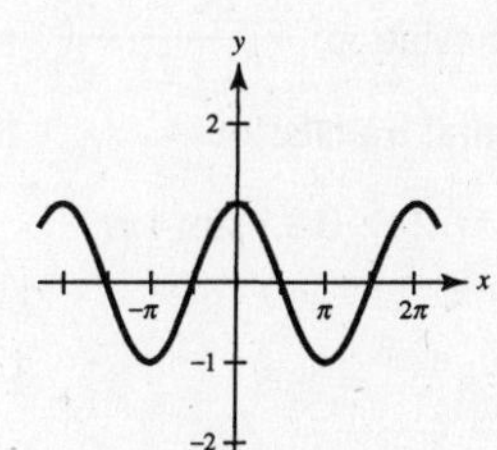

85. $f(x) = 5 \sin \frac{2x}{5}$

Amplitude: 5

Period: $\frac{2\pi}{\frac{2}{5}} = 5\pi$

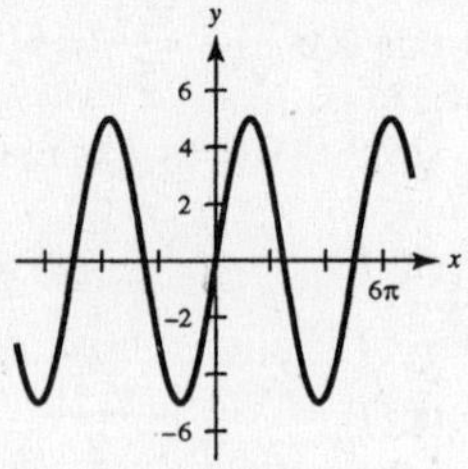

86. $f(x) = 8 \cos\left(-\frac{x}{4}\right)$

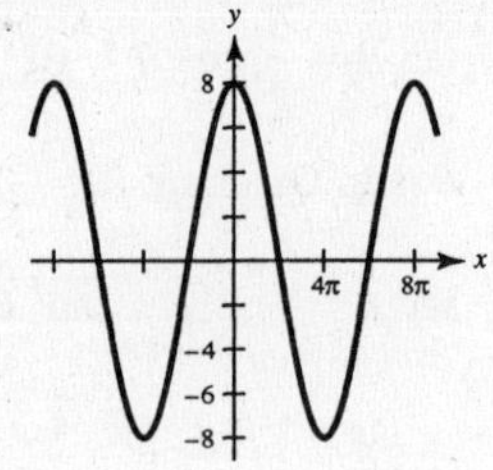

87. $y = 2 + \sin x$

Shift the graph of $y = \sin x$ two units upward.

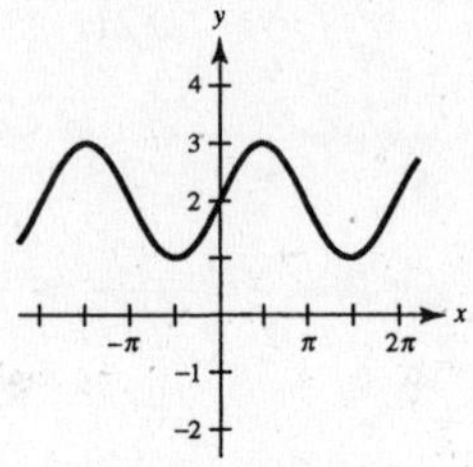

88. $y = -4 - \cos \pi x$

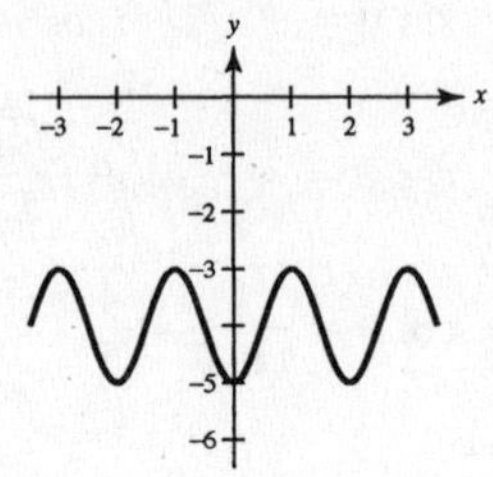

89. $g(t) = \frac{5}{2} \sin(t - \pi)$

Amplitude: $\frac{5}{2}$

Period: 2π

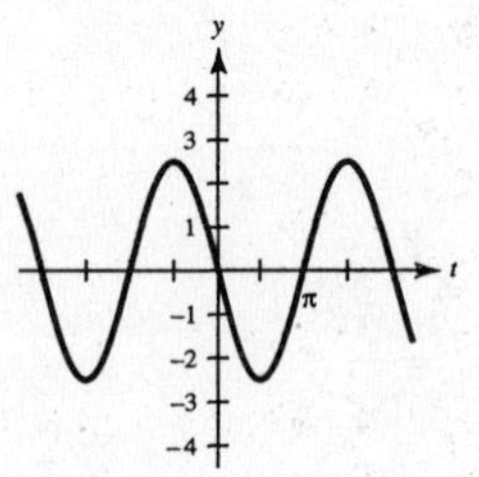

90. $g(t) = 3 \cos(t + \pi)$

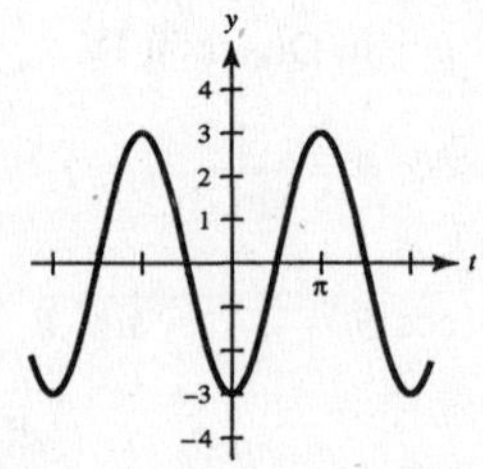

91. $y = 3 \sin x$

Amplitude: 3

Period: 2π

Matches graph (d).

92. $y = -3 \sin x$ matches graph (a).

Period: 2π

Amplitude: 3

93. $y = 2 \sin \pi x$

Amplitude: 2

Period: 2

Matches graph (b).

94. $y = 2 \sin \frac{x}{2}$ matches graph (c).

Period: 4π

Amplitude: 2

95. $y = a \sin bx$

(a) $a = 2, \frac{2\pi}{b} = \frac{1}{264} \Rightarrow b = 528\pi$

$y = 2 \sin(528\pi x)$

(b) $f = \frac{1}{\frac{1}{264}} = 264$ cycles per second.

96. (a) Amplitude: $\frac{1}{2}(2.35 - 1.65) = 0.35$

Period: $2(0.375 - 0.125) = 0.50$

(b) Vertical translation: $\frac{2.35 + 1.65}{2} = 2$

No horizontal translation.

Model: $y = 2 + 0.35 \sin(4\pi t)$

(c)

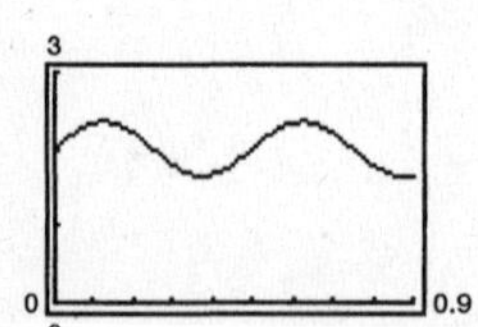

The model is a good fit to the data.

97. $f(x) = \tan x$

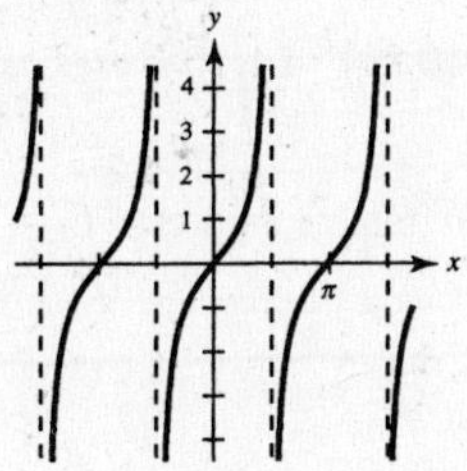

98. $f(t) = \tan\left(t - \dfrac{\pi}{4}\right)$

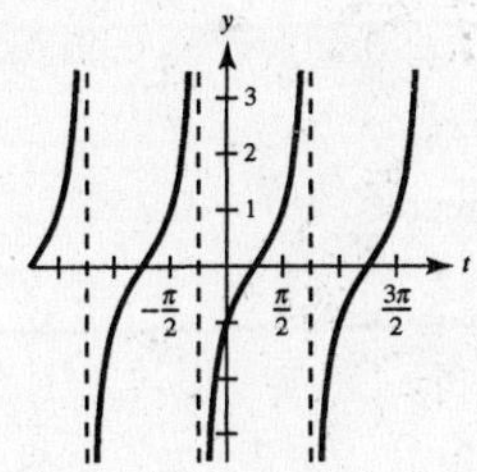

99. $f(x) = \cot x$

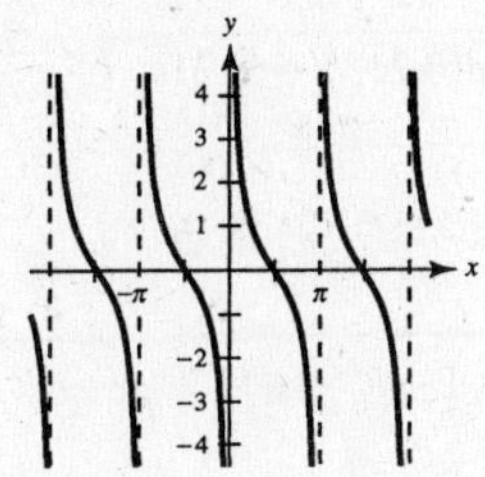

100. $g(t) = 2 \cot 2t$

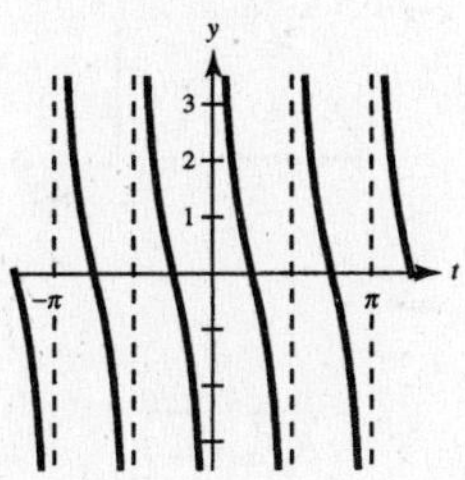

101. $f(x) = \sec x$

Graph $y = \cos x$ first.

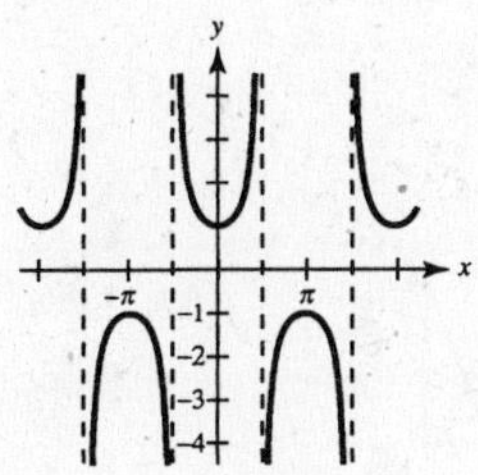

102. $h(t) = \sec\left(t - \dfrac{\pi}{4}\right)$

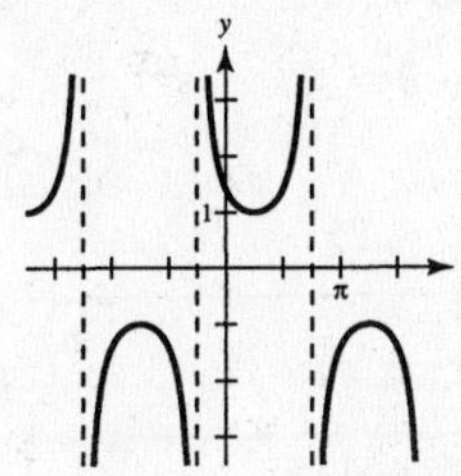

103. $f(x) = \csc x$

Graph $y = \sin x$ first.

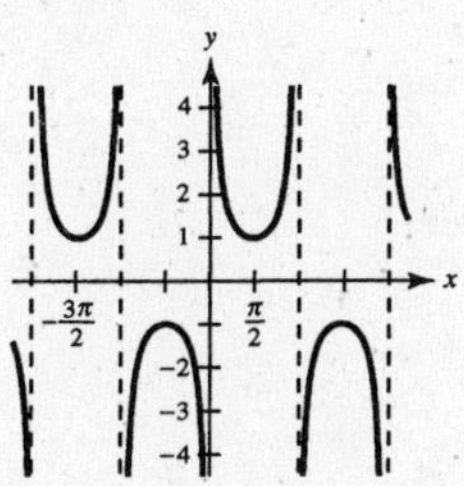

104. $f(t) = 3 \csc\left(2t + \dfrac{\pi}{4}\right)$

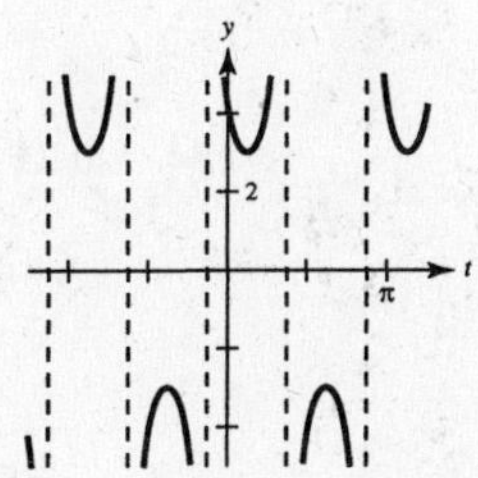

105. $f(x) = x \cos x$

Graph $y = x$ and $y = -x$ first.

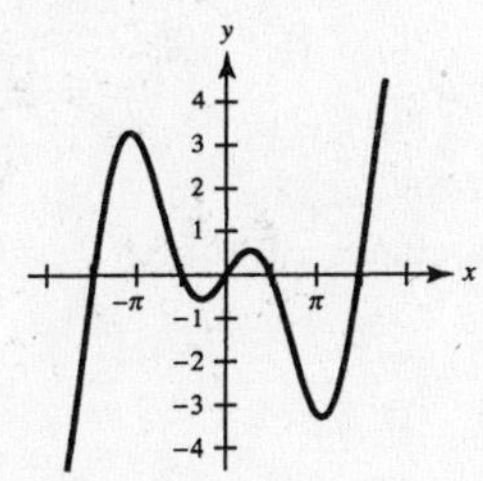

106. $g(x) = e^x \cos x$

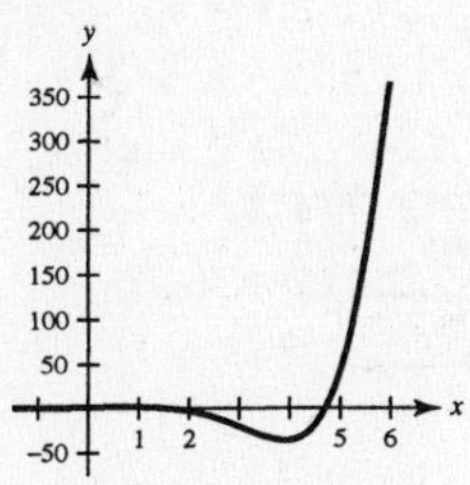

107. $\arcsin\left(-\dfrac{1}{2}\right) = -\arcsin\dfrac{1}{2} = -\dfrac{\pi}{6}$

108. $\arcsin(-1) = -\dfrac{\pi}{2}$

109. $\arccos\dfrac{\sqrt{3}}{2} = \dfrac{\pi}{6}$

110. $\arccos\left(\dfrac{\sqrt{2}}{2}\right) = \dfrac{\pi}{4}$

111. $\cos^{-1}(-1) = \pi$

112. $\cos^{-1}\left(\dfrac{\sqrt{3}}{2}\right) = \dfrac{\pi}{6}$

113. $\arccos 0.324 \approx 1.24$ radians

114. $\arccos(-0.888) \approx 2.66$

115. $\arctan 0.123 \approx 0.12$ radian

116. $\arctan(2.34) \approx 1.17$

117. $\arctan 5.783 \approx 1.40$ radians

118. $\arctan(99.1) \approx 1.56$

119. $\tan^{-1}(-1.5) \approx -0.98$ radian

120. $\tan^{-1}(8.2) \approx 1.45$

121. $\arcsin 0.4 \approx 0.41$ radian

122. $\arcsin(0.213) \approx 0.21$

123. $\sin^{-1}(-0.44) \approx -0.46$ radian

124. $\sin^{-1}(0.89) \approx 1.10$

125. $\sin(\arcsin 0.72) = 0.72$

126. $\cos(\arccos 0.25) = 0.25$

127. $\arctan(\tan \pi) = \arctan 0 = 0$

128. $\arccos[\cos(-5\pi)] - \arccos(\ 1) = \pi$

129. $\cos\left(\arctan \frac{3}{4}\right) = \frac{4}{5}$. Use a right triangle.

Let $\theta = \arctan \frac{3}{4}$

then $\tan \theta = \frac{3}{4}$

and $\cos \theta = \frac{4}{5}$.

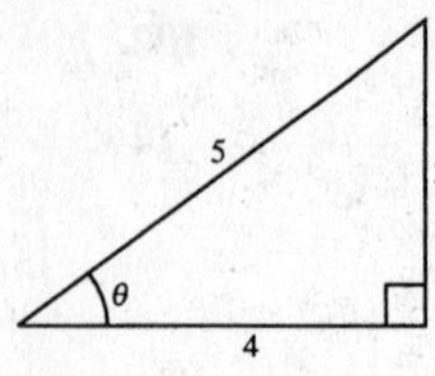

130. Let $u = \arccos \frac{3}{5}$.

$\tan\left(\arccos \frac{3}{5}\right) = \tan u = \frac{4}{3}$

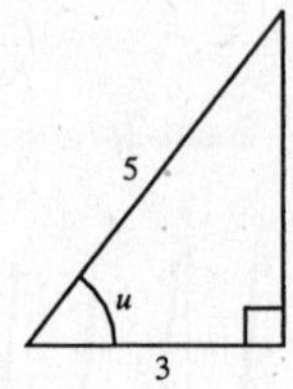

131. $\sec\left(\arctan \frac{12}{5}\right) = \frac{13}{5}$. Use a right triangle.

Let $\theta = \arctan \frac{12}{5}$

then $\tan \theta = \frac{12}{5}$

and $\sec \theta = \frac{13}{5}$.

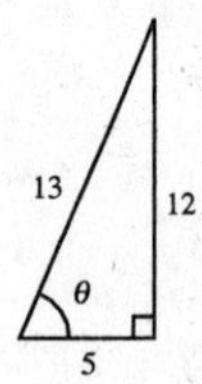

132. Let $u = \arcsin\left(-\frac{12}{13}\right)$.

$\cot\left[\arcsin\left(-\frac{12}{13}\right)\right] = \cot u$

$= -\frac{5}{12}$

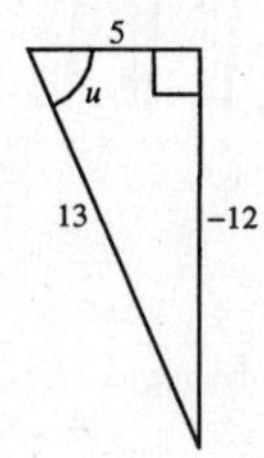

133. $\tan \theta = \dfrac{70}{30}$

$\theta = \arctan\left(\dfrac{70}{30}\right) \approx 66.8°$

134. $\tan 21° = \dfrac{h}{25}$

$h = 25 \tan 21° \approx 9.6$ feet

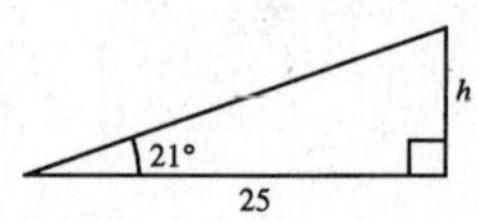

135. $\sin 48° = \dfrac{d_1}{650} \Rightarrow d_1 \approx 483$

$\cos 25° = \dfrac{d_2}{810} \Rightarrow d_2 \approx 734$

$d_1 + d_2 \approx 1217$

$\cos 48° = \dfrac{d_3}{650} \Rightarrow d_3 \approx 435$

$\sin 25° = \dfrac{d_4}{810} \Rightarrow d_4 \approx 342$

$d_3 - d_4 \approx 93$

$\tan \theta \approx \dfrac{93}{1217} \Rightarrow \theta \approx 4.4°$

$\sec 4.4° \approx \dfrac{D}{1217} \Rightarrow D \approx 1217 \sec 4.4° \approx 1221$

The distance is 1221 miles and the bearing is N 85.6° E.

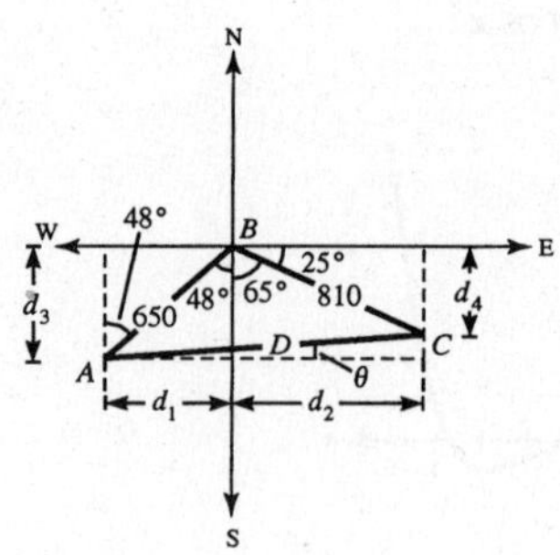

136. Amplitude: $\frac{1.5}{2} = 0.75$ inches

Period: 3 seconds

$d = a \cos bt$

$a = 0.75$

$b = \frac{2\pi}{3}$

$d = 0.75 \cos\left(\frac{2\pi t}{3}\right)$

Problem Solving for Chapter 9

1. The area of a circle is $A = \pi r^2 \Rightarrow \pi = \frac{A}{r^2}$.

The circumference of a circle is $C = 2\pi r$.

$$C = 2\left(\frac{A}{r^2}\right)r$$

$$C = \frac{2A}{r}$$

$$\frac{Cr}{2} = A$$

For a sector, $C = s = r\theta$

Thus, $A = \frac{(r\theta)r}{2} = \frac{1}{2}\theta r^2$ for a sector.

2. $f(x) = x - \cos x$

(a)

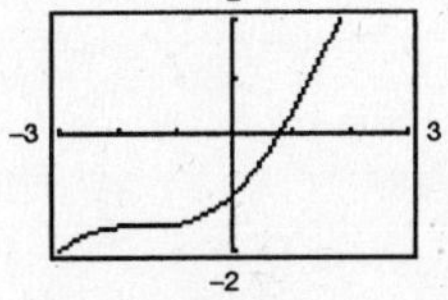

The zero between 0 and 1 appears to occur at $x \approx 0.739$.

(b) $x_n = \cos(x_{n-1})$

$x_0 = 1$

$x_1 = \cos 1 \approx 0.5403$

$x_2 = \cos 0.5403 \approx 0.8576$

$x_3 = \cos 0.8576 \approx 0.6543$

$x_4 = \cos 0.6543 \approx 0.7935$

$x_5 = \cos 0.7935 \approx 0.7014$

$x_6 = \cos 0.7014 \approx 0.7640$

$x_7 = \cos 0.7640 \approx 0.7221$

$x_8 = \cos 0.7221 \approx 0.7504$

$x_9 = \cos 0.7504 \approx 0.7314$

$\vdots$

This sequence appears to be approaching the zero of f: $x \approx 0.739$.

3. $y_2 = \frac{\pi}{2} - y_1$

$$\arctan x + \arctan\frac{1}{x} = y_1 + y_2$$

$$= y_1 + \left(\frac{\pi}{2} - y_1\right) = \frac{\pi}{2}$$

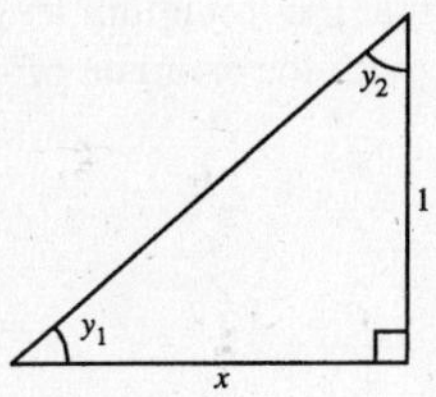

4. Let $\alpha = \arcsin x$ and $\beta = \arccos x$, then $\sin \alpha = x$ and $\cos \beta = x$. Thus, $\sin \alpha = \cos \beta$ which implies that α and β are complementary angles and we have

$$\alpha + \beta = \frac{\pi}{2}$$

$$\arcsin x + \arccos x = \frac{\pi}{2}.$$

5. $\arcsin x = \arcsin \dfrac{x}{1} = \arctan \dfrac{x}{\sqrt{1-x^2}},\ x > 0$

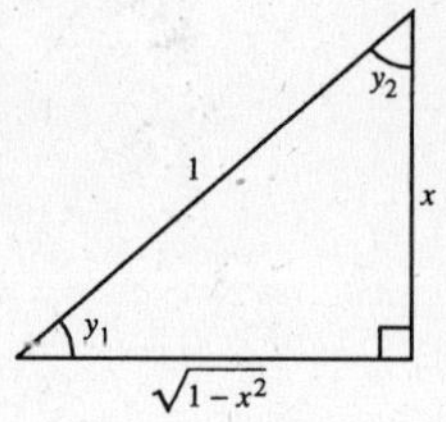

6. (a)

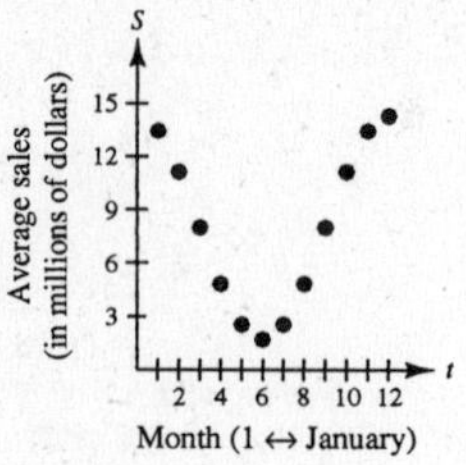

(b) $a = \dfrac{1}{2}(14.3 - 1.7) = 6.3$

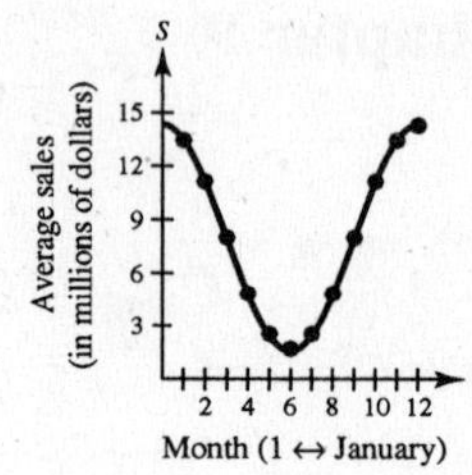

$\dfrac{2\pi}{b} = 12 \Rightarrow b = \dfrac{\pi}{6}$

Shift: $d = 14.3 - 6.3 = 8$

$S = d + a \cos bt$

$S = 8 + 6.3 \cos\left(\dfrac{\pi t}{6}\right)$

Note: Another model is $S = 8 + 6.3 \sin\left(\dfrac{\pi t}{6} + \dfrac{\pi}{2}\right)$.

The model is a good fit.

(c) Period: $\dfrac{2\pi}{\pi/6} = 12$

This corresponds to the 12 months in a year. Since the sales of outerwear is seasonal this is reasonable.

(d) The amplitude represents the maximum displacement from average sales of 8 million dollars. Sales are greatest in December (cold weather + Christmas) and least in June.

7. (a)

θ	L_1	L_2	$L_1 + L_2$
0.1	$\dfrac{2}{\sin 0.1}$	$\dfrac{3}{\cos 0.1}$	23.0
0.2	$\dfrac{2}{\sin 0.2}$	$\dfrac{3}{\cos 0.2}$	13.1
0.3	$\dfrac{2}{\sin 0.3}$	$\dfrac{3}{\cos 0.3}$	9.9
0.4	$\dfrac{2}{\sin 0.4}$	$\dfrac{3}{\cos 0.4}$	8.4

(b)

θ	L_1	L_2	$L_1 + L_2$
0.5	$\dfrac{2}{\sin 0.5}$	$\dfrac{3}{\cos 0.5}$	7.6
0.6	$\dfrac{2}{\sin 0.6}$	$\dfrac{3}{\cos 0.6}$	7.2
0.7	$\dfrac{2}{\sin 0.7}$	$\dfrac{3}{\cos 0.7}$	7.0
0.8	$\dfrac{2}{\sin 0.8}$	$\dfrac{3}{\cos 0.8}$	7.1

The minimum length of the elevator is 7.0 meters.

(c) $L = L_1 + L_2 = \dfrac{2}{\sin\theta} + \dfrac{3}{\cos\theta}$

(d)

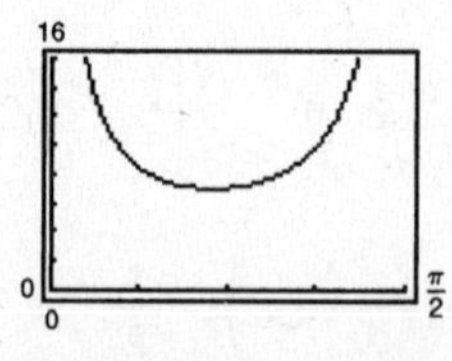

From the graph, it appears that the minimum length is 7.0 meters, which agrees with the estimate of part (b).

8. (a) $\arctan x \approx x - \dfrac{x^3}{3} + \dfrac{x^5}{5} - \dfrac{x^7}{7}$

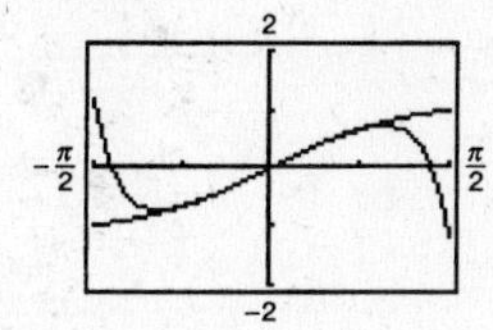

The approximation appears to be accurate over the interval $-0.5 < x < 0.5$.

(b) $\arctan x \approx x - \dfrac{x^3}{3} + \dfrac{x^5}{5} - \dfrac{x^7}{7} + \dfrac{x^9}{9}$

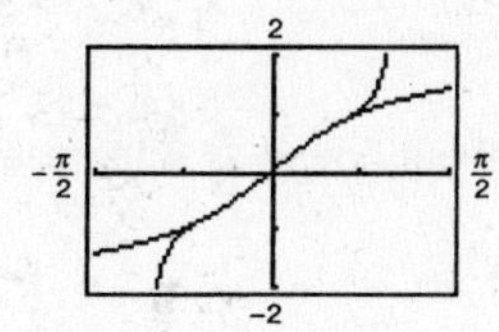

The accuracy improved.

9. $f(x) = \sqrt{x}$

$g(x) = 6 \arctan x$

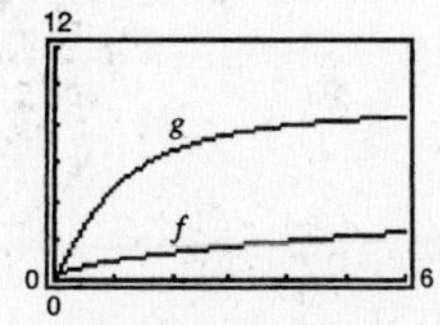

As x increases to infinity, g approaches 3π, but f has no maximum. Using the solve feature of the graphing utility, you find $a \approx 87.54$.

10. (a) & (b)

Base 1	*Base 2*	*Altitude*	*Area*
8	$8 + 16 \cos 10°$	$8 \sin 10°$	22.1
8	$8 + 16 \cos 20°$	$8 \sin 20°$	42.5
8	$8 + 16 \cos 30°$	$8 \sin 30°$	59.7
8	$8 + 16 \cos 40°$	$8 \sin 40°$	72.7
8	$8 + 16 \cos 50°$	$8 \sin 50°$	80.5
8	$8 + 16 \cos 60°$	$8 \sin 60°$	83.1
8	$8 + 16 \cos 70°$	$8 \sin 70°$	80.7

The maximum occurs when $\theta = 60°$ and is approximately 83.1 square feet.

(c) $A(\theta) = [8 + (8 + 16 \cos \theta)]\left[\dfrac{8 \sin \theta}{2}\right]$

$= (16 + 16 \cos \theta)(4 \sin \theta)$

$= 64\,(1 + \cos \theta)(\sin \theta)$

(d)

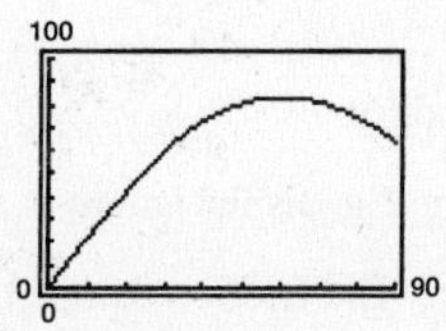

The maximum of 83.1 square feet occurs when $\theta = \dfrac{\pi}{3} = 60°$.

11. $d_1 + a_1 \sin(b_1 x + c_1) = d_2 + a_2 \cos(b_2 x + c_2)$. Assume that b_1 and b_2 are not both zero.

(a) $d_1 = d_2$ (b) $|a_1| = |a_2|$ (c) $|b_1| = |b_2|$

(d) $c_1 - c_2 = \dfrac{(2n + 1)\pi}{2}$, for some integer n

(e) Answers will vary. Examples:

$$1 + \sin x = 1 + \cos\left(x - \frac{\pi}{2}\right)$$

$$2 \sin(3x + \pi) = -2 \cos\left(-3x + \frac{\pi}{2}\right)$$

12. $f(x) = \sin^k x$

$f(-x) = \sin^k(-x) = (-\sin x)^k = (-1)^k \sin^k x$

If k is a positive even integer, then $f(-x) = \sin^k (x) = f(x)$ and f is an even function.

If k is a positive odd integer, then $f(-x) = -\sin^k x = -f(x)$ and f is an odd function.

The same is not true for $f(x) = \cos^k x$.

$f(-x) = \cos^k (-x) = \cos^k x = f(x)$, f is an even function for all positive integer values of k.

13. $d = \dfrac{(12\pi)(60)(24)(365)}{(5280)(12)} = \dfrac{1095\pi}{11} \approx 312.7$ miles

14. (a) $\tan\theta = \dfrac{x}{12}$

$x = 12\tan\theta$

Area = Area of triangle − Area of sector

$= \left(\dfrac{1}{2}bh\right) - \left(\dfrac{1}{2}r^2\theta\right)$

$= \dfrac{1}{2}(12)(12\tan\theta) - \dfrac{1}{2}(12^2)(\theta)$

$= 72\tan\theta - 72\theta$

$= 72(\tan\theta - \theta)$

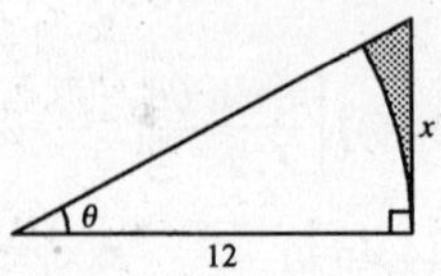

(b)

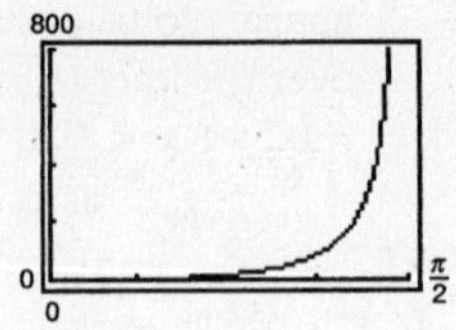

As $\theta \Rightarrow \dfrac{\pi}{2}$, $A \Rightarrow \infty$.

The area increases without bound as θ approaches $\dfrac{\pi}{2}$.

15. $y = Ae^{-kt}\cos bt = \frac{1}{5}e^{-t/10}\cos 6t$

(a) A is changed from $\frac{1}{5}$ to $\frac{1}{3}$: The displacement is increased.

(b) k is changed from $\frac{1}{10}$ to $\frac{1}{3}$: The friction damps the oscillations more rapidly.

(c) b is changed from 6 to 9: The frequency of oscillation is increased.

16. (a) $f(t - 2c) = f(t)$, t and $t - 2c$ differ by an integer multiple of the period.

(b) $f\left(t + \frac{1}{2}c\right) \neq f\left(\frac{1}{2}t\right)$, $\left(t + \frac{1}{2}c\right) - \left(\frac{1}{2}t\right) = \frac{1}{2}t + \frac{1}{2}c$ which is not an integer multiple of the period.

(c) $f\left(\frac{1}{2}[t + c]\right) \neq f\left(\frac{1}{2}t\right)$, $\left(\frac{1}{2}[t + c]\right) - \left(\frac{1}{2}t\right) = \frac{1}{2}c$ which is not an integer multiple of the period.

CHAPTER 10
Analytic Trigonometry

Section 10.1 Using Fundamental Trigonometric Identities **744**

Section 10.2 Verifying Trigonometric Identities **755**

Section 10.3 Solving Trigonometric Equations **761**

Section 10.4 Sum and Difference Formulas **774**

Section 10.5 Multiple-Angle and Product-to-Sum Formulas **788**

Review Exercises **809**

Problem Solving **819**

CHAPTER 10
Analytic Trigonometry

Section 10.1 Using Fundamental Trigonometric Identities

1. $\sin x = \dfrac{\sqrt{3}}{2}, \cos x = -\dfrac{1}{2} \Rightarrow x$ is in Quadrant II.

$$\tan x = \frac{\sin x}{\cos x} = \frac{\sqrt{3}/2}{-1/2} = -\sqrt{3}$$

$$\cot x = \frac{1}{\tan x} = -\frac{1}{\sqrt{3}} = -\frac{\sqrt{3}}{3}$$

$$\sec x = \frac{1}{\cos x} = \frac{1}{-1/2} = -2$$

$$\csc x = \frac{1}{\sin x} = \frac{1}{\sqrt{3}/2} = \frac{2}{\sqrt{3}} = \frac{2\sqrt{3}}{3}$$

2. $\tan x = \dfrac{\sqrt{3}}{3}, \cos x = -\dfrac{\sqrt{3}}{2}$

x is in Quadrant III.

$$\sin x = -\sqrt{1 - \left(-\frac{\sqrt{3}}{2}\right)^2} = -\sqrt{\frac{1}{4}} = -\frac{1}{2}$$

$$\csc x = \frac{1}{\sin x} = -2$$

$$\sec x = \frac{1}{\cos x} = -\frac{2}{\sqrt{3}} = -\frac{2\sqrt{3}}{3}$$

$$\cot x = \frac{1}{\tan x} = \frac{3}{\sqrt{3}} = \sqrt{3}$$

3. $\sec \theta = \sqrt{2}, \sin \theta = -\dfrac{\sqrt{2}}{2} \Rightarrow \theta$ is in Quadrant IV.

$$\cos \theta = \frac{1}{\sec \theta} = \frac{1}{\sqrt{2}} = \frac{\sqrt{2}}{2}$$

$$\tan \theta = \frac{\sin \theta}{\cos \theta} = \frac{-\sqrt{2}/2}{\sqrt{2}/2} = -1$$

$$\cot \theta = \frac{1}{\tan \theta} = -1$$

$$\csc \theta = \frac{1}{\sin \theta} = -\sqrt{2}$$

4. $\csc \theta = \dfrac{5}{3}, \tan \theta = \dfrac{3}{4}$

θ is in Quadrant I.

$$\sin \theta = \frac{1}{\csc \theta} = \frac{3}{5}$$

$$\cos \theta = \frac{\sin \theta}{\tan \theta} = \frac{3}{5} \cdot \frac{4}{3} = \frac{4}{5}$$

$$\sec \theta = \frac{1}{\cos \theta} = \frac{5}{4}$$

$$\cot \theta = \frac{1}{\tan \theta} = \frac{4}{3}$$

5. $\tan x = \dfrac{5}{12}, \sec x = -\dfrac{13}{12} \Rightarrow x$ is in

Quadrant III.

$$\cos x = \frac{1}{\sec x} = -\frac{12}{13}$$

$$\sin x = -\sqrt{1 - \cos^2 x} = -\sqrt{1 - \frac{144}{169}} = -\frac{5}{13}$$

$$\cot x = \frac{1}{\tan x} = \frac{12}{5}$$

$$\csc x = \frac{1}{\sin x} = -\frac{13}{5}$$

6. $\cot \phi = -3, \sin \phi = \dfrac{\sqrt{10}}{10}$

ϕ is in Quadrant II.

$$\cos \phi = \cot \phi \sin \phi = -\frac{3\sqrt{10}}{10}$$

$$\tan \phi = \frac{1}{\cot \phi} = -\frac{1}{3}$$

$$\csc \phi = \frac{1}{\sin \phi} = \sqrt{10}$$

$$\sec \phi = \frac{1}{\cos \phi} = -\frac{10}{3\sqrt{10}} = -\frac{\sqrt{10}}{3}$$

7. $\sec\phi = \frac{3}{2}$, $\csc\phi = -\frac{3\sqrt{5}}{5} \Rightarrow \phi$ is in Quadrant IV.

$\sin\phi = \frac{1}{\csc\phi} = \frac{1}{-3\sqrt{5}/5} = -\frac{\sqrt{5}}{3}$

$\cos\phi = \frac{1}{\sec\phi} = \frac{1}{3/2} = \frac{2}{3}$

$\tan\phi = \frac{\sin\phi}{\cos\phi} = \frac{-\sqrt{5}/3}{2/3} = -\frac{\sqrt{5}}{2}$

$\cot\phi = \frac{1}{\tan\phi} = \frac{1}{-\sqrt{5}/2} = -\frac{2}{\sqrt{5}} = -\frac{2\sqrt{5}}{5}$

8. $\cos\left(\frac{\pi}{2} - x\right) = \frac{3}{5}$, $\cos x = \frac{4}{5}$

x is in Quadrant I.

$\sin x = \sqrt{1 - \left(\frac{4}{5}\right)^2} = \frac{3}{5}$

$\tan x = \frac{\sin x}{\cos x} = \frac{3}{5} \cdot \frac{5}{4} = \frac{3}{4}$

$\csc x = \frac{1}{\sin x} = \frac{5}{3}$

$\sec x = \frac{1}{\cos x} = \frac{5}{4}$

$\cot x = \frac{1}{\tan x} = \frac{4}{3}$

9. $\sin(-x) = -\frac{1}{3} \Rightarrow \sin x = \frac{1}{3}$, $\tan x = -\frac{\sqrt{2}}{4} \Rightarrow x$ is in Quadrant II.

$\cos x = -\sqrt{1 - \sin^2 x} = -\sqrt{1 - \frac{1}{9}} = -\frac{2\sqrt{2}}{3}$

$\cot x = \frac{1}{\tan x} = \frac{1}{-\sqrt{2}/4} = -2\sqrt{2}$

$\sec x = \frac{1}{\cos x} = \frac{1}{-2\sqrt{2}/3} = -\frac{3\sqrt{2}}{4}$

$\csc x = \frac{1}{\sin x} = \frac{1}{1/3} = 3$

10. $\sec x = 4$, $\sin x > 0$

x is in Quadrant I.

$\cos x = \frac{1}{\sec x} = \frac{1}{4}$

$\sin x = \sqrt{1 - \left(\frac{1}{4}\right)^2} = \frac{\sqrt{15}}{4}$

$\tan x = \frac{\sin x}{\cos x} = \frac{\sqrt{15}}{4} \cdot \frac{4}{1} = \sqrt{15}$

$\csc x = \frac{1}{\sin x} = \frac{4}{\sqrt{15}} = \frac{4\sqrt{15}}{15}$

$\cot x = \frac{1}{\tan x} = \frac{1}{\sqrt{15}} = \frac{\sqrt{15}}{15}$

11. $\tan\theta = 2$, $\sin\theta < 0 \Rightarrow \theta$ is in Quadrant III.

$\sec\theta = -\sqrt{\tan^2\theta + 1} = -\sqrt{4 + 1} = -\sqrt{5}$

$\cos\theta = \frac{1}{\sec\theta} = -\frac{1}{\sqrt{5}} = -\frac{\sqrt{5}}{5}$

$\sin\theta = -\sqrt{1 - \cos^2\theta}$

$= -\sqrt{1 - \frac{1}{5}} = -\frac{2}{\sqrt{5}} = -\frac{2\sqrt{5}}{5}$

$\csc\theta = \frac{1}{\sin\theta} = -\frac{\sqrt{5}}{2}$

$\cot\theta = \frac{1}{\tan\theta} = \frac{1}{2}$

12. $\csc\theta = -5$, $\cos\theta < 0$

θ is in Quadrant III.

$\sin\theta = \frac{1}{\csc\theta} = -\frac{1}{5}$

$\cos\theta = -\sqrt{1 - \left(-\frac{1}{5}\right)^2} = \frac{-2\sqrt{6}}{5}$

$\tan\theta = \frac{\sin\theta}{\cos\theta} = -\frac{1}{5} \cdot -\frac{5}{2\sqrt{6}} = \frac{\sqrt{6}}{12}$

$\sec\theta = \frac{1}{\cos\theta} = -\frac{5}{2\sqrt{6}} = -\frac{5\sqrt{6}}{12}$

$\cot\theta = \frac{1}{\tan\theta} = \frac{12}{\sqrt{6}} = 2\sqrt{6}$

13. $\sin\theta = -1,\ \cot\theta = 0 \Rightarrow \theta = \frac{3\pi}{2}$

$\cos\theta = \sqrt{1 - \sin^2\theta} = 0$

$\sec\theta$ is undefined.

$\tan\theta$ is undefined.

$\csc\theta = -1$

14. $\tan\theta$ is undefined, $\sin\theta > 0$.

$\theta = \frac{\pi}{2}$

$\tan\theta = \frac{\sin\theta}{\cos\theta}$ is undefined $\Rightarrow \cos\theta = 0$

$\sin\theta = \sqrt{1 - 0^2} = 1$

$\csc\theta = \frac{1}{\sin\theta} = 1$

$\sec\theta = \frac{1}{\cos\theta}$ is undefined.

$\cot\theta = \frac{\cos\theta}{\sin\theta} = \frac{0}{1} = 0$

15. $\sec x\cos x = \sec x \cdot \frac{1}{\sec x} = 1$

The expression is matched with (d).

16. $\tan x\csc x = \frac{\sin x}{\cos x} \cdot \frac{1}{\sin x} = \frac{1}{\cos x} = \sec x$

Matches (a).

17. $\cot^2 x - \csc^2 x = \cot^2 x - (1 + \cot^2 x) = -1$

The expression is matched with (b).

18. $(1 - \cos^2 x)(\csc x) = (\sin^2 x)\frac{1}{\sin x} = \sin x$

Matches (f).

19. $\frac{\sin(-x)}{\cos(-x)} = \frac{-\sin x}{\cos x} = -\tan x$

The expression is matched with (e).

20. $\frac{\sin\left[\left(\frac{\pi}{2}\right) - x\right]}{\cos\left[\left(\frac{\pi}{2}\right) - x\right]} = \frac{\cos x}{\sin x} = \cot x$

Matches (c).

21. $\sin x\sec x = \sin x \cdot \frac{1}{\cos x} = \tan x$

The expression is matched with (b).

22. $\cos^2 x(\sec^2 x - 1) = \cos^2 x(\tan^2 x)$

$= \cos^2 x\left(\frac{\sin^2 x}{\cos^2 x}\right)$

$= \sin^2 x$

Matches (c).

23. $\sec^4 x - \tan^4 x = (\sec^2 x + \tan^2 x)(\sec^2 x - \tan^2 x)$

$= (\sec^2 x + \tan^2 x)(1) = \sec^2 x + \tan^2 x$

The expression is matched with (f).

24. $\cot x\sec x = \frac{\cos x}{\sin x} \cdot \frac{1}{\cos x} = \frac{1}{\sin x} = \csc x$

Matches (a).

25. $\frac{\sec^2 x - 1}{\sin^2 x} = \frac{\tan^2 x}{\sin^2 x} = \frac{\sin^2 x}{\cos^2 x} \cdot \frac{1}{\sin^2 x} = \sec^2 x$

The expression is matched with (e).

26. $\frac{\cos^2\left[\left(\frac{\pi}{2}\right) - x\right]}{\cos x} = \frac{\sin^2 x}{\cos x} = \frac{\sin x}{\cos x}\sin x = \tan x\sin x$

Matches (d).

27. $\cot\theta\sec\theta = \frac{\cos\theta}{\sin\theta} \cdot \frac{1}{\cos\theta} = \frac{1}{\sin\theta} = \csc\theta$

28. $\cos\beta\tan\beta = \cos\beta\frac{\sin\beta}{\cos\beta} = \sin\beta$

29. $\sin \phi (\csc \phi - \sin \phi) = (\sin \phi)\frac{1}{\sin \phi} - \sin^2 \phi$

$= 1 - \sin^2 \phi = \cos^2 \phi$

30. $\sec^2 x(1 - \sin^2 x) = \sec^2 x - \sec^2 x \sin^2 x$

$= \sec^2 x - \frac{1}{\cos^2 x} \cdot \sin^2 x$

$= \sec^2 x - \frac{\sin^2 x}{\cos^2 x}$

$= \sec^2 x - \tan^2 x = 1$

31. $\frac{\cot x}{\csc x} = \frac{\cos x/\sin x}{1/\sin x}$

$= \frac{\cos x}{\sin x} \cdot \frac{\sin x}{1} = \cos x$

32. $\frac{\csc \theta}{\sec \theta} = \frac{1/\sin \theta}{1/\cos \theta} = \frac{\cos \theta}{\sin \theta} = \cot \theta$

33. $\frac{1 - \sin^2 x}{\csc^2 x - 1} = \frac{\cos^2 x}{\cot^2 x} = \cos^2 x \tan^2 x = (\cos^2 x)\frac{\sin^2 x}{\cos^2 x}$

$= \sin^2 x$

34. $\frac{1}{\tan^2 x + 1} = \frac{1}{\sec^2 x} = \frac{1}{1/\cos^2 x} = \cos^2 x$

35. $\sec \alpha \frac{\sin \alpha}{\tan \alpha} = \frac{1}{\cos \alpha}(\sin \alpha) \cot \alpha = \frac{1}{\cos \alpha}(\sin \alpha)\left(\frac{\cos \alpha}{\sin \alpha}\right) = 1$

36. $\frac{\tan^2 \theta}{\sec^2 \theta} = \frac{\sin^2 \theta}{\cos^2 \theta} \cdot \frac{1}{\sec^2 \theta} = \frac{\sin^2 \theta}{\cos^2 \theta} \cdot \frac{1}{\frac{1}{\cos^2 \theta}} = \frac{\sin^2 \theta \cos^2 \theta}{\cos^2 \theta} = \sin^2 \theta$

37. $\cos\left(\frac{\pi}{2} - x\right)\sec x = (\sin x)(\sec x) = (\sin x)\left(\frac{1}{\cos x}\right) = \frac{\sin x}{\cos x} = \tan x$

38. $\cot\left(\frac{\pi}{2} - x\right) \cos x = \tan x \cos x = \frac{\sin x}{\cos x} \cdot \cos x = \sin x$

39. $\frac{\cos^2 y}{1 - \sin y} = \frac{1 - \sin^2 y}{1 - \sin y} = \frac{(1 + \sin y)(1 - \sin y)}{1 - \sin y} = 1 + \sin y$

40. $(\cos t)(1 + \tan^2 t) = (\cos t)(\sec^2 t) = \frac{\cos t}{\cos^2 t} = \frac{1}{\cos t} = \sec t$

41. $\sin \beta \tan \beta + \cos \beta = (\sin \beta)\frac{\sin \beta}{\cos \beta} + \cos \beta$

$= \frac{\sin^2 \beta}{\cos \beta} + \frac{\cos^2 \beta}{\cos \beta}$

$= \frac{\sin^2 \beta + \cos^2 \beta}{\cos \beta}$

$= \frac{1}{\cos \beta}$

$= \sec \beta$

42. $\csc \phi \tan \phi + \sec \phi = \frac{1}{\sin \phi} \cdot \frac{\sin \phi}{\cos \phi} + \sec \phi$

$= \frac{1}{\cos \phi} + \sec \phi = 2 \sec \phi$

43. $\cot u \sin u + \tan u \cos u = \frac{\cos u}{\sin u}(\sin u) + \frac{\sin u}{\cos u}(\cos u)$

$= \cos u + \sin u$

44. $\sin \theta \sec \theta + \cos \theta \csc \theta = \frac{\sin \theta}{\cos \theta} + \frac{\cos \theta}{\sin \theta}$

$= \frac{\sin^2 \theta + \cos^2 \theta}{\cos \theta \sin \theta}$

$= \frac{1}{\cos \theta \sin \theta}$

$= \sec \theta \csc \theta$

45. $\tan^2 x - \tan^2 x \sin^2 x = \tan^2 x(1 - \sin^2 x)$

$= \tan^2 x \cos^2 x$

$= \frac{\sin^2 x}{\cos^2 x} \cdot \cos^2 x$

$= \sin^2 x$

46. $\sin^2 x \csc^2 x - \sin^2 x = \sin^2 x(\csc^2 x - 1)$

$= \sin^2 x \cot^2 x$

$= \sin^2 x \cdot \frac{\cos^2 x}{\sin^2 x}$

$= \cos^2 x$

47. $\sin^2 x \sec^2 x - \sin^2 x = \sin^2 x(\sec^2 x - 1)$

$= \sin^2 x \tan^2 x$

48. $\cos^2 x + \cos^2 x \tan^2 x = \cos^2 x(1 + \tan^2 x)$

$= \cos^2 x + (\sec^2 x)$

$= \cos^2 x\left(\frac{1}{\cos^2 x}\right)$

$= 1$

49. $\frac{\sec^2 x - 1}{\sec x - 1} = \frac{(\sec x + 1)(\sec x - 1)}{\sec x - 1} = \sec x + 1$

50. $\frac{\cos^2 x - 4}{\cos x - 2} = \frac{(\cos x + 2)(\cos x - 2)}{\cos x - 2} = \cos x + 2$

51. $\tan^4 x + 2\tan^2 x + 1 = (\tan^2 x + 1)^2$

$= (\sec^2 x)^2$

$= \sec^4 x$

52. $1 - 2\cos^2 x + \cos^4 x = (1 - \cos^2 x)^2$

$= (\sin^2 x)^2$

$= \sin^4 x$

53. $\sin^4 x - \cos^4 x = (\sin^2 x + \cos^2 x)(\sin^2 x - \cos^2 x)$

$= (1)(\sin^2 x - \cos^2 x)$

$= \sin^2 x - \cos^2 x$

54. $\sec^4 x - \tan^4 x = (\sec^2 x + \tan^2 x)(\sec^2 x - \tan^2 x)$

$= (\sec^2 x + \tan^2 x)(1)$

$= \sec^2 x + \tan^2 x$

55. $\csc^3 x - \csc^2 x - \csc x + 1 = \csc^2 x(\csc x - 1) - 1(\csc x - 1)$

$= (\csc^2 x - 1)(\csc x - 1)$

$= \cot^2 x(\csc x - 1)$

56. $\sec^3 x - \sec^2 x - \sec x + 1 = \sec^2 x(\sec x - 1) - (\sec x - 1)$

$= (\sec^2 x - 1)(\sec x - 1)$

$= \tan^2 x(\sec x - 1)$

57. $(\sin x + \cos x)^2 = \sin^2 x + 2\sin x \cos x + \cos^2 x$

$= (\sin^2 x + \cos^2 x) + 2\sin x \cos x$

$= 1 + 2\sin x \cos x$

58. $(\cot x + \csc x)(\cot x - \csc x) = \cot^2 x - \csc^2 x$

$= -1$

59. $(2\csc x + 2)(2\csc x - 2) = 4\csc^2 x - 4 = 4(\csc^2 x - 1) = 4\cot^2 x$

60. $(3 - 3\sin x)(3 + 3\sin x) = 9 - 9\sin^2 x = 9(1 - \sin^2 x) = 9\cos^2 x$

61. $$\begin{aligned}\frac{1}{1+\cos x} + \frac{1}{1-\cos x} &= \frac{1 - \cos x + 1 + \cos x}{(1+\cos x)(1-\cos x)}\\ &= \frac{2}{1-\cos^2 x}\\ &= \frac{2}{\sin^2 x}\\ &= 2\csc^2 x\end{aligned}$$

62. $$\begin{aligned}\frac{1}{\sec x + 1} - \frac{1}{\sec x - 1} &= \frac{\sec x - 1 - (\sec x + 1)}{(\sec x + 1)(\sec x - 1)}\\ &= \frac{\sec x - 1 - \sec x - 1}{\sec^2 x - 1}\\ &= \frac{-2}{\tan^2 x}\\ &= -2\left(\frac{1}{\tan^2 x}\right) = -2\cot^2 x\end{aligned}$$

63. $$\begin{aligned}\frac{\cos x}{1+\sin x} + \frac{1+\sin x}{\cos x} &= \frac{\cos^2 x + (1+\sin x)^2}{\cos x(1+\sin x)}\\ &= \frac{\cos^2 x + 1 + 2\sin x + \sin^2 x}{\cos x(1+\sin x)}\\ &= \frac{2 + 2\sin x}{\cos x(1+\sin x)}\\ &= \frac{2(1+\sin x)}{\cos x(1+\sin x)}\\ &= \frac{2}{\cos x}\\ &= 2\sec x\end{aligned}$$

64. $$\begin{aligned}\tan x - \frac{\sec^2 x}{\tan x} &= \frac{\tan^2 x - \sec^2 x}{\tan x}\\ &= \frac{-1}{\tan x} = -\cot x\end{aligned}$$

65. $$\begin{aligned}\frac{\sin^2 y}{1-\cos y} &= \frac{1-\cos^2 y}{1-\cos y}\\ &= \frac{(1+\cos y)(1-\cos y)}{1-\cos y}\\ &= 1+\cos y\end{aligned}$$

66. $$\begin{aligned}\frac{5}{\tan x + \sec x}\cdot\frac{\tan x - \sec x}{\tan x - \sec x} &= \frac{5(\tan x - \sec x)}{\tan^2 x - \sec^2 x}\\ &= \frac{5(\tan x - \sec x)}{-1}\\ &= 5(\sec x - \tan x)\end{aligned}$$

67. $$\begin{aligned}\frac{3}{\sec x - \tan x}\cdot\frac{\sec x + \tan x}{\sec x + \tan x} &= \frac{3(\sec x + \tan x)}{\sec^2 x - \tan^2 x}\\ &= \frac{3(\sec x + \tan x)}{1}\\ &= 3(\sec x + \tan x)\end{aligned}$$

68. $$\begin{aligned}\frac{\tan^2 x}{\csc x + 1}\cdot\frac{\csc x - 1}{\csc x - 1} &= \frac{\tan^2 x(\csc x - 1)}{\csc^2 x - 1}\\ &= \frac{\tan^2 x(\csc x - 1)}{\cot^2 x}\\ &= \tan^2 x(\csc x - 1)\tan^2 x\\ &= \tan^4 x(\csc x - 1)\end{aligned}$$

69. $\cos\theta = \sqrt{1 - \sin^2\theta}$ is **not** an identity.

$\cos^2\theta + \sin^2\theta = 1 \implies \cos\theta = \pm\sqrt{1 - \sin^2\theta}$

70. The equation is **not** an identity.

$\cot\theta = \pm\sqrt{\csc^2\theta - 1}$

71. $\dfrac{\sin k\theta}{\cos k\theta} = \tan\theta$ is **not** an identity.

$\dfrac{\sin k\theta}{\cos k\theta} = \tan k\theta$

72. The equation is **not** an identity.

$\dfrac{1}{5\cos\theta} = \dfrac{1}{5}\left(\dfrac{1}{\cos\theta}\right) = \dfrac{1}{5}\sec\theta \neq 5\sec\theta$

73. $\sin\theta\csc\theta = 1$ is an identity.

$$\sin\theta\cdot\frac{1}{\sin\theta} = 1, \text{ provided } \sin\theta \neq 0.$$

74. The equation is **not** an identity because the angles may not be the same.

$$\sin\theta\csc\phi = \sin\theta\left(\frac{1}{\sin\phi}\right) \neq 1 \text{ unless } \theta = \phi$$

75. Since $\sin^2\theta + \cos^2\theta = 1$ and $\cos^2\theta = 1 - \sin^2\theta$:

$$\cos\theta = \pm\sqrt{1-\sin^2\theta}$$

$$\tan\theta = \frac{\sin\theta}{\cos\theta} = \pm\frac{\sin\theta}{\sqrt{1-\sin^2\theta}}$$

$$\cot\theta = \frac{1}{\tan\theta} = \pm\frac{\sqrt{1-\sin^2\theta}}{\sin\theta}$$

$$\sec\theta = \frac{1}{\cos\theta} = \pm\frac{1}{\sqrt{1-\sin^2\theta}}$$

$$\csc\theta = \frac{1}{\sin\theta}$$

76. $\cos\theta$

$$\sin\theta = \pm\sqrt{1-\cos^2\theta}$$

$$\tan\theta = \frac{\sin\theta}{\cos\theta} = \pm\frac{\sqrt{1-\cos^2\theta}}{\cos\theta}$$

$$\csc\theta = \frac{1}{\sin\theta} = \pm\frac{1}{\sqrt{1-\cos^2\theta}}$$

$$\sec\theta = \frac{1}{\cos\theta}$$

$$\cot\theta = \frac{1}{\tan\theta} = \pm\frac{\cos\theta}{\sqrt{1-\cos^2\theta}}$$

77. $y_1 = \cos\left(\frac{\pi}{2} - x\right)$, $y_2 = \sin x$

x	0.2	0.4	0.6	0.8	1.0	1.2	1.4
y_1	0.1987	0.3894	0.5646	0.7175	0.8414	0.9320	0.9854
y_2	0.1987	0.3894	0.5646	0.7175	0.8414	0.9320	0.9854

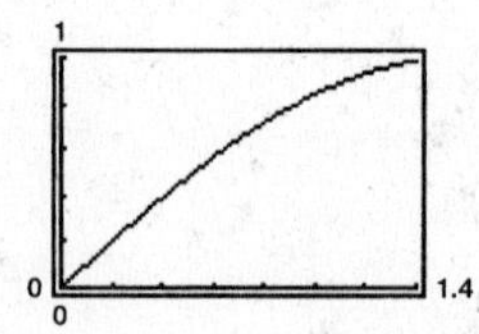

Conclusion: $y_1 = y_2$

78. $y_1 = \sec x - \cos x$, $y_2 = \sin x \tan x$

x	0.2	0.4	0.6	0.8	1.0	1.2	1.4
y_1	0.0403	0.1646	0.3863	0.7386	1.3105	2.3973	5.7135
y_2	0.0403	0.1646	0.3863	0.7386	1.3105	2.3973	5.7135

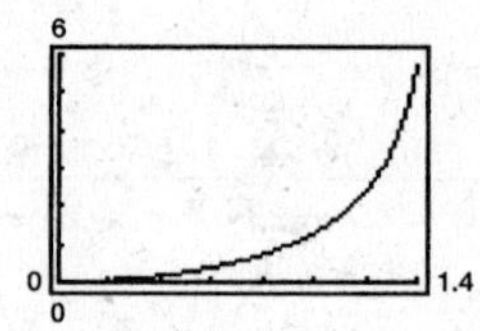

It appears that $y_1 = y_2$.

79. $y_1 = \dfrac{\cos x}{1 - \sin x}$, $y_2 = \dfrac{1 + \sin x}{\cos x}$

x	0.2	0.4	0.6	0.8	1.0	1.2	1.4
y_1	1.2230	1.5085	1.8958	2.4650	3.4082	5.3319	11.6814
y_2	1.2230	1.5085	1.8958	2.4650	3.4082	5.3319	11.6814

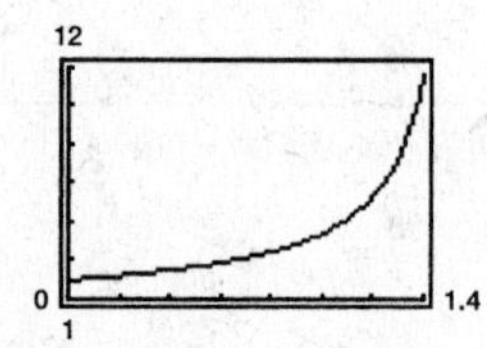

Conclusion: $y_1 = y_2$

80. $y_1 = \sec^4 x - \sec^2 x$, $y_2 = \tan^2 x + \tan^4 x$

x	0.2	0.4	0.6	0.8	1.0	1.2	1.4
y_1	0.0428	0.2107	0.6871	2.1841	8.3087	50.3869	1163.6143
y_2	0.0428	0.2107	0.6871	2.1841	8.3087	50.3869	1163.6143

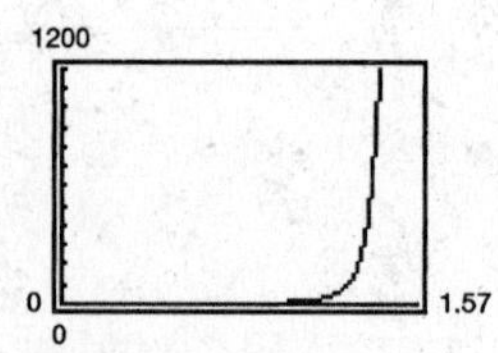

It appears that $y_1 = y_2$.

81. $\cos x \cot x + \sin x = \cos x\left(\dfrac{\cos x}{\sin x}\right) + \sin x$

$$= \frac{\cos^2 x}{\sin x} + \frac{\sin^2 x}{\sin x}$$

$$= \frac{\cos^2 x + \sin^2 x}{\sin x} = \frac{1}{\sin x} = \csc x$$

$y_1 = \cos x \cot x + \sin x$ and $y_2 = \csc x = \dfrac{1}{\sin x}$

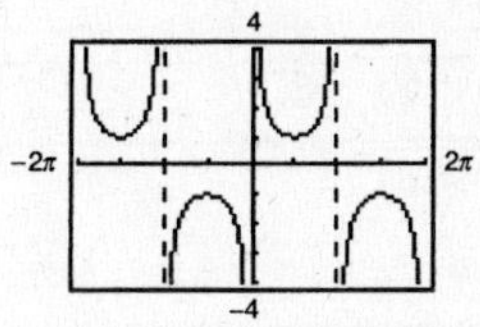

82. $\sec x \csc x - \tan x = \dfrac{1}{\cos x}\dfrac{1}{\sin x} - \dfrac{\sin x}{\cos x}$

$$= \frac{1}{\cos x \sin x} - \frac{\sin^2 x}{\cos x \sin x}$$

$$= \frac{\cos^2 x}{\cos x \sin x}$$

$$= \frac{\cos x}{\sin x}$$

$$= \cot x$$

$y_1 = \sec x \csc x - \tan x$ and $y_2 = \cot x = \dfrac{\cos x}{\sin x}$

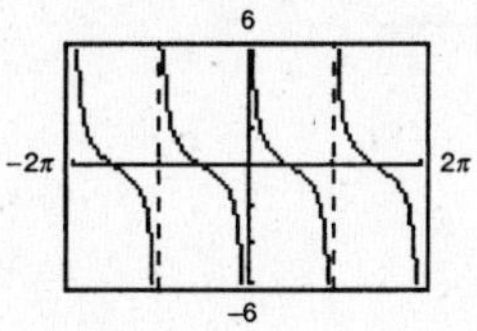

83. $\dfrac{1}{\sin x}\left(\dfrac{1}{\cos x} - \cos x\right) = \dfrac{1}{\sin x \cos x} - \dfrac{\cos x}{\sin x}$

$$= \frac{1 - \cos^2 x}{\sin x \cos x} = \frac{\sin^2 x}{\sin x \cos x} = \frac{\sin x}{\cos x} = \tan x$$

$y_1 = \dfrac{1}{\sin x}\left(\dfrac{1}{\cos x} - \cos x\right)$ and $y_2 = \tan x$

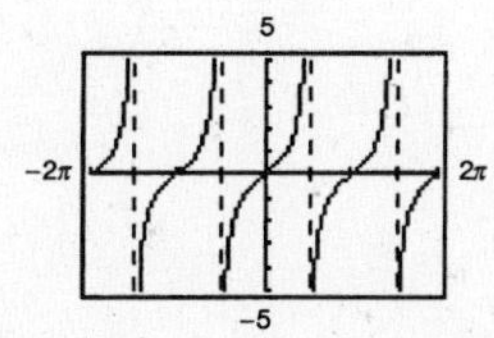

84. $\dfrac{1}{2}\left(\dfrac{1 + \sin \theta}{\cos \theta} + \dfrac{\cos \theta}{1 + \sin \theta}\right)$

$$= \frac{1}{2}\left[\frac{(1 + \sin \theta)^2 + \cos^2 \theta}{\cos \theta(1 + \sin \theta)}\right]$$

$$= \frac{1}{2}\left[\frac{1 + 2\sin \theta + \sin^2 \theta + \cos^2 \theta}{\cos \theta(1 + \sin \theta)}\right]$$

$$= \frac{1}{2}\left[\frac{2 + 2\sin \theta}{\cos \theta(1 + \sin \theta)}\right] = \frac{1}{2}\left[\frac{2(1 + \sin \theta)}{\cos \theta(1 + \sin \theta)}\right]$$

$$= \frac{1}{\cos \theta} = \sec \theta$$

$y_1 = \dfrac{1}{2}\left(\dfrac{1 + \sin \theta}{\cos \theta} + \dfrac{\cos \theta}{1 + \sin \theta}\right)$ and $y_2 = \sec \theta$

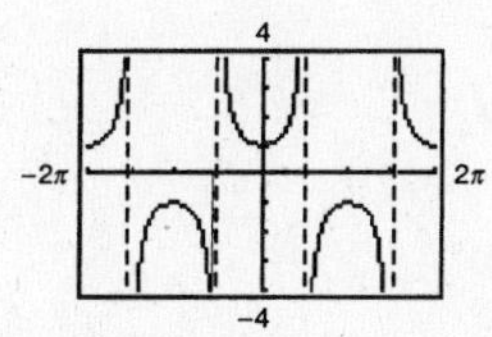

85. Let $x = 3\cos \theta$, then

$$\sqrt{9 - x^2} = \sqrt{9 - (3\cos \theta)^2} = \sqrt{9 - 9\cos^2 \theta} = \sqrt{9(1 - \cos^2 \theta)}$$

$$= \sqrt{9\sin^2 \theta} = 3\sin \theta.$$

86. Let $x = 2\cos\theta$, then

$$\sqrt{4 - x^2} = \sqrt{4 - (2\cos\theta)^2} = \sqrt{4 - 4\cos^2\theta} = \sqrt{4(1 - \cos^2\theta)} = \sqrt{4\sin^2\theta} = 2\sin\theta.$$

87. Let $x = 3\sec\theta$, then

$$\sqrt{x^2 - 9} = \sqrt{(3\sec\theta)^2 - 9} = \sqrt{9\sec^2\theta - 9} = \sqrt{9(\sec^2\theta - 1)} = \sqrt{9\tan^2\theta} = 3\tan\theta.$$

88. Let $x = 2\sec\theta$ then

$$\sqrt{x^2 - 4} = \sqrt{(2\sec\theta)^2 - 4} = \sqrt{4(\sec^2\theta - 1)} = \sqrt{4\tan^2\theta} = 2\tan\theta$$

89. Let $x = 5\tan\theta$, then

$$\sqrt{x^2 + 25} = \sqrt{(5\tan\theta)^2 + 25} = \sqrt{25\tan^2\theta + 25} = \sqrt{25(\tan^2\theta + 1)} = \sqrt{25\sec^2\theta} = 5\sec\theta$$

90. Let $x = 10\tan\theta$ then

$$\sqrt{x^2 + 100} = \sqrt{(10\tan\theta)^2 + 100} = \sqrt{100(\tan^2\theta + 1)} = \sqrt{100\sec^2\theta} = 10\sec\theta$$

91. Let $x = 3\sin\theta$, then $\sqrt{9 - x^2} = 3$ becomes

$$\sqrt{9 - (3\sin\theta)^2} = 3$$
$$\sqrt{9 - 9\sin^2\theta} = 3$$
$$\sqrt{9(1 - \sin^2\theta)} = 3$$
$$\sqrt{9\cos^2\theta} = 3$$
$$3\cos\theta = 3$$
$$\cos\theta = 1$$
$$\sin\theta = \sqrt{1 - \cos^2\theta} = \sqrt{1 - (1)^2} = 0.$$

92. $x = 6\sin\theta$

$$3 = \sqrt{36 - x^2} = \sqrt{36 - (6\sin\theta)^2} = \sqrt{36(1 - \sin^2\theta)} = \sqrt{36\cos^2\theta} = 6\cos\theta$$

$$\cos\theta = \frac{3}{6} = \frac{1}{2}$$

$$\sin\theta = \pm\sqrt{1 - \cos^2\theta} = \pm\sqrt{1 - \left(\frac{1}{2}\right)^2} = \pm\sqrt{\frac{3}{4}} = \pm\frac{\sqrt{3}}{2}$$

93. Let $x = 2\cos\theta$, then $\sqrt{4 - x^2} = \sqrt{2}$ becomes

$$\sqrt{4 - (2\cos\theta)^2} = \sqrt{2}$$
$$\sqrt{4 - 4\cos^2\theta} = \sqrt{2}$$
$$\sqrt{4(1 - \cos^2\theta)} = \sqrt{2}$$
$$\sqrt{4\sin^2\theta} = \sqrt{2}$$
$$2\sin\theta = \sqrt{2}, \text{ since } \sqrt{4 - x^2} > 0$$
$$\sin\theta = \frac{\sqrt{2}}{2}$$
$$\cos\theta = \sqrt{1 - \sin^2\theta} = \sqrt{1 - \frac{1}{2}} = \sqrt{\frac{1}{2}} = \frac{\sqrt{2}}{2}.$$

94. The equation $-5\sqrt{3} = \sqrt{100 - x^2}$ has no real solution since the left side is negative, and the right side is non-negative for $-10 \le x \le 10$.

95. $\sin \theta = \sqrt{1 - \cos^2 \theta}$

Let $y_1 = \sin x$ and $y_2 = \sqrt{1 - \cos^2 x},\ 0 \le x < 2\pi$.

$y_1 = y_2$ for $0 \le x \le \pi$, so we have

$\sin\theta = \sqrt{1 - \cos^2 \theta}$ for $0 \le \theta \le \pi$.

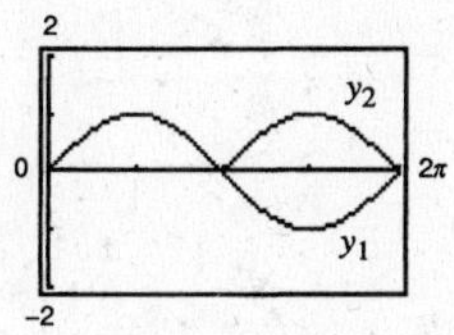

96. $\cos \theta = -\sqrt{1 - \sin^2 \theta}$

$\frac{\pi}{2} \le \theta \le \frac{3\pi}{2}$

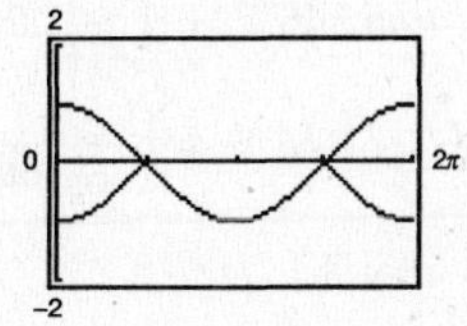

97. $\sec \theta = \sqrt{1 + \tan^2 \theta}$

Let $y_1 = \dfrac{1}{\cos x}$ and $y_2 = \sqrt{1 + \tan^2 x},\ 0 \le x < 2\pi$.

$y_1 = y_2$ for $0 \le x < \dfrac{\pi}{2}$ and $\dfrac{3\pi}{2} < x < 2\pi$, so we have

$\sec \theta = \sqrt{1 + \tan^2 \theta}$ for $0 \le \theta < \dfrac{\pi}{2}$ and $\dfrac{3\pi}{2} < \theta < 2\pi$.

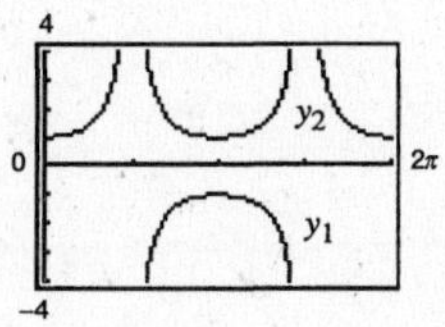

98. $\csc \theta = \sqrt{1 + \cot^2 \theta}$

$0 < \theta < \pi$

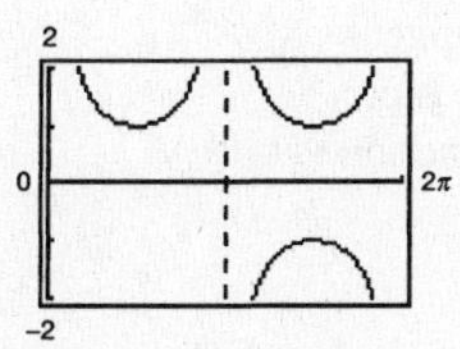

99. $\ln|\cos x| - \ln|\sin x| = \ln \dfrac{|\cos x|}{|\sin x|} = \ln|\cot x|$

100. $\ln|\sec x| + \ln|\sin x| = \ln|\sec x \sin x| = \ln\left|\dfrac{1}{\cos x} \cdot \sin x\right| = \ln|\tan x|$

101.
$$\begin{aligned}\ln|\cot t| + \ln(1 + \tan^2 t) &= \ln\big[|\cot t|(1 + \tan^2 t)\big]\\ &= \ln|\cot t \sec^2 t| = \ln\left|\frac{\cos t}{\sin t} \cdot \frac{1}{\cos^2 t}\right|\\ &= \ln\left|\frac{1}{\sin t \cos t}\right| = \ln|\csc t \sec t|\end{aligned}$$

102.
$$\begin{aligned}\ln(\cos^2 t) + \ln(1 + \tan^2 t) &= \ln[\cos^2 t(1 + \tan^2 t)]\\ &= \ln[\cos^2 t \sec^2 t]\\ &= \ln\left(\cos^2 t \cdot \frac{1}{\cos^2 t}\right)\\ &= \ln(1) = 0\end{aligned}$$

103. (a) $\csc^2 132° - \cot^2 132° \approx 1.8107 - 0.8107 = 1$

(b) $\csc^2 \dfrac{2\pi}{7} - \cot^2 \dfrac{2\pi}{7} \approx 1.6360 - 0.6360 = 1$

104. $\tan^2\theta + 1 = \sec^2\theta$

(a) $\theta = 346°$

$(\tan 346°)^2 + 1 \approx 1.0622$

$(\sec 346°)^2 = \left(\dfrac{1}{\cos 346°}\right)^2 \approx 1.0622$

(b) $\theta = 3.1$

$(\tan 3.1)^2 + 1 \approx 1.00173$

$(\sec 3.1)^2 = \left(\dfrac{1}{\cos 3.1}\right)^2 \approx 1.00173$

105. $\cos\left(\dfrac{\pi}{2} - \theta\right) = \sin\theta$

(a) $\theta = 80°$

$\cos(90° - 80°) = \sin 80°$

$0.9848 = 0.9848$

(b) $\theta = 0.8$

$\cos\left(\dfrac{\pi}{2} - 0.8\right) = \sin 0.8$

$0.7174 = 0.7174$

106. $\sin(-\theta) = -\sin\theta$

(a) $\theta = 250°$

$\sin(-250°) \approx 0.9397$

$-(\sin 250°) \approx 0.9397$

(b) $\theta = \frac{1}{2}$

$\sin\left(-\frac{1}{2}\right) \approx -0.4794$

$-\left(\sin\frac{1}{2}\right) \approx -0.4794$

107. $\mu W\cos\theta = W\sin\theta$

$\mu = \dfrac{W\sin\theta}{W\cos\theta} = \tan\theta$

108. $\sin^{-1/2}x\cos x = \dfrac{\cos x}{\sqrt{\sin x}}$

$= \dfrac{\cos x}{\sqrt{\sin x}} \cdot \dfrac{\sqrt{\sin x}}{\sqrt{\sin x}}$

$= \dfrac{\cos x\sqrt{\sin x}}{\sin x}$

$= \dfrac{\cos x}{\sin x}\sqrt{\sin x}$

$= \cot x\sqrt{\sin x}$

109. True.

For example, $\sin(-x) = -\sin(x)$ means the graph of $\sin(x)$ is symmetric about the origin.

110. False. A cofunction identity can be used to transform a tangent function so that it can be represented by a cotangent function.

111. True.

$1 + \tan^2\theta = \sec^2\theta$

$\tan^2\theta = \sec^2\theta - 1$

For $\dfrac{\pi}{2} < \theta \le \pi$, $\tan\theta \le 0$ so

$\tan\theta = -\sqrt{\sec^2\theta - 1}.$

112. True.

$\sin^2\theta + \cos^2\theta = 1$

$\sin^2\theta = 1 - \cos^2\theta$

For $\pi \le \theta \le 2\pi$, $\sin\theta \le 0$ so

$\sin\theta = -\sqrt{1 - \cos^2\theta}.$

113. (a) $\displaystyle\lim_{x\to(\pi/2)^-} \sin x = 1$

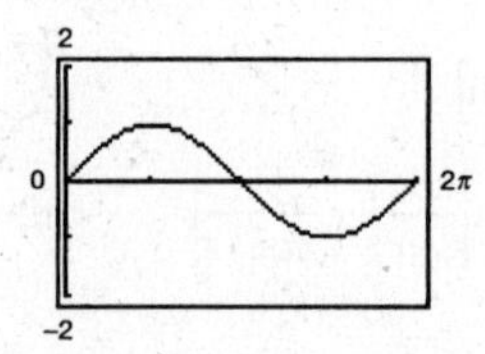

(b) $\displaystyle\lim_{x\to(\pi/2)^-} \csc x = 1$

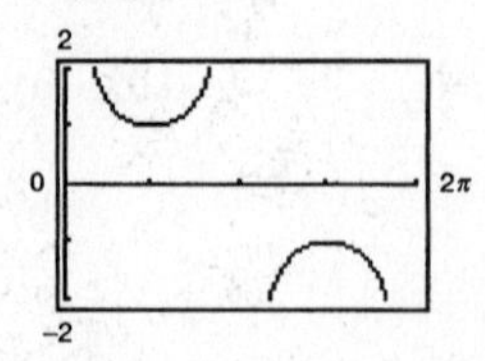

114. (a) $\displaystyle\lim_{x\to 0^+} \cos x = 1$

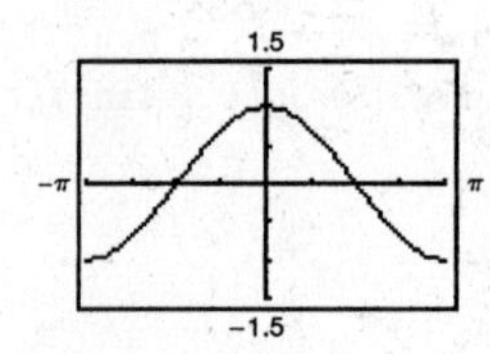

(b) $\displaystyle\lim_{x\to 0^+} \sec x = 1$

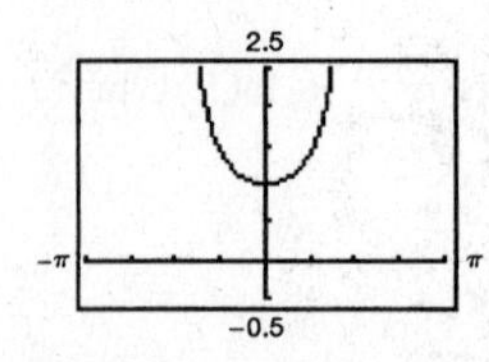

115. (a) $\displaystyle\lim_{x\to(\pi/2)^-} \tan x = \infty$

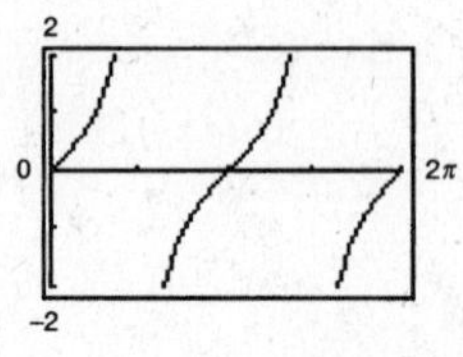

(b) $\displaystyle\lim_{x\to(\pi/2)^-} \cot x = 0$

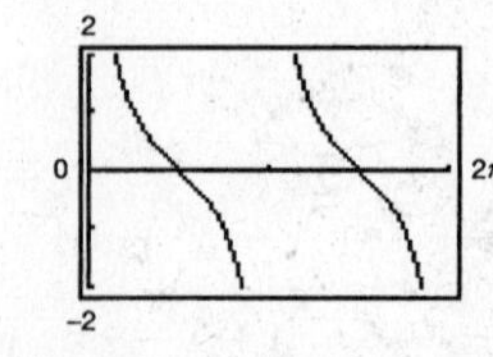

116. (a) $\lim_{x \to \pi^+} \sin x = 0$

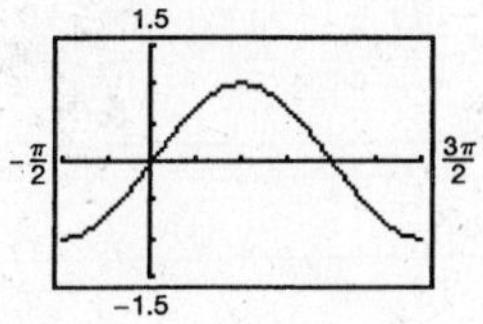

(b) $\lim_{x \to \pi^+} \csc x = -\infty$

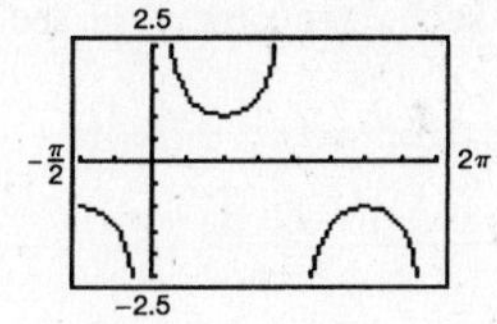

Section 10.2 Verifying Trigonometric Identities

1. $\sin t \csc t = \sin t\left(\dfrac{1}{\sin t}\right) = 1$

2. $\sec y \cos y = \left(\dfrac{1}{\cos y}\right)\cos y = 1$

3. $(1 + \sin \alpha)(1 - \sin \alpha) = 1 - \sin^2 \alpha = \cos^2 \alpha$

4. $\cot^2 y(\sec^2 y - 1) = \cot^2 y \tan^2 y = 1$

5. $$\begin{aligned}\cos^2 \beta - \sin^2 \beta &= (1 - \sin^2 \beta) - \sin^2 \beta \\ &= 1 - 2\sin^2 \beta\end{aligned}$$

6. $$\begin{aligned}\cos^2 \beta - \sin^2 \beta &= \cos^2 \beta - (1 - \cos^2 \beta) \\ &= 2\cos^2 \beta - 1\end{aligned}$$

7. $$\begin{aligned}\tan^2 \theta + 4 &= (\sec^2 \theta - 1) + 4 \\ &= \sec^2 \theta + 3\end{aligned}$$

8. $$\begin{aligned}2 - \sec^2 z &= 2 - (1 + \tan^2 z) \\ &= 1 - \tan^2 z\end{aligned}$$

9. $$\begin{aligned}\sin^2 \alpha - \sin^4 \alpha &= \sin^2 \alpha(1 - \sin^2 \alpha) \\ &= (1 - \cos^2 \alpha)(\cos^2 \alpha) \\ &= \cos^2 \alpha - \cos^4 \alpha\end{aligned}$$

10. $$\begin{aligned}\cos x + \sin x \tan x &= \cos x + \sin x\left(\frac{\sin x}{\cos x}\right) \\ &= \frac{\cos^2 x + \sin^2 x}{\cos x} \\ &= \frac{1}{\cos x} \\ &= \sec x\end{aligned}$$

11. $$\begin{aligned}\frac{\csc^2 \theta}{\cot \theta} &= \csc^2 \theta\left(\frac{1}{\cot \theta}\right) = \csc^2 \theta \tan \theta \\ &= \left(\frac{1}{\sin^2 \theta}\right)\left(\frac{\sin \theta}{\cos \theta}\right) = \left(\frac{1}{\sin \theta}\right)\left(\frac{1}{\cos \theta}\right) \\ &= \csc \theta \sec \theta\end{aligned}$$

12. $$\begin{aligned}\cos t(\csc^2 t - 1) &= \cos t \cot^2 t \\ &= \sin t\left(\frac{\cos t}{\sin t}\right)\cot^2 t \\ &= \frac{1}{\csc t}\cot^3 t \\ &= \frac{\cot^3 t}{\csc t}\end{aligned}$$

13. $$\begin{aligned}\frac{\cot^2 t}{\csc t} &= \frac{\cos^2 t}{\sin^2 t} \cdot \sin t \\ &= \frac{\cos^2 t}{\sin t} \\ &= \frac{1 - \sin^2 t}{\sin t} = \frac{1}{\sin t} - \frac{\sin^2 t}{\sin t} \\ &= \csc t - \sin t\end{aligned}$$

14. $$\begin{aligned}\frac{1}{\tan \beta} + \tan \beta &= \frac{1 + \tan^2 \beta}{\tan \beta} \\ &= \frac{\sec^2 \beta}{\tan \beta}\end{aligned}$$

15. $\sin^{1/2} x \cos x - \sin^{5/2} x \cos x = \sin^{1/2} x \cos x(1 - \sin^2 x) = \sin^{1/2} x \cos x \cdot \cos^2 x = \cos^3 x\sqrt{\sin x}$

16. $\sec^6 x(\sec x\tan x) - \sec^4 x(\sec x\tan x) = \sec^4 x(\sec x\tan x)(\sec^2 x - 1)$

$$= \sec^4 x(\sec x\tan x)\tan^2 x$$

$$= \sec^5 x\tan^3 x$$

17. $\dfrac{1}{\sec x\tan x} = \cos x\cot x = \cos x \cdot \dfrac{\cos x}{\sin x}$

$$= \frac{\cos^2 x}{\sin x}$$

$$= \frac{1 - \sin^2 x}{\sin x}$$

$$= \frac{1}{\sin x} - \sin x$$

$$= \csc x - \sin x$$

18. $\dfrac{\sec\theta - 1}{1 - \cos\theta} = \dfrac{\sec\theta - 1}{1 - (1/\sec\theta)} \cdot \dfrac{\sec\theta}{\sec\theta}$

$$= \frac{\sec\theta(\sec\theta - 1)}{\sec\theta - 1}$$

$$= \sec\theta$$

19. $\dfrac{1}{\sec x} - \dfrac{1}{\cos x} = \cos x - \sec x$ by the reciprocal identities.

20. $\sec x - \cos x = \dfrac{1}{\cos x} - \cos x$

$$= \frac{1 - \cos^2 x}{\cos x}$$

$$= \frac{\sin^2 x}{\cos x}$$

$$= \sin x \cdot \frac{\sin x}{\cos x}$$

$$= \sin x\tan x$$

21. $\sin x\cos x + \sin^3 x\sec x = \sin x\left[\cos x + \sin^2 x\left(\dfrac{1}{\cos x}\right)\right]$

$$= \sin x\left[\frac{\cos^2 x + \sin^2 x}{\cos x}\right]$$

$$= \sin x\left(\frac{1}{\cos x}\right)$$

$$= \frac{\sin x}{\cos x}$$

$$= \tan x$$

22. $\dfrac{\sec x + \tan x}{\sec x - \tan x} = \dfrac{\sec x + \tan x}{\sec x - \tan x} \cdot \dfrac{\sec x + \tan x}{\sec x + \tan x}$

$$= \frac{(\sec x + \tan x)^2}{\sec^2 x - \tan^2 x}$$

$$= \frac{(\sec x + \tan x)^2}{1}$$

$$= (\sec x + \tan x)^2$$

23. $\dfrac{1}{\tan x} + \dfrac{1}{\cot x} = \dfrac{\cot x + \tan x}{\tan x\cot x}$

$$= \frac{\cot x + \tan x}{1}$$

$$= \tan x + \cot x$$

24. $\dfrac{1}{\sin x} - \dfrac{1}{\csc x} = \dfrac{\csc x - \sin x}{\sin x\csc x}$

$$= \frac{\csc x - \sin x}{1}$$

$$= \csc x - \sin x$$

25. $$\frac{\cos\theta\cot\theta}{1-\sin\theta}-1=\frac{\cos\theta\cot\theta-(1-\sin\theta)}{1-\sin\theta}$$
$$=\frac{\cos\theta\left(\frac{\cos\theta}{\sin\theta}\right)-1+\sin\theta}{1-\sin\theta}\cdot\frac{\sin\theta}{\sin\theta}$$
$$=\frac{\cos^2\theta-\sin\theta+\sin^2\theta}{\sin\theta(1-\sin\theta)}$$
$$=\frac{1-\sin\theta}{\sin\theta(1-\sin\theta)}$$
$$=\frac{1}{\sin\theta}$$
$$=\csc\theta$$

26. $$\frac{1+\sin\theta}{\cos\theta}+\frac{\cos\theta}{1+\sin\theta}=\frac{(1+\sin\theta)^2+\cos^2\theta}{\cos\theta(1+\sin\theta)}$$
$$=\frac{1+2\sin\theta+\sin^2\theta+\cos^2\theta}{\cos\theta(1+\sin\theta)}$$
$$=\frac{2+2\sin\theta}{\cos\theta(1+\sin\theta)}$$
$$=\frac{2(1+\sin\theta)}{\cos\theta(1+\sin\theta)}=\frac{2}{\cos\theta}$$
$$=2\sec\theta$$

27. $$\frac{1}{\sin x+1}+\frac{1}{\csc x+1}=\frac{\csc x+1+\sin x+1}{(\sin x+1)(\csc x+1)}$$
$$=\frac{\sin x+\csc x+2}{\sin x\csc x+\sin x+\csc x+1}$$
$$=\frac{\sin x+\csc x+2}{1+\sin x+\csc x+1}$$
$$=\frac{\sin x+\csc x+2}{\sin x+\csc x+2}$$
$$=1$$

28. $$\cos x-\frac{\cos x}{1-\tan x}=\frac{\cos x(1-\tan x)-\cos x}{1-\tan x}$$
$$=\frac{-\cos x\tan x}{1-\tan x}$$
$$=\frac{-\cos x(\sin x/\cos x)}{1-(\sin x/\cos x)}\cdot\frac{\cos x}{\cos x}$$
$$=\frac{-\sin x\cos x}{\cos x-\sin x}$$
$$=\frac{\sin x\cos x}{\sin x-\cos x}$$

29. $$\tan\left(\frac{\pi}{2}-\theta\right)\tan\theta=\cot\theta\tan\theta=\left(\frac{1}{\tan\theta}\right)\tan\theta=1$$

30. $$\frac{\cos[(\pi/2)-x]}{\sin[(\pi/2)-x]}=\frac{\sin x}{\cos x}=\tan x$$

31. $$\frac{\csc(-x)}{\sec(-x)}=\frac{\frac{1}{\sin(-x)}}{\frac{1}{\cos(-x)}}=\frac{\cos(-x)}{\sin(-x)}=\frac{\cos x}{-\sin x}=-\cot x$$

32. $$(1+\sin y)[1+\sin(-y)]=(1+\sin y)(1-\sin y)$$
$$=1-\sin^2 y$$
$$=\cos^2 y$$

33. $$\frac{\cos(-\theta)}{1+\sin(-\theta)}=\frac{\cos\theta}{1-\sin\theta}\cdot\frac{1+\sin\theta}{1+\sin\theta}$$
$$=\frac{\cos\theta(1+\sin\theta)}{1-\sin^2\theta}$$
$$=\frac{\cos\theta(1+\sin\theta)}{\cos^2\theta}$$
$$=\frac{1+\sin\theta}{\cos\theta}$$
$$=\frac{1}{\cos\theta}+\frac{\sin\theta}{\cos\theta}$$
$$=\sec\theta+\tan\theta$$

34. $$\frac{\csc(-\theta)+1}{\cos(-\theta)+\cot(-\theta)}=\frac{-\csc\theta+1}{\cos\theta-\cot\theta}$$
$$=\frac{-\csc\theta+1}{\cos\theta\left[1-\frac{1}{\sin\theta}\right]}$$
$$=\frac{1-\csc\theta}{\cos\theta(1-\csc\theta)}$$
$$=\frac{1}{\cos\theta}$$
$$=\sec\theta$$

35. $$\frac{\sin x\cos y+\cos x\sin y}{\cos x\cos y-\sin x\sin y}=\frac{\dfrac{\sin x\cos y}{\cos x\cos y}+\dfrac{\cos x\sin y}{\cos x\cos y}}{\dfrac{\cos x\cos y}{\cos x\cos y}-\dfrac{\sin x\sin y}{\cos x\cos y}}$$
$$=\frac{\tan x+\tan y}{1-\tan x\tan y}$$

36. $$\frac{\tan x+\tan y}{1-\tan x\tan y}=\frac{\dfrac{1}{\cot x}+\dfrac{1}{\cot y}}{1-\dfrac{1}{\cot x}\cdot\dfrac{1}{\cot y}}\cdot\frac{\cot x\cot y}{\cot x\cot y}$$
$$=\frac{\cot y+\cot x}{\cot x\cot y-1}$$

37. $$\frac{\tan x+\cot y}{\tan x\cot y}=\frac{\dfrac{1}{\cot x}+\dfrac{1}{\tan y}}{\dfrac{1}{\cot x}\cdot\dfrac{1}{\tan y}}\cdot\frac{\cot x\tan y}{\cot x\tan y}=\tan y+\cot x$$

38. $$\frac{\cos x-\cos y}{\sin x+\sin y}+\frac{\sin x-\sin y}{\cos x+\cos y}=\frac{(\cos x-\cos y)(\cos x+\cos y)+(\sin x-\sin y)(\sin x+\sin y)}{(\sin x+\sin y)(\cos x+\cos y)}$$
$$=\frac{\cos^2 x-\cos^2 y+\sin^2 x-\sin^2 y}{(\sin x+\sin y)(\cos x+\cos y)}$$
$$=\frac{(\cos^2 x+\sin^2 x)-(\cos^2 y+\sin^2 y)}{(\sin x+\sin y)(\cos x+\cos y)}$$
$$=0$$

39. $$\sqrt{\frac{1+\sin\theta}{1-\sin\theta}}=\sqrt{\frac{1+\sin\theta}{1-\sin\theta}\cdot\frac{1+\sin\theta}{1+\sin\theta}}$$
$$=\sqrt{\frac{(1+\sin\theta)^2}{1-\sin^2\theta}}$$
$$=\sqrt{\frac{(1+\sin\theta)^2}{\cos^2\theta}}$$
$$=\frac{1+\sin\theta}{|\cos\theta|}$$

40. $$\sqrt{\frac{1-\cos\theta}{1+\cos\theta}}=\sqrt{\frac{1-\cos\theta}{1+\cos\theta}\cdot\frac{1-\cos\theta}{1-\cos\theta}}$$
$$=\sqrt{\frac{(1-\cos\theta)^2}{1-\cos^2\theta}}$$
$$=\sqrt{\frac{(1-\cos\theta)^2}{\sin^2\theta}}$$
$$=\frac{1-\cos\theta}{|\sin\theta|}$$

41. $\cos^2\beta+\cos^2\left(\frac{\pi}{2}-\beta\right)=\cos^2\beta+\sin^2\beta=1$

42. $\sec^2 y-\cot^2\left(\frac{\pi}{2}-y\right)=\sec^2 y-\tan^2 y=1$

43. $$\sin t\csc\left(\frac{\pi}{2}-t\right)=\sin t\sec t=\sin t\left(\frac{1}{\cos t}\right)$$
$$=\frac{\sin t}{\cos t}=\tan t$$

44. $\sec^2\left(\frac{\pi}{2}-x\right)-1=\csc^2 x-1=\cot^2 x$

45. $$2\sec^2 x-2\sec^2 x\sin^2 x-\sin^2 x-\cos^2 x=2\sec^2 x(1-\sin^2 x)-(\sin^2 x+\cos^2 x)$$
$$=2\sec^2 x(\cos^2 x)-1$$
$$=2\cdot\frac{1}{\cos^2 x}\cdot\cos^2 x-1$$
$$=2-1$$
$$=1$$

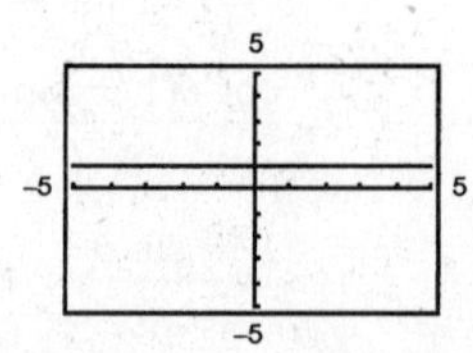

46. $\csc x(\csc x - \sin x) + \dfrac{\sin x - \cos x}{\sin x} + \cot x = \csc^2 x - \csc x \sin x + 1 - \dfrac{\cos x}{\sin x} + \cot x$

$= \csc^2 x - 1 + 1 - \cot x + \cot x$

$= \csc^2 x$

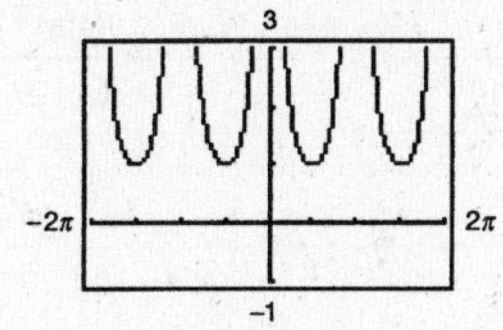

47. $2 + \cos^2 x - 3\cos^4 x = (1 - \cos^2 x)(2 + 3\cos^2 x)$

$= \sin^2 x(2 + 3\cos^2 x)$

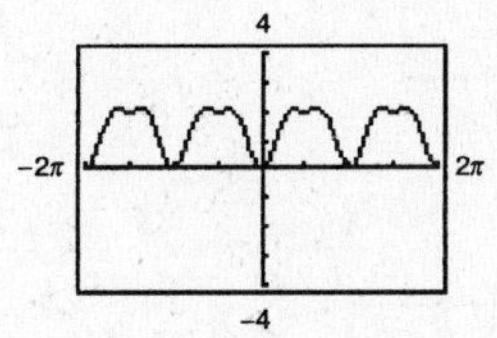

48. $4\tan^4 x + \tan^2 x - 3 = (\tan^2 x + 1)(4\tan^2 x - 3)$

$= \sec^2 x(4\tan^2 x - 3)$

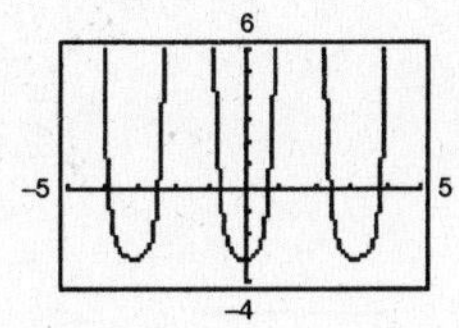

49. $\csc^4 x - 2\csc^2 x + 1 = (\csc^2 x - 1)^2$

$= (\cot^2 x)^2 = \cot^4 x$

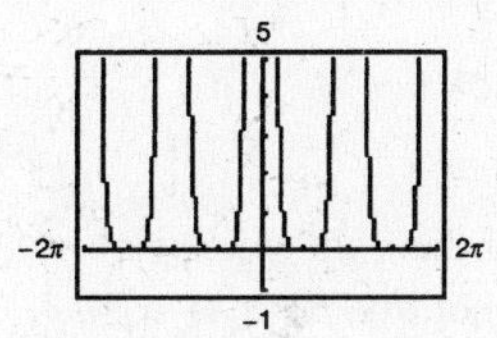

50. $(\sin^4 \beta - 2\sin^2 \beta + 1)\cos \beta = (\sin^2 \beta - 1)^2 \cos \beta$

$= (-\cos^2 \beta)^2 \cos \beta$

$= \cos^5 \beta$

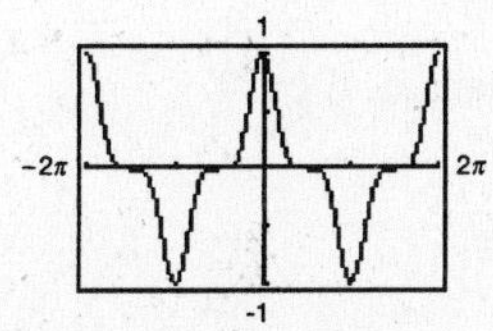

51. $\sec^4 \theta - \tan^4 \theta = (\sec^2 \theta + \tan^2 \theta)(\sec^2 \theta - \tan^2 \theta)$

$= (1 + \tan^2 \theta + \tan^2 \theta)(1)$

$= 1 + 2\tan^2 \theta$

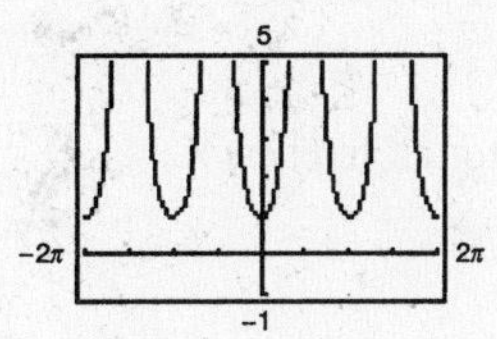

52. $\csc^4 \theta - \cot^4 \theta = (\csc^2 \theta - \cot^2 \theta)(\csc^2 \theta + \cot^2 \theta)$

$= \csc^2 \theta + \cot^2 \theta$

$= \csc^2 \theta + (\csc^2 \theta - 1)$

$= 2\csc^2 \theta - 1$

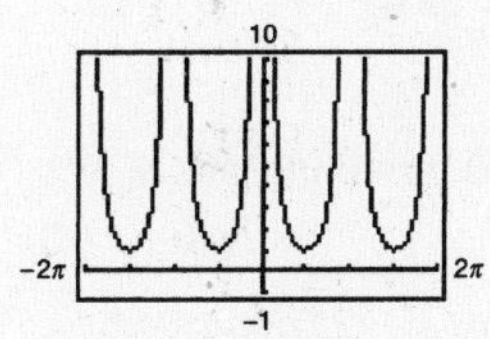

53. $\dfrac{\cos x}{1 + \sin x} = \dfrac{\cos x}{1 + \sin x} \cdot \dfrac{1 - \sin x}{1 - \sin x}$

$= \dfrac{\cos x(1 - \sin x)}{1 - \sin^2 x}$

$= \dfrac{\cos x(1 - \sin x)}{\cos^2 x}$

$= \dfrac{1 - \sin x}{\cos x}$

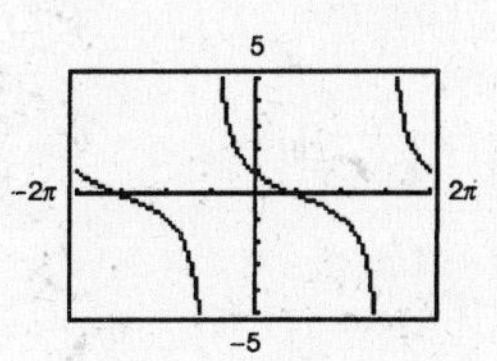

54. $\dfrac{\cot \alpha}{\csc \alpha - 1} \cdot \dfrac{\csc \alpha + 1}{\csc \alpha + 1} = \dfrac{\cot \alpha(\csc \alpha + 1)}{\csc^2 \alpha - 1}$

$= \dfrac{\cot \alpha(\csc \alpha + 1)}{\cot^2 \alpha}$

$= \dfrac{\csc \alpha + 1}{\cot \alpha}$

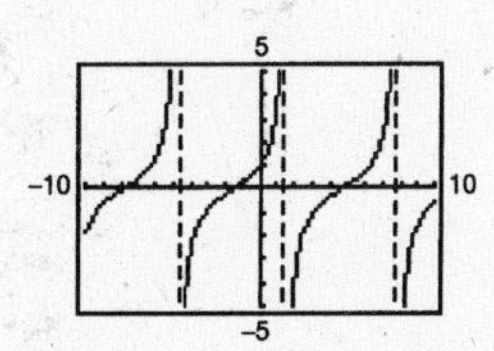

55. $\dfrac{\tan^3\alpha - 1}{\tan\alpha - 1} = \dfrac{(\tan\alpha - 1)(\tan^2\alpha + \tan\alpha + 1)}{\tan\alpha - 1} = \tan^2\alpha + \tan\alpha + 1$

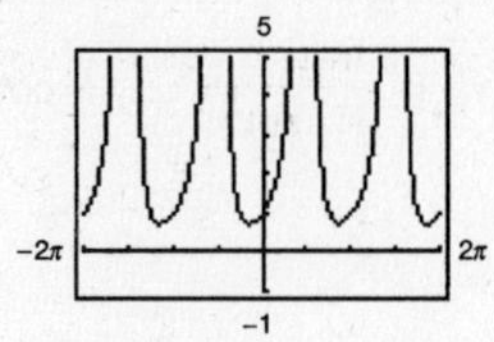

56. $\dfrac{\sin^3\beta + \cos^3\beta}{\sin\beta + \cos\beta} = \dfrac{(\sin\beta + \cos\beta)(\sin^2\beta - \sin\beta\cos\beta + \cos^2\beta)}{\sin\beta + \cos\beta}$

$= \sin^2\beta + \cos^2\beta - \sin\beta\cos\beta$

$= 1 - \sin\beta\cos\beta$

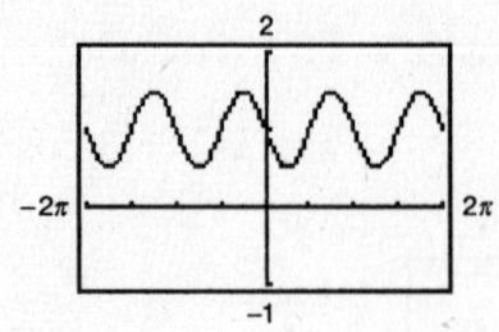

57. Since $\sin^2\theta = 1 - \cos^2\theta$, then $\sin\theta = \pm\sqrt{1 - \cos^2\theta}$; $\sin\theta \neq \sqrt{1 - \cos\theta}$ if θ lies in Quadrant III or IV.

One such angle is $\theta = \dfrac{7\pi}{4}$.

58. $\tan\theta = \sqrt{\sec^2\theta - 1}$

True identity: $\tan\theta = \pm\sqrt{\sec^2\theta - 1}$

$\tan\theta = \sqrt{\sec^2\theta - 1}$ is not true for $\pi/2 < \theta < \pi$ or $3\pi/2 < \theta < 2\pi$. Thus, the equation is not true for $\theta = 3\pi/4$.

59. $\sqrt{\tan^2 x} = |\tan x|$

$\sqrt{\tan^2 x} \neq \tan x$ if x lies in Quadrant II or IV.

One such angle is $x = \dfrac{3\pi}{4}$.

60. $\sqrt{\sin^2 x + \cos^2 x} = \sin x + \cos x$

$\sqrt{\sin^2 x + \cos^2 x} \neq \sin x + \cos x$

The left side is 1 for any x, but the right side is not necessarily 1. The equation is not true for $x = \pi/4$.

61. When n is even,

$$\cos\left[\frac{(2n+1)\pi}{2}\right] = \cos\frac{\pi}{2} = 0.$$

When n is odd,

$$\cos\left[\frac{(2n+1)\pi}{2}\right] = \cos\frac{3\pi}{2} = 0.$$

Thus, $\cos\left[\dfrac{(2n+1)\pi}{2}\right] = 0$ for all integers n.

62. $\sin\left[\dfrac{(12n+1)\pi}{6}\right] = \sin\left[\dfrac{1}{6}(12n\pi + \pi)\right]$

$= \sin\left(2n\pi + \dfrac{\pi}{6}\right)$

$= \sin\dfrac{\pi}{6} = \dfrac{1}{2}$

Thus, $\sin\left[\dfrac{(12n+1)\pi}{6}\right] = \dfrac{1}{2}$ for all integers n.

63. $\ln|\sec\theta| = \ln\left|\dfrac{1}{\cos\theta}\right|$

$= \ln 1 - \ln|\cos\theta|$

$= -\ln|\cos\theta|$

64. $-\ln|\sec\theta + \tan\theta| = -\ln\left|\dfrac{1}{\cos\theta} + \dfrac{\sin\theta}{\cos\theta}\right|$

$= \ln\left|\dfrac{1 + \sin\theta}{\cos\theta}\right|^{-1}$

$= \ln\left|\dfrac{\cos\theta}{1 + \sin\theta} \cdot \dfrac{1 - \sin\theta}{1 - \sin\theta}\right|$

$= \ln\left|\dfrac{\cos\theta - \cos\theta\sin\theta}{1 - \sin^2\theta}\right|$

$= \ln\left|\dfrac{\cos\theta - \cos\theta\sin\theta}{\cos^2\theta}\right|$

$= \ln\left|\dfrac{1}{\cos\theta} - \dfrac{\sin\theta}{\cos\theta}\right|$

$= \ln|\sec\theta - \tan\theta|$

65. $\sin^2 25° + \sin^2 65° = \sin^2 25° + \cos^2(90° - 65°) = \sin^2 25° + \cos^2 25° = 1$

66. $\cos^2 55° + \cos^2 35° = \cos^2 55° + \sin^2(90° - 35°) = \cos^2 55° + \sin^2 55° = 1$

67. $$\begin{aligned}\cos^2 20° + \cos^2 52° + \cos^2 38° + \cos^2 70° &= \cos^2 20° + \cos^2 52° + \sin^2(90° - 38°) + \sin^2(90° - 70°)\\ &= \cos^2 20° + \cos^2 52° + \sin^2 52° + \sin^2 20°\\ &= (\cos^2 20° + \sin^2 20°) + (\cos^2 52° + \sin^2 52°)\\ &= 1 + 1\\ &= 2\end{aligned}$$

68. $$\begin{aligned}\sin^2 12° + \sin^2 40° + \sin^2 50° + \sin^2 78° &= \sin^2 12° + \sin^2 78° + \sin^2 40° + \sin^2 50°\\ &= \cos^2(90° - 12°) + \sin^2 78° + \cos^2(90° - 40°) + \sin^2 50°\\ &= \cos^2 78° + \sin^2 78° + \cos^2 50° + \sin^2 50°\\ &= 1 + 1 = 2\end{aligned}$$

69. $$\begin{aligned}\cos x - \csc x \cot x &= \cos x - \left(\frac{1}{\sin x}\right)\left(\frac{\cos x}{\sin x}\right)\\ &= \cos x\left(1 - \frac{1}{\sin^2 x}\right)\\ &= \cos x(1 - \csc^2 x)\\ &= -\cos x(\csc^2 x - 1)\\ &= -\cos x \cot^2 x\end{aligned}$$

70. $$\begin{aligned}-\csc^2 x - \sec^2 x &= -\left(\frac{1}{\sin^2 x} + \frac{1}{\cos^2 x}\right)\\ &= -\left(\frac{\cos^2 x + \sin^2 x}{\sin^2 x \cos^2 x}\right)\\ &= -\frac{1}{\sin^2 x \cos^2 x}\\ &= -\frac{1}{\sin^2 x} \cdot \frac{1}{\cos^2 x}\\ &= -\csc^2 x \sec^2 x\end{aligned}$$

71. False. For the equation to be an identity, it must be true for all values of θ in the domain.

72. True. An identity is an equation that is true for all real values in the domain of the variable.

73. False

$$\begin{aligned}\frac{(\cos\theta + \sin\theta)^2}{\sin\theta} &= \frac{\cos^2\theta + 2\sin\theta\cos\theta + \sin^2\theta}{\sin\theta}\\ &= \frac{1 + 2\sin\theta\cos\theta}{\sin\theta}\\ &= \frac{1}{\sin\theta} + \frac{2\sin\theta\cos\theta}{\sin\theta}\\ &= \csc\theta + 2\cos\theta\end{aligned}$$

74. False

$$\begin{aligned}\frac{(\tan\theta + 1)^2}{\sec\theta} &= \frac{\tan^2\theta + 2\tan\theta + 1}{\sec\theta}\\ &= \frac{\sec^2\theta + 2\tan\theta}{\sec\theta}\\ &= \sec\theta + \frac{2\tan\theta}{\sec\theta}\\ &= \sec\theta + 2\sin\theta\end{aligned}$$

Section 10.3 Solving Trigonometric Equations

1. $2\cos x - 1 = 0$

(a) $2\cos\frac{\pi}{3} - 1 = 2\left(\frac{1}{2}\right) - 1 = 0$

(b) $2\cos\frac{5\pi}{3} - 1 = 2\left(\frac{1}{2}\right) - 1 = 0$

2. $\sec x - 2 = 0$

(a) $x = \dfrac{\pi}{3}$

$$\sec\frac{\pi}{3} - 2 = \frac{1}{\cos(\pi/3)} - 2$$
$$= \frac{1}{1/2} - 2$$
$$= 2 - 2 = 0$$

(b) $x = \dfrac{5\pi}{3}$

$$\sec\frac{5\pi}{3} - 2 = \frac{1}{\cos(5\pi/3)} - 2$$
$$= \frac{1}{1/2} - 2$$
$$= 2 - 2 = 0$$

3. $3\tan^2 2x - 1 = 0$

(a) $$3\left[\tan 2\left(\frac{\pi}{12}\right)\right]^2 - 1 = 3\tan^2\frac{\pi}{6} - 1$$
$$= 3\left(\frac{1}{\sqrt{3}}\right)^2 - 1$$
$$= 0$$

(b) $$3\left[\tan 2\left(\frac{5\pi}{12}\right)\right]^2 - 1 = 3\tan^2\frac{5\pi}{6} - 1$$
$$= 3\left(-\frac{1}{\sqrt{3}}\right)^2 - 1$$
$$= 0$$

4. $2\cos^2 4x - 1 = 0$

(a) $x = \dfrac{\pi}{16}$

$$2\cos^2\left[4\left(\frac{\pi}{16}\right)\right] - 1 = 2\cos^2\frac{\pi}{4} - 1$$
$$= 2\left(\frac{\sqrt{2}}{2}\right)^2 - 1$$
$$= 2\left(\frac{1}{2}\right) - 1 = 1 - 1 = 0$$

(b) $x = \dfrac{3\pi}{16}$

$$2\cos^2\left[4\left(\frac{3\pi}{16}\right)\right] - 1 = 2\cos^2\frac{3\pi}{4} - 1$$
$$= 2\left(-\frac{\sqrt{2}}{2}\right)^2 - 1$$
$$= 2\left(\frac{1}{2}\right) - 1 = 0$$

5. $2\sin^2 x - \sin x - 1 = 0$

(a) $$2\sin^2\frac{\pi}{2} - \sin\frac{\pi}{2} - 1 = 2(1)^2 - 1 - 1$$
$$= 0$$

(b) $$2\sin^2\frac{7\pi}{6} - \sin\frac{7\pi}{6} - 1 = 2\left(-\frac{1}{2}\right)^2 - \left(-\frac{1}{2}\right) - 1$$
$$= \frac{1}{2} + \frac{1}{2} - 1 = 0$$

6. $\csc^4 x - 4\csc^2 x = 0$

(a) $x = \dfrac{\pi}{6}$

$$\csc^4\frac{\pi}{6} - 4\csc^2\frac{\pi}{6} = \frac{1}{\sin^4\frac{\pi}{6}} - \frac{4}{\sin^2\frac{\pi}{6}}$$
$$= \frac{1}{\left(\frac{1}{2}\right)^4} - \frac{4}{\left(\frac{1}{2}\right)^2}$$
$$= 16 - 16 = 0$$

(b) $x = \dfrac{5\pi}{6}$

$$\csc^4\frac{5\pi}{6} - 4\csc\frac{5\pi}{6} = \frac{1}{\sin^4\frac{5\pi}{6}} - \frac{4}{\sin^2\frac{5\pi}{6}}$$
$$= \frac{1}{\left(\frac{1}{2}\right)^4} - \frac{4}{\left(\frac{1}{2}\right)^2}$$
$$= 16 - 16 = 0$$

7. $2\cos x + 1 = 0$

$2\cos x = -1$

$\cos x = -\frac{1}{2}$

$x = \frac{2\pi}{3} + 2n\pi$

or $x = \frac{4\pi}{3} + 2n\pi$

8. $2\sin x + 1 = 0$

$2\sin x = -1$

$\sin x = -\frac{1}{2}$

$x = \frac{7\pi}{6} + 2n\pi$

or $x = \frac{11\pi}{6} + 2n\pi$

9. $\sqrt{3}\csc x - 2 = 0$

$\sqrt{3}\csc x = 2$

$\csc x = \frac{2}{\sqrt{3}}$

$x = \frac{\pi}{3} + 2n\pi$

or $x = \frac{2\pi}{3} + 2n\pi$

10. $\tan x + \sqrt{3} = 0$

$\tan x = -\sqrt{3}$

$x = \frac{2\pi}{3} + n\pi$

11. $3\sec^2 x - 4 = 0$

$\sec^2 x = \frac{4}{3}$

$\sec x = \pm\frac{2}{\sqrt{3}}$

$x = \frac{\pi}{6} + n\pi$

or $x = \frac{5\pi}{6} + n\pi$

12. $3\cot^2 x - 1 = 0$

$\cot^2 x = \frac{1}{3}$

$\cot x = \pm\frac{1}{\sqrt{3}}$

$x = \frac{\pi}{3} + n\pi$

or $x = \frac{2\pi}{3} + n\pi$

13. $\sin x(\sin x + 1) = 0$

$\sin x = 0$ or $\sin x = -1$

$x = n\pi$ $\qquad x = \frac{3\pi}{2} + 2n\pi$

14. $(3\tan^2 x - 1)(\tan^2 x - 3) = 0$

$3\tan^2 x - 1 = 0$ or $\tan^2 x - 3 = 0$

$\tan x = \pm\frac{1}{\sqrt{3}}$

$x = \frac{\pi}{6} + n\pi$

or $x = \frac{5\pi}{6} + n\pi$

$\tan x = \pm\sqrt{3}$

$x = \frac{\pi}{3} + n\pi$

or $x = \frac{2\pi}{3} + n\pi$

15. $4\cos^2 x - 1 = 0$

$\cos^2 x = \frac{1}{4}$

$\cos^2 x = \pm\frac{1}{2}$

$x = \frac{\pi}{3} + n\pi$

or $x = \frac{2\pi}{3} + n\pi$

16. $\sin^2 x = 3\cos^2 x$

$\sin^2 x - 3(1 - \sin^2 x) = 0$

$4\sin^2 x = 3$

$\sin x = \pm\frac{\sqrt{3}}{2}$

$x = \frac{\pi}{3} + n\pi$

or $x = \frac{2\pi}{3} + n\pi$

17. $\cos^3 x = \cos x$

$\cos^3 x - \cos x = 0$

$\cos x(\cos^2 x - 1) = 0$

$\cos x = 0$ or $\cos^2 x - 1 = 0$

$x = \frac{\pi}{2}, \frac{3\pi}{2}$ $\qquad \cos x = \pm 1$

$x = 0, \pi$

18. $\sec^2 x - 1 = 0$

$\sec^2 x = 1$

$\sec x = \pm 1$

$x = 0$

or $x = \pi$

19. $3\tan^3 x - \tan x = 0$

$\tan x(3\tan^2 x - 1) = 0$

$\tan x = 0$ or $3\tan^2 x - 1 = 0$

$x = 0, \pi$ $\qquad \tan x = \pm\dfrac{\sqrt{3}}{3}$

$x = \dfrac{\pi}{6}, \dfrac{5\pi}{6}, \dfrac{7\pi}{6}, \dfrac{11\pi}{6}$

20. $2\sin^2 x = 2 + \cos x$

$2 - 2\cos^2 x = 2 + \cos x$

$2\cos^2 x + \cos x = 0$

$\cos x(2\cos x + 1) = 0$

$\cos x = 0$ or $2\cos x + 1 = 0$

$x = \dfrac{\pi}{2}, \dfrac{3\pi}{2}$ $\qquad 2\cos x = -1$

$\cos x = -\dfrac{1}{2}$

$x = \dfrac{2\pi}{3}, \dfrac{4\pi}{3}$

21. $\sec^2 x - \sec x - 2 = 0$

$(\sec x - 2)(\sec x + 1) = 0$

$\sec x - 2 = 0$ or $\sec x + 1 = 0$

$\sec x = 2$ $\qquad \sec x = -1$

$x = \dfrac{\pi}{3}, \dfrac{5\pi}{3}$ $\qquad x = \pi$

22. $\sec x \csc x = 2\csc x$

$\sec x \csc x - 2\csc x = 0$

$\csc x(\sec x - 2) = 0$

$\csc x = 0$ or $\sec x - 2 = 0$

No solution $\qquad \sec x = 2$

$x = \dfrac{\pi}{3}, \dfrac{5\pi}{3}$

23. $2\sin x + \csc x = 0$

$2\sin x + \dfrac{1}{\sin x} = 0$

$2\sin^2 x + 1 = 0$

$\sin^2 x = -\dfrac{1}{2} \implies$ No solution

24. $\sec x + \tan x = 1$

$\dfrac{1}{\cos x} + \dfrac{\sin x}{\cos x} = 1$

$1 + \sin x = \cos x$

$(1 + \sin x)^2 = \cos^2 x$

$1 + 2\sin x + \sin^2 x = \cos^2 x$

$1 + 2\sin x + \sin^2 x = 1 - \sin^2 x$

$2\sin^2 x + 2\sin x = 0$

$2\sin x(\sin x + 1) = 0$

$\sin x = 0$ or $\sin x + 1 = 0$

$x = 0, \pi$ $\qquad \sin x = -1$

$x = \dfrac{3\pi}{2}$

(π is extraneous) $\qquad \left(\dfrac{3\pi}{2} \text{ is extraneous}\right)$

$x = 0$ is the only solution.

25. $2\cos^2 x + \cos x - 1 = 0$

$(2\cos x - 1)(\cos x + 1) = 0$

$2\cos x - 1 = 0$ or $\cos x + 1 = 0$

$\cos x = \dfrac{1}{2}$ $\qquad \cos x = -1$

$x = \dfrac{\pi}{3}, \dfrac{5\pi}{3}$ $\qquad x = \pi$

26. $2\sin^2 x + 3\sin x + 1 = 0$

$(2\sin x + 1)(\sin x + 1) = 0$

$2\sin x + 1 = 0$ or $\sin x + 1 = 0$

$\sin x = -\dfrac{1}{2}$ $\qquad \sin x = -1$

$x = \dfrac{7\pi}{6}, \dfrac{11\pi}{6}$ $\qquad x = \dfrac{3\pi}{2}$

27.
$$2\sec^2 x + \tan^2 x - 3 = 0$$
$$2(\tan^2 x + 1) + \tan^2 x - 3 = 0$$
$$3\tan^2 x - 1 = 0$$
$$\tan x = \pm\frac{\sqrt{3}}{3}$$
$$x = \frac{\pi}{6}, \frac{5\pi}{6}, \frac{7\pi}{6}, \frac{11\pi}{6}$$

28.
$$\cos x + \sin x \tan x = 2$$
$$\cos x + \sin x\left(\frac{\sin x}{\cos x}\right) = 2$$
$$\frac{\cos^2 x + \sin^2 x}{\cos x} = 2$$
$$\frac{1}{\cos x} = 2$$
$$\cos x = \frac{1}{2}$$
$$x = \frac{\pi}{3}, \frac{5\pi}{3}$$

29. $\cos 2x = \frac{1}{2}$

$2x = \frac{\pi}{3} + 2n\pi$ or $2x = \frac{5\pi}{3} + 2n\pi$

$x = \frac{\pi}{6} + n\pi$ $\quad x = \frac{5\pi}{6} + n\pi$

30. $\sin 2x = -\frac{\sqrt{3}}{2}$

$2x = \frac{4\pi}{3} + 2n\pi$ or $2x = \frac{5\pi}{3} + 2n\pi$

$x = \frac{2\pi}{3} + n\pi$ $\quad x = \frac{5\pi}{6} + n\pi$

31. $\tan 3x = 1$

$3x = \frac{\pi}{4} + 2n\pi$ or $3x = \frac{5\pi}{4} + 2n\pi$

$x = \frac{\pi}{12} + \frac{2n\pi}{3}$ $\quad x = \frac{5\pi}{12} + \frac{2n\pi}{3}$

These can be combined as $x = \frac{\pi}{12} + \frac{n\pi}{3}$.

32. $\sec 4x = 2$

$4x = \frac{\pi}{3} + 2n\pi$ or $4x = \frac{5\pi}{3} + 2n\pi$

$x = \frac{\pi}{12} + \frac{n\pi}{2}$ $\quad x = \frac{5\pi}{12} + \frac{n\pi}{2}$

33. $\cos\left(\frac{x}{2}\right) = \frac{\sqrt{2}}{2}$

$\frac{x}{2} = \frac{\pi}{4} + 2n\pi$ or $\frac{x}{2} = \frac{7\pi}{4} + 2n\pi$

$x = \frac{\pi}{2} + 4n\pi$ $\quad x = \frac{7\pi}{2} + 4n\pi$

34. $\sin\frac{x}{2} = -\frac{\sqrt{3}}{2}$

$\frac{x}{2} = \frac{4\pi}{3} + 2n\pi$ or $\frac{x}{2} = \frac{5\pi}{3} + 2n\pi$

$x = \frac{8\pi}{3} + 4n\pi$ $\quad x = \frac{10\pi}{3} + 4n\pi$

35. $2\sin^2 2x = 1$

$\sin 2x = \pm\frac{1}{\sqrt{2}} = \pm\frac{\sqrt{2}}{2}$

$2x = \frac{\pi}{4} + \frac{n\pi}{2}$

Thus, $x = \frac{\pi}{8} + \frac{n\pi}{4}$

36. $\tan^2 3x = 3$

$\tan 3x = \pm\sqrt{3}$

$3x = \frac{\pi}{3} + n\pi \Rightarrow x = \frac{\pi}{9} + \frac{n\pi}{3}$

or

$3x = \frac{2\pi}{3} + n\pi \Rightarrow x = \frac{2\pi}{9} + \frac{n\pi}{3}$

37. $\tan 3x(\tan x - 1) = 0$

$\tan 3x = 0$ or $\tan x - 1 = 0$

$3x = n\pi$ $\quad$ $\tan x = 1$

$x = \frac{n\pi}{3}$ $\quad$ $x = \frac{\pi}{4} + n\pi$

38. $\cos 2x(2\cos x + 1) = 0$

$\cos 2x = 0$ or $2\cos x + 1 = 0$

$2x = \frac{\pi}{2} + n\pi$ $\quad$ $\cos x = -\frac{1}{2}$

$x = \frac{\pi}{4} + \frac{n\pi}{2}$ $\quad$ $x = \frac{2\pi}{3} + 2n\pi$

or

$x = \frac{4\pi}{3} + 2n\pi$

39. $y = \sin\frac{\pi x}{2} + 1$

$\sin\frac{\pi x}{2} + 1 = 0$

$\sin\frac{\pi x}{2} = -1$

$\frac{\pi x}{2} = \frac{3\pi}{2} + 2n\pi$

$x = 3 + 4n$

For $-2 < x < 4$ the intercepts are -1 and 3.

40. $y = \sin\pi x + \cos\pi x$

$\sin\pi x + \cos\pi x = 0$

$\sin\pi x = -\cos\pi x$

$\pi x = -\frac{\pi}{4} + n\pi$

$x = -\frac{1}{4} + n$

For $-1 < x < 3$ the intercepts are $-\frac{1}{4}, \frac{3}{4}, \frac{7}{4}, \frac{11}{4}$.

41. $y = \tan^2\left(\frac{\pi x}{6}\right) - 3$

$\tan^2\left(\frac{\pi x}{6}\right) - 3 = 0$

$\tan^2\left(\frac{\pi x}{6}\right) = 3$

$\tan\left(\frac{\pi x}{6}\right) = \pm\sqrt{3}$

$\frac{\pi x}{6} = \frac{\pi}{3} + n\pi$ or $\frac{\pi x}{6} = \frac{2\pi}{3} + n\pi$

$x = 2 + 6n$ or $x = 4 + 6n$

For $-3 < x < 3$ the intercepts are -2 and 2.

42. $y = \sec^4\left(\frac{\pi x}{8}\right) - 4$

$\sec^4\left(\frac{\pi x}{8}\right) - 4 = 0$

$\sec^4\left(\frac{\pi x}{8}\right) = 4$

$\sec\left(\frac{\pi x}{8}\right) = \pm\sqrt{2}$

$\frac{\pi x}{8} = \frac{\pi}{4} + \frac{\pi}{2}n$

$x = 2 + 4n$

For $-3 < x < 3$ the intercepts are -2 and 2.

43. $6y^2 - 13y + 6 = 0$

$(3y - 2)(2y - 3) = 0$

$3y - 2 = 0$ or $2y - 3 = 0$

$y = \frac{2}{3}$ $\quad$ $y = \frac{3}{2}$

$6\cos^2 x - 13\cos x + 6 = 0$

$(3\cos x - 2)(2\cos x - 3) = 0$

$3\cos x - 2 = 0$ or $2\cos x - 3 = 0$

$\cos x = \frac{2}{3}$ $\quad$ $\cos x = \frac{3}{2}$ (No solution)

$x \approx 0.8411 + 2n\pi, 5.4421 + 2n\pi$

44. $y^2 + y - 20 = 0$

$(y + 5)(y - 4) = 0$

$y + 5 = 0$ or $y - 4 = 0$

$y = -5$ $\qquad y = 4$

$\sin^2 x + \sin x - 20 = 0$

$(\sin x + 5)(\sin x - 4) = 0$

$\sin x = -5$ or $\sin x = 4$

No solution $\qquad$ No solution

45. $f(x) = \cos \dfrac{1}{x}$

(a) Domain: All real numbers except $x = 0$.

(b) Symmetry: y-axis; Asymptote: $y = 1$

(c) The function values oscillate between -1 and 1 more rapidly.

(d) Infinitely many

(e) $\cos \dfrac{1}{x} = 0$

$\dfrac{1}{x} = n\pi + \dfrac{\pi}{2} \Rightarrow x = \dfrac{1}{n\pi + \pi/2}$ x is the greatest when $n = 0$. Then $x = \dfrac{2}{\pi} \approx 0.6366$

46. $f(x) = \dfrac{\sin x}{x}$

(a) Domain: all real numbers except $x = 0$.

(b) Symmetry: y-axis; Asymptote: $y = 0$

(c) As $x \longrightarrow 0, f(x) \longrightarrow 1$.

(d) $\dfrac{\sin x}{x} = 0$ has four solutions in the interval $[-8, 8]$.

$(\sin x)\left(\dfrac{1}{x}\right) = 0$

$\sin x = 0$

$x = -2\pi, -\pi, \pi, 2\pi$

47. $2 \sin x + \cos x = 0$

$2 \sin x = -\cos x$

$2 = -\dfrac{\cos x}{\sin x}$

$2 = -\cot x$

$-2 = \cot x$

$-\dfrac{1}{2} = \tan x$

$x = \arctan\left(-\dfrac{1}{2}\right)$

$x = \pi - \arctan\left(\dfrac{1}{2}\right) \approx 2.6779$ or $x = 2\pi - \arctan\left(\dfrac{1}{2}\right) \approx 5.8195$

Graph $y_1 = 2 \sin x + \cos x$.

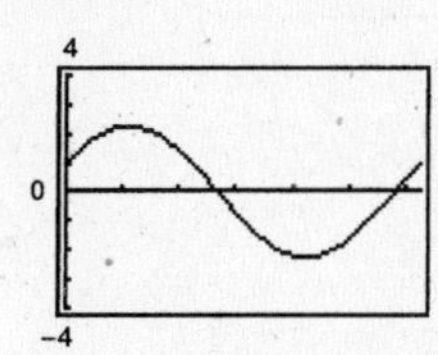

The x-intercepts occur at $x \approx 2.6779$ and $x \approx 5.8195$.

48. $\dfrac{\cos x \cot x}{1 - \sin x} = 3$

$$y_1 = \left(\frac{\dfrac{\cos x}{\tan x}}{1 - \sin x}\right) - 3$$

$x \approx 0.5236, 2.6180$

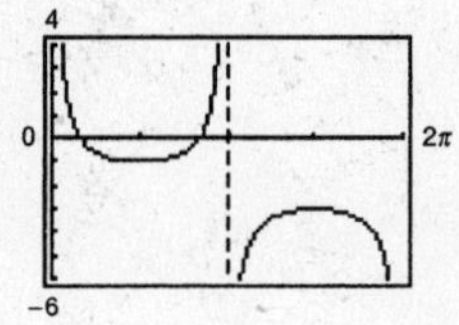

49.
$$\frac{1 + \sin x}{\cos x} + \frac{\cos x}{1 + \sin x} = 4$$
$$\frac{(1 + \sin x)^2 + \cos^2 x}{\cos x(1 + \sin x)} = 4$$
$$\frac{1 + 2\sin x + \sin^2 x + \cos^2 x}{\cos x(1 + \sin x)} = 4$$
$$\frac{2 + 2\sin x}{\cos x(1 + \sin x)} = 4$$
$$\frac{2}{\cos x} = 4$$
$$\cos x = \frac{1}{2}$$
$$x = \frac{\pi}{3}, \frac{5\pi}{3}$$

Graph $y_1 = \dfrac{1 + \sin x}{\cos x} + \dfrac{\cos x}{1 + \sin x} - 4.$

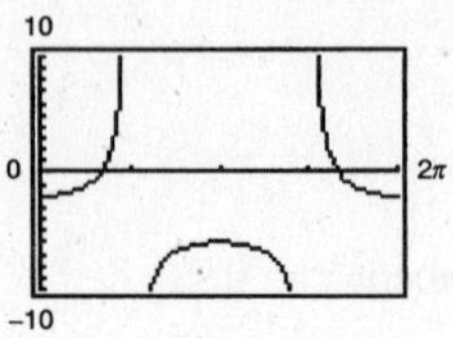

The x-intercepts occur at $x = \dfrac{\pi}{3} \approx 1.0472$ and $x = \dfrac{5\pi}{3} \approx 5.2360.$

50. $4\sin^3 x + 2\sin^2 x - 2\sin x - 1 = 0$

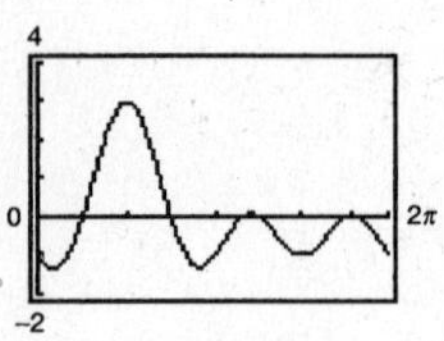

$x \approx 0.7854, 2.3562, 3.6652, 3.9270, 5.4978, 5.7596$

51. $x \tan x - 1 = 0$

Graph $y_1 = x \tan x - 1.$

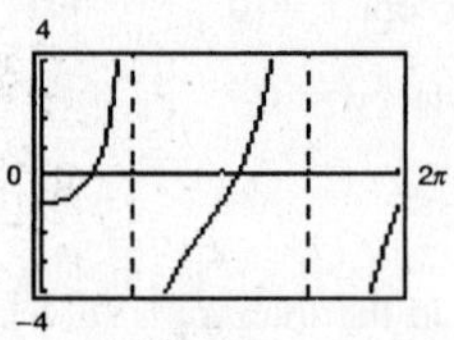

The x-intercepts occur at $x \approx 0.8603$ and $x \approx 3.4256$

52. $x \cos x - 1 = 0$

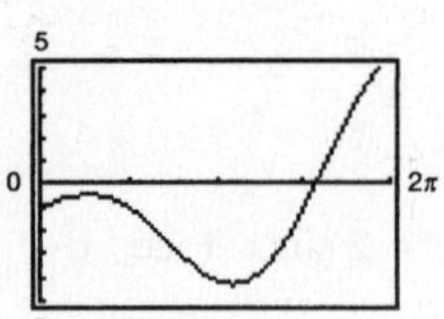

$x \approx 4.9172$

53. $\sec^2 x + 0.5 \tan x - 1 = 0$

Graph $y_1 = \dfrac{1}{(\cos x)^2} + 0.5 \tan x - 1.$

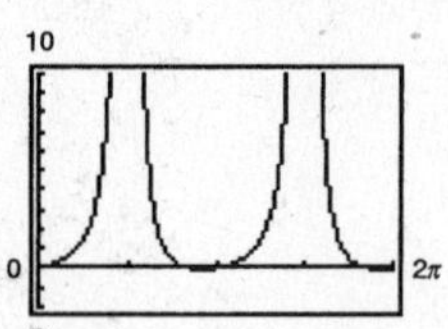

The x-intercepts occur at $x = 0$, $x \approx 2.6779$, $x = \pi \approx 3.1416$, and $x \approx 5.8195.$

54. $\csc^2 x + 0.5 \cot x - 5 = 0$

$$y_1 = \left(\frac{1}{\sin x}\right)^2 + \frac{1}{2\tan x} - 5$$

$x \approx 0.5153, 2.7259, 3.6569, 5.8675$

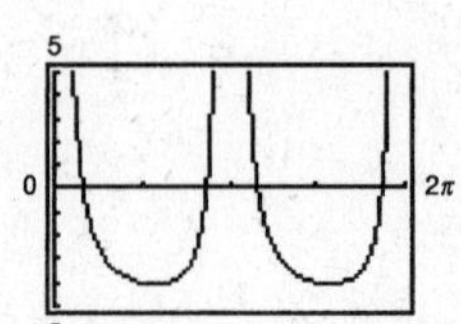

55. $2\tan^2 x + 7\tan x - 15 = 0$

$(2\tan x - 3)(\tan x + 5) = 0$

$2\tan x - 3 = 0$ or $\tan x + 5 = 0$

$\tan x = 1.5$ $\quad\quad$ $\tan x = -5$

$x \approx 0.9828, 4.1244$ $\quad\quad$ $x \approx 1.7682,\ 4.9098$

Graph $y_1 = 2\tan^2 x + 7\tan x - 15$.

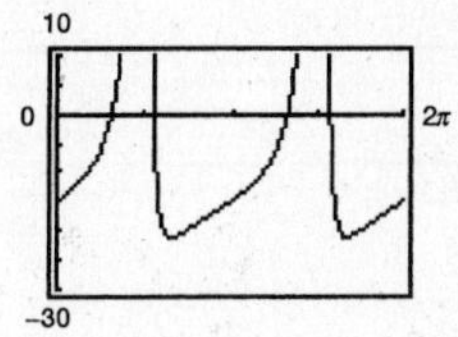

The x-intercepts occur at $x \approx 0.9828$, $x \approx 1.7682$, $x \approx 4.1244$, and $x \approx 4.9098$.

56. $6\sin^2 x - 7\sin x + 2 = 0$

$x \approx 0.5236\left(\text{or } \dfrac{\pi}{6}\right), 0.7297, 2.4119, 2.6180\left(\text{or } \dfrac{5\pi}{6}\right)$

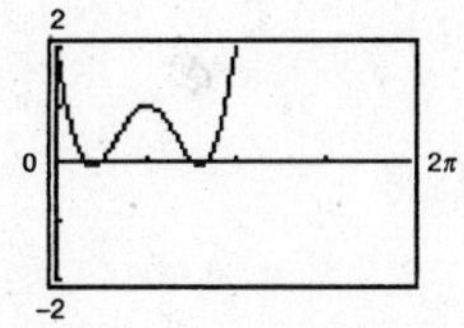

57. $12\sin^2 x - 13\sin x + 3 = 0$

$$\sin x = \frac{-(-13) \pm \sqrt{(-13)^2 - 4(12)(3)}}{2(12)}$$

$$= \frac{13 \pm 5}{24}$$

$\sin x = \dfrac{1}{3}$ or $\sin x = \frac{3}{4}$

$x \approx 0.3398,\ 2.8018$ $\quad\quad$ $x \approx 0.8481,\ 2.2935$

Graph $y_1 = 12\sin^2 x - 13\sin x + 3$.

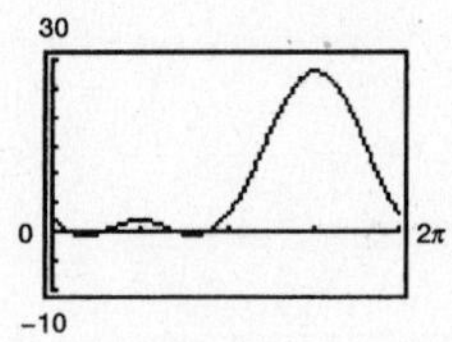

The x-intercepts occur at $x \approx 0.3398$, $x \approx 0.8481$, $x \approx 2.2935$,and $x \approx 2.8018$.

58. $3\tan^2 x + 4\tan x - 4 = 0$

$$\tan x = \frac{-4 \pm \sqrt{4^2 - 4(3)(-4)}}{2(3)} = \frac{-4 \pm \sqrt{64}}{6} = -2, \frac{2}{3}$$

$\tan x = -2$ $\quad\quad$ $\tan x = \dfrac{2}{3}$

$x = \arctan(-2) + n\pi$ $\quad\quad$ $x = \arctan\left(\dfrac{2}{3}\right) + n\pi$

$\approx -1.1071 + n\pi$ $\quad\quad$ $\approx 0.5880 + n\pi$

The values of x in $[0, 2\pi)$ are 0.5880, 3.7296, 2.0344, 5.1760.

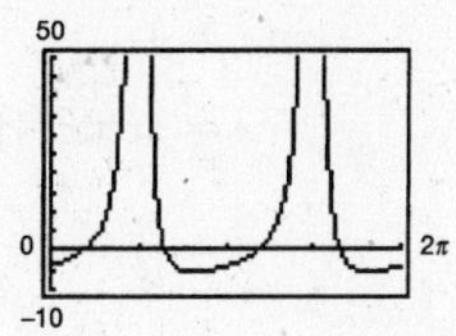

59. $\tan^2 x + 3\tan x + 1 = 0$

$$\tan x = \frac{-3 \pm \sqrt{3^2 - 4(1)(1)}}{2(1)} = \frac{-3 \pm \sqrt{5}}{2}$$

$\tan x = \dfrac{-3 - \sqrt{5}}{2}$ or $\tan x = \dfrac{-3 + \sqrt{5}}{2}$

$x \approx 1.9357,\ 5.0773$ $\quad\quad$ $x \approx 2.7767,\ 5.9183$

Graph $y_1 = \tan^2 x + 3\tan x + 1$.

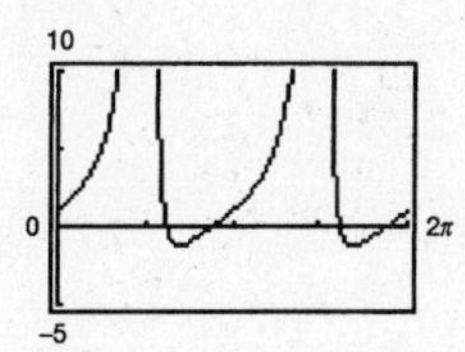

The x-intercepts occur at $x \approx 1.9357$, $x \approx 2.7767$, $x \approx 5.0773$, and $x \approx 5.9183$.

60. $4\cos^2 x - 4\cos x - 1 = 0$

$$\cos x = \frac{4 \pm \sqrt{(-4)^2 - 4(4)(-1)}}{2(4)}$$

$$= \frac{4 \pm \sqrt{32}}{8} = \frac{1 \pm \sqrt{2}}{2}$$

$\cos x = \dfrac{1 - \sqrt{2}}{2}$ $\qquad$ $\cos x = \dfrac{1 + \sqrt{2}}{2}$

$x = \arccos\left(\dfrac{1 - \sqrt{2}}{2}\right)$ $\qquad$ No solution

≈ 1.7794 $\qquad$ $\left(\dfrac{1 + \sqrt{2}}{2} > 1\right)$

Solutions in $[0, 2\pi)$ are $\arccos\left(\dfrac{1 - \sqrt{2}}{2}\right)$ and $2\pi - \arccos\left(\dfrac{1 - \sqrt{2}}{2}\right)$: 1.7794, 4.5038.

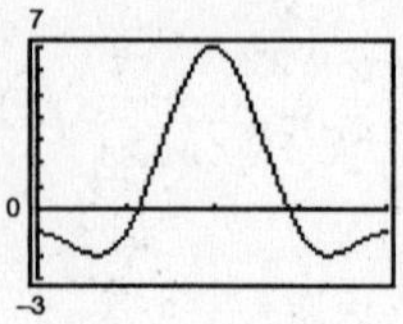

61. $\tan^2 x - 6\tan x + 5 = 0$

$(\tan x - 1)(\tan x - 5) = 0$

$\tan x - 1 = 0$ or $\tan x - 5 = 0$

$\tan x = 1$ $\qquad$ $\tan x = 5$

$x = \dfrac{\pi}{4}, \dfrac{5\pi}{4}$ $\qquad$ $x = \arctan 5, \arctan 5 + \pi$

62. $\sec^2 x + \tan x - 3 = 0$

$1 + \tan^2 x + \tan x - 3 = 0$

$\tan^2 x + \tan x - 2 = 0$

$(\tan x + 2)(\tan x - 1) = 0$

$\tan x + 2 = 0$ $\qquad$ $\tan x - 1 = 0$

$\tan x = -2$ $\qquad$ $\tan x = 1$

$x = \arctan(-2) + n\pi$ $\qquad$ $x = \arctan(1) + n\pi$

$\approx -1.1071 + n\pi$ $\qquad$ $= \dfrac{\pi}{4} + n\pi$

Solutions in $[0, 2\pi)$ are $\arctan(-2) + \pi, \arctan(-2) + 2\pi, \dfrac{\pi}{4}, \dfrac{5\pi}{4}$.

63. $2\cos^2 x - 5\cos x + 2 = 0$

$(2\cos x - 1)(\cos x - 2) = 0$

$2\cos x - 1 = 0$ or $\cos x - 2 = 0$

$\cos x = \dfrac{1}{2}$ $\qquad$ $\cos x = 2$

$x = \dfrac{\pi}{3}, \dfrac{5\pi}{3}$ $\qquad$ $x = \arccos 2,$

$2\pi - \arccos 2$

64. $2\sin^2 x - 7\sin x + 3 = 0$

$(\sin x - 3)(2\sin x - 1) = 0$

$\sin x - 3 = 0$ $\qquad$ $2\sin x - 1 = 0$

No solution $\qquad$ $\sin x = \dfrac{1}{2}$

$x = \dfrac{\pi}{6}, \dfrac{5\pi}{6}$

Solutions in $[0, 2\pi)$ are $\dfrac{\pi}{6}, \dfrac{5\pi}{6}$.

65. (a) $f(x) = \sin x + \cos x$

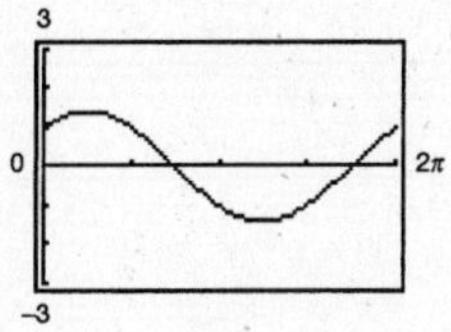

Maximum point: (0.7854, 1.4142)

Minimum point: (3.9270, −1.4142)

(b) $\cos x - \sin x = 0$

$$\cos x = \sin x$$

$$1 = \frac{\sin x}{\cos x}$$

$$\tan x = 1$$

$$x = \frac{\pi}{4}, \frac{5\pi}{4} \approx 0.7854, 3.9270$$

$$f\left(\frac{\pi}{4}\right) = \sin\frac{\pi}{4} + \cos\frac{\pi}{4} = \frac{\sqrt{2}}{2} + \frac{\sqrt{2}}{2} = \sqrt{2}$$

$$f\left(\frac{5\pi}{4}\right) = \sin\frac{5\pi}{4} + \cos\frac{5\pi}{4} = -\sin\frac{\pi}{4} + \left(-\cos\frac{\pi}{4}\right) = -\frac{\sqrt{2}}{2} - \frac{\sqrt{2}}{2} = -\sqrt{2}$$

Therefore, the maximum point in the interval $[0, 2\pi)$ is $(\pi/4, \sqrt{2})$ and the minimum point is $(5\pi/4, -\sqrt{2})$.

66. (a) $f(x) = \sin^2 x + \cos x$

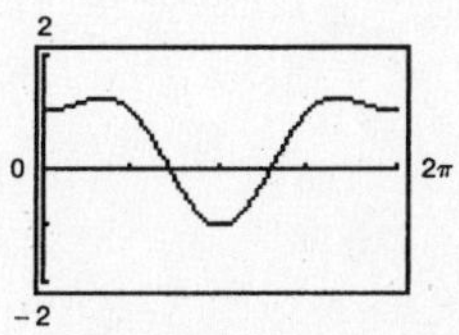

Maximum points: (1.0472, 1.25), (5.2360, 1.25)

Minimum points: (0, 1), (3.1416, −1)

(b) $2 \sin x \cos x - \sin x = 0$

$\sin x(2 \cos x - 1) = 0$

$\sin x = 0 \Rightarrow x = 0 \quad \text{or} \quad x = \pi \approx 3.1416$

$\cos x = \frac{1}{2} \Rightarrow x = \frac{\pi}{3} \approx 1.0472 \quad \text{or} \quad x = \frac{5\pi}{3} \approx 5.2360$

$f(0) = 1, f(\pi) = -1, f\left(\frac{\pi}{3}\right) = f\left(\frac{5\pi}{3}\right) = 1.25$

Maximum points: $\left(\frac{\pi}{3}, 1.25\right), \left(\frac{5\pi}{3}, 1.25\right)$

Minimum points: $(0, 1), (\pi, -1)$

67. (a) $f(x) = 2 \sin x + \cos 2x$

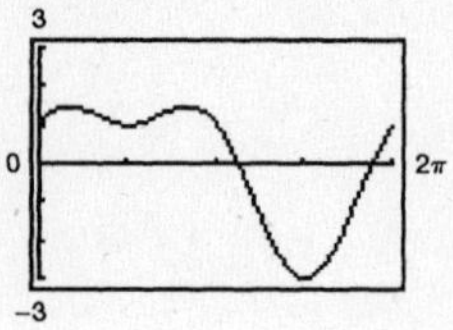

Maximum points: (0.5240, 1.5), (2.6180, 1.5)

Minimum points: (1.5708, 1.0), (4.7124, −3.0)

(b) $2 \cos x - 4 \sin x \cos x = 0$

$2 \cos x(1 - 2 \sin x) = 0$

$2 \cos x = 0$

$x = \frac{\pi}{2}, \frac{3\pi}{2}$

$\approx 1.5708, 4.7124$

$f\left(\frac{\pi}{2}\right) = 1$

$f\left(\frac{3\pi}{2}\right) = -3$

$1 - 2 \sin x = 0$

$\sin x = \frac{1}{2}$

$x = \frac{\pi}{6}, \frac{5\pi}{6}$

$\approx 0.5240, 2.6180$

$f\left(\frac{\pi}{6}\right) = f\left(\frac{5\pi}{6}\right) = 1.5$

Maximum points: $\left(\frac{\pi}{6}, 1.5\right), \left(\frac{5\pi}{6}, 1.5\right)$

Minimum points: $\left(\frac{\pi}{2}, 1\right), \left(\frac{3\pi}{2}, -3\right)$

68. (a) $f(x) = \cos 2x - 2 \sin x$

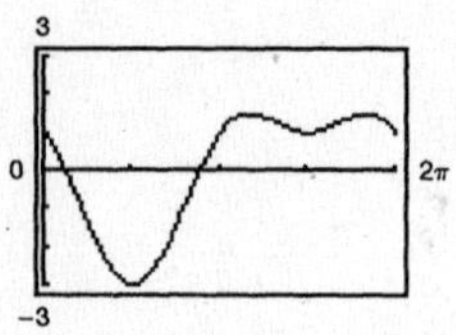

Maximum points: (3.6652, 1.5), (5.7596, 1.5)

Minimum points: (1.5708, −3.0), (4.7124, 1.0)

(b) $2 \sin x \cos x + \cos x = 0$

$\cos x(2 \sin x + 1) = 0$

$\cos x = 0 \Rightarrow x = \frac{\pi}{2} \approx 1.5708$ or $x = \frac{3\pi}{2} \approx 4.7124$

$\sin x = -\frac{1}{2} \Rightarrow x = \frac{7\pi}{6} \approx 3.6652$ or $x = \frac{11\pi}{6} \approx 5.7596$

$f\left(\frac{\pi}{2}\right) = -3, f\left(\frac{3\pi}{2}\right) = 1, f\left(\frac{7\pi}{6}\right) = f\left(\frac{11\pi}{6}\right) = 1.5$

Maximum points: $\left(\frac{7\pi}{6}, 1.5\right), \left(\frac{11\pi}{6}, 1.5\right)$

Minimum points: $\left(\frac{\pi}{2}, -3\right), \left(\frac{3\pi}{2}, 1\right)$

69. $y = \frac{1}{12}(\cos 8t - 3 \sin 8t)$

$$\frac{1}{12}(\cos 8t - 3 \sin 8t) = 0$$

$$\cos 8t = 3 \sin 8t$$

$$\frac{1}{3} = \tan 8t$$

$$8t \approx 0.32175 + n\pi$$

$$t \approx 0.04 + \frac{n\pi}{8}$$

In the interval $0 \le t \le 1$, $t \approx 0.04, 0.43$, and 0.83.

70. $y_1 = 1.56e^{-0.22t} \cos 4.9t$, $y_2 = 1$, $y_3 = -1$

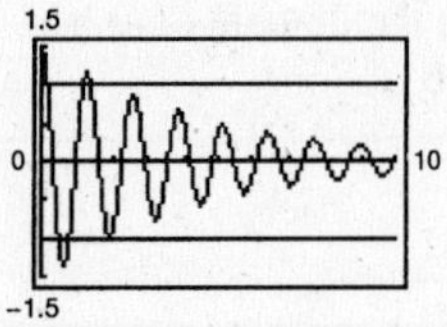

Right most point of intersection: $(1.9575, -1)$

The displacement does not exceed one inch from equilibrium after $t = 1.96$ seconds.

71. $S = 74.50 + 43.75 \sin \frac{\pi t}{6}$

t	1	2	3	4	5	6	7	8	9	10	11	12
S	96.4	112.4	118.3	112.4	96.4	74.5	52.6	36.6	30.8	36.6	52.6	74.5

Sales exceed 100,000 units during February, March, and April.

72. range $= 300$ feet

$$v_0 = 100 \text{ feet per second}$$

$$r = \frac{1}{32}v_0^2 \sin 2\theta$$

$$\frac{1}{32}(100)^2 \sin 2\theta = 300$$

$$\sin 2\theta = 0.96$$

$$2\theta = \arcsin(0.96) \approx 73.74°$$

$$\theta \approx 36.9°$$

or

$$2\theta = 180° - \arcsin(0.96) \approx 106.26°$$

$$\theta \approx 53.1°$$

73. Range $= 1000$ yards $= 3000$ feet

$$v_0 = 1200 \text{ feet per second}$$

$$r = \frac{1}{32} v_0^2 \sin 2\theta$$

$$3000 = \frac{1}{32}(1200)^2 \sin 2\theta$$

$$\sin 2\theta \approx 0.066667$$

$$2\theta \approx 3.8°$$

$$\theta \approx 1.9°$$

74. (a) and (b)

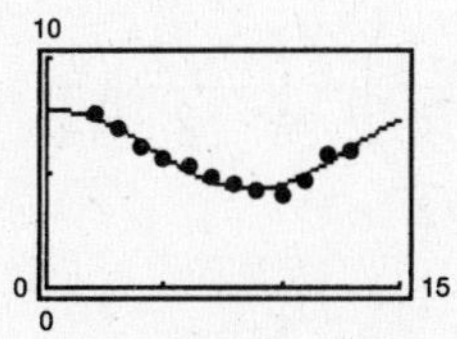

The model is a good fit to the data.

(c) The constant term gives the average rate of 5.99%.

(d) Period: $\frac{2\pi}{0.38} \approx 16.5$ years

(e) $r = 6$ when $t \approx 21$ which corresponds to 2011.

75. True. The period of $2 \sin 4t - 1$ is $\frac{\pi}{2}$ and the period of $2 \sin t - 1$ is 2π.

In the interval $[0, 2\pi)$ the first equation has four cycles whereas the second equation has only one cycle, thus the first equation has four times the x-intercepts (solutions) as the second equation.

76. False.

$\sin x = 3.4$ has no solution since 3.4 is outside the range of sin. Also, 3.4 is outside the domain of arcsin, so $x = \arcsin(3.4)$ is an invalid equation.

77. True

$\tan 4x = 0$

In general, $4x = 0 + n\pi$ so

$$x = \frac{n\pi}{4}, \text{ where } n \text{ is an integer.}$$

78. False

$\sin 2x = -1$

In general, $2x = \frac{3\pi}{2} + 2n\pi$ so

$$x = \frac{3\pi}{4} + n\pi, \text{ where } n \text{ is an integer.}$$

79. $y_1 = 2\sin x$

$y_2 = 3x + 1$

From the graph we see that there is only one point of intersection.

80. By inspecting the graphs of y_1 and y_2, it appears they intersect at three points.

Section 10.4 Sum and Difference Formulas

1. (a) $\cos\left(\frac{\pi}{4} + \frac{\pi}{3}\right) = \cos\frac{\pi}{4}\cos\frac{\pi}{3} - \sin\frac{\pi}{4}\sin\frac{\pi}{3}$

$$= \frac{\sqrt{2}}{2}\cdot\frac{1}{2} - \frac{\sqrt{2}}{2}\cdot\frac{\sqrt{3}}{2}$$

$$= \frac{\sqrt{2} - \sqrt{6}}{4}$$

(b) $\cos\frac{\pi}{4} + \cos\frac{\pi}{3} = \frac{\sqrt{2}}{2} + \frac{1}{2} = \frac{\sqrt{2} + 1}{2}$

2. (a) $\sin\left(\frac{3\pi}{4} + \frac{5\pi}{6}\right) = \sin\frac{3\pi}{4}\cos\frac{5\pi}{6} + \cos\frac{3\pi}{4}\sin\frac{5\pi}{6}$

$$= \left(\frac{\sqrt{2}}{2}\right)\left(-\frac{\sqrt{3}}{2}\right) + \left(-\frac{\sqrt{2}}{2}\right)\left(\frac{1}{2}\right)$$

$$= -\frac{\sqrt{6} + \sqrt{2}}{4}$$

(b) $\sin\frac{3\pi}{4} + \sin\frac{5\pi}{6} = \frac{\sqrt{2}}{2} + \frac{1}{2} = \frac{\sqrt{2} + 1}{2}$

3. (a) $\sin\left(\frac{7\pi}{6} - \frac{\pi}{3}\right) = \sin\frac{5\pi}{6} = \sin\frac{\pi}{6} = \frac{1}{2}$

(b) $\sin\frac{7\pi}{6} - \sin\frac{\pi}{3} = -\frac{1}{2} - \frac{\sqrt{3}}{2} = \frac{-1 - \sqrt{3}}{2}$

4. (a) $\cos\left(\frac{2\pi}{3} - \frac{\pi}{6}\right) = \cos\frac{2\pi}{3}\cos\frac{\pi}{6} + \sin\frac{2\pi}{3}\sin\frac{\pi}{6}$

$$= \left(-\frac{1}{2}\right)\left(\frac{\sqrt{3}}{2}\right) + \left(\frac{\sqrt{3}}{2}\right)\left(\frac{1}{2}\right) = 0$$

(b) $\cos\frac{2\pi}{3} + \cos\frac{\pi}{6} = -\frac{1}{2} + \frac{\sqrt{3}}{2} = \frac{\sqrt{3} - 1}{2}$

5. (a) $\cos(120° + 45°) = \cos 120°\cos 45° - \sin 120°\sin 45°$

$$= \left(-\frac{1}{2}\right)\left(\frac{\sqrt{2}}{2}\right) - \left(\frac{\sqrt{3}}{2}\right)\left(\frac{\sqrt{2}}{2}\right)$$

$$= \frac{-\sqrt{2} - \sqrt{6}}{4}$$

(b) $\cos 120° + \cos 45° = -\frac{1}{2} + \frac{\sqrt{2}}{2} = \frac{-1 + \sqrt{2}}{2}$

6. (a) $\sin(135° - 30°) = \sin 135°\cos 30° - \cos 135°\sin 30°$

$$= \left(\frac{\sqrt{2}}{2}\right)\left(\frac{\sqrt{3}}{2}\right) - \left(-\frac{\sqrt{2}}{2}\right)\left(\frac{1}{2}\right)$$

$$= \frac{\sqrt{6} + \sqrt{2}}{4}$$

(b) $\sin 135° - \cos 30° = \frac{\sqrt{2}}{2} - \frac{\sqrt{3}}{2}$

$$= \frac{\sqrt{2} - \sqrt{3}}{2}$$

7. $\sin 105° = \sin(60° + 45°)$

$= \sin 60° \cos 45° + \cos 60° \sin 45°$

$= \frac{\sqrt{3}}{2} \cdot \frac{\sqrt{2}}{2} + \frac{1}{2} \cdot \frac{\sqrt{2}}{2}$

$= \frac{\sqrt{2}}{4}(\sqrt{3} + 1)$

$\cos 105° = \cos(60° + 45°)$

$= \cos 60° \cos 45° - \sin 60° \sin 45°$

$= \frac{1}{2} \cdot \frac{\sqrt{2}}{2} - \frac{\sqrt{3}}{2} \cdot \frac{\sqrt{2}}{2}$

$= \frac{\sqrt{2}}{4}(1 - \sqrt{3})$

$\tan 105° = \tan(60° + 45°)$

$= \frac{\tan 60° + \tan 45°}{1 - \tan 60° \tan 45°}$

$= \frac{\sqrt{3} + 1}{1 - \sqrt{3}} = \frac{\sqrt{3} + 1}{1 - \sqrt{3}} \cdot \frac{1 + \sqrt{3}}{1 + \sqrt{3}}$

$= \frac{4 + 2\sqrt{3}}{-2} = -2 - \sqrt{3}$

8. $165° = 135° + 30°$

$\sin 165° = \sin(135° + 30°)$

$= \sin 135° \cos 30° + \sin 30° \cos 135°$

$= \sin 45° \cos 30° - \sin 30° \cos 45°$

$= \frac{\sqrt{2}}{2} \cdot \frac{\sqrt{3}}{2} - \frac{1}{2} \cdot \frac{\sqrt{2}}{2}$

$= \frac{\sqrt{2}}{4}(\sqrt{3} - 1)$

$\cos 165° = \cos(135° + 30°)$

$= \cos 135° \cos 30° - \sin 135° \sin 30°$

$= -\cos 45° \cos 30° - \sin 45° \sin 30°$

$= -\frac{\sqrt{2}}{2} \cdot \frac{\sqrt{3}}{2} - \frac{\sqrt{2}}{2} \cdot \frac{1}{2}$

$= -\frac{\sqrt{2}}{4}(\sqrt{3} + 1)$

$\tan 165° = \tan(135° + 30°)$

$= \frac{\tan 135° + \tan 30°}{1 - \tan 135° \tan 30°}$

$= \frac{-\tan 45° + \tan 30°}{1 + \tan 45° \tan 30°}$

$= \frac{-1 + \frac{\sqrt{3}}{3}}{1 + \frac{\sqrt{3}}{3}}$

$= -2 + \sqrt{3}$

9. $\sin 195° = \sin(225° - 30°)$

$= \sin 225° \cos 30° - \cos 225° \sin 30°$

$= -\sin 45° \cos 30° + \cos 45° \sin 30°$

$= -\frac{\sqrt{2}}{2} \cdot \frac{\sqrt{3}}{2} + \frac{\sqrt{2}}{2} \cdot \frac{1}{2}$

$= \frac{\sqrt{2}}{4}(1 - \sqrt{3})$

$\cos 195° = \cos(225° - 30°)$

$= \cos 225° \cos 30° + \sin 225° \sin 30°$

$= -\cos 45° \cos 30° - \sin 45° \sin 30°$

$= -\frac{\sqrt{2}}{2} \cdot \frac{\sqrt{3}}{2} - \frac{\sqrt{2}}{2} \cdot \frac{1}{2}$

$= -\frac{\sqrt{2}}{4}(\sqrt{3} + 1)$

$\tan 195° = \tan(225° - 30°)$

$= \frac{\tan 225° - \tan 30°}{1 + \tan 225° \tan 30°}$

$= \frac{\tan 45° - \tan 30°}{1 + \tan 45° \tan 30°}$

$= \frac{1 - \left(\frac{\sqrt{3}}{3}\right)}{1 + \left(\frac{\sqrt{3}}{3}\right)} = \frac{3 - \sqrt{3}}{3 + \sqrt{3}} \cdot \frac{3 - \sqrt{3}}{3 - \sqrt{3}}$

$= \frac{12 - 6\sqrt{3}}{6} = 2 - \sqrt{3}$

10. $255° = 300° - 45°$

$$\sin 255° = \sin(300° - 45°)$$
$$= \sin 300° \cos 45° - \sin 45° \cos 300°$$
$$= -\sin 60° \cos 45° - \sin 45° \cos 60°$$
$$= -\frac{\sqrt{3}}{2} \cdot \frac{\sqrt{2}}{2} - \frac{\sqrt{2}}{2} \cdot \frac{1}{2}$$
$$= -\frac{\sqrt{2}}{4}(\sqrt{3} + 1)$$

$$\cos 255° = \cos(300° - 45°)$$
$$= \cos 300° \cos 45° + \sin 300° \sin 45°$$
$$= \cos 60° \cos 45° - \sin 60° \sin 45°$$
$$= \frac{1}{2} \cdot \frac{\sqrt{2}}{2} - \frac{\sqrt{3}}{2} \cdot \frac{\sqrt{2}}{2}$$
$$= \frac{\sqrt{2}}{4}(1 - \sqrt{3})$$

$$\tan 255° = \tan(300° - 45°)$$
$$= \frac{\tan 300° - \tan 45°}{1 + \tan 300° \tan 45°}$$
$$= \frac{-\tan 60° - \tan 45°}{1 - \tan 60° \tan 45°}$$
$$= \frac{-\sqrt{3} - 1}{1 - \sqrt{3}} = 2 + \sqrt{3}$$

11. $\sin \frac{11\pi}{12} = \sin\left(\frac{3\pi}{4} + \frac{\pi}{6}\right)$

$$= \sin \frac{3\pi}{4} \cos \frac{\pi}{6} + \cos \frac{3\pi}{4} \sin \frac{\pi}{6}$$
$$= \frac{\sqrt{2}}{2} \cdot \frac{\sqrt{3}}{2} + \left(-\frac{\sqrt{2}}{2}\right)\frac{1}{2}$$
$$= \frac{\sqrt{2}}{4}(\sqrt{3} - 1)$$

$$\cos \frac{11\pi}{12} = \cos\left(\frac{3\pi}{4} + \frac{\pi}{6}\right)$$
$$= \cos \frac{3\pi}{4} \cos \frac{\pi}{6} - \sin \frac{3\pi}{4} \sin \frac{\pi}{6}$$
$$= -\frac{\sqrt{2}}{2} \cdot \frac{\sqrt{3}}{2} - \frac{\sqrt{2}}{2} \cdot \frac{1}{2}$$
$$= -\frac{\sqrt{2}}{4}(\sqrt{3} + 1)$$

$$\tan \frac{11\pi}{12} = \tan\left(\frac{3\pi}{4} + \frac{\pi}{6}\right)$$
$$= \frac{\tan \frac{3\pi}{4} + \tan \frac{\pi}{6}}{1 - \tan \frac{3\pi}{4} \tan \frac{\pi}{6}}$$
$$= \frac{-1 + \frac{\sqrt{3}}{3}}{1 - (-1)\frac{\sqrt{3}}{3}}$$
$$= \frac{-3 + \sqrt{3}}{3 + \sqrt{3}} \cdot \frac{3 - \sqrt{3}}{3 - \sqrt{3}}$$
$$= \frac{-12 + 6\sqrt{3}}{6} = -2 + \sqrt{3}$$

12. $\frac{7\pi}{12} = \frac{\pi}{3} + \frac{\pi}{4}$

$$\sin \frac{7\pi}{12} = \sin\left(\frac{\pi}{3} + \frac{\pi}{4}\right)$$
$$= \sin \frac{\pi}{3} \cos \frac{\pi}{4} + \sin \frac{\pi}{4} \cos \frac{\pi}{3}$$
$$= \frac{\sqrt{3}}{2} \cdot \frac{\sqrt{2}}{2} + \frac{\sqrt{2}}{2} \cdot \frac{1}{2}$$
$$= \frac{\sqrt{2}}{4}(\sqrt{3} + 1)$$

$$\cos \frac{7\pi}{12} = \cos\left(\frac{\pi}{3} + \frac{\pi}{4}\right)$$
$$= \cos \frac{\pi}{3} \cos \frac{\pi}{4} - \sin \frac{\pi}{3} \sin \frac{\pi}{4}$$
$$= \frac{1}{2} \cdot \frac{\sqrt{2}}{2} - \frac{\sqrt{3}}{2} \cdot \frac{\sqrt{2}}{2}$$
$$= \frac{\sqrt{2}}{4}(1 - \sqrt{3})$$

$$\tan \frac{7\pi}{12} = \tan\left(\frac{\pi}{3} + \frac{\pi}{4}\right)$$
$$= \frac{\tan \frac{\pi}{3} + \tan \frac{\pi}{4}}{1 - \tan \frac{\pi}{3} \tan \frac{\pi}{4}}$$
$$= \frac{\sqrt{3} + 1}{1 - \sqrt{3}}$$
$$= -2 - \sqrt{3}$$

13. $\sin\dfrac{17\pi}{12} = \sin\left(\dfrac{9\pi}{4} - \dfrac{5\pi}{6}\right)$

$= \sin\dfrac{9\pi}{4}\cos\dfrac{5\pi}{6} - \cos\dfrac{9\pi}{4}\sin\dfrac{5\pi}{6}$

$= \dfrac{\sqrt{2}}{2}\left(-\dfrac{\sqrt{3}}{2}\right) - \left(\dfrac{\sqrt{2}}{2}\right)\left(\dfrac{1}{2}\right)$

$= -\dfrac{\sqrt{2}}{4}(\sqrt{3} + 1)$

$\cos\dfrac{17\pi}{12} = \cos\left(\dfrac{9\pi}{4} - \dfrac{5\pi}{6}\right)$

$= \cos\dfrac{9\pi}{4}\cos\dfrac{5\pi}{6} + \sin\dfrac{9\pi}{4}\sin\dfrac{5\pi}{6}$

$= \dfrac{\sqrt{2}}{2}\left(-\dfrac{\sqrt{3}}{2}\right) + \dfrac{\sqrt{2}}{2}\left(\dfrac{1}{2}\right)$

$= \dfrac{\sqrt{2}}{4}(1 - \sqrt{3})$

$\tan\dfrac{17\pi}{12} = \tan\left(\dfrac{9\pi}{4} - \dfrac{5\pi}{6}\right)$

$= \dfrac{\tan(9\pi/4) - \tan(5\pi/6)}{1 + \tan(9\pi/4)\tan(5\pi/6)}$

$= \dfrac{1 - (-\sqrt{3}/3)}{1 + (-\sqrt{3}/3)}$

$= \dfrac{3 + \sqrt{3}}{3 - \sqrt{3}} \cdot \dfrac{3 + \sqrt{3}}{3 + \sqrt{3}}$

$= \dfrac{12 + 6\sqrt{3}}{6} = 2 + \sqrt{3}$

14. $-\dfrac{\pi}{12} = \dfrac{\pi}{6} - \dfrac{\pi}{4}$

$\sin\left(-\dfrac{\pi}{12}\right) = \sin\left(\dfrac{\pi}{6} - \dfrac{\pi}{4}\right)$

$= \sin\dfrac{\pi}{6}\cos\dfrac{\pi}{4} - \sin\dfrac{\pi}{4}\cos\dfrac{\pi}{6}$

$= \dfrac{1}{2} \cdot \dfrac{\sqrt{2}}{2} - \dfrac{\sqrt{2}}{2} \cdot \dfrac{\sqrt{3}}{2}$

$= \dfrac{\sqrt{2}}{4}(1 - \sqrt{3})$

$\cos\left(-\dfrac{\pi}{12}\right) = \cos\left(\dfrac{\pi}{6} - \dfrac{\pi}{4}\right)$

$= \cos\dfrac{\pi}{6}\cos\dfrac{\pi}{4} + \sin\dfrac{\pi}{6}\sin\dfrac{\pi}{4}$

$= \dfrac{\sqrt{3}}{2} \cdot \dfrac{\sqrt{2}}{2} + \dfrac{1}{2} \cdot \dfrac{\sqrt{2}}{2}$

$= \dfrac{\sqrt{2}}{4}(\sqrt{3} + 1)$

$\tan\left(-\dfrac{\pi}{12}\right) = \tan\left(\dfrac{\pi}{6} - \dfrac{\pi}{4}\right)$

$= \dfrac{\tan(\pi/6) - \tan(\pi/4)}{1 + \tan(\pi/6)\tan(\pi/4)}$

$= \dfrac{(\sqrt{3}/3) - 1}{1 + (\sqrt{3}/3)}$

$= -2 + \sqrt{3}$

15. $285° = 225° + 60°$

$\sin 285° = \sin(225° + 60°)$

$= \sin 225° \cos 60° + \cos 225° \sin 60°$

$= -\dfrac{\sqrt{2}}{2}\left(\dfrac{1}{2}\right) - \dfrac{\sqrt{2}}{2}\left(\dfrac{\sqrt{3}}{2}\right) = -\dfrac{\sqrt{2}}{4}(\sqrt{3} + 1)$

$\cos 285° = \cos(225° + 60°)$

$= \cos 225° \cos 60° - \sin 225° \sin 60°$

$= -\dfrac{\sqrt{2}}{2}\left(\dfrac{1}{2}\right) - \left(-\dfrac{\sqrt{2}}{2}\right)\left(\dfrac{\sqrt{3}}{2}\right) = \dfrac{\sqrt{2}}{4}(\sqrt{3} - 1)$

$\tan 285° = \tan(225° + 60°)$

$= \dfrac{\tan 225° + \tan 60°}{1 - \tan 225° \tan 60°} = \dfrac{1 + \sqrt{3}}{1 - \sqrt{3}} \cdot \dfrac{1 + \sqrt{3}}{1 + \sqrt{3}}$

$= \dfrac{4 + 2\sqrt{3}}{-2} = -2 - \sqrt{3} = -(2 + \sqrt{3})$

16. $-105 = 30° - 135°$

$\sin(30° - 135°) = \sin 30° \cos 135° - \cos 30° \sin 135°$

$= \sin 30°(-\cos 45°) - \cos 30° \sin 45°$

$= \left(\dfrac{1}{2}\right)\left(-\dfrac{\sqrt{2}}{2}\right) - \left(\dfrac{\sqrt{3}}{2}\right)\left(\dfrac{\sqrt{2}}{2}\right)$

$= -\dfrac{\sqrt{2}}{4}(1 + \sqrt{3})$

$\cos(30° - 135°) = \cos 30° \cos 135° + \sin 30° \sin 135°$

$= \cos 30°(-\cos 45°) + \sin 30° \sin 45°$

$= \left(\dfrac{\sqrt{3}}{2}\right)\left(-\dfrac{\sqrt{2}}{2}\right) + \left(\dfrac{1}{2}\right)\left(\dfrac{\sqrt{2}}{2}\right)$

$= \dfrac{\sqrt{2}}{4}(1 - \sqrt{3})$

$\tan(30° - 135°) = \dfrac{\tan 30° - \tan 135°}{1 + \tan 30° \tan 135°}$

$= \dfrac{\tan 30° - (-\tan 45°)}{1 + \tan 30°(-\tan 45°)}$

$= \dfrac{(\sqrt{3}/3) - (-1)}{1 + (\sqrt{3}/3)(-1)} = 2 + \sqrt{3}$

17. $-165° = -(120° + 45°)$

$$\sin(-165°) = \sin[-(120° + 45°)]$$
$$= -\sin(120° + 45°)$$
$$= -[\sin 120° \cos 45° + \cos 120° \sin 45°]$$
$$= -\left[\frac{\sqrt{3}}{2} \cdot \frac{\sqrt{2}}{2} - \frac{1}{2} \cdot \frac{\sqrt{2}}{2}\right]$$
$$= -\frac{\sqrt{2}}{4}(\sqrt{3} - 1)$$

$$\cos(-165°) = \cos[-(120° + 45°)]$$
$$= \cos(120° + 45°)$$
$$= \cos 120° \cos 45° - \sin 120° \sin 45°$$
$$= -\frac{1}{2} \cdot \frac{\sqrt{2}}{2} - \frac{\sqrt{3}}{2} \cdot \frac{\sqrt{2}}{2}$$
$$= -\frac{\sqrt{2}}{4}(1 + \sqrt{3})$$

$$\tan(-165°) = \tan[-(120° + 45°)]$$
$$= -\tan(120° + \tan 45°)$$
$$= -\frac{\tan 120° + \tan 45°}{1 - \tan 120° \tan 45°}$$
$$= -\frac{-\sqrt{3} + 1}{1 - (-\sqrt{3})(1)}$$
$$= -\frac{1 - \sqrt{3}}{1 + \sqrt{3}} \cdot \frac{1 - \sqrt{3}}{1 - \sqrt{3}}$$
$$= -\frac{4 - 2\sqrt{3}}{-2}$$
$$= 2 - \sqrt{3}$$

18. $15° = 45° - 30°$

$$\sin 15° = \sin(45° - 30°)$$
$$= \sin 45° \cos 30° - \cos 45° \sin 30°$$
$$= \left(\frac{\sqrt{2}}{2}\right)\left(\frac{\sqrt{3}}{2}\right) - \left(\frac{\sqrt{2}}{2}\right)\left(\frac{1}{2}\right)$$
$$= \frac{\sqrt{2}(\sqrt{3} - 1)}{4}$$
$$= \frac{\sqrt{2}}{4}(\sqrt{3} - 1)$$

$$\cos 15° = \cos(45° - 30°)$$
$$= \cos 45° \cos 30° + \sin 45° \sin 30°$$
$$= \left(\frac{\sqrt{2}}{2}\right)\left(\frac{\sqrt{3}}{2}\right) + \left(\frac{\sqrt{2}}{2}\right)\left(\frac{1}{2}\right)$$
$$= \frac{\sqrt{2}(\sqrt{3} + 1)}{4}$$
$$= \frac{\sqrt{2}}{4}(\sqrt{3} + 1)$$

$$\tan 15° = \tan(45° - 30°)$$
$$= \frac{\tan 45° - \tan 30°}{1 + \tan 45° \tan 30°}$$
$$= \frac{1 - (\sqrt{3}/3)}{1 + (1)(\sqrt{3}/3)}$$
$$= \frac{(3 - \sqrt{3})/3}{(3 + \sqrt{3})/3}$$
$$= \frac{3 - \sqrt{3}}{3 + \sqrt{3}} \cdot \frac{3 - \sqrt{3}}{3 - \sqrt{3}}$$
$$= \frac{12 - 6\sqrt{3}}{6}$$
$$= 2 - \sqrt{3}$$

19. $\dfrac{13\pi}{12} = \dfrac{3\pi}{4} + \dfrac{\pi}{3}$

$$\sin\frac{13\pi}{12} = \sin\left(\frac{3\pi}{4} + \frac{\pi}{3}\right)$$
$$= \sin\frac{3\pi}{4}\cos\frac{\pi}{3} + \cos\frac{3\pi}{4}\sin\frac{\pi}{3}$$
$$= \frac{\sqrt{2}}{2} \cdot \frac{1}{2} + \left(-\frac{\sqrt{2}}{2}\right)\left(\frac{\sqrt{3}}{2}\right)$$
$$= \frac{\sqrt{2}}{4}(1 - \sqrt{3})$$

$$\cos\frac{13\pi}{12} = \cos\left(\frac{3\pi}{4} + \frac{\pi}{3}\right)$$
$$= \cos\frac{3\pi}{4}\cos\frac{\pi}{3} - \sin\frac{3\pi}{4}\sin\frac{\pi}{3}$$
$$= -\frac{\sqrt{2}}{2} \cdot \frac{1}{2} - \frac{\sqrt{2}}{2} \cdot \frac{\sqrt{3}}{2} = -\frac{\sqrt{2}}{4}(1 + \sqrt{3})$$

$$\tan\frac{13\pi}{12} = \tan\left(\frac{3\pi}{4} + \frac{\pi}{3}\right)$$
$$= \frac{\tan\left(\frac{3\pi}{4}\right) + \tan\left(\frac{\pi}{3}\right)}{1 - \tan\left(\frac{3\pi}{4}\right)\tan\left(\frac{\pi}{3}\right)}$$
$$= \frac{-1 + \sqrt{3}}{1 - (-1)(\sqrt{3})}$$
$$= -\frac{1 - \sqrt{3}}{1 + \sqrt{3}} \cdot \frac{1 - \sqrt{3}}{1 - \sqrt{3}}$$
$$= -\frac{4 - 2\sqrt{3}}{-2}$$
$$= 2 - \sqrt{3}$$

20. $-\frac{7\pi}{12} = -\frac{\pi}{3} - \frac{\pi}{4}$

$$\sin\left(-\frac{7\pi}{12}\right) = \sin\left(-\frac{\pi}{3} - \frac{\pi}{4}\right)$$
$$= \sin\left(-\frac{\pi}{3}\right)\cos\left(\frac{\pi}{4}\right) - \cos\left(-\frac{\pi}{3}\right)\sin\left(\frac{\pi}{4}\right)$$
$$= \left(-\frac{\sqrt{3}}{2}\right)\left(\frac{\sqrt{2}}{2}\right) - \left(\frac{1}{2}\right)\left(\frac{\sqrt{2}}{2}\right)$$
$$= -\frac{\sqrt{2}}{4}(\sqrt{3} + 1)$$

$$\cos\left(-\frac{7\pi}{12}\right) = \cos\left(-\frac{\pi}{3} - \frac{\pi}{4}\right)$$
$$= \cos\left(-\frac{\pi}{3}\right)\cos\left(\frac{\pi}{4}\right) + \sin\left(-\frac{\pi}{3}\right)\sin\left(\frac{\pi}{4}\right)$$
$$= \left(\frac{1}{2}\right)\left(\frac{\sqrt{2}}{2}\right) + \left(-\frac{\sqrt{3}}{2}\right)\left(\frac{\sqrt{2}}{2}\right)$$
$$= \frac{\sqrt{2}}{4}(1 - \sqrt{3})$$

$$\tan\left(-\frac{7\pi}{12}\right) = \tan\left(-\frac{\pi}{3} - \frac{\pi}{4}\right)$$
$$= -\frac{\tan(-\pi/3) - \tan(\pi/4)}{1 + \tan(-\pi/3)\tan(\pi/4)}$$
$$= \frac{-\sqrt{3} - 1}{1 + (-\sqrt{3})(1)}$$
$$= 2 + \sqrt{3}$$

21. $-\frac{13\pi}{12} = -\left(\frac{3\pi}{4} + \frac{\pi}{3}\right)$

$$\sin\left[-\left(\frac{3\pi}{4} + \frac{\pi}{3}\right)\right] = -\sin\left(\frac{3\pi}{4} + \frac{\pi}{3}\right)$$
$$= -\left[\sin\frac{3\pi}{4}\cos\frac{\pi}{3} + \cos\frac{3\pi}{4}\sin\frac{\pi}{3}\right]$$
$$= -\left[\frac{\sqrt{2}}{2}\left(\frac{1}{2}\right) + \left(-\frac{\sqrt{2}}{2}\right)\left(\frac{\sqrt{3}}{2}\right)\right]$$
$$= -\frac{\sqrt{2}}{4}(1 - \sqrt{3}) = \frac{\sqrt{2}}{4}(\sqrt{3} - 1)$$

$$\cos\left[-\left(\frac{3\pi}{4} + \frac{\pi}{3}\right)\right] = \cos\left(\frac{3\pi}{4} + \frac{\pi}{3}\right)$$
$$= \cos\frac{3\pi}{4}\cos\frac{\pi}{3} - \sin\frac{3\pi}{4}\sin\frac{\pi}{3}$$
$$= -\frac{\sqrt{2}}{2}\left(\frac{1}{2}\right) - \frac{\sqrt{2}}{2}\left(\frac{\sqrt{3}}{2}\right)$$
$$= -\frac{\sqrt{2}}{4}(\sqrt{3} + 1)$$

$$\tan\left[-\left(\frac{3\pi}{4} + \frac{\pi}{3}\right)\right] = -\tan\left(\frac{3\pi}{4} + \frac{\pi}{3}\right)$$
$$= -\frac{\tan(3\pi/4) + \tan(\pi/3)}{1 - \tan(3\pi/4)\tan(\pi/3)}$$
$$= -\frac{-1 + \sqrt{3}}{1 - (-\sqrt{3})}$$
$$= \frac{1 - \sqrt{3}}{1 + \sqrt{3}} \cdot \frac{1 - \sqrt{3}}{1 - \sqrt{3}}$$
$$= \frac{4 - 2\sqrt{3}}{-2}$$
$$= -2 + \sqrt{3}$$

22. $\frac{5\pi}{12} = \frac{\pi}{4} + \frac{\pi}{6}$

$$\sin\left(\frac{\pi}{4} + \frac{\pi}{6}\right) = \sin\frac{\pi}{4}\cos\frac{\pi}{6} + \cos\frac{\pi}{4}\sin\frac{\pi}{6}$$
$$= \left(\frac{\sqrt{2}}{2}\right)\left(\frac{\sqrt{3}}{2}\right) + \left(\frac{\sqrt{2}}{2}\right)\left(\frac{1}{2}\right) = \frac{\sqrt{2}}{4}(\sqrt{3} + 1)$$

$$\cos\left(\frac{\pi}{4} + \frac{\pi}{6}\right) = \cos\frac{\pi}{4}\cos\frac{\pi}{6} - \sin\frac{\pi}{4}\sin\frac{\pi}{6}$$
$$= \left(\frac{\sqrt{2}}{2}\right)\left(\frac{\sqrt{3}}{2}\right) - \left(\frac{\sqrt{2}}{2}\right)\left(\frac{1}{2}\right) = \frac{\sqrt{2}}{2}(\sqrt{3} - 1)$$

$$\tan\left(\frac{\pi}{4} + \frac{\pi}{6}\right) = \frac{\tan\frac{\pi}{4} + \tan\frac{\pi}{6}}{1 - \tan\frac{\pi}{4}\tan\frac{\pi}{6}} = \frac{1 + \frac{\sqrt{3}}{3}}{1 - (1)\left(\frac{\sqrt{3}}{3}\right)} = \sqrt{3} + 2$$

23. $\cos 25° \cos 15° - \sin 25° \sin 15° = \cos(25° + 15°) = \cos 40°$

24. $\sin 140° \cos 50° + \cos 140° \sin 50° = \sin(140° + 50°) = \sin 190°$

25. $\dfrac{\tan 325° - \tan 86°}{1 + \tan 325° \tan 86°} = \tan(325° - 86°) = \tan 239°$

26. $\dfrac{\tan 140° - \tan 60°}{1 + \tan 140° \tan 60°} = \tan(140° - 60°) = \tan 80°$

27. $\sin 3 \cos 1.2 - \cos 3 \sin 1.2 = \sin(3 - 1.2) = \sin 1.8$

28. $\cos \dfrac{\pi}{7} \cos \dfrac{\pi}{5} - \sin \dfrac{\pi}{7} \sin \dfrac{\pi}{5} = \cos\left(\dfrac{\pi}{7} + \dfrac{\pi}{5}\right)$

$= \cos \dfrac{12\pi}{35}$

29. $\dfrac{\tan 2x + \tan x}{1 - \tan 2x \tan x} = \tan(2x + x) = \tan 3x$

30. $\cos 3x \cos 2y + \sin 3x \sin 2y = \cos(3x - 2y)$

31. $\sin 330° \cos 30° - \cos 330° \sin 30° = \sin(330° - 30°)$

$= \sin 300°$

$= -\dfrac{\sqrt{3}}{2}$

32. $\cos 15° \cos 60° + \sin 15° \sin 60° = \cos(15° - 60°)$

$= \cos(-45°) = \dfrac{\sqrt{2}}{2}$

33. $\sin \dfrac{\pi}{12} \cos \dfrac{\pi}{4} + \cos \dfrac{\pi}{12} \sin \dfrac{\pi}{4} = \sin\left(\dfrac{\pi}{12} + \dfrac{\pi}{4}\right)$

$= \sin \dfrac{\pi}{3}$

$= \dfrac{\sqrt{3}}{2}$

34. $\cos \dfrac{\pi}{16} \cos \dfrac{3\pi}{16} - \sin \dfrac{\pi}{16} \sin \dfrac{3\pi}{16} = \cos\left(\dfrac{\pi}{16} + \dfrac{3\pi}{16}\right)$

$= \cos \dfrac{\pi}{4} = \dfrac{\sqrt{2}}{2}$

35. $\dfrac{\tan 25° + \tan 110°}{1 - \tan 25° \tan 110°} = \tan(25° + 110°)$

$= \tan 135°$

$= -1$

36. $\dfrac{\tan\left(\dfrac{5\pi}{4}\right) - \tan\left(\dfrac{\pi}{12}\right)}{1 + \tan\left(\dfrac{5\pi}{4}\right)\tan\left(\dfrac{\pi}{12}\right)} = \tan\left(\dfrac{5\pi}{4} - \dfrac{\pi}{12}\right)$

$= \tan\left(\dfrac{7\pi}{6}\right)$

$= \tan\left(\dfrac{\pi}{6}\right) = \dfrac{\sqrt{3}}{3}$

For Exercises 37–44, we have:

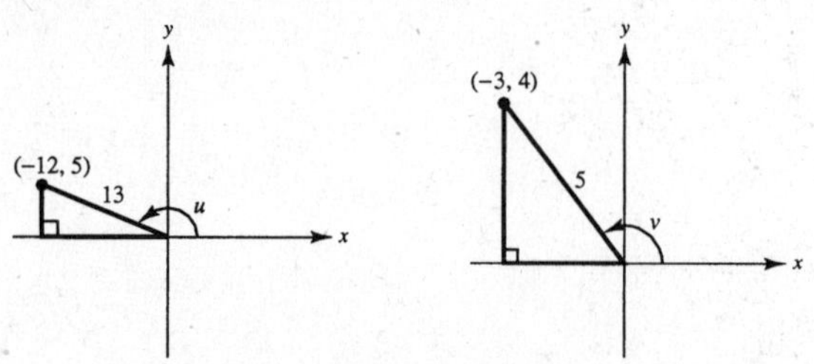

Figures for Exercises 37–44

$\sin u = \frac{5}{13}$, **u in Quadrant II** $\Rightarrow$ $\cos u = -\frac{12}{13}$, $\tan u = -\frac{5}{12}$

$\cos v = -\frac{3}{5}$, **v in Quadrant II** $\Rightarrow$ $\sin v = \frac{4}{5}$, $\tan v = -\frac{4}{3}$.

37. $\sin(u + v) = \sin u \cos v + \cos u \sin v$

$= \left(\frac{5}{13}\right)\left(-\frac{3}{5}\right) + \left(-\frac{12}{13}\right)\left(\frac{4}{5}\right)$

$= -\frac{63}{65}$

38. $\cos(u - v) = \cos u \cos v + \sin u \sin v$

$= \left(-\frac{12}{13}\right)\left(-\frac{3}{5}\right) - \left(\frac{5}{13}\right)\left(\frac{4}{5}\right)$

$= \frac{36}{65} + \frac{20}{65} = \frac{56}{65}$

39. $\cos(u + v) = \cos u \cos v - \sin u \sin v$

$$= \left(-\tfrac{12}{13}\right)\left(-\tfrac{3}{5}\right) - \left(\tfrac{5}{13}\right)\left(\tfrac{4}{5}\right)$$

$$= \tfrac{16}{65}$$

40. $\sin(v - u) = \sin v \cos u - \cos v \sin u$

$$= \left(\frac{4}{5}\right)\left(-\frac{12}{13}\right) - \left(-\frac{3}{5}\right)\left(\frac{5}{13}\right)$$

$$= -\frac{48}{65} + \frac{15}{65} = -\frac{33}{65}$$

41. $\tan(u + v) = \dfrac{\tan u + \tan v}{1 - \tan u \tan v} = \dfrac{-\frac{5}{12} + \left(-\frac{4}{3}\right)}{1 - \left(-\frac{5}{12}\right)\left(-\frac{4}{3}\right)} = \dfrac{-\frac{21}{12}}{1 - \frac{5}{9}}$

$$= \left(-\frac{7}{4}\right)\left(\frac{9}{4}\right) = -\frac{63}{16}$$

42. $\csc(u - v) = \dfrac{1}{\sin(u - v)} = \dfrac{1}{-\sin(v - u)}$

$$= \frac{1}{-\left(-\frac{33}{65}\right)} = \frac{65}{33}$$

43. $\sec(v - u) = \dfrac{1}{\cos(v - u)} = \dfrac{1}{\cos v \cos u + \sin v \sin u}$

$$= \frac{1}{\left(-\frac{3}{5}\right)\left(-\frac{12}{13}\right) + \left(\frac{4}{5}\right)\left(\frac{5}{13}\right)} = \frac{1}{\left(\frac{36}{65}\right) + \left(\frac{20}{65}\right)} = \frac{1}{\frac{56}{65}}$$

$$= \frac{65}{56}$$

44. $\cot(u + v) = \dfrac{1}{\tan(u + v)} = \dfrac{1}{-\frac{63}{16}} = -\dfrac{16}{63}$

$$\tan(u + v) = \frac{\tan u + \tan v}{1 - \tan u \tan v} = \frac{\left(-\frac{5}{12}\right) + \left(-\frac{4}{3}\right)}{1 - \left(-\frac{5}{12}\right)\left(-\frac{4}{3}\right)}$$

$$= \frac{-\frac{7}{4}}{\frac{4}{9}} = -\frac{63}{16}$$

For Exercises 45–50, we have:

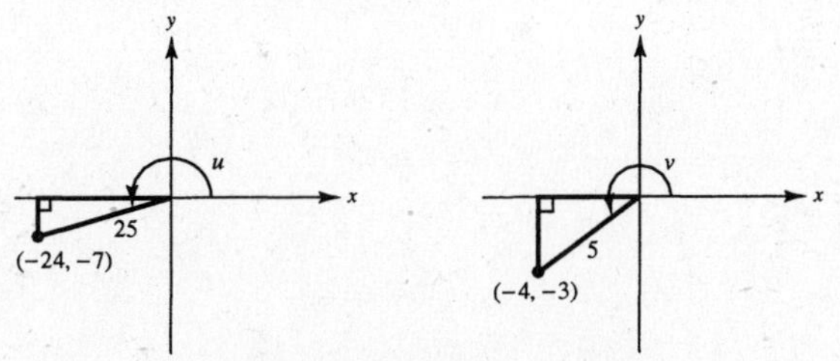

Figures for Exercises 45–50

$\sin u = -\frac{7}{25}$, u in Quadrant III $\Rightarrow$ $\cos u = -\frac{24}{25}$, $\tan u = \frac{7}{24}$

$\cos v = -\frac{4}{5}$, v in Quadrant III $\Rightarrow$ $\sin v = -\frac{3}{5}$, $\tan v = \frac{3}{4}$.

45. $\cos(u + v) = \cos u \cos v - \sin u \sin v$

$$= \left(-\frac{24}{25}\right)\left(-\frac{4}{5}\right) - \left(-\frac{7}{25}\right)\left(-\frac{3}{5}\right)$$

$$= \frac{3}{5}$$

46. $\sin(u + v) = \sin u \cos v + \cos u \sin v$

$$= \left(-\frac{7}{25}\right)\left(-\frac{4}{5}\right) + \left(-\frac{24}{25}\right)\left(-\frac{3}{5}\right)$$

$$= \frac{28}{125} + \frac{72}{125} = \frac{100}{125} = \frac{4}{5}$$

47. $\tan(u - v) = \dfrac{\tan u - \tan v}{1 + \tan u \tan v}$

$$= \frac{\frac{7}{24} - \frac{3}{4}}{1 + \left(\frac{7}{24}\right)\left(\frac{3}{4}\right)} = \frac{-\frac{11}{24}}{\frac{39}{32}} = -\frac{44}{117}$$

48. $\cot(v - u) = \dfrac{1}{\tan(v - u)} = \dfrac{1}{\frac{44}{117}} = \dfrac{117}{44}$

$$\tan(v - u) = \frac{\tan v - \tan u}{1 + \tan v \tan u} = \frac{\left(\frac{3}{4}\right) - \left(\frac{7}{24}\right)}{1 + \left(\frac{3}{4}\right)\left(\frac{7}{24}\right)}$$

$$= \frac{\frac{11}{24}}{\frac{39}{32}} = \frac{44}{117}$$

49. $\sec(u + v) = \dfrac{1}{\cos(u + v)} = \dfrac{1}{\frac{3}{5}} = \dfrac{5}{3}$

Use Exercise 45 for $\cos(u + v)$.

50. $\csc(u - v) = \dfrac{1}{\sin(u - v)} = \dfrac{1}{-\frac{44}{125}} = -\dfrac{125}{44}$

$\sin(u - v) = \sin u \cos v - \cos u \sin v = \left(-\frac{7}{25}\right)\left(-\frac{4}{5}\right) - \left(-\frac{24}{25}\right)\left(-\frac{3}{5}\right)$

$= \dfrac{28}{125} - \dfrac{72}{125} = -\dfrac{44}{125}$

51. $\sin(3\pi - x) = \sin 3\pi \cos x - \sin x \cos 3\pi$

$= (0)(\cos x) - (-1)(\sin x)$

$= \sin x$

52. $\sin\left(\dfrac{\pi}{2} + x\right) = \sin\dfrac{\pi}{2}\cos x + \sin x \cos\dfrac{\pi}{2}$

$= (1)(\cos x) + (\sin x)(0)$

$= \cos x$

53. $\sin\left(\dfrac{\pi}{6} + x\right) = \sin\dfrac{\pi}{6}\cos x + \cos\dfrac{\pi}{6}\sin x$

$= \dfrac{1}{2}\left(\cos x + \sqrt{3}\sin x\right)$

54. $\cos\left(\dfrac{5\pi}{4} - x\right) = \cos\dfrac{5\pi}{4}\cos x + \sin\dfrac{5\pi}{4}\sin x$

$= -\dfrac{\sqrt{2}}{2}(\cos x + \sin x)$

55. $\cos(\pi - \theta) + \sin\left(\dfrac{\pi}{2} + \theta\right) = \cos\pi\cos\theta + \sin\pi\sin\theta + \sin\dfrac{\pi}{2}\cos\theta + \cos\dfrac{\pi}{2}\sin\theta$

$= (-1)(\cos\theta) + (0)(\sin\theta) + (1)(\cos\theta) + (\sin\theta)(0)$

$= -\cos\theta + \cos\theta$

$= 0$

56. $\tan\left(\dfrac{\pi}{4} - \theta\right) = \dfrac{\tan\frac{\pi}{4} - \tan\theta}{1 + \tan\frac{\pi}{4}\tan\theta} = \dfrac{1 - \tan\theta}{1 + \tan\theta}$

57. $\cos\left(\dfrac{3\pi}{2} - x\right) = \cos\dfrac{3\pi}{2}\cos x + \sin\dfrac{3\pi}{2}\sin x$

$= (0)(\cos x) + (-1)(\sin x)$

$= -\sin x$

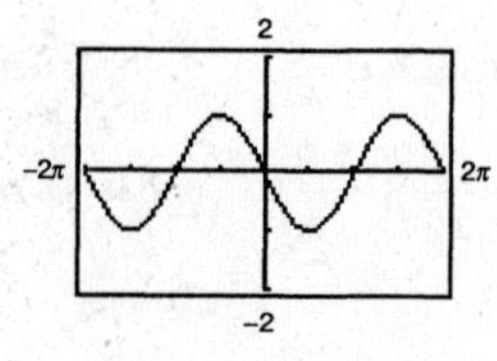

58. $\cos(\pi + x) = \cos\pi\cos x - \sin\pi\sin x$

$= (-1)\cos x - (0)\sin x$

$= -\cos x$

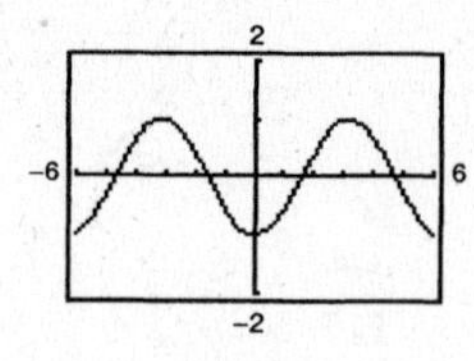

59. $\sin\left(\dfrac{3\pi}{2} + \theta\right) = \sin\dfrac{3\pi}{2}\cos\theta + \cos\dfrac{3\pi}{2}\sin\theta$

$= (-1)(\cos\theta) + (0)(\sin\theta)$

$= -\cos\theta$

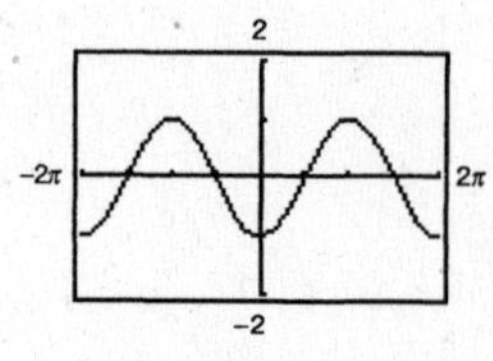

60. $\tan(\pi + \theta) = \dfrac{\tan \pi + \tan \theta}{1 - \tan \pi \tan \theta}$

$= \dfrac{0 + \tan \theta}{1 - (0) \tan \theta}$

$= \tan \theta$

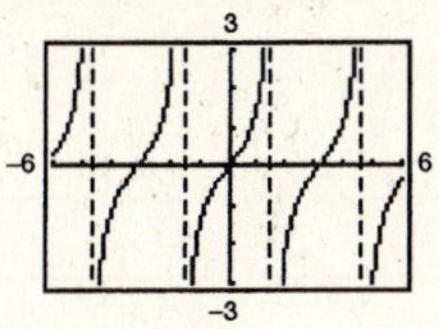

61. $\cos(x + y) \cos(x - y) = (\cos x \cos y - \sin x \sin y)(\cos x \cos y + \sin x \sin y)$

$= \cos^2 x \cos^2 y - \sin^2 x \sin^2 y$

$= \cos^2 x(1 - \sin^2 y) - \sin^2 x \sin^2 y$

$= \cos^2 x - \cos^2 x \sin^2 y - \sin^2 x \sin^2 y$

$= \cos^2 x - \sin^2 y(\cos^2 x + \sin^2 x)$

$= \cos^2 x - \sin^2 y$

62. $\sin(x + y) \sin(x - y) = (\sin x \cos y + \sin y \cos x)(\sin x \cos y - \sin y \cos x)$

$= \sin^2 x \cos^2 y - \sin^2 y \cos^2 x$

$= \sin^2 x(1 - \sin^2 y) - \sin^2 y \cos^2 x$

$= \sin^2 x - \sin^2 x \sin^2 y - \sin^2 y \cos^2 x$

$= \sin^2 x - \sin^2 y(\sin^2 x + \cos^2 x)$

$= \sin^2 x - \sin^2 y$

63. $\sin(x + y) + \sin(x - y) = \sin x \cos y + \cos x \sin y + \sin x \cos y - \cos x \sin y$

$= 2 \sin x \cos y$

64. $\cos(x + y) + \cos(x - y) = \cos x \cos y - \sin x \sin y + \cos x \cos y + \sin x \sin y$

$= 2 \cos x \cos y$

65. $\sin(\arcsin x + \arccos x) = \sin(\arcsin x) \cos(\arccos x) + \sin(\arccos x) \cos(\arcsin x)$

$= x \cdot x + \sqrt{1 - x^2} \cdot \sqrt{1 - x^2}$

$= x^2 + 1 - x^2$

$= 1$

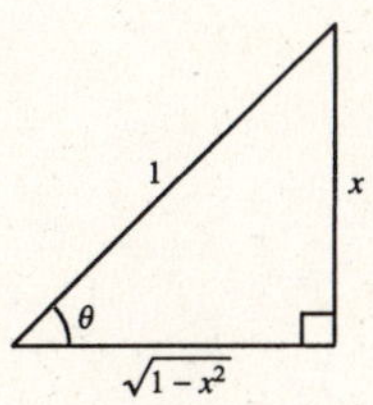

$\theta = \arcsin x$

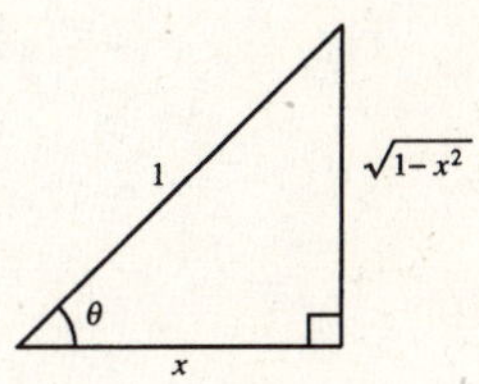

$\theta = \arccos x$

66. Let:

$u = \arctan 2x$ and $v = \arccos x$

$\tan u = 2x \qquad \cos v = x$

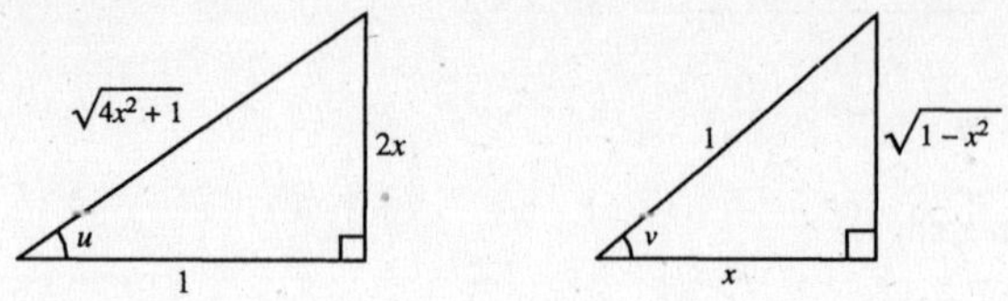

$$\begin{aligned}\sin(\arctan 2x - \arccos x) &= \sin(u - v)\\ &= \sin u \cos v - \cos u \sin v\\ &= \frac{2x}{\sqrt{4x^2+1}}(x) - \frac{1}{\sqrt{4x^2+1}}\left(\sqrt{1-x^2}\right)\\ &= \frac{2x^2 - \sqrt{1-x^2}}{\sqrt{4x^2+1}}\end{aligned}$$

67. $$\begin{aligned}\cos(\arccos x + \arcsin x) &= \cos(\arccos x)\cos(\arcsin x) - \sin(\arccos x)\sin(\arcsin x)\\ &= x \cdot \sqrt{1-x^2} - \sqrt{1-x^2} \cdot x\\ &= 0\end{aligned}$$

(Use the triangles in Exercise 65.)

68. Let:

$u = \arccos x$ and $v = \arctan x$

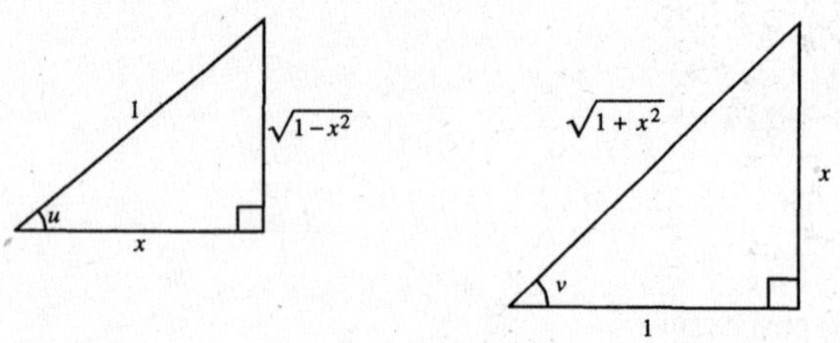

$$\begin{aligned}\cos(\arcos x - \arctan x) &= \cos(u - v)\\ &= \cos u \cos v + \sin u \sin v\\ &= (x)\left(\frac{1}{\sqrt{1+x^2}}\right) + \left(\sqrt{1-x^2}\right)\left(\frac{x}{\sqrt{1+x^2}}\right)\\ &= \frac{x + x\sqrt{1-x^2}}{\sqrt{1+x^2}}\end{aligned}$$

69. $$\begin{aligned}\sin\left(x + \frac{\pi}{3}\right) + \sin\left(x - \frac{\pi}{3}\right) &= 1\\ \sin x \cos\frac{\pi}{3} + \cos x \sin\frac{\pi}{3} + \sin x \cos\frac{\pi}{3} - \cos x \sin\frac{\pi}{3} &= 1\\ 2\sin x(0.5) &= 1\\ \sin x &= 1\\ x &= \frac{\pi}{2}\end{aligned}$$

70.

$$\sin\left(x + \frac{\pi}{6}\right) - \sin\left(x - \frac{\pi}{6}\right) = \frac{1}{2}$$

$$\sin x \cos \frac{\pi}{6} + \cos x \sin \frac{\pi}{6} - \left(\sin x \cos \frac{\pi}{6} - \cos x \sin \frac{\pi}{6}\right) = \frac{1}{2}$$

$$2 \cos x(0.5) = \frac{1}{2}$$

$$\cos x = \frac{1}{2}$$

$$x = \frac{\pi}{3}, \frac{5\pi}{3}$$

71.

$$\cos\left(x + \frac{\pi}{4}\right) - \cos\left(x - \frac{\pi}{4}\right) = 1$$

$$\cos x \cos \frac{\pi}{4} - \sin x \sin \frac{\pi}{4} - \left(\cos x \cos \frac{\pi}{4} + \sin x \sin \frac{\pi}{4}\right) = 1$$

$$-2 \sin x\left(\frac{\sqrt{2}}{2}\right) = 1$$

$$-\sqrt{2} \sin x = 1$$

$$\sin x = -\frac{1}{\sqrt{2}}$$

$$\sin x = -\frac{\sqrt{2}}{2}$$

$$x = \frac{5\pi}{4}, \frac{7\pi}{4}$$

72.

$$\cos\left(x + \frac{\pi}{4}\right) + \cos\left(x - \frac{\pi}{4}\right) = 1$$

$$\cos x \cos \frac{\pi}{4} - \sin x \sin \frac{\pi}{4} + \cos x \cos \frac{\pi}{4} + \sin x \sin \frac{\pi}{4} = 1$$

$$2 \cos x\left(\frac{\sqrt{2}}{2}\right) = 1$$

$$\sqrt{2} \cos x = 1$$

$$\cos x = \frac{1}{\sqrt{2}}$$

$$\cos x = \frac{\sqrt{2}}{2}$$

$$x = \frac{\pi}{4}, \frac{7\pi}{4}$$

73.
$$\tan(x+\pi)+2\sin(x+\pi)=0$$
$$\frac{\tan x+\tan\pi}{1-\tan x\tan\pi}+2(\sin x\cos\pi+\cos x\sin\pi)=0$$
$$\frac{\tan x+0}{1-\tan x(0)}+2[\sin x(-1)+\cos x(0)]=0$$
$$\frac{\tan x}{1}-2\sin x=0$$
$$\frac{\sin x}{\cos x}=2\sin x$$
$$\sin x=2\sin x\cos x$$
$$\sin x(1-2\cos x)=0$$
$$\sin x=0 \quad\text{or}\quad \cos x=\frac{1}{2}$$
$$x=0,\pi \qquad x=\frac{\pi}{3},\frac{5\pi}{3}$$

74.
$$\tan(x+\pi)-\cos\left(x+\frac{\pi}{2}\right)=0$$
$$\frac{\tan x+\tan\pi}{1-\tan x\tan\pi}-\left(\cos x\cos\frac{\pi}{2}-\sin x\sin\frac{\pi}{2}\right)=0$$
$$\frac{\tan x+0}{1-\tan x(0)}-\cos x(0)+\sin x(1)=0$$
$$\tan x+\sin x=0$$
$$\frac{\sin x}{\cos x}+\sin x=0$$
$$\sin x(\sec x+1)=0$$
$$\sin x=0 \quad\text{or}\quad \sec x=-1$$
$$x=0,\pi \qquad x=\pi$$

75. $\sin(x+\Delta x)=\sin x\cos\Delta x+\cos x\sin\Delta x$

76. $\cos(x+\Delta x)=\cos x\cos\Delta x-\sin x\sin\Delta x$

77. $\sin(\theta+\Delta\theta)=\sin\theta\cos\Delta\theta+\cos\theta\sin\Delta\theta$

78. $\cos(\theta+\Delta\theta)=\cos\theta\cos\Delta\theta-\sin\theta\sin\Delta\theta$

79. False. $\sin(u\pm v)=\sin u\cos v\pm\cos u\sin v$.

In Exercises 2–3, parts (a) and (b) are unequal.

80. False.

$\cos(u\pm v)=\cos u\cos v\mp\sin u\sin v$

81. False. $\cos\left(x-\frac{\pi}{2}\right)=\cos x\cos\frac{\pi}{2}+\sin x\sin\frac{\pi}{2}$

$=(\cos x)(0)+(\sin x)(1)$

$=\sin x$

82. True.

$\sin\left(x-\frac{\pi}{2}\right)=-\sin\left(\frac{\pi}{2}-x\right)=-\cos x$

83. $C=\arctan\frac{b}{a}\implies\sin C=\frac{b}{\sqrt{a^2+b^2}},\ \cos C=\frac{a}{\sqrt{a^2+b^2}}$

$$\sqrt{a^2+b^2}\sin(B\theta+C)=\sqrt{a^2+b^2}\left(\sin B\theta\cdot\frac{a}{\sqrt{a^2+b^2}}+\frac{b}{\sqrt{a^2+b^2}}\cdot\cos B\theta\right)=a\sin B\theta+b\cos B\theta$$

84. $C = \arctan\frac{a}{b} \Rightarrow \sin C = \frac{a}{\sqrt{a^2+b^2}},\ \cos C = \frac{b}{\sqrt{a^2+b^2}}$

$$\sqrt{a^2+b^2}\cos(B\theta - C) = \sqrt{a^2+b^2}\left(\cos B\theta \cdot \frac{b}{\sqrt{a^2+b^2}} + \sin B\theta \cdot \frac{a}{\sqrt{a^2+b^2}}\right)$$

$$= b\cos B\theta + a\sin B\theta$$

$$= a\sin B\theta + b\cos B\theta$$

85. $\sin\theta + \cos\theta$

$a = 1,\ b = 1,\ B = 1$

(a) $C = \arctan\frac{b}{a} = \arctan 1 = \frac{\pi}{4}$

$$\sin\theta + \cos\theta = \sqrt{a^2+b^2}\sin(B\theta + C)$$

$$= \sqrt{2}\sin\left(\theta + \frac{\pi}{4}\right)$$

(b) $C = \arctan\frac{a}{b} = \arctan 1 = \frac{\pi}{4}$

$$\sin\theta + \cos\theta = \sqrt{a^2+b^2}\cos(B\theta - C)$$

$$= \sqrt{2}\cos\left(\theta - \frac{\pi}{4}\right)$$

86. $3\sin 2\theta + 4\cos 2\theta$

$a = 3, b = 4, B = 2$

(a) $C = \arctan\frac{b}{a} = \arctan\frac{4}{3} \approx 0.9273$

$$3\sin 2\theta + 4\cos 2\theta = \sqrt{a^2+b^2}\sin(B\theta + C)$$

$$\approx 5\sin(2\theta + 0.9273)$$

(b) $C = \arctan\frac{a}{b} = \arctan\frac{3}{4} \approx 0.6435$

$$3\sin 2\theta + 4\cos 2\theta = \sqrt{a^2+b^2}\cos(B\theta - C)$$

$$\approx 5\cos(2\theta - 0.6435)$$

87. $12\sin 3\theta + 5\cos 3\theta$

$a = 12,\ b = 5,\ B = 3$

(a) $C = \arctan\frac{b}{a} = \arctan\frac{5}{12} \approx 0.3948$

$$12\sin 3\theta + 5\cos 3\theta = \sqrt{a^2+b^2}\sin(B\theta + C)$$

$$\approx 13\sin(3\theta + 0.3948)$$

(b) $C = \arctan\frac{a}{b} = \arctan\frac{12}{5} \approx 1.1760$

$$12\sin 3\theta + 5\cos 3\theta = \sqrt{a^2+b^2}\cos(B\theta - C)$$

$$\approx 13\cos(3\theta - 1.1760)$$

88. $\sin 2\theta - \cos 2\theta$

$a = 1, b = -1, B = 2$

(a) $C = \arctan\frac{b}{a} = \arctan(-1) = -\frac{\pi}{4}$

$$\sin 2\theta - \cos 2\theta = \sqrt{a^2+b^2}\sin(B\theta + C)$$

$$= \sqrt{2}\sin\left(2\theta - \frac{\pi}{4}\right)$$

(b) $C = \arctan\frac{a}{b} = \arctan(-1) = -\frac{\pi}{4}$

$$\sin 2\theta - \cos 2\theta = \sqrt{a^2+b^2}\cos(B\theta - C)$$

$$= \sqrt{2}\cos\left(2\theta + \frac{\pi}{4}\right)$$

89. $C = \arctan\frac{b}{a} = \frac{\pi}{2} \Rightarrow a = 0$

$\sqrt{a^2+b^2} = 2 \Rightarrow b = 2$

$B = 1$

$$2\sin\left(\theta + \frac{\pi}{2}\right) = (0)(\sin\theta) + (2)(\cos\theta) = 2\cos\theta$$

90. $C = \arctan\frac{a}{b} = -\frac{3\pi}{4} \Rightarrow a = b, a < 0, b < 0$

$\sqrt{a^2+b^2} = 5 \Rightarrow a = b = \frac{-5\sqrt{2}}{2}$

$B = 1$

$$5\cos\left(\theta + \frac{3\pi}{4}\right) = -\frac{5\sqrt{2}}{2}\sin\theta - \frac{5\sqrt{2}}{2}\cos\theta$$

91. $$\frac{\cos(x+\Delta x)-\cos x}{\Delta x} = \frac{\cos x\cos\Delta x - \sin x\sin\Delta x - \cos x}{\Delta x}$$

$$= \frac{-\sin x\sin\Delta x - \cos x + \cos x\cos\Delta x}{\Delta x}$$

$$= \frac{-\sin x\sin\Delta x}{\Delta x} - \frac{\cos x(1-\cos\Delta x)}{\Delta x}$$

$$= -\sin x\left(\frac{\sin\Delta x}{\Delta x}\right) - \cos x\left(\frac{1-\cos\Delta x}{\Delta x}\right)$$

Section 10.5 Multiple-Angle and Product-to-Sum Formulas

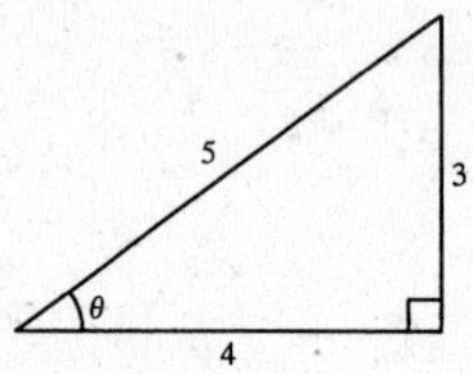

Figure for Exercises 1–8

1. $\sin\theta = \frac{3}{5}$

2. $\tan\theta = \frac{3}{4}$

3. $\cos 2\theta = 2\cos^2\theta - 1$

$$= 2\left(\frac{4}{5}\right)^2 - 1$$

$$= \frac{32}{25} - 1$$

$$= \frac{7}{25}$$

4. $\sin 2\theta = 2\sin\theta\cos\theta$

$$= 2\left(\frac{3}{5}\right)\left(\frac{4}{5}\right)$$

$$= \frac{24}{25}$$

5. $\tan 2\theta = \dfrac{2\tan\theta}{1-\tan^2\theta}$

$$= \frac{2(3/4)}{1-(3/4)^2}$$

$$= \frac{3/2}{1-9/16}$$

$$= \frac{3}{2}\cdot\frac{16}{7}$$

$$= \frac{24}{7}$$

6. $\sec 2\theta = \dfrac{1}{\cos 2\theta}$

$$= \frac{1}{7/25} \quad \text{From Exercise \#3}$$

$$= \frac{25}{7}$$

7. $\csc 2\theta = \dfrac{1}{\sin 2\theta}$

$$= \frac{1}{24/25} \quad \text{From Exercise \#4}$$

$$= \frac{25}{24}$$

8. $\cot 2\theta = \dfrac{1}{\tan 2\theta}$

$$= \frac{1}{24/7} \quad \text{From Exercise \#5}$$

$$= \frac{7}{24}$$

9. $\sin 2x - \sin x = 0$

$$2\sin x\cos x - \sin x = 0$$

$$\sin x(2\cos x - 1) = 0$$

$\sin x = 0$ or $2\cos x - 1 = 0$

$x = 0, \pi$ $\qquad \cos x = \frac{1}{2}$

$x = \frac{\pi}{3}, \frac{5\pi}{3}$

$x = 0, \frac{\pi}{3}, \pi, \frac{5\pi}{3}$

10. $\sin 2x + \cos x = 0$

$$2\sin x\cos x + \cos x = 0$$

$$\cos x(2\sin x + 1) = 0$$

$\cos x = 0$ or $2\sin x + 1 = 0$

$x = \frac{\pi}{2}, \frac{3\pi}{2}$ $\qquad \sin x = -\frac{1}{2}$

$x = \frac{7\pi}{6}, \frac{11\pi}{6}$

11. $4 \sin x \cos x = 1$

$$2 \sin 2x = 1$$

$$\sin 2x = \frac{1}{2}$$

$$2x = \frac{\pi}{6} + 2n\pi \quad \text{or} \quad 2x = \frac{5\pi}{6} + 2n\pi$$

$$x = \frac{\pi}{12} + n\pi \qquad x = \frac{5\pi}{12} + n\pi$$

$$x = \frac{\pi}{12}, \frac{13\pi}{12} \qquad x = \frac{5\pi}{12}, \frac{17\pi}{12}$$

12. $\sin 2x \sin x = \cos x$

$$2 \sin x \cos x \sin x - \cos x = 0$$

$$\cos x(2 \sin^2 x - 1) = 0$$

$$\cos x = 0 \quad \text{or} \quad 2 \sin^2 x - 1 = 0$$

$$x = \frac{\pi}{2}, \frac{3\pi}{2} \qquad \sin^2 x = \frac{1}{2}$$

$$\sin x = \pm\frac{\sqrt{2}}{2}$$

$$x = \frac{\pi}{4}, \frac{3\pi}{4}, \frac{5\pi}{4}, \frac{7\pi}{4}$$

13. $\cos 2x = \cos x$

$$\cos^2 x - \sin^2 x = \cos x$$

$$\cos^2 x - (1 - \cos^2 x) - \cos x = 0$$

$$2 \cos^2 x - \cos x - 1 = 0$$

$$(2 \cos x + 1)(\cos x - 1) = 0$$

$$2 \cos x + 1 = 0 \quad \text{or} \quad \cos x - 1 = 0$$

$$\cos x = -\frac{1}{2} \qquad \cos x = 1$$

$$x = \frac{2\pi}{3}, \frac{4\pi}{3} \qquad x = 0$$

14. $\cos 2x + \sin x = 0$

$$1 - 2 \sin^2 x + \sin x = 0$$

$$2 \sin^2 x - \sin x - 1 = 0$$

$$(2 \sin x + 1)(\sin x - 1) = 0$$

$$2 \sin x + 1 = 0 \quad \text{or} \quad \sin x - 1 = 0$$

$$\sin x = -\frac{1}{2} \qquad \sin x = 1$$

$$x = \frac{7\pi}{6}, \frac{11\pi}{6} \qquad x = \frac{\pi}{2}$$

15. $\tan 2x - \cot x = 0$

$$\frac{2 \tan x}{1 - \tan^2 x} = \cot x$$

$$2 \tan x = \cot x(1 - \tan^2 x)$$

$$2 \tan x = \cot x - \cot x \tan^2 x$$

$$2 \tan x = \cot x - \tan x$$

$$3 \tan x = \cot x$$

$$3 \tan x - \cot x = 0$$

$$3 \tan x - \frac{1}{\tan x} = 0$$

$$\frac{3 \tan^2 x - 1}{\tan x} = 0$$

$$\frac{1}{\tan x}(3 \tan^2 x - 1) = 0$$

$$\cot x(3 \tan^2 x - 1) = 0$$

$$\cot x = 0 \quad \text{or} \quad 3 \tan^2 x - 1 = 0$$

$$x = \frac{\pi}{2}, \frac{3\pi}{2} \qquad \tan^2 x = \frac{1}{3}$$

$$\tan x = \pm\frac{\sqrt{3}}{3}$$

$$x = \frac{\pi}{6}, \frac{5\pi}{6}, \frac{7\pi}{6}, \frac{11\pi}{6}$$

$$x = \frac{\pi}{6}, \frac{\pi}{2}, \frac{5\pi}{6}, \frac{7\pi}{6}, \frac{3\pi}{2}, \frac{11\pi}{6}$$

16.

$$\tan 2x - 2\cos x = 0$$

$$\frac{2\tan x}{1-\tan^2 x} = 2\cos x$$

$$2\tan x = 2\cos x(1-\tan^2 x)$$

$$2\tan x = 2\cos x - 2\cos x\tan^2 x$$

$$2\tan x = 2\cos x - 2\cos x\frac{\sin^2 x}{\cos^2 x}$$

$$2\tan x = 2\cos x - 2\frac{\sin^2 x}{\cos x}$$

$$\tan x = \cos x - \frac{\sin^2 x}{\cos x}$$

$$\frac{\sin x}{\cos x} = \cos x - \frac{\sin^2 x}{\cos x}$$

$$\frac{\sin x}{\cos x} + \frac{\sin^2 x}{\cos x} - \cos x = 0$$

$$\frac{\sin x + \sin^2 x - \cos^2 x}{\cos x} = 0$$

$$\frac{1}{\cos x}[\sin x + \sin^2 x - (1-\sin^2 x)] = 0$$

$$\sec x[2\sin^2 x + \sin x - 1] = 0$$

$$\sec x(2\sin x - 1)(\sin x + 1) = 0$$

$\sec x = 0$ or $2\sin x - 1 = 0$ or $\sin x + 1 = 0$

No solution; $\sin x = \frac{1}{2}$, $x = \frac{\pi}{6}, \frac{5\pi}{6}$; $\sin x = -1$, $x = \frac{3\pi}{2}$

Also, values for which $\cos x = 0$ need to be checked.

$\frac{\pi}{2}, \frac{3\pi}{2}$ are solutions.

$x = \frac{\pi}{6}, \frac{\pi}{2}, \frac{5\pi}{6}, \frac{3\pi}{2}$

17.

$$\sin 4x = -2\sin 2x$$

$$\sin 4x + 2\sin 2x = 0$$

$$2\sin 2x\cos 2x + 2\sin 2x = 0$$

$$2\sin 2x(\cos 2x + 1) = 0$$

$2\sin 2x = 0$ or $\cos 2x + 1 = 0$

$\sin 2x = 0$; $\cos 2x = -1$

$2x = n\pi$; $2x = \pi + 2n\pi$

$x = \frac{n}{2}\pi$; $x = \frac{\pi}{2} + n\pi$

$x = 0, \frac{\pi}{2}, \pi, \frac{3\pi}{2}$; $x = \frac{\pi}{2}, \frac{3\pi}{2}$

18.
$$(\sin 2x + \cos 2x)^2 = 1$$
$$\sin^2 2x + 2\sin 2x\cos 2x + \cos^2 2x = 1$$
$$2\sin 2x\cos 2x = 0$$
$$\sin 4x = 0$$
$$4x = n\pi$$
$$x = \frac{n\pi}{4}$$
$$x = 0, \frac{\pi}{4}, \frac{\pi}{2}, \frac{3\pi}{4}, \pi, \frac{5\pi}{4}, \frac{3\pi}{2}, \frac{7\pi}{4}$$

19. $6\sin x\cos x = 3(2\sin x\cos x)$
$= 3\sin 2x$

20. $6\cos^2 x - 3 = 3(2\cos^2 x - 1)$
$= 3\cos 2x$

21. $4 - 8\sin^2 x = 4(1 - 2\sin^2 x)$
$= 4\cos 2x$

22. $(\cos x + \sin x)(\cos x - \sin x) = \cos^2 x - \sin^2 x$
$= \cos 2x$

23. $\sin u = -\frac{4}{5}, \pi < u < \frac{3\pi}{2} \Rightarrow \cos u = -\frac{3}{5}$

$$\sin 2u = 2\sin u\cos u = 2\left(-\frac{4}{5}\right)\left(-\frac{3}{5}\right) = \frac{24}{25}$$
$$\cos 2u = \cos^2 u - \sin^2 u = \frac{9}{25} - \frac{16}{25} = -\frac{7}{25}$$
$$\tan 2u = \frac{2\tan u}{1 - \tan^2 u} = \frac{2\left(\frac{4}{3}\right)}{1 - \frac{16}{9}} = \frac{8}{3}\left(-\frac{9}{7}\right) = -\frac{24}{7}$$

24. $\cos u = -\frac{2}{3}, \frac{\pi}{2} < u < \pi$

$$\sin 2u = 2\sin u\cos u = 2\cdot\frac{\sqrt{5}}{3}\left(-\frac{2}{3}\right) = -\frac{4\sqrt{5}}{9}$$
$$\cos 2u = \cos^2 u - \sin^2 u = \frac{4}{9} - \frac{5}{9} = -\frac{1}{9}$$
$$\tan 2u = \frac{2\tan u}{1 - \tan^2 u} = \frac{2\left(-\frac{\sqrt{5}}{2}\right)}{1 - \frac{5}{4}} = 4\sqrt{5}$$

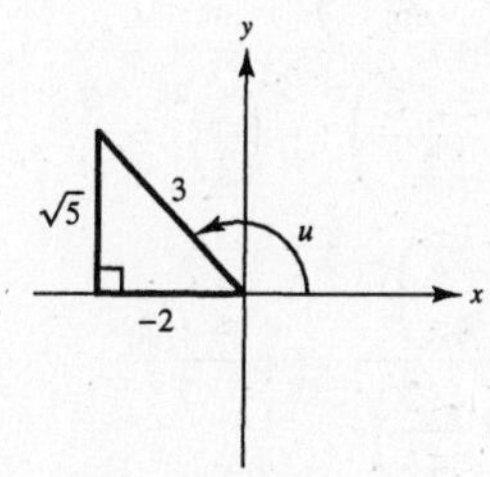

$\csc u = 3, \frac{\pi}{2} < u < \pi$

25. $\tan u = \frac{3}{4}, 0 < u < \frac{\pi}{2} \Rightarrow \sin u = \frac{3}{5}$ and $\cos u = \frac{4}{5}$

$$\sin 2u = 2\sin u\cos u = 2\left(\frac{3}{5}\right)\left(\frac{4}{5}\right) = \frac{24}{25}$$
$$\cos 2u = \cos^2 u - \sin^2 u = \frac{16}{25} - \frac{9}{25} = \frac{7}{25}$$
$$\tan 2u = \frac{2\tan u}{1 - \tan^2 u} = \frac{2\left(\frac{3}{4}\right)}{1 - \frac{9}{16}} = \frac{3}{2}\left(\frac{16}{7}\right) = \frac{24}{7}$$

26. $\cot u = -4, \frac{3\pi}{2} < u < 2\pi$

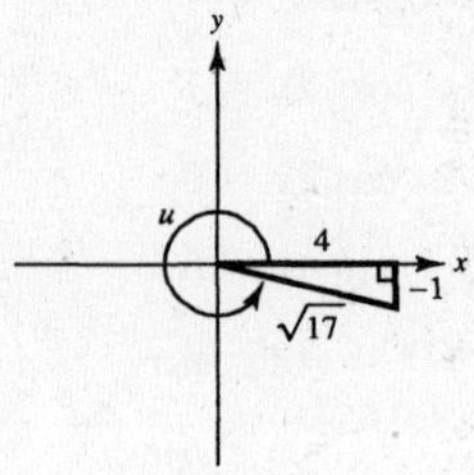

$$\sin 2u = 2\sin u\cos u = 2\left(-\frac{1}{\sqrt{17}}\right)\left(\frac{4}{\sqrt{17}}\right) = -\frac{8}{17}$$

$$\cos 2u = \cos^2 u - \sin^2 u$$

$$= \left(\frac{4}{\sqrt{17}}\right)^2 - \left(-\frac{1}{\sqrt{17}}\right)^2 = \frac{15}{17}$$

$$\tan 2u = \frac{2\tan u}{1 - \tan^2 u} = \frac{2\left(-\frac{1}{4}\right)}{1 - \left(-\frac{1}{4}\right)^2} = -\frac{8}{15}$$

27. $\sec u = -\frac{5}{2}, \frac{\pi}{2} < u < \pi \Rightarrow \sin u = \frac{\sqrt{21}}{5}$ and $\cos u = -\frac{2}{5}$

$$\sin 2u = 2\sin u\cos u = 2\left(\frac{\sqrt{21}}{5}\right)\left(-\frac{2}{5}\right) = -\frac{4\sqrt{21}}{25}$$

$$\cos 2u = \cos^2 u - \sin^2 u = \left(-\frac{2}{5}\right)^2 - \left(\frac{\sqrt{21}}{5}\right)^2 = -\frac{17}{25}$$

$$\tan 2u = \frac{2\tan u}{1 - \tan^2 u} = \frac{2\left(-\frac{\sqrt{21}}{2}\right)}{1 - \left(-\frac{\sqrt{21}}{2}\right)^2}$$

$$= \frac{-\sqrt{21}}{1 - \frac{21}{4}} = \frac{4\sqrt{21}}{17}$$

28. $\sin 2u = 2\sin u\cos u = 2\cdot\frac{1}{3}\left(-\frac{2\sqrt{2}}{3}\right) = -\frac{4\sqrt{2}}{9}$

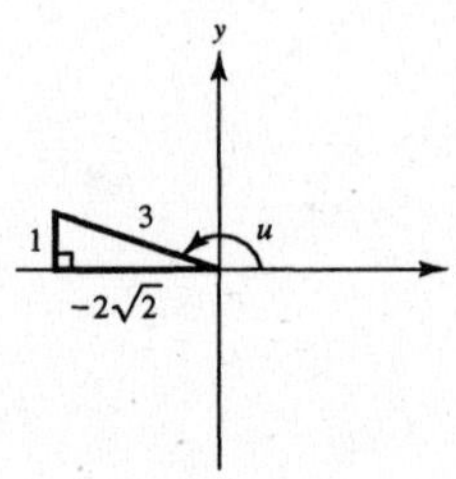

$$\cos 2u = \cos^2 u - \sin^2 u = \left(-\frac{2\sqrt{2}}{3}\right)^2 - \left(\frac{1}{3}\right)^2 = \frac{7}{9}$$

$$\tan 2u = \frac{2\tan u}{1 - \tan^2 u} = \frac{2\left(-\frac{\sqrt{2}}{4}\right)}{1 - \left(-\frac{\sqrt{2}}{4}\right)^2} = -\frac{4\sqrt{2}}{7}$$

29. $\cos^4 x = (\cos^2 x)(\cos^2 x) = \left(\frac{1 + \cos 2x}{2}\right)\left(\frac{1 + \cos 2x}{2}\right) = \frac{1 + 2\cos 2x + \cos^2 2x}{4}$

$$= \frac{1 + 2\cos 2x + \frac{1 + \cos 4x}{2}}{4}$$

$$= \frac{2 + 4\cos 2x + 1 + \cos 4x}{8}$$

$$= \frac{3 + 4\cos 2x + \cos 4x}{8}$$

$$= \frac{1}{8}(3 + 4\cos 2x + \cos 4x)$$

30. $\sin^8 x = \sin^4 x \sin^4 x = (\sin^2 x \sin^2 x)(\sin^2 x \sin^2 x)$

$$\sin^2 x = \frac{1 - \cos 2x}{2}$$

$$\sin^4 x = \left(\frac{1 - \cos 2x}{2}\right)\left(\frac{1 - \cos 2x}{2}\right)$$

$$= \frac{1}{4}(1 - 2\cos 2x + \cos^2 2x)$$

$$= \frac{1}{4}\left(1 - 2\cos 2x + \left(\frac{1 + \cos 4x}{2}\right)\right)$$

$$= \frac{1}{8}(3 - 4\cos 2x + \cos 4x)$$

$$\sin^8 x = \sin^4 x \sin^4 x$$

$$= \frac{1}{64}(3 - 4\cos 2x + \cos 4x)(3 - 4\cos 2x + \cos 4x)$$

$$= \frac{1}{64}(9 - 24\cos 2x + 16\cos^2 2x + 6\cos 4x - 8\cos 2x\cos 4x + \cos^2 4x)$$

$$= \frac{1}{64}\left[9 - 24\cos 2x + 16\left(\frac{1 + \cos 4x}{2}\right) + 6\cos 4x - 8\frac{1}{2}(\cos 6x + \cos 2x) + \left(\frac{1 + \cos 8x}{2}\right)\right]$$

$$= \frac{1}{64}\left[\frac{35}{2} - 28\cos 2x + 14\cos 4x - 4\cos 6x + \frac{1}{2}\cos 8x\right]$$

$$= \frac{1}{128}[35 - 56\cos 2x + 28\cos 4x - 8\cos 6x + \cos 8x]$$

In the above, we used $\cos 2x \cos 4x = \frac{1}{2}(\cos 6x + \cos 2x)$ which follows from $\cos(4x \pm 2x) = \cos 4x \cos 2x \mp \sin 4x \sin 2x$.

31. $(\sin^2 x)(\cos^2 x) = \left(\frac{1 - \cos 2x}{2}\right)\left(\frac{1 + \cos 2x}{2}\right)$

$$= \frac{1 - \cos^2 2x}{4}$$

$$= \frac{1}{4}\left(1 - \frac{1 + \cos 4x}{2}\right)$$

$$= \frac{1}{8}(2 - 1 - \cos 4x)$$

$$= \frac{1}{8}(1 - \cos 4x)$$

32. $\sin^4 x \cos^4 x = \sin^2 x \sin^2 x \cos^2 x \cos^2 x$

$$= (\sin^2 x \cos^2 x)(\sin^2 x \cos^2 x)$$

$$= \left(\frac{1}{4}\sin^2 2x\right)\left(\frac{1}{4}\sin^2 2x\right)$$

$$= \left[\frac{1}{4}\left(\frac{1 - \cos 4x}{2}\right)\right]\left[\frac{1}{4}\left(\frac{1 - \cos 4x}{2}\right)\right]$$

$$= \frac{1}{64}[1 - 2\cos 4x + \cos^2 4x]$$

$$= \frac{1}{64}\left[1 - 2\cos 4x + \left(\frac{1 + \cos 8x}{2}\right)\right]$$

$$= \frac{1}{64}\left[\frac{3}{2} - 2\cos 4x + \frac{1}{2}\cos 8x\right]$$

$$= \frac{1}{128}(3 - 4\cos 4x + \cos 8x)$$

33. $\sin^2 x \cos^4 x = \sin^2 x \cos^2 x \cos^2 x = \left(\frac{1-\cos 2x}{2}\right)\left(\frac{1+\cos 2x}{2}\right)\left(\frac{1+\cos 2x}{2}\right)$

$$= \frac{1}{8}(1-\cos 2x)(1+\cos 2x)(1+\cos 2x)$$

$$= \frac{1}{8}(1-\cos^2 2x)(1+\cos 2x)$$

$$= \frac{1}{8}(1+\cos 2x-\cos^2 2x-\cos^3 2x)$$

$$= \frac{1}{8}\left[1+\cos 2x-\left(\frac{1+\cos 4x}{2}\right)-\cos 2x\left(\frac{1+\cos 4x}{2}\right)\right]$$

$$= \frac{1}{16}[2+2\cos 2x-1-\cos 4x-\cos 2x-\cos 2x\cos 4x]$$

$$= \frac{1}{16}\left[1+\cos 2x-\cos 4x-\left(\frac{1}{2}\cos 2x+\frac{1}{2}\cos 6x\right)\right]$$

$$= \frac{1}{32}(2+2\cos 2x-2\cos 4x-\cos 2x-\cos 6x)$$

$$= \frac{1}{32}(2+\cos 2x-2\cos 4x-\cos 6x)$$

34. $\sin^4 x \cos^2 x = \sin^2 x \sin^2 x \cos^2 x$

$$= \left(\frac{1-\cos 2x}{2}\right)\left(\frac{1-\cos 2x}{2}\right)\left(\frac{1+\cos 2x}{2}\right)$$

$$= \frac{1}{8}(1-\cos 2x)(1-\cos^2 2x)$$

$$= \frac{1}{8}(1-\cos 2x-\cos^2 2x+\cos^3 2x)$$

$$= \frac{1}{8}\left[1-\cos 2x-\left(\frac{1+\cos 4x}{2}\right)+\cos 2x\left(\frac{1+\cos 4x}{2}\right)\right]$$

$$= \frac{1}{16}[2-2\cos 2x-1-\cos 4x+\cos 2x+\cos 2x\cos 4x]$$

$$= \frac{1}{16}\left[1-\cos 2x-\cos 4x+\frac{1}{2}\cos 2x+\frac{1}{2}\cos 6x\right]$$

$$= \frac{1}{32}[2-2\cos 2x-2\cos 4x+\cos 2x+\cos 6x]$$

$$= \frac{1}{32}[2-\cos 2x-2\cos 4x+\cos 6x]$$

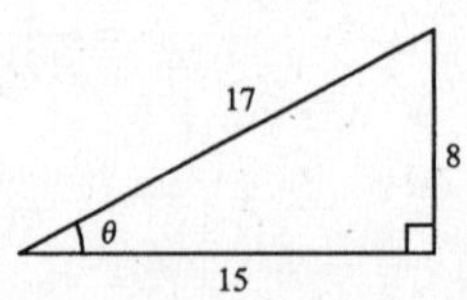

$\sin\theta = \frac{8}{17}$

$\cos\theta = \frac{15}{17}$

Figure for Exercises 35–39

35. $\cos\frac{\theta}{2} = \sqrt{\frac{1+\cos\theta}{2}} = \sqrt{\frac{1+\frac{15}{17}}{2}} = \sqrt{\frac{32}{34}} = \sqrt{\frac{16}{17}} = \frac{4\sqrt{17}}{17}$

36. $\sin\frac{\theta}{2} = \sqrt{\frac{1-\cos\theta}{2}} = \sqrt{\frac{1-\frac{15}{17}}{2}}$

$= \sqrt{\frac{\frac{2}{17}}{2}} = \frac{1}{\sqrt{17}} = \frac{\sqrt{17}}{17}$

37. $\tan\frac{\theta}{2} = \frac{\sin\theta}{1+\cos\theta} = \frac{\frac{8}{17}}{1+\frac{15}{17}} = \frac{8}{17}\cdot\frac{17}{32} = \frac{1}{4}$

38. $\sec\frac{\theta}{2} = \frac{1}{\cos\frac{\theta}{2}} = \frac{1}{\sqrt{\frac{1+\cos\theta}{2}}} = \frac{1}{\sqrt{\frac{1+\frac{15}{17}}{2}}}$

$= \frac{1}{\sqrt{\frac{16}{17}}} = \frac{\sqrt{17}}{4}$

39. $\csc\frac{\theta}{2} = \frac{1}{\sin\frac{\theta}{2}} = \frac{1}{\sqrt{\frac{(1-\cos\theta)}{2}}} = \frac{1}{\sqrt{\frac{1-\frac{15}{17}}{2}}}$

$= \frac{1}{\sqrt{\frac{1}{17}}} = \sqrt{17}$

40. $\cot\frac{\theta}{2} = \frac{1}{\tan\frac{\theta}{2}} = \frac{\sin\theta}{1-\cos\theta} = \frac{\frac{8}{17}}{1-\frac{15}{17}} = \frac{\frac{8}{17}}{\frac{2}{17}} = 4$

41. $\sin 75^\circ = \sin\left(\frac{1}{2}\cdot 150^\circ\right) = \sqrt{\frac{1-\cos 150^\circ}{2}} = \sqrt{\frac{1+\frac{\sqrt{3}}{2}}{2}} = \frac{1}{2}\sqrt{2+\sqrt{3}}$

$\cos 75^\circ = \cos\left(\frac{1}{2}\cdot 150^\circ\right) = \sqrt{\frac{1+\cos 150^\circ}{2}} = \sqrt{\frac{1-\frac{\sqrt{3}}{2}}{2}} = \frac{1}{2}\sqrt{2-\sqrt{3}}$

$\tan 75^\circ = \tan\left(\frac{1}{2}\cdot 150^\circ\right) = \frac{\sin 150^\circ}{1+\cos 150^\circ} = \frac{\frac{1}{2}}{1-\frac{\sqrt{3}}{2}} = \frac{1}{2-\sqrt{3}}\cdot\frac{2+\sqrt{3}}{2+\sqrt{3}} = \frac{2+\sqrt{3}}{4-3} = 2+\sqrt{3}$

42. $\sin 165^\circ = \sin\left(\frac{1}{2}\cdot 330^\circ\right) = \sqrt{\frac{1-\cos 330^\circ}{2}} = \sqrt{\frac{1-\frac{\sqrt{3}}{2}}{2}} = \frac{1}{2}\sqrt{2-\sqrt{3}}$

$\cos 165^\circ = \cos\left(\frac{1}{2}\cdot 330^\circ\right) = -\sqrt{\frac{1+\cos 330^\circ}{2}} = -\sqrt{\frac{1+\frac{\sqrt{3}}{2}}{2}} = -\frac{1}{2}\sqrt{2+\sqrt{3}}$

$\tan 165^\circ = \tan\left(\frac{1}{2}\cdot 330^\circ\right) = \frac{\sin 330^\circ}{1+\cos 330^\circ} = \frac{-\frac{1}{2}}{1+\frac{\sqrt{3}}{2}} = \frac{-1}{2+\sqrt{3}} = \sqrt{3}-2$

43. $\sin 112^\circ 30' = \sin\left(\frac{1}{2}\cdot 225^\circ\right) = \sqrt{\frac{1-\cos 225^\circ}{2}} = \sqrt{\frac{1+\frac{\sqrt{2}}{2}}{2}} = \frac{1}{2}\sqrt{2+\sqrt{2}}$

$\cos 112^\circ 30' = \cos\left(\frac{1}{2}\cdot 225^\circ\right) = -\sqrt{\frac{1+\cos 225^\circ}{2}} = -\sqrt{\frac{1-\frac{\sqrt{2}}{2}}{2}} = -\frac{1}{2}\sqrt{2-\sqrt{2}}$

$\tan 112^\circ 30' = \tan\left(\frac{1}{2}\cdot 225^\circ\right) = \frac{\sin 225^\circ}{1+\cos 225^\circ} = \frac{-\frac{\sqrt{2}}{2}}{1-\frac{\sqrt{2}}{2}} = -1-\sqrt{2}$

44. $\sin 67° 30' = \sin\left(\frac{1}{2} \cdot 135°\right) = \sqrt{\frac{1 - \cos 135°}{2}} = \sqrt{\frac{1 + \frac{\sqrt{2}}{2}}{2}} = \frac{1}{2}\sqrt{2 + \sqrt{2}}$

$\cos 67° 30' = \cos\left(\frac{1}{2} \cdot 135°\right) = \sqrt{\frac{1 + \cos 135°}{2}} = \sqrt{\frac{1 - \frac{\sqrt{2}}{2}}{2}} = \frac{1}{2}\sqrt{2 - \sqrt{2}}$

$\tan 67° 30' = \tan\left(\frac{1}{2} \cdot 135°\right) = \frac{\sin 135°}{1 + \cos 135°} = \frac{\frac{\sqrt{2}}{2}}{1 - \frac{\sqrt{2}}{2}} = 1 + \sqrt{2}$

45. $\sin\frac{\pi}{8} = \sin\left[\frac{1}{2}\left(\frac{\pi}{4}\right)\right] = \sqrt{\frac{1 - \cos\frac{\pi}{4}}{2}} = \frac{1}{2}\sqrt{2 - \sqrt{2}}$

$\cos\frac{\pi}{8} = \cos\left[\frac{1}{2}\left(\frac{\pi}{4}\right)\right] = \sqrt{\frac{1 + \cos\frac{\pi}{4}}{2}} = \frac{1}{2}\sqrt{2 + \sqrt{2}}$

$\tan\frac{\pi}{8} = \tan\left[\frac{1}{2}\left(\frac{\pi}{4}\right)\right] = \frac{\sin\frac{\pi}{4}}{1 + \cos\frac{\pi}{4}} = \frac{\frac{\sqrt{2}}{2}}{1 + \frac{\sqrt{2}}{2}} = \sqrt{2} - 1$

46. $\sin\frac{\pi}{12} = \sin\left[\frac{1}{2}\left(\frac{\pi}{6}\right)\right] = \sqrt{\frac{1 - \cos\frac{\pi}{6}}{2}} = \sqrt{\frac{1 - \frac{\sqrt{3}}{2}}{2}} = \frac{1}{2}\sqrt{2 - \sqrt{3}}$

$\cos\frac{\pi}{12} = \cos\left[\frac{1}{2}\left(\frac{\pi}{6}\right)\right] = \sqrt{\frac{1 + \cos\frac{\pi}{6}}{2}} = \frac{1}{2}\sqrt{2 + \sqrt{3}}$

$\tan\frac{\pi}{12} = \tan\left[\frac{1}{2}\left(\frac{\pi}{6}\right)\right] = \frac{\sin\frac{\pi}{6}}{1 + \cos\frac{\pi}{6}} = \frac{\frac{1}{2}}{1 + \frac{\sqrt{3}}{2}} = 2 - \sqrt{3}$

47. $\sin\frac{3\pi}{8} = \sin\left(\frac{1}{2} \cdot \frac{3\pi}{4}\right) = \sqrt{\frac{1 - \cos\frac{3\pi}{4}}{2}} = \sqrt{\frac{1 + \frac{\sqrt{2}}{2}}{2}} = \frac{1}{2}\sqrt{2 + \sqrt{2}}$

$\cos\frac{3\pi}{8} = \cos\left(\frac{1}{2} \cdot \frac{3\pi}{4}\right) = \sqrt{\frac{1 + \cos\frac{3\pi}{4}}{2}} = \sqrt{\frac{1 - \frac{\sqrt{2}}{2}}{2}} = \frac{1}{2}\sqrt{2 - \sqrt{2}}$

$\tan\frac{3\pi}{8} = \tan\left(\frac{1}{2} \cdot \frac{3\pi}{4}\right) = \frac{\sin\frac{3\pi}{4}}{1 + \cos\frac{3\pi}{4}} = \frac{\frac{\sqrt{2}}{2}}{1 - \frac{\sqrt{2}}{2}} = \frac{\frac{\sqrt{2}}{2}}{\frac{(2 - \sqrt{2})}{2}} = \frac{\sqrt{2}}{2 - \sqrt{2}} \cdot \frac{2 + \sqrt{2}}{2 + \sqrt{2}} = \sqrt{2} + 1$

48. $\sin\dfrac{7\pi}{12} = \sin\left[\dfrac{1}{2}\left(\dfrac{7\pi}{6}\right)\right] = \sqrt{\dfrac{1-\cos 7\pi/6}{2}} = \sqrt{\dfrac{1+\dfrac{\sqrt{3}}{2}}{2}} = \dfrac{1}{2}\sqrt{2+\sqrt{3}}$

$\cos\dfrac{7\pi}{12} = \cos\left[\dfrac{1}{2}\left(\dfrac{7\pi}{6}\right)\right] = -\sqrt{\dfrac{1+\cos\dfrac{7\pi}{6}}{2}} = -\sqrt{\dfrac{1-\dfrac{\sqrt{3}}{2}}{2}} = -\dfrac{1}{2}\sqrt{2-\sqrt{3}}$

$\tan\dfrac{7\pi}{12} = \tan\left[\dfrac{1}{2}\left(\dfrac{7\pi}{6}\right)\right] = \dfrac{\sin\dfrac{7\pi}{6}}{1+\cos\dfrac{7\pi}{6}} = \dfrac{-\dfrac{1}{2}}{1-\dfrac{\sqrt{3}}{2}} = -2-\sqrt{3}$

49. $\sin u = \dfrac{5}{13}, \dfrac{\pi}{2} < u < \pi \implies \cos u = -\dfrac{12}{13}$

$\sin\left(\dfrac{u}{2}\right) = \sqrt{\dfrac{1-\cos u}{2}} = \sqrt{\dfrac{1+\frac{12}{13}}{2}} = \dfrac{5\sqrt{26}}{26}$

$\cos\left(\dfrac{u}{2}\right) = \sqrt{\dfrac{1+\cos u}{2}} = \sqrt{\dfrac{1-\frac{12}{13}}{2}} = \dfrac{\sqrt{26}}{26}$

$\tan\left(\dfrac{u}{2}\right) = \dfrac{\sin u}{1+\cos u} = \dfrac{\frac{5}{13}}{1-\frac{12}{13}} = 5$

50. $\cos u = \dfrac{3}{5}, 0 < u < \dfrac{\pi}{2}$

$\sin\left(\dfrac{u}{2}\right) = \sqrt{\dfrac{1-\cos u}{2}} = \sqrt{\dfrac{1-\frac{3}{5}}{2}} = \dfrac{\sqrt{5}}{5}$

$\cos\left(\dfrac{u}{2}\right) = \sqrt{\dfrac{1+\cos u}{2}} = \sqrt{\dfrac{1+\frac{3}{5}}{2}} = \dfrac{2\sqrt{5}}{5}$

$\tan\left(\dfrac{u}{2}\right) = \dfrac{\sin u}{1+\cos u} = \dfrac{\frac{4}{5}}{1+\frac{3}{5}} = \dfrac{1}{2}$

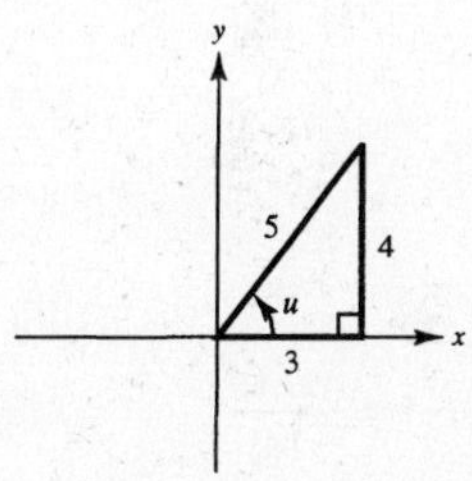

51. $\tan u = -\dfrac{5}{8}, \dfrac{3\pi}{2} < u < 2\pi \implies \sin u = -\dfrac{5}{\sqrt{89}}$ and $\cos u = \dfrac{8}{\sqrt{89}}$

$\sin\left(\dfrac{u}{2}\right) = \sqrt{\dfrac{1-\cos u}{2}} = \sqrt{\dfrac{1-\dfrac{8}{\sqrt{89}}}{2}}\sqrt{\dfrac{\sqrt{89}-8}{2\sqrt{89}}} = \sqrt{\dfrac{89-8\sqrt{89}}{178}}$

$\cos\left(\dfrac{u}{2}\right) = -\sqrt{\dfrac{1+\cos u}{2}} = -\sqrt{\dfrac{1+\dfrac{8}{\sqrt{89}}}{2}} = -\sqrt{\dfrac{\sqrt{89}+8}{2\sqrt{89}}} = -\sqrt{\dfrac{89+8\sqrt{89}}{178}}$

$\tan\left(\dfrac{u}{2}\right) = \dfrac{1-\cos u}{\sin u} = \dfrac{1-\dfrac{8}{\sqrt{89}}}{-\dfrac{5}{\sqrt{89}}} = \dfrac{8-\sqrt{89}}{5}$

52. $\cot u = 3,\ \pi < u < \dfrac{3\pi}{2}$

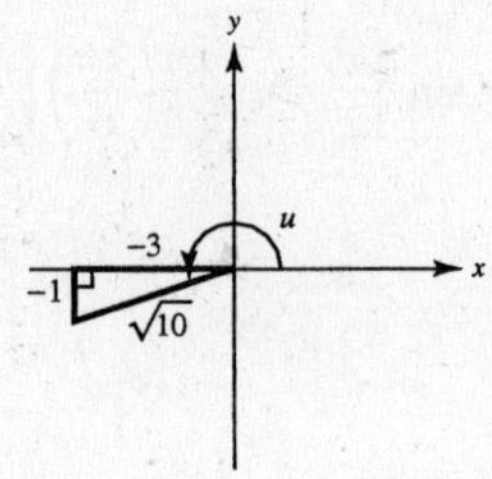

$$\sin\left(\frac{u}{2}\right) = \sqrt{\frac{1 - \cos u}{2}} = \sqrt{\frac{1 + \frac{3}{\sqrt{10}}}{2}} = \sqrt{\frac{10 + 3\sqrt{10}}{20}} = \frac{1}{2}\sqrt{\frac{10 + 3\sqrt{10}}{5}}$$

$$\cos\left(\frac{u}{2}\right) = -\sqrt{\frac{1 + \cos u}{2}} = -\sqrt{\frac{1 - \frac{3}{\sqrt{10}}}{2}} = -\sqrt{\frac{10 - 3\sqrt{10}}{20}} = -\frac{1}{2}\sqrt{\frac{10 - 3\sqrt{10}}{5}}$$

$$\tan\left(\frac{u}{2}\right) = \frac{1 - \cos u}{\sin u} = \frac{1 + \frac{3}{\sqrt{10}}}{-\frac{1}{\sqrt{10}}} = -3 - \sqrt{10}$$

53. $\csc u = -\dfrac{5}{3},\ \pi < u < \dfrac{3\pi}{2} \Rightarrow \sin u = -\dfrac{3}{5}$ and $\cos u = -\dfrac{4}{5}$

$$\sin\left(\frac{u}{2}\right) = \sqrt{\frac{1 - \cos u}{2}} = \sqrt{\frac{1 + \frac{4}{5}}{2}} = \frac{3\sqrt{10}}{10}$$

$$\cos\left(\frac{u}{2}\right) = -\sqrt{\frac{1 + \cos u}{2}} = -\sqrt{\frac{1 - \frac{4}{5}}{2}} = -\frac{\sqrt{10}}{10}$$

$$\tan\left(\frac{u}{2}\right) = \frac{1 - \cos u}{\sin u} = \frac{1 + \frac{4}{5}}{-\frac{3}{5}} = -3$$

54. $\sec u = -\dfrac{7}{2},\ \dfrac{\pi}{2} < u < \pi$

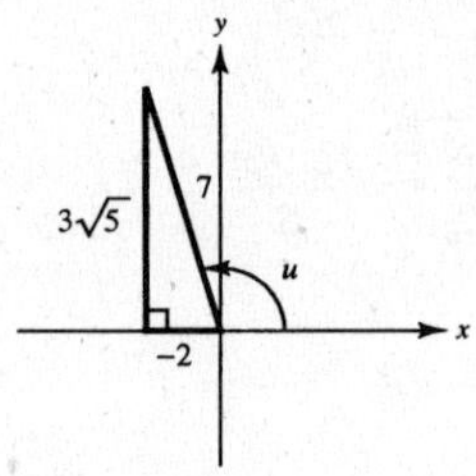

$$\sin\left(\frac{u}{2}\right) = \sqrt{\frac{1 - \cos u}{2}} = \sqrt{\frac{1 + \frac{2}{7}}{2}} = \frac{3\sqrt{14}}{14}$$

$$\cos\left(\frac{u}{2}\right) = \sqrt{\frac{1 + \cos u}{2}} = \sqrt{\frac{1 - \frac{2}{7}}{2}} = \frac{\sqrt{70}}{14}$$

$$\tan\left(\frac{u}{2}\right) = \frac{1 - \cos u}{\sin u} = \frac{1 + \frac{2}{7}}{\frac{3\sqrt{5}}{7}} = \frac{3\sqrt{5}}{5}$$

55. $\sqrt{\dfrac{1 - \cos 6x}{2}} = |\sin 3x|$

56. $\sqrt{\dfrac{1 + \cos 4x}{2}} = \left|\cos \dfrac{4x}{2}\right| = |\cos 2x|$

57.
$$-\sqrt{\frac{1 - \cos 8x}{1 + \cos 8x}} = -\frac{\sqrt{\frac{1 - \cos 8x}{2}}}{\sqrt{\frac{1 + \cos 8x}{2}}}$$
$$= -\frac{|\sin 4x|}{|\cos 4x|}$$
$$= -|\tan 4x|$$

58. $-\sqrt{\dfrac{1 - \cos(x - 1)}{2}} = -\left|\sin\left(\dfrac{x - 1}{2}\right)\right|$

59. $\sin\frac{x}{2} + \cos x = 0$

$$\pm\sqrt{\frac{1-\cos x}{2}} = -\cos x$$

$$\frac{1-\cos x}{2} = \cos^2 x$$

$$0 = 2\cos^2 x + \cos x - 1$$

$$= (2\cos x - 1)(\cos x + 1)$$

$\cos x = \frac{1}{2}$ or $\cos x = -1$

$x = \frac{\pi}{3}, \frac{5\pi}{3}$ $\quad$ $x = \pi$

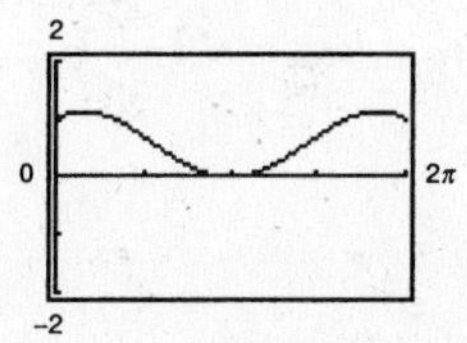

By checking these values in the original equation, we see that $x = \pi/3$ and $x = 5\pi/3$ are extraneous, and $x = \pi$ is the only solution.

60. $h(x) = \sin\frac{x}{2} + \cos x - 1$

$$\sin\frac{x}{2} + \cos x - 1 = 0$$

$$\pm\sqrt{\frac{1-\cos x}{2}} = 1 - \cos x$$

$$\frac{1-\cos x}{2} = 1 - 2\cos x + \cos^2 x$$

$$1 - \cos x = 2 - 4\cos x + 2\cos^2 x$$

$$2\cos^2 x - 3\cos x + 1 = 0$$

$$(2\cos x - 1)(\cos x - 1) = 0$$

$2\cos x - 1 = 0$ or $\cos x - 1 = 0$

$\cos x = \frac{1}{2}$ $\quad$ $\cos x = 1$

$x = \frac{\pi}{3}, \frac{5\pi}{3}$ $\quad$ $x = 0$

$0, \frac{\pi}{3}$, and $\frac{5\pi}{3}$ are all solutions to the equation.

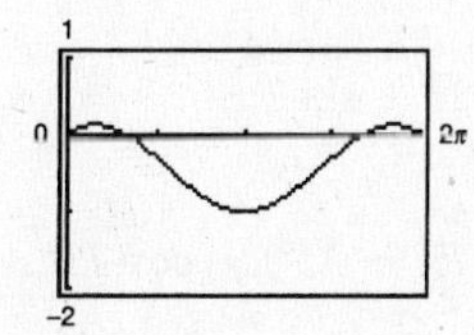

61. $\cos\frac{x}{2} - \sin x = 0$

$$\pm\sqrt{\frac{1+\cos x}{2}} = \sin x$$

$$\frac{1+\cos x}{2} = \sin^2 x$$

$$1 + \cos x = 2\sin^2 x$$

$$1 + \cos x = 2 - 2\cos^2 x$$

$$2\cos^2 x + \cos x - 1 = 0$$

$$(2\cos x - 1)(\cos x + 1) = 0$$

$2\cos x - 1 = 0$ or $\cos x + 1 = 0$

$\cos x = \frac{1}{2}$ $\quad$ $\cos x = -1$

$x = \frac{\pi}{3}, \frac{5\pi}{3}$ $\quad$ $x = \pi$

$x = \frac{\pi}{3}, \pi, \frac{5\pi}{3}$

$\pi/3$, π, and $5\pi/3$ are all solutions to the equation.

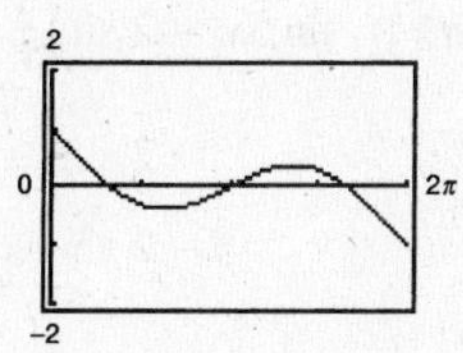

62. $g(x) = \tan\frac{x}{2} - \sin x$

$$\tan\frac{x}{2} - \sin x = 0$$

$$\frac{1-\cos x}{\sin x} = \sin x$$

$$1 - \cos x = \sin^2 x$$

$$1 - \cos x = 1 - \cos^2 x$$

$$\cos^2 x - \cos x = 0$$

$$\cos x(\cos x - 1) = 0$$

$\cos x = 0$ or $\cos x - 1 = 0$

$x = \frac{\pi}{2}, \frac{3\pi}{2}$ $\quad$ $\cos x = 1$

$x = 0$

0, $\pi/2$, and $3\pi/2$ are all solutions to the equation.

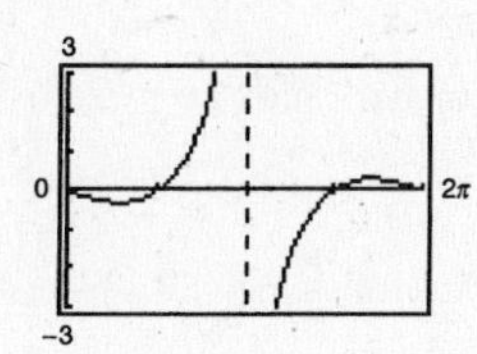

63. $6\sin\frac{\pi}{4}\cos\frac{\pi}{4} = 6\cdot\frac{1}{2}\left[\sin\left(\frac{\pi}{4}+\frac{\pi}{4}\right)+\sin\left(\frac{\pi}{4}-\frac{\pi}{4}\right)\right]$

$= 3\left(\sin\frac{\pi}{2}+\sin 0\right)$

64. $4\cos\frac{\pi}{3}\sin\frac{5\pi}{6} = 4\cdot\frac{1}{2}\left[\sin\left(\frac{\pi}{3}+\frac{5\pi}{6}\right)-\sin\left(\frac{\pi}{3}-\frac{5\pi}{6}\right)\right]$

$= 2\left[\sin\left(\frac{7\pi}{6}\right)-\sin\left(-\frac{\pi}{2}\right)\right]$

$= 2\left[\sin\left(\frac{7\pi}{6}\right)+\sin\left(\frac{\pi}{2}\right)\right]$

65. $\cos 4\theta\sin 6\theta = \frac{1}{2}[\sin(4\theta+6\theta)-\sin(4\theta-6\theta)]$

$= \frac{1}{2}[\sin 10\theta - \sin(-2\theta)]$

$= \frac{1}{2}(\sin 10\theta + \sin 2\theta)$

66. $3\sin 2\alpha\sin 3\alpha = 3\cdot\frac{1}{2}[\cos(2\alpha-3\alpha)-\cos(2\alpha+3\alpha)]$

$= \frac{3}{2}[\cos(-\alpha)-\cos 5\alpha]$

$= \frac{3}{2}(\cos\alpha - \cos 5\alpha)$

67. $5\cos(-5\beta)\cos 3\beta = 5\cdot\frac{1}{2}[\cos(-5\beta-3\beta)+\cos(-5\beta+3\beta)] = \frac{5}{2}[\cos(-8\beta)+\cos(-2\beta)]$

$= \frac{5}{2}(\cos 8\beta + \cos 2\beta)$

68. $\cos 2\theta\cos 4\theta = \frac{1}{2}[\cos(2\theta-4\theta)+\cos(2\theta+4\theta)] = \frac{1}{2}[\cos(-2\theta)+\cos 6\theta]$

$= \frac{1}{2}(\cos 2\theta + \cos 6\theta)$

69. $\sin(x+y)\sin(x-y) = \frac{1}{2}(\cos 2y - \cos 2x)$

70. $\sin(x+y)\cos(x-y) = \frac{1}{2}(\sin 2x + \sin 2y)$

71. $\cos(\theta-\pi)\sin(\theta+\pi) = \frac{1}{2}(\sin 2\theta + \sin 2\pi)$

72. $\sin(\theta+\pi)\sin(\theta-\pi) = \frac{1}{2}(\cos 2\pi - \cos 2\theta)$

73. $10\cos 75^\circ\cos 15^\circ = 10\left(\frac{1}{2}\right)[\cos(75^\circ-15^\circ)+\cos(75^\circ+15^\circ)] = 5[\cos 60^\circ + \cos 90^\circ]$

74. $6\sin 45^\circ\cos 15^\circ = 6\left(\frac{1}{2}\right)(\sin 60^\circ + \sin 30^\circ) = 3(\sin 60^\circ + \sin 30^\circ)$

75. $\sin 60^\circ + \sin 30^\circ = 2\sin\left(\frac{60^\circ+30^\circ}{2}\right)\cos\left(\frac{60^\circ-30^\circ}{2}\right)$

$= 2\sin 45^\circ\cos 15^\circ$

76. $\cos 120^\circ + \cos 30^\circ = 2\cos\left(\frac{120^\circ+30^\circ}{2}\right)\cos\left(\frac{120^\circ-30^\circ}{2}\right)$

$= 2\cos 75^\circ\cos 45^\circ$

77. $\cos\frac{3\pi}{4} - \cos\frac{\pi}{4} = -2\sin\left(\frac{\frac{3\pi}{4}+\frac{\pi}{4}}{2}\right)\sin\left(\frac{\frac{3\pi}{4}-\frac{\pi}{4}}{2}\right)$

$= -2\sin\frac{\pi}{2}\sin\frac{\pi}{4}$

78. $\sin\frac{5\pi}{4} - \sin\frac{3\pi}{4} = 2\cos\left(\frac{\frac{5\pi}{4}+\frac{3\pi}{4}}{2}\right)\sin\left(\frac{\frac{5\pi}{4}-\frac{3\pi}{4}}{2}\right)$

$= 2\cos\pi\sin\frac{\pi}{4}$

79. $\sin 5\theta - \sin 3\theta = 2\cos\left(\frac{5\theta+3\theta}{2}\right)\sin\left(\frac{5\theta-3\theta}{2}\right)$

$= 2\cos 4\theta\sin\theta$

80. $\sin 3\theta + \sin\theta = 2\sin\left(\frac{3\theta+\theta}{2}\right)\cos\left(\frac{3\theta-\theta}{2}\right)$

$= 2\sin 2\theta\cos\theta$

81. $\cos 6x + \cos 2x = 2\cos\left(\frac{6x+2x}{2}\right)\cos\left(\frac{6x-2x}{2}\right)$

$= 2\cos 4x\cos 2x$

82. $\sin x + \sin 5x = 2\sin\left(\frac{x+5x}{2}\right)\cos\left(\frac{x-5x}{2}\right)$

$= 2\sin 3x\cos(-2x)$

$= 2\sin 3x\cos 2x$

83. $\sin(\alpha+\beta) - \sin(\alpha-\beta) = 2\cos\left(\frac{\alpha+\beta+\alpha-\beta}{2}\right)\sin\left(\frac{\alpha+\beta-\alpha+\beta}{2}\right) = 2\cos\alpha\sin\beta$

84. $\cos(\phi + 2\pi) + \cos\phi = 2\cos\left(\frac{\phi + 2\pi + \phi}{2}\right)\cos\left(\frac{\phi + 2\pi - \phi}{2}\right)$

$= 2\cos(\phi + \pi)\cos(\pi)$

85. $\cos\left(\theta + \frac{\pi}{2}\right) - \cos\left(\theta - \frac{\pi}{2}\right) = -2\sin\left[\frac{\left(\theta + \frac{\pi}{2}\right) + \left(\theta - \frac{\pi}{2}\right)}{2}\right]\sin\left[\frac{\left(\theta + \frac{\pi}{2}\right) - \left(\theta - \frac{\pi}{2}\right)}{2}\right]$

$= -2\sin\theta\sin\frac{\pi}{2}$

86. $\sin\left(x + \frac{\pi}{2}\right) + \sin\left(x - \frac{\pi}{2}\right) = 2\sin\left(\frac{x + \frac{\pi}{2} + x - \frac{\pi}{2}}{2}\right)\cos\left(\frac{x + \frac{\pi}{2} - \left(x - \frac{\pi}{2}\right)}{2}\right)$

$= 2\sin x\cos\frac{\pi}{2}$

87. $\sin 6x + \sin 2x = 0$

$2\sin\left(\frac{6x + 2x}{2}\right)\cos\left(\frac{6x - 2x}{2}\right) = 0$

$2(\sin 4x)\cos 2x = 0$

$\sin 4x = 0$ or $\cos 2x = 0$

$4x = n\pi$ $\quad\quad 2x = \frac{\pi}{2} + n\pi$

$x = \frac{n\pi}{4}$ $\quad\quad x = \frac{\pi}{4} + \frac{n\pi}{2}$

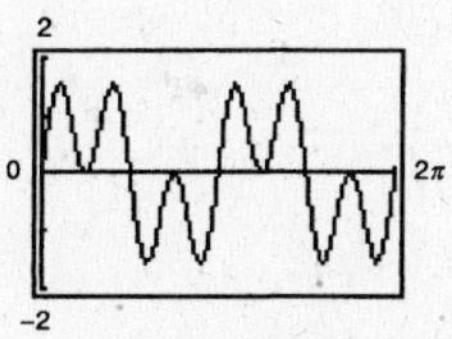

In the interval $[0, 2\pi)$ we have

$x = 0, \frac{\pi}{4}, \frac{\pi}{2}, \frac{3\pi}{4}, \pi, \frac{5\pi}{4}, \frac{3\pi}{2}, \frac{7\pi}{4}.$

88. $h(x) = \cos 2x - \cos 6x$

$\cos 2x - \cos 6x = 0$

$-2\sin 4x\sin(-2x) = 0$

$2\sin 4x\sin 2x = 0$

$\sin 4x = 0$ or $\sin 2x = 0$

$4x = n\pi$ $\quad\quad 2x = n\pi$

$x = \frac{n\pi}{4}$ $\quad\quad x = \frac{n\pi}{2}$

$x = 0, \frac{\pi}{4}, \frac{\pi}{2}, \frac{3\pi}{4}, \pi, \frac{5\pi}{4}, \frac{3\pi}{2}, \frac{7\pi}{4}$ $\quad\quad x = 0, \frac{\pi}{2}, \pi, \frac{3\pi}{2}$

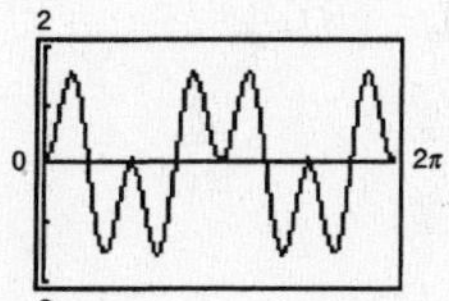

89. $\dfrac{\cos 2x}{\sin 3x - \sin x} - 1 = 0$

$$\frac{\cos 2x}{\sin 3x - \sin x} = 1$$

$$\frac{\cos 2x}{2\cos 2x \sin x} = 1$$

$$2\sin x = 1$$

$$\sin x = \frac{1}{2}$$

$$x = \frac{\pi}{6}, \frac{5\pi}{6}$$

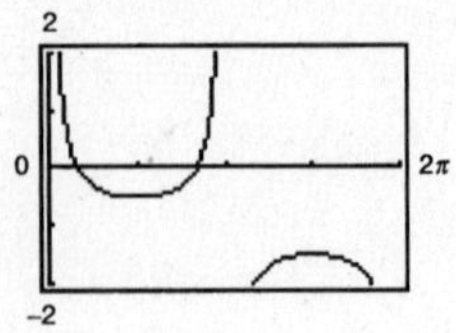

90. $f(x) = \sin^2 3x - \sin^2 x$

$$\sin^2 3x - \sin^2 x = 0$$

$$(\sin 3x + \sin x)(\sin 3x - \sin x) = 0$$

$$(2\sin 2x \cos x)(2\cos 2x \sin x) = 0$$

$\sin 2x = 0 \Rightarrow x = 0, \dfrac{\pi}{2}, \pi, \dfrac{3\pi}{2}$ or

$\cos x = 0 \Rightarrow x = \dfrac{\pi}{2}, \dfrac{3\pi}{2}$ or

$\cos 2x = 0 \Rightarrow x = \dfrac{\pi}{4}, \dfrac{3\pi}{4}, \dfrac{5\pi}{4}, \dfrac{7\pi}{4}$ or

$\sin x = 0 \Rightarrow x = 0, \pi$

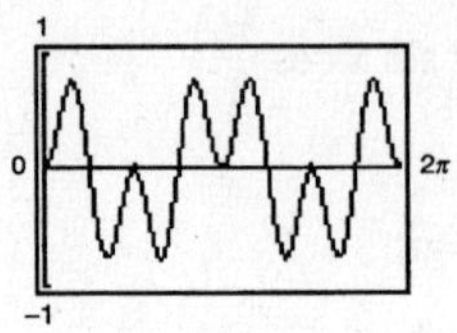

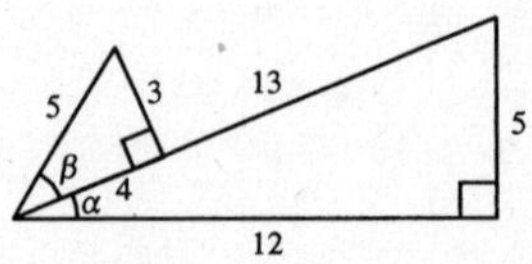

Figure for Exercises 91 and 94

91. $\sin^2 \alpha = \left(\dfrac{5}{13}\right)^2 = \dfrac{25}{169}$

$$\sin^2 \alpha = 1 - \cos^2 \alpha = 1 - \left(\frac{12}{13}\right)^2$$

$$= 1 - \frac{144}{169} = \frac{25}{169}$$

92. $\cos^2 \alpha = (\cos \alpha)^2 = \left(\dfrac{12}{13}\right)^2 = \dfrac{144}{169}$

$$\cos^2 \alpha = 1 - \sin^2 \alpha = 1 - \left(\frac{5}{13}\right)^2$$

$$= 1 - \frac{25}{169} = \frac{144}{169}$$

93. $\sin \alpha \cos \beta = \left(\dfrac{5}{13}\right)\left(\dfrac{4}{5}\right) = \dfrac{4}{13}$

$$\sin \alpha \cos \beta = \cos\left(\frac{\pi}{2} - \alpha\right) \sin\left(\frac{\pi}{2} - \beta\right)$$

$$= \left(\frac{5}{13}\right)\left(\frac{4}{5}\right) = \frac{4}{13}$$

94. $\cos \alpha \sin \beta = \left(\dfrac{12}{13}\right)\left(\dfrac{3}{5}\right) = \dfrac{36}{65}$

$$\cos \alpha \sin \beta = \sin\left(\frac{\pi}{2} - \alpha\right) \cos\left(\frac{\pi}{2} - \beta\right)$$

$$= \left(\frac{12}{13}\right)\left(\frac{3}{5}\right) = \frac{36}{65}$$

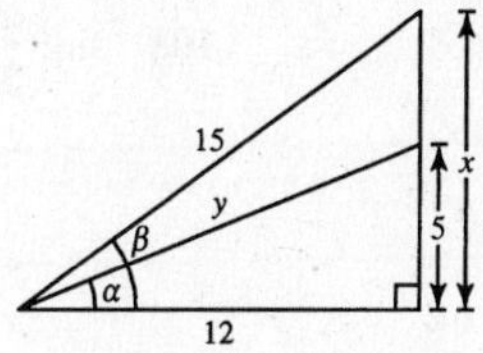

Figure for Exercises 95-102

$$x = \sqrt{15^2 - 12^2} = 9$$

$$y = \sqrt{12^2 + 5^2} = 13$$

$\sin \alpha = \frac{5}{13}$ $\quad \sin \beta = \frac{9}{15} = \frac{3}{5}$

$\cos \alpha = \frac{12}{13}$ $\quad \cos \beta = \frac{12}{15} = \frac{4}{5}$

$\tan \alpha = \frac{5}{12}$ $\quad \tan \beta = \frac{9}{12} = \frac{3}{4}$

95. $\sin \beta = \frac{9}{15} = \frac{3}{5}$

96. $\cos \alpha = \frac{12}{13}$

97. $\cos 2\beta = 2\cos^2 \beta - 1$
$= 2\left(\frac{4}{5}\right)^2 - 1$
$= \frac{32}{25} - 1$
$= \frac{7}{25}$

98. $\sin 2\alpha = 2 \sin \alpha \cos \alpha$
$= 2\left(\frac{5}{13}\right)\left(\frac{12}{13}\right)$
$= \frac{120}{169}$

99. $\sin \frac{\beta}{2} = \sqrt{\frac{1 - \cos \beta}{2}}$
$= \sqrt{\frac{1 - \frac{4}{5}}{2}}$
$= \sqrt{\frac{1}{10}}$
$= \frac{\sqrt{10}}{10}$

100. $\cos \frac{\beta}{2} = \sqrt{\frac{1 + \cos \beta}{2}}$
$= \sqrt{\frac{1 + \frac{4}{5}}{2}}$
$= \sqrt{\frac{9}{10}}$
$= \frac{3\sqrt{10}}{10}$

101. $\tan 2\alpha = \frac{2 \tan \alpha}{1 - \tan^2 \alpha}$
$= \frac{2\left(\frac{5}{12}\right)}{1 - \left(\frac{5}{12}\right)^2}$
$= \frac{\frac{5}{6}}{1 - \frac{25}{144}}$
$= \left(\frac{5}{6}\right)\left(\frac{144}{119}\right)$
$= \frac{120}{119}$

102. $\tan 2\beta = \frac{2 \tan \beta}{1 - \tan^2 \beta}$
$= \frac{2\left(\frac{3}{4}\right)}{1 - \left(\frac{3}{4}\right)^2}$
$= \frac{\frac{3}{2}}{1 - \frac{9}{16}}$
$= \left(\frac{3}{2}\right)\left(\frac{16}{7}\right)$
$= \frac{24}{7}$

103. $\csc 2\theta = \frac{1}{\sin 2\theta}$
$= \frac{1}{2 \sin \theta \cos \theta}$
$= \frac{1}{\sin \theta} \cdot \frac{1}{2 \cos \theta}$
$= \frac{\csc \theta}{2 \cos \theta}$

104. $\sec 2\theta = \frac{1}{\cos 2\theta} = \frac{1}{\cos^2 \theta - \sin^2 \theta}$
$= \frac{\frac{1}{\cos^2 \theta}}{1 - \frac{\sin^2 \theta}{\cos^2 \theta}}$
$= \frac{\sec^2 \theta}{1 - \tan^2 \theta}$
$= \frac{\sec^2 \theta}{1 - (\sec^2 \theta - 1)}$
$= \frac{\sec^2 \theta}{2 - \sec^2 \theta}$

105. $\cos^2 2\alpha - \sin^2 2\alpha = \cos [2(2\alpha)]$
$= \cos 4\alpha$

106. $\cos^4 x - \sin^4 x = (\cos^2 x - \sin^2 x)(\cos^2 x + \sin^2 x)$
$= (\cos 2x)(1) = \cos 2x$

107. $(\sin x + \cos x)^2 = \sin^2 x + 2\sin x\cos x + \cos^2 x$
$= (\sin^2 x + \cos^2 x) + 2\sin x\cos x$
$= 1 + \sin 2x$

108. $\sin\left(\frac{\alpha}{3}\right)\cos\left(\frac{\alpha}{3}\right) = \frac{1}{2}\left[2\left(\sin\left(\frac{\alpha}{3}\right)\cos\left(\frac{\alpha}{3}\right)\right)\right]$
$= \frac{1}{2}\sin\frac{2\alpha}{3}$

109. $1 + \cos 10y = 1 + \cos^2 5y - \sin^2 5y$
$= 1 + \cos^2 5y - (1 - \cos^2 5y)$
$= 2\cos^2 5y$

110. $\frac{\cos 3\beta}{\cos\beta} = \frac{\cos(2\beta + \beta)}{\cos\beta} = \frac{\cos 2\beta\cos\beta - \sin 2\beta\sin\beta}{\cos\beta}$
$= \frac{(\cos^2\beta - \sin^2\beta)\cos\beta - (2\sin\beta\cos\beta)\sin\beta}{\cos\beta}$
$= \frac{\cos^3\beta - 3\sin^2\beta\cos\beta}{\cos\beta}$
$= \cos^2\beta - 3\sin^2\beta = 1 - \sin^2\beta - 3\sin^2\beta$
$= 1 - 4\sin^2\beta$

111. $\sec\frac{u}{2} = \frac{1}{\cos\frac{u}{2}}$
$= \pm\sqrt{\frac{2}{1 + \cos u}}$
$= \pm\sqrt{\frac{2\sin u}{\sin u(1 + \cos u)}}$
$= \pm\sqrt{\frac{2\sin u}{\sin u + \sin u\cos u}}$
$= \pm\sqrt{\frac{\frac{2\sin u}{\cos u}}{\frac{\sin u}{\cos u} + \frac{\sin u\cos u}{\cos u}}}$
$= \pm\sqrt{\frac{2\tan u}{\tan u + \sin u}}$

112. $\tan\frac{u}{2} = \frac{1 - \cos u}{\sin u}$
$= \frac{1}{\sin u} - \frac{\cos u}{\sin u}$
$= \csc u - \cot u$

113. $\frac{\sin x \pm \sin y}{\cos x + \cos y} = \frac{2\sin\left(\frac{x \pm y}{2}\right)\cos\left(\frac{x \mp y}{2}\right)}{2\cos\left(\frac{x + y}{2}\right)\cos\left(\frac{x - y}{2}\right)}$
$= \tan\left(\frac{x \pm y}{2}\right)$

114. $\frac{\sin x + \sin y}{\cos x - \cos y} = \frac{2\sin\left(\frac{x + y}{2}\right)\cos\left(\frac{x - y}{2}\right)}{-2\sin\left(\frac{x + y}{2}\right)\sin\left(\frac{x - y}{2}\right)}$
$= -\cot\left(\frac{x - y}{2}\right)$

115. $\frac{\cos 4x + \cos 2x}{\sin 4x + \sin 2x} = \frac{2\cos\left(\frac{4x + 2x}{2}\right)\cos\left(\frac{4x - 2x}{2}\right)}{2\sin\left(\frac{4x + 2x}{2}\right)\cos\left(\frac{4x - 2x}{2}\right)}$
$= \frac{2\cos 3x\cos x}{2\sin 3x\cos x}$
$= \cot 3x$

116. $\frac{\cos t + \cos 3t}{\sin 3t - \sin t} = \frac{2\cos\left(\frac{4t}{2}\right)\cos\left(-\frac{2t}{2}\right)}{2\cos\left(\frac{4t}{2}\right)\sin\left(\frac{2t}{2}\right)}$
$= \frac{\cos(-t)}{\sin(t)} = \frac{\cos(t)}{\sin(t)} = \cot t$

117. $\sin\left(\frac{\pi}{6} + x\right) + \sin\left(\frac{\pi}{6} - x\right) = 2\sin\frac{\pi}{6}\cos x = 2 \cdot \frac{1}{2}\cos x = \cos x$

118. $\cos\left(\frac{\pi}{3} + x\right) + \cos\left(\frac{\pi}{3} - x\right) = 2\cos\left(\frac{\frac{\pi}{3} + x + \frac{\pi}{3} - x}{2}\right)\cos\left(\frac{\frac{\pi}{3} + x - \left(\frac{\pi}{3} - x\right)}{2}\right)$

$= 2\cos\left(\frac{\pi}{3}\right)\cos(x) = 2\left(\frac{1}{2}\right)\cos x = \cos x$

119. $\cos 3\beta = \cos(2\beta + \beta)$

$= \cos 2\beta \cos\beta - \sin 2\beta \sin \beta$

$= (\cos^2 \beta - \sin^2 \beta)\cos \beta - 2 \sin \beta \cos \beta \sin \beta$

$= \cos^3 \beta - \sin^2 \beta \cos \beta - 2 \sin^2 \beta \cos \beta$

$= \cos^3 \beta - 3 \sin^2 \beta \cos \beta$

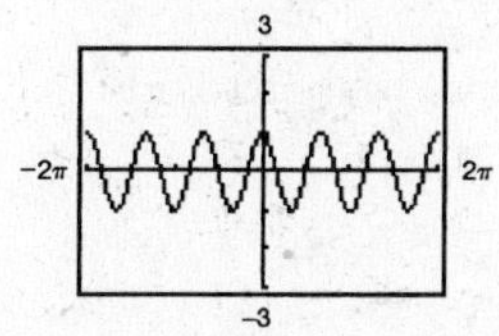

120. $\sin 4\beta = 2 \sin 2\beta \cos 2\beta$

$= 2(2 \sin \beta \cos \beta)(1 - 2 \sin^2 \beta)$

$= 4 \sin \beta \cos \beta(1 - 2 \sin^2 \beta)$

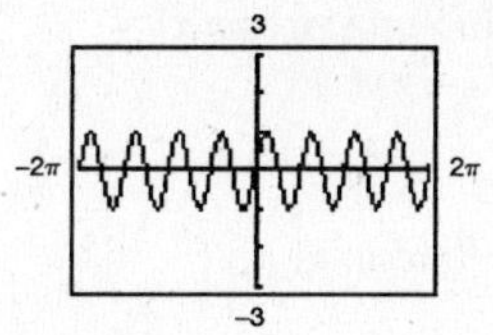

121. $\frac{\cos 4x - \cos 2x}{2 \sin 3x} = \frac{-2 \sin\left(\frac{4x + 2x}{2}\right)\sin\left(\frac{4x - 2x}{2}\right)}{2 \sin 3x}$

$= \frac{-2 \sin 3x \sin x}{2 \sin 3x}$

$= -\sin x$

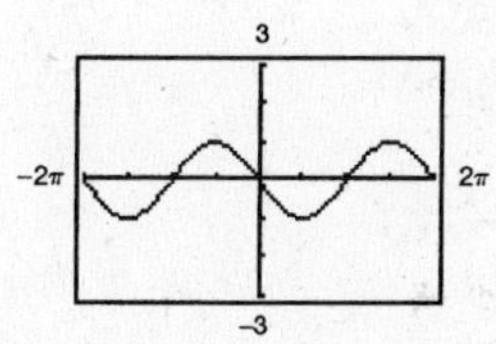

122. $\frac{\cos 3x - \cos x}{\sin 3x - \sin x} = \frac{-2 \sin\left(\frac{3x + x}{2}\right)\sin\left(\frac{3x - x}{2}\right)}{2 \cos\left(\frac{3x + x}{2}\right)\sin\left(\frac{3x - x}{2}\right)}$

$= \frac{-2 \sin 2x \sin x}{2 \cos 2x \sin x}$

$= -\tan 2x$

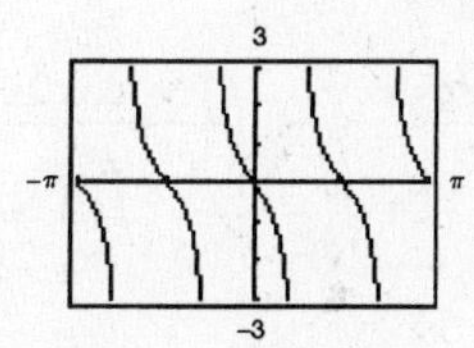

123. $\sin^2 x = \frac{1 - \cos 2x}{2} = \frac{1}{2} - \frac{\cos 2x}{2}$

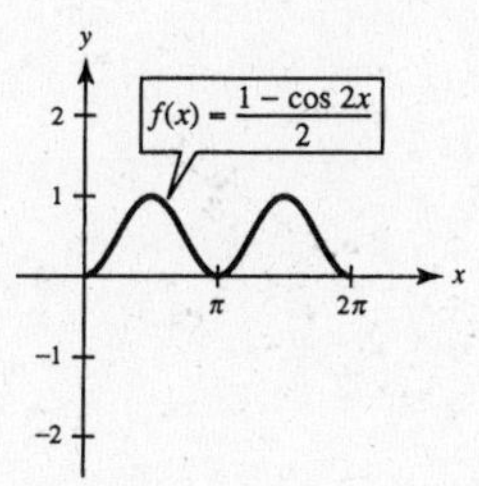

124. $f(x) = \cos^4 x = (\cos^2 x)(\cos^2 x)$

$= \left(\frac{1 + \cos 2x}{2}\right)\left(\frac{1 + \cos 2x}{2}\right)$

$= \frac{1}{4}(1 + 2 \cos 2x + \cos^2 2x)$

$= \frac{1}{4}\left[1 + 2 \cos 2x + \left(\frac{1 + \cos 4x}{2}\right)\right]$

$= \frac{1}{8}[2 + 4 \cos 2x + (1 + \cos 4x)]$

$= \frac{1}{8}(3 + 4 \cos 2x + \cos 4x)$

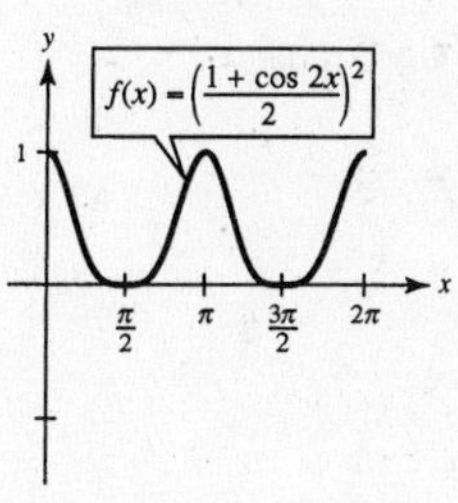

125. $f(x) = \cos^2 x = \dfrac{1 + \cos 2x}{2} = \dfrac{1}{2} + \dfrac{\cos 2x}{2}$

Shifted upward by $\dfrac{1}{2}$ unit.

Amplitude: $|a| = \dfrac{1}{2}$

Period: $\dfrac{2\pi}{2} = \pi$

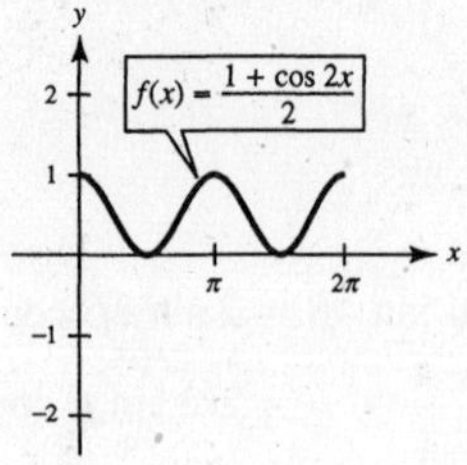

126. $f(x) = \sin^3 x = \sin x(\sin^2 x)$

$$= \sin x\left(\frac{1 - \cos 2x}{2}\right)$$

$$= \frac{1}{2}\sin x(1 - \cos 2x)$$

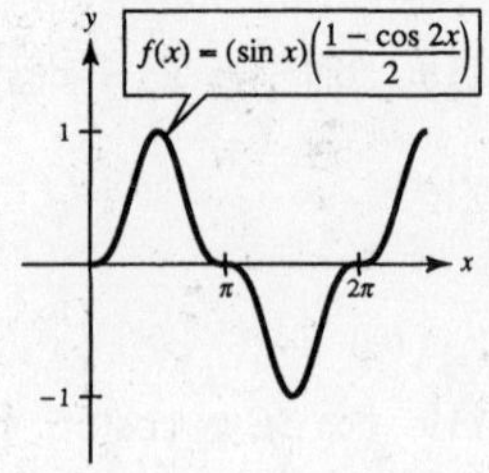

127. $\sin(2 \arcsin x) = 2 \sin(\arcsin x) \cos(\arcsin x)$

$$= 2x\sqrt{1 - x^2}$$

128. $\cos(2 \arccos x) = \cos^2(\arccos x) - \sin^2(\arccos x)$

$$= x^2 - (1 - x^2) = 2x^2 - 1$$

129. $\cos(2 \arcsin x) = 1 - 2 \sin^2(\arcsin x)$

$$= 1 - 2x^2$$

130. $\sin(2 \arccos x) = 2 \sin(\arccos x) \cos(\arccos x)$

$$= 2\left(\sqrt{1 - x^2}\right)(x)$$

$$= 2x\sqrt{1 - x^2}$$

131. $\cos(2 \arctan x) = \cos^2(\arctan x) - \sin^2(\arctan x)$

$$= \left(\frac{1}{\sqrt{1 + x^2}}\right)^2 - \left(\frac{x}{\sqrt{1 + x^2}}\right)^2$$

$$= \frac{1 - x^2}{1 + x^2}$$

132. $\sin(2 \arctan x) = 2 \sin(\arctan x) \cos(\arctan x)$

$$= 2\left(\frac{x}{\sqrt{1 + x^2}}\right)\left(\frac{1}{\sqrt{1 + x^2}}\right)$$

$$= \frac{2x}{1 + x^2}$$

133. $f(x) = \sin^4 x + \cos^4 x$

(a) $\sin^4 x + \cos^4 x = (\sin^2 x)^2 + (\cos^2 x)^2$

$$= \left(\frac{1 - \cos 2x}{2}\right)^2 + \left(\frac{1 + \cos 2x}{2}\right)^2$$

$$= \frac{1}{4}[(1 - \cos 2x)^2 + (1 + \cos 2x)^2]$$

$$= \frac{1}{4}(1 - 2 \cos 2x + \cos^2 2x + 1 + 2 \cos 2x + \cos^2 2x)$$

$$= \frac{1}{4}(2 + 2 \cos^2 2x)$$

$$= \frac{1}{4}\left[2 + 2\left(\frac{1 + \cos 2(2x)}{2}\right)\right]$$

$$= \frac{1}{4}(3 + \cos 4x)$$

(b) $\sin^4 x + \cos^4 x = (\sin^2 x)^2 + \cos^4 x$

$$= (1 - \cos^2 x)^2 + \cos^4 x$$

$$= 1 - 2 \cos^2 x + \cos^4 x + \cos^4 x$$

$$= 2 \cos^4 x - 2 \cos^2 x + 1$$

(c) $\sin^4 x + \cos^4 x = \sin^4 x + 2 \sin^2 x \cos^2 x + \cos^4 x - 2 \sin^2 x \cos^2 x$

$$= (\sin^2 x + \cos^2 x)^2 - 2 \sin^2 x \cos^2 x$$

$$= 1 - 2 \sin^2 x \cos^2 x$$

—CONTINUED—

133. —CONTINUED—

(d) $1 - 2\sin^2 x\cos^2 x = 1 - (2\sin x\cos x)(\sin x\cos x)$

$$= 1 - (\sin 2x)\left(\frac{1}{2}\sin 2x\right)$$

$$= 1 - \frac{1}{2}\sin^2 2x$$

(e) No, it does not mean that one of you is wrong. There is often more than one way to rewrite a trigonometric expression.

134. (a)

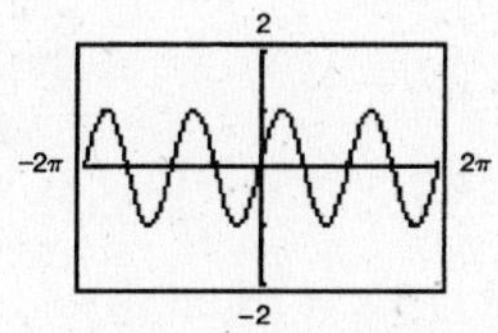

(b) The graph appears to be that of $\sin 2x$.

(c) $2\sin x\left[2\cos^2\left(\frac{x}{2}\right) - 1\right] = 2\sin x\cos x$

$$= \sin 2x$$

135. $r = \frac{1}{32}v_0^2\sin 2\theta$

$$= \frac{1}{32}v_0^2(2\sin\theta\cos\theta)$$

$$= \frac{1}{16}v_0^2\sin\theta\cos\theta$$

136. (a) $\sin\frac{\theta}{2} = \frac{1}{1}$

$$\frac{\theta}{2} = \arcsin 1$$

$$\theta = 2\left(\frac{\pi}{2}\right) = \pi$$

(b) $\sin\frac{\theta}{2} = \frac{1}{4.5}$

$$\frac{\theta}{2} = \arcsin\left(\frac{1}{4.5}\right)$$

$$\theta = 2\arcsin\left(\frac{1}{4.5}\right) \approx 0.4482$$

(c) $M = \frac{\text{speed}}{\text{speed of sound}} = \frac{\text{speed}}{760}$

speed $= 760M$

For $M = 1$: speed $= 760$ miles per hour

For $M = 4.5$: speed $= 760(4.5) = 3420$ miles per hour

(d) $\sin\frac{\theta}{2} = \frac{1}{M}$

$$\sqrt{\frac{1 - \cos\theta}{2}} = \frac{1}{M}$$

$$\frac{1 - \cos\theta}{2} = \frac{1}{M^2}$$

$$1 - \cos\theta = \frac{2}{M^2}$$

$$\cos\theta = 1 - \frac{2}{M^2} = \frac{M^2 - 2}{M^2}$$

137. False. For $u < 0$,

$$\sin 2u = -\sin(-2u)$$

$$= -2\sin(-u)\cos(-u)$$

$$= -2(-\sin u)\cos u$$

$$= 2\sin u\cos u.$$

138. False.

$$\sin\frac{u}{2} = -\sqrt{\frac{1 - \cos u}{2}}$$

when u is in the second quadrant.

139. True

$$\sqrt{\frac{1 - \cos(\pi/6)}{2}} = \sin\frac{(\pi/6)}{2}$$

$$= \sin\frac{\pi}{12} = \sin 2\left(\frac{\pi}{24}\right)$$

$$= 2\sin\frac{\pi}{24}\cos\frac{\pi}{24}$$

140. True

$$\sqrt{\frac{1 + \cos(\pi/4)}{2}} = \cos\frac{(\pi/4)}{2}$$

$$= \cos\frac{\pi}{8} = \cos 2\left(\frac{\pi}{16}\right)$$

$$= \cos^2\frac{\pi}{16} - \sin^2\frac{\pi}{16}$$

141. (a) $y = 4\sin\dfrac{x}{2} + \cos x$

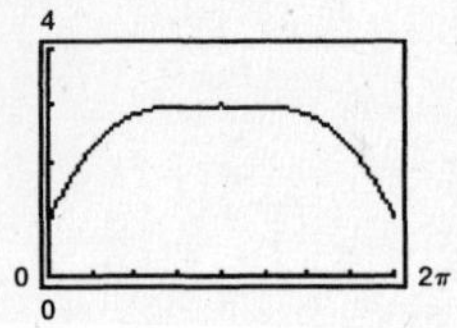

Maximum point: (3.1416, 3.0)

(b)
$$2\cos\frac{x}{2} - \sin x = 0$$
$$2\left(\pm\sqrt{\frac{1+\cos x}{2}}\right) = \sin x$$
$$4\left(\frac{1+\cos x}{2}\right) = \sin^2 x$$
$$2(1+\cos x) = 1 - \cos^2 x$$
$$\cos^2 x + 2\cos x + 1 = 0$$
$$(\cos x + 1)^2 = 0$$
$$\cos x = -1$$
$$x = \pi \approx 3.1416$$

142. (a) $f(x) = 2\cos\dfrac{x}{2} + \sin 2x$

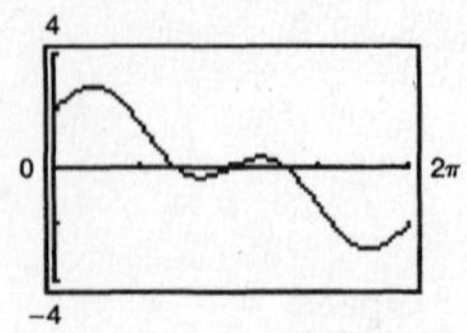

Maximum points: (0.6993, 2.8642), (3.6754, 0.3486)

Minimum points: (2.6078, −0.3486), (5.5839, −2.8642)

(b) $g(x) = 2\cos 2x - \sin\dfrac{x}{2}$

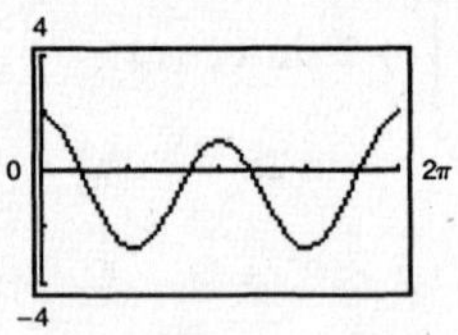

Zeros: $x \approx 0.6993, 2.6078, 3.6754, 5.5839$

143. $f(x) = \cos 2x - 2\sin x$

(a)

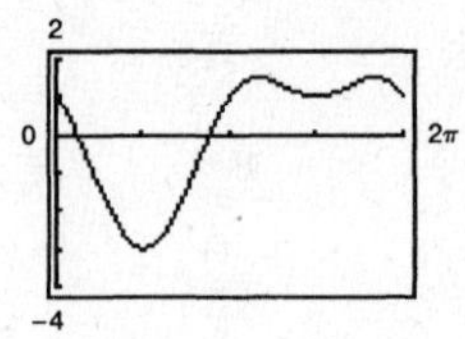

Maximum points: (3.6652, 1.5), (5.7596, 1.5)

Minimum points: (1.5708, −3), (4.7124, 1)

(b) $-2\cos x(2\sin x + 1) = 0$

$-2\cos x = 0$ or $2\sin x + 1 = 0$

$\cos x = 0$ $\qquad \sin x = -\dfrac{1}{2}$

$x = \dfrac{\pi}{2}, \dfrac{3\pi}{2}$ $\qquad x = \dfrac{7\pi}{6}, \dfrac{11\pi}{6}$

$\dfrac{\pi}{2} \approx 1.5708$ $\qquad \dfrac{7\pi}{6} \approx 3.6652$

$\dfrac{3\pi}{2} \approx 4.7124$ $\qquad \dfrac{11\pi}{6} \approx 5.7596$

144. $f(x) = \dfrac{1}{2}\sin 2x - \sin x$

(a)

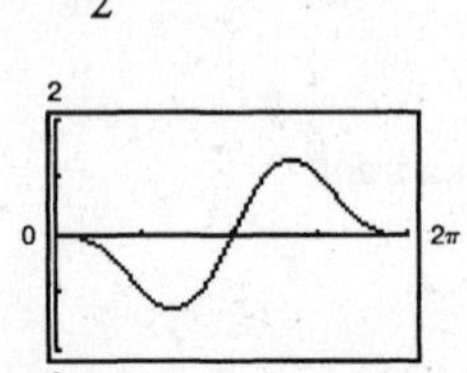

Maximum point: (4.1888, 1.2990)

Minimum point: (0, 0), (2.0944, −1.2990)

(b)
$$\cos 2x - \cos x = 0$$
$$2\cos^2 x - 1 - \cos x = 0$$
$$2\cos^2 x - \cos x - 1 = 0$$
$$(2\cos x + 1)(\cos x - 1) = 0$$

$2\cos x + 1 = 0$ or $\cos x - 1 = 0$

$\cos x = -\dfrac{1}{2}$ $\qquad \cos x = 1$

$x = \dfrac{2\pi}{3} \approx 2.0944$ $\qquad x = 0$

$x = \dfrac{4\pi}{3} \approx 4.1888$

145. $f(x) = \frac{1}{2}\cos 2x + \cos x$

(a)

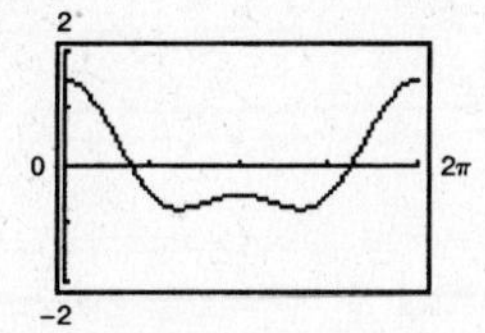

Maximum points: (0, 1.5), (3.1416, −0.5)

Minimum points: (2.0944, −0.75), (4.1888, −0.75)

(b)

$$-\sin 2x - \sin x = 0$$

$$-2\sin x\cos x - \sin x = 0$$

$$-\sin x(2\cos x + 1) = 0$$

$$-\sin x = 0 \quad \text{or} \quad 2\cos x + 1 = 0$$

$$\sin x = 0 \qquad \cos x = -\frac{1}{2}$$

$$x = 0,\ \pi \approx 3.1416 \qquad x = \frac{2\pi}{3}, \frac{4\pi}{3}$$

$$\approx 2.0944,\ 4.1888$$

Review Exercises for Chapter 10

1. $\dfrac{1}{\cos x} = \sec x$

2. $\dfrac{1}{\sin x} = \csc x$

3. $\dfrac{1}{\sec x} = \cos x$

4. $\dfrac{1}{\tan x} = \cot x$

5. $\dfrac{\cos x}{\sin x} = \cot x$

6. $\sqrt{1 + \tan^2 x} = \sqrt{\sec^2 x} = |\sec x|$

7. $\sin x = \frac{3}{5},\ \cos x = \frac{4}{5}$

$$\tan x = \frac{\sin x}{\cos x} = \frac{\frac{3}{5}}{\frac{4}{5}} = \frac{3}{4}$$

$$\cot x = \frac{1}{\tan x} = \frac{4}{3}$$

$$\sec x = \frac{1}{\cos x} = \frac{5}{4}$$

$$\csc x = \frac{1}{\sin x} = \frac{5}{3}$$

8. $\tan\theta = \frac{2}{3},\ \sec\theta = \frac{\sqrt{13}}{3}$

θ is in Quadrant I.

$$\cos\theta = \frac{1}{\sec\theta} = \frac{3}{\sqrt{13}} = \frac{3\sqrt{13}}{13}$$

$$\sin\theta = \sqrt{1 - \cos^2\theta} = \sqrt{1 - \frac{9}{13}} = \sqrt{\frac{4}{13}} = \frac{2\sqrt{13}}{13}$$

$$\csc\theta = \frac{1}{\sin\theta} = \frac{\sqrt{13}}{2}$$

$$\cot\theta = \frac{1}{\tan\theta} = \frac{3}{2}$$

9. $\sin\left(\frac{\pi}{2} - x\right) = \frac{\sqrt{2}}{2} \Rightarrow \cos x = \frac{1}{\sqrt{2}} = \frac{\sqrt{2}}{2},\ \sin x = -\frac{\sqrt{2}}{2}$

$$\tan x = \frac{\sin x}{\cos x} = \frac{-\frac{1}{\sqrt{2}}}{\frac{1}{\sqrt{2}}} = -1$$

$$\cot x = \frac{1}{\tan x} = -1$$

$$\sec x = \frac{1}{\cos x} = \sqrt{2}$$

$$\csc x = \frac{1}{\sin x} = -\sqrt{2}$$

10. $\csc\left(\frac{\pi}{2} - \theta\right) = \sec\theta = 9, \sin\theta = \frac{4\sqrt{5}}{9}$

θ is in Quadrant I.

$$\cos\theta = \frac{1}{\sec\theta} = \frac{1}{9}$$

$$\tan\theta = \frac{\sin\theta}{\cos\theta} = \frac{\frac{4\sqrt{5}}{9}}{\frac{1}{9}} = 4\sqrt{5}$$

$$\csc\theta = \frac{1}{\sin\theta} = \frac{9}{4\sqrt{5}} = \frac{9\sqrt{5}}{20}$$

$$\cot\theta = \frac{1}{\tan\theta} = \frac{1}{4\sqrt{5}} = \frac{\sqrt{5}}{20}$$

11. $\frac{1}{\cot^2 x + 1} = \frac{1}{\csc^2 x} = \sin^2 x$

12. $\frac{\tan\theta}{1 - \cos^2\theta} = \frac{\frac{\sin\theta}{\cos\theta}}{\sin^2\theta} = \frac{1}{\sin\theta\cos\theta}$

$= \csc\theta\sec\theta$

13. $\tan^2 x(\csc^2 x - 1) = \tan^2 x(\cot^2 x) = \tan^2 x\left(\frac{1}{\tan^2 x}\right) = 1$

14. $\cot^2 x(\sin^2 x) = \frac{\cos^2 x}{\sin^2 x}\sin^2 x = \cos^2 x$

15. $\frac{\sin\left(\frac{\pi}{2} - \theta\right)}{\sin\theta} = \frac{\cos\theta}{\sin\theta} = \cot\theta$

16. $\frac{\cot\left(\frac{\pi}{2} - u\right)}{\cos u} = \frac{\tan u}{\cos u} = \tan u\sec u$

17. $\cos^2 x + \cos^2 x\cot^2 x = \cos^2 x(1 + \cot^2 x) = \cos^2 x(\csc^2 x)$

$= \cos^2 x\left(\frac{1}{\sin^2 x}\right) = \frac{\cos^2 x}{\sin^2 x} = \cot^2 x$

18. $\tan^2\theta\csc^2\theta - \tan^2\theta = \tan^2\theta(\csc^2\theta - 1) = \tan^2\theta\cot^2\theta = 1$

19. $(\tan x + 1)^2\cos x = (\tan^2 x + 2\tan x + 1)\cos x$

$= (\sec^2 x + 2\tan x)\cos x$

$= \sec^2 x\cos x + 2\left(\frac{\sin x}{\cos x}\right)\cos x$

$= \sec x + 2\sin x$

20. $(\sec x - \tan x)^2 = \sec^2 x - 2\sec x\tan x + \tan^2 x$

$= 1 + \tan^2 x - 2\sec x\tan x + \tan^2 x$

$= 1 - 2\sec x\tan x + 2\tan^2 x$

21. $\frac{1}{\csc\theta + 1} - \frac{1}{\csc\theta - 1} = \frac{(\csc\theta - 1) - (\csc\theta + 1)}{(\csc\theta + 1)(\csc\theta - 1)}$

$= \frac{-2}{\csc^2\theta - 1}$

$= \frac{-2}{\cot^2\theta}$

$= -2\tan^2\theta$

22. $\frac{\cos^2 x}{1 - \sin x} = \frac{\cos^2 x}{(1 - \sin x)} \cdot \frac{(1 + \sin x)}{(1 + \sin x)}$

$= \frac{\cos^2 x(1 + \sin x)}{1 - \sin^2 x}$

$= 1 + \sin x$

23. $\cos x(\tan^2 x + 1) = \cos x\sec^2 x$

$= \frac{1}{\sec x}\sec^2 x$

$= \sec x$

24. $\sec^2 x\cot x - \cot x = \cot x(\sec^2 x - 1) = \cot x\tan^2 x$

$= \left(\frac{1}{\tan x}\right)\tan^2 x$

$= \tan x$

25. $\cos\left(x + \frac{\pi}{2}\right) = \cos x \cos \frac{\pi}{2} - \sin x \sin \frac{\pi}{2}$

$= (\cos x)(0) - (\sin x)(1)$

$= -\sin x$

26. $\cot\left(\frac{\pi}{2} - x\right) = \tan x$ by the Cofunction Identity

27. $\dfrac{1}{\tan \theta \csc \theta} = \dfrac{1}{\dfrac{\sin \theta}{\cos \theta} \cdot \dfrac{1}{\sin \theta}} = \cos \theta$

28. $\dfrac{1}{\tan x \csc x \sin x} = \dfrac{1}{(\tan x)\left(\dfrac{1}{\sin x}\right)(\sin x)} = \dfrac{1}{\tan x}$

$= \cot x$

29. $\sin^5 x \cos^2 x = \sin^4 x \cos^2 x \sin x$

$= (1 - \cos^2 x)^2 \cos^2 x \sin x$

$= (1 - 2\cos^2 x + \cos^4 x) \cos^2 x \sin x$

$= (\cos^2 x - 2\cos^4 x + \cos^6 x)\sin x$

30. $\cos^3 x \sin^2 x = \cos x \cos^2 x \sin^2 x$

$= \cos x(1 - \sin^2 x)\sin^2 x$

$= \cos x(\sin^2 x - \sin^4 x)$

$= (\sin^2 x - \sin^4 x)\cos x$

31. $\sin x = \sqrt{3} - \sin x$

$\sin x = \dfrac{\sqrt{3}}{2}$

$x = \dfrac{\pi}{3} + 2\pi n, \dfrac{2\pi}{3} + 2\pi n$

32. $4 \cos \theta = 1 + 2 \cos \theta$

$2 \cos \theta = 1$

$\cos \theta = \dfrac{1}{2}$

$\theta = \dfrac{\pi}{3} + 2n\pi$ or $\dfrac{5\pi}{3} + 2n\pi$

33. $3\sqrt{3} \tan u = 3$

$\tan u = \dfrac{1}{\sqrt{3}}$

$u = \dfrac{\pi}{6} + n\pi$

34. $\dfrac{1}{2} \sec x - 1 = 0$

$\dfrac{1}{2} \sec x = 1$

$\sec x = 2$

$\cos x = \dfrac{1}{2}$

$x = \dfrac{\pi}{3} + 2n\pi$ or $\dfrac{5\pi}{3} + 2n\pi$

35. $3 \csc^2 x = 4$

$\csc^2 x = \dfrac{4}{3}$

$\sin x = \pm\dfrac{\sqrt{3}}{2}$

$x = \dfrac{\pi}{3} + 2\pi n, \dfrac{2\pi}{3} + 2\pi n, \dfrac{4\pi}{3} + 2\pi n, \dfrac{5\pi}{3} + 2\pi n$

36. $4 \tan^2 u - 1 = \tan^2 u$

$3 \tan^2 u - 1 = 0$

$\tan^2 u = \dfrac{1}{3}$

$\tan u = \pm\dfrac{1}{\sqrt{3}} = \pm\dfrac{\sqrt{3}}{3}$

$u = \dfrac{\pi}{6} + n\pi$ or $\dfrac{5\pi}{6} + n\pi$

37.
$$2\cos^2 x - \cos x = 1$$
$$2\cos^2 x - \cos x - 1 = 0$$
$$(2\cos x + 1)(\cos x - 1) = 0$$
$$2\cos x + 1 = 0 \qquad \cos x - 1 = 0$$
$$\cos x = -\frac{1}{2} \qquad \cos x = 1$$
$$x = \frac{2\pi}{3}, \frac{4\pi}{3} \qquad x = 0$$

38.
$$2\sin^2 x - 3\sin x = -1$$
$$2\sin^2 x - 3\sin x + 1 = 0$$
$$(2\sin x - 1)(\sin x - 1) = 0$$
$$2\sin x - 1 = 0 \quad \text{or} \quad \sin x - 1 = 0$$
$$\sin x = \frac{1}{2} \qquad \sin x = 1$$
$$x = \frac{\pi}{6}, \frac{5\pi}{6} \qquad x = \frac{\pi}{2}$$

39.
$$\cos^2 x + \sin x = 1$$
$$1 - \sin^2 x + \sin x - 1 = 0$$
$$-\sin x(\sin x - 1) = 0$$
$$\sin x = 0 \qquad \sin x - 1 = 0$$
$$x = 0, \pi \qquad \sin x = 1$$
$$x = \frac{\pi}{2}$$

40.
$$\sin^2 x + 2\cos x = 2$$
$$1 - \cos^2 x + 2\cos x = 2$$
$$0 = \cos^2 x - 2\cos x + 1$$
$$0 = (\cos x - 1)^2$$
$$\cos x - 1 = 0$$
$$\cos x = 1$$
$$x = 0$$

41. $2\sin 2x - \sqrt{2} = 0$
$$\sin 2x = \frac{\sqrt{2}}{2}$$
$$2x = \frac{\pi}{4} + 2\pi n, \frac{3\pi}{4} + 2\pi n$$
$$x = \frac{\pi}{8} + \pi n, \frac{3\pi}{8} + \pi n$$
$$x = \frac{\pi}{8}, \frac{3\pi}{8}, \frac{9\pi}{8}, \frac{11\pi}{8}$$

42. $\sqrt{3}\tan 3x = 0$
$$\tan 3x = 0$$
$$3x = 0, \pi, 2\pi, 3\pi, 4\pi, 5\pi$$
$$x = 0, \frac{\pi}{3}, \frac{2\pi}{3}, \pi, \frac{4\pi}{3}, \frac{5\pi}{3}$$

43. $\cos 4x(\cos x - 1) = 0$
$$\cos 4x = 0 \qquad \cos x - 1 = 0$$
$$4x = \frac{\pi}{2} + 2\pi n, \frac{3\pi}{2} + 2\pi n \qquad \cos x = 1$$
$$x = \frac{\pi}{8} + \frac{\pi}{2}n, \frac{3\pi}{8} + \frac{\pi}{2}n \qquad x = 0$$
$$x = 0, \frac{\pi}{8}, \frac{3\pi}{8}, \frac{5\pi}{8}, \frac{7\pi}{8}, \frac{9\pi}{8}, \frac{11\pi}{8}, \frac{13\pi}{8}, \frac{15\pi}{8}$$

44. $3\csc^2 5x = -4$
$$\csc^2 5x = -\frac{4}{3}$$
$$\csc 5x = \pm\sqrt{-\frac{4}{3}}$$
No real solution

45. $\sin^2 x - 2\sin x = 0$
$$\sin x(\sin x - 2) = 0$$
$$\sin x = 0 \qquad \sin x - 2 = 0$$
$x = 0, \pi$ No solution

46. $2\cos^2 x + 3\cos x = 0$
$$\cos x(2\cos x + 3) = 0$$
$$\cos x = 0 \quad \text{or} \quad 2\cos x + 3 = 0$$
$$x = \frac{\pi}{2}, \frac{3\pi}{2} \qquad 2\cos x = -3$$
$$\cos x = -\frac{3}{2}$$
No solution

47. $\tan^2\theta + \tan\theta - 12 = 0$

$(\tan\theta + 4)(\tan\theta - 3) = 0$

$\tan\theta + 4 = 0 \qquad \tan\theta - 3 = 0$

$\theta = \arctan(-4) + n\pi \qquad \theta = \arctan 3 + n\pi$

$\theta = \arctan(-4) + \pi, \arctan(-4) + 2\pi, \arctan 3, \arctan 3 + \pi$

48. $\sec^2 x + 6\tan x + 4 = 0$

$1 + \tan^2 x + 6\tan x + 4 = 0$

$\tan^2 x + 6\tan x + 5 = 0$

$(\tan x + 5)(\tan x + 1) = 0$

$\tan x + 5 = 0$ or $\tan x + 1 = 0$

$\tan x = -5 \qquad \tan x = -1$

$x = \arctan(-5) + \pi \qquad x = \dfrac{3\pi}{4}, \dfrac{7\pi}{4}$

$x = \arctan(-5) + 2\pi$

49.
$$\begin{aligned}\sin 285^\circ &= \sin(315^\circ - 30^\circ)\\ &= \sin 315^\circ\cos 30^\circ - \cos 315^\circ \sin 30^\circ\\ &= \left(-\frac{\sqrt{2}}{2}\right)\left(\frac{\sqrt{3}}{2}\right) - \left(\frac{\sqrt{2}}{2}\right)\left(\frac{1}{2}\right)\\ &= -\frac{\sqrt{2}}{4}(\sqrt{3}+1)\end{aligned}$$

$$\begin{aligned}\cos 285^\circ &= \cos(315^\circ - 30^\circ)\\ &= \cos 315^\circ\cos 30^\circ + \sin 315^\circ \sin 30^\circ\\ &= \left(\frac{\sqrt{2}}{2}\right)\left(\frac{\sqrt{3}}{2}\right) + \left(-\frac{\sqrt{2}}{2}\right)\left(\frac{1}{2}\right)\\ &= \frac{\sqrt{2}}{4}(\sqrt{3}-1)\end{aligned}$$

$$\begin{aligned}\tan 285^\circ &= \tan(315^\circ - 30^\circ) = \frac{\tan 315^\circ - \tan 30^\circ}{1 + \tan 315^\circ \tan 30^\circ}\\ &= \frac{(-1) - \left(\frac{\sqrt{3}}{3}\right)}{1 + (-1)\left(\frac{\sqrt{3}}{3}\right)} = -2 - \sqrt{3}\end{aligned}$$

50.
$$\begin{aligned}\sin(345^\circ) &= \sin(300^\circ + 45^\circ)\\ &= \sin 300^\circ \cos 45^\circ + \cos 300^\circ \sin 45^\circ\\ &= -\frac{\sqrt{3}}{2}\cdot\frac{\sqrt{2}}{2} + \frac{1}{2}\cdot\frac{\sqrt{2}}{2}\\ &= \frac{\sqrt{2}}{4}(-\sqrt{3}+1) = \frac{\sqrt{2}}{4}(1-\sqrt{3})\end{aligned}$$

$$\begin{aligned}\cos(345^\circ) &= \cos(300^\circ + 45^\circ)\\ &= \cos 300^\circ\cos 45^\circ - \sin 300^\circ \sin 45^\circ\\ &= \frac{1}{2}\cdot\frac{\sqrt{2}}{2} - \left(-\frac{\sqrt{3}}{2}\right)\frac{\sqrt{2}}{2}\\ &= \frac{\sqrt{2}}{4}(1+\sqrt{3})\end{aligned}$$

$$\begin{aligned}\tan(345^\circ) &= \tan(300^\circ + 45^\circ)\\ &= \frac{\tan 300^\circ + \tan 45^\circ}{1 - \tan 300^\circ \tan 45^\circ} = \frac{-\sqrt{3}+1}{1+\sqrt{3}(1)}\cdot\frac{1-\sqrt{3}}{1-\sqrt{3}}\\ &= \frac{4-2\sqrt{3}}{-2} = -2+\sqrt{3}\end{aligned}$$

51.
$$\begin{aligned}\sin\frac{25\pi}{12} &= \sin\left(\frac{11\pi}{6} + \frac{\pi}{4}\right) = \sin\frac{11\pi}{6}\cos\frac{\pi}{4} + \cos\frac{11\pi}{6}\sin\frac{\pi}{4}\\ &= \left(-\frac{1}{2}\right)\left(\frac{\sqrt{2}}{2}\right) + \left(\frac{\sqrt{3}}{2}\right)\left(\frac{\sqrt{2}}{2}\right) = \frac{\sqrt{2}}{4}(\sqrt{3}-1)\end{aligned}$$

$$\begin{aligned}\cos\frac{25\pi}{12} &= \cos\left(\frac{11\pi}{6} + \frac{\pi}{4}\right) = \cos\frac{11\pi}{6}\cos\frac{\pi}{4} - \sin\frac{11\pi}{6}\sin\frac{\pi}{4}\\ &= \left(\frac{\sqrt{3}}{2}\right)\left(\frac{\sqrt{2}}{2}\right) - \left(-\frac{1}{2}\right)\left(\frac{\sqrt{2}}{2}\right) = \frac{\sqrt{2}}{4}(\sqrt{3}+1)\end{aligned}$$

$$\begin{aligned}\tan\frac{25\pi}{12} &= \tan\left(\frac{11\pi}{6} + \frac{\pi}{4}\right) = \frac{\tan\frac{11\pi}{6} + \tan\frac{\pi}{4}}{1 - \tan\frac{11\pi}{6}\tan\frac{\pi}{4}}\\ &= \frac{\left(-\frac{\sqrt{3}}{3}\right) + 1}{1 - \left(-\frac{\sqrt{3}}{3}\right)(1)} = 2 - \sqrt{3}\end{aligned}$$

52. $\sin\left(\frac{19\pi}{12}\right) = \sin\left(\frac{11\pi}{6} - \frac{\pi}{4}\right)$

$$= \sin\frac{11\pi}{6}\cos\frac{\pi}{4} - \cos\frac{11\pi}{6}\sin\frac{\pi}{4}$$

$$= -\frac{1}{2}\cdot\frac{\sqrt{2}}{2} - \frac{\sqrt{3}}{2}\cdot\frac{\sqrt{2}}{2}$$

$$= -\frac{\sqrt{2}}{4}(1 + \sqrt{3}) = -\frac{\sqrt{2}}{4}(\sqrt{3} + 1)$$

$\cos\left(\frac{19\pi}{12}\right) = \cos\left(\frac{11\pi}{6} - \frac{\pi}{4}\right)$

$$= \cos\frac{11\pi}{6}\cos\frac{\pi}{4} + \sin\frac{11\pi}{6}\sin\frac{\pi}{4}$$

$$= \frac{\sqrt{3}}{2}\cdot\frac{\sqrt{2}}{2} + \left(-\frac{1}{2}\right)\frac{\sqrt{2}}{2}$$

$$= \frac{\sqrt{2}}{4}(\sqrt{3} - 1)$$

$\tan\left(\frac{19\pi}{12}\right) = \tan\left(\frac{11\pi}{6} - \frac{\pi}{4}\right)$

$$= \frac{\tan\frac{11\pi}{6} - \tan\frac{\pi}{4}}{1 + \tan\frac{11\pi}{6}\tan\frac{\pi}{4}}$$

$$= \frac{-\frac{\sqrt{3}}{3} - 1}{1 + \left(-\frac{\sqrt{3}}{3}\right)(1)} = \frac{-\sqrt{3} - 3}{3 - \sqrt{3}}\cdot\frac{3 + \sqrt{3}}{3 + \sqrt{3}}$$

$$= \frac{-(12 + 6\sqrt{3})}{6} = -2 - \sqrt{3}$$

53. $\sin 60^\circ \cos 45^\circ - \cos 60^\circ \sin 45^\circ = \sin(60^\circ - 45^\circ)$

$$= \sin 15^\circ$$

54. $\cos 45^\circ \cos 120^\circ - \sin 45^\circ \sin 120^\circ = \cos(45^\circ + 120^\circ)$

$$= \cos 165^\circ$$

55. $\frac{\tan 25^\circ + \tan 10^\circ}{1 - \tan 25^\circ \tan 10^\circ} = \tan(25^\circ + 10^\circ)$

$$= \tan 35^\circ$$

56. $\frac{\tan 68^\circ - \tan 115^\circ}{1 + \tan 68^\circ \tan 115^\circ} = \tan(68^\circ - 115^\circ) = \tan(-47^\circ)$

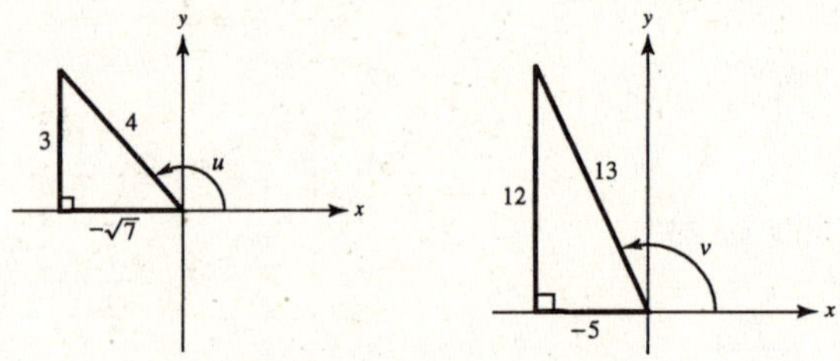

Figures for Exercises 57–62

57. $\sin(u + v) = \sin u \cos v + \cos u \sin v$

$$= \left(\frac{3}{4}\right)\left(-\frac{5}{13}\right) + \left(-\frac{\sqrt{7}}{4}\right)\left(\frac{12}{13}\right)$$

$$= -\frac{3}{52}(5 + 4\sqrt{7})$$

58. $\tan(u + v) = \frac{\tan u + \tan v}{1 - \tan u \tan v} = \frac{\left(-\frac{3}{\sqrt{7}}\right) + \left(-\frac{12}{5}\right)}{1 - \left(-\frac{3}{\sqrt{7}}\right)\left(-\frac{12}{5}\right)}$

$$= \frac{15 + 12\sqrt{7}}{36 - 5\sqrt{7}}\cdot\frac{36 + 5\sqrt{7}}{36 + 5\sqrt{7}} = \frac{960 + 507\sqrt{7}}{1121}$$

59. $\cos(u - v) = \cos u \cos v + \sin u \sin v$

$$= \left(-\frac{\sqrt{7}}{4}\right)\left(-\frac{5}{13}\right) + \left(\frac{3}{4}\right)\left(\frac{12}{13}\right)$$

$$= \frac{1}{52}(5\sqrt{7} + 36)$$

60. $\sin(u - v) = \sin u \cos v - \cos u \sin v$

$$= \left(\frac{3}{4}\right)\left(-\frac{5}{13}\right) - \left(-\frac{\sqrt{7}}{4}\right)\left(\frac{12}{13}\right)$$

$$= \frac{-15 + 12\sqrt{7}}{52} = \frac{12\sqrt{7} - 15}{52}$$

61. $\cos(u + v) = \cos u \cos v - \sin u \sin v$

$$= \left(-\frac{\sqrt{7}}{4}\right)\left(-\frac{5}{13}\right) - \left(\frac{3}{4}\right)\left(\frac{12}{13}\right)$$

$$= \frac{1}{52}(5\sqrt{7} - 36)$$

62. $\tan(u - v) = \dfrac{\tan u - \tan v}{1 + \tan u \tan v} = \dfrac{\left(-\frac{3}{\sqrt{7}}\right) - \left(-\frac{12}{5}\right)}{1 + \left(-\frac{3}{\sqrt{7}}\right)\left(-\frac{12}{5}\right)}$

$$= \frac{-15 + 12\sqrt{7}}{36 + 5\sqrt{7}} \cdot \frac{36 - 5\sqrt{7}}{36 - 5\sqrt{7}}$$

$$= \frac{-960 + 507\sqrt{7}}{1121}$$

63. $\sin\left(x + \frac{\pi}{4}\right) - \sin\left(x - \frac{\pi}{4}\right) = 1$

$$2 \cos x \sin \frac{\pi}{4} = 1$$

$$\cos x = \frac{\sqrt{2}}{2}$$

$$x = \frac{\pi}{4}, \frac{7\pi}{4}$$

64. $\cos\left(x + \frac{\pi}{6}\right) - \cos\left(x - \frac{\pi}{6}\right) = 1$

$$\left(\cos x \cos \frac{\pi}{6} - \sin x \sin \frac{\pi}{6}\right) - \left(\cos x \cos \frac{\pi}{6} + \sin x \sin \frac{\pi}{6}\right) = 1$$

$$-2 \sin x \sin \frac{\pi}{6} = 1$$

$$-2 \sin x\left(\frac{1}{2}\right) = 1$$

$$\sin x = -1$$

$$x = \frac{3\pi}{2}$$

65. $\sin 4x = 2 \sin 2x \cos 2x$

$$= 2[2 \sin x \cos x(\cos^2 x - \sin^2 x)]$$

$$= 4 \sin x \cos x(2 \cos^2 x - 1)$$

$$= 8 \cos^3 x \sin x - 4 \cos x \sin x$$

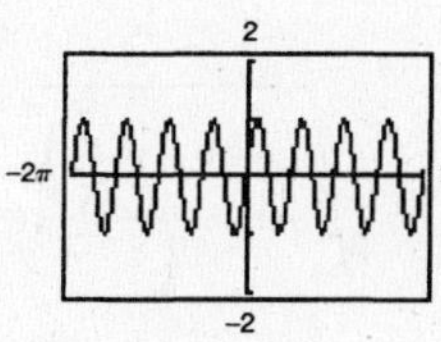

66. $\dfrac{1 - \cos 2x}{1 + \cos 2x} = \dfrac{1 - (1 - 2\sin^2 x)}{1 + (2\cos x^2 - 1)}$

$$= \frac{2 \sin^2 x}{2 \cos^2 x}$$

$$= \tan^2 x$$

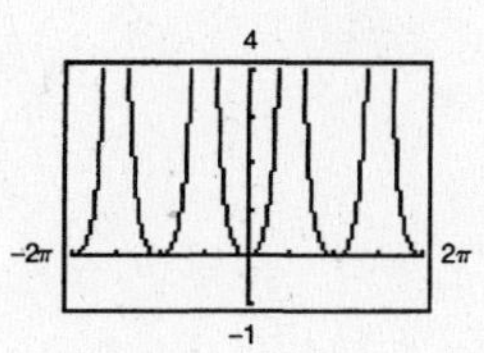

67. $\sin u = -\frac{4}{5}, \pi < u < \frac{3\pi}{2}$

$$\cos u = -\sqrt{1 - \sin^2 u} = -\frac{3}{5}$$

$$\tan u = \frac{\sin u}{\cos u} = \frac{4}{3}$$

$$\sin 2u = 2 \sin u \cos u = 2\left(-\frac{4}{5}\right)\left(-\frac{3}{5}\right) = \frac{24}{25}$$

$$\cos 2u = \cos^2 u - \sin^2 u = \left(-\frac{3}{5}\right)^2 - \left(-\frac{4}{5}\right)^2 = -\frac{7}{25}$$

$$\tan 2u = \frac{2 \tan u}{1 - \tan^2 u} = \frac{2\left(\frac{4}{3}\right)}{1 - \left(\frac{4}{3}\right)^2} = -\frac{24}{7}$$

68. $\cos u = -\dfrac{2}{\sqrt{5}}, \dfrac{\pi}{2} < u < \pi \Rightarrow \sin u = \dfrac{1}{\sqrt{5}}$ and $\tan u = -\dfrac{1}{2}$

$$\sin 2u = 2\sin u \cos u = 2\left(\frac{1}{\sqrt{5}}\right)\left(-\frac{2}{\sqrt{5}}\right) = -\frac{4}{5}$$

$$\cos 2u = \cos^2 u - \sin^2 u = \left(-\frac{2}{\sqrt{5}}\right)^2 - \left(\frac{1}{\sqrt{5}}\right)^2 = \frac{3}{5}$$

$$\tan 2u = \frac{2\tan u}{1 - \tan^2 u} = \frac{2\left(-\frac{1}{2}\right)}{1 - \left(-\frac{1}{2}\right)^2} = \frac{-1}{\frac{3}{4}} = -\frac{4}{3}$$

69. $\tan^2 2x = \dfrac{\sin^2 2x}{\cos^2 2x} = \dfrac{\frac{1 - \cos 4x}{2}}{\frac{1 + \cos 4x}{2}} = \dfrac{1 - \cos 4x}{1 + \cos 4x}$

70. $\cos^2 3x = \dfrac{1 + \cos 6x}{2}$

71. $\sin^2 x \tan^2 x = \sin^2 x\left(\dfrac{\sin^2 x}{\cos^2 x}\right) = \dfrac{\sin^4 x}{\cos^2 x}$

$$= \frac{\left(\frac{1 - \cos 2x}{2}\right)^2}{\frac{1 + \cos 2x}{2}} = \frac{\frac{1 - 2\cos 2x + \cos^2 2x}{4}}{\frac{1 + \cos 2x}{2}}$$

$$= \frac{1 - 2\cos 2x + \frac{1 + \cos 4x}{2}}{2(1 + \cos 2x)}$$

$$= \frac{2 - 4\cos 2x + 1 + \cos 4x}{4(1 + \cos 2x)}$$

$$= \frac{3 - 4\cos 2x + \cos 4x}{4(1 + \cos 2x)}$$

72. $\cos^2 x \tan^2 x = \sin^2 x = \dfrac{1 - \cos 2x}{2}$

73. $\sin(-75°) = -\sqrt{\dfrac{1 - \cos 150°}{2}} = -\sqrt{\dfrac{1 - \left(-\frac{\sqrt{3}}{2}\right)}{2}} = -\dfrac{\sqrt{2 + \sqrt{3}}}{2}$

$$= -\frac{1}{2}\sqrt{2 + \sqrt{3}}$$

$$\cos(-75°) = \sqrt{\frac{1 + \cos 150°}{2}} = \sqrt{\frac{1 + \left(-\frac{\sqrt{3}}{2}\right)}{2}} = \frac{\sqrt{2 - \sqrt{3}}}{2}$$

$$= \frac{1}{2}\sqrt{2 - \sqrt{3}}$$

$$\tan(-75°) = -\left(\frac{1 - \cos 150°}{\sin 150°}\right) = -\left(\frac{1 - \left(-\frac{\sqrt{3}}{2}\right)}{\frac{1}{2}}\right) = -(2 + \sqrt{3})$$

$$= -2 - \sqrt{3}$$

74. $\sin 15° = \sin\left(\frac{30°}{2}\right) = \sqrt{\frac{1 - \cos 30°}{2}} = \sqrt{\frac{1 - \frac{\sqrt{3}}{2}}{2}} = \frac{1}{2}\sqrt{2 - \sqrt{3}}$

$\cos 15° = \cos\left(\frac{30°}{2}\right) = \sqrt{\frac{1 + \cos 30°}{2}} = \sqrt{\frac{1 + \frac{\sqrt{3}}{2}}{2}} = \frac{1}{2}\sqrt{2 + \sqrt{3}}$

$\tan 15° = \tan\left(\frac{30°}{2}\right) = \frac{1 - \cos 30°}{\sin 30°} = \frac{1 - \frac{\sqrt{3}}{2}}{\frac{1}{2}} = 2 - \sqrt{3}$

75. $\sin\left(\frac{19\pi}{12}\right) = -\sqrt{\frac{1 - \cos\frac{19\pi}{6}}{2}} = -\sqrt{\frac{1 - \left(-\frac{\sqrt{3}}{2}\right)}{2}} = -\frac{\sqrt{2 + \sqrt{3}}}{2}$

$= -\frac{1}{2}\sqrt{2 + \sqrt{3}}$

$\cos\left(\frac{19\pi}{12}\right) = \sqrt{\frac{1 + \cos\frac{19\pi}{6}}{2}} = \sqrt{\frac{1 + \left(-\frac{\sqrt{3}}{2}\right)}{2}} = \frac{\sqrt{2 - \sqrt{3}}}{2}$

$= \frac{1}{2}\sqrt{2 - \sqrt{3}}$

$\tan\left(\frac{19\pi}{12}\right) = \frac{1 - \cos\frac{19\pi}{6}}{\sin\frac{19\pi}{6}} = \frac{1 - \left(-\frac{\sqrt{3}}{2}\right)}{-\frac{1}{2}} = -2 - \sqrt{3}$

76. $\sin\left(-\frac{17\pi}{12}\right) = \sin\left(-\frac{\frac{17\pi}{6}}{2}\right) = \sqrt{\frac{1 - \cos\left(-\frac{17\pi}{6}\right)}{2}} = \sqrt{\frac{1 - \left(-\frac{\sqrt{3}}{2}\right)}{2}}$

$= \frac{1}{2}\sqrt{2 + \sqrt{3}}$

$\cos\left(-\frac{17\pi}{12}\right) = \cos\left(\frac{-\frac{17\pi}{6}}{2}\right) = -\sqrt{\frac{1 + \cos\left(-\frac{17\pi}{6}\right)}{2}} = -\sqrt{\frac{1 + \left(-\frac{\sqrt{3}}{2}\right)}{2}}$

$= -\frac{1}{2}\sqrt{2 - \sqrt{3}}$

$\tan\left(-\frac{17\pi}{12}\right) = \tan\left(\frac{-\left(\frac{17\pi}{6}\right)}{2}\right) = \frac{1 - \cos\left(-\frac{17\pi}{6}\right)}{\sin\left(-\frac{17\pi}{6}\right)} = \frac{1 - \left(-\frac{\sqrt{3}}{2}\right)}{-\frac{1}{2}} = -2 - \sqrt{3}$

77. $-\sqrt{\frac{1 + \cos 10x}{2}} = -\left|\cos\frac{10x}{2}\right| = -|\cos 5x|$

78. $\frac{\sin 6x}{1 + \cos 6x} = \tan 3x$

79. $r = \frac{1}{32} v_0^2 \sin 2\theta$

range = 100 feet

$v_0 = 80$ feet per second

$r = \frac{1}{32}(80)^2 \sin 2\theta = 100$

$\sin 2\theta = 0.5$

$2\theta = 30°, 150°$

$\theta = 15° \text{ or } 75° = \frac{\pi}{12} \text{ or } \frac{5\pi}{12}$

80. Volume V of the trough will be the area A of the isosceles triangle times the length l of the trough.

$V = A \cdot l$

(a) $A = \frac{1}{2}bh$

$\cos\frac{\theta}{2} = \frac{h}{0.5} \Rightarrow h = 0.5 \cos\frac{\theta}{2}$

$\sin\frac{\theta}{2} = \frac{\frac{b}{2}}{0.5} \Rightarrow \frac{b}{2} = 0.5 \sin\frac{\theta}{2}$

$A = 0.5 \sin\frac{\theta}{2} 0.5 \cos\frac{\theta}{2}$

$= (0.5)^2 \sin\frac{\theta}{2} \cos\frac{\theta}{2}$

$= 0.25 \sin\frac{\theta}{2} \cos\frac{\theta}{2}$ square meters

$V = (0.25)(4) \sin\frac{\theta}{2} \cos\frac{\theta}{2}$ cubic meters

$= \sin\frac{\theta}{2} \cos\frac{\theta}{2}$ cubic meters

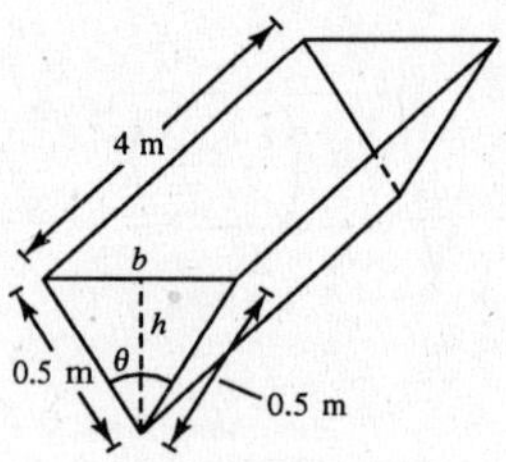

Not drawn to scale

(b) $V = \sin\frac{\theta}{2} \cos\frac{\theta}{2}$

$= \frac{1}{2}\left(2 \sin\frac{\theta}{2} \cos\frac{\theta}{2}\right)$

$= \frac{1}{2} \sin\theta$ cubic meters

Volume is maximum when $\theta = \frac{\pi}{2}$.

81. $\cos\frac{\pi}{6} \sin\frac{\pi}{6} = \frac{1}{2}\left[\sin\frac{\pi}{3} - \sin 0\right]$

82. $6 \sin 15° \sin 45° = 6\left(\frac{1}{2}\right)[\cos(15° - 45°) - \cos(15° + 45°)]$

$= 3[\cos(-30°) - \cos 60°]$

$= 3(\cos 30° - \cos 60°)$

83. $\cos 5\theta \cos 3\theta = \frac{1}{2}[\cos 2\theta + \cos 8\theta]$

84. $4 \sin 3\alpha \cos 2\alpha = 4\left(\frac{1}{2}\right)[\sin(3\alpha + 2\alpha) + \sin(3\alpha - 2\alpha)]$

$= 2(\sin 5\alpha + \sin\alpha)$

85. $\sin 60° + \sin 90° = 2 \sin 75° \cos 15°$

86. $\cos 3\theta + \cos 2\theta = 2\cos\left(\dfrac{3\theta + 2\theta}{2}\right)\cos\left(\dfrac{3\theta - 2\theta}{2}\right)$

$$= 2\cos\frac{5\theta}{2}\cos\frac{\theta}{2}$$

87. $\cos\left(x + \dfrac{\pi}{6}\right) - \cos\left(x - \dfrac{\pi}{6}\right) = -2\sin(x)\sin\left(\dfrac{\pi}{6}\right)$

88. $\sin\left(x + \dfrac{\pi}{4}\right) - \sin\left(x - \dfrac{\pi}{4}\right) = 2\cos\left[\dfrac{\left(x + \frac{\pi}{4}\right) + \left(x - \frac{\pi}{4}\right)}{2}\right]\sin\left[\dfrac{\left(x + \frac{\pi}{4}\right) - \left(x - \frac{\pi}{4}\right)}{2}\right]$

$$= 2\cos x \sin\frac{\pi}{4}$$

89. $y = \sqrt{x + 3} + 4\cos x$

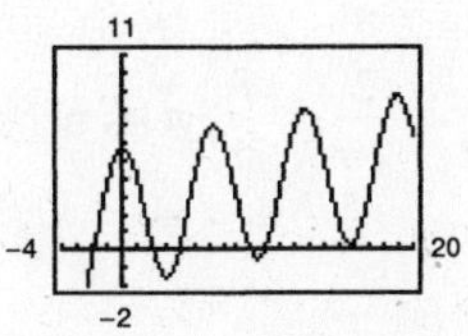

Zeros: $x \approx -1.8431, 2.1758, 3.9903, 8.8935, 9.8820$

90. $y = 2 - \dfrac{1}{2}x^2 + 3\sin\dfrac{\pi x}{2}$

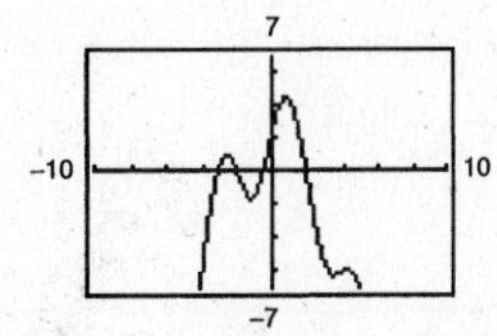

Approximate roots: $-3.1395, -2.0000, -0.4378, 2.0000$

91. (a) $y = 1.5\sin 8t - 0.5\cos 8t = \dfrac{1}{2}(3\sin 8t - 1\cos 8t)$

Using the identity

$a\sin B\theta + b\cos B\theta = \sqrt{a^2 + b^2}\sin(B\theta + C),$

$C = \arctan\dfrac{b}{a}, a > 0$

(Exercise 83, Section 10.4), we have

$$y = \frac{1}{2}\sqrt{(3)^2 + (-1)^2}\sin\left(8t + \arctan\left(-\frac{1}{3}\right)\right)$$

$$= \frac{\sqrt{10}}{2}\sin\left(8t - \arctan\left(\frac{1}{3}\right)\right).$$

(b) Amplitude $= \dfrac{\sqrt{10}}{2}$ feet

(c) Frequency $= \dfrac{1}{\frac{2\pi}{8}} = \dfrac{4}{\pi}$ cycles per second

Problem Solving for Chapter 10

1. $f(x) = \tan\dfrac{\pi x}{4}$

Since $\tan \pi/4 = 1$, $x = 1$ is the smallest nonnegative fixed point.

2. Graph $y = \cos x$ and $y = x$ on the same set of axes. Their point of intersection gives the value of c such that $f(c) = c \Rightarrow \cos c = c$.

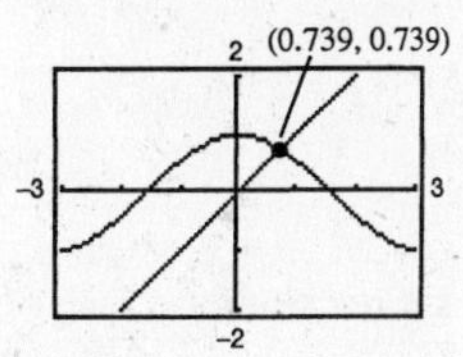

$c \approx 0.739$

3. $A = 2x \cos x,\ 0 < x < \dfrac{\pi}{2}$

(a)

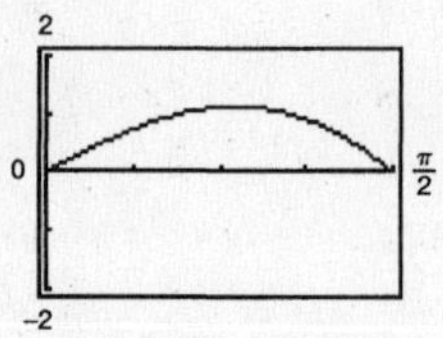

The maximum area of $A \approx 1.12$ occurs when $x \approx 0.86$.

(b) $A \geq 1$ for $0.6 < x < 1.1$

4. $y = \dfrac{1}{3}\sin 2t + \dfrac{1}{4}\cos 2t$

(a) $a = \dfrac{1}{3},\ b = \dfrac{1}{4},\ B = 2$

$$C = \arctan\frac{b}{a} = \arctan\frac{3}{4} \approx 0.6435$$

$$y \approx \sqrt{\left(\frac{1}{3}\right)^2 + \left(\frac{1}{4}\right)^2}\sin(2t + 0.6435)$$

$$= \frac{5}{12}\sin(2t + 0.6435)$$

(b) Amplitude: $\dfrac{5}{12}$ foot

(c) Frequency: $\dfrac{1}{\text{period}} = \dfrac{B}{2\pi} = \dfrac{2}{2\pi} = \dfrac{1}{\pi}$ cycles per second

5. $f(x) = 3\sin(0.6x - 2)$

(a) Zero:
$$\sin(0.6x - 2) = 0$$
$$0.6x - 2 = 0$$
$$0.6x = 2$$
$$x = \frac{2}{0.6} = \frac{10}{3}$$

(b) $g(x) = -0.45x^2 + 5.52x - 13.70$

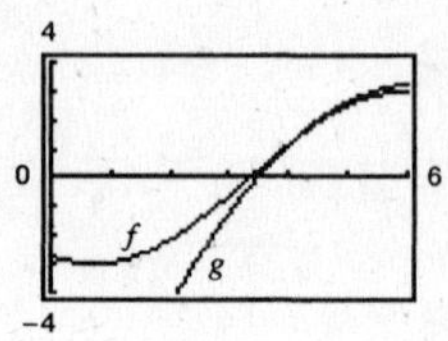

For $3.5 \leq x \leq 6$ the approximation appears to be good.

(c) $-0.45x^2 + 5.52x - 13.70 = 0$

$$x = \frac{-5.52 \pm \sqrt{(5.52)^2 - 4(-0.45)(-13.70)}}{2(-0.45)}$$

$x \approx 3.46, 8.81$

The zero of g on $[0, 6]$ is 3.46. The zero is close to the zero $\frac{10}{3} \approx 3.33$ of f.

6.
$$y_1 = A\cos 2\pi\left(\frac{t}{T} - \frac{x}{\lambda}\right)$$
$$y_2 = A\cos 2\pi\left(\frac{t}{T} + \frac{x}{\lambda}\right)$$
$$y_1 + y_2 = A\cos 2\pi\left(\frac{t}{T} - \frac{x}{\lambda}\right) + A\cos 2\pi\left(\frac{t}{T} + \frac{x}{\lambda}\right)$$
$$y_1 + y_2 = A\left[\cos 2\pi\frac{t}{T}\cos 2\pi\frac{x}{\lambda} + \sin 2\pi\frac{t}{T}\sin 2\pi\frac{x}{\lambda}\right] + A\left[\cos 2\pi\frac{t}{T}\cos 2\pi\frac{x}{\lambda} - \sin 2\pi\frac{t}{T}\sin 2\pi\frac{x}{\lambda}\right]$$
$$= 2A\cos 2\pi\frac{t}{T}\cos 2\pi\frac{x}{\lambda}$$

7.
$$\cos(n\pi + \theta) = \cos n\pi\cos\theta - \sin n\pi\sin\theta$$
$$= (-1)^n(\cos\theta) - (0)(\sin\theta)$$
$$= (-1)^n(\cos\theta), \text{ where } n \text{ is an integer.}$$

8.
$$\sin(n\pi + \theta) = \sin n\pi\cos\theta + \sin\theta\cos n\pi$$
$$= (0)(\cos\theta) + (\sin\theta)(-1)^n$$
$$= (-1)^n(\sin\theta), \text{ where } n \text{ is an integer.}$$

9. 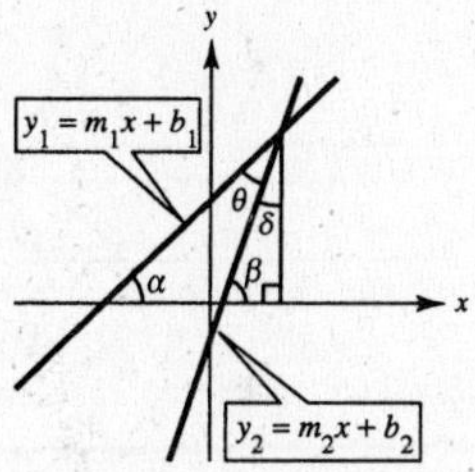

Let $m_2 > m_1 > 0$.

$m_1 = \tan \alpha$ and $m_2 = \tan \beta$

$\beta + \delta = 90° \implies \delta = 90° - \beta$

$\alpha + \theta + \delta = 90° \implies \alpha + \theta + (90° - \beta) = 90° \implies \theta = \beta - \alpha$

Therefore, $\theta = \arctan m_2 - \arctan m_1$, $m_2 > m_1 > 0$.

10. (a) For $y = x$ and $y = \sqrt{3}x$ we have $m_1 = 1$ and $m_2 = \sqrt{3}$.

$$\begin{aligned}\theta &= \arctan\sqrt{3} - \arctan 1\\ &= 60° - 45°\\ &= 15°\end{aligned}$$

(b) For $y = x$ and $y = \frac{1}{\sqrt{3}}x$ we have $m_2 = 1$ and $m_1 = \frac{1}{\sqrt{3}}$.

$$\begin{aligned}\theta &= \arctan 1 - \arctan\frac{1}{\sqrt{3}}\\ &= 45° - 30°\\ &= 15°\end{aligned}$$

11. $$\begin{aligned}\sin^2\left(\theta + \frac{\pi}{4}\right) + \sin^2\left(\theta - \frac{\pi}{4}\right) &= \left[\sin\theta\cos\frac{\pi}{4} + \cos\theta\sin\frac{\pi}{4}\right]^2 + \left[\sin\theta\cos\frac{\pi}{4} - \cos\theta\sin\frac{\pi}{4}\right]^2\\ &= \left[\frac{\sin\theta}{\sqrt{2}} + \frac{\cos\theta}{\sqrt{2}}\right]^2 + \left[\frac{\sin\theta}{\sqrt{2}} - \frac{\cos\theta}{\sqrt{2}}\right]^2\\ &= \frac{\sin^2\theta}{2} + \sin\theta\cos\theta + \frac{\cos^2\theta}{2} + \frac{\sin^2\theta}{2} - \sin\theta\cos\theta + \frac{\cos^2\theta}{2}\\ &= \sin^2\theta + \cos^2\theta\\ &= 1\end{aligned}$$

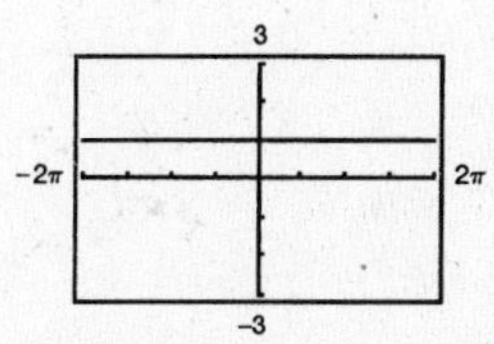

12. From the figure, it appears that $u + v = w$. Assume that u, v, and w are all in Quadrant I. From the figure:

$\tan u = \frac{s}{3s} = \frac{1}{3}$

$\tan v = \frac{s}{2s} = \frac{1}{2}$

$\tan w = \frac{s}{s} = 1$

$$\tan(u + v) = \frac{\tan u + \tan v}{1 - \tan u \tan v} = \frac{1/3 + 1/2}{1 - (1/3)(1/2)} = \frac{5/6}{1 - (1/6)} = 1 = \tan w.$$

Thus, $\tan(u + v) = \tan w$. Because u, v, and w are all in Quadrant I, we have

$$\begin{aligned}\arctan[\tan(u + v)] &= \arctan[\tan w]\\ u + v &= w.\end{aligned}$$

13. To prove the identity for $\sin(u + v)$ we first need to prove the identity for $\cos(u - v)$. Assume $0 < v < u < 2\pi$ and locate u, v, and $u - v$ on the unit circle.

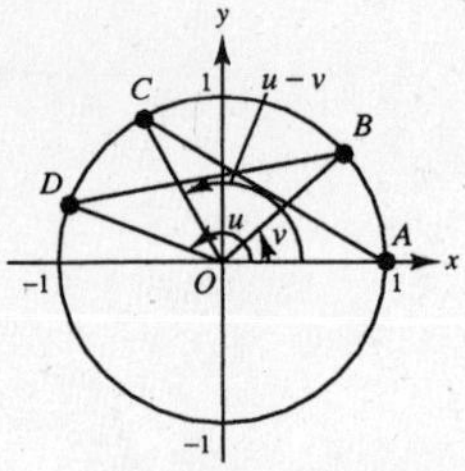

The coordinates of the points on the circle are:

$A = (1, 0)$, $B = (\cos v, \sin v)$, $C = (\cos(u - v), \sin(u - v))$, and $D = (\cos u, \sin u)$.

Since ∠DOB = ∠COA, chords AC and BD are equal. By the distance formula we have:

$$\sqrt{[\cos(u - v) - 1]^2 + [\sin(u - v) - 0]^2} = \sqrt{(\cos u - \cos v)^2 + (\sin u - \sin v)^2}$$

$$\cos^2(u - v) - 2\cos(u - v) + 1 + \sin^2(u - v) = \cos^2 u - 2\cos u\cos v + \cos^2 v + \sin^2 u - 2\sin u\sin v + \sin^2 v$$

$$[\cos^2(u - v) + \sin^2(u - v)] + 1 - 2\cos(u - v) = (\cos^2 u + \sin^2 u) + (\cos^2 v + \sin^2 v) - 2\cos u\cos v - 2\sin u\sin v$$

$$2 - 2\cos(u - v) = 2 - 2\cos u\cos v - 2\sin u\sin v$$

$$-2\cos(u - v) = -2(\cos u\cos v + \sin u\sin v)$$

$$\cos(u - v) = \cos u\cos v + \sin u\sin v$$

Now, to prove the identity for $\sin(u + v)$, use cofunction identities.

$$\sin(u + v) = \cos\left[\frac{\pi}{2} - (u + v)\right] = \cos\left[\left(\frac{\pi}{2} - u\right) - v\right] = \cos\left(\frac{\pi}{2} - u\right)\cos v + \sin\left(\frac{\pi}{2} - u\right)\sin v = \sin u\cos v + \cos u\sin v$$

14. First, prove $\cos(u - v) = \cos u\cos v + \sin u\sin v$ using the figure containing points

$A(1, 0)$

$B(\cos(u - v), \sin(u - v))$

$C(\cos v, \sin v)$

$D(\cos u, \sin u)$

on the unit circle.

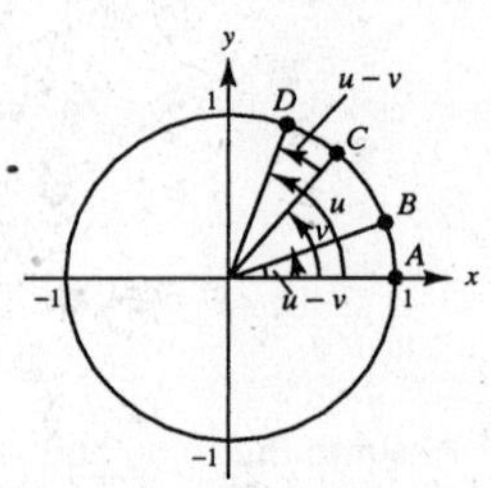

Since chords AB and CD are each subtended by angle $u - v$, their lengths are equal. Equating $[d(A, B)]^2 = [d(C, D)]^2$ we have $(\cos(u - v) - 1)^2 + \sin^2(u - v) = (\cos u - \cos v)^2 + (\sin u - \sin v)^2$. Simplifying and solving for $\cos(u - v)$, we have $\cos(u - v) = \cos u\cos v + \sin u\sin v$.

Using $\sin\theta = \cos\left(\frac{\pi}{2} - \theta\right)$ we have

$$\sin(u - v) = \cos\left[\frac{\pi}{2} - (u - v)\right] = \cos\left[\left(\frac{\pi}{2} - u\right) - (-v)\right]$$

$$= \cos\left(\frac{\pi}{2} - u\right)\cos(-v) + \sin\left(\frac{\pi}{2} - u\right)\sin(-v)$$

$$= \sin u\cos v - \cos u\sin v.$$

15. (a) $A = \frac{1}{2}bh$

$$\cos\frac{\theta}{2} = \frac{h}{10} \Rightarrow h = 10\cos\frac{\theta}{2}$$

$$\sin\frac{\theta}{2} = \frac{(1/2)b}{10} \Rightarrow \frac{1}{2}b = 10\sin\frac{\theta}{2}$$

$$A = 10\sin\frac{\theta}{2}10\cos\frac{\theta}{2} \Rightarrow A = 100\sin\frac{\theta}{2}\cos\frac{\theta}{2}$$

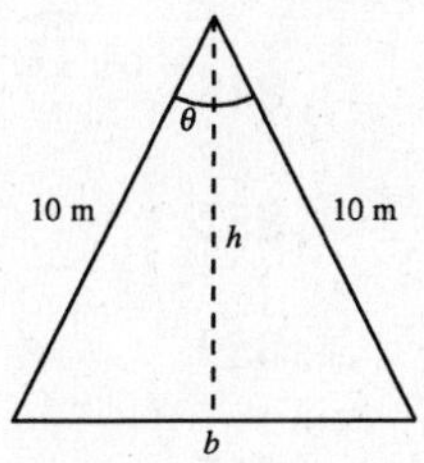

(b) $A = 100\sin\frac{\theta}{2}\cos\frac{\theta}{2}$

$$A = 50\left(2\sin\frac{\theta}{2}\cos\frac{\theta}{2}\right)$$

$$A = 50\sin\theta$$

When $\theta = \pi/2$, $\sin\theta = 1 \Rightarrow$ the area is a maximum.

$$A = 50\sin\frac{\pi}{2} = 50(1) = 50 \text{ square feet}$$

16. $\frac{x}{2} = 2r\sin^2\frac{\theta}{2} = 2r\left(\frac{1-\cos\theta}{2}\right)$

$$= r(1 - \cos\theta)$$

$$x = 2r(1 - \cos\theta)$$

17. $\lim_{x\to 0^-} f(x) = \lim_{x\to 0^-} (a^2 - 2) = a^2 - 2$

$$\lim_{x\to 0^+} f(x) = \lim_{x\to 0^+} \frac{ax}{\tan x} = a \left(\text{because } \lim_{x\to 0}\frac{\tan x}{x} = 1\right)$$

Thus,

$$a^2 - 2 = a$$

$$a^2 - a - 2 = 0$$

$$(a - 2)(a + 1) = 0$$

$$a = -1, 2$$

18.

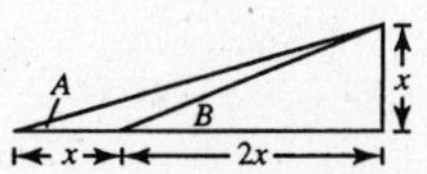

$\tan A = \frac{x}{3x} = \frac{1}{3}$ and $\tan B = \frac{x}{2x} = \frac{1}{2}$

$$\tan(A + B) = \frac{\tan A + \tan B}{1 - \tan A \tan B}$$

$$= \frac{\frac{1}{3} + \frac{1}{2}}{1 - \left(\frac{1}{3}\right)\left(\frac{1}{2}\right)} = \frac{\frac{5}{6}}{\frac{5}{6}}$$

$$= 1$$

Thus, $A + B = \arctan 1 = \frac{\pi}{4}$

CHAPTER 11
Trigonometric Functions and Calculus

Section 11.1 Limits of Trigonometric Functions 825

Section 11.2 Trigonometric Functions: Differentiation 831

Section 11.3 Trigonometric Functions: Integration 845

Section 11.4 Inverse Trigonometric Functions: Differentiation 852

Section 11.5 Inverse Trigonometric Functions: Integration 859

Section 11.6 Hyperbolic Functions 865

Review Exercises 873

Problem Solving 881

CHAPTER 11
Trigonometric Functions and Calculus

Section 11.1 Limits of Trigonometric Functions

1.

x	-0.1	-0.01	-0.001	0.001	0.01	0.1
$f(x)$	0.9983	0.99998	1.0000	1.0000	0.99998	0.9983

$\lim_{x\to 0} \frac{\sin x}{x} \approx 1.0000$ (Actual limit is 1.) (Make sure you use radian mode.)

2.

x	-0.1	-0.01	-0.001	0.001	0.01	0.1
$f(x)$	0.0500	0.0050	0.0005	-0.0005	-0.0050	-0.0500

$\lim_{x\to 0} \frac{\cos x - 1}{x} \approx 0.0000$ (Actual limit is 0.)(Make sure you use radian mode.)

3. $\lim_{x\to\pi/2} \tan x$ does not exist since the function increases and decreases without bound as x approaches $\pi/2$.

4. $\lim_{x\to 0} \sec x = 1$

5. $\lim_{x\to 0} \cos(1/x)$ does not exist since the function oscillates between -1 and 1 as x approaches 0.

6. $\lim_{x\to 1} \sin(\pi x) = 0$

7. $\lim_{x\to\pi/2} \sin x = \sin\frac{\pi}{2} = 1$

8. $\lim_{x\to\pi} \tan x = \tan \pi = 0$

9. $\lim_{x\to 2} \cos\frac{\pi x}{3} = \cos\frac{\pi 2}{3} = -\frac{1}{2}$

10. $\lim_{x\to 1} \sin\frac{\pi x}{2} = \sin\frac{\pi}{2} = 1$

11. $\lim_{x\to 0} \sec 2x = \sec 0 = 1$

12. $\lim_{x\to\pi} \cos 3x = \cos 3\pi = -1$

13. $\lim_{x\to 5\pi/6} \sin x = \sin\frac{5\pi}{6} = \frac{1}{2}$

14. $\lim_{x\to 5\pi/3} \cos x = \cos\frac{5\pi}{3} = \frac{1}{2}$

15. $\lim_{x\to 3} \tan\left(\frac{\pi x}{4}\right) = \tan\frac{3\pi}{4} = -1$

16. $\lim_{x\to 7} \sec\left(\frac{\pi x}{6}\right) = \sec\frac{7\pi}{6} = \frac{-2\sqrt{3}}{3}$

17. $f(x) = 3x - \cos x$ is continuous for all real x.

18. Continuous for all real x

19. $f(x) = \begin{cases} \tan\frac{\pi x}{4}, & |x| < 1 \\ x, & |x| \ge 1 \end{cases} = \begin{cases} \tan\frac{\pi x}{4}, & -1 < x < 1 \\ x, & x \le -1 \text{ or } x \ge 1 \end{cases}$

has **possible** discontinuities at $x = -1$, $x = 1$.

1. $f(-1) = -1$ $\qquad f(1) = 1$

2. $\lim_{x\to -1} f(x) = -1$ $\qquad \lim_{x\to 1} f(x) = 1$

3. $f(-1) = \lim_{x\to -1} f(x)$ $\qquad f(1) = \lim_{x\to 1} f(x)$

f is continuous at $x = \pm 1$, therefore, f is continuous for all real x.

20. Continuous for all real x

21. $f(x) = \csc 2x$ has nonremovable discontinuities at integer multiples of $\pi/2$.

22. Nonremovable discontinuities at each odd integer k

23. $f(x) = \dfrac{\sin x}{x}$

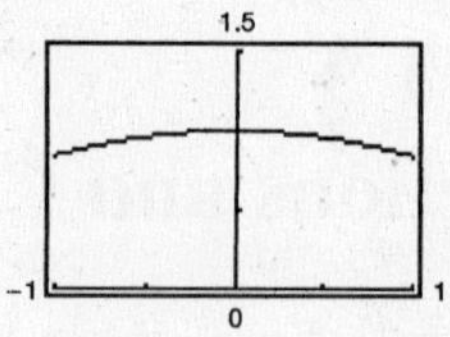

Although the graph appears continuous on $[-1, 1]$, there is a discontinuity at $x = 0$. Examining a function graphically and analytically ensures that you will find all points where the function is not defined.

24. The graph appears to be continuous on $[-1, 1]$. But, the graph has a removable discontinuity at $x = 0$. This hole at $(0, 1)$ does not appear on the graph.

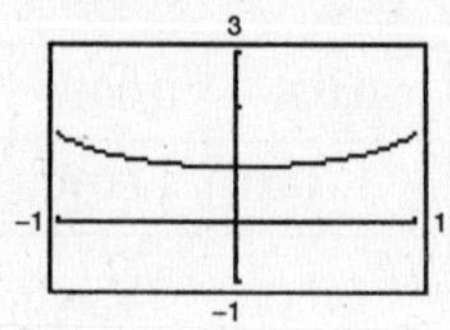

25. $f(x) = \tan 2x = \dfrac{\sin 2x}{\cos 2x}$ has vertical asymptotes at

$$x = \frac{(2n + 1)\pi}{4} = \frac{\pi}{4} + \frac{n\pi}{2},\ n \text{ any integer.}$$

26. $f(x) = \sec \pi x = \dfrac{1}{\cos \pi x}$ has vertical asymptotes at

$$x = \frac{2n + 1}{2},\ n \text{ any integer.}$$

27. $s(t) = \dfrac{t}{\sin t}$ has vertical asymptotes at $t = n\pi$, n a nonzero integer. There is no vertical asymptote at $t = 0$ since

$$\lim_{t\to 0} \frac{t}{\sin t} = 1.$$

28. $g(\theta) = \dfrac{\tan \theta}{\theta} = \dfrac{\sin \theta}{\theta \cos \theta}$ has vertical asymptotes at

$$\theta = \frac{(2n + 1)\pi}{2} = \frac{\pi}{2} + n\pi,\ n \text{ any integer.}$$

There is no vertical asymptote at $\theta = 0$ since

$$\lim_{\theta\to 0} \frac{\tan \theta}{\theta} = 1.$$

29. $f(t) = \dfrac{\sin 3t}{t}$

t	-0.1	-0.01	-0.001	0	0.001	0.01	0.1
$f(t)$	2.96	2.9996	3	?	3	2.9996	2.96

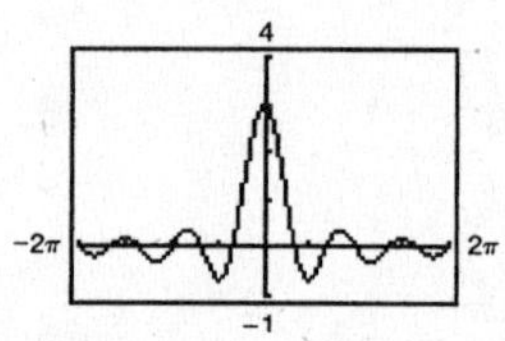

Analytically, $\displaystyle\lim_{t\to 0} \frac{\sin 3t}{t} = \lim_{t\to 0}\left[3\left(\frac{\sin 3t}{3t}\right)\right] = 3(1) = 3.$

30. $\displaystyle\lim_{h\to 0} (1 + \cos 2h) = 1 + 1 = 2$

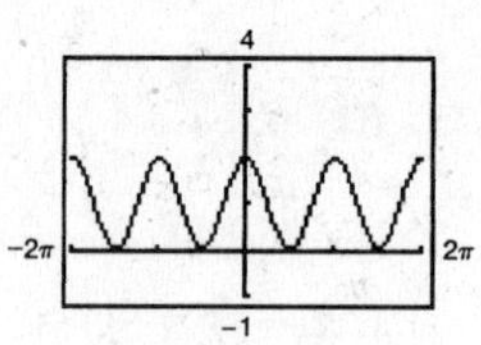

31. $f(x) = \dfrac{\sin x^2}{x}$

x	-0.1	-0.01	-0.001	0	0.001	0.01	0.1
$f(x)$	-0.099998	-0.01	-0.001	?	0.001	0.01	0.099998

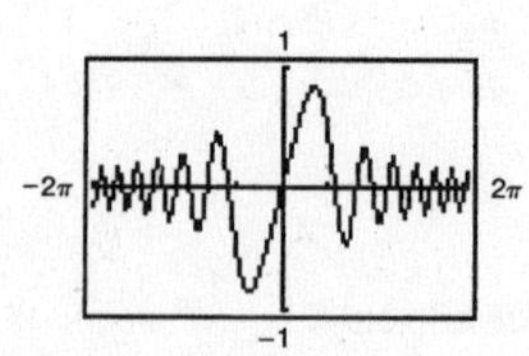

Analytically, $\displaystyle\lim_{x\to 0} \frac{\sin x^2}{x} = \lim_{x\to 0} x\left(\frac{\sin x^2}{x^2}\right) = 0(1) = 0.$

32. $f(x) = \dfrac{\sin x}{\sqrt[3]{x}}$

x	-0.1	-0.01	-0.001	0	0.001	0.01	0.1
$f(x)$	0.215	0.0464	0.01	?	0.01	0.0464	0.215

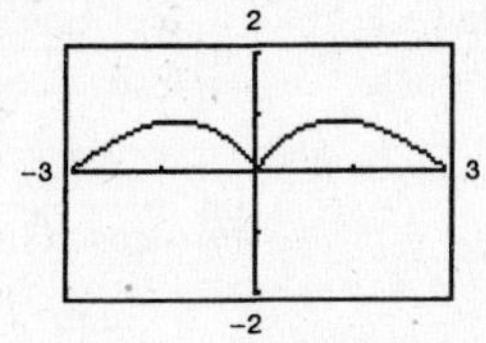

The limit appears to equal 0.

Analytically, $\displaystyle\lim_{x\to 0}\frac{\sin x}{\sqrt[3]{x}} = \lim_{x\to 0}\sqrt[3]{x^2}\left(\frac{\sin x}{x}\right) = (0)(1) = 0.$

33. $f(x) = x\cos x$

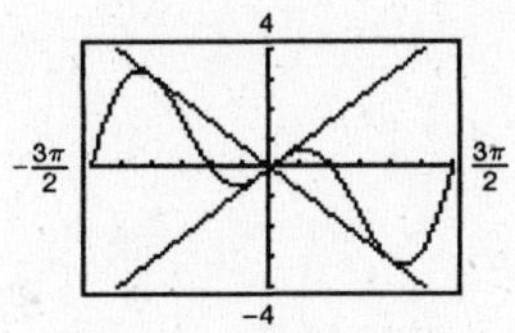

$\displaystyle\lim_{x\to 0} x\cos x = 0$

34. $f(x) = |x\sin x|$

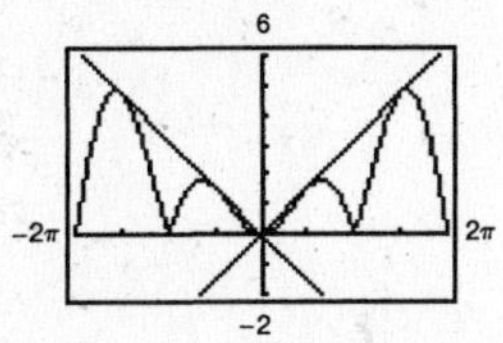

$\displaystyle\lim_{x\to 0} |x\sin x| = 0$

35. $f(x) = |x|\sin x$

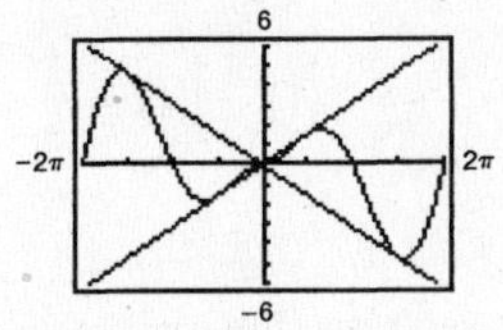

$\displaystyle\lim_{x\to 0} |x|\sin x = 0$

36. $f(x) = |x|\cos x$

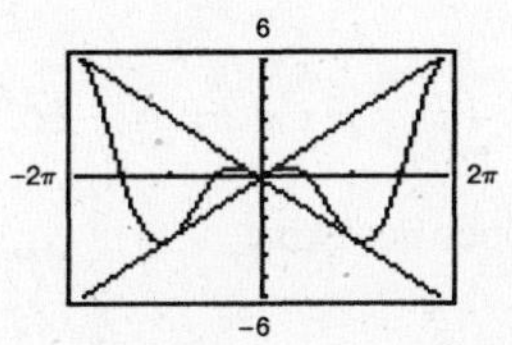

$\displaystyle\lim_{x\to 0} |x|\cos x = 0$

37. $f(x) = x\sin\dfrac{1}{x}$

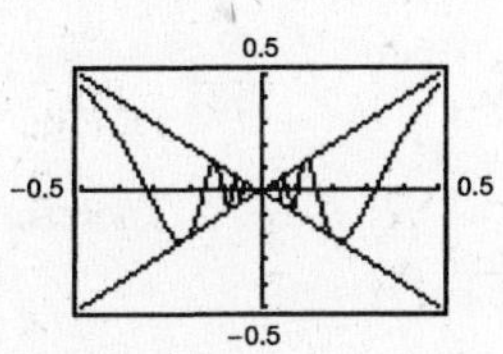

$\displaystyle\lim_{x\to 0}\left(x\sin\frac{1}{x}\right) = 0$

38. $h(x) = x\cos\dfrac{1}{x}$

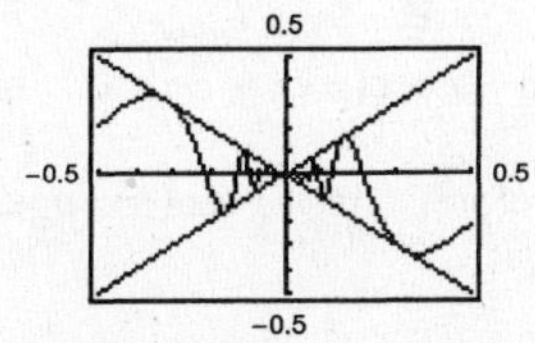

$\displaystyle\lim_{x\to 0}\left(x\cos\frac{1}{x}\right) = 0$

39. $\displaystyle\lim_{x\to 0}\frac{\sin x}{5x} = \lim_{x\to 0}\left[\left(\frac{\sin x}{x}\right)\left(\frac{1}{5}\right)\right] = (1)\left(\frac{1}{5}\right) = \frac{1}{5}$

40. $\displaystyle\lim_{x\to 0}\frac{3(1-\cos x)}{x} = \lim_{x\to 0}\left[3\left(\frac{1-\cos x}{x}\right)\right]$

$= (3)(0) = 0$

41. $\displaystyle\lim_{x\to 0}\frac{\sin x(1-\cos x)}{2x^2} = \lim_{x\to 0}\left[\frac{1}{2}\cdot\frac{\sin x}{x}\cdot\frac{1-\cos x}{x}\right]$

$= \dfrac{1}{2}(1)(0) = 0$

42. $\displaystyle\lim_{\theta\to 0}\frac{\cos\theta\tan\theta}{\theta} = \lim_{\theta\to 0}\frac{\sin\theta}{\theta} = 1$

43. $\displaystyle\lim_{\theta\to 0}\frac{\sin^2 x}{x} = \lim_{\theta\to 0}\left[\frac{\sin x}{x}\sin x\right] = (1)\sin 0 = 0$

44. $\displaystyle\lim_{x\to 0}\frac{\tan^2 x}{x} = \lim_{x\to 0}\frac{\sin^2 x}{x\cos^2 x} = \lim_{x\to 0}\left[\frac{\sin x}{x}\cdot\frac{\sin x}{\cos^2 x}\right]$

$= (1)(0) = 0$

45. $\displaystyle\lim_{h\to 0}\frac{(1-\cos h)^2}{h} = \lim_{h\to 0}\left[\frac{1-\cos h}{h}(1-\cos h)\right]$

$= (0)(0) = 0$

46. $\displaystyle\lim_{\phi\to\pi}\phi\sec\phi = -\pi$

47. $\displaystyle\lim_{x\to(\pi/2)}\frac{\cos x}{\cot x} = \lim_{x\to(\pi/2)}\sin x = 1$

48. $\lim\limits_{x\to(\pi/4)} \dfrac{1-\tan x}{\sin x - \cos x} = \lim\limits_{x\to(\pi/4)} \dfrac{\cos x - \sin x}{\sin x \cos x - \cos^2 x}$

$= \lim\limits_{x\to(\pi/4)} \dfrac{-(\sin x - \cos x)}{\cos x(\sin x - \cos x)}$

$= \lim\limits_{x\to(\pi/4)} \dfrac{-1}{\cos x}$

$= \lim\limits_{x\to(\pi/4)} (-\sec x) = -\sqrt{2}$

49. $\lim\limits_{t\to 0} \dfrac{\sin 3t}{2t} = \lim\limits_{t\to 0} \left(\dfrac{\sin 3t}{3t}\right)\left(\dfrac{3}{2}\right) = (1)\left(\dfrac{3}{2}\right) = \dfrac{3}{2}$

50. $\lim\limits_{x\to 0} \dfrac{\sin 2x}{\sin 3x} = \underset{x\to 0}{\text{im}} \left[2\left(\dfrac{\sin 2x}{2x}\right)\left(\dfrac{1}{3}\right)\left(\dfrac{3x}{\sin 3x}\right)\right]$

$= 2(1)\left(\dfrac{1}{3}\right)(1) = \dfrac{2}{3}$

51. $\lim\limits_{x\to\pi} \cot x$

Does not exist.

52. $\lim\limits_{x\to\pi/2} \sec x = \lim\limits_{x\to\pi/2} \dfrac{1}{\cos x}$

Does not exist.

53. $\lim\limits_{x\to 0^+} \dfrac{2}{\sin x} = \infty$

54. $\lim\limits_{x\to\pi/2^+} \dfrac{-2}{\cos x} = \infty$

55. $\lim\limits_{x\to\pi} \dfrac{\sqrt{x}}{\csc x} = \lim\limits_{x\to\pi} (\sqrt{x}\sin x) = 0$

56. $\lim\limits_{x\to 0} \dfrac{(x+2)}{\cot x} = \lim\limits_{x\to 0} [(x+2)\tan x] = 0$

57. $\lim\limits_{x\to(1/2)^-} x\sec(\pi x) = \infty$ and $\lim\limits_{x\to(1/2)^+} x\sec(\pi x) = -\infty.$

Therefore, $\lim\limits_{x\to(1/2)} x\sec(\pi x)$ does not exist.

58. $\lim\limits_{x\to(1/2)^-} x^2\tan\pi x = \infty$ and $\lim\limits_{x\to(1/2)^+} x^2 \tan \pi x = -\infty.$

Therefore, $\lim\limits_{x\to(1/2)} x^2 \tan \pi x$ does not exist.

59. $\lim\limits_{x\to\pi/6} \sin x = \dfrac{1}{2}$

means that as x gets closer to $\dfrac{\pi}{6}$, $\sin x$ gets closer to $\dfrac{1}{2}$.

60. No. $\lim\limits_{x\to 3} f(x)$ could equal any number, or not exist at all.

61. No. $f(\pi)$ could equal any number, or not exist at all.

62. Let $f(x) = \dfrac{1}{\sin\left(\frac{1}{2}\pi x\right)}$.

63. $f(x) = x,\ g(x) = \sin x,\ h(x) = \dfrac{\sin x}{x}$

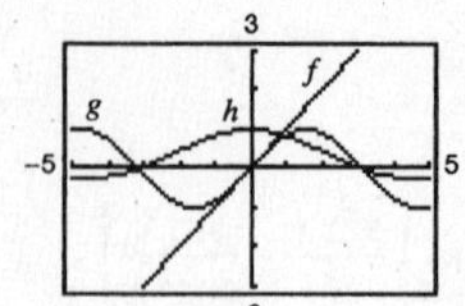

When you are "close to" 0 the magnitude of f is approximately equal to the magnitude of g. Thus, $|g|/|f| \approx 1$ when x is "close to" 0.

64. $f(x) = x,\ g(x) = \sin^2 x,\ h(x) = \dfrac{\sin^2 x}{x}$

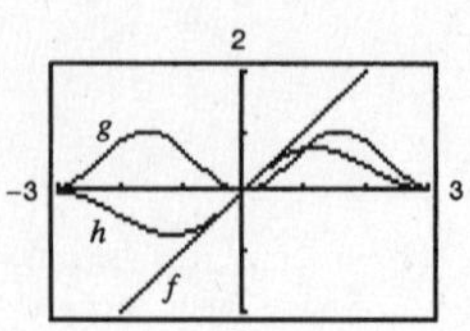

When you are "close to" 0 the magnitude of g is "smaller" than the magnitude of f and the magnitude of g is approaching zero "faster" than the magnitude of f. Thus, $|g|/|f| \approx 0$ when x is "close to" 0.

65. (a)

x	1	0.5	0.2	0.1	0.01	0.001	0.0001
$f(x)$	0.1585	0.0411	0.0067	0.0017	≈ 0	≈ 0	≈ 0

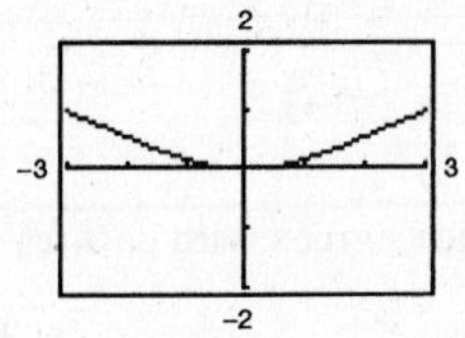

$$\lim_{x\to 0^+} \frac{x - \sin x}{x} = 0$$

(b)

x	1	0.5	0.2	0.1	0.01	0.001	0.0001
$f(x)$	0.1585	0.0823	0.0333	0.0167	0.0017	≈ 0	≈ 0

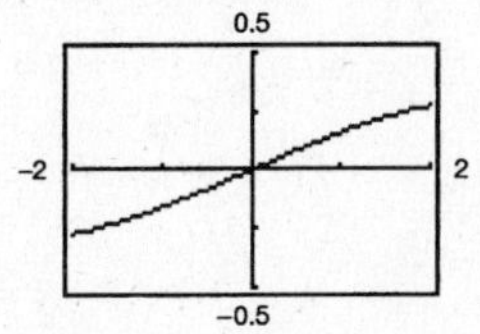

$$\lim_{x\to 0^+} \frac{x - \sin x}{x^2} = 0$$

(c)

x	1	0.5	0.2	0.1	0.01	0.001	0.0001
$f(x)$	0.1585	0.1646	0.1663	0.1666	0.1667	0.1667	0.1667

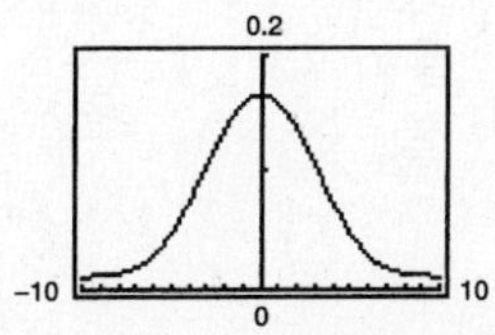

$$\lim_{x\to 0^+} \frac{x - \sin x}{x^3} = 0.1167 \ (1/6)$$

(d)

x	1	0.5	0.2	0.1	0.01	0.001	0.0001
$f(x)$	0.1585	0.3292	0.8317	1.6658	16.67	166.7	1667.0

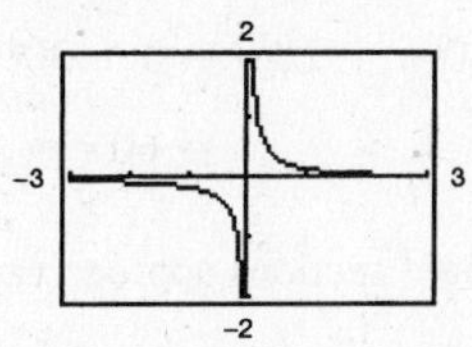

$$\lim_{x\to 0^+} \frac{x - \sin x}{x^4} = \infty$$

For $n \geq 3$, $\displaystyle\lim_{x\to 0^+} \frac{x - \sin x}{x^n} = \infty$.

66. $f(x) = \dfrac{\sec x - 1}{x^2}$

(a) The domain of f is all $x \neq 0, \pi/2 + n\pi$.

(b)

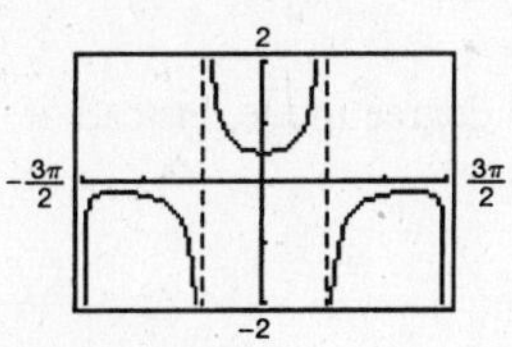

The domain is not obvious. The hole at $x = 0$ is not apparent.

(c) $\displaystyle\lim_{x\to 0} f(x) = \frac{1}{2}$

(d) $$\frac{\sec x - 1}{x^2} = \frac{\sec x - 1}{x^2} \cdot \frac{\sec x + 1}{\sec x + 1} = \frac{\sec^2 x - 1}{x^2(\sec x + 1)}$$

$$= \frac{\tan^2 x}{x^2(\sec x + 1)} = \frac{1}{\cos^2 x}\left(\frac{\sin^2 x}{x^2}\right)\frac{1}{\sec x + 1}$$

Hence, $$\lim_{x\to 0} \frac{\sec x - 1}{x^2} = \lim_{x\to 0} \frac{1}{\cos^2 x}\left(\frac{\sin^2 x}{x^2}\right)\frac{1}{\sec x + 1}$$

$$= 1(1)\left(\frac{1}{2}\right) = \frac{1}{2}.$$

67. (a) $\lim_{x\to 0} \frac{1-\cos x}{x^2} = \lim_{x\to 0} \frac{1-\cos x}{x^2} \cdot \frac{1+\cos x}{1+\cos x}$

$= \lim_{x\to 0} \frac{1-\cos^2 x}{x^2(1+\cos x)}$

$= \lim_{x\to 0} \frac{\sin^2 x}{x^2} \cdot \frac{1}{1+\cos x}$

$= (1)\left(\frac{1}{2}\right) = \frac{1}{2}$

(b) Thus, $\frac{1-\cos x}{x^2} \approx \frac{1}{2} \Rightarrow 1 - \cos x \approx \frac{1}{2}x^2$

$\Rightarrow \cos x \approx 1 - \frac{1}{2}x^2$ for $x \approx 0$.

(c) $\cos(0.1) \approx 1 - \frac{1}{2}(0.1)^2 = 0.995$

(d) $\cos(0.1) \approx 0.9950$, which agrees with part (c).

68. (a) $r = 50\pi \sec^2 \frac{\pi}{6} = \frac{200\pi}{3}$ ft/sec

(b) $r = 50\pi \sec^2 \frac{\pi}{3} = 200\pi$ ft/sec

(c) $\lim_{\theta\to(\pi/2)^-} [50\pi \sec^2 \theta] = \infty$

69. (a) $A = \frac{1}{2}bh - \frac{1}{2}r^2\theta = \frac{1}{2}(10)(10 \tan \theta) - \frac{1}{2}(10)^2\theta$

$= 50 \tan \theta - 50\,\theta$

Domain: $\left(0, \frac{\pi}{2}\right)$

(b)

ϕ	0.3	0.6	0.9	1.2	1.5
$f(\theta)$	0.47	4.21	18.0	68.6	630.1

(c)

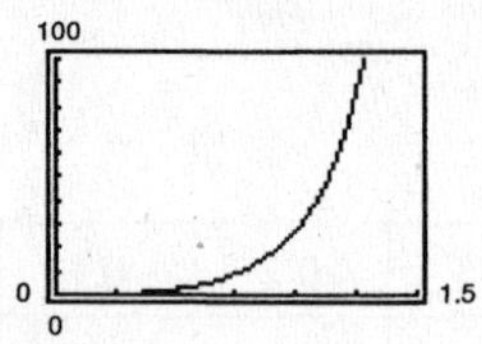

(d) $\lim_{\theta\to\pi/2^-} A = \infty$

70. (a) Because the circumference of the motor is half that of the saw arbor, the saw makes $1700/2 = 850$ revolutions per minute.

(b) The belt and saw rotate in opposite directions.

(c) $2(20 \cot \phi) + 2(10 \cot \phi)$: straight sections. The angle subtended in each circle is

$$2\pi - \left(2\left(\frac{\pi}{2} - \phi\right)\right) = \pi + 2\phi.$$

Thus, the length of the belt around the pulleys is

$$20(\pi + 2\phi) + 10(\pi + 2\phi) = 30(\pi + 2\phi).$$

Total length $= 60 \cot \phi + 30(\pi + 2\phi)$

Domain: $\left(0, \frac{\pi}{2}\right)$

(d)

ϕ	0.3	0.6	0.9	1.2	1.5
L	306.2	217.9	195.9	189.6	188.5

(e)

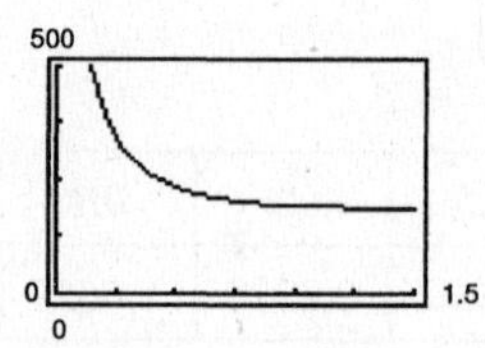

(f) $\lim_{\phi\to(\pi/2)^-} [60 \cot \phi + 30(\pi + 2\phi)] = 0 + 30(\pi + \pi)$

$= 60\pi \approx 188.4956$

As $\phi \to \left(\frac{\pi}{2}\right)^-$ the straight sections approach zero in length so the total length approaches the sum of the pulley circumferences or 60π.

(g) $\lim_{\phi\to 0^+} [60 \cot \phi + 30(\pi + 2\phi)] = \infty$

71. The calculator was set in degree mode instead of radian mode.

72. The calculator was set in degree mode, instead of radian mode.

73. $\lim_{x\to 0}\frac{1-\cos x}{x} = \lim_{x\to 0}\frac{1-\cos x}{x}\cdot\frac{1+\cos x}{1+\cos x}$

$= \lim_{x\to 0}\frac{1-\cos^2 x}{x(1+\cos x)} = \lim_{x\to 0}\frac{\sin^2 x}{x(1+\cos x)}$

$= \lim_{x\to 0}\frac{\sin x}{x}\cdot\frac{\sin x}{1+\cos x}$

$= \left[\lim_{x\to 0}\frac{\sin x}{x}\right]\left[\lim_{x\to 0}\frac{\sin x}{1+\cos x}\right]$

$= (1)(0) = 0$

74. Let y be a real number. If $y = 0$, then $x = 0$. If $y > 0$, then let $0 < x_0 < \pi/2$ such that $M = \tan x_0 > y$ (this is possible since the tangent function increases without bound on $[0, \pi/2)$). By the Intermediate Value Theorem, $f(x) = \tan x$ is continuous on $[0, x_0]$ and $0 < y < M$, which implies that there exists x between 0 and x_0 such that $\tan x = y$. The argument is similar if $y < 0$.

75. False. Tangent, cotangent, secant and cosecant are not continuous at all real numbers.

76. False. $\lim_{x\to\pi/2}\tan x \neq \tan\frac{\pi}{2}$ Since $\tan\frac{\pi}{2}$ is not defined.

Section 11.2 Trigonometric Functions: Differentiation

1. $f(x) = 2\sin x + 3\cos x$

$f'(x) = 2\cos x - 3\sin x$

2. $g(t) = \pi\cos t$

$g'(t) = -\pi\sin t$

3. $f(x) = 6\sqrt{x} + 5\cos x = 6x^{1/2} + 5\cos x$

$f'(x) = 3x^{-1/2} - 5\sin x = \frac{3}{\sqrt{x}} - 5\sin x$

4. $g(t) = \sqrt[4]{t} + 8\sec t = t^{1/4} + 8\sec t$

$g'(t) = \frac{1}{4}t^{-3/4} + 8\sec t\tan t = \frac{1}{4t^{3/4}} + 8\sec t\tan t$

5. $f(x) = -x + \tan x$

$f'(x) = -1 + \sec^2 x = \tan^2 x$

6. $y = x + \cot x$

$y' = 1 - \csc^2 x = -\cot^2 x$

7. $y = \frac{1}{x} - 3\sin x$

$y' = \frac{-1}{x^2} - 3\cos x$

8. $h(s) = \frac{1}{s} - 10\csc s$

$h'(s) = -\frac{1}{s^2} + 10\csc s\cot s$

9. $f(x) = x^3\cos x$

$f'(x) = x^3(-\sin x) + \cos x(3x^2)$

$= 3x^2\cos x - x^3\sin x$

10. $g(x) = \sqrt{x}\sin x$

$g'(x) = \sqrt{x}\cos x + \sin x\left(\frac{1}{2\sqrt{x}}\right)$

$= \sqrt{x}\cos x + \frac{1}{2\sqrt{x}}\sin x$

11. $f(t) = t^2\sin t$

$f'(t) = t^2\cos t + 2t\sin t$

$= t(t\cos t + 2\sin t)$

12. $f(\theta) = (\theta + 1)\cos\theta$

$f'(x) = (\theta + 1)(-\sin\theta) + (\cos\theta)(1)$

$= \cos\theta - (\theta + 1)\sin\theta$

13. $f(t) = \frac{\cos t}{t}$

$f'(t) = \frac{-t\sin t - \cos t}{t^2} = -\frac{t\sin t + \cos t}{t^2}$

14. $f(x) = \frac{\sin x}{x}$

$f'(x) = \frac{x\cos x - \sin x}{x^2}$

15. $g(x) = \frac{\sin x}{x^2}$

$g'(x) = \frac{x^2\cos x - \sin x(2x)}{x^4} = \frac{x\cos x - 2\sin x}{x^3}$

16. $f(t) = \frac{\cos t}{t^3}$

$f'(t) = \frac{t^3(-\sin t) - \cos t(3t^2)}{(t^3)^2} = -\frac{t\sin t + 3\cos t}{t^4}$

17. $g(\theta) = \dfrac{\theta}{1 - \sin\theta}$

$g'(\theta) = \dfrac{1 - \sin\theta + \theta\cos\theta}{(\sin\theta - 1)^2}$ (form of answer may vary)

18. $f(\theta) = \dfrac{\sin\theta}{1 - \cos\theta}$

$f'(\theta) = \dfrac{(1 - \cos\theta)\cos\theta - \sin\theta(\sin\theta)}{(1 - \cos\theta)^2}$

$= \dfrac{\cos\theta - 1}{(1 - \cos\theta)^2} = \dfrac{1}{\cos\theta - 1}$

19. $y = \dfrac{3(1 - \sin x)}{2\cos x} = \dfrac{3}{2}(\sec x - \tan x)$

$y' = \dfrac{3}{2}(\sec x\tan x - \sec^2 x) = \dfrac{3}{2}\sec x(\tan x - \sec x)$

$= \dfrac{3}{2}(\sec x\tan x - \tan^2 x - 1)$

20. $y = \dfrac{\sec x}{x}$

$y' = \dfrac{x\sec x\tan x - \sec x}{x^2}$

$= \dfrac{\sec x(x\tan x - 1)}{x^2}$

21. $y = -\csc x - \sin x$

$y' = \csc x\cot x - \cos x$

$= \dfrac{\cos x}{\sin^2 x} - \cos x$

$= \cos x(\csc^2 x - 1)$

$= \cos x\cot^2 x$

22. $f(x) = x^2\tan x$

$f'(x) = x^2\sec^2 x + 2x\tan x$

$= x(x\sec^2 x + 2\tan x)$

23. $y = 2x\sin x + x^2\cos x$

$y' = 2x\cos x + 2\sin x + x^2(-\sin x) + 2x\cos x$

$= 4x\cos x + 2\sin x - x^2\sin x$

24. $h(\theta) = 5\theta\sec\theta + \theta\tan\theta$

$h'(\theta) = 5\theta\sec\theta\tan\theta + 5\sec\theta + \theta\sec^2\theta + \tan\theta$

25. (a) $y = \sin x$

$y' = \cos x$

$y'(0) = 1$

1 cycle in $[0, 2\pi]$.

(b) $y = \sin 2x$

$y' = 2\cos 2x$

$y'(0) = 2$

2 cycles in $[0, 2\pi]$.

26. (a) $y = \sin 3x$

$y' = 3\cos 3x$

$y'(0) = 3$

3 cycles in $[0, 2\pi]$.

(b) $y = \sin\left(\frac{1}{2}\right)x$

$y' = \left(\frac{1}{2}\right)\cos\left(\frac{1}{2}\right)x$

$y'(0) = \frac{1}{2}$

$\frac{1}{2}$ cycle in $[0, 2\pi]$.

27. $f(x) = \sin x\cos x, \left(\dfrac{\pi}{2}, 0\right)$

$f'(x) = \cos^2 x - \sin^2 x$

$f'\left(\dfrac{\pi}{2}\right) = -1$

Tangent line: $y - 0 = -1\left(x - \dfrac{\pi}{2}\right)$

$y = -x + \dfrac{\pi}{2}$

28. $f(x) = x\sin x + \cos x, (\pi, -1)$

$f'(x) = x\cos x + \sin x - \sin x = x\cos x$

$f'(\pi) = -\pi$

Tangent line: $y + 1 = -\pi(x - \pi)$

$y = -\pi x + \pi^2 - 1$

29. $f(x) = e^{\sin x}, (0, 1)$

$f'(x) = e^{\sin x} \cdot \cos x$

$f'(0) = 1$

Tangent line: $y - 1 = 1(x - 0)$

$y = x + 1$

30. $f(x) = \sin(\cos x), \left(\frac{3\pi}{2}, 0\right)$

$f'(x) = \cos(\cos x)(-\sin x)$

$f'\left(\frac{3\pi}{2}\right) = (1)(1) = 1$

Tangent line: $y - 0 = 1\left(x - \frac{3\pi}{2}\right)$

$$y = x - \frac{3\pi}{2}$$

31. $y = \cos 3x$

$\frac{dy}{dx} = -3\sin 3x$

32. $y = \sin \pi x$

$\frac{dy}{dx} = \pi \cos \pi x$

33. $g(x) = 3\tan 4x$

$g'(x) = 12\sec^2 4x$

34. $h(x) = \sec(x^2)$

$h'(x) = 2x\sec(x^2)\tan(x^2)$

35. $y = \sin(\pi x)^2 = \sin(\pi^2 x^2)$

$y' = \cos(\pi^2 x^2)[2\pi^2 x] = 2\pi^2 x\cos(\pi^2 x^2)$

36. $y = \cos(1 - 2x)^2 = \cos((1 - 2x)^2)$

$y' = -\sin(1 - 2x)^2(2(1 - 2x)(-2))$

$= 4(1 - 2x)\sin(1 - 2x)^2$

37. $h(x) = \sin 2x \cos 2x$

$h'(x) = \sin 2x(-2\sin 2x) + \cos 2x(2\cos 2x)$

$= 2\cos^2 2x - 2\sin^2 2x$

$= 2\cos 4x$

Alternate solution: $h(x) = \frac{1}{2}\sin 4x$

$h'(x) = \frac{1}{2}\cos 4x(4) = 2\cos 4x$

38. $g(\theta) = \sec\left(\frac{1}{2}\theta\right)\tan\left(\frac{1}{2}\theta\right)$

$g'(\theta) = \sec\left(\frac{1}{2}\theta\right)\sec^2\left(\frac{1}{2}\theta\right)\frac{1}{2} + \tan\left(\frac{1}{2}\theta\right)\sec\left(\frac{1}{2}\theta\right)\tan\left(\frac{1}{2}\theta\right)\frac{1}{2}$

$= \frac{1}{2}\sec\left(\frac{1}{2}\theta\right)\left[\sec^2\left(\frac{1}{2}\theta\right) + \tan^2\left(\frac{1}{2}\theta\right)\right]$

39. $f(x) = \frac{\cot x}{\sin x} = \frac{\cos x}{\sin^2 x}$

$f'(x) = \frac{\sin^2 x(-\sin x) - \cos x(2\sin x\cos x)}{\sin^4 x}$

$= \frac{-\sin^2 x - 2\cos^2 x}{\sin^3 x} = \frac{-1 - \cos^2 x}{\sin^3 x}$

40. $g(v) = \frac{\cos v}{\csc v} = \cos v \cdot \sin v$

$g'(v) = \cos v(\cos v) + \sin v(-\sin v) = \cos^2 v - \sin^2 v = \cos 2v$

41. $y = 4\sec^2 x$

$y' = 8\sec x \cdot \sec x\tan x = 8\sec^2 x\tan x$

42. $y = 2\tan^3 x$

$y' = 6\tan^2 x \cdot \sec^2 x$

43. $f(\theta) = \frac{1}{4}\sin^2 2\theta = \frac{1}{4}(\sin 2\theta)^2$

$f'(\theta) = 2\left(\frac{1}{4}\right)(\sin 2\theta)(\cos 2\theta)(2)$

$= \sin 2\theta\cos 2\theta = \frac{1}{2}\sin 4\theta$

44. $g(t) = 5\cos^2 \pi t = 5(\cos \pi t)^2$

$g'(t) = 10\cos \pi t(-\sin \pi t)(\pi)$

$= -10\pi(\sin \pi t)(\cos \pi t) = -5\pi\sin 2\pi t$

45. $f(x) = 3\sec^2(\pi t - 1)$

$f'(t) = 6\sec(\pi t - 1)\sec(\pi t - 1)\tan(\pi t - 1)(\pi)$

$= 6\pi\sec^2(\pi t - 1)\tan(\pi t - 1) = \dfrac{6\pi\sin(\pi t - 1)}{\cos^3(\pi t - 1)}$

46. $h(t) = 2\cot^2(\pi t + 2)$

$h'(t) = 4\cot(\pi t + 2)(-\csc^2(\pi t + 2)(\pi))$

$= -4\pi\cot(\pi t + 2)\csc^2(\pi t + 2)$

47. $y = \sqrt{x} + \dfrac{1}{4}\sin(2x)^2$

$= \sqrt{x} + \dfrac{1}{4}\sin(4x^2)$

$\dfrac{dy}{dx} = \dfrac{1}{2}x^{-1/2} + \dfrac{1}{4}\cos(4x^2)(8x)$

$= \dfrac{1}{2\sqrt{x}} + 2x\cos(2x)^2$

48. $y = 3x - 5\cos(\pi x)^2$

$= 3x - 5\cos(\pi^2 x^2)$

$\dfrac{dy}{dx} = 3 + 5\sin(\pi^2 x^2)(2\pi^2 x)$

$= 3 + 10\pi^2 x\sin(\pi x)^2$

49. $y = \sin x^{1/3} + (\sin x)^{1/3}$

$y' = \cos x^{1/3}\left(\dfrac{1}{3}x^{-2/3}\right) + \dfrac{1}{3}(\sin x)^{-2/3}\cos x$

$= \dfrac{1}{3}\left[\dfrac{\cos x^{1/3}}{x^{2/3}} + \dfrac{\cos x}{(\sin x)^{2/3}}\right]$

50. $y = e^x(\sin x + \cos x)$

$\dfrac{dy}{dx} = e^x(\cos x - \sin x) + (\sin x + \cos x)(e^x)$

$= e^x(2\cos x) = 2e^x\cos x$

51. $y = \ln|\sin x|$

$\dfrac{dy}{dx} = \dfrac{\cos x}{\sin x} = \cot x$

52. $y = \ln|\csc x|$

$y' = \dfrac{-\csc x \cdot \cot x}{\csc x} = -\cot x$

53. $y = \ln|\csc x - \cot x|$

$y' = \dfrac{-\csc x\cot x + \csc^2 x}{\csc x - \cot x}$

$= \dfrac{\csc x(\csc x - \cot x)}{\csc x - \cot x} = \csc x$

54. $y = \ln|\sec x + \tan x|$

$\dfrac{dy}{dx} = \dfrac{\sec x\tan x + \sec^2 x}{\sec x + \tan x}$

$= \dfrac{\sec x(\sec x + \tan x)}{\sec x + \tan x} = \sec x$

55. $y = \ln\left|\dfrac{\cos x}{1 - \sin x}\right| = \ln|\cos x| - \ln|1 - \sin x|$

$y' = \dfrac{1}{\cos x}(-\sin x) - \dfrac{1}{1 - \sin x}(-\cos x)$

$= \dfrac{-\sin x}{\cos x} + \dfrac{\cos x}{1 - \sin x} = \dfrac{-\sin x + \sin^2 x + \cos^2 x}{\cos x(1 - \sin x)}$

$= \dfrac{-\sin x + 1}{\cos x(1 - \sin x)}$

$= \dfrac{1}{\cos x} = \sec x$

56. $y = \ln\sqrt{1 + \sin^2 x} = \dfrac{1}{2}\ln(1 + \sin^2 x)$

$\dfrac{dy}{dx} = \left(\dfrac{1}{2}\right)\dfrac{2\sin x\cos x}{1 + \sin^2 x} = \dfrac{\sin x\cos x}{1 + \sin^2 x}$

57. Yes. f is continuous on $[0, \pi]$ and differentiable on $(0, \pi)$.

$f'(c) = \dfrac{f(b) - f(a)}{b - a}$

$\cos(c) = \dfrac{0 - 0}{\pi - 0} = 0$

$c = \dfrac{\pi}{2}$

58. Yes. f is continuous on $[0, \pi]$ and differentiable on $(0, \pi)$.

$$f'(x) = 2\cos x + 2\cos 2x = \frac{0-0}{\pi - 0} = 0$$

$$2\cos^2 x + \cos x - 1 = 0$$

$$(2\cos x - 1)(\cos x + 1) = 0$$

$$\cos x = \frac{1}{2} \Rightarrow x = \frac{\pi}{3}$$

$$\cos x = -1 \Rightarrow x = \pi$$

$$c = \frac{\pi}{3}$$

59.

$$y = \frac{1 + \csc x}{1 - \csc x}$$

$$y' = \frac{(1 - \csc x)(-\csc x \cot x) - (1 + \csc x)(\csc x \cot x)}{(1 - \csc x)^2}$$

$$= \frac{-2\csc x \cot x}{(1 - \csc x)^2}$$

$$y'\left(\frac{\pi}{6}\right) = \frac{-2(2)(\sqrt{3})}{(1-2)^2} = -4\sqrt{3}$$

60. $f(x) = \tan x \cot x = 1$

$f'(x) = 0$

$f'(1) = 0$

61.

$$h(t) = \frac{\sec t}{t}$$

$$h'(t) = \frac{t(\sec t \tan t) - (\sec t)(1)}{t^2}$$

$$= \frac{\sec t(t \tan t - 1)}{t^2}$$

$$h'(\pi) = \frac{\sec \pi(\pi \tan \pi - 1)}{\pi^2} = \frac{1}{\pi^2}$$

62.

$$f(x) = \sin x(\sin x + \cos x)$$

$$f'(x) = \sin x(\cos x - \sin x) + (\sin x + \cos x)\cos x$$

$$= \sin x \cos x - \sin^2 x + \sin x \cos x + \cos^2 x$$

$$= \sin 2x + \cos 2x$$

$$f'\left(\frac{\pi}{4}\right) = \sin\frac{\pi}{2} + \cos\frac{\pi}{2} = 1$$

63. $y = e^x(\cos\sqrt{2}\,x + \sin\sqrt{2}\,x)$

$$y' = e^x(-\sqrt{2}\sin\sqrt{2}\,x + \sqrt{2}\cos\sqrt{2}\,x) + e^x(\cos\sqrt{2}x + \sin\sqrt{2}\,x)$$

$$y'' = e^x(-2\cos\sqrt{2}\,x - 2\sin\sqrt{2}\,x) + 2e^x(-\sqrt{2}\sin\sqrt{2}\,x + \sqrt{2}\cos\sqrt{2}\,x) + e^x(\cos\sqrt{2}\,x + \sin\sqrt{2}\,x)$$

$$= -e^x(\cos\sqrt{2}\,x + \sin\sqrt{2}\,x) + 2e^x(-\sqrt{2}\sin\sqrt{2}\,x + \sqrt{2}\cos\sqrt{2}\,x)$$

Therefore, $y'' - 2y' + 3y = 0$.

64. (a) $y = e^x(3\cos 2x - 4\sin 2x)$

$$y' = e^x(-6\sin 2x - 8\cos 2x) + e^x(3\cos 2x - 4\sin 2x)$$

$$= e^x(-5\cos 2x - 10\sin 2x)$$

$$y'' = e^x(10\sin 2x - 20\cos 2x) + e^x(-5\cos 2x - 10\sin 2x)$$

$$= -25e^x\cos 2x$$

$$y'' - 2y' + 5y = -25e^x\cos 2x + e^x(10\cos 2x + 20\sin 2x) + e^x(15\cos 2x - 20\sin 2x) = 0$$

65. $f(x) = x^n \sin x$

$$f'(x) = x^n\cos x + nx^{n-1}\sin x$$

$$= x^{n-1}(x\cos x + n\sin x)$$

When $n = 1$: $f'(x) = x\cos x + \sin x$.

When $n = 2$: $f'(x) = x(x\cos x + 2\sin x)$.

When $n = 3$: $f'(x) = x^2(x\cos x + 3\sin x)$.

When $n = 4$: $f'(x) = x^3(x\cos x + 4\sin x)$.

For general n, $f'(x) = x^{n-1}(x\cos x + n\sin x)$.

66. $f(x) = \dfrac{\cos x}{x^n} = x^{-n}\cos x$

$$f'(x) = -x^{-n}\sin x - nx^{-n-1}\cos x$$
$$= -x^{-n-1}(x\sin x + n\cos x)$$
$$= -\frac{x\sin x + n\cos x}{x^{n+1}}$$

When $n = 1$: $f'(x) = -\dfrac{x\sin x + \cos x}{x^2}$.

When $n = 2$: $f'(x) = -\dfrac{x\sin x + 2\cos x}{x^3}$.

When $n = 3$: $f'(x) = -\dfrac{x\sin x + 3\cos x}{x^4}$.

When $n = 4$: $f'(x) = -\dfrac{x\sin x + 4\cos x}{x^5}$.

For general n, $f'(x) = -\dfrac{x\sin x + n\cos x}{x^{n+1}}$.

67.
$$\sin x + 2\cos 2y = 1$$
$$\cos x - 4(\sin 2y)y' = 0$$
$$y' = \frac{\cos x}{4\sin 2y}$$

68.
$$(\sin \pi x + \cos \pi y)^2 = 2$$
$$2(\sin \pi x + \cos \pi y)[\pi\cos \pi x - \pi(\sin \pi y)y'] = 0$$
$$\pi\cos \pi x - \pi(\sin \pi y)y' = 0$$
$$y' = \frac{\cos \pi x}{\sin \pi y}$$

69.
$$\sin x = x(1 + \tan y)$$
$$\cos x = x(\sec^2 y)y' + (1 + \tan y)(1)$$
$$y' = \frac{\cos x - \tan y - 1}{x\sec^2 y}$$

70.
$$\cot y = x - y$$
$$(-\csc^2 y)y' = 1 - y'$$
$$y' = \frac{1}{1 - \csc^2 y}$$
$$= \frac{1}{-\cot^2 y} = -\tan^2 y$$

71.
$$y = \sin(xy)$$
$$y' = [xy' + y]\cos(xy)$$
$$y' - x\cos(xy)y' = y\cos(xy)$$
$$y' = \frac{y\cos(xy)}{1 - x\cos(xy)}$$

72.
$$x = \sec\frac{1}{y}$$
$$1 = -\frac{y'}{y^2}\sec\frac{1}{y}\tan\frac{1}{y}$$
$$y' = \frac{-y^2}{\sec(1/y)\tan(1/y)} = -y^2\cos\left(\frac{1}{y}\right)\cot\left(\frac{1}{y}\right)$$

73. (a) $f(x) = \cos x \qquad f\left(\dfrac{\pi}{3}\right) = \dfrac{1}{2}$

$f'(x) = -\sin x \qquad f'\left(\dfrac{\pi}{3}\right) = -\dfrac{\sqrt{3}}{2}$

$f''(x) = -\cos x \qquad f''\left(\dfrac{\pi}{3}\right) = -\dfrac{1}{2}$

$$P_1(x) = -\frac{\sqrt{3}}{2}\left(x - \frac{\pi}{3}\right) + \frac{1}{2}$$
$$P_2(x) = -\frac{1}{4}\left(x - \frac{\pi}{3}\right)^2 - \frac{\sqrt{3}}{2}\left(x - \frac{\pi}{3}\right) + \frac{1}{2}$$

(b)

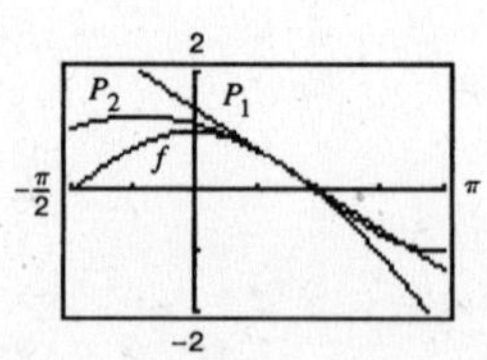

(c) P_2 is a better approximation than P_1.

(d) The accuracy worsens as you move away from $x = a = \dfrac{\pi}{3}$.

74. $f(x) = \sin x \qquad f\left(\frac{\pi}{2}\right) = 1$

$f'(x) = \cos x \qquad f'\left(\frac{\pi}{2}\right) = 0$

$f''(x) = -\sin x \qquad f''\left(\frac{\pi}{2}\right) = -1$

(a) $P_1(x) = f'(a)(x - a) + f(a) = 0\left(x - \frac{\pi}{2}\right) + 1 = 1$

$P_2(x) = \frac{1}{2}f''(a)(x - a)^2 + f'(a)(x - a) + f(a) = \frac{1}{2}(-1)\left(x - \frac{\pi}{2}\right)^2 + 1$

$= 1 - \frac{1}{2}\left(x - \frac{\pi}{2}\right)^2$

(b)

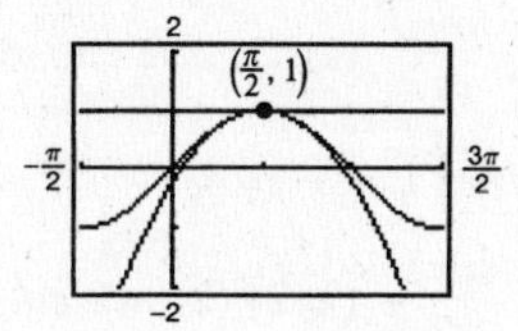

(c) P_2 is a better approximation than P_1.

(d) The accuracy worsens as you move farther away from $x = a = \frac{\pi}{2}$.

75. (a) $f(x) = \tan\frac{\pi x}{4} \qquad f(1) = 1$

$f'(x) = \frac{\pi}{4}\sec^2\frac{\pi x}{4} \qquad f'(1) = \frac{\pi}{4}(2) = \frac{\pi}{2}$

$f''(x) = \frac{\pi}{2}\sec^2\frac{\pi x}{4} \cdot \tan\frac{\pi x}{4}\left(\frac{\pi}{4}\right) \qquad f''(1) = \frac{\pi^2}{8}(2)(1) = \frac{\pi^2}{4}$

$P_1(x) = f'(1)(x - 1) + f(1) = \frac{\pi}{2}(x - 1) + 1$

$P_2(x) = \frac{1}{2}\left(\frac{\pi^2}{4}\right)(x - 1)^2 + f'(1)(x - 1) + f(1) = \frac{\pi^2}{8}(x - 1)^2 + \frac{\pi}{2}(x - 1) + 1$

(b)

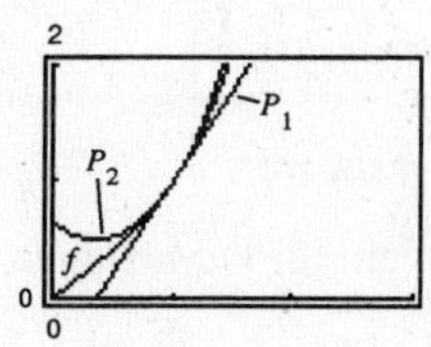

(c) P_2 is a better approximation than P_1.

(d) The accuracy worsens as you move away from $x = a = 1$.

76. (a) $f(x) = \sec(2x)$

$f'(x) = 2(\sec 2x)(\tan 2x)$

$f''(x) = 2[2(\sec 2x)(\tan 2x)]\tan 2x + 2(\sec 2x)(\sec^2 2x)(2)$

$= 4[(\sec 2x)(\tan^2 2x) + \sec^3 2x]$

$f\left(\frac{\pi}{6}\right) = \sec\left(\frac{\pi}{3}\right) = 2$

$f'\left(\frac{\pi}{6}\right) = 2\sec\left(\frac{\pi}{3}\right)\tan\left(\frac{\pi}{3}\right) = 4\sqrt{3}$

$f''\left(\frac{\pi}{6}\right) = 4[2(3) + 2^3] = 56$

$P_1(x) = 4\sqrt{3}\left(x - \frac{\pi}{6}\right) + 2$

$P_2(x) = \frac{1}{2}(56)\left(x - \frac{\pi}{6}\right)^2 + 4\sqrt{3}\left(x - \frac{\pi}{6}\right) + 2$

$= 28\left(x - \frac{\pi}{6}\right)^2 + 4\sqrt{3}\left(x - \frac{\pi}{6}\right) + 2$

(b)

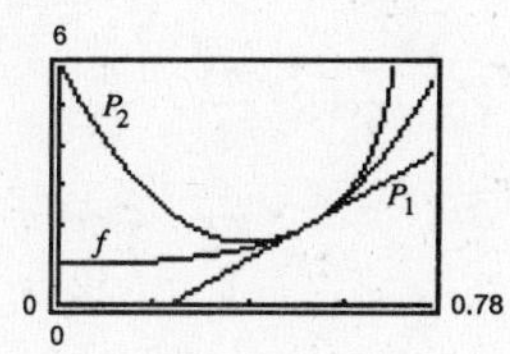

(c) P_2 is a better approximation than P_1.

(d) The accuracy worsens as you move away from $x = a = \frac{\pi}{6}$.

77. $y = \tan x$

$\dfrac{dx}{dt} = 2$

$\dfrac{dy}{dt} = \sec^2 x \dfrac{dx}{dt}$

(a) When $x = -\pi/3$,

$$\frac{dy}{dt} = (2)^2(2) = 8 \text{ cm/sec.}$$

(b) When $x = -\pi/4$,

$$\frac{dy}{dt} = (\sqrt{2})^2(2) = 4 \text{ cm/sec.}$$

(c) When $x = 0$,

$$\frac{dy}{dt} = (1)^2(2) = 2 \text{ cm/sec.}$$

78. $y = \sin x$

$\dfrac{dx}{dt} = 2$

$\dfrac{dy}{dt} = \cos x \dfrac{dx}{dt}$

(a) When $x = \pi/6$,

$$\frac{dy}{dt} = \left(\cos\frac{\pi}{6}\right)(2) = \sqrt{3} \text{ cm/sec.}$$

(b) When $x = \pi/4$,

$$\frac{dy}{dt} = \left(\cos\frac{\pi}{4}\right)(2) = \sqrt{2} \text{ cm/sec.}$$

(c) When $x = \pi/3$,

$$\frac{dy}{dt} = \left(\cos\frac{\pi}{3}\right)(2) = 1 \text{ cm/sec.}$$

79. $f(x) = \cos \pi x, \left[0, \dfrac{1}{6}\right]$

$f'(x) = -\pi \sin \pi x$

Left endpoint: $(0, 1)$ Maximum

Right endpoint: $\left(\dfrac{1}{6}, \dfrac{\sqrt{3}}{2}\right)$ Minimum

80. $g(x) = \sec x, \left[-\dfrac{\pi}{6}, \dfrac{\pi}{3}\right]$

$g'(x) = \sec x \tan x$

Left endpoint: $\left(-\dfrac{\pi}{6}, \dfrac{2}{\sqrt{3}}\right) \approx \left(-\dfrac{\pi}{6}, 1.1547\right)$

Right endpoint: $\left(\dfrac{\pi}{3}, 2\right)$ Maximum

Critical number: $(0, 1)$ Minimum

81. $y = \dfrac{4}{x} + \tan\dfrac{\pi x}{8}, \left[1, \dfrac{7}{2}\right]$

$$y' = \frac{-4}{x^2} + \frac{\pi}{8}\sec^2\frac{\pi x}{8} = 0$$

$$\frac{\pi}{8}\sec^2\frac{\pi x}{8} = \frac{4}{x^2}$$

$$\cos^2\frac{\pi x}{8} = \frac{\pi x^2}{32}$$

Using a graphing utility, $x \approx 2.1345$ and $f(x) \approx 2.9856$.

Endpoints: $\left(1, \sqrt{2} + 3\right) \approx (1, 4.4142)$

$\left(\dfrac{7}{2}, 6.1702\right)$.

Minimum: $(2.1345, 2.9856)$

Maximum: $(3.5, 6.1702)$

82. $y = x^2 - 2 - \cos x, [-1, 3]$

$y' = 2x + \sin x$

Left endpoint: $(-1, -1.5403)$

Right endpoint: $(3, 7.99)$ Maximum

Critical number: $(0, -3)$ Minimum

83. $f(x) = \sin x, [0, 2\pi]$

$f(0) = f(2\pi) = 0$

f is continuous on $[0, 2\pi]$. f is differentiable on $(0, 2\pi)$. Rolle's Theorem applies.

$f'(x) = \cos x$

c values: $\dfrac{\pi}{2}, \dfrac{3\pi}{2}$

84. $f(x) = \cos x, [0, 2\pi]$

$f(0) = f(2\pi) = 1$

f is continuous on $[0, 2\pi]$. f is differentiable on $(0, 2\pi)$. Rolle's Theorem applies.

$f'(x) = -\sin x$

c value: π

85. $f(x) = \cos 2x, \left[-\frac{\pi}{12}, \frac{\pi}{6}\right]$

$$f\left(-\frac{\pi}{12}\right) = \frac{\sqrt{3}}{2}$$

$$f\left(\frac{\pi}{6}\right) = \frac{1}{2}$$

$$f\left(-\frac{\pi}{12}\right) \neq f\left(\frac{\pi}{6}\right)$$

Rolle's Theorem does not apply.

86. $f(x) = \frac{6x}{\pi} - 4\sin^2 x, \left[0, \frac{\pi}{6}\right]$

$$f(0) = f\left(\frac{\pi}{6}\right) = 0$$

f is continuous on $[0, \pi/6]$. f is differentiable on $(0, \pi/6)$. Rolle's Theorem applies.

$$f'(x) = \frac{6}{\pi} - 8\sin x\cos x = 0$$

$$\frac{6}{\pi} = 8\sin x\cos x$$

$$\frac{3}{4\pi} = \frac{1}{2}\sin 2x$$

$$\frac{3}{2\pi} = \sin 2x$$

$$\frac{1}{2}\arcsin\left(\frac{3}{2\pi}\right) = x$$

$$x \approx 0.2489$$

c value: 0.2489

87. $f(x) = \tan x, [0, \pi]$

$f(0) = f(\pi) = 0$

f is not continuous on $[0, \pi]$ since $f(\pi/2)$ does not exist. Rolle's Theorem does not apply.

88. $f(x) = \sec x, \left[-\frac{\pi}{4}, \frac{\pi}{4}\right]$

$$f\left(-\frac{\pi}{4}\right) = f\left(\frac{\pi}{4}\right) = \sqrt{2}$$

f is continuous on $[-\pi/4, \pi/4]$. f is differentiable on $(-\pi/4, \pi/4)$. Rolle's Theorem applies.

$$f'(x) = \sec x\tan x$$

$$\sec x\tan x = 0$$

$$x = 0$$

c value: 0

89. $f(x) = 2\sin x + \sin 2x$

$f'(x) = 2\cos x + 2\cos 2x = 0$

$2(2\cos^2 x + \cos x - 1) = 0$

$2(\cos x + 1)(2\cos x - 1) = 0$ when $\cos x = \frac{1}{2}, -1$ or $x = \frac{\pi}{3}, \pi, \frac{5\pi}{3}$.

$f''(x) = -2\sin x - 4\sin 2x$

$= -2\sin x(1 + 4\cos x) = 0$ when $x = 0, 1.823, \pi, 4.460$.

Relative maximum: $\left(\frac{\pi}{3}, 2.598\right)$

Relative minimum: $\left(\frac{5\pi}{3}, -2.598\right)$

Points of inflection: $(1.823, 1.452), (\pi, 0,) (4.46, -1.452)$

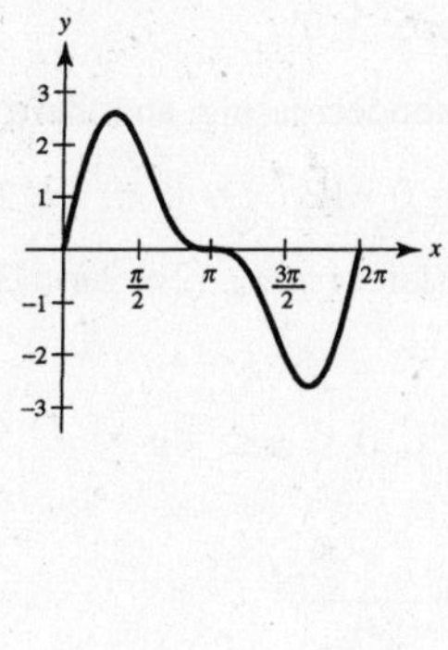

90. $f(x) = 2\sin x + \cos 2x,\ 0 \le x \le 2\pi$

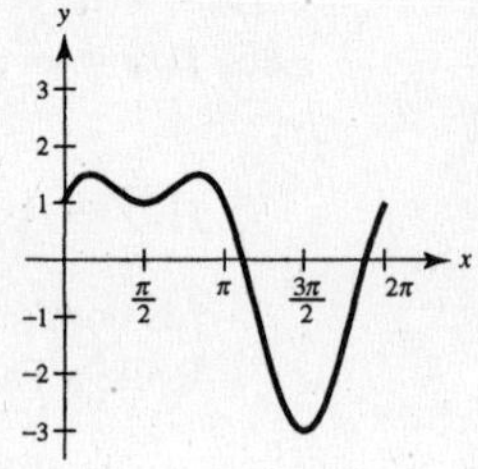

91. $y = \sin x - \dfrac{1}{18}\sin 3x,\ 0 \le x \le 2\pi$

$y' = \cos x - \dfrac{1}{6}\cos 3x = 0$ when $x = \dfrac{\pi}{2}, \dfrac{3\pi}{2}$.

$y'' = -\sin x + \dfrac{1}{2}\sin 3x = 0$ when $x = 0, \dfrac{\pi}{6}, \dfrac{5\pi}{6}, \pi, \dfrac{7\pi}{6}, \dfrac{11\pi}{6}$.

Relative maximum: $\left(\dfrac{\pi}{2}, \dfrac{19}{18}\right)$

Relative minimum: $\left(\dfrac{3\pi}{2}, -\dfrac{19}{18}\right)$

Inflection points: $\left(\dfrac{\pi}{6}, \dfrac{4}{9}\right), \left(\dfrac{5\pi}{6}, \dfrac{4}{9}\right), (\pi, 0), \left(\dfrac{7\pi}{6}, -\dfrac{4}{9}\right), \left(\dfrac{11\pi}{6}, -\dfrac{4}{9}\right)$

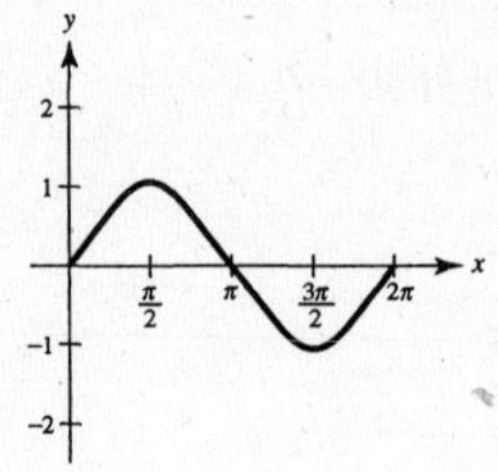

92. $y = \cos x - \dfrac{1}{2}\cos 2x,\ 0 \le x \le 2\pi$

$y' = -\sin x + \sin 2x = -\sin x(1 - 2\cos x) = 0$ when $x = 0, \pi, \dfrac{\pi}{3}, \dfrac{5\pi}{3}$.

$y'' = -\cos x + 2\cos 2x = -\cos x + 2(2\cos^2 x - 1)$

$= 4\cos^2 x - \cos x - 2 = 0$ when $\cos x = \dfrac{1 \pm \sqrt{33}}{8} \approx 0.8431, -0.5931$.

Therefore, $x \approx 0.5678$ or 5.7154, $x \approx 2.2057$ or 4.0775.

Relative maxima: $\left(\dfrac{\pi}{3}, \dfrac{3}{4}\right), \left(\dfrac{5\pi}{3}, \dfrac{3}{4}\right)$

Relative minimum: $\left(\pi, -\dfrac{3}{2}\right)$

Inflection points: $(0.5678, 0.6323), (2.2057, -0.4448), (5.7154, 0.6323), (4.0775, -0.4448)$

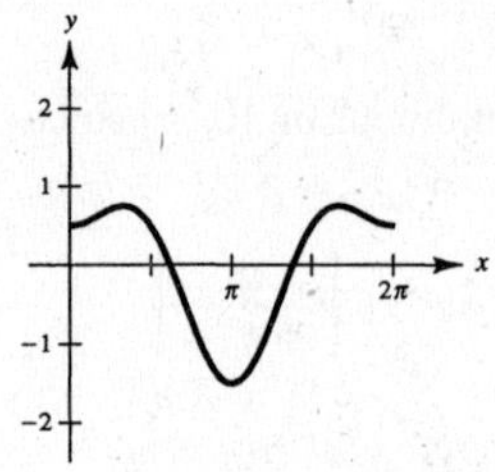

93. $f(x) = x - \sin x$

$f'(x) = 1 - \cos x \ge 0$

Therefore, f is nondecreasing, and there are no relative extrema.

$f''(x) = \sin x = 0$ when $x = 0, \pi, 2\pi, 3\pi, 4\pi$.

Points of inflection: $(\pi, \pi), (2\pi, 2\pi), (3\pi, 3\pi)$

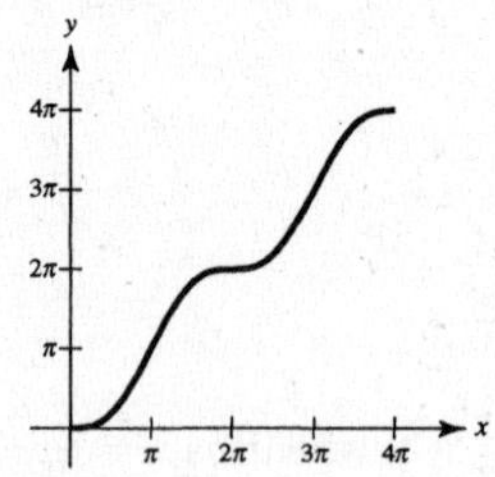

94. $f(x) = \cos x - x,\ 0 \le x \le 4\pi$

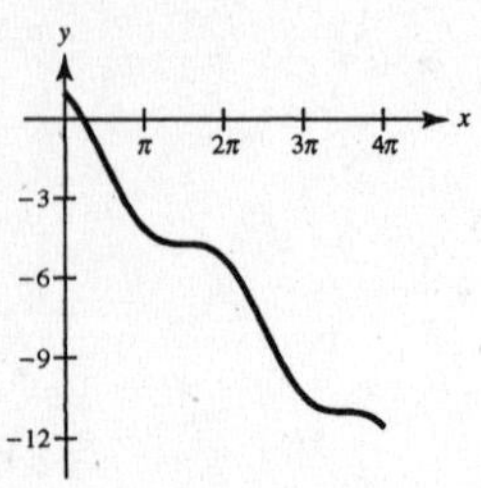

95. $y = 2(\csc x + \sec x),\ 0 < x < \dfrac{\pi}{2}$

$y' = 2(\sec x \tan x - \csc x \cot x) = 0 \Rightarrow x = \pi/4$

Relative minimum: $\left(\dfrac{\pi}{4}, 4\sqrt{2}\right)$

Vertical asymptotes: $x = 0,\ x = \dfrac{\pi}{2}$

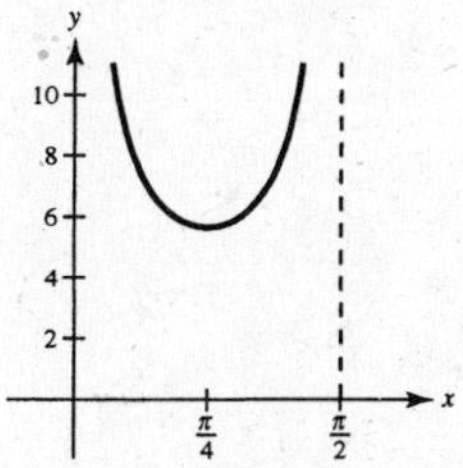

96. $g(x) = x \tan x,\ -\dfrac{3\pi}{2} < x < \dfrac{3\pi}{2}$

$g'(x) = \dfrac{x + \sin x \cos x}{\cos^2 x} = 0$ when $x = 0$

$g''(x) = \dfrac{2(\cos x + x \sin x)}{\cos^3 x}$

Vertical asymptotes: $x = -\dfrac{3\pi}{2}, -\dfrac{\pi}{2}, \dfrac{\pi}{2}, \dfrac{3\pi}{2}$

Intercepts: $(-\pi, 0), (0, 0), (\pi, 0)$

Symmetric with respect to y-axis.

Increasing on

$\left(0, \dfrac{\pi}{2}\right)$ and $\left(\dfrac{\pi}{2}, \dfrac{3\pi}{2}\right)$.

Points of inflection:
$(\pm 2.80, -1)$

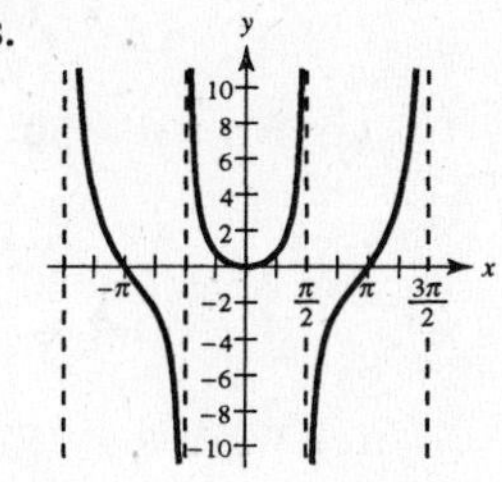

97. Since $\dfrac{\pi}{2} < 2 < 4 < \dfrac{3\pi}{2}$ and $f(x) = \sin x$ is decreasing on $\left(\dfrac{\pi}{2}, \dfrac{3\pi}{2}\right)$, $f(2) > f(4)$.

98. Least value: $4(3) + 5 = 17$

Greatest value: $6(3) + 5 = 23$

99. Answers will vary.
One example is

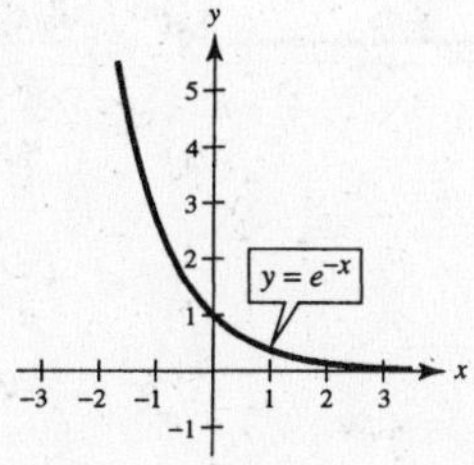

100. $g(x) = f(1 - 2x)$

$g'(x) = f'(1 - 2x)(-2)$

101. $y = A \cos \omega t$

(a) Amplitude: $A = \dfrac{3.5}{2} = 1.75$

$y = 1.75 \cos \omega t$

Period: $10 \Rightarrow \omega = \dfrac{2\pi}{10} = \dfrac{\pi}{5}$

$y = 1.75 \cos \dfrac{\pi t}{5}$

(b) $v = y' = 1.75\left[-\dfrac{\pi}{5} \sin \dfrac{\pi t}{5}\right]$

$= -0.35\pi \sin \dfrac{\pi t}{5}$

102. (a) Using a graphing utility, or by trial and error, you obtain a model of the form

$T(t) = 64.18 - 22.15 \sin\left(\dfrac{\pi t}{6} + 1\right).$

(b)

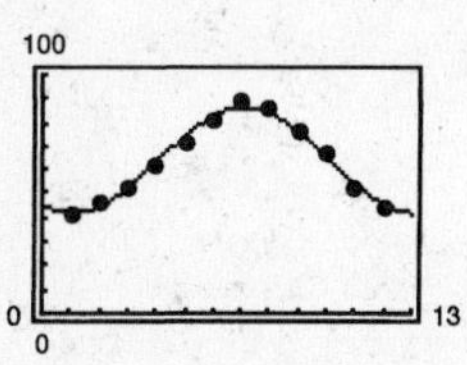

(c) $T'(t) = -22.15 \cos\left(\dfrac{\pi t}{6} + 1\right)\left(\dfrac{\pi}{6}\right)$

$= -11.60 \cos\left(\dfrac{\pi t}{6} + 1\right)$

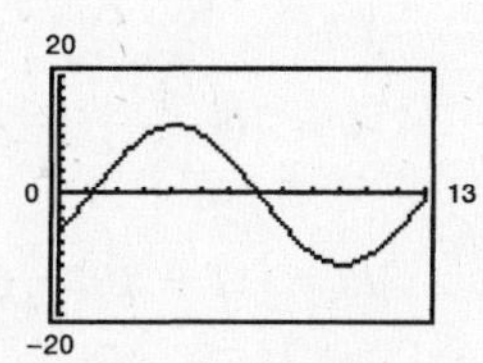

(d) The temperature changes most rapidly when $t \approx 4.1$ (April) and $t \approx 10.1$ (October). The temperature changes most slowly ($T'(t) = 0$) when $t \approx 1.1$ (January) and $t \approx 7.1$ (July).

103. $\tan\theta = \dfrac{y}{x}$

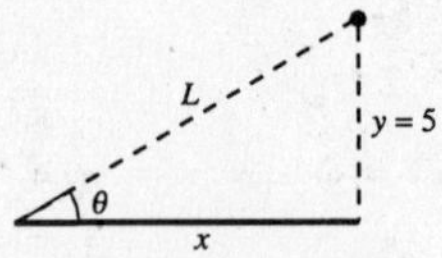

$$\frac{dx}{dt} = -600 \text{ mi/hr}$$

$$(\sec^2\theta)\frac{d\theta}{dt} = -\frac{y}{x^2}\cdot\frac{dx}{dt}$$

$$\frac{d\theta}{dt} = \cos^2\theta\left(-\frac{y}{x^2}\right)\frac{dx}{dt} - \frac{x^2}{L^2}\left(-\frac{y}{x^2}\right)\frac{dx}{dt}$$

$$= \left(-\frac{y^2}{L^2}\right)\left(\frac{1}{y}\right)\frac{dx}{dt} = (-\sin^2\theta)\left(\frac{1}{5}\right)(-600) = 120\sin^2\theta$$

(a) When $\theta = 30°$, $\dfrac{d\theta}{dt} = \dfrac{120}{4} = 30 \text{ rad/hr} = \dfrac{1}{2} \text{ rad/min}$.

(b) When $\theta = 60°$, $\dfrac{d\theta}{dt} = 120\left(\dfrac{3}{4}\right) = 90 \text{ rad/hr} = \dfrac{3}{2} \text{ rad/min}$.

(c) When $\theta = 75°$, $\dfrac{d\theta}{dt} = 120\sin^2 75° \approx 111.96 \text{ rad/hr} \approx 1.87 \text{ rad/min}$.

104. $\tan\theta = \dfrac{x}{50}$

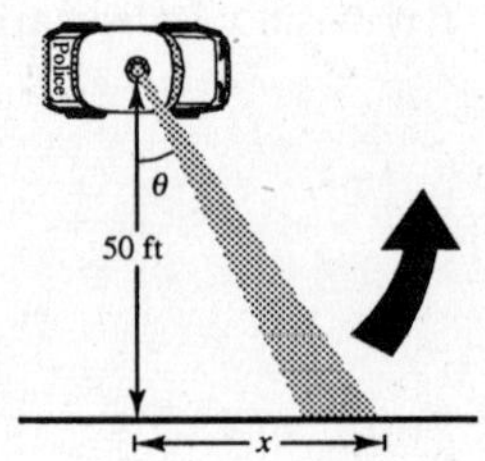

$$\frac{d\theta}{dt} = 30(2\pi) = 60\pi \text{ rad/min} = \pi \text{ rad/sec}$$

$$\sec^2\theta\left(\frac{d\theta}{dt}\right) = \frac{1}{50}\left(\frac{dx}{dt}\right)$$

$$\frac{dx}{dt} = 50\sec^2\theta\left(\frac{d\theta}{dt}\right)$$

(a) When $\theta = 30°$, $\dfrac{dx}{dt} = \dfrac{200\pi}{3}$ ft/sec.

(b) When $\theta = 60°$, $\dfrac{dx}{dt} = 200\pi$ ft/sec.

(c) When $\theta = 70°$, $\dfrac{dx}{dt} \approx 427.43\pi$ ft/sec.

105. $r = \dfrac{v_0^{\,2}}{32}(\sin 2\theta)$

$v_0 = 2200$ ft/sec

θ changes from 10° to 11°.

$$dr = \frac{(2200)^2}{16}(\cos 2\theta)\,d\theta$$

$$\theta = 10\left(\frac{\pi}{180}\right)$$

$$d\theta = (11-10)\frac{\pi}{180}$$

$$\Delta r \approx dr$$

$$= \frac{(2200)^2}{16}\cos\left(\frac{20\pi}{180}\right)\left(\frac{\pi}{180}\right) \approx 4961 \text{ feet}$$

$$\approx 4961 \text{ feet}$$

106. $h = 50\tan\theta$

$\theta = 71.5° = 1.2479$ radians

$dh = 50\sec^2\theta\cdot d\theta$

$$\left|\frac{dh}{h}\right| = \left|\frac{50\sec^2(1.2479)}{50\tan(1.2479)}d\theta\right| \le 0.06$$

$$\left|\frac{9.9316}{2.9886}d\theta\right| \le 0.06$$

$$|d\theta| \le 0.018$$

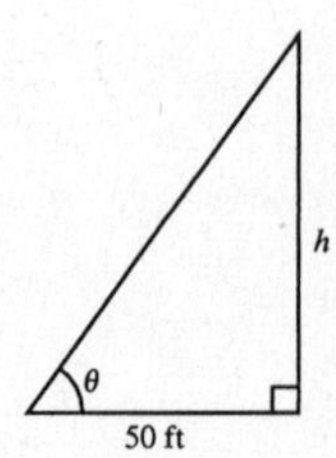

107. $F\cos\theta = k(W - F\sin\theta)$

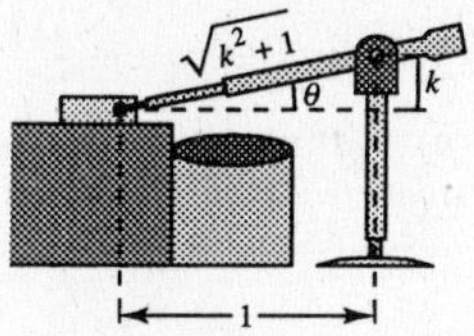

$$F = \frac{kW}{\cos\theta + k\sin\theta}$$

$$\frac{dF}{d\theta} = \frac{-kW(k\cos\theta - \sin\theta)}{(\cos\theta + k\sin\theta)^2} = 0$$

$k\cos\theta = \sin\theta \Rightarrow k = \tan\theta \Rightarrow \theta = \arctan k$

Since $\cos\theta + k\sin\theta = \dfrac{1}{\sqrt{k^2+1}} + \dfrac{k^2}{\sqrt{k^2+1}} = \sqrt{k^2+1}$,

the minimum force is $F = \dfrac{kW}{\cos\theta + k\sin\theta} = \dfrac{kW}{\sqrt{k^2+1}}$.

108. $V = \frac{1}{3}\pi r^2 h = \frac{1}{3}\pi r^2\sqrt{144 - r^2}$

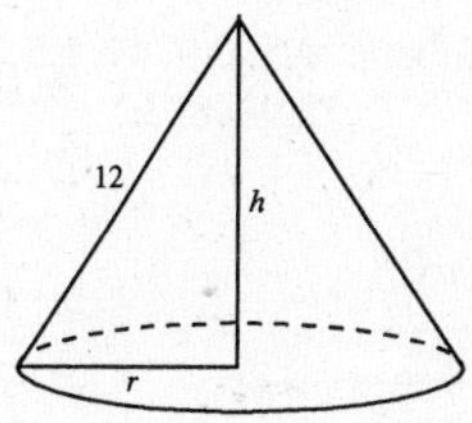

$$\frac{dV}{dr} = \frac{1}{3}\pi\left[r^2\left(\frac{1}{2}\right)(144 - r^2)^{-1/2}(-2r) + 2r\sqrt{144 - r^2}\right]$$

$$= \frac{1}{3}\pi\left[\frac{288r - 3r^3}{\sqrt{144 - r^2}}\right] = \pi\left[\frac{r(96 - r^2)}{\sqrt{144 - r^2}}\right] = 0 \text{ when } r = 0, 4\sqrt{6}.$$

By the First Derivative Test, V is maximum when $r = 4\sqrt{6}$ and $h = 4\sqrt{3}$.

Area of circle: $A = \pi(12)^2 = 144\pi$

Lateral surface area of cone: $S = \pi(4\sqrt{6})\sqrt{(4\sqrt{6})^2 + (4\sqrt{3})^2} = 48\sqrt{6}\pi$

Area of sector: $144\pi - 48\sqrt{6}\pi = 72\theta$, $\theta = \dfrac{144\pi - 48\sqrt{6}\pi}{72} \approx 1.153$ radians or 66°

109. (a)

Base 1	Base 2	Altitude	Area
8	8 + 16 cos 10°	8 sin 10°	≈ 22.1
8	8 + 16 cos 20°	8 sin 20°	≈ 42.5
8	8 + 16 cos 30°	8 sin 30°	≈ 59.7
8	8 + 16 cos 40°	8 sin 40°	≈ 72.7
8	8 + 16 cos 50°	8 sin 50°	≈ 80.5
8	8 + 16 cos 60°	8 sin 60°	≈ 83.1

(b)

Base 1	Base 2	Altitude	Area
8	8 + 16 cos 10°	8 sin 10°	≈ 22.1
8	8 + 16 cos 20°	8 sin 20°	≈ 42.5
8	8 + 16 cos 30°	8 sin 30°	≈ 59.7
8	8 + 16 cos 40°	8 sin 40°	≈ 72.7
8	8 + 16 cos 50°	8 sin 50°	≈ 80.5
8	8 + 16 cos 60°	8 sin 60°	≈ 83.1
8	8 + 16 cos 70°	8 sin 70°	≈ 80.7
8	8 + 16 cos 80°	8 sin 80°	≈ 74.0
8	8 + 16 cos 90°	8 sin 90°	≈ 64.0

The maximum cross-sectional area is approximately 83.1 square feet.

(c) $A = (a + b)\dfrac{h}{2}$

$$= [8 + (8 + 16\cos\theta)]\frac{8\sin\theta}{2}$$

$$= 64(1 + \cos\theta)\sin\theta,\ 0° < \theta < 90°$$

(d) $\dfrac{dA}{d\theta} = 64(1 + \cos\theta)\cos\theta + (-64\sin\theta)\sin\theta$

$$= 64(\cos\theta + \cos^2\theta - \sin^2\theta)$$

$$= 64(2\cos^2\theta + \cos\theta - 1)$$

$$= 64(2\cos\theta - 1)(\cos\theta + 1)$$

$$= 0 \text{ when } \theta = 60°, 180°, 300°.$$

The maximum occurs when $\theta = 60°$.

(e)

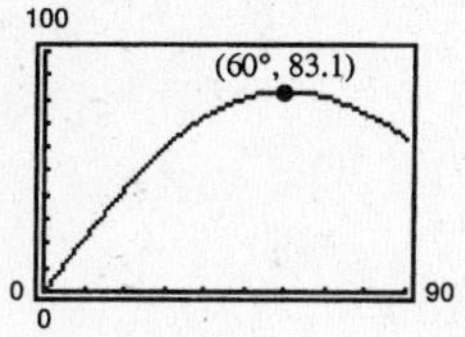

110. $f(x + p) = f(x)$ for all x.

(a) Yes, $f'(x + p) = f'(x)$, which shows that f' is periodic as well.

(b) Yes, let $g(x) = f(2x)$, so $g'(x) = 2f'(2x)$. Since f' is periodic, so is g'.

111. $f(x) = \dfrac{\cos^2 \pi x}{\sqrt{x^2 + 1}}, \quad 0 < x < 4$

(a)

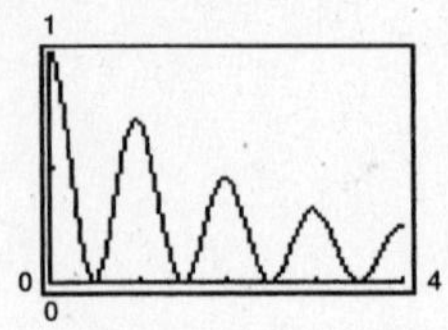

Critical numbers: 0.5, 0.9747, 1.5, 1.9796, 2.5, 2.9848, 3.5

(b) $f'(x) = \dfrac{-\cos \pi x[x \cos \pi x + 2\pi(x^2 + 1)\sin \pi x]}{(x^2 + 1)^{3/2}}$

$f'(x) = 0$ when $x \approx 0.5, 0.9747, 1.5, 1.9796, 2.5, 2.9848, 3.5$

They are the same values as in part (a).

112. $f(x) = \tan(\sin \pi x)$

(a)

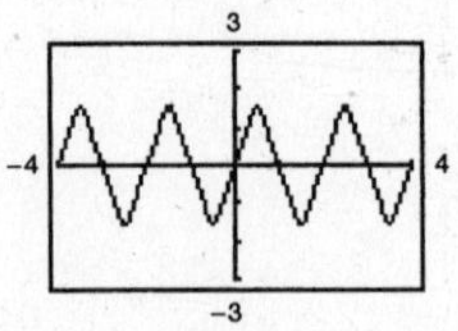

(b) The graph is symmetric about the origin:
$f(-x) = \tan(\sin(-\pi x)) = \tan(-\sin(\pi x)) = -\tan(\sin \pi x) = -f(x)$

(c) Yes, the function is periodic. $f(x + 2) = f(x)$. Period 2.

(d) Absolute maximum at $(0.5, 1.5574)$.

Absolute minimum at $(-0.5, -1.5574)$.

(e) Concave upward on $(0, 0.2837)$ and $(0.7163, 1)$.

Concave downward on $(0.2837, 0.7163)$.

113. False. If $y = (1 - x)^{1/2}$, then $y' = \frac{1}{2}(1 - x)^{-1/2}(-1)$.

114. False. If $f(x) = \sin^2 2x$, then $f'(x) = 2(\sin 2x)(2 \cos 2x)$.

115. False. First apply the Product Rule.

116. False.

$f(x) = 3 \sin x + 2 \cos x$

$f'(x) = 3 \cos x - 2 \sin x$

$3 \cos x - 2 \sin x = 0$

$3 \cos x = 2 \sin x$

$\frac{3}{2} = \tan x$

Critical number: $x = \tan^{-1}\left(\frac{3}{2}\right)$

$f\left(\tan^{-1} \frac{3}{2}\right) \approx 3.60555$ is the maximum value of y.

117. True. $y = \sin(bx)$

Slope: $y' = b \cos(bx)$

$-b \le y' \le b$ (Assume $b > 0$)

118. (a) $\sec x = \dfrac{1}{\cos x}$

$$\frac{d}{dx}[\sec x] = \frac{d}{dx}\left[\frac{1}{\cos x}\right] = \frac{(\cos x)(0) - (1)(-\sin x)}{(\cos x)^2} = \frac{\sin x}{\cos x \cos x} = \frac{1}{\cos x} \cdot \frac{\sin x}{\cos x} = \sec x \tan x$$

(b) $\csc x = \dfrac{1}{\sin x}$

$$\frac{d}{dx}[\csc x] = \frac{d}{dx}\left[\frac{1}{\sin x}\right] = \frac{(\sin x)(0) - (1)(\cos x)}{(\sin x)^2} = -\frac{\cos x}{\sin x \sin x} = -\frac{1}{\sin x} \cdot \frac{\cos x}{\sin x} = -\csc x \cot x$$

(c) $\cot x = \dfrac{\cos x}{\sin x}$

$$\frac{d}{dx}[\cot x] = \frac{d}{dx}\left[\frac{\cos x}{\sin x}\right] = \frac{\sin x(-\sin x) - (\cos x)(\cos x)}{(\sin x)^2} = -\frac{\sin^2 x + \cos^2 x}{\sin^2 x} = -\frac{1}{\sin^2 x} = -\csc^2 x$$

Section 11.3 Trigonometric Functions: Integration

1. $\displaystyle\int (2\sin x + 3\cos x)\,dx = -2\cos x + 3\sin x + C$

2. $\displaystyle\int (t^2 - \sin t)\,dt = \frac{1}{3}t^3 + \cos t + C$

3. $\displaystyle\int (1 - \csc t \cot t)\,dt = t + \csc t + C$

4. $\displaystyle\int (\theta^2 + \sec^2 \theta)\,d\theta = \frac{1}{3}\theta^3 + \tan\theta + C$

5. $\displaystyle\int (\sec^2\theta - \sin\theta)\,d\theta = \tan\theta + \cos\theta + C$

6. $\displaystyle\int \sec y(\tan y - \sec y)\,dy = \int (\sec y \tan y - \sec^2 y)\,dy$

$$= \sec y - \tan y + C$$

7. $\displaystyle\int (\tan^2 y + 1)\,dy = \int \sec^2 y\,dy = \tan y + C$

Check: $\dfrac{d}{dy}(\tan y + C) = \sec^2 y = \tan^2 y + 1$

8. $\displaystyle\int \frac{\cos x}{1 - \cos^2 x}\,dx = \int \frac{\cos x}{\sin^2 x}\,dx = \int \left(\frac{1}{\sin x}\right)\left(\frac{\cos x}{\sin x}\right)dx$

$$= \int \csc x \cot x\,dx = -\csc x + C$$

Check: $\dfrac{d}{dx}[-\csc x + C] = \csc x \cot x = \dfrac{1}{\sin x} \cdot \dfrac{\cos x}{\sin x}$

$$= \frac{\cos x}{1 - \cos^2 x}$$

9. $\displaystyle\int \pi \sin \pi x\,dx = -\cos \pi x + C$

10. $\displaystyle\int 4x^3 \sin x^4\,dx = \int \sin x^4(4x^3)\,dx = -\cos x^4 + C$

11. $\displaystyle\int \sin 2x\,dx = \frac{1}{2}\int (\sin 2x)(2x)\,dx = -\frac{1}{2}\cos 2x + C$

12. $\displaystyle\int \cos 6x\,dx = \frac{1}{6}\int (\cos 6x)(6)\,dx$

$$= \frac{1}{6}\sin 6x + C$$

13. $\displaystyle\int \frac{1}{\theta^2}\cos\frac{1}{\theta}\,d\theta = -\int \cos\frac{1}{\theta}\left(-\frac{1}{\theta^2}\right)d\theta = -\sin\frac{1}{\theta} + C$

14. $\displaystyle\int x \sin x^2\,dx = \frac{1}{2}\int (\sin x^2)(2x)\,dx$

$$= -\frac{1}{2}\cos x^2 + C$$

15. $\int \sin 2x \cos 2x\,dx = \frac{1}{2}\int (\sin 2x)(2\cos 2x)\,dx = \frac{1}{2}\frac{(\sin 2x)^2}{2} + C = \frac{1}{4}\sin^2 2x + C$ OR

$\int \sin 2x \cos 2x\,dx = -\frac{1}{2}\int (\cos 2x)(-2\sin 2x)\,dx = -\frac{1}{2}\frac{(\cos 2x)^2}{2} + C_1 = -\frac{1}{4}\cos^2 2x + C_1$ OR

$\int \sin 2x \cos 2x\,dx = \frac{1}{2}\int 2\sin 2x \cos 2x\,dx = \frac{1}{2}\int \sin 4x\,dx = -\frac{1}{8}\cos 4x + C_2$

16. $\int \sec(1-x)\tan(1-x)\,dx = -\int [\sec(1-x)\tan(1-x)](-1)\,dx = -\sec(1-x) + C$

17. $\int \tan^4 x \sec^2 x\,dx = \frac{\tan^5 x}{5} + C = \frac{1}{5}\tan^5 x + C$

18. $\int \sqrt{\tan x}\sec^2 x\,dx = \frac{(\tan x)^{3/2}}{3/2} + C = \frac{2}{3}(\tan x)^{3/2} + C$

19. $\int \frac{\csc^2 x}{\cot^3 x}\,dx = -\int (\cot x)^{-3}(-\csc^2 x)\,dx$

$= -\frac{(\cot x)^{-2}}{-2} + C = \frac{1}{2\cot^2 x} + C = \frac{1}{2}\tan^2 x + C = \frac{1}{2}(\sec^2 x - 1) + C = \frac{1}{2}\sec^2 x + C_1$

20. $\int \frac{\sin x}{\cos^3 x}\,dx = -\int (\cos x)^{-3}(-\sin x)\,dx = -\frac{(\cos x)^{-2}}{-2} + C = \frac{1}{2\cos^2 x} + C = \frac{1}{2}\sec^2 x + C$

21. $\int \cot^2 x\,dx = \int (\csc^2 x - 1)\,dx = -\cot x - x + C$

22. $\int \csc^2\left(\frac{x}{2}\right)dx = 2\int \csc^2\left(\frac{x}{2}\right)\left(\frac{1}{2}\right)dx = -2\cot\left(\frac{x}{2}\right) + C$

23. Let $u = e^x$, $du = e^x dx$.

$\int e^x \cos(e^x)dx = \int \cos u\,du = \sin u + C = \sin(e^x) + C$

24. Let $u = \sin x$, $du = \cos dx$.

$\int \cos x e^{\sin x}\,dx = e^{\sin x} + C$

25. $\int \frac{\cos\theta}{\sin\theta}\,d\theta = \ln|\sin\theta| + C$

$(u = \sin\theta, du = \cos\theta\,d\theta)$

26. $\int \tan 5\theta\,d\theta = \frac{1}{5}\int \frac{5\sin 5\theta}{\cos 5\theta}\,d\theta$

$= -\frac{1}{5}\ln|\cos 5\theta| + C$

27. $\int \csc 2x\,dx = \frac{1}{2}\int (\csc 2x)(2)\,dx$

$= -\frac{1}{2}\ln|\csc 2x + \cot 2x| + C$

28. $\int \sec\frac{x}{2}\,dx = 2\int \sec\frac{x}{2}\left(\frac{1}{2}\right)dx$

$= 2\ln\left|\sec\frac{x}{2} + \tan\frac{x}{2}\right| + C$

29. $\int \frac{\cos t}{1+\sin t}\,dt = \ln|1 + \sin t| + C$

30. $u = \cot t$, $du = -\csc^2 t\,dt$

$\int \frac{\csc^2 t}{\cot t}dt = -\ln|\cot t| + C$

31. $\int \frac{\sec x \tan x}{\sec x - 1}\,dx = \ln|\sec x - 1| + C$

32. $\int (\sec t + \tan t)\,dt = \ln|\sec t + \tan t| - \ln|\cos t| + C$

$= \ln\left|\frac{\sec t + \tan t}{\cos t}\right| + C$

$= \ln|\sec t(\sec t + \tan t)| + C$

33. $$\int_0^{\pi} (1 + \sin x)\,dx = \Big[x - \cos x\Big]_0^{\pi}$$
$$= (\pi + 1) - (0 - 1)$$
$$= 2 + \pi$$

34. $$\int_0^{\pi/4} \frac{1 - \sin^2 \theta}{\cos^2 \theta}\,d\theta = \int_0^{\pi/4} d\theta = \Big[\theta\Big]_0^{\pi/4} = \frac{\pi}{4}$$

35. $$\int_{-\pi/6}^{\pi/6} \sec^2 x\,dx = \Big[\tan x\Big]_{-\pi/6}^{\pi/6} = \frac{\sqrt{3}}{3} - \left(-\frac{\sqrt{3}}{3}\right) = \frac{2\sqrt{3}}{3}$$

36. $$\int_{\pi/4}^{\pi/2} (2 - \csc^2 x)\,dx = \Big[2x + \cot x\Big]_{\pi/4}^{\pi/2} = (\pi + 0) - \left(\frac{\pi}{2} + 1\right) = \frac{\pi}{2} - 1 = \frac{\pi - 2}{2}$$

37. $$\int_{-\pi/3}^{\pi/3} 4 \sec \theta \tan \theta\,d\theta = \Big[4 \sec \theta\Big]_{-\pi/3}^{\pi/3} = 4(2) - 4(2) = 0$$

38. $$\int_{-\pi/2}^{\pi/2} (2t + \cos t)\,dt = \Big[t^2 + \sin t\Big]_{-\pi/2}^{\pi/2} = \left(\frac{\pi^2}{4} + 1\right) - \left(\frac{\pi^2}{4} - 1\right) = 2$$

39.

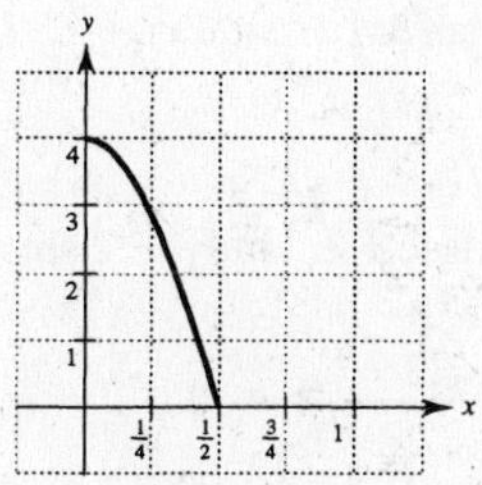

b. $A \approx \frac{4}{3}$ square units

40.

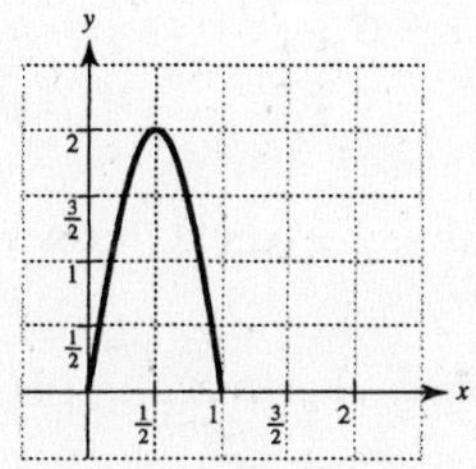

d. $\int_0^1 2 \sin \pi x\,dx \approx \frac{5}{4}$

41. (a)

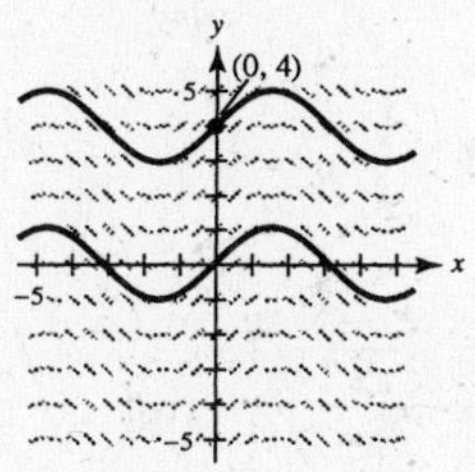

(b) $y = \int \cos x\,dx = \sin x + C$

$4 = \sin 0 + C \implies C = 4$

$y = \sin x + 4$

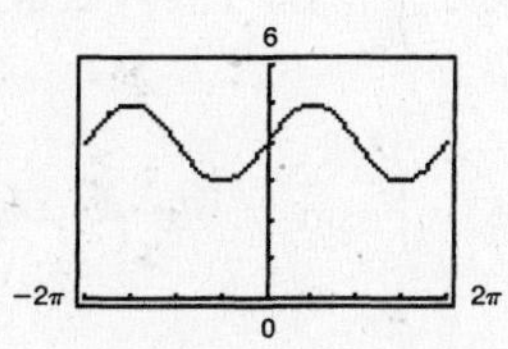

42. (a)

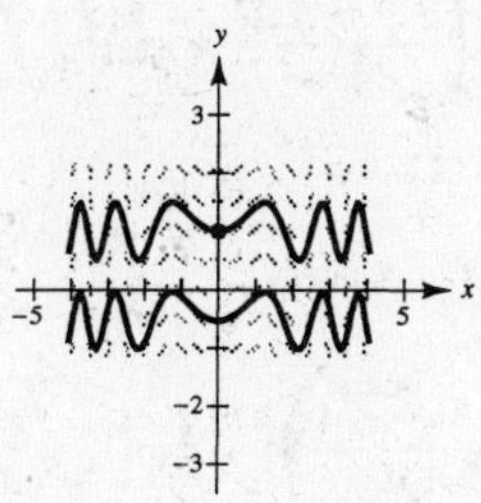

(b) $y = \int x \cos x^2\,dx = \frac{1}{2}\int 2x \cos x^2\,dx$

$= \frac{1}{2} \sin x^2 + C$

$1 = \frac{1}{2} \sin 0 + C = C$

$y = \frac{1}{2} \sin x^2 + 1$

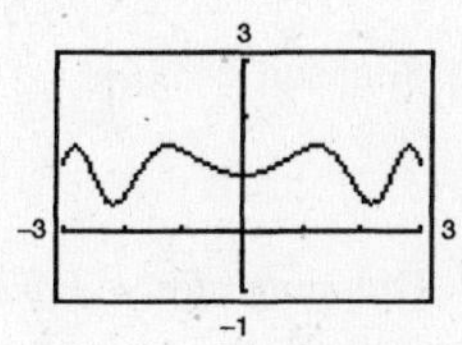

43. $s = \int \tan(2\theta)\, d\theta$

$= \frac{1}{2}\int \tan(2\theta)(2\, d\theta)$

$= -\frac{1}{2}\ln|\cos 2\theta| + C$

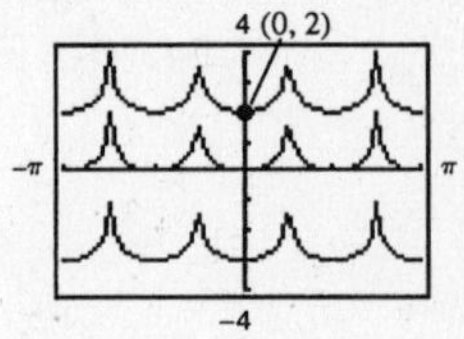

$(0, 2)$: $2 = -\frac{1}{2}\ln|\cos(0)| + C \Rightarrow C = 2$

$s = -\frac{1}{2}\ln|\cos 2\theta| + 2$

44. $r = \int \frac{\sec^2 t}{\tan t + 1}\, dt$

$= \ln|\tan t + 1| + C$

$(\pi, 4)$: $4 = \ln|0 + 1| + C \Rightarrow C = 4$

$r = \ln|\tan t + 1| + 4$

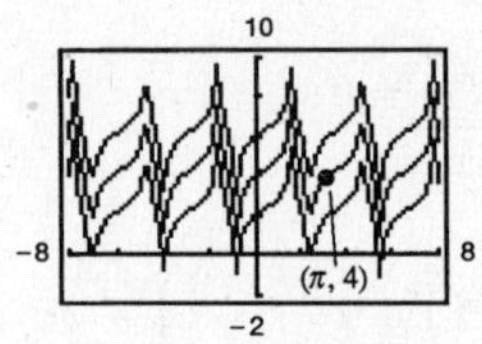

45. $\int \cos(1 - x)\, dx = -\sin(1 - x) + C$

46. $\int \frac{\tan^2 2x}{\sec 2x}\, dx = \frac{1}{2}[\ln|\sec 2x + \tan 2x| - \sin 2x] + C$

47. $\int_{\pi/4}^{\pi/2} (\csc x - \sin x)\, dx = \Big[-\ln|\csc x + \cot x| + \cos x\Big]_{\pi/4}^{\pi/2}$

$= \ln(\sqrt{2} + 1) - \frac{\sqrt{2}}{2} \approx 0.174$

48. $\int_{-\pi/4}^{\pi/4} \frac{\sin^2 x - \cos^2 x}{\cos x}\, dx = \Big[\ln|\sec x + \tan x| - 2\sin x\Big]_{-\pi/4}^{\pi/4}$

$= \ln\left(\frac{\sqrt{2} + 1}{\sqrt{2} - 1}\right) - 2\sqrt{2} \approx -1.066$

49. $\int_0^{\pi/2} \sin^2 x\, dx \approx 0.7854 \left(= \frac{\pi}{4}\right)$

50. $\int_0^3 x \sin x\, dx \approx 3.1111$

51. $\int_0^4 \sin \sqrt{x}\, dx \approx 3.4832$

52. $\int_1^2 \frac{\sin x}{x}\, dx \approx 0.6593$

53. Trapezoidal: $\int_0^{\sqrt{\pi/2}} \cos(x^2)\, dx \approx \frac{\sqrt{\pi/2}}{8}\cos 0 + 2\cos\left(\frac{\sqrt{\pi/2}}{4}\right)^2 + 2\cos\left(\frac{\sqrt{\pi/2}}{2}\right)^2 + 2\cos\left(\frac{3\sqrt{\pi/2}}{4}\right)^2 + \cos\left(\sqrt{\frac{\pi}{2}}\right)^2$

≈ 0.957

Simpson's: $\int_0^{\sqrt{\pi/2}} \cos(x^2)\, dx \approx \frac{\sqrt{\pi/2}}{12}\left[\cos 0 + 4\cos\left(\frac{\sqrt{\pi/2}}{4}\right)^2 + 2\cos\left(\frac{\sqrt{\pi/2}}{2}\right)^2 + 4\cos\left(\frac{3\sqrt{\pi/2}}{4}\right)^2 + \cos\left(\sqrt{\frac{\pi}{2}}\right)^2\right]$

≈ 0.978

Graphing utility: 0.977

54. Trapezoidal: $\int_0^{\sqrt{\pi/4}} \tan(x^2)\, dx \approx \frac{\sqrt{\pi/4}}{8}\left[\tan 0 + 2\tan\left(\frac{\sqrt{\pi/4}}{4}\right)^2 + 2\tan\left(\frac{\sqrt{\pi/4}}{2}\right)^2 + 2\tan\left(\frac{3\sqrt{\pi/4}}{4}\right)^2 + \tan\left(\sqrt{\frac{\pi}{4}}\right)^2\right]$

≈ 0.271

Simpson's: $\int_0^{\sqrt{\pi/4}} \tan(x^2)\, dx \approx \frac{\sqrt{\pi/4}}{12}\left[\tan 0 + 4\tan\left(\frac{\sqrt{\pi/4}}{4}\right)^2 + 2\tan\left(\frac{\sqrt{\pi/4}}{2}\right)^2 + 4\tan\left(\frac{3\sqrt{\pi/4}}{4}\right)^2 + \tan\left(\sqrt{\frac{\pi}{4}}\right)^2\right]$

≈ 0.257

Graphing utility: 0.256

55. Trapezoidal: $\int_1^{1.1} \sin x^2\,dx \approx \frac{1}{80}[\sin(1) + 2\sin(1.025)^2 + 2\sin(1.05)^2 + 2\sin(1.075)^2 + \sin(1.1)^2] \approx 0.089$

Simpson's: $\int_1^{1.1} \sin x^2\,dx \approx \frac{1}{120}[\sin(1) + 4\sin(1.025)^2 + 2\sin(1.05)^2 + 4\sin(1.075)^2 + \sin(1.1)^2] \approx 0.089$

Graphing utility: 0.089

56. Trapezoidal: $\int_0^{\pi/2} \sqrt{1+\cos^2 x}\,dx \approx \frac{\pi}{16}\left[\sqrt{2} + 2\sqrt{1+\cos^2(\pi/8)} + 2\sqrt{1+\cos^2(\pi/4)} + 2\sqrt{1+\cos^2(3\pi/8)} + 1\right] \approx 1.910$

Simpson's: $\int_0^{\pi/2} \sqrt{1+\cos^2 x}\,dx \approx \frac{\pi}{24}\left[\sqrt{2} + 4\sqrt{1+\cos^2(\pi/8)} + 2\sqrt{1+\cos^2(\pi/4)} + 4\sqrt{1+\cos^2(3\pi/8)} + 1\right] \approx 1.910$

Graphing utility: 1.910

57. Trapezoidal: $\int_0^{\pi/4} x\tan x\,dx \approx \frac{\pi}{32}\left[0 + 2\left(\frac{\pi}{16}\right)\tan\left(\frac{\pi}{16}\right) + 2\left(\frac{2\pi}{16}\right)\tan\left(\frac{2\pi}{16}\right) + 2\left(\frac{3\pi}{16}\right)\tan\left(\frac{3\pi}{16}\right) + \frac{\pi}{4}\right] \approx 0.194$

Simpson's: $\int_0^{\pi/4} x\tan x\,dx \approx \frac{\pi}{48}\left[0 + 4\left(\frac{\pi}{16}\right)\tan\left(\frac{\pi}{16}\right) + 2\left(\frac{2\pi}{16}\right)\tan\left(\frac{2\pi}{16}\right) + 4\left(\frac{3\pi}{16}\right)\tan\left(\frac{3\pi}{16}\right) + \frac{\pi}{4}\right] \approx 0.186$

Graphing utility: 0.186

58. Trapezoidal: $\int_0^{\pi} \frac{\sin x}{x}\,dx \approx \frac{\pi}{8}\left[1 + \frac{2\sin(\pi/4)}{\pi/4} + \frac{2\sin(\pi/2)}{\pi/2} + \frac{2\sin(3\pi/4)}{3\pi/4} + 0\right] \approx 1.836$

Simpson's: $\int_0^{\pi} \frac{\sin x}{x}\,dx \approx \frac{\pi}{12}\left[1 + \frac{4\sin(\pi/4)}{\pi/4} + \frac{2\sin(\pi/2)}{\pi/2} + \frac{4\sin(3\pi/4)}{3\pi/4} + 0\right] \approx 1.852$

Graphing utility: 1.852

59. $A = \int_0^{\pi/2} \cos x\,dx = \left[\sin x\right]_0^{\pi/2} = 1$

60. $A = \int_0^{\pi} (x + \sin x)\,dx$

$= \left[\frac{x^2}{2} - \cos x\right]_0^{\pi} = \frac{\pi^2}{2} + 2 = \frac{\pi^2 + 4}{2}$

61. $A = \int_0^{\pi} (2\sin x + \sin 2x)dx$

$= -\left[2\cos x + \frac{1}{2}\cos 2x\right]_0^{\pi} = 4$

62. $A = \int_0^{\pi} (\sin x + \cos 2x)\,dx$

$= \left[-\cos x + \frac{1}{2}\sin 2x\right]_0^{\pi} = 2$

63. Area $= \int_{\pi/2}^{2\pi/3} \sec^2\left(\frac{x}{2}\right)dx = 2\int_{\pi/2}^{2\pi/3} \sec^2\left(\frac{x}{2}\right)\left(\frac{1}{2}\right)dx = \left[2\tan\left(\frac{x}{2}\right)\right]_{\pi/2}^{2\pi/3} = 2(\sqrt{3} - 1)$

64. Let $u = 2x$, $du = 2\,dx$.

Area $= \int_{\pi/12}^{\pi/4} \csc 2x\cot 2x\,dx = \frac{1}{2}\int_{\pi/12}^{\pi/4} \csc 2x\cot 2x(2)\,dx = \left[-\frac{1}{2}\csc 2x\right]_{\pi/12}^{\pi/4} = \frac{1}{2}$

65. $\int_{-\pi/4}^{\pi/4} \sin x\,dx = 0$ since $\sin x$ is symmetric to the origin.

66. $\int_{-\pi/4}^{\pi/4} \cos x\,dx = 2\int_0^{\pi/4} \cos x\,dx = 2\left[\sin x\right]_0^{\pi/4} = \sqrt{2}$, since $\cos x$ is symmetric to the y-axis.

67. $f(x) = \sin^2 x \cos x$ is even.

$$\int_{-\pi/2}^{\pi/2} \sin^2 x \cos x \, dx = 2\int_0^{\pi/2} \sin^2 x(\cos x) \, dx = 2\left[\frac{\sin^3 x}{3}\right]_0^{\pi/2} = \frac{2}{3}$$

68. $\int_{-\pi/2}^{\pi/2} \sin x \cos x \, dx = 0$ since $\sin(-x)\cos(-x) = -\sin x \cos x$ and hence is symmetric to the origin.

69. $\int_{-\pi/4}^{\pi/4} 2\sec^2 x \, dx = \left[2\tan x\right]_{-\pi/4}^{\pi/4} = 2(1) - 2(-1) = 4$

$$f(c)\left[\frac{\pi}{4} - \left(-\frac{\pi}{4}\right)\right] = 4$$

$$2\sec^2 c = \frac{8}{\pi}$$

$$\sec^2 c = \frac{4}{\pi}$$

$$\sec c = \pm\frac{2}{\sqrt{\pi}}$$

$$c = \pm\operatorname{arcsec}\left(\frac{2}{\sqrt{\pi}}\right)$$

$$= \pm\arccos\frac{\sqrt{\pi}}{2} \approx \pm 0.4817$$

70. $\int_{-\pi/3}^{\pi/3} \cos x \, dx = \left[\sin x\right]_{-\pi/3}^{\pi/3} = \sqrt{3}$

$$f(c)\left[\frac{\pi}{3} - \left(-\frac{\pi}{3}\right)\right] = \sqrt{3}$$

$$\cos c = \frac{3\sqrt{3}}{2\pi}$$

$$c \approx \pm 0.5971$$

71. $\frac{1}{\pi - 0}\int_0^{\pi} \sin x \, dx = \left[-\frac{1}{\pi}\cos x\right]_0^{\pi} = \frac{2}{\pi}$

$$\text{Average value} = \frac{2}{\pi}$$

$$\sin x = \frac{2}{\pi}$$

$$x \approx 0.690, 2.451$$

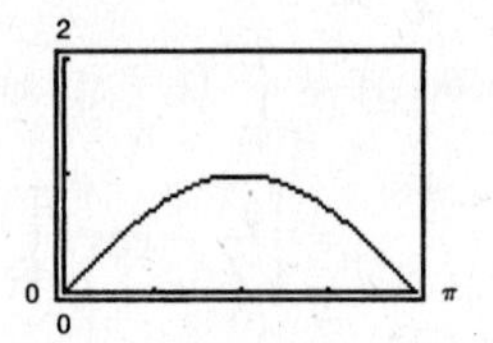

72. $\frac{1}{(\pi/2) - 0}\int_0^{\pi/2} \cos x \, dx = \left[\frac{2}{\pi}\sin x\right]_0^{\pi/2} = \frac{2}{\pi}$

$$\text{Average value} = \frac{2}{\pi}$$

$$\cos x = \frac{2}{\pi}$$

$$x \approx 0.881$$

73. $F(x) = \int_0^x t\cos t \, dt$

$F'(x) = x\cos x$

74. $F(x) = \int_0^x \sin t^2 \, dt$

$F'(x) = \sin x^2$

75. $F(x) = \int_{\pi/4}^x \sec^2 t \, dt$

$F'(x) = \sec^2 x$

76. $F(x) = \int_{\pi/3}^x \sec t \tan t \, dt$

$F'(x) = \sec x \tan x$

77. No, $\tan x$ is not continuous at $x = \frac{\pi}{2}$.

78. If $u = \pi x$, $du = \pi \, dx$ and $\int \sec^2(\pi x) \, dx = \int \sec^2(u) \frac{du}{\pi}$.

79. $f(x) = x\cos x$ is odd and $[-\pi, \pi]$ is centered at 0.

80. The result will be less than $\int_0^{\pi} \sin x \, dx = 2$ because the graph is concave downward.

81. (a)

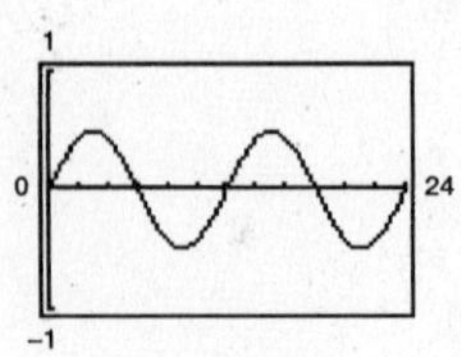

The area above the x-axis equals the area below the x-axis. Thus, the average value is zero.

(b)

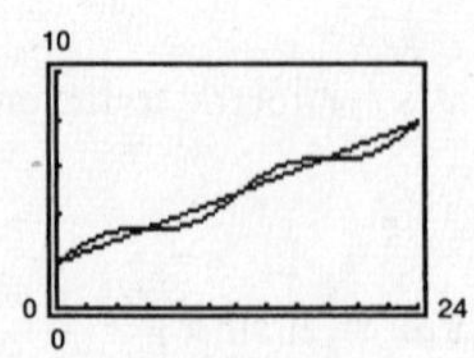

The average value of S appears to be g.

82. $R = 3.121 + 2.399\sin(0.524t + 1.377)$

(a) 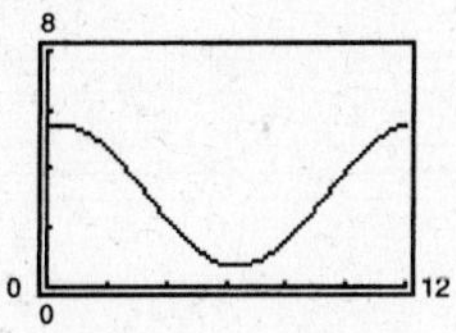

Relative minimum: (6.4, 0.7) or June

Relative maximum: (0.4, 5.5) or January

(b) $\int_0^{12} R(t)\,dt \approx 37.47$ inches

(c) $\frac{1}{3}\int_9^{12} R(t)\,dt \approx \frac{1}{3}(13) = 4.33$ inches

83. $\frac{1}{b-a}\int_a^b \left[74.50 + 43.75\sin\frac{\pi t}{6}\right] dt = \frac{1}{b-a}\left[74.50t - \frac{262.5}{\pi}\cos\frac{\pi t}{6}\right]_a^b$

(a) $\frac{1}{3}\left[74.50t - \frac{262.5}{\pi}\cos\frac{\pi t}{6}\right]_0^3 = \frac{1}{3}\left(223.5 + \frac{262.5}{\pi}\right) \approx 102.352$ thousand units

(b) $\frac{1}{3}\left[74.50t - \frac{262.5}{\pi}\cos\frac{\pi t}{6}\right]_3^6 = \frac{1}{3}\left(447 + \frac{262.5}{\pi} - 223.5\right) \approx 102.352$ thousand units

(c) $\frac{1}{12}\left[74.50t - \frac{262.5}{\pi}\cos\frac{\pi t}{6}\right]_0^{12} = \frac{1}{12}\left(894 - \frac{262.5}{\pi} + \frac{262.5}{\pi}\right) = 74.5$ thousand units

84. (a) 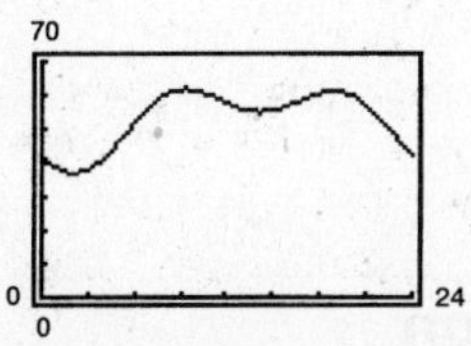

Maximum flow: $R \approx 61.713$ at $t = 9.36$.
[(18.861, 61.177) is a relative maximum.]

(b) Volume $= \int_0^{24} R(t)\,dt \approx 1272$ (thousands of gallons)

85. $\frac{1}{b-a}\int_a^b [2\sin(60\pi t) + \cos(120\pi t)]\,dt = \frac{1}{b-a}\left[-\frac{1}{30\pi}\cos(60\pi t) + \frac{1}{120\pi}\sin(120\pi t)\right]_a^b$

(a) $\frac{1}{(1/60) - 0}\left[-\frac{1}{30\pi}\cos(60\pi t) + \frac{1}{120\pi}\sin(120\pi t)\right]_0^{1/60} = 60\left[\left(\frac{1}{30\pi} + 0\right) - \left(-\frac{1}{30\pi}\right)\right] = \frac{4}{\pi} \approx 1.273$ amps

(b) $\frac{1}{(1/240) - 0}\left[-\frac{1}{30\pi}\cos(60\pi t) + \frac{1}{120\pi}\sin(120\pi t)\right]_0^{1/240} = 240\left[\left(-\frac{1}{30\sqrt{2}\pi} + \frac{1}{120\pi}\right) - \left(-\frac{1}{30\pi}\right)\right]$

$= \frac{2}{\pi}(5 - 2\sqrt{2}) \approx 1.382$ amps

(c) $\frac{1}{(1/30) - 0}\left[-\frac{1}{30\pi}\cos(60\pi t) + \frac{1}{120\pi}\sin(120\pi t)\right]_0^{1/30} = 30\left[\left(\frac{-1}{30\pi}\right) - \left(-\frac{1}{30\pi}\right)\right] = 0$ amps

86. $\int_0^t \sin\sqrt{x}\,dx = 2,\ n = 10$

By trial and error, we obtain $t \approx 2.477$.

87. $-\ln|\cos x| + C = \ln\left|\frac{1}{\cos x}\right| + C = \ln|\sec x| + C$

88. $\ln|\sin x| + C = \ln\left|\frac{1}{\csc x}\right| + C = -\ln|\csc x| + C$

89. $\ln|\sec x + \tan x| + C = \ln\left|\dfrac{(\sec x + \tan x)(\sec x - \tan x)}{(\sec x - \tan x)}\right| + C$

$$= \ln\left|\frac{\sec^2 x - \tan^2 x}{\sec x - \tan x}\right| + C = \ln\left|\frac{1}{\sec x - \tan x}\right| + C = -\ln|\sec x - \tan x| + C$$

90. $-\ln|\csc x + \cot x| + C = -\ln\left|\dfrac{(\csc x + \cot x)(\csc x - \cot x)}{(\csc x - \cot x)}\right| + C$

$$= -\ln\left|\frac{\csc^2 x - \cot^2 x}{\csc x - \cot x}\right| + c = -\ln\left|\frac{1}{\csc x - \cot x}\right| + C = \ln|\csc x - \cot x| + C$$

91. True

$$\int_a^b \sin x\,dx = \Big[-\cos x\Big]_a^b = -\cos b + \cos a = -\cos(b + 2\pi) + \cos a = \int_a^{b+2\pi} \sin x\,dx$$

92. True

$$4\int \sin x \cos x\,dx = 2\int \sin 2x\,dx = -\cos 2x + C$$

93. False

$$\int \sin^2 2x \cos 2x\,dx = \frac{1}{2}\int (\sin 2x)^2(2\cos 2x)\,dx = \frac{1}{2}\frac{(\sin 2x)^3}{3} + C = \frac{1}{6}\sin^3 2x + C$$

Section 11.4 Inverse Trigonometric Functions: Differentiation

1. $y = \arcsin x$

$$y' = \frac{1}{\sqrt{1 - x^2}}$$

At $x = 0$, $y' = 1$.

2. $y = \arcsin 2x$

$$y' = \frac{2}{\sqrt{1 - 4x^2}}$$

At $x = 0$, $y' = 2$.

3. $y = \arcsin \dfrac{x}{2}$

$$y' = \frac{1/2}{\sqrt{1 - x^2/4}}$$

At $x = 0$, $y' = \dfrac{1}{2}$.

4. $y = \arcsin \dfrac{x}{3}$

$$y' = \frac{1/3}{\sqrt{1 - x^2/9}}$$

At $x = 0$, $y' = \dfrac{1}{3}$.

5. $f(x) = 2\arcsin(x - 1)$

$$f'(x) = \frac{2}{\sqrt{1 - (x - 1)^2}} = \frac{2}{\sqrt{2x - x^2}}$$

6. $f(t) = \arcsin t^2$

$$f'(t) = \frac{2t}{\sqrt{1 - t^4}}$$

7. $g(x) = 3\arccos \dfrac{x}{2}$

$$g'(x) = \frac{-3(1/2)}{\sqrt{1 - (x^2/4)}} = \frac{-3}{\sqrt{4 - x^2}}$$

8. $f(x) = \operatorname{arcsec} 2x$

$$f'(x) = \frac{2}{|2x|\sqrt{4x^2 - 1}} = \frac{1}{|x|\sqrt{4x^2 - 1}}$$

9. $f(x) = \arctan \dfrac{x}{a}$

$$f'(x) = \frac{1/a}{1 + (x^2/a^2)} = \frac{a}{a^2 + x^2}$$

10. $f(x) = \arctan \sqrt{x}$

$$f'(x) = \left(\frac{1}{1 + x}\right)\left(\frac{1}{2\sqrt{x}}\right) = \frac{1}{2\sqrt{x}(1 + x)}$$

11. $g(x) = \dfrac{\arcsin 3x}{x}$

$$g'(x) = \frac{x\left(3/\sqrt{1-9x^2}\right) - \arcsin 3x}{x^2}$$

$$= \frac{3x - \sqrt{1-9x^2}\arcsin 3x}{x^2\sqrt{1-9x^2}}$$

12. $h(x) = x^2 \arctan x$

$$h'(x) = 2x \arctan x + \frac{x^2}{1+x^2}$$

13. $h(t) = \sin(\arccos t) = \sqrt{1-t^2}$

$$h'(t) = \frac{1}{2}(1-t^2)^{-1/2}(-2t) = \frac{-t}{\sqrt{1-t^2}}$$

14. $f(x) = \arcsin x + \arccos x = \dfrac{\pi}{2}$

$$f'(x) = 0$$

15. $y = x \arccos x - \sqrt{1-x^2}$

$$y' = \arccos x - \frac{x}{\sqrt{1-x^2}} - \frac{1}{2}(1-x^2)^{-1/2}(-2x)$$

$$= \arccos x$$

16. $y = \ln(t^2+4) - \dfrac{1}{2}\arctan\dfrac{t}{2}$

$$y' = \frac{2t}{t^2+4} - \frac{1}{2}\cdot\frac{1}{1+\left(\frac{t}{2}\right)^2}\left(\frac{1}{2}\right)$$

$$= \frac{2t}{t^2+4} - \frac{1}{t^2+4} = \frac{2t-1}{t^2+4}$$

17. $y = \dfrac{1}{2}\left(\dfrac{1}{2}\ln\dfrac{x+1}{x-1} + \arctan x\right)$

$$= \frac{1}{4}[\ln(x+1) - \ln(x-1)] + \frac{1}{2}\arctan x$$

$$\frac{dy}{dx} = \frac{1}{4}\left(\frac{1}{x+1} - \frac{1}{x-1}\right) + \frac{1/2}{1+x^2} = \frac{1}{1-x^4}$$

18. $y = \dfrac{1}{2}\left[x\sqrt{4-x^2} + 4\arcsin\left(\dfrac{x}{2}\right)\right]$

$$y' = \frac{1}{2}\left[x\frac{1}{2}(4-x^2)^{-1/2}(-2x) + \sqrt{4-x^2} + 2\frac{1}{\sqrt{1-(x/2)^2}}\right]$$

$$= \frac{1}{2}\left[\frac{-x^2}{\sqrt{4-x^2}} + \sqrt{4-x^2} + \frac{4}{\sqrt{4-x^2}}\right]$$

$$= \sqrt{4-x^2}$$

19. $y = x\arcsin x + \sqrt{1-x^2}$

$$\frac{dy}{dx} = x\left(\frac{1}{\sqrt{1-x^2}}\right) + \arcsin x - \frac{x}{\sqrt{1-x^2}} = \arcsin x$$

20. $y = x\arctan 2x - \dfrac{1}{4}\ln(1+4x^2)$

$$\frac{dy}{dx} = \frac{2x}{1+4x^2} + \arctan(2x) - \frac{1}{4}\left(\frac{8x}{1+4x^2}\right) = \arctan(2x)$$

21. $y = 8\arcsin\dfrac{x}{4} - \dfrac{x\sqrt{16-x^2}}{2}$

$$y' = 2\frac{1}{\sqrt{1-(x/4)^2}} - \frac{\sqrt{16-x^2}}{2} - \frac{x}{4}(16-x^2)^{-1/2}(-2x)$$

$$= \frac{8}{\sqrt{16-x^2}} - \frac{\sqrt{16-x^2}}{2} + \frac{x^2}{2\sqrt{16-x^2}}$$

$$= \frac{16-(16-x^2)+x^2}{2\sqrt{16-x^2}} = \frac{x^2}{\sqrt{16-x^2}}$$

22. $y = 25\arcsin\dfrac{x}{5} - x\sqrt{25-x^2}$

$$y' = 5\frac{1}{\sqrt{1-(x/5)^2}} - \sqrt{25-x^2} - x\frac{1}{2}(25-x^2)^{-1/2}(-2x)$$

$$= \frac{25}{\sqrt{25-x^2}} - \frac{(25-x^2)}{\sqrt{25-x^2}} + \frac{x^2}{\sqrt{25-x^2}}$$

$$= \frac{2x^2}{\sqrt{25-x^2}}$$

23. $y = \arctan x + \dfrac{x}{1+x^2}$

$$y' = \frac{1}{1+x^2} + \frac{(1+x^2) - x(2x)}{(1+x^2)^2}$$
$$= \frac{(1+x^2) + (1-x^2)}{(1+x^2)^2}$$
$$= \frac{2}{(1+x^2)^2}$$

24. $y = \arctan\dfrac{x}{2} - \dfrac{1}{2(x^2+4)}$

$$y' = \frac{1}{2}\frac{1}{1+(x/2)^2} + \frac{1}{2}(x^2+4)^{-2}(2x)$$
$$= \frac{2}{x^2+4} + \frac{x}{(x^2+4)^2}$$
$$= \frac{2x^2+8+x}{(x^2+4)^2}$$

25. $f(x) = 2\arcsin x, \left(\dfrac{1}{2}, \dfrac{\pi}{3}\right)$

$$f'(x) = \frac{2}{\sqrt{1-x^2}}$$
$$f'\left(\frac{1}{2}\right) = \frac{2}{\sqrt{1-1/4}} = \frac{4}{\sqrt{3}} = \frac{4\sqrt{3}}{3}$$

Tangent line: $y - \dfrac{\pi}{3} = \dfrac{4\sqrt{3}}{3}\left(x - \dfrac{1}{2}\right)$

$$y = \frac{4\sqrt{3}}{3}x + \frac{\pi}{3} - \frac{2\sqrt{3}}{3}$$

26. $f(x) = \dfrac{1}{2}\arccos x, \left(\dfrac{-\sqrt{2}}{2}, \dfrac{3\pi}{8}\right)$

$$f'(x) = \frac{-1}{2\sqrt{1-x^2}}$$
$$f'\left(\frac{-\sqrt{2}}{2}\right) = \frac{-1}{2\sqrt{1/2}} = \frac{-\sqrt{2}}{2}$$

Tangent line: $y - \dfrac{3\pi}{8} = \dfrac{-\sqrt{2}}{2}\left(x + \dfrac{\sqrt{2}}{2}\right)$

$$y = \frac{-\sqrt{2}}{2}x + \frac{3\pi}{8} - \frac{1}{2}$$

27. $f(x) = \arcsin x, a = \dfrac{1}{2}$

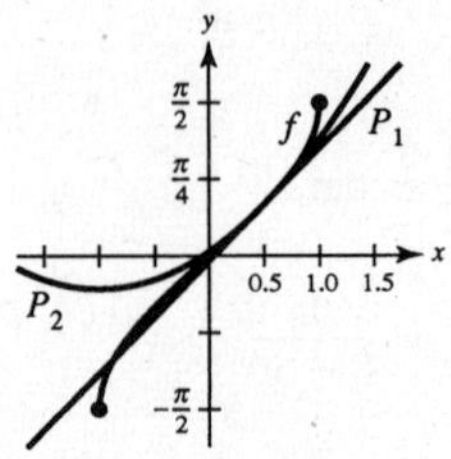

$$f'(x) = \frac{1}{\sqrt{1-x^2}}$$
$$f''(x) = \frac{x}{(1-x^2)^{3/2}}$$
$$P_1(x) = f\left(\frac{1}{2}\right) + f'\left(\frac{1}{2}\right)\left(x - \frac{1}{2}\right) = \frac{\pi}{6} + \frac{2\sqrt{3}}{3}\left(x - \frac{1}{2}\right)$$
$$P_2(x) = f\left(\frac{1}{2}\right) + f'\left(\frac{1}{2}\right)\left(x - \frac{1}{2}\right) + \frac{1}{2}f''\left(\frac{1}{2}\right)\left(x - \frac{1}{2}\right)^2 = \frac{\pi}{6} + \frac{2\sqrt{3}}{3}\left(x - \frac{1}{2}\right) + \frac{2\sqrt{3}}{9}\left(x - \frac{1}{2}\right)^2$$

28. $f(x) = \arctan x, a = 1$

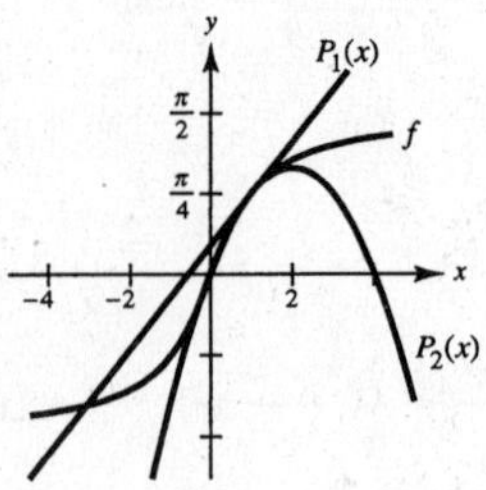

$$f'(x) = \frac{1}{1+x^2}$$
$$f''(x) = \frac{-2x}{(1+x^2)^2}$$
$$P_1(x) = f(1) + f'(1)(x-1) = \frac{\pi}{4} + \frac{1}{2}(x-1)$$
$$P_2(x) = f(1) + f'(1)(x-1) + \frac{1}{2}f''(1)(x-1)^2 = \frac{\pi}{4} + \frac{1}{2}(x-1) - \frac{1}{4}(x-1)^2$$

29. $f(x) = \text{arcsec } x - x$

$f'(x) = \dfrac{1}{|x|\sqrt{x^2 - 1}} - 1$

$= 0$ when $|x|\sqrt{x^2 - 1} = 1$.

$x^2(x^2 - 1) = 1$

$x^4 - x^2 - 1 = 0$ when $x^2 = \dfrac{1 + \sqrt{5}}{2}$ or

$x = \pm\sqrt{\dfrac{1 + \sqrt{5}}{2}} = \pm 1.272.$

Relative maximum: $(1.272, -0.606)$

Relative minimum: $(-1.272, 3.747)$

30. $f(x) = \arcsin x - 2x$

$f'(x) = \dfrac{1}{\sqrt{1 - x^2}} - 2 = 0$ when $\sqrt{1 - x^2} = \dfrac{1}{2}$ or

$x = \pm\dfrac{\sqrt{3}}{2}.$

$f''(x) = \dfrac{x}{(1 - x^2)^{3/2}}$

$f''\left(\dfrac{\sqrt{3}}{2}\right) > 0$

Relative minimum: $\left(\dfrac{\sqrt{3}}{2}, -0.68\right)$

$f''\left(-\dfrac{\sqrt{3}}{2}\right) < 0$

Relative maximum: $\left(-\dfrac{\sqrt{3}}{2}, 0.68\right)$

31. $f(x) = \arctan x - \arctan(x - 4)$

$f'(x) = \dfrac{1}{1 + x^2} - \dfrac{1}{1 + (x - 4)^2} = 0$

$1 + x^2 = 1 + (x - 4)^2$

$0 = -8x + 16$

$x = 2$

By the First Derivative Test, $(2, 2.214)$ is a relative maximum.

32. $h(x) = \arcsin x - 2 \arctan x$

$h'(x) = \dfrac{1}{\sqrt{1 - x^2}} - \dfrac{2}{1 + x^2} = 0$

$1 + x^2 = 2\sqrt{1 - x^2}$

$1 + 2x^2 + x^4 = 4(1 - x^2)$

$x^4 + 6x^2 - 3 = 0$

$x = \pm 0.681$

By the First Derivative Test, $(-0.681, 0.447)$ is a relative maximum and $(0.681, -0.447)$ is a relative minimum.

33. $f(x) = \arcsin x$ is an increasing function.

34. The graph of $y = \text{arccot } x$ is decreasing.

35. Since $\tan(\arctan x) = x$, $\dfrac{d}{dx}[\tan(\arctan x)] = \dfrac{d}{dx}(x) = 1.$

36. The derivatives are algebraic. See Theorem 11.6.

37. (a) $\cot \theta = \dfrac{x}{5}$

$\theta = \text{arccot}\left(\dfrac{x}{5}\right)$

(b) $\dfrac{d\theta}{dt} = \dfrac{-\frac{1}{5}}{1 + \left(\frac{x}{5}\right)^2}\dfrac{dx}{dt} = \dfrac{-5}{x^2 + 25}\dfrac{dx}{dt}$

If $\dfrac{dx}{dt} = -400$ and $x = 10$, $\dfrac{d\theta}{dt} = 16$ rad/hr.

If $\dfrac{dx}{dt} = -400$ and $x = 3$, $\dfrac{d\theta}{dt} \approx 58.824$ rad/hr.

38. (a) $\cot \theta = \dfrac{x}{3}$

$\theta = \text{arccot}\left(\dfrac{x}{3}\right)$

(b) $\dfrac{d\theta}{dt} = \dfrac{-3}{x^2 + 9}\dfrac{dx}{dt}$

If $x = 10$, $\dfrac{d\theta}{dt} \approx 11.01$ rad/hr.

If $x = 3$, $\dfrac{d\theta}{dt} \approx 66.667$ rad/hr.

A lower altitude results in a greater rate of change of θ.

39. (a) $h(t) = -16t^2 + 256$

$-16t^2 + 256 = 0$ when $t = 4$ sec.

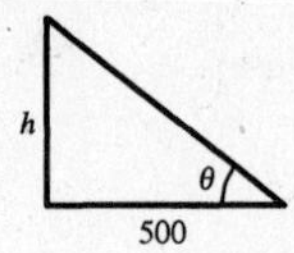

(b) $\tan\theta = \dfrac{h}{500} = \dfrac{-16t^2 + 256}{500}$

$$\theta = \arctan\left[\frac{16}{500}(-t^2 + 16)\right]$$

$$\frac{d\theta}{dt} = \frac{-8t/125}{1 + [(4/125)(-t^2 + 16)]^2} = \frac{-1000t}{15{,}625 + 16(16 - t^2)^2}$$

When $t = 1$, $d\theta/dt \approx -0.0520$ rad/sec.

When $t = 2$, $d\theta/dt \approx -0.1116$ rad/sec.

40. $\cos\theta = \dfrac{750}{s}$

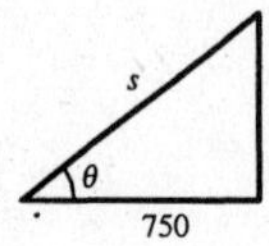

$$\theta = \arccos\left(\frac{750}{s}\right)$$

$$\frac{d\theta}{dt} = \frac{d\theta}{ds}\cdot\frac{ds}{dt} = \frac{-1}{\sqrt{1 - (750/s)^2}}\left(\frac{-750}{s^2}\right)\frac{ds}{dt}$$

$$= \frac{750}{s\sqrt{s^2 - 750^2}}\,\frac{ds}{dt}$$

41. $\tan(\arctan x + \arctan y) = \dfrac{\tan(\arctan x) + \tan(\arctan y)}{1 - \tan(\arctan x)\tan(\arctan y)} = \dfrac{x + y}{1 - xy},\ xy \neq 1$

Therefore,

$$\arctan x + \arctan y = \arctan\left(\frac{x + y}{1 - xy}\right),\ xy \neq 1.$$

Let $x = \frac{1}{2}$ and $y = \frac{1}{3}$.

$$\arctan\left(\frac{1}{2}\right) + \arctan\left(\frac{1}{3}\right) = \arctan\frac{(1/2) + (1/3)}{1 - [(1/2)\cdot(1/3)]} = \arctan\frac{5/6}{1 - (1/6)} = \arctan\frac{5/6}{5/6} = \arctan 1 = \frac{\pi}{4}$$

42. (a) Let $y = \arcsin u$. Then

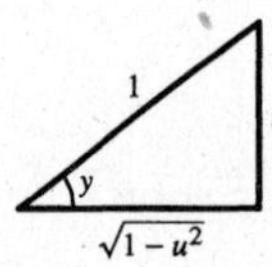

$$\sin y = u$$

$$\cos y \cdot y' = u'$$

$$\frac{dy}{dx} = \frac{u'}{\cos y} = \frac{u'}{\sqrt{1 - u^2}}.$$

(b) Let $y = \arctan u$. Then

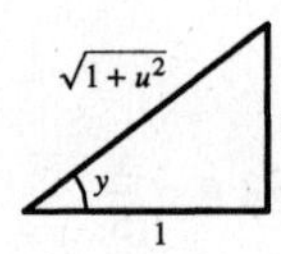

$$\tan y = u$$

$$\sec^2 y\,\frac{dy}{dx} = u'$$

$$\frac{dy}{dx} = \frac{u'}{\sec^2 y} = \frac{u'}{1 + u^2}.$$

(c) Let $y = \operatorname{arcsec} u$. Then

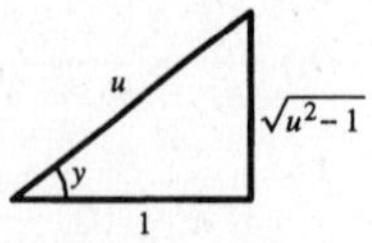

$$\sec y = u$$

$$\sec y \tan y\,\frac{dy}{dx} = u'$$

$$\frac{dy}{dx} = \frac{u'}{\sec y \tan y} = \frac{u'}{|u|\sqrt{u^2 - 1}}.$$

Note: The absolute value sign in the formula for the derivative of arcsec u is necessary because the inverse secant function has a positive slope at every value in its domain.

(d) Let $y = \arccos u$. Then

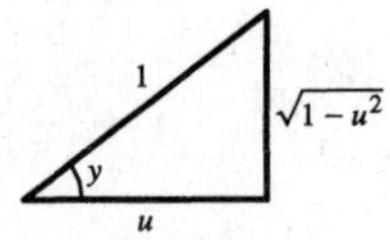

$$\cos y = u$$

$$-\sin y\,\frac{dy}{dx} = u'$$

$$\frac{dy}{dx} = -\frac{u'}{\sin y} = -\frac{u'}{\sqrt{1 - u^2}}.$$

—CONTINUED—

42. —CONTINUED—

(e) Let $y = \operatorname{arccot} u$. Then

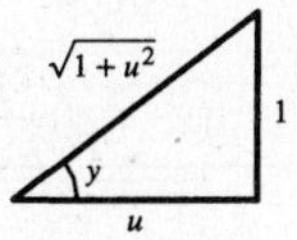

$$\cot y = u$$

$$-\csc^2 y \frac{dy}{dx} = u'$$

$$\frac{dy}{dx} = \frac{u'}{-\csc^2 y} = -\frac{u'}{1+u^2}.$$

(f) Let $y = \operatorname{arccsc} u$. Then

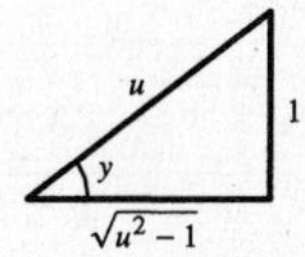

$$\csc y = u$$

$$-\csc y \cot y \frac{dy}{dx} = u'$$

$$\frac{dy}{dx} = \frac{u'}{-\csc y \cot y} = -\frac{u'}{|u|\sqrt{u^2-1}}.$$

Note: The absolute value sign in the formula for the derivative of arccsc u is necessary because the inverse cosecant function has a negative slope at every value in its domain.

43. $f(x) = \arcsin\left(\frac{x-2}{2}\right) - 2\arcsin\frac{\sqrt{x}}{2}$

$$f'(x) = \frac{\frac{1}{2}}{\sqrt{1-\left(\frac{x-2}{2}\right)^2}} - 2\frac{\frac{1}{4}x^{-1/2}}{\sqrt{1-\frac{x}{4}}}$$

$$= \frac{\frac{1}{2}}{\sqrt{1-\left(\frac{x^2-4x+4}{4}\right)}} - \frac{\frac{1}{2}}{\sqrt{x}\sqrt{1-\frac{x}{4}}}$$

$$= \frac{1}{\sqrt{4x-x^2}} - \frac{1}{\sqrt{4x-x^2}} = 0$$

Since $f'(x) = 0$, f is constant.

44. $f(x) = \sin x$

$g(x) = \arcsin(\sin x)$

(a) The range of $y = \arcsin x$ is $-\pi/2 \le y \le \pi/2$.

(b) Maximum: $\pi/2$

Minimum: $-\pi/2$

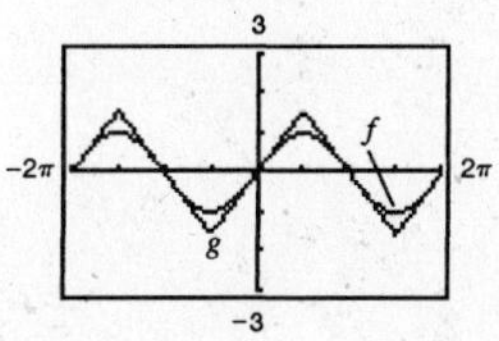

45. True.

$$\frac{d}{dx}[\arctan x] = \frac{1}{1+x^2} > 0 \text{ for all } x.$$

46. True.

$$\frac{d}{dx}[\arctan(\tan x)] = \frac{\sec^2 x}{1+\tan^2 x} = \frac{\sec^2 x}{\sec^2 x} = 1$$

47. False.

$$\arccos\left(\frac{1}{2}\right) = \frac{\pi}{3}$$

48. False.

$$\arcsin\left(\frac{\sqrt{2}}{2}\right) = \frac{\pi}{4}$$

49. $y = (7-4x)^2$

$y' = 2(7-4x)(-4) = -8(7-4x) = -56 + 32x$

50. $y = x\sqrt{4-x^2}$

$$y' = x\frac{1}{2}(4-x^2)^{-1/2}(-2x) + (4-x^2)^{1/2}$$

$$= \frac{-x^2}{\sqrt{4-x^2}} + \sqrt{4-x^2}$$

$$= \frac{4-2x^2}{\sqrt{4-x^2}}$$

51. $y = \dfrac{x}{\sqrt{x+5}}$

$$y' = \frac{\sqrt{x+5} - x\frac{1}{2}(x+5)^{-1/2}}{x+5}$$

$$= \frac{x+5-\frac{1}{2}x}{(x+5)^{3/2}}$$

$$= \frac{10+x}{2(x+5)^{3/2}}$$

52. $y = \dfrac{\sqrt{x}+2}{x^2+4}$

$$y' = \frac{(x^2+4)\frac{1}{2\sqrt{x}} - (\sqrt{x}+2)(2x)}{(x^2+4)^2}$$

$$= \frac{x^{3/2} + 4x^{-1/2} - 8x - 4x^{3/2}}{2(x^2+4)^2}$$

$$= \frac{-3x^2 - 8x^{3/2} + 4}{2\sqrt{x}(x^2+4)^2}$$

53. $f(t) = 4\sin^2(2t)$

$$f'(t) = 8\sin(2t)\cos(2t)(2)$$

$$= 16\sin(2t)\cos(2t)$$

$$= 8\sin(4t)$$

54. $f(x) = \sec(\pi x)\tan(\pi x)$

$$f'(x) = \pi\sec(\pi x)\sec^2(\pi x) + \pi\tan^2(\pi x)\sec(\pi x)$$

$$= \pi\sec(\pi x)[\sec^2(\pi x) + \tan^2(\pi x)]$$

55. $f(x) = \dfrac{\cos(4x)}{x}$

$$f'(x) = \frac{-4\sin(4x)x - \cos(4x)}{x^2}$$

56. $g(t) = \sin\sqrt{kt}$

$$g'(t) = \cos\left(\sqrt{kt}\right)\frac{1}{2}(kt)^{-1/2}k$$

$$= \frac{\sqrt{k}\cos\sqrt{kt}}{2\sqrt{t}}$$

57. $g(x) = \ln\dfrac{\sqrt{x+5}}{x^2+1} = \dfrac{1}{2}\ln(x+5) - \ln(x^2+1)$

$$g'(x) = \frac{1}{2(x+5)} - \frac{2x}{x^2+1}$$

$$= \frac{-3x^2 - 20x + 1}{2(x+5)(x^2+1)}$$

58. $h(x) = \dfrac{\ln(x+1)}{x}$

$$h'(x) = \frac{x\frac{1}{x+1} - \ln(x+1)}{x^2}$$

$$= \frac{x - (x+1)\ln(x+1)}{x^2(x+1)}$$

59. $s = t\ln(t^2+2)$

$$s' = t\frac{2t}{t^2+2} + \ln(t^2+2)$$

$$= \frac{2t^2}{t^2+2} + \ln(t^2+2)$$

60. $h = \ln[t^2(t+3)]$

$$= \ln t^2 + \ln(t+3)$$

$$h' = \frac{2}{t} + \frac{1}{t+3}$$

$$= \frac{3t+6}{t(t+3)}$$

61. $f(x) = e^{x/2}\sin x$

$$f'(x) = e^{x/2}\cos x + \frac{1}{2}e^{x/2}\sin x$$

$$= \frac{1}{2}e^{x/2}[2\cos x + \sin x]$$

62. $g(x) = \dfrac{e^{2x}}{x^2}$

$$g'(x) = \frac{2x^2e^{2x} - 2xe^{2x}}{x^4}$$

$$= \frac{2e^{2x}(x-1)}{x^3}$$

63. $y = e^{-x}\ln x^2 = 2e^{-x}\ln x$

$$y' = -2e^{-x}\ln x + \frac{2}{x}e^{-x}$$

$$= 2e^{-x}\left[\frac{1}{x} - \ln x\right]$$

$$= e^{-x}\left[\frac{2}{x} - \ln x^2\right]$$

64. $y = \ln e^{x^2} = x^2$

$$y' = 2x$$

65. $y = 5^x$

$y' = 5^x \ln 5$

66. $y = \log_3 (2x)$

$$y' = \frac{1}{2x \ln 3}(2) = \frac{1}{x \ln 3}$$

Section 11.5 Inverse Trigonometric Functions: Integration

1. $\displaystyle\int \frac{5}{\sqrt{9 - x^2}}\,dx = 5 \arcsin\left(\frac{x}{3}\right) + C$

2. $\displaystyle\int \frac{3}{\sqrt{1 - 4x^2}}\,dx = \frac{3}{2}\int \frac{2}{\sqrt{1 - 4x^2}}\,dx = \frac{3}{2}\arcsin(2x) + C$

3. Let $u = 3x$, $du = 3\,dx$.

$$\int_0^{1/6} \frac{1}{\sqrt{1 - 9x^2}}\,dx = \frac{1}{3}\int_0^{1/6} \frac{1}{\sqrt{1 - (3x)^2}}(3)\,dx = \left[\frac{1}{3}\arcsin(3x)\right]_0^{1/6} = \frac{\pi}{18}$$

4. $\displaystyle\int_0^1 \frac{1}{\sqrt{4 - x^2}}\,dx = \left[\arcsin\frac{x}{2}\right]_0^1 = \frac{\pi}{6}$

5. $\displaystyle\int \frac{7}{16 + x^2}\,dx = \frac{7}{4}\arctan\left(\frac{x}{4}\right) + C$

6. $\displaystyle\int \frac{4}{1 + 9x^2}\,dx = \frac{4}{3}\int \frac{3}{1 + 9x^2}\,dx = \frac{4}{3}\arctan(3x) + C$

7. Let $u = 2x$, $du = 2\,dx$.

$$\int_0^{\sqrt{3}/2} \frac{1}{1 + 4x^2}\,dx = \frac{1}{2}\int_0^{\sqrt{3}/2} \frac{2}{1 + (2x)^2}\,dx = \left[\frac{1}{2}\arctan(2x)\right]_0^{\sqrt{3}/2} = \frac{\pi}{6}$$

8. $\displaystyle\int_{\sqrt{3}}^3 \frac{1}{9 + x^2}\,dx = \left[\frac{1}{3}\arctan\frac{x}{3}\right]_{\sqrt{3}}^3 = \frac{\pi}{36}$

9. $\displaystyle\int \frac{1}{x\sqrt{4x^2 - 1}}\,dx = \int \frac{2}{2x\sqrt{(2x)^2 - 1}}\,dx = \operatorname{arcsec}|2x| + C$

10. $\displaystyle\int \frac{1}{4 + (x - 1)^2}\,dx = \frac{1}{2}\arctan\left(\frac{x - 1}{2}\right) + C$

11. $\displaystyle\int \frac{x^3}{x^2 + 1}\,dx = \int \left[x - \frac{x}{x^2 + 1}\right]dx = \int x\,dx - \frac{1}{2}\int \frac{2x}{x^2 + 1}\,dx = \frac{1}{2}x^2 - \frac{1}{2}\ln(x^2 + 1) + C$ (Use long division.)

12. $\displaystyle\int \frac{x^4 - 1}{x^2 + 1}\,dx = \int (x^2 - 1)\,dx = \frac{1}{3}x^3 - x + C$

13. Let $u = t^2$, $du = 2t\,dt$.

$$\int \frac{t}{\sqrt{1 - t^4}}\,dt = \frac{1}{2}\int \frac{1}{\sqrt{1 - (t^2)^2}}(2t)\,dt = \frac{1}{2}\arcsin(t^2) + C$$

14. Let $u = x^2$, $du = 2x\,dx$.

$$\int \frac{1}{x\sqrt{x^4 - 4}}\,dx = \frac{1}{2}\int \frac{1}{x^2\sqrt{(x^2)^2 - 2^2}}(2x)\,dx$$

$$= \frac{1}{4}\operatorname{arcsec}\frac{x^2}{2} + C$$

15. Let $u = \arcsin x$, $du = \dfrac{1}{\sqrt{1 - x^2}}\,dx$.

$$\int_0^{1/\sqrt{2}} \frac{\arcsin x}{\sqrt{1 - x^2}}\,dx = \left[\frac{1}{2}\arcsin^2 x\right]_0^{1/\sqrt{2}} = \frac{\pi^2}{32} \approx 0.308$$

16. Let $u = \arccos x$, $du = -\dfrac{1}{\sqrt{1 - x^2}}\,dx$.

$$\int_0^{1/\sqrt{2}} \frac{\arccos x}{\sqrt{1 - x^2}}\,dx = -\int_0^{1/\sqrt{2}} \frac{-\arccos x}{\sqrt{1 - x^2}}\,dx = \left[-\frac{1}{2}\arccos^2 x\right]_0^{1/\sqrt{2}} = \frac{3\pi^2}{32} \approx 0.925$$

17. Let $u = 1 - x^2$, $du = -2x\,dx$.

$$\int_{-1/2}^{0} \frac{x}{\sqrt{1-x^2}}\,dx = -\frac{1}{2}\int_{-1/2}^{0} (1-x^2)^{-1/2}(-2x)\,dx$$
$$= \left[-\sqrt{1-x^2}\right]_{-1/2}^{0} = \frac{\sqrt{3}-2}{2}$$
$$\approx -0.134$$

18. Let $u = 1 + x^2$, $du = 2x\,dx$.

$$\int_{-\sqrt{3}}^{0} \frac{x}{1+x^2}\,dx = \frac{1}{2}\int_{-\sqrt{3}}^{0} \frac{1}{1+x^2}(2x)\,dx$$
$$= \left[\frac{1}{2}\ln(1+x^2)\right]_{-\sqrt{3}}^{0} = -\ln 2$$

19. Let $u = e^{2x}$, $du = 2e^{2x}\,dx$.

$$\int \frac{e^{2x}}{4+e^{4x}}\,dx = \frac{1}{2}\int \frac{2e^{2x}}{4+(e^{2x})^2}\,dx = \frac{1}{4}\arctan\frac{e^{2x}}{2} + C$$

20. $$\int_1^2 \frac{1}{3+(x-2)^2}\,dx = \int_1^2 \frac{1}{(\sqrt{3})^2+(x-2)^2}\,dx$$
$$= \left[\frac{1}{\sqrt{3}}\arctan\left(\frac{x-2}{\sqrt{3}}\right)\right]_1^2 = \frac{\sqrt{3}\pi}{18}$$

21. Let $u = \cos x$, $du = -\sin x\,dx$.

$$\int_{\pi/2}^{\pi} \frac{\sin x}{1+\cos^2 x}\,dx = -\int_{\pi/2}^{\pi} \frac{-\sin x}{1+\cos^2 x}\,dx$$
$$= \left[-\arctan(\cos x)\right]_{\pi/2}^{\pi} = \frac{\pi}{4}$$

22. $$\int_0^{\pi/2} \frac{\cos x}{1+\sin^2 x}\,dx = \arctan(\sin x)\Big]_0^{\pi/2} = \frac{\pi}{4}$$

23. $\displaystyle\int \frac{1}{\sqrt{x}\sqrt{1-x}}\,dx.$ $u = \sqrt{x}$, $x = u^2$, $dx = 2u\,du$

$$\int \frac{1}{u\sqrt{1-u^2}}(2u\,du) = 2\int \frac{du}{\sqrt{1-u^2}} = 2\arcsin u + C$$
$$= 2\arcsin\sqrt{x} + C$$

24. $\displaystyle\int \frac{3}{2\sqrt{x}(1+x)}\,dx.$ $u = \sqrt{x}$, $du = \dfrac{1}{2\sqrt{x}}\,dx$, $dx = 2u\,du$

$$\frac{3}{2}\int \frac{2u\,du}{u(1+u^2)} = 3\int \frac{du}{1+u^2} = 3\arctan u + C$$
$$= 3\arctan\sqrt{x} + C$$

25. $$\int \frac{x-3}{x^2+1}\,dx = \frac{1}{2}\int \frac{2x}{x^2+1}\,dx - 3\int \frac{1}{x^2+1}\,dx$$
$$= \frac{1}{2}\ln(x^2+1) - 3\arctan x + C$$

26. $$\int \frac{4x+3}{\sqrt{1-x^2}}\,dx = (-2)\int \frac{-2x}{\sqrt{1-x^2}}\,dx + 3\int \frac{1}{\sqrt{1-x^2}}\,dx$$
$$= -4\sqrt{1-x^2} + 3\arcsin x + C$$

27. $$\int \frac{x+5}{\sqrt{9-(x-3)^2}}\,dx = \int \frac{(x-3)}{\sqrt{9-(x-3)^2}}\,dx + \int \frac{8}{\sqrt{9-(x-3)^2}}\,dx$$
$$= -\sqrt{9-(x-3)^2} + 8\arcsin\left(\frac{x-3}{3}\right) + C$$
$$= -\sqrt{6x-x^2} + 8\arcsin\left(\frac{x}{3}-1\right) + C$$

28. $$\int \frac{x-2}{(x+1)^2+4}\,dx = \frac{1}{2}\int \frac{2x+2}{(x+1)^2+4}\,dx - \int \frac{3}{(x+1)^2+4}\,dx$$
$$= \frac{1}{2}\ln(x^2+2x+5) - \frac{3}{2}\arctan\left(\frac{x+1}{2}\right) + C$$

29. $$\int_0^2 \frac{1}{x^2-2x+2}\,dx = \int_0^2 \frac{1}{1+(x-1)^2}\,dx = \left[\arctan(x-1)\right]_0^2 = \frac{\pi}{2}$$

30. $$\int_{-2}^{2} \frac{dx}{x^2+4x+13} = \int_{-2}^{2} \frac{dx}{(x+2)^2+9} = \left[\frac{1}{3}\arctan\left(\frac{x+2}{3}\right)\right]_{-2}^{2} = \frac{1}{3}\arctan\left(\frac{4}{3}\right)$$

31. $\displaystyle\int \frac{2x}{x^2 + 6x + 13}\,dx = \int \frac{2x + 6}{x^2 + 6x + 13}\,dx - 6\int \frac{1}{x^2 + 6x + 13}\,dx = \int \frac{2x + 6}{x^2 + 6x + 13}\,dx - 6\int \frac{1}{4 + (x + 3)^2}\,dx$

$$= \ln|x^2 + 6x + 13| - 3\arctan\left(\frac{x + 3}{2}\right) + C$$

32. $\displaystyle\int \frac{2x - 5}{x^2 + 2x + 2}\,dx = \int \frac{2x + 2}{x^2 + 2x + 2}\,dx - 7\int \frac{1}{1 + (x + 1)^2}\,dx = \ln|x^2 + 2x + 2| - 7\arctan(x + 1) + C$

33. $\displaystyle\int \frac{1}{\sqrt{-x^2 - 4x}}\,dx = \int \frac{1}{\sqrt{4 - (x + 2)^2}}\,dx$

$$= \arcsin\left(\frac{x + 2}{2}\right) + C$$

34. $\displaystyle\int \frac{2}{\sqrt{-x^2 + 4x}}\,dx = \int \frac{2}{\sqrt{4 - (x^2 - 4x + 4)}}\,dx$

$$= \int \frac{2}{\sqrt{4 - (x - 2)^2}}\,dx$$

$$= 2\arcsin\left(\frac{x - 2}{2}\right) + C$$

35. Let $u = -x^2 - 4x$, $du = (-2x - 4)\,dx$.

$$\int \frac{x + 2}{\sqrt{-x^2 - 4x}}\,dx = -\frac{1}{2}\int (-x^2 - 4x)^{-1/2}(-2x - 4)\,dx$$

$$= -\sqrt{-x^2 - 4x} + C$$

36. Let $u = x^2 - 2x$, $du = (2x - 2)\,dx$.

$$\int \frac{x - 1}{\sqrt{x^2 - 2x}}\,dx = \frac{1}{2}\int (x^2 - 2x)^{-1/2}(2x - 2)\,dx$$

$$= \sqrt{x^2 - 2x} + C$$

37. $\displaystyle\int_2^3 \frac{2x - 3}{\sqrt{4x - x^2}}\,dx = \int_2^3 \frac{2x - 4}{\sqrt{4x - x^2}}\,dx + \int_2^3 \frac{1}{\sqrt{4x - x^2}}\,dx = -\int_2^3 (4x - x^2)^{-1/2}(4 - 2x)\,dx + \int_2^3 \frac{1}{\sqrt{4 - (x - 2)^2}}\,dx$

$$= \left[-2\sqrt{4x - x^2} + \arcsin\left(\frac{x - 2}{2}\right)\right]_2^3 = 4 - 2\sqrt{3} + \frac{\pi}{6} \approx 1.059$$

38. $\displaystyle\int \frac{1}{(x - 1)\sqrt{x^2 - 2x}}\,dx = \int \frac{1}{(x - 1)\sqrt{(x - 1)^2 - 1}}\,dx = \operatorname{arcsec}|x - 1| + C$

39. Let $u = x^2 + 1$, $du = 2x\,dx$.

$$\int \frac{x}{x^4 + 2x^2 + 2}\,dx = \frac{1}{2}\int \frac{2x}{(x^2 + 1)^2 + 1}\,dx$$

$$= \frac{1}{2}\arctan(x^2 + 1) + C$$

40. Let $u = x^2 - 4$, $du = 2x\,dx$.

$$\int \frac{x}{\sqrt{9 + 8x^2 - x^4}}\,dx = \frac{1}{2}\int \frac{2x}{\sqrt{25 - (x^2 - 4)^2}}\,dx$$

$$= \frac{1}{2}\arcsin\left(\frac{x^2 - 4}{5}\right) + C$$

41. Let $u = \sqrt{e^t - 3}$. Then $u^2 + 3 = e^t$, $2u\,du = e^t\,dt$, and $\dfrac{2u\,du}{u^2 + 3} = dt$.

$$\int \sqrt{e^t - 3}\,dt = \int \frac{2u^2}{u^2 + 3}\,du = \int 2\,du - \int 6\frac{1}{u^2 + 3}\,du$$

$$= 2u - 2\sqrt{3}\arctan\frac{u}{\sqrt{3}} + C = 2\sqrt{e^t - 3} - 2\sqrt{3}\arctan\sqrt{\frac{e^t - 3}{3}} + C$$

42. Let $u = \sqrt{x - 2}$, $u^2 + 2 = x$, $2u\,du = dx$.

$$\int \frac{\sqrt{x - 2}}{x + 1}\,dx = \int \frac{2u^2}{u^2 + 3}\,du = \int \frac{2u^2 + 6 - 6}{u^2 + 3}\,du = 2\int du - 6\int \frac{1}{u^2 + 3}\,du$$

$$= 2u - \frac{6}{\sqrt{3}}\arctan\frac{u}{\sqrt{3}} + C = 2\sqrt{x - 2} - 2\sqrt{3}\arctan\sqrt{\frac{x - 2}{3}} + C$$

43. A perfect square trinomial is an expression in x with three terms that factors as a perfect square.

Example: $x^2 + 6x + 9 = (x + 3)^2$

44. The term is $\left(\frac{3}{2}\right)^2 = \frac{9}{4}$: $x^2 + 3x = x^2 + 3x + \frac{9}{4} - \frac{9}{4} = \left(x + \frac{3}{2}\right)^2 - \frac{9}{4}$

45. (a) $\displaystyle\int \frac{1}{\sqrt{1 - x^2}}\,dx = \arcsin x + C,\ u = x$

(b) $\displaystyle\int \frac{x}{\sqrt{1 - x^2}}\,dx = -\sqrt{1 - x^2} + C,\ u = 1 - x^2$

(c) $\displaystyle\int \frac{1}{x\sqrt{1 - x^2}}\,dx$ cannot be evaluated using the basic integration rules.

46. (a) $\displaystyle\int e^{x^2}\,dx$ cannot be evaluated using the basic integration rules.

(b) $\displaystyle\int xe^{x^2}\,dx = \frac{1}{2}e^{x^2} + C,\ u = x^2$

(c) $\displaystyle\int \frac{1}{x^2}e^{1/x}\,dx = -e^{1/x} + C,\ u = \frac{1}{x}$

47. (a)

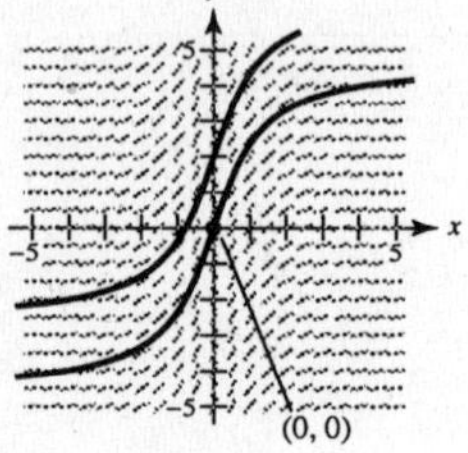

(b) $\displaystyle\frac{dy}{dx} = \frac{3}{1 + x^2},\ (0, 0)$

$$y = 3\int \frac{dx}{1 + x^2} = 3 \arctan x + C$$

$(0, 0)$: $0 = 3\arctan(0) + C \Rightarrow C = 0$

$y = 3 \arctan x$

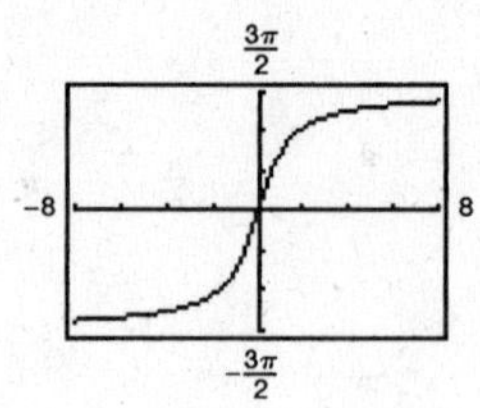

48. (a)

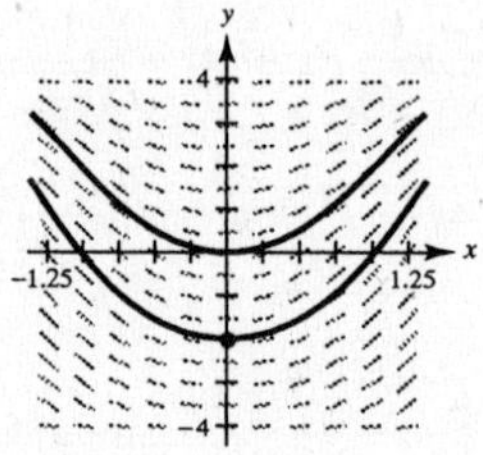

(b) $\displaystyle\frac{dy}{dx} = x\sqrt{16 - y^2},\ (0, -2)$

$$\frac{dy}{\sqrt{16 - y^2}} = x\,dx$$

$$\arcsin\left(\frac{y}{4}\right) = \frac{x^2}{2} + C$$

$$(0, -2)\text{: } \arcsin\left(-\frac{2}{4}\right) = C \Rightarrow C = -\frac{\pi}{6}$$

$$\arcsin\left(\frac{y}{4}\right) = \frac{x^2}{2} - \frac{\pi}{6}$$

$$\frac{y}{4} = \sin\left(\frac{x^2}{2} - \frac{\pi}{6}\right)$$

$$y = 4\sin\left(\frac{x^2}{2} - \frac{\pi}{6}\right)$$

49. $\displaystyle\frac{dy}{dx} = \frac{10}{x\sqrt{x^2 - 1}},\ y(3) = 0$

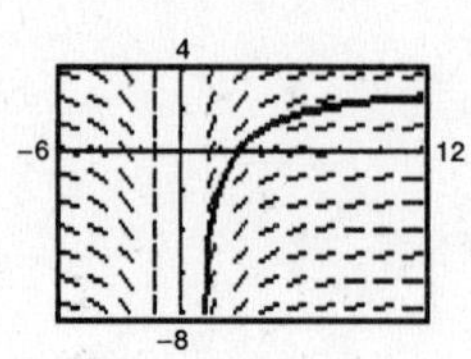

50. $\displaystyle\frac{dy}{dx} = \frac{2y}{\sqrt{16 - x^2}},\ y(0) = 2$

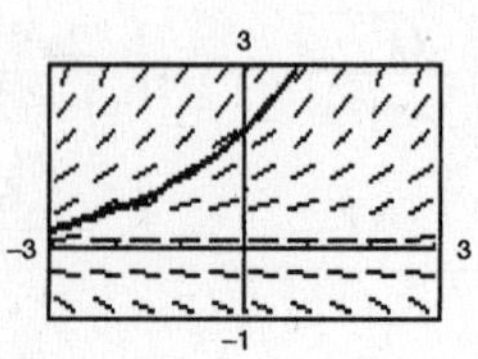

51. $A = \int_1^3 \frac{1}{x^2 - 2x + 1 + 4}\,dx = \int_1^3 \frac{1}{(x-1)^2 + 2^2}\,dx$

$= \left[\frac{1}{2}\arctan\left(\frac{x-1}{2}\right)\right]_1^3 = \frac{1}{2}\arctan(1) = \frac{\pi}{8} \approx 0.3927$

52. $A = \int_0^1 \frac{1}{\sqrt{4 - x^2}}\,dx = \left[\arcsin\frac{x}{2}\right]_0^1 = \frac{\pi}{6}$

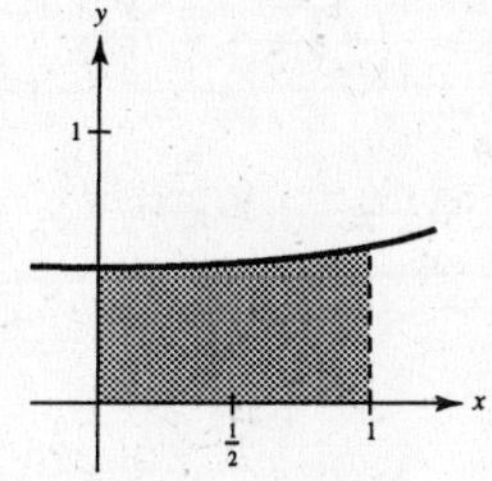

53. (a) $\int_0^1 \frac{4}{1 + x^2}\,dx = \left[4\arctan x\right]_0^1 = 4\arctan 1 - 4\arctan 0 = 4\left(\frac{\pi}{4}\right) - 4(0) = \pi$

(b) Let $n = 6$.

$4\int_0^1 \frac{1}{1 + x^2}\,dx \approx 4\left(\frac{1}{18}\right)\left[1 + \frac{4}{1 + (1/36)} + \frac{2}{1 + (1/9)} + \frac{4}{1 + (1/4)} + \frac{2}{1 + (4/9)} + \frac{4}{1 + (25/36)} + \frac{1}{2}\right] \approx 3.1415918$

(c) 3.1415927

54. $F(x) = \frac{1}{2}\int_x^{x+2} \frac{2}{t^2 + 1}\,dt$

(a) $F(x)$ represents the average value of $f(x)$ over the interval $[x, x + 2]$. Maximum at $x = -1$, since the graph is greatest on $[-1, 1]$.

(b) $F(x) = \left[\arctan t\right]_x^{x+2} = \arctan(x + 2) - \arctan x$

$F'(x) = \frac{1}{1 + (x+2)^2} - \frac{1}{1 + x^2} = \frac{(1 + x^2) - (x^2 + 4x + 5)}{(x^2 + 1)(x^2 + 4x + 5)} = \frac{-4(x + 1)}{(x^2 + 1)(x^2 + 4x + 5)} = 0$ when $x = -1$.

55. (a) $\frac{d}{dx}\left[\arcsin\left(\frac{u}{a}\right) + C\right] = \frac{1}{\sqrt{1 - (u^2/a^2)}}\left(\frac{u'}{a}\right) = \frac{u'}{\sqrt{a^2 - u^2}}$

Thus, $\int \frac{du}{\sqrt{a^2 - u^2}} = \arcsin\left(\frac{u}{a}\right) + C.$

(b) $\frac{d}{dx}\left[\frac{1}{a}\arctan\frac{u}{a} + C\right] = \frac{1}{a}\left[\frac{u'/a}{1 + (u/a)^2}\right] = \frac{1}{a^2}\left[\frac{u'}{(a^2 + u^2)/a^2}\right] = \frac{u'}{a^2 + u^2}$

Thus, $\int \frac{du}{a^2 + u^2} = \int \frac{u'}{a^2 + u^2}\,dx = \frac{1}{a}\arctan\frac{u}{a} + C.$

(c) Assume $u > 0$.

$\frac{d}{dx}\left[\frac{1}{a}\operatorname{arcsec}\frac{u}{a} + C\right] = \frac{1}{a}\left[\frac{u'/a}{(u/a)\sqrt{(u/a)^2 - 1}}\right] = \frac{1}{a}\left[\frac{u'}{u\sqrt{(u^2 - a^2)/a^2}}\right] = \frac{u'}{u\sqrt{u^2 - a^2}}.$

The case $u < 0$ is handled in a similar manner.

Thus, $\int \frac{du}{u\sqrt{u^2 - a^2}} = \int \frac{u'}{u\sqrt{u^2 - a^2}}\,dx = \frac{1}{a}\operatorname{arcsec}\frac{|u|}{a} + C.$

56. Let $f(x) = \arctan x - \dfrac{x}{1 + x^2}$

$$f'(x) = \frac{1}{1 + x^2} - \frac{1 - x^2}{(1 + x^2)^2} = \frac{2x^2}{(1 + x^2)^2} > 0 \text{ for } x > 0.$$

Since $f(0) = 0$ and f is increasing for $x > 0$, $\arctan x - \dfrac{x}{1 + x^2} > 0$ for $x > 0$. Thus,

$$\arctan x > \frac{x}{1 + x^2}.$$

Let $g(x) = x - \arctan x$

$$g'(x) = 1 - \frac{1}{1 + x^2} = \frac{x^2}{1 + x^2} > 0 \text{ for } x > 0.$$

Since $g(0) = 0$ and g is increasing for $x > 0$, $x - \arctan x > 0$ for $x > 0$. Thus, $x > \arctan x$. Therefore,

$$\frac{x}{1 + x^2} < \arctan x < x.$$

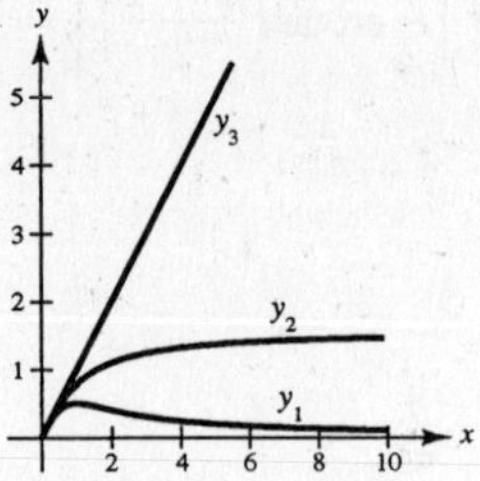

57. $\displaystyle\int \frac{1}{\sqrt{6x - x^2}}\, dx$

(a) $6x - x^2 = 9 - (x^2 - 6x + 9) = 9 - (x - 3)^2$

$$\int \frac{1}{\sqrt{6x - x^2}}\, dx = \int \frac{dx}{\sqrt{9 - (x - 3)^2}} = \arcsin\left(\frac{x - 3}{3}\right) + C$$

(b) $u = \sqrt{x}$, $u^2 = x$, $2u\, du = dx$

$$\int \frac{1}{\sqrt{6u^2 - u^4}}(2u\, du) = \int \frac{2}{\sqrt{6 - u^2}}\, du = 2\arcsin\left(\frac{u}{\sqrt{6}}\right) + C$$

$$= 2\arcsin\left(\frac{\sqrt{x}}{\sqrt{6}}\right) + C$$

(c)

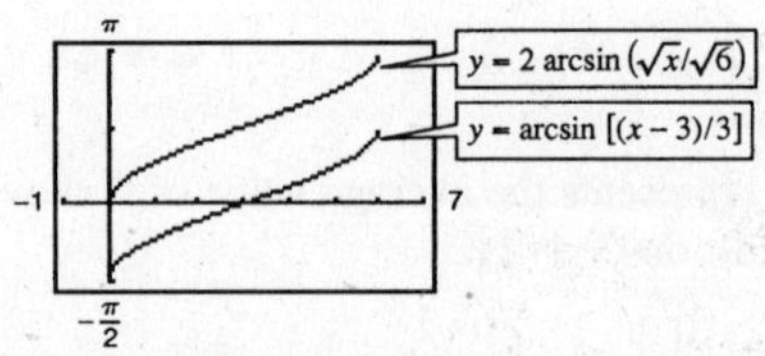

The antiderivatives differ by a constant, $\pi/2$.

Domain: $[0, 6]$

58. (a) $v(t) = -32t + 500$

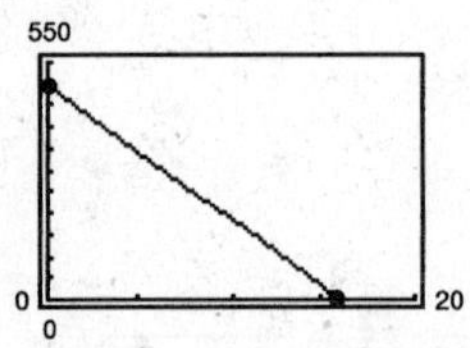

(b) $s(t) = \displaystyle\int v(t)\, dt = \int (-32t + 500)\, dt$

$= -16t^2 + 500t + C$

$s(0) = -16(0) + 500(0) + C = 0 \Rightarrow C = 0$

$s(t) = -16t^2 + 500t$

When the object reaches its maximum height, $v(t) = 0$.

$$v(t) = -32t + 500 = 0$$

$$-32t = -500$$

$$t = 15.625$$

$$s(15.625) = -16(15.625)^2 + 500(15.625)$$

$$= 3906.25 \text{ ft (Maximum height)}$$

—CONTINUED—

58. —CONTINUED—

(c) $$\int \frac{1}{32 + kv^2}\,dv = -\int dt$$

$$\frac{1}{\sqrt{32k}}\arctan\left(\sqrt{\frac{k}{32}}v\right) = -t + C_1$$

$$\arctan\left(\sqrt{\frac{k}{32}}v\right) = -\sqrt{32k}t + C$$

$$\sqrt{\frac{k}{32}}v = \tan\left(C - \sqrt{32k}\,t\right)$$

$$v = \sqrt{\frac{32}{k}}\tan\left(C - \sqrt{32k}\,t\right)$$

When $t = 0$, $v = 500$, $C = \arctan\left(500\sqrt{k/32}\right)$, and we have

$$v(t) = \sqrt{\frac{32}{k}}\tan\left[\arctan\left(500\sqrt{\frac{k}{32}}\right) - \sqrt{32k}\,t\right].$$

(d) When $k = 0.001$, $v(t) = \sqrt{32{,}000}\tan\left[\arctan\left(500\sqrt{0.00003125}\right) - \sqrt{0.032}\,t\right]$.

$v(t) = 0$ when $t_0 \approx 6.86$ sec.

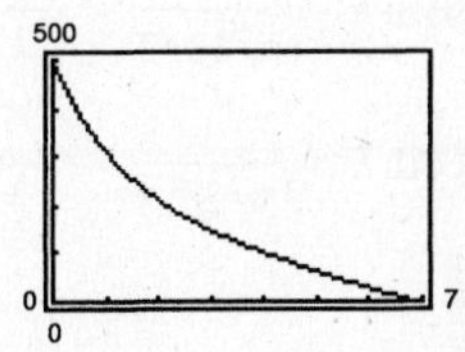

(e) $$h = \int_0^{6.86} \sqrt{32{,}000}\tan\left[\arctan\left(500\sqrt{0.00003125}\right) - \sqrt{0.032}\,t\right] dt$$

Simpson's Rule: $n = 10$; $h \approx 1088$ feet

(f) Air resistance lowers the maximum height.

Section 11.6 Hyperbolic Functions

1. (a) $\sinh 3 = \dfrac{e^3 - e^{-3}}{2} \approx 10.018$

(b) $\tanh(-2) = \dfrac{\sinh(-2)}{\cosh(-2)} = \dfrac{e^{-2} - e^2}{e^{-2} + e^2} \approx -0.964$

2. (a) $\cosh(0) = \dfrac{e^0 + e^0}{2} = 1$

(b) $\operatorname{sech}(1) = \dfrac{2}{e + e^{-1}} \approx 0.648$

3. (a) $\cosh^{-1}(2) = \ln\left(2 + \sqrt{3}\right) \approx 1.317$

(b) $\operatorname{sech}^{-1}\left(\dfrac{2}{3}\right) = \ln\left(\dfrac{1 + \sqrt{1 - (4/9)}}{2/3}\right) \approx 0.962$

4. (a) $\sinh^{-1}(0) = 0$

(b) $\tanh^{-1}(0) = 0$

5. $$\tanh^2 x + \operatorname{sech}^2 x = \left(\frac{e^x - e^{-x}}{e^x + e^{-x}}\right)^2 + \left(\frac{2}{e^x + e^{-x}}\right)^2 = \frac{e^{2x} - 2 + e^{-2x} + 4}{(e^x + e^{-x})^2} = \frac{e^{2x} + 2 + e^{-2x}}{e^{2x} + 2 + e^{-2x}} = 1$$

6. $$\frac{1 + \cosh 2x}{2} = \frac{1 + (e^{2x} + e^{-2x})/2}{2} = \frac{e^{2x} + 2 + e^{-2x}}{4} = \left(\frac{e^x + e^{-x}}{2}\right)^2 = \cosh^2 x$$

7. $$\sinh x \cosh y + \cosh x \sinh y = \left(\frac{e^x - e^{-x}}{2}\right)\left(\frac{e^y + e^{-y}}{2}\right) + \left(\frac{e^x + e^{-x}}{2}\right)\left(\frac{e^y - e^{-y}}{2}\right)$$

$$= \frac{1}{4}\left[e^{x+y} - e^{-x+y} + e^{x-y} - e^{-(x+y)} + e^{x+y} + e^{-x+y} - e^{x-y} - e^{-(x+y)}\right]$$

$$= \frac{1}{4}\left[2\left(e^{x+y} - e^{-(x+y)}\right)\right] = \frac{e^{(x+y)} - e^{-(x+y)}}{2} = \sinh(x + y)$$

8. $2\cosh\left(\frac{x+y}{2}\right)\cosh\left(\frac{x-y}{2}\right) = 2\left[\frac{e^{(x+y)/2}+e^{-(x+y)/2}}{2}\right]\left[\frac{e^{(x-y)/2}+e^{-(x-y)/2}}{2}\right]$

$$= 2\left[\frac{e^x+e^y+e^{-y}+e^{-x}}{4}\right] = \frac{e^x+e^{-x}}{2} + \frac{e^y+e^{-y}}{2}$$

$$= \cosh x + \cosh y$$

9. $\sinh x = \frac{3}{2}$

$$\cosh^2 x - \left(\frac{3}{2}\right)^2 = 1 \Rightarrow \cosh^2 x = \frac{13}{4} \Rightarrow \cosh x = \frac{\sqrt{13}}{2}$$

$$\tanh x = \frac{3/2}{\sqrt{13}/2} = \frac{3\sqrt{13}}{13}$$

$$\operatorname{csch} x = \frac{1}{3/2} = \frac{2}{3}$$

$$\operatorname{sech} x = \frac{1}{\sqrt{13}/2} = \frac{2\sqrt{13}}{13}$$

$$\coth x = \frac{1}{3/\sqrt{13}} = \frac{\sqrt{13}}{3}$$

10. $\tanh x = \frac{1}{2}$

$$\left(\frac{1}{2}\right)^2 + \operatorname{sech}^2 x = 1 \Rightarrow \operatorname{sech}^2 x = \frac{3}{4} \Rightarrow \operatorname{sech} x = \frac{\sqrt{3}}{2}$$

$$\cosh x = \frac{1}{\sqrt{3}/2} = \frac{2\sqrt{3}}{3}$$

$$\coth x = \frac{1}{1/2} = 2$$

$$\sinh x = \tanh x \cosh x = \left(\frac{1}{2}\right)\left(\frac{2\sqrt{3}}{3}\right) = \frac{\sqrt{3}}{3}$$

$$\operatorname{csch} x = \frac{1}{\sqrt{3}/3} = \sqrt{3}$$

Putting these in order:

$\sinh x = \frac{\sqrt{3}}{3}$ $\quad$ $\operatorname{csch} x = \sqrt{3}$

$\cosh x = \frac{2\sqrt{3}}{3}$ $\quad$ $\operatorname{sech} x = \frac{\sqrt{3}}{2}$

$\tanh x = \frac{1}{2}$ $\quad$ $\coth x = 2$

11. $y = \sinh(1 - x^2)$

$y' = -2x\cosh(1 - x^2)$

12. $y = \coth(3x)$

$y' = -3\operatorname{csch}^2(3x)$

13. $f(x) = \ln(\sinh x)$

$$f'(x) = \frac{1}{\sinh x}(\cosh x) = \coth x$$

14. $g(x) = \ln(\cosh x)$

$$g'(x) = \frac{1}{\cosh x}(\sinh x) = \tanh x$$

15. $y = x\cosh x - \sinh x$

$y' = x\sinh x + \cosh x - \cosh x = x\sinh x$

16. $h(t) = t - \coth t$

$h'(t) = 1 + \operatorname{csch}^2 t = \coth^2 t$

17. $f(t) = \arctan(\sinh t)$

$$f'(t) = \frac{1}{1+\sinh^2 t}(\cosh t)$$

$$= \frac{\cosh t}{\cosh^2 t} = \operatorname{sech} t$$

18. $g(x) = \operatorname{sech}^2 3x$

$g'(x) = -2\operatorname{sech}(3x)\operatorname{sech}(3x)\tanh(3x)(3)$

$= -6\operatorname{sech}^2 3x \tanh 3x$

19. Let $y = g(x)$.

$$y = x^{\cosh x}$$

$$\ln y = \cosh x \ln x$$

$$\frac{1}{y}\left(\frac{dy}{dx}\right) = \frac{\cosh x}{x} + \sinh x \ln x$$

$$\frac{dy}{dx} = \frac{y}{x}[\cosh x + x(\sinh x) \ln x]$$

$$= \frac{x^{\cosh x}}{x}[\cosh x + x(\sinh x) \ln x]$$

20. $f(x) = e^{\sinh x}$

$$f'(x) = (\cosh x)(e^{\sinh x})$$

21. $y = (\cosh x - \sinh x)^2$

$$y' = 2(\cosh x - \sinh x)(\sinh x - \cosh x)$$

$$= -2(\cosh x - \sinh x)^2 = -2e^{-2x}$$

22. $y = \operatorname{sech}(x + 1)$

$$y' = -\operatorname{sech}(x + 1)\tanh(x + 1)$$

23. $f(x) = \sin x \sinh x - \cos x \cosh x, \ -4 \le x \le 4$

$$f'(x) = \sin x \cosh x + \cos x \sinh x - \cos x \sinh x + \sin x \cosh x$$

$$= 2 \sin x \cosh x = 0 \text{ when } x = 0, \pm\pi.$$

Relative maxima: $(\pm\pi, \cosh \pi)$

Relative minimum: $(0, -1)$

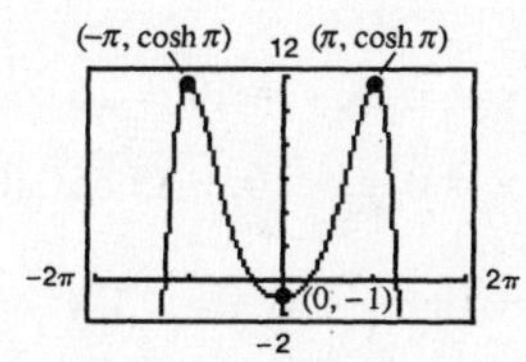

24. $f(x) = x \sinh(x - 1) - \cosh(x - 1)$

$$f'(x) = x \cosh(x - 1) + \sinh(x - 1) - \sinh(x - 1) = x \cosh(x - 1)$$

$f'(x) = 0$ for $x = 0$. By the First Derivative Test, $(0, -\cosh(-1)) \approx (0, -1.543)$ is a relative minimum.

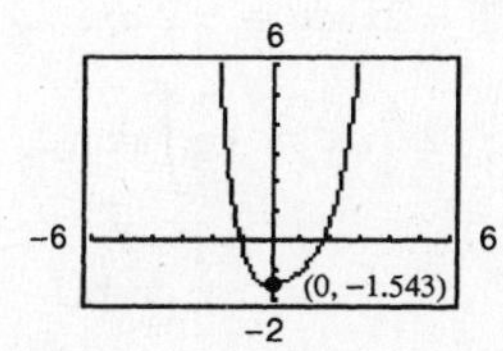

25. $g(x) = x \operatorname{sech} x = \dfrac{x}{\cosh x}$

Relative maximum: $(1.20, 0.66)$

Relative minimum: $(-1.20, -0.66)$

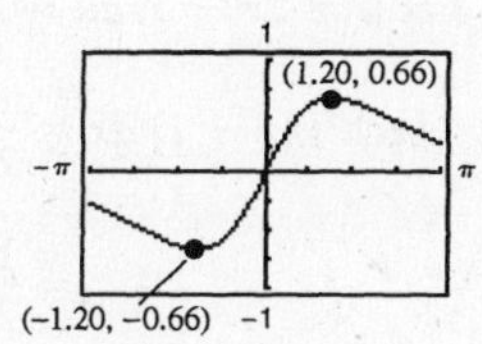

26. $h(x) = 2 \tanh x - x$

Relative maximum: $(0.88, 0.53)$

Relative minimum: $(-0.88, -0.53)$

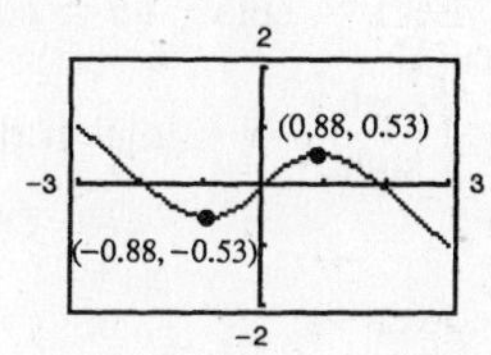

27. $y = a \sinh x$

$$y' = a \cosh x$$

$$y'' = a \sinh x$$

$$y''' = a \cosh x$$

Therefore, $y''' - y' = 0$.

28. $y = a \cosh x$

$$y' = a \sinh x$$

$$y'' = a \cosh x$$

Therefore, $y'' - y = 0$.

29. $f(x) = \tanh x \qquad f(0) = 0$

$f'(x) = \operatorname{sech}^2 x \qquad f'(0) = 1$

$f''(x) = -2 \operatorname{sech}^2 x \tanh x \qquad f''(0) = 0$

$P_1(x) = P_2(x) = x$

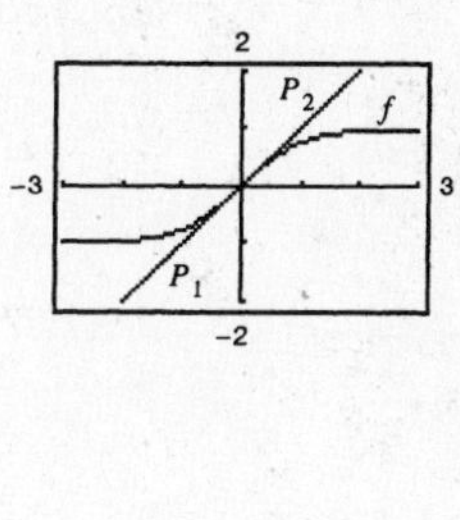

30. $f(x) = \cosh x \qquad f(0) = \cosh(0) = 1$

$f'(x) = \sinh x \qquad f'(0) = \sinh(0) = 0$

$f''(x) = \cosh x \qquad f''(0) = \cosh(0) = 1$

$P_1(x) = f(0) + f'(0)(x - 0) = 1$

$P_2(x) = 1 + \frac{1}{2}x^2$

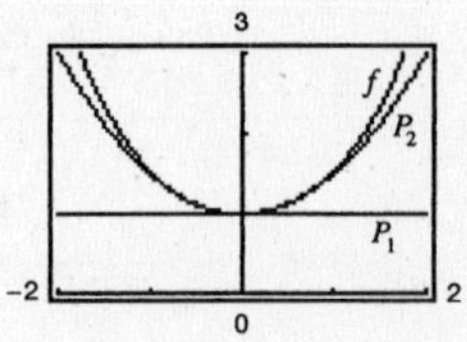

31. (a) $y = 10 + 15 \cosh \dfrac{x}{15}, \; -15 \le x \le 15$

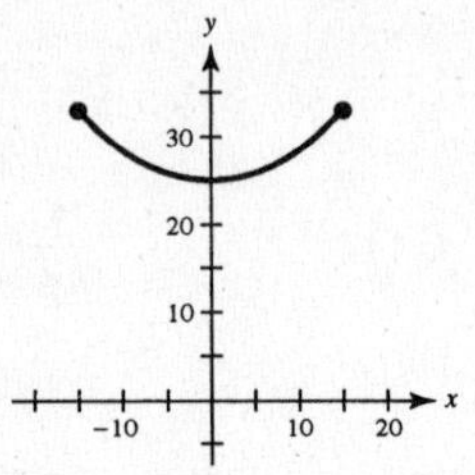

(b) At $x = \pm 15$, $y = 10 + 15 \cosh(1) \approx 33.146$.

At $x = 0$, $y = 10 + 15 \cosh(0) = 25$.

(c) $y' = \sinh \dfrac{x}{15}$. At $x = 15$, $y' = \sinh(1) \approx 1.175$

32. (a) $y = 18 + 25 \cosh \dfrac{x}{25}, \; -25 \le x \le 25$

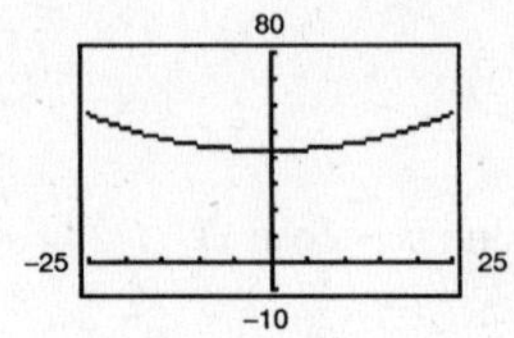

(b) At $x = \pm 25$, $y = 18 + 25 \cosh(1) \approx 56.577$.

At $x = 0$, $y = 18 + 25 = 43$.

(c) $y' = \sinh \dfrac{x}{25}$. At $x = 25$, $y' = \sinh(1) \approx 1.175$

33. Let $u = 1 - 2x$, $du = -2\,dx$.

$$\int \sinh(1 - 2x)\,dx = -\frac{1}{2}\int \sinh(1 - 2x)(-2)\,dx = -\frac{1}{2}\cosh(1 - 2x) + C$$

34. Let $u = \sqrt{x}$, $du = \dfrac{1}{2\sqrt{x}}\,dx$.

$$\int \frac{\cosh \sqrt{x}}{\sqrt{x}}\,dx = 2\int \cosh \sqrt{x}\left(\frac{1}{2\sqrt{x}}\right)dx = 2\sinh\sqrt{x} + C$$

35. Let $u = \sinh x$, $du = \cosh x\,dx$.

$$\int \frac{\cosh x}{\sinh x}\,dx = \ln|\sinh x| + C$$

36. Let $u = 2x - 1$, $du = 2\,dx$.

$$\int \operatorname{sech}^2(2x - 1)\,dx = \frac{1}{2}\int \operatorname{sech}^2(2x - 1)(2)\,dx = \frac{1}{2}\tanh(2x - 1) + C$$

37. Let $u = \dfrac{x^2}{2}$, $du = x\,dx$.

$$\int x \operatorname{csch}^2 \frac{x^2}{2}\,dx = \int \left(\operatorname{csch}^2 \frac{x^2}{2}\right)x\,dx = -\coth \frac{x^2}{2} + C$$

38. Let $u = \operatorname{sech} x$, $du = -\operatorname{sech} x \tanh x\,dx$.

$$\int \operatorname{sech}^3 x \tanh x\,dx = -\int \operatorname{sech}^2 x(-\operatorname{sech} x \tanh x)\,dx = -\frac{1}{3}\operatorname{sech}^3 x + C$$

39. Let $u = \dfrac{1}{x}$, $du = -\dfrac{1}{x^2}\,dx$.

$$\int \frac{\operatorname{csch}(1/x)\coth(1/x)}{x^2}\,dx = -\int \operatorname{csch}\frac{1}{x}\coth\frac{1}{x}\left(-\frac{1}{x^2}\right)dx = \operatorname{csch}\frac{1}{x} + C$$

40. $$\int \cosh^2 x\,dx = \int \frac{1 + \cosh 2x}{2}\,dx = \frac{1}{2}\left[x + \frac{\sinh 2x}{2}\right] + C = \frac{1}{2}x + \frac{1}{4}\sinh 2x + C$$

41. $\displaystyle\int_0^4 \frac{1}{25 - x^2}\,dx = \left[\frac{1}{10}\ln\left|\frac{5+x}{5-x}\right|\right]_0^4 = \frac{1}{10}\ln 9 = \frac{1}{5}\ln 3$

42. $\displaystyle\int_0^4 \frac{1}{\sqrt{25 - x^2}}\,dx = \left[\arcsin\frac{x}{5}\right]_0^4 = \arcsin\frac{4}{5}$

43. Let $u = 2x$, $du = 2\,dx$.

$$\int_0^{\sqrt{2}/4} \frac{1}{\sqrt{1-(2x)^2}}(2)\,dx = \left[\arcsin(2x)\right]_0^{\sqrt{2}/4} = \frac{\pi}{4}$$

44. $\displaystyle\int \frac{2}{x\sqrt{1+4x^2}}\,dx = 2\int \frac{1}{(2x)\sqrt{1+(2x)^2}}(2)\,dx$

$$= -2\ln\left(\frac{1+\sqrt{1+4x^2}}{|2x|}\right) + C$$

45. Let $u = x^2$, $du = 2x\,dx$.

$$\int \frac{x}{x^4+1}\,dx = \frac{1}{2}\int \frac{2x}{(x^2)^2+1}\,dx = \frac{1}{2}\arctan(x^2) + C$$

46. Let $u = \sinh x$, $du = \cosh x\,dx$.

$$\int \frac{\cosh x}{\sqrt{9-\sinh^2 x}}\,dx = \arcsin\left(\frac{\sinh x}{3}\right) + C$$

$$= \arcsin\left(\frac{e^x - e^{-x}}{6}\right) + C$$

47. $y = \cosh^{-1}(3x)$

$$y' = \frac{3}{\sqrt{9x^2-1}}$$

48. $y = \tanh^{-1}\left(\dfrac{x}{2}\right)$

$$y' = \frac{1}{1-(x/2)^2}\left(\frac{1}{2}\right) = \frac{2}{4-x^2}$$

49. $y = \sinh^{-1}(\tan x)$

$$y' = \frac{1}{\sqrt{\tan^2 x + 1}}(\sec^2 x) = |\sec x|$$

50. $y = \operatorname{sech}^{-1}(\cos 2x)$, $0 < x < \dfrac{\pi}{4}$

$$y' = \frac{-1}{\cos 2x\sqrt{1-\cos^2 2x}}(-2\sin 2x) = \frac{2\sin 2x}{\cos 2x|\sin 2x|} = \frac{2}{\cos 2x} = 2\sec 2x,$$

since $\sin 2x \ge 0$ for $0 < x < \pi/4$.

51. $y = \coth^{-1}(\sin 2x)$

$$y' = \frac{1}{1-\sin^2 2x}(2\cos 2x) = 2\sec 2x$$

52. $y = (\operatorname{csch}^{-1} x)^2$

$$y' = 2\operatorname{csch}^{-1} x\left(\frac{-1}{|x|\sqrt{1+x^2}}\right) = \frac{-2\operatorname{csch}^{-1} x}{|x|\sqrt{1+x^2}}$$

53. $y = 2x\sinh^{-1}(2x) - \sqrt{1+4x^2}$

$$y' = 2x\left(\frac{2}{\sqrt{1+4x^2}}\right) + 2\sinh^{-1}(2x) - \frac{4x}{\sqrt{1+4x^2}}$$

$$= 2\sinh^{-1}(2x)$$

54. $y = x\tanh^{-1} x + \ln\sqrt{1-x^2} = x\tanh^{-1} x + \dfrac{1}{2}\ln(1-x^2)$

$$y' = x\left(\frac{1}{1-x^2}\right) + \tanh^{-1} x + \frac{-x}{1-x^2} = \tanh^{-1} x$$

55. $y = a\operatorname{sech}^{-1}\left(\dfrac{x}{a}\right) - \sqrt{a^2-x^2}$

$$\frac{dy}{dx} = \frac{-1}{(x/a)\sqrt{1-(x^2/a^2)}} + \frac{x}{\sqrt{a^2-x^2}} = \frac{-a^2}{x\sqrt{a^2-x^2}} + \frac{x}{\sqrt{a^2-x^2}} = \frac{x^2-a^2}{x\sqrt{a^2-x^2}} = \frac{-\sqrt{a^2-x^2}}{x}$$

56. Equation of tangent line through $P = (x_0, y_0)$:

$$y - a\,\text{sech}^{-1}\frac{x_0}{a} + \sqrt{a^2 - x_0^2} = -\frac{\sqrt{a^2 - x_0^2}}{x_0}(x - x_0)$$

When $x = 0$,

$$y = a\,\text{sech}^{-1}\frac{x_0}{a} - \sqrt{a^2 - x_0^2} + \sqrt{a^2 - x_0^2} = a\,\text{sech}^{-1}\frac{x_0}{a}.$$

Hence, Q is the point $[0, a\,\text{sech}^{-1}(x_0/a)]$.

Distance from P to Q: $d = \sqrt{x_0^2 + \left(-\sqrt{a^2 - x_0^2}\right)^2} = a$

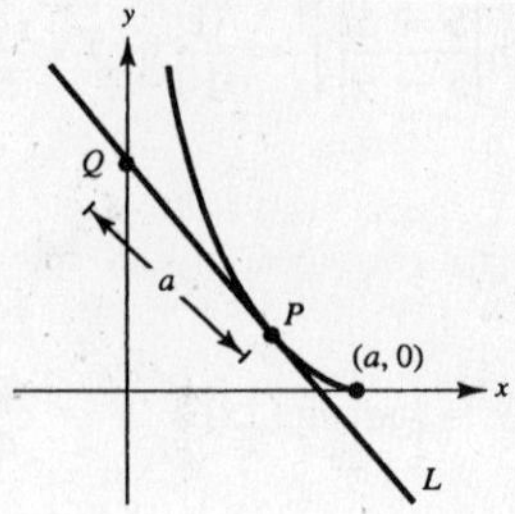

57. See "Definition of the Hyperbolic Functions" on page 709.

58.

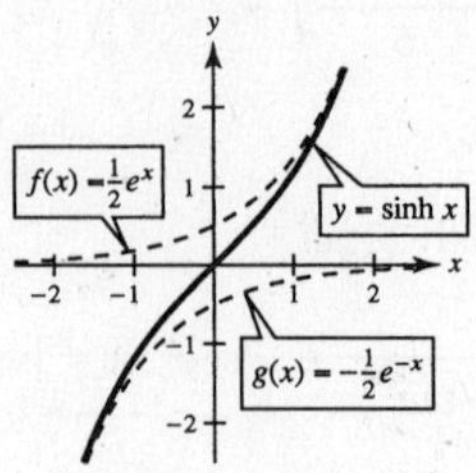

Domain: $(-\infty, \infty)$

Range: $(-\infty, \infty)$

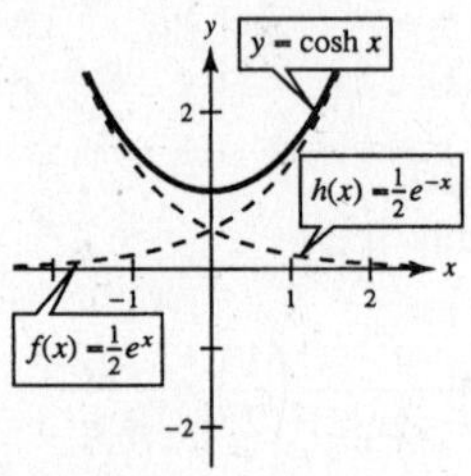

Domain: $(-\infty, \infty)$

Range: $[1, \infty)$

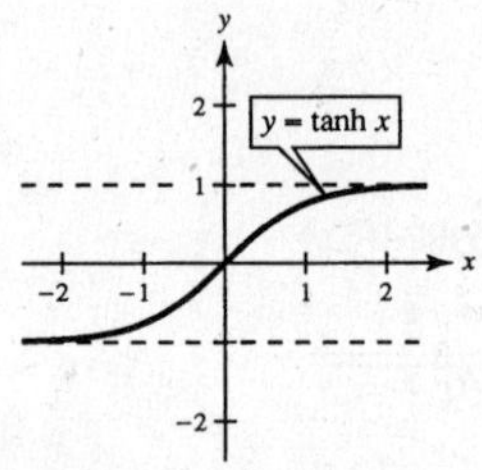

Domain: $(-\infty, \infty)$

Range: $(-1, 1)$

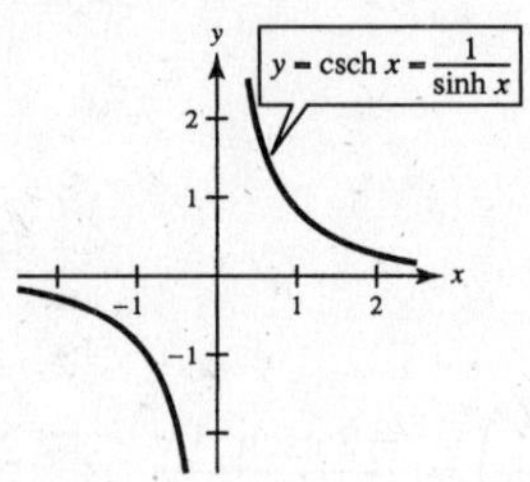

Domain: $(-\infty, 0) \cup (0, \infty)$

Range: $(-\infty, 0) \cup (0, \infty)$

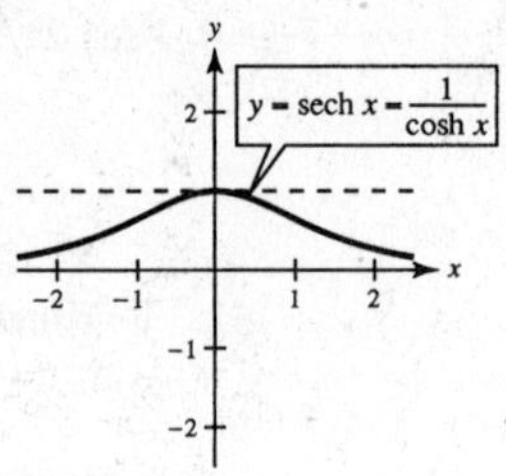

Domain: $(-\infty, \infty)$

Range: $(0, 1]$

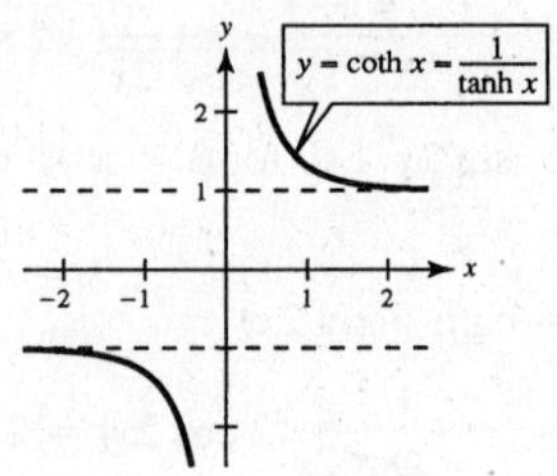

Domain: $(-\infty, 0) \cup (0, \infty)$

Range: $(-\infty, -1) \cup (1, \infty)$

59. $$\int \frac{1}{\sqrt{1 + e^{2x}}}\,dx = \int \frac{e^x}{e^x\sqrt{1 + (e^x)^2}}\,dx = -\text{csch}^{-1}(e^x) + C = -\ln\left(\frac{1 + \sqrt{1 + e^{2x}}}{e^x}\right) + C$$

60. $$\int \frac{x}{9 - x^4}\,dx = -\frac{1}{2}\int \frac{-2x}{9 - (x^2)^2}\,dx = -\frac{1}{2}\left(\frac{1}{6}\right)\ln\left|\frac{3 - x^2}{3 + x^2}\right| + C$$

$$= -\frac{1}{12}\ln\left|\frac{3 - x^2}{3 + x^2}\right| + C$$

61. Let $u = \sqrt{x}$, $du = \dfrac{1}{2\sqrt{x}}\,dx$.

$$\int \frac{1}{\sqrt{x}\sqrt{1 + x}}\,dx = 2\int \frac{1}{\sqrt{1 + \left(\sqrt{x}\right)^2}}\left(\frac{1}{2\sqrt{x}}\right)dx = 2\sinh^{-1}\sqrt{x} + C = 2\ln\left(\sqrt{x} + \sqrt{1 + x}\right) + C$$

62. Let $u = x^{3/2}$, $du = \frac{3}{2}\sqrt{x}\,dx$.

$$\int \frac{\sqrt{x}}{\sqrt{1+x^3}}\,dx = \frac{2}{3}\int \frac{1}{\sqrt{1+(x^{3/2})^2}}\left(\frac{3}{2}\sqrt{x}\right)dx = \frac{2}{3}\sinh^{-1}(x^{3/2}) + C = \frac{2}{3}\ln\left(x^{3/2} + \sqrt{1+x^3}\right) + C$$

63. $$\int \frac{-1}{4x - x^2}\,dx = \int \frac{1}{(x-2)^2 - 4}\,dx = \frac{1}{4}\ln\left|\frac{(x-2)-2}{(x-2)+2}\right| = \frac{1}{4}\ln\left|\frac{x-4}{x}\right| + C$$

64. $$\int \frac{dx}{(x+2)\sqrt{x^2+4x+8}} = \int \frac{dx}{(x+2)\sqrt{(x+2)^2+4}}$$
$$= -\frac{1}{2}\ln\left(\frac{2 + \sqrt{(x+2)^2+4}}{|x+2|}\right) + C$$

65. $$\int \frac{1}{1 - 4x - 2x^2}\,dx = \int \frac{1}{3 - 2(x+1)^2}\,dx = \frac{-1}{\sqrt{2}}\int \frac{\sqrt{2}}{\left[\sqrt{2}(x+1)\right]^2 - \left(\sqrt{3}\right)^2}\,dx$$
$$= \frac{-1}{2\sqrt{6}}\ln\left|\frac{\sqrt{2}(x+1) - \sqrt{3}}{\sqrt{2}(x+1) + \sqrt{3}}\right| + C = \frac{1}{2\sqrt{6}}\ln\left|\frac{\sqrt{2}(x+1) + \sqrt{3}}{\sqrt{2}(x+1) - \sqrt{3}}\right| + C$$

66. $$\int \frac{1}{(x+1)\sqrt{2x^2+4x+8}}\,dx = \int \frac{1}{(x+1)\sqrt{2(x+1)^2+6}}\,dx$$
$$= \frac{1}{\sqrt{2}}\int \frac{1}{(x+1)\sqrt{(x+1)^2 + \left(\sqrt{3}\right)^2}}\,dx = -\frac{1}{\sqrt{6}}\ln\left(\frac{\sqrt{3} + \sqrt{(x+1)^2+3}}{x+1}\right) + C$$

67. $$y = \int \frac{1}{\sqrt{16x^2 - 8x + 80}}\,dx$$
$$= \int \frac{1}{\sqrt{(4x-1)^2 + 79}}\,dx \quad (u = 4x - 1,\ a^2 = 79)$$
$$= \frac{1}{4}\ln\left[4x - 1 + \sqrt{(4x-1)^2 + 79}\right] + C$$
$$= \frac{1}{4}\ln\left[4x - 1 + \sqrt{16x^2 - 8x + 80}\right] + C$$
$$= \frac{\ln\left[2\sqrt{2}\sqrt{2x^2 - x + 10} + 4x - 1\right]}{4} + C$$

68. Let $u = 2(x-1)$, $du = 2\,dx$.

$$y = \int \frac{1}{(x-1)\sqrt{-4x^2 + 8x - 1}}\,dx$$
$$= \int \frac{2}{2(x-1)\sqrt{\left(\sqrt{3}\right)^2 - [2(x-1)]^2}}\,dx$$
$$= -\frac{1}{\sqrt{3}}\ln\left|\frac{\sqrt{3} + \sqrt{-4x^2 + 8x - 1}}{2(x-1)}\right| + C$$

69. $$y = \int \frac{x^3 - 21x}{5 + 4x - x^2}\,dx$$
$$= \int \left(-x - 4 + \frac{20}{5 + 4x - x^2}\right)dx$$
$$= \int (-x - 4)\,dx + 20\int \frac{1}{3^2 - (x-2)^2}\,dx$$
$$= -\frac{x^2}{2} - 4x + \frac{20}{6}\ln\left|\frac{(x-2)+3}{(x-2)-3}\right| + C$$
$$= -\frac{x^2}{2} - 4x + \frac{10}{3}\ln\left|\frac{x+1}{x-5}\right| + C$$
$$= \frac{-x^2}{2} - 4x - \frac{10}{3}\ln\left|\frac{x-5}{x+1}\right| + C$$

70. $$y = \int \frac{1 - 2x}{4x - x^2}\,dx$$
$$= \int \frac{4 - 2x}{4x - x^2}\,dx + 3\int \frac{1}{(x-2)^2 - 4}\,dx$$
$$= \ln|4x - x^2| + \frac{3}{4}\ln\left|\frac{(x-2)-2}{(x-2)+2}\right| + C$$
$$= \ln|4x - x^2| + \frac{3}{4}\ln\left|\frac{x-4}{x}\right| + C$$

71. $A = \int_0^4 \operatorname{sech}\frac{x}{2}\tanh\frac{x}{2}\,dx$

$= -2\operatorname{sech}\frac{x}{2}\Big]_0^4$

$= -2[\operatorname{sech} 2 - \operatorname{sech} 0]$

$= \frac{-4}{e^2 + e^{-2}} + 2$

72. $A = \int_0^2 \tanh 2x\,dx$

$= \int_0^2 \frac{e^{2x} - e^{-2x}}{e^{2x} + e^{-2x}}\,dx$

$= \frac{1}{2}\int_0^2 \frac{1}{e^{2x} + e^{-2x}}(2)(e^{2x} - e^{-2x})\,dx$

$= \left[\frac{1}{2}\ln(e^{2x} + e^{-2x})\right]_0^2$

$= \frac{1}{2}\ln(e^4 + e^{-4}) - \frac{1}{2}\ln 2$

$= \ln\sqrt{\frac{e^4 + e^{-4}}{2}} \approx 1.654$

73. $A = \int_0^2 \frac{5x}{\sqrt{x^4 + 1}}\,dx$

$= \frac{5}{2}\int_0^2 \frac{2x}{\sqrt{(x^2)^2 + 1}}\,dx$

$= \left[\frac{5}{2}\ln\left(x^2 + \sqrt{x^4 + 1}\right)\right]_0^2$

$= \frac{5}{2}\ln\left(4 + \sqrt{17}\right) \approx 5.237$

74. $A = \int_3^5 \frac{6}{\sqrt{x^2 - 4}}\,dx$

$= \left[6\ln\left(x + \sqrt{x^2 - 4}\right)\right]_3^5$

$= 6\ln\left(5 + \sqrt{21}\right) - 6\ln\left(3 + \sqrt{5}\right)$

$= 6\ln\left(\frac{5 + \sqrt{21}}{3 + \sqrt{5}}\right) \approx 3.626$

75. $y = \cosh x = \frac{e^x + e^{-x}}{2}$

$y' = \frac{e^x - e^{-x}}{2} = \sinh x$

76. $y = \operatorname{sech}^{-1} x$

$\operatorname{sech} y = x$

$-\operatorname{sech} y \tanh y(y') = 1$

$y' = \frac{-1}{\operatorname{sech} y \tanh y} = \frac{-1}{\operatorname{sech} y\sqrt{1 - \operatorname{sech}^2 y}} = \frac{-1}{x\sqrt{1 - x^2}}$

77. $y = \cosh^{-1} x$

$\cosh y = x$

$(\sinh y)(y') = 1$

$y' = \frac{1}{\sinh y} = \frac{1}{\sqrt{\cosh^2 y - 1}} = \frac{1}{\sqrt{x^2 - 1}}$

78. $\frac{d}{dx}[\operatorname{sech} x] = \frac{d}{dx}\left[\frac{2}{e^x + e^{-x}}\right] = \frac{-2}{(e^x + e^{-x})^2}(e^x - e^{-x})$

$= \frac{-2}{e^x + e^{-x}} \cdot \frac{e^x - e^{-x}}{e^x + e^{-x}} = -\operatorname{sech} x \tanh x$

79. $\int \frac{3k}{16}\,dt = \int \frac{1}{x^2 - 12x + 32}\,dx$

$\frac{3kt}{16} = \int \frac{1}{(x - 6)^2 - 4}\,dx = \frac{1}{2(2)}\ln\left|\frac{(x - 6) - 2}{(x - 6) + 2}\right| + C = \frac{1}{4}\ln\left|\frac{x - 8}{x - 4}\right| + C$

When $x = 0$: $t = 0$

$C = -\frac{1}{4}\ln(2)$

When $x = 1$: $t = 10$

$\frac{30k}{16} = \frac{1}{4}\ln\left|\frac{-7}{-3}\right| - \frac{1}{4}\ln(2) = \frac{1}{4}\ln\left(\frac{7}{6}\right)$

$k = \frac{2}{15}\ln\left(\frac{7}{6}\right)$

When $t = 20$: $\left(\frac{3}{16}\right)\left(\frac{2}{15}\right)\ln\left(\frac{7}{6}\right)(20) = \frac{1}{4}\ln\frac{x - 8}{2x - 8}$

$\ln\left(\frac{7}{6}\right)^2 = \ln\frac{x - 8}{2x - 8}$

$\frac{49}{36} = \frac{x - 8}{2x - 8}$

$62x = 104$

$x = \frac{104}{62} = \frac{52}{31} \approx 1.677$ kg

80. (a) $v(t) = -32t$

(b) $s(t) = \int v(t)\,dt = \int(-32t)\,dt = -16t^2 + C$

$s(0) = -16(0)^2 + C = 400 \Rightarrow C = 400$

$s(t) = -16t^2 + 400$

(c) $\dfrac{dv}{dt} = -32 + kv^2$

$$\int \frac{dv}{kv^2 - 32} = \int dt$$

$$\int \frac{dv}{32 - kv^2} = -\int dt$$

Let $u = \sqrt{k}\,v$, then $du = \sqrt{k}\,dv$.

$$\frac{1}{\sqrt{k}} \cdot \frac{1}{2\sqrt{32}} \ln\left|\frac{\sqrt{32} + \sqrt{k}\,v}{\sqrt{32} - \sqrt{k}\,v}\right| = -t + C$$

Since $v(0) = 0$, $C = 0$.

$$\ln\left|\frac{\sqrt{32} + \sqrt{k}\,v}{\sqrt{32} - \sqrt{k}\,v}\right| = -2\sqrt{32k}\,t$$

$$\frac{\sqrt{32} + \sqrt{k}\,v}{\sqrt{32} - \sqrt{k}\,v} = e^{-2\sqrt{32k}\,t}$$

$$\sqrt{32} + \sqrt{k}\,v = e^{-2\sqrt{32k}\,t}\left(\sqrt{32} - \sqrt{k}\,v\right)$$

$$v\left(\sqrt{k} + \sqrt{k}\,e^{-2\sqrt{32k}\,t}\right) = \sqrt{32}\left(e^{-2\sqrt{32k}\,t} - 1\right)$$

$$v = \frac{\sqrt{32}\left(e^{-2\sqrt{32k}\,t} - 1\right)}{\sqrt{k}\left(e^{-2\sqrt{32k}\,t} + 1\right)} \cdot \frac{e^{\sqrt{32k}\,t}}{e^{\sqrt{32k}\,t}}$$

$$= \frac{\sqrt{32}}{\sqrt{k}}\left[\frac{-\left(e^{\sqrt{32k}\,t} - e^{-\sqrt{32k}\,t}\right)}{e^{\sqrt{32k}\,t} + e^{-\sqrt{32k}\,t}}\right]$$

$$= -\frac{\sqrt{32}}{\sqrt{k}}\tanh\left(\sqrt{32k}\,t\right)$$

(d) $\displaystyle\lim_{t\to\infty}\left[-\frac{\sqrt{32}}{\sqrt{k}}\tanh\left(\sqrt{32k}\,t\right)\right] = -\frac{\sqrt{32}}{\sqrt{k}}.$

The velocity is bounded by $-\sqrt{32}/\sqrt{k}$.

(e) Since $\int\tanh(ct)\,dt = (1/c)\ln\cosh(ct)$ (which can be verified by differentiation), then

$$s(t) = \int -\frac{\sqrt{32}}{\sqrt{k}}\tanh\left(\sqrt{32k}\,t\right)dt$$

$$= -\frac{\sqrt{32}}{\sqrt{k}}\frac{1}{\sqrt{32k}}\ln\left[\cosh\left(\sqrt{32k}\,t\right)\right] + C$$

$$= -\frac{1}{k}\ln\left[\cosh\left(\sqrt{32k}\,t\right)\right] + C.$$

When $t = 0$,

$s(0) = C$

$= 400 \Rightarrow 400 - (1/k)\ln\left[\cosh\left(\sqrt{32k}\,t\right)\right]$.

When $k = 0.01$,

$s_2(t) = 400 - 100\ln\left(\cosh\sqrt{0.32}\,t\right)$

$s_1(t) = -16t^2 + 400.$

$s_1(t) = 0$ when $t = 5$ seconds.

$s_2(t) = 0$ when $t \approx 8.3$ seconds

When air resistance is not neglected, it takes approximately 3.3 more seconds to reach the ground.

81. As k increases, the time required for the object to reach the ground increases.

82. Let $y = \arcsin(\tanh x)$. Then,

$$\sin y = \tanh x = \frac{e^x - e^{-x}}{e^x + e^{-x}} \text{ and}$$

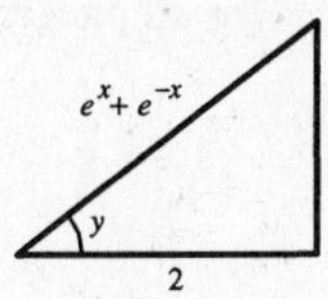

$$\tan y = \frac{e^x - e^{-x}}{2} = \sinh x.$$

Thus, $y = \arctan(\sinh x)$. Therefore,

$\arctan(\sinh x) = \arcsin(\tanh x)$.

Review Exercises for Chapter 11

1. $\displaystyle\lim_{x\to 3}\sec(\pi x) = \sec(3\pi) = -1$

2. $\displaystyle\lim_{\theta\to -1/2}\tan(2\pi\theta) = \tan(-\pi) = 0$

3. $\displaystyle\lim_{x\to\pi/2}\cot x = \cot\left(\frac{\pi}{2}\right) = 0$

4. $\displaystyle\lim_{x\to\pi/3}\sec x = \sec\frac{\pi}{3} = \frac{1}{\cos\frac{\pi}{3}} = 2$

5. $\lim_{\alpha\to 0} \dfrac{\sin 5\alpha}{3\alpha} = \lim_{\alpha\to 0} \dfrac{\sin 5\alpha}{5\alpha} \cdot \dfrac{5}{3} = \dfrac{5}{3}$

6. $\lim_{t\to 0} \dfrac{2(1-\cos t)}{10t} = \lim_{t\to 0} \dfrac{1}{5}\dfrac{(1-\cos t)}{t} = \dfrac{1}{5}(0) = 0$

7. $$\lim_{\Delta x\to 0} \frac{\sin[(\pi/6)+\Delta x]-(1/2)}{\Delta x} = \lim_{\Delta x\to 0} \frac{\sin(\pi/6)\cos\Delta x + \cos(\pi/6)\sin\Delta x - (1/2)}{\Delta x}$$
$$= \lim_{\Delta x\to 0} \frac{1}{2}\cdot\frac{(\cos\Delta x - 1)}{\Delta x} + \lim_{\Delta x\to 0} \frac{\sqrt{3}}{2}\cdot\frac{\sin\Delta x}{\Delta x}$$
$$= 0 + \frac{\sqrt{3}}{2}(1) = \frac{\sqrt{3}}{2}$$

8. $$\lim_{\Delta x\to 0} \frac{\cos(\pi+\Delta x)+1}{\Delta x} = \lim_{\Delta x\to 0} \frac{\cos\pi\cos\Delta x - \sin\pi\sin\Delta x + 1}{\Delta x}$$
$$= \lim_{\Delta x\to 0}\left[-\frac{(\cos\Delta x - 1)}{\Delta x}\right] - \lim_{\Delta x\to 0}\left[\sin\pi\frac{\sin\Delta x}{\Delta x}\right]$$
$$= -0 - (0)(1) = 0$$

9. $\lim_{x\to 0^+} \dfrac{\sin 4x}{5x} = \lim_{x\to 0^+}\left[\dfrac{4}{5}\left(\dfrac{\sin 4x}{4x}\right)\right] = \dfrac{4}{5}$

10. $\lim_{x\to 0^+} \dfrac{\sec x}{x} = \infty$

11. $\lim_{x\to 0^+} \dfrac{\csc 2x}{x} = \lim_{x\to 0^+} \dfrac{1}{x\sin 2x} = \infty$

12. $\lim_{x\to 0^-} \dfrac{\cos^2 x}{x} = -\infty$

13. $f(x) = \csc \dfrac{\pi x}{2}$

Nonremovable discontinuities at each even integer.

Continuous on

$(2k, 2k+2)$

for all integers k.

14. $f(x) = \tan 2x$

Nonremovable discontinuities when

$$x = \frac{(2n+1)\pi}{4}.$$

Continuous on

$$\left(\frac{(2n-1)\pi}{4}, \frac{(2n+1)\pi}{4}\right)$$

for all integers n.

15. $f(x) = \dfrac{-4}{x} + \tan\dfrac{\pi x}{8}$ is continuous on $[1, 3]$.

$f(1) = -4 + \tan\dfrac{\pi}{8} < 0$ and $f(3) = -\dfrac{4}{3} + \tan\dfrac{3\pi}{8} > 0$.

By the Intermediate Value Theorem, $f(1) = 0$ for at least one value of c between 1 and 3.

16. $f(x) = \dfrac{\tan 2x}{x}$

(a)

x	-0.1	-0.01	-0.001	0.001	0.01	0.1
$f(x)$	2.0271	2.0003	2.0000	2.0000	2.0003	2.0271

$\lim_{x\to 0} \dfrac{\tan 2x}{x} = 2$

(b) Yes, define

$$f(x) = \begin{cases} \dfrac{\tan 2x}{x}, & x \neq 0 \\ 2, & x = 0 \end{cases}.$$

Now $f(x)$ is continuous at $x = 0$.

17. $f(\theta) = 2\theta - 3\sin\theta$

$f'(\theta) = 2 - 3\cos\theta$

18. $g(\alpha) = 4\cos\alpha + 6$

$g'(\alpha) = -4\sin\alpha$

19. $h(x) = \sqrt{x}\sin x = x^{1/2}\sin x$

$h'(x) = \dfrac{1}{2\sqrt{x}}\sin x + \sqrt{x}\cos x$

20. $f(t) = t^3\cos t$

$f'(t) = t^3(-\sin t) + \cos t(3t^2)$

$= -t^3\sin t + 3t^2\cos t$

21. $y = \dfrac{x^2}{\cos x}$

$y' = \dfrac{\cos x\,(2x) - x^2(-\sin x)}{\cos^2 x} = \dfrac{2x\cos x + x^2\sin x}{\cos^2 x}$

22. $y = \dfrac{\sin x}{x^2}$

$y' = \dfrac{(x^2)\cos x - (\sin x)(2x)}{x^4} = \dfrac{x\cos x - 2\sin x}{x^3}$

23. $y = 3\sec x$

$y' = 3\sec x\tan x$

24. $y = 2x - \tan x$

$y' = 2 - \sec^2 x$

25. $y = \dfrac{1}{2}\csc 2x$

$y' = \dfrac{1}{2}(-\csc 2x\cot 2x)(2)$

$= -\csc 2x\cot 2x$

26. $y = \csc 3x + \cot 3x$

$y' = -3\csc 3x\cot 3x - 3\csc^2 3x$

$= -3\csc 3x(\cot 3x + \csc 3x)$

27. $y = \dfrac{x}{2} - \dfrac{\sin 2x}{4}$

$y' = \dfrac{1}{2} - \dfrac{1}{4}\cos 2x(2)$

$= \dfrac{1}{2}(1 - \cos 2x) = \sin^2 x$

28. $y = \dfrac{1 + \sin x}{1 - \sin x}$

$y' = \dfrac{(1 - \sin x)\cos x - (1 + \sin x)(-\cos x)}{(1 - \sin x)^2}$

$= \dfrac{2\cos x}{(1 - \sin x)^2}$

29. $y = \dfrac{2}{3}\sin^{3/2} x - \dfrac{2}{7}\sin^{7/2} x$

$y' = \sin^{1/2} x\cos x - \sin^{5/2} x\cos x$

$= (\cos x)\sqrt{\sin x}(1 - \sin^2 x)$

$= (\cos^3 x)\sqrt{\sin x}$

30. $y = \dfrac{\sec^7 x}{7} - \dfrac{\sec^5 x}{5}$

$y' = \sec^6 x(\sec x\tan x) - \sec^4 x(\sec x\tan x)$

$= \sec^5 x\tan x(\sec^2 x - 1)$

$= \sec^5 x\tan^3 x$

31. $y = -x\tan x$

$y' = -x\sec^2 x - \tan x$

32. $y = x\cos x - \sin x$

$y' = -x\sin x + \cos x - \cos x = -x\sin x$

33. $y = \dfrac{\sin\pi x}{x + 2}$

$y' = \dfrac{(x + 2)\pi\cos\pi x - \sin\pi x}{(x + 2)^2}$

34. $y = \dfrac{\cos(x - 1)}{x - 1}$

$y' = \dfrac{-(x - 1)\sin(x - 1) - \cos(x - 1)(1)}{(x - 1)^2}$

$= -\dfrac{1}{(x - 1)^2}[(x - 1)\sin(x - 1) + \cos(x - 1)]$

35. $y = \ln|\tan\theta|$

$y' = \dfrac{1}{\tan\theta} \cdot \sec^2\theta$

$= \dfrac{\cos\theta}{\sin\theta} \cdot \dfrac{1}{\cos^2\theta}$

$= \csc\theta \cdot \sec\theta$

or, $y' = \dfrac{1}{\tan\theta}(\tan^2\theta + 1)$

$= \tan\theta + \cot\theta$

36. $y = e^{-x}\cos(\pi x)$

$y' = e^{-x}(-\pi\sin(\pi x)) - e^{-x}\cos(\pi x)$

$= -e^{-x}(\pi\sin\pi x + \cos\pi x)$

37. $f(x) = \cot x$

$f'(x) = -\csc^2 x$

$f'' = -2\csc x(-\csc x \cdot \cot x)$

$= 2\csc^2 x\cot x$

38. $y = \sin^2 x$

$y' = 2\sin x\cos x = \sin 2x$

$y'' = 2\cos 2x$

39. $f(\theta) = 3\tan\theta$

$f'(\theta) = 3\sec^2\theta$

$f''(\theta) = 6\sec\theta(\sec\theta\tan\theta)$

$= 6\sec^2\theta\tan\theta$

40. $h(t) = 4\sin t - 5\cos t$

$h'(t) = 4\cos t + 5\sin t$

$h''(t) = -4\sin t + 5\cos t$

41. $g(x) = \csc x, \left[\dfrac{\pi}{6}, \dfrac{\pi}{3}\right]$

$g'(x) = -\csc x\cot x$

Left endpoint: $\left(\dfrac{\pi}{6}, 2\right)$ Maximum

Right endpoint: $\left(\dfrac{\pi}{3}, \dfrac{2\sqrt{3}}{3}\right)$ Minimum

42. $f(x) = 2x - \tan x, \left[-\dfrac{\pi}{2}, \dfrac{\pi}{4}\right]$

$f'(x) = 2 - \sec^2 x = 0$

$\sec x = \pm\sqrt{2}$

$x = \pm\dfrac{\pi}{4}$

$f\left(-\dfrac{\pi}{2}\right)$ undefined

$f\left(\dfrac{\pi}{4}\right) = \dfrac{\pi}{2} - 1 \approx 0.5708$

$f\left(-\dfrac{\pi}{4}\right) = -\dfrac{\pi}{2} + 1 \approx -0.5708$ minimum

Since $\lim_{x\to(-\pi/2)^+} f(x) = \infty$, there is no maximum.

43. $y = 2\sin x + 3\cos x$

$y' = 2\cos x - 3\sin x$

$y'' = -2\sin x - 3\cos x$

$y'' + y = -(2\sin x + 3\cos x) + (2\sin x + 3\cos x)$

$= 0$

44. $y = \dfrac{(10 - \cos x)}{x}$

$xy + \cos x = 10$

$xy' + y - \sin x = 0$

$xy' = \sin x - y$

$xy' + y = (\sin x - y) + y = \sin x$

45. $\tan(x + y) = x$

$(1 + y')\sec^2(x + y) = 1$

$y' = \dfrac{1 - \sec^2(x + y)}{\sec^2(x + y)}$

$= \dfrac{-\tan^2(x + y)}{\tan^2(x + y) + 1}$

$= -\dfrac{x^2}{x^2 + 1}$

At $(0, 0)$: $y' = 0$.

46. $x\cos y = 1$

$x[-y'\sin y] + \cos y = 0$

$y' = \dfrac{\cos y}{x\sin y}$

$= \dfrac{1}{x}\cot y = \dfrac{\cot y}{x}$

At $\left(2, \dfrac{\pi}{3}\right)$: $y' = \dfrac{1}{2\sqrt{3}}$.

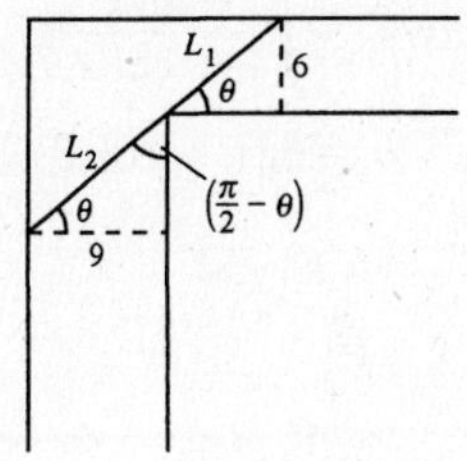

47. $\csc\theta = \dfrac{L_1}{6}$ or $L_1 = 6\csc\theta$ (see figure)

$$\csc\left(\frac{\pi}{2}-\theta\right) = \frac{L_2}{9} \text{ or } L_2 = 9\csc\left(\frac{\pi}{2}-\theta\right)$$

$$L = L_1 + L_2 = 6\csc\theta + 9\csc\left(\frac{\pi}{2}-\theta\right) = 6\csc\theta + 9\sec\theta$$

$$\frac{dL}{d\theta} = -6\csc\theta\cot\theta + 9\sec\theta\tan\theta = 0$$

$$\tan^3\theta = \frac{2}{3} \Longrightarrow \tan\theta = \frac{\sqrt[3]{2}}{\sqrt[3]{3}}$$

$$\sec\theta = \sqrt{1+\tan^2\theta} = \sqrt{1+\left(\frac{2}{3}\right)^{2/3}} = \frac{\sqrt{3^{2/3}+2^{2/3}}}{3^{1/3}}$$

$$\csc\theta = \frac{\sec\theta}{\tan\theta} = \frac{\sqrt{3^{2/3}+2^{2/3}}}{2^{1/3}}$$

$$L = 6\frac{(3^{2/3}+2^{2/3})^{1/2}}{2^{1/3}} + 9\frac{(3^{2/3}+2^{2/3})^{1/2}}{3^{1/3}} = 3(3^{2/3}+2^{2/3})^{3/2}\text{ ft} \approx 21.07\text{ ft (Compare to Exercise 48 using } a = 6 \text{ and } b = 9.)$$

48. Using Exercise 73 as a guide we have $L_1 = a\csc\theta$ and $L_2 = b\sec\theta$. Then $dL/d\theta = -a\csc\theta\cot\theta + b\sec\theta\tan\theta = 0$ when

$$\tan\theta = \sqrt[3]{a/b},\ \sec\theta = \frac{\sqrt{a^{2/3}+b^{2/3}}}{b^{1/3}},\ \csc\theta = \frac{\sqrt{a^{2/3}+b^{2/3}}}{a^{1/3}} \text{ and}$$

$$L = L_1 + L_2 = a\csc\theta + b\sec\theta = a\frac{(a^{2/3}+b^{2/3})^{1/2}}{a^{1/3}} + b\frac{(a^{2/3}+b^{2/3})^{1/2}}{b^{1/3}} = (a^{2/3}+b^{2/3})^{3/2}.$$

This matches the result of Exercise 47.

49. $\displaystyle\int (4x - 3\sin x)\,dx = 2x^2 + 3\cos x + C$

50. $\displaystyle\int (5\cos x - 2\sec^2 x)\,dx = 5\sin x - 2\tan x + C$

51. $\displaystyle\int \sin^3 x\cos x\,dx = \frac{1}{4}\sin^4 x + C$

52. $\displaystyle\int x\sin 3x^2\,dx = \frac{1}{6}\int(\sin 3x^2)(6x)\,dx = -\frac{1}{6}\cos 3x^2 + C$

53. $\displaystyle\int \frac{\sin\theta}{\sqrt{1-\cos\theta}}\,dx = \int(1-\cos\theta)^{-1/2}\sin\theta\,d\theta = 2(1-\cos\theta)^{1/2} + C = 2\sqrt{1-\cos\theta} + C$

54. $\displaystyle\int \frac{\cos x}{\sqrt{\sin x}}\,dx = \int(\sin x)^{-1/2}\cos x\,dx = 2(\sin x)^{1/2} + C = 2\sqrt{\sin x} + C$

55. $\displaystyle\int \tan^n x\sec^2 x\,dx = \frac{\tan^{n+1} x}{n+1} + C,\ n \neq -1$

56. $\displaystyle\int (1+\sec\pi x)^2\sec\pi x\tan\pi x\,dx = \frac{1}{\pi}\int(1+\sec\pi x)^2(\pi\sec\pi x\tan\pi x)\,dx = \frac{1}{3\pi}(1+\sec\pi x)^3 + C$

57. $\displaystyle\int \sec 2x\tan 2x\,dx = \frac{1}{2}\int(\sec 2x\tan 2x)(2)\,dx = \frac{1}{2}\sec 2x + C$

58. $\displaystyle\int \cot^4\alpha\csc^2\alpha\,d\alpha = -\int(\cot\alpha)^4(-\csc^2\alpha)\,d\alpha = -\frac{1}{5}\cot^5\alpha + C$

59. $\displaystyle\int \frac{\sin x}{1+\cos x}\,dx = -\int \frac{-\sin x}{1+\cos x}\,dx$

$= -\ln|1+\cos x| + C$

60. Let $u = \tan x$, $du = \sec^2 x\,dx$

$\displaystyle\int e^{\tan x}\sec^2 x\,dx = \int e^u\,du$

$= e^u + C$

$= e^{\tan x} + C$

61. $f(x) = \displaystyle\int \cos\frac{x}{2}\,dx = 2\sin\frac{x}{2} + C$

Since $f(0) = 3 = 2\sin 0 + C$, $C = 3$. Thus,

$f(x) = 2\sin\dfrac{x}{2} + 3.$

62. $f(x) = \displaystyle\int \pi\sec\pi x\tan\pi x\,dx = \sec\pi x + C$

Since $f(1/3) = 1 = \sec(\pi/3) + C$, $C = -1$. Thus

$f(x) = \sec\pi x - 1.$

63. $\displaystyle\int_0^{\pi}\cos\left(\frac{x}{2}\right)dx = 2\int_0^{\pi}\cos\left(\frac{x}{2}\right)\frac{1}{2}\,dx = \left[2\sin\left(\frac{x}{2}\right)\right]_0^{\pi} = 2$

64. $\displaystyle\int_{-\pi/4}^{\pi/4}\sin 2x\,dx = 0$ since $\sin 2x$ is an odd function.

65. $\displaystyle\int_0^{\pi/3}\sec\theta\,d\theta = \Big[\ln|\sec\theta + \tan\theta|\Big]_0^{\pi/3} = \ln\left(2+\sqrt{3}\right)$

66. $\displaystyle\int_0^{\pi/4}\tan\left(\frac{\pi}{4} - x\right)dx = \left[\ln\left|\cos\left(\frac{\pi}{4} - x\right)\right|\right]_0^{\pi/4}$

$= 0 - \ln\left(\dfrac{1}{\sqrt{2}}\right) = \dfrac{1}{2}\ln 2$

67.

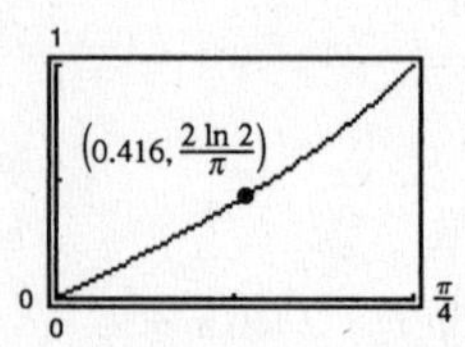

$\text{Average value} = \dfrac{1}{\frac{\pi}{4} - 0}\displaystyle\int_0^{\pi/4}\tan x\,dx$

$= \dfrac{-4}{\pi}\ln|\cos x|\Big]_0^{\pi/4}$

$= \dfrac{-4}{\pi}\ln\left(\dfrac{\sqrt{2}}{2}\right)$

$= \dfrac{2}{\pi}\ln 2 \approx 0.4413$

$\tan x = \dfrac{2}{\pi}\ln 2 \Rightarrow x \approx 0.4156$

68.

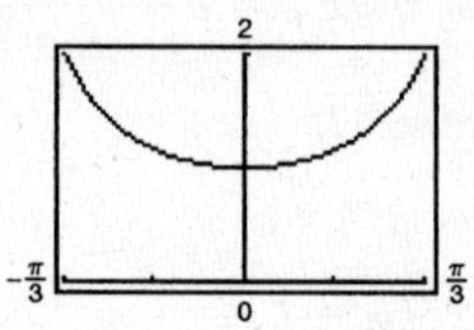

$\text{Average value} = \dfrac{1}{\frac{\pi}{3} - \left(-\frac{\pi}{3}\right)}\displaystyle\int_{-\pi/3}^{\pi/3}\sec x\,dx$

$= \dfrac{3}{2\pi}\ln|\sec x + \tan x|\Big]_{-\pi/3}^{\pi/3}$

$= \dfrac{3}{2\pi}\left[\ln(2+\sqrt{3}) - \ln(2-\sqrt{3})\right]$

$= \dfrac{3}{2\pi}\ln\left(\dfrac{2+\sqrt{3}}{2-\sqrt{3}}\right) \approx 1.2576$

$\sec x = \dfrac{3}{2\pi}\ln\left(\dfrac{2+\sqrt{3}}{2-\sqrt{3}}\right) \Rightarrow x \approx \pm 0.6515$

69. $\text{Area} = \displaystyle\int_0^{\pi/3}\sec^2 x\,dx = \tan x\Big]_0^{\pi/3} = \sqrt{3}$

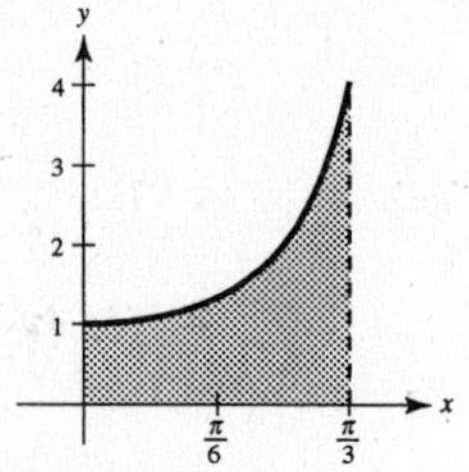

70. $\displaystyle\int_{-\pi/4}^{\pi/4}\cos x\,dx = 2\int_0^{\pi/4}\cos x\,dx = \Big[2\sin x\Big]_0^{\pi/4} = \sqrt{2}$

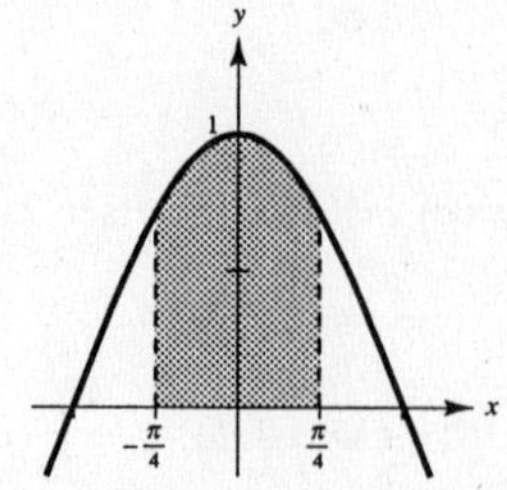

71. $F(x) = \int_0^x \tan^4 t\, dt$

$F'(x) = \tan^4 x$

72. $F(x) = \int_{\pi/2}^{x^3} \cos t\, dt$

$F'(x) = \cos(x^3)(3x^2) = 3x^2 \cos(x^3)$

73. (a) $C = 0.1\int_8^{20}\left[12 \sin \frac{\pi(t-8)}{12}\right] dt = \left[-\frac{14.4}{\pi} \cos \frac{\pi(t-8)}{12}\right]_8^{20} = \frac{-14.4}{\pi}(-1-1) \approx \$9.17$

74. $\int_0^2 1.75 \sin \frac{\pi t}{2}\, dt = -\frac{2}{\pi}\left[1.75 \cos \frac{\pi t}{2}\right]_0^2 = -\frac{2}{\pi}(1.75)(-1-1) = \frac{7}{\pi} \approx 2.2282\text{liters}$

Increase is

$$\frac{7}{\pi} - \frac{5.1}{\pi} = \frac{1.9}{\pi} \approx 0.6048 \text{ liters.}$$

75. $\int_0^\pi \sqrt{x} \sin x\, dx,\ n = 4$

Trapezoidal Rule: 2.3290

Simpson's Rule: 2.4491

76. $\int_0^{\pi/2} \sqrt{1 + \cos^2 x}\, dx,\ n = 2$

Trapezoidal Rule: 1.9100

Simpson's Rule: 1.9146

77. $y = \tan(\arcsin x) = \frac{x}{\sqrt{1-x^2}}$

$y' = \frac{(1-x^2)^{1/2} + x^2(1-x^2)^{-1/2}}{1-x^2} = (1-x^2)^{-3/2}$

78. $y = \arctan(x^2 - 1)$

$y' = \frac{2x}{1 + (x^2-1)^2} = \frac{2x}{x^4 - 2x^2 + 2}$

79. $y = x \operatorname{arcsec} x$

$y' = \frac{x}{|x|\sqrt{x^2-1}} + \operatorname{arcsec} x$

80. $y = \frac{1}{2} \arctan e^{2x}$

$y' = \frac{1}{2}\left(\frac{1}{1+e^{4x}}\right)(2e^{2x}) = \frac{e^{2x}}{1+e^{4x}}$

81. $y = x(\arcsin x)^2 - 2x + 2\sqrt{1-x^2} \arcsin x$

$y' = \frac{2x \arcsin x}{\sqrt{1-x^2}} + (\arcsin x)^2 - 2 + \frac{2\sqrt{1-x^2}}{\sqrt{1-x^2}} - \frac{2x}{\sqrt{1-x^2}} \arcsin x = (\arcsin x)^2$

82. $y = \sqrt{x^2-4} - 2 \operatorname{arcsec} \frac{x}{2},\ 2 < x < 4$

$y' = \frac{x}{\sqrt{x^2-4}} - \frac{1}{(|x|/2)\sqrt{(x/2)^2-1}} = \frac{x}{\sqrt{x^2-4}} - \frac{4}{|x|\sqrt{x^2-4}} = \frac{x^2-4}{|x|\sqrt{x^2-4}} = \frac{\sqrt{x^2-4}}{x}$

83. Let $u = e^{2x}$, $du = 2e^{2x}\, dx$.

$$\int \frac{1}{e^{2x}+e^{-2x}}\, dx = \int \frac{e^{2x}}{1+e^{4x}}\, dx = \frac{1}{2}\int \frac{1}{1+(e^{2x})^2}(2e^{2x})\, dx = \frac{1}{2}\arctan(e^{2x}) + C$$

84. Let $u = 5x$, $du = 5\, dx$.

$$\int \frac{1}{3+25x^2}\, dx = \frac{1}{5}\int \frac{1}{(\sqrt{3})^2 + (5x)^2}(5)\, dx$$

$$= \frac{1}{5\sqrt{3}} \arctan \frac{5x}{\sqrt{3}} + C$$

85. Let $u = 16 + x^2$, $du = 2x\, dx$.

$$\int \frac{x}{16+x^2}\, dx = \frac{1}{2}\int \frac{1}{16+x^2}(2x)\, dx = \frac{1}{2}\ln(16+x^2) + C$$

86. $\int \frac{4-x}{\sqrt{4-x^2}}\,dx = 4\int \frac{1}{\sqrt{4-x^2}}\,dx + \frac{1}{2}\int (4-x^2)^{-1/2}(-2x)\,dx = 4\arcsin\frac{x}{2} + \sqrt{4-x^2} + C$

87. Let $u = \arctan\left(\frac{x}{2}\right)$, $du = \frac{2}{4+x^2}\,dx$.

$$\int \frac{\arctan(x/2)}{4+x^2}\,dx = \frac{1}{2}\int \left(\arctan\frac{x}{2}\right)\left(\frac{2}{4+x^2}\right)dx$$
$$= \frac{1}{4}\left(\arctan\frac{x}{2}\right)^2 + C$$

88. Let $u = \arcsin x$, $du = \frac{1}{\sqrt{1-x^2}}\,dx$.

$$\int \frac{\arcsin x}{\sqrt{1-x^2}}\,dx = \frac{1}{2}(\arcsin x)^2 + C$$

89. $y' = \sqrt{1-y^2}$

(a)

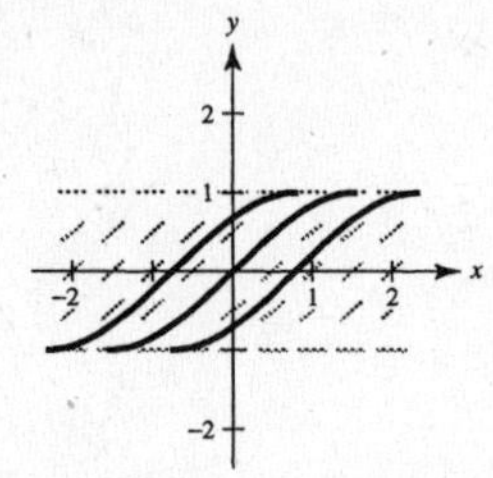

(b) Rate of change is greatest when $y = 0$. (slope lines are slope 1)

Rate of change is least when $y = \pm 1$. (slope lines are horizontal)

(c) $\int \frac{dy}{\sqrt{1-y^2}} = \int dx$

$\arcsin y = x + C$

$y = \sin(x + C)$, $-\frac{\pi}{2} \le x + C \le \frac{\pi}{2}$

90. $\int \frac{dy}{\sqrt{A^2 - y^2}} = \int \sqrt{\frac{k}{m}}\,dt$

$\arcsin\left(\frac{y}{A}\right) = \sqrt{\frac{k}{m}}\,t + C$

Since $y = 0$ when $t = 0$, you have $C = 0$. Thus,

$$\sin\left(\sqrt{\frac{k}{m}}\,t\right) = \frac{y}{A}$$
$$y = A\sin\left(\sqrt{\frac{k}{m}}\,t\right).$$

91. $y = 2x - \cosh\sqrt{x}$

$y' = 2 - \frac{1}{2\sqrt{x}}(\sinh\sqrt{x}) = 2 - \frac{\sinh\sqrt{x}}{2\sqrt{x}}$

92. $y = x\tanh^{-1} 2x$

$y' = x\left(\frac{2}{1-4x^2}\right) + \tanh^{-1} 2x = \frac{2x}{1-4x^2} + \tanh^{-1} 2x$

93. Let $u = x^2$, $du = 2x\,dx$.

$$\int \frac{x}{\sqrt{x^4-1}}\,dx = \frac{1}{2}\int \frac{1}{\sqrt{(x^2)^2-1}}(2x)\,dx$$
$$= \frac{1}{2}\ln\left(x^2 + \sqrt{x^4-1}\right) + C$$

94. Let $u = x^3$, $du = 3x^2\,dx$.

$$\int x^2(\operatorname{sech} x^3)^2\,dx = \frac{1}{3}\int (\operatorname{sech} x^3)^2(3x^2)\,dx = \frac{1}{3}\tanh x^3 + C$$

Problem Solving for Chapter 11

1. $f(x) = a + b \cos cx$

$f'(x) = -bc \sin cx$

At $(0, 1)$: $a + b = 1$ Equation 1

At $\left(\frac{\pi}{4}, \frac{3}{2}\right)$: $a + b \cos\left(\frac{c\pi}{4}\right) = \frac{3}{2}$ Equation 2

$-bc \sin\left(\frac{c\pi}{4}\right) = 1$ Equation 3

From Equation 1, $a = 1 - b$. Equation 2 becomes $(1 - b) + b \cos\left(\frac{c\pi}{4}\right) = \frac{3}{2} \Rightarrow -b + b \cos \frac{c\pi}{4} = \frac{1}{2}$.

From Equation 3, $b = \dfrac{-1}{c \sin\left(\frac{c\pi}{4}\right)}$. Thus $\dfrac{1}{c \sin\left(\frac{c\pi}{4}\right)} + \dfrac{-1}{c \sin\left(\frac{c\pi}{4}\right)} \cos\left(\frac{c\pi}{4}\right) = \frac{1}{2}$

$1 - \cos\left(\frac{c\pi}{4}\right) = \frac{1}{2} c \sin\left(\frac{c\pi}{4}\right)$.

Graphing the equation $g(c) = \frac{1}{2} c \sin\left(\frac{c\pi}{4}\right) + \cos\left(\frac{c\pi}{4}\right) - 1$, you see that many values of c will work.

One answer: $c = 2$, $b = -\frac{1}{2}$, $a = \frac{3}{2} \Rightarrow f(x) = \frac{3}{2} - \frac{1}{2} \cos 2x$

2. (a)

z (degrees)	0.1	0.01	0.0001
$\frac{\sin z}{z}$	0.0174533	0.0174533	0.0174533

(b) $\lim_{z \to 0} \frac{\sin z}{z} \approx 0.0174533$

In fact, $\lim_{z \to 0} \frac{\sin z}{z} = \frac{\pi}{180}$.

(c) $$\frac{d}{dz}(\sin z) = \lim_{\Delta z \to 0} \frac{\sin(z + \Delta z) - \sin z}{\Delta z}$$

$$= \lim_{\Delta z \to 0} \frac{\sin z \cdot \cos \Delta z + \sin \Delta z \cdot \cos z - \sin z}{\Delta z}$$

$$= \lim_{\Delta z \to 0} \left[\sin z\left(\frac{\cos \Delta z - 1}{\Delta z}\right)\right] + \lim_{\Delta z \to 0} \left[\cos z\left(\frac{\sin \Delta z}{\Delta z}\right)\right]$$

$$= (\sin z)(0) + (\cos z)\left(\frac{\pi}{180}\right) = \frac{\pi}{180} \cos z$$

(d) $S(90) = \sin\left(\frac{\pi}{180} 90\right) = \sin \frac{\pi}{2} = 1$; $C(180) = \cos\left(\frac{\pi}{180} 180\right) = -1$

$$\frac{d}{dz}S(z) = \frac{d}{dz} \sin(cz) = c \cdot \cos(cz) = \frac{\pi}{180}C(z)$$

(e) The formulas for the derivatives are more complicated in degrees.

3. $E(\phi) = \dfrac{\tan\phi(1 - 0.1\tan\phi)}{0.1 + \tan\phi} = \dfrac{10\tan\phi - \tan^2\phi}{1 + 10\tan\phi}$

$$E'(\phi) = \frac{(1 + 10\tan\phi)(10\sec^2\phi - 2\tan\phi\sec^2\phi) - (10\tan\phi - \tan^2\phi)10\sec^2\phi}{(1 + 10\tan\phi)^2} = 0$$

$\Rightarrow (1 + 10\tan\phi)(10\sec^2\phi - 2\tan\phi\sec^2\phi) = (10\tan\phi - \tan^2\phi)10\sec^2\phi$

$\Rightarrow 10\sec^2\phi - 2\tan\phi\sec^2\phi + 100\tan\phi\sec^2\phi - 20\tan^2\phi\sec^2\phi = 100\tan\phi\sec^2\phi - 10\tan^2\phi\sec^2\phi$

$\Rightarrow 10 - 2\tan\phi = 10\tan^2\phi$

$\Rightarrow 10\tan^2\phi + 2\tan\phi - 10 = 0$

$$\tan\phi = \frac{-2 \pm \sqrt{4 + 400}}{20} \approx 0.90499, -1.10499$$

Using the positive value, $\phi \approx 0.736$, or $42.1°$.

4. (a)

x	0.1	0.2	0.3	0.4	0.5	1.0
$\sin x$	0.09983	0.19867	0.29552	0.38942	0.47943	0.84147

$\sin x < x$

(b) Let $f(x) = \sin x$. For $x \ge 1$, $\sin x < 1$ is obviously true, so assume $0 < x < 1$. Then $f'(x) = \cos x$ and on $[0, x]$ you have by the Mean Value Theorem,

$$f'(c) = \frac{f(x) - f(0)}{x - 0}, \quad 0 < c < x$$

$$\cos(c) = \frac{\sin x}{x}.$$

Hence, $\left|\dfrac{\sin x}{x}\right| = |\cos(c)| < 1$

$\Rightarrow |\sin x| < |x|$

$\Rightarrow \sin x < x.$

5. $d = \sqrt{13^2 + x^2}$, $\sin\theta = \dfrac{x}{d}$

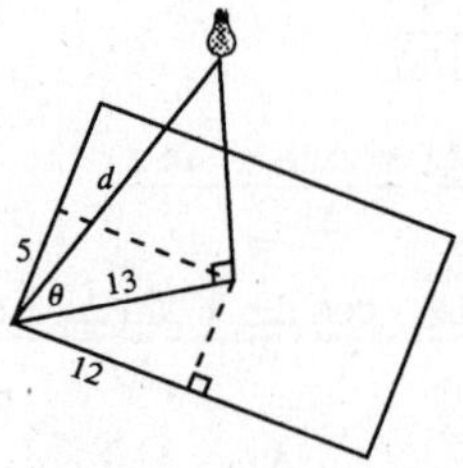

Let A be the amount of illumination at one of the corners, as indicated in the figure. Then

$$A = \frac{kI}{(13^2 + x^2)}\sin\theta = \frac{kIx}{(13^2 + x^2)^{3/2}}$$

$$A'(x) = kI\frac{(x^2 + 169)^{3/2}(1) - x\left(\frac{3}{2}\right)(x^2 + 169)^{1/2}(2x)}{(169 + x^2)^3} = 0$$

$\Rightarrow (x^2 + 169)^{3/2} = 3x^2(x^2 + 169)^{1/2}$

$x^2 + 169 = 3x^2$

$2x^2 = 169$

$$x = \frac{13}{\sqrt{2}} \approx 9.19 \text{ feet.}$$

By the First Derivative Test, this is a maximum.

6. (a) Let $A = \int_0^b \frac{f(x)}{f(x) + f(b - x)}\,dx.$

Let $u = b - x$, $du = -dx$.

$$A = \int_b^0 \frac{f(b - u)}{f(b - u) + f(u)}(-du) = \int_0^b \frac{f(b - u)}{f(b - u) + f(u)}\,du = \int_0^b \frac{f(b - x)}{f(b - x) + f(x)}\,dx$$

Then,

$$2A = \int_0^b \frac{f(x)}{f(x) + f(b - x)}\,dx + \int_0^b \frac{f(b - x)}{f(b - x) + f(x)}\,dx = \int_0^b 1\,dx = b.$$

Thus, $A = \frac{b}{2}$.

(b) $b = 1 \Rightarrow \int_0^1 \frac{\sin x}{\sin(1 - x) + \sin x}\,dx = \frac{1}{2}$

7. Since $-|f(x)| \le f(x) \le |f(x)|$,

$$-\int_a^b |f(x)|\,dx \le \int_a^b f(x)\,dx \le \int_a^b |f(x)|\,dx \Rightarrow \left|\int_a^b f(x)\,dx\right| \le \int_a^b |f(x)|\,dx.$$

8. By Theorem 6.8, $0 < f(x) \le M \Rightarrow \int_a^b f(x)\,dx \le \int_a^b M\,dx = M(b - a).$

Similarly, $m \le f(x) \Rightarrow m(b - a) = \int_a^b m\,dx \le \int_a^b f(x)\,dx.$

Thus, $m(b - a) \le \int_a^b f(x)\,dx \le M(b - a)$. On the interval $[0, 1]$, $1 \le \sqrt{1 + x^4} \le \sqrt{2}$ and $b - a = 1$.

Thus, $1 \le \int_0^1 \sqrt{1 + x^4}\,dx \le \sqrt{2}$. $\left(\textbf{Note: } \int_0^1 \sqrt{1 + x^4}\,dx \approx 1.0894\right)$

9. (a) The area of the sector of a circle of radius r is $A = \frac{tr^2}{2} = \frac{1}{2}r^2 t$. Since $r = 1$, $A = \frac{1}{2}(1)^2 t = \frac{1}{2}t$ or $t = 2A$.

(b) Let A = Area of AOP.

$$\begin{aligned}
A &= \frac{1}{2}\text{Base} \cdot \text{Height} - \int_1^{\cosh t} \sqrt{x^2 - 1}\,dx \\
&= \frac{1}{2}\cosh t \cdot \sinh t - \int_1^{\cosh t} \sqrt{x^2 - 1}\,dx \\
&= \frac{1}{2}\cosh t \cdot \sinh t - \left(\frac{-\ln\left|\sqrt{x^2 - 1} + x\right|}{2} + \frac{x\sqrt{x^2 - 1}}{2}\right)_1^{\cosh t} \\
&= \frac{1}{2}\cosh t \cdot \sinh t - \frac{1}{2}\left[x\sqrt{x^2 - 1} - \ln\left|\sqrt{x^2 - 1} + x\right|\right]_1^{\cosh t} \\
&= \frac{1}{2}\cosh t \cdot \sinh t - \frac{1}{2}(\cosh t \sinh t - \ln|\sinh t + \cosh t|) \\
&= \frac{1}{2}\ln e^t \\
&= \frac{1}{2}t
\end{aligned}$$

Therefore, $t = 2A$.

10. $f(x) = \sin(\ln x)$

(a) Domain: $x > 0$ or $(0, \infty)$

(b) $f(x) = 1 = \sin(\ln x) \Rightarrow \ln x = \frac{\pi}{2} + 2k\pi.$

Two values are $x = e^{\pi/2}, e^{(\pi/2)+2\pi}.$

(c) $f(x) = -1 = \sin(\ln x) \Rightarrow \ln x = \frac{3\pi}{2} + 2k\pi.$

Two values are $x = e^{-\pi/2}, e^{3\pi/2}.$

(d) Since the range of the sine function is $[-1, 1]$, parts (b) and (c) show that the range of f is $[-1, 1]$.

(e) $f'(x) = \frac{1}{x}\cos(\ln x)$

$$f'(x) = 0 \Rightarrow \cos(\ln x) = 0 \Rightarrow \ln x = \frac{\pi}{2} + k\pi \Rightarrow$$

$$x = e^{\pi/2} \text{ on } [1, 10]$$

$$\left.\begin{aligned} f(e^{\pi/2}) &= 1 \\ f(1) &= 0 \\ f(10) &\approx 0.7440 \end{aligned}\right\} \text{ Maximum is 1 at } x = e^{\pi/2} \approx 4.8105$$

(f) 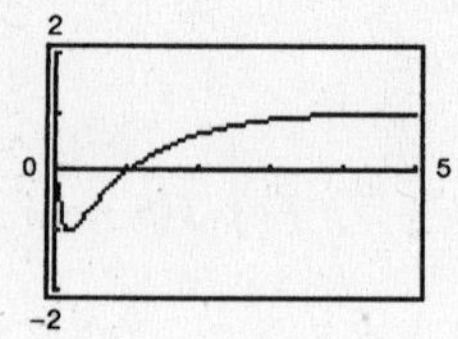

$\lim_{x\to 0^+} f(x)$ seems to be $-\frac{1}{2}$. (This is incorrect.)

(g) For the points $x = e^{\pi/2}, e^{-3\pi/2}, e^{-7\pi/2}, \ldots$ we have $f(x) = 1$.

For the points $x = e^{-\pi/2}, e^{-5\pi/2}, e^{-9\pi/2}, \ldots$ we have $f(x) = -1$.

That is, as $x \to 0^+$, there is an infinite number of points where $f(x) = 1$, and an infinite number where $f(x) = -1$. Thus $\lim_{x\to 0+} \sin(\ln x)$ does not exist.

You can verify this by graphing $f(x)$ on small intervals close to the origin.

11. $\tan\theta_1 = \frac{3}{x}$

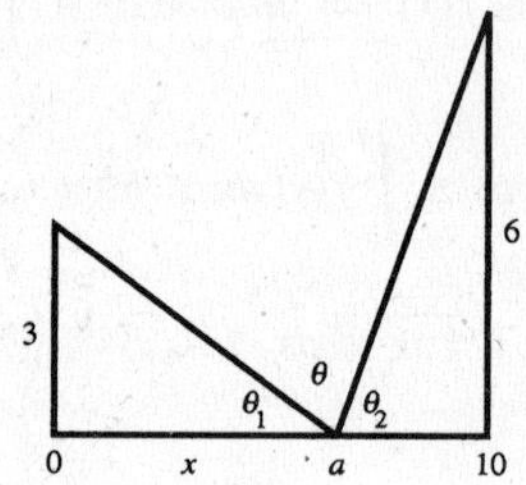

$\tan\theta_2 = \frac{6}{10 - x}$

Minimize $\theta_1 + \theta_2$:

$$f(x) = \theta_1 + \theta_2 = \arctan\left(\frac{3}{x}\right) + \arctan\left(\frac{6}{10 - x}\right)$$

$$f'(x) = \frac{1}{1 + \frac{9}{x^2}}\left(\frac{-3}{x^2}\right) + \frac{1}{1 + \frac{36}{(10 - x)^2}}\left(\frac{6}{(10 - x)^2}\right) = 0$$

$$\frac{3}{x^2 + 9} = \frac{6}{(10 - x)^2 + 36}$$

$$(10 - x)^2 + 36 = 2(x^2 + 9)$$

$$100 - 20x + x^2 + 36 = 2x^2 + 18$$

$$x^2 + 20x - 118 = 0$$

$$x = \frac{-20 \pm \sqrt{20^2 - 4(-118)}}{2} = -10 \pm \sqrt{218}$$

$a = -10 + \sqrt{218} \approx 4.7648 \quad f(a) \approx 1.4153$

$\theta = \pi - (\theta_1 + \theta_2) \approx 1.7263$ or $98.9°$

Endpoints: $a = 0$: $\theta \approx 1.0304$

$a = 10$: $\theta \approx 1.2793$

Maximum is 1.7263 at $a = -10 + \sqrt{218} \approx 4.7648$.

12. Let $u = \tan x$, $du = \sec^2 x\, dx$.

$$\text{Area} = \int_0^{\pi/4} \frac{1}{\sin^2 x + 4\cos^2 x}\, dx = \int_0^{\pi/4} \frac{\sec^2 x}{\tan^2 x + 4}\, dx$$

$$= \int_0^1 \frac{du}{u^2 + 4}$$

$$= \left[\frac{1}{2}\arctan\left(\frac{u}{2}\right)\right]_0^1$$

$$= \frac{1}{2}\arctan\left(\frac{1}{2}\right)$$

13. (a)

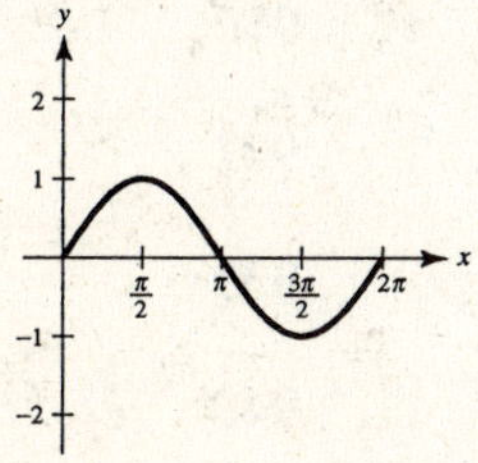

$$\int_0^{\pi} \sin x\, dx = -\int_{\pi}^{2\pi} \sin x\, dx \implies \int_0^{2\pi} \sin x\, dx = 0$$

(b)

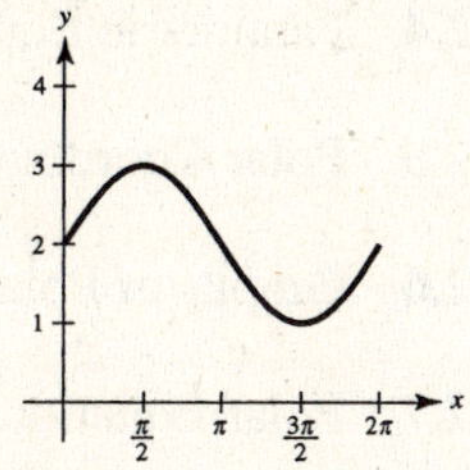

$$\int_0^{2\pi} (\sin x + 2)\, dx = 2(2\pi) = 4\pi$$

(c)

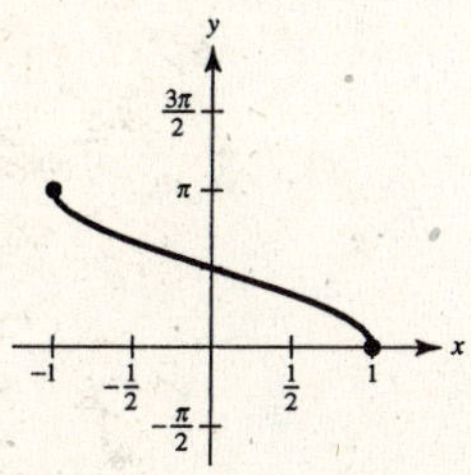

$$\int_{-1}^{1} \arccos x\, dx = \frac{1}{2}(2\pi) = \pi$$

(d)

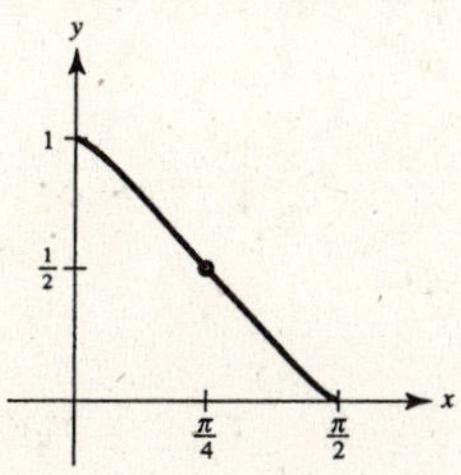

$y = \dfrac{1}{1 + (\tan x)^{\sqrt{2}}}$ is symmetric with respect to the point $\left(\dfrac{\pi}{4}, \dfrac{1}{2}\right)$. $\displaystyle\int_0^{\pi/2} \frac{1}{1 + (\tan x)^{\sqrt{2}}}\, dx = \frac{\pi}{2}\left(\frac{1}{2}\right) = \frac{\pi}{4}$

14. $f(x) = \sin \beta x$

(a) $f'(x) = \beta \cos \beta x$

$f''(x) = -\beta^2 \sin \beta x$

$f'''(x) = -\beta^3 \cos \beta x$

$f^{(4)}(x) = \beta^4 \sin \beta x$

(b) $f''(x) + \beta^2 f(x) = -\beta^2 \sin \beta x + \beta^2(\sin \beta x) = 0$

(c) $f^{(2k)}(x) = (-1)^k \beta^{2k} \sin \beta x$

$f^{(2k-1)}(x) = (-1)^{k+1} \beta^{2k-1} \cos \beta x$

CHAPTER 12
Topics in Analytic Geometry

Section 12.1 Introduction to Conics: Parabolas 887

Section 12.2 Ellipses and Implicit Differentiation 895

Section 12.3 Hyperbolas and Implicit Differentiation 906

Section 12.4 Parametric Equations and Calculus 916

Section 12.5 Polar Coordinates and Calculus 931

Section 12.6 Graphs of Polar Equations 939

Section 12.7 Polar Equations of Conics 948

Review Exercises 954

Problem Solving 968

CHAPTER 12
Topics in Analytic Geometry

Section 12.1 Introduction to Conics: Parabolas

1. $y^2 = -4x$
Vertex: (0, 0)
Opens to the left since p is negative.
Matches graph (e).

2. $x^2 = 2y$
Vertex: (0, 0)
$p = \frac{1}{2} > 0$
Opens upward.
Matches graph (b).

3. $x^2 = -8y$
Vertex: (0, 0)
Opens downward since p is negative.
Matches graph (d).

4. $y^2 = -12x$
Vertex: (0, 0)
$p = -3 < 0$
Opens to the left.
Matches graph (f).

5. $(y - 1)^2 = 4(x - 3)$
Vertex: (3, 1)
Opens to the right since p is positive.
Matches graph (a).

6. $(x + 3)^2 = -2(y - 1)$
Vertex: $(-3, 1)$
$p = -\frac{1}{2} < 0$
Opens downward.
Matches graph (c).

7. $y = \frac{1}{2}x^2$
$x^2 = 2y$
$x^2 = 4\left(\frac{1}{2}\right)y \Rightarrow h = 0, k = 0, p = \frac{1}{2}$
Vertex: (0, 0)
Focus: $\left(0, \frac{1}{2}\right)$
Directrix: $y = -\frac{1}{2}$

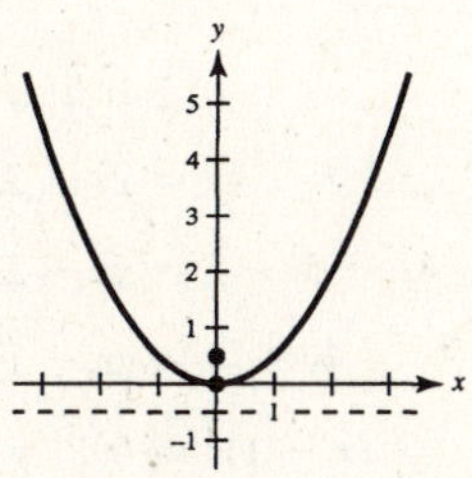

8. $y = -2x^2 \Rightarrow x^2 = 4\left(-\frac{1}{8}\right)y$
Vertex: (0, 0)
Focus: $\left(0, -\frac{1}{8}\right)$
Directrix: $y = \frac{1}{8}$

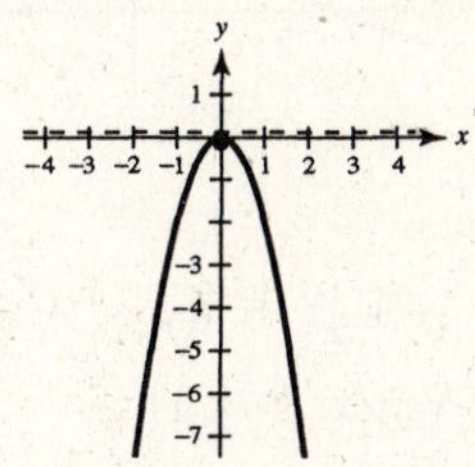

9. $y^2 = -6x$
$y^2 = 4\left(-\frac{3}{2}\right)x \Rightarrow h = 0, k = 0, p = -\frac{3}{2}$
Vertex: (0, 0)
Focus: $\left(-\frac{3}{2}, 0\right)$
Directrix: $x = \frac{3}{2}$

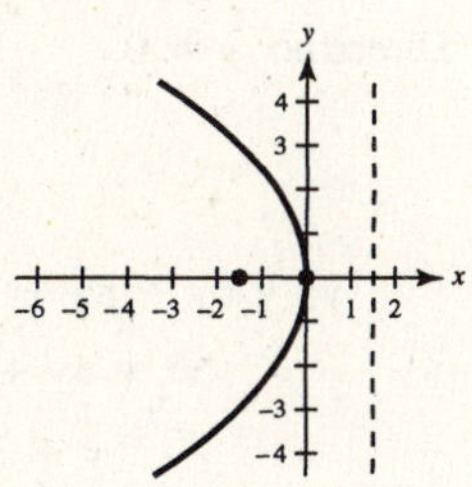

10. $y^2 = 3x \Rightarrow y^2 = 4\left(\frac{3}{4}\right)x$
Vertex: (0, 0)
Focus: $\left(\frac{3}{4}, 0\right)$
Directrix: $x = -\frac{3}{4}$

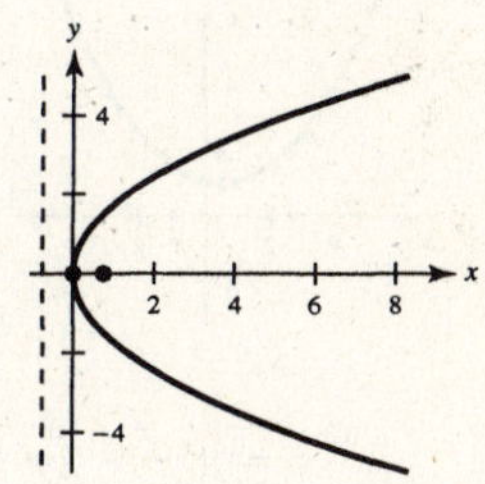

11. $x^2 + 6y = 0$
$x^2 = -6y = 4\left(-\frac{3}{2}\right)y \Rightarrow h = 0, k = 0, p = -\frac{3}{2}$
Vertex: (0, 0)
Focus: $\left(0, -\frac{3}{2}\right)$
Directrix: $y = \frac{3}{2}$

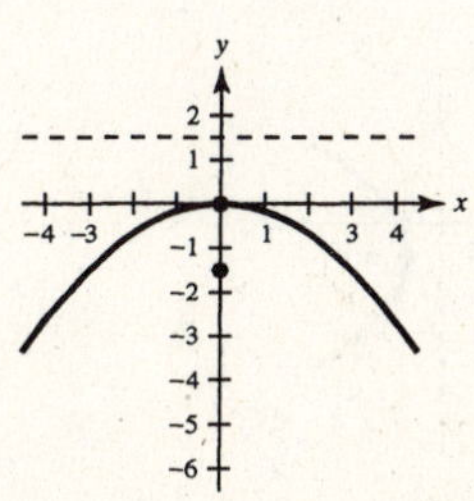

12. $x + y^2 = 0$

$y^2 = -x = 4\left(-\frac{1}{4}\right)x$

Vertex: $(0, 0)$

Focus: $\left(-\frac{1}{4}, 0\right)$

Directrix: $x = \frac{1}{4}$

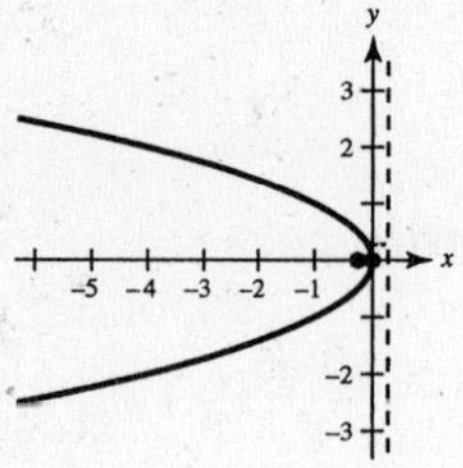

13. $(x - 1)^2 + 8(y + 2) = 0$

$(x - 1)^2 = 4(-2)(y + 2)$

$h = 1, k = -2, p = -2$

Vertex: $(1, -2)$

Focus: $(1, -4)$

Directrix: $y = 0$

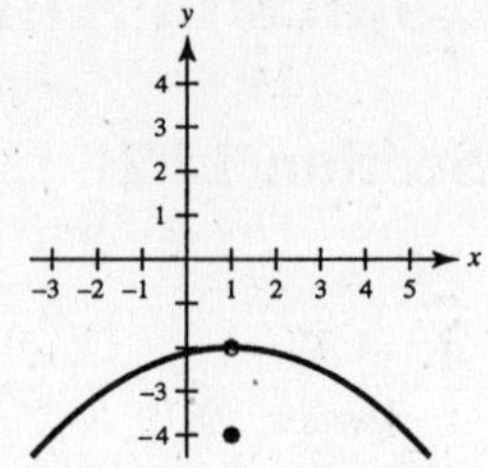

14. $(x + 5) + (y - 1)^2 = 0$

$(y - 1)^2 = 4\left(-\frac{1}{4}\right)(x + 5)$

Vertex: $(-5, 1)$

Focus: $\left(-5 + \left(-\frac{1}{4}\right), 1\right) \Rightarrow \left(-\frac{21}{4}, 1\right)$

Directrix: $x = -5 - \left(-\frac{1}{4}\right) = -\frac{19}{4}$

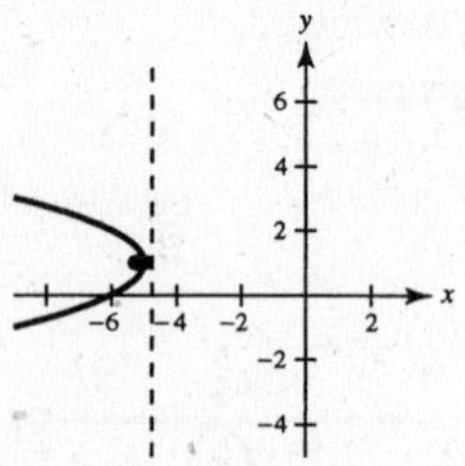

15. $\left(x + \frac{3}{2}\right)^2 = 4(y - 2)$

$\left(x + \frac{3}{2}\right)^2 = 4(1)(y - 2)$

$h = -\frac{3}{2}, k = 2, p = 1$

Vertex: $\left(-\frac{3}{2}, 2\right)$

Focus: $\left(-\frac{3}{2}, 3\right)$

Directrix: $y = 1$

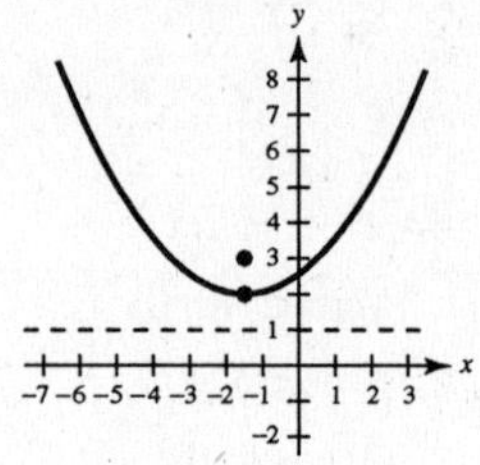

16. $\left(x + \frac{1}{2}\right)^2 = 4(y - 1) = 4(1)(y - 1)$

Vertex: $\left(-\frac{1}{2}, 1\right)$

Focus: $\left(-\frac{1}{2}, 1 + 1\right) \Rightarrow \left(-\frac{1}{2}, 2\right)$

Directrix: $y = 1 - 1 = 0$

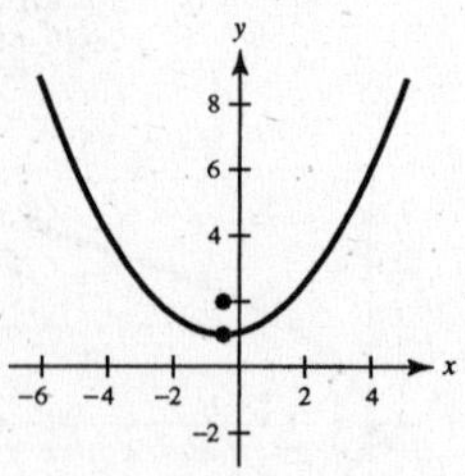

17.

$$y = \tfrac{1}{4}(x^2 - 2x + 5)$$
$$4y = x^2 - 2x + 5$$
$$4y - 5 + 1 = x^2 - 2x + 1$$
$$4y - 4 = (x - 1)^2$$
$$(x - 1)^2 = 4(1)(y - 1)$$

$h = 1, k = 1, p = 1$

Vertex: $(1, 1)$

Focus: $(1, 2)$

Directrix: $y = 0$

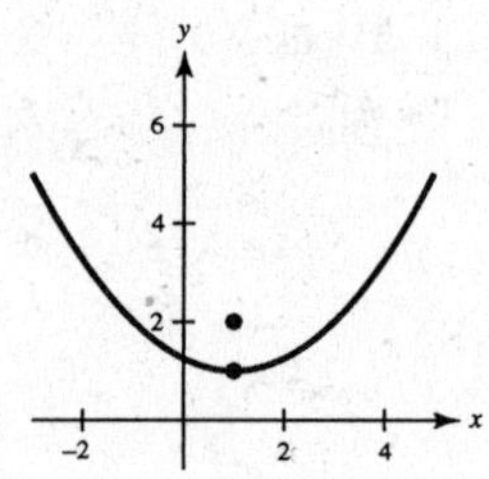

18.

$$x = \tfrac{1}{4}(y^2 + 2y + 33)$$
$$4x = y^2 + 2y + 1 - 1 + 33 = (y + 1)^2 + 32$$
$$(y + 1)^2 = 4(1)(x - 8)$$

Vertex: $(8, -1)$

Focus: $(9, -1)$

Directrix: $x = 7$

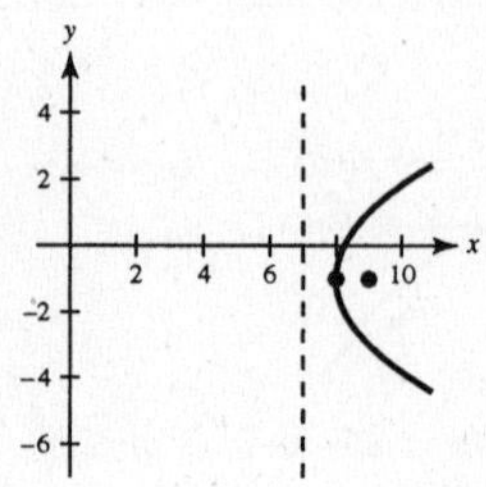

19.

$$y^2 + 6y + 8x + 25 = 0$$
$$y^2 + 6y + 9 = -8x - 25 + 9$$
$$(y + 3)^2 = 4(-2)(x + 2)$$

$h = -2, k = -3, p = -2$

Vertex: $(-2, -3)$

Focus: $(-4, -3)$

Directrix: $x = 0$

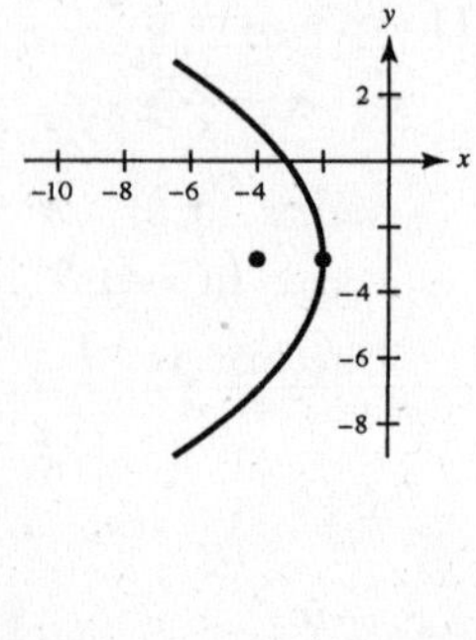

20. $y^2 - 4y - 4x = 0$

$y^2 - 4y + 4 = 4x + 4$

$(y - 2)^2 = 4(1)(x + 1)$

Vertex: $(-1, 2)$
Focus: $(0, 2)$
Directrix: $x = -2$

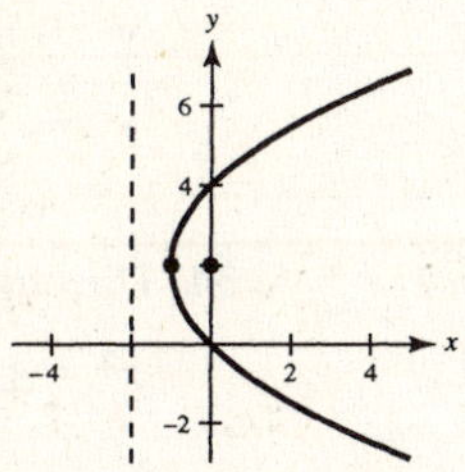

21. $x^2 + 4x + 6y - 2 = 0$

$x^2 + 4x = -6y + 2$

$x^2 + 4x + 4 = -6y + 2 + 4$

$(x + 2)^2 = -6(y - 1)$

$(x + 2)^2 = 4\left(-\frac{3}{2}\right)(y - 1)$

$h = -2, k = 1, p = -\frac{3}{2}$

Vertex: $(-2, 1)$

Focus: $\left(-2, -\frac{1}{2}\right)$

Directrix: $y = \frac{5}{2}$

On the graphing calculator, enter:

$y_1 = -\frac{1}{6}(x^2 + 4x - 2)$

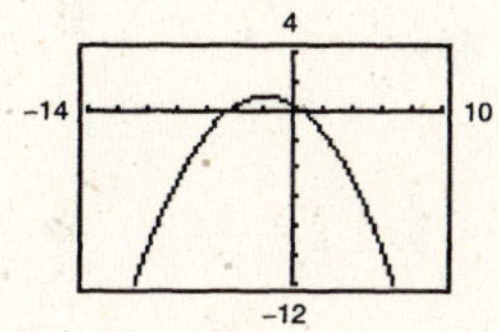

22. $x^2 - 2x + 8y + 9 = 0$

$x^2 - 2x + 1 = -8y - 9 + 1$

$(x - 1)^2 = -8(y + 1) = 4(-2)(y + 1)$

Vertex: $(1, -1)$
Focus: $(1, -3)$
Directrix: $y = 1$

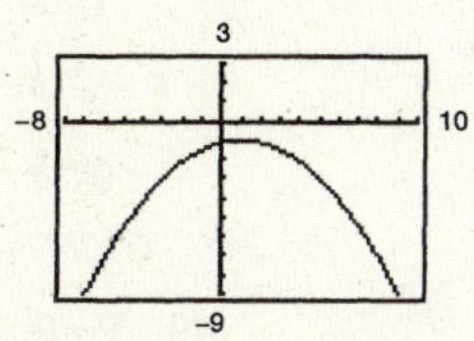

23. $y^2 + x + y = 0$

$y^2 + y + \frac{1}{4} = -x + \frac{1}{4}$

$\left(y + \frac{1}{2}\right)^2 = 4\left(-\frac{1}{4}\right)\left(x - \frac{1}{4}\right)$

$h = \frac{1}{4}, k = -\frac{1}{2}, p = -\frac{1}{4}$

Vertex: $\left(\frac{1}{4}, -\frac{1}{2}\right)$

Focus: $\left(0, -\frac{1}{2}\right)$

Directrix: $x = \frac{1}{2}$

To use a graphing calculator, enter:

$y_1 = -\frac{1}{2} + \sqrt{\frac{1}{4} - x}$

$y_2 = -\frac{1}{2} - \sqrt{\frac{1}{4} - x}$

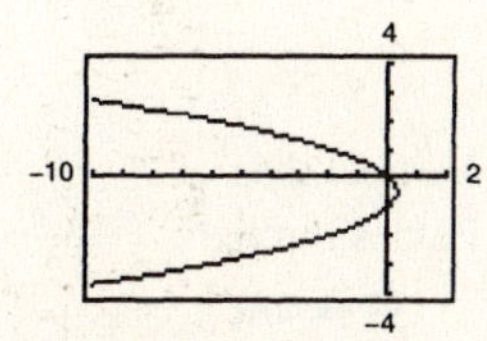

24. $y^2 - 4x - 4 = 0$

$y^2 = 4x + 4 = 4(1)(x + 1)$

Vertex: $(-1, 0)$
Focus: $(0, 0)$
Directrix: $x = -2$

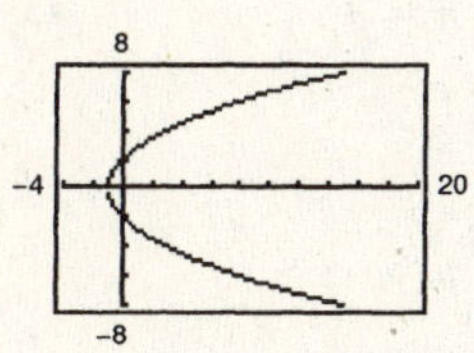

25. $y^2 - 8x = 0 \Rightarrow y = \pm\sqrt{8x}$

$x - y + 2 = 0 \Rightarrow y = x + 2$

The point of tangency is (2, 4).

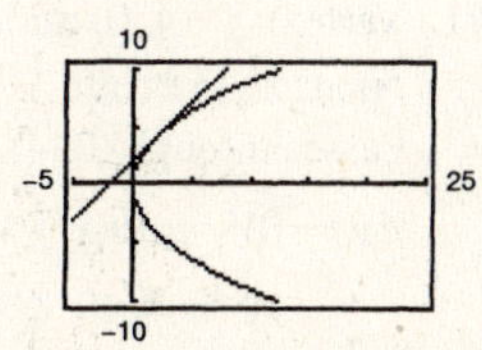

26. $x^2 + 12y = 0 \Rightarrow y_1 = -\frac{1}{12}x^2$

$x + y - 3 = 0 \Rightarrow y_2 = 3 - x$

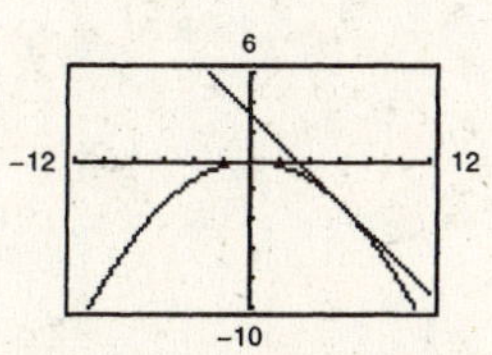

Using the trace or intersect feature, the point of tangency is $(6, -3)$.

27. Vertex: $(0, 0) \Rightarrow h = 0, k = 0$

Graph opens upward.

$x^2 = 4py$

Point on graph: (3, 6)

$3^2 = 4p(6)$

$9 = 24p$

$\frac{3}{8} = p$

Thus, $x^2 = 4\left(\frac{3}{8}\right)y \Rightarrow x^2 = \frac{3}{2}y$.

28. Point: $(-2, 6)$

$y^2 = 4px$

$6^2 = 4p(-2)$

$p = -\frac{9}{2}$

$y^2 = -18x$

29. Vertex: $(0, 0) \Rightarrow h = 0, k = 0$

Focus: $\left(0, -\frac{3}{2}\right) \Rightarrow p = -\frac{3}{2}$

$x^2 = 4py$

$x^2 = 4\left(-\frac{3}{2}\right)y$

$x^2 = -6y$

30. Focus: $(2, 0) \Rightarrow p = 2$

$y^2 = 4px$

$y^2 = 8x$

31. Vertex: $(0, 0) \Rightarrow h = 0, k = 0$

Focus: $(-2, 0) \Rightarrow p = -2$

$y^2 = 4px$

$y^2 = 4(-2)x$

$y^2 = -8x$

32. Focus: $(0, -2) \Rightarrow p = -2$

$x^2 = 4py$

$x^2 = -8y$

33. Vertex: $(0, 0) \Rightarrow h = 0, k = 0$

Directrix: $y = -1 \Rightarrow p = 1$

$x^2 = 4py$

$x^2 = 4(1)y$

$x^2 = 4y$

34. Directrix: $y = 3 \Rightarrow p = -3$

$x^2 = 4py$

$x^2 = -12y$

35. Vertex: $(0, 0) \Rightarrow h = 0, k = 0$

Directrix: $x = 2 \Rightarrow p = -2$

$y^2 = 4px$

$y^2 = 4(-2)x$

$y^2 = -8x$

36. Directrix: $x = -3 \Rightarrow p = 3$

$y^2 = 4px$

$y^2 = 12x$

37. Vertex: $(0, 0) \Rightarrow h = 0, k = 0$

Horizontal axis and passes through the point $(4, 6)$

$y^2 = 4px$

$6^2 = 4p(4)$

$36 = 16p \Rightarrow p = \frac{9}{4}$

$y^2 = 4\left(\frac{9}{4}\right)x$

$y^2 = 9x$

38. Vertical axis

Passes through: $(-3, -3)$

$x^2 = 4py$

$(-3)^2 = 4p(-3)$

$9 = -12p$

$p = -\frac{3}{4}$

$x^2 = -3y$

39. Vertex: $(3, 1)$ and opens downward. Passes through $(2, 0)$ and $(4, 0)$.

$y = -(x - 2)(x - 4)$

$= -x^2 + 6x - 8$

$= -(x - 3)^2 + 1$

$(x - 3)^2 = -(y - 1)$

40. Vertex: $(5, 3) \Rightarrow h = 5, k = 3$

Passes through: $(4.5, 4)$

$(y - k)^2 = 4p(x - h)$

$(y - 3)^2 = 4p(x - 5)$

$1 = 4p(4.5 - 5)$

$p = -\frac{1}{2}$

$(y - 3)^2 = -2(x - 5)$

41. Vertex: $(-4, 0)$ and opens to the right. Passes through $(0, 4)$.

$(y - 0)^2 = 4p(x + 4)$

$4^2 = 4p(0 + 4)$

$16 = 16p$

$1 = p$

$y^2 = 4(x + 4)$

42. Vertex: $(3, -3) \Rightarrow h = 3, k = -3$

Passes through: $(0, 0)$

$(x - h)^2 = 4p(y - k)$

$(x - 3)^2 = 4p(y + 3)$

$9 = 12p$

$p = \frac{3}{4}$

$(x - 3)^2 = 3(y + 3)$

43. Vertex: $(5, 2)$

Focus: $(3, 2)$

Horizontal axis

$p = 3 - 5 = -2$

$(y - 2)^2 = 4(-2)(x - 5)$

$(y - 2)^2 = -8(x - 5)$

44. Vertex: $(-1, 2) \Rightarrow h = -1, k = 2$

Focus: $(-1, 0) \Rightarrow p = -2$

$(x - h)^2 = 4p(y - k)$

$(x + 1)^2 = 4(-2)(y - 2)$

$(x + 1)^2 = -8(y - 2)$

45. Vertex: $(0, 4)$

Directrix: $y = 2$

Vertical axis

$p = 4 - 2 = 2$

$(x - 0)^2 = 4(2)(y - 4)$

$x^2 = 8(y - 4)$

46. Vertex: $(-2, 1) \Rightarrow h = -2, k = 1$

Directrix: $x = 1 \Rightarrow p = -3$

$(y - k)^2 = 4p(x - h)$

$(y - 1)^2 = 4(-3)(x + 2)$

$(y - 1)^2 = -12(x + 2)$

47. Focus: $(2, 2)$

Directrix: $x = -2$

Horizontal axis

Vertex: $(0, 2)$

$p = 2 - 0 = 2$

$(y - 2)^2 = 4(2)(x - 0)$

$(y - 2)^2 = 8x$

48. Focus: $(0, 0)$

Directrix: $y = 8 \Rightarrow p = -4 \Rightarrow h = 0, k = 4$

$(x - h)^2 = 4p(y - k)$

$x^2 = 4(-4)(y - 4)$

$x^2 = -16(y - 4)$

49. $(y - 3)^2 = 6(x + 1)$

For the upper half of the parabola:

$y - 3 = +\sqrt{6(x + 1)}$

$y = \sqrt{6(x + 1)} + 3$

50. $(y + 1)^2 = 2(x - 4)$

$y + 1 = \pm\sqrt{2(x - 4)}$

$y = -1 \pm \sqrt{2(x - 4)}$

Lower half of parabola: $y = -1 - \sqrt{2(x - 4)}$

51. $x^2 = 4y$

$2x = 4\dfrac{dy}{dx}$

$\dfrac{dy}{dx} = \dfrac{x}{2}$

52. $x^2 = \dfrac{1}{4}y$

$y = 4x^2$

$\dfrac{dy}{dx} = 8x$

53. $y^2 = 6x$

$2y\dfrac{dy}{dx} = 6$

$\dfrac{dy}{dx} = \dfrac{3}{y}$

54. $y^2 = -8x$

$2y\dfrac{dy}{dx} = -8$

$\dfrac{dy}{dx} = -\dfrac{4}{y}$

55. $(x - 2)^2 = 6(y + 3)$

$2(x - 2) = 6\dfrac{dy}{dx}$

$\dfrac{dy}{dx} = \dfrac{x - 2}{3}$

56. $(x + 4)^2 = -3(y - 1)$

$2(x + 4) = -3\dfrac{dy}{dx}$

$\dfrac{dy}{dx} = -\dfrac{2(x + 4)}{3}$

57. $(y + 3)^2 = -8(x - 2)$

$2(y + 3)\dfrac{dy}{dx} = -8$

$\dfrac{dy}{dx} = -\dfrac{4}{y + 3}$

58. $\left(y - \dfrac{3}{2}\right)^2 = 4(x + 4)$

$2\left(y - \dfrac{3}{2}\right)\dfrac{dy}{dx} = 4$

$\dfrac{dy}{dx} = \dfrac{4}{2y - 3}$

59. $x^2 = 2y$

$2x = 2\dfrac{dy}{dx}$

$\dfrac{dy}{dx} = x$

At $(4, 8)$, $m = \left.\dfrac{dy}{dx}\right]_{(4,\,8)} = 4.$

Tangent line: $y - 8 = 4(x - 4)$

$4x - y - 8 = 0$

60. $x^2 = 2y$

$2x = 2\dfrac{dy}{dx}$

$\dfrac{dy}{dx} = x$

At $\left(-3, \dfrac{9}{2}\right)$, $m = \dfrac{dy}{dx}(-3) = -3.$

Tangent line: $y - \dfrac{9}{2} = -3(x + 3)$

$y - \dfrac{9}{2} = -3x - 9$

$3x + y + \dfrac{9}{2} = 0$

$6x + 2y + 9 = 0$

61. $y = -2x^2$

$\dfrac{dy}{dx} = -4x$

At $(-1, -2)$, $m = \left.\dfrac{dy}{dx}\right]_{(-1,\,-2)} = 4.$

Tangent line: $y + 2 = 4(x + 1)$

$4x - y + 2 = 0$

62. $y = -2x^2$

$\dfrac{dy}{dx} = -4x$

At $(2, -8)$, $m = \dfrac{dy}{dx}(2) = -8$.

Tangent line: $y + 8 = -8(x - 2)$

$8x + y - 8 = 0$

63. $y^2 = 2(x - 3)$

$2y\dfrac{dy}{dx} = 2$

$\dfrac{dy}{dx} = \dfrac{1}{y}$

At $(5, 2)$, $m = \left.\dfrac{dy}{dx}\right]_{(5, 2)} = \dfrac{1}{2}$.

Tangent line: $y - 2 = \dfrac{1}{2}(x - 5)$

$2y - 4 = x - 5$

$x - 2y - 1 = 0$

64. $y^2 = 2(x - 3)$

$2y\dfrac{dy}{dx} = 2$

$\dfrac{dy}{dx} = \dfrac{1}{y}$

At $(11, 4)$, $m = \left.\dfrac{dy}{dx}\right]_{(11, 4)} = \dfrac{1}{4}$.

Tangent line: $y - 4 = \dfrac{1}{4}(x - 11)$

$4y - 16 = x - 11$

$0 = x - 4y + 5$

65. $(x - 1)^2 = 6(y + 2)$

$2(x - 1) = 6\dfrac{dy}{dx}$

$\dfrac{dy}{dx} = \dfrac{x - 1}{3}$

At $(-5, 4)$, $m = \left.\dfrac{dy}{dx}\right]_{(-5, 4)} = -2$.

Tangent line: $y - 4 = -2(x + 5)$

$2x + y + 6 = 0$

66. $(x - 1)^2 = 6(y + 2)$

$2(x - 1) = 6\dfrac{dy}{dx}$

$\dfrac{dy}{dx} = \dfrac{x - 1}{3}$

At $(10, 11.5)$, $m = \left.\dfrac{dy}{dx}\right]_{(10, 11.5)} = 3$.

Tangent line: $y - 11.5 = 3(x - 10)$

$2y - 23 = 6x - 60$

$6x - 2y - 37 = 0$

67. $R = 265x - \frac{5}{4}x^2$

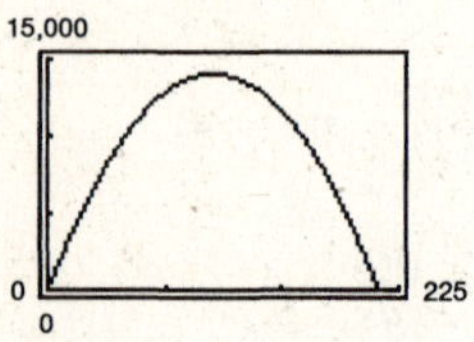

The revenue is maximum when $x = 106$ units.

68. Maximum revenue occurs at $x = 135$.

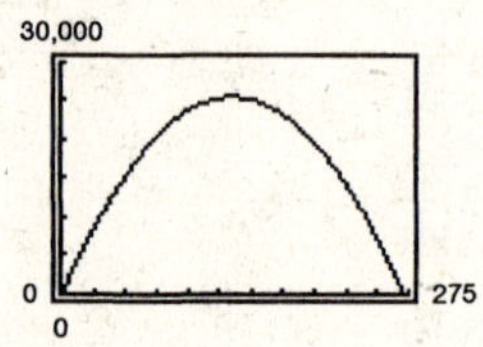

69. A circle is formed when a plane intersects the top or bottom half of a double-napped cone and is perpendicular to the axis through the center of the cone.

70. An ellipse is formed when a plane intersects only the top or bottom half of a double–napped cone but is not perpendicular to the axis of the cone, not parallel to the side of the cone, and does not intersect the vertex.

71. A parabola is formed when a plane intersects the top or bottom half of a double-napped cone, is parallel to the side of the cone, and does not intersect the vertex.

72. A hyperbola is formed when a plane intersects both halves of a double–napped cone, is parallel to the axis of the cone, and does not intersect the vertex.

73. $x^2 = 4py$

(a) As p increases, the graph becomes wider.

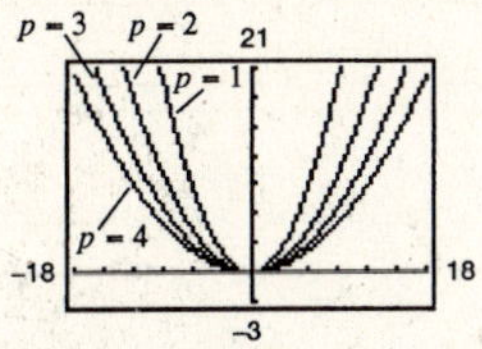

(b)

Value of p	*Focus*
$p = 1$	$(0, 1)$
$p = 2$	$(0, 2)$
$p = 3$	$(0, 3)$
$p = 4$	$(0, 4)$

(c) The length of the focal chord is $4|p|$.

When $p = 1,\ 4|p| = 4.$

When $p = 2,\ 4|p| = 8.$

When $p = 3,\ 4|p| = 12.$

When $p = 4,\ 4|p| = 16.$

(d) The endpoints of the focal chord give us two easily determined additional points on the graph.

74. $y - y_1 = \dfrac{x_1}{2p}(x - x_1)$

Slope: $m = \dfrac{x_1}{2p}$

75. Vertex: $(0, 0) \implies h = 0, k = 0$

Focus: $(0, 4.5) \implies p = 4.5$

$(x - h)^2 = 4p(y - k)$

$(x - 0)^2 = 4(4.5)(y - 0)$

$x^2 = 18y$ or $y = \dfrac{1}{18}x^2$

76. (a)

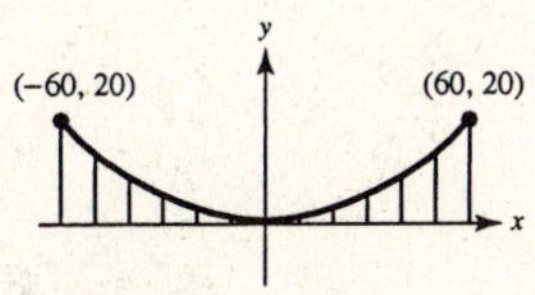

(b) $(x - 0)^2 = 4p(y - 0)$

$x^2 = 4py$

At $(60, 20)$: $60^2 = 4p(20) \implies p = 45$

$x^2 = 4(45)y$

$y = \dfrac{x^2}{180}$

(c)

x	0	20	40	60
y	0	$2\frac{2}{9}$	$8\frac{8}{9}$	20

77. (a) Vertex: $(0, 0) \implies h = 0, k = 0$

Points on the parabola: $(\pm 16, -0.4)$

$x^2 = 4py$

$(\pm 16)^2 = 4p(-0.4)$

$256 = -1.6p$

$-160 = p$

$x^2 = 4(-160y)$

$x^2 = -640y$

$y = -\dfrac{1}{640}x^2$

(b) When $y = -0.1$ we have, $-0.1 = -\dfrac{1}{640}x^2$

$64 = x^2$

$\pm 8 = x.$

Thus, 8 feet away from the center of the road, the road surface is 0.1 foot lower than in the middle.

78. $y = -0.08x^2 + x + 4$

(a)

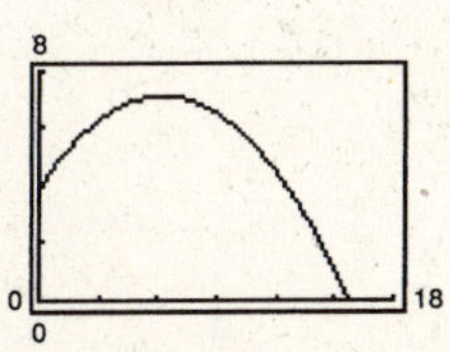

(b) Maximum height: 7.125 feet

Range: 15.687 feet

(c) $y = -0.08(x^2 - 12.5x) + 4$

$= -0.08(x^2 - 12.5x + 39.0625) + 3.125 + 4$

$= -0.08(x - 6.25)^2 + 7.125$

Vertex: $(6.25, 7.125)$

Maximum height: 7.125 feet

79. The position of the target is on the x-axis, so first, let $y = 0$.

$$0 = 30{,}000 - \frac{x^2}{39{,}204}$$

$$\frac{x^2}{39{,}204} = 30{,}000$$

$$x^2 = 30{,}000(39{,}204)$$

$$x = \sqrt{30{,}000(39{,}204)}$$

$$x \approx 34{,}294.606 \text{ feet}$$

Since the bomber is flying at 792 feet per second,

$$\frac{34{,}294.606 \text{ feet}}{792 \text{ feet per second}} \approx 43.3 \text{ seconds.}$$

The bomb should be released 43.3 seconds prior to being over the target.

80. The point on the graph of $y^2 = 6x$ that is closest to the focus of the parabola would be the vertex $(0, 0)$.

81. $x^2 = 2y, y = 3$

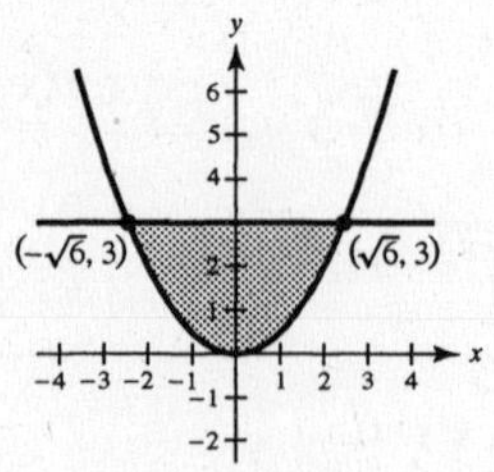

$$A = 2\int_0^{\sqrt{6}} \left(3 - \frac{x^2}{2}\right) dx$$

$$= 2\left[3x - \frac{x^3}{6}\right]_0^{\sqrt{6}}$$

$$= 4\sqrt{6} \text{ square units}$$

82. $x^2 = 4(y - 1), y = 10$

$$y = \frac{1}{4}x^2 + 1 = 10$$

$$x^2 = 36$$

$$x = \pm 6$$

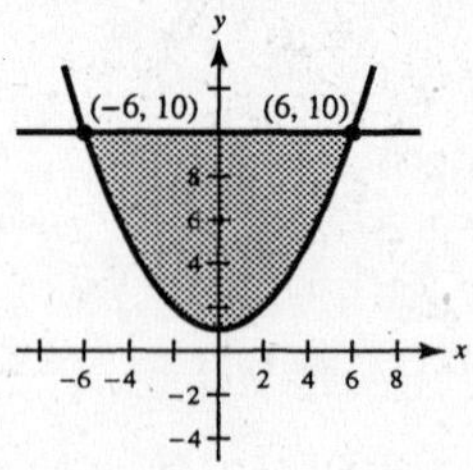

$$A = 2\int_0^6 \left(10 - \frac{1}{4}x^2 - 1\right) dx$$

$$= 2\int_0^6 \left(9 - \frac{1}{4}x^2\right) dx$$

$$= 2\left[9x - \frac{x^3}{12}\right]_0^6$$

$$= 2[54 - 18] = 72$$

83. $y^2 = 4x, x = 5$

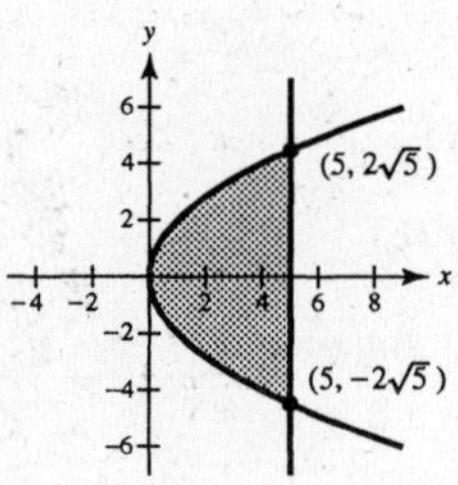

$$A = 2\int_0^{2\sqrt{5}} \left(5 - \frac{y^2}{4}\right) dy$$

$$= 2\left[5y - \frac{y^3}{12}\right]_0^{2\sqrt{5}}$$

$$= \frac{40\sqrt{5}}{3} \text{ square units}$$

84. $y^2 = -4(x + 1), x = -5 \Rightarrow y = \pm 4$

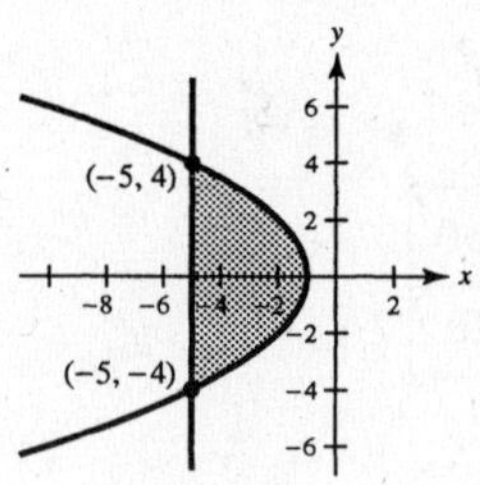

$$A = 2\int_0^4 \left(-\frac{1}{4}y^2 - 1 + 5\right) dy$$

$$= 2\int_0^4 \left(-\frac{1}{4}y^2 + 4\right) dy$$

$$= 2\left[\frac{-1}{12}y^3 + 4y\right]_0^4$$

$$= 2\left[\frac{-16}{3} + 16\right] = \frac{64}{3}$$

85. $(x - 2)^2 = 4y, x = 0, x = 4, y = 0$

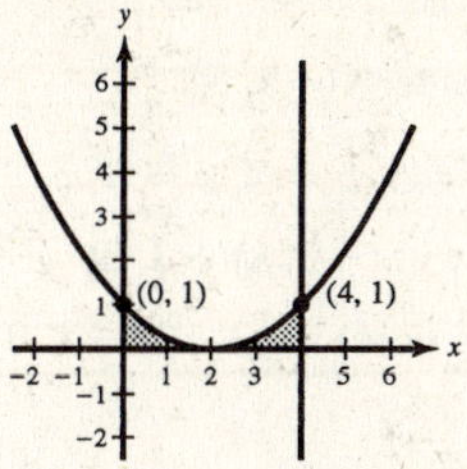

By symmetry $A = 2\int_0^2 \frac{1}{4}(x - 2)^2\,dx$

$$= \frac{1}{2}\frac{(x-2)^3}{3}\Bigg]_0^2$$

$$= \frac{4}{3} \text{ square units}$$

86. $(x + 1)^2 = -8(y - 2), y = 0$

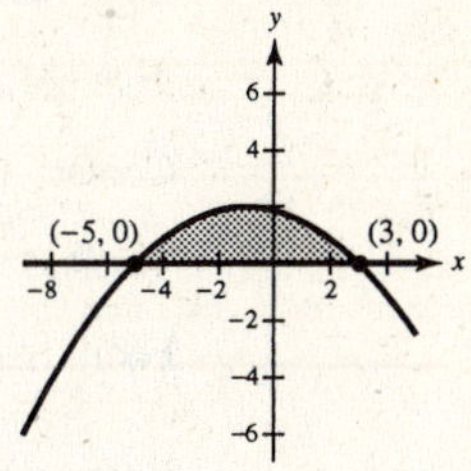

$$A = \int_{-5}^{3}\left[-\frac{1}{8}(x + 1)^2 + 2\right]dx$$

$$= -\frac{1}{24}(x + 1)^3 + 2x\Bigg]_{-5}^{3}$$

$$= \frac{32}{3} \text{ square units}$$

87. False. It is not possible for a parabola to intersect its directrix. If the graph crossed the directrix there would exist points nearer the directrix than the focus.

88. True. If the axis (line connecting the vertex and focus) is horizontal, then the directrix must be vertical.

Section 12.2 Ellipses and Implicit Differentiation

1. $\frac{x^2}{4} + \frac{y^2}{9} = 1$

Center: $(0, 0)$

$a = 3, b = 2$

Vertical major axis

Matches graph (b).

2. $\frac{x^2}{9} + \frac{y^2}{4} = 1$

Center: $(0, 0)$

$a = 3, b = 2$

Horizontal major axis

Matches graph (c).

3. $\frac{x^2}{4} + \frac{y^2}{25} = 1$

Center: $(0, 0)$

$a = 5, b = 2$

Vertical major axis

Matches graph (d).

4. $\frac{y^2}{4} + \frac{x^2}{4} = 1$

Center: $(0, 0)$

Circle of radius: 2

Matches graph (f).

5. $\frac{(x - 2)^2}{16} + (y + 1)^2 = 1$

Center: $(2, -1)$

$a = 4, b = 1$

Horizontal major axis

Matches graph (a).

6. $\frac{(x + 2)^2}{9} + \frac{(y + 2)^2}{4} = 1$

Center: $(-2, -2)$

$a = 3, b = 2$

Horizontal major axis

Matches graph (e).

7. $\frac{x^2}{25} + \frac{y^2}{16} = 1$

Center: $(0, 0)$

$a = 5, b = 4, c = 3$

Foci: $(\pm 3, 0)$

Vertices: $(\pm 5, 0)$

$e = \frac{3}{5}$

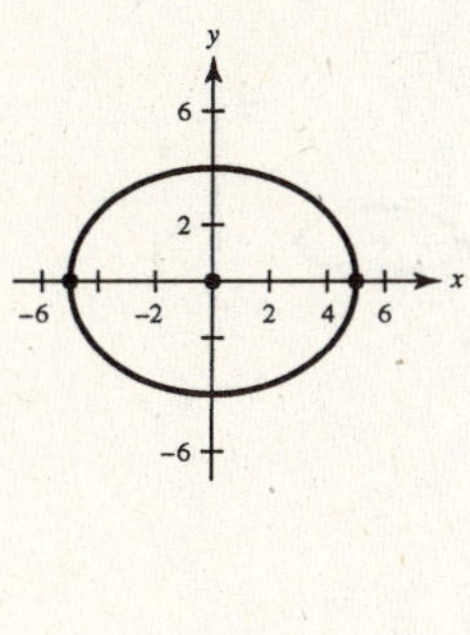

8. $\frac{x^2}{81} + \frac{y^2}{144} = 1$

$a^2 = 144, b^2 = 81, c^2 = 63$

Center: $(0, 0)$

Vertices: $(0, \pm 12)$

Foci: $\left(0, \pm 3\sqrt{7}\right)$

Eccentricity: $e = \frac{c}{a} = \frac{\sqrt{7}}{4}$

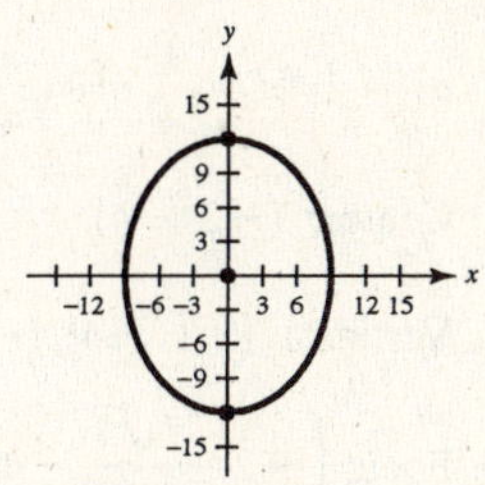

9. $\frac{x^2}{5} + \frac{y^2}{9} = 1$

$a = 3, b = \sqrt{5}, c = 2$

Center: $(0, 0)$

Foci: $(0, \pm 2)$

Vertices: $(0, \pm 3)$

$e = \frac{2}{3}$

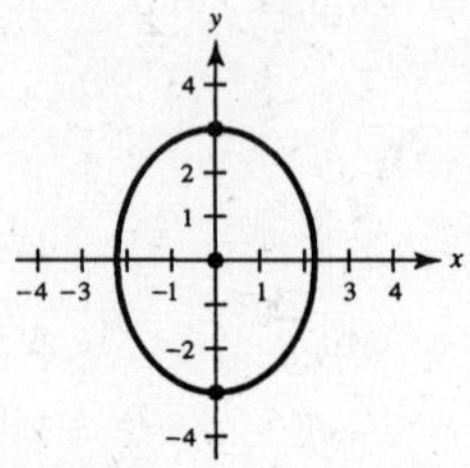

10. $\frac{x^2}{64} + \frac{y^2}{28} = 1$

$a^2 = 64, b^2 = 28, c^2 = 36$

Center: $(0, 0)$

Vertices: $(\pm 8, 0)$

Foci: $(\pm 6, 0)$

Eccentricity: $e = \frac{c}{a} = \frac{3}{4}$

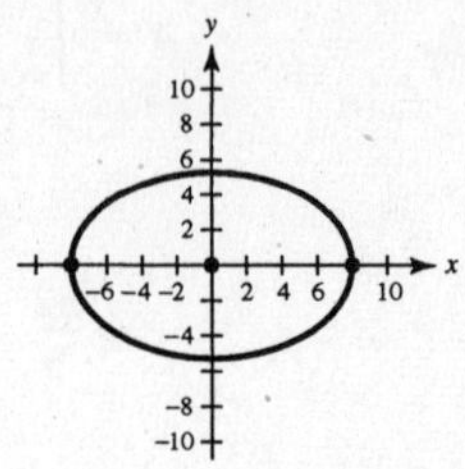

11. $\frac{(x + 3)^2}{16} + \frac{(y - 5)^2}{25} = 1$

Center: $(-3, 5)$

$a = 5, b = 4, c = 3$

Foci: $(-3, 8)(-3, 2)$

Vertices: $(-3, 10)(-3, 0)$

$e = \frac{3}{5}$

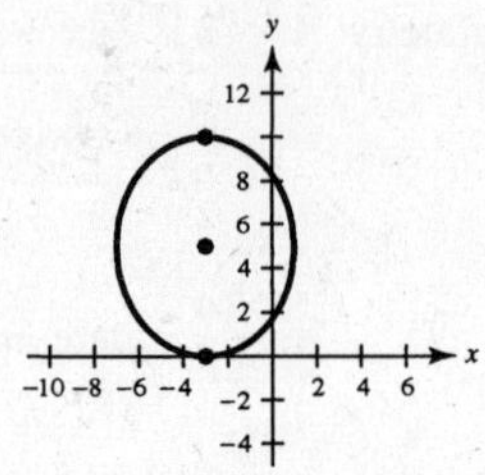

12. $\frac{(x - 4)^2}{12} + \frac{(y + 3)^2}{16} = 1$

$a^2 = 16, b^2 = 12, c^2 = 4$

Center: $(4, -3)$

Vertices: $(4, 1), (4, -7)$

Foci: $(4, -1), (4, -5)$

Eccentricity: $e = \frac{c}{a} = \frac{1}{2}$

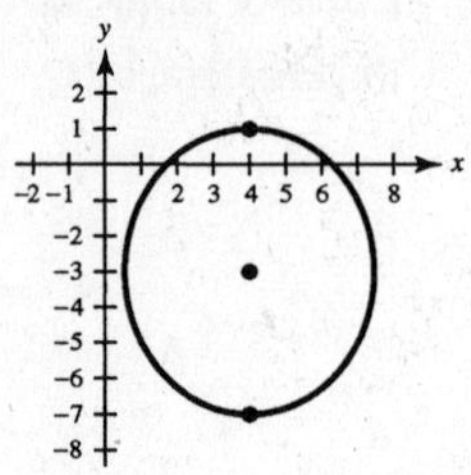

13. $\frac{(x + 5)^2}{\frac{9}{4}} + (y - 1)^2 = 1$

Center: $(-5, 1)$

$a = \frac{3}{2}, b = 1, c = \frac{\sqrt{5}}{2}$

Foci: $\left(-5 + \frac{\sqrt{5}}{2}, 1\right), \left(-5 - \frac{\sqrt{5}}{2}, 1\right)$

Vertices: $\left(-\frac{7}{2}, 1\right), \left(-\frac{13}{2}, 1\right)$

$e = \frac{\frac{\sqrt{5}}{2}}{\frac{3}{2}} = \frac{\sqrt{5}}{3}$

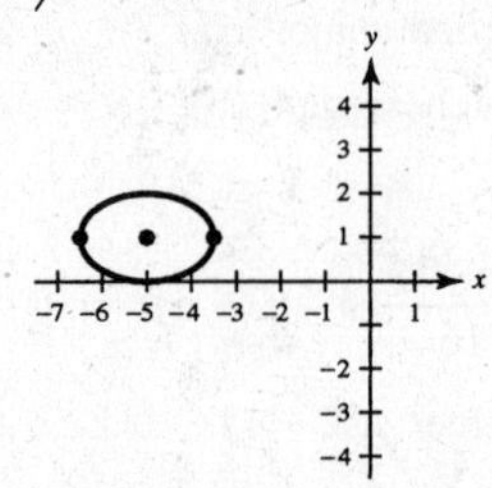

14. $\frac{(x + 2)^2}{1} + \frac{(y + 4)^2}{\frac{1}{4}} = 1$

$a^2 = 1, b^2 = \frac{1}{4}, c^2 = \frac{3}{4}$

Center: $(-2, -4)$

Vertices: $(-1, -4), (-3, -4)$

Foci: $\left(-2 \pm \frac{\sqrt{3}}{2}, -4\right) \Rightarrow \left(\frac{-4 \pm \sqrt{3}}{2}, -4\right)$

$e = \frac{\sqrt{3}}{2}$

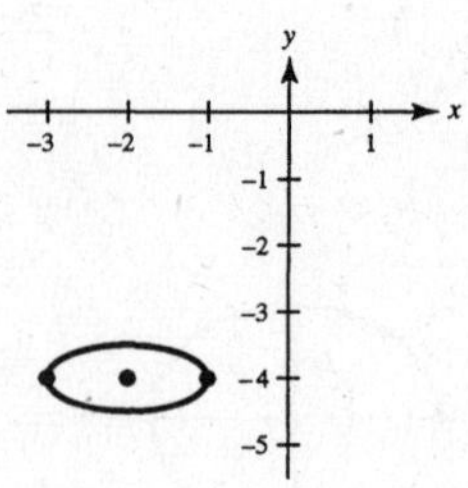

15. $9x^2 + 4y^2 + 36x - 24y + 36 = 0$

$$9(x^2 + 4x + 4) + 4(y^2 - 6y + 9) = -36 + 36 + 36$$

$$9(x + 2)^2 + 4(y - 3) = 36$$

$$\frac{(x + 2)^2}{4} + \frac{(y - 3)^2}{9} = 1$$

$a = 3, b = 2, c = \sqrt{5}$

Center: $(-2, 3)$

Foci: $\left(-2, 3 \pm \sqrt{5}\right)$

Vertices: $(-2, 6), (-2, 0)$

$e = \dfrac{\sqrt{5}}{3}$

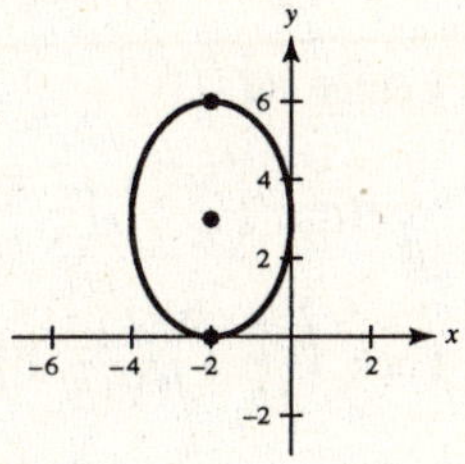

16. $9x^2 + 4y^2 - 54x + 40y + 37 = 0$

$$9(x^2 - 6x + 9) + 4(y^2 + 10y + 25) = -37 + 81 + 100$$

$$\frac{(x - 3)^2}{16} + \frac{(y + 5)^2}{36} = 1$$

$a^2 = 36, b^2 = 16, c^2 = 20$

Center: $(3, -5)$

Vertices: $(3, 1), (3, -11)$

Foci: $\left(3, -5 \pm 2\sqrt{5}\right)$

Eccentricity: $e = \dfrac{c}{a} = \dfrac{\sqrt{5}}{3}$

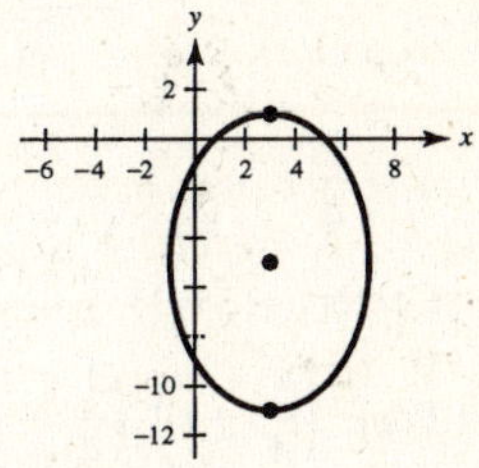

17. $x^2 + 5y^2 - 8x - 30y - 39 = 0$

$$(x^2 - 8x + 16) + 5(y^2 - 6y + 9) = 39 + 16 + 45$$

$$(x - 4)^2 + 5(y - 3)^2 = 100$$

$$\frac{(x - 4)^2}{100} + \frac{(y - 3)^2}{20} = 1$$

Center: $(4, 3)$

$a = 10, b = 2\sqrt{5}, c = 4\sqrt{5}$

Foci: $\left(4 \pm 4\sqrt{5}, 3\right)$

Vertices: $(14, 3), (-6, 3)$

$e = \dfrac{4\sqrt{5}}{10} = \dfrac{2\sqrt{5}}{5}$

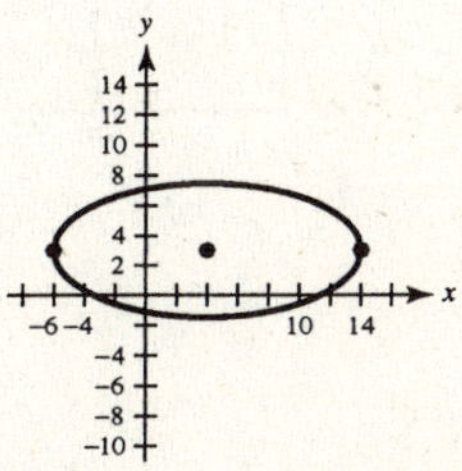

18. $x^2 + 4y^2 - 6x + 20y - 2 = 0$

$$(x^2 - 6x + 9) + 4\left(y^2 + 5y + \frac{25}{4}\right) = 2 + 9 + 25$$

$$\frac{(x - 3)^2}{36} + \frac{\left(y + \frac{5}{2}\right)^2}{9} = 1$$

$a^2 = 36, b^2 = 9, c^2 = 27$

Center: $\left(3, -\dfrac{5}{2}\right)$

Vertices: $\left(9, -\dfrac{5}{2}\right), \left(-3, -\dfrac{5}{2}\right)$

Foci: $\left(3 \pm 3\sqrt{3}, -\dfrac{5}{2}\right)$

Eccentricity: $e = \dfrac{c}{a} = \dfrac{\sqrt{3}}{2}$

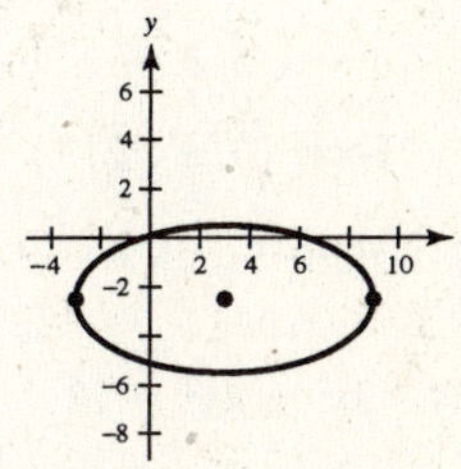

19. $3x^2 + y^2 + 9x - 5y + 1 = 0$

$$3(x^2 + 3x) + (y^2 - 5y) = -1$$

$$3\left(x^2 + 3x + \frac{9}{4}\right) + \left(y^2 - 5y + \frac{25}{4}\right) = -1 + \frac{27}{4} + \frac{25}{4}$$

$$3\left(x + \frac{3}{2}\right)^2 + \left(y - \frac{5}{2}\right)^2 = 12$$

$$\frac{\left(x + \frac{3}{2}\right)^2}{4} + \frac{\left(y - \frac{5}{2}\right)^2}{12} = 1$$

$a = \sqrt{12} = 2\sqrt{3}, b = 2, c = \sqrt{a^2 - b^2} = \sqrt{8} = 2\sqrt{2}$

Vertical major axis

Center: $\left(-\dfrac{3}{2}, \dfrac{5}{2}\right)$

Vertices: $\left(-\dfrac{3}{2}, \dfrac{5}{2} \pm 2\sqrt{3}\right)$

Foci: $\left(-\dfrac{3}{2}, \dfrac{5}{2} \pm 2\sqrt{2}\right)$

Eccentricity: $e = \dfrac{2\sqrt{2}}{2\sqrt{3}} = \dfrac{\sqrt{6}}{3}$

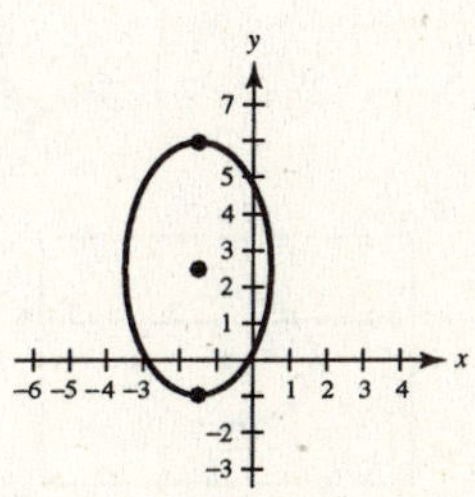

20. $3x^2 + y^2 + 18x - 2y - 8 = 0$

$$3(x^2 + 6x + 9) + (y^2 - 2y + 1) = 8 + 27 + 1$$

$$\frac{(x + 3)^2}{12} + \frac{(y - 1)^2}{36} = 1$$

$a^2 = 36, b^2 = 12, c^2 = 24$

Center: $(-3, 1)$

Vertices: $(-3, 7), (-3, -5)$

Foci: $\left(-3, 1 \pm 2\sqrt{6}\right)$

Eccentricity: $e = \dfrac{c}{a} = \dfrac{\sqrt{6}}{3}$

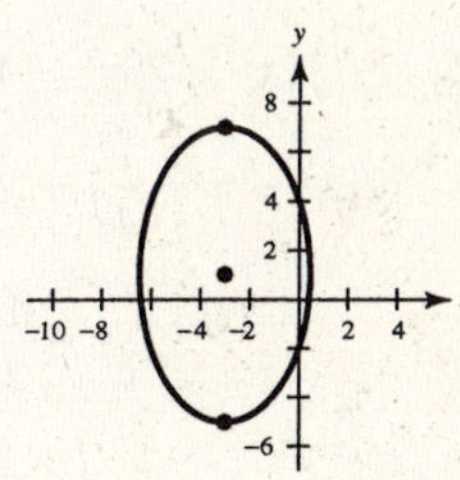

21. $16x^2 + 25y^2 - 32x + 50y + 16 = 0$

$$16(x^2 - 2x + 1) + 25(y^2 + 2y + 1) = -16 + 16 + 25$$

$$16(x - 1)^2 + 25(y + 1)^2 = 25$$

$$\frac{(x - 1)^2}{\frac{25}{16}} + (y + 1)^2 = 1$$

$a = \frac{5}{4}, b = 1, c = \frac{3}{4}$

Center: $(1, -1)$

Foci: $\left(\frac{7}{4}, -1\right), \left(\frac{1}{4}, -1\right)$

Vertices: $\left(\frac{9}{4}, -1\right), \left(-\frac{1}{4}, -1\right)$

$e = \frac{3}{5}$

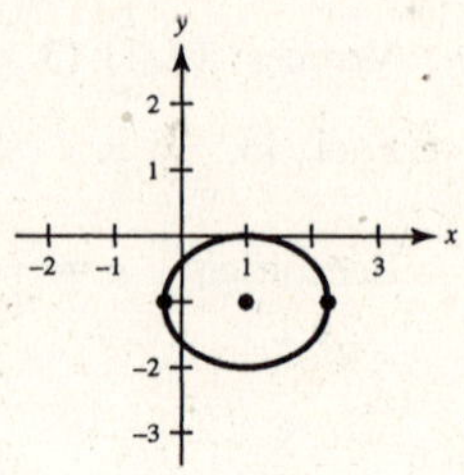

22. $9x^2 + 25y^2 - 36x - 50y + 60 = 0$

$$9(x^2 - 4x + 4) + 25(y^2 - 2y + 1) = -60 + 36 + 25$$

$$\frac{(x - 2)^2}{\frac{1}{9}} + \frac{(y - 1)^2}{\frac{1}{25}} = 1$$

$a^2 = \frac{1}{9}, b^2 = \frac{1}{25}, c^2 = \frac{16}{225}$

Center: $(2, 1)$

Vertices: $\left(\frac{5}{3}, 1\right), \left(\frac{7}{3}, 1\right)$

Foci: $\left(\frac{34}{15}, 1\right), \left(\frac{26}{15}, 1\right)$

Eccentricity: $e = \frac{c}{a} = \frac{4}{5}$

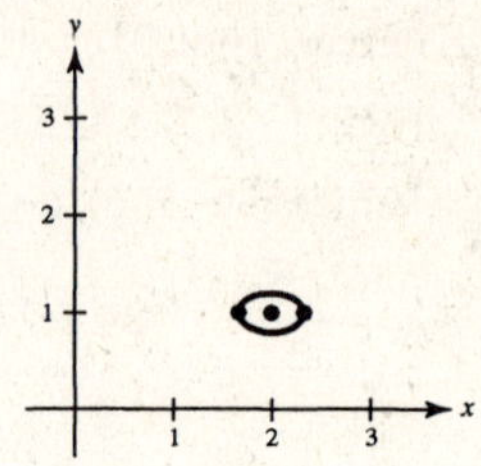

23. $5x^2 + 3y^2 = 15$

$$\frac{x^2}{3} + \frac{y^2}{5} = 1$$

Center: $(0, 0)$

$a = \sqrt{5}, b = \sqrt{3}, c = \sqrt{2}$

Foci: $(0, \pm\sqrt{2})$

Vertices: $(0, \pm\sqrt{5})$

To graph, solve for y.

$$y^2 = \frac{15 - 5x^2}{3}$$

$$y_1 = \sqrt{\frac{15 - 5x^2}{3}}$$

$$y_2 = -\sqrt{\frac{15 - 5x^2}{3}}$$

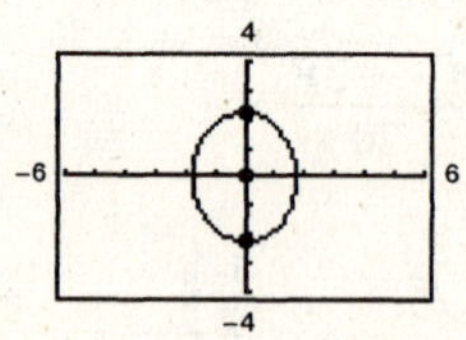

24. $3x^2 + 4y^2 = 12$

$$\frac{x^2}{4} + \frac{y^2}{3} = 1$$

$a^2 = 4, b^2 = 3, c^2 = 1$

Center: $(0, 0)$

Vertices: $(\pm 2, 0)$

Foci: $(\pm 1, 0)$

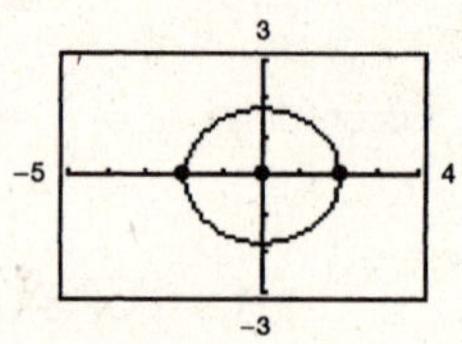

25. $12x^2 + 20y^2 - 12x + 40y - 37 = 0$

$$12\left(x^2 - x + \frac{1}{4}\right) + 20(y^2 + 2y + 1) = 37 + 3 + 20$$

$$12\left(x - \frac{1}{2}\right)^2 + 20(y + 1)^2 = 60$$

$$\frac{\left(x - \frac{1}{2}\right)^2}{5} + \frac{(y + 1)^2}{3} = 1$$

$a = \sqrt{5}, b = \sqrt{3}, c = \sqrt{2}$

Center: $\left(\frac{1}{2}, -1\right)$

Foci: $\left(\frac{1}{2} \pm \sqrt{2}, -1\right)$

Vertices: $\left(\frac{1}{2} \pm \sqrt{5}, -1\right)$

To graph, solve for y.

$$(y + 1)^2 = 3\left[1 - \frac{(x - 0.5)^2}{5}\right]$$

$$y_1 = -1 + \sqrt{3\left[1 - \frac{(x - 0.5)^2}{5}\right]}$$

$$y_2 = -1 - \sqrt{3\left[1 - \frac{(x - 0.5)^2}{5}\right]}$$

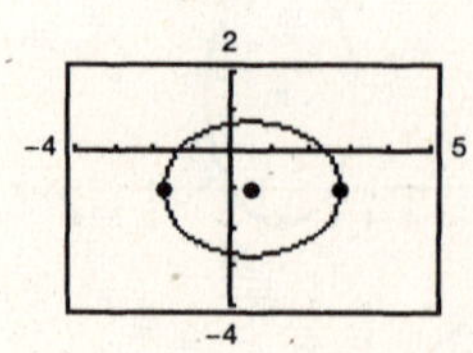

26. $36x^2 + 9y^2 + 48x - 36y - 72 = 0$

$$36\left(x^2 + \frac{4}{3}x + \frac{4}{9}\right) + 9(y^2 - 4y + 4) = 72 + 16 + 36$$

$$\frac{\left(x + \frac{2}{3}\right)^2}{\frac{31}{9}} + \frac{(y-2)^2}{\frac{124}{9}} = 1$$

$$a^2 = \frac{124}{9}, 6^2 = \frac{31}{9}, c^2 = \frac{31}{3}$$

Center: $\left(-\frac{2}{3}, 2\right)$

Vertices: $\left(-\frac{2}{3}, 2 \pm \frac{2\sqrt{31}}{3}\right)$

Foci: $\left(-\frac{2}{3}, 2 \pm \frac{\sqrt{93}}{3}\right)$

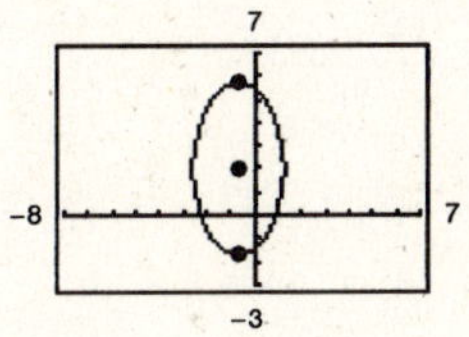

27. Center: $(0, 0)$

$a = 4, b = 2$

Vertical major axis

$$\frac{(x-h)^2}{b^2} + \frac{(y-k)^2}{a^2} = 1$$

$$\frac{x^2}{4} + \frac{y^2}{16} = 1$$

28. Vertices: $(\pm 2, 0) \Rightarrow a = 2$

Endpoints of minor axis: $\left(0, \pm\frac{3}{2}\right) \Rightarrow b = \frac{3}{2}$

$$\frac{x^2}{a^2} + \frac{y^2}{b^2} = 1$$

$$\frac{x^2}{2^2} + \frac{y^2}{\left(\frac{3}{2}\right)^2} = 1$$

$$\frac{x^2}{4} + \frac{4y^2}{9} = 1$$

29. Vertices: $(\pm 6, 0)$

$a = 6, c = 2 \Rightarrow b = \sqrt{32} = 4\sqrt{2}$

Foci: $(\pm 2, 0)$

Horizontal major axis

Center: $(0, 0)$

$$\frac{(x-h)^2}{a^2} + \frac{(y-k)^2}{b^2} = 1$$

$$\frac{x^2}{36} + \frac{y^2}{32} = 1$$

30. Vertices: $(0, \pm 8) \Rightarrow a = 8$

Foci: $(0, \pm 4) \Rightarrow c = 4$

$b^2 = a^2 - c^2 = 64 - 16 = 48$

$$\frac{x^2}{b^2} + \frac{y^2}{a^2} = 1$$

$$\frac{x^2}{48} + \frac{y^2}{64} = 1$$

31. Foci: $(\pm 5, 0) \Rightarrow c = 5$

Center: $(0, 0)$

Horizontal major axis

Major axis of length 12 $\Rightarrow 2a = 12$

$a = 6$

$6^2 - b^2 = 5^2 \Rightarrow b^2 = 11$

$$\frac{(x-h)^2}{a^2} + \frac{(y-k)^2}{b^2} = 1$$

$$\frac{x^2}{36} + \frac{y^2}{11} = 1$$

32. Foci: $(\pm 2, 0) \Rightarrow c = 2$

Major axis length: $8 \Rightarrow a = 4$

$b^2 = a^2 - c^2 = 16 - 4 = 12$

$$\frac{x^2}{a^2} + \frac{y^2}{b^2} = 1$$

$$\frac{x^2}{16} + \frac{y^2}{12} = 1$$

33. Vertices: $(0, \pm 5) \Rightarrow a = 5$

Center: $(0, 0)$

Vertical major axis

$$\frac{(x-h)^2}{b^2} + \frac{(y-k)^2}{a^2} = 1$$

$$\frac{x^2}{b^2} + \frac{y^2}{25} = 1$$

Point: $(4, 2)$

$$\frac{4^2}{b^2} + \frac{2^2}{25} = 1$$

$$\frac{16}{b^2} = 1 - \frac{4}{25} = \frac{21}{25}$$

$$400 = 21b^2$$

$$\frac{400}{21} = b^2$$

$$\frac{x^2}{\frac{400}{21}} + \frac{y^2}{25} = 1$$

$$\frac{21x^2}{400} + \frac{y^2}{25} = 1$$

34. Major axis vertical

Passes through: $(0, 4)$ and $(2, 0)$

$a = 4, b = 2$

$$\frac{x^2}{b^2} + \frac{y^2}{a^2} = 1$$

$$\frac{x^2}{4} + \frac{y^2}{16} = 1$$

35. Center: $(2, 3)$

$a = 3, \quad b = 1$

Vertical major axis

$$\frac{(x-h)^2}{b^2} + \frac{(y-k)^2}{a^2} = 1$$

$$\frac{(x-2)^2}{1} + \frac{(y-3)^2}{9} = 1$$

36. Vertices: $(4, \pm 4) \Rightarrow a = 4$

Center: $(4, 0) \Rightarrow h = 4, k = 0$

Endpoints of minor axis: $(1, 0), (7, 0) \Rightarrow b = 3$

$$\frac{(x-h)^2}{b^2} + \frac{(y-k)^2}{a^2} = 1$$

$$\frac{(x-4)^2}{9} + \frac{y^2}{16} = 1$$

37. Center: $(-2, 3)$

$a = 4, \quad b = 3$

Horizontal major axis

$$\frac{(x-h)^2}{a^2} + \frac{(y-k)^2}{b^2} = 1$$

$$\frac{(x+2)^2}{16} + \frac{(y-3)^2}{9} = 1$$

38. Vertices: $(0, -1), (4, -1) \Rightarrow a = 2$

Center: $(2, -1) \Rightarrow h = 2, k = -1$

Endpoints of minor axis: $(2, 0), (2, -2) \Rightarrow b = 1$

$$\frac{(x-h)^2}{a^2} + \frac{(y-k)^2}{b^2} = 1$$

$$\frac{(x-2)^2}{4} + \frac{(y+1)^2}{1} = 1$$

39. Vertices: $(0, 4), (4, 4) \Rightarrow a = 2$

Minor axis of length 2 $\Rightarrow b = 1$

Center: $(2, 4) = (h, k)$

$$\frac{(x-h)^2}{a^2} + \frac{(y-k)^2}{b^2} = 1$$

$$\frac{(x-2)^2}{4} + \frac{(y-4)^2}{1} = 1$$

40. Foci: $(0, 0), (4, 0) \Rightarrow c = 2, h = 2, k = 0$

Major axis length: $8 \Rightarrow a = 4$

$b^2 = a^2 - c^2 = 16 - 4 = 12$

$$\frac{(x-h)^2}{a^2} + \frac{(y-k)^2}{b^2} = 1$$

$$\frac{(x-2)^2}{16} + \frac{y^2}{12} = 1$$

41. Foci: $(0, 0), (0, 8) \Rightarrow c = 4$

Major axis of length 16 $\Rightarrow a = 8$

$b^2 = a^2 - c^2 = 64 - 16 = 48$

Center: $(0, 4) = (h, k)$

$$\frac{(x-h)^2}{b^2} + \frac{(y-k)^2}{a^2} = 1$$

$$\frac{x^2}{48} + \frac{(y-4)^2}{64} = 1$$

42. Center: $(2, -1) \Rightarrow h = 2, k = -1$

Vertex: $\left(2, \frac{1}{2}\right) \Rightarrow a = \frac{3}{2}$

Minor axis length: $2 \Rightarrow b = 1$

$$\frac{(x-h)}{b^2} + \frac{(y-k)^2}{a^2} = 1$$

$$\frac{(x-2)^2}{1} + \frac{(y+1)^2}{\left(\frac{3}{2}\right)^2} = 1$$

$$(x-2)^2 + \frac{4(y+1)^2}{9} = 1$$

43. Vertices: $(3, 1), (3, 9) \Rightarrow a = 4$

Center: $(3, 5)$

Minor axis of length $6 \Rightarrow b = 3$

Vertical major axis

$$\frac{(x-h)^2}{b^2} + \frac{(y-k)^2}{a^2} = 1$$

$$\frac{(x-3)^2}{9} + \frac{(y-5)^2}{16} = 1$$

44. Center: $(3, 2) \Rightarrow h = 3, k = 2$

$a = 3c$

Foci: $(1, 2), (5, 2) \Rightarrow c = 2, a = 6$

$b^2 = a^2 - c^2 = 36 - 4 = 32$

$$\frac{(x-h)^2}{a^2} + \frac{(y-k)^2}{b^2} = 1$$

$$\frac{(x-3)^2}{36} + \frac{(y-2)^2}{32} = 1$$

45. Center: $(0, 4)$

Vertices: $(-4, 4), (4, 4) \Rightarrow a = 4$

$a = 2c \Rightarrow 4 = 2c \Rightarrow c = 2$

$2^2 = 4^2 - b^2 \Rightarrow b^2 = 12$

Horizontal major axis

$$\frac{(x-h)^2}{a^2} + \frac{(y-k)^2}{b^2} = 1$$

$$\frac{x^2}{16} + \frac{(y-4)^2}{12} = 1$$

46. Vertices: $(5, 0), (5, 12) \Rightarrow a = 6$

Endpoints of the minor axis:

$(1, 6), (9, 6) \Rightarrow b = 4$

Center: $(5, 6) \Rightarrow h = 5, k = 6$

$$\frac{(x-h)^2}{b^2} + \frac{(y-k)^2}{a^2} = 1$$

$$\frac{(x-5)^2}{16} + \frac{(y-6)^2}{36} = 1$$

47. Vertices: $(\pm 5, 0) \Rightarrow a = 5$

Eccentricity: $\frac{3}{5} \Rightarrow c = \frac{3}{5}a = 3$

$b^2 = a^2 - c^2 = 25 - 9 = 16$

Center: $(0, 0) = (h, k)$

$$\frac{(x-h)^2}{a^2} + \frac{(y-k)^2}{b^2} = 1$$

$$\frac{x^2}{25} + \frac{y^2}{16} = 1$$

48. Vertices: $(0, \pm 8) \Rightarrow a = 8$

$e = \frac{1}{2} \Rightarrow \frac{c}{a} = \frac{1}{2}, c = 4$

$b^2 = a^2 - c^2 = 64 - 16 = 48$

Center: $(0, 0)$

$$\frac{x^2}{b^2} + \frac{y^2}{a^2} = 1$$

$$\frac{x^2}{48} + \frac{y^2}{64} = 1$$

49. (a) The length of the string is $2a$.

(b) The path is an ellipse since the sum of the distances from the two fixed points is constant.

50. If $2a = 10$ then b must be less than 5. As b approaches 5 the ellipse becomes more circular in shape.

51. $\frac{x^2}{9} + \frac{y^2}{4} = 1$

$$\frac{2}{9}x + \frac{1}{2}y\frac{dy}{dx} = 0$$

$$\frac{dy}{dx} = \frac{-\frac{2}{9}x}{\frac{1}{2}y} = -\frac{4x}{9y}$$

52. $\frac{x^2}{25} + \frac{y^2}{64} = 1$

$$\frac{2x}{25} + \frac{y}{32}\frac{dy}{dx} = 0$$

$$\frac{dy}{dx} = \frac{-2x/25}{y/32} = -\frac{64x}{25y}$$

53. $\dfrac{(x-4)^2}{4} + \dfrac{(y+2)^2}{16} = 1$

$4(x-4)^2 + (y+2)^2 = 16$

$8(x-4) + 2(y+2)\dfrac{dy}{dx} = 0$

$\dfrac{dy}{dx} = -\dfrac{4(x-4)}{y+2}$

54. $\dfrac{(x+5)^2}{36} + \dfrac{(y-3)^2}{9} = 1$

$(x+5)^2 + 4(y-3)^2 = 36$

$2(x+5) + 8(y-3)\dfrac{dy}{dx} = 0$

$\dfrac{dy}{dx} = -\dfrac{x+5}{4(y-3)}$

55. $9x^2 + 4y^2 - 36x + 8y + 31 = 0$

$18x + 8y\dfrac{dy}{dx} - 36 + 8\dfrac{dy}{dx} = 0$

$\dfrac{dy}{dx}(8y+8) = -18x + 36$

$\dfrac{dy}{dx} = \dfrac{-18(x-2)}{8(y+1)} = -\dfrac{9(x-2)}{4(y+1)}$

56. $3x^2 + 25y^2 - 216x - 300y + 324 = 0$

$6x + 50y\dfrac{dy}{dx} - 216 - 300\dfrac{dy}{dx} = 0$

$\dfrac{dy}{dx}(50y - 300) = -6x + 216$

$\dfrac{dy}{dx} = \dfrac{-6(x-36)}{50(y-6)} = -\dfrac{3(x-36)}{25(y-6)}$

57. $\dfrac{(x-2)^2}{16} + \dfrac{y^2}{12} = 1$

(a) $\dfrac{(x-2)}{8} + \dfrac{y}{6}\dfrac{dy}{dx} = 0$

$\dfrac{dy}{dx} = -\dfrac{3(x-2)}{4y}$

At (0, 3), $\dfrac{dy}{dx} = \dfrac{1}{2}$.

Tangent line: $y - 3 = \dfrac{1}{2}(x - 0)$

$\dfrac{1}{2}x - y + 3 = 0$

(b) The center of the ellipse is (2, 0). The point (0, 3) is two units to the left and three units above the center. By symmetry, the point two units to the right and three unit below the center, $(4, -3)$, also has a slope of $\frac{1}{2}$.

$y + 3 = \dfrac{1}{2}(x - 4)$

$0 = \dfrac{1}{2}x - y - 5$ or $x - 2y - 10 = 0$

(c)

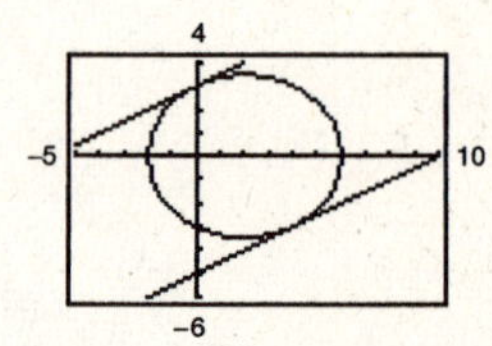

Graph: $y_1 = \sqrt{12 - \frac{3}{4}(x-2)^2}$

$y_2 = -\sqrt{12 - \frac{3}{4}(x-2)^2}$

$y_3 = \dfrac{1}{2}x + 3$

$y_4 = \dfrac{1}{2}x - 5$

58. $\dfrac{(x-2)^2}{4} + (y+1)^2 = 1$

(a) $\dfrac{(x-2)}{2} + 2(y+1)\dfrac{dy}{dx} = 0$

$\dfrac{dy}{dx} = -\dfrac{(x-2)}{4(y+1)}$

At $\left(3, -\dfrac{2+\sqrt{3}}{2}\right), \dfrac{dy}{dx} = \dfrac{\sqrt{3}}{6}$

Tangent line: $y + \dfrac{2+\sqrt{3}}{2} = \dfrac{\sqrt{3}}{6}(x-3)$

$y = \dfrac{\sqrt{3}}{6}x - \sqrt{3} - 1$

$\sqrt{3}x - 6y - 6\sqrt{3} - 6 = 0$

(b) The center of the ellipse is $(2, -1)$. The point $\left(3, -\dfrac{2+\sqrt{3}}{2}\right)$ is one unit to the right $\dfrac{\sqrt{3}}{2}$ unit below the center. By symmetry, the point one unit to the left and $\dfrac{\sqrt{3}}{2}$ unit above the center, $\left(1, \dfrac{-2+\sqrt{3}}{2}\right)$, also has a slope of $\dfrac{\sqrt{3}}{6}$.

$y - \left(\dfrac{-2+\sqrt{3}}{2}\right) = \dfrac{\sqrt{3}}{6}(x-1)$

$y = \dfrac{\sqrt{3}}{6}x + \dfrac{\sqrt{3}}{3} - 1$

$\sqrt{3}x - 6y + 2\sqrt{3} - 6 = 0$

(c) 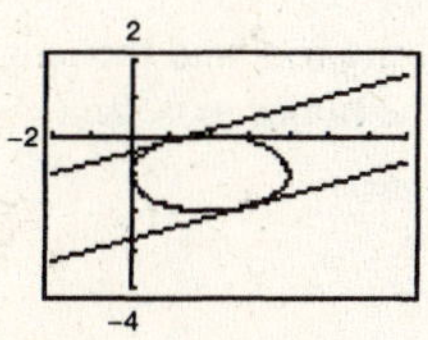

Graph: $y_1 = -1 + \sqrt{1 - \frac{1}{4}(x-2)^2}$

$y_2 = -1 - \sqrt{1 - \frac{1}{4}(x-2)^2}$

$y_3 = \dfrac{\sqrt{3}}{6}x - \sqrt{3} - 1$

$y_4 = \dfrac{\sqrt{3}}{6}x + \dfrac{\sqrt{3}}{3} - 1$

59. $x^2 + 4y^2 + 6x - 16y + 9 = 0$

$2x + 8y\dfrac{dy}{dx} + 6 - 16\dfrac{dy}{dx} = 0$

$\dfrac{dy}{dx}(8y - 16) = -2x - 6$

$\dfrac{dy}{dx} = \dfrac{-2(x+3)}{8(y-2)} = -\dfrac{x+3}{4(y-2)}$

Horizontal tangents when $\dfrac{dy}{dx} = 0$: $x = -3$

$(-3)^2 + 4y^2 + 6(-3) - 16y + 9 = 0$

$4y^2 - 16y = 0$

$4y(y-4) = 0$

$y = 0$ or $y = 4$

The endpoints of the minor axis are $(-3, 0)$ and $(-3, 4)$.

Vertical tangents when $\dfrac{dy}{dx}$ does not exist: $y = 2$

$x^2 + 4(2)^2 + 6x - 16(2) + 9 = 0$

$x^2 + 6x - 7 = 0$

$(x+7)(x-1) = 0$

$x = -7$ or $x = 1$

The endpoints of the major axis are $(-7, 2)$ and $(1, 2)$.

60. $9x^2 + y^2 - 90x + 2y + 190 = 0$

$$18x + 2y\frac{dy}{dx} - 90 + 2\frac{dy}{dx} = 0$$

$$\frac{dy}{dx}(2y + 2) = -18x + 90$$

$$\frac{dy}{dx} = \frac{-18(x - 5)}{2(y + 1)}$$

Horizontal tangents when $\frac{dy}{dx} = 0$: $x = 5$

$$9(5)^2 + y^2 - 90(5) + 2y + 190 = 0$$

$$y^2 + 2y - 35 = 0$$

$$(y - 5)(y + 7) = 0$$

$$y = 5 \quad \text{or} \quad y = -7$$

The endpoints of the major axis are $(5, 5)$ and $(5, -7)$.

Vertical tangents when $\frac{dy}{dx}$ does not exist: $y = -1$

$$9x^2 + (-1)^2 - 90x + 2(-1) + 190 = 0$$

$$9x^2 - 90x + 189 = 0$$

$$9(x - 7)(x - 3) = 0$$

$$x = 7 \quad \text{or} \quad x = 3$$

The endpoints of the minor axis are $(7, -1)$ and $(3, -1)$.

61. $\frac{x^2}{4} + \frac{y^2}{1} = 1$

$a = 2, b = 1$

Area $= \pi ab = 2\pi$ square units.

62. $\frac{x^2}{16} + \frac{y^2}{9} = 1$

$a = 4, b = 3$

Area $= \pi ab = 12\pi$ square units.

63. $3x^2 + 2y^2 = 6$

$$\frac{x^2}{2} + \frac{y^2}{3} = 1$$

$a = \sqrt{3}, b = \sqrt{2}$

Area $= \pi ab = \sqrt{6}\pi$ square units.

64. $5x^2 + 7y^2 = 70$

$$\frac{x^2}{14} + \frac{y^2}{10} = 1$$

$a = \sqrt{14}, b = \sqrt{10}$

Area $= \pi ab = \sqrt{140}\pi = 2\sqrt{35}\pi$ square units.

65. The tacks should be placed at the foci and the length of the string is the length of the major axis, $2a$.

Center: $(0, 0)$

$a = 3, b = 2, c = \sqrt{5}$

Foci (Positions of the tacks): $(\pm\sqrt{5}, 0)$

Length of string: 6 feet

66. (a)

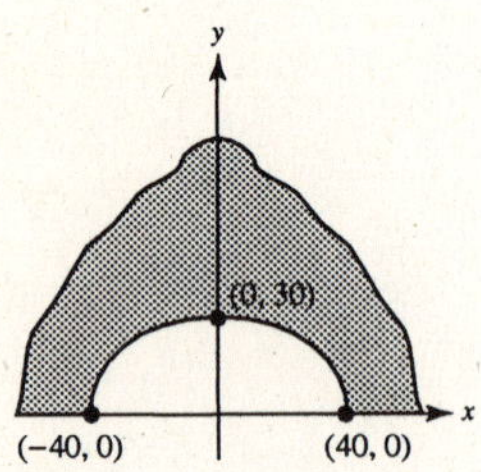

(b) $a = 40, b = 30$

$$\frac{x^2}{a^2} + \frac{y^2}{b^2} = 1$$

$$\frac{x^2}{1600} + \frac{y^2}{900} = 1$$

(c) When $x = \pm 35$, you are five feet from the edge of the tunnel.

$$\frac{35^2}{1600} + \frac{y^2}{900} = 1$$

$$y^2 = 900 - 900\left(\frac{1225}{1600}\right) = \frac{3375}{16}$$

$$y = \pm\sqrt{\frac{3375}{16}} \approx \pm 14.5$$

The height of the tunnel five feet from its edge is 14.5 feet.

67. Area of circle: $\pi r^2 = 100\pi$

Area of ellipse: $\pi(a)(10)$

$10a\pi = 2(100\pi)$

$10a\pi = 200\pi$

$a = 20$

Length of major axis: $2a = 40$

68. $a = \dfrac{36.18}{2} = 18.09$

$e = \dfrac{c}{a} = 0.97$

$c = ea \approx 17.55$

$b^2 = a^2 - c^2 \approx 19.34$

$\dfrac{x^2}{a^2} + \dfrac{y^2}{b^2} = 1$

$\dfrac{x^2}{327.25} + \dfrac{y^2}{19.34} = 1$

69. $a - c = 0.34$

$a + c = 4.08$

$2a = 4.42 \Rightarrow a = 2.21$

$c = 4.08 - a \Rightarrow c = 1.87$

$b^2 = a^2 - c^2 = 2.21^2 - 1.87^2 = 1.3872 \Rightarrow b \approx 1.18$

$\dfrac{x^2}{a^2} + \dfrac{y^2}{b^2} = 1$

$\dfrac{x^2}{4.88} + \dfrac{y^2}{1.39} = 1$

70. $a + c = 6378 + 938 = 7316$

$a - c = 6378 + 212 = 6590$

Solving this system for a and c gives

$a = 6953,\ c = 363$

$e = \dfrac{c}{a} = \dfrac{363}{6953} \approx 0.052.$

71. $\dfrac{x^2}{25} + \dfrac{y^2}{16} = 1 \Rightarrow y = \pm\dfrac{4}{5}\sqrt{25 - x^2}$

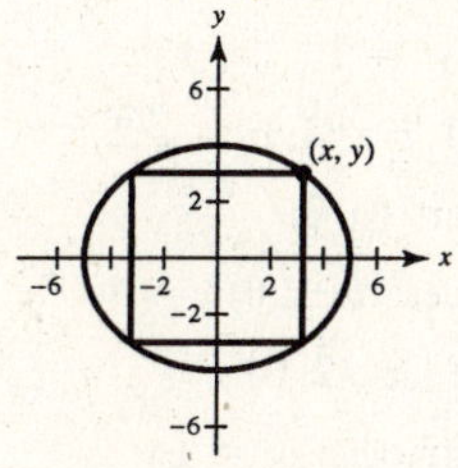

Let (x, y) be the corner of the rectangle in Quadrant I. Then the area to be maximized is:

$A = (2x)(2y) = 4x\left(\dfrac{4}{5}\sqrt{25 - x^2}\right) = \dfrac{16}{5}x\sqrt{25 - x^2}$

$\dfrac{dA}{dx} = \dfrac{16}{5}\left[\dfrac{-x^2}{\sqrt{25 - x^2}} + \sqrt{25 - x^2}\right] = \dfrac{16}{5}\left[\dfrac{25 - 2x^2}{\sqrt{25 - x^2}}\right]$

The critical numbers are $x = \pm 5$ and $x = \pm\dfrac{5\sqrt{2}}{2}$.

Choosing x in Quadrant I yields $x = \dfrac{5\sqrt{2}}{2}$ and $y = \dfrac{4}{5}\sqrt{25 - \left(\dfrac{5\sqrt{2}}{2}\right)^2} = 2\sqrt{2}.$

The length is $2x = 5\sqrt{2}$ units.

The width is $2y = 4\sqrt{2}$ units.

72. $\dfrac{x^2}{a^2} + \dfrac{y^2}{b^2} = 1$

(a) $a + b = 20 \Rightarrow b = 20 - a$

$A = \pi ab = \pi a(20 - a)$

(b) $264 = \pi a(20 - a)$

$0 = -\pi a^2 + 20\pi a - 264$

$0 = \pi a^2 - 20\pi a + 264$

By the Quadratic Formula: $a \approx 14$ or $a \approx 6$. Choosing the larger value of a, we have $a \approx 14$ and $b \approx 6$.

The equation of an ellipse with an area of 264 is

$\dfrac{x^2}{196} + \dfrac{y^2}{36} = 1.$

(c)

a	8	9	10	11	12	13
A	301.6	311.0	314.2	311.0	301.6	285.9

The area is maximum when $a = 10$ and the ellipse is a circle.

(d)

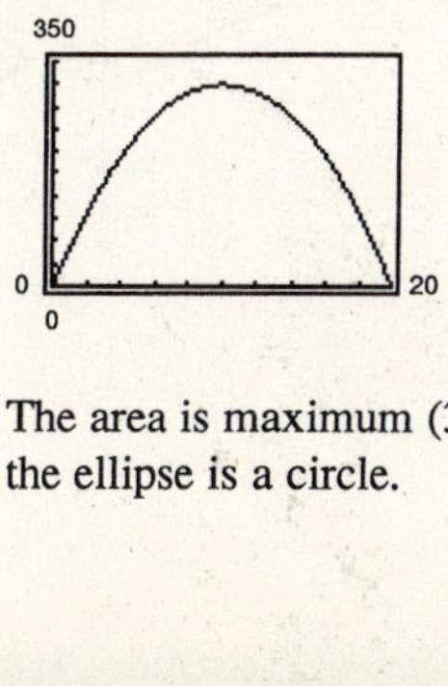

The area is maximum (314.16) when $a = b = 10$ and the ellipse is a circle.

73. False. The graph of $\frac{x^2}{4} + y^4 = 1$ is not an ellipse. The degree on y is 4, not 2.

74. True. If e is close to 1, the ellipse is elongated and the foci are close to the vertices.

75. True. The area of the circle is 16π. The area of the ellipse is $4\pi b$, where $b < 4$ since $2a = 8$ is the major axis length.

76. False. The foci of an ellipse cannot occur outside the ellipse because $0 < c < a$ and both the foci and vertices are on the major axis.

77. Let $A = C = 0$ and $D = E = F = 1$. Then $x + y + 1 = 0$ is a line, not an ellipse.

78. True.

Section 12.3 Hyperbolas and Implicit Differentiation

1. $\frac{y^2}{9} - \frac{x^2}{25} = 1$

Center: $(0, 0)$

$a = 3, b = 5$

Vertical transverse axis

Matches graph (b).

2. $\frac{y^2}{25} - \frac{x^2}{9} = 1$

Center: $(0, 0)$

$a = 5, b = 3$

Vertical transverse axis

Matches graph (c).

3. $\frac{(x-1)^2}{16} - \frac{y^2}{4} = 1$

Center: $(1, 0)$

$a = 4, b = 2$

Horizontal transverse axis

Matches graph (a).

4. $\frac{(x+1)^2}{16} - \frac{(y-2)^2}{9} = 1$

Center: $(-1, 2)$

$a = 4, b = 3$

Horizontal transverse axis

Matches graph (d).

5. $x^2 - y^2 = 1$

$a = 1, b = 1, c = \sqrt{2}$

Center: $(0, 0)$

Vertices: $(\pm 1, 0)$

Foci: $(\pm\sqrt{2}, 0)$

Asymptotes: $y = \pm x$

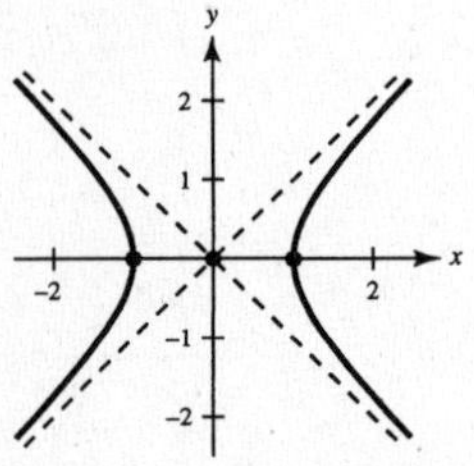

6. $\frac{x^2}{9} - \frac{y^2}{25} = 1$

$a = 3, b = 5$

$c = \sqrt{3^2 + 5^2} = \sqrt{34}$

Center: $(0, 0)$

Vertices: $(\pm 3, 0)$

Foci: $(\pm\sqrt{34}, 0)$

Asymptotes: $y = \pm\frac{5}{3}x$

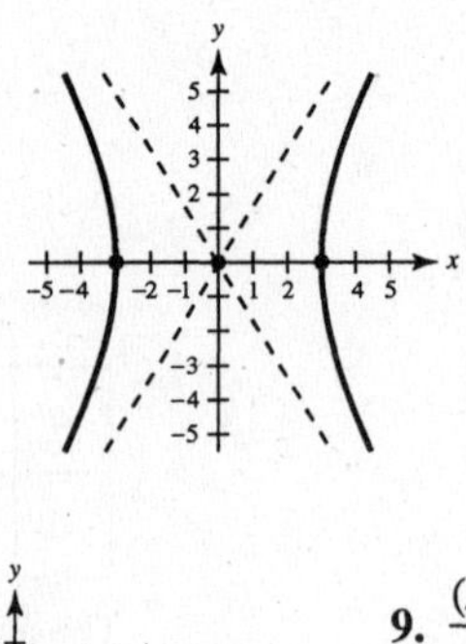

7. $\frac{y^2}{25} - \frac{x^2}{81} = 1$

$a = 5, b = 9, c = \sqrt{106}$

Center: $(0, 0)$

Vertices: $(0, \pm 5)$

Foci: $(0, \pm\sqrt{106})$

Asymptotes: $y = \pm\frac{5}{9}x$

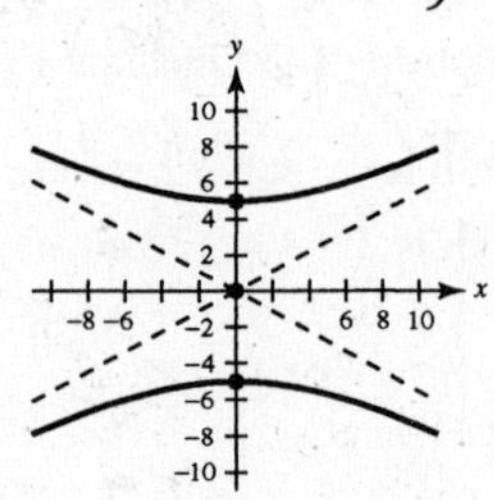

8. $\frac{x^2}{36} - \frac{y^2}{4} = 1$

$a = 6, b = 2,$

$c = \sqrt{36 + 4} = 2\sqrt{10}$

Center: $(0, 0)$

Vertices: $(\pm 6, 0)$

Foci: $(\pm 2\sqrt{10}, 0)$

Asymptotes: $y = \pm\frac{1}{3}x$

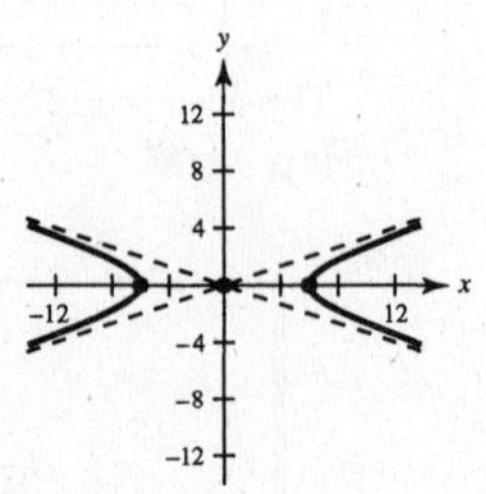

9. $\frac{(x-1)^2}{4} - \frac{(y+2)^2}{1} = 1$

$a = 2, b = 1, c = \sqrt{5}$

Center: $(1, -2)$

Vertices: $(-1, -2), (3, -2)$

Foci: $(1 \pm \sqrt{5}, -2)$

Asymptotes: $y = -2 \pm \frac{1}{2}(x - 1)$

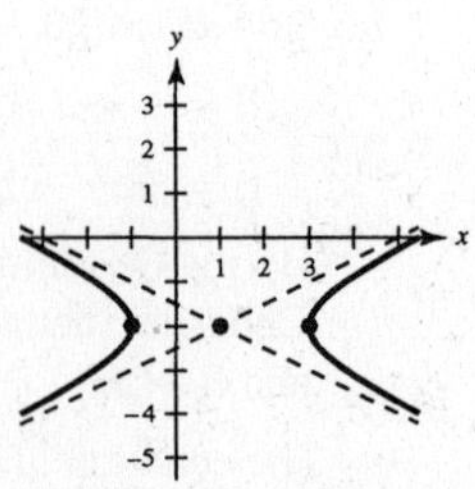

10. $\dfrac{(x+3)^2}{144} - \dfrac{(y-2)^2}{25} = 1$

$a = 12, b = 5$

$c = \sqrt{144 + 25} = 13$

Center: $(-3, 2)$

Vertices: $(9, 2), (-15, 2)$

Foci: $(10, 2), (-16, 2)$

Asymptotes: $y = 2 \pm \dfrac{5}{12}(x + 3)$

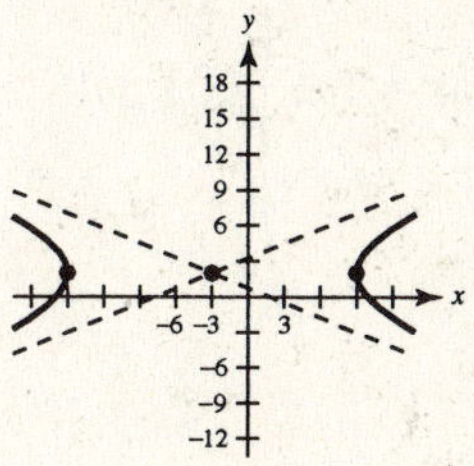

11. $\dfrac{(y+6)^2}{\frac{1}{9}} - \dfrac{(x-2)^2}{\frac{1}{4}} = 1$

$a = \dfrac{1}{3}, b = \dfrac{1}{2}, c = \dfrac{\sqrt{13}}{6}$

Center: $(2, -6)$

Vertices: $\left(2, -\dfrac{17}{3}\right), \left(2, -\dfrac{19}{3}\right)$

Foci: $\left(2, -6 \pm \dfrac{\sqrt{13}}{6}\right)$

Asymptotes: $y = -6 \pm \dfrac{2}{3}(x - 2)$

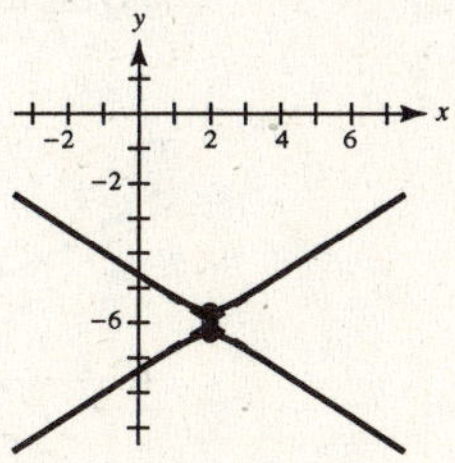

12. $\dfrac{(y-1)^2}{\frac{1}{4}} - \dfrac{(x+3)^2}{\frac{1}{16}} = 1$

$a = \dfrac{1}{2}, b = \dfrac{1}{4}$

$c = \sqrt{\dfrac{1}{4} + \dfrac{1}{16}} = \dfrac{\sqrt{5}}{4}$

Center: $(-3, 1)$

Vertices: $\left(-3, \dfrac{3}{2}\right), \left(-3, \dfrac{1}{2}\right)$

Foci: $\left(-3, 1 \pm \dfrac{\sqrt{5}}{4}\right)$

Asymptotes:
$y = 1 \pm 2(x + 3)$

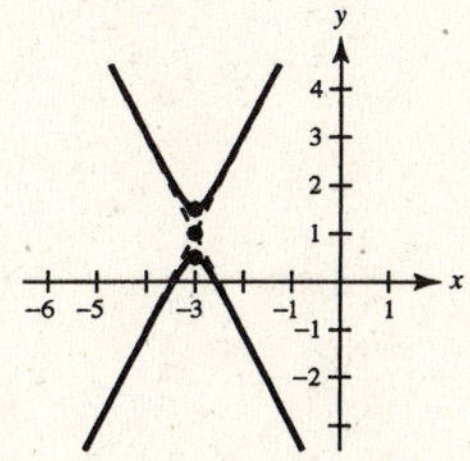

13.

$$9x^2 - y^2 - 36x - 6y + 18 = 0$$
$$9(x^2 - 4x + 4) - (y^2 + 6y + 9) = -18 + 36 - 9$$
$$9(x-2)^2 - (y+3)^2 = 9$$
$$\frac{(x-2)^2}{1} - \frac{(y+3)^2}{9} = 1$$

$a = 1, b = 3, c = \sqrt{10}$

Center: $(2, -3)$

Vertices: $(1, -3), (3, -3)$

Foci: $(2 \pm \sqrt{10}, -3)$

Asymptotes: $y = -3 \pm 3(x - 2)$

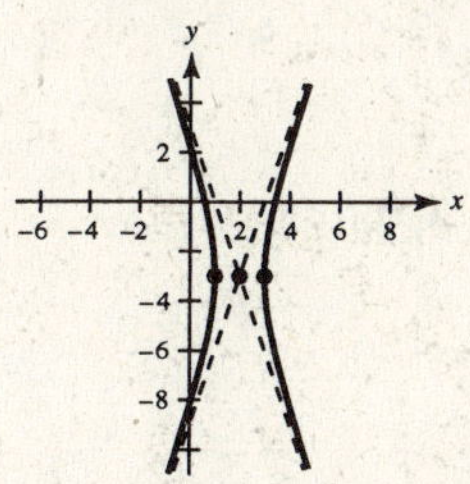

14.

$$x^2 - 9y^2 + 36y - 72 = 0$$
$$x^2 - 9(y^2 - 4y + 4) = 72 - 36$$
$$x^2 - 9(y-2)^2 = 36$$
$$\frac{x^2}{36} - \frac{(y-2)^2}{4} = 1$$

$a = 6, b = 2, c = \sqrt{36 + 4} = 2\sqrt{10}$

Center: $(0, 2)$

Vertices: $(\pm 6, 2)$

Foci: $(\pm 2\sqrt{10}, 2)$

Asymptotes: $y = 2 \pm \dfrac{1}{3}x$

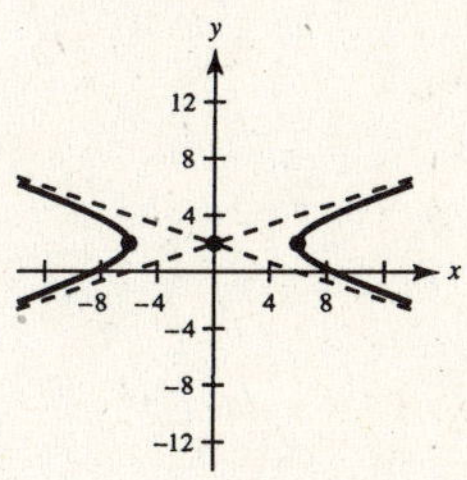

15.

$$x^2 - 9y^2 + 2x - 54y - 80 = 0$$
$$(x^2 + 2x + 1) - 9(y^2 + 6y + 9) = 80 + 1 - 81$$
$$(x+1)^2 - 9(y+3)^2 = 0$$
$$y + 3 = \pm\frac{1}{3}(x + 1)$$

Degenerate hyperbola is two lines intersecting at $(-1, -3)$.

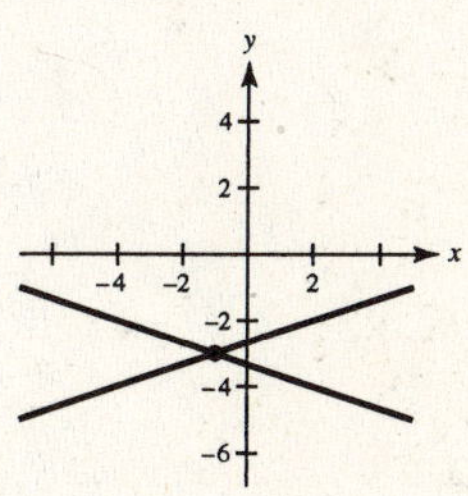

16.
$$16y^2 - x^2 + 2x + 64y + 63 = 0$$
$$16(y^2 + 4y + 4) - (x^2 - 2x + 1) = -63 + 64 - 1$$
$$16(y + 2)^2 - (x - 1)^2 = 0$$
$$y + 2 = \pm\tfrac{1}{4}(x - 1)$$

Degenerate hyperbola: The graph is two lines intersecting at $(1, -2)$.

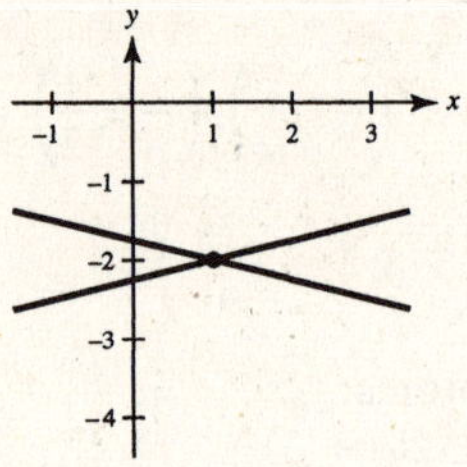

17.
$$16y^2 - 9x^2 - 18x - 32y = 137$$
$$16(y^2 - 2y + 1) - 9(x^2 + 2x + 1) = 137 + 16 - 9$$
$$16(y - 1)^2 - 9(x + 1)^2 = 144$$
$$\frac{(y - 1)^2}{9} - \frac{(x + 1)^2}{16} = 1$$

Center: $(-1, 1)$

$a = 3, b = 4 \implies c = 5$

Vertices: $(-1, 4), (-1, -2)$

Foci: $(-1, 6), (-1, -4)$

Asymptotes: $y = k \pm \frac{a}{b}(x - h) = 1 \pm \frac{3}{4}(x + 1)$

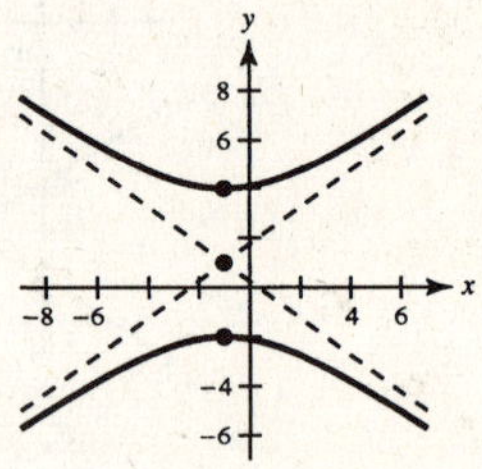

18.
$$9x^2 - y^2 + 72x - 2y = -134$$
$$9(x^2 + 8x + 16) - (y^2 + 2y + 1) = -134 + 144 - 1$$
$$9(x + 4)^2 - (y + 1)^2 = 9$$
$$(x + 4)^2 - \frac{(y + 1)^2}{9} = 1$$

Center: $(-4, -1)$

$a = 1, b = 3, c = \sqrt{10}$

Vertices: $(-3, -1), (-5, -1)$

Foci: $\left(-4 \pm \sqrt{10}, -1\right)$

Asymptotes: $y = k \pm \frac{b}{a}(x - h) = -1 \pm 3(x + 4)$

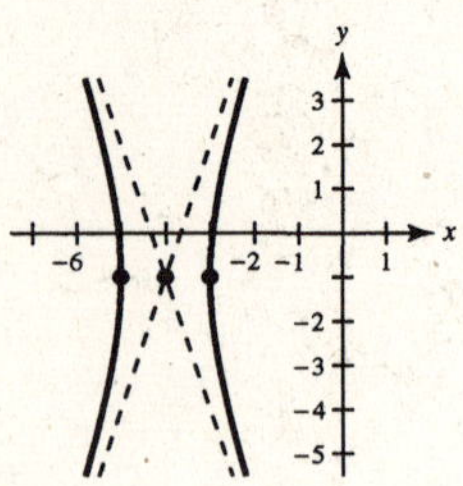

19. $2x^2 - 3y^2 = 6$

$$\frac{x^2}{3} - \frac{y^2}{2} = 1$$

$a = \sqrt{3}, b = \sqrt{2}, c = \sqrt{5}$

Center: $(0, 0)$

Vertices: $\left(\pm\sqrt{3}, 0\right)$

Foci: $\left(\pm\sqrt{5}, 0\right)$

Asymptotes: $y = \pm\sqrt{\frac{2}{3}}x = \pm\frac{\sqrt{6}}{3}x$

To use a graphing calculator, solve for y first.

$$y^2 = \frac{2x^2 - 6}{3}$$

$$\left.\begin{aligned} y_1 &= \sqrt{\frac{2x^2 - 6}{3}} \\ y_2 &= -\sqrt{\frac{2x^2 - 6}{3}} \end{aligned}\right\} \text{ Hyperbola}$$

$$\left.\begin{aligned} y_3 &= \frac{\sqrt{6}}{3}x \\ y_4 &= -\frac{\sqrt{6}}{3}x \end{aligned}\right\} \text{ Asymptotes}$$

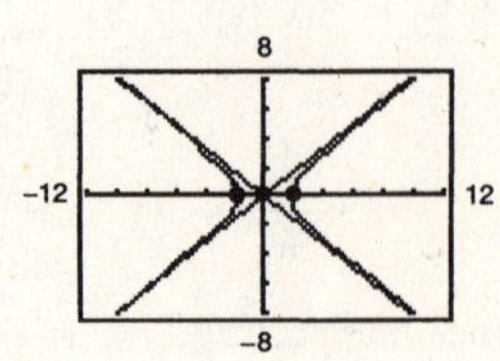

20. $6y^2 - 3x^2 = 18$

$$\frac{y^2}{3} - \frac{x^2}{6} = 1$$

$a = \sqrt{3}, b = \sqrt{6}, c = \sqrt{3+6} = 3$

Center: $(0, 0)$

Vertices: $(0, \pm\sqrt{3})$

Foci: $(0, \pm 3)$

Asymptotes: $y = \pm\frac{\sqrt{2}}{2}x$

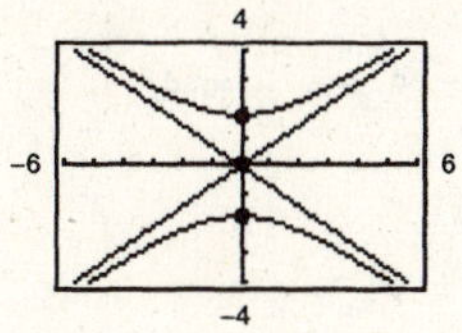

21.
$$9y^2 - x^2 + 2x + 54y + 62 = 0$$
$$9(y^2 + 6y + 9) - (x^2 - 2x + 1) = -62 - 1 + 81$$
$$9(y+3)^2 - (x-1)^2 = 18$$
$$\frac{(y+3)^2}{2} - \frac{(x-1)^2}{18} = 1$$

$a = \sqrt{2}, b = 3\sqrt{2}, c = 2\sqrt{5}$

Center: $(1, -3)$

Vertices: $(1, -3 \pm \sqrt{2})$

Foci: $(1, -3 \pm 2\sqrt{5})$

Asymptotes: $y = -3 \pm \frac{1}{3}(x - 1)$

To use a graphing calculator, solve for y first.

$$9(y+3)^2 = 18 + (x-1)^2$$
$$y = -3 \pm \sqrt{\frac{18 + (x-1)^2}{9}}$$

$$\left.\begin{aligned} y_1 &= -3 + \frac{1}{3}\sqrt{18 + (x-1)^2} \\ y_2 &= -3 - \frac{1}{3}\sqrt{18 + (x-1)^2} \end{aligned}\right\} \text{Hyperbola}$$

$$\left.\begin{aligned} y_3 &= -3 + \frac{1}{3}(x-1) \\ y_4 &= -3 - \frac{1}{3}(x-1) \end{aligned}\right\} \text{Asymptotes}$$

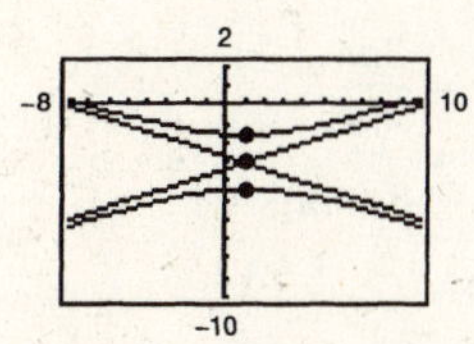

22.
$$9x^2 - y^2 + 54x + 10y + 55 = 0$$
$$9(x^2 + 6x + 9) - (y^2 - 10y + 25) = -55 + 81 - 25$$
$$\frac{(x+3)^2}{\frac{1}{9}} - \frac{(y-5)^2}{1} = 1$$

$a = \frac{1}{3}, b = 1, c = \frac{\sqrt{10}}{3}$

Center: $(-3, 5)$

Vertices: $\left(-3 \pm \frac{1}{3}, 5\right) \Rightarrow \left(-\frac{10}{3}, 5\right), \left(-\frac{8}{3}, 5\right)$

Foci: $\left(-3 \pm \frac{\sqrt{10}}{3}, 5\right)$

Asymptotes: $y = 5 \pm 3(x + 3)$

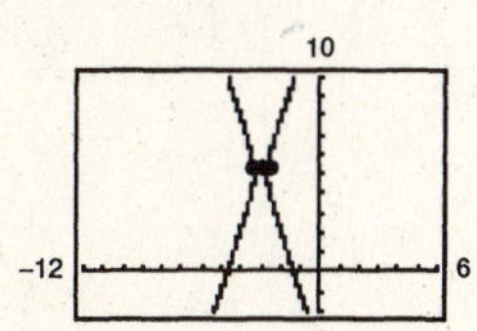

23. Vertices: $(0, \pm 2) \Rightarrow a = 2$

Foci: $(0, \pm 4) \Rightarrow c = 4$

$b^2 = c^2 - a^2 = 16 - 4 = 12$

Center: $(0, 0) = (h, k)$

$$\frac{(y-k)^2}{a^2} - \frac{(x-h)^2}{b^2} = 1$$

$$\frac{y^2}{4} - \frac{x^2}{12} = 1$$

24. Vertices: $(\pm 4, 0) \implies a = 4$

Foci: $(\pm 6, 0) \implies c = 6$

$b^2 = c^2 - a^2 = 36 - 16 = 20 \implies b = 2\sqrt{5}$

$$\frac{x^2}{a^2} - \frac{y^2}{b^2} = 1$$

$$\frac{x^2}{16} - \frac{y^2}{20} = 1$$

25. Vertices: $(\pm 1, 0) \implies a = 1$

Asymptotes: $y = \pm 5x \implies \dfrac{b}{a} = 5, b = 5$

Center: $(0, 0) = (h, k)$

$$\frac{(x-h)^2}{a^2} - \frac{(y-k)^2}{b^2} = 1$$

$$\frac{x^2}{1} - \frac{y^2}{25} = 1$$

26. Vertices: $(0, \pm 3) \implies a = 3$

Asymptotes: $y = \pm 3x \implies \dfrac{a}{b} = 3, b = 1$

$$\frac{y^2}{a^2} - \frac{x^2}{b^2} = 1$$

$$\frac{y^2}{9} - x^2 = 1$$

27. Foci: $(0, \pm 8) \implies c = 8$

Asymptotes: $y = \pm 4x \implies \dfrac{a}{b} = 4 \implies a = 4b$

Center: $(0, 0) = (h, k)$

$c^2 = a^2 + b^2 \implies 64 = 16b^2 + b^2$

$$\frac{64}{17} = b^2 \implies a^2 = \frac{1024}{17}$$

$$\frac{(y-k)^2}{a^2} - \frac{(x-h)^2}{b^2} = 1$$

$$\frac{y^2}{\frac{1024}{17}} - \frac{x^2}{\frac{64}{17}} = 1$$

$$\frac{17y^2}{1024} - \frac{17x^2}{64} = 1$$

28. Foci: $(\pm 10, 0) \implies c = 10$

Asymptotes: $y = \pm \dfrac{3}{4}x \implies \dfrac{b}{a} = \dfrac{3m}{4m}$

$c^2 = a^2 + b^2 \implies 100 = (3m)^2 + (4m)^2$

$100 = 25m^2$

$2 = m$

$a = 4(2) = 8$

$b = 3(2) = 6$

$$\frac{x^2}{a^2} - \frac{y^2}{b^2} = 1$$

$$\frac{x^2}{64} - \frac{y^2}{36} = 1$$

29. Vertices: $(2, 0), (6, 0) \implies a = 2$

Foci: $(0, 0), (8, 0) \implies c = 4$

$b^2 = c^2 - a^2 = 16 - 4 = 12$

Center: $(4, 0) = (h, k)$

$$\frac{(x-h)^2}{a^2} - \frac{(y-k)^2}{b^2} = 1$$

$$\frac{(x-4)^2}{4} - \frac{y^2}{12} = 1$$

30. Vertices: $(2, 3), (2, -3) \implies a = 3$

Center: $(2, 0)$

Foci: $(2, 6), (2, -6) \implies c = 6$

$b^2 = c^2 - a^2 = 36 - 9 = 27$

$$\frac{(y-k)^2}{a^2} - \frac{(x-h)^2}{b^2} = 1$$

$$\frac{y^2}{9} - \frac{(x-2)^2}{27} = 1$$

31. Vertices: $(4, 1), (4, 9) \implies a = 4$

Foci: $(4, 0), (4, 10) \implies c = 5$

$b^2 = c^2 - a^2 = 25 - 16 = 9$

Center: $(4, 5) = (h, k)$

$$\frac{(y-k)^2}{a^2} - \frac{(x-h)^2}{b^2} = 1$$

$$\frac{(y-5)^2}{16} - \frac{(x-4)^2}{9} = 1$$

32. Vertices: $(-2, 1), (2, 1) \Rightarrow a = 2$

Center: $(0, 1)$

Foci: $(-3, 1), (3, 1) \Rightarrow c = 3$

$b^2 = c^2 - a^2 = 9 - 4 = 5$

$$\frac{(x-h)^2}{a^2} - \frac{(y-k)^2}{b^2} = 1$$

$$\frac{x^2}{4} - \frac{(y-1)^2}{5} = 1$$

33. Vertices: $(2, 3), (2, -3) \Rightarrow a = 3$

Passes through the point: $(0, 5)$

Center: $(2, 0) = (h, k)$

$$\frac{(y-k)^2}{a^2} - \frac{(x-h)^2}{b^2} = 1$$

$$\frac{y^2}{9} - \frac{(x-2)^2}{b^2} = 1 \Rightarrow \frac{(x-2)^2}{b^2} = \frac{y^2}{9} - 1 = \frac{y^2-9}{9} \Rightarrow b^2 = \frac{9(x-2)^2}{y^2-9} = \frac{9(-2)^2}{25-9} = \frac{36}{16} = \frac{9}{4}$$

$$\frac{y^2}{9} - \frac{(x-2)^2}{9/4} = 1$$

$$\frac{y^2}{9} - \frac{4(x-2)^2}{9} = 1$$

34. Vertices: $(-2, 1), (2, 1) \Rightarrow a = 2$

Center: $(0, 1)$

Point on curve: $(5, 4)$

$$\frac{(x-h)^2}{a^2} - \frac{(y-k)^2}{b^2} = 1$$

$$\frac{x^2}{4} - \frac{(y-1)^2}{b^2} = 1$$

$$\frac{25}{4} - \frac{9}{b^2} = 1$$

$$b^2 = \frac{12}{7}$$

$$\frac{x^2}{4} - \frac{(y-1)^2}{\frac{12}{7}} = 1$$

$$\frac{x^2}{4} - \frac{7(y-1)^2}{12} = 1$$

35. Vertices: $(0, 4), (0, 0) \Rightarrow a = 2$

Passes through the point $\left(\sqrt{5}, -1\right)$

Center: $(0, 2) = (h, k)$

$$\frac{(y-k)^2}{a^2} - \frac{(x-h)^2}{b^2} = 1$$

$$\frac{(y-2)^2}{4} - \frac{x^2}{b^2} = 1 \Rightarrow \frac{x^2}{b^2} = \frac{(y-2)^2}{4} - 1 = \frac{(y-2)^2 - 4}{4}$$

$$\Rightarrow b^2 = \frac{4x^2}{(y-2)^2 - 4} = \frac{4\left(\sqrt{5}\right)^2}{(-1-2)^2 - 4} = \frac{20}{5} = 4$$

$$\frac{(y-2)^2}{4} - \frac{x^2}{4} = 1$$

36. Vertices: $(1, \pm 2) \implies a = 2$

Center: $(1, 0)$

Point on curve: $\left(0, \sqrt{5}\right)$

$$\frac{(y-k)^2}{a^2} - \frac{(x-h)^2}{b^2} = 1$$

$$\frac{y^2}{4} - \frac{(x-1)^2}{b^2} = 1$$

$$\frac{5}{4} - \frac{1}{b^2} = 1$$

$$b^2 = 4$$

$$\frac{y^2}{4} - \frac{(x-1)^2}{4} = 1$$

37. Vertices: $(1, 2), (3, 2) \implies a = 1$

Asymptotes: $y = x, y = 4 - x$

$\frac{b}{a} = 1 \implies \frac{b}{1} = 1 \implies b = 1$

Center: $(2, 2) = (h, k)$

$$\frac{(x-h)^2}{a^2} - \frac{(y-k)^2}{b^2} = 1$$

$$\frac{(x-2)^2}{1} - \frac{(y-2)^2}{1} = 1$$

38. Vertices: $(3, 0), (3, 6) \implies a = 3$

Center: $(3, 3)$

Asymptotes: $y = 6 - x, y = x$

$\frac{a}{b} = 1 \implies b = 3$

$$\frac{(y-k)^2}{a^2} - \frac{(x-h)^2}{b^2} = 1$$

$$\frac{(y-3)^2}{9} - \frac{(x-3)^2}{9} = 1$$

39. Vertices: $(0, 2), (6, 2) \implies a = 3$

Asymptotes: $y = \frac{2}{3}x, y = 4 - \frac{2}{3}x$

$\frac{b}{a} = \frac{2}{3} \implies b = 2$

Center: $(3, 2) = (h, k)$

$$\frac{(x-h)^2}{a^2} - \frac{(y-k)^2}{b^2} = 1$$

$$\frac{(x-3)^2}{9} - \frac{(y-2)^2}{4} = 1$$

40. Vertices: $(3, 0), (3, 4) \implies a = 2$

Asymptotes: $y = \frac{2}{3}x, y = 4 - \frac{2}{3}x$

$\frac{a}{b} = \frac{2}{3} \implies b = 3$

Center: $(3, 2) = (h, k)$

$$\frac{(y-k)^2}{a^2} - \frac{(x-h)^2}{b^2} = 1$$

$$\frac{(y-2)^2}{4} - \frac{(x-3)^2}{9} = 1$$

41. Foci: $(2, 2)$ and $(10, 2)$

From the foci we see that the transverse axis is horizontal, the center is $(6, 2)$ and $c = 4$.

Also, $2a = 6 \implies a = 3$.

$c^2 = a^2 + b^2 \implies 16 = 9 + b^2 \implies b = \sqrt{7}$

$$\frac{(x-6)^2}{9} - \frac{(y-2)^2}{7} = 1$$

42. Foci: $(-3, 0)$ and $(-3, 3)$

From the foci we see that the transverse axis is vertical, the center is $(-3, 1.5)$ and $c = 1.5$.

Also, $2a = 2 \implies a = 1$.

$(1.5)^2 = 1^2 + b^2 \implies b = \sqrt{1.25}$

$$\frac{(y-1.5)^2}{1.25} - \frac{(x+3)^2}{1} = 1$$

43. Center $(0, 0)$

Horizontal transverse axis.

Foci: $(\pm c, 0)$

$2a = d_1 - d_2$

$$
\begin{aligned}
2a &= \left|\sqrt{(x+c)^2 + y^2} - \sqrt{(x-c)^2 + y^2}\right| \\
2a + \sqrt{(x-c)^2 + y^2} &= \sqrt{(x+c)^2 + y^2} \\
4a^2 + 4a\sqrt{(x-c)^2 + y^2} + (x-c)^2 + y^2 &= (x+c)^2 + y^2 \\
4a\sqrt{(x-c)^2 + y^2} &= 4cx - 4a^2 \\
a\sqrt{(x-c)^2 + y^2} &= cx - 2a \\
a^2(x^2 - acx + c^2 + y^2) &= c^2x^2 - 2a^2cx + a^4 \\
a^2(c^2 - a^2) &= (c^2 - a^2)x^2 - a^2y^2
\end{aligned}
$$

Let $b^2 = c^2 - a^2$. Then, $a^2b^2 = b^2x^2 - a^2y^2 \implies 1 = \dfrac{x^2}{a^2} - \dfrac{y^2}{b^2}$.

44. The asymptotes of a hyperbola can be drawn by drawing the diagonal lines through the corners of its central rectangle.

45. $\dfrac{x^2}{64} - \dfrac{y^2}{36} = 1$

$$\frac{x}{32} - \frac{y}{18}\frac{dy}{dx} = 0$$

$$\frac{dy}{dx} = \frac{9x}{16y}$$

46. $\dfrac{y^2}{16} - \dfrac{x^2}{25} = 1$

$$\frac{y}{8}\frac{dy}{dx} - \frac{2x}{25} = 0$$

$$\frac{dy}{dx} = \frac{16x}{25y}$$

47. $\dfrac{(y-3)^2}{9} - \dfrac{(x+1)^2}{9} = 1$

$$\frac{2(y-3)}{9}\frac{dy}{dx} - \frac{2(x+1)}{9} = 0$$

$$\frac{dy}{dx} = \frac{x+1}{y-3}$$

48. $(x+1)^2 - \dfrac{(y-2)^2}{9} = 1$

$$2(x+1) - \frac{2(y-2)}{9}\frac{dy}{dx} = 0$$

$$\frac{dy}{dx} = \frac{9(x+1)}{y-2}$$

49. $x^2 - 2y^2 + 8y - 17 = 0$

$$2x - 4y\frac{dy}{dx} + 8\frac{dy}{dx} = 0$$

$$\frac{dy}{dx}(-4y + 8) = -2x$$

$$\frac{dy}{dx} = \frac{x}{2(y-2)}$$

50. $x^2 - 5y^2 + 20x + 2y - 35 = 0$

$$2x - 10y\frac{dy}{dx} + 20 + 2\frac{dy}{dx} = 0$$

$$\frac{dy}{dx}(-10y + 2) = -2x - 20$$

$$\frac{dy}{dx} = \frac{x+10}{5y-1}$$

51. $x^2 - 4y^2 + 2x + 16y - 19 = 0$

$$2x - 8y\frac{dy}{dx} + 2 + 16\frac{dy}{dx} = 0$$

$$(16 - 8y)\frac{dy}{dx} = -2x - 2$$

$$\frac{dy}{dx} = \frac{2x+2}{8y-16} = \frac{x+1}{4y-8}$$

52. $4y^2 - x^2 + 4x - 5 = 0$

$$8y\frac{dy}{dx} - 2x + 4 = 0$$

$$8y\frac{dy}{dx} = 2x - 4$$

$$\frac{dy}{dx} = \frac{x-2}{4y}$$

53. $\dfrac{(x-2)^2}{16} - \dfrac{y^2}{12} = 1$

(a) $\dfrac{(x-2)}{8} - \dfrac{y}{6}\dfrac{dy}{dx} = 0$

$$\frac{dy}{dx} = \frac{3(x-2)}{4y}$$

At $(10, 6)$, $\dfrac{dy}{dx} = 1$.

Tangent line: $y = x - 4$ or $x - y - 4 = 0$

(b) The center of the hyperbola is $(2, 0)$. The point $(10, 6)$ is eight units to the right and six units above the center. By symmetry, the point eight units to the left and six units below the center, $(-6, -6)$ also has a slope of 1.

$y = x$ or $x - y = 0$

(c)

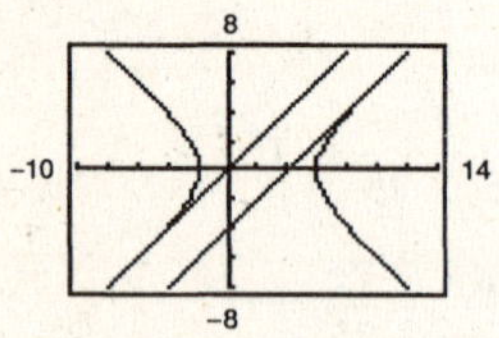

Graph: $y_1 = \sqrt{\frac{3}{4}(x - 2(^2 - 12}$

$y_2 = -\sqrt{\frac{3}{4}(x - 2)^2 - 12}$

$y_3 = x - 4$

$y_4 = x$

54. $\dfrac{(y-3)^2}{4} - \dfrac{(x-1)^2}{9} = 1$

(a) $\dfrac{(y-3)}{2}\dfrac{dy}{dx} - \dfrac{2(x-1)}{9} = 0$

$$\frac{dy}{dx} = \frac{4(x-1)}{9(y-3)}$$

At $\left(5, \dfrac{19}{3}\right)$, $\dfrac{dy}{dx} = \dfrac{8}{15}$.

Tangent line: $y - \dfrac{19}{3} = \dfrac{8}{15}(x - 5)$

$$y = \frac{8}{15}x + \frac{11}{3}$$

$$8x - 15y + 55 = 0$$

(b) The center of the hyperbola is $(1, 3)$. The point $\left(5, \frac{19}{3}\right)$ is four units to the right and $\frac{10}{3}$ units above the center. By symmetry, the point four units to the left and $\frac{10}{3}$ units below the center, $\left(-3, -\frac{1}{3}\right)$, also has a slope of $\frac{8}{15}$.

$$y + \frac{1}{3} = \frac{8}{15}(x + 3)$$

$$y = \frac{8}{15}x + \frac{19}{15}$$

$$8x - 15y + 19 = 0$$

(c)

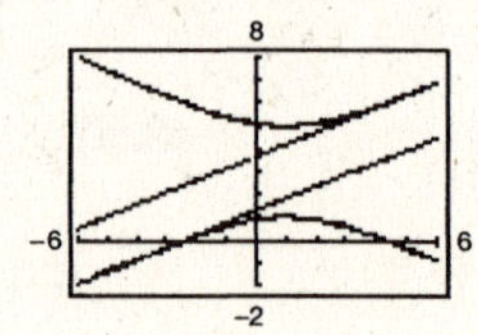

Graph: $y_1 = 3 + \sqrt{4 + \frac{4}{9}(x-1)^2}$

$y_2 = 3 - \sqrt{4 + \frac{4}{9}(x-1)^2}$

$y_3 = \dfrac{8}{15}x + \dfrac{11}{3}$

$y_4 = \dfrac{8}{15}x + \dfrac{19}{15}$

55. $4y^2 - x^2 + 6x + 40y + 75 = 0$

$$8y\frac{dy}{dx} - 2x + 6 + 40\frac{dy}{dx} = 0$$

$$\frac{dy}{dx}(8y + 40) = 2x - 6$$

$$\frac{dy}{dx} = \frac{x-3}{4(y+5)}$$

$$\frac{dy}{dx} = 0 \text{ when } x = 3.$$

$\dfrac{dy}{dx}$ does not exist when $y = -5$.

When $x = 3$: $4y^2 - 3^2 + 6(3) + 40y + 75 = 0$

$$4y^2 + 40y + 84 = 0$$

$$4(y + 7)(y + 3) = 0$$

$$y = -7 \text{ or } y = -3$$

When $y = -5$: $4(-5)^2 - x^2 + 6x + 40(-5) + 75 = 0$

$$x^2 - 6x + 25 = 0$$

No real solutions

The vertices are $(3, -7)$ and $(3, -3)$.

56. $16x^2 - 9y^2 + 64x + 18y + 19 = 0$

$$32x - 18y\frac{dy}{dx} + 64 + 18\frac{dy}{dx} = 0$$

$$\frac{dy}{dx}(-18y + 18) = -32x - 64$$

$$\frac{dy}{dx} = \frac{16(x+2)}{9(y-1)}$$

$\frac{dy}{dx} = 0$ when $x = -2$.

$\frac{dy}{dx}$ does not exist when $y = 1$.

When $x = -2$: $16(-2)^2 - 9y^2 + 64(-2) + 18y + 19 = 0$

$$-9y^2 + 18y - 45 = 0$$

$$-9(y^2 - 2y + 5) = 0$$

No real solutions

When $y = 1$: $16x^2 - 9(1)^2 + 64x + 18(1) + 19 = 0$

$$16x^2 + 64x + 28 = 0$$

$$4(2x+1)(2x+7) = 0$$

$x = -\frac{1}{2}$ OR $x = -\frac{7}{2}$

The vertices are $(-\frac{1}{2}, 1)$ and $(-\frac{7}{2}, 1)$.

57. $x^2 + y^2 - 6x + 4y + 9 = 0$

$A = 1, C = 1$

$A = C \Longrightarrow$ Circle

58. $x^2 + 4y^2 - 6x + 16y + 21 = 0$

$AC = (1)(4) > 0$ and $A \neq C$

The graph is an ellipse.

59. $4x^2 - y^2 - 4x - 3 = 0$

$A = 4, C = -1$

$AC = (4)(-1) = -4 < 0 \Longrightarrow$ Hyperbola

60. $y^2 - 6y - 4x + 21 = 0$

$AC = (0)(1) = 0$

The graph is a parabola.

61. $4x^2 + 3y^2 + 8x - 24y + 51 = 0$

$A = 4, C = 3$

$AC = 4(3) = 12 > 0 \Longrightarrow$ Ellipse

62. $4y^2 - 2x^2 - 4y - 8x - 15 = 0$

$AC = (-2)(4) < 0$

The graph is a hyperbola.

63. $25x^2 - 10x - 200y - 119 = 0$

$A = 25, C = 0$

$AC = 25(0) = 0 \Longrightarrow$ Parabola

64. $4y^2 + 4x^2 - 24x + 35 = 0$

$A = C$

The graph is a circle.

65. Since $\overline{AB} = 1100$ feet and the sound takes one second longer to reach B than A, the explosion must occur on the vertical line through A and B below A.

Foci: $(\pm 3300, 0) \Longrightarrow c = 3300$

Center: $(0, 0) = (h, k)$

$$\frac{\overline{CE}}{1100} - \frac{\overline{AE}}{1100} = 4 \Longrightarrow 2a = 4400, a = \frac{4400}{2} = 2200$$

$$b^2 = c^2 - a^2 = (3300)^2 - (2200)^2 = 6{,}050{,}000$$

$$\frac{x^2}{(2200)^2} - \frac{y^2}{6{,}050{,}000} = 1$$

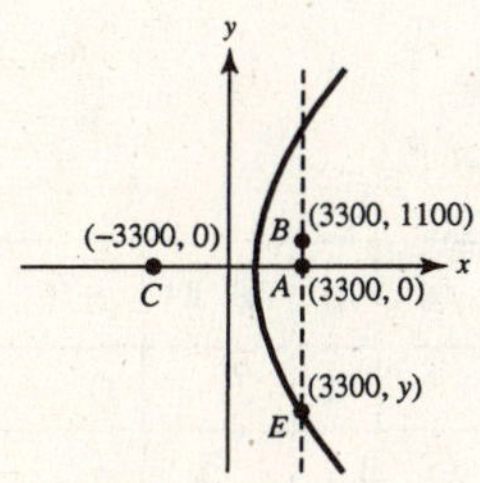

$$y^2 = 6{,}050{,}000\left(\frac{x^2}{(2200)^2} - 1\right)$$

$$y^2 = 6{,}050{,}000\left(\frac{(3300)^2}{(2200)^2} - 1\right) = 7{,}562{,}500$$

$$y = -2750$$

The explosion occurs at $(3300, -2750)$.

66. Foci: $(\pm 150, 0) \Rightarrow c = 150$

Center: $(0, 0) = (h, k)$

$$\frac{d_2}{186{,}000} - \frac{d_1}{186{,}000} = 0.001 \Rightarrow 2a = 186, a = 93$$

$$b^2 = c^2 - a^2 = 150^2 - 93^2 = 13{,}851$$

$$\frac{x^2}{93^2} - \frac{y^2}{13{,}851} = 1$$

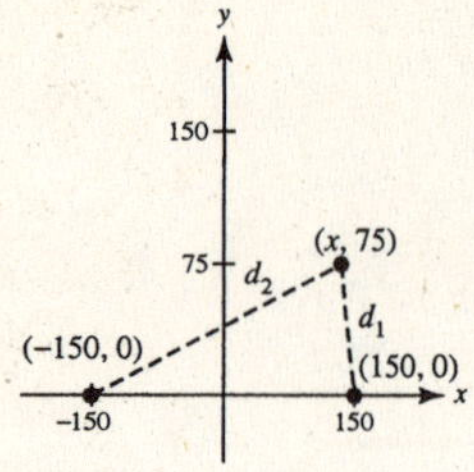

$$x^2 = 93^2\left(1 + \frac{75^2}{13{,}851}\right) \approx 12{,}161$$

$$x \approx 110.3 \text{ miles}$$

67. True. For a hyperbola, $c^2 = a^2 + b^2$ or $e^2 = \dfrac{c^2}{a^2} = 1 + \dfrac{b^2}{a^2}$.

The larger the ratio of b to a, the larger the eccentricity $e = \dfrac{c}{a}$ of the hyperbola.

68. False. For the trivial solution of two intersecting lines to occur, the standard form of the equation of the hyperbola would be equal to zero.

$$\frac{(x-h)^2}{a^2} - \frac{(y-k)^2}{b^2} = 0 \quad \text{or} \quad \frac{(y-k)^2}{a^2} - \frac{(x-h)^2}{b^2} = 0$$

Section 12.4 Parametric Equations and Calculus

1. $x = \sqrt{t}, y = 3 - t$

(a)

t	0	1	2	3	4
x	0	1	$\sqrt{2}$	$\sqrt{3}$	2
y	3	2	1	0	-1

(b)

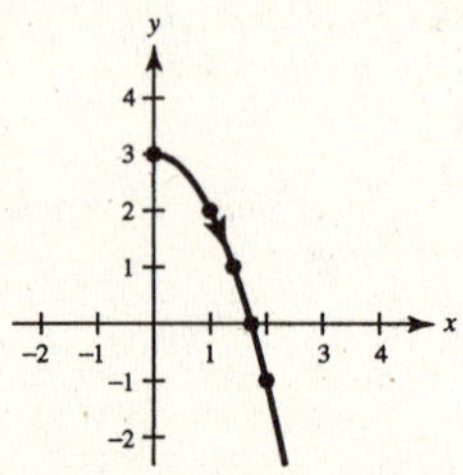

(c) $x^2 = t$

$y = 3 - x^2$

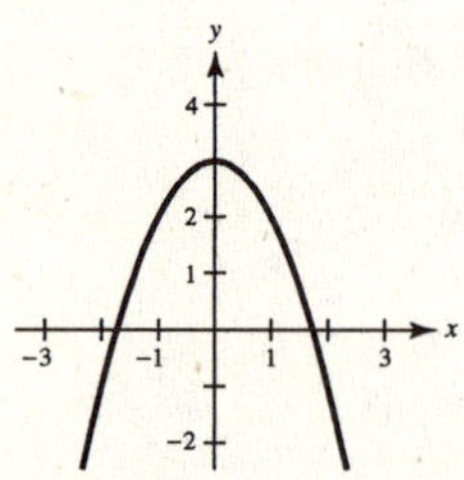

The graph of the rectangular equation shows the entire parabola rather than just the right half.

2. $x = 4\cos^2\theta, y = 2\sin\theta$

(a)

θ	$-\pi/2$	$-\pi/4$	0	$\pi/4$	$\pi/2$
x	0	2	4	2	0
y	-2	$-\sqrt{2}$	0	$\sqrt{2}$	2

(b)

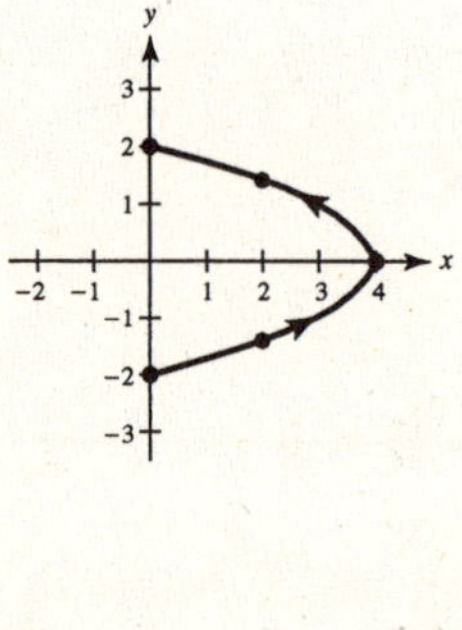

(c) $\cos^2\theta + \sin^2\theta = 1$

$$\frac{x}{4} + \frac{y^2}{4} = 1$$

$$x + y^2 = 4$$

$$y^2 = 4 - x$$

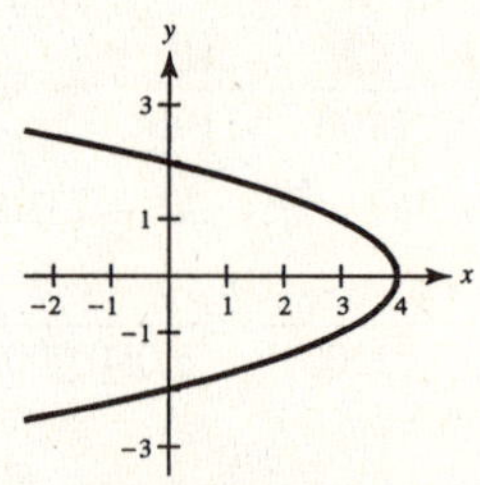

The graph of the rectangular equation shows the entire parabola rather than the restricted portion in Quadrants I and IV.

3. $x = 3t - 3 \Rightarrow t = \dfrac{x + 3}{3}$

$y = 2t + 1 \Rightarrow y = \dfrac{2}{3}(x + 3) + 1 = \dfrac{2}{3}x + 3$

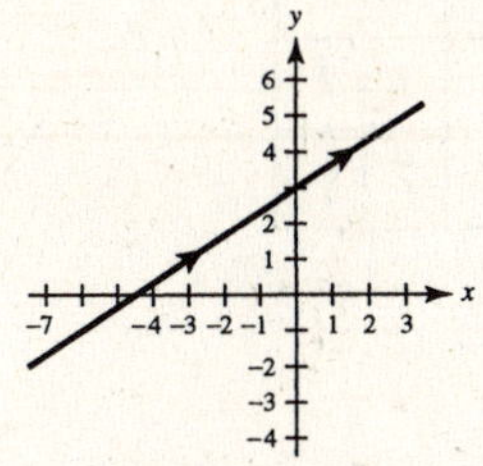

4. $x = 3 - 2t \Rightarrow t = -\frac{1}{2}x + \frac{3}{2}$

$y = 2 + 3t$

t	-3	-2	-1	0	1	2	3
x	9	7	5	3	1	-1	-3
y	-7	-4	-1	2	5	8	11

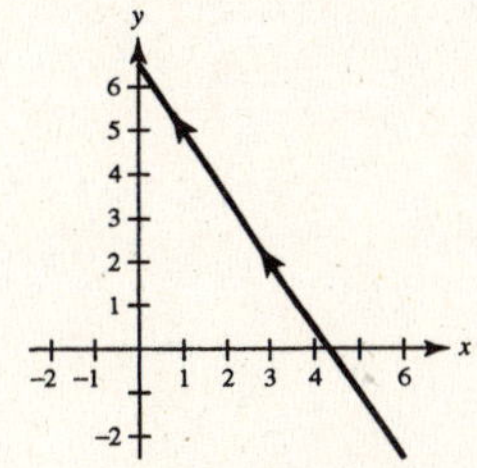

$$y = 2 + 3t$$
$$y = 2 + 3\left(-\tfrac{1}{2}x + \tfrac{3}{2}\right)$$
$$y = 2 - \tfrac{3}{2}x + \tfrac{9}{2}$$
$$2y = 4 - 3x + 9$$
$$3x + 2y - 13 = 0$$

5. $x = \frac{1}{4}t \Rightarrow t = 4x$

$y = t^2 \Rightarrow y = 16x^2$

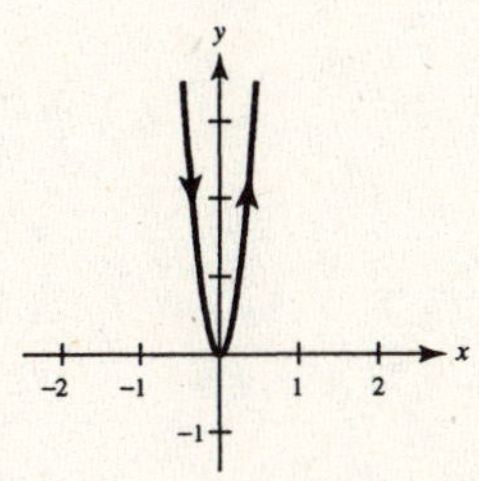

6. $x = t$

$y = t^3$

t	-3	-2	-1	0	1	2	3
x	-3	-2	-1	0	1	2	3
y	-27	-8	-1	0	1	8	27

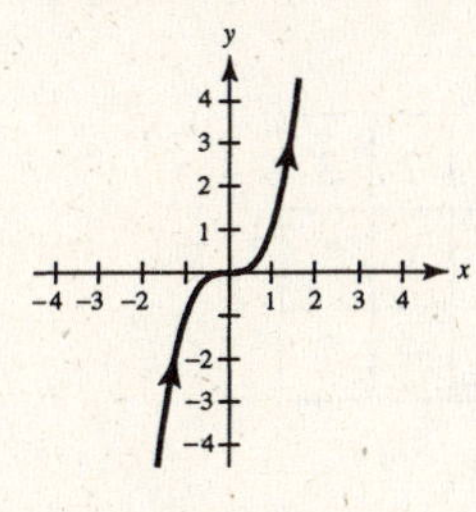

$y = t^3$

$y = x^3$

7. $x = t + 2 \Rightarrow t = x - 2$

$y = t^2 \Rightarrow y = (x - 2)^2$

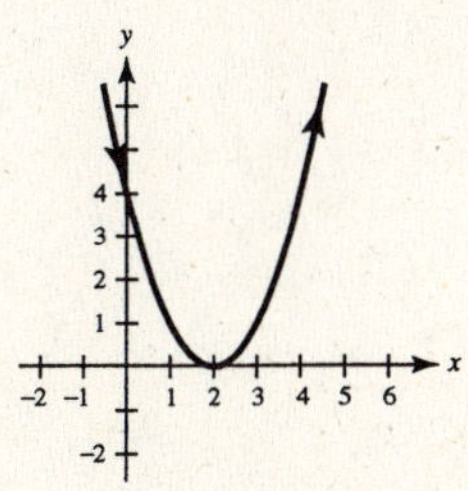

8. $x = \sqrt{t} \Rightarrow x^2 = t, t \geq 0$

$y = 1 - t$

t	0	1	2	3
x	0	1	$\sqrt{2}$	$\sqrt{3}$
y	1	0	-1	-2

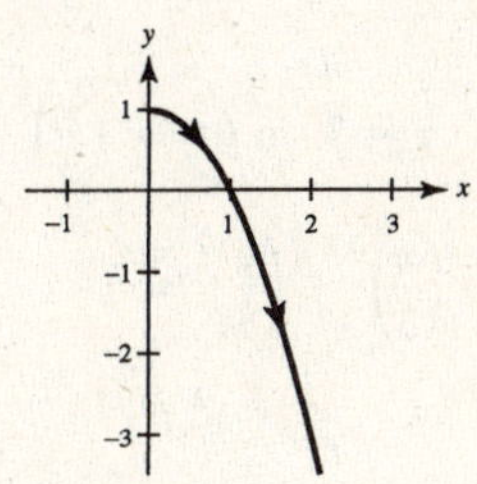

$y = 1 - t$

$y = 1 - x^2, x \geq 0$

9. $x = t + 1 \Rightarrow t = x - 1$

$$y = \frac{t}{t+1} \Rightarrow y = \frac{x-1}{x} = 1 - \frac{1}{x}$$

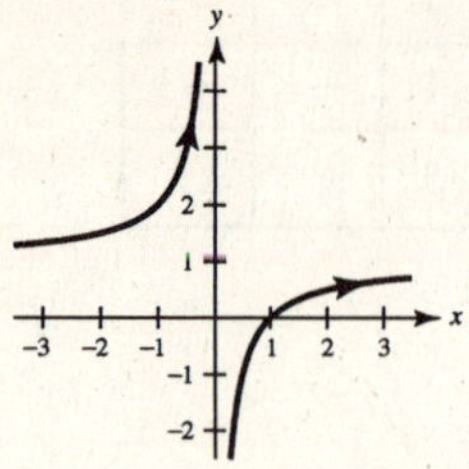

10. $x = t - 1 \Rightarrow t = x + 1$

$$y = \frac{t}{t-1}$$

t	-3	-2	-1	0	2	3
x	-4	-3	-2	-1	1	2
y	$\frac{3}{4}$	$\frac{2}{3}$	$\frac{1}{2}$	0	2	$\frac{3}{2}$

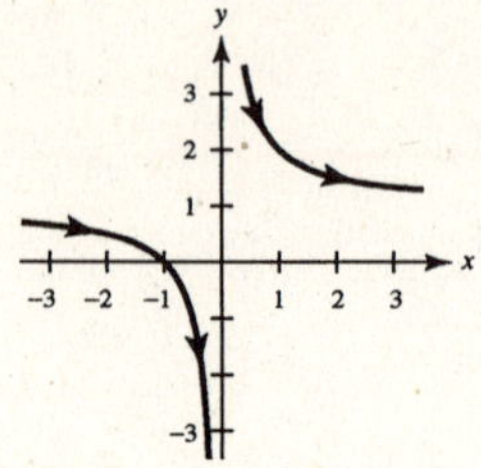

$$y = \frac{t}{t-1}$$

$$y = \frac{x+1}{x+1-1} = \frac{x+1}{x}$$

11. $x = 2(t+1) \Rightarrow \frac{x}{2} - 1 = t$ or $t = \frac{x-2}{2}$

$$y = |t - 2| \Rightarrow \quad y = \left|\frac{x}{2} - 1 - 2\right| = \left|\frac{x}{2} - 3\right| = \left|\frac{x-6}{2}\right|$$

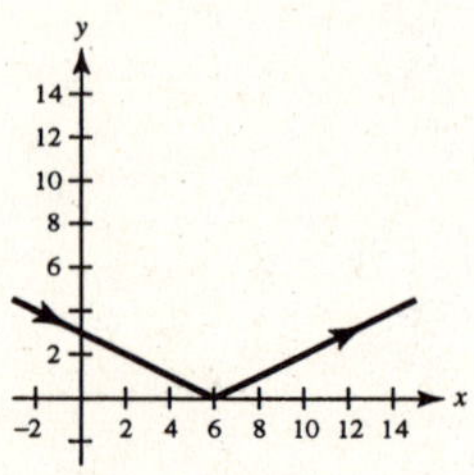

12. $x = |t - 1|$

$y = t + 2 \Rightarrow t = y - 2$

t	-3	-2	-1	0	1	2	3
x	4	3	2	1	0	1	2
y	-1	0	1	2	3	4	5

$x = |t - 1|$

$x = |y - 2 - 1| \Rightarrow x = |y - 3|$

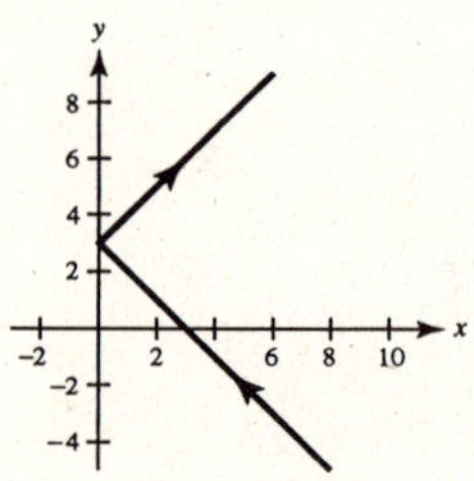

13. $x = 3\cos\theta \Rightarrow \left(\frac{x}{3}\right)^2 = \cos^2\theta$

$$y = 3\sin\theta \Rightarrow \left(\frac{y}{3}\right)^2 = \sin^2\theta$$

$$\left(\frac{x}{3}\right)^2 + \left(\frac{y}{3}\right)^2 = 1$$

$$x^2 + y^2 = 9$$

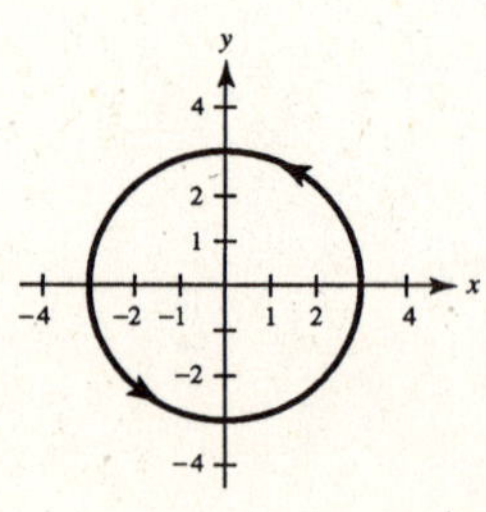

14. $x = 2\cos\theta$

$y = 3\sin\theta$

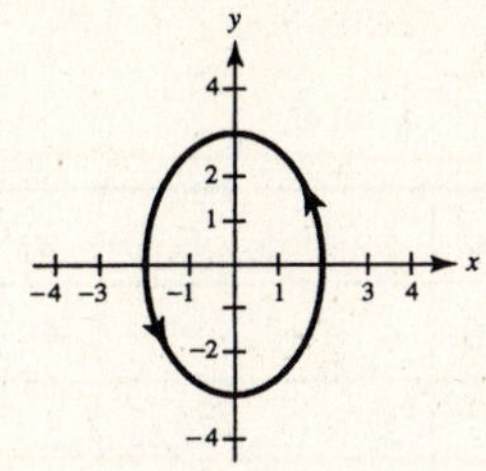

t	0	$\pi/4$	$\pi/2$	$3\pi/4$	π	$5\pi/4$	$3\pi/2$	$7\pi/4$	2π
x	2	$\sqrt{2}$	0	$-\sqrt{2}$	-2	$-\sqrt{2}$	0	$\sqrt{2}$	2
y	0	$3\sqrt{2}/2$	3	$3\sqrt{2}/2$	0	$-3\sqrt{2}/2$	-3	$-3\sqrt{2}/2$	0

$$\cos^2\theta + \sin^2\theta = 1$$

$$\left(\frac{x}{2}\right)^2 + \left(\frac{y}{3}\right)^2 = 1$$

$$\frac{x^2}{4} + \frac{y^2}{9} = 1$$

15. $x = 4\sin 2\theta \Rightarrow \left(\frac{x}{4}\right)^2 = \sin^2 2\theta$

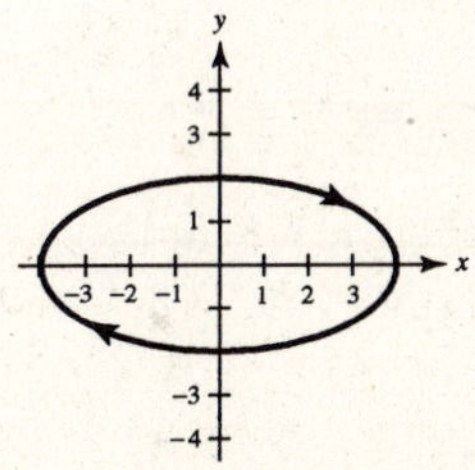

$y = 2\cos 2\theta \Rightarrow \left(\frac{y}{2}\right)^2 = \cos^2 2\theta$

$$\left(\frac{x}{4}\right)^2 + \left(\frac{y}{2}\right)^2 = 1$$

$$\frac{x^2}{16} + \frac{y^2}{4} = 1$$

16. $x = \cos\theta$

$y = 2\sin 2\theta$

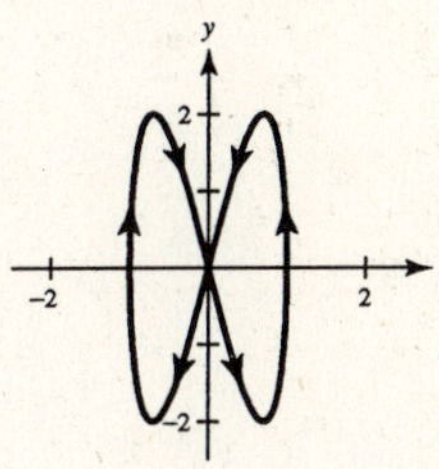

t	0	$\pi/4$	$\pi/2$	$3\pi/4$	π	$5\pi/4$	$3\pi/2$	$7\pi/4$	2π
x	1	$\sqrt{2}/2$	0	$-\sqrt{2}/2$	-1	$-\sqrt{2}/2$	0	$\sqrt{2}/2$	1
y	0	2	0	-2	0	2	0	-2	0

$$y = 2\sin 2\theta$$
$$y = 2(2\sin\theta\cos\theta)$$
$$y^2 = 16\sin^2\theta\cos^2\theta$$
$$y^2 = 16(1 - x^2)x^2$$
$$y^2 = 16x^2(1 - x^2)$$

17. $x = 4 + 2\cos\theta \Rightarrow \left(\frac{x-4}{2}\right)^2 = \cos^2\theta$

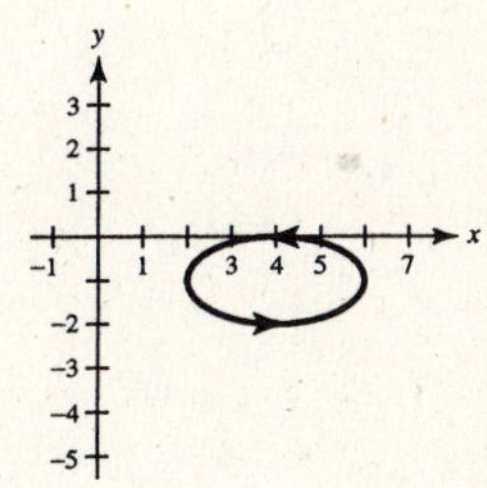

$y = -1 + \sin\theta \Rightarrow (y+1)^2 = \sin^2\theta$

$$\frac{(x-4)^2}{4} + \frac{(y+1)^2}{1} = 1$$

18. $x = 4 + 2\cos\theta$

$y = 2 + 3\sin\theta$

t	0	$\pi/4$	$\pi/2$	$3\pi/4$	π	$5\pi/4$	$3\pi/2$	$7\pi/4$	2π
x	6	$4 + \sqrt{2}$	4	$4 - \sqrt{2}$	2	$4 - \sqrt{2}$	4	$4 + \sqrt{2}$	6
y	2	$2 + (3\sqrt{2}/2)$	5	$2 + (3\sqrt{2}/2)$	2	$2 - (3\sqrt{2}/2)$	-1	$2 - (3\sqrt{2}/2)$	2

$\cos^2\theta + \sin^2\theta = 1$

$$\frac{(x-4)^2}{4} + \frac{(y-2)^2}{9} = 1$$

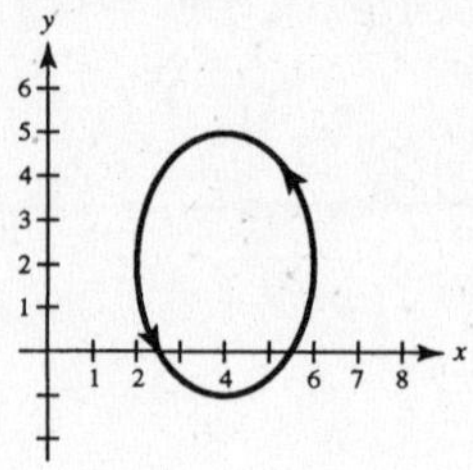

19. $x = e^{-t} \Rightarrow \dfrac{1}{x} = e^t$

$y = e^{3t} \Rightarrow y = (e^t)^3$

$y = \left(\dfrac{1}{x}\right)^3$

$y = \dfrac{1}{x^3},\ x > 0, y > 0$

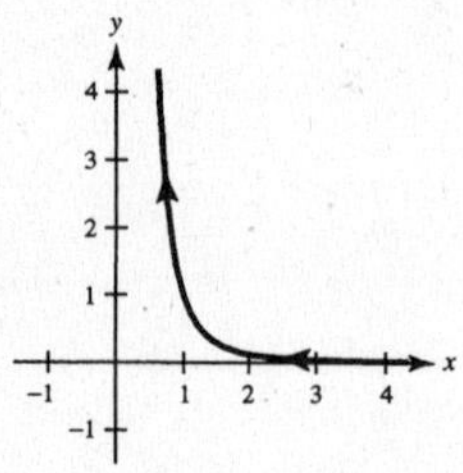

20. $x = e^{2t}$

$y = e^t \Rightarrow y^2 = e^{2t}$

t	-3	-2	-1	0	1	2
x	0.0025	0.0183	0.1353	1	7.3891	54.5982
y	0.0498	0.1353	0.3679	1	2.7183	7.3891

$x = e^{2t} = y^2$

$y^2 = x, y > 0$

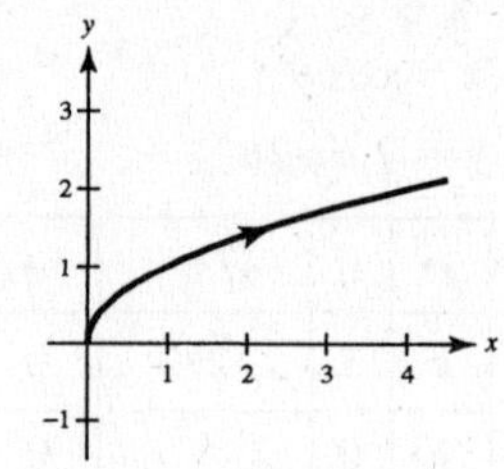

21. $x = t^3 \Rightarrow x^{1/3} = t$

$y = 3\ln t \Rightarrow y = \ln t^3$

$y = \ln(x^{1/3})^3$

$y = \ln x$

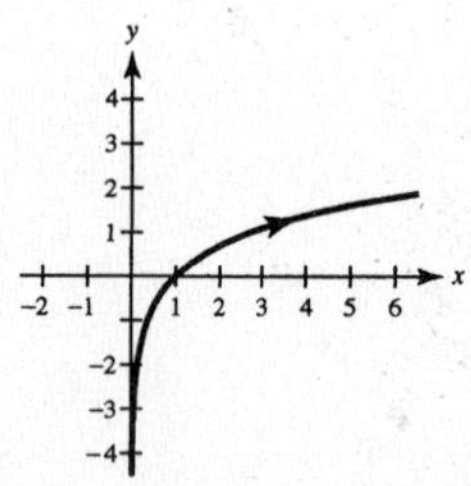

22. $x = \ln 2t \Rightarrow t = \frac{1}{2}e^x$

$y = 2t^2$

t	1	2	3	4
x	0.6931	1.3863	1.7918	2.0794
y	2	8	18	32

$y = 2t^2 = 2\left(\frac{1}{2}e^x\right)^2 = \frac{1}{2}e^{2x}$

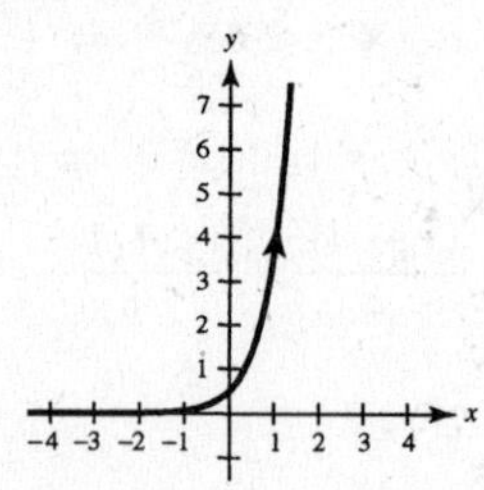

23. By eliminating the parameters in (a) to (d), we get $y = 2x + 1$. They differ from each other in orientation and in restricted domains. These curves are all smooth except for (b).

(a) $x = t, y = 2t + 1$

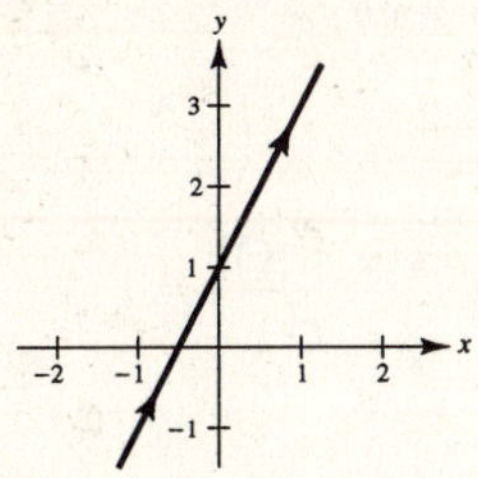

(b) $x = \cos\theta \qquad y = 2\cos\theta + 1$

$-1 \le x \le 1 \quad -1 \le y \le 3$

$\dfrac{dx}{d\theta} = \dfrac{dy}{d\theta} = 0$ when $\theta = 0, \pm\pi, \pm 2\pi, \ldots$

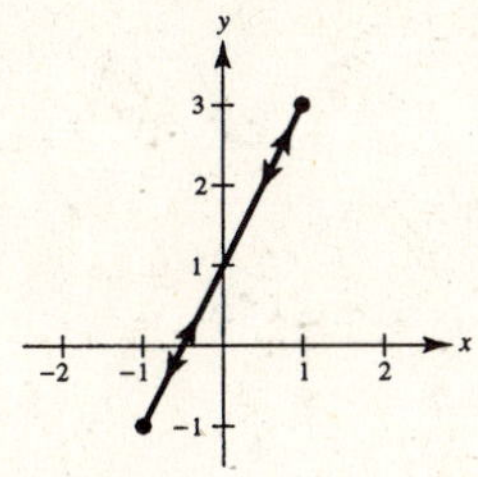

(c) $x = e^{-t} \qquad y = 2e^{-t} + 1$

$x > 0 \qquad y > 1$

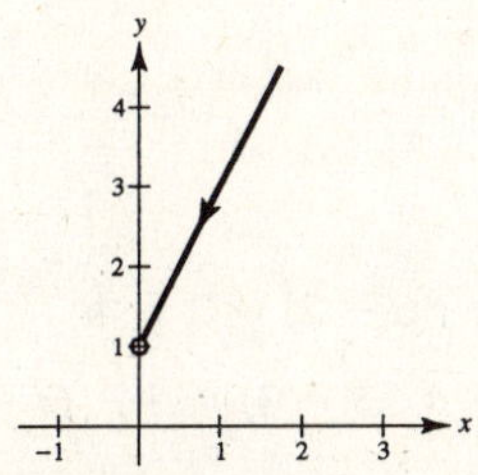

(d) $x = e^{t} \qquad y = 2e^{t} + 1$

$x > 0 \qquad y > 1$

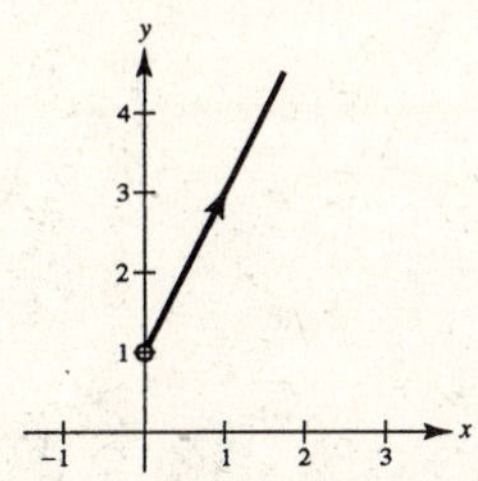

24. By eliminating the parameters in (a) to (d), we get $x^2 + y^2 = 4$. They differ from each other in orientation and in restricted domains. These curves are all smooth.

(a) $x = 2\cos\theta, y = 2\sin\theta$

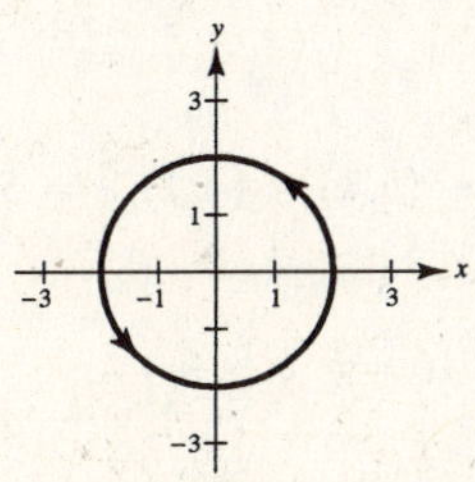

(b) $x = \dfrac{\sqrt{4t^2 - 1}}{|t|} = \sqrt{4 - \dfrac{1}{t^2}} \qquad y = \dfrac{1}{t}$

$x \ge 0, x \ne 2 \qquad y \ne 0$

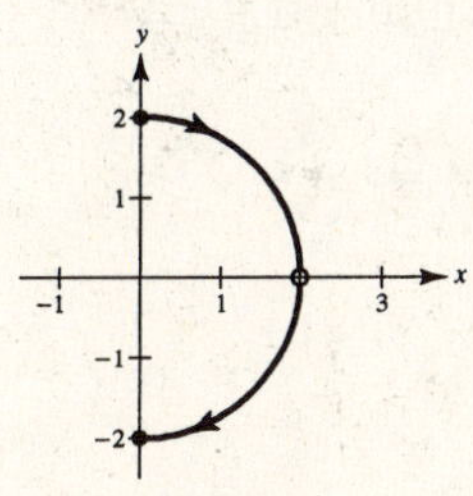

(c) $x = \sqrt{t} \qquad y = \sqrt{4 - t}$

$x \ge 0 \qquad y \ge 0$

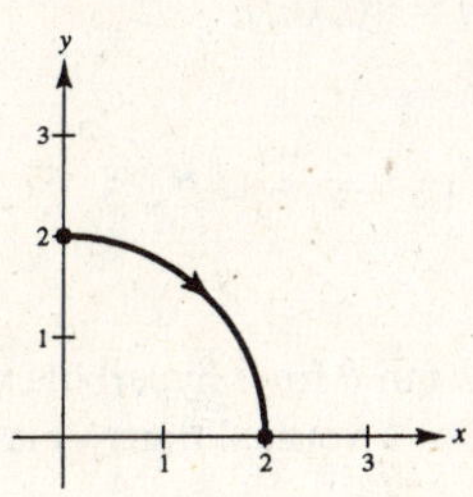

(d) $x = -\sqrt{4 - e^{2t}} \qquad y = e^{t}$

$-2 < x \le 0 \qquad y > 0$

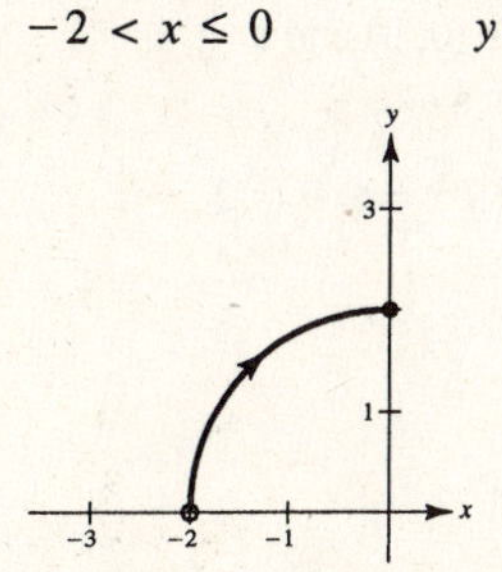

25. The curves are identical on $0 < \theta < \pi$.
Represent $y = 2(1 - x^2)$.

26. The orientations are reversed. The graphs are the same.

27. $x = x_1 + t(x_2 - x_1), y = y_1 + t(y_2 - y_1)$

$$\frac{x - x_1}{x_2 - x_1} = t$$

$$y = y_1 + \left(\frac{x - x_1}{x_2 - x_1}\right)(y_2 - y_1)$$

$$y - y_1 = \frac{y_2 - y_1}{x_2 - x_1}(x - x_1) = m(x - x_1)$$

28. $x = h + r \cos \theta, y = k + r \sin \theta$

$$\cos \theta = \frac{x - h}{r}, \sin \theta = \frac{y - k}{r}$$

$$\cos^2 \theta + \sin^2 \theta = \frac{(x - h)^2}{r^2} + \frac{(y - k)^2}{r^2} = 1$$

$$(x - h)^2 + (y - k)^2 = r^2$$

29. $x = h + a \cos \theta, y = k + b \sin \theta$

$$\frac{x - h}{a} = \cos \theta, \frac{y - k}{b} = \sin \theta$$

$$\frac{(x - h)^2}{a^2} + \frac{(y - k)^2}{b^2} = 1$$

30. $x = h + a \sec \theta, y = k + b \tan \theta$

$$\frac{x - h}{a} = \sec \theta, \frac{y - k}{b} = \tan \theta$$

$$\frac{(x - h)^2}{a^2} - \frac{(y - k)^2}{b^2} = 1$$

31. From Exercise 27 we have:
$x = 0 + t(6 - 0) = 6t$
$y = 0 + t(-3 - 0) = -3t$

32. Line through $(2, 3)$ and $(6, -3)$
From Exercise 27 we have:
$x = x_1 + t(x_2 - x_1) = 2 + t(6 - 2) = 2 + 4t$
$y = y_1 + t(y_2 - y_1) = 3 + t(-3 - 3) = 3 - 6t$

33. From Exercise 28 we have:
$x = 3 + 4 \cos \theta$
$y = 2 + 4 \sin \theta$

34. Circle with center $(-3, 2)$; radius: 5
From Exercise 28 we have:
$x = h + r \cos \theta = -3 + 5 \cos \theta$
$y = k + r \sin \theta = 2 + 5 \sin \theta$

35. Vertices: $(\pm 4, 0) \Rightarrow (h, k) = (0, 0)$ and $a = 4$
Foci: $(\pm 3, 0) \Rightarrow c = 3$
$c^2 = a^2 - b^2 \Rightarrow 9 = 16 - b^2 \Rightarrow b = \sqrt{7}$
From Exercise 29 we have:
$x = 4 \cos \theta$
$y = \sqrt{7} \sin \theta$

36. Ellipse
Vertices: $(4, 7), (4, -3) \Rightarrow (h, k) = (4, 2), a = 5$
Foci: $(4, 5), (4, -1) \Rightarrow c = 3$
$b^2 = a^2 - c^2 = 25 - 9 = 16 \Rightarrow b = 4$
From Exercise 29 we have:
$x = h + a \cos \theta = 4 + 4 \cos \theta$
$y = k + b \sin \theta = 2 + 5 \sin \theta$

37. Vertices: $(\pm 4, 0) \Rightarrow (h, k) = (0, 0)$ and $a = 4$
Foci: $(\pm 5, 0) \Rightarrow c = 5$
$c^2 = a^2 + b^2 \Rightarrow 25 = 16 + b^2 \Rightarrow b = 3$
From Exercise 30 we have:
$x = 4 \sec \theta$
$y = 3 \tan \theta$

38. Hyperbola
Vertices: $(0, \pm 2) \Rightarrow (h, k) = (0, 0), a = 2$
Foci: $(0, \pm 4) \Rightarrow c = 4$
$b^2 = c^2 - a^2 = 16 - 4 = 12 \Rightarrow b = 2\sqrt{3}$
From Exercise 30 we have:
$x = h + a \sec \theta, y = k + b \tan \theta$ for a hyperbola with horizontal transverse axis. For a vertical transverse axis we have:
$x = h + b \tan \theta = 2\sqrt{3} \tan \theta$
$y = k + a \sec \theta = 2 \sec \theta$

39. $y = 3x - 2$

(a) $t = x \Rightarrow x = t$ and $y = 3t - 2$

(b) $t = 2 - x \Rightarrow x = -t + 2$ and $y = 3(-t + 2) - 2 = -3t + 4$

40. $x = 3y - 2$

(a) $t = x$

$x = t$

$y = \frac{1}{3}(t + 2)$

(b) $t = 2 - x$

$x = 2 - t$

$y = \frac{1}{3}(x + 2) = \frac{1}{3}(4 - t)$

41. $y = x^2$

(a) $t = x \Rightarrow x = t$ and $y = t^2$

(b) $t = 2 - x \Rightarrow x = -t + 2$ and $y = (-t + 2)^2 = t^2 - 4t + 4$

42. $y = x^3$

(a) $t = x$

$x = t$

$y = t^3$

(b) $t = 2 - x$

$x = 2 - t$

$y = (2 - t)^3$

43. $y = x^2 + 1$

(a) $t = x \Rightarrow x = t$ and $y = t^2 + 1$

(b) $t = 2 - x \Rightarrow x = -t + 2$ and $y = (-t + 2)^2 + 1 = t^2 - 4t + 5$

44. $y = 2 - x$

(a) $t = x$

$x = t$

$y = 2 - t$

(b) $t = 2 - x$

$x = 2 - t$

$y = 2 - (2 - t) = t$

45. $y = \dfrac{1}{x}$

(a) $t = x \Rightarrow x = t$ and $y = \dfrac{1}{t}$

(b) $t = 2 - x \Rightarrow x = -t + 2$ and $y = \dfrac{1}{-t + 2} = \dfrac{-1}{t - 2}$

46. $y = \dfrac{1}{2x}$

(a) $t = x$

$x = t$

$y = \dfrac{1}{2t}$

(b) $t = 2 - x$

$x = 2 - t$

$y = \dfrac{1}{2(2 - t)} = \dfrac{1}{4 - 2t}$

47. $x = 2 \sec t,\ y = 3 \tan t$

$\sec^2 t - \tan^2 t = 1$

$\dfrac{x^2}{4} - \dfrac{y^2}{9} = 1$

$x = 2 \sec(-t) = 2 \sec t$

$y = 3 \tan(-t) = -3 \tan t$

$\sec^2 t - \tan^2 t = 1$

$\dfrac{x^2}{4} - \dfrac{y^2}{9} = 1$

The graph would still be the same except the orientation of the curve would be reversed.

48. $x = 4\cos t,\ y = 3\sin t$

$\dfrac{x^2}{16} + \dfrac{y^2}{9} = 1$

(a) $x = 4\cos 2t,\ y = 3\sin 2t$

The object is moving twice as fast around the ellipse.

(b) $x = 5\cos t,\ y = 3\sin t$

The object now moves around the ellipse.

$\dfrac{x^2}{25} + \dfrac{y^2}{9} = 1$

49. $x = 2\cos\theta \Rightarrow -2 \le x \le 2$

$y = \sin 2\theta \Rightarrow -1 \le y \le 1$

Matches graph (b).

Domain: $[-2, 2]$

Range: $[-1, 1]$

50. $x = 4\cos^3\theta \Rightarrow -4 \le x \le 4$

$y = 6\sin^3\theta \Rightarrow -6 \le y \le 6$

Matches graph (c).

Domain: $[-4, 4]$

Range: $[-6, 6]$

51. $x = \frac{1}{2}(\cos\theta + \theta\sin\theta)$

$y = \frac{1}{2}(\sin\theta - \theta\cos\theta)$

Matches graph (d).

Domain: $(-\infty, \infty)$

Range: $(-\infty, \infty)$

52. $x = \frac{1}{2}\cot\theta \Rightarrow -\infty < x < \infty$

$y = 4\sin\theta\cos\theta \Rightarrow -2 \le y \le 2$

Matches graph (a).

Domain: $(-\infty, \infty)$

Range: $[-2, 2]$

53. $x = 4(\theta - \sin\theta)$

$y = 4(1 - \cos\theta)$

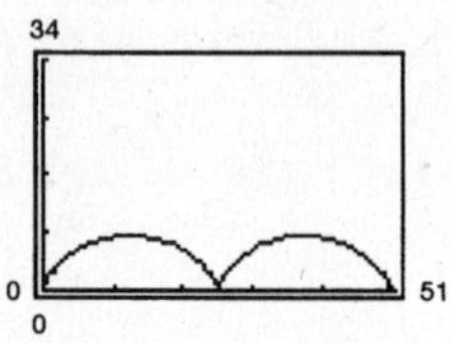

54. $x = \theta + \sin\theta$

$y = 1 - \cos\theta$

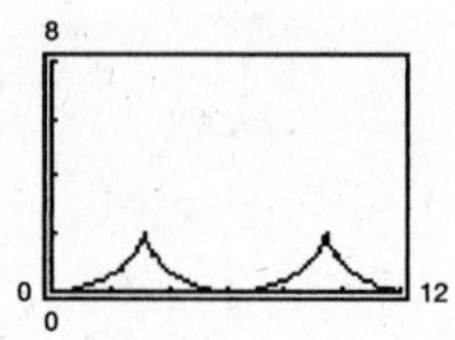

55. $x = \theta - \frac{3}{2}\sin\theta$

$y = 1 - \frac{3}{2}\cos\theta$

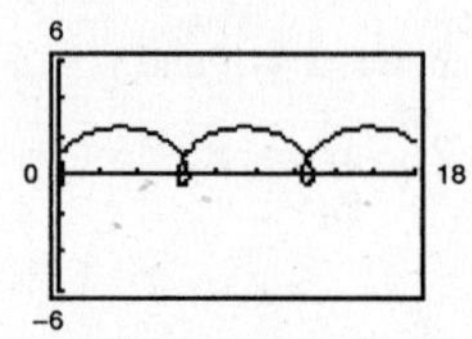

56. $x = 2\theta - 4\sin\theta$

$y = 2 - 4\cos\theta$

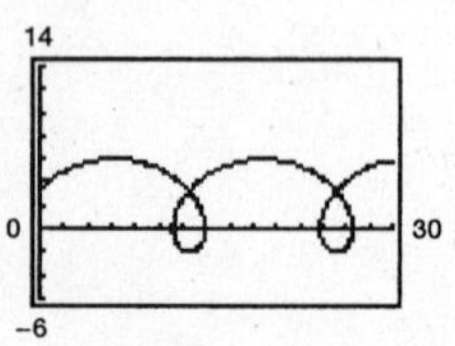

57. $x = 3\cos^3\theta$

$y = 3\sin^3\theta$

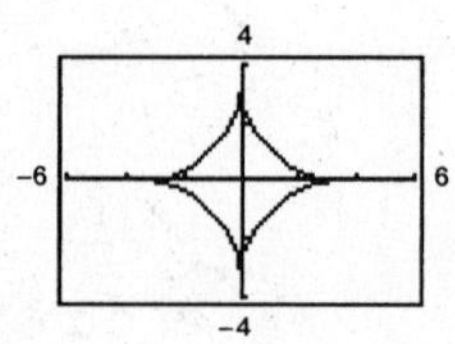

58. $x = 8\theta - 4\sin\theta$

$y = 8 - 4\cos\theta$

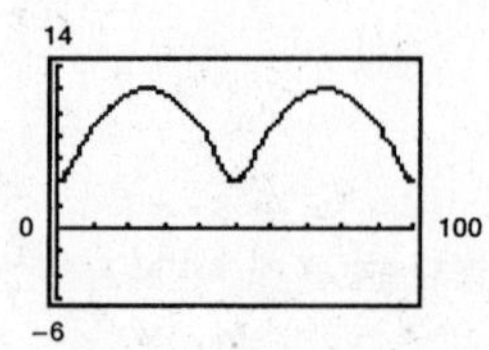

59. $x = 2\cot\theta$

$y = 2\sin^2\theta$

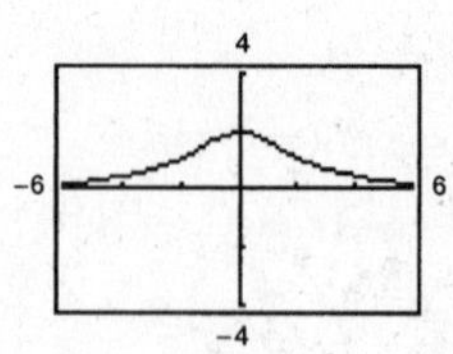

60. $x = \dfrac{3t}{1 + t^3}$

$y = \dfrac{3t^2}{1 + t^3}$

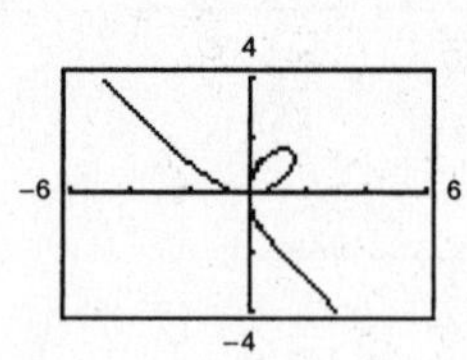

61. (a)

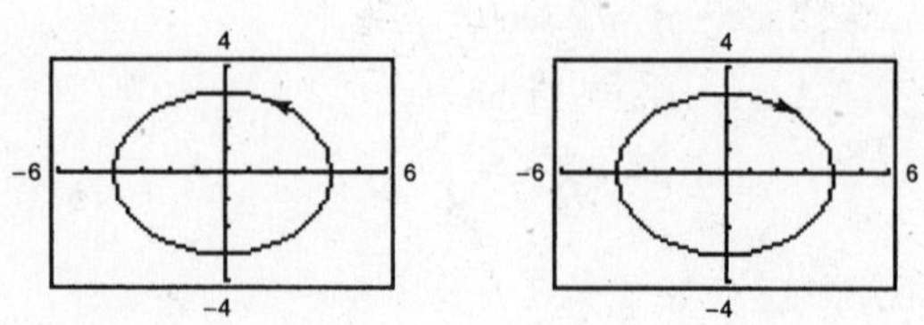

(b) The orientation of the second curve is the reverse of the first curve.

(c) The orientation will be reversed.

(d) Many answers possible. For example, $x = 1 + t$, $y = 1 + 2t$, and $x = 1 - t$, $y = 1 - 2t$.

62. The set of points (x, y) corresponding to the rectangular equation of a set of parametric equations does not show the orientation of the curve nor any restriction on the domain of the original parametric equations.

63. See definition on page 749.

64. Each point (x, y) in the plane is determined by the plane curve $x = f(t)$, $y = g(t)$. For each t, plot (x, y). As t increases, the curve is traced out in a specific direction called the orientation of the curve.

65. $x = (v_0 \cos \theta)t$ and $y = h + (v_0 \sin \theta)t - 16t^2$

(a) $\theta = 60°,\ v_0 = 88$ ft/sec

$x = (88 \cos 60°)t$ and $y = (88 \sin 60°)t - 16t^2$

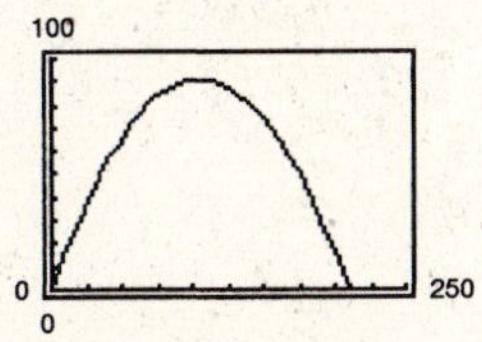

Maximum height: 90.8 feet

Range: 209.6 feet

(b) $\theta = 60°,\ v_0 = 132$ ft/sec

$x = (132 \cos 60°)t$ and $y = (132 \sin 60°)t - 16t^2$

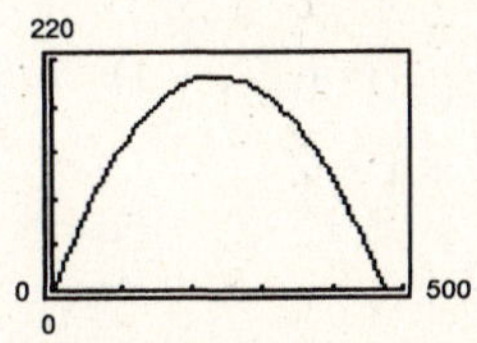

Maximum height: 204.2 feet

Range: 471.6 feet

(c) $\theta = 45°,\ v_0 = 88$ ft/sec

$x = (88 \cos 45°)t$ and $y = (88 \sin 45°)t - 16t^2$

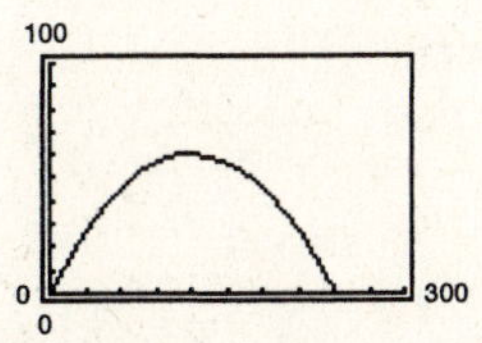

Maximum height: 60.5 ft

Range: 242.0 ft

(d) $\theta = 45°,\ v_0 = 132$ ft/sec

$x = (132 \cos 45°)t$ and $y = (132 \sin 45°)t - 16t^2$

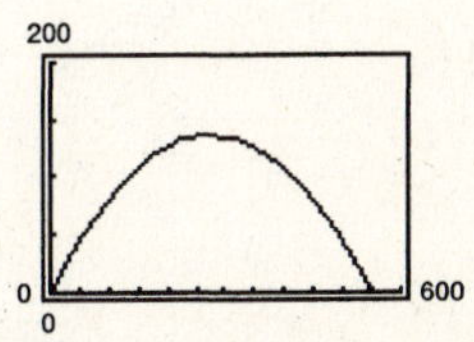

Maximum height: 136.1 ft

Range: 544.5 ft

66. $x = (v_0 \cos \theta)t$

$y = h + (v_0 \sin \theta)t - 16t^2$

(a) $\theta = 15°, v_0 = 60$ ft/sec

Maximum height: 3.8 feet
Range: 56.3 feet

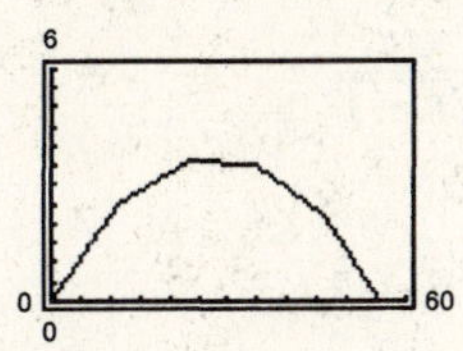

(b) $\theta = 15°, v_0 = 100$ ft/sec

Maximum height: 10.5 feet
Range: 156.3 feet

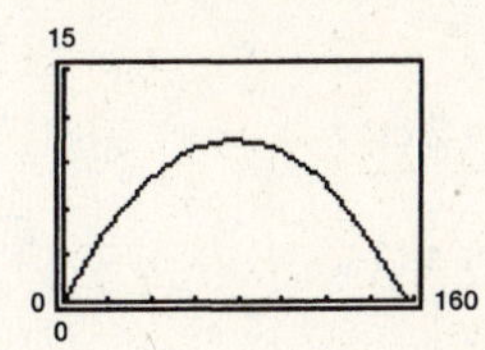

(c) $\theta = 30°, v_0 = 60$ ft/sec

Maximum height: 14.1 feet
Range: 97.4 feet

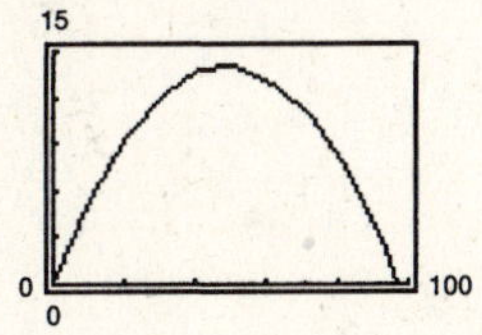

(d) $\theta = 30°, v_0 = 100$ ft/sec

Maximum height: 39.1 feet
Range: 270.6 feet

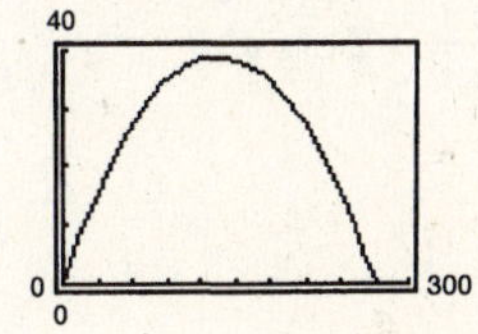

67. $x = 2t,\ y = 3t - 1$

(a) $\dfrac{dy}{dx} = \dfrac{3}{2}$

(b) $t = \dfrac{x}{2}$

$y = \dfrac{3x}{2} - 1$

$\dfrac{dy}{dx} = \dfrac{3}{2}$

68. $x = \sqrt{t},\ y = 3t - 1$

(a) $\dfrac{dy}{dx} = \dfrac{3}{\dfrac{1}{2\sqrt{t}}} = \dfrac{3}{\dfrac{1}{2x}} = 6x$

(b) $x^2 = t \Rightarrow y = 3x^2 - 1$

$\dfrac{dy}{dx} = 6x$

69. $x = t + 1,\ y = t^2 + 3t$

(a) $\dfrac{dy}{dx} = 2t + 3$

(b) $x - 1 = t \Rightarrow y = (x - 1)^2 + 3(x - 1)$

$= x^2 + x - 2$

$\dfrac{dy}{dx} = 2x + 1$

$= 2(t + 1) + 1$

$= 2t + 3$

70. $x = t^2 + 3t + 2,\ y = 2t$

(a) $\dfrac{dy}{dx} = \dfrac{2}{2t + 3}$

(b) $x = \left(\dfrac{y}{2}\right)^2 + 3\left(\dfrac{y}{2}\right) + 2 = \dfrac{1}{4}y^2 + \dfrac{3}{2}y + 2$

$1 = \left(\dfrac{1}{2}y + \dfrac{3}{2}\right)\dfrac{dy}{dx}$

$\dfrac{dy}{dx} = \dfrac{2}{y + 3} = \dfrac{2}{2t + 3}$

71. $x = 2\cos t,\ y = 2\sin t$

(a) $\dfrac{dy}{dx} = \dfrac{2\cos t}{-2\sin t} = -\cot t$

(b) $\cos^2 t + \sin^2 t = 1$

$\left(\dfrac{x}{2}\right)^2 + \left(\dfrac{y}{2}\right)^2 = 1$

$x^2 + y^2 = 4$

$2x + 2y\dfrac{dy}{dx} = 0$

$\dfrac{dy}{dx} = \dfrac{-x}{y} = \dfrac{-2\cos t}{2\sin t} = -\cot t$

72. $x = \cos t,\ y = 3\sin t$

(a) $\dfrac{dy}{dx} = \dfrac{3\cos t}{-\sin t} = -3\cot t$

(b) $\cos^2 t + \sin^2 t = 1$

$(x)^2 + \left(\dfrac{y}{3}\right)^2 = 1$

$9x^2 + y^2 = 9$

$18x + 2y\dfrac{dy}{dx} = 0$

$\dfrac{dy}{dx} = \dfrac{-9x}{y} = -\dfrac{9\cos t}{3\sin t} = -3\cot t$

73. $x = 2 + \sec t,\ y = 1 + 2\tan t$

(a) $\dfrac{dy}{dx} = \dfrac{2\sec^2 t}{\sec t\tan t} = 2\sec t\cot t$

$= 2\left(\dfrac{1}{\cos t}\right)\left(\dfrac{\cos t}{\sin t}\right) = 2\csc t$

(b) $\sec^2 t - \tan^2 t = 1$

$(x - 2)^2 - \dfrac{(y - 1)^2}{4} = 1$

$2(x - 2) - \dfrac{1}{2}(y - 1)\dfrac{dy}{dx} = 0$

$\dfrac{dy}{dx} = \dfrac{4(x - 2)}{y - 1} = \dfrac{4\sec t}{2\tan t} = 2\csc t$

74. $x = \sqrt{t},\ y = \sqrt{t - 1}$

(a) $\dfrac{dy}{dx} = \dfrac{\dfrac{1}{2}(t - 1)^{-1/2}}{\dfrac{1}{2}t^{-1/2}} = \dfrac{\sqrt{t}}{\sqrt{t - 1}}$

(b) $x^2 = t,\ y^2 = t - 1$

$y^2 = x^2 - 1$

$2y\dfrac{dy}{dx} = 2x$

$\dfrac{dy}{dx} = \dfrac{x}{y} = \dfrac{\sqrt{t}}{\sqrt{t - 1}}$

75. $x = \cos^3 t,\ y = \sin^3 t$

(a) $\dfrac{dy}{dx} = \dfrac{3\sin^2 t\cos t}{-3\cos^2 t\sin t} = -\tan t$

(b) $\cos^2 t + \sin^2 t = 1$

$x^{2/3} + y^{2/3} = 1$

$\dfrac{2}{3}x^{-1/3} + \dfrac{2}{3}y^{-1/3}\dfrac{dy}{dx} = 0$

$\dfrac{dy}{dx} = -\dfrac{y^{1/3}}{x^{1/3}} = -\dfrac{\sin t}{\cos t} = -\tan t$

76. $x = t - \sin t, y = 1 - \cos t$

(a) $\dfrac{dy}{dx} = \dfrac{\sin t}{1 - \cos t} = \dfrac{\sin t(1 + \cos t)}{\sin^2 t} = \csc t + \cot t$

(b) $\cos t = 1 - y \implies t = \arccos(1 - y)$

$x = \arccos(1 - y) - \sin(\arccos(1 - y))$

$x = \arccos(1 - y) - \sqrt{1 - (1 - y)^2}$

$1 = \left[\dfrac{1}{\sqrt{1 - (1 - y)^2}} - \dfrac{1 - y}{\sqrt{1 - (1 - y)^2}}\right]\dfrac{dy}{dx} = \dfrac{y}{\sqrt{1 - (1 - y)^2}}\dfrac{dy}{dx}$

$\sqrt{1 - (1 - y)^2} = y\dfrac{dy}{dx}$

$\sin(\arccos(1 - y)) = y\dfrac{dy}{dx}$

$\sin t = (1 - \cos t)\dfrac{dy}{dx}$

$\dfrac{dy}{dx} = \dfrac{\sin t}{1 - \cos t} = \csc t + \cot t$

77. $\dfrac{dy}{dx} = \dfrac{dy/dt}{dx/dt} = \dfrac{-4}{2t} = \dfrac{-2}{t}$

78. $\dfrac{dy}{dx} = \dfrac{dy/dt}{dx/dt} = \dfrac{-1}{(1/3)t^{-2/3}} = -3t^{2/3}$

79. $\dfrac{dy}{dx} = \dfrac{dy/d\theta}{dx/d\theta} = \dfrac{-2\cos\theta\sin\theta}{2\sin\theta\cos\theta} = -1$

$\left[\textbf{Note: } x + y = 1 \implies y = 1 - x \text{ and } \dfrac{dy}{dx} = -1\right]$

80. $\dfrac{dy}{dx} = \dfrac{dy/d\theta}{dx/d\theta} = \dfrac{(-1/2)e^{-\theta/2}}{2e^{\theta}} = -\dfrac{1}{4}e^{-3\theta/2} = \dfrac{-1}{4e^{3\theta/2}}$

81. $x = 2t,\ y = 3t - 1$

$\dfrac{dy}{dx} = \dfrac{dy/dt}{dx/dt} = \dfrac{3}{2}$

$\dfrac{d^2y}{dx^2} = 0$ Line

82. $x = \sqrt{t},\ y = 3t - 1$

$\dfrac{dy}{dx} = \dfrac{3}{1/(2\sqrt{t})} = 6\sqrt{t} = 6$ when $t = 1$.

$\dfrac{d^2y}{dx^2} = \dfrac{3/\sqrt{t}}{1/(2\sqrt{t})} = 6$ concave upwards

83. $x = t + 1,\ y = t^2 + 3t$

$\dfrac{dy}{dx} = \dfrac{2t + 3}{1} = 1$ when $t = -1$.

$\dfrac{d^2y}{dx^2} = 2$ concave upwards

84. $x = t^2 + 3t + 2, y = 2t$

$\dfrac{dy}{dx} = \dfrac{2}{2t + 3} = \dfrac{2}{3}$ when $t = 0$.

$\dfrac{d^2y}{dx^2} = \dfrac{-2(2)/(2t + 3)^2}{2t + 3} = \dfrac{-4}{(2t + 3)^3} = \dfrac{-4}{27}$ when $t = 0$.

concave downward

85. $x = 2\cos\theta,\ y = 2\sin\theta$

$$\frac{dy}{dx} = \frac{2\cos\theta}{-2\sin\theta} = -\cot\theta = -1 \text{ when } \theta = \frac{\pi}{4}.$$

$$\frac{d^2y}{dx^2} = \frac{\csc^2\theta}{-2\sin\theta} = \frac{-\csc^3\theta}{2} = -\sqrt{2} \text{ when } \theta = \frac{\pi}{4}.$$

concave downward

86. $x = \cos\theta,\ y = 3\sin\theta$

$$\frac{dy}{dx} = \frac{3\cos\theta}{-\sin\theta} = -3\cot\theta;\ \frac{dy}{dx} \text{ is undefined when } \theta = 0.$$

$$\frac{d^2y}{dx^2} = \frac{3\csc^2\theta}{-\sin\theta} = \frac{-3}{\sin^3\theta};\ \frac{d^2y}{dx^2} \text{ is undefined when } \theta = 0.$$

87. $x = 2 + \sec\theta,\ y = 1 + 2\tan\theta$

$$\frac{dy}{dx} = \frac{2\sec^2\theta}{\sec\theta\tan\theta}$$

$$= \frac{2\sec\theta}{\tan\theta} = 2\csc\theta = 4 \text{ when } \theta = \frac{\pi}{6}.$$

$$\frac{d^2y}{dx^2} = \frac{d\left[\dfrac{dy}{dx}\right]}{\dfrac{dx}{d\theta}} = \frac{-2\csc\theta\cot\theta}{\sec\theta\tan\theta}$$

$$= -2\cot^3\theta = -6\sqrt{3} \text{ when } \theta = \frac{\pi}{6}.$$

concave downward

88. $x = \sqrt{t},\ y = \sqrt{t-1}$

$$\frac{dy}{dx} = \frac{1/\left(2\sqrt{t-1}\right)}{1/\left(2\sqrt{t}\right)}$$

$$= \frac{\sqrt{t}}{\sqrt{t-1}} = \sqrt{2} \text{ when } t = 2.$$

$$\frac{d^2y}{dx^2} = \frac{\left[\sqrt{t-1}/\left(2\sqrt{t}\right) - \sqrt{t}\left(1/2\sqrt{t-1}\right)\right]/(t-1)}{1/\left(2\sqrt{t}\right)}$$

$$= \frac{-1}{(t-1)^{3/2}} = -1 \text{ when } t = 2.$$

concave downward

89. $x = \cos^3\theta,\ y = \sin^3\theta$

$$\frac{dy}{dx} = \frac{3\sin^2\theta\cos\theta}{-3\cos^2\theta\sin\theta}$$

$$= -\tan\theta = -1 \quad \text{when } \theta = \frac{\pi}{4}.$$

$$\frac{d^2y}{dx^2} = \frac{-\sec^2\theta}{-3\cos^2\theta\sin\theta} = \frac{1}{3\cos^4\theta\sin\theta}$$

$$= \frac{\sec^4\theta\csc\theta}{3} = \frac{4\sqrt{2}}{3} \text{ when } \theta = \frac{\pi}{4}.$$

concave upward

90. $x = \theta - \sin\theta,\ y = 1 - \cos\theta$

$$\frac{dy}{dx} = \frac{\sin\theta}{1-\cos\theta} = 0 \text{ when } \theta = \pi.$$

$$\frac{d^2y}{dx^2} = \frac{\dfrac{[(1-\cos\theta)\cos\theta - \sin^2\theta]}{(1-\cos\theta)^2}}{(1-\cos\theta)}$$

$$= \frac{-1}{(1-\cos\theta)^2} = -\frac{1}{4} \text{ when } \theta = \pi.$$

concave downward

91. $x = 1 - t,\ y = t^2$

Horizontal tangents: $\dfrac{dy}{dt} = 2t = 0$ when $t = 0$.

Point: $(1, 0)$

Vertical tangents: $\dfrac{dx}{dt} = -1 \neq 0$; none

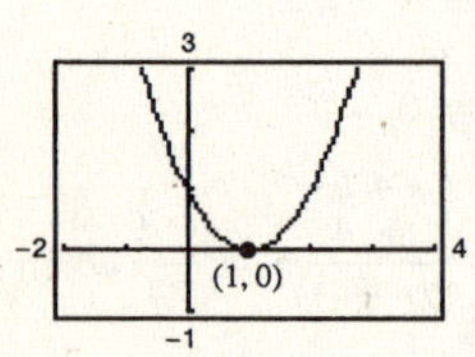

92. $x = t + 1,\ y = t^2 + 3t$

Horizontal tangents: $\dfrac{dy}{dt} = 2t + 3 = 0$ when $t = -\dfrac{3}{2}$.

Point: $\left(-\dfrac{1}{2}, -\dfrac{9}{4}\right)$

Vertical tangents: $\dfrac{dx}{dt} = 1 \neq 0$; none

93. $x = 1 - t,\ y = t^3 - 3t$

Horizontal tangents: $\dfrac{dy}{dt} = 3t^2 - 3 = 0$ when $t = \pm 1$.

Points: $(0, -2), (2, 2)$

Vertical tangents: $\dfrac{dx}{dt} = -1 \neq 0$; none

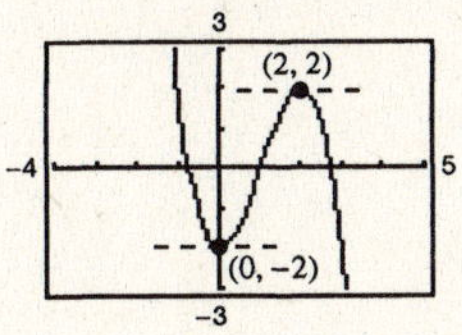

94. $x = t^2 - t + 2,\ y = t^3 - 3t$

Horizontal tangents: $\dfrac{dy}{dt} = 3t^2 - 3 = 0$ when $t = \pm 1$.

Points: $(2, -2), (4, 2)$

Vertical tangents: $\dfrac{dx}{dt} = 2t - 1 = 0$ when $t = \dfrac{1}{2}$.

Point: $\left(\dfrac{7}{4}, -\dfrac{11}{8}\right)$

95. $x = 3\cos\theta,\ y = 3\sin\theta$

Horizontal tangents: $\dfrac{dy}{d\theta} = 3\cos\theta = 0$ when $\theta = \dfrac{\pi}{2}, \dfrac{3\pi}{2}$.

Points: $(0, 3), (0, -3)$

Vertical tangents: $\dfrac{dx}{d\theta} = -3\sin\theta = 0$ when $\theta = 0, \pi$.

Points: $(3, 0), (-3, 0)$

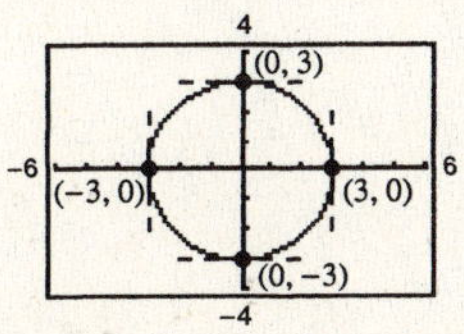

96. $x = \cos\theta,\ y = 2\sin 2\theta$

Horizontal tangents: $\dfrac{dy}{d\theta} = 4\cos 2\theta = 0$ when

$\theta = \dfrac{\pi}{4}, \dfrac{3\pi}{4}, \dfrac{5\pi}{4}, \dfrac{7\pi}{4}$.

Points: $\left(\dfrac{\sqrt{2}}{2}, 2\right), \left(-\dfrac{\sqrt{2}}{2}, -2\right), \left(-\dfrac{\sqrt{2}}{2}, 2\right), \left(\dfrac{\sqrt{2}}{2}, -2\right)$

Vertical tangents: $\dfrac{dx}{d\theta} = -\sin\theta = 0$ when $\theta = 0, \pi$.

Points: $(1, 0), (-1, 0)$

97. $x = 4 + 2\cos\theta,\ y = -1 + \sin\theta$

Horizontal tangents: $\dfrac{dy}{d\theta} = \cos\theta = 0$ when $\theta = \dfrac{\pi}{2}, \dfrac{3\pi}{2}$.

Points: $(4, 0), (4, -2)$

Vertical tangents: $\dfrac{dx}{d\theta} = -2\sin\theta = 0$ when $\theta = 0, \pi$.

Points: $(6, -1), (2, -1)$

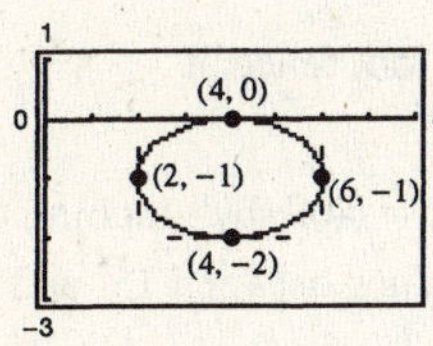

98. $x = 4\cos^2\theta,\ y = 2\sin\theta$

Horizontal tangents: $\dfrac{dy}{d\theta} = 2\cos\theta = 0$ when $\theta = \dfrac{\pi}{2}, \dfrac{3\pi}{2}$.

Since $dx/d\theta = 0$ at $\pi/2$ and $3\pi/2$, exclude them.

Vertical tangents: $\dfrac{dx}{d\theta} = -8\cos\theta\sin\theta = 0$ when

$\theta = 0, \pi$.

Point: $(4, 0)$

99. $x = \sec\theta,\ y = \tan\theta$

Horizontal tangents: $\dfrac{dy}{d\theta} = \sec^2\theta \neq 0$; none

Vertical tangents: $\dfrac{dx}{d\theta} = \sec\theta\tan\theta = 0$ when $x = 0, \pi$.

Points: $(1, 0), (-1, 0)$

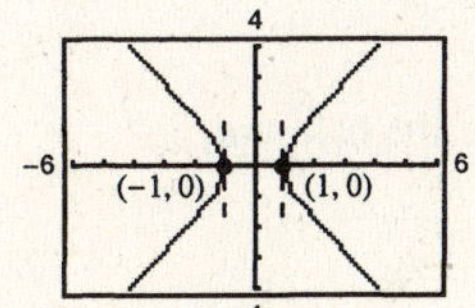

100. $x = \cos^2 \theta,\ y = \cos \theta$

Horizontal tangents: $\dfrac{dy}{d\theta} = -\sin \theta = 0$ when $x = 0, \pi$.

Since $dx/d\theta = 0$ at these values, exclude them.

Vertical tangents: $\dfrac{dx}{d\theta} = -2 \cos \theta \sin \theta = 0$ when

$$\theta = \frac{\pi}{2}, \frac{3\pi}{2}.$$

(Exclude 0, π.)

Point: $(0, 0)$

101. (a) $x = (v_0 \cos \theta)t$

$y = h + (v_0 \sin \theta)t - 16t^2$

$h = 5,\ v_0 = 240,\ \theta = 10°$

$x = (240 \cos 10°)t$

$y = 5 + (240 \sin 10°)t - 16t^2$

(b) $5 + (240 \sin 10°)t - 16t^2 = 0 \Rightarrow t = \dfrac{240 \sin 10° + \sqrt{(240 \sin 10°)^2 + 320}}{32}$

≈ 2.7196 seconds

Distance traveled $= x = (240 \cos 10°)(2.7196) \approx 643$ ft

(c)

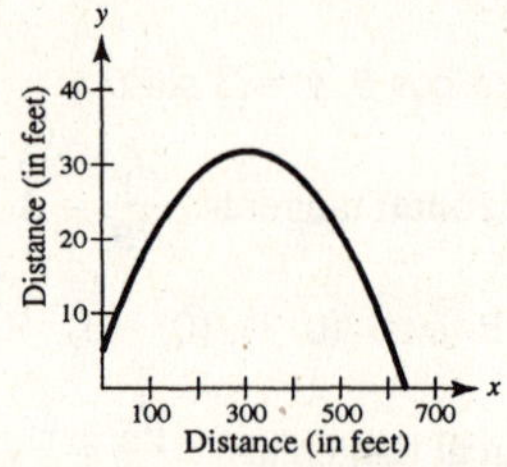

Maximum height: 32.1 feet

(d) From (b), $t \approx 2.72$ seconds

102. (a) 100 miles per hour $= 100\left(\dfrac{5280}{3600}\right)$ ft/sec $= \dfrac{440}{3}$ ft/sec

$x = \left(\dfrac{440}{3} \cos \theta\right)t \approx (146.67 \cos \theta)t$

$y = 3 + \left(\dfrac{440}{3} \sin \theta\right)t - 16t^2$

$\approx 3 + (146.67 \sin \theta)t - 16t^2$

(b) For $\theta = 15°$, we have:

$x = \left(\dfrac{440}{3} \cos 15°\right)t \approx 141.7t$

$y = 3 + \left(\dfrac{440}{3} \sin 15°\right)t - 16t^2 \approx 3 + 38.0t - 16t^2$

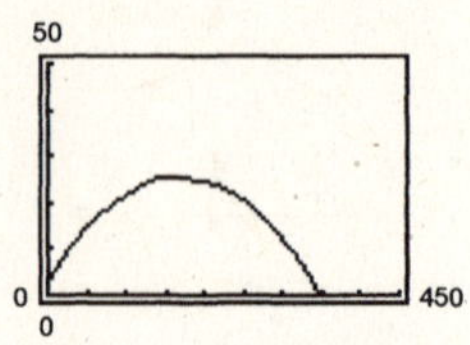

The ball hits the ground inside the ballpark, so it is not a home run.

(c) For $\theta = 23°$, we have:

$x = \left(\dfrac{440}{3} \cos 23°\right)t \approx 135.0t$

$y = 3 + \left(\dfrac{440}{3} \sin 23°\right)t - 16t^2 \approx 3 + 57.3t - 16t^2$

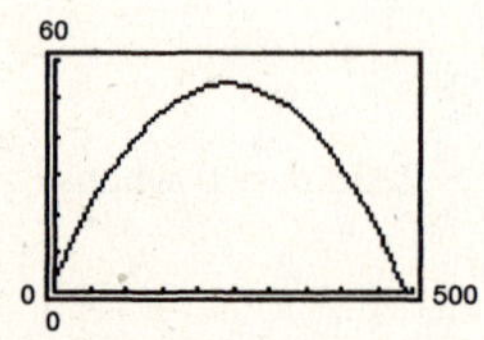

The ball easily clears the 10-foot fence at 400 feet so it is a home run.

(d) Find θ so that $y = 10$ when $x = 400$ by graphing the parametric equations for θ values between 15° and 23°. This occurs when $\theta \approx 19.4°$.

103. $x = (v_0 \cos\theta)t \Rightarrow t = \dfrac{x}{v_0 \cos\theta}$

$$y = h + (v_0 \sin\theta)t - 16t^2$$
$$= h + (v_0 \sin\theta)\left(\frac{x}{v_0 \cos\theta}\right) - 16\left(\frac{x}{v_0 \cos\theta}\right)^2$$
$$= h + (\tan\theta)x - \frac{16x^2}{v_0^2 \cos^2\theta}$$
$$= -\frac{16 \sec^2\theta}{v_0^2}x^2 + (\tan\theta)x + h$$

104. $y = 7 + x - 0.02x^2$

(a) Exercise 103 result: $y = -\dfrac{16 \sec^2\theta}{v_0^2}x^2 + (\tan\theta)x + h$

$h = 7$

$\tan\theta = 1 \Rightarrow \theta = 45°$

$\dfrac{16 \sec^2 45°}{v_0^2} = 0.02 \Rightarrow v_0 = 40$

$x = (v_0 \cos\theta)t = (40 \cos 45°)t$

$y = h + (v_0 \sin\theta)t - 16t^2 \approx 7 + (40 \cos 45°)t - 16t^2$

(b)

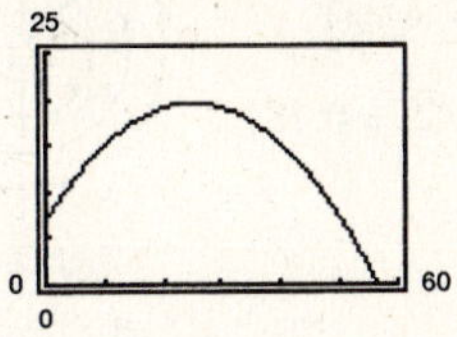

(c) Maximum height: 19.5 feet

Range: 56.2 feet

105. True.

$x = t$

$y = t^2 + 1 \Rightarrow y = x^2 + 1$

$x = 3t$

$y = 9t^2 + 1 \Rightarrow y = x^2 + 1$

106. False. Since $t^2 \geq 0$, the graph is that of $y = x$ for $x \geq 0$.

107. False.

A single rectangular equation may have many different parametric representations.

$y = 3 - 2x$

Let $x = t$ and $y = 3 - 2t$.

Let $x = t + 1$ and $y = 3 - 2(t + 1) = 1 - 2t$.

108. True. The graph is a hyperbola.

$x = 2 \tan t, y = \sec t$

$\sec^2 t - \tan^2 t = 1$

$y^2 - \dfrac{x^2}{4} = 1$

Section 12.5 Polar Coordinates and Calculus

1. Polar Coordinates: $\left(4, -\dfrac{\pi}{3}\right)$

Additional representations

$\left(4, -\dfrac{\pi}{3} + 2\pi\right) = \left(4, \dfrac{5\pi}{3}\right)$

$\left(-4, -\dfrac{\pi}{3} - \pi\right) = \left(-4, -\dfrac{4\pi}{3}\right)$

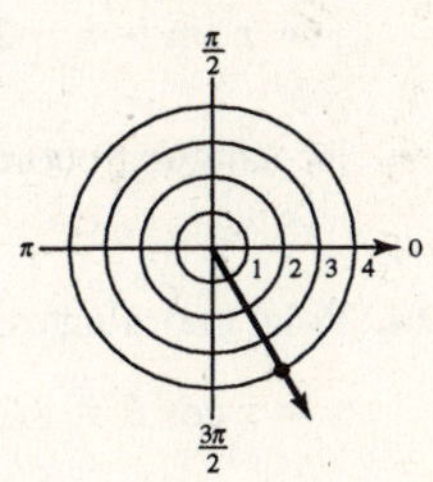

2. $\left(-1, -\dfrac{3\pi}{4}\right)$

$\left(1, \dfrac{\pi}{4}\right)$

$\left(-1, \dfrac{5\pi}{4}\right)$

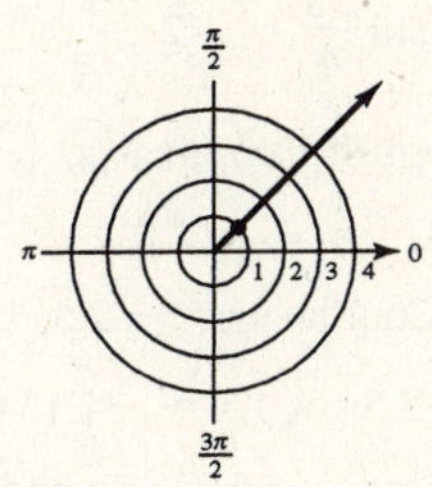

3. Polar Coordinates: $\left(0, -\frac{7\pi}{6}\right)$

Additional representations

$$\left(0, -\frac{7\pi}{6} + 2\pi\right) = \left(0, \frac{5\pi}{6}\right)$$

$$\left(0, -\frac{7\pi}{6} + \pi\right) = \left(0, -\frac{\pi}{6}\right)$$

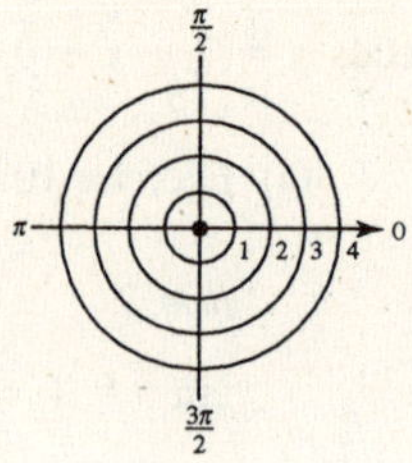

4. $\left(16, \frac{5\pi}{2}\right)$

$\left(16, \frac{\pi}{2}\right)$

$\left(-16, \frac{3\pi}{2}\right)$

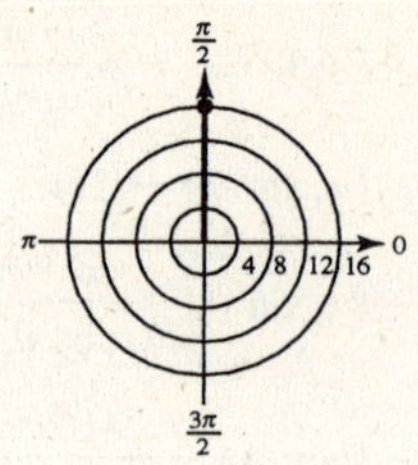

5. Polar Coordinates: $\left(\sqrt{2}, 2.36\right)$

Additional representations

$\left(\sqrt{2}, 2.36 + 2\pi\right) \approx \left(\sqrt{2}, 8.64\right)$

$\left(-\sqrt{2}, 2.36 - \pi\right) \approx \left(-\sqrt{2}, -0.78\right)$

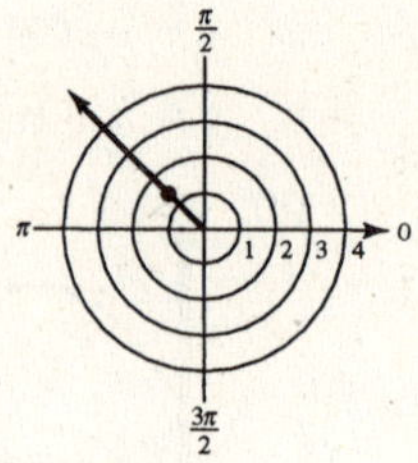

6. $(-3, -1.57)$

$(3, 1.5716)$

$(-3, 4.7132)$

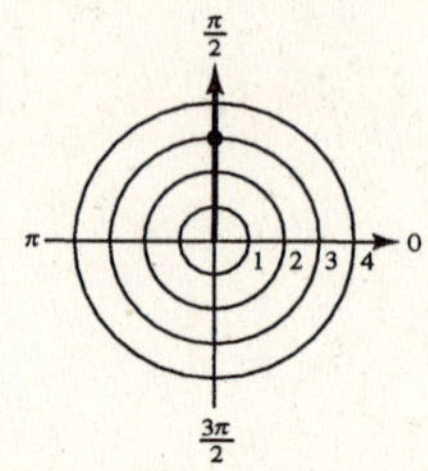

7. Polar Coordinates: $\left(3, \frac{\pi}{2}\right)$

$x = 3 \cos \frac{\pi}{2} = 0$

$y = 3 \sin \frac{\pi}{2} = 3$

Rectangular Coordinates: $(0, 3)$

8. Polar coordinates: $\left(3, \frac{3\pi}{2}\right) = (r, \theta)$

$x = r \cos \theta = 3 \cos \frac{3\pi}{2} = 0$

$y = r \sin \theta = 3 \sin \frac{3\pi}{2} = -3$

Rectangular coordinates: $(0, -3)$

9. Polar Coordinates: $\left(-1, \frac{5\pi}{4}\right)$

$x = -1 \cos\left(\frac{5\pi}{4}\right) = \frac{\sqrt{2}}{2}, y = -1 \sin\left(\frac{5\pi}{4}\right) = \frac{\sqrt{2}}{2}$

Rectangular Coordinates: $\left(\frac{\sqrt{2}}{2}, \frac{\sqrt{2}}{2}\right)$

10. Polar coordinates: $(0, -\pi) = (r, \theta)$

$x = r \cos \theta = 0$

$y = r \sin \theta = 0$

Rectangular coordinates: $(0, 0)$

11. Polar Coordinates: $\left(2, \frac{3\pi}{4}\right)$

$x = 2 \cos \frac{3\pi}{4} = -\sqrt{2}$

$y = 2 \sin \frac{3\pi}{4} = \sqrt{2}$

Rectangular Coordinates: $\left(-\sqrt{2}, \sqrt{2}\right)$

12. Polar coordinates:

$x = r \cos \theta = -2 \cos \frac{7\pi}{6} = \sqrt{3}$

$y = r \sin \theta = -2 \sin \frac{7\pi}{6} = 1$

Rectangular coordinates: $\left(\sqrt{3}, 1\right)$

13. Polar Coordinates: $(-2.5, 1.1)$

$x = -2.5 \cos 1.1 \approx -1.134$

$y = -2.5 \sin 1.1 \approx -2.228$

Rectangular Coordinates: $(-1.134, -2.228)$

14. Polar coordinates: $(8.25, 3.5) = (r, \theta)$

$x = r \cos \theta = 8.25 \cos 3.5 \approx -7.7258$

$y = r \sin \theta = 8.25 \sin 3.5 \approx -2.8940$

Rectangular coordinates: $(-7.7258, -2.8940)$

15. Rectangular Coordinates: $(1, 1)$

$r = \pm\sqrt{2}$, $\tan\theta = 1$, $\theta = \dfrac{\pi}{4}$ or $\dfrac{5\pi}{4}$

Polar Coordinates: $\left(\sqrt{2}, \dfrac{\pi}{4}\right), \left(-\sqrt{2}, \dfrac{5\pi}{4}\right)$

16. Rectangular coordinates: $(-3, -3)$

$r = \pm 3\sqrt{2}$, $\tan\theta = 1$, $\theta = \dfrac{\pi}{4}$ or $\theta = \dfrac{5\pi}{4}$

Polar coordinates: $\left(3\sqrt{2}, \dfrac{5\pi}{4}\right), \left(-3\sqrt{2}, \dfrac{\pi}{4}\right)$

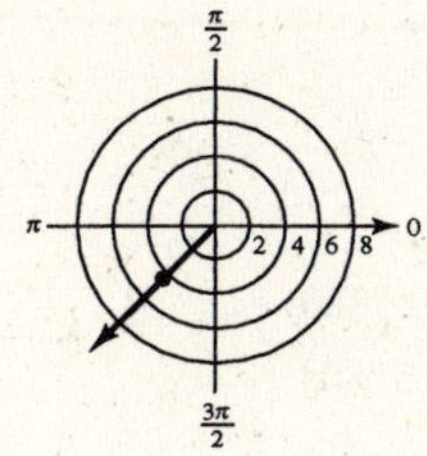

17. Rectangular Coordinates: $(-6, 0)$

$r = \pm 6$, $\tan\theta = 0$, $\theta = 0$ or π

Polar Coordinates: $(6, \pi), (-6, 0)$

18. Rectangular coordinates: $(0, -5)$

$r = \pm 5$, $\tan\theta$ undefined, $\theta = \dfrac{\pi}{2}$ or $\theta = \dfrac{3\pi}{2}$

Polar coordinates: $\left(5, \dfrac{3\pi}{2}\right), \left(-5, \dfrac{\pi}{2}\right)$

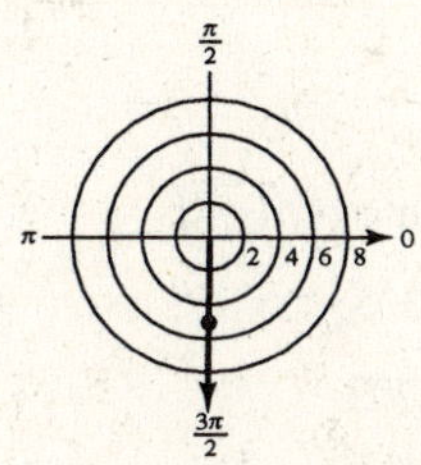

19. Rectangular Coordinates: $(-3, 4)$

$r = \pm\sqrt{9 + 16} = \pm 5$, $\tan\theta = -\dfrac{4}{3}$, $\theta \approx 2.2143, 5.3559$

Polar Coordinates: $(5, 2.2143), (-5, 5.3559)$

20. Rectangular coordinates: $(3, -1)$

$r = \sqrt{9 + 1} = \pm\sqrt{10}$, $\tan\theta = -\frac{1}{3}$, $\theta \approx -0.322 \approx 5.961$ or $\theta \approx 2.820$

Polar coordinates: $\left(-\sqrt{10}, 2.820\right), \left(\sqrt{10}, 5.961\right)$

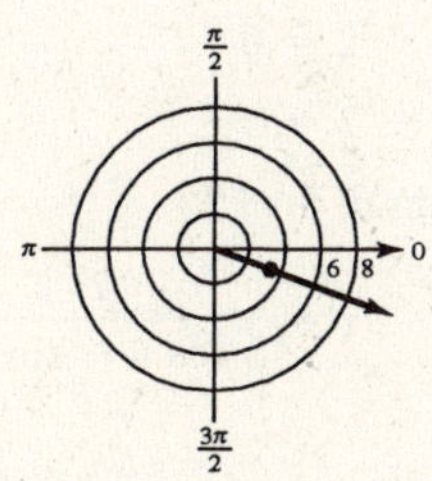

21. Rectangular Coordinates: $\left(-\sqrt{3}, -\sqrt{3}\right)$

$r = \pm\sqrt{3 + 3} = \pm\sqrt{6}$, $\tan\theta = 1$, $\theta = \dfrac{\pi}{4}$ or $\dfrac{5\pi}{4}$

Polar Coordinates: $\left(\sqrt{6}, \dfrac{5\pi}{4}\right), \left(-\sqrt{6}, \dfrac{\pi}{4}\right)$

22. Rectangular coordinates: $(5, 12)$

$r = \sqrt{25 + 144} = \pm 13$, $\tan\theta = \frac{12}{5}$, $\theta \approx 1.176$ or $\theta \approx 4.318$

Polar coordinates: $(13, 1.176)$, $(-13, 4.318)$

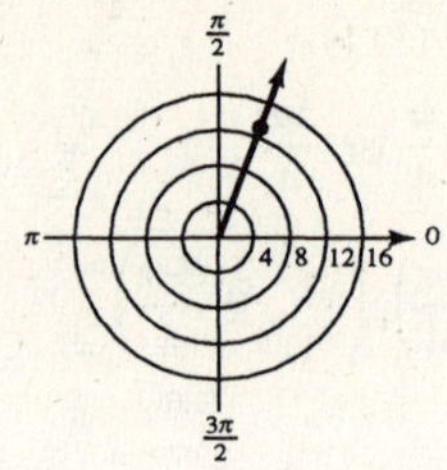

23. Rectangular: $(3, -2)$

$(3, -2) \blacktriangleright$ Pol

$\approx (3.606, -0.5880)$

or $\left(\sqrt{13}, -0.5880\right)$

24. Rectangular coordinates: $(-5, 2)$

$R \blacktriangleright Pr(-5, 2) \approx 5.385$

$R \blacktriangleright P\theta(-5, 2) \approx 2.761$

$\approx (5.385, 2.761)$

25. Rectangular: $\left(\sqrt{3}, 2\right)$

$\left(\sqrt{3}, 2\right) \blacktriangleright$ Pol

$\approx (2.646, 0.8571)$

or $\left(\sqrt{7}, 0.8571\right)$

26. Rectangular coordinates: $\left(3\sqrt{2}, 3\sqrt{2}\right)$

$R \blacktriangleright Pr\left(3\sqrt{2}, 3\sqrt{2}\right) = 6$

$R \blacktriangleright P\theta\left(3\sqrt{2}, 3\sqrt{2}\right) \approx 0.785$

$= \left(6, \frac{\pi}{4}\right)$

27. Rectangular: $\left(\frac{5}{2}, \frac{4}{3}\right)$

$\left(\frac{5}{2}, \frac{4}{3}\right) \blacktriangleright$ Pol

$\approx (2.833, 0.4900)$ or $\left(\frac{17}{6}, 0.4900\right)$

28. Rectangular coordinates: $\left(\frac{7}{4}, \frac{3}{2}\right)$

$R \blacktriangleright Pr\left(\frac{7}{4}, \frac{3}{2}\right) \approx 2.305$

$R \blacktriangleright P\theta\left(\frac{7}{4}, \frac{3}{2}\right) \approx 0.709$

$\approx (2.305, 0.709)$

29. $x^2 + y^2 = 9$

$r = 3$

30. $x^2 + y^2 = 16$

$r = 4$

31.

$$y = 4$$
$$r\sin\theta = 4$$
$$r = 4\csc\theta$$

32.

$$x = 10$$
$$r\cos\theta = 10$$
$$r = 10\sec\theta$$

33.

$$3x - y + 2 = 0$$
$$3r\cos\theta - r\sin\theta + 2 = 0$$
$$r(3\cos\theta - \sin\theta) = -2$$
$$r = \frac{-2}{3\cos\theta - \sin\theta}$$

34.

$$3x + 5y - 2 = 0$$
$$3r\cos\theta + 5r\sin\theta - 2 = 0$$
$$r(3\cos\theta + 5\sin\theta) = 2$$
$$r = \frac{2}{3\cos\theta + 5\sin\theta}$$

35.

$$xy = 16$$
$$(r\cos\theta)(r\sin\theta) = 16$$
$$r^2 = 16\sec\theta\csc\theta = 32\csc 2\theta$$

36.

$$y = x$$
$$r\cos\theta = r\sin\theta$$
$$1 = \tan\theta$$
$$\theta = \frac{\pi}{4}$$

37. $y^2 - 8x - 16 = 0$

$r^2 \sin^2 \theta - 8r \cos \theta - 16 = 0$

By the Quadratic Formula, we have:

$$r = \frac{-(-8\cos\theta) \pm \sqrt{(-8\cos\theta)^2 - 4(\sin^2\theta)(-16)}}{2\sin^2\theta}$$

$$= \frac{8\cos\theta \pm \sqrt{64\cos^2\theta + 64\sin^2\theta}}{2\sin^2\theta}$$

$$= \frac{8\cos\theta \pm \sqrt{64(\cos^2\theta + \sin^2\theta)}}{2\sin^2\theta}$$

$$= \frac{8\cos\theta \pm 8}{2\sin^2\theta}$$

$$= \frac{4(\cos\theta \pm 1)}{1 - \cos^2\theta}$$

$$r = \frac{4(\cos\theta + 1)}{(1 + \cos\theta)(1 - \cos\theta)} = \frac{4}{1 - \cos\theta}$$

or

$$r = \frac{4(\cos\theta - 1)}{(1 + \cos\theta)(1 - \cos\theta)} = \frac{-4}{1 + \cos\theta}$$

38. $(x^2 + y^2)^2 = 9(x^2 - y^2)$

$(r^2)^2 = 9(r^2 \cos^2 \theta - r^2 \sin^2 \theta)$

$= 9r^2(\cos^2 \theta - \sin^2 \theta)$

$r^2 = 9 \cos 2\theta$

39. $x^2 + y^2 = a^2$

$r^2 = a^2$

$r = a$

40. $x^2 + y^2 = 9a^2$

$r = 3a$

41. $y = b$

$r \sin \theta = b$

$r = b \csc \theta$

42. $x = 4a$

$r \cos \theta = 4a$

$r = 4a \sec \theta$

43. $x^2 + y^2 - 2ax = 0$

$r^2 - 2a\, r \cos \theta = 0$

$r(r - 2a \cos \theta) = 0$

$r - 2a \cos \theta = 0$

$r = 2a \cos \theta$

44. $x^2 + y^2 - 2ay = 0$

$r^2 - 2ar \sin \theta = 0$

$r = 2a \sin \theta$

45. $r = 4 \sin \theta$

$r^2 = 4r \sin \theta$

$x^2 + y^2 = 4y$

$x^2 + y^2 - 4y = 0$

46. $r = 3 \cos \theta$

$r^2 = 3r \cos \theta$

$x^2 + y^2 = 3x$

$x^2 + y^2 - 3x = 0$

47. $\theta = \frac{2\pi}{3}$

$\tan \theta = \tan \frac{2\pi}{3}$

$\frac{y}{x} = -\sqrt{3}$

$y = -\sqrt{3}x$

$\sqrt{3}x + y = 0$

48. $r = 4$

$r^2 = 16$

$x^2 + y^2 = 16$

49. $r = 2 \csc \theta$

$r \sin \theta = 2$

$y = 2$

50. $r^2 = \sin 2\theta = 2 \sin \theta \cos \theta$

$r^2 = 2\left(\frac{y}{r}\right)\left(\frac{x}{r}\right) = \frac{2xy}{r^2}$

$r^4 = 2xy$

$(x^2 + y^2)^2 = 2xy$

51.

$$r = 2\sin 3\theta$$
$$r = 2\sin(\theta + 2\theta)$$
$$r = 2[\sin\theta\cos 2\theta + \cos\theta\sin 2\theta]$$
$$r = 2[\sin\theta(1 - 2\sin^2\theta) + \cos\theta(2\sin\theta\cos\theta)]$$
$$r = 2[\sin\theta - 2\sin^3\theta + 2\sin\theta\cos^2\theta]$$
$$r = 2[\sin\theta - 2\sin^3\theta + 2\sin\theta(1 - \sin^2\theta)]$$
$$r = 2(3\sin\theta - 4\sin^3\theta)$$
$$r^4 = 6r^3\sin\theta - 8r^3\sin^3\theta$$
$$(x^2 + y^2)^2 = 6(x^2 + y^2)y - 8y^3$$
$$(x^2 + y^2)^2 = 6x^2y - 2y^3$$

52.

$$r = \frac{1}{1 - \cos\theta}$$
$$r - r\cos\theta = 1$$
$$\sqrt{x^2 + y^2} - x = 1$$
$$x^2 + y^2 = 1 + 2x + x^2$$
$$y^2 = 2x + 1$$

53.

$$r = \frac{6}{2 - 3\sin\theta}$$
$$r(2 - 3\sin\theta) = 6$$
$$2r = 6 + 3r\sin\theta$$
$$2\left(\pm\sqrt{x^2 + y^2}\right) = 6 + 3y$$
$$4(x^2 + y^2) = (6 + 3y)^2$$
$$4x^2 + 4y^2 = 36 + 36y + 9y^2$$
$$4x^2 - 5y^2 - 36y - 36 = 0$$

54.

$$r = \frac{6}{2\cos\theta - 3\sin\theta}$$
$$r = \frac{6}{2(x/r) - 3(y/r)}$$
$$r = \frac{6r}{2x - 3y}$$
$$1 = \frac{6}{2x - 3y}$$
$$2x - 3y = 6$$

55.

$$r = 6$$
$$r^2 = 36$$
$$x^2 + y^2 = 36$$

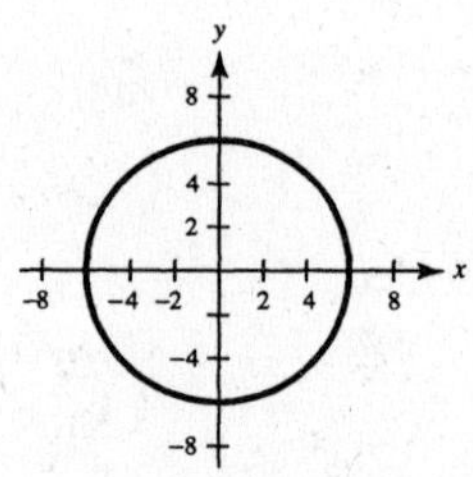

56.

$$r = 8$$
$$r^2 = 64$$
$$x^2 + y^2 = 64$$

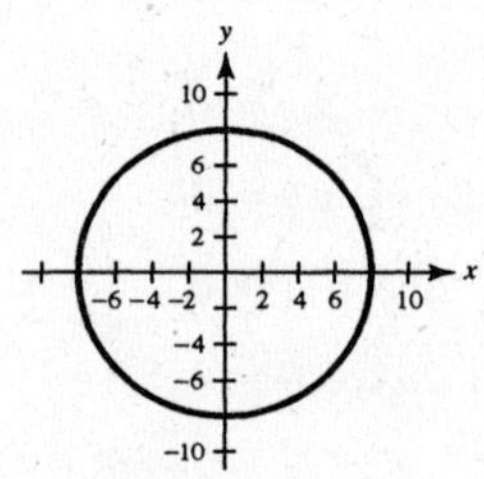

57.

$$\theta = \frac{\pi}{6}$$
$$\tan\theta = \tan\frac{\pi}{6}$$
$$\frac{y}{x} = \frac{\sqrt{3}}{3}$$
$$y = \frac{\sqrt{3}}{3}x$$
$$3y = \sqrt{3}x$$
$$-\sqrt{3}x + 3y = 0$$

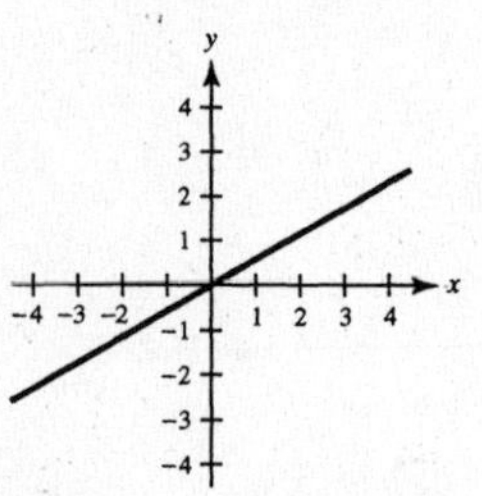

58.

$$\theta = \frac{3\pi}{4}$$
$$\tan\theta = \tan\frac{3\pi}{4}$$
$$\frac{y}{x} = -1$$
$$y = -x$$
$$x + y = 0$$

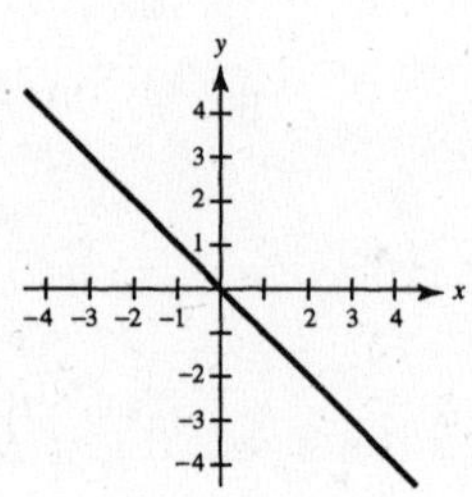

59.

$$r = 3\sec\theta$$
$$r\cos\theta = 3$$
$$x = 3$$
$$x - 3 = 0$$

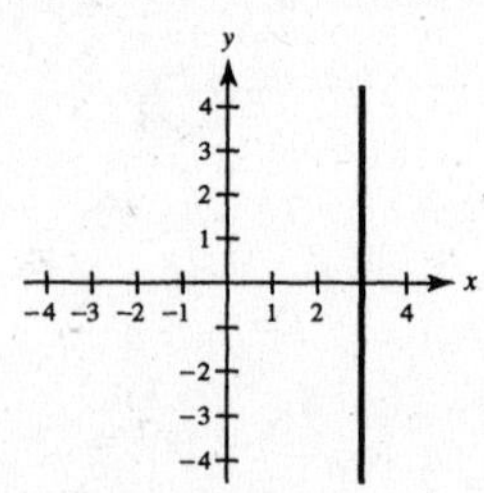

60. $r = 2 \csc \theta$

$r \sin \theta = 2$

$y = 2$

$y - 2 = 0$

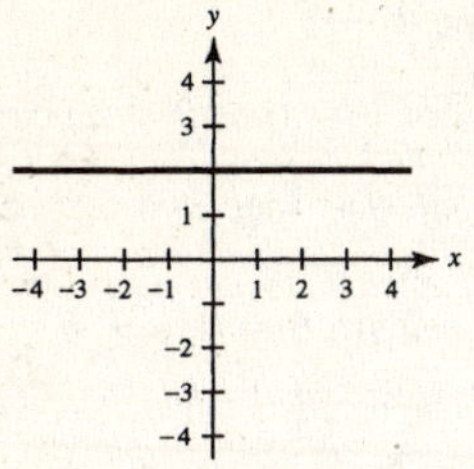

61. $r = 1 + \sin \theta, 0 \le \theta < 2\pi$

$x = r \cos \theta = (1 + \sin \theta) \cos \theta$

$y = r \sin \theta = (1 + \sin \theta) \sin \theta$

$$\frac{dx}{d\theta} = (1 + \sin \theta)(-\sin \theta) + \cos^2 \theta = 0$$

$$-\sin \theta - \sin^2 \theta + (1 - \sin^2 \theta) = 0$$

$$2 \sin^2 \theta + \sin \theta - 1 = 0$$

$$(2 \sin \theta - 1)(\sin \theta + 1) = 0$$

$$\sin \theta = \frac{1}{2} \implies \theta = \frac{\pi}{6}, \frac{5\pi}{6}$$

$$\sin \theta = -1 \implies \theta = \frac{3\pi}{2}$$

$$\frac{dy}{d\theta} = (1 + \sin \theta)\cos \theta + \cos \theta \sin \theta = 0$$

$$\cos \theta (2 \sin \theta + 1) = 0$$

$$\cos \theta = 0 \implies \theta = \frac{\pi}{2}, \frac{3\pi}{2}$$

$$\sin \theta = -\frac{1}{2} \implies \theta = \frac{7\pi}{6}, \frac{11\pi}{6}$$

Vertical tangents: $\left(\frac{3}{2}, \frac{\pi}{6}\right), \left(\frac{3}{2}, \frac{5\pi}{6}\right)$

Horizontal tangents: $\left(2, \frac{\pi}{2}\right), \left(\frac{1}{2}, \frac{7\pi}{6}\right), \left(\frac{1}{2}, \frac{11\pi}{6}\right)$

(**Note:** $\theta = \frac{3\pi}{2}$ corresponds to (0, 0), which is a cusp.)

62. $r = 1 - 2 \cos \theta$

$x = r \cos \theta = (1 - 2 \cos \theta) \cos \theta$

$y = r \sin \theta = (1 - 2 \cos \theta) \sin \theta$

$$\frac{dx}{d\theta} = (1 - 2 \cos \theta)(-\sin \theta) + 2 \sin \theta \cos \theta = 0$$

$$\sin \theta(2 \cos \theta - 1 + 2 \cos \theta) = 0$$

$$\sin \theta = 0 \implies \theta = 0, \pi$$

$$\cos \theta = \frac{1}{4} \implies \theta = 1.3181, 4.9651$$

$$\frac{dy}{d\theta} = (1 - 2 \cos \theta) \cos \theta + 2 \sin \theta \sin \theta = 0$$

$$\cos \theta - 2 \cos^2 \theta + 2(1 - \cos^2 \theta) = 0$$

$$4 \cos^2 \theta - \cos \theta - 2 = 0$$

$$\cos \theta = \frac{1}{8}(1 \pm \sqrt{33})$$

$$\theta = 0.5678, 2.2057, 4.0775, 5.7154$$

Vertical tangents: $(-1, 0), (3, \pi), \left(\frac{1}{2}, 1.3181\right), \left(\frac{1}{2}, 4.9651\right)$

Horizontal tangents: $(-0.6861, 0.5678), (2.1861, 2.2057)$
$(2.1861, 4.0775), (-0.6861, 5.7154)$

63. $r = 2\csc\theta + 3$

$x = r\cos\theta = (2\csc\theta + 3)\cos\theta = 2\cot\theta + 3\cos\theta$

$y = r\sin\theta = (2\csc\theta + 3)\sin\theta = 2 + 3\sin\theta$

$\frac{dx}{d\theta} = -2\csc^2\theta - 3\sin\theta = 0$

$$\frac{2}{\sin^2\theta} = -3\sin\theta$$

$$\sin^3\theta = \frac{-2}{3}$$

$$\sin\theta \approx -0.87358$$

$$\theta \approx 4.2041, 5.2207$$

$\frac{dy}{d\theta} = 3\cos\theta = 0 \Rightarrow \theta = \frac{\pi}{2}, \frac{3\pi}{2}$

Vertical tangents: $(0.7106, 4.2041), (0.7106, 5.2207)$

Horizontal tangents: $\left(5, \frac{\pi}{2}\right), \left(1, \frac{3\pi}{2}\right)$

64. $r = \sec\theta + 2$

$x = r\cos\theta = (\sec\theta + 2)\cos\theta = 1 + 2\cos\theta$

$y = r\sin\theta = (\sec\theta + 2)\sin\theta = \tan\theta + 2\sin\theta$

$\frac{dx}{d\theta} = -2\sin\theta = 0 \Rightarrow \theta = 0, \pi$

$\frac{dy}{d\theta} = \sec^2\theta + 2\cos\theta = 0$

$$2\cos\theta = -\frac{1}{\cos^2\theta}$$

$$\cos^3\theta = -\frac{1}{2}$$

$$\cos\theta \approx -0.79370$$

$$\theta \approx 2.4877, 3.7955$$

Vertical tangents: $(3, 0), (1, \pi)$

Horizontal tangents: $(0.7401, 2.4877), (0.7401, 3.7955)$

65. $r = 2(h\cos\theta + k\sin\theta)$

$r^2 = 2r(h\cos\theta + k\sin\theta)$

$x^2 + y^2 = 2hx + 2ky$

$(x^2 - 2hx) + (y^2 - 2ky) = 0$

$(x^2 - 2hx + h^2) + (y^2 - 2ky + k^2) = h^2 + k^2$

$(x - h)^2 + (y - k)^2 = h^2 + k^2$

Radius: $\sqrt{h^2 + k^2}$

Center: (h, k)

66.

$$r = \cos\theta + 3\sin\theta$$

$$r = \frac{x}{r} + \frac{3y}{r}$$

$$r^2 = x + 3y$$

$$x^2 + y^2 = x + 3y$$

$$x^2 - x + y^2 - 3y = 0$$

$$\left(x - \frac{1}{2}\right)^2 + \left(y - \frac{3}{2}\right)^2 = \frac{5}{2}$$

The graph is a circle.

67. Given point: $\left(4, \frac{\pi}{6}\right)$

(a) $\left(-4, \frac{\pi}{6}\right)$: Symmetric to the pole

(b) $\left(4, -\frac{\pi}{6}\right)$: Symmetric to the polar axis

(c) $\left(-4, -\frac{\pi}{6}\right)$: Symmetric to the line $\theta = \frac{\pi}{2}$

68. (a) $x^2 + y^2 = 25 \Rightarrow r^2 = 25 \Rightarrow r = 5$

The graphs are the same.

(b) $x - y = 0 \Rightarrow r\cos\theta - r\sin\theta = 0 \Rightarrow \sin\theta = \cos\theta$

The graphs are the same.

69. True. Because r is a directed distance, then the point (r, θ) can be represented as $(r, \theta + 2n\pi)$.

70. False.

(r_1, θ) and (r_2, θ) represent the same point only if $r_1 = r_2$.

71. False.

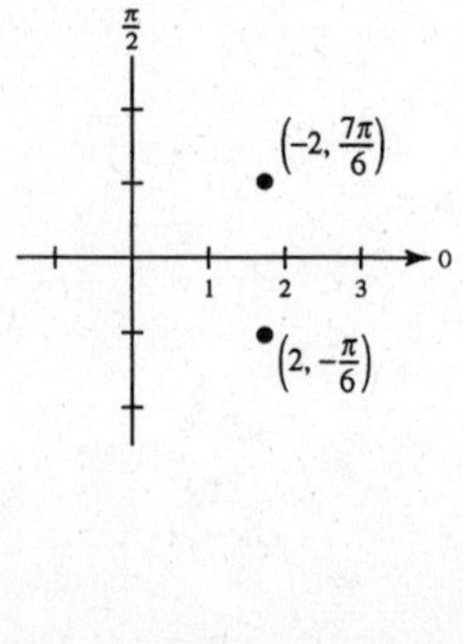

Section 12.6 Graphs of Polar Equations

1. $r = 3\cos 2\theta$

Rose curve with 4 petals

2. $r = 5 - 5\sin\theta$

Cardioid

3. $r = 3(1 - 2\cos\theta)$

Limaçon with inner loop

4. $r^2 = 16\cos 2\theta$

Lemniscate

5. $r = 6\sin 2\theta$

Rose curve with 4 petals

6. $r = 3\cos\theta$

Circle

7. $r = 5 + 4\cos\theta$

$\theta = \dfrac{\pi}{2}$: $-r = 5 + 4\cos(-\theta)$

$-r = 5 + 4\cos\theta$

Not an equivalent equation

Polar axis: $r = 5 + 4\cos(-\theta)$

$r = 5 + 4\cos\theta$

Equivalent equation

Pole: $-r = 5 + 4\cos\theta$

Not an equivalent equation

Answer: Symmetric with respect to polar axis

8. $r = 16\cos 3\theta$

$\theta = \dfrac{\pi}{2}$: $-r = 16\cos(3(-\theta))$

$-r = 16\cos(-3\theta)$

$-r = 16\cos 3\theta$

Not an equivalent equation

Polar axis: $r = 16\cos(3(-\theta))$

$r = 16\cos(-3\theta)$

$r = 16\cos 3\theta$

Equivalent equation

Pole: $-r = 16\cos 3\theta$

Not an equivalent equation

Answer: Symmetric with respect to polar axis

9. $r = \dfrac{2}{1 + \sin\theta}$

$\theta = \dfrac{\pi}{2}$: $r = \dfrac{2}{1 + \sin(\pi - \theta)}$

$r = \dfrac{2}{1 + \sin\pi\cos\theta - \cos\pi\sin\theta}$

$r = \dfrac{2}{1 + \sin\theta}$

Equivalent equation

Polar axis: $r = \dfrac{2}{1 + \sin(-\theta)}$

$r = \dfrac{2}{1 - \sin\theta}$

Not an equivalent equation

Pole: $-r = \dfrac{2}{1 + \sin\theta}$

Answer: Symmetric with respect to $\theta = \pi/2$

10. $r = \dfrac{3}{2 + \cos\theta}$

$\theta = \dfrac{\pi}{2}$: $-r = \dfrac{3}{2 + \cos(-\theta)}$

Not an equivalent equation

Polar axis: $r = \dfrac{3}{2 + \cos(-\theta)}$

Equivalent equation

Pole: $-r = \dfrac{3}{2 + \cos\theta}$

Not an equivalent equation

Answer: Symmetric with respect to polar axis

11. $r^2 = 16\cos 2\theta$

$\theta = \frac{\pi}{2}$: $(-r)^2 = 16\cos 2(-\theta)$

$r^2 = 16\cos 2\theta$

Equivalent equation

Polar axis: $r^2 = 16\cos 2(-\theta)$

$r^2 = 16\cos 2\theta$

Equivalent equation

Pole: $(-r)^2 = 16\cos 2\theta$

$r^2 = 16\cos 2\theta$

Equivalent equation

Answer: Symmetric with respect to $\theta = \frac{\pi}{2}$, the polar axis, and the pole

12. $r^2 = 36\sin 2\theta$

$\theta = \frac{\pi}{2}$: $(-r)^2 = 36\sin(-2\theta)$

Not an equivalent equation

Polar axis: $r^2 = 36\sin(-2\theta)$

Not an equivalent equation

Pole: $(-r)^2 = 36\sin 2\theta$

Equivalent equation

Answer: Symmetric with respect to pole

13. $|r| = |10(1 - \sin\theta)| = 10|1 - \sin\theta| \le 10(2) = 20$

$|1 - \sin\theta| = 2$

$1 - \sin\theta = 2$ or $1 - \sin\theta = -2$

$\sin\theta = -1$ $\quad$ $\sin\theta = 3$

$\theta = \frac{3\pi}{2}$ $\quad$ Not possible

Maximum: $|r| = 20$ when $\theta = \frac{3\pi}{2}$.

$0 = 10(1 - \sin\theta)$

$\sin\theta = 1$

$\theta = \frac{\pi}{2}$

Zero: $r = 0$ when $\theta = \frac{\pi}{2}$.

14. $|r| = |6 + 12\cos\theta| \le |6| + |12\cos\theta|$

$= 6 + 12|\cos\theta| \le 18$

$\cos\theta = 1$

$\theta = 0$

Maximum: $|r| = 18$ when $\theta = 0$.

$0 = 6 + 12\cos\theta$

$\cos\theta = -\frac{1}{2}$

$\theta = \frac{2\pi}{3}, \frac{4\pi}{3}$

Zero: $r = 0$ when $\theta = \frac{2\pi}{3}, \frac{4\pi}{3}$.

15. $|r| = |4\cos 3\theta| = 4|\cos 3\theta| \le 4$

$|\cos 3\theta| = 1$

$\cos 3\theta = \pm 1$

$\theta = 0, \frac{\pi}{3}, \frac{2\pi}{3}$

Maximum: $|r| = 4$ when $\theta = 0, \frac{\pi}{3}, \frac{2\pi}{3}$.

$0 = 4\cos 3\theta$

$\cos 3\theta = 0$

$\theta = \frac{\pi}{6}, \frac{\pi}{2}, \frac{5\pi}{6}$

Zero: $r = 0$ when $\theta = \frac{\pi}{6}, \frac{\pi}{2}, \frac{5\pi}{6}$.

16. $|r| = |3\sin 2\theta| = 3|\sin 2\theta| \le 3$

$|\sin 2\theta| = 1$

$\sin 2\theta = \pm 1$

$\theta = \frac{\pi}{4}, \frac{3\pi}{4}, \frac{5\pi}{4}, \frac{7\pi}{4}$

Maximum: $|r| = 3$ when $\theta = \frac{\pi}{4}, \frac{3\pi}{4}, \frac{5\pi}{4}, \frac{7\pi}{4}$.

$0 = 3\sin 2\theta$

$\sin 2\theta = 0$

$\theta = 0, \frac{\pi}{2}, \pi, \frac{3\pi}{2}$

Zero: $r = 0$ when $\theta = 0, \frac{\pi}{2}, \pi, \frac{3\pi}{2}, 2\pi$.

17. Circle: $r = 5$

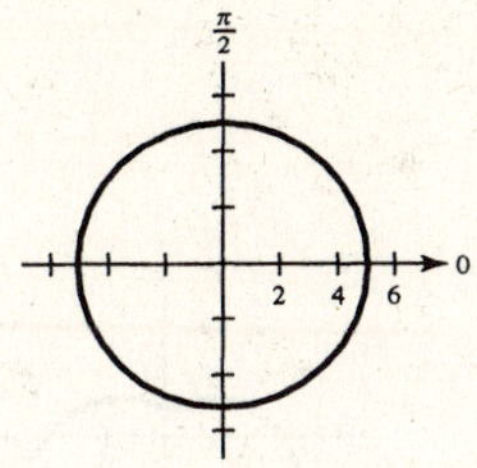

18. Circle: $r = 2$

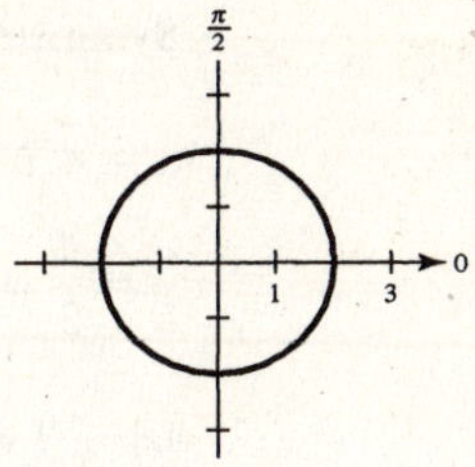

19. Circle: $r = \dfrac{\pi}{6}$

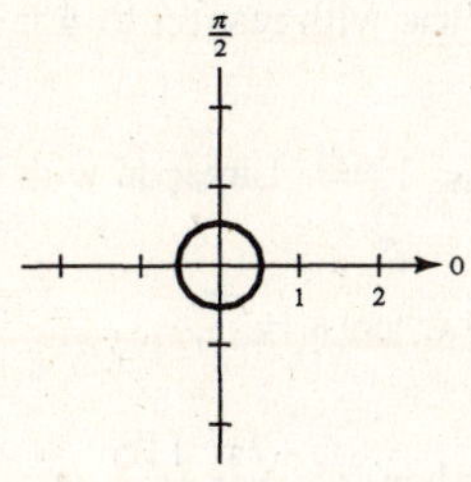

20. $r = -\dfrac{3\pi}{4}$

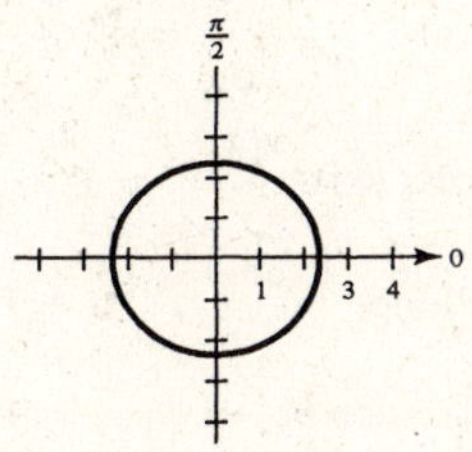

21. $r = 3 \sin \theta$

Symmetric with respect to $\theta = \dfrac{\pi}{2}$.

Circle with a radius of $\dfrac{3}{2}$.

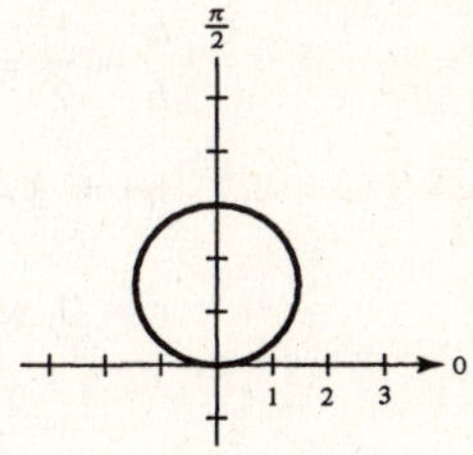

22. $r = 4 \cos \theta$

Symmetric with respect to polar axis.

Circle with radius 2.

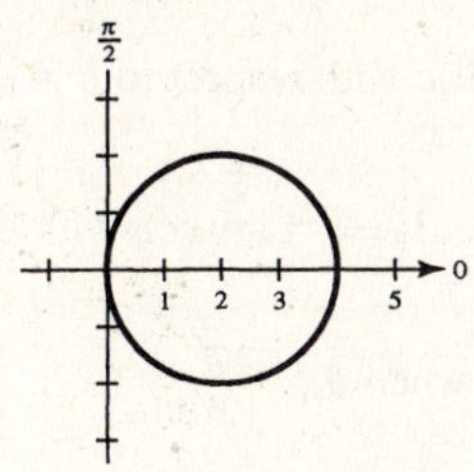

23. $r = 3(1 - \cos \theta)$

Symmetric with respect to the polar axis.

$\dfrac{a}{b} = \dfrac{3}{3} = 1 \Rightarrow$ Cardioid

$|r| = 6$ when $\theta = \pi$.

$r = 0$ when $\theta = 0$.

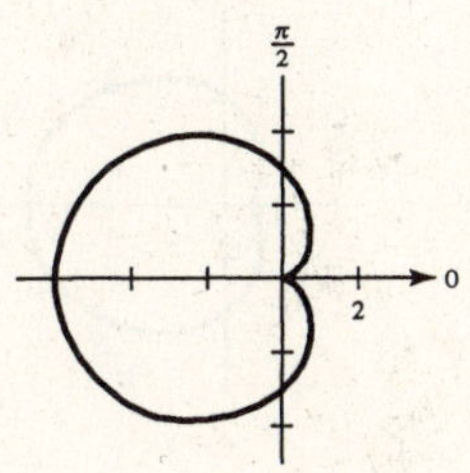

24. $r = 4(1 - \sin \theta)$

Symmetric with respect to $\dfrac{\pi}{2}$.

$\dfrac{a}{b} = \dfrac{4}{4} = 1 \Rightarrow$ Cardioid

$|r| = 8$ when $\theta = \dfrac{3\pi}{2}$.

$r = 0$ when $\theta = \dfrac{\pi}{2}$.

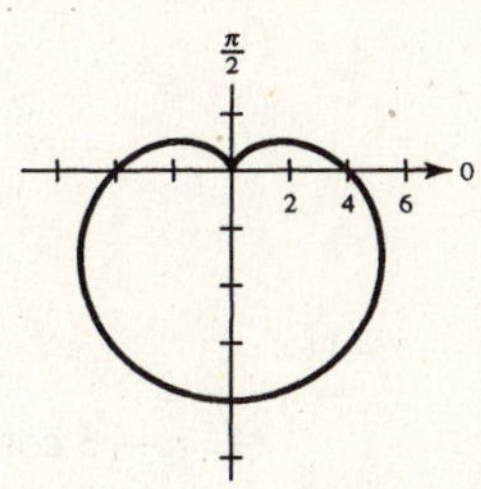

25. $r = 4(1 + \sin \theta)$

Symmetric with respect to $\theta = \dfrac{\pi}{2}$.

$\dfrac{a}{b} = \dfrac{4}{4} = 1 \Rightarrow$ Cardioid

$|r| = 8$ when $\theta = \dfrac{\pi}{2}$.

$r = 0$ when $\theta = \dfrac{3\pi}{2}$.

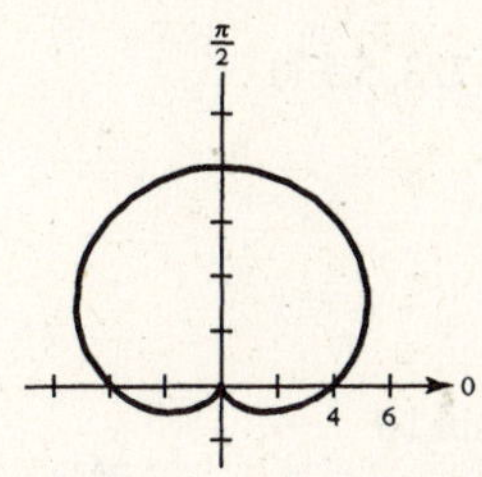

26. $r = 2(1 + \cos \theta)$

Symmetric with respect to polar axis.

$\dfrac{a}{b} = \dfrac{2}{2} = 1 \Rightarrow$ Cardioid

$|r| = 4$ when $\theta = 0$.

$r = 0$ when $\theta = \pi$.

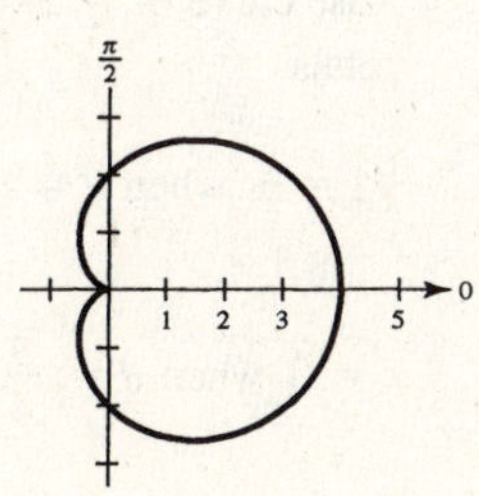

27. $r = 3 + 6 \sin \theta$

Symmetric with respect to $\theta = \dfrac{\pi}{2}$.

$\dfrac{a}{b} = \dfrac{3}{6} < 1 \Longrightarrow$ Limaçon with inner loop

$|r| = 9$ when $\theta = \dfrac{\pi}{2}$.

$r = 0$ when $\theta = \dfrac{7\pi}{6}, \dfrac{11\pi}{6}$.

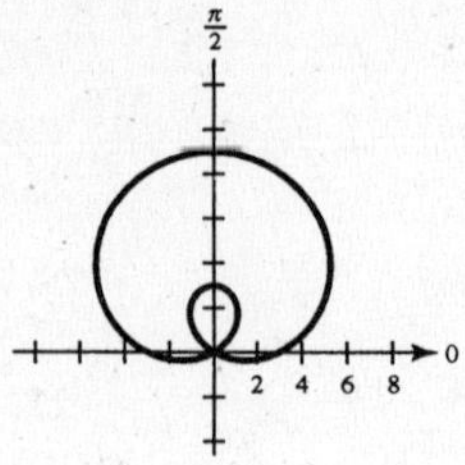

28. $r = 4 - 3 \sin \theta$

Symmetric with respect to $\dfrac{\pi}{2}$.

$a = 4, b = 3$

$\dfrac{a}{b} = \dfrac{4}{3} \Longrightarrow$ Dimpled limaçon

$|r| = 7$ when $\theta = \dfrac{3\pi}{2}$.

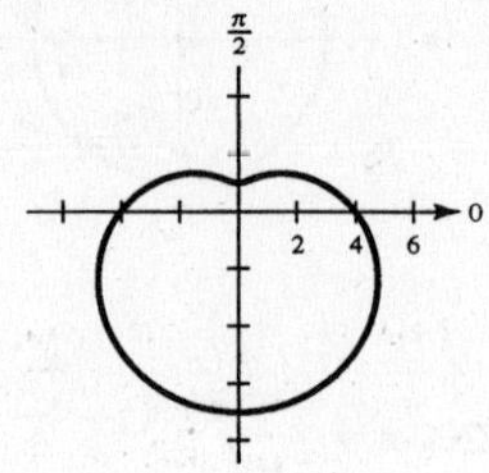

29. $r = 1 - 2 \sin \theta$

Symmetric with respect to $\theta = \dfrac{\pi}{2}$.

$\dfrac{a}{b} = \dfrac{1}{2} < 1 \Longrightarrow$ Limaçon with inner loop

$|r| = 3$ when $\theta = \dfrac{3\pi}{2}$.

$r = 0$ when $\theta = \dfrac{\pi}{6}, \dfrac{5\pi}{6}$.

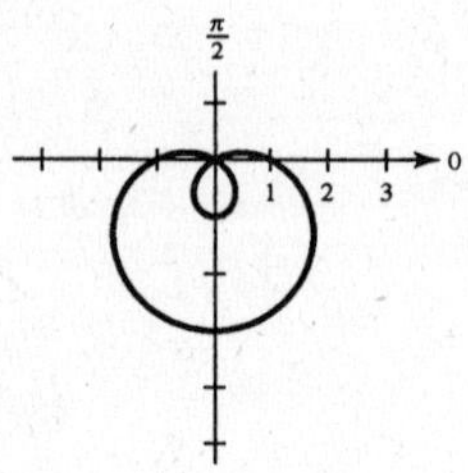

30. $r = 1 - 2 \cos \theta$

Symmetric with respect to the polar axis.

$\dfrac{a}{b} = \dfrac{1}{2} \Longrightarrow$ Limaçon with inner loop

$|r| = 3$ when $\theta = \pi$.

$r = 0$ when $\theta = \dfrac{\pi}{3}, \dfrac{5\pi}{3}$.

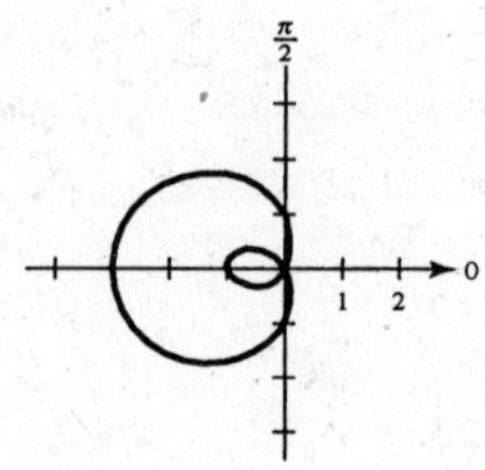

31. $r = 3 - 4 \cos \theta$

Symmetric with respect to the polar axis.

$\dfrac{a}{b} = \dfrac{3}{4} < 1 \Longrightarrow$ Limaçon with inner loop

$|r| = 7$ when $\theta = \pi$.

$r = 0$ when $\cos \theta = \dfrac{3}{4}$ or

$\theta \approx 0.723, 5.560$.

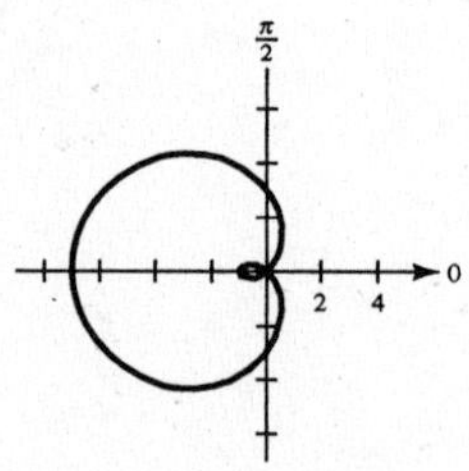

32. $r = 4 + 3 \cos \theta$

Symmetric with respect to the polar axis.

$\dfrac{a}{b} = \dfrac{4}{3} > 1 \Longrightarrow$ Dimpled limaçon

$|r| = 7$ when $\theta = 0$.

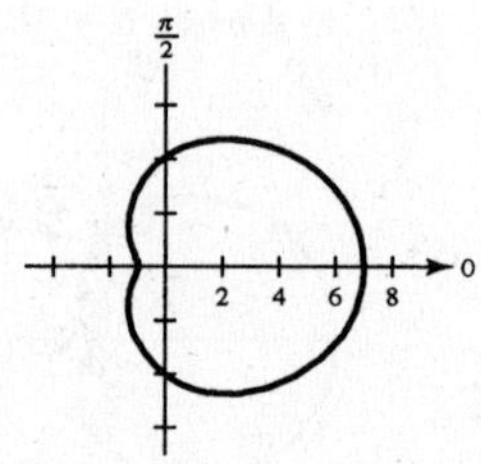

33. $r = 5 \sin 2\theta$

Symmetric with respect to $\theta = \dfrac{\pi}{2}$, the polar axis and the pole.

Rose curve ($n = 2$) with 4 petals

$|r| = 5$ when $\theta = \dfrac{\pi}{4}, \dfrac{3\pi}{4}, \dfrac{5\pi}{4}, \dfrac{7\pi}{4}$.

$r = 0$ when $\theta = 0, \dfrac{\pi}{2}, \pi$.

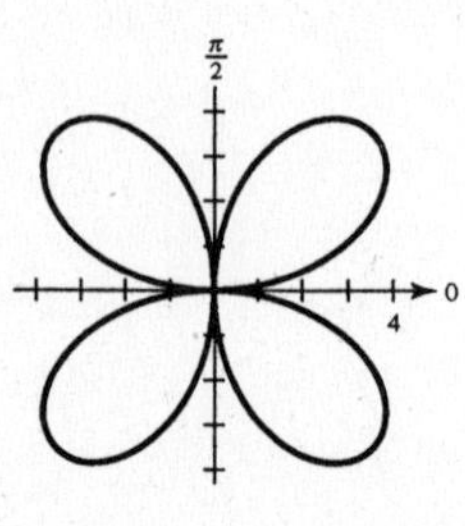

34. $r = 3 \cos 2\theta$

Symmetric with respect to the polar axis; the pole, and $\theta = \dfrac{\pi}{2}$.

Rose curve ($n = 2$) with four petals

$|r| = 3$ when $\theta = 0, \dfrac{\pi}{2}, \pi, \dfrac{3\pi}{2}$.

$r = 0$ when $\theta = \dfrac{\pi}{4}, \dfrac{3\pi}{4}, \dfrac{5\pi}{4}, \dfrac{7\pi}{4}$.

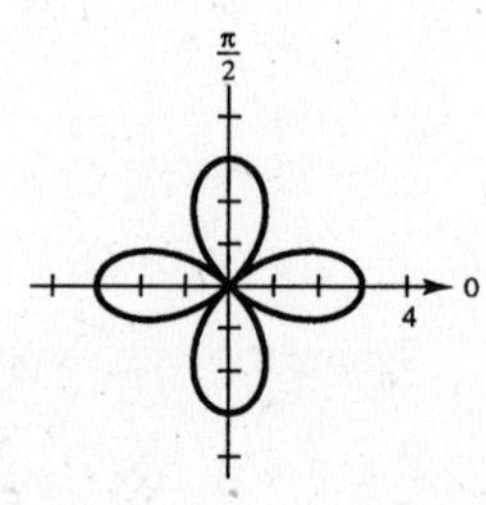

35. $r = 2 \sec \theta$

$r = \dfrac{2}{\cos \theta}$

$r \cos \theta = 2$

$x = 2 \Rightarrow$ Line

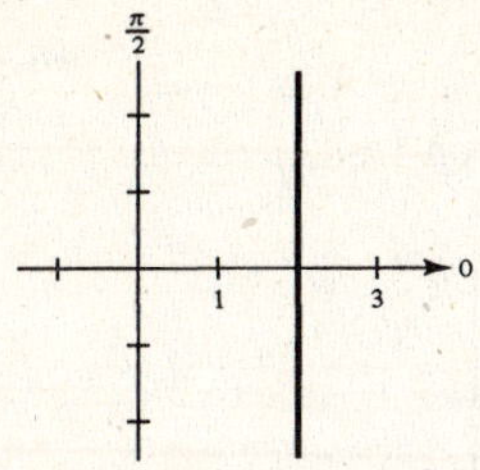

36. $r = 5 \csc \theta$

$r \sin \theta = 5$

$y = 5 \Rightarrow$ Line

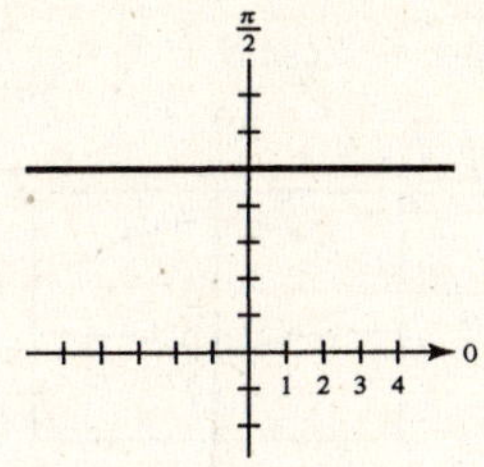

37. $r = \dfrac{3}{\sin \theta - 2 \cos \theta}$

$r(\sin \theta - 2 \cos \theta) = 3$

$y - 2x = 3$

$y = 2x + 3 \Rightarrow$ Line

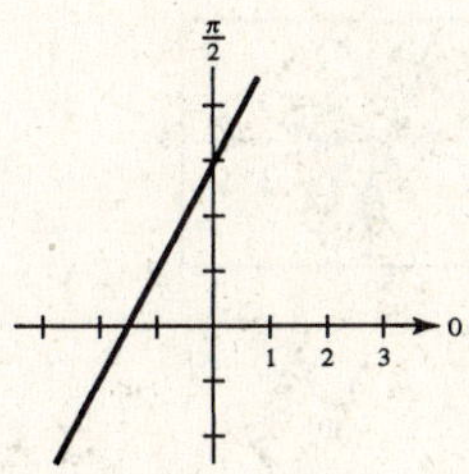

38. $r = \dfrac{6}{2 \sin \theta - 3 \cos \theta}$

$r(2 \sin \theta - 3 \cos \theta) = 6$

$2y - 3x = 6$

$y = \frac{3}{2}x + 3 \Rightarrow$ Line

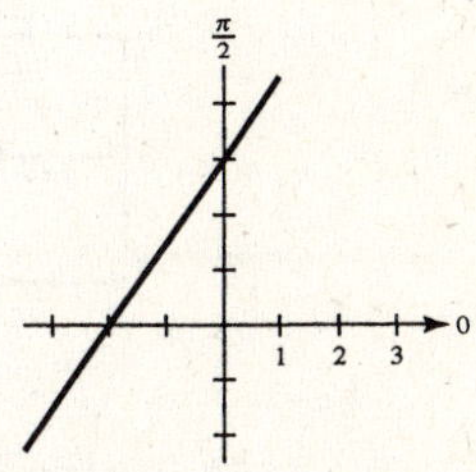

39. $r^2 = 9 \cos 2\theta$

Symmetric with respect to the polar axis, $\theta = \dfrac{\pi}{2}$, and the pole.

Lemniscate

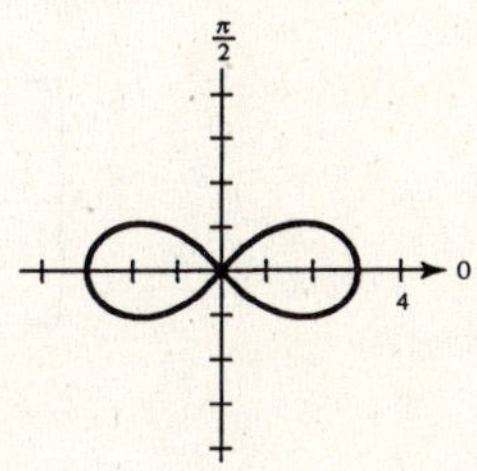

40. $r^2 = 4 \sin \theta$

$r = 2\sqrt{\sin \theta}$

$r = -2\sqrt{\sin \theta}$

$0 \le \theta \le \pi$

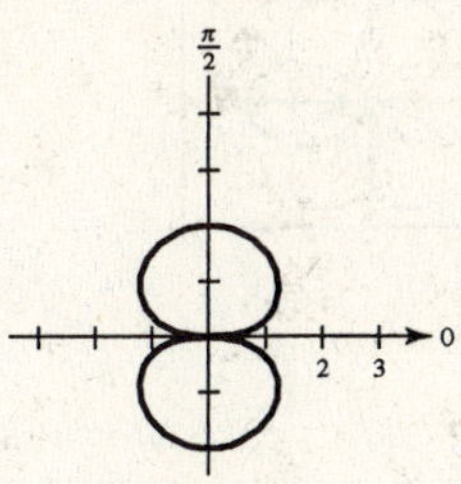

41. $r = 8 \cos \theta$

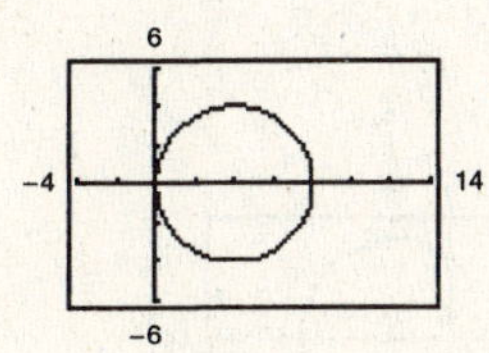

42. $r = \cos 2\theta$

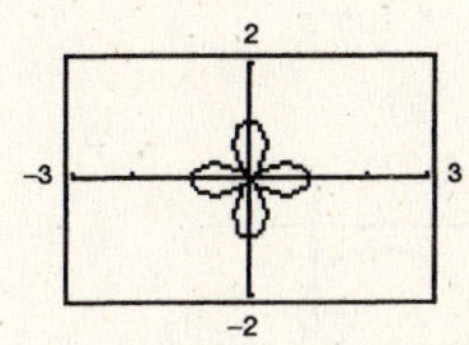

43. $r = 3(2 - \sin \theta)$

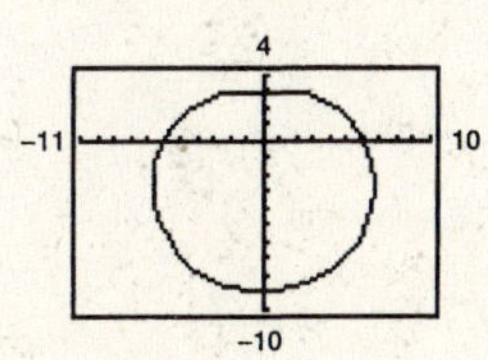

44. $r = 2 \cos(3\theta - 2)$

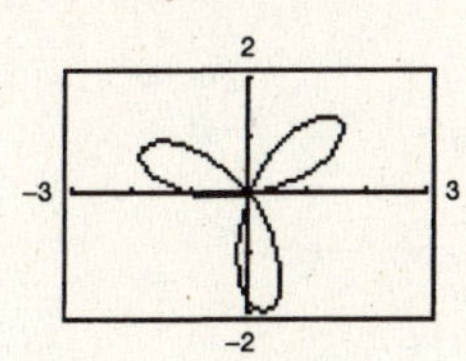

45. $r = 8 \sin \theta \cos^2 \theta$

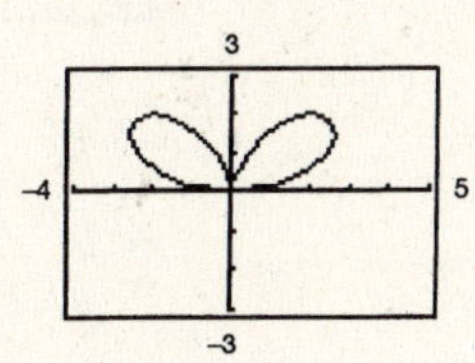

46. $r = 2 \csc \theta + 5 = \dfrac{2}{\sin \theta} + 5$

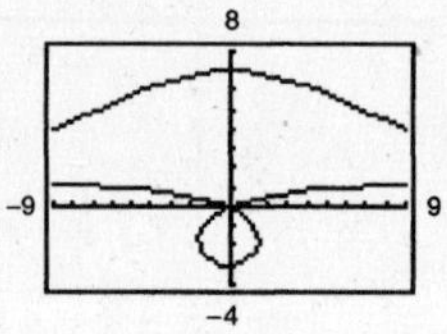

47. $r = 1 + 2 \cos \theta$

Limaçon with inner loop

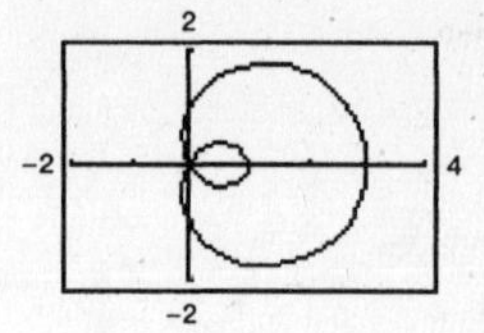

48. $r = 1 - 2 \sin \theta$

$0 \le \theta \le 2\pi$

Limaçon with inner loop

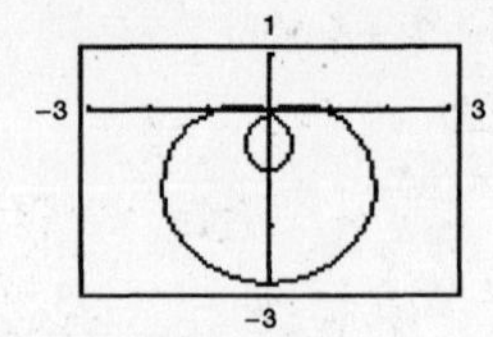

49. $r = 1 + \sin \theta$

Cardioid

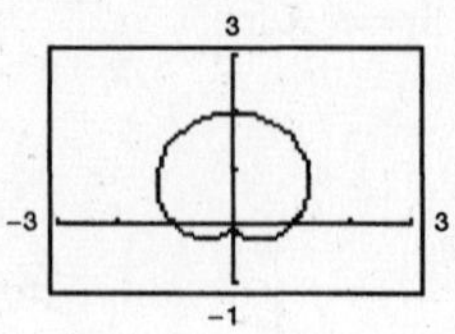

50. $r = 1 - \cos \theta$

$0 \le \theta \le 2\pi$

Cardioid

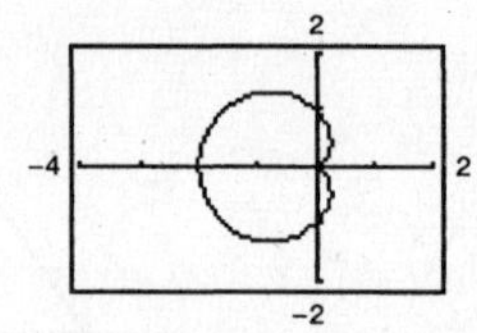

51. $r = 2 \sin 4\theta$

Rose curve

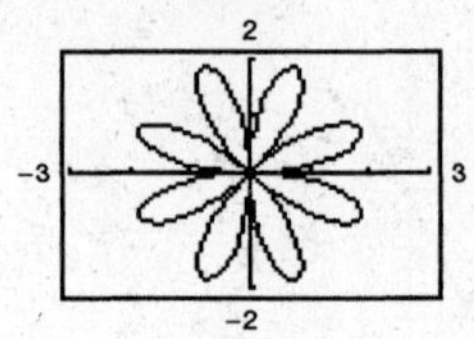

52. $r = 4 \cos 2\theta$

$0 \le \theta \le 2\pi$

Rose curve

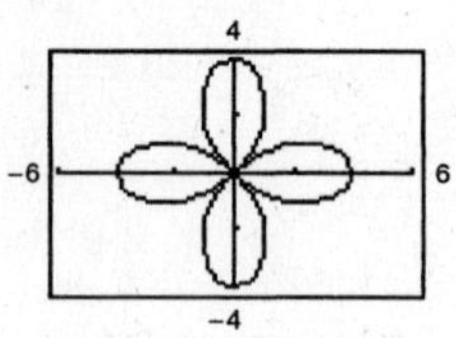

53. $r = \sin 5\theta$

Rose curve

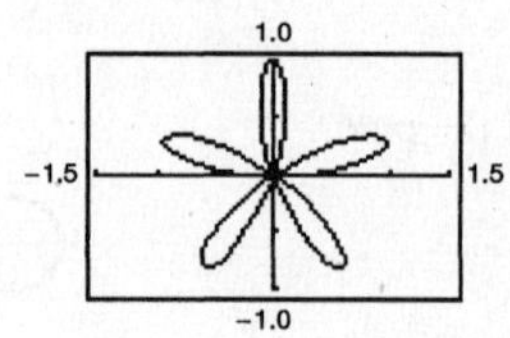

54. $r = \cos 3\theta$

$0 \le \theta \le 2\pi$

Rose curve

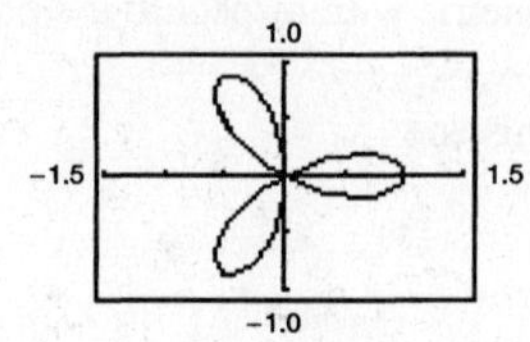

55. $r = 3 - 4 \cos \theta$

$0 \le \theta < 2\pi$

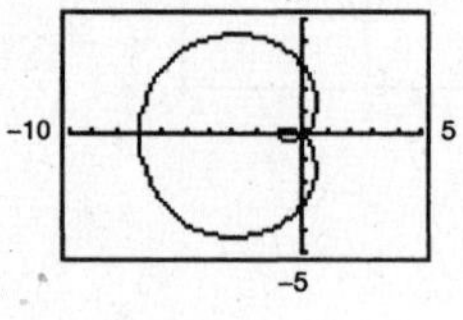

56. $r = 5 + 4 \cos \theta$

$0 \le \theta < 2\pi$

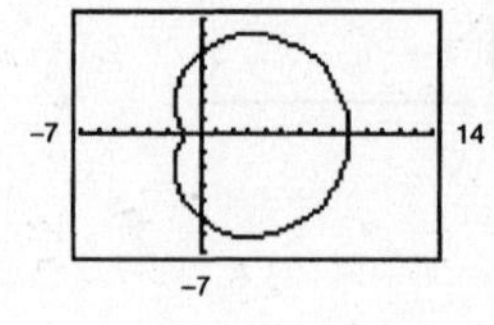

57. $r = 2 \cos\left(\dfrac{3\theta}{2}\right)$

$0 \le \theta < 4\pi$

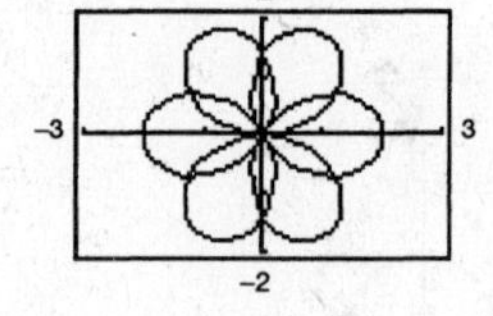

58. $r = 3 \sin\left(\dfrac{5\theta}{2}\right)$

$0 \le \theta < 4\pi$

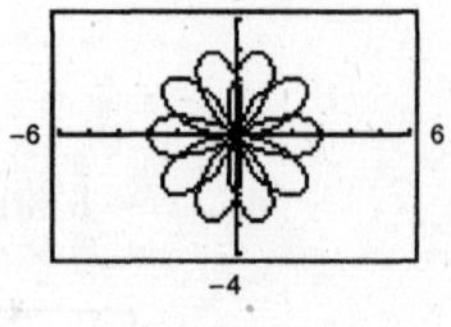

59. $r^2 = 9 \sin 2\theta$

Graph as $r_1 = 3\sqrt{\sin 2\theta}$ and $r_2 = -3\sqrt{\sin 2\theta}$. It is traced out once for $0 \le \theta < \dfrac{\pi}{2}$.

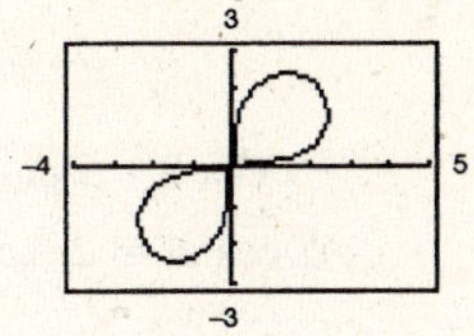

60. $r^2 = 4 \cos 2\,\theta$

Graph as $r_1 = 2\sqrt{\cos 2\theta}$ and

$r_2 = -2\sqrt{\cos 2\theta}$, $0 \le \pi$

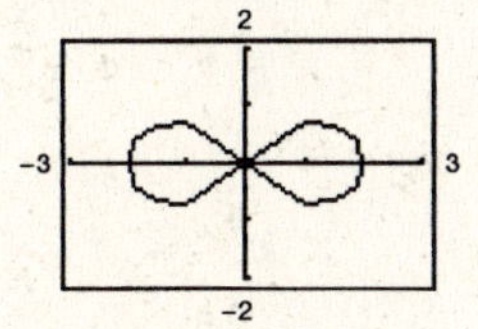

61. $r^2 = \dfrac{1}{\theta}$

$0 < \theta < \infty$

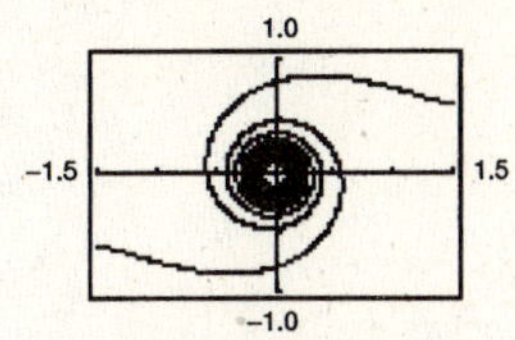

62. $r^2 = \dfrac{4}{\theta}$

Graph as $r_1 = \dfrac{2}{\sqrt{\theta}}$ and $r_2 = \dfrac{-2}{\sqrt{\theta}}$.

$0 < \theta < \infty$

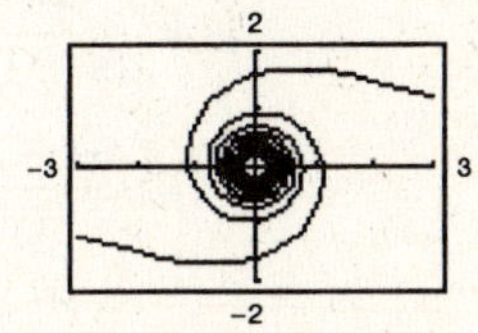

63. $r = 6 \cos \theta$

(a) $0 \le \theta \le \dfrac{\pi}{2}$

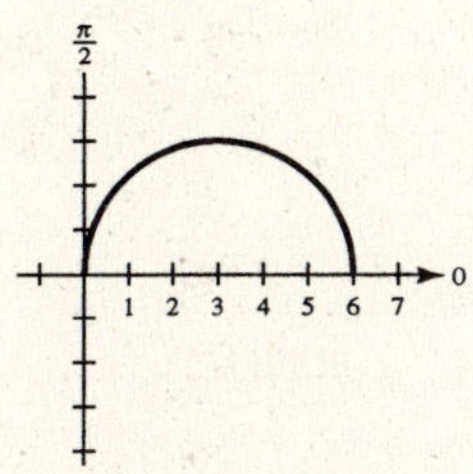

Upper half of circle

(b) $\dfrac{\pi}{2} \le \theta \le \pi$

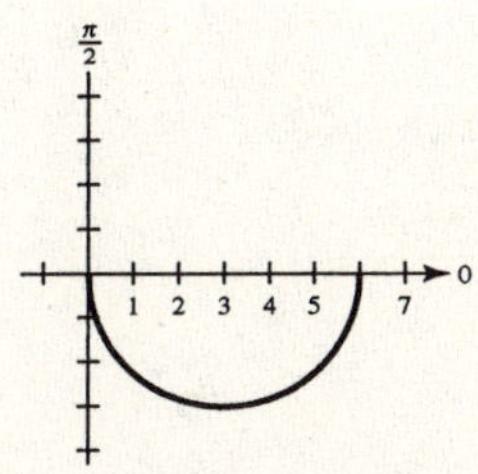

Lower half of circle

(c) $-\dfrac{\pi}{2} \le \theta \le \dfrac{\pi}{2}$

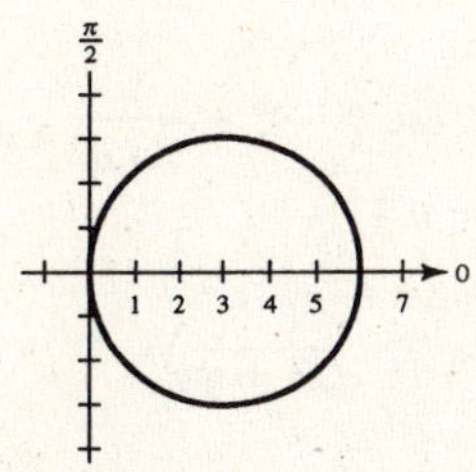

Full Circle

(d) $\dfrac{\pi}{4} \le \theta \le \dfrac{3\pi}{4}$

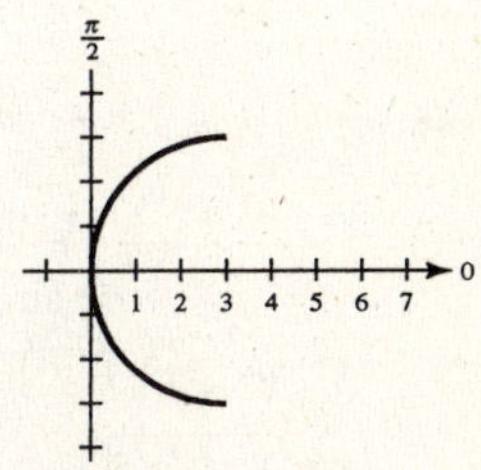

Left half of circle

64. $r = 2 + k \sin \theta$

$k = 0$: $r = 2$

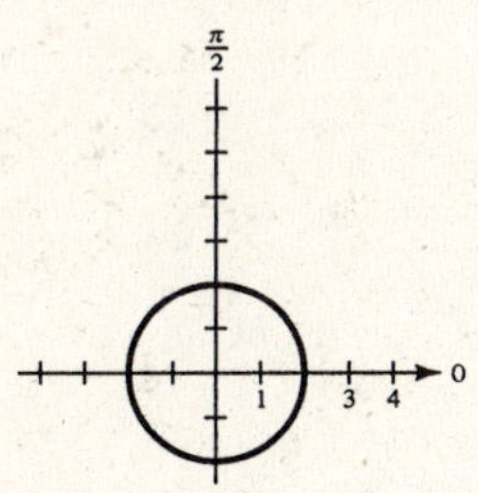

Circle

$k = 1$: $r = 2 + \sin \theta$

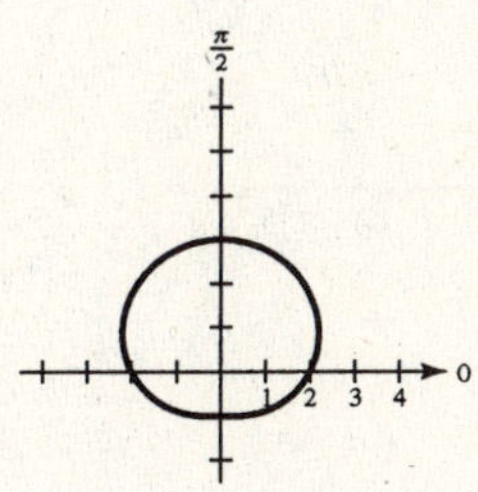

Convex Limaçon

$k = 2$: $r = 2 + 2 \sin \theta$

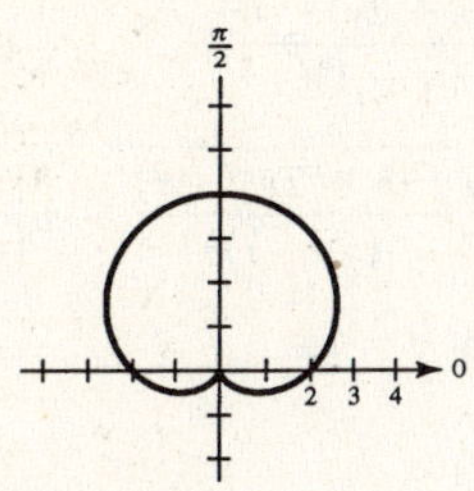

Cardioid

$k = 3$: $r = 2 + 3 \sin \theta$

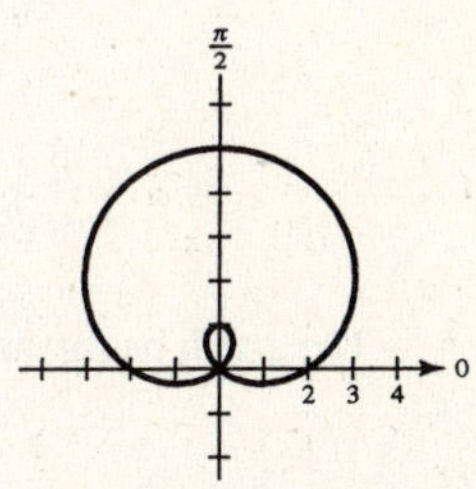

Limaçon with inner loop

65.

$$r = 2 - \sec\theta = 2 - \frac{1}{\cos\theta}$$

$$r\cos\theta = 2\cos\theta - 1$$

$$r(r\cos\theta) = 2r\cos\theta - r$$

$$\left(\pm\sqrt{x^2+y^2}\right)x = 2x - \left(\pm\sqrt{x^2+y^2}\right)$$

$$\left(\pm\sqrt{x^2+y^2}\right)(x+1) = 2x$$

$$\left(\pm\sqrt{x^2+y^2}\right) = \frac{2x}{x+1}$$

$$x^2 + y^2 = \frac{4x^2}{(x+1)^2}$$

$$y^2 = \frac{4x^2}{(x+1)^2} - x^2$$

$$= \frac{4x^2 - x^2(x+1)^2}{(x+1)^2} = \frac{4x^2 - x^2(x^2+2x+1)}{(x+1)^2}$$

$$= \frac{-x^4 - 2x^3 + 3x^2}{(x+1)^2} = \frac{-x^2(x^2+2x-3)}{(x+1)^2}$$

$$y = \pm\sqrt{\frac{x^2(3-2x-x^2)}{(x+1)^2}} = \pm\left|\frac{x}{x+1}\right|\sqrt{3-2x-x^2}$$

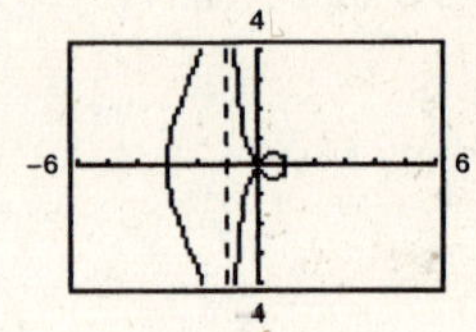

The graph has an asymptote at $x = -1$.

66.

$$r = 2 + \csc\theta = 2 + \frac{1}{\sin\theta}$$

$$r\sin\theta = 2\sin\theta + 1$$

$$r(r\sin\theta) = 2r\sin\theta + r$$

$$\left(\pm\sqrt{x^2+y^2}\right)(y) = 2y + \left(\pm\sqrt{x^2+y^2}\right)$$

$$\left(\pm\sqrt{x^2+y^2}\right)(y-1) = 2y$$

$$\left(\pm\sqrt{x^2+y^2}\right) = \frac{2y}{y-1}$$

$$x^2 + y^2 = \frac{4y^2}{(y-1)^2}$$

$$x^2 = \frac{y^2(3+2y-y^2)}{(y-1)^2}$$

$$x = \sqrt{\frac{y^2(3+2y-y^2)}{(y-1)^2}} = \pm\left|\frac{y}{y-1}\right|\sqrt{3+2y-y^2}$$

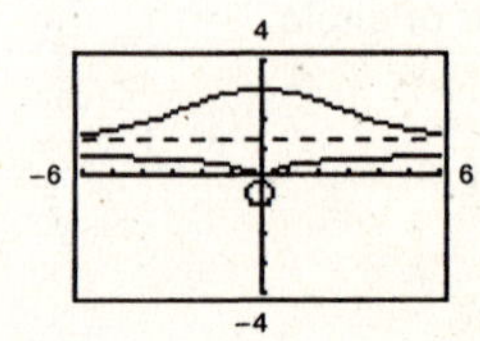

The graph has an asymptote at $y = 1$.

67. $r = \dfrac{3}{\theta}$

$$\theta = \frac{3}{r} = \frac{3\sin\theta}{r\sin\theta} = \frac{3\sin\theta}{y}$$

$$y = \frac{3\sin\theta}{\theta}$$

As $\theta \to 0, y \to 3$.

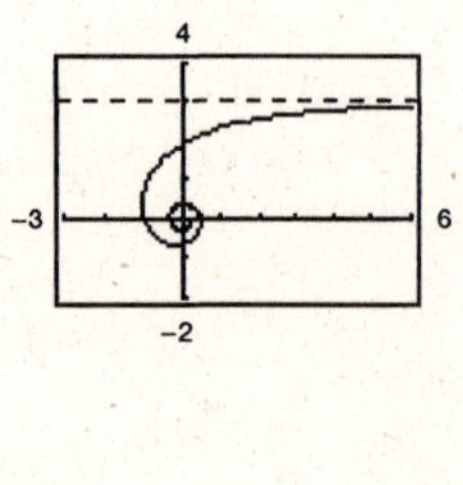

68.

$$r = 2\cos 2\theta\sec\theta = \frac{2\cos 2\theta}{\cos\theta}$$

$$r = \frac{2(\cos^2\theta - \sin^2\theta)}{\cos\theta}$$

$$r\cos\theta = 2(\cos^2\theta - \sin^2\theta)$$

$$x = 2(\cos^2\theta - \sin^2\theta)$$

As $\theta \to \dfrac{\pi}{2}, x \to -2$.

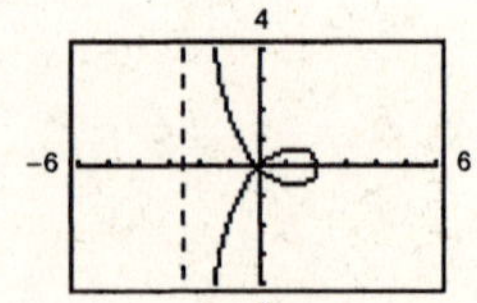

69. True. For a graph to have polar axis symmetry, replace (r, θ) by $(r, -\theta)$ or $(-r, \pi - \theta)$.

70. False. For a graph to be symmetric about the pole, one portion of the graph coincides with the other portion when rotated π radians about the pole.

71. True. Since $\frac{a}{b} = \frac{1}{2} < 1$, the graph is a limaçon with an inner loop.

72. False. The graph of $r = 3 - 2\sin\theta$ is a dimpled limaçon.

73. True

74. True

75. (a) $r = 1 - \sin\theta$

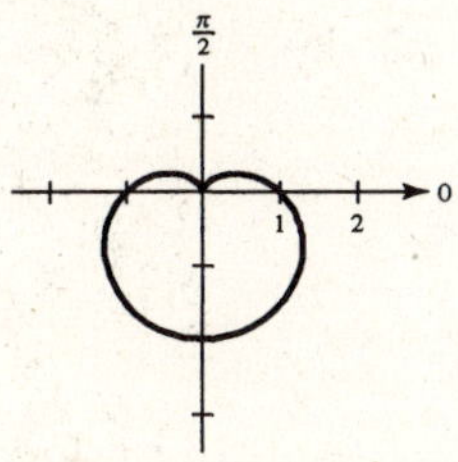

(b) $r = 1 - \sin\left(\theta - \frac{\pi}{4}\right)$

Rotate the graph of $r = 1 - \sin\theta$ through the angle $\frac{\pi}{4}$.

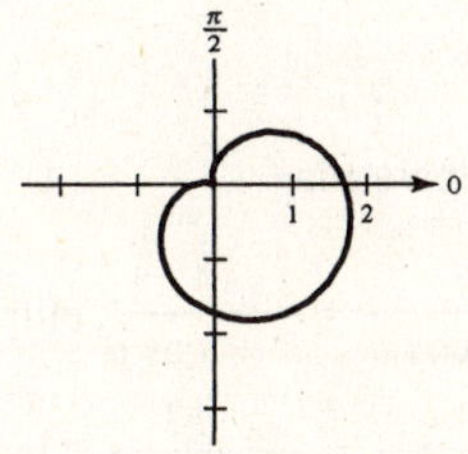

76. (a)

$$r = 3\sec\theta$$

$$r = \frac{3}{\cos\theta}$$

$$r\cos\theta = 3 \Rightarrow x = 3$$

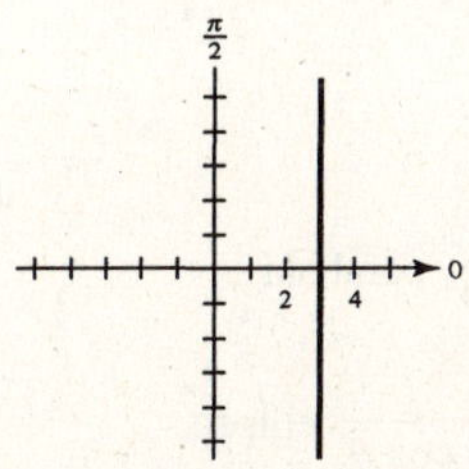

(b)

$$r = 3\sec\left(\theta - \frac{\pi}{4}\right)$$

$$r = \frac{3}{\cos(\theta - (\pi/4))}$$

$$r = \frac{3}{\cos\theta\cos(\pi/4) + \sin\theta\sin(\pi/4)}$$

$$\frac{\sqrt{2}}{2}r\cos\theta + \frac{\sqrt{2}}{2}r\sin\theta = 3$$

$$\frac{\sqrt{2}}{2}x + \frac{\sqrt{2}}{2}y = 3$$

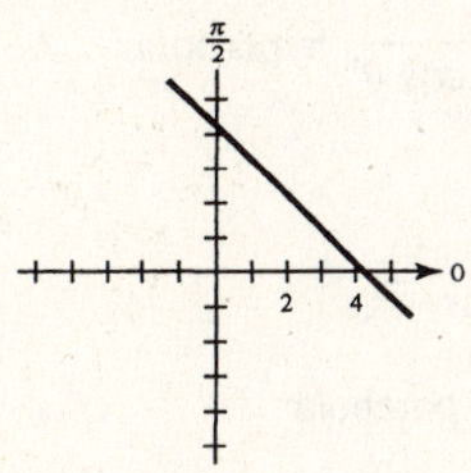

(c)

$$r = 3\sec\left(\theta + \frac{\pi}{3}\right)$$

$$r = \frac{3}{\cos(\theta + (\pi/3))}$$

$$r = \frac{3}{\cos\theta\cos(\pi/3) - \sin\theta\sin(\pi/3)}$$

$$\frac{1}{2}r\cos\theta - \frac{\sqrt{3}}{2}r\sin\theta = 3$$

$$\frac{1}{2}x - \frac{\sqrt{3}}{2}y = 3$$

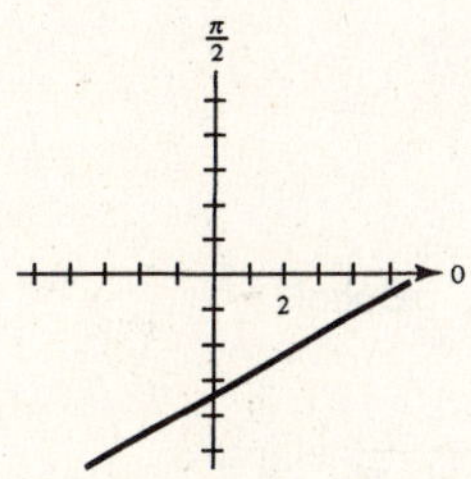

—CONTINUED—

76. —CONTINUED—

(d) $r = 3\sec\left(\theta - \dfrac{\pi}{2}\right)$

$$r = \frac{3}{\cos(\theta - (\pi/2))}$$

$$r = \frac{3}{\cos\theta\cos(\pi/2) + \sin\theta\sin(\pi/2)}$$

$r\sin\theta = 3 \Rightarrow y = 3$

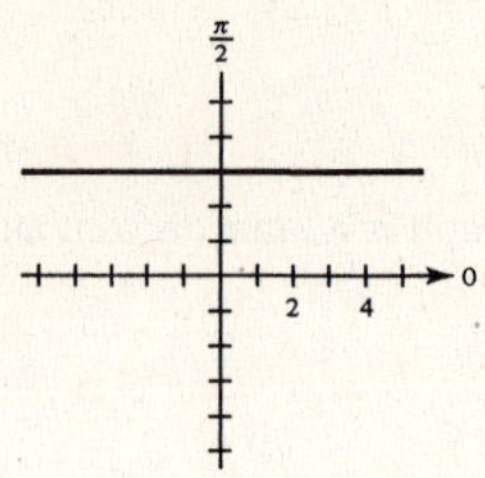

Section 12.7 Polar Equations of Conics

1. $r = \dfrac{4e}{1 + e\cos\theta}$

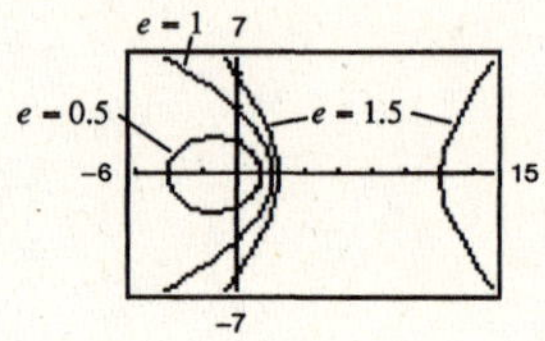

(a) $e = 1, r = \dfrac{4}{1 + \cos\theta}$, parabola

(b) $e = 0.5, r = \dfrac{2}{1 + 0.5\cos\theta} = \dfrac{4}{2 + \cos\theta}$, ellipse

(c) $e = 1.5, r = \dfrac{6}{1 + 1.5\cos\theta} = \dfrac{12}{2 + 3\cos\theta}$, hyperbola

2. $r = \dfrac{4e}{1 - e\cos\theta}$

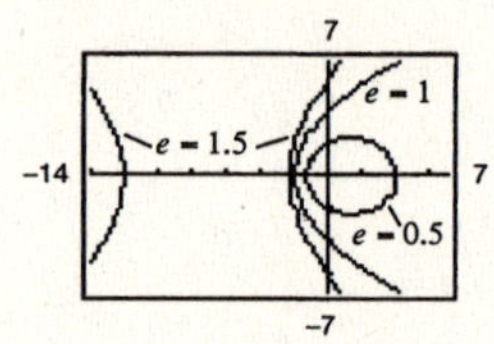

(a) $e = 1, r = \dfrac{4}{1 - \cos\theta}$, parabola

(b) $e = 0.5, r = \dfrac{2}{1 - 0.5\cos\theta}$, ellipse

(c) $e = 1.5, r = \dfrac{6}{1 - 1.5\cos\theta}$, hyperbola

3. $r = \dfrac{4e}{1 - e\sin\theta}$

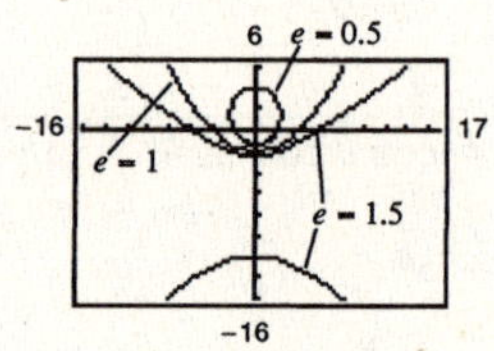

(a) $e = 1, r = \dfrac{4}{1 - \sin\theta}$, parabola

(b) $e = 0.5, r = \dfrac{2}{1 - 0.5\sin\theta} = \dfrac{4}{2 - \sin\theta}$, ellipse

(c) $e = 1.5, r = \dfrac{6}{1 - 1.5\sin\theta} = \dfrac{12}{2 - 3\sin\theta}$, hyperbola

4. $r = \dfrac{4e}{1 + e\sin\theta}$

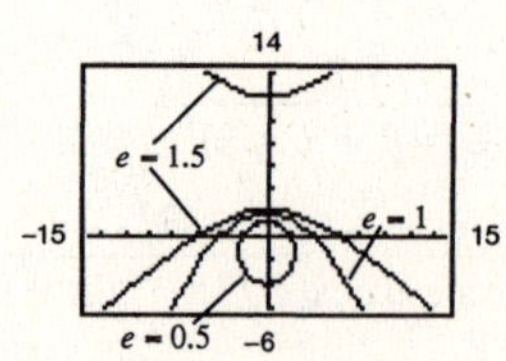

(a) $e = 1, r = \dfrac{4}{1 + \sin\theta}$, parabola

(b) $e = 0.5, r = \dfrac{2}{1 + 0.5\sin\theta}$, ellipse

(c) $e = 1.5, r = \dfrac{6}{1 + 1.5\sin\theta}$, hyperbola

5. $r = \dfrac{2}{1 + \cos\theta}$

$e = 1 \Rightarrow$ Parabola

Vertical directrix to the right of the pole

Matches graph (f).

6. $r = \dfrac{3}{2 - \cos\theta}$

$e = \dfrac{1}{2} \Rightarrow$ Ellipse

Vertical directrix to the left of the pole

Matches graph (c).

7. $r = \dfrac{3}{1 + 2\sin\theta}$

$e = 2 \Rightarrow$ Hyperbola

Matches graph (d).

8. $r = \dfrac{2}{1 - \sin\theta}$

$e = 1 \Rightarrow$ Parabola

Horizontal directrix below pole

Matches graph (e).

9. $r = \dfrac{4}{2 + \cos\theta}$

$= \dfrac{2}{1 + 0.5\cos\theta}$

$e = 0.5 \Rightarrow$ Ellipse

Vertical directrix to the right of the pole

Matches graph (a).

10. $r = \dfrac{4}{1 - 3\sin\theta}$

$e = 3 \Rightarrow$ Hyperbola

Horizontal directrix below pole

Matches graph (b).

11. $r = \dfrac{2}{1 - \cos\theta}$

$e = 1$, the graph is a parabola.

Vertex: $(1, \pi)$

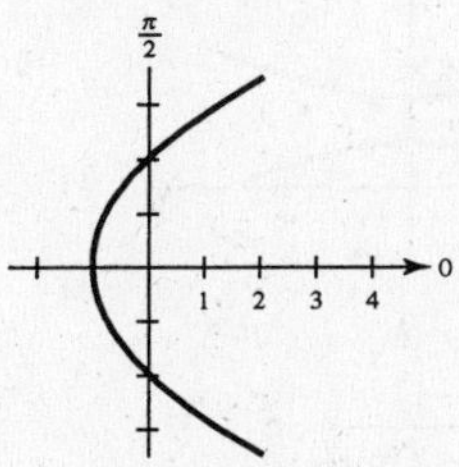

12. $r = \dfrac{3}{1 + \sin\theta}$

$e = 1 \Rightarrow$ Parabola

Vertex: $\left(\dfrac{3}{2}, \dfrac{\pi}{2}\right)$

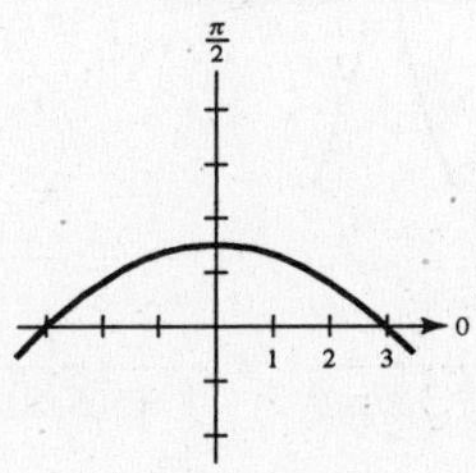

13. $r = \dfrac{5}{1 + \sin\theta}$

$e = 1$, the graph is a parabola.

Vertex: $\left(\dfrac{5}{2}, \dfrac{\pi}{2}\right)$

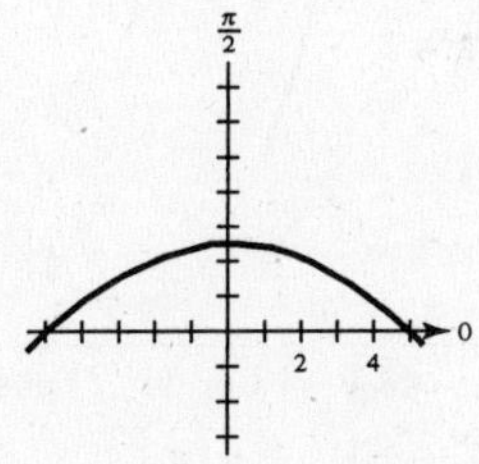

14. $r = \dfrac{6}{1 + \cos\theta}$

$e = 1 \Rightarrow$ Parabola

Vertex: $(3, 0)$

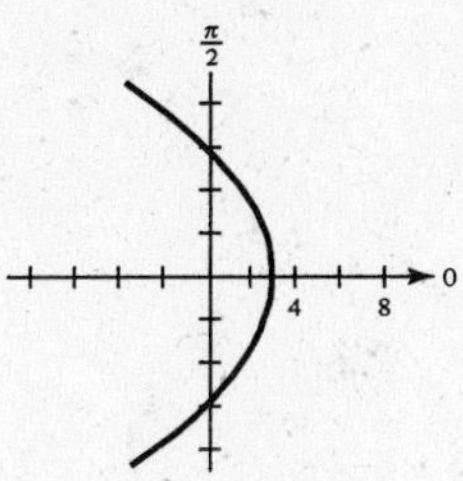

15. $r = \dfrac{2}{2 - \cos\theta} = \dfrac{1}{1 - (1/2)\cos\theta}$

$e = \dfrac{1}{2}1$, the graph is an ellipse.

Vertices: $(2, 0), \left(\dfrac{2}{3}, \pi\right)$

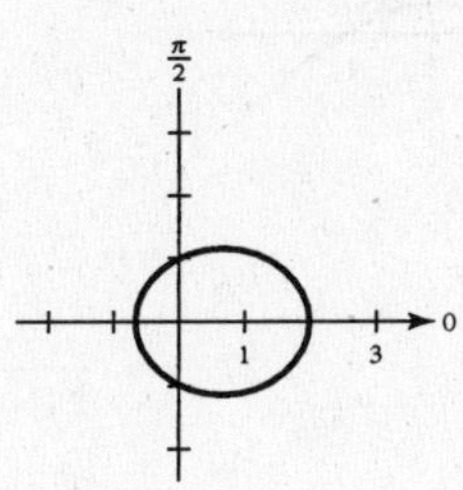

16. $r = \dfrac{3}{3 + \sin\theta} = \dfrac{1}{1 + (1/3)\sin\theta}$

$e = \dfrac{1}{3} < 1 \Rightarrow$ Ellipse

Vertices: $\left(\dfrac{3}{4}, \dfrac{\pi}{2}\right), \left(\dfrac{3}{2}, \dfrac{3\pi}{2}\right)$

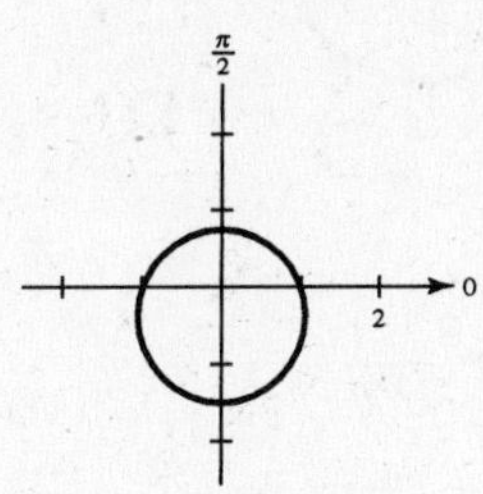

17. $r = \dfrac{6}{2 + \sin\theta} = \dfrac{3}{1 + (1/2)\sin\theta}$

$e = \dfrac{1}{2} < 1$, the graph is an ellipse.

Vertices: $\left(2, \dfrac{\pi}{2}\right), \left(6, \dfrac{3\pi}{2}\right)$

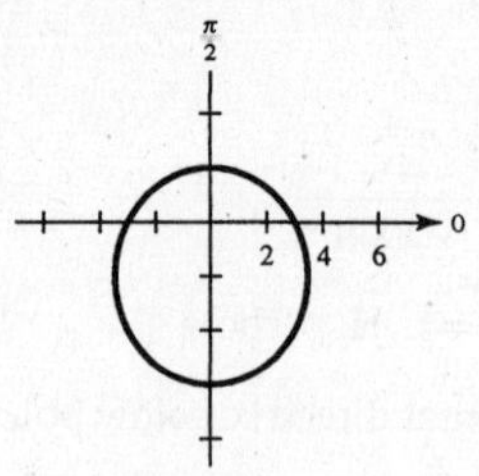

18. $r = \dfrac{9}{3 - 2\cos\theta} = \dfrac{3}{1 - (2/3)\cos\theta}$

$e = \dfrac{2}{3} < 1 \Rightarrow$ Ellipse

Vertices: $(9, 0), \left(\dfrac{9}{5}, \pi\right)$

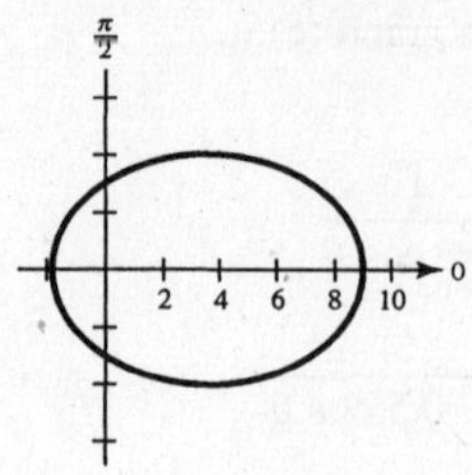

19. $r = \dfrac{3}{2 + 4\sin\theta} = \dfrac{3/2}{1 + 2\sin\theta}$

$e = 2 > 1$, the graph is a hyperbola.

Vertices: $\left(\dfrac{1}{2}, \dfrac{\pi}{2}\right), \left(-\dfrac{3}{2}, \dfrac{3\pi}{2}\right)$

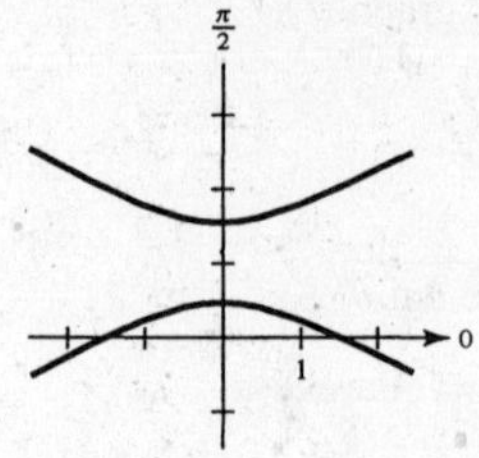

20. $r = \dfrac{5}{-1 + 2\cos\theta} = \dfrac{-5}{1 - 2\cos\theta}$

$e = 2 > 1 \Rightarrow$ Hyperbola

Vertices: $(5, 0), \left(-\dfrac{5}{3}, \pi\right)$

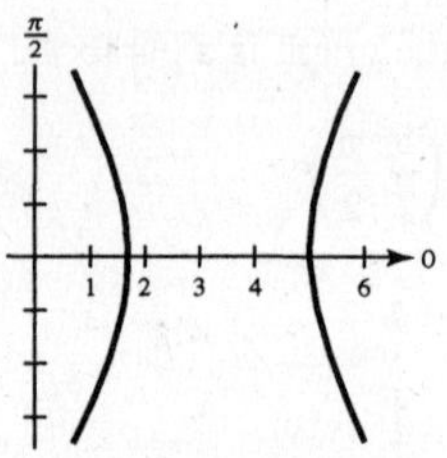

21. $r = \dfrac{3}{2 - 6\cos\theta} = \dfrac{3/2}{1 - 3\cos\theta}$

$e = 3 > 1$, the graph is a hyperbola.

Vertices: $\left(-\dfrac{3}{4}, 0\right), \left(\dfrac{3}{8}, \pi\right)$

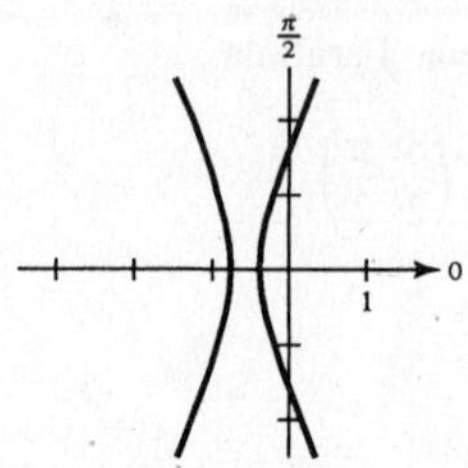

22. $r = \dfrac{3}{2 + 6\sin\theta} = \dfrac{3/2}{1 + 3\sin\theta}$

$e = 3 > 1 \Rightarrow$ Hyperbola

Vertices: $\left(\dfrac{3}{8}, \dfrac{\pi}{2}\right), \left(-\dfrac{3}{4}, \dfrac{3\pi}{2}\right)$

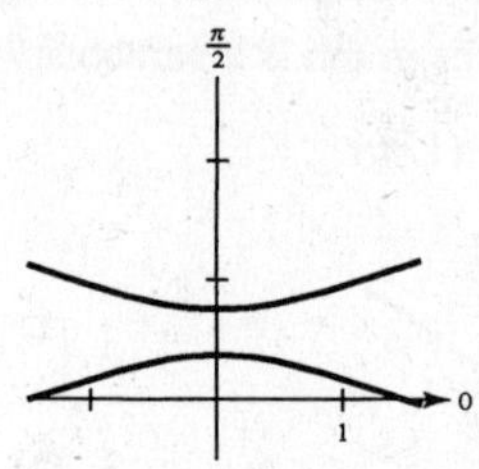

23. $r = \dfrac{4}{2 - \cos\theta} = \dfrac{2}{1 - (1/2)\cos\theta}$

$e = \dfrac{1}{2} < 1$, the graph is an ellipse.

Vertices: $(4, 0), \left(\dfrac{4}{3}, \pi\right)$

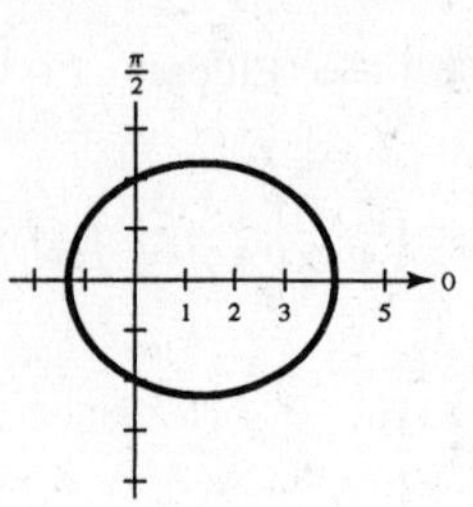

24. $r = \dfrac{2}{2 + 3\sin\theta} = \dfrac{1}{1 + (3/2)\sin\theta}$

$e = \dfrac{3}{2} > 1 \Rightarrow$ Hyperbola

Vertices: $\left(\dfrac{2}{5}, \dfrac{\pi}{2}\right), \left(-2, \dfrac{3\pi}{2}\right)$

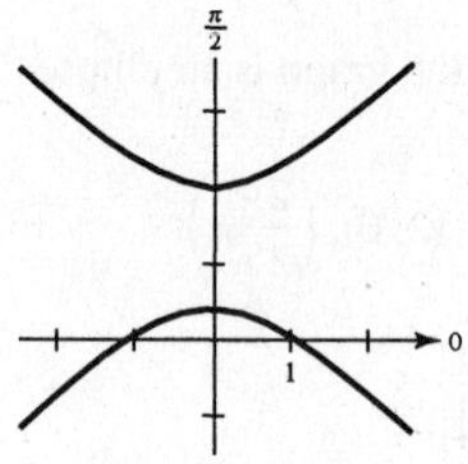

25. $r = \dfrac{-1}{1 - \sin\theta}$

$e = 1 \Rightarrow$ Parabola

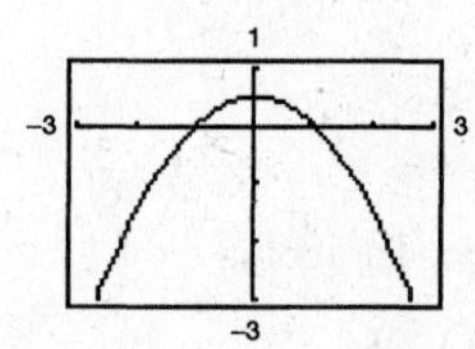

26. $r = \dfrac{2}{1 + \sin\theta}$

$e = 1 \Rightarrow$ Parabola

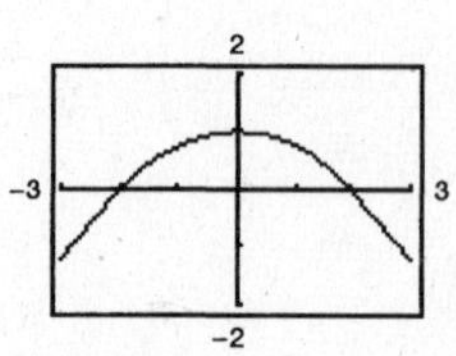

27. $r = \dfrac{-5}{2 + 4\sin\theta} = \dfrac{-(5/2)}{1 + 2\sin\theta}$

$e = 2 \Rightarrow$ Hyperbola

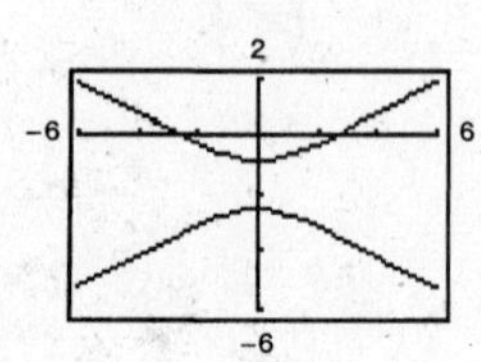

28. $r = \dfrac{6}{3 - 2\cos\theta} = \dfrac{2}{1 - (2/3)\cos\theta}$

$e = \dfrac{2}{3} \Rightarrow$ Ellipse

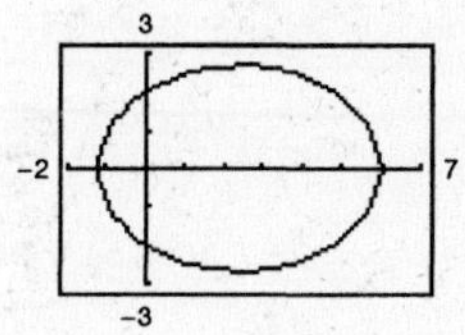

29. $r = \dfrac{3}{-4 + 2\cos\theta}$

$e = \dfrac{1}{2} \Rightarrow$ Ellipse

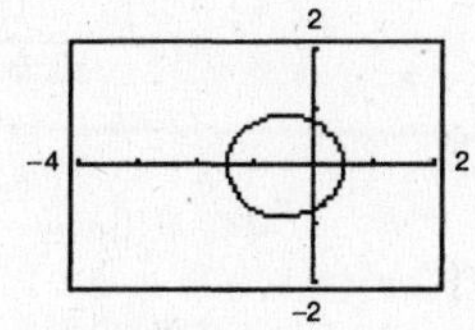

30. $r = \dfrac{5}{-1 + 2\cos\theta} = \dfrac{-5}{1 - 2\cos\theta}$

$e = 2 \Rightarrow$ Hyperbola

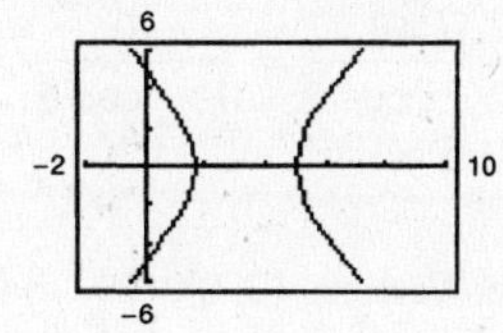

31. $r = \dfrac{4}{1 - 2\cos\theta}$

$e = 2 \Rightarrow$ Hyperbola

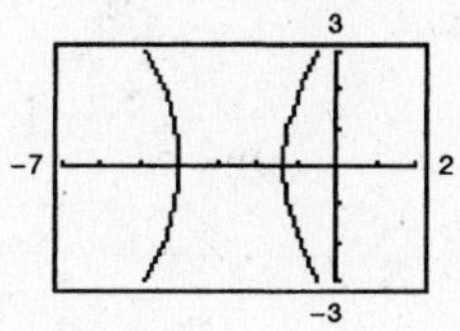

32. $r = \dfrac{12}{2 - \cos\theta} = \dfrac{6}{1 - (1/2)\cos\theta}$

$e = \dfrac{1}{2} \Rightarrow$ Ellipse

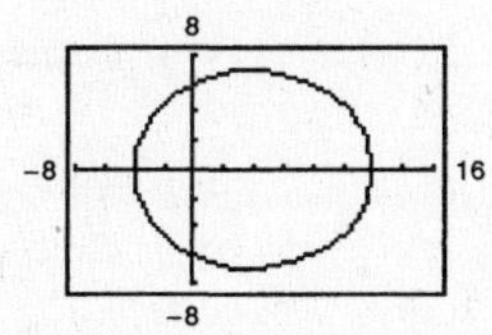

33. Parabola: $e = 1$

Directrix: $x = -1$

Vertical directrix to the left of the pole

$r = \dfrac{1(1)}{1 - 1\cos\theta} = \dfrac{1}{1 - \cos\theta}$

34. Parabola: $e = 1$

Directrix: $y = -2$

$p = 2$

Horizontal directrix below the pole

$r = \dfrac{1(2)}{1 - 1\sin\theta} = \dfrac{2}{1 - \sin\theta}$

35. Ellipse: $e = \dfrac{1}{2}$

Directrix: $y = 1$

$p = 1$

Horizontal directrix above the pole

$r = \dfrac{(1/2)(1)}{1 + (1/2)\sin\theta} = \dfrac{1}{2 + \sin\theta}$

36. Ellipse: $e = \dfrac{3}{4}$

Directrix: $y = -3$

$p = 3$

Horizontal directrix below the pole

$r = \dfrac{(3/4)(3)}{1 - (3/4)\sin\theta} = \dfrac{9}{4 - 3\sin\theta}$

37. Hyperbola: $e = 2$

Directrix: $x = 1$

$p = 1$

Vertical directrix to the right of the pole

$r = \dfrac{2(1)}{1 + 2\cos\theta} = \dfrac{2}{1 + 2\cos\theta}$

38. Hyperbola: $e = \dfrac{3}{2}$

Directrix: $x = -1$

$p = 1$

Vertical directrix to the left of the pole

$r = \dfrac{(3/2)(1)}{1 - (3/2)\cos\theta} = \dfrac{3}{2 - 3\cos\theta}$

39. Parabola

Vertex: $\left(1, -\dfrac{\pi}{2}\right) \Rightarrow e = 1, p = 2$

Horizontal directrix below the pole

$r = \dfrac{1(2)}{1 - 1\sin\theta} = \dfrac{2}{1 - \sin\theta}$

40. Parabola

Vertex: $(6, 0) \Rightarrow e = 1, p = 12$

Vertical directrix to the right of the pole

$r = \dfrac{1(12)}{1 + 1\cos\theta} = \dfrac{12}{1 + \cos\theta}$

41. Parabola

Vertex: $(5, \pi) \Rightarrow e = 1, p = 10$

Vertical directrix to the left of the pole

$$r = \frac{1(10)}{1 - 1\cos\theta} = \frac{10}{1 - \cos\theta}$$

42. Parabola

Vertex: $\left(10, \frac{\pi}{2}\right) \Rightarrow e = 1, p = 20$

Horizontal directrix above the pole

$$r = \frac{1(20)}{1 + 1\sin\theta} = \frac{20}{1 + \sin\theta}$$

43. Ellipse: Vertices $(2, 0), (10, \pi)$

Center: $(4, \pi)$; $c = 4, a = 6, e = \frac{2}{3}$

Vertical directrix to the right of the pole

$$r = \frac{(2/3)p}{1 + (2/3)\cos\theta} = \frac{2p}{3 + 2\cos\theta}$$

$$2 = \frac{2p}{3 + 2\cos 0}$$

$p = 5$

$$r = \frac{2(5)}{3 + 2\cos\theta} = \frac{10}{3 + 2\cos\theta}$$

44. Ellipse

Vertices: $\left(2, \frac{\pi}{2}\right), \left(4, \frac{3\pi}{2}\right)$

Center: $\left(1, \frac{3\pi}{2}\right)$; $c = 1, a = 3, e = \frac{1}{3}$

Horizontal directrix above the axis

$$r = \frac{1/3p}{1 + (1/3)\sin\theta} = \frac{p}{3 + \sin\theta}$$

$$2 = \frac{p}{3 + \sin(\pi/2)}$$

$p = 8$

$$r = \frac{8}{3 + \sin\theta}$$

45. Ellipse: Vertices $(20, 0), (4, \pi)$

Center: $(8, 0)$; $c = 8, a = 12, e = \frac{2}{3}$

Vertical directrix to the left of the pole

$$r = \frac{(2/3)p}{1 - (2/3)\cos\theta} = \frac{2p}{3 - 2\cos\theta}$$

$$20 = \frac{2p}{3 - 2\cos 0}$$

$p = 10$

$$r = \frac{2(10)}{3 - 2\cos\theta} = \frac{20}{3 - 2\cos\theta}$$

46. Hyperbola

Vertices: $(2, 0), (8, 0)$

Center: $(5, 0)$; $c = 5, a = 3, e = \frac{5}{3}$

Vertical directrix to the right of the pole

$$r = \frac{(5/3)p}{1 + (5/3)\cos\theta} = \frac{5p}{3 + 5\cos\theta}$$

$$2 = \frac{5p}{3 + 5\cos 0}$$

$p = \frac{16}{5}$

$$r = \frac{5(16/5)}{3 + 5\cos\theta} = \frac{16}{3 + 5\cos\theta}$$

47. Hyperbola: Vertices $\left(1, \frac{3\pi}{2}\right), \left(9, \frac{3\pi}{2}\right)$

Center: $\left(5, \frac{3\pi}{2}\right)$; $c = 5, a = 4, e = \frac{5}{4}$

Horizontal directrix below the pole

$$r = \frac{(5/4)p}{1 - (5/4)\sin\theta} = \frac{5p}{4 - 5\sin\theta}$$

$$1 = \frac{5p}{4 - 5\sin(3\pi/2)}$$

$p = \frac{9}{5}$

$$r = \frac{5(9/5)}{4 - 5\sin\theta} = \frac{9}{4 - 5\sin\theta}$$

48. Hyperbola

Vertices: $\left(4, \frac{\pi}{2}\right), \left(-1, \frac{3\pi}{2}\right)$

Center: $\left(\frac{5}{2}, \frac{\pi}{2}\right)$; $c = \frac{5}{2}, a = \frac{3}{2}, e = \frac{5}{3}$

Horizontal directrix above the pole

$$r = \frac{5/3p}{1 + (5/3)\sin\theta} = \frac{5p}{3 + 5\sin\theta}$$

$$1 = \frac{5p}{3 + 5\sin(-3\pi/2)}$$

$p = \frac{8}{5}$

$$r = \frac{5(8/5)}{3 + 5\sin\theta} = \frac{8}{3 + 5\sin\theta}$$

49.
$$\frac{x^2}{a^2} + \frac{y^2}{b^2} = 1$$
$$x^2b^2 + y^2a^2 = a^2b^2$$
$$b^2r^2\cos^2\theta + a^2r^2\sin^2\theta = a^2b^2$$
$$r^2[b^2\cos^2\theta + a^2(1 - \cos^2\theta] = a^2b^2$$
$$r^2[a^2 + \cos^2\theta(b^2 - a^2)] = a^2b^2$$
$$r^2 = \frac{a^2b^2}{a^2 + (b^2 - a^2)\cos^2\theta} = \frac{a^2b^2}{a^2 - c^2\cos^2\theta}$$
$$= \frac{b^2}{1 - (c/a)^2\cos^2\theta} = \frac{b^2}{1 - e^2\cos^2\theta}$$

50.
$$\frac{x^2}{a^2} - \frac{y^2}{b^2} = 1$$
$$x^2b^2 - y^2a^2 = a^2b^2$$
$$b^2r^2\cos^2\theta - a^2r^2\sin^2\theta = a^2b^2$$
$$r^2[b^2\cos^2\theta - a^2(1 - \cos^2\theta) = a^2b^2$$
$$r^2[-a^2 + \cos^2\theta(a^2 + b^2)] = a^2b^2$$
$$r^2 = \frac{a^2b^2}{-a^2 + c^2\cos^2\theta} = \frac{b^2}{-1 + (c^2/a^2)\cos^2\theta}$$
$$= \frac{-b^2}{1 - e^2\cos^2\theta}$$

51. $\frac{x^2}{169} + \frac{y^2}{144} = 1$

$a = 13, b = 12, c = 5, e = \frac{5}{13}$

$$r^2 = \frac{144}{1 - \frac{25}{169}\cos^2\theta} = \frac{24{,}336}{169 - 25\cos^2\theta}$$

52. $\frac{x^2}{25} + \frac{y^2}{16} = 1$

$a = 5, b = 4, c = 3, e = \frac{3}{5}$

$$r^2 = \frac{400}{25 - 9\cos^2\theta}$$

53. $\frac{x^2}{9} - \frac{y^2}{16} = 1$

$a = 3, b = 4, c = 5, e = \frac{5}{3}$

$$r^2 = \frac{-16}{1 - \frac{25}{9}\cos^2\theta} = \frac{144}{25\cos^2\theta - 9}$$

54. $\frac{x^2}{36} - \frac{y^2}{4} = 1$

$a = 6, b = 2, c = 2\sqrt{10}, e = \frac{\sqrt{10}}{3}$

$$r^2 = \frac{-4}{1 - (10/9)\cos^2\theta} = \frac{-36}{9 - 10\cos^2\theta}$$
$$= \frac{36}{10\cos^2\theta - 9}$$

55. Hyperbola

One focus: $\left(5, \frac{\pi}{2}\right)$

Vertices: $\left(4, \frac{\pi}{2}\right), \left(4, -\frac{\pi}{2}\right)$

$c = 5, a = 4, b = 3, e = \frac{5}{4}$

If the vertices were $(4, 0)$ and $(4, \pi)$, we would have:

$$r^2 = \frac{-9}{1 - \frac{25}{16}\cos^2\theta} = \frac{144}{25\cos^2\theta - 16}$$

The equation for the given hyperbola can be found by rotating through an angle of $\pi/2$, thus:

$$r^2 = \frac{144}{25\cos^2\left(\frac{\pi}{2} - \theta\right) - 16} = \frac{144}{25\sin^2\theta - 16}$$

56. Ellipse

One focus: $(4, 0)$

Vertices: $(5, 0), (5, \pi)$

$a = 5, c = 4, b = 3, e = \frac{4}{5}$

$$r^2 = \frac{9}{1 - (16/25)\cos^2\theta} = \frac{225}{25 - 16\cos^2\theta}$$

57. Vertex: $\left(4100, \frac{\pi}{2}\right)$

Focus: $(0, 0)$

$e = 1, p = 8200$

$$r = \frac{ep}{1 + e\sin\theta} = \frac{8200}{1 + \sin\theta}$$

When $\theta = 30°$, $r = 8200/1.5 \approx 5466.67$.

Distance between the surface of the earth and the satellite is $r - 4000 \approx 1467$ miles.

58. From Exercise 57, $r = \dfrac{8200}{1 + \sin\theta}$.

$\theta = 60°$: $r = \dfrac{8200}{1 + \sin 60°} \approx 4394.4$

Distance between surface of Earth and satellite:

$4394.4 - 4000 = 394.4$ miles

59. False. If e remains fixed, and p changes, then the lengths of the major axis and minor axis change.

For example, graph $r = \dfrac{5}{1 - \frac{2}{3}\sin\theta}$, with $e = \dfrac{2}{3}$ and $p = \dfrac{15}{2}$ and $r = \dfrac{6}{1 - \frac{2}{3}\sin\theta}$, with $e = \dfrac{2}{3}$ and $p = 9$, on the same set of coordinate axes.

60. True. The graphs are identical, although the graphs are not traced out in the same order as θ goes from 0 to 2π.

61. True.

$$r = \frac{10}{3 + 2\cos\theta} = \frac{\frac{10}{3}}{1 + \frac{2}{3}\cos\theta} \Rightarrow e = \frac{2}{3} \text{ and } p = 5$$

Ellipse with a horizontal major axis.
Center: $(-4, 0)$

$a = 6, c = 4, b = \sqrt{20}$

$$\frac{(x + 4)^2}{36} + \frac{y^2}{20} = 1$$

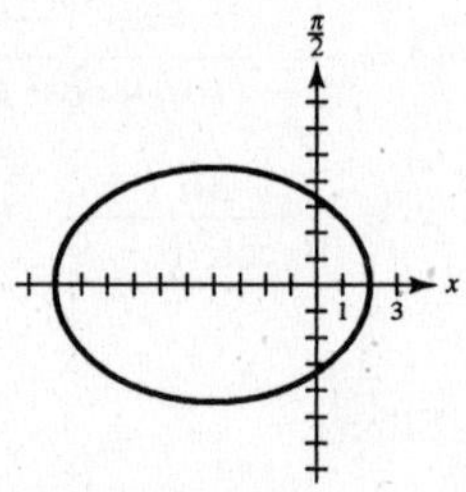

62. False.

$$r = \frac{10}{2 + \sin\theta} = \frac{5}{1 + \frac{1}{2}\sin\theta}$$

$e = \dfrac{1}{2} < 1$ so the graph is an ellipse, not a hyperbola.

Review Exercises for Chapter 12

1. Hyperbola

2. A parabola is formed.

3. Vertex: $(4, 2) = (h, k)$

Focus: $(4, 0) \Rightarrow p = -2$

$(x - h)^2 = 4p(y - k)$

$(x - 4)^2 = -8(y - 2)$

4. Vertex: $(2, 0) = (h, k)$

Focus: $(0, 0) \Rightarrow p = -2$

$(y - k)^2 = 4p(x - h)$

$y^2 = -8(x - 2)$

5. Vertex: $(0, 2) = (h, k)$

Directrix: $x = -3 \Rightarrow p = 3$

$(y - k)^2 = 4p(x - h)$

$(y - 2)^2 = 12x$

6. Vertex: $(2, 2) = (h, k)$

Directrix: $y = 0 \Rightarrow p = 2$

$(x - h)^2 = 4p(y - k)$

$(x - 2)^2 = 8(y - 2)$

7. $x^2 = -2y \Rightarrow p = -\frac{1}{2}$

Focus: $\left(0, -\frac{1}{2}\right)$

$d_1 = b + \frac{1}{2}$

$d_2 = \sqrt{(2-0)^2 + \left(-2 + \frac{1}{2}\right)^2}$

$= \sqrt{4 + \frac{9}{4}} = \frac{5}{2}$

$d_1 = d_2$

$b + \frac{1}{2} = \frac{5}{2}$

$b = 2$

The slope of the line is $m = \frac{-2-2}{2-0} = -2$

Tangent line: $y = -2x + 2$

x-intercept: $(1, 0)$

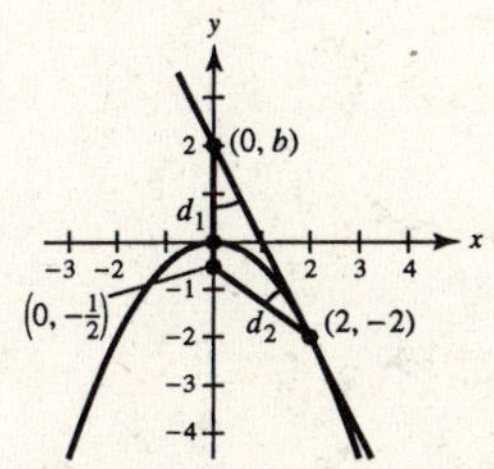

8. $x^2 = -2y$

$p = -\frac{1}{2}$

Focus: $\left(0, -\frac{1}{2}\right)$

Tangent line through point $(-4, -8)$:

Slope: m

y-intercept: $(0, b)$

$d_1 = b + \frac{1}{2}$

$d_2 = \sqrt{(-4-0)^2 + \left(-8 + \frac{1}{2}\right)^2} = \frac{17}{2}$

$d_1 = d_2 \Rightarrow b = 8$

$m = \frac{-8-8}{-4-0} = 4$

$y = 4x + 8$

x-intercept of tangent line: $(-2, 0)$

9. Parabola

Opens downward

Vertex: $(0, 12)$

$(x-h)^2 = 4p(y-k)$

$x^2 = 4p(y-12)$

Solution points: $(\pm 4, 10)$

$16 = 4p(10-12)$

$16 = -8p$

$-2 = p$

$x^2 = -8(y-12)$

To find the x-intercepts, let $y = 0$.

$x^2 = 96$

$x = \pm\sqrt{96} = \pm 4\sqrt{6}$

At the base z, the archway is $2(4\sqrt{6}) = 8\sqrt{6}$ meters wide.

10. $y^2 = 4px$

$p = 1.5$

$y^2 = 6x$

11. Vertices: $(-3, 0), (7, 0) \Rightarrow a = 5$
$(h, k) = (2, 0)$

Foci: $(0, 0), (4, 0) \Rightarrow c = 2$

$b^2 = a^2 - c^2 = 25 - 4 = 21$

$$\frac{(x-h)^2}{a^2} + \frac{(y-k)^2}{b^2} = 1$$

$$\frac{(x-2)^2}{25} + \frac{y^2}{21} = 1$$

12. Vertices: $(2, 0), (2, 4) \Rightarrow a = 2, (h, k) = (2, 2)$

Foci: $(2, 1), (2, 3) \Rightarrow c = 1$

$b^2 = a^2 - c^2 = 4 - 1 = 3$

$$\frac{(x-h)^2}{b^2} + \frac{(y-k)^2}{a^2} = 1$$

$$\frac{(x-2)^2}{3} + \frac{(y-2)^2}{4} = 1$$

13. Vertices: $(0, \pm 6) \Rightarrow a = 6, (h, k) = (0, 0)$

Passes through $(2, 2)$

$$\frac{(x-h)^2}{b^2} + \frac{(y-k)^2}{a^2} = 1$$

$$\frac{x^2}{b^2} + \frac{y^2}{36} = 1 \Rightarrow b^2 = \frac{36x^2}{36 - y^2} = \frac{36(4)}{36 - 4} = \frac{9}{2}$$

$$\frac{x^2}{9/2} + \frac{y^2}{36} = 1$$

$$\frac{2x^2}{9} + \frac{y^2}{36} = 1$$

14. Vertices: $(0, 1), (4, 1) \Rightarrow a = 2, (h, k) = (2, 1)$

Endpoints of minor axis: $(2, 0), (2, 2) \Rightarrow b = 1$

$$\frac{(x-h)^2}{a^2} + \frac{(y-k)^2}{b^2} = 1$$

$$\frac{(x-2)^2}{4} + (y-1)^2 = 1$$

15. $2a = 10 \Rightarrow a = 5$

$b = 4$

$c^2 = a^2 - b^2 = 25 - 16 = 9 \Rightarrow c = 3$

The foci occur 3 feet from the center of the arch on a line connecting the tops of the pillars.

16. $\frac{x^2}{324} + \frac{y^2}{196} = 1$

$a = \sqrt{324} = 18, b = \sqrt{196} = 14$

$c = \sqrt{a^2 - b^2} = \sqrt{128} = 8\sqrt{2}$

Longest distance: $2a = 36$ feet

Shortest distance: $2b = 28$ feet

Distance between foci: $2c = 16\sqrt{2}$ feet

17. $16x^2 + 9y^2 - 32x + 72y + 16 = 0$

$16(x^2 - 2x + 1) + 9(y^2 + 8y + 16) = -16 + 16 + 144$

$16(x-1)^2 + 9(y+4)^2 = 144$

$$\frac{(x-1)^2}{9} + \frac{(y+4)^2}{16} = 1$$

$a = 4, b = 3, c = \sqrt{7}$

Center: $(1, -4)$

Vertices: $(1, 0)$ and $(1, -8)$

Foci: $\left(1, -4 \pm \sqrt{7}\right)$

Eccentricity: $e = \frac{\sqrt{7}}{4}$

18. $4x^2 + 25y^2 + 16x - 150y + 141 = 0$

$4(x^2 + 4x + 4) + 25(y^2 - 6y + 9) = -141 + 16 + 225$

$$\frac{(x+2)^2}{25} + \frac{(y-3)^2}{4} = 1$$

Center: $(-2, 3)$

$a = 5, b = 2, c = \sqrt{a^2 - b^2} = \sqrt{21}$

Vertices: $(3, 3), (-7, 3)$

Foci: $\left(-2 \pm \sqrt{21}, 3\right)$

Eccentricity: $e = \frac{c}{a} = \frac{\sqrt{21}}{5}$

19. $\dfrac{(x+2)^2}{81} + \dfrac{(y-1)^2}{100} = 1$

$a = 10, b = 9, c = \sqrt{19}$

Center: $(-2, 1)$

Vertices: $(-2, 11)$ and $(-2, -9)$

Foci: $\left(-2, 1 \pm \sqrt{19}\right)$

Eccentricity: $e = \dfrac{\sqrt{19}}{10}$

20. $\dfrac{(x-5)^2}{1} + \dfrac{(y+3)^2}{36} = 1$

Center: $(5, -3)$

$a = 6, b = 1, c = \sqrt{a^2 - b^2} = \sqrt{35}$

Vertices: $(5, 3), (5, -9)$

Foci: $\left(5, -3 \pm \sqrt{35}\right)$

Eccentricity: $e = \dfrac{c}{a} = \dfrac{\sqrt{35}}{6}$

21. $\dfrac{(x-1)^2}{9} + \dfrac{(y-5)^2}{25} = 1$

$\dfrac{2(x-1)}{9} + \dfrac{2(y-5)}{25}\dfrac{dy}{dx} = 0$

$\dfrac{dy}{dx} = -\dfrac{25(x-1)}{9(y-5)}$

At $(4, 5)$, $\dfrac{dy}{dx}$ is undefined. The tangent line is vertical.

$x = 4$

$x - 4 = 0$

22. $\dfrac{(x+2)^2}{4} + \dfrac{(y+2)^2}{25} = 1$

$\dfrac{(x+2)}{2} + \dfrac{2(y+2)}{25}\dfrac{dy}{dx} = 0$

$\dfrac{dy}{dx} = \dfrac{-25(x+2)}{4(y+2)}$

At $(-2, 3)$, $\dfrac{dy}{dx} = 0$.

The tangent line is horizontal.

$y = 3$

$y - 3 = 0$

23. $x^2 + 5y^2 = 10$

$\dfrac{x^2}{10} + \dfrac{y^2}{2} = 1$

$a = \sqrt{10}, b = \sqrt{2}$

Area $= \pi ab = \sqrt{20}\,\pi = 2\sqrt{5}\,\pi$ square units.

24. $4x^2 + y^2 = 4$

$\dfrac{x^2}{1} + \dfrac{y^2}{4} = 1$

$a = 2, b = 1$

Area $= \pi ab = 2\pi$ square units

25. $(x^2 - 2x + 1) + 4(y^2 - 4y + 4) = -13 + 1 + 16$

$(x-1)^2 + 4(y-2)^2 = 4$

$\dfrac{(x-1)^2}{4} + \dfrac{(y-2)^2}{1} = 1$

Area $= \pi ab = \pi(2)(1) = 2\pi$

26. $16(x^2 - 2x + 1) + 4(y^2 - 2y + 1) = 44 + 16 + 4$

$16(x-1)^2 + 4(y-1)^2 = 64$

$\dfrac{(x-1)^2}{4} + \dfrac{(y-1)^2}{16} = 1$

Area $= \pi ab = \pi(2)(4) = 8\pi$

27. Vertices: $(0, \pm 1) \Rightarrow a = 1, (h, k) = (0, 0)$

Foci: $(0, \pm 3) \Rightarrow c = 3$

$b^2 = c^2 - a^2 = 9 - 1 = 8$

$\dfrac{(y-k)^2}{a^2} - \dfrac{(x-h)^2}{b^2} = 1$

$y^2 - \dfrac{x^2}{8} = 1$

28. Vertices: $(2, 2), (-2, 2) \Rightarrow a = 2, (h, k) = (0, 2)$

Foci: $(4, 2), (-4, 2) \Rightarrow c = 4$

$b^2 = c^2 - a^2 = 16 - 4 = 12$

$\dfrac{(x-h)^2}{a^2} - \dfrac{(y-k)^2}{b^2} = 1$

$\dfrac{x^2}{4} - \dfrac{(y-2)^2}{12} = 1$

29. Foci: $(0, 0), (8, 0) \Rightarrow c = 4, (h, k) = (4, 0)$

Asymptotes: $y = \pm 2(x - 4) \Rightarrow \dfrac{b}{a} = 2, b = 2a$

$b^2 = c^2 - a^2 \Rightarrow 4a^2 = 16 - a^2\ 130 \Rightarrow a^2 = \dfrac{16}{5},$

$b^2 = \dfrac{64}{5}$

$$\frac{(x-h)^2}{a^2} - \frac{(y-k)^2}{b^2} = 1$$

$$\frac{(x-4)^2}{\frac{16}{5}} - \frac{y^2}{\frac{64}{5}} = 1$$

$$\frac{5(x-4)^2}{16} - \frac{5y^2}{64} = 1$$

30. Foci: $(3, \pm 2) \Rightarrow c = 2, (h, k) = (3, 0)$

Asymptotes: $y = \pm 2(x - 3) \Rightarrow \dfrac{a}{b} = 2, a = 2b$

$b^2 = c^2 - a^2 = 4 - 4b^2 \Rightarrow b^2 = \dfrac{4}{5}, a^2 = \dfrac{16}{5}$

$$\frac{(y-k)^2}{a^2} - \frac{(x-h)^2}{b^2} = 1$$

$$\frac{y^2}{16/5} - \frac{(x-3)^2}{4/5} = 1 \Rightarrow \frac{5y^2}{16} - \frac{5(x-3)^2}{4} = 1$$

31. $9x^2 - 16y^2 - 18x - 32y - 151 = 0$

$$9(x^2 - 2x + 1) - 16(y^2 + 2y + 1) = 151 + 9 - 16$$

$$9(x-1)^2 - 16(y+1)^2 = 144$$

$$\frac{(x-1)^2}{16} - \frac{(y+1)^2}{9} = 1$$

$a = 4, b = 3, c = 5$

Center: $(1, -1)$

Vertices: $(5, -1)$ and $(-3, -1)$

Foci: $(6, -1)$ and $(-4, -1)$

Asymptotes: $y = -1 \pm \dfrac{3}{4}(x - 1)$

$$y = \frac{3}{4}x - \frac{7}{4} \quad \text{or} \quad y = -\frac{3}{4}x - \frac{1}{4}$$

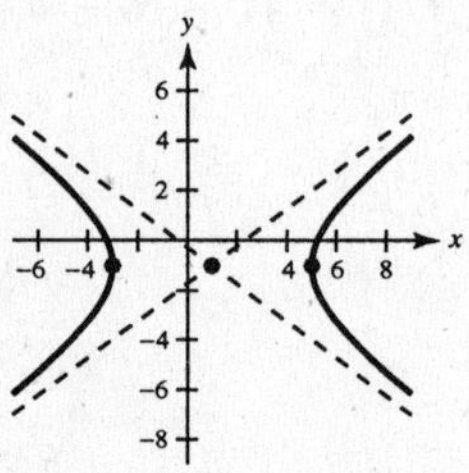

32. $-4x^2 + 25y^2 - 8x + 150y + 121 = 0$

$$-4(x^2 + 2x + 1) + 25(y^2 + 6y + 9) = -121 - 4 + 225$$

$$\frac{(y+3)^2}{4} - \frac{(x+1)^2}{25} = 1$$

Center: $(-1, -3)$

$a = 2, b = 5, c = \sqrt{a^2 + b^2} = \sqrt{29}$

Vertices: $(-1, -1), (-1, -5)$

Foci: $\left(-1, -3 \pm \sqrt{29}\right)$

Asymptotes: $y = -3 \pm \dfrac{2}{5}(x + 1)$

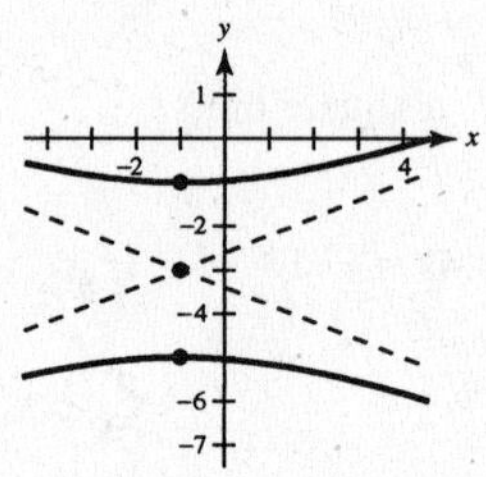

33. $\dfrac{(x-3)^2}{16} - \dfrac{(y+5)^2}{4} = 1$

$a = 4, b = 2, c = \sqrt{20} = 2\sqrt{5}$

Center: $(3, -5)$

Vertices: $(7, -5)$ and $(-1, -5)$

Foci: $\left(3 \pm 2\sqrt{5}, -5\right)$

Asymptotes: $y = -5 \pm \dfrac{1}{2}(x - 3)$

$$y = \frac{1}{2}x - \frac{13}{2} \quad \text{or} \quad y = -\frac{1}{2}x - \frac{7}{2}$$

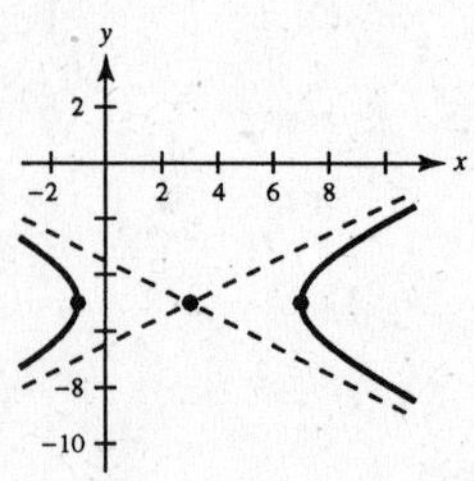

34. $\dfrac{(y-1)^2}{4} - x^2 = 1$

Center: $(0, 1)$

$a = 2, b = 1,$

$c = \sqrt{a^2 + b^2} = \sqrt{5}$

Vertices: $(0, 3), (0, -1)$

Foci: $\left(0, 1 \pm \sqrt{5}\right)$

Asymptotes: $y = 1 \pm 2x$

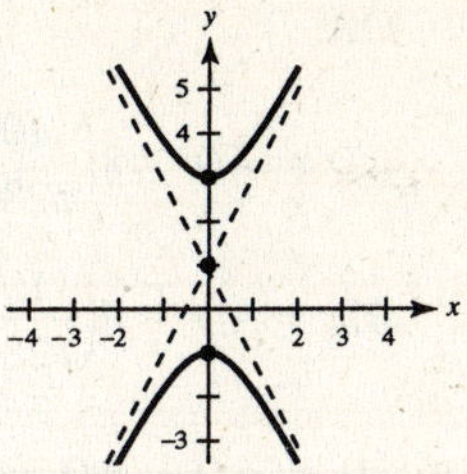

35. $\dfrac{x^2}{9} - \dfrac{y^2}{3} = 1$

$\dfrac{2x}{9} - \dfrac{2y}{3}\dfrac{dy}{dx} = 0$

$\dfrac{dy}{dx} = \dfrac{x}{3y}$

At $(6, 3)$, $\dfrac{dy}{dx} = \dfrac{6}{9} = \dfrac{2}{3}$.

Tangent line: $y - 3 = \dfrac{2}{3}(x - 6)$

$3y - 9 = 2x - 12$

$2x - 3y - 3 = 0$

36. $\dfrac{y^2}{4} - \dfrac{x^2}{2} = 1$

$\dfrac{y}{2}\dfrac{dy}{dx} - x = 0$

$\dfrac{dy}{dx} = \dfrac{2x}{y}$

At $(4, 6)$, $\dfrac{dy}{dx} = \dfrac{8}{6} = \dfrac{4}{3}$.

Tangent line: $y - 6 = \dfrac{4}{3}(x - 4)$

$3y - 18 = 4x - 16$

$4x - 3y + 2 = 0$

37. Foci: $(\pm 100, 0) \Longrightarrow c = 100$

Center: $(0, 0)$

$\dfrac{d_2}{186{,}000} - \dfrac{d_1}{186{,}000} = 0.0005 \Longrightarrow d_2 - d_1 = 93 = 2a \Longrightarrow a = 46.5$

$b^2 = c^2 - a^2 = 100^2 - 46.5^2 = 7837.75$

$\dfrac{x^2}{2162.25} - \dfrac{y^2}{7837.75} = 1$

$y^2 = 7837.75\left(\dfrac{60^2}{2162.25} - 1\right) \approx 5211.5736$

$y \approx 72$ miles

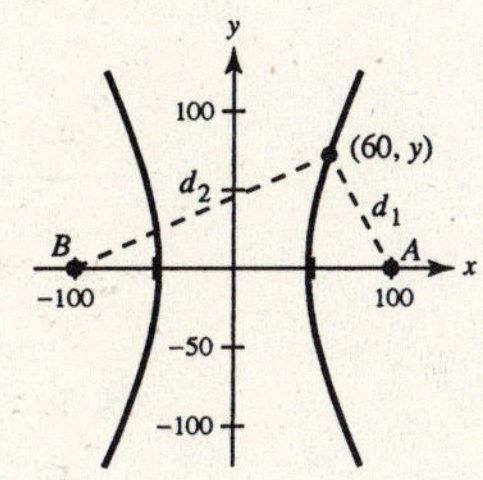

38. $BD = AD + 6\left(\frac{1100}{5280}\right)$

$CD = AD + 8\left(\frac{1100}{5280}\right)$

$2a = CD - BD = 2\left(\frac{1100}{5280}\right)$

$a = \frac{5}{24}, c = 2 \Rightarrow b^2 = \frac{2279}{576}$

Thus, we have $\frac{576x^2}{25} - \frac{576y^2}{2279} = 1.$

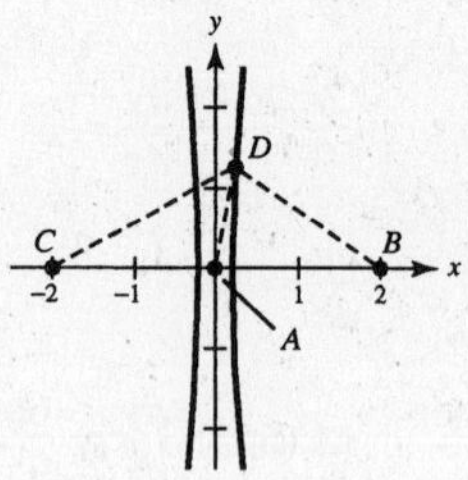

OR:

$CD = AD + 8\left(\frac{1100}{5280}\right)$

$BD = AD + 6\left(\frac{1100}{5280}\right)$

$2a = BD - AD = 6\left(\frac{1100}{5280}\right)$

$a = 3\left(\frac{5}{24}\right) = \frac{5}{8}, c = 1 \Rightarrow b^2 = \frac{39}{64}$

Center: $(1, 0)$

$\frac{64(x-1)^2}{25} - \frac{64y^2}{39} = 1$

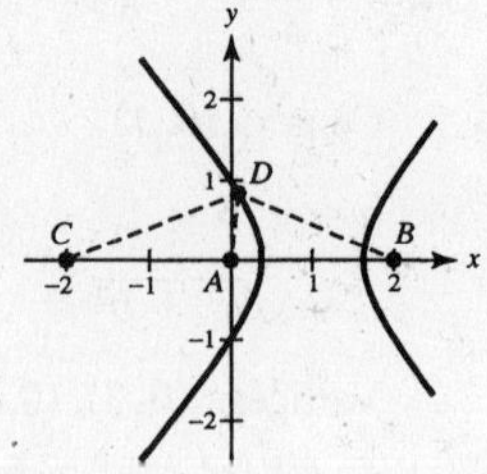

39. $5x^2 - 2y^2 + 10x - 4y + 17 = 0$

$AC = 5(-2) = -10 < 0$

The graph is a hyperbola.

40. $-4y^2 + 5x + 3y + 7 = 0$

$AC = (0)(-4) = 0 \Rightarrow$ Parabola

41. $4x^2 + y^2 - 16x = 0$

$AC = 4(1) = 4 > 0$

Ellipse

42. $4x^2 + 4y^2 - 8x + 16y - 19 = 0$

$A = C = 4$

Circle

43. $x = 3 \cos 0 = 3$

$y = 2 \sin^2 0 = 0$

44. $x = 3 \cos \theta, y = 2 \sin^2 \theta$

$\theta = \frac{\pi}{3}$

$x = 3 \cos \frac{\pi}{3} = \frac{3}{2}, y = 2 \sin^2 \frac{\pi}{3} = \frac{3}{2}$

45. $x = 3 \cos \frac{\pi}{6} = \frac{3\sqrt{3}}{2}$

$y = 2 \sin^2 \frac{\pi}{6} = \frac{1}{2}$

46. $x = 3 \cos \theta, y = 2 \sin^2 \theta$

$\theta = -\frac{\pi}{4}$

$x = 3 \cos\left(-\frac{\pi}{4}\right) = \frac{3\sqrt{2}}{2}, y = 2 \sin^2\left(-\frac{\pi}{4}\right) = 1$

47. $x = 2t \Rightarrow \frac{x}{2} = t$

$y = 4t \Rightarrow y = 4\left(\frac{x}{2}\right) = 2x$

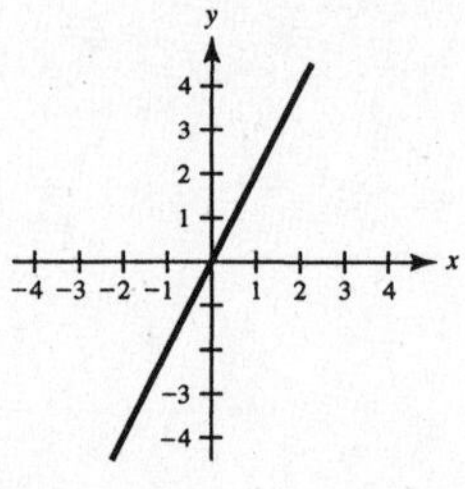

48. $x = 1 + 4t, y = 2 - 3t$

$t = \frac{x-1}{4}$

$y = 2 - 3\left(\frac{x-1}{4}\right)$

$3x + 4y = 11$

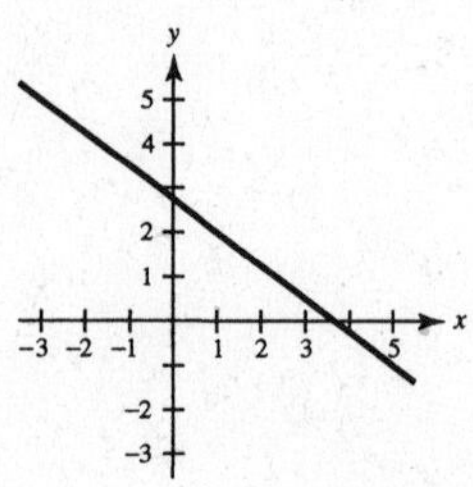

49. $x = t^2, x \geq 0$

$y = \sqrt{t} \Rightarrow y^2 = t$

$x = (y^2)^2 \Rightarrow x = y^4 \Rightarrow y = \sqrt[4]{x}$

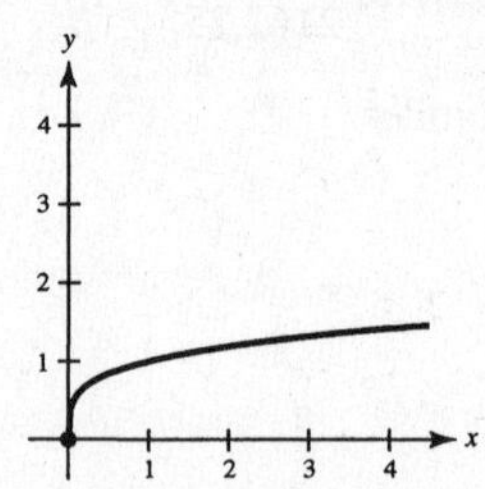

50. $x = t + 4, y = t^2$

$t = x - 4$

$y = (x - 4)^2$

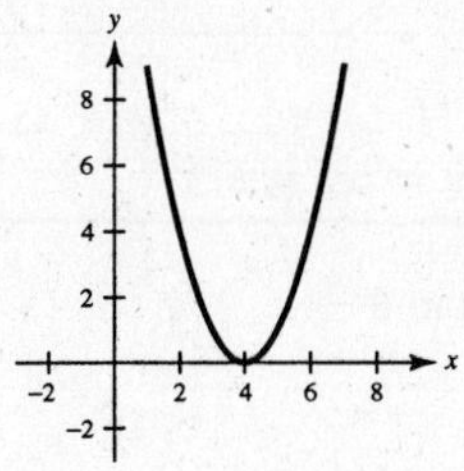

51. $x = 6 \cos \theta, y = 6 \sin \theta$

$\cos \theta = \frac{x}{6}, \sin \theta = \frac{y}{6}$

$\frac{x^2}{36} + \frac{y^2}{36} = 1$

$x^2 + y^2 = 36$

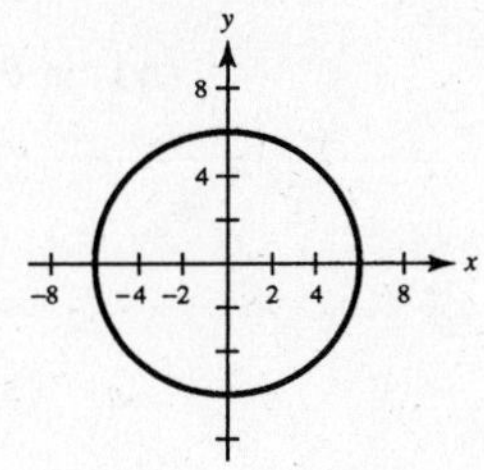

52. $x = 3 + 3 \cos \theta, y = 2 + 5 \sin \theta$

$\cos \theta = \frac{x - 3}{3}, \sin \theta = \frac{y - 2}{5}$

$\frac{(x - 3)^2}{9} + \frac{(y - 2)^2}{25} = 1$

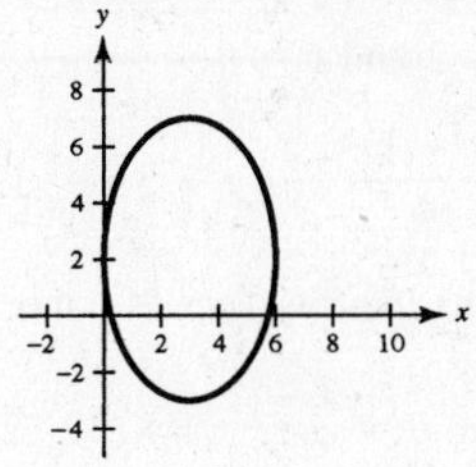

53. $x = t^2 - 4t$

$y = t^3$

$\frac{dx}{dt} = 2t - 4 = 0 \Rightarrow t = 2$

$\frac{dy}{dt} = 3t^2 \Rightarrow t = 0$

Vertical tangent: $(-4, 8), (t = 2)$

Horizontal tangent: $(0, 0)$ $(t = 0)$

54. $x = 2t^3 - 3t^2$

$y = t^2 + 4t$

$\frac{dx}{dt} = 6t^2 - 6t = 6t(t - 1) = 0 \Rightarrow t = 0,1$

$\frac{dy}{dt} = 2t + 4 = 0 \Rightarrow t = -2$

Vertical tangents: $(0, 0), (-1, 5)$ $(t = 0, 1)$

Horizontal tangent: $(-28, -4)$ $(t = -2)$

55. $x = 2 \sin t$

$y = -\cos t$

$\frac{dx}{dt} = 2 \cos t = 0 \Rightarrow t = \frac{\pi}{2}, \frac{3\pi}{2}$

$\frac{dy}{dt} = \sin t = 0 \Rightarrow t = 0, \pi$

Vertical tangents: $(2, 0)$ $(-2, 0)$

Horizontal tangents: $(0, -1), (0, 1)$

56. $x = 1 + \cos t$

$y = 1 - 2 \sin t$

$\frac{dx}{dt} = -\sin t = 0 \Rightarrow t = 0, \pi$

$\frac{dy}{dt} = -2 \cos t = 0 \Rightarrow t = \frac{\pi}{2}, \frac{3\pi}{2}$

Vertical tangents: $(2, 1), (0, 1)$

Horizontal tangents: $(1, -1), (1, 3)$

57. $(h, k) = (-3, 4)$

$2a = 8 \Rightarrow a = 4$

$2b = 6 \Rightarrow b = 3$

$\frac{(x + 3)^2}{16} + \frac{(y - 4)^2}{9} = 1$

$x = -3 + 4 \cos \theta$

$y = 4 + 3 \sin \theta$

This solution is not unique.

58. Hyperbola

Vertices: $(0, \pm 4)$

Foci: $(0, \pm 5)$

Center: $(0, 0)$

$a = 4, c = 5, b = \sqrt{c^2 - a^2} = 3$

$x = 3 \tan \theta, y = 4 \sec \theta$

59. Hyperbola

Asymptotes: $y = 3 \pm \frac{1}{2}(x + 1)$

Center: $(-1, 3)$

If the transverse axis is *horizontal*, then $\pm\frac{b}{a} = \pm\frac{1}{2}$.

Let $a = 2$ and $b = 1$.

$$\frac{(x + 1)^2}{4} - (y - 3)^2 = 1$$

Using the identity $\sec^2 \theta - \tan^2 \theta = 1$, we have

$$\sec \theta = \frac{x + 1}{2} \implies x = 2 \sec \theta - 1$$

$$\tan \theta = y - 3 \implies y = \tan \theta + 3.$$

If the transverse axis is *vertical*, then $\pm\frac{a}{b} = \pm\frac{1}{2}$.

Let $a = 1$ and $b = 2$.

$$(y - 3)^2 - \frac{(x + 1)^2}{4} = 1$$

This time, $\sec \theta = y - 3 \implies y = \sec \theta + 3$

and $\tan \theta = \frac{x + 1}{2} \implies x = 2 \tan \theta - 1$.

60. $x = \cos 3\theta + 5 \cos \theta$

$y = \sin 3\theta + 5 \sin \theta$

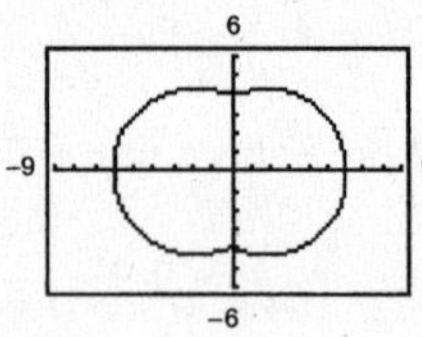

61. Polar coordinates: $\left(2, \frac{\pi}{4}\right)$

$x = 2 \cos \frac{\pi}{4} = \sqrt{2}$

$y = 2 \sin \frac{\pi}{4} = \sqrt{2}$

Rectangular coordinates: $\left(\sqrt{2}, \sqrt{2}\right)$

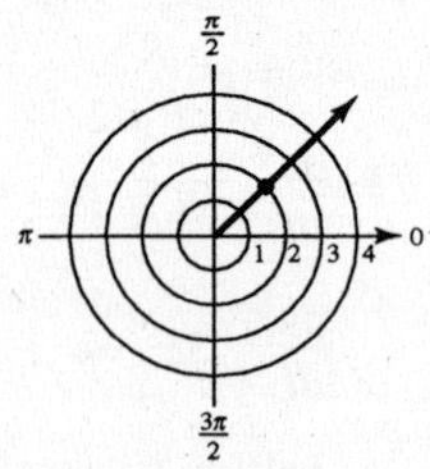

62. Polar coordinates: $\left(-5, -\frac{\pi}{3}\right) = \left(5, \frac{2\pi}{3}\right)$

Rectangular coordinates:

$$\left(5 \cos \frac{2\pi}{3}, 5 \sin \frac{2\pi}{3}\right) = \left(-\frac{5}{2}, \frac{5\sqrt{3}}{2}\right)$$

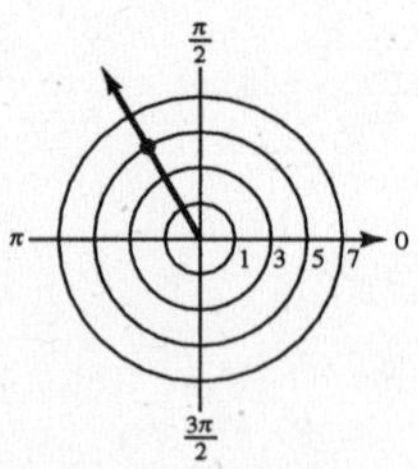

63. Polar coordinates: $(-7, 4.19)$

$x = -7 \cos 4.19 \approx 3.4927$

$y = -7 \sin 4.19 \approx 6.0664$

Rectangular coordinates: $(3.4927, 6.0664)$

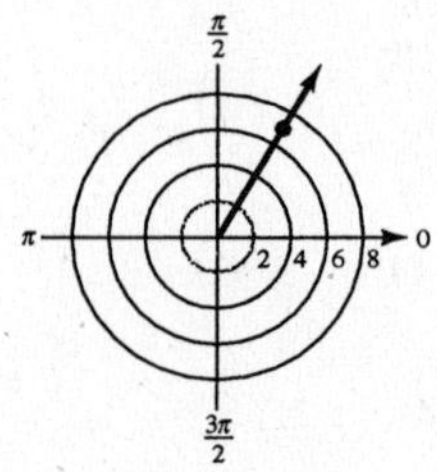

64. Polar coordinates: $\left(\sqrt{3}, 2.62\right)$

Rectangular coordinates: $\left(\sqrt{3} \cos 2.62, \sqrt{3} \sin 2.62\right)$

$\approx (-1.5017, 0.8630)$

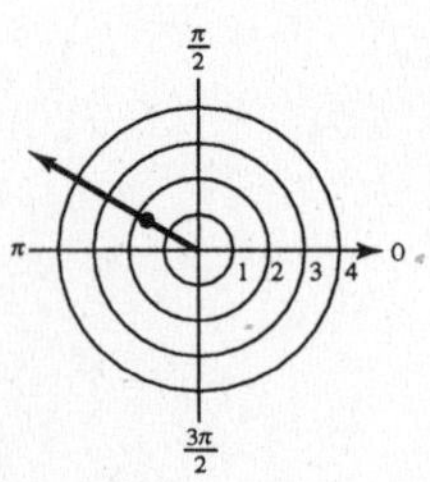

65. Rectangular coordinates: $(0, 2)$

$r = \pm\sqrt{0^2 + 2^2} = \pm 2$

$\tan \theta$ is undefined $\implies \theta = \frac{\pi}{2}, \frac{3\pi}{2}$

Polar coordinates: $\left(2, \frac{\pi}{2}\right), \left(-2, \frac{3\pi}{2}\right), \left(2, \frac{5\pi}{2}\right)$

66. Rectangular coordinates: $\left(-\sqrt{5}, \sqrt{5}\right)$

Polar coordinates:

$r = \sqrt{\left(-\sqrt{5}\right)^2 + \left(\sqrt{5}\right)^2} = \sqrt{10}$

$\tan\theta = -1, \theta = \dfrac{3\pi}{4}$

$\left(\sqrt{10}, \dfrac{3\pi}{4}\right), \left(-\sqrt{10}, \dfrac{7\pi}{4}\right)$

67. Rectangular coordinates: (4, 6)

$r = \pm\sqrt{4^2 + 6^2} = \pm\sqrt{52} \approx \pm 7.2111$

$\tan\theta = \dfrac{6}{4} \implies \theta \approx 0.9828, 4.1244$

Polar coordinates: (7.2111, 0.9828), (−7.2111, 4.1244), (7.2111, 7.2660)

68. Rectangular coordinates: (3, −4)

Polar coordinates:

$r = \sqrt{3^2 + (-4)^2} = 5$

$\tan\theta = -\dfrac{4}{3}, \theta \approx -0.9273$

(5, 5.356), (−5, 2.214)

69.
$$r = 3\cos\theta$$
$$r^2 = 3r\cos\theta$$
$$x^2 + y^2 = 3x$$

70.
$$r = 10$$
$$r^2 = 100$$
$$x^2 + y^2 = 100$$

71. $r = \dfrac{2}{1 + \sin\theta}$
$$r(1 + \sin\theta) = 2$$
$$r + r\sin\theta = 2$$
$$\pm\sqrt{x^2 + y^2} + y = 2$$
$$\pm\sqrt{x^2 + y^2} = 2 - y$$
$$x^2 + y^2 = (2 - y)^2$$
$$x^2 + y^2 = 4 - 4y + y^2$$
$$x^2 + 4y - 4 = 0$$

72.
$$r^2 = \cos 2\theta$$
$$r^2 = \left(\frac{x}{r}\right)^2 - \left(\frac{y}{r}\right)^2$$
$$(x^2 + y^2)^2 = x^2 - y^2$$

73. $(x^2 + y^2)^2 = ax^2y$
$$(r^2)^2 = ar^2\cos^2\theta\, r\sin\theta$$
$$r = a\cos^2\theta\sin\theta$$

74. $x^2 + y^2 - 4x = 0$
$$r^2 - 4r\cos\theta = 0$$
$$r = 4\cos\theta$$

75. $xy = 5$
$$(r\cos\theta)(r\sin\theta) = 5$$
$$r^2 = \frac{5}{\cos\theta\sin\theta}$$

76. $xy = -2$
$$(r\cos\theta)(r\sin\theta) = -2$$
$$r^2 = \frac{-2}{\cos\theta\sin\theta}$$

77. $r = 4$

Circle of radius 4 centered at the pole

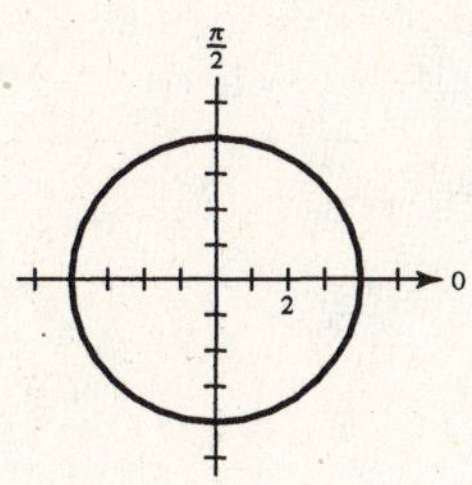

78. $r = 2\theta$

Symmetric with respect to $\theta = \dfrac{\pi}{2}$

Spiral

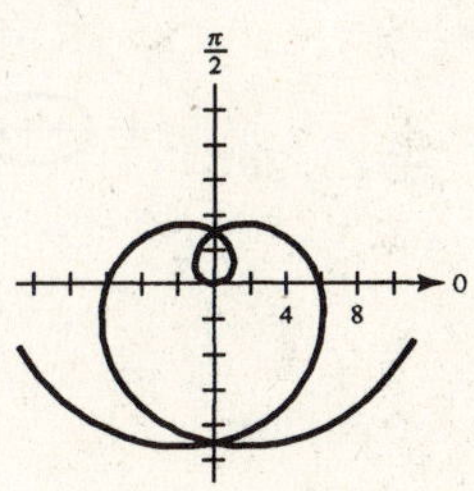

79. $r = 4\sin 2\theta$

Symmetric with respect to $\theta = \pi/2$, the polar axis, and the pole.

Rose curve ($n = 2$) with 4 petals

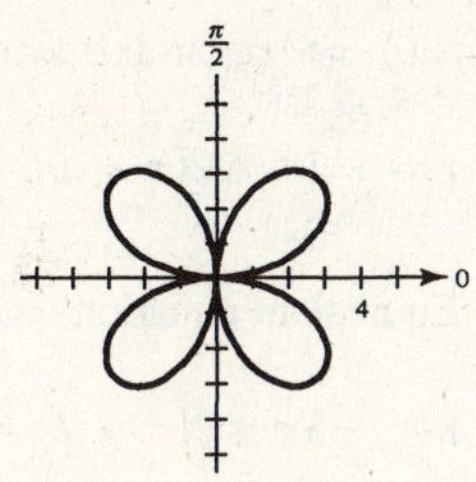

80. $r = \cos 5\theta$

Symmetric with respect to polar axis

Rose curve $(n = 5)$ with five petals

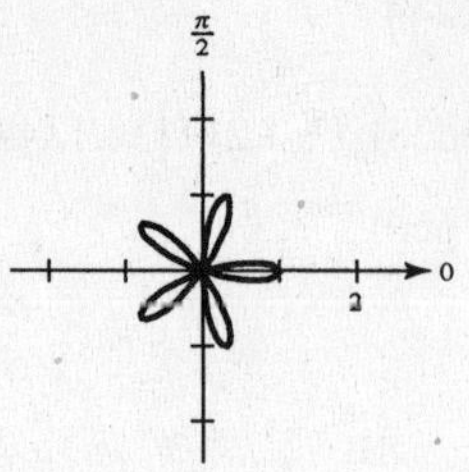

81. $r = -2(1 + \cos\theta)$

Symmetric with respect to the polar axis

$\frac{a}{b} = \frac{2}{2} = 1 \Rightarrow$ Cardioid

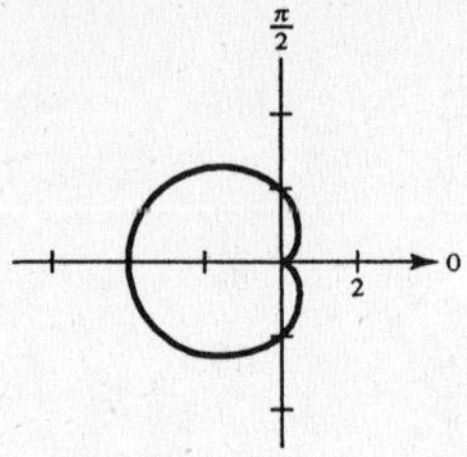

82. $r = 3 - 4\cos\theta$

Symmetric with respect to polar axis

$\frac{a}{b} = \frac{3}{4} < 0 \Rightarrow$ Limaçon with inner loop

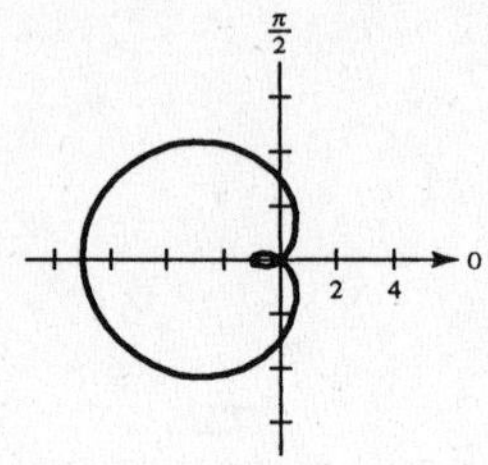

83. $r = 2 + 6\sin\theta$

Limaçon with inner loop

$r = f(\sin\theta) \Rightarrow \theta = \frac{\pi}{2}$ symmetry

Maximum value: $|r| = 8$ when $\theta = \frac{\pi}{2}$.

Zeros: $2 + 6\sin\theta = 0 \Rightarrow \sin\theta = -\frac{1}{3} \Rightarrow \theta \approx 3.4814, 5.9433$

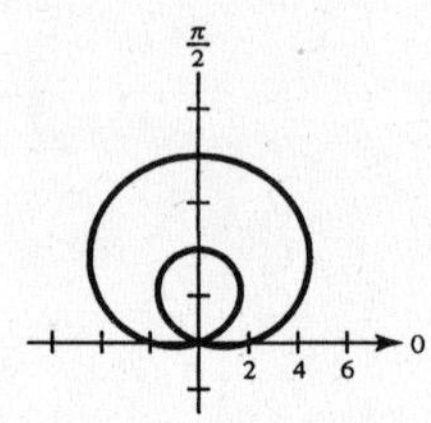

84. $r = 5 - 5\cos\theta$

$r = 5(1 - \cos\theta)$

Symmetry: polar axis

Maximum values of $|r|$: $|r| = 10$ when $\theta = \pi$.

Zeros of r: $r = 0$ when $\theta = 0, 2\pi$.

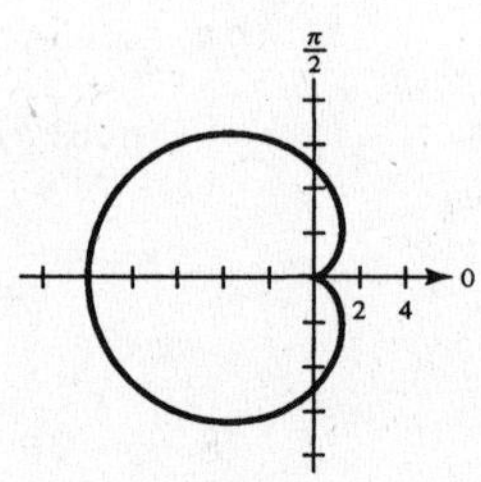

85. $r = -3\cos 2\theta$

Rose curve with 4 petals

$r = f(\cos\theta) \Rightarrow$ polar axis symmetry

$\theta = \frac{\pi}{2}$: $r = -3\cos 2(\pi - \theta) = -3\cos(2\pi - 2\theta) = -3\cos 2\theta$

Equivalent equation $\Rightarrow \theta = \frac{\pi}{2}$ symmetry

Pole: $r = -3\cos 2(\pi + \theta) = -3\cos(2\pi + 2\theta) = -3\cos 2\theta$

Equivalent equation $\Rightarrow$ pole symmetry

Maximum value: $|r| = 3$ when $\theta = 0, \frac{\pi}{2}, \pi, \frac{3\pi}{2}$.

Zeros: $-3\cos 2\theta = 0$ when $\cos 2\theta = 0 \Rightarrow \theta = \frac{\pi}{4}, \frac{3\pi}{4}, \frac{5\pi}{4}, \frac{7\pi}{4}$.

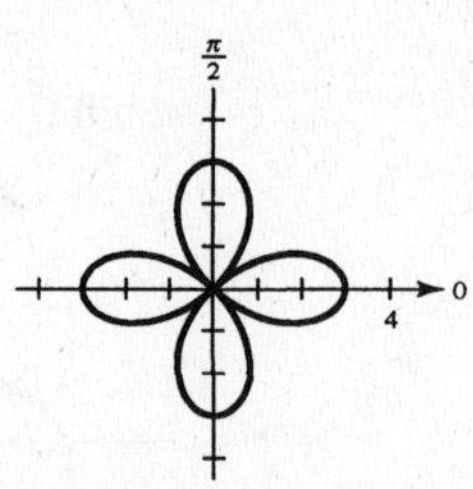

86. $r^2 = \cos 2\theta$

Symmetry: polar axis, $\theta = \dfrac{\pi}{2}$, pole

Maximum values of $|r|$: $|r| = 1$ when $\theta = 0, \pi$.

Zeros of r: $r = 0$ when $\theta = \dfrac{\pi}{4}, \dfrac{3\pi}{4}$.

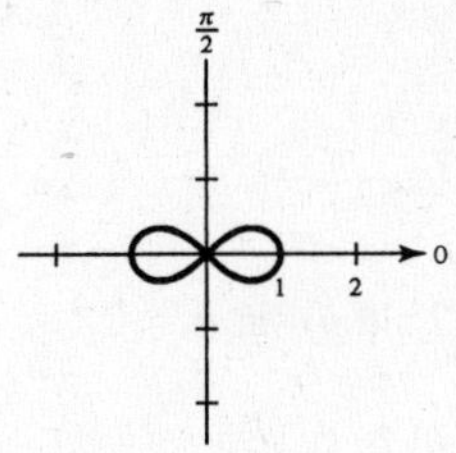

87. $r = 3(2 - \cos \theta)$

$= 6 - 3 \cos \theta$

$\dfrac{a}{b} = \dfrac{6}{3} = 2$

The graph is a convex limaçon.

88. $r = 3(1 - 2 \cos \theta)$

$r = 3 - 6 \cos \theta$

Limaçon with inner loop.

89. $r = 4 \cos 3\theta$

The graph is a rose curve with 3 petals.

90. $r^2 = 9 \cos 2\theta$

Lemniscate

91. $r = 2(1 + \sin \theta),\ \theta = \dfrac{\pi}{2}$

$$\frac{dy}{dx} = \frac{2(1 + \sin \theta) \cos \theta + (2 \cos \theta) \sin \theta}{-2(1 + \sin \theta) \sin \theta + (2 \cos \theta) \cos \theta}$$

$$= \frac{2 \sin \theta \cos \theta + \cos \theta}{\cos^2 \theta - \sin^2 \theta - \sin \theta}$$

At $\theta = \dfrac{\pi}{2}, \dfrac{dy}{dx} = 0$.

92. $r = 2(1 - \cos \theta),\ \theta = \dfrac{3\pi}{2}$

$$\frac{dy}{dx} = \frac{2(1 - \cos \theta) \cos \theta + (2 \sin \theta) \sin \theta}{-2(1 - \cos \theta) \sin \theta + 2 \sin \theta \cos \theta}$$

$$= \frac{\sin^2 \theta - \cos^2 \theta + \cos \theta}{2 \sin \theta \cos \theta - \sin \theta}$$

At $\theta = \dfrac{3\pi}{2}, \dfrac{dy}{dx} = \dfrac{1}{1} = 1$.

93. $r = 3 + 2 \cos \theta,\ \theta = \pi$

$$\frac{dy}{dx} = \frac{(3 + 2 \cos \theta) \cos \theta + (-2 \sin \theta) \sin \theta}{-(3 + 2 \cos \theta) \sin \theta + (-2 \sin \theta) \cos \theta}$$

$$= \frac{2 \cos^2 \theta - 2 \sin^2 \theta + 3 \cos \theta}{-4 \sin \theta \cos \theta - 3 \sin \theta}$$

At $\theta = \pi, \dfrac{dy}{dx}$ is undefined.

94. $r = 3(1 - \cos \theta),\ \theta = 0$

$$\frac{dy}{dx} = \frac{3(1 - \cos \theta) \cos \theta + (3 \sin \theta) \sin \theta}{-3(1 - \cos \theta) \sin \theta + (3 \sin \theta) \cos \theta}$$

At $\theta = 0, \dfrac{dy}{dx}$ is undefined.

95. $r = 1 - 2\sin\theta$

$x = r\cos\theta = (1 - 2\sin\theta)\cos\theta = \cos\theta - 2\sin\theta\cos\theta$

$y = r\sin\theta = (1 - 2\sin\theta)\sin\theta = \sin\theta - 2\sin^2\theta$

$$\frac{dx}{d\theta} = -\sin\theta + 2\sin^2\theta - 2\cos^2\theta$$

$$= -\sin\theta + 2\sin^2\theta - 2(1 - \sin^2\theta)$$

$$= 4\sin^2\theta - \sin\theta - 2 = 0$$

$$\sin\theta = \frac{1 \pm \sqrt{33}}{8} \approx 0.84307, -0.59307$$

$\theta \approx 1.0030, 2.1386, 3.7765, 5.6483$

$$\frac{dy}{d\theta} = \cos\theta - 4\sin\theta\cos\theta = \cos\theta(1 - 4\sin\theta) = 0$$

$$\cos\theta = 0 \Rightarrow \theta = \frac{\pi}{2}, \frac{3\pi}{2}$$

$$\sin\theta = \frac{1}{4} \Rightarrow \theta = 0.2527, 2.8889$$

Vertical tangents: $(-0.6861, 1.0030), (-0.6861, 2.1386),$
$(2.1861, 3.7765), (2.1861, 5.6483)$

Horizontal tangents: $\left(-1, \frac{\pi}{2}\right), \left(3, \frac{3\pi}{2}\right),$
$\left(\frac{1}{2}, 0.2527\right), \left(\frac{1}{2}, 2.8889\right)$

96. $r = 1 + 2\cos\theta$

$x = r\cos\theta = (1 + 2\cos\theta)\cos\theta$

$y = r\sin\theta = (1 + 2\cos\theta)\sin\theta$

$$\frac{dx}{d\theta} = (1 + 2\cos\theta)(-\sin\theta) + (-2\sin\theta)\cos\theta = 0$$

$$= \sin\theta(-4\cos\theta - 1) = 0$$

$\sin\theta = 0 \Rightarrow \theta = 0, \pi$

$$\cos\theta = -\frac{1}{4} \Rightarrow 1.8235, 4.4597$$

$$\frac{dy}{d\theta} = (1 + 2\cos\theta)\cos\theta + (-2\sin\theta)\sin\theta = 0$$

$\cos\theta + 2\cos^2\theta - 2(1 - \cos^2\theta) = 0$

$4\cos^2\theta + \cos\theta - 2 = 0$

$$\cos\theta = \frac{1}{8}(-1 \pm \sqrt{33})$$

$\theta \approx 0.9359, 5.3473, 3.7094, 2.5738$

Vertical tangents: $(3, 0), (-1, \pi), \left(\frac{1}{2}, 1.8235\right), \left(\frac{1}{2}, 4.4597\right)$

Horizontal tangents: $(2.1861, 0.9359), (2.1861, 5.3473)$
$(-0.6861, 3.7094), (-0.6861, 2.5738)$

97. $r = \cos\theta, 0 \le \theta \le \pi$

$x = r\cos\theta = \cos^2\theta$

$y = r\sin\theta = \cos\theta\sin\theta$

$$\frac{dx}{d\theta} = -2\cos\theta\sin\theta = 0 \Rightarrow \theta = 0, \frac{\pi}{2}$$

$$\frac{dy}{d\theta} = \cos^2\theta - \sin^2\theta = 0 \Rightarrow \theta = \frac{\pi}{4}, \frac{3\pi}{4}$$

Vertical tangents: $(1, 0), \left(0, \frac{\pi}{2}\right)$

Horizontal tangents: $\left(\frac{\sqrt{2}}{2}, \frac{\pi}{4}\right)\left(-\frac{\sqrt{2}}{2}, \frac{3\pi}{4}\right)$

98. $r = 1 + \cos\theta$

$x = r\cos\theta = (1 + \cos\theta)\cos\theta$

$y = r\sin\theta = (1 + \cos\theta)\sin\theta$

$$\frac{dx}{d\theta} = (-\sin\theta)\cos\theta + (1 + \cos\theta)(-\sin\theta) = 0$$

$\sin\theta(1 + 2\cos\theta) = 0$

$\sin\theta = 0 \Rightarrow \theta = 0, \pi$

$$1 + 2\cos\theta = 0 \Rightarrow \cos\theta = -\frac{1}{2} \Rightarrow \theta = \frac{2\pi}{3}, \frac{4\pi}{3}$$

$$\frac{dy}{d\theta} = (-\sin\theta)\sin\theta + (1 + \cos\theta)\cos\theta = 0$$

$\cos^2\theta + \cos\theta - (1 - \cos^2\theta) = 0$

$2\cos^2\theta + \cos\theta - 1 = 0$

$(2\cos\theta - 1)(\cos\theta + 1) = 0$

$$\cos\theta = \frac{1}{2} \Rightarrow \theta = \frac{\pi}{3}, \frac{5\pi}{3}$$

$\cos\theta = -1 \Rightarrow \theta = \pi$

Vertical tangents: $(2, 0), \left(\frac{1}{2}, \frac{2\pi}{3}\right), \left(\frac{1}{2}, \frac{4\pi}{3}\right)$

Horizontal tangents: $\left(\frac{3}{2}, \frac{\pi}{3}\right), \left(\frac{3}{2}, \frac{5\pi}{3}\right)$

[**Note:** $(0, \pi)$ is a cusp.]

99. $r = \dfrac{1}{1 + 2\sin\theta}, e = 2$

Hyperbola symmetric with respect to $\theta = \dfrac{\pi}{2}$ and having vertices at $\left(\dfrac{1}{3}, \dfrac{\pi}{2}\right)$ and $\left(-1, \dfrac{3\pi}{2}\right)$.

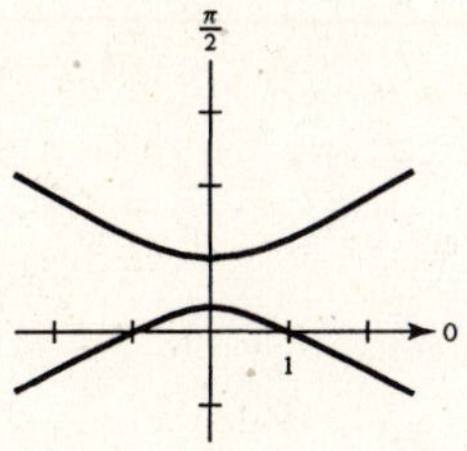

100. $r = \dfrac{2}{1 - \sin\theta}, e = 1$

Parabola symmetric with $\theta = \pi/2$ and the vertex at $(1, 3\pi/2)$.

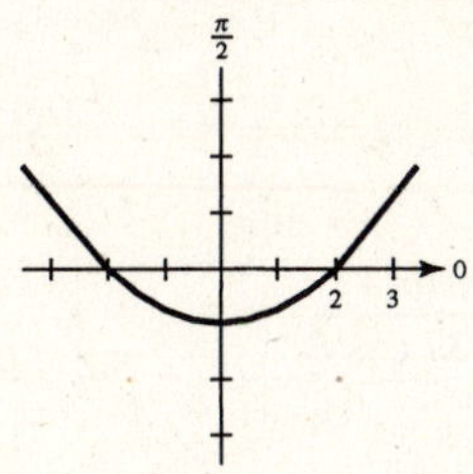

101. $r = \dfrac{4}{5 - 3\cos\theta}$

$r = \dfrac{\frac{4}{5}}{1 - \left(\frac{3}{5}\right)\cos\theta}, e = \dfrac{3}{5}$

Ellipse symmetric with respect to the polar axis and having vertices at $(2, 0)$ and $\left(\dfrac{1}{2}, \pi\right)$.

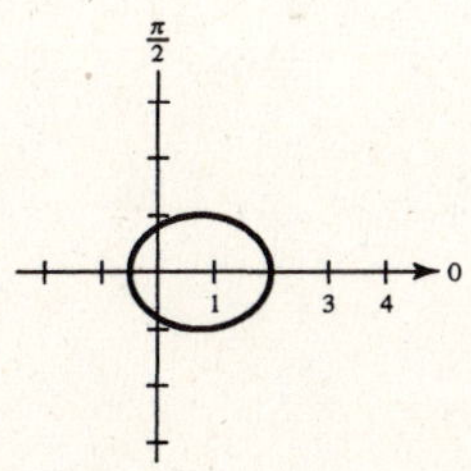

102. $r = \dfrac{16}{4 + 5\cos\theta}$

$r = \dfrac{4}{1 + (5/4)\cos\theta}$

$e = \dfrac{5}{4}$

Hyperbola

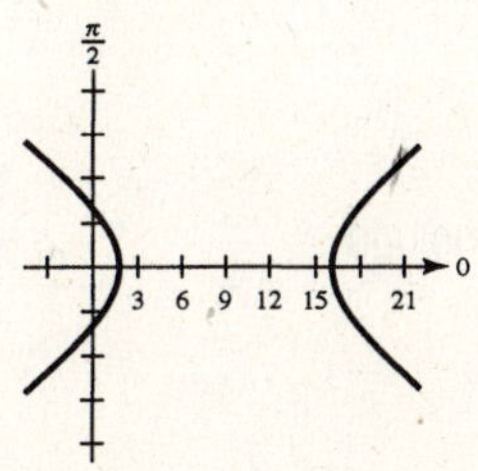

103. Parabola: $r = \dfrac{ep}{1 - e\cos\theta}, e = 1$

Vertex: $(2, \pi)$

Focus: $(0, 0) \implies p = 4$

$r = \dfrac{4}{1 - \cos\theta}$

104. Parabola: $r = \dfrac{ep}{1 + e\sin\theta}, \quad e = 1$

Vertex: $\left(2, \dfrac{\pi}{2}\right)$

Focus: $(0, 0) \implies p = 4$

$r = \dfrac{4}{1 + \sin\theta}$

105. Ellipse: $r = \dfrac{ep}{1 - e\cos\theta}$

Vertices: $(5, 0), (1, \pi) \implies a = 3$

One focus: $(0, 0) \implies c = 2$

$e = \dfrac{c}{a} = \dfrac{2}{3}, p = \dfrac{5}{2}$

$r = \dfrac{(2/3)(5/2)}{1 - (2/3)\cos\theta} = \dfrac{5/3}{1 - (2/3)\cos\theta} = \dfrac{5}{3 - 2\cos\theta}$

106. Hyperbola: $r = \dfrac{ep}{1 + e\cos\theta}$

Vertices: $(1, 0), (7, 0) \implies a = 3$

One focus: $(0, 0) \implies c = 4$

$e = \dfrac{c}{a} = \dfrac{4}{3}, p = \dfrac{7}{4}$

$r = \dfrac{(4/3)(7/4)}{1 + (4/3)\cos\theta} = \dfrac{7/3}{1 + (4/3)\cos\theta} = \dfrac{7}{3 + 4\cos\theta}$

107. $a + c = 122{,}800 + 4000 = 126{,}800$

$a - c = \quad 119 + 4000 = \quad 4{,}119$

$2a = 130{,}919$

$a = 65{,}459.5$

$c = 61{,}340.5$

$$e = \frac{c}{a} = \frac{61{,}340.5}{65{,}459.5} \approx 0.937$$

$$r = \frac{ep}{1 - e\cos\theta} \approx \frac{0.937p}{1 - 0.937\cos\theta}$$

$r = 126{,}800$ when $\theta = 0$

$$126{,}800 = \frac{ep}{1 - e\cos 0} = \frac{ep}{1 - e}$$

$ep = 126{,}800(1 - e) \approx 7978.81$

Thus, $r = \dfrac{7978.81}{1 - 0.937\cos\theta}$

When $\theta = \dfrac{\pi}{3}$, $r \approx 15{,}011.9$

The distance from the surface of the earth and the satellite is $15{,}011.9 - 4000 \approx 11{,}011.9$ miles.

108. Parabola: $r = \dfrac{ep}{1 + e\sin\theta}$, $e = 1$

Vertex: $\left(6{,}000{,}000, \dfrac{\pi}{2}\right)$

Focus: $(0, 0) \Rightarrow p = 12{,}000{,}000$

$$r = \frac{12{,}000{,}000}{1 + \sin\theta}$$

$$\theta = -\frac{\pi}{3}$$

$r \approx 89{,}600{,}000$ miles

Problem Solving for Chapter 12

1. (a) $A = 2\displaystyle\int_0^b 2\sqrt{py}\,dy = 4\sqrt{p}\left(\frac{2}{3}y^{3/2}\right)\Big]_0^b = \frac{8}{3}p^{1/2}b^{3/2}$

(b) If $p = 2$ and $b = 4$, then $A = \dfrac{8}{3}\sqrt{2}(4)^{3/2} = \dfrac{64\sqrt{2}}{3}$ square units.

(c) As p approaches zero, the parabola becomes narrower and narrower, thus the area becomes smaller and smaller.

2. For $\dfrac{x^2}{a^2} + \dfrac{y^2}{b^2} = 1$, we have $c^2 = a^2 - b^2$

When $x = c$: $\dfrac{c^2}{a^2} + \dfrac{y^2}{b^2} = 1$

$$y^2 = b^2\left(1 - \frac{a^2 - b^2}{a^2}\right) = \frac{b^4}{a^2}$$

$$y = \frac{b^2}{a}$$

Length of latus rectum: $2y = \dfrac{2b^2}{a}$

3. $x = \dfrac{4t}{1+t^3},\ y = \dfrac{4t^2}{1+t^3}$

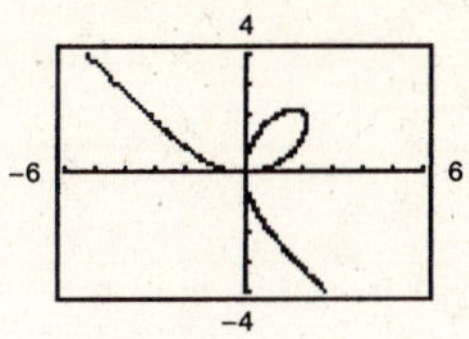

(a) $x^3 + y^3 = 4xy$

(b) $\dfrac{dy}{dt} = \dfrac{(1+t^3)(8t) - 4t^2(3t^2)}{(1+t^3)^2}$

$= \dfrac{4t(2-t^3)}{(1+t^3)^2} = 0$ when $t = 0$ or $t = \sqrt[3]{2}$.

Points: $(0, 0), \left(\dfrac{4\sqrt[3]{2}}{3}, \dfrac{4\sqrt[3]{4}}{3}\right) \approx (1.6799, 2.1165)$

(c) Since $\dfrac{dy}{dx} = \dfrac{dy/dt}{dx/dt}$, the points of horizontal tangency occur when $\dfrac{dy}{dt} = 0$ (but $\dfrac{dx}{dt} \neq 0$).

$\dfrac{dx}{dt} = \dfrac{4(1-2t^3)}{(1+t^3)^2}$

$\dfrac{dy}{dt} = \dfrac{-4t(t^3-2)}{(t^3+1)^2} = 0$ when $t = 0$ or $t = \sqrt[3]{2}$.

When $t = 0$: $x = 0$, $y = 0$, and $\dfrac{dx}{dt} = 4$.

When $t = \sqrt[3]{2}$: $x = \dfrac{4\sqrt[3]{2}}{3}$, $y = \dfrac{4\sqrt[3]{4}}{3}$, and $\dfrac{dx}{dt} = -\dfrac{4}{3}$.

4. Center: $(0, 0) = (h, k)$

Focus: $(24, 0) \Rightarrow c = 24$

Solution point: $(24, 24)$

$24^2 = a^2 + b^2 \Rightarrow b^2 = 24^2 - a^2$

$\dfrac{(x-h)^2}{a^2} - \dfrac{(y-k)^2}{b^2} = 1$

$\dfrac{x^2}{a^2} - \dfrac{y^2}{24^2 - a^2} = 1 \Rightarrow \dfrac{24^2}{a^2} - \dfrac{24^2}{24^2 - a^2} = 1$

Solving yields $a^2 = \dfrac{(3 - \sqrt{5})24^2}{2} \approx 220.0124$ and

$b^2 \approx 355.9876$.

Thus, we have $\dfrac{x^2}{220.0124} - \dfrac{y^2}{355.9876} = 1$.

The right vertex is at $(a, 0) \approx (14.83, 0)$.

5. The coordinates of point (x, y) can be thought of as the sum of two vectors:

From origin to center of small circle:

$\langle 3\cos\theta, 3\sin\theta \rangle$

From center of small circle to point (x, y):

$\langle \cos\beta, \sin\beta \rangle$

Because the small circle rotates by 2θ when its center has rotated by θ, we have $\beta = \pi + 3\theta$.

$x = 3\cos\theta + \cos(\pi + 3\theta) = 3\cos\theta - \cos 3\theta$

$y = 3\sin\theta + \sin(\pi + 3\theta) = 3\sin\theta - \sin 3\theta$

6. When the circle has rolled θ radians, the center is at $(a\theta, a)$.

$\sin\theta = \sin(\pi - \theta)$

$= \dfrac{|AC|}{b} = \dfrac{|BD|}{b} \Rightarrow |BD| = b\sin\theta$

$\cos\theta = -\cos(\pi - \theta)$

$= \dfrac{|AP|}{-b} \Rightarrow |AP| = -b\cos\theta$

Therefore, $x = a\theta - b\sin\theta$ and $y = a - b\cos\theta$.

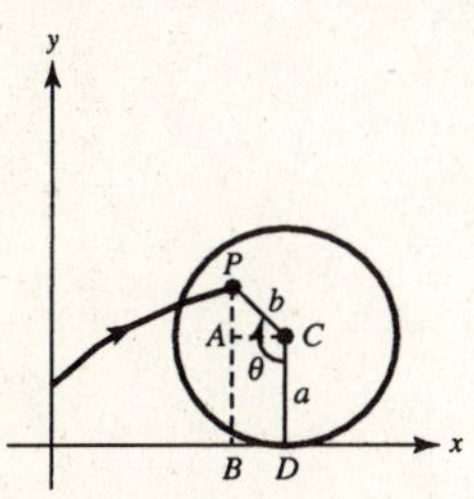

7. (a) $(r_1, \theta_1) = (x_1, y_1)$ where $x_1 = r_1 \cos \theta_1$ and $y_1 = r_1 \sin \theta_1$.

$(r_2, \theta_2) = (x_2, y_2)$ where $x_2 = r_2 \cos \theta_2$ and $y_2 = r_2 \sin \theta_2$.

$$\begin{aligned} d &= \sqrt{(x_1 - x_2)^2 + (y_1 - y_2)^2} \\ &= \sqrt{x_1^2 - 2x_1x_2 + x_2^2 + y_1^2 - 2y_1y_2 + y_2^2} \\ &= \sqrt{(x_1^2 + y_1^2) + (x_2^2 + y_2^2) - 2(x_1x_2 + y_1y_2)} \\ &= \sqrt{r_1^2 + r_2^2 - 2(r_1r_2 \cos \theta_1 \cos \theta_2 + r_1r_2 \sin \theta_1 \sin \theta_2)} \\ &= \sqrt{r_1^2 + r_2^2 - 2r_1r_2 \cos(\theta_1 - \theta_2)} \end{aligned}$$

(b) If $\theta_1 = \theta_2$, then

$$\begin{aligned} d &= \sqrt{r_1^2 + r_2^2 - 2r_1r_2} \\ &= \sqrt{(r_1 - r_2)^2} \\ &= |r_1 - r_2|. \end{aligned}$$

This represents the distance between two points on the line $\theta = \theta_1 = \theta_2$.

(c) If $\theta_1 - \theta_2 = 90°$, then

$d = \sqrt{r_1^2 + r_2^2}$.

This is the result of the Pythagorean Theorem.

(d) The results should be the same. For example, use the points

$$\left(3, \frac{\pi}{6}\right) \text{ and } \left(4, \frac{\pi}{3}\right).$$

The distance is $d \approx 2.053$.

Now use the representations

$$\left(-3, \frac{7\pi}{6}\right) \text{ and } \left(-4, \frac{4\pi}{3}\right).$$

The distance is still $d \approx 2.053$.

8. $r = 6[1 + \cos(\theta - \phi)]$

(a)

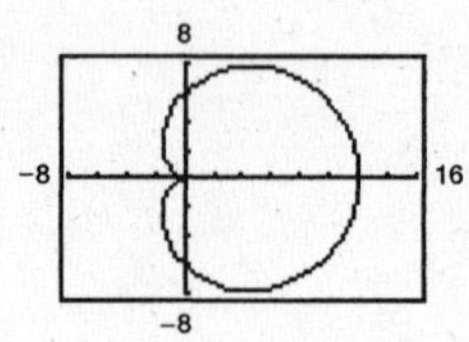

(b)

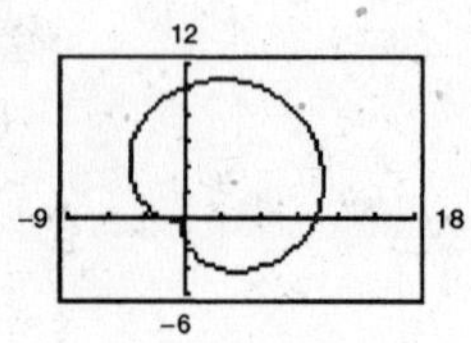

(c)

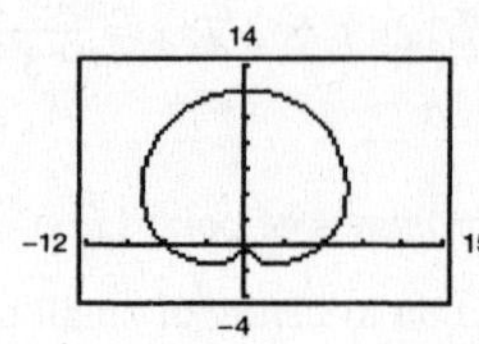

The angle ϕ has the effect of rotating the graph by the angle ϕ. For part (c),

$$\begin{aligned} r &= 6\left[1 + \cos\left(\theta - \frac{\pi}{2}\right)\right] \\ &= 6(1 + \sin \theta). \end{aligned}$$

9. $r = 3 \sin k\theta$

(a) $r = 3 \sin 1.5\theta$

$0 \le \theta < 4\pi$

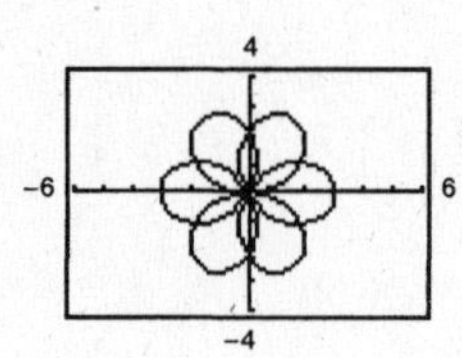

(b) $r = 3 \sin 2.5\theta$

$0 \le \theta < 4\pi$

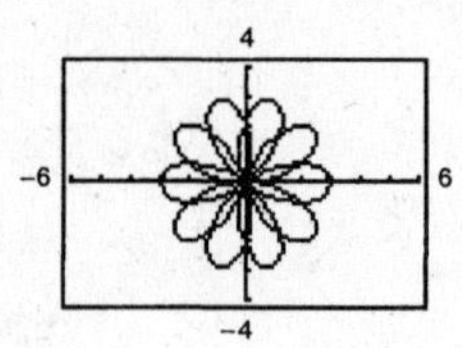

(c) Yes. $r = 3 \sin(k\theta)$.

Find the minimum value of $\theta (\theta > 0)$ that is a multiple of 2π that makes $k\theta$ a multiple of 2π.

10. $y = \overline{QB} - \overline{QA}$

$\overline{QP} = \text{arc } QC = r\theta$

$\overline{QA} = r\theta \sin(90° - \theta)$

$= r\theta \cos \theta$

$\overline{QB} = r \sin \theta$

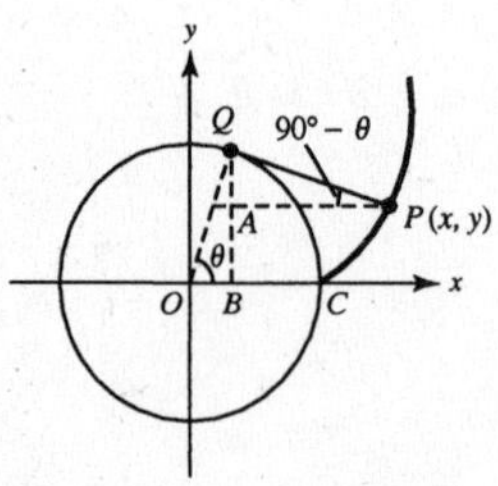

Therefore, $y = r \sin \theta - r\theta \cos \theta = r(\sin \theta - \theta \cos \theta)$.

Similarly, $x = \overline{OB} + \overline{AP}$. Therefore, $x = r \cos \theta + r\theta \sin \theta = r(\cos \theta + \theta \sin \theta)$.

11. $r = e^{\cos\theta} - 2\cos 4\theta + \sin^5\left(\frac{\theta}{12}\right)$

(a) No, the graph appears to have a period of 2π but does not. For example, $r(\pi) \neq r(3\pi)$.

(b) By using the table feature of the calculator we have

$r \approx 4.077$ when $\theta \approx 5.54$ for $0 \le \theta \le 2\pi$ and

$r \approx 4.46$ when $\theta \approx 11.83$ for $0 \le \theta \le 4\pi$.

The graph is not periodic. As θ increases the maximum value of r changes.

12.

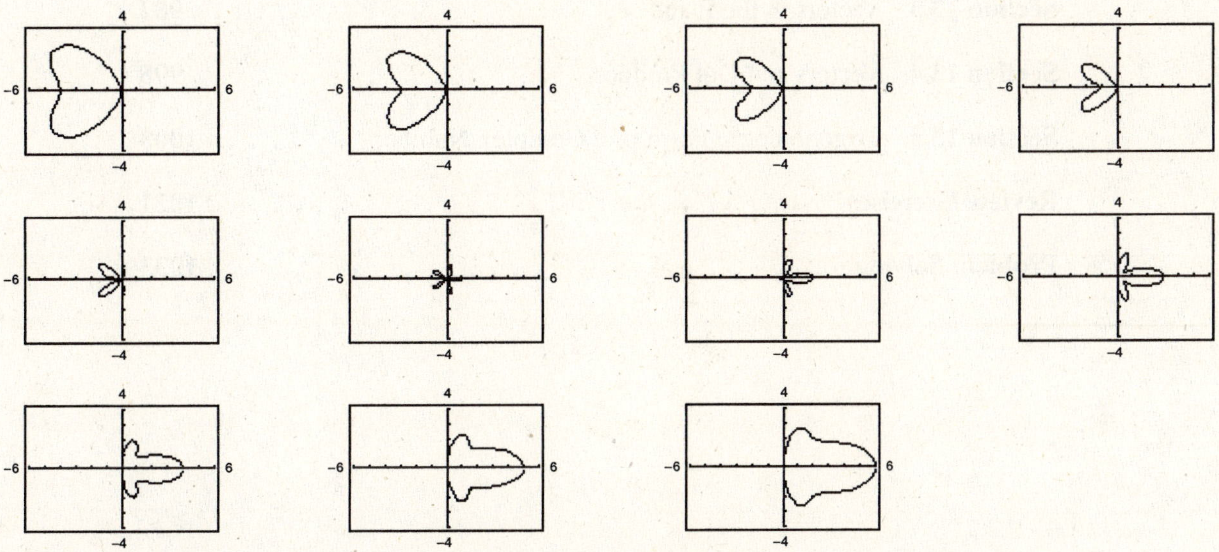

$n = 1, 2, 3, 4, 5$ produce "bells"; $n = -1, -2, -3, -4, -5$ produce "hearts".

13. Since the axis of symmetry is the x-axis, the vertex is $(h, 0)$ and $y^2 = 4p(x - h)$.

Also, since the focus is $(0, 0)$, $0 - h = p \Rightarrow h = -p$ and $y^2 = 4p(x + p)$.

14. Let (x, x) be the corner of the square in Quadrant I.

$A = 4x^2$

$\frac{x^2}{a^2} + \frac{x^2}{b^2} = 1 \Rightarrow x^2 = \frac{a^2b^2}{a^2 + b^2}$

Thus, $A = \frac{4a^2b^2}{a^2 + b^2}$.

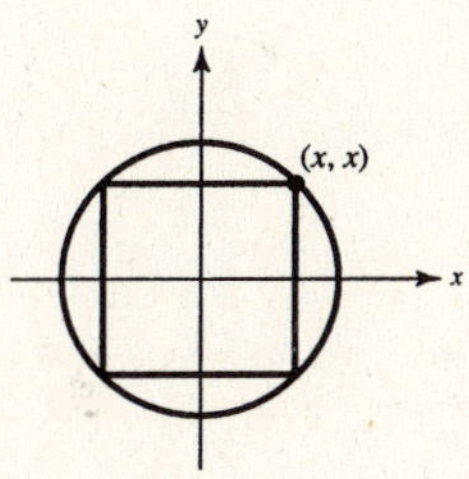

CHAPTER 13
Additional Topics in Trigonometry

Section 13.1 Law of Sines **973**

Section 13.2 Law of Cosines **979**

Section 13.3 Vectors in the Plane **987**

Section 13.4 Vectors and Dot Products **998**

Section 13.5 Trigonometric Form of a Complex Number **1003**

Review Exercises **1021**

Problem Solving **1033**

CHAPTER 13
Additional Topics in Trigonometry

Section 13.1 Law of Sines

1. Given: $A = 30°$, $B = 45°$, $a = 20$

$C = 180° - A - B = 105°$

$$b = \frac{a}{\sin A}(\sin B) = \frac{20 \sin 45°}{\sin 30°} = 20\sqrt{2} \approx 28.28$$

$$c = \frac{a}{\sin A}(\sin C) = \frac{20 \sin 105°}{\sin 30°} \approx 38.64$$

2.

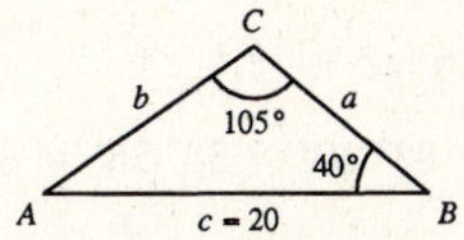

Given: $B = 40°$, $C = 105°$, $c = 20$

$A = 180° - B - C = 35°$

$$a = \frac{c}{\sin C}(\sin A) = \frac{20 \sin 35°}{\sin 105°} \approx 11.88$$

$$b = \frac{c}{\sin C}(\sin B) = \frac{20 \sin 40°}{\sin 105°} \approx 13.31$$

3. Given: $A = 25°$, $B = 35°$, $a = 3.5$

$C = 180° - A - B = 120°$

$$b = \frac{a}{\sin A}(\sin B) = \frac{3.5}{\sin 25°}(\sin 35°) \approx 4.8$$

$$c = \frac{a}{\sin A}(\sin C) = \frac{3.5}{\sin 25°}(\sin 120°) \approx 7.2$$

4.

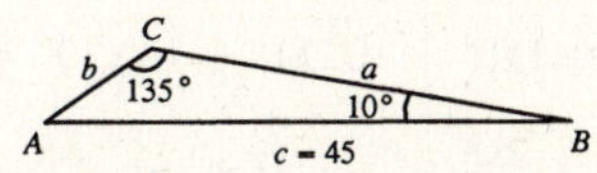

Given: $B = 10°$, $C = 135°$, $c = 45$

$A = 180° - B - C = 35°$

$$a = \frac{c}{\sin C}(\sin A) = \frac{45 \sin 35°}{\sin 135°} \approx 36.50$$

$$b = \frac{c}{\sin C}(\sin B) = \frac{45 \sin 10°}{\sin 135°} \approx 11.05$$

5. Given: $A = 36°$, $a = 8$, $b = 5$

$$\sin B = \frac{b \sin A}{a} = \frac{5 \sin 36°}{8} \approx 0.36737 \implies B \approx 21.55°$$

$C = 180° - A - B \approx 180° - 36° - 21.55° = 122.45°$

$$c = \frac{a}{\sin A}(\sin C) = \frac{8}{\sin 36°}(\sin 122.45°) \approx 11.49$$

6. Given: $A = 60°$, $a = 9$, $c = 10$

$$\sin C = \frac{c \sin A}{a} = \frac{10 \sin 60°}{9} \approx 0.9623 \implies C \approx 74.21° \text{ or } C \approx 105.79°$$

Case 1

$C \approx 74.21°$

$B = 180° - A - C \approx 45.79°$

$$b = \frac{a}{\sin A}(\sin B) \approx \frac{9 \sin 45.79°}{\sin 60°} \approx 7.45$$

Case 2

$C \approx 105.79°$

$B = 180° - A - C \approx 14.21°$

$$b = \frac{a}{\sin A}(\sin B) \approx \frac{9 \sin 14.21°}{\sin 60°} \approx 2.55$$

7. Given: $A = 102.4°, C = 16.7°, a = 21.6$

$B = 180° - A - C = 60.9°$

$$b = \frac{a}{\sin A}(\sin B) = \frac{21.6}{\sin 102.4°}(\sin 60.9°) \approx 19.3$$

$$c = \frac{a}{\sin A}(\sin C) = \frac{21.6}{\sin 102.4°}(\sin 16.7°) \approx 6.4$$

8. Given: $A = 24.3°, C = 54.6°, c = 2.68$

$B = 180° - A - C = 101.1°$

$$a = \frac{c}{\sin C}(\sin A) = \frac{2.68 \sin 24.3°}{\sin 54.6°} \approx 1.35$$

$$b = \frac{c}{\sin C}(\sin B) = \frac{2.68 \sin 101.1°}{\sin 54.6°} \approx 3.23$$

9. Given: $A = 83° 20', \ C = 54.6°, \ c = 18.1$

$B = 180° - A - C = 180° - 83° 20' - 54° 36' = 42° 4'$

$$a = \frac{c}{\sin C}(\sin A) = \frac{18.1}{\sin 54.6°}(\sin 83° 20') \approx 22.05$$

$$b = \frac{c}{\sin C}(\sin B) = \frac{18.1}{\sin 54.6°}(\sin 42° 4') \approx 14.88$$

10. Given: $A = 5° 40', B = 8° 15', b = 4.8$

$C = 180° - A - B = 166° 5'$

$$a = \frac{b}{\sin B}(\sin A) = \frac{4.8 \sin 5° 40'}{\sin 8° 15'} \approx 3.30$$

$$c = \frac{b}{\sin B}(\sin C) = \frac{4.8 \sin 166° 5'}{\sin 8° 15'} \approx 8.05$$

11. Given: $B = 15° 30', \ a = 4.5, \ b = 6.8$

$$\sin A = \frac{a \sin B}{b} = \frac{4.5 \sin 15° 30'}{6.8} \approx 0.17685 \implies A \approx 10° 11'$$

$C = 180° - A - B \approx 180° - 10° 11' - 15° 30' = 154° 19'$

$$c = \frac{b}{\sin B}(\sin C) = \frac{6.8}{\sin 15° 30'}(\sin 154° 19') \approx 11.03$$

12. Given: $B = 2° 45', b = 6.2, c = 5.8$

$$\sin C = \frac{c \sin B}{b} = \frac{5.8 \sin 2° 45'}{6.2} \approx 0.04488 \implies C \approx 2.57°$$

$A = 180° - B - C \approx 174.68°$

$$a = \frac{b}{\sin B}(\sin A) \approx \frac{6.2 \sin 174.68°}{\sin 2° 45'} \approx 11.98$$

13. Given: $C = 145°, \ b = 4, \ c = 14$

$$\sin B = \frac{b \sin C}{c} = \frac{4 \sin 145°}{14} \approx 0.16387 \implies B \approx 9.43°$$

$A = 180° - B - C \approx 180° - 9.43° - 145° = 25.57°$

$$a = \frac{c}{\sin C}(\sin A) \approx \frac{14}{\sin 145°}(\sin 25.57°) \approx 10.53$$

14. Given: $A = 100°, a = 125, c = 10$

$$\sin C = \frac{c \sin A}{a} = \frac{10 \sin 100°}{125} \approx 0.07878 \implies C \approx 4.52°$$

$B = 180° - A - C \approx 75.48°$

$$b = \frac{a}{\sin A}(\sin B) \approx \frac{125 \sin 75.48°}{\sin 100°} \approx 122.87$$

15. Given: $A = 110° 15', \ a = 48, \ b = 16$

$$\sin B = \frac{b \sin A}{a} = \frac{16 \sin 110° 15'}{48} \approx 0.31273 \implies B \approx 18° 13'$$

$C = 180° - A - B \approx 180° - 110° 15' - 18° 13' = 51° 32'$

$$c = \frac{a}{\sin A}(\sin C) = \frac{48}{\sin 110° 15'}(\sin 51° 32') \approx 40.06$$

16. Given: $C = 85° 20', a = 35, c = 50$

$$\sin A = \frac{a \sin C}{c} = \frac{35 \sin 85° 20'}{50} \approx 0.6977 \implies A \approx 44.24°$$

$$B = 180° - A - C \approx 50.43°$$

$$b = \frac{c \sin B}{\sin C} \approx \frac{50 \sin 50.43°}{\sin 85° 20'} \approx 38.67$$

17. Given: $A = 55°, B = 42°, c = \frac{3}{4}$

$$C = 180° - A - B = 83°$$

$$a = \frac{c}{\sin C}(\sin A) = \frac{0.75}{\sin 83°}(\sin 55°) \approx 0.62$$

$$b = \frac{c}{\sin C}(\sin B) = \frac{0.75}{\sin 83°}(\sin 42°) \approx 0.51$$

18. Given: $B = 28°, C = 104°, a = 3\frac{5}{8}$

$$A = 180° - B - C = 48°$$

$$b = \frac{a \sin B}{\sin A} = \frac{3\frac{5}{8} \sin 28°}{\sin 48°} \approx 2.29$$

$$c = \frac{a \sin C}{\sin A} = \frac{3\frac{5}{8} \sin 104°}{\sin 48°} \approx 4.73$$

19. Given: $A = 58°, a = 11.4, b = 12.8$

$$\sin B = \frac{b \sin A}{a} = \frac{12.8 \sin 58°}{11.4} \approx 0.9522 \implies B \approx 72.2° \text{ or } B \approx 107.8°$$

Case 1

$$B \approx 72.2°$$

$$C = 180° - A - B \approx 49.8°$$

$$c = \frac{a}{\sin A}(\sin C) \approx \frac{11.4 \sin 49.8°}{\sin 58°} \approx 10.27$$

Case 2

$$B \approx 107.8°$$

$$C = 180° - A - B \approx 14.2°$$

$$c = \frac{a}{\sin A}(\sin C) \approx \frac{11.4 \sin 14.2°}{\sin 58°} \approx 3.30$$

20. Given: $a = 4.5,\ b = 12.8,\ A = 58°$

$h = 12.8 \sin 58° \approx 10.86$

Since $a < h$, no triangle is formed.

21. Given: $a = 18, b = 20, A = 76°$

$h = 20 \sin 76° \approx 19.41$

Since $a < h$, no triangle is formed.

22. Given: $A = 76°, a = 34, b = 21$

$$\sin B = \frac{b \sin A}{a} = \frac{21 \sin 76°}{34} \approx 0.5993 \implies B \approx 36.82°$$

$$C = 180° - A - B \approx 67.18°$$

$$c = \frac{a \sin C}{\sin A} \approx \frac{34 \sin 67.18°}{\sin 76°} \approx 32.30$$

23. Given: $a = 125,\ b = 200,\ A = 110°$

No triangle is formed because A is obtuse and $a < b$.

24. Given: $A = 110°, a = 125, b = 100$

$$\sin B = \frac{b \sin A}{a} = \frac{100 \sin 110°}{125} \approx 0.75175 \implies B \approx 48.74°$$

$$C = 180° - A - B \approx 21.26°$$

$$c = \frac{a \sin C}{\sin A} \approx \frac{125 \sin 21.26°}{\sin 110°} \approx 48.23$$

25. Given: $A = 22°, a = \frac{5}{7}, b = \frac{5}{7}$

$$\sin B = \frac{b \sin A}{a} = \frac{\frac{5}{7} \sin 22°}{\frac{5}{7}} = \sin 22° \implies B = 22°$$

$$C = 180° - A - B = 136°$$

$$c = \frac{a \sin C}{\sin A} = \frac{\frac{5}{7} \sin 136°}{\sin 22°} \approx 1.32$$

26. Given: $a = \frac{5}{12}, b = 1\frac{3}{8}, A = 22°$

$h = \left(1\frac{3}{8}\right) \sin 22° \approx 0.52$

Since $a < h$, no triangle is formed.

27. $\text{Area} = \frac{1}{2}ab \sin C = \frac{1}{2}(4)(6) \sin 120° \approx 10.4$

28. $\text{Area} = \frac{1}{2}ac \sin B = \frac{1}{2}(62)(20) \sin 130° \approx 474.9$

29. $\text{Area} = \frac{1}{2}bc \sin A = \frac{1}{2}(57)(85) \sin 43° 45' \approx 1675.2$

30. $A = 5° 15', b = 4.5, c = 22$

$\text{Area} = \frac{1}{2}bc \sin A = \left(\frac{1}{2}\right)(4.5)(22) \sin 5.25° \approx 4.529$

31. $\text{Area} = \frac{1}{2}ac \sin B = \frac{1}{2}(105)(64) \sin(72°30') \approx 3204.5$

32. $C = 84° 30', a = 16, b = 20$

$\text{Area} = \frac{1}{2}ab \sin C = \left(\frac{1}{2}\right)(16)(20) \sin 84.5° \approx 159.3$

33. Given: $A = 36°, \ a = 5$

(a) One solution if $b \le 5$ or $b = \dfrac{5}{\sin 36°}$

(b) Two solutions if $5 < b < \dfrac{5}{\sin 36°}$

(c) No solution if $b > \dfrac{5}{\sin 36°}$

34. Given: $A = 60°, a = 10$

(a) One solution if $b \le 10$ or $b = \dfrac{10}{\sin 60°}$.

(b) Two solutions if $10 < b < \dfrac{10}{\sin 60°}$.

(c) No solutions if $b > \dfrac{10}{\sin 60°}$.

35. Given: $A = 10°, a = 10.8$

(a) One solution if $b \le 10.8$ or $b = \dfrac{10.8}{\sin 10°}$

(b) Two solutions if $10.8 < b < \dfrac{10.8}{\sin 10°}$

(c) No solution if $b > \dfrac{10.8}{\sin 10°}$

36. Given: $A = 88°, a = 315.6$

(a) One solution if $b \le 315.6$ or $b = \dfrac{315.6}{\sin 88°}$

(b) Two solutions if $315.6 < b < \dfrac{315.6}{\sin 88°}$

(c) No solutions if $b > \dfrac{315.6}{\sin 88°}$

37. Law of Sines

If ABC is a triangle with sides a, b, and c, then

$$\frac{a}{\sin A} = \frac{b}{\sin B} = \frac{c}{\sin C}.$$

38. To use the Law of Sines to solve a right triangle you would need to know the measure of at least one side, and either another side or one of the other angles.

39. $C = 180° - 23° - 94° = 63°$

$h = \dfrac{35}{\sin 63°}(\sin 23°) \approx 15.3$ meters

40. (a)

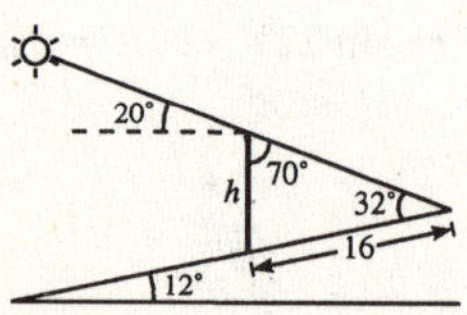

(b) $\dfrac{h}{\sin 32°} = \dfrac{16}{\sin 70°}$

(c) $h = \dfrac{16 \sin 32°}{\sin 70°} \approx 9.0$ meters

41. $\dfrac{\sin(42° - \theta)}{10} = \dfrac{\sin 48°}{17}$

$\sin(42° - \theta) \approx 0.43714$

$42° - \theta \approx 25.9°$

$\theta \approx 16.1°$

42.

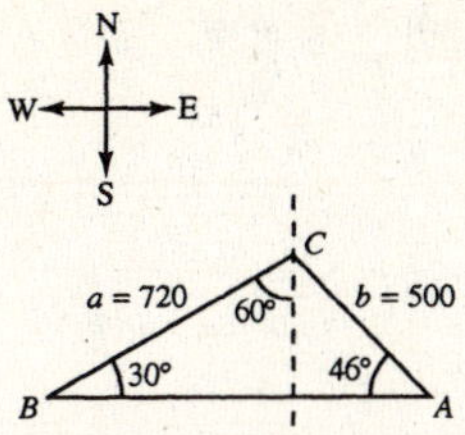

Given: $A = 46°, a = 720, b = 500$

$$\sin B = \frac{b \sin A}{a} = \frac{500 \sin 46°}{720} \approx 0.50 \implies B \approx 30°$$

The bearing from C to B is $S\ 60°\ W$.

43. Given: $c = 100$

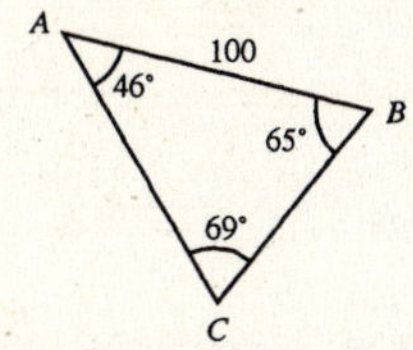

$A = 74° - 28° = 46°,$

$B = 180° - 41° - 74° = 65°,$

$C = 180° - 46° - 65° = 69°$

$$a = \frac{c}{\sin C}(\sin A) = \frac{100}{\sin 69°}(\sin 46°) \approx 77 \text{ meters}$$

44. (a)

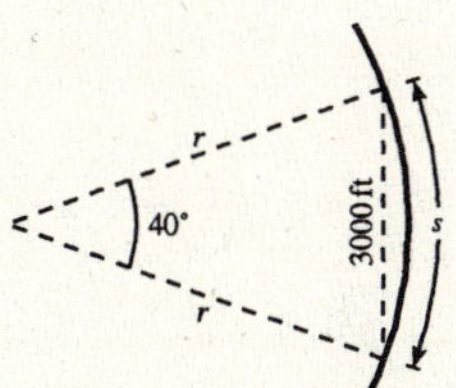

(b) $r = \dfrac{3000 \sin\left[\frac{1}{2}(180° - 40°)\right]}{\sin 40°} \approx 4385.71$ feet

(c) $s \approx 40°\left(\dfrac{\pi}{180°}\right)4385.71 \approx 3061.80$ feet

45. (a)

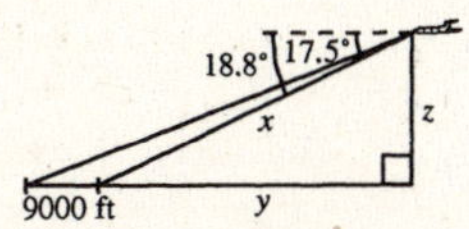

(b) $\dfrac{x}{\sin 17.5°} = \dfrac{9000}{\sin 1.3°}$

$x \approx 119{,}289.1261$ feet ≈ 22.6 miles

(c) $\dfrac{y}{\sin 71.2°} = \dfrac{x}{\sin 90°}$

$y = x \sin 71.2° \approx 119{,}289.1261 \sin 71.2°$

$\approx 112{,}924.963$ feet ≈ 21.4 miles

(d) $z = x \sin 18.8° \approx 119{,}289.1261 \sin 18.8° \approx 38{,}443$ feet

46.

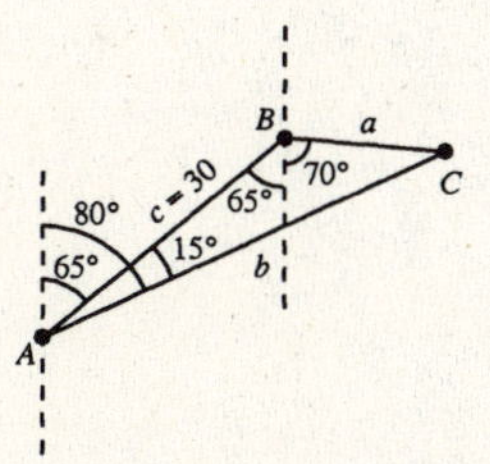

Given: $A = 15°, B = 135°, c = 30$

$C = 180° - A - B = 30°$

From Pine Knob:

$$b = \frac{c \sin B}{\sin C} = \frac{30 \sin 135°}{\sin 30°} \approx 42.4 \text{ kilometers}$$

From Colt Station:

$$a = \frac{c \sin A}{\sin C} = \frac{30 \sin 15°}{\sin 30°} \approx 15.5 \text{ kilometers}$$

47.

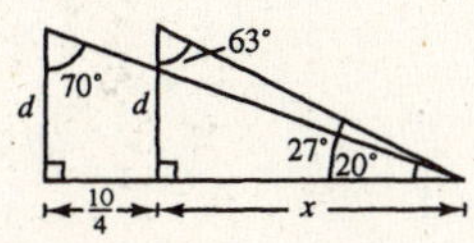

In 15 minutes the boat has traveled

$$(10 \text{ mph})\left(\frac{1}{4} \text{ hr}\right) = \frac{10}{4} \text{ miles.}$$

$\tan 63° = \dfrac{x}{d} \qquad \implies d \tan 63° = x$

$\tan 70° = \dfrac{x + (10/4)}{d} \implies d \tan 70° = x + \dfrac{10}{4}$

$\implies d \tan 70° - \dfrac{10}{4} = x$

$d \tan 70° - \dfrac{10}{4} = d \tan 63°$

$d \tan 70° - d \tan 63° = \dfrac{10}{4}$

$d(\tan 70° - \tan 63°) = 2.5$

$$d = \frac{2.5}{\tan 70° - \tan 63°} \approx 3.2 \text{ miles}$$

48. $A = 90° - 62° = 28°$

$B = 90° + 38° = 128°,\ c = 5$

$C = 180° - 128° - 28° = 24°$

$$a = \frac{c}{\sin C}(\sin A) = \frac{5}{\sin 24°}(\sin 28°) \approx 5.77$$

$d = a \sin(90° - 38°) \approx 5.77 \sin 52° \approx 4.55$ miles

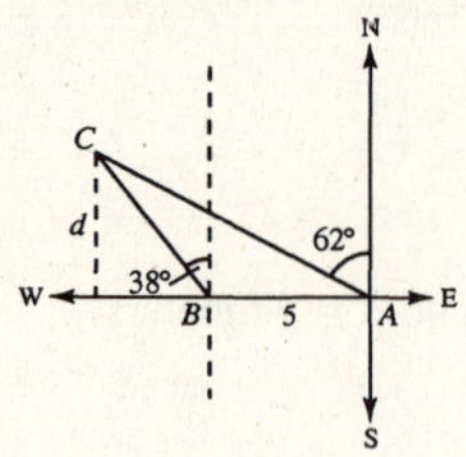

49. $\alpha = 180° - (180° - \phi + \theta) = \phi - \theta$

$$\frac{d}{\sin \theta} = \frac{2}{\sin \alpha}$$

$$d = \frac{2 \sin \theta}{\sin(\phi - \theta)}$$

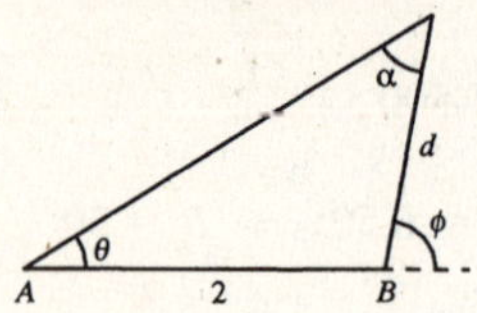

50. (a) $\sin \alpha = \dfrac{5.45}{58.36}$

$\alpha \approx 5.36°$

(b) $\beta = 180° - \theta - (\alpha + 90°) \approx 84.64° - \theta$

(c) $\dfrac{d}{\sin \beta} = \dfrac{58.36}{\sin \theta}$

$$d = \frac{58.36 \sin \beta}{\sin \theta} = \frac{58.36 \sin(84.64° - \theta)}{\sin \theta}$$

(d)

θ	d(meters)
10°	324.08
20°	154.19
30°	95.19
40°	63.80
50°	43.30
60°	28.10

51. True. The longest side of a triangle is always opposite the largest angle.

52. False. Two sides and one opposite angle do not necessarily determine a unique triangle.

53. True.

Assume $C = 90°$. Then $\sin C = 1$

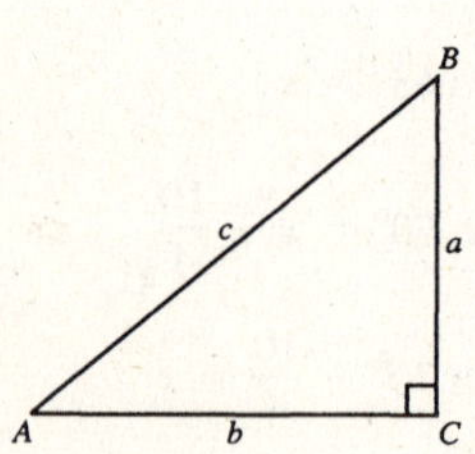

$$\sin A = \frac{a}{c} \Rightarrow c = \frac{a}{\sin A}$$

$$\Rightarrow \frac{a}{\sin A} = \frac{c}{\sin C}$$

$$\sin B = \frac{b}{c} \Rightarrow c = \frac{b}{\sin B}$$

$$\Rightarrow \frac{b}{\sin B} = \frac{c}{\sin C}.$$

Thus, $\dfrac{a}{\sin A} = \dfrac{b}{\sin B} = \dfrac{c}{\sin C}.$

54. False. The area of an oblique triangle is

Area $= \frac{1}{2}ab \sin C$, *not* $\frac{1}{2}ab \sin A$.

55. (a) $A = 30°, a = 15, c = 25$

$$\sin C = \frac{25 \sin 30°}{15} \implies C \approx 56.4427°$$

$$B = 180° - A - C \implies B \approx 93.5573°$$

$$\text{Area} = \frac{1}{2}ac \sin B \approx \frac{1}{2}(15)(25) \sin 93.5573°$$

$$\approx 187.14 \text{ square units}$$

(b) $b = \dfrac{a \sin B}{\sin A} \approx \dfrac{15 \sin 93.5573°}{\sin 30°} \approx 29.94$

The coordinates for point C are as follows:

$x = b \cos A \approx 29.94 \cos 30° \approx 25.93$

$y = b \sin A \approx 29.94 \sin 30° \approx 14.97$

The line between $(0, 0)$ and C is: $y \approx 0.577x$

The line between $(25, 0)$ and C is: $y \approx 16.097(x - 25)$

Solving for x in each equation yields $x \approx 1.733y$ and $x \approx 25 + 0.062y$.

$$\text{Area} \approx \int_0^{14.97} [(25 + 0.062y) - 1.733y]\, dy$$

$$= \int_0^{14.97} (25 - 1.671y) dy$$

$$= (25y - 0.835y^2)\Big]_0^{14.97}$$

$$\approx 187.13 \text{ square units}$$

Note: The answers are slightly different due to rounding errors.

Section 13.2 Law of Cosines

1. Given: $a = 7, b = 10, c = 15$

$$\cos C = \frac{a^2 + b^2 - c^2}{2ab} = \frac{49 + 100 - 225}{2(7)(10)} \approx -0.5429 \implies C \approx 122.88°$$

$$\sin B = \frac{b \sin C}{c} = \frac{10 \sin 122.88°}{15} \approx 0.5599 \implies B \approx 34.05°$$

$$A \approx 180° - 34.05° - 122.88° \approx 23.07°$$

2. Given: $a = 8, b = 3, c = 9$

$$\cos A = \frac{b^2 + c^2 - a^2}{2bc} = \frac{3^2 + 9^2 - 8^2}{2(3)(9)} \approx 0.4815 \implies A \approx 61.2°$$

$$\cos C = \frac{a^2 + b^2 - c^2}{2ab} = \frac{8^2 + 3^2 - 9^2}{2(8)(3)} \approx -0.1667 \implies C \approx 99.6°$$

$$B \approx 180° - 61.2° - 99.6° \approx 19.2°$$

3. Given: $A = 30°,\ b = 15,\ c = 30$

$$a^2 = b^2 + c^2 - 2bc \cos A = 225 + 900 - 2(15)(30) \cos 30° \approx 345.5771$$

$$a \approx 18.6$$

$$\cos B = \frac{a^2 + c^2 - b^2}{2ac} \approx \frac{(18.6)^2 + 900 - 225}{2(18.6)(30)} \approx 0.9148$$

$$B \approx 23.8°$$

$$C \approx 180° - 30° - 23.8° = 126.2°$$

4. Given: $C = 105°, a = 10, b = 4.5$

$$c^2 = a^2 + b^2 - 2ab\cos C = 10^2 + 4.5^2 - 2(10)(4.5)\cos 105° \approx 143.5437 \Rightarrow c \approx 12.0$$

$$\cos B = \frac{a^2 + c^2 - b^2}{2ac} \approx \frac{10^2 + (12.0)^2 - (4.5)^2}{2(10)(12.0)} \approx 0.93229 \Rightarrow B \approx 21.2°$$

$$A = 180° - 105° - 21.3° \approx 53.7°$$

5. $a = 11, b = 14, c = 20$

$$\cos C = \frac{a^2 + b^2 - c^2}{2ab} = \frac{121 + 196 - 400}{2(11)(14)} \approx -0.2695 \Rightarrow C \approx 105.63°$$

$$\sin B = \frac{b\sin C}{c} = \frac{14\sin 105.63°}{20} \approx 0.6741 \Rightarrow B \approx 42.39°$$

$$A \approx 180° - 42.39° - 105.63° \approx 31.98°$$

6. Given: $a = 55, b = 25, c = 72$

$$\cos C = \frac{a^2 + b^2 - c^2}{2ab} = \frac{55^2 + 25^2 - 72^2}{2(55)(25)} \approx -0.5578 \Rightarrow C \approx 123.90°$$

$$\cos A = \frac{b^2 + c^2 - a^2}{2bc} = \frac{25^2 + 72^2 - 55^2}{2(25)(72)} \approx 0.7733 \Rightarrow A \approx 39.35°$$

$$B = 180° - 123.90° - 39.35° \approx 16.75°$$

7. Given: $a = 75.4,\ b = 52,\ c = 52$

$$\cos A = \frac{b^2 + c^2 - a^2}{2bc} = \frac{52^2 + 52^2 - 75.4^2}{2(52)(52)} = -0.05125 \Rightarrow A \approx 92.94°$$

$$\sin B = \frac{b\sin A}{a} \approx \frac{52(0.9987)}{75.4} \approx 0.68875 \Rightarrow B \approx 43.53°$$

$$C = B \approx 43.53°$$

8. Given: $a = 1.42, b = 0.75, c = 1.25$

$$\cos A = \frac{b^2 + c^2 - a^2}{2bc} = \frac{(0.75)^2 + (1.25)^2 - (1.42)^2}{2(0.75)(1.25)} = 0.05792 \Rightarrow A \approx 86.7°$$

$$\cos B = \frac{a^2 + c^2 - b^2}{2ac} = \frac{(1.42)^2 + (1.25)^2 - (0.75)^2}{2(1.42)(1.25)} \approx 0.8497 \Rightarrow B \approx 31.8°$$

$$C = 180° - 86.7° - 31.8° \approx 61.5°$$

9. Given: $A = 135°, b = 4, c = 9$

$$a^2 = b^2 + c^2 - 2bc\cos A = 16 + 81 - 2(4)(9)\cos 135° \approx 147.9117 \Rightarrow a \approx 12.16$$

$$\sin B = \frac{b\sin A}{a} = \frac{4\sin 135°}{12.16} \approx 0.2326 \Rightarrow B \approx 13.45°$$

$$C \approx 180° - 135° - 13.45° \approx 31.55°$$

10. Given: $A = 55°, b = 3, c = 10$

$$a^2 = b^2 + c^2 - 2bc\cos A = 3^2 + 10^2 - 2(3)(10)\cos 55° \approx 74.585 \Rightarrow a \approx 8.64$$

$$\sin B = \frac{b\sin A}{a} \approx \frac{3\sin 55°}{8.64} \approx 0.2844 \Rightarrow A \approx 16.5°$$

$$C \approx 180° - 16.5° - 55° \approx 108.5°$$

11. Given: $B = 10° 35', a = 40, c = 30$

$$b^2 = a^2 + c^2 - 2ac\cos B = 1600 + 900 - 2(40)(30)\cos 10° 35' \approx 140.8268 \implies b \approx 11.9$$

$$\sin C = \frac{c\sin B}{b} = \frac{30\sin 10° 35'}{11.9} \approx 0.4630 \implies C \approx 27.58° \approx 27° 35'$$

$$A \approx 180° - 10° 35' - 27° 35' = 141° 50'$$

12. Given: $B = 75° 20', a = 6.2, c = 9.5$

$$b^2 = a^2 + c^2 - 2ac\cos B = (6.2)^2 + (9.5)^2 - 2(6.2)(9.5)\cos 75\ 20' \approx 98.8636 \implies b \approx 9.94$$

$$\sin A = \frac{a\sin B}{b} \approx \frac{6.2\sin 75° 20'}{9.94} \approx 0.6034 \implies A \approx 37.1°$$

$$C \approx 180° - 75° 20' - 37.1° \approx 67.6°$$

13. Given: $B = 125° 40', \ a = 32, \ c = 32$

$$b^2 = a^2 + c^2 - 2ac\cos B \approx 32^2 + 32^2 - 2(32)(32)(-0.5831) \approx 3242.1 \implies b \approx 56.9$$

$$A = C \implies 2A = 180° - 125° 40' = 54° 20' \implies A = C = 27° 10'$$

14. Given: $C = 15° 15', a = 6.25, b = 2.15$

$$c^2 = a^2 + b^2 - 2ab\cos C = (6.25)^2 + (2.15)^2 - 2(6.25)(2.15)\cos 15° 15' \approx 17.7563 \implies c \approx 4.21$$

$$\cos A = \frac{b^2 + c^2 - a^2}{2bc} \approx \frac{(2.15)^2 + (4.21)^2 - (6.25)^2}{2(2.15)(4.21)} \approx -0.92338 \implies A \approx 157.4°$$

$$B \approx 180 - 15\ 15' - 157.4 \approx 7.4°$$

15. $C = 43°, a = \frac{4}{9}, b = \frac{7}{9}$

$$c^2 = a^2 + b^2 - 2ab\cos C = \left(\frac{4}{9}\right)^2 + \left(\frac{7}{9}\right)^2 - 2\left(\frac{4}{9}\right)\left(\frac{7}{9}\right)\cos 43° \approx 0.2968 \implies c \approx 0.5448$$

$$\sin A = \frac{a\sin C}{c} = \frac{(4/9)\sin 43°}{0.5448} \approx 0.5564 \implies A \approx 33.8°$$

$$B \approx 180° - 43° - 33.8° \approx 103.2°$$

16. Given: $C = 103°, a = \frac{3}{8}, b = \frac{3}{4}$

$$c^2 = a^2 + b^2 - 2ab\cos C = \left(\frac{3}{8}\right)^2 + \left(\frac{3}{4}\right)^2 - 2\left(\frac{3}{8}\right)\left(\frac{3}{4}\right)\cos 103° \approx 0.8297 \implies c \approx 0.91$$

$$\cos A = \frac{b^2 + c^2 - a^2}{2bc} \approx \frac{\left(\frac{3}{4}\right)^2 + (0.91)^2 - \left(\frac{3}{8}\right)^2}{2\left(\frac{3}{4}\right)(0.91)} \approx 0.9157 \implies A \approx 23.7°$$

$$B \approx 180° - 23.7° - 103° \approx 53.3°$$

17. $a = 5, \ b = 7, \ c = 10 \implies s = \frac{a + b + c}{2} = 11$

$$\text{Area} = \sqrt{s(s-a)(s-b)(s-c)} = \sqrt{11(6)(4)(1)} \approx 16.25$$

18. $a = 12, \ b = 15, \ c = 9 \implies s = \frac{12 + 15 + 9}{2} = 18$

$$\text{Area} = \sqrt{18(6)(3)(9)} = 54$$

19. $a = 2.5, b = 10.2, c = 9 \Rightarrow s = \dfrac{a+b+c}{2} = 10.85$

$\text{Area} = \sqrt{s(s-a)(s-b)(s-c)} = \sqrt{10.85(8.35)(0.65)(1.85)} \approx 10.44$

20. Given: $a = 75.4, b = 52, c = 52$

$s = \dfrac{75.4 + 52 + 52}{2} = 89.7$

$\text{Area} = \sqrt{s(s-a)(s-b)(s-c)} = \sqrt{89.7(14.3)(37.7)(37.7)} \approx 1350$

21. $a = 12.32, b = 8.46, c = 15.05 \Rightarrow s = \dfrac{a+b+c}{2} = 17.915$

$\text{Area} = \sqrt{s(s-a)(s-b)(s-c)} = \sqrt{17.915(5.595)(9.455)(2.865)} \approx 52.11$

22. Given: $a = 3.05, b = 0.75, c = 2.45$

$s = \dfrac{3.05 + 0.75 + 2.45}{2} = 3.125$

$\text{Area} = \sqrt{s(s-a)(s-b)(s-c)}$

$= \sqrt{3.125(0.075)(2.375)(0.675)}$

≈ 0.61

23. Law of Cosines

If ABC is a triangle with sides a, b, and c, then

Standard Form	*Alternate Form*
$a^2 = b^2 + c^2 - 2bc\cos A$	$\cos A = \dfrac{b^2 + c^2 - a^2}{2\,bc}$
$b^2 = a^2 + c^2 - 2ac\cos B$	$\cos B = \dfrac{a^2 + c^2 - b^2}{2\,ac}$
$c^2 = a^2 + b^2 - 2ab\cos C$	$\cos C = \dfrac{a^2 + b^2 - c^2}{2\,ab}$

24. Four Cases:

1. Two angles and any side (AAS or ASA)
2. Two sides and any angle opposite one of them (SSA)
3. Three sides (SSS)
4. Two sides and their included angle (SAS)

The first two cases can be solved by the Law of Sines. The last two cases require the Law of Cosines.

25.

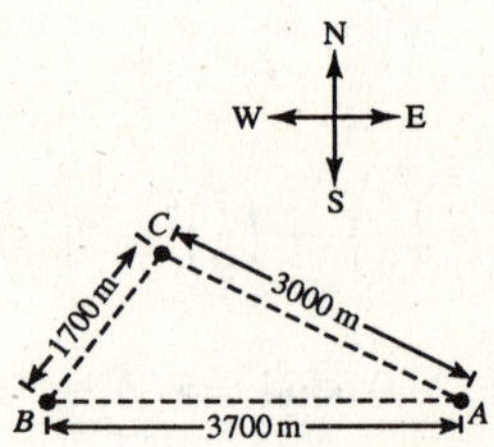

$\cos B = \dfrac{1700^2 + 3700^2 - 3000^2}{2(1700)(3700)} \Rightarrow B \approx 52.9°$

Bearing: $90° - 52.9° = $ N 37.1° E

$\cos C = \dfrac{1700^2 + 3000^2 - 3700^2}{2(1700)(3000)} \Rightarrow C \approx 100.2°$

Bearing: $A = 180° - 52.9° - 100.2° = 26.9° \Rightarrow$ S 63.1° E

26.

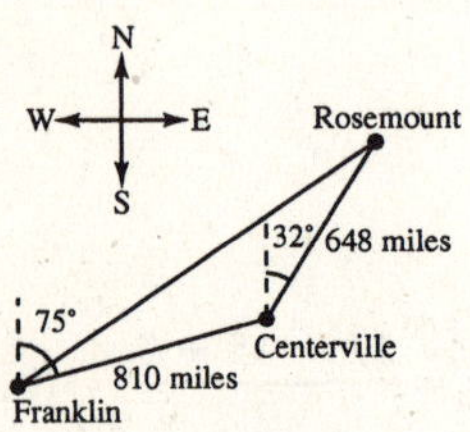

Distance from Franklin to Rosemount:

$d = \sqrt{810^2 + 648^2 - 2(810)(648)\cos(137°)} \approx 1357.8$ miles

Bearing from Franklin to Rosemount:

N $(75° - \theta)$ E

$\cos\theta \approx \dfrac{(1357.8)^2 + 810^2 - 648^2}{2(1357.8)(810)} \approx 0.9455$

$\theta \approx 19.0°$

N 56.0° E

27. 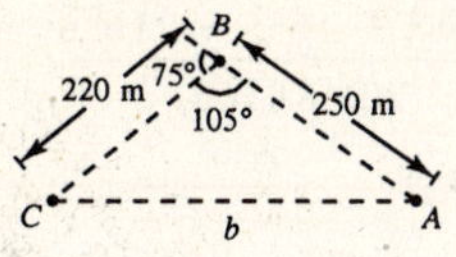

$b^2 = 220^2 + 250^2 - 2(220)(250)\cos 105° \Rightarrow b \approx 373.3$ meters

28. 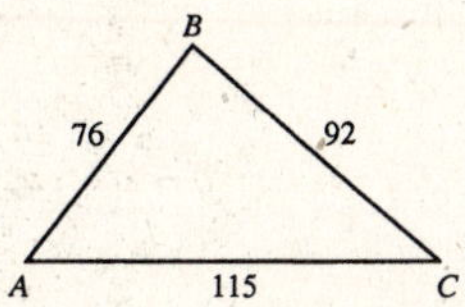

$$\cos A = \frac{115^2 + 76^2 - 92^2}{2(115)(76)} \approx 0.6028 \Rightarrow A \approx 52.9°$$

$$\cos C = \frac{115^2 + 92^2 - 76^2}{2(115)(92)} \approx 0.75203 \Rightarrow c \approx 41.2°$$

29. 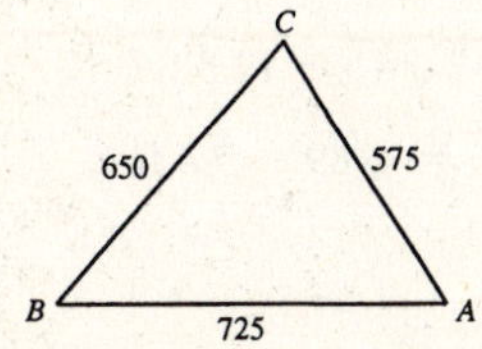

The largest angle is across from the largest side.

$$\cos C = \frac{650^2 + 575^2 - 725^2}{2(650)(575)}$$

$C \approx 72.3°$

30. $\cos\theta = \dfrac{2^2 + 3^2 - (4.5)^2}{2(2)(3)} \approx -0.60417$

$\theta \approx 127.2°$

31. $C = 180° - 53° - 67° = 60°$

$c^2 = a^2 + b^2 - 2ab\cos C$

$= 36^2 + 48^2 - 2(36)(48)(0.5)$

$= 1872$

$c \approx 43.3$ mi

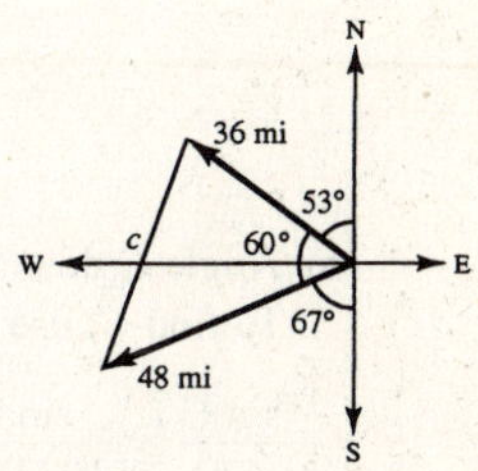

32. $d^2 = 60.5^2 + 90^2 - 2(60.5)(90)\cos 45° \approx 4059.8572 \Rightarrow d \approx 63.7$ ft

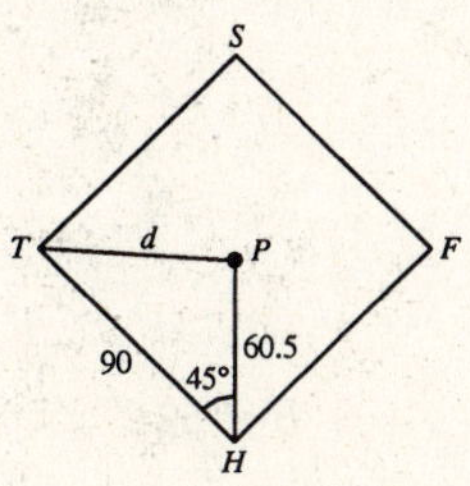

33. The angles at the base of the tower are 96° and 84°. The longer guy wire g_1 is given by:

$g_1{}^2 = 75^2 + 100^2 - 2(75)(100)\cos 96° \approx 17{,}192.9 \Rightarrow g_1 \approx 131.1$ feet

The shorter guy wire g_2 is given by:

$g_2{}^2 = 75^2 + 100^2 - 2(75)(100)\cos 84° \approx 14{,}057.1 \Rightarrow g_2 \approx 118.6$ feet

34. (a) $\cos\theta = \dfrac{273^2 + 178^2 - 235^2}{2(273)(178)}$

$\theta \approx 58.4°$

Bearing: N 58.4° W

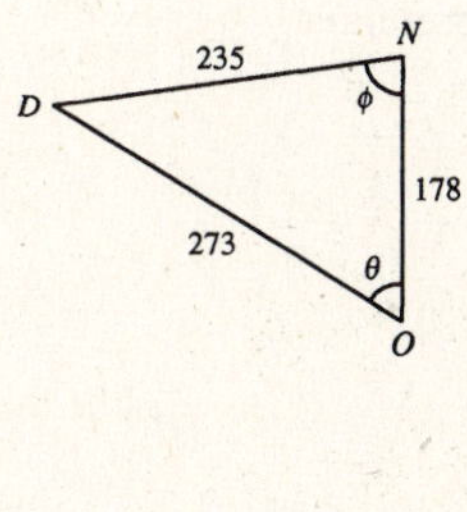

(b) $\cos\phi = \dfrac{235^2 + 178^2 - 273^2}{2(235)(178)}$

$\phi \approx 81.5°$

Bearing: S 81.5° W

35. (a) $\cos \theta = \dfrac{165^2 + 216^2 - 368^2}{2(165)(216)}$

$\theta \approx 149.7°$

$\theta - 90° \approx 59.7°$

Bearing: N 59.7° E

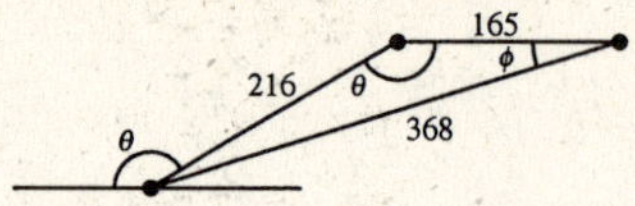

(b) $\cos \phi = \dfrac{368^2 + 165^2 - 216^2}{2(368)(165)}$

$\phi \approx 17.2°$

$90° - \phi \approx 72.8°$

Bearing: N 72.8° E

36. $\overline{RS} = \sqrt{8^2 + 10^2} = \sqrt{164} = 2\sqrt{41} \approx 12.8$ ft

$\overline{PQ} = \frac{1}{2}\sqrt{16^2 + 10^2} = \frac{1}{2}\sqrt{356} = \sqrt{89} \approx 9.4$ ft

$\tan P = \frac{10}{16}$

$P = \arctan \frac{5}{8} \approx 32.0°$

$\overline{QS} \approx \sqrt{8^2 + 9.4^2 - 2(8)(9.4) \cos 32°} \approx \sqrt{24.81} \approx 5.0$ ft

37. (a) $x^2 = 330^2 + 420^2 - 2(330)(420) \cos 8°$

$\approx 10{,}797.69134$

$x \approx 103.9119 \approx 103.9$ feet

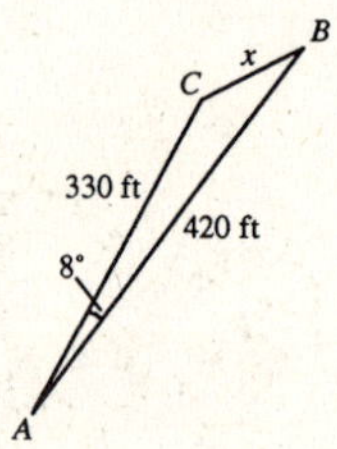

(b) Angle C does not change as θ changes. To find C, use the Law of Sines.

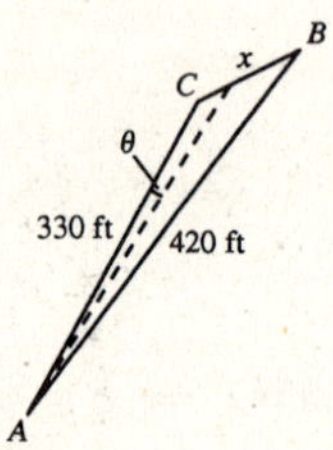

$\dfrac{\sin C}{420} = \dfrac{\sin 8°}{103.9119}$

$\sin C = \dfrac{420 \sin 8°}{103.9119} \approx 0.56252$

Since C is obtuse, $C \approx 180° - \sin^{-1} 0.5652$

$\approx 145.77°$

$B = 180° - C - \theta \approx 34.23° - \theta$

$\dfrac{\sin(34.23° - \theta)}{330} = \dfrac{\sin \theta}{x}$

$x = \dfrac{330 \sin \theta}{\sin(34.23° - \theta)}$

By the Quotient Rule,

$$\frac{dx}{dt} = \frac{330\left[\sin(34.23° - \theta) \cos \theta \dfrac{d\theta}{dt} - \sin \theta \cos(34.23° - \theta)\left(-\dfrac{d\theta}{dt}\right)\right]}{\sin^2(34.23° - \theta)}$$

$$= \frac{330\dfrac{d\theta}{dt} \sin 34.23°}{\sin^2(34.23° - \theta)} \quad \text{(by the sum and difference formula)}$$

When $\theta = 3°$, $\dfrac{d\theta}{dt} = \dfrac{1.4\pi}{180}$ rad/sec, and we have

$$\frac{dx}{dt} = \frac{330\left(\dfrac{1.4\pi}{180}\right) \sin 34.23°}{\sin^2 31.23°} \approx 16.87 \text{ feet per second}$$

38. (a) $a^2 = 35^2 + 20^2 - 2(35)(20)\cos 42° \implies a \approx 24.2$ miles

(b) Given: $A = 42°$, $b = 35$ miles, $c = 20$ miles, $a \approx 24.2$ miles,

$\frac{dc}{dt} = 300$ miles per hour, $\frac{db}{dt} = 375$ miles per hour

$$a^2 = b^2 + c^2 - 2bc\cos 42°$$

$$2a\frac{da}{dt} = 2b\frac{db}{dt} + 2c\frac{dc}{dt} - 2\cos 42°\left(b\frac{dc}{dt} + c\frac{db}{dt}\right)$$

$$\frac{da}{dt} = \frac{b\frac{db}{dt} + c\frac{dc}{dt} - \cos 42°\left(b\frac{dc}{dt} + c\frac{db}{dt}\right)}{a}$$

$$= \frac{(35)(375) + (20)(300) - \cos 42°[(35)(300) + (20)(375)]}{24.2}$$

$$\approx 238 \text{ miles per hour}$$

39. $a^2 = 20^2 + 20^2 - 2(20)(20)\cos 11° \implies a \approx 3.8$ miles

40. 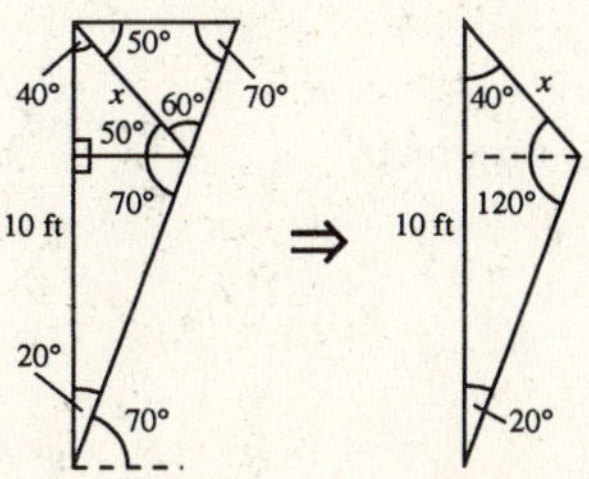

$$\frac{x}{\sin 20°} = \frac{10}{\sin 120°}$$

$$x = \frac{10\sin 20°}{\sin 120°}$$

$$x \approx 3.95 \text{ feet}$$

41. $d^2 = 10^2 + 7^2 - 2(10)(7)\cos\theta$

$$\theta = \arccos\left[\frac{10^2 + 7^2 - d^2}{2(10)(7)}\right]$$

$$s = \frac{360° - \theta}{360°}(2\pi r) = \frac{(360° - \theta)\pi}{45°}$$

d (inches)	9	10	12	13	14	15	16
θ (degrees)	60.9°	69.5°	88.0°	98.2°	109.6°	122.9°	139.8°
s (inches)	20.88	20.28	18.99	18.28	17.48	16.55	15.37

42. area $= 2\left[\frac{1}{2}(70)(100)\sin 70°\right]$

≈ 6577.8 square meters

(The area of the parallelogram is the sum of the areas of two triangles.)

43. $a = 200, b = 500, c = 600 \implies s = \frac{200 + 500 + 600}{2} = 650$

Area $= \sqrt{650(450)(150)(50)} \approx 46{,}837.5$ square feet

44. (a) $7^2 = 1.5^2 + x^2 - 2(1.5)(x\cos\theta)$

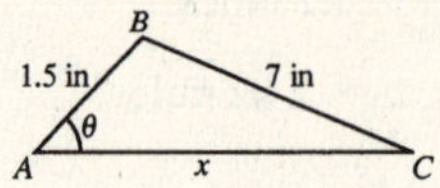

$49 = 2.25 + x^2 - 3x\cos\theta$

$x^2 - 3x\cos\theta - 46.75 = 0$

$$x = \frac{3\cos\theta \pm \sqrt{9\cos^2\theta + 187}}{2}$$

Choosing the positive value for x yields

$$x = \frac{1}{2}\left(3\cos\theta + \sqrt{9\cos^2\theta + 187}\right)$$

(b) $d = 8.5 - x = 8.5 - 0.5\left(3\cos\theta + \sqrt{9\cos^2\theta + 187}\right)$

(c)

θ	0	45°	90°	135°	180°
d	0″	0.5202″	1.6626″	2.6415″	3″

(d) $d(5°) \approx 0.0069$ inch

(e) $$\frac{dd}{d\theta} = \frac{3\sin\theta}{2} + \frac{9\sin\theta\cos\theta}{2\sqrt{9\cos^2\theta + 187}}$$

$$\frac{d^2d}{d\theta^2} = \frac{3\cos\theta}{2} + \frac{9\left[18\cos^4\theta - 9\sin^4\theta - 356\sin^2\theta + 178\right]}{2(9\cos^2\theta + 187)^{3/2}}$$

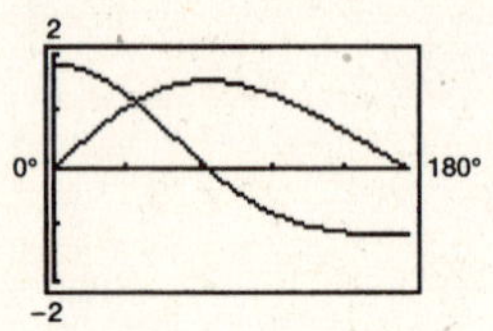

The speed is 0 for $\theta = 0°$ and $\theta = 180°$. The speed is maximum at 78.4° (1.368 radians).

(f) $\frac{d\theta}{dt} = 2500(2\pi) = 5000\pi$ radians per minute

$7^2 = 1.5^2 + x^2 - 2(1.5)(x)\cos\theta$

$x = \frac{1}{2}\left(3\cos\theta + \sqrt{9\cos^2\theta + 187}\right)$ from part (a)

$$0 = 2x\frac{dx}{dt} - 3\left(-x\sin\theta\frac{d\theta}{dt} + \cos\theta\frac{dx}{dt}\right)$$

$$0 = \frac{dx}{dt}(2x - 3\cos\theta) + 3x\sin\theta\frac{d\theta}{dt}$$

$$\frac{dx}{dt} = \frac{(3x\sin\theta)\left(\frac{d\theta}{dt}\right)}{3\cos\theta - 2x} = \frac{15000\pi x\sin\theta}{3\cos\theta - 2x}$$

θ	0°	30°	90°	150°
x	8.5 in.	8.259 in.	6.837 in.	5.661 in.
$\frac{dx}{dt}$	0 in./min	−13,980 in./min	−23,562 in./min	−9582 in./min

—CONTINUED—

44. —CONTINUED—

(g) $$\frac{d^2x}{dt^2} = (15{,}000\pi)\frac{\left[(3\cos\theta - 2x)\left(x\cos\theta\frac{d\theta}{dt} + \sin\theta\frac{dx}{dt}\right) - (x\sin\theta)\left(-3\sin\theta\frac{d\theta}{dt} - 2\frac{dx}{dt}\right)\right]}{(3\cos\theta - 2x)^2}$$

$$= \frac{(15{,}000\pi)\left[(3\cos\theta - 2x)\left(5000\pi x\cos\theta + \sin\theta\frac{dx}{dt}\right) + (x\sin\theta)\left(15{,}000\pi\sin\theta + 2\frac{dx}{dt}\right)\right]}{(3\cos\theta - 2x)^2}$$

θ	0°	180°
x	8.5 in.	5.5 in.
$\frac{dx}{dt}$	0 in./min	0 in./min
$\frac{d^2x}{dt^2}$	$-449{,}419{,}486$ in./min^2	290,800,844 in./min^2

The acceleration is not the same.

45. False. The average of the three sides of a triangle is $\frac{a+b+c}{3}$, not $\frac{a+b+c}{2}$.

46. False. To solve an SSA triangle, the Law of Sines is needed.

47. True. Assume $C = 90°$, then $\cos C = 0$ and $c^2 = a^2 + b^2 - 2ab\cos C$ becomes $c^2 = a^2 + b^2$.

The Law of Cosines simplifies to the Pythagorean Theorem.

48. False. Assume $\cos C$ is negative.

Then $\frac{\pi}{2} < C < \pi$ and C is obtuse.

Section 13.3 Vectors in the Plane

1. Initial point: $(0, 0)$

Terminal point: $(3, 2)$

$\mathbf{v} = \langle 3 - 0, 2 - 0\rangle = \langle 3, 2\rangle$

$\|\mathbf{v}\| = \sqrt{3^2 + 2^2} = \sqrt{13}$

2. Initial point: $(0, 0)$

Terminal point: $(-4, -2)$

$\mathbf{v} = \langle -4 - 0, -2 - 0\rangle = \langle -4, -2\rangle$

$\|\mathbf{v}\| = \sqrt{(-4)^2 + (-2)^2} = \sqrt{20} = 2\sqrt{5}$

3. Initial point: $(2, 2)$

Terminal point: $(-1, 4)$

$\mathbf{v} = \langle -1 - 2, 4 - 2\rangle = \langle -3, 2\rangle$

$\|\mathbf{v}\| = \sqrt{(-3)^2 + 2^2} = \sqrt{13}$

4. Initial point: $(-1, -1)$

Terminal point: $(3, 5)$

$\mathbf{v} = \langle 3 - (-1), 5 - (-1)\rangle = \langle 4, 6\rangle$

$\|\mathbf{v}\| = \sqrt{4^2 + 6^2} = \sqrt{52} = 2\sqrt{13}$

5. Initial point: $(3, -2)$

Terminal point: $(3, 3)$

$\mathbf{v} = \langle 3 - 3, 3 - (-2)\rangle = \langle 0, 5\rangle$

$\|\mathbf{v}\| = \sqrt{0^2 + 5^2} = \sqrt{25} = 5$

6. Initial point: $(-4, -1)$

Terminal point: $(3, -1)$

$\mathbf{v} = \langle 3 - (-4), -1 - (-1)\rangle = \langle 7, 0\rangle$

$\|\mathbf{v}\| = \sqrt{7^2 + 0^2} = 7$

7. Initial point: $(-1, 5)$

Terminal point: $(15, 12)$

$\mathbf{v} = \langle 15 - (-1), 12 - 5 \rangle = \langle 16, 7 \rangle$

$\|\mathbf{v}\| = \sqrt{16^2 + 7^2} = \sqrt{305}$

8. Initial point: $(1, 11)$

Terminal point: $(9, 3)$

$\mathbf{v} = \langle 9 - 1, 3 - 11 \rangle = \langle 8, -8 \rangle$

$\|\mathbf{v}\| = \sqrt{8^2 + (-8)^2} = 8\sqrt{2}$

9. Initial point: $(-3, -5)$

Terminal point: $(5, 1)$

$\mathbf{v} = \langle 5 - (-3), 1 - (-5) \rangle = \langle 8, 6 \rangle$

$\|\mathbf{v}\| = \sqrt{8^2 + 6^2} = \sqrt{100} = 10$

10. Initial point: $(-3, 11)$

Terminal point: $(9, 40)$

$\mathbf{v} = \langle 9 - (-3), 40 - 11 \rangle = \langle 12, 29 \rangle$

$\|\mathbf{v}\| = \sqrt{12^2 + 29^2} = \sqrt{985}$

11. Initial point: $(1, 3)$

Terminal point: $(-8, -9)$

$\mathbf{v} = \langle -8 - 1, -9 - 3 \rangle = \langle -9, -12 \rangle$

$\|\mathbf{v}\| = \sqrt{(-9)^2 + (-12)^2} = \sqrt{225} = 15$

12. Initial point: $(-2, 7)$

Terminal point: $(5, -17)$

$\mathbf{v} = \langle 5 - (-2), -17 - 7 \rangle = \langle 7, -24 \rangle$

$\|\mathbf{v}\| = \sqrt{7^2 + (-24)^2} = 25$

13.

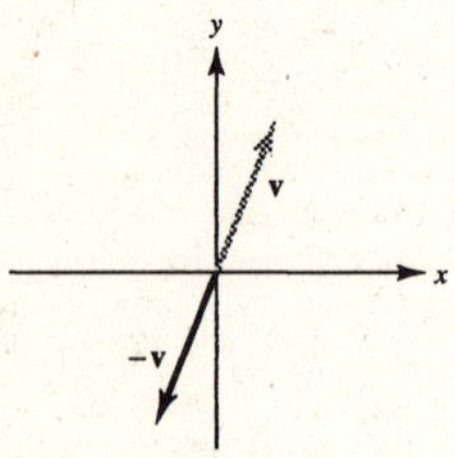

14. $5\mathbf{v}$

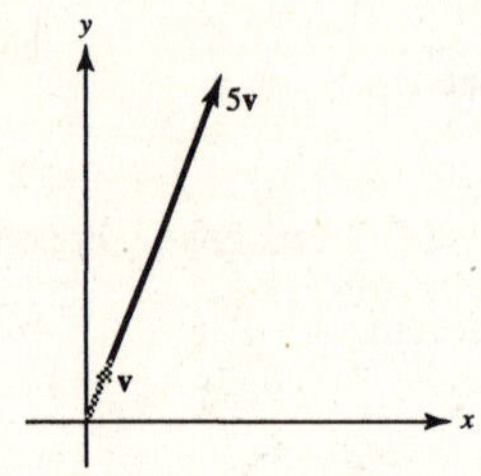

15.

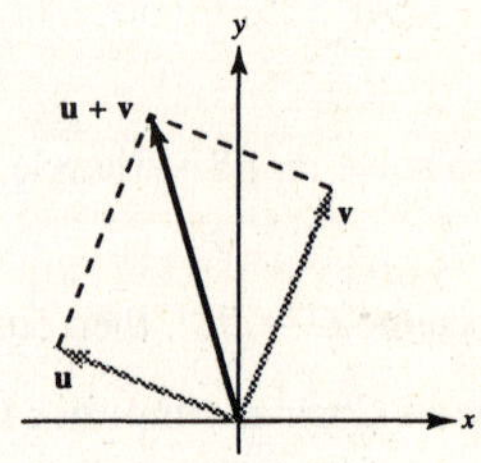

16. $\mathbf{u} - \mathbf{v}$

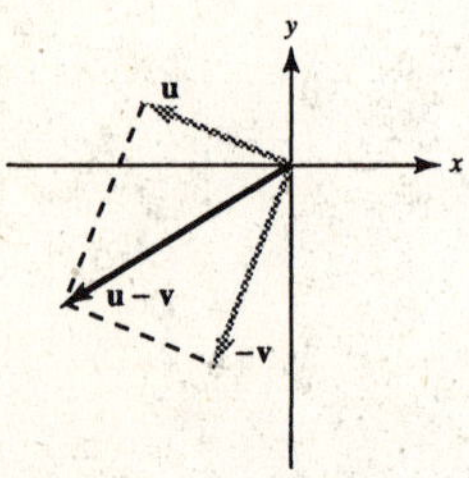

17. $\mathbf{u} + 2\mathbf{v}$

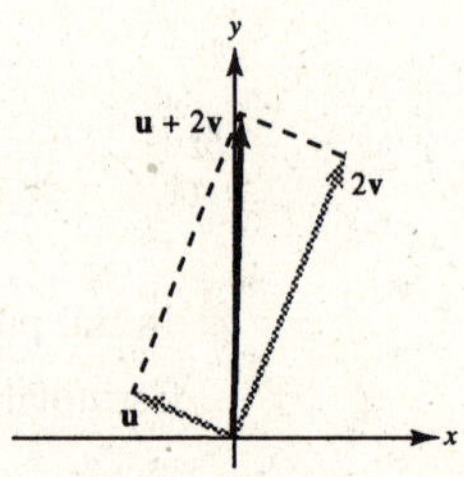

18. $\mathbf{v} - \frac{1}{2}\mathbf{u}$

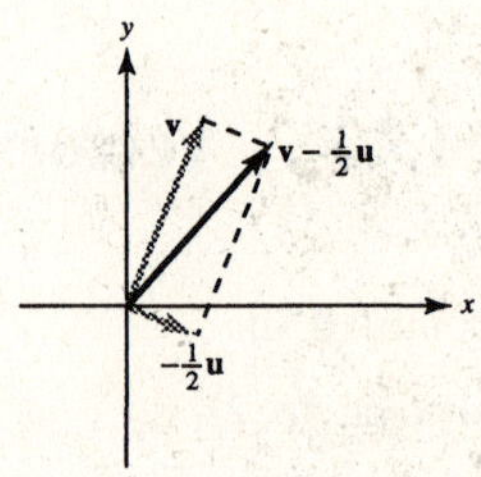

19. $\mathbf{u} = \langle 2, 1 \rangle, \mathbf{v} = \langle 1, 3 \rangle$

(a) $\mathbf{u} + \mathbf{v} = \langle 3, 4 \rangle$

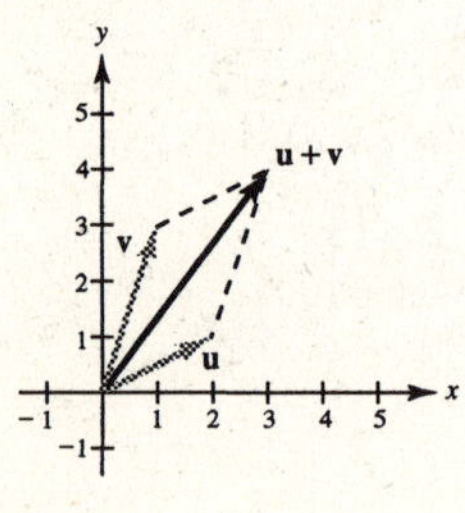

(b) $\mathbf{u} - \mathbf{v} = \langle 1, -2 \rangle$

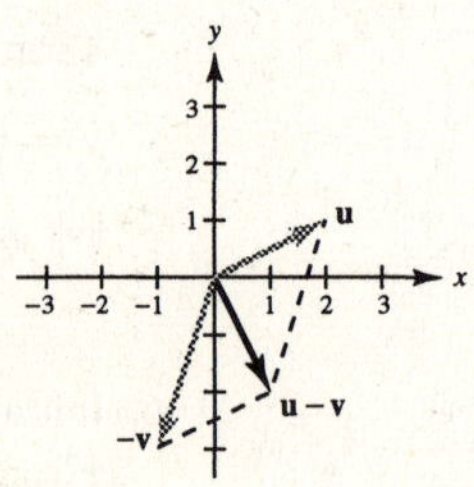

(c) $2\mathbf{u} - 3\mathbf{v} = \langle 4, 2 \rangle - \langle 3, 9 \rangle$

$= \langle 1, -7 \rangle$

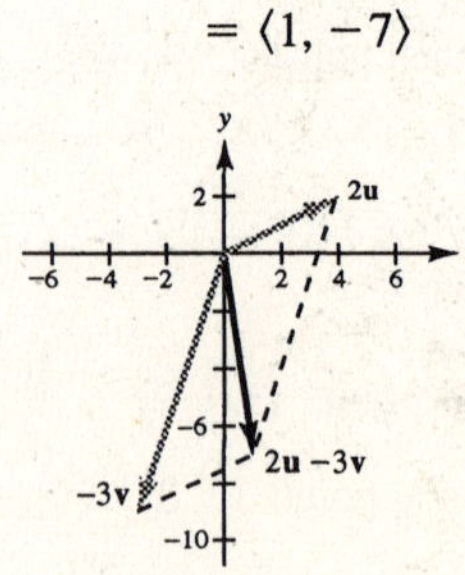

20. $\mathbf{u} = \langle 2, 3\rangle, \mathbf{v} = \langle 4, 0\rangle$

(a) $\mathbf{u} + \mathbf{v} = \langle 6, 3\rangle$

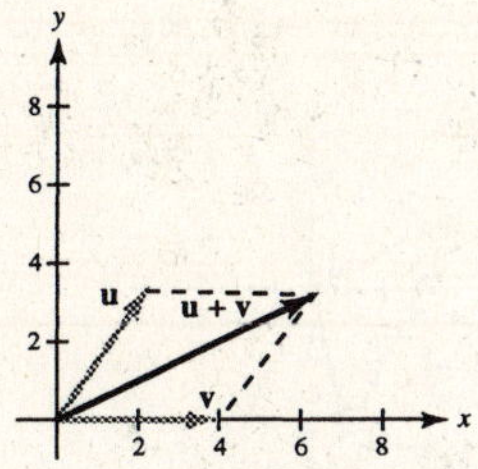

(b) $\mathbf{u} - \mathbf{v} = \langle -2, 3\rangle$

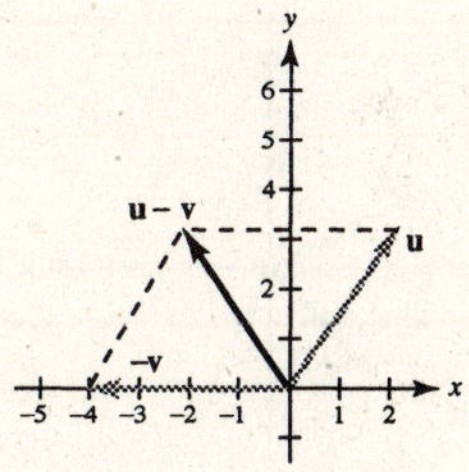

(c) $2\mathbf{u} - 3\mathbf{v} = \langle 4, 6\rangle - \langle 12, 0\rangle$

$= \langle -8, 6\rangle$

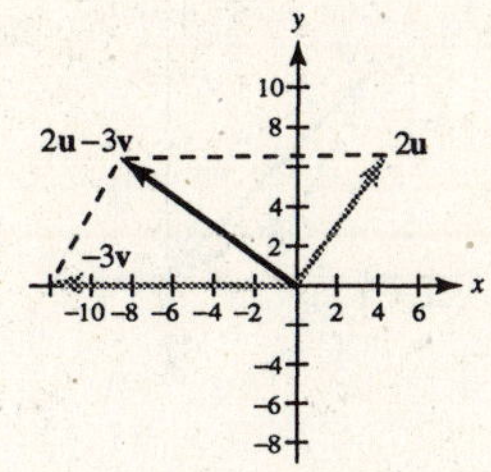

21. $\mathbf{u} = \langle -5, 3\rangle, \mathbf{v} = \langle 0, 0\rangle$

(a) $\mathbf{u} + \mathbf{v} = \langle -5, 3\rangle = \mathbf{u}$

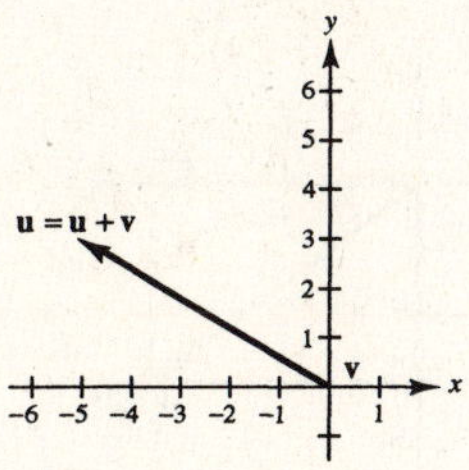

(b) $\mathbf{u} - \mathbf{v} = \langle -5, 3\rangle = \mathbf{u}$

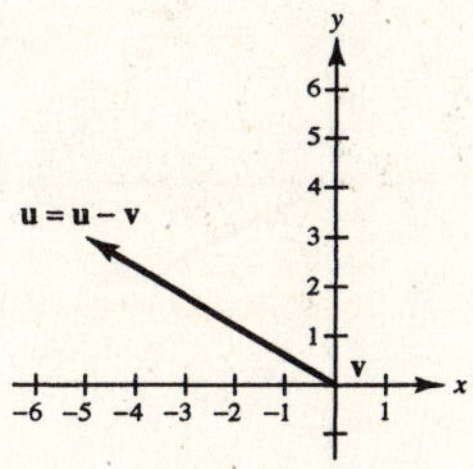

(c) $2\mathbf{u} - 3\mathbf{v} = 2\mathbf{u} = \langle -10, 6\rangle$

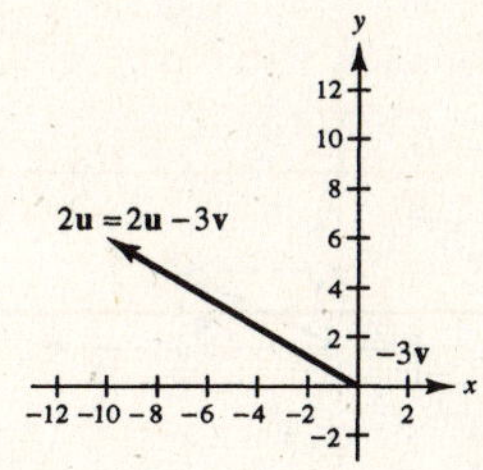

22. $\mathbf{u} = \langle 0, 0\rangle, \mathbf{v} = \langle 2, 1\rangle$

(a) $\mathbf{u} + \mathbf{v} = \langle 2, 1\rangle$

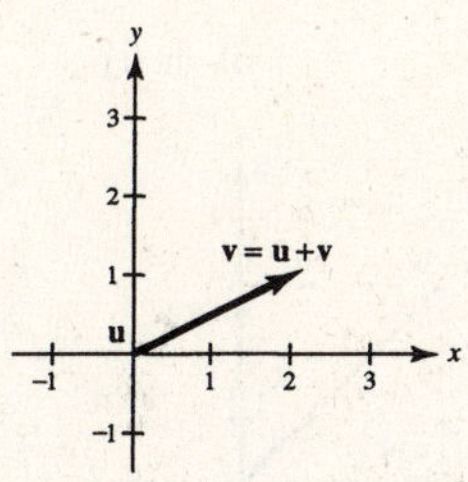

(b) $\mathbf{u} - \mathbf{v} = \langle -2, -1\rangle$

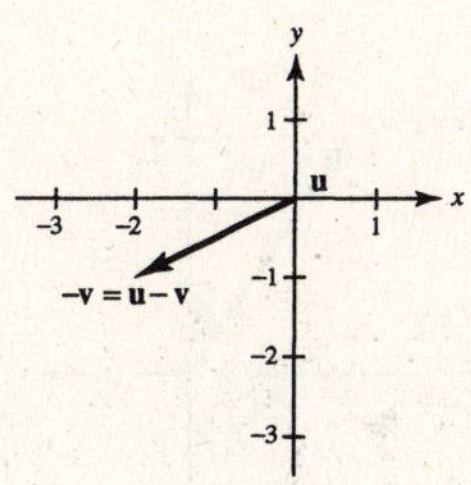

(c) $2\mathbf{u} - 3\mathbf{v} = \langle 0, 0\rangle - \langle 6, 3\rangle$

$= \langle -6, -3\rangle$

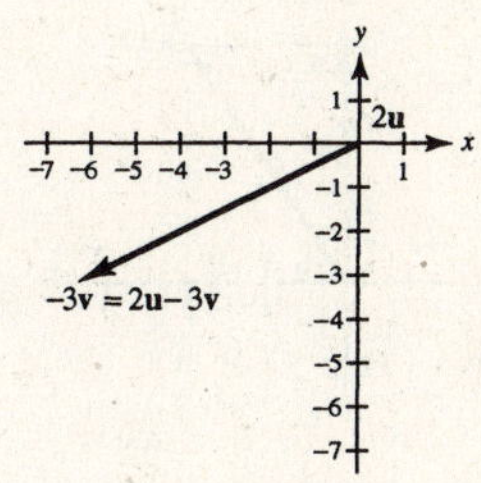

23. $\mathbf{u} = \mathbf{i} + \mathbf{j}, \mathbf{v} = 2\mathbf{i} - 3\mathbf{j}$

(a) $\mathbf{u} + \mathbf{v} = 3\mathbf{i} - 2\mathbf{j}$

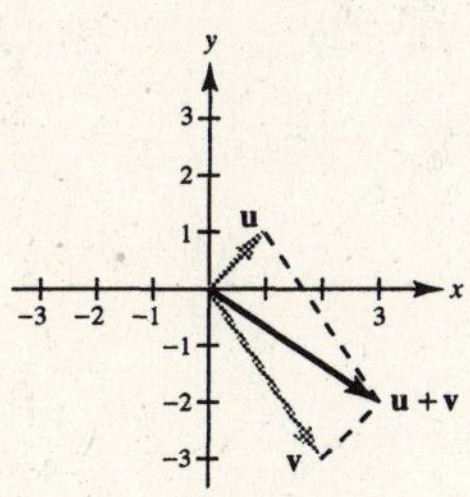

(b) $\mathbf{u} - \mathbf{v} = -\mathbf{i} + 4\mathbf{j}$

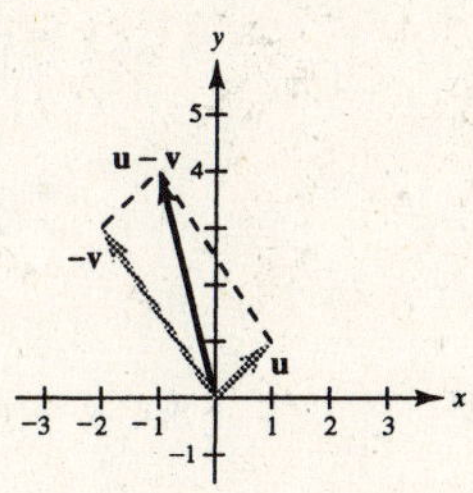

(c) $2\mathbf{u} - 3\mathbf{v} = (2\mathbf{i} + 2\mathbf{j}) - (6\mathbf{i} - 9\mathbf{j})$

$= -4\mathbf{i} + 11\mathbf{j}$

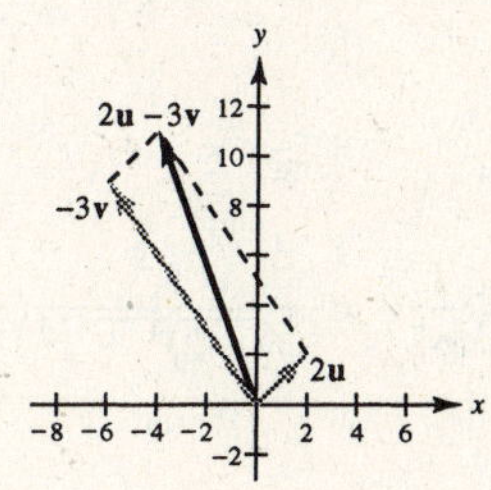

24. $\mathbf{u} = -2\mathbf{i} + \mathbf{j}, \mathbf{v} = -\mathbf{i} + 2\mathbf{j}$

(a) $\mathbf{u} + \mathbf{v} = -3\mathbf{i} + 3\mathbf{j}$

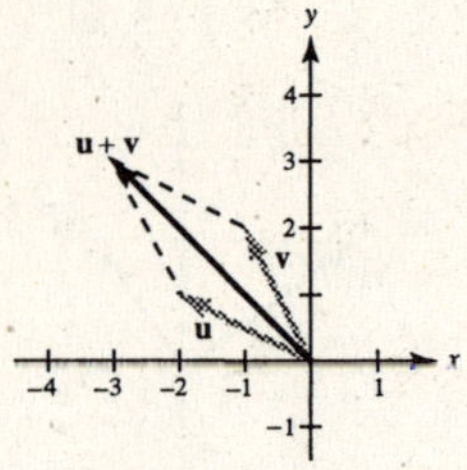

(b) $\mathbf{u} - \mathbf{v} = -\mathbf{i} - \mathbf{j}$

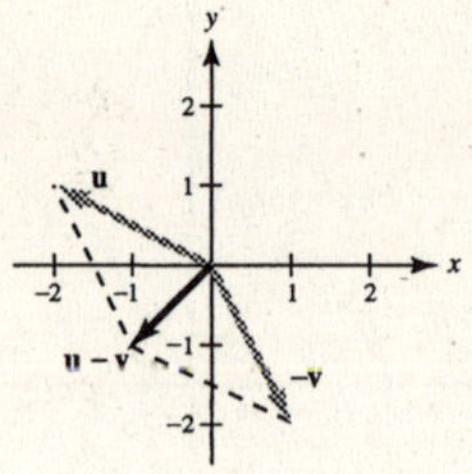

(c) $2\mathbf{u} - 3\mathbf{v} = (-4\mathbf{i} + 2\mathbf{j}) - (-3\mathbf{i} + 6\mathbf{j})$

$= -\mathbf{i} - 4\mathbf{j}$

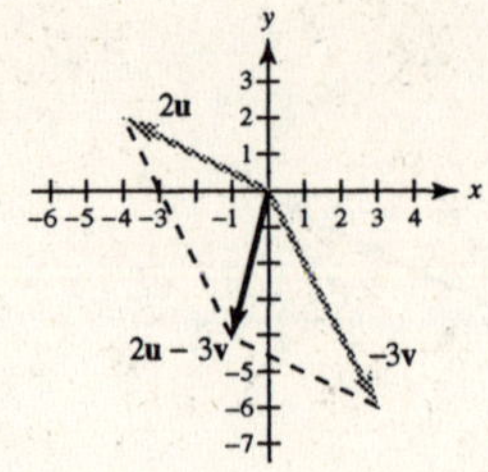

25. $\mathbf{u} = 2\mathbf{i}, \mathbf{v} = \mathbf{j}$

(a) $\mathbf{u} + \mathbf{v} = 2\mathbf{i} + \mathbf{j}$

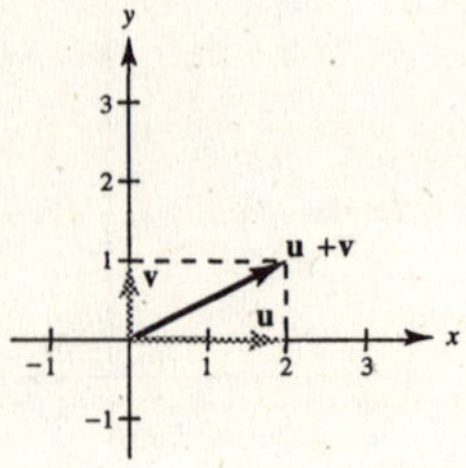

(b) $\mathbf{u} - \mathbf{v} = 2\mathbf{i} - \mathbf{j}$

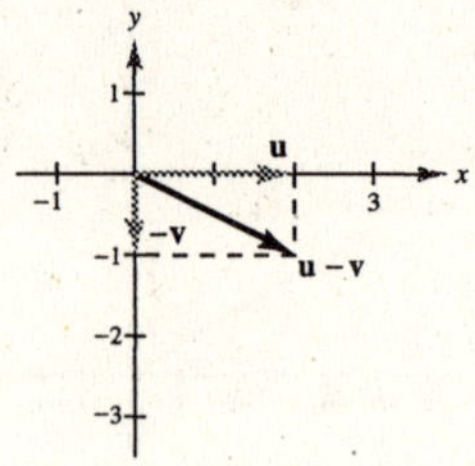

(c) $2\mathbf{u} - 3\mathbf{v} = 4\mathbf{i} - 3\mathbf{j}$

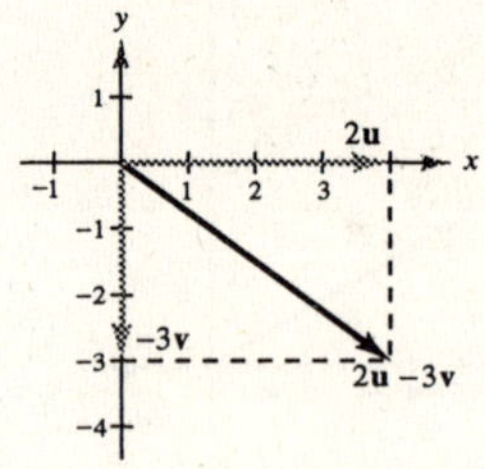

26. $\mathbf{u} = 3\mathbf{j}, \mathbf{v} = 2\mathbf{i}$

(a) $\mathbf{u} + \mathbf{v} = 2\mathbf{i} + 3\mathbf{j}$

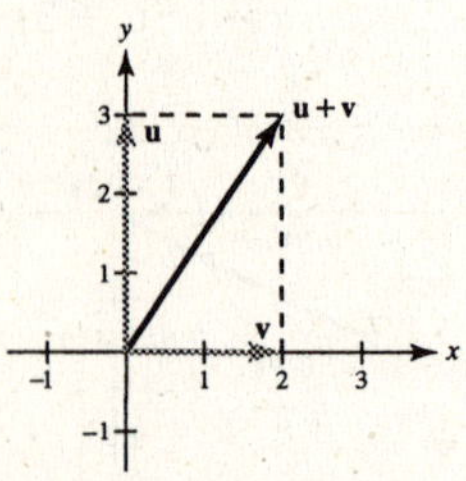

(b) $\mathbf{u} - \mathbf{v} = -2\mathbf{i} + 3\mathbf{j}$

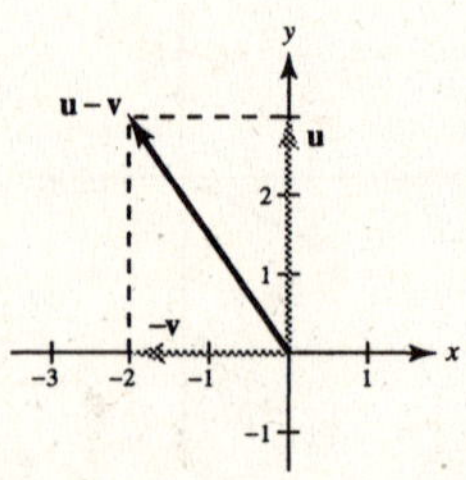

(c) $2\mathbf{u} - 3\mathbf{v} = 6\mathbf{j} - 6\mathbf{i}$

$= -6\mathbf{i} + 6\mathbf{j}$

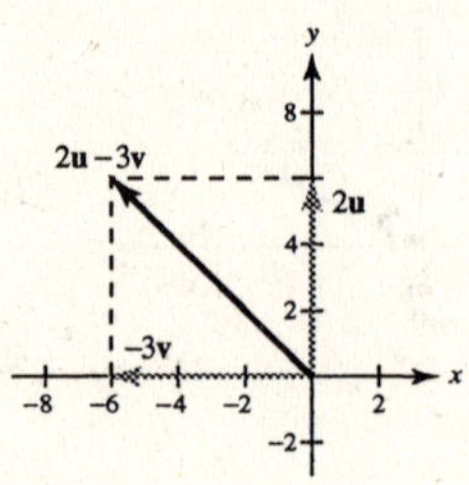

27. $\mathbf{v} = \dfrac{1}{\|\mathbf{u}\|}\mathbf{u} = \dfrac{1}{\sqrt{3^2 + 0^2}}\langle 3, 0\rangle = \dfrac{1}{3}\langle 3, 0\rangle = \langle 1, 0\rangle$

28. $\mathbf{u} = \langle 0, -2\rangle$

$\mathbf{v} = \dfrac{1}{\|\mathbf{u}\|}\mathbf{u} = \dfrac{1}{\sqrt{0^2 + (-2)^2}}\langle 0, -2\rangle$

$= \dfrac{1}{2}\langle 0, -2\rangle = \langle 0, -1\rangle$

29. $\mathbf{u} = \dfrac{1}{\|\mathbf{v}\|}\mathbf{v} = \dfrac{1}{\sqrt{(-2)^2 + 2^2}}\langle -2, 2\rangle = \dfrac{1}{2\sqrt{2}}\langle -2, 2\rangle$

$= \left\langle -\dfrac{1}{\sqrt{2}}, \dfrac{1}{\sqrt{2}}\right\rangle$

30. $\mathbf{v} = \langle 5, -12\rangle$

$\mathbf{u} = \dfrac{1}{\|\mathbf{v}\|}\mathbf{v} = \dfrac{1}{\sqrt{5^2 + (-12)^2}}\langle 5, -12\rangle$

$= \dfrac{1}{13}\langle 5, -12\rangle$

$= \left\langle \dfrac{5}{13}, -\dfrac{12}{13}\right\rangle$

31. $\mathbf{u} = \frac{1}{\|\mathbf{v}\|}\mathbf{v} = \frac{1}{\sqrt{6^2 + (-2)^2}}(6\mathbf{i} - 2\mathbf{j}) = \frac{1}{\sqrt{40}}(6\mathbf{i} - 2\mathbf{j})$

$= \frac{1}{2\sqrt{10}}(6\mathbf{i} - 2\mathbf{j}) = \frac{3}{\sqrt{10}}\mathbf{i} - \frac{1}{\sqrt{10}}\mathbf{j}$

32. $\mathbf{v} = \mathbf{i} + \mathbf{j}$

$\mathbf{u} = \frac{1}{\|\mathbf{v}\|}\mathbf{v} = \frac{1}{\sqrt{1^2 + 1^2}}(\mathbf{i} + \mathbf{j})$

$= \frac{1}{\sqrt{2}}(\mathbf{i} + \mathbf{j}) = \frac{\sqrt{2}}{2}\mathbf{i} + \frac{\sqrt{2}}{2}\mathbf{j}$

33. $\mathbf{u} = \frac{1}{\|\mathbf{w}\|}\mathbf{w} = \frac{1}{4}(4\mathbf{j}) = \mathbf{j}$

34. $\mathbf{w} = -6\mathbf{i}$

$\mathbf{v} = \frac{1}{\|\mathbf{w}\|}\mathbf{w} = \frac{1}{\sqrt{(-6)^2 + 0^2}}(-6\mathbf{i})$

$= \frac{1}{6}(-6\mathbf{i}) = -\mathbf{i}$

35. $\mathbf{u} = \frac{1}{\|\mathbf{w}\|}\mathbf{w} = \frac{1}{\sqrt{1^2 + (-2)^2}}(\mathbf{i} - 2\mathbf{j}) = \frac{1}{\sqrt{5}}(\mathbf{i} - 2\mathbf{j})$

$= \frac{1}{\sqrt{5}}\mathbf{i} - \frac{2}{\sqrt{5}}\mathbf{j}$

36. $\mathbf{w} = 7\mathbf{j} - 3\mathbf{i}$

$\mathbf{v} = \frac{1}{\|\mathbf{w}\|}\mathbf{w} = \frac{1}{\sqrt{(-3)^2 + 7^2}}(-3\mathbf{i} + 7\mathbf{j})$

$= -\frac{3}{\sqrt{58}}\mathbf{i} + \frac{7}{\sqrt{58}}\mathbf{j}$

37. $5\left(\frac{1}{\|\mathbf{u}\|}\mathbf{u}\right) = 5\left(\frac{1}{\sqrt{3^2 + 3^2}}\langle 3, 3\rangle\right) = \frac{5}{3\sqrt{2}}\langle 3, 3\rangle$

$= \left\langle \frac{5}{\sqrt{2}}, \frac{5}{\sqrt{2}} \right\rangle$

38. $\mathbf{v} = 6\left(\frac{1}{\|\mathbf{u}\|}\mathbf{u}\right) = 6\left(\frac{1}{\sqrt{(-3)^2 + 3^2}}\langle -3, 3\rangle\right)$

$= 6\left(\frac{1}{3\sqrt{2}}\langle -3, 3\rangle\right) = \left\langle -\frac{6}{\sqrt{2}}, \frac{6}{\sqrt{2}} \right\rangle$

39. $9\left(\frac{1}{\|\mathbf{u}\|}\mathbf{u}\right) = 9\left(\frac{1}{\sqrt{2^2 + 5^2}}\langle 2, 5\rangle\right) = \frac{9}{\sqrt{29}}\langle 2, 5\rangle$

$= \left\langle \frac{18}{\sqrt{29}}, \frac{45}{\sqrt{29}} \right\rangle$

40. $\mathbf{v} = 10\left(\frac{1}{\|\mathbf{u}\|}\mathbf{u}\right) = 10\left(\frac{1}{\sqrt{0^2 + (-10)^2}}\langle -10, 0\rangle\right)$

$= 10\left(\frac{1}{10}\langle -10, 0\rangle\right) = \langle -10, 0\rangle$

41. $\mathbf{v} = \frac{3}{2}\mathbf{u} = \frac{3}{2}(2\mathbf{i} - \mathbf{j})$

$= 3\mathbf{i} - \frac{3}{2}\mathbf{j} = \left\langle 3, -\frac{3}{2} \right\rangle$

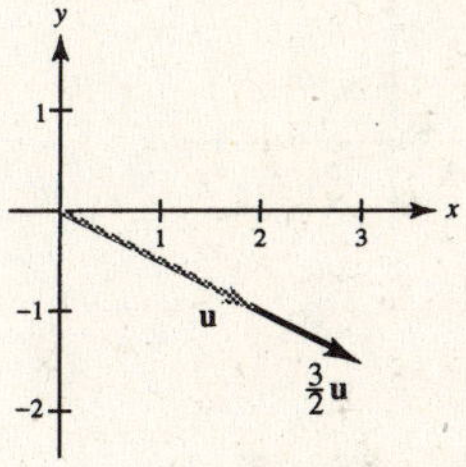

42. $\mathbf{v} = \frac{3}{4}\mathbf{w} = \frac{3}{4}(\mathbf{i} + 2\mathbf{j})$

$= \frac{3}{4}\mathbf{i} + \frac{3}{2}\mathbf{j} = \left\langle \frac{3}{4}, \frac{3}{2} \right\rangle$

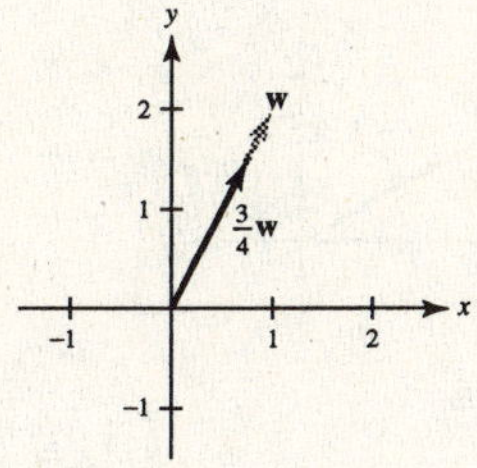

43. $\mathbf{v} = \mathbf{u} + 2\mathbf{w}$

$= (2\mathbf{i} - \mathbf{j}) + 2(\mathbf{i} + 2\mathbf{j})$

$= 4\mathbf{i} + 3\mathbf{j} = \langle 4, 3\rangle$

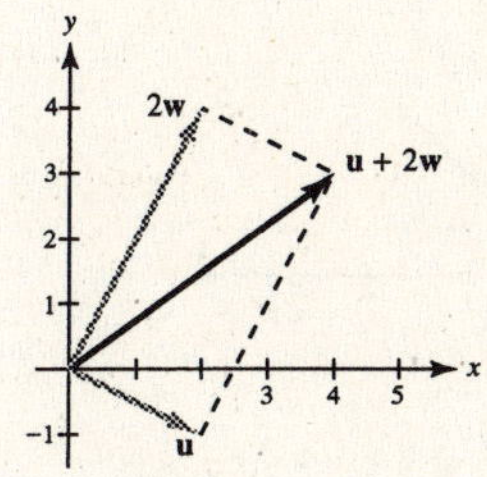

44. $\mathbf{v} = -\mathbf{u} + \mathbf{w}$

$= -(2\mathbf{i} - \mathbf{j}) + (\mathbf{i} + 2\mathbf{j})$

$= -\mathbf{i} + 3\mathbf{j} = \langle -1, 3\rangle$

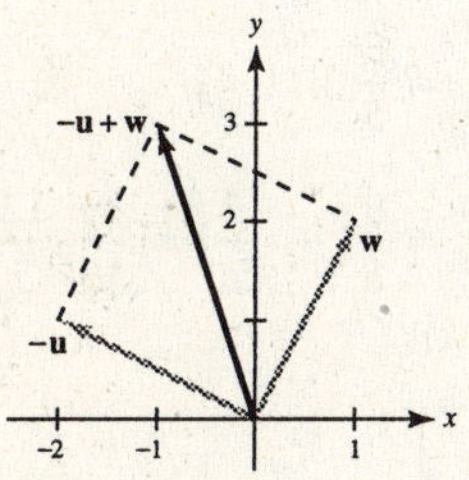

45. $\mathbf{v} = \frac{1}{2}(3\mathbf{u} + \mathbf{w})$

$= \frac{1}{2}(6\mathbf{i} - 3\mathbf{j} + \mathbf{i} + 2\mathbf{j})$

$= \frac{7}{2}\mathbf{i} - \frac{1}{2}\mathbf{j} = \left\langle \frac{7}{2}, -\frac{1}{2} \right\rangle$

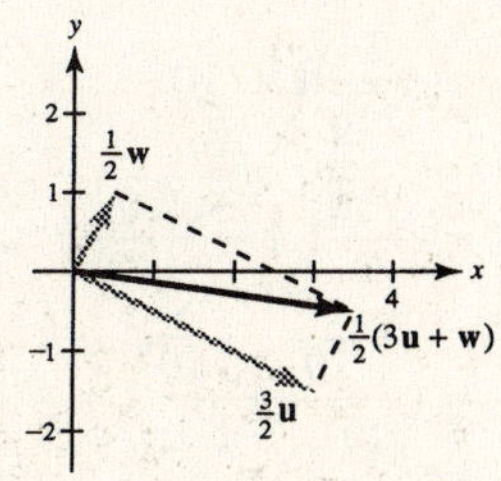

46. $\mathbf{v} = \mathbf{u} - 2\mathbf{w}$

$= (2\mathbf{i} - \mathbf{j}) - 2(\mathbf{i} + 2\mathbf{j})$

$= -5\mathbf{j} = \langle 0, -5 \rangle$

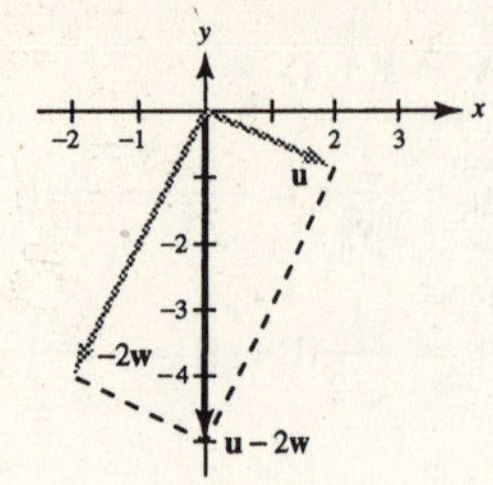

47. $\mathbf{v} = 3(\cos 60°\mathbf{i} + \sin 60°\mathbf{j})$

$\|\mathbf{v}\| = 3, \ \theta = 60°$

48. $\mathbf{v} = 8(\cos 135°\,\mathbf{i} + \sin 135°\,\mathbf{j})$

$\|\mathbf{v}\| = 8, \theta = 135°$

49. $\mathbf{v} = 6\mathbf{i} - 6\mathbf{j}$

$\|\mathbf{v}\| = \sqrt{6^2 + (-6)^2} = \sqrt{72}$

$= 6\sqrt{2}$

$\tan \theta = \dfrac{-6}{6} = -1$

Since **v** lies in Quadrant IV, $\theta = 315°$.

50. $\mathbf{v} = -5\mathbf{i} + 4\mathbf{j}$

$\|\mathbf{v}\| = \sqrt{(-5)^2 + 4^2} = \sqrt{41}$

$\tan \theta = -\dfrac{4}{5}$

Since **v** lies in Quadrant II, $\theta = 141.3°$.

51. $\|\mathbf{v}\| = 5 \quad \theta = 0°$

$\mathbf{v} = \langle 5 \cos 0°, 5 \sin 0° \rangle$

$= \langle 5, 0 \rangle$

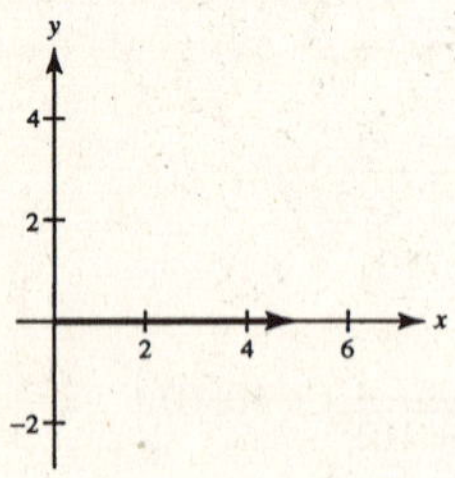

52. $\mathbf{v} = \langle \cos 45°, \sin 45° \rangle$

$= \left\langle \dfrac{\sqrt{2}}{2}, \dfrac{\sqrt{2}}{2} \right\rangle$

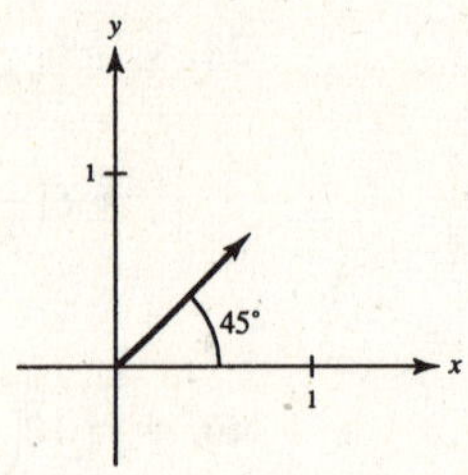

53. $\mathbf{v} = \left\langle \dfrac{7}{2} \cos 150°, \dfrac{7}{2} \sin 150° \right\rangle$

$= \left\langle -\dfrac{7\sqrt{3}}{4}, \dfrac{7}{4} \right\rangle$

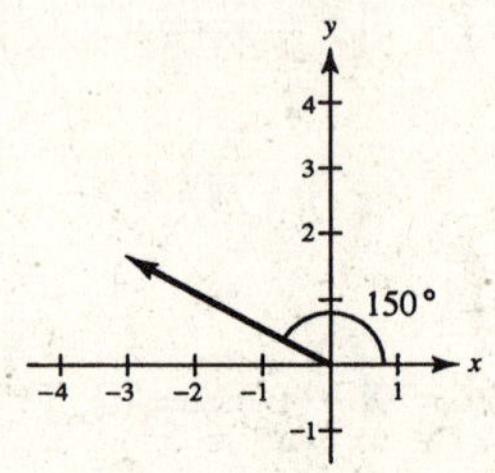

54. $\mathbf{v} = \left\langle \dfrac{5}{2} \cos 45°, \dfrac{5}{2} \sin 45° \right\rangle$

$= \left\langle \dfrac{5\sqrt{2}}{4}, \dfrac{5\sqrt{2}}{4} \right\rangle$

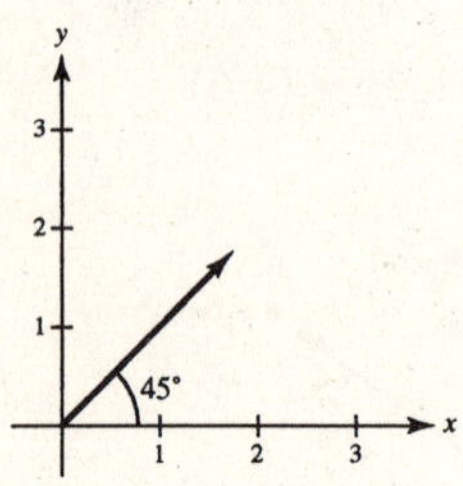

55. $\mathbf{v} = \left\langle 3\sqrt{2} \cos 150°, 3\sqrt{2} \sin 150° \right\rangle$

$= \left\langle -\dfrac{3\sqrt{6}}{2}, \dfrac{3\sqrt{2}}{2} \right\rangle$

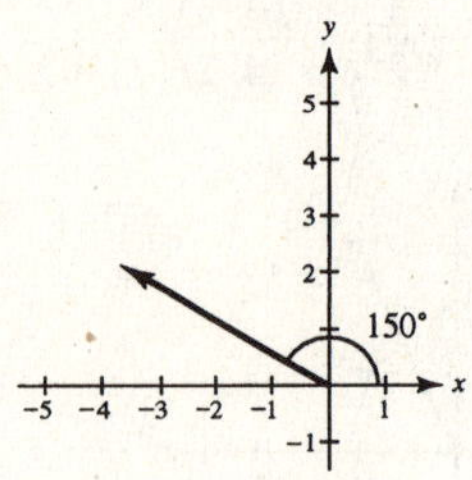

56. $\mathbf{v} = \left\langle 4\sqrt{3} \cos 90°, 4\sqrt{3} \sin 90° \right\rangle$

$= \left\langle 0, 4\sqrt{3} \right\rangle$

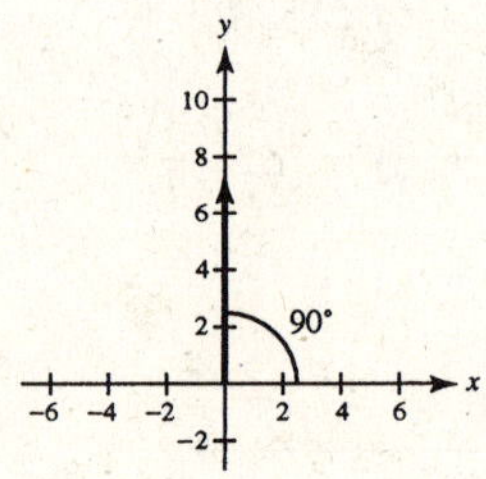

57. $\mathbf{v} = 2\left(\dfrac{1}{\sqrt{1^2 + 3^2}}\right)(\mathbf{i} + 3\mathbf{j})$

$= \dfrac{2}{\sqrt{10}}(\mathbf{i} + 3\mathbf{j})$

$= \dfrac{\sqrt{10}}{5}\mathbf{i} + \dfrac{3\sqrt{10}}{5}\mathbf{j}$

$= \left\langle \dfrac{\sqrt{10}}{5}, \dfrac{3\sqrt{10}}{5} \right\rangle$

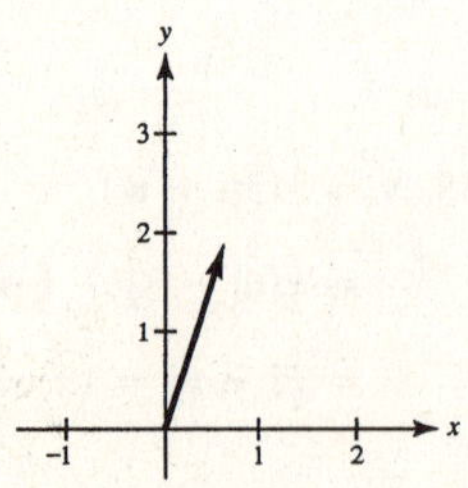

58. $\mathbf{v} = 3\left(\dfrac{1}{\sqrt{3^2 + 4^2}}\right)(3\mathbf{i} + 4\mathbf{j})$

$= \dfrac{3}{5}(3\mathbf{i} + 4\mathbf{j})$

$= \dfrac{9}{5}\mathbf{i} + \dfrac{12}{5}\mathbf{j} = \left\langle \dfrac{9}{5}, \dfrac{12}{5} \right\rangle$

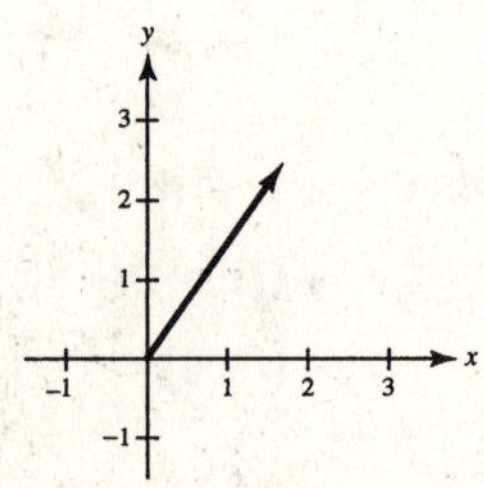

59. $\mathbf{u} = \langle 5\cos 0°, 5\sin 0°\rangle = \langle 5, 0\rangle$

$\mathbf{v} = \langle 5\cos 90°, 5\sin 90°\rangle = \langle 0, 5\rangle$

$\mathbf{u} + \mathbf{v} = \langle 5, 5\rangle$

60. $\mathbf{u} = \langle 4\cos 60°, 4\sin 60°\rangle = \langle 2, 2\sqrt{3}\rangle$

$\mathbf{v} = \langle 4\cos 90°, 4\sin 90°\rangle = \langle 0, 4\rangle$

$\mathbf{u} + \mathbf{v} = \langle 2, 4 + 2\sqrt{3}\rangle$

61. $\mathbf{u} = \langle 20\cos 45°, 20\sin 45°\rangle = \langle 10\sqrt{2}, 10\sqrt{2}\rangle$

$\mathbf{v} = \langle 50\cos 180°, 50\sin 180°\rangle = \langle -50, 0\rangle$

$\mathbf{u} + \mathbf{v} = \langle 10\sqrt{2} - 50, 10\sqrt{2}\rangle$

62. $\mathbf{u} = \langle 50\cos 30°, 50\sin 30°\rangle = \langle 25\sqrt{3}, 25\rangle$

$\approx \langle 43.301, 25\rangle$

$\mathbf{v} = \langle 30\cos 110°, 30\sin 110°\rangle \approx \langle -10.261, 28.191\rangle$

$\mathbf{u} + \mathbf{v} \approx \langle 33.04, 53.19\rangle$

63. $\mathbf{v} = \mathbf{i} + \mathbf{j}$

$\mathbf{w} = 2\mathbf{i} - 2\mathbf{j}$

$\mathbf{u} = \mathbf{v} - \mathbf{w} = -\mathbf{i} + 3\mathbf{j}$

$\|\mathbf{v}\| = \sqrt{2}$

$\|\mathbf{w}\| = 2\sqrt{2}$

$\|\mathbf{v} - \mathbf{w}\| = \sqrt{10}$

$$\cos\alpha = \frac{\|\mathbf{v}\|^2 + \|\mathbf{w}\|^2 - \|\mathbf{v} - \mathbf{w}\|^2}{2\|\mathbf{v}\|\,\|\mathbf{w}\|} = \frac{2 + 8 - 10}{2\sqrt{2}\cdot 2\sqrt{2}} = 0$$

$\alpha = 90°$

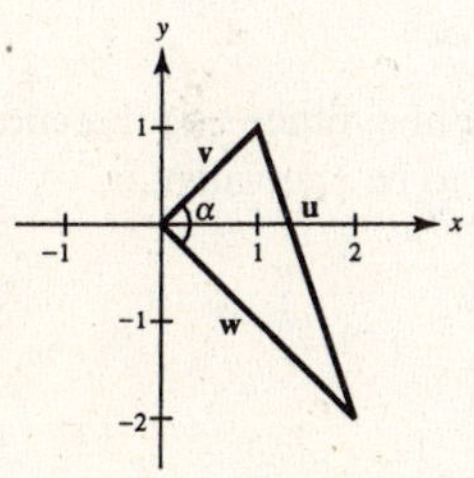

64. $\mathbf{v} = 3\mathbf{i} - 2\mathbf{j}$

$\mathbf{w} = 2\mathbf{i} + 2\mathbf{j}$

$\mathbf{u} = \mathbf{v} - \mathbf{w} = \mathbf{i} - 4\mathbf{j}$

$$\cos\theta = \frac{\|\mathbf{v}\|^2 + \|\mathbf{w}\|^2 - \|\mathbf{v} - \mathbf{w}\|^2}{2\|\mathbf{v}\|\,\|\mathbf{w}\|} = \frac{13 + 8 - 17}{2\sqrt{13}\sqrt{8}}$$

$$= \frac{1}{\sqrt{13}\sqrt{2}}$$

$\theta \approx 78.7°$

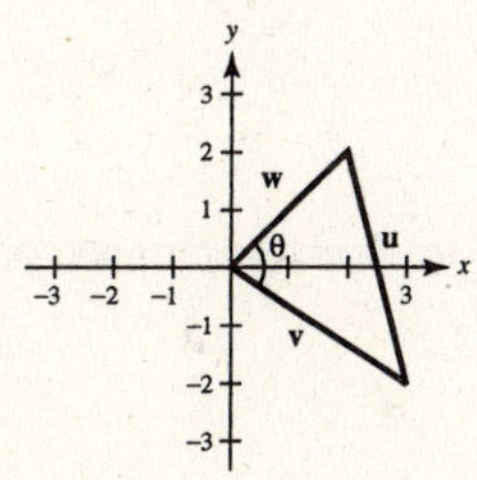

65. $\mathbf{v} = \mathbf{i} + \mathbf{j}$

$\mathbf{w} = 3\mathbf{i} - \mathbf{j}$

$\mathbf{u} = \mathbf{v} - \mathbf{w} = -2\mathbf{i} + 2\mathbf{j}$

$$\cos\alpha = \frac{\|\mathbf{v}\|^2 + \|\mathbf{w}\|^2 - \|\mathbf{v} - \mathbf{w}\|^2}{2\|\mathbf{v}\|\,\|\mathbf{w}\|} = \frac{2 + 10 - 8}{2\sqrt{2}\,\sqrt{10}} \approx 0.4472$$

$\alpha = 63.4°$

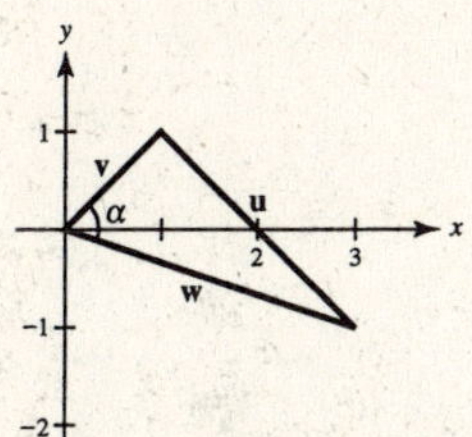

66. $\mathbf{v} = \mathbf{i} + 2\mathbf{j}$

$\mathbf{w} = 2\mathbf{i} - \mathbf{j}$

$\mathbf{u} = \mathbf{v} - \mathbf{w} = -\mathbf{i} + 3\mathbf{j}$

$$\cos\theta = \frac{\|\mathbf{v}\|^2 + \|\mathbf{w}\|^2 - \|\mathbf{v} - \mathbf{w}\|^2}{2\|\mathbf{v}\|\,\|\mathbf{w}\|} = \frac{5 + 5 - 10}{2\sqrt{5}\sqrt{5}} = 0$$

$\theta = 90°$

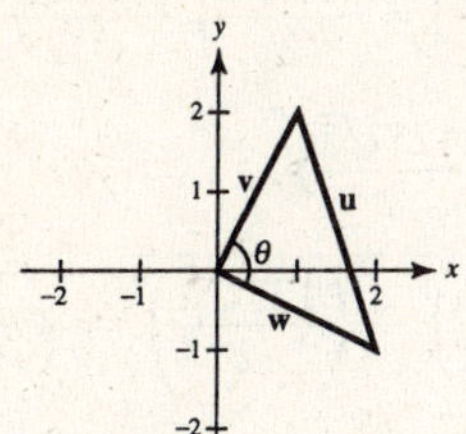

67. Force One: $\mathbf{u} = 45\mathbf{i}$

Force Two: $\mathbf{v} = 60 \cos \theta \mathbf{i} + 60 \sin \theta \mathbf{j}$

Resultant Force: $\mathbf{u} + \mathbf{v} = (45 + 60 \cos \theta)\mathbf{i} + 60 \sin \theta \mathbf{j}$

$$\|\mathbf{u} + \mathbf{v}\| = \sqrt{(45 + 60 \cos \theta)^2 + (60 \sin \theta)^2} = 90$$

$$2025 + 5400 \cos \theta + 3600 = 8100$$

$$5400 \cos \theta = 2475$$

$$\cos \theta = \frac{2475}{5400} \approx 0.4583$$

$$\theta \approx 62.7^\circ$$

68. Force One: $\mathbf{u} = 3000\mathbf{i}$

Force Two: $\mathbf{v} = 1000 \cos \theta \mathbf{i} + 1000 \sin \theta \mathbf{j}$

Resultant Force: $\mathbf{u} + \mathbf{v} = (3000 + 1000 \cos \theta)\mathbf{i} + 1000 \sin \theta \mathbf{j}$

$$\|\mathbf{u} + \mathbf{v}\| = \sqrt{(3000 + 1000 \cos \theta)^2 + (1000 \sin \theta)^2} = 3750$$

$$9{,}000{,}000 + 6{,}000{,}000 \cos \theta + 1{,}000{,}000 = 14{,}062,500$$

$$6{,}000{,}000 \cos \theta = 4{,}062,500$$

$$\cos \theta = \frac{4{,}062{,}500}{6{,}000{,}000} \approx 0.6771$$

$$\theta \approx 47.4^\circ$$

69. The difference $\mathbf{u} - \mathbf{v}$ is the vector from the terminal point of $\mathbf{v}$ to the terminal point of $\mathbf{u}$.

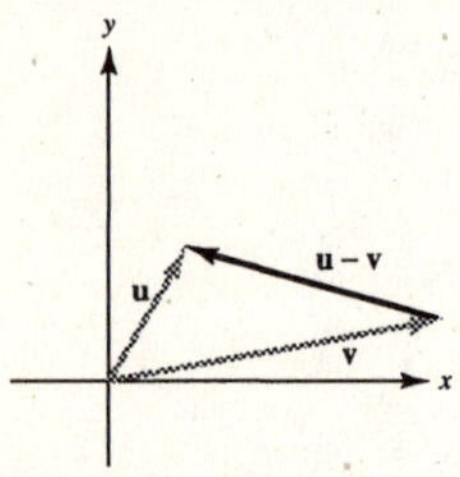

70. Two vectors that have the same magnitude and direction are equivalent. *A* and *C* appear to be equivalent.

71. The sum of $\mathbf{u}$ and $\mathbf{v}$ will be greater in figure (a). The angle between $\mathbf{u}$ and $\mathbf{v}$ is acute.

72. $\mathbf{u} = 300 \cos 0^\circ \mathbf{i} + 300 \sin 0^\circ \mathbf{j} = 300\mathbf{i}$

$$\mathbf{v} = 125 \cos 45^\circ \mathbf{i} + 125 \sin 45^\circ \mathbf{j} = \frac{125\sqrt{2}}{2}\mathbf{i} + \frac{125\sqrt{2}}{2}\mathbf{j}$$

$$\mathbf{u} + \mathbf{v} = \left(300 + \frac{125\sqrt{2}}{2}\right)\mathbf{i} + \frac{125\sqrt{2}}{2}\mathbf{j}$$

$$\|\mathbf{u} + \mathbf{v}\| = \sqrt{\left(300 + \frac{125\sqrt{2}}{2}\right)^2 + \left(\frac{125\sqrt{2}}{2}\right)^2} \approx 398.32 \text{ newtons}$$

$$\tan \theta = \frac{\frac{125\sqrt{2}}{2}}{300 + \frac{125\sqrt{2}}{2}} = \frac{125\sqrt{2}}{600 + 125\sqrt{2}}$$

$$\theta \approx 12.82^\circ$$

73.

$$\mathbf{u} = (2000\cos 30°)\mathbf{i} + (2000\sin 30°)\mathbf{j}$$
$$\approx 1732.05\mathbf{i} + 1000\mathbf{j}$$
$$\mathbf{v} = (900\cos(-45°))\mathbf{i} + (900\sin(-45°))\mathbf{j}$$
$$\approx 636.4\mathbf{i} + -636.4\mathbf{j}$$
$$\mathbf{u} + \mathbf{v} \approx 2368.45\mathbf{i} + 363.6\mathbf{j}$$
$$\|\mathbf{u} + \mathbf{v}\| \approx \sqrt{(2368.45)^2 + (363.6)^2} \approx 2396.20 \text{ newtons}$$
$$\tan\theta = \frac{363.6}{2368.45} \approx 0.1535 \implies \theta \approx 8.7°$$

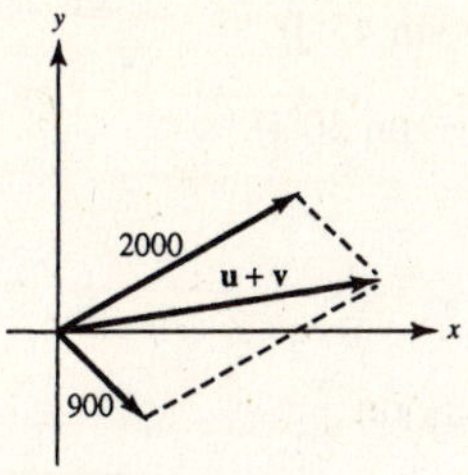

74. $\mathbf{u} = 75\cos 45°\mathbf{i} + 75\sin 45°\mathbf{j} = \dfrac{75\sqrt{2}}{2}\mathbf{i} + \dfrac{75\sqrt{2}}{2}\mathbf{j}$

$\mathbf{v} = 100\cos 60°\mathbf{i} + 100\sin 60°\mathbf{j} = 50\mathbf{i} + 50\sqrt{3}\mathbf{j}$

$\mathbf{w} = 125\cos 120°\mathbf{i} + 125\sin 120°\mathbf{j} = -\dfrac{125}{2}\mathbf{i} + \dfrac{125\sqrt{3}}{2}\mathbf{j}$

$\mathbf{u} + \mathbf{v} + \mathbf{w} \approx 40.533\mathbf{i} + 247.889\mathbf{j}$

$\|\mathbf{u} + \mathbf{v} + \mathbf{w}\| \approx 251.18$ pounds

$\tan\theta \approx \dfrac{247.889}{40.533}$

$\theta \approx 80.71°$

75.

$$\mathbf{u} = (70\cos 60°)\mathbf{i} - (70\sin 60°)\mathbf{j} \approx 35\mathbf{i} - 60.62\mathbf{j}$$
$$\mathbf{v} = (40\cos 45°)\mathbf{i} + (40\sin 45°)\mathbf{j} \approx 28.28\mathbf{i} + 28.28\mathbf{j}$$
$$\mathbf{w} = (60\cos 120°)\mathbf{i} + (60\sin 120°)\mathbf{j} \approx -30\mathbf{i} + 51.96\mathbf{j}$$
$$\mathbf{u} + \mathbf{v} + \mathbf{w} = 33.28\mathbf{i} + 19.62\mathbf{j}$$
$$\|\mathbf{u} + \mathbf{v} + \mathbf{w}\| \approx 38.6 \text{ pounds}$$
$$\tan\theta \approx \frac{19.62}{33.28} \approx 0.5895$$
$$\theta \approx 30.5°$$

76. Horizontal component of velocity: $70\cos 35° \approx 57.34$ feet per second

Vertical component of velocity: $70\sin 35° \approx 40.15$ feet per second

77. Horizontal component of velocity: $1100\cos 5° \approx 1095.8$ feet per second

Vertical component of velocity: $1100\sin 5° \approx 95.9$ feet per second

78. (a)

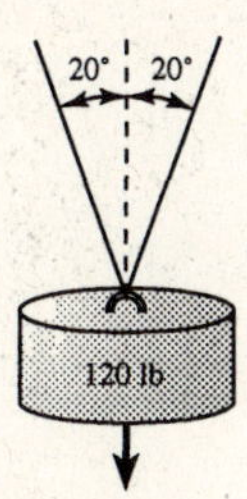

(b) Rope 1: $\mathbf{u} = \|\mathbf{u}\|(\cos 70°\mathbf{i} - \sin 70°\mathbf{j})$

Rope 2: $\mathbf{v} = \|\mathbf{u}\|(-\cos 70°\mathbf{i} - \sin 70°\mathbf{j})$

Resultant: $\mathbf{u} + \mathbf{v} = -120\mathbf{j}$

$$-\|\mathbf{u}\|\sin 70° - \|\mathbf{u}\|\sin 70° = -120$$
$$\|\mathbf{u}\| \approx 63.9$$

Therefore, the tension of each rope is $\|\mathbf{u}\| \approx 63.9$ pounds.

79. Cable $\overrightarrow{AC}$: $\mathbf{u} = \|\mathbf{u}\|(\cos 45°\mathbf{i} - \sin 45°\mathbf{j})$

Cable $\overrightarrow{BC}$: $\mathbf{v} = \|\mathbf{v}\|(-\cos 30°\mathbf{i} - \sin 30°\mathbf{j})$

Resultant: $\mathbf{u} + \mathbf{v} = -3000\mathbf{j}$

$\|\mathbf{u}\| \cos 45° - \|\mathbf{v}\| \cos 30° = 0$

$-\|\mathbf{u}\| \sin 45° - \|\mathbf{v}\| \sin 30° = -3000$

Solving this system yields:

$T_{AC} = \|\mathbf{u}\| \approx 2689.7$ pounds

$T_{BC} = \|\mathbf{v}\| \approx 2196.2$ pounds

80. Rope $\overrightarrow{AC}$: $\mathbf{u} = 15\mathbf{i} - 36\mathbf{j}$

The vector lies in Quadrant IV and its reference angle is $\arctan\left(\frac{12}{5}\right)$.

$\mathbf{u} = \|\mathbf{u}\|\left[\cos\left(\arctan \frac{12}{5}\right)\mathbf{i} - \sin\left(\arctan \frac{12}{5}\right)\mathbf{j}\right]$

Rope $\overrightarrow{BC}$: $\mathbf{v} = -30\mathbf{i} - 36\mathbf{j}$

The vector lies in Quadrant III and its reference angle is $\arctan\left(\frac{6}{5}\right)$.

$\mathbf{v} = \|\mathbf{v}\|\left[-\cos\left(\arctan \frac{6}{5}\right)\mathbf{i} - \sin\left(\arctan \frac{6}{5}\right)\mathbf{j}\right]$

Resultant: $\mathbf{u} + \mathbf{v} = -4000\mathbf{j}$

$\|\mathbf{u}\| \cos\left(\arctan \frac{12}{5}\right) - \|\mathbf{v}\| \cos\left(\arctan \frac{6}{5}\right) = 0$

$-\|\mathbf{u}\| \sin\left(\arctan \frac{12}{5}\right) - \|\mathbf{v}\| \sin\left(\arctan \frac{6}{5}\right) = -4000$

Solving this system of equations yields: $T_{AC} = \|\mathbf{u}\| \approx 2888.9$ pounds

$T_{BC} = \|\mathbf{v}\| \approx 1735.6$ pounds

81. Towline 1: $\mathbf{u} = \|\mathbf{u}\|(\cos 18°\mathbf{i} + \sin 18°\mathbf{j})$

Towline 2: $\mathbf{v} = \|\mathbf{u}\|(\cos 18°\mathbf{i} - \sin 18°\mathbf{j})$

Resultant: $\mathbf{u} + \mathbf{v} = 6000\mathbf{i}$

$\|\mathbf{u}\| \cos 18° + \|\mathbf{u}\| \cos 18° = 6000$

$\|\mathbf{u}\| \approx 3154.4$

Therefore, the tension on each towline is

$\|\mathbf{u}\| \approx 3154.4$ pounds.

82. Airspeed: $\mathbf{u} = (875 \cos 58°)\mathbf{i} - (875 \sin 58°)\mathbf{j}$

Groundspeed: $\mathbf{v} = (800 \cos 50°)\mathbf{i} - (800 \sin 50°)\mathbf{j}$

Wind: $\mathbf{w} = \mathbf{v} - \mathbf{u} = (800 \cos 50° - 875 \cos 58°)\mathbf{i} + (-800 \sin 50° + 875 \sin 58°)\mathbf{j}$

$\approx 50.5507\mathbf{i} + 129.2065\mathbf{j}$

Wind speed: $\|\mathbf{w}\| \approx \sqrt{(50.5077)^2 + (129.2065)^2}$

≈ 138.7 kilometers per hour

Wind direction: $\tan \theta \approx \dfrac{129.2065}{50.5507}$

$\theta \approx 68.6°$ Bearing: $90° - 68.6° = 21.4°$

N 21.4° E

83. (a)

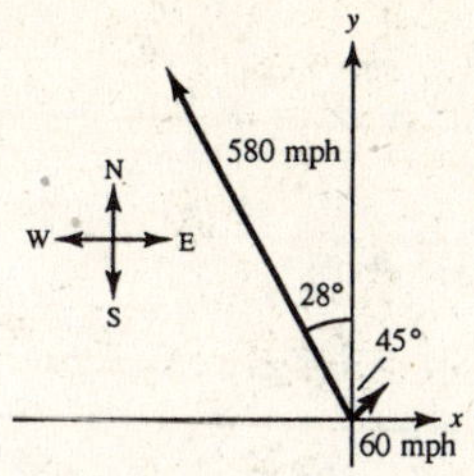

(b) Plane: $\mathbf{u} = (580 \cos 118°)\mathbf{i} + (580 \sin 118°)\mathbf{j} \approx -272.3\mathbf{i} + 512.1\mathbf{j}$

Wind: $\mathbf{v} = (60 \cos 45°)\mathbf{i} + (60 \sin 45°)\mathbf{j} \approx 42.4\mathbf{i} + 42.4\mathbf{j}$

$$\mathbf{u} + \mathbf{v} \approx -229.9\mathbf{i} + 554.5\mathbf{j}$$

$$\|\mathbf{u} + \mathbf{v}\| \approx \sqrt{(-229.9)^2 + (554.5)^2} \approx 600.3$$

$$\tan\theta \approx -\frac{554.5}{229.9} \approx -2.4119 \Rightarrow \theta \approx 112.5$$

The ground speed is 600.3 miles per hour and the heading is N 22.5° W.

84. $W = FD = (100 \cos 50°)(30) = 1928.4$ foot–pounds

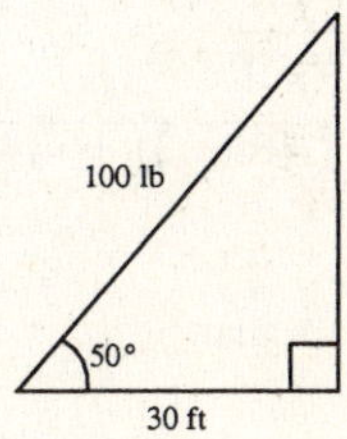

85. Horizontal force: $\mathbf{u} = \|\mathbf{u}\|\mathbf{i}$

Weight: $\mathbf{w} = -\mathbf{j}$

Rope: $\mathbf{t} = \|\mathbf{t}\|(\cos 135°\mathbf{i} + \sin 135°\mathbf{j})$

$$\mathbf{u} + \mathbf{w} + \mathbf{t} = \mathbf{0} \Rightarrow \|\mathbf{u}\| + \|\mathbf{t}\|\cos 135° = 0$$

$$-1 + \|\mathbf{t}\|\sin 135° = 0$$

$\|\mathbf{t}\| \approx \sqrt{2}$ pounds

$\|\mathbf{u}\| \approx 1$ pound

86. Program.

Answers will vary depending on the model of the graphing utility.

The following program is written for a *TI-82* or *TI-83* or *TI-83 Plus* graphing calculator. The program sketches two vectors $\mathbf{u} = a\mathbf{i} + b\mathbf{j}$ and $\mathbf{v} = c\mathbf{i} + d\mathbf{j}$ in standard position, and then sketches the vector difference $\mathbf{u} - \mathbf{v}$ using the parallelogram law.

```
PROGRAM: SUBVECT
:Input "ENTER A", A
:Input "ENTER B", B
:Input "ENTER C", C
:Input "ENTER D", D
:Line (0, 0, A, B)
:Line (0, 0, C, D)
:Pause
:A-C→E
:B-D→F
:Line (A, B, C, D)
:Line (A, B, E, F)
:Line (0, 0, E, F)
:Pause
:ClrDraw
:Stop
```

87. $\mathbf{u} = \langle 5 - 1, 2 - 6\rangle = \langle 4, -4\rangle$

$\mathbf{v} = \langle 9 - 4, 4 - 5\rangle = \langle 5, -1\rangle$

$\mathbf{u} - \mathbf{v} = \langle -1, -3\rangle$ or $\mathbf{v} - \mathbf{u} = \langle 1, 3\rangle$

88. $\mathbf{u} = \langle 80 - 10, 80 - 60\rangle = \langle 70, 20\rangle$

$\mathbf{v} = \langle -20 - (-100), 70 - 0\rangle = \langle 80, 70\rangle$

$\mathbf{u} - \mathbf{v} = \langle 70 - 80, 20 - 70\rangle = \langle -10, -50\rangle$

$\mathbf{v} - \mathbf{u} = \langle 80 - 70, 70 - 20\rangle = \langle 10, 50\rangle$

89. True. See Example 1.

90. True, by the definition of a unit vector.

$$\mathbf{u} = \frac{\mathbf{v}}{\|\mathbf{v}\|} \Rightarrow \mathbf{v} = \|\mathbf{v}\|\mathbf{u}$$

91. False, $a = b = 0$.

92. True.

$\|\mathbf{u}\| = \sqrt{a^2 + b^2} = 1 \Rightarrow a^2 + b^2 = 1$

Section 13.4 Vectors and Dot Products

1. $\mathbf{u} = \langle 6, 1 \rangle,\ \mathbf{v} = \langle -2, 3 \rangle$

$\mathbf{u} \cdot \mathbf{v} = 6(-2) + 1(3) = -9$

2. $\mathbf{u} = \langle 5, 12 \rangle, \mathbf{v} = \langle -3, 2 \rangle$

$\mathbf{u} \cdot \mathbf{v} = 5(-3) + 12(2) = 9$

3. $\mathbf{u} - 4\mathbf{i} - 2\mathbf{j},\ \mathbf{v} = \mathbf{i} - \mathbf{j}$

$\mathbf{u} \cdot \mathbf{v} = 4(1) + (-2)(-1) = 6$

4. $\mathbf{u} = 3\mathbf{i} + 4\mathbf{j}, \mathbf{v} = 7\mathbf{i} - 2\mathbf{j}$

$\mathbf{u} \cdot \mathbf{v} = 3(7) + 4(-2) = 13$

5. $\mathbf{u} = \langle 2, 2 \rangle$

$\mathbf{u} \cdot \mathbf{u} = 2(2) + 2(2) = 8$

The result is a scalar.

6. $\mathbf{v} = \langle -3, 4 \rangle$

$$\|\mathbf{v}\| + 3 = \sqrt{\mathbf{v} \cdot \mathbf{v}} + 3 = \sqrt{(-3)(-3) + 4(4)} + 3 = 8$$

The result is a scalar.

7. $\mathbf{u} = \langle 2, 2 \rangle,\ \mathbf{v} = \langle -3, 4 \rangle$

$$(\mathbf{u} \cdot \mathbf{v})\mathbf{v} = [(2)(-3) + 2(4)]\langle -3, 4 \rangle = 2\langle -3, 4 \rangle = \langle -6, 8 \rangle$$

The result is a vector.

8. $\mathbf{u} = \langle 2, 2 \rangle, \mathbf{v} = \langle -3, 4 \rangle$

$3\mathbf{u} \cdot \mathbf{v} = 3[(2)(-3) + (2)(4)] = 3(2) = 6$

The result is a scalar.

9. $\mathbf{u} = \langle -5, 12 \rangle$

$\|\mathbf{u}\| = \sqrt{\mathbf{u} \cdot \mathbf{u}} = \sqrt{(-5)^2 + 12^2} = 13$

10. $\mathbf{u} = \langle 2, -4 \rangle$

$\|\mathbf{u}\| = \sqrt{\mathbf{u} \cdot \mathbf{u}} = \sqrt{2(2) + (-4)(-4)} = \sqrt{20} = 2\sqrt{5}$

11. $\mathbf{u} = 20\mathbf{i} + 25\mathbf{j}$

$\|\mathbf{u}\| = \sqrt{(20)^2 + (25)^2} = \sqrt{1025} = 5\sqrt{41}$

12. $\mathbf{u} = 12\mathbf{i} - 16\mathbf{j}$

$\|\mathbf{u}\| = \sqrt{\mathbf{u} \cdot \mathbf{u}} = \sqrt{12(12) + (-16)(-16)} = \sqrt{400} = 20$

13. $\mathbf{u} = 6\mathbf{j}$

$\|\mathbf{u}\| = \sqrt{(0)^2 + (6)^2} = \sqrt{36} = 6$

14. $\mathbf{u} = -21\mathbf{i}$

$\|\mathbf{u}\| = \sqrt{\mathbf{u} \cdot \mathbf{u}} = \sqrt{(-21)(-21) + 0(0)} = \sqrt{21^2} = 21$

15. $\mathbf{u} = \langle 1, 0 \rangle,\ \mathbf{v} = \langle 0, -2 \rangle$

$$\cos\theta = \frac{\mathbf{u} \cdot \mathbf{v}}{\|\mathbf{u}\|\,\|\mathbf{v}\|} = \frac{0}{(1)(2)} = 0$$

$\theta = 90°$

16. $\mathbf{u} = \langle 3, 2 \rangle, \mathbf{v} = \langle 4, 0 \rangle$

$$\cos\theta = \frac{\mathbf{u} \cdot \mathbf{v}}{\|\mathbf{u}\|\,\|\mathbf{v}\|} = \frac{3(4) + 2(0)}{\sqrt{13}(4)} = \frac{3}{\sqrt{13}} \approx 0.8321$$

$\theta \approx 33.7°$

17. $\mathbf{u} = 3\mathbf{i} + 4\mathbf{j},\ \mathbf{v} = -2\mathbf{j}$

$$\cos\theta = \frac{\mathbf{u} \cdot \mathbf{v}}{\|\mathbf{u}\|\,\|\mathbf{v}\|} = -\frac{8}{(5)(2)}$$

$\theta = \arccos\left(-\frac{4}{5}\right)$

$\theta \approx 143.13°$

18. $\mathbf{u} = 2\mathbf{i} - 3\mathbf{j}, \mathbf{v} = \mathbf{i} - 2\mathbf{j}$

$$\cos\theta = \frac{\mathbf{u} \cdot \mathbf{v}}{\|\mathbf{u}\|\,\|\mathbf{v}\|} = \frac{2(1) + (-3)(-2)}{\sqrt{2^2 + 3^2}\sqrt{1^2 + 2^2}} = \frac{8}{\sqrt{65}} \approx 0.992278$$

$\theta \approx 7.13°$

19. $\mathbf{u} = 2\mathbf{i} - \mathbf{j}, \mathbf{v} = 6\mathbf{i} + 4\mathbf{j}$

$$\cos\theta = \frac{\mathbf{u} \cdot \mathbf{v}}{\|\mathbf{u}\|\,\|\mathbf{v}\|} = \frac{8}{\sqrt{5}\sqrt{52}} \Rightarrow \theta \approx 60.26°$$

20. $\mathbf{u} = -6\mathbf{i} - 3\mathbf{j}, \mathbf{v} = -8\mathbf{i} + 4\mathbf{j}$

$$\cos \mathbf{u} = \frac{\mathbf{u} \cdot \mathbf{v}}{\|\mathbf{u}\|\,\|\mathbf{v}\|} = \frac{-6(-8) + (-3)(4)}{\sqrt{45}\sqrt{80}}$$

$$= \frac{36}{60} = 0.6$$

$$\theta \approx 53.13^\circ$$

21. $\mathbf{u} = 5\mathbf{i} + 5\mathbf{j},\ \mathbf{v} = -6\mathbf{i} + 6\mathbf{j}$

$$\cos \theta = \frac{\mathbf{u} \cdot \mathbf{v}}{\|\mathbf{u}\|\,\|\mathbf{v}\|} = 0 \Rightarrow \theta = 90^\circ$$

22. $\mathbf{u} = 2\mathbf{i} - 3\mathbf{j}, \mathbf{v} = 4\mathbf{i} + 3\mathbf{j}$

$$\cos \theta = \frac{\mathbf{u} \cdot \mathbf{v}}{\|\mathbf{u}\|\,\|\mathbf{v}\|} = \frac{2(4) + (-3)(3)}{\sqrt{13}\sqrt{25}} \approx -0.0555$$

$$\theta \approx 93.18^\circ$$

23. $\mathbf{u} = \left(\cos \frac{\pi}{3}\right)\mathbf{i} + \left(\sin \frac{\pi}{3}\right)\mathbf{j} = \frac{1}{2}\mathbf{i} + \frac{\sqrt{3}}{2}\mathbf{j}$

$$\mathbf{v} = \left(\cos \frac{3\pi}{4}\right)\mathbf{i} + \left(\sin \frac{3\pi}{4}\right)\mathbf{j} = -\frac{\sqrt{2}}{2}\mathbf{i} + \frac{\sqrt{2}}{2}\mathbf{j}$$

$$\|\mathbf{u}\| = \|\mathbf{v}\| = 1$$

$$\cos \theta = \frac{\mathbf{u} \cdot \mathbf{v}}{\|\mathbf{u}\|\,\|\mathbf{v}\|} = \mathbf{u} \cdot \mathbf{v} = \left(\frac{1}{2}\right)\left(-\frac{\sqrt{2}}{2}\right) + \left(\frac{\sqrt{3}}{2}\right)\left(\frac{\sqrt{2}}{2}\right) = \frac{-\sqrt{2} + \sqrt{6}}{4}$$

$$\theta = \arccos\left(\frac{-\sqrt{2} + \sqrt{6}}{4}\right) = 75^\circ = \frac{5\pi}{12}$$

24. $\mathbf{u} = \cos\left(\frac{\pi}{4}\right)\mathbf{i} + \sin\left(\frac{\pi}{4}\right)\mathbf{j} = \frac{\sqrt{2}}{2}\mathbf{i} + \frac{\sqrt{2}}{2}\mathbf{j}$

$$\mathbf{v} = \cos\left(\frac{\pi}{2}\right)\mathbf{i} + \sin\left(\frac{\pi}{2}\right)\mathbf{j} = \mathbf{j}$$

$$\cos \theta = \frac{\mathbf{u} \cdot \mathbf{v}}{\|\mathbf{u}\|\,\|\mathbf{v}\|} = \frac{\frac{\sqrt{2}}{2}(0) + \frac{\sqrt{2}}{2}(1)}{1 \cdot 1} = \frac{\sqrt{2}}{2}$$

$$\theta = \frac{\pi}{4}$$

25. $P = (1, 2),\ Q = (3, 4),\ R = (2, 5)$

$$\overrightarrow{PQ} = \langle 2, 2\rangle,\ \overrightarrow{PR} = \langle 1, 3\rangle,\ \overrightarrow{QR} = \langle -1, 1\rangle$$

$$\cos \alpha = \frac{\overrightarrow{PQ} \cdot \overrightarrow{PR}}{\|\overrightarrow{PQ}\|\,\|\overrightarrow{PR}\|} = \frac{8}{(2\sqrt{2})(\sqrt{10})} \Rightarrow \alpha = \arccos \frac{2}{\sqrt{5}} \approx 26.6^\circ$$

$$\cos \beta = \frac{\overrightarrow{PQ} \cdot \overrightarrow{QR}}{\|\overrightarrow{PQ}\|\,\|\overrightarrow{QR}\|} = 0 \Rightarrow \beta = 90^\circ.\text{ Thus, } \gamma = 180^\circ - 26.6^\circ - 90^\circ = 63.4^\circ.$$

26. $P = (-3, -4), Q = (1, 7), R = (8, 2)$

$\overrightarrow{PQ} = \langle 4, 11\rangle, \overrightarrow{QR} = \langle 7, -5\rangle,$

$\overrightarrow{PR} = \langle 11, 6\rangle, \overrightarrow{QP} = \langle -4, -11\rangle$

$$\cos\alpha = \frac{\overrightarrow{PQ}\cdot\overrightarrow{PR}}{\|\overrightarrow{PQ}\|\,\|\overrightarrow{PR}\|} = \frac{110}{(\sqrt{137})(\sqrt{157})} \Rightarrow \alpha \approx 41.4^\circ$$

$$\cos\beta = \frac{\overrightarrow{QR}\cdot\overrightarrow{QP}}{\|\overrightarrow{QR}\|\,\|\overrightarrow{QP}\|} = \frac{27}{(\sqrt{74})(\sqrt{137})} \Rightarrow \beta \approx 74.4^\circ$$

$\alpha \approx 180^\circ - 41.4^\circ - 74.4^\circ = 64.2^\circ$

27. $P = (-3, 0), Q = (2, 2), R = (0, 6)$

$\overrightarrow{QP} = \langle -5, -2\rangle, \overrightarrow{PR} = \langle 3, 6\rangle, \overrightarrow{QR} = \langle -2, 4\rangle$

$$\cos\alpha = \frac{\overrightarrow{PQ}\cdot\overrightarrow{PR}}{\|\overrightarrow{PQ}\|\,\|\overrightarrow{PR}\|} = \frac{27}{\sqrt{29}\sqrt{45}} \Rightarrow \alpha \approx 41.6^\circ$$

$$\cos\beta = \frac{\overrightarrow{QP}\cdot\overrightarrow{QR}}{\|\overrightarrow{QP}\|\,\|\overrightarrow{GR}\|} = \frac{2}{\sqrt{29}\sqrt{20}} \Rightarrow \beta \approx 85.2^\circ$$

$\delta = 180^\circ - 41.6^\circ - 85.2^\circ = 53.2^\circ$

28. $P = (-3, 5), Q = (-1, 9), R = (7, 9)$

$\overrightarrow{PQ} = \langle 2, 4\rangle, \overrightarrow{QR} = \langle 8, 0\rangle,$

$\overrightarrow{PR} = \langle 10, 4\rangle, \overrightarrow{QP} = \langle -2, -4\rangle$

$$\cos\alpha = \frac{\overrightarrow{PQ}\cdot\overrightarrow{PR}}{\|\overrightarrow{PQ}\|\,\|\overrightarrow{PR}\|} = \frac{36}{(\sqrt{20})(\sqrt{116})}$$

$\Rightarrow \alpha \approx 41.6^\circ$

$$\cos\beta = \frac{\overrightarrow{QR}\cdot\overrightarrow{QP}}{\|\overrightarrow{QR}\|\,\|\overrightarrow{QP}\|} = \frac{-16}{8(\sqrt{20})}$$

$\Rightarrow \beta \approx 116.6^\circ$

$y \approx 180^\circ - 41.6^\circ - 116.6^\circ = 21.8^\circ$

29. $\mathbf{u}\cdot\mathbf{v} = \|\mathbf{u}\|\,\|\mathbf{v}\|\cos\theta$

$= (4)(10)\cos\frac{2\pi}{3}$

$= 40\left(-\frac{1}{2}\right)$

$= -20$

30. $\|\mathbf{u}\| = 100, \|\mathbf{v}\| = 250, \theta = \frac{\pi}{6}$

$\mathbf{u}\cdot\mathbf{v} = \|\mathbf{u}\|\,\|\mathbf{v}\|\cos\theta$

$= (100)(250)\cos\frac{\pi}{6}$

$= 25{,}000\cdot\frac{\sqrt{3}}{2}$

$= 12{,}500\sqrt{3}$

31. $\mathbf{u}\cdot\mathbf{v} = \|\mathbf{u}\|\,\|\mathbf{v}\|\cos\theta$

$= (81)(64)\cos\frac{\pi}{4}$

$= 5184\left(\frac{\sqrt{2}}{2}\right)$

$= 2592\sqrt{2}$

32. $\|\mathbf{u}\| = 9, \|\mathbf{v}\| = 144, \theta = \frac{\pi}{2}$

$\mathbf{u}\cdot\mathbf{v} = \|\mathbf{u}\|\,\|\mathbf{v}\|\cos\theta$

$= (9)(144)\cos\frac{\pi}{2}$

$= 0$

33. $\mathbf{u} = \langle -12, 30\rangle,\ \mathbf{v} = \left\langle \frac{1}{2}, -\frac{5}{4}\right\rangle$

$\mathbf{u} = -24\mathbf{v} \Rightarrow$ **u** and **v** are parallel.

34. $\mathbf{u} = \langle 3, 15\rangle, \mathbf{v} = \langle -1, 5\rangle$

$\mathbf{u} \neq k\mathbf{v} \Rightarrow$ Not parallel

$\mathbf{u}\cdot\mathbf{v} \neq 0 \Rightarrow$ Not orthogonal

Neither

35. $\mathbf{u} = \frac{1}{4}(3\mathbf{i} - \mathbf{j}),\ \mathbf{v} = 5\mathbf{i} + 6\mathbf{j}$

$\mathbf{u} \neq k\mathbf{v} \Rightarrow$ Not parallel

$\mathbf{u}\cdot\mathbf{v} \neq 0 \Rightarrow$ Not orthogonal

Neither

36. $\mathbf{u} = i, \mathbf{v} = -2\mathbf{i} + 2\mathbf{j}$

$\mathbf{u} \neq k\mathbf{v} \Rightarrow$ Not parallel

$\mathbf{u}\cdot\mathbf{v} \neq 0 \Rightarrow$ Not orthogonal

Neither

37. $\mathbf{u} = 2\mathbf{i} - 2\mathbf{j},\ \mathbf{v} = -\mathbf{i} - \mathbf{j}$

$\mathbf{u}\cdot\mathbf{v} = 0 \Rightarrow$ **u** and **v** are orthogonal.

38. $\mathbf{u} = \langle \cos\theta, \sin\theta\rangle$

$\mathbf{v} = \langle \sin\theta, -\cos\theta\rangle$

$\mathbf{u}\cdot\mathbf{v} = 0 \Rightarrow$ **u** and **v** are orthogonal.

39. $\mathbf{u} = \langle 2, 2\rangle,\ \mathbf{v} = \langle 6, 1\rangle$

$$\mathbf{w}_1 = \text{proj}_{\mathbf{v}}\mathbf{u} = \left(\frac{\mathbf{u}\cdot\mathbf{v}}{\|\mathbf{v}\|^2}\right)\mathbf{v} = \frac{14}{37}\mathbf{v} = \frac{14}{37}\langle 6, 1\rangle$$

$$\mathbf{w}_2 = \mathbf{u} - \mathbf{w}_1 = \langle 2, 2\rangle - \frac{14}{37}\langle 6, 1\rangle = \left\langle -\frac{10}{37}, \frac{60}{37}\right\rangle = \frac{10}{37}\langle -1, 6\rangle$$

40. $\mathbf{u} = \langle 4, 2\rangle, \mathbf{v} = \langle 1, -2\rangle$

$$\mathbf{w}_1 = \text{proj}_{\mathbf{v}}\mathbf{u} = \left(\frac{\mathbf{u}\cdot\mathbf{v}}{\|\mathbf{v}\|^2}\right)\mathbf{v} = 0\langle 1, -2\rangle = \langle 0, 0\rangle$$

$$\mathbf{w}_2 = \mathbf{u} - \mathbf{w}_1 = \langle 4, 2\rangle - \langle 0, 0\rangle = \langle 4, 2\rangle$$

41. $\mathbf{u} = \langle 0, 3\rangle,\ \mathbf{v} = \langle 2, 15\rangle$

$$\mathbf{w}_1 = \text{proj}_{\mathbf{v}}\mathbf{u} = \left(\frac{\mathbf{u}\cdot\mathbf{v}}{\|\mathbf{v}\|^2}\right)\mathbf{v} = \frac{45}{229}\langle 2, 15\rangle$$

$$\mathbf{w}_2 = \mathbf{u} - \mathbf{w}_1 = \langle 0, 3\rangle - \frac{45}{229}\langle 2, 15\rangle = \left\langle -\frac{90}{229}, \frac{12}{229}\right\rangle = \frac{6}{229}\langle -15, 2\rangle$$

42. $\mathbf{u} = \langle -3, -2\rangle, \mathbf{v} = \langle -4, -1\rangle$

$$\mathbf{w}_1 = \text{proj}_{\mathbf{v}}\mathbf{u} = \left(\frac{\mathbf{u}\cdot\mathbf{v}}{\|\mathbf{v}\|^2}\right)\mathbf{v} = \left(\frac{14}{17}\right)\langle -4, -1\rangle$$

$$\mathbf{w}_2 = \mathbf{u} - \mathbf{w}_1 = \langle -3, -2\rangle - \frac{14}{17}\langle -4, -1\rangle$$

$$= \frac{5}{17}\langle 1, -4\rangle$$

43. $\mathbf{u} = \langle 3, 5\rangle$

For $\mathbf{v}$ to be orthogonal to $\mathbf{u}$, $\mathbf{u}\cdot\mathbf{v}$ must equal 0.

Two possibilities: $\langle -5, 3\rangle$ and $\langle 5, -3\rangle$

44. $\mathbf{u} = \langle -6, 3\rangle$

For $\mathbf{v}$ to be orthogonal to $\mathbf{u}$, $\mathbf{u}\cdot\mathbf{v}$ must be equal to 0.

Two possibilities: $\langle 3, 6\rangle, \langle -3, -6\rangle$

45. $\mathbf{u} = \frac{2}{3}\mathbf{i} - \frac{1}{2}\mathbf{j}$

For $\mathbf{u}$ and $\mathbf{v}$ to be orthogonal, $\mathbf{u}\cdot\mathbf{v}$ must equal 0.

Two possibilities: $\frac{1}{2}\mathbf{i} + \frac{2}{3}\mathbf{j}$ and $-\frac{1}{2}\mathbf{i} - \frac{2}{3}\mathbf{j}$

46. $\mathbf{u} = -\frac{5}{2}\mathbf{i} - 3\mathbf{j}$

For $\mathbf{v}$ to be orthogonal to $\mathbf{u}$, $\mathbf{u}\cdot\mathbf{v}$ must be equal to 0.

Two possibilities: $\mathbf{v} = 3\mathbf{i} - \frac{5}{2}\mathbf{j}$

$$\mathbf{v} = -3\mathbf{i} + \frac{5}{2}\mathbf{j}$$

47. $\mathbf{w} = \|\text{proj}_{\overrightarrow{PQ}}\mathbf{v}\|\,\|\overrightarrow{PQ}\|$ where $\overrightarrow{PQ} = \langle 4, 7\rangle$ and $\mathbf{v} = \langle 1, 4\rangle$.

$$\text{proj}_{\overrightarrow{PQ}}\mathbf{v} = \left(\frac{\mathbf{v}\cdot\overrightarrow{PQ}}{\|\overrightarrow{PQ}\|^2}\right)\overrightarrow{PQ} = \left(\frac{32}{65}\right)\langle 4, 7\rangle$$

$$\mathbf{w} = \|\text{proj}_{\overrightarrow{PQ}}\mathbf{v}\|\,\|\overrightarrow{PQ}\| = \left(\frac{32\sqrt{65}}{65}\right)\left(\sqrt{65}\right) = 32$$

48. $P = (1, 3), Q = (-3, 5), \mathbf{v} = -2\mathbf{i} + 3\mathbf{j}$

$$\text{work} = \mathbf{v}\cdot\overrightarrow{PQ}$$

$$= (-2\mathbf{i} + 3\mathbf{j})\cdot(-4\mathbf{i} + 2\mathbf{j})$$

$$= (-2)(-4) + 3(2) = 14$$

49. Since $\frac{\mathbf{u}\cdot\mathbf{v}}{\|\mathbf{u}\|\,\|\mathbf{v}\|} = \cos\theta$, we have $\mathbf{u}\cdot\mathbf{v} = \|\mathbf{u}\|\,\|\mathbf{v}\|\cos\theta$.

The dot product of $\mathbf{u}$ and $\mathbf{v}$ equals the product of their lengths (magnitudes) when $\cos\theta = 1$. This happens when the angle between them is 0.

50. $W = (\cos\theta)\,\|F\|\,\|\overrightarrow{PQ}\|$ and $\|F_1\| = \|F_2\|$

(a) If $\theta_1 = -\theta_2$ then the work is the same since

$\cos(-\theta) = \cos\theta$.

(b) If $\theta_1 = 60°$ then $W_1 = \dfrac{1}{2}\|F_1\|\,\|\overrightarrow{PQ}\|$

If $\theta_2 = 30°$ then $W_2 = \dfrac{\sqrt{3}}{2}\|F_2\|\,\|\overrightarrow{PQ}\|$

$W_2 = \sqrt{3}\,W_1$.

51. $\mathbf{u} = \langle 1650, 3200\rangle$, $\mathbf{v} = \langle 15.25, 10.50\rangle$

$\mathbf{u}\cdot\mathbf{v} = 1650(15.25) + 3200(10.50) = \$58{,}762.50$

This gives the total revenue that can be earned by selling all of the units.

52. $\mathbf{u} = \langle 1650, 3200\rangle$

$\mathbf{v} = \langle 15.25, 10.50\rangle$

Increase prices by 5%: $1.05\mathbf{v}$ Operation: Scalar multiplication of a vector.

$$\begin{aligned}\mathbf{u}\cdot 1.05\mathbf{v} &= 1.05\mathbf{u}\cdot\mathbf{v}\\ &= 1.05[1650(15.25) + 3200(10.50)]\\ &= 1.05(58{,}762.50)\\ &= 61{,}700.63\end{aligned}$$

53. (a) $\mathbf{F} = -30{,}000\mathbf{j}$

$\mathbf{v} = (\cos d°)\mathbf{i} + (\sin d°)\mathbf{j}$

$$\mathbf{w}_1 = \text{proj}_{\mathbf{v}}\mathbf{F} = \left(\frac{\mathbf{F}\cdot\mathbf{v}}{\|\mathbf{v}\|^2}\right)\mathbf{v} = (\mathbf{F}\cdot\mathbf{v})\mathbf{v} = -30{,}000\,(\sin d°)\mathbf{v}$$

The magnitude of this force is $30{,}000(\sin d°)$ and this force, in pounds, is needed to keep the truck from rolling down the hill.

(b)

d(degrees)	Force(pounds)
0°	0
1°	523.6
2°	1047.0
3°	1570.1
4°	2092.7
5°	2614.7
6°	3135.9
7°	3656.1
8°	4175.2
9°	4693.0
10°	5209.4

(c) $\mathbf{w}_2 = \mathbf{F} - \mathbf{w}_1 = -30{,}000\mathbf{j} + 30{,}000\sin 5°(\cos 5°\mathbf{i} + \sin 5°\mathbf{j})$

$\approx 2604.72\mathbf{i} - 29{,}772.12\mathbf{j}$

$\|\mathbf{w}_2\| \approx 29{,}885.8$ pounds

54. $\mathbf{F} = -30{,}000\mathbf{j}$ Gravitational force

$\mathbf{v} = (\cos 8°)\mathbf{i} + (\sin 8°)\mathbf{j}$

$$\begin{aligned}\mathbf{w}_1 = \text{proj}_{\mathbf{v}}\mathbf{F} &= \left(\frac{\mathbf{F}\cdot\mathbf{v}}{\|\mathbf{v}\|^2}\right)\mathbf{v}\\ &= (\mathbf{F}\cdot\mathbf{v})\mathbf{v} \approx -4175.2\mathbf{v}\end{aligned}$$

The magnitude of this force is 4175.2; therefore, a force of 4175.2 pounds is needed to keep the truck from rolling down the hill.

$$\begin{aligned}\mathbf{w}_2 = \mathbf{F} - \mathbf{w}_1 &= -30{,}000\mathbf{j} + 4175.2[(\cos 8°)\mathbf{i} + (\sin 8°)\mathbf{j}]\\ &= (4175.2\cos 8°)\mathbf{i} + (-30{,}000 + 4175.2\sin 8°)\mathbf{j}\end{aligned}$$

$\|\mathbf{w}_2\| \approx 29{,}708.0$ pounds

55. $\mathbf{w} = (245)(3) = 735$ Newton-meters

56. work $= (2400)(5) = 12{,}000$ foot-pounds

57. $\mathbf{w} = (\cos 30°)(45)(20) \approx 779.4$ foot-pounds

58. work $= (\cos 35°)(15,691)(800)$

$\approx 10{,}282{,}652$ newton $-$ meters

59. False. The dot product of two vectors is a scalar which can be positive, zero, or negative.

60. True.

By definition, $\mathbf{u}$ and $\mathbf{v}$ are orthogonal if $\mathbf{u} \cdot \mathbf{v} = 0$.

61. False. Work is represented by a scalar.

62. True.

$W = \mathbf{F} \cdot \overrightarrow{PQ} = 0$ if $\mathbf{F}$ and $\overrightarrow{PQ}$ are orthogonal.

63. Let $\mathbf{u} = \langle u_1, u_2 \rangle$, $\mathbf{v} = \langle v_1, v_2 \rangle$, $\mathbf{w} = \langle w_1, w_2 \rangle$ and $\mathbf{0} = \langle 0, 0 \rangle$.

(a) $$\begin{aligned} \mathbf{0} \cdot \mathbf{v} &= \langle 0, 0 \rangle \cdot \langle v_1, v_2 \rangle \\ &= 0(v_1) + 0(v_2) \\ &= 0 + 0 \\ &= 0 \end{aligned}$$

(b) $$\begin{aligned} \mathbf{u} \cdot (\mathbf{v} + \mathbf{w}) &= \langle u_1, u_2 \rangle \cdot (\langle v_1, v_2 \rangle + w_1, w_2 \rangle) \\ &= \langle u_1, u_2 \rangle \cdot \langle v_1 + w_1, v_2 + w_2 \rangle \\ &= u_1(v_1 + w_1) + u_2(v_2 + w_2) \\ &= u_1v_1 + u_1w_1 + u_2v_2 + u_2w_2 \\ &= (u_1v_1 + u_2v_2) + (u_1w_1 + u_2w_2) \\ &= \mathbf{u} \cdot \mathbf{v} + \mathbf{u} \cdot \mathbf{w} \end{aligned}$$

(c) $$\begin{aligned} c(\mathbf{u} \cdot \mathbf{v}) &= c(\langle u_1, u_2 \rangle \cdot \langle v_1, v_2 \rangle) \\ &= c(u_1v_1 + u_2v_2) \\ &= u_1(cv_1) + u_2(cv_2) \\ &= \langle u_1, u_2 \rangle \cdot \langle cv_1, cv_2 \rangle \\ &= \mathbf{u} \cdot (c\mathbf{v}) \end{aligned}$$

64. 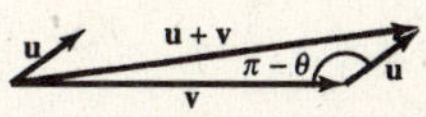

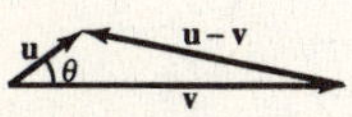

By the Law of Cosines we have the following:

$$\begin{aligned} \|\mathbf{u} + \mathbf{v}\|^2 &= \|\mathbf{u}\|^2 + \|\mathbf{v}\|^2 - 2\|\mathbf{u}\|\,\|\mathbf{v}\| \cos(\pi - \theta) \\ &= \|\mathbf{u}\|^2 + \|\mathbf{v}\|^2 + 2\|\mathbf{u}\|\,\|\mathbf{v}\| \cos \theta \end{aligned}$$

$$\|\mathbf{u} - \mathbf{v}\|^2 = \|\mathbf{u}\|^2 + \|\mathbf{v}\|^2 - 2\|\mathbf{u}\|\,\|\mathbf{v}\| \cos \theta$$

$$\begin{aligned} \|\mathbf{u} + \mathbf{v}\|^2 - \|\mathbf{u} - \mathbf{v}\|^2 &= 4\|\mathbf{u}\|\,\|\mathbf{v}\| \cos \theta \\ &= 4(\mathbf{u} \cdot \mathbf{v}) \end{aligned}$$

Alternate Solution:

$$\begin{aligned} &\|\mathbf{u} + \mathbf{v}\|^2 - \|\mathbf{u} - \mathbf{v}\|^2 \\ &= (\mathbf{u} + \mathbf{v}) \cdot (\mathbf{u} + \mathbf{v}) - (\mathbf{u} - \mathbf{v}) \cdot (\mathbf{u} - \mathbf{v}) \\ &= (\mathbf{u} \cdot \mathbf{u} + 2\mathbf{u} \cdot \mathbf{v} + \mathbf{v} \cdot \mathbf{v}) - (\mathbf{u} \cdot \mathbf{u} - 2\mathbf{u} \cdot \mathbf{v} + \mathbf{v} \cdot \mathbf{v}) \\ &= 4\mathbf{u} \cdot \mathbf{v} \end{aligned}$$

Section 13.5 Trigonometric Form of a Complex Number

1. $|-7i| = \sqrt{0^2 + (-7)^2}$

$= \sqrt{49} = 7$

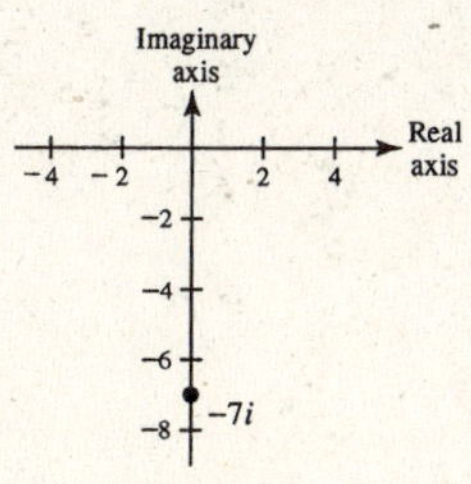

2. $|-7| = \sqrt{(-7)^2 + 0^2} = \sqrt{49} = 7$

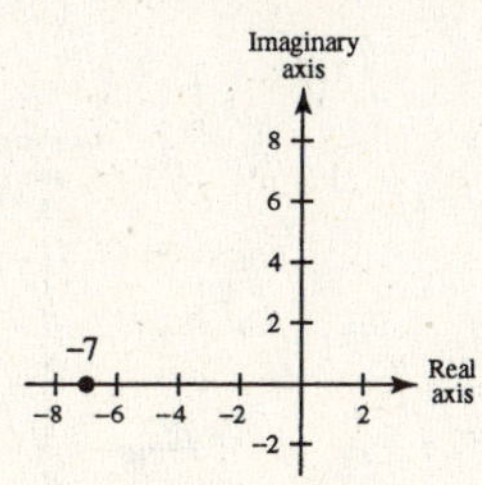

3. $|-4 + 4i| = \sqrt{(-4)^2 + (4)^2}$

$= \sqrt{32} = 4\sqrt{2}$

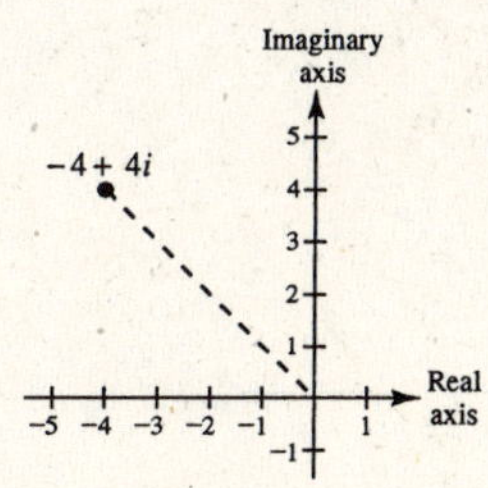

4. $|5 - 12i| = \sqrt{5^2 + (-12)^2}$

$= \sqrt{169} = 13$

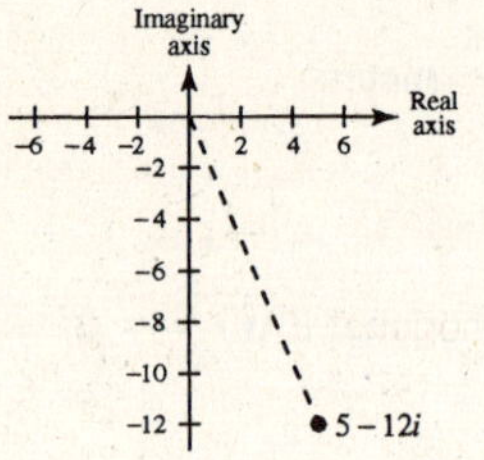

5. $|6 - 7i| = \sqrt{6^2 + (-7)^2}$

$= \sqrt{85}$

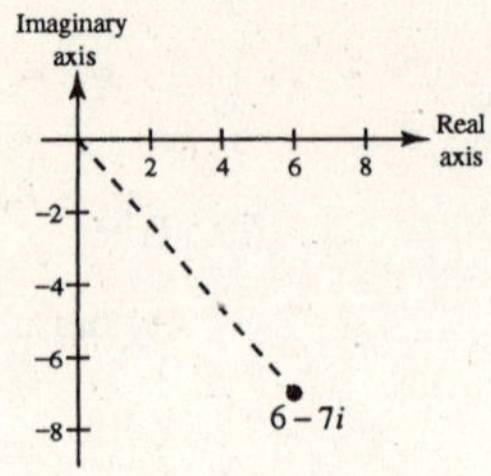

6. $|-8 + 3i| = \sqrt{(-8)^2 + (3)^2}$

$= \sqrt{73}$

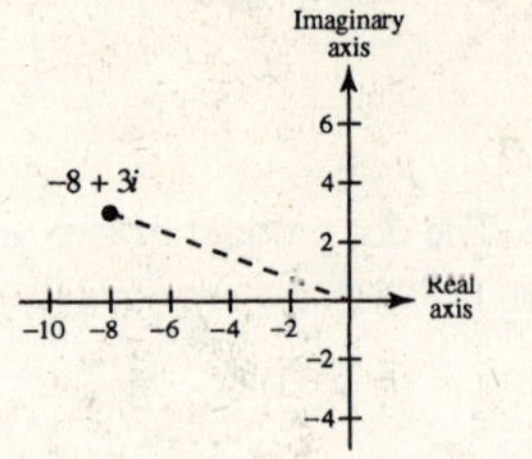

7. $z = 3i$

$r = \sqrt{0^2 + 3^2} = \sqrt{9} = 3$

$\tan\theta = \dfrac{3}{0}$, undefined $\Rightarrow \theta = \dfrac{\pi}{2}$

$z = 3\left(\cos\dfrac{\pi}{2} + i\sin\dfrac{\pi}{2}\right)$

8. $z = -2$

$r = \sqrt{(-2)^2 + 0^2} = \sqrt{4} = 2$

$\tan\theta = \dfrac{0}{-2} \Rightarrow \theta = \pi$

$z = 2(\cos\pi + i\sin\pi)$

9. $z = 3 - i$

$r = \sqrt{(3)^2 + (-1)^2} = \sqrt{10}$

$\tan\theta = -\dfrac{1}{3}$, θ is in Quadrant IV.

$\theta \approx 5.96$ radians

$z \approx \sqrt{10}(\cos 5.96 + i\sin 5.96)$

10. $z = -1 + \sqrt{3}i$

$r = \sqrt{(-1)^2 + (\sqrt{3})^2} = \sqrt{4} = 2$

$\tan\theta = \dfrac{\sqrt{3}}{-1} = -\sqrt{3} \Rightarrow \theta = \dfrac{2\pi}{3}$

$z = 2\left(\cos\dfrac{2\pi}{3} + i\sin\dfrac{2\pi}{3}\right)$

11. $z = 3 - 3i$

$r = \sqrt{3^2 + (-3)^2} = \sqrt{18} = 3\sqrt{2}$

$\tan\theta = \dfrac{-3}{3} = -1$, θ is in Quadrant IV $\Rightarrow \theta = \dfrac{7\pi}{4}$.

$z = 3\sqrt{2}\left(\cos\dfrac{7\pi}{4} + i\sin\dfrac{7\pi}{4}\right)$

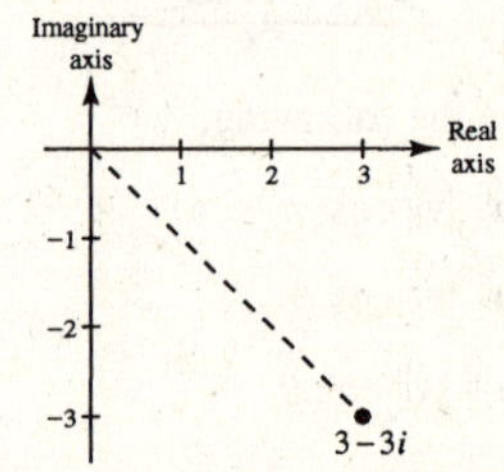

12. $z = 2 + 2i$

$r = \sqrt{2^2 + 2^2} = \sqrt{8} = 2\sqrt{2}$

$\tan\theta = \dfrac{2}{2} = 1 \Rightarrow \theta = \dfrac{\pi}{4}$

$z = 2\sqrt{2}\left(\cos\dfrac{\pi}{4} + i\sin\dfrac{\pi}{4}\right)$

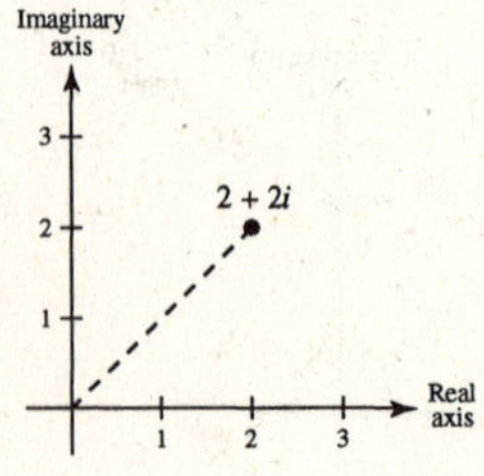

13. $z = \sqrt{3} + i$

$r = \sqrt{(\sqrt{3})^2 + 1^2} = \sqrt{4} = 2$

$\tan\theta = \dfrac{1}{\sqrt{3}} = \dfrac{\sqrt{3}}{3} \Rightarrow \theta = \dfrac{\pi}{6}$

$z = 2\left(\cos\dfrac{\pi}{6} + i\sin\dfrac{\pi}{6}\right)$

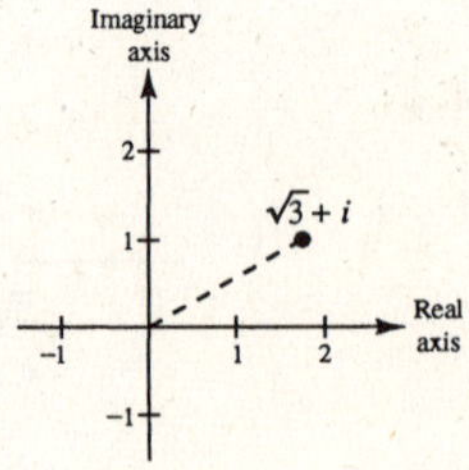

14. $z = 4 - 4\sqrt{3}i$

$r = \sqrt{4^2 + (-4\sqrt{3})^2} = 8$

$\tan \theta = \dfrac{-4\sqrt{3}}{4} = -\sqrt{3} \Rightarrow \theta = \dfrac{5\pi}{3}$

$z = 8\left(\cos \dfrac{5\pi}{3} + i \sin \dfrac{5\pi}{3}\right)$

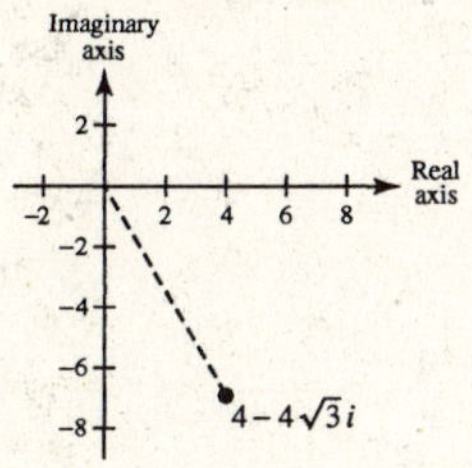

15. $z = -2(1 + \sqrt{3}i)$

$r = \sqrt{(-2)^2 + (-2\sqrt{3})^2} = \sqrt{16} = 4$

$\tan \theta = \dfrac{\sqrt{3}}{1} = \sqrt{3}$, θ is in Quadrant III $\Rightarrow \theta = \dfrac{4\pi}{3}$.

$z = 4\left(\cos \dfrac{4\pi}{3} + i \sin \dfrac{4\pi}{3}\right)$

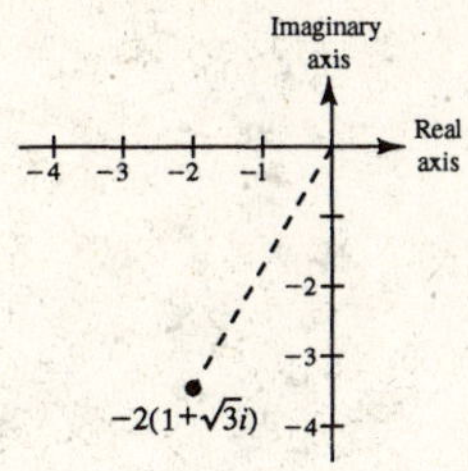

16. $z = \dfrac{5}{2}(\sqrt{3} - i)$

$r = \sqrt{\left(\dfrac{5}{2}\sqrt{3}\right)^2 + \left(\dfrac{5}{2}(-1)\right)^2} = \sqrt{\dfrac{100}{4}} = \sqrt{25} = 5$

$\tan \theta = \dfrac{-1}{\sqrt{3}} = \dfrac{-\sqrt{3}}{3} \Rightarrow \theta = \dfrac{11\pi}{6}$

$z = 5\left(\cos \dfrac{11\pi}{6} + i \sin \dfrac{11\pi}{6}\right)$

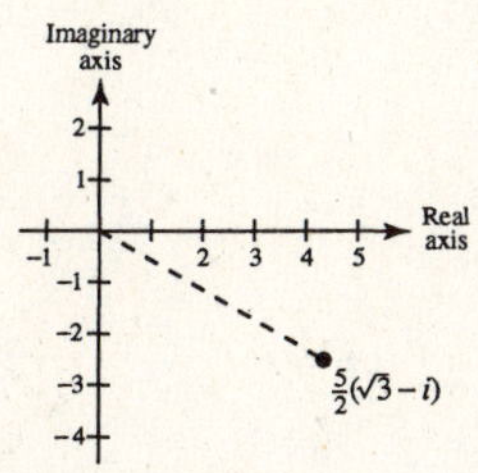

17. $z = -5i$

$r = \sqrt{0^2 + (-5)^2} = \sqrt{25} = 5$

$\tan \theta = \dfrac{-5}{0}$, undefined $\Rightarrow \theta = \dfrac{3\pi}{2}$

$z = 5\left(\cos \dfrac{3\pi}{2} + i \sin \dfrac{3\pi}{2}\right)$

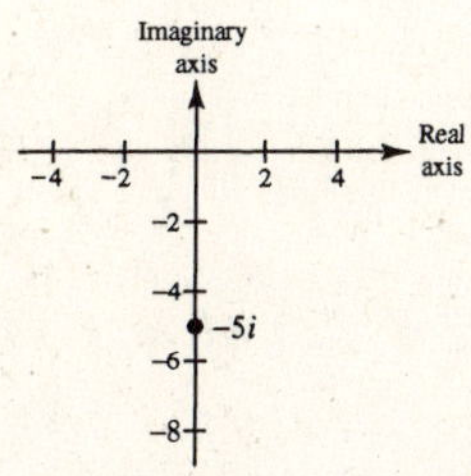

18. $z = 4i$

$r = \sqrt{0^2 + 4^2} = \sqrt{16^2} = 4$

$\tan \theta = \dfrac{4}{0}$, undefined $\Rightarrow \theta = \dfrac{\pi}{2}$

$z = 4\left(\cos \dfrac{\pi}{2} + i \sin \dfrac{\pi}{2}\right)$

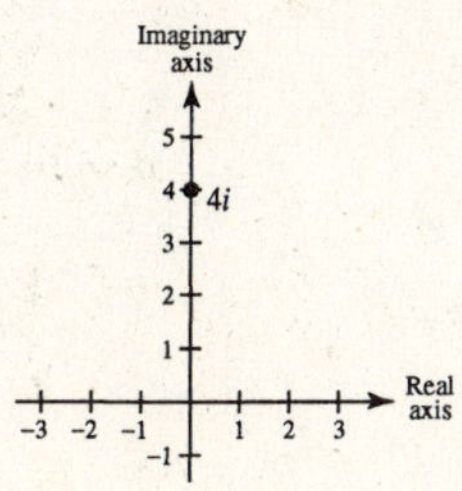

19. $z = -7 + 4i$

$r = \sqrt{(-7)^2 + (4)^2} = \sqrt{65}$

$\tan \theta = \dfrac{4}{-7}$, θ is in Quadrant II $\Rightarrow \theta \approx 2.62$.

$z \approx \sqrt{65}(\cos 2.62 + i \sin 2.62)$

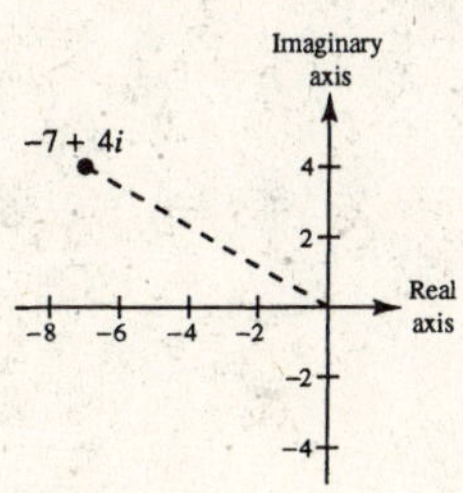

20. $z = 3 - i$

$r = \sqrt{(3)^2 + (-1)^2} = \sqrt{10}$

$\tan\theta = \dfrac{-1}{3}$ or $\theta \approx 5.96$, θ is in Quadrant IV

$z = \sqrt{10}(\cos(5.96) + i\sin(5.96))$

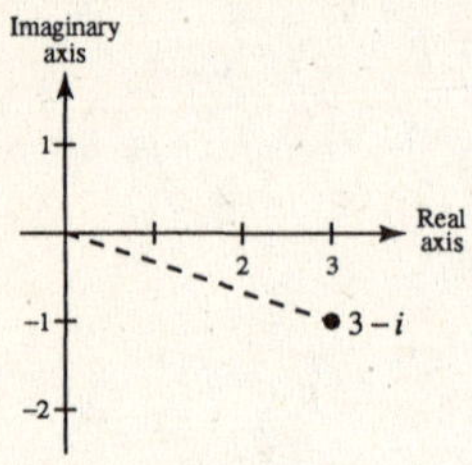

21. $z = 7 + 0i$

$r = \sqrt{(7)^2 + (0)^2} = \sqrt{49} = 7$

$\tan\theta = \dfrac{0}{7} = 0 \Rightarrow \theta = 0$

$z = 7(\cos 0 + i\sin 0)$

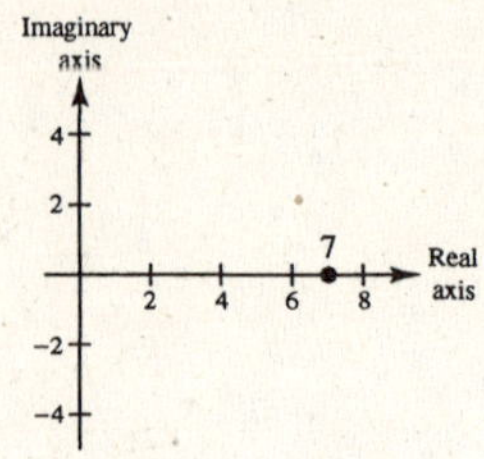

22. $z = 4$

$r = \sqrt{4^2 + 0^2} = \sqrt{16^2} = 4$

$\tan\theta = \dfrac{0}{4} = 0 \Rightarrow \theta = 0$

$z = 4(\cos 0 + i\sin 0)$

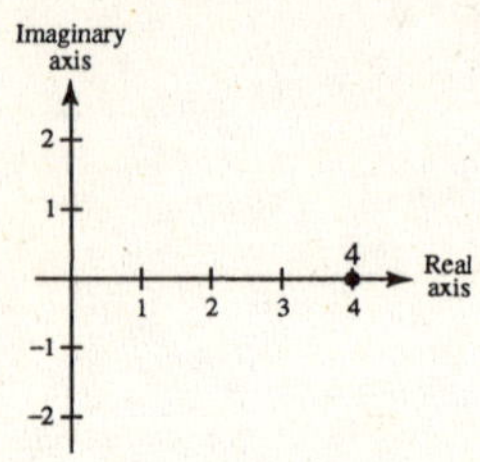

23. $z = 3 + \sqrt{3}i$

$r = \sqrt{(3)^2 + (\sqrt{3})^2} = \sqrt{12}$

$= 2\sqrt{3}$

$\tan\theta = \dfrac{\sqrt{3}}{3} \Rightarrow \theta = \dfrac{\pi}{6}$

$z = 2\sqrt{3}\left(\cos\dfrac{\pi}{6} + i\sin\dfrac{\pi}{6}\right)$

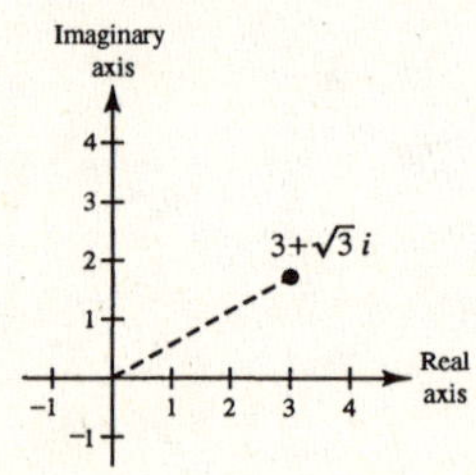

24. $z = 2\sqrt{2} - i$

$r = \sqrt{(2\sqrt{2})^2 + (-1)^2} = \sqrt{9} = 3$

$\tan\theta = \dfrac{-1}{2\sqrt{2}} = -\dfrac{\sqrt{2}}{4} \Rightarrow \theta \approx 5.94$

$z = 3(\cos(5.94) + i\sin(5.94))$

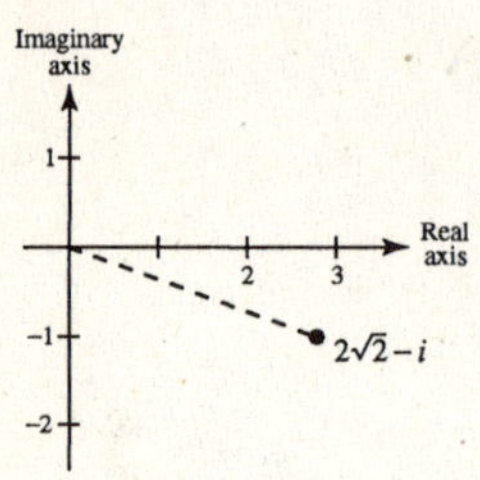

25. $z = -3 - i$

$r = \sqrt{(-3)^2 + (-1)^2} = \sqrt{10}$

$\tan\theta = \dfrac{-1}{-3} = \dfrac{1}{3}$, θ is in Quadrant III $\Rightarrow \theta \approx 3.46$.

$z \approx \sqrt{10}\,(\cos 3.46 + i\sin 3.46)$

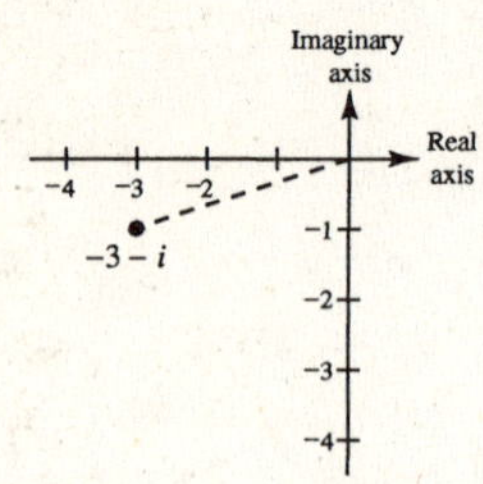

26. $z = 1 + 3i$

$r = \sqrt{1^2 + 3^2} = \sqrt{10}$

$\tan\theta = \frac{3}{1} = 3 \Rightarrow \theta \approx 1.25$

$z \approx \sqrt{10}(\cos 1.25 + i\sin 1.25)$

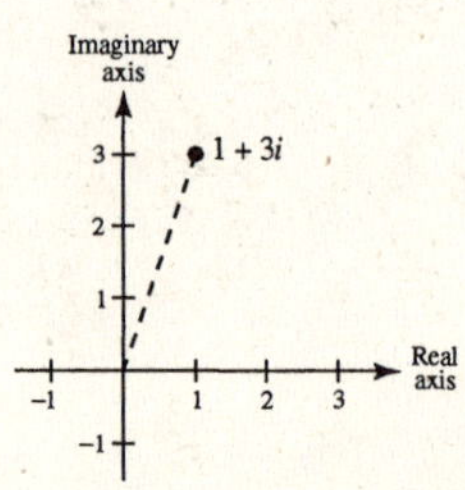

27. $z = 5 + 2i$

$r \approx 5.39$

$\theta \approx 0.38$

$z \approx 5.39(\cos 0.38 + i \sin 0.38)$

28. $z = 8 + 3i$

$r \approx 8.54$

$\theta \approx 0.36$

$z \approx 8.54(\cos 0.36 + i \sin 0.36)$

29. $z = -3 + i$

$r \approx 3.16$

$\theta \approx 2.82$

$z \approx 3.16(\cos 2.82 + i \sin 2.82)$

30. $z = -5 - i$

$r \approx 5.10$

$\theta \approx 3.34$

$z \approx 5.10(\cos 3.34 + i \sin 3.34)$

31. $z = 3\sqrt{2} - 7i$

$r \approx 8.19$

$\theta \approx 5.26$

$z \approx 8.19(\cos 5.26 + i \sin 5.26)$

32. $z = 4\sqrt{5} - 4i$

$r \approx 9.80$

$\theta \approx 5.86$

$z \approx 9.80(\cos 5.86 + i \sin 5.86)$

33. $z = -8 - 5\sqrt{3}i$

$r \approx 11.79$

$\theta \approx 3.97$

$z \approx 11.79(\cos 3.97 + i \sin 3.97)$

34. $z = -9 - 2\sqrt{10}i$

$r = 11.0$

$\theta \approx 3.75$

$z \approx 11.0(\cos 3.75 + i \sin 3.75)$

35. $3(\cos 120° + i \sin 120°) = 3\left(-\frac{1}{2} + \frac{\sqrt{3}}{2}i\right)$

$= -\frac{3}{2} + \frac{3\sqrt{3}}{2}i$

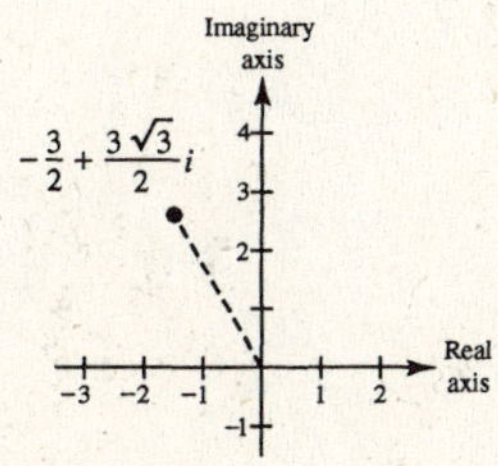

36. $5(\cos 135° + i \sin 135°) = 5\left[-\frac{\sqrt{2}}{2} + i\left(\frac{\sqrt{2}}{2}\right)\right]$

$= -\frac{5\sqrt{2}}{2} + \frac{5\sqrt{2}}{2}i$

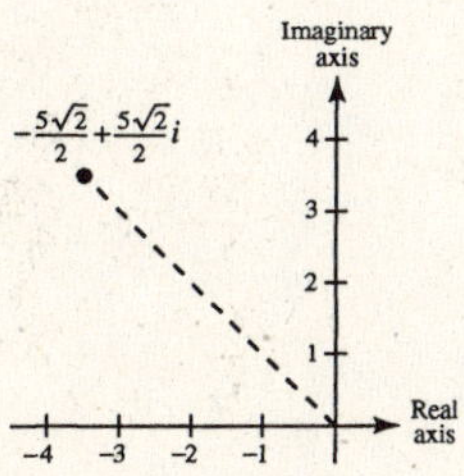

37. $\frac{3}{2}(\cos 300° + i \sin 300°) = \frac{3}{2}\left[\frac{1}{2} + i\left(-\frac{\sqrt{3}}{2}\right)\right]$

$= \frac{3}{4} - \frac{3\sqrt{3}}{4}i$

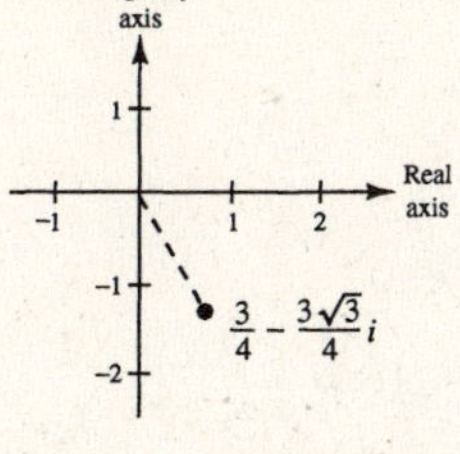

38. $\frac{1}{4}(\cos 225° + i \sin 225°) = \frac{1}{4}\left(-\frac{\sqrt{2}}{2} + i\left(-\frac{\sqrt{2}}{2}\right)\right)$

$= -\frac{\sqrt{2}}{8} - i\frac{\sqrt{2}}{8}$

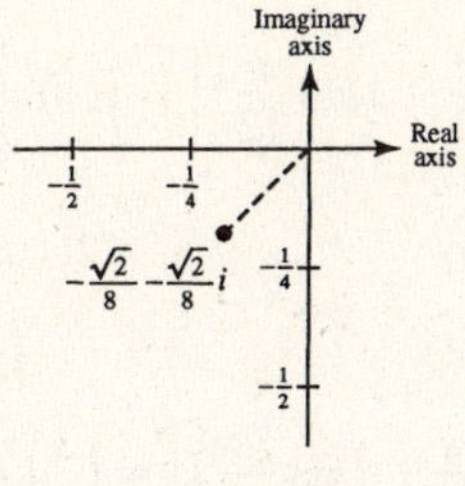

39. $3.75\left(\cos \frac{3\pi}{4} + i \sin \frac{3\pi}{4}\right) = -\frac{15\sqrt{2}}{8} + \frac{15\sqrt{2}}{8}i$

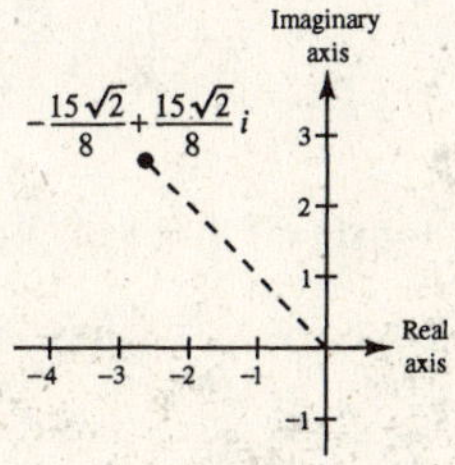

40. $6\left(\cos\frac{5\pi}{12} + i\sin\frac{5\pi}{12}\right) \approx 1.5529 + i\,5.7956$

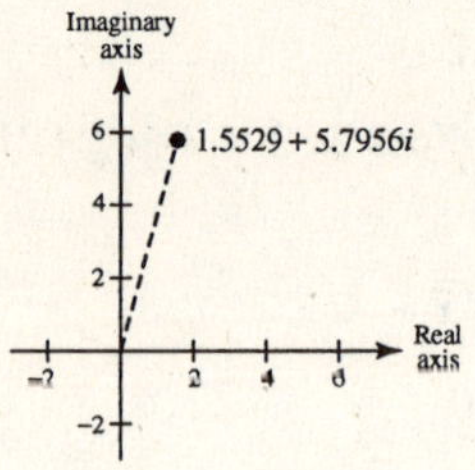

41. $8\left(\cos\frac{\pi}{2} + i\sin\frac{\pi}{2}\right) = 8(0 + i) = 8i$

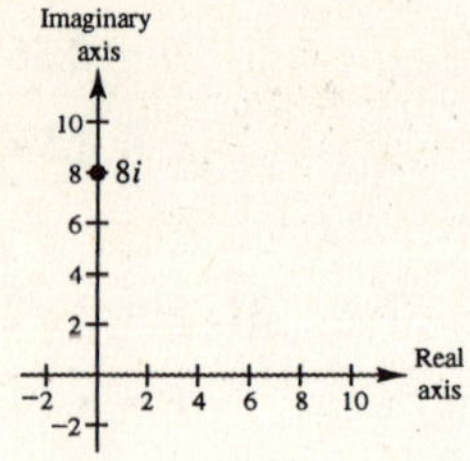

42. $7(\cos 0 + i\sin 0) = 7$

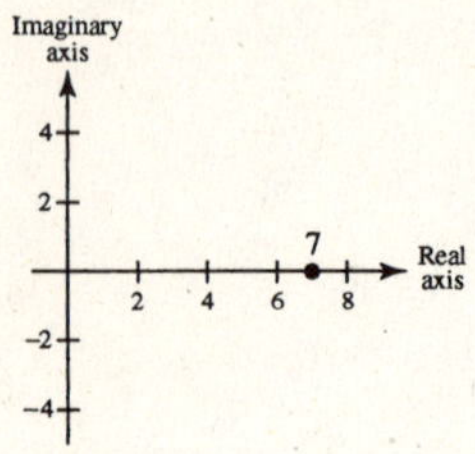

43. $3[\cos(18°45') + i\sin(18°45')] \approx 2.8408 + 0.9643i$

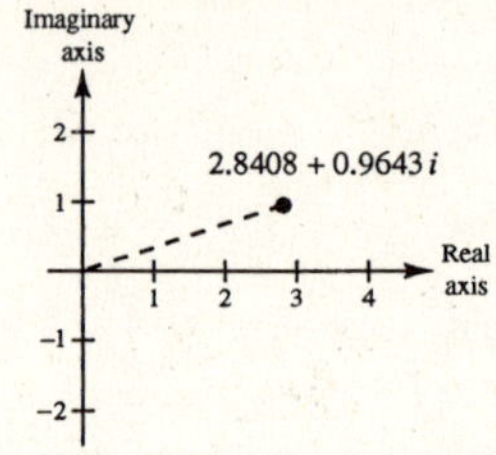

44. $6[\cos(230°\,30') + i\sin(230°\,30')] \approx -3.8165 - 4.6297i$

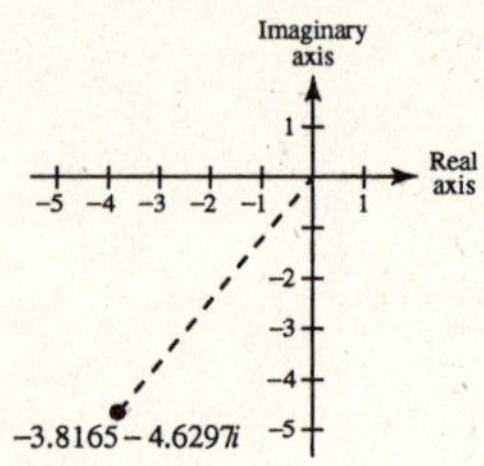

45. $5\left(\cos\frac{\pi}{9} + i\sin\frac{\pi}{9}\right) \approx 4.70 + 1.71i$

46. $10\left(\cos\frac{2\pi}{5} + i\sin\frac{2\pi}{5}\right) \approx 3.09 + 9.51i$

47. $3(\cos 165.5° + i\sin 165.5°) \approx -2.90 + 0.75i$

48. $9(\cos 58° + i\sin 58°) \approx 4.77 + 7.63i$

49. $$\left[2\left(\cos\frac{\pi}{4} + i\sin\frac{\pi}{4}\right)\right]\left[6\left(\cos\frac{\pi}{12} + i\sin\frac{\pi}{12}\right)\right] = (2)(6)\left[\cos\left(\frac{\pi}{4} + \frac{\pi}{12}\right) + i\sin\left(\frac{\pi}{4} + \frac{\pi}{12}\right)\right]$$
$$= 12\left(\cos\frac{\pi}{3} + i\sin\frac{\pi}{3}\right)$$

50. $$\left[\frac{3}{4}\left(\cos\frac{\pi}{3} + i\sin\frac{\pi}{3}\right)\right]\left[4\left(\cos\frac{3\pi}{4} + i\sin\frac{3\pi}{4}\right)\right] = \left(\frac{3}{4}\right)(4)\left[\cos\left(\frac{\pi}{3} + \frac{3\pi}{4}\right) + i\sin\left(\frac{\pi}{3} + \frac{3\pi}{4}\right)\right]$$
$$= 3\left(\cos\frac{13\pi}{12} + i\sin\frac{13\pi}{12}\right)$$

51. $$\left[\tfrac{5}{3}(\cos 140° + i\sin 140°)\right]\left[\tfrac{2}{3}(\cos 60° + i\sin 60°)\right] = \left(\tfrac{5}{3}\right)\left(\tfrac{2}{3}\right)[\cos(140° + 60°) + i\sin(140° + 60°)]$$
$$= \tfrac{10}{9}(\cos 200° + i\sin 200°)$$

52. $[0.5(\cos 100° + i \sin 100°)][0.8(\cos 300° + i \sin 300°)] = (0.5)(0.8)[\cos(100° + 300°) + i \sin(100° + 300°)]$

$$= 0.4(\cos 400° + i \sin 400°)$$

$$= 0.4(\cos 40° + i \sin 40°)$$

53. $[0.45(\cos 310° + i \sin 310°)][0.60(\cos 200° + i \sin 200°)] = (0.45)(0.60)[\cos(310° + 200°) + i \sin(310° + 200°)]$

$$= 0.27(\cos 510° + i \sin 510°)$$

$$= 0.27(\cos 150° + i \sin 150°)$$

54. $(\cos 5° + i \sin 5°)(\cos 20° + i \sin 20°) = \cos(5° + 20°) + i \sin(5° + 20°)$

$$= \cos 25° + i \sin 25°$$

55. $\dfrac{\cos 50° + i \sin 50°}{\cos 20° + i \sin 20°} = \cos(50° - 20°) + i \sin(50° - 20°)$

$$= \cos 30° + i \sin 30°$$

56. $\dfrac{2(\cos 120° + i \sin 120°)}{4(\cos 40° + i \sin 40°)} = \dfrac{2}{4}[\cos(120° - 40°) + i \sin(120° - 40°)]$

$$= \frac{1}{2}(\cos 80° + i \sin 80°)$$

57. $\dfrac{\cos\dfrac{5\pi}{3} + i \sin\dfrac{5\pi}{3}}{\cos \pi + i \sin \pi} = \cos\left(\dfrac{5\pi}{3} - \pi\right) + i \sin\left(\dfrac{5\pi}{3} - \pi\right) = \cos\left(\dfrac{2\pi}{3}\right) + i \sin\left(\dfrac{2\pi}{3}\right)$

58. $\dfrac{5[\cos(4.3) + i \sin(4.3)]}{4[\cos(2.1) + i \sin(2.1)]} = \dfrac{5}{4}[\cos(4.3 - 2.1) + i \sin(4.3 - 2.1)$

$$= \frac{5}{4}[\cos(2.2) + i \sin(2.2)]$$

59. $\dfrac{12(\cos 52° + i \sin 52°)}{3(\cos 110° + i \sin 110°)} = 4[\cos(52° - 110°) + i \sin(52° - 110°)]$

$$= 4[\cos(-58°) + i \sin(-58°)]$$

60. $\dfrac{6[\cos 40° + i \sin 40°]}{7[\cos 100° + i \sin 100°]} = \dfrac{6}{7}[\cos(40° - 100°) + i \sin(40° - 100°)]$

$$= \frac{6}{7}[\cos 300° + i \sin 300°]$$

61. (a) $2 + 2i = 2\sqrt{2}\left(\cos\dfrac{\pi}{4} + i \sin\dfrac{\pi}{4}\right)$

$1 - i = \sqrt{2}\left[\cos\left(-\dfrac{\pi}{4}\right) + i \sin\left(-\dfrac{\pi}{4}\right)\right]$

(b) $(2 + 2i)(1 - i) = \left[2\sqrt{2}\left(\cos\dfrac{\pi}{4} + i \sin\dfrac{\pi}{4}\right)\right]\left[\sqrt{2}\left(\cos\left(-\dfrac{\pi}{4}\right) + i \sin\left(-\dfrac{\pi}{4}\right)\right)\right] = 4(\cos 0 + i \sin 0) = 4$

(c) $(2 + 2i)(1 - i) = 2 - 2i + 2i - 2i^2 = 2 + 2 = 4$

62. (a) $\sqrt{3} + i = 2(\cos 30° + i \sin 30°)$

$1 + i = \sqrt{2}(\cos 45° + i \sin 45°)$

(b) $$\begin{aligned}(\sqrt{3} + i)(1 + i) &= [2(\cos 30° + i \sin 30°)][\sqrt{2}(\cos 45° + i \sin 45°)]\\ &= 2\sqrt{2}(\cos 75° + i \sin 75°)\\ &= 2\sqrt{2}\left[\left(\frac{\sqrt{6} - \sqrt{2}}{4}\right) + \left(\frac{\sqrt{6} + \sqrt{2}}{4}\right)i\right]\\ &= (\sqrt{3} - 1) + (\sqrt{3} + 1)i\end{aligned}$$

(c) $(\sqrt{3} + i)(1 + i) = \sqrt{3} + (\sqrt{3} + 1)i + i^2 = (\sqrt{3} - 1) + (\sqrt{3} + 1)i$

63. (a) $-2i = 2\left[\cos\left(-\frac{\pi}{2}\right) + i \sin\left(-\frac{\pi}{2}\right)\right]$

$1 + i = \sqrt{2}\left(\cos\frac{\pi}{4} + i \sin\frac{\pi}{4}\right)$

(b) $$\begin{aligned}-2i(1 + i) &= 2\left[\cos\left(-\frac{\pi}{2}\right) + i \sin\left(-\frac{\pi}{2}\right)\right]\left[\sqrt{2}\left(\cos\frac{\pi}{4} + i \sin\frac{\pi}{4}\right)\right]\\ &= 2\sqrt{2}\left[\cos\left(-\frac{\pi}{4}\right) + i \sin\left(-\frac{\pi}{4}\right)\right]\\ &= 2\sqrt{2}\left[\frac{1}{\sqrt{2}} - \frac{1}{\sqrt{2}}i\right] = 2 - 2i\end{aligned}$$

(c) $-2i(1 + i) = -2i - 2i^2 = -2i + 2 = 2 - 2i$

64. (a) $4 = 4(\cos 0 + i \sin 0)$

$1 - \sqrt{3}i = 2\left(\cos\left(-\frac{\pi}{3}\right) + i \sin\left(\frac{-\pi}{3}\right)\right)$

(b) $$\begin{aligned}4(1 - \sqrt{3}i) &= 8\left(\cos\left(-\frac{\pi}{3}\right) + i \sin\left(-\frac{\pi}{3}\right)\right)\\ &= 8\left(\frac{1}{2} + i\left(-\frac{\sqrt{3}}{2}\right)\right)\\ &= 4 - 4\sqrt{3}i\end{aligned}$$

(c) $4(1 - \sqrt{3}i) = 4 - 4\sqrt{3}i$

65. (a) $3 + 4i \approx 5(\cos 0.9273 + i \sin 0.9273)$

$1 - \sqrt{3}i = 2\left(\cos\frac{5\pi}{3} + i \sin\frac{5\pi}{3}\right)$

(b) $$\begin{aligned}\frac{3 + 4i}{1 - \sqrt{3}i} &\approx \frac{5(\cos 0.9273 + i \sin 0.9273)}{2\left(\cos\frac{5\pi}{3} + i \sin\frac{5\pi}{3}\right)}\\ &\approx 2.5[\cos(-4.3087) + i \sin(-4.3087)]\\ &\approx -0.982 + 2.299i\end{aligned}$$

(c) $$\begin{aligned}\frac{3 + 4i}{1 - \sqrt{3}i} &= \frac{3 + 4i}{1 - \sqrt{3}i} \cdot \frac{1 + \sqrt{3}i}{1 + \sqrt{3}i}\\ &= \frac{3 + (4 + 3\sqrt{3})i + 4\sqrt{3}i^2}{1 + 3}\\ &= \frac{3 - 4\sqrt{3}}{4} + \frac{4 + 3\sqrt{3}}{4}i\\ &\approx -0.982 + 2.299i\end{aligned}$$

66. (a) $1 + \sqrt{3}i = 2\left(\cos\frac{\pi}{3} + i \sin\frac{\pi}{3}\right)$

$6 - 3i \approx 3\sqrt{5}[\cos(-0.464) + i \sin(-0.464)]$

(b) $$\begin{aligned}\frac{1 + \sqrt{3}i}{6 - 3i} &\approx \frac{2}{3\sqrt{5}}\left[\cos\left(\frac{\pi}{3} + 0.464\right) + i \sin\left(\frac{\pi}{3} + 0.464\right)\right]\\ &\approx \frac{2\sqrt{5}}{15}[\cos 1.51 + i \sin 1.51]\\ &\approx 0.018 + 0.298i\end{aligned}$$

(c) $$\begin{aligned}\frac{1 + \sqrt{3}i}{6 - 3i} \cdot \frac{6 + 3i}{6 + 3i} &= \frac{(6 - 3\sqrt{3}) + i(3 + 6\sqrt{3})}{45}\\ &= \frac{2 - \sqrt{3}}{15} + i\frac{1 + 2\sqrt{3}}{15}\\ &\approx 0.018 + 0.298i\end{aligned}$$

67. (a) $5 = 5(\cos 0 + i \sin 0)$

$2 + 3i \approx \sqrt{13}(\cos 0.9828 + i \sin 0.9828)$

(b) $\dfrac{5}{2 + 3i} \approx \dfrac{5(\cos 0 + i \sin 0)}{\sqrt{13}(\cos 0.9828 + i \sin 0.9828)} = \dfrac{5\sqrt{13}}{13}[\cos(-0.9828) + i \sin(-0.9828)] \approx 0.769 - 1.154i$

(c) $\dfrac{5}{2 + 3i} = \dfrac{5}{2 + 3i} \cdot \dfrac{2 - 3i}{2 - 3i} = \dfrac{10 - 15i}{13} = \dfrac{10}{13} - \dfrac{15}{13}i \approx 0.769 - 1.154i$

68. (a) $4i = 4(\cos 90° + i \sin 90°)$

$-4 + 2i = 2\sqrt{5}(\cos 153.4° + i \sin 153.4°)$

(b) $\dfrac{4i}{-4 + 2i} = \dfrac{4(\cos 90° + i \sin 90°)}{2\sqrt{5}(\cos 153.4° + i \sin 153.4°)}$

$= \dfrac{2\sqrt{5}}{5}(\cos 296.6° + i \sin 296.6°)$

$\approx 0.400 - 0.800i$

(c) $\dfrac{4i}{-4 + 2i} = \dfrac{4i}{-4 + 2i} \cdot \dfrac{-4 - i}{-4 - 2i}$

$= \dfrac{8 - 16i}{20}$

$= \dfrac{2}{5} - \dfrac{4}{5}i$

$= 0.400 - 0.800i$

69. Let $z = x + iy$ such that:

$|z| = 2 \Rightarrow 2 = \sqrt{x^2 + y^2}$

$\Rightarrow 4 = x^2 + y^2$: circle with radius of 2

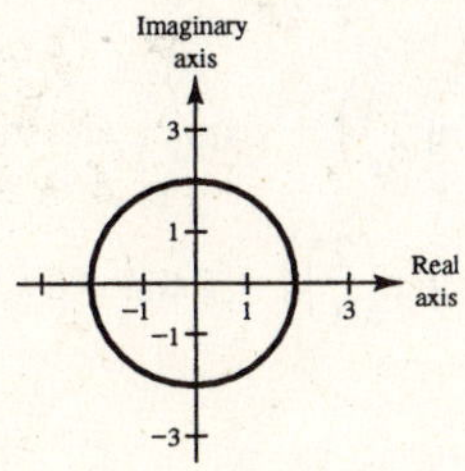

70. $|z| = 3$

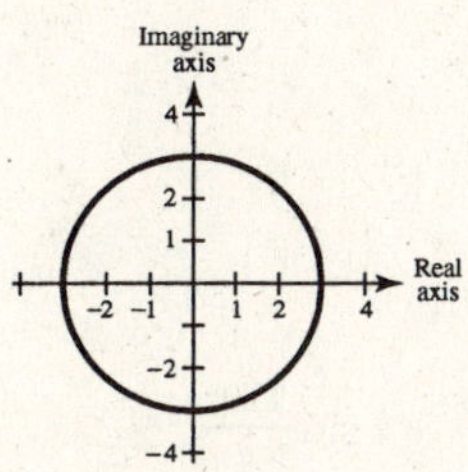

71. Let $\theta = \dfrac{\pi}{6}$.

Let $z = x + iy$ such that:

$\tan \dfrac{\pi}{6} = \dfrac{y}{x}$; line $y = \dfrac{\sqrt{3}}{3}x$

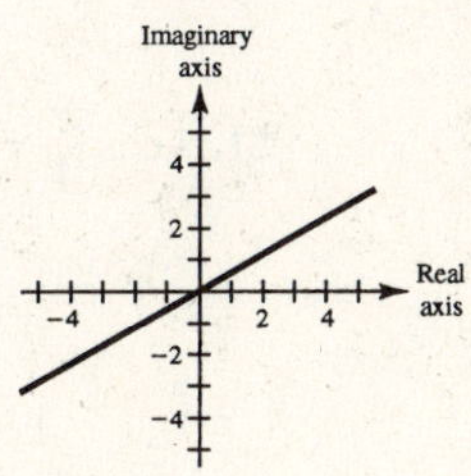

72. $\theta = \dfrac{5\pi}{4}$

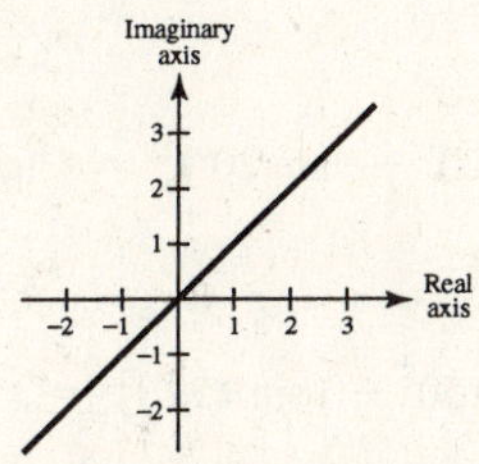

73. $(1 + i)^5 = \left[\sqrt{2}\left(\cos \dfrac{\pi}{4} + i \sin \dfrac{\pi}{4}\right)\right]^5$

$= (\sqrt{2})^5\left(\cos \dfrac{5\pi}{4} + i \sin \dfrac{5\pi}{4}\right)$

$= 4\sqrt{2}\left(-\dfrac{\sqrt{2}}{2} - \dfrac{\sqrt{2}}{2}i\right)$

$= -4 - 4i$

74. $(2 + 2i)^6 = \left[2\sqrt{2}\left(\cos \dfrac{\pi}{4} + i \sin \dfrac{\pi}{4}\right)\right]^6$

$= (2\sqrt{2})^6\left(\cos \dfrac{6\pi}{4} + i \sin \dfrac{6\pi}{4}\right)$

$= 512\left(\cos \dfrac{3\pi}{2} + i \sin \dfrac{3\pi}{2}\right)$

$= -512i$

75. $(-1+i)^{10} = \left[\sqrt{2}\left(\cos\frac{3\pi}{4} + i\sin\frac{3\pi}{4}\right)\right]^{10}$

$= (\sqrt{2})^{10}\left(\cos\frac{30\pi}{4} + i\sin\frac{30\pi}{4}\right)$

$= 32\left[\cos\left(\frac{3\pi}{2} + 6\pi\right) + i\sin\left(\frac{3\pi}{2} + 6\pi\right)\right]$

$= 32\left(\cos\frac{3\pi}{2} + i\sin\frac{3\pi}{2}\right)$

$= 32[0 + i(-1)]$

$= -32i$

76. $(3-2i)^8 = \left[\sqrt{13}\left(\cos\left(-\arctan\left(\frac{2}{3}\right)\right) + i\sin\left(-\arctan\left(\frac{2}{3}\right)\right)\right)\right]^8$

$= (\sqrt{13})^8\left[\cos\left(-8\arctan\left(\frac{2}{3}\right)\right) + i\sin\left(-8\arctan\left(\frac{2}{3}\right)\right)\right]$

$= -239 + 28{,}560i$

77. $2(\sqrt{3}+i)^7 = 2\left[2\left(\cos\frac{\pi}{6} + i\sin\frac{\pi}{6}\right)\right]^7$

$= 2\left[2^7\left(\cos\frac{7\pi}{6} + i\sin\frac{7\pi}{6}\right)\right]$

$= 256\left(-\frac{\sqrt{3}}{2} - \frac{1}{2}i\right)$

$= -128\sqrt{3} - 128i$

78. $4(1-\sqrt{3}i)^3 = 4\left[2\left(\cos\frac{5\pi}{3} + i\sin\frac{5\pi}{3}\right)\right]^3$

$= 4[2^3(\cos 5\pi + i\sin 5\pi)]$

$= 32(-1)$

$= -32$

79. $[5(\cos 20° + i\sin 20°)]^3 = 5^3(\cos 60° + i\sin 60°) = \frac{125}{2} + \frac{125\sqrt{3}}{2}i$

80. $[3(\cos 150° + i\sin 150°)]^4 = 3^4(\cos 600° + i\sin 600°)$

$= 81(\cos 240° + i\sin 240°)$

$= 81(-\cos 60° - i\sin 60°)$

$= -\frac{81}{2} - \frac{81\sqrt{3}}{2}i$

81. $\left(\cos\frac{\pi}{4} + i\sin\frac{\pi}{4}\right)^{12} = \cos\frac{12\pi}{4} + i\sin\frac{12\pi}{4}$

$= \cos 3\pi + i\sin 3\pi$

$= -1$

82. $\left[2\left(\cos\frac{\pi}{2} + i\sin\frac{\pi}{2}\right)\right]^8 = 2^8(\cos 4\pi + i\sin 4\pi)$

$= 256(\cos 0 + i\sin 0)$

$= 256$

83. $[5(\cos 3.2 + i\sin 3.2)]^4 = 5^4(\cos 12.8 + i\sin 12.8)$

$\approx 608.02 + 144.69i$

84. $(\cos 0 + i\sin 0)^{20} = \cos 0 + i\sin 0$

$= 1$

85. $(3-2i)^5 \approx [3.6056[\cos(-0.588) + i\sin(-0.588)]]^5$

$\approx (3.6056)^5[\cos(-2.94) + i\sin(-2.94)]$

$\approx -597 - 122i$

86. $(\sqrt{5} - 4i)^3 = -43\sqrt{5} + 4i$

87. $[3(\cos 15° + i \sin 15°)]^4 = 81(\cos 60° + i \sin 60°)$

$= \dfrac{81}{2} + \dfrac{81\sqrt{3}}{2}i$

88. $[2(\cos 10° + i \sin 10°)]^8 = 256(\cos 80° + i \sin 80°)$

$\approx 44.4539 + 252.1108i$

89. $\left[2\left(\cos \dfrac{\pi}{10} + i \sin \dfrac{\pi}{10}\right)\right]^5 = 2^5\left(\cos \dfrac{\pi}{2} + i \sin \dfrac{\pi}{2}\right)$

$= 32i$

90. $\left[2\left(\cos \dfrac{\pi}{8} + i \sin \dfrac{\pi}{8}\right)\right]^6 = 64\left(\cos \dfrac{3\pi}{4} + i \sin \dfrac{3\pi}{4}\right) = -32\sqrt{2} + 32\sqrt{2}i$

91. (a) Square roots of $5(\cos 120° + i \sin 120°)$:

$$\sqrt{5}\left[\cos\left(\frac{120° + 360°k}{2}\right) + i \sin\left(\frac{120° + 360°k}{2}\right)\right],\ k = 0,\ 1$$

$k = 0$: $\sqrt{5}(\cos 60° + i \sin 60°)$

$k = 1$: $\sqrt{5}(\cos 240° + i \sin 240°)$

(b)

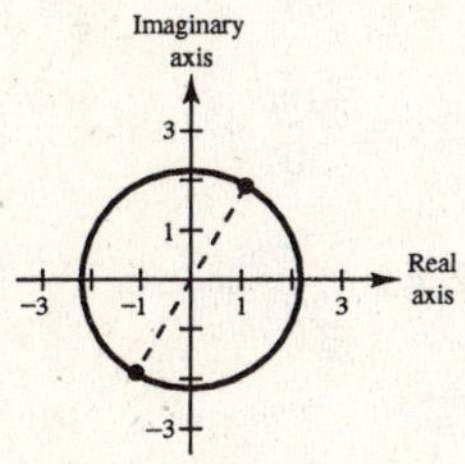

(c) $\dfrac{\sqrt{5}}{2} + \dfrac{\sqrt{15}}{2}i,\ -\dfrac{\sqrt{5}}{2} - \dfrac{\sqrt{15}}{2}i$

92. (a) Square roots of $16(\cos 60° + i \sin 60°)$:

$$\sqrt{16}\left[\cos\left(\frac{60° + k\,360°}{2}\right) + i \sin\left(\frac{60° + k\,360°}{2}\right)\right]$$

$k = 0, 1$

$k = 0$: $4(\cos 30° + i \sin 30°)$

$k = 1$: $4(\cos 210° + i \sin 210°)$

(b)

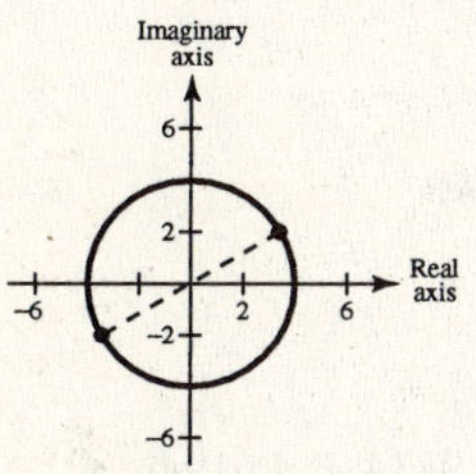

(c) $2\sqrt{3} + 2i,\ -2\sqrt{3} - 2i$

93. (a) Cube roots of $8\left(\cos \dfrac{2\pi}{3} + i \sin \dfrac{2\pi}{3}\right)$:

$$\sqrt[3]{8}\left[\cos\left(\frac{\frac{2\pi}{3} + 2\pi k}{3}\right) + i \sin\left(\frac{\frac{2\pi}{3} + 2\pi k}{3}\right)\right],\ k = 0, 1, 2$$

$k = 0$: $2\left(\cos \dfrac{2\pi}{9} + i \sin \dfrac{2\pi}{9}\right)$

$k = 1$: $2\left(\cos \dfrac{8\pi}{9} + i \sin \dfrac{8\pi}{9}\right)$

$k = 2$: $2\left(\cos \dfrac{14\pi}{9} + i \sin \dfrac{14\pi}{9}\right)$

(b)

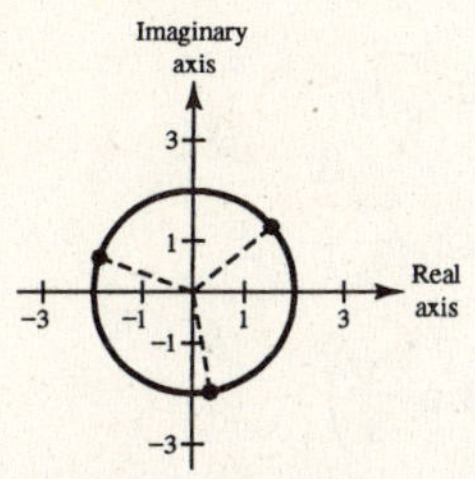

(c) $1.5321 + 1.2856i$

$-1.8794 + 0.6840i$

$0.3473 - 1.9696i$

94. (a) Fifth roots of $32\left(\cos\frac{5\pi}{6} + i\sin\frac{5\pi}{6}\right)$:

$$\sqrt[5]{32}\left[\cos\left(\frac{\frac{5\pi}{6} + 2k\pi}{5}\right) + i\sin\left(\frac{\frac{5\pi}{6} + 2k\pi}{5}\right)\right]$$

$k = 0, 1, 2, 3, 4$

$k = 0$: $2\left(\cos\frac{\pi}{6} + i\sin\frac{\pi}{6}\right)$

$k = 1$: $2\left(\cos\frac{17\pi}{30} + i\sin\frac{17\pi}{30}\right)$

$k = 2$: $2\left(\cos\frac{29\pi}{30} + i\sin\frac{29\pi}{30}\right)$

$k = 3$: $2\left(\cos\frac{41\pi}{30} + i\sin\frac{41\pi}{30}\right)$

$k = 4$: $2\left(\cos\frac{53\pi}{30} + i\sin\frac{53\pi}{30}\right)$

(b)

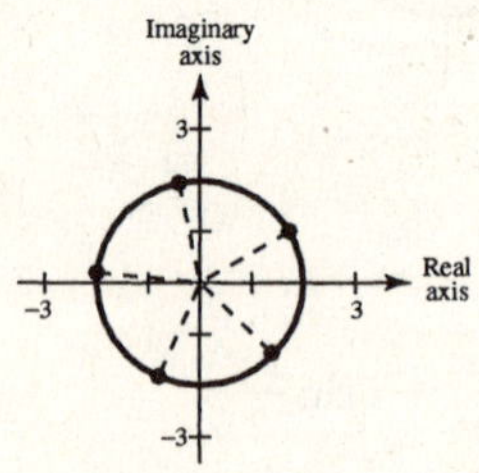

(c) $\sqrt{3} + i, -0.4158 + 1.9563i, -1.9890 + 0.2091i,$
$-0.8135 - 1.8271i, 1.4863 - 1.3383i$

95. (a) Square roots of $-25i = 25\left(\cos\frac{3\pi}{2} + i\sin\frac{3\pi}{2}\right)$:

$$\sqrt{25}\left[\cos\left(\frac{\frac{3\pi}{2} + 2k\pi}{2}\right) + i\sin\left(\frac{\frac{3\pi}{2} + 2k\pi}{2}\right)\right], k = 0, 1$$

$k = 0$: $5\left(\cos\frac{3\pi}{4} + i\sin\frac{3\pi}{4}\right)$

$k = 1$: $5\left(\cos\frac{7\pi}{4} + i\sin\frac{7\pi}{4}\right)$

(b)

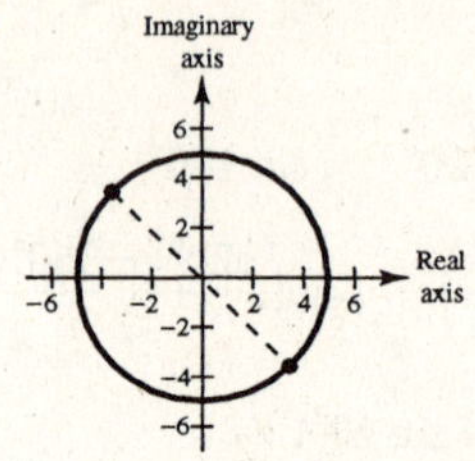

(c) $-\frac{5\sqrt{2}}{2} + \frac{5\sqrt{2}}{2}i, \frac{5\sqrt{2}}{2} - \frac{5\sqrt{2}}{2}i$

96. (a) Fourth roots of $625i = 625\left(\cos\frac{\pi}{2} + i\sin\frac{\pi}{2}\right)$:

$$\sqrt[4]{625}\left[\cos\left(\frac{\frac{\pi}{2} + 2k\pi}{4}\right) + i\sin\left(\frac{\frac{\pi}{2} + 2k\pi}{4}\right)\right]$$

$k = 0, 1, 2, 3$

$k = 0$: $5\left(\cos\frac{\pi}{8} + i\sin\frac{\pi}{8}\right)$

$k = 1$: $5\left(\cos\frac{5\pi}{8} + i\sin\frac{5\pi}{8}\right)$

$k = 2$: $5\left(\cos\frac{9\pi}{8} + i\sin\frac{9\pi}{8}\right)$

$k = 3$: $5\left(\cos\frac{13\pi}{8} + i\sin\frac{13\pi}{8}\right)$

(b)

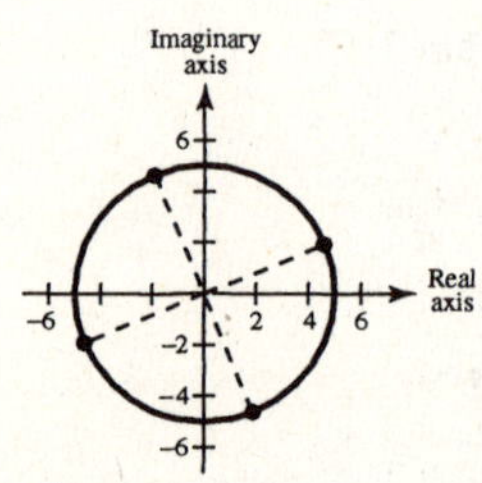

(c) $4.6194 + 1.9134i, -1.9134 + 4.6194i,$
$-4.6194 - 1.9134i, 1.9134 - 4.6194i$

97. (a) Cube roots of $-\frac{125}{2}(1 + \sqrt{3}i) = 125\left(\cos\frac{4\pi}{3} + i\sin\frac{4\pi}{3}\right)$:

$$\sqrt[3]{125}\left[\cos\left(\frac{\frac{4\pi}{3} + 2k\pi}{3}\right) + i\sin\left(\frac{\frac{4\pi}{3} + 2k\pi}{3}\right)\right],\ k = 0, 1, 2$$

$k = 0$: $5\left(\cos\frac{4\pi}{9} + i\sin\frac{4\pi}{9}\right)$

$k = 1$: $5\left(\cos\frac{10\pi}{9} + i\sin\frac{10\pi}{9}\right)$

$k = 2$: $5\left(\cos\frac{16\pi}{9} + i\sin\frac{16\pi}{9}\right)$

(b)

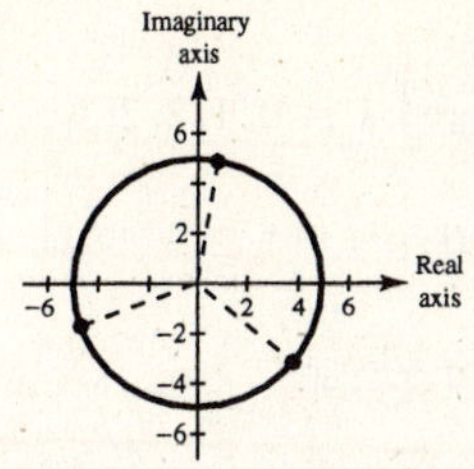

(c) $0.8682 + 4.924i,\ -4.6985 - 1.710i,\ 3.8302 - 3.214i$

98. (a) Cube roots of $-4\sqrt{2}(1 - i) = 8\left(\cos\frac{3\pi}{4} + i\sin\frac{3\pi}{4}\right)$:

$$\sqrt[3]{8}\left[\cos\left(\frac{\frac{3\pi}{4} + 2k\pi}{3}\right) + i\sin\left(\frac{\frac{3\pi}{4} + 2k\pi}{3}\right)\right]$$

$k = 0, 1, 2$

$k = 0$: $2\left(\cos\frac{\pi}{4} + i\sin\frac{\pi}{4}\right)$

$k = 1$: $2\left(\cos\frac{11\pi}{12} + i\sin\frac{11\pi}{12}\right)$

$k = 2$: $2\left(\cos\frac{19\pi}{12} + i\sin\frac{19\pi}{12}\right)$

(b)

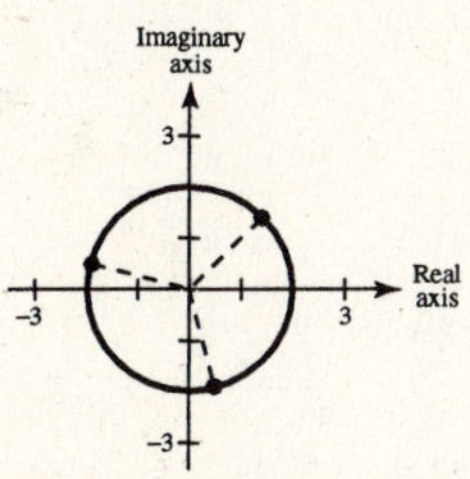

(c) $1.414 + 1.414i,\ -1.932 + 0.5176i,\ 0.5176 - 1.9319i$

99. (a) Fourth roots of $16 = 16(\cos 0 + i\sin 0)$:

$$\sqrt[4]{16}\left[\cos\frac{0 + 2\pi k}{4} + i\sin\frac{0 + 2\pi k}{4}\right],\ k = 0, 1, 2, 3$$

$k = 0$: $2(\cos 0 + i\sin 0)$

$k = 1$: $2\left(\cos\frac{\pi}{2} + i\sin\frac{\pi}{2}\right)$

$k = 2$: $2(\cos\pi + i\sin\pi)$

$k = 3$: $2\left(\cos\frac{3\pi}{2} + i\sin\frac{3\pi}{2}\right)$

(b)

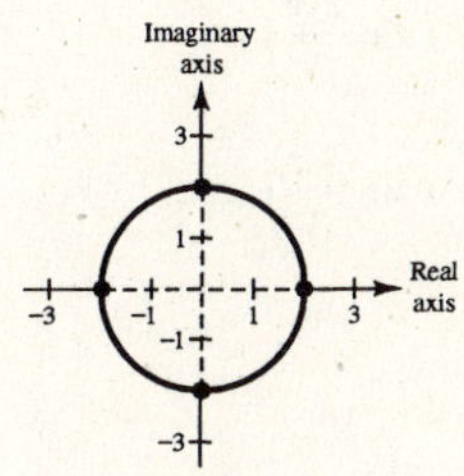

(c) $2,\ 2i,\ -2,\ -2i$

100. (a) Fourth roots of $i = \cos\frac{\pi}{2} + i\sin\frac{\pi}{2}$:

$$\sqrt[4]{1}\left[\cos\left(\frac{\frac{\pi}{2} + 2k\pi}{4}\right) + i\sin\left(\frac{\frac{\pi}{2} + 2k\pi}{4}\right)\right]$$

$k = 0, 1, 2, 3$

$k = 0$: $\cos\frac{\pi}{8} + i\sin\frac{\pi}{8}$

$k = 1$: $\cos\frac{5\pi}{8} + i\sin\frac{5\pi}{8}$

$k = 2$: $\cos\frac{9\pi}{8} + i\sin\frac{9\pi}{8}$

$k = 3$: $\cos\frac{13\pi}{8} + i\sin\frac{13\pi}{8}$

(b)

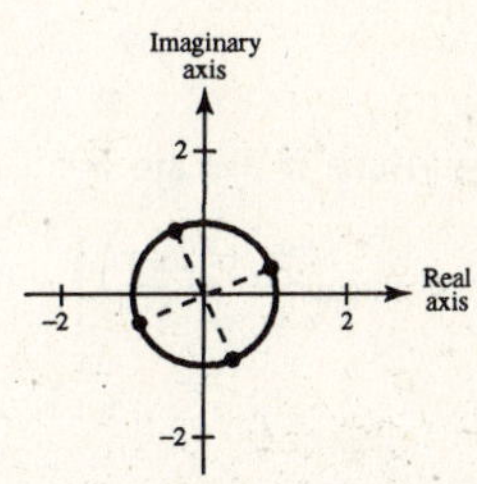

(c) $0.9239 + 0.3827i,\ -0.3827 + 0.9239i,$
$-0.9239 - 0.3827i,\ 0.3827 - 0.9239i$

101. (a) Fifth roots of $1 = \cos 0 + i \sin 0$:

$$\cos\left(\frac{2k\pi}{5}\right) + i \sin\left(\frac{2k\pi}{5}\right), k = 0, 1, 2, 3, 4$$

$k = 0$: $\cos 0 + i \sin 0$

$k = 1$: $\cos \frac{2\pi}{5} + i \sin \frac{2\pi}{5}$

$k = 2$: $\cos \frac{4\pi}{5} + i \sin \frac{4\pi}{5}$

$k = 3$: $\cos \frac{6\pi}{5} + i \sin \frac{6\pi}{5}$

$k = 4$: $\cos \frac{8\pi}{5} + i \sin \frac{8\pi}{5}$

(b)

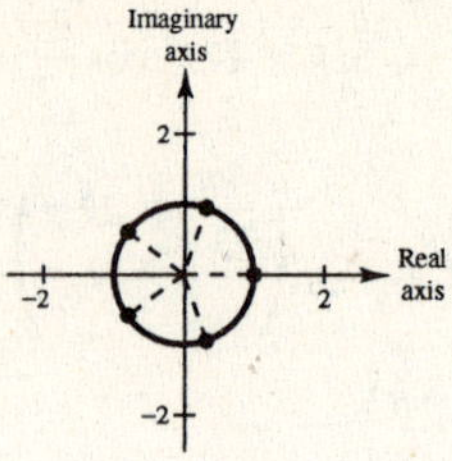

(c) $1, 0.3090 + 0.9511i, -0.8090 + 0.5878i, -0.8090 - 0.5878i, 0.3090 - 0.9511i$

102. (a) Cube roots of $1000 = 1000(\cos 0 + i \sin 0)$:

$$\sqrt[3]{1000}\left(\cos \frac{2k\pi}{3} + i \sin \frac{2k\pi}{3}\right)$$

$k = 0, 1, 2$

$k = 0$: $10(\cos 0 + i \sin 0)$

$k = 1$: $10\left(\cos \frac{2\pi}{3} + i \sin \frac{2\pi}{3}\right)$

$k = 2$: $10\left(\cos \frac{4\pi}{3} + i \sin \frac{4\pi}{3}\right)$

(b)

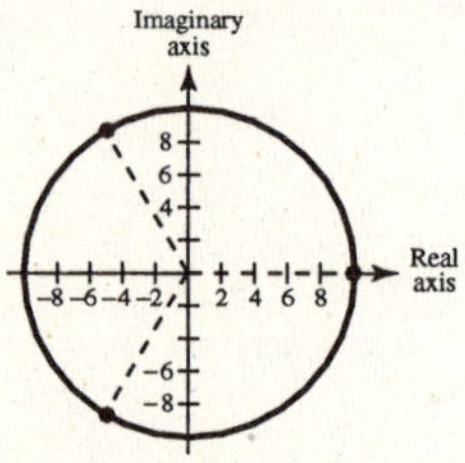

(c) $10, -5 + 5\sqrt{3}i, -5 - 5\sqrt{3}i$

103. (a) Cube roots of $-125 = 125(\cos \pi + i \sin \pi)$:

$$\sqrt[3]{125}\left[\cos\left(\frac{\pi + 2\pi k}{3}\right) + i \sin\left(\frac{\pi + 2\pi k}{3}\right)\right], k = 0, 1, 2$$

$k = 0$: $5\left(\cos \frac{\pi}{3} + i \sin \frac{\pi}{3}\right)$

$k = 1$: $5(\cos \pi + i \sin \pi)$

$k = 2$: $5\left(\cos \frac{5\pi}{3} + i \sin \frac{5\pi}{3}\right)$

(b)

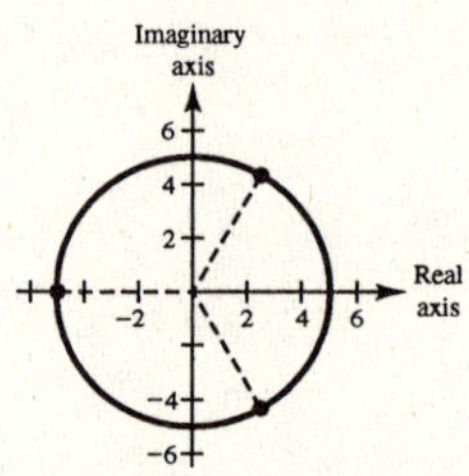

(c) $\frac{5}{2} + \frac{5\sqrt{3}}{2}i, -5, \frac{5}{2} - \frac{5\sqrt{3}}{2}i$

104. (a) Fourth roots of $-4 = 4(\cos \pi + i \sin \pi)$:

$$\sqrt[4]{4}\left[\cos\left(\frac{\pi + 2k\pi}{4}\right) + i \sin\left(\frac{\pi + 2k\pi}{4}\right)\right]$$

$k = 0, 1, 2, 3$

$k = 0$: $\sqrt{2}\left(\cos \frac{\pi}{4} + i \sin \frac{\pi}{4}\right)$

$k = 1$: $\sqrt{2}\left(\cos \frac{3\pi}{4} + i \sin \frac{3\pi}{4}\right)$

$k = 2$: $\sqrt{2}\left(\cos \frac{5\pi}{4} + i \sin \frac{5\pi}{4}\right)$

$k = 3$: $\sqrt{2}\left(\cos \frac{7\pi}{4} + i \sin \frac{7\pi}{4}\right)$

(b)

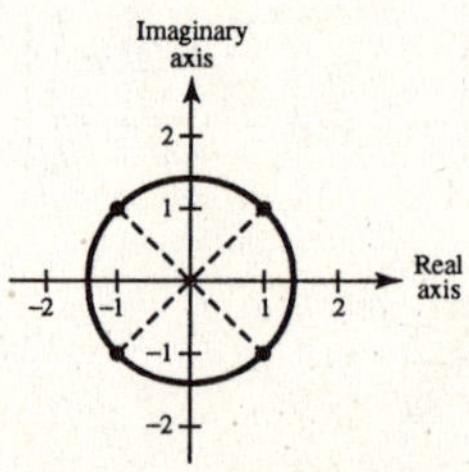

(c) $1 + i, -1 + i, -1 - i, 1 - i$

105. (a) Fifth roots of $128(-1 + i) = 128\sqrt{2}\left(\cos\frac{3\pi}{4} + i\sin\frac{3\pi}{4}\right)$:

$$\sqrt[5]{128\sqrt{2}}\left[\cos\left(\frac{\frac{3\pi}{4} + 2\pi k}{5}\right) + i\sin\left(\frac{\frac{3\pi}{4} + 2\pi k}{5}\right)\right],\ k = 0, 1, 2, 3, 4;\ \sqrt[5]{128\sqrt{2}} = 2\sqrt[5]{4\sqrt{2}} = 2(2^{5/2})^{1/5}$$

$$= 2(2^{1/2}) = 2\sqrt{2}$$

$k = 0$: $2\sqrt{2}\left(\cos\frac{3\pi}{20} + i\sin\frac{3\pi}{20}\right)$

$k = 1$: $2\sqrt{2}\left(\cos\frac{11\pi}{20} + i\sin\frac{11\pi}{20}\right)$

$k = 2$: $2\sqrt{2}\left(\cos\frac{19\pi}{20} + i\sin\frac{19\pi}{20}\right)$

$k = 3$: $2\sqrt{2}\left(\cos\frac{27\pi}{20} + i\sin\frac{27\pi}{20}\right)$

$k = 4$: $2\sqrt{2}\left(\cos\frac{7\pi}{4} + i\sin\frac{7\pi}{4}\right)$

(b)

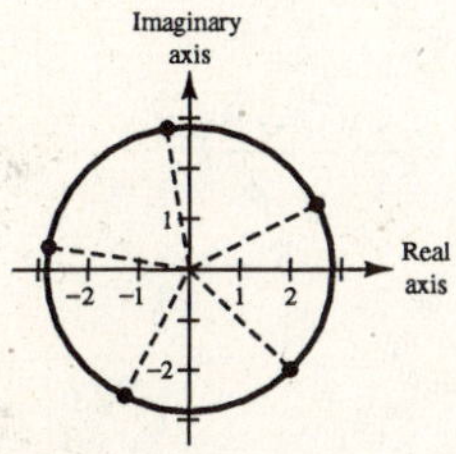

(c) $2.5201 + 1.2841i, -0.4425 + 2.7936i, -2.7936 + 0.4425i, -1.2841 - 2.5201i, 2 - 2i$

106. (a) Sixth roots of $64i = 64\left(\cos\frac{\pi}{2} + i\sin\frac{\pi}{2}\right)$:

$$\sqrt[6]{64}\left[\cos\left(\frac{\frac{\pi}{2} + 2k\pi}{6}\right) + i\sin\left(\frac{\frac{\pi}{2} + 2k\pi}{6}\right)\right]$$

$k = 0, 1, 2, 3, 4, 5$

$k = 0$: $2\left(\cos\frac{\pi}{12} + i\sin\frac{\pi}{12}\right)$

$k = 1$: $2\left(\cos\frac{5\pi}{12} + i\sin\frac{5\pi}{12}\right)$

$k = 2$: $2\left(\cos\frac{3\pi}{4} + i\sin\frac{3\pi}{4}\right)$

$k = 3$: $2\left(\cos\frac{13\pi}{12} + i\sin\frac{13\pi}{12}\right)$

$k = 4$: $2\left(\cos\frac{17\pi}{12} + i\sin\frac{17\pi}{12}\right)$

$k = 5$: $2\left(\cos\frac{7\pi}{4} + i\sin\frac{7\pi}{4}\right)$

(b)

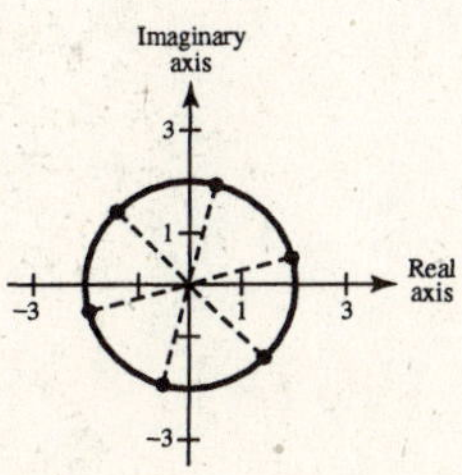

(c) $1.9319 + 0.5176i, 0.5176 + 1.9319i, -\sqrt{2} + \sqrt{2}i,$
$-1.9319 - 0.5176i, -0.5176 - 1.9319i, \sqrt{2} - \sqrt{2}i$

107. A complex number z has exactly n distinct n^{th} roots. Since one fourth root is shown, that means that there are **three** more fourth roots which are not shown.

108. The other three fourth roots all have the same magnitude as the shown root. They are also equally spaced about the circle, so successive roots are each 90° apart. From the figure it appears that the roots are:

$2(\cos 30° + i \sin 30°) = \sqrt{3} + i$

$2(\cos 120° + i \sin 120°) = -1 + \sqrt{3}i$

$2(\cos 210° + i \sin 210°) = -\sqrt{3} - i$

$2(\cos 300° + i \sin 300°) = 1 - \sqrt{3}i$

109. $x^4 - i = 0$

$x^4 = i$

The solutions are the fourth roots of

$i = \cos\frac{\pi}{2} + i \sin\frac{\pi}{2}$:

$$\sqrt[4]{1}\left[\cos\left(\frac{\frac{\pi}{2} + 2k\pi}{4}\right) + i \sin\left(\frac{\frac{\pi}{2} + 2k\pi}{4}\right)\right], k = 0, 1, 2, 3$$

$k = 0: \cos\frac{\pi}{8} + i \sin\frac{\pi}{8}$

$k = 1: \cos\frac{5\pi}{8} + i \sin\frac{5\pi}{8}$

$k = 2: \cos\frac{9\pi}{8} + i \sin\frac{9\pi}{8}$

$k = 3: \cos\frac{13\pi}{8} + i \sin\frac{13\pi}{8}$

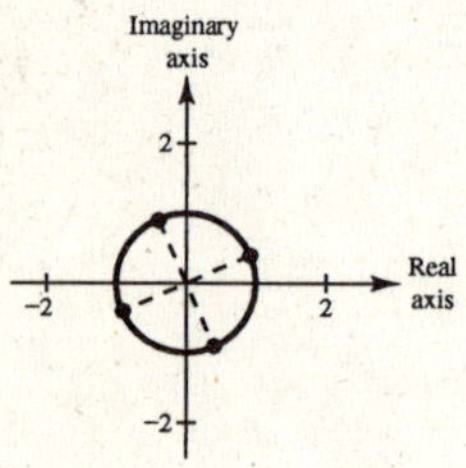

110. $x^3 + 1 = 0$

$x^3 = -1$

The solutions are the cube roots of $-1 = \cos \pi + i \sin \pi$:

$$\cos\left(\frac{\pi + 2k\pi}{3}\right) + i \sin\left(\frac{\pi + 2k\pi}{3}\right)$$

$k = 0, 1, 2$

$k = 0: \cos\frac{\pi}{3} + i \sin\frac{\pi}{3} = \frac{1}{2} + \frac{\sqrt{3}}{2}i$

$k = 1: \cos \pi + i \sin \pi = -1$

$k = 2: \cos\frac{5\pi}{3} + i \sin\frac{5\pi}{3} = \frac{1}{2} - \frac{\sqrt{3}}{2}i$

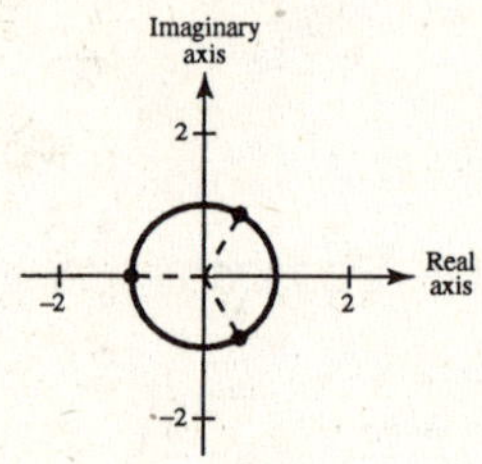

111. $x^5 + 243 = 0$

$x^5 = -243$

The solutions are the fifth roots of

$-243 = 243(\cos \pi + i \sin \pi)$:

$$\sqrt[5]{243}\left[\cos\left(\frac{\pi + 2k\pi}{5}\right) + i \sin\left(\frac{\pi + 2k\pi}{5}\right)\right], k = 0, 1, 2, 3, 4$$

$k = 0: 3\left(\cos\frac{\pi}{5} + i \sin\frac{\pi}{5}\right)$

$k = 1: 3\left(\cos\frac{3\pi}{5} + i \sin\frac{3\pi}{5}\right)$

$k = 2: 3(\cos \pi + i \sin \pi) = -3$

$k = 3: 3\left(\cos\frac{7\pi}{5} + i \sin\frac{7\pi}{5}\right)$

$k = 4: 3\left(\cos\frac{9\pi}{5} + i \sin\frac{9\pi}{5}\right)$

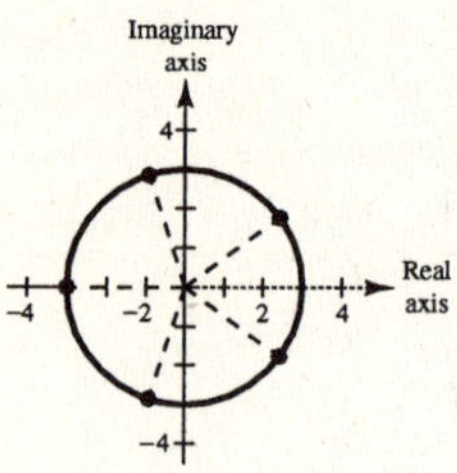

112. $x^3 - 27 = 0$

$$x^3 = 27$$

The solutions are the cube roots of $27 = 27(\cos 0 + i \sin 0)$:

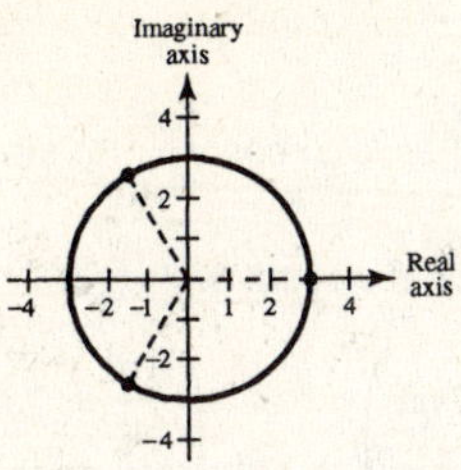

$$\sqrt[3]{27}\left[\cos\left(\frac{2k\pi}{3}\right) + i \sin\left(\frac{2k\pi}{3}\right)\right]$$

$k = 0, 1, 2$

$k = 0$: $3(\cos 0 + i \sin 0) = 3$

$k = 1$: $3\left(\cos \frac{2\pi}{3} + i \sin \frac{2\pi}{3}\right) = -\frac{3}{2} + \frac{3\sqrt{3}}{2}$

$k = 2$: $3\left(\cos \frac{4\pi}{3} + i \sin \frac{4\pi}{3}\right) = -\frac{3}{2} - \frac{3\sqrt{3}}{2}$

113. $x^4 + 16i = 0$

$$x^4 = -16i$$

The solutions are the fourth roots of $-16i = 16\left(\cos \frac{3\pi}{2} + i \sin \frac{3\pi}{2}\right)$:

$$\sqrt[4]{16}\left[\cos \frac{\frac{3\pi}{2} + 2\pi k}{4} + i \sin \frac{\frac{3\pi}{2} + 2\pi k}{4}\right], k = 0, 1, 2, 3$$

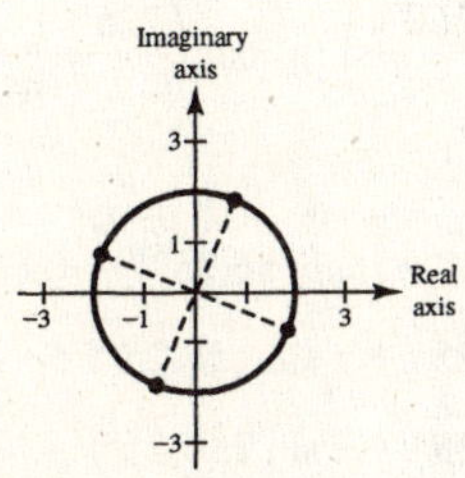

$k = 0$: $2\left(\cos \frac{3\pi}{8} + i \sin \frac{3\pi}{8}\right)$

$k = 1$: $2\left(\cos \frac{7\pi}{8} + i \sin \frac{7\pi}{8}\right)$

$k = 2$: $2\left(\cos \frac{11\pi}{8} + i \sin \frac{11\pi}{8}\right)$

$k = 3$: $2\left(\cos \frac{15\pi}{8} + i \sin \frac{15\pi}{8}\right)$

114. $x^6 - 64i = 0$

$$x^6 = 64i$$

The solutions are the sixth roots of $64i$:

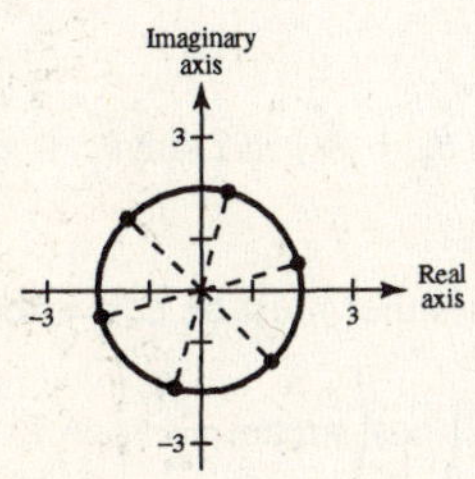

$$\sqrt[6]{64}\left[\cos\left(\frac{\frac{\pi}{2} + 2k\pi}{6}\right) + i \sin\left(\frac{\frac{\pi}{2} + 2k\pi}{6}\right)\right]$$

$k = 0, 1, 2, 3, 4, 5$

$k = 0$: $2\left(\cos \frac{\pi}{12} + i \sin \frac{\pi}{12}\right) \approx 1.932 + 0.5176i$

$k = 1$: $2\left(\cos \frac{5\pi}{12} + i \sin \frac{5\pi}{12}\right) \approx 0.5176 + 1.932i$

$k = 2$: $2\left(\cos \frac{3\pi}{4} + i \sin \frac{3\pi}{4}\right) \approx -1.414 + 1.414i$

$k = 3$: $2\left(\cos \frac{13\pi}{12} + i \sin \frac{13\pi}{12}\right) \approx -1.932 - 0.5176i$

$k = 4$: $2\left(\cos \frac{17\pi}{12} + i \sin \frac{17\pi}{12}\right) \approx -0.5176 - 1.932i$

$k = 5$: $2\left(\cos \frac{7\pi}{4} + i \sin \frac{7\pi}{4}\right) \approx 1.414 - 1.414i$

115. $x^3 - (1 - i) = 0$

$$x^3 = 1 - i = \sqrt{2}\left(\cos\frac{7\pi}{4} + i\sin\frac{7\pi}{4}\right)$$

The solutions are the cube roots of $1 - i$:

$$\sqrt[3]{\sqrt{2}}\left[\cos\left(\frac{\frac{7\pi}{4} + 2\pi k}{3}\right) + i\sin\left(\frac{\frac{7\pi}{4} + 2\pi k}{3}\right)\right],\ k = 0, 1, 2$$

$k = 0$: $\sqrt[6]{2}\left(\cos\frac{7\pi}{12} + i\sin\frac{7\pi}{12}\right)$

$k = 1$: $\sqrt[6]{2}\left(\cos\frac{5\pi}{4} + i\sin\frac{5\pi}{4}\right)$

$k = 2$: $\sqrt[6]{2}\left(\cos\frac{23\pi}{12} + i\sin\frac{23\pi}{12}\right)$

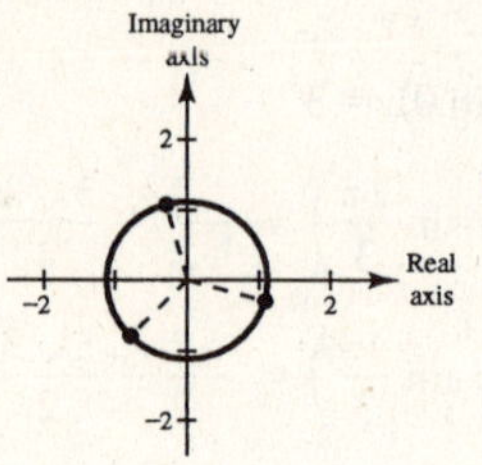

116. $x^4 + (1 + i) = 0$

$$x^4 = -1 - i = \sqrt{2}(\cos 225° + i\sin 225°)$$

The solutions are the fourth roots of $-1 - i$:

$$\sqrt[4]{\sqrt{2}}\left[\cos\left(\frac{225° + 360°k}{4}\right) + i\sin\left(\frac{225° + 360°k}{4}\right)\right]$$

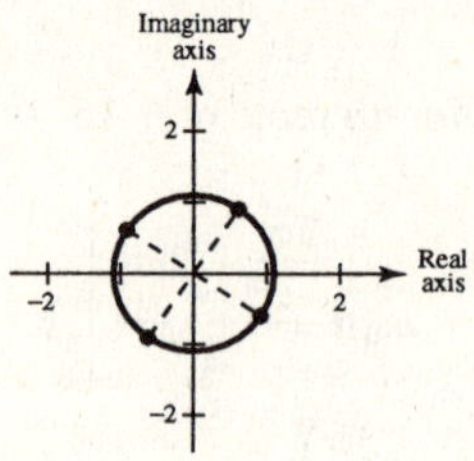

$k = 0, 1, 2, 3$

$k = 0$: $\sqrt[8]{2}(\cos 56.25° + i\sin 56.25°) \approx 0.6059 + 0.9067i$

$k = 1$: $\sqrt[8]{2}(\cos 146.25° + i\sin 146.25°) \approx -0.9067 + 0.6059i$

$k = 2$: $\sqrt[8]{2}(\cos 236.25° + i\sin 236.25°) \approx -0.6059 - 0.9067i$

$k = 3$: $\sqrt[8]{2}(\cos 326.25° + i\sin 326.25°) \approx 0.9067 - 0.6059i$

117. True, by the definition of the absolute value of a complex number.

118. True, because successive nth roots have arguments that differ by $2\pi/n$.

119. True. $z_1z_2 = r_1r_2[\cos(\theta_1 + \theta_2) + i\sin(\theta_1 + \theta_2)]$ and $z_1z_2 = 0$ if and only if $r_1 = 0$ and/or $r_2 = 0$.

120. False. The complex number must be converted to trigonometric form before applying DeMoivre's Theorem.

$$(4 + \sqrt{6}i)^8 = \left[\sqrt{22}\left(\cos\left(\arctan\frac{\sqrt{6}}{4}\right) + i\sin\left(\arctan\frac{\sqrt{6}}{4}\right)\right)\right]^8$$

121. True.

$$2 - 2\sqrt{3}i = 4\left(\cos\frac{5\pi}{3} + i\sin\frac{5\pi}{3}\right)$$

$$(2 - 2\sqrt{3}i)^3 = 64(\cos 5\pi + i\sin 5\pi)$$

$$= 64(\cos\pi + i\sin\pi)$$

122. $-\frac{1}{2}(1 + \sqrt{3}i) = -\left(\cos\frac{4\pi}{3} + i\sin\frac{4\pi}{3}\right)$

$$\left[-\frac{1}{2}(1 + \sqrt{3}i)\right]^6 = \left[-\left(\cos\frac{4\pi}{3} + i\sin\frac{4\pi}{3}\right)\right]^6$$

$$= \cos 8\pi + i\sin 8\pi$$

$$= 1$$

123. $$2^{-1/4}(1 - i) = 2^{-1/4}\left[\sqrt{2}\left(\cos\frac{7\pi}{4} + i\sin\frac{7\pi}{4}\right)\right]$$

$$= 2^{1/4}\left(\cos\frac{7\pi}{4} + i\sin\frac{7\pi}{4}\right)$$

$$\left[2^{1/4}\left(\cos\frac{7\pi}{4} + i\sin\frac{7\pi}{4}\right)\right]^4 = (2^{1/4})^4(\cos 7\pi + i\sin 7\pi)$$

$$= 2(\cos\pi + i\sin\pi)$$

$$= -2$$

124. $$\frac{z_1}{z_2} = \frac{r_1(\cos\theta_1 + i\sin\theta_1)}{r_2(\cos\theta_2 + i\sin\theta_2)} \cdot \frac{\cos\theta_2 - i\sin\theta_2}{\cos\theta_2 - i\sin\theta_2}$$

$$= \frac{r_1}{r_2(\cos^2\theta_2 + \sin^2\theta_2)}[\cos\theta_1\cos\theta_2 + \sin\theta_1\sin\theta_2 + i(\sin\theta_1\cos\theta_2 - \sin\theta_2\cos\theta_1)]$$

$$= \frac{r_1}{r_2}[\cos(\theta_1 - \theta_2) + i\sin(\theta_1 - \theta_2)]$$

125. (a) $2(\cos 30° + i\sin 30°)$

$2(\cos 150° + i\sin 150°)$

$2(\cos 270° + i\sin 270°)$

(b) These are the cube roots of $8i$.

126. (a) $3(\cos 45° + i\sin 45°)$

$3(\cos 135° + i\sin 135°)$

$3(\cos 225° + i\sin 225°)$

$3(\cos 315° + i\sin 315°)$

(b) These are the fourth roots of -81.

(c) The fourth roots of -81:

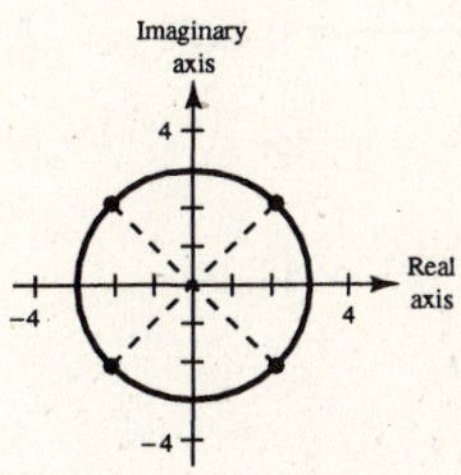

Review Exercises for Chapter 13

1. Given: $A = 35°, B = 71°, a = 8$

$C = 180° - 35° - 71° = 74°$

$$b = \frac{a\sin B}{\sin A} = \frac{8\sin 71°}{\sin 35°} \approx 13.19$$

$$c = \frac{a\sin C}{\sin A} = \frac{8\sin 74°}{\sin 35°} \approx 13.41$$

2. Given: $A = 22°, B = 121°, a = 17$

$C = 180° - A - B = 37°$

$$b = \frac{a\sin B}{\sin A} = \frac{17\sin 121°}{\sin 22°} \approx 38.90$$

$$c = \frac{a\sin C}{\sin A} = \frac{17\sin 37°}{\sin 22°} \approx 27.31$$

3. Given: $B = 72°, C = 82°, b = 54$

$A = 180° - 72° - 82° = 26°$

$$a = \frac{b\sin A}{\sin B} = \frac{54\sin 26°}{\sin 72°} \approx 24.89$$

$$c = \frac{b\sin C}{\sin B} = \frac{54\sin 82°}{\sin 72°} \approx 56.23$$

4. Given: $B = 10°, C = 20°, c = 33$

$A = 180° - B - C = 150°$

$$a = \frac{c\sin A}{\sin C} = \frac{33\sin 150°}{\sin 20°} \approx 48.24$$

$$b = \frac{c\sin B}{\sin C} = \frac{33\sin 10°}{\sin 20°} \approx 16.75$$

5. Given: $A = 16°, B = 98°, c = 8.4$

$C = 180° - 16° - 98° = 66°$

$$a = \frac{c \sin A}{\sin C} = \frac{8.4 \sin 16°}{\sin 66°} \approx 2.53$$

$$b = \frac{c \sin B}{\sin C} = \frac{8.4 \sin 98°}{\sin 66°} \approx 9.11$$

6. Given: $A = 95°, B = 45°, c = 104.8$

$C = 180° - A - B = 40°$

$$a = \frac{c \sin A}{\sin C} = \frac{104.8 \sin 95°}{\sin 40°} \approx 162.42$$

$$b = \frac{c \sin B}{\sin C} = \frac{104.8 \sin 45°}{\sin 40°} \approx 115.29$$

7. Given: $A = 24°, C = 48°, b = 27.5$

$B = 180° - 24° - 48° = 108°$

$$a = \frac{b \sin A}{\sin B} = \frac{27.5 \sin 24°}{\sin 108°} \approx 11.76$$

$$c = \frac{b \sin C}{\sin B} = \frac{27.5 \sin 48°}{\sin 108°} \approx 21.49$$

8. Given: $B = 64°, C = 36°, a = 367$

$A = 180° - B - C = 80°$

$$b = \frac{a \sin B}{\sin A} = \frac{367 \sin 64°}{\sin 80°} \approx 334.95$$

$$c = \frac{a \sin C}{\sin A} = \frac{367 \sin 36°}{\sin 80°} \approx 219.04$$

9. Given: $B = 150°, b = 30, c = 10$

$$\sin C = \frac{c \sin B}{b} = \frac{10 \sin 150°}{30} \approx 0.1667 \implies C \approx 9.59°$$

$A \approx 180° - 150° - 9.59° = 20.41°$

$$a = \frac{b \sin A}{\sin B} = \frac{30 \sin 20.41°}{\sin 150°} \approx 20.92$$

10. Given: $B = 150°, a = 10, b = 3$

$$\sin A = \frac{a \sin B}{b} = \frac{10 \sin 150°}{3} \approx 1.67 > 1$$

No solution

11. $A = 75°, a = 51.2, b = 33.7$

$$\sin B = \frac{b \sin A}{a} = \frac{33.7 \sin 75°}{51.2} \approx 0.6358 \implies B \approx 39.48°$$

$C \approx 180° - 75° - 39.48° = 65.52°$

$$c = \frac{a \sin C}{\sin A} = \frac{51.2 \sin 65.52°}{\sin 75°} \approx 48.24$$

12. Given: $B = 25°, \ a = 6.2, \ b = 4$

$$\sin A = \frac{a \sin B}{b} \approx 0.6551 \implies A \approx 40.9° \text{ or } 139.1°$$

Case 1: $A \approx 40.9°$

$C \approx 180° - 25° - 40.9° = 114.1°$

$c \approx 8.6$

Case 2: $A \approx 139.1°$

$C \approx 180° - 25° - 139.1° = 15.9°$

$c \approx 2.6$

13. Area $= \frac{1}{2}bc \sin A = \frac{1}{2}(5)(7)\sin 27° \approx 7.945$

14. $B = 80°, a = 4, c = 8$

Area $= \frac{1}{2}ac \sin B = \frac{1}{2}(4)(8)(0.9848) = 15.76$

15. Area $= \frac{1}{2}ab \sin C = \frac{1}{2}(16)(5)\sin 123° \approx 33.547$

16. $A = 11°, b = 22, c = 21$

Area $= \frac{1}{2}bc \sin A \approx \frac{1}{2}(22)(21)(0.1908)$

$= 44.08$

17. $\tan 17° = \dfrac{h}{x + 50} \implies h = (x + 50)\tan 17°$

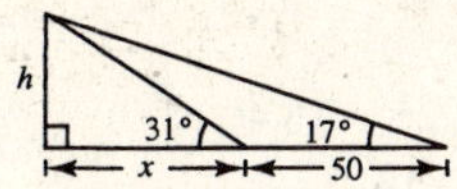

$$h = x \tan 17° + 50 \tan 17°$$

$\tan 31° = \dfrac{h}{x} \implies h = x \tan 31°$

$$x \tan 17° + 50 \tan 17° = x \tan 31°$$

$$50 \tan 17° = x(\tan 31° - \tan 17°)$$

$$\frac{50 \tan 17°}{\tan 31° - \tan 17°} = x$$

$$x \approx 51.7959$$

$$h = x \tan 31°$$

$$\approx 51.7959 \tan 31°$$

$$\approx 31.1 \text{ meters}$$

18. $16^2 = w^2 + 12^2 - 2w(12) \cos 140°$

$w^2 - (24 \cos 140°)w - 112 = 0 \implies w \approx 4.83$

19.

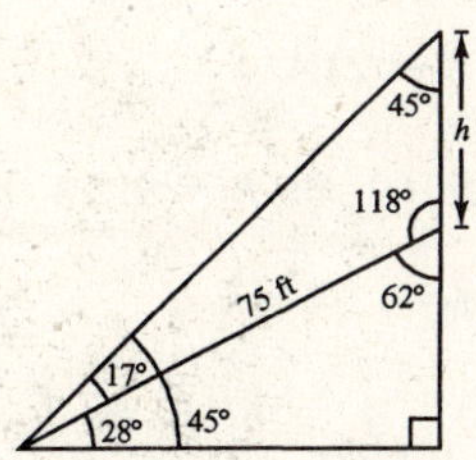

$$\frac{h}{\sin 17°} = \frac{75}{\sin 45°}$$

$$h = \frac{75 \sin 17°}{\sin 45°}$$

$$h \approx 31.01 \text{ feet}$$

20. Given: $a = 5, b = 8, c = 10$

$\cos C = \dfrac{a^2 + b^2 - c^2}{2ab} - 0.1375 \implies C \approx 97.90°$

$\cos B = \dfrac{a^2 + c^2 - b^2}{2ac} = 0.61 \implies B \approx 52.41°$

$A = 180° - B - C \approx 29.69°$

21. Given: $a = 80, b = 60, c = 100$

$\cos C = \dfrac{a^2 + b^2 - c^2}{2ab} = \dfrac{6400 + 3600 - 10{,}000}{2(80)(60)} = 0 \implies C = 90°$

$\sin A = \dfrac{80}{100} = 0.8 \implies A \approx 53.13°$

$\sin B = \dfrac{60}{100} = 0.6 \implies B \approx 36.87°$

22. Given: $a = 2.5, b = 5.0, c = 4.5$

$\cos B = \dfrac{a^2 + c^2 - b^2}{2ac} = 0.0667 \implies B \approx 86.18°$

$\cos C = \dfrac{a^2 + b^2 - c^2}{2ab} = 0.44 \implies C \approx 63.90°$

$A = 180° - B - C \approx 29.92°$

23. Given: $a = 16.4, b = 8.8, c = 12.2$

$$\cos A = \frac{b^2 + c^2 - a^2}{2bc} = \frac{8.8^2 + 12.2^2 - 16.4^2}{2(8.8)(12.2)} \approx -0.1988 \Rightarrow A \approx 101.47^\circ$$

$$\sin B = \frac{b \sin A}{a} \approx \frac{8.8 \sin 101.47^\circ}{16.4} \approx 0.5259 \Rightarrow B \approx 31.73^\circ$$

$$C \approx 180^\circ - 101.47^\circ - 31.73^\circ = 46.80^\circ$$

24. Given: $B = 110^\circ, a = 4, c = 4$

$$b = \sqrt{a^2 + c^2 - 2ac \cos B} \approx 6.55$$

$$A = C = \tfrac{1}{2}(180^\circ - 110^\circ) = 35^\circ$$

25. Given: $B = 150^\circ, a = 10, c = 20$

$$b^2 = 10^2 + 20^2 - 2(10)(20) \cos 150^\circ \Rightarrow b \approx 29.09$$

$$\sin A = \frac{a \sin B}{b} \approx \frac{10 \sin 150^\circ}{29.09} \Rightarrow A \approx 9.90^\circ$$

$$C \approx 180^\circ - 150^\circ - 9.90^\circ = 20.10^\circ$$

26. Given: $C = 43^\circ, a = 22.5, b = 31.4$

$$c = \sqrt{a^2 + b^2 - 2ab \cos C} \approx 21.42$$

$$\cos B = \frac{a^2 + c^2 - b^2}{2ac} \approx -0.02169 \Rightarrow B \approx 91.24^\circ$$

$$A = 180^\circ - B - C \approx 45.76^\circ$$

27. Given: $A = 62^\circ, b = 11.34, c = 19.52$

$$a^2 = 11.34^2 + 19.52^2 - 2(11.34)(19.52) \cos 62^\circ \Rightarrow a \approx 17.37$$

$$\sin B = \frac{b \sin A}{a} \approx \frac{11.34 \sin 62^\circ}{17.37} \Rightarrow B \approx 35.20^\circ$$

$$C \approx 180^\circ - 62^\circ - 35.20^\circ = 82.80^\circ$$

28. Length of AC $= \sqrt{300^2 + 425^2 - 2(300)(425) \cos 115^\circ}$

≈ 615.1 meters

29. (a)

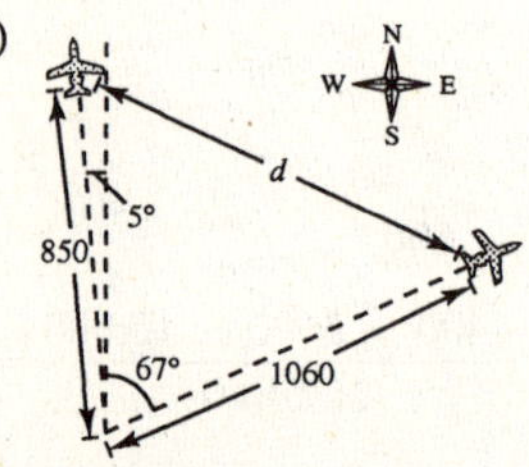

(b) $d^2 = 850^2 + 1060^2 - 2(850)(1060) \cos 72^\circ$

$\approx 1{,}289{,}251$

$d \approx 1135$ miles

30. $a = 4,\ b = 5,\ c = 7$

$$s = \frac{a + b + c}{2} = \frac{4 + 5 + 7}{2} = 8$$

$$\text{Area} = \sqrt{s(s - a)(s - b)(s - c)}$$

$$= \sqrt{8(4)(3)(1)} \approx 9.798$$

31. $a = 15, b = 8, c = 10$

$$s = \frac{15 + 8 + 10}{2} = 16.5$$

$$\text{Area} = \sqrt{16.5(1.5)(8.5)(6.5)} \approx 36.979$$

32. $a = 12.3, b = 15.8, c = 3.7$

$$s = \frac{a+b+c}{2} = \frac{12.3 + 15.8 + 3.7}{2} = 15.9$$

$$\begin{aligned}\text{Area} &= \sqrt{s(s-a)(s-b)(s-c)}\\ &= \sqrt{15.9(3.6)(0.1)(12.2)}\\ &= 8.4\end{aligned}$$

33. $a = 38.1, b = 26.7, c = 19.4$

$$s = \frac{38.1 + 26.7 + 19.4}{2} = 42.1$$

$$\text{Area} = \sqrt{42.1(4)(15.4)(22.7)} \approx 242.630$$

34.

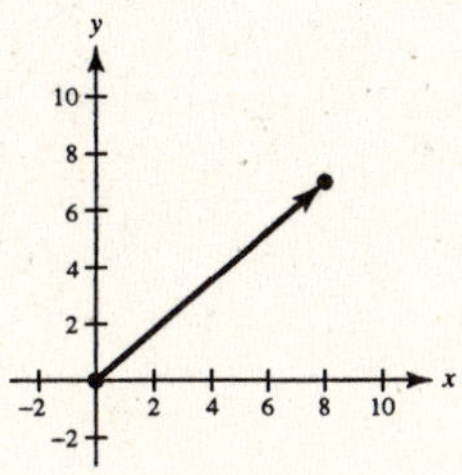

35. Initial point: $(3, 4)$

Terminal point: $(-5, -7)$

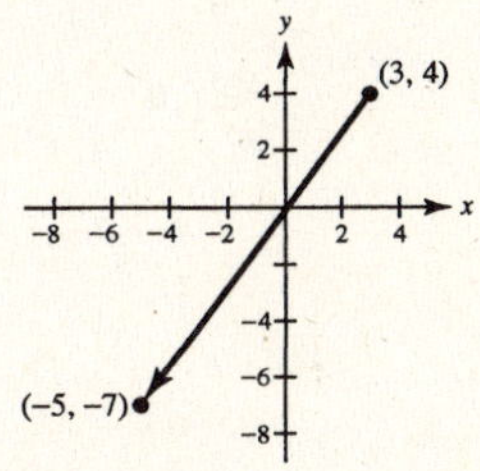

36.

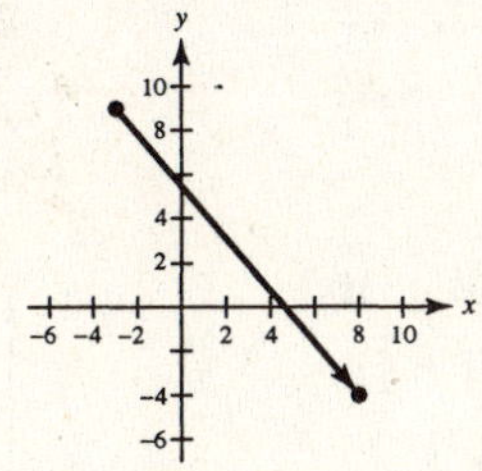

37. Initial point: $(-6, -8)$

Terminal point: $(8, 3)$

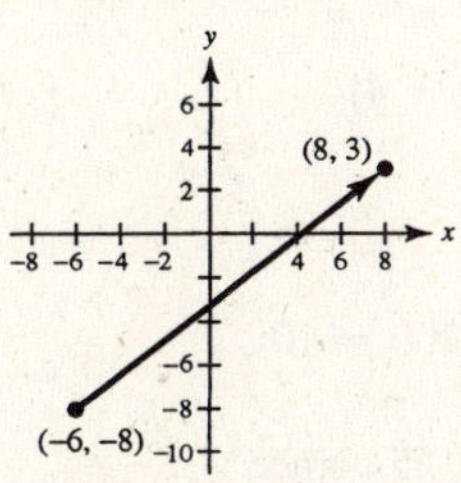

38. Initial point: $(-5, 4)$

Terminal point: $(2, -1)$

$\mathbf{v} = \langle 2 - (-5), -1 - 4\rangle = \langle 7, -5\rangle$

39. Initial point: $(0, 1)$

Terminal point: $\left(6, \frac{7}{2}\right)$

$\mathbf{v} = \left\langle 6 - 0, \frac{7}{2} - 1\right\rangle = \left\langle 6, \frac{5}{2}\right\rangle$

40. Initial point: $(0, 10)$

Terminal point: $(7, 3)$

$\mathbf{v} = \langle 7 - 0, 3 - 10\rangle = \langle 7, -7\rangle$

41. Initial point: $(1, 5)$

Terminal point: $(15, 9)$

$\mathbf{v} = \langle 15 - 1, 9 - 5\rangle = \langle 14, 4\rangle$

42. $\|\mathbf{v}\| = 6, \theta = 135°$

$$\begin{aligned}\mathbf{v} &= \langle 6\cos 135°, 6\sin 135°\rangle\\ &= \left\langle -3\sqrt{2}, 3\sqrt{2}\right\rangle\end{aligned}$$

43. $\|\mathbf{v}\| = \dfrac{1}{3}, \theta = 210°$

$$\left\langle \frac{1}{3}\cos 210°, \frac{1}{3}\sin 210°\right\rangle = \left\langle -\frac{\sqrt{3}}{6}, -\frac{1}{6}\right\rangle$$

44. $\mathbf{u} = 6\mathbf{i} - 5\mathbf{j}, \mathbf{v} = 10\mathbf{i} + 3\mathbf{j}$

$$\begin{aligned}2\mathbf{u} + \mathbf{v} &= 2(6\mathbf{i} - 5\mathbf{j}) + (10\mathbf{i} + 3\mathbf{j})\\ &= 22\mathbf{i} - 7\mathbf{j}\end{aligned}$$

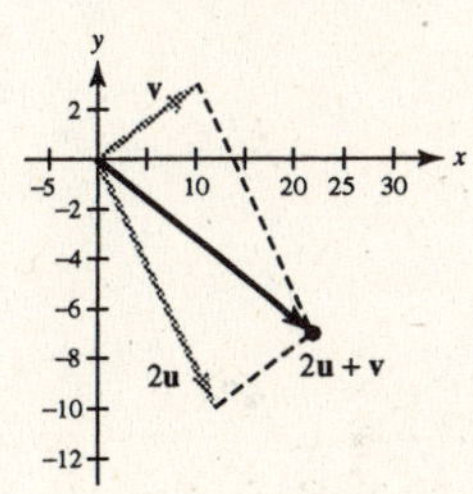

45. $\mathbf{u} = 6\mathbf{i} - 5\mathbf{j},\ \mathbf{v} = 10\mathbf{i} + 3\mathbf{j}$

$4\mathbf{u} - 5\mathbf{v} = (24\mathbf{i} - 20\mathbf{j}) - (50\mathbf{i} + 15\mathbf{j}) = -26\mathbf{i} - 35\mathbf{j}$

$= \langle -26, -35 \rangle$

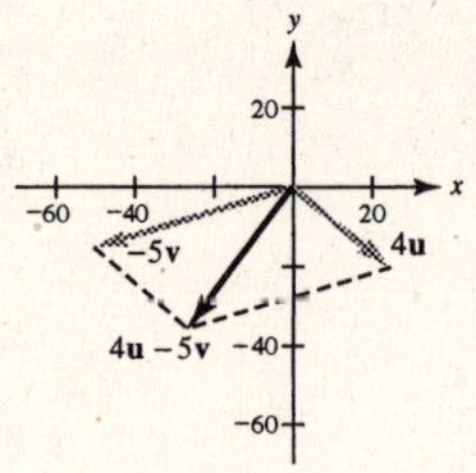

46. $\mathbf{v} = 10\mathbf{i} + 3\mathbf{j}$

$3\mathbf{v} = 3(10\mathbf{i} + 3\mathbf{j}) = 30\mathbf{i} + 9\mathbf{j}$

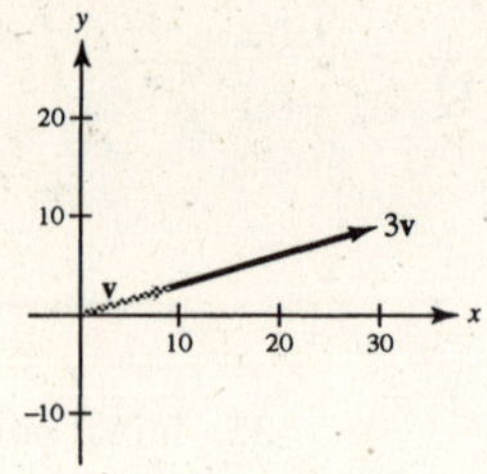

47. $\mathbf{v} = 10\mathbf{i} + 3\mathbf{j}$

$\frac{1}{2}\mathbf{v} = 5\mathbf{i} + \frac{3}{2}\mathbf{j} = \langle 5, \frac{3}{2} \rangle$

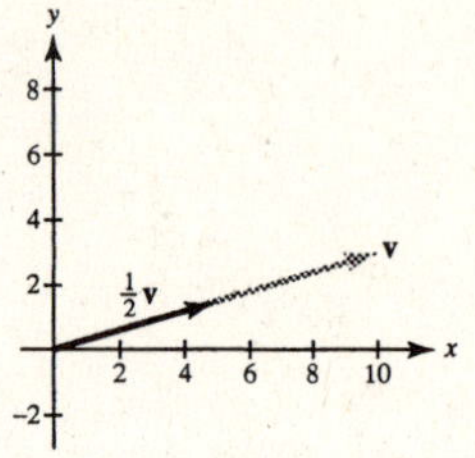

48. $\mathbf{u} = \langle -3, 4 \rangle = -3\mathbf{i} + 4\mathbf{j}$

49. $\mathbf{u} = \langle -6, -8 \rangle = -6\mathbf{i} - 8\mathbf{j}$

50. Initial point: $(3, 4)$

Terminal point: $(9, 8)$

$\mathbf{u} = (9 - 3)\mathbf{i} + (8 - 4)\mathbf{j} = 6\mathbf{i} + 4\mathbf{j}$

51. Initial point: $(-2, 7)$

Terminal point: $(5, -9)$

$\mathbf{u} = \langle 5 - (-2), -9 - 7 \rangle = \langle 7, -16 \rangle = 7\mathbf{i} - 16\mathbf{j}$

52. $\mathbf{v} = -10\mathbf{i} + 10\mathbf{j}$

$\|\mathbf{v}\| = \sqrt{(-10)^2 + (10)^2} = \sqrt{200} = 10\sqrt{2}$

$\tan\theta = \dfrac{10}{-10} = -1 \Rightarrow \theta = 135°$ since

$\mathbf{v}$ is in Quadrant II.

$\mathbf{v} = 10\sqrt{2}(\mathbf{i}\cos 135° + \mathbf{j}\sin 135°)$

53. $\mathbf{v} = 4\mathbf{i} - \mathbf{j}$

$\|\mathbf{v}\| = \sqrt{4^2 + (-1)^2} = \sqrt{17}$

$\tan\theta = \dfrac{-1}{4}$, θ in Quadrant IV $\Rightarrow \theta \approx 346°$

$\mathbf{v} \approx \sqrt{17}(\cos 346°\,\mathbf{i} + \sin 346°\,\mathbf{j})$

54. $\mathbf{v} = 7(\cos 60°\,\mathbf{i} + \sin 60°\,\mathbf{j})$

$\|\mathbf{v}\| = 7$

$\theta = 60°$

55. $\mathbf{v} = 3(\cos 150°\mathbf{i} + \sin 150°\,\mathbf{j})$

$\|\mathbf{v}\| = 3,\ \theta = 150°$

56. $\mathbf{v} = 5\mathbf{i} + 4\mathbf{j}$

$\|\mathbf{v}\| = \sqrt{5^2 + 4^2} = \sqrt{41}$

$\tan\theta = \dfrac{4}{5} \Rightarrow \theta \approx 38.7°$

57. $\mathbf{v} = -4\mathbf{i} + 7\mathbf{j}$

$\|\mathbf{v}\| = \sqrt{(-4)^2 + 7^2} = \sqrt{65}$

$\tan\theta = \dfrac{7}{-4}$, θ in Quadrant II $\Rightarrow \theta \approx 119.7°$

58. $\mathbf{v} = -3\mathbf{i} - 3\mathbf{j}$

$\|\mathbf{v}\| = \sqrt{(-3)^2 + (-3)^2} = 3\sqrt{2}$

$\tan\theta = \dfrac{-3}{-3} = 1 \Rightarrow \theta = 225°$

59. $\mathbf{v} = 8\mathbf{i} - \mathbf{j}$

$\|\mathbf{v}\| = \sqrt{8^2 + (-1)^2} = \sqrt{65}$

$\tan\theta = \dfrac{-1}{8}$, θ in Quadrant IV $\Rightarrow \theta \approx 352.9°$

60. Magnitude of resultant:

$c = \sqrt{85^2 + 50^2 - 2(85)(50)\cos 165°}$

≈ 133.92 pounds

Let θ be the angle between the resultant and the 85-pound force.

$\cos\theta \approx \dfrac{(133.92)^2 + 85^2 - 50^2}{2(133.92)(85)} \approx 0.9953 \Rightarrow \theta \approx 5.6°$

61. Rope One: $\mathbf{u} = \|\mathbf{u}\|(\cos 30°\mathbf{i} - \sin 30°\mathbf{j}) = \|\mathbf{u}\|\left(\dfrac{\sqrt{3}}{2}\mathbf{i} - \dfrac{1}{2}\mathbf{j}\right)$

Rope Two: $\mathbf{v} = \|\mathbf{u}\|(-\cos 30°\mathbf{i} - \sin 30°\mathbf{j}) = \|\mathbf{u}\|\left(-\dfrac{\sqrt{3}}{2}\mathbf{i} - \dfrac{1}{2}\mathbf{j}\right)$

Resultant: $\mathbf{u} + \mathbf{v} = -\|\mathbf{u}\|\mathbf{j} = -180\mathbf{j}$

$\|\mathbf{u}\| = 180$

Therefore, the tension on each rope is $\|\mathbf{u}\| = 180$ lbs.

62. Airspeed: $\mathbf{u} = 724(\cos 60°\mathbf{i} + \sin 60°\mathbf{j})$

$= 362(\mathbf{i} + \sqrt{3}\mathbf{j})$

Wind: $\mathbf{w} = 32\mathbf{i}$

Groundspeed $= \mathbf{u} + \mathbf{w} = (394\mathbf{i} + 362\sqrt{3}\mathbf{j})$

$\|\mathbf{u} + \mathbf{w}\| = \sqrt{(394)^2 + (362\sqrt{3})^2} \approx 740.5$ km/hr

$\tan\theta = \dfrac{362\sqrt{3}}{394} \Rightarrow \theta \approx 57.9°$

Bearing: N 32.1° E

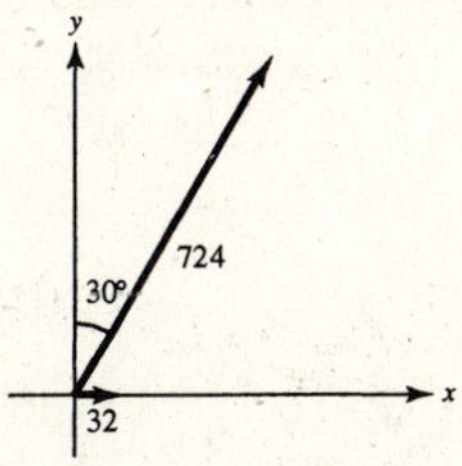

63. $\mathbf{u} = \langle 6, 7\rangle$

$\mathbf{v} = \langle -3, 9\rangle$

$\mathbf{u}\cdot\mathbf{v} = 6(-3) + 7(9) = 45$

64. $\mathbf{u} = \langle -7, 12\rangle, \mathbf{v} = \langle -4, -14\rangle$

$\mathbf{u}\cdot\mathbf{v} = -7(-4) + 12(-14) = -140$

65. $\mathbf{u} = 3\mathbf{i} + 7\mathbf{j}$

$\mathbf{v} = 11\mathbf{i} - 5\mathbf{j}$

$\mathbf{u}\cdot\mathbf{v} = 3(11) + 7(-5) = -2$

66. $\mathbf{u} = -7\mathbf{i} + 2\mathbf{j}, \mathbf{v} = 16\mathbf{i} - 12\mathbf{j}$

$\mathbf{u}\cdot\mathbf{v} = -7(16) + 2(-12) = -136$

67. $\mathbf{u} = \langle -3, 4\rangle$

$2\mathbf{u} = \langle -6, 8\rangle$

$2\mathbf{u}\cdot\mathbf{u} = (-6)(-3) + 8(4) = 50$

The result is a scalar.

68. $\mathbf{u} = \langle -3, 4\rangle$

$2\mathbf{u} = \langle -6, 8\rangle$

$\mathbf{v} = \langle 2, 1\rangle$

$2\mathbf{u}\cdot\mathbf{v} = (-6)(2) + (8)(1) = -4$

The result is a scalar.

69. $\mathbf{u} = \langle -3, 4\rangle, \mathbf{v} = \langle 2, 1\rangle$

$\mathbf{u}\cdot\mathbf{v} = (-3)(2) + 4(1) = -2$

$\mathbf{u}(\mathbf{u}\cdot\mathbf{v}) = \mathbf{u}(-2) = -2\mathbf{u} = \langle 6, -8\rangle$

The result is a vector.

70. $\mathbf{u} = \langle -3, 4\rangle, \mathbf{v} = \langle 2, 1\rangle$

$3\mathbf{u}\cdot\mathbf{v} = 3(-3(2) + 4(1)) = 3(-2) = -6$; scalar

71. $\mathbf{u} = \langle -3, 4 \rangle$

$\|\mathbf{u}\|^2 = \mathbf{u} \cdot \mathbf{u} = (-3)^2 + (4)^2 = 25$

The result is a scalar.

72. $\mathbf{v} = \langle 2, 1 \rangle$

$\|\mathbf{v}\|^2 = \mathbf{v} \cdot \mathbf{v} = 2^2 + 1^2 = 5$; scalar

73. $\mathbf{u} = \cos\frac{7\pi}{4}\mathbf{i} + \sin\frac{7\pi}{4}\mathbf{j} = \left\langle \frac{1}{\sqrt{2}}, -\frac{1}{\sqrt{2}} \right\rangle$

$\mathbf{v} = \cos\frac{5\pi}{6}\mathbf{i} + \sin\frac{5\pi}{6}\mathbf{j} = \left\langle -\frac{\sqrt{3}}{2}, \frac{1}{2} \right\rangle$

$\cos\theta = \frac{\mathbf{u} \cdot \mathbf{v}}{\|\mathbf{u}\|\,\|\mathbf{v}\|} = \frac{-\sqrt{3} - 1}{2\sqrt{2}} \Rightarrow \theta = \frac{11\pi}{12}$

74. $\mathbf{u} = \cos 45^\circ\,\mathbf{i} + \sin 45^\circ\,\mathbf{j}$

$\mathbf{v} = \cos 300^\circ\,\mathbf{i} + \sin 300^\circ\,\mathbf{j}$

Angle between $\mathbf{u}$ and $\mathbf{v}$: $60^\circ + 45^\circ = 105^\circ$

75. $\mathbf{u} = \langle 2\sqrt{2}, -4 \rangle, \mathbf{v} = \langle -\sqrt{2}, 1 \rangle$

$\cos\theta = \frac{\mathbf{u} \cdot \mathbf{v}}{\|\mathbf{u}\|\,\|\mathbf{v}\|} = \frac{-8}{(\sqrt{24})(\sqrt{3})} \Rightarrow \theta \approx 160.5^\circ$

76. $\mathbf{u} = \langle 3, \sqrt{3} \rangle, \mathbf{v} = \langle 4, 3\sqrt{3} \rangle$

$\cos\theta = \frac{\mathbf{u} \cdot \mathbf{v}}{\|\mathbf{u}\|\,\|\mathbf{v}\|} = \frac{21}{\sqrt{12}\sqrt{43}} \Rightarrow \theta \approx 22.4^\circ$

77. $\mathbf{u} = \langle -3, 8 \rangle$

$\mathbf{v} = \langle 8, 3 \rangle$

$\mathbf{u} \cdot \mathbf{v} = -3(8) + 8(3) = 0$

$\mathbf{u}$ and $\mathbf{v}$ are orthogonal.

78. $\mathbf{u} = \langle \frac{1}{4}, -\frac{1}{2} \rangle, \mathbf{v} = \langle -2, 4 \rangle$

$\mathbf{v} = -8\mathbf{u} \Rightarrow$ Parallel

79. $\mathbf{u} = -\mathbf{i}$

$\mathbf{v} = \mathbf{i} + 2\mathbf{j}$

$\mathbf{u} \cdot \mathbf{v} \neq 0 \Rightarrow$ Not orthogonal

$\mathbf{v} \neq k\mathbf{u} \Rightarrow$ Not parallel

Neither

80. $\mathbf{u} = -2\mathbf{i} + \mathbf{j}, \mathbf{v} = 3\mathbf{i} + 6\mathbf{j}$

$\mathbf{u} \cdot \mathbf{v} = 0 \Rightarrow$ Orthogonal

81. $\mathbf{u} = \langle -4, 3 \rangle,\ \mathbf{v} = \langle -8, -2 \rangle$

$\mathbf{w}_1 = \text{proj}_{\mathbf{v}}\mathbf{u} = \left(\frac{\mathbf{u} \cdot \mathbf{v}}{\|\mathbf{v}\|^2}\right)\mathbf{v} = \left(\frac{26}{68}\right)\langle -8, -2 \rangle$

$= -\frac{13}{17}\langle 4, 1 \rangle$

$\mathbf{w}_2 = \mathbf{u} - \mathbf{w}_1 = \langle -4, 3 \rangle - \left(-\frac{13}{17}\right)\langle 4, 1 \rangle$

$= \frac{16}{17}\langle -1, 4 \rangle$

82. $\mathbf{u} = \langle 5, 6 \rangle, \mathbf{v} = \langle 10, 0 \rangle$

$\text{proj}_{\mathbf{v}}\mathbf{u} = \left(\frac{\mathbf{u} \cdot \mathbf{v}}{\|\mathbf{v}\|^2}\right)\mathbf{v} = \frac{50}{100}\langle 10, 0 \rangle = \langle 5, 0 \rangle$

$\mathbf{u} - \text{proj}_{\mathbf{v}}\mathbf{u} = \langle 0, 6 \rangle$

83. $\mathbf{u} = \langle 2, 7 \rangle,\ \mathbf{v} = \langle 1, -1 \rangle$

$\mathbf{w}_1 = \text{proj}_{\mathbf{v}}\mathbf{u} = \left(\frac{\mathbf{u} \cdot \mathbf{v}}{\|\mathbf{v}\|^2}\right)\mathbf{v} = -\frac{5}{2}\langle 1, -1 \rangle$

$= \frac{5}{2}\langle -1, 1 \rangle$

$\mathbf{w}_2 = \mathbf{u} - \mathbf{w}_1 = \langle 2, 7 \rangle - \left(\frac{5}{2}\right)\langle -1, 1 \rangle$

$= \frac{9}{2}\langle 1, 1 \rangle$

84. $\mathbf{u} = \langle -3, 5\rangle, \mathbf{v} = \langle -5, 2\rangle$

$$\text{proj}_{\mathbf{v}}\mathbf{u} = \left(\frac{\mathbf{u} \cdot \mathbf{v}}{\|\mathbf{v}\|^2}\right)\mathbf{v} = \frac{25}{29}\langle -5, 2\rangle$$

$$\mathbf{u} - \text{proj}_{\mathbf{v}}\mathbf{u} = \langle -3, 5\rangle - \frac{25}{29}\langle -5, 2\rangle$$

$$= \frac{19}{29}\langle 2, 5\rangle$$

85. $P = (5, 3), Q = (8, 9) \Rightarrow \overrightarrow{PQ} = \langle 3, 6\rangle$

$$W = \mathbf{F} \cdot \overrightarrow{PQ} = \langle 2, 7\rangle \cdot \langle 3, 6\rangle = 48$$

86. $\text{work} = \mathbf{v} \cdot \overrightarrow{PQ}$

$$= (3\mathbf{i} - 6\mathbf{j}) \cdot (-10\mathbf{i} + 17\mathbf{j})$$

$$= -30 - 102$$

$$= -132$$

87. $|7i| = \sqrt{0^2 + 7^2} = 7$

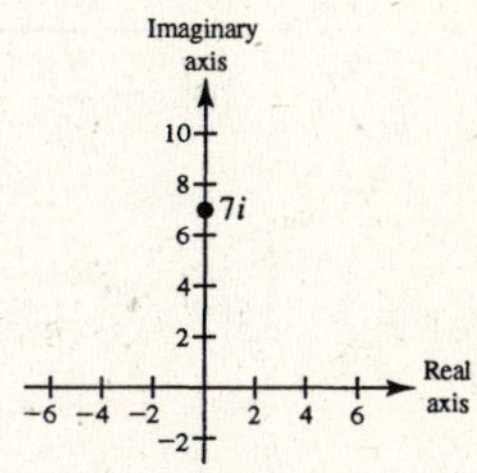

88. $|-6i| = 6$

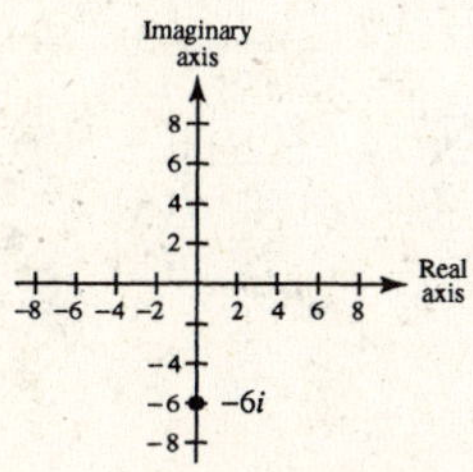

89. $|5 + 3i| = \sqrt{5^2 + 3^2} = \sqrt{34}$

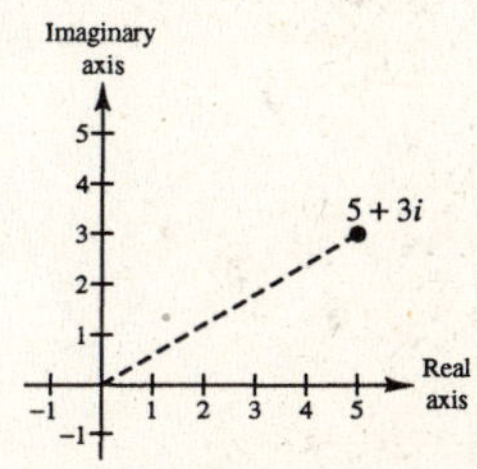

90. $|-10 - 4i| = \sqrt{(-10)^2 + (-4)^2}$

$$= 2\sqrt{29}$$

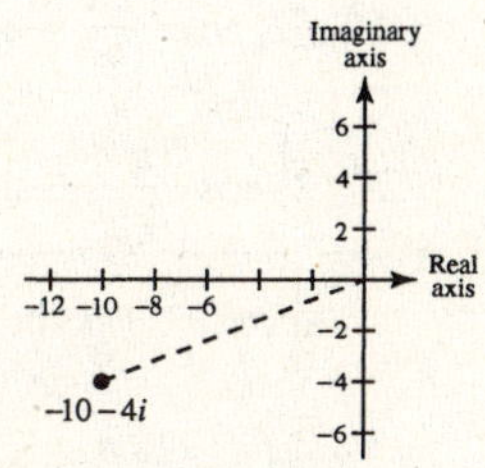

91. $5 - 5i$

$$r = \sqrt{5^2 + (-5)^2} = \sqrt{50} = 5\sqrt{2}$$

$\tan\theta = \dfrac{-5}{5} = -1 \Rightarrow \theta = \dfrac{7\pi}{4}$ since the complex number is in Quadrant IV.

$$5 - 5i = 5\sqrt{2}\left(\cos\frac{7\pi}{4} + i\sin\frac{7\pi}{4}\right)$$

92. $z = 5 + 12i$

$$|z| = \sqrt{5^2 + 12^2} = 13$$

$$\tan\theta = \frac{12}{5} \Rightarrow \theta \approx 1.176$$

$$z \approx 13(\cos 1.176 + i\sin 1.176)$$

93. $-3\sqrt{3} + 3i$

$$r = \sqrt{(-3\sqrt{3})^2 + 3^2} = \sqrt{36} = 6$$

$\tan\theta = \dfrac{3}{-3\sqrt{3}} = -\dfrac{1}{\sqrt{3}} \Rightarrow \theta = \dfrac{5\pi}{6}$ since the complex number is in Quadrant II.

$$-3\sqrt{3} + 3i = 6\left(\cos\frac{5\pi}{6} + i\sin\frac{5\pi}{6}\right)$$

94. $z = -7$

$$|z| = 7$$

$$\tan\theta = \frac{0}{-7} = 0 \Rightarrow \theta = \pi$$

$$z = 7(\cos\pi + i\sin\pi)$$

95. (a) $z_1 = 2\sqrt{3} - 2i = 4\left(\cos\frac{11\pi}{6} + i\sin\frac{11\pi}{6}\right)$

$z_2 = -10i = 10\left(\cos\frac{3\pi}{2} + i\sin\frac{3\pi}{2}\right)$

(b) $z_1z_2 = \left[4\left(\cos\frac{11\pi}{6} + i\sin\frac{11\pi}{6}\right)\right]\left[10\left(\cos\frac{3\pi}{2} + i\sin\frac{3\pi}{2}\right)\right] = 40\left(\cos\frac{10\pi}{3} + i\sin\frac{10\pi}{3}\right)$

$$\frac{z_1}{z_2} = \frac{4\left(\cos\frac{11\pi}{6} + i\sin\frac{11\pi}{6}\right)}{10\left(\cos\frac{3\pi}{2} + i\sin\frac{3\pi}{2}\right)} = \frac{2}{5}\left(\cos\frac{\pi}{3} + i\sin\frac{\pi}{3}\right)$$

96. (a) $z_1 = -3(1 + i) = 3\sqrt{2}\left(\cos\frac{5\pi}{4} + i\sin\frac{5\pi}{4}\right)$

$z_2 = 2(\sqrt{3} + i) = 4\left(\cos\frac{\pi}{6} + i\sin\frac{\pi}{6}\right)$

(b) $z_1z_2 = \left[3\sqrt{2}\left(\cos\frac{5\pi}{4} + i\sin\frac{5\pi}{4}\right)\right]\left[4\left(\cos\frac{\pi}{6} + i\sin\frac{\pi}{6}\right)\right]$

$= 12\sqrt{2}\left(\cos\frac{17\pi}{12} + i\sin\frac{17\pi}{12}\right)$

$$\frac{z_1}{z_2} = \frac{3\sqrt{2}\left[\cos\frac{5\pi}{4} + i\sin\frac{5\pi}{4}\right]}{4\left[\cos\frac{\pi}{6} + i\sin\frac{\pi}{6}\right]}$$

$= \frac{3\sqrt{2}}{4}\left(\cos\frac{13\pi}{12} + i\sin\frac{13\pi}{12}\right)$

97. $\left[5\left(\cos\frac{\pi}{12} + i\sin\frac{\pi}{12}\right)\right]^4 = 5^4\left(\cos\frac{4\pi}{12} + i\sin\frac{4\pi}{12}\right)$

$= 625\left(\cos\frac{\pi}{3} + i\sin\frac{\pi}{3}\right)$

$= 625\left(\frac{1}{2} + \frac{\sqrt{3}}{2}i\right)$

$= \frac{625}{2} + \frac{625\sqrt{3}}{2}i$

98. $\left[2\left(\cos\frac{4\pi}{15} + i\sin\frac{4\pi}{15}\right)\right]^5 = 2^5\left(\cos\frac{4\pi}{3} + i\sin\frac{4\pi}{3}\right)$

$= 32\left(-\frac{1}{2} - \frac{\sqrt{3}}{2}i\right)$

$= -16 - 16\sqrt{3}i$

99. $(2 + 3i)^6 \approx [\sqrt{13}(\cos 56.3° + i\sin 56.3°)]^6$

$= 13^3(\cos 337.9° + i\sin 337.9°)$

$\approx 13^3(0.9263 - 0.3769i)$

$\approx 2035 - 828i$

100. $(1 - i)^8 = \left[\sqrt{2}(\cos 315° + i\sin 315°)\right]^8$

$= 16(\cos 2520° + i\sin 2520°)$

$= 16(\cos 0° + i\sin 0°)$

$= 16$

101. (a) The trigonometric form of the three roots shown is:

$4(\cos 60° + i\sin 60°)$

$4(\cos 180° + i\sin 180°)$

$4(\cos 300° + i\sin 300°)$

(b) Since there are three evenly spaced roots on the circle of radius 4, they are cube roots of a complex number of modulus $4^3 = 64$. Cubing them yields -64.

$[4(\cos 60° + i\sin 60°)]^3 = -64$

$[4(\cos 180° + i\sin 180°)]^3 = -64$

$[4(\cos 300° + i\sin 300°)]^3 = -64$

102. (a) The trigonometric forms of the four roots shown are:

$4(\cos 60° + i \sin 60°)$

$4(\cos 150° + i \sin 150°)$

$4(\cos 240° + i \sin 240°)$

$4(\cos 330° + i \sin 330°)$

(b) Since there are four evenly spaced roots on the circle of radius 4, they are fourth roots of a complex number of modulus 4^4. In this case, raising them to the fourth power yields $-128 - 128\sqrt{3}i$.

103. Sixth roots of $-729i = 729\left(\cos\frac{3\pi}{2} + i \sin\frac{3\pi}{2}\right)$:

$$\sqrt[6]{729}\left[\cos\left(\frac{\frac{3\pi}{2} + 2k\pi}{6}\right) + i\sin\left(\frac{\frac{3\pi}{2} + 2k\pi}{6}\right)\right], k = 0, 1, 2, 3, 4, 5$$

$k = 0: 3\left(\cos\frac{\pi}{4} + i\sin\frac{\pi}{4}\right)$

$k = 1: 3\left(\cos\frac{7\pi}{12} + i\sin\frac{7\pi}{12}\right)$

$k = 2: 3\left(\cos\frac{11\pi}{12} + i\sin\frac{11\pi}{12}\right)$

$k = 3: 3\left(\cos\frac{5\pi}{4} + i\sin\frac{5\pi}{4}\right)$

$k = 4: 3\left(\cos\frac{19\pi}{12} + i\sin\frac{19\pi}{12}\right)$

$k = 5: 3\left(\cos\frac{23\pi}{12} + i\sin\frac{23\pi}{12}\right)$

104. Fourth roots of $256 = 256(\cos 0 + i \sin 0)$:

$$\sqrt[4]{256}\left(\cos\frac{2\pi k}{4} + i\sin\frac{2\pi k}{4}\right)$$

$k = 0, 1, 2, 3$

$k = 0:\ 4(\cos 0 + i\sin 0) = 4$

$k = 1:\ 4\left(\cos\frac{\pi}{2} + i\sin\frac{\pi}{2}\right) = 4i$

$k = 2:\ 4(\cos\pi + i\sin\pi) = -4$

$k = 3:\ 4\left(\cos\frac{3\pi}{2} + i\sin\frac{3\pi}{2}\right) = -4i$

105. Cube roots of $64 = 64(\cos 0 + i \sin 0)$:

$$\sqrt[3]{64}\left[\cos\left(\frac{0 + 2\pi k}{3}\right) + i\sin\left(\frac{0 + 2\pi k}{3}\right)\right],$$

$k = 0, 1, 2$

$k = 0:\ 4(\cos 0 + i\sin 0) = 4$

$k = 1:\ 4\left(\cos\frac{2\pi}{3} + i\sin\frac{2\pi}{3}\right) = -2 + 2\sqrt{3}i$

$k = 2:\ 4\left(\cos\frac{4\pi}{3} + i\sin\frac{4\pi}{3}\right) = -2 - 2\sqrt{3}i$

106. Fifth roots of $32 = 32(\cos 0 + i \sin 0)$

$$\sqrt[5]{32}\left[\cos\left(\frac{0 + 2\pi k}{5}\right) + i\sin\left(\frac{0 + 2\pi k}{5}\right)\right],$$

$k = 0, 1, 2, 3, 4$

$k = 0:\ 2(\cos 0 + i\sin 0) = 2$

$k = 1:\ 2\left(\cos\frac{2\pi}{5} + i\sin\frac{2\pi}{5}\right)$

$k = 2:\ 2\left(\cos\frac{4\pi}{5} + i\sin\frac{4\pi}{5}\right)$

$k = 3:\ 2\left(\cos\frac{6\pi}{5} + i\sin\frac{6\pi}{5}\right)$

$k = 4:\ 2\left(\cos\frac{8\pi}{5} + i\sin\frac{8\pi}{5}\right)$

107. $x^4 + 81 = 0$

$x^4 = -81$ Solve by finding the fourth roots of -81.

$-81 = 81(\cos \pi + i \sin \pi)$

$$\sqrt[4]{-81} = \sqrt[4]{81}\left[\cos\left(\frac{\pi + 2\pi k}{4}\right) + i \sin\left(\frac{\pi + 2\pi k}{4}\right)\right],\ k = 0, 1, 2, 3$$

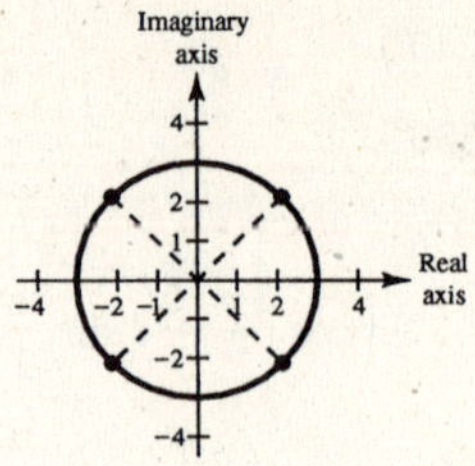

$$k = 0{:}\ 3\left(\cos\frac{\pi}{4} + i \sin\frac{\pi}{4}\right) = \frac{3\sqrt{2}}{2} + \frac{3\sqrt{2}}{2}i$$

$$k = 1{:}\ 3\left(\cos\frac{3\pi}{4} + i \sin\frac{3\pi}{4}\right) = -\frac{3\sqrt{2}}{2} + \frac{3\sqrt{2}}{2}i$$

$$k = 2{:}\ 3\left(\cos\frac{5\pi}{4} + i \sin\frac{5\pi}{4}\right) = -\frac{3\sqrt{2}}{2} - \frac{3\sqrt{2}}{2}i$$

$$k = 3{:}\ 3\left(\cos\frac{7\pi}{4} + i \sin\frac{7\pi}{4}\right) = \frac{3\sqrt{2}}{2} - \frac{3\sqrt{2}}{2}i$$

108. $x^5 - 32 = 0$

$x^5 = 32$

$32 = 32(\cos 0 + i \sin 0)$

$$\sqrt[3]{32} = \sqrt[5]{32}\left[\cos\left(0 + \frac{2\pi k}{5}\right) + i \sin\left(0 + \frac{2\pi k}{5}\right)\right]$$

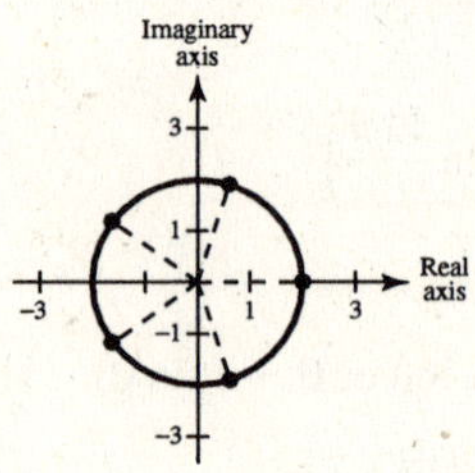

$k = 0, 1, 2, 3, 4$

$k = 0{:}\ 2(\cos 0 + i \sin 0) = 2$

$$k = 1{:}\ 2\left(\cos\frac{2\pi}{5} + i \sin\frac{2\pi}{5}\right) = 0.6180 + 1.9021i$$

$$k = 2{:}\ 2\left(\cos\frac{4\pi}{5} + i \sin\frac{4\pi}{5}\right) = -1.6180 + 1.1756i$$

$$k = 3{:}\ 2\left(\cos\frac{6\pi}{5} + i \sin\frac{6\pi}{5}\right) = -1.6180 - 1.1756i$$

$$k = 4{:}\ 2\left(\cos\frac{8\pi}{5} + i \sin\frac{8\pi}{5}\right) = 0.6180 - 1.9021i$$

109. $x^3 + 8i = 0$

$x^3 = -8i$ Solve by finding the cube roots of $-8i$.

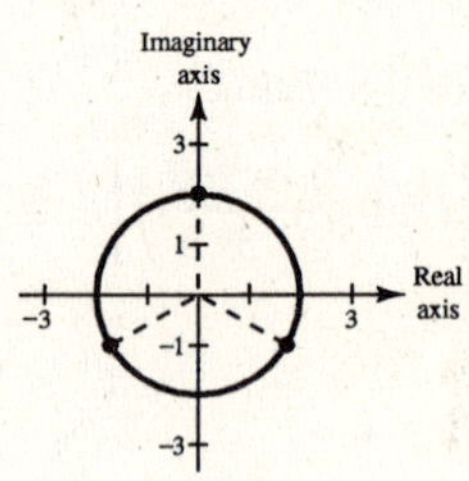

$$-8i = 8\left(\cos\frac{3\pi}{2} + i \sin\frac{3\pi}{2}\right)$$

$$\sqrt[3]{-8i} = \sqrt[3]{8}\left[\cos\left(\frac{\frac{3\pi}{2} + 2\pi k}{3}\right) + i \sin\left(\frac{\frac{3\pi}{2} + 2\pi k}{3}\right)\right],\ k = 0, 1, 2$$

$$k = 0{:}\ 2\left(\cos\frac{\pi}{2} + i \sin\frac{\pi}{2}\right) = 2i$$

$$k = 1{:}\ 2\left(\cos\frac{7\pi}{6} + i \sin\frac{7\pi}{6}\right) = -\sqrt{3} - i$$

$$k = 2{:}\ 2\left(\cos\frac{11\pi}{6} + i \sin\frac{11\pi}{6}\right) = \sqrt{3} - i$$

110. $(x^3 - 1)(x^2 + 1) = 0$

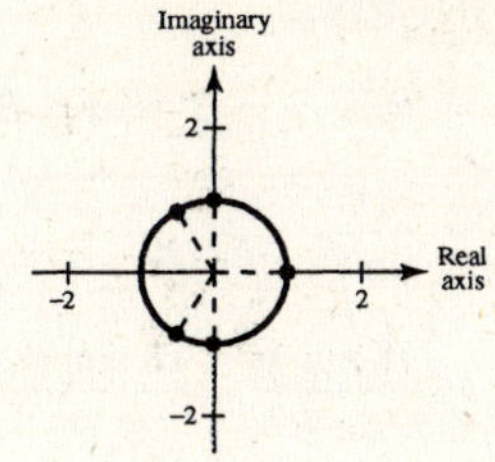

$$x^3 - 1 = 0$$
$$x^2 + 1 = 0$$
$$x^3 = 1$$
$$1 = 1(\cos 0 + i \sin 0)$$
$$\sqrt[3]{1} = \sqrt[3]{1}\left[\cos\left(\frac{0 + 2\pi k}{3}\right) + i \sin\left(\frac{0 + 2\pi k}{3}\right)\right],\ k = 0, 1, 2$$
$$1(\cos 0 + i \sin 0) = 1$$
$$1\left(\cos\frac{2\pi}{3} + i \sin\frac{2\pi}{3}\right) = -\frac{1}{2} + \frac{\sqrt{3}}{2}i$$
$$1\left(\cos\frac{4\pi}{3} + i \sin\frac{4\pi}{3}\right) = -\frac{1}{2} - \frac{\sqrt{3}}{2}i$$
$$x^2 + 1 = 0$$
$$x^2 = -1$$
$$-1 = 1(\cos \pi + i \sin \pi)$$
$$\sqrt{-1} = \sqrt{1}\left[\cos\left(\frac{\pi + 2\pi k}{2}\right) + i \sin\left(\frac{\pi + 2\pi k}{2}\right),\ k = 0, 1\right.$$
$$k = 0, 1$$
$$1\left(\cos\frac{\pi}{2} + i \sin\frac{\pi}{2}\right) = i$$
$$1\left(\cos\frac{3\pi}{2} + i \sin\frac{3\pi}{2}\right) = -i$$

Problem Solving for Chapter 13

1. (a) $\dfrac{\sin \alpha}{9} = \dfrac{\sin \beta}{18}$

$\sin \alpha = 0.5 \sin \beta$

$\alpha = \arcsin(0.5 \sin \beta)$; Domain: $0 < \beta < \pi$

(b) $$\frac{d\alpha}{d\beta} = \frac{0.5 \cos \beta}{\sqrt{1 - (0.5 \sin \beta)^2}} = \frac{0.5 \cos \beta}{\sqrt{1 - \dfrac{\sin^2\beta}{4}}} \cdot \frac{2}{2} = \frac{\cos \beta}{\sqrt{4 - \sin^2\beta}}$$

Maximum: $\left(\dfrac{\pi}{2}, \dfrac{\pi}{6}\right)$

Range: $0 < \alpha \le \dfrac{\pi}{6}$

(c)

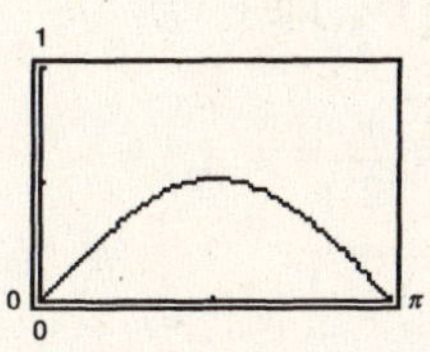

(d) $$\frac{d\alpha}{dt} = \frac{\cos \beta}{\sqrt{4 - \sin^2\beta}}\left(\frac{d\beta}{dt}\right)$$

When $\dfrac{d\beta}{dt} = 0.2$ and $\beta = \dfrac{\pi}{4}$ we have $\dfrac{d\alpha}{dt} = \dfrac{\frac{\sqrt{2}}{2}(0.2)}{\sqrt{4 - \frac{1}{2}}} = \dfrac{\sqrt{7}}{35} \approx 0.0756$ radians per second.

—CONTINUED—

1. —CONTINUED—

(e) $\gamma = \pi - \alpha - \beta = \pi - \beta - \arcsin(0.5 \sin \beta)$

$$\frac{c}{\sin \gamma} = \frac{18}{\sin \beta}$$

$$c = \frac{18 \sin \gamma}{\sin \beta} = \frac{18 \sin[\pi - \beta - \arcsin(0.5 \sin \beta)]}{\sin \beta}; \text{ Domain: } 0 < \beta < \pi$$

(f)

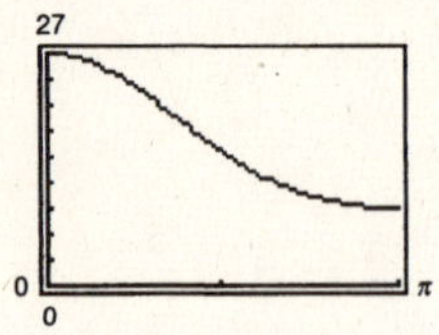

Range: $9 < c < 27$

(g) $$\frac{dc}{dt} = \frac{18\left[\sin \beta \cos[\pi - \beta - \arcsin(0.5 \sin \beta)]\left(-1 - \frac{\cos \beta}{\sqrt{4 - \sin^2 \beta}}\right) - \sin[\pi - \beta - \arcsin(0.5 \sin \beta)] \cos \beta\right]}{\sin^2 \beta} \cdot \frac{d\beta}{dt}$$

When $\frac{d\beta}{dt} = 0.2$ and $\beta = \frac{\pi}{4}$, $\frac{dc}{dt} \approx -1.7539$ radians per second.

(h)

β	0	0.4	0.8	1.2	1.6	2.0	2.4	2.8
α	0	0.1960	0.3669	0.4848	0.5234	0.4720	0.3445	0.1683
c	Undef.	25.95	23.07	19.19	15.33	12.29	10.31	9.27

(i) When $\beta \to 0$, $c \to 27$. On the graph it is undefined. The function obtained by using the Law of Sines is not valid when $\beta = 0$ since we would no longer have a triangle.

2. $\mathbf{F}_1 = \langle 10, 0 \rangle$, $\mathbf{F}_2 = 5\langle \cos \theta, \sin \theta \rangle$

(a) $\mathbf{F}_1 + \mathbf{F}_2 = \langle 10 + 5 \cos \theta, 5 \sin \theta \rangle$

$$\begin{aligned} \|\mathbf{F}_1 + \mathbf{F}_2\| &= \sqrt{(10 + 5\cos\theta)^2 + (5 \sin \theta)^2} \\ &= \sqrt{100 + 100 \cos \theta + 25 \cos^2 \theta + 25 \sin^2 \theta} \\ &= 5\sqrt{4 + 4 \cos \theta + \cos^2 \theta + \sin^2 \theta} \\ &= 5\sqrt{4 + 4 \cos \theta + 1} \\ &= 5\sqrt{5 + 4 \cos \theta} \end{aligned}$$

(b)

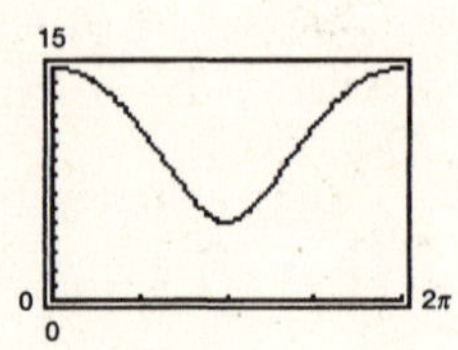

(c) Range: $[5, 15]$
Maximum is 15 when $\theta = 0$.
Minimum is 5 when $\theta = \pi$.

(d) The magnitude of the resultant is never 0 because the magnitudes of $\mathbf{F}_1$ and $\mathbf{F}_2$ are not the same.

3. (a) Initial point: $(0, 0)$

Terminal point: $\left(\frac{u_1 + v_1}{2}, \frac{u_2 + v_2}{2}\right)$

$$\mathbf{w} = \left\langle \frac{u_1 + v_1}{2}, \frac{u_2 + v_2}{2} \right\rangle = \frac{1}{2}(\mathbf{u} + \mathbf{v})$$

(b) Initial point: (u_1, u_2)

Terminal point: $\frac{1}{2}(u_1 + v_1, u_2 + v_2)$

$$\begin{aligned} \mathbf{w} &= \left\langle \frac{u_1 + v_1}{2} - u_1, \frac{u_2 + v_2}{2} - u_2 \right\rangle \\ &= \left\langle \frac{v_1 - u_1}{2}, \frac{v_2 - u_2}{2} \right\rangle = \frac{1}{2}(\mathbf{v} - \mathbf{u}) \end{aligned}$$

4. $\frac{1}{2}bc(1 + \cos A) = \frac{1}{2}bc\left[1 + \frac{b^2 + c^2 - a^2}{2bc}\right]$

$$= \frac{1}{2}bc\left[\frac{2bc + b^2 + c^2 - a^2}{2bc}\right]$$

$$= \frac{1}{4}[(b + c)^2 - a^2]$$

$$= \frac{1}{4}[(b + c) + a][(b + c) - a]$$

$$= \frac{b + c + a}{2} \cdot \frac{b + c - a}{2}$$

$$= \frac{a + b + c}{2} \cdot \frac{-a + b + c}{2}$$

5. $\frac{1}{2}bc(1 - \cos A) = \frac{1}{2}bc\left[1 + \frac{a^2 - (b^2 + c^2)}{2bc}\right]$

$$= \frac{1}{2}bc\left[\frac{2bc + a^2 - b^2 - c^2}{2bc}\right]$$

$$= \frac{a^2 - (b^2 - 2bc + c^2)}{4}$$

$$= \frac{a^2 - (b - c)^2}{4}$$

$$= \left(\frac{a - (b - c)}{2}\right)\left(\frac{a + (b - c)}{2}\right)$$

$$= \frac{a - b + c}{2} \cdot \frac{a + b - c}{2}$$

6. (a) Working with $\triangle OBC$, we have $\cos \alpha = \frac{a/2}{R}$.

This implies that $2R = \frac{a}{\cos \alpha}$.

Since we know that

$$\frac{a}{\sin A} = \frac{b}{\sin B} = \frac{c}{\sin C},$$

we can complete the proof by showing that $\cos \alpha = \sin A$. The solution of the system

$A + B + C = 180°$

$\alpha - C + A = \beta$

$\alpha + \beta = B$ is $\alpha = 90° - A$. Therefore:

$$2R = \frac{a}{\cos \alpha} = \frac{a}{\cos(90° - A)} = \frac{a}{\sin A}.$$

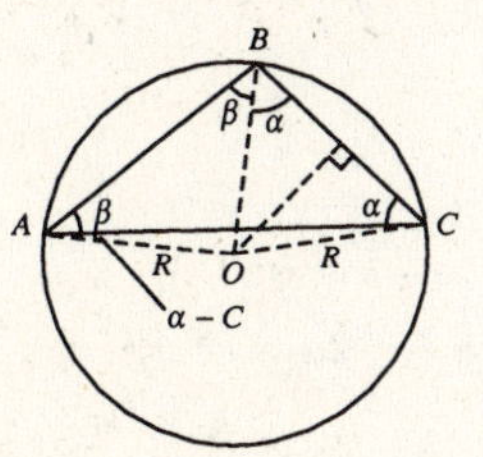

(b) By Heron's Formula, the area of the triangle is

$$\text{Area} = \sqrt{s(s - a)(s - b)(s - c)}.$$

We can also find the area by dividing the area into six triangles and using the fact that the area is $\frac{1}{2}$ the base times the height. Using the figure as given, we have

$$\text{Area} = \frac{1}{2}xr + \frac{1}{2}xr + \frac{1}{2}yr + \frac{1}{2}yr + \frac{1}{2}zr + \frac{1}{2}zr$$

$$= r(x + y + z)$$

$$= rs.$$

Therefore: $rs = \sqrt{s(s - a)(s - b)(s - c)} \Rightarrow$

$$r = \sqrt{\frac{(s - a)(s - b)(s - c)}{s}}.$$

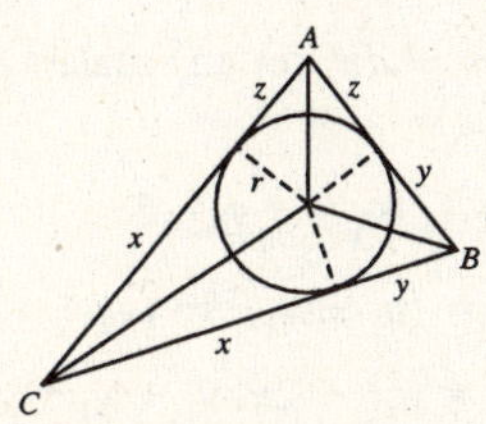

(c) $a = 25, b = 55, c = 72$

(i) area of triangle: $s = \frac{1}{2}(25 + 55 + 72) = 76$

area $= \sqrt{76(51)(21)(4)} \approx 570.60$

(ii) area of circumscribed circle:

$$\cos C = \frac{25^2 + 55^2 - 72^2}{2(25)(55)} \approx -0.5578 \Rightarrow C \approx 123.9°$$

$$R = \frac{1}{2}\left(\frac{c}{\sin C}\right) \approx 43.37$$

area $= \pi R^2 \approx 5909.2$

(iii) area of inscribed circle:

$$r = \sqrt{\frac{(s - a)(s - b)(s - c)}{s}} = \sqrt{\frac{(51)(21)(4)}{76}} \approx 7.51$$

area $= \pi r^2 \approx 177.09$

(d) Given: $a = 200$ ft, $b = 250$ ft, $c = 325$ ft

$$s = \frac{200 + 250 + 325}{2} \approx 387.5$$

Radius of the inscribed circle: $r = \sqrt{\frac{(s - a)(s - b)(s - c)}{s}} = \sqrt{\frac{(187.5)(137.5)(62.5)}{387.5}} \approx 64.5$ ft

Circumference of an inscribed circle: $C = 2\pi r \approx 2\pi(64.5) \approx 405.3$ ft

7. (a) $A = \frac{1}{2}(30)(20)\sin\left(\theta + \frac{\theta}{2}\right) - \frac{1}{2}(8)(20)\sin\frac{\theta}{2} - \frac{1}{2}(8)(30)\sin\theta$

$$= 300\sin\frac{3\theta}{2} - 80\sin\frac{\theta}{2} - 120\sin\theta$$

$$= 20\left[15\sin\frac{3\theta}{2} - 4\sin\frac{\theta}{2} - 6\sin\theta\right]$$

(b) 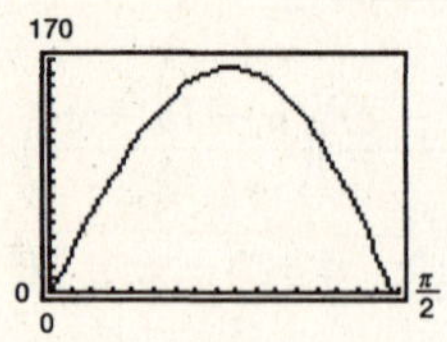

(c) Domain: $0 \le \theta \le 1.6690$

The area would increase and the domain would increase at the right endpoint of its interval.

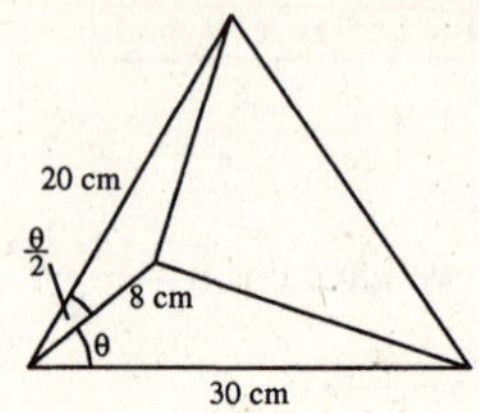

(d) $\frac{dA}{d\theta} = 20\left[15\left(\frac{3}{2}\right)\cos\frac{3\theta}{2} - 4\left(\frac{1}{2}\right)\cos\frac{\theta}{2} - 6\cos\theta\right]$

$$= 20\left[\frac{45}{2}\cos\frac{3\theta}{2} - 2\cos\frac{\theta}{2} - 6\cos\theta\right]$$

$$= 10\left[45\cos\frac{3\theta}{2} - 4\cos\frac{\theta}{2} - 12\cos\theta\right]$$

Critical number: $\theta \approx 0.8782$

8. Since $\mathbf{u}$ is orthogonal to $\mathbf{v}$ and $\mathbf{w}$, $\mathbf{u}\cdot\mathbf{v} = 0$ and $\mathbf{u}\cdot\mathbf{w} = 0$.

$$\mathbf{u}\cdot(c\mathbf{v} + d\mathbf{w}) = \mathbf{u}\cdot(c\mathbf{v}) + \mathbf{u}\cdot(d\mathbf{w})$$
$$= c(\mathbf{u}\cdot\mathbf{v}) + d(\mathbf{u}\cdot\mathbf{w})$$
$$= c(0) + d(0)$$
$$= 0$$

Thus, $\mathbf{u}$ is orthogonal to $c\mathbf{v} + d\mathbf{w}$ for any scalars c and d.

9. (a) Let $\mathbf{u} = \langle u_1, u_2\rangle$ and $\mathbf{v} = \langle v_1, v_2\rangle$, then

$$\|\mathbf{u}+\mathbf{v}\|^2 + \|\mathbf{u}-\mathbf{v}\|^2 = (u_1+v_1)^2 + (u_2+v_2)^2 + (u_1-v_1)^2 + (u_2-v_2)^2$$
$$= u_1^2 + 2u_1v_1 + v_1^2 + u_2^2 + 2u_2v_2 + v_2^2 + u_1^2 - 2u_1v_1 + v_1^2 + u_2^2 - 2u_2v_2 + v_2^2$$
$$= 2(u_1^2 + u_2^2) + 2(v_1^2 + v_2^2)$$
$$= 2\|\mathbf{u}\|^2 + 2\|\mathbf{v}\|^2.$$

(b) The sum of the squares of the lengths of the diagonals of a parallelagram is equal to the sum of the squares of the lengths of all four sides.

10. $\bar{z} = r[\cos(-\theta) + i\sin(-\theta)]$

$$= r[\cos\theta + -i\sin\theta]$$
$$= r\cos\theta - ir\sin\theta$$

which is the complex conjugate of $r(\cos\theta + i\sin\theta) = r\cos\theta + ir\sin\theta$.

11. (a) $z\bar{z} = [r(\cos\theta + i\sin\theta)][r(\cos(-\theta) + i\sin(-\theta))]$

$$= r^2[\cos(\theta - \theta) + i\sin(\theta - \theta)]$$
$$= r^2[\cos 0 + i\sin 0]$$
$$= r^2$$

(b) $\frac{z}{\bar{z}} = \frac{r(\cos\theta + i\sin\theta)}{r[\cos(-\theta) + i\sin(-\theta)]}, z \neq 0$

$$= \frac{r}{r}[\cos(\theta - (-\theta)) + i\sin(\theta - (-\theta))]$$
$$= \cos 2\theta + i\sin 2\theta$$

12. $z = r(\cos\theta + i\sin\theta)$

$-z = -r(\cos\theta + i\sin\theta)$

$= r(-\cos\theta + -i\sin\theta)$

$= r(\cos(\theta + \pi) + i\sin(\theta + \pi))$

13. Euler's Formula: $e^{a+bi} = e^{a}(\cos b + i\sin b)$

$e^{\pi i} + 1 = e^{0}(\cos\pi + i\sin\pi) + 1 = -1 + i(0) + 1 = 0$

14. (a)

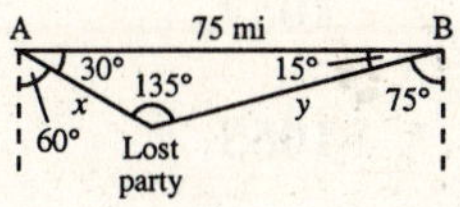

$\dfrac{x}{\sin 15°} = \dfrac{75}{\sin 135°}$ and $\dfrac{y}{\sin 30°} = \dfrac{75}{\sin 135°}$

$x \approx 27.452$ miles $\qquad y \approx 53.033$ miles

(b)

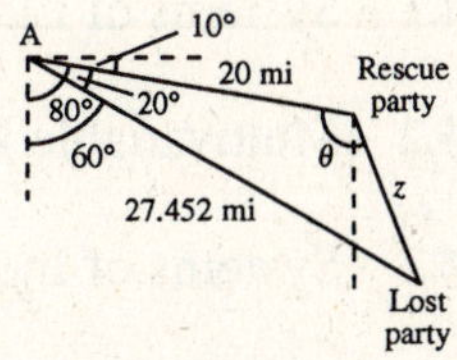

$z^2 = (27.452)^2 + (20)^2 - 2(27.452)(20)\cos 20°$

$z \approx 11.034$ miles

$\dfrac{\sin\theta}{27.452} = \dfrac{\sin 20°}{11.034}$

$\sin\theta \approx 0.8509$

$\theta = 180° - \sin^{-1}(0.8509)$

$\theta \approx 121.69°$

To find the bearing, we have $\theta - 10° - 90° \approx 21.69$

Bearing: S 21.69° E

15. $z_1 = 2(\cos\theta + i\sin\theta)$

$z_2 = 2(\cos(\pi - \theta) + i\sin(\pi - \theta))$

$z_1z_2 = (2)(2)[\cos(\theta + (\pi - \theta)) + i\sin(\theta + (\pi - \theta))]$

$= 4(\cos\pi + i\sin\pi)$

$= -4$

$\dfrac{z_1}{z_2} = \dfrac{2(\cos\theta + i\sin\theta)}{2(\cos(\pi - \theta) + i\sin(\pi - \theta))}$

$= 1[\cos(\theta - (\pi - \theta)) + i\sin(\theta - (\pi - \theta))]$

$= \cos(2\theta - \pi) + i\sin(2\theta - \pi)$

$= \cos 2\theta\cos\pi + \sin 2\theta\sin\pi + i(\sin 2\theta\cos\pi - \cos 2\theta\sin\pi)$

$= -\cos 2\theta - i\sin 2\theta$

16. 3/4 mile = 1320 yards

$x^2 = 1320^2 + 300^2 - 2(1320)(300)\cos 10°$

$x \approx 1025.881$ yards ≈ 0.58 mile

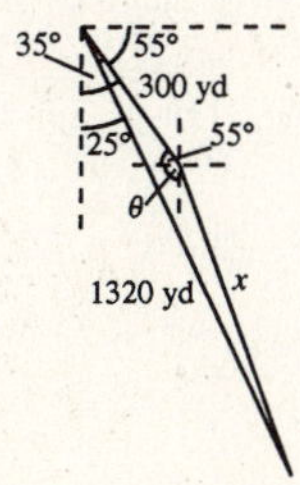

$\dfrac{\sin\theta}{1320} = \dfrac{\sin 10°}{1025.881}$

$\sin\theta \approx 0.2234$

$\theta = 180° - \sin^{-1}(0.2234)$

$\theta \approx 167.09°$

Bearing: $\theta - 55° - 90° \approx 22.09°$

S 22.09° E

17. The base is an equilateral triangle

$\Rightarrow B = \dfrac{\sqrt{3}}{4}x^2 \tan 35° = \dfrac{h}{x} \Rightarrow h = x\tan 35°$

Thus, $V = \dfrac{1}{3}Bh$

$= \dfrac{1}{3}\left(\dfrac{\sqrt{3}}{4}x^2\right)(x\tan 35°)$

$= \dfrac{\sqrt{3}}{12}x^3\tan 35°$

CHAPTER 14
Systems of Equations and Matrices

Section 14.1 Systems of Linear Equations in Two Variables 1039

Section 14.2 Multivariable Linear Systems 1055

Section 14.3 Systems of Inequalities 1070

Section 14.4 Matrices and Systems of Equations 1080

Section 14.5 Operations with Matrices 1094

Section 14.6 The Inverse of a Square Matrix 1103

Section 14.7 The Determinant of a Square Matrix 1117

Review Exercises 1126

Problem Solving 1150

CHAPTER 14
Systems of Equations and Matrices

Section 14.1 Systems of Linear Equations in Two Variables

1. $\begin{cases} 4x - y = 1 \\ 6x + y = -6 \end{cases}$

(a) $4(0) - (-3) \neq 1$

$(0, -3)$ is **not** a solution.

(b) $4(-1) - (-4) \neq 1$

$(-1, -4)$ is **not** a solution.

(c) $4\left(-\frac{3}{2}\right) - (-2) \neq 1$

$\left(-\frac{3}{2}, -2\right)$ is **not** a solution.

(d) $4\left(-\frac{1}{2}\right) - (-3) = 1$

$6\left(-\frac{1}{2}\right) + (-3) = -6$

$\left(-\frac{1}{2}, -3\right)$ **is** a solution.

2. $\begin{cases} 4x^2 + y = 3 \\ -x - y = 11 \end{cases}$

(a) $4(2)^2 + (-13) \stackrel{?}{=} 3$

$16 - 13 = 3$

$-2 - (-13) \stackrel{?}{=} 11$

$-2 + 13 = 11$

$(2, -13)$ **is** a solution.

(b) $4(2)^2 + (-9) \stackrel{?}{=} 3$

$16 - 9 \neq 3$

$(2, -9)$ is **not** a solution.

(c) $4\left(-\frac{3}{2}\right)^2 + \left(-\frac{31}{3}\right) \stackrel{?}{=} 3$

$\frac{36}{4} - \frac{31}{3} \neq 3$

$\left(-\frac{3}{2}, -\frac{31}{3}\right)$ is **not** a solution.

(d) $4\left(-\frac{7}{4}\right)^2 + \left(-\frac{37}{4}\right) \stackrel{?}{=} 3$

$\frac{49}{4} - \frac{37}{4} = 3$

$-\left(-\frac{7}{4}\right) - \left(-\frac{37}{4}\right) \stackrel{?}{=} 11$

$\frac{7}{4} + \frac{37}{4} = 11$

$\left(-\frac{7}{4}, -\frac{37}{4}\right)$ **is** a solution.

3. $\begin{cases} y = -2e^x \\ 3x - y = 2 \end{cases}$

(a) $0 \neq -2e^{-2}$

$(-2, 0)$ is **not** a solution.

(b) $-2 = -2e^0$

$3(0) - (-2) = 2$

$(0, -2)$ **is** a solution.

(c) $-3 \neq -2e^0$

$(0, -3)$ is **not** a solution.

(d) $2 \neq -2e^{-1}$

$(-1, 2)$ is **not** a solution.

4. $\begin{cases} -\log x + 3 = y \\ \frac{1}{9}x + y = \frac{28}{9} \end{cases}$

(a) $-\log 9 + 3 \neq \frac{37}{9}$

$\left(9, \frac{37}{9}\right)$ is **not** a solution.

(b) $-\log 10 + 3 = 2$

$\frac{1}{9}(10) + 2 = \frac{28}{9}$

$(10, 2)$ **is** a solution.

(c) $-\log(1) + 3 = 3$

$\frac{1}{9}(1) + 3 = \frac{28}{9}$

$(1, 3)$ **is** a solution.

(d) $-\log 2 + 3 \neq 4$

$(2, 4)$ is **not** a solution.

5. $\begin{cases} x - y = 0 & \text{Equation 1} \\ 5x - 3y = 10 & \text{Equation 2} \end{cases}$

Solve for y in Equation 1: $y = x$

Substitute for y in Equation 2: $5x - 3x = 10$

Solve for x: $2x = 10 \Rightarrow x = 5$

Back-substitute in Equation 1: $y = x = 5$

Solution: $(5, 5)$

6. $\begin{cases} x + 2y = 1 & \text{Equation 1} \\ 5x - 4y = -23 & \text{Equation 2} \end{cases}$

Solve for x in Equation 1: $x = 1 - 2y$

Substitute for x in Equation 2: $5(1 - 2y) - 4y = -23$

Solve for y: $-14y = -28 \Rightarrow y = 2$

Back-substitute $y = 2$: $x = 1 - 2y = 1 - 2(2) = -3$

Solution: $(-3, 2)$

7. $\begin{cases} 2x - y + 2 = 0 & \text{Equation 1} \\ 4x + y - 5 = 0 & \text{Equation 2} \end{cases}$

Solve for y in Equation 1: $y = 2x + 2$

Substitute for y in Equation 2: $4x + (2x + 2) - 5 = 0$

Solve for x: $6x - 3 = 0 \Rightarrow x = \frac{1}{2}$

Back-substitute $x = \frac{1}{2}$: $y = 2x + 2 = 2\left(\frac{1}{2}\right) + 2 = 3$

Solution: $\left(\frac{1}{2}, 3\right)$

8. $\begin{cases} 6x - 3y - 4 = 0 & \text{Equation 1} \\ x + 2y - 4 = 0 & \text{Equation 2} \end{cases}$

Solve for x in Equation 2: $x = 4 - 2y$

Substitute for x in Equation 1: $6(4 - 2y) - 3y - 4 = 0$

Solve for y: $24 - 12y - 3y - 4 = 0 \Rightarrow -15y = -20$

$\Rightarrow y = \frac{4}{3}$

Back-substitute $y = \frac{4}{3}$: $x = 4 - 2y = 4 - 2\left(\frac{4}{3}\right) = \frac{4}{3}$

Solution: $\left(\frac{4}{3}, \frac{4}{3}\right)$

9. $\begin{cases} 1.5x + 0.8y = 2.3 & \text{Equation 1} \\ 0.3x - 0.2y = 0.1 & \text{Equation 2} \end{cases}$

Multiply the equations by 10.

$15x + 8y = 23$ Revised Equation 1

$3x - 2y = 1$ Revised Equation 2

Solve for y in revised Equation 2: $y = \frac{3}{2}x - \frac{1}{2}$

Substitute for y in revised Equation 1: $15x + 8\left(\frac{3}{2}x - \frac{1}{2}\right) = 23$

Solve for x: $15x + 12x - 4 = 23 \Rightarrow 27x = 27 \Rightarrow x = 1$

Back-substitute $x = 1$: $y = \frac{3}{2}(1) - \frac{1}{2} = 1$

Solution: $(1, 1)$

10. $\begin{cases} 0.5x + 3.2y = 9.0 & \text{Equation 1} \\ 0.2x - 1.6y = -3.6 & \text{Equation 2} \end{cases}$

Multiply the equations by 10.

$5x + 32y = 90$ Revised Equation 1

$2x - 16y = -36$ Revised Equation 2

Solve for x in revised Equation 2: $x = 8y - 18$

Substitute for x in revised Equation 1: $5(8y - 18) + 32y = 90$

Solve for y: $40y - 90 + 32y = 90 \Rightarrow 72y = 180$

$\Rightarrow y = \frac{5}{2}$

Back-substitute $y = \frac{5}{2}$: $x = 8\left(\frac{5}{2}\right) - 18 = 2$

Solution: $\left(2, \frac{5}{2}\right)$

11. $\begin{cases} \frac{1}{5}x + \frac{1}{2}y = 8 & \text{Equation 1} \\ x + y = 20 & \text{Equation 2} \end{cases}$

Solve for x in Equation 2: $x = 20 - y$

Substitute for x in Equation 1: $\frac{1}{5}(20 - y) + \frac{1}{2}y = 8$

Solve for y: $4 + \frac{3}{10}y = 8 \Rightarrow y = \frac{40}{3}$

Back-substitute $y = \frac{40}{3}$: $x = 20 - y = 20 - \frac{40}{3} = \frac{20}{3}$

Solution: $\left(\frac{20}{3}, \frac{40}{3}\right)$

12. $\begin{cases} \frac{1}{2}x + \frac{3}{4}y = 10 & \text{Equation 1} \\ \frac{3}{4}x - y = 4 & \text{Equation 2} \end{cases}$

Solve for y in Equation 2: $y = \frac{3}{4}x - 4$

Substitute for y in Equation 1: $\frac{1}{2}x + \frac{3}{4}\left(\frac{3}{4}x - 4\right) = 10$

Solve for x: $\frac{1}{2}x + \frac{9}{16}x - 3 = 10 \Rightarrow \frac{17}{16}x = 13 \Rightarrow x = \frac{208}{17}$

Back-substitute $x = \frac{208}{17}$: $y = \frac{3}{4}\left(\frac{208}{17}\right) - 4 = \frac{88}{17}$

Solution: $\left(\frac{208}{17}, \frac{88}{17}\right)$

13. $\begin{cases} 6x + 5y = -3 & \text{Equation 1} \\ -x - \frac{5}{6}y = -7 & \text{Equation 2} \end{cases}$

Solve for x in Equation 2: $x = 7 - \frac{5}{6}y$

Substitute for x in Equation 1: $6\left(7 - \frac{5}{6}y\right) + 5y = -3$

Solve for y: $42 - 5y + 5y = -3 \Rightarrow 42 = -3$ (False)

No solution

14. $\begin{cases} -\frac{2}{3}x + y = 2 & \text{Equation 1} \\ 2x - 3y = 6 & \text{Equation 2} \end{cases}$

Solve for y in Equation 1: $y = \frac{2}{3}x + 2$

Substitute for y in Equation 2: $2x - 3\left(\frac{2}{3}x + 2\right) = 6$

Solve for x: $2x - 2x - 6 = 6 \Rightarrow 0 = 12$ Inconsistent

No solution

15. $\begin{cases} x^2 - y = 0 & \text{Equation 1} \\ 2x + y = 0 & \text{Equation 2} \end{cases}$

Solve for y in Equation 2: $y = -2x$

Substitute for y in Equation 1: $x^2 - (-2x) = 0$

Solve for x:
$x^2 + 2x = 0 \Rightarrow x(x + 2) = 0 \Rightarrow x = 0, -2$

Back-substitute $x = 0$: $y = -2(0) = 0$

Back-substitute $x = -2$: $y = -2(-2) = 4$

Solutions: $(0, 0), (-2, 4)$

16. $\begin{cases} x - 2y = 0 & \text{Equation 1} \\ 3x - y^2 = 0 & \text{Equation 2} \end{cases}$

Solve for x in Equation 1: $x = 2y$

Substitute for x in Equation 2: $3(2y) - y^2 = 0$

Solve for y:
$6y - y^2 = 0 \Rightarrow y(6 - y) = 0 \Rightarrow y = 0, 6$

Back-substitute $y = 0$: $x = 2(0) = 0$

Back-substitute $y = 6$: $x = 2(6) = 12$

Solutions: $(0, 0), (12, 6)$

17. $\begin{cases} x^3 - y = 0 & \text{Equation 1} \\ x - y = 0 & \text{Equation 2} \end{cases}$

Solve for y in Equation 2: $y = x$

Substitute for y in Equation 1: $x^3 - x = 0$

Solve for x: $x(x + 1)(x - 1) = 0 \Rightarrow x = 0, \pm 1$

Back-substitute $x = 0$: $y = 0$

Back-substitute $x = 1$: $y = 1$

Back-substitute $x = -1$: $y = -1$

Solutions: $(0, 0), (1, 1), (-1, -1)$

18. $\begin{cases} y = -x & \text{Equation 1} \\ y = x^3 + 3x^2 + 2x & \text{Equation 2} \end{cases}$

Substitute for y in Equation 2: $-x = x^3 + 3x^2 + 2x$

Solve for x: $x^3 + 3x^2 + 3x = 0 \Rightarrow x(x^2 + 3x + 3) = 0$

$$\Rightarrow x = 0, \frac{-3 \pm i\sqrt{3}}{2}$$

Back-substitute $x = 0$: $y = 0$

The only real solution is $(0, 0)$.

19. $\begin{cases} y = x^3 - 2x + 1 & \text{Equation 1} \\ y = x^2 - 1 & \text{Equation 2} \end{cases}$

Substitute for y in Equation 2: $x^3 - 2x + 1 = x^2 - 1$

Solve for x: $x^3 - x^2 - 2x + 2 = 0 \Rightarrow x^2(x - 1) - 2(x - 1) = 0$

$$\Rightarrow (x - 1)(x^2 - 2) = 0$$

$$\Rightarrow x = 1, \pm\sqrt{2}$$

Back-substitute $x = 1 \Rightarrow y = 0$

$x = \pm\sqrt{2} \Rightarrow y = 1$

Solutions: $(1, 0), \left(\pm\sqrt{2}, 1\right)$

20. $\begin{cases} y = x^3 - x & \text{Equation 1} \\ y = x^2 - 1 & \text{Equation 2} \end{cases}$

Substitute for y in Equation 2: $x^3 - x = x^2 - 1$

Solve for x: $x^3 - x^2 - x + 1 = 0 \Rightarrow x^2(x - 1) - 1(x - 1) = 0$

$$\Rightarrow (x^2 - 1)(x - 1) = 0$$

$$\Rightarrow (x + 1)(x - 1)^2 = 0$$

$$\Rightarrow x = \pm 1$$

Back-substitute $x = -1 \Rightarrow y = 0$

$x = 1 \Rightarrow y = 0$

Solutions: $(\pm 1, 0)$

21. $\begin{cases} 2x + y = 5 & \text{Equation 1} \\ x - y = 1 & \text{Equation 2} \end{cases}$

Add Equation 1 to Equation 2: $3x = 6 \Rightarrow x = 2$

Back-substitute $x = 2 \Rightarrow y = 1$

Solution: $(2, 1)$

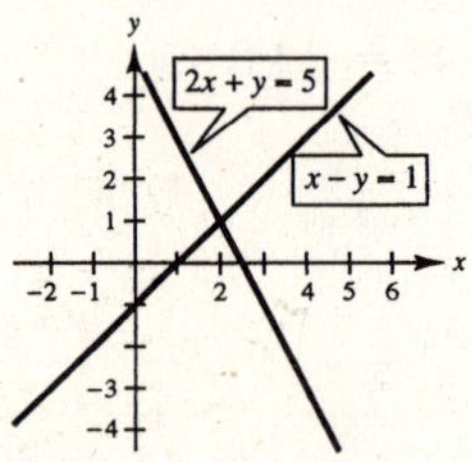

22. $\begin{cases} x + 3y = 1 & \text{Equation 1} \\ -x + 2y = 4 & \text{Equation 2} \end{cases}$

Add Equation 1 to Equation 2: $5y = 5 \Rightarrow y = 1$

Back-substitute $y = 1 \Rightarrow x = -2$

Solution: $(-2, 1)$

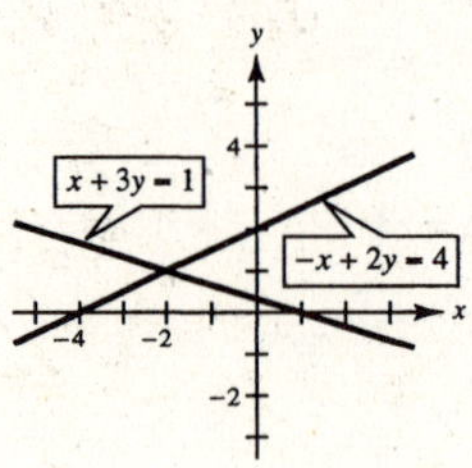

23. $\begin{cases} x + y = 0 & \text{Equation 1} \\ 3x + 2y = 1 & \text{Equation 2} \end{cases}$

Multiply Equation 1 by -2: $-2x - 2y = 0$

Add this to Equation 2 to eliminate y: $x = 1$

Substitute $x = 1$ in Equation 1: $1 + y = 0 \Rightarrow y = -1$

Solution: $(1, -1)$

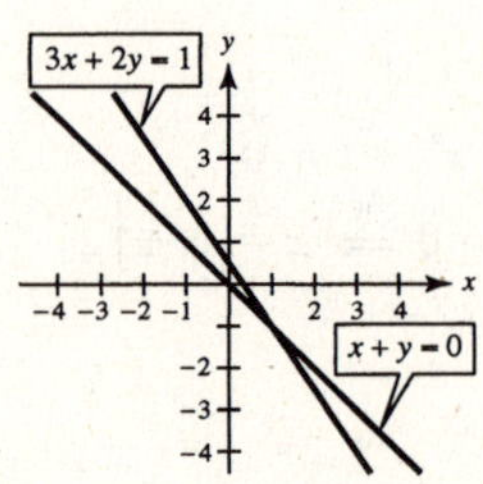

24. $\begin{cases} 2x - y = 3 & \text{Equation 1} \\ 4x + 3y = 21 & \text{Equation 2} \end{cases}$

Multiply Equation 1 by 3: $6x - 3y = 9$

Add this to Equation 2 to eliminate y:

$$\begin{array}{rrl} 6x & - 3y & = 9 \\ 4x & + 3y & = 21 \\ \hline 10x & & = 30 \\ & \Rightarrow x & = 3 \end{array}$$

Substitute $x = 3$ in Equation 1: $2(3) - y = 3 \Rightarrow y = 3$

Solution: $(3, 3)$

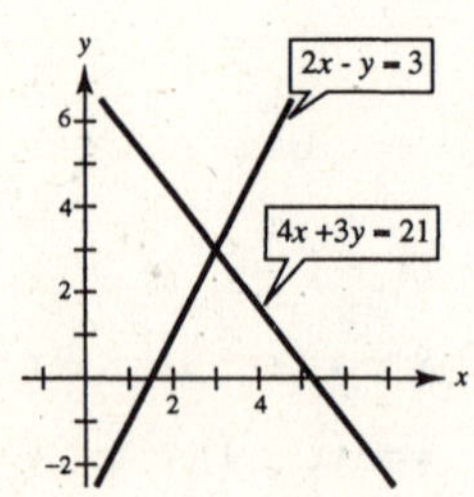

25. $\begin{cases} x - y = 2 & \text{Equation 1} \\ -2x + 2y = 5 & \text{Equation 2} \end{cases}$

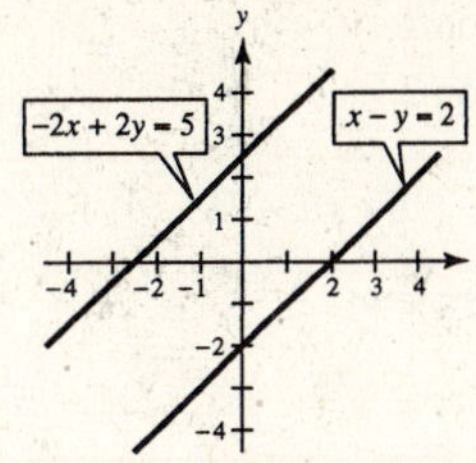

Multiply Equation 1 by 2: $2x - 2y = 4$

Add this to Equation 2: $0 = 9$

There are no solutions.

26. $\begin{cases} 3x + 2y = 3 & \text{Equation 1} \\ 6x + 4y = 14 & \text{Equation 2} \end{cases}$

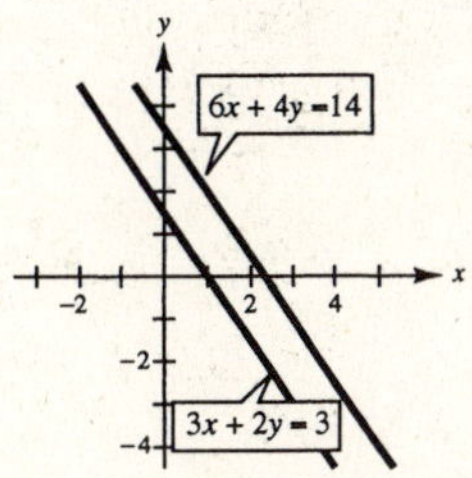

Multiply Equation 1 by -2: $-6x - 4y = -6$

Add this to Equation 2:
$$\begin{aligned} -6x - 4y &= -6 \\ 6x + 4y &= 14 \\ \hline 0 &= 8 \end{aligned}$$

There are no solutions.

27. $\begin{cases} 3x - 2y = 5 & \text{Equation 1} \\ -6x + 4y = -10 & \text{Equation 2} \end{cases}$

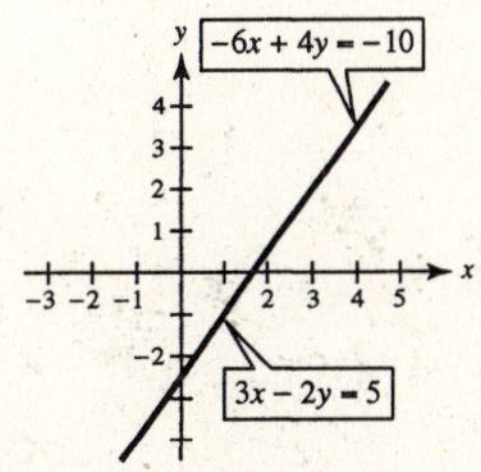

Multiply Equation 1 by 2 and add to Equation 2: $0 = 0$

The equations are dependent. There are infinitely many solutions. Let $x = a$.

$3a - 2y = 5 \implies y = \frac{1}{2}(3a - 5)$

Solution: $\left(a, \frac{1}{2}(3a - 5)\right)$ where a is any real number.

28. $\begin{cases} 9x - 3y = -15 & \text{Equation 1} \\ -3x + y = 5 & \text{Equation 2} \end{cases}$

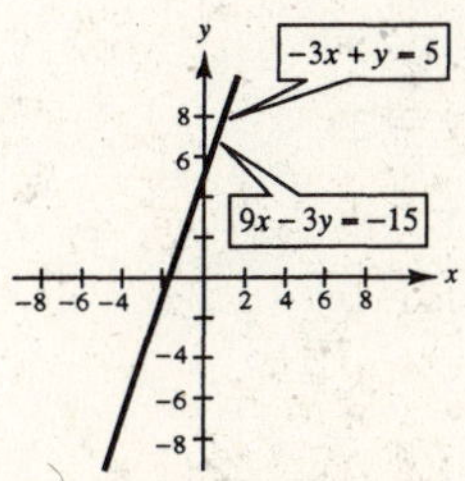

Multiply Equation 2 by 3: $-9x + 3y = 15$

Add this to Equation 1:
$$\begin{aligned} 9x - 3y &= -15 \\ -9x + 3y &= 15 \\ \hline 0 &= 0 \end{aligned}$$

There are infinitely many solutions. Let $x = a$

$-3a + y = 5 \implies y = 3a + 5$

Solution: $(a, 3a + 5)$, where a is any real number.

29. $\begin{cases} 9x + 3y = 1 & \text{Equation 1} \\ 3x - 6y = 5 & \text{Equation 2} \end{cases}$

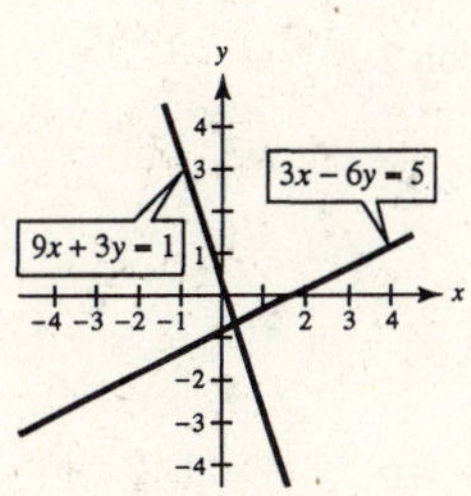

Multiply Equation 2 by (-3):
$$\begin{aligned} 9x + 3y &= 1 \\ -9x + 18y &= -15 \end{aligned}$$

Add to eliminate x: $21y = -14 \implies y = -\frac{2}{3}$

Substitute $y = -\frac{2}{3}$ in Equation 1: $9x + 3\left(-\frac{2}{3}\right) = 1$

$$x = \frac{1}{3}$$

Solution: $\left(\frac{1}{3}, -\frac{2}{3}\right)$

30. $\begin{cases} 5x + 3y = -18 & \text{Equation 1} \\ 2x - 6y = 1 & \text{Equation 2} \end{cases}$

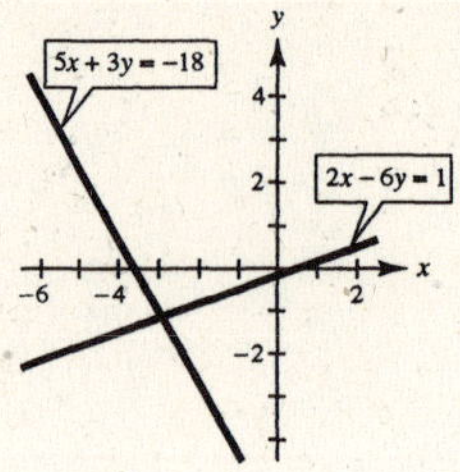

Multiply Equation 1 by 2: $10x + 6y = -36$

Add this to Equation 2 to eliminate y:

$$\begin{aligned} 10x + 6y &= -36 \\ 2x - 6y &= 1 \\ \hline 12x \quad &= -35 \Rightarrow x = \frac{-35}{12} \end{aligned}$$

Substitute $x = -\frac{35}{12}$ in Equation 2:

$2\left(-\frac{35}{12}\right) - 6y = 1 \Rightarrow y = -\frac{41}{36}$

Solution: $\left(-\frac{35}{12}, -\frac{41}{36}\right)$

31. $\begin{cases} x + 2y = 4 & \text{Equation 1} \\ x - 2y = 1 & \text{Equation 2} \end{cases}$

Add to eliminate y:

$2x = 5$

$x = \frac{5}{2}$

Substitute $x = \frac{5}{2}$ in Equation 1:

$\frac{5}{2} + 2y = 4 \Rightarrow y = \frac{3}{4}$

Solution: $\left(\frac{5}{2}, \frac{3}{4}\right)$

32. $\begin{cases} 3x - 5y = 2 & \text{Equation 1} \\ 2x + 5y = 13 & \text{Equation 2} \end{cases}$

Add to eliminate y:

$$\begin{aligned} 3x - 5y &= 2 \\ 2x + 5y &= 13 \\ \hline 5x \quad &= 15 \Rightarrow x = 3 \end{aligned}$$

Substitute $x = 3$ in Equation 1: $3(3) - 5y = 2 \Rightarrow y = \frac{7}{5}$

Solution: $\left(3, \frac{7}{5}\right)$

33. $\begin{cases} 2x + 3y = 18 & \text{Equation 1} \\ 5x - y = 11 & \text{Equation 2} \end{cases}$

Multiply Equation 2 by 3: $15x - 3y = 33$

Add this to Equation 1 to eliminate y:

$17x = 51 \Rightarrow x = 3$

Substitute $x = 3$ in Equation 1:

$6 + 3y = 18 \Rightarrow y = 4$

Solution: $(3, 4)$

34. $\begin{cases} x + 7y = 12 & \text{Equation 1} \\ 3x - 5y = 10 & \text{Equation 2} \end{cases}$

Multiply Equation 1 by -3: $-3x - 21y = -36$

Add this to Equation 2 to eliminate x:

$$\begin{aligned} -3x - 21y &= -36 \\ 3x - 5y &= 10 \\ \hline -26y &= -26 \\ \Rightarrow y &= 1 \end{aligned}$$

Substitute $y = 1$ in Equation 1: $x + 7 = 12 \Rightarrow x = 5$

Solution: $(5, 1)$

35. $\begin{cases} 3x + 2y = 10 & \text{Equation 1} \\ 2x + 5y = 3 & \text{Equation 2} \end{cases}$

Multiply Equation 1 by 2 and Equation 2 by (-3):

$$\begin{cases} 6x + 4y = 20 \\ -6x - 15y = -9 \end{cases}$$

Add to eliminate x: $-11y = 11 \Rightarrow y = -1$

Substitute $y = -1$ in Equation 1:

$3x - 2 = 10 \Rightarrow x = 4$

Solution: $(4, -1)$

36. $\begin{cases} 2r + 4s = 5 & \text{Equation 1} \\ 16r + 50s = 55 & \text{Equation 2} \end{cases}$

Multiply Equation 1 by (-8): $-16r - 32s = -40$

Add this to Equation 2 to eliminate r:

$$\begin{aligned} -16r - 32s &= -40 \\ 16r + 50s &= 55 \\ \hline 18s &= 15 \\ \Rightarrow s &= \frac{5}{6} \end{aligned}$$

Substitute $s = \frac{5}{6}$ in Equation 1: $2r + 4\left(\frac{5}{6}\right) = 5 \Rightarrow r = \frac{5}{6}$

Solution: $\left(\frac{5}{6}, \frac{5}{6}\right)$

37. $\begin{cases} 5u + 6v = 24 & \text{Equation 1} \\ 3u + 5v = 18 & \text{Equation 2} \end{cases}$

Multiply Equation 1 by 5 and Equation 2 by -6:

$$\begin{cases} 25u + 30v = 120 \\ -18u - 30v = -108 \end{cases}$$

Add to eliminate v: $7u = 12 \Rightarrow u = \frac{12}{7}$

Substitute $u = \frac{12}{7}$ in Equation 1:

$$5\left(\tfrac{12}{7}\right) + 6v = 24 \Rightarrow 6v = \tfrac{108}{7} \Rightarrow v = \tfrac{18}{7}$$

Solution: $\left(\frac{12}{7}, \frac{18}{7}\right)$

38. $\begin{cases} 3x + 11y = 4 & \text{Equation 1} \\ -2x - 5y = 9 & \text{Equation 2} \end{cases}$

Multiply Equation 1 by 2 and Equation 2 by 3:

$$\begin{cases} 6x + 22y = 8 \\ -6x - 15y = 27 \end{cases}$$

Add to eliminate x:

$$\begin{aligned} 6x + 22y &= 8 \\ -6x - 15y &= 27 \\ \hline 7y &= 35 \Rightarrow y = 5 \end{aligned}$$

Substitute $y = 5$ in Equation 1: $3x + 11(5) = 4$

$\Rightarrow x = -17$

Solution: $(-17, 5)$

39. $\begin{cases} 1.8x + 1.2y = 4 & \text{Equation 1} \\ 9x + 6y = 3 & \text{Equation 2} \end{cases}$

Multiply Equation 1 by 10 and Equation 2 by -2:

$$\begin{cases} 18x + 12y = 40 \\ -18x - 12y = -6 \end{cases}$$

Add to eliminate x and y: $0 = 34$

Inconsistent

No solution

40. $\begin{cases} 3.1x - 2.9y = -10.2 & \text{Equation 1} \\ 15.5x - 14.5y = 21 & \text{Equation 2} \end{cases}$

Multiply Equation 1 by 10 and Equation 2 by (-2):

$$\begin{cases} 31x - 29y = -102 \\ -31x + 29y = -42 \end{cases}$$

Add to eliminate x:

$$\begin{aligned} 31x - 29y &= -102 \\ -31x + 29y &= -42 \\ \hline 0 &= -144 \end{aligned}$$

There are no solutions.

41. $\begin{cases} \dfrac{x}{4} + \dfrac{y}{6} = 1 & \text{Equation 1} \\ x - y = 3 & \text{Equation 2} \end{cases}$

Multiply Equation 1 by 6: $\frac{3}{2}x + y = 6$

Add this to Equation 2 to eliminate y: $\frac{5}{2}x = 9 \Rightarrow x = \frac{18}{5}$

Substitute $x = \frac{18}{5}$ in Equation 2:

$$\frac{18}{5} - y = 3$$

$$y = \frac{3}{5}$$

Solution: $\left(\dfrac{18}{5}, \dfrac{3}{5}\right)$

42. $\begin{cases} \dfrac{2}{3}x + \dfrac{1}{6}y = \dfrac{2}{3} & \text{Equation 1} \\ 4x + y = 4 & \text{Equation 2} \end{cases}$

Multiply Equation 1 by (-6): $-4x - y = -4$

Add this to Equation 2:

$$\begin{aligned} -4x - y &= -4 \\ 4x + y &= 4 \\ \hline 0 &= 0 \end{aligned}$$

There are infinitely many solutions. Let $x = a$

$$4a + y = 4 \Rightarrow y = 4 - 4a$$

Solution: $(a_1, 4 - 4a)$ where a is any real number

43. $\begin{cases} 2.5x - 3y = 1.5 & \text{Equation 1} \\ 2x - 2.4y = 1.2 & \text{Equation 2} \end{cases}$

Multiply Equation 1 by 20 and Equation 2 by -25:

$$\begin{aligned} 50x - 60y &= 30 \\ -50x + 60y &= -30 \end{aligned}$$

Add to eliminate x and y: $0 = 0$

The equations are dependent.

There are infinitely many solutions.

Let $x = a$, then $2.5a - 3y = 1.5 \Rightarrow y = \dfrac{2.5a - 1.5}{3} = \dfrac{5}{6}a - \dfrac{1}{2}$.

Solution: $\left(a, \dfrac{5}{6}a - \dfrac{1}{2}\right)$ where a is any real number.

44. $\begin{cases} 6.3x + 7.2y = 5.4 & \text{Equation 1} \\ 5.6x + 6.4y = 4.8 & \text{Equation 2} \end{cases}$

Multiply Equation 1 by 56 and Equation 2 by -63:

$$\begin{cases} 352.8x + 403.2y = 302.4 \\ -352.8x - 403.2y = -302.4 \end{cases}$$

Add to eliminate x and y:

$$\begin{array}{r} 352.8x + 403.2y = 302.4 \\ -352.8x - 403.2y = -302.4 \\ \hline 0 = 0 \end{array}$$

There are infinitely many solutions. Let $x = a$

$$5.6a + 6.4y = 4.8 \implies y = -0.875a + 0.75$$

Solution: $(a, -0.875a + 0.75)$, where a is any real number

45. $\begin{cases} 0.05x - 0.03y = 0.21 & \text{Equation 1} \\ 0.07x + 0.02y = 0.16 & \text{Equation 2} \end{cases}$

Multiply Equation 1 by 200 and Equation 2 by 300.

$$\begin{cases} 10x - 6y = 42 \\ 21x + 6y = 48 \end{cases}$$

Add to eliminate y: $31x = 90 \implies x = \frac{90}{31}$

Back-substitute $x = \frac{90}{31}$ into the revised Equation 1

$$10\left(\tfrac{90}{31}\right) - 6y = 42 \implies y = -\tfrac{1}{6}\left(42 - \tfrac{900}{31}\right) = -\tfrac{67}{31}$$

Solution: $\left(\frac{90}{31}, -\frac{67}{31}\right)$

46. $\begin{cases} 0.2x - 0.5y = -27.8 & \text{Equation 1} \\ 0.3x + 0.4y = 68.7 & \text{Equation 2} \end{cases}$

Multiply Equation 1 by 40 and Equation 2 by 50.

$$\begin{cases} 8x - 20y = -1112 \\ 15x + 20y = 3435 \end{cases}$$

Add to eliminate y: $23x = 2323 \implies x = 101$

Back-substitute $x = 101$ into the revised Equation 2.

$$15(101) + 20y = 3435 \implies y = \tfrac{1}{20}(3435 - 1515) = 96$$

Solution: $(101, 96)$

47. $\begin{cases} 4b + 3m = 3 & \text{Equation 1} \\ 3b + 11m = 13 & \text{Equation 2} \end{cases}$

Multiply Equation 1 by 11 and Equation 2 by -3.

$$\begin{cases} 44b + 33m = 33 \\ -9b - 33m = -39 \end{cases}$$

Add to eliminate y: $35b = -6 \implies b = -\frac{6}{35}$

Back-substitute $b = -\frac{6}{35}$ into Equation 1.

$$4\left(-\tfrac{6}{35}\right) + 3m = 3 \implies m = \tfrac{1}{3}\left(3 + \tfrac{24}{35}\right) = \tfrac{43}{35}$$

Solution: $\left(-\frac{6}{35}, \frac{43}{35}\right)$

48. $\begin{cases} 2x + 5y = 8 & \text{Equation 1} \\ 5x + 8y = 10 & \text{Equation 2} \end{cases}$

Multiply Equation 1 by 5 and Equation 2 by -2.

$$\begin{cases} 10x + 25y = 40 \\ -10x - 16y = -20 \end{cases}$$

Add to eliminate x: $9y = 20 \implies y = \frac{20}{9}$

Back-substitute $y = \frac{20}{9}$ into Equation 1.

$$2x + 5\left(\tfrac{20}{9}\right) = 8 \implies x = \tfrac{1}{2}\left(8 - \tfrac{100}{9}\right) = -\tfrac{14}{9}$$

Solution: $\left(-\frac{14}{9}, \frac{20}{9}\right)$

49. $\begin{cases} \dfrac{x+3}{4} + \dfrac{y-1}{3} = 1 & \text{Equation 1} \\ 2x - y = 12 & \text{Equation 2} \end{cases}$

Multiply Equation 1 by 12 to eliminate the fractions.

$$3(x+3) + 4(y-1) = 12 \implies 3x + 4y = 12 - 9 + 4 = 7$$

$$\begin{cases} 3x + 4y = 7 \\ 2x - y = 12 \end{cases}$$

Multiply Equation 2 by 4.

$$\begin{cases} 3x + 4y = 7 \\ 8x - 4y = 48 \end{cases}$$

Add to eliminate y: $11x = 55 \implies x = 5$

Back-substitute $x = 5$ into Equation 2.

$$2(5) - y = 12 \implies y = -(12 - 10) = -2$$

Solution: $(5, -2)$

50. $\begin{cases} \dfrac{x-1}{2} + \dfrac{y+2}{3} = 4 & \text{Equation 1} \\ x - 2y = 5 & \text{Equation 2} \end{cases}$

Multiply Equation 1 by 6 to eliminate the fractions.

$$3(x-1) + 2(y+2) = 24 \implies 3x + 2y = 24 + 3 - 4 = 23$$

$$\begin{cases} 3x + 2y = 23 \\ x - 2y = 5 \end{cases}$$

Add to eliminate y: $4x = 28 \implies x = 7$

Back-substitute $x = 7$ into Equation 2.

$$7 - 2y = 5 \implies y = -\frac{1}{2}(5 - 7) = 1$$

Solution: $(7, 1)$

51. $\begin{cases} 2x - 5y = 0 \implies y = \frac{2}{5}x \\ x - y = 3 \implies y = x - 3 \end{cases}$

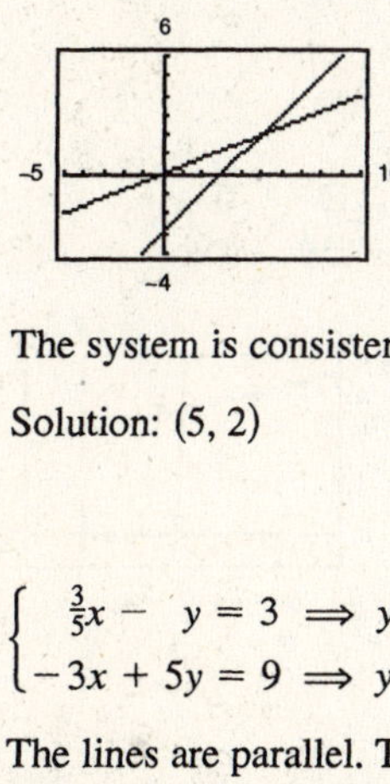

The system is consistent. There is one solution.

Solution: $(5, 2)$

52. $\begin{cases} 2x + y = 5 \implies y = -2x + 5 \\ x - 2y = -1 \implies y = \dfrac{x+1}{2} \end{cases}$

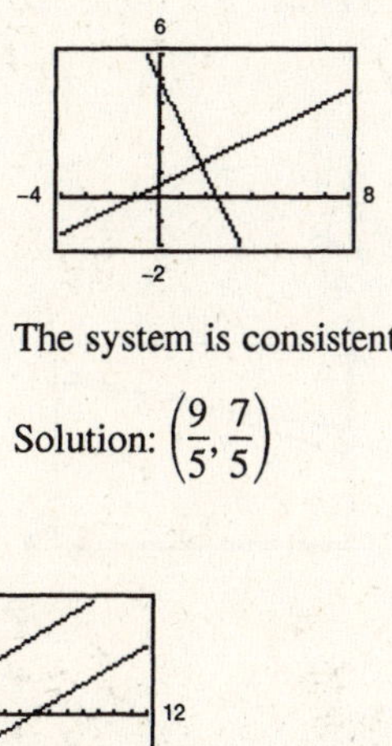

The system is consistent. There is one solution.

Solution: $\left(\frac{9}{5}, \frac{7}{5}\right)$

53. $\begin{cases} \frac{3}{5}x - y = 3 \implies y = \frac{3}{5}x - 3 \\ -3x + 5y = 9 \implies y = \frac{3}{5}x + \frac{9}{5} \end{cases}$

The lines are parallel. The system is inconsistent.

There are no solutions.

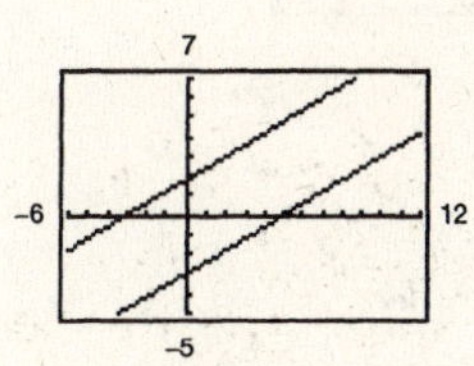

54. $\begin{cases} 4x - 6y = 9 \Rightarrow y = \frac{2}{3}x - \frac{3}{2} \\ \frac{16}{3}x - 8y = 12 \Rightarrow y = \frac{2}{3}x - \frac{3}{2} \end{cases}$

The system is consistent.

There are infinitely many solutions.

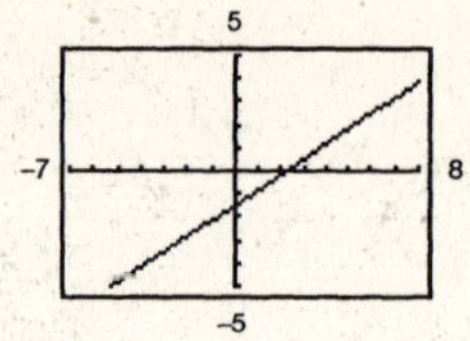

55. $\begin{cases} x + 7y = 2 \Rightarrow y = -\frac{1}{7}x + \frac{2}{7} \\ 4x - y = 9 \Rightarrow y = 4x - 9 \end{cases}$

The system is consistent.

There is one solution.

Solution: $\left(\frac{65}{29}, -\frac{1}{29}\right)$

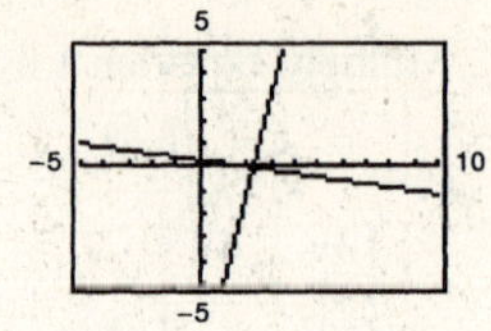

56. $\begin{cases} 8x - 14y = 5 \Rightarrow y = \frac{4}{7}x - \frac{5}{14} \\ 2x - 3.5y = 1.25 \Rightarrow y = \frac{4}{7}x - \frac{5}{14} \end{cases}$

The system is consistent.

There are infinitely many solutions.

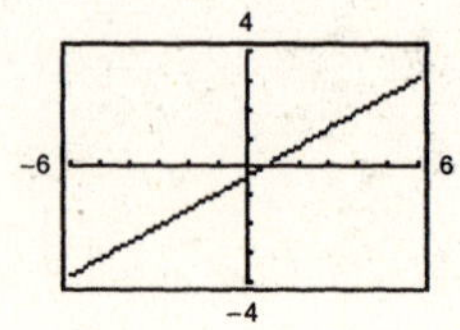

57. $\begin{cases} -x + 7y = 3 \Rightarrow y = \frac{1}{7}x + \frac{3}{7} \\ -\frac{1}{7}x + y = 5 \Rightarrow y = \frac{1}{7}x + 5 \end{cases}$

The lines are parallel.

The system is inconsistent.

There are no solutions.

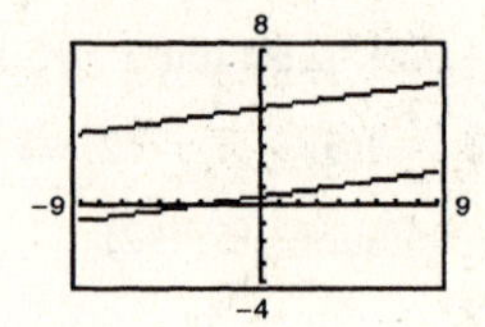

58. $\begin{cases} -7x + 6y = -4 \Rightarrow y = \frac{7}{6}x - \frac{2}{3} \\ y + \frac{7}{6}x = -1 \Rightarrow y = -\frac{7}{6}x - 1 \end{cases}$

The system is consistent.

There is one solution.

Solution: $\left(-\frac{1}{7}, -\frac{5}{6}\right)$

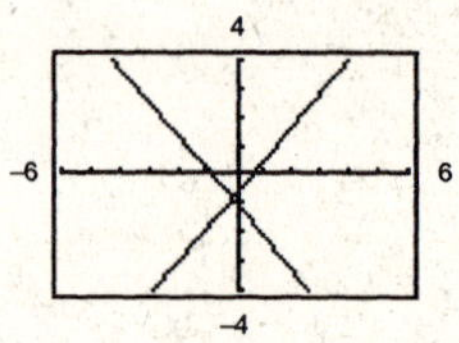

59. $\begin{cases} 8x + 9y = 42 \Rightarrow y = -\frac{8}{9}x + \frac{14}{3} \\ 6x - y = 16 \Rightarrow y = 6x - 16 \end{cases}$

The system is consistent.

There is one solution.

Solution: $(3, 2)$

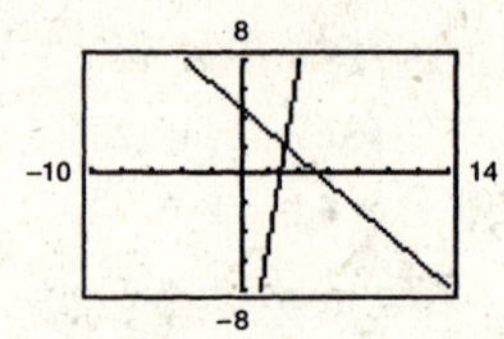

60. $\begin{cases} 4y = -8 \Rightarrow y = -2 \\ 7x - 2y = 25 \Rightarrow y = \frac{7}{2}x - \frac{25}{2} \end{cases}$

The system is consistent.

There is one solution.

Solution: $(3, -2)$

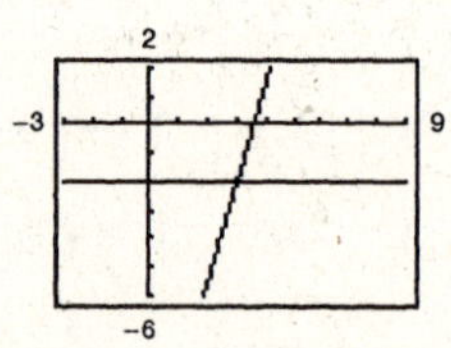

61. $\begin{cases} \frac{3}{2}x - \frac{1}{5}y = 8 \Rightarrow y = \frac{15}{2}x - 40 \\ -2x + 3y = 3 \Rightarrow y = \frac{2}{3}x + 1 \end{cases}$

The system is consistent.

There is one solution.

Solution: $(6, 5)$

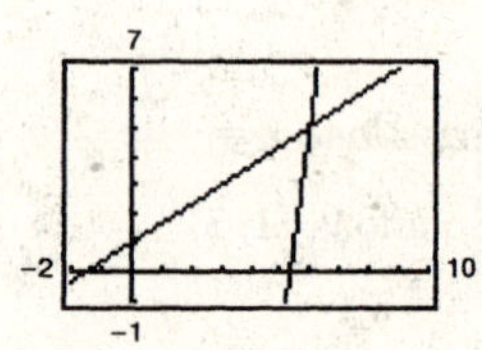

62. $\begin{cases} \frac{3}{4}x - \frac{5}{2}y = -9 \Rightarrow y = \frac{3}{10}x + \frac{18}{5} \\ -x + 6y = 28 \Rightarrow y = \frac{1}{6}x + \frac{14}{3} \end{cases}$

The system is consistent.

There is one solution.

Solution: $(8, 6)$

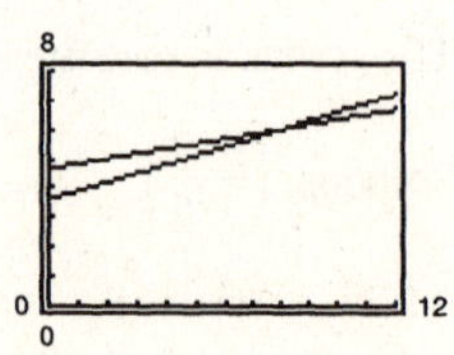

63. $\begin{cases} 0.5x + 2.2y = 9 \Rightarrow y = -\frac{5}{22}x + \frac{45}{11} \\ 6x + 0.4y = -22 \Rightarrow y = -15x - 55 \end{cases}$

The system is consistent.

There is one solution.

Solution: $(-4, 5)$

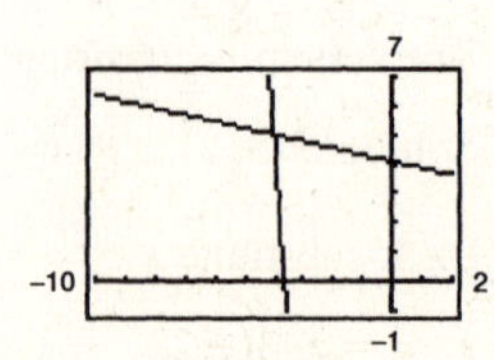

64. $\begin{cases} 2.4x + 3.8y = -17.6 \Rightarrow y = -\frac{12}{19}x - \frac{88}{19} \\ 4x - 0.2y = -3.2 \Rightarrow y = 20x + 16 \end{cases}$

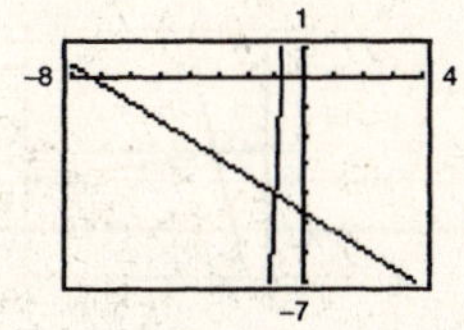

The system is consistent.

There is one solution.

Solution: $(-1, -4)$

65. $\begin{cases} 3x - 5y = 7 & \text{Equation 1} \\ 2x + y = 9 & \text{Equation 2} \end{cases}$

Multiply Equation 2 by 5:

$$10x + 5y = 45$$

Add this to Equation 1:

$$13x = 52 \Rightarrow x = 4$$

Back-substitute $x = 4$ into Equation 2:

$$2(4) + y = 9 \Rightarrow y = 1$$

Solution: $(4, 1)$

66. $\begin{cases} -x + 3y = 17 & \text{Equation 1} \\ 4x + 3y = 7 & \text{Equation 2} \end{cases}$

Subtract Equation 2 from Equation 1 to eliminate y:

$$\begin{aligned} -x + 3y &= 17 \\ \underline{-4x - 3y} &= \underline{-7} \\ -5x \quad &= 10 \Rightarrow x = -2 \end{aligned}$$

Substitute $x = -2$ in Equation 1:

$$-(-2) + 3y = 17 \Rightarrow y = 5$$

Solution: $(-2, 5)$

67. $\begin{cases} y = 2x - 5 & \text{Equation 1} \\ y = 5x - 11 & \text{Equation 2} \end{cases}$

Since both equations are solved for y, set them equal to one another and solve for x.

$$\begin{aligned} 2x - 5 &= 5x - 11 \\ 6 &= 3x \\ 2 &= x \end{aligned}$$

Back-substitute $x = 2$ into Equation 1:

$$y = 2(2) - 5 = -1$$

Solution: $(2, -1)$

68. $\begin{cases} 7x + 3y = 16 & \text{Equation 1} \\ y = x + 2 & \text{Equation 2} \end{cases}$

Substitute for y in Equation 1:

$$\begin{aligned} 7x + 3(x + 2) &= 16 \\ 7x + 3x + 6 &= 16 \\ 10x &= 10 \Rightarrow x = 1 \end{aligned}$$

Substitute $x = 1$ in Equation 2: $y = 1 + 2 = 3$

Solution: $(1, 3)$

69. $\begin{cases} x - 5y = 21 & \text{Equation 1} \\ 6x + 5y = 21 & \text{Equation 2} \end{cases}$

Add the equations: $7x = 42 \Rightarrow x = 6$

Back-substitute $x = 6$ into Equation 1:

$$6 - 5y = 21 \Rightarrow -5y = 15 \Rightarrow y = -3$$

Solution: $(6, -3)$

70. $\begin{cases} y = -3x - 8 & \text{Equation 1} \\ y = 15 - 2x & \text{Equation 2} \end{cases}$

Since both equations are solved for y, set them equal to one another and solve for x:

$$\begin{aligned} -3x - 8 &= 15 - 2x \\ -x &= 23 \\ x &= -23 \end{aligned}$$

Back-substitute $x = -23$ into Equation 1:

$$y = -3(-23) - 8 = 61$$

Solution: $(-23, 61)$

71. $\begin{cases} -2x + 8y = 19 & \text{Equation 1} \\ y = x - 3 & \text{Equation 2} \end{cases}$

Substitute the expression for y from Equation 2 into Equation 1.

$-2x + 8(x - 3) = 19 \Rightarrow -2x + 8x - 24 = 19 \Rightarrow 6x = 43 \Rightarrow x = \frac{43}{6}$

Back-substitute $x = \frac{43}{6}$ into Equation 2:

$y = \frac{43}{6} - 3 \Rightarrow y = \frac{25}{6}$

Solution: $\left(\frac{43}{6}, \frac{25}{6}\right)$

72. $\begin{cases} 4x - 3y = 6 & \text{Equation 1} \\ -5x + 7y = -1 & \text{Equation 2} \end{cases}$

Multiply Equation 1 by 5 and Equation 2 by 4:

$\begin{cases} 20x - 15y = 30 \\ -20x + 28y = -4 \end{cases}$

Add to eliminate x:

$$\begin{aligned} 20x - 15y &= 30 \\ -20x + 28y &= -4 \\ \hline 13y &= 26 \Rightarrow y = 2 \end{aligned}$$

Back-substitute $y = 2$ into Equation 1:

$4x - 3(2) = 6 \Rightarrow x = 3$

Solution: $(3, 2)$

73. Demand = Supply

$50 - 0.5x = 0.125x$

$50 = 0.625x$

$x = 80$ units

$p = \$10$

Point of Equilibrium: $(80, 10)$

74. Demand = Supply

$100 - 0.05x = 25 + 0.1x$

$75 = 0.15x$

$x = 500$ units

$p = \$75$

Point of Equilibrium: $(500, 75)$

75. Demand = Supply

$140 - 0.00002x = 80 + 0.00001x$

$60 = 0.00003x$

$x = 2{,}000{,}000$ units

$p = \$100$

Point of Equilibrium: $(2{,}000{,}000, 100)$

76. Demand = Supply

$400 - 0.0002x = 225 + 0.0005x$

$175 = 0.0007x$

$x = 250{,}000$ units

$p = \$350$

Point of Equilibrium: $(250{,}000, 350)$

77. A solution of a system is an ordered pair that satisfies each equation in the system.

78. For a linear system the result will be a contradictory equation such as $0 = N$, where N is a nonzero real number. For a nonlinear system there may be an equation with imaginary solutions.

79. The advantage of the method of substitution over the graphical method is that substitution gives exact solutions but graphical solutions, may only be approximate.

80. $y = x^2$

(a) Line with two points of intersection.

$y = 2x$

(0, 0) and (2, 4)

(b) Line with one point of intersection.

$y = 0$

(0, 0)

(c) Line with no points of intersection.

$y = x - 2$

81. $\begin{cases} 100y - x = 200 & \text{Equation 1} \\ 99y - x = -198 & \text{Equation 2} \end{cases}$

Subtract Equation 2 from Equation 1 to eliminate x:

$$\begin{array}{r} 100y - x = 200 \\ -\ 99y + x = 198 \\ \hline y \quad = 398 \end{array}$$

Substitute $y = 398$ into Equation 1: $100(398) - x = 200 \implies x = 39{,}600$

Solution: (39,600, 398)

The lines are not parallel. The scale on the axes must be changed to see the point of intersection.

82. $\begin{cases} 21x - 20y = 0 & \text{Equation 1} \\ 13x - 12y = 120 & \text{Equation 2} \end{cases}$

Multiply Equation 2 by $\left(-\frac{5}{3}\right)$: $-\frac{65}{3}x + 20y = -200$

Add this to Equation 1 to eliminate y:

$-\frac{2}{3}x = -200 \implies x = 300$

Back-substitute $x = 300$ in Equation 1:

$21(300) - 20y = 0 \implies y = 315$

Solution: (300, 315)

The lines are not parallel. It is necessary to change the scale on the axes to see the point of intersection.

83. No, it is not possible for a consistent system of linear equations to have exactly two solutions. Either the lines will intersect once or they will coincide and then the system would have infinite solutions.

84. Answers will vary.

(a) No solution

$\begin{cases} x + y = 10 \\ x + y = 20 \end{cases}$

(b) Infinite number of solutions

$\begin{cases} x + y = 3 \\ 2x + 2y = 6 \end{cases}$

85. (a) $\begin{cases} x + y = 25{,}000 \\ 0.06x + 0.085y = 2000 \end{cases}$

As the amount at 6% increases, the amount at 8.5% decreases. The amount of interest is fixed at \$2000.

(b) The point of intersection occurs when $x = 5000$, so the most that can be invested at 6% and still earn \$2000 per year in interest is \$5000.

86. (a) $\begin{cases} x + y = 20{,}000 \\ 0.065x + 0.085y = 1600 \end{cases}$

(b) The solution to the system is $x = 5000$ and $y = 15{,}000$. In order to meet the requirement of \$1600 per year in interest, \$5000 must be invested at 6.5%.

87. $0.06x = 0.03x + 350$

$0.03x = 350$

$x \approx \$11{,}666.67$

To make the straight commission offer the better offer, you would have to sell more than \$11,666.67 per week.

88. $35{,}000 + 0.02x = 30{,}000 + 0.03x$

$5000 = 0.01x$

$500{,}000 = x$

For the second offer to be better, you would have to sell more than \$500,000 per year.

89. $2l + 2w = 40 \implies l + w = 20 \implies w = 20 - l$

$lw = 96 \implies l(20 - l) = 96$

$20l - l^2 = 96$

$0 = l^2 - 20l + 96$

$0 = (l - 8)(l - 12)$

$l = 8$ or $l = 12$

If $l = 8$, then $w = 12$.

If $l = 12$, then $w = 8$.

Since the length is supposed to be greater than the width, we have $l = 12$ kilometers and $w = 8$ kilometers.
Dimensions: 8×12 kilometers

90. $A = \frac{1}{2}bh$

$1 = \frac{1}{2}a^2$

$a^2 = 2$

$a = \sqrt{2}$

The dimensions are $\sqrt{2} \times \sqrt{2}$ inches.

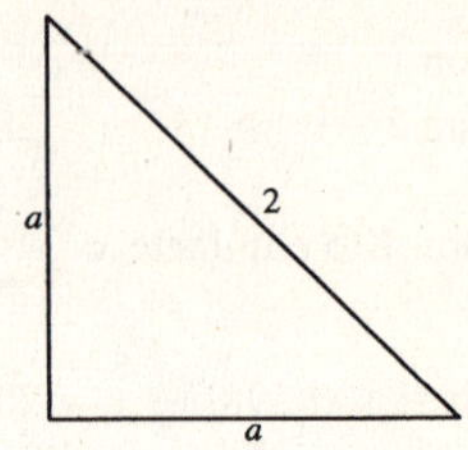

91. Let $r_1 =$ the air speed of the plane and $r_2 =$ the wind air speed.

$3.6(r_1 - r_2) = 1800$ Equation 1

$3(r_1 + r_2) = 1800$ Equation 2

$$\begin{aligned} r_1 - r_2 &= 500 \\ r_1 + r_2 &= 600 \\ 2r_1 &= 1100 \\ r_1 &= 550 \\ 550 + r_2 &= 600 \\ r_2 &= 50 \end{aligned}$$

The air speed of the plane is 550 mph and the speed of the wind is 50 mph.

92. Let $x =$ the speed of the plane that leaves first and $y =$ the speed of the plane that leaves second.

$$\begin{cases} y - x = 80 & \text{Equation 1} \\ 2x + \frac{3}{2}y = 3200 & \text{Equation 2} \end{cases}$$

$$\begin{aligned} -2x + 2y &= 160 \\ 2x + \tfrac{3}{2}y &= 3200 \\ \hline \tfrac{7}{2}y &= 3360 \\ y &= 960 \\ 960 - x &= 80 \\ x &= 880 \end{aligned}$$

Solution: First plane: 880 kilometers per hour; Second plane: 960 kilometers per hour

93. Let $x =$ number of deodorant containers produced by Machine 1

$y =$ number of deodorant containers produced by Machine 2

$x + y = 1764$ Equation 1

$x = 1.8y$ Equation 2

Substitute $1.8y$ for x in Equation 1.

$1.8y + y = 1764 \implies 2.8y = 1764 \implies y = 630$

Back-substitute $y = 630$ into Equation 2.

$x = 1.8(630) \implies x = 1134$

Solution: Machine 1 produces 1134 deodorant containers and Machine 2 produces 630 deodorant containers.

94. Let x = the number of balloons inflated by the father; y = the number of balloons inflated by the child

$$\begin{cases} x + y = 80 & \text{Equation 1} \Rightarrow (-2)\text{ Equation 1} & -2x - 2y = -160 \\ 2x = 3y & \text{Equation 2} \Rightarrow \text{Equation 2} & 2x - 3y = 0 \end{cases}$$

$$\begin{aligned} -5y &= -160 \\ y &= 32 \\ x + 32 &= 80 \\ x &= 48 \end{aligned}$$

The father inflated 48 balloons and the child inflated 32 balloons.

95. Let x = the number of liters at 20%

Let y = the number of liters at 50%.

(a) $\begin{cases} x + y = 10 \\ 0.2x + 0.5y = 0.3(10) \end{cases}$

(b)

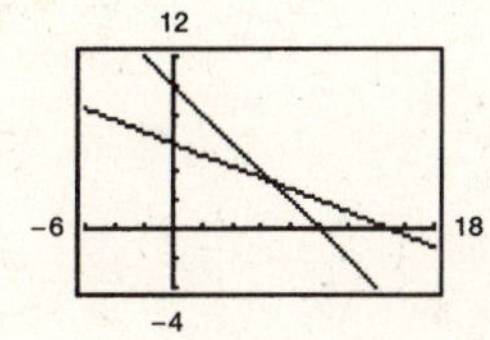

As x increases, y decreases.

(c) $-2 \cdot$ Equation 1 $\quad -2x - 2y = -20$

$10 \cdot$ Equation 2 $\quad 2x + 5y = 30$

$$\begin{aligned} 3y &= 10 \\ y &= \tfrac{10}{3} \\ x + \tfrac{10}{3} &= 10 \\ x &= \tfrac{20}{3} \end{aligned}$$

In order to obtain the specified concentration of the final mixture, $6\frac{2}{3}$ liters of the 20% solution and $3\frac{1}{3}$ liters of the 50% solution are required.

96. Let x = the number of gallons of 87 octane gasoline; y = the number of gallons of 92 octane gasoline

(a) $\begin{cases} x + y = 500 & \text{Equation 1} \\ 87x + 92y = 44{,}500 & \text{Equation 2} \end{cases}$

(b)

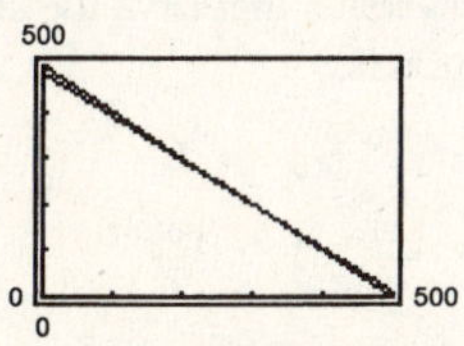

As the amount of 87 octane gasoline increases, the amount of 92 octane gasoline decreases.

(c) (-87) Equation 1 $\quad -87x - 87y = -43{,}500$

Equation 2 $\quad 87x + 92y = 44{,}500$

$$\begin{aligned} 5y &= 1000 \\ y &= 200 \\ x + 200 &= 500 \\ x &= 300 \end{aligned}$$

Solution: 87 octane: 300 gallons;
92 octane: 200 gallons

97. Let x = the number of student tickets; y = the number of adult tickets

$$\begin{cases} x + y = 1435 & \text{Equation 1} \\ 1.50x + 5.00y = 3552.50 & \text{Equation 2} \end{cases}$$

Multiply Equation 1 by -1.50.

$$\begin{cases} -1.50x - 1.50y = -2152.50 \\ 1.50x + 5.00y = 3552.50 \end{cases} \quad \text{Add the equations}$$

$$\begin{aligned} 3.50y &= 1400.00 \\ y &= 400 \\ x + 400 = 1435 \Rightarrow x &= 1035 \end{aligned}$$

Solution: 1035 student tickets and 400 adult tickets were sold.

98. Let $x =$ the number of jackets sold before noon;
$y =$ the number of jackets sold after noon

$$\begin{cases} x + y = 214 & \text{Equation 1} \\ 31.95x + 18.95y = 5108.30 & \text{Equation 2} \end{cases}$$

Multiply Equation 1 by -31.95.

$$\begin{cases} -31.95x - 31.95y = -6837.30 \\ 31.95x + 18.95y = 5108.30 \quad \text{Add the equations} \end{cases}$$

$$-13y = -1729$$

$$y = 133$$

$$x + 133 = 214 \Rightarrow x = 81$$

Solution: 81 jackets were sold before noon and 133 jackets were sold after noon.

99. (1.00, 450), (1.25, 375), (1.50, 330)

(a) $\begin{cases} 3b + 3.75a = 1155 \\ 3.75b + 4.8125a = 1413.75 \end{cases}$

By elimination we have $a = -240$ and $b = 685$.

(b) Least squares regression line:
$y = -240x + 685$

(c)

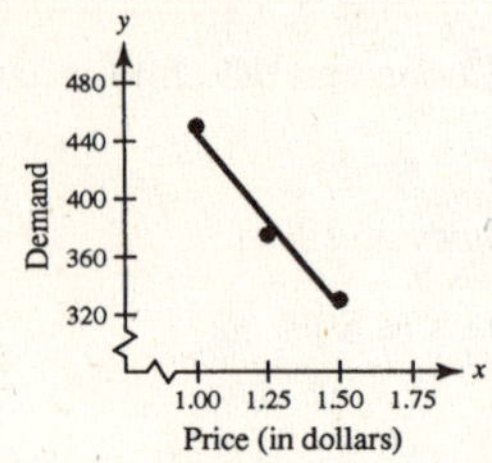

(d) $y = -240(1.40) + 685$

$= 349$ units

100. (a) (1.0, 32), (1.5, 41), (2.0, 48), (2.5, 53)

$$\begin{cases} 4b + 7a = 174 \Rightarrow & -7b - 12.25a = -304.5 \\ 7b + 13.5a = 322 \Rightarrow & 7b + 13.5a = 322 \end{cases}$$

$$1.25a = 17.5$$

$$a = 14$$

$$4b + 98 = 174$$

$$b = 19$$

Least squares regression line: $y = 14x + 19$

(b) $y = 14x + 19$

(c)

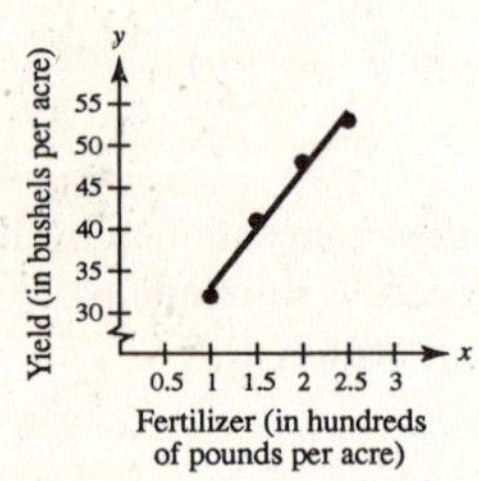

(d) When $x = 1.6$: $y = 14(1.6) + 19 = 41.4$ bushels per acre.

101. There are infinitely many systems that have the solution (6, 3). One possible system is:

$$\begin{cases} x + y = 9 \\ 3x - 2y = 12 \end{cases}$$

102. There are infinitely many systems that have the solution $(8, -2)$. One possible system is:

$$8 - 2 = 6 \Rightarrow x + y = 6$$

$$2(8) - (-2) = 18 \Rightarrow 2x - y = 18$$

103. There are infinitely many systems that have the solution $\left(3, \frac{5}{2}\right)$. One possible system is:

$$\begin{cases} 2x + 2y = 11 \\ x - 4y = -7 \end{cases}$$

104. There are infinitely many systems that have the solution $\left(-\frac{2}{3}, -10\right)$. One possible system is:

$$3\left(-\tfrac{2}{3}\right) + 3(-10) = -32 \Rightarrow 3x + 3y = -32$$

$$3\left(-\tfrac{2}{3}\right) + 6(-10) = -62 \Rightarrow 3x + 6y = -62$$

105. $\begin{cases} 4x - 8y = -3 & \text{Equation 1} \\ 2x + ky = 16 & \text{Equation 2} \end{cases}$

Multiply Equation 2 by -2: $-4x - 2ky = -32$

Add this to Equation 1:
$$\begin{aligned} 4x - 8y &= -3 \\ -4x - 2ky &= -32 \\ \hline -8y - 2ky &= -35 \end{aligned}$$

The system is inconsistent if $-8y - 2ky = 0$. This occurs when $k = -4$.

106. $\begin{cases} 15x + 3y = 6 \Rightarrow & 30x + 6y = 12 \\ -10x + ky = 9 \Rightarrow & -30x + 3ky = 27 \end{cases}$

$$(6 + 3k)y = 39$$

If $k = -2$, then we would have $0 = 39$ and the system would be inconsistent.

107. False. To solve a system of equations by substitution, you can solve for either variable in one of the two equations and then back-substitute.

108. False. The system can have at most 4 solutions because a parabola and a circle can intersect at most 4 times.

109. False. Two lines that coincide have infinitely many points of intersection.

110. False. Solving a system of equations algebraically will always give an exact solution.

111. True. If there are no points of intersection (solutions) then the lines must be parallel.

112. False.

The system $\begin{cases} x^2 + y^2 = 25 \\ x^2 - y^2 = 25 \end{cases}$ is not linear.

By elimination we have two solutions, $(5, 0)$ and $(-5, 0)$.

Section 14.2 Multivariable Linear Systems

1. $\begin{cases} 3x - y + z = 1 \\ 2x - 3z = -14 \\ 5y + 2z = 8 \end{cases}$

(a) $3(2) - (0) + (-3) \neq 1$

$(2, 0, -3)$ is **not** a solution.

(b) $3(-2) - (0) + 8 \neq 1$

$(-2, 0, 8)$ is **not** a solution.

(c) $3(0) - (-1) + 3 \neq 1$

$(0, -1, 3)$ is **not** a solution.

(d) $\begin{aligned} 3(-1) - (0) + 4 &= 1 \\ 2(-1) - 3(4) &= -14 \\ 5(0) + 2(4) &= 8 \end{aligned}$

$(-1, 0, 4)$ **is** a solution.

2. $\begin{cases} 3x + 4y - z = 17 \\ 5x - y + 2z = 2 \\ 2x - 3y + 7z = -21 \end{cases}$

(a) $3(3) + 4(-1) - 2 \neq 17$

$(3, -1, 2)$ is **not** a solution.

(b) $\begin{aligned} 3(1) + 4(3) - (-2) &= 17 \\ 5(1) - 3 + 2(-2) &= -2 \\ 2(1) - 3(3) + 7(-2) &= -21 \end{aligned}$

$(1, 3, -2)$ **is** a solution.

(c) $3(4) + 4(1) - (-3) \neq 17$

$(4, 1, -3)$ is **not** a solution.

(d) $3(1) + 4(-2) - 2 \neq 17$

$(1, -2, 2)$ is **not** a solution.

3. $\begin{cases} 4x + y - z = 0 \\ -8x - 6y + z = -\frac{7}{4} \\ 3x - y = -\frac{9}{4} \end{cases}$

(a) $4\left(\frac{1}{2}\right) + \left(-\frac{3}{4}\right) - \left(-\frac{7}{4}\right) \neq 0$

$\left(\frac{1}{2}, -\frac{3}{4}, -\frac{7}{4}\right)$ is **not** a solution.

(b) $4\left(-\frac{3}{2}\right) + \left(\frac{5}{4}\right) - \left(-\frac{5}{4}\right) \neq 0$

$\left(-\frac{3}{2}, \frac{5}{4}, -\frac{5}{4}\right)$ is **not** a solution.

(c) $\begin{aligned} 4\left(-\frac{1}{2}\right) + \left(\frac{3}{4}\right) - \left(-\frac{5}{4}\right) &= 0 \\ -8\left(-\frac{1}{2}\right) - 6\left(\frac{3}{4}\right) + \left(-\frac{5}{4}\right) &= -\frac{7}{4} \\ 3\left(-\frac{1}{2}\right) - \left(\frac{3}{4}\right) &= -\frac{9}{4} \end{aligned}$

$\left(-\frac{1}{2}, \frac{3}{4}, -\frac{5}{4}\right)$ **is** a solution.

(d) $4\left(-\frac{1}{2}\right) + \left(\frac{1}{6}\right) - \left(-\frac{3}{4}\right) \neq 0$

$\left(-\frac{1}{2}, \frac{1}{6}, -\frac{3}{4}\right)$ is **not** a solution.

4. $\begin{cases} -4x - y - 8z = -6 \\ y + z = 0 \\ 4x - 7y = 6 \end{cases}$

(a) $\begin{aligned} -4(-2) - (-2) - 8(2) &= -6 \\ -2 + 2 &= 0 \\ 4(-2) - 7(-2) &= 6 \end{aligned}$

$(-2, -2, 2)$ **is** a solution.

(b) $-4\left(-\frac{33}{2}\right) - (-10) - 8(10) \neq -6$

$\left(-\frac{33}{2}, -10, 10\right)$ is **not** a solution.

(c) $-4\left(\frac{1}{8}\right) - \left(-\frac{1}{2}\right) - 8\left(\frac{1}{2}\right) \neq -6$

$\left(\frac{1}{8}, -\frac{1}{2}, \frac{1}{2}\right)$ is **not** a solution.

(d) $\begin{aligned} -4\left(-\frac{11}{2}\right) - (-4) - 8(4) &= -6 \\ -4 + 4 &= 0 \\ 4\left(-\frac{11}{2}\right) - 7(-4) &= 6 \end{aligned}$

$\left(-\frac{11}{2}, -4, 4\right)$ **is** a solution.

5. $\begin{cases} 2x - y + 5z = 24 & \text{Equation 1} \\ \quad\; y + 2z = 6 & \text{Equation 2} \\ \qquad\quad z = 4 & \text{Equation 3} \end{cases}$

Back-substitute $z = 4$ into Equation 2.

$$y + 2(4) = 6$$
$$y = -2$$

Back-substitute $y = -2$ and $z = 4$ into Equation 1.

$$2x - (-2) + 5(4) = 24$$
$$2x + 22 = 24$$
$$x = 1$$

Solution: $(1, -2, 4)$

6. $\begin{cases} 4x - 3y - 2z = 21 & \text{Equation 1} \\ \quad\; 6y - 5z = -8 & \text{Equation 2} \\ \qquad\quad z = -2 & \text{Equation 3} \end{cases}$

Back-substitute $z = -2$ in Equation 2:

$$6y - 5(-2) = -8$$
$$y = -3$$

Back-substitute $z = -2$ and $y = -3$ in Equation 1:

$$4x - 3(-3) - 2(-2) = 21$$
$$4x + 13 = 21$$
$$x = 2$$

Solution: $(2, -3, -2)$

7. $\begin{cases} 2x + y - 3z = 10 & \text{Equation 1} \\ \quad\; y = 2 & \text{Equation 2} \\ \quad\; y - z = 4 & \text{Equation 3} \end{cases}$

Back-substitute $y = 2$ into Equation 3.

$$2 - z = 4$$
$$z = -2$$

Back-substitute $y = 2$ and $z = -2$ into Equation 1.

$$2x + 2 - 3(-2) = 10$$
$$2x + 8 = 10$$
$$x = 1$$

Solution: $(1, 2, -2)$

8. $\begin{cases} x = 8 & \text{Equation 1} \\ 2x + 3y = 10 & \text{Equation 2} \\ x - y + 2z = 22 & \text{Equation 3} \end{cases}$

Back-substitute $x = 8$ in Equation 2:

$$2(8) + 3y = 10$$
$$y = -2$$

Back-substitute $x = 8$ and $y = -2$ in Equation 3:

$$8 - (-2) + 2z = 22$$
$$z = 6$$

Solution: $(8, -2, 6)$

9. $\begin{cases} 4x - 2y + z = 8 & \text{Equation 1} \\ \qquad\quad 2z = 4 & \text{Equation 2} \\ \quad -y + z = 4 & \text{Equation 3} \end{cases}$

From Equation 2 we have $z = 2$. Back-substitute $z = 2$ into Equation 3.

$$-y + 2 = 4$$
$$y = -2$$

Back-substitute $y = -2$ and $z = 2$ into Equation 1.

$$4x - 2(-2) + 2 = 8$$
$$4x + 6 = 8$$
$$x = \frac{1}{2}$$

Solution: $\left(\frac{1}{2}, -2, 2\right)$

10. $\begin{cases} 5x - 8z = 22 \\ \quad 3y - 5z = 10 \\ \qquad\quad z = -4 \end{cases}$

Back-substitute $z = -4$ in Equation 2:

$3y - 5(-4) = 10 \Rightarrow y = -\frac{10}{3}$

Back-substitute $z = -4$ in Equation 1:

$5x - 8(-4) = 22 \Rightarrow x = -2$

Solution: $\left(-2, -\frac{10}{3}, -4\right)$

11. $\begin{cases} x + y + z = 6 & \text{Equation 1} \\ 2x - y + z = 3 & \text{Equation 2} \\ 3x - z = 0 & \text{Equation 3} \end{cases}$

$\begin{cases} x + y + z = 6 \\ -3y - z = -9 & -2\text{Eq.1} + \text{Eq.2} \\ -3y - 4z = -18 & -3\text{Eq.1} + \text{Eq.3} \end{cases}$

$\begin{cases} x + y + z = 6 \\ -3y - z = -9 \\ -3z = -9 & -\text{Eq.2} + \text{Eq.3} \end{cases}$

$\begin{cases} x + y + z = 6 \\ -3y - z = -9 \\ z = 3 & -\frac{1}{3}\text{Eq.3} \end{cases}$

$-3y - 3 = -9 \Rightarrow y = 2$

$x + 2 + 3 = 6 \Rightarrow x = 1$

Solution: $(1, 2, 3)$

12. $\begin{cases} x + y + z = 3 & \text{Equation 1} \\ x - 2y + 4z = 5 & \text{Equation 2} \\ 3y + 4z = 5 & \text{Equation 3} \end{cases}$

$\begin{cases} x + y + z = 3 \\ -3y + 3z = 2 & (-1)\text{Eq.1} + \text{Eq.2} \\ 3y + 4z = 5 \end{cases}$

$\begin{cases} x + y + z = 3 \\ -3y + 3z = 2 \\ 7z = 7 & \text{Eq.2} + \text{Eq.3} \end{cases}$

$\begin{cases} x + y + z = 3 \\ y - z = -\frac{2}{3} & \left(-\frac{1}{3}\right)\text{Eq.2} \\ z = 1 & \left(\frac{1}{7}\right)\text{Eq.3} \end{cases}$

$y - 1 = -\frac{2}{3} \Rightarrow y = \frac{1}{3}$

$x + \frac{1}{3} + 1 = 3 \Rightarrow x = \frac{5}{3}$

Solution: $\left(\frac{5}{3}, \frac{1}{3}, 1\right)$

13. $\begin{cases} 2x + 2z = 2 \\ 5x + 3y = 4 \\ 3y - 4z = 4 \end{cases}$

$\begin{cases} x + z = 1 & \frac{1}{2}\text{Eq.1} \\ 5x + 3y = 4 \\ 3y - 4z = 4 \end{cases}$

$\begin{cases} x + z = 1 \\ 3y - 5z = -1 & -5\text{Eq.1} + \text{Eq.2} \\ 3y - 4z = 4 \end{cases}$

$\begin{cases} x + z = 1 \\ 3y - 5z = -1 \\ z = 5 & -\text{Eq.2} + \text{Eq.3} \end{cases}$

$3y - 5(5) = -1 \Rightarrow y = 8$

$x + 5 = 1 \Rightarrow x = -4$

Solution: $(-4, 8, 5)$

14. $\begin{cases} x + y - z = -1 & \text{Interchange equations} \\ 2x + 4y + z = 1 \\ x - 2y - 3z = 2 \end{cases}$

$\begin{cases} x + y - z = -1 \\ 2y + 3z = 3 & (-2)\text{Eq.1} + \text{Eq.2} \\ -3y - 2z = 3 & (-1)\text{Eq.1} + \text{Eq.3} \end{cases}$

$\begin{cases} x + y - z = -1 \\ 2y + 3z = 3 \\ -6y - 4z = 6 & 2\text{ Eq.3} \end{cases}$

$\begin{cases} x + y - z = -1 \\ 2y + 3z = 3 \\ 5z = 15 & 3\text{Eq.2} + \text{Eq.3} \end{cases}$

$\begin{cases} x + y - z = -1 \\ y + \frac{3}{2}z = \frac{3}{2} & \left(\frac{1}{2}\right)\text{Eq.2} \\ z = 3 & \left(\frac{1}{5}\right)\text{Eq.3} \end{cases}$

$y + \frac{3}{2}(3) = \frac{3}{2} \Rightarrow y = -3$

$x - 3 - 3 = -1 \Rightarrow x = 5$

Solution: $(5, -3, 3)$

15. $\begin{cases} 3x + 3y = 9 & \text{Interchange Equations.} \\ 2x - 3z = 10 \\ 6y + 4z = -12 \end{cases}$

$\begin{cases} x + y = 3 & \frac{1}{3}\text{Eq.1} \\ 2x - 3z = 10 \\ 6y + 4z = -12 \end{cases}$

$\begin{cases} x + y = 3 \\ -2y - 3z = 4 & -2\text{Eq.1} + \text{Eq.2} \\ 6y + 4z = -12 \end{cases}$

$\begin{cases} x + y = 3 \\ -2y - 3z = 4 \\ -5z = 0 & 3\text{Eq.2} + \text{Eq.3} \end{cases}$

$\begin{cases} x + y = 3 \\ -2y - 3z = 4 \\ z = 0 & -\frac{1}{5}\text{Eq.3} \end{cases}$

$-2y - 3(0) = 4 \Rightarrow y = -2$

$x - 2 = 3 \Rightarrow x = 5$

Solution: $(5, -2, 0)$

16. $$\begin{cases} x + 4y + z = 0 \\ 2x + 4y - z = 7 \\ 2x - 4y + 2z = -6 \end{cases} \quad \text{Interchange equations}$$

$$\begin{cases} x + 4y + z = 0 & \\ -4y - 3z = 7 & (-2)\text{Eq.1} + \text{Eq.2} \\ -12y = -6 & (-2)\text{Eq.1} + \text{Eq.3} \end{cases}$$

$$\begin{cases} x + 4y + z = 0 & \\ -4y - 3z = 7 & \\ 9z = -27 & (-3)\text{Eq.2} + \text{Eq.3} \end{cases}$$

$$\begin{cases} x + 4y + z = 0 & \\ y + \frac{3}{4}z = -\frac{7}{4} & \left(-\frac{1}{4}\right)\text{Eq.2} \\ z = -3 & \left(\frac{1}{9}\right)\text{Eq.3} \end{cases}$$

$$y + \tfrac{3}{4}(-3) = -\tfrac{7}{4} \Rightarrow y = \tfrac{1}{2}$$

$$x + 4\left(\tfrac{1}{2}\right) + (-3) = 0 \Rightarrow x = 1$$

Solution: $\left(1, \frac{1}{2}, -3\right)$

17. $$\begin{cases} x - 2y + 2z = -9 \\ 2x + y - z = 7 \\ 3x - y + z = 5 \end{cases} \quad \text{Interchange Equations.}$$

$$\begin{cases} x - 2y + 2z = -9 & \\ 5y - 5z = 25 & -2\text{Eq.1} + \text{Eq.2} \\ 5y - 5z = 32 & -3\text{Eq.1} + \text{Eq.3} \end{cases}$$

$$\begin{cases} x - 2y + 2z = -9 & \\ 5y - 5z = 25 & \\ 0 = 7 & -\text{Eq.2} + \text{Eq.3} \end{cases}$$

Inconsistent, no solution.

18. $$\begin{cases} x - 11y + 4z = 3 \\ 5x - 3y + 2z = 3 \\ 2x + 4y - z = 7 \end{cases} \quad \text{Interchange equations}$$

$$\begin{cases} x - 11y + 4z = 3 & \\ 52y - 18z = -12 & (-5)\text{Eq.1} + \text{Eq.2} \\ 26y - 9z = 1 & (-2)\text{Eq.1} + \text{Eq.3} \end{cases}$$

$$\begin{cases} x - 11y + 4z = 3 & \\ 52y - 18z = -12 & \\ 0 = 7 & \left(-\frac{1}{2}\right)\text{Eq.2} + \text{Eq.3} \end{cases}$$

Inconsistent; no solution

19. $$\begin{cases} 3x - 5y + 5z = 1 \\ 5x - 2y + 3z = 0 \\ 7x - y + 3z = 0 \end{cases}$$

$$\begin{cases} 6x - 10y + 10z = 2 & 2\text{Eq.1} \\ 5x - 2y + 3z = 0 & \\ 7x - y + 3z = 0 & \end{cases}$$

$$\begin{cases} x - 8y + 7z = 2 & -\text{Eq.2} + \text{Eq.1} \\ 5x - 2y + 3z = 0 & \\ 7x - y + 3z = 0 & \end{cases}$$

$$\begin{cases} x - 8y + 7z = 2 & \\ 38y - 32z = -10 & -5\text{Eq.1} + \text{Eq.2} \\ 55y - 46z = -14 & -7\text{Eq.1} + \text{Eq.3} \end{cases}$$

$$\begin{cases} x - 8y + 7z = 2 & \\ 2090y - 1760z = -550 & 55\text{Eq.2} \\ -2090y + 1748z = 532 & -38\text{Eq.3} \end{cases}$$

$$\begin{cases} x - 8y + 7z = 2 & \\ 2090y - 1760z = -550 & \\ -12z = -18 & \text{Eq.2} + \text{Eq.3} \end{cases}$$

$$-12z = -18 \Rightarrow z = \tfrac{3}{2}$$

$$38y - 32\left(\tfrac{3}{2}\right) = -10 \Rightarrow y = 1$$

$$x - 8(1) + 7\left(\tfrac{3}{2}\right) = 2 \Rightarrow x = -\tfrac{1}{2}$$

Solution: $\left(-\frac{1}{2}, 1, \frac{3}{2}\right)$

20. $$\begin{cases} 2x + y + 3z = 1 & \text{Equation 1} \\ 2x + 6y + 8z = 3 & \text{Equation 2} \\ 6x + 8y + 18z = 5 & \text{Equation 3} \end{cases}$$

$$\begin{cases} 2x + y + 3z = 1 & \\ 5y + 5z = 2 & (-1)\text{Eq.1} + \text{Eq.2} \\ 5y + 9z = 2 & (-3)\text{Eq.1} + \text{Eq.3} \end{cases}$$

$$\begin{cases} 2x + y + 3z = 1 & \\ 5y + 5z = 2 & \\ 4z = 0 & (-1)\text{Eq.2} + \text{Eq.3} \end{cases}$$

$$\begin{cases} x + \frac{1}{2}y + \frac{3}{2}z = \frac{1}{2} & \left(\frac{1}{2}\right)\text{Eq.1} \\ y + z = \frac{2}{5} & \left(\frac{1}{5}\right)\text{Eq.2} \\ z = 0 & \left(\frac{1}{4}\right)\text{Eq.3} \end{cases}$$

$$y + 0 = \tfrac{2}{5} \Rightarrow y = \tfrac{2}{5}$$

$$x + \tfrac{1}{2}\left(\tfrac{2}{5}\right) + \tfrac{3}{2}(0) = \tfrac{1}{2} \Rightarrow x = \tfrac{3}{10}$$

Solution: $\left(\frac{3}{10}, \frac{2}{5}, 0\right)$

21. $\begin{cases} x + 2y - 7z = -4 \\ 2x + y + z = 13 \\ 3x + 9y - 36z = -33 \end{cases}$

$\begin{cases} x + 2y - 7z = -4 \\ -3y + 15z = 21 & -2\text{Eq.}1 + \text{Eq.}2 \\ 3y - 15z = -21 & -3\text{Eq.}1 + \text{Eq.}3 \end{cases}$

$\begin{cases} x + 2y - 7z = -4 \\ -3y + 15z = 21 \\ 0 = 0 & \text{Eq.}2 + \text{Eq.}3 \end{cases}$

$\begin{cases} x + 2y - 7z = -4 \\ y - 5z = -7 & -\frac{1}{3}\text{Eq.}2 \end{cases}$

$\begin{cases} x + 3z = 10 & -2\text{Eq.}2 + \text{Eq.}1 \\ y - 5z = -7 \end{cases}$

Let $z = a$, then:

$y = 5a - 7$

$x = -3a + 10$

Solution: $(-3a + 10, 5a - 7, a)$

22. $\begin{cases} 2x + y - 3z = 4 & \text{Equation 1} \\ 4x + 2z = 10 & \text{Equation 2} \\ -2x + 3y - 13z = -8 & \text{Equation 3} \end{cases}$

$\begin{cases} 2x + y - 3z = 4 \\ -2y + 8z = 2 & (-2)\text{Eq.}1 + \text{Eq.}2 \\ 4y - 16z = -4 & \text{Eq.}1 + \text{Eq.}3 \end{cases}$

$\begin{cases} 2x + y - 3z = 4 \\ -2y + 8z = 2 \\ 0 = 0 & 2\text{Eq.}2 + \text{Eq.}3 \end{cases}$

$\begin{cases} 2x + z = 5 & (\frac{1}{2})\text{Eq.}2 + \text{Eq.}1 \\ -2y + 8z = 2 \end{cases}$

$\begin{cases} x + z/2 = \frac{5}{2} & (\frac{1}{2})\text{Eq.}1 \\ y - 4z = -1 & (-\frac{1}{2})\text{Eq. }2 \end{cases}$

$z = a$

$y - 4a = -1 \Rightarrow y = 4a - 1$

$x + \frac{1}{2}a = \frac{5}{2} \Rightarrow x = -\frac{1}{2}a + \frac{5}{2}$

Solution: $\left(-\frac{1}{2}a + \frac{5}{2}, 4a - 1, a\right)$

23. $\begin{cases} 3x - 3y + 6z = 6 \\ x + 2y - z = 5 \\ 5x - 8y + 13z = 7 \end{cases}$

$\begin{cases} x - y + 2z = 2 & \frac{1}{3}\text{Eq.}1 \\ x + 2y - z = 5 \\ 5x - 8y + 13z = 7 \end{cases}$

$\begin{cases} x - y + 2z = 2 \\ 3y - 3z = 3 & -\text{Eq.}1 + \text{Eq.}2 \\ -3y + 3z = -3 & -5\text{Eq.}1 + \text{Eq.}3 \end{cases}$

$\begin{cases} x - y + 2z = 2 \\ y - z = 1 & \frac{1}{3}\text{Eq.}2 \\ 0 = 0 & \text{Eq.}2 + \text{Eq.}3 \end{cases}$

$\begin{cases} x + z = 3 & \text{Eq.}2 + \text{Eq.}1 \\ y - z = 1 \end{cases}$

Let $z = a$, then:

$y = a + 1$

$x = -a + 3$

Solution: $(-a + 3, a + 1, a)$

24. $\begin{cases} x + 4z = 13 & \text{Equation 1} \\ 4x - 2y + z = 7 & \text{Equation 2} \\ 2x - 2y - 7z = -19 & \text{Equation 3} \end{cases}$

$\begin{cases} x + 4z = 13 \\ -2y - 15z = -45 & (-4)\text{Eq.}1 + \text{Eq.}2 \\ -2y - 15z = -45 & (-2)\text{Eq.}1 + \text{Eq.}3 \end{cases}$

$\begin{cases} x + 4z = 13 \\ -2y - 15z = -45 \\ 0 = 0 & (-1)\text{Eq.}2 + \text{Eq.}3 \end{cases}$

$\begin{cases} x + 4z = 13 \\ y + \frac{15}{2}z = \frac{45}{2} & (-\frac{1}{2})\text{Eq.}2 \end{cases}$

$z = a$

$y + \frac{15}{2}a = \frac{45}{2} \Rightarrow y = -\frac{15}{2}a + \frac{45}{2}$

$x + 4a = 13 \Rightarrow x = -4a + 13$

Solution: $\left(-4a + 13, -\frac{15}{2}a + \frac{45}{2}, a\right)$

25. $\begin{cases} x - 2y + 5z = 2 \\ 4x - z = 0 \end{cases}$

Let $z = a$, then $x = \frac{1}{4}a$.

$\frac{1}{4}a - 2y + 5a = 2$

$a - 8y + 20a = 8$

$-8y = -21a + 8$

$y = \frac{21}{8}a - 1$

Answer: $\left(\frac{1}{4}a, \frac{21}{8}a - 1, a\right)$

To avoid fractions, we could go back and let $z = 8a$, then $4x - 8a = 0 \Rightarrow x = 2a$.

$2a - 2y + 5(8a) = 2$

$-2y + 42a = 2$

$y = 21a - 1$

Solution: $(2a, 21a - 1, 8a)$

26. $\begin{cases} x - 3y + 2z = 18 & \text{Equation 1} \\ 5x - 13y + 12z = 80 & \text{Equation 2} \end{cases}$

$\begin{cases} x - 3y + 2z = 18 \\ 2y + 2z = -10 & (-5)\text{Eq.1} + \text{Eq.2} \end{cases}$

$\begin{cases} x - 3y + 2z = 18 \\ y + z = -5 & \left(\frac{1}{2}\right)\text{Eq.2} \end{cases}$

$\begin{cases} x + 5z = 3 & 3\text{Eq.2} + \text{Eq.1} \\ y + z = -5 \end{cases}$

Let $z = a$, then: $y + a = -5 \Rightarrow y = -a - 5$

$x + 5a = 3 \Rightarrow x = -5a + 3$

Solution: $(-5a + 3, -a - 5, a)$

27. $\begin{cases} 2x - 3y + z = -2 \\ -4x + 9y = 7 \end{cases}$

$\begin{cases} 2x - 3y + z = -2 \\ 3y + 2z = 3 & 2\text{Eq.1} + \text{Eq.2} \end{cases}$

$\begin{cases} 2x + 3z = 1 & \text{Eq.2} + \text{Eq.1} \\ 3y + 2z = 3 \end{cases}$

Let $x = a$, then:

$y = -\frac{2}{3}a + 1$

$x = -\frac{3}{2}a + \frac{1}{2}$

Solution: $\left(-\frac{3}{2}a + \frac{1}{2}, -\frac{2}{3}a + 1, a\right)$

28. $\begin{cases} 2x + 3y + 3z = 7 & \text{Equation 1} \\ 4x + 18y + 15z = 44 & \text{Equation 2} \end{cases}$

$\begin{cases} 2x + 3y + 3z = 7 \\ 12y + 9z = 30 & (-2)\text{Eq.1} + \text{Eq.2} \end{cases}$

$\begin{cases} 2x + \frac{3}{4}z = -\frac{1}{2} & \left(-\frac{1}{4}\right)\text{Eq.2} + \text{Eq.1} \\ 12y + 9z = 30 \end{cases}$

$\begin{cases} x + \frac{3}{8}z = -\frac{1}{4} & \left(\frac{1}{2}\right)\text{Eq.1} \\ y + \frac{3}{4}z = \frac{5}{2} & \left(\frac{1}{12}\right)\text{Eq.2} \end{cases}$

Let $z = a$, then:

$y + \frac{3}{4}a = \frac{5}{2} \Rightarrow y = -\frac{3}{4}a + \frac{5}{2}$

$x + \frac{3}{8}a = -\frac{1}{4} \Rightarrow x = -\frac{3}{8}a - \frac{1}{4}$

Solution: $\left(-\frac{3}{8}a - \frac{1}{4}, -\frac{3}{4}a + \frac{5}{2}, a\right)$

29. $\begin{cases} x + 3w = 4 \\ 2y - z - w = 0 \\ 3y - 2w = 1 \\ 2x - y + 4z = 5 \end{cases}$

$\begin{cases} x + 3w = 4 \\ 2y - z - w = 0 \\ 3y - 2w = 1 \\ -y + 4z - 6w = -3 & -2\text{Eq.1} + \text{Eq.4} \end{cases}$

$\begin{cases} x + 3w = 4 & -\text{Eq.4 and} \\ y - 4z + 6w = 3 & \text{interchange} \\ 2y - z - w = 0 & \text{the equations.} \\ 3y - 2w = 1 \end{cases}$

$\begin{cases} x + 3w = 4 \\ y - 4z + 6w = 3 \\ 7z - 13w = -6 & -2\text{Eq.2} + \text{Eq.3} \\ 12z - 20w = -8 & -3\text{Eq.2} + \text{Eq.4} \end{cases}$

$\begin{cases} x + 3w = 4 \\ y - 4z + 6w = 3 \\ z - 3w = -2 & -\frac{1}{2}\text{Eq.4} + \text{Eq.3} \\ 12z - 20w = -8 \end{cases}$

$\begin{cases} x + 3w = 4 \\ y - 4z + 6w = 3 \\ z - 3w = -2 \\ 16w = 16 & -12\text{Eq.3} + \text{Eq.4} \end{cases}$

$16w = 16 \Rightarrow w = 1$

$z - 3(1) = -2 \Rightarrow z = 1$

$y - 4(1) + 6(1) = 3 \Rightarrow x = 1$

$x + 3(1) = 4 \Rightarrow x = 1$

Solution: $(1, 1, 1, 1)$

30.
$$\begin{cases} x + y + z + w = 6 & \text{Equation 1} \\ 2x + 3y - w = 0 & \text{Equation 2} \\ -3x + 4y + z + 2w = 4 & \text{Equation 3} \\ x + 2y - z + w = 0 & \text{Equation 4} \end{cases}$$

$$\begin{cases} x + y + z + w = 6 & \\ y - 2z - 3w = -12 & (-2)\text{Eq.1} + \text{Eq.2} \\ 7y + 4z + 5w = 22 & 3\text{Eq.1} + \text{Eq.3} \\ y - 2z = -6 & (-1)\text{Eq.1} + \text{Eq.4} \end{cases}$$

$$\begin{cases} x + y + z + w = 6 & \\ y - 2z - 3w = -12 & \\ 18z + 26w = 106 & (-7)\text{Eq.2} + \text{Eq.3} \\ 3w = 6 & (-1)\text{Eq.2} + \text{Eq.4} \end{cases}$$

$$\begin{cases} x + y + z + w = 6 & \\ y - 2z - 3w = -12 & \\ z + \frac{13}{9}w = \frac{53}{9} & \left(\frac{1}{18}\right)\text{Eq.3} \\ w = 2 & \left(\frac{1}{3}\right)\text{Eq.4} \end{cases}$$

$$z + \tfrac{13}{9}(2) = \tfrac{53}{9} \Rightarrow z = 3$$
$$y - 2(3) - 3(2) = -12 \Rightarrow y = 0$$
$$x + 0 + 3 + 2 = 6 \Rightarrow x = 1$$

Solution: $(1, 0, 3, 2)$

31.
$$\begin{cases} x + 4z = 1 \\ x + y + 10z = 10 \\ 2x - y + 2z = -5 \end{cases}$$

$$\begin{cases} x + 4z = 1 & \\ y + 6z = 9 & -\text{Eq.1} + \text{Eq.2} \\ -y - 6z = -7 & -2\text{Eq.1} + \text{Eq.3} \end{cases}$$

$$\begin{cases} x + 4z = 1 & \\ y + 6z = 9 & \\ 0 = 2 & \text{Eq.2} + \text{Eq.3} \end{cases}$$

No solution, inconsistent

32.
$$\begin{cases} 2x - 2y - 6z = -4 & \text{Equation 1} \\ -3x + 2y + 6z = 1 & \text{Equation 2} \\ x - y - 5z = -3 & \text{Equation 3} \end{cases}$$

$$\begin{cases} x - y - 5z = -3 & \\ -3x + 2y + 6z = 1 & \text{Interchange equations} \\ 2x - 2y - 6z = -4 & \end{cases}$$

$$\begin{cases} x - y - 5z = -3 & \\ -y - 9z = -8 & 3\text{Eq.1} + \text{Eq.2} \\ 4z = 2 & (-2)\text{Eq.1} + \text{Eq.3} \end{cases}$$

$$\begin{cases} x - y - 5z = -3 & \\ y + 9z = 8 & (-1)\text{Eq.2} \\ z = \frac{1}{2} & \left(\frac{1}{4}\right)\text{Eq.3} \end{cases}$$

$$y + 9\left(\tfrac{1}{2}\right) = 8 \Rightarrow y = \tfrac{7}{2}$$
$$x - \tfrac{7}{2} - 5\left(\tfrac{1}{2}\right) = -3 \Rightarrow x = 3$$

Solution: $\left(3, \frac{7}{2}, \frac{1}{2}\right)$

33.
$$\begin{cases} 2x + 3y = 0 \\ 4x + 3y - z = 0 \\ 8x + 3y + 3z = 0 \end{cases}$$

$$\begin{cases} 2x + 3y = 0 & \\ -3y - z = 0 & -2\text{Eq.1} + \text{Eq.2} \\ -9y + 3z = 0 & -4\text{Eq.1} + \text{Eq.3} \end{cases}$$

$$\begin{cases} 2x + 3y = 0 & \\ -3y - z = 0 & \\ 6z = 0 & -3\text{Eq.2} + \text{Eq.3} \end{cases}$$

$$6z = 0 \Rightarrow z = 0$$
$$-3y - 0 = 0 \Rightarrow y = 0$$
$$2x + 3(0) = 0 \Rightarrow x = 0$$

Solution: $(0, 0, 0)$

34.
$$\begin{cases} 4x + 3y + 17z = 0 \\ 5x + 4y + 22z = 0 \\ 4x + 2y + 19z = 0 \end{cases}$$

$$\begin{cases} 5x + 4y + 22z = 0 & \\ 4x + 3y + 17z = 0 & \text{Interchange} \\ 4x + 2y + 19z = 0 & \text{equations} \end{cases}$$

$$\begin{cases} x + y + 5z = 0 & (-1)\text{Eq.2} + \text{Eq.1} \\ 4x + 3y + 17z = 0 & \\ 4x + 2y + 19z = 0 & \end{cases}$$

$$\begin{cases} x + y + 5z = 0 & \\ -y - 3z = 0 & (-4)\text{Eq.1} + \text{Eq.2} \\ -2y - z = 0 & (-4)\text{Eq.1} + \text{Eq.3} \end{cases}$$

$$\begin{cases} x + y + 5z = 0 & \\ y + 3z = 0 & (-1)\text{Eq.2} \\ 5z = 0 & (-2)\text{Eq.2} + \text{Eq.3} \end{cases}$$

$$\begin{cases} x + y + 5z = 0 & \\ y + 3z = 0 & \\ z = 0 & \left(\frac{1}{5}\right)\text{Eq.3} \end{cases}$$

$$y + 3(0) = 0 \Rightarrow y = 0$$
$$x + 0 + 5(0) = 0 \Rightarrow x = 0$$

Solution: $(0, 0, 0)$

35. $\begin{cases} 12x + 5y + z = 0 \\ 23x + 4y - z = 0 \end{cases}$

$\begin{cases} 24x + 10y + 2z = 0 \\ 23x + 4y - z = 0 \end{cases}$ 2Eq.1

$\begin{cases} x + 6y + 3z = 0 \\ 23x + 4y - z = 0 \end{cases}$ −Eq.2 + Eq.1

$\begin{cases} x + 6y + 3z = 0 \\ -134y - 70z = 0 \end{cases}$ −23Eq.1 + Eq.2

$\begin{cases} x + 6y + 3z = 0 \\ -67y - 35z = 0 \end{cases}$ $\frac{1}{2}$Eq.2

To avoid fractions, let $z = 67a$, then:

$$-67y - 35(67a) = 0$$
$$y = -35a$$
$$x + 6(-35a) + 3(67a) = 0$$
$$x = 9a$$

Solution: $(9a, -35a, 67a)$

36. $\begin{cases} 2x - y - z = 0 & \text{Equation 1} \\ -2x + 6y + 4z = 2 & \text{Equation 2} \end{cases}$

$\begin{cases} 2x - y - z = 0 \\ 5y + 3z = 2 & \text{Eq.1 + Eq.2} \end{cases}$

$\begin{cases} x - \frac{1}{2}y - \frac{1}{2}z = 0 & (\frac{1}{2})\text{Eq.1} \\ y + \frac{3}{5}z = \frac{2}{5} & (\frac{1}{5})\text{Eq.2} \end{cases}$

Let $z = a$, then:

$$y + \tfrac{3}{5}a = \tfrac{2}{5} \Rightarrow y = -\tfrac{3}{5}a + \tfrac{2}{5}$$
$$x - \tfrac{1}{2}\left(-\tfrac{3}{5}a + \tfrac{2}{5}\right) - \tfrac{1}{2}a = 0 \Rightarrow x = \tfrac{1}{5}a + \tfrac{1}{5}$$

Solution: $\left(\frac{1}{5}a + \frac{1}{5}, -\frac{3}{5}a + \frac{2}{5}, a\right)$

37. $\begin{cases} x - 2y + 3z = 5 & \text{Equation 1} \\ -x + 3y - 5z = 4 & \text{Equation 2} \\ 2x \quad\quad - 3z = 0 & \text{Equation 3} \end{cases}$

Add Equation 1 to Equation 2.

$\begin{cases} x - 2y + 3z = 5 \\ y - 2z = 9 \\ 2x \quad\quad - 3z = 0 \end{cases}$

This is a first step in putting the system in row-echelon form.

38. $\begin{cases} x - 2y + 3z = 5 & \text{Equation 1} \\ -x + 3y - 5z = 4 & \text{Equation 2} \\ 2x \quad\quad - 3z = 0 & \text{Equation 3} \end{cases}$

Add −2 times Equation 1 to Equation 3.

$\begin{cases} x - 2y + 3z = 5 \\ -x + 3y - 5z = 4 \\ 4y - 9z = -10 \end{cases}$

This is a first step in putting the system in row-echelon form.

39. No, they are not equivalent. For the second system to be equivalent to the first, the constant in the second equation should be −11 and the coefficient of z in the third equation should be 2.

40. The first system is inconsistent because 4 times Equation 1 added to Equation 2 yields $0 = 20$.

41. When using Gaussian elimination to solve a system of linear equations, a system has no solution when there is a row representing a contradictory equation such as $0 = N$, where N is a nonzero real number.

For instance: $x + y = 3$ Equation 1
$-x - y = 3$ Equation 2

$x + y = 0$
$0 = 6$ Eq.1 + Eq.2

No solution

42. If a system of three linear equations with three unknowns has a unique solution, then graphically we have three planes that intersect at only one point.

43. $y = ax^2 + bx + c$ passing through $(0, 0), (2, -2), (4, 0)$

$(0, 0)$: $0 = c$

$(2, -2)$: $-2 = 4a + 2b + c \Rightarrow -1 = 2a + b$

$(4, 0)$: $0 = 16a + 4b + c \Rightarrow 0 = 4a + b$

Solution: $a = \frac{1}{2}, b = -2, c = 0$

The equation of the parabola is $y = \frac{1}{2}x^2 - 2x$.

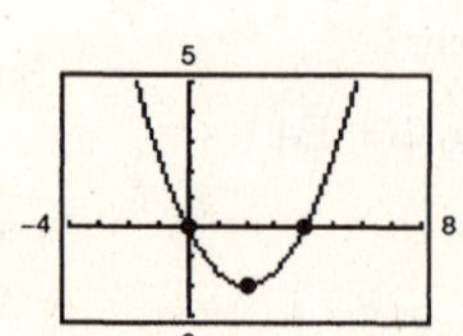

44. $y = ax^2 + bx + c$ passing through

$(0, 3), (1, 4), (2, 3)$

$(0, 3)$: $3 = c$

$(1, 4)$: $4 = a + b + c \Rightarrow 1 = a + b$

$(2, 3)$: $3 = 4a + 2b + c \Rightarrow 0 = 2a + b$

Solution: $a = -1, b = 2, c = 3$

The equation of the parabola is $y = -x^2 + 2x + 3$.

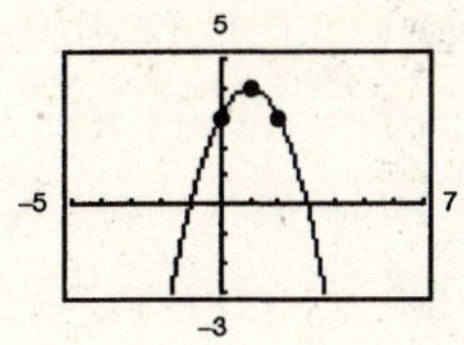

45. $y = ax^2 + bx + c$ passing through $(2, 0), (3, -1), (4, 0)$

$(2, 0)$: $0 = 4a + 2b + c$

$(3, -1)$: $-1 = 9a + 3b + c$

$(4, 0)$: $0 = 16a + 4b + c$

$$\begin{cases} 0 = 4a + 2b + c & \\ -1 = 5a + b & -\text{Eq.1} + \text{Eq.2} \\ 0 = 12a + 2b & -\text{Eq.1} + \text{Eq.3} \end{cases}$$

$$\begin{aligned} 0 &= 4a + 2b + c & \\ -1 &= 5a + b & \\ 2 &= 2a & -2\text{Eq.2} - 3\text{Eq.} \end{aligned}$$

Solution: $a = 1, b = -6, c = 8$

The equation of the parabola is $y = x^2 - 6x + 8$.

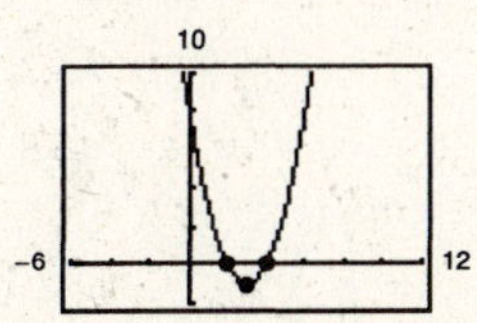

46. $y = ax^2 + bx + c$ passing through

$(1, 3), (2, 2), (3, -3)$

$(1, 3)$: $3 = a + b + c$

$(2, 2)$: $2 = 4a + 2b + c$

$(3, -3)$: $-3 = 9a + 3b + c$

$$\begin{cases} a + b + c = 3 & \\ 3a + b = -1 & (-1)\text{Eq.1} + \text{Eq.2} \\ 8a + 2b = -6 & (-1)\text{Eq.1} + \text{Eq.3} \end{cases}$$

$$\begin{cases} a + b + c = 3 & \\ 3a + b = -1 & \\ 2a = -4 & (-2)\text{Eq.2} + \text{Eq.3} \end{cases}$$

Solution: $a = -2, b = 5, c = 0$

The equation of the parabola is $y = -2x^2 + 5x$.

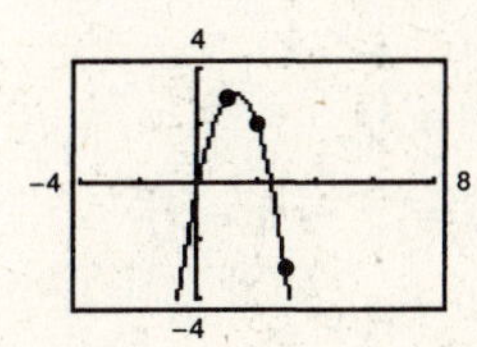

47. $x^2 + y^2 + Dx + Ey + F = 0$ passing through $(0, 0), (2, 2), (4, 0)$

$(0, 0)$: $F = 0$

$(2, 2)$: $8 + 2D + 2E + F = 0 \Rightarrow D + E = -4$

$(4, 0)$: $16 + 4D + F = 0 \Rightarrow D = -4$ and $E = 0$

The equation of the circle is $x^2 + y^2 - 4x = 0$.

To graph, let $y_1 = \sqrt{4x - x^2}$ and $y_2 = -\sqrt{4x - x^2}$.

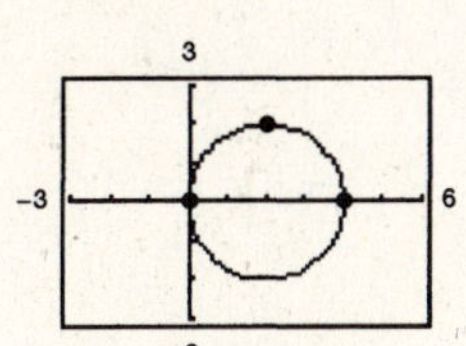

48. $x^2 + y^2 + Dx + Ey + F = 0$ passing through $(0, 0), (0, 6), (3, 3)$

$(0, 0)$: $F = 0$

$(0, 6)$: $36 + 6E + F = 0 \Rightarrow E = -6$

$(3, 3)$: $18 + 3D + 3E + F = 0 \Rightarrow D = 0$

The equation of the circle is $x^2 + y^2 - 6y = 0$.

To graph, complete the square first, then solve for y.

$$\begin{aligned} x^2 + (y^2 - 6y + 9) &= 9 \\ x^2 + (y - 3)^2 &= 9 \\ (y - 3)^2 &= 9 - x^2 \\ y - 3 &= \pm\sqrt{9 - x^2} \\ y &= 3 \pm \sqrt{9 - x^2} \end{aligned}$$

Let $y_1 = 3 + \sqrt{9 - x^2}$

and $y_2 = 3 - \sqrt{9 - x^2}$.

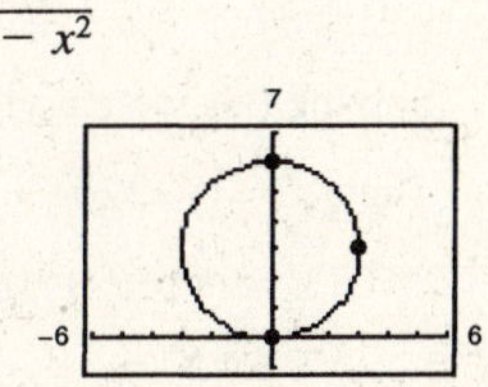

49. $x^2 + y^2 + Dx + Ey + F = 0$ passing through $(-3, -1), (2, 4), (-6, 8)$

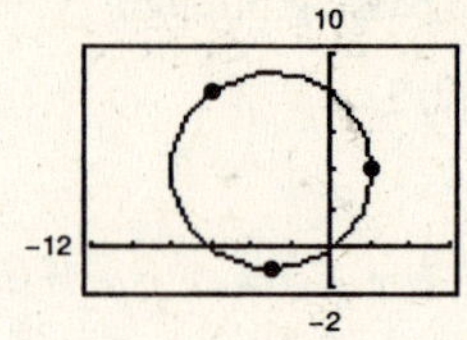

$(-3, -1)$: $10 - 3D - E + F = 0 \Rightarrow 10 = 3D + E - F$

$(2, 4)$: $20 + 2D + 4E + F = 0 \Rightarrow 20 = -2D - 4E - F$

$(-6, 8)$: $100 - 6D + 8E + F = 0 \Rightarrow 100 = 6D - 8E - F$

Solution: $D = 6, E = -8, F = 0$

The equation of the circle is $x^2 + y^2 + 6x - 8y = 0$.

To graph, complete the squares first, then solve for y.

$$(x^2 + 6x + 9) + (y^2 - 8y + 16) = 0 + 9 + 16$$
$$(x + 3)^2 + (y - 4)^2 = 25$$
$$(y - 4)^2 = 25 - (x + 3)^2$$
$$y - 4 = \pm\sqrt{25 - (x + 3)^2}$$
$$y = 4 \pm \sqrt{25 - (x + 3)^2}$$

Let $y_1 = 4 + \sqrt{25 - (x + 3)^2}$ and $y_2 = 4 - \sqrt{25 - (x + 3)^2}$.

50. $x^2 + y^2 + Dx + Ey + F = 0$ passing through $(0, 0), (0, -2), (3, 0)$

$(0, 0)$: $F = 0$

$(0, -2)$: $4 - 2E + F = 0 \Rightarrow E = 2$

$(3, 0)$: $9 + 3D + F = 0 \Rightarrow D = -3$

The equation of the circle is $x^2 + y^2 - 3x + 2y = 0$.

To graph, complete the squares first, then solve for y.

$$\left(x^2 - 3x + \tfrac{9}{4}\right) + (y^2 + 2y + 1) = \tfrac{9}{4} + 1$$
$$\left(x - \tfrac{3}{2}\right)^2 + (y + 1)^2 = \tfrac{13}{4}$$
$$(y + 1)^2 = \tfrac{13}{4} - \left(x - \tfrac{3}{2}\right)^2$$
$$y + 1 = \pm\sqrt{\tfrac{13}{4} - \left(x - \tfrac{3}{2}\right)^2}$$
$$y = -1 \pm \sqrt{\tfrac{13}{4} - \left(x - \tfrac{3}{2}\right)^2}$$

Let $y_1 = -1 + \sqrt{\tfrac{13}{4} - \left(x - \tfrac{3}{2}\right)^2}$ and $y_2 = -1 - \sqrt{\tfrac{13}{4} - \left(x - \tfrac{3}{2}\right)^2}$.

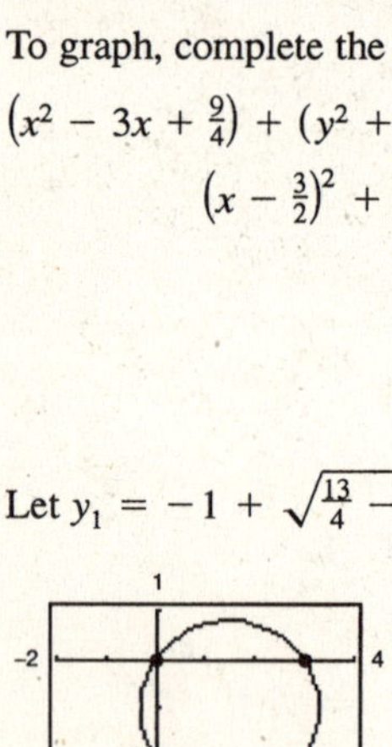

51. $s = \frac{1}{2}at^2 + v_0t + s_0$

$(1, 128), (2, 80), (3, 0)$

$128 = \frac{1}{2}a + v_0 + s_0 \Rightarrow a + 2v_0 + 2s_0 = 256$

$80 = 2a + 2v_0 + s_0 \Rightarrow 2a + 2v_0 + s_0 = 80$

$0 = \frac{9}{2}a + 3v_0 + s_0 \Rightarrow 9a + 6v_0 + 2s_0 = 0$

Solving this system yields $a = -32, v_0 = 0, s_0 = 144$.

Thus, $s = \frac{1}{2}(-32)t^2 + (0)t + 144$

$= -16t^2 + 144$.

52. $s = \frac{1}{2}at^2 + v_0t + s_0$

(1, 48), (2, 64), (3, 48)

$$\begin{cases} 48 = \frac{1}{2}a + v_0 + s_0 \\ 64 = 2a + 2v_0 + s_0 \\ 48 = \frac{9}{2}a + 3v_0 + s_0 \end{cases}$$

$$\begin{cases} a + 2v_0 + 2s_0 = 96 & \text{2Eq. 1} \\ 2a + 2v_0 + s_0 = 64 \\ 9a + 6v_0 + 2s_0 = 96 & \text{2Eq. 3} \end{cases}$$

$$\begin{cases} a + 2v_0 + 2s_0 = 96 \\ -2v_0 - 3s_0 = -128 & (-2)\text{Eq.1} + \text{Eq.2} \\ -12v_0 - 16s_0 = -768 & (-9)\text{Eq.1} + \text{Eq.3} \end{cases}$$

$$\begin{cases} a + 2v_0 + 2s_0 = 96 \\ -2v_0 - 3s_0 = -128 \\ 2s_0 = 0 & (-6)\text{Eq.2} + \text{Eq.3} \end{cases}$$

$$\begin{cases} a + 2v_0 + 2s_0 = 96 \\ v_0 + 1.5s_0 = 64 & (-0.5)\text{Eq.2} \\ s_0 = 0 & (0.5)\text{Eq.3} \end{cases}$$

$v_0 + 1.5(0) = 64 \Rightarrow v_0 = 64$

$a + 2(64) + 2(0) = 96 \Rightarrow a = -32$

Thus, $s = \frac{1}{2}(-32)t^2 + 64t + 0$

$= -16t^2 + 64t.$

53. $s = \frac{1}{2}at^2 + v_0t + s_0$

(1, 452), (2, 372), (3, 260)

$452 = \frac{1}{2}a + v_0 + s_0 \Rightarrow a + 2v_0 + 2s_0 = 904$

$372 = 2a + 2v_0 + s_0 \Rightarrow 2a + 2v_0 + s_0 = 372$

$260 = \frac{9}{2}a + 3v_0 + s_0 \Rightarrow 9a + 6v_0 + 2s_0 = 520$

Solving this system yields $a = -32$, $v_0 = -32$, $s_0 = 500$.

Thus, $s = \frac{1}{2}(-32)t^2 + (-32)t + 500$

$= -16t^2 - 32t + 500.$

54. $s = \frac{1}{2}at^2 + v_0t + s_0$

(1, 132), (2, 100), (3, 36)

$$\begin{cases} 132 = \frac{1}{2}a + v_0 + s_0 \\ 100 = 2a + 2v_0 + s_0 \\ 36 = \frac{9}{2}a + 3v_0 + s_0 \end{cases}$$

$$\begin{cases} a + 2v_0 + 2s_0 = 264 & \text{2Eq. 1} \\ 2a + 2v_0 + s_0 = 100 \\ 9a + 6v_0 + 2s_0 = 72 & \text{2Eq. 3} \end{cases}$$

$$\begin{cases} a + 2v_0 + 2s_0 = 264 \\ -2v_0 - 3s_0 = -428 & (-2)\text{Eq.1} + \text{Eq.2} \\ -12v_0 - 16s_0 = -2304 & (-9)\text{Eq.1} + \text{Eq.3} \end{cases}$$

$$\begin{cases} a + 2v_0 + 2s_0 = 264 \\ -2v_0 - 3s_0 = -428 \\ 2s_0 = 264 & (-6)\text{Eq.2} + \text{Eq.3} \end{cases}$$

$$\begin{cases} a + 2v_0 + 2s_0 = 264 \\ v_0 + 1.5s_0 = 214 & (-0.5)\text{Eq.2} \\ s_0 = 132 & (0.5)\text{Eq.3} \end{cases}$$

$v_0 + 1.5(132) = 214 \Rightarrow v_0 = 16$

$a + 2(16) + 2(132) = 264 \Rightarrow a = -32$

Thus, $s = \frac{1}{2}(-32)t^2 + 16t + 132$

$= -16t^2 + 16t + 132.$

55. Let x = number of touchdowns.
Let y = number of extra-point kicks.
Let z = number of field goals.

$$\begin{cases} x + y + z = 9 \\ 6x + y + 3z = 30 \\ x - y = 0 \end{cases}$$

$$\begin{cases} x + y + z = 9 \\ -5y - 3z = -24 & -6\text{Eq.1} + \text{Eq.2} \\ -2y - z = -9 & -\text{Eq.1} + \text{Eq.3} \end{cases}$$

$$\begin{cases} x + y + z = 9 \\ y + \frac{3}{5}z = \frac{24}{5} & -\frac{1}{5}\text{Eq.2} \\ -2y - z = -9 \end{cases}$$

$$\begin{cases} x + y + z = 9 \\ y + \frac{3}{5}z = \frac{24}{5} \\ \frac{1}{5}z = \frac{3}{5} & 2\text{Eq.2} + \text{Eq.3} \end{cases}$$

$\frac{1}{5}z = \frac{3}{5} \Rightarrow z = 3$

$y + \frac{3}{5}(3) = \frac{24}{5} \Rightarrow y = 3$

$x + 3 + 3 = 9 \Rightarrow x = 3$

Thus, 3 touchdowns, 3 extra-kick points, and 3 field goals were scored.

56. Let x = number of three-point baskets.
Let y = number of two-point baskets.
Let z = number of one-point free throws.

$$\begin{cases} 3x + 2y + z = 90 \\ 15x - y = 0 \\ 12x - z = 0 \end{cases}$$

$$\begin{cases} 3x + 2y + z = 90 \\ -11y - 5z = -450 & -5\text{Eq.1} + \text{Eq.2} \\ -8y - 5z = -360 & -4\text{Eq.1} + \text{Eq.3} \end{cases}$$

$$\begin{cases} 3x + 2y + z = 90 \\ -11y - 5z = -450 \\ 3y = 90 & -\text{Eq.2} + \text{Eq.3} \end{cases}$$

$$3y = 90 \implies y = 30$$

$$-11(30) - 5z = -450 \implies z = -\tfrac{1}{5}(-450 + 330) = 24$$

$$3x + 2(30) + 24 = 90 \implies x = \tfrac{1}{3}(90 - 60 - 24) = 2$$

Thus, 2 three-point baskets, 30 two-point baskets, and 24 one-point free throws were scored.

57. Let x = amount at 8%.
Let y = amount at 9%.
Let z = amount at 10%.

$$\begin{cases} x + y + z = 775{,}000 \\ 0.08x + 0.09y + 0.10z = 67{,}500 \\ x = 4z \end{cases}$$

$$y + 5z = 775{,}000 \qquad \text{Substitute } 4z \text{ for } x \text{ in Eq.1 and Eq.2.}$$

$$0.09y + 0.42z = 67{,}500$$

$z = 75{,}000$

$y = 775{,}000 - 5z = 400{,}000$

$x = 4z = 300{,}000$

$300,000 was borrowed at 8%.

$400,000 was borrowed at 9%.

$75,000 was borrowed at 10%.

58. Let x = amount at 8%.
Let y = amount at 9%.
Let z = amount at 10%.

$$\begin{cases} x + y + z = 800{,}000 \\ 0.08x + 0.09y + 0.10z = 67{,}000 \\ x = 5z \end{cases}$$

$$\begin{cases} y + 6z = 800{,}000 \\ 0.09y + 0.5z = 67{,}000 \end{cases}$$

$z = 125{,}000$

$y = 800{,}000 - 6(125{,}000) = 50{,}000$

$x = 5(125{,}000) = 625{,}000$

Solution: x = $625,000 at 8%
y = $50,000 at 9%
z = $125,000 at 10%

59. Let C = amount in certificates of deposit.
Let M = amount in municipal bonds.
Let B = amount in blue-chip stocks.
Let G = amount in growth or speculative stocks.

$$\begin{cases} C + M + B + G = 500{,}000 \\ 0.10C + 0.08M + 0.12B + 0.13G = 0.10(500{,}000) \\ B + G = \tfrac{1}{4}(500{,}000) \end{cases}$$

This system has infinitely many solutions.

Let $G = s$, then $B = 125{,}000 - s$

$M = 125{,}000 + \tfrac{1}{2}s$

$C = 250{,}000 - \tfrac{1}{2}s.$

One possible solution is to let $s = 50{,}000$.

Certificates of deposit: $225,000

Municipal bonds: $150,000

Blue-chip stocks: $75,000

Growth or speculative stocks: $50,000

60. Let C = amount in certificates of deposit.
Let M = amount in municipal bonds.
Let B = amount in blue-chip stocks.
Let G = amount in growth or speculative stocks.

$$\begin{cases} C + M + B + G = 500{,}000 \\ 0.09C + 0.05M + 0.12B + 0.14G = 0.10(500{,}000) \\ B + G = \tfrac{1}{4}(500{,}000) \end{cases}$$

This system has infinitely many solutions.

Let $G = s$, then $B = 125{,}000 - s$

$M = \tfrac{1}{2}s - 31{,}250$

$C = 406{,}250 - \tfrac{1}{2}s.$

Solution:

$406{,}250 - \tfrac{1}{2}s$ in certificates of deposit,

$-31{,}250 + \tfrac{1}{2}s$ in municipal bonds,

$125{,}000 - s$ in blue-chip stocks,

s in growth stocks

One possible solution is to let s = $100,000.

Certificates of deposit: $356,250
Municipal bonds: $18,750
Blue-chip stocks: $25,000
Growth or speculative stocks: $100,000

61. Let x = liters of spray X.

Let y = liters of spray Y.

Let z = liters of spray Z.

$$\left.\begin{array}{l}\text{Chemical A: } \frac{1}{5}x + \frac{1}{2}z = 12\\ \text{Chemical B: } \frac{2}{5}x + \frac{1}{2}z = 16\end{array}\right\} \Rightarrow x = 20, z = 16$$

Chemical C: $\frac{2}{5}x + y = 26 \quad \Rightarrow y = 18$

20 liters of spray X, 18 liters of spray Y, and 16 liters of spray Z are needed to get the desired mixture.

62.
$$\begin{cases} I_1 - I_2 + I_3 = 0 & \text{Equation 1}\\ 3I_1 + 2I_2 = 7 & \text{Equation 2}\\ 2I_2 + 4I_3 = 8 & \text{Equation 3}\end{cases}$$

$$\begin{cases} I_1 - I_2 + I_3 = 0 & \\ 5I_2 - 3I_3 = 7 & (-3)\text{Eq.1} + \text{Eq.2}\\ 2I_2 + 4I_3 = 8 & \end{cases}$$

$$\begin{cases} I_1 - I_2 + I_3 = 0 & \\ 10I_2 - 6I_3 = 14 & 2\text{Eq.2}\\ 10I_2 + 20I_3 = 40 & 5\text{Eq.3}\end{cases}$$

$$\begin{cases} I_1 - I_2 + I_3 = 0 & \\ 10I_2 - 6I_3 = 14 & \\ 26I_3 = 26 & (-1)\text{Eq.2} + \text{Eq.3}\end{cases}$$

$$26I_3 = 26 \Rightarrow I_3 = 1$$
$$10I_2 - 6(1) = 14 \Rightarrow I_2 = 2$$
$$I_1 - 2 + 1 = 0 \Rightarrow I_1 = 1$$

Solution: $I_1 = 1, I_2 = 2, I_3 = 1$

63.

	Product	
Truck	A	B
Large	6	3
Medium	4	4
Small	0	3

Let x = number of large trucks.

Let y = number of medium trucks.

Let z = number of small trucks.

$$\begin{cases} 6x + 4y \geq 15\\ 3x + 4y + 3z \geq 16\end{cases}$$

Possible solutions:

(1) 4 medium trucks

(2) 2 large trucks, 1 medium truck, 2 small trucks

(3) 3 large trucks, 1 medium truck, 1 small truck

(4) 3 large trucks, 3 small trucks

64. Let x = number of touchdowns.
Let y = number of extra-point kicks.
Let z = number of field goals.

$$\begin{cases} x + y + z = 17\\ 6x + y + 3z = 59\\ x - z = 0\end{cases}$$

$$\begin{cases} x + y + z = 17 & \\ -5y - 3z = -43 & -6\text{Eq.1} + \text{Eq.2}\\ -y - 2z = -17 & -\text{Eq.1} + \text{Eq.3}\end{cases}$$

$$\begin{cases} x + y + z = 17 & \\ -y - 2z = -17 & \text{Interchange Eq.2 + Eq.3}\\ -5y - 3z = -43 & \end{cases}$$

$$\begin{cases} x + y + z = 17 & \\ y + 2z = 17 & -\text{Eq.2}\\ 7z = 42 & -5\text{Eq.2} + \text{Eq.3}\end{cases}$$

$$7z = 42 \Rightarrow z = 6$$
$$y + 2(6) = 17 \Rightarrow y = 5$$
$$x + 5 + 6 = 17 \Rightarrow x = 6$$

Thus, 6 touchdowns, 5 extra-kick points, and 6 field goals were scored.

65. (a) To use as little of the 50% solution as possible, the chemist should use no 10% solution.

$$x(0.20) + (10 - x)(0.50) = 10(0.25)$$
$$x(0.20) + 5 - 0.50x = 2.5$$
$$-0.30x = -2.5$$
$$x = 8\tfrac{1}{3} \text{ liters of 20\% solution}$$
$$10 - x = 1\tfrac{2}{3} \text{ liters of 50\% solution}$$

(b) To use as much 50% solution as possible, the chemist should use no 20% solution.

$$x(0.10) + (10 - x)0.50 = 10(0.25)$$
$$0.10x + 5 - 0.50x = 2.5$$
$$-0.40x = -2.5$$
$$x = 6\tfrac{1}{4} \text{ liters of 10\% solution}$$
$$10 - x = 3\tfrac{3}{4} \text{ liters of 50\% solution}$$

—CONTINUED—

65. —CONTINUED—

(c) To use 2 liters of 50% solution we let x = the number of liters at 10% and y = the number of liters at 20%.

$0.10x + 0.20y + 2(0.50) = 10(0.25)$ Equation 1

$x + y = 8$ Equation 2

Solution: $y = 7$ liters of 20% solution; $x = 1$ liter of 10% solution

66. $\begin{cases} t_1 - 2t_2 = 0 \\ t_1 - 2a = 128 \Rightarrow 2t_2 - 2a = 128 \\ t_2 + a = 32 \Rightarrow -2t_2 - 2a = -64 \end{cases}$

$$-4a = 64$$
$$a = -16$$
$$t_2 = 48$$
$$t_1 = 96$$

So, $t_1 = 96$ pounds

$t_2 = 48$ pounds

$a = -16$ feet per second squared.

67. $\begin{cases} t_1 - 2t_2 = 0 & \text{Equation 1} \\ t_1 - 2a = 128 & \text{Equation 2} \\ t_2 + 2a = 64 & \text{Equation 3} \end{cases}$

$\begin{cases} t_1 - 2t_2 = 0 \\ 2t_2 - 2a = 128 & (-1)\text{Eq.1} + \text{Eq.2} \\ t_2 + 2a = 64 \end{cases}$

$\begin{cases} t_1 - 2t_2 = 0 \\ 2t_2 - 2a = 128 \\ 3a = 0 & (-\frac{1}{2})\text{Eq.2} + \text{Eq.3} \end{cases}$

$3a = 0 \Rightarrow a = 0$

$2t_2 - 2(0) = 128 \Rightarrow t_2 = 64$

$t_1 - 2(64) = 0 \Rightarrow t_1 = 128$

Solution: $a = 0$ ft/sec^2

$t_1 = 128$ lb

$t_2 = 64$ lb

68. (a) (30, 55), (40, 105), (50, 188)

Quadratic model: $y \approx 0.165x^2 - 6.55x + 103$

(b)

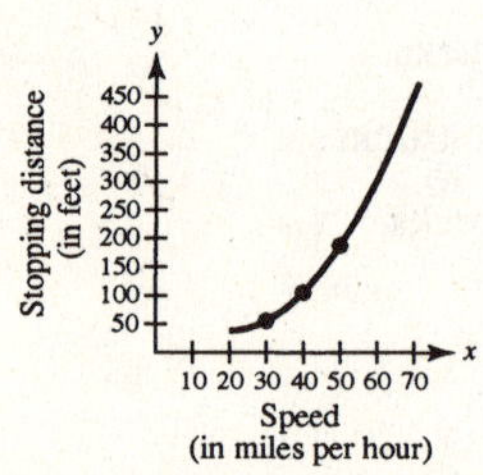

(c) When $x = 70$, $y = 453$ feet.

69. (a) (100, 75), (120, 68), (140, 55)

$\begin{cases} c + 100b + 10{,}000a = 75 & \text{Equation 1} \\ c + 120b + 14{,}400a = 68 & \text{Equation 2} \\ c + 140b + 19{,}600a = 55 & \text{Equation 3} \end{cases}$

$\begin{cases} c + 100b + 10{,}000a = 75 \\ 20b + 4400a = -7 & (-1)\text{Eq.1} + \text{Eq.2} \\ 40b + 9600a = -20 & (-1)\text{Eq.1} + \text{Eq.3} \end{cases}$

$\begin{cases} c + 100b + 10{,}000a = 75 \\ 20b + 4400a = -7 \\ 800a = -6 & (-2)\text{Eq.2} + \text{Eq.3} \end{cases}$

$\begin{cases} c + 100b + 10{,}000a = 75 \\ b + 220a = -\frac{7}{20} & (\frac{1}{20})\text{Eq.2} \\ a = -\frac{6}{800} & (\frac{1}{800})\text{Eq.2} \end{cases}$

$a = -\frac{6}{800} = -0.0075$

$b + 220(-0.0075) = -\frac{7}{20} \Rightarrow b = \frac{26}{20} = 1.3$

$c + 100(1.3) + 10{,}000(-0.0075) = 75 \Rightarrow c = 20$

So, $y = -0.0075x^2 + 1.3x + 20$

(b)

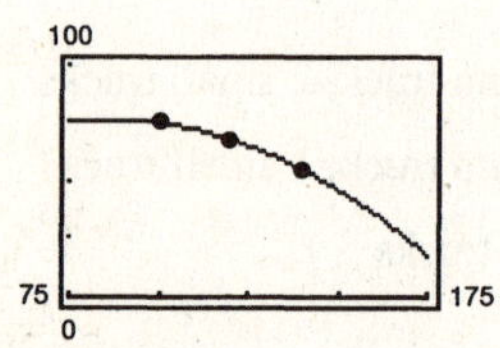

(c) For $x = 170$:

$y = -0.0075(170)^2 + 1.3(170) + 20$

$= 24.25\%$

70. (a) (15, 22.3), (40, 30), (65, 27.4)

Quadratic model: $y \approx -0.00824x^2 + 0.7612x + 12.736$

(b)

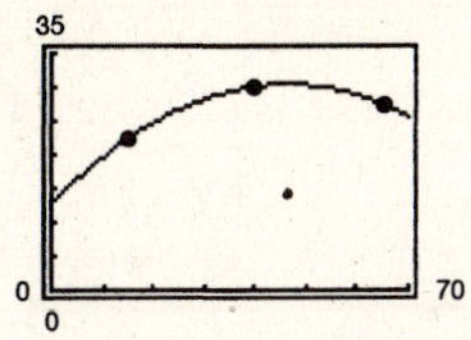

(c) $y(55) \approx 29.68$ miles per gallon

71. There are an infinite number of linear systems that have $(4, -1, 2)$ as their solution. One such system is as follows:

$$\begin{cases} 3x + y - z = 9 \\ x + 2y - z = 0 \\ -x + y + 3z = 1 \end{cases}$$

72. There are an infinite number of linear systems that have $(-5, -2, 1)$ as their solution. One such system is:

$$\begin{cases} x + y + z = -6 \\ -2x - y + 3z = 15 \\ x + 4y - z = -14 \end{cases}$$

73. There are an infinite number of linear systems that have $\left(3, -\frac{1}{2}, \frac{7}{4}\right)$ as their solution. One such system is as follows:

$$\begin{cases} x + 2y - 4z = -5 \\ -x - 4y + 8z = 13 \\ x + 6y + 4z = 7 \end{cases}$$

74. There are an infinite number of linear systems that have $\left(2, -\frac{1}{2}, -1\right)$ as their solution. One such system is:

$$\begin{cases} x + 2y + z = 0 \\ x + 2z = 0 \\ 3x + 4y - 5z = 9 \end{cases}$$

75.

$$\begin{cases} y + \lambda = 0 \\ x + \lambda = 0 \\ x + y - 10 = 0 \end{cases} \quad \left.\begin{matrix} y + \lambda = 0 \\ x + \lambda = 0 \end{matrix}\right\} \Rightarrow x = y = -\lambda$$

$$x + y - 10 = 0 \Rightarrow 2x - 10 = 0$$

$$x = 5$$

$$y = 5$$

$$\lambda = -5$$

76.

$$\begin{cases} 2x + \lambda = 0 \\ 2y + \lambda = 0 \\ x + y - 4 = 0 \end{cases} \quad \left.\begin{matrix} 2x + \lambda = 0 \\ 2y + \lambda = 0 \end{matrix}\right\} x = y = -\frac{\lambda}{2}$$

$$x + y - 4 = 0 \Rightarrow 2x - 4 = 0$$

$$2x = 4$$

$$x = 2$$

$$y = 2$$

$$\lambda = -4$$

77.

$$\begin{cases} 2x - 2x\lambda = 0 \Rightarrow 2x(1 - \lambda) = 0 \Rightarrow \lambda = 1 \text{ or } x = 0 \\ -2y + \lambda = 0 \\ y - x^2 = 0 \end{cases}$$

If $\lambda = 1$:

$$2y = \lambda \Rightarrow y = \frac{1}{2}$$

$$x^2 = y \Rightarrow x = \pm\sqrt{\frac{1}{2}} = \pm\frac{\sqrt{2}}{2}$$

If $x = 0$:

$$x^2 = y \Rightarrow y = 0$$

$$2y = \lambda \Rightarrow \lambda = 0$$

Solution: $x = \pm\frac{\sqrt{2}}{2}$ or $x = 0$

$y = \frac{1}{2}$ $\quad y = 0$

$\lambda = 1$ $\quad \lambda = 0$

78. $\begin{cases} 2 + 2y + 2\lambda = 0 \\ 2x + 1 + \lambda = 0 \Rightarrow \lambda = -2x - 1 \\ 2x + y - 100 = 0 \end{cases}$

$$2 + 2y + 2(-2x - 1) = 0 \Rightarrow -4x + 2y = 0 \Rightarrow -4x + 2y = 0$$

$$2x + y - 100 = 0 \Rightarrow 2x + y = 100 \Rightarrow \quad \underline{4x + 2y = 200}$$

$$4y = 200$$

$$y = 50$$

$$x = 25$$

$$\lambda = -2(25) - 1 = -51$$

79. False. Equation 2 does not have a leading coefficient of 1.

80. True. If a system of three linear equations is inconsistent, then it has no points in common to all three equations.

81. False. A linear system of three equations with three unknowns can have exactly one solution, infinitely many solutions, or no solution. In the case of no solution, the system would be inconsistant.

82. False. There are infinitely many solutions. Equation 2 equals three times Equation 1.

Section 14.3 Systems of Inequalities

1. $x \geq 2$

Using a solid line, graph the vertical line $x = 2$ and shade to the right of this line.

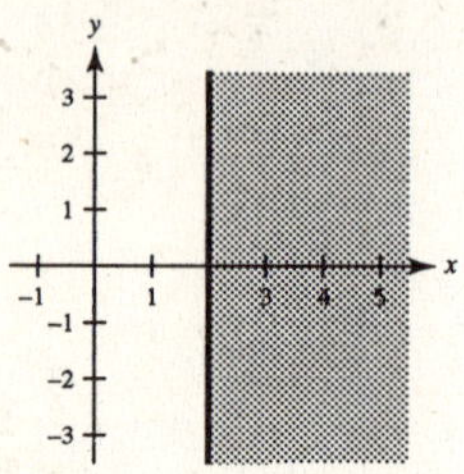

2. $x \leq 4$

Using a solid line, graph the vertical line $x = 4$, and shade to the left of this line.

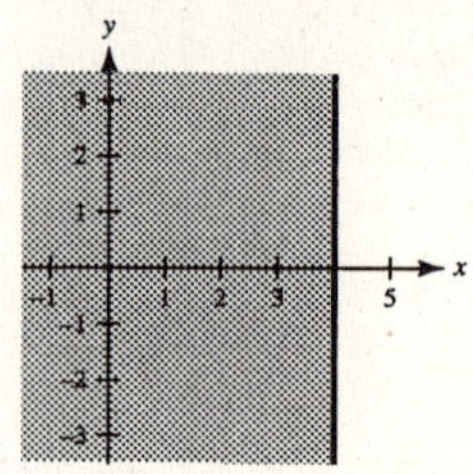

3. $y \geq -1$

Using a solid line, graph the horizontal line $y = -1$ and shade above this line.

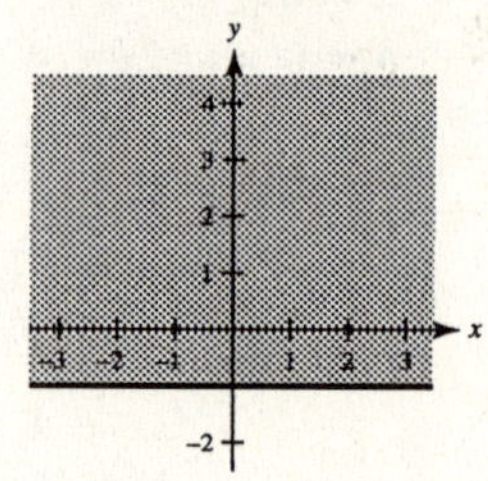

4. $y \leq 3$

Using a solid line, graph the horizontal line $y = 3$, and shade below this line.

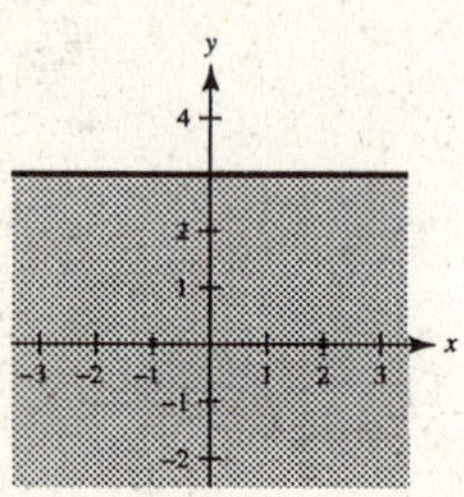

5. $y < 2 - x$

Using a dashed line, graph $y = 2 - x$, and then shade below the line.
(Use (0, 0) as a test point.)

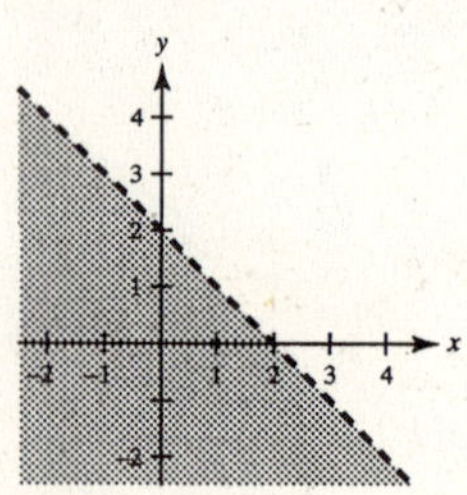

6. $y > 2x - 4$

Using a dashed line, graph $y = 2x - 4$, and shade above the line.
(Use (0, 0) as a test point.)

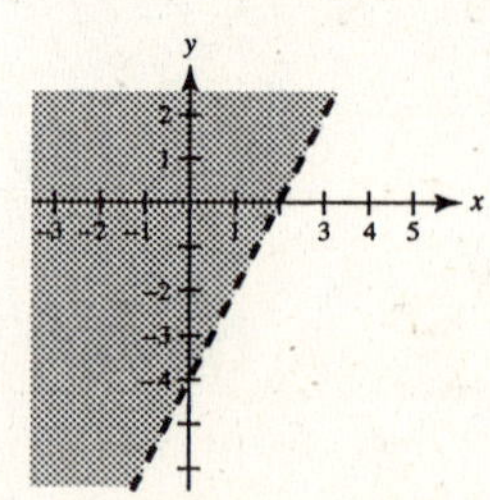

7. $2y - x \geq 4$

Using a solid line, graph $2y - x = 4$, and then shade above the line. (Use $(0, 0)$ as a test point.)

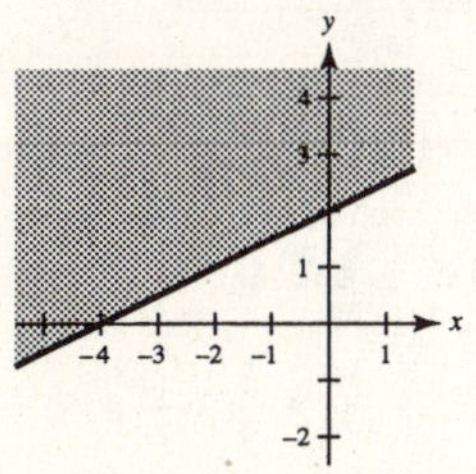

8. $5x + 3y \geq -15$

Using a solid line, graph $5x + 3y = -15$, and shade above the line. (Use $(0, 0)$ as a test point.)

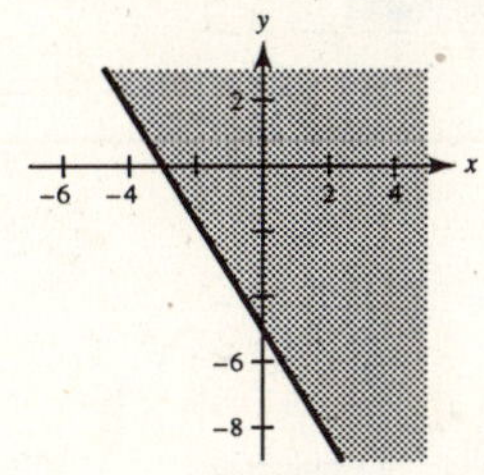

9. $(x + 1)^2 + (y - 2)^2 < 9$

Using a dashed line, sketch the circle $(x + 1)^2 + (y - 2)^2 = 9$.

Center: $(-1, 2)$

Radius: 3

Test point: $(0, 0)$. Shade the inside of the circle.

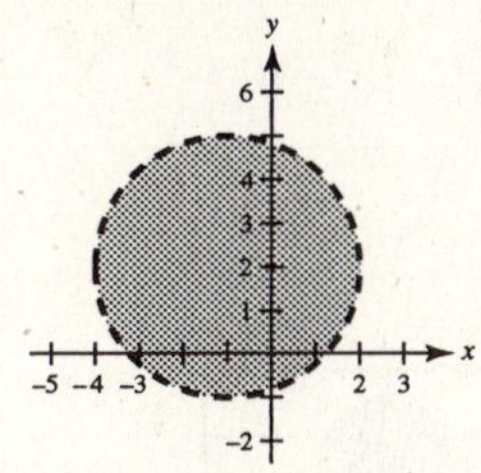

10. $y^2 - x < 0$

Using a dashed line, graph the parabola $y^2 - x = 0$, and shade the region inside this parabola. (Use $(1, 0)$ as a test point.)

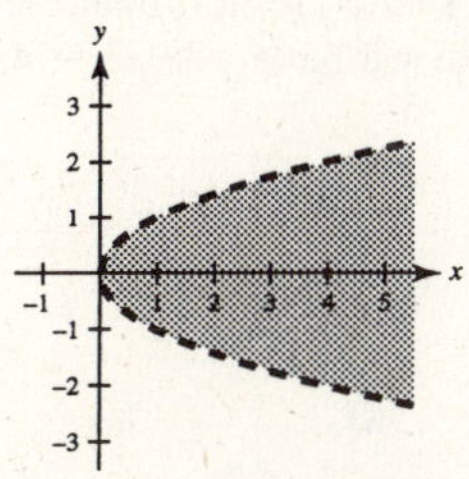

11. $y \leq \dfrac{1}{1 + x^2}$

Using a solid line, graph $y = \dfrac{1}{1 + x^2}$, and then shade below the curve. (Use $(0, 0)$ as a test point.)

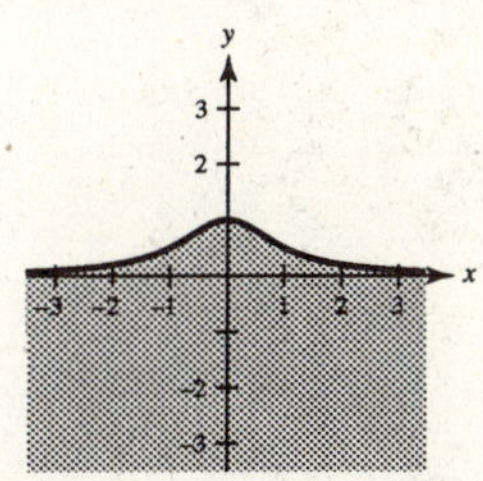

12. $y > \dfrac{-15}{x^2 + x + 4}$

Using a dashed line, graph $y = \dfrac{-15}{x^2 + x + 4}$ and then shade above the curve. (Use $(0, 0)$ as a test point.)

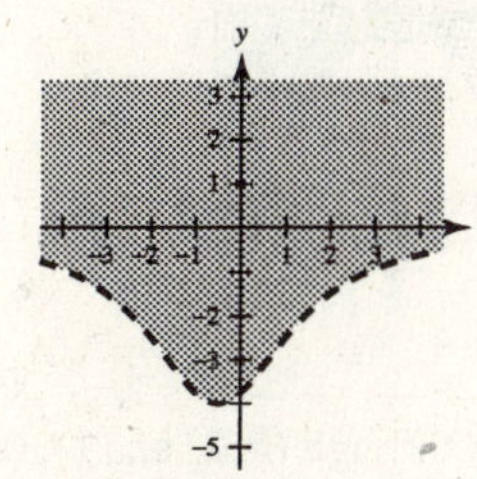

13. $y < \ln x$

Using a dashed line, sketch $y = \ln x$, and then shade below the curve. (Use $(e, 0)$ as a test point.)

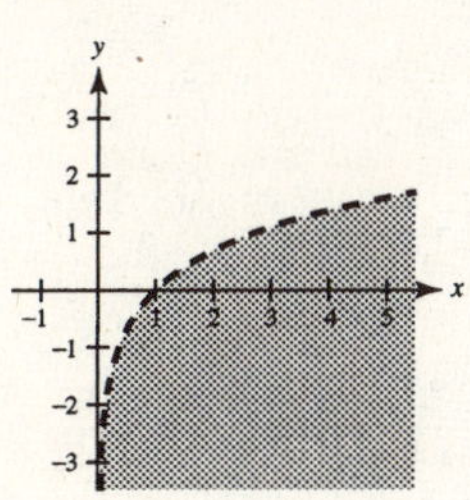

14. $y \geq 6 - \ln(x + 5)$

Using a solid line, graph $y = 6 - \ln(x + 5)$ and shade above the curve. (Use $(0, 6)$ as a test point.)

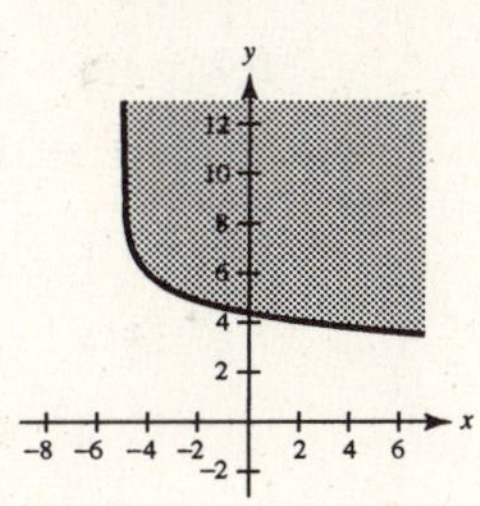

15. $y < 3^{-x-4}$

Using a dashed line, sketch $y = 3^{-x-4}$, and then shade below the curve. (Use $(0, 0)$ as a test point.)

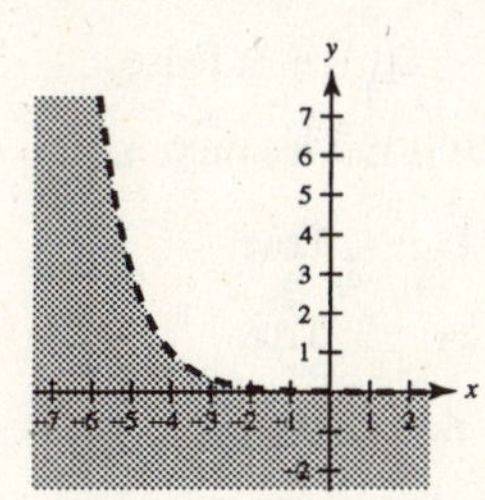

16. $y \leq 2^{2x-0.5} - 7$

Using a solid line, graph $y = 2^{2x-0.5} - 7$ and shade below the curve. (Use $(0, 0)$ as a test point.)

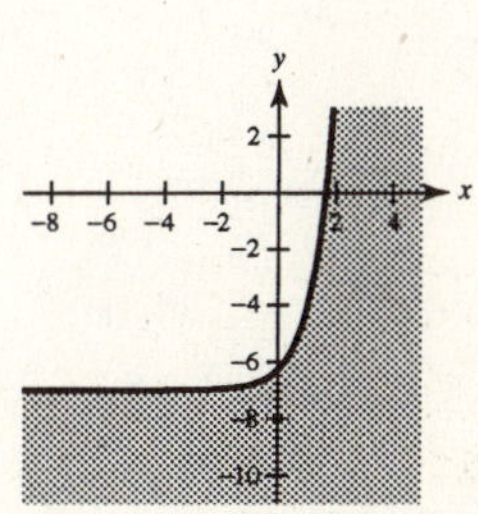

17. $y \geq \dfrac{2}{3}x - 1$

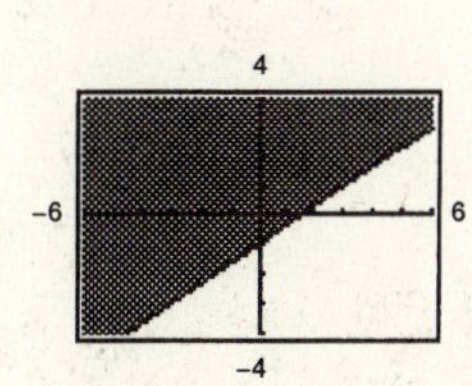

18. $y \le 6 - \frac{3}{2}x$

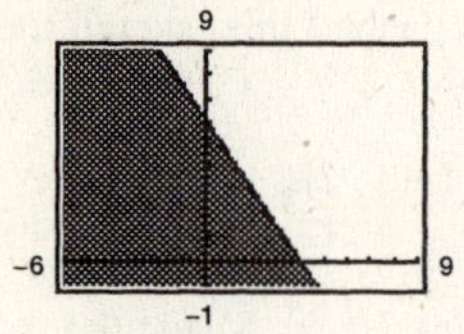

19. $y < -3.8x + 1.1$

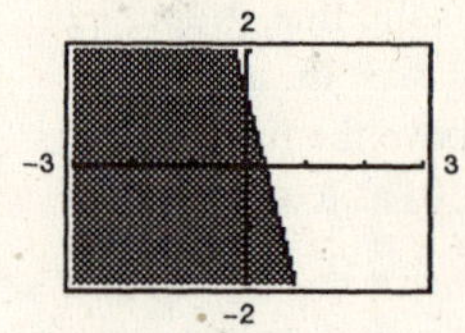

20. $y \ge -20.74 + 2.66x$

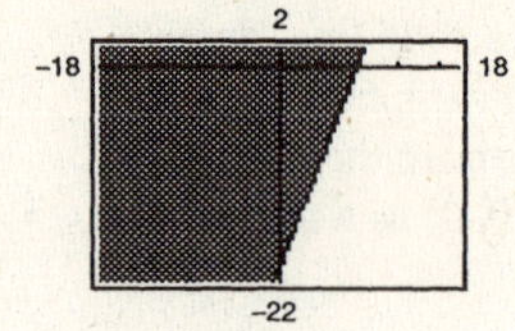

21. $x^2 + 5y - 10 \le 0$

$$y \le 2 - \frac{x^2}{5}$$

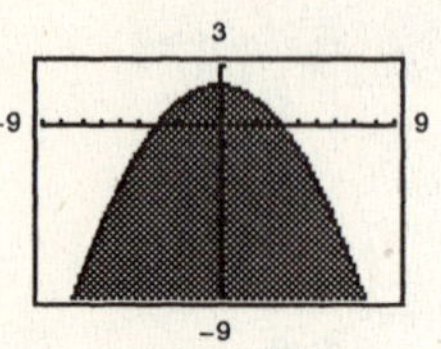

22. $2x^2 - y - 3 > 0$

$$y < 2x^2 - 3$$

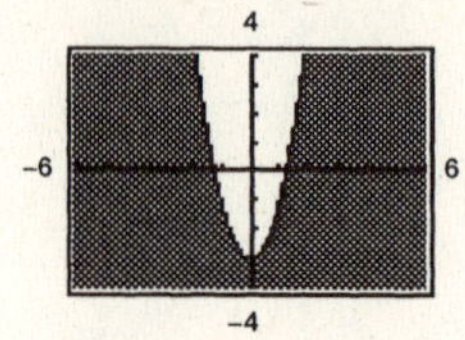

23. $\frac{5}{2}y - 3x^2 - 6 \ge 0$

$$y \ge \frac{2}{5}(3x^2 + 6)$$

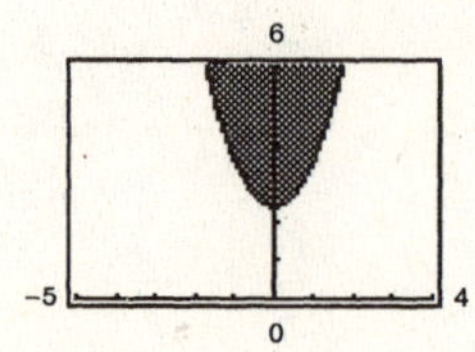

24. $-\frac{1}{10}x^2 - \frac{3}{8}y < -\frac{1}{4}$

$$y > \frac{2}{3} - \frac{4}{15}x^2$$

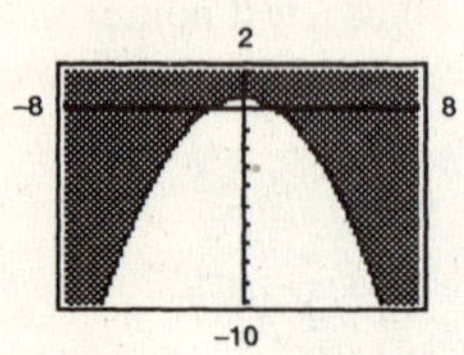

25. The line through $(-4, 0)$ and $(0, 2)$ is $y = \frac{1}{2}x + 2$. For the shaded region below the line, we have $y \le \frac{1}{2}x + 2$.

26. The parabola through $(-2, 0)$, $(0, -4)$, $(2, 0)$ is $y = x^2 - 4$. For the shaded region inside the parabola, we have $y \ge x^2 - 4$.

27. The line through $(0, 2)$ and $(3, 0)$ is $y = -\frac{2}{3}x + 2$. For the shaded region above the line, we have $y \ge -\frac{2}{3}x + 2$.

28. The circle shown is $x^2 + y^2 = 9$. For the shaded region inside the circle, we have $x^2 + y^2 \le 9$.

29. $\begin{cases} x \ge -4 \\ y > -3 \\ y \le -8x - 3 \end{cases}$

(a) $0 \le -8(0) - 3$, False

$(0, 0)$ is **not** a solution.

(b) $-3 > -3$, False

$(-1, -3)$ is **not** a solution.

(c) $-4 \ge -4$, True

$0 > -3$, True

$0 \le -8(-4) - 3$, True

$(-4, 0)$ **is** a solution.

(d) $-3 \ge -4$, True

$11 > -3$, True

$11 \le -8(-3) - 3$, True

$(-3, 11)$ **is** a solution.

30. $\begin{cases} -2x + 5y \geq 3 \\ y < 4 \\ -4x + 2y < 7 \end{cases}$

(a) $-2(0) + 5(2) \geq 3$, True

$2 < 4$, True

$-4(0) + 2(2) < 7$, True

$(0, 2)$ **is** a solution.

(b) $-2(-6) + 5(4) \geq 3$, True

$4 < 4$, False

$(-6, 4)$ is **not** a solution.

(c) $-2(-8) + 5(-2) \geq 3$, True

$-2 < 4$, True

$-4(-8) + 2(-2) < 7$, False

$(-8, -2)$ is **not** a solution.

(d) $-2(-3) + 5(2) \geq 3$, True

$2 < 4$, True

$-4(-3) + 2(2) < 7$, False

$(-3, 2)$ is **not** a solution.

31. $\begin{cases} 3x + y > 1 \\ -y - \frac{1}{2}x^2 \leq -4 \\ -15x + 4y > 0 \end{cases}$

(a) $3(0) + (10) > 1$, True

$-10 - \frac{1}{2}(0)^2 \leq -4$, True

$-15(0) + 4(10) > 0$, True

$(0, 10)$ **is** a solution.

(b) $3(0) + (-1) > 1$, False $\Rightarrow$ $(0, -1)$ is **not** a solution.

(c) $3(2) + (9) > 1$, True

$-9 - \frac{1}{2}(2)^2 \leq -4$, True

$-15(2) + 4(9) > 0$, True

$(2, 9)$ **is** a solution.

(d) $3(-1) + 6 > 1$, True

$-6 - \frac{1}{2}(-1)^2 \leq -4$, True

$-15(-1) + 4(6) > 0$, True

$(-1, 6)$ **is** a solution.

32. $\begin{cases} x^2 + y^2 \geq 36 \\ -3x + y \leq 10 \\ \frac{2}{3}x - y \geq 5 \end{cases}$

(a) $(-1)^2 + 7^2 \geq 36$, True

$-3(-1) + 7 \leq 10$, True

$\frac{2}{3}(-1) - 7 \geq 5$, False

$(-1, 7)$ is **not** a solution.

(b) $(-5)^2 + 1^2 \geq 36$, False

$(-5, 1)$ is **not** a solution.

(c) $6^2 + 0^2 \geq 36$, True

$-3(6) + 0 \leq 0$, True

$\frac{2}{3}(6) - 0 \geq 5$, False

$(6, 0)$ is **not** a solution.

(d) $4^2 + (-8)^2 \geq 36$, True

$-3(4) - 8 \leq 10$, True

$\frac{2}{3}(4) - (-8) \geq 5$, True

$(4, -8)$ **is** a solution.

33. $\begin{cases} x + y \leq 1 \\ -x + y \leq 1 \\ y \geq 0 \end{cases}$

First, find the points of intersection of each pair of equations.

Vertex A	Vertex B	Vertex C
$x + y = 1$	$x + y = 1$	$-x + y = 1$
$-x + y = 1$	$y = 0$	$y = 0$
$(0, 1)$	$(1, 0)$	$(-1, 0)$

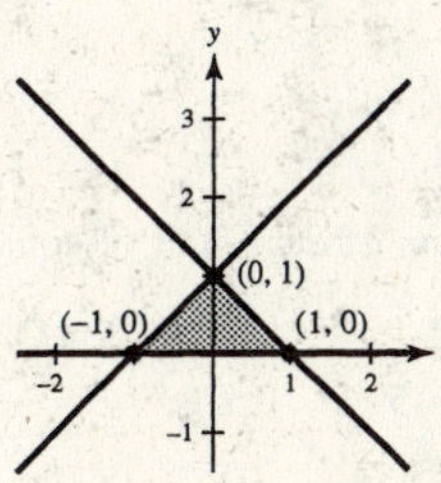

34. $\begin{cases} 3x + 2y < 6 \\ x \quad\quad > 0 \\ \quad\quad y > 0 \end{cases}$

First, find the points of intersection of each pair of equations.

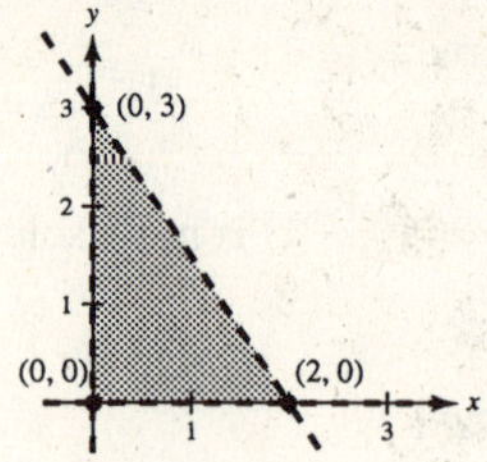

Vertex A	Vertex B	Vertex C
$3x + 2y = 6$	$x = 0$	$3x + 2y = 6$
$x = 0$	$y = 0$	$y = 0$
$(0, 3)$	$(0, 0)$	$(2, 0)$

35. $\begin{cases} x^2 + y \le \;\;5 \\ x \quad\quad \ge -1 \\ \quad\quad y \ge \;\;0 \end{cases}$

First, find the points of intersection of each pair of equations.

Vertex A	Vertex B	Vertex C
$x^2 + y = 5$	$x^2 + y = 5$	$x = -1$
$x = -1$	$y = 0$	$y = 0$
$(-1, 4)$	$(\pm\sqrt{5}, 0)$	$(-1, 0)$

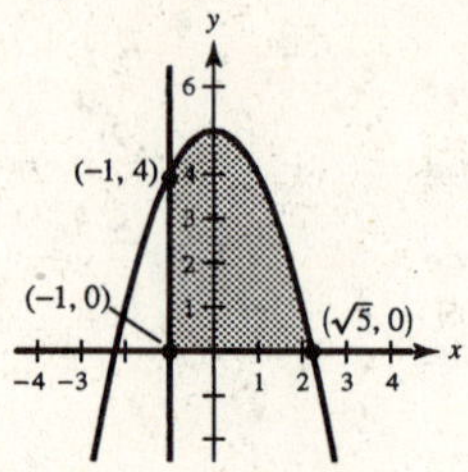

36. $\begin{cases} 2x^2 + y \ge 2 \\ x \quad\quad \le 2 \\ \quad\quad y \le 1 \end{cases}$

First, find the points of intersection of each pair of equations.

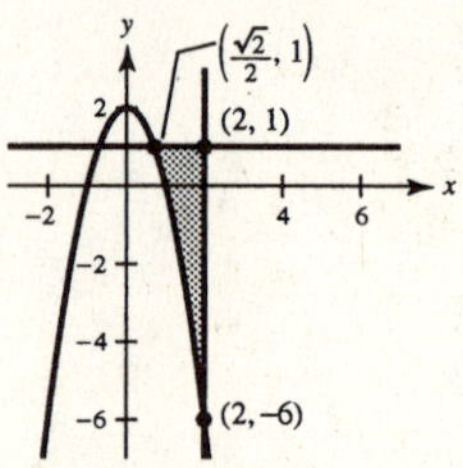

Vertex A	Vertex B	Vertex C
$2x^2 + y = 2$	$x = 2$	$2x^2 + y = 2$
$x = 2$	$y = 1$	$y = 1$
$(2, -6)$	$(2, 1)$	$\left(\frac{\sqrt{2}}{2}, 1\right)$

37. $\begin{cases} -3x + 2y < \;\;6 \\ \;\;x - 4y > -2 \\ \;\;2x + \;y < \;\;3 \end{cases}$

First, find the points of intersection of each pair of equations.

Vertex A	Point B	Vertex C
$-3x + 2y = 6$	$-3x + 2y = 6$	$x - 4y = -2$
$x - 4y = -2$	$2x + y = 3$	$2x + y = 3$
$(-2, 0)$	$(0, 3)$	$\left(\frac{10}{9}, \frac{7}{9}\right)$

Note that B is not a vertex of the solution region.

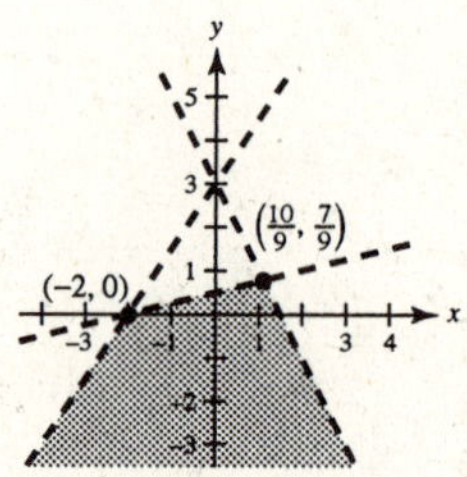

38. $\begin{cases} \;\;x - 7y > -36 \\ 5x + 2y > \;\;\;5 \\ 6x - 5y > \;\;\;6 \end{cases}$

First, find the points of intersection of each pair of equations.

Vertex A	Vertex B	Vertex C
$x - 7y = -36$	$5x + 2y = 5$	$x - 7y = -36$
$5x + 2y = 5$	$6x - 5y = 6$	$6x - 5y = 6$
$(-1, 5)$	$(1, 0)$	$(6, 6)$

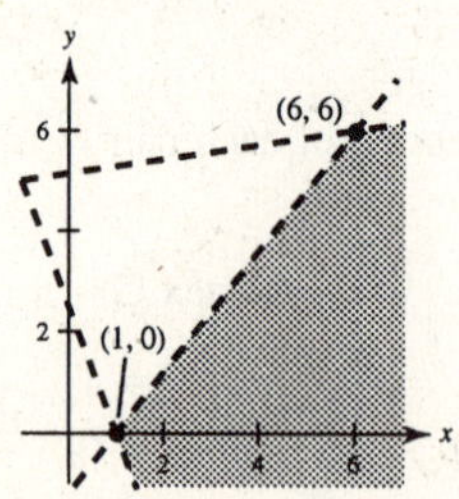

39. $\begin{cases} 2x + y > 2 \\ 6x + 3y < 2 \end{cases}$

The lines are parallel. There are no points of intersection. There is no region common to both inequalities.

The system has no solution.

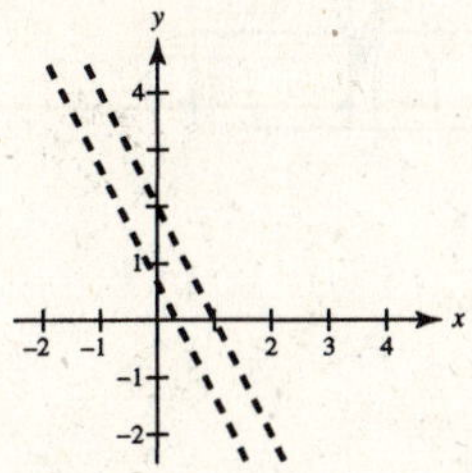

40. $\begin{cases} x - 2y < -6 \\ 5x - 3y > -9 \end{cases}$

Point of intersection: $(0, 3)$

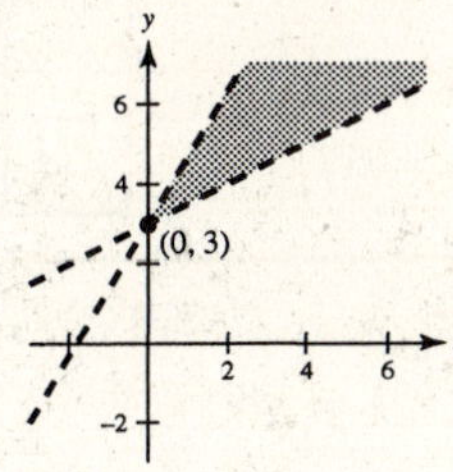

41. $\begin{cases} x > y^2 \\ x < y + 2 \end{cases}$

Points of intersection:

$$y^2 = y + 2$$
$$y^2 - y - 2 = 0$$
$$(y + 1)(y - 2) = 0$$
$$y = -1, 2$$

$(1, -1), (4, 2)$

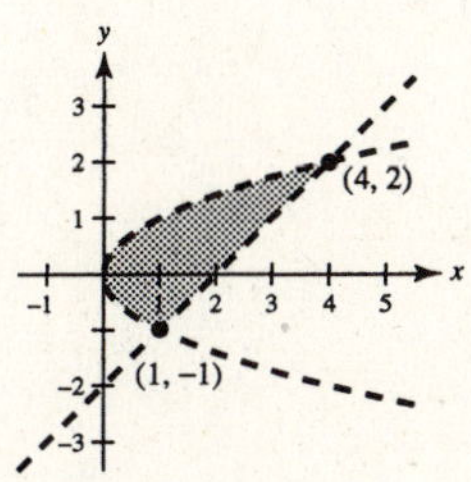

42. $\begin{cases} x - y^2 > 0 \\ x - y > 2 \end{cases}$

Points of intersection:

$$y^2 = y + 2$$
$$y^2 - y - 2 = 0$$
$$(y + 1)(y - 2) = 0$$
$$y = -1, 2$$

$(1, -1), (4, 2)$

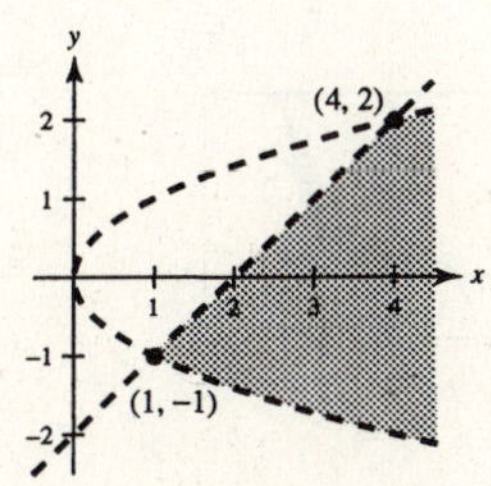

43. $\begin{cases} x^2 + y^2 \le 9 \\ x^2 + y^2 \ge 1 \end{cases}$

There are no points of intersection. The region common to both inequalities is the region between the circles.

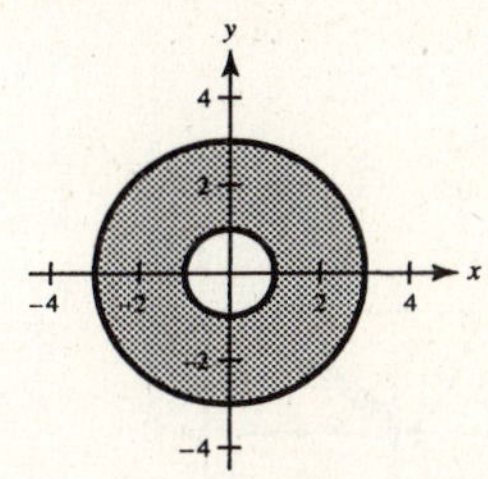

44. $\begin{cases} x^2 + y^2 \le 25 \\ 4x - 3y \le 0 \end{cases}$

Points of intersection:

$$x^2 + \left(\tfrac{4}{3}x\right)^2 = 25$$
$$\tfrac{25}{9}x^2 = 25$$
$$x = \pm 3$$

$(-3, -4), (3, 4)$

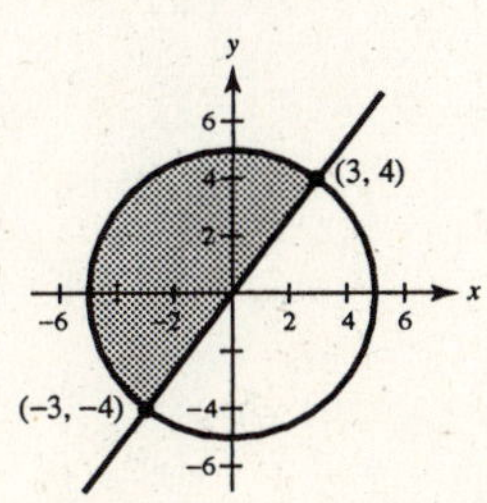

45. $3x + 4 \ge y^2$

$x - y < 0$

Points of intersection:

$$x - y = 0 \Rightarrow y = x$$
$$3y + 4 = y^2$$
$$0 = y^2 - 3y - 4$$
$$0 = (y - 4)(y + 1)$$
$$y = 4 \text{ or } y = -1$$
$$x = 4 \quad x = -1$$

$(4, 4)$ and $(-1, -1)$

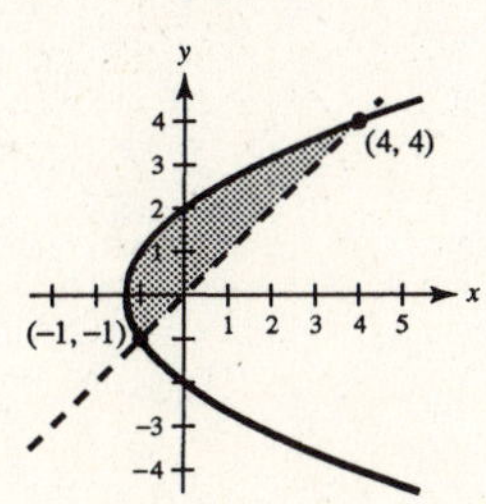

46. $\begin{cases} x < 2y - y^2 \\ 0 < x + y \end{cases}$

Points of intersection:

$$-y = 2y - y^2$$
$$y^2 - 3y = 0$$
$$y(y - 3) = 0$$
$$y = 0, 3$$

$(0, 0), (-3, 3)$

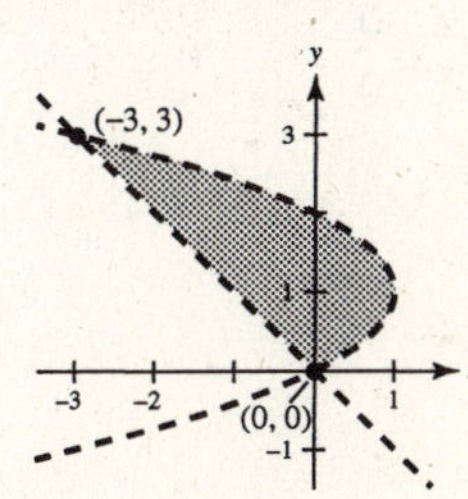

47. $\begin{cases} y \le \sqrt{3x} + 1 \\ y \ge x^2 + 1 \end{cases}$

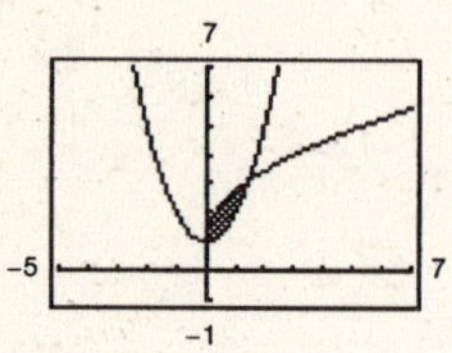

48. $\begin{cases} y < -x^2 + 2x + 3 \\ y > x^2 - 4x + 3 \end{cases}$

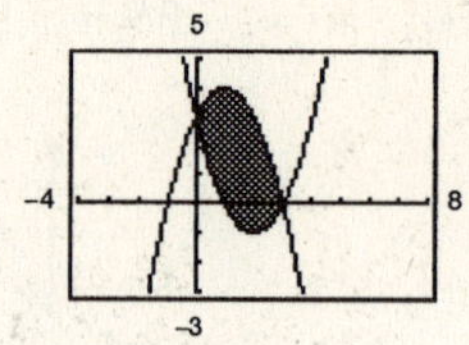

49. $\begin{cases} y < x^3 - 2x + 1 \\ y > -2x \\ x \le 1 \end{cases}$

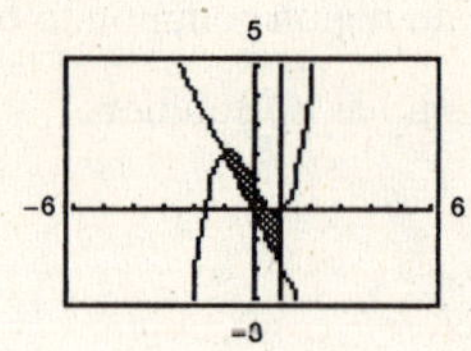

50. $\begin{cases} y \ge x^4 - 2x^2 + 1 \\ y \le 1 - x^2 \end{cases}$

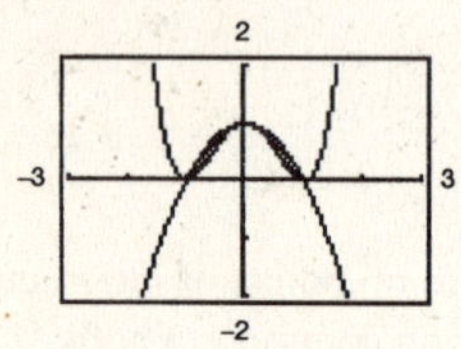

51. $\begin{cases} x^2y \ge 1 \Rightarrow y \ge \dfrac{1}{x^2} \\ 0 < x \le 4 \\ y \le 4 \end{cases}$

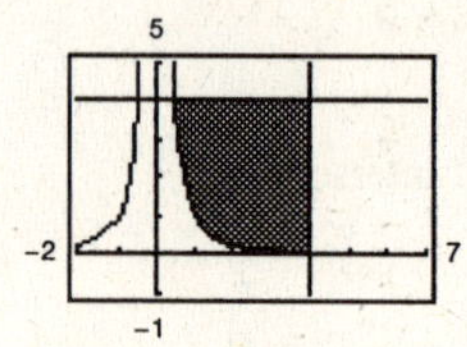

52. $\begin{cases} y \le e^{-x^2/2} \\ y \ge 0 \\ -2 \le x \le 2 \end{cases}$

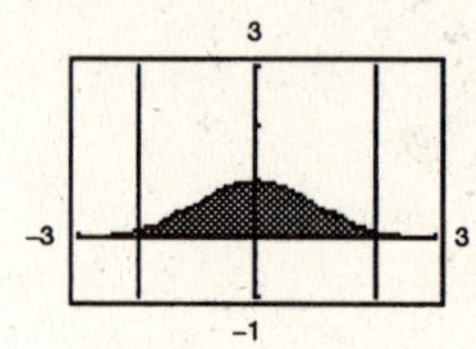

53. $\begin{cases} y \le \sqrt{x} \\ y \ge x^2 \end{cases}$

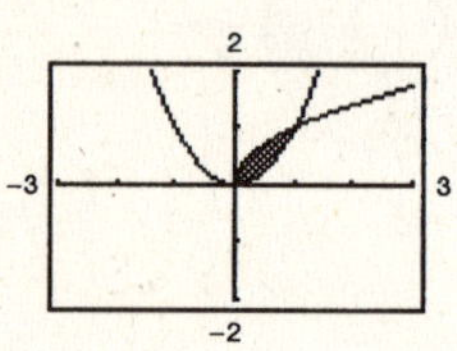

54. $\begin{cases} y \ge x^3 - 3x \\ y \le x^2 + 2x + 3 \\ x \ge -1 \end{cases}$

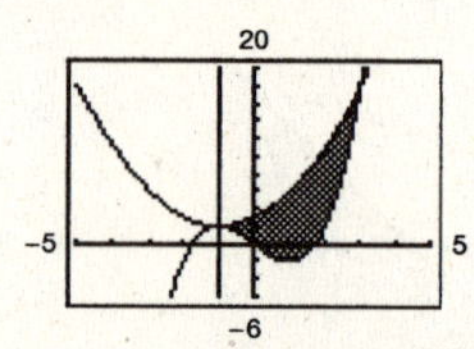

55. $\begin{cases} y \le 4 - x \\ x \ge 0 \\ y \ge 0 \end{cases}$

56. $(0, 6), (3, 0), (0, -3)$

$\begin{cases} y < 6 - 2x \\ y \ge x - 3 \\ x \ge 1 \end{cases}$

57. Line through points $(0, 4)$ and $(4, 0)$: $y = 4 - x$

Line through points $(0, 2)$ and $(8, 0)$: $y = 2 - \frac{1}{4}x$

$\begin{cases} y \ge 4 - x \\ y \ge 2 - \frac{1}{4}x \\ x \ge 0 \\ y \ge 0 \end{cases}$

58. Circle: $x^2 + y^2 > 4$

59. $\begin{cases} x^2 + y^2 \le 16 \\ x \ge 0 \\ y \ge 0 \end{cases}$

60. $(0, 0), (0, 4), (\sqrt{8}, \sqrt{8})$

$\begin{cases} x^2 + y^2 \le 16 \\ x \le y \\ x \ge 0 \end{cases}$

61. Rectangular region with vertices at $(2, 1)$, $(5, 1)$, $(5, 7)$, and $(2, 7)$

$$\begin{cases} x \geq 2 \\ x \leq 5 \\ y \geq 1 \\ y \leq 7 \end{cases}$$

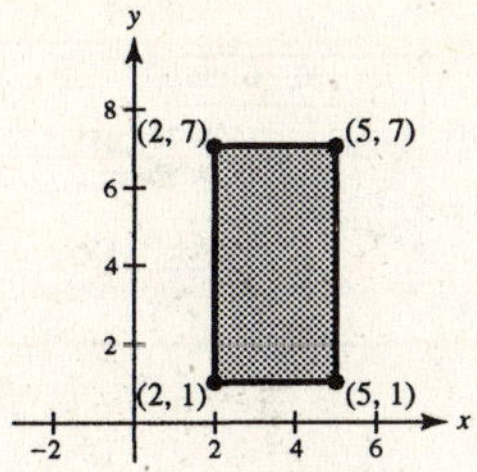

This system may be written as:

$$\begin{cases} 2 \leq x \leq 5 \\ 1 \leq y \leq 7 \end{cases}$$

62. Parallelogram with vertices at $(0, 0)$, $(4, 0)$, $(1, 4)$, $(5, 4)$

$(0, 0), (4, 0)$: $y \geq 0$

$(4, 0), (5, 4)$: $4x - y \leq 16$

$(1, 4), (5, 4)$: $y \leq 4$

$(0, 0), (1, 4)$: $4x - y \geq 0$

$$\begin{cases} 4x - y \geq 0 \\ 4x - y \leq 16 \\ 0 \leq y \leq 4 \end{cases}$$

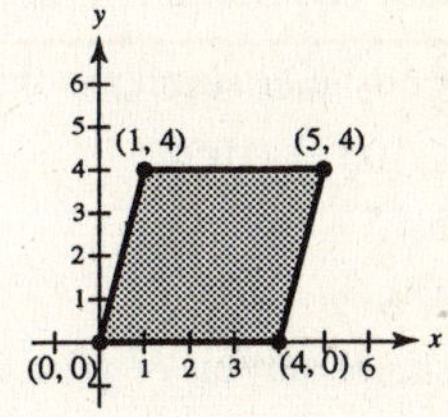

63. Triangle with vertices at $(0, 0)$, $(5, 0)$, $(2, 3)$

$(0, 0), (5, 0)$ Line: $y = 0$

$(0, 0), (2, 3)$ Line: $y = \frac{3}{2}x$

$(2, 3), (5, 0)$ Line: $y = -x + 5$

$$\begin{cases} y \leq \frac{3}{2}x \\ y \leq -x + 5 \\ y \geq 0 \end{cases}$$

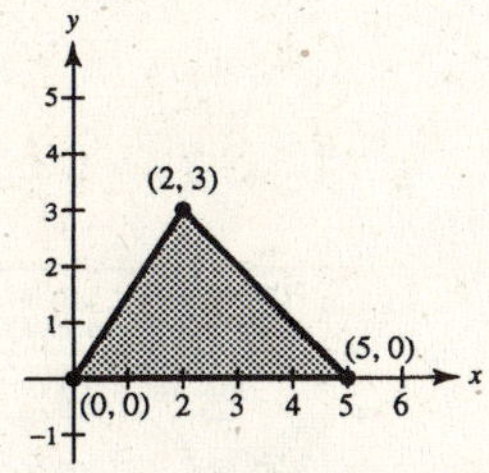

64. Triangle with vertices at $(-1, 0)$, $(1, 0)$, $(0, 1)$

$(-1, 0), (1, 0)$: $y \geq 0$

$(-1, 0), (0, 1)$: $y \leq x + 1$

$(0, 1), (1, 0)$: $y \leq -x + 1$

$$\begin{cases} y \leq x + 1 \\ y \leq -x + 1 \\ y \geq 0 \end{cases}$$

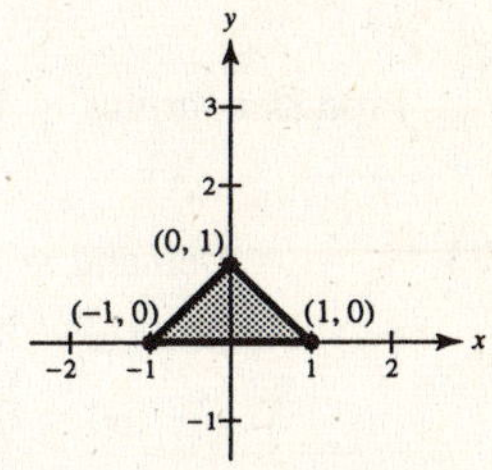

65. $\begin{cases} x^2 + y^2 \leq 16 \Rightarrow \text{region inside the circle} \\ x + y \geq 4 \Rightarrow \text{region above the line} \end{cases}$

Matches graph (d).

66. $x^2 + y^2 \leq 16 \Rightarrow$ region inside the circle
$x + y \leq 4 \Rightarrow$ region below the line

Matches graph (b).

67. $\begin{cases} x^2 + y^2 \geq 16 \Rightarrow \text{region outside the circle} \\ x + y \geq 4 \Rightarrow \text{region above the line} \end{cases}$

Matches graph (c).

68. $x^2 + y^2 \geq 16 \Rightarrow$ region outside the circle
$x + y \leq 4 \Rightarrow$ region below the line

Matches graph (a).

69. (a) The solution of $x + 2y \leq 6$ would include the points on the line $x + 2y = 6$. The line would be drawn solid in the graph and shading remains the same.

(b) Shading would be above the dashed line in the graph of $x + 2y > 6$.

70. Test a point on either side of the boundary.

71. Demand = Supply

$$50 - 0.5x = 0.125x$$
$$50 = 0.625x$$
$$80 = x$$
$$10 = p$$

Point of equilibrium: $(80, 10)$

The consumer surplus is the area of the triangular region defined by

$$\begin{cases} p \leq 50 - 0.5x \\ p \geq 10 \\ x \geq 0. \end{cases}$$

Consumer surplus $= \frac{1}{2}(\text{base})(\text{height}) = \frac{1}{2}(80)(40) = \$1600$

The producer surplus is the area of the triangular region defined by

$$\begin{cases} p \geq 0.125x \\ p \leq 10 \\ x \geq 0. \end{cases}$$

Producer surplus $= \frac{1}{2}(\text{base})(\text{height}) = \frac{1}{2}(80)(10) = \$400$

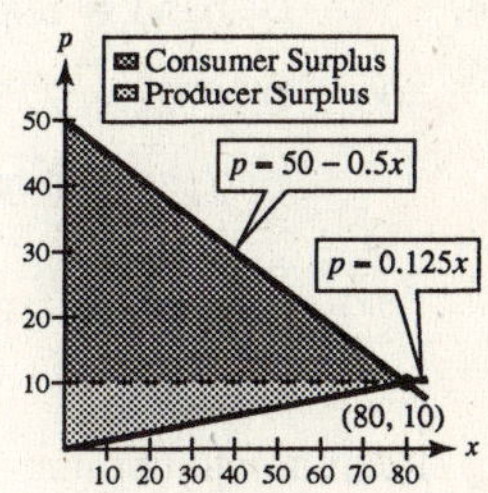

72. Demand = Supply

$$100 - 0.05x = 25 + 0.1x$$
$$75 = 0.15x$$
$$500 = x$$
$$75 = p$$

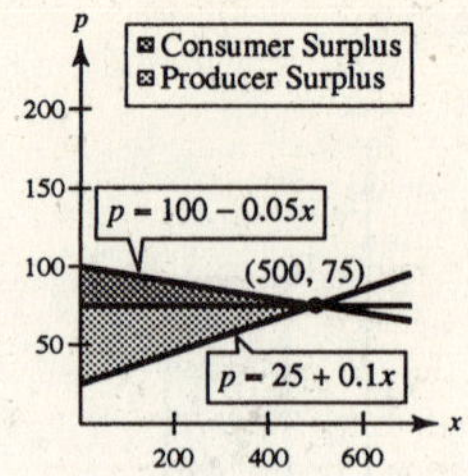

Point of equilibrium: (500, 75)

The consumer surplus is the area of the triangular region defined by

$$\begin{cases} p \leq 100 - 0.05x \\ p \geq 75 \\ x \geq 0. \end{cases}$$

Consumer surplus $= \frac{1}{2}(\text{base})(\text{height}) = \frac{1}{2}(500)(25) = 6250$

The producer surplus is the area of the triangular region defined by

$$\begin{cases} p \geq 25 + 0.1x \\ y \leq 75 \\ x \geq 0. \end{cases}$$

Producer surplus $= \frac{1}{2}(\text{base})(\text{height}) = \frac{1}{2}(500)(50) = 12{,}500$

73. Demand = Supply

$$140 - 0.00002x = 80 + 0.00001x$$
$$60 = 0.00003x$$
$$2{,}000{,}000 = x$$
$$100 = p$$

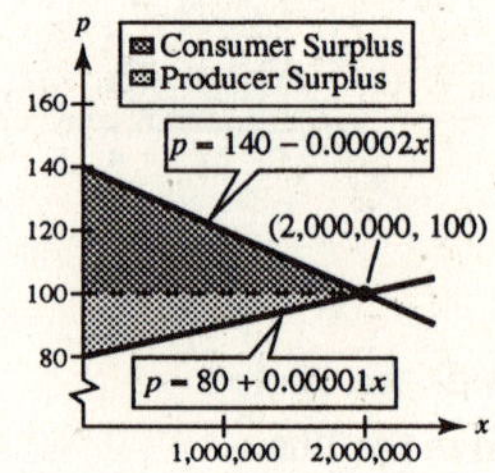

Point of equilibrium: (2,000,000, 100)

The consumer surplus is the area of the triangular region defined by

$$\begin{cases} p \leq 140 - 0.00002x \\ p \geq 100 \\ x \geq 0. \end{cases}$$

Consumer surplus $= \frac{1}{2}(\text{base})(\text{height}) = \frac{1}{2}(2{,}000{,}000)(40) = \$40{,}000{,}000$ or \$40 million

The producer surplus is the area of the triangular region defined by

$$\begin{cases} p \geq 80 + 0.00001x \\ p \leq 100 \\ x \geq 0. \end{cases}$$

Producer surplus $= \frac{1}{2}(\text{base})(\text{height}) = \frac{1}{2}(2{,}000{,}000)(20) = \$20{,}000{,}000$ or \$20 million

74. Demand = Supply

$$400 - 0.0002x = 225 + 0.0005x$$
$$175 = 0.0007x$$
$$250{,}000 = x$$
$$350 = p$$

Point of equilibrium: (250,000, 350)

The consumer surplus is the area of the triangular region defined by

$$\begin{cases} p \leq 400 - 0.0002x \\ p \geq 350 \\ x \geq 0. \end{cases}$$

Consumer surplus
$= \frac{1}{2}(\text{base})(\text{height}) = \frac{1}{2}(250{,}000)(50) = 6{,}250{,}000$

The producer surplus is the area of the triangular region defined by

$$\begin{cases} p \geq 225 + 0.0005x \\ p \leq 350 \\ x \geq 0. \end{cases}$$

Producer surplus $= \frac{1}{2}(\text{base})(\text{height}) = \frac{1}{2}(250{,}000)(125) = 15{,}625{,}000$

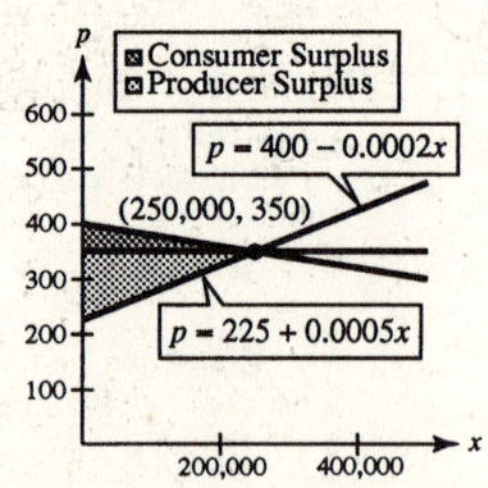

75. x = number of tables

y = number of chairs

$$\begin{cases} x + \frac{3}{2}y \le 12 & \text{Assembly center} \\ \frac{4}{3}x + \frac{3}{2}y \le 15 & \text{Finishing center} \\ x \ge 0 \\ y \ge 0 \end{cases}$$

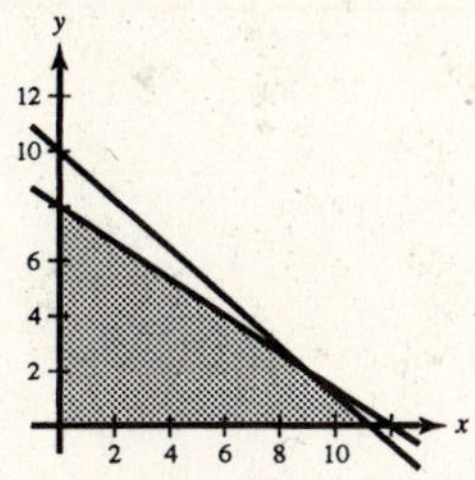

76. x = number of model A

y = number of model B

$$\begin{cases} x \ge 2y \\ 8x + 12y \le 200 \\ x \ge 4 \\ y \ge 2 \end{cases}$$

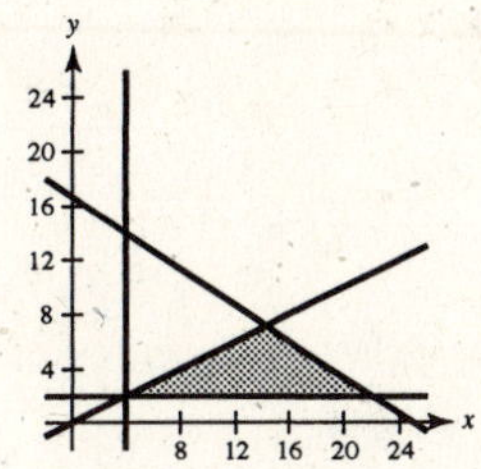

77. x = amount in smaller account

y = amount in larger account

Account constraints:

$$\begin{cases} x + y \le 20{,}000 \\ y \ge 2x \\ x \ge 5{,}000 \\ y \ge 5{,}000 \end{cases}$$

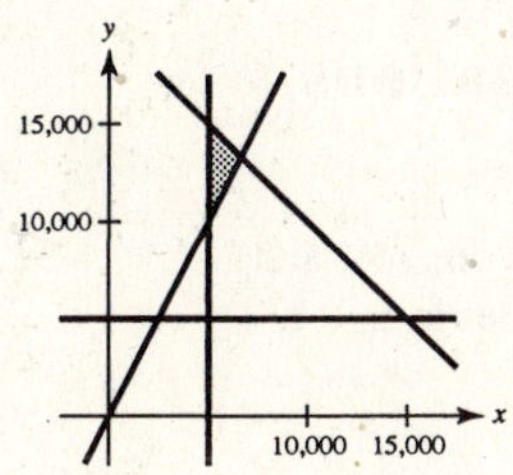

78. x = number of \$30 tickets

y = number of \$20 tickets

$$\begin{cases} x + y \le 3000 \\ 30x + 20y \ge 75{,}000 \\ x \le 2000 \\ x \ge 0 \\ y \ge 0 \end{cases}$$

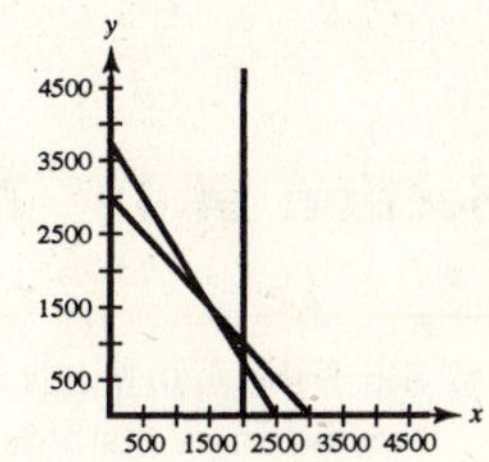

79. x = number of packages of gravel

y = number of bags of stone

$$\begin{cases} 55x + 70y \le 7500 & \text{Weight} \\ x \ge 50 \\ y \ge 40 \end{cases}$$

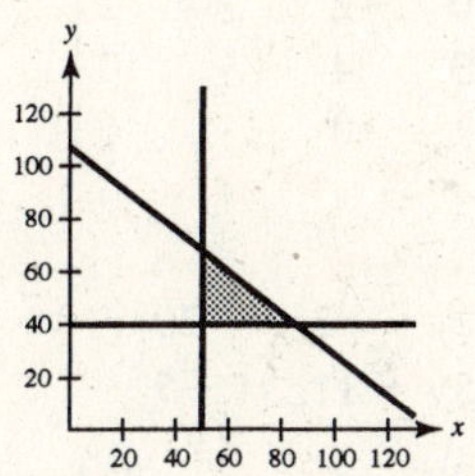

80. x = number of ounces of food X

y = number of ounces of food Y

$$\begin{cases} 20x + 10y \ge 300 & \text{(calcium)} \\ 15x + 10y \ge 150 & \text{(iron)} \\ 10x + 20y \ge 200 & \text{(vitamin B)} \\ x \ge 0 \\ y \ge 0 \end{cases}$$

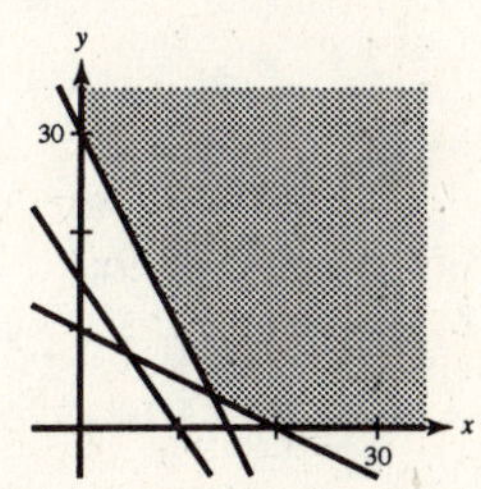

81.
$$\begin{cases} xy \ge 500 & \text{Body = building space} \\ 2x + \pi y \ge 125 & \text{Track (Two semi–circles and two lengths)} \\ x \ge 0 \\ y \ge 0 \end{cases}$$

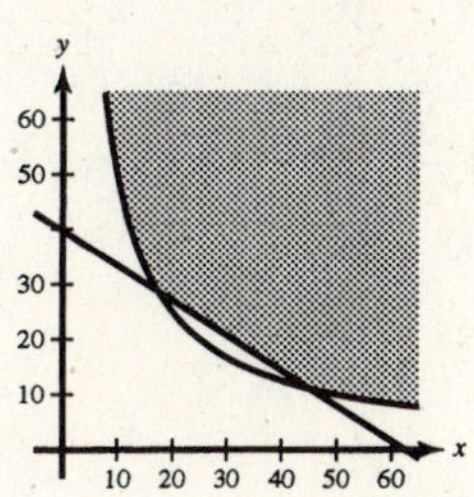

82. The graph is a half-line on the real number line; on the rectangular coordinate system, the graph is a half-plane.

83. True. The figure is a rectangle with length of 9 units and width of 11 units.

84. False. The graph shows the solution of the system

$$\begin{cases} y < 6 \\ -4x - 9y < 6 \\ 3x + y^2 \geq 2. \end{cases}$$

85. False.

The graph shows the solution to the following system:

$$\begin{cases} y \geq x^2 \\ x \geq 0 \\ y \leq 2 \end{cases}$$

86. True.

The graph of the system $\begin{cases} x^2 + y^2 \leq 4 \\ x \geq 0 \\ y \geq 0 \end{cases}$ is a quarter circle of radius 2. The area is $A = \frac{1}{4}\pi(2)^2 = \pi$ square miles.

Section 14.4 Matrices and Systems of Equations

1. Since the matrix has one row and two columns, its order is 1×2.

2. Since the matrix has one row and four columns, its order is 1×4.

3. Since the matrix has three rows and one column, its order is 3×1.

4. Since the matrix has three rows and four columns, its order is 3×4.

5. Since the matrix has two rows and two columns, its order is 2×2.

6. Since the matrix has two rows and three columns, its order is 2×3.

7. $\begin{cases} 4x - 3y = -5 \\ -x + 3y = 12 \end{cases}$

$$\begin{bmatrix} 4 & -3 & \vdots & -5 \\ -1 & 3 & \vdots & 12 \end{bmatrix}$$

8. $\begin{cases} 7x + 4y = 22 \\ 5x - 9y = 15 \end{cases}$

$$\begin{bmatrix} 7 & 4 & \vdots & 22 \\ 5 & -9 & \vdots & 15 \end{bmatrix}$$

9. $\begin{cases} x + 10y - 2z = 2 \\ 5x - 3y + 4z = 0 \\ 2x + y \quad = 6 \end{cases}$

$$\begin{bmatrix} 1 & 10 & -2 & \vdots & 2 \\ 5 & -3 & 4 & \vdots & 0 \\ 2 & 1 & 0 & \vdots & 6 \end{bmatrix}$$

10. $\begin{cases} -x - 8y + 5z = 8 \\ -7x \quad - 15z = -38 \\ 3x - y + 8z = 20 \end{cases}$

$$\begin{bmatrix} -1 & -8 & 5 & \vdots & 8 \\ -7 & 0 & -15 & \vdots & -38 \\ 3 & -1 & 8 & \vdots & 20 \end{bmatrix}$$

11. $\begin{cases} 7x - 5y + z = 13 \\ 19x \quad - 8z = 10 \end{cases}$

$$\begin{bmatrix} 7 & -5 & 1 & \vdots & 13 \\ 19 & 0 & -8 & \vdots & 10 \end{bmatrix}$$

12. $\begin{cases} 9x + 2y - 3z = 20 \\ -25y + 11z = -5 \end{cases}$

$$\begin{bmatrix} 9 & 2 & -3 & \vdots & 20 \\ 0 & -25 & 11 & \vdots & -5 \end{bmatrix}$$

13. $\begin{bmatrix} 1 & 2 & \vdots & 7 \\ 2 & -3 & \vdots & 4 \end{bmatrix}$

$$\begin{cases} x + 2y = 7 \\ 2x - 3y = 4 \end{cases}$$

14. $\begin{bmatrix} 7 & -5 & \vdots & 0 \\ 8 & 3 & \vdots & -2 \end{bmatrix}$

$$\begin{cases} 7x - 5y = 0 \\ 8x + 3y = -2 \end{cases}$$

15. $\begin{bmatrix} 2 & 0 & 5 & \vdots & -12 \\ 0 & 1 & -2 & \vdots & 7 \\ 6 & 3 & 0 & \vdots & 2 \end{bmatrix}$

$$\begin{cases} 2x \quad + 5z = -12 \\ y - 2z = 7 \\ 6x + 3y \quad = 2 \end{cases}$$

16. $\begin{bmatrix} 4 & -5 & -1 & \vdots & 18 \\ -11 & 0 & 6 & \vdots & 25 \\ 3 & 8 & 0 & \vdots & -29 \end{bmatrix}$

$\begin{cases} 4x - 5y - z = 18 \\ -11x + 6z = 25 \\ 3x + 8y = -29 \end{cases}$

17. $\begin{bmatrix} 9 & 12 & 3 & 0 & \vdots & 0 \\ -2 & 18 & 5 & 2 & \vdots & 10 \\ 1 & 7 & -8 & 0 & \vdots & -4 \\ 3 & 0 & 2 & 0 & \vdots & -10 \end{bmatrix}$

$\begin{cases} 9x + 12y + 3z = 0 \\ -2x + 18y + 5z + 2w = 10 \\ x + 7y - 8z = -4 \\ 3x + 2z = -10 \end{cases}$

18. $\begin{bmatrix} 6 & 2 & -1 & -5 & \vdots & -25 \\ -1 & 0 & 7 & 3 & \vdots & 7 \\ 4 & -1 & -10 & 6 & \vdots & 23 \\ 0 & 8 & 1 & -11 & \vdots & -21 \end{bmatrix}$

$\begin{cases} 6x + 2y - z - 5w = -25 \\ -x + 7z + 3w = 7 \\ 4x - y - 10z + 6w = 23 \\ 8y + z - 11w = -21 \end{cases}$

19. $\begin{bmatrix} 1 & 0 & 0 & 0 \\ 0 & 1 & 1 & 5 \\ 0 & 0 & 0 & 0 \end{bmatrix}$

This matrix is in reduced row–echelon form.

20. $\begin{bmatrix} 1 & 3 & 0 & 0 \\ 0 & 0 & 1 & 8 \\ 0 & 0 & 0 & 0 \end{bmatrix}$

This matrix is in reduced row-echelon form.

21. $\begin{bmatrix} 2 & 0 & 4 & 0 \\ 0 & -1 & 3 & 6 \\ 0 & 0 & 1 & 5 \end{bmatrix}$

The first nonzero entries in rows one and two are not one. The matrix is not in row–echelon form.

22. $\begin{bmatrix} 1 & 0 & 2 & 1 \\ 0 & 1 & -3 & 10 \\ 0 & 0 & 1 & 0 \end{bmatrix}$

This matrix is in row-echelon form.

23. $\begin{bmatrix} 1 & 4 & 3 \\ 2 & 10 & 5 \end{bmatrix}$

$-2R_1 + R_2 \rightarrow \begin{bmatrix} 1 & 4 & 3 \\ 0 & \boxed{2} & -1 \end{bmatrix}$

24. $\begin{bmatrix} 3 & 6 & 8 \\ 4 & -3 & 6 \end{bmatrix}$

$\frac{1}{3}R_1 \rightarrow \begin{bmatrix} 1 & \boxed{2} & \frac{8}{3} \\ 4 & -3 & 6 \end{bmatrix}$

25. $\begin{bmatrix} 1 & 1 & 4 & -1 \\ 3 & 8 & 10 & 3 \\ -2 & 1 & 12 & 6 \end{bmatrix}$

$\begin{matrix} \\ -3R_1 + R_2 \rightarrow \\ 2R_1 + R_3 \rightarrow \end{matrix} \begin{bmatrix} 1 & 1 & 4 & -1 \\ 0 & 5 & \boxed{-2} & \boxed{6} \\ 0 & 3 & \boxed{20} & \boxed{4} \end{bmatrix}$

$\begin{matrix} \\ \frac{1}{5}R_2 \rightarrow \\ \\ \end{matrix} \begin{bmatrix} 1 & 1 & 4 & -1 \\ 0 & 1 & -\frac{2}{5} & \frac{6}{5} \\ 0 & 3 & \boxed{20} & \boxed{4} \end{bmatrix}$

26. $\begin{bmatrix} 2 & 4 & 8 & 3 \\ 1 & -1 & -3 & 2 \\ 2 & 6 & 4 & 9 \end{bmatrix}$

$\begin{matrix} \frac{1}{2}R_1 \rightarrow \\ \\ \\ \end{matrix} \begin{bmatrix} 1 & \boxed{2} & \boxed{4} & \boxed{\frac{3}{2}} \\ 1 & -1 & -3 & 2 \\ 2 & 6 & 4 & 9 \end{bmatrix}$

$\begin{matrix} \\ -R_1 + R_2 \rightarrow \\ -2R_1 + R_3 \rightarrow \end{matrix} \begin{bmatrix} 1 & 2 & 4 & \frac{3}{2} \\ 0 & \boxed{-3} & -7 & \frac{1}{2} \\ 0 & 2 & \boxed{-4} & \boxed{6} \end{bmatrix}$

27. $\begin{bmatrix} -2 & 5 & 1 \\ 3 & -1 & -8 \end{bmatrix} \rightarrow \begin{bmatrix} 13 & 0 & -39 \\ 3 & -1 & -8 \end{bmatrix}$

Add five times Row 2 to Row 1.

28. $\begin{bmatrix} 3 & -1 & -4 \\ -4 & 3 & 7 \end{bmatrix} \rightarrow \begin{bmatrix} 3 & -1 & -4 \\ 5 & 0 & -5 \end{bmatrix}$

Add 3 times Row 1 to Row 2.

29. $\begin{bmatrix} 0 & -1 & -5 & 5 \\ -1 & 3 & -7 & 6 \\ 4 & -5 & 1 & 3 \end{bmatrix} \rightarrow \begin{bmatrix} -1 & 3 & -7 & 6 \\ 0 & -1 & -5 & 5 \\ 0 & 7 & -27 & 27 \end{bmatrix}$

Interchange Row 1 and Row 2. Then add four times the new Row 1 to Row 3.

30. $\begin{bmatrix} -1 & -2 & 3 & -2 \\ 2 & -5 & 1 & -7 \\ 5 & 4 & -7 & 6 \end{bmatrix} \rightarrow \begin{bmatrix} -1 & -2 & 3 & -2 \\ 0 & -9 & 7 & -11 \\ 0 & -6 & 8 & -4 \end{bmatrix}$

Add 2 times Row 1 to Row 2.
Add 5 times Row 1 to Row 3.

31. $\begin{bmatrix} 1 & 2 & 3 \\ 2 & -1 & -4 \\ 3 & 1 & -1 \end{bmatrix}$

(a) $\begin{bmatrix} 1 & 2 & 3 \\ 0 & -5 & -10 \\ 3 & 1 & -1 \end{bmatrix}$ (b) $\begin{bmatrix} 1 & 2 & 3 \\ 0 & -5 & -10 \\ 0 & -5 & -10 \end{bmatrix}$ (c) $\begin{bmatrix} 1 & 2 & 3 \\ 0 & -5 & -10 \\ 0 & 0 & 0 \end{bmatrix}$

(d) $\begin{bmatrix} 1 & 2 & 3 \\ 0 & 1 & 2 \\ 0 & 0 & 0 \end{bmatrix}$ (e) $\begin{bmatrix} 1 & 0 & -1 \\ 0 & 1 & 2 \\ 0 & 0 & 0 \end{bmatrix}$ This system is in reduced row-echelon form.

32. $\begin{bmatrix} 7 & 1 \\ 0 & 2 \\ -3 & 4 \\ 4 & 1 \end{bmatrix}$ (a) $\begin{bmatrix} 7 & 1 \\ 0 & 2 \\ -3 & 4 \\ 1 & 5 \end{bmatrix}$ (b) $\begin{bmatrix} 1 & 5 \\ 0 & 2 \\ -3 & 4 \\ 7 & 1 \end{bmatrix}$ (c) $\begin{bmatrix} 1 & 5 \\ 0 & 2 \\ 0 & 19 \\ 7 & 1 \end{bmatrix}$

(d) $\begin{bmatrix} 1 & 5 \\ 0 & 2 \\ 0 & 19 \\ 0 & -34 \end{bmatrix}$ (e) $\begin{bmatrix} 1 & 5 \\ 0 & 1 \\ 0 & 19 \\ 0 & -34 \end{bmatrix}$ (f) $\begin{bmatrix} 1 & 0 \\ 0 & 1 \\ 0 & 0 \\ 0 & 0 \end{bmatrix}$ This matrix is in reduced row-echelon form.

33. $\begin{bmatrix} 1 & 1 & 0 & 5 \\ -2 & -1 & 2 & -10 \\ 3 & 6 & 7 & 14 \end{bmatrix}$

$\begin{matrix} \\ 2R_1 + R_2 \rightarrow \\ -3R_1 + R_3 \rightarrow \end{matrix} \begin{bmatrix} 1 & 1 & 0 & 5 \\ 0 & 1 & 2 & 0 \\ 0 & 3 & 7 & -1 \end{bmatrix}$

$\begin{matrix} \\ \\ -3R_2 + R_3 \rightarrow \end{matrix} \begin{bmatrix} 1 & 1 & 0 & 5 \\ 0 & 1 & 2 & 0 \\ 0 & 0 & 1 & -1 \end{bmatrix}$

34. $\begin{bmatrix} 1 & 2 & -1 & 3 \\ 3 & 7 & -5 & 14 \\ -2 & -1 & -3 & 8 \end{bmatrix}$

$\begin{matrix} \\ -3R_1 + R_2 \rightarrow \\ 2R_1 + R_3 \rightarrow \end{matrix} \begin{bmatrix} 1 & 2 & -1 & 3 \\ 0 & 1 & -2 & 5 \\ 0 & 3 & -5 & 14 \end{bmatrix}$

$\begin{matrix} \\ \\ -3R_2 + R_3 \rightarrow \end{matrix} \begin{bmatrix} 1 & 2 & -1 & 3 \\ 0 & 1 & -2 & 5 \\ 0 & 0 & 1 & -1 \end{bmatrix}$

35. $\begin{bmatrix} 1 & -1 & -1 & 1 \\ 5 & -4 & 1 & 8 \\ -6 & 8 & 18 & 0 \end{bmatrix}$

$\begin{matrix} \\ -5R_1 + R_2 \rightarrow \\ 6R_1 + R_3 \rightarrow \end{matrix} \begin{bmatrix} 1 & -1 & -1 & 1 \\ 0 & 1 & 6 & 3 \\ 0 & 2 & 12 & 6 \end{bmatrix}$

$\begin{matrix} \\ \\ -2R_2 + R_3 \rightarrow \end{matrix} \begin{bmatrix} 1 & -1 & -1 & 1 \\ 0 & 1 & 6 & 3 \\ 0 & 0 & 0 & 0 \end{bmatrix}$

36. $\begin{bmatrix} 1 & -3 & 0 & -7 \\ -3 & 10 & 1 & 23 \\ 4 & -10 & 2 & -24 \end{bmatrix}$

$\begin{matrix} \\ 3R_1 + R_2 \rightarrow \\ -4R_1 + R_3 \rightarrow \end{matrix} \begin{bmatrix} 1 & -3 & 0 & -7 \\ 0 & 1 & 1 & 2 \\ 0 & 2 & 2 & 4 \end{bmatrix}$

$\begin{matrix} \\ \\ -2R_2 + R_3 \rightarrow \end{matrix} \begin{bmatrix} 1 & -3 & 0 & -7 \\ 0 & 1 & 1 & 2 \\ 0 & 0 & 0 & 0 \end{bmatrix}$

37. Use the reduced row-echelon form feature of a graphing utility.

$$\begin{bmatrix} 3 & 3 & 3 \\ -1 & 0 & -4 \\ 2 & 4 & -2 \end{bmatrix} \Rightarrow \begin{bmatrix} 1 & 0 & 0 \\ 0 & 1 & 0 \\ 0 & 0 & 1 \end{bmatrix}$$

38. Use the reduced row-echelon form feature of a graphing utility.

$$\begin{bmatrix} 1 & 3 & 2 \\ 5 & 15 & 9 \\ 2 & 6 & 10 \end{bmatrix} \Rightarrow \begin{bmatrix} 1 & 3 & 0 \\ 0 & 0 & 1 \\ 0 & 0 & 0 \end{bmatrix}$$

39. Use the reduced row–echelon form feature of a graphing utility.

$$\begin{bmatrix} 1 & 2 & 3 & -5 \\ 1 & 2 & 4 & -9 \\ -2 & -4 & -4 & 3 \\ 4 & 8 & 11 & -14 \end{bmatrix} \Rightarrow \begin{bmatrix} 1 & 2 & 0 & 0 \\ 0 & 0 & 1 & 0 \\ 0 & 0 & 0 & 1 \\ 0 & 0 & 0 & 0 \end{bmatrix}$$

40. Use the reduced row-echelon form feature of a graphing utility.

$$\begin{bmatrix} -2 & 3 & -1 & -2 \\ 4 & -2 & 5 & 8 \\ 1 & 5 & -2 & 0 \\ 3 & 8 & -10 & -30 \end{bmatrix} \Rightarrow \begin{bmatrix} 1 & 0 & 0 & 0 \\ 0 & 1 & 0 & 0 \\ 0 & 0 & 1 & 0 \\ 0 & 0 & 0 & 1 \end{bmatrix}$$

41. Use the reduced row–echelon form feature of a graphing utility.

$$\begin{bmatrix} -3 & 5 & 1 & 12 \\ 1 & -1 & 1 & 4 \end{bmatrix} \Rightarrow \begin{bmatrix} 1 & 0 & 3 & 16 \\ 0 & 1 & 2 & 12 \end{bmatrix}$$

42. Use the reduced row-echelon form feature of a graphing utility.

$$\begin{bmatrix} 5 & 1 & 2 & 4 \\ -1 & 5 & 10 & -32 \end{bmatrix} \Rightarrow \begin{bmatrix} 1 & 0 & 0 & 2 \\ 0 & 1 & 2 & -6 \end{bmatrix}$$

43. $\begin{cases} x - 2y = 4 \\ y = -3 \end{cases}$

$x - 2(-3) = 4$

$x = -2$

Solution: $(-2, -3)$

44. $\begin{cases} x + 5y = 0 \\ y = -1 \end{cases}$

$x + 5(-1) = 0$

$x = 5$

Solution: $(5, -1)$

45. $\begin{cases} x - y + 2z = 4 \\ y - z = 2 \\ z = -2 \end{cases}$

$y - (-2) = 2$

$y = 0$

$x - 0 + 2(-2) = 4$

$x = 8$

Solution: $(8, 0, -2)$

46. $\begin{cases} x + 2y - 2z = -1 \\ y + z = 9 \\ z = -3 \end{cases}$

$y + (-3) = 9$

$y = 12$

$x + 2(12) - 2(-3) = -1$

$x = -31$

Solution: $(-31, 12, -3)$

47. $\begin{bmatrix} 1 & 0 & \vdots & 3 \\ 0 & 1 & \vdots & -4 \end{bmatrix}$

$x = 3$

$y = -4$

Solution: $(3, -4)$

48. $\begin{bmatrix} 1 & 0 & \vdots & -6 \\ 0 & 1 & \vdots & 10 \end{bmatrix}$

$x = -6$

$y = 10$

Solution: $(-6, 10)$

49. $\begin{bmatrix} 1 & 0 & 0 & \vdots & -4 \\ 0 & 1 & 0 & \vdots & -10 \\ 0 & 0 & 1 & \vdots & 4 \end{bmatrix}$

$x = -4$

$y = -10$

$z = 4$

Solution: $(-4, -10, 4)$

50. $\begin{bmatrix} 1 & 0 & 0 & \vdots & 5 \\ 0 & 1 & 0 & \vdots & -3 \\ 0 & 0 & 1 & \vdots & 0 \end{bmatrix}$

$x = 5$

$y = -3$

$z = 0$

Solution: $(5, -3, 0)$

51. $\begin{cases} x + 2y = 7 \\ 2x + y = 8 \end{cases}$

$$\begin{bmatrix} 1 & 2 & \vdots & 7 \\ 2 & 1 & \vdots & 8 \end{bmatrix}$$

$$-2R_1 + R_2 \rightarrow \begin{bmatrix} 1 & 2 & \vdots & 7 \\ 0 & -3 & \vdots & -6 \end{bmatrix}$$

$$-\tfrac{1}{3}R_2 \rightarrow \begin{bmatrix} 1 & 2 & \vdots & 7 \\ 0 & 1 & \vdots & 2 \end{bmatrix}$$

$\begin{cases} x + 2y = 7 \\ y = 2 \end{cases}$

$y = 2$

$x + 2(2) = 7 \Rightarrow x = 3$

Solution: $(3, 2)$

52. $\begin{cases} 2x + 6y = 16 \\ 2x + 3y = 7 \end{cases}$

$$\begin{bmatrix} 2 & 6 & \vdots & 16 \\ 2 & 3 & \vdots & 7 \end{bmatrix}$$

$$-R_1 + R_2 \rightarrow \begin{bmatrix} 2 & 6 & \vdots & 16 \\ 0 & -3 & \vdots & -9 \end{bmatrix}$$

$$\begin{matrix} \tfrac{1}{2}R_1 \rightarrow \\ -\tfrac{1}{3}R_2 \rightarrow \end{matrix} \begin{bmatrix} 1 & 3 & \vdots & 8 \\ 0 & 1 & \vdots & 3 \end{bmatrix}$$

$\begin{cases} x + 3y = 8 \\ y = 3 \end{cases}$

$y = 3$

$x + 3(3) = 8 \Rightarrow x = -1$

Solution: $(-1, 3)$

53. $\begin{cases} 3x - 2y = -27 \\ x + 3y = 13 \end{cases}$

$$\begin{bmatrix} 3 & -2 & \vdots & -27 \\ 1 & 3 & \vdots & 13 \end{bmatrix}$$

$$\begin{matrix} R_1 \\ R_2 \end{matrix} \begin{bmatrix} 1 & 3 & \vdots & 13 \\ 3 & -2 & \vdots & -27 \end{bmatrix}$$

$$-3R_1 + R_2 \rightarrow \begin{bmatrix} 1 & 3 & & 13 \\ 0 & -11 & & -66 \end{bmatrix}$$

$$-\tfrac{1}{11}R_2 \rightarrow \begin{bmatrix} 1 & 3 & \vdots & 13 \\ 0 & 1 & \vdots & 6 \end{bmatrix}$$

$\begin{cases} x + 3y = 13 \\ y = 6 \end{cases}$

$y = 6$

$x + 3(6) = 13 \Rightarrow x = -5$

Solution: $(-5, 6)$

54. $\begin{cases} -x + y = 4 \\ 2x - 4y = -34 \end{cases}$

$$\begin{bmatrix} -1 & 1 & \vdots & 4 \\ 2 & -4 & \vdots & -34 \end{bmatrix}$$

$$\begin{matrix} (-1)R_1 \rightarrow \\ (\tfrac{1}{2})R_2 \rightarrow \end{matrix} \begin{bmatrix} 1 & -1 & \vdots & -4 \\ 1 & -2 & \vdots & -17 \end{bmatrix}$$

$$-R_1 + R_2 \rightarrow \begin{bmatrix} 1 & -1 & \vdots & -4 \\ 0 & -1 & \vdots & -13 \end{bmatrix}$$

$$(-1)R_2 \rightarrow \begin{bmatrix} 1 & -1 & \vdots & -4 \\ 0 & 1 & \vdots & 13 \end{bmatrix}$$

$\begin{cases} x - y = -4 \\ y = 13 \end{cases}$

$y = 13$

$x - 13 = -4 \Rightarrow x = 9$

Solution: $(9, 13)$

55. $\begin{cases} -2x + 6y = -22 \\ x + 2y = -9 \end{cases}$

$$\begin{bmatrix} -2 & 6 & \vdots & -22 \\ 1 & 2 & \vdots & -9 \end{bmatrix}$$

$$\begin{matrix} R_1 \\ R_2 \end{matrix} \begin{bmatrix} 1 & 2 & \vdots & -9 \\ -2 & 6 & \vdots & -22 \end{bmatrix}$$

$$2R_1 + R_2 \rightarrow \begin{bmatrix} 1 & 2 & \vdots & -9 \\ 0 & 10 & \vdots & -40 \end{bmatrix}$$

$$\tfrac{1}{10}R_2 \rightarrow \begin{bmatrix} 1 & 2 & \vdots & -9 \\ 0 & 1 & \vdots & -4 \end{bmatrix}$$

$\begin{cases} x + 2y = -9 \\ y = -4 \end{cases}$

$y = -4$

$x + 2(-4) = -9 \Rightarrow x = -1$

Solution: $(-1, -4)$

56. $\begin{cases} 5x - 5y = -5 \\ -2x - 3y = 7 \end{cases}$

$$\begin{bmatrix} 5 & -5 & \vdots & -5 \\ -2 & -3 & \vdots & 7 \end{bmatrix}$$

$$\tfrac{1}{5}R_1 \rightarrow \begin{bmatrix} 1 & -1 & \vdots & -1 \\ -2 & -3 & \vdots & 7 \end{bmatrix}$$

$$2R_1 + R_2 \rightarrow \begin{bmatrix} 1 & -1 & \vdots & -1 \\ 0 & -5 & \vdots & 5 \end{bmatrix}$$

$$-\tfrac{1}{5}R_2 \rightarrow \begin{bmatrix} 1 & -1 & \vdots & -1 \\ 0 & 1 & \vdots & -1 \end{bmatrix}$$

$\begin{cases} x - y = -1 \\ y = -1 \end{cases}$

$y = -1$

$x - (-1) = -1 \Rightarrow x = -2$

Solution: $(-2, -1)$

57. $\begin{cases} -x + 2y = 1.5 \\ 2x - 4y = 3.0 \end{cases}$

$$\begin{bmatrix} -1 & 2 & \vdots & 1.5 \\ 2 & -4 & \vdots & 3.0 \end{bmatrix}$$

$$2R_1 + R_2 \rightarrow \begin{bmatrix} -1 & 2 & \vdots & 1.5 \\ 0 & 0 & \vdots & 6.0 \end{bmatrix}$$

The system is inconsistent and there is no solution.

58. $\begin{cases} x - 3y = 5 \\ -2x + 6y = -10 \end{cases}$

$$\begin{bmatrix} 1 & -3 & \vdots & 5 \\ -2 & 6 & \vdots & -10 \end{bmatrix}$$

$$2R_1 + R_2 \rightarrow \begin{bmatrix} 1 & -3 & \vdots & 5 \\ 0 & 0 & \vdots & 0 \end{bmatrix}$$

$x - 3y = 5$

$y = a$

$x = 3a + 5$

Solution: $(3a + 5, a)$, where a is a real number

59. $\begin{cases} x - 3z = -2 \\ 3x + y - 2z = 5 \\ 2x + 2y + z = 4 \end{cases}$

$$\begin{bmatrix} 1 & 0 & -3 & \vdots & -2 \\ 3 & 1 & -2 & \vdots & 5 \\ 2 & 2 & 1 & \vdots & 4 \end{bmatrix}$$

$$\begin{matrix} \\ -3R_1 + R_2 \rightarrow \\ -2R_1 + R_3 \rightarrow \end{matrix} \begin{bmatrix} 1 & 0 & -3 & \vdots & -2 \\ 0 & 1 & 7 & \vdots & 11 \\ 0 & 2 & 7 & \vdots & 8 \end{bmatrix}$$

$$\begin{matrix} \\ \\ -2R_2 + R_3 \rightarrow \end{matrix} \begin{bmatrix} 1 & 0 & -3 & \vdots & -2 \\ 0 & 1 & 7 & \vdots & 11 \\ 0 & 0 & -7 & \vdots & -14 \end{bmatrix}$$

$$\begin{matrix} \\ \\ -\frac{1}{7}R_3 \rightarrow \end{matrix} \begin{bmatrix} 1 & 0 & -3 & \vdots & -2 \\ 0 & 1 & 7 & \vdots & 11 \\ 0 & 0 & 1 & \vdots & 2 \end{bmatrix}$$

$\begin{cases} x - 3z = -2 \\ y + 7z = 11 \\ z = 2 \end{cases}$

$z = 2$

$y + 7(2) = 11 \Rightarrow y = -3$

$x - 3(2) = -2 \Rightarrow x = 4$

Solution: $(4, -3, 2)$

60. $\begin{cases} 2x - y + 3z = 24 \\ 2y - z = 14 \\ 7x - 5y = 6 \end{cases}$

$$\begin{bmatrix} 2 & -1 & 3 & \vdots & 24 \\ 0 & 2 & -1 & \vdots & 14 \\ 7 & -5 & 0 & \vdots & 6 \end{bmatrix}$$

$$R_3 + (-3)R_1 \rightarrow \begin{bmatrix} 1 & -2 & -9 & \vdots & -66 \\ 0 & 2 & -1 & \vdots & 14 \\ 7 & -5 & 0 & \vdots & 6 \end{bmatrix}$$

$$\begin{matrix} \\ \\ -7R_1 + R_3 \rightarrow \end{matrix} \begin{bmatrix} 1 & -2 & -9 & \vdots & -66 \\ 0 & 2 & -1 & \vdots & 14 \\ 0 & 9 & 63 & \vdots & 468 \end{bmatrix}$$

$$\begin{matrix} \\ 4R_2 \rightarrow \\ \\ \end{matrix} \begin{bmatrix} 1 & -2 & -9 & \vdots & -66 \\ 0 & 8 & -4 & \vdots & 56 \\ 0 & 9 & 63 & \vdots & 468 \end{bmatrix}$$

$$\begin{matrix} \\ -R_3 + R_2 \rightarrow \\ \\ \end{matrix} \begin{bmatrix} 1 & -2 & -9 & \vdots & -66 \\ 0 & -1 & -67 & \vdots & -412 \\ 0 & 9 & 63 & \vdots & 468 \end{bmatrix}$$

$$\begin{matrix} \\ \\ 9R_2 + R_3 \rightarrow \end{matrix} \begin{bmatrix} 1 & -2 & -9 & \vdots & -66 \\ 0 & -1 & -67 & \vdots & -412 \\ 0 & 0 & -540 & \vdots & -3240 \end{bmatrix}$$

$$\begin{matrix} \\ -R_2 \rightarrow \\ -\frac{1}{540}R_3 \rightarrow \end{matrix} \begin{bmatrix} 1 & -2 & -9 & \vdots & -66 \\ 0 & 1 & 67 & \vdots & 412 \\ 0 & 0 & 1 & \vdots & 6 \end{bmatrix}$$

$\begin{cases} x - 2y - 9z = -66 \\ y + 67z = 412 \\ z = 6 \end{cases}$

$z = 6$

$y + 67(6) = 412 \Rightarrow y = 10$

$x - 2(10) - 9(6) = -66 \Rightarrow x = 8$

Solution: $(8, 10, 6)$

61. $\begin{cases} -x + y - z = -14 \\ 2x - y + z = 21 \\ 3x + 2y + z = 19 \end{cases}$

$$\begin{bmatrix} -1 & 1 & -1 & \vdots & -14 \\ 2 & -1 & 1 & \vdots & 21 \\ 3 & 2 & 1 & \vdots & 19 \end{bmatrix}$$

$$\begin{matrix} -R_1 \rightarrow \\ \\ \\ \end{matrix} \begin{bmatrix} 1 & -1 & 1 & \vdots & 14 \\ 2 & -1 & 1 & \vdots & 21 \\ 3 & 2 & 1 & \vdots & 19 \end{bmatrix}$$

$$\begin{matrix} \\ -2R_1 + R_2 \rightarrow \\ -3R_1 + R_3 \rightarrow \end{matrix} \begin{bmatrix} 1 & -1 & 1 & \vdots & 14 \\ 0 & 1 & -1 & \vdots & -7 \\ 0 & 5 & -2 & \vdots & -23 \end{bmatrix}$$

$$\begin{matrix} \\ \\ -5R_2 + R_3 \rightarrow \end{matrix} \begin{bmatrix} 1 & -1 & 1 & \vdots & 14 \\ 0 & 1 & -1 & \vdots & -7 \\ 0 & 0 & 3 & \vdots & 12 \end{bmatrix}$$

$$\begin{matrix} \\ \\ \frac{1}{3}R_3 \rightarrow \end{matrix} \begin{bmatrix} 1 & -1 & 1 & \vdots & 14 \\ 0 & 1 & -1 & \vdots & -7 \\ 0 & 0 & 1 & \vdots & 4 \end{bmatrix}$$

$\begin{cases} x - y + z = 14 \\ y - z = -7 \\ z = 4 \end{cases}$

$z = 4$

$y - 4 = -7 \Rightarrow y = -3$

$x - (-3) + 4 = 14 \Rightarrow x = 7$

Solution: $(7, -3, 4)$

62. $\begin{cases} 2x + 2y - z = 2 \\ x - 3y + z = -28 \\ -x + y = 14 \end{cases}$

$$\begin{bmatrix} 2 & 2 & -1 & \vdots & 2 \\ 1 & -3 & 1 & \vdots & -28 \\ -1 & 1 & 0 & \vdots & 14 \end{bmatrix}$$

$$\begin{matrix} R_2 \\ R_1 \\ \\ \end{matrix} \begin{bmatrix} 1 & -3 & 1 & \vdots & -28 \\ 2 & 2 & -1 & \vdots & 2 \\ -1 & 1 & 0 & \vdots & 14 \end{bmatrix}$$

$$\begin{matrix} \\ R_3 \\ R_2 \end{matrix} \begin{bmatrix} 1 & -3 & 1 & \vdots & -28 \\ -1 & 1 & 0 & \vdots & 14 \\ 2 & 2 & -1 & \vdots & 2 \end{bmatrix}$$

$$\begin{matrix} \\ R_1 + R_2 \rightarrow \\ -2R_1 + R_3 \rightarrow \end{matrix} \begin{bmatrix} 1 & -3 & 1 & \vdots & -28 \\ 0 & -2 & 1 & \vdots & -14 \\ 0 & 8 & -3 & \vdots & 58 \end{bmatrix}$$

$$\begin{matrix} \\ \\ 4R_2 + R_3 \rightarrow \end{matrix} \begin{bmatrix} 1 & -3 & 1 & \vdots & -28 \\ 0 & -2 & 1 & \vdots & -14 \\ 0 & 0 & 1 & \vdots & 2 \end{bmatrix}$$

$$\begin{matrix} \\ -\frac{1}{2}R_2 \rightarrow \\ \\ \end{matrix} \begin{bmatrix} 1 & -3 & 1 & -28 \\ 0 & 1 & -\frac{1}{2} & 7 \\ 0 & 0 & 1 & 2 \end{bmatrix}$$

$\begin{cases} x - 3y + z = -28 \\ y - \frac{1}{2}z = 7 \\ z = 2 \end{cases}$

$z = 2$

$y - \frac{1}{2}(2) = 7 \Rightarrow y = 8$

$x - 3(8) + 2 = -28 \Rightarrow x = -6$

Solution: $(-6, 8, 2)$

63. $\begin{cases} x + 2y - 3z = -28 \\ 4y + 2z = 0 \\ -x + y - z = -5 \end{cases}$

$$\begin{bmatrix} 1 & 2 & -3 & \vdots & -28 \\ 0 & 4 & 2 & \vdots & 0 \\ -1 & 1 & -1 & \vdots & -5 \end{bmatrix}$$

$$\begin{matrix} \\ \frac{1}{4}R_2 \rightarrow \\ R_1 + R_3 \rightarrow \end{matrix} \begin{bmatrix} 1 & 2 & -3 & \vdots & -28 \\ 0 & 1 & \frac{1}{2} & \vdots & 0 \\ 0 & 3 & -4 & \vdots & -33 \end{bmatrix}$$

$$\begin{matrix} \\ \\ -3R_2 + R_3 \rightarrow \end{matrix} \begin{bmatrix} 1 & 2 & -3 & -28 \\ 0 & 1 & \frac{1}{2} & 0 \\ 0 & 0 & -\frac{11}{2} & -33 \end{bmatrix}$$

$$\begin{matrix} \\ \\ -\frac{2}{11}R_3 \rightarrow \end{matrix} \begin{bmatrix} 1 & 2 & -3 & \vdots & -28 \\ 0 & 1 & \frac{1}{2} & \vdots & 0 \\ 0 & 0 & 1 & \vdots & 6 \end{bmatrix}$$

$\begin{cases} x + 2y - 3z = -28 \\ y + \frac{1}{2}z = 0 \\ z = 6 \end{cases}$

$z = 6$

$y + \frac{1}{2}(6) = 0 \Rightarrow y = -3$

$x + 2(-3) - 3(6) = -28 \Rightarrow x = -4$

Solution: $(-4, -3, 6)$

64. $\begin{cases} 3x - 2y + z = 15 \\ -x + y + 2z = -10 \\ x - y - 4z = 14 \end{cases}$

$$\begin{bmatrix} 3 & -2 & 1 & \vdots & 15 \\ -1 & 1 & 2 & \vdots & -10 \\ 1 & -1 & -4 & \vdots & 14 \end{bmatrix}$$

$$\begin{matrix} R_3 \\ \\ R_1 \end{matrix} \begin{bmatrix} 1 & -1 & -4 & \vdots & 14 \\ -1 & 1 & 2 & \vdots & -10 \\ 3 & -2 & 1 & \vdots & 15 \end{bmatrix}$$

$$\begin{matrix} \\ R_1 + R_2 \rightarrow \\ -3R_1 + R_3 \rightarrow \end{matrix} \begin{bmatrix} 1 & -1 & -4 & \vdots & 14 \\ 0 & 0 & -2 & \vdots & 4 \\ 0 & 1 & 13 & \vdots & -27 \end{bmatrix}$$

$$\begin{matrix} \\ R_3 \\ R_2 \end{matrix} \begin{bmatrix} 1 & -1 & -4 & 14 \\ 0 & 1 & 13 & -27 \\ 0 & 0 & -2 & 4 \end{bmatrix}$$

$$\begin{matrix} \\ \\ -\frac{1}{2}R_3 \rightarrow \end{matrix} \begin{bmatrix} 1 & -1 & -4 & \vdots & 14 \\ 0 & 1 & 13 & \vdots & -27 \\ 0 & 0 & 1 & \vdots & -2 \end{bmatrix}$$

$\begin{cases} x - y - 4z = 14 \\ y + 13z = -27 \\ z = -2 \end{cases}$

$z = -2$

$y + 13(-2) = -27 \implies y = -1$

$x - (-1) - 4(-2) = 14 \implies x = 5$

Solution: $(5, -1, -2)$

65. $\begin{cases} x + y - 5z = 3 \\ x - 2z = 1 \\ 2x - y - z = 0 \end{cases}$

$$\begin{bmatrix} 1 & 1 & -5 & \vdots & 3 \\ 1 & 0 & -2 & \vdots & 1 \\ 2 & -1 & -1 & \vdots & 0 \end{bmatrix}$$

$$\begin{matrix} \\ -R_1 + R_2 \rightarrow \\ -2R_1 + R_3 \rightarrow \end{matrix} \begin{bmatrix} 1 & 1 & -5 & \vdots & 3 \\ 0 & -1 & 3 & \vdots & -2 \\ 0 & -3 & 9 & \vdots & -6 \end{bmatrix}$$

$$\begin{matrix} \\ -R_2 \rightarrow \\ \\ \end{matrix} \begin{bmatrix} 1 & 1 & -5 & \vdots & 3 \\ 0 & 1 & -3 & \vdots & 2 \\ 0 & -3 & 9 & \vdots & -6 \end{bmatrix}$$

$$\begin{matrix} -R_2 + R_1 \rightarrow \\ \\ 3R_2 + R_3 \rightarrow \end{matrix} \begin{bmatrix} 1 & 0 & -2 & \vdots & 1 \\ 0 & 1 & -3 & \vdots & 2 \\ 0 & 0 & 0 & \vdots & 0 \end{bmatrix}$$

$\begin{cases} x - 2z = 1 \\ y - 3z = 2 \end{cases}$

$z = a$

$y - 3a = 2 \implies y = 3a + 2$

$x - 2a = 1 \implies x = 2a + 1$

Solution: $(2a + 1, 3a + 2, a)$

66. $\begin{cases} 2x + 3z = 3 \\ 4x - 3y + 7z = 5 \\ 8x - 9y + 15z = 9 \end{cases}$

$$\begin{bmatrix} 2 & 0 & 3 & \vdots & 3 \\ 4 & -3 & 7 & \vdots & 5 \\ 8 & -9 & 15 & \vdots & 9 \end{bmatrix}$$

$$\begin{matrix} \\ -2R_1 + R_2 \rightarrow \\ -4R_1 + R_3 \rightarrow \end{matrix} \begin{bmatrix} 2 & 0 & 3 & \vdots & 3 \\ 0 & -3 & 1 & \vdots & -1 \\ 0 & -9 & 3 & \vdots & -3 \end{bmatrix}$$

$$\begin{matrix} \\ \\ -3R_2 + R_3 \rightarrow \end{matrix} \begin{bmatrix} 2 & 0 & 3 & \vdots & 3 \\ 0 & -3 & 1 & \vdots & -1 \\ 0 & 0 & 0 & \vdots & 0 \end{bmatrix}$$

$$\begin{matrix} \frac{1}{2}R_1 \rightarrow \\ -\frac{1}{3}R_2 \rightarrow \\ \\ \end{matrix} \begin{bmatrix} 1 & 0 & \frac{3}{2} & \vdots & \frac{3}{2} \\ 0 & 1 & -\frac{1}{3} & \vdots & \frac{1}{3} \\ 0 & 0 & 0 & \vdots & 0 \end{bmatrix}$$

$z = a$

$y = \frac{1}{3}a + \frac{1}{3}$

$x = -\frac{3}{2}a + \frac{3}{2}$

Solution: $\left(-\frac{3}{2}a + \frac{3}{2}, \frac{1}{3}a + \frac{1}{3}, a\right)$, where a is a real number.

67. $\begin{cases} x + 2y + z + 2w = 8 \\ 3x + 7y + 6z + 9w = 26 \end{cases}$

$$\begin{bmatrix} 1 & 2 & 1 & 2 & \vdots & 8 \\ 3 & 7 & 6 & 9 & \vdots & 26 \end{bmatrix}$$

$$\begin{matrix} \\ -3R_1 + R_2 \rightarrow \end{matrix} \begin{bmatrix} 1 & 2 & 1 & 2 & \vdots & 8 \\ 0 & 1 & 3 & 3 & \vdots & 2 \end{bmatrix}$$

$$\begin{matrix} -2R_2 + R_1 \rightarrow \\ \\ \end{matrix} \begin{bmatrix} 1 & 0 & -5 & -4 & \vdots & 4 \\ 0 & 1 & 3 & 3 & \vdots & 2 \end{bmatrix}$$

$\begin{cases} x - 5z - 4w = 4 \\ y + 3z + 3w = 2 \end{cases}$

$w = a, z = b$

$y + 3b + 3a = 2 \implies y = 2 - 3b - 3a$

$x - 5b - 4a = 4 \implies x = 4 + 5b + 4a$

Solution: $(4 + 5b + 4a, 2 - 3b - 3a, b, a)$, where a and b are real numbers.

68. $\begin{cases} 4x + 12y - 7z - 20w = 22 \\ 3x + 9y - 5z - 28w = 30 \end{cases}$

$$\begin{bmatrix} 4 & 12 & -7 & -20 & \vdots & 22 \\ 3 & 9 & -5 & -28 & \vdots & 30 \end{bmatrix}$$

$$\begin{matrix} -R_2 + R_1 \rightarrow \\ \end{matrix} \begin{bmatrix} 1 & 3 & -2 & 8 & \vdots & -8 \\ 3 & 9 & -5 & -28 & \vdots & 30 \end{bmatrix}$$

$$\begin{matrix} \\ -3R_1 + R_2 \rightarrow \end{matrix} \begin{bmatrix} 1 & 3 & -2 & 8 & \vdots & -8 \\ 0 & 0 & 1 & -52 & \vdots & 54 \end{bmatrix}$$

$$\begin{matrix} 2R_2 + R_1 \rightarrow \\ \end{matrix} \begin{bmatrix} 1 & 3 & 0 & -96 & \vdots & 100 \\ 0 & 0 & 1 & -52 & \vdots & 54 \end{bmatrix}$$

$w = a$

$z = 52a + 54$

$y = b$

$x = -3b + 96a + 100$

Solution: $(-3b + 96a + 100, b, 52a + 54, a)$ where a and b are real numbers.

69. $\begin{cases} -x + y = -22 \\ 3x + 4y = 4 \\ 4x - 8y = 32 \end{cases}$

$$\begin{bmatrix} -1 & 1 & \vdots & -22 \\ 3 & 4 & \vdots & 4 \\ 4 & -8 & \vdots & 32 \end{bmatrix}$$

$$\begin{matrix} -R_1 \rightarrow \\ \\ \end{matrix} \begin{bmatrix} 1 & -1 & \vdots & 22 \\ 3 & 4 & \vdots & 4 \\ 4 & -8 & \vdots & 32 \end{bmatrix}$$

$$\begin{matrix} \\ -3R_1 + R_2 \rightarrow \\ -4R_1 + R_3 \rightarrow \end{matrix} \begin{bmatrix} 1 & -1 & \vdots & 22 \\ 0 & 7 & \vdots & -62 \\ 0 & -4 & \vdots & -56 \end{bmatrix}$$

$$\begin{matrix} \\ \frac{1}{7}R_2 \rightarrow \\ -\frac{1}{4}R_3 \rightarrow \end{matrix} \begin{bmatrix} 1 & -1 & \vdots & 22 \\ 0 & 1 & \vdots & -\frac{62}{7} \\ 0 & 1 & \vdots & 14 \end{bmatrix}$$

$$\begin{matrix} \\ \\ -R_2 + R_3 \rightarrow \end{matrix} \begin{bmatrix} 1 & -1 & 22 \\ 0 & 1 & -\frac{62}{7} \\ 0 & 0 & \frac{160}{7} \end{bmatrix}$$

The system is inconsistent and there is no solution.

70. $\begin{cases} x + 2y = 0 \\ x + y = 6 \\ 3x - 2y = 8 \end{cases}$

$$\begin{bmatrix} 1 & 2 & \vdots & 0 \\ 1 & 1 & \vdots & 6 \\ 3 & -2 & \vdots & 8 \end{bmatrix}$$

$$\begin{matrix} \\ -R_1 + R_2 \rightarrow \\ -3R_1 + R_3 \rightarrow \end{matrix} \begin{bmatrix} 1 & 2 & \vdots & 0 \\ 0 & -1 & \vdots & 6 \\ 0 & -8 & \vdots & 8 \end{bmatrix}$$

$$\begin{matrix} \\ \\ -8R_2 + R_3 \rightarrow \end{matrix} \begin{bmatrix} 1 & 2 & \vdots & 0 \\ 0 & -1 & \vdots & 6 \\ 0 & 0 & \vdots & -40 \end{bmatrix}$$

The system in inconsistent and there is no solution.

71. Use the reduced row–echelon form feature of a graphing utility.

$\begin{cases} 3x + 3y + 12z = 6 \\ x + y + 4z = 2 \\ 2x + 5y + 20z = 10 \\ -x + 2y + 8z = 4 \end{cases}$ $\begin{bmatrix} 3 & 3 & 12 & \vdots & 6 \\ 1 & 1 & 4 & \vdots & 2 \\ 2 & 5 & 20 & \vdots & 10 \\ -1 & 2 & 8 & \vdots & 4 \end{bmatrix} \Rightarrow \begin{bmatrix} 1 & 0 & 0 & \vdots & 0 \\ 0 & 1 & 4 & \vdots & 2 \\ 0 & 0 & 0 & \vdots & 0 \\ 0 & 0 & 0 & \vdots & 0 \end{bmatrix}$

$z = a$

$y = 2 - 4a$

$x = 0$

$\begin{cases} x = 0 \\ y + 4z = 2 \end{cases}$

Solution: $(0, 2 - 4a, a)$

72. Use the reduced row-echelon form feature of a graphing utility.

$$\begin{cases} 2x + 10y + 2z = 6 \\ x + 5y + 2z = 6 \\ x + 5y + z = 3 \\ -3x - 15y - 3z = -9 \end{cases}$$

$$\begin{bmatrix} 2 & 10 & 2 & \vdots & 6 \\ 1 & 5 & 2 & \vdots & 6 \\ 1 & 5 & 1 & \vdots & 3 \\ -3 & -15 & -3 & \vdots & -9 \end{bmatrix} \Rightarrow \begin{bmatrix} 1 & 5 & 0 & \vdots & 0 \\ 0 & 0 & 1 & \vdots & 3 \\ 0 & 0 & 0 & \vdots & 0 \\ 0 & 0 & 0 & \vdots & 0 \end{bmatrix}$$

$$\begin{cases} z = 3 \\ x + 5y = 0 \end{cases}$$

$z = 3$

$y = a$

$x + 5a = 0 \Rightarrow x = -5a$

Solution: $(-5a, a, 3)$, where a is a real number.

73. Use the reduced row–echelon form feature of a graphing utility.

$$\begin{cases} 2x + y - z + 2w = -6 \\ 3x + 4y + w = 1 \\ x + 5y + 2z + 6w = -3 \\ 5x + 2y - z - w = 3 \end{cases} \quad \begin{bmatrix} 2 & 1 & -1 & 2 & \vdots & -6 \\ 3 & 4 & 0 & 1 & \vdots & 1 \\ 1 & 5 & 2 & 6 & \vdots & -3 \\ 5 & 2 & -1 & -1 & \vdots & 3 \end{bmatrix} \Rightarrow \begin{bmatrix} 1 & 0 & 0 & 0 & \vdots & 1 \\ 0 & 1 & 0 & 0 & \vdots & 0 \\ 0 & 0 & 1 & 0 & \vdots & 4 \\ 0 & 0 & 0 & 1 & \vdots & -2 \end{bmatrix}$$

$x = 1$

$y = 0$

$z = 4$

$w = -2$

Solution: $(1, 0, 4, -2)$

74. Use the reduced row-echelon form feature of a graphing utility.

$$\begin{cases} x + 2y + 2z + 4w = 11 \\ 3x + 6y + 5z + 12w = 30 \\ x + 3y - 3z + 2w = -5 \\ 6x - y - z + w = -9 \end{cases}$$

$$\begin{bmatrix} 1 & 2 & 2 & 4 & \vdots & 11 \\ 3 & 6 & 5 & 12 & \vdots & 30 \\ 1 & 3 & -3 & 2 & \vdots & -5 \\ 6 & -1 & -1 & 1 & \vdots & -9 \end{bmatrix} \Rightarrow \begin{bmatrix} 1 & 0 & 0 & 0 & \vdots & -1 \\ 0 & 1 & 0 & 0 & \vdots & 1 \\ 0 & 0 & 1 & 0 & \vdots & 3 \\ 0 & 0 & 0 & 1 & \vdots & 1 \end{bmatrix}$$

$$\begin{cases} x = -1 \\ y = 1 \\ z = 3 \\ w = 1 \end{cases}$$

$w = 1$

$z = 3$

$y = 1$

$x = -1$

Solution: $(-1, 1, 3, 1)$

75. Use the reduced row–echelon form feature of a graphing utility.

$$\begin{cases} x + y + z + w = 0 \\ 2x + 3y + z - 2w = 0 \\ 3x + 5y + z = 0 \end{cases} \quad \begin{bmatrix} 1 & 1 & 1 & 1 & \vdots & 0 \\ 2 & 3 & 1 & -2 & \vdots & 0 \\ 3 & 5 & 1 & 0 & \vdots & 0 \end{bmatrix} \Rightarrow \begin{bmatrix} 1 & 0 & 2 & 0 & \vdots & 0 \\ 0 & 1 & -1 & 0 & \vdots & 0 \\ 0 & 0 & 0 & 1 & \vdots & 0 \end{bmatrix}$$

$$\begin{cases} x + 2z = 0 \\ y - z = 0 \\ w = 0 \end{cases}$$

Let $z = a$. Then $x = -2a$ and $y = a$.

Solution: $(-2a, a, a, 0)$, where a is a real number.

76. $\begin{cases} x + 2y + z + 3w = 0 \\ x - y + w = 0 \\ y - z + 2w = 0 \end{cases}$

$$\begin{bmatrix} 1 & 2 & 1 & 3 & \vdots & 0 \\ 1 & -1 & 0 & 1 & \vdots & 0 \\ 0 & 1 & -1 & 2 & \vdots & 0 \end{bmatrix} \Rightarrow \begin{bmatrix} 1 & 0 & 0 & 2 & \vdots & 0 \\ 0 & 1 & 0 & 1 & \vdots & 0 \\ 0 & 0 & 1 & -1 & \vdots & 0 \end{bmatrix}$$

$\begin{cases} x + 2w = 0 \\ y + w = 0 \\ z - w = 0 \end{cases}$

$w = a, z = a, y = -a, x = -2a$

Solution: $(-2a, -a, a, a)$, where a is a real number.

77. (a) $\begin{cases} x - 2y + z = -6 \\ y - 5z = 16 \\ z = -3 \end{cases}$

$y - 5(-3) = 16$

$y = 1$

$x - 2(1) + (-3) = -6$

$x = -1$

Solution: $(-1, 1, -3)$

(b) $\begin{cases} x + y - 2z = 6 \\ y + 3z = -8 \\ z = -3 \end{cases}$

$y + 3(-3) = -8$

$y = 1$

$x + (1) - 2(-3) = 6$

$x = -1$

Solution: $(-1, 1, -3)$

Both systems yield the same solution, namely $(-1, 1, -3)$.

78. (a) $\begin{cases} x - 3y + 4z = -11 \\ y - z = -4 \\ z = 2 \end{cases}$

$y - 2 = -4$

$y = -2$

$x - 3(-2) + 4(2) = -11$

$x = -25$

(b) $\begin{cases} x + 4y = -11 \\ y + 3z = 4 \\ z = 2 \end{cases}$

$y + 3(2) = 4$

$y = -2$

$x + 4(-2) = -11$

$x = -3$

The systems do **not** yield the same solution.

79. (a) $\begin{cases} x - 4y + 5z = 27 \\ y - 7z = -54 \\ z = 8 \end{cases}$

$y - 7(8) = -54$

$y = 2$

$x - 4(2) + 5(8) = 27$

$x = -5$

Solution: $(-5, 2, 8)$

(b) $\begin{cases} x - 6y + z = 15 \\ y + 5z = 42 \\ z = 8 \end{cases}$

$y + 5(8) = 42$

$y = 2$

$x - 6(2) + (8) = 15$

$x = 19$

Solution: $(19, 2, 8)$

The systems do **not** yield the same solution.

80. (a) $\begin{cases} x + 3y - z = 19 \\ y + 6z = -18 \\ z = -4 \end{cases}$

$y + 6(-4) = -18$

$y = 6$

$x + 3(6) - (-4) = 19$

$x = -3$

(b) $\begin{cases} x - y + 3z = -15 \\ y - 2z = 14 \\ z = -4 \end{cases}$

$y - 2(-4) = 14$

$y = 6$

$x - 6 + 3(-4) = -15$

$x = 3$

The systems do **not** yield the same solution.

81. (a) In the row-echelon form of an augmented matrix that corresponds to an inconsistent sytem of linear equations, there exists a row consisting of all zeros except for the entry in the last column.

(b) In the row-echelon form of an augmented matrix that corresponds to a system with an infinte number of solutions, there are fewer rows with nonzero entries than there are variables.

82. 1. Interchange two rows.

2. Multiply a row by a nonzero constant.

3. Add a multiple of one row to another row.

83. They are the same.

84. A matrix in row–echelon form is in reduced row–echelon form if every column that has a leading 1 has zeros in every position above and below its leading 1.

85. $\begin{cases} I_1 - I_2 + I_3 = 0 \\ 3I_1 + 4I_2 = 18 \\ I_2 + 3I_3 = 6 \end{cases}$

$$\begin{bmatrix} 1 & -1 & 1 & \vdots & 0 \\ 3 & 4 & 0 & \vdots & 18 \\ 0 & 1 & 3 & \vdots & 6 \end{bmatrix}$$

$$-3R_1 + R_2 \rightarrow \begin{bmatrix} 1 & -1 & 1 & \vdots & 0 \\ 0 & 7 & -3 & \vdots & 18 \\ 0 & 1 & 3 & \vdots & 6 \end{bmatrix}$$

$$\begin{matrix} \\ R_3 \\ R_2 \end{matrix} \begin{bmatrix} 1 & -1 & 1 & \vdots & 0 \\ 0 & 1 & 3 & \vdots & 6 \\ 0 & 7 & -3 & \vdots & 18 \end{bmatrix}$$

$$-7R_2 + R_3 \rightarrow \begin{bmatrix} 1 & -1 & 1 & \vdots & 0 \\ 0 & 1 & 3 & \vdots & 6 \\ 0 & 0 & -24 & \vdots & -24 \end{bmatrix}$$

$$-\tfrac{1}{24}R_3 \rightarrow \begin{bmatrix} 1 & -1 & 1 & \vdots & 0 \\ 0 & 1 & 3 & \vdots & 6 \\ 0 & 0 & 1 & \vdots & 1 \end{bmatrix}$$

$\begin{cases} I_1 - I_2 + I_3 = 0 \\ I_2 + 3I_3 = 6 \\ I_3 = 1 \end{cases}$

$I_3 = 1$

$I_2 + 3(1) = 6 \Rightarrow I_2 = 3$

$I_1 - 3 + 1 = 0 \Rightarrow I_1 = 2$

86. x = amount at 7%, y = amount at 8%, z = amount at 10%.

$z = 4x \quad \Rightarrow \quad -4x + z = 0$

$\begin{cases} x + y + z = 1{,}500{,}000 \\ 0.07x + 0.08y + 0.10z = 130{,}500 \\ -4x + z = 0 \end{cases}$

$$\begin{bmatrix} 1 & 1 & 1 & \vdots & 1{,}500{,}000 \\ 0.07 & 0.08 & 0.10 & \vdots & 130{,}500 \\ -4 & 0 & 1 & \vdots & 0 \end{bmatrix}$$

$$\begin{matrix} \\ -0.07R_1 + R_2 \rightarrow \\ 4R_1 + R_3 \rightarrow \end{matrix} \begin{bmatrix} 1 & 1 & 1 & \vdots & 1{,}500{,}000 \\ 0 & 0.01 & 0.03 & \vdots & 25{,}500 \\ 0 & 4 & 5 & \vdots & 6{,}000{,}000 \end{bmatrix}$$

$$100R_2 \rightarrow \begin{bmatrix} 1 & 1 & 1 & \vdots & 1{,}500{,}000 \\ 0 & 1 & 3 & \vdots & 2{,}550{,}000 \\ 0 & 4 & 5 & \vdots & 6{,}000{,}000 \end{bmatrix}$$

$$-4R_2 + R_3 \rightarrow \begin{bmatrix} 1 & 1 & 1 & \vdots & 1{,}500{,}000 \\ 0 & 1 & 3 & \vdots & 2{,}550{,}000 \\ 0 & 0 & -7 & \vdots & -4{,}200{,}000 \end{bmatrix}$$

$$-\tfrac{1}{7}R_3 \rightarrow \begin{bmatrix} 1 & 1 & 1 & \vdots & 1{,}500{,}000 \\ 0 & 1 & 3 & \vdots & 2{,}550{,}000 \\ 0 & 0 & 1 & \vdots & 600{,}000 \end{bmatrix}$$

$\begin{cases} x + y + z = 1{,}500{,}000 \\ y + 3z = 2{,}550{,}000 \\ z = 600{,}000 \end{cases}$

$y + 3(600{,}000) = 2{,}550{,}000 \Rightarrow y = 750{,}000$

$x + 750{,}000 + 600{,}000 = 1{,}500{,}000 \Rightarrow x = 150{,}000$

Solution: \$150,000 at 7%, \$750,000 at 8%, and \$600,000 at 10%

87. $x =$ amount at 9%, $y =$ amount at 10%, $z =$ amount at 12%

$$\begin{aligned} x + y \quad z &= 500{,}000 \\ 0.09x + 0.10y + 0.12z &= 52{,}000 \\ 2.5x - y &= 0 \end{aligned}$$

$$\left[\begin{array}{ccc:c} 1 & 1 & 1 & 500{,}000 \\ 0.09 & 0.10 & 0.12 & 52{,}000 \\ 2.5 & -1 & 0 & 0 \end{array}\right]$$

$$\begin{array}{r} \\ -0.09R_1 + R_2 \rightarrow \\ -2.5R_1 + R_3 \rightarrow \end{array} \left[\begin{array}{ccc:c} 1 & 1 & 1 & 500{,}000 \\ 0 & 0.01 & 0.03 & 7{,}000 \\ 0 & -3.5 & -2.5 & -1{,}250{,}000 \end{array}\right]$$

$$\begin{array}{r} \\ 100R_2 \rightarrow \\ 2R_3 \rightarrow \end{array} \left[\begin{array}{ccc:c} 1 & 1 & 1 & 500{,}000 \\ 0 & 1 & 3 & 700{,}000 \\ 0 & -7 & -5 & -2{,}500{,}000 \end{array}\right]$$

$$\begin{array}{r} -R_2 + R_1 \rightarrow \\ \\ 7R_2 + R_3 \rightarrow \end{array} \left[\begin{array}{ccc:c} 1 & 0 & -2 & -200{,}000 \\ 0 & 1 & 3 & 700{,}000 \\ 0 & 0 & 16 & 2{,}400{,}000 \end{array}\right]$$

$$\begin{array}{r} \\ \\ \frac{1}{16}R_3 \rightarrow \end{array} \left[\begin{array}{ccc:c} 1 & 0 & -2 & -200{,}000 \\ 0 & 1 & 3 & 700{,}000 \\ 0 & 0 & 1 & 150{,}000 \end{array}\right]$$

$$\begin{cases} x - 2z = -200{,}000 \\ y + 3z = 700{,}000 \\ z = 150{,}000 \end{cases}$$

$y + 3(150{,}000) = 700{,}000 \Rightarrow y = 250{,}000$

$x - 2(150{,}000) = -200{,}000 \Rightarrow x = 100{,}000$

Solution: $(100{,}000, 250{,}000, 150{,}000)$

Answer: \$100,000 at 9%, \$250,000 at 10%, \$150,000 at 12%

88. $y = ax^2 + bx + c$

$$\begin{cases} a + b + c = 8 \\ 4a + 2b + c = 13 \\ 9a + 3b + c = 20 \end{cases}$$

$$\left[\begin{array}{ccc:c} 1 & 1 & 1 & 8 \\ 4 & 2 & 1 & 13 \\ 9 & 3 & 1 & 20 \end{array}\right]$$

$$\begin{array}{r} \\ -4R_1 + R_2 \rightarrow \\ -9R_1 + R_3 \rightarrow \end{array} \left[\begin{array}{ccc:c} 1 & 1 & 1 & 8 \\ 0 & -2 & -3 & -19 \\ 0 & -6 & -8 & -52 \end{array}\right]$$

$$\begin{array}{r} \\ -\frac{1}{2}R_2 \rightarrow \\ -3R_2 + R_3 \rightarrow \end{array} \left[\begin{array}{ccc:c} 1 & 1 & 1 & 8 \\ 0 & 1 & \frac{3}{2} & \frac{19}{2} \\ 0 & 0 & 1 & 5 \end{array}\right]$$

$$\begin{cases} a + b + c = 8 \\ b + \frac{3}{2}c = \frac{19}{2} \\ c = 5 \end{cases}$$

$c = 5$

$b + \frac{3}{2}(5) = \frac{19}{2} \Rightarrow b = 2$

$a + 2 + 5 = 8 \Rightarrow a = 1$

Equation of parabola: $y = x^2 + 2x + 5$

89. $f(x) = ax^2 + bx + c$

$f(1) = a + b + c = 9$

$f(2) = 4a + 2b + c = 8$

$f(3) = 9a + 3b + c = 5$

$$\begin{bmatrix} 1 & 1 & 1 & \vdots & 9 \\ 4 & 2 & 1 & \vdots & 8 \\ 9 & 3 & 1 & \vdots & 5 \end{bmatrix}$$

$$\begin{matrix} \\ -4R_1 + R_2 \rightarrow \\ -9R_1 + R_3 \rightarrow \end{matrix} \begin{bmatrix} 1 & 1 & 1 & \vdots & 9 \\ 0 & -2 & -3 & \vdots & -28 \\ 0 & -6 & -8 & \vdots & -76 \end{bmatrix}$$

$$\begin{matrix} \\ -\frac{1}{2}R_2 \rightarrow \\ \\ \end{matrix} \begin{bmatrix} 1 & 1 & 1 & \vdots & 9 \\ 0 & 1 & \frac{3}{2} & \vdots & 14 \\ 0 & -6 & -8 & \vdots & -76 \end{bmatrix}$$

$$\begin{matrix} \\ \\ 6R_2 + R_3 \rightarrow \end{matrix} \begin{bmatrix} 1 & 1 & 1 & \vdots & 9 \\ 0 & 1 & \frac{3}{2} & \vdots & 14 \\ 0 & 0 & 1 & \vdots & 8 \end{bmatrix}$$

$$\begin{cases} a + b + c = 9 \\ b + \frac{3}{2}c = 14 \\ c = 8 \end{cases}$$

$c = 8$

$b + \frac{3}{2}(8) = 14 \Rightarrow b = 2$

$a + (2) + (8) = 9 \Rightarrow a = -1$

Equation of parabola: $y = -x^2 + 2x + 8$

90. (a) $(0, 5.0),\ (15, 9.6),\ (30, 12.4)$

$y = ax^2 + bx + c$

$$\begin{cases} c = 5 \\ 225a + 15b + c = 9.6 \Rightarrow 225a + 15b = 4.6 \\ 900a + 30b + c = 12.4 \Rightarrow 900a + 30b = 7.4 \end{cases}$$

$$\begin{bmatrix} 225 & 15 & \vdots & 4.6 \\ 900 & 30 & \vdots & 7.4 \end{bmatrix}$$

$$\begin{matrix} \\ -4R_1 + R_2 \rightarrow \end{matrix} \begin{bmatrix} 225 & 15 & \vdots & 4.6 \\ 0 & -30 & \vdots & -11 \end{bmatrix}$$

$$\begin{matrix} \frac{1}{225}R_1 \rightarrow \\ \left(-\frac{1}{30}\right)R_2 \rightarrow \end{matrix} \begin{bmatrix} 1 & \frac{1}{15} & \vdots & \frac{23}{1125} \\ 0 & 1 & \vdots & \frac{11}{30} \end{bmatrix}$$

$$\begin{cases} a + \frac{1}{15}b = \frac{23}{1125} \\ b = \frac{11}{30} \end{cases}$$

$a + \frac{1}{15}\left(\frac{11}{30}\right) = \frac{23}{1125} \Rightarrow a = -\frac{1}{250} = -0.004$

Equation of parabola: $y = -0.004x^2 + 0.367x + 5.$

(b)

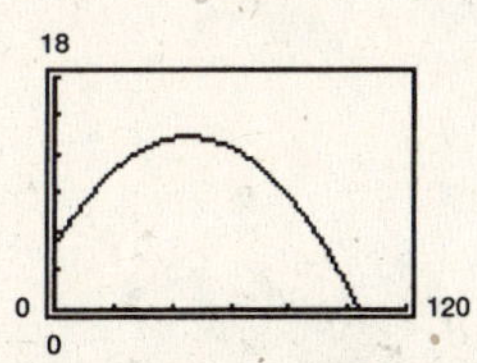

The maximum height is approximately 13 feet and the ball strikes the ground at approximately 104 feet.

(c) The maximum occurs at the vertex.

$$x = -\frac{b}{2a} = \frac{-0.367}{2(-0.004)} = 45.875$$

$y = -0.004(45.875)^2 + 0.367(45.875) + 5 = 13.418$ feet

The ball strikes the ground when $y = 0$.

$-0.004x^2 + 0.367x + 5 = 0$

By the Quadratic Formula and using the positive value for x we have $x \approx 103.793$ feet.

91. (a) $(9, 121.1), (10, 130.9), (11, 141.5)$

$f(t) = at^2 + bt + c$

$f(9) = 81a + 9b + c = 121.1$

$f(10) = 100a + 10b + c = 130.9$

$f(11) = 121a + 11b + c = 141.5$

$$\begin{bmatrix} 81 & 9 & 1 & \vdots & 121.1 \\ 100 & 10 & 1 & \vdots & 130.9 \\ 121 & 11 & 1 & \vdots & 141.5 \end{bmatrix} \Rightarrow \begin{bmatrix} 1 & 0 & 0 & \vdots & 0.4 \\ 0 & 1 & 0 & \vdots & 2.2 \\ 0 & 0 & 1 & \vdots & 68.9 \end{bmatrix}$$

$f(t) = 0.4t^2 + 2.2t + 68.9$

(b)

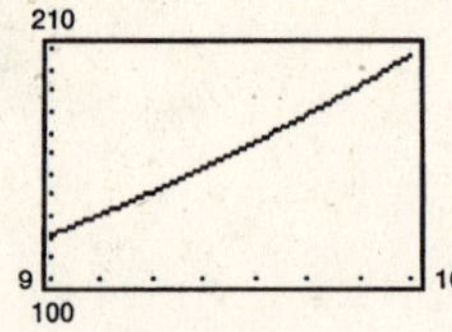

(c) $f(12) = \$152.9$ billion

This is close to the actual value of \$153.0 billion for 2002 sales.

(d) $f(16) = \$206.5$ billion

The parabola grows without bound as t increases. Even though drug store sales are also growing, this figure represents an increase in sales of nearly 35% in four years, which seems excessive.

92. False. It is a 2×4 matrix.

93. False. The rows are in the wrong order. To change this matrix to reduced row–echelon form, interchange Row 1 and Row 4, and interchange Row 2 and Row 3.

94. False. Gaussian elimination reduces a matrix until a row-echelon form is obtained and Gauss-Jordan elimination reduces a matrix until a reduced row-echelon form is obtained.

95. False. See Example 8 in this section.

Section 14.5 Operations with Matrices

1. $x = -4,\ y = 22$

2. $x = 13, y = 12$

3. $x + 1 = 2x - 1 \Rightarrow x = 2$

$-y = y + 1 \Rightarrow y = -\frac{1}{2}$

4. $-x = 2x - 3 \Rightarrow 3 = 3x \Rightarrow x = 1$

$2y + 1 = 5 \Rightarrow 2y = 4 \Rightarrow y = 2$

5. $2x + 1 = 5,\ 3x = 6,\ 3y - 5 = 4$

$x = 2,\ y = 3$

6. $x + 2 = 2x + 6, \quad y + 2 = 11$

$-4 = x \quad y = 9$

7. (a) $A + B = \begin{bmatrix} 1 & -1 \\ 2 & -1 \end{bmatrix} + \begin{bmatrix} 2 & -1 \\ -1 & 8 \end{bmatrix} = \begin{bmatrix} 1+2 & -1-1 \\ 2-1 & -1+8 \end{bmatrix} = \begin{bmatrix} 3 & -2 \\ 1 & 7 \end{bmatrix}$

(b) $A - B = \begin{bmatrix} 1 & -1 \\ 2 & -1 \end{bmatrix} - \begin{bmatrix} 2 & -1 \\ -1 & 8 \end{bmatrix} = \begin{bmatrix} 1-2 & -1+1 \\ 2+1 & -1-8 \end{bmatrix} = \begin{bmatrix} -1 & 0 \\ 3 & -9 \end{bmatrix}$

(c) $3A = 3\begin{bmatrix} 1 & -1 \\ 2 & -1 \end{bmatrix} = \begin{bmatrix} 3(1) & 3(-1) \\ 3(2) & 3(-1) \end{bmatrix} = \begin{bmatrix} 3 & -3 \\ 6 & -3 \end{bmatrix}$

(d) $3A - 2B = \begin{bmatrix} 3 & -3 \\ 6 & -3 \end{bmatrix} - 2\begin{bmatrix} 2 & -1 \\ -1 & 8 \end{bmatrix} = \begin{bmatrix} 3 & -3 \\ 6 & -3 \end{bmatrix} + \begin{bmatrix} -4 & 2 \\ 2 & -16 \end{bmatrix} = \begin{bmatrix} -1 & -1 \\ 8 & -19 \end{bmatrix}$

8. (a) $A + B = \begin{bmatrix} 1 & 2 \\ 2 & 1 \end{bmatrix} + \begin{bmatrix} -3 & -2 \\ 4 & 2 \end{bmatrix} = \begin{bmatrix} 1-3 & 2-2 \\ 2+4 & 1+2 \end{bmatrix} = \begin{bmatrix} -2 & 0 \\ 6 & 3 \end{bmatrix}$

(b) $A - B = \begin{bmatrix} 1 & 2 \\ 2 & 1 \end{bmatrix} - \begin{bmatrix} -3 & -2 \\ 4 & 2 \end{bmatrix} = \begin{bmatrix} 1+3 & 2+2 \\ 2-4 & 1-2 \end{bmatrix} = \begin{bmatrix} 4 & 4 \\ -2 & -1 \end{bmatrix}$

(c) $3A = 3\begin{bmatrix} 1 & 2 \\ 2 & 1 \end{bmatrix} = \begin{bmatrix} 3(1) & 3(2) \\ 3(2) & 3(1) \end{bmatrix} = \begin{bmatrix} 3 & 6 \\ 6 & 3 \end{bmatrix}$

(d) $3A - 2B = \begin{bmatrix} 3 & 6 \\ 6 & 3 \end{bmatrix} - 2\begin{bmatrix} -3 & -2 \\ 4 & 2 \end{bmatrix} = \begin{bmatrix} 3+6 & 6+4 \\ 6-8 & 3-4 \end{bmatrix} = \begin{bmatrix} 9 & 10 \\ -2 & -1 \end{bmatrix}$

9. $A = \begin{bmatrix} 6 & -1 \\ 2 & 4 \\ -3 & 5 \end{bmatrix}$, $B = \begin{bmatrix} 1 & 4 \\ -1 & 5 \\ 1 & 10 \end{bmatrix}$

(a) $A + B = \begin{bmatrix} 7 & 3 \\ 1 & 9 \\ -2 & 15 \end{bmatrix}$ (b) $A - B = \begin{bmatrix} 5 & -5 \\ 3 & -1 \\ -4 & -5 \end{bmatrix}$ (c) $3A = \begin{bmatrix} 18 & -3 \\ 6 & 12 \\ -9 & 15 \end{bmatrix}$

(d) $3A - 2B = \begin{bmatrix} 18 & -3 \\ 6 & 12 \\ -9 & 15 \end{bmatrix} - \begin{bmatrix} 2 & 8 \\ -2 & 10 \\ 2 & 20 \end{bmatrix} = \begin{bmatrix} 16 & -11 \\ 8 & 2 \\ -11 & -5 \end{bmatrix}$

10. (a) $A + B = \begin{bmatrix} 2 & 1 & 1 \\ -1 & -1 & 4 \end{bmatrix} + \begin{bmatrix} 2 & -3 & 4 \\ -3 & 1 & -2 \end{bmatrix} = \begin{bmatrix} 2+2 & 1-3 & 1+4 \\ -1-3 & -1+1 & 4-2 \end{bmatrix} = \begin{bmatrix} 4 & -2 & 5 \\ -4 & 0 & 2 \end{bmatrix}$

(b) $A - B = \begin{bmatrix} 2 & 1 & 1 \\ -1 & -1 & 4 \end{bmatrix} - \begin{bmatrix} 2 & -3 & 4 \\ -3 & 1 & -2 \end{bmatrix} = \begin{bmatrix} 2-2 & 1-(-3) & 1-4 \\ -1-(-3) & -1-1 & 4-(-2) \end{bmatrix} = \begin{bmatrix} 0 & 4 & -3 \\ 2 & -2 & 6 \end{bmatrix}$

(c) $3A = 3\begin{bmatrix} 2 & 1 & 1 \\ -1 & -1 & 4 \end{bmatrix} = \begin{bmatrix} 3(2) & 3(1) & 3(1) \\ 3(-1) & 3(-1) & 3(4) \end{bmatrix} = \begin{bmatrix} 6 & 3 & 3 \\ -3 & -3 & 12 \end{bmatrix}$

(d) $3A - 2B = \begin{bmatrix} 6 & 3 & 3 \\ -3 & -3 & 12 \end{bmatrix} - 2\begin{bmatrix} 2 & -3 & 4 \\ -3 & 1 & -2 \end{bmatrix} = \begin{bmatrix} 6 & 3 & 3 \\ -3 & -3 & 12 \end{bmatrix} + \begin{bmatrix} -4 & 6 & -8 \\ 6 & -2 & 4 \end{bmatrix}$

$= \begin{bmatrix} 2 & 9 & -5 \\ 3 & -5 & 16 \end{bmatrix}$

11. $A = \begin{bmatrix} 2 & 2 & -1 & 0 & 1 \\ 1 & 1 & -2 & 0 & -1 \end{bmatrix}$, $B = \begin{bmatrix} 1 & 1 & -1 & 1 & 0 \\ -3 & 4 & 9 & -6 & -7 \end{bmatrix}$

(a) $A + B = \begin{bmatrix} 3 & 3 & -2 & 1 & 1 \\ -2 & 5 & 7 & -6 & -8 \end{bmatrix}$

(b) $A - B = \begin{bmatrix} 1 & 1 & 0 & -1 & 1 \\ 4 & -3 & -11 & 6 & 6 \end{bmatrix}$

(c) $3A = \begin{bmatrix} 6 & 6 & -3 & 0 & 3 \\ 3 & 3 & -6 & 0 & -3 \end{bmatrix}$

(d) $3A - 2B = \begin{bmatrix} 6 & 6 & -3 & 0 & 3 \\ 3 & 3 & -6 & 0 & -3 \end{bmatrix} - \begin{bmatrix} 2 & 2 & -2 & 2 & 0 \\ -6 & 8 & 18 & -12 & -14 \end{bmatrix} = \begin{bmatrix} 4 & 4 & -1 & -2 & 3 \\ 9 & -5 & -24 & 12 & 11 \end{bmatrix}$

12. (a) $A + B = \begin{bmatrix} -1 & 4 & 0 \\ 3 & -2 & 2 \\ 5 & 4 & -1 \\ 0 & 8 & -6 \\ -4 & -1 & 0 \end{bmatrix} + \begin{bmatrix} -3 & 5 & 1 \\ 2 & -4 & -7 \\ 10 & -9 & -1 \\ 3 & 2 & -4 \\ 0 & 1 & -2 \end{bmatrix}$

$= \begin{bmatrix} -1-3 & 4+5 & 0+1 \\ 3+2 & -2-4 & 2-7 \\ 5+10 & 4-9 & -1-1 \\ 0+3 & 8+2 & -6-4 \\ -4+0 & -1+1 & 0-2 \end{bmatrix}$

$= \begin{bmatrix} -4 & 9 & 1 \\ 5 & -6 & -5 \\ 15 & -5 & -2 \\ 3 & 10 & -10 \\ -4 & 0 & -2 \end{bmatrix}$

(b) $A - B = \begin{bmatrix} -1 & 4 & 0 \\ 3 & -2 & 2 \\ 5 & 4 & -1 \\ 0 & 8 & -6 \\ -4 & -1 & 0 \end{bmatrix} - \begin{bmatrix} -3 & 5 & 1 \\ 2 & -4 & -7 \\ 10 & -9 & -1 \\ 3 & 2 & -4 \\ 0 & 1 & -2 \end{bmatrix}$

$= \begin{bmatrix} -1+3 & 4-5 & 0-1 \\ 3-2 & -2+4 & 2+7 \\ 5-10 & 4+9 & -1+1 \\ 0-3 & 8-2 & -6+4 \\ -4-0 & -1-1 & 0+2 \end{bmatrix}$

$= \begin{bmatrix} 2 & -1 & -1 \\ 1 & 2 & 9 \\ -5 & 13 & 0 \\ -3 & 6 & -2 \\ -4 & -2 & 2 \end{bmatrix}$

(c) $3A = 3\begin{bmatrix} -1 & 4 & 0 \\ 3 & -2 & 2 \\ 5 & 4 & -1 \\ 0 & 8 & -6 \\ -4 & -1 & 0 \end{bmatrix} = \begin{bmatrix} -3 & 12 & 0 \\ 9 & -6 & 6 \\ 15 & 12 & -3 \\ 0 & 24 & -18 \\ -12 & -3 & 0 \end{bmatrix}$

(d) $3A - 2B = \begin{bmatrix} -3 & 12 & 0 \\ 9 & -6 & 6 \\ 15 & 12 & -3 \\ 0 & 24 & -18 \\ -12 & -3 & 0 \end{bmatrix} - 2\begin{bmatrix} -3 & 5 & 1 \\ 2 & -4 & -7 \\ 10 & -9 & -1 \\ 3 & 2 & -4 \\ 0 & 1 & -2 \end{bmatrix}$

$= \begin{bmatrix} -3 & 12 & 0 \\ 9 & -6 & 6 \\ 15 & 12 & -3 \\ 0 & 24 & -18 \\ -12 & -3 & 0 \end{bmatrix} + \begin{bmatrix} 6 & -10 & -2 \\ -4 & 8 & 14 \\ -20 & 18 & 2 \\ -6 & -4 & 8 \\ 0 & -2 & 4 \end{bmatrix}$

$= \begin{bmatrix} 3 & 2 & -2 \\ 5 & 2 & 20 \\ -5 & 30 & -1 \\ -6 & 20 & -10 \\ -12 & -5 & 4 \end{bmatrix}$

13. $A = \begin{bmatrix} 6 & 0 & 3 \\ -1 & -4 & 0 \end{bmatrix}, B = \begin{bmatrix} 8 & -1 \\ 4 & -3 \end{bmatrix}$

(a) $A + B$ is not possible. A and B do not have the same order.

(b) $A - B$ is not possible. A and B do not have the same order.

(c) $3A = \begin{bmatrix} 18 & 0 & 9 \\ -3 & -12 & 0 \end{bmatrix}$

(d) $3A - 2B$ is not possible. A and B do not have the same order.

14. (a) $A + B$ is not possible. A and B do not have the same order.

(b) $A - B$ is not possible. A and B do not have the same order.

(c) $3A = 3\begin{bmatrix} 3 \\ 2 \\ -1 \end{bmatrix} = \begin{bmatrix} 9 \\ 6 \\ -3 \end{bmatrix}$

(d) $3A - 2B$ is not possible. A and B do not have the same order.

15. $\begin{bmatrix} -5 & 0 \\ 3 & -6 \end{bmatrix} + \begin{bmatrix} 7 & 1 \\ -2 & -1 \end{bmatrix} + \begin{bmatrix} -10 & -8 \\ 14 & 6 \end{bmatrix} = \begin{bmatrix} -5+7+(-10) & 0+1+(-8) \\ 3+(-2)+14 & -6+(-1)+6 \end{bmatrix} = \begin{bmatrix} -8 & -7 \\ 15 & -1 \end{bmatrix}$

16. $\begin{bmatrix} 6 & 8 \\ -1 & 0 \end{bmatrix} + \begin{bmatrix} 0 & 5 \\ -3 & -1 \end{bmatrix} + \begin{bmatrix} -11 & -7 \\ 2 & -1 \end{bmatrix} = \begin{bmatrix} 6+0+(-11) & 8+5+(-7) \\ -1+(-3)+2 & 0+(-1)+(-1) \end{bmatrix} = \begin{bmatrix} -5 & 6 \\ -2 & -2 \end{bmatrix}$

17. $4\left(\begin{bmatrix} -4 & 0 & 1 \\ 0 & 2 & 3 \end{bmatrix} - \begin{bmatrix} 2 & 1 & -2 \\ 3 & -6 & 0 \end{bmatrix}\right) = 4\begin{bmatrix} -6 & -1 & 3 \\ -3 & 8 & 3 \end{bmatrix} = \begin{bmatrix} -24 & -4 & 12 \\ -12 & 32 & 12 \end{bmatrix}$

18. $\frac{1}{2}([5 \quad -2 \quad 4 \quad 0] + [14 \quad 6 \quad -18 \quad 9]) = \frac{1}{2}[5 + 14 \quad -2 + 6 \quad 4 + (-18) \quad 0 + 9]$

$= \frac{1}{2}[19 \quad 4 \quad -14 \quad 9]$

$= \left[\frac{19}{2} \quad 2 \quad -7 \quad \frac{9}{2}\right]$

19. $-3\left(\begin{bmatrix} 0 & -3 \\ 7 & 2 \end{bmatrix} + \begin{bmatrix} -6 & 3 \\ 8 & 1 \end{bmatrix}\right) - 2\begin{bmatrix} 4 & -4 \\ 7 & -9 \end{bmatrix} = -3\begin{bmatrix} -6 & 0 \\ 15 & 3 \end{bmatrix} - \begin{bmatrix} 8 & -8 \\ 14 & -18 \end{bmatrix} = \begin{bmatrix} 18 & 0 \\ -45 & -9 \end{bmatrix} - \begin{bmatrix} 8 & -8 \\ 14 & -18 \end{bmatrix} = \begin{bmatrix} 10 & 8 \\ -59 & 9 \end{bmatrix}$

20. $-1\begin{bmatrix} 4 & 11 \\ -2 & -1 \\ 9 & 3 \end{bmatrix} + \frac{1}{6}\left(\begin{bmatrix} -5 & -1 \\ 3 & 4 \\ 0 & 13 \end{bmatrix} + \begin{bmatrix} 7 & 5 \\ -9 & -1 \\ 6 & -1 \end{bmatrix}\right) = \begin{bmatrix} -4 & -11 \\ 2 & 1 \\ -9 & -3 \end{bmatrix} + \frac{1}{6}\begin{bmatrix} -5 + 7 & -1 + 5 \\ 3 + (-9) & 4 + (-1) \\ 0 + 6 & 13 + (-1) \end{bmatrix}$

$= \begin{bmatrix} -4 & -11 \\ 2 & 1 \\ -9 & -3 \end{bmatrix} + \frac{1}{6}\begin{bmatrix} 2 & 4 \\ -6 & 3 \\ 6 & 12 \end{bmatrix}$

$= \begin{bmatrix} -4 & -11 \\ 2 & 1 \\ -9 & -3 \end{bmatrix} + \begin{bmatrix} \frac{1}{3} & \frac{2}{3} \\ -1 & \frac{1}{2} \\ 1 & 2 \end{bmatrix}$

$= \begin{bmatrix} -4 + \frac{1}{3} & -11 + \frac{2}{3} \\ 2 + (-1) & 1 + \frac{1}{2} \\ -9 + 1 & -3 + 2 \end{bmatrix} = \begin{bmatrix} -\frac{11}{3} & -\frac{31}{3} \\ 1 & \frac{3}{2} \\ -8 & -1 \end{bmatrix}$

21. $\frac{3}{7}\begin{bmatrix} 2 & 5 \\ -1 & -4 \end{bmatrix} + 6\begin{bmatrix} -3 & 0 \\ 2 & 2 \end{bmatrix} \approx \begin{bmatrix} -17.143 & 2.143 \\ 11.571 & 10.286 \end{bmatrix}$

22. $55\left(\begin{bmatrix} 14 & -11 \\ -22 & 19 \end{bmatrix} + \begin{bmatrix} -22 & 20 \\ 13 & 6 \end{bmatrix}\right) = 55\begin{bmatrix} 14 + (-22) & -11 + 20 \\ -22 + 13 & 19 + 6 \end{bmatrix}$

$= 55\begin{bmatrix} -8 & 9 \\ -9 & 25 \end{bmatrix} = \begin{bmatrix} -440 & 495 \\ -495 & 1375 \end{bmatrix}$

23. $-\begin{bmatrix} 3.211 & 6.829 \\ -1.004 & 4.914 \\ 0.055 & -3.889 \end{bmatrix} - \begin{bmatrix} -1.630 & -3.090 \\ 5.256 & 8.335 \\ -9.768 & 4.251 \end{bmatrix} = \begin{bmatrix} -1.581 & -3.739 \\ -4.252 & -13.249 \\ 9.713 & -0.362 \end{bmatrix}$

24. $-12\left(\begin{bmatrix} 6 & 20 \\ 1 & -9 \\ -2 & 5 \end{bmatrix} + \begin{bmatrix} 14 & -15 \\ -8 & -6 \\ 7 & 0 \end{bmatrix} + \begin{bmatrix} -31 & -19 \\ 16 & 10 \\ 24 & -10 \end{bmatrix}\right) = -12\begin{bmatrix} 6 + 14 + (-31) & 20 + (-15) + (-19) \\ 1 + (-8) + 16 & -9 + (-6) + 10 \\ -2 + 7 + 24 & 5 + 0 + (-10) \end{bmatrix}$

$= -12\begin{bmatrix} -11 & -14 \\ 9 & -5 \\ 29 & -5 \end{bmatrix} = \begin{bmatrix} 132 & 168 \\ -108 & 60 \\ -348 & 60 \end{bmatrix}$

25. $X = 3\begin{bmatrix} -2 & -1 \\ 1 & 0 \\ 3 & -4 \end{bmatrix} - 2\begin{bmatrix} 0 & 3 \\ 2 & 0 \\ -4 & -1 \end{bmatrix} = \begin{bmatrix} -6 & -3 \\ 3 & 0 \\ 9 & -12 \end{bmatrix} - \begin{bmatrix} 0 & 6 \\ 4 & 0 \\ -8 & -2 \end{bmatrix} = \begin{bmatrix} -6 & -9 \\ -1 & 0 \\ 17 & -10 \end{bmatrix}$

26. $2X = 2A - B$

$$X = A - \tfrac{1}{2}B = \begin{bmatrix} -2 & -1 \\ 1 & 0 \\ 3 & -4 \end{bmatrix} - \tfrac{1}{2}\begin{bmatrix} 0 & 3 \\ 2 & 0 \\ -4 & -1 \end{bmatrix} = \begin{bmatrix} -2 & -1 \\ 1 & 0 \\ 3 & -4 \end{bmatrix} - \begin{bmatrix} 0 & \frac{3}{2} \\ 1 & 0 \\ -2 & -\frac{1}{2} \end{bmatrix} = \begin{bmatrix} -2 & -\frac{5}{2} \\ 0 & 0 \\ 5 & -\frac{7}{2} \end{bmatrix}$$

27. $$X = -\tfrac{3}{2}A + \tfrac{1}{2}B = -\tfrac{3}{2}\begin{bmatrix} -2 & -1 \\ 1 & 0 \\ 3 & -4 \end{bmatrix} + \tfrac{1}{2}\begin{bmatrix} 0 & 3 \\ 2 & 0 \\ -4 & -1 \end{bmatrix} = \begin{bmatrix} 3 & \frac{3}{2} \\ -\frac{3}{2} & 0 \\ -\frac{9}{2} & 6 \end{bmatrix} + \begin{bmatrix} 0 & \frac{3}{2} \\ 1 & 0 \\ -2 & -\frac{1}{2} \end{bmatrix} = \begin{bmatrix} 3 & 3 \\ -\frac{1}{2} & 0 \\ -\frac{13}{2} & \frac{11}{2} \end{bmatrix}$$

28. $2A + 4B = -2X$

$$X = -A - 2B = -1\begin{bmatrix} -2 & -1 \\ 1 & 0 \\ 3 & -4 \end{bmatrix} - 2\begin{bmatrix} 0 & 3 \\ 2 & 0 \\ -4 & -1 \end{bmatrix} = \begin{bmatrix} 2 & 1 \\ -1 & 0 \\ -3 & 4 \end{bmatrix} + \begin{bmatrix} 0 & -6 \\ -4 & 0 \\ 8 & 2 \end{bmatrix} = \begin{bmatrix} 2 & -5 \\ -5 & 0 \\ 5 & 6 \end{bmatrix}$$

29. (a) $AB = \begin{bmatrix} 1 & 2 \\ 4 & 2 \end{bmatrix}\begin{bmatrix} 2 & -1 \\ -1 & 8 \end{bmatrix} = \begin{bmatrix} (1)(2) + (2)(-1) & (1)(-1) + (2)(8) \\ (4)(2) + (2)(-1) & (4)(-1) + (2)(8) \end{bmatrix} = \begin{bmatrix} 0 & 15 \\ 6 & 12 \end{bmatrix}$

(b) $BA = \begin{bmatrix} 2 & -1 \\ -1 & 8 \end{bmatrix}\begin{bmatrix} 1 & 2 \\ 4 & 2 \end{bmatrix} = \begin{bmatrix} (2)(1) + (-1)(4) & (2)(2) + (-1)(2) \\ (-1)(1) + (8)(4) & (-1)(2) + (8)(2) \end{bmatrix} = \begin{bmatrix} -2 & 2 \\ 31 & 14 \end{bmatrix}$

(c) $A^2 = \begin{bmatrix} 1 & 2 \\ 4 & 2 \end{bmatrix}\begin{bmatrix} 1 & 2 \\ 4 & 2 \end{bmatrix} = \begin{bmatrix} (1)(1) + (2)(4) & (1)(2) + (2)(2) \\ (4)(1) + (2)(4) & (4)(2) + (2)(2) \end{bmatrix} = \begin{bmatrix} 9 & 6 \\ 12 & 12 \end{bmatrix}$

30. (a) $AB = \begin{bmatrix} 2 & -1 \\ 1 & 4 \end{bmatrix}\begin{bmatrix} 0 & 0 \\ 3 & -3 \end{bmatrix} = \begin{bmatrix} 2(0) + (-1)3 & 2(0) + (-1)(-3) \\ 1(0) + 4(3) & 1(0) + 4(-3) \end{bmatrix} = \begin{bmatrix} -3 & 3 \\ 12 & -12 \end{bmatrix}$

(b) $BA = \begin{bmatrix} 0 & 0 \\ 3 & -3 \end{bmatrix}\begin{bmatrix} 2 & -1 \\ 1 & 4 \end{bmatrix} = \begin{bmatrix} 0(2) + (0)1 & 0(-1) + (0)(4) \\ 3(2) + (-3)(1) & 3(-1) + (-3)4 \end{bmatrix} = \begin{bmatrix} 0 & 0 \\ 3 & -15 \end{bmatrix}$

(c) $A^2 = \begin{bmatrix} 2 & -1 \\ 1 & 4 \end{bmatrix}\begin{bmatrix} 2 & -1 \\ 1 & 4 \end{bmatrix} = \begin{bmatrix} 2(2) + (-1)(1) & 2(-1) + (-1)4 \\ 1(2) + 4(1) & 1(-1) + 4(4) \end{bmatrix} = \begin{bmatrix} 3 & -6 \\ 6 & 15 \end{bmatrix}$

31. (a) $AB = \begin{bmatrix} 3 & -1 \\ 1 & 3 \end{bmatrix}\begin{bmatrix} 1 & -3 \\ 3 & 1 \end{bmatrix} = \begin{bmatrix} (3)(1) + (-1)(3) & (3)(-3) + (-1)(1) \\ (1)(1) + (3)(3) & (1)(-3) + (3)(1) \end{bmatrix} = \begin{bmatrix} 0 & -10 \\ 10 & 0 \end{bmatrix}$

(b) $BA = \begin{bmatrix} 1 & -3 \\ 3 & 1 \end{bmatrix}\begin{bmatrix} 3 & -1 \\ 1 & 3 \end{bmatrix} = \begin{bmatrix} (1)(3) + (-3)(1) & (1)(-1) + (-3)(3) \\ (3)(3) + (1)(1) & (3)(-1) + (1)(3) \end{bmatrix} = \begin{bmatrix} 0 & -10 \\ 10 & 0 \end{bmatrix}$

(c) $A^2 = \begin{bmatrix} 3 & -1 \\ 1 & 3 \end{bmatrix}\begin{bmatrix} 3 & -1 \\ 1 & 3 \end{bmatrix} = \begin{bmatrix} (3)(3) + (-1)(1) & (3)(-1) + (-1)(3) \\ (1)(3) + (3)(1) & (1)(-1) + (3)(3) \end{bmatrix} = \begin{bmatrix} 8 & -6 \\ 6 & 8 \end{bmatrix}$

32. (a) $AB = \begin{bmatrix} 1 & -1 \\ 1 & 1 \end{bmatrix}\begin{bmatrix} 1 & 3 \\ -3 & 1 \end{bmatrix} = \begin{bmatrix} 1(1) + (-1)(-3) & 1(3) + (-1)(1) \\ 1(1) + 1(-3) & 1(3) + 1(1) \end{bmatrix} = \begin{bmatrix} 4 & 2 \\ -2 & 4 \end{bmatrix}$

(b) $BA = \begin{bmatrix} 1 & 3 \\ -3 & 1 \end{bmatrix}\begin{bmatrix} 1 & -1 \\ 1 & 1 \end{bmatrix} = \begin{bmatrix} 1(1) + (3)1 & 1(-1) + 3(1) \\ -3(1) + (1)(1) & -3(-1) + 1(1) \end{bmatrix} = \begin{bmatrix} 4 & 2 \\ -2 & 4 \end{bmatrix}$

(c) $A^2 = \begin{bmatrix} 1 & -1 \\ 1 & 1 \end{bmatrix}\begin{bmatrix} 1 & -1 \\ 1 & 1 \end{bmatrix} = \begin{bmatrix} 1(1) + (-1)(1) & 1(-1) + (-1)(1) \\ 1(1) + (1)(1) & 1(-1) + 1(1) \end{bmatrix} = \begin{bmatrix} 0 & -2 \\ 2 & 0 \end{bmatrix}$

33. (a) $AB = \begin{bmatrix} 7 \\ 8 \\ -1 \end{bmatrix} \begin{bmatrix} 1 & 1 & 2 \end{bmatrix} = \begin{bmatrix} 7(1) & 7(1) & 7(2) \\ 8(1) & 8(1) & 8(2) \\ -1(1) & -1(1) & -1(2) \end{bmatrix} = \begin{bmatrix} 7 & 7 & 14 \\ 8 & 8 & 16 \\ -1 & -1 & -2 \end{bmatrix}$

(b) $BA = \begin{bmatrix} 1 & 1 & 2 \end{bmatrix} \begin{bmatrix} 7 \\ 8 \\ -1 \end{bmatrix} = [(1)(7) + (1)(8) + (2)(-1)] = [13]$

(c) A^2 is not possible.

34. (a) $AB = \begin{bmatrix} 3 & 2 & 1 \end{bmatrix} \begin{bmatrix} 2 \\ 3 \\ 0 \end{bmatrix} = [3(2) + 2(3) + 1(0)] = [12]$

(b) $BA = \begin{bmatrix} 2 \\ 3 \\ 0 \end{bmatrix} \begin{bmatrix} 3 & 2 & 1 \end{bmatrix} = \begin{bmatrix} 2(3) & 2(2) & 2(1) \\ 3(3) & 3(2) & 3(1) \\ 0(3) & 0(2) & 0(1) \end{bmatrix} = \begin{bmatrix} 6 & 4 & 2 \\ 9 & 6 & 3 \\ 0 & 0 & 0 \end{bmatrix}$

(c) The number of columns of A does not equal the number of rows of A; the multiplication is not possible.

35. A is 3×2 and B is 3×3. AB is not possible.

36. A is 2×4, B is 2×2. AB is not possible.

37. A is 3×3, B is $3 \times 2 \implies AB$ is 3×2.

$$\begin{bmatrix} 0 & -1 & 0 \\ 4 & 0 & 2 \\ 8 & -1 & 7 \end{bmatrix} \begin{bmatrix} 2 & 1 \\ -3 & 4 \\ 1 & 6 \end{bmatrix} = \begin{bmatrix} (0)(2) + (-1)(-3) + (0)(1) & (0)(1) + (-1)(4) + (0)(6) \\ (4)(2) + (0)(-3) + (2)(1) & (4)(1) + (0)(4) + (2)(6) \\ (8)(2) + (-1)(-3) + (7)(1) & (8)(1) + (-1)(4) + (7)(6) \end{bmatrix} = \begin{bmatrix} 3 & -4 \\ 10 & 16 \\ 26 & 46 \end{bmatrix}$$

38. A is 3×2, B is $2 \times 2 \implies AB$ is 3×2.

$$AB = \begin{bmatrix} -1 & 3 \\ 4 & -5 \\ 0 & 2 \end{bmatrix} \begin{bmatrix} 1 & 2 \\ 0 & 7 \end{bmatrix} = \begin{bmatrix} -1 & 19 \\ 4 & -27 \\ 0 & 14 \end{bmatrix}$$

39. A is 3×3, B is $3 \times 3 \implies AB$ is 3×3.

$$\begin{bmatrix} 1 & 0 & 0 \\ 0 & 4 & 0 \\ 0 & 0 & -2 \end{bmatrix} \begin{bmatrix} 3 & 0 & 0 \\ 0 & -1 & 0 \\ 0 & 0 & 5 \end{bmatrix} = \begin{bmatrix} (1)(3) + (0)(0) + (0)(0) & (1)(0) + (0)(-1) + (0)(0) & (1)(0) + (0)(0) + (0)(5) \\ (0)(3) + (4)(0) + (0)(0) & (0)(0) + (4)(-1) + (0)(0) & (0)(0) + (4)(0) + (0)(5) \\ (0)(3) + (0)(0) + (-2)(0) & (0)(0) + (0)(-1) + (-2)(0) & (0)(0) + (0)(0) + (-2)(5) \end{bmatrix}$$

$$= \begin{bmatrix} 3 & 0 & 0 \\ 0 & -4 & 0 \\ 0 & 0 & -10 \end{bmatrix}$$

40. A is 3×3, B is $3 \times 3 \implies AB$ is 3×3.

$$AB = \begin{bmatrix} 5 & 0 & 0 \\ 0 & -8 & 0 \\ 0 & 0 & 7 \end{bmatrix} \begin{bmatrix} \frac{1}{5} & 0 & 0 \\ 0 & -\frac{1}{8} & 0 \\ 0 & 0 & \frac{1}{2} \end{bmatrix} = \begin{bmatrix} 1 & 0 & 0 \\ 0 & 1 & 0 \\ 0 & 0 & \frac{7}{2} \end{bmatrix}$$

41. A is 3×3, B is $3 \times 3 \implies AB$ is 3×3.

$$\begin{bmatrix} 0 & 0 & 5 \\ 0 & 0 & -3 \\ 0 & 0 & 4 \end{bmatrix} \begin{bmatrix} 6 & -11 & 4 \\ 8 & 16 & 4 \\ 0 & 0 & 0 \end{bmatrix} =$$

$$\begin{bmatrix} (0)(6) + (0)(8) + (5)(0) & (0)(-11) + (0)(16) + (5)(0) & (0)(4) + (0)(4) + (5)(0) \\ (0)(6) + (0)(8) + (-3)(0) & (0)(-11) + (0)(16) + (-3)(0) & (0)(4) + (0)(4) + (-3)(0) \\ (0)(6) + (0)(8) + (4)(0) & (0)(-11) + (0)(16) + (4)(0) & (0)(4) + (0)(4) + (4)(0) \end{bmatrix} = \begin{bmatrix} 0 & 0 & 0 \\ 0 & 0 & 0 \\ 0 & 0 & 0 \end{bmatrix}$$

42. A is 2×1, B is $1 \times 4 \implies AB$ is 2×4.

$$\begin{bmatrix} 10 \\ 12 \end{bmatrix}\begin{bmatrix} 6 & -2 & 1 & 6 \end{bmatrix} = \begin{bmatrix} 60 & -20 & 10 & 60 \\ 72 & -24 & 12 & 72 \end{bmatrix}$$

43. $$\begin{bmatrix} 5 & 6 & -3 \\ -2 & 5 & 1 \\ 10 & -5 & 5 \end{bmatrix}\begin{bmatrix} 1 & -1 & 2 \\ 8 & 1 & 4 \\ 4 & -2 & 9 \end{bmatrix} = \begin{bmatrix} 41 & 7 & 7 \\ 42 & 5 & 25 \\ -10 & -25 & 45 \end{bmatrix}$$

44. $$\begin{bmatrix} 11 & -12 & 4 \\ 14 & 10 & 12 \\ 6 & -2 & 9 \end{bmatrix}\begin{bmatrix} 12 & 10 \\ -5 & 12 \\ 15 & 16 \end{bmatrix} = \begin{bmatrix} 252 & 30 \\ 298 & 452 \\ 217 & 180 \end{bmatrix}$$

45. $$\begin{bmatrix} -3 & 8 & -6 & 8 \\ -12 & 15 & 9 & 6 \\ 5 & -1 & 1 & 5 \end{bmatrix}\begin{bmatrix} 3 & 1 & 6 \\ 24 & 15 & 14 \\ 16 & 10 & 21 \\ 8 & -4 & 10 \end{bmatrix} = \begin{bmatrix} 151 & 25 & 48 \\ 516 & 279 & 387 \\ 47 & -20 & 87 \end{bmatrix}$$

46. A is 3×3, B is 4×2. AB is not possible.

47. A is 2×4 and B is $2 \times 4 \implies AB$ is not possible.

48. $$\begin{bmatrix} 15 & -18 \\ -4 & 12 \\ -8 & 22 \end{bmatrix}\begin{bmatrix} -7 & 22 & 1 \\ 8 & 16 & 24 \end{bmatrix} = \begin{bmatrix} -249 & 42 & -417 \\ 124 & 104 & 284 \\ 232 & 176 & 520 \end{bmatrix}$$

49. $$\begin{bmatrix} 3 & 1 \\ 0 & -2 \end{bmatrix}\begin{bmatrix} 1 & 0 \\ -2 & 2 \end{bmatrix}\begin{bmatrix} 1 & 0 \\ 2 & 4 \end{bmatrix} = \begin{bmatrix} 5 & 8 \\ -4 & -16 \end{bmatrix}$$

50. $$-3\left(\begin{bmatrix} 6 & 5 & -1 \\ 1 & -2 & 0 \end{bmatrix}\begin{bmatrix} 0 & 3 \\ -1 & -3 \\ 4 & 1 \end{bmatrix}\right) = \begin{bmatrix} 27 & -6 \\ -6 & -27 \end{bmatrix}$$

51. $$\begin{bmatrix} 0 & 2 & -2 \\ 4 & 1 & 2 \end{bmatrix}\left(\begin{bmatrix} 4 & 0 \\ 0 & -1 \\ -1 & 2 \end{bmatrix} + \begin{bmatrix} -2 & 3 \\ -3 & 5 \\ 0 & -3 \end{bmatrix}\right) = \begin{bmatrix} -4 & 10 \\ 3 & 14 \end{bmatrix}$$

52. $$\begin{bmatrix} 3 \\ -1 \\ 5 \\ 7 \end{bmatrix}\left(\begin{bmatrix} 5 & -6 \end{bmatrix} + \begin{bmatrix} 7 & -1 \end{bmatrix} + \begin{bmatrix} -8 & 9 \end{bmatrix}\right) = \begin{bmatrix} 12 & 6 \\ -4 & -2 \\ 20 & 10 \\ 28 & 14 \end{bmatrix}$$

53. (a) $$\begin{bmatrix} -1 & 1 \\ -2 & 1 \end{bmatrix}\begin{bmatrix} x_1 \\ x_2 \end{bmatrix} = \begin{bmatrix} 4 \\ 0 \end{bmatrix}$$

(b) $$\left[\begin{array}{cc:c} -1 & 1 & 4 \\ -2 & 1 & 0 \end{array}\right]$$

$$-R_2 + R_1 \rightarrow \left[\begin{array}{cc:c} 1 & 0 & 4 \\ -2 & 1 & 0 \end{array}\right]$$

$$2R_1 + R_2 \rightarrow \left[\begin{array}{cc:c} 1 & 0 & 4 \\ 0 & 1 & 8 \end{array}\right]$$

$$X = \begin{bmatrix} 4 \\ 8 \end{bmatrix}$$

54. (a) $$\begin{bmatrix} 2 & 3 \\ 1 & 4 \end{bmatrix}\begin{bmatrix} x_1 \\ x_2 \end{bmatrix} = \begin{bmatrix} 5 \\ 10 \end{bmatrix}$$

(b) $$\begin{matrix} R_2 \\ R_1 \end{matrix}\left[\begin{array}{cc:c} 1 & 4 & 10 \\ 2 & 3 & 5 \end{array}\right]$$

$$-2R_1 + R_2 \rightarrow \left[\begin{array}{cc:c} 1 & 4 & 10 \\ 0 & -5 & -15 \end{array}\right]$$

$$-\tfrac{1}{5}R_2 \rightarrow \left[\begin{array}{cc:c} 1 & 4 & 10 \\ 0 & 1 & 3 \end{array}\right]$$

$$-4R_2 + R_1 \rightarrow \left[\begin{array}{cc:c} 1 & 0 & -2 \\ 0 & 1 & 3 \end{array}\right]$$

$$X = \begin{bmatrix} -2 \\ 3 \end{bmatrix}$$

55. (a) $$\begin{bmatrix} -2 & -3 \\ 6 & 1 \end{bmatrix}\begin{bmatrix} x_1 \\ x_2 \end{bmatrix} = \begin{bmatrix} -4 \\ -36 \end{bmatrix}$$

(b) $$\left[\begin{array}{cc:c} -2 & -3 & -4 \\ 6 & 1 & -36 \end{array}\right]$$

$$3R_1 + R_2 \rightarrow \left[\begin{array}{cc:c} -2 & -3 & -4 \\ 0 & -8 & -48 \end{array}\right]$$

$$\begin{matrix} -\tfrac{1}{2}R_1 \rightarrow \\ -\tfrac{1}{8}R_2 \rightarrow \end{matrix}\left[\begin{array}{cc:c} 1 & \tfrac{3}{2} & 2 \\ 0 & 1 & 6 \end{array}\right]$$

$$-\tfrac{3}{2}R_2 + R_1 \rightarrow \left[\begin{array}{cc:c} 1 & 0 & -7 \\ 0 & 1 & 6 \end{array}\right]$$

$$X = \begin{bmatrix} -7 \\ 6 \end{bmatrix}$$

56. (a) $$\begin{bmatrix} -4 & 9 \\ 1 & -3 \end{bmatrix}\begin{bmatrix} x_1 \\ x_2 \end{bmatrix} = \begin{bmatrix} -13 \\ 12 \end{bmatrix}$$

(b) $$\begin{array}{r} R_1 \\ R_2 \end{array}\left[\begin{array}{rr|r} 1 & -3 & 12 \\ -4 & 9 & -13 \end{array}\right]$$

$$4R_1 + R_2 \to \left[\begin{array}{rr|r} 1 & -3 & 12 \\ 0 & -3 & 35 \end{array}\right]$$

$$-\tfrac{1}{3}R_2 \to \left[\begin{array}{rr|r} 1 & -3 & 12 \\ 0 & 1 & -\frac{35}{3} \end{array}\right]$$

$$3R_2 + R_1 \to \left[\begin{array}{rr|r} 1 & 0 & -23 \\ 0 & 1 & -\frac{35}{3} \end{array}\right]$$

$$X = \begin{bmatrix} -23 \\ -\frac{35}{3} \end{bmatrix}$$

57. (a) $$\begin{bmatrix} 1 & -2 & 3 \\ -1 & 3 & -1 \\ 2 & -5 & 5 \end{bmatrix}\begin{bmatrix} x_1 \\ x_2 \\ x_3 \end{bmatrix} = \begin{bmatrix} 9 \\ -6 \\ 17 \end{bmatrix}$$

(b) $$\left[\begin{array}{rrr|r} 1 & -2 & 3 & 9 \\ -1 & 3 & -1 & -6 \\ 2 & -5 & 5 & 17 \end{array}\right]$$

$$\begin{array}{r} \\ R_1 + R_2 \to \\ -2R_2 + R_3 \to \end{array}\left[\begin{array}{rrr|r} 1 & -2 & 3 & 9 \\ 0 & 1 & 2 & 3 \\ 0 & -1 & -1 & -1 \end{array}\right]$$

$$\begin{array}{r} 2R_2 + R_1 \to \\ \\ R_2 + R_3 \to \end{array}\left[\begin{array}{rrr|r} 1 & 0 & 7 & 15 \\ 0 & 1 & 2 & 3 \\ 0 & 0 & 1 & 2 \end{array}\right]$$

$$\begin{array}{r} -7R_3 + R_1 \to \\ -2R_3 + R_2 \to \\ \\ \end{array}\left[\begin{array}{rrr|r} 1 & 0 & 0 & 1 \\ 0 & 1 & 0 & -1 \\ 0 & 0 & 1 & 2 \end{array}\right]$$

$$X = \begin{bmatrix} 1 \\ -1 \\ 2 \end{bmatrix}$$

58. (a) $$\begin{bmatrix} 1 & 1 & -3 \\ -1 & 2 & 0 \\ 1 & -1 & 1 \end{bmatrix}\begin{bmatrix} x_1 \\ x_2 \\ x_3 \end{bmatrix} = \begin{bmatrix} 9 \\ 6 \\ -5 \end{bmatrix}$$

(b) $$\left[\begin{array}{rrr|r} 1 & 1 & -3 & 9 \\ -1 & 2 & 0 & 6 \\ 1 & -1 & 1 & -5 \end{array}\right]$$

$$\begin{array}{r} \\ R_1 + R_2 \to \\ -R_1 + R_3 \to \end{array}\left[\begin{array}{rrr|r} 1 & 1 & -3 & 9 \\ 0 & 3 & -3 & 15 \\ 0 & -2 & 4 & -14 \end{array}\right]$$

$$\begin{array}{r} \\ \tfrac{1}{3}R_2 \to \\ \\ \end{array}\left[\begin{array}{rrr|r} 1 & 1 & -3 & 9 \\ 0 & 1 & -1 & 5 \\ 0 & -2 & 4 & -14 \end{array}\right]$$

$$\begin{array}{r} -R_2 + R_1 \to \\ \\ 2R_2 + R_3 \to \end{array}\left[\begin{array}{rrr|r} 1 & 0 & -2 & 4 \\ 0 & 1 & -1 & 5 \\ 0 & 0 & 2 & -4 \end{array}\right]$$

$$\begin{array}{r} \\ \\ \tfrac{1}{2}R_3 \to \end{array}\left[\begin{array}{rrr|r} 1 & 0 & -2 & 4 \\ 0 & 1 & -1 & 5 \\ 0 & 0 & 1 & -2 \end{array}\right]$$

$$\begin{array}{r} 2R_3 + R_1 \to \\ R_3 + R_2 \to \\ \\ \end{array}\left[\begin{array}{rrr|r} 1 & 0 & 0 & 0 \\ 0 & 1 & 0 & 3 \\ 0 & 0 & 1 & -2 \end{array}\right]$$

$$X = \begin{bmatrix} 0 \\ 3 \\ -2 \end{bmatrix}$$

59. (a) $$\begin{bmatrix} 1 & -5 & 2 \\ -3 & 1 & -1 \\ 0 & -2 & 5 \end{bmatrix}\begin{bmatrix} x_1 \\ x_2 \\ x_3 \end{bmatrix} = \begin{bmatrix} -20 \\ 8 \\ -16 \end{bmatrix}$$

(b) $$\left[\begin{array}{rrr|r} 1 & -5 & 2 & -20 \\ -3 & 1 & -1 & 8 \\ 0 & -2 & 5 & -16 \end{array}\right]$$

$$\begin{array}{r} \\ 3R_1 + R_2 \to \\ \\ \end{array}\left[\begin{array}{rrr|r} 1 & -5 & 2 & -20 \\ 0 & -14 & 5 & -52 \\ 0 & -2 & 5 & -16 \end{array}\right]$$

$$\begin{array}{r} \\ -R_3 + R_2 \to \\ \\ \end{array}\left[\begin{array}{rrr|r} 1 & -5 & 2 & -20 \\ 0 & -12 & 0 & -36 \\ 0 & -2 & 5 & -16 \end{array}\right]$$

$$\begin{array}{r} \\ -\tfrac{1}{12}R_2 \to \\ \\ \end{array}\left[\begin{array}{rrr|r} 1 & -5 & 2 & -20 \\ 0 & 1 & 0 & 3 \\ 0 & -2 & 5 & -16 \end{array}\right]$$

$$\begin{array}{r} 5R_2 + R_1 \to \\ \\ 2R_2 + R_3 \to \end{array}\left[\begin{array}{rrr|r} 1 & 0 & 2 & -5 \\ 0 & 1 & 0 & 3 \\ 0 & 0 & 5 & -10 \end{array}\right]$$

$$\begin{array}{r} \\ \\ \tfrac{1}{5}R_3 \to \end{array}\left[\begin{array}{rrr|r} 1 & 0 & 2 & -5 \\ 0 & 1 & 0 & 3 \\ 0 & 0 & 1 & -2 \end{array}\right]$$

$$\begin{array}{r} -2R_3 + R_1 \to \\ \\ \\ \end{array}\left[\begin{array}{rrr|r} 1 & 0 & 0 & -1 \\ 0 & 1 & 0 & 3 \\ 0 & 0 & 1 & -2 \end{array}\right]$$

$$X = \begin{bmatrix} -1 \\ 3 \\ -2 \end{bmatrix}$$

60. (a) $\begin{bmatrix} 1 & -1 & 4 \\ 1 & 3 & 0 \\ 0 & -6 & 5 \end{bmatrix}\begin{bmatrix} x_1 \\ x_2 \\ x_3 \end{bmatrix} = \begin{bmatrix} 17 \\ -11 \\ 40 \end{bmatrix}$

(b) $\begin{bmatrix} 1 & -1 & 4 & \vdots & 17 \\ 1 & 3 & 0 & \vdots & -11 \\ 0 & -6 & 5 & \vdots & 40 \end{bmatrix}$

$\begin{matrix} \\ -R_1 + R_2 \rightarrow \\ \\ \end{matrix}\begin{bmatrix} 1 & -1 & 4 & \vdots & 17 \\ 0 & 4 & -4 & \vdots & -28 \\ 0 & -6 & 5 & \vdots & 40 \end{bmatrix}$

$\begin{matrix} \\ \frac{1}{4}R_2 \rightarrow \\ \\ \end{matrix}\begin{bmatrix} 1 & -1 & 4 & \vdots & 17 \\ 0 & 1 & -1 & \vdots & -7 \\ 0 & -6 & 5 & \vdots & 40 \end{bmatrix}$

$\begin{matrix} \\ \\ 6R_2 + R_3 \rightarrow \end{matrix}\begin{bmatrix} 1 & -1 & 4 & \vdots & 17 \\ 0 & 1 & -1 & \vdots & -7 \\ 0 & 0 & -1 & \vdots & -2 \end{bmatrix}$

$\begin{matrix} R_2 + R_1 \rightarrow \\ \\ -R_3 \rightarrow \end{matrix}\begin{bmatrix} 1 & 0 & 3 & \vdots & 10 \\ 0 & 1 & -1 & \vdots & -7 \\ 0 & 0 & 1 & \vdots & 2 \end{bmatrix}$

$\begin{matrix} -3R_3 + R_1 \rightarrow \\ R_3 + R_2 \rightarrow \\ \\ \end{matrix}\begin{bmatrix} 1 & 0 & 0 & \vdots & 4 \\ 0 & 1 & 0 & \vdots & -5 \\ 0 & 0 & 1 & \vdots & 2 \end{bmatrix}$

$X = \begin{bmatrix} 4 \\ -5 \\ 2 \end{bmatrix}$

For 61–70, *A* is of order 2 × 3, *B* is of order 2 × 3, *C* is of order 3 × 2, and *D* is of order 2 × 2.

61. $A + 2C$ is not possible. A and C are not of the same order.

62. $B - 3C$ is not possible. B and C are not of the same order.

63. AB is not possible. The number of columns of A does not equal the number of rows of B.

64. BC is possible. The resulting order is 2×2.

65. $BC - D$ is possible. The resulting order is 2×2.

66. $CB - D$ is not possible. The order of CB is 3×3, but the order of D is 2×2.

67. (CA) is 3×3 so $(CA)D$ is not possible.

68. $(BC)D$ is possible. The resulting order is 2×2.

69. $D(A - 3B)$ is possible. The resulting order is 2×3.

70. $(BC - D)A$ is possible. The resulting order is 2×3.

71. The product of two diagonal matrices of the same order is a diagonal matrix whose entries are the products of the corresponding diagonal entries of A and B.

72. $\begin{bmatrix} a_{11} & a_{12} & a_{13} \\ a_{21} & a_{22} & a_{23} \\ a_{31} & a_{32} & a_{33} \end{bmatrix} + \begin{bmatrix} a_{11} & a_{12} & a_{13} \\ a_{21} & a_{22} & a_{23} \\ c_1 & c_2 & c_3 \end{bmatrix} = \begin{bmatrix} 2a_{11} & 2a_{12} & 2a_{13} \\ 2a_{21} & 2a_{22} & 2a_{23} \\ a_{31} + c_1 & a_{32} + c_2 & a_{33} + c_3 \end{bmatrix}$

The error is that the corresponding entries in rows 1 and 2 were not added in the text.

73. $1.2\begin{bmatrix} 70 & 50 & 25 \\ 35 & 100 & 70 \end{bmatrix} = \begin{bmatrix} 84 & 60 & 30 \\ 42 & 120 & 84 \end{bmatrix}$

74. $1.10\begin{bmatrix} 100 & 90 & 70 & 30 \\ 40 & 20 & 60 & 60 \end{bmatrix} = \begin{bmatrix} 110 & 99 & 77 & 33 \\ 44 & 22 & 66 & 66 \end{bmatrix}$

75. $BA = [3.50 \quad 6.00]\begin{bmatrix} 125 & 100 & 75 \\ 100 & 175 & 125 \end{bmatrix} = [\$1037.50 \quad \$1400.00 \quad \$1012.50]$

The entries in the matrix represent the profits for both crops at the three outlets.

76. $BA = [\$39.50 \quad \$44.50 \quad \$56.50]\begin{bmatrix} 5{,}000 & 4{,}000 \\ 6{,}000 & 10{,}000 \\ 8{,}000 & 5{,}000 \end{bmatrix} = [\$916{,}500 \quad \$885{,}500]$

The entries represent the costs of the three models of the product at the two warehouses.

77. $ST = \begin{bmatrix} 3 & 2 & 2 & 3 & 0 \\ 0 & 2 & 3 & 4 & 3 \\ 4 & 2 & 1 & 3 & 2 \end{bmatrix} \begin{bmatrix} 840 & 1100 \\ 1200 & 1350 \\ 1450 & 1650 \\ 2650 & 3000 \\ 3050 & 3200 \end{bmatrix} = \begin{bmatrix} \$15,770 & \$18,300 \\ \$26,500 & \$29,250 \\ \$21,260 & \$24,150 \end{bmatrix}$

The entries represent the wholesale and retail inventory values of the inventories at the three outlets.

78. $ST = \begin{bmatrix} 1.0 & 0.5 & 0.2 \\ 1.6 & 1.0 & 0.2 \\ 2.5 & 2.0 & 0.4 \end{bmatrix} \begin{bmatrix} \$12 & \$10 \\ \$9 & \$8 \\ \$8 & \$7 \end{bmatrix} = \begin{bmatrix} \$18.10 & \$15.40 \\ \$29.80 & \$25.40 \\ \$51.20 & \$43.80 \end{bmatrix}$

The columns of ST represent the total cost for labor to produce one small, medium and large boat at plants A and B respectively.

79. True. The sum of two matrices of different orders is undefined.

80. False. For most matrices, $AB \neq BA$.

81. False. $\begin{bmatrix} -2 & 4 \\ -3 & 0 \\ 6 & 1 \end{bmatrix} \begin{bmatrix} 1 & 1 \\ 1 & 1 \end{bmatrix} = \begin{bmatrix} 2 & 2 \\ -3 & -3 \\ 7 & 7 \end{bmatrix}$

82. True. An $n \times m$ matrix times an $m \times p$ matrix *does* yield an $n \times p$ matrix.

Section 14.6 The Inverse of a Square Matrix

1. $AB = \begin{bmatrix} 2 & 1 \\ 5 & 3 \end{bmatrix} \begin{bmatrix} 3 & -1 \\ -5 & 2 \end{bmatrix} = \begin{bmatrix} 6-5 & -2+2 \\ 15-15 & -5+6 \end{bmatrix} = \begin{bmatrix} 1 & 0 \\ 0 & 1 \end{bmatrix}$

$BA = \begin{bmatrix} 3 & -1 \\ -5 & 2 \end{bmatrix} \begin{bmatrix} 2 & 1 \\ 5 & 3 \end{bmatrix} = \begin{bmatrix} 6-5 & 3-3 \\ -10+10 & -5+6 \end{bmatrix} = \begin{bmatrix} 1 & 0 \\ 0 & 1 \end{bmatrix}$

2. $AB = \begin{bmatrix} 1 & -1 \\ -1 & 2 \end{bmatrix} \begin{bmatrix} 2 & 1 \\ 1 & 1 \end{bmatrix} = \begin{bmatrix} 2-1 & 1-1 \\ -2+2 & -1+2 \end{bmatrix} = \begin{bmatrix} 1 & 0 \\ 0 & 1 \end{bmatrix}$

$BA = \begin{bmatrix} 2 & 1 \\ 1 & 1 \end{bmatrix} \begin{bmatrix} 1 & -1 \\ -1 & 2 \end{bmatrix} = \begin{bmatrix} 2-1 & -2+2 \\ 1-1 & -1+2 \end{bmatrix} = \begin{bmatrix} 1 & 0 \\ 0 & 1 \end{bmatrix}$

3. $AB = \begin{bmatrix} 1 & 2 \\ 3 & 4 \end{bmatrix} \begin{bmatrix} -2 & 1 \\ \frac{3}{2} & -\frac{1}{2} \end{bmatrix} = \begin{bmatrix} -2+3 & 1-1 \\ -6+6 & 3-2 \end{bmatrix} = \begin{bmatrix} 1 & 0 \\ 0 & 1 \end{bmatrix}$

$BA = \begin{bmatrix} -2 & 1 \\ \frac{3}{2} & -\frac{1}{2} \end{bmatrix} \begin{bmatrix} 1 & 2 \\ 3 & 4 \end{bmatrix} = \begin{bmatrix} -2+3 & -4+4 \\ \frac{3}{2}-\frac{3}{2} & 3-2 \end{bmatrix} = \begin{bmatrix} 1 & 0 \\ 0 & 1 \end{bmatrix}$

4. $AB = \begin{bmatrix} 1 & -1 \\ 2 & 3 \end{bmatrix} \begin{bmatrix} \frac{3}{5} & \frac{1}{5} \\ -\frac{2}{5} & \frac{1}{5} \end{bmatrix} = \begin{bmatrix} \frac{3}{5}+\frac{2}{5} & \frac{1}{5}-\frac{1}{5} \\ \frac{6}{5}-\frac{6}{5} & \frac{2}{5}+\frac{3}{5} \end{bmatrix} = \begin{bmatrix} 1 & 0 \\ 0 & 1 \end{bmatrix}$

$BA = \begin{bmatrix} \frac{3}{5} & \frac{1}{5} \\ -\frac{2}{5} & \frac{1}{5} \end{bmatrix} \begin{bmatrix} 1 & -1 \\ 2 & 3 \end{bmatrix} = \begin{bmatrix} \frac{3}{5}+\frac{2}{5} & -\frac{3}{5}+\frac{3}{5} \\ -\frac{2}{5}+\frac{2}{5} & \frac{2}{5}+\frac{3}{5} \end{bmatrix} = \begin{bmatrix} 1 & 0 \\ 0 & 1 \end{bmatrix}$

5. $AB = \begin{bmatrix} 2 & -17 & 11 \\ -1 & 11 & -7 \\ 0 & 3 & -2 \end{bmatrix} \begin{bmatrix} 1 & 1 & 2 \\ 2 & 4 & -3 \\ 3 & 6 & -5 \end{bmatrix} = \begin{bmatrix} 2-17+33 & 2-68+66 & 4+51-55 \\ -1+22-21 & -1+44-42 & -2-33+35 \\ 6-6 & 12-12 & -9+10 \end{bmatrix} = \begin{bmatrix} 1 & 0 & 0 \\ 0 & 1 & 0 \\ 0 & 0 & 1 \end{bmatrix}$

$BA = \begin{bmatrix} 1 & 1 & 2 \\ 2 & 4 & -3 \\ 3 & 6 & -5 \end{bmatrix} \begin{bmatrix} 2 & -17 & 11 \\ -1 & 11 & -7 \\ 0 & 3 & -2 \end{bmatrix} = \begin{bmatrix} 2-1 & -17+11+6 & 11-7-4 \\ 4-4 & -34+44-9 & 22-28+6 \\ 6-6 & -51+66-15 & 33-42+10 \end{bmatrix} = \begin{bmatrix} 1 & 0 & 0 \\ 0 & 1 & 0 \\ 0 & 0 & 1 \end{bmatrix}$

6. $AB = \begin{bmatrix} -4 & 1 & 5 \\ -1 & 2 & 4 \\ 0 & -1 & -1 \end{bmatrix}\begin{bmatrix} -\frac{1}{2} & 1 & \frac{3}{2} \\ \frac{1}{4} & -1 & -\frac{11}{4} \\ -\frac{1}{4} & 1 & \frac{7}{4} \end{bmatrix}$

$= \begin{bmatrix} 2 + \frac{1}{4} - \frac{5}{4} & -4 - 1 + 5 & -6 - \frac{11}{4} + \frac{35}{4} \\ \frac{1}{2} + \frac{1}{2} - 1 & -1 - 2 + 4 & -\frac{3}{2} - \frac{11}{2} + 7 \\ -\frac{1}{4} + \frac{1}{4} & 1 - 1 & \frac{11}{4} - \frac{7}{4} \end{bmatrix}$

$= \begin{bmatrix} 1 & 0 & 0 \\ 0 & 1 & 0 \\ 0 & 0 & 1 \end{bmatrix}$

$BA = \begin{bmatrix} -\frac{1}{2} & 1 & \frac{3}{2} \\ \frac{1}{4} & -1 & -\frac{11}{4} \\ -\frac{1}{4} & 1 & \frac{7}{4} \end{bmatrix}\begin{bmatrix} -4 & 1 & 5 \\ -1 & 2 & 4 \\ 0 & -1 & -1 \end{bmatrix}$

$= \begin{bmatrix} 2 - 1 & -\frac{1}{2} + 2 - \frac{3}{2} & -\frac{5}{2} + 4 - \frac{3}{2} \\ -1 + 1 & \frac{1}{4} - 2 + \frac{11}{4} & \frac{5}{4} - 4 + \frac{11}{4} \\ 1 - 1 & -\frac{1}{4} + 2 - \frac{7}{4} & -\frac{5}{4} + 4 - \frac{7}{4} \end{bmatrix}$

$= \begin{bmatrix} 1 & 0 & 0 \\ 0 & 1 & 0 \\ 0 & 0 & 1 \end{bmatrix}$

7. $AB = \begin{bmatrix} 2 & 0 & 1 & 1 \\ 3 & 0 & 0 & 1 \\ -1 & 1 & -2 & 1 \\ 4 & -1 & 1 & 0 \end{bmatrix}\begin{bmatrix} -1 & 2 & -1 & -1 \\ -4 & 9 & -5 & -6 \\ 0 & 1 & -1 & -1 \\ 3 & -5 & 3 & 3 \end{bmatrix}$

$= \begin{bmatrix} -2 + 3 & 4 + 1 - 5 & -2 - 1 + 3 & -2 - 1 + 3 \\ 0 & 6 - 5 & 0 & 0 \\ 1 - 4 + 3 & -2 + 9 - 2 - 5 & 1 - 5 + 2 + 3 & 1 - 6 + 2 + 3 \\ 0 & 8 - 9 + 1 & -4 + 5 - 1 & -4 + 6 - 1 \end{bmatrix} = \begin{bmatrix} 1 & 0 & 0 & 0 \\ 0 & 1 & 0 & 0 \\ 0 & 0 & 1 & 0 \\ 0 & 0 & 0 & 1 \end{bmatrix}$

$BA = \begin{bmatrix} -1 & 2 & -1 & -1 \\ -4 & 9 & -5 & -6 \\ 0 & 1 & -1 & -1 \\ 3 & -5 & 3 & 3 \end{bmatrix}\begin{bmatrix} 2 & 0 & 1 & 1 \\ 3 & 0 & 0 & 1 \\ -1 & 1 & -2 & 1 \\ 4 & -1 & 1 & 0 \end{bmatrix}$

$= \begin{bmatrix} -2 + 6 + 1 - 4 & 0 & -1 + 2 - 1 & -1 + 2 - 1 \\ -8 + 27 + 5 - 24 & -5 + 6 & -4 + 10 - 6 & -4 + 9 - 5 \\ 3 + 1 - 4 & 0 & 2 - 1 & 0 \\ 6 - 15 - 3 + 12 & 0 & 3 - 6 + 3 & 3 - 5 + 3 \end{bmatrix} = \begin{bmatrix} 1 & 0 & 0 & 0 \\ 0 & 1 & 0 & 0 \\ 0 & 0 & 1 & 0 \\ 0 & 0 & 0 & 1 \end{bmatrix}$

8. $AB = \begin{bmatrix} -2 & 0 & 1 & 0 \\ 1 & -1 & -3 & 0 \\ -2 & -1 & 0 & -2 \\ 0 & 1 & 3 & -1 \end{bmatrix}\begin{bmatrix} -3 & -3 & 1 & -2 \\ 12 & 14 & -5 & 10 \\ -5 & -6 & 2 & -4 \\ -3 & -4 & 1 & -3 \end{bmatrix}$

$= \begin{bmatrix} 6 - 5 & 6 - 6 & -2 + 2 & 4 - 4 \\ -3 - 12 + 15 & -3 - 14 + 18 & 1 + 5 - 6 & -2 - 10 + 12 \\ 6 - 12 + 6 & 6 - 14 + 8 & -2 + 5 - 2 & 4 - 10 + 6 \\ 12 - 15 + 3 & 14 - 18 + 4 & -5 + 6 - 1 & 10 - 12 + 3 \end{bmatrix}$

$= \begin{bmatrix} 1 & 0 & 0 & 0 \\ 0 & 1 & 0 & 0 \\ 0 & 0 & 1 & 0 \\ 0 & 0 & 0 & 1 \end{bmatrix}$

$BA = \begin{bmatrix} -3 & -3 & 1 & -2 \\ 12 & 14 & -5 & 10 \\ -5 & -6 & 2 & -4 \\ -3 & -4 & 1 & -3 \end{bmatrix}\begin{bmatrix} -2 & 0 & 1 & 0 \\ 1 & -1 & -3 & 0 \\ -2 & -1 & 0 & -2 \\ 0 & 1 & 3 & -1 \end{bmatrix}$

$= \begin{bmatrix} 6 - 3 - 2 & 3 - 1 - 2 & -3 + 9 - 6 & -2 + 2 \\ -24 + 14 + 10 & -14 + 5 + 10 & 12 - 42 + 30 & 10 - 10 \\ 10 - 6 - 4 & 6 - 2 - 4 & -5 + 18 - 12 & -4 + 4 \\ 6 - 4 - 2 & 4 - 1 - 3 & -3 + 12 - 9 & -2 + 3 \end{bmatrix}$

$= \begin{bmatrix} 1 & 0 & 0 & 0 \\ 0 & 1 & 0 & 0 \\ 0 & 0 & 1 & 0 \\ 0 & 0 & 0 & 1 \end{bmatrix}$

9. $AB = \frac{1}{3}\begin{bmatrix} -2 & 2 & 3 \\ 1 & -1 & 0 \\ 0 & 1 & 4 \end{bmatrix}\begin{bmatrix} -4 & -5 & 3 \\ -4 & -8 & 3 \\ 1 & 2 & 0 \end{bmatrix} = \frac{1}{3}\begin{bmatrix} -8+8+3 & 10-16+6 & -6+6 \\ -4+4 & -5+8 & 3-3 \\ -4+4 & -8+8 & 3 \end{bmatrix}$

$= \frac{1}{3}\begin{bmatrix} 3 & 0 & 0 \\ 0 & 3 & 0 \\ 0 & 0 & 3 \end{bmatrix} = \begin{bmatrix} 1 & 0 & 0 \\ 0 & 1 & 0 \\ 0 & 0 & 1 \end{bmatrix}$

$BA = \frac{1}{3}\begin{bmatrix} -4 & -5 & 3 \\ -4 & -8 & 3 \\ 1 & 2 & 0 \end{bmatrix}\begin{bmatrix} -2 & 2 & 3 \\ 1 & -1 & 0 \\ 0 & 1 & 4 \end{bmatrix} = \frac{1}{3}\begin{bmatrix} 8-5 & -8+5+3 & -12+12 \\ 8-8 & -8+8+3 & -12+12 \\ -2+2 & 2-2 & 3 \end{bmatrix} = \begin{bmatrix} 1 & 0 & 0 \\ 0 & 1 & 0 \\ 0 & 0 & 1 \end{bmatrix}$

10. $AB = \frac{1}{3}\begin{bmatrix} -1 & 1 & 0 & -1 \\ 1 & -1 & 1 & 0 \\ -1 & 1 & 2 & 0 \\ 0 & -1 & 1 & 1 \end{bmatrix}\begin{bmatrix} -3 & 1 & 1 & -3 \\ -3 & -1 & 2 & -3 \\ 0 & 1 & 1 & 0 \\ -3 & -2 & 1 & 0 \end{bmatrix} = \frac{1}{3}\begin{bmatrix} 3-3+3 & -1-1+2 & -1+2-1 & 3-3 \\ -3+3 & 1+1+1 & 1-2+1 & -3+3 \\ 3-3 & -1-1+2 & -1+2+2 & 3-3 \\ 3-3 & 1+1-2 & -2+1+1 & 3 \end{bmatrix}$

$= \begin{bmatrix} 1 & 0 & 0 & 0 \\ 0 & 1 & 0 & 0 \\ 0 & 0 & 1 & 0 \\ 0 & 0 & 0 & 1 \end{bmatrix}$

$BA = \frac{1}{3}\begin{bmatrix} -3 & 1 & 1 & -3 \\ -3 & -1 & 2 & -3 \\ 0 & 1 & 1 & 0 \\ -3 & -2 & 1 & 0 \end{bmatrix}\begin{bmatrix} -1 & 1 & 0 & -1 \\ 1 & -1 & 1 & 0 \\ -1 & 1 & 2 & 0 \\ 0 & -1 & 1 & 1 \end{bmatrix} = \frac{1}{3}\begin{bmatrix} 3+1-1 & -3-1+1+3 & 1+2-3 & 3-3 \\ 3-1-2 & -3+1+2+3 & -1+4-3 & 3-3 \\ 1-1 & -1+1 & 1+2 & 0 \\ 3-2-1 & -3+2+1 & -2+2 & 3 \end{bmatrix}$

$= \begin{bmatrix} 1 & 0 & 0 & 0 \\ 0 & 1 & 0 & 0 \\ 0 & 0 & 1 & 0 \\ 0 & 0 & 0 & 1 \end{bmatrix}$

11. $[A \ \vdots \ I] = \left[\begin{array}{cc:cc} 2 & 0 & 1 & 0 \\ 0 & 3 & 0 & 1 \end{array}\right]$

$\begin{array}{l} \frac{1}{2}R_1 \rightarrow \\ \frac{1}{3}R_2 \rightarrow \end{array} \left[\begin{array}{cc:cc} 1 & 0 & \frac{1}{2} & 0 \\ 0 & 1 & 0 & \frac{1}{3} \end{array}\right] = [I \ \vdots \ A^{-1}]$

$A^{-1} = \begin{bmatrix} \frac{1}{2} & 0 \\ 0 & \frac{1}{3} \end{bmatrix}$

12. $[A \ \vdots \ I] = \left[\begin{array}{cc:cc} 1 & 2 & 1 & 0 \\ 3 & 7 & 0 & 1 \end{array}\right]$

$-3R_1 + R_2 \rightarrow \left[\begin{array}{cc:cc} 1 & 2 & 1 & 0 \\ 0 & 1 & -3 & 1 \end{array}\right]$

$-2R_2 + R_1 \rightarrow \left[\begin{array}{cc:cc} 1 & 0 & 7 & -2 \\ 0 & 1 & -3 & 1 \end{array}\right] = [I \ \vdots \ A^{-1}]$

$A^{-1} = \begin{bmatrix} 7 & -2 \\ -3 & 1 \end{bmatrix}$

13. $[A \ \vdots \ I] = \left[\begin{array}{cc:cc} 1 & -2 & 1 & 0 \\ 2 & -3 & 0 & 1 \end{array}\right]$

$-2R_1 + R_2 \rightarrow \left[\begin{array}{cc:cc} 1 & -2 & 1 & 0 \\ 0 & 1 & -2 & 1 \end{array}\right]$

$2R_2 + R_1 \rightarrow \left[\begin{array}{cc:cc} 1 & 0 & -3 & 2 \\ 0 & 1 & -2 & 1 \end{array}\right] = [I \ \vdots \ A^{-1}]$

$A^{-1} = \begin{bmatrix} -3 & 2 \\ -2 & 1 \end{bmatrix}$

14. $[A \ \vdots \ I] = \left[\begin{array}{cc:cc} -7 & 33 & 1 & 0 \\ 4 & -19 & 0 & 1 \end{array}\right]$

$2R_2 + R_1 \rightarrow \left[\begin{array}{cc:cc} 1 & -5 & 1 & 2 \\ 4 & -19 & 0 & 1 \end{array}\right]$

$-4R_1 + R_2 \rightarrow \left[\begin{array}{cc:cc} 1 & -5 & 1 & 2 \\ 0 & 1 & -4 & -7 \end{array}\right]$

$5R_2 + R_1 \rightarrow \left[\begin{array}{cc:cc} 1 & 0 & -19 & -33 \\ 0 & 1 & -4 & -7 \end{array}\right] = [I \ \vdots \ A^{-1}]$

$A^{-1} = \begin{bmatrix} -19 & -33 \\ -4 & -7 \end{bmatrix}$

15. $[A \ \vdots \ I] = \begin{bmatrix} -1 & 1 & \vdots & 1 & 0 \\ -2 & 1 & \vdots & 0 & 1 \end{bmatrix}$

$-R_2 + R_1 \rightarrow \begin{bmatrix} 1 & 0 & \vdots & 1 & -1 \\ -2 & 1 & \vdots & 0 & 1 \end{bmatrix}$

$2R_1 + R_2 \rightarrow \begin{bmatrix} 1 & 0 & \vdots & 1 & -1 \\ 0 & 1 & \vdots & 2 & -1 \end{bmatrix} = [I \ \vdots \ A^{-1}]$

$A^{-1} = \begin{bmatrix} 1 & -1 \\ 2 & -1 \end{bmatrix}$

16. $[A \ \vdots \ I] = \begin{bmatrix} 11 & 1 & \vdots & 1 & 0 \\ -1 & 0 & \vdots & 0 & 1 \end{bmatrix}$

$10R_2 + R_1 \rightarrow \begin{bmatrix} 1 & 1 & \vdots & 1 & 10 \\ -1 & 0 & \vdots & 0 & 1 \end{bmatrix}$

$R_1 + R_2 \rightarrow \begin{bmatrix} 1 & 1 & \vdots & 1 & 10 \\ 0 & 1 & \vdots & 1 & 11 \end{bmatrix}$

$-R_2 + R_1 \rightarrow \begin{bmatrix} 1 & 0 & \vdots & 0 & -1 \\ 0 & 1 & \vdots & 1 & 11 \end{bmatrix} = [I \ \vdots \ A^{-1}]$

$A^{-1} = \begin{bmatrix} 0 & -1 \\ 1 & 11 \end{bmatrix}$

17. $[A \ \vdots \ I] = \begin{bmatrix} 2 & 4 & \vdots & 1 & 0 \\ 4 & 8 & \vdots & 0 & 1 \end{bmatrix}$

$-2R_1 + R_2 \rightarrow \begin{bmatrix} 2 & 4 & \vdots & 1 & 0 \\ 0 & 0 & \vdots & -2 & 1 \end{bmatrix}$

The two zeros in the second row imply that the inverse does not exist.

18. $[A \ \vdots \ I] = \begin{bmatrix} 2 & 3 & \vdots & 1 & 0 \\ 1 & 4 & \vdots & 0 & 1 \end{bmatrix}$

$\begin{matrix} R_2 \\ R_1 \end{matrix} \begin{bmatrix} 1 & 4 & \vdots & 0 & 1 \\ 2 & 3 & \vdots & 1 & 0 \end{bmatrix}$

$-2R_1 + R_2 \rightarrow \begin{bmatrix} 1 & 4 & \vdots & 0 & 1 \\ 0 & -5 & \vdots & 1 & -2 \end{bmatrix}$

$-\frac{1}{5}R_2 \rightarrow \begin{bmatrix} 1 & 4 & \vdots & 0 & 1 \\ 0 & 1 & \vdots & -\frac{1}{5} & \frac{2}{5} \end{bmatrix}$

$-4R_2 + R_1 \rightarrow \begin{bmatrix} 1 & 0 & \vdots & \frac{4}{5} & -\frac{3}{5} \\ 0 & 1 & \vdots & -\frac{1}{5} & \frac{2}{5} \end{bmatrix} = [I \ \vdots \ A^{-1}]$

$A^{-1} = \frac{1}{5}\begin{bmatrix} 4 & -3 \\ -1 & 2 \end{bmatrix}$

19. $A = \begin{bmatrix} 2 & 7 & 1 \\ -3 & -9 & 2 \end{bmatrix}$ *A* has no inverse because it is not square.

20. $A = \begin{bmatrix} -2 & 5 \\ 6 & -15 \\ 0 & 1 \end{bmatrix}$ *A* has no inverse because it is not square.

21. $[A \ \vdots \ I] = \begin{bmatrix} 1 & 1 & 1 & \vdots & 1 & 0 & 0 \\ 3 & 5 & 4 & \vdots & 0 & 1 & 0 \\ 3 & 6 & 5 & \vdots & 0 & 0 & 1 \end{bmatrix}$

$\begin{matrix} \\ -3R_1 + R_2 \rightarrow \\ -3R_1 + R_3 \rightarrow \end{matrix} \begin{bmatrix} 1 & 1 & 1 & \vdots & 1 & 0 & 0 \\ 0 & 2 & 1 & \vdots & -3 & 1 & 0 \\ 0 & 3 & 2 & \vdots & -3 & 0 & 1 \end{bmatrix}$

$\begin{matrix} \\ \frac{1}{2}R_2 \rightarrow \\ \\ \end{matrix} \begin{bmatrix} 1 & 1 & 1 & \vdots & 1 & 0 & 0 \\ 0 & 1 & \frac{1}{2} & \vdots & -\frac{3}{2} & \frac{1}{2} & 0 \\ 0 & 3 & 2 & \vdots & -3 & 0 & 1 \end{bmatrix}$

$\begin{matrix} -R_2 + R_1 \rightarrow \\ \\ -3R_2 + R_3 \rightarrow \end{matrix} \begin{bmatrix} 1 & 0 & \frac{1}{2} & \vdots & \frac{5}{2} & -\frac{1}{2} & 0 \\ 0 & 1 & \frac{1}{2} & \vdots & -\frac{3}{2} & \frac{1}{2} & 0 \\ 0 & 0 & \frac{1}{2} & \vdots & \frac{3}{2} & -\frac{3}{2} & 1 \end{bmatrix}$

$\begin{matrix} -R_3 + R_1 \rightarrow \\ -R_3 + R_2 \rightarrow \\ \\ \end{matrix} \begin{bmatrix} 1 & 0 & 0 & \vdots & 1 & 1 & -1 \\ 0 & 1 & 0 & \vdots & -3 & 2 & -1 \\ 0 & 0 & \frac{1}{2} & \vdots & \frac{3}{2} & -\frac{3}{2} & 1 \end{bmatrix}$

$\begin{matrix} \\ \\ 2R_3 \rightarrow \end{matrix} \begin{bmatrix} 1 & 0 & 0 & \vdots & 1 & 1 & -1 \\ 0 & 1 & 0 & \vdots & -3 & 2 & -1 \\ 0 & 0 & 1 & \vdots & 3 & -3 & 2 \end{bmatrix} = [I \ \vdots \ A^{-1}]$

$A^{-1} = \begin{bmatrix} 1 & 1 & -1 \\ -3 & 2 & -1 \\ 3 & -3 & 2 \end{bmatrix}$

22. $$[A \;\vdots\; I] = \left[\begin{array}{rrr|rrr} 1 & 2 & 2 & 1 & 0 & 0 \\ 3 & 7 & 9 & 0 & 1 & 0 \\ -1 & -4 & -7 & 0 & 0 & 1 \end{array}\right]$$

$$\begin{array}{r} \\ -3R_1 + R_2 \rightarrow \\ R_1 + R_3 \rightarrow \end{array} \left[\begin{array}{rrr|rrr} 1 & 2 & 2 & 1 & 0 & 0 \\ 0 & 1 & 3 & -3 & 1 & 0 \\ 0 & -2 & -5 & 1 & 0 & 1 \end{array}\right]$$

$$\begin{array}{r} -2R_2 + R_1 \rightarrow \\ \\ 2R_2 + R_3 \rightarrow \end{array} \left[\begin{array}{rrr|rrr} 1 & 0 & -4 & 7 & -2 & 0 \\ 0 & 1 & 3 & -3 & 1 & 0 \\ 0 & 0 & 1 & -5 & 2 & 1 \end{array}\right]$$

$$\begin{array}{r} 4R_3 + R_1 \rightarrow \\ -3R_3 + R_2 \rightarrow \\ \\ \end{array} \left[\begin{array}{rrr|rrr} 1 & 0 & 0 & -13 & 6 & 4 \\ 0 & 1 & 0 & 12 & -5 & -3 \\ 0 & 0 & 1 & -5 & 2 & 1 \end{array}\right] = [I \;\vdots\; A^{-1}]$$

$$A^{-1} = \begin{bmatrix} -13 & 6 & 4 \\ 12 & -5 & -3 \\ -5 & 2 & 1 \end{bmatrix}$$

23. $$[A \;\vdots\; I] = \left[\begin{array}{rrr|rrr} 1 & 0 & 0 & 1 & 0 & 0 \\ 3 & 4 & 0 & 0 & 1 & 0 \\ 2 & 5 & 5 & 0 & 0 & 1 \end{array}\right]$$

$$\begin{array}{r} \\ -3R_1 + R_2 \rightarrow \\ -2R_1 + R_3 \rightarrow \end{array} \left[\begin{array}{rrr|rrr} 1 & 0 & 0 & 1 & 0 & 0 \\ 0 & 4 & 0 & -3 & 1 & 0 \\ 0 & 5 & 5 & -2 & 0 & 1 \end{array}\right]$$

$$\begin{array}{r} \\ \\ -\frac{5}{4}R_2 + R_3 \rightarrow \end{array} \left[\begin{array}{rrr|rrr} 1 & 0 & 0 & 1 & 0 & 0 \\ 0 & 4 & 0 & -3 & 1 & 0 \\ 0 & 0 & 5 & \frac{7}{4} & -\frac{5}{4} & 1 \end{array}\right]$$

$$\begin{array}{r} \\ \frac{1}{4}R_2 \rightarrow \\ \frac{1}{5}R_3 \rightarrow \end{array} \left[\begin{array}{rrr|rrr} 1 & 0 & 0 & 1 & 0 & 0 \\ 0 & 1 & 0 & -\frac{3}{4} & \frac{1}{4} & 0 \\ 0 & 0 & 1 & \frac{7}{20} & -\frac{1}{4} & \frac{1}{5} \end{array}\right] = [I \;\vdots\; A^{-1}]$$

$$A^{-1} = \begin{bmatrix} 1 & 0 & 0 \\ -\frac{3}{4} & \frac{1}{4} & 0 \\ \frac{7}{20} & -\frac{1}{4} & \frac{1}{5} \end{bmatrix}$$

24. $$[A \;\vdots\; I] = \left[\begin{array}{rrr|rrr} 1 & 0 & 0 & 1 & 0 & 0 \\ 3 & 0 & 0 & 0 & 1 & 0 \\ 2 & 5 & 5 & 0 & 0 & 1 \end{array}\right] \begin{array}{r} \\ -3R_1 + R_2 \rightarrow \\ -2R_1 + R_3 \rightarrow \end{array} \left[\begin{array}{rrr|rrr} 1 & 0 & 0 & 1 & 0 & 0 \\ 0 & 0 & 0 & -3 & 1 & 0 \\ 0 & 5 & 5 & -2 & 0 & 1 \end{array}\right]$$

Since the first three entries of row 2 are all zeros, the inverse of A does not exist.

25. $$[A \;\vdots\; I] = \left[\begin{array}{rrrr|rrrr} -8 & 0 & 0 & 0 & 1 & 0 & 0 & 0 \\ 0 & 1 & 0 & 0 & 0 & 1 & 0 & 0 \\ 0 & 0 & 4 & 0 & 0 & 0 & 1 & 0 \\ 0 & 0 & 0 & -5 & 0 & 0 & 0 & 1 \end{array}\right]$$

$$\begin{array}{r} -\frac{1}{8}R_1 \rightarrow \\ \\ \frac{1}{4}R_3 \rightarrow \\ -\frac{1}{5}R_4 \rightarrow \end{array} \left[\begin{array}{rrrr|rrrr} 1 & 0 & 0 & 0 & -\frac{1}{8} & 0 & 0 & 0 \\ 0 & 1 & 0 & 0 & 0 & 1 & 0 & 0 \\ 0 & 0 & 1 & 0 & 0 & 0 & \frac{1}{4} & 0 \\ 0 & 0 & 0 & 1 & 0 & 0 & 0 & -\frac{1}{5} \end{array}\right] = [I \;\vdots\; A^{-1}]$$

$$A^{-1} = \begin{bmatrix} -\frac{1}{8} & 0 & 0 & 0 \\ 0 & 1 & 0 & 0 \\ 0 & 0 & \frac{1}{4} & 0 \\ 0 & 0 & 0 & -\frac{1}{5} \end{bmatrix}$$

26. $[A \; \vdots \; I] = \left[\begin{array}{cccc:cccc} 1 & 3 & -2 & 0 & 1 & 0 & 0 & 0 \\ 0 & 2 & 4 & 6 & 0 & 1 & 0 & 0 \\ 0 & 0 & -2 & 1 & 0 & 0 & 1 & 0 \\ 0 & 0 & 0 & 5 & 0 & 0 & 0 & 1 \end{array}\right]$

$$\begin{array}{r} \\ \frac{1}{2}R_2 \rightarrow \\ \\ \frac{1}{5}R_4 \rightarrow \end{array} \left[\begin{array}{cccc:cccc} 1 & 3 & -2 & 0 & 1 & 0 & 0 & 0 \\ 0 & 1 & 2 & 3 & 0 & \frac{1}{2} & 0 & 0 \\ 0 & 0 & -2 & 1 & 0 & 0 & 1 & 0 \\ 0 & 0 & 0 & 1 & 0 & 0 & 0 & \frac{1}{5} \end{array}\right]$$

$$\begin{array}{r} -3R_2 + R_1 \rightarrow \\ R_3 + R_2 \rightarrow \\ -R_4 + R_3 \rightarrow \\ \\ \end{array} \left[\begin{array}{cccc:cccc} 1 & 0 & -8 & -9 & 1 & -\frac{3}{2} & 0 & 0 \\ 0 & 1 & 0 & 4 & 0 & \frac{1}{2} & 1 & 0 \\ 0 & 0 & -2 & 0 & 0 & 0 & 1 & -\frac{1}{5} \\ 0 & 0 & 0 & 1 & 0 & 0 & 0 & \frac{1}{5} \end{array}\right]$$

$$\begin{array}{r} -4R_3 + R_1 \rightarrow \\ -4R_4 + R_2 \rightarrow \\ -\frac{1}{2}R_3 \rightarrow \\ \\ \end{array} \left[\begin{array}{cccc:cccc} 1 & 0 & 0 & -9 & 1 & -\frac{3}{2} & -4 & \frac{4}{5} \\ 0 & 1 & 0 & 0 & 0 & \frac{1}{2} & 1 & -\frac{4}{5} \\ 0 & 0 & 1 & 0 & 0 & 0 & -\frac{1}{2} & \frac{1}{10} \\ 0 & 0 & 0 & 1 & 0 & 0 & 0 & \frac{1}{5} \end{array}\right]$$

$$\begin{array}{r} 9R_4 + R_1 \rightarrow \\ \\ \\ \\ \end{array} \left[\begin{array}{cccc:cccc} 1 & 0 & 0 & 0 & 1 & -\frac{3}{2} & -4 & \frac{13}{5} \\ 0 & 1 & 0 & 0 & 0 & \frac{1}{2} & 1 & -\frac{4}{5} \\ 0 & 0 & 1 & 0 & 0 & 0 & -\frac{1}{2} & \frac{1}{10} \\ 0 & 0 & 0 & 1 & 0 & 0 & 0 & \frac{1}{5} \end{array}\right] = [I \; \vdots \; A^{-1}]$$

$$A^{-1} = \frac{1}{10}\begin{bmatrix} 10 & -15 & -40 & 26 \\ 0 & 5 & 10 & -8 \\ 0 & 0 & -5 & 1 \\ 0 & 0 & 0 & 2 \end{bmatrix}$$

27. $A = \begin{bmatrix} 1 & 2 & -1 \\ 3 & 7 & -10 \\ -5 & -7 & -15 \end{bmatrix}$

$A^{-1} = \begin{bmatrix} -175 & 37 & -13 \\ 95 & -20 & 7 \\ 14 & -3 & 1 \end{bmatrix}$

28. $A = \begin{bmatrix} 10 & 5 & -7 \\ -5 & 1 & 4 \\ 3 & 2 & -2 \end{bmatrix}$

$A^{-1} = \begin{bmatrix} -10 & -4 & 27 \\ 2 & 1 & -5 \\ -13 & -5 & 35 \end{bmatrix}$

29. $A = \begin{bmatrix} 1 & 1 & 2 \\ 3 & 1 & 0 \\ -2 & 0 & 3 \end{bmatrix}$

$A^{-1} = \frac{1}{2}\begin{bmatrix} -3 & 3 & 2 \\ 9 & -7 & -6 \\ -2 & 2 & 2 \end{bmatrix} = \begin{bmatrix} -1.5 & 1.5 & 1 \\ 4.5 & -3.5 & -3 \\ -1 & 1 & 1 \end{bmatrix}$

30. $A = \begin{bmatrix} 3 & 2 & 2 \\ 2 & 2 & 2 \\ -4 & 4 & 3 \end{bmatrix}$

$A^{-1} = \begin{bmatrix} 1 & -1 & 0 \\ 7 & -8.5 & 1 \\ -8 & 10 & -1 \end{bmatrix}$

31. $A = \begin{bmatrix} -\frac{1}{2} & \frac{3}{4} & \frac{1}{4} \\ 1 & 0 & -\frac{3}{2} \\ 0 & -1 & \frac{1}{2} \end{bmatrix}$

$A^{-1} = \begin{bmatrix} -12 & -5 & -9 \\ -4 & -2 & -4 \\ -8 & -4 & -6 \end{bmatrix}$

32. $\begin{bmatrix} -\frac{5}{6} & \frac{1}{3} & \frac{11}{6} \\ 0 & \frac{2}{3} & 2 \\ 1 & -\frac{1}{2} & -\frac{5}{2} \end{bmatrix}$

A^{-1} does not exist.

33. $A = \begin{bmatrix} 0.1 & 0.2 & 0.3 \\ -0.3 & 0.2 & 0.2 \\ 0.5 & 0.4 & 0.4 \end{bmatrix}$

$A^{-1} = \frac{5}{11}\begin{bmatrix} 0 & -4 & 2 \\ -22 & 11 & 11 \\ 22 & -6 & -8 \end{bmatrix} = \begin{bmatrix} 0 & -1.\overline{81} & 0.\overline{90} \\ -10 & 5 & 5 \\ 10 & -2.\overline{72} & -3.\overline{63} \end{bmatrix}$

34. $A = \begin{bmatrix} 0.6 & 0 & -0.3 \\ 0.7 & -1 & 0.2 \\ 1 & 0 & -0.9 \end{bmatrix}$

$A^{-1} = \begin{bmatrix} 3.75 & 0 & -1.25 \\ 3.458\overline{3} & -1 & -1.375 \\ 4.1\overline{6} & 0 & -2.5 \end{bmatrix}$

35. $A = \begin{bmatrix} 1 & 0 & 3 & 0 \\ 0 & 2 & 0 & 4 \\ 1 & 0 & 3 & 0 \\ 0 & 2 & 0 & 4 \end{bmatrix}$

A^{-1} does not exist.

36. $A = \begin{bmatrix} 4 & 8 & -7 & 14 \\ 2 & 5 & -4 & 6 \\ 0 & 2 & 1 & -7 \\ 3 & 6 & -5 & 10 \end{bmatrix}$

$A^{-1} = \begin{bmatrix} 27 & -10 & 4 & -29 \\ -16 & 5 & -2 & 18 \\ -17 & 4 & -2 & 20 \\ -7 & 2 & -1 & 8 \end{bmatrix}$

37. $A = \begin{bmatrix} -1 & 0 & 1 & 0 \\ 0 & 2 & 0 & -1 \\ 2 & 0 & -1 & 0 \\ 0 & -1 & 0 & 1 \end{bmatrix}$

$A^{-1} = \begin{bmatrix} 1 & 0 & 1 & 0 \\ 0 & 1 & 0 & 1 \\ 2 & 0 & 1 & 0 \\ 0 & 1 & 0 & 2 \end{bmatrix}$

38. $A = \begin{bmatrix} 1 & -2 & -1 & -2 \\ 3 & -5 & -2 & -3 \\ 2 & -5 & -2 & -5 \\ -1 & 4 & 4 & 11 \end{bmatrix}, \quad A^{-1} = \begin{bmatrix} -24 & 7 & 1 & -2 \\ -10 & 3 & 0 & -1 \\ -29 & 7 & 3 & -2 \\ 12 & -3 & -1 & 1 \end{bmatrix}$

39. $A = \begin{bmatrix} a & b \\ c & d \end{bmatrix}, A^{-1} = \dfrac{1}{ad - bc}\begin{bmatrix} d & -b \\ -c & a \end{bmatrix}$

$A = \begin{bmatrix} 5 & -2 \\ 2 & 3 \end{bmatrix}$

$ad - bc = (5)(3) - (-2)(2) = 19$

$A^{-1} = \frac{1}{19}\begin{bmatrix} 3 & 2 \\ -2 & 5 \end{bmatrix} = \begin{bmatrix} \frac{3}{19} & \frac{2}{19} \\ -\frac{2}{19} & \frac{5}{19} \end{bmatrix}$

40. $A = \begin{bmatrix} 7 & 12 \\ -8 & -5 \end{bmatrix}$

$ad - bc = 7(-5) - 12(-8) = -35 + 96 = 61$

$A^{-1} = \frac{1}{61}\begin{bmatrix} -5 & -12 \\ 8 & 7 \end{bmatrix} = \begin{bmatrix} -\frac{5}{61} & -\frac{12}{61} \\ \frac{8}{61} & \frac{7}{61} \end{bmatrix}$

41. $A = \begin{bmatrix} -4 & -6 \\ 2 & 3 \end{bmatrix}$

$ad - bc = (-4)(3) - (2)(-6) = 0$

Since $ad - bc = 0$, A^{-1} does not exist.

42. $A = \begin{bmatrix} -12 & 3 \\ 5 & -2 \end{bmatrix}$

$ad - bc = (-12)(-2) - 3(5) = 24 - 15 = 9$

$A^{-1} = \frac{1}{9}\begin{bmatrix} -2 & -3 \\ -5 & -12 \end{bmatrix} = \begin{bmatrix} -\frac{2}{9} & -\frac{1}{3} \\ -\frac{5}{9} & -\frac{4}{3} \end{bmatrix}$

43. $A = \begin{bmatrix} \frac{7}{2} & -\frac{3}{4} \\ \frac{1}{5} & \frac{4}{5} \end{bmatrix}$

$ad - bc = \left(\frac{7}{2}\right)\left(\frac{4}{5}\right) - \left(-\frac{3}{4}\right)\left(\frac{1}{5}\right) = \frac{28}{10} + \frac{3}{20} = \frac{59}{20}$

$A^{-1} = \dfrac{1}{\frac{59}{20}}\begin{bmatrix} \frac{4}{5} & \frac{3}{4} \\ -\frac{1}{5} & \frac{7}{2} \end{bmatrix} = \frac{20}{59}\begin{bmatrix} \frac{4}{5} & \frac{3}{4} \\ -\frac{1}{5} & \frac{7}{2} \end{bmatrix} = \begin{bmatrix} \frac{16}{59} & \frac{15}{59} \\ -\frac{4}{59} & \frac{70}{59} \end{bmatrix}$

44. $A = \begin{bmatrix} -\frac{1}{4} & \frac{9}{4} \\ \frac{5}{3} & \frac{8}{9} \end{bmatrix}$

$ad - bc = \left(-\frac{1}{4}\right)\left(\frac{8}{9}\right) - \left(\frac{9}{4}\right)\left(\frac{5}{3}\right) = -\frac{143}{36}$

$A^{-1} = -\frac{36}{143}\begin{bmatrix} \frac{8}{9} & -\frac{9}{4} \\ -\frac{5}{3} & -\frac{1}{4} \end{bmatrix} = \begin{bmatrix} -\frac{32}{143} & \frac{81}{143} \\ \frac{60}{143} & \frac{9}{143} \end{bmatrix}$

45. $\begin{bmatrix} x \\ y \end{bmatrix} = \begin{bmatrix} -3 & 2 \\ -2 & 1 \end{bmatrix}\begin{bmatrix} 5 \\ 10 \end{bmatrix} = \begin{bmatrix} 5 \\ 0 \end{bmatrix}$

Solution: $(5, 0)$

46. $\begin{bmatrix} x \\ y \end{bmatrix} = \begin{bmatrix} -3 & 2 \\ -2 & 1 \end{bmatrix}\begin{bmatrix} 0 \\ 3 \end{bmatrix} = \begin{bmatrix} 6 \\ 3 \end{bmatrix}$

Solution: $(6, 3)$

47. $\begin{bmatrix} x \\ y \end{bmatrix} = \begin{bmatrix} -3 & 2 \\ -2 & 1 \end{bmatrix}\begin{bmatrix} 4 \\ 2 \end{bmatrix} = \begin{bmatrix} -8 \\ -6 \end{bmatrix}$

Solution: $(-8, -6)$

48. $\begin{bmatrix} x \\ y \end{bmatrix} = \begin{bmatrix} -3 & 2 \\ -2 & 1 \end{bmatrix}\begin{bmatrix} 1 \\ -2 \end{bmatrix} = \begin{bmatrix} -7 \\ -4 \end{bmatrix}$

Solution: $(-7, -4)$

49. $\begin{bmatrix} x \\ y \\ z \end{bmatrix} = \begin{bmatrix} 1 & 1 & -1 \\ -3 & 2 & -1 \\ 3 & -3 & 2 \end{bmatrix}\begin{bmatrix} 0 \\ 5 \\ 2 \end{bmatrix} = \begin{bmatrix} 3 \\ 8 \\ -11 \end{bmatrix}$

Solution: $(3, 8, -11)$

50. $\begin{bmatrix} x \\ y \\ z \end{bmatrix} = \begin{bmatrix} 1 & 1 & -1 \\ -3 & 2 & -1 \\ 3 & -3 & 2 \end{bmatrix}\begin{bmatrix} -1 \\ 2 \\ 0 \end{bmatrix} = \begin{bmatrix} 1 \\ 7 \\ -9 \end{bmatrix}$

Solution: $(1, 7, -9)$

51. $\begin{bmatrix} x_1 \\ x_2 \\ x_3 \\ x_4 \end{bmatrix} = \begin{bmatrix} -24 & 7 & 1 & -2 \\ -10 & 3 & 0 & -1 \\ -29 & 7 & 3 & -2 \\ 12 & -3 & -1 & 1 \end{bmatrix}\begin{bmatrix} 0 \\ 1 \\ -1 \\ 2 \end{bmatrix} = \begin{bmatrix} 2 \\ 1 \\ 0 \\ 0 \end{bmatrix}$

Solution: $(2, 1, 0, 0)$

52. $\begin{bmatrix} x_1 \\ x_2 \\ x_3 \\ x_4 \end{bmatrix} = \begin{bmatrix} -24 & 7 & 1 & -2 \\ -10 & 3 & 0 & -1 \\ -29 & 7 & 3 & -2 \\ 12 & -3 & -1 & 1 \end{bmatrix}\begin{bmatrix} 1 \\ -2 \\ 0 \\ -3 \end{bmatrix} = \begin{bmatrix} -32 \\ -13 \\ -37 \\ 15 \end{bmatrix}$

Solution: $(-32, -13, -37, 15)$

53. $A = \begin{bmatrix} 3 & 4 \\ 5 & 3 \end{bmatrix}$

$A^{-1} = \dfrac{1}{9 - 20}\begin{bmatrix} 3 & -4 \\ -5 & 3 \end{bmatrix}$

$\begin{bmatrix} x \\ y \end{bmatrix} = -\dfrac{1}{11}\begin{bmatrix} 3 & -4 \\ -5 & 3 \end{bmatrix}\begin{bmatrix} -2 \\ 4 \end{bmatrix} = -\dfrac{1}{11}\begin{bmatrix} -22 \\ 22 \end{bmatrix} = \begin{bmatrix} 2 \\ -2 \end{bmatrix}$

Solution: $(2, -2)$

54. $A = \begin{bmatrix} 18 & 12 \\ 30 & 24 \end{bmatrix}$

$A^{-1} = \dfrac{1}{432 - 360}\begin{bmatrix} 24 & -12 \\ -30 & 18 \end{bmatrix}$

$\begin{bmatrix} x \\ y \end{bmatrix} = \dfrac{1}{72}\begin{bmatrix} 24 & -12 \\ -30 & 18 \end{bmatrix}\begin{bmatrix} 13 \\ 23 \end{bmatrix}$

$= \dfrac{1}{72}\begin{bmatrix} 36 \\ 24 \end{bmatrix} = \begin{bmatrix} \frac{1}{2} \\ \frac{1}{3} \end{bmatrix}$

Solution: $\left(\frac{1}{2}, \frac{1}{3}\right)$

55. $A = \begin{bmatrix} -0.4 & 0.8 \\ 2 & -4 \end{bmatrix}$

$A^{-1} = \dfrac{1}{1.6 - 1.6}\begin{bmatrix} -4 & -0.8 \\ -2 & -0.4 \end{bmatrix}$

A^{-1} does not exist.

This implies that there is no unique solution; that is, either the system is inconsistent *or* there are infinitely many solutions.

Find the reduced row–echelon form of the matrix corresponding to the system.

$\begin{bmatrix} -0.4 & 0.8 & \vdots & 1.6 \\ 2 & -4 & \vdots & 5 \end{bmatrix}$

$-2.5R_1 \rightarrow \begin{bmatrix} 1 & -2 & \vdots & -4 \\ 2 & -4 & \vdots & 5 \end{bmatrix}$

$-2R_1 + R_2 \rightarrow \begin{bmatrix} 1 & -2 & \vdots & -4 \\ 0 & 0 & \vdots & 13 \end{bmatrix}$

The given system is inconsistent and there is no solution.

56. $A = \begin{bmatrix} 0.2 & -0.6 \\ -1 & 1.4 \end{bmatrix}$

$A^{-1} = \dfrac{1}{0.28 - 0.6}\begin{bmatrix} 1.4 & 0.6 \\ 1 & 0.2 \end{bmatrix}$

$\begin{bmatrix} x \\ y \end{bmatrix} = -\dfrac{1}{0.32}\begin{bmatrix} 1.4 & 0.6 \\ 1 & 0.2 \end{bmatrix}\begin{bmatrix} 2.4 \\ -8.8 \end{bmatrix}$

$= -\dfrac{1}{0.32}\begin{bmatrix} -1.92 \\ 0.64 \end{bmatrix} = \begin{bmatrix} 6 \\ -2 \end{bmatrix}$

Solution: $(6, -2)$

57. $A = \begin{bmatrix} 3 & 6 \\ 6 & 14 \end{bmatrix}$

$A^{-1} = \dfrac{1}{42 - 36}\begin{bmatrix} 14 & -6 \\ -6 & 3 \end{bmatrix}$

$\begin{bmatrix} x \\ y \end{bmatrix} = \dfrac{1}{6}\begin{bmatrix} 14 & -6 \\ -6 & 3 \end{bmatrix}\begin{bmatrix} 6 \\ 11 \end{bmatrix} = \dfrac{1}{6}\begin{bmatrix} 18 \\ -3 \end{bmatrix} = \begin{bmatrix} 3 \\ -\frac{1}{2} \end{bmatrix}$

Solution: $\left(3, -\frac{1}{2}\right)$

58. $A = \begin{bmatrix} 3 & 2 \\ 2 & 10 \end{bmatrix}$

$A^{-1} = \dfrac{1}{30-4}\begin{bmatrix} 10 & -2 \\ -2 & 3 \end{bmatrix}$

$\begin{bmatrix} x \\ y \end{bmatrix} = \dfrac{1}{26}\begin{bmatrix} 10 & -2 \\ -2 & 3 \end{bmatrix}\begin{bmatrix} 1 \\ 6 \end{bmatrix} = \dfrac{1}{26}\begin{bmatrix} -2 \\ 16 \end{bmatrix} = \begin{bmatrix} -\frac{1}{13} \\ \frac{8}{13} \end{bmatrix}$

Solution: $\left(-\frac{1}{13}, \frac{8}{13}\right)$

59. $A = \begin{bmatrix} -\frac{1}{4} & \frac{3}{8} \\ \frac{3}{2} & \frac{3}{4} \end{bmatrix}$

$A^{-1} = \dfrac{1}{-\frac{3}{16} - \frac{9}{16}}\begin{bmatrix} \frac{3}{4} & -\frac{3}{8} \\ -\frac{3}{2} & -\frac{1}{4} \end{bmatrix} = -\frac{4}{3}\begin{bmatrix} \frac{3}{4} & -\frac{3}{8} \\ -\frac{3}{2} & -\frac{1}{4} \end{bmatrix} = \begin{bmatrix} -1 & \frac{1}{2} \\ 2 & \frac{1}{3} \end{bmatrix}$

$\begin{bmatrix} x \\ y \end{bmatrix} = \begin{bmatrix} -1 & \frac{1}{2} \\ 2 & \frac{1}{3} \end{bmatrix}\begin{bmatrix} -2 \\ -12 \end{bmatrix} = \begin{bmatrix} -4 \\ -8 \end{bmatrix}$

Solution: $(-4, -8)$

60. $A = \begin{bmatrix} \frac{5}{6} & -1 \\ \frac{4}{3} & -\frac{7}{2} \end{bmatrix}$

$A^{-1} = \dfrac{1}{-\frac{35}{12} + \frac{4}{3}}\begin{bmatrix} -\frac{7}{2} & 1 \\ -\frac{4}{3} & \frac{5}{6} \end{bmatrix}$

$\begin{bmatrix} x \\ y \end{bmatrix} = -\dfrac{12}{19}\begin{bmatrix} -\frac{7}{2} & 1 \\ -\frac{4}{3} & \frac{5}{6} \end{bmatrix}\begin{bmatrix} -20 \\ -51 \end{bmatrix}$

$= -\dfrac{12}{19}\begin{bmatrix} 19 \\ -\frac{95}{6} \end{bmatrix} = \begin{bmatrix} -12 \\ 10 \end{bmatrix}$

Solution: $(-12, 10)$

61. $A = \begin{bmatrix} 4 & -1 & 1 \\ 2 & 2 & 3 \\ 5 & -2 & 6 \end{bmatrix}$

Find A^{-1}.

$$[A \;\vdots\; I] = \left[\begin{array}{rrr|rrr} 4 & -1 & 1 & 1 & 0 & 0 \\ 2 & 2 & 3 & 0 & 1 & 0 \\ 5 & -2 & 6 & 0 & 0 & 1 \end{array}\right]$$

$$\begin{matrix} R_1 \\ \\ R_3 \end{matrix}\left[\begin{array}{rrr|rrr} 5 & -2 & 6 & 0 & 0 & 1 \\ 2 & 2 & 3 & 0 & 1 & 0 \\ 4 & -1 & 1 & 1 & 0 & 0 \end{array}\right]$$

$$\begin{matrix} -R_3 + R_1 \rightarrow \\ \\ \\ \end{matrix}\left[\begin{array}{rrr|rrr} 1 & -1 & 5 & -1 & 0 & 1 \\ 2 & 2 & 3 & 0 & 1 & 0 \\ 4 & -1 & 1 & 1 & 0 & 0 \end{array}\right]$$

$$\begin{matrix} \\ -2R_1 + R_2 \rightarrow \\ -4R_1 + R_3 \rightarrow \end{matrix}\left[\begin{array}{rrr|rrr} 1 & -1 & 5 & -1 & 0 & 1 \\ 0 & 4 & -7 & 2 & 1 & -2 \\ 0 & 3 & -19 & 5 & 0 & -4 \end{array}\right]$$

$$\begin{matrix} \\ -R_3 + R_2 \rightarrow \\ \\ \end{matrix}\left[\begin{array}{rrr|rrr} 1 & -1 & 5 & -1 & 0 & 1 \\ 0 & 1 & 12 & -3 & 1 & 2 \\ 0 & 3 & -19 & 5 & 0 & -4 \end{array}\right]$$

$$\begin{matrix} R_2 + R_1 \rightarrow \\ \\ -3R_2 + R_3 \rightarrow \end{matrix}\left[\begin{array}{rrr|rrr} 1 & 0 & 17 & -4 & 1 & 3 \\ 0 & 1 & 12 & -3 & 1 & 2 \\ 0 & 0 & -55 & 14 & -3 & -10 \end{array}\right]$$

$$\begin{matrix} \\ \\ -\frac{1}{55}R_3 \rightarrow \end{matrix}\left[\begin{array}{rrr|rrr} 1 & 0 & 17 & -4 & 1 & 3 \\ 0 & 1 & 12 & -3 & 1 & 2 \\ 0 & 0 & 1 & -\frac{14}{55} & \frac{3}{55} & \frac{2}{11} \end{array}\right]$$

$$\begin{matrix} -17R_3 + R_1 \rightarrow \\ -12R_3 + R_2 \rightarrow \\ \\ \end{matrix}\left[\begin{array}{rrr|rrr} 1 & 0 & 0 & \frac{18}{55} & \frac{4}{55} & -\frac{1}{11} \\ 0 & 1 & 0 & \frac{3}{55} & \frac{19}{55} & -\frac{2}{11} \\ 0 & 0 & 1 & -\frac{14}{55} & \frac{3}{55} & \frac{2}{11} \end{array}\right] = [I \;\vdots\; A^{-1}]$$

$$A^{-1} = \frac{1}{55}\begin{bmatrix} 18 & 4 & -5 \\ 3 & 19 & -10 \\ -14 & 3 & 10 \end{bmatrix}$$

$$\begin{bmatrix} x \\ y \\ z \end{bmatrix} = \frac{1}{55}\begin{bmatrix} 18 & 4 & -5 \\ 3 & 19 & -10 \\ -14 & 3 & 10 \end{bmatrix}\begin{bmatrix} -5 \\ 10 \\ 1 \end{bmatrix} = \frac{1}{55}\begin{bmatrix} -55 \\ 165 \\ 110 \end{bmatrix} = \begin{bmatrix} -1 \\ 3 \\ 2 \end{bmatrix}$$

Solution: $(-1, 3, 2)$

62. $A = \begin{bmatrix} 4 & -2 & 3 \\ 2 & 2 & 5 \\ 8 & -5 & -2 \end{bmatrix}$

$A^{-1} = \frac{1}{82}\begin{bmatrix} -21 & 19 & 16 \\ -44 & 32 & 14 \\ 26 & -4 & -12 \end{bmatrix}$

$\begin{bmatrix} x \\ y \\ z \end{bmatrix} = \frac{1}{82}\begin{bmatrix} -21 & 19 & 16 \\ -44 & 32 & 14 \\ 26 & -4 & -12 \end{bmatrix}\begin{bmatrix} -2 \\ 16 \\ 4 \end{bmatrix} = \frac{1}{82}\begin{bmatrix} 410 \\ 656 \\ -164 \end{bmatrix} = \begin{bmatrix} 5 \\ 8 \\ -2 \end{bmatrix}$

Solution: $(5, 8, -2)$

63. $A = \begin{bmatrix} 5 & -3 & 2 \\ 2 & 2 & -3 \\ 1 & -7 & 8 \end{bmatrix}$

A^{-1} does not exist. This implies that there is no unique solution; that is, either the system is inconsistent *or* the system has infinitely many solution. Use a graphing utility to find the reduced row–echelon form of the matrix corresponding to the system.

$\begin{bmatrix} 5 & -3 & 2 & \vdots & 2 \\ 2 & 2 & -3 & \vdots & 3 \\ 1 & -7 & 8 & \vdots & -4 \end{bmatrix}$

$\begin{bmatrix} 1 & 0 & -\frac{5}{16} & \vdots & \frac{13}{16} \\ 0 & 1 & -\frac{19}{16} & \vdots & \frac{11}{16} \\ 0 & 0 & 0 & \vdots & 0 \end{bmatrix}$

$\begin{cases} x - \frac{5}{16}z = \frac{13}{16} \\ y - \frac{19}{16}z = \frac{11}{16} \end{cases}$

Let $z = a$. Then $x = \frac{5}{16}a + \frac{13}{16}$ and $y = \frac{19}{16}a + \frac{11}{16}$.

Solution: $\left(\frac{5}{16}a + \frac{13}{16}, \frac{19}{16}a + \frac{11}{16}, 16a\right)$, where a is a real number.

64. $A = \begin{bmatrix} 3 & -2 & 1 \\ -4 & 1 & -3 \\ 1 & -5 & 1 \end{bmatrix}$

$A^{-1} = \begin{bmatrix} 0.56 & 0.12 & -0.2 \\ -0.04 & -0.08 & -0.2 \\ -0.76 & -0.52 & 0.2 \end{bmatrix}$

$\begin{bmatrix} x \\ y \\ z \end{bmatrix} = \begin{bmatrix} 0.56 & 0.12 & -0.2 \\ -0.04 & -0.08 & -0.2 \\ -0.76 & -0.52 & 0.2 \end{bmatrix}\begin{bmatrix} -29 \\ 37 \\ -24 \end{bmatrix} = \begin{bmatrix} -7 \\ 3 \\ -2 \end{bmatrix}$

Solution: $(-7, 3, -2)$

65. $A = \begin{bmatrix} 2 & 3 & 5 \\ 3 & 5 & 9 \\ 5 & 9 & 17 \end{bmatrix}$

A^{-1} does not exist. This implies that there is no unique solution; that is, either the system is inconsistent *or* the system has infinitely many solution. Use a graphing utility to find the reduced row–echelon form of the matrix corresponding to the system.

$\begin{bmatrix} 2 & 3 & 5 & \vdots & 4 \\ 3 & 5 & 9 & \vdots & 7 \\ 5 & 9 & 17 & \vdots & 13 \end{bmatrix}$

$\begin{bmatrix} 1 & 0 & -2 & \vdots & -1 \\ 0 & 1 & 3 & \vdots & 2 \\ 0 & 0 & 0 & \vdots & 0 \end{bmatrix}$

$\begin{cases} x - 2z = -1 \\ y + 3z = 2 \end{cases}$

Let $z = a$. Then $x = 2a - 1$ and $y = -3a + 2$.

Solution: $(2a - 1, -3a + 2, a)$, where a is a real number.

66. $A = \begin{bmatrix} -8 & 7 & -10 \\ 12 & 3 & -5 \\ 15 & -9 & 2 \end{bmatrix}$

$A^{-1} \approx \begin{bmatrix} -0.034 & 0.066 & -0.004 \\ -0.086 & 0.117 & -0.139 \\ -0.133 & 0.029 & -0.094 \end{bmatrix}$

$\begin{bmatrix} x \\ y \\ z \end{bmatrix} \approx \begin{bmatrix} -0.034 & 0.066 & -0.004 \\ -0.086 & 0.117 & -0.139 \\ -0.133 & 0.029 & -0.094 \end{bmatrix}\begin{bmatrix} -151 \\ 86 \\ 187 \end{bmatrix} \approx \begin{bmatrix} 10 \\ -3 \\ 5 \end{bmatrix}$

Solution: $(10, -3, 5)$

67. $A = \begin{bmatrix} 7 & -3 & 0 & 2 \\ -2 & 1 & 0 & -1 \\ 4 & 0 & 1 & -2 \\ -1 & 1 & 0 & -1 \end{bmatrix}$

$A^{-1} = \begin{bmatrix} 0 & -1 & 0 & 1 \\ -1 & -5 & 0 & 3 \\ -2 & -4 & 1 & -2 \\ -1 & -4 & 0 & 1 \end{bmatrix}$

$$\begin{bmatrix} x \\ y \\ z \\ w \end{bmatrix} = \begin{bmatrix} 0 & -1 & 0 & 1 \\ -1 & -5 & 0 & 3 \\ -2 & -4 & 1 & -2 \\ -1 & -4 & 0 & 1 \end{bmatrix} \begin{bmatrix} 41 \\ -13 \\ 12 \\ -8 \end{bmatrix} = \begin{bmatrix} 5 \\ 0 \\ -2 \\ 3 \end{bmatrix}$$

Solution: $(5, 0, -2, 3)$

68. $A = \begin{bmatrix} 2 & 5 & 0 & 1 \\ 1 & 4 & 2 & -2 \\ 2 & -2 & 5 & 1 \\ 1 & 0 & 0 & -3 \end{bmatrix}$

$A^{-1} \approx \begin{bmatrix} 0.338 & -0.352 & 0.141 & 0.394 \\ 0.042 & 0.164 & -0.066 & -0.117 \\ -0.141 & 0.230 & 0.108 & -0.164 \\ 0.113 & -0.117 & 0.047 & -0.202 \end{bmatrix}$

$$\begin{bmatrix} x \\ y \\ z \\ w \end{bmatrix} \approx \begin{bmatrix} 0.338 & -0.352 & 0.141 & 0.394 \\ 0.042 & 0.164 & -0.066 & -0.117 \\ -0.141 & 0.230 & 0.108 & -0.164 \\ 0.113 & -0.117 & 0.047 & -0.202 \end{bmatrix} \begin{bmatrix} 11 \\ -7 \\ 3 \\ -1 \end{bmatrix} \approx \begin{bmatrix} 6.21 \\ -0.77 \\ -2.67 \\ 2.40 \end{bmatrix}$$

Solution: $(6.21, -0.77, -2.67, 2.40)$

69. The inverse matrix can be calculated once and used for more than one exercise.

70. If A is a square matrix and if there exists matrix B such that $AB = BA =$ identity matrix, then B is called the inverse of A and is denoted by A^{-1}.

71. A matrix is singular if it does not have an inverse.

72. (a) Given $A = \begin{bmatrix} a_{11} & 0 \\ 0 & a_{22} \end{bmatrix}$, $A^{-1} = \begin{bmatrix} \frac{1}{a_{11}} & 0 \\ 0 & \frac{1}{a_{22}} \end{bmatrix}$.

Given $A = \begin{bmatrix} a_{11} & 0 & 0 \\ 0 & a_{22} & 0 \\ 0 & 0 & a_{33} \end{bmatrix}$,

$$A^{-1} = \begin{bmatrix} \frac{1}{a_{11}} & 0 & 0 \\ 0 & \frac{1}{a_{22}} & 0 \\ 0 & 0 & \frac{1}{a_{33}} \end{bmatrix}.$$

(b) In general, the inverse of a matrix in the form of A is

$$\begin{bmatrix} \frac{1}{a_{11}} & 0 & 0 & \cdots & 0 \\ 0 & \frac{1}{a_{22}} & 0 & \cdots & 0 \\ 0 & 0 & \frac{1}{a_{33}} & \cdots & 0 \\ \vdots & \vdots & \vdots & \cdots & \vdots \\ 0 & 0 & 0 & \cdots & \frac{1}{a_{nn}} \end{bmatrix}.$$

73. $A = \begin{bmatrix} 1 & 1 & 1 \\ 0.065 & 0.07 & 0.09 \\ 0 & 2 & -1 \end{bmatrix}$ $\quad [A \;\vdots\; I] = \left[\begin{array}{ccc|ccc} 1 & 1 & 1 & 1 & 0 & 0 \\ 0.065 & 0.07 & 0.09 & 0 & 1 & 0 \\ 0 & 2 & -1 & 0 & 0 & 1 \end{array}\right]$

$$200R_2 \rightarrow \left[\begin{array}{ccc|ccc} 1 & 1 & 1 & 1 & 0 & 0 \\ 13 & 14 & 18 & 0 & 200 & 0 \\ 0 & 2 & -1 & 0 & 0 & 1 \end{array}\right]$$

$$-13R_1 + R_2 \rightarrow \left[\begin{array}{ccc|ccc} 1 & 1 & 1 & 1 & 0 & 0 \\ 0 & 1 & 5 & -13 & 200 & 0 \\ 0 & 2 & -1 & 0 & 0 & 1 \end{array}\right]$$

$$\begin{matrix} -R_2 + R_1 \rightarrow \\ \\ -2R_2 + R_3 \rightarrow \end{matrix} \left[\begin{array}{ccc|ccc} 1 & 0 & -4 & 14 & -200 & 0 \\ 0 & 1 & 5 & -13 & 200 & 0 \\ 0 & 0 & -11 & 26 & -400 & 1 \end{array}\right]$$

$$-\tfrac{1}{11}R_3 \rightarrow \left[\begin{array}{ccc|ccc} 1 & 0 & -4 & 14 & -200 & 0 \\ 0 & 1 & 5 & -13 & 200 & 0 \\ 0 & 0 & 1 & -\frac{26}{11} & \frac{400}{11} & -\frac{1}{11} \end{array}\right]$$

$$\begin{matrix} 4R_3 + R_1 \rightarrow \\ -5R_3 + R_2 \rightarrow \\ \\ \end{matrix} \left[\begin{array}{ccc|ccc} 1 & 0 & 0 & \frac{50}{11} & -\frac{600}{11} & -\frac{4}{11} \\ 0 & 1 & 0 & -\frac{13}{11} & \frac{200}{11} & \frac{5}{11} \\ 0 & 0 & 1 & -\frac{26}{11} & \frac{400}{11} & -\frac{1}{11} \end{array}\right] = [I \;\vdots\; A^{-1}]$$

$$X = A^{-1}B = \frac{1}{11}\begin{bmatrix} 50 & -600 & -4 \\ -13 & 200 & 5 \\ -26 & 400 & -1 \end{bmatrix}\begin{bmatrix} 10{,}000 \\ 705 \\ 0 \end{bmatrix} = \begin{bmatrix} 7000 \\ 1000 \\ 2000 \end{bmatrix}$$

Answer: \$7000 in AAA–rated bonds, \$1000 in A–rated bonds, \$2000 in B–rated bonds

74. $A = \begin{bmatrix} 1 & 1 & 1 \\ 0.065 & 0.07 & 0.09 \\ 0 & 2 & -1 \end{bmatrix}$

$$[A \;\vdots\; I] = \left[\begin{array}{ccc|ccc} 1 & 1 & 1 & 1 & 0 & 0 \\ 0.065 & 0.07 & 0.09 & 0 & 1 & 0 \\ 0 & 2 & -1 & 0 & 0 & 1 \end{array}\right]$$

$$200R_2 \rightarrow \left[\begin{array}{ccc|ccc} 1 & 1 & 1 & 1 & 0 & 0 \\ 13 & 14 & 18 & 0 & 200 & 0 \\ 0 & 2 & -1 & 0 & 0 & 1 \end{array}\right]$$

$$-13R_1 + R_2 \rightarrow \left[\begin{array}{ccc|ccc} 1 & 1 & 1 & 1 & 0 & 0 \\ 0 & 1 & 5 & -13 & 200 & 0 \\ 0 & 2 & -1 & 0 & 0 & 1 \end{array}\right]$$

$$\begin{matrix} -R_2 + R_1 \rightarrow \\ \\ -2R_2 + R_3 \rightarrow \end{matrix} \left[\begin{array}{ccc|ccc} 1 & 0 & -4 & 14 & -200 & 0 \\ 0 & 1 & 5 & -13 & 200 & 0 \\ 0 & 0 & -11 & 26 & -400 & 1 \end{array}\right]$$

$$-\tfrac{1}{11}R_3 \rightarrow \left[\begin{array}{ccc|ccc} 1 & 0 & -4 & 14 & -200 & 0 \\ 0 & 1 & 5 & -13 & 200 & 0 \\ 0 & 0 & 1 & -\frac{26}{11} & \frac{400}{11} & -\frac{1}{11} \end{array}\right]$$

$$\begin{matrix} 4R_3 + R_1 \rightarrow \\ -5R_3 + R_2 \rightarrow \\ \\ \end{matrix} \left[\begin{array}{ccc|ccc} 1 & 0 & 0 & \frac{50}{11} & -\frac{600}{11} & -\frac{4}{11} \\ 0 & 1 & 0 & -\frac{13}{11} & \frac{200}{11} & \frac{5}{11} \\ 0 & 0 & 1 & -\frac{26}{11} & \frac{400}{11} & -\frac{1}{11} \end{array}\right] = [I \;\vdots\; A^{-1}]$$

$$X = A^{-1}B = \frac{1}{11}\begin{bmatrix} 50 & -600 & -4 \\ -13 & 200 & 5 \\ -26 & 400 & -1 \end{bmatrix}\begin{bmatrix} 10{,}000 \\ 760 \\ 0 \end{bmatrix} = \begin{bmatrix} 4000 \\ 2000 \\ 4000 \end{bmatrix}$$

Solution: \$4000 in AAA-rated bonds, \$2000 in A-rated bonds, \$4000 in B-rated bonds.

75. Use the inverse matrix A^{-1} from Exercise 73.

$$X = A^{-1}B = \frac{1}{11}\begin{bmatrix} 50 & -600 & -4 \\ -13 & 200 & 5 \\ -26 & 400 & -1 \end{bmatrix}\begin{bmatrix} 12{,}000 \\ 835 \\ 0 \end{bmatrix} = \begin{bmatrix} 9000 \\ 1000 \\ 2000 \end{bmatrix}$$

Answer: \$9000 in AAA–rated bonds, \$1000 in A–rated bonds, \$2000 in B–rated bonds

76. Use the inverse matrix A^{-1} from Exercise 74.

$$X = A^{-1}B = \frac{1}{11}\begin{bmatrix} 50 & -600 & -4 \\ -13 & 200 & 5 \\ -26 & 400 & -1 \end{bmatrix}\begin{bmatrix} 500{,}000 \\ 38{,}000 \\ 0 \end{bmatrix} = \begin{bmatrix} 200{,}000 \\ 100{,}000 \\ 200{,}000 \end{bmatrix}$$

Solution: \$200,000 in AAA-rated bonds, \$100,000 in A-rated bonds, and \$200,000 in B-rated bonds.

77. $A = \begin{bmatrix} 2 & 0 & 4 \\ 0 & 1 & 4 \\ 1 & 1 & -1 \end{bmatrix}$

$$[A \;\vdots\; I] = \left[\begin{array}{ccc|ccc} 2 & 0 & 4 & 1 & 0 & 0 \\ 0 & 1 & 4 & 0 & 1 & 0 \\ 1 & 1 & -1 & 0 & 0 & 1 \end{array}\right]$$

$$\begin{matrix} R_1 \\ \\ R_3 \end{matrix}\left[\begin{array}{ccc|ccc} 1 & 1 & -1 & 0 & 0 & 1 \\ 0 & 1 & 4 & 0 & 1 & 0 \\ 2 & 0 & 4 & 1 & 0 & 0 \end{array}\right]$$

$$\begin{matrix} \\ \\ -2R_1 + R_3 \rightarrow \end{matrix}\left[\begin{array}{ccc|ccc} 1 & 1 & -1 & 0 & 0 & 1 \\ 0 & 1 & 4 & 0 & 1 & 0 \\ 0 & -2 & 6 & 1 & 0 & -2 \end{array}\right]$$

$$\begin{matrix} -R_2 + R_1 \rightarrow \\ \\ 2R_2 + R_3 \rightarrow \end{matrix}\left[\begin{array}{ccc|ccc} 1 & 0 & -5 & 0 & -1 & 1 \\ 0 & 1 & 4 & 0 & 1 & 0 \\ 0 & 0 & 14 & 1 & 2 & -2 \end{array}\right]$$

$$\begin{matrix} \\ \\ \frac{1}{14}R_3 \rightarrow \end{matrix}\left[\begin{array}{ccc|ccc} 1 & 0 & -5 & 0 & -1 & 1 \\ 0 & 1 & 4 & 0 & 1 & 0 \\ 0 & 0 & 1 & \frac{1}{14} & \frac{1}{7} & -\frac{1}{7} \end{array}\right]$$

$$\begin{matrix} 5R_3 + R_1 \rightarrow \\ -4R_3 + R_2 \rightarrow \\ \\ \end{matrix}\left[\begin{array}{ccc|ccc} 1 & 0 & 0 & \frac{5}{14} & -\frac{2}{7} & \frac{2}{7} \\ 0 & 1 & 0 & -\frac{2}{7} & \frac{3}{7} & \frac{4}{7} \\ 0 & 0 & 1 & \frac{1}{14} & \frac{1}{7} & -\frac{1}{7} \end{array}\right] = [I \;\vdots\; A^{-1}]$$

$$A^{-1} = \frac{1}{14}\begin{bmatrix} 5 & -4 & 4 \\ -4 & 6 & 8 \\ 1 & 2 & -2 \end{bmatrix}$$

$$\begin{bmatrix} I_1 \\ I_2 \\ I_3 \end{bmatrix} = \frac{1}{14}\begin{bmatrix} 5 & -4 & 4 \\ -4 & 6 & 8 \\ 1 & 2 & -2 \end{bmatrix}\begin{bmatrix} 14 \\ 28 \\ 0 \end{bmatrix} = \begin{bmatrix} -3 \\ 8 \\ 5 \end{bmatrix}$$

Answer: $I_1 = -3$ amperes, $I_2 = 8$ amperes, $I_3 = 5$ amperes

78. $A = \begin{bmatrix} 2 & 0 & 4 \\ 0 & 1 & 4 \\ 1 & 1 & -1 \end{bmatrix}$

$$[A \ \vdots \ I] = \left[\begin{array}{ccc|ccc} 2 & 0 & 4 & 1 & 0 & 0 \\ 0 & 1 & 4 & 0 & 1 & 0 \\ 1 & 1 & -1 & 0 & 0 & 1 \end{array}\right]$$

$$\begin{matrix} R_3 \\ \\ R_1 \end{matrix}\left[\begin{array}{ccc|ccc} 1 & 1 & -1 & 0 & 0 & 1 \\ 0 & 1 & 4 & 0 & 1 & 0 \\ 2 & 0 & 4 & 1 & 0 & 0 \end{array}\right]$$

$$\begin{matrix} \\ \\ -2R_1 + R_3 \rightarrow \end{matrix}\left[\begin{array}{ccc|ccc} 1 & 1 & -1 & 0 & 0 & 1 \\ 0 & 1 & 4 & 0 & 1 & 0 \\ 0 & -2 & 6 & 1 & 0 & -2 \end{array}\right]$$

$$\begin{matrix} -R_2 + R_1 \rightarrow \\ \\ 2R_2 + R_3 \rightarrow \end{matrix}\left[\begin{array}{ccc|ccc} 1 & 0 & -5 & 0 & -1 & 1 \\ 0 & 1 & 4 & 0 & 1 & 0 \\ 0 & 0 & 14 & 1 & 2 & -2 \end{array}\right]$$

$$\begin{matrix} \\ \\ \frac{1}{14}R_3 \rightarrow \end{matrix}\left[\begin{array}{ccc|ccc} 1 & 0 & -5 & 0 & -1 & 1 \\ 0 & 1 & 4 & 0 & 1 & 0 \\ 0 & 0 & 1 & \frac{1}{14} & \frac{1}{7} & -\frac{1}{7} \end{array}\right]$$

$$\begin{matrix} 5R_3 + R_1 \rightarrow \\ -4R_3 + R_2 \rightarrow \\ \\ \end{matrix}\left[\begin{array}{ccc|ccc} 1 & 0 & 0 & \frac{5}{14} & -\frac{2}{7} & \frac{2}{7} \\ 0 & 1 & 0 & -\frac{2}{7} & \frac{3}{7} & \frac{4}{7} \\ 0 & 0 & 1 & \frac{1}{14} & \frac{1}{7} & -\frac{1}{7} \end{array}\right] = [I \ \vdots \ A^{-1}]$$

$$A^{-1} = \frac{1}{14}\begin{bmatrix} 5 & -4 & 4 \\ -4 & 6 & 8 \\ 1 & 2 & -2 \end{bmatrix}$$

$$\begin{bmatrix} I_1 \\ I_2 \\ I_3 \end{bmatrix} = \frac{1}{14}\begin{bmatrix} 5 & -4 & 4 \\ -4 & 6 & 8 \\ 1 & 2 & -2 \end{bmatrix}\begin{bmatrix} 24 \\ 23 \\ 0 \end{bmatrix} = \begin{bmatrix} 2 \\ 3 \\ 5 \end{bmatrix}$$

Solution: $I_1 = 2$ amperes, $I_2 = 3$ amperes, $I_3 = 5$ amperes

79. The sum of two invertible matrices is not necessarily invertible. For example, let

$$A = \begin{bmatrix} 1 & 0 \\ 0 & 1 \end{bmatrix} \text{ and } B = \begin{bmatrix} -1 & 0 \\ 0 & -1 \end{bmatrix}.$$

Then $A + B = \begin{bmatrix} 0 & 0 \\ 0 & 0 \end{bmatrix}$ which is clearly singular.

80. $A = \begin{bmatrix} 3 & x \\ -2 & -3 \end{bmatrix} \Rightarrow A^{-1} = \dfrac{1}{-9 + 2x}\begin{bmatrix} -3 & -x \\ 2 & 3 \end{bmatrix}$

If $A^{-1} = A$ then $\begin{bmatrix} \dfrac{-3}{-9 + 2x} & \dfrac{-x}{-9 + 2x} \\ \dfrac{2}{-9 + 2x} & \dfrac{3}{-9 + 2x} \end{bmatrix} = \begin{bmatrix} 3 & x \\ -2 & -3 \end{bmatrix}$.

Equating the first entry in row 1 yields

$$\frac{-3}{-9 + 2x} = 3 \Rightarrow -3 = -27 + 6x \Rightarrow x = 4.$$

We then check and see that $x = 4$ works for the other three entries.

$$\frac{-4}{-9 + 2(4)} = 4$$

$$\frac{2}{-9 + 2(4)} = -2$$

$$\frac{3}{-9 + 2(4)} = -3$$

Thus, $x = 4$.

81. If $A = \begin{bmatrix} 4 & x \\ -2 & -3 \end{bmatrix}$ is singular then
$ad - bc = -12 + 2x = 0$. Thus, $x = 6$.

82. If $A = \begin{bmatrix} \sin\theta & \cos\theta \\ -\cos\theta & \sin\theta \end{bmatrix}$ then $ad - bc = \sin^2\theta + \cos^2\theta = 1$ for all θ. Since $ad - bc \neq 0$ for any value of θ, A is invertible

and $A^{-1} = \begin{bmatrix} \sin\theta & -\cos\theta \\ \cos\theta & \sin\theta \end{bmatrix}$.

83. True. If B is the inverse of A, then $AB = I = BA$.

84. True. If A and B are both square matrices and $AB = I_N$, it can be shown that $BA = I_N$.

85. True. If A is of order $m \times n$ and B is of order $n \times m$ (where $m \neq n$), the products AB and BA are of different orders and so cannot be equal to each other.

86. True. The inverse of a nonsingular matrix **is** unique.

87. True. If A can be row reduced to the identity matrix, then A has an inverse and is nonsingular.

88. False. The matrix $\begin{bmatrix} a & b \\ c & d \end{bmatrix}$ is invertible if $ad - bc \neq 0$.

Section 14.7 The Determinant of a Square Matrix

1. 5

2. -8

3. $\begin{vmatrix} 2 & 1 \\ 3 & 4 \end{vmatrix} = 2(4) - 1(3) = 8 - 3 = 5$

4. $\begin{vmatrix} -3 & 1 \\ 5 & 2 \end{vmatrix} = (-3)(2) - (5)(1) = -11$

5. $\begin{vmatrix} 5 & 2 \\ -6 & 3 \end{vmatrix} = 5(3) - 2(-6) = 15 + 12 = 27$

6. $\begin{vmatrix} 2 & -2 \\ 4 & 3 \end{vmatrix} = (2)(3) - (4)(-2) = 14$

7. $\begin{vmatrix} -7 & 0 \\ 3 & 0 \end{vmatrix} = -7(0) - 0(3) = 0$

8. $\begin{vmatrix} 4 & -3 \\ 0 & 0 \end{vmatrix} = (4)(0) - (0)(-3) = 0$

9. $\begin{vmatrix} 2 & -3 \\ -6 & 9 \end{vmatrix} = (2)(9) - (-6)(-3) = 0$

10. $\begin{vmatrix} -3 & -2 \\ -6 & -1 \end{vmatrix} = (-3)(-1) - (-2)(-6) = 3 - 12 = -9$

11. $\begin{vmatrix} 4 & 7 \\ -2 & 5 \end{vmatrix} = (4)(5) - (-2)(7) = 34$

12. $\begin{vmatrix} 9 & 0 \\ 7 & 8 \end{vmatrix} = 9(8) - 0(7) = 72 - 0 = 72$

13. $\begin{vmatrix} -\frac{1}{2} & \frac{1}{3} \\ -6 & \frac{1}{3} \end{vmatrix} = \left(-\frac{1}{2}\right)\left(\frac{1}{3}\right) - \left(\frac{1}{3}\right)(-6) = \frac{11}{6}$

14. $\begin{vmatrix} \frac{2}{3} & \frac{4}{3} \\ -1 & -\frac{1}{3} \end{vmatrix} = \left(\frac{2}{3}\right)\left(-\frac{1}{3}\right) - (-1)\left(\frac{4}{3}\right) = \frac{10}{9}$

15. $\begin{vmatrix} 0.3 & 0.2 & 0.2 \\ 0.2 & 0.2 & 0.2 \\ -0.4 & 0.4 & 0.3 \end{vmatrix} = -0.002$

16. $\begin{vmatrix} 0.1 & 0.2 & 0.3 \\ -0.3 & 0.2 & 0.2 \\ 0.5 & 0.4 & 0.4 \end{vmatrix} = -0.022$

17. $\begin{vmatrix} 0.9 & 0.7 & 0 \\ -0.1 & 0.3 & 1.3 \\ -2.2 & 4.2 & 6.1 \end{vmatrix} = -4.842$

18. $\begin{vmatrix} 0.1 & 0.1 & -4.3 \\ 7.5 & 6.2 & 0.7 \\ 0.3 & 0.6 & -1.2 \end{vmatrix} = -11.217$

19. $\begin{bmatrix} 3 & 4 \\ 2 & -5 \end{bmatrix}$

(a) $M_{11} = -5$, $M_{12} = 2$, $M_{21} = 4$, $M_{22} = 3$

(b) $C_{11} = M_{11} = -5$, $C_{12} = -M_{12} = -2$, $C_{21} = -M_{21} = -4$, $C_{22} = M_{22} = 3$

20. $\begin{bmatrix} 11 & 0 \\ -3 & 2 \end{bmatrix}$

(a) $M_{11} = 2$, $M_{12} = -3$, $M_{21} = 0$, $M_{22} = 11$

(b) $C_{11} = M_{11} = 2$, $C_{12} = M_{12} = 3$, $C_{21} = M_{21} = 0$, $C_{22} = M_{22} = 11$

21. $\begin{bmatrix} 3 & 1 \\ -2 & -4 \end{bmatrix}$

(a) $M_{11} = -4$, $M_{12} = -2$, $M_{21} = 1$, $M_{22} = 3$

(b) $C_{11} = M_{11} = -4$, $C_{12} = -M_{12} = 2$, $C_{21} = -M_{21} = -1$, $C_{22} = M_{22} = 3$

22. $\begin{bmatrix} -6 & 5 \\ 7 & -2 \end{bmatrix}$

(a) $M_{11} = -2$, $M_{12} = 7$, $M_{21} = 5$, $M_{22} = -6$

(b) $C_{11} = M_{11} = -2$, $C_{12} = -M_{12} = -7$, $C_{21} = -M_{21} = -5$, $C_{22} = M_{22} = -6$

23. $\begin{bmatrix} 4 & 0 & 2 \\ -3 & 2 & 1 \\ 1 & -1 & 1 \end{bmatrix}$

(a) $M_{11} = \begin{vmatrix} 2 & 1 \\ -1 & 1 \end{vmatrix} = 2 - (-1) = 3$

$M_{12} = \begin{vmatrix} -3 & 1 \\ 1 & 1 \end{vmatrix} = -3 - 1 = -4$

$M_{13} = \begin{vmatrix} -3 & 2 \\ 1 & -1 \end{vmatrix} = 3 - 2 = 1$

$M_{21} = \begin{vmatrix} 0 & 2 \\ -1 & 1 \end{vmatrix} = 0 - (-2) = 2$

$M_{22} = \begin{vmatrix} 4 & 2 \\ 1 & 1 \end{vmatrix} = 4 - 2 = 2$

$M_{23} = \begin{vmatrix} 4 & 0 \\ 1 & -1 \end{vmatrix} = -4 - 0 = -4$

$M_{31} = \begin{vmatrix} 0 & 2 \\ 2 & 1 \end{vmatrix} = 0 - 4 = -4$

$M_{32} = \begin{vmatrix} 4 & 2 \\ -3 & 1 \end{vmatrix} = 4 - (-6) = 10$

$M_{33} = \begin{vmatrix} 4 & 0 \\ -3 & 2 \end{vmatrix} = 8 - 0 = 8$

(b) $C_{11} = (-1)^2 M_{11} = 3$

$C_{12} = (-1)^3 M_{12} = 4$

$C_{13} = (-1)^4 M_{13} = 1$

$C_{21} = (-1)^3 M_{21} = -2$

$C_{22} = (-1)^4 M_{22} = 2$

$C_{23} = (-1)^5 M_{23} = 4$

$C_{31} = (-1)^4 M_{31} = -4$

$C_{32} = (-1)^5 M_{32} = -10$

$C_{33} = (-1)^6 M_{33} = 8$

24. $\begin{bmatrix} 1 & -1 & 0 \\ 3 & 2 & 5 \\ 4 & -6 & 4 \end{bmatrix}$

(a) $M_{11} = \begin{vmatrix} 2 & 5 \\ -6 & 4 \end{vmatrix} = 8-(-30) = 38$

$M_{12} = \begin{vmatrix} 3 & 5 \\ 4 & 4 \end{vmatrix} = 12 - 20 = -8$

$M_{13} = \begin{vmatrix} 3 & 2 \\ 4 & -6 \end{vmatrix} = -18 - 8 = -26$

$M_{21} = \begin{vmatrix} -1 & 0 \\ -6 & 4 \end{vmatrix} = -4 - 0 = -4$

$M_{22} = \begin{vmatrix} 1 & 0 \\ 4 & 4 \end{vmatrix} = 4 - 0 = 4$

$M_{23} = \begin{vmatrix} 1 & -1 \\ 4 & -6 \end{vmatrix} = -6 - (-4) = -2$

$M_{31} = \begin{vmatrix} -1 & 0 \\ 2 & 5 \end{vmatrix} = -5 - 0 = -5$

$M_{32} = \begin{vmatrix} 1 & 0 \\ 3 & 5 \end{vmatrix} = 5 - 0 = 5$

$M_{33} = \begin{vmatrix} 1 & -1 \\ 3 & 2 \end{vmatrix} = 2 - (-3) = 5$

(b) $C_{11} = (-1)^2M_{11} = 38$

$C_{12} = (-1)^3M_{12} = 8$

$C_{13} = (-1)^4M_{13} = -26$

$C_{21} = (-1)^3M_{21} = 4$

$C_{22} = (-1)^4M_{22} = 4$

$C_{23} = (-1)^5M_{23} = 2$

$C_{31} = (-1)^4M_{31} = -5$

$C_{32} = (-1)^5M_{32} = -5$

$C_{33} = (-1)^6M_{33} = 5$

25. $\begin{bmatrix} 3 & -2 & 8 \\ 3 & 2 & -6 \\ -1 & 3 & 6 \end{bmatrix}$

(a) $M_{11} = \begin{vmatrix} 2 & -6 \\ 3 & 6 \end{vmatrix} = 12 + 18 = 30$

$M_{12} = \begin{vmatrix} 3 & -6 \\ -1 & 6 \end{vmatrix} = 18 - 6 = 12$

$M_{13} = \begin{vmatrix} 3 & 2 \\ -1 & 3 \end{vmatrix} = 9 + 2 = 11$

$M_{21} = \begin{vmatrix} -2 & 8 \\ 3 & 6 \end{vmatrix} = -12 - 24 = -36$

$M_{22} = \begin{vmatrix} 3 & 8 \\ -1 & 6 \end{vmatrix} = 18 + 8 = 26$

$M_{23} = \begin{vmatrix} 3 & -2 \\ -1 & 3 \end{vmatrix} = 9 - 2 = 7$

$M_{31} = \begin{vmatrix} -2 & 8 \\ 2 & -6 \end{vmatrix} = 12 - 16 = -4$

$M_{32} = \begin{vmatrix} 3 & 8 \\ 3 & -6 \end{vmatrix} = -18 - 24 = -42$

$M_{33} = \begin{vmatrix} 3 & -2 \\ 3 & 2 \end{vmatrix} = 6 + 6 = 12$

(b) $C_{11} = (-1)^2M_{11} = 30$

$C_{12} = (-1)^3M_{12} = -12$

$C_{13} = (-1)^4M_{13} = 11$

$C_{21} = (-1)^3M_{21} = 36$

$C_{22} = (-1)^4M_{22} = 26$

$C_{23} = (-1)^5M_{23} = -7$

$C_{31} = (-1)^4M_{31} = -4$

$C_{32} = (-1)^5M_{32} = 42$

$C_{33} = (-1)^6M_{33} = 12$

26. $\begin{bmatrix} -2 & 9 & 4 \\ 7 & -6 & 0 \\ 6 & 7 & -6 \end{bmatrix}$

(a) $M_{11} = \begin{vmatrix} -6 & 0 \\ 7 & -6 \end{vmatrix} = 36$

$M_{12} = \begin{vmatrix} 7 & 0 \\ 6 & -6 \end{vmatrix} = -42$

$M_{13} = \begin{vmatrix} 7 & -6 \\ 6 & 7 \end{vmatrix} = 85$

$M_{21} = \begin{vmatrix} 9 & 4 \\ 7 & -6 \end{vmatrix} = -82$

$M_{22} = \begin{vmatrix} -2 & 4 \\ 6 & -6 \end{vmatrix} = -12$

$M_{23} = \begin{vmatrix} -2 & 9 \\ 6 & 7 \end{vmatrix} = -68$

$M_{31} = \begin{vmatrix} 9 & 4 \\ -6 & 0 \end{vmatrix} = 24$

$M_{32} = \begin{vmatrix} -2 & 4 \\ 7 & 0 \end{vmatrix} = -28$

$M_{33} = \begin{vmatrix} -2 & 9 \\ 7 & -6 \end{vmatrix} = -51$

(b) $C_{11} = (-1)^2M_{11} = 36$

$C_{12} = (-1)^3M_{12} = 42$

$C_{13} = (-1)^4M_{13} = 85$

$C_{21} = (-1)^3M_{21} = 82$

$C_{22} = (-1)^4M_{22} = -12$

$C_{23} = (-1)^5M_{23} = 68$

$C_{31} = (-1)^4M_{31} = 24$

$C_{32} = (-1)^5M_{32} = 28$

$C_{33} = (-1)^6M_{33} = -51$

27. (a) $\begin{vmatrix} -3 & 2 & 1 \\ 4 & 5 & 6 \\ 2 & -3 & 1 \end{vmatrix} = -3\begin{vmatrix} 5 & 6 \\ -3 & 1 \end{vmatrix} - 2\begin{vmatrix} 4 & 6 \\ 2 & 1 \end{vmatrix} + \begin{vmatrix} 4 & 5 \\ 2 & -3 \end{vmatrix} = -3(23) - 2(-8) - 22 = -75$

(b) $\begin{vmatrix} -3 & 2 & 1 \\ 4 & 5 & 6 \\ 2 & -3 & 1 \end{vmatrix} = -2\begin{vmatrix} 4 & 6 \\ 2 & 1 \end{vmatrix} + 5\begin{vmatrix} -3 & 1 \\ 2 & 1 \end{vmatrix} + 3\begin{vmatrix} -3 & 1 \\ 4 & 6 \end{vmatrix} = -2(-8) + 5(-5) + 3(-22) = -75$

28. (a) $\begin{vmatrix} -3 & 4 & 2 \\ 6 & 3 & 1 \\ 4 & -7 & -8 \end{vmatrix} = -6\begin{vmatrix} 4 & 2 \\ -7 & -8 \end{vmatrix} + 3\begin{vmatrix} -3 & 2 \\ 4 & -8 \end{vmatrix} - 1\begin{vmatrix} -3 & 4 \\ 4 & -7 \end{vmatrix} = -6(-18) + 3(16) - (5) = 151$

(b) $\begin{vmatrix} -3 & 4 & 2 \\ 6 & 3 & 1 \\ 4 & -7 & -8 \end{vmatrix} = 2\begin{vmatrix} 6 & 3 \\ 4 & -7 \end{vmatrix} - \begin{vmatrix} -3 & 4 \\ 4 & -7 \end{vmatrix} - 8\begin{vmatrix} -3 & 4 \\ 6 & 3 \end{vmatrix} = 2(-54) - (5) - 8(-33) = 151$

29. (a) $\begin{vmatrix} 5 & 0 & -3 \\ 0 & 12 & 4 \\ 1 & 6 & 3 \end{vmatrix} = 0\begin{vmatrix} 0 & -3 \\ 6 & 3 \end{vmatrix} + 12\begin{vmatrix} 5 & -3 \\ 1 & 3 \end{vmatrix} - 4\begin{vmatrix} 5 & 0 \\ 1 & 6 \end{vmatrix} = 0(18) + 12(18) - 4(30) = 96$

(b) $\begin{vmatrix} 5 & 0 & -3 \\ 0 & 12 & 4 \\ 1 & 6 & 3 \end{vmatrix} = 0\begin{vmatrix} 0 & 4 \\ 1 & 3 \end{vmatrix} + 12\begin{vmatrix} 5 & -3 \\ 1 & 3 \end{vmatrix} - 6\begin{vmatrix} 5 & -3 \\ 0 & 4 \end{vmatrix} = 0(-4) + 12(18) - 6(20) = 96$

30. (a) $\begin{vmatrix} 10 & -5 & 5 \\ 30 & 0 & 10 \\ 0 & 10 & 1 \end{vmatrix} = 0\begin{vmatrix} -5 & 5 \\ 0 & 10 \end{vmatrix} - 10\begin{vmatrix} 10 & 5 \\ 30 & 10 \end{vmatrix} + \begin{vmatrix} 10 & -5 \\ 30 & 0 \end{vmatrix} = 0(-50) - 10(-50) + 150 = 650$

(b) $\begin{vmatrix} 10 & -5 & 5 \\ 30 & 0 & 10 \\ 0 & 10 & 1 \end{vmatrix} = 10\begin{vmatrix} 0 & 10 \\ 10 & 1 \end{vmatrix} - 30\begin{vmatrix} -5 & 5 \\ 10 & 1 \end{vmatrix} + 0\begin{vmatrix} -5 & 5 \\ 0 & 10 \end{vmatrix} = 10(-100) - 30(-55) + 0(-50) = 650$

31. (a) $\begin{vmatrix} 6 & 0 & -3 & 5 \\ 4 & 13 & 6 & -8 \\ -1 & 0 & 7 & 4 \\ 8 & 6 & 0 & 2 \end{vmatrix} = -4\begin{vmatrix} 0 & -3 & 5 \\ 0 & 7 & 4 \\ 6 & 0 & 2 \end{vmatrix} + 13\begin{vmatrix} 6 & -3 & 5 \\ -1 & 7 & 4 \\ 8 & 0 & 2 \end{vmatrix} - 6\begin{vmatrix} 6 & 0 & 5 \\ -1 & 0 & 4 \\ 8 & 6 & 2 \end{vmatrix} - 8\begin{vmatrix} 6 & 0 & -3 \\ -1 & 0 & 7 \\ 8 & 6 & 0 \end{vmatrix}$

$= -4(-282) + 13(-298) - 6(-174) - 8(-234) = 170$

(b) $\begin{vmatrix} 6 & 0 & -3 & 5 \\ 4 & 13 & 6 & -8 \\ -1 & 0 & 7 & 4 \\ 8 & 6 & 0 & 2 \end{vmatrix} = 0\begin{vmatrix} 4 & 6 & -8 \\ -1 & 7 & 4 \\ 8 & 0 & 2 \end{vmatrix} + 13\begin{vmatrix} 6 & -3 & 5 \\ -1 & 7 & 4 \\ 8 & 0 & 2 \end{vmatrix} + 0\begin{vmatrix} 6 & -3 & 5 \\ 4 & 6 & -8 \\ 8 & 0 & 2 \end{vmatrix} + 6\begin{vmatrix} 6 & -3 & 5 \\ 4 & 6 & -8 \\ -1 & 7 & 4 \end{vmatrix}$

$= 0 + 13(-298) + 0 + 6(674) = 170$

32. (a) $\begin{vmatrix} 10 & 8 & 3 & -7 \\ 4 & 0 & 5 & -6 \\ 0 & 3 & 2 & 7 \\ 1 & 0 & -3 & 2 \end{vmatrix} = 0\begin{vmatrix} 8 & 3 & -7 \\ 0 & 5 & -6 \\ 0 & -3 & 2 \end{vmatrix} - 3\begin{vmatrix} 10 & 3 & -7 \\ 4 & 5 & -6 \\ 1 & -3 & 2 \end{vmatrix} + 2\begin{vmatrix} 10 & 8 & -7 \\ 4 & 0 & -6 \\ 1 & 0 & 2 \end{vmatrix} - 7\begin{vmatrix} 10 & 8 & 3 \\ 4 & 0 & 5 \\ 1 & 0 & -3 \end{vmatrix}$

$= 0(-64) - 3(-3) + 2(-112) - 7(136) = -1167$

(b) $\begin{vmatrix} 10 & 8 & 3 & -7 \\ 4 & 0 & 5 & -6 \\ 0 & 3 & 2 & 7 \\ 1 & 0 & -3 & 2 \end{vmatrix} = 10\begin{vmatrix} 0 & 5 & -6 \\ 3 & 2 & 7 \\ 0 & -3 & 2 \end{vmatrix} - 4\begin{vmatrix} 8 & 3 & -7 \\ 3 & 2 & 7 \\ 0 & -3 & 2 \end{vmatrix} + 0\begin{vmatrix} 8 & 3 & -7 \\ 0 & 5 & -6 \\ 0 & -3 & 2 \end{vmatrix} - 1\begin{vmatrix} 8 & 3 & -7 \\ 0 & 5 & -6 \\ 3 & 2 & 7 \end{vmatrix}$

$= 10(24) - 4(245) + 0(-64) - 1(427) = -1167$

33. Expand along Column 1.

$$\begin{vmatrix} 2 & -1 & 0 \\ 4 & 2 & 1 \\ 4 & 2 & 1 \end{vmatrix} = 2\begin{vmatrix} 2 & 1 \\ 2 & 1 \end{vmatrix} - 4\begin{vmatrix} -1 & 0 \\ 2 & 1 \end{vmatrix} + 4\begin{vmatrix} -1 & 0 \\ 2 & 1 \end{vmatrix} = 2(0) - 4(-1) + 4(-1) = 0$$

34. Expand along Row 3.

$$\begin{vmatrix} -2 & 2 & 3 \\ 1 & -1 & 0 \\ 0 & 1 & 4 \end{vmatrix} = 0\begin{vmatrix} 2 & 3 \\ -1 & 0 \end{vmatrix} - 1\begin{vmatrix} -2 & 3 \\ 1 & 0 \end{vmatrix} + 4\begin{vmatrix} -2 & 2 \\ 1 & -1 \end{vmatrix}$$

$$= 0(3) - 1(-3) + 4(0) = 3$$

35. Expand along Row 2.

$$\begin{vmatrix} 6 & 3 & -7 \\ 0 & 0 & 0 \\ 4 & -6 & 3 \end{vmatrix} = 0\begin{vmatrix} 3 & -7 \\ -6 & 3 \end{vmatrix} - 0\begin{vmatrix} 6 & -7 \\ 4 & 3 \end{vmatrix} + 0\begin{vmatrix} 6 & 3 \\ 4 & -6 \end{vmatrix} = 0$$

36. Expand along Column 3.

$$\begin{vmatrix} 1 & 1 & 2 \\ 3 & 1 & 0 \\ -2 & 0 & 3 \end{vmatrix} = 2\begin{vmatrix} 3 & 1 \\ -2 & 0 \end{vmatrix} - 0\begin{vmatrix} 1 & 1 \\ -2 & 0 \end{vmatrix} + 3\begin{vmatrix} 1 & 1 \\ 3 & 1 \end{vmatrix}$$

$$= 2(2) - 0(2) + 3(-2) = -2$$

37. $\begin{vmatrix} -1 & 2 & 5 \\ 0 & 3 & 4 \\ 0 & 0 & 3 \end{vmatrix} = -1\begin{vmatrix} 3 & 4 \\ 0 & 3 \end{vmatrix} = -1(9) = -9$

38. Expand along Row 1.

$$\begin{vmatrix} 1 & 0 & 0 \\ -4 & -1 & 0 \\ 5 & 1 & 5 \end{vmatrix} = 1\begin{vmatrix} -1 & 0 \\ 1 & 5 \end{vmatrix} - 0\begin{vmatrix} -4 & 0 \\ 5 & 5 \end{vmatrix} + 0\begin{vmatrix} -4 & -1 \\ 5 & 1 \end{vmatrix}$$

$$= 1(-5) - 0(-20) + 0(1) = -5$$

39. Expand along Column 3.

$$\begin{vmatrix} 1 & 4 & -2 \\ 3 & 2 & 0 \\ -1 & 4 & 3 \end{vmatrix} = -2\begin{vmatrix} 3 & 2 \\ -1 & 4 \end{vmatrix} + 3\begin{vmatrix} 1 & 4 \\ 3 & 2 \end{vmatrix} = -2(14) + 3(-10) = -58$$

40. Expand along Row 3.

$$\begin{vmatrix} 2 & -1 & 3 \\ 1 & 4 & 4 \\ 1 & 0 & 2 \end{vmatrix} = 1\begin{vmatrix} -1 & 3 \\ 4 & 4 \end{vmatrix} - 0\begin{vmatrix} 2 & 3 \\ 1 & 4 \end{vmatrix} + 2\begin{vmatrix} 2 & -1 \\ 1 & 4 \end{vmatrix} = 1(-16) - 0(5) + 2(9) = 2$$

41. Expand along Column 3.

$$\begin{vmatrix} 2 & 6 & 6 & 2 \\ 2 & 7 & 3 & 6 \\ 1 & 5 & 0 & 1 \\ 3 & 7 & 0 & 7 \end{vmatrix} = 6\begin{vmatrix} 2 & 7 & 6 \\ 1 & 5 & 1 \\ 3 & 7 & 7 \end{vmatrix} - 3\begin{vmatrix} 2 & 6 & 2 \\ 1 & 5 & 1 \\ 3 & 7 & 7 \end{vmatrix} = 6(-20) - 3(16) = -168$$

42. Expand along Row 2.

$$\begin{vmatrix} 3 & 6 & -5 & 4 \\ -2 & 0 & 6 & 0 \\ 1 & 1 & 2 & 2 \\ 0 & 3 & -1 & -1 \end{vmatrix} = -(-2)\begin{vmatrix} 6 & -5 & 4 \\ 1 & 2 & 2 \\ 3 & -1 & -1 \end{vmatrix} - 6\begin{vmatrix} 3 & 6 & 4 \\ 1 & 1 & 2 \\ 0 & 3 & -1 \end{vmatrix} = 2(-63) - 6(-3) = -108$$

43. Expand along Column 1.

$$\begin{vmatrix} 5 & 3 & 0 & 6 \\ 4 & 6 & 4 & 12 \\ 0 & 2 & -3 & 4 \\ 0 & 1 & -2 & 2 \end{vmatrix} = 5\begin{vmatrix} 6 & 4 & 12 \\ 2 & -3 & 4 \\ 1 & -2 & 2 \end{vmatrix} - 4\begin{vmatrix} 3 & 0 & 6 \\ 2 & -3 & 4 \\ 1 & -2 & 2 \end{vmatrix} = 5(0) - 4(0) = 0$$

44. Expand along Row 3.

$$\begin{vmatrix} 1 & 4 & 3 & 2 \\ -5 & 6 & 2 & 1 \\ 0 & 0 & 0 & 0 \\ 3 & -2 & 1 & 5 \end{vmatrix} = 0$$

45. Expand along Column 2, then along Column 4.

$$\begin{vmatrix} 3 & 2 & 4 & -1 & 5 \\ -2 & 0 & 1 & 3 & 2 \\ 1 & 0 & 0 & 4 & 0 \\ 6 & 0 & 2 & -1 & 0 \\ 3 & 0 & 5 & 1 & 0 \end{vmatrix} = -2\begin{vmatrix} -2 & 1 & 3 & 2 \\ 1 & 0 & 4 & 0 \\ 6 & 2 & -1 & 0 \\ 3 & 5 & 1 & 0 \end{vmatrix} = (-2)(-2)\begin{vmatrix} 1 & 0 & 4 \\ 6 & 2 & -1 \\ 3 & 5 & 1 \end{vmatrix} = 4(103) = 412$$

46. Expand along Column 1.

$$\begin{vmatrix} 5 & 2 & 0 & 0 & -2 \\ 0 & 1 & 4 & 3 & 2 \\ 0 & 0 & 2 & 6 & 3 \\ 0 & 0 & 3 & 4 & 1 \\ 0 & 0 & 0 & 0 & 2 \end{vmatrix} = 5\begin{vmatrix} 1 & 4 & 3 & 2 \\ 0 & 2 & 6 & 3 \\ 0 & 3 & 4 & 1 \\ 0 & 0 & 0 & 2 \end{vmatrix} = 5 \cdot 1\begin{vmatrix} 2 & 6 & 3 \\ 3 & 4 & 1 \\ 0 & 0 & 2 \end{vmatrix} = 5(-20) = -100$$

47. $\begin{vmatrix} 3 & 8 & -7 \\ 0 & -5 & 4 \\ 8 & 1 & 6 \end{vmatrix} = -126$

48. $\begin{vmatrix} 5 & -8 & 0 \\ 9 & 7 & 4 \\ -8 & 7 & 1 \end{vmatrix} = 223$

49. $\begin{vmatrix} 7 & 0 & -14 \\ -2 & 5 & 4 \\ -6 & 2 & 12 \end{vmatrix} = 0$

50. $\begin{vmatrix} 3 & 0 & 0 \\ -2 & 5 & 0 \\ 12 & 5 & 7 \end{vmatrix} = 105$

51. $\begin{vmatrix} 1 & -1 & 8 & 4 \\ 2 & 6 & 0 & -4 \\ 2 & 0 & 2 & 6 \\ 0 & 2 & 8 & 0 \end{vmatrix} = -336$

52. $\begin{vmatrix} 0 & -3 & 8 & 2 \\ 8 & 1 & -1 & 6 \\ -4 & 6 & 0 & 9 \\ -7 & 0 & 0 & 14 \end{vmatrix} = 7441$

53. (a) $\begin{vmatrix} -1 & 0 \\ 0 & 3 \end{vmatrix} = -3$

(b) $\begin{vmatrix} 2 & 0 \\ 0 & -1 \end{vmatrix} = -2$

(c) $\begin{bmatrix} -1 & 0 \\ 0 & 3 \end{bmatrix}\begin{bmatrix} 2 & 0 \\ 0 & -1 \end{bmatrix} = \begin{bmatrix} -2 & 0 \\ 0 & -3 \end{bmatrix}$

(d) $\begin{vmatrix} -2 & 0 \\ 0 & -3 \end{vmatrix} = 6$

54. (a) $|A| = \begin{vmatrix} -2 & 1 \\ 4 & -2 \end{vmatrix} = 0$

(b) $|B| = \begin{vmatrix} 1 & 2 \\ 0 & -1 \end{vmatrix} = -1$

(c) $AB = \begin{bmatrix} -2 & 1 \\ 4 & -2 \end{bmatrix}\begin{bmatrix} 1 & 2 \\ 0 & -1 \end{bmatrix} = \begin{bmatrix} -2 & -5 \\ 4 & 10 \end{bmatrix}$

(d) $|AB| = \begin{vmatrix} -2 & -5 \\ 4 & 10 \end{vmatrix} = 0$

55. (a) $\begin{vmatrix} 0 & 1 & 2 \\ -3 & -2 & 1 \\ 0 & 4 & 1 \end{vmatrix} = -21$

(b) $\begin{vmatrix} 3 & -2 & 0 \\ 1 & -1 & 2 \\ 3 & 1 & 1 \end{vmatrix} = -19$

(c) $\begin{bmatrix} 0 & 1 & 2 \\ -3 & -2 & 1 \\ 0 & 4 & 1 \end{bmatrix}\begin{bmatrix} 3 & -2 & 0 \\ 1 & -1 & 2 \\ 3 & 1 & 1 \end{bmatrix} = \begin{bmatrix} 7 & 1 & 4 \\ -8 & 9 & -3 \\ 7 & -3 & 9 \end{bmatrix}$

(d) $\begin{vmatrix} 7 & 1 & 4 \\ -8 & 9 & -3 \\ 7 & -3 & 9 \end{vmatrix} = 399$

56. (a) $|A| = \begin{vmatrix} 3 & 2 & 0 \\ -1 & -3 & 4 \\ -2 & 0 & 1 \end{vmatrix} = -23$

(b) $|B| = \begin{vmatrix} -3 & 0 & 1 \\ 0 & 2 & -1 \\ -2 & -1 & 1 \end{vmatrix} = 1$

(c) $AB = \begin{bmatrix} 3 & 2 & 0 \\ -1 & -3 & 4 \\ -2 & 0 & 1 \end{bmatrix}\begin{bmatrix} -3 & 0 & 1 \\ 0 & 2 & -1 \\ -2 & -1 & 1 \end{bmatrix} = \begin{bmatrix} -9 & 4 & 1 \\ -5 & -10 & 6 \\ 4 & -1 & -1 \end{bmatrix}$

(d) $|AB| = \begin{vmatrix} -9 & 4 & 1 \\ -5 & -10 & 6 \\ 4 & -1 & -1 \end{vmatrix} = -23$

57. (a) $\begin{vmatrix} -1 & 2 & 1 \\ 1 & 0 & 1 \\ 0 & 1 & 0 \end{vmatrix} = 2$

(b) $\begin{vmatrix} -1 & 0 & 0 \\ 0 & 2 & 0 \\ 0 & 0 & 3 \end{vmatrix} = -6$

(c) $\begin{bmatrix} -1 & 2 & 1 \\ 1 & 0 & 1 \\ 0 & 1 & 0 \end{bmatrix}\begin{bmatrix} -1 & 0 & 0 \\ 0 & 2 & 0 \\ 0 & 0 & 3 \end{bmatrix} = \begin{bmatrix} 1 & 4 & 3 \\ -1 & 0 & 3 \\ 0 & 2 & 0 \end{bmatrix}$

(d) $\begin{vmatrix} 1 & 4 & 3 \\ -1 & 0 & 3 \\ 0 & 2 & 0 \end{vmatrix} = -12$

58. (a) $|A| = \begin{vmatrix} 2 & 0 & 1 \\ 1 & -1 & 2 \\ 3 & 1 & 0 \end{vmatrix} = 0$

(b) $|B| = \begin{vmatrix} 2 & -1 & 4 \\ 0 & 1 & 3 \\ 3 & -2 & 1 \end{vmatrix} = -7$

(c) $AB = \begin{bmatrix} 2 & 0 & 1 \\ 1 & -1 & 2 \\ 3 & 1 & 0 \end{bmatrix}\begin{bmatrix} 2 & -1 & 4 \\ 0 & 1 & 3 \\ 3 & -2 & 1 \end{bmatrix} = \begin{bmatrix} 7 & -4 & 9 \\ 8 & -6 & 3 \\ 6 & -2 & 15 \end{bmatrix}$

(d) $|AB| = \begin{vmatrix} 7 & -4 & 9 \\ 8 & -6 & 3 \\ 6 & -2 & 15 \end{vmatrix} = 0$

59. $\begin{vmatrix} w & x \\ y & z \end{vmatrix} = wz - xy$

$-\begin{vmatrix} y & z \\ w & x \end{vmatrix} = -(xy - wz) = wz - xy$

Thus, $\begin{vmatrix} w & x \\ y & z \end{vmatrix} = -\begin{vmatrix} y & z \\ w & x \end{vmatrix}$.

60. $\begin{vmatrix} w & cx \\ y & cz \end{vmatrix} = cwz - cxy = c(wz - xy)$

$c\begin{vmatrix} w & x \\ y & z \end{vmatrix} = c(wz - xy)$

Thus, $\begin{vmatrix} w & cx \\ y & cz \end{vmatrix} = c\begin{vmatrix} w & x \\ y & z \end{vmatrix}$.

61. $\begin{vmatrix} 1 & x & x^2 \\ 1 & y & y^2 \\ 1 & z & z^2 \end{vmatrix} = \begin{vmatrix} y & y^2 \\ z & z^2 \end{vmatrix} - \begin{vmatrix} x & x^2 \\ z & z^2 \end{vmatrix} + \begin{vmatrix} x & x^2 \\ y & y^2 \end{vmatrix}$

$$= (yz^2 - y^2z) - (xz^2 - x^2z) + (xy^2 - x^2y)$$
$$= yz^2 - xz^2 - y^2z + x^2z + xy(y - x)$$
$$= z^2(y - x) - z(y^2 - x^2) + xy(y - x)$$
$$= z^2(y - x) - z(y - x)(y + x) + xy(y - x)$$
$$= (y - x)[z^2 - z(y + x) + xy]$$
$$= (y - x)[z^2 - zy - zx + xy]$$
$$= (y - x)[z^2 - zx - zy + xy]$$
$$= (y - x)[z(z - x) - y(z - x)]$$
$$= (y - x)(z - x)(z - y)$$

62. $\begin{vmatrix} a+b & a & a \\ a & a+b & a \\ a & a & a+b \end{vmatrix} = (a + b)\begin{vmatrix} a+b & a \\ a & a+b \end{vmatrix} - a\begin{vmatrix} a & a \\ a & a+b \end{vmatrix} + a\begin{vmatrix} a & a \\ a+b & a \end{vmatrix}$

$$= (a + b)[(a + b)^2 - a^2] - a[a(a + b) - a^2] + a[a^2 - a(a + b)]$$
$$= (a + b)^3 - a^2(a + b) - a^2(a + b) + a^3 + a^3 - a^2(a + b)$$
$$= (a + b)^3 - 3a^2(a + b) + 2a^3$$
$$= a^3 + 3a^2b + 3ab^2 + b^3 - 3a^3 - 3a^2b + 2a^3$$
$$= 3ab^2 + b^3 = b^2(3a + b)$$

63. $\begin{vmatrix} x-1 & 2 \\ 3 & x-2 \end{vmatrix} = 0$

$$(x - 1)(x - 2) - 6 = 0$$
$$x^2 - 3x - 4 = 0$$
$$(x + 1)(x - 4) = 0$$
$$x = -1 \text{ or } x = 4$$

64. $\begin{vmatrix} x-2 & -1 \\ -3 & x \end{vmatrix} = 0$

$$x(x - 2) - (-3)(-1) = 0$$
$$x^2 - 2x - 3 = 0$$
$$(x + 1)(x - 3) = 0$$
$$x = -1 \text{ or } x = 3$$

65. $\begin{vmatrix} x+3 & 2 \\ 1 & x+2 \end{vmatrix} = 0$

$$(x + 3)(x + 2) - 2 = 0$$
$$x^2 + 5x + 4 = 0$$
$$(x + 1)(x + 4) = 0$$
$$x = -1 \text{ or } x = -4$$

66. $\begin{vmatrix} x+4 & -2 \\ 7 & x-5 \end{vmatrix} = 0$

$$(x + 4)(x - 5) - 7(-2) = 0$$
$$x^2 - x - 6 = 0$$
$$(x + 2)(x - 3) = 0$$
$$x = -2 \text{ or } x = 3$$

67. A square matrix is a square array of numbers. The determinant of a square matrix is a real number.

68. The cofactor C_{ij} of a square matrix is the minor M_{ij} of the matrix multiplied by $(-1)^{i+j}$.

69. Points: $(-4, 0)$ and $(4, 4)$

Equation: $\begin{vmatrix} x & y & 1 \\ -4 & 0 & 1 \\ 4 & 4 & 1 \end{vmatrix} = 0$

$x\begin{vmatrix} 0 & 1 \\ 4 & 1 \end{vmatrix} - y\begin{vmatrix} -4 & 1 \\ 4 & 1 \end{vmatrix} + 1\begin{vmatrix} -4 & 0 \\ 4 & 4 \end{vmatrix} = 0$

$-4x + 8y - 16 = 0 \Rightarrow x - 2y + 4 = 0$

70. Points: $(2, 5)$ and $(6, -1)$

Equation: $\begin{vmatrix} x & y & 1 \\ 2 & 5 & 1 \\ 6 & -1 & 1 \end{vmatrix} = 0$

$x\begin{vmatrix} 5 & 1 \\ -1 & 1 \end{vmatrix} - y\begin{vmatrix} 2 & 1 \\ 6 & 1 \end{vmatrix} + 1\begin{vmatrix} 2 & 5 \\ 6 & -1 \end{vmatrix} = 0$

$6x + 4y - 32 = 0 \Rightarrow 3x + 2y - 16 = 0$

71. Points: $\left(-\frac{5}{2}, 3\right)$ and $\left(\frac{7}{2}, 1\right)$

Equation: $\begin{vmatrix} x & y & 1 \\ -\frac{5}{2} & 3 & 1 \\ \frac{7}{2} & 1 & 1 \end{vmatrix}$

$x\begin{vmatrix} 3 & 1 \\ 1 & 1 \end{vmatrix} - y\begin{vmatrix} -\frac{5}{2} & 1 \\ \frac{7}{2} & 1 \end{vmatrix} + 1\begin{vmatrix} -\frac{5}{2} & 3 \\ \frac{7}{2} & 1 \end{vmatrix} = 0$

$2x + 6y - 13 = 0$

72. Points: $(-0.8, 0.2)$ and $(0.7, 3.2)$

Equation: $\begin{vmatrix} x & y & 1 \\ -0.8 & 0.2 & 1 \\ 0.7 & 3.2 & 1 \end{vmatrix}$

$x\begin{vmatrix} 0.2 & 1 \\ 3.2 & 1 \end{vmatrix} - y\begin{vmatrix} -0.8 & 1 \\ 0.7 & 1 \end{vmatrix} + 1\begin{vmatrix} -0.8 & 0.2 \\ 0.7 & 3.2 \end{vmatrix} = 0$

$-3x + 1.5y - 2.7 = 0 \Rightarrow 10x - 5y + 9 = 0$

73. $\begin{vmatrix} 4u & -1 \\ -1 & 2v \end{vmatrix} = 8uv - 1$

74. $\begin{vmatrix} 3x^2 & -3y^2 \\ 1 & 1 \end{vmatrix} = 3x^2 - (-3y^2) = 3x^2 + 3y^2$

75. $\begin{vmatrix} e^{2x} & e^{3x} \\ 2e^{2x} & 3e^{3x} \end{vmatrix} = 3e^{5x} - 2e^{5x} = e^{5x}$

76. $\begin{vmatrix} e^{-x} & xe^{-x} \\ -e^{-x} & (1-x)e^{-x} \end{vmatrix} = (1-x)e^{-2x} - (-xe^{-2x}) = e^{-2x} - xe^{-2x} + xe^{-2x} = e^{-2x}$

77. True. If an entire row is zero, then each cofactor in the expansion is multiplied by zero.

78. True. If a square matrix has two columns that are equal, then elementary column operations can be used to create a column with all zeros.

79. Let $A = \begin{bmatrix} x_{11} & x_{12} & x_{13} \\ x_{21} & x_{22} & x_{23} \\ x_{31} & x_{32} & x_{33} \end{bmatrix}$ and $|A| = 5$.

$2A = \begin{bmatrix} 2x_{11} & 2x_{12} & 2x_{13} \\ 2x_{21} & 2x_{22} & 2x_{23} \\ 2x_{31} & 2x_{32} & 2x_{33} \end{bmatrix}$

$$
\begin{aligned}
|2A| &= 2x_{11}\begin{vmatrix} 2x_{22} & 2x_{23} \\ 2x_{32} & 2x_{33} \end{vmatrix} - 2x_{12}\begin{vmatrix} 2x_{21} & 2x_{23} \\ 2x_{31} & 2x_{33} \end{vmatrix} + 2x_{13}\begin{vmatrix} 2x_{21} & 2x_{22} \\ 2x_{31} & 2x_{32} \end{vmatrix} \\
&= 2[x_{11}(4x_{22}x_{33} - 4x_{32}x_{23}) - x_{12}(4x_{21}x_{33} - 4x_{31}x_{23}) + x_{13}(4x_{21}x_{32} - 4x_{31}x_{22})] \\
&= 8[x_{11}(x_{22}x_{33} - x_{32}x_{23}) - x_{12}(x_{21}x_{33} - x_{31}x_{23}) + x_{13}(x_{21}x_{32} - x_{31}x_{22})] \\
&= 8|A|
\end{aligned}
$$

Thus, $|2A| = 8|A| = 8(5) = 40$.

Review Exercises for Chapter 14

1. $\begin{cases} x^2 - y^2 = 9 \\ x - y = 1 \Rightarrow x = y + 1 \end{cases}$

$(y+1)^2 - y^2 = 9$

$2y + 1 = 9$

$y = 4$

$x = 5$

Solution: $(5, 4)$

2. $\begin{cases} x^2 + y^2 = 169 \\ 3x + 2y = 39 \Rightarrow x = \frac{1}{3}(39 - 2y) \end{cases}$

$\left[\frac{1}{3}(39 - 2y)\right]^2 + y^2 = 169$

$\frac{1}{9}(1521 - 156y + 4y^2) + y^2 = 169$

$1521 - 156y + 4y^2 + 9y^2 = 1521$

$13y^2 - 156y = 0$

$13y(y - 12) = 0 \Rightarrow y = 0, 12$

$y = 0$: $x = \frac{1}{3}(39 - 2(0)) = 13$

$y = 12$: $x = \frac{1}{3}(39 - 2(12)) = 5$

Solution: $(13, 0), (5, 12)$

3. $\begin{cases} y = 2x^2 \\ y = x^4 - 2x^2 \Rightarrow 2x^2 = x^4 - 2x^2 \end{cases}$

$0 = x^4 - 4x^2$

$0 = x^2(x^2 - 4)$

$0 = x^2(x + 2)(x - 2)$

$x = 0, x = -2, x = 2$

$y = 0, y = 8, y = 8$

Solutions: $(0, 0), (-2, 8), (2, 8)$

4. $\begin{cases} x = y + 3 \\ x = y^2 + 1 \end{cases}$

$y + 3 = y^2 + 1$

$0 = y^2 - y - 2$

$0 = (y - 2)(y + 1) \Rightarrow y = 2, -1$

$y = 2$: $x = 2 + 3 = 5$

$y = -1$: $x = -1 + 3 = 2$

Solution: $(5, 2), (2, -1)$

5. $\begin{cases} 2x - y = 10 \\ x + 5y = -6 \end{cases}$

Point of intersection: $(4, -2)$

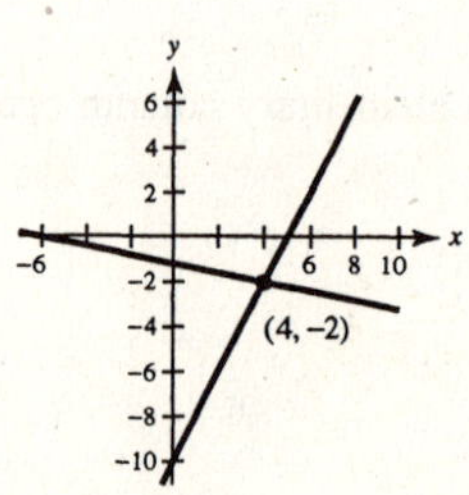

6. $y^2 - 2y + x = 0 \Rightarrow (y - 1)^2 = 1 - x \Rightarrow y = 1 \pm \sqrt{1 - x}$

$x + y = 0 \Rightarrow y = -x$

Points of intersection: $(0, 0)$ and $(-3, 3)$

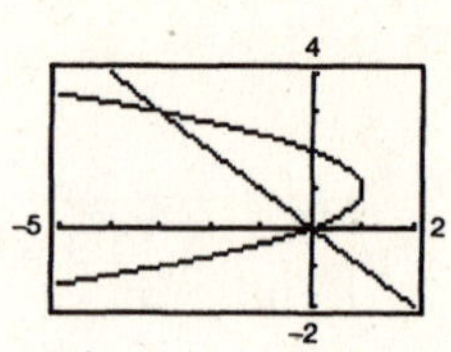

7. $\begin{cases} y = -2e^{-x} \\ 2e^x + y = 0 \Rightarrow y = -2e^x \end{cases}$

Point of intersection: $(0, -2)$

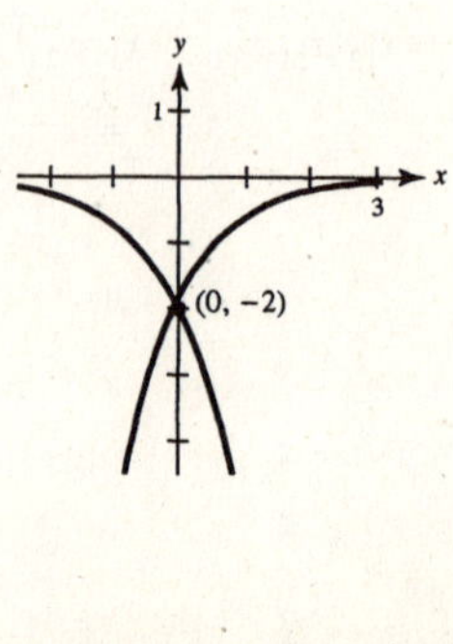

8. $y = 2(6 - x)$

$y = 2^{x-2}$

Point of intersection: $(4, 4)$

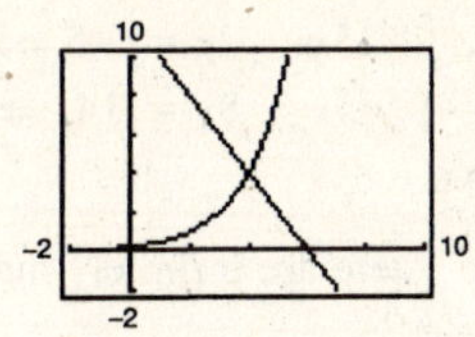

9. $\begin{cases} y = 2x^2 - 4x + 1 \\ y = x^2 - 4x + 3 \end{cases}$

Point of intersection:
$(1.41, -0.66), (-1.41, 10.66)$

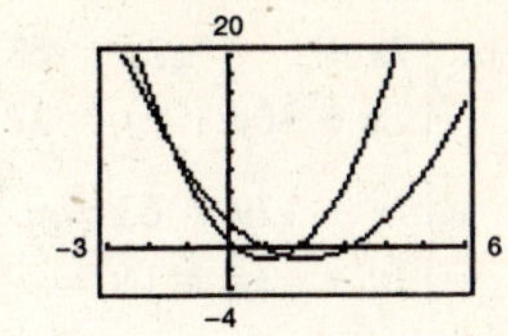

10. $y = \ln(x - 1) - 3$

$y = 4 - \frac{1}{2}x$

Point of intersection:
$(9.68, -0.84)$

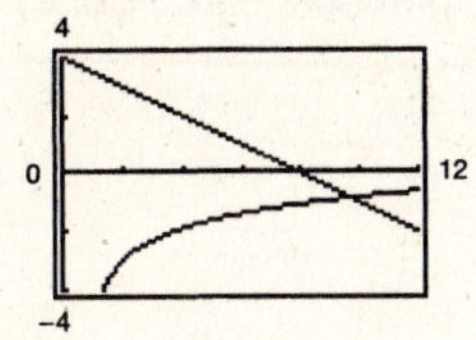

11. $\begin{cases} 2x - y = 2 \Rightarrow 16x - 8y = 16 \\ 6x + 8y = 39 \Rightarrow 6x + 8y = 39 \end{cases}$

$$22x = 55$$

$$x = \frac{55}{22} = \frac{5}{2}$$

Back-substitute $x = \frac{5}{2}$ into Equation 1

$$2\left(\frac{5}{2}\right) - y = 2$$

$$y = 3$$

Solution: $\left(\frac{5}{2}, 3\right)$

12. $\begin{cases} 40x + 30y = 24 \Rightarrow 40x + 30y = 24 \\ 20x - 50y = -14 \Rightarrow -40x + 100y = 28 \end{cases}$

$$130y = 52$$

$$y = \frac{2}{5}$$

Back-substitute $y = \frac{2}{5}$ in Equation 1.

$$40x + 30\left(\frac{2}{5}\right) = 24$$

$$40x = 12$$

$$x = \frac{3}{10}$$

Solution: $\left(\frac{3}{10}, \frac{2}{5}\right)$

13. $\begin{cases} 0.2x + 0.3y = 0.14 \Rightarrow 20x + 30y = 14 \Rightarrow 20x + 30y = 14 \\ 0.4x + 0.5y = 0.20 \Rightarrow 4x + 5y = 2 \Rightarrow -20x - 25y = -10 \end{cases}$

$$5y = 4$$

$$y = \frac{4}{5}$$

Back-substitute $y = \frac{4}{5}$ into Equation 2.

$$4x + 5\left(\frac{4}{5}\right) = 2$$

$$4x = -2$$

$$x = -\frac{1}{2}$$

Solution: $\left(-\frac{1}{2}, \frac{4}{5}\right) = (-0.5, 0.8)$

14. $\begin{cases} 12x + 42y = -17 \Rightarrow 36x + 126y = -51 \\ 30x - 18y = 19 \Rightarrow 210x - 126y = 133 \end{cases}$

$$246x = 82$$

$$x = \frac{1}{3}$$

Back-substitute $x = \frac{1}{3}$ in Equation 1.

$$12\left(\frac{1}{3}\right) + 42y = -17$$

$$42y = -21$$

$$y = -\frac{1}{2}$$

Solution: $\left(\frac{1}{3}, -\frac{1}{2}\right)$

15. $\begin{cases} 3x - 2y = 0 \Rightarrow 3x - 2y = 0 \\ 3x + 2(y + 5) = 10 \Rightarrow 3x + 2y = 0 \end{cases}$

$$6x = 0$$

$$x = 0$$

Back-substitute $x = 0$ into Equation 1.

$$3(0) - 2y = 0$$

$$2y = 0$$

$$y = 0$$

Solution: $(0, 0)$

16. $\begin{cases} 7x + 12y = 63 \\ 2x + 3(y+2) = 21 \end{cases}$

$$\begin{cases} 7x + 12y = 63 \implies -7x - 12y = -63 \\ 2x + 3y = 15 \implies \underline{8x + 12y = 60} \end{cases}$$
$$x = -3$$

Back-substitute $x = -3$ in Equation 1.

$7(-3) + 12y = 63$

$12y = 84$

$y = 7$

Solution: $(-3, 7)$

17. $\begin{cases} 1.25x - 2y = 3.5 \implies 5x - 8y = 14 \\ 5x - 8y = 14 \implies \underline{-5x + 8y = -14} \end{cases}$

$$0 = 0$$

There are infinitely many solutions.

Let $y = a$, then $5x - 8a = 14 \implies x = \frac{8}{5}a + \frac{14}{5}a.$

Solution: $\left(\frac{8}{5}a + \frac{14}{5}, a\right)$

18. $\begin{cases} 1.5x + 2.5y = 8.5 \implies 3x + 5y = 17 \\ 6x + 10y = 24 \implies \underline{-3x - 5y = -12} \end{cases}$

$$0 = 5$$

The system is inconsistent. There is no solution.

19. $\begin{cases} -3x - 5y = -1 \\ 6x + y = 4 \end{cases}$

The system is consistent. The lines intersect at one point so there is one solution.

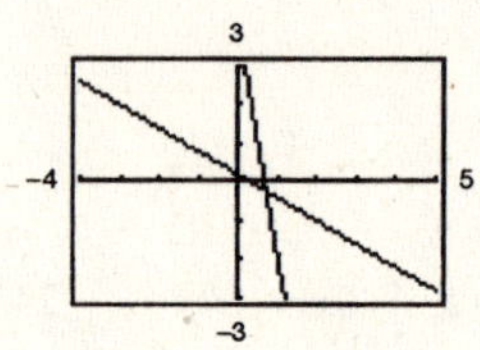

20. $\begin{cases} \frac{1}{5}x = -4 + y \implies y = \frac{1}{5}x + 4 \\ 5y = x \implies y = \frac{1}{5}x \end{cases}$

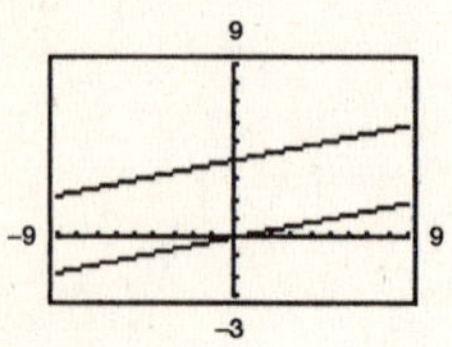

The system is inconsistent. The lines are parallel. No solution.

21. $\begin{cases} 6x - 14.4y = 1.8 \\ 1.2x - 2.88y = 0.36 \end{cases}$

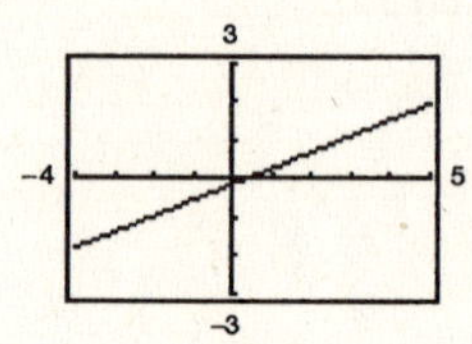

The system is consistent. The lines are identical. There are infinitely many solutions.

22. $\begin{cases} \frac{8}{5}x - y = 3 \implies y = \frac{8}{5}x - 3 \\ -5y + 8x = -2 \implies y = \frac{8}{5}x + \frac{2}{5} \end{cases}$

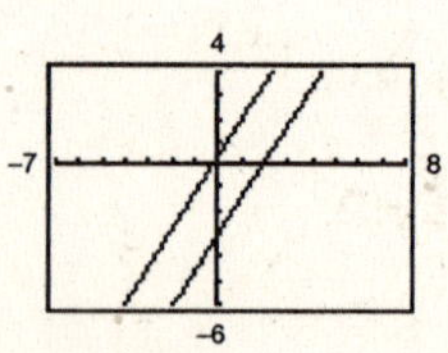

The system is inconsistent. The lines are parallel. No solution.

23. $C = 2.15x + 50{,}000$

$R = 6.95x$

Break-Even: $R = C$

$6.95x = 2.15x + 50{,}000$

$4.80x = 50{,}000$

$x \approx 10{,}417$ units

24. Let $x =$ the amount of 90% solution.

Let $y =$ the amount of 50% solution.

$$\begin{cases} x + y = 200 \implies y = 200 - x \\ 0.90x + 0.50y = 0.75(200) \end{cases}$$

$0.90x + 0.50(200 - x) = 150$

$0.90x + 100 - 0.50x = 150$

$0.40x = 50$

$x = 125$

$y = 200 - 125 = 75$

Solution: 125 liters of 90% solution,
75 liters of 50% solution

25. $x =$ number of \$9.95 compact discs

$y =$ number of \$14.95 compact discs

$x + y = 650 \implies y = 650 - x$

$9.95x + 14.95y = 7717.50$

$9.95x + 14.95(650 - x) = 7717.50$

$-5x = -2000$

$x = 400$

$y = 250$

Solution: 400 at \$9.95 and 250 at \$14.95

26. Let x = speed of the slower plane.

Let y = speed of the faster plane.

Then, distance of first plane + distance of second plane = 275 miles

(rate of first plane)(time) + (rate of second plane)(time) = 275 miles

$$\begin{cases} x\left(\frac{40}{60}\right) + y\left(\frac{40}{60}\right) = 275 \\ y = x + 25 \end{cases}$$

$$\tfrac{2}{3}x + \tfrac{2}{3}(x + 25) = 275$$

$$4x + 50 = 825$$

$$4x = 775$$

$$x = 193.75 \text{ mph}$$

$$y = x + 25 = 218.75 \text{ mph}$$

27. $37 - 0.0002x = 22 + 0.00001x$

$$15 = 0.00021x$$

$$x = \frac{500{,}000}{7},\ p = \frac{159}{7}$$

Point of equilibrium: $\left(\frac{500{,}000}{7}, \frac{159}{7}\right)$

28. $45 + 0.0002x = 120 - 0.0001x$

$$0.0003x = 75$$

$$x = 250{,}000 \text{ units}$$

$$p = \$95.00$$

Point of equilibrium: (250,000, 95)

29. $\begin{cases} x - 4y + 3z = 3 \\ -y + z = -1 \\ z = -5 \end{cases}$

$$-y + (-5) = -1 \Rightarrow y = -4$$

$$x - 4(-4) + 3(-5) = 3 \Rightarrow x = 2$$

Solution: $(2, -4, -5)$

30. $\begin{cases} x - 7y + 8z = 85 \\ y - 9z = -35 \\ z = 3 \end{cases}$

$$y - 9(3) = -35 \Rightarrow y = -8$$

$$x - 7(-8) + 8(3) = 85 \Rightarrow x = 5$$

Solution: $(5, -8, 3)$

31. $\begin{cases} x + 2y + 6z = 4 \\ -3x + 2y - z = -4 \\ 4x + 2z = 16 \end{cases}$

$$\begin{cases} x + 2y + 6z = 4 & \\ 8y + 17z = 8 & \text{3Eq.1 + Eq.2} \\ -8y - 22z = 0 & -\text{4Eq.1 + Eq.3} \end{cases}$$

$$\begin{cases} x + 2y + 6z = 4 & \\ 8y + 17z = 8 & \\ -5z = 8 & \text{Eq.2 + Eq.3} \end{cases}$$

$$\begin{cases} x + 2y + 6z = 4 & \\ 8y + 17z = 8 & \\ z = -\frac{8}{5} & -\tfrac{1}{5}\text{Eq.3} \end{cases}$$

$$8y + 17\left(-\tfrac{8}{5}\right) = 8 \Rightarrow y = \tfrac{22}{5}$$

$$x + 2\left(\tfrac{22}{5}\right) + 6\left(-\tfrac{8}{5}\right) = 4 \Rightarrow x = \tfrac{24}{5}$$

Solution: $\left(\frac{24}{5}, \frac{22}{5}, -\frac{8}{5}\right)$

32. $\begin{cases} x + 3y - z = 13 & \text{Equation 1} \\ 2x - 5z = 23 & \text{Equation 2} \\ 4x - y - 2z = 14 & \text{Equation 3} \end{cases}$

$$\begin{cases} x + 3y - z = 13 & \\ -6y - 3z = -3 & (-2)\text{Eq.1 + Eq.2} \\ -13y + 2z = -38 & (-4)\text{Eq.1 + Eq.3} \end{cases}$$

$$\begin{cases} x + 3y - z = 13 & \\ -6y - 3z = -3 & \\ \frac{17}{2}z = -\frac{63}{2} & \left(-\frac{13}{6}\right)\text{Eq.2 + Eq.3} \end{cases}$$

$$\begin{cases} x + 3y - z = 13 & \\ y + \frac{1}{2}z = \frac{1}{2} & \left(-\frac{1}{6}\right)\text{Eq.2} \\ z = -\frac{63}{17} & \left(\frac{2}{17}\right)\text{Eq.3} \end{cases}$$

$$y + \tfrac{1}{2}\left(-\tfrac{63}{17}\right) = \tfrac{1}{2} \Rightarrow y = \tfrac{40}{17}$$

$$x + 3\left(\tfrac{40}{17}\right) - \left(-\tfrac{63}{17}\right) = 13 \Rightarrow x = \tfrac{38}{17}$$

Solution: $\left(\frac{38}{17}, \frac{40}{17}, -\frac{63}{17}\right)$

33. $\begin{cases} x - 2y + z = -6 \\ 2x - 3y = -7 \\ -x + 3y - 3z = 11 \end{cases}$

$\begin{cases} x - 2y + z = -6 & \\ y - 2z = 5 & -2\text{Eq.1} + \text{Eq.2} \\ y - 2z = 5 & \text{Eq.1} + \text{Eq.3} \end{cases}$

$\begin{cases} x - 2y + z = -6 & \\ y - 2z = 5 & \\ 0 = 0 & -\text{Eq.2} + \text{Eq.3} \end{cases}$

Let $z = a$, then:

$$y = 2a + 5$$

$$x - 2(2a + 5) + a = -6$$

$$x - 3a - 10 = -6$$

$$x = 3a + 4$$

Solution: $(3a + 4, 2a + 5, a)$ where a is any real number.

34. $\begin{cases} 2x + 6z = -9 & \text{Equation 1} \\ 3x - 2y + 11z = -16 & \text{Equation 2} \\ 3x - y + 7z = -11 & \text{Equation 3} \end{cases}$

$\begin{cases} -x + 2y - 5z = 7 & (-1)\text{Eq.2} + \text{Eq.1} \\ 3x - 2y + 11z = -16 & \\ 3x - y + 7z = -11 & \end{cases}$

$\begin{cases} -x + 2y - 5z = 7 & \\ 4y - 4z = 5 & 3\text{Eq.1} + \text{Eq.2} \\ 5y - 8z = 10 & 3\text{Eq.1} + \text{Eq.3} \end{cases}$

$\begin{cases} -x + 2y - 5z = 7 & \\ 4y - 4z = 5 & \\ -3y = 0 & (-2)\text{Eq.2} + \text{Eq.3} \end{cases}$

$\begin{cases} -x + 2y - 5z = 7 & \\ y - z = \frac{5}{4} & (\frac{1}{4})\text{Eq.2} \\ y = 0 & (-\frac{1}{3})\text{Eq.3} \end{cases}$

$$(0) - z = \tfrac{5}{4} \Rightarrow z = -\tfrac{5}{4}$$

$$-x + 2(0) - 5\left(-\tfrac{5}{4}\right) = 7 \Rightarrow x = -\tfrac{3}{4}$$

Solution: $\left(-\frac{3}{4}, 0, -\frac{5}{4}\right)$

35. $\begin{cases} 5x - 12y + 7z = 16 \Rightarrow \\ 3x - 7y + 4z = 9 \Rightarrow \end{cases} \begin{cases} 15x - 36y + 21z = 48 \\ -15x + 35y - 20z = -45 \end{cases}$

$$-y + z = 3$$

Let $y = a$.

Then $z = a + 3$
and $5x - 12a + 7(a + 3) = 16 \Rightarrow x = a - 1$.

Solution: $(a - 1, a, a + 3)$ where a is any real number.

36. $\begin{cases} 2x + 5y - 19z = 34 \Rightarrow \\ 3x + 8y - 31z = 54 \Rightarrow \end{cases} \begin{matrix} 6x + 15y - 57z = 102 \\ -6x - 16y + 62z = -108 \end{matrix}$

$$-y + 5z = -6$$

Let $z = a$. Then:

$$-y + 5a = -6 \Rightarrow y = 5a + 6$$

$$2x + 5(5a + 6) - 19a = 34 \Rightarrow x = -3a + 2$$

Solution: $(-3a + 2, 5a + 6, a)$ where a is any real number.

37. $y = ax^2 + bx + c$ through $(0, -5)$, $(1, -2)$, and $(2, 5)$.

$(0, -5)\colon -5 = + c \Rightarrow \quad c = -5$

$(1, -2)\colon -2 = a + b + c \Rightarrow \begin{cases} a + b = 3 \\ 2a + b = 5 \end{cases}$

$(2, 5)\colon 5 = 4a + 2b + c \Rightarrow$

$\begin{cases} 2a + b = 5 \\ -a - b = -3 \end{cases}$

$$a = 2$$

$$b = 1$$

The equation of the parabola is $y = 2x^2 + x - 5$.

38. $y = ax^2 + bx + c$ through $(-5, 6)$, $(1, 0)$, $(2, 20)$.

$(-5, 6)\colon 6 = 25a - 5b + c$

$(1, 0)\colon 0 = a + b + c \Rightarrow c = -a - b$

$(2, 20)\colon 20 = 4a + 2b + c$

$\begin{cases} 24a - 6b = 6 \Rightarrow \\ 3a + b = 20 \Rightarrow \end{cases} \begin{matrix} 24a - 6b = 6 \\ -24a - 8b = -160 \end{matrix}$

$$-14b = -154$$

$$b = 11$$

$$3a + 11 = 20 \Rightarrow a = 3$$

$$c = -3 - 11 \Rightarrow c = -14$$

The equation of the parabola is $y = 3x^2 + 11x - 14$.

39. $x^2 + y^2 + Dx + Ey + F = 0$ through $(-1, -2)$, $(5, -2)$ and $(2, 1)$.

$$\begin{aligned}(-1,-2)&: 5 - D - 2E + F = 0 \Rightarrow \\ (5,-2)&: 29 + 5D - 2E + F = 0 \Rightarrow \\ (2, 1)&: 5 + 2D + 2E + F = 0 \Rightarrow\end{aligned}\begin{cases} D + 2E - F = 5 \\ 5D - 2E + F = -29 \\ 2D + E + F = -5\end{cases}$$

From the first two equations we have

$$6D = -24$$
$$D = -4.$$

Substituting $D = -4$ into the second and third equations yields:

$$\begin{aligned}-20 - 2E + F = -29 &\Rightarrow \\ -8 + E + F = -5 &\Rightarrow\end{aligned}\begin{cases}-2E + F = -9 \\ -E - F = -3\end{cases}$$

$$\begin{aligned}-3E &= -12 \\ E &= 4 \\ F &= -1\end{aligned}$$

The equation of the circle is $x^2 + y^2 - 4x + 4y - 1 = 0$.

40. $x^2 + y^2 + Dx + Ey + F = 0$ through $(1, 4)$, $(4, 3)$, $(-2, -5)$.

$(1, 4)$: $17 + D + 4E + F = 0$

$(4, 3)$: $25 + 4D + 3E + F = 0$

$(-2, -5)$: $29 - 2D - 5E + F = 0$

$$\begin{cases} D + 4E + F = -17 & \text{Equation 1} \\ 4D + 3E + F = -25 & \text{Equation 2} \\ 2D + 5E - F = 29 & \text{Equation 3}\end{cases}$$

$$\begin{cases} D + 4E + F = -17 & \\ -13E - 3F = 43 & (-4)\text{Eq.}1 + \text{Eq.}2 \\ -3E - 3F = 63 & (-2)\text{Eq.}1 + \text{Eq.}3\end{cases}$$

$$\begin{cases} D + 4E + F = -17 & \\ -3E - 3F = 63 & \text{Interchange equations} \\ -13E - 3F = 43 & \end{cases}$$

$$\begin{cases} D + 4E + F = -17 & \\ -3E - 3F = 63 & \\ 10F = -230 & (-\tfrac{13}{3})\text{Eq.}2 + \text{Eq.}3\end{cases}$$

$F = -23, E = 2, D = -2$

The equation of the circle is $x^2 + y^2 - 2x + 2y - 23 = 0$.

41. From the following chart we obtain our system of equations.

	A	B	C
Mixture X	$\frac{1}{5}$	$\frac{2}{5}$	$\frac{2}{5}$
Mixture Y	0	0	1
Mixture Z	$\frac{1}{3}$	$\frac{1}{3}$	$\frac{1}{3}$
Desired Mixture	$\frac{6}{27}$	$\frac{8}{27}$	$\frac{13}{27}$

$$\left.\begin{aligned}\tfrac{1}{5}x + \tfrac{1}{3}z &= \tfrac{6}{27} \\ \tfrac{2}{5}x + \tfrac{1}{3}z &= \tfrac{8}{27}\end{aligned}\right\} x = \tfrac{10}{27}, z = \tfrac{12}{27}$$

$$\tfrac{2}{5}x + y + \tfrac{1}{3}z = \tfrac{13}{27} \Rightarrow y = \tfrac{5}{27}$$

To obtain the desired mixture, use 10 gallons of spray X, 5 gallons of spray Y, and 12 gallons of spray Z.

42. Let x = amount invested at 7%
y = amount invested at 9%
z = amount invested at 11%.

$y = x - 3000$ and $z = x - 5000 \Rightarrow y + z = 2x - 8000$

$$\begin{cases} x + y + z = 40{,}000 \\ 0.07x + 0.09y + 0.11z = 3500 \\ y + z = 2x - 8000 \end{cases}$$

$$x + (2x - 8000) = 40{,}000 \Rightarrow x = 16{,}000$$
$$y = 16{,}000 - 3000 \Rightarrow y = 13{,}000$$
$$z = 16{,}000 - 5000 \Rightarrow z = 11{,}000$$

Thus, \$16,000 was invested at 7%, \$13,000 at 9% and \$11,000 at 11%.

43. $s(t) = \frac{1}{2}at^2 + v_0t + s_0$

$(1, 52), (2, 72), (3, 60)$

$52 = \frac{1}{2}a + v_0 + s_0 \Rightarrow a + 2v_0 + 2s_0 = 104$

$72 = 2a + 2v_0 + s_0 \Rightarrow 2a + 2v_0 + s_0 = 72$

$60 = \frac{9}{2}a + 3v_0 + s_0 \Rightarrow 9a + 6v_0 + 2s_0 = 120$

Solving this system yields $a = -32$, $v_0 = 68$, $s_0 = 0$.

Thus, $s(t) = \frac{1}{2}(-32)t^2 + 68t + 0 = -16t^2 + 68t$.

44. $s(t) = \frac{1}{2}at^2 + v_0t + s_0$

$(1, 36), (2, 48), (3, 36)$

$36 = \frac{1}{2}a + v_0 + s_0 \Rightarrow a + 2v_0 + 2s_0 = 72$

$48 = 2a + 2v_0 + s_0 \Rightarrow 2a + 2v_0 + s_0 = 48$

$36 = \frac{9}{2}a + 3v_0 + s_0 \Rightarrow 9a + 6v_0 + 2s_0 = 72$

Solving this system yields $a = -24$, $v_0 = 48$, $s_0 = 0$.

Thus, $s(t) = \frac{1}{2}(-24)t^2 + 48t + 0 = -12t^2 + 48t$.

45. $y \le 5 - \frac{1}{2}x$

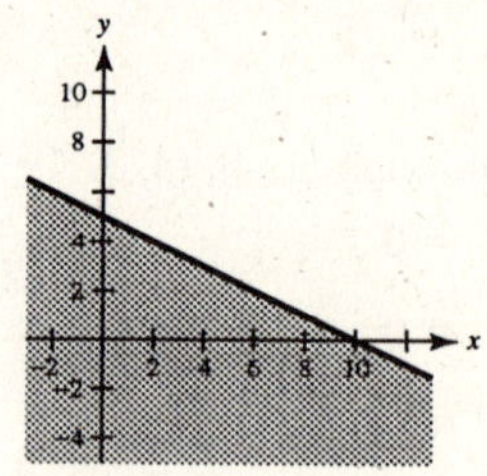

46. $3y - x \ge 7$

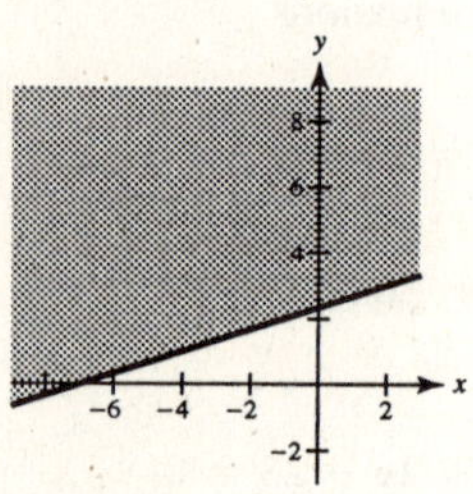

47. $y - 4x^2 > -1$

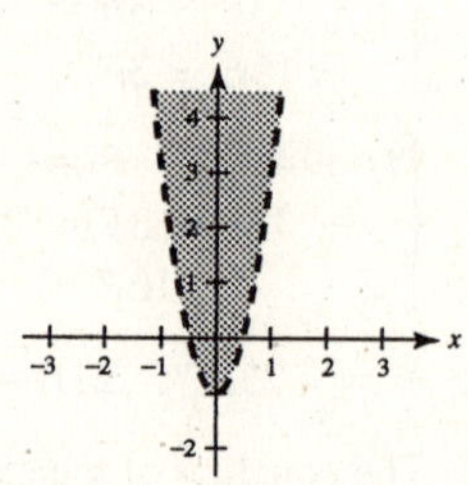

48. $y \le 2 \ln x - 6$

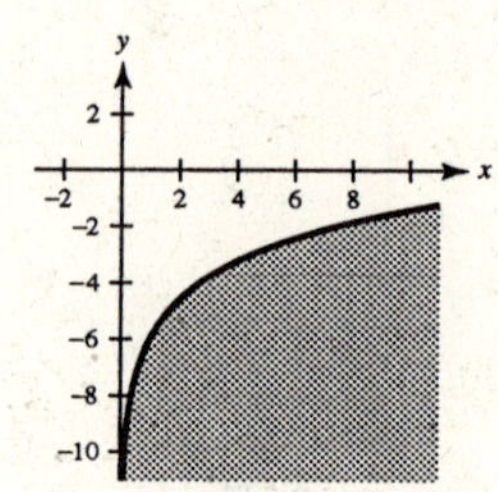

49. $\begin{cases} x + 2y \le 160 \\ 3x + y \le 180 \\ x \ge 0 \\ y \ge 0 \end{cases}$

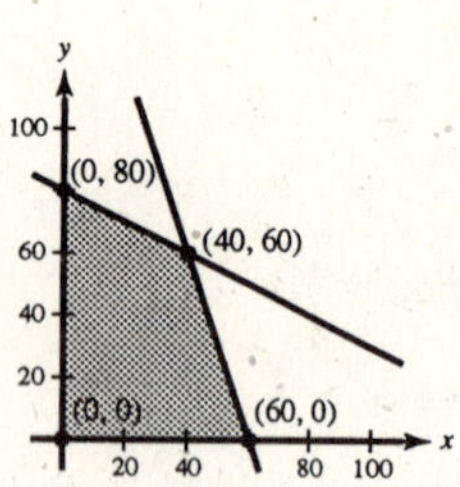

Vertex A	Vertex B	Vertex C	Vertex D
$x + 2y = 160$	$x + 2y = 160$	$3x + y = 180$	$x = 0$
$3x + y = 180$	$x = 0$	$y = 0$	$y = 0$
$(40, 60)$	$(0, 80)$	$(60, 0)$	$(0, 0)$

50. $\begin{cases} 2x + 3y \le 24 \\ 2x + y \le 16 \\ x \ge 0 \\ y \ge 0 \end{cases}$

Vertices:
$(0, 0), (0, 8), (6, 4), (8, 0)$

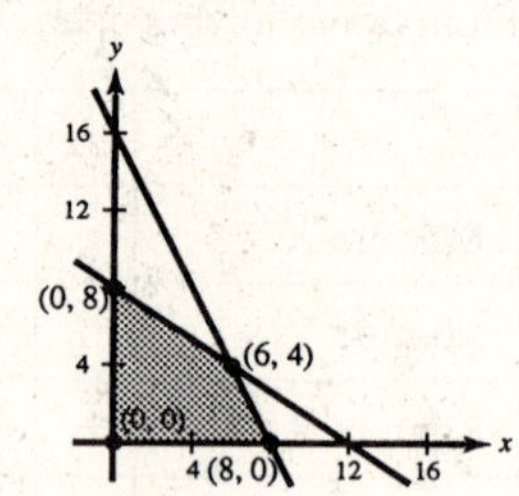

51. $\begin{cases} 3x + 2y \geq 24 \\ x + 2y \geq 12 \\ 2 \leq x \leq 15 \\ y \leq 15 \end{cases}$

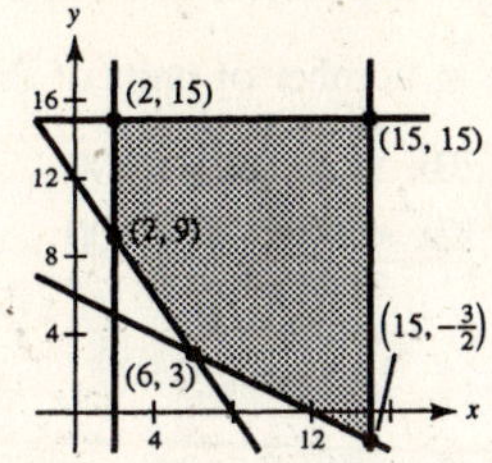

Vertex A	Vertex B
$3x + 2y = 24$	$3x + 2y = 24$
$x + 2y = 12$	$x = 2$
$(6, 3)$	$(2, 9)$

Vertex C	Vertex D	Vertex E
$x = 2$	$x = 15$	$x + 2y = 12$
$y = 15$	$y = 15$	$x = 15$
$(2, 15)$	$(15, 15)$	$\left(15, -\frac{3}{2}\right)$

52. $\begin{cases} 2x + y \geq 16 \\ x + 3y \geq 18 \\ 0 \leq x \leq 25 \\ 0 \leq y \leq 25 \end{cases}$

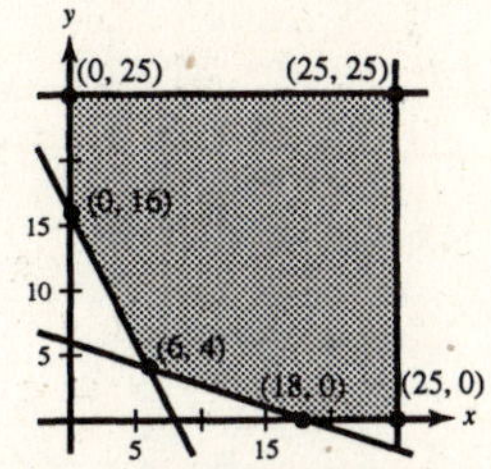

Vertices: $(6, 4), (0, 16), (0, 25), (25, 25), (25, 0), (18, 0)$

53. $\begin{cases} y < x + 1 \\ y > x^2 - 1 \end{cases}$

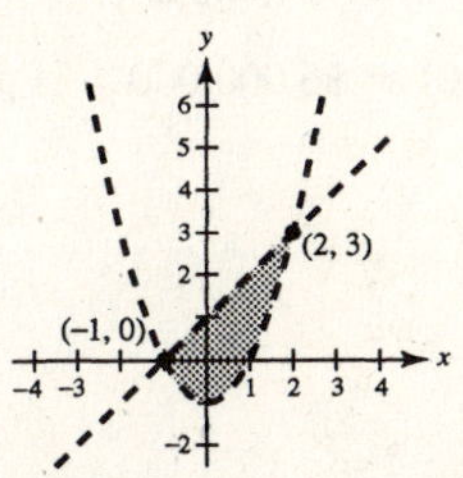

Vertices:

$x + 1 = x^2 - 1$

$0 = x^2 - x - 2 = (x + 1)(x - 2)$

$x = -1$ or $x = 2$

$y = 0 \qquad y = 3$

$(-1, 0) \qquad (2, 3)$

54. $\begin{cases} y \leq 6 - 2x - x^2 \\ y \geq x + 6 \end{cases}$

Vertices: $x + 6 = 6 - 2x - x^2$

$x^2 + 3x = 0$

$x(x + 3) = 0 \implies x = 0, -3$

$(0, 6), (-3, 3)$

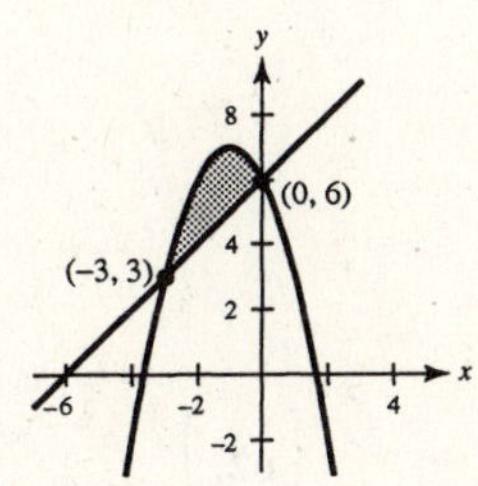

55. $\begin{cases} 2x - 3y \geq 0 \\ 2x - y \leq 8 \\ y \geq 0 \end{cases}$

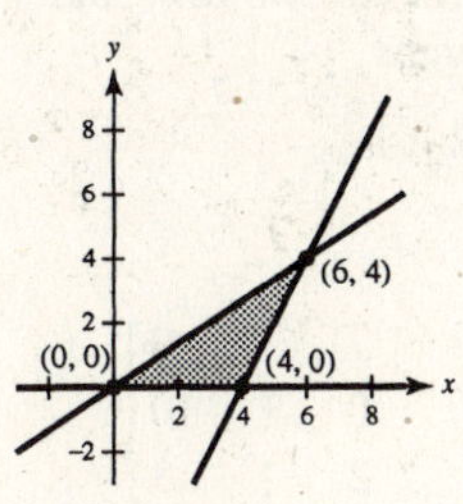

Vertex A	Vertex B	Vertex C
$2x - 3y = 0$	$2x - 3y = 0$	$2x - y = 8$
$2x - y = 8$	$y = 0$	$y = 0$
$(6, 4)$	$(0, 0)$	$(4, 0)$

56. $\begin{cases} x^2 + y^2 \leq 9 \implies y^2 \leq 9 - x^2 \\ (x - 3)^2 + y^2 \leq 9 \implies y^2 \leq 9 - (x - 3)^2 \end{cases}$

Vertices: $9 - x^2 = 9 - (x - 3)^2$

$(x - 3)^2 - x^2 = 0$

$x^2 - 6x + 9 - x^2 = 0$

$x = \dfrac{3}{2}$

$\left(\dfrac{3}{2}, \pm\dfrac{3\sqrt{3}}{2}\right)$

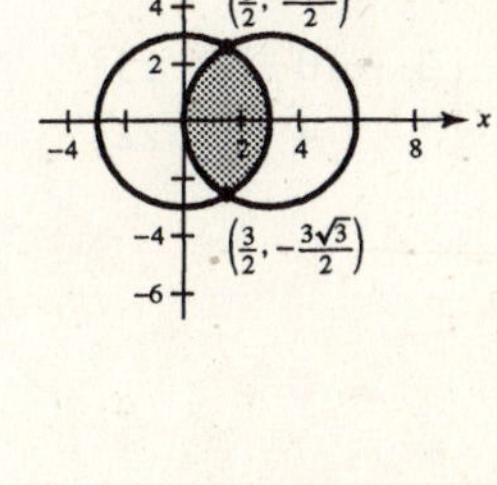

57. Let x = the number of bushels for Harrisburg, and
y = the number of bushels for Philadelphia.

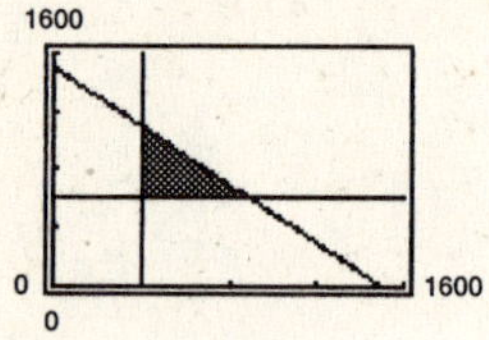

$$\begin{cases} x \geq 400 \\ y \geq 600 \\ x + y \leq 1500 \end{cases}$$

58. x = number of units of Product I

y = number of units of Product II

$$\begin{cases} 20x + 30y \leq 24{,}000 \\ 12x + 8y \leq 12{,}400 \\ x \geq 0 \\ y \geq 0 \end{cases}$$

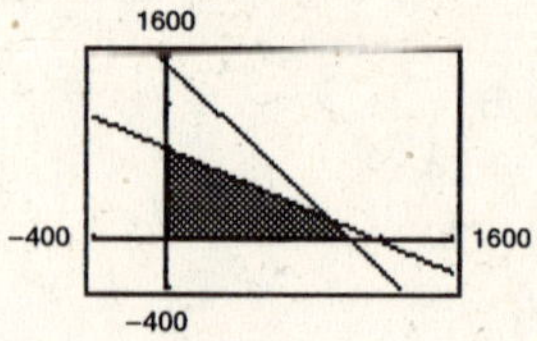

59. $160 - 0.0001x = 70 + 0.0002x$

$$90 = 0.0003x$$

$$x = 300{,}000 \text{ units}$$

$$p = \$130$$

Point of equilibrium: (300,000, 130)

Consumer surplus: $\frac{1}{2}(300{,}000)(30) = \$4{,}500{,}000$

Producer surplus: $\frac{1}{2}(300{,}000)(60) = \$9{,}000{,}000$

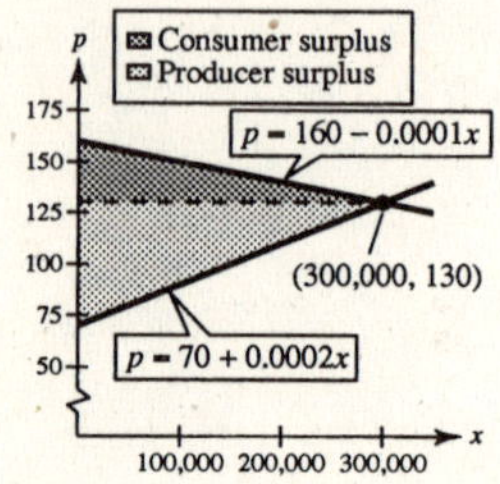

60. $130 - 0.0002x = 30 + 0.0003x$

$$100 = 0.0005x$$

$$x = 200{,}000 \text{ units}$$

$$p = \$90$$

Point of equilibrium: (200,000, 90)

Consumer surplus: $\frac{1}{2}(200{,}000)(40) = \$4{,}000{,}000$

Producer surplus: $\frac{1}{2}(200{,}000)(60) = \$6{,}000{,}000$

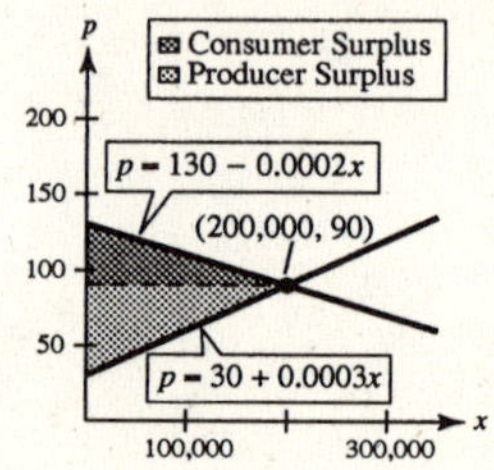

61. $\begin{bmatrix} -4 \\ 0 \\ 5 \end{bmatrix}$

Order: 3×1

62. $\begin{bmatrix} 3 & -1 & 0 & 6 \\ -2 & 7 & 1 & 4 \end{bmatrix}$

Since the matrix has two rows and four columns, its order is 2×4.

63. $[3]$

Order: 1×1

64. $[6 \quad 2 \quad -5 \quad 8 \quad 0]$

Since the matrix has one row and five columns, its order is 1×5.

65. $\begin{bmatrix} 2 & 3 & -2 \\ 1 & 0 & 1 \end{bmatrix}$

The matrix has two rows and three columns.

Order: 2×3

66. $\begin{bmatrix} 2 & 1 & 4 & -1 & 0 \\ 1 & -2 & 0 & -2 & 1 \end{bmatrix}$

The matrix has two rows and five columns.

Order: 2×5

67. $\begin{cases} 3x - 10y = 15 \\ 5x + 4y = 22 \end{cases}$

$$\left[\begin{array}{cc:c} 3 & -10 & 15 \\ 5 & 4 & 22 \end{array}\right]$$

68. $\begin{cases} 8x - 7y + 4z = 12 \\ 3x - 5y + 2z = 20 \\ 5x + 3y - 3z = 26 \end{cases}$

$$\left[\begin{array}{ccc:c} 8 & -7 & 4 & 12 \\ 3 & -5 & 2 & 20 \\ 5 & 3 & -3 & 26 \end{array}\right]$$

69. $\left[\begin{array}{ccc:c} 5 & 1 & 7 & -9 \\ 4 & 2 & 0 & 10 \\ 9 & 4 & 2 & 3 \end{array}\right]$

$$\begin{cases} 5x + y + 7z = -9 \\ 4x + 2y = 10 \\ 9x + 4y + 2z = 3 \end{cases}$$

70. $\begin{bmatrix} 13 & 16 & 7 & 3 & \vdots & 2 \\ 1 & 21 & 8 & 5 & \vdots & 12 \\ 4 & 10 & -4 & 3 & \vdots & -1 \end{bmatrix}$ $\quad \begin{cases} 13x + 16y + 7z + 3w = 2 \\ x + 21y + 8z + 5w = 12 \\ 4x + 10y - 4z + 3w = -1 \end{cases}$

71.

$$\begin{bmatrix} 0 & 1 & 1 \\ 1 & 2 & 3 \\ 2 & 2 & 2 \end{bmatrix}$$

$$\begin{matrix} R_1 \\ R_2 \\ \\ \end{matrix}\begin{bmatrix} 1 & 2 & 3 \\ 0 & 1 & 1 \\ 2 & 2 & 2 \end{bmatrix}$$

$$\begin{matrix} \\ \\ -2R_1 + R_3 \rightarrow \end{matrix}\begin{bmatrix} 1 & 2 & 3 \\ 0 & 1 & 1 \\ 0 & -2 & -4 \end{bmatrix}$$

$$\begin{matrix} -2R_2 + R_1 \rightarrow \\ \\ 2R_2 + R_3 \rightarrow \end{matrix}\begin{bmatrix} 1 & 0 & 1 \\ 0 & 1 & 1 \\ 0 & 0 & -2 \end{bmatrix}$$

$$\begin{matrix} \\ \\ -\frac{1}{2}R_3 \rightarrow \end{matrix}\begin{bmatrix} 1 & 0 & 1 \\ 0 & 1 & 1 \\ 0 & 0 & 1 \end{bmatrix}$$

$$\begin{matrix} -R_3 + R_1 \rightarrow \\ -R_3 + R_2 \rightarrow \\ \\ \end{matrix}\begin{bmatrix} 1 & 0 & 0 \\ 0 & 1 & 0 \\ 0 & 0 & 1 \end{bmatrix}$$

72.

$$\begin{bmatrix} 1 & 1 & 1 & 0 \\ 1 & 1 & 0 & 1 \\ 1 & 0 & 1 & 1 \\ 0 & 1 & 1 & 1 \end{bmatrix}$$

$$\begin{matrix} \\ -R_1 + R_2 \rightarrow \\ -R_1 + R_3 \rightarrow \\ \\ \end{matrix}\begin{bmatrix} 1 & 1 & 1 & 0 \\ 0 & 0 & -1 & 1 \\ 0 & -1 & 0 & 1 \\ 0 & 1 & 1 & 1 \end{bmatrix}$$

$$\begin{matrix} R_3 + R_1 \rightarrow \\ \\ \\ R_3 + R_4 \rightarrow \end{matrix}\begin{bmatrix} 1 & 0 & 1 & 1 \\ 0 & 0 & -1 & 1 \\ 0 & -1 & 0 & 1 \\ 0 & 0 & 1 & 2 \end{bmatrix}$$

$$\begin{matrix} \\ -R_2 \rightarrow \\ -R_3 \rightarrow \\ R_2 + R_4 \rightarrow \end{matrix}\begin{bmatrix} 1 & 0 & 1 & 1 \\ 0 & 0 & 1 & -1 \\ 0 & 1 & 0 & -1 \\ 0 & 0 & 0 & 3 \end{bmatrix}$$

$$\begin{matrix} \\ R_3 \\ R_2 \\ \frac{1}{3}R_4 \rightarrow \end{matrix}\begin{bmatrix} 1 & 0 & 1 & 1 \\ 0 & 1 & 0 & -1 \\ 0 & 0 & 1 & -1 \\ 0 & 0 & 0 & 1 \end{bmatrix}$$

$$\begin{matrix} -R_4 + R_1 \rightarrow \\ R_4 + R_2 \rightarrow \\ R_4 + R_3 \rightarrow \\ \\ \end{matrix}\begin{bmatrix} 1 & 0 & 1 & 0 \\ 0 & 1 & 0 & 0 \\ 0 & 0 & 1 & 0 \\ 0 & 0 & 0 & 1 \end{bmatrix}$$

$$\begin{matrix} -R_3 + R_1 \rightarrow \\ \\ \\ \\ \end{matrix}\begin{bmatrix} 1 & 0 & 0 & 0 \\ 0 & 1 & 0 & 0 \\ 0 & 0 & 1 & 0 \\ 0 & 0 & 0 & 1 \end{bmatrix}$$

73. $\begin{bmatrix} 1 & 2 & 3 & \vdots & 9 \\ 0 & 1 & -2 & \vdots & 2 \\ 0 & 0 & 0 & \vdots & 0 \end{bmatrix}$

Consistent

Infinitely many solutions

74. $\begin{bmatrix} 1 & 2 & 3 & \vdots & 9 \\ 0 & 1 & -2 & \vdots & 2 \\ 0 & 0 & 0 & \vdots & 8 \end{bmatrix}$

Inconsistent

75. $\begin{bmatrix} 1 & 2 & 3 & \vdots & 9 \\ 0 & 1 & -2 & \vdots & 2 \\ 0 & 0 & 1 & \vdots & -3 \end{bmatrix}$

Consistent

One solution

76. $\begin{bmatrix} 1 & 2 & 3 & 10 & 6 & \vdots & 0 \\ 0 & 1 & -5 & -2 & 0 & \vdots & 5 \\ 0 & 0 & 1 & 12 & 0 & \vdots & -2 \\ 0 & 0 & 0 & 1 & 1 & \vdots & 0 \end{bmatrix}$

Consistent
Infinitely many solutions

77. $\begin{bmatrix} 5 & 4 & \vdots & 2 \\ -1 & 1 & \vdots & -22 \end{bmatrix}$

$4R_2 + R_1 \rightarrow \begin{bmatrix} 1 & 8 & \vdots & -86 \\ -1 & 1 & \vdots & -22 \end{bmatrix}$

$R_1 + R_2 \rightarrow \begin{bmatrix} 1 & 8 & \vdots & -86 \\ 0 & 9 & \vdots & -108 \end{bmatrix}$

$\frac{1}{9}R_2 \rightarrow \begin{bmatrix} 1 & 8 & \vdots & -86 \\ 0 & 1 & \vdots & -12 \end{bmatrix}$

$\begin{cases} x + 8y = -86 \\ y = -12 \end{cases}$

$y = -12$

$x + 8(-12) = -86 \Rightarrow x = 10$

Solution: $(10, -12)$

78. $\begin{bmatrix} 2 & -5 & \vdots & 2 \\ 3 & -7 & \vdots & 1 \end{bmatrix}$

$\frac{1}{2}R_1 \rightarrow \begin{bmatrix} 1 & -\frac{5}{2} & \vdots & 1 \\ 3 & -7 & \vdots & 1 \end{bmatrix}$

$-3R_1 + R_2 \rightarrow \begin{bmatrix} 1 & -\frac{5}{2} & \vdots & 1 \\ 0 & \frac{1}{2} & \vdots & -2 \end{bmatrix}$

$2R_3 \rightarrow \begin{bmatrix} 1 & -\frac{5}{2} & \vdots & 1 \\ 0 & 1 & \vdots & -4 \end{bmatrix}$

$\begin{cases} x - \frac{5}{2}y = 1 \\ y = -4 \end{cases}$

$y = -4$

$x - \frac{5}{2}(-4) = 1 \Rightarrow x = -9$

Solution: $(-9, -4)$

79. $\begin{bmatrix} 0.3 & -0.1 & \vdots & -0.13 \\ 0.2 & -0.3 & \vdots & -0.25 \end{bmatrix}$

$\begin{matrix} 10R_1 \rightarrow \\ 10R_2 \rightarrow \end{matrix} \begin{bmatrix} 3 & -1 & \vdots & -1.3 \\ 2 & -3 & \vdots & -2.5 \end{bmatrix}$

$-R_2 + R_1 \rightarrow \begin{bmatrix} 1 & 2 & \vdots & 1.2 \\ 2 & -3 & \vdots & -2.5 \end{bmatrix}$

$-2R_1 + R_2 \rightarrow \begin{bmatrix} 1 & 2 & \vdots & 1.2 \\ 0 & -7 & \vdots & -4.9 \end{bmatrix}$

$-\frac{1}{7}R_2 \rightarrow \begin{bmatrix} 1 & 2 & \vdots & 1.2 \\ 0 & 1 & \vdots & 0.7 \end{bmatrix}$

$\begin{cases} x + 2y = 1.2 \\ y = 0.7 \end{cases}$

$y = 0.7$

$x + 2(0.7) = 1.2 \Rightarrow x = -0.2$

Solution: $(-0.2, 0.7)$

80. $\begin{bmatrix} 0.2 & -0.1 & \vdots & 0.07 \\ 0.4 & -0.5 & \vdots & -0.01 \end{bmatrix}$

$\begin{matrix} 5R_1 \rightarrow \\ -2R_1 + R_2 \rightarrow \end{matrix} \begin{bmatrix} 1 & -0.5 & \vdots & 0.35 \\ 0 & -0.3 & \vdots & -0.15 \end{bmatrix}$

$-\frac{1}{0.3}R_2 \rightarrow \begin{bmatrix} 1 & -0.5 & \vdots & 0.35 \\ 0 & 1 & \vdots & 0.5 \end{bmatrix}$

$\begin{cases} x - 0.5y = 0.35 \\ y = 0.5 \end{cases}$

$y = 0.5$

$x - 0.5(0.5) = 0.35 \Rightarrow x = 0.6$

81. $\begin{bmatrix} 2 & 3 & 1 & \vdots & 10 \\ 2 & -3 & -3 & \vdots & 22 \\ 4 & -2 & 3 & \vdots & -2 \end{bmatrix}$

$\begin{matrix} \\ -R_1 + R_2 \rightarrow \\ -2R_1 + R_3 \rightarrow \end{matrix} \begin{bmatrix} 2 & 3 & 1 & \vdots & 10 \\ 0 & -6 & -4 & \vdots & 12 \\ 0 & -8 & 1 & \vdots & -22 \end{bmatrix}$

$\begin{matrix} \frac{1}{2}R_1 \rightarrow \\ -\frac{1}{6}R_2 \rightarrow \\ \\ \end{matrix} \begin{bmatrix} 1 & \frac{3}{2} & \frac{1}{2} & \vdots & 5 \\ 0 & 1 & \frac{2}{3} & \vdots & -2 \\ 0 & -8 & 1 & \vdots & -22 \end{bmatrix}$

$8R_2 + R_3 \rightarrow \begin{bmatrix} 1 & \frac{3}{2} & \frac{1}{2} & 5 \\ 0 & 1 & \frac{2}{3} & -2 \\ 0 & 0 & \frac{19}{3} & -38 \end{bmatrix}$

$\frac{3}{19}R_3 \rightarrow \begin{bmatrix} 1 & \frac{3}{2} & \frac{1}{2} & \vdots & 5 \\ 0 & 1 & \frac{2}{3} & \vdots & -2 \\ 0 & 0 & 1 & \vdots & -6 \end{bmatrix}$

$z = -6$

$y + \frac{2}{3}(-6) = -2 \Rightarrow y = 2$

$x + \frac{3}{2}(2) + \frac{1}{2}(-6) = 5 \Rightarrow x = 5$

Solution: $(5, 2, -6)$

82. $\begin{bmatrix} 2 & 3 & 3 & \vdots & 3 \\ 6 & 6 & 12 & \vdots & 13 \\ 12 & 9 & -1 & \vdots & 2 \end{bmatrix}$

$$\begin{matrix} \\ -3R_1 + R_2 \rightarrow \\ -6R_1 + R_3 \rightarrow \end{matrix} \begin{bmatrix} 2 & 3 & 3 & \vdots & 3 \\ 0 & -3 & 3 & \vdots & 4 \\ 0 & -9 & -19 & \vdots & -16 \end{bmatrix}$$

$$\begin{matrix} R_2 + R_1 \rightarrow \\ \\ -3R_2 + R_3 \rightarrow \end{matrix} \begin{bmatrix} 2 & 0 & 6 & \vdots & 7 \\ 0 & -3 & 3 & \vdots & 4 \\ 0 & 0 & -28 & \vdots & -28 \end{bmatrix}$$

$$\begin{matrix} \frac{1}{2}R_1 \rightarrow \\ -\frac{1}{3}R_2 \rightarrow \\ -\frac{1}{28}R_3 \rightarrow \end{matrix} \begin{bmatrix} 1 & 0 & 3 & \vdots & \frac{7}{2} \\ 0 & 1 & -1 & \vdots & -\frac{4}{3} \\ 0 & 0 & 1 & \vdots & 1 \end{bmatrix}$$

$$\begin{cases} x + 3z = \frac{7}{2} \\ y - z = -\frac{4}{3} \\ z = 1 \end{cases}$$

$z = 1$

$y - 1 = -\frac{4}{3} \Rightarrow y = -\frac{1}{3}$

$x + 3(1) = \frac{7}{2} \Rightarrow x = \frac{1}{2}$

Solution: $\left(\frac{1}{2}, -\frac{1}{3}, 1\right)$

83. $\begin{bmatrix} 2 & 1 & 2 & \vdots & 4 \\ 2 & 2 & 0 & \vdots & 5 \\ 2 & -1 & 6 & \vdots & 2 \end{bmatrix}$

$$\begin{matrix} \\ -R_1 + R_2 \rightarrow \\ -R_1 + R_3 \rightarrow \end{matrix} \begin{bmatrix} 2 & 1 & 2 & \vdots & 4 \\ 0 & 1 & -2 & \vdots & 1 \\ 0 & -2 & 4 & \vdots & -2 \end{bmatrix}$$

$$\begin{matrix} -R_2 + R_1 \rightarrow \\ \\ 2R_2 + R_3 \rightarrow \end{matrix} \begin{bmatrix} 2 & 0 & 4 & \vdots & 3 \\ 0 & 1 & -2 & \vdots & 1 \\ 0 & 0 & 0 & \vdots & 0 \end{bmatrix}$$

$$\begin{matrix} \frac{1}{2}R_1 \rightarrow \\ \\ \\ \end{matrix} \begin{bmatrix} 1 & 0 & 2 & \vdots & \frac{3}{2} \\ 0 & 1 & -2 & \vdots & 1 \\ 0 & 0 & 0 & \vdots & 0 \end{bmatrix}$$

Let $z = a$, then:

$y - 2a = 1 \Rightarrow y = 2a + 1$

$x + 2a = \frac{3}{2} \Rightarrow x = -2a + \frac{3}{2}$

Solution: $\left(-2a + \frac{3}{2}, 2a + 1, a\right)$

84. $\begin{bmatrix} 1 & 2 & 6 & & 1 \\ 2 & 5 & 15 & & 4 \\ 3 & 1 & 3 & & -6 \end{bmatrix}$

$$\begin{matrix} \\ -2R_1 + R_2 \rightarrow \\ -3R_1 + R_3 \rightarrow \end{matrix} \begin{bmatrix} 1 & 2 & 6 & \vdots & 1 \\ 0 & 1 & 3 & \vdots & 2 \\ 0 & -5 & -15 & \vdots & -9 \end{bmatrix}$$

$$\begin{matrix} \\ \\ 5R_2 + R_3 \rightarrow \end{matrix} \begin{bmatrix} 1 & 2 & 6 & \vdots & 1 \\ 0 & 1 & 3 & \vdots & 2 \\ 0 & 0 & 0 & \vdots & 1 \end{bmatrix}$$

Because the last row consists of all zeros except for the last entry, the system is inconsistent and there is no solution.

85. $\begin{bmatrix} 2 & 1 & 1 & 0 & \vdots & 6 \\ 0 & -2 & 3 & -1 & \vdots & 9 \\ 3 & 3 & -2 & -2 & \vdots & -11 \\ 1 & 0 & 1 & 3 & \vdots & 14 \end{bmatrix}$

$$\begin{matrix} -R_4 + R_1 \\ \\ \\ \\ \end{matrix} \begin{bmatrix} 1 & 1 & 0 & -3 & \vdots & -8 \\ 0 & -2 & 3 & -1 & \vdots & 9 \\ 3 & 3 & -2 & -2 & \vdots & -11 \\ 1 & 0 & 1 & 3 & \vdots & 14 \end{bmatrix}$$

$$\begin{matrix} \\ \\ -3R_1 + R_3 \rightarrow \\ -R_1 + R_4 \rightarrow \end{matrix} \begin{bmatrix} 1 & 1 & 0 & -3 & \vdots & -8 \\ 0 & -2 & 3 & -1 & \vdots & 9 \\ 0 & 0 & -2 & 7 & \vdots & 13 \\ 0 & -1 & 1 & 6 & \vdots & 22 \end{bmatrix}$$

—CONTINUED—

85. —CONTINUED—

$$-3R_4 + R_2 \rightarrow \begin{bmatrix} 1 & 1 & 0 & -3 & \vdots & -8 \\ 0 & 1 & 0 & -19 & \vdots & -57 \\ 0 & 0 & -2 & 7 & \vdots & 13 \\ 0 & -1 & 1 & 6 & \vdots & 22 \end{bmatrix}$$

$$R_2 + R_4 \rightarrow \begin{bmatrix} 1 & 1 & 0 & -3 & \vdots & -8 \\ 0 & 1 & 0 & -19 & \vdots & -57 \\ 0 & 0 & -2 & 7 & \vdots & 13 \\ 0 & 0 & 1 & -13 & \vdots & -35 \end{bmatrix}$$

$$\begin{matrix} \\ \\ R_4 \\ R_3 \end{matrix} \begin{bmatrix} 1 & 1 & 0 & -3 & \vdots & -8 \\ 0 & 1 & 0 & -19 & \vdots & -57 \\ 0 & 0 & 1 & -13 & \vdots & -35 \\ 0 & 0 & -2 & 7 & \vdots & 13 \end{bmatrix}$$

$$2R_3 + R_4 \rightarrow \begin{bmatrix} 1 & 1 & 0 & -3 & \vdots & -8 \\ 0 & 1 & 0 & -19 & \vdots & -57 \\ 0 & 0 & 1 & -13 & \vdots & -35 \\ 0 & 0 & 0 & -19 & \vdots & -57 \end{bmatrix}$$

$$\tfrac{1}{19}R_4 \rightarrow \begin{bmatrix} 1 & 1 & 0 & -3 & \vdots & -8 \\ 0 & 1 & 0 & -19 & \vdots & -57 \\ 0 & 0 & 1 & -13 & \vdots & -35 \\ 0 & 0 & 0 & 1 & \vdots & 3 \end{bmatrix}$$

$w = 3$

$z - 13(3) = -35 \Rightarrow z = 4$

$y - 19(3) = -57 \Rightarrow y = 0$

$x + 0 - 3(3) = -8 \Rightarrow x = 1$

Solution: $(1, 0, 4, 3)$

86.

$$\begin{bmatrix} 1 & 2 & 0 & 1 & \vdots & 3 \\ 0 & -3 & 3 & 0 & \vdots & 0 \\ 4 & 4 & 1 & 2 & \vdots & 0 \\ 2 & 0 & 1 & 0 & \vdots & 3 \end{bmatrix}$$

$$\begin{matrix} \\ -\tfrac{1}{3}R_2 \rightarrow \\ -4R_1 + R_3 \rightarrow \\ -2R_1 + R_4 \rightarrow \end{matrix} \begin{bmatrix} 1 & 2 & 0 & 1 & \vdots & 3 \\ 0 & 1 & -1 & 0 & \vdots & 0 \\ 0 & -4 & 1 & -2 & \vdots & -12 \\ 0 & -4 & 1 & -2 & \vdots & -3 \end{bmatrix}$$

$$-R_3 + R_4 \rightarrow \begin{bmatrix} 1 & 2 & 0 & 1 & \vdots & 3 \\ 0 & 1 & -1 & 0 & \vdots & 0 \\ 0 & -4 & 1 & -2 & \vdots & -12 \\ 0 & 0 & 0 & 0 & \vdots & 9 \end{bmatrix}$$

Because the last row consists of all zeros except for the last entry, the system is inconsistent and there is no solution.

87.

$$\left[\begin{array}{rrrcr} -1 & 1 & 2 & \vdots & 1 \\ 2 & 3 & 1 & \vdots & -2 \\ 5 & 4 & 2 & \vdots & 4 \end{array}\right]$$

$$\begin{array}{r} -R_1 \rightarrow \\ \\ \\ \end{array}\left[\begin{array}{rrrcr} 1 & -1 & -2 & \vdots & -1 \\ 2 & 3 & 1 & \vdots & -2 \\ 5 & 4 & 2 & \vdots & 4 \end{array}\right]$$

$$\begin{array}{r} \\ -2R_1 + R_2 \rightarrow \\ -5R_1 + R_3 \rightarrow \end{array}\left[\begin{array}{rrrcr} 1 & -1 & -2 & \vdots & -1 \\ 0 & 5 & 5 & \vdots & 0 \\ 0 & 9 & 12 & \vdots & 9 \end{array}\right]$$

$$\begin{array}{r} \\ \frac{1}{5}R_2 \rightarrow \\ \\ \end{array}\left[\begin{array}{rrrcr} 1 & -1 & -2 & \vdots & -1 \\ 0 & 1 & 1 & \vdots & 0 \\ 0 & 9 & 12 & \vdots & 9 \end{array}\right]$$

$$\begin{array}{r} R_2 + R_1 \rightarrow \\ \\ -9R_2 + R_3 \rightarrow \end{array}\left[\begin{array}{rrrcr} 1 & 0 & -1 & \vdots & -1 \\ 0 & 1 & 1 & \vdots & 0 \\ 0 & 0 & 3 & \vdots & 9 \end{array}\right]$$

$$\begin{array}{r} \\ \\ \frac{1}{3}R_3 \rightarrow \end{array}\left[\begin{array}{rrrcr} 1 & 0 & -1 & \vdots & -1 \\ 0 & 1 & 1 & \vdots & 0 \\ 0 & 0 & 1 & \vdots & 3 \end{array}\right]$$

$$\begin{array}{r} R_3 + R_1 \rightarrow \\ -R_3 + R_2 \rightarrow \\ \\ \end{array}\left[\begin{array}{rrrcr} 1 & 0 & 0 & \vdots & 2 \\ 0 & 1 & 0 & \vdots & -3 \\ 0 & 0 & 1 & \vdots & 3 \end{array}\right]$$

$x = 2, y = -3, z = 3$

Solution: $(2, -3, 3)$

88. $\begin{cases} 4x + 4y + 4z = 5 \\ 4x - 2y - 8z = 1 \\ 5x + 3y + 8z = 6 \end{cases}$

$$\left[\begin{array}{rrrcr} 4 & 4 & 4 & \vdots & 5 \\ 4 & -2 & -8 & \vdots & 1 \\ 5 & 3 & 8 & \vdots & 6 \end{array}\right]$$

$$\begin{array}{r} \frac{1}{4}R_1 \rightarrow \\ \\ \\ \end{array}\left[\begin{array}{rrrcr} 1 & 1 & 1 & \vdots & \frac{5}{4} \\ 4 & -2 & -8 & \vdots & 1 \\ 5 & 3 & 8 & \vdots & 6 \end{array}\right]$$

$$\begin{array}{r} \\ -4R_1 + R_2 \rightarrow \\ -5R_1 + R_3 \rightarrow \end{array}\left[\begin{array}{rrrcr} 1 & 1 & 1 & \vdots & \frac{5}{4} \\ 0 & -6 & -12 & \vdots & -4 \\ 0 & -2 & 3 & \vdots & -\frac{1}{4} \end{array}\right]$$

$$\begin{array}{r} \\ -\frac{1}{6}R_2 \rightarrow \\ \\ \end{array}\left[\begin{array}{rrrcr} 1 & 1 & 1 & \vdots & \frac{5}{4} \\ 0 & 1 & 2 & \vdots & \frac{2}{3} \\ 0 & -2 & 3 & \vdots & -\frac{1}{4} \end{array}\right]$$

$$\begin{array}{r} -R_2 + R_1 \rightarrow \\ \\ 2R_2 + R_3 \rightarrow \end{array}\left[\begin{array}{rrrcr} 1 & 0 & -1 & \vdots & \frac{7}{12} \\ 0 & 1 & 2 & \vdots & \frac{2}{3} \\ 0 & 0 & 7 & \vdots & \frac{13}{12} \end{array}\right]$$

$$\begin{array}{r} \\ \\ \frac{1}{7}R_3 \rightarrow \end{array}\left[\begin{array}{rrrcr} 1 & 0 & -1 & \vdots & \frac{7}{12} \\ 0 & 1 & 2 & \vdots & \frac{2}{3} \\ 0 & 0 & 1 & \vdots & \frac{13}{84} \end{array}\right]$$

$$\begin{array}{r} R_3 + R_1 \rightarrow \\ -2R_3 + R_2 \rightarrow \\ \\ \end{array}\left[\begin{array}{rrrcr} 1 & 0 & 0 & \vdots & \frac{31}{42} \\ 0 & 1 & 0 & \vdots & \frac{5}{14} \\ 0 & 0 & 1 & \vdots & \frac{13}{84} \end{array}\right]$$

$x = \frac{31}{42}$

$y = \frac{5}{14}$

$z = \frac{13}{84}$

Solution: $\left(\frac{31}{42}, \frac{5}{14}, \frac{13}{84}\right)$

89.

$$\left[\begin{array}{rrrcr} 2 & -1 & 9 & \vdots & -8 \\ -1 & -3 & 4 & \vdots & -15 \\ 5 & 2 & -1 & \vdots & 17 \end{array}\right]$$

$$\begin{array}{r} R_2 + R_1 \rightarrow \\ \\ \\ \end{array}\left[\begin{array}{rrrcr} 1 & -4 & 13 & \vdots & -23 \\ -1 & -3 & 4 & \vdots & -15 \\ 5 & 2 & -1 & \vdots & 17 \end{array}\right]$$

$$\begin{array}{r} \\ R_1 + R_2 \rightarrow \\ -5R_1 + R_3 \rightarrow \end{array}\left[\begin{array}{rrrcr} 1 & -4 & 13 & \vdots & -23 \\ 0 & -7 & 17 & \vdots & -38 \\ 0 & 22 & -66 & \vdots & 132 \end{array}\right]$$

$$\begin{array}{r} \\ R_3 \\ R_2 \end{array}\left[\begin{array}{rrrcr} 1 & -4 & 13 & \vdots & -23 \\ 0 & 22 & -66 & \vdots & 132 \\ 0 & -7 & 17 & \vdots & 38 \end{array}\right]$$

$$\begin{array}{r} \\ \frac{1}{22}R_2 \rightarrow \\ \\ \end{array}\left[\begin{array}{rrrcr} 1 & -4 & 13 & \vdots & -23 \\ 0 & 1 & -3 & \vdots & 6 \\ 0 & -7 & 17 & \vdots & -38 \end{array}\right]$$

$$\begin{array}{r} \\ \\ 7R_2 + R_3 \rightarrow \end{array}\left[\begin{array}{rrrcr} 1 & -4 & 13 & \vdots & -23 \\ 0 & 1 & -3 & \vdots & 6 \\ 0 & 0 & -4 & \vdots & 4 \end{array}\right]$$

$$\begin{array}{r} \\ \\ -\frac{1}{4}R_3 \end{array}\left[\begin{array}{rrrcr} 1 & -4 & 13 & \vdots & -23 \\ 0 & 1 & -3 & \vdots & 6 \\ 0 & 0 & 1 & \vdots & -1 \end{array}\right]$$

$$\begin{array}{r} 4R_2 + R_1 \rightarrow \\ \\ \\ \end{array}\left[\begin{array}{rrrcr} 1 & 0 & 1 & \vdots & 1 \\ 0 & 1 & -3 & \vdots & 6 \\ 0 & 0 & 1 & \vdots & -1 \end{array}\right]$$

$$\begin{array}{r} -R_3 + R_1 \rightarrow \\ 3R_3 + R_2 \rightarrow \\ \\ \end{array}\left[\begin{array}{rrrcr} 1 & 0 & 0 & \vdots & 2 \\ 0 & 1 & 0 & \vdots & 3 \\ 0 & 0 & 1 & \vdots & -1 \end{array}\right]$$

$x = 2, y = 3, z = -1$

Solution: $(2, 3, -1)$

90. $\begin{cases} -3x + y + 7z = -20 \\ 5x - 2y - z = 34 \\ -x + y + 4z = -8 \end{cases}$

$$\begin{bmatrix} -3 & 1 & 7 & \vdots & -20 \\ 5 & -2 & -1 & \vdots & 34 \\ -1 & 1 & 4 & \vdots & -8 \end{bmatrix}$$

$$\begin{matrix} R_3 \\ \\ R_1 \end{matrix}\begin{bmatrix} -1 & 1 & 4 & \vdots & -8 \\ 5 & -2 & -1 & \vdots & 34 \\ -3 & 1 & 7 & \vdots & -20 \end{bmatrix}$$

$$-1R_1\begin{bmatrix} 1 & -1 & -4 & \vdots & 8 \\ 5 & -2 & -1 & \vdots & 34 \\ -3 & 1 & 7 & \vdots & -20 \end{bmatrix}$$

$$\begin{matrix} \\ -5R_1 + R_2 \rightarrow \\ 3R_1 + R_3 \rightarrow \end{matrix}\begin{bmatrix} 1 & -1 & -4 & \vdots & 8 \\ 0 & 3 & 19 & \vdots & -6 \\ 0 & -2 & -5 & \vdots & 4 \end{bmatrix}$$

$$\begin{matrix} \\ \frac{1}{3}R_2 \rightarrow \\ \\ \end{matrix}\begin{bmatrix} 1 & -1 & -4 & \vdots & 8 \\ 0 & 1 & \frac{19}{3} & \vdots & -2 \\ 0 & -2 & -5 & \vdots & 4 \end{bmatrix}$$

$$\begin{matrix} R_2 + R_1 \rightarrow \\ \\ 2R_2 + R_3 \rightarrow \end{matrix}\begin{bmatrix} 1 & 0 & \frac{7}{3} & \vdots & 6 \\ 0 & 1 & \frac{19}{3} & \vdots & -2 \\ 0 & 0 & \frac{23}{3} & \vdots & 0 \end{bmatrix}$$

$$\begin{matrix} \\ \\ \frac{3}{23}R_3 \rightarrow \end{matrix}\begin{bmatrix} 1 & 0 & \frac{7}{3} & \vdots & 6 \\ 0 & 1 & \frac{19}{3} & \vdots & -2 \\ 0 & 0 & 1 & \vdots & 0 \end{bmatrix}$$

$$\begin{matrix} -\frac{7}{3}R_3 + R_1 \rightarrow \\ -\frac{19}{3}R_3 + R_2 \rightarrow \\ \\ \end{matrix}\begin{bmatrix} 1 & 0 & 0 & \vdots & 6 \\ 0 & 1 & 0 & \vdots & -2 \\ 0 & 0 & 1 & \vdots & 0 \end{bmatrix}$$

$x = 6, y = -2, z = 0$

Solution: $(6, -2, 0)$

91. Use the reduced row–echelon form feature of a graphing utility.

$$\begin{bmatrix} 3 & -1 & 5 & -2 & \vdots & -44 \\ 1 & 6 & 4 & -1 & \vdots & 1 \\ 5 & -1 & 1 & 3 & \vdots & -15 \\ 0 & 4 & -1 & -8 & \vdots & 58 \end{bmatrix} \Rightarrow \begin{bmatrix} 1 & 0 & 0 & 0 & \vdots & 2 \\ 0 & 1 & 0 & 0 & \vdots & 6 \\ 0 & 0 & 1 & 0 & \vdots & -10 \\ 0 & 0 & 0 & 1 & \vdots & -3 \end{bmatrix}$$

$x = 2, y = 6, z = -10, w = -3$

Solution: $(2, 6, -10, -3)$

92. Use the reduced row-echelon form feature of the graphing utility.

$$\begin{bmatrix} 4 & 12 & 2 & \vdots & 20 \\ 1 & 6 & 4 & \vdots & 12 \\ 1 & 6 & 1 & \vdots & 8 \\ -2 & -10 & -2 & \vdots & -10 \end{bmatrix} \Rightarrow \begin{bmatrix} 1 & 0 & 0 & \vdots & 0 \\ 0 & 1 & 0 & \vdots & 0 \\ 0 & 0 & 1 & \vdots & 0 \\ 0 & 0 & 0 & \vdots & 1 \end{bmatrix}$$

The system is inconsistent and there is no solution.

93. $\begin{bmatrix} -1 & x \\ y & 9 \end{bmatrix} = \begin{bmatrix} -1 & 12 \\ -7 & 9 \end{bmatrix} \Rightarrow x = 12$ and $y = -7$

94. $\begin{bmatrix} x+3 & 4 & -4y \\ 0 & -3 & 2 \\ -2 & y+5 & 6x \end{bmatrix} = \begin{bmatrix} 5x-1 & 4 & -44 \\ 0 & -3 & 2 \\ -2 & 16 & 6 \end{bmatrix}$

$\left.\begin{aligned} x + 3 &= 5x - 1 \\ -4y &= -44 \\ y + 5 &= 16 \\ 6x &= 6 \end{aligned}\right\} x = 1$ and $y = 11$

95. Since A and B are both of order 2×2, $A + 3B$ can be performed.

96. Since A and B are not of the same order, $A + 3B$ cannot be performed.

97. $$\begin{bmatrix} 7 & 3 \\ -1 & 5 \end{bmatrix} + \begin{bmatrix} 10 & -20 \\ 14 & -3 \end{bmatrix} = \begin{bmatrix} 7+10 & 3-20 \\ -1+14 & 5-3 \end{bmatrix} = \begin{bmatrix} 17 & -17 \\ 13 & 2 \end{bmatrix}$$

98. Since the matrices are not of the same order, the operation cannot be performed.

99. $$-2\begin{bmatrix} 1 & 2 \\ 5 & -4 \\ 6 & 0 \end{bmatrix} + 8\begin{bmatrix} 7 & 1 \\ 1 & 2 \\ 1 & 4 \end{bmatrix} = \begin{bmatrix} -2 & -4 \\ -10 & 8 \\ -12 & 0 \end{bmatrix} + \begin{bmatrix} 56 & 8 \\ 8 & 16 \\ 8 & 32 \end{bmatrix} = \begin{bmatrix} 54 & 4 \\ -2 & 24 \\ -4 & 32 \end{bmatrix}$$

100. $$-\begin{bmatrix} 8 & -1 & 8 \\ -2 & 4 & 12 \\ 0 & -6 & 0 \end{bmatrix} - 5\begin{bmatrix} -2 & 0 & -4 \\ 3 & -1 & 1 \\ 6 & 12 & -8 \end{bmatrix}$$

$$= \begin{bmatrix} -8 & 1 & -8 \\ 2 & -4 & -12 \\ 0 & 6 & 0 \end{bmatrix} + \begin{bmatrix} 10 & 0 & 20 \\ -15 & 5 & -5 \\ -30 & -60 & 40 \end{bmatrix}$$

$$= \begin{bmatrix} -8+10 & 1+0 & -8+20 \\ 2-15 & -4+5 & -12-5 \\ 0-30 & 6-60 & 0+40 \end{bmatrix} = \begin{bmatrix} 2 & 1 & 12 \\ -13 & 1 & -17 \\ -30 & -54 & 40 \end{bmatrix}$$

101. $$X = 3A - 2B = 3\begin{bmatrix} -4 & 0 \\ 1 & -5 \\ -3 & 2 \end{bmatrix} - 2\begin{bmatrix} 1 & 2 \\ -2 & 1 \\ 4 & 4 \end{bmatrix}$$

$$= \begin{bmatrix} -14 & -4 \\ 7 & -17 \\ -17 & -2 \end{bmatrix}$$

102. $$X = \frac{1}{6}(4A + 3B) = \frac{1}{6}\left(4\begin{bmatrix} -4 & 0 \\ 1 & -5 \\ -3 & 2 \end{bmatrix} + 3\begin{bmatrix} 1 & 2 \\ -2 & 1 \\ 4 & 4 \end{bmatrix}\right) = \frac{1}{6}\left(\begin{bmatrix} -16 & 0 \\ 4 & -20 \\ -12 & 8 \end{bmatrix} + \begin{bmatrix} 3 & 6 \\ -6 & 3 \\ 12 & 12 \end{bmatrix}\right) = \frac{1}{6}\begin{bmatrix} -16+3 & 0+6 \\ 4-6 & -20+3 \\ -12+12 & 8+12 \end{bmatrix}$$

$$= \frac{1}{6}\begin{bmatrix} -13 & 6 \\ -2 & -17 \\ 0 & 20 \end{bmatrix} = \begin{bmatrix} -\frac{13}{6} & 1 \\ -\frac{1}{3} & -\frac{17}{6} \\ 0 & \frac{10}{3} \end{bmatrix}$$

103. $$X = \frac{1}{3}[B - 2A] = \frac{1}{3}\left(\begin{bmatrix} 1 & 2 \\ -2 & 1 \\ 4 & 4 \end{bmatrix} - 2\begin{bmatrix} -4 & 0 \\ 1 & -5 \\ -3 & 2 \end{bmatrix}\right)$$

$$= \frac{1}{3}\begin{bmatrix} 9 & 2 \\ -4 & 11 \\ 10 & 0 \end{bmatrix}$$

104. $X = \frac{1}{3}(2A - 5B) = \frac{1}{3}\left(2\begin{bmatrix} -4 & 0 \\ 1 & -5 \\ -3 & 2 \end{bmatrix} - 5\begin{bmatrix} 1 & 2 \\ -2 & 1 \\ 4 & 4 \end{bmatrix}\right) = \frac{1}{3}\left(\begin{bmatrix} -8 & 0 \\ 2 & -10 \\ -6 & 4 \end{bmatrix} + \begin{bmatrix} -5 & -10 \\ 10 & -5 \\ -20 & -20 \end{bmatrix}\right) = \frac{1}{3}\begin{bmatrix} -8-5 & 0-10 \\ 2+10 & -10-5 \\ -6-20 & 4-20 \end{bmatrix}$

$= \frac{1}{3}\begin{bmatrix} -13 & -10 \\ 12 & -15 \\ -26 & -16 \end{bmatrix} = \begin{bmatrix} -\frac{13}{3} & -\frac{10}{3} \\ 4 & -5 \\ -\frac{26}{3} & -\frac{16}{3} \end{bmatrix}$

105. Not possible because the number of columns of A does not equal the number of rows of B.

106. Since A is 3×2 and B is 2×2, AB can be performed.

107. $\begin{bmatrix} 1 & 2 \\ 5 & -4 \\ 6 & 0 \end{bmatrix}\begin{bmatrix} 6 & -2 & 8 \\ 4 & 0 & 0 \end{bmatrix} = \begin{bmatrix} 1(6)+2(4) & 1(-2)+2(0) & 1(8)+2(0) \\ 5(6)+(-4)(4) & 5(-2)+(-4)(0) & 5(8)+(-4)(0) \\ 6(6)+(0)(4) & 6(-2)+(0)(0) & 6(8)+(0)(0) \end{bmatrix}$

$= \begin{bmatrix} 14 & -2 & 8 \\ 14 & -10 & 40 \\ 36 & -12 & 48 \end{bmatrix}$

108. Not possible because the number of columns of the first matrix does not equal the number of rows of the second matrix.

109. $\begin{bmatrix} 1 & 5 & 6 \\ 2 & -4 & 0 \end{bmatrix}\begin{bmatrix} 6 & 4 \\ -2 & 0 \\ 8 & 0 \end{bmatrix} = \begin{bmatrix} 1(6)+5(-2)+6(8) & 1(4)+5(0)+6(0) \\ 2(6)-4(-2)+0(8) & 2(4)-4(0)+0(0) \end{bmatrix}$

$= \begin{bmatrix} 44 & 4 \\ 20 & 8 \end{bmatrix}$

110. $\begin{bmatrix} 1 & 3 & 2 \\ 0 & 2 & -4 \\ 0 & 0 & 3 \end{bmatrix}\begin{bmatrix} 4 & -3 & 2 \\ 0 & 3 & -1 \\ 0 & 0 & 2 \end{bmatrix} = \begin{bmatrix} 1(4) & 1(-3)+3(3) & 1(2)+3(-1)+2(2) \\ 0 & 2(3) & 2(-1)+(-4)(2) \\ 0 & 0 & 3(2) \end{bmatrix}$

$= \begin{bmatrix} 4 & 6 & 3 \\ 0 & 6 & -10 \\ 0 & 0 & 6 \end{bmatrix}$

111. $\begin{bmatrix} 4 \\ 6 \end{bmatrix}\begin{bmatrix} 6 & -2 \end{bmatrix} = \begin{bmatrix} 4(6) & 4(-2) \\ 6(6) & 6(-2) \end{bmatrix} = \begin{bmatrix} 24 & -8 \\ 36 & -12 \end{bmatrix}$

112. $\begin{bmatrix} 4 & -2 & 6 \end{bmatrix}\begin{bmatrix} -2 & 1 \\ 0 & -3 \\ 2 & 0 \end{bmatrix} = \begin{bmatrix} 4(-2)-2(0)+6(2) & 4(1)-2(-3)+6(0) \end{bmatrix}$

$= \begin{bmatrix} 4 & 10 \end{bmatrix}$

113. $\begin{bmatrix} 2 & 1 \\ 6 & 0 \end{bmatrix}\left(\begin{bmatrix} 4 & 2 \\ -3 & 1 \end{bmatrix} + \begin{bmatrix} -2 & 4 \\ 0 & 4 \end{bmatrix}\right) = \begin{bmatrix} 2 & 1 \\ 6 & 0 \end{bmatrix}\begin{bmatrix} 2 & 6 \\ -3 & 5 \end{bmatrix}$

$= \begin{bmatrix} 2(2)+1(-3) & 2(6)+1(5) \\ 6(2)+0 & 6(6)+0 \end{bmatrix}$

$= \begin{bmatrix} 1 & 17 \\ 12 & 36 \end{bmatrix}$

114. $-3\begin{bmatrix}1 & -1\\4 & 2\end{bmatrix}\left(\begin{bmatrix}0 & 3\\1 & 2\end{bmatrix}\begin{bmatrix}1 & 0\\5 & -3\end{bmatrix}\right) = \begin{bmatrix}-3 & 3\\-12 & -6\end{bmatrix}\begin{bmatrix}0(1)+3(5) & 0(0)+3(-3)\\1(1)+2(5) & 1(0)+2(-3)\end{bmatrix}$

$$= \begin{bmatrix}-3 & 3\\-12 & -6\end{bmatrix}\begin{bmatrix}15 & -9\\11 & -6\end{bmatrix}$$

$$= \begin{bmatrix}-3(15)+3(11) & -3(-9)+3(-6)\\-12(15)-6(11) & -12(-9)-6(-6)\end{bmatrix}$$

$$= \begin{bmatrix}-12 & 9\\-246 & 144\end{bmatrix}$$

115. $\begin{bmatrix}4 & 1\\11 & -7\\12 & 3\end{bmatrix}\begin{bmatrix}3 & -5 & 6\\2 & -2 & -2\end{bmatrix} = \begin{bmatrix}14 & -22 & 22\\19 & -41 & 80\\42 & -66 & 66\end{bmatrix}$

116. $\begin{bmatrix}-2 & 3 & 10\\4 & -2 & 2\end{bmatrix}\begin{bmatrix}1 & 1\\-5 & 2\\3 & 2\end{bmatrix} = \begin{bmatrix}13 & 24\\20 & 4\end{bmatrix}$

117. $\begin{cases}5x + 4y = 2\\-x + y = -22\end{cases}$

(a) $\begin{bmatrix}5 & 4\\-1 & 1\end{bmatrix}\begin{bmatrix}x\\y\end{bmatrix} = \begin{bmatrix}2\\-22\end{bmatrix}$

(b) $\left[\begin{array}{cc:c}5 & 4 & 2\\-1 & 1 & -22\end{array}\right]$ $\begin{array}{c}4R_2 + R_1\rightarrow\end{array}\left[\begin{array}{cc:c}1 & 8 & -86\\-1 & 1 & -22\end{array}\right]$ $\begin{array}{c}\\R_1 + R_2\rightarrow\end{array}\left[\begin{array}{cc:c}1 & 8 & -86\\0 & 9 & -108\end{array}\right]$

$\begin{array}{c}\\\frac{1}{9}R_2\rightarrow\end{array}\left[\begin{array}{cc:c}1 & 8 & -86\\0 & 1 & -12\end{array}\right]$ $\begin{array}{c}-8R_2 + R_1\rightarrow\end{array}\left[\begin{array}{cc:c}1 & 0 & 10\\0 & 1 & -12\end{array}\right]$

Solution: $x = 10, y = -12$

118. $\begin{cases}2x + 3y + z = 10\\2x - 3y - 3z = 22\\4x - 2y + 3z = -2\end{cases}$ $\begin{bmatrix}2 & 3 & 1\\2 & -3 & -3\\4 & -2 & 3\end{bmatrix}\begin{bmatrix}x\\y\\z\end{bmatrix} = \begin{bmatrix}10\\22\\-2\end{bmatrix}$

119. $BA = [10.25 \quad 14.50 \quad 17.75]\begin{bmatrix}8200 & 7400\\6500 & 9800\\5400 & 4800\end{bmatrix} = [\$274{,}150 \quad \$303{,}150]$

The merchandise shipped to warehouse 1 is worth $274,150, and the merchandise shipped to warehouse 2 is worth $303,150.

120. $A_n = 1.25A = \begin{bmatrix}10{,}250 & 9250\\8125 & 12{,}250\\6750 & 6000\end{bmatrix}$

$BA_n = [\$342{,}687.50 \quad \$378{,}937.50]$

121. $AB = \begin{bmatrix}-4 & -1\\7 & 2\end{bmatrix}\begin{bmatrix}-2 & -1\\7 & 4\end{bmatrix} = \begin{bmatrix}-4(-2)+(-1)(7) & -4(-1)+(-1)(4)\\7(-2)+2(7) & 7(-1)+2(4)\end{bmatrix}$

$$= \begin{bmatrix}1 & 0\\0 & 1\end{bmatrix} = I$$

$BA = \begin{bmatrix}-2 & -1\\7 & 4\end{bmatrix}\begin{bmatrix}-4 & -1\\7 & 2\end{bmatrix} = \begin{bmatrix}-2(-4)+(-1)(7) & -2(-1)+(-1)(2)\\7(-4)+4(7) & 7(-1)+4(2)\end{bmatrix}$

$$= \begin{bmatrix}1 & 0\\0 & 1\end{bmatrix} = I$$

122. $AB = \begin{bmatrix} 1 & -1 & 0 \\ -1 & 0 & -1 \\ 8 & -4 & 2 \end{bmatrix} \begin{bmatrix} -2 & 1 & \frac{1}{2} \\ -3 & 1 & \frac{1}{2} \\ 2 & -2 & -\frac{1}{2} \end{bmatrix} = \begin{bmatrix} 1 & 0 & 0 \\ 0 & 1 & 0 \\ 0 & 0 & 1 \end{bmatrix} = I$

$BA = \begin{bmatrix} -2 & 1 & \frac{1}{2} \\ -3 & 1 & \frac{1}{2} \\ 2 & -2 & -\frac{1}{2} \end{bmatrix} \begin{bmatrix} 1 & -1 & 0 \\ -1 & 0 & -1 \\ 8 & -4 & 2 \end{bmatrix} = \begin{bmatrix} 1 & 0 & 0 \\ 0 & 1 & 0 \\ 0 & 0 & 1 \end{bmatrix} = I$

123. $[A \ \vdots \ I] = \left[\begin{array}{cc|cc} -6 & 5 & 1 & 0 \\ -5 & 4 & 0 & 1 \end{array}\right]$

$-\frac{1}{6}R_1 \rightarrow \left[\begin{array}{cc|cc} 1 & -\frac{5}{6} & -\frac{1}{6} & 0 \\ -5 & 4 & 0 & 1 \end{array}\right]$

$5R_1 + R_2 \rightarrow \left[\begin{array}{cc|cc} 1 & -\frac{5}{6} & -\frac{1}{6} & 0 \\ 0 & -\frac{1}{6} & -\frac{5}{6} & 1 \end{array}\right]$

$-6R_2 \rightarrow \begin{bmatrix} 1 & -\frac{5}{6} & -\frac{1}{6} & 0 \\ 0 & 1 & 5 & -6 \end{bmatrix}$

$\frac{5}{6}R_2 + R_1 \rightarrow \left[\begin{array}{cc|cc} 1 & 0 & 4 & -5 \\ 0 & 1 & 5 & -6 \end{array}\right] = [I \ \vdots \ A^{-1}]$

$A^{-1} = \begin{bmatrix} 4 & -5 \\ 5 & -6 \end{bmatrix}$

124. $[A \ \vdots \ I] = \left[\begin{array}{ccc|ccc} 0 & -2 & 1 & 1 & 0 & 0 \\ -5 & -2 & -3 & 0 & 1 & 0 \\ 7 & 3 & 4 & 0 & 0 & 1 \end{array}\right]$

$R_3 \leftrightarrow R_1 \quad \left[\begin{array}{ccc|ccc} 7 & 3 & 4 & 0 & 0 & 1 \\ -5 & -2 & -3 & 0 & 1 & 0 \\ 0 & -2 & 1 & 1 & 0 & 0 \end{array}\right]$

$R_2 + R_1 \rightarrow \left[\begin{array}{ccc|ccc} 2 & 1 & 1 & 0 & 1 & 1 \\ -5 & -2 & -3 & 0 & 1 & 0 \\ 0 & -2 & 1 & 1 & 0 & 0 \end{array}\right]$

$5R_1 + 2R_2 \rightarrow \left[\begin{array}{ccc|ccc} 2 & 1 & 1 & 0 & 1 & 1 \\ 0 & 1 & -1 & 0 & 7 & 5 \\ 0 & -2 & 1 & 1 & 0 & 0 \end{array}\right]$

$\begin{matrix} -R_2 + R_1 \rightarrow \\ \\ 2R_2 + R_3 \rightarrow \end{matrix} \left[\begin{array}{ccc|ccc} 2 & 0 & 2 & 0 & -6 & -4 \\ 0 & 1 & -1 & 0 & 7 & 5 \\ 0 & 0 & -1 & 1 & 14 & 10 \end{array}\right]$

$\begin{matrix} \frac{1}{2}R_1 \rightarrow \\ \\ -R_3 \rightarrow \end{matrix} \left[\begin{array}{ccc|ccc} 1 & 0 & 1 & 0 & -3 & -2 \\ 0 & 1 & -1 & 0 & 7 & 5 \\ 0 & 0 & 1 & -1 & -14 & -10 \end{array}\right]$

$\begin{matrix} -R_3 + R_1 \rightarrow \\ R_3 + R_2 \rightarrow \\ \end{matrix} \left[\begin{array}{ccc|ccc} 1 & 0 & 0 & 1 & 11 & 8 \\ 0 & 1 & 0 & -1 & -7 & -5 \\ 0 & 0 & 1 & -1 & -14 & -10 \end{array}\right] = [I \ \vdots \ A^{-1}]$

$A^{-1} = \begin{bmatrix} 1 & 11 & 8 \\ -1 & -7 & -5 \\ -1 & -14 & -10 \end{bmatrix}$

125. $\begin{bmatrix} 2 & 0 & 3 \\ -1 & 1 & 1 \\ 2 & -2 & 1 \end{bmatrix}^{-1} = \begin{bmatrix} \frac{1}{2} & -1 & -\frac{1}{2} \\ \frac{1}{2} & -\frac{2}{3} & -\frac{5}{6} \\ 0 & \frac{2}{3} & \frac{1}{3} \end{bmatrix}$

126. $A = \begin{bmatrix} 1 & 4 & 6 \\ 2 & -3 & 1 \\ -1 & 18 & 16 \end{bmatrix}$

A^{-1} does not exist.

127. $A = \begin{bmatrix} -7 & 2 \\ -8 & 2 \end{bmatrix}$

$$A^{-1} = \frac{1}{-7(2) - 2(-8)}\begin{bmatrix} 2 & -2 \\ 8 & -7 \end{bmatrix} = \frac{1}{2}\begin{bmatrix} 2 & -2 \\ 8 & -7 \end{bmatrix} = \begin{bmatrix} 1 & -1 \\ 4 & -\frac{7}{2} \end{bmatrix}$$

128. $A = \begin{bmatrix} 10 & 4 \\ 7 & 3 \end{bmatrix}$

$ad - bc = (10)(3) - (4)(7) = 2$

$$A^{-1} = \frac{1}{10(3) - 4(7)}\begin{bmatrix} 3 & -4 \\ -7 & 10 \end{bmatrix} = \frac{1}{2}\begin{bmatrix} 3 & -4 \\ -7 & 10 \end{bmatrix} = \begin{bmatrix} \frac{3}{2} & -2 \\ -\frac{7}{2} & 5 \end{bmatrix}$$

129. $A = \begin{bmatrix} -\frac{1}{2} & 20 \\ \frac{3}{10} & -6 \end{bmatrix}$

$$A^{-1} = \frac{1}{-\frac{1}{2}(-6) - 20\left(\frac{3}{10}\right)}\begin{bmatrix} -6 & -20 \\ -\frac{3}{10} & -\frac{1}{2} \end{bmatrix} = -\frac{1}{3}\begin{bmatrix} -6 & -20 \\ -\frac{3}{10} & -\frac{1}{2} \end{bmatrix}$$

$$= \begin{bmatrix} 2 & \frac{20}{3} \\ \frac{1}{10} & \frac{1}{6} \end{bmatrix}$$

130. $A = \begin{bmatrix} -\frac{3}{4} & \frac{5}{2} \\ -\frac{4}{5} & -\frac{8}{3} \end{bmatrix}$

$ad - bc = \left(-\frac{3}{4}\right)\left(-\frac{8}{3}\right) - \left(\frac{5}{2}\right)\left(-\frac{4}{5}\right) = 2 + 2 = 4$

$$A^{-1} = \frac{1}{4}\begin{bmatrix} -\frac{8}{3} & -\frac{5}{2} \\ \frac{4}{5} & -\frac{3}{4} \end{bmatrix} = \begin{bmatrix} -\frac{2}{3} & -\frac{5}{8} \\ \frac{1}{5} & -\frac{3}{16} \end{bmatrix}$$

131. $\begin{cases} -x + 4y = 8 \\ 2x - 7y = -5 \end{cases}$

$$\begin{bmatrix} x \\ y \end{bmatrix} = \begin{bmatrix} -1 & 4 \\ 2 & -7 \end{bmatrix}^{-1}\begin{bmatrix} 8 \\ -5 \end{bmatrix} = \begin{bmatrix} 7 & 4 \\ 2 & 1 \end{bmatrix}\begin{bmatrix} 8 \\ -5 \end{bmatrix}$$

$$= \begin{bmatrix} 7(8) + 4(-5) \\ 2(8) + 1(-5) \end{bmatrix} = \begin{bmatrix} 36 \\ 11 \end{bmatrix}$$

Solution: $(36, 11)$

132. $\begin{cases} 5x - y = 13 \\ -9x + 2y = -24 \end{cases}$

$$\begin{bmatrix} x \\ y \end{bmatrix} = \begin{bmatrix} 5 & -1 \\ -9 & 2 \end{bmatrix}^{-1}\begin{bmatrix} 13 \\ -24 \end{bmatrix} = \begin{bmatrix} 2 & 1 \\ 9 & 5 \end{bmatrix}\begin{bmatrix} 13 \\ -24 \end{bmatrix} = \begin{bmatrix} 2 \\ -3 \end{bmatrix}$$

Solution: $(2, -3)$

133. $\begin{cases} -3x + 10y = 8 \\ 5x - 17y = -13 \end{cases}$

$$\begin{bmatrix} x \\ y \end{bmatrix} = \begin{bmatrix} -3 & 10 \\ 5 & -17 \end{bmatrix}^{-1}\begin{bmatrix} 8 \\ -13 \end{bmatrix} = \begin{bmatrix} -17 & -10 \\ -5 & -3 \end{bmatrix}\begin{bmatrix} 8 \\ -13 \end{bmatrix}$$

$$= \begin{bmatrix} -17(8) + (-10)(-13) \\ -5(8) + (-3)(-13) \end{bmatrix} = \begin{bmatrix} -6 \\ -1 \end{bmatrix}$$

Solution: $(-6, -1)$

134. $\begin{cases} 4x - 2y = -10 \\ -19x + 9y = 47 \end{cases}$

$$\begin{bmatrix} x \\ y \end{bmatrix} = \begin{bmatrix} 4 & -2 \\ -19 & 9 \end{bmatrix}^{-1}\begin{bmatrix} -10 \\ 47 \end{bmatrix} = \begin{bmatrix} -\frac{9}{2} & -1 \\ -\frac{19}{2} & -2 \end{bmatrix}\begin{bmatrix} -10 \\ 47 \end{bmatrix} = \begin{bmatrix} -2 \\ 1 \end{bmatrix}$$

Solution: $(-2, 1)$

135. $\begin{cases} 3x + 2y - z = 6 \\ x - y + 2z = -1 \\ 5x + y + z = 7 \end{cases}$

$$\begin{bmatrix} x \\ y \\ z \end{bmatrix} = \begin{bmatrix} 3 & 2 & -1 \\ 1 & -1 & 2 \\ 5 & 1 & 1 \end{bmatrix}^{-1} \begin{bmatrix} 6 \\ -1 \\ 7 \end{bmatrix} = \begin{bmatrix} -1 & -1 & 1 \\ 3 & \frac{8}{3} & -\frac{7}{3} \\ 2 & \frac{7}{3} & -\frac{5}{3} \end{bmatrix} \begin{bmatrix} 6 \\ -1 \\ 7 \end{bmatrix}$$

$$= \begin{bmatrix} -1(6) - 1(-1) + 1(7) \\ 3(6) + \frac{8}{3}(-1) - \frac{7}{3}(7) \\ 2(6) + \frac{7}{3}(-1) - \frac{5}{3}(7) \end{bmatrix} = \begin{bmatrix} 2 \\ -1 \\ -2 \end{bmatrix}$$

Solution: $(2, -1, -2)$

136. $\begin{cases} -x + 4y - 2z = 12 \\ 2x - 9y + 5z = -25 \\ -x + 5y - 4z = 10 \end{cases}$

$$\begin{bmatrix} x \\ y \\ z \end{bmatrix} = \begin{bmatrix} -1 & 4 & -2 \\ 2 & -9 & 5 \\ -1 & 5 & -4 \end{bmatrix}^{-1} \begin{bmatrix} 12 \\ -25 \\ 10 \end{bmatrix} = \begin{bmatrix} -11 & -6 & -2 \\ -3 & -2 & -1 \\ -1 & -1 & -1 \end{bmatrix} \begin{bmatrix} 12 \\ -25 \\ 10 \end{bmatrix} = \begin{bmatrix} -2 \\ 4 \\ 3 \end{bmatrix}$$

Solution: $(-2, 4, 3)$

137. $\begin{cases} -2x + y + 2z = -13 \\ -x - 4y + z = -11 \\ -y - z = 0 \end{cases}$

$$\begin{bmatrix} x \\ y \\ z \end{bmatrix} = \begin{bmatrix} -2 & 1 & 2 \\ -1 & -4 & 1 \\ 0 & -1 & -1 \end{bmatrix}^{-1} \begin{bmatrix} -13 \\ -11 \\ 0 \end{bmatrix} = \begin{bmatrix} -\frac{5}{9} & \frac{1}{9} & -1 \\ \frac{1}{9} & -\frac{2}{9} & 0 \\ -\frac{1}{9} & \frac{2}{9} & -1 \end{bmatrix} \begin{bmatrix} -13 \\ -11 \\ 0 \end{bmatrix}$$

$$= \begin{bmatrix} -\frac{5}{9}(-13) + \frac{1}{9}(-11) - 1(0) \\ \frac{1}{9}(-13) - \frac{2}{9}(-11) + 0(0) \\ -\frac{1}{9}(-13) + \frac{2}{9}(-11) - 1(0) \end{bmatrix} = \begin{bmatrix} 6 \\ 1 \\ -1 \end{bmatrix}$$

Solution: $(6, 1, -1)$

138. $\begin{cases} 3x - y + 5z = -14 \\ -x + y + 6z = 8 \\ -8x + 4y - z = 44 \end{cases}$

$$\begin{bmatrix} x \\ y \\ z \end{bmatrix} = \begin{bmatrix} 3 & -1 & 5 \\ -1 & 1 & 6 \\ -8 & 4 & -1 \end{bmatrix}^{-1} \begin{bmatrix} -14 \\ 8 \\ 44 \end{bmatrix} = \begin{bmatrix} \frac{25}{6} & -\frac{19}{6} & \frac{11}{6} \\ \frac{49}{6} & -\frac{37}{6} & \frac{23}{6} \\ -\frac{2}{3} & \frac{2}{3} & -\frac{1}{3} \end{bmatrix} \begin{bmatrix} -14 \\ 8 \\ 44 \end{bmatrix} = \begin{bmatrix} -3 \\ 5 \\ 0 \end{bmatrix}$$

Solution: $(-3, 5, 0)$

139. $\begin{cases} x + 2y = -1 \\ 3x + 4y = -5 \end{cases}$

$$\begin{bmatrix} x \\ y \end{bmatrix} = \begin{bmatrix} 1 & 2 \\ 3 & 4 \end{bmatrix}^{-1} \begin{bmatrix} -1 \\ -5 \end{bmatrix} = \begin{bmatrix} -2 & 1 \\ \frac{3}{2} & -\frac{1}{2} \end{bmatrix} \begin{bmatrix} -1 \\ -5 \end{bmatrix} = \begin{bmatrix} -3 \\ 1 \end{bmatrix}$$

Solution: $(-3, 1)$

140. $\begin{cases} x + 3y = 23 \\ -6x + 2y = -18 \end{cases}$

$$\begin{bmatrix} x \\ y \end{bmatrix} = \begin{bmatrix} 1 & 3 \\ -6 & 2 \end{bmatrix}^{-1} \begin{bmatrix} 23 \\ -18 \end{bmatrix} = \begin{bmatrix} 0.1 & -0.15 \\ 0.3 & 0.05 \end{bmatrix} \begin{bmatrix} 23 \\ -18 \end{bmatrix} = \begin{bmatrix} 5 \\ 6 \end{bmatrix}$$

$x = 5, y = 6$

Solution: $(5, 6)$

141. $\begin{cases} -3x - 3y - 4z = 2 \\ y + z = -1 \\ 4x + 3y + 4z = -1 \end{cases}$

$$\begin{bmatrix} x \\ y \\ z \end{bmatrix} = \begin{bmatrix} -3 & -3 & -4 \\ 0 & 1 & 1 \\ 4 & 3 & 4 \end{bmatrix}^{-1} \begin{bmatrix} 2 \\ -1 \\ -1 \end{bmatrix} = \begin{bmatrix} 1 & 0 & 1 \\ 4 & 4 & 3 \\ -4 & -3 & -3 \end{bmatrix} \begin{bmatrix} 2 \\ -1 \\ -1 \end{bmatrix} = \begin{bmatrix} 1 \\ 1 \\ -2 \end{bmatrix}$$

Solution: $(1, 1, -2)$

142. $\begin{cases} x - 3y - 2z = 8 \\ -2x + 7y + 3z = -19 \\ x - y - 3z = 3 \end{cases}$

$$\begin{bmatrix} x \\ y \\ z \end{bmatrix} = \begin{bmatrix} 1 & -3 & -2 \\ -2 & 7 & 3 \\ 1 & -1 & -3 \end{bmatrix}^{-1} \begin{bmatrix} 8 \\ -19 \\ 3 \end{bmatrix} = \begin{bmatrix} -18 & -7 & 5 \\ -3 & -1 & 1 \\ -5 & -2 & 1 \end{bmatrix} \begin{bmatrix} 8 \\ -19 \\ 3 \end{bmatrix} = \begin{bmatrix} 4 \\ -2 \\ 1 \end{bmatrix}$$

$x = 4, y = -2, z = 1$

Solution: $(4, -2, 1)$

143. $\begin{vmatrix} 8 & 5 \\ 2 & -4 \end{vmatrix} = 8(-4) - 5(2) = -42$

144. $\begin{vmatrix} -9 & 11 \\ 7 & -4 \end{vmatrix} = (-9)(-4) - (11)(7) = -41$

145. $\begin{vmatrix} 50 & -30 \\ 10 & 5 \end{vmatrix} = 50(5) - (-30)(10) = 550$

146. $\begin{vmatrix} 14 & -24 \\ 12 & -15 \end{vmatrix} = (14)(-15) - (-24)(12) = 78$

147. $\begin{bmatrix} 2 & -1 \\ 7 & 4 \end{bmatrix}$

(a) $M_{11} = 4$

$M_{12} = 7$

$M_{21} = -1$

$M_{22} = 2$

(b) $C_{11} = M_{11} = 4$

$C_{12} = -M_{12} = -7$

$C_{21} = -M_{21} = 1$

$C_{22} = M_{22} = 2$

148. $\begin{bmatrix} 3 & 6 \\ 5 & -4 \end{bmatrix}$

(a) $M_{11} = -4$

$M_{12} = 5$

$M_{21} = 6$

$M_{22} = 3$

(b) $C_{11} = M_{11} = -4$

$C_{12} = -M_{12} = -5$

$C_{21} = -M_{21} = -6$

$C_{22} = M_{22} = 3$

149. $\begin{bmatrix} 3 & 2 & -1 \\ -2 & 5 & 0 \\ 1 & 8 & 6 \end{bmatrix}$

(a) $M_{11} = \begin{vmatrix} 5 & 0 \\ 8 & 6 \end{vmatrix} = 30$

$M_{12} = \begin{vmatrix} -2 & 0 \\ 1 & 6 \end{vmatrix} = -12$

$M_{13} = \begin{vmatrix} -2 & 5 \\ 1 & 8 \end{vmatrix} = -21$

$M_{21} = \begin{vmatrix} 2 & -1 \\ 8 & 6 \end{vmatrix} = 20$

$M_{22} = \begin{vmatrix} 3 & -1 \\ 1 & 6 \end{vmatrix} = 19$

$M_{23} = \begin{vmatrix} 3 & 2 \\ 1 & 8 \end{vmatrix} = 22$

$M_{31} = \begin{vmatrix} 2 & -1 \\ 5 & 0 \end{vmatrix} = 5$

$M_{32} = \begin{vmatrix} 3 & -1 \\ -2 & 0 \end{vmatrix} = -2$

$M_{33} = \begin{vmatrix} 3 & 2 \\ -2 & 5 \end{vmatrix} = 19$

(b) $C_{11} = M_{11} = 30$

$C_{12} = -M_{12} = 12$

$C_{13} = M_{13} = -21$

$C_{21} = -M_{21} = -20$

$C_{22} = M_{22} = 19$

$C_{23} = -M_{23} = -22$

$C_{31} = M_{31} = 5$

$C_{32} = -M_{32} = 2$

$C_{33} = M_{33} = 19$

150. $\begin{bmatrix} 8 & 3 & 4 \\ 6 & 5 & -9 \\ -4 & 1 & 2 \end{bmatrix}$

(a) $M_{11} = \begin{vmatrix} 5 & -9 \\ 1 & 2 \end{vmatrix} = 19$

$M_{12} = \begin{vmatrix} 6 & -9 \\ -4 & 2 \end{vmatrix} = -24$

$M_{13} = \begin{vmatrix} 6 & 5 \\ -4 & 1 \end{vmatrix} = 26$

$M_{21} = \begin{vmatrix} 3 & 4 \\ 1 & 2 \end{vmatrix} = 2$

$M_{22} = \begin{vmatrix} 8 & 4 \\ -4 & 2 \end{vmatrix} = 32$

$M_{23} = \begin{vmatrix} 8 & 3 \\ -4 & 1 \end{vmatrix} = 20$

$M_{31} = \begin{vmatrix} 3 & 4 \\ 5 & -9 \end{vmatrix} = -47$

$M_{32} = \begin{vmatrix} 8 & 4 \\ 6 & -9 \end{vmatrix} = -96$

$M_{33} = \begin{vmatrix} 8 & 3 \\ 6 & 5 \end{vmatrix} = 22$

(b) $C_{11} = M_{11} = 19$

$C_{12} = -M_{12} = 24$

$C_{13} = M_{13} = 26$

$C_{21} = -M_{21} = -2$

$C_{22} = M_{22} = 32$

$C_{23} = -M_{23} = -20$

$C_{31} = M_{31} = -47$

$C_{32} = -M_{32} = 96$

$C_{33} = M_{33} = 22$

151. Expand using Column 2.

$$\begin{vmatrix} -2 & 4 & 1 \\ -6 & 0 & 2 \\ 5 & 3 & 4 \end{vmatrix} = -4\begin{vmatrix} -6 & 2 \\ 5 & 4 \end{vmatrix} - 3\begin{vmatrix} -2 & 1 \\ -6 & 2 \end{vmatrix}$$

$$= -4(-34) - 3(2) = 130$$

152. Expand using Row 3.

$$\begin{vmatrix} 4 & 7 & -1 \\ 2 & -3 & 4 \\ -5 & 1 & -1 \end{vmatrix} = -5\begin{vmatrix} 7 & -1 \\ -3 & 4 \end{vmatrix} - 1\begin{vmatrix} 4 & -1 \\ 2 & 4 \end{vmatrix} - 1\begin{vmatrix} 4 & 7 \\ 2 & -3 \end{vmatrix}$$

$$= -5(25) - (18) - (-26) = -117$$

153. Expand along Row 1.

$$\begin{vmatrix} 3 & 0 & -4 & 0 \\ 0 & 8 & 1 & 2 \\ 6 & 1 & 8 & 2 \\ 0 & 3 & -4 & 1 \end{vmatrix} = 3\begin{vmatrix} 8 & 1 & 2 \\ 1 & 8 & 2 \\ 3 & -4 & 1 \end{vmatrix} + (-4)\begin{vmatrix} 0 & 8 & 2 \\ 6 & 1 & 2 \\ 0 & 3 & 1 \end{vmatrix}$$

$$= 3[8(8 - (-8)) - 1(1 - 6) + 2(-4 - 24)] - 4[0 - 6(8 - 6) + 0]$$

$$= 3[128 + 5 - 56] - 4[-12]$$

$$= 279$$

154. Expand using Row 1, then use Row 3 of each 3×3 matrix.

$$\begin{vmatrix} -5 & 6 & 0 & 0 \\ 0 & 1 & -1 & 2 \\ -3 & 4 & -5 & 1 \\ 1 & 6 & 0 & 3 \end{vmatrix} = -5\begin{vmatrix} 1 & -1 & 2 \\ 4 & -5 & 1 \\ 6 & 0 & 3 \end{vmatrix} - 6\begin{vmatrix} 0 & -1 & 2 \\ -3 & -5 & 1 \\ 1 & 0 & 3 \end{vmatrix}$$

$$= -5[6(-1 + 10) + 3(-5 + 4)] - 6[(-1 + 10) + 3(0 - 3)]$$

$$= -5(54 - 3) - 6(9 - 9)$$

$$= -255$$

155. Points: $(0, 0), (-2, 2)$

Equation: $\begin{vmatrix} x & y & 1 \\ 0 & 0 & 1 \\ -2 & 2 & 1 \end{vmatrix} - \begin{vmatrix} x & y \\ -2 & 2 \end{vmatrix} = -(2x + 2y) = 0$ or $x + y = 0$

156. Points: $(-4, 3), (2, 1)$

Equation: $\begin{vmatrix} x & y & 1 \\ -4 & 3 & 1 \\ 2 & 1 & 1 \end{vmatrix} = x\begin{vmatrix} 3 & 1 \\ 1 & 1 \end{vmatrix} - y\begin{vmatrix} -4 & 1 \\ 2 & 1 \end{vmatrix} + \begin{vmatrix} -4 & 3 \\ 2 & 1 \end{vmatrix} = 2x + 6y - 10 = 0 \Longrightarrow x + 3y - 5 = 0$

157. Points: $(10, 7), (-2, -7)$

Equation:

$$\begin{vmatrix} x & y & 1 \\ 10 & 7 & 1 \\ -2 & -7 & 1 \end{vmatrix} = \begin{vmatrix} 10 & 7 \\ -2 & -7 \end{vmatrix} - \begin{vmatrix} x & y \\ -2 & -7 \end{vmatrix} + \begin{vmatrix} x & y \\ 10 & 7 \end{vmatrix} = -70 + 14 - (-7x + 2y) + 7x - 10y = 0 \text{ or } 7x - 6y - 28 = 0$$

158. Points: $\left(-\frac{1}{2}, 3\right), \left(\frac{5}{2}, 1\right)$

Equation: $\begin{vmatrix} x & y & 1 \\ -\frac{1}{2} & 3 & 1 \\ \frac{5}{2} & 1 & 1 \end{vmatrix} = x\begin{vmatrix} 3 & 1 \\ 1 & 1 \end{vmatrix} - y\begin{vmatrix} -\frac{1}{2} & 1 \\ \frac{5}{2} & 1 \end{vmatrix} + \begin{vmatrix} -\frac{1}{2} & 3 \\ \frac{5}{2} & 1 \end{vmatrix} = 2x + 3y - 8 = 0$

159. Points: $(-3, 10), (1, -2)$

Equation: $\begin{vmatrix} x & y & 1 \\ -3 & 10 & 1 \\ 1 & -2 & 1 \end{vmatrix} = x\begin{vmatrix} 10 & 1 \\ -2 & 1 \end{vmatrix} - y\begin{vmatrix} -3 & 1 \\ 1 & 1 \end{vmatrix} + \begin{vmatrix} -3 & 10 \\ 1 & -2 \end{vmatrix} = 12x + 4y - 4 = 0 \Longrightarrow 3x + y - 1 = 0$

160. Points: $\left(-2, \frac{11}{2}\right), \left(1, -\frac{1}{2}\right)$

Equation: $\begin{vmatrix} x & y & 1 \\ -2 & \frac{11}{2} & 1 \\ 1 & -\frac{1}{2} & 1 \end{vmatrix} = x\begin{vmatrix} \frac{11}{2} & 1 \\ -\frac{1}{2} & 1 \end{vmatrix} - y\begin{vmatrix} -2 & 1 \\ 1 & 1 \end{vmatrix} + \begin{vmatrix} -2 & \frac{11}{2} \\ 1 & -\frac{1}{2} \end{vmatrix} = 6x + 3y - \frac{9}{2} = 0 \Rightarrow 4x + 2y - 3 = 0$

Problem Solving for Chapter 14

1. (a)

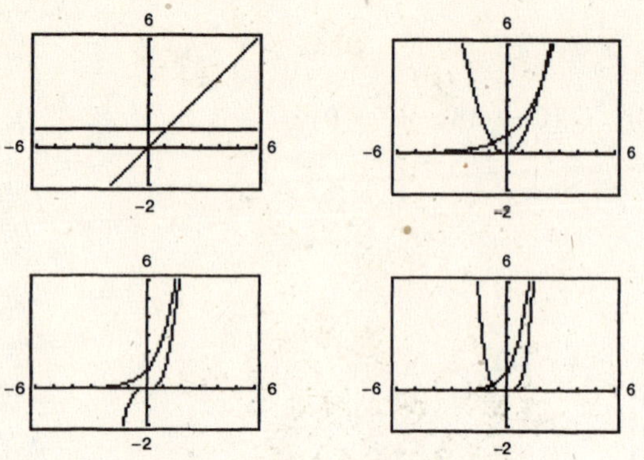

(b) If b is even, there are three points of intersection.
If b is odd, there are two points of intersection.

2. $x =$ radius of smaller circle

$y =$ radius of larger circle

(a) Constraints on circles: $\pi y^2 - \pi x^2 \geq 10$

$y > x$

$x > 0$

(b)

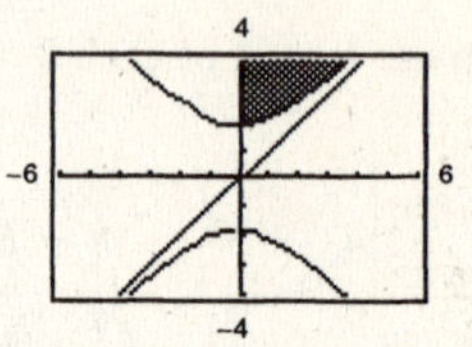

(c) The line is an asymptote to the boundary. The larger the circles, the closer the radii can be and the constraint still be satisfied.

3.

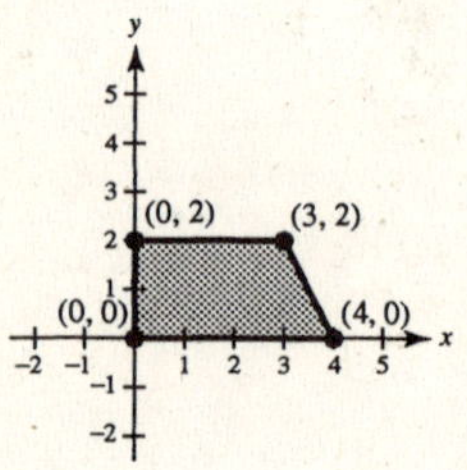

System of Inequalities

$$\begin{cases} 0 \leq y \leq 2 \\ x \geq 0 \\ y \leq -2x + 8 \end{cases}$$

Find the equations of the lines through each pair of adjacent vertices.

(0, 0), (0, 2): $x = 0$

(0, 0), (4, 0): $y = 0$

(0, 2), (3, 2): $y = 2$

(3, 2), (4, 0): $y = -2x + 8$

4. Let $A = \begin{bmatrix} 1 & 3 \\ -2 & 4 \end{bmatrix}$ and $B = \begin{bmatrix} -4 & 0 \\ 3 & 5 \end{bmatrix}$.

$|A| = \begin{vmatrix} 1 & 3 \\ -2 & 4 \end{vmatrix} = 10$, $|B| = \begin{vmatrix} -4 & 0 \\ 3 & 5 \end{vmatrix} = -20$, $|A| + |B| = -10$

$A + B = \begin{bmatrix} -3 & 3 \\ 1 & 9 \end{bmatrix}$, $|A + B| = \begin{vmatrix} -3 & 3 \\ 1 & 9 \end{vmatrix} = -30$

Thus, $|A + B| \neq |A| + |B|$. Your answer may differ, depending on how you choose A and B.

5. $z = a$

$y = -4a + 1$

$x = -3a - 2$

One possible system is:

$$\begin{cases} x + y + 7z = (-3a - 2) + (-4a + 1) + 7a = -1 \\ x + 2y + 11z = (-3a - 2) + 2(-4a + 1) + 11a = 0 \\ 2x + y + 10z = 2(-3a - 2) + (-4a + 1) + 10a = -3 \end{cases}$$

or

$$\begin{cases} x + y + 7z = -1 \\ x + 2y + 11z = 0 \\ 2x + y + 10z = -3 \end{cases}$$

(Note that the coefficients of x, y, and z have been chosen so that the a–terms cancel.)

6. (a) $P^2 = \begin{bmatrix} 0.6 & 0.1 & 0.1 \\ 0.2 & 0.7 & 0.1 \\ 0.2 & 0.2 & 0.8 \end{bmatrix}\begin{bmatrix} 0.6 & 0.1 & 0.1 \\ 0.2 & 0.7 & 0.1 \\ 0.2 & 0.2 & 0.8 \end{bmatrix} = \begin{bmatrix} 0.40 & 0.15 & 0.15 \\ 0.28 & 0.53 & 0.17 \\ 0.32 & 0.32 & 0.68 \end{bmatrix}$

The P^2 matrix gives the proportion of the voting population that changed parties or remained loyal to their party from the first election to the third.

(b) $P^3 = P^2P = \begin{bmatrix} 0.40 & 0.15 & 0.15 \\ 0.28 & 0.53 & 0.17 \\ 0.32 & 0.32 & 0.68 \end{bmatrix}\begin{bmatrix} 0.6 & 0.1 & 0.1 \\ 0.2 & 0.7 & 0.1 \\ 0.2 & 0.2 & 0.8 \end{bmatrix} = \begin{bmatrix} 0.300 & 0.175 & 0.175 \\ 0.308 & 0.433 & 0.217 \\ 0.392 & 0.392 & 0.608 \end{bmatrix}$

$$P^4 = P^3P = \begin{bmatrix} 0.300 & 0.175 & 0.175 \\ 0.308 & 0.433 & 0.217 \\ 0.392 & 0.392 & 0.608 \end{bmatrix}\begin{bmatrix} 0.6 & 0.1 & 0.1 \\ 0.2 & 0.7 & 0.1 \\ 0.2 & 0.2 & 0.8 \end{bmatrix} = \begin{bmatrix} 0.250 & 0.188 & 0.188 \\ 0.315 & 0.377 & 0.248 \\ 0.435 & 0.435 & 0.565 \end{bmatrix}$$

$$P^5 = P^4P = \begin{bmatrix} 0.250 & 0.188 & 0.188 \\ 0.315 & 0.377 & 0.248 \\ 0.435 & 0.435 & 0.565 \end{bmatrix}\begin{bmatrix} 0.6 & 0.1 & 0.1 \\ 0.2 & 0.7 & 0.1 \\ 0.2 & 0.2 & 0.8 \end{bmatrix} = \begin{bmatrix} 0.225 & 0.194 & 0.194 \\ 0.314 & 0.345 & 0.267 \\ 0.461 & 0.461 & 0.539 \end{bmatrix}$$

$$P^6 = \begin{bmatrix} 0.213 & 0.197 & 0.197 \\ 0.311 & 0.326 & 0.280 \\ 0.477 & 0.477 & 0.523 \end{bmatrix}$$

$$P^7 = \begin{bmatrix} 0.206 & 0.198 & 0.198 \\ 0.308 & 0.316 & 0.288 \\ 0.486 & 0.486 & 0.514 \end{bmatrix}$$

$$P^8 = \begin{bmatrix} 0.203 & 0.199 & 0.199 \\ 0.305 & 0.309 & 0.292 \\ 0.492 & 0.492 & 0.508 \end{bmatrix}$$

As P is raised to higher and higher powers, the resulting matrices appear to be approaching the matrix

$$\begin{bmatrix} 0.2 & 0.2 & 0.2 \\ 0.3 & 0.3 & 0.3 \\ 0.5 & 0.5 & 0.5 \end{bmatrix}.$$

7. $AC = \begin{bmatrix} 0 & 1 \\ 0 & 1 \end{bmatrix}\begin{bmatrix} 2 & 3 \\ 2 & 3 \end{bmatrix} = \begin{bmatrix} 2 & 3 \\ 2 & 3 \end{bmatrix}$

$BC = \begin{bmatrix} 1 & 0 \\ 1 & 0 \end{bmatrix}\begin{bmatrix} 2 & 3 \\ 2 & 3 \end{bmatrix} = \begin{bmatrix} 2 & 3 \\ 2 & 3 \end{bmatrix}$

Thus, $AC = BC$ even though $A \neq B$.

8. $AB = \begin{bmatrix} 3 & 3 \\ 4 & 4 \end{bmatrix}\begin{bmatrix} 1 & -1 \\ -1 & 1 \end{bmatrix} = \begin{bmatrix} 0 & 0 \\ 0 & 0 \end{bmatrix}$

$AB = O$ and neither A nor B is O.

9. $A = \begin{bmatrix} 1 & 2 \\ -2 & 1 \end{bmatrix}$

(a) $A^2 - 2A + 5I = \begin{bmatrix} 1 & 2 \\ -2 & 1 \end{bmatrix}\begin{bmatrix} 1 & 2 \\ -2 & 1 \end{bmatrix} - 2\begin{bmatrix} 1 & 2 \\ -2 & 1 \end{bmatrix} + 5\begin{bmatrix} 1 & 0 \\ 0 & 1 \end{bmatrix}$

$= \begin{bmatrix} -3 & 4 \\ -4 & -3 \end{bmatrix} + \begin{bmatrix} -2 & -4 \\ 4 & -2 \end{bmatrix} + \begin{bmatrix} 5 & 0 \\ 0 & 5 \end{bmatrix}$

$= \begin{bmatrix} 0 & 0 \\ 0 & 0 \end{bmatrix} = 0$

(b) $A^{-1} = \dfrac{1}{(1) - (-4)}\begin{bmatrix} 1 & -2 \\ 2 & 1 \end{bmatrix} = \dfrac{1}{5}\begin{bmatrix} 1 & -2 \\ 2 & 1 \end{bmatrix}$

$\dfrac{1}{5}(2I - A) = \dfrac{1}{5}\left[\begin{bmatrix} 2 & 0 \\ 0 & 2 \end{bmatrix} - \begin{bmatrix} 1 & 2 \\ -2 & 1 \end{bmatrix}\right] = \dfrac{1}{5}\begin{bmatrix} 1 & -2 \\ 2 & 1 \end{bmatrix}$

Thus, $A^{-1} = \dfrac{1}{5}(2I - A)$.

(c) $A^2 - 2A + 5I = 0$

$A^2 - 2A = -5I$

$(A - 2I)A = -5I$

$-\dfrac{1}{5}(A - 2I)A = -5I$

$\dfrac{1}{5}(2I - A)A = I$

Thus, $A^{-1} = \dfrac{1}{5}(2I - A)$.

10. $A = \begin{bmatrix} 1 & 2 \\ 1 & 3 \end{bmatrix}, B = \begin{bmatrix} 2 & -1 \\ 1 & -1 \end{bmatrix}$

$AB = \begin{bmatrix} 4 & -3 \\ 5 & -4 \end{bmatrix}$

$(AB)^{-1} = \begin{bmatrix} 4 & -3 \\ 5 & -4 \end{bmatrix} = AB$

$A^{-1}B^{-1} = \begin{bmatrix} 3 & -2 \\ -1 & 1 \end{bmatrix}\begin{bmatrix} 1 & -1 \\ 1 & -2 \end{bmatrix} = \begin{bmatrix} 1 & 1 \\ 0 & -1 \end{bmatrix}$

$B^{-1}A^{-1} = \begin{bmatrix} 1 & -1 \\ 1 & -2 \end{bmatrix}\begin{bmatrix} 3 & -2 \\ -1 & 1 \end{bmatrix} = \begin{bmatrix} 4 & -3 \\ 5 & -4 \end{bmatrix} = (AB)^{-1}$

Conjecture: $(AB)^{-1} = B^{-1}A^{-1}$

11. $AA^{-1} = \begin{bmatrix} a & b \\ c & d \end{bmatrix}\left(\frac{1}{ad - bc}\right)\begin{bmatrix} d & -b \\ -c & a \end{bmatrix} = \frac{1}{ad - bc}\begin{bmatrix} a & b \\ c & d \end{bmatrix}\begin{bmatrix} d & -b \\ -c & a \end{bmatrix}$

$= \frac{1}{ad - bc}\begin{bmatrix} ad - bc & 0 \\ 0 & ad - bc \end{bmatrix} = \begin{bmatrix} 1 & 0 \\ 0 & 1 \end{bmatrix}$

$A^{-1}A = \frac{1}{ad - bc}\begin{bmatrix} d & -b \\ -c & a \end{bmatrix}\begin{bmatrix} a & b \\ c & d \end{bmatrix} = \frac{1}{ad - bc}\begin{bmatrix} ad - bc & 0 \\ 0 & ad - bc \end{bmatrix} = \begin{bmatrix} 1 & 0 \\ 0 & 1 \end{bmatrix}$

12. $\begin{cases} x + 3y + z = 3 \\ x + 5y + 5z = 1 \\ 2x + 6y + 3z = 8 \end{cases}$

$$\begin{bmatrix} 1 & 3 & 1 & \vdots & 3 \\ 1 & 5 & 5 & \vdots & 1 \\ 2 & 6 & 3 & \vdots & 8 \end{bmatrix}$$

$$\begin{matrix} \\ -R_1 + R_2 \to \\ -2R_1 + R_3 \to \end{matrix} \begin{bmatrix} 1 & 3 & 1 & \vdots & 3 \\ 0 & 2 & 4 & \vdots & -2 \\ 0 & 0 & 1 & \vdots & 2 \end{bmatrix}$$

$$\frac{1}{2}R_2 \to \begin{bmatrix} 1 & 3 & 1 & \vdots & 3 \\ 0 & 1 & 2 & \vdots & -1 \\ 0 & 0 & 1 & \vdots & 2 \end{bmatrix}$$

This is a matrix in row-echelon form.

$$\begin{bmatrix} 1 & 3 & \frac{3}{2} & \vdots & 4 \\ 0 & 1 & \frac{7}{4} & \vdots & -\frac{3}{2} \\ 0 & 0 & 1 & \vdots & 2 \end{bmatrix}$$

The row-echelon form feature of a graphing utility yields ∴.

There are infinitely many matrices in row–echelon form that correspond to the original system of equations. All such matrices will yield the same solution, namely $(16, -5, 2)$.

13. (a) $\begin{vmatrix} 4 & 5 & 6 \\ 7 & 8 & 9 \\ 10 & 11 & 12 \end{vmatrix} = 0$

$\begin{vmatrix} 33 & 34 & 35 \\ 36 & 37 & 38 \\ 39 & 40 & 41 \end{vmatrix} = 0$ $\quad \begin{vmatrix} -5 & -4 & -3 \\ -2 & -1 & 0 \\ 1 & 2 & 3 \end{vmatrix} = 0$

$\begin{vmatrix} 19 & 20 & 21 & 22 \\ 23 & 24 & 25 & 26 \\ 27 & 28 & 29 & 30 \\ 31 & 32 & 33 & 34 \end{vmatrix} = 0$ $\quad \begin{vmatrix} 57 & 58 & 59 & 60 \\ 61 & 62 & 63 & 64 \\ 65 & 66 & 67 & 68 \\ 69 & 70 & 71 & 72 \end{vmatrix} = 0$

For an $n \times n$ matrix $(n > 2)$ with consecutive integer entries, the determinant appears to be 0.

—CONTINUED—

13. —CONTINUED—

(b) For the 3 × 3 case we have the following:

$$\begin{vmatrix} x & x+1 & x+2 \\ x+3 & x+4 & x+5 \\ x+6 & x+7 & x+8 \end{vmatrix} = x\begin{vmatrix} x+4 & x+5 \\ x+7 & x+8 \end{vmatrix} - (x+1)\begin{vmatrix} x+3 & x+5 \\ x+6 & x+8 \end{vmatrix} + (x+2)\begin{vmatrix} x+3 & x+4 \\ x+6 & x+7 \end{vmatrix}$$

$$= x[(x+4)(x+8) - (x+7)(x+5)] - (x+1)[(x+3)(x+8)$$

$$- (x+6)(x+5)] + (x+2)[(x+3)(x+7) - (x+6)(x+4)]$$

$$= x[(x^2+12x+32) - (x^2+12x+35)] - (x+1)[(x^2+11x+24)$$

$$- (x^2+11x+30)] + (x+2)[(x^2+10x+21) - (x^2+10x+24)]$$

$$= -3x - (x+1)(-6) + (x+2)(-3)$$

$$= -3x + 6x + 6 - 3x - 6 = 0$$

14. $\begin{cases} 3x - 5y = 8 \\ 2x + k_1y = k_2 \end{cases}$

$$6x - 10y = 16$$

$$-6x - 3k_1y = -3k_2$$

$$(-10 - 3k_1)y = 16 - 3k_2$$

For this system to have an infinite number of solutions, this last equation should be 0 = 0. Thus,

$-10 - 3k_1 = 0$ and $16 - 3k_2 = 0$

$-10 = 3k_1$ $\qquad 16 = 3k_2$

$-\frac{10}{3} = k_1$ $\qquad \frac{16}{3} = k_2$

15. There are a finite number of solutions.

(a) If both equations are linear, then the maximum number of solutions to a finite system is *one.*

(b) If one equation is linear and the other is quadratic, the the maximum number of solutions is *two.*

(c) If both equations are quadratic, then the maximum number of solutions is *four.*

16. No. Each matrix is in row-echelon form, but the third matrix cannot be achieved from the first or second matrix with elementary row operations.

17. $6 = \pm\frac{1}{2}\begin{vmatrix} -2 & -3 & 1 \\ 1 & -1 & 1 \\ -8 & x & 1 \end{vmatrix}$

$$\pm 12 = \begin{vmatrix} 1 & -1 \\ -8 & x \end{vmatrix} - \begin{vmatrix} -2 & -3 \\ -8 & x \end{vmatrix} + \begin{vmatrix} -2 & -3 \\ 1 & -1 \end{vmatrix}$$

$$\pm 12 = (x - 8) - (-2x - 24) + 5$$

$$\pm 12 = 3x + 21$$

$$x = \frac{-21 \pm 12}{3} = -7 \pm 4$$

$$x = -3 \text{ or } x = -11$$

18. Vertices: (0, 25), (10, 0), (28, 5)

$$\text{Area} = \frac{1}{2}\begin{vmatrix} 0 & 25 & 1 \\ 10 & 0 & 1 \\ 28 & 5 & 1 \end{vmatrix} = 250 \text{ square miles}$$

19. Answers will vary. One example is:

$$A = \begin{bmatrix} 1 & 0 \\ -1 & 0 \end{bmatrix}$$

20. $(a - b)(b - c)(c - a) = -a^2b + a^2c + ab^2 - ac^2 - b^2c + bc^2$

$$\begin{vmatrix} 1 & 1 & 1 \\ a & b & c \\ a^2 & b^2 & c^2 \end{vmatrix} = \begin{vmatrix} b & c \\ b^2 & c^2 \end{vmatrix} - \begin{vmatrix} a & c \\ a^2 & c^2 \end{vmatrix} + \begin{vmatrix} a & b \\ a^2 & b^2 \end{vmatrix}$$

$$= bc^2 - b^2c - ac^2 + a^2c + ab^2 - a^2b$$

Thus, $\begin{vmatrix} 1 & 1 & 1 \\ a & b & c \\ a^2 & b^2 & c^2 \end{vmatrix} = (a - b)(b - c)(c - a).$

21. $(a - b)(b - c)(c - a)(a + b + c) = -a^3b + a^3c + ab^3 - ac^3 - b^3c + bc^3$

$$\begin{vmatrix} 1 & 1 & 1 \\ a & b & c \\ a^3 & b^3 & c^3 \end{vmatrix} = \begin{vmatrix} b & c \\ b^3 & c^3 \end{vmatrix} - \begin{vmatrix} a & c \\ a^3 & c^3 \end{vmatrix} + \begin{vmatrix} a & b \\ a^3 & b^3 \end{vmatrix}$$

$$= bc^3 - b^3c - ac^3 + a^3c + ab^3 - a^3b$$

Thus, $\begin{vmatrix} 1 & 1 & 1 \\ a & b & c \\ a^3 & b^3 & c^3 \end{vmatrix} = (a - b)(b - c)(c - a)(a + b + c).$

22. $\begin{vmatrix} x & 0 & c \\ -1 & x & b \\ 0 & -1 & a \end{vmatrix} = x\begin{vmatrix} x & b \\ -1 & a \end{vmatrix} + c\begin{vmatrix} -1 & x \\ 0 & -1 \end{vmatrix}$

$$= x(ax + b) + c(1 - 0)$$

$$= ax^2 + bx + c$$

23. $\begin{vmatrix} x & 0 & 0 & d \\ -1 & x & 0 & c \\ 0 & -1 & x & b \\ 0 & 0 & -1 & a \end{vmatrix} = x\begin{vmatrix} x & 0 & c \\ -1 & x & b \\ 0 & -1 & a \end{vmatrix} - d\begin{vmatrix} -1 & x & 0 \\ 0 & -1 & x \\ 0 & 0 & -1 \end{vmatrix}$

$$= \underbrace{x(ax^2 + bx + c)}_{\text{From Exercise 22}} - d\left(-\begin{vmatrix} -1 & x \\ 0 & -1 \end{vmatrix}\right)$$

$$= ax^3 + bx^2 + cx + d$$

24. $\begin{vmatrix} 2 & -5 & 1 \\ 4 & x & 1 \\ 5 & -2 & 1 \end{vmatrix} = 0$

$$2\begin{vmatrix} x & 1 \\ -2 & 1 \end{vmatrix} + 5\begin{vmatrix} 4 & 1 \\ 5 & 1 \end{vmatrix} + \begin{vmatrix} 4 & x \\ 5 & -2 \end{vmatrix} = 0$$

$$2(x + 2) + 5(-1) + (-8 - 5x) = 0$$

$$-3x - 9 = 0$$

$$x = -3$$